COMBUSTION
FOSSIL POWER

A Reference Book on Fuel Burning and Steam Generation

Editor
JOSEPH G. SINGER, P.E.
Fellow, American Society of Mechanical Engineers

Fourth Edition

Published by

COMBUSTION ENGINEERING, INC.
1000 Prospect Hill Road
Windsor, Connecticut 06095
1991

ASEA BROWN BOVERI

Disclaimer

The information in this book has been obtained by Combustion Engineering, Inc. from sources that are believed to be reliable. Neither Combustion Engineering, Inc., nor any of the authors employed thereby, makes any warranty, express or implied, or assumes any legal liability or responsibility for the accuracy, completeness, or usefulness of any information, apparatus, product or process disclosed in this book, or represents that the use of any such information, apparatus, product or process would not infringe privately owned rights. This book is published with the understanding that Combustion Engineering, Inc. and the authors employed thereby are supplying information but are not attempting to render engineering or other professional services. If such services are required, the assistance of an appropriate professional should be sought.

Layout, Artwork and Mechanicals
Joseph P. Amenta

RAND McNALLY
Printed in U.S.A.

Table of Contents

APPENDICES

ACKNOWLEDGEMENT

We dedicate this edition of COMBUSTION, with thanks, to the many people who contributed their talent and time, in great or small measure, to revising and updating its technical base, and encouraging and supporting its publication.

A. A. Arstikaitis	D. Frabotta	W. S. Mikus
G. F. Barcikowski	F. Gabrielli	N. C. Mohn
T. J. Barker	D. E. Gelbar	K. W. Morris
T. K. Benton	D. E. Gorski	J. J. Moskal
J. D. Bianca	J. Grusha	J. M. Niziolek
R. D. Blodgett	G. P. Hammer	A. L. Plumley
M. P. Borden	R. W. Hanson	E. A. Ramspeck
R. W. Borio	J. P. Harmon	J. H. Ratcliffe
C. R. Bozzuto	D. J. Hart	J. A. Ray
G. E. Bresowar	C. F. Horlitz, Jr.	J. W. Regan
H. E. Burbach	G. D. Jukkola	E. K. Rickard
G. M. Chase	S. E. Kmiotek	J. R. Rode
I. M. Clark	R. F. Konopacki	G. Sailer
M. B. Cohen	R. C. Kunkel	A. J. Seibert
J. R. Comparato	A. A. Levasseur	R. S. Skowyra
K. E. Crooke	V. Llinares, Jr.	R. T. Smuda
D. P. Davies	G. R. Lovejoy	P. L. Stanwicks
M. A. Delin	R. E. Mahoney	E. A. Steen
R. E. Donais	J. A. Makuch	G. R. Strich
R. E. Drake, II	T. V. Malizewski	W. R. Sylvester
W. B. Ferguson	J. F. Mangold	G. W. Thimot
J. D. Fishburn	J. L. Marion	W. F. Walsh
J. C. Flynn, Jr.	O. Martinez, Jr.	C. H. Wells, Jr.
A. E. Fournier	T. P. Mastronarde	J. E. Wilmoth
J. D. Fox	M. S. McCartney	E. A. Zielinski

To Mrs. Judith C. Gorski, who served as copy editor for the Third and Fourth Editions, we owe a debt of gratitude. Her concern for logic and clarity is reflected throughout the text.

To T. F. Gawlicki, we say thanks for expediting this edition to completion. Without him, it never would have been printed.

To my wife, Alice, I submit an overdue apology for the three years of creeping disarray.

Note: In addition to the technical editing, rewriting, and proofreading of the entire book, the Editor wrote *much* of Chapters 7, 15, and 24; *most* of Chapters 1 through 6, and 14; and *all* of Chapter 16, Appendices A and E, and the Index.

Preface to the Fourth Edition

The book COMBUSTION is intended to provide a thorough insight into the workings of the steam-producing side of a modern high-efficiency power plant burning a wide variety of fossil fuels. We have revised or added to over three-quarters of the material in the 1981 edition. We have made several chapters even more generic than in the last edition, which should make it even more useful to the engineers, developers, and operating people who regularly refer to those sections. It appears that the principal value of this text to academia is as a supplemental reference for both class and project work (which it is) and a source of information for writing textbooks on overall power generation, which it is not.

Because of the maturation of the fluidized-bed boiler since 1981, we have built an entire chapter around that technology and its power-producing applications. We have not placed specific emphasis on this or any other way of burning coal, because economics and federal regulations will dictate their usage. We have also included a new chapter on the C-E laboratory capabilities for research and development in the fuel-burning, steam-generating, and emission-control fields. The chapter illustrates the scope of equipment and talented personnel needed to back up the design and operation of today's large, high-pressure steam-generating equipment.

Engineers throughout the world have found the previous edition of this book useful in their day-to-day activities to do with power plants. In revising and updating it, we anticipate that it will be of even greater use in the unprecedented construction of fossil-fueled power-generation installations that will take place in the next 10 to 15 years. This edition *looks forward* to that design and building program, and presents the total range of concepts and equipment that will be used to take the universe of steam power generation into the twenty-first century.

The world community is making steady progress in converting to the SI-metric measurement system (Système International des Unités). America will take a major step in 1992 when all government procurement will be done in SI units only. However, the many duplicate 300- to 600-megawatt units that are going to be ordered by American and other utilities in the next several years will be built in U. S. Customary (USC) units, as were their prototypes. Consequently, spare and replacement parts will continue to be furnished in USC dimensions well into the middle of the next century. "Bi-lingualism" of measurement will be with us a long time; hence the substantial expansion that we have made to the appendix on engineering conversions.

Note that the six chapters that are either new or that have been substantially rewritten have both SI-metric and USC units. In the body of the text, numbers given in parentheses following a mathematical value will be either close approximations or exact equalities, depending on the context. As in the previous edition, the results of empirical testing are reported in the units in which the data was generated.

It surprised me, during the 3 years that I have worked on this revision, that so much has changed since 1980. The fact that very few major purchases of large boilers were made in that time did not stop the brilliant engineering minds of the industry from refining, optimizing, and, in some areas, completely changing designs for use in the upcoming campaign of power-plant construction. We can only surmise that all competitive industries do the same—to continue to invent and grow and update when things are slow—when management provides the encouragement and backing to do so. I am proud to have worked for 45 years with such dedicated and ingenious people.

As we said in the Preface to the previous edition:

"Please note that throughout this book many illustrations and figures are used to explain physical and chemical phenomena. The values in these figures are approximate and are for illustration only. None should be used for design standards or other purposes where exact values are required. Although extensive, the text is limited to introductory material on many of the subjects and is only the 'tip of a large iceberg' that represents the accumulated experience of the technical personnel of a major steam-generator supplier such as C-E."

Joseph G. Singer
Easter, 1991

Combustion Engineering—An Introduction

The world today faces a critical challenge as nations strive to satisfy the basic human needs of food, shelter, and clothing which are so dependent on adequate supplies of energy. The great increase in the use of energy has come at the expense of our fossil fuels—coal, oil and gas—and people are more aware that natural resources have finite limits.

As we search for the ultimate solutions to the energy crisis, we must at the same time continue to look for interim ways of meeting the immediate growth in demand for energy. Technological improvements in mining, moving, processing and using fossil fuels can, of course, stretch energy-resource reserves, as can a determined effort at conservation. Similarly, application of technologies such as coal gasification or fluidized-bed combustion can widen the use of the world's vast coal resources. And finding new ways to make the best use of our energy supplies is what Combustion Engineering's tradition is all about.

A NEW NAME AND RAPID GROWTH

To tell the story of Combustion Engineering is to trace the development of fossil power technology in the United States. The year 1912 marks a milestone for C-E. The organization bearing the Combustion Engineering name was formed in that year, although its roots had been firmly planted in the 1880's.

The merger of the Grieve Grate Company and the American Stoker Company brought together two well-known manufacturers of fuel-burning equipment to become Combustion Engineering Corp. The Grieve grate was a specialized grate capable of burning anthracite-coal screenings; the American Stoker Company manufactured a screw-type stoker to burn bituminous coal. Soon, the new organization acquired a license to build a single-retort underfeed stoker of English design. Known as the Type E stoker, it remained popular for small boilers for many years.

During the twenties, other stoker technology was acquired as were manufacturing plants along the Monongahela River south of Pittsburgh, Pennsylvania. C-E fabricated all its stokers in this plant which was built about the turn of the century. In addition to the Coxe traveling grate and the Green chain grate, C-E offered several types of underfeed stokers including the Type E, the Low Ram Stoker, the Skelly Stoker, and the C-E Multiple-Retort Stoker.

As the Company grew, new stoker designs were developed such as the Lloyd Traveling

Grate and the C-E Continuous Ash Discharge Spreader Stoker. But as grate surfaces grew to 24-ft wide by 28-ft long in size, the traveling grate stoker reached a practical limit—and a unit this size burning anthracite could generate only about 200,000 pounds of steam per hour. Despite its limitations, however, the continuous ash discharge spreader stoker remains in use today on boilers burning coal, wood, and bark.

A QUANTUM JUMP IN COMBUSTION TECHNOLOGY

In 1918, C-E launched into the developing technology of pulverized-coal firing. Overcoming the limitation placed on boiler capacity by the size restrictions of stokers, pulverized-coal firing opened the door to a large expansion in capacity of coal-fired boilers.

Capitalizing on a pulverized-coal firing system for steam locomotives, Combustion Engineering successfully adapted this technology to stationary utility boilers at the Oneida Street and Lakeside Stations of the Milwaukee Electric Railway and Light Co. This pioneering work, done jointly with Milwaukee Electric's engineers and the Bureau of Mines, marks the most important single development in the art of solid-fuel combustion in this century. As a result, in 1980, the Oneida Street Station, where pulverized coal was first put to use by a utility, was named an Historic Engineering Landmark by the American Society of Mechanical Engineers.

MANUFACTURING CAPABILITY EXPANDS

It was not until 1925 that C-E entered the steam boiler business, beginning with a bent-tube boiler design installed in the early twenties at the Ford Motor Company's River Rouge plant. For this same installation, Combustion had supplied the pulverizers and fuel-burning equipment. The Ford Company's 200,000 lb/hr boilers were among the first large units specifi-

cally designed for pulverized-coal firing.

Acquisition of two boiler companies augmented C-E's manufacturing capabilities. At the former Heine Boiler plant site in St. Louis is located one of C-E's manufacturing plants today, while the present major manufacturing facility in Chattanooga is on the former Hedges-Walsh and Weidner Boiler Co. site.

Until this time, steam-generating units were an assembly of separate components bought from different manufacturers. With the addition of these two boiler companies, C-E was able to combine all components (boiler, superheater, water-cooled furnace, fuel-burning equipment and heat-recovery equipment) into one coordinated design.

THE MOVE TOWARD MERGER

Paralleling the growth of Combustion Engineering in this period was the Locomotive Superheater Company, founded in 1910 to further the use of superheated steam in the boilers of American locomotives. This company had established a manufacturing facility in East Chicago, Indiana, where today Combustion Engineering assembles coal pulverizing equipment. With the development in 1914 of a patented forged return bend for joining two tubes, the Locomotive Superheater Co. produced heat exchangers that very quickly were accepted as a standard in the railroad industry. Virtually all of the steam locomotives subsequently built in the U.S. were equipped with these superheaters. In 1917, The Superheater Company, as it became known, began designing and building superheaters for use in stationary boilers and the superheater design achieved prominence in the stationary boiler field. It was here that the paths of Combustion Engineering and the Superheater Company crossed.

During the Great Depression, Combustion Engineering was reorganized as a subsidiary of The Superheater Company. Close ties developed between these two organizations, leading to the eventual merger of the two in 1948 under the name of Combustion Engineering-Superheater, Inc. In 1953 the name Superheater was eliminated and the Company took the name by which it's known throughout the world today—Combustion Engineering Inc.

C-E: PIONEER IN BOILER DESIGN

After its entry in 1925 into the steam-generating equipment business, Combustion made many significant contributions to furnace design and firing techniques.

Traditionally, boilers had refractory-lined furnaces. But with the advent of pulverized-coal firing, refractory-lined furnaces could not tolerate the high heat-release rates without severe ash-fouling and erosion problems. Consequently, it became necessary to cool the furnace, at the same time protecting the refractory, by lining its inside surfaces with tubing through which water flowed as part of the boiler circulating system. Combustion was in the forefront of this development.

Initially, bottom water screens shielded only the furnace floor as in the Oneida and Lakeside Stations of Milwaukee Electric in 1920. In 1923 at the Springdale Station of West Penn Power Co., water-tube use was extended to the rear furnace walls. The first installation of water tubes on furnace side walls was also in 1923 at Consolidated Edison's Hell Gate Station. In 1925 the Cahokia Station of Union Electric became the first boiler with bottom, rear and side-wall water tubing.

Water-cooling technology matured from bare tubes on wide spacing to close-spaced tubes with welded fins, then to tangent tubes presenting a completely water-cooled metal surface to the flame. Today's modern boilers have fusion-welded furnace walls, a C-E development in 1950 at the Kearny Station of Public Service Electric and Gas of New Jersey. During this era steam generator sizes expanded rapidly, until in 1929 Combustion put into service at the East River Station of New York Edison Co., (now Consolidated Edison) the first unit to develop steam at 1,000,000 lb/hr.

As the search for higher efficiency of utility plants continued, operating pressures and temperatures also rose. In the mid-1920's most utility boilers supplied steam to turbines at 400 psig, with a few as high as 600 psig. By the late 20's, some boilers were sold for throttle pressures of 1200 pounds per square inch, including C-E boilers supplied to Milwaukee Electric and Kansas City Power and Light. In this same period, steam temperatures increased gradually from between 500°F and 600°F to the 700°F to 750°F range, a substantial advancement considering the materials available at the time.

PULVERIZED FIRING DEVELOPS

The pioneering work on pulverized-coal firing at Milwaukee was followed in 1927 by the introduction of tangential firing in which fuel and air are introduced into a furnace from its four corners instead of through front or rear walls. First installed commercially at the U.S. Rubber Co. plant in Detroit, Michigan, this firing system provides the optimum in rapid and intimate mixing of fuel and oxygen to promote efficient combustion. Later, in 1940, a major modification to the basic tangential firing idea was applied to a boiler at Duke Power Co's., Buck Station—the fuel and air nozzles in the corners were made vertically tiltable. The new design permitted the flame envelope created by the tangential action to be moved up and down within the furnace, thereby changing the heat-absorption pattern of the furnace wall. The resulting variation in temperature of the hot gases entering the superheater surfaces provided a means of controlling steam temperature. C-E has had such tangential firing as its preferred method of suspension firing of most fuels for over 50 years.

Another equipment design to become a standard in the industry is the C-E Bowl Mill, introduced commercially in 1933. Originally based on the Raymond technology, this pulverizer design has been modified and increased in size over the years from a coal capacity of 3500 lb/hr coal to more than 100 tons/hr to keep pace with the requirement for supplying fuel to larger and larger steam generators.

WELDED DRUMS LEAD TO HIGHER PRESSURES, TEMPERATURES

A major contribution to the development of modern steam generators was the welded boiler drum. Until the late twenties, heavy pressure vessels were made of formed plates riveted together with the joints caulked to prevent leakage. This construction physically limited operating steam pressure.

At its Chattanooga plant, Combustion developed and perfected processes, techniques, and machines for fusion welding of heavy plate to form cylindrical shells for drums. The first such welded drum was dedicated as a National Historic Landmark by ASME on May 2, 1980—50 years after the drum was successfully tested. Welded construction permitted steam pressures to increase and, in 1931, Combustion installed the first boiler to operate at 1800 psig at the Phillip Carey Co. in Lockland, Ohio. By 1953, C-E had designed the first 2650 psig unit for the Kearny Station of Public Service Electric and Gas of New Jersey.

With improved endurance of materials that could tolerate higher operating temperatures and pressures, outlet steam temperatures began to climb from the 750°F level. By 1939, steam temperatures reached 925°F when C-E installed another pioneering unit at the River Rouge plant of Ford Motor Co.

The first unit to exceed 1000°F total steam temperature was sold to the Public Service Electric and Gas of New Jersey for its Sewaren Station in 1949. This boiler was designed to operate at 1050°F at the superheater outlet. Steam temperatures continued to inch upward; the Kearny unit of the same utility, purchased in 1953, was designed to operate at 1100°F and 2376 psig at the superheater outlet and 1050°F at the reheater outlet. In 1954, the C-E supercritical pressure unit for Philadelphia Electric Co's. Eddystone Station was designed for outlet conditions of 1210°F at 5300 psig.

In the search for higher efficiencies, the utility industry adopted the reheat cycle. This concept involved reheating steam in a section of the boiler after some of its energy had been extracted through expansion in the initial section of the turbine. The reheated steam, now at or near the initial steam temperature, was then returned to the final section of the turbine.

Although a few reheat units had been built in the 1920's, it was not until after World War II that the surge toward reheat in the utility industry began in earnest. Combustion sold the first post-war reheat unit to Boston Edison Co. for its Edgar Station in 1947. In the following years, reheat designs became the norm, with C-E supplying more than 630 reheat units to utility users around the world.

CIRCULATING SYSTEM ADVANCES

Combustion Engineering designed and installed its first Controlled Circulation® steam generator at Montaup Electric Co. in Somerset, Massachusetts in 1942. In this design, controlled circulation of water is assured by using pumps to provide a positive flow through the heat-absorbing tubes in the furnace walls; the available pump head permits the use of distribution orifices in the tube circuits. The Controlled Circulation steam generator overcomes the problem of decreasing thermal-circulation effect as the operating pressure approaches the 3208 psia critical point. This positive-circulation design has been well accepted in the utility industry around the world.

A license agreement was signed in 1953 with Sulzer Brothers of Switzerland for the rights to build and sell their Monotube steam-generator design in the United States. (The Monotube design is a once-through type steam generator in which the water that is introduced at the inlet of the unit passes continuously through the tubing, where it is heated to the desired outlet temperature with no internal recirculation.) In 1954, The Philadelphia Electric Company purchased the first unit of this design for its Eddystone Plant. This unit was designed to operate at 5300 psig and 1210°F at the superheater outlet, the highest pressure and temperature for a unit of commercial size ever. The unit went into operation in 1959, followed by several other supercritical and subcritical pressure installations of the Monotube type.

Combustion's next advancement in high-pressure technology area was the Combined Circulation® design which uses the principle of Controlled Circulation in the furnace walls (recirculation of the fluid by pumps) superimposed on the once-through flow of the Monotube design. This design eliminated the Monotube-boiler requirement for a high through-flow in the furnace walls during start-up and the necessity for a correspondingly large turbine bypass system. The first unit of this design was installed at the Tennessee Valley Authority's Bull Run Station and placed in operation in 1966. It was designed for 3650 psig superheater outlet pressure at 1003°F superheater and 1003°F reheater outlet temperatures. The steam capacity of 6,400,000 lb/hr served a turbine capable of generating 900 megawatts. Many similar units of this type followed.

BROADENED CAPABILITIES

Combustion Engineering developed and produced air preheaters of the plate type in the early 1920's to provide heat recovery from the hot gases leaving boilers, thereby improving boiler efficiency. During the same time, the Ljungstrom continuous regenerative-type air preheater was developed, in connection with the first turbine locomotive. In the U.S., the first Ljungstrom installations for industrial power plant boilers were made in 1923.

In 1925, the Air Preheater Corporation was founded, jointly owned by the Ljungstrom Turbine Manufacturing Co. of Sweden and the James Howden Co., Ltd. The Superheater Company acquired this company after the Depression, along with the exclusive license to manufacture the Ljungstrom® Air Preheater in the United States and Canada. The Ljungstrom Air Preheater utilizes a rotating heating element made up of closely spaced metal plates packed in baskets through which alternately pass the hot gases leaving the boiler and the cold combustion air going to the furnace.

This air preheater design, manufactured in C-E's Wellsville, New York facility, has been universally accepted in the industry and has been applied to steam generators of all manufacturers.

Combustion Engineering pioneered flue-gas scrubbers for sulfur-dioxide removal with research and development work begun in 1963. A pilot facility was constructed in its Windsor, Connecticut, development laboratory, followed in 1966 and 1967 by experimental field work on a Detroit Edison Co. unit at the St. Clair Station. C-E made the first commercial scrubber installations in the United States in 1968 at Union Electric Co's Meramec Station and Kansas Power and Light Co's. Lawrence Station.

In 1977, C-E Walther, Inc., a subsidiary of C-E acquired Pollution Control-Walther, an American company producing the Walther (Cologne, Germany) design of rigid-frame electrostatic precipitators. With this acquisition, Combustion broadened its scope of supply in environmental control systems, which now includes fabric filters, dry sulfur-dioxide removal systems, wet electrostatic precipitators, rod scrubbers and wet scrubbers.

It was also in 1977 that C-E re-entered the ash-handling field, a move aimed toward providing full-scope ash systems for pulverized coal-fired boilers and associated precipitators, fabric filters, or dry scrubbers. Beginning with the Combusco water-sealed ash drag conveyor designed in 1919 by the Underfeed Stoker Company, Ltd., Combustion furnished many bottom-ash hoppers and removal systems, and pyrite hoppers for C-E pulverizers. Today, it is a prime supplier of large submerged scraper conveyors, under license from its associated company EVT of Stuttgart, Germany.

Meanwhile in 1955, C-E's nuclear power activity had begun with the design and construction of a prototype marine nuclear propulsion system. Going into operation in 1959, the S1C prototype operated for more than 10 years, both as an R&D and as a Naval training facility. In 1967, another C-E reactor, the BONUS plant, achieved full operation as the first nuclear power plant in the United States with an integral superheating core. Also in 1967, the Company sold its first nuclear steam supply system to the utility industry with Consumers Power's purchase of its Palisades nuclear station.

During the years, C-E continued to diversify into new markets while strengthening its position in traditional ones. A case in point is the Company's activities at the international level which began in 1923. Since that time, Combustion—through direct sales, associated companies, and licensees—has gained world-wide reputation as a leading supplier of steam-generating equipment with installations in more than 70 countries. About 40 percent of the free world's electric power currently is generated by equipment of C-E design.

THE MODERN C-E: GROWTH AND DIVERSIFICATION

Industrial expansion and general business growth characterized the U.S. national economy during the 1950's and 60's. Reflecting confidence in the strength of its position in the utility and industrial power fields, C-E took steps to expand its range of products and services for a changing and growing marketplace.

During these eventful years, C-E transformed from a manufacturer of steam supply systems into a multifaceted organization deeply involved in building products, oil and gas exploration, production, refining, and petrochemical plants. Starting with The Lummus Company acquisition in 1957, Combustion Engineering has operated with the policy that successful diversification is built upon technology or markets familiar to C-E.

In 1990, C-E became, by acquisition, a wholly-owned subsidiary of Asea Brown Boveri Inc., a Delaware corporation. Thus, C-E is now part of the ABB Group, one of the world's largest electrical engineering companies. Backed by some 215,000 employees in 140 countries worldwide, the ABB Group is helping to provide affordable electricity, industrial competitiveness, environmental protection, mass transportation, and a variety of industrial services.

C-E, as a member of the ABB Group, is well positioned to meet the world's energy challenges head-on by offering to energy-related industries, both nationally and internationally, equipment and services in a blend of technologies.

The combination of Asea Brown Boveri Inc. and C-E brings together two companies with complementary capabilities and reputations as industry and market leaders. The combined U. S. operations of Asea Brown Boveri Inc. and C-E represent a total of 30,000 employees.

THE YEARS AHEAD

We have reviewed the history of C-E as the company came of age and matured in the utility and industrial markets, and have examined its many significant contributions. Advances made over the last three quarters of a century in the development of energy conversion processes from fossil fuels have been great. Those individuals who have worked in the industry during some part of that time have witnessed not only tremendous growth in the art and science of fuel burning and steam generation, but also remarkable changes in the lifestyles of people throughout the world brought about in part by that growth. There are underway today new developments which will open new vistas for fossil fuels. That is what much of this book is all about.

As a member of the ABB Group, C-E continues to embrace the values that have kept it at the cutting edge of technology. A dedication to service and the courage to lead—attributes which have been major influences throughout its history—still guide the company's operations today. As it faces the technical challenges of our energy future, C-E, as a member of the ABB Group will continue to build on its reputation by supplying only the highest quality steam-generating and other energy equipment.

CHAPTER 1

Steam Power-Plant Design

The design of equipment for a fossil-fueled steam supply system starts with the application of certain basic fundamentals governing the relationship between properties of matter that define the conversion of energy from one form to another. Known as the first and second laws of thermodynamics, these cornerstones of power-plant design provide a quantitative method of looking at the sequential processes of working fluids as a function of temperature, pressure, enthalpy, and entropy. The design of a modern-day power plant represents, however, more than the application of thermodynamic data. It is a synthesis of economic considerations with thermal performance criteria that govern the selection of steam-generating equipment, whether the installation is for producing steam at an industrial site or utility power plant.

In this first chapter, which presents an overview of power-plant design, the opening section explores basic thermodynamic cycles: Carnot, Rankine regenerative reheat and cogenerative cycles are covered. A second section describes certain general design approaches relating to industrial installations. The third section deals with the principal factors involved in the selection of steam-generating equipment for electric utilities. Finally, the concluding section examines power-plant economics.

THERMODYNAMIC CYCLES FOR POWER PLANTS

Historically, the roots of thermodynamics go back to attempts to quantify instinctive concepts of systems designed to produce work from various energy sources. For continuously converting heat into work, the first practical power systems used steam as a working fluid.

By definition, a thermodynamic cycle is a series of processes combined in such a way that the thermodynamic states at which the working fluid exists are repeated periodically. Customarily, in an electric generating station, the fluid is cycled through a sequence of processes in a closed loop designed to maximize generation of electric power from the fuel consumed consistent with plant economics.

The design of a specific power plant represents an optimization of thermodynamic and economic considerations, the latter including initial, production, and distribution costs. In this first chapter, basic thermodynamics and economics applicable to large utility and industrial power plants will be presented. These form the basis for the selection of plant cycles and equipment. In this connection, it may be mentioned that an outstanding treatment of the subject is given in the Zerban and Nye book *Power*

Plants, 1960 edition, particularly with reference to Chapter 2.

LAWS OF THERMODYNAMICS

The first and second laws of thermodynamics govern the thermodynamic analysis of fluid cycles. These laws are stated in equation form as follows:

FIRST LAW

$$\Delta E = Q - W + \sum_i (\pm h_i \pm e_{xi}) m_i$$

(1)

where ΔE is change in energy content of system, Q is heat transferred *to* system, W is work transferred *from* system and $(h_i + e_{xi})m_i$ is energy convected into or out of system by mass, m_i, with enthalpy, h_i, and extrinsic energy, e_{xi}.

Extrinsic energy, e_{xi}, is dependent on the frame of reference. For a fluid system, e_{xi} = kinetic energy + potential energy = $V_i^2/2_{gj} + z_i$, in which V_i is velocity and z is elevation above datum.

This equation applies equally to processes and cycles, steady- and transient-flow situations. For example, in a closed system where fluid streams do not cross the boundary, $m_i = 0$, and if the process is cyclic, then $\Delta E = 0$ and Eq. 1 becomes

$$\sum_{cycle} Q = \sum_{cycle} W$$

(2)

This implies 100 percent efficiency.

As another example, if the steady-state adiabatic expansion in a turbine is being analyzed, then $\Delta E = 0$, $Q = 0$, and Eq. 1 reduces to:

$$W = - \sum_i (h_i + e_{xi})m_i$$

$$= m[(h + e_x)_{in} - (h + e_x)_{out}]$$

$$W = m\left[\left(h + \frac{V^2}{2_{gj}} + z\right)_{in} - \left(h + \frac{V^2}{2_{gj}} + z\right)_{out}\right]$$

(3)

When changes in kinetic energy and elevation of the fluid stream may be neglected, Eq. 3 reduces to the familiar $W \simeq m(h_{in} - h_{out})$.

SECOND LAW

$$\Delta S = \left(\frac{Q}{T}\right) + I + \sum_i S_i$$

(4)

where ΔS is change in entropy of system; $(Q/T) = \Sigma_i (Q_i/T_i)$ i.e. the sum over the system boundaries of the heat transferred, Q_i, at a position on the boundary where the local temperature is T_i, and I is irreversibility. (For consistency with other second-law statements, $I \geqq 0$. For reversible processes or cycles, $I = 0$; for irreversible processes or cycles, $I > 0$.) $\Sigma_i S_i$ is the entropy flow into and out of system associated with mass flow, m_i, into and out of system. For a reversible cyclic process involving a closed system, $I = 0$, $\Delta S = 0$, $\Sigma_i S_i = 0$ and Eq. 4 reduces to

$$\left(\frac{Q}{T}\right) = 0$$

(5)

The steady-flow, adiabatic expansion of a fluid through a turbine is governed by the following equation.

$$I = - \sum_i m_i s_i = -m(s_{in} - s_{out})$$

$$= m(s_{out} - s_{in})$$

(6)

For reversible, adiabatic expansion, $I = 0$ and the process is characterized by the familiar isentropic property, $s_{out} = s_{in}$. Turbine expansion is not wholly isentropic. This nonisentropicity is taken into account by defining the turbine efficiency. The primary advantage of writing the second law in an equation (Eq. 4) rather than the usual inequality is its usefulness in analyzing processes and cycles in a direct, quantitative, manner, similar to the first-law analysis.[1]

Eqs. 1 and 4 thus provide quantitative means of examining all processes encountered in power-plant analysis regardless of the fluids used or the specific cycle employed.

CARNOT CYCLE

In 1824, Sadi Carnot, a French engineer, published a small, moderately technical book, *Reflections on the Motive Power of Fire.*[2] Carnot made here three important contributions: the concept of reversibility, the concept of a cycle, and the specification of a heat engine producing maximum work when operating cyclically between two reservoirs each at a fixed temperature.

The Carnot cycle consists of several reversible isothermal and isentropic processes, which may be viewed as occurring in either the nonflow or flow device shown in Figs. 1A and 1B. In the first instance, the heat source and sink are placed in contact with the device to accomplish the required isothermal heat addition (a-b) and rejection (c-d) respectively. The insulation shown replaces the heat reservoirs for executing the reversible adiabatic processes involving expansion (b-c) and compression (d-a). The process characteristics for good heat transfer and work transfer are not the same and are partially in conflict; Fig. 1B, therefore, shows a flow system for executing the Carnot cycle with the work and heat-transfer processes assigned to separate devices. For both nonflow and flow systems, the state changes experienced by the working fluid are shown in the temperature-entropy diagram of Fig. 1C.

The classic Carnot cycle is such, then, that no other can have a better efficiency between the specified temperature limits than the Carnot value. Other cycles may equal it, but none can exceed it.

$$\text{Carnot efficiency} = \frac{T_h - T_l}{T_h} = 1 - \frac{T_l}{T_h,} \tag{7}$$

where T_h is the temperature of the heat source and T_l is the temperature of the heat sink, all in terms of absolute temperature.

Practical attempts to attain the Carnot cycle encounter irreversibilities in the form of finite temperature differences during the heat-transfer processes and fluid friction during work-transfer processes. The compression process *d-a*, moreover, is difficult to perform on a two-phase mixture and requires an input of work ranging from a fifth to a third of the turbine output. When realistic irreversibilities are introduced, the Carnot cycle net work is reduced; the size and cost of equipment increase. Consequently, other cycles appear more attractive as practical models.[3]

In relation to the Carnot efficiency, the high-temperature heat source cannot be defined in terms of maximum temperature. Instead, the weighted average of the temperature of the

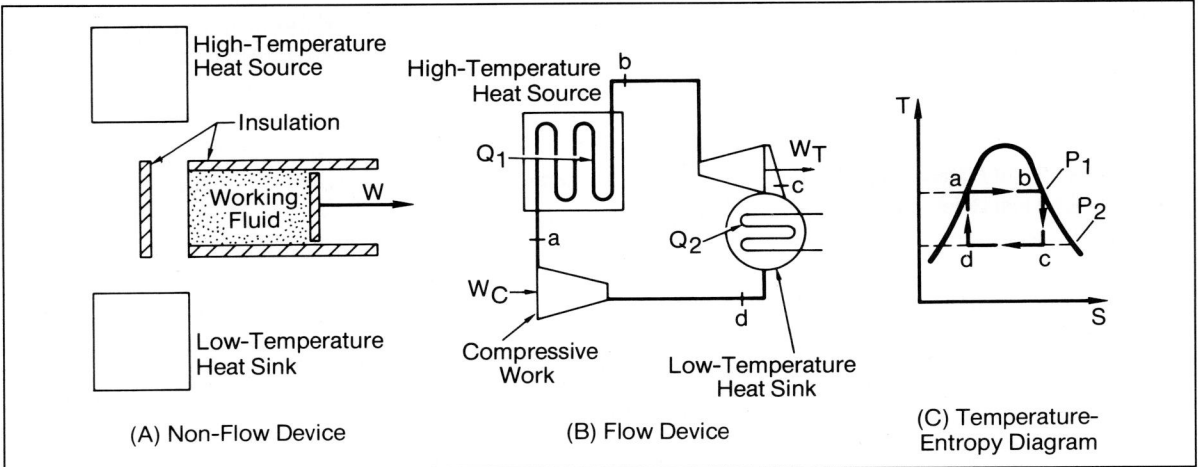

Fig. 1 **Carnot cycle**

(A) Non-Flow Device   (B) Flow Device   (C) Temperature-Entropy Diagram working fluid must be calculated, involving the heating of the feedwater as it leaves the last heater, and the evaporation, superheating, and reheating processes.

RANKINE CYCLE

The cornerstone of the modern steam power plant is a modification of the Carnot cycle proposed by W. J. M. Rankine[4], a distinguished Scottish engineering professor of thermodynamics and applied mechanics. The elements comprising the Rankine cycle are the same as those appearing in Fig. 1B with one exception: because the condensation process accompanying the heat-rejection process continues until the saturated liquid state is reached, a simple liquid pump replaces the two-phase compressor. The temperature-entropy and enthalpy-entropy diagrams of Fig. 2 illustrate the state changes for the Rankine cycle. With the exception that compression terminates (state a) at boiling pressure rather than the boiling temperature (state a'), the cycle resembles a Carnot cycle. The triangle bounded by a-a' and the line connecting to the temperature-entropy curve in Fig. 2A signify the loss of cycle work because of the irreversible heating of the liquid from state a to saturated liquid. The lower pressure at state a, compared to a', makes possible a much smaller work of compression between d-a. For operating plants it amounts to 1 percent or less of the turbine output.

This modification eliminates the two-phase vapor compression process, reduces compression work to a negligible amount, and makes the Rankine cycle less sensitive than the Carnot cycle to the irreversibilities bound to occur in an actual plant. As a result, when compared with a Carnot cycle operating between the same temperature limits and with realistic component efficiencies, the Rankine cycle has a larger net work output per unit mass of fluid circulated, smaller size and lower cost of equipment. In addition, due to its relative insensitivity to irreversibilities, its operating plant thermal efficiencies will exceed those of the Carnot cycle.

REGENERATIVE RANKINE CYCLE

Refinements in component design soon brought power plants based on the Rankine cycle to their peak thermal efficiencies, with further increases realized by modifying the basic cycle. This occurred through increasing the temperature of saturated steam supplied to the turbine, by increasing the turbine inlet temperature through constant-pressure superheat, by reducing the sink temperature, and by reheating the working vapor after partial expansion followed by continued expansion to the final sink temperature. In practice, all of these are employed with yet another important modification. In the previous section, the irreversibility associated with the heating of the compressed liquid to saturation by a finite temperature difference was cited as the primary thermodynamic cause of lower thermal efficiency for the Rankine cycle. The regenerative cycle attempts to eliminate this irreversibility by using as heat

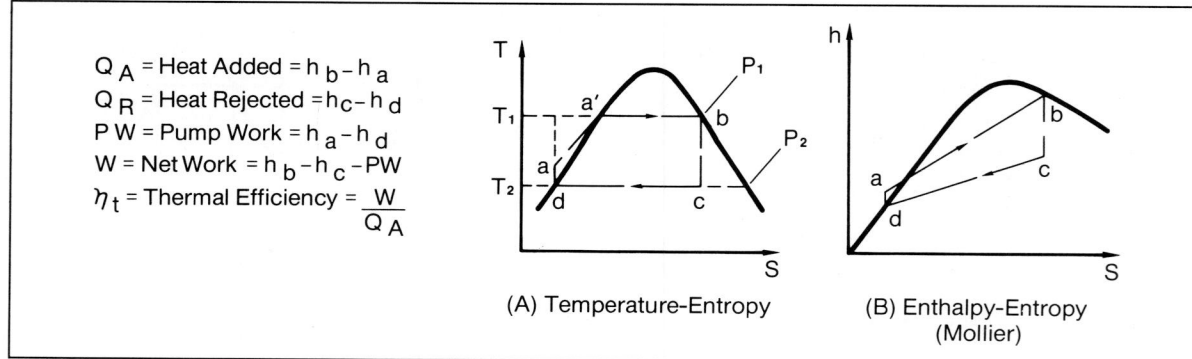

Q_A = Heat Added = $h_b - h_a$
Q_R = Heat Rejected = $h_c - h_d$
PW = Pump Work = $h_a - h_d$
W = Net Work = $h_b - h_c - PW$
η_t = Thermal Efficiency = $\dfrac{W}{Q_A}$

(A) Temperature-Entropy

(B) Enthalpy-Entropy
(Mollier)

Fig. 2 Simple Rankine cycle (without superheat)

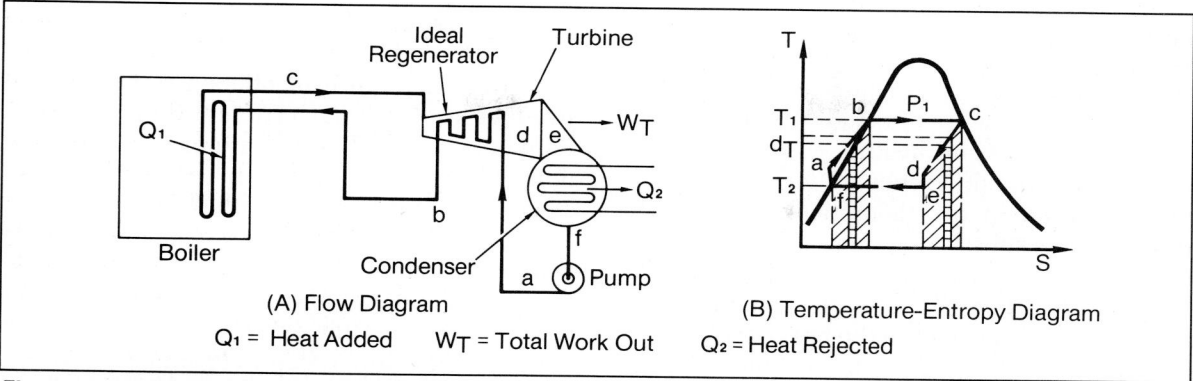

Fig. 3 Idealized regenerative Rankine cycle

(A) Flow Diagram
Q₁ = Heat Added   W_T = Total Work Out   Q₂ = Heat Rejected

sources other parts of the cycle with temperatures slightly above that of the compressed liquid being heated. Fig. 3 is an idealized form of such a procedure.

The condensed liquid at f is pumped to the pressure P_1, passes through coils around the turbine, and receives heat from the fluid expanding in the turbine. The liquid and vapor flow counter to one another and, by reversible heat transfer over the infinitesimal temperature difference d_T, the liquid is brought to the saturated state at T_1 (process b-c) and then rejects heat at the constant temperature T_2 (process e-f). Such a system, by the second law, will have a thermal efficiency equal to a Carnot cycle operating between the same temperatures.

This procedure of transferring heat from one part of a cycle to another in order to eliminate or reduce external irreversibilities is called "re-

generative heating" which is basic to all regenerative cycles. Though thermodynamically desirable, the idealized regenerative cycle just described has several features which preclude its use in practice. Locating the heat exchanger around the turbine increases design difficulties and cost. Even if these problems were solved, heat transfer could not be accomplished reversibly in the time available; further, cooling as described causes the vapor to reach excessive moisture content.

The scheme shown in Fig. 4 permits a practical approach to regeneration without encountering these problems. Extraction or "bleeding" of steam at state c for use in the "open" heater avoids excessive cooling of the vapor during turbine expansion; in the heater, liquid from the condenser increases in temperature by ΔT. (Regenerative cycle heaters are called "open" or

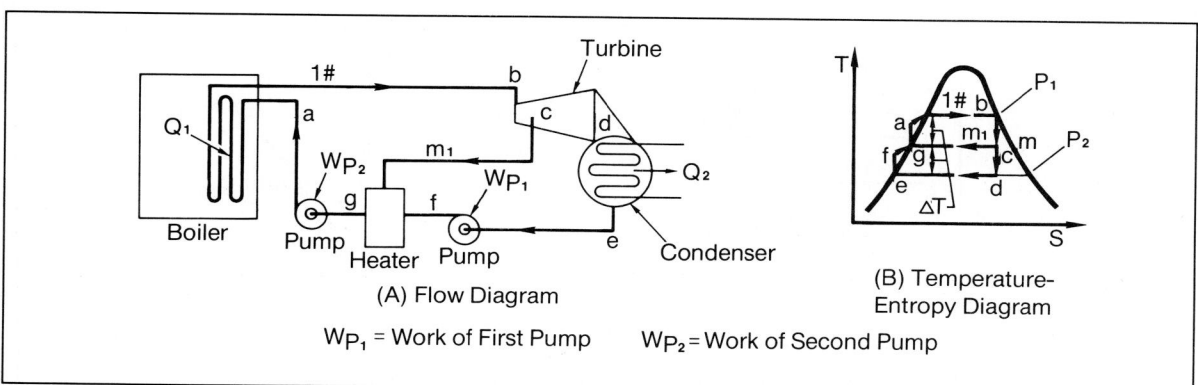

Fig. 4 Practical single-extraction regenerative cycle

(A) Flow Diagram
W_P₁ = Work of First Pump   W_P₂ = Work of Second Pump

"closed" depending on whether hot and cold fluids are mixed directly to share energy or kept separate with energy exchange occurring indirectly through the use of metal coils.)

The extraction and heating substitute the finite temperature difference ΔT for the infinitesimal dT used in the theoretical regeneration process. This substitution, while failing to realize the full potential of regeneration, halves the temperature difference through which the condensate must be heated in the basic Rankine cycle. Additional extractions and heaters permit a closer approximation to the maximum efficiency of the idealized regenerative cycle, with further improvement over the simple Rankine cycle shown in Fig. 2.

Reducing the temperature difference between the liquid entering the boiler and that of the saturated fluid increases the cycle thermal efficiency. The price paid is a decrease in net work produced per pound of vapor entering the turbine and an increase in the size, complexity, and initial cost of the plant. Additional improvements in cycle performance may be realized by continuing to accept the consequences of increasing the number of feedwater heating stages. Balancing cycle thermal efficiency against plant size, complexity, and cost for production of power at minimum cost determines the optimum number of heaters.[5]

REHEAT CYCLE

The use of *superheat* offers a simple way to improve the thermal efficiency of the basic Rankine cycle and reduce vapor moisture content to acceptable levels in the low-pressure stages of the turbine. But with the continued increase of steam temperatures and pressures to achieve better cycle efficiency, in some situations available superheat temperatures are insufficient to prevent excessive moisture from forming in the low-pressure turbine stages.

The solution to this problem is to interrupt the expansion process, remove the vapor for *reheat* at constant pressure, and return it to the turbine for continued expansion to condenser pressure. The thermodynamic cycle using this modification of the Rankine cycle is called the "reheat cycle." Reheating may be carried out in a section of the boiler supplying primary steam, in a separately fired heat exchanger, or in a steam-to-steam heat exchanger. Most present-day utility units combine superheater and reheater in the same boiler.

Usual central-station practice combines both regenerative and reheat modifications to the basic Rankine cycle. Fig. 5 is a temperature-entropy diagram of a single-reheat regenerative cycle with two stages of feedwater heating. For large installations, reheat makes possible an

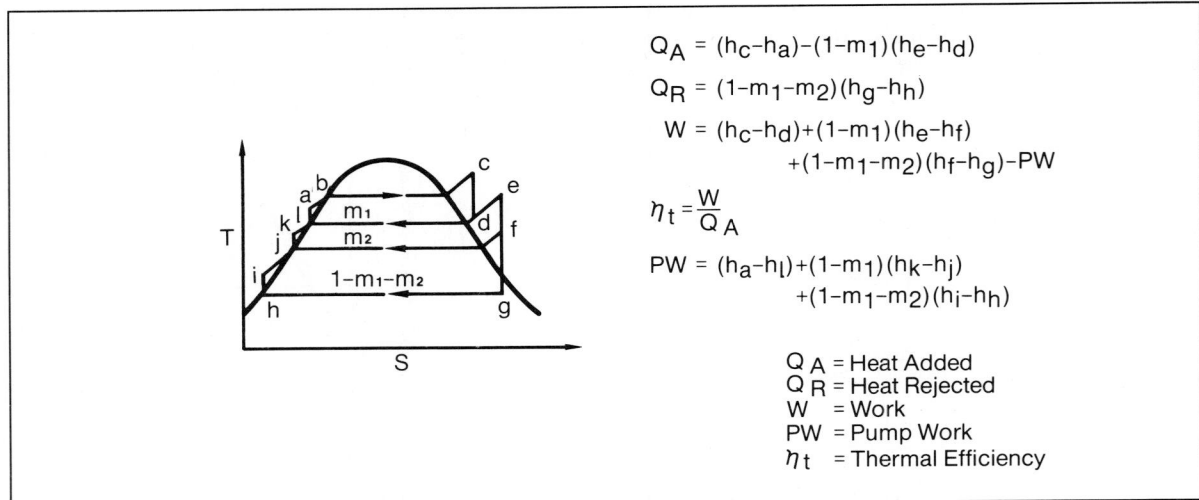

$$Q_A = (h_c - h_a) - (1 - m_1)(h_e - h_d)$$

$$Q_R = (1 - m_1 - m_2)(h_g - h_h)$$

$$W = (h_c - h_d) + (1 - m_1)(h_e - h_f)$$
$$+ (1 - m_1 - m_2)(h_f - h_g) - PW$$

$$\eta_t = \frac{W}{Q_A}$$

$$PW = (h_a - h_l) + (1 - m_1)(h_k - h_j)$$
$$+ (1 - m_1 - m_2)(h_i - h_h)$$

Q_A = Heat Added
Q_R = Heat Rejected
W = Work
PW = Pump Work
η_t = Thermal Efficiency

Fig. 5 **Regenerative cycle with single reheat and two feedwater heaters**

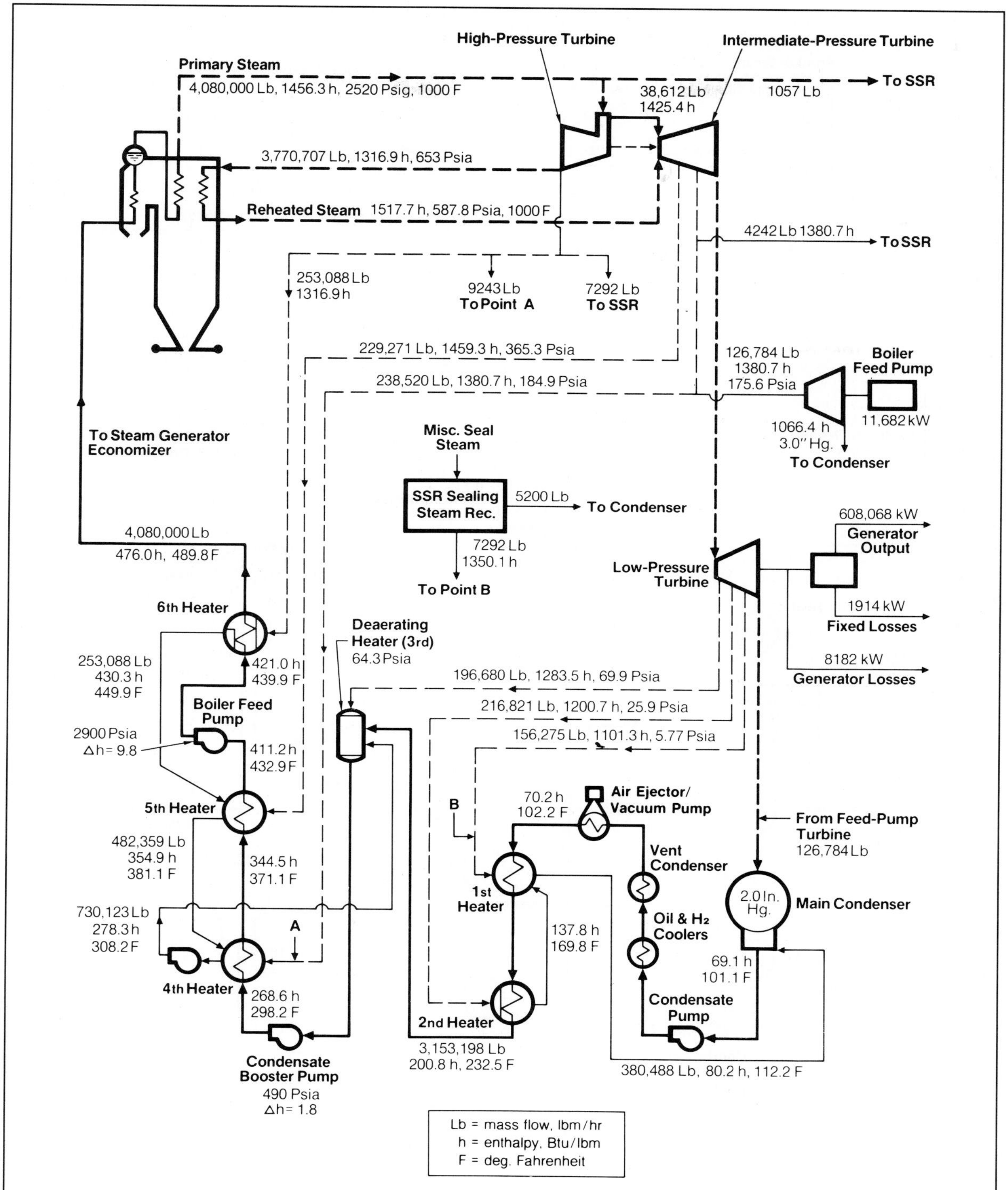

High-Pressure Turbine **Intermediate-Pressure Turbine**

Primary Steam
4,080,000 Lb, 1456.3 h, 2520 Psig, 1000 F

38,612 Lb
1425.4 h   1057 Lb   ➤ **To SSR**

3,770,707 Lb, 1316.9 h, 653 Psia

Reheated Steam 1517.7 h, 587.8 Psia, 1000 F

4242 Lb 1380.7 h   ➤**To SSR**

253,088 Lb
1316.9 h

9243 Lb   7292 Lb
To Point A   **To SSR**

229,271 Lb, 1459.3 h, 365.3 Psia

238,520 Lb, 1380.7 h, 184.9 Psia

126,784 Lb   **Boiler**
1380.7 h   **Feed Pump**
175.6 Psia

To Steam Generator Economizer

1066.4 h   11,682 kW
3.0" Hg.
To Condenser

Misc. Seal Steam

SSR Sealing Steam Rec.   5200 Lb   **To Condenser**

7292 Lb
1350.1 h

608,068 kW
Generator Output

To Point B

Low-Pressure Turbine

4,080,000 Lb
476.0 h, 489.8 F

1914 kW
Fixed Losses

6th Heater

Deaerating Heater (3rd)
64.3 Psia

196,680 Lb, 1283.5 h, 69.9 Psia

8182 kW
Generator Losses

253,088 Lb
430.3 h
449.9 F

421.0 h
439.9 F

216,821 Lb, 1200.7 h, 25.9 Psia

156,275 Lb, 1101.3 h, 5.77 Psia

Boiler Feed Pump

2900 Psia
Δh= 9.8

411.2 h
432.9 F

70.2 h
102.2 F

Air Ejector/ Vacuum Pump

5th Heater

B

Vent Condenser

From Feed-Pump Turbine
126,784 Lb

482,359 Lb
354.9 h
381.1 F

344.5 h
371.1 F

1st Heater

137.8 h
169.8 F

Oil & H₂ Coolers

2.0 In. Hg.   **Main Condenser**

730,123 Lb
278.3 h
308.2 F

A

4th Heater

268.6 h
298.2 F

2nd Heater

69.1 h
101.1 F

Condensate Pump

Condensate Booster Pump
490 Psia
Δh= 1.8

3,153,198 Lb
200.8 h, 232.5 F

380,488 Lb, 80.2 h, 112.2 F

Lb = mass flow, lbm/hr
h = enthalpy, Btu/lbm
F = deg. Fahrenheit

Fig. 6 **Reheat regenerative cycle, 600-MW subcritical-pressure fossil power plant (U.S. units)**

ABB
ASEA BROWN BOVERI

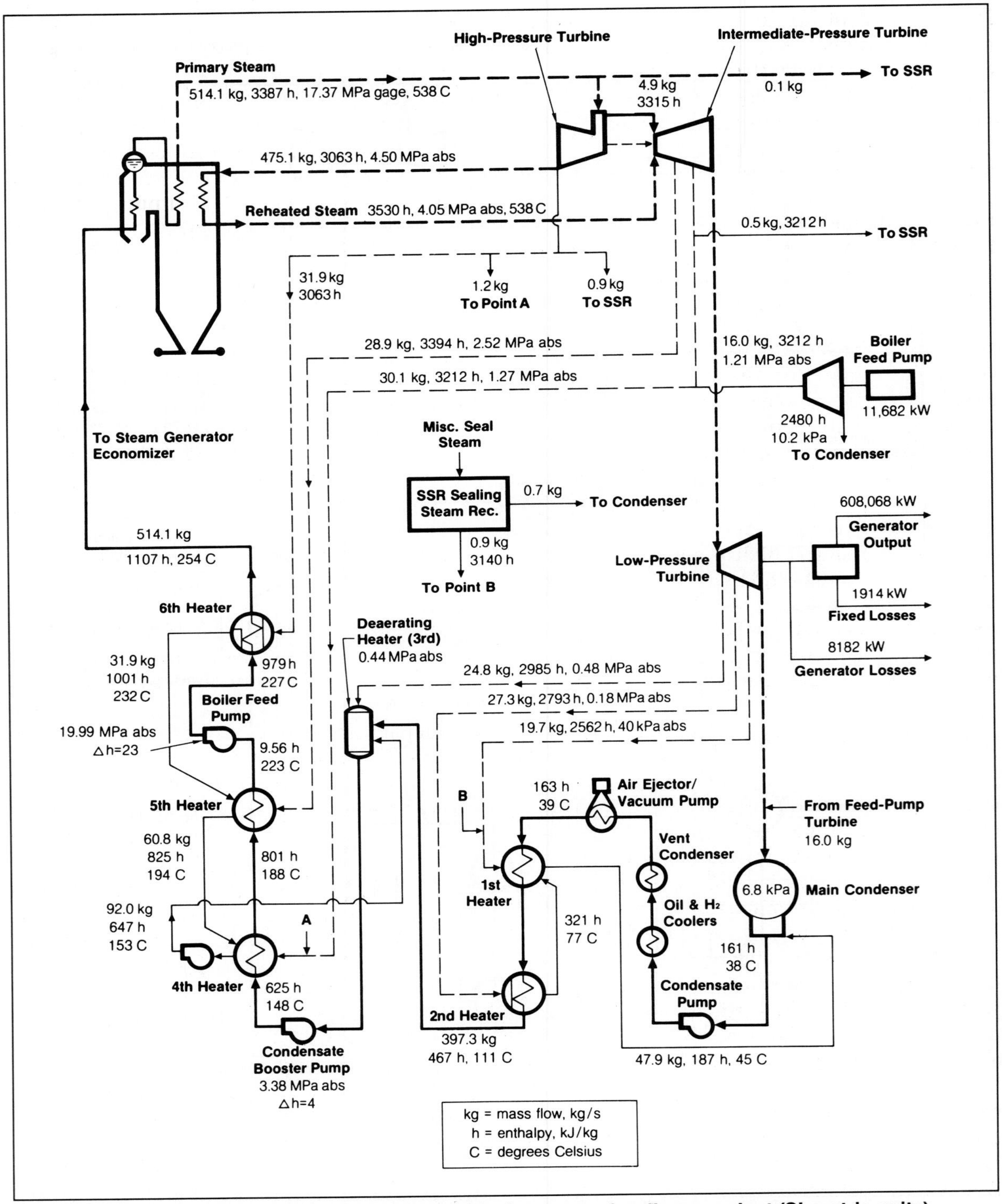

Fig. 7 Reheat regenerative cycle, 600-MW subcritical-pressure fossil power plant (SI-metric units)

improvement of approximately 5 percent in thermal efficiency and substantially reduces the heat rejected to the condenser cooling water.[6] The operating characteristics and economics of modern plants justify the installation of only one stage of reheat except for units operating at supercritical pressure.

Figs. 6 and 7 show a flow diagram for a 600-MW fossil-fueled reheat cycle designed for initial turbine conditions of 2520 psig and 1000°F (17.4 MPa gage and 538°C) steam. Six feedwater heaters are supplied by exhaust steam from the high-pressure turbine and extraction steam from the intermediate and low-pressure turbines. Except for the deaerating heater (third), all heaters shown are closed heaters. Three pumps are shown: (1) the condensate pump which pumps the condensate through oil and hydrogen gas coolers, vent condenser, air ejector, or vacuum pump, first and second heaters, and deaerating heater; (2) the condensate booster pump which pumps the condensate through fourth and fifth heaters; and (3) the boiler feed pump which pumps the condensate through the sixth heater to the economizer and boiler. The mass flows noted on the diagram are at the prescribed conditions for full-load operation.

SUPERCRITICAL-PRESSURE CYCLE

A definite relationship exists between operating temperature and optimum pressure of a cycle. The supercritical-pressure cycle is in use worldwide to obtain the highest possible thermodynamic efficiencies with fossil-fuel steam-generation equipment.

A regenerative-reheat cycle is used with 6 to 8 stages of feedwater heating and, because of the high inlet temperature and pressure, two stages of reheat can be justified and have been used in several installations. Figs. 8 and 9 show a typical supercritical-pressure steam turbine heat balance. Seven feedwater heaters are shown with steam extractions from the various turbines supplying energy for the regenerative heating processes. For such supercritical pressure plants, cycle efficiencies exceeding 40 percent have been reached.

Philadelphia Electric Company's Eddystone Unit 1, a double-reheat supercritical-pressure unit with original outlet steam conditions of 5300 psig, 1210°F (36.5 MPa gage, 654°C), was the high-water mark of cycle efficiency for conventional (non-binary cycle) steam plants. Although later operating at lower steam conditions, it produced electrical output with an input of less than 8200 Btu per kilowatt hour, corresponding to an overall thermal efficiency of about 42 percent.

COGENERATIVE STEAM CYCLES

In many instances, power and thermal-energy needs may be combined in a single power plant which will operate at a high annual load factor and higher thermal efficiency than obtainable by addressing the needs separately. Such a combined heat-and-power arrangement is termed a cogeneration plant, which will be discussed in detail in the following sections.

For heating service (such as in a district-heating system), the steam is generated at a pressure and temperature sufficiently high that the exhaust from the steam turbine is at steam conditions suitable for delivery to the steam mains or heat exchangers (for a high-temperature water system) and distribution to the users. The needs of industrial plants for process steam may be met by similar arrangements using either turbine exhaust or extracted steam from an appropriate turbine stage. The selection of optimum exhaust or extraction conditions will depend on the proportions of power and process heat required of the particular plant. Process-steam requirements are usually in the low-pressure range with modest superheat required if at all; consequently, initial steam pressure and temperature will be well within the limits of current technology.

Cogenerative steam cycles have been applied over a wide range of capacities extending from small industrial and institutional power plants to large central stations serving the electrical and heating needs of a metropolitan center or the power/process requirements of a major petro-chemical installation. The food-process-

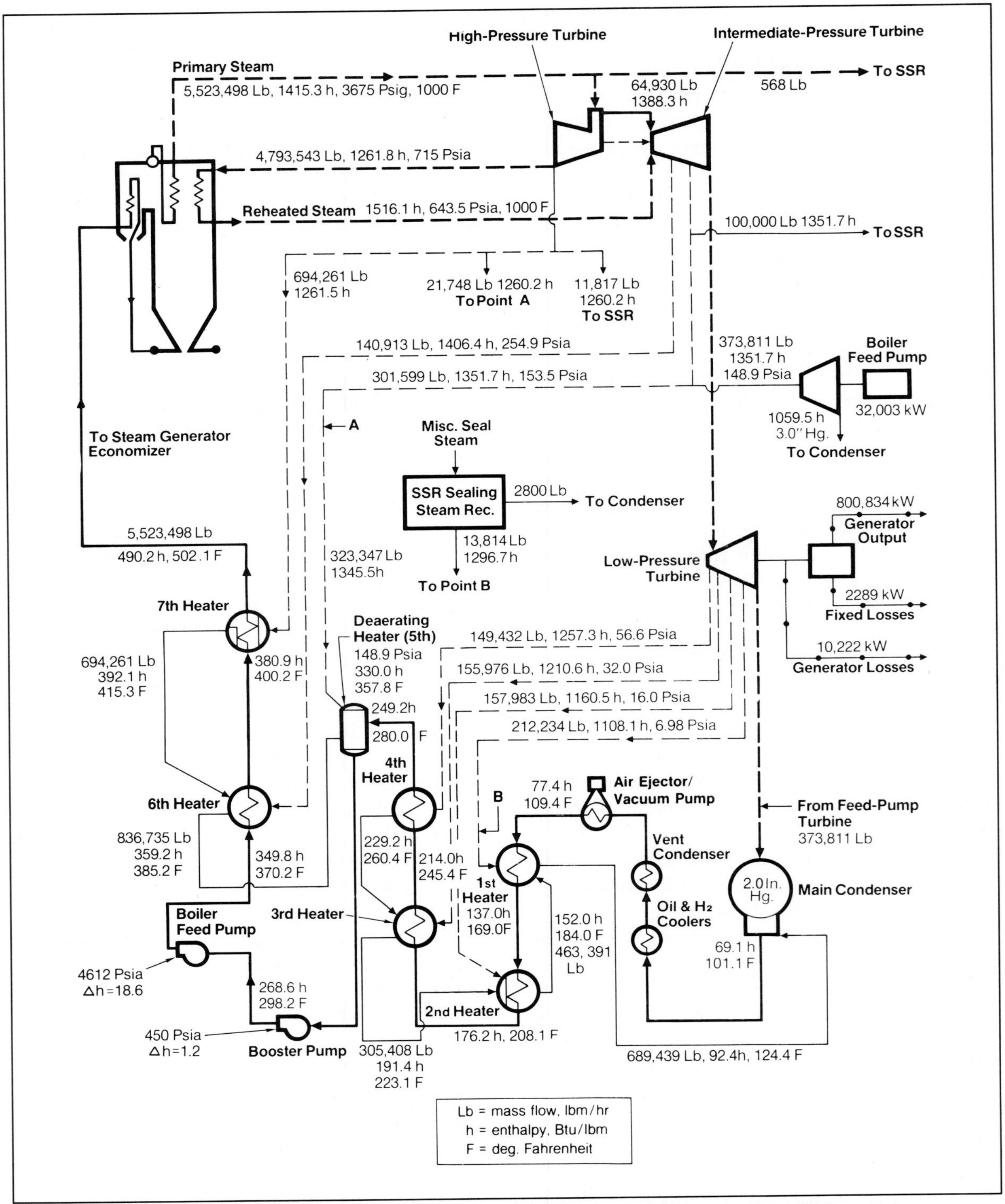

Fig. 8 Reheat regenerative cycle, 800-MW supercritical-pressure fossil power plant (U.S. units)

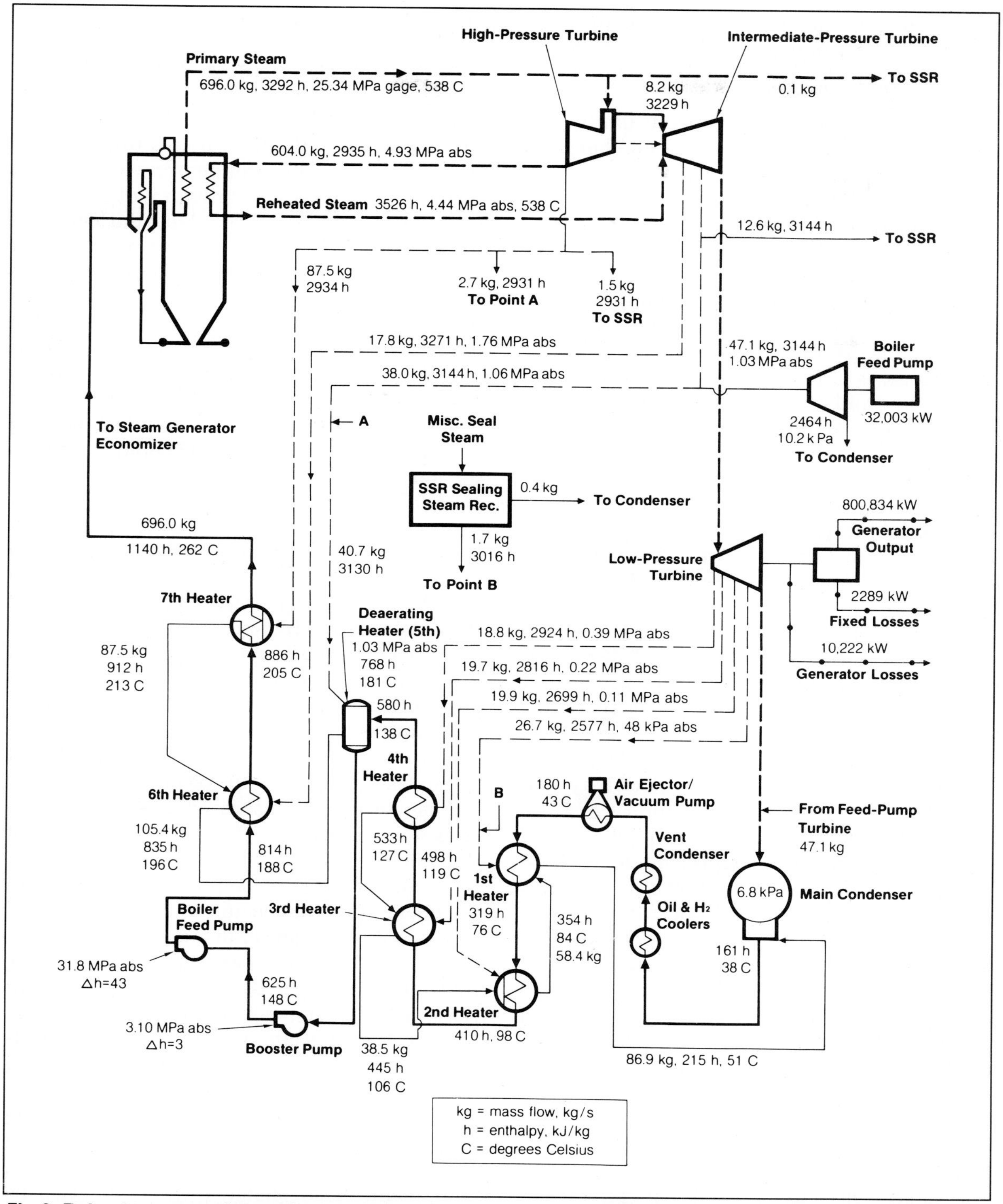

Primary Steam
696.0 kg, 3292 h, 25.34 MPa gage, 538 C

High-Pressure Turbine

Intermediate-Pressure Turbine

8.2 kg
3229 h

0.1 kg → **To SSR**

604.0 kg, 2935 h, 4.93 MPa abs

Reheated Steam 3526 h, 4.44 MPa abs, 538 C

12.6 kg, 3144 h → **To SSR**

87.5 kg
2934 h

2.7 kg, 2931 h
To Point A

1.5 kg
2931 h
To SSR

47.1 kg, 3144 h
1.03 MPa abs

Boiler Feed Pump

17.8 kg, 3271 h, 1.76 MPa abs

38.0 kg, 3144 h, 1.06 MPa abs

2464 h
10.2 k Pa
To Condenser

32,003 kW

To Steam Generator Economizer

A

Misc. Seal Steam

SSR Sealing Steam Rec. 0.4 kg → **To Condenser**

40.7 kg
3130 h

1.7 kg
3016 h
To Point B

800,834 kW

Generator Output

696.0 kg
1140 h, 262 C

Low-Pressure Turbine

2289 kW
Fixed Losses

10,222 kW
Generator Losses

7th Heater

87.5 kg
912 h
213 C

886 h
205 C

Deaerating Heater (5th)
1.03 MPa abs
768 h
181 C

18.8 kg, 2924 h, 0.39 MPa abs

19.7 kg, 2816 h, 0.22 MPa abs

19.9 kg, 2699 h, 0.11 MPa abs

26.7 kg, 2577 h, 48 kPa abs

580 h
138 C

4th Heater

6th Heater

105.4 kg
835 h
196 C

814 h
188 C

533 h
127 C

498 h
119 C

1st Heater
319 h
76 C

B

180 h
43 C

Air Ejector/ Vacuum Pump

Vent Condenser

354 h
84 C
58.4 kg

Oil & H₂ Coolers

From Feed-Pump Turbine
47.1 kg

6.8 kPa **Main Condenser**

3rd Heater

Boiler Feed Pump

31.8 MPa abs
△h=43

625 h
148 C

3.10 MPa abs
△h=3

Booster Pump

38.5 kg
445 h
106 C

2nd Heater
410 h, 98 C

161 h
38 C

86.9 kg, 215 h, 51 C

kg = mass flow, kg/s
h = enthalpy, kJ/kg
C = degrees Celsius

Fig.9 **Reheat regenerative cycle, 800-MW supercritical-pressure fossil power plant (SI-metric units)**

ing, chemical, and paper industries are three among the many industrial users of such cycles.

COGENERATION: THE DEFINITION

Cogeneration is the simultaneous production of power (usually electricity) and another form of useful thermal energy (usually steam or hot water) from a single fuel-consuming process. It is also referred to by the terms combined heat and power (CHP), combined power and process, or simply power/process. The fuel can be burned in a boiler, a gas turbine, or a diesel engine. It is important to realize that cogeneration, as a practical and time-honored technology, is a matter of design of cycles and their piping and valving systems, and not of basic differences in the boilers, turbines (steam or gas) and heat exchangers that are made part of the cycles.

A typical steam-based cogeneration plant burns fossil fuel (coal, oil, or natural gas, for example) or a renewable resource such as plant or animal waste material (biomass) to generate steam which is piped to a non-condensing turbine to drive a generator. Since the steam leaving the turbine retains much of its energy, it can be used further for heating or other heat-absorbing applications. For instance, every naval or merchant ship that uses heat for cooking, space-heating, or laundry operation, in addition to generating electricity and driving propellers, is a cogenerator.

The efficiencies of cogenerating steam and electricity can result in significant energy-cost saving for facilities as varied as a manufacturing or processing plant, an office or apartment complex, or a hospital. Steam or hot water can be used for heating or cooling of buildings, or for industrial processes. The power can be to compress air or gas for a variety of industrial purposes. Cogenerated steam can be injected into the ground in steam-flood (enhanced oil recovery) projects, or can be used as the heat source in absorption-type refrigeration plants. If the amount of energy produced exceeds the cogenerator's own needs, either steam/ hot water, electricity, or both, can be sold for use by others.

REGENERATION VERSUS COGENERATION: THE THERMODYNAMICS

The principal purpose of a *regenerative power-plant cycle* is the high-efficiency generation of electrical power, accomplished by extracting steam at multiple points from a steam turbine, ahead of the condenser. As shown in the diagrams of Figs. 6 through 9 of this chapter, the extracted steam-flows heat the feedwater in successive steps as it returns to the boiler from the condenser. As described earlier, regenerative cycles are essentially closed cycles, as the working fluid constantly recycles through the heat-exchange equipment. Makeup water is added only to replace steam or water lost in gland sealing or other outward leakage. As also described earlier, single or double reheating can be added to the basic regenerative cycle to further enhance thermal efficiency. In large electric-utility plants, 6 to 8 stages of feedwater heating are used to minimize heat rejection to the condenser, and there is a minimum of one stage of reheat.

Although a *cogenerative power-plant cycle* uses a similarly efficient arrangement of cycle elements (steam generator, turbine or turbines, feedwater heaters and other heat exchangers, and some modified form of condenser) it has a dual purpose. Not only does a cogeneration cycle produce electrical or shaft power, it also generates a stream or streams of thermal energy for use in a heat-absorbing or heated-fluid-consuming process, all from the combustion of the same fuel.

By minimizing, or completely eliminating, the energy loss that occurs in the water-cooled or air-cooled condenser of a regenerative cycle, cogenerative cycles can have significantly higher overall thermal efficiencies than conventional condensing regenerative systems. The level of efficiency attained depends on the heat-to-power ratio and the type of process forming the basis for the cycle.

Like a regenerative cycle, a cogenerative cycle can be "closed", with the heat energy being added to and taken from it in various pieces of equipment. A cogenerative cycle can also be semi-closed: in such a semi-closed or "exportive" cycle, some of the extracted heat leaves the cycle with fluid that is never returned from the heat-using process. Substantial amounts of makeup water, then, may be needed in the semi-closed cycle to allow for the loss of the treated water during the process.

Regenerative cycles are somewhat standardized, differing only slightly from one cycle designer to another. In contrast, cogeneration systems are unlimited in thermodynamic possibilities, and thus in the configuration and sizing of all the cycle elements. To a significant extent, then, the *process* determines the design, arrangement, and cost of components, as well as their pressure and temperature levels. Since in many geographical areas cogenerators are unregulated by public agencies in many aspects of their operation, they are similarly unlimited in economic possibilities.[7]

Achieving High Thermal Efficiency in a Fossil Power Plant

A principal thrust of the study of the thermodynamics of fossil-fired power plants is the maximization of thermal efficiency: that is, the most efficient production of power (usually electrical) from a supply of fuel that has chemical energy. In attempting to achieve such high thermal efficiency, the designer of a conventional Rankine cycle raises the temperature of heat supply to a turbine and lowers the rejection temperature, the latter being done by dropping the condenser pressure. The temperature of heat to the turbine is increased by

- raising boiler pressure
- reheating between turbine stages, and
- regeneratively heating the boiler feedwater,

all as described in previous sections.

Whether such modifications to achieve higher efficiency justify the extra capital required is a matter for economic study; thermodynamics is seldom the sole criterion. In maximizing the work output (W) for a given amount of fuel fired (F), the designer minimizes the heat rejection from the plant (Q_R). The designer of a *cogeneration* cycle is interested in using some of the heat rejected usefully

(as Q_U), so that $Q_R = Q_U + Q_{NU}$, where Q_{NU} is "non-useful" heat rejected; there is, then, less emphasis on maximizing thermal efficiency. But because electric power is a valued product (of greater value than useful heat rejected), a cogenerator still may want to produce that electricity at minimum cost, so it is important to be aware of the steps taken to improve conventional plant efficiency.[8]

EFFICACY OF COGENERATION PLANTS

There are many approaches to mathematically defining the true value of any given cogenerative cycle. The energy utilization factor (EUF) of Porter and Mastanaiah[9] is expressed as the sum of the work output (W) and the useful rejected heat (Q_U) divided by the fuel fired (F), all in consistent energy units. But since the work (or electrical power) produced and the useful heat will usually be of different monetary value, the EUF is not satisfactory as a criterion of *economic* performance. Horlock and others (references 8 through 12) have written extensively on the subject of comparing the economic values of various cogenerative cycles.

CLASSIFICATION OF COGENERATION PLANTS

Cogeneration projects can be classified as either *topping* or *bottoming* systems. In a topping system, the steam produces electricity first and all or a part of the exhausted thermal energy is then used in an industrial process, or to provide space heating or cooling. Bottoming cogeneration systems use the waste heat from industrial processes or other high-temperature thermal processes to generate electricity by generating steam in a waste-heat (heat-recovery) boiler.

Fig. 10 shows a schematic diagram of a basic closed power/process cycle using a backpressure turbine. After superheated steam is generated at a suitably high pressure, it is admitted to the turbine, does useful work (W_T), and emerges usually in the superheated state, c. After desuperheating, saturated steam, d, enters the heater and is entirely condensed. Because the steam required for power generation will not equal at all times that required for process

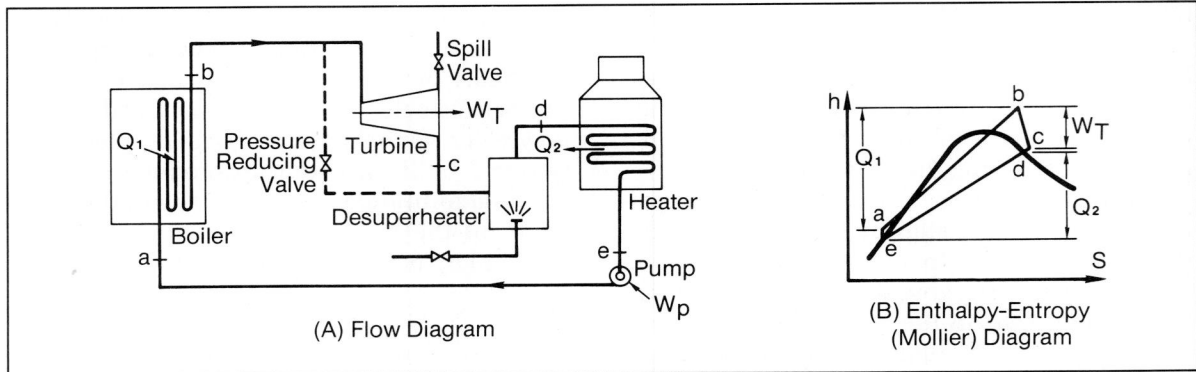

Fig. 10 Closed power/process cogenerative cycle using a backpressure turbine

work, some means of controlling the exhaust steam pressure must be employed to avoid variations in the pressure and, therefore, the steam saturation temperature. The control method depends on the circumstances. An ordinary centrifugal governor fitted to the backpressure turbine will cause the quantity of available exhaust steam to be controlled by the load on the turbine. Should the available exhaust be too small, live steam may be passed through a reducing valve into the desuperheater. If the quantity of exhaust steam exceeds requirements, then the excess steam may be blown to atmosphere, into an accumulator, or into a feed tank through the spill valve.

Figs. 11 and 12 illustrate a typical cogeneration system based on the steam topping cycle. It uses controlled-pressure automatic steam extraction and a controlled-backpressure turbine. This example has three closed feedwater heaters plus a deaerating heater.[13]

Such a steam-topping cogenerative cycle is suitable when large quantities of process steam and electricity are called for. This type of plant provides great potential for large industrial and power-producing installations, and incorporates steam generators of the types described in Chapter 8 of this text.

STEAM-GENERATOR DESIGN FOR A SPECIFIC CYCLE

The design of a steam-generating unit is independent of the regenerative or cogenerative cycle that it is part of. That is, the configuration and sizing of a boiler are based on

- generating a given amount of primary steam and reheat steam (if any) at a specified temperature and pressure.
- beginning with feedwater entering at a stated temperature
- with the reheat flow to the boiler being at a given temperature and heat content
- burning a particular fuel or fuels

The quantity of makeup water required by the cycle will of course affect the amount of chemicals needed for internal treatment of the boiler water. But it is the cycle or process equipment ahead of and following the boiler that places the generating plant in one category or the other. The actual regeneration or cogeneration, then, takes place *after* the boiler steam-outlet valves, resulting from the engineered ways in which the steam or hot-water flows are used.

The two types of cycles are, therefore, not mutually exclusive: there are many regenerative-cogenerative cycles in long-term operation throughout the world, from small shipboard installations to major district-heating/power-producing plants in Europe and America.

COMBINED GAS-TURBINE/STEAM TURBINE POWER CYCLES

A combined cycle is understood to be (at least in America) any one of a number of configurations of gas turbines, steam generators (or other heat-recovery equipment), and steam turbines

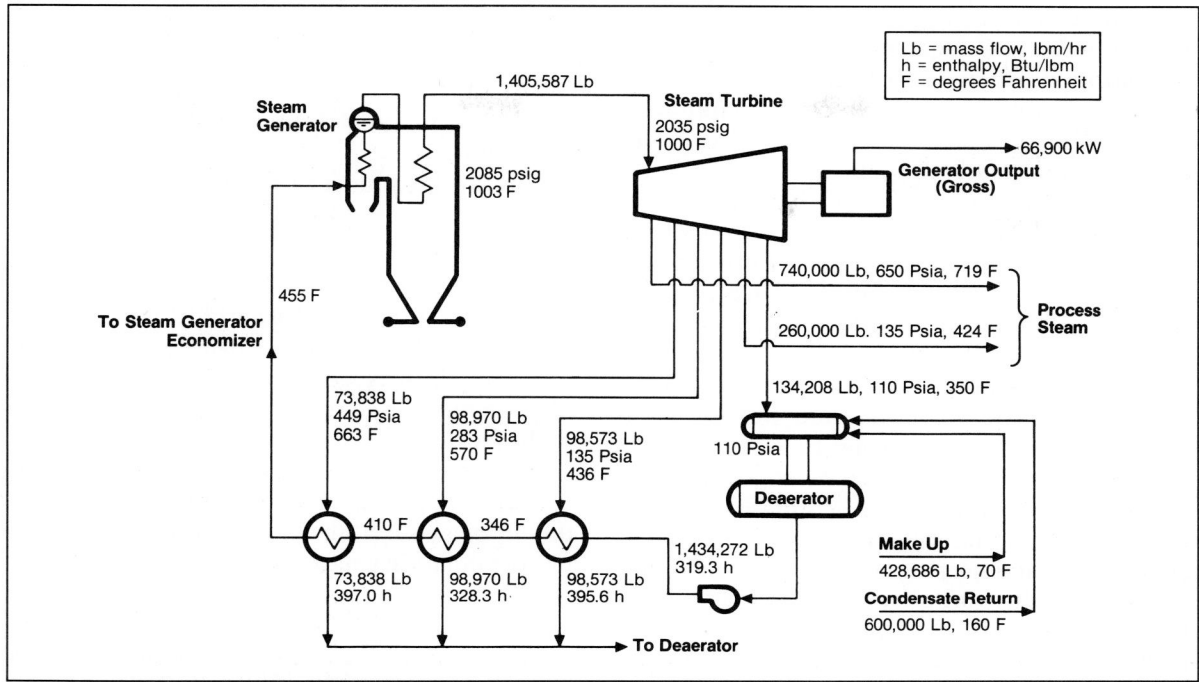

Fig. 11 Heat balance of a steam-topping cogenerative cycle (U.S. units).

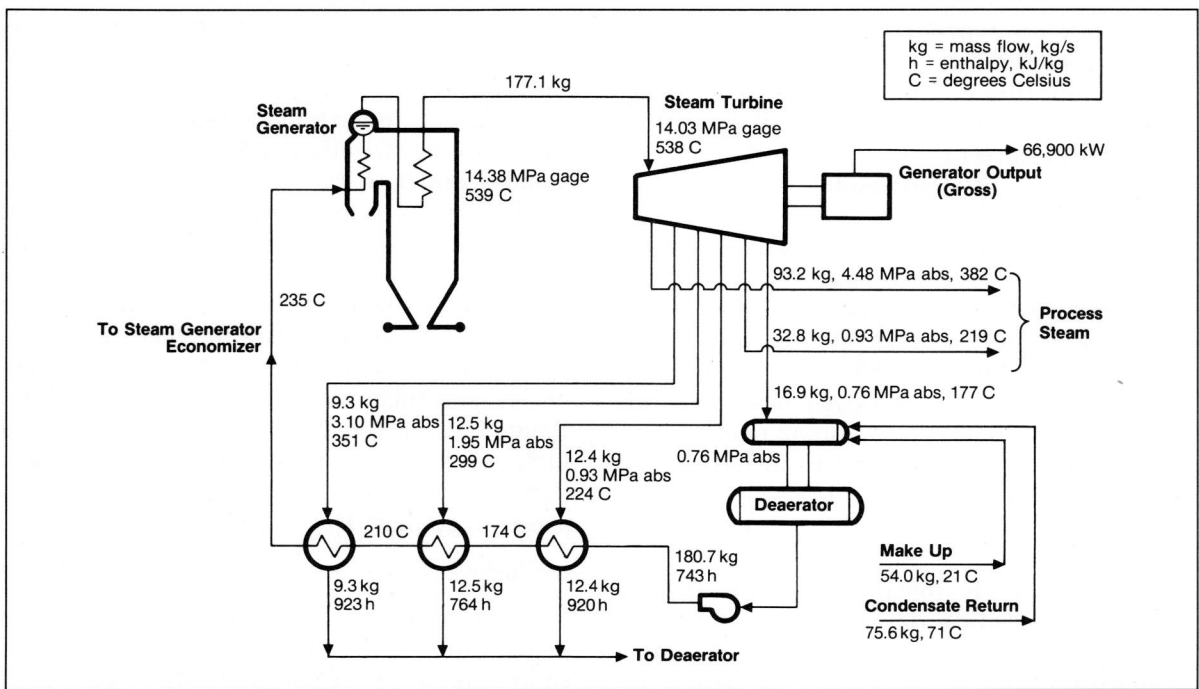

Fig. 12 Heat balance of a steam-topping cogenerative cycle (SI–metric units).

assembled for the improvement of cycle efficiency in the power-generation process.[14]

Analogous to mercury/steam, potassium/steam, or steam/ammonia cycles, the combined gas/steam power cycle is a binary cycle, although not normally thought of as such.[15,16] As a kind of marriage between the Brayton and Rankine cycles, it has the usual intent of increasing the working-fluid temperature through the high-temperature capabilities of the internal-combustion (gas) turbine while using the "external-combustion" (steam) turbine to reduce the temperature of heat rejection. The penalty involved is the required work of compression performed by the gas turbine.

Note that such a cycle can be cogenerative if the steam turbine is a non-condensing back-pressure type that further uses the exhaust steam for space heating, oil-field steam flooding, or an industrial process. A cycle is *not* cogenerative if the final output of both turbines is electrical power only, with no usable or used thermal byproduct.[17]

Gas turbines with integral combustion systems fire natural gas, jet fuel, light distillate oil or, with proper prefiring treatment, a wide range of residual and crude oils. *Open-cycle* gas turbines commonly satisfy the peaking and reserve requirements of the power-producing industry because of their quick-starting capability and low installed capital cost. Their low thermal efficiency and use of prime fuels, however, place them at economic disadvantage when operated for long periods of time. Depending on size, pressure ratio, and turbine inlet temperature, heat rates ranging from 10,500 to over 14,000 Btu/kWhr (HHV) are obtained firing oil fuel in simple open-cycle machines; this corresponds to between 24 and 32 percent thermal efficiency on a high-heat-value basis.

TYPES OF COMBINED CYCLES

Combined-cycle installations can fall into four broad classifications, primarily dependent on how the steam generator is used in conjunction with the gas turbine. There are both near-atmospheric-pressure and high-positive-

pressure (supercharged or turbocharged) arrangements.

1. Gas turbine plus unfired steam generator

2. Gas turbine plus supplementary-fired steam generator.

3. Gas turbine plus furnace-fired high-pressure, high-temperature steam generator.

4. Supercharged furnace-fired steam generator plus gas turbine.

The first three are essentially atmospheric-pressure designs, which use the heat energy, and sometimes the oxygen, contained in the exhaust of a combustion gas turbine; they take advantage of the fact that the gas turbine is otherwise penalized by the substantial heat loss associated with elevated stack-gas temperatures (in excess of 800°F or about 430°C). They are described in greater detail in Chapter 8.

Gas Turbine Plus Unfired Steam Generator

In this concept, a heat-recovery steam generator (HRSG) is installed at the discharge of a gas turbine to recover the energy in the gas-turbine exhaust and supply steam to a steam turbine (Fig. 13). All the fuel is fired in the gas turbine, and the steam generator depends entirely on the gas turbine as its heat source. The steam generator is often designed for low steam pressures and temperatures, usually limited to approximately 1500 psi and 900°F.

Because the power from the steam turbine will be produced without any additional fuel input and because there is only a small decrease in gas-turbine efficiency because of the backpressure (draft loss) of the steam generator, the overall plant thermal efficiency will be improved over that of the open-cycle gas turbines. Depending on the particular steam cycle and fuel, net plant heat rates below 9,000 Btu/kWhr (HHV) can be obtained (thermal efficiencies above 38 percent, HHV basis).

Gas Turbine Plus Supplementary-Fired Steam Generator

Gas-turbine exhaust contains as much as 18 percent oxygen and may be used as an oxygen source to support further combustion. Therefore, a modification of the simple waste-heat applica-

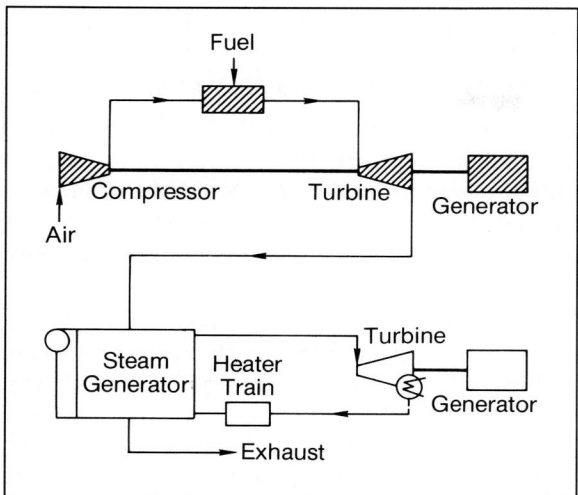

Fig. 13 **Gas turbine plus unfired heat-recovery steam generator (HRSG)**

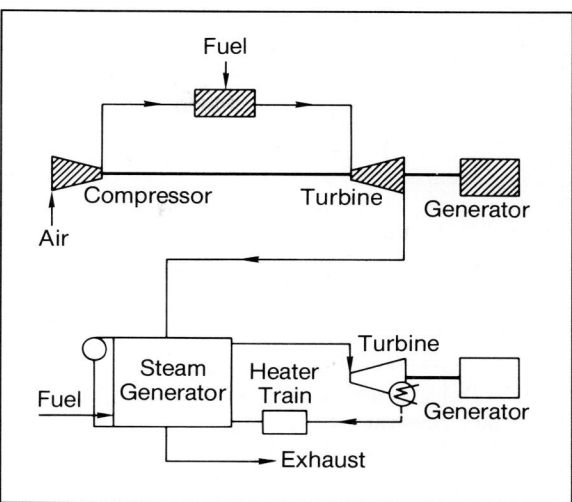

Fig. 14 **Gas turbine plus furnace-fired steam generator**

tion is the use of a supplementary firing system located in the connecting duct between the gas-turbine outlet and the inlet to the steam generator. The firing system will utilize a portion of the oxygen contained in the gas-turbine exhaust and be selected to limit the maximum gas temperature entering the steam generator to approximately 1400°F.

With a given gas-turbine size and this gas-temperature limit, the steam generation can double that of a simple waste-heat application, and the steam turbine will supply a greater proportion of the plant load. The higher steam-generator inlet gas temperature will allow steam conditions to be increased to levels of 2600 psig and 1000°F. The steam turbine designs are nonreheat or reheat, and may be either condensing or noncondensing. For most arrangements the final steam conditions are primarily based on the steam-turbine economics. Net plant heat rates of below 8,000 Btu/kWhr (HHV) are possible with this combination of equipment (for thermal efficiencies above 42 percent, HHV).

Gas Turbine Plus Furnace-Fired Steam Generator

The previous cycle used only a small portion of the available oxygen in the gas-turbine exhaust. Another adaptation is the design of a plant which

utilizes essentially all of the oxygen in the turbine exhaust to support further combustion. Typical gas turbines operate with 250 to 400 percent excess air and thus can support the combustion of approximately three to four times as much fuel in a downstream boiler as was burned in the gas turbine. The majority of fuel can now be fired in the boiler and 60 to 80 percent of the total plant power generation will be supplied by the steam turbine, with the remaining portion by the gas turbine (Fig. 14). The gas turbine may be considered as both an independent power supplier and a forced-draft fan for the boiler. Any of the high-pressure, high-temperature steam conditions utilized by modern steam turbines can be incorporated into this combined cycle.

And although the gas-turbine fuel in the previous arrangement is presently limited to gas and oil, this cycle allows the use of any fossil fuel in the steam-generator.

Plant heat rates of 9,530 Btu/kWhr (HHV) have been obtained in a U.S. installation[18]; thermal efficiencies higher than 36 percent (HHV) are now possible.

Supercharged Furnace-Fired Steam Generator Plus Gas Turbine

Another configuration for a combined cycle

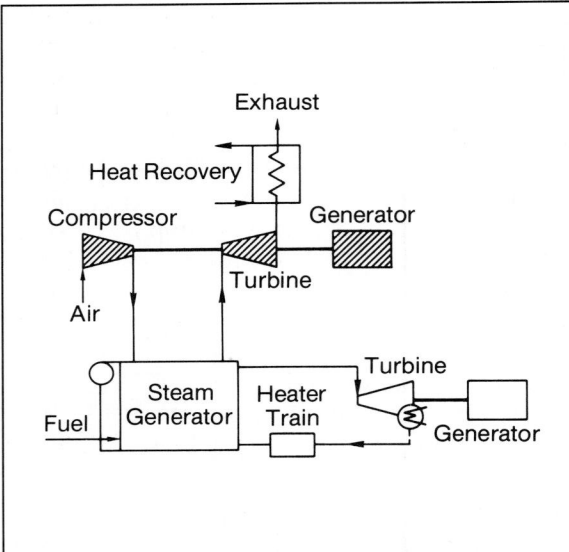

**Fig. 15 Supercharged furnace-fired
steam generator plus gas turbine**

includes the installation of a steam generator between the air compressor and the gas turbine
(Fig. 15). The air compressor serves as a forced-
draft fan and pressurizes the boiler, where all the
fuel is fired; the products of combustion, having
been partially cooled within the boiler complex,
are then discharged through a gas turbine. Additional heat is recovered by heat exchangers installed at the exhaust of the gas turbine. They are
used as economizers or feedwater heaters.

Although conventionally fired gas turbines
require high excess-air quantities, the firing
system for the supercharged furnace is designed to operate with excess-air levels commensurate with conventional units and the
air compressor is selected for such an air
capacity. The steam turbine supplies the majority of the plant electrical generation, with
the gas turbine either selected to provide sufficient power to drive the air compressor
(Velox or turbocharged cycle) or sized to supply
additional power generation. Chapter 9 includes a description of the pressurized fluidized-bed combustor (PFBC) employed in this
type of cycle, which is in active development
in several countries.[19]

INDUSTRIAL
POWER-PLANT DESIGN

This section focuses attention on the overall
design elements of the industrial or institutional power plant, exclusive of the marine propulsion plant. Many of these elements are
involved also in the design of more complex
utility power plants, which is the subject of the
next section. In addition, other chapters of the
book will cover specifics of the design of boilers
together with their various components and
supporting systems.

The most elementary type of industrial power
plant incorporates a boiler as a heat source and
a heating system as a load to dissipate the thermal energy released by fuel fired to the boiler.
From this concept onward, the industrial or institutional power plant may encompass various
degrees of complexity, as discussed in the previous section on cogenerative steam plants. Cycles may range from the very simple, with low
thermal efficiency, to the most complex and efficient arrangements as proposed for central
stations, including combinations of steam and
gas turbines. Boiler size may extend from the
generation of a few thousand pounds of steam
per hour to several million in large installations, and the same spread is true for steam turbines for industrial power generation. In short,
the industrial power plant can be used to illustrate virtually every aspect of the thermal engineering involved with a fossil-fueled plant.

TRENDS IN POWER-PLANT DESIGN

Up to this point the term *industrial power
plant* has been used without definition. In engineering terms, there is no physical difference
between industrial and electric-utility central-
station equipment, as is demonstrated in Chapter 8: Steam Generators for Process Use/Power
Production. At one time, power generation was
a part of virtually every industrial power plant,
but this changed for several reasons:

1. The public-utility systems are securely and
effectively interconnected electrically, and provide highly reliable service.

2. The cost of purchased electric power has resisted the effects of inflation better than almost any other commodity, largely through the economies-of-scale in the large centralized power plant.

3. The demands for additional electrical power have generally grown faster than the demands for thermal energy for space heating or process use, thereby exceeding the capabilities of backpressure generation.

The end result was that the majority of industrial power plants built after World War II and until recently (with the enactment of the Public Utility Regulatory Policies Act of 1978[20]) were for space heating and process steam. The exceptions were in situations such as the following:

1. Coordinated demands for steam and power, accompanied by the availability of waste fuels suitable for combustion in boilers. Examples are found in the pulp and paper, chemicals, petroleum-refining, food-processing, and steel industries. These five industries account for almost three-quarters of all manufacturing steam demand; in them, steam use is usually the limiting factor in the technical potential of such cogeneration.[13]

2. A balanced growth of electrical and steam-heating requirements. This is found in many institutional settings, such as universities, hospitals, penal institutions, and some district-heating schemes in metropolitan areas. Here, backpressure cogeneration can have marked advantages.

It is estimated that, in the late 1970's, less than 5 percent of American electrical generating capacity was in the form of cogeneration. It is now recognized that, where it is applicable, cogeneration is an important means of more effectively using all kinds of fuels in industrial-power situations.

POWER-PLANT STUDIES

The starting point of an industrial power plant is an engineering study. Power and steam loads must be ascertained and costs estimated before construction can be considered. In some instances, the organization for whom the power plant is being built may have an engineering staff of sufficient size and experience to make preliminary studies, develop the detailed design, evaluate bids, award contracts and supervise construction. But most often, a consulting engineer is called upon to perform one or more of these functions.

In any case, a preliminary report must be written to obtain authorization of capital funds. Not only does this report consider the total investment, but it also evaluates outlays for such items as operation, maintenance, depreciation, insurance, interest, and taxes. With its primary emphasis on economic factors, the preliminary report must be written so as to be intelligible to those whose background may be in finance or law rather than engineering.

A salient part of the preliminary report is the charting of anticipated loads for different conditions, such as daily load curves for winter and summer, weekdays, Saturdays, Sundays and holidays. Special consideration must be given to any unusual operating conditions and to the time of peak loads. If a manufacturing operation is involved and if it incorporates some process equipment with marked swings in demand, a detailed study should be made of the nature and frequency of the operation and its steam and power demands. As a part of load studies of existing plants, the peaks and valleys should be investigated with the objective of determining whether corrective measures might be taken to level out the load to increase effective output.

Typical load curves, as shown in Figs. 16, 17, and 18, help to determine the size of such equipment as boilers, turbines and auxiliaries. Studies must then be made of the capability of the equipment to meet not only the conditions plotted on the basis of past experience but also those forecast for a limited time in the future. Load curves should be made for both power and steam requirements, and due consideration should be given to the relative growth of power and steam in the future.

The engineer with the assignment to study the power and steam requirements must become thoroughly acquainted with the operating

Steam Power-Plant Design

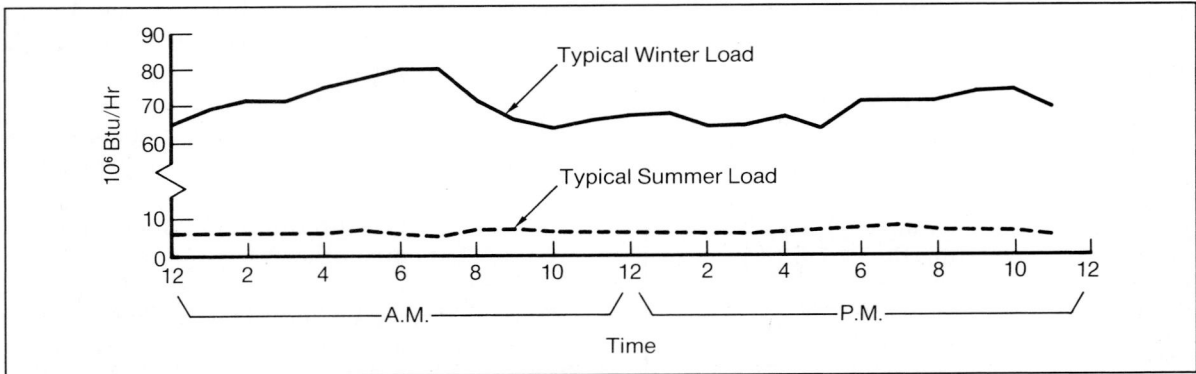

Fig. 16 Summer and winter heating loads in a high-temperature water installation

characteristics of the proposed installation or the extension to existing facilities. In the case of a hospital supplied by an isolated power plant, dependability of service is of paramount importance. The same may be true of an industrial process of a continuous nature, where an emergency may be very costly and possibly hazardous as well. On the other hand, some installations may have firm electrical connections to outside sources. Here, the design emphasis might be on continuity of service despite outage of generating plant equipment. In other situations, power and heating interruptions may be inconvenient, but may not cause severe problems or losses.

Federal, state and local regulations affect the installation and operation of all industrial facilities. In the case of a cogeneration plant or independent power producer, the greatest regulatory influences on project feasibility will most likely stem from regulations in the areas of electric power exchanges, fuel use, and environmental quality. Of these areas, electric power exchanges are subject to a combination of state and federal regulation; fuel use is regulated primarily by the federal government; and environmental quality is likely to involve regulation at the federal, state, and local levels.

Based on the definition of the problem and the detailed technical analysis, then, the preliminary report should offer conclusions and recommendations. These should include estimates of the required capital investment along with operating costs and fixed charges. In most reports of this nature there will be a number of analyses of economic alternatives, to aid man-

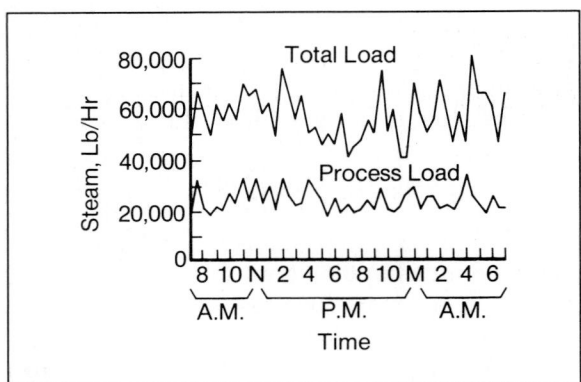

Fig. 17 Widely fluctuating processing load curve

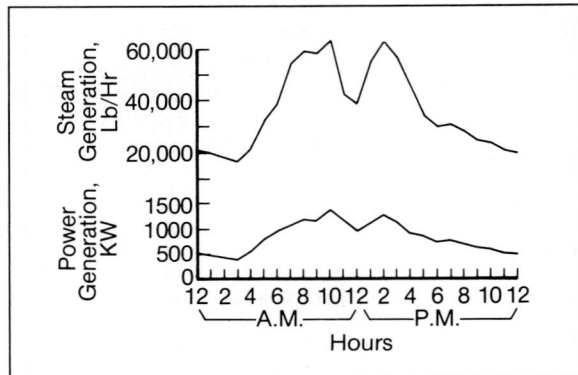

Fig. 18 Load curves for backpressure steam and power generation

agement in making decisions prior to authorization of construction.

DESIGN INVESTIGATIONS

Detailed design follows acceptance of the preliminary engineering study and authorization to proceed. Some topics of the preliminary report are investigated more thoroughly as part of the design process. The studies are coordinated with the purchase of materials and equipment and the making of detailed architectural and construction drawings.

The number and scope of design investigations will vary from plant to plant. As a very minimum, decisions must be made as to choice of cycle, selection of fuel, number and types of auxiliaries, extent of instrumentation and automatic control, and plans for isolated or interconnected operation.

CHOICE OF CYCLE

In the design of a new plant, careful consideration must be given to the choice of thermal cycle. If space-heating and process use constitute the entire load, then it is possible to make a choice between a steam and a high-temperature water cycle. Large industrial power plants with substantial process and electric loads are likely to make use of more advanced cogenerative cycles, as covered above.

CHOICE OF FUEL

Selection of fuel is based on a combined investigation of availability, cost, and operating requirements. Most industrial power plants use solid, liquid or gaseous fuels, either singly or in combination. Although generally these are commercially available fuels, sometimes they are byproducts of manufacturing and processing.

When comparing fuel costs, the following factors should be tabulated:

- base price of fuel
- cost of fuel, delivered
- cost of handling and reclaiming
- cost of labor, including firing and disposition of refuse

- cost of plant services, including power
- fixed charges on fuel-handling and burning equipment

Careful study of these outlays, adjusted to an annual basis, will provide a foundation for fuel choice. Other less tangible factors should also be evaluated, such as reliability of supply, future availability, individual fuel cost trends, ease of conversion from one fuel to another, and extent of future plant expansion.

Charges for byproduct fuels depend upon the method of accounting for process and power-plant costs. In some industries, such as steel and petroleum refining, byproduct fuels must be either burned in power-plant boilers as fast as they are produced or consumed as atmospheric flares. Other byproduct fuels, such as some types of wood and plant refuse, may be stored for limited periods. The economics of such byproduct fuels can become very complex or be as simple as the necessity of disposing of them immediately by some form of combustion. In many instances, conventional solid, liquid, or gaseous fuels must be fired to supplement byproduct fuels.

CHOICE OF AUXILIARIES

No steam power plant is complete with merely a boiler and a turbine. Auxiliary equipment is required in the form of fans, pumps, heaters, tanks and piping. In some instances, heat-recovery equipment is added to boilers, and generally some form of water-conditioning equipment will be required. Both fuel and ash handling systems are necessary if solid fuel is fired. If all electrical requirements cannot be satisfied through backpressure operation, a condenser is a necessity.

The designer of an industrial power plant must investigate to some degree each of these auxiliaries. In the case of tanks and piping, the investigation may be to determine optimum sizes and selection from among alternative piping and equipment arrangements. In other cases, the designer may decide only whether to use the auxiliary equipment.

Motor versus Steam-Turbine Drives

In most industrial power plants, the designer has a choice of steam-turbine or electric-motor drive for rotating auxiliaries. Electric drives are used commonly for such items as forced- and induced-draft fans and pumps for boiler feed, condensate and fuel oil. Yet the possibilities and advantages of turbine drives should not be overlooked. These may contribute to improvement of the plant heat balance and assure continuity of boiler plant operation in the event of power failure. Where more than one fan and pump are installed for each plant auxiliary service, the combination of motor and turbine drives becomes attractive, with provision for automatic starting of the steam-driven auxiliary in the event of power failure.

Because continuity of service is absolutely essential for the boiler feed pump, it is common practice to provide a steam-driven pump for use in the event of electric power failure. In addition, provision may be made to start this pump automatically if there is a marked drop of pressure in the boiler feed line.

Unlike the case of the boiler feed pump, the question of turbine drives for forced- and induced-draft fans does not have a clear-cut answer. Consideration should be given to the probability of electric-power outage and the necessity for maintaining full-load output in the event of such outage. The economics of operation must be weighed against the requirements for service continuity.

INSTRUMENTATION AND AUTOMATIC CONTROL

Industrial power plants vary widely in their use of instrumentation and automatic control. Instrumentation may be installed as an aid to operation, as a means of keeping records of use of fuel, steam and electricity, or for both purposes. Automatic controls may be specified to reduce operating personnel to a minimum and to assist in maintaining operation at a high level of efficiency.

Instruments assist in the operation of a power plant as well as in the collection of information on the cost of steam and power consumption. The designer must keep in mind the type of data which must be available at all times to the operator and must select the correct instruments to provide this information. Some of these instruments will be of the indicating type, while others will record data over prolonged periods. Both types are essential; some overlapping and duplication of readings can be justified because of the differing uses for operating and recorded information.

Every industrial power plant will incorporate instrumentation to indicate boiler and turbine loading. It is also common practice to provide instruments to show various steam and flue-gas temperatures, air and steam flow, feedwater and steam pressures, and electrical outputs. Chapter 13 describes the types of instrumentation and controls currently being installed in industrial power plants.

DRAWINGS AND SPECIFICATIONS

To build a power plant, drawings and specifications must be prepared to obtain bids and exercise proper supervision over purchasing and construction. The objective of such drawings and specifications is to describe the work to be done, primarily from the point of view of the results to be achieved. The engineer must use extreme care to see that they are clear, concise, and capable of but one interpretation. The drawings describe the work graphically and dimensionally while the specifications represent verbal descriptions.

It is well to consider the specifications as the rule book which governs the entire project. The drawings should indicate the location of equipment, including interconnecting piping and wiring. Between the two, the work to be done should be clearly set forth, with nothing essential omitted and with a determined effort to avoid inconsistencies and unnecessary overlap. The specifications should set forth the functions and limits of each item shown on the drawings. By their very nature, specifications are intended to be very detailed about what is to be done, even though they may include some general stipulation or conditions which relate to the work as a whole.

As an example, consider a piping system for

an industrial power plant. The drawings will show a piping layout, including detailed dimensions, and will indicate the pressure characteristics of the system. The specifications should list all design requirements: design pressure and temperature; operating pressure and temperature; the grade designation, sometimes called the schedule, of piping, fittings and valves; the type and pressure characteristics of joints, including any special requirements. These piping specifications should also contain provisions covering quality of materials and workmanship, including inspection and acceptance requirements.

PERFORMANCE SPECIFICATIONS

In the specifying of power-plant equipment, it is sometimes expedient to write a performance specification, with only such descriptions or physical limitations as are necessary to provide for the desired quality of materials and workmanship. Within this framework, the manufacturer is given as much freedom as possible to provide equipment which will best fulfill the functional requirements of the installation. By reason of specialized experience, a manufacturer frequently can best determine the detailed design of the particular equipment it supplies, given all the factors for performing economic analyses of available alternates.

ELECTRIC UTILITY POWER-PLANT DESIGN

The objective of this section is to show the steps that must be taken and the decisions that must be made in the preliminary design of a central station for the generation of electric power. Most of the information applies equally to installations being made by unregulated independent power producers.

Electric utilities are continuously thinking about generating facilities that must be in operation as much as ten or more years in the future. The *final* permitting, design, and construction of new capacity will typically require not less than four years. For these reasons, all basic de-

cisions must be made at least that long before the capacity is needed.

Management must find the best practicable answers to four fundamental questions about the provision of additional capacity: When? How big? Where? What kind? In more formal language, the following steps must be undertaken.

1. Forecast of loads to determine timing and size of additions.
2. Selection of plant location.
3. Selection of types of equipment.

Each step is the subject of a subsequent discussion and involves the careful weighing of both economic and technical alternatives.

FORECAST OF LOADS

The answers to When? and How big? are approached by making a forecast of future electric demands. Such activity is a continuous function in a major utility and is, necessarily, a matter of judgment in interpreting company records, economic trends, population shifts, technological changes and regulatory factors.

There are two significant characteristics of electric demand (usually called "load") in this connection: the peak load, which is the maximum demand on the generating plant for a relatively short period, as 15 minutes or one hour; and the average demand over a longer period, usually a month or a year. The ratio of the two is called the *load factor* and is typically in the range of 40 to 60 percent. A system with a relatively high load factor is fortunate in that it uses, on the average, more of its installed capacity profitably; it spreads its investment costs over a greater production than would be true of another system of the same size, having the same peak load, but with a lower load factor. The high-load-factor system can therefore afford to spend more money for more efficient equipment that will reduce operating costs. The chief of these is the cost of fuel, expenditures for which vary directly with station output. Improvements in station thermal efficiency are one way to reduce the outlay for fuel.

The records of electric demands are usually

kept in three major classifications: residential, commercial, and industrial. Special records applying to unusually large users of power are also available.

The forecaster studies the history of each type of load but chiefly gathers all available data on the growth of the area which is being studied. The forecaster watches population shifts, growth of suburban shopping centers, trends toward air conditioning and electric heating, changing processes in industry, development programs to attract new employers, statistics on per capita use of electricity; in short, every factor that bears on the future use of electrical energy in the area under investigation.

The result of this process is a graph showing the expected peak loads and load factors for a period of years into the future. An example of a peak-load forecast is the line X—X in Fig. 19.

To be reasonably sure of carrying the peak load of any given year, the total generating capacity of the system must exceed the expected peak load by a margin for reserve. The necessary reserve is of two kinds.

■ Spinning reserve is the excess generating capacity that is in operation and on the line at the time of peak load. It must be at least equal to the capacity of the largest single generating unit in use at that time on the system or its interconnections. It is necessary to provide spinning reserve to guard against the possibility of a mishap to the largest unit causing loss of

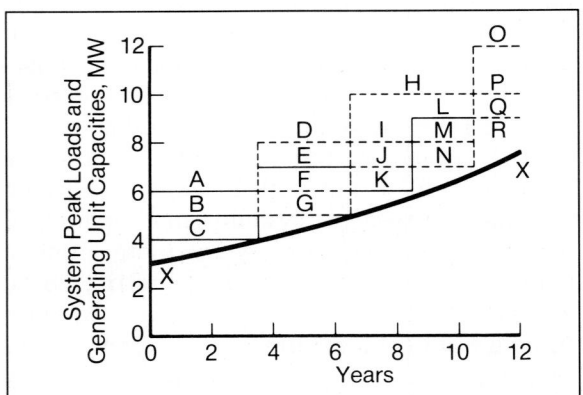

Fig. 19 Relationship between peak loads and installed capacity

its generating capacity, even if only for a few minutes.

■ Reserve for scheduled outages is generating capacity that is installed on the system but is not necessarily in operation at the time of peak load. Such reserve is needed because equipment must be inspected and overhauled at regular intervals. The amount of this reserve must be not less than the size of the largest unit that will be overhauled during the peak-load season.

DETERMINING SYSTEM RESERVE REQUIREMENTS

A simplified example will illustrate the point. Assume a system composed of twelve identical generating units, each capable of generating a continuous output designated as M kW, where M is any number. The total installed capacity is 12 M kW. The spinning reserve is one unit, and the reserve for scheduled outages is another unit because an overhaul typically takes about a month. This system can be expected to carry a peak load of 10 M kW, its firm capacity.

If the system consists of six units each of 2 M kW capacity, the installed capacity is the same as before, but deduction of the spinning reserve will bring its capacity down to 10 M kW, and the further deduction of reserve for scheduled outages will leave a firm capacity of only 8 M kW. However, it may be possible to schedule the overhauls of six units wholly outside the peak-load season, depending on the character of the system load variations through the year; this would make the firm capacity of the system 10 M kW as before.

Another example is indicated in Fig. 19. The system is assumed to consist of six identical units each of M kW capacity. Line A indicates present installed capacity. Deduction of spinning reserve gives line B, and further deduction of reserve for scheduled outages leaves the firm capacity, line C. Another unit must be in operation before the peak-load season of the fourth year. If this unit is of M kW capacity, the new situation is indicated by lines E, F and G. Similarly, other M kW units must be added, one before the seventh year, one before the ninth year, and so on.

If the seventh unit is twice the size of the earlier ones, the installed capacity in the fourth year is shown by line D. Spinning reserve must now be 2M kW, the size of the largest unit. Reserve for scheduled outages will, however, be M kW, making the firm capacity line G, the same as if unit seven were of the smaller size. It follows that installing a large unit increases the firm capacity of a system only by an amount equal to the capacity of some smaller unit.

When the second larger unit is installed, as is indicated for years seven to ten, the firm capacity will increase by 2M kW if it is possible to schedule outages of the two larger units out of the peak-load season (lines H, I and J). Otherwise line K will give the firm capacity.

Further additions of capacity are also indicated on Fig. 19. Solid lines apply to M kW units and dash lines to the larger 2M kW size.

The foregoing is known as the block system of generating capacity addition. The objective is to install the equivalent of two blocks more than the peak load, one to provide spinning reserve and the other to permit scheduled outages for maintenance and repair.

EVALUATING SYSTEM CAPACITY ADDITIONS

Several factors have led to modifications and deviations from this simplified method of capacity addition. Knowledge of system operation has become more complete, and reliability of components is somewhat better understood. This means that entire systems may be computer-simulated by mathematical models and that unit outages may be predicted by probability methods.[21] Using this approach the necessary system reserve requirements may be less than under the block system, but the largest single unit should not exceed 5 to 10 percent of the total system capacity.

In practice, these considerations for system capacity additions are far from clear-cut for a number of different reasons.

■ The units will be of different types, sizes, ages and vulnerability to accidental shutdown.

■ The actual capacity of a plant varies with condensing water temperature, and at times is subject to variations caused by changes in the fuel.

■ A system usually includes transmission lines which are subject to interruption of service from weather and other such causes quite different from those that affect generating units. Outages from such interruptions may be greater than the loss of a single unit.

■ There is the possibility that not only one unit, but two or more, may suffer unexpected interruptions simultaneously or following closely upon one another.

The graph of Fig. 19 plots predicted annual peak loads for certain future years. Another informative graph is constructed by getting from system records the number of hours per year in which the load equalled or exceeded a given amount. The system minimum load will correspond to 8760 hours per year, and the maximum load will appear, in all probability, for only one hour. Such a graph is called a load-duration curve. Fig. 20 represents an idealized example. The areas are energy units expressed as kWhr, and the total area under the curve, divided by the area of the rectangle enclosing it, is the annual load factor.

The chief value of this curve is to emphasize the brief period during which peak loads must be generated in the typical system. Even though the curve is drawn for a past year, its shape will not change markedly in the future unless the character of the load changes materially. Load-duration curves can, therefore, be drawn for future years, subject to the uncertainty of the predicted peak loads and load factors discussed earlier in this chapter.

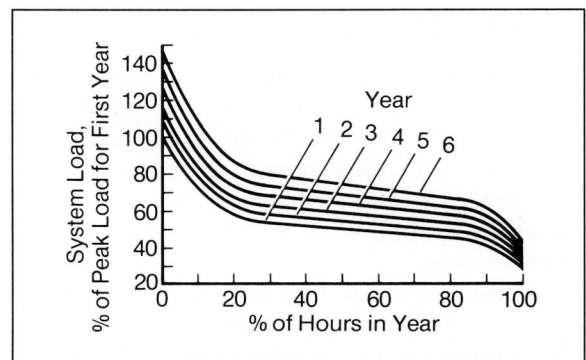

Fig. 20 **Load-duration curves**

It is apparent from the series of curves on a single set of axes in Fig. 20 that the system already has enough equipment to generate all but a relatively small part of the predicted loads. These peaks last a short time and involve only a small amount of generation above a horizontal line representing the system firm capacity. Assume for the moment that all equipment in the system is in good condition and can be expected to operate successfully throughout the period under study. On this assumption, the problem of meeting the future growth in peak load can be solved simply by providing the cheapest possible generating capacity large enough to take care of the growth in load; its fuel economy is not of prime importance because it will be used only a few hours per year, whereas investment costs go on continuously and must be kept to a minimum.[22]

CHOICE OF POWER-PLAN ADDITIONS

Generally, load-distribution requirements for most utility systems follow the pattern of Fig. 20. A typical utility may have three distinct load requirements—base, intermediate and peak. Prior to the mid-1970's, the traditional methods of meeting these load requirements were to have large nuclear and high-efficiency fossil-fuel steam plants designed for 6000 to 8000 hours of operation per year supplying base load. Specially designed fossil-fuel steam plants and former base-loaded fossil plants would supply intermediate load. Low-efficiency installations required to operate for 500 to 2000 hours per year would supply the peak-load and reserve requirements of the system. Gas turbines, old steam plants, and hydro-pumped storage plants would supply this latter capacity. (It should be noted that there is no clean-cut line of demarcation between load categories. Plants designed for one mode of operation may, in practice, be required to operate in other modes which were not originally contemplated.)

Traditionally, utilities purchased new additions of the largest and most efficient steam cycle available. These new units were assigned to base-load service and expected to operate at a capacity factor equal to their availability for approximately the first 10 years of commercial operation. This relegated previously base-loaded plants within the system to cyclic or lower load-factor operation.

Currently, several factors are operating to change this pattern. Improvements in heat rate have diminished. Utilities with a mix of nuclear and new fossil may require the fossil plants to enter cycling service directly. Finally, although the new fossil units may be designed for an efficient thermodynamic cycle, tail-end emission control systems may have high parasitic power requirements. Similarly, the new plant may be fueled with low-sulfur coal transported great distances at considerable expense. Such factors may result in the new fossil unit's having higher generation costs than some of the older plants, with a corresponding influence on the selection and loading of equipment for the most economic dispatch.

Final decisions as to size and timing may well be tied in with decisions as to location and type of equipment, as well as licensing and regulatory requirements that can result in exceedingly long construction lead times. As has been often demonstrated since the advent of longer plant-realization times resulting from the current extended permitting process, the risk of error in installing new facilities increases directly with the time of construction. This mitigates against the 12- to 15-year total time span of nuclear plants, and makes short-installation-time generating equipment (such as gas turbines), as well as power-purchase scenarios, very attractive to system planners. The analysis and study described to this point will disclose a relatively small number of practical possibilities, each of which must be studied in some detail to set up comparative investment and operating costs over a period of years.

SELECTION OF PLANT LOCATION

The location of an electric-generating station is determined by analysis of many factors that influence the selection in diverse ways. During the study and forecasting of future loads, it will usually become apparent that only a few attrac-

tive sites are available. Each of these must be studied to determine the effect of individual factors on its desirability. The object is to make an engineering economic analysis that will disclose the best choice, which in general is the location that will result in the lowest total cost to the owner in the long run.

CONDENSING WATER SUPPLY

As is well known, a steam turbine requires a relatively great quantity of condensing water for its economical operation. Under average conditions it is estimated that about 800 tons of water are required for each ton of coal burned. The supply of water should not vary through the year, its temperature should be as low as possible, the water should not be corrosive to the usual materials, and it should not contain suspended material that will interfere with the flow through pumps and tubes. These considerations suggest a river, a lake or an ocean, and central stations are located whenever possible on the bank of some such body of water.

The alternate to such a location is to provide cooling towers in which forced or natural-draft circulation of air cools the condensing water nearly to the prevailing wet-bulb temperature. Although the investment in cooling towers is high compared to once-through cooling, this higher cost is somewhat offset by the smaller waterfront construction and tunnels.

The extent to which condensing water is available sets a limit on the amount of power that can be generated at the site. Advanced steam conditions not only achieve greater cycle efficiency but also result in less rejection of heat to the condenser. For example, for the same initial steam conditions and a limited amount of condensing water, selection of a reheat regenerative cycle permits greater power output than a nonreheat cycle, everything else being equal. This can be of considerable importance to central stations which are being redeveloped on sites having limited amounts of cooling water. Fig. 21 shows the reduction of heat rejection to condenser cooling water as cycles of increased thermal efficiency and decreased net plant heat rate are incorporated in power plants.[23]

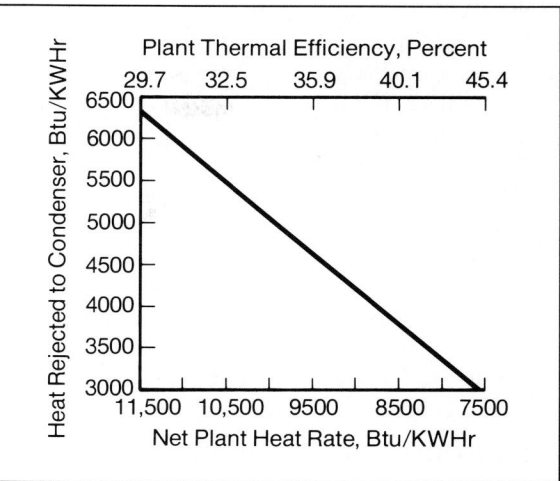

Fig. 21 Heat rejected versus net plant heat rate based on 1.5″ Hg backpressure

PUBLIC OPINION

The effect of public opinion on the choice of a site can be profound. If the public thinks of a power plant as an ugly, noisy, dirty place, spreading dust and ash around the neighborhood, it instantly objects to admitting one to genteel surroundings. The remedy is to design and build central stations that have good architectural treatment, that are clean and as quiet as possible. Many utilities have done this quite successfully.

OTHER CONSIDERATIONS

Additional concerns include availability of municipal services and suitable labor, appropriate zoning, convenient access, climate, tax situation and the like. Room for growth and freedom from undue risk of flood and earthquake are ordinarily essential.

The nature of surrounding installations may play a part in the selection. For example, the height of stacks will be subject to Federal Aeronautics Administration regulations and would be restricted near an airport.

In many site selection problems, one possibility is that of adding to an existing station. To do so has obvious economies in making use of existing facilities, no matter how much new construction may have to be provided. If the existing station site is large enough and has enough condensing water, and especially if it was designed for expansion, it is unlikely that a new site can compete successfully. Each case must be studied on its merits; the danger is in assuming that the "obvious" answer is the best.

Finally, in all studies and analyses, it is essential to remember that future conditions determine the best solution. All available data apply to past years, or at best to the present, and must be reviewed critically to ensure that the most likely future conditions have been derived or projected from them. This statement applies to every engineering economic study.

SELECTION OF TYPES OF EQUIPMENT

Having determined the size, timing, and location of the installation by the steps previously described, we must tackle the question, *What kind?* In practice, the steps are not taken separately and in succession; they overlap considerably but may follow the sequence described in the following text.

The primary choice is between two plant fuel cycles; that is, coal fuel or internal combustion (diesel or gas turbine). The possible consideration of nuclear, hydroelectric, or purchased power would take place before site selection, and is outside the scope of this book. Strictly on the basis of total economy, steam with coal fuel frequently will be preferred in the larger sizes, with internal combustion having an advantage for smaller units.

The size range within which each type of plant fits best is reasonably well defined by experience and will not be the subject of intensive study in the typical instance. Capacity installed for peaking purposes may be subject to very different economic evaluation than is the case for base-load generating equipment. Heat rate economies dominate the latter, while availability of peak-shaving capacity for short periods is of most importance in the former.

Assuming that a conventional steam cycle has been selected, studies will be made to find the best conditions as to steam and reheat pressures and temperatures, number and location of extraction stages, condenser pressure, and a host of

similar characteristics.[24] Only with experience and judgment as a guide is it possible to keep the combinations of variables within practical bounds. Even with the help of advanced methods of analysis, the basic method of solution is to assume a reasonable set of conditions and calculate the required investment in the central station and the resulting cost of operation under future load conditions. In some instances, the influence of a variable can be isolated and analyzed by itself, but all too often a change in assumed design conditions will be reflected throughout the cycle, resulting in an unexpected and perhaps undesired change at another point.[25]

The sizes of the units will have been fairly well determined by the studies of load growth and estimated capital cost. The unit system, under which a single boiler serves a single turbine, each with its own auxiliaries, is rather generally accepted, but there may well be exceptions, especially when a station has a process steam load in addition to an electric load.

At this point, a number of subsidiary economic studies will be required; for example, heater and condenser surface, turbine versus motor-driven auxiliaries, voltage of electric auxiliary drives and extent of building enclosure. The object is always to reduce the initial investment without increasing maintenance and operating costs.

Investment costs are initially determined by making a preliminary design of the feature in question and estimating from quotations and past experience the cost of its purchase and installation, keeping in mind the fact that such matters as building volume, foundations, steel work, steam and water piping, electrical wiring, and controls cannot be ignored.

Operating costs are estimated by calculating plant performance over the necessary range of loads under the assumed conditions and applying load duration data to find annual costs. It is generally necessary to include maintenance costs for a complete comparison.

The several assumed combinations are compared by standard methods of engineering economy to find the one that promises the lowest overall cost over the life of the station, giving due regard to the time value of money.

SOME POWER-GENERATION TERMS

CAPACITY FACTOR: a measure of the output of the plant over some time period. Capacity factor is the ratio of the energy generated by the unit during the time period to the energy that could have been generated had the unit run at its full rating over the entire period. (See Chapter 24.)

NET PLANT HEAT RATE (NPHR): the fuel-heat input required to generate a kWhr and deliver the generated power to the transmission line leaving the plant. The term "net" implies that energy needed to supply auxiliary equipment must be deducted prior to calculating NPHR. Plant efficiency can be calculated by dividing 3412 Btu/kWhr by NPHR.

REPLACEMENT POWER COSTS: the cost of supplying replacement energy in the event of a forced or scheduled outage. Replacement power may be available within the utility's system by bringing into operation units with high production costs, or may have to be purchased from other utilities.

AUXILIARY-POWER CHARGES: charges assessed to account for the effect of auxiliary-power requirements on the net output from the plant to the utility's grid. A *demand* charge, proportional to the cost of new generating capacity, applies regardless of the duration such auxiliaries are expected to run. An *energy* charge is applied against the time the auxiliaries will be in operation. The magnitude of the energy charge depends on fuel costs and other operating costs in the utility system.

See reference 24 for a typical engineering economy study performed for application to a large utility system.

The result of this engineering process will be to produce a set of preliminary arrangement drawings showing the station as it has been conceived; a set of abbreviated specifications covering the important equipment, structures and systems; a reasonably reliable construction cost estimate; and a report covering the alternates considered, advantages and disadvantages of each, comparative investment and operating costs, and other data individual to each case. Management will be able to base its decision on this material if it is complete and well presented. Then and only then does the detailed procurement and design process begin. The steam-turbine/generator and the boiler are purchased first, followed by other major components as dictated by their respective design and fabrication lead times.

As the equipment is purchased, hundreds of studies of comparative economics of the vendor offerings are made, based on capital costs, energy costs, and probably costs of operation and maintenance. The following section describes how such conventional power-plant economic studies are made.

POWER-PLANT ECONOMICS

Costs associated with owning and operating power-plant equipment are a basic concern of both the design engineer and the user. Often, the initial cost of the equipment must balance against the operating cost over the plant lifetime. Before discussing how such comparisons are made, it is necessary to consider the two general categories of power-plant costs.

Much of the money spent by a power-plant owner is for goods and services consumed within a relatively short time after acquisition. This category includes outlays for wages and salaries, operating and maintenance supplies, and fuel. Called *expenses*, such payments are normally made from revenue.

In contrast, other owner expenditures are for items whose usefulness continues for an extended period and which produce revenue in the future; money spent to construct a power plant is a typical example. Termed *capital investments*, expenditures of this nature are not ordinarily paid directly from revenue; this is because current revenue most often would be insufficient to cover large capital expenditures, and because the equipment is expected to provide service well into the future.

POWER-GENERATING COSTS

Total power-generating costs, or bus-bar costs, are the sum of fixed charges, fuel costs, and operation and maintenance (O&M) costs. The fixed charges are associated with the capital investment, whereas the fuel and operation and maintenance (O&M) costs are generally treated as expenses.

FIXED CHARGES

Capital costs are translated into annual costs by calculating the fixed charges on the plant or equipment.

Fixed charges are those costs incurred during each year of the lifetime of a plant, independent of how much energy production takes place in each year. Included in the fixed charges are usually all of the costs that are proportional to the capital investment in the installation, which will include depreciation, required return on investment, property insurance, federal income taxes, and state and local taxes.

Some components of the fixed-charge rate, such as depreciation, are bookkeeping expenses which do not represent cash outflow during the operation period. Other fixed-charge components, such as property insurance and *ad valorem* taxes, represent annual outlays which are in direct proportion to part or all of the initial investment. The required return on the investment and the income taxes associated with this return are a major part of the annual fixed charges. Also included is an interim replacement allowance, which provides for any intermittent replacement of equipment required before the end of the scheduled plant life. *Not included* in annual fixed charges are relatively constant fixed costs, such as liability insurance or plant staffing expenses, which bear no direct relation to the plant investment and are somewhat independent of the extent to which power is generated. (Such expenses are included in the plant operation and maintenance category.)

The proportionality constant that converts the initial capital cost of the plant to annual fixed charges is the fixed-charge rate. Therefore, the fixed charges on the plant are

$$C = IC \times FCR/100$$

where C = fixed charges, \$/yr.
IC = initial capital cost, \$, and
FCR = annual fixed-charge rate, %

(8)

or, on a per-unit-of-energy-generated basis:

$$c = \frac{IC \times 10^3 \times FCR/100}{G} = \frac{10 \times IC \times FCR}{G}$$

(9)

where c = fixed charges, mills/kWhr, and
G = energy generation, kWhr/yr.

In utility accounting, fixed-charge percentages are applied annually to the depreciated (book) value of the investment, and consequently, the amount charged to a specific plant decreases from year to year. When this procedure is followed for evaluations, it is necessary to calculate fixed charges and all other costs year-by-year over the life of the plant, to discount each future year's total costs to obtain the corresponding present worth, and to sum all these present worths to obtain the grand-total "present-worthed" cost of the plant being evaluated. This gives a single figure by which the relative economics of proposed plants may be judged.

A simpler procedure giving the same result is to use levelized annual fixed charges; that is, to apply a *constant* fixed-charge rate each year to the initial investment in such a manner that the present-worthed grand total of the constant charges equals the corresponding total of variable charges. The use of levelized fixed charges does not yield the year-by-year forecast of costs given by the more detailed approach, but such annual breakdowns are not usually required for evaluations. A conceptual approach to developing a fixed-charge rate is shown in Table I.

FUEL COSTS

For most economic evaluations, converting fuel costs on an as-received basis at the plant to

fuel costs per kilowatt-hour of power produced is sufficient. The relationships to do this are

$$F = \frac{FC \times NPHR}{10^5}$$

(10)

where   F = fuel cost, mills/kWhr
          FC = fuel cost, ¢/10⁶ (million)/Btu
      NPHR = net plant heat rate, Btu/kWhr
or

$$F = \frac{FC \times NTHR}{10^3 \times B\,E\,(1 - APR)}$$

(11)

where NTHR = net turbine heat rate, Btu/kWhr
          BE = boiler efficiency, %
       APR = auxiliary power requirements
                   (as a fraction of gross plant output)

Thus, if fuel costs for a coal-fired unit are projected at 300¢ per million Btu, and the net plant heat rate is 9300 Btu/kWhr, the fuel cost, F, is 28 mills/kWhr.

Table I. Typical Fixed-Charge Rate for Depreciable Portion of Plant

Component	Percent
Interest or return on investment	8.6 *
Depreciation * *	3.3
Interim replacements	0.4
Property insurance	0.5
Federal income taxes	3.0 * * *
State and local taxes	2.4 * * *
Total	18.2

*Levelized value: (50% debt @ 10% + 50% equity @ 14%) × .72 levelized average investment

* *Straight-line for 30 years

* * *For investor-owned companies

TIME-DEPENDENT COSTS

The time it takes to complete a project affects present-day direct costs (such as material and labor) and indirect costs (such as construction facilities and services) in two ways. Escalation takes place between project initiation and actual delivery of equipment and services. The utility also incurs interest charges on payments that must be made before the plant is placed into service. The magnitude of these escalation and interest-during-construction (IDC) costs on the steam-generating equipment depends, of course, on the vendor's escalation provisions and terms of payment. These charges can typically add 50 percent to the original cost of long-lead-time equipment, such as boilers and turbine-generators. Therefore, changes in the present-day price of new equipment, either upwards or downwards, are highly leveraged in their impact on the evaluated cost of the equipment installed in the plant.

Although each equipment payment is subject to different escalation and IDC charges as a result of the timing of the payment, the net effect of these charges can be represented by single factors. Each factor is a weighted-average of the charges on individual payments. The final, fully escalated cost of the equipment is greater than the present day cost of equipment by the factor F, where:

$$F = E_d \times E_c \times I_c$$

(12)

and where;

F = total factor to be applied to present-day costs to account for escalation and interest during construction

E_d = escalation factor during design period to start of construction

E_c = escalation factor during construction period (See Fig. 22)

I_c = interest factor during construction period (See Fig. 23)

Figs. 22 and 23 depict graphically the values of escalation and interest multipliers during the construction period based on typical cash-flow patterns for power plants.

Consider a unit with a present-day cost of $800 million. With construction beginning two years from now and commercial operation six years from now, escalation is expected to be 7 percent per year, and interest during construction, 8 percent per year:

$E_d = (1.07)^2 = 1.145$
$E_c = 1.135$ (See Fig. 22 for 48 months and 7%/yr.)
$I_c = 1.16$ (See Fig. 23 for 48 months and 8%/yr.)
$F = 1.145 \times 1.135 \times 1.16 = 1.51$

This plant then would have a capital cost at the commercial operation date of:
$1.51 \times \$800 \times 10^6 = \$1,208$ million

ECONOMIES OF SCALE

Many of the cost components that make up total power-plant costs are fixed; that is, they are independent of plant size, or else they increase slowly as plant size increases. If total plant costs increase at a slower rate, percentage-wise, than the rate at which design plant output increases, economies of scale are present. Power-plant construction experience indicates that economies of scale exist in the steam-generating plant, the turbine-generator, the condenser, and switchgear, and consequently, in total plant costs.

This type of cost behavior is often represented by some form of power relationship, less than unity, and usually between 0.7 and 0.85. With a power relationship "k," this would be expressed as follows:

$$\frac{IC_1}{IC_2} = \left(\frac{PO_1}{PO_2}\right)^k$$

where: IC_1 = capital cost of plant 1
IC_2 = capital cost of plant 2
PO_1 = output of plant 1
PO_2 = output of plant 2   **(13)**

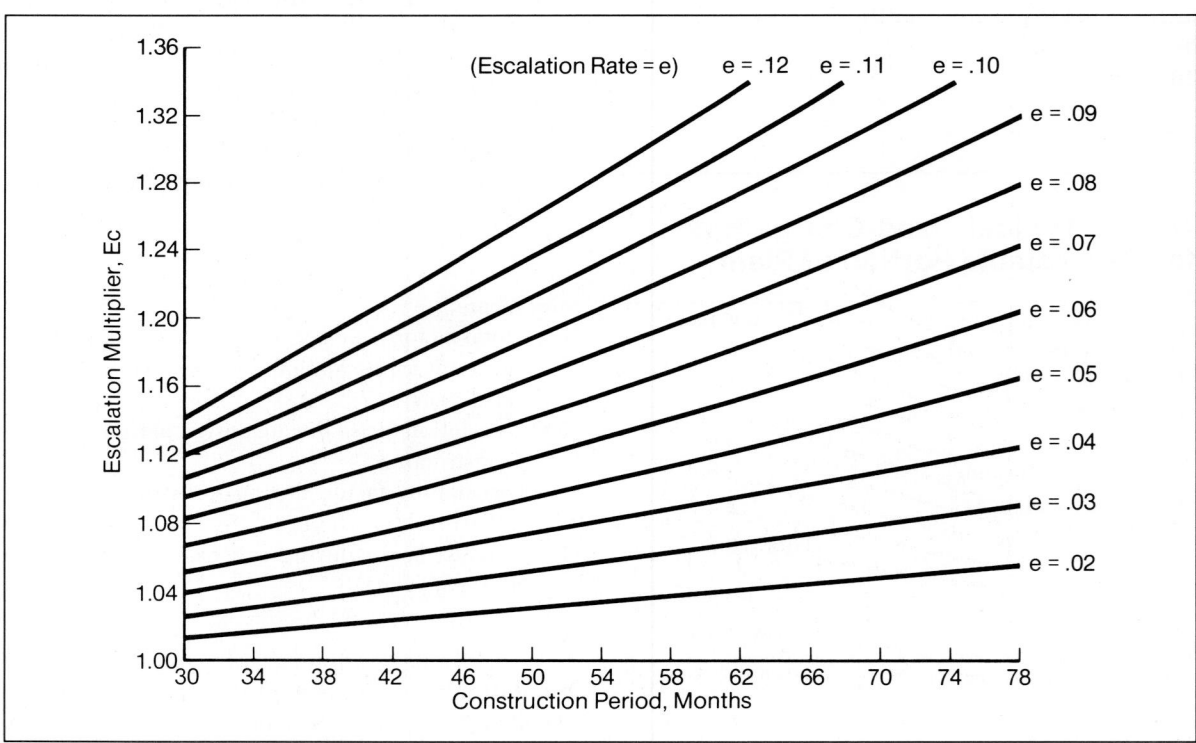

Fig. 22 Escalation multiplier

Thus, if the cost of a 300-MW generating unit is expected to be $2,000/kW, the total cost for a 500-MW unit will be, with k = 0.8

$$\left(\frac{500}{300}\right)^{0.8} \times \$2,000/kW \times 300,000\ kW$$

$$= \$903,000,000$$

Economies of scale are only one factor involved in the choice of unit size. Consideration must also be made of such factors as reliability, mode of operation, financing requirements, availability of sites, and regulations.

PRESENT-WORTH AND LEVELIZED COSTS

Present-worth and *levelized costs* are terms often used in discussing utility planning, evaluations, and engineering economics in general. The concept of present-worth costs is frequently explained by the statement that "a dollar today is worth more than a dollar tomorrow." While this sounds reasonable, it is important to understand the basis for the statement. It has nothing to do with inflation or escalation of costs. It simply means that a dollar received today can be invested and, by "tomorrow," will be equal to a dollar plus interest.

The concept of present worth or present value is a corollary. The present worth of some future expense (or revenue) is the amount that would have to equal that future expense (or revenue) eventually. Total present-worth costs are simply a way of combining payments made at different times to account for the value of interest. All utility expenditures are in this category.

Present-worth costs are calculated by simply reversing the process of calculating compound interest. Therefore, C_o is the present worth of some cost, C_1 occurring at time t in the future

$$C_o = \frac{C_1}{(1+r)^t}$$

(14)

where r is the interest rate or discount rate.

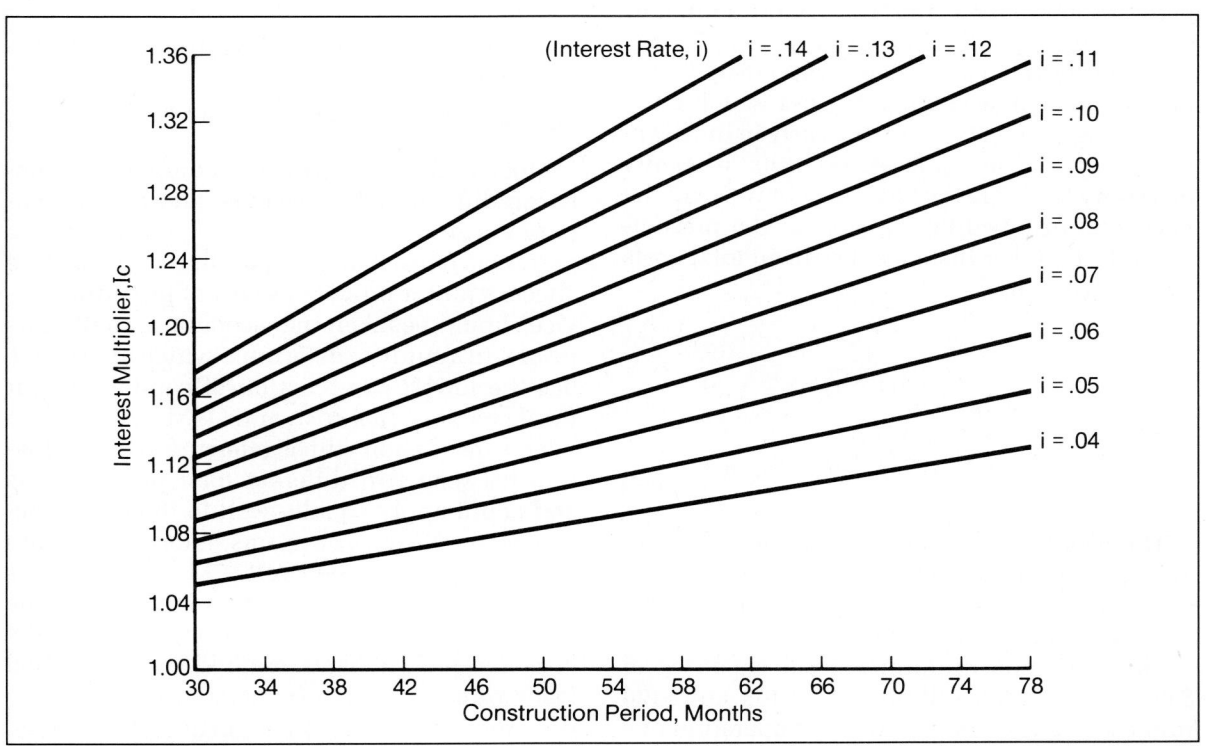

Fig. 23 Interest multiplier

The expression $\dfrac{1}{(1 + r)^t}$ is called the discount factor or present-worth factor.

Thus, the present worth of a payment of $100,000 to be made three years later, assuming a discount rate of 9 percent per year, is

$$\$100,000 \times \frac{1}{(1.09)^3} = \$77,200.$$

In other words, a payment of $77,200 now is equivalent to a payment of $100,000 three years later, assuming the money can be invested to generate a rate of return of 9%/yr.

The total present-worth costs (PWC) are the sum of all costs associated with a project properly discounted or present-worthed to a common point in time.

$$PWC = \sum_i \frac{C_i}{(1 + r)^t}$$

(15)

Total present-worth costs can be used to compare two alternatives that involve different cost expenditures over the life of a project.

A closely related concept is levelized costs. Levelizing is an averaging process which gives more weight to costs occurring early in the life of the project. To obtain levelized costs, it is only necessary to divide the total present worth of the payments involved by the sum of the present-worth factors. For instance, levelized total costs (LC) are given by:

$$LC = \frac{\sum \dfrac{C_i}{(1 + r)^t}}{\sum \dfrac{1}{(1 + r)^t}}$$

(16)

The resulting levelized cost is one, constant, annual value which is equivalent (i.e. has the same total present worth) as the non-uniform series of actual annual costs.

Thus, for non-uniform annual expenses of $157 million; $165 million; $167, $163, and $170 million over the first five years of operation of a new generating unit, the present-worth costs and

levelized costs, using a 9%/yr discount rate, are shown in Table II.

ECONOMIC ANALYSES

Most evaluations involve comparisons between different plant designs, alternative components or various modes of operation. To perform a consistent evaluation, everything that affects a power producer's cost of providing service should be taken into account. Usually this results in a study to determine which alternative has the lowest combination of capital cost (or fixed charges), fuel cost, and O&M cost.

This discussion will be referring to the evaluation between two different total power plants and could apply equally well to the choice between steam-supply systems, alternative pump designs, or operating cycles. Most evaluations can be classified into one of two approaches. The first of these is what is often referred to as the "revenue requirements method." This method leads to a comparison of generating costs (or revenue requirements) on a mills/kWhr basis, between alternatives. An equivalent-revenue-

requirements comparison can also be made by contrasting the total present worth of all costs, i.e. present worth of fixed charges on the capital investment plus the present worth of fuel and O&M costs. A comparison of present-worth totals, instead of mills/kWhr is only valid in the case where each alternative produces the same energy.

The revenue-requirements method, therefore, necessitates that all capital costs first be converted to annual fixed charges and then be combined with annual revenue requirements for fuel and O&M. It is then possible to calculate total present-worth revenue requirements for each alternative or levelized annual revenue requirements. The alternative with the lowest revenue requirements is the preferred choice in the evaluation.

The second method of performing an economic evaluation is what is often known as the "capitalized-cost" method. With the revenue-requirements approach, capital costs were converted to annual costs by use of the fixed-charge rate. The fixed charges could then be combined with fuel and O&M costs which were already on an annual basis. Conversely, with the capital-

ized-cost method, just the opposite is done. Annual costs such as fuel and O&M are divided by the fixed-charge rate (i.e. capitalized) so that the result can be combined with capital costs. This gives a very quick way of comparing the difference in operating costs with the difference in capital costs between various alternatives.

The system planner is still faced with the traditional problem of balancing plant capital cost, expected capacity factor, and fuel cost in determining which steam cycle to choose. A curve system (Fig. 24), using the capitalized-cost method described previously, is useful in performing such comparisons.

OTHER ASPECTS OF THE SELECTION OF POWER-PLANT EQUIPMENT

Intensive economic and cost analyses of the type described in the preceding section are very important in choosing the size, type, and equipment manufacturer for power and steam generation. But there are other surprisingly potent variables influencing the eventual generation economics of a given installation that the evaluating engineer has to take into account. Among these variables are:

Table II. Levelized Costs and Present-Worth Costs of a New Generating Unit

Year	Present-Worth Factor	Expenses 10⁶$	Present-Worth Expenses, 10⁶$
1	0.917	157	143.9
2	0.842	165	138.9
3	0.772	167	128.9
4	0.708	163	115.4
5	0.650	170	110.5
Totals	3.889		637.6

$$\text{Levelized Expenses} = \frac{\$637.6 \times 10^6}{3.889}$$
$$= \$164 \times 10^6/\text{yr}$$

PRESENT WORTH

Present worth of a fixed annual cost, FC, for n years at r %/yr. (If FC is equal to 1, this expression is equal to the sum of the present-worth factors for n years.)

1. $\text{PW} = \dfrac{\text{FC}}{r}\left[1 - \left(\dfrac{1}{1+r}\right)^{n}\right]$

Present worth of an annual cost, AC, which is escalating at e %/yr:

2. $\text{PW} = \text{AC}\left(\dfrac{1}{r-e}\right)\left[1 - \left(\dfrac{1+e}{1+r}\right)^{n}\right]$

where r is again the discount rate.

3. To obtain levelized annual costs (when escalation is taking place), divide the results obtained in expression 2 by the sum of the PW factors from expression 1.

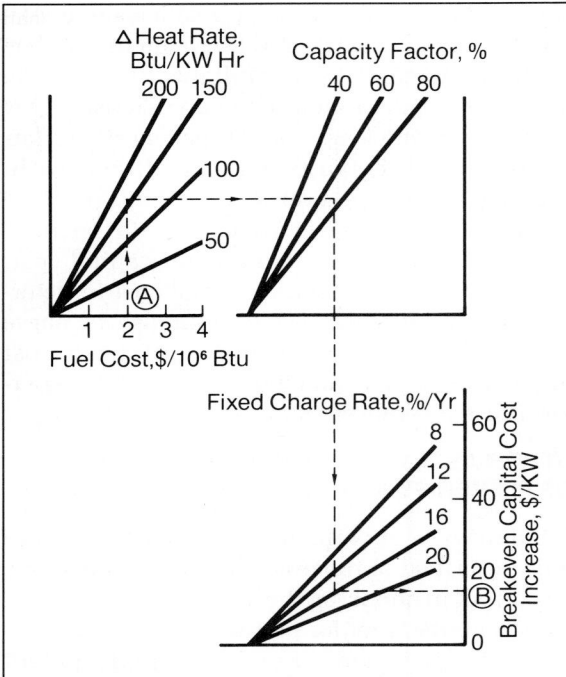

Fig. 24 Breakeven capital cost changes versus improvements in heat rate. EXAMPLE: Going from an 1800 psig, 1000°/1000° cycle to a 2400 psig, 1000°/1000° cycle results in an improvement in cycle efficiency of 160 Btu/kWhr. What is the breakeven increase in capital cost? Given: 70 percent capacity factor, $2/10⁶ Btu fuel cost, and 16 percent/yr fixed charge rate. Solution: Enter at point A; read the result at point B, about $15/kW.

a. the reliability of the many critical components associated directly with the generating process;

b. the reliability of fuel supply of a specific type and heating value during the lifetime of the plant;

c. regulatory obligations, present or future, that a plant is bound to observe during operation or, in the event of failure of equipment having an actual or alleged effect on public health or safety, shut down the generating operation;

d. compatibility with *future* regulatory requirements, which have to be anticipated during the siting and equipment-selection process; and

e. compatibility with future *social* and *political* requirements.

So, in addition to the effects of inflation (escalation and changes in interest rates) presented above, there is no question but that we must anticipate large but uncertain capital-cost increases in the construction of all power-producing plants because of federal and state regulatory impact. Laws yet to be enacted in the areas of clean air, clean water, conservation, resource recovery, public-safety, public health, aesthetics (plant visual impact), and product liability can be expected to affect decision making even more than purely economic and equipment- availability considerations.

REFERENCES

[1] Edward Arthur Bruges, *Available Energy and the Second Law Analysis.* New York: Academic Press, 1959.

J. H. Keenan, "A Steam Chart for Second Law Analysis." *Mechanical Engineering,* 54:195–204, 1932.

M. W. Thring, "The Virtue of Energy, Its Meaning and Practical Significance." *Institute of Fuel. Journal,* 17:116–123, 1944.

C. Birnie and E. F. Obert, "Evaluation and Location of the Losses in a 60,000 KW Power Station," *Proceedings of the Midwest Power Conference,* 11:187–193, 1949. Chicago: Illinois Institute of Technology, 1949.

Allen Keller, "Evaluation of Steam-Power-Plant Losses by Means of the Entropy-Balance Diagram." *Transactions of the ASME,* 72:949–953, October 1950.

C. A. Meyer, et al., "Availability Balance of Steam Power Plants." *Transactions of the ASME. Journal of Engineering for Power,* 81, Series A:35–42, January 1959.

[2] E. Clapeyron and R. Clausius, "Memoir on the Motive Power of Heat" in Sadi Nicolas Leonard Carnot, E. Clapeyron and R. Clausius, *Reflections on the Motive Power of Fire; and other papers on the 2nd law of Thermodynamics.* Gloucester, Ma.: Peter Smith, 1962. Also published as a Dover paperback.

[3] Edward F. Obert and Richard A. Gaggioli, *Thermodynamics,* latest edition. New York: McGraw-Hill.

Newman A. Hall and W. E. Ibele, *Engineering Thermodynamics.* Englewood Cliffs, N.J.: Prentice-Hall, 1960, pp. 447–520.

[4] William J. M. Rankine, *A Manual of the Steam Engine and Other Prime Movers*, revised by W. J. Millar with a section entitled "Gas, Oil, and Air Engines" by Brian Duncan. London: Griffin and Co., 1908.

[5] C. D. Weir, "Optimization of Heater Enthalpy Rises in Feed-Heating Trains." *Institution of Mechanical Engineers. Proeedings*, 174:769–796, 1960. Discussion by R. W. Haywood, pp. 784–787.

   G. Chiantore, et al., "Optimizing A Regenerative Steam-Turbine Cycle." *Transactions of the ASME. Journal of Engineering for Power*, 83, Series A: 433–443, October 1961.

   J. Kenneth Salisbury, *Steam Turbines and Their Cycles.* Huntington, N.Y.: Robert E. Krieger, 1974, Part 3: Cycle Analysis.

[6] Anon., "The Reheat Cycle—A Re-Evaluation." *Combustion*, 21(12): 38–40, June 1950. Papers given at the Symposium on the Reheat Cycle sponsored by the ASME and held in New York, November 29–December 3, 1948. *Transactions of the ASME*, 71:673–749, 1949.

   J. K. Salisbury, "Analysis of the Steam-Turbine Reheat Cycle." *Transactions of the ASME* 80: 1629–1642, November 1958.

   J. K. Salisbury, "Power-Plant Performance Monitoring." *Transactions of the ASME Journal of Engineering for Power*, 83, Series A: 409–422, October 1961.

[7] *Cogeneration: Special Section; Power*, June, 1987, Vol. 131, No. 6.

[8] J. H. Horlock, *Cogeneration—Combined Heat and Power (CHP): Thermodynamics and Economics*, Pergamon Press, Oxford (England), 1987.

[9] R. W. Porter and K. Mastanaiah, "Thermal-Economic Analysis of Heat-Matched Industrial Cogeneration Systems," *Energy*, 7, 2, 1982.

[10] A. R. J. Timmermans, "Combined Cycles and Their Possibilities," Lecture Series, *Combined Cycles for Power Generation*, Von Karman Institute for Fluid Dynamics, Rhode Saint Genese, Belgium, 1978.

[11] I. Oliker, "Steam Turbines for Cogeneration Power Plants," *Trans. ASME—Journal of Engineering for Power*, 102, 482–485, 1980.

[12] M. P. Polsky, "Fuel Effectiveness of Cogeneration," Joint Power Generation Conference, ASME Paper 80-JPGC/Pwr-8, 1980.

[13] M. S. Reddy, F. Afshar, and R. J. Hollmeier, "Evaluation of Alternative System Designs in a Cogeneration Plant," *Proceedings of the American Power Conference*, 44, 1982. Chicago: Illinois Institute of Technology, 1982.

[14] Henry J. Blaskowski and Joseph G. Singer, "Gas Turbine Boiler Applications." *Combustion*, 28(11): 38–44, May 1957.

   W. H. Clayton and Joseph G. Singer, "Steam Generator Designs for Combined Cycle Applications." *Combustion*, 44(10): 26–32, April 1973.

[15] "The Mercury Power Plant from South Meadow to Schiller," General Electric Co. GER-246. Reprinted from *Power Generation*, March 1950.

[16] A. P. Fraas, "A Potassium-Steam Binary Vapor Cycle for Better Fuel Economy and Reduced Thermal Pollution," ASME Paper No. 71-WA/Ener-9, Nov.–Dec. 1971.

[17] D. H. Cooke, "Combined Cycle Thermodynamic Inquiries and Options," ASME/IEEE Joint Power Generation Conference, Paper 87-JPGC/Pwr-61, 1987.

[18] J. B. Stout, et al., "A Large Combined Gas Turbine-Steam Turbine Generating Unit." *Proceedings of the American Power Conference*, 24:404–411, 1962. Chicago: Illinois Institute of Technology, 1962.

[19] S. R. Wysk, H. H. Ropers, K. Janssen, and S. G. Drenker, "A Pressurized Circulating Fluidized Bed for Utility Applications," Eighth International Conference on Fluidized Bed Combustion, Houston, Texas, March 1985.

   "PFBC Turbocharged Boiler Design and Economic Study," EPRI Research Project RP-2428-2, December 1985.

   "PFBC Turbocharged Boiler Design and Economic Study," Cost and Economic Data Package prepared by Fluor Engineers for EPRI Research Project RP-2428-1, February 1986.

[20] This Act is Public Law 95-617, Nov. 9, 1978, 92 Stat. 3117, as amended; it is commonly referred to as PURPA.

[21] AIEE Probability Applications Working Group, *Application of Probability Methods to Generating Capacity Problems*. AIEE Paper CP 60–37. New York: American Institute of Electrical Engineers, 1960. (Unpublished).

   C. J. Baldwin, "Modern Scientific Tools Used in the Power Industry for Tomorrow's Problems." *Proceedings of the American Power Conference*, 24: 94–105, 1962. Chicago: Illinois Institute of Technology, 1962.

[22] R. D. Brown and D. A. Harris, *Large Coal-Fired Cycling Units*. Paper given at the ASME-IEEE-ASCE Joint Power Generation Conference, Portland, Oregon, Sept. 28–Oct. 2, 1975.

   Peter H. Benziger and Joseph G. Singer, "Design for Cycling at Chalk Point." *Proceedings of the American Power Conference*, 34:415–423, 1972. Chicago: Illinois Institute of Technology, 1972.

[23] Stanley Moyer, "Industry's Water Problems." *ASME Paper No. 6 WA-141*. New York: American Society of Mechanical Engineers, 1961. Also in condensed form in *Mechanical Engineering*, 84(3): 46–49, March 1962.

[24] James W. Lyons, "Optimizing Designs of Fossil-Fired Generating Units." *Power Engineering*, 83(2):50–56, February 1979.

[25] W. A. Wilson, "An Analytic Procedure for Optimizing the Selection of Power-Plant Components." *Transactions of the ASME*, 79:1120–1128, July 1957.

CHAPTER 2

Fossil Fuels

Fossil fuels used for steam generation in utility and industrial power plants may be classified into solid, liquid, and gaseous fuels as in Table I. Each fuel may be further classified as a natural, manufactured, or byproduct fuel. Not mutually exclusive, these classifications necessarily overlap in some areas.

Obvious examples of natural fuels are coal, crude oil and natural gas. Residual oils which are fired in boilers might be considered as a byproduct of the refining of crude oil. Wood, although a natural fuel, is rarely burned in boilers except in the form of sawdust, shavings, slabs and bark which remain as a byproduct after lumbering and pulping operations. Coal is the natural fuel from which coke, coke-oven gas, char, tars, chemicals and industrial gases may be converted by carbonization. Coal may also be gasified to obtain industrial gases for heating, chemical reduction, and hydrogenation and synthesis reactions.[1]

Of all the fossil fuels used for steam generation in electric-utility and industrial power plants today, coal is the most important. It is widely available throughout much of the world, and the quantity and quality of coal reserves are better known than those of other fuels. Many studies have been made of coal availability and utilization and should be consulted for more detailed information.[2]

FORMATION OF COAL

Coal forms as the result of a natural chemical process in which plants absorb carbon dioxide from the atmosphere. Sunlight, moisture and other factors convert carbon dioxide into compounds containing carbon, hydrogen and oxygen, such as sugars, starch, cellulose, lignin and other complex substances that make up the plant structure. Under favorable conditions, vegetation is converted into some of the many forms of coal now known to mankind.

When organic matter begins to change to coal, peat is the first product. In a block of peat one can often see, with the naked eye, woody fragments of stems, roots, and bark. As peat is buried, it is cut off from the oxygen in air, and rapid decay of its organic matter is prevented by slowing bacterial action. The weight of more vegetation falling on the peat helps to compress and solidify it, as does the weight of water when the deposit sinks below a lake or sea, as has often happened. Sometimes, mineral sediments have settled from muddy flood waters while vegetable matter was accumulating and formed "partings" or layers of shale in the coal vein. At the end of coal-forming periods, swamps remain flooded for a long time, and earthy sediments are deposited in thick beds over the peat, further compressing it and starting "coalification," the coal-making process.

Many kinds of coal are found in a natural deposit. In some, transitions from the extremes of lignite to anthracite exist in a single bed. It is generally assumed that differences in rank of coal are not caused by different source materials but by the agencies of coal formation.

THE FORMATION CYCLE

Geologists usually name time, pressure and temperature as the agents that change peat into different types of coal. Chemists include micro-organisms as an additional important factor.

Time, it must be remembered, is in itself no agent; it is merely the duration or period during which an agent has an opportunity to act. It becomes crucial, however, because many organic chemical reactions are slow. In fact, it is believed that reactions are going on slowly in most complex organic substances. In a short period the results of such reactions would be negligible, but over millions of years the total results must be large. Although the greatest single factor in the process of formation was probably time, other factors must also be sought, because coals of the same age and, in fact, in the same bed may be of different rank.

Generally, pressure is considered to be the factor of next importance, because coal of higher rank is generally found in regions that have been under high pressure. Anthracite, for example, is associated with earth-folding or mountain formation, which processes bring about great internal pressure. Experimentally, it has not been determined that pressure alone can change organic substances chemically.

Likewise, heat has played an important part in this great natural chemical industry. The temperature need not be high, for time brings about a relatively great change even at the low temperatures prevailing in the earth's crust.

Man completes the cycle by burning the various products of the natural process to carbon dioxide, and then nature starts all over again.

There is no satisfactory definition of coal. It is a mixture of organic chemical and mineral materials produced by a natural process of growth and decay, accumulation of debris both vegetal and mineral, with some sorting and stratification, and accomplished by chemical, biological, bacteriological and metamorphic action. The organic chemical materials produce heat when burned; the mineral matter remains as the residue called ash.

CLASSIFICATION OF COAL

Coals are grouped according to rank, the degree of progressive alteration in the transformation from lignite to anthracite.

Table I. Classification of Fuels

Natural Fuels	Manufactured or Byproduct Fuels
Solid	
Coal	Coke and coke breeze
	Coal tar
Lignite	Lignite char
Peat	
Wood	Charcoal
	Bark, saw dust and wood waste
	Petroleum coke
	Bagasse
	Refuse
Liquid	
Petroleum	Gasoline
	Kerosene
	Fuel oil
	Gas oil
	Shale oil
	Petroleum fractions and residues
Gaseous	
Natural gas	Refinery gas
Liquefied petroleum gases (LPG)	Coke-oven gas
	Blast-furnace gas
	Producer gas
	Water gas
	Carburetted water gas
	Coal gas
	Regenerator waste gas

For the purposes of the power-plant operator, there are several suitable ranks of coal:

- anthracite
- bituminous
- subbituminous
- lignite

Table II. Classification of Coals by Rank[a]

Class and Group	Fixed Carbon Limits, % (Dry, Mineral-Matter-Free Basis)		Volatile Matter Limits, % (Dry, Mineral-Matter-Free Basis)		Calorific Value Limits, Btu/lb (Moist,[b] Mineral-Matter-Free Basis)		Agglomerating Character
	Equal or Greater Than	Less Than	Equal or Greater Than	Less Than	Equal or Greater Than	Less Than	
I. Anthracitic							
1. Meta-anthracite	98	. . .	. . .	2	. . .	. . .	nonagglom-erating
2. Anthracite	92	98	2	8	. . .	. . .	
3. Semianthracite[c]	86	92	8	14	. . .	. . .	
II. Bituminous							
1. Low-volatile bituminous coal	78	86	14	22	. . .	. . .	commonly agglomerating[e]
2. Medium volatile bituminous coal	69	78	22	31	. . .	. . .	
3. High-volatile A bituminous coal	. . .	69	31	. . .	14,000[d]	. . .	
4. High-volatile B bituminous coal	. . .	. . .	. . .	. . .	13,000[d]	14,000	
5. High-volatile C bituminous coal	. . .	. . .	. . .	. . .	11,500	13,000	
					10,500	11,500	agglomerating
III. Subbituminous							
1. Subbituminous A coal	. . .	. . .	. . .	. . .	10,500	11,500	nonagglom-erating
2. Subbituminous B coal	. . .	. . .	. . .	. . .	9,500	10,500	
3. Subbituminous C coal	. . .	. . .	. . .	. . .	8,300	9,500	
IV. Lignitic							
1. Lignite A	. . .	. . .	. . .	. . .	6,300	8,300	
2. Lignite B	. . .	. . .	. . .	. . .	. . .	6,300	

[a] This classification does not include a few coals, principally nonbanded varieties, which have unusual physical and chemical properties and which come within the limits of fixed carbon or calorific value of the high-volatile bituminous and subbituminous ranks. All of these coals either contain less than 48% dry, mineral-matter-free fixed carbon or have more than 15,500 moist, mineral-matter-free Btu per pound.

[b] Moist refers to coal containing its natural inherent moisture but not including visible water on the surface of the coal.

[c] If agglomerating, classify in low-volatile group of the bituminous class.

[d] Coals having 69% or more fixed carbon on the dry, mineral-matter-free basis shall be classified by fixed carbon, regardless of calorific value.

[e] It is recognized that there may be nonagglomerating varieties in these groups of the bituminous class, and there are notable exceptions in high-volatile C bituminous group.

Reprinted from *ASTM Standards* D 388, Classification of Coals by Rank.

Being extremely broad, these terms fail to define rank completely. Many investigators have attempted to set up some scientific system of classification to accurately define the boundary lines of variation. Some better known bases for classification are those of Persifor Frazer, Jr., who made the earliest published classification of American coals in 1877; M. R. Campbell and S. W. Parr, who both published papers on the subject in 1906; David White, with publications in 1909 and 1913; and O. C. Ralston, whose graphic studies of some 3000 coal analyses were made in 1915. Each of these will serve to type a coal within narrow limits, but the suitability of a given coal for a specific purpose is best established by actual trial in the equipment for which it is selected.[3]

The American Society for Testing and Materials (ASTM) has established perhaps the most universally applicable basis for classifying coal according to fixed carbon and heating value (calorific value) calculated to a mineral-matter-free basis. As shown in Table II, this scheme represents a further development of the proposals of S. W. Parr. The high-rank coals are classified according to fixed carbon on the dry basis, and the low-rank coals according to Btu on the moist basis. Agglomerating indices differentiate between certain adjacent groups.

In commercial practice, it frequently suffices to calculate to a dry, ash-free basis. The ash, however, does not correspond to the mineral matter in coal. Thus, if the ash-free basis is used when classifying coal according to rank, significant errors may be introduced. Remarkably uniform results, however, are obtained from heating values of coals of a given rank and source when calculated to a dry mineral-matter-free basis.

In ASTM classification by rank, the agglomerating index, or caking quality of a coal, indicates the dividing line between noncaking coals and those having weakly caking properties. The noncaking designation applies only to coals that produce a noncoherent residue which can be poured out of the crucible as a powder or flakes that will pulverize easily with thumb and finger pressure.

The transformation of vegetal matter through wood and peat to lignite and finally to anthracite results in a reduction of volatile matter and oxygen content, with a simultaneous increase in carbon content. This is illustrated graphically in Fig. 1 in which moisture-and-ash-free volatile matter, fixed carbon, oxygen content, and high heating value are given for various ranks of coals as well as wood and peat.

RANKS OF COAL

The method of "proximate analysis" identifies the degree of coalification of the higher

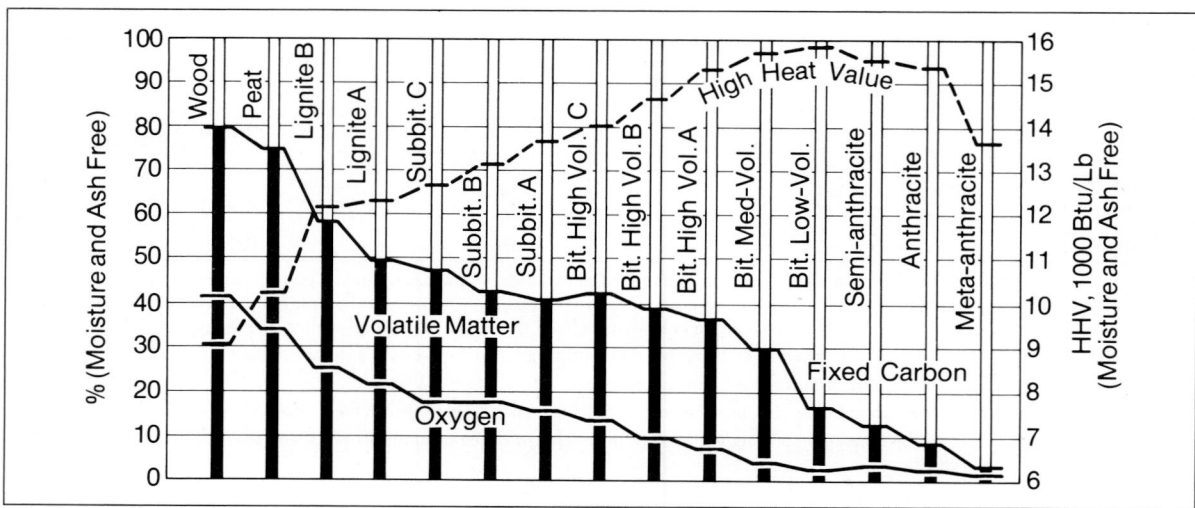

Fig. 1. **Progressive stages of transformation of vegetal matter into coal**

rank coals. In this method, a chemical analysis determines four constituents in coal: (1) water, called moisture; (2) mineral impurity, called ash, left when the coal is completely burned; (3) volatile matter, consisting of gases driven out when coal is heated to certain temperatures; and (4) fixed carbon, the coke-like residue that burns at higher temperatures after volatile matter has been driven off. For the lower rank coals, heating value and caking and weathering properties determine rank. The coal rank increases as the amount of fixed carbon increases and the amounts of moisture and volatile matter decrease. (Moisture and volatile matter were squeezed and distilled from coal during its formation by pressure and heat, raising the proportion of fixed carbon.) On an ash-free basis, the difference in constituents between a typical lignite, the lowest rank of coal, and an anthracite, the highest rank, is clearly shown in the following analysis:

	Lignite	Anthracite
Fixed Carbon (FC)	30%	92%
Volatile Matter (VM)	33%	5%
Moisture (H_2O)	37%	3%

The following description of coals by rank gives some of their physical characteristics.

ANTHRACITE

Hard and very brittle, anthracite is dense, shiny black, and homogeneous with no marks of layers. Unlike the lower rank coals, it has a high percentage of fixed carbon and a low percentage of volatile matter. Anthracites include a variety of slow-burning fuels merging into graphite at one end and into bituminous coal at the other. They are the hardest coals on the market, consisting almost entirely of fixed carbon, with the little volatile matter present in them chiefly as methane, CH_4. Anthracite is usually graded into small sizes before being burned on stokers: the "meta-anthracites" burn so slowly as to require mixing with other coals, while the "semianthracites," which have more volatile matter, are burned with relative ease if properly fired. Most anthracites have a lower heating value than the highest grade bitumi-

nous coals. Anthracite is used principally for heating homes and in gas production.

Some semianthracites are dense, but softer than anthracite, shiny gray, and somewhat granular in structure. The grains have a tendency to break off in handling the lump, and produce a coarse, sandlike slack.

Other semianthracites are dark gray and distinctly granular. The grains break off easily in handling and produce a coarse slack. The granular structure has been produced by small vertical cracks in horizontal layers of comparatively pure coal separated by very thin partings. The cracks are the result of heavy downward pressure, and probably shrinkage of the pure coal because of a drop in temperature.

BITUMINOUS

By far the largest group, bituminous coals derive their name from the fact that on being heated they are often reduced to a cohesive, binding, sticky mass. Their carbon content is less than that of anthracites, but they have more volatile matter. The character of their volatile matter is more complex than that of anthracites and they are higher in calorific value. They burn easily, especially in pulverized form, and their high volatile content makes them good for producing gas. Their binding nature enables them to be used in the manufacture of coke, while the nitrogen in them is utilized in processing ammonia.

The low-volatile bituminous coals are grayish black and distinctly granular in structure. The grain breaks off very easily, and handling reduces the coal to slack. Any lumps that remain are held together by thin partings. Because the grains consist of comparatively pure coal, the slack is usually lower in ash content than are the lumps.

Medium-volatile bituminous coals are the transition from high-volatile to low-volatile coal and, as such, have the characteristics of both. Many have a granular structure, are soft, and crumble easily. Some are homogeneous with very faint indications of grains or layers. Others are of more distinct laminar structure, are hard, and stand handling well.

High-volatile A bituminous coals are mostly homogeneous with no indication of grains, but some show distinct layers. They are hard and stand handling with little breakage. The moisture, ash and sulfur contents are low, and the heating value high.

High-volatile B bituminous coals are of distinct laminar structure; thin layers of black, shiny coal alternate with dull, charcoal-like layers. They are hard and stand handling well. Breakage occurs generally at right angles and parallel to the layers, so that the lumps generally have a cubical shape.

High-volatile C bituminous coals are of distinct laminar structure, are hard and stand handling well. They generally have high moisture, ash, and sulfur content and they are considered to be free-burning coals.

SUBBITUMINOUS

These coals are brownish black or black. Most are homogeneous with smooth surfaces, and with no indication of layers. They have high moisture content, as much as 15 to 30 percent, although appearing dry. When exposed to air they lose part of the moisture and crack with an audible noise. On long exposure to air, they disintegrate. They are free-burning, entirely noncoking, coals.

LIGNITE

Lignites are brown and of a laminar structure in which the remnants of woody fibers may be quite apparent. The word *lignite* comes from the Latin word *lignum* meaning wood. Their origin is mostly from plants rich in resin, so they are high in volatile matter. Freshly mined lignite is tough, although not hard, and it requires a heavy blow with a hammer to break the large lumps. But on exposure to air it loses moisture rapidly and disintegrates. Even when it appears quite dry, the moisture content may be as high as 30 percent. Owing to the high moisture and low heating value, it is not economical to transport it long distances.

Unconsolidated lignite (B in Table II) is also known as "brown coal." (Further differentiation between the various lignites and brown coals is given in Appendix A.) Brown coals are generally found close to the surface, contain more than 45 percent moisture, and are readily won by strip mining.

PEAT

Peat is not yet a commercial fuel in many countries where it is found, because of its very high moisture content and low heating value. It is a heterogeneous material of partially decomposed organic matter (plant material) and inorganic minerals that have accumulated in a water-saturated environment over a period of time. Its color can vary from yellow to brownish black, depending upon the degree of biological decay, mechanical disintegration of the plant fibers, and the presence of sediment. A water-saturated environment inhibits active biological decomposition of the plant material and promotes the retention of carbon and oxygen that would normally be released as gaseous products of the biological activity.

The universal problem in utilizing peat is its high moisture content. Even after draining a peat bog and solidifying the peat, it can still contain 70 to 95 percent water.

The sulfur content of peat is normally quite low, varying from negligible to less than 1 percent in dried peat. On the other hand, the ash content of peat can vary from 2 to 70 percent in assays of dry peat from a variety of sources.

COAL MINING

The two general methods of mining are stripping (open pit) and underground. In underground mining the coal is undercut, top-cut, or sheared and then blasted. Loading may be by hand or machine. If hand-loaded, the miners remove most of the larger visible impurities while shoveling the coal into mine cars or onto conveyors. With machine loading no such removal of impurities in the mine is possible. In strip mining the overburden is removed by large power shovels, and the coal by smaller shovels. If the coal is hard it is blasted prior to loading.

Most coal seams contain interstratified bands of impurities which must be removed in a coal-cleaning plant. Few coal seams are clean

enough to be mined by mechanical means and the coal shipped directly to market without mechanical cleaning.

Small strip mines frequently load coal of inferior quality. Outcrop coal from such mines is weathered, frequently mixed with impurities, and of low calorific value.

On the other hand, the better strip mines do not mine coal near the outcrop. Coal seams firm up and become equal to deep-mined coal when the overburden reaches a depth of 10 to 30 ft, depending on rank of coal and character of overburden. In general, coal resists weathering with increase of rank. Some anthracites can be mined to the outcrop, while low-rank bituminous coals usually are not acceptable if the overburden is less than 20 ft.

Practically all large, mechanized mines (underground and strip) have preparation plants in which the coal is sized, mechanically cleaned, and otherwise prepared to give a high-quality fuel conforming to size specifications.

COAL CLEANING

The oldest methods of cleaning coal use picking tables, which consist of a horizontal conveyor over which the coal moves slowly in comparatively thin layers. Workers on both sides of the conveyor pick the impurities from the coal. This is probably the best way to clean lump coal.

Most mechanical cleaning processes depend on differences in specific gravity of coal and associated impurities to effect a separation. Other physical properties of coal and refuse utilized to a minor extent are shape, resiliency, coefficient of sliding friction, electrical conductivity, and froth-flotation differences. Magnetic separators are used widely to remove tramp iron and thus prevent such material from getting into pulverizer mechanisms.

Coal-cleaning processes may be wet or dry, depending on whether water or air is used as the medium. In general, wet processes are more efficient than dry. With coal for which dry processes are suitable, the product is dry and highly desirable. Coal from wet-process cleaners must be dried by drainage, centrifuges, filters, or heat dryers to avoid excessive water in the final product.

COAL SAMPLING

A sample must represent the bulk of the coal from which it is taken. The items that should be most representative are ash and moisture content. The weight of the gross sample and the method of collecting and handling it depend on the size of the coal, the moisture and ash content, and the purpose for which the sample is collected. After collection, the sample must be handled so that the moisture content does not change. If the coal is very wet, considerable moisture may be lost during the handling. It may be necessary to stabilize the moisture in the gross sample by air drying before crushing and quartering. In such cases, the air-drying moisture loss must be determined.

The standard methods for sampling coals and preparing them for analysis are ANSI/ *ASTM Standards* D 2234, Collection of a Gross Sample of Coal, and D 2013, Preparing Coal Samples for Analysis.

COAL ANALYSIS

Two types of coal analyses are in general use: the proximate and the ultimate analysis, both expressed in percent by weight.

The proximate analysis gives information on the behavior of coal when it is heated; that is, how much of the coal goes off as gas and tar vapors, called the volatile matter, and how much remains as fixed carbon. The proximate analysis is easy and supplies useful information to assist in the selection of coal for steam generation. Along with the determination of volatile matter and fixed carbon also given are the moisture and ash contents and the heating value in Btu per pound or MJ per kilogram. Sulfur is given as a separate determination. ANSI/

ASTM Standards D 3172 is the basic method for proximate analysis of coal and coke.

The ultimate analysis gives the elements of which the coal substance is composed. These elements include carbon, hydrogen, nitrogen, oxygen and sulfur. Ash content is determined as a whole, and, when desirable, separate analysis is made on the ash. *ASTM Standards* D 3176 is the standard method for ultimate analysis of coal and coke.

Coal analysis may be given on several bases, and it is customary to select the basis to suit the application. Thus, for the purposes of classification, the dry or moist and mineral-matter-free bases are generally used. In combustion calculations the as-received basis is applicable.

AS-RECEIVED BASIS

The as-received analysis of a fuel represents the actual proportions of the constituents in the fuel sample as received at the laboratory. The sample may be fuel as fired, as mined, or in any other given condition.

MOISTURE-FREE (DRY) BASIS

Moisture content is variable, even in the same coal, under different conditions of handling and exposure. For example, coal as received at a plant may contain an amount of moisture different from that received at the laboratory for analysis, and both may vary with weather conditions. Also, in a plant burning pulverized fuel, the coal may carry one percentage of moisture as delivered to the raw-coal bunker, another as delivered to the pulverizer, another as delivered to the pulverized-fuel bunker (in a storage system), and still another as fired. Furthermore, when a laboratory determines an ultimate analysis as wet, as-received, or as-fired, the moisture can be reported as hydrogen and oxygen and added to the hydrogen and oxygen of the coal itself.

DRY MINERAL-MATTER-FREE BASIS

As mentioned previously, because the ash does not correspond in percentage to the mineral matter in the coal, errors are introduced which become significant in problems of classifying coals according to rank. Two formulas available for making such calculations from the as-received basis, are the Parr and approximation formulas. (See box.)

PARR FORMULAS:

Dry, Mm-free FC =
$$\frac{FC - 0.15S}{100 - (M + 1.08A + 0.55S)} \times 100$$

Dry, Mm-free VM = 100 − Dry, Mm-free FC

Moist, Mm-free Btu =
$$\frac{Btu - 50S}{100 - (1.08A + 0.55S)} \times 100$$

Note: The above formula for fixed carbon is derived from the Parr formula for volatile matter.[4]

APPROXIMATION FORMULAS:

Dry, Mm-free FC =
$$\frac{FC}{100 - (M + 1.1A + 0.1S)} \times 100$$

Dry, Mm-free VM = 100 − Dry, Mm-free FC

Moist, Mm-free Btu =
$$\frac{Btu}{100 - (1.1A + 0.1S)} \times 100$$

where
- Mm = mineral matter
- Btu = British thermal units per pound (calorific value)
- FC = percent of fixed carbon
- VM = percent of volatile matter
- M = percent of moisture
- A = percent of ash
- S = percent of sulfur

The formulas can be used to check analyses and frequently to identify the source and rank of the fuel. For example, the heating value of coals of a given rank and source are remarkably uniform when calculated on a dry, Mm-free basis.

In commercial practice it frequently suffices to calculate to a dry, ash-free basis as follows:

Dry, ash-free FC = $\dfrac{FC}{100 - (M + A)} \times 100$

Dry, ash-free VM = 100 − Dry, ash-free FC

ITEMS OF PROXIMATE ANALYSIS

Coal as mined and shipped contains varying amounts of water. Accurate determination of this water is not as simple as one would expect, since the sample frequently can lose moisture on exposure to the atmosphere. This is particularly true during the reduction of the sample for analysis. To prevent this loss, it is customary to air-dry the entire sample under specified conditions before fine grinding for analysis. The moisture is then determined by a standard procedure of drying in an oven, and the loss in weight is corrected for the air-drying loss. The moisture in coal does not represent all of the water present, since water of decomposition (combined water) and water of hydration are not given off under the conditions of test. The moisture content of coals varies widely. In the high-rank low-volatile bituminous coals it is frequently under 5 percent. High-volatile bituminous coals may have as much as 12 percent and lignite, as high as 45 percent as mined. The finer sizes will often retain more moisture than the coarse sizes of the same coal subjected to rainfall or wet-washing.

ASTM defines total moisture in coal as that moisture determined as the loss in weight in an air atmosphere under rigidly controlled conditions of temperature, time and airflow as established in *ASTM Standards* D 3302.

INHERENT MOISTURE

Inherent moisture in coal is that moisture existing as a quality of the coal seam in its natural state of deposition and includes only that water considered to be a part of the deposit, and not that moisture which exists as a surface addition. There are a number of other terms relating to moisture in coal, including: bed moisture, equilibrium moisture, air-dry moisture loss, free moisture, water of hydration and others of lesser consequence.

CONDITION OF WATER IN COAL

Like many other substances of vegetable origin, coal contains water in the ordinary condition, which may give the coal a moist appearance. Coal also contains water that is frequently but erroneously called "water of combination," this water being concealed, as in hydrated salts, by the dry appearance of the material. The condition in which the so-called "water of combination" in coal exists is problematical; in the true chemical sense the water is probably not "combined," that is, united with another substance in definite proportions, as determined by its relative molecular weight and the number of molecules.

There is no sharp line of demarcation; the "free" water cannot be separated with exactness from the "combined" water by chemical analysis. Under all ordinary drying conditions, there is a tendency for more than the "free" water to leave the coal. But the "combined" or, "inherent" water, is distinctly different from the "free" water in its properties, as it has a subnormal vapor pressure and an energy change is involved in its combination with, or separation from, the coal substance. In other words, the inherent water is held more tenaciously and cannot be entirely removed by drying the coal in ordinary air which contains relatively large quantities of moisture.

PRACTICAL EFFECTS OF WATER

As ordinarily found in the market, coal, when superficially dry, contains water; an anthracitic or low-volatile coal in this condition contains possibly 2 percent water; a high-volatile bituminous coal, 6 to 10 percent; and a subbituminous coal, 14 to 18 percent. Even finely crushed or powdered coal does not become dry in ordinary air and requires an artificially and thoroughly dried atmosphere to remove all of its water content.

Humidity conditions may appreciably alter the weight of exposed coal, independently of the actual fall of rain or snow. In other words, coal is frequently hygroscopic. After being transferred from storage in an atmosphere of 40 percent humidity to one of 75 percent humidity, a bituminous coal, for example, may gain 3 percent and a subbituminous 7 percent in weight by absorption of moisture.

FREE SUPERFICIAL WATER

In wet coal, the water in excess of a certain percentage (which depends on the kind of coal) is mechanically held in the free state, its vapor pressure and other properties being, for all practical purposes, normal. This excess moisture may be termed superficial or accidental moisture. All water in the coal above approximately 3 percent in the anthracite coals, 12 percent in bituminous, and 22 percent in the subbituminous coals may be so classed, although this line of demarcation is more or less variable even among coals of the same kind.

The percentage of superficial water retained by coal that has been wetted and the water drained off is greatly affected by the size of the coal. In screenings or slack, which contains pieces of many different sizes, the particles pack closely together and form small interstices which retain water by capillarity. Screened lump, on the other hand, or run-of-mine with a large proportion of lump, has much larger interstices and relatively less surface to be wet, and therefore retains less water. See Table III.

There is no ready means of separating sharply the superficial and the inherent water of coal. Unless the air is saturated, the superficial water can all be removed by air drying, the rate of evaporation depending on the temperature, the humidity of the air, and the fineness of the coal. But at the same time, part of the inherent water leaves the coal and continues to do so as long as its vapor pressure is higher than that of the air.

By the expression "inherent water" in coal or in any material, therefore, is meant the water which exists as such but which has a vapor pressure less than the normal.

VOLATILE MATTER

The volatile matter is that portion which, exclusive of water vapor, is driven off in gas or vapor form when the coal is subjected to a standardized temperature test. It consists of hydrocarbons and other gases resulting from distillation and decomposition.

Table III. Water Retention After Draining[1]

Kind of Coal	* %	** %	*** %
Bituminous, New River, WV.:			
Screenings *a*	22.27	2.52	1.03
Sized "buckwheat" *b*	3.17	2.35	1.00
Bituminous, Pittsburgh bed, PA.:			
Screenings *a*	18.73	2.68	1.15
Sized "buckwheat" *b*	2.62	2.73	1.25
Bituminous, Macoupin County, IL.:			
Screenings *a*	28.60	6.84	.32
Sized "buckwheat" *b*	16.71	6.63	3.40
Subbituminous, Big Horn County, WY.:			
Screenings *a*	32.58	14.24	6.38
Sized "buckwheat" *b*	18.70	13.79	6.65

[1] From H. C. Porter and O. C. Ralston, "Some Properties of the Water in Coal," U.S. Bureau of Mines, Technical Paper 113. Washington; U.S. Bureau of Mines, 1916.
  *Total water content of coal after soaking and draining.
 **Normal water content in commercially "dry" coal.
***Water retained in coal after "air-drying" in laboratory at 35°C.
  a Through ½ inch mesh screen.
  b From ¼ inch to 1 inch in diameter.

Volatile matter is determined by prescribed methods which may vary according to the nature of the material, but in the case of coal and coke Method D 3175 is used.

Temperature and time are a vital concern in this test, since they actually determine the definition of volatile matter. Temperature must be 950°C ± 20°C, and heating time must be exactly 7 minutes.

The main constituents of volatile matter in all ranks of coal are hydrogen, oxygen, carbon monoxide, methane and other hydrocarbons, and that portion of moisture that is formed by chemical combination during thermal decomposition of the coal substance. The composition of volatile matter varies greatly for different ranks of coal.

Volatile matter is used to establish the rank of coals, to indicate coke yield on carbonization processes, to provide the basis for purchasing and selling, and to establish burning characteristics.

Because of the arbitrary nature of this test, caution must be used in comparing the results of volatile analyses with those obtained from tests run in other countries. Details of the various national standards vary greatly.

FIXED CARBON

The fixed carbon is the combustible residue left after driving off the volatile matter. It is not all carbon, and its form and hardness are an indication of the coking properties of a coal, and therefore, a guide in the choice of fuel-firing equipment. In general, the fixed carbon represents that portion of the fuel that must be burned in solid state, either in the fuel bed on a stoker, or as solid particles in the pulverized-fuel furnace.

The fixed carbon in a proximate analysis is a calculated figure obtained by subtracting from 100 the sum of the percentages of moisture, volatile matter and ash.

ASH

Ash is the noncombustible residue after complete combustion of the coal. The weight of ash is usually slightly less than that of the mineral matter originally present before burning.

For high-calcium-content coals, however, the ash can be higher than the mineral matter due to retention of the oxides of sulfur.[5]

The ASTM definition for coal ash is the inorganic residue remaining after ignition of combustible substances, determined by definite prescribed methods. This definition is followed by two notes, one of which states that ash may not be identical—in composition or quantity—to inorganic substances present in the coal before ignition. The second note specifies that, in the case of coal and coke, the methods shall be those prescribed by *ASTM Standards* D 3174. This method determines the ash content by weighing the residue remaining after the coal is burned under rigidly controlled conditions of sample weight, temperature, time and atmosphere, oxidizing or reducing.

Although the definition of ash content used in other countries is similar to the American definition, the ashing conditions may be different. Therefore, it is desirable to spell out the conditions used, particularly when the results from laboratories in different countries are to be compared.

During the burning process, various chemical and physical changes take place. The conditions of oxidation determine the number and extent of such changes; thus, a great degree of variability can be expected in separate determinations of ash content even on portions of the same coal sample unless standardized procedures are closely followed. In particular, this is true of coals with relatively large amounts of carbonates or pyrite.

Ash is usually considered the product of complete oxidation of coal. It is composed of the oxides formed from the mineral constituents of coal. However, these minerals may be present in two forms in coal: as visible impurities, or as minute impurities so finely divided and so intimately mixed that they may be considered a part of the coal structure.

The term *inherent* or *fixed ash content* is used to designate that portion of the ash content that is structurally part of the coal and cannot be separated from it by mechanical means. This is a relative term, however, and will have different values, since mechanical separation is accomplished at varying levels according to the size of the coal.

In general, when the term *inherent ash* is used, it is assumed it refers to the residue that remains after coal has been broken to the size at which it is to be used, cleaned by a mechanical process, and incinerated. Probably the first test developed for coal, the determination of ash content continues to be one of the most important of the tests performed.

ITEMS OF ULTIMATE ANALYSIS

Ultimate analysis is needed for the computation of air requirements, weight of products of combustion, and heat losses, on boiler tests. The air requirements and the weight of products of combustion determine fan sizes. The following are items of ultimate analysis as determined by *ASTM Standards* D 3176.

TOTAL CARBON

Total carbon includes both the carbon in the fixed carbon and in the volatile matter, and will be proportionately greater than the fixed carbon as the volatile content of the coal increases. All this carbon appears in the products of combustion as CO_2 when the fuel is completely burned.

HYDROGEN

All hydrogen in the fuel is burned to water and, together with the moisture in fuel, appears as water vapor in the waste gas. In the publications of the U.S. Bureau of Mines, ultimate analysis of coal on as-received basis includes moisture in the hydrogen and oxygen items. The weight of the water vapor in the products of combustion is nine times the weight of the hydrogen item.

NITROGEN

Nitrogen in most solid fuels is relatively low and of little importance in combustion calculations. However, a portion of the fuel nitrogen may react during combustion to form nitrogen oxides.

SULFUR

Sulfur content of a fuel is useful in judging the corrosiveness of the products of combustion. Combustion of sulfur forms oxides, which combine with water to form acids that may be deposited when the combustion gas is cooled below its dew-point temperature.

Sulfur in coal occurs mainly in three forms. It can be present in organic combination as part of the coal substance; it can be present as the sulfide ion in pyrites and marcasite; or it can be present as the sulfate ion. Rarely, sulfur may occur as elemental sulfur, but in insufficient quantities to be appreciable. ANSI/*ASTM Standards* D 2492 gives a standard test method for the determination of the forms of sulfur in coal; sulfate sulfur and pyritic sulfur are first determined, and organic sulfur is found by determining the difference from the total sulfur.

Two methods are generally accepted for total sulfur determination. Eschka's method converts all the sulfur present in the coal to the sulfate ion, which is then precipitated as $BaSO_4$. The Eschka method requires a time period— including sample preparation—of up to 24 hours, so a more rapid method of high-temperature combustion has been adopted. Details and analytical procedures for both of these methods, as well as the bomb-washing method, an alternate third method, may be found in *ASTM Standards* D 3177.

OXYGEN

The oxygen content of fuels is a guide to the rank of the fuel. The amount of oxygen is high in low-rank fuels like lignite. Oxygen in fuels is in combination with carbon or hydrogen and, therefore, represents a reduction in the potential heat of a fuel. High-oxygen fuels have low heating values.

As there is no direct ASTM method of determining oxygen, it is calculated by subtracting from 100 the sum of the other components of the ultimate analysis. Any errors incurred in the other determinations are placed on the oxygen, and a material balance cannot be obtained to aid in checking the accuracy of the ultimate analysis.[6]

HEATING VALUE

The calorific or heating value of a solid fuel is expressed in Btu per lb (or MJ per kg) of fuel on as-received, dry, or moisture- and ash-free basis. It is the amount of heat recovered when the products of complete combustion of a unit quantity of a fuel are cooled to the initial temperature of the air and fuel.

Heating values as determined in calorimeters are termed high or gross heating values, and include the latent heat of the water vapor in the products of combustion. The most common type of calorimeter in use today is the adiabatic bomb calorimeter, and *ASTM Standards* D 2015 covers this test. In actual operation of boilers, the water vapor in the combustion gas leaving is not cooled below its dew point, and this latent heat is not available for making steam. The latent heat can be subtracted from the high, or gross, heating value to give the low, or net, heating value.

This deduction in Btu per lb of fuel is equal to the total pounds of water vapor per lb of fuel (moisture in the fuel, plus vapor formed by combustion of hydrogen of the coal substance) multiplied by the latent heat of evaporation at the partial pressure of the vapor in the exit gas. The value used varies from 1030 to 1080 Btu per lb of water vapor. Low or net heating values are standard in European practice, and high heating values are standard in American practice.[7]

For anthracite and bituminous coal, gross heating value in Btu per pound of coal can be calculated approximately by a formula of the Dulong type. (See box below.)

DULONG-TYPE FORMULA

$$HHV = 14{,}600\ C + 62{,}000\left(H - \frac{O}{8}\right) + 4050\ S$$

where: HHV is in Btu/lb and C, H_2, O_2 and S are carbon, hydrogen, oxygen, and sulfur in the coal, respectively, expressed in fraction of a pound, usually as fired. The results are approximate because the formula does not take into account the heats of dissociation and similar phenomena occurring during combustion. For low-rank coals, the heating values obtained by this formula are generally too low.

ASH-FUSION TEMPERATURE

When coal ash is heated, it becomes soft and sticky and, as the temperature continues to rise, it becomes fluid. *ASTM Standards* D 1857 is the method for measuring fusibility of coal ash. This test is an observation of the temperatures at which triangular pyramids or cones prepared from coal ash and coke ash attain and pass through certain defined stages of fusing and flow when heated at a specific rate in controlled, mildly reducing, and where desired, oxidizing atmosphere.

This method is empirical, and strict observance of the requirements and conditions is necessary to obtain reproducible temperatures and to enable different laboratories to obtain concordant results.

Four stages of fusion temperature are usually reported in American practice:

1. Initial, or the first rounding of the cone
2. Softening, when the height has diminished until it is equal to the width at the base
3. Hemispherical, when the height of the lump equals one-half the width of the base
4. Fluid, when the mass is no higher than one-sixteenth inch

The test for fusion temperatures may be run in either a reducing or an oxidizing atmosphere. The ash-fusion temperature given in most existing tables is the softening temperature (H = W) in a reducing atmosphere. This subject is expanded in Chapter 3.

GRINDABILITY

This test determines the relative ease of pulverization of coal in comparison with coals chosen as standards. The Hardgrove method has been accepted as the standard, and *ASTM Standards* D 409 is the Grindability of Coal by the Hardgrove-Machine Method.

Each Hardgrove machine is calibrated by use of standard reference samples of coal, having grindability indexes of approximately 40, 60, 80 and 100. Standard coals may be obtained from the U.S. Bureau of Mines in Pittsburgh for the purpose of calibration.

The Hardgrove-index number reported by

the laboratory is based on an original soft coal chosen as a standard coal whose grindability index was set at 100. Therefore, the harder the coal, the lower the index number.

Since the grindability index varies, not only from seam to seam but within the same seam, grindability data are of utmost economic importance to the users of commercial grinding and pulverizing equipment.

The results of grindability measurements by the Hardgrove machine are affected by several factors, among them the ash and moisture content, temperature, and the presence of different petrographic constituents (organic components distinguishable by microscopic inspection). Chapter 11 has more information.

TYPICAL COAL ANALYSES

U.S. Bureau of Mines publications contain the proximate and ultimate analyses and softening temperatures of ash of several coals from various states. The coals are identified by country, mine, coal bed, and rank.[8]

Table IV (Pg. 2-16) gives analyses of several typical U.S. coals and coke, classified by the ASTM system of Table II. Besides proximate and ultimate analyses, high heating value and theoretical combustion-air requirements "A" (on a pounds-per-million-Btu basis) are given. Table V (Pg. 2-18) presents an overview of coals from outside the continental U.S.

Appendix A gives extensive data on both United States coals and coals from all the major producing areas of the world. Also, in Appendix A, the International Systems for Classification of Hard Coals, Brown Coals and Lignites are discussed in detail. Additional analyses of American, European, and Asian coals are given, along with international classifications.

BURNING CHARACTERISTICS OF COALS

Coal is burned in power plants in either the crushed and sized form, or in the crushed and subsequently pulverized form. Previous editions of this text have concentrated on the burn-

ing characteristics exhibited when used on different types of stokers. Of particular interest in this edition is the behavior of coal when burned in pulverized condition.

The combustion of pulverized coal is a very complex process consisting of a number of overlapping steps, including heating, ignition, devolatization and char (carbonaceous residue) burnout. The sequence of these events is depicted in Figure 2. Char burnout occupies the majority of the time required for complete combustion. This step, therefore, constitutes a bottleneck which dictates the overall combustion efficiency of pulverized coal. Each of these steps is briefly described below.

HEATING

The cloud of relatively cool pulverized coal particles entering a hot furnace is heated by flame radiation and by mixing with recirculated hot combustion products. This initial phase in the coal combustion process results in the loss of moisture from and minor devolatization of coal particles.

IGNITION/DEVOLATIZATION

Upon further heating of the particle cloud, a point is reached where ignition occurs. The temperature at which ignition occurs is controlled by a balance between the rates of heat generation and heat loss; hence it may vary depending upon the furnace design and operating conditions. Ignition stability exists when the rate of heat generation is equal to the rate of heat loss.

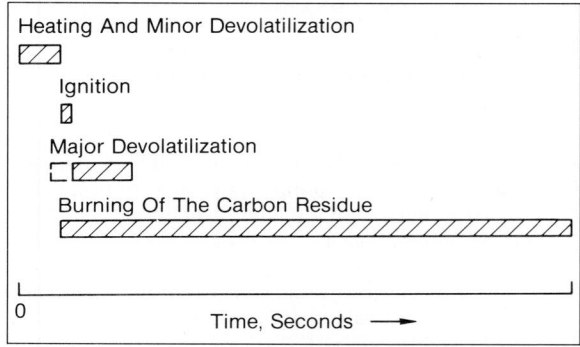

Fig. 2 Chronology of the combustion process

Table VI. Typical Flammability Indices

Coal Rank	Flammability Index (°F)
Anthracite	1450–1750
Bituminous	1050–1250
Subbituminous	900–1100
Lignite	800–1050

After ignition occurs, the devolatization rapidly proceeds, typically requiring approximately 10 percent of the overall combustion time beyond ignition. Studies have indicated that the volatile matter yields during the rapid heating rates (10^4 to 10^5°C) characteristic of dispersed particles in coal flames are generally significantly higher than those observed during slow heating rates encountered in ASTM proximate volatile-matter determination.[9,10] The rate of devolatization, the total quantity of volatile matter released, and the composition/calorific value of the volatile matter play significant roles during the early stages of combustion. Therefore, these characteristics have a strong impact on ignition stability.

C-E has developed a simple bench-scale test to directly assess the relative ignition stability characteristics of a pulverized coal.[11] This test determines a Flammability Index which is the ignition temperature of a suspension of pulverized fuel in an oxygen atmosphere under specific conditions. Typical ranges of flammability indices determined for coals of various rank are shown in Table VI. The flammability index of a fuel is useful in classifying relative ignition behavior by virtue of comparison with those of fuels with known commercial performance.

CHAR REACTIVITY

In trying to understand the burning characteristics of coal, char reactivity is focused upon because its burnout takes much longer than the volatile matter release (i.e., devolatilization) and burnout in the gas phase by approximately one order of magnitude (Figure 2). Char reactivity is affected by the quantity and rate of volatile matter released as well as other fuel aspects (swelling, agglomerating tendency and mineral matter).

Swelling and agglomeration affect the nature of the pore structure of the residual char, thus influencing the burning characteristics of this char. The mineral matter, depending on its chemical nature and degree of dispersion in the coal matrix, may catalyze or hinder the chemical reactivity of the coal particular char.

A procedure has been developed at C-E for preparing coal chars under specific conditions and characterizing them from the standpoints of their specific pore structures and reactivities.[12]

Char Preparation. C-E's Drop-Tube Furnace System (Figure 3), which is a 2-inch inside diameter laminar–flow reactor, is used to prepare

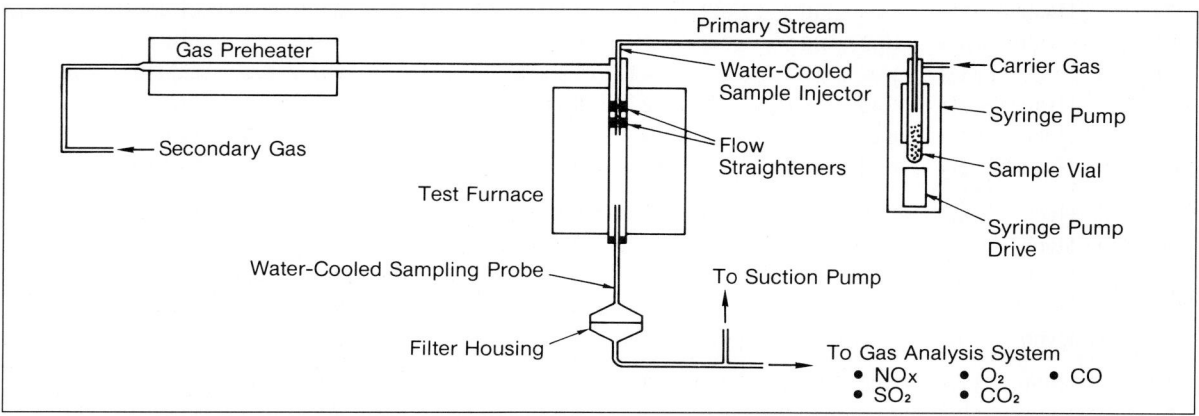

Fig. 3 Schematic of the Combustion Engineering Drop-Tube Furnace System

Table IV. Analyses of Typical U.S. Coals, as Mined

State	Rank	% Proximate Analysis				% Ultimate Analysis						HHV, Btu/Lb	A at Zero Excess Air, Lb/10⁶ Btu*
		H₂O	VM	FC	ASH	H₂O	C	H₂	S	O₂	N₂		
RI	Ma	13.3	2.5	65.3	18.9	13.3	64.2	0.4	0.3	2.7	0.2	9,313	808
CO	A	2.5	5.7	83.8	8.0	2.5	83.9	2.9	0.7	0.7	1.3	13,720	787
NM	A	2.9	5.5	82.7	8.9	2.9	82.3	2.6	0.8	1.3	1.2	13,340	786
PA **	A	5.4	3.8	77.1	13.7	5.4	76.1	1.8	0.6	1.8	0.6	11,950	791
***	A	2.3	3.1	87.7	6.9	2.3	86.7	1.9	0.5	0.9	0.8	13,540	794
****	A	4.9	3.7	82.2	9.2	4.9	81.6	1.8	0.5	1.3	0.7	12,820	788
AR	Sa	2.1	9.8	78.8	9.3	2.1	80.3	3.4	1.7	1.7	1.5	13,700	770
PA	Sa	3.0	8.4	78.9	9.7	3.0	80.2	3.3	0.7	2.0	1.1	13,450	777
VA	Sa	3.1	10.6	66.7	19.6	3.1	70.5	3.2	0.6	2.2	0.8	11,850	782
AR	Lvb	3.4	16.2	71.8	8.6	3.4	79.6	3.9	1.0	1.8	1.7	13,700	774
MD	Lvb	3.2	18.2	70.4	8.2	3.2	79.0	4.1	1.0	2.9	1.6	13,870	761
OK	Lvb	2.6	16.5	72.2	8.7	2.6	80.1	4.0	1.0	1.9	1.7	13,800	775
WV	Lvb	2.7	17.2	76.1	4.0	2.7	84.7	4.3	0.6	2.2	1.5	14,730	767
PA	Mvb	3.3	20.5	70.0	6.2	3.3	80.7	4.5	1.8	2.4	1.1	14,310	765
VA	Mvb	3.1	21.8	67.9	7.2	3.1	80.1	4.7	1.0	2.4	1.5	14,030	778
AL	Hvab	5.5	30.8	60.9	2.8	5.5	80.3	4.9	0.6	4.2	1.7	14,210	768
CO	Hvab	1.4	32.6	54.3	11.7	1.4	73.4	5.1	0.6	6.5	1.3	13,210	763
KS	Hvab	7.4	31.8	52.4	8.4	7.4	70.7	4.6	2.6	5.0	1.3	12,670	769
KY	Hvab	3.1	35.0	58.9	3.0	3.1	79.2	5.4	0.6	7.2	1.5	14,290	758
MO	Hvab	5.4	32.1	53.5	9.0	5.4	71.6	4.8	3.6	4.2	1.4	12,990	769
NM	Hvab	2.0	33.5	50.6	13.9	2.0	70.6	4.8	1.3	6.2	1.2	12,650	766
OH	Hvab	4.9	36.6	51.2	7.3	4.9	71.9	4.9	2.6	7.0	1.4	12,990	762
OK	Hvab	2.1	35.0	57.0	5.9	2.1	76.7	4.9	0.5	7.9	2.0	13,630	757
PA	Hvab	2.6	30.0	58.3	9.1	2.6	76.6	4.9	1.3	3.9	1.6	13,610	773
TN	Hvab	1.8	35.9	56.1	6.2	1.8	77.7	5.2	1.2	6.0	1.9	13,890	767
TX	Hvab	4.0	48.9	34.9	12.2	4.0	65.5	5.9	2.0	9.1	1.3	12,230	767
UT	Hvab	4.3	37.2	51.8	6.7	43	72.2	5.1	1.1	9.0	1.6	12,990	758
VA	Hvab	2.2	36.0	58.0	3.8	2.2	80.6	5.5	0.7	5.9	1.3	14,510	764
WA	Hvab	4.3	37.7	47.1	10.9	4.3	68.9	5.4	0.5	8.5	1.5	12,610	758
WV	Hvab	2.4	33.0	60.0	4.6	2.4	80.8	5.1	0.7	4.8	1.6	14,350	768
IL	Hvcb	8.0	33.0	50.6	8.4	8.0	68.7	4.5	1.2	7.6	1.6	12,130	766
KY	Hvcb	7.5	37.7	45.3	9.5	7.5	66.9	4.8	3.5	6.4	1.4	12,080	774
MO	Hvcb	10.5	32.0	44.6	12.9	10.5	63.4	4.2	2.5	5.2	1.3	11,300	773
OH	Hvcb	8.2	36.1	48.7	7.0	8.2	68.4	4.7	1.2	9.1	1.4	12,160	762
WY	Hvcb	5.1	40.5	49.8	4.6	5.1	73.0	5.0	0.5	10.6	1.2	12,960	757
IL	Hvbb	12.1	40.2	39.1	8.6	12.1	62.8	4.6	4.3	6.6	1.0	11,480	769
IN	Hvbb	12.4	36.6	42.3	8.7	12.4	63.4	4.3	2.3	7.6	1.3	11,420	758
IA	Hvbb	14.1	35.6	39.3	11.0	14.1	58.5	4.0	4.3	7.2	0.9	10,720	754

Table IV. **Analyses of Typical U.S. Coals, as Mined — *Continued***

State	Rank	% Proximate Analysis				% Ultimate Analysis						HHV, Btu/Lb	A at Zero Excess Air, Lb/10⁶ Btu*
		H₂O	VM	FC	ASH	H₂O	C	H₂	S	O₂	N₂		
MI	Hvbb	12.4	35.0	47.0	5.6	12.4	65.8	4.5	2.9	7.4	1.4	11,860	762
CO	Sub	19.6	30.5	45.9	4.0	19.6	58.8	3.8	0.3	12.2	1.3	10,130	756
WY	Sub	23.2	33.3	39.7	3.8	23.2	54.6	3.8	0.4	13.2	1.0	9,420	757
ND	Lig A	34.8	28.2	30.8	6.2	34.8	42.4	2.8	0.7	12.4	0.7	7,210	750
TX	Lig A	33.7	29.3	29.7	7.3	33.7	42.5	3.1	0.5	12.1	0.8	7,350	752

*A is the air required for combustion under stoichiometric conditions (no excess air), with 0.013 lb H₂0 per lb dry air.

**Orchard Bed, ***Mammoth Bed, ****Holmes Bed, RANK KEY: Ma-Meta-anthracite, A-Anthracite, Sa-Semi-anthracite, Lvb-Low-Vol. Bituminous, Mvb-Med.-Vol. Bituminous, Hvab-High-Vol. Bituminous A, Hvcb-High-Vol. Bituminous B, Hvbb-High-Vol. Bituminous C, Sub-Subbituminous, Lig A-Lignite A

each coal char. This entails pyrolyzing a $70 \pm 2\%$ through 200 mesh representative coal sample in nitrogen atmosphere at 2650°F to drive off the volatile matter under heating rates which are commensurate with those encountered in pulverized coal-fired boilers. The resulting char is size-graded to 200x400 mesh.

Char Characterization. This 200x400 mesh, volatile matter-free, char is subjected to pore structural analysis through measurements of its BET surface area in nitrogen at -196°C[13] and mercury and helium densities under specific conditions[14], which can be used to determine its total open porosity. This same char is also burned in a thermo-gravimetric analysis appa-

ratus (TGA) in air at 1290°F (700°C) to determine its burn-off under isothermal conditions as a function of time. These types of characterization give quick classifications of the pore nature and reactivity of a given coal char by virtue of comparison with the information in the data bank.

Typical Results. TGA burn-off curves of chars prepared from coals ranging in rank from lignite to anthracite are given in Figure 4 along with corresponding BET surface areas. In general, the reactivity of a char increases with increasing BET surface area. Fuels whose chars have BET surface areas greater than 50 m²/g are typically very reactive, and, hence, do not

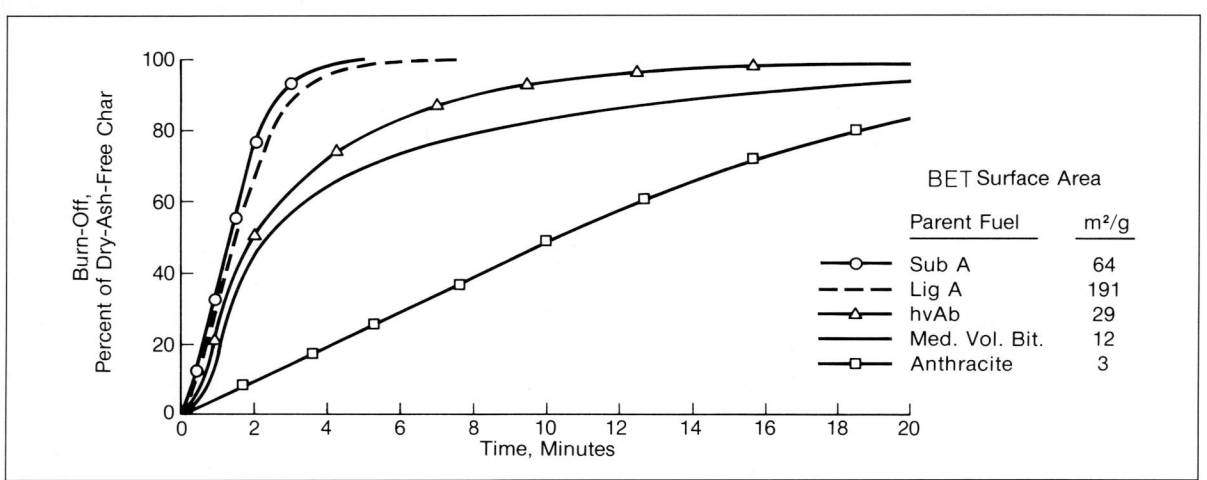

Fig. 4 **Thermogravimetric burn-off of 200x400 mesh DTFS chars at 700°C**

Table V. Typical Analyses of Coals of the World

Country	District or Mine	H₂0	Vol	FC	Ash	Sul	MJ/kg	HHV, Btu/Lb
			TYPICAL ANALYSIS—AS RECEIVED					
Argentina	Turbio River	8.6	34.8	40.9	15.7	0.9	24.1	10,360
Australia	New South Wales							
	Western Field	2.6	30.0	52.8	14.6	0.7	27.0	11,620
	Southern Field	0.6	23.29	65.08	11.0	3.7	30.4	13,090
	Queensland	1.5	37.0	49.9	11.6	. . .	28.6	12,300
	Victoria (Brown Coal)	66.3	17.7	15.3	0.7	0.1	8.6	3,700
Belgium	Batterie and Vidette	1.0	8.5	63.5	27.0	. . .	24.9	10,710
Brazil	Sao Jeronimo	13.8	24.7	27.1	34.4	3.0	15.3	6,600
	Sao Jeronimo-Washed	16.0	23.1	32.9	28.0	0.9	17.6	7,570
	Butia	11.5	32.0	42.9	13.6	1.3		. . .
Canada	Alberta-Drumheller	20.0	28.0	41.0	11.0	. . .	21.0	9,020
	Saskatchewan-Souris	35.0	23.4	34.6	8.0	. . .	17.0	7,290
	British Col.-Crows Nest	1.4	24.5	61.8	12.3	0.5	31.1	13,360
	Nova Scotia-Emery	4.0	33.7	51.6	10.7	2.5	29.3	12,600
Chile	Schwager	2.9	41.3	52.2	3.6	0.9	33.3	14,310
	Lota	3.4	39.6	55.4	1.6	0.7	33.1	14,220
	Mafil	12.6	35.6	40.0	11.8	0.6	23.5	10,120
China	Kailin	2.5	29.9	44.4	25.7	0.66	24.0	10,300
	Kew Loong Kieng	5.4	29.7	45.0	19.9	. . .	25.3	10,870
	Kiaping	3.5	24.4	41.1	31.0	. . .	21.6	9,290
Colombia	Bogota	5.3	23.4	63.7	7.6	0.8	33.1	14,220
France	Bethune	8.0	18.0	39.0	35.0	. . .	19.6	8,410
	Anzin	1.6	9.2	44.3	44.9	. . .	17.9	7,690
Germany	Frimmersdorf-Westfield	60.7	20.0	16.7	2.6	0.2	9.9	4,240
	Saar	9.7	31.8	48.7	9.8	. . .	25.7	11,040
	Saxony	53.1	25.3	18.1	3.5	0.9	10.5	4,500
	Westphalia	1.7	23.7	69.2	5.4	. . .	32.5	13,970
	Lower Silesia	4.5	25.6	56.1	13.8	. . .	27.4	11,770
United Kingdom								
Wales	Cardiff	1.5	11.0	85.5	2.0	0.8	33.5	14,400
	Arley	1.1	36.3	59.0	3.6	2.3	34.1	14,680
England	Durham	1.5	34.7	60.0	3.8	0.87	31.0	13,340
Scotland	Lanark	7.5	31.8	56.7	4.0	0.2	31.8	13,680
Greece	Aliveri	31.0	30.0	21.0	18.0	. . .	13.1	5,640

Table V. Typical Analyses of Coals of the World—*Continued*

Country	District or Mine	H₂0	TYPICAL ANALYSIS—AS RECEIVED				MJ/kg	HHV, Btu/Lb
			Vol	FC	Ash	Sul		
India	Bermo Seam	1.9	21.1	50.4	26.6	0.9	25.2	10,820
	Damodar Valley	4.0	12.8	41.1	42.1	0.3	18.3	7,850
	Trombay	7.2	20.8	44.7	27.3	1.0	22.2	9,560
	Umaria Field	5.3	27.1	47.8	19.8	. . .	25.5	10,980
	Palana	41.4	29.2	23.8	5.6	. . .	15.6	6,710
Italy	Sardinia	3.6	39.8	33.0	23.6	6.5	23.7	10,210
Japan	Hiyoshi-Anthracite	2.7	5.8	75.0	16.5	. . .	29.2	12,540
	Hukuho	8.6	33.8	44.5	13.1	. . .	28.6	12,300
	Niiura	19.0	24.8	31.2	25.0	. . .	19.8	8,520
Mexico	Palu	1.3	21.0	59.0	18.7	0.7	28.4	12,200
Peru	Goyllarisquisga	4.0	35.3	29.5	31.2	. . .	20.7	8,910
	Quishuarcancha	2.5	35.8	34.8	26.9		22.6	9,730
Poland	Katowice	17.0	21.0	40.0	22.0	. . .	18.8	8,100
	Upper Silesia	4.0	31.6	58.4	6.0	. . .	31.4	13,500
Russia	Donetz Anthracite	2.0	3.5	83.0	11.5	0.8	30.0	12,910
	Donetz Med. Vol.	3.5	21.0	71.0	4.5	3.6	33.8	14,530
	Donetz Long Flame	7.0	40.0	31.5	21.5	5.7	26.3	11,300
Spain	Asturias	10.2	11.9	47.6	30.3	0.8	20.8	8,930
	Asturias	5.9	20.0	60.5	14.0	. . .	25.6	11,020
	Zaragoza	19.5	34.6	25.4	20.5	. . .	16.7	7,200
Turkey	Raihenburg	17.1	34.9	24.8	23.2	0.5	15.5	6,660
South Africa	Natal	4.2	16.6	70.5	8.7	4.18	. . .	. . .
	Orange Free State	5.6	28.4	50.4	15.6	1.5	24.7	10,640
	Transvaal	2.2	27.0	57.5	13.3	0.7	28.2	12,120
Venezuela	Barcelona	3.0	36.1	57.5	3.4	1.4	31.8	13,680
Yugoslavia	Anatolia	1.4	29.4	51.3	17.9	1.0	27.3	11,720

necessarily show a correlation with TGA reactivities. Both the BET and TGA apparatus offer quick means for classifying char surface properties and preliminary reactivity assessment, respectively.

C-E has also developed an advanced methodology for more accurate prediction of carbon loss in commercial pulverized coal-fired boilers.[12] This methodology is described in Chapter 6.

SOLID FUELS OTHER THAN COAL

Many solid or semisolid fuels other than coal are used in the steam-generating process. As described in this section, the majority of these fuels result from the production of primary industrial goods; some are waste fuels that depend upon local economic conditions for their use as sources of energy. They occur in many different forms (most of them not capable of being pulverized like coal), have a wide range of calorific content, and many require special equipment for efficient burning.

COKE AND COKE BREEZE

Coke is the fused solid residue left when certain coals, petroleum, or tar pitch are heated in an atmosphere excluding oxygen, so as to expel their volatile content. The process of thus decomposing these fuels into their gaseous and solid fractions is known as destructive distillation or carbonization.

When anthracite is rapidly heated out of contact with air to, say, 1600°F, it will evolve some volatile matter, but remain essentially unchanged in form. Lignites, when subjected to the same treatment, give off a large amount of gas, while their solid, charry residue cracks and shrinks, but does not fuse together to form

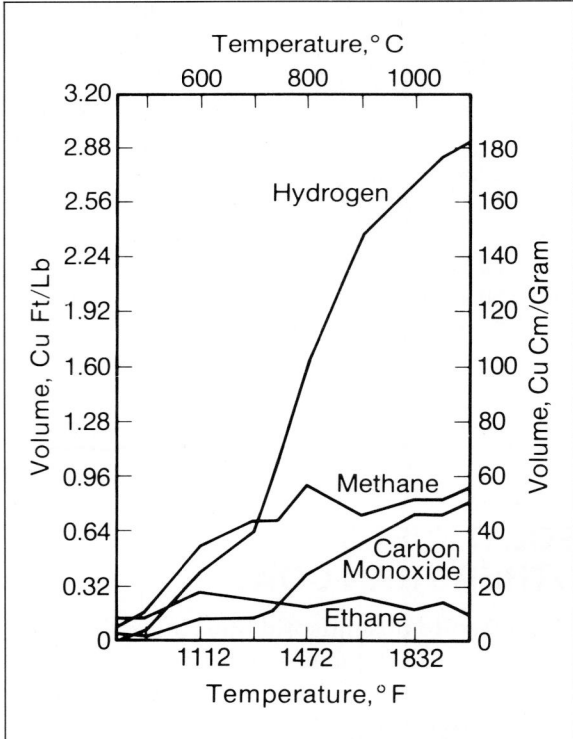

***Fig. 5* Gaseous products of distillation of Silkstone coal (Great Britain) at high temperatures**

"coke." Many varieties of bituminous coals, on the other hand, become plastic and "melt" when heated to only 500–700°F. Due to the resinous substances in these coals, the particles of the molten mass stick together, forming a porous coke which varies in color from dull gray to light silvery gray.

The character of the coke residue is dependent on the kind of fuel distilled, the temperature and pressure to which the fuel is exposed, and the type of oven used.

When the distillation process is carried on in temperatures over 700–900°F, it is called high-temperature carbonization, whereas low-temperature carbonization takes place below 700–900°F. Both the character and relative quantity of solid and gaseous distillation products depend on the temperature employed. This is best illustrated by a comparison between Fig. 5 which shows the volume and kind of gas evolved by an English coal at high temperatures and Fig. 6 which shows the same relation for a U.S. coal at low temperatures. With an increase in temperature, there is a corresponding increase in the evolution of hydrogen and carbon monoxide, at the expense of methane and the higher hydrocarbons. Because of its greater yield of gas, the high-temperature process is preferred for fuel-gas-making purposes and for metallurgical coke. Low-temperature distillation, on the other hand, requires less heat per ton of coke produced and is, therefore, more economical for making domestic coke.

Good coke from this process is shiny, hard, and uniformly porous. It has little volatile matter and is smokeless. If used for metallurgical purposes, it must not have too much sulfur; in foundry work, not much moisture; for domestic use, as little ash as possible; and for making blue-water gas, as high an ash-fusion temperature as possible. The proportion of ash in coke is seen from Fig. 6 to be higher than the ash content of the original coal.

For firing steam generators, good coke is normally considered uneconomical. But "coke breeze," degraded-size coke with as much as 40 percent passing through a ⅛-inch mesh screen, is advantageously burned on certain types of

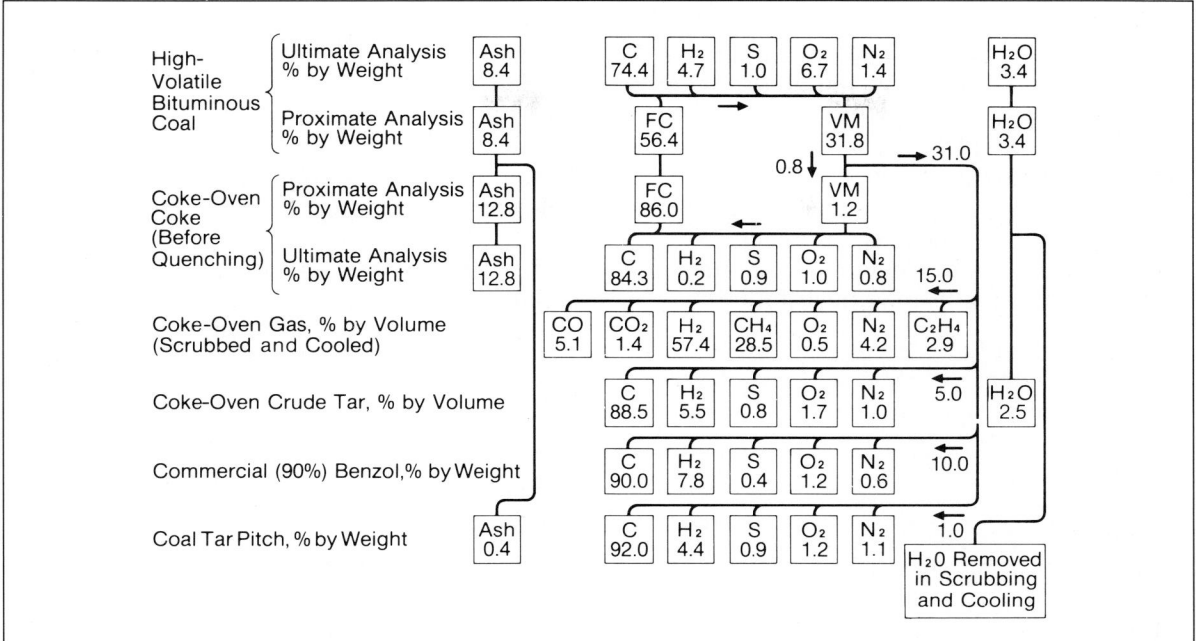

| High-Volatile Bituminous Coal | Ultimate Analysis % by Weight | Ash 8.4 | | C 74.4 | H₂ 4.7 | S 1.0 | O₂ 6.7 | N₂ 1.4 | | H₂O 3.4 |
| | Proximate Analysis % by Weight | Ash 8.4 | | FC 56.4 | | | VM 31.8 | | | H₂O 3.4 |

Fig. 6 Gaseous products from distillation of Pittsburgh coal at low temperatures

stokers. As the typical analyses of Table VII indicate, coke breeze usually has a higher ash content than the rest of the coke.

Nearly all coke produced now is carbonized at high temperatures in byproduct ovens of the slot type; these have largely displaced the older "beehive" ovens. The volatile products of the process, consisting essentially of gas, ammonia, tar and light oil, are recovered and separated into their various fractions. Fig. 7 shows diagrammatically the fuels taking part in and resulting from the coke-making process, together with their respective typical analyses.

PETROLEUM COKE

Petroleum coke is a byproduct of a process in which residual hydrocarbons are converted to lighter, more highly valued distillates. Two processes that produce these byproducts are in use: delayed coking and fluid coking.

DELAYED COKING

In this process, the reduced crude oil is heated rapidly and flows to isolated coking

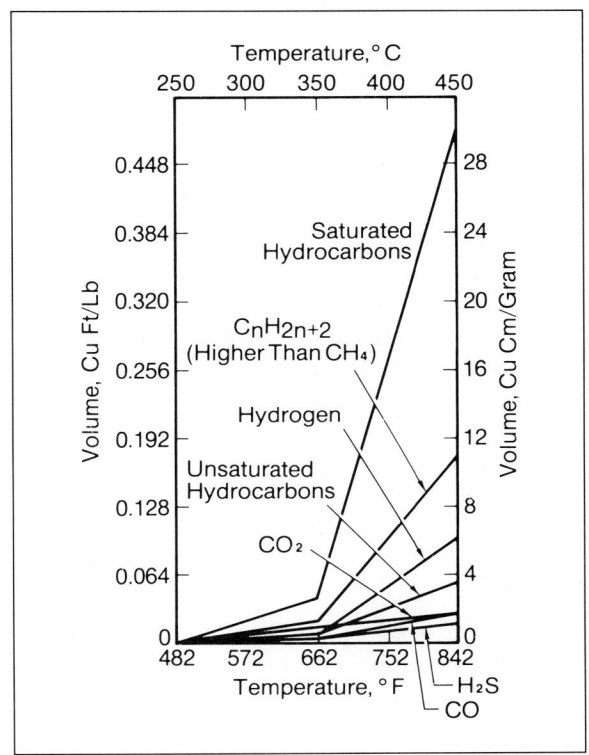

Fig. 7 Typical analyses of fuels involved in the manufacture of byproduct coke

drums where it is coked by its own contained heat. The process requires several drums to permit removal of the coke in one drum while the others remain on stream. The residual product which solidifies in these drums is termed *delayed coke*. When first removed from the drum, it has the appearance of run-of-mine coal, except that the coke is dull black.

The analysis of the coke varies with the crude from which it is made, ranging as follows:

Moisture	3–12%
Volatile matter	10–20%
Fixed carbon	71–88%
Ash	0.2–3.0%
Sulfur	2.9–5.4%
Btu/lb, dry	14,100–15,600 (32.8 to 36.3 MJ/kg)

FLUID COKING

Two large vessels are used in fluid coking. One is known as a reactor vessel, and the other, a burner vessel. In this process, fluid coke is both the catalyst and secondary product. The seed coke is first heated in the burner vessel, either by adding air and burning a portion of the coke, or by burning an extraneous fuel such as oil. The heated seed coke then flows into the reactor vessel where it comes in contact with the preheated residual oil and the lighter fractions of the oil are flashed off. The coke which is produced both deposits in uniform layers on the seed coke and forms new seed coke. Thus there is a constantly accumulating coke reservoir which is tapped off and is available as a boiler feed.

The coke thus formed is a hard, dry, spherical solid resembling black sand. It is composed of over 90-percent carbon with varying percentages of sulfur and ash, depending on the source of the crude oil. Typical analyses are

Fixed carbon	90–95%
Volatile matter	3–6.5%
Ash	0.2–0.5%
Sulfur	4.0–7.5%
HHV, Btu/lb, dry	14,100–14,600 (32.8 to 34.0 MJ/kg)

COAL TAR

Coal tar is a byproduct in the carbonization of coal. The tar compounds are extremely complex and number in the hundreds. The solid material which is insoluble in benzene is contained as colloidal and coarse dispersed particles and is known as "free carbon." The composition of the tar is dependent on the temperature of carbonization and, to a lesser extent, on the nature of the coking coal. The following is an example of a coal tar analysis.

Carbon	89.9%
Hydrogen	6.0%
Sulfur	1.2%
Oxygen	1.8%
Nitrogen	0.4%
Moisture	0.7%
Gravity, Baumé	1.18°
Viscosity at 122°F, SSF	900
Flash point	156°F (69°C)
Heating Value, Btu/lb, dry	16,750 (39.0 MJ/kg)

Coal tar is burned in boilers only when it cannot be sold for other purposes at a higher price than its equivalent fuel value. At ordinary temperature, the viscosity of tar is very high and must be heated for pumping. It burns like a fuel oil and the same equipment can be used.

COAL TAR PITCH

Coal tar pitch is used to a small extent for generation of steam. It is the residue resulting from the distillation and refining of coal tar. The pitch is solid at ordinary atmospheric temperature but becomes liquid at about 300°F. Generally, it is burned in pulverized form; in a few cases, it is melted and burned like oil or coal tar. When it is burned in pulverized form, it must be kept cool during pulverization and during delivery to the furnace. In some cases, it is preferred to coal. Because of its very low ash content, the stack gas is practically free from dust, and therefore there is no flue dust nuisance. When burned in liquid form, it must be kept very hot to prevent congealing. Table VIII shows analyses of coal tar pitch.

Table VII. Analyses of Typical Cokes, as Fired

	% Proximate Analysis					% Ultimate Analysis					HHV, Btu/Lb	A at Zero Excess Air, Lb/10⁶ Btu
	H₂O	VM	FC	ASH	H₂O	C	H₂	S	O₂	N₂		
Kind												
High-temperature coke	5.0	1.3	83.7	10.0	5.0	82.0	0.5	0.8	0.7	10.0	12,200	798
Low-temperature coke	2.8	15.1	72.1	10.0	2.8	74.5	3.2	1.8	6.1	10.0	12,600	763
Beehive coke	0.5	1.8	86.0	11.7	0.5	84.4	0.7	1.0	0.5	11.7	12,527	807
Byproduct coke	0.8	1.4	87.1	10.7	0.8	85.0	0.7	1.0	0.5	10.7	12,690	802
High-temperature coke breeze	12.0	4.2	65.8	18.0	12.0	66.8	1.2	0.6	0.5	18.0	10,200	805
Gas Works Coke:												
Horizontal retorts	0.8	1.4	88.0	9.8	0.8	86.8	0.6	0.7	0.2	9.8	12,820	808
Vertical retorts	1.3	2.5	86.3	9.9	1.3	85.4	1.0	0.7	0.3	9.9	12,770	809
Narrow coke ovens	0.7	2.0	85.3	12.0	0.7	84.6	0.5	0.7	0.3	12.0	12,550	802

WOOD

Wood is a complex vegetable tissue composed principally of cellulose, an organic compound having a definite chemical composition. It would, therefore, seem reasonable to assume that equal weights of different dry wood species will have practically the same heat content. However, owing to the presence of resins, gums and other substances in varying amounts, heat content is not uniform.

Ultimate analyses showing the chemical composition of several different wood species are given in Table IX. Because the substances making up these fuels are complex organic chemical compounds, and complex thermodynamic changes take place when they are burned in a furnace, it is not possible to make use of formulas such as the Dulong type to predict their heating values. These analyses do not indicate the amount of resins or similar substances present. But note that the heat content, on the dry basis, is greatest in the cases of highly resinous woods as fir and pine.

The moisture content of freshly cut wood varies from 30 to 50 percent. After air drying for approximately a year, this is reduced to between 18 and 25 percent.

Table VIII. Coal Tar Pitch

Proximate Analysis

Moisture	2.2%
VM	48.2%
FC	48.9%
Ash	0.7%

Ultimate Analysis

Carbon	90.1%
Hydrogen	4.9%
Sulfur	0.9%
Oxygen	0.6%
Nitrogen	0.6%
Moisture	2.2%
Ash	0.7%

HHV, dry	16,200 (37.7 MJ/kg)

Most wood as commercially available for steam generation is usually the waste product resulting from some manufacturing process. Its moisture content as received at the furnace will depend on (1) extraneous water from source or storage or handling in the rain and (2) whether it is "sap wood" or "heart wood," as well as on the species and on the time of year it is cut.

The use of wet refuse wood for steam generation falls into three broad classifications, each with its own specific combustion problems. As both producer and user of byproduct fuel, sawmills are in the first category. Byproduct fuel for in-plant use is usually made up of sawdust, shavings, bark and other wood waste in varying percentages dependent on the nature of the mill operation, the methods of storage and disposal of waste wood as well as the arrangement of available fuel burning equipment.

In the second classification are those plants which purchase their wood fuel supply, usually composed of chips with little or no sawdust and shavings.

Paper mills are in the third classification, as they must dispose of the wet wood refuse produced in their wood preparation plants. Their principal problem is one of wet bark disposal.

BARK

A common waste product in paper mills, bark results from debarking tree trunks used in making paper. The bark is peeled off the trunks in long rope-like strips. This shape and size, combined with the high moisture content, make the handling of the fuel difficult.

Bark as received from the barking drums contains 80 percent or more moisture, and in this condition is of no value as a fuel. This fact is best illustrated by Table X, the data for which have been prepared on the basis of a dry bark

Table IX. Typical Analyses of Dry Wood

	C	H₂	S	O₂	N₂	Ash	HHV, Btu/Lb	A at Zero Excess Air, Lb/10⁶ Btu
Softwoods†								
Cedar, white	48.80	6.37	. . .	44.46	. . .	0.37	8400*	709
Cypress	54.98	6.54	. . .	38.08	. . .	0.40	9870*	711
Fir, Douglas	52.3	6.3	. . .	40.5	0.1	0.8	9050	720
Hemlock, western	50.4	5.8	0.1	41.4	0.1	2.2	8620	706
Pine, pitch	59.00	7.19	. . .	32.68	. . .	1.13	11320*	702
white	52.55	6.08	. . .	41.25	. . .	0.12	8900*	723
yellow	52.60	7.02	. . .	40.07	. . .	0.31	9610*	710
Redwood	53.5	5.9	. . .	40.3	0.1	0.2	9040	722
Hardwoods†								
Ash, white	49.73	6.93	. . .	43.04	. . .	0.30	8920*	709
Beech	51.64	6.26	. . .	41.45	. . .	0.65	8760*	729
Birch, white	49.77	6.49	. . .	43.45	. . .	0.29	8650*	712
Elm	50.35	6.57	. . .	42.34	. . .	0.74	8810*	715
Hickory	49.67	6.49	. . .	43.11	. . .	0.73	8670*	711
Maple	50.64	6.02	. . .	41.74	0.25	1.35	8580	719
Oak, black	48.78	6.09	. . .	44.98	. . .	0.15	8180*	714
red	49.49	6.62	. . .	43.74	. . .	0.15	8690*	709
white	50.44	6.59	. . .	42.73	. . .	0.24	8810*	715
Poplar	51.64	6.26	. . .	41.45	. . .	0.65	8920*	716

*Calculated from reported high heating value of kiln-dried wood assumed to contain 8-percent moisture.

†The terms "hard" and "soft" wood, contrary to popular conception, have no reference to the actual hardness of the wood. According to the *Wood Handbook,* prepared by the Forest Products Laboratory of the U.S. Department of Agriculture, hardwoods belong to the botanical group of trees that are broad-leaved whereas softwoods belong to the group that have needle or scalelike leaves, such as evergreens: cypress, larch, and tamarack are exceptions.

heating value of 8750 Btu per lb. Thus, at 80 percent moisture the heating value is only 1750 Btu per lb (as received), and for every pound of dry substance there are 4 lbs of water which must be evaporated before any heat is available. Under these conditions, the bark will not support its own combustion, and it will be necessary to supply heat from some other source if the wet material is to be disposed of in the furnace of any steam generating unit.

HOG FUEL

In the manufacture of lumber, the amount of material removed from the log to produce sound lumber is approximately as follows: 18 percent in the form of slabs, edging and trimming; 10 percent as bark; and 20 percent as sawdust and shavings. While the total waste material will usually average 50 percent, distribution of different types of waste may vary widely from the approximations given above, owing to mill conditions as well as to the ultimate finished product.

The mills frequently use either the sawdust or a mixture of sawdust and shavings for steam production because these can be burned without further processing. The remainder of the so-called waste products requires further size reduction in a "hog" to facilitate feeding, rapid combustion, transportation and storage. These newly sized products, together with varying percentages of sawdust and shavings present, constitute *hog fuel*.

The percentage of sawdust and shavings present may be quite high if the fuel is to be burned at the sawmill. Table XI shows typical analyses of hog fuel.

A large paper producer conducted a series of tests to determine the moisture content, as well as the unit weight, of waste wood produced in its mill. The results of these tests, shown in Table XII, illustrate the wide variation in weight that will exist between units of hog fuel because of moisture and different wood-species content.

Hog fuel, as normally delivered to the furnace, contains variable amounts of moisture, averaging approximately 50 percent, most of which is in the cellular structure of the wood. Storage of logs in the mill pond, water lubrication of saws, and exposure to rain because of outdoor storage of the hog fuel all contribute to the high total moisture content. In addition, the hog fuel, on dry-wood basis, contains approximately 81 percent volatile matter and somewhat less than 18 percent fixed carbon. The noncombustible residue, in the form of wood ash, is only a small fraction of 1 percent.

Table X. Relationship of Bark Moisture to Heat Content

% H_2O	HHV, Btu/Lb	Lb water/ lb dry Substance
0	8750	0.00
20	7000	0.25
40	5250	0.67
50	4375	1.00
60	3500	1.50
70	2625	2.30
80	1750	4.00
90	875	9.00

Table XI. Analyses of Hog Fuels

Kind of Fuel	Western Hemlock	Douglas Fir	Pine Sawdust
H_2O As Received	57.9	35.9	. . .
H_2O Air Dried	7.3	6.5	6.3
Proximate Analysis, Dry Fuel			
VM	74.2	82.0	79.4
FC	23.6	17.2	20.1
Ash	2.2	0.8	0.5
Ultimate Analysis, Dry Fuel			
Hydrogen	5.8	6.3	6.3
Carbon	50.4	52.3	51.8
Nitrogen	0.1	0.1	0.1
Oxygen	41.4	40.5	41.3
Sulfur	0.1	0	0
Ash	2.2	0.8	0.5
HHV, Btu/lb, dry	8620	9050	9130

BAGASSE

Bagasse is a waste fuel produced when sugar is extracted from cane. The cane stalks grow in the field from 10 to 12 ft high. In Latin America, it is customary to cut the stalks again into smaller pieces, 2 to 3 ft long, before sending them through shredding machines or rotary knives. The shredding, or cutting operation, is called "disintegration" or "defibration," and serves to open up the hard rind and thereby facilitate the squeezing out of juice in the cane.

As in the case of wood, sugar cane consists of cellulose fiber which makes up the tissue enclosing such sugars as sucrose ($C_{12}H_{22}O_{11}$) and glucose ($C_6H_{12}O_6$). Water and small quantities of mineral ash are also present. On a moisture- and ash-free basis, the cane contains from 10 to 17 percent cellulose fiber and from 83 to 90 percent sugar.

To separate the juice from the fiber, the disintegrated cane is crushed between rollers. After this primary pressing, the issuing fiber, or "bagasse," is sent through more roller mills for further extraction. At the same time that the bagasse undergoes additional squeezing, it is sprayed, or "imbibed," with plain water or diluted juice to help dissolve more of the sugar. When hot water, generally taken from the boiler feed line, is used to soak the bagasse, the operation is known as "maceration." On leaving the last mill, a typical bagasse might have 40 percent fiber, 2.5 percent sugar, 55 percent moisture and 2.5 percent ash. The relatively high ash content in many bagasse samples is because of trash and dirt picked up in harvesting.

Table XIII shows typical ultimate analyses of bagasse from different countries, on a dry basis. (As-received bagasse contains 40 to 60 percent moisture by weight.) Carbon content is slightly lower; oxygen a little higher than in wood, possibly due to the extraction of sugar.

The size of bagasse pieces depends on the machinery employed to disintegrate the cane. A simple crusher will produce coarser bagasse than is obtained by adding a shredder ahead of the crusher. Aside from the special case where grindstones defibrate the cane, the finest size is obtained with rotating knives replacing both shredder and crusher.

FOOD-PROCESSING WASTES

The production of food from some fruits and vegetables results in burnable solids that are available for energy production. Among these are nut hulls, with a fuel value of about 7700 Btu/lb (18 MJ/kg) as fired; rice hulls, with

Table XII. Unit Weight and Moisture Content of Wood (as-received basis)

Type of Wood	% H₂0	Weight- lb/cu ft
Drum Barker (pressed)	63.4	19.3
Regular paper mill waste wood	56.9	21.5
Hemlock	53.7	19.4
Douglas fir	44.4	17.4

Table XIII. Typical Analyses of Bagasse, Dry

	% by Weight					HHV, Btu/Lb	A at Zero Excess Air, Lb/10⁶ Btu
	C	H₂	O₂	N₂	Ash		
Cuba	43.15	6.00	47.95	. . .	2.90	7985	629
Hawaii	46.20	6.40	45.90	. . .	1.50	8160	687
Java	46.03	6.56	45.55	0.18	1.68	8681	651
Mexico	47.30	6.08	45.30	. . .	1.32	8740	646
Peru	49.00	5.89	43.36	. . .	1.75	8380	700
Puerto Rico	44.21	6.31	47.72	0.41	1.35	8386	627

5200 to 6500 Btu/lb (12 to 15 MJ/kg); corn cobs, with 7500 to 8300 Btu/lb (17 to 19 MJ/kg); and coffee grounds, with between 4900 and 6500 Btu/lb (11 to 15 MJ/kg).

MUNICIPAL AND INDUSTRIAL REFUSE

Solid residential, commercial, and industrial waste, or "municipal solid waste" (MSW), is a fuel having some of the characteristics of wood, bagasse, and other "young" fossil fuels. It derives most of its heating value from its cellulosic content. When shredded and separated into light and heavy fractions, a fuel can be produced having thermal energy equal to lignitic coals, with high ash, but low sulfur, content. This beneficiated product is known as "refuse-derived fuel" (RDF).

The technical aspects of recovery of the energy from this resource, as opposed to its simple disposal, are well documented. (See references 15 through 25.)

Refuse is a highly time-, geography-, and weather-dependent fuel. Its character varies widely with the economic status of the people generating it. It is extremely heterogeneous but, in general, has increased in paper and plastic content and, therefore, heating value, through the years. The garbage fraction has decreased, as has the quantity of ash from coal-fired household furnaces. In suburban areas, it will have greater or lesser content of leaves and grass as a function of the day in the week, and more or less moisture content as a function of the weather. Urban refuse is characterized by large quantities of paper bags, cardboard, and similar dry combustible material.

Similarly, industrial refuse has higher thermal value today than earlier, because of the increased use of plastics and other synthetic materials, many of which have high heating values with little or no moisture or ash. Typical high-calorific-value components of industrial waste are rubber, wood, sawdust, fats, oils, waxes, solvents, paints, and other organic materials, many of them with heating values of 10,000 to 19,000 Btu per lb (23 to 44 MJ/kg).

Raw municipal refuse will be typically from one-third to one-half paper; 2 to 12 percent textiles, rubber, plastics, and leather; 7 to 24 percent food waste; 1 to 7 percent wood; 5 to 12 percent glass and ceramics; 7 to 11 percent metal; and 1 to 33 percent grass, leaves and dirt. Proximate and ultimate analyses of residential solid waste with magnetic metals removed are given in Table XIV. Further separation of heavy-density materials (nonferrous metals, stone, masonry, and glass) can increase refuse heating value to between 7,000 and 8,000 Btu per pound (16 to 19 MJ/kg).

LIQUID FUELS

Fuel oil used for steam generation purposes may be defined as petroleum or any of its liquid residues remaining after the more volatile constituents have been removed.

Petroleum is sometimes burned in its crude form. In this condition, most of it will contain lighter gasoline fractions which lower the flash point and present a fire hazard. Through limited fractional distillation, or topping, the lighter gasoline can be removed and a safe fuel oil can be produced. If the refining process is carried through extended fractional distilla-

Table XIV. Analyses of Typical Residential Solid Waste With Magnetic Metals Removed

Proximate Analysis	% As Received
H_2O	19.7–31.3
Ash	9.4–26.8
VM	36.8–56.2
FC	0.6–14.6
HHV, Btu/lb	3100–6500 (7 to 15 MJ/kg)

Ultimate Analysis	
H_2O	19.7–31.3
Ash	9.4–26.8
Carbon	23.4–42.8
Hydrogen	3.4–6.3
Nitrogen	0.2–0.4
Chlorine	0.1–0.9
Sulfur	0.1–0.4
Oxygen	15.4–31.9

Table XV. Detailed Requirements for Fuel Oils[A]

Grade of Fuel Oil	Flash Point, °C (°F) Min	Pour Point, °C (°F) Max	Water and Sediment Vol % Max	Carbon Residue on 10 % Bottoms % Max	Ash Weight % Max	Distillation Temp. C (°F) 10% Point Max	90% Point Min	90% Point Max
No. 1 Distillate oil intended for vaporizing pot-type burners and other burners requiring this grade	38 or legal (100)	−18[C] (0)	0.05	0.15	. . .	215 (420)	. . .	288 (550)
No. 2 Distillate oil for general purpose heating for use in burners not requiring No. 1	38 or legal (100)	−6[C] (20)	0.05	0.35	. . .	. . .	282[C] (540)	338 (640)
No. 4 Preheating not usually required for handling or burning	55 or legal (130)	−6[C] (20)	0.50	. . .	0.10	. . .	. . .	. . .
No. 5 (Light) Preheating may be required depending on climate and equipment	55 or legal (130)	. . .	1.00	. . .	0.10	. . .	. . .	. . .
No. 5 (Heavy) Preheating may be required for burning and, in cold climates, may be required for handling	55 or legal (130)	. . .	1.00	. . .	0.10	. . .	. . .	. . .
No. 6 Preheating required for burning and handling	60 (140)	[G]	2.00[E]	. . .	. . .	. . .	. . .	. . .

[A] It is the intent of these classifications that failure to meet any requirement of a given grade does not automatically place an oil in the next lower grade unless in fact it meets all requirements of the lower grade.

[B] In countries outside the United States other sulfur limits may apply.

[C] Lower or higher pour points may be specified whenever required by conditions of storage or use. When pour point less than −18°C (0°F) is specified, the minimum viscosity for Grade No. 2 shall be 1.8 cSt (32.0 SUS) and the minimum 90 % point shall be waived.

[D] Viscosity values in parentheses are for information only and not necessarily limiting.

[E] The amount of water by distillation plus the sediment by extraction shall not exceed 2.00 %. The amount of sediment by extraction shall not exceed 0.50 %. A deduction in quantity shall be made for all water and sediment in excess of 1.0 %.

[F] Where low sulfur fuel is required, fuel oil falling in the viscosity range of a lower numbered grade down to and including No. 4 may be supplied by agreement between purchaser and supplier. The viscosity range of the initial shipment shall be identified and advance notice shall be required when changing from one viscosity range to another. This notice shall be in sufficient time to permit the user to make the necessary adjustments.

[G] Where low sulfur fuel oil is required, Grade 6 fuel oil will be classified as low pour +15°C (60°F) max or high pour (no max.) Low pour fuel oil should be used unless all tanks and lines are heated.

Excerpted from *ASTM Standards* D 396, Specifications for Fuel Oils.

Table XV. Detailed Requirements for Fuel Oils[A] — *Continued*

Saybolt Viscosity, s[D]				Kinematic Viscosity, cSt[D]				Specific Gravity 60/60°F (deg API)	Copper Strip Corrosion	Sulfur, %
Universal at 38°C (100°F)		Furol at 50°C (122°F)		At 38°C (100°F)		At 50°C (122°F)				
Min	Max	Min	Max	Min	Max	Min	Max	Max	Max	Max
. . .	. . .	. . .	. . .	1.4	2.2	. . .	. . .	0.8499 (35 min)	No. 3	0.5 or legal
(32.6)	(37.9)	. . .	. . .	2.0[C]	3.6	. . .	. . .	0.8762 (30 min)	No. 3	0.5[B] or legal
(45)	(125)	. . .	. . .	5.8	26.4[F]	. . .	. . .	. . .	. . .	legal
(>125)	(300)	. . .	. . .	>26.4	65[F]	. . .	. . .	. . .	. . .	legal
(>300)	(900)	(23)	(40)	>65	194[F]	(42)	(81)	. . .	. . .	legal
(>900)	(9000)	(>45)	(300)	>92	638[F]	50°C	(122°F)	. . .	. . .	legal

tion and cracking, such fuels as gasoline, kerosene, gas oil, light fuel oils, lubricating oil, heavy fuel oil, residual tar, pitch and petroleum coke are produced.

PROPERTIES OF FUEL OIL

The term *fuel oil* may conveniently cover a wide range of petroleum products. It may be applied to crude petroleum, to a light petroleum fraction similar to kerosene or gas oil, or to a heavy residue left after distilling off the fixed gases, the gasoline, and more or less of the kerosene and gas oil. To provide standardization, specifications have been established, Table XV, for several grades of fuel oil.

Sometimes designated as light and medium domestic fuel oils, Grades No. 1 and No. 2 are specified mainly by the temperature of the distillation range. Grade No. 6, designated as heavy industrial fuel oil and sometimes known as Bunker C oil, is specified mainly by viscosity. The specific gravities of Grades 4, 5, and 6 are not specified because they will vary with the source of the crude petroleum and the extent of the refinery operation in cracking and distilling.

Despite the multiplicity of chemical compounds found in fuel oils, the typical analyses of these fuels, as shown in Table XVI, are fairly constant.

SPECIFIC GRAVITY

This is the ratio between the weight of any volume of oil at 60°F and the weight of an equal volume of water at 60°F. The common designation is *Sp Gr 60/60°F* and is expressed as a decimal carried to four places.

Gravity determinations are readily made by immersing a hydrometer into the sample and reading the scale at the point to which the instrument sinks in the oil. The specific gravity is either read direct or the gravity is measured in degrees API.

Formerly, the gravity of oil was measured in degrees Baumé but confusion developed over the use of the two so-called Baumé scales for light liquids. To overcome this, the American Petroleum Institute, the U.S. Bureau of Mines, and the U.S. Bureau of Standards agreed to recommend, in 1921, that only one scale be used in the petroleum industry, and that it be known as the API scale. The relation between specific gravity and °API is expressed by

$$\text{Sp Gr } 60/60°F = \frac{141.5}{131.5 + °API} \tag{1}$$

HEATING VALUE

This may be expressed in either Btu per gallon at 60°F or Btu per pound. The heating value per gallon increases with specific gravity, because there is more weight per gallon, and ranges from about 135,000 to 150,000 Btu. The heating value per pound of fuel oil varies inversely with the specific gravity, because lighter oil contains more hydrogen; it ranges from 18,300 to 19,500 Btu.

The exact determination of the heat content of fuel oil is made in a bomb calorimeter. However, there is an approximate relationship between specific gravity and the high heating value. For an uncracked distilate or residue,

$$\text{Btu/lb (high)} \approx 17,660 + (69 \times API \text{ gravity}) \tag{2}$$

For a cracked distillate,

$$\text{Btu/lb (High)} \approx 17,780 + (54 \times API \text{ gravity}) \tag{3}$$

VISCOSITY

This is defined as the measure of the resistance to flow. The greater this resistance, the longer it takes a given volume of oil to flow through a fixed orifice. The Saybolt Universal viscosity is expressed in seconds of time that it takes to run 60 cc through a standard size orifice at any desired temperature. Viscosity is commonly measured at 100°F, 150°F and 210°F. The oil is held at constant temperature within ± 0.25°F during the test period.

Fuel oil is very viscous, and it takes a long time to make a determination with the Saybolt Universal viscosimeter. For this reason, the viscosity of fuel oil is usually measured with a Saybolt Furol viscosimeter, which is the same as the Saybolt Universal except that the orifice is larger. Viscosity of 62 seconds Saybolt Furol is 600 seconds Saybolt Universal.[26]

Another measure of viscosity sometimes used is the so-called Engler degree or specific viscosity. The Engler degree is the quotient of time of outflow of 200 cc of oil, divided by the time of flow of 200 cc of water; that is, the viscosity of oil is compared with that of water.

Viscosity of fuel oil decreases as the temperature rises and becomes nearly constant above about 250°F. Therefore, when fuel oil is heated to reduce the viscosity for good atomization, there is little gain in heating the oil beyond 250°F. Moreover, since burners operate most efficiently with oil of constant viscosity, it is desirable to operate in the viscosity range where temperature variations have the least effect. This is illustrated in Fig. 8 for No. 6 Bunker C oil. Fig. 9 shows the viscosity-temperature relationships for the several grades of fuel. Table XVI shows typical analyses.

The relationship among specific gravity, deg

API, density in lb per gal, Btu per lb, and Btu per gal for petroleum products is graphically shown in Fig. 10. Also included are the ranges in deg API for gasoline, kerosene, gas oil and fuel oils. Knowing the value of any one of these characteristics, it is possible to determine all the others quickly. For example, assume the deg API to be 75, then the intersection of this value with the deg API curves is at a point A, through which a horizontal line is drawn to intersect the remaining curves. Then, by referring to their respective scales, it is possible to read the specific gravity B as 0.685, the density C as 5.675 lb per gal, and the higher heating value at D as 20,550 Btu per lb, or at E as 116,800 Btu per gal. Of particular interest is the fact that, although the high specific gravity fuel oils (15°API) have a lower heating value per pound than the lower specific gravity gasoline (60°API), the total heat per gallon, the basis on which they are purchased commercially, is considerably greater.

FLASH AND FIRE POINT

Flash point of fuel oil is the lowest temperature at which sufficient vapor is given off to

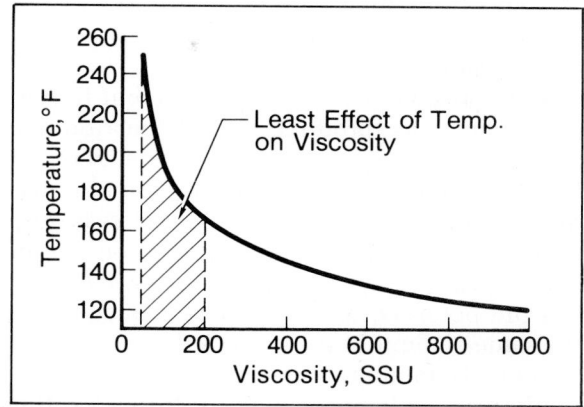

Fig. 8 **Viscosity versus temperature, No. 6 fuel oil**

Table XVI. Typical Analyses and Properties of Fuel Oils*

Grade	No. 1 Fuel Oil	No. 2 Fuel Oil	No. 4 Fuel Oil	No. 5 Fuel Oil	No. 6 Fuel Oil
Type	Distillate (Kerosene)	Distillate	Very Light Residual	Light Residual	Residual
Color	Light	Amber	Black	Black	Black
API gravity, 60°F	40	32	21	17	12
Specific gravity, 60/60°F	0.8251	0.8654	0.9279	0.9529	0.9861
Lb/U.S. gallon, 60°F	6.870	7.206	7.727	7.935	8.212
Viscos., Centistokes, 100°F	1.6	2.68	15.0	50.0	360.0
Viscos., Saybolt Univ., 100°F	31	35	77	232	. . .
Viscos., Saybolt Furol, 122°F	. . .	. . .	. . .	. . .	170
Pour point, °F	Below zero	Below zero	10	30	65
Temp. for pumping, °F	Atmospheric	Atmospheric	15 min.	35 min.	100
Temp. for atomizing, °F	Atmospheric	Atmospheric	25 min.	130	200
Carbon residue, %	Trace	Trace	2.5	5.0	12.0
Sulfur, %	0.1	0.4–0.7	0.4–1.5	2.0 max.	2.8 max.
Oxygen and nitrogen, %	0.2	0.2	0.48	0.70	0.92
Hydrogen, %	13.2	12.7	11.9	11.7	10.5
Carbon, %	86.5	86.4	86.10	85.55	85.70
Sediment and water, %	Trace	Trace	0.5 max.	1.0 max.	2.0 max.
Ash, %	Trace	Trace	0.02	0.05	0.08
Btu/gallon	137,000	141,000	146,000	148,000	150,000

* Courtesy Exxon Corporation

form a momentary flash when flame is brought near the surface. The values vary with the apparatus and procedure; both must be specified. Flash point specifications for the several grades of fuel oil are given in Table XV. Fire point is the lowest temperature at which the oil gives off enough vapor to burn continuously.

SULFUR AND ASH

Sulfur is a very undesirable element in fuel oil because its products of combustion are acidic and cause corrosion in economizers, air heaters and gas ducts. Because of the high hydrogen content in fuel oil, and the resulting high water vapor content in the products of combustion, a given amount of sulfur in fuel oil has the potential of doing more damage than the same amount of sulfur in coal.

Fuel oil contains all the solid impurities originally present in the crude oil. If these solids contain a large proportion of salt, they are very fusible and can cause considerable trouble.

OIL REFINERY REFUSE FUEL

Byproducts from refinery operation consist of a wide variety of refuse fuels. There are solids, such as asphaltic pitch and petroleum coke. The liquids, termed *sludge*, often have a high specific gravity and contain variable amounts of solid matter in suspension.

The characteristics of the sludge are governed by those of the crude oil used and the manner in which it is processed. Many of the suspended solids may be carbonaceous, in the form of small particles of oil coke.

Perhaps the most troublesome of sludges, because of its frequent and widely varying characteristics, is *acid sludge*. Its gravity may range between 5 and 14°API, and its viscosity is indeterminate. It contains changing quantities of weak sulfuric acid that may run as high as 40 percent, and this, together with the suspended carbonaceous material and flux, which must be added to make the sludge flow, causes heat

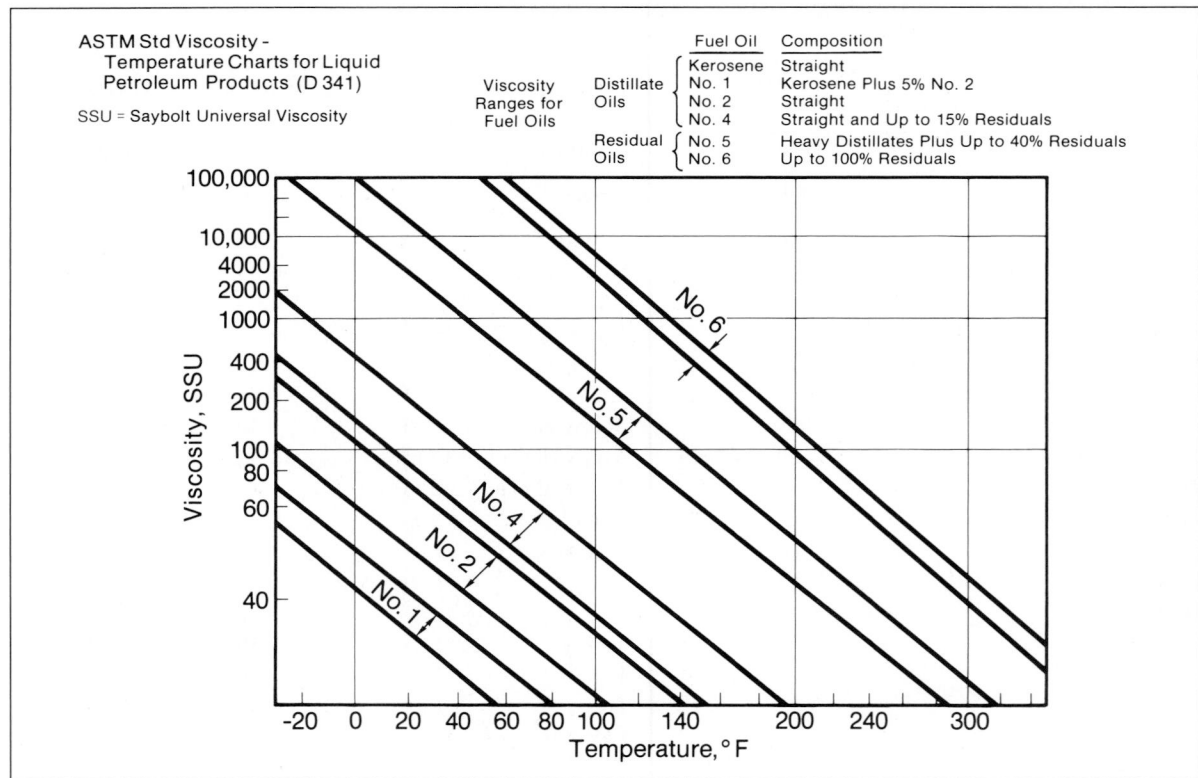

Fig. 9 **Viscosity ranges for fuel oils**

values to vary between 8,000 and 17,500 Btu/lb.

Alkaline sludges, such as soda tar and neutralized sludge, are less troublesome to fire since they are less variable in character than acid sludge.

GASEOUS FUELS

Gaseous fuels, when available, are ideal for steam generation because of the ease of control, the presence of little or no solid residue, and the low excess-air requirement, which contributes to high efficiency.

Properties of fuel gas considered to be of prime importance are composition, heating value, and specific gravity.

ANALYSIS

The analysis of fuel gas is expressed in terms of volume percentages of the component gases.

Determinations can be made by selective absorption in chemical solutions, by separation of components through distillation, by infrared or mass spectrometry, or by means of gas chromatography. Typical analyses of various gases are given under their specific headings.

HEATING VALUE

The heating value refers to the quantity of heat released during combustion of a unit amount of fuel gas. Determinations are made with a continuous flow (constant pressure) gas calorimeter. The heating value as determined in calorimeters is termed *high heating value* and is the quantity of heat evolved when the products of combustion are cooled to 60°F and the water vapor produced is completely condensed to a liquid at that temperature. The *low heating value* differs from the high heating value by the latent heat of evaporation of water

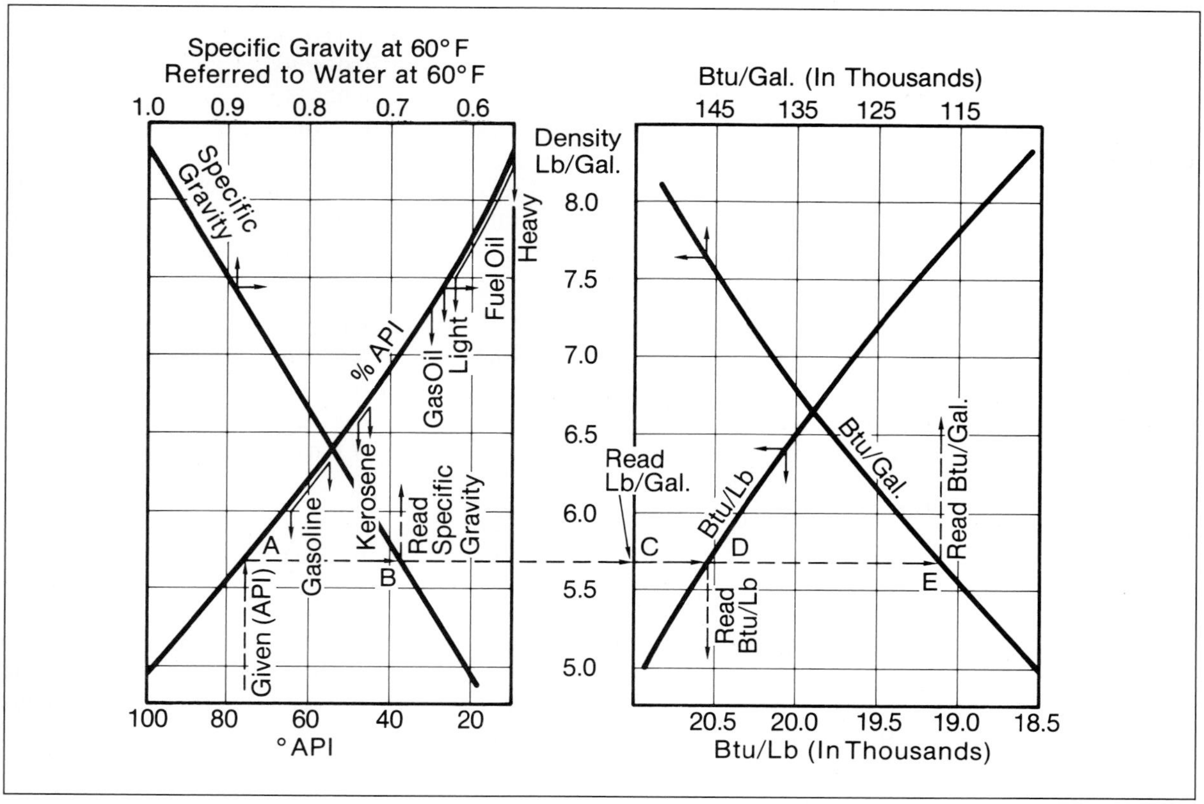

Fig. 10 **Chart showing the relationships of important characteristics of liquid fuels**

formed in the combustion process, as described on Page 2–12.

The heating value of manufactured gas is expressed as Btu per cu ft when measured at 60°F and 30 in. Hg, saturated with water vapor. The values for natural gas, however, are commonly reported at a pressure of 14.7 psia (pounds per square inch absolute) or 30 in. Hg, at a temperature of 80°F, and generally on a dry basis.

The heating value of gaseous fuels varies considerably, depending on the constituents present. When not obtainable by test, H_f can be calculated by summing up the heat evolved by the individual combustible fractions of the gas.

In Table XVII are shown the principal components, together with their properties at 60°F and 30 in. Hg, moisture-free. When present in different proportions, these make up various fuel gases.

ANSI/ASTM *Standards* D 3588 gives a method for calculating calorific value and specific gravity of gaseous fuels and includes a method for determining the repeatability and reproducibility of the calculated values.

SPECIFIC GRAVITY

Various methods for determining the specific gravity of a fuel gas are available but three

Table XVII. Combustion Constants of Dry Gases at 60°F and 30 In. Hg

	Chemical Formula	O₂ Reqd./ Cu Ft of Dry Gas, Cu Ft	CO₂ Formed/ Cu Ft of Dry Gas, Cu Ft	H₂O Formed/ Cu Ft of Dry Gas, Cu Ft	Density of Dry Gas, Lb/Cu Ft	HHV of Dry Gas Btu/Cu Ft*	HHV of Dry Gas Btu/Lb
Gas							
Oxygen	O_2	. . .	. . .	. . .	0.08461	. . .	. . .
Nitrogen (atmospheric)	N_2	. . .	. . .	. . .	0.07439	. . .	. . .
Air	. . .	. . .	. . .	. . .	0.07655	. . .	. . .
Carbon dioxide	CO_2	. . .	. . .	. . .	0.1170	. . .	. . .
Water vapor	H_2O	. . .	. . .	. . .	0.04758	. . .	. . .
Hydrogen	H_2	0.5	. . .	1.0	0.005327	325	60,991
Hydrogen sulfide	H_2S	1.5	1.0**	1.0	0.09109	647	7,100
Carbon monoxide	CO	0.5	1.0	. . .	0.07404	321	4,323
Saturated Hydrocarbons							
Methane	CH_4	2.0	1.0	2.0	0.04246	1014	23,896
Ethane	C_2H_6	3.5	2.0	3.0	0.08029	1789	22,282
Propane	C_3H_8	5.0	3.0	4.0	0.1196	2573	21,523
Butane	C_4H_{10}	6.5	4.0	5.0	0.1582	3392	21,441
Pentane	C_5H_{12}	8.0	5.0	6.0	0.1904	4200	22,058
Unsaturated Hydrocarbons or Illuminants							
Ethylene	C_2H_4	3.0	2.0	2.0	0.07421	1614	21,647
Propylene	C_3H_6	4.5	3.0	3.0	0.1110	2383	21,464
Butylene	C_4H_8	6.0	4.0	4.0	0.1480	3190	21,552
Pentylene	C_5H_{10}	7.5	5.0	5.0	0.1852	4000	21,600
Acetylene	C_2H_2	2.5	2.0	1.0	0.06971	1488	21,344
Benzene	C_6H_6	7.5	6.0	3.0	0.2060	3930	19,068
Toluene	C_7H_8	9.0	7.0	4.0	0.2431	4750	19,537

* If gas is saturated with moisture at 60°F and 30.0 in. Hg, reduce by 1.74%.
** SO_2 rather than CO_2

methods have been adopted as *ASTM Standards* D 1070, Test for Gravity, Specific, (Relative Density) of Gaseous Fuels.

DIRECT WEIGHING METHOD

This involves the determination of the weight differential between two equal volumes of gas and air, both at identical conditions of temperature and pressure.

PRESSURE BALANCE METHOD

In this method, a flask containing the gas is counterbalanced on a beam enclosed in a container. The beam is brought to balance by adjusting the air pressure within the container which varies the buoyancy of the flask. The procedure is repeated with air in the flask and the specific gravity determined from the ratio of the required absolute pressures.

DISPLACEMENT BALANCE METHOD

The instrument consists of a balance beam on each end of which two bells are suspended in a sealing liquid. One bell, containing air, is open to the atmosphere through holes in its top; the other bell, containing the gas, is open to the atmosphere through a 59-in. gas column connected to the space under the bell. An unbalanced force is produced which is equal to the pressure differential above and within the gas-filled bulb. The magnitude of this force is an indication of the specific gravity.

NATURAL GAS

Natural gas is perhaps the closest approach to an ideal fuel because it is practically free from noncombustible gas or solid residue. It is found compressed in porous rock and shale formations, or cavities, which are sealed between strata of close-textured rocks under the earth's surface. When these so-called gas-bearing sands of a pool are tapped by drilling wells, the gas is found to be under rock pressure, which may be as high as 2000 psig (pounds per square inch gage). As gas is withdrawn, this pressure gradually decreases until the field must be abandoned.

Natural-gas fields frequently exist in the neighborhood of oil deposits. Usually natural gas occupies the space above the oil and the oil, in turn, lies over salt water. The relative density of these substances accounts for their segregation, natural gas being lightest, water heaviest, and oil in between. But not all natural gas is associated with oil. At times it is found by itself, or directly in contact with salt water, hermetically sealed by the rock.

Although whenever possible natural gas is delivered at the required destination under well pressure, it is also frequently transported over long distances by means of pipelines and compressors. Thus, the furnace which uses this gas may be close to the well or at the end of a pipeline and the conditions in the pipelines may change the composition of natural gas going to the furnace. For instance, aside from the effect of compression, which liquefies the heavy members of the hydrocarbon family, water and oil are sometimes sprayed into the gas to keep it "moist." For this reason, any natural-gas analysis, as fired, should be taken at the point of use rather than at the well.

COMPOSITION OF NATURAL GAS

The characteristics of natural gas, as it comes out of the earth's surface, depend to some extent on its underground conditions. Generally, it is odorless and colorless. It burns with a blue flame and is highly explosive when mixed with air in the correct proportions. The range in its chemical composition is indicated by Table XVIII, where it is seen that methane, CH_4, and ethane, C_2H_6, are its principal combustible components. This is because in natural-gas analyses it is customary to report the heavier hydrocarbons in terms of CH_4 and C_2H_6, or if this grouping is not satisfactory, in terms of C_2H_6 and C_3H_8 (propane).

When sulfur is present in the oil deposit, the analysis of natural gas associated with this oil often includes hydrogen sulfide. This hydrogen sulfide is removed in most instances before distributing the gas because it is a potential source of pipeline corrosion.

In addition to its combustible constituents, natural gas may contain considerable amounts of carbon dioxide, CO_2, or nitrogen, N_2. It sometimes happens that gas is drawn from wells under suction and that, because of this, air will leak into the lines. The analysis of natural gas will then show the presence of oxygen.

DRY AND WET NATURAL GAS

If natural gas has been in contact with oil, it will be impregnated with varying amounts of heavy hydrocarbon vapors, such as pentane, C_5H_{12}, and hexane, C_6H_{14}, which are liquid at ordinary pressure and temperature. Known as "wet" natural gas, it is usually economical to dry it by liquefying the heavy vapors, which are then collected and called *casing-head gasoline*. In a complete chemical analysis of natural gas, a certain fraction of a heavy hydrocarbon is sometimes preceded by the letter "n," such as "n-butane," which means "normal" butane, and the remainder uses the prefix "iso," meaning "equal," as, for example, "iso-butane." The chemical formula is the same in both cases as the prefix merely denotes a different arrangement of the atoms in the hydrocarbon molecule. However, the physical and chemical properties of these so-called "isomers" are not usually the same.

"Dry" natural gas comes from wells away from oil deposits and is, therefore, comparatively devoid of heavy hydrocarbons.

Natural gases are also classified as either "sweet" or "sour." The *sour* gas is one which contains some mercaptans and a high percentage of hydrogen sulfide while the *sweet* gas is one in which these objectionable constituents have been removed. Analysis of sour and sweet gases are shown in Table XIX.

HEATING VALUE

The higher or gross heating value of natural gas is usually about 1000 Btu per cu ft, and it can be computed by adding together the heat contributed by volumetric percentages of the various component gases. This method will usually result in a lower value per cubic foot than that obtained by calorimetric determinations, because the unsaturated hydrocarbons are frequently grouped and reported with C_2H_6. For the same reason, the corresponding density, under standard conditions of 60°F and 30 in. Hg, will also be lower. The calculated Btu per pound, however, will be close to its actual value, because of the compensating effect of the lower calculated density.

LIQUEFIED PETROLEUM GAS (LPG)

The term *liquefied petroleum gas, LPG*, is applied to certain hydrocarbons which are gaseous under normal atmospheric conditions, but can be liquefied under moderate pressure at normal temperatures. LPG is derived from natural gas and from various petroleum-refinery sources such as crude distillation and cracking. The hydrocarbons in LPG are mainly of the paraffinic (saturated) series, principally propane, isobutane and normal butane.

The California Natural Gasoline Association has divided LPG mixtures into six standard

Table XVIII. Characteristics of Typical Natural Gases at 60° F and 30 In. Hg, Dry

			% by Volume					Density, Lb/Cu Ft	HHV		A at Zero Excess Air, Lb/10⁶ Btu
CO₂	N₂	H₂S	CH₄	C₂H₆	C₃H₈	C₄H₁₀	C₅H₁₂		Btu/Cu Ft**	Btu/Lb	
5.50	...	7.00	77.73	5.56	2.40	1.18	0.63*	0.05621	1061	18,880	738
3.51	32.00	0.50	52.54	3.77	2.22	2.02	3.44*	0.06610	874	13,220	729
26.2	0.7	...	59.2	13.9	...	...	...	0.06747	849	12,580	732
0.17	87.69	...	10.50	1.64	...	...	...	0.07120	136	1,907	732
0.20	0.60	...	99.20	...	...	...	...	0.04491	1006	22,410	732
...	0.60	...	...	79.40	20.00	...	...	0.08812	1935	21,960	735
...	0.50	...	...	21.80	77.70	...	...	0.11079	2389	21,560	738

*All hydrocarbons heavier than C_5H_{12} are assumed to be C_5H_{12}.
**If gas is saturated with moisture at 60°F and 30.0 in. Hg, reduce by 1.74%.

grades based on physical properties such as vapor pressure and specific gravity, as shown in Table XX.

The greatest use of LPG is as a domestic fuel with an appreciable amount being consumed in synthetic rubber production and chemical industries. It has a limited use in steam-generating units as an ignition and warm-up fuel but is generally prohibited for economic reasons from being a primary fuel.

REFINERY AND OIL GAS

Both refinery gas and oil gas may be defined as gaseous fuels composed of uncondensed distillation fractions. Aside from differences in composition, refinery gas is distinguished commercially as a byproduct, whereas oil gas is a main product of distillation.

A brief review of the methods of treating crudes will give a better understanding of the nature of oil and refinery gases.

Like coal, petroleum is the source of various solid, liquid, and gaseous products which are used as fuels and for other purposes. Among the solid fuels derived from it are the heavy tars, pitches, and cokes which are collected at the bottom of stills; the liquid products comprise gasoline, kerosene, gas oils, and various

Table XIX. Natural Gas from Arkansas Fields

Constituents	Raw-sour gas*	Sweetened-sour gas**
Carbon dioxide	5.50	0.00
Methane	77.73	88.83
Ethane	5.56	6.35
Propane	2.41	2.75
Butane	1.17	1.34
Pentane	0.39	0.45
Hexane and higher	0.24	0.28
Hydrogen sulfide	7.0	0.0004
HHV	. . .	. . .
Btu/cu ft at standard conditions	. . .	1159

*From Big Creek field
**From McKamie field

grades of fuel oils, as well as semifluid sludges left over from refinery operations. Finally, petroleum yields a variety of gaseous fuels, some of which are purposely manufactured for sale, such as oil gas and carburetted water gas, and others in the nature of waste products, known as refinery gases.

There are two basically different ways of processing crudes. The first and older is distillation; the second and more recent is cracking. Frequently, in modern refinery practice, both methods are combined into the same or successive operations.

DISTILLATION VERSUS CRACKING

In a typical continuous distillation process, the crude oil is first preheated in a tube still to a temperature of about 600°F. The charge is then sent to a fractionating tower, or column, where it is flashed at atmospheric pressure and separated into its various components according to their specific gravities.

In addition to distillation being carried out at atmospheric pressure or under vacuum, it may take place in the presence of steam or other inactive gas. Both vacuum and steam distillation permit fractionation at lower temperature.

Cracking is the thermal decomposition of petroleum. In a typical cracking process, the crude oil is heated in tube coils to a temperature of 850 to 1000°F and discharged, under a pressure of perhaps 750 psi, first into separators which eliminate the tar and heavy oil, and then into a fractionating tower.

PRODUCTION OF REFINERY FUELS

As a fuel, refinery gas is withdrawn from whatever point in the refining system it is found economical. It may come from raw gasoline tanks, from stabilizers, from cracking stills, or from adsorbers used in purifying operations. Likewise, depending on the process and the point of origin in the system, the fuel gas may be at, or below, or above atmospheric pressure; if it comes from a condenser, it may be at room temperature, but its temperature may be much higher if taken directly from a cracking still or fractionating tower. Often, refinery gases from various sources in the plant are blended together before being burned.

Because it is specifically manufactured for commercial purposes, oil gas deserves special mention, as the heat value standard for the locality where it is sold governs its final composition.

Oil gas is produced by spraying oil into a retort heated to about 1300°F, causing thermal decomposition. Cracking of the oil is then completed by passing the gas through a second retort, which converts the unstable hydrocarbons of low molecular weight into permanently gaseous hydrocarbons. In this manner about 85 percent of the oil is recovered as gas.

The wide range of characteristic analyses of these gases is given in Table XXI. The last two analyses are of oil gases; the others represent refinery gases.

COKE-OVEN GAS

As its name implies, coke-oven gas is a by-product of the destructive distillation of bituminous coal in the manufacture of coke. The raw coal is placed in ovens, which are externally heated until practically all the volatile matter is driven out. The coke-oven gas is made up chiefly of hydrogen, methane, ethylene, and carbon monoxide, with small percentages of carbon dioxide, nitrogen, oxygen, and heavy hydrocarbons or illuminants. The character of

the raw coal and the length of time taken in the roasting operation influence the exact composition of coke-oven gas.

As the coke-oven gas comes from the oven at high temperatures, it frequently contains various impurities in the form of tarry particles, dust, benzol, and hydrogen sulfide. With rare exceptions, however, it is always washed and cooled before being sent to a storage tank and distributed for burning.

Because of a breaking down of the heavy hydrocarbons, carbonization at high temperatures generally results in calorific values varying from 400 to 600 Btu per cu ft. On the other hand, low-temperature carbonization gives a smaller yield of coke-oven gas, but a higher heating value, ranging from 600 to 1000 Btu per cu ft. Table XXII indicates a spread between 477 and 829 Btu per cu ft in the high heating value of the analyses selected. The heating value of a coke-oven gas, when not given, may be accurately determined by a summation of the heats evolved by the individual combustible constituents of the gas.

PRODUCER GAS

Producer gas is formed through the partial combustion of coal or coke by passing less than stoichiometric air through a hot fuel bed. Because of the use of air, the nitrogen content of the gas is high, generally 50 to 55 percent by

Table XX. CNGA Standard Grade for LPG Mixture

CNGA Standard Grade	Max Vapor Pressure Psig at 100°F	Range of Allowable Sp Gr 60/60°F $H_2O = 1.0$	Composition
A	80	0.585–0.555	Predominantly butanes
B	100	0.560–0.545	Butane-propane mixture, largely butanes
C	125	0.550–0.535	Butane-propane mixture, proportions approximately equal
D	150	0.540–0.525	Butane-propane mixture, propane exceeds butane
E	175	0.530–0.510	Propane-butane mixture, largely propane
F	200	0.520–0.504	Predominantly propane

volume, and the heating value is rather low, ranging from about 140 to 180 Btu per cu ft. The ranges of component gases in this fuel are

CO	20–30%
H_2	8–20%
CH_4	0.5–3 %
CO_2	3–9 %
N_2	50–56%
O_2	0.1–0.3%

WATER GAS (BLUE GAS)

Water gas is made in a cyclic process, in which coke is "blown" with air to raise its temperature and then "blasted" with steam. The steam reacts with the hot carbon endothermally as follows:

$$C + H_2O \rightarrow CO + H_2 \tag{4}$$

The gas is sometimes called "blue gas" because of the characteristic blue flame with which it burns, the color resulting from the high percentage of hydrogen and carbon monoxide. A typical analysis shows this gas consists of the following components:

Table XXI. Characteristics of Typical Refinery and Oil Gases at 60°F and 30 In. Hg, Dry

% by volume								
O_2	...	...	...	...	...	2.3	0.9	0.1
N_2	...	...	...	...	...	8.7	8.4	2.7
CO_2	...	...	...	...	3.3	...	2.2	1.0
CO	...	...	...	...	1.5	...	14.3	6.8
H_2	...	...	...	...	5.6	...	50.9	59.2
H_2S	2.18	...	...	...	...	...	...	...
CH_4	41.62	4.30	92.10	5.0	30.9	30.3	15.9	25.4
C_2H_6	20.91	82.70	1.90	12.0	19.8	13.4	5.0	...
C_3H_8	19.72	13.00	4.50	30.0	38.1	19.1	...	...
C_4H_{10}*	9.05	...	1.30	34.0	0.0	14.7	...	...
C_5H_{12}†	6.52	...	0.02	19.0	...	1.8	...	...
$C_3H_3H_6$‡	...	...	...	...	0.2	9.7	2.4	4.8
Density Lb/Cu Ft	0.08676	0.08377	0.04845	0.13760	0.08102	0.09232	0.03631	0.02756
HHV Btu/Cu Ft**	1898	1858	1136	2988	1696	1844	519	586
Btu/Lb	21,880	22,170	23,460	21,720	20,930	19,970	14,300	21,270
A at Zero Excess Air Lb/10^6 Btu	722	725	723	717	725	715	648	657
CO_2 at Zero Excess Air %	13.3	13.4	12.1	13.9	13.4	13.6	12.2	10.6

*Includes both iso-C_4H_{10} and n-C_4H_{10}.
†Includes all saturated hydrocarbons heavier than C_5H_{12}, also both iso-C_5H_{12} and n-C_5H_{12}.
‡Includes all illuminants.
**If gas is saturated with moisture at 60°F and 30.0 in. Hg, reduce by 1.74%.

ABB
ASEA BROWN BOVERI

Table XXII. Characteristics of Typical Coke-Oven Gases at 60°F and 30 In. Hg, Dry

CO₂	O₂	N₂	CO	H₂	CH₂	C₂H₄	C₆H₆	Density, Lb/Cu Ft	HHV Btu/Cu Ft*	Btu/Lb	A at Zero Excess Air, Lb/10⁶ Btu
1.8	0.2	3.4	6.3	53.0	31.6	2.7	1.0	0.0298	596	20,010	674
1.4	0.5	4.2	5.1	57.4	28.5	2.9	...	0.0263	539	20,490	664
2.6	0.6	3.7	6.1	47.9	33.9	5.2	...	0.0316	603	19,070	676
3.13	...	...	11.93	42.16	37.14	4.76	0.88	0.0359	663	18,500	684
0.1	...	2.4	6.8	27.7	50.0	13.0	...	0.0393	829	21,100	700
0.75	...	12.1	6.0	53.0	28.15	...	...	0.0291	477	16,390	668

*If gas is saturated with moisture at 60°F and 30.0 in. Hg, reduce by 1.74%.

CO₂	5.6%
CO	37.0%
H₂	47.3%
CH₄	1.3%
N₂	8.8%
HHV, Btu/cu ft	287

CARBURETTED WATER GAS

Water gas enriched with fuel oil to raise its heating value is called *carburetted water gas*. The process involves the injection of oil into the carburetting chamber during the steam blow; the water gas, passing into the carburetor, picks up the oil vapor which is then cracked into gases. The composition and heating value of the carburetted water gas varies with the amount of oil used, the latter property varying between 400 and 700 Btu per cu ft. Table XXIII shows typical analyses of this gas.

REGENERATOR WASTE GAS

The lighter hydrocarbons or gas oils produced in the petroleum coking processes are further refined in catalytic-cracking units. These units fall into two general types: fluid units, in which fine powdered catalyst flows through the equipment with flow characteristics resembling a liquid, and moving-bed units, which use either spherical or pelleted catalyst circulated by elevators or gas lifts.

The cracking of the feed takes place in a chemical reactor vessel. The preheated catalyst is maintained in a fluid state, and during the cracking reaction the catalyst becomes coated with a coke deposit. To maintain catalyst activity, this material must be removed.

The spent catalyst is continuously removed from the reactor and transported to a regenerator vessel. In this unit compressed air fluidizes the catalyst and burns off the carbon. To keep compression costs to a minimum, as well as to keep the temperature inside the regenerator restricted to a level which will not destroy catalyst activity, the smallest amount of air is used that will effectively clean the catalyst.

Table XXIII. Analyses of Carburetted Water Gas

Component	From Coke %	From Anthracite %
Carbon dioxide and hydrogen sulfide	0.9	3.3
Nitrogen	6.8	4.2
Hydrogen	37.4	38.4
Carbon monoxide	35.0	31.0
Methane	8.1	12.7
Ethane	1.30	1.05
Ethylene	6.7	6.9
Propane	0.25	0.08
Propylene	1.5	0.87
Butane	0.0	0.0
Butylene	0.75	0.45
Liquid hydrocarbons	1.30	1.05

This combustion process, therefore, normally produces an appreciable percentage of carbon monoxide.

The gas, although at temperatures as high as 1125°F, has a heating value of not over 40 Btu per cu ft. A typical analysis of the constituents in regenerator waste gas follows:

CO	6.9%
CO_2	8.1%
O_2	0.8%
N_2	65.8%
H_2O	18.4%

BLAST-FURNACE GAS

Since blast-furnace gas is one of the most important byproducts in the smelting of iron ore to obtain pig iron, its characteristics are best understood by examining what takes place in a blast furnace when in operation.

Iron ore, coke, and limestone are alternately charged into the furnace from the top at the same time that air, previously heated to between 1100 and 1300°F, is blown into it through openings at the bottom. In moving down through the furnace, the mass of solid raw materials is vigorously scrubbed by air which sweeps upward at velocities up to 450 ft per sec. The process of extracting iron from its ore begins when carbon in the coke unites with oxygen in the air to form carbon monoxide, CO, with the liberation of heat. Next, the presence of more carbon together with this heat reduces the iron oxide and converts some of the carbon monoxide to carbon dioxide, CO_2. To facilitate removal of impurities from the molten mass, limestone is added and more CO_2 is formed.

Besides carbon monoxide and carbon dioxide, blast-furnace gas contains nitrogen, hydrogen, and sometimes, small amounts of methane (Table XXIV). The last two are the result of high-temperature dissociation of moisture which enters the furnace with the materials charged, whereas the nitrogen in the analyses is almost exclusively derived from the air blast.

While at one time iron and steel producers concentrated their entire attention on getting the highest yield of pig iron per ton of coke burned, now they are much more aware of the economic value of a quality blast-furnace gas.

Present-day furnaces burn this fuel in any one of three different states: (1) as raw gas with some of the dust and dirt removed; (2) after the gas has had a "primary" cleaning or washing; and (3) after it has been thoroughly dedusted or washed by means of "final" as well as "primary" cleaners.

DUST

The high velocity at which blasts of air sweep through the alternate layers of coke, iron ore, and limestone causes many fine particles, and some fairly large lumps, to be carried along in suspension as dust. In some cases, as much as 50 grains of dust per standard cubic foot (at 60°F and 30 in. Hg) have been measured in the gas leaving the top of the blast furnace. Burned in this state, the gas would quickly plug the firing equipment and boiler passages, and require almost continuous shutdown for cleaning. Generally, therefore, many of the suspended solids are removed, before firing, by passing the blast-furnace gas through a dry dust catcher. On exiting from this dust catcher as "raw gas," it still contains 3 to 15 grains of dust per standard cu ft, and, in this condition, it is sometimes burned under steam-generating units.

More often, however, before being delivered to a boiler furnace, the gas is given an additional, or primary, cleaning by means of either a dry separator or a washer, to reduce the dust content to 0.1–0.8 grain per standard cu ft. As desired, following the "primary" cleaners, either dry or wet final cleaners further remove remaining dust content to 0.005–0.05 grain per standard cu ft. When washed, blast-furnace gas can carry a considerable amount of water in suspension as fine droplets as well as water vapor.

TEMPERATURE

The temperature at which blast-furnace gas reaches the boiler depends on the kind of cleaning apparatus employed. As raw gas coming from the dust catcher, its temperature may be anywhere from 300 to 800°F. If the subsequent cleaning is done with dry separators, the drop

Table XXIV. Characteristics of Typical Blast-Furnace Gases at 60°F and 30 In. Hg, Dry

| % by Volume | | | | | Density, Lb/Cu Ft | HHV | | A at Zero Excess Air, Lb/10⁶ Btu |
CO₂	N₂	CO	H₂	CH₄		Btu/Cu Ft*	Btu/Lb	
14.5	57.5	25.0	3.0	. . .	0.0779	90.1	1150	576
13.0	57.6	26.2	3.2	. . .	0.0771	94.6	1219	575
15.59	59.28	23.35	1.7	0.08	0.0792	81.4	1021	577
8.7	56.5	32.8	1.8	0.2	0.0762	113.3	1478	579
5.7	59.0	34.0	1.3	. . .	0.0753	113.6	1498	576
6.0	60.0	27.0	2.0	5.0	0.0734	144.0	1950	631

* If gas is saturated with moisture at 60°F and 30.0 in. Hg, reduce by 1.74%

in gas temperature through each separator may be only 30 to 50°F. On the other hand, washers, using sprays of water to precipitate the dust, lower the temperature of the gas to 70–120°F.

Furthermore, in some cases the distance from blast furnace to boiler is sufficiently great to cause a drop in temperature from radiation losses in the connecting piping.

REFERENCES AND FOOTNOTES

[1] George Frederick Gebhardt, *Steam Power Plant Engineering*, Chapter II "Fuels," 6th ed. New York: John Wiley and Sons, 1925.

James A. Kent, ed. *Riegel's Handbook of Industrial Chemistry*, 7th ed. New York: Van Nostrand Reinhold Co. 1974.

[2] Sam H. Schurr and Bruce C. Netschert, *Energy in the American Economy, 1850–1975: An Economic Study of Its History and Prospects.* Originally published by John Hopkins Press, 1960. Reprinted 1977 by Greenwood Press, Inc., Westport, CT; pp 57–83 and 302–346.

A. J. Johnson and G. H. Auth, ed., *Fuels and Combustion Handbook.* New York: McGraw-Hill, 1951.

National Coal Association, *Coal Facts.* Washington: National Coal Association, 1979.

[3] William Kent, "Classification and Heating Value of American Coals," *Transactions of the ASME,* 36: 189–209, 1914.

A. C. Fieldner, "Constitution and Classification of Coal," *Transactions of the ASME,* Paper FSP 50–51: 49–50, 1927–1928.

Henry Kreisinger and B. J. Cross, "Burning Characteristics of Different Coals," *Transactions of the ASME,* Paper FSP 50–52: 49–50, 1927–1928.

[4] Illinois University at Urbana-Champagne. Engineering Experiment Station. Bulletin Series no. 37, "Unit Coal and the Composition of Coal Ash," by S. W. Parr and W. F. Wheeler, 1909.

Illinois University at Urbana-Champagne. Engineering Experiment Station Bulletin Series no. 180, "The Classification of Coal," by S. W. Parr, 1928.

[5] W. H. Ode and F. H. Gibson, "Effect of Sulfur Retention on Determined Ash in Lower Rank Coals," U. S. Bureau of Mines, Report of Investigation 5931. Washington: U. S. Bureau of Mines, 1962.

[6] National Research Council. Committee on Chemistry of Coal. *Chemistry of Coal Utilization,* Supplementary Volume, ed. by H. H. Lowry. New York: John Wiley, 1963. Chapter 5, "Coal Analysis and Mineral Matter," by W. H. Ode.

[7] H. Kreisinger, "High and Net Heat Values of Fuel," *Combustion,* 2(2):25–26 and 47, Aug. 1930.

[8] Roy F. Abernethy and Elsie M. Cochrane, "Fusibility of Ash of United States Coals," U. S. Bureau of Mines, Information Circular 7923. Washington: U.S. Bureau of Mines, 1960.

Roy F. Abernethy and Elsie M. Cochrane, "Free-Swelling and Grindability Indexes of United States Coals," U. S. Bureau of Mines, Information Circular 8025. Washington: U. S. Bureau of Mines, 1961.

[9] Badzioch, S., and Hawksley, P.G.W., "Kinetics of Thermal Decomposition of Pulverized Coal Particles," *Ind. Eng. Chem. Process Des. Develop.,* 9, 521 (1970).

[10] Nsakala, N., Essenhigh, R. H., and Walker, Jr., P. L., "Characteristics of Chars Produced from Lignites by Pyrolysis at 808°C following Rapid Heating," *Fuel*, 57, 605 (1978).

[11] Pollock, W. H., Goetz,, G. J. and Park, E. D., "Advancing the Art of Boiler Design by Combining Operating Experience and Advanced Coal Evaluation Techniques." Presented at the *American Power Conference* at Chicago, Illinois, April 18–20, 1983; also as Combustion Engineering publication TIS-7382.

[12] Nsakala, N., Patel, R. L. and Borio, R. W., "An Advanced Methodology for Prediction of Carbon Loss in Commercial Pulverized Coal-Fired Boilers," Presented at 1986 *ASME/IEEE Joint Power Generation Conference* at Portland, Oregon, October 19-23, 1986; also as Combustion Engineering publication TIS-8211.

[13] Brunauer, S., Emmett, P.H., and Teller, E., "Adsorption of Gases in Multi-layers," *J. Am. Chem, Soc.*, 60, 309 (1938).

[14] Nsakala, N., Patel, R. L., and Lao, T. C., "Combustion Characterization of Coals for Industrial Applications," Final Technical Report DOE/PC/40267-5, Contract AC22-81PC40267, March 1985.

[15] F. Nowak, "Considerations in the Construction of Large Refuse Incinerators," *Proceedings of the 1970 National Incinerator Conference*, Cincinnati, May 17–20, 1970. Sponsored by ASME Incinerator Division. New York: American Society of Mechanical Engineers, 1970, pp 86–97.

[16] J. W. Regan, "Generating Steam from Prepared Refuse," *Proceedings of the 1970 National Incinerator Conference*, Cincinnati, May 17–20, 1970. Sponsored by ASME Incinerator Division. New York: American Society of Mechanical Engineers, 1970, pp 216–223.

[17] ASME 71–WA/Inc–1

by A. D. Konopka—Systems Evaluation of Refuse as a Low Sulfur Fuel: Pt. 3–Air Pollution Aspects.

/Inc–2
by R. E. Sommerlad, et. al.—Systems Evaluation of Refuse as a Low Sulfur Fuel: Pt. 2—Steam Generator Aspects.

/Inc–3
by A. M. Roberts and E. M. Wilson—Systems Evaluation of Refuse as a Low Sulfur Fuel: Pt. 1—The Value of Refuse Energy and the Cost of Its Recovery.

Presented at the ASME Winter Annual Meeting, Washington, D. C., Nov. 28–Dec. 2, 1971. New York: American Society of Mechanical Engineers, 1971.

[18] John T. Pfeffer, *Reclamation of Energy from Organic Waste: Final Report*. Report No. PB-231-176 (EPA-670/2-74-016), prepared under EPA grant No. EPA-1P-80766, Department of Civil Engineering, University of Illinois, Urbana, Illinois, Mar. 1974. Springfield, Va.: National Technical Information Service.

[19] W. E. Franklin, D. Rendersky, L. J. Shanon, and W. R. Park, *Resource Recovery: Catalogue of Processes*, Report No. PB-214 148, Midwest Research Institute, Kansas City, Mo., Feb. 1973. Springfield, Va.: National Technical Information Service.

[20] W. E. Franklin, D. Bendersky, L. J. Shannon, and W. R. Park, *Resource Recovery: The State of Technology*, Report No. PB-214 149, Midwest Research Institute, Kansas City, February, 1973. Springfield, Va.: National Technical Information Service.

[21] Gerald E. Dreifke, David L. Klumb, and Jerrel D. Smith, "Solid Waste as a Utility Fuel," *Proceedings of the American Power Conference*, Vol. 35: 1198–1206. Chicago: Illinois Institute of Technology, 1973.

[22] D. Joseph Hagerty, et al., *Solid Waste Management*. New York: Van Nostrand Reinhold Co., 1973.

[23] Joseph G. Singer and Joseph F. Mullen, *Closing the Refuse Power Cycle*, presented at the ASME-IEEE Joint Power Generation Conference, New Orleans, Sept. 16–19, 1973, ASME Paper-73-PWR-18; also as Combustion Engineering Publication TIS 3612.

[24] Robert A. Lowe, *Energy Recovery from Waste: Solid Waste as Supplementary Fuel in Power Plant Boilers*, Report No. PB-256 494 (EPA-SW-36d. ii), Grant No. S-802255, prepared by the City of St. Louis, Mo. in cooperation with Union Electric Co., St. Louis, Mo., 1973.

[25] John H. Fernandes and R. C. Shenk, "The Place of Incineration in Resource Recovery of Solid Waste," *Proceedings of the 1974 National Incinerator Conference*, Miami, May 12–15, 1974: Resource Recovery Through Incineration, 1–10, sponsored by the ASME Incinerator Division. New York: American Society of Mechanical Engineers, 1974.

[26] *ASTM Standards*, Parts 15, 24, and 140, D 2161, "Standard Method for Conversion of Kinematic Viscosity to Saybolt Universal Viscosity or to Saybolt Furol Viscosity." Philadelphia: American Society for Testing and Materials, latest edition.

CHAPTER 3

Properties of Coal Ash

As mined, coal contains varying quantities of mineral matter which, when the coal is burned, results in the incombustible residue known as ash. Most of this mineral matter was carried in by wind and water; only a small fraction was inherent in the original vegetation. The mineral matter from which ash is formed is an undesirable substance in the coal. An inert material on which freight must be paid, it must be removed from the furnace as fast as it accumulates; otherwise, it may cause interruption in operation. Also, provision must be made for handling and disposing of the ash, both functions representing items of cost.

The mineral matter of coal cannot be completely separated by any physical method to identify fully the individual minerals or to determine them quantitatively. Therefore, to assess the amounts of the major elements present in coal, an analysis is made to determine the percentages of the *oxides* of these elements present in the coal ash.

The ash analysis does not indicate the nature or distribution of the mineral matter but, nevertheless, provides information which is useful in both the study and the utilization of coal. Examination of many coal-ash analyses shows, in almost all cases, that oxides of silicon, aluminum, iron, calcium and magnesium account for 95 percent or more of the ash. Of the other components present, oxides of sodium and potassium are important only as they act as fluxes in fusing the ash.

ASH IN COAL: PRINCIPAL DESIGN FACTOR

The management of this coal ash is one of the major considerations in the design of a coal-fired steam generator. The behavior of the mineral matter in a coal, as it influences furnace-wall *slagging* during the combustion process, is a significant factor in furnace sizing in terms of volume, plan area, and fuel-burning zone. Mineral matter formed as tenacious deposits on waterwall surfaces insulates the heat-transfer portions of the furnace—thus, the ash properties affect the disposition of both radiant and convective heating surface, as well as the number of furnace-wall cleaning devices (usually known as sootblowers).

The behavior of this mineral matter, as it influences *fouling* of convection-type surfaces, affects the amount of such surfaces, their spacing, and the quantity of sootblowers for cleaning these heat transfer surfaces.

And the *percentage* of this mineral matter in the coal is significant because of the amount of slagging and fouling that it can cause, the burden that it places on ash-handling equipment and disposal facilities, the frequency and length of sootblowing, and the rate of wear of pulverizer parts. Thus, for any given chemical and physical properties of coal ash, their net effect on the boiler during a given time period will be directly proportional to the amount fired in the furnace.

A corollary of this observation on the management of ash is that completely de-ashed coal of any type or rank can be burned in a furnace designed specifically for oil or gas firing. That is to say, ash is what gives a coal its "character." Without ash, *all* furnaces could easily be designed on the basis of heat transfer only.

Students, designers, and users of large pulverized-coal boilers must realize that since the 1920's, substantial work has been done to establish "scientific" methods for the sizing of furnaces. Further, during these years, much has been learned about the ways in which certain coals (and their ash) act in different sizes and types of furnaces.

But it must be understood and accepted that, as the art exists today, engineering judgment, based upon what we know about the ash in a coal (based on ASTM analyses and other parametric data) and how that ash has "behaved" in actual furnace operation, is the only guide to sizing new and larger furnaces.

In this chapter, then, most of the useful physical and chemical properties of ash are given. Useful parameters for categorizing ash also are

presented and discussed—all as they relate to slagging and fouling of boiler surfaces. Methods of ash analysis that have been developed recently are described; and the mechanism of fire-side metal loss by either corrosion or erosion caused by fuel ash is covered. Appendix B contains additional engineering data on other significant ash properties. All such information is of vital concern to engineers making decisions about furnace size and heating-surface arrangements.

COMPOSITION OF COAL ASH

Coal ash consists almost entirely of metal oxides. The composition varies over a wide range, and there is no "typical" ash analysis.

CHEMICAL ANALYSES

Chemical analyses of coal ash provide data from which to estimate coal-ash and slag characteristics, as well as calculate various correlation parameters. Analyses are generally performed on ash prepared to *ASTM Standards* D 3174, Ash in the Analyses Sample of Coal and Coke. Pulverized coal is burned in a ceramic crucible, in air, at 1290–1380°F, to completion (except for the ash residue).

Chemical analyses of coal ash are reported as the mass percent of each equivalent oxide and are generally expressed as:

$$SiO_2 + Al_2O_3 + Fe_2O_3 + CaO + MgO + Na_2O + K_2O + TiO_2 + P_2O_5 + SO_3 = 100\%$$

(1)

Because it is usually present in insignificant quantities, P_2O_5 is sometimes omitted.

Table I gives the ranges of the chemical composition of most coal ash.[1]

Research has shown that the chemical composition of coal ash prepared to *ASTM Standards* D 3174 is different from that of flyash, bottom ash, or waterwall slag as they occur in actual boiler furnaces. However, because few adequate alternatives have been established, coal ash and slag behavior commonly are estimated using the chemical composition of the

Table I. Coal-Ash Chemical Composition Ranges

Oxide Component	Percentage
SiO_2	10–70
Al_2O_3	8–38
Fe_2O_3	2–50
CaO	0.5–30
MgO	0.3–8
Na_2O	0.1–8
K_2O	0.1–3
TiO_2	0.4–3.5
SO_3	0.1–30

ash. Methods involving gravity separation of pulverized-coal fractions can be used in an effort to simulate the segregation effects in an actual furnace. (See the section of this chapter on Selective Deposition of Ash Constituents.)

MINERALOGICAL ANALYSES

Mineralogical analyses of coal ash attempt to identify the original mineral-matter forms in coal. The thermal behavior of coal ash in an operating furnace largely depends on the reactions between the ash-forming minerals, which cannot be detected by the ordinary chemical analyses. Accordingly, it can be argued that coal-ash and slag behavior should be characterized by its mineralogical composition and not by its chemical composition. However, to date, neither a standard procedure for the low-temperature ashing technique nor a criterion for mineral detection has been established.

Mineralogical analyses of coal ash can be performed either on ash prepared according to *ASTM Standards D 3174* or, preferably, on ash prepared according to the low-temperature ashing technique. Low-temperature ash is produced by placing pulverized-coal samples in an oxygen plasma at approximately 400°F. The oxygen will gradually break down the organic matter without excessive thermal decomposition of the inorganic mineral species.

At least a hundred mineral species are associated with coal. Table II lists the more abundant mineral forms.

Thermal reactions among ash-forming minerals are an important, although very complex, subject. Fig. 1 gives an overview of important steps during the heating of mineral matter.

CHEMICAL REACTIONS DURING COMBUSTION

Most ash constituents will not be in the form shown in Table I when the coal goes to the fur-

Table II. Typical Mineral Species Found in Coal

Mineral Species	Formula
Kaolinite	$Al_2O_3 \cdot 2SiO_2 \cdot H_2O$
Illite	$K_2O \cdot 3Al_2O_3 \cdot 6SiO_2 \cdot 2H_2O$
Muscovite	$K_2O \cdot 3Al_2O_3 \cdot 6SiO_2 \cdot 2H_2O$
Biotite	$K_2O \cdot MgO \cdot Al_2O_3 \cdot 3SiO_2 \cdot H_2O$
Orthoclase	$K_2O \cdot Al_2O_3 \cdot 6SiO_2$
Albite	$Na_2O \cdot Al_2O_3 \cdot 6SiO_2$
Calcite	$CaCO_3$
Dolomite	$CaCO_3 \cdot MgCO_3$
Siderite	$FeCO_3$
Pyrite	FeS_2
Gypsum	$CaSO_4 \cdot 2H_2O$
Quartz	SiO_2
Hematite	Fe_2O_3
Magnetite	Fe_3O_4
Rutile	TiO_2
Halite	$NaCl$
Sylvite	KCl

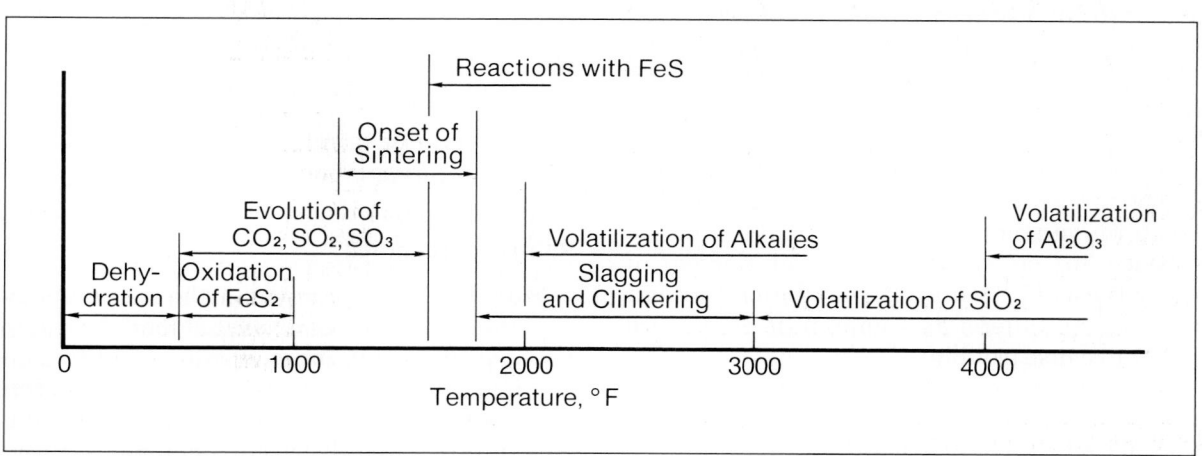

Fig. 1. Effect of heating on coal mineral matter

nace. As has been pointed out, the bulk of the ash is mineral matter brought in by water, and is largely a mixture of the various compounds. These compounds change considerably as they pass through the furnace and are subjected to high furnace temperature. Chemical reactions take place among these substances, and also between them and the reducing gases and hot carbon. They form new and more complex compounds, usually containing less oxygen than the original constituents, because the oxygen is taken away by the reducing gases and hot carbon. This is particularly true of the ferric oxide, Fe_2O_3, which occurs in the coal largely as pyrites, FeS_2. This may be reduced to FeS or to the lower oxides, Fe_3O_4 and FeO, or even metallic iron, Fe. The lower sulfide and oxides have lower softening and fusion temperatures. While going through the fusion state, the new compounds form solutions, and this change further lowers the fusion temperature.

As a rule, ash high in SiO_2 or alumina has high softening temperature, and this temperature is not greatly affected by reducing atmosphere. The alkali metal compounds Na_2O and K_2O, as well as the alkaline earth, magnesium oxide, tend to lower the softening temperature of the ash.

IRON IN COAL ASH

Iron compounds are responsible for much of the misbehavior of coal ash. Therefore, coals with ash high in iron are always suspected of causing trouble. If the ash, and particularly the iron, is uniformly distributed through the coal, difficulties are more likely to occur than if the iron compounds are in large pieces, separated from the coal. When the coal is pulverized, the larger and heavier pieces of ash are rejected by some pulverizers and do not go through the furnace. In screenings with a high percent of fines, the ash is uniformly distributed through the coal, and therefore compounds of iron are likely to cause trouble.

In pulverized-coal firing, there may be a considerable separation of the different constituents of the ash. The ash deposited at the furnace bottom will have chemical composition and physical properties different from those of the ash carried by the gases.

In slagging- or wet-bottom pulverized-coal furnaces, a considerable part of the iron compounds in the slag may be reduced to metallic iron, which sinks to the bottom of the molten slag and is difficult to remove in either hot or cold state.

In some coals, the iron is in the form of pyrite, FeS_2, which, while passing through the furnace, undergoes various changes. Both the iron and the sulfur may combine with oxygen, iron forming the lower oxides and sulfur, SO_2 or SO_3. Sulfur may also combine with the alkaline metals, Na and K, to form sulfur compounds, which have very low fusion temperature. In slagging-bottom furnaces, where the molten ash remains in a liquid state over extended periods of time, some of these sodium and potassium sulfur compounds are vaporized, then condensed on watercooled furnace walls, and may become an important factor in external corrosion of furnace-wall tubes.

COAL-ASH SLAGGING AND DEPOSITION

A number of parameters are used for evaluating coal-ash behavior as they affect furnace slagging and deposition on both the furnace walls and convection surfaces. Some of these parameters are
- ash-fusibility temperatures
- base/acid ratio
- iron/calcium ratio
- silica/alumina ratio
- iron/dolomite ratio
- dolomite percentage
- ferric percentage.

Such indices help organize the ash composition into "building blocks" that lend insight to expected behavior under burning conditions.

Some confusion exists between the terms *slag* and *deposits* (or deposition). Definitions as they apply to pulverized-coal firing are:
1. *Slag*—fused deposits or resolidified molten

material that forms primarily on furnace walls or other surfaces exposed predominately to radiant heat or high gas temperatures.

2. *Deposits*—bonded (sintered or cemented) ash buildup that forms primarily on such convection surfaces as superheater and reheater tubes but also on furnace walls at lower than slag-producing temperatures.

A summary of ash properties in terms of slagging and deposition is given in the following sections of this chapter.

MEASURING ASH-FUSIBILITY TEMPERATURES

Ash fusibility has long been recognized as a tool for measuring the performance of coals related to slagging and deposit buildup. It is still perhaps the most basic means for predicting coal-ash performance. Other parameters are used primarily to explain or amplify ash-fusibility temperatures. In general, high fusion temperatures result in low slagging potential in dry-bottom furnaces, while low fusion temperatures are considered mandatory for wet-bottom (slag-tap) furnaces.

Many experimental methods have been developed for measuring the fusion temperatures of coal ash. Most tests are based on the gradual thermal deformation of a coal-ash sample, and the characteristic fusion temperatures are reported when the sample fuses into certain distinctive shapes or heights. Most test methods are strictly empirical and have little theoretical basis. The following descriptions characterize some of the methods.

ASTM STANDARDS METHOD

ASTM Standards D 1857, Fusibility of Coal Ash, gives the experimental procedure for one method of testing. The test is based on the gradual thermal deformation of a pyramid-shaped ash sample, 3/4 inch in height and 1/4 inch in equilateral-triangular base width. Mounted on a refractory substrate, the sample is heated at a prescribed rate in a gas or electric furnace. A controlled atmosphere (reducing or oxidizing, depending upon the tests to be per-

formed) is maintained inside the furnace. During the heating process, changes in the shape of the pyramid are observed, Fig. 2, and the following four characteristic deformation temperatures are reported:

1. Initial Deformation Temperature (IT): The temperature at which the tip of the ash pyramid begins to show any evidence of deformation. Shrinkage of the cone is ignored if the tip remains sharp.

2. Softening Temperature (ST), H = W: The temperature at which the ash sample has fused into a spherical shape in which the height is equal to the width at the base. The H = W softening temperature in a reducing atmosphere frequently is referred to as the "fusion temperature."

3. Hemispherical Temperature (HT), H = 1/2W: The temperature at which the ash sample has fused into a hemispherical shape where the height is equal to 1/2 the width at the base.

4. Fluid Temperature (FT): The temperature at which the ash sample has fused down into a nearly flat layer with a maximum height of 1/16 inch.

Prior to the revision of *ASTM Standards* D 1857, only three characteristic deformation temperatures (initial deformation, softening, and fluid) were observed with no dimensional (height-to-width) specifications for the softening and fluid points. Because of such differ-

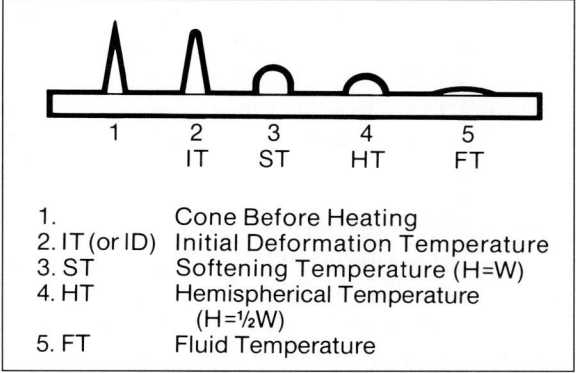

1.	Cone Before Heating
2. IT (or ID)	Initial Deformation Temperature
3. ST	Softening Temperature (H=W)
4. HT	Hemispherical Temperature (H=½W)
5. FT	Fluid Temperature

Fig. 2. Critical temperature points as defined in *ASTM Standards* D 1857

ences in definitions, earlier investigations have reported the softening temperatures without distinguishing between their being spherical or hemispherical. Unless specified otherwise, the softening temperature may be assumed to represent the spherical point, where H = W.

Although the earlier standard procedure used only reducing atmospheres for the tests, many studies give the ash-fusion temperatures without specifying the atmosphere; it thus must be assumed that the tests were carried out in a reducing atmosphere.

BRITISH STANDARD METHOD

In principle, the British standard method is similar to the ASTM method, but differs from it in several experimental specifications. B.S. 1016, Part 15, gives the procedure for this method of testing. The British method specifies that the ash sample be a pyramid of 8 to 13 mm in height with the height 2 to 3 times the width of the triangular base. Mounted on a refractory substrate, the sample is heated at a prescribed rate in an electric furnace with a controlled atmosphere (either reducing or oxidizing) maintained inside the furnace. During the heating process, changes in the sample shape are observed through a magnifying optical instrument and the following three characteristic deformation temperatures are reported:

1. Initial Deformation Temperature: The temperature at which the tip of the ash pyramid begins to show any evidence of deformation.

2. Hemispherical Temperature: The temperature at which the ash sample has fused into a hemispherical shape where the height is equal to ½ the width.

3. Flow Temperature: The temperature at which the ash sample has fused into a flattened disk shape such that the height is ⅙ the width.

GERMAN STANDARD METHOD

German Standard DIN 51 730 describes the procedure for this method of testing. The test is based on the gradual thermal deformation of a cylindric sample of coal ash, 3 mm in height and 3 mm in diameter, or a cubic sample of coal ash with 3 mm edges. The sample is mounted

on a platinum-covered refractory substrate, which is placed on an electric heating element. The heating element, together with the ash sample, is inserted into a small furnace and heated at a prescribed rate. A controlled atmosphere (either reducing or oxidizing) is maintained inside the furnace. An image of the sample is projected onto a graduated screen so that changes in the shape of the ash can be observed. The following three characteristic temperatures are reported:

1. Softening Temperature: The temperature at which the first indication of softening appears, mainly in the form of rounding of the edges or swelling of the sample.

2. Hemispherical Temperature: The temperature at which the ash sample has fused into a hemispherical shape where the height is equal to ½ the base width.

3. Flow Temperature: The temperature at which the ash sample has fused into a flattened disk such that the height is reduced to ⅓ the original sample height.

OTHER METHODS

Similar to the German method in experimental procedure and specifications, the International Standard Method, ISO 540, requires the observation of three characteristic temperatures—initial deformation, hemispherical, and flow. However, a greater latitude is allowed in the shapes and dimensions of the ash sample, because it has been shown that geometrical variations in the sample, within certain limits, have little influence on the observed results.

The Russian experimental procedure (GOST 2057–68) is similar to the German; characteristic deformation temperatures reported during the heating of the sample are the deformation temperature, the softening (hemispherical) temperature, and the temperature of beginning of liquid fused state.

RELATION BETWEEN CHEMICAL AND MINERALOGICAL COMPOSITION

Barrett[2], Estep *et al*[3], and Thiessen *et al*[4] have shown that the fusion behavior of coal ashes is

quite similar to that of the corresponding oxide mixtures. Barrett concluded that, if coal ashes are finely ground, intimately mixed, and brought to the same standard state before the fusion tests are made, then, (1) the fusion temperatures depend only on the composition and not on the manner in which the composition is obtained; and (2) the softening temperature is closely related to the equilibrium phase diagram of the corresponding oxide components which make up the ashes.

ASH-SOFTENING TEMPERATURE RELATIONSHIP TO DEPOSIT REMOVAL

If ash arrives at a heat-absorbing surface at a temperature near its softening temperature, the resulting deposit is likely to be porous in structure. Depending on the strength of the bond, it may fall off the metal surfaces from its own weight or it can be removed by sootblowing. If such a deposit is permitted to build up in a zone of high gas temperature, its surface (because of the insulating properties of the ash) can reach the melting point and then run down the wall surfaces in the furnace.

If ash particles arrive at the heat-absorbing surface at temperatures below the softening temperature, they will not form a bonded structure, but instead, will settle out as dust. As such, removal is comparatively simple.

On the other hand, if ash particles have been subjected to temperatures higher than the softening temperature for a sufficient time to become plastic or liquid, the resulting deposit will be a coarse, fused mass condensed on the cooler metal surface. The resolidified material is tightly bonded and difficult to remove. As this material builds up on itself, its insulating properties continuously increase the surface temperatures until the fluid temperature is reached and slag runoff results.

INITIAL DEFORMATION-FLUID TEMPERATURE DIFFERENTIAL

The temperature differential between initial deformation and fluid temperatures gives an insight as to the type of deposit formation to expect on furnace tube surfaces. A small tem-perature spread from initial deformation to fluid temperatures indicates that the wall slag will be thin, running and tenacious. This type of slag is extremely difficult to control by sootblowing. As the range from initial deformation to fluid temperature increases, the resulting slag deposit will build up to thicker proportions before the surface becomes sufficiently liquid to run. The tube-ash bond is less adhesive and therefore responds better to removal by sootblowing.

PROPERTIES OF COAL-ASH COMPONENTS

The acidic oxide constituents SiO_2, Al_2O_3, and TiO_2 found in coal ash are generally considered to produce high melting temperatures. Temperatures will be lowered proportionally by the relative amounts of basic oxides, Fe_2O_3, CaO, MgO, Na_2O, and K_2O available in the ash for reaction. Minerals associated with coals consist of multiple combinations of elements listed in the first column of Table III. If these elements were each exposed to air under the proper heat conditions, each would be converted to its highest form of oxide. Individual melting temperatures of these oxides are listed in the third column, and, assuming no interaction between them, the melting temperature of the ash would be represented by the weighted average of the oxides. But an interaction takes place at elevated temperatures that produces complex salts having entirely different physical constants. Among other changes, the melting temperatures of the compounds are usually lowered. The table lists some of the combinations, with melting temperatures.

The sum of the noted acidic components can range from 20 to 90 percent of the ash. Generally, highly acid coal ashes have high ash-fusion temperatures and high slag viscosities.

The sum of the basic components can be between 5 and 80 percent. Either high or low basic values can result in high fusion temperatures. Minimum fusion temperatures occur at about 30 to 40 percent basic component.

BASE/ACID RATIO

This index is defined as:

$$\frac{B}{A} = \frac{Fe_2O_3 + CaO + MgO + Na_2O + K_2O}{SiO_2 + Al_2O_3 + TiO_2} \tag{2}$$

Because most elements combine with each other according to their acidic or basic properties, the base/acid ratio reflects a potential for ash-containing metals to combine in the combustion process to produce low melting salts. Extremes at either end of the base/acid ratio indicate a minimum potential for forming combinations with low fusibility temperatures. Note that the alkaline metals, sodium and potassium, are exceptions in that they form low-melting ash regardless of the base/acid combinations: therefore, their presence in an ash composition in sufficient concentrations significantly alters base/acid influences.

Although the base/acid ratio is an indication of the fusion characteristics and slagging potential of a coal ash, it should not be used as the sole criterion of evaluation. A simple ratio such as the base/acid ratio does not take into account differences in fluxing action among the several bases nor does it recognize the degrees of interaction with the acidic components. (*Fluxing property,* in this connotation, is defined as a property of the component or substance which lowers the fusion temperature of the mixture; the terms *flux* and *fluxing property* are also used to describe the lowering in viscosity of a liquid mixture.)

Also, an ash in which the major constituents are alumina and iron will have different slagging characteristics from an ash in which the components include higher percentages of silica and calcium and yet both can have the same base/acid ratio.

In any case, the base/acid ratio is a valuable supplement to ash-fusibility temperatures and other data in predicting the relative performance of coal ash in the furnace. With most ash, a base/acid ratio in the 0.4 to 0.7 range manifests low ash-fusibility temperatures and a higher slagging potential.

SILICA/ALUMINA RATIO

The silica/alumina ratio is defined as:

$$\frac{SiO_2}{Al_2O_3} \tag{3}$$

The general range of values is between 0.8 and 4.0.

The silica/alumina ratio can provide additional information relating to ash fusibility. As both of these constituents are acidic and are, therefore, considered high melting as oxides, silica is more likely to form lower melting species (silicates) with basic constituents than is alumina. For two coals having equal base/acid ratios, the one with a higher silica/alumina ratio should have lower fusibility temperatures.

Table III. Properties of Coal-Ash Components

Element	Oxide	Melting Temp (°F)	Chemical Property	Compound	Melting Temp (°F)
Si	SiO_2	3120	Acidic	Na_2SiO_3	1610
Al	Al_2O_3	3710	Acidic	K_2SiO_3	1790
Ti	TiO_2	3340	Acidic	$Al_2O_3 \cdot Na_2O \cdot 6SiO_2$	2010
Fe	Fe_2O_3	2850	Basic	$Al_2O_3 \cdot K_2O \cdot 6SiO_2$	2100
Ca	CaO	4570	Basic	$FeSiO_3$	2090
Mg	MgO	5070	Basic	$CaO \cdot Fe_2O_3$	2280
Na	Na_2O	Sublimes at 2330	Basic	$CaO \cdot MgO \cdot 2SiO_2$	2535
K	K_2O	Decomposes at 660	Basic	$CaSiO_3$	2804

Effects of increasing amounts of alumina on the formation of the silicates depend on the amount and the combination of bases present, temperature and time of reaction, and other factors. A finite evaluation is virtually impossible because of the complex interactions. Fig. 3 illustrates the effect of silica/alumina ratios on ash-fusibility temperatures. In laboratory investigations, varying proportions of silica and alumina were mixed with an Illinois coal and then burned in a muffle furnace used to prepare ash for fusibility determinations. Fusibility temperatures were determined in a reducing atmosphere. According to these data, no appreciable change is noted between silica/alumina ratios of 1.7 and 2.8. At ratios less than 1.7, the softening and fluid temperatures both increase; at ratios greater than 2.8, the fluid temperature decreases. Initial deformation temperatures were not affected by any wide changes in ratios.

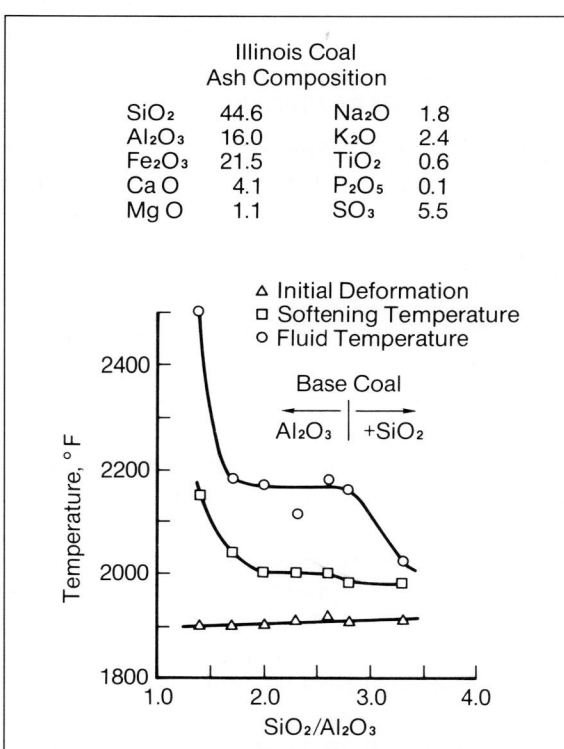

Fig. 3. Ash-fusibility temperatures as affected by silica/alumina ratios (C-E laboratory data)

IRON/CALCIUM RATIO

This parameter, also known as the iron-oxide/calcium-oxide ratio, is defined as:

$$\frac{Fe_2O_3}{CaO}$$

$$(4)$$

As indicated, the base/acid ratio does not account for differences in fluxing action. Among the five basic-oxide components, iron and calcium are the most important, primarily because they make up the largest amount of the basic constituents. Iron oxides comprise 5 to 40 percent of the ash and calcium oxides make up 2 to 30 percent. The following statements, extracted from ref. 5, relate the iron-calcium relationship to the fusibility of ash.

"1. In the absence of lime, iron oxides do not make the most fusible slags."

"2. Ferric oxide, up to about 20 percent, has a fluxing effect which increases with the percentage, but above 20 and up to 40 percent it does not materially increase the fusibility."

"3. The combined fluxing effect of iron oxides and lime has a complex relation to the percentage of the two fluxes present. For ashes in which the percentage of iron oxide is over 14 percent, the addition of a given percentage of lime produces a lowering of the fusion temperatures greater than the same additional percentage of iron oxide."

"4. The quantity of magnesia present in coal ashes is usually smaller than the lime; its fluxing effect with lime present is greater than that of an equal additional amount of lime in about the ratio of 3:2."

Comparisons such as the following confirm this decrease.

Coal	Fe_2O_3	CaO	Fe_2O_3/CaO Ratio	Softening Temperature, °F
(1)	31.8	0.3	106.0	2360
(2)	24.8	2.0	12.4	2270
(3)	21.3	4.8	4.4	2130

The iron/calcium ratios of the three examples are 106, 12.4 and 4.4 respectively. Fig. 4 shows the effect of limestone addition to fusibility temperatures on an Eastern U.S. coal. Based on these two data sources as well as observation of other fusibility-ash composition relationships, Fe_2O_3/CaO ratios between 10 and 0.2 have a marked effect on lowering the fusibility temperatures of coal ash. Extreme effects are evident between ratios of 3 and 0.3.

IRON/DOLOMITE RATIO

The iron-oxide/dolomite ratio is defined as:

$$\frac{Fe_2O_3}{CaO + MgO}$$

(5)

This parameter is essentially identical to the Fe_2O_3/CaO ratio. CaO and MgO have been

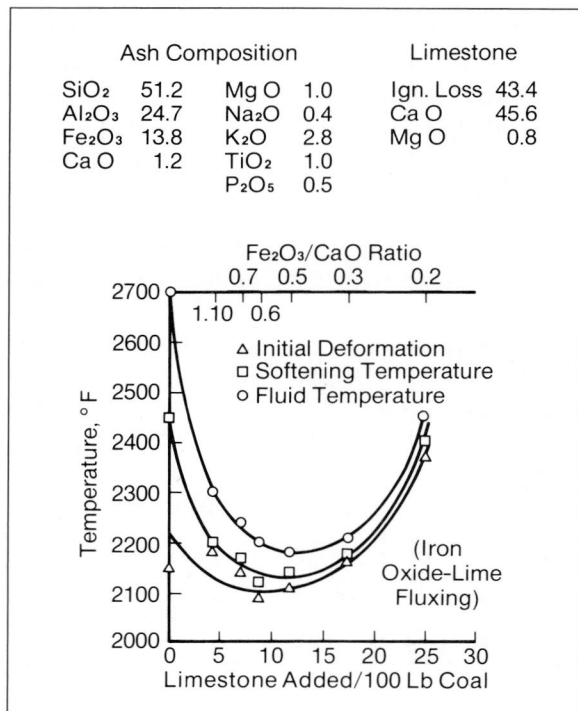

Ash Composition				Limestone	
SiO_2	51.2	MgO	1.0	Ign. Loss	43.4
Al_2O_3	24.7	Na_2O	0.4	CaO	45.6
Fe_2O_3	13.8	K_2O	2.8	MgO	0.8
CaO	1.2	TiO_2	1.0		
		P_2O_5	0.5		

Fig. 4. **Ash-fusibility temperatures of an Eastern coal as affected by limestone addition**

found to have similar fluxing properties, and the use of this ratio is recommended when the MgO content of the ash is high.

Bituminous-type ash is characterized by an iron-oxide/dolomite ratio greater than one, while lignitic ash has $Fe_2O_3/CaO + MgO$ in a ratio that is less than one.

DOLOMITE PERCENTAGE (DP)

This is defined as:

$$DP = \left(\frac{CaO + MgO}{Fe_2O_3 + CaO + MgO + Na_2O + K_2O} \right) \times 100$$

(6)

This parameter is used primarily for coal ashes with a basic-oxide content over 40 percent. It has been empirically related to the viscosity of coal-ash slags.[6] Its general range is from 40 to 98 percent. At a given percent basic, a higher DP usually results in higher fusion temperatures and higher slag viscosities.

EQUIVALENT Fe_2O_3 AND FERRIC PERCENTAGE (FP)

These parameters describe the degree of iron oxidation in coal-ash slags. They are defined as:

$$Equiv. \; Fe_2O_3 = Fe_2O_3 + 1.11 FeO + 1.43 Fe, \; and$$

$$FP = \left(\frac{Fe_2O_3}{Equiv. \; Fe_2O_3} \right) \times 100$$

(7)

These terms were introduced when it was observed that the ash-fusion temperatures and slag-crystallization temperatures were lower in a reducing atmosphere than in an oxidizing atmosphere. Experiments showed that this deviation was due to the different forms of iron oxide that could exist depending on the surrounding atmosphere. It was concluded that FeO and Fe are stronger fluxing agents than Fe_2O_3 and result in depression of ash-fusion and slag-crystallization temperatures. Thus, equivalent Fe_2O_3 and FP depend on combus-

tion conditions and are usually determined by chemical analysis of actual furnace slags.

In *ASTM Standards* D 1857, it has been specified that a reducing atmosphere should contain a mixture of oxidizing and reducing gases in which the content of the oxidizing gas is within the limits 20 to 80 percent by volume. In this range of gaseous composition, the iron is mainly in the ferrous state, FeO. Under stronger oxidizing conditions, the iron is mainly in the ferric state, Fe_2O_3; under more reducing conditions, the iron is mainly in the metallic state, Fe.

SILICA PERCENTAGE (SP)

This relationship is defined as:

$$SP = \left(\frac{SiO_2}{SiO_2 + Equiv. \ Fe_2O_3 + CaO + MgO} \right) \times 100$$

(8)

SP, which has a range of 35 to 90 percent, has been empirically correlated with the viscosity of coal-ash slags. As SP increases, the slag viscosity increases. (Silica percentage is also occasionally referred to as silica ratio or equivalent silica content.)

TOTAL ALKALIES (Na₂O + K₂O)

As previously noted, sodium and potassium metals join with other elements in combinations that produce low fusibility temperatures. The influence of alkalies on fusibility and slagging potential in the furnace is proportional to the quantity in the ash.

With sodium-containing coals, the rate of buildup and the ability to control deposits effectively seem to be related to the amount of sodium present. This is particularly true of coals with lignitic-type ash (low iron, high alkali, and high-alkaline earths) as found primarily in U.S. coals west of the Mississippi. The ash of such coals will contain CaO as high as 40 percent and the sodium oxide may range from 1 to 15 percent, as extremes.

Many sodium compounds melt at temperatures below 900°C (1650°F). In addition, sodium solids manifest a property beyond melting temperatures alone that contributes to deposit building. Besides low melting temperatures, some sodium compounds volatilize at relatively low temperatures. These disperse throughout the gas stream and subsequently condense on other ash particles and on metal surfaces as heat is absorbed and gas temperatures are lowered. The condensed sodium provides a binding matrix for ash particles to fuse together and build up on tube surfaces. [7,8,9]

Sodium oxide is transformed directly from a solid to a gaseous state at 2330°F. Because certain lignites and low-rank coals may have sodium that is organically bound, they are particularly susceptible to the formation of sodium oxide and consequent vaporization in the combustion zone. The sodium is released to a free state as the organics are consumed in the flame and are immediately available for vaporization into the gas stream. On the other hand, sodium tied up as mineral salts must be first decomposed to a free state before it is available for vaporization and, therefore, may not be as proportionally active.

A widespread distribution of sodium vapor in the combusion gases assures contact with ash particles and metal surfaces as the gas cools and the sodium condenses. Thus, particle surfaces are "conditioned" for bonding and the resulting deposit forms rapidly. The rate of buildup is proportional to the amount of active sodium in the coal. This phenomenon applies particularly to high-alkali ash defined as Western U.S. type ash. Because deposit buildup is rapid with such coals, tube spacing in convective passes must be maintained on wide centers to minimize bridging.

FORMS OF ALKALIES IN COAL

Alkalies can be present in coal in various forms. Alkalies which vaporize during combustion are often classified as active alkalies, because they are free to react or condense subsequently in the boiler. Active alkalies consist primarily of simple inorganic salts and organically bound alkalies. More stable forms of alkalies exist in impurities such as the clay and shale minerals (silicates) which remain relatively inert during combustion, and therefore are less influential in the ash deposition process. Fig. 5 summarizes the different categories in which various alkali forms can be associated.

H. E. Crossley in his Melchett lecture to the Institute of Fuels in 1962 had this to say about alkalies in coal:

"It was known that alkalies existed in coal in two forms: chlorides and complex silicates. The chlorides could be expected to volatilize readily during combustion but the silicates could be expected to retain alkali in the ashes. Thus, if an attempt was made to relate the forms of alkali-rich deposits with alkali in coal, there could be presence of the complex silicates. It was decided instead to use the amount of chlorine as the index of volatile alkalies and this has been standard practice ever since. In consequence, the determination of chlorine has become an important item in the ultimate analysis of coals."[10] (This statement does not include or describe the mechanism of volatile sodium as found in lignitic-type coals.)

The chlorine content itself can be used as an indication of the fouling tendency of a coal, a fact which has been recognized not only in this country, but in England. Crossley classifies coals as follows:

Fouling Potential	Chlorine %, Coal Basis
High	Greater than 0.3
Medium	0.3 to 0.15
Low	Less than 0.15

American experience supports these conclusions, indicating that tightly bonded deposits in the convection areas can be expected when the chlorine level exceeds 0.30. Rather than depending on the measurement in the coal as the yardstick for evaluating the ash fouling potential, another approach is to measure the water-soluble sodium in the coal. The relationship of water-soluble sodium as measured and the chlorine, interpreted as sodium, is shown in the following table:

Fouling Potential	Cl%, Coal Basis	Equiv. Na %
High	Greater than 0.5	Greater than 0.33
Medium	0.3	0.2
Low	Less than 0.1	Less than 0.07

Although the water-soluble sodium in a coal is indicative of the sodium existing in an active form, it may not represent all of the active sodium in the coal. For many coals, in particular those of low rank, a high percentage of the sodium is present in an organically associated form; such organically bound alkalies are not soluble in water.

A TECHNIQUE FOR DISTINGUISHING ACTIVE FROM INACTIVE ALKALI

To segregate the different categories of alkalies, a technique has been developed utilizing a dilute acetic-acid medium. Leaching coal samples in a dilute acetic-acid solution exposes the coal to a source of hydrogen ions. It appears that an ion exchange possibility exists in which the hydrogen ions (H^+) of the acid displace or exchange with metallic elements associated with the organic coal molecules. These elements include sodium, potassium, calcium, and magnesium.

The weak acid used during leaching provides the hydrogen-ion source necessary to displace organically bound elements, but does not have the strong dissolving powers which would be necessary to break down the complex extraneous minerals. The constituents (in particular alkalies) removed from the coal essentially consist of the organically bound material along with the simple inorganic compounds. Alkalies present either in organic fractions or in simple inorganic salts can be expected to volatilize during combustion, and are therefore of primary interest.

RESULTS OF WEAK-ACID ANALYSIS

Analysis of acetic-acid-soluble alkalies has been conducted on a cross-section of U.S. coals for which performance characteristics are well established through commercial operation.

Detailed data comparing the alkali contents of the ash of seven U.S. coals are summarized in Table IV. The ash varies in fouling tendencies from severe to low or minimal. Values presented for sodium in the ash were determined by the conventional ashing and lithium tetraborate fusion procedure.

In general, the coals containing high alkali contents and, in particular, high sodium contents had the most severe fouling tendencies. This is the expected trend. However, after studying the acetic-acid-soluble alkaline contents of the coals, one is better able to relate their fouling behavior. Past work has primarily focused on sodium as the bad-acting constituent responsible for fouling. It is possible to understand why sodium is frequently more influential than potassium by comparing the relative quantities of acetic-acid-soluble sodium and potassium in Table IV.

What is believed to be the primary reason that sodium has a greater influence is that sodium is frequently present in an active form, whereas potassium is typically contained in a less active form. Values for acetic-acid-soluble sodium in U.S. coals frequently range up to 10 times more than the soluble-potassium values. In many U.S. coals, much of the potassium appears to be tied up in extraneous clay minerals which do not break down readily when heated, making the potassium in this form less reactive. Although the potassium contents of these coals range from 0.5 to 1.9 percent K_2O in the ash, the acetic-acid-soluble potassium generally is equivalent to less than 0.1 percent K_2O in the ash. Conversely, the acetic-acid-sodium content of the high-fouling coals represents virtually all of the Na_2O detected in the ash.

The coals containing the most soluble alkali were low-rank coals (subbituminous coals and lignites). It is believed that such coals can contain substantial quantities of alkali (as well as alkaline-earth constituents) present as organically bound humates. It appears that the potential for having organically bound metallic compounds decreases as the coal rank increases. The Utah high-volatile bituminous illustrates this tendency. This fuel contains a relatively high sodium content (3.8 percent Na_2O in ash). But the soluble sodium content is only about 40 percent of the "total" sodium and the overall fouling potential is low to moderate.

In the case of the particular coals selected, all

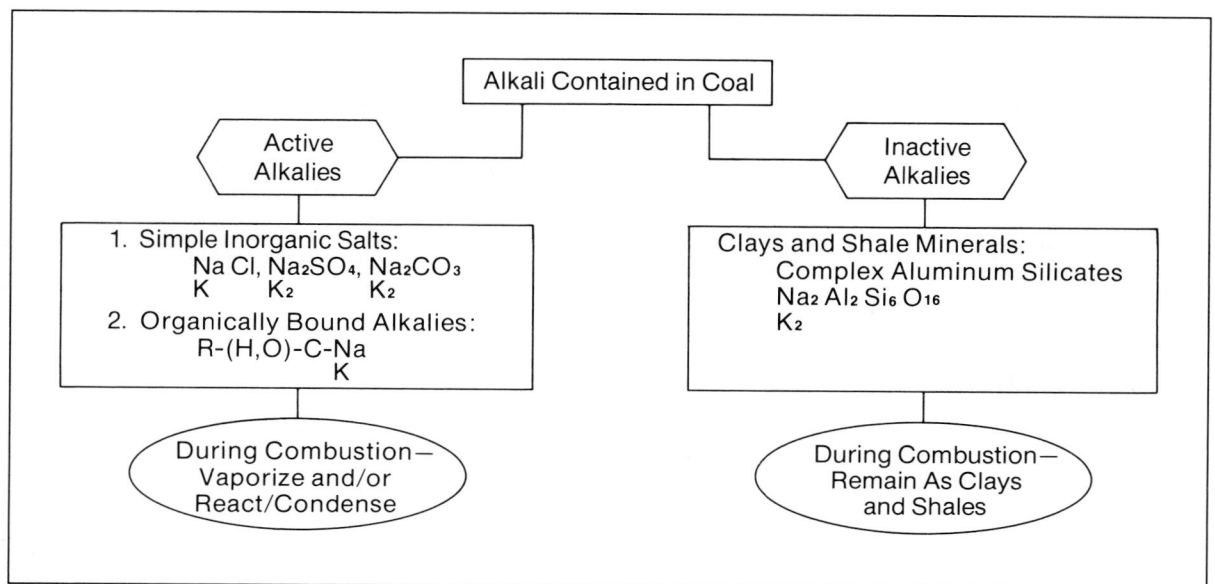

Fig. 5. **Manner in which alkalies are contained in coal**

of the high-fouling coals had quantities of soluble sodium greater than the "total" quantity of sodium detected in the ash by the conventional technique. The lower values for "total" sodium may be the result of a slight loss of organically bound material during the ashing process.

RELATION OF FOULING POTENTIAL TO ALKALI FORM

The data in Table IV indicate that fouling potential of coal ash is related to both total quantity of acetic-acid-soluble alkali and to the

Table IV. Comparative Alkali Content of U.S. Coals

Rank Region	Lignite ND	Sub B MT	Lignite TX (Yegua)	Lignite TX (Wilcox)	hvBb UT	hvAb PA	Lignite TX (Wilcox)
HHV, Btu/lb, Dry Basis	10640	12130	7750	9710	12870	13200	8420
Ash Composition (%)							
SiO_2	20.0	33.9	62.1	52.3	52.5	51.1	57.9
Al_2O_3	9.1	11.4	15.1	17.4	18.9	30.7	21.8
Fe_2O_3	10.3	10.8	3.5	5.3	1.1	10.0	3.9
CaO	22.4	21.0	6.2	9.4	13.2	1.6	7.1
MgO	6.4	2.7	0.7	3.2	1.3	0.9	2.1
Na_2O	5.0	5.8	3.6	0.9	3.8	0.4	0.7
K_2O	0.5	1.6	1.9	1.2	0.9	1.7	0.8
TiO	0.4	0.7	0.9	1.2	1.2	2.0	1.1
SO_3	21.9	12.0	6.1	9.6	6.2	1.4	4.4
Fouling Potential	Severe	High	High	Moderate	Moderate	Low	Low
Lb Ash/10^6 Btu, Dry Basis	9.0	4.6	43.3	20.1	7.9	10.2	34.4
Acetic-Acid-Soluble							
Sodium (Na, ppm)	3980	2680	9650	1030	1120	250	340
Potassium (K, ppm)	. . .	. . .	1230	85	85	. . .	110
Alkali in Ash, % Wt.							
Na_2O	5.0	5.8	3.6	0.9	3.8	0.4	0.7
K_2O	0.5	1.6	1.9	1.2	0.9	1.7	0.8
Equiv. Sol. Alkali in Ash, % Wt. of Ash							
Na_2O	5.58	6.45	3.88	0.71	1.49	0.15	0.16
K_2O	. . .	. . .	0.44	0.04	0.08	. . .	0.05
% Sol. Alkali of Total (Equiv. Sol. Na_2O)/(Na_2O in Ash)	112	111	108	79	39	38	23
(Lbs Sol. Na)/(10^6 Btu Fired)	0.374	0.221	1.245	0.106	0.087	0.018	.040
(Lbs Sol. Na)/(lb Ash/10^6 Btu Fired)	0.044	0.048	0.223	0.005	0.014	0.002	0.001

equivalent percentage of soluble alkali in ash. The total volatile alkali is believed to be reflected by furnace input of soluble alkali listed in Table IV as pounds of soluble sodium per million Btu fired. It is postulated that the deposition mechanism could involve the condensation of the gaseous alkali species which provides the sticky, bonding matrix to build convection-pass deposits. The greater the quantity of this "glue," the more severe the deposition. Conversely, the greater the quantity of ash particles, the thinner this "glue" would be spread. Because this bonding matrix is a surface phenomenon, it would depend on both the total quantity of vapor species and the diluting influence of the ash. Therefore, a combination of pounds of soluble sodium per million Btu fired and pounds of soluble sodium per pound of ash per million Btu fired should be used when evaluating the fouling potential of a fuel. (See Table IV, bottom two entries.)

VARIATION OF ASH PROPERTIES WITHIN A COAL SEAM

Note that it is possible to experience vastly different ash properties within the same coal seam. This point is illustrated with the two comparative lignites mined in Eastern Texas from the Wilcox formation. The sodium contents of the two lignites are very similar (0.9 percent and 0.7 percent Na_2O in ash respectively). However, the ash-fouling behavior of these fuels is significantly different, as vividly illustrated by the acetic-acid-soluble sodium contents of the coal. Roughly 80 percent of the sodium in the moderate-fouling Wilcox lignite could be considered active compared to only 23 percent for the low-fouling lignite. On a concentration basis (pounds of sodium per pound of ash per million Btu fired), the moderate-fouling lignite has more than 10 times that of the low-fouling lignite because of the ash dilution.

The acetic-acid-soluble technique is relatively new, such that the limited data thus far accumulated neither allow development of quantitative relationships nor define the reliability of the analysis. But the major value of the technique is the apparent sensitivity to the "active alkalies" present, which seem to be a predominant factor in high-fouling behavior.

SELECTIVE DEPOSITION OF ASH CONSTITUENTS

Several slagging and fouling indexes to predict the behavior of coal ash in furnaces during the combustion of coal have been reviewed in preceding text. These analyses consider a homogeneous distribution of the mineral matter throughout the coal matrix. Such an assumption may be applicable to stoker firing, where the coal particles, as fired, are relatively large, but may not be applicable in pulverized coal where the composition of mineral matter of individual particles can be quite different.[11, 12, 13] Because the behavior of an individual coal-ash particle depends upon its composition, *average* properties may not be an accurate index of the slagging and fouling potential of a coal.

Most evaluations of coal ash, including the determination of fusibility temperatures, do not consider the phenomenon of selective deposition when attempting to predict slagging and fouling behavior. The fractionation of a coal prior to a determination of fusibility temperatures can provide a better insight into what occurs within the furnace; i.e., individual particles of ash can act independently of one another with respect to their physical state and reactions in the flame.[14-17]

THE DISTRIBUTION OF MINERAL MATTER IN THE COAL MATRIX

Ash-fusibility temperatures as determined by ASTM procedures are the temperatures at which the composite ash sample initially deforms, softens, and becomes fluid. These values, as well as such ratios as base/acid, iron/calcium, and silica/alumina, provide a guide to the slagging and fouling characteristics of the coal being burned. But they do not predict how the various minerals behave during the combustion process. Ash deposits on furnace walls differ significantly from the ash of the as-fired coal. Use of the coal-ash fusibility values and the coal/ash ratios does not explain the selective deposition of ash constituents noted on furnace walls.

Table V shows the composition of deposits collected from a boiler burning a high-volatile bituminous coal and compares it with the composition of the as-fired ash. The most notable difference in the composition of these two samples is the iron content, the ash deposit having an iron content over three times as great as the coal ash as-fired. The amount of difference between as-fired ash and that taken from tube surfaces is a function of the distribution of mineral matter in the coal matrix.

These data suggest a mechanism that can explain why some coals slag more than conventional analysis would indicate. Coal is a heterogenous substance in terms of its organic and inorganic content. The coal-preparation engineer makes use of this knowledge in most beneficiation processes. Density differences or surface property differences are exploited when separating clean coal from the refuse or high-ash-containing parts of the coal. Also the potential for liberating the inorganic material increases as the coal is ground finer.

The pulverization of coal, then, increases the potential for liberating inorganic particles. It has been estimated by Moody and Langan[18] that 40 to 70 percent of the particles are pure coal (containing inherent mineral matter only), 20 to 40 percent occur as coal and mineral matter (extraneous as well as inherent), and 10

Table V. Comparison of Ash Deposit Composition with As-Fired Coal Ash Composition

Ash Composition	As-Fired Coal Ash	Lower Furnace Ash Deposit
SiO_2	47.0	33.3
Al_2O_3	26.7	18.0
Fe_2O_3	14.6	43.5
CaO	2.2	1.2
MgO	0.7	0.5
Na_2O	0.4	0.2
K_2O	2.3	1.6
TiO_2	1.3	0.8
P_2O_5	0.9	0.4
SO_3	1.1	0.5

to 40 percent are pure mineral matter (extraneous only).

Comparing the composition of particles that result from pulverization can show an extreme difference in constituents. Because these particles are fired in suspension, the behavior of the mineral matter in a given particle will depend on the composition of that particle alone and, of course, the makeup of the surrounding gases. Each particle will reach a physical state (sticky, sintered, or molten) related to its *own* composition and reaction with surrounding gases; it will not be affected by the composition or physical state of other particles in suspension.

Fig. 6 illustrates a hypothetical breakdown of a lump of coal during pulverization. Certain

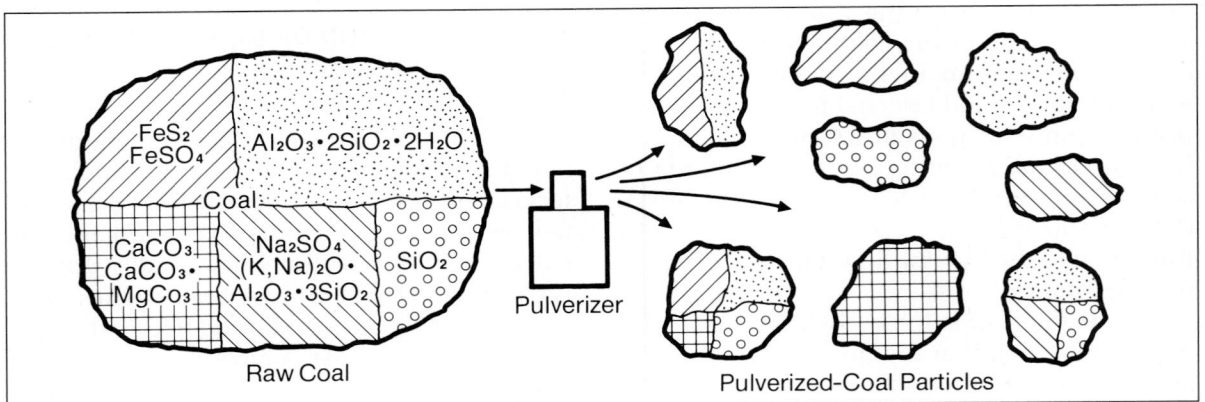

Fig. 6. **Segregation of mineral matter during grinding**

portions of the coal may be enriched with specific minerals. When the parent lump has been pulverized, the composition of any given particle will depend on the local composition of the lump and is not representative of the average composition of the lump. When these particles are fired in a furnace, their propensity for deposition will be a function of their individual physical states, aerodynamic properties, and adhesive properties (or wettability). Ash particles with a high propensity for deposition are probably molten, high-density, spherical particles that do not follow gas stream-lines, and that readily wet the heat-transfer surfaces they contact. Ash particles with a low propensity for deposition are probably dry (high-melting temperature), irregularly shaped particles that are more likely to follow gas stream-lines than to contact the furnace walls.

As discussed earlier, iron compounds are among the materials having the lowest fusing temperatures found in coal ash.[5, 19] It is postulated that particles high in iron form high-density molten spheres that have low drag coefficients, thereby facilitating their penetration through the gas stream to the tube wall. Ash particles that have very high fusing temperatures do not melt to form spheres; their irregular shapes give higher drag coefficients and allow them to follow gas stream-lines. Some particles having low melting temperatures contain carbonate compounds whose gaseous evolution of CO_2 generates cenospheres, which, being hollow, are low in density and also follow gas stream-lines.

COAL ANALYSIS AS A FUNCTION OF PARTICLE SIZE

In predicting behavioral tendencies of mineral matter during combustion, it is important to analyze coal in the size range at which it is being used, because segregation of mineral matter is a function of particle size. As each particle is different, it would be ideal to analyze and quantify individual particles; however, this could prove to be difficult and costly. As a practical alternative to analyzing individual

particles, a gravity fractionation technique has been developed.[20]

PROCEDURE FOR GRAVITY FRACTIONATION

The procedure requires that the coal be pulverized to the same fineness as that during firing; the coal particles then are separated by density, using various mixtures of organic liquids having specific gravities of 1.3, 1.5, 1.7, 1.9, 2.1, 2.5, and 2.9. Fewer separations might be adequate, but as the procedure was being developed a finer resolution of gravity fractions was found desirable. A 1-pound sample is normally used and is well stirred into the 1.7 gravity liquid using conventional separatory funnels. After separating the 1-pound sample into the fraction that sinks and the fraction that floats in the 1.7 liquid, each fraction is filtered and washed with a low-density organic solvent. The float fraction is then taken to the next lighter gravity liquid; the sink fraction is taken to the next heavier gravity liquid as shown in Fig. 7. After separation, the eight fractions are air-dried at 250°F and submitted for the follow-

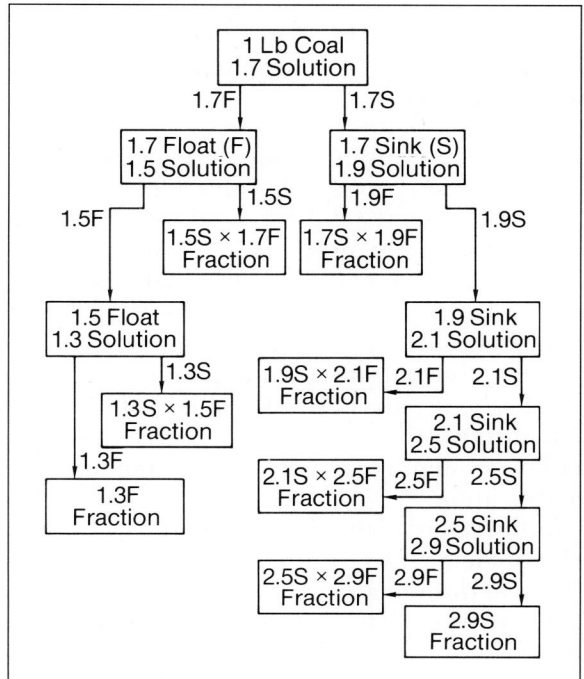

Fig. 7. Gravity separation procedure

ing analyses: percent ash, ash composition, ash fusibility, and, optionally, X-ray diffraction.

RESULTS OF ANALYSES

Table VI shows the analyses of three coals selected for the gravity fractionation technique.

Tables VII through IX present data for each of the coal gravity fractions. Compositional differences among the gravity fractions are thought to represent the most important data obtained; fusibility temperatures of ash contained in the gravity fractions would be controlled by composition, for instance. Results here primarily compare composition and behavior of the ash deposit.

SLAGGING INFLUENCED BY IRON CONTENT OF ASH

The role of iron is one of the more important factors in assessing slagging potential. Frequently, investigators use the total iron concentration in a coal ash to assess the slagging potential; however, this iron value could be misleading as it does not account for distribution of iron compounds in the coal matrix. In addition to the importance of the distribution of mineral matter within the coal matrix, the type of minerals also plays a role in the ash deposition process. For example, iron is frequently present as pyrites (FeS_2) and carbonates ($FeCO_3$) and/or as an impurity in calcite and dolomite, and less frequently as prochlorite ($Fe_2Mg_2Al_2Si_2O_{11}$), hornblende ($CaFe_3Si_4O_{12}$), hematite (Fe_2O_3), and others.[21] Pyrites will more likely exist as discrete particles within the heavier gravity coal fractions whereas carbonates are more likely to be disseminated throughout the coal matrix and not show a concentration with gravity fractionation. The occurrence of these and other iron-containing minerals and their degree of segregation in the coal matrix are functions of coalification.

Initial stages of the reaction of mineral matter in the flame can influence ash deposition. Physical differences (such as density and size) and, more importantly, melting temperatures (and thus shapes of particles), are determined by the products of transient reactions in the flame, depending on the type of mineral:

	FeS₂	FeS	FeCO₃	FeO
Specific Gravity	5.02	4.62	3.96	5.70
Melting Temp (°F)	1382(d)	2182	1112(d)	2511

Where d indicates that the compound decomposes at the temperature shown. The compounds that reach the walls control the type of deposit formed. The reactions below illustrate this point:

$$FeS_2 (c) + O_2 \xrightarrow[\text{oxidation}]{\text{initial}} FeS(l) + SO_2(g)$$

$$(9)$$

$$FeCO_3 (c) \xrightarrow[\text{reaction}]{\text{initial}} FeO(c, l) + CO_2(g)$$

where c = crystalline, g = gaseous, and l = liquid

$$(10)$$

Table VI. ASTM Properties of Test Coals

Coal*	A	B	C
ASTM RANK	hvA	hvC	Subbit C
Proximate Analysis			
Moisture	0.7	2.7	10.2
Volatile Matter	32.3	41.7	31.5
Fixed Carbon	54.4	46.4	42.8
Ash	12.6	9.2	11.9
Fusibility Temperatures (°F)			
IT	2430	2080	2130
ST	2510	2140	2170
HT	2560	2170	2210
FT	2660	2300	2310
Ash Composition			
SiO_2	50.7	52.0	50.2
Al_2O_3	30.2	17.5	16.9
TiO_2	1.9	1.3	0.8
Fe_2O_3	9.5	15.5	8.9
CaO	1.6	4.5	11.5
MgO	0.9	1.1	3.5
Na_2O	1.0	0.6	1.8
K_2O	2.7	2.8	1.7
P_2O_5	< 0.1	< 0.1	...
SO_3	1.4	4.2	4.3

*Reporting Basis: As-fired (pulverized)

Table VII. Gravity Separations—Coal A

Gravity Fraction:	1.3F	1.3 × 1.5	1.5 × 1.7	1.7 × 1.9	1.9 × 2.1	2.1 × 2.5	2.5 × 2.9	2.9S
Wt% of Coal	11.32	56.92	9.79	13.33	1.08	2.05	4.05	1.46
% Ash	2.3	7.4	20.4	14.6	51.1	79.1	88.2	71.0
Fusibility Temperatures (°F)								
IT	2420	2700+	2700+	2700+	2520	2320	2420	2010
ST	2520	2700+	2700+	2700+	2670	2400	2510	2040
HT	2690	2700+	2700+	2700+	2700+	2470	2660	2070
FT	2700+	2700+	2700+	2700+	2700+	2610	2700+	2130
Ash Composition								
SiO_2	44.2	52.8	53.7	52.5	53.7	50.6	58.0	26.6
Al_2O_3	36.3	29.9	27.1	27.9	27.1	24.9	22.4	8.6
TiO_2	1.7	1.6	1.1	1.3	1.0	0.8	0.9	0.8
Fe_2O_3	7.0	6.4	6.8	7.2	7.2	5.3	6.7	59.3
CaO	4.7	2.2	1.7	1.6	1.4	5.1	1.3	1.2
MgO	1.2	1.1	1.1	1.1	1.2	1.1	1.1	0.6
Na_2O	0.7	1.2	1.1	1.2	1.2	1.0	1.0	0.8
K_2O	2.9	0.4	3.1	3.1	3.3	3.0	3.1	1.2
SO_3	1.0	0.4	0.4	0.4	0.4	6.2	0.8	0.8

Table VIII. Gravity Separations—Coal B

Gravity Fraction:	1.3F *	1.3 × 1.5	1.5 × 1.7	1.7 × 1.9	1.9 × 2.1	2.1 × 2.5	2.5 × 2.9	2.9S
Wt% of Coal	0.13	45.45	45.52	3.87	2.83	1.20	0.44	0.56
% Ash	1.0	3.2	6.9	32.5	41.3	53.4	78.7	64.3
Fusibility Temperatures (°F)								
IT		2030	2150	1950	1900	1930	2050	2020
ST		2110	2240	2060	1950	1960	2100	2040
HT		2230	2420	2150	1970	2010	2150	2050
FT		2470	2500	2230	2140	2150	2210	2210
Ash Composition								
SiO_2		55.1	57.7	50.1	44.4	39.3	22.4	12.7
Al_2O_3		19.0	21.5	15.2	13.3	11.8	6.8	4.3
TiO_2		2.1	1.2	1.0	0.6	0.6	0.4	0.5
Fe_2O_3		13.2	13.5	24.9	30.5	26.0	22.0	57.3
CaO		2.3	1.9	1.7	3.7	9.5	20.1	9.7
MgO		1.1	1.2	0.9	0.7	0.5	0.3	0.2
Na_2O		0.5	0.4	0.4	0.3	0.3	0.3	0.3
K_2O		2.1	2.2	1.8	1.4	1.1	0.4	0.3
SO_3		0.6	0.3	0.7	2.7	9.3	26.1	12.3
P_2O_5		< 0.1	< 0.1	. . .	. . .	. . .	. . .	< 0.1

*Insufficient sample

ABB
ASEA BROWN BOVERI

The first reaction, the oxidation of iron pyrite initially to pyrrhotite and SO_2, is one of the more significant reactions the mineral matter experiences in the combustion zone. The product FeS forms molten spheres which, through less drag and higher density (compared to most other particles in the fireball) are more likely to contact the walls of the furnace. Depending on the composition of ash deposits already present on the furnace wall, subsequent reactions involving FeS may form a material that melts at a relatively low temperature. The reactions in (11), (12A) and (12B) illustrate two possible products with significantly different melting temperatures. Reaction 11 shows the formation of magnetite (Fe_3O_4) which will tolerate a rela-

$$FeS(l) + Fe_2O_3\,(c, l) + 3/2\;O_2(g)$$
$$\rightarrow Fe_3O_4\,(c, l) + SO_2\,(g) \tag{11}$$

$$2FeS(l) + 3O_2(g) \rightarrow 2FeO(c, l) + 2SO_2(g) \tag{12A}$$

$$FeO(l) + SiO_2(c, l) \rightarrow FeSiO_3(c, l)$$
$$\text{(melting temp 2095°F)} \tag{12B}$$

tively high melting temperature before decomposing to FeO and O_2. The presence of Fe_3O_4 in the deposit would contribute toward a "dry" deposit. Reaction 12A shows the formation of FeO as an intermediate product which can react with silica to form a low-melting iron silicate (shown in 12B). The presence of iron silicate would contribute toward a molten deposit.

Iron compounds in the "2.9 sink" are given special emphasis as they often contain mostly pyrites, which can cause slagging. The bar charts depicted in Fig. 8 show that iron compounds in coal B are more uniformly distributed throughout the coal matrix than in coal C. Thus, coal C has the high potential for slagging. This agrees with observations made in the field; a furnace burning coal C had more slagging problems than a unit burning coal B.

Table IX. Gravity Separations—Coal C

Gravity Fraction:	1.3F *	1.3 × 1.5	1.5 × 1.7	1.7 × 1.9	1.9 × 2.1	2.1 × 2.5	2.5 × 2.9	2.9S
Wt% of Coal	0.2	62.76	32.28	1.56	1.09	0.62	0.45	1.05
% Ash		6.1	6.4	21.8	39.9	82.6	84.0	68.3
Fusibility Temperatures (°F)								
IT		2230	2260	2240	2310	2180	2350	2210
ST		2290	2370	2340	2380	2230	2440	2250
HT		2310	2390	2400	2430	2280	2490	2380
FT		2340	2420	2630	2550	2420	2530	2530
Ash Composition								
SiO_2		41.1	43.7	54.5	53.4	51.0	37.0	6.8
Al_2O_3		18.5	24.2	25.7	20.7	10.3	3.0	0.7
TiO_2		0.8	0.8	0.6	0.8	0.4	0.3	0.3
Fe_2O_3		3.2	2.0	7.1	5.9	3.5	3.7	79.0
CaO		19.6	16.1	7.0	8.2	11.9	22.8	3.9
MgO		5.1	4.3	1.4	1.0	0.6	0.8	0.3
Na_2O		0.3	0.5	0.3	0.2	0.2	0.2	0.2
K_2O		0.3	0.3	0.4	0.2	0.4	0.4	0.2
SO_3		9.5	7.7	2.7	6.5	9.6	11.0	6.2
P_2O_5		< 0.1	< 0.1	< 0.1	< 0.1	0.2	0.3	0.2

*Insufficient sample

Table X shows that the degree of segregation of iron compounds within the coal matrix has no bearing on the total percentage of iron in the raw coal. If the coals are ranked solely by iron content, there is a significant difference between the orders of ranking, influenced by measurements of total or segregated iron.

The gravity fractionation technique has been used to estimate the slagging potential of a wide variety of coals. Table XI is a summary of tests on sixteen U.S. coals, listing them in order from severe to low potential.[22]

The slagging potential was established based on the potential of a fuel to form molten ash deposits that would be uncontrollable by conventional steam or air sootblowing. An examination of the iron in the ash (expressed as percent Fe_2O_3) indicates poor correlation with slagging potential. Several coals with low to moderate (4 to 9 percent) iron contents (coals 2, 5, 6) exhibit severe slagging potential. Several coals with medium to high (13 to 16 percent) iron contents (coals 10, 11, 12) exhibit less severe, although still high, slagging potentials. Examination of the ash fusion temperatures also reveals poor correlation.

An examination of the gravity fractionation data, however, shows excellent correlation with slagging potential. The iron in the ash of the 2.9 sink fraction correlates very well with slagging potential. The 2.9 sink gravity fraction provides an excellent indication of segregated, reduced (largely pyrite) iron. The low- to moderate-iron coals that exhibited severe slagging potentials had high amounts of segregated iron (expressed as percent Fe_2O_3 in ash of 2.9 sink

Table X. Total Iron Compared with Segregated Iron

Coal	% Iron in Entire Coal Ash	Total % Iron in 2.9 Sink Fraction
A	9.5	39.9
B	15.5	11.9
C	8.9	69.7

fraction) varying between 74 to 79 percent. The medium to high iron coals that exhibited lower (but still high) slagging potentials had less segregated iron, varying between 57 to 60 percent. The wide applicability of this test is indicated by the inclusion of several Western coals.

Correlation of slagging potential with segregated iron content (from gravity fractionation) is shown in Figure 9. This relationship has a very high level of confidence for fuels that exhibit 50 percent or more Fe_2O_3 in the ash of the 2.9 sink fraction. In these cases the impact of segregated, reduced iron plays a major role in slagging. Correlation of slagging potential with this segregated iron has a much higher confidence level than correlation with iron in the ash.

CORROSION BY FUEL ASH

During normal operation of utility boilers, metal wastage has occurred primarily (Fig. 10)

- on waterwalls near the firing zone
- in the high-temperature superheater and reheater sections
- in the low-temperature gas passes and air heaters.

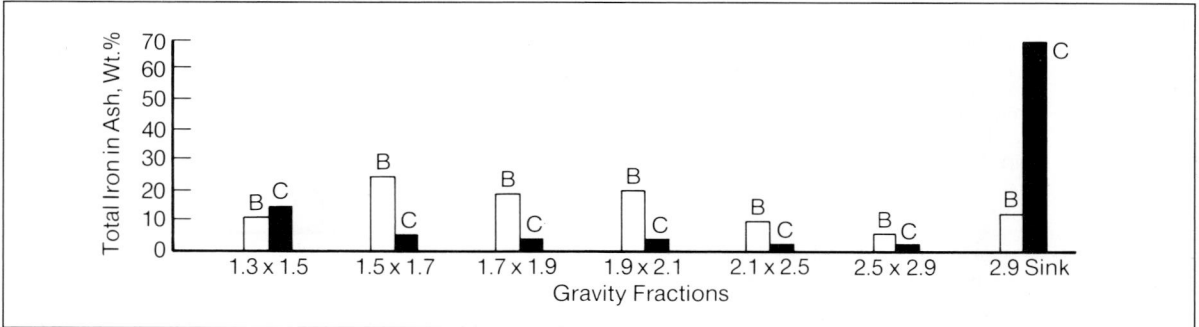

Fig. 8. **Iron content of various gravity fractions, coal B vs coal C**

Table XI. **Ash Slagging Potential of U.S. Coals as a Function of Iron in 2.9 Sink Fraction**

Coal	Coal Rank	Geographical Region	Ash Content, %	Ash Fusibility Temps. (Red. Atm.), %			Fe$_2$O$_3$ in Ash, %	Fe$_2$O$_3$ in Ash 2.9 Sink Fraction, %	Slagging Potential
				I.D.	S.T.	F.T.			
1	Hvb	Midwest	13.0	2010	2100	2390	22.7	85	9.7 Severe
2	Sub	Montana	11.7	2040	2120	2310	8.2	75	9.5 Severe
3	Mvb	Penn.	16.8	2110	2350	2640	19.0	79	9.2 Severe
4	Hvb	W. Kentucky	13.5	1980	2060	2270	27.2	78	8.8 Severe
5	Sub	Montana	11.0	2130	2170	2310	8.9	79	8.7 Severe
6	Hvb	Illinois	7.7	2330	2440	2570	4.0	74	8.4 Severe
7	Hvb	Midwest	13.4	1940	2050	2250	22.9	72	8.3 Severe
8	Hvb	Ohio	15.4	1970	2050	2370	22.6	62	6.6 High
9	Hvb	Penn.	17.1	2310	2530	2700	9.5	59	6.5 High
10	Hvb	Illinois	12.3	2050	2050	2140	13.0	59	6.5 High
11	Hvb	Penn.	16.6	2360	2460	2700	12.7	60	6.3 High
12	Hvb	Illinois	10.5	2080	2140	2300	15.9	57	6.1 High
13	Hvb	Penn.	16.6	2570	2700	2700	9.8	58	5.2 High
14	Hvb	Ky. & Tenn.	15.7	2700	2700	2700	4.2	50	4.5 Moderate
15	Hvb	Arizona	13.3	2570	2700	2700	5.8	43	3.1 Moderate
16	Hvb	Virginia	13.9	2350	2550	2700	8.3	40	1.4 Low

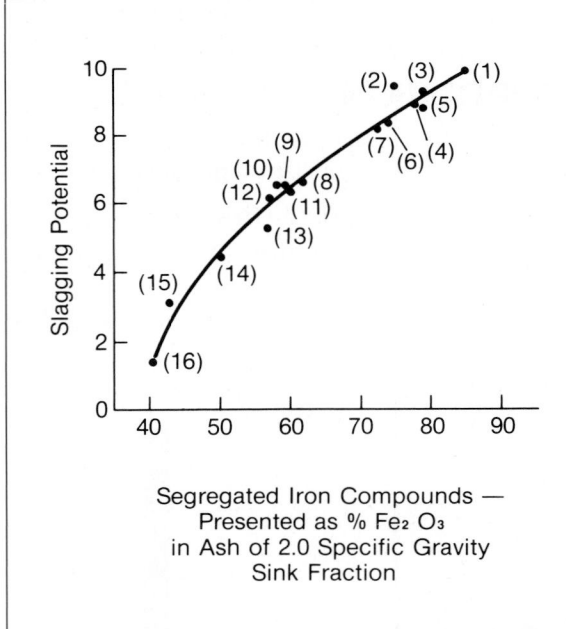

Fig. 9: **Slagging potential vs percent iron in 2.9 specific gravity sink fraction (Numbers in parentheses correspond to coal numbers in Table XI)**

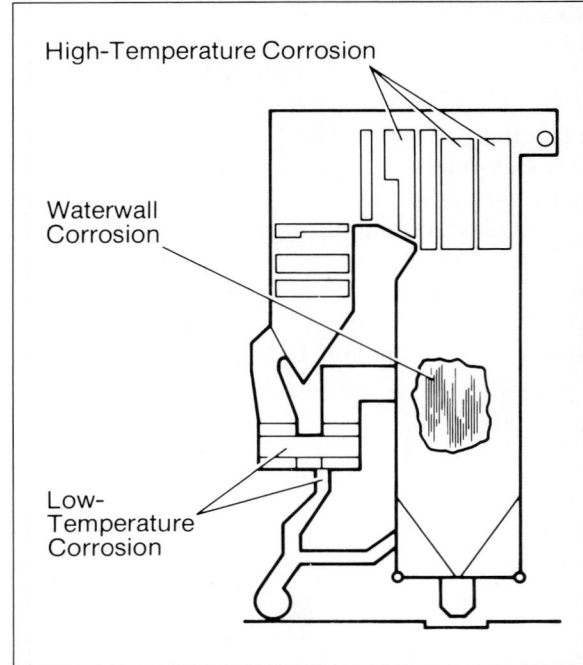

Fig. 10. **Steam-generator side elevation showing areas where wastage has occurred**

To reduce failures in these areas, C-E has established the underlying causes of major types of corrosion and has developed practical procedures for the control of waterwall and low-temperature wastage.

In recent years the most common temperature for the main steam and reheat on larger steam-generating units has been 1005°F. As a result, fundamental studies of high-temperature corrosion begun in the 1960's have been discontinued. Except for some very promising work in identifying the corrosive constituents in coals,[23] little new information concerning high-temperature coal ash corrosion is available.

CONCLUSIONS FROM SERVICE EXPERIENCE OF COAL-FIRED UNITS

Although the U.S. utility industry has, in general, abandoned the 1050°F and higher steam temperatures, there are a number of steam generators operating at those temperatures from which the following has been established:

1. All coals do not produce an ash that is "corrosive." Only a small percentage of coal-fired units experienced serious corrosion requiring major operating or maintenance corrections. Although the percentage is small, the actual numbers of units that experienced serious corrosion problems were substantial, and the concern and attention given to the coal-ash corrosion problem are certainly warranted. But it is important to know that many coal-fired units operate at steam temperatures of 1050°F with essentially no corrosion, or at acceptably low corrosion rates.

2. For coals that are "corrosive," metal temperature is an important corrosion-rate variable. As surface metal temperatures increase to about 1200°F, corrosion rates increase significantly. A unit delivering steam at temperatures of 1050°F or higher will have a greater corrosion problem than a 1005°F steam unit *if* the coal is corrosive. The corrosion rates will be higher, and the amount of material undergoing corrosion will be significantly increased. For the types of coal that can be clas-

sified as noncorrosive, however, temperature does not appear to be an important variable with regard to corrosion. If the coals are noncorrosive, there need not be a greater corrosion concern as steam temperatures are increased above 1005°F.

3. The molten ash produced by corrosive coals is highly aggressive and corrosion is not easily prevented.

a. All potentially viable pressure-part materials corrode at unacceptable rates. Currently, no tube material has an adequate high-temperature strength to resist corrosion if the coal ash is corrosive.

b. Various tube coatings have been tried. Electroplated chromium has proven effective in preventing corrosion, but it is an expensive solution with an indeterminate adherence life. Bi-metal tubes with an outer corrosion-resistant clad high in chromium, such as alloy IN 671 (50 Cr-50 Ni), show promise of having adequate corrosion resistance, but the cost of such tubing is high. In addition, these alloying elements are expected to be in critically short supply in the future. Other less expensive corrosion-resistant coatings have been tried without success.

The high-temperature corrosion that did occur was a liquid-phase coal-ash attack under tightly bonded deposits that was observed when surface metal temperatures exceeded 1100°F. This attack affected superheaters and reheaters made of austenitic (stainless) and ferritic steel.[23-39]

CONCLUSIONS FROM RESEARCH ON COAL-ASH CORROSION

Research programs have attempted to determine causes and suggest remedies for this type of wastage. Both austenitic (stainless) and ferritic steel are subject to this wastage, which takes place under tightly bonded deposits. While this excessive deposit-type wastage of ferritic alloys increases with increasing temperature, wastage of austenitic alloys is characterized as a function of temperature by the skewed "bell-shaped" curve shown in Fig. 11.

Careful inspection at various temperatures reveals a thin white or yellow layer next to the tube surface. This layer, which is molten within the 1000° to 1300°F temperature range, contains a high concentration of alkali and sulfur (Fig. 12).

Simultaneous field and laboratory investigations, including chemical and X-ray diffraction analyses, have shown that complex alkali-iron-trisulfates are major constituents of the white inner layers associated with wastage on both superheater and reheater tubes and on corrosion probes, as well as in controlled laboratory wastage experiments. By means of laboratory weight-loss tests, the bell-shaped corrosion curve has been reproduced under the same conditions needed for synthesis of the complex sulfate.[28] As noted earlier, this same complex sulfate was formed as a byproduct of waterwall wastage.[24,33]

With the principal corrosive defined, efforts turned toward establishing the mechanism of complex sulfate formation to assist in recommendation of adequate protective measures. It is significant that the concentration of compounds thought to be responsible for corrosion is considerably greater in probe and tube deposits, as a result of selective deposition, than in the coal ash or flyash from which these compounds originate.

An important explanation advanced by early investigators is that the initial deposits may be a powderlike material containing alkalies and iron oxide which react with sulfur trioxide to form alkali-iron-trisulfates.[40] Concentration of alkalies on the tube surface may also occur by thermal migration of molten material through the deposits to the tube surface.[28]

On the basis of the selective deposition observed, it was felt that individual particles of flyash vary in composition and, therefore, have different fusion temperatures. Some of the particles that are molten or semimolten at relatively low temperatures continue to stick to the tubes. Sodium and potassium compounds released during the combustion process in a form capable of reaction with SO_3 in the flue gas, may condense or deposit on the tubes as the initial layer. This explains the formation of bonded deposits in regions where the gas temperature is significantly lower than the fusion temperature of the total coal ash.

Fig. 11 also shows a comparison of the equilibrium curves and the "bell-shaped" wastage curve for austenitic stainless steel. It has been suggested that the potassium/sodium ratio in the coal ash or tube deposit is significant. The molar ratio of alkalies determines the temperature range over which the complex sul-

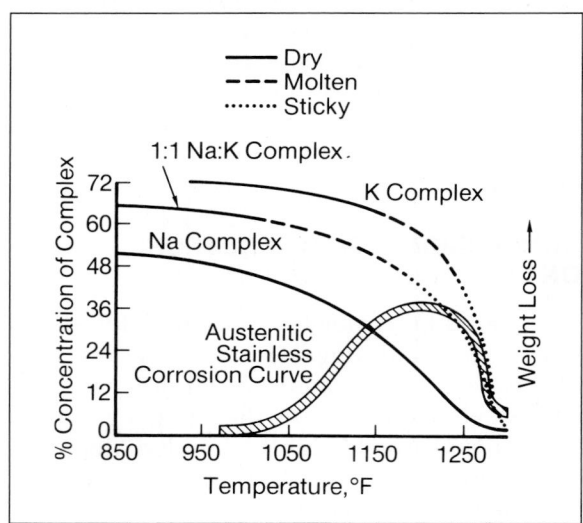

Fig. 11. Physical state of complex alkali-iron-trisulfates as a function of temperature and the corrosion of austenitic alloys

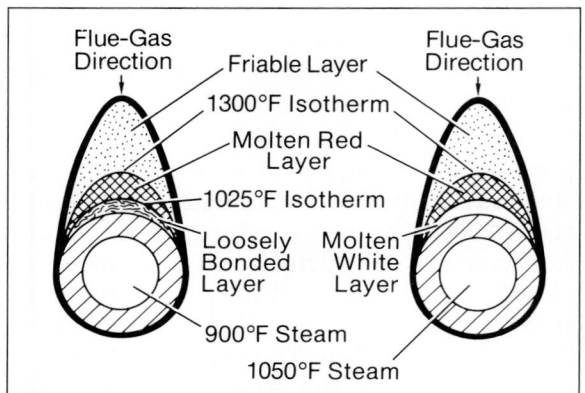

Fig. 12. Effect of steam temperature on deposit structure

fates are molten. The equilibrium concentration of sodium-iron-trisulfate was not molten from 1150°F to 1275°F. The greatest fluid range appears to be the 1:1 molar sodium/potassium ratio. This corresponds to the 1030° to 1275°F region where corrosion was observed. This is to be expected because a molten corrosive is generally more reactive than its solid counterpart. Consequently, an alkaline molar ratio of about 1:1 in a deposit is corrosive over a wider temperature range than any other combination because of the greater range of fluidity of the complex sulfate.

The roles of the various compounds necessary to form a corrosive deposit are placed in proper perspective by the equilibrium curve. The steps that influence formation and growth of deposits from combustion products of coal may be summarized as follows:

1. During combustion, pyrite (FeS_2) and organic sulfur react with oxygen:

$$2FeS_2 + 5\frac{1}{2}O_2 \rightarrow Fe_2O_3 + 4SO_2$$
$$RS \text{ (organic sulfide)} + O_2 \rightarrow SO_2$$
$$SO_2 + \frac{1}{2}O_2 \rightarrow SO_3$$

2. In the high flame temperature, the Na and K in the clays and shales react to form Na_2O and K_2O.

3. The Na_2O and K_2O then react with SO_3 either in the gas stream or after deposition on the tube:

$$(Na_2 \text{ or } K_2)O + SO_3 \rightarrow (Na_2 \text{ or } K_2) SO_4$$

4. The alkali sulfates, iron oxide, and SO_3 then react to form the complex sulfate:

$$3(K_2 \text{ or } Na_2)SO_4 + Fe_2O_3 + 3SO_3 \rightarrow 2(K_3 \text{ or } Na_3)$$
$$Fe(SO_4)_3$$

When the tube deposit contains a sufficiently high potassium-sulfate/sodium-sulfate ratio, formation of liquid pyrosulfate could occur. SO_3 reacts directly with the alkali-sulfate mixture forming pyrosulfate, which, in turn, attacks the oxide to form complex sulfate at a higher rate than in the case of the dry reaction.

The melting point of a mixed alkali pyrosulfate ($K_{1.5}Na_{0.5}S_2O_7$) is 535°F when SO_3 is above 7 ppm. The low-melting pyrosulfates, as well as the complex sulfates, are considered primary agents in superheater corrosion, as was previously noted.

5. The complex sulfate in the molten phase then reacts with the tube metal:

$$2(K_3 \text{ or } Na_3) Fe(SO_4)_3 + 6Fe \rightarrow$$
$$\frac{3}{2}FeS + \frac{3}{2}Fe_3O_4 +$$
$$Fe_2O_3 + 3(K_2 \text{ or } Na_2) SO_4 + \frac{3}{2}SO_2$$

The reaction products shown in item 5 above tend to slow down the corrosion rate, but if the deposit spalls, the reaction begins anew. The rate of spalling, then, affects the overall rate of corrosion.

PREDICTION OF COAL-ASH CORROSIVITY

Based on an understanding of the corrosion process outlined above, work has been done to quantify the effects of specific coal-ash constituents on high-temperature corrosion.[41, 42]

Certain relationships were developed between coal-ash constituents and the degree to which they affect corrosion rates. The study has shown that the alkalies, alkaline earths, iron, and sulfur are the most significant constituents relative to high-temperature corrosion. It is significant that the alkali concentration had been measured by an acid leaching process. Such a procedure gives a better measure of the simpler forms of sodium and potassium than total alkali measurements. These simpler forms of the alkalies are thought to be more available for reaction in forming alkali-iron-trisulfate, K_3 or $Na_3 Fe(SO_4)_3$, the species directly responsible for tube wastage according to previous investigation by C-E and others. Because the alkalies are generally the least plentiful of the four constituents mentioned, they are usually the most sensitive corrosion indicators.

The alkaline-earth materials are important as they may preferentially retain the alkalies as sulfates or double salts of the type $K_2Ca_2(SO_4)_3$, thus preventing formation of the aggressive compound $K_3Fe(SO_4)_3$.[43]

SHORT-TERM PROBES AS CORROSION INDICATORS

During the study noted, probes were inserted for periods of five minutes for purposes of measuring the acid-soluble sodium and potassium during each test firing.[23] These probes were inserted during tests in which raw coals and prepared coals were burned as well as during tests in which additives were used. Acid-soluble sodium and potassium values were expressed as a ratio and plotted against measured corrosion rate and they showed a remarkably good correlation (Fig. 13). It was noted that the acid-soluble Na_2O/K_2O ratio measured from probe deposits is probably a closer approximation of the relative concentration of sodium and potassium-iron-trisulfate than that measured from the coal, for the following reason:

The many physical and chemical phenomena that cause changes in the coal constituents during combustion have already taken place when the resultant probe deposits are measured for sodium and potassium content. Therefore, an acid-soluble measure of the alkalies in the *deposit* rather than in the *coal* would automatically consider the effects of combustion on the mineral matter in the coal.

This short-term acid-soluble-alkalies meas-

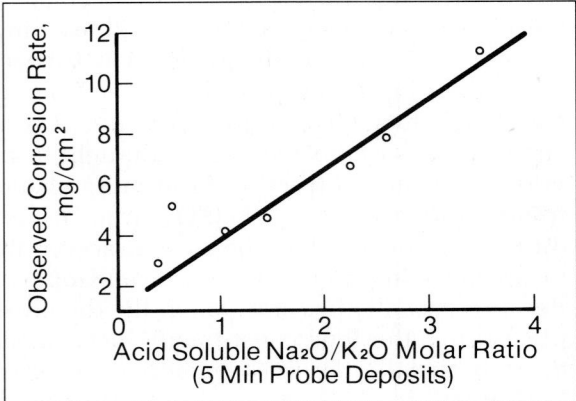

Fig. 13. Acid soluble Na_2O/K_2O molar ratio vs observed corrosion rate

urement approach is a simple and accurate way of assessing the corrosive potential of coal ash.

METHODS OF MINIMIZING THE CORROSIVE POTENTIAL OF COAL ASH

Addition or removal of certain constituents may modify a potentially corrosive coal. The entire operating system at the supplier facility should be examined from the seam face, where mining is begun, to the point of loading for shipment. These operations may be summarized as follows:

1. Mining: the coal is removed from the ground by one of several mining methods and transported from the mine.

2. Coal preparation: the run-of-mine (raw) coal may now be processed by any combination of many preparation procedures.

3. Coal additives and blending: additives or other coals may be blended with the original coal prior to shipment.

MINING

Analysis of channel samples of coal removed from a seam enclosed on two sides by "barren" material showed that the corrosion-affecting minerals, as well as other mineral matter, originated primarily from the roof, floor, or partings in the seam.[23] Therefore, the corrosiveness of a coal may be reduced by mining so that the concentration of alkaline-earth materials is increased with respect to the alkalies. This can be accomplished by inclusion or exclusion of seam roof material, floor material, and material contained in partings.

COAL PREPARATION

The procedures used to prepare coals for market should be examined. The results of bench-scale chemical and physical tests revealed that the individual chemical constituents of coal mineral matter often concentrate in specific size and/or weight fractions of the coal. Thus, preparation practices which employ sizing and gravity concentrating techniques can be used to alter the chemical composition of the coal product. Because high-temperature corrosion can be related to

coal mineral-matter properties, it is possible to reduce corrosive potential of many coals through the judicious application of conventional coal-preparation methods. In one case, the corrosiveness of the coal was reduced from 10.2 mg/cm²-300 hr to 2.9 mg/cm²-300 hr because of the benefits derived from a beneficiation process.[23]

COAL ADDITIVES AND BLENDING

Adding materials such as dolomite and limestone, or blending two or more coals, can counteract the corrosive nature of a coal.

The use of additives results in the preferential formation of sulfates or double salts of the type $K_2Ca_2(SO_4)_3$. Additives can also have beneficial effects in reducing problems from slagging and fouling. Tests have showed that, in most cases, ash deposits become more friable when an additive is used and deposits become easier to remove.[23]

CHLORIDE AS A FACTOR IN CORROSION

The research establishing the mechanism of the liquid-phase-deposit high-temperature corrosion has not shown any significant corrosion at the chloride levels of 0.1 to 0.2 percent normally encountered in coal firing.[44]

Investigations at C-E showed that the probe corrosion rate for clean coal increased by a factor of 3.3 when firing raw and clean coals from the same mine source. In this case, the coal had been cleaned by a gravity separation in carbon tetrachloride. Therefore, the residual chloride in the cleaned coal was 1.3 percent as opposed to a 0.05-percent level in the raw coal. This corrosion rate was confirmed by subsequent testing.[23]

Chloride concentrations in American coals range from 0 to 0.6 percent while coals from the United Kingdom peak at 0.8 percent. The 0.6 percent high for U.S. chloride levels is somewhat misleading in that the majority of U.S. coals have chloride levels less than 0.2 percent. Because of this, chlorine thus far has not been implicated by U.S. investigators as a major contributing factor to tube wastage.[45]

The role of chloride in increased corrosion may take several paths. Chloride naturally occurs in coal as an inorganic material and may be carried over with the flyash or released by strong sulfur acid as HCl. It can also be volatilized as NaCl. In addition, chloride can enter with refuse in either an organic or inorganic form. The resulting chloride compounds usually are concentrated on cooled tube surfaces where additional reactions may occur.

It is thought that two major adverse mechanisms are involved in the potential increase in corrosion rate in incinerators and/or fossil fuel-fired boilers in the presence of chloride. One involves formation of eutectics or complexes resulting in a lowering of the melting point of deposits. This phenomenon is of particular concern in waterwall wastage where the temperature of molten salts would increase.

The second mechanism, probably of greater concern in the superheater, involves the reaction of gas-phase sulfuric acid with deposit chloride to release HCl near the heated tube surface. Subsequent reaction may involve stepwise formation of volatile ferric chloride [46, 47] and/or unstable chloride or oxychlorides of other alloy components.

Heat-transfer surfaces installed in incinerators in both Europe and the U.S. have excessive corrosion rates.[48] It has been judged that combinations of sulfur, chlorine, lead, tin and zinc have been the corrosive agents. In these studies, where raw unprepared refuse is fired, it has been thought that the concentration of lead, tin, and zinc deposits results in a lowering of the deposit melting point with an acceleration of liquid-phase corrosion.

Battelle researchers have conducted investigations that show opposing effects of chlorine and sulfur in refuse.[49, 50] An increase in chlorine increases the corrosion rates of low-alloy steels while an increase of sulfur decreases the corrosion rates of all the steels investigated. They reported that the major contribution to the corrosion reaction is in the type of compounds that are deposited on the corrosion probe. Their investigation confirms the importance of the chloride reaction as reported by other investigators of the corrosion process.

CORROSION PROBLEMS FROM OIL ASH

Although this chapter focuses primarily on the physical and chemical properties of coal ash as they relate to slagging and fouling of boiler surfaces, fuel oil can also cause problems of corrosion.

Burning residual fuel oil in utility steam generators has not resulted in wastage of *waterwall* tubes. The principal wastage problems associated with boilers firing high-sulfur oils are (a) high-temperature corrosion of superheater and reheater tubes by low-melting-ash deposits, and (b) low-temperature corrosion of air heater, ducts, and dust-collector equipment by condensed sulfuric acid in the flue gas.

Analyses of typical residual oil ash as recorded in Table XII show that the composition of oil ash varies widely, and indicate that many elements may be found in the ash. However, the bulk of the ash contains compounds of vanadium, iron, nickel, sulfur, and alkali metals. The compounds of particular interest are those of vanadium, sodium, and sulfur which are associated with deposit formation and accelerated corrosion at high temperatures. The ash content of heavy-residual oil varies considerably, but is usually not over 0.1 percent. Concentration of vanadium in the ash is related to the source of the crude.

Vanadium content of residual-oil ash may vary from a trace to 30 percent or more. Similar variations occur in sodium content. Sulfur contents in oil have been reported from 0.7 to more than 5 percent.

HIGH-TEMPERATURE CORROSION BY OIL ASH

Opinion varies considerably on the exact mechanism of oil-ash corrosion, but most investigators agree that accelerated corrosion occurs by the fluxing action of molten sodium-vanadium complexes on the protective oxide scale on the tube. Various compounds containing different sodium/vanadium ratios frequently considered oxygen carriers, have been mentioned as participating in such corrosion.

Both melting and liquid temperatures on heating and cooling have been determined for several sodium-vanadium compounds and sodium sulfate.[51, 52, 53] The data in Table XIII show differences between the two liquidus curves for vanadium compounds, but not sodium sulfate. These differences are considered to be results of the absorption and evolution of oxygen by the complexes when heated to the melting point or cooled to the solidification point. These temperature differences could be of importance both to the formation of deposits and subsequent high-temperature corrosion by such deposits. Sulfur trioxide and nascent oxygen have also been suggested as corrodents.

A review of the work on this subject results in a curve relating metal wastage to the phase diagram of Na_2O versus V_2O_5.[54] Fig. 14 plots the results for an iron-base alloy. It can be seen that the greatest wastage occurs at a Na_2O/V_2O_5 ratio of about 1:5. Molten-temperature range of this ratio is 1150° to 1200°F. However, as was noted previously, eutectics occur at temperatures considerably lower; metal wastage would be possible at lower temperatures if the corresponding ratio of Na_2O/V_2O_5 is present. In ac-

Table XII. Analyses of Ash from Heavy Fuel Oils

	% by Weight			
	Mid-Cont.	Texas	Iran	Venezuela
SiO_2	31.7	1.6	12.1	2.4
Fe_2O_3				
Al_2O_3	31.8	8.9	18.1	18.4
TiO_2				
CaO	12.6	5.3	12.7	2.9
MgO	4.2	2.5	0.2	2.6
MnO	0.4	0.3	TR	. . .
V_2O_5	TR	1.4	38.5	50.5
NiO	0.5	1.5	10.7	. . .
Na_2O	6.9	30.8	. . .	12.7
K_2O	. . .	1.0	. . .	0.5
SO_3	10.8	42.1	7.0	8.6
Chloride	. . .	4.6	. . .	. . .

tual practice, diluents in the ash usually raise the melting point to 1100°F or higher; corrosion of metal surfaces below 1100°F is not usually a problem.

Boilers operating at steam temperatures of 1005°F or higher will have some superheater and reheater tubes operating at temperatures above 1100°F and will, therefore, be subject to corrosion. Field measurements of superheater and reheater tubes wastage have indicated corrosion rates of as much as 0.030 in. per year.

Many approaches have been taken to prevent high-temperature oil-ash corrosion. These include oil treatment to remove vanadium, sodium, and sulfur from the oil[55, 56]; the use of corrosion-resistant alloys and protective coatings; low-excess air operation; and the use of magnesium metal, magnesium oxide, or dolomite additives.

LOW-TEMPERATURE CORROSION FROM OIL ASH

The greater part of the sulfur released in the combustion of fuel oil appears in the products of combustion as SO_2. A small percentage is oxidized to SO_3 which, in the cooled flue gas, reacts with water in the vapor phase to form sulfuric acid. The formation of SO_3 in boilers is a complex process which is not thoroughly understood. Three possible mechanisms have received the most attention: (1) oxidation of SO_2 by molecular oxygen, (2) oxidation of SO_2 in the flame by atomic oxygen, and (3) catalytic oxidation of SO_2. Various authors have considered these mechanisms,[57–59] but extensive field tests have indicated that catalytic oxidation in the convection pass is a major source of SO_3.

If metal temperature is below the acid dew point, the sulfuric acid condenses on and corrodes it. If the gas temperature falls below the dew point, the sulfuric acid condenses on flyash particles which can agglomerate to form acid smuts. It follows that low-temperature corrosion and acid smuts will be eliminated by maintaining both metal and gas temperatures above the acid dew point or by the reduction of SO_3 in the flue gas.

The relation of SO_3 concentration to acid dew point temperature was established in a laboratory investigation.[60] In this study a direct determination was made of SO_3 in the gas stream. By maintaining a specially designed condenser at a temperature between the acid dew point

Table XIII. Liquidus/Melting Points Sodium and Vanadium Oil-Ash Constituents

Compound	Melting Point	Liquidus Temp		Ref.
		Heating	Cooling	
V_2O_5	1247, 1274°F	1220°F	1165°F	51, 52
$Na_2O \cdot 6V_2O_5$	1255	185	1060	52
$5\,Na_2O \cdot V_2O_4$				
$\cdot 11V_2O_5$	995	984	922	51, 53
$Na_2O \cdot 3V_2O_5$	1220	1040	1020	52
$NaVO_3$				
$(Na_2O \cdot V_2O_5)$	1140	1040	1020	52
Na_2SO_4	1625	1625	1625	52

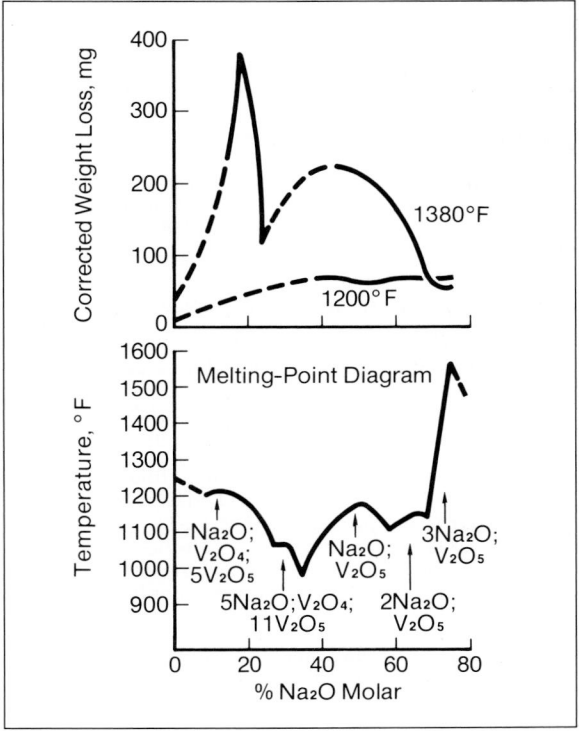

Fig. 14. Effect of *Na_2O/V_2O_5* mixtures on oil-ash corrosion

and the water dew point, sulfuric acid is condensed while all other flue-gas constituents remain in the gas phase.

Acid dew point is defined as "the temperature at which the combustion gases are saturated with sulfuric acid" and the dewpoint/ acid concentration relationship was determined using the condensation sampling method. Fig. 15 graphically illustrates the agreement of dew-point determination of this type with thermodynamically calculated data, and shows electrical dew-point meter measurements. In view of the uncertainties involved with using dew-point meters, it is preferable to use the condenser analytical method together with dew-point curves to predict the temperatures at which the corrosive sulfuric condensate will first appear.

Although the low-temperature sections of a boiler can be protected against acid corrosion, and acid-smut fallout eliminated by maintaining metal and gas temperatures above the dew point, it is readily recognized that a more positive solution to these problems is the reduction

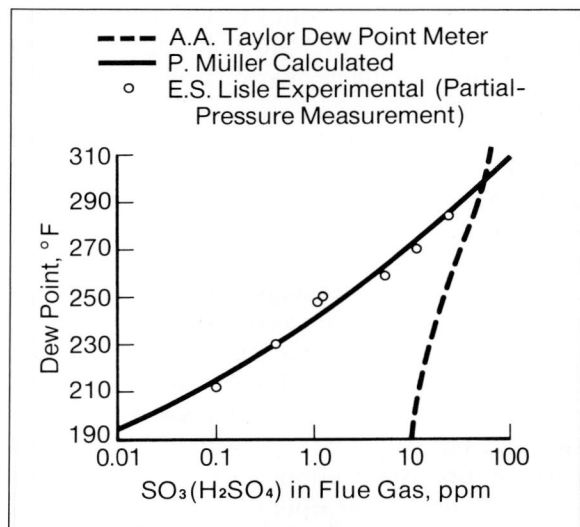

Fig. 15. Dew point as a function of H_2SO_4 /(SO_3) concentration

of sulfur-trioxide in the flue gas. Firing with low excess air plus alkaline additives to reduce SO_3 concentration has been done extensively in the United States and Europe.[61]

REFERENCES AND FOOTNOTES

[1] W. A. Selvig and F. H. Gibson, "Analyses of Ash from United States Coals," *U. S. Bureau of Mines Bulletin 567.* Washington: U.S. Bureau of Mines, 1956.

W. A. Selvig and F. H. Gibson, "Analyses of Ash from Coal of the United States," *U. S. Bureau of Mines Technical Paper 679.* Washington: U. S. Bureau of Mines, 1945.

[2] Elliott P. Barrett, "The Fusion, Flow, and Clinkering of Coal Ash: A Survey of the Chemical Background," Chapter 15 of *Chemistry of Coal Utilization,* Vol. I, edited by H. H. Lowry, pp. 496–571. New York: John Wiley and Sons and London: Chapman and Hall, Ltd., 1945.

[3] Thomas G. Estep, Harry Seltz, and William J. Osborn, "Determination of the Effect of Oxides of Sodium, Calcium, and Magnesium on the Ash Fusion Temperatures by the Use of Synthetic Coal Ash," *Mining and Metallurgical Investigations, Carnegie Institute of Technology, Mining and Metallurgical Advisory Boards, Bulletin 74,* 1937.

[4] G. Thiessin, C. G. Ball, and P. E. Grotts, "Coal Ash and Coal Mineral Matter," *Industrial and Engineering Chemistry,* 28: 335–361, 1936.

[5] P. Nicholls and W. A. Selvig, "Clinker Formation as Related to the Fusibility of Coal Ash," *U. S. Bureau of Mines Bulletin 364.* Washington: U. S. Bureau of Mines, 1932.

[6] A. F. Duzy, "Fusibility-Viscosity of Lignite-Type Ash," *ASME Paper No. 65—WA/FU-7.* New York: American Society of Mechanical Engineers, 1965.

[7] G. H. Gronhovd, et al., "Study of Factors Affecting Ash Deposition from Lignite and Other Coals," *ASME Paper No. 69–WA/C1-1.* New York: American Society of Mechanical Engineers, 1969.

[8] W. Beckering, et al., "Examination of Coal and Coal Ash by X-Ray Techniques," *Technology and Use of Lignite, Proceedings;* by James L. Elder and Wayne R. Kube. Co-sponsored by the U.S. Bureau of Mines and the University of North Dakota, Grand Forks, N.D., May 1–2, 1969, pp. 89–102. (U.S. Bureau of Mines Information Circular 8471). Washington: U.S. Bureau of Mines, 1970.

[9] J. R. Partington, *Textbook of Inorganic Chemistry,* 6th edition, p. 689. London: MacMillan Co., 1950.

[10] H. E. Crossley, "The Melchett Lecture for 1962: A Contribution to the Development of Power Stations," *Institute of Fuel. Journal,* 36 (269): 228–239, June 1963.

[11] James V. O'Gorman and Philip L. Walker, Jr., "Mineral Matter Characteristics of Some American Coals," *Fuel,* 50 (2): 135–151, April 1971.

[12] James V. O'Gorman and Philip L. Walker, Jr., "Thermal Behavior of Mineral Fractions Separated from Selected American Coals," *Fuel*, 52 (1): 71–79, January 1973.

[13] J. E. Payner and W. G. Marskell, "The Distribution of Mineral Matter in Pulverized Fuel and Solid Products of Combustion," *The Mechanism of Corrosion by Fuel Impurities*, proceedings of the International Conference held at the Marchwood Engineering Laboratories, Marchwood, near Southampton, Hampshire, England, May 20–24, 1963, H. R. Johnson and D. J. Littler, general editors. London: Butterworths, 1963, pp. 113–136.

[14] G. B. Gould and H. L. Brunjes, "Proportions of Free Fusible Material in Coal Ash, as an Index of Clinker and Slag Formation," *American Institute of Mining and Metallurgical Engineers Technical Publication 1175*; New York: American Institute of Mining and Metallurgical Engineers, 1940.

[15] R. W. Borio and R. R. Narciso, Jr., "The Use of Gravity Fractionation Techniques for Assessing Slagging and Fouling Potential of Coal Ash," *Transactions of the ASME Journal of Engineering for Power*, 101, Series A (4): 500–505, October 1979; also in Combustion Engineering publication TIS-5823; and in *ASME Paper No. 78–WA/CD-3*. New York: American Society of Mechanical Engineers, 1978.

[16] R. W. Bryers, "Influence of the Distribution of Mineral Matter in Coal on Fireside Ash Deposition," *Transactions of the ASME Journal of Engineering for Power*, 101, Series A (4), 506–515, October 1979; also in *ASME Paper No. 78–WA/CD-4*. New York: American Society of Mechanical Engineers, 1978.

[17] H. R. Hazard, et al., "Coal Mineral Matter and Furnace Slagging," *Proceedings of the American Power Conference*, 41: 610–617, 1979. Chicago: Illinois Institute of Technology, 1979.

[18] A. H. Moody and D. D. Langan, Jr., "Fusion Characteristics of Fractionated Coal Ashes," *Combustion*, 5 (4): 15–17, October 1933.

[19] R. W. Bryers, "The Physical and Chemical Characteristics of Pyrites and Their Influence on Fireside Problems in Steam Generators," *ASME Paper No. 75–WA/CD-2*. New York: American Society of Mechanical Engineers, 1975.

[20] See Reference 15.

[21] William T. Reid, *External Corrosion and Deposits; Boilers and Gas Turbines*. New York: American Elsevier Pub. Co., 1971.

[22] W. H. Pollock, G. J. Goetz, and E. D. Park, "Advancing the Art of Boiler Design by Combining Operating Experience and Advanced Coal Evaluation Techniques," *Proceedings of the American Power Conference*, 45: also as Combustion Engineering publication TIS-7382.

[23] R. W. Borio, et al., "The Control of High-Temperature Fire-Side Corrosion in Utility Coal-Fired Boilers," *U. S. Department of the Interior, Office of Coal Research, Research and Development Report No. 41*, April 1969.

Washington: U. S. Department of Interior, Office of Coal Research, 1969.

[24] A. C. Corey, H. A. Grabowski, and B. J. Cross, "External Corrosion of Furnace-Wall Tubes—III. Further Data on Sulfate Deposits and the Significance of Iron Sulfide Deposits," *Transactions of the ASME*, 71: 951–963 November 1949.

[25] A. L. Plumley, J. Jonakin, and R. E. Vuia, "A Review Study of Fireside Corrosion in Utility and Industrial Boilers. Part I—How to Lower Corrosion in Boilers," *Canadian Chemical Processing*, 51:52, June, 1967. Based on Combustion Engineering publication TIS-2775.

A. L. Plumley, J. Jonakin, and R. E. Vuia, "A Review Study of Fireside Corrosion in Utility and Industrial Boilers. Part II—M_GO. Injection to Lower Boiler Corrosion," *Canadian Chemical Processing*, 51:70, July, 1967. Based on Combustion Engineering publication TIS-2775.

[26] J. G. Koopman, et al., "Development and Use of a Probe for Studying Corrosion in Superheaters and Reheaters," *Proceedings of the American Power Conference*, 21: 236–245, 1959.

[27] James Jonakin, G. A. Rice, and J. T. Reese, "Fireside Corrosion of Superheater and Reheater Tubing," *ASME Paper No. 59-FU-5*. New York: American Society of Mechanical Engineers, 1959.

[28] Wharton Nelson and Carl Cain, Jr., "Corrosion of Superheaters and Reheaters of Pulverized-Coal-Fired Boilers," *Transactions of the ASME. Journal of Engineering for Power*, 82, Series A: 192–204, 1960

[29] J. T. Reese, James Jonakin, and J. G. Koopman, "How Coal Properties Relate to Corrosion of High Temperature Boiler Surfaces," *Proceedings of the American Power Conference*, 23: 391–399, 1961.

[30] Carl Cain, Jr. and Wharton Nelson, "Corrosion of Superheaters and Reheaters of Pulverized-Coal-Fired Boilers II." *Transactions of the ASME. Journal of Engineering for Power*, 83, Series A: 468–474, October 1961.

[31] Wharton Nelson and E. S. Lisle, "A Laboratory Evaluation of Catalyst Poisons for Reducing High-Temperature Gas-Side Corrosion and Ash Bonding in Coal-Fired Boilers," *Journal of the Institute of Fuel*, 37 (284): 378–384, September 1964.

[32] Wharton Nelson and E. S. Lisle, "High Temperature External Corrosion on Coal-Fired Boilers: Siliceous Inhibitors," *Journal of the Institute of Fuel*, 38 (291): 179–186, April 1965.

[33] R. C. Corey, et al., "External Corrosion of Furnace-Wall Tubes—II. Significance of Sulfate Deposits and Sulfur Trioxide in Corrosion Mechanism," *Transactions of the ASME*, 67: 289–302, 1945.

[34] H. R. Johnson and D. J. Littler, general editors, *The Mechanism of Corrosion by Fuel Impurities*, proceedings of the International Conference held at the Engineering Laboratories, Marchwood, near Southampton, Hampshire, England, May 20–24, 1963. London: Butterworths, 1963.

[35] S. A. Goldberg, J. J. Gallagher, and A. A. Orning, "A Laboratory Study of High Temperature Corrosion on Fireside Surfaces of Coal-Fired Steam Generators," *Transactions of the ASME. Journal of Engineering for Power,* 90, Series A: 193–212, 1968.

[36] M. Weintraub, S. Goldberg, and A. A. Orning, "A Study of Sulfur Reactions in Furnace Deposits," *Transactions of the ASME. Journal of Engineering for Power,* 83, Series A: 444–450, October 1961.

[37] P. Sedor, E. K. Diehl, and D. H. Barnhart, "External Corrosion of Superheaters in Boilers Firing High-Alkali Coals," *Transactions of the ASME. Journal of Engineering for Power,* 82, Series A: 181–193, July, 1960.

[38] H. E. Crossley, "External Boiler Deposits," *Journal of Institute of Fuel,* 25 (145): 221–225, September, 1952.

[39] P. H. Crumley, A. W. Fletcher, and D. S. Wikon, "The Formation of Bonded Deposits in Pulverized-Fuel-Fired Boilers," *Journal of the Institute of Fuel,* 28 (170): 117–120, March 1955.

[40] C. H. Anderson and E. Diehl, "Bonded Fireside Deposits in Coal-Fired Boilers; A Progress Report on Manner of Formation," *ASME Paper No. 55-A-200.* New York: American Society of Mechanical Engineers, 1955. Abstracted in *Mechanical Engineering,* 78: 271, 1956.

[41] R. W. Borio and R. P. Hensel, "Coal-Ash Composition as Related to High-Temperature Fireside Corrosion and Sulfur Oxides Emission Control," *Transactions of the ASME. Journal of Engineering for Power,* 94, Series A: 142–148, April 1972; also in. *ASME Paper No. 71-WA/CD-4.* New York; American Society of Mechanical Engineers 1971; and in Combustion Engineering publication TIS-3019.

[42] R. W. Borio, et al., "Control of High-Temperature Metal Wastage in Pulverized Coal-Fired Steam Generators," Combustion Engineering publication TIS-5055.

[43] A. Rahmel, "Influence of Calcium and Magnesium Sulfates on High Temperature Oxidation of Austenitic Chrome-Nickel Steels in the Presence of Alkali Sulfates and Sulfur Trioxides," *Mechanism of Corrosion by Fuel Impurities;* proceedings of the International Conference held at the Marchwood Engineering Laboratories, Marchwood near Southhampton, Hampshire, England, May 20–24, 1963. H. R. Johnson and D. J. Littler, general editors. London: Butterworths, 1963.

[44] A. L. Flumley, "Incinerator Corrosion Potential," *ASME Incinerator Division Corrosion Symposium.* New York: American Society of Mechanical Engineers, 1970.

[45] A. J. B. Cutler, et al., "The Role of Chloride in Corrosion Caused by Flue Gases and Their Deposits," *ASME Paper No. 70 WA/CD-1.* New York: American Society of Mechanical Engineers, 1970.

[46] K. Fassler, H. Leib, and H. Spaun, "Corrosion in Refuse Incinerators," *Mittelungen der VGB,* 48 (2): 126–139, 1968.

[47] R. Baum and C. H. Parker, "Incinerator Corrosion in the Presence of Polyvinyl Chloride and Other Acid-Releasing Constituents," *Plastecology 1972,* Society of Plastic Engineers, Regular Technical Conference, Rosemont, Illinois, October 11–12, 1972, pp. 44–51.

[48] R. E. Sommerlad, R. W. Bryers, and J. D. Shenker, "Systems Evaluation of Refuse as a Low Sulfur Fuel: Part 2—Steam Generator Aspects," *ASME Paper No. 71-WA/INC/2.* New York: American Society of Mechanical Engineers, 1971.

[49] H. H. Krause, D. A. Vaughan, and W. K. Boyd, "Corrosion and Deposits from Combustion on Solid Waste. Part III—Effects of Sulfur on Boiler Tube Metal," *Transactions of the ASME. Journal of Engineering for Power,* 97 (3), Series A: 448–452, July 1975. Also published as *ASME Paper No. 74-WA/CD-5.* New York: American Society of Mechanical Engineers, 1974.

[50] H. H. Krause, D. A. Vaughan, and W. K. Boyd, "Corrosion and Deposits from Combustion of Solid Wastes. Part III—Effects of Sulfur on Boiler Tube Metals," *Transactions of the ASME. Journal of Engineering for Power,* 97 (3), Series A: 448–452, July 1975.

[51] A. T. Bouden, P. Draper, and H. Rowling, "The Problem of Fuel Oil Ash Deposition in Open Cycle Gas Turbines," *Proceedings of the Institution of Mechanical Engineers,* 167: 291–300, 1953.

[52] Norman D. Phillips and Charles L. Wagoner, "Use of Differential Thermal Analysis in Exploring Minimum Temperature Limits of Oil Ash Corrosion," *Corrosion,* 17 (8): 102–106, August 1961.

[53] T. Widell and I. Juhasz, "Softening Temperature of Residual Fuel Oil Ash," *Combustion,* 22 (11): 51, May 1951.

[54] J. J. MacFarlane and N. Stephenson, "Communication," pp. 30T–31T of article by S. H. Frederick and T. F. Eden, "Corrosion Aspects of the Vanadium Problem in Gas Turbines," *Corrosion,* 11: 19T–33T, 1955.

[55] A. Voorhies, Jr., et al., "Improvement in Fuel Oil Quality, I—Demetalization of Residual Fuels II—Desulfurization of Residual Fuels." *Mechanism of Corrosion by Fuel Impurities,* proceedings of the International Conference held at the Marchwood Engineering Laboratories, Marchwood, near Southampton, Hampshire, England, May 20–24, 1963. H. R. Johnson and D. J. Littler, general editors. London: Butterworths, 1963.

[56] C. J. A. Edwards, *Literature Survey Primarily Covering Fireside Deposits and Corrosion Associated with Boiler Superheater Tubes in Oil-Fired Naval Steam-Raising Installation with Some References Also to Similar Problems in Land-Based Installations and Gas Turbines.* British Petroleum Research Centre, 1964.

[57] G. Whittingham, "The Oxidation of Sulfur Dioxide in the Combustion Process," *Third Symposium on Combustion, Flame, and Explosion Phenomena,* sponsored by the Combustion Institute and held at the University of Wisconsin, 1949, pp. 453–459. Pittsburgh: The Combustion Institute, 1971.

[58] A. B. Hedley, "Sulfur Trioxide in Combustion Gases," *Fuel Society Journal,* 43: 45–54, 1962.

[59] Arthur Levy and E. L. Merryman, "SO_3 Formation in H_2S Flames," *Transactions of the ASME. Journal of Engineering for Power*, 87, Series A: 116–123, January 1965.

Ibid., 374–378, October 1965.

[60] E. S. Lisle and J. D. Sensenbaugh, "The Determination of Sulfur Trioxide and Acid Dew Point in Flue Gases," *Combustion*, 36 (7): 12–16, January 1965.

[61] J. T. Reese, J. Jonakin, and V. Z. Caracristi, "Prevention of Residual Oil Combustion Problems by Use of Low Excess Air and Magnesium Additive," *ASME Paper No. 64-PWR-3*. New York: American Society of Mechanical Engineers, 1964.

BIBLIOGRAPHY

Bassa, G. and B. Bator, "The Influence of Distribution of Minerals in Pulverized Coals with High Mineral Contents on the Working Conditions of Steam Generators," *The Mechanism of Corrosion by Fuel Impurities*, proceedings of the International Conference held at the Marchwood Engineering Laboratories, Marchwood, near Southampton, Hampshire, England, May 20–24, 1963, H. R. Johnson and D. J. Littler, general editors. London: Butterworths, 1963, pp. 90–101.

Bishop, R. J. and J. A. C. Samms, "Pilot Scale Investigations of the Formation of Bonded Deposits," *The Mechanism of Corrosion by Fuel Impurities*, proceedings of the International Conference held at the Marchwood Engineering Laboratories, Marchwood, near Southampton, Hampshire, England, May 20–24, 1963, H. R. Johnson and D. J. Littler, general editors. London: Butterworths, 1963, pp. 155–172.

Boow, J., "Sodium/Ash Reactions in the Formation of Fireside Deposits in Pulverized-Fuel-Fired Boilers," *Fuel* 51(3): 170–173, July 1972.

Borio, R. W., G. J. Goetz, and A. A. Levasseur, "Slagging and Fouling Properties of Coal Ash Deposits as Determined in a Laboratory Test Facility," paper presented at the ASME Winter Annual Meeting, December 1977; also as Combustion Engineering publication TIS-5155.

Bryers, R. W., "The Physical and Chemical Characteristics of Pyrites and Their Influence on Fireside Problems in Steam Generators," *ASME Paper No. 75-WA/CD-2*. New York: American Society of Mechanical Engineers, 1975.

Bryers, R. W., and T. E. Taylor, "An Examination of the Relationship Between Ash Chemistry and Ash Fusion Temperatures in Various Coal Size and Gravity Fractions Using Polynomial Regression Analysis," *Transactions of the ASME. Journal of Engineering for Power*, 98, Series A: 528–539, October 1976.

Burbach, H. E. and E. A. Ramspeck, "Steam Generation Design for High-Sodium Subbituminous Coals," presented at ASME-IEEE-ASCE Joint Power Generation Conference, St. Louis, MO, October 4–8, 1981; also as Combustion Engineering publication TIS-6867.

Crossley, H. E., "External Boiler Deposits. Paper 4—Special Study of Ash and Clinker in Industry," *Journal of the Institute of Fuel*, 25(145): 221–225, September 1952.

Dundersdale, J., et al., "Studies Relating to the Behaviour of Sodium During the Combustion of Solid Fuels," *The Mechanism of Corrosion by Fuel Impurities*, proceedings of the International Conference held at the Marchwood Engineering Laboratories, Marchwood, near Southampton, Hampshire, England, May 20–24, 1963, H. R. Johnson and D. J. Littler, general editors. London: Butterworths, 1963, pp. 139–144.

Garner, L. J., "The Formation of Boiler Deposits from the Combustion of Victorian Brown Coals," *Journal of the Institute of Fuel*, 40(314): 107–116, March 1967.

Goetz, G. J., N. Y. Nsakala, and R. W. Borio, "Development of Method for Determining Emissivities and Absorptivities of Coal Ash Deposits," paper presented at the 1978 Winter Annual ASME Meeting, December, 1978; also as Combustion Engineering publication TIS-5890.

Goetz, G. J., N. Y. Nsakala, R. L. Patel, and T. C. Lao, "Combustion and Gasification Characteristics of Chars from Four Commercially Significant Coals of Different Rank," EPRI Report AP-2601, September 1982; condensed version presented at EPRI's Second Coal Gasification Conference, October 20–20, 1982, Palo Alto, CA.

Hale, G. L., A. A. Levasseur, A. L. Tyler and R. P. Hensel, "The Alkali Metals in Coal: A Study of Their Nature and Their Impact on Ash Fouling," paper presented at Coal Technology '80, Houston, TX, November 1980; also as Combustion Engineering publication TIS-6645.

Hedley, A. B., et al., "Available Mechanisms for Deposition from a Combustion Gas Stream," *ASME Paper No. 65-WA/CD-4*. New York: American Society of Mechanical Engineers, 1965.

Khrustalev, B. A., "Spectral Radiation Properties of Some Materials at High Temperatures and Their Influence on the Integral Absorption and Radiation Properties," *Heat Transfer-Soviet Research*, 5(2): 60–64, March/April 1973.

Krzhizhanovskii, R. E., et al., "Influence of Particle Size of Ash on the Structure and Effective Thermal Conductivity of Loose Deposits," *Thermal Engineering* (translation of *Teploenergetika*), 19(10): 36–39, 1972.

Kuleshova, I. A., et al., "Investigation of Fraction Composition and Properties of Pulverized Coal," *Thermal Engineering* (translation of *Teploenergetika*), 17(6):33–38, 1970.

Littlejohn, R. F., "Mineral Matter and Ash Distribution in 'As-Fired' Samples of Pulverized Fuels," *Journal of the Institute of Fuel*, 39(301): 59–67, February 1966.

Littlejohn, R. F. and J. D. Watt, "The Distribution of Mineral Matter in Pulverized Fuel," *The Mechanism of Corrosion by Fuel Impurities*, proceedings of the International Conference held at the Marchwood Engineering Laboratories, Marchwood, near Southampton, Hampshire, England, May 20–24, 1963, H. R. Johnson and D. J. Littler, general editors. London: Butterworths, 1963, pp. 102–112.

Marskell, W. G. and J. M. Miller, "Some Aspects of Deposit Formation in Pilot-Scale Pulverized-Fuel-Fired Installations." *Journal of the Institute of Fuel*, 29(188):380–387, September 1956.

Moody, A. H. and D. D. Langan, Jr., "Fusion Characteristics of Fractionated Coal Ashes," *Combustion*, 5(4):15–17, October 1933.

Moody, A. H. and D. D. Langan, Jr., "Fusion Temperature of Coal Ash as Related to Composition," *Combustion*, 6(8):13–20, February 1935.

Mulcahy, M. F., et al. "Fireside Deposits and Their Effect on Heat Transfer in a Pulverized-Fuel-Fired Boiler, Part I: The Radiant Emittance and Effective Thermal Conductance of the Deposits," *Journal of the Institute of Fuel*, 39(308):385–394, September 1966.

National Research Council. Committee on Chemical Utilization of Coal. Chemistry of Coal Utilization, prepared by the Committee on Chemical Utilization of Coal, Division of Chemistry and Chemical Technology, National Research Council, H. H. Lowry, chairman, Supplementary volume. New York: John Wiley and Sons, Inc., and London: Chapman and Hall, Ltd., 1963.

Raask, E., "Reactions of Coal Impurities During Combustion and Deposition of Ash Constituents on Cooled Surfaces," *The Mechanism of Corrosion by Fuel Impurities*, proceedings of the International Conference held at the Marchwood Engineering Laboratories, Marchwood, near Southampton, Hampshire, England, May 20–24, 1963, H. R. Johnson and D. J. Littler, general editors. London: Butterworths, 1963, pp. 145–154.

Regan, J. W., "Impact of Coal Characteristics on Boiler Design," presented at Coal Technology '82, December 7–9, 1982, Houston, TX; also published as Combustion Engineering publication TIS-7291.

Smith, R. A. and L. R. Glicksman, "Radiation Properties of Slag," *ASME Paper No. 69-WA/PWR-7*. New York: American Society of Mechanical Engineers, 1969.

Tufte, Philip H., et al., "Ash Fouling Potentials of Western Subbituminous Coals Determined in a Pilot-Plant Test Furnace," *Proceedings of the American Power Conference*, 38:661–671, 1976. Chicago: Ill. Institute of Technology, 1976.

CHAPTER 4

Combustion Processes

This chapter includes the basic equations used to describe how fuel and air injected into a furnace are transformed into gaseous products of combustion. A major section is devoted to the mathematical background and experiential basis for the combustion reactions in furnaces. There is a summary discussion of the physical mechanisms currently used for burning coal and other solid fossil fuels. The material concludes with the theoretical and empirical criteria for the formation of nitrogen oxides during combustion, and methods for reducing their emission from a furnace.

HISTORICAL INTRODUCTION

Although from his earliest existence on earth man has been fascinated by fire, he did not achieve a quantitative understanding of the combustion process until about the year 1880. Prior to that date, one can trace the development of many hypotheses concerning the nature and properties of fire, including some that were expressed in supernatural terms of fear and awe. However, even the existence of the now discredited phlogiston theory of combustion did not prevent enterprising engineers from designing and constructing boilers to generate steam for the earliest steam engines.

Phlogiston was a hypothetical mysterious substance which sometimes was presumed to have the property of negative weight and which combined with a body to render it combustible. First proposed by G. E. Stahl in 1697, the phlogiston theory dominated the chemical thought of the 18th century. Even such a perceptive observer as Joseph Priestly, who in 1774 discovered the unique power of oxygen for supporting combustion, accepted the phlogiston theory. In the years between 1775 and 1781, Antoine L. Lavoisier substituted for it the theory of oxygenation and provided experimental evidence that combustion was the union of the substance burned with the oxygen of the atmosphere.

In 1755 Joseph Black discovered carbon dioxide, and in 1781 Henry Cavendish demonstrated the compound nature of water. At about this same time Lavoisier made the precise measurements and formulated the volume and weight relationships that underlie the modern theory of combustion.

Beyond this, in 1811 Amedeo Avogadro established that the number of molecules in a unit volume under standard conditions is the same for all gases. During this same period John Dalton enunciated the law of partial pressures, and in 1803 his study of the physical properties of gases led to formulation of the atomic theory, including the law of combining weights. A related observation was made by Gay-Lussac in 1808 that gases always combine in volumes that bear simple ratios to each other.

COMBUSTION FUNDAMENTALS

To the engineer concerned with boiler design and performance, combustion may be considered as the chemical union of the combustible of a fuel and the oxygen of the air, controlled at such a rate as to produce useful heat energy. The principal combustible constituents are elemental carbon, hydrogen, and their compounds. In the combustion process, the compounds and elements are burned to carbon dioxide and water vapor. Small quantities of sulfur are present in most fuels. Although sulfur is a combustible and contributes slightly to the heating value of the fuel, its presence is generally detrimental because of the corrosive nature of its compounds.

Air, the usual source of oxygen for combustion in boilers, is a mixture of oxygen, nitrogen and small amounts of water vapor, carbon dioxide, argon and other elements. The compositions of dry and wet atmospheric air are given in Table I.

In an ideal situation, the combustion process would occur with the exact proportions of oxygen and a combustible that are called for in theory (the stoichiometric quantities). But it is impracticable to operate a boiler at the theoretical level of zero percent excess oxygen. In practice this condition is approached by providing an excess of oxygen in the form of excess air from the atmosphere. The amount of excess air varies with the fuel, boiler load, and the type of firing equipment.

Table I. Composition of Combustion Air

Dry Atmospheric Air

The volumetric composition of dry atmospheric air given in NACA Report 1235 (Standard Atmosphere—Tables and Data for Altitudes to 65,800 feet, November 20, 1952),[1] and the molecular weights of the gases constituting dry air are as follows:

	Volume %	Mol. Wgt.
Nitrogen	78.09	28.016
Oxygen	20.95	32.000
Argon	0.93	39.944
Carbon dioxide	0.03	44.010

(Neon, helium, krypton, hydrogen, xenon, ozone, and radon, combined, are less than 0.003%.)

Dry air with this composition has an apparent molecular weight of 28.97 lb/lb mol and a density at 32°F and 14.7 psia of $28.97 \div 359 = 0.0807$ lb/ft³. The oxygen content is 23.14% by weight. The lb dry air/lb oxygen $= 1 \div 0.2314 = 4.32$.

Wet Atmospheric Air

Wet atmospheric air is defined in this text as the above air plus 0.013 lb of water vapor/lb of dry air. (Air at 80°F, 60% relative humidity, and 14.7 psia pressure contains 1.3% water vapor by weight. See Fig. 1.)

Wet air with this amount of water vapor has an apparent molecular weight of 28.74 lb/lb mol and a density at 32°F and 14.7 psia of 0.0801 lb/ft³. The oxygen content is 22.84% by weight. The lb wet atmospheric air/lb oxygen $= 1 \div 0.2284 = 4.38$.

The mass of nitrogen, argon, carbon dioxide and water/lb oxygen $= 77.16/22.84 = 3.38$ lb.

COMBUSTION EQUATIONS

For combustion calculations, however, it is customary to write the combustion reaction equations on the basis of theoretical oxygen only, notwithstanding the presence of excess air and nitrogen. A partial list of these combustion equations and the approximate heat released in the reactions are given in Table II.

All combustion calculations are based on fundamental chemical reactions shown in Table II. Not only do the equations indicate what substances are involved in the reaction, but they also show the molecular proportions in which they take part.

Each molecule has a numerical value that represents its relative weight or molecular weight. This molecular weight is the sum of the atomic weights of the atoms composing the molecule. For example, carbon, C, has a molecular weight of 12; oxygen, O_2, has a molecular weight of $2 \times 16 = 32$; and carbon dioxide, CO_2, has a molecular weight of $12 + (2 \times 16) = 44$. These molecular weights are only relative values and may be expressed in any units. Note that the molecular weights in Table II are the whole-number values of the main isotopes of each substance.

It has been established that a molecular weight of any substance in the gaseous state and under the same conditions of temperature and pressure will occupy the same volume. This relationship is very significant. The volume will, of course, vary numerically for different units of weight and for different conditions of temperature and pressure. For combustion calculations, the pound and the cubic foot are the units commonly used in the U.S. and, unless otherwise stated, the temperature and pressure are understood to be 32°F and 14.7 psia. Thus a molecular weight of 32 lb of oxygen at 32°F and atmospheric pressure will have

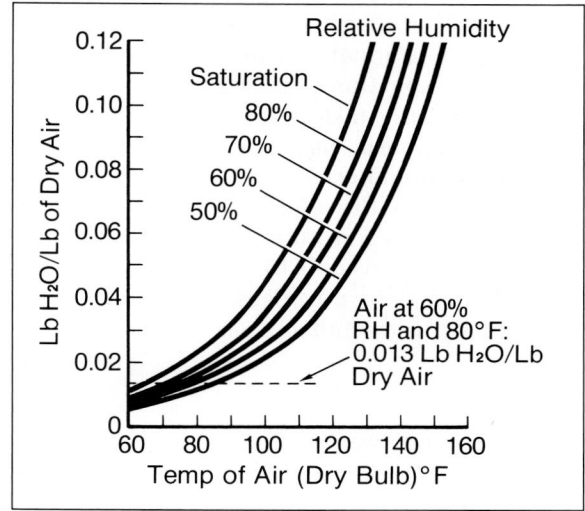

Fig. 1 **Moisture content of dry air as a function of dry-bulb temperature and relative humidity**

Table II. Combustion Equations

Combustible	Molecular Weight	Reaction	Heat Release,* Btu/lb
Carbon	12	$C + O_2 \rightarrow CO_2$	14,100
Hydrogen	2	$H_2 + 0.5\,O_2 \rightarrow H_2O$	61,000
Sulfur	32	$S + O_2 \rightarrow SO_2$	4,000
Hydrogen sulfide	34	$H_2S + 1.5\,O_2 \rightarrow SO_2 + H_2O$	7,100
Methane	16	$CH_4 + 2\,O_2 \rightarrow CO_2 + 2H_2O$	23,900
Ethane	30	$C_2H_6 + 3.5\,O_2 \rightarrow 2CO_2 + 3H_2O$	22,300
Propane	44	$C_3H_8 + 5\,O_2 \rightarrow 3CO_2 + 4H_2O$	21,500
Butane	58	$C_4H_{10} + 6.5\,O_2 \rightarrow 4CO_2 + 5H_2O$	21,300
Pentane	72	$C_5H_{12} + 8\,O_2 \rightarrow 5CO_2 + 6H_2O$	22,000

*Higher heating value/lb of combustible

the same volume as a molecular weight of 44 lb of carbon dioxide under the same conditions. This volume is 359 cu ft.

CONCEPT OF THE MOLE

A molecular weight expressed in pounds is called a *pound mole,* or simply a *mole,* and the volume that it occupies is called a *molal volume.* Molal volume varies with changes in temperature and pressure according to Boyle's and Charles' laws and may be corrected to any desired conditions. Volume is directly proportional to the absolute temperature and inversely proportional to the absolute pressure. Because combustion processes in steam boiler furnaces usually take place at practically constant atmospheric pressure, pressure corrections are seldom necessary.

Returning to the combustion equation for carbon and oxygen and applying these concepts, it is now possible to write this reaction in several ways. For purposes of molar analysis carbon may be treated as a gas.

$$(1)\ C\ +\ (1)\ O_2\ =\ (1)\ CO_2 \tag{1}$$

$$1\ mol\ C\ +\ 1\ mol\ O_2\ =\ 1\ mol\ CO_2 \tag{2}$$

$$12\ lb\ C\ +\ 32\ lb\ O_2\ =\ 44\ lb\ CO_2 \tag{3}$$

Dividing through by 12,

$$1\ lb\ C\ +\ 2.67\ lb\ O_2\ =\ 3.67\ lb\ CO_2 \tag{4}$$

$$1\ volume\ C\ +\ 1\ volume\ O_2\ =\ 1\ volume\ CO_2 \tag{5}$$

Because there are 4.32 lbs of dry air/lb of oxygen, the stoichiometric combustion of 1 lb of carbon requires 11.52 lb of dry air, or 11.68 lb of wet air (with 1.3 percent water vapor).

Each equation balances; there are the same number of atoms of each element and the same weight of reacting substances on each side of the arrow but not necessarily the same number of molecules, moles or volumes. Thus, one atom of carbon combined with one molecule of oxygen gives only one molecule of carbon dioxide; two moles of hydrogen plus one mole of oxygen yield two moles of water vapor.

It will be evident from a consideration of the mole-volume relationship that percent by volume is numerically the same as mole percent.

Because a mole represents a definite weight as well as a definite volume, it is a means of converting analyses by weight into analyses by volume and vice versa. Volumetric fractions of the several constituents of a gas can be multiplied by their respective molecular weights, with the sum of the products then being equal to the apparent molecular weight of one mole of gas. The percent by weight of each component can then be determined. Finally, the density of any gas at any temperature is found by dividing the molecular weight of the gas by the molal volume at that temperature, Fig. 2.

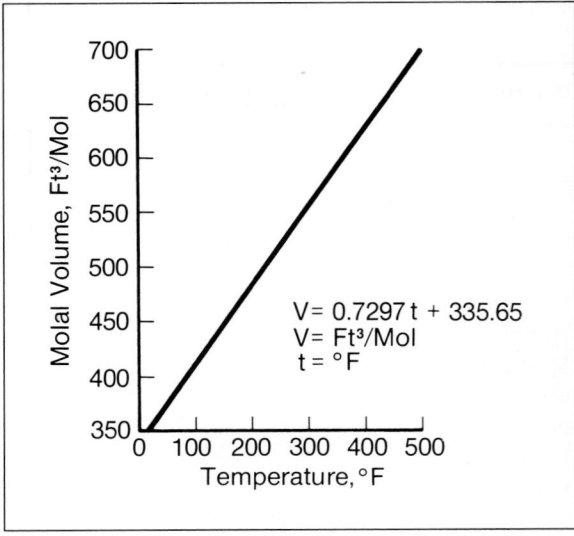

Fig. 2 Gas density determination

METHODS OF COMBUSTION CALCULATIONS

Two methods of combustion calculations are presented in this text. The first is known as the mole method and is based on the chemical relationships previously explained. The second method uses the firing of a million Btu as a basis for calculation.

THE MOLE METHOD

Table II gives the basic combustion reactions for the carbon, hydrogen, and sulfur in coal.

Assume a high-volatile bituminous coal of the following analysis, burned at 23 percent excess air; perform calculations on the basis of 100 lb of as-fired fuel. The fuel analysis as fired is the following

	% by Weight	Mole Weight
C	63.50	12
H_2	4.07	2
S	1.53	32
O_2	7.46	32
N_2	1.28	28
H_2O	15.00	18
Ash	7.16	
HHV	11,200 Btu/lb	

The calculation of air weight for combustion must be made on the basis of an oxygen balance because oxygen is the only element common to all oxidizing reactions. Oxygen contained in the fuel must be deducted from the calculated quantity needed because it is already combined with carbon, hydrogen, or other combustible constituents of the coal.

The molar relations are as given in Table II:

Mol	Mol O_2	Mol CO_2	Mol H_2O	Mol SO_2
C	1	1		
H_2	0.5		1	
S	1			1

AIR FOR COMBUSTION

O_2 for C	63.5/12 =	5.29 lb mol
O_2 for H_2	$(4.07 \times 0.5)/2$ =	1.02 lb mol
O_2 for S	1.53/32 =	0.05 lb mol
Total for 100 lb fuel		6.36 lb mol

Less O_2 in fuel	7.46/32 =	0.23 lb mol
O_2 required		= 6.13 lb mol
O_2 in excess air	6.13×0.23 =	1.41 lb mol
Total O_2 required/100 lb fuel =		7.54 lb mol

Dry air required $= [(7.54 \text{ lb mol } O_2) \times$

$$\frac{(100 \text{ lb mol air})}{(20.95 \text{ lb mol } O_2)}] = 36.0 \text{ lb mol}/100 \text{ lb fuel}$$

$36.0 \times 28.97 = 1043$ lb dry air/100 lb fuel

WEIGHT OF DRY PRODUCTS OF COMBUSTION

The weight of gaseous products of combustion can be calculated from the volumetric analysis of flue gas. Not only the weight of the flue gas/100 lb of coal but its analysis and volume can be calculated from the information given in the preceding example.

To obtain the wet products of combustion, or total wet gas when a fuel burns completely, the weight of the fuel is added to the weight of atmospheric air supplied for its combustion. If some of the fuel is ash or, if because of incomplete combustion, some of the fuel does not leave the furnace with the gases, then there will be less burned-out fuel in the products.

The wet products of combustion in the above example, then, are the fuel (100 lb − 7.16 lb ash = 92.84 lb/100 lb) plus the air required for combustion or (rounded) 93 + 1043 = 1136 lb/100 lb of fuel.

THE MILLION-BTU METHOD

This method for combustion calculations is based on the concept that the weight of air required in the combustion of a unit weight of any commercial fuel is more nearly proportional to the unit heat value than to the unit weight of the fuel. Consequently, the weights of air, dry gas, moisture, wet gas, and other quantities are expressed in pounds per million Btu fired.

In connection with this calculation method, the following items will be discussed:

1. Fuel in products, F
2. Atmospheric air for combustion, A
3. Effect of unburned combustible
4. Products of combustion, P
5. Moisture in the combustion air, W_a

6. Moisture from fuel in products of combustion, W_f

7. Dry gas content of combustion products, P_d

The first four items are necessary for the calculation of the gas and air quantities. Items 5 to 7 form the basis of heat balance calculations, in either the design or testing of a steam generating unit.

FUEL IN PRODUCTS, F

As defined earlier, F is that portion of the fuel fired which appears in the gaseous products of combustion. Because all quantities are to be those required for, or resulting from, the firing of 1,000,000 Btu, F must be calculated on that basis. If a fuel contains no ash, F is obtained by dividing 1,000,000 by the as-fired heating value of the fuel. For solid fuels where ash and/or solid combustible loss must be considered,

$$F = \frac{10^4(100 - \% \text{ ash} - \% \text{ solid combustible loss})}{\text{fuel heat value}}$$

where:

F = $lb/10^6$ Btu fired

% ash = percent by weight in fuel as fired

% solid combustible loss = percent by weight in fuel as fired

fuel heat value (HHV) = high heat value as fired, Btu/lb

(6)

ATMOSPHERIC AIR FOR COMBUSTION, A

In accordance with the molar method, the theoretical weight of dry air (zero excess) may be calculated from the fuel analysis and the formula:

$$A_{dry} = \left[\frac{11.52 (\%C) + 34.57 (\%H - \%O/8) + 4.32 (\%S)}{\text{HHV}} \right] \times 10^4$$

(7)

in which numerator and denominator are on the same basis—as-fired, moisture-free, or

moisture-and-ash free—and A_{dry} is in $lb/10^6$ Btu fired.

For air with 1.3-percent moisture by weight (80°F and 60-percent relative humidity), the formula becomes

$$A_{wet} = \left[\frac{11.68 (\%C) + 35.03 (\%H - \%O/8) + 4.38 (\%S)}{\text{HHV}} \right] \times 10^4$$

(8)

Values of A_{wet} range from 570 $lb/10^6$ Btu for pure hydrogen to above 800 $lb/10^6$ Btu for certain cokes and meta-anthracite coals, as shown in Fig. 3 for various fuels burned in steam generators. Any calculated values of wet air for combustion differing substantially from these values should lead to a cross-verification of the ultimate analysis and the observed high heating value. The analysis and HHV of the fuel have to be from the same sample to avoid errors in air and gas weight determinations.

EFFECT OF UNBURNED COMBUSTIBLE

In the combustion of solid fuels, even in pulverized form, it is not feasible to burn the available combustible completely. Thus, the

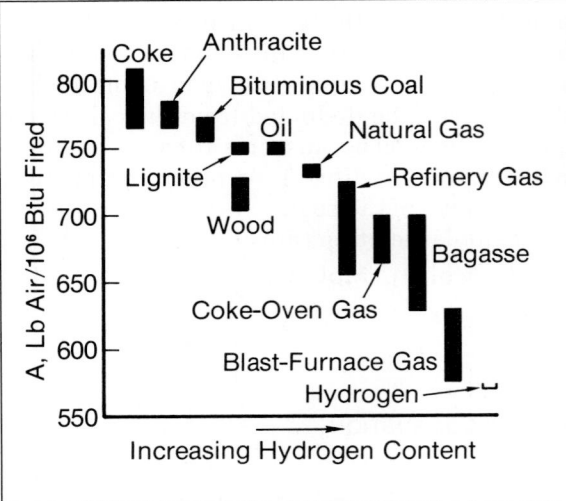

***Fig. 3* Combustion-air requirements for various fuels at zero excess-air—a range of values as an approximate function of hydrogen content**

air requirement per million Btu *fired* has to be reduced to the air required per million Btu *burned*. This is done by multiplying the combustion air A by the combustible-loss correction factor C.

The unburned combustible loss can be expressed either as percent carbon heat loss or the percent combustible weight loss. These are related by the expression:

$$\% \text{ Carbon heat loss} = \frac{14,600}{\text{HHV}} \times \% \text{ solid comb. wgt. loss}$$ (9)

in which 14,600 is the heat value for combustibles in refuse recommended by *A.S.M.E. Performance Test Code*, and HHV is the high heating value of the as-fired fuel.

If a fuel has carbon as its only combustible constituent, the factor C equals

$$1 - \frac{\% \text{ solid combustible weight loss}}{100}$$ (10)

If, however, all heat in the fuel does not come from carbon alone (so that the air is not strictly proportional to carbon burned), the factor C

will not be exact. For high-carbon, low-volatile fuels it will be nearly exact and will result in only a small error even for fuels low in fixed carbon and high in hydrogen. The error involved by using Eq. 10 in all cases is quite within the limits of accuracy of all other combustion calculations.

Finally, C can be expressed as a function of percent heat loss by combining the above relationships:

$$C = 1 - \frac{\% \text{ carbon heat loss}}{100} \times \frac{\text{HHV}}{14,600}$$ (11)

Fig. 4 is a graphical solution of this equation.

PRODUCTS OF COMBUSTION, P

The total gaseous products of combustion, P, become the sum of F and A (as corrected for combustible loss). Thus,

$$P = F + CA$$ (12)

where:
P = total gaseous products of combustion, lb/10^6 Btu fired
F = fuel fired exclusive of ash or solid carbon loss, lb/10^6 Btu fired
A = atmospheric air consumed, lb/10^6 Btu fired
C = combustible loss correction factor

MOISTURE IN COMBUSTION AIR, W_a

For heat-balance calculations, the moisture in air will be 1.3 percent of the air weight/10^6 Btu, or $W_a = 0.013A$, for ambient conditions of 80°F and 60-percent relative humidity. For air at a higher or lower temperature or relative humidity, the moisture content will be as shown in Fig. 1; air per million Btu, A, and W_a must be adjusted appropriately.

MOISTURE FROM FUEL, W_f

This item is separately reported both in an *ASME Performance Test Codes* heat balance

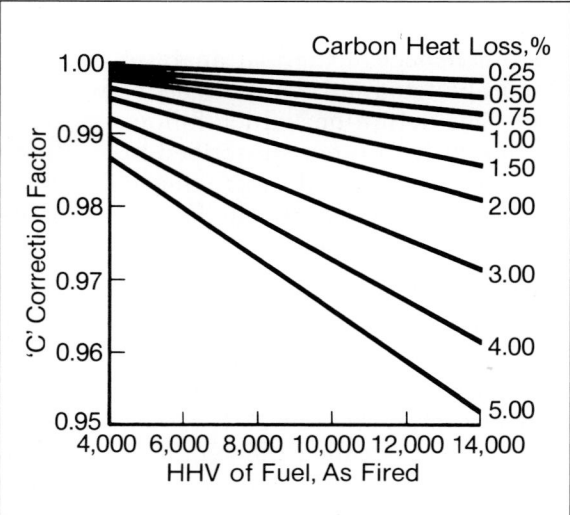

Fig. 4 Graphical solution of equation 11

and in a predicted heat balance. In the case of some fuels (such as natural and refinery gases), the heat loss because of this moisture may be the largest single item in the heat balance. W_f includes the combined surface and inherent moisture, W_c, from a fuel, plus the moisture formed by the combustion of hydrogen, W_h. W_c will vary from zero, or a mere trace in fuel oil, to over 115 lb/10^6 Btu fired in the case of green wood; W_h will vary from zero or a trace in lamp-black to 100 lb/10^6 Btu fired in the case of some refinery gases. Thus,

$$W_f = W_c + W_h, \text{ where}$$
$$W_c = H_2O \times \frac{10^4}{HHV}$$
$$\text{and } W_h = 9 \times H \times \frac{10^4}{HHV}$$

$$(13)$$

W_c and W_h are in lb/10^6 Btu fired, H_2O is the percent water by weight in the as-fired fuel, H is the hydrogen in the fuel (percent by weight as fired) and HHV is the high heating value of the as-fired fuel, Btu/lb.

DRY GAS, P_d

The dry-gas content of the combustion products is used in the calculation of the dry-gas-loss item of a boiler heat balance.

The dry gas may be determined by subtracting the water vapor from the total products, thus,

$$P_d = P - (W_a + W_f) \qquad (14)$$

where

P_d = dry gas, lb/10^6 Btu fired

P = total products of combustion

W_a = moisture in air

W_f = moisture from fuel

COMBUSTION CALCULATIONS BY GRAPHICAL METHODS

In Appendix C, we present a convenient graphical system for solving the equations of this million Btu method, for coal and most other power plant fuels. Examples of all

the significant combustion calculations are included to reinforce the above information.

In the event that an ultimate analysis of a given coal is not available, the curves and data in Appendix C can be used to arrive at close approximate solutions.

One of the principal purposes of Appendix C is to enable an engineer required to perform combustion calculations to obtain a "visual feel" for the order of magnitude of the calculated values. Engineering judgments about such concepts are very difficult to make based on calculations done on electronic computing equipment. When used in commercial problem solving, the answers obtained from computers stand alone, with no reference to physical input parameters. Thus, there is a lack of the judgmental development that can be gained from using graphical devices and one's skills may be limited to computer inputting.

THERMOCHEMISTRY

Energy is associated with the forces which bind atoms together to form a molecule. A rearrangement of the atoms to form new molecules, such as occurs in a chemical reaction, entails the liberation or absorption of energy. The branch of thermodynamics which deals with this subject is referred to as *thermochemistry*. Unfortunately, the working definitions regarding this aspect of thermal analysis are often obscurely stated and contradictory between authors. The definitions which follow are therefore not universal nor subscribed to by all authors in this field. In the main they will rely on the authoritative treatises by Wark[2], Reynolds and Perkins[3], and Van Wylen and Sonntag[4].

DEFINITIONS OF TERMS

Before proceeding, it is necessary to remark here that the reference datum for chemical thermodynamic tabulations is usually taken to be 1 atm pressure and 25°C (298°K, 77°F, or 537°R). This is called the *standard reference state* and, by convention, the enthalpy of every elemental substance (such as H_2, O_2, N_2, Cl) is defined to be zero at the standard reference state. We will briefly review several other defi-nitions concerned with the emission or con-sumption of energy accompanying a thermo-chemical reaction.

INTERNAL ENERGY OF REACTION

If a chemical reaction occurs at constant vol-ume and at 25°C initial temperature and the products of reaction are returned to 25°C, then the energy excess (liberated) or deficit (absorp-tion) required to meet the conditions of this system is known as the internal energy of reac-tion. The symbol assigned this energy is ΔU_R. Heat of reaction at constant volume is an alter-nate for the internal energy of reaction. By first law analysis, we can write for this item Q = ΔU.

ENTHALPY OF REACTION

If a chemical reaction is initiated at 25°C and the products are returned to 25°C and this pro-cess is carried out at constant pressure, then the energy supplied or disposed of is referred to as the enthalpy of reaction. The symbol assigned this energy is ΔH_R. Heat of reaction at constant pressure is an alternate term for the enthalpy of reaction. Although both ΔU_R and ΔH_R can be found tabularly listed, ΔH_R is the more com-monly encountered as it is conveniently meas-ured in a steady-flow device. Heat of reaction is a short form for heat of reaction at constant pressure, and is tabulated on a basis of Btu per mol of fuel burned. Heating value is synony-mous with the heat of reaction, except that it is always taken as being positive.

ENTHALPY OF COMBUSTION

This is defined, sometimes, as the negative of the heating value. The symbol assigned the en-thalpy of combustion is ΔH_C.

HIGHER HEATING VALUE; LOWER HEATING VALUE

If water in the products of combustion of a fossil fuel is taken as being condensed and low-ered in temperature to 77°F at 1 atm, the con-densation heat release h_{fg} is added to ΔH_C and this is called the higher heating value (HHV). If the H_2O remains as a vapor, then h_{fg} (1050.3 Btu/lb$_m$ at 25°C) is not added to ΔH_C depending

on whether the enthalpy of combustion tabula-tion is on a liquid or vapor (H_2O) basis.

EXOTHERMIC HEAT

A chemical reaction which liberates energy (heat) is called exothermic and is denoted by ΔH_R (see heat of reaction at constant pressure). The sign convention adopted for the liberated energy is negative. This sign convention is not universal (not used in all textbooks).

Use of the negative convention invariably is justified by the argument that heat is removed from the system in order to return it to 25°C. The negative convention for exothermic reac-tions is, in fact, physically correct and can be justified on a more convincing basis. For example, in thermochemical terms:

H_2 (gas) + ½ O_2 (gas) → H_2O (liquid):
ΔH_R, 25°C = −122,970 Btu/lb mol
ΔH_R, 25°C = − 68.137 kcal/g mol

a reaction which is also the HHV for the com-bustion of hydrogen. Fig. 5 demonstrates the principles introduced, on the basis of the for-mation of a lb-mol of H_2O, where one mole of a substance whose molecular weight is M is de-fined as M unit masses of that material.[5]

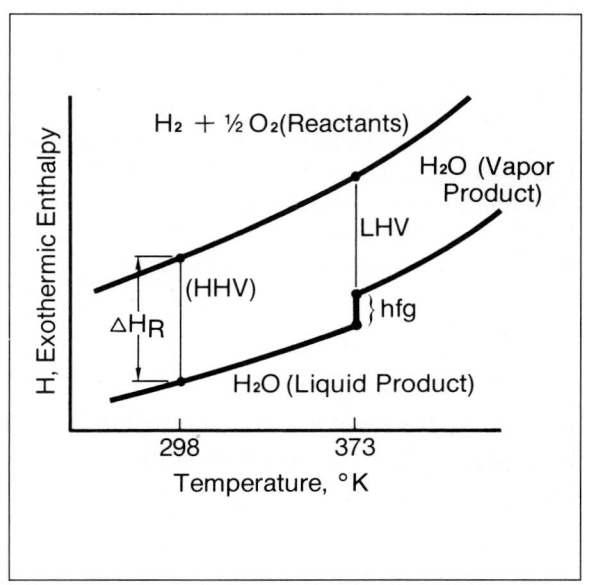

Fig. 5 Exothermic enthalpy of reaction for formation of H_2O at 1 atm, constant pressure

ENTHALPY OF FORMATION

Although the internal energy and enthalpy of reaction are basic quantities in the solution of thermochemical problems, the myriad chemical reactions effectively prohibit the overwhelming task of tabulating ΔU_R and ΔH_R values. Therefore, the concept of the enthalpy of formation was introduced. Applying the first law to a chemical reaction in a closed system at constant pressure, we write

$$\underset{i \text{ reactants}}{\Sigma(\nu_i H_i)} + \Delta H_R = \underset{i \text{ products}}{\Sigma(\nu_i H_i)}$$

where H_i is the molar enthalpy of any product or reactant at the pressure and temperature of the reaction, and ν_i again is the stoichiometric coefficient for each, based on the balanced chemical equation. It is clear that many sensible enthalpy tables for various gases account for changes in temperature at constant composition, whereas the thermochemical process being considered is defined in terms of constant temperature with variable composition. It follows that sensible enthalpy tables are not sufficient for evaluating energy transformations resulting from chemical reactions.

To circumvent this difficulty, the concept of an enthalpy of formation ΔH_f, is introduced. This is defined as the energy released or absorbed when a specific chemical compound, such as CO_2 or H_2O, is formed from its elements. The reactions, again, are such that initial reactant temperature and final product temperature are the same; written symbolically

$$Q = \Delta H_f = H_{compound} - \underset{i}{\Sigma(\nu_i H_i)}_{elements}$$

In this case the reactants are taken arbitrarily as the stable form of the elements at the given initial state. For example, the stable forms of gases such as hydrogen, nitrogen, and oxygen at 1 atm pressure and normal temperatures are $H_2(g)$, and $O_2(g)$. On the other hand, the stable form of carbon under these same conditions is $C(s)$. Because a value of zero has been assigned to the enthalpy of all stable elements at 1 atm and 25°C, the enthalphy H_i of any compound at the same state is simply the enthalpy of formation of that substance, or

$$\Delta H_f = H_{compound} - \text{zero} = H_{compound}$$

Because the chemical enthalpy of a stable element is always zero, by definition, the enthalpy of any compound is composed of two parts: that associated with its formation from elements and the sensible enthalpy associated with a change of state at constant composition. The sum of these two parts is called the absolute enthalpy of a substance.

The foregoing is illustrated by the reaction

$$C(s) + O_2(g) \rightarrow CO_2(g)$$

As solid carbon and gaseous diatomic oxygen are the stable forms of these elements at 25°C and 1 atm, their chemical enthalpies are chosen to be zero. The enthalpy of reaction is easily measured experimentally for this reaction, and it is found to be —169,290 Btu/lb mol, or —94.054 kcal/g mol. These values are also the enthalpy of formation of carbon dioxide, from which it follows that ΔH_R, ΔH_c, and ΔH_f are related but not synonymous. This is shown somewhat more clearly by Table III.

The usefulness of ΔH_f values is based on the assumption that the enthalpy of any compound is equal, arbitrarily, to its enthalpy of formation from stable elements. We may, therefore, substitute ΔH_f values for H_i values:

$$\Delta H_r = \underset{i}{\Sigma(\nu_i \Delta H_f)}_{products} - \underset{i}{\Sigma(\nu_i \Delta H_f)}_{reactants}$$

This equation is completely general for any reaction as long as the enthalpies of formation for all products and reactants are measured at the same temperature and pressure. This equation is widely used in thermochemistry. Its principal merit is, instead of requiring tabulations of thousands of values of ΔH_R for all conceivable reactions of interest, it is necessary only to tabulate ΔH_f values for compounds of interest. From the knowledge of ΔH_f values,

applying this equation leads directly to any desired enthalpy of reaction.

We have briefly presented several different definitions or labels; however, giving something a label does not necessarily constitute a satisfactory explanation. If, in addition, we literally interpret the words of Van Wylen[4] ''The justification of this procedure of arbitrarily assigning the value of zero to the enthalpy of the elements at 25°C, 1 atm rests on the fact that, in the absence of nuclear reactions, the mass of each element is conserved in a chemical reaction. No conflicts or ambiguities arise with this choice of reference state, and it proves to be very convenient in studying chemical reactions from a thermodynamic point of view'', then we are in a quandary regarding the origin of the energy we are tabulating.

Obviously, no one is going to abandon the law of conservation of mass as it is far too useful. On the other hand, we do not create energy out of thin air. The problem is resolved if we recognize that the law of conservation of mass is a near exact approximation which has been replaced by the law of conservation of mass-energy. There is nothing in Einstein's equation that prohibits or invalidates it outside the world of nuclear transition. For the HHV to be extracted from the reaction and do work, some-

thing else must disappear . . . that something is a minute amount of mass. We can quite readily calculate a conversion constant on a per Btu and on a per kcal basis. We write the well known Einstein relation as:

$$\delta m = \Delta E/C^2$$

where[5]

$$C = 2.99793 \times 10^5 \, km/sec$$
$$= 1.86282 \times 10^5 \, miles/sec$$

For a ΔE of 1 Btu, C of 9.8357×10^8 ft/sec, the gravitational constant of 32.174 ft-lb_m/lb_f sec^2 and the relationship for the mechanical-equivalent of heat, 1 Btu = 778.16 ft-lb_f[ref 2]

$$(1 \, Btu) \times \frac{778.16 \, ft\text{-}lb_f}{Btu} \times$$

$$\frac{32.174 \, ft\text{-}lb_m}{lb_f\text{-}sec^2 \, (96.741 \times 10^{16} \frac{ft^2}{sec^2})}$$

yields $2.588 \times 10^{-14} \, lb_m/Btu$

Using the more recent value of 777.649 for J (see ref 6), we get $2.586 \times 10^{-14} \, lb_m/Btu$; on a metric basis, we get 4.655×10^{-14} kg/kcal. It follows that we adopt $2.586 \times 10^{-14} \, lb_m/Btu$ and 4.655×10^{-14} kg/kcal as the mass equivalency

Table III. Enthalpy of Formation vs. Enthalpy of Combustion[1] at 25°C and at 1 atm Pressure

Substance	Chemical Formula	ΔH_f(25°C) kcal/g mol	ΔH_f(25°C) Btu/lb mol	ΔH_c(25°C) kcal/g mol	ΔH_c(25°C) Btu/lb mol
Carbon	C (solid)	0	0	−94.054	−169,290
Hydrogen	H₂ (gas)	0	0	−68.317	−122,970
Oxygen	O₂ (gas)	0	0	0	0
Carbon Monoxide	CO (gas)	−26.417	−47,540	−67.636	−121,750
Carbon Dioxide	CO₂ (gas)	−94.054	−169,290	0	0
Water	H₂O (vapor)	−57.798	−104,040	0	0
Water	H₂O (liquid)	−68.317	−122,970	0	0
Methane	CH₄ (gas)	−17.895	−32,210	−212.80	−383,040
Propane	C₃H₈ (gas)	−24.82	−44,680	−530.60	−955,070
Methyl Alcohol	CH₃OH (gas)	−48.08	−86,540	−182.61	−328,700

For ΔH_C the water in products of combustion is assumed to be liquid. The ΔH_C values are therefore also HHV values (except that HHV values are always given as positive).

constants. In the case of a power plant which averages a heat fired of, say, 3000×10^6 Btu/hr then, over a 25-year operating lifetime we find that

$$3 \times 10^9 \times 24 \times 365 \times 25 \times 2.586 \times 10^{-14} \cong 17 \text{ lb}$$

So about 17 lb of matter have been converted into energy, of which roughly 6.8 lb became electrical energy. This loss of mass is true only, of course, in the case where we transfer energy out of the system. In an adiabatic process where the system is not cooled down to the standard reference state, then no mass is lost because no energy is transferred.

THE GIBBS AND HELMHOLTZ FUNCTIONS

The enthalpy of formation concept we have just discussed implies that the reaction under consideration goes to completion. The disruptive energies available in a flame as, for example, temperature, are sufficiently high that this is not strictly true; thus, the reaction goes to completion in a somewhat cooler region adjoining the bulk flame volume. This is commonly referred to as residual combustion. The bulk flame composition in terms of the established product constituents (O_2, O, H_2O, OH, CO, CO_2, N_2, N, SO_2, SO, NH, NH_3, CN, HCN) attains some equilibrium composition consistent with the prevailing temperature.

By thermodynamic calculations it is possible to predict how far a reaction need go to reach equilibrium at a specified temperature by considering the thermodynamic potential of the reaction according to the Gibbs Free Energy Change ΔG.[7,8] Because this includes entropy in its determination, it is appropriate at this point to dispel some of the confusion that historically has attended this quantity. In physics a system is said to be in its most probable configuration when the number of accessible states is a maximum.

This is defined as follows[9]:

$$u = \text{system energy}$$
$$N = \text{number of atoms in the system}$$

The number of accessible states is called the degeneracy function g (not to be confused with the Gibbs function g) and is written as g(N, U). For any meaningful system this is a very large number and therefore the physicist defines for convenience a smaller number σ such that

$$\sigma(N,u) = \ln g(N,u)$$

This definition is called the *Entropy* or *Fundamental Entropy* and is related to the thermodynamic entropy S by the definition

$$S = K_B \sigma$$

where
$K_B = $ the Boltzmann constant $= 1.381 \times 10^{-16}$ erg/°K

Taking two systems, not in contact, we define system 1 to have accessible states g_1 and system 2 to have accessible states g_2. Then the combined accessible states are $g_1 g_2$. The physicist's entropy of the combined system is

$$\sigma = \ln(g_1 g_2) = \ln g_1 + \ln g_2 = \sigma_1 + \sigma_2$$

The total entropy is, thus, the sum of the entropies of the separate systems. From statistical mechanics we find that a quantity called the *Fundamental Temperature* is defined in terms of the fundamental entropy as

$$\frac{1}{\tau} = \left(\frac{\partial \sigma(N,u)}{\partial u} \right)$$

Performing the operation,

$$\frac{1}{\tau} = \frac{\partial \sigma(N,u)}{\partial u} = \frac{\partial \ln g(N,u)}{\partial u} = \frac{1}{g} \frac{\partial g}{\partial u}$$

$$= \frac{1}{g(N,u)} \lim_{\Delta u \to 0} \frac{g(N,u + \Delta u) - g(N,u)}{\Delta u}$$

As $N \to 0$, we conclude that $g(N,u) \to 0$ because for zero atoms we have zero accessible states, or $\frac{1}{\tau} \to \infty$ from which $\tau \to \infty$. It follows that

$$\frac{K_B}{\tau} = K_B \frac{\partial \sigma}{\partial u} = \frac{\partial K_B \sigma}{\partial u} = \frac{\partial S}{\partial u}$$

We now define a quantity T such that $\tau = K_B T$. Then

$$\frac{1}{T} = \frac{\partial S}{\partial u}$$

which we write in engineering form as

$$dS = \frac{du}{T}$$

In a more restricted sense, this is generally encountered as

$$S = \int \frac{dq}{T}$$

and is the entropy associated with the Gibbs equation. We can write the first law in the form (per unit mass)

$$dq = dU + Pdv \tag{15A}$$

$$dq = dh - vdP \tag{15B}$$

where h is the usual classical thermodynamic enthalpy function, defined as U + Pv, but is evaluated in terms of T and P. By taking h = h(T,P), we write

$$dh = \left(\frac{\partial h}{\partial T}\right)_P dT + \left(\frac{\partial h}{\partial P}\right)_T dP$$

which introduces the definition for the specific heat at constant pressure as:

$$C_P = \left(\frac{\partial h}{\partial T}\right)_P$$

We can write the second law in the Clausius form as:

$$dq = TdS \tag{16}$$

Combining (16) and (Eq. 15A), we get

$$dS = \frac{1}{T}dU + \frac{P}{T}dv \tag{17A}$$

which is called the Gibbs equation. Writing Eq. 17A as:

$$TdS = dU + Pdv \tag{17B}$$

and assuming a system at constant volume (dv = 0), we get the definition

$$T = \left(\frac{\partial u}{\partial S}\right)_V \tag{17C}$$

Writing Eq. 15B as

$$TdS = dh - vdP \tag{18}$$

and taking a system at constant pressure (dP = 0) we also get the definition

$$T = \left(\frac{\partial h}{\partial S}\right)_P \tag{19}$$

Writing Eq. 15A as

$$TdS = dU + Pdv$$

Then at constant u (du = 0) it follows that

$$\frac{P}{T} = \left(\frac{\partial S}{\partial V}\right)_u \tag{20}$$

We can then show the Gibbs equation from

$$S = S(u, v)$$

$$dS = \left(\frac{\partial S}{\partial u}\right)_v du + \left(\frac{\partial S}{\partial v}\right)_u dV \tag{21}$$

Substituting Eqs. 17C and 20 into 21, we get

$$dS = \frac{1}{T} du + \frac{P}{T} dV$$

Now differentiate TS

$$d(TS) = TdS + SdT$$

or

$$TdS = -SdT + d(TS)$$

Since (15B) can be written as

$$TdS = dh - vdP$$

we get

$$dh - vdP = d(TS) - SdT$$

then

$$dh - d(TS) = d(h - TS) = VdP - SdT = d(g)$$

The definition

$$g = h - TS \tag{22}$$

is known as the Gibbs function (Fig. 6) and is the *Free Enthalpy* but is also commonly called the Gibbs free energy. The equation

$$d(h - TS) = VdP - SdT \tag{23}$$

gives

$$\left(\frac{\partial(h - TS)}{\partial P}\right)_T = v \text{ at constant T} \tag{24}$$

and

$$\left(\frac{\partial(h - TS)}{\partial T}\right)_P = -S \text{ at constant P} \tag{25}$$

Taking $\frac{\partial}{\partial T}$ of Eq. 24 and $\frac{\partial}{\partial P}$ of Eq. 25, we get

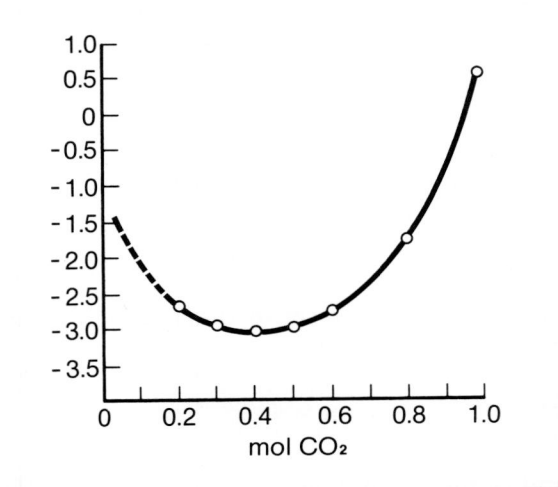

Fig. 6 The Gibbs function calculation shows that equilibrium is established at about 0.4 mol CO_2

$$\frac{\partial(h - TS)}{\partial T \partial P} = \left(\frac{\partial v}{\partial T}\right)_P , \quad \frac{\partial(h - TS)}{\partial P \partial T} = -\left(\frac{\partial S}{\partial P}\right)_T$$

from which

$$\left(\frac{\partial S}{\partial P}\right)_T = -\left(\frac{\partial v}{\partial T}\right)_P \tag{26}$$

Returning to Eq. 15B, we write the classic definition of C_P

$$C_P = \left(\frac{\partial h}{\partial T}\right)_P \text{ or } dq = C_P (dT)_P = dh$$

Then Eq. 15B becomes

$$C_P (dT)_P = Tds - \cancel{vdP}_0$$

which yields an alternate definition for C_P as

$$C_P = T \left(\frac{\partial S}{\partial T}\right)_P \tag{27}$$

For engineering applications the entropy S is given as a function of temperature and pressure. (See Properties of Steam, Appendix D.)

Taking $\quad S = S(T,P)$, then

$$dS = \left(\frac{\partial S}{\partial T}\right)_P dT + \left(\frac{\partial S}{\partial P}\right)_T dP$$

(28)

Substituting Eqs. 26 and 27, we get

$$dS = \frac{C_P}{T} dT - \left(\frac{\partial v}{\partial T}\right)_P dP$$

(29)

where Eq. 29 is a convenient form for calculating change in entropy at constant pressure.

If we combine Eq. 15A with the differential of TS, we get

$$d(TS) - SdT = du + Pdv$$

or

$$d(u - TS) = -Pdv - SdT = d(f)$$

(30)

In Eq. 30 the definition

$$f = u - TS$$

(31)

is called the Helmholtz function and is also known as the free energy property of the system. To distinguish this from Gibbs free energy, it is customary to call this the Helmholtz free energy.

From the classical definition of enthalpy as $h = u + Pv$, we write the Gibbs function as

$$g = h - TS = u + Pv - TS$$

(32)

Combining Eqs. 32 with 31, we find that the Gibbs and Helmholtz functions are related as

$$g = f + Pv$$

(33)

Eq. 33 possibly accounts for the fact that the Gibbs function is also referred to as the Gibbs-Helmholtz equation.

Differentiating the enthalpy, $h = u + Pv$, we get

$$dh = du + Pdv + vdP$$

(34)

and writing the Gibbs Eq. 17A as

$$du = TdS - Pdv$$

(35)

we get upon combining Eqs. 34 and 35

$$dh = TdS + vdP$$

(36)

Differentiating the Gibbs function Eq. 22

$$dg = dh - TdS - SdT$$

(37)

and combining Eqs. 36 and 37, we get a differential equation of state

$$dg = vdP - SdT$$

(38)

At constant pressure we write Eq. 38 as

$$dg = -SdT$$

(39)

In the JANAF[10] form, this is written as

$$\frac{d(\Delta G°)}{dT} = -\Delta S°$$

(40)

where the circular superscript ° indicates the thermodynamic standard of 1 atm and 25°C.

The Gibbs and Helmholtz functions have negative values since TS is greater than u or h.

On a T-S diagram for steam, this is clearly illustrated as shown (g = h − TS) (Fig. 7).

The equilibrium composition of a mixture reacting in an isothermal constant volume enclosure is the composition with the least absolute value of the Helmholtz function (f = u − TS) (Fig. 8).

A chemical reaction proceeding at constant temperature and pressure has the Gibbs function of the mixture decreasing to a minimum at the equilibrium composition. The Gibbs function (Fig. 9) allows us to determine the equilibrium composition of any reactive mixture of known pressure and temperature, regardless of whether these were constant during the reaction or not.

The Equilibrium Coefficient and the van't Hoff Equation

Eq. 38 is written as

$$dg = vdP$$

(41)

for a constant-temperature process. We are interested in this, as the Gibbs function criterion for equilibrium is based on constant temperature and total pressure.

Assuming ideal gases, we replace v in Eq. 41 by v = RT/P, giving

$$dg = RT \frac{dP}{P}$$

(42)

Integrating from standard state T° to T for the Gibbs differential and from standard state pressure of 1 atm to partial pressure P_i at Gibbs condition g_T at temperature T, and for any i^{th} component of a multicomponent gas, we get

$$g_T - g_{T°} = RT \ln \frac{P_i \, atm}{1 \, atm}$$

Taking R to be the universal gas constant, ft-lb/lb-mol-°R, and taking J as 778 ft-lb = 1 Btu (mechanical equivalent of heat), then

$$g_T - g_{T°} = \frac{RT}{J} \ln \frac{P_i}{1} \quad \text{Btu/lb-mol}$$

Writing this in terms of component i of a mixture

$$g_{i,T} = g_{i,T°} + \frac{RT}{J} \ln P_i$$

(43)

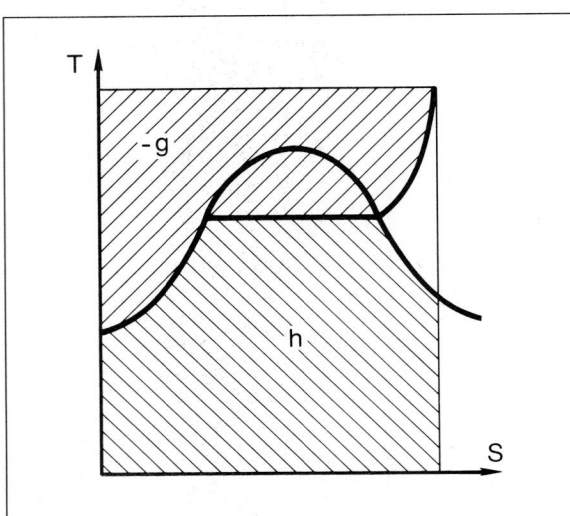

Fig. 7 Enthalpy h and free enthalpy g (Gibbs function).

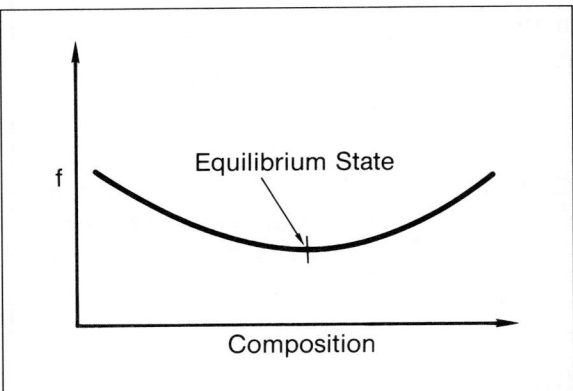

Fig. 8 Equilibrium at isothermal and constant volume conditions (Helmholtz function)

In Eq. 43, the value of g_i of the ith component at temperature T is found from the Gibbs function of the component at temperature T° (25°C) and its partial pressure P_i (atm.) in the mixture.

STOICHIOMETRIC COMBUSTION OF GAS

To introduce the next step, we consider the combustion of natural gas and assume it to be 100 percent CH_4.

$$CH_4 + 2O_2 \rightarrow CO_2 + 2H_2O \tag{44}$$

or in symbolic form

$$\nu_A A + \nu_B B \rightarrow \nu_E E + \nu_F F \tag{45}$$

where the ν_i are called the stoichiometric coefficients. Because we normally burn with air, we can assume air to be 21 percent by volume oxygen and take the balance as nitrogen.

$$21\% \; O_2 + 79\% \; N_2 \rightarrow 100\% \; air$$

or

$$1 \; mol \; O_2 + 3.76 \; mols \; N_2 = 4.76 \; mols \; air$$

We rewrite Eq. 44 as

$$CH_4 + 2O_2 + 7.52\,N_2 \rightarrow CO_2 + 2H_2O + 7.52\,N_2$$

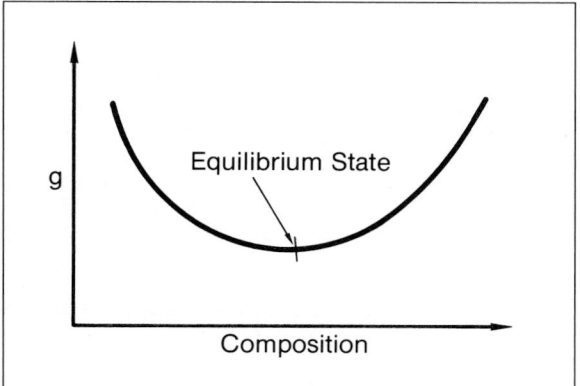

Fig. 9 **Equilibrium at isothermal and constant pressure conditions (Gibbs function)**

The values of ν_i are, of course, always known when the reaction is known. For an ideal gas the properties of the components are independent of one another (Gibbs-Dalton Law), wherefore the Gibbs function of the mixture is just the sum of the Gibbs functions of the i components. Then

$$G_{mix} = \sum N_i \, g_{i,T} = \sum N_i \left(g_{i,T°} + \frac{RT}{J} \ln P_i \right) \tag{46}$$

where N_i is the number of moles of each constituent present in the enclosure. If a differential amount of reactants A,B are combusted in time $d\gamma$ to form products E,F (see Eq. 45) and writing these mole amounts as dN_A, dN_B, dN_E, dN_F, then the change in the Gibbs function according to Eq. 46 is

$$dG_M = \left(g_{E,T°} + \frac{RT}{J} \ln P_E \right) dN_E + \left(g_{F,T°} \right.$$
$$\left. + \frac{RT}{J} \ln P_F \right) dN_F - \left(g_{A,T°} + \frac{RT}{J} \ln P_A \right)$$
$$dN_A - \left(g_{B,T°} + \frac{RT}{J} \ln P_B \right) dN_B \tag{47A}$$

$$= g_{E,T°} \, dN_E + g_{F,T°} \, dN_F - g_{A,T°} \, dN_A - g_{B,T°} \, dN_B$$
$$+ \frac{RT}{J} \ln \frac{P_E{}^{dN_E} P_F{}^{dN_F}}{P_A{}^{dN_A} P_B{}^{dN_B}} \tag{47B}$$

From Eq. 45, we realize that the stoichiometric coefficients specify the number of moles of reactants and products that are involved in the process. For differential amounts of reaction, we write

$$dN_A + dN_B \rightarrow dN_E + dN_F \tag{48}$$

which implies that instead of ν_A moles disappearing, we have $k\nu_A$ moles disappearing. Or

$$k\nu_A + k\nu_B \rightarrow k\nu_E + k\nu_F \tag{49}$$

Here k is referred to as a proportionality constant. Substituting Eq. 49 (that is, $k\nu_E = dN_E$ and so on) into Eq. 47A, we get

$$k\left[\nu_E\left(g_{E,T^\circ} + \frac{RT}{J}\ln P_E\right) + \nu_F\left(g_{F,T^\circ} + \frac{RT}{J}\ln P_F\right)\right.$$
$$- \nu_A\left(g_{A,T^\circ} + \frac{RT}{J}\ln P_A\right)$$
$$\left. - \nu_B\left(g_{B,T^\circ} + \frac{RT}{J}\ln P_B\right)\right] = dG_M \tag{50}$$

We are interested in the case of chemical equilibrium (Fig. 9) and assume an infinitesimal reaction around equilibrium to result in a net Gibbs function change of zero. For this event we write Eq. 50 as

$$O = \nu_E\,g_{E,T^\circ} + \nu_F\,g_{F,T^\circ} - \nu_A\,g_{A,T^\circ} - \nu_B\,g_{B,T^\circ}$$
$$+ \frac{RT}{J}\ln\frac{P_E^{\nu E}\,P_F^{\nu F}}{P_A^{\nu A}\,P_B^{\nu B}} \tag{51A}$$

The quantity $\nu_E\,g_{E,T^\circ} + \nu_F\,g_{F,T^\circ} - \nu_A\,g_{A,T^\circ} - \nu_B\,g_{B,T^\circ}$ is known as the standard-state Gibbs function change for a reaction and is usually written as ΔG_{T°. Or, in general,

$$\Delta G_{T^\circ} = \sum_{i=1}^{n} \nu_i\,g_{i,T^\circ}\Big/products$$
$$- \sum_{j=1}^{m} \nu_i\,g_{j,T^\circ}\Big/reactants \tag{51B}$$

Eq. 51A now is therefore written as

$$- \Delta G_{T^\circ} = \frac{RT}{J}\ln\frac{P_E^{\nu E}\,P_F^{\nu F}}{P_A^{\nu A}\,P_B^{\nu B}} \tag{52A}$$

EQUILIBRIUM CONSTANT

For ideal gas reactions, it is customary to define an *equilibrium constant*, K_P, such that

$$K_P = \frac{P_E^{\nu E}\,P_F^{\nu F}}{P_A^{\nu A}\,P_B^{\nu B}} \tag{52B}$$

Or, in general (i refers to products, j to reactants)

$$K_P = \frac{\prod_1^n P_i^{\nu_i}}{\prod_1^m P_j^{\nu_j}} \tag{52C}$$

Dropping the conversion constant J to obtain the usual textbook form, Eq. 52A then becomes

$$\Delta G_{T^\circ} = - RT\ln K_P \tag{53}$$

We now recast Eq. 22, taking into account the constant temperature assumption, in the form

$$\Delta G_{T^\circ} = \Delta H_{T^\circ} - T\Delta S_{T^\circ} \tag{54}$$

Substituting Eq. 54 into Eq. 53, we get

$$- R\ln K_P = \frac{\Delta H_{T^\circ}}{T} - \Delta S_{T^\circ} \tag{55}$$

Differentiating Eq. 55 with respect to T, we get

$$- R\frac{d(\ln K_P)}{dT} = - \frac{\Delta H_{T^\circ}}{T^2} + \frac{1}{T}\frac{d(\Delta H_{T^\circ})}{dT} - \frac{d(\Delta S_{T^\circ})}{dT} \tag{56}$$

Now Eq. 18 for the case of constant pressure is

$$TdS = dh \tag{57}$$

Writing Eq. 57 as $T\Delta S_{T°} = \Delta H_{T°}$ and differentiating with respect to T, keeping in mind that T itself is held constant, we get

$$\frac{Td(\Delta S_{T°})}{dT} = \frac{d(\Delta H_{T°})}{dT}$$

As these terms in Eq. 56 cancel, we get

$$\frac{d(\ln K_P)}{dT} = \frac{\Delta H_{T°}}{RT^2}$$

$$(58)$$

Since $d(1/T) = -dT/T^2$, Eq. 58 can also be written as

$$\frac{d(\ln K_P)}{d(1/T)} = -\frac{\Delta H_{T°}}{R}$$

$$(59)$$

as well as

$$\Delta H_{T°} = RT^2 \frac{d(\ln K_P)}{dT}$$

$$(60)$$

Eqs. 58, 59, and 60 are known as the van't Hoff equation. Generally the enthalpy of reaction is approximately independent of temperature. If such is the case, then Eq. 59 can be integrated as

$$\int_{K_{P_1}}^{K_{P_2}} d \ln (K_P) = \int_{1/T_1}^{1/T_2} -\frac{\Delta H_{T°}}{R} d(1/T)$$

or

$$\ln \frac{K_{P_2}}{K_{P_1}} = -\frac{\Delta H_{T°}}{R}\left(\frac{1}{T_2} - \frac{1}{T_1}\right)$$

$$(61)$$

For exothermic reactions $\Delta H_{T°}$ is negative by convention (heat or radiation is carried away from the system). By second law considerations one expects $T_2 > T_1$. Therefore the right-hand side of Eq. 61 is negative. This requires that $K_{P2} < K_{P1}$. Thus for exothermic reactions the equilibrium constant K_P decreases with increasing temperature. For endothermic reactions one expects the opposite to be the case. Eq. 61 also tells us, as well as the van't Hoff equation in general, that K_P is a function of temperature, but that this is physically meaningless unless the reaction is also specified. How the stoichiometric equation is specified is, unfortunately, also important. This is illustrated by the following identical reactions

(a) $CO + \frac{1}{2}O_2 \rightarrow CO_2$

(b) $2CO + O_2 \rightarrow 2CO_2$

The standard state Gibbs function change for the preceding reaction is

$$\Delta G_{T°} = \nu_{CO2}\, g_{CO2} - (\nu_{CO}\, g_{CO} + \nu_{O2}\, g_{O2})$$

or

$$\Delta G_{T°}/_{(a)} = 1g_{CO_2} - 1g_{CO} - \tfrac{1}{2}g_{O_2}$$

$$\Delta G_{T°}/_{(b)} = 2g_{CO_2} - 2g_{CO} - 1g_{O_2}$$

Thus $\Delta G_{T°}/_{(b)} = 2\Delta G_{T°}/_{(a)}$. Writing Eq. 53 as

$$K_P = e^{-\Delta G_{T°}/RT}$$

$$(62)$$

and taking the ratio of K_P's, we get

$$\frac{(K_P)_b}{(K_P)_a} = \frac{e^{-2\Delta G_{T°}/_{(a)}/RT}}{e^{-\Delta G_{T°}/_{(b)}/RT}} = \frac{e^{-2x}}{e^{-x}}$$

If we now square $(K_P)_a$, we get the ratio to be equal to unity:

$$\frac{(K_P)_b}{[(K_P)_a]^2} = \frac{e^{-2x}}{(e^{-x})^2} = 1$$

$$(63)$$

or

$$(K_P)_b = (K_P)_a{}^2$$

Thus by doubling the stoichiometric equation, we have squared the equilibrium coefficient. If we take the "backward-rate" or dissociation reaction for (a), we write

$$\text{(c)} \quad CO_2 \rightarrow CO + \tfrac{1}{2} O_2$$

and

$$\Delta G_{T°/(c)} = 1g_{CO} + \tfrac{1}{2}g_{O_2} - 1g_{CO_2} = -\Delta G_{T°/(a)}$$

Similarly, we take

$$\frac{(K_P)_c}{(K_P)_a^{-1}} = \frac{e^{-(-x)}}{(e^{-x})^{-1}} = 1 \tag{64}$$

or

$$(K_P)_c = 1/(K_P)_a$$

with the useful result that the backward-rate equilibrium coefficient is simply the reciprocal of the forward-rate equilibrium coefficient. By use of the perfect gas law, the equilibrium constant, K_P, can be put in somewhat more useful form than is given by Eqs. 52B and 52C. Realistically speaking, if one had a cylinder containing a mixture of gases at a total pressure of 1000 psig, it would be difficult to invent a pressure gage that would sequentially read the partial pressures and identify the corresponding components. An Orsat analysis identifying the components and the volumetric composition is a more practical method.

For component i in volume v, we write by the perfect gas law

$$P_i = \frac{M_i R_i T}{v} \frac{(\text{lb-ft/°R}) \, °R}{\text{ft}^3}$$

$$= \frac{M_i \dfrac{\mu_i R_i}{\mu_i} T}{v} \frac{\text{lb}}{\text{ft}^2} = \frac{M_i}{\mu_i} R_o T}{v} \frac{\text{lb}}{\text{ft}^2}$$

Since $P_{total} = \sum P_i$,

then $\dfrac{M_i \, \text{lb}}{\mu_i \, \text{lb/mol}} = N_i$

where:

N_i is the number of moles of each constituent present in volume v

Taking the ratio P_i/P_T, we get

$$\frac{P_i}{P_T} = \frac{M_i/\mu_i R_o T/v}{\sum (M_i/\mu_i R_o T)/v} = \frac{N_i}{\sum N_i} = \frac{N_i}{N_{mT}}$$

where

M_T = mols total (products)

Then

$$P_i = \frac{N_i}{N_{mT}} P_T = x_i P_T \tag{65}$$

Substituting Eq. 65 into Eq. 52C, we get

$$K_p = \frac{\prod\limits_1^n \left(\dfrac{N_i}{N_{mT}} P_T \right)^{\nu i}}{\prod\limits_1^m \left(\dfrac{N_j}{N_{mT}} \right)^{\nu j}}$$

$$= \frac{\prod\limits^n N_i^{\nu i} (P_T/N_{mT})^{\nu i}}{\prod\limits^m N_j^{\nu j} (P_T/N_{mT})^{\nu j}}$$

$$= \frac{\prod\limits^n N_i^{\nu i}}{\prod\limits^m N_j^{\nu j}} \left(\frac{P_T}{N_{mT}} \right)^{\Sigma \nu i - \Sigma \nu j}$$

$$= \frac{\prod\limits^n N_i^{\nu i}}{\prod\limits^m N_j^{\nu j}} \left(\frac{P_T}{N_{mT}} \right)^{\Delta \nu} \tag{66}$$

Using reaction 45 as an example, we would write Eq. 66 as

$$K_{P/(31)} = \frac{N_E^{\nu E} N_F^{\nu F}}{N_A^{\nu A} N_B^{\nu B}} \left(\frac{P_T}{N_{mT}} \right)^{\nu E + \nu F - \nu A - \nu B} \tag{67}$$

APPLICATION TO COMBUSTION CALCULATIONS

The following calculations demonstrate the application of the foregoing discussion.

BURNING PROPANE

Burn propane (C_3H_8) stoichiometrically (0% excess air). The reaction on a molar basis is, assuming dry air,

$$C_3H_8 + 5O_2 + 18.8N_2 \rightarrow 3CO_2 + 4H_2O + 18.8N_2$$
(68)

The air-to-fuel ratio therefore is

$$\frac{5 \text{ mol } O_2 \times \dfrac{32 \text{ lb}}{\text{mol}} + 18.8 \text{ mol } N_2 \times \dfrac{28 \text{ lb}}{\text{mol}}}{1 \text{ mol } C_3H_8 \times \dfrac{44 \text{ lb}}{\text{mol}}} = 15.7$$

or, equivalently

$$\frac{23.8 \text{ mol air} \times 29 \text{ lb air/mol air}}{1 \text{ mol } C_3H_8 \times 44 \text{ lb } C_3H_8/\text{mol } C_3H_8}$$
$$= 15.7 \text{ lb air/lb } C_3H_8$$

Now suppose that we burn the propane sub-stoichiometrically at 80 percent theoretical dry air. To solve the problem of insufficient air, we make the assumption that H_2 will always go to H_2O in favor of C going to CO_2. (On the basis of reaction kinetics this is a valid assumption.) Reaction 68 now becomes

$$1C_3H_8 + 4O_2 + 15.04N_2 \rightarrow$$
$$aCO + bCO_2 + 4H_2O + 15.04N_2$$
(69A)

This reaction solves as

$$C_3H_8 + 4O_2 + 15.04N_2 \rightarrow$$
$$2CO + 1CO_2 + 4H_2O + 15.04N_2$$
(69B)

Reaction 69B is an approximation as we have implicitly assumed that the reaction goes to completion. Continuing with the combustion of propane, we may assume, say, 150 percent theoretical dry air and consequently write reaction 68 as

$$C_3H_8 + 7.5O_2 + 28.2N_2 \rightarrow$$
$$3CO_2 + 4H_2O + 2.5O_2 + 28.2N_2$$
(70)

Products $= 3 + 4 + 2.5 + 28.2 = 37.7$ mols. The product mole fractions are

$$CO_2 = \frac{3}{37.7} \cong 0.08 \quad H_2O = \frac{4}{37.7} \cong 0.106$$

$$O_2 = \frac{2.5}{37.7} \cong 0.066 \quad N_2 = \frac{28.2}{37.7} \cong 0.748$$

Then $P_{H_2O} = 0.106 \times 14.7 = 1.56$ psia with a corresponding dew point (saturation temperature) of 117°F. Converting this to a wet-air basis of 80°F, 1 atm, 1.013 lb wet air/lb dry air we calculate

$$\frac{0.013 \text{ lb } H_2O}{\text{lb dry air}} \times \frac{29 \text{ lb dry air}}{\text{mol air}} \times \frac{1 \text{ mol } H_2O}{18 \text{ lb } H_2O}$$
$$= 0.021 \frac{\text{mol } H_2O}{\text{mol air}}$$

From the left-hand side of (70), we have 35.7 mols air used in the reaction or 35.7 mols air $\times$ 0.021 mol H_2O/mol air $= 0.75$ mol H_2O introduced with the air. Wherefore the products [right-hand side of (70)] now contain $4 + 0.75 = 4.75$ mol H_2O. Thus,

$$P_{H_2O} = \frac{4 + .75}{37.7 + .75} = 0.1235 \text{ atm.} = 1.816 \text{ psia}$$

Now suppose that we have a volumetric (Orsat) analysis of the products ($CO_2 = 11.5$ percent, $O_2 = 2.7$ percent, and $CO = 0.7$ percent) and we know we are burning propane. We then write reaction (67) as

$$xC_3H_8 + aO_2 + a3.76N_2 \rightarrow 11.5CO_2 +$$
$$0.7CO + 2.7O_2 + 85.1N_2 + bH_2O$$
(71A)

The carbon balance for reaction 71A is

$$3X = 11.5 + 0.7 \quad X = 4.0667 \quad \cong 4.07$$

The hydrogen balance for reaction 71A is

$$8X = 8 \times 4.0667 = 2b \quad b \cong 16.25$$

The oxygen balance for reaction 71A is

$$2a = 2 \times 11.5 + 0.7 + 5.4 + 16.25$$
$$a \cong 22.65$$

Substituting for x, a, b in reaction 71A and dividing through by 4.07, then we get

$$C_3H_8 + 5.56O_2 + \frac{85.1}{4.07}N_2 \rightarrow \frac{11.5}{4.07}CO_2$$

$$+ \frac{0.7}{4.07}CO + \frac{2.7}{4.07}O_2 + \frac{85.1}{4.07}N_2 + \frac{16.25}{4.07}H_2O$$

(71B)

Comparing reaction 71B with the stoichiometric reaction (71A), it follows that the excess air determined by Orsat analysis is

$$\frac{5.56}{5} \times 100 \cong 111 \text{ percent theoretical air}$$

or 11 percent excess air.

BURNING HYDROCARBON FUEL

Next we suppose that we have an Orsat analysis ($CO_2 - 12.1\%$, $O_2 - 3.8\%$, $CO - 0.9\%$) and we suspect we are burning some hydrocarbon fuel. We then write

$$C_xH_y + aO_2 + a3.76N_2 \rightarrow 12.1CO_2 + 0.9CO +$$

$$83.2N_2 + 3.8O_2 + bH_2O$$

(72A)

"a" from the nitrogen component is

$$a\,3.76N_2 = 83.2N_2, a = 83.2/3.76 = 22.1$$

Then, from the oxygen balance

$$2 \times 22.1 = 2 \times 12.1 + 0.9 + 2 \times 3.8 + b$$

$$b = 11.5$$

From the carbon balance, we write

$$x = 12.1 + 0.9 = 13.0$$

For the hydrogen balance, we write

$$y = 2b = 2 \times 11.5 = 23.0$$

wherefore reaction (72A) becomes

$$C_{13}H_{23} + 22.1O_2 + 83.2N_2 \rightarrow 12.1CO_2 + 0.9CO +$$

$$83.2N_2 + 3.8O_2 + 11.5H_2O$$

(72B)

Here the formula $C_{13}H_{23}$ is an average composition of the fuel components.

From reaction 72B the combustion air is $22.1 + 83.2 = 105.3$ mols per mol fuel. To convert reaction (72B) to a stoichiometric basis we realize that, ideally, 0.9 CO should be 0.9 CO_2 and that 3.8 O_2 goes to zero; then the 0.9 oxygen supplies 0.45 O_2 to the 0.9 CO_2 so an additional 0.45 O_2 must be supplied by the left-hand side.

Therefore, the O_2 requirement on the left-hand side of reaction 72B is

$$22.1\,O_2 - 3.8\,O_2 + 0.45\,O_2 = 18.75\,O_2$$

Hence, reaction 72B becomes

$$C_{13}H_{23} + 18.75\,O_2 + 3.76 \times 18.75N_2$$

$$\rightarrow 13\,CO_2 + 11.5\,H_2O + 3.76 \times 18.75\,N_2$$

(73)

from which it follows that the excess air percent is

$$\frac{22.1 - 18.75}{18.75} \times 100 \cong 17.9\%$$

GIBBS FUNCTION CALCULATION

Consider the water-gas reaction

$$CO + H_2O \rightarrow CO_2 + H_2$$

and apply the Gibbs function to the total system. First, we assume the following reactions

$$1\,CO + 1\,H_2O \rightarrow 1\,CO_2 + 1\,H_2$$

$$1\,CO + 1\,H_2O \rightarrow 0.8\,CO_2 + 0.8\,H_2 +$$
$$0.2\,CO + 0.2\,H_2O$$

$$1\,CO + 1\,H_2O \rightarrow 0.6\,CO_2 + 0.6\,H_2 +$$
$$0.4\,CO + 0.4\,H_2O$$

$$1\,CO + 1\,H_2O \rightarrow 0.5\,CO_2 + 0.5\,H_2 +$$
$$0.5\,CO + 0.5\,H_2O$$

The Gibbs function for the total system is Σ products $- \Sigma$ reactants

$$= \sum_i (h_i - T\Delta S_i)_{products} - \sum_j (h_j - T\Delta S_j)_{reactants}$$

To apply the above to the postulated reactions, we use Eq. 29. From $Pv = R_o T/\mu = RT$, we write

$$v = \frac{R}{P} T$$

then

$$\left(\frac{\partial v}{\partial T}\right)_P = \frac{R}{P}$$

Substituting in Eq. 29

$$dS = \frac{C_p}{T} dT - \frac{R}{P} dP \tag{74}$$

Then, for component i,

$$(S_i)_{T,P} = S_i^o + \int_{To}^{T} C_P \frac{dT}{T} - R \ln \frac{P_i}{P_o} \tag{75}$$

In the JANAF tables, the entropy at temperature T for component i is tabulated at 1 atmosphere and we therefore make a "correction" for component i at partial pressure P_i starting at pressure $P_o = 1$ atm. Hence,

$$(S_i)_{T,Pi} = S_{i,1}^o + \int_{To}^{T} C_P \frac{dT}{T} - R \ln \frac{P_i}{1}$$

$$= S_{i,T,1} = R \ln P_i \tag{76}$$

From the JANAF tables, we tabulate the following values at 1000°K and 1 atm, Table IV. We assume the reaction to proceed at 1000°K for convenience, to simplify the calculations.

HEAT GAS-PHASE REACTION

Consider the gas-phase reaction

$$1CO + \frac{1}{2}O_2 \rightleftharpoons 1CO_2 \tag{77}$$

and assume the reaction is in equilibrium at 1 atm and 3000°K. Assume negligible tempera-

Table IV. Gibbs Function Calculation—Step 1

Component	$H^o_{1000} - H^o_{298}$ kcal/mol I	$H^o_{298} - H^o_0$ kcal/mol II	$H^o_{1000} - H^o_0$ Σ(I + II) Sensible Enthalpy	ΔH^o_F (1000°K) Enthalpy of Formation kcal/mol	S^o_{1000} Entropy at 1 atm. cal/mol-°K
CO_2	7.984	2.238	10.222	−94.321	63.444
H_2	4.944	2.024	6.968	0.000	39.702
CO	5.183	2.072	7.255	−26.771	56.028
H_2O	6.209	2.367	8.576	−59.246	55.592

Table V. Gibbs Function Calculation—Step 2

$H_{products}$ at 1 mol each

10.222	CO_2
6.968	H_2
17.190	kcal/mol

$H_{reactants}$ at 1 mol each

7.255	CO
8.576	H_2O
15.831	kcal/mol

$H_{sensible} = H_{at\ 1\ mol} \times$ fractional mol

Table V. Gibbs Function Calculation—Step 2 — *(Continued)*

Mol	0.8	0.6	0.5	0.4	0.3	0.2
CO_2	8.17	6.133	5.111	4.089	3.067	2.044
H_2	5.574	4.181	3.484	2.787	2.090	1.394
Σ	13.752	10.314	8.595	6.876	5.157	3.438
Mol	0.2	0.4	0.5	0.6	0.7	0.8
CO	1.451	2.902	3.627	4.353	5.078	5.804
H_2O	1.715	3.430	4.288	5.145	6.003	6.861
Σ	3.166	6.332	7.915	9.498	11.081	12.665
$\Sigma_{Products}$	16.918	16.646	16.510	16.374	16.238	16.103

$$H_{products} - H_{reactants} \text{ vs. mols } CO_2$$

Mol CO_2	1.0	1.08	0.6	0.5	0.4	0.3	0.2
Products	17.190	16.918	16.646	16.510	16.374	16.238	16.103
Reactants	−15.831	−15.831	−15.831	−15.831	−15.831	−15.831	−15.831
Σ	1.359	1.087	0.815	0.679	0.543	0.407	0.272

Table VI. Gibbs Function Calculation—Step 3

Calculating entropies according to Eq. 76, we get (at 1000°K)

S_{CO_2}	=	$1.0(63.444 - 1.986 \ln 0.5)$	=	64.821	
S_{H_2}	=	$1.0(39.702 - 1.986 \ln 0.5)$	=	41.079	
Total		$S_{products}$ at 1 atm (& 1 mol CO_2)	=	105.900	cal/mol-°K

S_{CO}	=	$1.0(56.028 - 1.986 \ln 0.5)$	=	57.405	
S_{H_2O}	=	$1.0(55.592 - 1.986 \ln 0.5)$	=	56.969	
Total		$S_{reactants}$ at 1 atm	=	114.374	cal/mol-°K

S_{CO_2}	=	$0.8(63.444 - 1.986 \ln 0.4)$	=	52.211	
S_{H_2}	=	$0.8(39.702 - 1.986 \ln 0.4)$	=	33.218	
S_{CO}	=	$0.2(56.028 - 1.986 \ln 0.1)$	=	12.120	
S_{H_2O}	=	$0.2(55.592 - 1.986 \ln 0.1)$	=	12.033	
Total		$S_{products}$ at 0.8 mol CO_2	=	109.582	cal/mol-°K

S_{CO_2}	=	$0.6(63.444 - 1.986 \ln 0.3)$	=	39.501	
S_{H_2}	=	$0.6(39.702 - 1.986 \ln 0.3)$	=	25.256	
S_{CO}	=	$0.4(56.028 - 1.986 \ln 0.2)$	=	23.690	
S_{H_2O}	=	$0.4(55.592 - 1.986 \ln 0.2)$	=	23.515	
Total		$S_{products}$ at 0.6 mol CO_2	=	111.962	cal/mol-°K

ABB
ASEA BROWN BOVERI

Table VI. Gibbs Function Calculation—Step 3 — *(Continued)*

Calculating entropies according to Eq. 76, we get (at 1000°K)

S_{CO_2}	=	0.5(63.444−1.986 ln 0.25)	=	33.099
S_{H_2}	=	0.5(39.702−1.986 ln 0.25)	=	21.228
S_{CO}	=	0.5(56.028−1.986 ln 0.25)	=	29.390
S_{H_2O}	=	0.5(55.592−1.986 ln 0.25)	=	29.173
Total		$S_{products}$ at 0.5 mol CO_2	=	112.890 cal/mol-°K
S_{CO_2}	=	0.4(63.444−1.986 ln 0.2)	=	26.656
S_{H_2}	=	0.4(39.702−1.986 ln 0.2)	=	17.159
S_{CO}	=	0.6(56.028−1.986 ln 0.3)	=	35.051
S_{H_2O}	=	0.6(55.592−1.986 ln 0.3)	=	34.790
Total		$S_{products}$ at 0.4 mol CO_2	=	113.656 cal/mol-°K
S_{CO_2}	=	0.3(63.444−1.986 ln 0.15)	=	20.163
S_{H_2}	=	0.3(39.702−1.986 ln 0.15)	=	13.041
S_{CO}	=	0.7(56.028−1.986 ln 0.35)	=	40.679
S_{H_2}	=	0.7(55.592−1.986 ln 0.35)	=	40.374
Total		$S_{products}$ at 0.3 mol CO_2	=	114.257 cal/mol-°K
S_{CO_2}	=	0.2(63.444−1.986 ln 0.1)	=	13.603
S_{H_2}	=	0.2(39.702−1.986 ln 0.1)	=	8.855
S_{CO}	=	0.8(56.028−1.986 ln 0.4)	=	46.278
S_{H_2O}	=	0.8(55.592−1.986 ln 0.4)	=	45.930
Total		$S_{products}$ at 0.2 mol CO_2	=	114.666 cal/mol-°K

$$S_{products} \times 1000°K = S_{products} \text{ in kcal/mol}$$

$$(-S_{products} + S_{reactants})T = -T\,\Delta S$$

$$\Delta = \Sigma \text{ products} - \Sigma \text{ reactants}$$

Mol CO_2	1.0	0.8	0.6	0.5	0.4	0.3	0.2
$S°_{products}$	−105.900	−109.582	−111.962	−112.890	−113.656	−114.257	−114.666
$S°_{reactants}$	114.374	114.374	114.374	114.374	114.374	114.374	114.374
Σ	+ 8.474	+ 4.792	+ 2.412	+ 1.484	+ 0.718	+ 0.117	− 0.292

Table VII. Gibbs Function Calculation—Step 4

For the heat of formation (exothermic), we get, at 1000°K & 1 atm total pressure

Mol CO_2	1.0	0.8	0.6	0.5	0.4	0.3	0.2
H_{FCO_2}	−94.321	−75.457	−56.593	−47.160	−37.728	−28.296	−18.864
H_{FH_2}	0	0	0	0	0	0	0
Σ	−94.321	−75.457	−56.593	−47.160	−37.728	−28.296	−18.864

ABB
ASEA BROWN BOVERI

Table VII. Gibbs Function Calculation—Step 4 — *(Continued)*

For the heat of formation (exothermic), we get, at 1000°K & 1 atm. total pressure

Mol_{H_2O}	0	0.2	0.4	0.5	0.6	0.7	0.8
H_{FCO}	0	-5.354	-10.708	-13.385	-16.062	-18.740	-21.417
H_{FCO_2}	0	-11.849	-23.698	-29.623	-35.548	-41.472	-47.397
Σ	0	-17.203	-34.406	-43.008	-51.610	-60.212	-68.814
$\Sigma H_F{}^{prod}$	-94.321	-92.660	-90.999	-90.168	-89.338	-88.508	-87.678
$\Sigma H_F{}^{reac}$	$+85.017$	$+85.017$	$+85.017$	$+85.017$	$+85.017$	$+85.017$	$+85.017$
Σ	-9.304	-7.643	-5.982	-5.151	-4.321	-3.491	-2.661

Finally

$$(\underset{products}{\Sigma\ H_S + H_F - TS}) - (\underset{reactants}{\Sigma\ H_S + H_F - TS}) \text{ is tabulated in Table VIII}$$

as

$$\Sigma\left[H_{S_{products}} - H_{S_{reactants}}\right] + \Sigma\left[H_{F_{products}} - H_{F_{reactants}}\right] - \Sigma\left[TS_{products} - TS_{reactants}\right]$$

where Δ means Σ products $-\ \Sigma$ reactants

Table VIII. Gibbs Function Calculation—Final Step

Mol CO_2	1.0	0.8	0.6	0.5	0.4	0.3	0.2
$\Delta S°$	$+8.474$	$+4.792$	$+2.412$	$+1.484$	$+0.718$	$+0.117$	-0.292
$\Delta H_S°$	$+1.359$	$+1.087$	$+0.815$	$+0.679$	$+0.543$	$+0.407$	$+0.272$
$\Delta H_F°$	-9.304	-7.643	-5.982	-5.151	-4.321	-3.491	-2.661
Gibbs	$+0.529$	-1.764	-2.755	-2.987	-3.060	-2.967	-2.681

ture dissociation for O_2. For the reaction

$$CO_2 \rightleftharpoons 1CO + \tfrac{1}{2}O_2$$

at 1 atm and 3000°K, Ref. 2 gives the equilibrium coefficient as

$$\text{Log}_{10} K_P = -0.485$$

The equilibrium coefficient we need, according to Eq. 64, is therefore

$$(K_P)_a = {}^1\!/(K_P)_b$$

$$(K_P)_a = \frac{1}{10^{-0.485}} = 10^{0.485} = 3.055$$

From Eq. 66, we have for (a)

$$K_P = \frac{\prod\limits_{m}^{n} N_i{}^{\nu_i}}{\prod N_j{}^{\nu_i}} \left(\frac{P_T}{N_{mt}}\right)^{\Sigma \nu_i - \Sigma \nu_j}$$

or

$$K_P = 3.055 = \frac{(N_{CO_2})^1}{(N_{CO})^1 (N_{O^2})^{\frac{1}{2}}}\left(\frac{1\ \text{atm.}}{N_{mT}}\right)^{1-(1+\frac{1}{2})}$$

(78)

As we anticipated that the reaction does not go to completion, we rewrite reaction (77) with CO and O_2 also on the right-hand side. We note in Eq. 66 that the stoichiometric coefficients ap-

pear as exponents only; there are therefore no restrictions on the moles. We can therefore consider the case of an equimolal reaction (for convenience) and write

$$1CO + 1O_2 \xrightarrow[\text{EQUIL.}]{} x\,CO + y\,O_2 + z\,CO_2$$

$$(79)$$

In this reaction the product moles are

$$N_{mT} = x + y + z$$

Substituting the mixture [the right-hand side of reaction 79] into reaction 78 we get

$$\frac{z^1}{x^1 y^{\frac{1}{2}}} \left(\frac{1}{x+y+z}\right)^{-\frac{1}{2}} = \frac{z\,(x+y+z)^{\frac{1}{2}}}{x y^{\frac{1}{2}}}$$

$$(80)$$

In addition, the carbon balance demands that $1 = x + z$ and the oxygen balance requires that $3 = x + 2y + 2z$, then

$$z = 1 - x \text{ and } y = \tfrac{1}{2}(3 - x - 2z)$$
$$= \tfrac{1}{2}(3 - x - 2 + 2x) = \frac{1 + x}{2}$$

Thus

$$N_{mT} = x + y + z = x + \frac{1+x}{2} + 1 - x = \frac{3+x}{2}$$

Substituting into Eq. 80, we get

$$3.055 = \frac{(1-x)\left(\dfrac{3+x}{2}\right)^{\frac{1}{2}}}{x\left(\dfrac{1+x}{2}\right)^{\frac{1}{2}}} = \frac{(1-x)(3+x)^{\frac{1}{2}}}{x(1+x)^{\frac{1}{2}}}$$

Squaring both sides and simplifying

$$8.3x^3 + 8.3x^2 + 5x - 3 = 0$$

By inspection $x = \tfrac{1}{3}$ is an approximate solution. By additional calculation we find that x is close to 0.34. Then

$$z = 1 - x = 1 - .34 = .66$$

and

$$y = \frac{1+x}{2} = \frac{1.34}{2} = .67$$

and it follows that the equilibrium reaction is

$$1\,CO + 1\,O_2 \rightarrow 0.34\,CO + 0.67\,O_2 + 0.66\,CO_2$$

which is the equilibrium composition we would expect for the specified reaction.

CALCULATION WITH AIR

Now let us do a more realistic calculation using air instead of oxygen. We write Eq. 79 as

$$CO + O_2 + 3.76\,N_2 \rightarrow x\,CO + y\,O_2 + z\,CO_2 + 3.76\,N_2$$

The moles of mixture at equilibrium are

$$N_{mT} = x + y + z + 3.76$$

We also note that N_2, carried along as an inert, has no effect in Eq. 66, that is,

$$K_P = \frac{\prod N_i{}^{\nu_i} N_{N_2}{}^{3.76}}{\prod N_j{}^{\nu_j} N_{N_2}{}^{3.76}} \left(\frac{P_T}{N_{mT}}\right)^{\nu_i + 3.76 - \nu_j - 3.76}$$

and, as before,

$$3.055 = \frac{z^1}{x^1 y^{1/2}} \left(\frac{1 \text{ atm.}}{x + y + z + 3.76}\right)^{-\frac{1}{2}}$$

The carbon and oxygen balances remain the same

$$1 = x + z, \quad 3 = x + 2y + 2z$$

and

$$z = 1 - x, \quad y = \frac{1+x}{2}$$

then

$$3.055 = \frac{(1-x)(1)^{-\frac{1}{2}}}{x\left(\dfrac{1+x}{2}\right)^{\frac{1}{2}}(x+y+z+3.76)^{-\frac{1}{2}}}$$

Substituting for y and z

$$3.055 = \frac{(1-x)\left(\dfrac{10.52+x}{2}\right)^{\frac{1}{2}}}{x\left(\dfrac{1+x}{2}\right)^{\frac{1}{2}}}$$

squaring both sides and simplifying

$$8.3x^3 + .78x^2 + 20.04x - 10.52 = 0$$

By inspection, it is clear that $\tfrac{1}{3} < x < \tfrac{1}{2}$; by successive approximations we determine that

$$x = 0.46, \; z = 1 - x = 0.54, \; y = \frac{1+x}{2} = 0.73$$

$$1CO + 1O_2 + 3.76N_2 \rightarrow 0.46CO + 0.73O_2 + 0.54CO_2 + 3.76N_2$$

Comparing this with the previous solution

with the inert nitrogen constituent dropped

$$1\ CO + 1\ O_2 \rightarrow 0.34\ CO + 0.67\ O_2 + 0.66\ CO_2$$

We see that the presence of an inert (nitrogen) decreases the amount of CO_2 formed. If we now take a less drastic temperature, say 1000°K, and perform the equilibrium calculation, we find that for the reaction

$$CO_2 \rightarrow CO + \tfrac{1}{2}\ O_2$$

Reference (2) gives the equilibrium constant as $\text{Log}_{10} K_P = -10.221$. Then

$$(K_P)_a = \frac{1}{(K_P)_b} = \frac{1}{10^{-10.221}} = 10^{10.221}$$

Taking K_P as 10^{10} (for convenience) we get

$$(10^{10})^2 = 10^{20} = \frac{(1-x)^2\,(3+x)}{x^2 + x^3}$$

Hence

$$10^{20}\,x^3 + 10^{20}\,x^2 = 3 - 5x + x^2 + x^3$$

which can be rewritten closely as

$$10^{20}\,x^3 + 10^{20}\,x^2 + 5x - 3 = 0$$

We can conclude that the x value is effectively zero. Thus,

$$1\ CO + 1\ O_2 \rightarrow \tfrac{1}{2}\ O_2 + 1\ CO_2$$

At 1000°K, therefore, the reaction effectively goes to completion which is far from being the case at 3000°K.

We used a K_P for the reaction $1\ CO + \tfrac{1}{2}\ O_2 \rightarrow 1\ CO_2$ of 3.055 at 3000°K. To estimate the K_P for the same reaction at 2000°K, we use the van't Hoff isobar Eq. 61.

$$\ln \frac{K_{P_2}}{K_{P_1}} = -\frac{\Delta H_{T_o}}{R}\left(\frac{1}{T_2} - \frac{1}{T_1}\right)$$

The universal gas constant, R, has units of

$$1.986\ ^{cal}\!/_{g-mol \times k} \quad \text{or} \quad 1.986\ ^{Btu}\!/_{lb-mol-\,°R}$$

The enthalpy of reaction, $\Delta H_T°$, is given by reference (2) as $-119,500$ Btu/mol at 2000°K and $-117,200$ Btu/mol at 3000°K. Taking a numerical average of $-118,350$ Btu/mol, we get

$$\ln \frac{K_{P_2}}{K_{P_1}} = \frac{118350}{1.986}\left(\frac{1}{3600} - \frac{1}{5400}\right) \cong 5.54, \text{ then}$$

$$e^{\ln \frac{K_{P_2}}{K_{P_1}}} = e^{5.54} \cong 255 = \frac{K_{P_2}}{3.005}, \text{ or}$$

$$K_{P_2} = 778$$

which compares reasonably well with the value of 766 given by Wark[2]. The van't Hoff equation therefore gives a fair value in the absence of experimental data.

METHODS OF BURNING SOLID FUELS

Of particular interest in the practical combustion of fuels are the physical processes employed. Chapter 12 treats the firing of liquid and gaseous fuels. Here we briefly discuss four principal mechanisms currently used for burning coal and other solid fossil fuels. These mechanisms differ in

- the place where fuel is introduced into the furnace
- the way in which the fuel is injected
- the size of the fuel required by the process
- the temperature of any transport or carrier air, also called "primary air"
- the relative upward velocity of the combustion air and the fuel particles in the furnace, while burning is taking place
- the amount of recirculation (if any) of solid particles of ash, or fuel plus ash, from a point beyond the furnace back into the furnace; this is turn affects
- the residence time of the fuel and carbon-bearing ash particles in the furnace; and, very importantly,
- the average temperature of the fuel as combustion proceeds.

The following overview will summarize these differences as they relate to firing solid fuel either in suspension firing of pulverized fuel (called open-furnace firing or entrained flow); on a slowly moving or fixed grate (generally referred to as stoker firing); in a bubbling type of fluidized bed; or in a circulating fluidized bed.

Stoker firing, historically the oldest way to burn both coal and waste solid fuels, is still

very much in use. Pulverized-fuel suspension firing, a very efficient and practical way to burn very large quantities of coal and lignite, often has the liability of requiring expensive flue-gas cleanup systems for sulfur-oxide reduction. Fluidized-bed firing is ideal for handling low-grade fuels where sulfur removal is accomplished in-bed instead of in a post-combustion scrubbing system.

STOKER FIRING

Chapter 12 of this text describes in detail the equipment for burning coal as well as other types of fuel on a grate. In the majority of coal-fired applications, the grate or fuel bed is continually in slow motion, usually from the rear to the front of the furnace bottom. (Grate motion is from front-to-rear for mass burning of coal or waste fuels.) Metering feeders either deposit the fuel directly on the grate, or spread it mechanically or pneumatically from points usually 10 to 20 feet above the grate. With the latter method, significant burning occurs while the fuel is in suspension, with the remainder taking place on the grate; the proportions will vary with fuel size, fuel reactivity, and the mode of injection.

In stoker firing, heated air passes upward through apertures in the grate, with dampers often positioned in undergrate zones in order to achieve proper biasing of airflow. Over-grate air (termed overfire air) adds turbulence to the combustible-containing gases coming from the grate and supplies the required air for the portion of fuel that burns in suspension. As described in Chapter 12, a relatively small amount of recirculation of solid material increases the residence time of the larger-size particles. The principal mode of heat transfer in the furnace is radiation from both the hot fuel bed and the burning gas and char in the open combustion chamber.

The average temperature of a stoker fuel bed, which cannot be calculated, is a function of:
- the grate speed (for a moving grate)
- the temperature of the air flowing upward through the grate
- the excess-air percentage

- the volatile-to-fixed-carbon ratio of the fuel
- the as-fired fuel moisture content
- the size of the fuel.

SUSPENSION FIRING

Most large steam generators producing power through the burning of solid fuels are of the entrained-flow reactor type, most frequently called suspension fired; much of the material in Chapters 6, 7, and 8 concerns this combustion process. Chapters 11 and 12 cover the equipment for pulverizing and firing the fuel for such boilers; the flue-gas emission control and ash-handling systems of Chapters 15 and 16, as well as the discussion on the formation of nitrogen oxides in the next section of this chapter, are oriented in large toward this type of firing.

In pulverized firing, coal is ground to the fineness of face powder, which a stream of primary air transports into the furnace; it ignites as the fuel-air mixture enters the furnace, where it is joined by the bulk of combustion air (called the secondary air) which is heated to temperatures between 500 to 800°F (260 to 425°C). Residence time of the burning fuel in the furnace is measured in seconds; solid-particle recirculation is not used; and radiation of heat to the furnace-wall tubing is the principal mode of transfer.

The following variables affect the average temperature of both gas and solid particles in a suspension-fired or entrained-flow furnace:
- the type of firing (corner/tangential/vortex or multiple wall burners)
- excess-air percentage
- fuel reactivity and moisture content
- air distribution in the furnace
- firing density (either heat released in the active firing volume, or per square foot or square meter of furnace plan area)
- furnace geometry
- preheated-air temperature
- the dirtiness of the furnace wall, partially a function of the soot-blowing cycle.

Although the maximum attainable temperature (the adiabatic flame temperature) in such a furnace can be calculated, it will not be achieved. The immediate long-beam radiative cooling that takes place in a water-cooled chamber during the combustion process makes it impossible to attain the adiabatic temperature. Furnace exit-gas temperature *can* be measured. Thus, some approximation to an arithmetic or logarithmic *average* temperature of the gas and ash particles passing through the furnace can be calculated.

In order to vary furnace exit temperature with load, change in fuel, or furnace-wall dirtiness, staged combustion, biased firing, or fuel-nozzle tilt (in tangential firing) are used in large open combustion chambers; such arrangements allow control of superheater or reheater outlet steam temperature, or furnace nitrogen-oxide production, or both.

FLUIDIZED-BED FIRING

Chapter 9 covers in detail the technology and design of equipment for the two distinct mechanisms used for fluidized-bed firing in power plants: bubbling beds and circulating (sometimes called recirculating) beds. In a bubbling-bed combustor, fuel size and vertical air velocity are regulated to establish a discrete horizontal plane which divides the active bed from the entrained-flow "open furnace" above. In a circulating type fluid-bed unit, smaller fuel size and higher air velocities extend the zone of high solids-to-gas density up to the top of the furnace, and beyond, into the process separator, normally a cyclone.

With the bubbling fluid bed (BFB), the conduction/convection heat transfer is to furnace wall tubes and other heating surface that may be immersed in the bed; radiation transfer occurs above the active bed. With the circulating fluid bed (CFB), the bulk of conduction/convection heat transfer is to the combustor wall tubes. Further heat absorption can be by tubular heating surface outside the combustor, depending upon the specific application.

The basic mechanism for the control of bed temperature and heat transfer to the walls of the combustor, and to any immersed heating surface in the bed of a fluidized-bed boiler, is the variation in total solids inventory, the distribu-

tion of combustion air, and the solids recycle rate. Many of the variable described above for stoker and suspension firing also have an influence. But, in contrast to those methods of firing, the temperature in a fluidized-bed combustor is controllable in a narrow range, as further discussed in Chapter 9. Such control helps to minimize nitrogen-oxide formation and allows the optimum temperature level for calcining limestone for the capture of sulfur oxides.

THE FORMATION AND CONTROL OF NO$_x$ IN STEAM-GENERATING EQUIPMENT

Nitrogen monoxide (NO) and nitrogen dioxide (NO$_2$) are byproducts of the combustion process of virtually all fossil fuels. Historically, the quantity of these inorganic compounds in the products of combustion was not sufficient to affect boiler performance, and their presence was largely ignored. In recent years, oxides of nitrogen have been shown to be key constituents in the complex photochemical oxidant reaction with sunlight to form smog. Today, the emission of NO$_2$ and NO (collectively referred to as NO$_x$) is regulated by both state and federal authorities and has become an important consideration in the design of fuel-firing equipment.

THERMAL NO$_x$

The formation of NO$_x$ in the combustion process is often explained in terms of the source of nitrogen required for the reaction. The N$_2$ can originate from the atmospheric air, in which case the product is referred to as "thermal NO$_x$" or from the organically bound nitrogen components found in all coals and fuel oils which are termed "fuel NO$_x$." It is important to note that even though NO$_x$ consists usually of 95 percent NO and only 5 percent NO$_2$, the normal practice is to calculate concentrations of NO$_x$ as 100 percent NO$_2$.

The mechanisms involving thermal NO_x were first described by Zeldovich[11] and later modified to what is referred to as the extended Zeldovich mechanism.

$$N_2 + O \rightleftharpoons NO + N \tag{81}$$

$$N + O_2 \rightleftharpoons NO + O \tag{82}$$

$$N + OH \rightleftharpoons NO + H \tag{83}$$

As the equilibrium values predicted by this mechanism are higher than those actually measured, it is generally assumed that Eq. 81 is rate-determining due to its high activation energy of 317 kJ/mol. A better understanding of thermal NO_x can be derived from bench-scale tests that measure NO_x in a heated mixture of N_2, O_2, and argon [12]. These tests show that thermal NO_x can be predicted by the following equation:

$$[NO] = K_1 e^{(-K_2/T)}[N_2][O_2]^{1/2}t$$
$$\text{where } [\ ] = \text{mol fraction}$$
$$T = \text{temperature}$$
$$t = \text{time}$$
$$K_1, K_2 = \text{constants} \tag{84}$$

From this equation, it can be seen that thermal NO can be decreased by reducing time, temperature, and concentrations of N_2 and O_2. The fact that temperature in this equation is an exponential clearly demonstrates its importance in the control of thermal NO_x. In practice, the Zeldovich mechanisms demonstrated by Eq. 84 are sufficient for predicting NO_x only in regions that are downstream of the flame front.

Fenimore[13] coined the phase "prompt NO" to describe the NO generated within the flame front region for which, because of the very short residence time, the Zeldovich reactions proved inadequate. Subsequent investigation by DSoeto[14] showed that, as in Zeldovich's reactions, temperature in all cases reduced prompt NO; however, additional O_2 increased NO for fuel-rich flame fronts but decreased NO for fuel-lean flame fronts.

The effects of fuel-rich vs. fuel-lean flame fronts on prompt NO can be seen from the formation of NO from diffusion vs. premixed gas flame. As shown in Fig. 10, the stoichiometry at the premixed flame front is the same as the overall stoichiometry of the flame, since all the air is mixed with all the fuel. In the diffusion flame, fuel and air are introduced separately and mixed by burner-induced turbulence. The flame front forms before mixing is complete and is, therefore, fuel-rich even though the total fuel/air mixture may be air-rich. Tests by Takahashi[15] et al., Fig. 11, show the bell-

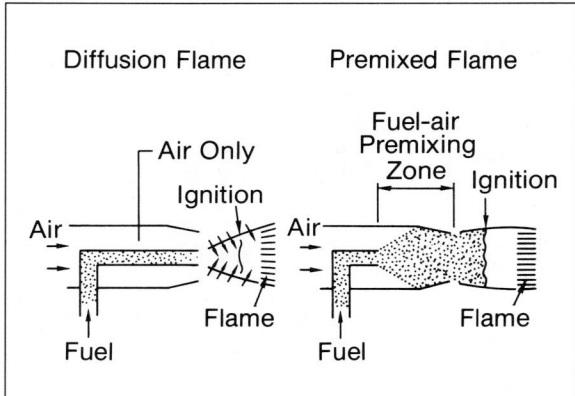

Fig. 10 Diffusion-flame and premixed-flame mechanisms

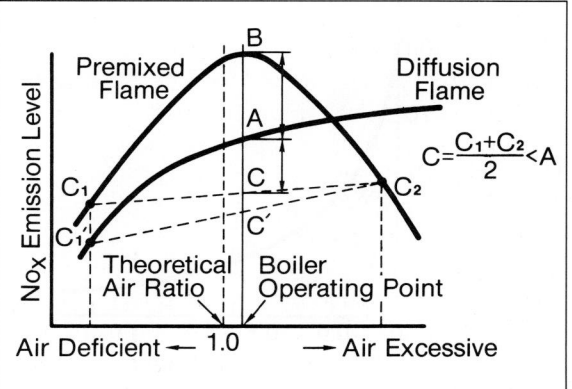

Fig. 11 NO_x versus stoichiometry for premixed and diffusion flames

shaped curve characteristic of prompt NO for premixed flames, but the diffusion flame does not show decreasing NO with increasing O_2 because the actual flame front is never air-rich.

It is interesting to note that Takahashi has shown that two separate flames, one fuel-lean premixed flame, plus one fuel-rich diffusion flame (points C_1 and C_2, Fig. 11) can be combined to produce a lower NO(at C') than either type flame with the same overall stoichiometry (points A or B).

FUEL NO$_x$

Although the kinetics involved in the conversion of organically bound nitrogen compounds found in fossil fuels are not yet well understood, numerous investigators have shown fuel NO_x to be an important mechanism in NO_x formation from fuel oil, and the dominant mechanism in NO_x generated from the

combustion of coal. Bench-scale tests, Fig. 12, burning fuel oils in a mixture of oxygen and carbon dioxide (to exclude thermal NO_x) have shown a remarkable correlation between the percent N_2 in the fuel oil versus NO_x.[16] Fig. 12 also illustrates the fact that the percent of fuel-nitrogen conversion is not constant, but decreases with increasing fuel nitrogen.

Similar bench-scale tests run for various coals have not produced similar results. Fig. 13 illustrates the large contribution of fuel NO_x, yet there is no apparent correlation between the quantity of fuel-bound nitrogen and fuel NO_x. Clearly, the fuel-nitrogen conversion rate is not constant, but will vary widely depending more on coal rank than on actual nitrogen content.

One explanation of the fuel-nitrogen conversion rate is shown in Fig. 14 which was obtained by subtracting a calculated Zeldovich thermal-NO_x value from total NO_x measured in large tangentially fired utility units. The resultant fuel-NO_x correlated well with the fuel nitrogen/oxygen ratio which suggests that fuel-bound oxygen, or some other fuel property that correlates with fuel oxygen, influences the percent conversion of fuel nitrogen to fuel NO_x.

Although there is little doubt that fuel-bound nitrogen is an important contributor to total NO_x emissions from coal and oil, the mechanisms involved in the transformation of fuel nitrogen to NO appear to be every bit as complex as the combustion process itself. On the

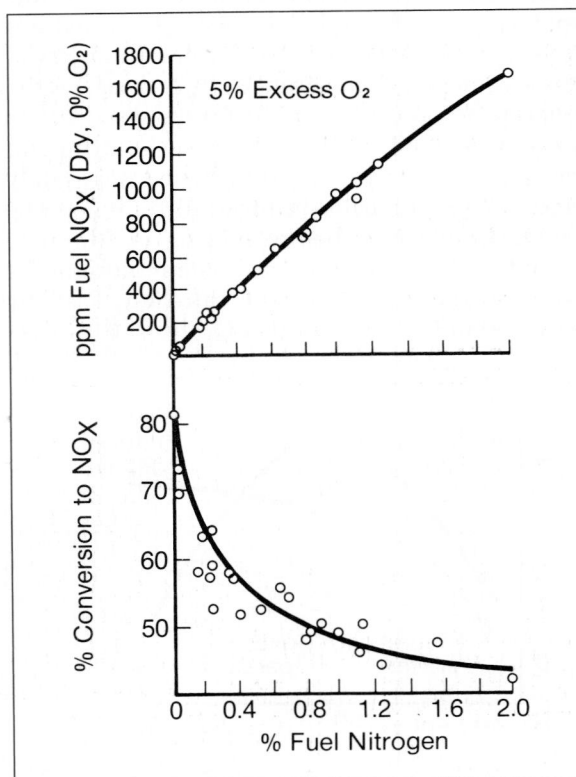

Fig. 12 Fuel NO and percent conversion of fuel nitrogen to Fuel NO—liquid fuels

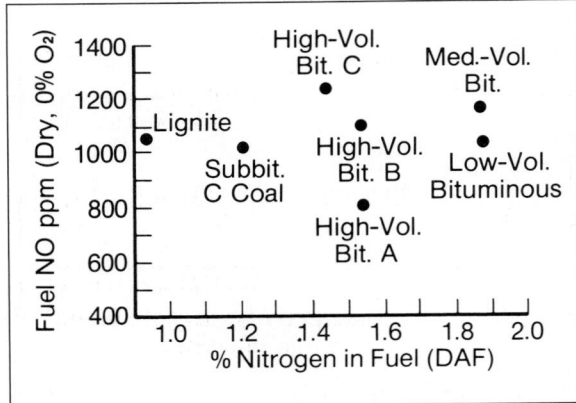

Fig. 13 Fuel NO and percent conversion of fuel nitrogen to fuel NO—pulverized coal, premixed

other hand, several investigators have made quantitative measurements in laboratory tests that provide valuable insight to the potential control of fuel NO_x.[17]

Pershing[18] isolated fuel NO_x by burning coal in a synthetic oxidant mixture that had a specific heat similar to air, containing 21 percent O_2, 18 percent CO_2, and 61 percent argon. On the basis of four different coals, the studies showed that fuel NO_x, unlike thermal NO_x, was relatively insensitive to the range of temperatures found in most flames. Levy[19] et al. found that, under pyrolysis, 65 percent of the fuel nitrogen remained in the char after all the volatile matter had evolved at the 1380°F used in the ASTM proximate-analysis test, and 90 percent had evolved by 2400°F. These tests, then, suggest that the fuel-bound nitrogen begins to evolve in the later stages of devolatization of coal and that subsequent high temperatures do not significantly affect the conversion to NO.

A most significant property of fuel nitrogen conversion that affects the design of fuel-firing equipment relates to the availability of oxygen to react with the fuel-nitrogen compounds in their gaseous state. Simply stated, the compounds that evolve from a coal particle such as NCH and NH_3, are relatively unstable and will reduce to harmless N_2 under fuel-rich conditions, or to NO under air-rich conditions. The technique used to demonstrate this is staging.

In staging, a portion of the total air required to complete combustion is withheld initially and the balance of the air is mixed with the incomplete products of combustion only after the oxygen content of the first-stage air is consumed. By varying the amount of first-stage air, the suspension-fired combustion of a coal particle or oil droplet can be interrupted at different stages of the reaction because of lack of oxygen, and allowed to proceed further at such time as the balance of air (second-stage air) is introduced. For NO_x control, the ideal quantity of first-stage air would be that which is sufficient to only generate the temperatures necessary to drive the gaseous state, but insufficient to provide sufficient oxidant to complete a reaction to NO.

Fig. 15 shows the effect of first-stage stoichiometric ratio on outlet NO_x levels for liquid fuels with various nitrogen levels. Even though

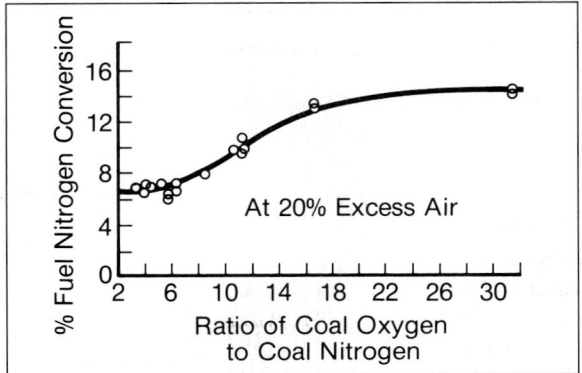

Fig. 14 Fuel-bound nitrogen conversion to NO_x in pulverized-coal combustion; the influence of the coal-oxygen to coal-nitrogen ratio

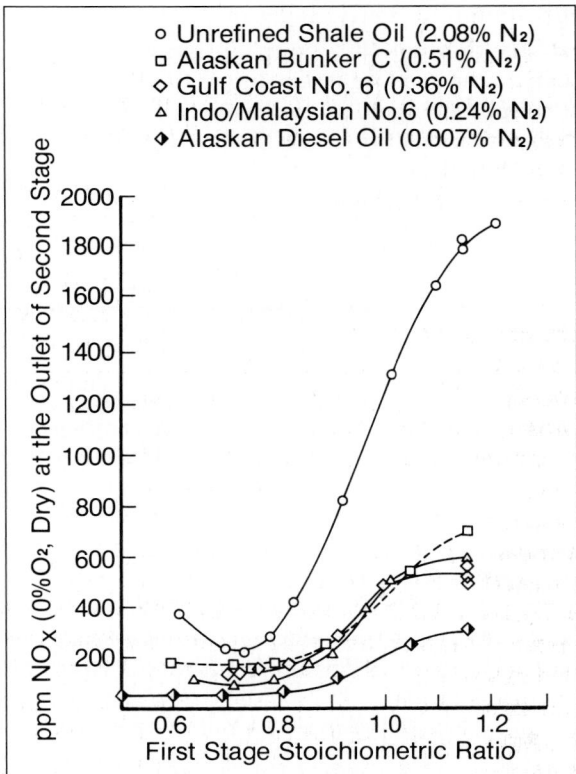

Fig. 15 The effect of liquid fuel composition on emissions—staged suspension firing

the tests used air for an oxidant and the total NO_x levels represent both thermal and fuel NO_x, the increased effectiveness of staging with higher fuel-nitrogen content is clear. Fig. 16 shows a similar but less dramatic effect on a Utah bituminous coal. This and other tests have shown that for coal, unlike oil, decreasing the first-stage stoichiometric air below an optimum level will increase NO_x levels, which suggests the possible requirement of an oxidant to reduce the intermediate HCN, NH_3, and other radicals.

The most important design criteria relevant to the control of both thermal and fuel NO_x for coal firing can be summarized as follows:

■ Coals with the lowest fuel-nitrogen and the lowest fuel oxygen/nitrogen ratios generally will produce the lowest NO_x.

■ The fuel NO_x can be minimized by controlling the quantity of air permitted to mix with the fuel in the early stages of combustion.

■ The thermal-NO_x contribution to total NO_x can be reduced by operating at the lowest practical percentages, as well as by minimizing gas temperatures throughout the furnace through the use of low-turbulence diffusion flames and large water-cooled furnaces.

Chapter 12 describes the equipment used in large suspension-fired steam generators to minimize the formation of NO_x in the fuel-burning process in a furnace. In Chapter 15, post-combustion (after-furnace) methods are given for chemically or catalytically reducing the quantity of nitrogen oxides passing to the atmosphere. An optimum system for overall NO_x reduction will result from the integration of both in-furnace and post-combustion techniques.

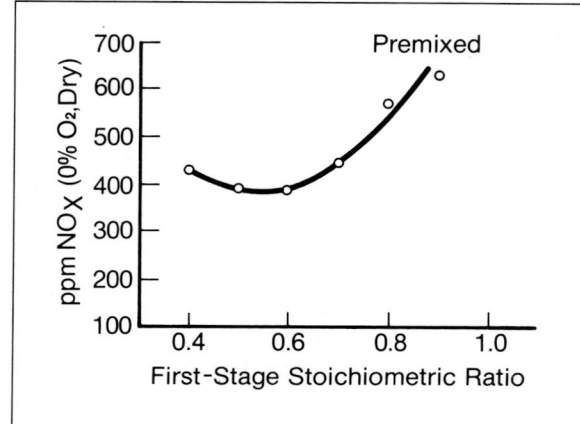

Fig. 16 The effect of initial fuel/air mixing—staged, Utah bituminous coal

---

REFERENCES

[1] "Standard Atmosphere Tables and Data for altitudes to 65,800 feet," National Advisory Committee for Aeronautics Report 1235 International Civil Aviation Organization (ICAO), Montreal, Canada and Langley Aeronautical Laboratory, Langley Field, VA. 1952.

[2] Kenneth Wark, *Thermodynamics*, 2nd ed., McGraw-Hill, 1971.

[3] William C. Reynolds and Henry C. Perkins, *Engineering Thermodynamics*, 1st ed., McGraw-Hill, 1970.

[4] Gordon J. Van Wylen and Richard E. Sonntag, *Fundamentals of Classical Thermodynamics*, 2nd ed., John Wiley & Sons, Inc., 1973.

[5] E. U. Condon and Hugh Odishaw, *Handbook of Physics*, 2nd ed., New York: McGraw-Hill, 1967.

[6] R. E. Bolz and G. L. Tuve, eds., *Handbook (of Tables for) Applied Engineering Science*, Publ. by the Chemical Rubber Co., Cleveland, Ohio, Copyright 1970.

[7] It may be noted that the term *Gibbs Free Energy* is in common use. Strictly and historically speaking it is incorrect. The Gibbs function of the system is termed the "Free Enthalpy" property. The Helmholtz function of the system is more properly called the "Free Energy" property. See, for example, E. Schmidt[8].

[8] E. Schmidt, *Thermodynamics*, translated from 3rd German edition, Oxford University Press, London, England, 1st English ed., 1947.

[9] C. Kittel, *Thermal Physics*, John Wiley & Sons., New York, 1969.

[10] JANAF (Joint Army, Navy, Air Force Project), *Thermochemical Tables*, 2nd ed., 1970 issued 1971, NSRDS-NBS 37, U.S. Government Printing Office, Washington, D.C.

[11] Ya. B. Zel'dovich, "The Oxidation of Nitrogen in Combustion and Explosions," *Acta Physicochimica U.S.S.R.*, 21: 577–628, 1946.

[12] D. J. MacKinnon, "Nitric Oxide Formation at High Temperature," *Air Pollution Control Association Journal*, 24(3): 237–239, March 1974.

[13] C. P. Fenimore, "Formation of Nitric Oxide From Fuel Nitrogen in Ethylene Flames," *Combustion and Flame*, 19(2): 289–296, October 1972.

[14] G. De Soete, "Mechanism of Nitric Oxide Formation From Ammonia and Amines in Hydrocarbon Flames," *Revue de L'Institut Francais du Petrole et Annuales Des Combustible Liquides*, 28(1): 95–108, 1973.

[15] Yasuro Takahashi, et al., "Development of Super-Low NO_x PM Burner," *Mitsubishi Technical Bulletin No. 134.* Mitsubishi Heavy Industries, Ltd., 1979.

[16] W. W. Habelt, "The Influence of Coal Oxygen to Coal Nitrogen Ratio on NO_x Formation," Presented at the 70th Annual AIChE Meeting, New York, November 13–17, 1977.

[17] W. W. Habelt and B. M. Howell, "Control of NO_x Formation in Tangentially Coal-Fired Steam Generators" *Proceedings of the NO_x Control Technology Seminar*, San Francisco, February 5–6, 1976. Special Report. Report no. PB-253 661 (EPRI SR-39). Springfield, VA: National Technical Information Service, 1976.

[18] D. W. Pershing and J. O. L. Wendt, "The Effect of Coal Composition on Thermal and Fuel NO_x Production from Pulverized Coal Combustion," *Central States Section, The Combustion Institute Spring Meeting*, Columbus, Ohio, April 5–6, 1976, Pittsburgh: The Combustion Institute, 1976.

[19] J. M. Levy, et al., *Combustion Research on the Fate of Fuel-Nitrogen Under Conditions Of Pulverized Coal Combustion*. Final Task Report. Report no. PB-286 208 (EPA-600/7-78/165). Springfield, VA: National Technical Information Service, 1978.

CHAPTER 5

Steam Generation

revious chapters have covered steam power-plant cycles, and general engineering and economic decisions related to the purchase of a steam-generating unit. Also discussed were fuels and their evaluation for use in a boiler furnace. The characteristics of coal ash were covered in great detail to emphasize the significant influence of mineral matter in solid fuels on the sizing and satisfactory operation of a large boiler. This earlier material, which also "burned" the fuel and determined the products of combustion, as well as presenting some of the principles of combustion in a furnace, established the theory behind the hardware.

This chapter will cover *steam generation* by considering the rôles of the different components that make up a steam generator.

Included are the *properties of steam* used to perform basic heat-balance and sizing calculations. Also examined are the criteria involved in the *configuration of boiler pressure parts and auxiliaries* as a function of system pressures and temperatures. Generically introduced are the various parts of a reheat-cycle steam generator and an explanation of their functioning as heat-absorbing apparatus. (More specific information in the following four chapters will supplement and amplify these descriptions.) Also presented is the vital function of *water-steam circulation* in a modern

high-pressure furnace, along with the basic principles of thermally induced and pumped circulation. The important aspects of the *separation of steam* from the steam-water mixture entering the upper drum of a subcritical-pressure boiler and the general types of mechanical drum internals are covered.

STEAM AS A WORKING FLUID

The theoretical amount of work that can be obtained from steam used in a prime mover is equivalent to the change in its total heat content from its condition at the entering state to that at the exhaust state.

The efficiency of the prime mover in converting the heat energy to mechanical effort governs the actual work obtained.

For economic studies involved in the selection and design of all steam- and power-generation equipment, it is necessary to understand thoroughly the properties of steam, the use of steam tables and the use of superheat. A brief review of these fundamentals that apply to the generation of steam will be helpful.

PROPERTIES OF STEAM

Steam results from adding sufficient heat to water to cause it to vaporize. This vaporization occurs in two steps: first, by adding heat to the water to raise it to the boiling temperature; second, by continuing the addition of heat to change the state from water to steam.

When heated at average atmospheric pressure at sea level, each pound of water increases in temperature about 1°F for each Btu added until 212°F is reached. Additional heat does not cause the temperature to rise but, if continued, results in boiling, and the water changes its state from a liquid to a vapor.

When water is heated to the boiling point in a closed vessel, the vapor released causes the pressure in the vessel to rise. With the rise in pressure, the temperature at which the water boils also will rise. It has been determined experimentally that during the change in state of any substance from a liquid to a vapor at constant pressure, the vapor in contact with its liquid will remain at constant temperature until the vaporization has been completed. Thus, the temperature at which boiling occurs for any given pressure is constant and is called the *saturation temperature*. This temperature is the same for the water as it is for the vapor with which it is in contact.

DEFINITIONS

The heat of the liquid is the heat used to raise a unit weight of water, normally one pound, from 32°F to the saturation temperature corresponding to a given pressure. Also called the enthalpy of the saturated liquid, this is stated in Btu per lb or kilojoules per kilogram.

The latent heat of vaporization is the heat added to a unit weight of water at saturation temperature to vaporize it completely and produce dry saturated steam. This is the enthalpy of evaporation or vaporization.

Dry saturated steam contains no moisture, and is at saturated temperature for the given pressure. Its total heat content, or enthalpy of the saturated vapor, is equal to the heat of the liquid plus the heat of vaporization.

Steam which contains water in any form, either as minute droplets, mist, or fog, is called wet steam. Wet steam may result from the entrainment of water in boiling or from partial condensation. In either case, the total heat content of the mixture is less than that of dry saturated steam, because vaporization is incomplete. The percentage of dry vapor by weight in the mixture is known as the quality of the steam. Thus, with 3-percent moisture in steam, the quality is 97 percent. The total heat of wet steam is equal to the heat of the liquid plus the percentage of latent heat of vaporization represented by the steam quality. The temperature of wet steam is the same as dry steam at the corresponding pressure.

Steam with a temperature higher than that of saturated steam at the same pressure is called superheated. Accomplished by maintaining the pressure and adding heat to the steam after its removal from contact with water, superheating results in an increase in temperature and volume. The total heat of superheated steam, or its enthalpy, is equal to the total heat of dry saturated steam plus the heat of superheating. The term total heat is the general engineering expression applicable to any steam condition, whether wet, dry saturated, or superheated. Also known as the enthalpy of steam, it is the amount of heat that must be added to a unit weight of water at 32°F to produce the end state under consideration.

The properties of steam have been the subject of considerable research in many of the countries of the world for many years, as stated in Appendix D. The Mollier diagram, constructed as an enthalpy-entropy chart to show the steam tables graphically, is particularly useful in the analysis of power-plant cycles. It visualizes the process of expansion of steam through the various sections of a steam turbine and helps in the quick but approximate solution of many other thermodynamic problems. Appendix D tells how to obtain a C-E Mollier chart based on steam properties calculated by C-E.

SUPERHEATED AND SATURATED STEAM

The properties of superheated steam approximate those of a perfect gas. One important characteristic is the dependence of internal energy on temperature; thus, the closer steam approaches a perfect gas, the better it does its work. In addition, it contains no moisture, nor can it condense until its temperature has been

lowered to that of saturated steam at the same pressure. This particular characteristic is of considerable value because, with the correct amount of superheat, it is possible to eliminate condensation in steam lines and to decrease the moisture in the steam turbine exhaust.

With saturated steam, the heat available depends entirely upon pressure, while with superheated steam there is additional heat, proportionate to the degree of superheat. This additional heat is obtained through increased expenditure of fuel, but the economic benefits derived result in a net efficiency gain of considerable magnitude (see Chapter 1). By using a comparatively small amount of superheat, it is possible to reduce moisture at exhaust conditions and to effect an increase in the percentage of heat utilized.

LIMITS OF STEAM TEMPERATURE AND PRESSURE

The materials of superheater construction govern the practical limits of steam tempera-ture and pressure. There has been considerable development in the metallurgy of steel alloys and in the manufacturing of both the tubing and the finished sections of the superheater. These have made possible the design of super-heaters and reheaters for high-temperature and high-pressure boiler installations, as have improvements in the art of welding. While those installations which border on the limitations of available materials may be more interesting, and perhaps more sensational, the large majority is in the 750° to 1050°F temperature range. Although commonplace, they represent the places where the greatest savings are made.

A study of steam temperatures accompanying installations using 2400 psig throttle pressures discloses that in the 1950's, 1050°F was overwhelmingly selected as the primary temperature. The trend since that time has been in the direction of 1000°F primary temperature, and is a well established—but surely not exclusive—pattern. Again, experience has proved to be the arbiter. First, the initial cost of

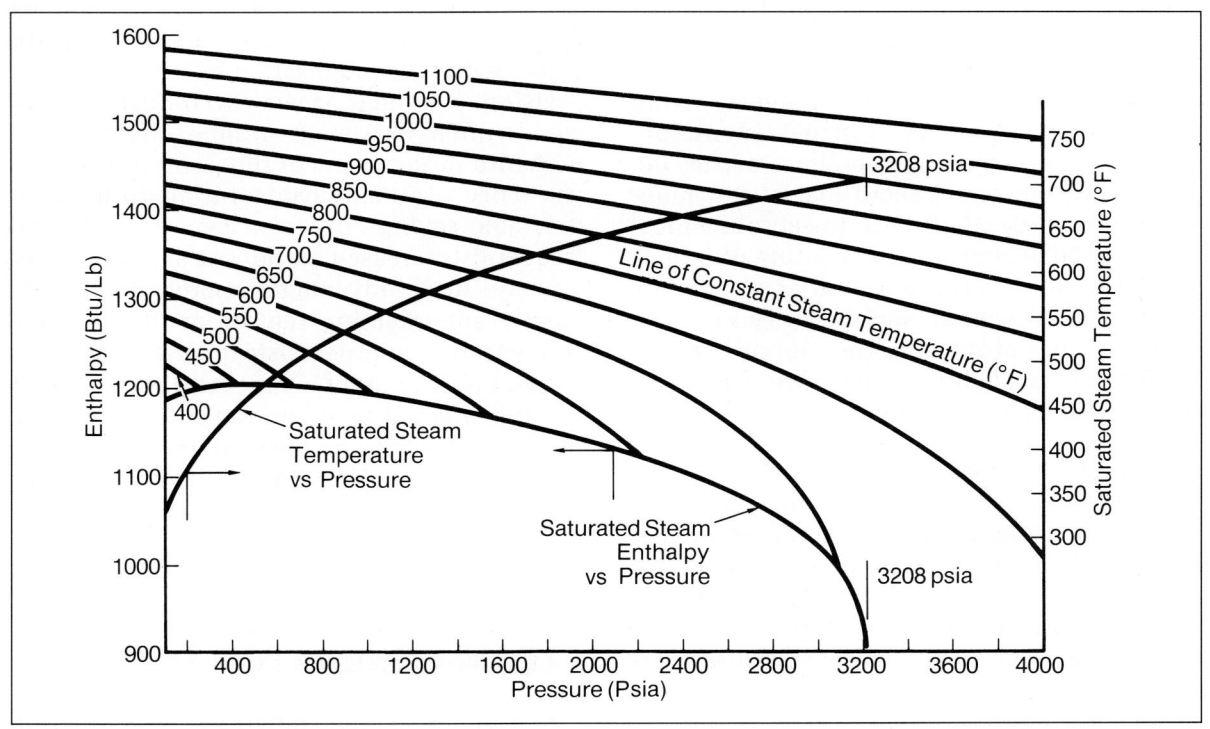

Fig. 1 **Variation in heat content (enthalpy) of steam as a function of pressure and temperature**

boilers and turbines for the higher temperature level proved to be less attractive as experience accumulated. In addition, even though accurate and conservative methods of superheater metal selection are used, problems do arise during sustained elevated temperature operation because of the adverse effects of certain fuel constituents on unit availability. Some owners therefore decided to reduce primary steam temperatures while continuing to lower capital cost through larger units. Other utilities chose to take advantage of the thermal gain from higher primary steam temperature and maintained the established 1050°F level.

Units employing the nominal 3500-psi cycle most commonly use 1000°F superheat and 1000°F reheat. About 30 percent of existing supercritical-pressure units have a second reheat stage at 1000°F, justified by high fuel cost and high anticipated load factor.

The curves in Fig. 1 show the variation in heat content of steam at different pressures. Also included are the heat of the liquid and the temperature of saturated steam. The values used in plotting these curves are from the 1967 ASME Steam Tables in Appendix D. When using the steam tables, or charts prepared from them, note that pressure is stated as absolute (psia). The steam gage on a boiler indicates the pressure in the vessel above that of the atmosphere. For these conditions, the absolute pressure is equal to atmospheric pressure plus gage pressure. Condensers and similar equipment operate under vacuum, and their gages usually read in inches of mercury (in. Hg) or kPa. For these conditions, absolute pressure is equal to atmospheric pressure minus vacuum.

REDUCING STEAM CONSUMPTION BY SUPERHEATING

Previous paragraphs have touched on the value of superheated as compared with saturated steam. A further discussion will serve to illustrate more clearly the manner in which superheating reduces steam consumption, particularly in regard to industrial and marine-boiler installations.

In the case of a steam turbine, high superheat is of utmost importance as the absence of moisture in steam decreases friction losses and erosion of turbine blades. If dry saturated steam is used in a turbine, condensation rate increases as the steam passes through the succeeding stages. The friction losses increase rapidly, because the condensate is actually inert material which acts to check the speed of the turbine rotors. By using superheated steam, the condensation can be limited to a relatively few stages in the turbine discharge end, thus reducing windage loss and friction between rotor and vapor because of lower density and absence of moisture in the initial stages.

Pipelines carrying steam lose heat by radiation. Thus, if steam entering a line is dry saturated, any loss in heat will immediately cause condensation which is usually discharged from the line through traps, and is frequently wasted. So, in addition to the heat loss from radiation, there is also loss of the heat in the condensate. Of course, if the condensate is returned to the hotwell, a portion of the heat in the liquid is recovered. By adding a sufficient amount of superheat to the steam, it can be piped without condensation loss.

Using saturated steam in industrial processes often results in minute quantities of moisture in the steam at the point of use. The use of dry saturated steam translates into a considerable savings in steam consumption and frequently increased output. As explained previously, these advantages accrue from sufficiently superheating the steam to overcome condensation in process-steam pipelines.

The installation of a superheater in a boiler has the effect of reducing the amount of work that must be done by the evaporative surfaces to produce the same power. In other words, installing a superheater has the effect of increasing plant capacity.

FUNDAMENTALS OF STEAM-GENERATOR DESIGN

Water-tube boilers range in capacity from small low-pressure heating units generating a

few thousand pounds of steam per hour to large reheat steam generators operating in the super-critical pressure region and serving turbine-generators in the million kilowatt range. In slightly different terms, capacity may be magnified more than a thousand times from the smallest to the largest, pressure may extend from just above atmospheric to values as high as 5000 psig, and steam temperatures may vary from the boiling point to a highly superheated condition at 1050°F or above.

What are the common elements in boilers having such a diversity of design parameters? To answer this question it may be well to define the primary function of a boiler which is simply *to generate steam at pressures above the atmospheric.* Steam is generated by the absorption of heat produced in the combustion of fuel. In some instances, such as waste-heat boilers, heated fluids serve as the heat source.

Generation of steam by heat absorption from products of combustion suggests that a boiler must have a pressure-parts system to convert incoming feedwater into steam; a structure within which the combustion reaction may take place, at the same time facilitating heat transfer and supporting boiler components; a means of introducing fuel and removing waste products; and controls and instruments to regulate and monitor operation. A boiler designer has to work with such things as drums, headers and tubing, which make up the pressure-parts system and enclose the furnace in which combustion takes place; burners and related fuel and ash-handling equipment; and fans to supply combustion air and exhaust waste gases. Various types of instruments and controls link these elements together in a physical and an operational sense.

A BRIEF HISTORY OF STEAM GENERATION

A furnace for firing a fossil fuel is a device for generating controlled heat with the objective of doing useful work. The work application may be *direct,* as with rotary kilns, or *indirect,* as in boilers for industrial or marine use, or for the generation of electric power. A further differentiation is whether the furnace enclosure is cooled, such as with *waterwalls,* or uncooled, with a refractory lining.

Furnaces developed originally from a need to fire pottery (4000 B.C.) and to smelt copper (3000 B.C.). Hastening and improving combustion by the use of bellows to blow air into the furnace were used about 2000 B.C.

Closely associated with the furnace is the corresponding steam boiler. Such boilers appear to be of Greek and Roman origin and were employed for household services. The Pompeiian water boiler, incorporating the water-tube principle, is one of the earliest recorded instances of boilers doing mechanical work (130 B.C.). It sent steam to Hero's engine, a hollow sphere mounted and revolving on trunnions, one of which permitted the passage of steam, which was exhausted through two right-angled nozzles that caused the sphere to rotate. This was the world's first reaction turbine.

For the next 1600 years, furnaces in general and waterwall furnaces in particular were essentially a neglected technology. This can be partly ascribed to the fact that steam as a working fluid had no application until the invention of the first commercially successful steam engine by Thomas Savery in 1698. In 1705, Newcomen's engine followed and by 1711, this engine was in general use for pumping water out of coal mines. Self-regulating steam valves were incorporated in 1713.

Many varieties of firetube boilers were invented in the second half of the 18th century, culminating with the Scotch marine boiler. As the name implies, in the firetube boiler the tubes may be considered to be a component part of the furnace, with the combustion process completed within the tube bundles. But such units were limited to operating pressure of about 150 psig, because of available steel-plate thicknesses. The development of the modern water-tube furnace for steam generation at higher pressures and in larger sizes than available with firetube boilers is the subject of the various chapters of this book which cover central-station steam generators, industrial-type boilers, and fluidized-bed and marine boilers, respectively.

BOILER OUTPUT

The output or capacity of a steam generator is either expressed in pounds of steam per hour or in the power output of a turbine generator in those cases where a single boiler provides the entire steam supply for an electric generating unit. Neither term is a true measure of the thermal energy supplied by the boiler.

Actual boiler output in terms of heat energy depends on several factors other than quantity of steam. These include temperature of feedwater entering the economizer, steam pressure and steam temperature at the superheater outlet, and the quantity, temperature and pressure of steam entering and leaving the reheater. Similarly, because turbine and generator efficiencies affect the boiler output, the generator output in kilowatts is not entirely a true measure of the energy output of the boiler alone.

These elements vary with the size and purpose of the power plant in which the boiler is installed. A large central station in which high thermal efficiency is a primary requisite has many more refinements and auxiliaries than a small heating plant in which minimum capital investment may be an important criterion. No matter how many of these elements may be present, however, the boiler designer must integrate them so that the boiler as a whole can function as a carefully adjusted, complex system which is capable of efficient operation over a wide load range.

STEAM-GENERATOR FUNCTIONS

In addition to its primary function of evaporating water to steam at high pressure, the modern boiler has to

■ produce that steam at exceptionally high purity, using stationary mechanical devices to remove impurities in the boiling water

■ superheat the steam generated in the unit to a specified temperature, and maintain that temperature over a designated range of load

■ re-superheat (reheat) the steam which is returned to the boiler after expanding through the high-pressure stages of the turbine, and maintain that reheat temperature constant over a specified range of load

■ reduce the gas temperature leaving the unit to a level that satisfies the requirement for high thermal efficiency and at the same time is suitable for processing in the emission-control equipment downstream of the boiler

BOILER EFFICIENCY

A diagram is one of the best ways to comprehend the significance of boiler efficiency as it is affected by the heat absorption pattern in a steam generator. Fig. 2 shows the distribution of heat energy in a reheat boiler for utility use. It is apparent that the primary source of heat is the fuel (supplemented by thermal energy from fans, pumps, and pulverizers), but that the preheated air for combustion adds directly to the total heat in the furnace. The amount of heat in the preheated air corresponds to that extracted from the exhaust gases by the air heater. This concept is discussed further in both this chapter and Chapter 6.

Of the total heat entering the furnace, the major portion is absorbed as sensible and latent heat of vaporization in the heating surfaces of the economizer, furnace, superheater and reheater. This absorbed heat represents the boiler output in the form of superheated and reheated steam. Losses which account for the remainder of the heat supplied to the furnace consist of the heat contained in the flue gas leaving the air heater (principally the sensible heat of the gas and latent heat in the moisture from the fuel), smaller losses from less-than-perfect combustion, and radiation from the boiler and its ancillaries. A description of these losses and how they are calculated is given in Chapter 6.

HEAT-ABSORBING SURFACES

The objective of the boiler designer is to arrange heat-transfer surface and fuel-burning equipment to optimize thermal efficiency and economic investment. One may select from waterwalls, superheaters, and reheaters, each of which absorbs heat from the furnace gas as it performs its respective function of heating water to the saturation point and of superheating and resuperheating steam. Also to be chosen are air heaters and economizers to recover heat from the furnace exit gases to preheat combustion air and increase the temperature of incoming feedwater.

The boiler designer must proportion heat-absorbing and heat-recovery surfaces to make best use of the heat released by the fuel. Waterwalls, superheaters, and reheaters are exposed to convection and radiant heat, whereas convection heat transfer predominates in air heaters and economizers.

The relative amounts of such surfaces vary with the size and operating conditions of the boiler. A small low-pressure heating plant with no heat-recovery equipment has quite a different arrangement from a large high-pressure unit operating on a reheat regenerative cycle and incorporating heat-recovery equipment.

Fig. 3 shows how the proportion of energy absorbed varies with different types of boilers.

In a heating-plant boiler operating with a minimum of feedwater heating and no superheater it is evident that most of the heat absorbed is utilized in evaporating water to steam. In a large reheat unit with feedwater heaters and heat-recovery equipment, heat for evaporation is comparatively small, whereas heat for superheating and reheating accounts for more than half of the total input.

Fig. 4 is a plot of temperature vs. heat absorption for four different high-pressure steam cycles: the single-reheat 1800 psig cycle at 5-percent overpressure; the 2400 psig, 1000°F cycle with single reheat to 1000°F, again at 5-percent overpressure (2620 psig at the superheater outlet); and a single-reheat and a double-reheat supercritical cycle, both at 1000°F at the turbine throttle. Such heat-absorption profiles will vary with entering feedwater temperature, cold-reheat temperature, and relative size of furnace and economizer.

FACTORS INFLUENCING BOILER DESIGN

In addition to the basics of unit size, steam pressure, and steam temperature, the designer must consider other factors that influence the overall design of the steam generator.

FUELS

Coal, although the most common fuel, is the most difficult one to burn. The many types of coal and their characteristics are covered thoroughly in Chapter 2 and Appendix A.

As described in Chapter 3, the ash in coal consists of a number of objectionable chemical elements and compounds. The high percentage of ash that can occur in coal has a serious effect on furnace performance.

At the high temperatures resulting from the burning of fuel in the furnace, fractions of ash can become partially fused and sticky. Depend-

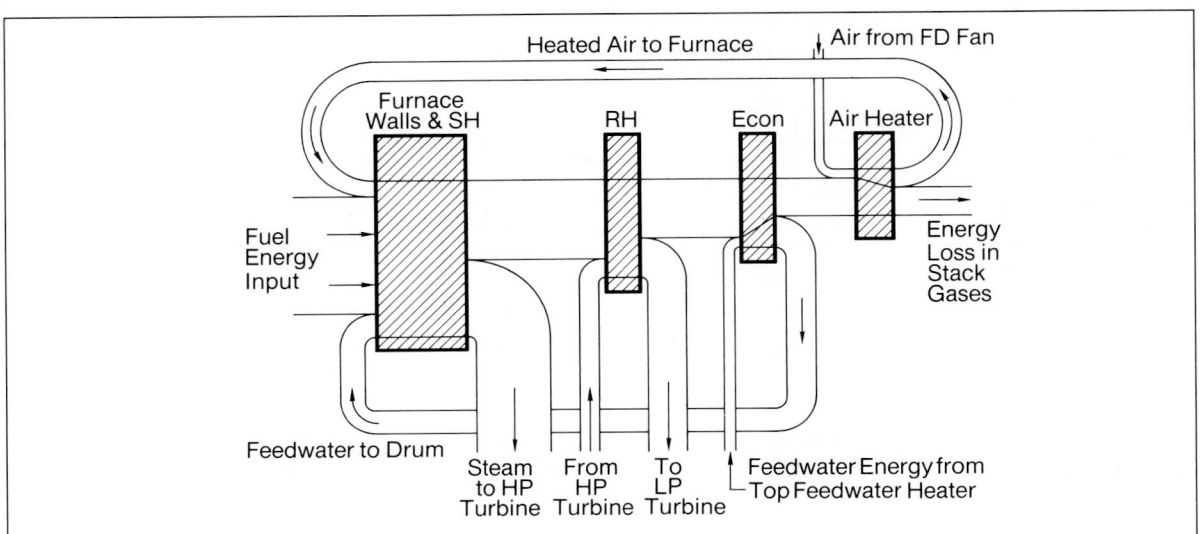

Fig. 2 Steam generator energy flow

ing on the quantity and fusion temperature, the partially fused ash may adhere to surfaces contacted by the ash-containing combustion gases, causing objectionable buildup of slag on or bridging between tubes. Chemicals in the ash may attack materials such as the alloy steel used in superheaters and reheaters.

In addition to the deposits in the high-temperature sections, the air heater (the coolest part of a boiler) may be subject to corrosion and plugging of gas passages from sulfur compounds in the fuel acting in combination with moisture present in the flue gas.

THE FURNACE

Heat generated in the combustion process appears as furnace radiation and sensible heat in the products of combustion. Water circulating through tubes that form the furnace wall lining absorbs as much as 50 percent of this heat which, in turn, generates steam by the evaporation of part of the circulated water.

Furnace design must consider water-heating and steam generation in the wall tubes as well as the processes of combustion. Practically all large modern boilers have walls comprised of water-cooled tubes to form complete metal coverage of the furnace enclosure. Similarly, areas outside of the furnace which form enclosures for sections of superheaters, reheaters and economizers also use either water- or steam-cooled tube surfaces. Present practice is to use tube arrangements and configurations which permit essentially complete elimination of refractories in all areas that are exposed to high-temperature gases.

Waterwalls usually consist of tangent or nearly tangent vertical tubes connected at top and bottom to headers. These tubes receive their water supply from the boiler drum by means of downcomer tubes connected between the bottom of the drum and the lower headers. The steam, along with a substantial quantity of water, is discharged from the top of the water-wall tubes into the upper waterwall headers and then passes through riser tubes to the boiler drum. Here the steam is separated from the water, which together with the incoming feedwater, is returned to the waterwalls through the downcomers.

Tube diameter and thickness are of concern from the standpoints of circulation and metal

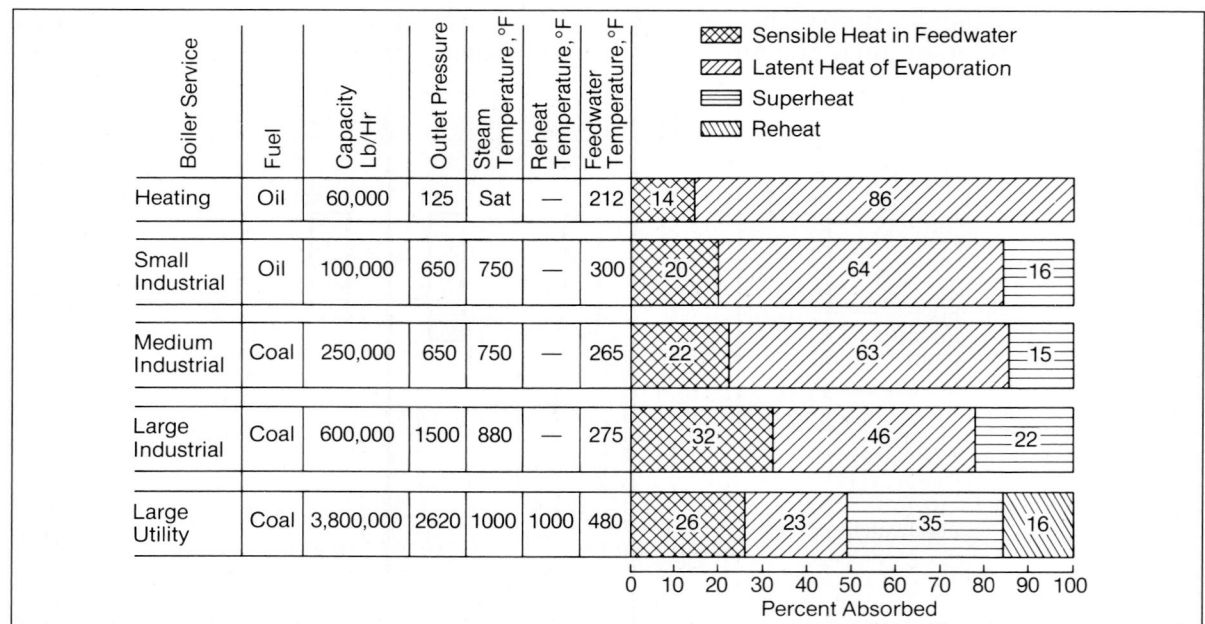

Fig. 3 **Heat absorption by various types of subcritical-pressure boilers**

Boiler Service	Fuel	Capacity Lb/Hr	Outlet Pressure	Steam Temperature, °F	Reheat Temperature, °F	Feedwater Temperature, °F	Percent Absorbed
Heating	Oil	60,000	125	Sat	—	212	Sensible Heat in Feedwater 14 / Latent Heat of Evaporation 86
Small Industrial	Oil	100,000	650	750	—	300	20 / 64 / Superheat 16
Medium Industrial	Coal	250,000	650	750	—	265	22 / 63 / 15
Large Industrial	Coal	600,000	1500	880	—	275	32 / 46 / 22
Large Utility	Coal	3,800,000	2620	1000	1000	480	26 / 23 / 35 / Reheat 16

temperatures. Thermo-syphonic (also called thermal or natural) circulation boilers generally use larger diameter tubes than positive (pumped) circulation or once-through boilers. This practice is dictated largely by the need for more liberal flow area to provide the lower velocities necessary with the limited head available. The use of small-diameter tubes is an advantage in high-pressure boilers because the lesser tube thicknesses required result in lower outside tube-metal temperatures. Such small-diameter tubes are used in recirculation boilers in which pumps provide an adequate head for circulation and maintain the desired velocities. Circulation of water and steam in both sub-critical and supercritical boilers is discussed further in other sections of this chapter.

SUPERHEATERS AND REHEATERS

The function of a superheater is to raise the boiler steam temperature above the saturated temperature level. As steam enters the super-heater in an essentially dry condition, further absorption of heat sensibly increases the steam temperature.

The reheater receives superheated steam which has partly expanded through the turbine. As described earlier, the rôle of the re-heater in the boiler is to re-superheat this steam to a desired temperature.

Superheater and reheater design depends on the specific duty to be performed. For relatively low final outlet temperatures, superheaters solely of the convection type are generally used. For higher final temperatures, surface requirements are larger and, of necessity, superheater elements are located in very high gas-temperature zones. Wide-spaced platens or panels, or wall-type superheaters or reheaters

of the radiant type, can then be used. Fig. 5 shows an arrangement of such platen and panel surfaces. A relatively small number of panels are located on horizontal centers of 5 to 8 feet to permit substantial radiant heat absorption. Platen sections, on 14″ to 28″ centers, are placed downstream of the panel elements; such spacing provides high heat absorption by both radiation and convection. (These modes of heat transfer are discussed further in Chapter 6.) Convection sections are arranged for essentially pure counterflow of steam and gas, with steam entering at the bottom and leaving at the top of the pass, while gas flow is opposite. As explained in Chapter 6, this arrangement allows a maximum mean temperature difference between the two media and minimizes the heating surface in the primary sections.

Metallurgy

Superheaters and reheaters designed for high pressures and temperatures above 1000°F require high-strength alloy tubing. Besides selection of materials for strength and oxidation resistance, the use of high steam pressure requires very thick walls in all tubing subject to steam pressure. Such thick tubes will have high outside metal temperatures and, because chemical action is accelerated at high temperatures, the tube metal is more likely to experience external corrosion. This is of particular concern when burning fuels containing objectionable impurities. The designer takes account of such conditions in selecting material and tube sizes. See Chapters 3, 6, and 17 for information relating to corrosion from coal ash, selection of superheater and reheater surface, and materials used for commercial construction of this equipment.

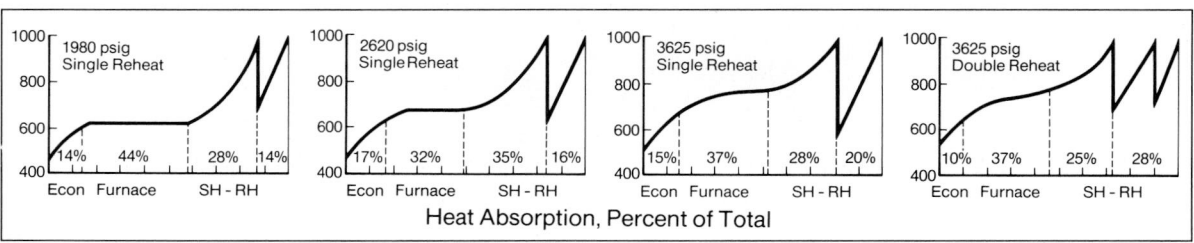

Fig. 4 **Heat absorption variation with cycle pressure**

ECONOMIZERS

Economizers help to improve boiler efficiency by extracting heat from flue gases discharged from the final superheater section of a radiant/reheat unit, or the evaporative bank of a non-reheat boiler. In the economizer, heat is transferred to the feedwater, which enters at a temperature appreciably lower than that of saturated steam. Generally, economizers are arranged for downward flow of gas and upward flow of water.

Water enters from a lower header and flows through horizontal tubing comprising the heating surface. Return bends at the ends of the tubing provide continuous tube elements, whose upper ends connect to an outlet header that is in turn connected to the boiler drum by means of tubes or large pipes. Tubes that form the heating surface may be plain or provided with extended surface such as fins. Frequently, they are arranged in staggered relationship in the gas pass to obtain high heat transfer and to lessen the space requirements.

Designing the economizer for counterflow of gas and water results in maximum mean temperature difference for heat transfer. (See Chapter 6 for more on this temperature difference.) Upward flow of water helps avoid water hammer, which may occur under some operating conditions. To avoid generating steam in the economizer, the design ordinarily provides exiting water temperatures below that of saturated steam during normal operation.

As shown in Fig. 5, economizers of a typical utility-type boiler are located in the same pass as the primary or horizontal sections of the superheater, or superheater and reheater, depending on the arrangement of the surface.

Tubing forming the heating surface is generally low-carbon steel. Because steel is subject to corrosion in the presence of even extremely low concentrations of oxygen, it is necessary to provide water that is practically 100 percent oxygen-free. In central stations and other large plants it is common practice to use deaerators for oxygen removal.

Small low-pressure boilers may have economizers made of cast iron, which is not as sub-

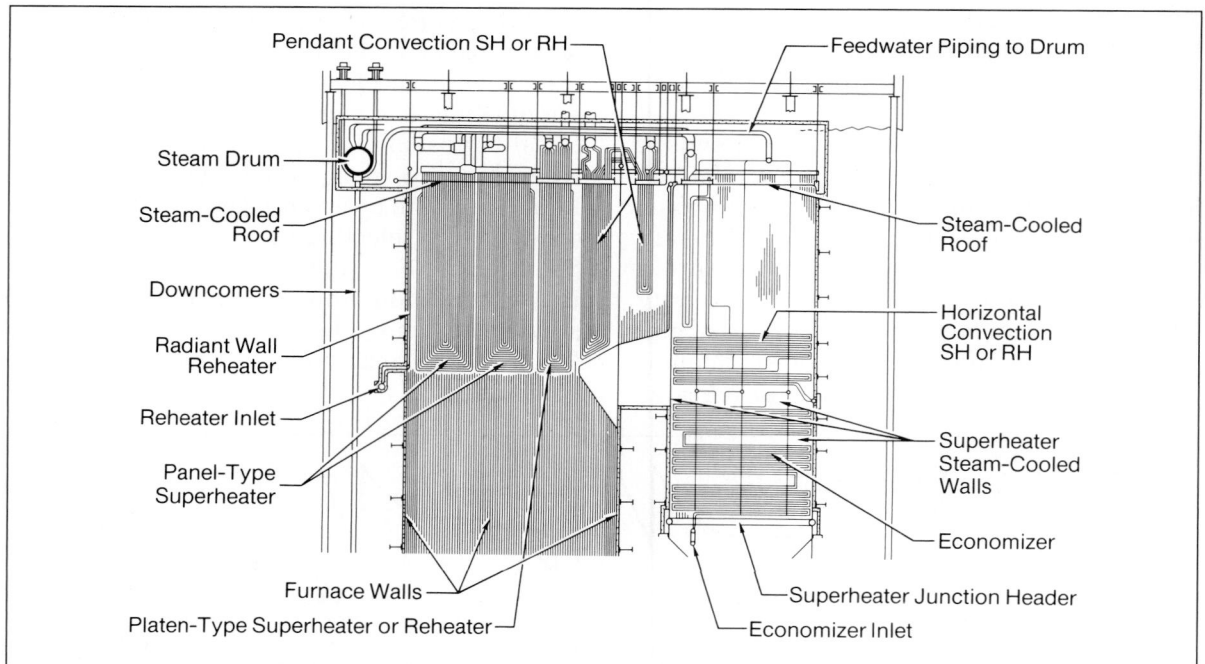

Fig. 5 **Arrangement of superheater, reheater and economizer of a pulverized-coal-fired steam-generator**

ject to oxygen corrosion. However, the design pressure for this material is limited to approximately 250 psig. Whereas cast-iron tubes find little application today, cast-iron fins shrunk on steel tubes are practical and can be used at any boiler pressure.

AIR HEATERS

Steam-generator air heaters have two important and concomitant functions: they cool the gases before they pass to the atmosphere, thereby increasing fuel-firing efficiency; at the same time, they raise the temperature of the incoming air of combustion. Depending on the pressure and temperature cycle, the type of fuel, and the type of boiler involved, one of the two functions will have prime importance.

For instance, in a low-pressure gas- or oil-fired industrial or marine boiler, combustion-gas temperature can be lowered in several ways—by a boiler bank, by an economizer, or by an air heater. Here, an air heater has principally a gas-cooling function, as no preheating is required to burn the oil or gas.

If the boiler is a high-pressure reheat unit burning a high-moisture subbituminous or lignitic coal, high preheated-air temperatures are needed to evaporate the moisture in the coal before ignition can take place. Here, the air-heating function becomes primary. Without exception, then, large pulverized-coal boilers either for industry or electric-power generation use air heaters to reduce the temperature of the combustion products from the 600°-800°F level to final exit-gas temperatures of 275°F to 350°F. In these units the combustion air is heated from about 80°F to between 500°F to 750°F, depending on coal calorific value and moisture content.

In theory, only the primary air—that used to actually dry the coal in the pulverizers—must be heated. Ignited fuel can burn without preheating the secondary and tertiary air. But there is considerable advantage to the furnace heat-transfer process from heating *all* the combustion air; it increases the rate of burning and helps raise flame adiabatic temperature.

Chapter 11 presents the various uses of air

heaters in the pulverization process. Some of the available types of air heaters are described in Chapter 14. Establishing inlet air and outlet gas temperatures to minimize corrosion of air-heater surface is considered in Chapter 6.

In sizing air heaters, note that, while the cost of pressure parts of a unit increase as the steam pressure increases, the air-heater cost is independent of pressure. Consequently, it is not difficult to justify the cost of larger air heaters on boilers operating at higher pressure levels.

CONTROL OF STEAM TEMPERATURE

Maintaining turbine efficiency over a wide load range and avoiding fluctuations in turbine metal temperatures require constant primary steam and reheat temperatures over the anticipated operating load range. To satisfy this requirement, a boiler must be equipped with means for controlling and maintaining such steam temperatures over the desired range. If uncontrolled, steam temperatures will rise as the steam output increases. This is particularly characteristic of convection-type superheaters, which account for the major share of heat absorbed by the superheater.

To provide an economical installation that operates at minimum metal temperatures, superheaters and reheaters should be designed to provide exactly the specified steam temperature at maximum output. An optimum design is one in which *all* the gas leaving the furnace passes over *all* the installed superheater and reheater surface at 100 percent boiler rating, without need of either superheater or reheater spray-water. To satisfy this the *means of control* must maintain full steam and reheat temperatures over the total control range.

TECHNIQUES OF STEAM-TEMPERATURE CONTROL

Steam-temperature control devices must be incorporated in the original design of a boiler firing system, in the superheater or reheater circuitry, or in arrangements of dampers for gas bypass. The following are the most frequently used means of control.

■ Firing-system manipulation, in which the effective release of heat from the fuel-burning process is made to occur at a higher or lower portion of the furnace; this affects the heat absorption pattern in the furnace and, consequently, the furnace exit-gas temperature.

■ Desuperheating by water sprayed into piping ahead of, in between, or following superheater or reheater sections.

■ Recirculation of gas, in which a portion of the combustion gases are brought back to the furnace and are added to the "once-through" flow of gas passing over superheater and reheater.

■ Gas bypass around some of the installed heating surface that is "excess" in certain parts of the load range; the purpose is to prevent such surfaces from absorbing heat from the bypassed gas so that the desired steam temperature is achieved without using any other means.

Subsequent sections describe in more detail these different methods of control, used in one form or another by all manufacturers.

Firing-System Manipulation

There are two common ways to vertically displace the zone of highest heat release in a furnace to achieve a change in the outlet gas temperature. The first, often used with front- or rear-wall fixed burners, is to insert or withdraw levels of burners as a function of load; removing lower levels and firing through the remaining upper levels effectively moves the high-heat-release zone higher in the furnace. This method necessitates backup by spray desuperheating for vernier control.

Tilting fuel and air nozzles, used in corner (tangential) firing systems are a positive means of controlling outlet gas temperature smoothly without cycling equipment in and out of service. Superheater or reheater steam temperatures are regulated by changes in nozzle position. Directing the flame toward the upper part of the furnace maintains a higher gas outlet temperature during light load—or when the walls are clean—than is the case if the flame were directed horizontally into the furnace. At higher loads, or when the walls are coated with ash or slag, nozzles can be positioned horizontally or angled downward.

Desuperheating Control

Desuperheating is the reduction of temperature of superheated steam accomplished by spraying water into the piping either ahead of or behind a superheater or reheater section. To minimize the amount of water used in the process, most large boiler installations use desuperheating in combination with one of the other temperature-control methods. This is because of water-treatment and pumping-power considerations.

Rising temperature with increasing output characterizes the performance of a superheater or reheater which receives its heat by convection from gas flowing over it. With desuperheating control *only*, the superheater must be designed for full temperature at some partial load. As a result, there is excessive heat-transfer surface, with corresponding excess temperatures, at higher loads. A desuperheater can be used to remove this excess in temperature. If located beyond the outlet of the superheater, a desuperheater will condition the steam before it is passed along to the turbine. Although this arrangement is practical for steam temperatures of 825°F and lower, the preferred location of the desuperheater, especially for temperatures above 850°F, is *between* sections of the superheater. In such interstage installations, the steam is first passed through one or more primary superheating sections, where it is raised to some intermediate temperature. It is then passed through the desuperheater and its temperature controlled so that, after continuing through the secondary of final stage of superheating, the required constant outlet temperature is maintained.

Desuperheating of *reheat* steam is generally not desirable because of the deleterious effect on plant heat rate. To evaporate the water that is added to the reheat flow (usually ahead of the point where the reheat steam is introduced into the steam generator) requires firing additional fuel in the boiler. There also results an increased condenser loss for the same primary steam flow. Consequently, reheat outlet temperature is best controlled using some means other than water spray, unless it is unavoidable.

The heat given up by the steam during a temperature reduction is picked up by the cooling water in three steps. First, its temperature is raised to that of saturated water; then the water is evaporated; and finally, the steam thus generated is raised to the final condition of temperature at the desuperheater outlet. By setting up a simple heat balance, it is possible to determine quickly the quantity of water required to desuperheat for any given set of conditions.

Desuperheaters are either the indirect or direct (mixing) type. The water available for use as the temperature-regulating medium governs the selection for any specific installation. This is not so important with the indirect (noncontact) type, where a tubular heat exchanger is used, since the steam to be desuperheated is separated from the cooling medium, and the heat is transferred through the separating tube wall. In the direct type, however, the cooling medium is injected into, and mixed with, the superheated steam to reduce the temperature. To be used for this purpose, the desuperheating water must be of condensate quality, containing very few solids.

In industrial or marine practice, to help reduce costs, non-contact desuperheating coils may be installed in one of the boiler drums. The principal objection to this design is its relatively low capacity. Through necessity, all parts must be sufficiently small to be conveniently handled through the boiler-drum access opening. The coils must therefore be compact, and made in sections to permit assembly and installation within the drum space limitations. If large or bulky, and if the number of parts are many, the assembly of the coils may become almost impossible.

Gas Recirculation

In this temperature-control means, described further in Chapter 6, a portion of the flue gas is diverted from the main stream at a point following the superheater and reheater (usually between the economizer outlet and the air-heater inlet) and is recirculated to the furnace. The gas passes through a recirculating fan and mixes with the gas in the furnace, causing a reduction in heat absorption. The heat available to the superheater and reheater increases, as does the quantity of gas passing over the surfaces. Both of these factors increase steam temperature.

Bypass-Damper Control

An arrangement of bypass dampers in a relatively cool gas zone downstream of superheater or reheater sections provides an acceptable means for maintaining constant steam temperature; automatic controllers adjust the dampers to provide the required degree of temperature control. If load changes are abrupt, frequent, and of considerable magnitude, there is likely to be some hunting in positioning the damper. As there is also some lag in the response to temperature changes, final temperature varies over a plus or minus 10°F range. This variation is characteristic where the regulation of steam temperature depends solely on control of gas flow through bypass-damper operation.

During the early damper-opening periods, the gas flow rapidly increases and then falls off as the full-open position is approached (see Fig. 6). The data were taken from an operational test during which the capacity of the unit was held constant and the effect of variable bypass-damper opening against steam temperature was recorded. There is an increase of 15°F in steam temperature for a change in damper from 100 percent to 40 percent open. However, the temperature increase from 40 percent open to fully-closed damper is 50°F. It is apparent

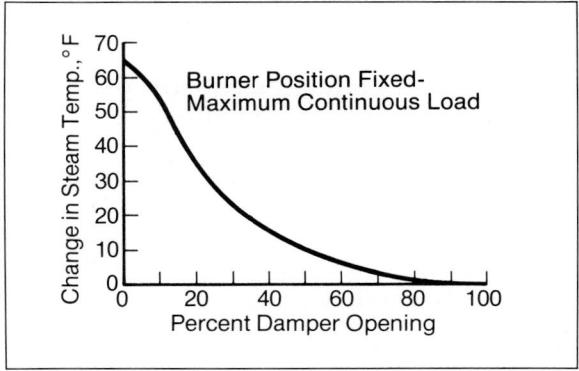

Fig. 6 Chart illustrating steam temperature change with respect to damper opening

that sensitive responses, resulting in more uni-form temperatures, are obtainable when the regulator operation is confined to the early stages of damper opening. In other words, any changes in capacity that would normally re-quire the bypass damper to operate above 50 percent open should also incorporate another means of control so as to keep within the sensi-tive control range.

WATER/STEAM CIRCULATION

The term circulation, as applied to steam generators, is the movement of water, steam, or a mixture of the two, through heated tubes. The tubes can be in furnace walls, boiler banks, economizers, superheaters, or reheaters. Ade-quate circulation results in the cooling fluid absorbing heat from the tube metal at a rate which maintains the tube temperature at or below design conditions. Adequate circulation also keeps the tube within the other physical and chemical limitations required by the inside and outside environment. In boilers, circula-tion through the varied systems of heated tubes can involve just the flow entering and leaving the system (called the once-through flow), a means of recirculating the fluid, or some com-bination of these two circulation concepts.

Steam generators of all manufacturers have similar pressure-part systems. For any given steam-power cycle, the economizer contains fluid in the lowest temperature range, with its inlet temperature being that leaving the top feedwater heater. The superheater contains fluid in the highest temperature range, with its outlet temperature essentially fixed. The evaporator contains the middle range of fluid temperatures.

Circulation in economizers, superheaters and reheaters is most commonly of the once-through type. The furnace-wall system of high-pressure units uses either once-through or recirculation flow, or some combination. By convention, these modes of circulation have become terms of reference for the complete steam generating unit.

At pressures below the 3208-psia critical point, the major portion of the evaporator oper-ates at a saturation temperature established by the pressure of the furnace-wall system. At supercritical pressures, the wall system has no fixed fluid temperature; a continuous tempera-ture increase occurs in the furnace-cooling fluid between the furnace-wall inlet and outlet.

The most critical circulating system in a large boiler is that of the furnace walls; they are at the same time the area of highest heat-absorption rates and a major structural component of the unit. This present section, and Chapter 7, dis-cuss the various types of furnace-wall circula-tion that the designer can use, their differences, and the advantages of each.

CIRCULATION IN FURNACE WALLS

There are four types of furnace-wall circula-tion systems used for present-day steam genera-tors. Their application depends on pressure, unit size and planned operating mode, required maneuverability, and the manufacturer's de-sign philosophy. Fig. 7 shows the four types diagrammatically; they are:

■ thermally induced, also called thermal, thermo-syphonic, or "natural", with inherent recirculation

■ thermally induced, pump-assisted, with recirculation

■ once-through, with no recirculation

■ once-through, with superimposed pumped recirculation

Because of the influence that pressure has on the behavior of water in the steam-generating process, it is necessary to distinguish between the different circulation systems at subcritical and supercritical pressures.

THERMAL CIRCULATION

The first type, thermal circulation, is a recir-culating system at subcritical pressure. It de-pends on the static-head difference between the water in the downcomers (usually un-heated) and the steam-water mixture in the heated generating tubes to produce circulation and maintain sufficient mass velocity and mix-ture quality for adequate furnace-wall cooling. Circulation begins only after heat is applied to

the vertical generating tubes and, when once begun, is proportional to the amount of heat locally supplied. To maintain the overall system pressure drop as low as possible, boilers using this principle for cooling the furnace enclosure need large flow areas in downcomers, supply tubes, generating tubes, and relief tubes to the drum.

Thermal-circulation boilers are the dominant choice of all manufacturers for low and medium subcritical-pressure operation, as they have been since the early development of the water-tube boiler. Theoretically, thermal circulation can be used at any waterwall operating pressure below the critical pressure of 3208 psia, as long as there is some finite density difference between the water in the downcomers and the saturated steam-water mixture in the heated circuits. Practically, the maximum furnace-wall pressure that has been used is about 400 psi below the critical pressure.

Later in this chapter are discussed the principles of thermal circulation including some of the mathematical bases for subcritical-circulation calculations in this mode. Also covered is the design background for different types of drum internals used to separate steam from the recirculating water so that it can proceed on to the superheater as a heatable vapor.

PUMP-ASSISTED THERMAL CIRCULATION

This type of subcritical-boiler recirculation system uses an external mechanical force, produced by one or more low-pressure-differential pumps located in the downtake system, to supplement the thermal head. The thermal-circulation circuitry may now include pump-isolating valves and, in many designs in this category, waterwall-circuit flow-control orifices. The orifices are varied in size to insure uniform circulation under changing furnace conditions. In general, waterwall tubes of boilers having positive-circulation systems of the type described are smaller in diameter than those of thermal circulation units; in the thermal units, large tubes minimize friction losses and take maximum advantage of the limited head provided by density difference.

Pump-assisted thermal circulation units find particular application at the high subcritical-pressure levels, where there is reduced fluid static-head energy available to recirculate the furnace-wall fluid. The reduced driving force is of concern where a very large number of parallel tube circuits makes even heat distribution a problem, such that some tubes can overheat under certain operating conditions; also, if there is less circulation energy available, it limits the boiler maneuverability.

Note that the only difference between the basic thermally induced boiler and the pump-assisted is in the circulating systems. Otherwise, both types share the same kind of firing methods, means of superheating and reheating and controlling superheater and reheater outlet temperatures, heat-recovery

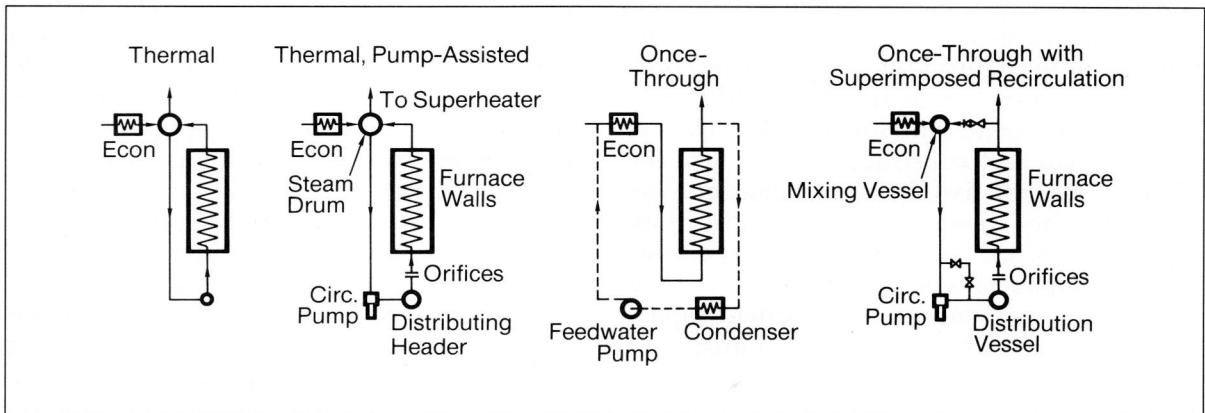

Fig. 7 **Steam-generator circulation systems**

equipment and structural support. A C-E developed positive-circulation design is described in Chapter 7.

ONCE-THROUGH CIRCULATION

In a once-through boiler, there is no recirculation of water within the unit. In elemental form, the boiler is merely a length of tubing through which water is pumped. Heat is applied, and the water flowing through the tube is converted into steam and superheated to the desired temperature at the outlet. In actual practice, the single tube is replaced by numerous small tubes arranged to provide effective heat transfer similar to the arrangement in drum type boilers. Note that the economizer and superheater operate on the once-through principle even in drum-type boilers. The fundamental differences lie only in the heat-absorbing circuits or evaporating portion of the unit. The word *evaporating* includes boiling at subcritical pressures and heating into the compressible vapor region at supercritical pressures.

In the typical reheat unit, the evaporating surface consists of waterwalls which form the major part, if not all, of the furnace enclosure. The principal distinguishing features of a once-through boiler, then, are related to the design and operation of the waterwalls.

The distinctive design requirements stem partly from the temperatures that may exist in the waterwalls of a once-through unit, in which there is a considerable rise in temperature from waterwall inlet to outlet. Even greater differences may exist during start-up or upset conditions. And there can be significant differences between adjacent individual tubes. The furnace design and construction must take into account the relative thermal expansions produced by these temperature differences in various parts of the furnace and between tubes, casing, and buckstays. The first distinguishing characteristic of once-through boilers, then, is their furnace-wall temperature patterns.

The second principal characteristic concerns means of preserving furnace tubing integrity. Unlike a drum-type boiler, in which circula-

tion of the working fluid is provided whenever the unit is fired, the once-through cooling flow is essentially proportional to the firing rate and is inadequate to provide waterwall tube protection during startup and at low loads. That is, in an elemental once-through unit, the design does not permit internal recirculation; nevertheless, a through flow has to be established at least in the highly heated portions of the circuitry. The operation of the power-plant feedwater system, including the boiler feed pump, is needed to produce the once-through flow, and suitable means must be provided to dispose of the circulated flow without incurring loss of heat or working fluid. This is normally done with a steam-turbine bypass system.

The once-through design has been successfully applied to both the high-subcritical and supercritical pressure ranges. In this connection, it is useful to point out some differences in boiler operation in the two pressure regimes.

Supercritical Versus Subcritical Once-Through Operation

When water is heated at a pressure above 3208 psia it does not boil. Therefore, it does not have a saturation temperature, nor does it produce a two-phase mixture of water and steam. Instead, the fluid undergoes a transition in the enthalpy range between 850 and 1050 Btu/lb (Fig. 1). In this range its physical properties (including density, compressibility and viscosity) change continuously from those of a liquid (water) to that of a vapor (steam). The temperature rises steadily; the specific heat and rate of rise varies considerably during the process. The nature of supercritical steam generation rules out the use of a boiler drum to separate steam from water; drumless units are universal for supercritical operation.

During the boiling process at subcritical pressure, individual molecules break out of the dense liquid clusters and, as the physical surroundings permit, form a separate vapor phase. At supercritical pressure, as heat is added to the liquid, the clusters gradually divide into smaller clusters and the spacing of the molecules gradually becomes less dense until the transition to the wide-spaced, random arrangement of a vapor is completed.

The most important result of this situation is that all impurities in the feedwater must either be deposited in the furnace and superheater surfaces or must be carried over into the turbine. As described in Chapter 20 only the best possible feedwater treatment can be considered acceptable. Because blowdown from the drum and furnace walls (as is possible with drum-type units) is not available to remove impurities from the system, not only the makeup water but also a portion of the condensate must be purified. Otherwise the concentration of impurities in the system will build up gradually until intolerable conditions or tube failures force an outage for cleaning or repairs.

More information on the development of the once-through boiler and its operation at both subcritical and supercritical pressures can be found in the 1967 edition of this text.[1]

ONCE-THROUGH FLOW
WITH SUPERIMPOSED RECIRCULATION

This fourth common type of boiler circulation, used for supercritical service, involves the superposition of pump-assisted circulation, as previously described, on the once-through flow. *Superposition* means that the pumps are permanently installed, but with a local bypass system, to allow them to be in-line or in the bypassed position as a function of load. The pumps then operate to assist circulation as required, on start-up and at low loads.

By adding to the once-through flow as necessary, the boiler-water circulating pumps act to protect the furnace-wall system from excessive metal temperatures. At loads that are between about 60 percent and 100 percent of maximum continuous rating, the pumps are no longer needed for wall-tube protection, and the unit operates on once-through flow.

With this type of unit, it is usual to supply a start-up system that contains shutoff and throttling valves (arranged between the furnace wall and the superheater) which can be bypassed through a low-pressure system with a water separator. This arrangement permits the furnace wall to operate at a pressure above the 3300 psig level while allowing the superheater to operate at a subcritical pressure level during start-up. The system also lends itself to variable-pressure operation through the superheater and steam turbine over any specified load range. Chapter 7 presents more detail on such start-up equipment.

COMPARISON OF BOILER DESIGNS FOR SUBCRITICAL STEAM CYCLES

Turbines for 2400 psig operation are usually designed for steam pressures of 2520 psig at the turbine throttle—a condition of 5 percent overpressure. A boiler-drum *operating* pressure of between 2750 and 2850 psig is required to allow for pressure drop through the superheater and the main steam line. As indicated earlier, with the densities of steam and water rapidly approaching each other above this pressure level, it represents a reasonable limit for a drum-type boiler incorporating steam separation and recirculation.

An advantage of drum-type units over once-through designs is their ability to operate with marginal quality feedwater. Steam separation permits elimination of undesirable solids by blowdown without affecting steam purity. The tendency of solid constituents to remain with the water phase makes this possible.

Because water is converted completely to steam in a once-through unit, solids in the feedwater must ultimately deposit in either the boiler, the superheater, or the turbine. Although high-quality makeup water can be provided consistently, the condenser always represents a possible threat of contamination. Demineralization of the condensate is necessary in systems with once-through boilers if power production is not to be curtailed during condenser leaks.

The furnace-wall tubes of the majority of coal-fired, drum-type, steam generators are virtually always some type of carbon steel. Its use is possible because the waterwalls are used exclusively for steam generation (evaporation). As a result, the fluid temperature in these tubes is constant and predictable. Even with very high furnace heat-transfer rates, mean-wall tube-metal temperatures seldom exceed 800°F.

TWO-PHASE FLOW

The design of boiler furnaces in the subcritical pressure range requires close attention to the adequacy of coolant flow necessary to keep the evaporative process in waterwall tubes within the nucleate boiling range. The importance of this aspect of two-phase flow has been brought about by the elevated steam pressures and better steam qualities for which boilers are being designed. Conditions are aggravated by the increased and varying absorption rates to which some furnaces are subjected.[2]

In simplest terms, the design problem centers on the choice of a physical arrangement that provides the coolest waterwall at minimum pump power. To accomplish this the designer must know the conditions under which the evaporative process passes from nucleate to transitional and finally to film boiling. Also necessary is information on the manner in which waterwall metal temperature can be expected to change as these phases occur.

An understanding of what is involved in two-phase flow under these conditions is possible from observing the changes that are produced in the film conductance, as indicated most directly by the inside tube-metal temperature when the heat flux, coolant mass flow, enthalpy and pressure are varied.

If a tube were heated uniformly along its length while water at a sufficiently high mass flow were passed through it at a given pressure, the inside metal temperature would follow a plot as described by the solid line in Fig. 8. If the flow rate were reduced, a point would be reached along the mixed or quality phase section of the plot where the metal temperature rises, as indicated by the dashed line in the illustration. The point of departure from the horizontal is known as DNB or the point of departure from nucleate boiling. The maximum point on the curve can be referred to as the point of film boiling or the minimum conductance point. Between DNB and the film-boiling point is the region of transitional boiling.

At DNB, the bubbles of steam forming on the hot tube surface *begin* to interfere with the flow of water to the tube surface, and eventually coalesce into a film of steam which effectively blankets the hot tube surface. The indication of this departure is defined as tube metal temperature fluctuations of 20°F or greater at frequencies of 10 seconds per cycle or slower. The transition from this point of departure from nucleate boiling to the point of steady-state film boiling is unstable because of the sweeping away of the coalescing bubble groups. This unstable phenomenon exhibits itself first with cyclical tube-metal temperature fluctuations of 10° to 20°F at a frequency appreciably slower than that of nucleate boiling bubble generation (something in the order of seconds per cycle versus tens of cycles per second for nucleate boiling). As this unsteady transition approaches full film boiling, metal temperature fluctuations of 50° to 100°F at even slower frequencies are observed. Fig. 9 is another way of illustrating the phenomenon.

CIRCULATION IN A SUBCRITICAL WATER-TUBE CIRCUIT

In recirculation-type boilers, the difference in density between steam and water is utilized to provide (or assist in providing) water circulation. (See Fig. 10) Generally, the downtakes

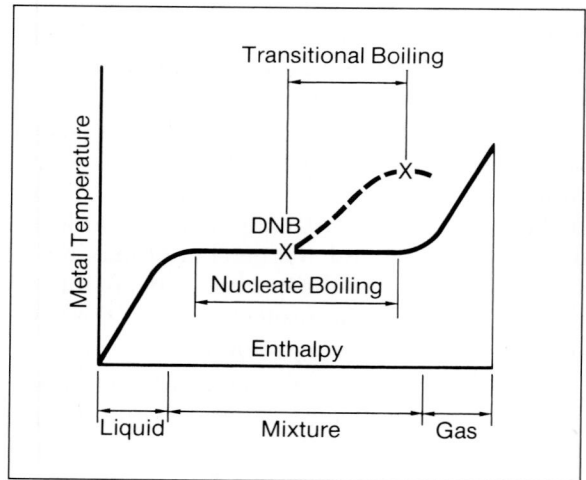

Fig. 8 Inside tube-metal temperature as a function of two-phase flow conditions

external to the furnace serve as the high-density leg of the system. The heated furnace walls or the forward portion of a boiler bank, containing a mixture of steam and water, constitutes the low-density leg. The available head to promote circulation is reduced by frictional and entrance and exit losses in the several circuits. The "adequate circulation" referred to earlier can be achieved only if the losses are low and make available a sufficient circulating head.

Before proceeding to the analytical and quantitative calculations of boiler circulation, consider what happens in a water-tube of a subcritical boiler. A steam-generating circuit (Fig. 11), for the velocity and resistance involved, will circulate as much water as the differential head will allow. The resistance is the sum of that due to steam flow and that due to water flow. Because the downcomer head is constant, the downcomer flow is a function of the sum of the downcomer head and the riser tube losses. Steam flow is a function of the firing rate, and the water-to-steam ratio in the riser circuit will vary for each operating condition.

The water in the mixture does not necessarily flow uniformly or at the same velocity as the steam, as most riser tubes are more or less verti-

cal, and the flow is against gravity. Because the water is heavier, it tends to separate and recirculate. The degree of such steam-water slippage and consequent recirculation depends on: the relative densities of steam and water, the relative amounts of steam and water in the mixture, the steam flow velocity, and the internal area of the riser tube. As might be expected, maximum slippage is found in boilers having large-diameter tubes, high ratios of water to steam, and low flow velocities.[3]

CIRCULATION RATIO

Circulation ratio is defined as the mass rate of water fed to the steam-generating tubes divided by the mass rate of steam generated. If steam is condensed in the drum by coming in direct contact with the feedwater, the mass of steam for the above calculation will be greater than the net output of the steam drum. In a boiler with drum internals capable of separating steam from water without condensing steam in the drum, the circulation ratio is the total circulation mass rate divided by the total steam mass rate leaving the steam drum.

The average mass of water per pound of steam leaving the steam-generator circuits and entering the steam drum, then, is the circulation ratio minus one.

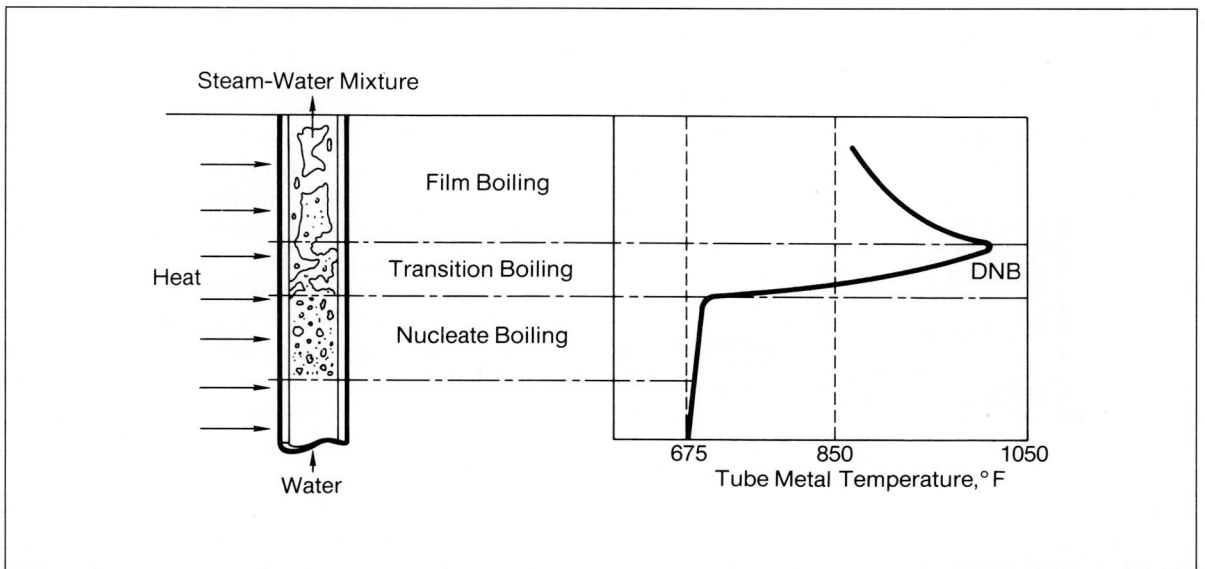

Fig. 9 Effect of DNB on tube metal temperature

Fig. 10 Chart showing relationship of steam and water mixtures

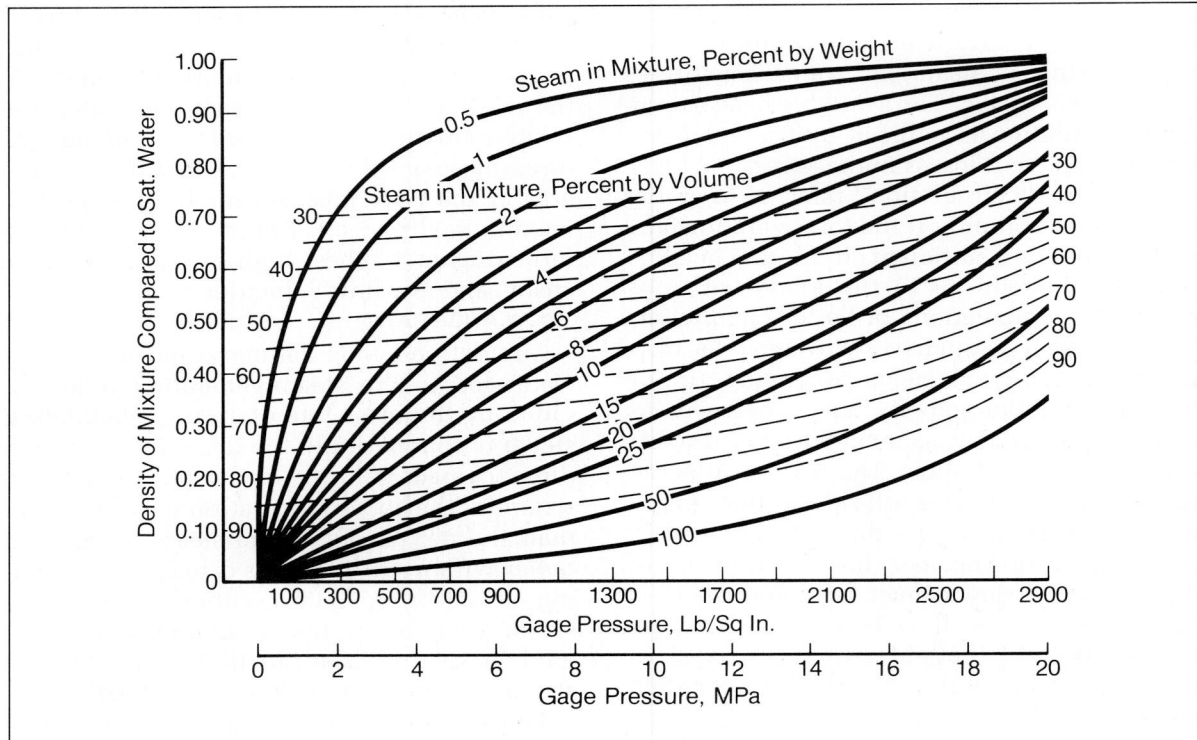

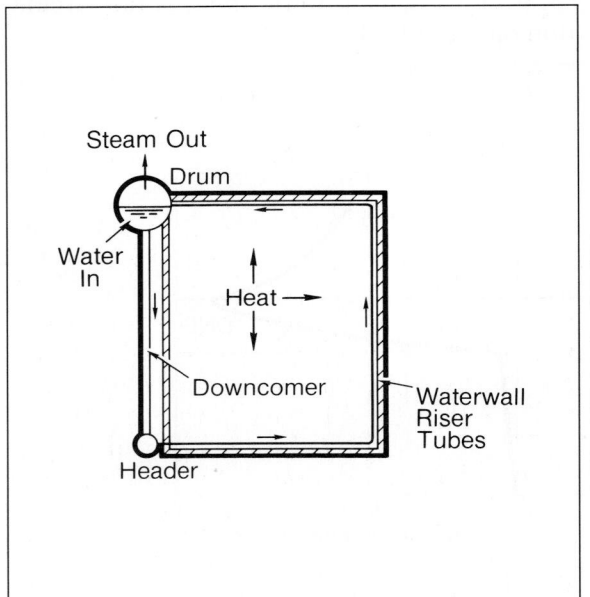

Fig. 11 Typical steam generating circuit in a subcritical-pressure water-tube boiler

GENERAL BOILER CIRCULATION

There are two distinct but related ways to consider boiler circulation. The primary view sees circulating fluid transmitting heat from an external source to a machine for conversion of thermal energy into mechanical energy. In this view the boiler is regarded as a thermal transformer in which heat-transfer functions are of first importance and are controlled essentially through fluid circulation.

A second viewpoint emphasizes circulation as a means of protecting the structure of the boiler by moving fluid rapidly through its drums, tubes and headers to prevent overheating. This places circulation also in the important role of a structural-protective mechanism.

Ideally these viewpoints are combined in a design in which efficiency of heat transfer is maximized while cost is minimized.

From a physical standpoint, there are three principles involved in the study of thermally induced circulation of fluid:

- differential pressure from expansion of a fluid by the application of heat to some portion of a circuit causes circulation
- with equilibrium established, the work available from expansion balances the work done against resistances to flow in the circuit
- mass flow of fluid in a closed circuit is constant throughout the circuit and unaffected by volume expansion resulting from a steady-state heat input

A unified analysis of subcritical-pressure boiler circulation proposed by R. W. Haywood in 1951 incorporated revisions of the hydraulic theory proposed by W. Y. Lewis and S. A. Robertson in 1940, the thermodynamic approach proposed by R. S. Silver in 1943, and the expansion approach set forth by R. F. Davis in 1947. Following is the Haywood analysis with changes in wording and terminology.[4]

FLUID FLOW IN HEATED PIPES

The classical analysis of frictional flow of a homogeneous fluid along a heated pipe must be understood before it can be applied to the special case of circulation, without bubble slip, in a subcritical boiler U-tube circuit.

The general energy equation relating to a unit mass of a homogeneous fluid in steady flow in a heated pipe is:

$$J dq = J dh + d\left(\frac{V^2}{2g_o}\right) + \frac{g}{g_o} dZ \tag{1}$$

$$= J dE + p dv + v dp + \frac{V dV}{g_o} + \frac{g}{g_o} dZ \tag{2}$$

where q is heat flow into the fluid; h is enthalpy of the fluid; V is velocity of the fluid; Z is height above a datum level; J is mechanical equivalent of heat; g_o is standard gravitational acceleration; g is acceleration; E is internal energy; p is pressure and v is specific volume.

This equation applies whether the flow is frictional or frictionless. To develop from it the equation for frictional flow, the equation of state for the fluid is needed, namely,

$$dE + \frac{p dv}{J} = T dS \tag{3}$$

and the fact that for frictional flow

$$T dS = dq + \frac{dW_f}{J} \tag{4}$$

where S is entropy of the unit, and W_f is the mechanical work dissipated by friction per unit mass of fluid.

The substitution of Eq. 3 and Eq. 4 in Eq. 2 gives

$$dW_f + v dp + \frac{V dV}{g_o} + \frac{g}{g_o} dZ = o \tag{5}$$

All the terms in Eq. 5 represent energy quantities per unit mass of fluid, and dW_f is that part of the available mechanical energy which is dissipated by friction. In the simple case of skin friction in a straight pipe

$$dW_f = \frac{4f}{d} \frac{V^2}{2g_o} dl \tag{6}$$

where f is a dimensionless friction coefficient; d is the inside diameter of the pipe and l is length of pipe.

The substitution of this in Eq. 5 would give the momentum equation for frictional flow in a straight pipe, but this is only a special case of the more general equation. For other "losses", such as in bends, etc., dW_f may usually be expressed as some function of the kinetic energy at the point, namely,

$$dW_f = K \frac{V^2}{2g_o} \tag{7}$$

where K is the friction loss coefficient for a particular point.

It is important to note that the units of dW_f are ft-lb-weight of energy/lb mass of fluid.

The Simple U-Circuit

To show the equivalence of the different theories of circulation, the idealized simple U-circuit shown in Fig. 12 is considered. In these circumstances the following simplifying assumptions are justified where in other circumstances they might be questionable:

The pressures at points A and F, just above the tubes, are assumed to be the same, and equal to p_s, and the kinetic energy of the fluid is assumed to be negligible at A and F.

D-E is a heated riser, in which the steam and water are assumed to move together as a homogeneous mixture, and in which there is a uniform increase in specific volume of the mixture with distance along the tube from D to E. Thus in DE, $dZ = dL = a \, dv$, where $a = l/[v_1(v_2/v_1 - 1)]$.

Frictional effects in the portion of the circuit from C to D are neglected.

HYDRODYNAMIC PRINCIPLES OF CIRCULATION

The hydrodynamic analysis is concerned

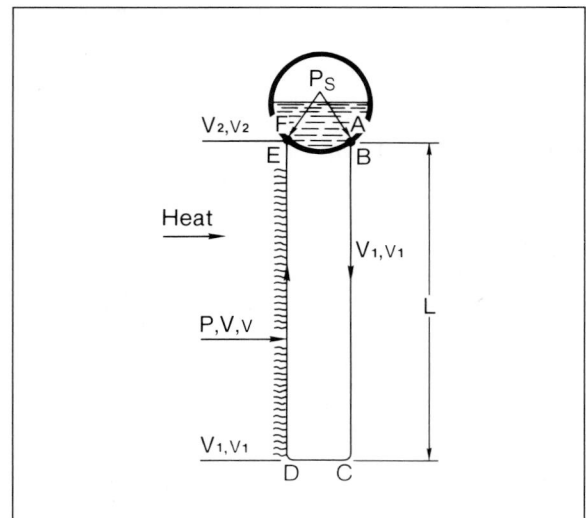

Fig. 12 **Simple U-circuit, subcritical drum-type boiler**

with the variation of *pressure*, not "head", the condition to be fulfilled being that

$$\int_A^F dp = 0$$

around the circuit. Hence Eq. 5 is applied in the form

$$- \int dp = \int \frac{g}{g_o} \frac{dz}{v} + \int \frac{V dV}{g_o v} + \int \frac{dW_f}{v}$$

(8)

from which

$$\int_C^B \frac{g}{g_o} \frac{dz}{v} - \int_D^E \frac{g}{g_o} \frac{dz}{v}$$
$$= \int_A^F \frac{dW_f}{v} + \int_A^F \frac{V dV}{g_o v}$$

(9)

All the terms in Eq. 9 represent pressures (e.g. lb_f/sq ft), so that the hydrodynamic analysis of circulation may be verbally expressed as:

Circulation rate adjusts itself until a state of equilibrium has been reached such that the difference between the hydrostatic pressures at the feet of the downcomer and riser legs is equal to the sum of the total pressure drops due to friction and acceleration in the circuit.

Hydrostatic pressure is the pressure which would exist at the point if there were a stationary column of liquid above the point, whose density varied in the same way as it would under the actual dynamic conditions of circulation. Pressure rise due to deceleration is included within the meaning of the term *pressure drop due to acceleration*.

By carrying out these integrations for the circuit of Fig. 12, the respective terms of Eq. 9 are readily shown to be

$$\int_B^C \frac{g}{g_o} \frac{dz}{v} - \int_D^E \frac{g}{g_o} \frac{dz}{v}$$
$$= \frac{g}{g_o} \frac{l}{v_1} \left[1 - \frac{1}{v_2/v_1 - 1} \log_e v_2/v_1 \right]$$

(10)

$$\int_A^F \frac{dW_f}{v} = \frac{V_1^2}{2g_o v_1}$$

$$\times [K_A + (v_2/v_1) K_F + (2f \, l/d) (v_2/v_1 + 3)]$$

(11)

$$\int_A^F \frac{V dV}{g_o v} = \frac{V_1^2}{2g_o v_1} (v_2/v_1 - 1)$$

(12)

from which the velocity of circulation is given nondimensionally by the expression

$$\frac{V_1^2}{2gl} = \frac{1 - \dfrac{1}{v_2/v_1 - 1} \log_e v_2/v_1}{C}$$

(13)

where

$$C = \lfloor (K_A + (v_2/v_1) K_F)$$
$$+ (2fl/d) (v_2/v_1 + 3) + (v_2/v_1 - 1)]$$

(14)

THERMODYNAMIC (EXPANSION) METHOD OF CIRCULATION

For the case in which fluid is assumed to flow as a homogeneous mixture, the thermodynamic (or expansion) analysis of circulation is of interest. Resulting from an essentially thermodynamic approach, it is at the same time very expressive of the physical nature of the process to describe it as an expansion theory. Its development depends on an understanding of the significance of the expansion work term, pdv, in the frictional steady-flow process. This is found by substituting in Eq. 5 the relation

$$vdp = d(pv) - pdv$$

(15)

giving

$$pdv = dW_f + d(pv) + d\left(\frac{V^2}{2g_o}\right) + \frac{g}{g_o} dz$$

(16)

All the terms in Eq. 16 represent energy quantities (e.g. ft lb force of energy/lb$_m$ of fluid).

For integration around the steam generation circuit, Eq. 16 may be written

$$\int pdv = \int dW_f + \Delta FE + \Delta KE + \Delta PE$$

(17)

where ΔFE is the increase in flow energy pv; ΔKE is the increase in kinetic energy and ΔPE is the increase in potential energy.

Applying this to the circuit of Fig. 12

$$\Delta FE - p_s (v_2 - v_1); \quad \Delta KE = 0; \quad \Delta PE = 0$$

Also, $p = p_s + p'$, where p' is the local excess of pressure at any point above the drum pressure. Substitution of these relations in Eq. 17 gives

$$\int_A^F p' dv = \int_A^F dW_f$$

(18)

The thermodynamic or expansion analysis of circulation may thus be expressed as:

When unit mass of fluid flowing around the circuit expands by an amount dv at a point where the pressure is $(p_s + p')$, it does an amount of work $(p_s + p')$ dv against the surrounding fluid, but the work done at the boundaries of the system in consequence of this expansion is only p$_s$dv.

Thus an excess of mechanical work equal to p'dv is available for overcoming friction (or which, in the more general case, would also be available for imparting increased kinetic and potential energy to the fluid). This excess of mechanical work may be termed the *work available for circulation,* and the rate of circulation adjusts itself until a state of equilibrium is established between the work available for circulation and the mechanical energy dissipated by friction in the circuit.

As they are derived from a common equation, it is evident that both thermodynamic and hydrodynamic analyses give the same answer for the rate of circulation. But the hydrodynamic

approach is simpler to apply because the thermodynamic approach requires a knowledge of the pressure at all points in the circuit where expansion is taking place, and this pressure must first be calculated by the procedure used in the hydrodynamic method. The steps involved are outlined as follows:

In evaluating $\int_A^F p'dv$, the integration is necessary only over that part of the circuit with a change in volume, namely, in the heated riser. Thus

$$\int_A^F p'dv = \int_D^E p''dv + (p_E - p_3)(v_2 - v_1)$$

(19)

where $p'' = (p - p_E)$, and is evaluated from Eq. 8 to give

$$p'' = \frac{g}{g_o}a \log_e \frac{v_2}{v} + \frac{G^2}{g_o}(v_2 - v)$$
$$+ \frac{aB}{2}(v_2^2 - v^2)$$

(20)

where $B = 2fG^2/g_od$ and G is mass velocity, V/v. Also

$$(p_E - p_s) = -v_2/v_1(1 - K_F)\frac{V_1^2}{2g_ov_1}$$

(21)

Substituting Eqs. (20) and (21) in Eq. (19) and integrating,

$$\text{"work available"} = \int_A^F p'dv = \frac{g}{g_o}L$$
$$\times \left[1 - \frac{1}{v_2/v_1 - 1}\log_e v_2/v_1\right] + \frac{V_1^2}{2g_o}$$
$$\times \left[(v_2/v_1 - 1)^2 + \frac{2fL/d}{3}2(v_2/v_1) + 1\right)$$
$$\times (v_2/v_1 - 1) \times v_2/v_1$$
$$\times (v_2/v_1 - 1)(1 - K_F)\right]$$

(22)

The mechanical energy dissipated in friction in the whole circuit is given by

$$\int_A^F dW_f = K_A \frac{V_1^2}{2g_o} + 4fL/d\frac{V_1^2}{2g_o}$$
$$+ aB\int_{r1}^{r2} v^2dv + K_F \frac{V_2^2}{2g_o}$$

where $a = \frac{1}{(v_2 - v_1)}$.

Thus

$$W_f = \frac{V_1^2}{2g_o}\left[K_A + (v_2/v_1)^2K_F + \frac{4fL/d}{3}\right.$$
$$\left. \times (\, (v_2/v_1)^2 + (v_2/v_1) + 4)\right]$$

(23)

Equating (22) and (23) results again in Eqs. (13) and (14) thus giving the same answer for the rate of circulation as was given by the hydrodynamic method. Although the thermodynamic analysis may clarify the physical processes involved in circulation, evaluating the rate of circulation by this means is considerably more complicated than by the hydrodynamic method, and the latter is therefore much preferred for practical use.

SEPARATION OF STEAM AND WATER

Water technology for steam generators is not limited to conditioning or treatment of makeup and, feedwater, as discussed in detail in Chapter 20. It is also concerned with the phase transformation of water to steam, and the separation of liquid and gaseous constituents. The study of these phenomena considers physical chemistry, fluid flow and mechanical design.

Despite many theoretical analyses of steam and water separation and a great number of hypotheses to explain these phenomena, steam and water separation in boilers retains many aspects of an engineering art and has thus far defied completely rational understanding. Experimental work on both model and full-scale

ABB
ASEA BROWN BOVERI

apparatus continues to provide useful information. Engineering criteria for the design of separators for subcritical-pressure steam and water separation, require correlation of these laboratory results with widely varied experience, as evidenced by the successful operation for many years of effective and efficient separation devices. But the challenge remains to develop a verifiable theoretical structure that provides an analytically rigorous explanation of the phenomena of steam and water separation in terms of the mass flows and physical equipment size found in large boilers.

THE REQUIREMENT FOR STEAM SEPARATION

Steam generated in a subcritical-pressure recirculation-type boiler is intimately mixed with large and variable amounts of circulating boiler water. Before the steam leaves the boiler and enters the superheater, practically all of this associated boiler water must be separated from the steam. This separation must be done within a limited space in the drum, within a matter of seconds, and under a variety of velocity, pressure and other operating conditions. The pressure drop across the steam and water separators must not be sufficient to affect boiler circulation or water-level control.

Nearly all of the liquid and solid impurities in the steam and water mixture must be separated from the steam before it is suitable for use. Any unseparated liquid in the steam contains dissolved and suspended boiler-water salts which appear as a solids impurity in the steam when the moisture is evaporated in the superheater or directed to a turbine or other steam driven apparatus.

The moisture content in a saturated steam is defined by its quality, which is a measure of the percent by weight of dry steam in a steam-water mixture. The solids content in a saturated steam is defined by its purity, a measure of the parts per million of solids impurity in the steam. Solids content in steam from high subcritical-pressure or supercritical boilers is measured in parts per billion.

The relationship between moisture carry-over, solids carryover and boiler-water concentration is best illustrated by an example. For a 0.01 percent moisture carryover of boiler water having 1000 ppm concentration, the solids in the steam would be 1000×0.0001 or 0.1 ppm. Conversely, the moisture carryover can be calculated if the boiler-water concentration and the steam impurity content are known.

BOILER DRUM

The drum of a subcritical boiler serves two functions, the first being that of separating steam from the mixture of water and steam discharged into it. Second, the drum houses equipment to "dry" steam after being separated from the water.

As the quantity of water contained in the boiler below the water level is relatively small compared to the total steam output, the matter of water storage is not significant. Primarily, the space required to accommodate steam-separating and purifying equipment determines drum size. Drum diameter and length should be sufficient to provide accessibility for installation and inspection. Length generally depends on furnace width; in the case of high-capacity units, it is controlled by the space required for the steam-separating devices.

The weight of water in the mixture delivered to the drum for separation depends on the circulation and may range from less than two to over 25 times the weight of steam. To reduce this water to the small fraction found in the steam requires a high efficiency of water separation. The equation indicating the percent of water separation necessary to give a steam impurity may be expressed as:

$$\text{Percent of water separation} = 100 - \frac{100\,P_s}{NC_b} \quad (24)$$

where
P_s = ppm of impurity in steam
N = circulation ratio, lb of water/lb of steam
C_b = boiler-water concentration

DRUM INTERNALS

Drum internals in subcritical-pressure boil-

ers separate water from steam and direct the flow of water and steam to establish an optimum distribution of drum metal temperature during boiler operation. Such apparatus may consist of baffles which change the direction of flow of a steam and water mixture, separators which use a spinning action for removing water from steam, or steam purifiers such as washers and screen dryers. These devices are used singly or in consort to remove impurities from the steam leaving the boiler drum.

Numerous factors affect the separation of water from steam in a boiler, among which are

■ the density of water with respect to the steam

■ the available pressure drop for drum-internal design

■ the amount of water in the mixture delivered to the steam drum

■ the quantity or total throughput of water and steam to be separated

■ viscosity, surface tension and other such factors affected by pressure

■ water level in the drum

■ the concentration of boiler-water solids

There is a considerable difference in the density of water and steam as pressure increases toward the critical point. This relationship is shown in Fig. 13 which is a plot of the ratio of the density of water to the density of

steam as a function of pressure. The density of water at 1200 psia is approximately 16 times that of steam. At 2800 psia, the density of water is about three times that of steam. Thus, as pressure increases, separating water from steam with simple devices becomes more difficult and requires more efficient apparatus to achieve primary separation in a confined area.

WATER SEPARATION STAGES

The stages of water separation are designated as primary separation, secondary separation, and drying. The devices used are primary separators, secondary separators and dryers. The term dryer was established during the course of development of water-tube boilers on the basis that the final stage of separation delivered a "dry" and saturated steam.

PRIMARY AND SECONDARY SEPARATORS

Primary and secondary separation reduces the water content of the steam from the boiler tube circuits to a moisture level that a final-stage dryer can handle. Frequently, the design of equipment for primary and secondary separation is influenced as much by the boiler design as by the basic function of separating the water from the steam.

Practically all drum internals consist of plate baffles, banks of screens, arrangements of corrugated or bent plate, and devices using radial acceleration to disengage water from steam.

Baffle plates are generally used to change or reverse a flow pattern to assist gravity separation in the open drum space. Fig. 14 illustrates

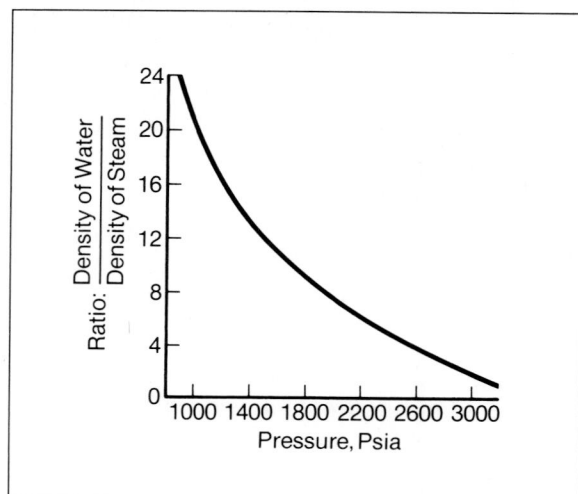

Fig. 13 **Water/steam density relationship**

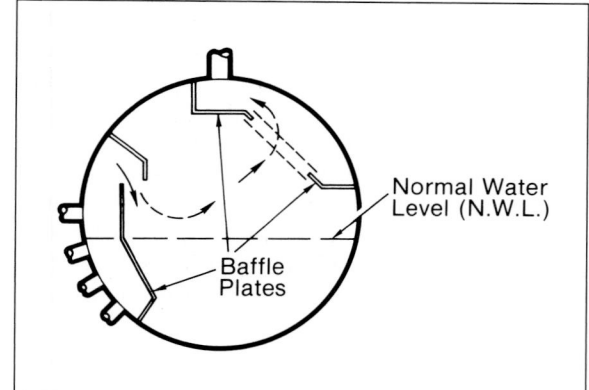

Fig. 14 **Baffle plates in a steam drum**

an example of simple baffle arrangements. Baffle plates change flow direction of water and steam and act as impact plates. Water separating out on such plates normally will drain off through or adjacent to the steam flow, and a controlling factor in design and operation is the steam flow velocity through such drainage. Areas under and around baffles must be sufficient to prevent excessive re-entrainment of spray. Limited in their impact-separating capacity, the chief purpose of plate baffles is to direct flow to make maximum use of the gravity separating capacity available from any low-velocity steam space in the drum.

Screens of wire mesh material are generally satisfactory for secondary separators or dryers. The limiting factor for screen separators is the steam flow velocity through the free area of the screen and the water drainage capacity of the screen. Screens are effective in separating spray. Riser mixtures normally have too much water in them for screens to handle as primary separators. Other factors affecting performance of screens as primary separators are circulation ratio, size of screen, and steam flow velocity through the screen. Examples of screen arrangements as primary or secondary separators in boiler drums are shown in Fig. 15.

Bent and corrugated plates are commonly used for all three stages of separation. Fig. 16 illustrates an example of a corrugated plate assembly which has the advantage of a higher ratio of free area to projected area than do

screens. This equipment can be relatively small for the same steam flow velocity.

Bulk separators, Fig. 17, deflect large quantities of water directed into the drum from active risers. The apparatus reduces the bulk water entrained with the steam and directs it below the water level. This design has allowed higher evaporation rates per unit of drum length since it satisfactorily removes the bulk of the spray water from the steam. The bulk separator reduces moisture by providing impact surface, a change of direction for the mixture, a drain trough, and a screen layer for reduction of fine spray.

Reversing hoods, shown in Fig. 18, combine all the desirable design features of baffle and change-of-direction principles. As indicated, steam and water from the active generating tubes are directed behind a baffle into the slotted reversing hoods. These primary separators are simply an arrangement of baffle plates to guide steam and water in a manner to give maximum utilization of gravity separation in open drum space. By accelerating the gravity flow of water and by reversing the flow of steam, normal gravity potential is increased and separation enhanced.

At higher pressures, water and steam are separated most efficiently in a drum internal utilizing radial acceleration to disengage the entrained particle from the steam. Guide vanes impart spinning action to the turbulent mixture, and the circular motion causes the heavier

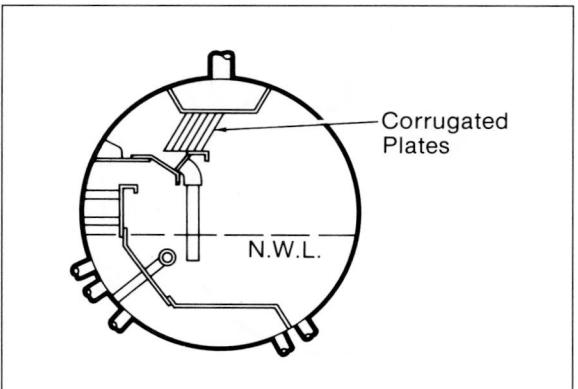

Fig. 15 Use of screens for secondary separation   **Fig. 16 Corrugated-plate assembly**

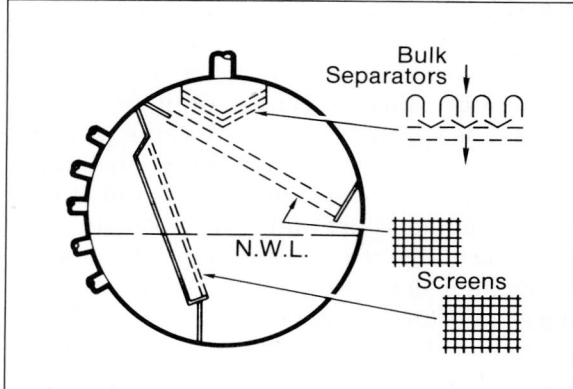

Fig. 17 Bulk separators for removal of spray water from steam

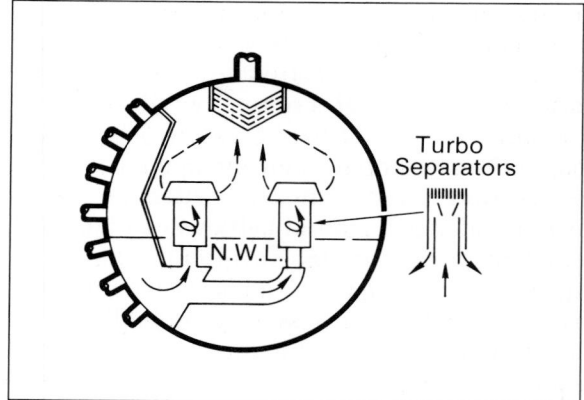

Fig. 19 Turbo separators followed by corrugated-plate assembly

water particles to move radially outward through the steam and to impinge upon the outer wall of the separator where they can be collected. The efficiency of separation is related to pressure in that the relative densities of water and steam determine the resistance to particle motion due to buoyant effects. As operating pressure is increased, and as the density of saturated steam approaches that of water, relative motion is more difficult to achieve and separation efficiency is decreased.

The turbo separator, shown in Fig. 19 uses the radial acceleration principle. Equipped with a corrugated plate assembly at the outlet of the separator, it provides primary and secondary separation of water from steam at pressures up to the critical operating point.

As the water and steam mixture circulated from the waterwalls is introduced into the drum, it sweeps the drum shell on its path to the bottom. The confining baffle concentric with the drum shell creates effective velocities and rapid heat transfer. The mixture enters the separators arranged along the length of the drum. Vanes spin the mixture as it travels upward through the separators and thus create a separating force. The concentrated layer of water flowing upward along the surface of the primary tube is skimmed off and directed downward through an outer concentric tube for discharge below the waterline with minimum disturbance to the water level.

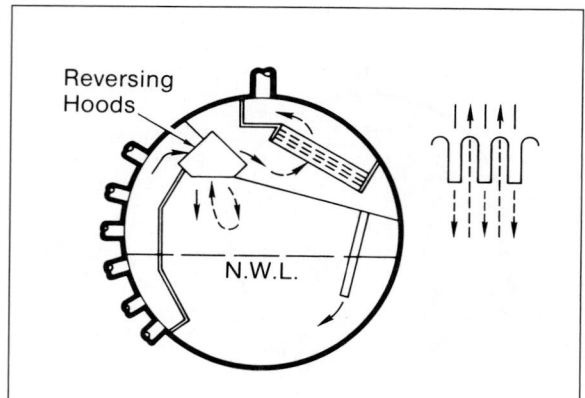

Fig. 18 Reversing-hood primary separators

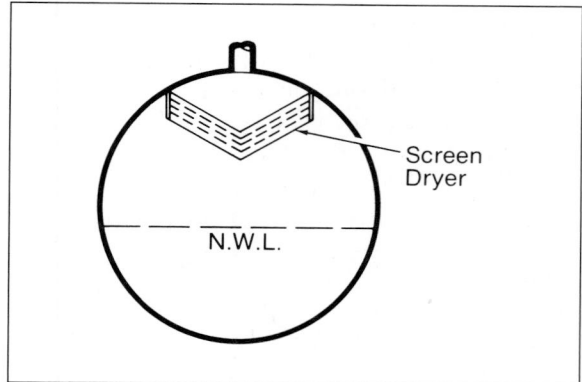

Fig. 20 Screen dryer for removal of residual moisture from steam

The steam and the remaining entrained water continue upward through a steam collector nozzle and turn horizontally into the secondary separator. The velocity at this point is low and water cannot be re-entrained from wetted surfaces and runs off the plates. Leaving the separator, the steam flows upward into the final dryer which facilitates handling of spray and dehydration of foam. The turbo separator has no inherent capacity limit because there is no water seal under a high differential pressure. Changes in water level do not affect the efficiency of the turbo separator[5].

DRYERS

A component of drum internals, dryers function to remove residual moisture from steam after primary and secondary separators have eliminated most of the circulating water. They are designed to have a large surface area on which moisture can deposit and from which it can drain back into the drum by gravity. Flow velocity through the free area of a dryer must be restricted to limits above which deposition or drainage may be inhibited.

Closely spaced corrugated or bent plates, screens or mats of woven wire mesh can be used as dryer surface materials. The screen dryers are a practical compromise of performance and drainability that have given satisfactory service for many years.

The design of dryers requires consideration of a number of factors. Space limitation in the drum restricts the dryer size. Other factors as sturdiness, leak-proof installation, drainage facilities and provision for cleaning due to possible plugging of dryer free areas must also be considered when selecting a steam dryer. The pressure drop across a dryer is normally low because of the low flow velocities and relatively small amounts of water involved.

Dryers operate on a low-velocity deposition principle, not on a velocity separation principle. Formation of insoluble residues on the dryers from boiler water entrained with the steam decreases the free area, increases the local velocity, and promotes carryover. Similar results are noted by the filming action of foamy boiler water in the dryer. The free surface areas are reduced significantly, increasing the local velocity and facilitating re-entrainment of boiler water and carryover. Dryers in boilers operating with foamy boiler water or large amounts of suspended matter in the boiler water should be inspected periodically and cleaned as necessary. Fig. 20 is a simplified representation of a screen dryer for final drying of steam.

AUXILIARY DRUM INTERNALS

Feedwater lines, blowdown lines and chemical feed lines are also installed in the boiler drum. In some marine boilers, desuperheaters may also be installed. Although feedwater, blowdown and chemical feed lines normally do not take up much drum space, their location can be a minor complication in the overall internals arrangement. To give satisfactory distribution of flow, these lines are usually run to the center of the drum where they feed other branch lines that are perforated.

Feedwater lines are submerged in the drum water but must be arranged so as to avoid discharge of cold feedwater against the bare drum shell, as temperature variations can cause severe thermal stress in the thick drum shells. In some cases, it is desirable to concentrate the colder feedwater flow into the downcomer tubes to condense steam entrained in the downcomer flow and improve boiler circulation. It is also necessary to prevent water hammer which can occur readily if steam leaks back into the feedwater system. The inflow of cold water may suddenly condense the steam and create a vacuum. Water hammer results when water rushing into the vacuum stops suddenly.

To control scale, sludge, and corrosion, chemical feed lines introduce chemicals in a manner which insures rapid mixing with the feedwater. The chemicals are generally added in a concentrated form, and it is necessary to flush the lines periodically with clean water to prevent plugging due to reaction and deposition.

Blowdown lines periodically or continuously remove a portion of water from the boiler. Sufficient high-solids boiler water is re-

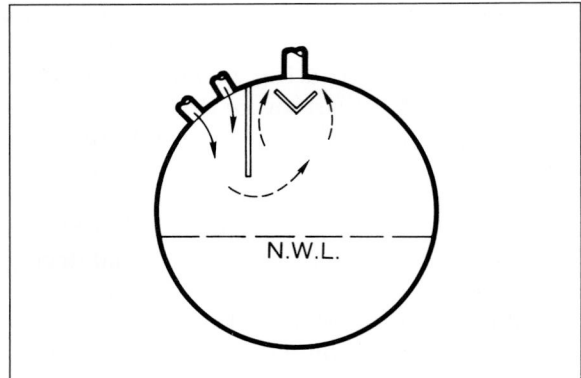

Fig. 21 Dry-box construction

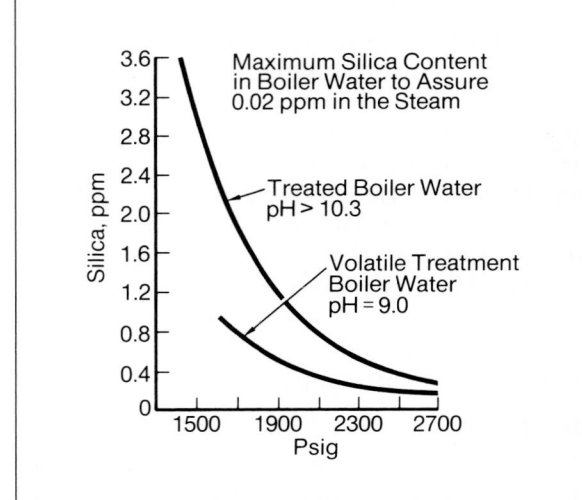

Fig. 23 Relationship of silica in boiler water to operating pressure

moved and replaced by low-concentration feedwater to maintain a desired concentration in the boiler water. These lines are located to minimize the occlusion of feedwater and chemical feed, and are designed to prevent entrainment of steam.

Perforated plates, or tapered-plate restrictions at the top of the drum, distribute steam flow. Not a steam dryer, this dry-box arrangement distributes the flow to allow a minimum of steam outlets to assure a satisfactory velocity distribution to a steam dryer. Fig. 21 shows a dry box construction.

STEAM WASHERS

Steam washers are special drum internals in which low-concentration feedwater is sprayed into the steam space to dilute the solids content in the moisture being carried over to the steam dryer. Originally used in multiple-drum boil-

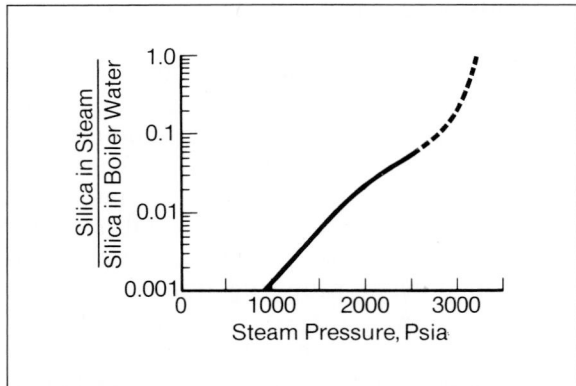

Fig. 22 Distribution of silica in steam and water

ers at pressures from 600 to 1200 psig, they became obsolete as water treatment improved and pressures increased. Interest in steam washers has been renewed however because of their potential for reducing the amount of silica vaporized with the steam. This reduction is not possible with mechanical devices used for steam separation.

Vaporization of silica increases with pressure, as illustrated in Fig. 22. At 2500 psia, the silica in the steam is about 10 times the quantity at 1500 psia, for the same concentration in the boiler water.[6] Boiler-water pH affects the vaporization of silica, as shown in Fig. 23. Controlling silica at recommended values such as these will assure less than 20 ppb in the steam.

Laboratory results and field data conclusively indicate that silica reduction by steam washing in boilers above 1800 psig is ineffective. Improving the quality of makeup water, then, is the only effective means of controlling the silica problem above 1800 psig.[7]

VAPOROUS CARRYOVER

As operating pressures increase, the steam phase exhibits greater solvent capabilities for the salts that may be present in the water phase. These salts will be partitioned in an equilib-

Table I. Summary of Laboratory Results

Drum pressure (psig)	2600	2800	3000
Concentration in boiler water, ppm	15	15	15
Sodium sulfate, %	0.02	0.04	0.28
Disodium phosphate, %	0.01	0.03	0.41
Trisodium phosphate, %	0.02	0.04	0.35
Sodium chloride, %	0.04	0.09	0.39
Sodium hydroxide, %	0.02	0.08	0.55

rium between the steam and water. Known as vaporous carryover, this phenomenon will contribute additional boiler-water solids directly to the steam, independent of the efficiency of steam-water separation components.[8]

Although silica was the first material found to exhibit significant vaporous carryover, it is now realized that such carryover contributes a major proportion of total solids in the steam as drum operating pressures increase above 2600 psig. And it is recognized that controlling the amount of vaporous solids in the boiler water will be required for these other salts as it was for silica.

Table I shows results of a laboratory experiment conducted in 1973-1974 to study vaporous carryover.[9] For various salts, it lists percent vaporous carryover, which is the ratio of the salt concentration in the steam and boiler water. In each case, sodium was measured and then converted to the concentration value.

REFERENCES

[1] Glenn R. Fryling, ed., *Combustion Engineering: A Reference Book on Fuel Burning and Steam Generation*, Rev. ed. New York: Combustion Engineering, Inc., 1966, pp. 25-1–25-27.

[2] Warren M. Rohsenow and H. Choi, *Heat, Mass and Momentum Transfer*. Englewood Cliffs, NJ: Prentice-Hall, 1961, pp. 211–236.

Samson S. Kutateladze, *Fundamentals of Heat Transfer*, 2nd rev. and augm. ed. Translated by Scripta Technica, Inc., and edited by Robert D. Cess. New York: Academic Press, 1963, pp. 342–379 and 380–398.

[3] G. A. Nothman and R. C. Binder, "Slip Velocity in Boiler-Tube Circuits," *Combustion*, 14(12): 40–42, June 1943.

[4] R. W. Haywood, "Research into Fundamentals of Boiler Circulation Theory," *Proceedings of the General Discussion on Heat Transfer*, jointly sponsored by the American Society of Mechanical Engineers and the Institution of Mechanical Engineers, London, Sept. 11–13, 1951, and Atlantic City, Nov. 26–28, 1951, pp. 63–65. London: Institution of Mechanical Engineers, 1951.

W. Yorath Lewis and S. A. Robertson, "The Circulation of Water and Steam in Water Tube Boilers and the Rational Simplification of Boiler Design," *Proceedings of the Institution of Mechanical Engineers*, 143: 147–178, 1940.

S. A. Robertson, "Communication on the Circulation of Water and Steam in Water Tube Boilers and the Rational Simplification of Boiler Design," *Proceedings of the Institution of Mechanical Engineers*, 144: 184–190, 1940.

R. S. Silver, "A Thermodynamic Theory of Circulation in Water Tube Boilers," *Proceedings of the Institution of Mechanical Engineers*, 153: 261–281, 1945.

R. F. Davis, "Expansion Theory of Circulation in Water Tube Boilers," *Engineering*, 163: 145–148, 1947.

[5] T. Ravese, "The Application and Development of the Turbo Steam Separator," *Combustion*, 26(1): 45–47, July 1954.

E. M. Powell and H. A. Grabowski, "Drum Internals and High-Pressure Boiler Design," *ASME Paper No. 54-A-242*. New York: American Society of Mechanical Engineers, 1954.

[6] F. G. Straub and H. A. Grabowski, "Silica Deposition in Steam Turbines," *Transactions of the ASME*, 67: 309–316, May 1945.

[7] H. A. Klein, "Evaluation of Steam Washers in Power-Plant Boilers," *Transactions of the ASME. Journal of Engineering for Power*, 83, Series A: 343–353, October 1961.

[8] F. Gabrielli and H. A. Grabowski, "Steam Purity at High Pressure," Presented at the ASME-IEEE-ASCE Joint Power Generation Conference, Charlotte, NC, Oct. 8–10, 1979.

[9] S. L. Goodstine, "Vaporous Carryover of Sodium Salts in High-Pressure Steam," *Proceedings of the American Power Conference*, 36: 784–789, 1974. Chicago: Illinois Institute of Technology, 1974.

CHAPTER 6

Designing for Boiler Performance

To design a large steam generator, manufacturers follow a step-by-step process that begins with establishing basic output-steam conditions and systematically building to the structural design of furnace walls, headers and supporting steel.

So that each manufacturer can develop a competitive bid that will give reasonably similar steam- and water-side performance, owners issue a set of design criteria that form the basic ground rules. Generally, the scope-of-supply specifications issued to a boiler manufacturer cover the pressure parts, the pulverizing and fuel-burning equipment, the air heaters, the sootblowers, the furnace containment and ductwork, and the furnace safeguard supervisory system. "Standard versions" of specifications include

- primary-steam and reheat-steam pressures and temperatures
- feedwater temperature
- complete fuel and ash analyses
- ambient-air and outlet-gas temperatures

Often, the mechanical-draft fans, structural steel, and furnace-bottom ash-handling equipment are also part of the scope of supply. But even if these items are not included, their required performance and physical characteristics must be defined because of their important role in performance, accessibility, and maintenance of the boiler and because they directly interface with the primary steam-generating equipment. Only when the owner asks for expanded-scope "boiler islands" will the manufacturer design such peripherals to the steam generator as walkways and stairs, steam and feedwater piping, the boiler building, and the coal silos or bins. The steam-generator supplier can also be responsible for the design of precipitators, flue-gas desulfurization systems, and complete ash-handling equipment.

AN OVERVIEW OF DESIGN BASICS

One of the first steps in the boiler design process is to calculate boiler efficiency by using the required excess-air percentage for the fuel to be burned. Working in a systematic fashion, the designer next determines fuel, air and gas quantities, selects pulverizers, and establishes a preliminary configuration of furnace windboxes.

From analysis of pulverizing-air requirements, the feedwater temperature, and the variation of these and other factors over the normal operating range of the unit, the designer arrives at air-heater inlet and outlet gas and air temperatures. In turn, this determination allows calculation of net heat input, furnace size, and outlet gas temperature.

SELECTING HEAT-ABSORBING EQUIPMENT

Having settled on the basic "box size" and gas temperature leaving the furnace, the designer now calculates the gas-temperature drops for all of the superheater, reheater and economizer sections up to the air-heater inlet. A graphic representation of the design calculations to this point gives the skeleton unit shown in Fig. 1.

By an iterative process, the several high-pressure heat-transfer components are "surfaced." That is, the designer assumes a total physical configuration, including tubing diameters and spacings, then defines the net area for gas flow between the tubes (called the *free gas area*) and determines heat-transfer rates for each section. When a satisfactory arrangement of surface is determined, the designer calculates the steam- and water-side pressure drops, and gas-side pressure losses (also referred to as *draft losses*). Finally, metal temperatures are calculated and tubing diameters, materials, and thicknesses are chosen.

With the overall steam-generating unit shape now developed, the cut-and-try strategy can be applied to the circulation-system design for which a general approach was outlined in Chapter 5. Usually, the waterwall tubing diameter, spacing and thickness are predetermined from design experience with units of the same size and pressure level. To arrive at the desired system performance, therefore, the designer's effort centers on proper sizing of downtakes, supply tubes, risers, and circulating pumps.

SELECTING NON-HEAT-ABSORBING EQUIPMENT

At this point the design of the basic heat-absorbing circuitry is complete and the design engineer proceeds with specifications for

■ drums, headers, desuperheaters, and connecting piping (discussed later in this chapter)

■ the ignition system (see Chapter 12) air heaters, pumps, and sootblowers (see Chapter 14)

■ ductwork and fans (the sizing of which are covered in this chapter and the mechanical features of which are described in Chapter 14)

■ the furnace-wall structure, which has to consider both explosive and implosive forces (see Chapter 13)

■ ash-handling equipment (see Chapter 16)

■ the structural support system, including hanger rods, columns, girders, and beams

■ platforms, walkways, and stairs

■ other balance-of-boiler-island items, such as specialized chemical-recovery-boiler auxiliaries, emission-control equipment, and stacks.

Except in the case of marine boilers and some retrofit situations, an owner usually does not severely restrict the physical space made available for a major boiler installation. Although the designer has a free hand to establish an optimum configuration by using existing company's standards, costs of building volume and auxiliary-equipment power consumption will often lead to consideration of several options in both basic pressure-part and pulverizer arrangement as well as the selection of the boiler-related auxiliaries. The process of selecting the equipment will be described in greater detail in subsequent sections.

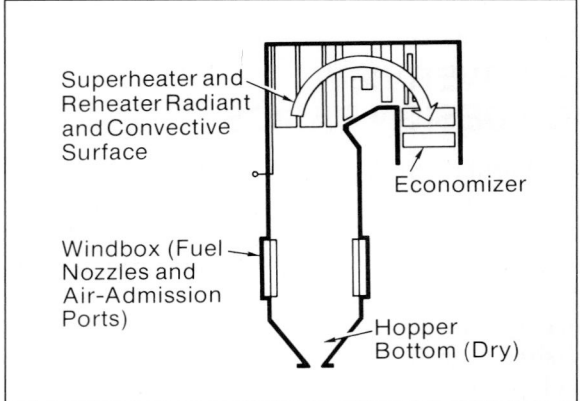

Fig. 1. Skeleton arrangement of radiant-reheat unit showing superheater, reheater and economizer downstream of furnace.

Superheater and Reheater Radiant and Convective Surface

Economizer

Windbox (Fuel Nozzles and Air-Admission Ports)

Hopper Bottom (Dry)

This introductory look at the design process has viewed its basis, its scope, and its sequence. The next section explains step-by-step how to calculate unit heat absorption, efficiency, and fuel fired.

HEAT ABSORPTION AND EFFICIENCY CALCULATIONS

The first calculation in the boiler-design process is that of the heat absorbed by the working fluids—the primary steam flow and the reheated-steam flow. The purchaser establishes these flows at one or more load points, and gives the steam-generator supplier the pressures and temperatures at the inlets and outlets of the primary and reheat headers.

Boiler capacity is given in terms of the mass flow of primary steam at maximum continuous rating (MCR). In this and other texts, nominal boiler size is commonly expressed in "megawatts", an approximation used for convenience—steam-turbine generators are rated in electrical output on the basis of stated condensing conditions. Boilers are properly rated on their steam output(s) at certain pressures and temperatures.

HEAT ABSORBED

With the outlet steam conditions identified, the heat content or enthalpy, h, in Btu per lb, is read from steam tables (Appendix F) by interpolating as necessary. The heat in the feedwater, h_f, is read from tables or charts for compressed water such as Fig. 2. For a nonreheat unit, if W_s represents the steam produced in pounds per hour, the total boiler output or heat absorbed, Q_{abs}, will be

$$Q_{abs} = W_s(h - h_f)$$

(1)

For a unit with one or more stages of steam reheating, the heat absorbed in the reheater(s) is added to the primary-steam heat absorption to arrive at the total Q_{abs}. Fig. 3 emphasizes that this total heat absorption takes place between point A, where the fuel and preheated combustion air enter the furnace and burn, and point B, where the gases leave the economizer, which is usually the final primary-working-fluid heat absorber. The heat of the products of combus-

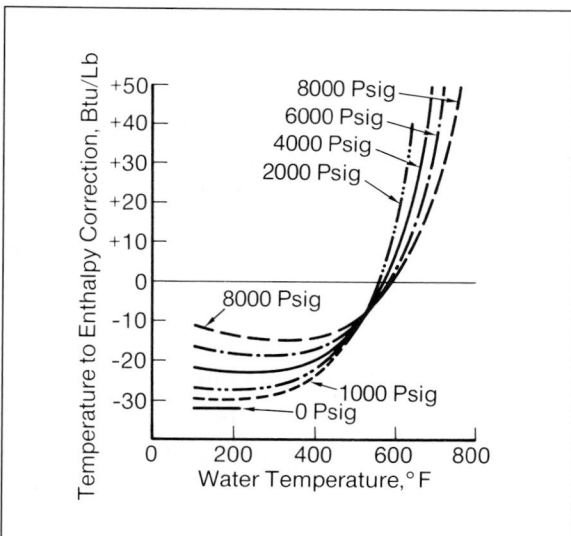

Fig. 2. Enthalpy of Compressed Water (to obtain enthalpy in Btu/lb (above 32°F) for any water temperature and pressure, add or subtract the value on the ordinate from the water temperature. Example: 400°F water at 4000 PSIG, read −20 from curve. $h_f = (400 − 20) = 380$ Btu/lb)

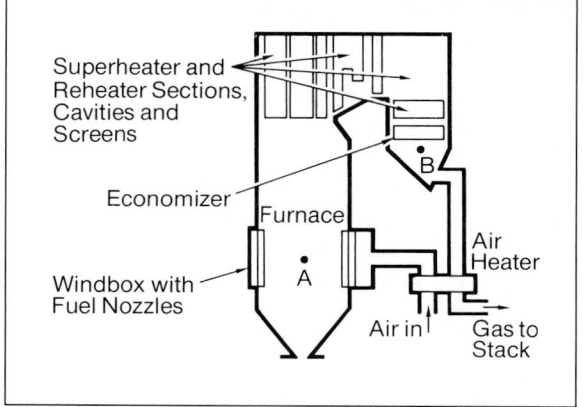

Fig. 3. Total heat absorption by the primary working fluids in a steam generator—the water and steam—takes place between points A and B. The theoretical or adiabatic temperature at point A includes the heat added to the combustion air in the air heater. (Heat is not *absorbed* in the air heater, but is transferred only.)

tion passing through the air heater is picked up by the air and transferred immediately to the furnace, essentially without loss.

HEAT FIRED

The fuel fired, Q_f, in Btu's per hour, is given by the relation

$$Q_f = \frac{Q_{abs}}{\eta}$$

(2)

where η is the overall chemical efficiency of the steam generator. (See Chapter 21, Testing and Measurements, for more information on the definition of chemical-heat efficiency.)

The quantity of fuel to be burned, the amount of air to be supplied, and the products formed are all calculated from Q_f. These three quantities largely determine the size of the furnace and the other components.

The amount of fuel fired, sometimes called the combustion rate, is found by dividing Q_f by the high heating value of the fuel (to give the mass of fuel fired per hour) or, for gaseous fuels, the volumetric rate which is the cubic feet of gas fired per hour.

STEAM-GENERATOR EFFICIENCY

The efficiency of a boiler is most frequently taken to be the ratio of the heat absorbed by the water and steam to the chemical heat in the fuel fired. It is not a measure of the efficiency with which the coal or other fuel is fired into the combustion chamber or onto a grate, although the carbon heat loss item in a heat balance will account for this. Modern suspension-burning systems reduce this carbon heat loss inefficiency to minimal amounts. Nor is boiler efficiency the net rate at which an electrical generating plant produces power, called the net plant heat rate in Btu/kWhr. But boiler efficiency must be known to determine the "Btu" consumed so that the net plant heat rate can be established for a given boiler turbine unit.

Simplistically, a boiler absorbs most of the heat in the fuel fired but is unable to "capture" the heat in the gases going to the stack. Included in that lost heat to the stack is the latent heat of vaporization of the water in the fuel and the water formed by the combustion of hydrogen, which is impossible for the boiler to absorb. Thus, efficiency becomes a function of the type and analysis of fuel fired, the excess air at which it is burned, the air temperature entering and the gas temperature leaving the air preheater, and several other factors described later. Based on the fuel and the above temperatures, heat losses and credits are calculated as a percent of the heat input, and the efficiency becomes

$$\eta = 100 - \%\text{Losses} + \%\text{Credits}$$

(3)

The process of accounting for all the heat losses, as well as the heat available in the steam, is known as a *heat balance*. The heat balance method is frequently used to test for the efficiency of an operating unit. Unlike the test engineer who can measure these losses accurately, the designer must assign values to some of them based on experience.

The following losses are those which must be known before the efficiency can be determined.

- loss in dry products of combustion
- loss due to moisture in air
- loss due to moisture from fuel
- loss due to water vapor in gaseous fuels
- loss due to moisture from hydrogen in fuel
- sensible heat loss in ash
- loss due to unburned combustible
- loss due to radiation and convection
- other losses and heat credits

COMBUSTION CALCULATIONS

With the fuel analysis given, the excess-air percentage decided on (see subsection Excess Air for Combustion, which follows), and the moisture content of the air stated or assumed, the designer can perform the combustion calculations as delineated in Chapter 4. By using the combustion equations or the graphical

methods presented in Appendix C, the air for combustion, the dry weight of products, the weight of water resulting from burning the fuel, and the other items required for heat-balance work are determined. Also, a check is made of the ultimate fuel analysis supplied versus the high heating value, to discover early any discrepancy in this correlation (see Chapter 4 for more on that aspect of the calculation).

EXCESS AIR FOR COMBUSTION

Perfect or stoichiometric combustion is the complete oxidation of all the combustible constituents of a fuel, consuming exactly 100 percent of the oxygen contained in the combustion air. Excess air is any amount above that theoretical quantity.

Commercial fuels can be burned satisfactorily only when the amount of air supplied to them exceeds that which is theoretically calculated as required from equations showing the chemical reactions involved. The quantity of excess air provided in any particular case depends on

- the physical state of the fuel in the combustion chamber
- fuel particle size, or oil viscosity
- the proportion of inert matter present
- the design of furnace and fuel burning equipment

For complete combustion, solid fuels require the greatest, and gaseous fuels the least, quantity of excess air. Fuels that are finely subdivided on entering the furnace burn more easily and require less excess air than those induced in large lumps or masses. Burners, stokers and furnaces having design features producing a high degree of turbulence and mixing of the fuel with the combustion air require less excess air.

Table I indicates the range in values for the excess-air percentage commonly employed by the designer. These are expressed in percent of theoretical air, and are understood to be at the design load condition of the boiler. (At lower loads, both in design and in operation, higher percentages are sometimes used.)

Table I. Excess Air at Furnace Outlet

Fuels		% Excess air
Solid	Coal (pulverized)	15–30
	Coke	20–40
	Wood	25–50
	Bagasse	25–45
Liquid	Oil	3–15
Gaseous	Natural gas	5–10
	Refinery gas	8–15
	Blast-furnace gas	15–25
	Coke-oven gas	5–10

AIR TEMPERATURE FOR COMBUSTION, t_a

Efficiency calculations are based on the air temperature that will actually be entering the air heater, not on the ambient temperature (t_{aa}) nor the specified reference air temperature (t_{ra}). The increase in temperature of the air as it passes through the forced-draft (FD) fan and the steam or hot-water air heater must be added to t_{aa} to arrive at t_a.

A function of the power consumed by the fan, air temperature rise through an FD fan can be calculated from

$$\Delta T_{fd} \simeq \frac{10,300(HP)}{W_a}$$

(4)

where ΔT_{fd} is in °F, HP is the brake horsepower to the fan shaft, and W_a is the weight of air being handled in lb/hr. The relation is based on conversion to heat of the air horsepower and the horsepower corresponding to the inefficiency of the fan. This heat must be stored in the air. The formula includes a small correction required for velocity pressure at the fan outlet but the formula cannot be used to predict the temperature rise at the no-flow or shut-off condition.

AIR TEMPERATURE RISE BY STEAM OR HOT-WATER HEATING

It is common in power plants to provide heating coils ahead of the main air heater inlets to preheat the air entering those heaters. Although primarily used to reduce the potential for corrosion on the heating surfaces of the air heater, preheating air with steam or hot water can substantially improve the overall plant efficiency. Preheating increases the heat content of the incoming combustion air which helps increase unit efficiency and thereby decrease the fuel that is fired in the boiler.

The economic and practical limits to improved boiler efficiency by lowering the unit's exit-gas temperature, have essentially been reached. Corrosion and/or plugging of air heaters and dust-collection equipment generally determine the lower temperature limit, which depends largely on the sulfur content of the fuel being burned. Efforts to design and operate modern units for sustained periods with 250°F final flue-gas temperature at full load generally have been unsuccessful with many fuels, and usually the design objective[1] is at least 25°F higher, or 275°F. Extraction-steam air preheating has both increased the efficiency and protected air heater surfaces because the steam

flow through the high-pressure turbine blading can be greater for a given electrical output. Thus, steam-turbine efficiency is improved by extracting steam for air heating just as it is improved with feedwater heating.

AIR HEATER COLD-END PROTECTION

The part of an air heater called the *cold end* is the section in which the incoming "cold" air meets the exiting cooled products of combustion. The sum of these temperatures divided by 2 is called the average cold-end temperature (ACET). Since cold-end fouling and corrosion can be related to the ACET level, it is usual practice to work to such a temperature to establish the exit-gas temperature and the required air temperature leaving the steam or hot-water heater. A guide for setting the cold-end temperature is given in Fig. 4.[2]

For subbituminous and lignitic coals having the sum of the lime (CaO) and magnesia (MgO) greater than the ferric oxide (Fe_2O_3) in the ash, the sulfur content used for reading the curve should be adjusted to *equivalent sulfur*, ES.

$$ES = 14,000 \times S/HHV$$

(5)

where S is the percent sulfur in the coal as fired and HHV is the high heating value of the coal as fired.

Thus, for a low-rank coal with an HHV of 7,000 Btu/lb and an as-fired sulfur content of 1 percent, the curve must be read at an ES of 2 percent, for an ACET of 163°F.

EXIT GAS TEMPERATURE— CORRECTED VS. UNCORRECTED

Note that the ACET in Fig. 4 is termed "uncorrected". That means that the gas temperature used in the averaging calculation is not adjusted for the leakage from air- to gas-side of the air heater. Because of the differential pressure between the air entering and the gases exiting an air heater (the latter being below atmospheric pressure in most instances), there is a flow of cooling and diluting air over to the gas-side of the heater.

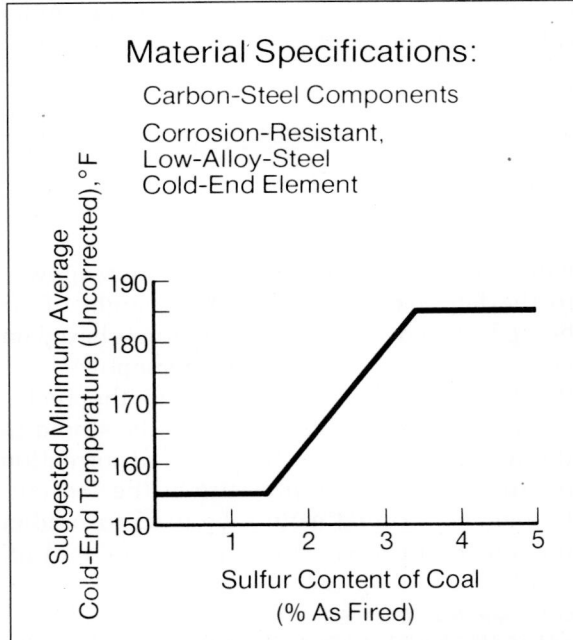

Fig. 4. Recommended minimum average cold-end temperature for coal firing

The dilution raises the excess-air percentage that is measured in the exiting flue gases. The measured exit temperature at the preheater is lower than the uncorrected (actually un-measurable) temperature because of this air leakage. In efficiency calculations, designers use the uncorrected (theoretical) temperature and the excess-air percentage *entering* the air preheater as the basis for the heat losses. The result would be the same if the corrected temperature and the corrected excess air were to be used in the calculations.

It is important to realize that the more input parameters specified, the less the potential for differences in fuel fired and air and gas quantities when bids are taken. The preceding groundwork is fundamental to the calculation of the several heat losses.

LOSS IN THE DRY PRODUCTS OF COMBUSTION

Frequently the largest of all, the loss in dry products of combustion represents the difference between the heat content of the dry exhaust gases and the heat content these gases would have at the temperature of the ambient air. The dry gas (P_d) is calculated by subtracting the water vapor in the products from the total products of combustion (P) with all quantities expressed in pounds per million Btu input (see Chapter 4). Knowing P_d, the percent dry-gas loss is found from

$$L_{dg} = \frac{P_d \times 0.24\,(T_g - T_a)}{10^4}$$

(6)

LOSS DUE TO MOISTURE IN AIR

The moisture in the combustion air (W_a) is determined for specified conditions of temperature and relative humidity, as described in Chapter 4. For an atmospheric air requirement (A) in pounds per million Btu fired and typical conditions of 80°F and 60-percent relative humidity, W_a will be 0.013 CA, where C is the combustible-loss correction factor.

The heat loss, in percent of heat input, equals

$$L_a = \frac{W_a\,(0.46)\,(T_g - T_a)}{10^4}$$

(7)

LOSS DUE TO MOISTURE IN FUEL

This loss represents the difference in the heat content of the moisture in the exit gases and that at the temperature of the ambient air. If H_2O includes both the surface and hygroscopic moisture in percent by weight of fuel fired, then

$$W_c = H_2O\,(10^4)/\text{HHV}$$

(8)

For T_g higher than 575°F, the loss due to fuel moisture equals

$$L_f = \frac{W_c\,(1066 + 0.5T_g - T_f)}{10^4}$$

(9)

and for T_g lower than 575°F,

$$L_f = \frac{W_c\,(1089 + 0.46T_g - T_f)}{10^4}$$

(10)

in which T_f corresponds to the temperature of the fuel received at the pulverizers or the oil or gas guns. This temperature is ordinarily taken as 80°F or the reference air temperature if the actual temperature is not known.

It should be understood that W_c does not include water vapor as found in gaseous fuels.

LOSS DUE TO WATER VAPOR IN GASEOUS FUELS

Moisture is also present in many gaseous fuels, particularly in blast-furnace gas and coke-oven gas, which are frequently dedusted by passing them through sprays of water. This moisture exists in the gas in two separate forms, each requiring different treatment in calculating the heat balance.

Washed gas contains *entrained* water in the form of suspended globules. Where exceptionally clean gas is not required, the entrained moisture, W_c, averages 7 lb per million Btu. No entrained water is in unwashed gas.

In addition to the visible moisture in a liquid state, nearly all gaseous fuels contain some water vapor. In natural gas, the water vapor is there because of salt water that has been in contact with the gas in the ground, or because of *rehydration*. In refinery gas, blast-furnace gas, and coke-oven gas it is present either owing to the nature of the process of which these gases are byproducts, or because of subsequent cleaning operations.

The water vapor (W_s) in pounds per million Btu is of such magnitude that it can usually be neglected in heat-balance calculations for natural gas, refinery gas, or coke-oven gas. For blast-furnace gas saturated with moisture at 60°F, W_s may be taken as 8 lb per million Btu without serious error.

In a heat balance for gaseous fuels, W_s must be considered separately from W_n and W_c, since it exists as a vapor which already has the latent heat of vaporization, and therefore requires no heat evolved by the fuel to vaporize it. In every respect it is similar to W_a so that the percent loss due to its presence is

$$L_s = \frac{W_s (0.46) (T_g - T_a)}{10^4}$$

(11)

LOSS DUE TO MOISTURE FROM HYDROGEN IN FUEL

This loss includes the sensible and latent heats of the moisture from the combustion of the hydrogen in the fuel above the ambient or reference air temperature. The water from hydrogen (W_h) equals $9 \times H \times 10^4/HHV$, as previously defined. When only the proximate analysis is available, it is necessary to resort to an empirical relation between the hydrogen in the fuel and the volatile matter, such as is found in Appendix C. But remember that, in deriving the hydrogen moisture from a proximate analysis, the loss due to the water loss may be in error as much as 1.5 percent—equal to or greater than some other items in the heat balance.

For T_g higher than 575°F, the loss due to moisture from hydrogen in fuel equals

$$L_h = W_h (1066 + 0.5 \, T_g - T_a)/10^4$$

(12)

and for T_g lower than 575°F, the loss equals

$$L_h = W_h (1089 + 0.46 \, T_g - T_a)/10^4$$

(13)

in which T_a is the air temperature entering the air preheater; that is, the ambient air temperature plus the FD fan and steam-air-heater temperature rises. The above relationships apply to solid, liquid and gaseous fuels.

SENSIBLE HEAT LOSS IN ASH

Hot ash from the burning of a solid fuel can represent a substantial heat loss, dependent on the quantity and the temperature at which ash is rejected from the steam generator or auxiliary equipment. The significant locations of ash extraction from most boilers are the furnace bottom and the hoppers under the economizer, the air heater(s) and the flyash-collection device, usually an electrostatic precipitator or a fabric filter. Flyash leaving an air heater at a temperature such as 300°F, for example, represents a heat loss whether it is collected in a precipitator or not. Further, even if it is cooled below 300°F in the collection process, such cooling has to be accounted for.

The *ASME Performance Test Code* PTC 4.1 provides for the measurement of the sensible heat lost in the ash leaving a boiler. The only specific method given (at this writing) is for solid-fuel-fired units in which the ash or slag

falling from the walls of a furnace is caught in a wet receiver. The heat loss of this ash/slag is determined by measuring the temperature increase and flow of the cooling water in and out of the bottom-ash hopper, and making the indicated corrections for evaporation and sluicegate leakage. Flyash heat is Code-designated as L_d, the heat loss due to sensible heat in the flue dust; this is different and distinct from any loss attributed to unburned carbon in the ash.

For the majority of pulverized-fuel boilers firing low- to medium-ash coals, the heat loss is well below 1 percent of the fuel fired, and is not accounted for explicitly either in a predicted heat balance or in an ASME Short Form efficiency test. Since it is both difficult and expensive to measure, the ash heat loss on dry-bottom pulverized-fuel boilers with wet ash receivers usually is considered part of the unmeasured and unaccounted-for losses described below.

For stoker-fired and fluidized-bed boilers, which maintain as much of the ash as possible in a high-temperature moving or circulating bed, the hot ash and other solids leaving the bottom of the unit, for high-ash fuels, may contain a large amount of heat representing well over 2 percent of the heating value of the fuel fired. See Figure 5, which indicates the magnitude of this heat loss as a function of the pounds of ash per million Btu fired and the rejection temperature. The correlation is based on specific heats varying with temperature, for ash of an assumed composition. (The specific-heat variation does not consider the effect of phase changes of certain ash constituents which can occur in the range of 800 to 1200°F.) For contract efficiency determinations, the boiler supplier and owner must agree both on the specific heats to be used and the means of determining reject-ash/solids quantity and distribution.

Note that the ash-removal temperature, for heat-balance purposes, is the temperature below which no heat is added to any of the steam-generator working fluids—fuel, air, water, or steam. Also, for fluid-bed combustion, there is usually an additive solid, such as limestone, to facilitate operation and/or capture sulfur, that adds to the effective ash quantity per million

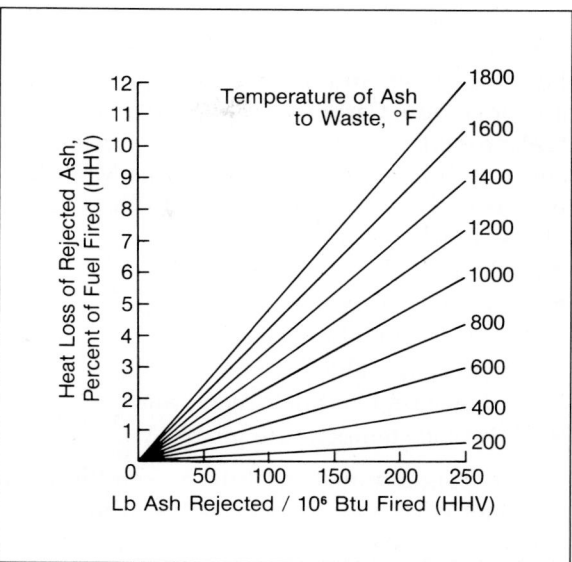

Fig. 5 Heat loss of ash rejected from stoker grate or fluidized bed above 80°F reference temperature. Ash here includes fuel ash plus rejected limestone or other additive/sorbent.

Btu fired. Thus, to determine an overall value for the heat loss attributable to the hot ash leaving a boiler, the temperatures at all rejection points, as well as the quantities at all points, have to be established by calculation or estimation. A summation of the weighted values of the rejects will give the total heat loss due to the sensible heat in ash-plus-additive.

LOSS DUE TO UNBURNED COMBUSTIBLE

This loss comprises unliberated thermal energy due to incomplete oxidation of the combustible matter in the fuel. The loss can be from unburned carbon that is trapped in the flyash or the furnace-bottom ash deposited in the ash receiver of a pulverized-fuel fired or fluid-bed boiler, or it can be from the incomplete combustion of the carbon in the fuel as evidenced by the presence of carbon monoxide (CO) in the gaseous products leaving the furnace. For design calculations, liquid and gaseous fuels are generally presumed to burn with zero combustible loss; CO loss is assumed to be zero, regardless of the fuel.

For suspension firing of pulverized coal and other solid fuels, control of unburned carbon heat loss is an important objective in design and operation. Carbon burnout is dependent on fuel properties, furnace and firing system design, and unit operating conditions. Major factors influencing complete combustion of carbon are:

- fuel reactivity
- fuel fineness/particle size
- efficiency of fuel-air mixing
- excess air available for combustion
- residence time in the furnace, and
- the furnace temperature profile.

The designer must carefully match these parameters with fuel characteristics to minimize carbon heat loss. In most instances, the loss due to unburned carbon is controllable to below $1/2$ percent of the fuel fired.

C-E has developed a methodology for predicting solid carbon loss during the suspension firing of pulverized coal using advanced laboratory characterization and mathematical furnace modelling techniques. The methodology is described in Reference 3; it involved a comprehensive reactivity study of a wide range of solid fuels in C-E's Drop Tube Furnace System (DTFS), which incorporates an entrained laminar-flow reactor. The DTFS is described in further detail in Chapter 3 of this text, and in References 4, 5, and 6 of this chapter.

The fundamental information obtained from this study comprises physical and chemical characteristics and combustion kinetics of many coals, chars, coal-derived synfuels, and refuse-derived fuels. In devising this improved method for predicting combustion efficiency, it was determined that better accuracy could be achieved if coal pyrolysis and volatile-matter combustion were separated from the dominant heterogeneous char-burning phase of coal combustion. An accurate simulation of temperature/time history to which coal particles are exposed was of primary importance; this was accomplished in the Drop Tube Furnace. Finally, it was recognized that meaningful calibration of a mathematical model would be necessary, achieved through acquisition of field data taken during combustion of the same fuels as tested in the laboratory. The validity of the predictive technique was tested by comparison between carbon-loss values generated by the laboratory and those directly measured in operating full-scale units using the same fuel.

LOSS DUE TO RADIATION AND CONVECTION

This loss is a comprehensive term used in a heat balance calculation to account for heat losses to the air through conduction, radiation, and convection. The heat emanates from the boiler, ductwork, and pulverizers. Because it is very difficult to measure on large operational boilers, this is the only significant loss for which computation is not based on test measurements in the *Performance Test Code* for boilers.

Fig. 6, used for both design work and performance tests on fully watercooled furnaces, was extrapolated by the American Boiler Manufacturers Association from data obtained from some relatively small boilers. The curve values were later checked by the ASME Test Code committee against actual measured losses on several large boilers and were found to be conservatively high.[7] Note that this item does not account for or include radiation, convection, or infiltration losses from a precipitator located ahead of the air heater(s).

For boilers having ratios of external surface to volume, or surface temperatures, that vary considerably from those used in establishing these curves, direct calculation of radiation/convection loss can be done.

UNACCOUNTED-FOR LOSSES

This item represents unclassified, contentious, and difficult-to-measure losses that are included in a heat balance to arrive at a guaranteed contractual efficiency.

The value used consists of three components, of unstated weight. First, there are the unmeasured losses which, because of the difficulty or the great expense of measurement, are best included in a convenient margin. Typical is the

sensible-heat-in-ash loss for pulverized-coal boilers covered in the previous section. Approximation of such a loss by calculation frequently will be a more satisfactory approach than setting up and operating the very complicated testing equipment and instrumentation to obtain meaningful results.

Second, the loss includes some unstated tolerance for instrument error. Whether the short-form performance test is made, or the fully detailed and very costly long-form procedure is followed, it is important to arrive at an efficiency that can be accepted both by the owner and the supplier. Chapter 22 on field testing and measurement covers the probable, and rather large, measurement error that can be expected in performance testing. Having some "room for error" will often eliminate the need for retests on the basis of faulty instrumentation, poor placement or operation of probes,

and other potential areas of disagreement.

The third component, grouped with instrument-error, is the manufacturer's tolerance in making an efficiency guarantee.

HEAT CREDITS

A heat credit is any energy obtained from outside the envelope of the steam generator (see Fig. 7, Chapter 22) and added to the working fluid or fluids, the air, or the gas inside the envelope boundary. This is over and above the chemical heat in the fuel fired and is usually in the form of electrical energy.

As a practical matter only major credits are considered in most heat-balances. These include the heat from the power conversion in pulverizer motors, primary-air-fan motors, gas-recirculating-fan motors, and circulating-pump motors.

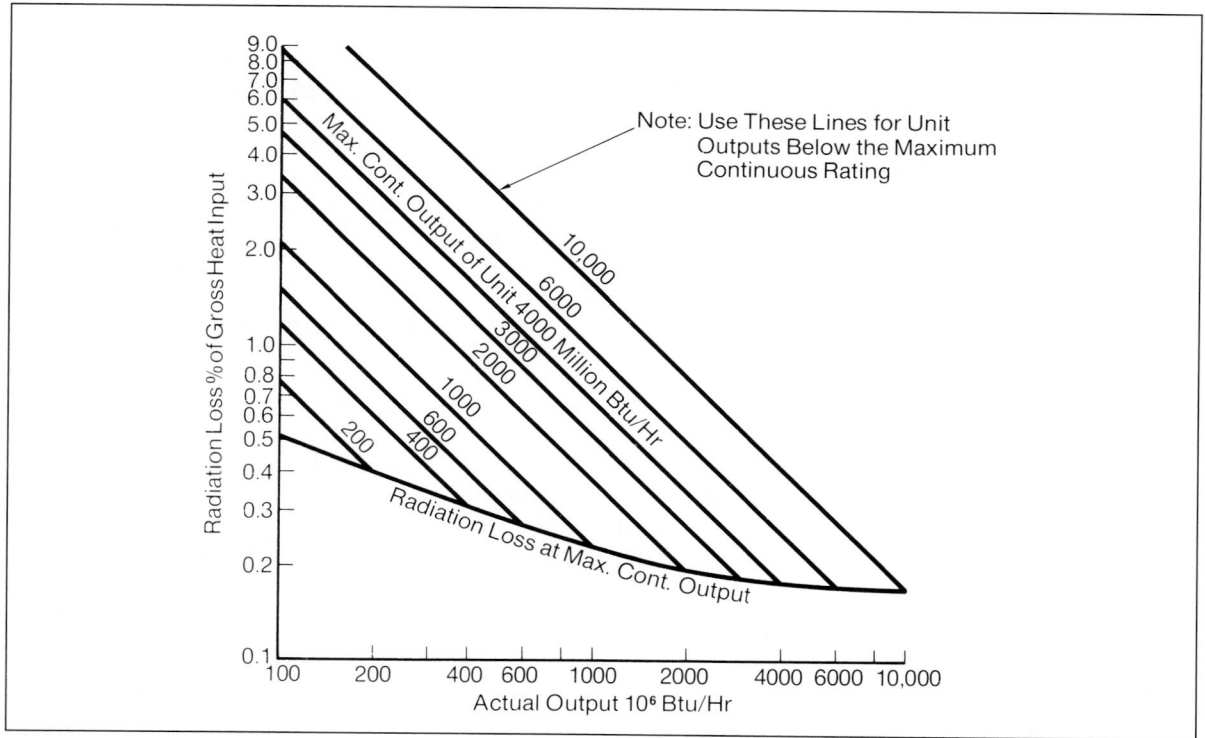

Fig. 6. Radiation loss for completely water cooled furnaces. The radiation loss values obtained from this curve are for a differential of 50°F between surface and ambient temperature and for an air velocity of 100 feet per minute over the surface. The curve is based on an analysis done by the American Boiler Manufacturers Association (ABMA), and is published in the original format in the *ASME Performance Test Code* PTC-4.1.

Credits are calculated by determining the power input to the motor (or steam turbine if the steam is supplied from outside the envelope) and dividing by the heat fired (Q_f) if the power is applied to the air or gas, or by the heat absorbed (Q_{abs}) if the power is applied directly to the working fluid; the quotient is then expressed in percent of fuel fired for use in the heat-balance calculation.

HEAT LOSS IN INTEGRAL PRECIPITATOR

It has been demonstrated that there is a substantial gas-temperature drop in the ductwork and through the walls of a precipitator located between the boiler economizer and the air heater(s). This temperature drop results from radiation, convection, and air inleakage. No tests have been performed to categorize these contributors to the problem.

Logically, there is a loss of energy which has to be charged someplace. Because the precipitator is not part of the steam-generator circuitry and because the heat is not returned to the air or gas, the loss should not be charged to the boiler.

The most satisfactory approach is to consider the integral precipitator as another "heat absorber", similar to a second reheater, and take the energy absorbed as an additional output. Using the higher Q_{abs} will give the correct Q_f and the correct air and gas weight without penalizing the steam-generator efficiency.

The heat "lost" or "absorbed" in such a precipitator can be calculated by

$$L_{ip} = \frac{W_g \times c_p \times \Delta T_g}{Q_f} \times 100$$

(14)

where
  W_g is the gas flow through the precipitator, lb/hr
  c_p is the specific heat of the gas, Btu/lb-°F
  ΔT_g is the gas-temperature drop through the precipitator, °F, and
  Q_f is the heat fired, Btu/hr

Note that this method does not take into account the air leaking into the precipitator. If

such leakage can be anticipated and quantified, it should be given special consideration because of its effect on air-heater heat transfer (the lower gas temperature entering the air heater reduces the heat head for thermal transfer) and its effect on the ID-fan performance and power consumption.

HEAT-BALANCE CALCULATION

The heat balance of Table II for a boiler to serve a 500-MW turbine is based on firing a bituminous coal of the analysis given in Chapter 4, page 4–5, at an excess-air percentage of 23 percent. The carbon loss, FD-fan temperature rise, and heat credits are assumed for this example, and will vary from case to case. Combustion air has 0.013 lb of water vapor per lb of air (80°F and 60 percent relative humidity). Fuel temperature is taken as equal to ambient air temperature.

FIRING SYSTEM DESIGN

With combustion and heat-balance calculations completed, and the overall steam-generator heat absorption fixed, the designer can now arrive at the weight of fuel fired.

$$\text{Fuel fired, lb/hr} = \frac{Q_f}{\text{HHV}}$$

(15)

where Q_f is in Btu/hr and the high heating value, HHV, is in Btu/lb.

PULVERIZER SELECTION

It is customary in pulverizer selection to provide one or two "spare" or standby mills to allow for on-line maintenance at boiler full load. (See Chapter 23 for more comments on pulverizer maintenance with the unit in operation.) If there are "n" number of pulverizers installed then (n – 1) mills will be capable of full boiler load grinding the worst fuel that the unit must handle; with an average more

Table II. Heat-Balance Calculations

Exit gas temperature, uncorrected for leakage, t_g, °F	285
Ambient air temperature, t_{aa}, °F	80
Temperature rise, FD fan, Δt_{fd}, °F	10
Air temperature entering air heater, T_a, °F	90
Solid combustible loss, % by weight	0.20
Combustible loss correction factor, C	0.9985
Fuel in products, F, lb/10^6 Btu	83
Air for combustion, A, lb/10^6 Btu	936
Air in products, C × A, lb/10^6 Btu	934
Total products, P = (F + CA) lb/10^6 Btu	1017
Water in air, W_a lb/10^6 Btu	12.1
Water in fuel, W_c lb/10^6 Btu	13.4
Water formed in combustion, W_h lb/10^6 Btu	32.7
Dry gas in products, P_d lb/10^6 Btu	959

Heat loss in dry gas	%	4.49
Heat loss from water in fuel	%	1.53
Heat loss from combustion of H_2	%	3.70
Heat loss from moisture in air	%	0.11
Heat loss from unburned carbon	%	0.20
Radiation and convection loss	%	0.18
Unmeasured losses*	%	1.50
Total losses,	%	11.71
Less heat credits**	%	0.25
Net losses,	%	11.46
Efficiency (100 − net losses)	%	88.54

*Unmeasured losses include tolerance for instrument error and manufacturer's margin.

**Heat credits here are the sum of motor inputs to pulverizers, primary-air fans, and boiler-water circulating pumps, determined in a separate calculation.

favorable fuel, (n − 2) mills can accommodate the maximum load. In either case, pulverizer selection is done using capacity curves similar to those given in Chapter 11, allowing for the wearing parts being in worn condition.

The capacity of a given size pulverizer varies significantly with coal moisture, the grindability of the coal (its "ability to be ground"), and the fineness required for satisfactory, low-carbon-loss burning. Chapter 11, Pulverizers and Pulverized–Coal Systems, covers these parameters and their effects on pulverizer performance. The wider the range in coal calorific value, moisture content and grindability, the greater must be the overall pulverizer capacity.

The statistical probability of all the unfavorable characteristics occurring at the same time, and the total possible period of operation at full boiler load with the worst fuel, have to be weighed against the cost and complexity of multiple installed-spare pulverizers.

WINDBOX DESIGN

With tangential or corner firing, each pulverizer installed delivers ground coal to a row of fuel nozzles located at different levels in the furnace. A row will contain 4 or 8 nozzles depending upon whether the furnace is of the *open type* or of the divided type; the latter usually has a full or partial centerwall between the two firing chambers, with 4 nozzles per horizontal row in each of the two chambers. In an open furnace with six pulverizers, there will be four fuel nozzles arranged in six horizontal rows for a total of 24. Further detail about tangential windboxes and associated ignition equipment is given in Chapter 12. Fuel-Firing Systems.

With front- or rear-wall firing using fixed round burners, the piping from the pulverizers to the burners is configured in various ways depending upon the manufacturer's preference, steam-temperature control means, and nitrogen-oxide reduction strategy.

Windbox sizing for either tangential or wall firing must recognize the effect of plant altitude on airflow velocity, both in primary-air/fuel piping and in secondary-air compartments. Air volume and corresponding velocity vary linearly with the *elevation correction factor* (barometric pressure at sea level divided by the barometric pressure at altitude; see Fig. 7). Pressure drop through a given windbox opening or burner throat varies as the square of the actual velocity divided by the elevation correction factor; this must be taken into account in fan sizing, covered later in this chapter.

AIR-HEATER DESIGN

Having designed the windbox and selected the pulverizers to accommodate the least fa-

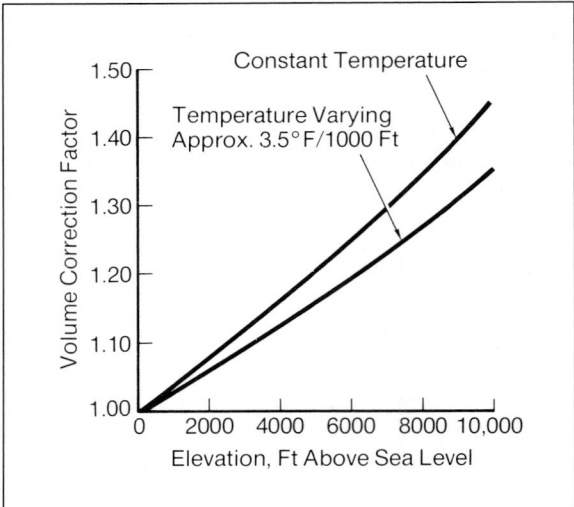

Fig. 7 Volumetric correction for elevation above sea level (Calculated from the data of Standard Atmosphere—Tables and Data to 65,800 Feet, NACA Report 1235)

vorable present or future fuel, the designer now has sufficient information to define the air preheater requirements.

The gas flow is the primary determinant in sizing an air heater, since the pressure drop through the gas side varies predictably with load (being essentially proportional to the square of the gas flow and linear with absolute gas temperature). Primary and secondary airflows are additive to the total airflow required, but each will change with fuel moisture, load, and the number of pulverizers in operation. The maximum primary airflow and pressure required by a certain size air heater will be important factors in primary-air fan and motor selection.

Air heater design is covered in detail in Chapter 14. The use of the trisector and other air-heater configurations is treated in connection with pulverizing systems in Chapter 11.

FURNACE PERFORMANCE

The following section deals with the thermal performance of a large steam-generator furnace. Although the determination of furnace

exit-gas temperature is the key to the performance of all the very expensive superheater and reheater surfaces installed behind it, that temperature is not amenable to calculation. Only through field experience with similar fuels and generally the same mode of furnace operation can such temperatures be determined with confidence. And despite using the most sophisticated computer techniques in the analysis of the upper and lower furnace regions as heat-transfer entities, the data input to those computers is based on a subjective judgment of the slagging character of every coal burned.

We will discuss heat transfer in general as a brief introduction to the subject of furnace heat transfer. We will then show a portion of the analytical approach we take to furnace performance, and will refer to other published material for the interested reader.

Finally, we leave the reader with the perception that furnace outlet temperature determination, while not a secret process, is certainly and necessarily a proprietary one.

HEAT TRANSFER

A necessary step in the design of boilers is a detailed engineering analysis of heat transfer to estimate the cost, practicability and size of equipment to interchange a specified amount of heat in a given time. A steam-generating-unit furnace wall, for example, can be operated successfully over extended periods only if surface temperatures can be maintained within acceptable limits by removing heat continuously and at rapid rates. The size and geometry of a boiler furnace depend not only upon the amount of heat to be transmitted but also on the rate at which it can be absorbed or transferred.

MODES OF HEAT TRANSFER

Most textbooks on heat transfer define three modes of heat flow: *convection, conduction* and *radiation*. These three encompass much of the field of physics including mechanics, heat, acoustics, optics and electricity. For gases, conduction takes place by elastic impact; for solid nonconductors, in longitudinal vibration; and for metals, in electronic movement.

The laws of aerodynamics and hydrodynamics govern convection which involves the transportation and exchange of heat due to the mixing motion of different parts of a fluid.

Radiative heat transfer is an electromagnetic event which in fossil-fuel-fired furnaces occurs mainly in the infrared range.

Usually, heat is transferred simultaneously by more than one of these modes.

Convection

Authorities differ as to whether Sir Isaac Newton formulated a definition of convection or a law of cooling when he first expressed the following relationship in 1701:

$$q = hA_s\Delta T$$

(16)

where q denotes the time rate of heat flow, h is the surface or film coefficient of heat transfer, A_s is the heat-transfer-surface area and ΔT is a temperature difference between a surface and a fluid in contact with it.

Conduction

More than a century later, in 1828, the French physicist and mathematician J. B. J. Fourier formulated the basic law of heat conduction, which may be expressed as follows:

$$q = -kA_c\frac{dT}{dx}$$

(17)

where k is the thermal conductivity, A_c is a constant cross-sectional area, and dT/dx is the temperature gradient at the section. The minus sign signifies that the heat is flowing in the direction of decreasing temperature, in accordance with the second law of thermodynamics.

Radiation

In 1879, Joseph Stefan empirically discovered the third fundamental law of heat transfer, the basic equation for total thermal radiation

from an ideal radiator or black body. It was derived theoretically by Ludwig Boltzmann in 1884, and may be expressed as follows:

$$q = \sigma A_s T^4$$

(18)

where σ is the Stefan-Boltzmann constant for total black radiation, A_s is the heat-transfer-surface area, and T is the absolute temperature. (See equations 23 and 23A.)

The Stefan-Boltzmann equation represents the total radiant energy emitted by a black body in all directions, but it does not reveal the distribution of energy in the spectrum. The distribution of emissive power among the different wave lengths was derived by Max Planck in 1900, using the quantum concept.[8]

Heat transfer in boilers will be covered in more detail after furnace performance is discussed. In a later section this chapter covers fluid-to-fluid transfer rates and convection heat-transfer in detail.

FURNACE DESIGN CONSIDERATIONS

The furnace of a suspension-fired steam generator is a large water-cooled chamber in which fuel and air are mixed and burned. Its purpose is to generate heat under controlled conditions with the objective of doing useful work in a steam turbine or other machine. In a thermal-circulation (non-pumped circulation) furnace, some of the heat absorbed is used to move the water-steam mixture against the internal frictional resistance of the 1-in. to 3-in. diameter furnace-wall tubing.

FURNACES FOR PULVERIZED COAL FIRING

Furnaces for burning coal are more liberally sized than those for oil or gas firing. This is necessary to complete combustion within the confines of the furnace and to prevent the formation of objectionable ash or slag deposits. Furnace-wall heat-absorption rates are low enough that tubing metal temperatures do not greatly influence the furnace size.

The furnace dimensions of a unit to fire pulverized fuel have a major influence on the arrangement of firing equipment, the location and quantity of convective heat-transfer surfaces, the quantity and length of soot blowing equipment, the amount of supporting structural steel, the extent of platforms and stairways, and the arrangement of ductwork. A furnace must be properly proportioned, with adequate height between the top row of coal nozzles and the furnace arch, to assure adequate retention time for the gaseous combustion products. The upper furnace area uses widely spaced superheater or reheater panels and platens to further cool the gases. These measures insure that the furnace outlet temperature at the entrance to the close-spaced convection surfaces will be sufficiently low to avoid excessive ash accumulations.

A designer determines the furnace plan area by carefully studying all of the coals anticipated to be burned and arriving at an appropriate value of net heat input to plan area (NHI/PA). Values presently used for a wide range of coals are from 1.4 to 2.0 million Btu/hr-sq ft.

FURNACES FOR OIL AND NATURAL-GAS FIRING

Oil does not require, at what has been normal excess-air requirements, as large a furnace volume as coal to achieve complete combustion. However, the rapid burning of and high radiation rate from oil results in high heat-absorption rates in the active burning zone of the furnace. The furnace size must, therefore, be increased above the minimum required for complete combustion, to reduce heat-absorption rates and avoid excessive furnace-wall metal temperatures.

Natural-gas firing permits the selection of smaller furnaces than for oil firing primarily because a more uniform heat-absorption pattern is obtained.

This brief discussion of furnace sizing relates primarily to tangential firing. Some of the statements made do not necessarily apply to furnace and firing-system combinations employed by manufacturers other than C-E.

Once the furnace size has been established, (based principally on experience with similar fuels burned in other units), and the windboxes have been located in relation to the panels and platens in the upper furnace, an analysis is made to arrive at the outlet gas temperature.

FURNACE ANALYSIS

Analysis here means the engineering solution to yield the information required by an engineer for the design of tangentially fired utility furnaces. The gas-side requirement is two-fold: (1) selection of the requisite grade of water-wall steel requires, quantitatively, the axial absorption profile, and (2) the furnace outlet temperature must be known to within about ±50°F to correctly size the sequence of superheaters, reheaters, economizers and air heater.

Since analysis of radiative transfer between parallel planes is not trivial, study of a vortex-fired utility furnace becomes a project of fair magnitude. This may be verified by a scan of the family of "Parallel Plate" papers [9-45]. Thus, the analysis of industrial furnaces has developed on an overall phenomenon basis. An early example of this is the Rosin Equation (1925) for coal dust flames, discussed by Essenhigh in References 24, 26, 27 and 32.

The engineering approach to various aspects of furnace behavior, as distinguished from the theoretical approach, is reasonably well characterized by [24-31] and also well represented by Thekdi et al.[32] An excellent survey of the current state-of-the-theoretical-art is given by Beer and Siddall.[33]

A promising approach is the Hottel zoning method,[34-36] which is intended as a realistic scheme for calculating radiant heat exchange with respect to volume-distributed heat release. Although from the literature it is clear that the zoning method has gained widespread acceptance, rigorous solutions for utility furnaces are not yet available.[37]

EMISSIVITY OF THE COMBUSTION PRODUCTS

Furnace heat-transfer calculations must deal on a quantitative basis with the emissivity of the gaseous products of combustion, gas-to-gas absorptivity, and metallic-surface-to-gas absorptivity.

Because about 95 percent of the heat transfer in large combustion chambers is by radiation, it is important to evaluate the radiative power of the gas/fuel/flame media. If heat transfer requires quantification, the emissivity of the radiating components in the furnace must be known. To make this identification, however, demands the capability to predict the formation, space-dependent concentration, and lifetime of transient species such as OH, C, CO, CN, HCN, N, NO, HCO, H, C_2, CH and other radicals which may be present. Not the least problem in the determination is predicting soot formation (size, distribution, and density) in hydrocarbon flames or particle radiation in coal flames.

Similarly, the radiation contribution by the unburned fuel itself, if the ignition temperatures are high, may not be insignificant either. At this time, neither the chemical and reaction kinetics nor the individual spectra are established or even known completely.

Only the main products of combustion, CO_2 and H_2O, have accepted emissivity values. An engineering approach based on CO_2 and H_2O radiation with an allowance for other contributors constitutes an acceptable alternative because the design engineer is interested in the effective overall gas emissivity as it results from the interacting radiating components. This requires establishing a relation by which the total effect can be expressed in terms of known components.

DEFINITIONS USED IN FURNACE DESIGN

Effective Projected Radiant Surface (EPRS)

Effective projected radiant surface is the total projected area of the planes which pass through the centers of all wall tubes, plus the area of a plane which passes perpendicular to the gas flow where the furnace gases reach the first convection superheater or reheater surface. In calculating the EPRS, the surfaces of both sides of the superheater and reheater platens extending into the furnace may be included.

Furnace Volume

The cubage of the furnace within the walls and planes defined under EPRS.

Volumetric Heat Release Rate

The total quantity of thermal energy above fixed datum introduced into a furnace by the fuel, considered to be the product of the hourly fuel rate and its high heat value, and expressed in Btu per hour per cubic foot of furnace volume. This value, does not include the heat added by preheated air nor the heat unavailable through the evaporation of moisture in fuel and that from the combustion of hydrogen.

Heat Available or Net Heat Input

The thermal energy above a fixed datum that is capable of being absorbed for useful work. In boiler practice, the heat available in the furnace is usually taken to be the higher heating value of the fuel corrected by subtracting radiation losses, unburned combustible, latent heat of the water in the fuel or formed by the burning of hydrogen, and adding sensible heat in the air (and recirculated gas if used) for combustion, all above an ambient or reference temperature.

Furnace Release Rate

Furnace release rate is the heat available per sq ft of heat absorbing surface in the furnace (the EPRS).

Furnace Plan Heat-Release Rate

Furnace plan heat-release rate is usually based on the net heat input at a horizontal cross-sectional plane of the furnace through the firing zone, expressed in million Btu/hr-sq ft. The area of the plan is calculated from the horizontal length and width of the furnace taken from the centerline of the waterwall tubes.

MODIFIED EMISSIVITY AND ABSORPTIVITY FACTORS

An assumption that fuel and air are ideally and instantaneously converted to CO_2 and H_2O seriously underestimates the waterwall heat-absorption rates. That is, the Hottel emissivities[34] used for calculating radiative transfer, even though they have been confirmed time and time again (see Ref. 39), constitute only a partial contributor to the emissivity in a tangentially fired furnace. It is clear that an additional emissivity factor is necessary. Bueters[40,41] showed that the source condition required "blackening", which led to the development and use of the "F_E operator".

$$\epsilon = \frac{\text{energy actually radiated by the system}}{\text{energy radiated if the system were black,}}$$

the combustion-product emissivity is related to Hottel emissivity according to the equation

$$\bar{\epsilon} = \frac{F_E - 1 + \epsilon_H}{F_E}, \ 1 \leq F_E < \infty \tag{19}$$

It follows that the wall-to-gas re-radiation absorptivity is blackened accordingly, or

$$\alpha_{w \to g} = \frac{F_E - 1 + \alpha_H}{F_E} \tag{20}$$

where,

$\bar{\epsilon}$ = combustion products emissivity

F_E = a value which modifies Hottel emissivities to yield combustion-products emissivity

ϵ_H = the combined emissivity for CO_2 and H_2O as given by Hottel[38]

$\alpha_{w \to g}$ = gas absorptivity with respect to bounding-surface radiation

α_H = gas absorptivity as given by Hottel[38]

The gas-to-gas absorptivity, $\alpha_{g \to g}$, is analogously defined as

$$\alpha_{g \to g} = \frac{F_A - 1 + \alpha_H}{F_A} \tag{21}$$

where F_A is the gas-to-gas radiation factor for combustion products.

The F_E factor, as defined, is not related to factors published elsewhere.[42] The behavior of F_E is physically what one would expect for a transition region from a Hottel radiation, $F_E = 1$, dependent on beam length, pressure, and temperature, to the black-body limit, $F_E \to \infty$, where emissivity is independent of beam length, temperature, and pressure. This is illustrated by Fig. 8 where $\epsilon(F_E = 1)$ is taken at typical Hottel values for industrial furnaces of 0.2, 0.3, and 0.4, all at temperature T. The F_E "operator" has therefore been adopted as being fundamentally preferable to using a multiplying factor or corrective addition on ϵ_H.[40]

To determine and correlate source descriptors such as F_E, the heat-absorption profile and furnace outlet temperature are measured on operating steam-generating units. The resulting data is processed by a computer which uses it as a boundary condition to solve for source descriptors.[41]

F_E values for large natural-gas-fired furnaces are about 1.2. Applying this to a Hottel emissiv-

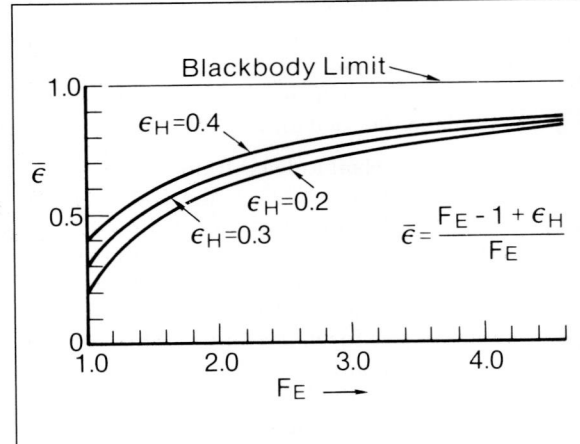

Fig. 8 Combustion products emissivity ($\bar{\epsilon}$) (for radiating components) vs F_E

ity of about 0.34 for gas temperatures in the range of 3000°F, one arrives at $\epsilon = 0.45$. This accounts for additional contributions by convection, transient species, and soot. An altered Hottel emissivity of about 0.39 would be estimated from the work of Ludwig and Boyton on water vapor.[43] Assuming that this is a more exact value and that convective transfer is 4 to 5 percent, we can postulate roughly

$$0.45\sigma T^4 = \text{Hottel radiation} + \text{convection} + \text{additional radiation}$$

$$= 0.39\sigma T^4 + 0.02\sigma T^4 + 0.04\sigma T^4 \tag{22}$$

For a rather typical temperature of $3500°R_1$ we would arrive at a convective coefficient of

$$0.02\sigma T^4 = h\,\Delta T, \quad \Delta T \simeq 2000°F$$

$$h = \frac{.02 \times .17134 \times 10^{-8} \times 3500^4}{2000}$$

$$\cong 2.6\ \text{Btu/ft}^2\text{-hr-°F}$$

This value is in the range of values for combustion chambers given by Steward and Cannon.[44]

When quantitative convective correlations for tangentially-fired furnaces have been determined along the lines established by Hottel, the F_E value for fuels on which no operating experience is available may be predetermined.

RADIATION FORMULATION

Hottel and Egbert[45] give a Stefan-Boltzmann form of equation for net radiation as

$$q = 0.1723\ A\ \acute{\epsilon}_{surf}\left[\epsilon_g\left(\frac{T_g}{100}\right)^4 - \alpha\left(\frac{T_s}{100}\right)^4\right] \tag{23}$$

Modified by F_E, this equation becomes

$$q = 0.17134\ A_s\ \acute{\epsilon}_s\left[\epsilon_g\left(\frac{T_g}{100}\right)^4 - \alpha_{w\to g}\left(\frac{T_w}{100}\right)^4\right] \tag{23A}$$

where

q = heat given up by flame Btu/hr

0.17134×10^{-8} Btu/ft²-hr-(°R)⁴ is the Stefan-Boltzmann constant,

$\acute{\epsilon}_s$ = effective exchange factor to account for infinite number of reflections and re-radiations; written in terms of surface emissivity (dimensionless)

ϵ_g = radiative (combustion products) emissivity, dimensionless

A_s = effective surface area of flame envelope and receiving surface when equal, sq ft

$\alpha_{w\to g}$ = fraction of bounding-surface radiation which is absorbed by the gas, as above

T_g = gas temperature, °R

T_w = cold-surface temperature, °R

This equation is used to formulate the gas-to-gas radiation exchange and gas-temperature distribution inside a furnace by means of a horizontal slicing technique.[41]

In Eq. 23A, the area of the flame will vary substantially with furnace design and fuel fired. The type of firing used affects the rate at which gas sweeps over heating surface, and the area of the flame varies with the rate of firing, other things being constant. Flame emissivity depends on many factors including

■ flame luminosity due to burning particles

■ the concentration of water vapor and carbon dioxide in the flame

■ the volume of flame and its temperature.

The temperature of the flame envelope will be neither constant nor uniform in a watercooled furnace of a steam-generating unit. Combustion characteristics of different firing equipment vary widely. Some produce a short, but intense and highly turbulent flame, while combustion with others is relatively slow, thereby producing a comparatively long flame of greater volume.

Although for any given furnace the area of cold surface is fixed, slag accumulation may influence its effectiveness. The emissivity of the watercooling installed depends on the material used and its condition. For an oxidized steel furnace, however, the emissivity will approach unity at high temperatures.

Generally, the temperature of the cold surface may be taken as that of the water and steam inside the tube, without introducing an appreciable error. But depending on its area, thickness and density, slag accumulation will affect not only the area but also the temperature of the cold or receiving surface. (See Appendix B, Determination of Coal-Ash Properties.)

DESIGN APPROACH

From the foregoing discussion of the variables affecting the temperature field and radiant heat transfer in the boiler furnace, the reasons why the designer does not usually approach the problem in a purely theoretical manner should be apparent. In practice, the designer resorts to the use and interpretation of empirical data collected from boilers of similar design operating under similar conditions. Known theories of heat transfer are applied to the empirical data so that they may be used for conditions differing from those tested. This is necessary to improve and advance the art of designing boiler furnaces. Furthermore, to predict the temperature of flue gas leaving a given furnace in advance of actual operation requires a background of accumulated data, experience and judgment generally possessed only by manufacturers of the equipment involved.

In the simplified empirical design approach, traverses are made of the gas stream leaving the furnace to determine the weighted average temperature. Such measurement of boiler furnace performance is simple in principle but difficult to attain in practice. There are many problems in taking measurements at high temperature in a complex equipment configuration of large size.

The average outlet temperature, along with the calculated mass flow of the products of combustion and the mean specific heat above an arbitrary datum, will give the heat content of the gas leaving the furnace above the datum. This may be done according to the following equation:

$$q = Mc_p(T_t - 70) \tag{24}$$

where c_p is the mean specific heat at constant pressure from 70°F to the leaving temperature, T_t is the gas temperature leaving the furnace, and M is the mass flow rate of gas.

The heat absorbed by the radiant heating surface exposed to the furnace gas is the difference between the net heat liberated in the furnace available for absorption and the heat content of the gas leaving the furnace.

TYPICAL FIELD TEST RESULTS

Reference 46 reports a full scale field study of the foregoing design approach. The study involved a 600-MW tangentially-fired boiler burning a highly slagging high-volatile bituminous coal in pulverized form.

The boiler furnace, shown in outline in Fig. 9, is 56-ft wide, 49-ft deep, and 183-ft high. During the tests the net heat input on the plan area was 2.1×10^6 Btu/sq ft-hr. Furnace temperatures were measured using aspirating, single-shield high-velocity thermocouples on 15-ft water-cooled probes extending 14 ft into the furnace.

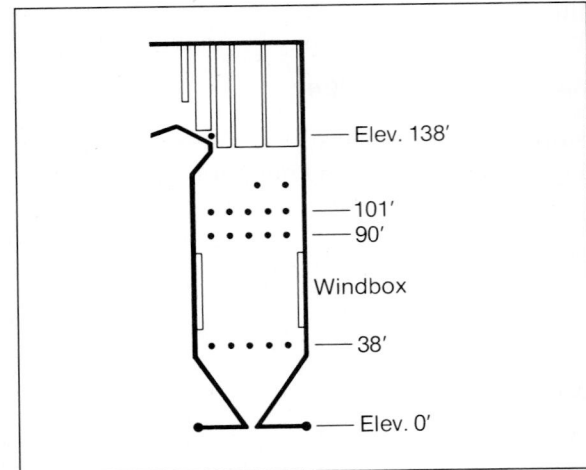

Fig. 9 **Test furnace outline, showing test-probe elevation 138 ft.**

Table III summarizes the gas temperatures measured at the 138-ft level through 7 sighting ports opposite the furnace nose. The average gas temperature reported in this test program for the 7 traverse lines is 2232°F, with a maximum variation of ± 150°F. (Temperatures measured in ports 2, 3, and 4 were done on one day, and are higher than those measured in ports 5, 6, 7, and 8 on the next day.) As is to be expected, temperature levels anywhere in a large furnace firing pulverized coal will vary throughout a day, and through a week. This occurs as a function of the thickness of slag and ash on the furnace walls, the sootblowing pattern, slag falls during the night, and burner flow and mixing patterns.[46]

ELEMENTS OF BOILER HEAT TRANSFER

Earlier, this chapter introduced the concepts of heat transfer by various models. Ref. 47 gives much of the historical background of their development. From an understanding of these concepts, it is apparent that the calculation of heat transfer occuring in superheater, reheater, and economizer sections downstream of the furnace has a first-order dependence on the furnace-outlet gas temperature. And, after the different types of surfaces are installed and the unit goes into operation, the furnace leaving temperature has a similarly significant effect on how they perform.

FLOW OF HEAT THROUGH FURNACE TUBES

The heat which is given up by the furnace gas, and which results in the generation of steam within the tube exposed to that gas, must pass through a series of resistances: the gas film adjacent to the tube; slag on the tube surface; tube wall; internal scale deposited by the water evaporated; the water and steam film. If the slag and scale are not bonded intimately to the tube surface, two more resistances may be added: the gas film between the slag and tube, and the steam film between the internal scale and the tube wall. Fig. 10 illustrates the temperature gradient resulting from the flow of heat through some of these series resistances.

Heat transfer through the gas film and the evaporating film does not follow the simple equation for conduction. The temperature gradient through the other resistances, however, is inversely proportional to the thermal conductivity. For conductance through a body having flat parallel plane surfaces, the following equation may be used:

$$\Delta t = \frac{(q/A_s)\,x}{k}$$

(25)

where

Δt is the temperature drop across any single thermal resistance,

q/A_s is the heat flux per unit of surface area,

x is the thickness of the resistance, and

Table III. Gas temperatures at elevation 138 ft

Traverse Distance[a]	Temperature,°F						
	2[b]	3	4	5	6	7	8
2	2239	2249	2313	2225	2130	2170	2165
4	2263	2296	2290	2165	2090	2080	2200
6	2290	2315	2305	2165	2115	2080	2260
8	2340	2340	2310	2155	2100	2100	2260
10	2366	2379	2330	2165	2130	2145	2220
12	2384	2374	2318	2240	2100	2155	2255
14	2377	2375	2300	2230	2060	2180	2290

(a) Distance in feet from boiler inside wall.
(b) Port number (typical).

k is the thermal conductivity for unit thickness,

all expressed in consistent units.

For radial heat flow in a tube wall we write:

$$\Delta t = \frac{q/A_s}{k} r_2 \left[\log_e \frac{r_2}{r_1} \right]$$

(26)

where r_2 is the outside radius, r_1 is the inside radius, and q/A_s is the heat flux at the outside tube surface, again in consistent units.

The temperature gradient shown in Fig. 10 is typical of furnace tubes exposed to radiant heat. From inspection, the resistance having the dominating influence on heat-transfer rate is that through the gas film. Thus, for many practical purposes, the temperature of the receiving surface may be taken as equal to that of the water and steam within the tube. This approximation can be closer to fact when studying transfer in convective zones than it is for furnace-wall tubes. Accumulations of slag or ash on the outside surface of furnace tubing, or scale on the inside, may easily become the controlling resistance to any heat transfer if allowed to form unchecked.

As described in Chapter 20, Power Plant Water Technology, inside scale can cause sufficient increase in tube metal temperature to produce overheating and rupture.

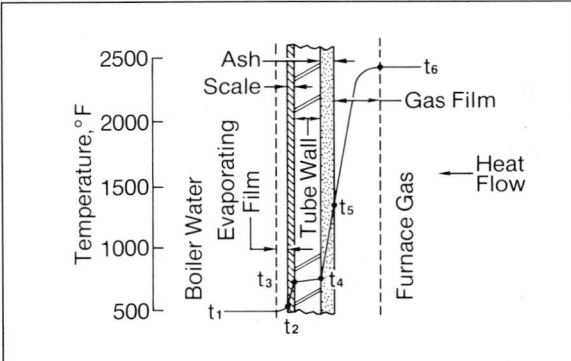

Fig. 10 Temperature gradient for furnace-wall tubes exposed to radiant heat, with slag accumulation

FLUID-TO-FLUID HEAT-TRANSFER RATES

The fundamental theory of heat-transfer in the steady state is analogous to, and based on, the same concept as Ohm's law, namely: *That flow varies directly as the potential and inversely as the resistance.* Thus in Fig. 10, the potential between the two fluids is the temperature difference $(t_6 - t_1)$, and there are five resistances in series—the hot fluid (gas) film, the slag or ash layer, the metal separating the fluids, the internal scale or sludge layer, and the cold-fluid evaporating film.

Applying the Ohm's law analogy, the flow of heat, Q, will equal the temperature difference divided by the resistance. In each instance, as with Ohm's law the resistance is the reciprocal of the thermal conductance.

In a boiler bank or economizer the temperature difference, Δt_s, between cold fluid and clean heating surface will be very small compared to Δt_g, because the thermal conductance of the water film is high. But in a superheater or reheater the dry steam becomes the cold fluid, and the temperature gradient across the film will be higher because the thermal conductance of these films is lower.

The designers of boiler banks and economizers, therefore, are interested only in the overall thermal conductance or heat-transfer rate, and are not greatly concerned with the intermediate rates or temperatures, because the metal temperature will be close to that of the cold fluid. On the other hand, designers of superheaters, furnace waterwalls, and reheaters are also concerned about the metal temperatures.

EFFECT OF GAS-FLOW DIRECTIONALITY

Whether the gas flows parallel or transverse to the tube-bank axis has a major effect on the rate of heat transfer. When gas flows through tubes, the flow is parallel to the axes for their entire length. When flowing outside tubes, however, there is usually a combination of parallel-flow convection and cross-flow convection, in addition to local radiation.

Equations for heat transfer with parallel flow contain a factor for inside tube diameter, but the actual tube diameter can be used only when

the gas actually flows through the tubes. For flow outside the tubes, an *equivalent diameter* is used:

$$d_e = \frac{4A_c}{P}$$

(27)

where d_e is the equivalent diameter, A_c is the free gas passage area and P is the gas-touched perimeter of tubes, all in consistent units.

When calculating the heat transfer in a parallel-flow pass of a boiler, coefficients for flow parallel to the axis of the tubes should be applied only to those portions where the gas is confined between parallel baffles and forced to travel with minimal turbulence along the axis of the tubes. For the portions at the ends of such passes, cross-flow transfer rates may be used for the effective surface opposite the entrance and exit openings. Consideration should be given to the variation of mass velocities in those portions, if any.

In general, the convective heat-transfer coefficient with flow *across* tube banks is considerably higher than with parallel flow. Under some conditions it is twice as much for comparable velocities. Fig. 11 shows the approximate variation. The upper limit represents flow across small diameter tubes, and the lower limit, flow parallel to tubes that are spaced on relatively wide centers.

FACTORS AFFECTING CONVECTION HEAT TRANSFER FROM GAS

Among most designers of steam-generating equipment, the convection heat-transfer rate in those parts of the apparatus not exposed to luminous radiation is the overall rate between the fluid giving up and the fluid receiving the heat. It has been shown that the gas film conductance may be used instead of the overall rate without much error in certain parts of the equipment. But even the gas film conductance must include a factor or an addition to allow for nonluminous radiation from hot gases, such as carbon dioxide and water vapor, both of which are present in the products of combustion of all common fuels. Therefore, when we speak of the

convection rate, it will be understood to include this radiation, and the expression *pure convection* will be used in connection with all data or curves not including this allowance.

An analysis of pure convection will show that the heat-transfer rate per unit area of heating surface will be affected by numerous variables given in the following expression.

$$R_c = f(D, V, \rho, \mu, c_p, k, a)$$

where:

R_c = film conductance, fluid to solid, for pure convection

D = linear dimension of solid surface

V = linear velocity of fluid stream

ρ = density of fluid

μ = viscosity of fluid

c_p = specific heat of fluid at constant pressure

k = conductivity of fluid

a = geometric relation ratio or combined ratios to cover the effect of spacing, width, depth and length

Investigators have found that good correlation of test data has resulted when these variables are combined into dimensionless groups or ratios, as follows:

Nusselt number or group = Nu = $R_c D/k$
Reynolds number or group = Re = $DV \rho/\mu$
Prandtl number or group = Pr = $c_p \mu/k$

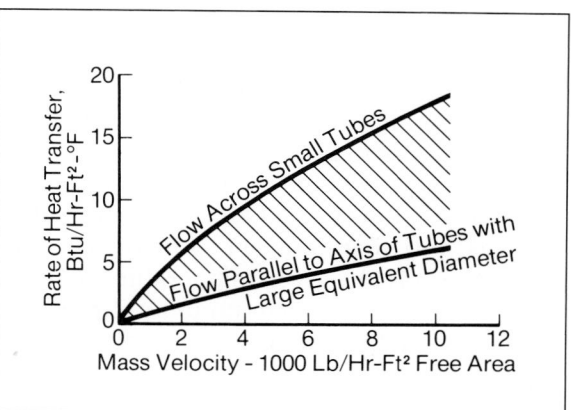

Fig. 11 **Chart showing the range of convective heat-transfer rate between conditions of parallel- and cross-flow of gas**

For forced convection it may, therefore, be written that

$$R_cD/k = f[(DV \; \rho/\mu), (c_p \; \mu/k), (a)]$$

(28)

Expressed in terms of power functions, this becomes

$$R_cD/k = k_pF_a \; (DM/\mu)^n(C_p\mu/k)^m$$

(29)

in which

k_p = constant proportionality factor to be experimentally determined

F_a = arrangement factor replacing a, to be experimentally determined

M   = V × ρ = mass velocity of fluid

m, n   = power functions to be experimentally determined

The factor, a, and F_a in the two foregoing equations could probably be eliminated if all parts and dimensions of all heat-transfer apparatus were geometrically similar. This is impossible in commercial practice, and has not always been done in designing apparatus for research purposes. Therefore, where D in these equations is merely taken as tube diameter, complete agreement between data from different sources is not to be expected. This may partly explain the disagreement in values of m, n and k as determined by different investigators. Higher values for n have been found for flow parallel to the axis of tubes than for flow transverse to the axis.

The numerical value of the Prandtl group, c_p μ/k, is nearly the same for all the constituents of flue gas, and, therefore, is practically constant for all flue gas at a given temperature. There is little opportunity to determine by experiment to what extent this group enters into heat transfer from gas, and some investigators do not include it in their equations. However, there is considerable variation in the specific heat of gases at high temperature, and there is no certainty that proper correction for gas temperature has been allowed when this group is omitted. In general, heat-transfer equations containing no temperature term, or containing none of the fluid properties which vary with temperature, are to be considered only for limited application in the range for which the heat-transfer equations are derived.

In regard to tube arrangement and spacing, it has been definitely established that, in banks of tubes, the depth of the bank affects the heat-transfer rate. There is a marked increase for each row up to about the fifth, and gradually less and less increase to the tenth. This is caused by increased turbulence from row to row. For one particular Reynolds number, and other variables kept constant, Pierson[48] found approximately 25 percent lower cross-flow heat-transfer rate for two rows than for ten rows for both *in-line-* and *staggered-tube* arrangements, but the difference between five rows and ten rows was about 9 percent. Experiments of other would indicate that a very turbulent gas stream, on approaching the bank, may affect these comparisons and produce higher heat-transfer rates for shallow banks than indicated in previous text.

It is obviously incorrect to expect staggered tube rows to give better cross-flow heat transfer than tubes in line without considering the effect of depth spacing in the direction of flow.

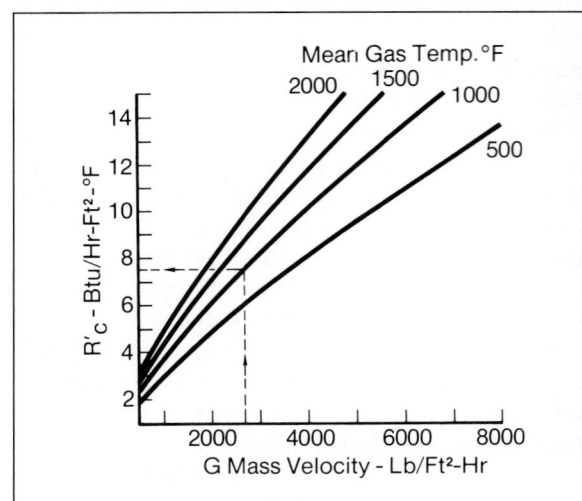

Fig. 12 **Convection transfer rate**

When the depth spacing is great in both cases, so that the stream has a chance to expand fully after passing each row, there can be no possible advantage with staggered rows, and the two should be equal. The method used by Grimison[49] for correlation of Pierson's and Huge's[50] data indicates this to be true under certain conditions, but it is also affected by the gas velocity when tube diameter and temperature are maintained constant.

CONVECTIVE HEAT-TRANSFER RATES

In day-to-day calculations of superheater and reheater heat transfer, it is convenient to work from curves that synthesize the above factors. For convection transfer, the general equation forming the basis for either curves or computational programs is

$$R_c = k_c c_p M^x / D^y \qquad (30)$$

where

R_c = pure convection heat transfer, gas to metal

k_c = a constant

c_p = specific heat of gas at constant pressure

M = gas mass velocity through the free area between the tubes

D = tube diameter, all being in consistent units

and x and y are exponents; the constant and the exponents must be determined empirically and depend on tube arrangement and spacing.

From tests on large steam-generating units or other heat-recovery equipment, it is difficult to ascertain the precise effect on heat transfer of variations in velocity, tube diameter, distribution and direction of flow, tube spacing, or gas temperature. But it is, nevertheless, necessary to adjust the various coefficients and exponents in such basic relations as Eq. 30 to reflect operational experience.

Typical parametric curves based on test data are shown in Figs. 12, 13, and 14. The overall convection rate is $R_c = R'_c \times F_d \times F_a$.

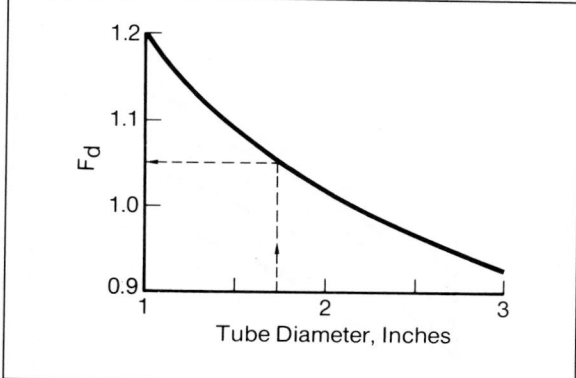

Fig. 13 Convection transfer-rate correction factor for tube geometry

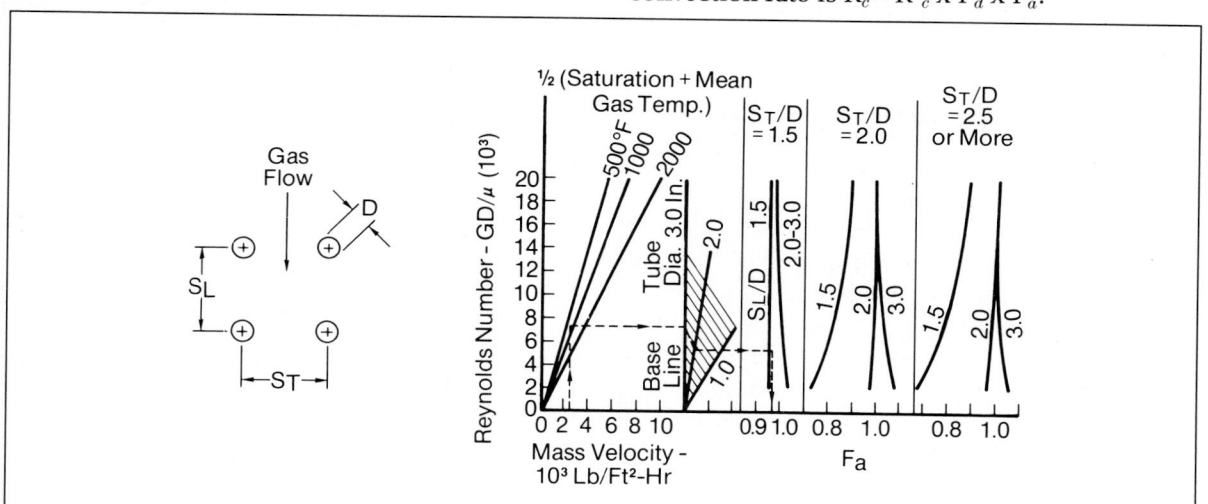

Fig. 14 Convection transfer-rate correction factors, Reynolds number and tube geometry

THE EFFECT OF NONLUMINOUS RADIATION

Carbon dioxide and water vapor are the principal radiating components of boiler flue gas; their combined radiating effect has historically been referred to as nonluminous radiation. Transfer rates and radiant beam lengths for nonluminous radiation are to a great extent based on work done by Hottel and others previously referenced.

In all cases where the flue-gas temperature is high and the tube spacing relatively great, the nonluminous radiation will be of considerable magnitude and should be added to the pure convection rate. Figs. 15 through 18 show typical rates and correction factors. The radiant beam length is calculated according to Fig. 17 and is entered into Fig. 18 to obtain the beam-length correction factor, F_b. The basic nonluminous rate from Fig. 15 is multiplied by F_f (Fig. 16) and F_b to get a corrected rate to be used in subsequent calculations.

EFFECT OF VARIATION IN TEMPERATURE DIFFERENCE

The temperature difference between gas and steam, water, or air produces the flow of heat; the logarithmic temperature difference (LMTD) represents the effective difference between two fluids. The LMTD depends on the relative directions of flow of the fluids as they pass through or over the surface. That is, do

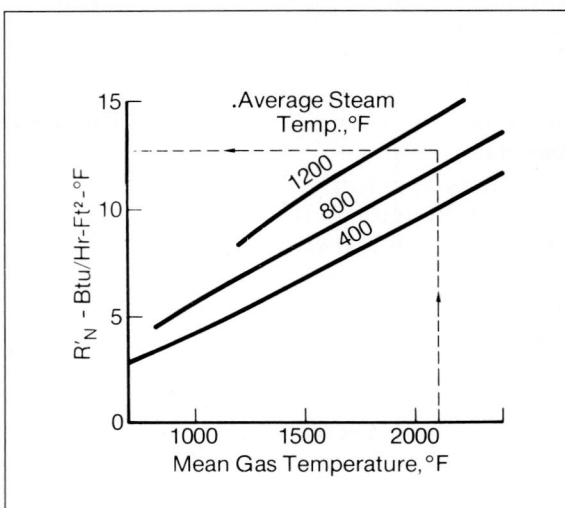

Fig. 15 Transfer rate — nonluminous radiation

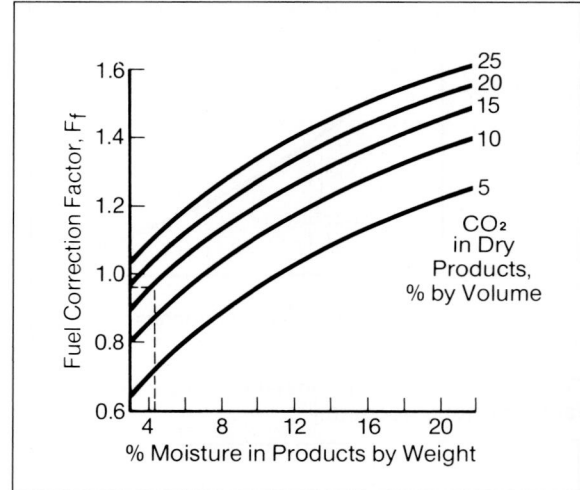

Fig. 16 Fuel correction factor for nonluminous radiation

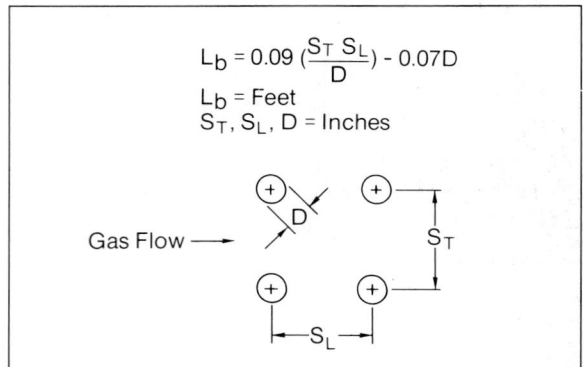

$$L_b = 0.09 \left(\frac{S_T \; S_L}{D}\right) - 0.07D$$

L_b = Feet
S_T, S_L, D = Inches

Fig. 17 Radiant beam length, in-line tubes

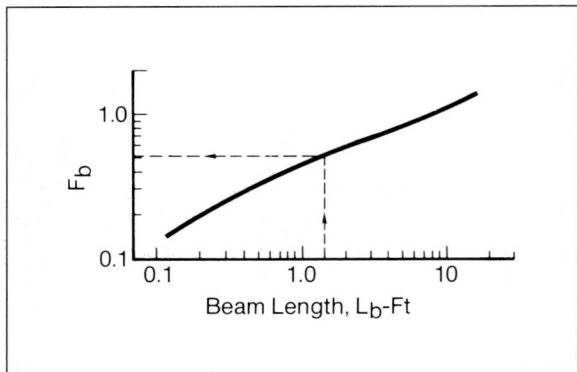

Fig. 18 Correction factor, radiant beam length

they flow counter to, parallel to, or across one another. For parallel or counter flow

$$LMTD = \frac{\text{greatest difference} - \text{least difference}}{\log_e\left(\frac{\text{greatest difference}}{\text{least difference}}\right)}$$

or,

$$LMTD = \frac{\Delta T_1 - \Delta T_2}{\log_e\left(\frac{\Delta T_1}{\Delta T_2}\right)}$$

(31)

where ΔT_1 and ΔT_2 are given in Fig. 19. These figures show that for a given heat recovery the greatest temperature difference will be obtained, and the least heating surface required, when the two fluids flow counter to each other. Furthermore, with parallel flow, the highest temperature of the heated fluid can only approach but not equal the lowest temperature of the heating fluid.

In waterwalls or economizer surface, the temperature difference between cold fluid and clean heating surface will be very small compared to the temperature difference between the clean surface and the mean gas temperature because the thermal conductance of the water film is high. In a superheater or reheater, on the other hand, the temperature gradient across the cold film will be higher because the thermal conductance of these films is lower.

In the design and proportioning of waterwall and economizer surface, interest is primarily in the overall thermal conductance or heat-transfer rate. The intermediate rates or temperatures are of secondary concern because the metal temperature will be close to that of the cold fluid. Design of superheaters, reheaters and waterwalls must also consider the metal temperatures. In the case of superheaters, knowledge of metal temperature is necessary for the most economical use of alloy material. The designer of a heat-transfer device such as a tubular air heater, on the other hand, must know the temperature of the metal separating the two fluids to combat condensation of

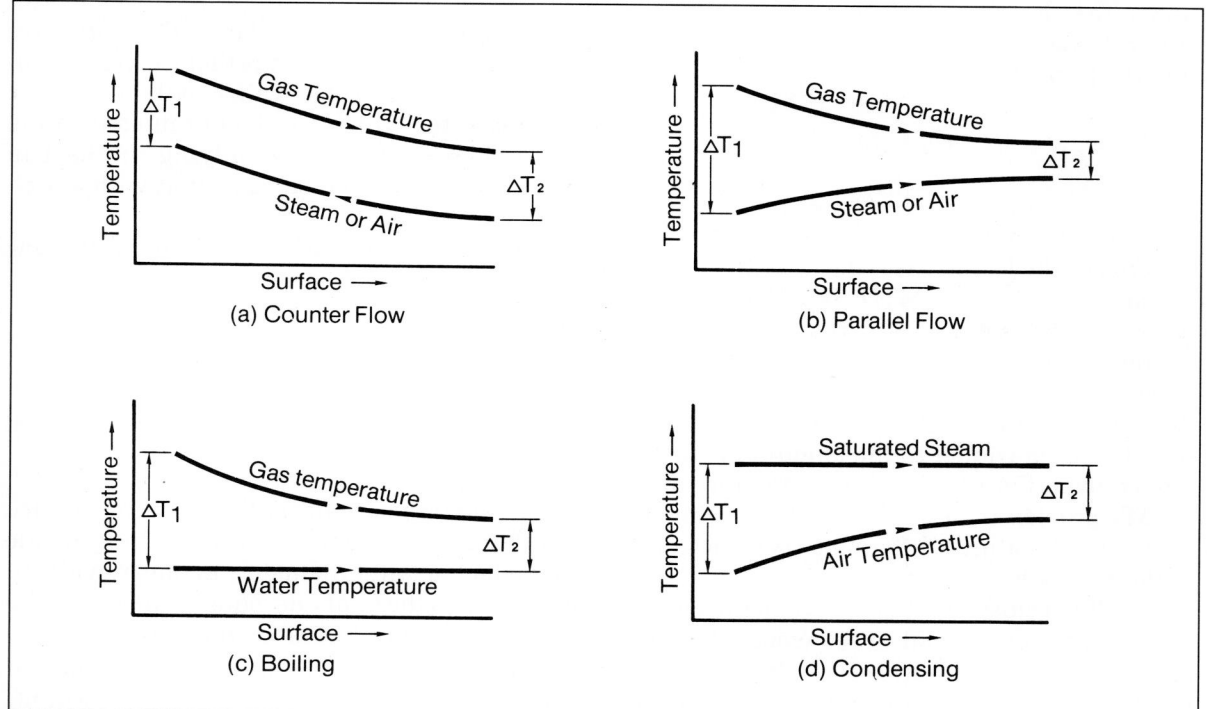

Fig. 19 Fluid temperatures in heat exchangers

moisture and acid vapors, or to minimize the effect by selection of noncorrosive materials.

APPLICATION OF HEAT-TRANSFER RATES TO BOILER CALCULATIONS

As discussed, superheater and reheater tubular surfaces in the upper furnace and beyond, (loosely termed the "convection pass") absorb heat in three ways: by direct radiation from the flame; by nonluminous radiation, chiefly from the carbon dioxide and water vapor in the combustion gases; and by heat transfer through pure convection.

The contributions from these sources are additive and directly affect both heat absorption and metal temperatures. That is, a superheater in a high-temperature gas zone that "sees" the fireball may have the same convection rate, R_c, as a low-temperature superheater section in a 1000°F zone. But the effective overall transfer rate, in this instance including the long-beam-length-radiation, may be double the R_c.

From the second law of thermodynamics, we know that a given surface will absorb heat proportionally to the temperature difference and the transfer rate.

$$q/A_s = R_t \quad (LMTD) \tag{32}$$

where q is the heat absorbed per unit time, A_s is the area of the heating surface, R_t is the overall transfer rate, and LMTD is as previously defined—all in consistent units.

The logarithmic mean temperature difference must be used in Eq. 32 because the reduction in temperature between T_1 and T_2 is not directly proportional to the amount of surface passed over. For small surfaces, and for values of ΔT_2 greater than half ΔT_1, the arithmetic mean temperature difference can be used without appreciable error.

Also, for a closed system in which the weight of gas entering is equal to the weight leaving, and in which losses due to radiation and convection from the boundary walls can be neglected, the quantity of heat absorbed by the surface will equal the heat given up by the gas:

$$q = M_g c_p \, \Delta T \tag{33}$$

where M_g is the mass flow-rate of gas, c_p is the mean specific heat at constant pressure, and ΔT is the gas temperature drop, all in consistent units. Figs. 20 and 21 give specific heats at constant pressure of the products of combustion of several solid, liquid, and gaseous fuels, burned at usual excess-air percentages. Fig. 22 with specific heats of air at varying moisture contents is useful for air-heater calculations.

When heat transfer in commercial apparatus is calculated from changes in fluid temperature determined by test measurements on similar apparatus, several additional errors may be introduced. In taking the temperature readings, the gas stream is large, and simultaneous readings at many points are necessary. If there is stratification or nonuniform flow, the arithmetic average of these readings may introduce an error which may be much larger than any error in the temperature readings themselves. An incorrect specific heat value will introduce another error which can be eliminated in the research laboratory by supplying the heat in some way which can be measured with a high degree of accuracy.

For any heat-transfer surface, then, the gas temperature drop is

$$\Delta T_g = \frac{R_t A_s (LMTD)}{M_g c_p} \tag{34}$$

where the transfer rate and the heating surface are defined on the same basis—tubing inside diameter, outside diameter, or mean-wall diameter. R_t is the total rate, equal to pure convection rate plus nonluminous transfer rate.

Where luminous or "long-beam" radiation from the fireball is to be taken into account, curves such as in Fig. 23 are used to give

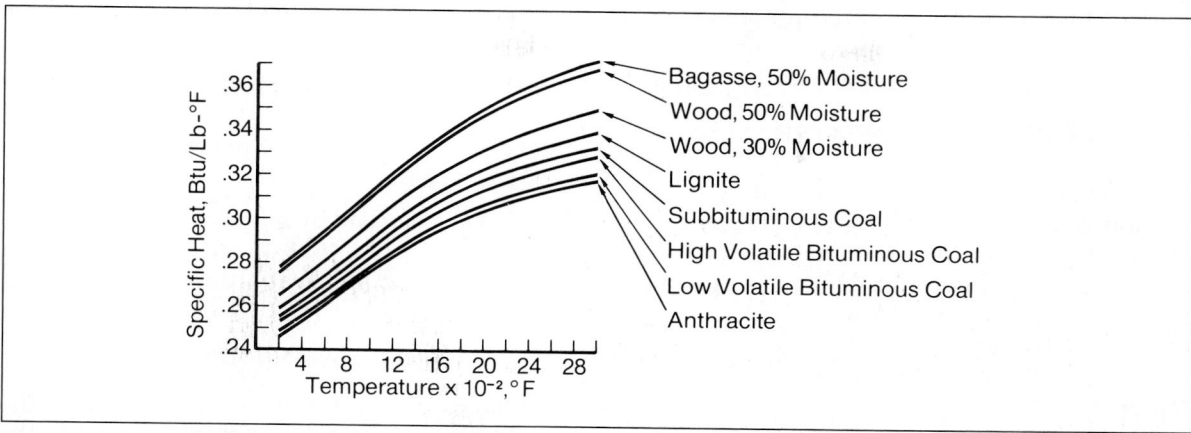

Fig. 20 **Specific heat at constant pressure for products of combustion of solid fuels**

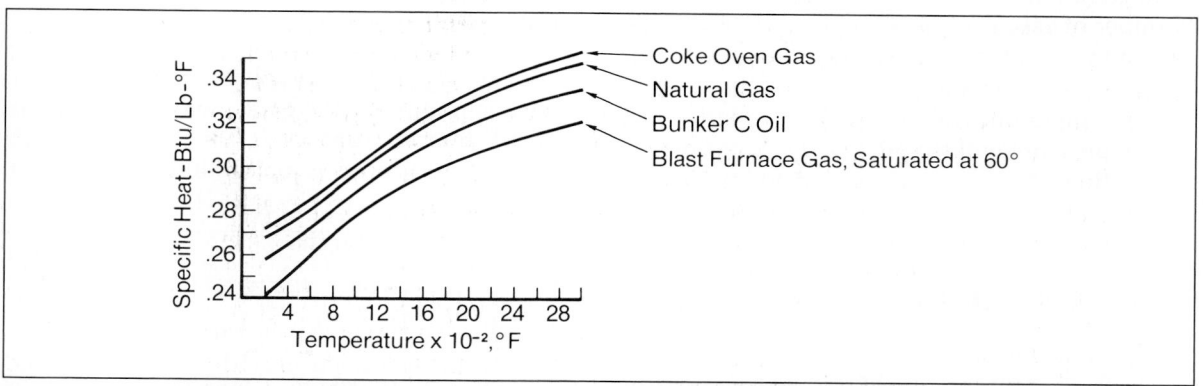

Fig. 21 **Specific heat at constant pressure for products of combustion of liquid and gaseous fuels**

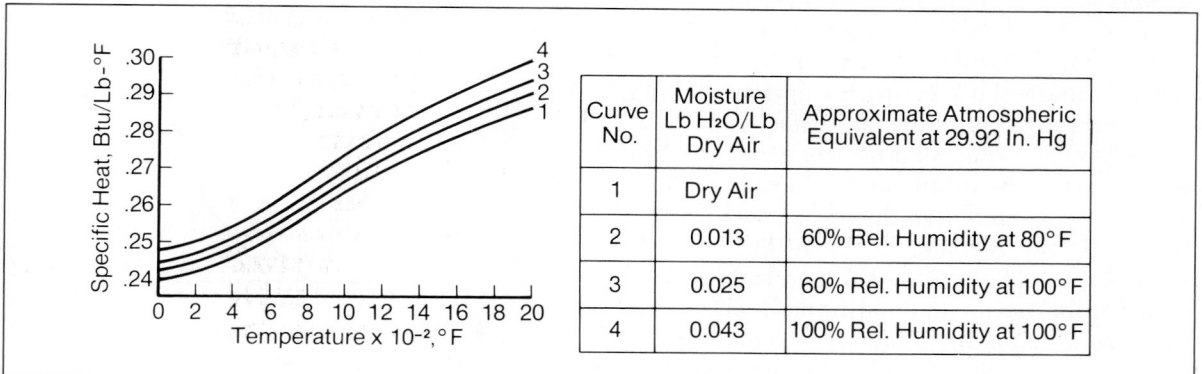

Curve No.	Moisture Lb H_2O/Lb Dry Air	Approximate Atmospheric Equivalent at 29.92 In. Hg
1	Dry Air	
2	0.013	60% Rel. Humidity at 80° F
3	0.025	60% Rel. Humidity at 100° F
4	0.043	100% Rel. Humidity at 100° F

Fig. 22 **Specific heat at constant pressure of air at varying moisture contents**

radiant-heat absorption on flat projected areas of banks of tubing; they are to be corrected for view factor and depth of tube bank.

CALCULATION OF A REHEATER

An over-simplified example of a single-section reheater will tie together much of the preceding discussion. Assume the section is located in a moderately high gas-temperature zone, with both inlet and outlet steam conditions known. It is then possible to calculate in one step without assumption of intermediate temperatures.

KNOWN DATA

Tube diameter	2⅛ in. OD
Free gas area, FGA	1675 sq. ft.
Number of assemblies, N	90
Steam temperature leaving, t_2	1005 °F
Steam pressure leaving, P_2	488 psig
Steam temperature entering, t_1	680°F
Steam pressure entering P_1	500 psig
Steam flow, W_s	3,500,000 lb/hr
Gas mass flow, M_g	4,600,000 lb/hr
Gas temperature entering T_1	2000°F

CALCULATION OF HEATING SURFACE

The reheater section is counterflow and its entering gas temperature, T_1, is the same as the gas temperature leaving the superheater surface upstream, previously calculated.

To find the gas temperature leaving the reheater, T_2, and the heating surface, S, required to raise the steam temperature from 680°F to 1005°F, the following steps are required.

The amount of direct furnace radiation which passes through superheater surfaces ahead of the reheater is first determined using the projected area of the tubing and a correlation such as Fig. 23. Assume that the direct radiation absorbed by the reheater is 25 million Btu per hour. The total heat to be absorbed is now established from the steam conditions in and out of the reheater. The direct radiation is deducted to obtain the convective plus non-luminous heat transferred, q_c.

Enthalpy of steam at 1005°F and 488 psig = H_2 = 1523 Btu/lb

Enthalpy of steam at 680°F and 500 psig = H_1 = 1345 Btu/lb

Heat absorbed by reheater per lb of steam = ΔH = 178 Btu

Total heat absorbed by reheater = $W_s\Delta H$ = 3,500,000 × 178 = 623 × 10⁶ Btu/hr. Subtracting the radiant heat absorbed (25 million Btu/hr.),

$$q_c = 598 \times 10^6 \, \text{Btu/h}$$

Since q_c is also the heat absorbed from the gas, the following relationship, leading to the solution of T_2, may be expressed.

$$q_c = M_g \, c_p \, \Delta T_n$$

$$\Delta T_g = \frac{q_c}{M_g c_p} = T_1 - T_2$$

For solution of T_2 all values are known in the above equations except c_p which for an assumed temperature of 1790°F, firing high-volatile bituminous coal, is 0.308 (from Fig. 19)

$$\Delta T_g = \frac{598 \times 10^6}{4,600,000 \times 0.308} = 422°F$$

$$T_2 = 2000 - 422 = 1578°F$$

Heating surface, A_s, is the only remaining unknown in connection with this reheater section. This can be determined from

$$A_s = \frac{M_g c_p \Delta T_g}{R_t \, (\text{LMTD})}$$

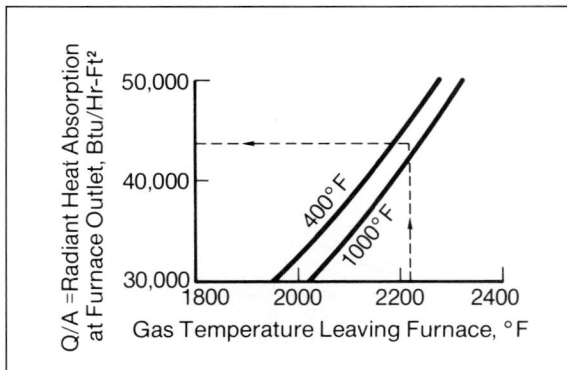

Fig. 23 **Radiant heat absorption at furnace outlet**

From Fig. 24, which shows the temperature gradient across the reheater, we calculate the LMTD.

$$\text{LMTD} = \frac{(T_1 - t_2) - (T_2 - t_1)}{\text{Log}_e \dfrac{T_1 - t_2}{T_2 - t_1}} =$$

$$\frac{995 - 898}{\text{Log}_e \dfrac{995}{898}} = 946°F$$

$$R_t = R_n + R_c$$
$$R_n = R'_n \times F_b \times F_f$$
$$T_m = t_{avg} + \text{LMTD}$$
$$T_m = \frac{1005 + 680}{2} + 946 = 1789°F$$

Assume a radiant beam length of 1.2 ft

$$R'_n = 10.5 \text{ Btu/hr-sq ft-°F}$$
   (from Fig. 15)
$$F_b = 0.45 \text{ (from Fig. 18)}$$
$$F_f = \text{Use } 1.05$$
$$R_n = 10.5 \times 0.45 \times 1.05 = 5.0 \text{ Btu/hr-sq ft-°F}$$
$$R_c = R'_c \times F_d \times F_a$$

Mass velocity, G, is required to evaluate R'_c

$$G = \frac{W_2}{\text{FGA}} = \frac{4,600,000}{1675}$$
   $$= 2750 \text{ lb/hr-sq ft}$$
$$R'_c = 9.5 \text{ Btu/hr-sq ft-°F}$$
   (from Fig. 11 based on $T_m = 1789°F$)
$$F_d = 1.0 \text{ (from Fig. 13)}$$

Use 0.95 for F_a (Fig. 14 would be used for actual tube spacings.)

$$R_c = R'_c \times F_d \times F_a$$
   $$= 9.5 \times 1.0 \times 0.95$$
   $$= 9.0 \text{ Btu/hr-sq ft-°F}$$
$$R_t = R_n + R_c = 5.0 + 9.0$$
   $$= 14.0 \text{ Btu/hr-sq ft-°F}$$
$$S = \frac{M_g \times c_g \times \Delta T_g}{R_t \times \text{LMTD}}$$
   $$= \frac{(4,600,000) (0.308) (422)}{(14.0) \qquad (946)}$$
   $$= 45,100 \text{ sq ft}$$

DRAFT LOSSES AND THE SPECIFICATION OF FANS

Intimately tied to the physical arrangement of tubular heating surfaces and the determination of convective and nonluminous heat-transfer rates is the draft loss, or pressure drop, of the gases passing over them.

Although there is no direct mathematical relationship between heat transfer and draft loss, it can be said in general that convection transfer rates and draft losses are directly proportional, but to an unknown power function. That is, the rate of heat transfer can seldom be increased without producing an increase in pressure differential.

The engineer designing superheater, reheater, boiler bank, economizer, and air heater surfaces has to balance the cost of heating surface against the power penalty for pressure drop and draft loss to arrive at an economic configuration. And because the several different transfer surfaces (depending on their design pressure level and operating temperature) have widely varying costs per unit of surface area, they too have to be evaluated carefully to arrive at the optimum mix. It is because of this relationship of transfer rate and draft loss that it is necessary to include here terminology and calculation methods for draft losses through tube banks and power-plant stacks.

DRAFT

Although often used in many ways, the term *draft* is defined here as pressure difference *below* that of the atmosphere. The absolute pressure of the atmosphere at mean sea level is 29.9 in. Hg, or 14.7 psia under conditions referred to as *normal barometer*. Without any change in elevation above sea level, barometric pressure may vary as much as 3 in. Hg from *normal* conditions. Thus, as the absolute pressure corresponding to zero, or balanced draft, is quite variable, it is customary, and more convenient, to express draft values in terms of gage pressure rather than absolute pressure.

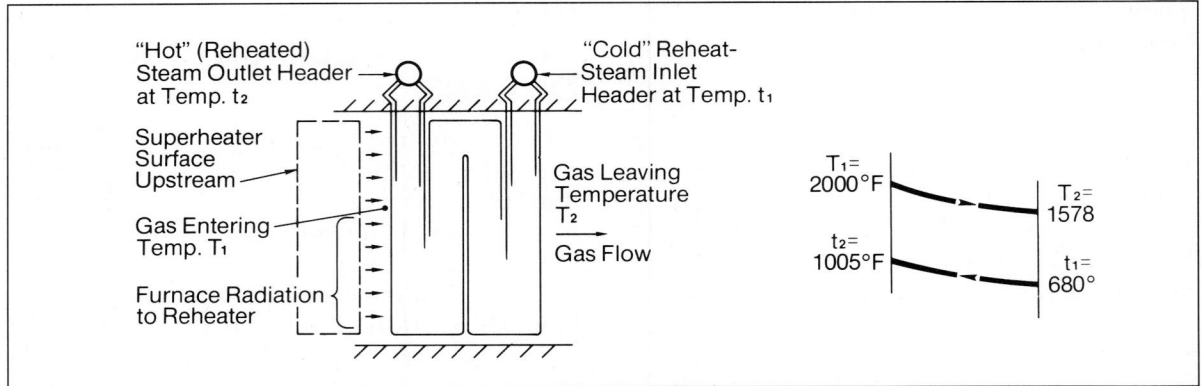

Fig. 24 **Schematic arrangement and temperature gradient across reheater**

For example, a simple U-tube half filled with water may be used as a draft gage, by connecting one leg to a duct or furnace in which the pressure of draft is to be measured, and leaving the other leg open to atmospheric pressure. If the absolute pressure in the furnace is exactly equal to the absolute atmospheric pressure outside the furnace, the water level in the two legs of the U-tube will remain at the same elevation. This is known as a balanced-draft condition, and the draft is said to be zero. On the other hand, if the absolute pressure in the furnace is less than that of the outside atmosphere, the level in the tube leg exposed to the atmosphere will move down, and the level in the other leg will move up. The difference in level is then the draft measured in inches of water, and generally written inches water gage (″WG).

In recording draft readings it is not necessary to place a minus sign ahead of the numerical value of draft when the pressure at point of measurement is lower than that of the atmosphere. But if the absolute pressure at the point of measurement is greater than that of the atmosphere, it should preferably be called *pressure*. If it is called draft, the numerical values should be preceded by a plus sign.

FORCED AND INDUCED DRAFT

The term *forced draft* is indicative of an absolute pressure higher than that of the atmosphere, and this expression is therefore not consistent with the word *draft* as previously defined. The term *induced draft* is indicative of an absolute pressure lower than that of the atmosphere, but the word induced is somewhat superfluous when using the previous definition of draft. Nevertheless, these expressions are in common usage and are used in this text, despite these inconsistencies.

PRESSURE DROP AND DRAFT LOSS

The difference between pressure gage readings in parts of a system operating with a positive pressure relative to that of the atmosphere, is generally called *pressure drop;* whereas the difference between gage readings in parts of a system operating with a negative pressure relative to that of the atmosphere, is generally called *draft loss*. The actual gage readings are referred to as pressure or draft, as the case may be, except that in reporting data taken at a point where there is normally a draft, it is called draft even when there is a pressure, but a plus sign should be used as a prefix to the figures given.

PRESSURE

When the fluid to be measured has motion, it has a *velocity pressure* in addition to its *static pressure*. The *total pressure* is the algebraic sum of velocity pressure and static pressure. Static pressure is the force per unit area exerted on a wall by adjacent fluid which is at rest, or which is flowing without disturbance along the wall of a conduit. To obtain the static pressure of a fluid at rest, use any pipe or tube having an

opening at the point in question. Generally, a tube with a special construction at the end, called a "static tube," is used to determine the static pressure of fluid moving in a straight conduit of sufficient length. Sometimes, a static hole in the wall of the conduit is used.

In the case of fluids flowing in curved paths, such as near elbows, centrifugal force of the fluid affects the pressure. Care must be taken to locate static holes and impact tubes so that they are not affected by curved paths, eddies, or turbulence; otherwise the determination of static pressure or total pressure will be inaccurate.

MEASUREMENT OF TOTAL PRESSURE

Total pressure is measured with an open-ended tube (known as an impact tube) pointed upstream. The excess of total pressure as shown by the impact tube over the static pressure is the exact equivalent of the energy represented by the velocity. This difference is known as velocity pressure or velocity head. When static pressure is below atmospheric, and therefore negative, the total pressure will be a numerically smaller negative number, or may even be positive, depending on the magnitude of the draft and the fluid velocity.

Any statement of pressure should be accompanied by an indication of whether it is static or total, whether it is gage or absolute, and at what base plant altitude, if greater than 1000 ft above sea level. After correction for difference in local elevation, the difference in total pressure between two points in a closed system is a measure of the pressure loss between them; the difference in static pressure between the same points will give the same result only if the fluid velocities at both points are the same. When the velocity of a fluid stream in a horizontal conduit is decreased by a diverging section, the velocity head is decreased and the static head increased. Thus it is quite possible to have a downstream static pressure that is higher than upstream, even though there is a loss in total pressure in the direction of flow.

See Chapter 22 on testing and measurements for draft-loss observations in the field.

STACK EFFECT

Flow of hot gas or hot air in a vertical duct introduces a factor of correction for elevation. Because the fluid in the duct has a lower density than the outside air, *stack or chimney effect (or thermal static draft)* exists. The pressure drop for downward flow increases, and it decreases for upward flow, compared to what the pressure drop would normally be under the same conditions in a horizontal duct. This stack-effect correction may be of considerable magnitude when determining the true pressure drop or draft loss in a tall piece of apparatus, and can be of particular significance in a series of boiler passages and flues, with the direction of the gas flow alternately up and down.

The static draft, as this difference in pressure can be properly called, must be calculated, since simple draft-gage readings indicate the combined values of stack-effect and frictional flow losses. The following formula is used:

$$\text{Static Draft} = 27.7\, Z_c \left(\frac{P_a}{R_a T_a} - \frac{P_c}{R_c T_c} \right)$$

(35)

where static draft is in inches of water, the height under consideration is Z_c feet, p_a and p_c are the absolute pressures (psia) of the ambient atmosphere and the combustion gas, respectively, T_a and T_c are the average absolute temperatures, and R_a and R_c are the respective gas constants. The value of R_a is 53.3, while that of R_c depends upon the composition of the gases. At usual excess-air percentages, R_c is 52.2 for bituminous coal, 53.2 for fuel oil, 55.6 for natural gas, and 49.3 for blast-furnace gas.[51]

Note that the equation for static draft involves only one dimension, the height of the gas pass or stack, and that the static draft is the maximum that can be produced by any such pass of a given height when filled with stationary gases of a given density and surrounded by a still atmosphere of a given density.

DRAFT-LOSS CALCULATIONS

The variables in the pressure differential across the convection surface of a steam-generating unit include friction due to flow across tubes; loss of head in turns; friction due to flow through, or parallel to, tubes; and stack effect. Named in order of magnitude, they occur in practically all steam generating units.

Gas and air velocities in steam-generating units will always be such as to produce turbulent flow, particularly at high unit ratings. Under these conditions friction loss, when measured in terms of inches of water, will vary directly as the square of the mass velocity, and inversely as density, which also means directly with the absolute temperature. Eqs. 36, 37, and 38 show the mechanics of friction loss for the three types itemized above.

| Flow across tubes | $PD = fN\,H_v$ | **(36)** |

| Turn loss | $PD = K_t \times H_v$ | **(37)** |

| Flow along tubes | $PD = f\dfrac{L}{D} \times H_v$ | **(38)** |

where:

PD = pressure drop, "WG

 f = friction factor, dimensionless

 N = number of restrictions

K_t = constant, depending on type of turns

 L = length of tube in ft

 D = inside diameter, or equivalent diameter in ft

H_v = velocity head, "WG

The velocity head is calculated from Eq. 39.

$$H_v = 0.0002307\,\frac{(G/1000)^2}{d}$$

(39)

where:

 G = mass velocity, lb/hr-sq ft free area

 d = density lb/cu ft

For flow across tubes, the friction factor will depend on the tube diameter and arrangement, and on the Reynolds number, which is in turn directly proportional to velocity and tube di-

ameter, and inversely proportional to the viscosity of the gas. Of the two main variables, the tube arrangement is by far the more important. Generally the friction loss from flow across staggered tubes is considerably greater than for tubes in a line. When tubes are in line, however, the friction loss will increase as the spacing in the direction of gas flow is increased. When that spacing reaches approximately four times the tube diameter, the friction loss will be the same, whether the tubes are in line or staggered. It is not within the scope of this text to discuss in detail the variation of friction factor with tube arrangement. A close approximation, however, is 0.24 when the tubes are in line, and 0.36 for staggered tubes.

The number of restrictions will be equal to the number of tube rows crossed over, unless the tubes are so staggered that the minimum free area for the passage of air or gas is determined by the diagonal clearance between adjacent tubes. If that is the case, the number of restrictions will be the number of tube rows minus one.

A calculation of draft loss across 22 rows of tubes in a convection bank, with the tubes on a square pitch of 3½" is as follows:

$$\frac{S_T}{D} = \frac{\text{Transverse Spacing}}{\text{Outside diameter of tube}} = \frac{3\frac{1}{2}''}{2\frac{1}{2}''} = 1.4$$

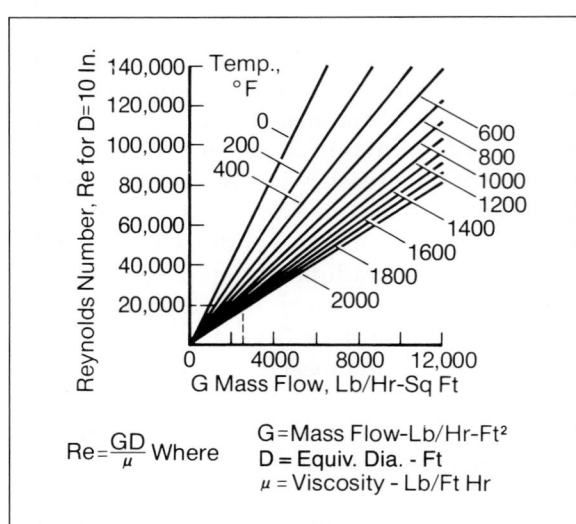

$$Re = \frac{GD}{\mu} \quad \text{Where} \quad \begin{array}{l} G = \text{Mass Flow-Lb/Hr-Ft}^2 \\ D = \text{Equiv. Dia. - Ft} \\ \mu = \text{Viscosity - Lb/Ft Hr} \end{array}$$

Fig. 25 Mass flow versus Reynolds number.

$$\frac{S_L}{D} = \frac{\text{Longitudinal Spacing}}{\text{Outside diameter of tube}} = \frac{3\frac{1}{2}''}{2\frac{1}{2}''} = 1.4$$

Reynolds number (from Fig. 25) =

$$20,000 \times \frac{2\frac{1}{2}''}{10''} = 5000$$

Friction factor, f (from Fig. 26)   = 0.092

Draft loss per restriction (from Fig. 27)   = 0.027

Draft loss through bank =

$$(0.027)\,\frac{(0.092)}{(0.100)}\,(22) = 0.55'' \text{ WG}$$

DRAFT LOSSES IN DUCTWORK

Ductwork systems form a substantial part of a pulverized-coal steam generator, both in physical size and in cost. The principal elements are (a) the main cold- and hot-air ducts from the forced-draft fans to the air preheaters, and then to the furnace; (b) the hot gas ducts from the economizer to the air heater(s) and then to the emission-control equipment and on to the induced-draft fans; and (c) the high-pressure primary-air ducts from the primary-air fans to the air heaters and on to the pulverizers. Smaller ducts are needed for sealing air to pulverizers and other equipment, and ignitor and scanner cooling air.

To determine the size and performance of these ducts, the designer must first establish the predicted maximum flows, the allowable duct velocities (based on company standards and parasitical-power costs), and both the ambient barometric pressure and the effect of localized pressures in the ducts above or below the ambient barometric pressure.

AIR AND GAS VOLUME CALCULATIONS

Air and gas weights result from the combustion calculations plus considerations of tempering air, air-heater leakage, and casing infiltration. To convert gas weight to cubic feet per minute (CFM) at sea level (14.7 psia), use this relation.

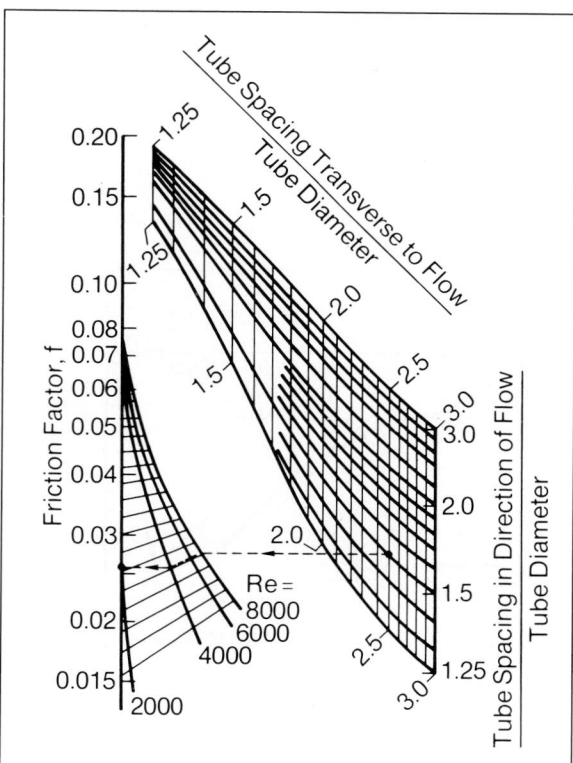

Fig. 26 Friction factors for in-line tube banks

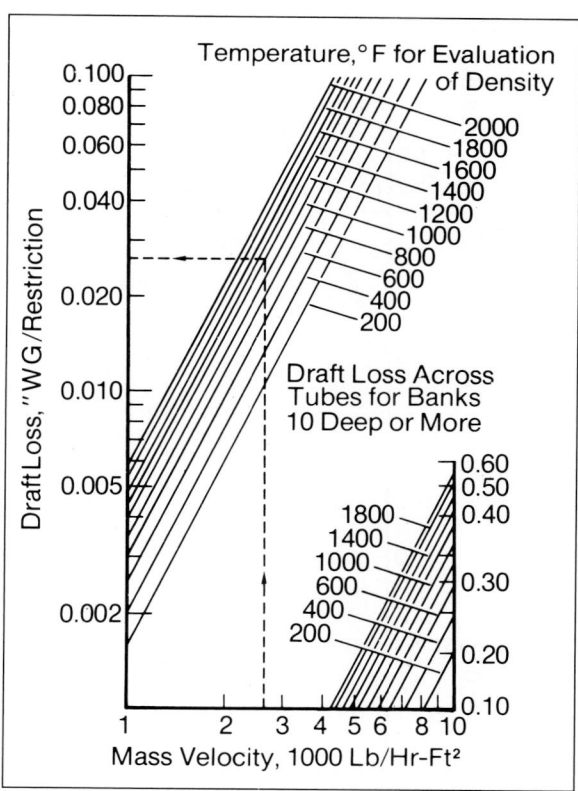

Fig. 27 Draft loss across tube banks

$$CFM = \frac{W_g (T + 460)}{(MW)_g \, 82.2}$$

(40)

where CFM is the cu ft/min of air or gas at temperature T in °F, W_g is the air or gas weight in lb/hr, and MW is the apparent molecular weight of the air or gas mixture in pounds.

This volumetric flow rate at 14.7 psia has to be corrected for plant elevation above sea level using Fig. 7 and for local pressure in the ductwork system. Fig. 28 illustrates the variation in pressure that exists in a ductwork system from FD-fan inlet to ID-fan outlet. Actual gas volumes at each point of interest must be calculated using a correction curve like Fig. 29.

VELOCITIES IN DUCTS

As a general rule, the following velocities are used in arriving at the cross-sectional flow areas of boiler ducts.

Cold-air ducts	2000 to 2500 ft/min
Hot-air ducts	3000 to 3500 ft/min
Gas ducts	3500 to 4000 ft/min

Velocities can be higher at higher temperatures because the gas is less dense and therefore has less impact energy; lower static-pressure loss results for a given temperature.

DUCT-LOSS DETERMINATION

Air pressure and gas draft losses are a power function of velocity (approximately the square) and an inverse function of the specific volume of the fluid. The most significant losses occur in turns and at abrupt changes in flow area. Vaning is often necessary to reduce losses where there are tight turns or large variations in cross-section, such as between an air-heater outlet and a precipitator inlet (where the gas velocity drops below 10 ft/sec).

Curves such as those in Figs. 30 and 31 are used to proportion ductwork for desired pres-

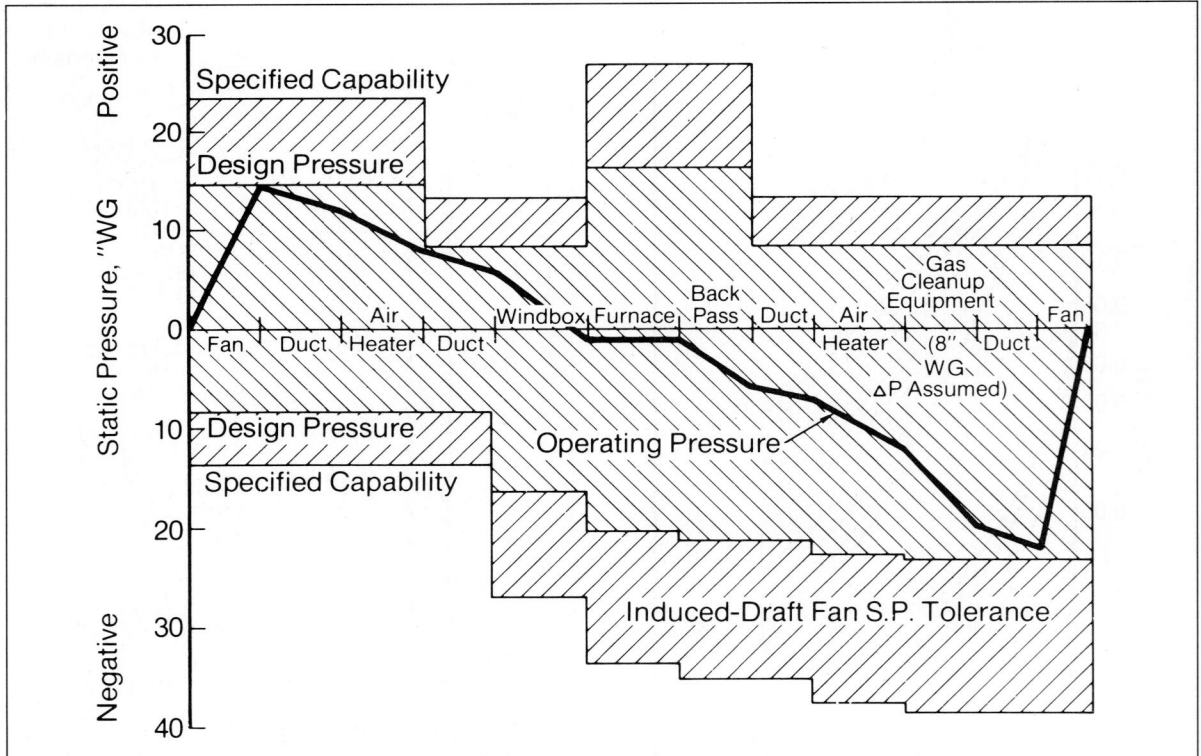

Fig. 28 Profile of air and gas pressures through a steam generating unit

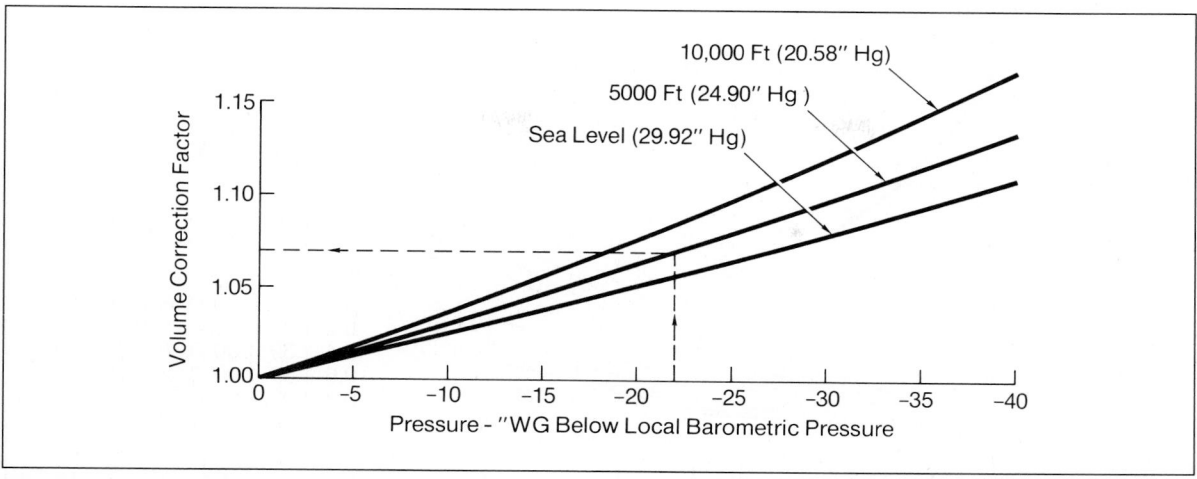

Fig. 29 Volumetric correction factor for negative pressure below the local barometric pressure. This provides an additional correction to the basic elevation correction curve, Fig. 6, to adjust for pressures in a steam generator that are below the local atmospheric pressure. Calculated from the data of Standard Atmosphere—Tables and Data to 65,800 feet, NACA Report 1235

sure or draft losses. They show that, for very large ducts, the resistance may be practically independent of length and almost wholly dependent on the design of bends.

DRAFT LOSS IN STACKS

A stack can operate to produce a net static draft or flow of gas because it contains a column of heated gas and is connected at its base, indirectly through the breeching and boiler, to the cooler outside air. To do so, the acceleration loss of the gas and the friction loss from boiler, ductwork, and stack must be overcome by the theoretical differential static head between the hot and cold column, which in turn depends on the height of the hot column and the density of the gas. (See Eq. 35.)

Stack performance curves are usually plotted for sea-level gas density, and corrections are made for elevations above sea level. It is obvious, however, that weather conditions which reduce barometric pressure have the same effect in reducing the capacity of a stack as if it were located at a higher altitude. Stack selection, or calculation of its performance, requires the determination of:

- gas weight to be handled
- mean gas density within stack at elevation
- air density outside the stack, at elevation

The gas quantity calculation must make allowance for the number of boilers connected to the stack, the maximum excess air at which they may be operated, and the air leakage through idle units as well as through the many duct connections.

The gas-density calculation involves a knowledge of the gas temperature from each unit, and the effect of air leakage on the total mixture temperature. Heat losses in emission-control equipment, ductwork, and the stack itself must also be considered, because it is the *mean* density in the stack that must be calculated.

The barometric pressure at plant elevation must be taken into account, because it affects the density of both the heated column inside the stack and the cold column of outside air.

Fig. 32 is a convenient graphical method of analyzing a stack of given diameter and height for the net effect of thermal draft versus friction loss. It is plotted for an 80°F ambient temperature and sea level pressure of 29.92 in. Hg. The net gain or loss must be divided by the elevation correction factor (Fig. 7) for elevation above sea level, or for a depressed barometer from atmospheric conditions that have to be taken into account.

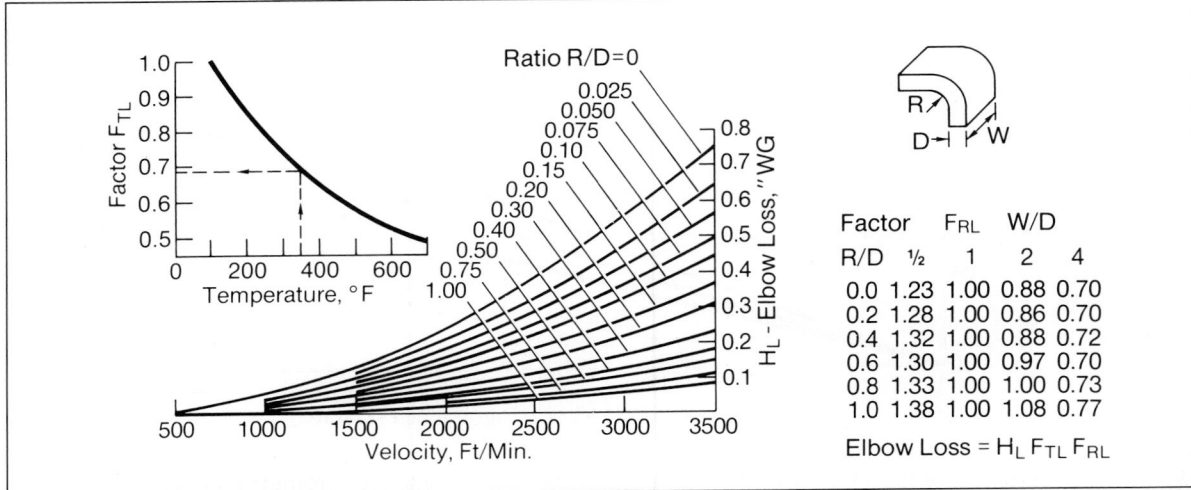

Fig. 30 Chart for determining draft/pressure loss for 90° elbows

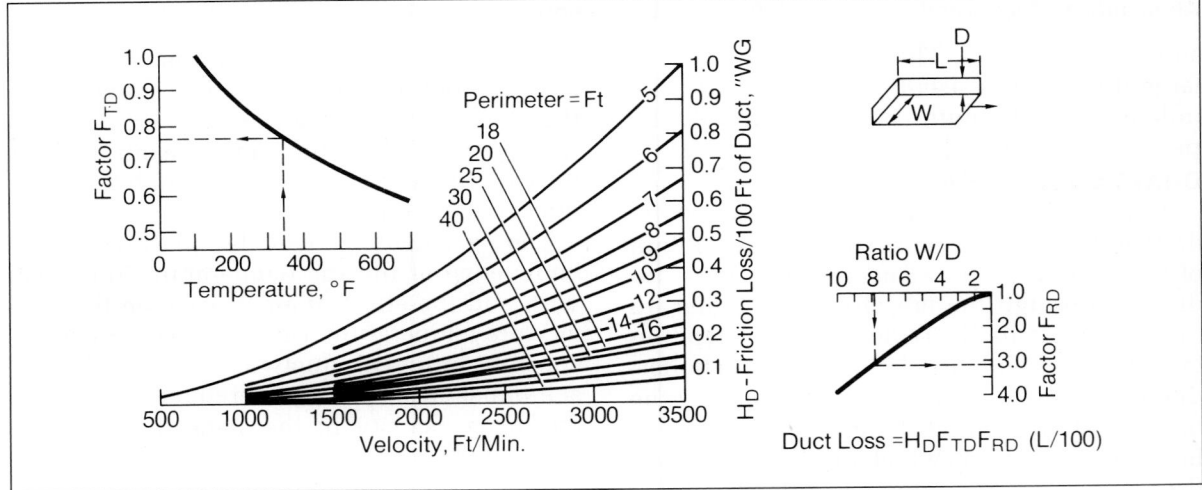

Fig. 31 Chart for determining friction loss for straight runs of ductwork

The effective height of the stack, rather than the actual height, is used with Fig. 32 as in any stack-effect calculation. If allowance has been made for static draft in all apparatus and connections between the boiler and the stack, then the effective height is the distance between the top of the stack and the centerline of the opening in the stack where the gas enters.

SPECIFYING POWER-PLANT FANS

Specifying the required operating conditions for fans at various load points is generally left to the boiler manufacturer. The fan manufacturer assumes responsibility for producing the volume and pressure specified although limited by the fact that standard fan test codes specify certain conditions which can be maintained in a formal testing setup but which are seldom if ever duplicated in an actual installation. It is thus practically impossible, or at least very expensive, to determine comparable performance of laboratory and field conditions. Furthermore, the connections made to a fan may affect its performance to such an extent

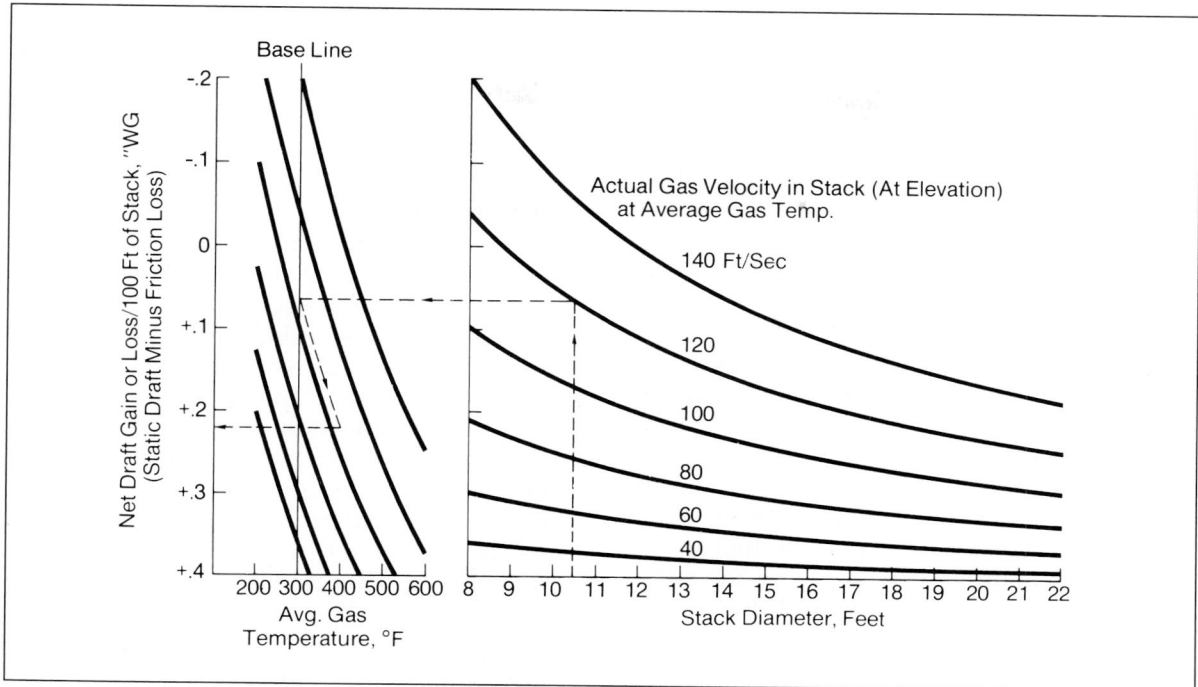

Fig. 32 **Net draft (thermal static draft minus friction loss) versus gas velocity and stack diameter at 80°F ambient temperature and 29.92″ Hg**

that it would be unwise to select it without reasonably large pressure and volume tolerances unless it were shop tested with inlet and outlet connections identical to those to be used under service conditions. In this connection it is always desirable to submit the layout of ductwork connected to fans to the fan manufacturer for evaluation and comments.

FORCED DRAFT FANS

The requirements of FD fans are calculated from the fuel fired and the excess air. For pressurized firing, the calculation is quite accurate as all air passes through the FD fans. For balanced-draft units, the calculation is more complicated as the excess air measurement does not usually correspond to the air through the FD fans. Other sources of airflow into the unit are setting infiltration and tempering air.

Infiltration and Tempering

Determined by tightness of expansion joints, observation and access doors and other penetrations of the boiler, casing infiltration varies

with life, operation and maintenance of the unit. Moisture in the fuel determines tempering air which is high with dry coal and low with wet coal. In assigning values for infiltration and tempering note that too high an estimate of airflow through fan and air preheater results in low predicted exit-gas temperature.

Effect of Air Temperature

The temperature and elevation of the fan above sea level both affect the density of the air, which in turn affects the capacity of the fan. The moisture content of the air, especially for tropical conditions, also has to be taken into account (see Fig. 1, Chapter 4).

If air preheater protection such as a steam coil is provided *ahead* of the fan, the air temperature leaving must be considered. If hot-air recirculation is used for air preheater protection, then both temperature and extra volume must be considered.

Pressure Specification

On balanced-draft units the required static

head is the sum of all series resistances in the secondary air system, including cold air duct, steam air heater, air preheater, air metering device, hot air duct, dampers and burner pressure drop. On pressure-fired units the additional loss from the furnace to the stack outlet must also be included in determining total system resistance. Where stack effect is present, it must also be considered in determining total head requirement of the fan.

Selection

The volume and static pressure calculated according to the foregoing give the actual required fan capacity under predicted operating conditions. The calculation assumes excess air, steady load, tight casing, normal fuel and commercially clean surfaces as might be expected during an acceptance test. These actual conditions are needed to evaluate power consumption at the various loads, select the control equipment and provide a fan that will operate at maximum efficiency at the desired normal output of the steam-generating unit.

These actual calculated operating conditions should not be used as maximum requirements in purchasing the fan. At times it may be necessary and desirable to operate with more excess air, or the actual temperature at the fan may be higher than anticipated, or the fan may not be up to expectations because of poor inlet and discharge conditions. Therefore, when specifying conditions for fan selection, liberal excess factors such as the following are generally applied to both volume and pressure:

- balanced-draft unit, 25 percent excess volume, 50 percent excess static pressure
- pressure-fired unit, 20 percent excess volume, 25 percent excess static pressure

These values may, of course, be modified if extreme conservatism was used in arriving at the so-called actual requirements, or if they were calculated for an unusual peak load well above the desired continuous rating.

INDUCED-DRAFT FAN

The induced-draft fan is usually positioned at the outlet of the terminal heat recovery apparatus. It may be located either ahead of, or after, the dust collector. Sometimes it is an integral part of, and built into, the stack base. This fan therefore must handle the gas resulting from combustion of the fuel as well as all infiltration occurring up to the fan inlet, including leakage in the air preheater.

Volume Specifications

The volume requirements of ID fans are figured at the calculated density existing at the fan inlet. Based on the flue-gas temperature entering the fan, density should also be corrected for elevation above sea level and, of course, be based on the actual specific volume of the products of combustion at the excess air percentage leaving the last piece of heat-transfer or emission-control equipment.

Flue-gas weight on which the volume is based is calculated from the fuel requirements and excess air. This gas weight should include moisture from fuel, from combustion of hydrogen, and from the air or any other source.

The flyash carried in the flue gas from a pulverized-coal fired unit does not appreciably increase the total volume of the mixture handled, even though it does increase density. The presence of ash may increase the power requirements of the drive for a given speed and capacity. But since ash will generally be less than five grains/cu ft, no allowance need be made other than for fan excess factors recommended in "Selection" subsection below.

Pressure Specification

The ID-fan must provide a static head equal to the series resistance from the furnace to stack outlet, including resistance of superheater, reheater, boiler bank, economizer, air preheater, dust collector and all ductwork. Besides resistances, the net stack effect must be included together with required furnace draft in arriving at the total draft requirement of the ID fan.

Selection

The volume and static pressure calculated in accordance with the foregoing give the actual required fan capacity under good operating conditions, with assumed excess air, commercially clean surfaces and normal leakage values. As with the FD fan, such conditions are

needed for evaluating power, setting controls and providing a fan that will operate at maximum efficiency at desired normal output of the boiler, but the actual calculated operating conditions should not be used as maximum requirements in purchasing the fan. Operation at higher excess air may be advisable; higher amounts of gas recirculation may be used; leakage and infiltration may be higher than estimated; surfaces may be dirty, which will increase temperature as well as resistance; and the fan may not perform to expectations because of poor inlet and discharge connections. Therefore, in specifications for ID fans, liberal excess factors are generally applied to both volume and pressure, as follows: 20 percent excess volume and 30 percent excess static pressure. As with the FD fan, these figures may be modified if extreme conservatism was used in arriving at the so-called actual requirements, or if they were calculated for an unusual peak load well above the desired maximum continuous rate of operation.

GAS RECIRCULATION FAN

For the same size unit and the same amount of heat released by the fuel, oil firing requires much larger heating surfaces in the superheater and reheater than for coal firing. Much of this relatively expensive surface can be eliminated by increasing the gas mass flow through the convection passes, using gas recirculation. This is common on straight oil-fired units and is a good solution for combination coal and oil units. The gas recirculating duct system is shown in Fig. 33. Gas is tapped off the main gas duct at point A, passes through the gas recirculating fan and enters the bottom of the furnace at B. Damper C controls the flow and is part of the steam temperature control system.

The gas recirculation fan is necessary to overcome the pressure differential between the furnace and the economizer outlet. If for any reason the fan is shut down while the boiler is in operation, high temperature furnace gas would flow backward through the fan and ducts, which are not designed to withstand such temperatures. Therefore shutoff damper

D, automatically closes when the fan stops. However, even the best shutoff damper has some leakage, so higher pressure air is tapped off the main duct at point E. The shutoff damper F is ordinarily shut, but opens automatically with closing of the other shutoff damper. There will then be a small flow of air in both directions from point G thus protecting all equipment that might be damaged.

Volume Specifications

The amount of recirculation needed to make steam temperature at the control load determines the volume requirements of the GR fan. Specifying the flue-gas volume to be handled together with the gas temperature is a matter for the superheater designer to determine. However, it must be given at both the upper and lower end of the operating range of the gas-recirculation requirement.

Pressure Specifications

Gas-recirculation flow results in pressure drops throughout the GR system, including ducts, dampers and dust collector. Because the amount of gas recirculated adds to and is common with the main gas flow through the superheater, reheater and economizer, it also increases the drops through these. It is these latter resistances which generally dominate in establishing the total head requirements of the recirculation fan.

At control load or some other partial boiler load, GR requirements are maximum. But because the main gas flow is low, the total head requirements of the recirculation fan are moderate. At full boiler load, little or no gas recirculation may be required. However, since the main gas flow through the boiler is high, the pressure differential across the superheater, reheater and economizer is high and the recirculation fan must produce a high static head to balance this differential. In selecting the fan, the relationship of fan head to the differential is a critical factor to consider.

If at any point over its operating range the fan could not develop enough head to overcome the differential developed by the boiler,

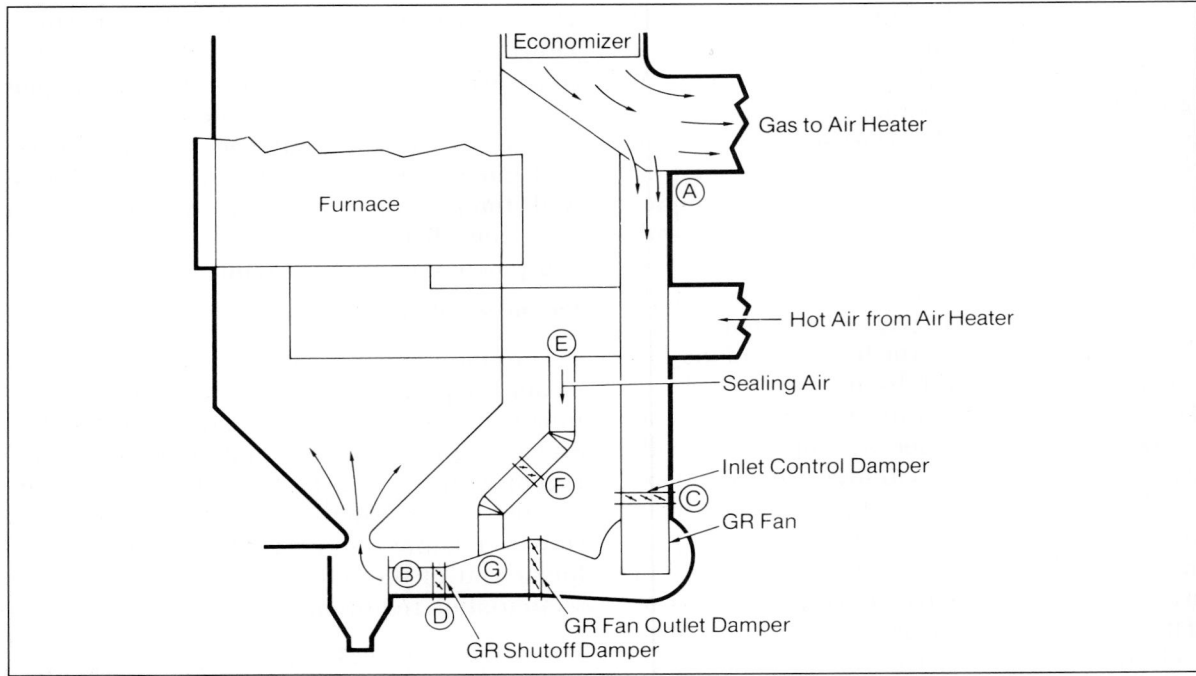

Fig. 33 Typical gas-recirculating system

backflow of furnace gas will result through the fan with very serious consequences.

To protect against this condition, it is usual to specify as a minimum point on the fan characteristic a pressure at least 2 in. wg. above the maximum superheater-reheater-economizer differential at full load on the boiler. Normally, control-load requirements dictate fan size. The following are usual excess percentages applied to the calculated requirements at this load: 15 to 20 percent volume excess, with the static pressure boost adjusted to accommodate the increased volume.

STRESS ANALYSIS AND STRUCTURAL DESIGN OF STEAM GENERATOR COMPONENTS

In a steam generator design, the structural integrity of the pressure parts and their supports is of major concern. Not only must the drums,

headers, and tubing withstand all the pressure stresses under hostile conditions of high temperature and, frequently, corrosive atmospheres, but they must also retain their positions, contour, and roundness under a variety of externally imposed loads. Extremely sophisticated analyses of the involved stresses must be made of both the pressure-retaining parts and their supporting hanger rods, U-bolts, lugs, and structural members.

This section describes the important aspect of the boiler design engineer's work that can be called "designing for structural performance."

MATERIALS SELECTION FOR SUPERHEATERS AND REHEATERS

From the view point of thermodynamics, the power-plant cycle efficiency is related to primary and reheat steam temperatures. A steam temperature increase improves the cycle efficiency, everything else being the same. However, superheater and reheater outlet conditions are limited by the ability of materials to

withstand physical stress and by economic breakeven points.

In the high-temperature region of boilers, allowable stress goes down as materials are exposed to high flue-gas temperatures necessary to superheat and reheat steam to high temperatures. At the same time, more costly grades of alloy steel are required to withstand the physical stress imposed by elevated temperatures and pressures.

As an example of temperature increase effect, the allowable stress for a material known as SA-213 T-11 drops from 11,000 psi at 950°F to 6600 psi at 1000°F. Up to 850°F it is usually possible to specify carbon steel, but beyond this point more costly alloys are required.

In view of these facts, selection of materials is a most important consideration in superheater and reheater design. Proper material selection requires knowledge not only of metal temperature conditions but also of the economics of manufacture and fabrication of various steel alloys. These aspects are amplified in Chapters 17 and 18 on materials and metallurgy, and steam-generator manufacture. A systematic procedure to calculate the metal temperatures of superheater and reheater tubing is explained in the following subsection.

SUPERHEATER AND REHEATER METAL TEMPERATURES

Before proceeding to make metal temperature calculations, the designer assumes the following data to be known for the point in question:

- gas temperature, T_g
- steam temperature at the point in question, t_s
- mass velocity of gas, G_g
- mass velocity of steam, G_s
- tubing size and estimated thickness
- tubing material
- steam pressure
- heat absorbed, q, Btu/hr.

(1) Total heat-transfer rate, R_t (as defined in previous section on boiler calculations):

$$R_t = R_c + R_n \tag{41}$$

(2) Total heat absorption rate per unit area, $\dfrac{q}{A_o}$ is calculated from

$$\frac{q}{A_o} = R_t (T_g - t_s) \tag{42}$$

By knowing R_t, t_s and T_g the localized heat absorption rate per unit area can be obtained.

If the superheater or reheater tube in question is subject to direct furnace radiation, this should be taken into consideration in Eq. 42. Fig. 34 is a polar plot of a reheater tube facing the furnace, showing the higher metal temperature rise on the front face of the tube compared to the back or downstream side.

(3) Steam side film conductance of heat transfer h_c:

After the heat absorption per unit is determined, the next step is to establish steam-side film conductance of heat transfer, h_c. By dimensional analysis and experimental work, a general relationship is developed to calculate the film conductance for forced convection in fluids. This may be evaluated from the equation $(Nu) = a(Re)^b(Pr)^c$, where $a = 0.023$, $b = 0.8$, $c = 0.4$ and the terms are respectively Nusselt, Reynolds and Prandtl numbers. The correlation may then be expressed as follows.

$$\frac{h_c D}{k} = 0.023 \left(\frac{DG}{\mu}\right) 0.8 \left(\frac{c_p \mu}{k}\right) 0.4 \tag{43}$$

where D is an expression of applicable diameter, k is the thermal conductivity of the fluid, G is the mass velocity, μ is the absolute viscosity of the fluid and c_p is the specific heat at constant pressure. The values of a, b and c are determined experimentally and should be used

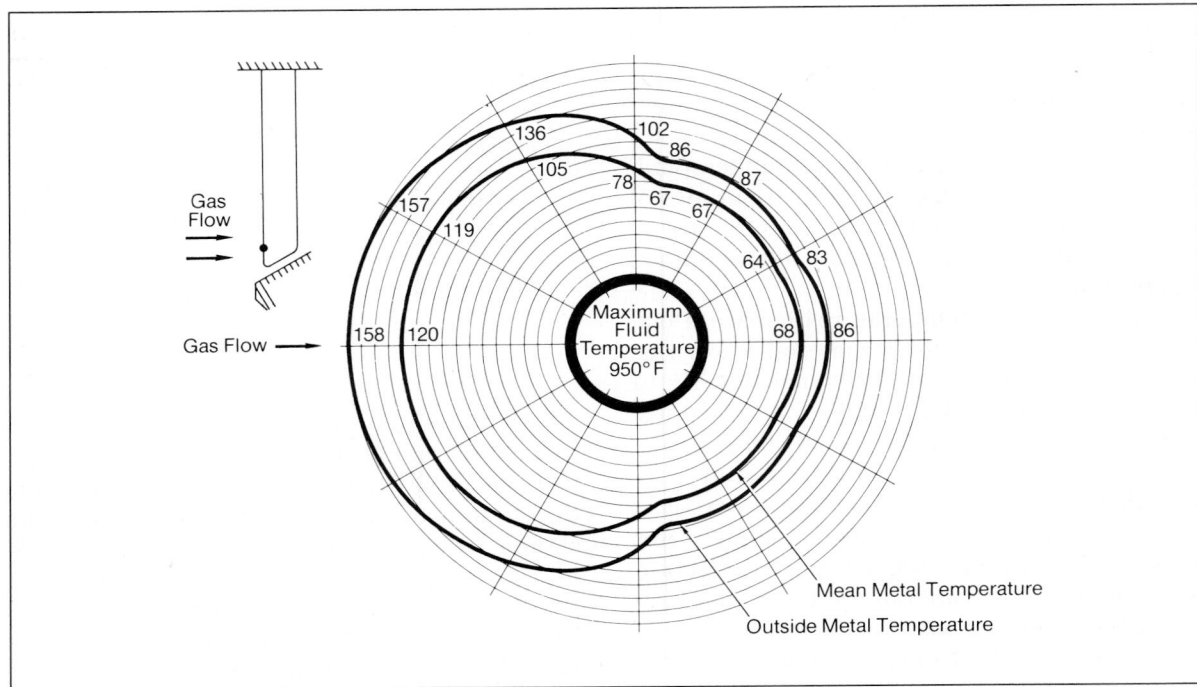

Fig. 34. Metal temperature distribution in a reheater tube

with discretion, taking into consideration results available in heat-transfer literature.

(4) Steam-side film temperature drop, Δt_s:

By knowing the heat absorption rate per unit
area and steam-side film conductance of
heat transfer, the steam-side film temperature
drop can be obtained from the following
relationship.

$$\Delta t_s = \frac{q}{A_o} \frac{1}{h_c} \frac{D_o}{D_i}$$

(44)

where D_o is the outside diameter of tube and D_i
is the inside diameter of tube. The D_o/D_i term
appears in the above formula because q/A_o was
based on outside diameter of tube.

(5) Tube wall temperature drop, ΔT:

Temperature drop through a tube can be calculated from the conduction equation:

$$\Delta T = \frac{q}{A_o} D_o \log_e \frac{D_o}{D_i} \frac{1}{2k}$$

(45)

where k is the thermal conductivity of the tube
for which the temperature is being calculated.

The tube metal temperature at any given
point in the tubing will be the sum of the steam
temperature, the film temperature drop and the
tube wall temperature drop.

METAL TEMPERATURE TOLERANCES

The preceding represents an ideal case based
on the assumptions that the steam flow rate and
temperature and gas flow rate and temperature
are uniform for all the elements. These conditions actually are never experienced in the performance of the steam generator. It is necessary
therefore to provide so called "imbalances" or
tolerances that are anticipated. These include
the following:

■ Desuperheater allowance which is the number
of degrees that the steam temperature will rise
at the particular point upstream of the superheater when the full use of the desuperheater is
made. Usually the design is such that little or
no desuperheating is required. However, when
a desuperheater is installed all pressure parts

ahead of the desuperheater are designed to include the desuperheater allowance.

■ Steam temperature imbalance which allows for the increase in expected steam temperature at a given point caused by the unbalanced distribution in steam flows, gas flows and gas temperatures.

■ Film and metal drop imbalance which is the increase in the film and metal temperature drop caused by the imbalance distribution in the gas flows and gas temperatures.

The maximum metal temperature at any given point in the tubing will be the sum of the steam temperature, the film and metal temperature drops, the desuperheater allowance — if any, the steam temperature unbalance and the film and metal drop imbalance.

After the metal temperature is established, the selection of material is made in accordance with the *ASME* Boiler and Pressure Vessel Code. Section I of the Code covering power boilers gives details for calculating tube wall thickness and the allowable stress values for the various materials. In superheater and reheater design it has been found desirable to limit the use of various materials according to the oxidation-resistance limits in Table IV, which are in all cases lower than the maximum permissable metal temperatures established by the *ASME* Boiler and Pressure Vessel Code.

PRESSURE-DROP CALCULATIONS

Another aspect of the design of superheaters, reheaters, and economizers that relates to the selection of materials in previous paragraphs is the steam or water pressure drop that occurs in operation. Just as there is a semi-proportional relationship between transfer rate and draft loss, as previously explained, there is a similar connection between pressure drop and metal temperature. That is, the fluid film thickness inside a heated tube is affected by the mass flow of fluid through the tube. And pressure drop is roughly a square function of the mass flow.

In evaluating whether a given pressure drop is too high through any proposed circuitry, the principal concerns most often are (a) the design pressure of upstream pressure parts, and (b) the power penalty for any increment of pressure drop versus the cost of using more expensive alloys in superheater or reheater tubing to reduce wall thickness and, thereby, the pressure drop. An example of the first situation is the influence on boiler drum and waterwall tubes and headers of the superheater pressure drop: the superheater drop is added to the required superheater-outlet pressure to arrive at the operating pressure in the drum, the furnace walls, the risers, and the downtakes. With tolerances specified by the *ASME Code,* the actual design or working pressure of the steam generator is established. The higher the superheater pressure drop, the higher the ASME design pressure for the entire unit, which has a direct effect on the thickness and cost of the pressure parts.

Basic Pressure-Drop Equations

The following equations are used for water

Table IV. Maximum Outside-Surface Metal Temperatures

ASME Code Alloy Specification	Nominal Composition	Oxidation Limit, °F
SA-213	Carbon Steel	850
SA-213 T-1	Carbon- ½ Mo	900
SA-213 T-11	1¼ Cr- ½ Mo	1025
SA-213 T-22	2¼ Cr- 1% Mo	1100
SA-213 T-9	9 Cr- 1 Mo	1175
SA-213 Type 304H	18 Cr- 8 Ni	1300
SA-213 Type 347H	18 Cr-10Ni	1300

and steam pressure-drop calculations on sub-critical-pressure boilers, and are similar to those for supercritical units.

Water flowing through pipes and tubing:

$$\Delta P = \frac{1.4 \times F^{1.85} L}{10^3 D^{4.85}}$$

(46)

where

F = flow per path, 1000 lb/hr

L = length of tube plus equivalent length of bends, ft

D = inside diameter in inches

Steam flowing through tubing, up to 6″ O.D.:

$$\Delta P = \frac{4.11 \times F^{1.85} V L}{10^4 D^{4.97}}$$

(47)

Steam flow through pipes:

$$\Delta P = 1.306 \left(1 + \frac{3.6}{D}\right) \frac{L V F^2}{D^5}$$

(48)

where

F = flow per path, lb/min

V = specific volume, cu ft/lb

and L and D are as defined above.

For determining the total equivalent length of a circuit, the following factors are used, where the equivalent length is equal to the factor multiplied by the inside diameter (ID) of the tubing. Thus, a 180° bend in a tube with 2 in. ID has an equivalent length (for use in Eq. 46, 47, 48) of 30 × 2 = 60 in. = 5 ft.

Equivalent Lengths

	MULTIPLYING FACTOR ON ID
Entrance loss	30
Exit loss	30
Squeezed return bend	50
180° bend	30
90° bend	25
45° bend	15
90° short radius elbow	24
90° long radius elbow	12

Entrance and exit losses are applied to the junctions between tubes and headers. A squeezed

return bend is one with a bend radius less than the tube outside diameter (OD). A 90° short-radius elbow is defined as having a bend radius less than 3 times the pipe OD. Anything greater is considered a long-radius elbow.

MECHANICAL STRESSES IN DRUMS, HEADERS AND PIPING

Industrial safety codes, such as the *ASME Code*, provide the designer with acceptable formulas for sizing various components of a pressure vessel. Typical formulas making use of equations from the basic theory of elasticity for specific shells of revolution presented here are taken from Section I, Rules for Construction of Power Boilers, of the *ASME Code*.[52]

SPHERICAL SHELL

The minimum thickness, t, of a spherical shell of inside radius, R, for a design maximum allowable working pressure, P, and a maximum allowable working stress, S, all in consistent units, is

$$t = \frac{P R}{2 \times S - 0.2 P} \text{ or } \frac{P R}{1.65}$$

(49)

on an elastic shell basis.

Eq. 49 gives the correct thickness for thin-walled shells, but when the thickness of the shell is greater than about 0.36 R, the elastic theory requires that thickness be calculated on a thick-walled basis. The formula for thickness is

$$t = R \left[\sqrt[3]{\frac{(S + P)}{2S - P}} - 1 \right]$$

(50)

CYLINDRICAL SHELL

The minimum thickness for a thin-wall cylindrical shell for piping, drums, and headers in the steam generator is

$$t = \frac{PR}{SE - (1 - y) P} + C$$

(51)

where C is an additional allowance for threading and structural stability and y is a tempera-

ture coefficient, as in Table V.

For thick-wall cylinders, in which the thickness is greater than one-half the inside radius, the elastic theory requires that the shell wall thickness be determined from

$$t = R \left(\sqrt{\frac{SE + P}{SE - P}} - 1 \right)$$

(52)

In this equation, E is the efficiency, expressed as a decimal, and called the ligament efficiency. It is based on the pitch, diametral size, and configuration of the holes in a drum or header. For openings that form a definite pattern in pressure parts, the efficiency of ligament between the tube holes is determined per paragraph PG-52 of Section I of the Code.

When the pitch of the tubes on every tube row is equal, as in Fig. 35, the formula is

$$E = \frac{p - d}{p}$$

(53)

where p is the longitudinal pitch of adjacent openings and d is the diameter of the openings, in consistent dimensions, and E is the efficiency of ligament, as a decimal. For a pitch of $5\frac{1}{4}$ in. and tube-hole diameter of $3\frac{9}{32}$ in., efficiency is $(5.25 - 3.28)/5.25 = 0.375$.

When the pitch of the tube holes on any one row is unequal, the formula is

$$E = \frac{p_1 - n d}{p_1}$$

(54)

where p_1 is the pitch between corresponding openings in a series of symmetrical groups of openings and n is the number of openings in length p_1. The pitch in these equations is measured either on the flat plate before rolling or on the median surface after rolling.

For the spacing shown in Fig. 35 B, the efficiency is $(12 - 2 \times 3.281) / 12 = 0.453$. For the spacing in Fig. 35 C, ligament efficiency is $(29.25 - 5 \times 3.281)/29.25 = 0.439$.

The strength of those ligaments between tube holes that are subjected to a longitudinal stress shall be at least one-half the required strength of the ligaments coming between the tube holes which are subjected to a circumferential stress. When tubes or holes are arranged in a drum or shell in symmetrical groups along lines parallel to the axis and the same spacing is used for each group, no group efficiency shall be less than that on which the maximum allowable working pressure is based.

HEMISPHERICAL HEAD

The formulas for a hemispherical head are identical to those for a spherical shell. Either industrial-code rules or the theory of elasticity give the required thicknesses for these shells.[53] If a pressure vessel consists of a cylindrical shell capped with a hemispherical head, the

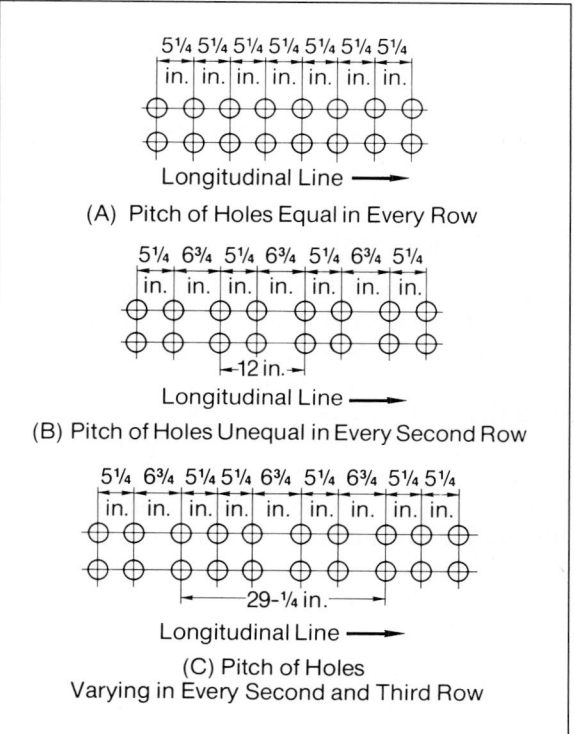

Fig. 35 **Examples of tube spacings in drums or headers**

Table V. Temperature Coefficient y used in Design Calculations

Design Temperature	Up to 900°F	901° to 950°F	Above 950°F
y factor for ferritic steels	0.40	0.50	0.70

From Section I, *ASME* Boiler Code. Between temperatures listed, determine y by interpolation.

component sizes can be determined by codes or by the theory of elasticity using the simple membrane formulas.

SEMIELLIPSOIDAL HEAD

The standard semiellipsoidal head recognized by the *ASME* Code has half the minor axis, or the depth of the head, at least equal to one-quarter the inside diameter of the head. By elastic theory this head requires a thickness of

$$t = \frac{PR}{S - (1 - y)P} + C$$

(55)

The R in the above equation is either the ½ length of the major axis, or the radius of the mating cylinder.

TORISPHERICAL HEAD

A torispherical head is a shell which combines part of a torus with a spherical cap. In the *ASME* Code the knuckle, or torus, radius is required to be 3 times the head thickness, but not less than 0.06 times the diameter of the shell, and the crown radius may not exceed the outside diameter of the head flange. There are no directly applicable formulas based on elastic theory but any given head shape can readily be calculated from basic equations. The *ASME* Code specifies the thickness to be

$$t = \frac{5PL}{4.8S}$$

(56)

where L is the radius to which the head is dished, measured on the concave side.

FLAT HEAD

A flat head can be considered as a circular plate which is uniformly loaded by pressure. The amount of end fixity depends on the method of mounting and the stiffness of the cylindrical shell to which the flat head is attached. If the plate is assumed to be simply supported, then no moments are transferred to the shell, and the maximum plate stress occurs at the center of the plate. The stress is:

$$S_{max} = \frac{3(3 + v)P R^2}{8t^2}$$

(57)

The degree of end fixity is determined by applied edge moments,

$$M_o = C_1P R^2$$

and superposed stresses,

$$S_s = \frac{6M_o}{t^2}$$

By substituting allowable stresses in these equations, the required thickness of the plate can be expressed as:

$$t = d \sqrt{\frac{C_2P}{S}}$$

(58)

where d is the inside diameter of the header at the flat end plate, and C_2 is a factor that depends on the method of attachment of the head and other items in Code paragraph PG-31.4.

Eq. 58 gives the thickness of a flat head as required by the *ASME* Code. The constant C_2 in the equation is a factor depending on the method of attachment of the head, and on the shell, pipe or header dimensions. It must be found from appropriate tables in the Code, and these tables usually specify the end fixity of the flat plate. In the industrial code, the constant is usually greater than that required by elastic theory, thereby adding another factor of safety to the code design.

DISCONTINUITY STRESSES

When a pressure vessel is subjected to internal pressure, redundant forces and moments

are induced in areas of structural discontinuity such as the intersection of a cylindrical shell and a hemispherical head. This can best be visualized, Fig. 36, by assuming that the shell and head act as separate units with internal pressure applied to the individual, simply supported parts. Because the deflections of the components differ, shear forces and moments must be applied at the edges to rejoin the components for a compatible structure. The resulting stresses (usually bending stresses) induced by these shear forces and moments are called discontinuity stresses. The bending stresses reach a maximum at, or near, the discontinuity, but they attenuate with distance from the joint.[54] A commonly accepted approach to the derivation of these discontinuity stresses is the theory of beams on an elastic foundation.[53,55]

If a pressure vessel with a hemispherical head is designed for minimum thickness of the mating components, then the maximum stress in the pressure vessel will be 27 percent higher than the circumferential stress in the cylindrical shell away from all discontinuities. On the other hand, for a vessel of uniform thickness, the maximum stress is only 3 percent higher than in the same size cylindrical shell.

A pressure vessel with a standard semiellipsoidal head of the same thickness as a mating cylindrical shell exhibits a maximum stress 13 percent higher than the circumferential membrane stress in the cylinder.[56] Unfortunately, membrane stresses and discontinuity stresses are very sensitive to changes in curvature.

REINFORCED OPENINGS

For very small unreinforced circular holes, the theory of elasticity shows that a stress concentration factor of three will exist around the hole under an uniaxial tension. The stress concentration is reduced to two for the state of biaxial tension. In a cylindrical shell under pressure, the state of stress is such that the circumferential stress is twice the longitudinal stress. Under this condition the stress concentration at an unreinforced penetration would reach a peak of 2.5. The stress concentration ef-

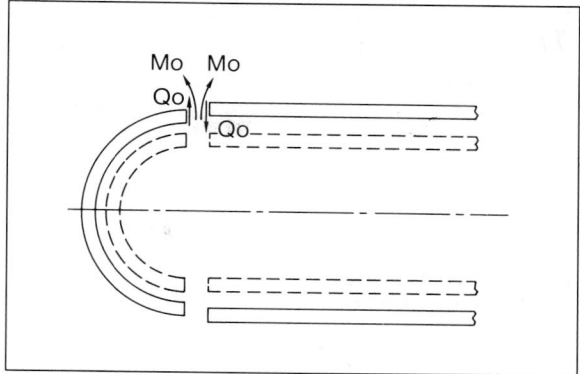

Fig. 36. Discontinuity forces at a head-cylinder junction under internal pressure

fect would be localized at the edge of the hole and could cause some plastic deformation with an inherent redistribution of stresses.

The *ASME* Code requires that, for holes larger than a certain diameter, reinforcement to the opening should be added by the nozzle attachment, increased shell thickness, or the pad for a cover plate if the opening is to be used for future servicing operations.

The Code's rules specify that the area removed for the penetration in a given cross section be reinforced within specified boundaries. At best this is a crude method; however, experience has shown that the stresses do not cause failure at nozzles if the level of stress in the shell is at a reasonable limit.[57] The best type of reinforcement is one in which the reinforcement is equally distributed on each side of the shell, that is, equal amounts of reinforcement extend inside and outside the shell.[58] This method of reinforcement is not always practical; most commonly, then, it is necessary that all the reinforcement must be on the outside of the shell.

STRESS ANALYSIS AND STRUCTURAL DESIGN OF FURNACE WALLS

Welded furnace walls are one of the most important components of boilers. They may be incorporated into the design of small shop-assembled boilers as well as in the largest utility-type units. Furnace walls, also frequently referred to as waterwalls, provide the enclosure around the major parts of a boiler and are integrally concerned with all of its essential functions. As a link in the steam generating cycle, they are components of the pressure parts system. The support of the furnace-wall system is linked to that of other components and will be discussed in the final section of this chapter dealing with structural design of boilers.

Welded furnace walls must be safe and reliable over a range of operating conditions. The furnace-wall tubes have the highest heat-flux rates in the boiler and are subjected to the most extreme fluctuations in temperature and pressure. During boiler startup the walls experience considerable overfiring and variable pressure conditions are encountered as the unit is brought up to its normal operating level. The furnace-wall enclosure must be designed to withstand both positive and negative inside pressures. In addition, special conditions sometimes occur in which the enclosure must withstand wind and earthquake loadings.

The crucial feature in the protection of a furnace-wall tube against failure is the assurance of adequate circulation of fluid within the tube under all operating conditions. For this reason, the thermodynamic requirement for the particular type of furnace determines the internal diameter of the wall tube. For some furnaces the wall will contain small-diameter tubing with thick walls while for other furnaces the tubing will be large diameter and relatively thin walls. Thus, mechanical behavior of furnace walls will vary greatly for different types of furnaces.

MECHANICAL LOADING OF FURNACE WALLS

Analytical and experimental techniques make it possible to determine the stresses in the furnace wall in considerable detail. With this knowledge and the realization that different stresses have different degrees of significance, the furnace walls are designed to give the desired thermal behavior, at the same time being safe and reliable.[59]

As a starting point, the tube thickness is determined to give the required internal diameter for proper thermal performance of the furnace. Since the tubes are a part of the pressure system of the steam generator, the determination of wall thickness is within the jurisdiction of Section I of the *ASME* Code.[52] Thus the equation used for tube thickness is

$$t = \frac{PD}{2S + P} + 0.005D$$

(59)

where t is thickness, P is fluid pressure, D is outside tube diameter and S is the allowable stress according to Table PG 23.1 of Section I, *ASME* Boiler and Pressure Vessel Code.

Following this procedure satisfies the *ASME* Code requirements for minimum tube wall thickness and gives a wall panel with the required flexibility to keep the thermal stresses within reasonable limits on severe thermal transients. Except under special conditions, a thicker wall tube does not improve the overall stress condition. As the pressure stress becomes smaller because of thicker tube, the thermal stress increases rather rapidly with greater wall thickness. A change to a better material is the most satisfactory method of gaining improvement in strength requirements.

Other mechanical loads also act on the furnace wall and must be analyzed to assure its reliability. Some of these additional loads are furnace firing pressure, dead-weight loads, wind loads, earthquake loads and reactive loads that exist between the furnace wall and the rigid buckstay system.

DETERMINATION OF FURNACE STRESSES

The straight analytical determination of stresses in the furnace walls is difficult because of the complex geometry of the panel. To begin the calculation, it is necessary to assume a fin size. For a furnace panel with normal heat-absorption rates, the fin width will be about one half inch. With this close tube spacing, the wall can be considered a stiffened orthogonal plate. Using the notation of Timoshenko,[53] the differential equation of equilibrium for small displacements of an orthotropic plate is

$$D_x \frac{\partial^4 W}{\partial x^4} + 2(D_1 + 2D_{xy}) \frac{\partial^4 W}{\partial x^2 \partial y^2}$$

$$+ D_y \frac{\partial^4 W}{\partial y^4} = q$$

(60)

where W is the lateral displacement,

$$D_x = \frac{E'_x h^3}{12}, \quad D_y = \frac{E'_y h^3}{12},$$

$$D_1 = \frac{E'' h^3}{12}, \quad D_{xy} = \frac{G h^3}{12},$$

q is uniform lateral load, E is the modulus of elasticity and h is wall thickness.

The numerical computer techniques of finite elements are used to solve Eq. 60. The tube and fin geometry of the furnace wall is transformed into an equivalent orthotropic plate by the use of equivalent load functions as developed by C-E engineers. The mechanical loads such as furnace pressure, dead weight, buckstay loading, wind and earthquake and the thermal loads from temperature distributions are applied to an orthotropic finite element plate model of the furnace walls. The transformed equivalent orthotropic plate elements are used to model the furnace wall and beam elements are used for the buckstay system. The finite element mathematical model of the furnace is solved using one of the finite element computer programs such as "MARC-CE Nonlinear Finite Element Program" or the "STRUDL Finite Element Program". The computer program findings are resultant forces and moments acting on the various equivalent orthotropic plate elements of the furnace-wall mode (Fig. 37).

BALANCED THERMAL LOADING

Balanced thermal loading is a condition which results in a local variation in temperature that produces thermal stresses in tubes and fins; this condition occurs because of balanced heat absorption from one side of the wall. The local metal temperatures existing in a tube and fin depend on the panel geometry, the heat

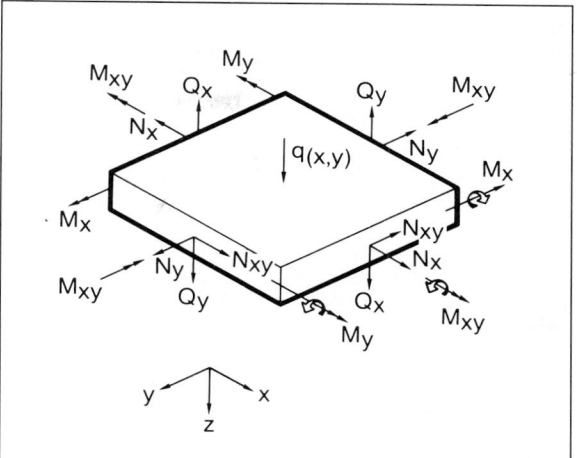

Fig. 37 Orthotropic plate subjected to in-plane and lateral loads

flux, inside film coefficient and the type of boiler. Fig. 38 gives an example for a Controlled Circulation® boiler of the performance of the four critical temperatures: at the outside, mean and inside diameter on the crown of the tube, and at the fin tip, as a function of inside film coefficient. The curves illustrate the importance of maintaining a minimum inside film coefficient to keep the tube cool. Using the relationship between heat flux and nucleate boiling coefficient shown in Fig. 39, the actual maximum temperatures in the tube and fin above the saturation temperature are plotted against flux rate in Fig. 40. This relationship is essentially proportional and demonstrates the range of absorption to which tubing and fin material may be exposed.

Temperature Distribution

The first step in determining the thermal stress in the fin and tube is to obtain a temperature distribution, T. The tube and fin respond with sufficient speed so that, at any given condition, this relationship can be approximated by the steady state.

Thus $\nabla^2 T = 0$

where $\nabla^2 T = \frac{\partial T^2}{\partial x^2} + \frac{\partial T^2}{\partial y^2}$

(61)

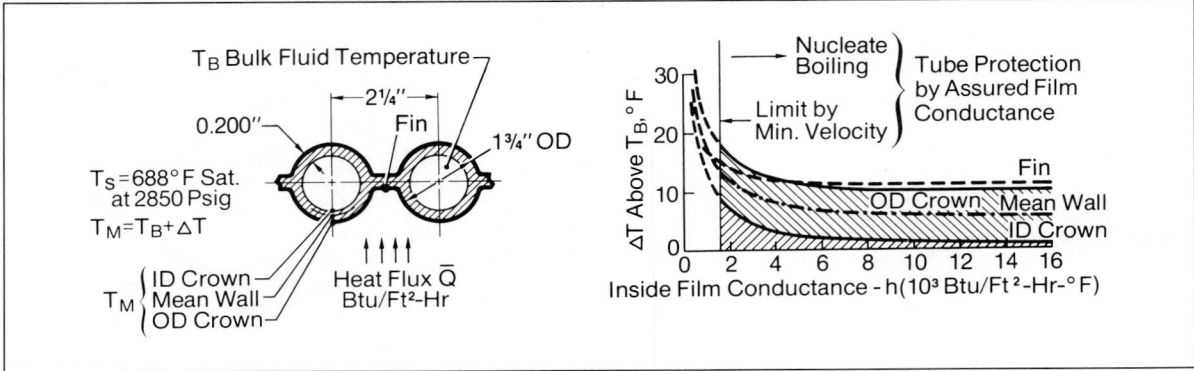

Fig. 38 Furnace-wall tube temperature plot

Numerical computer techniques are used to solve this differential equation. The CEWIND computer program for calculating temperature distributions is based on the finite difference method and provides directly the surface temperature and the internal temperature distributions. Fig. 41 shows some thermal isotherms generated by this program. Often a special computer tape will be made of the thermal results so that temperatures may be directly input to computer thermal stress programs.

The MARC-HEAT Computer Program will solve the same equation with a finite element type of analysis and has the advantage that the same finite element model may be used for both the thermal and stress analysis programs. However, stresses or strains require an extrapolation procedure.

The CE-NOAXCYL Nonlinear Axisymmetric Finite Element Analysis Program is a special purpose computer program which will provide a transient heat transfer analysis with non-axisymmetric loading on a symmetric tube and fin cross section. This program has the capability of making plastic and creep strain calculations if they are required.

Thermal Stresses

Thermal stresses occur in the tube and fin because the elements cannot deform freely. Some of the thermal stress results from differences between the mean temperature of the section and the local temperature. The remaining thermal stress is imposed on the tube and fin by the comparatively stiff buckstay system.

A finite element model is made of the tube and fin geometry as shown in Fig. 42. If the temperature distributions of the tube and fin (Fig. 41) are input into one of the finite element computer programs such as "MARC-CE Non-

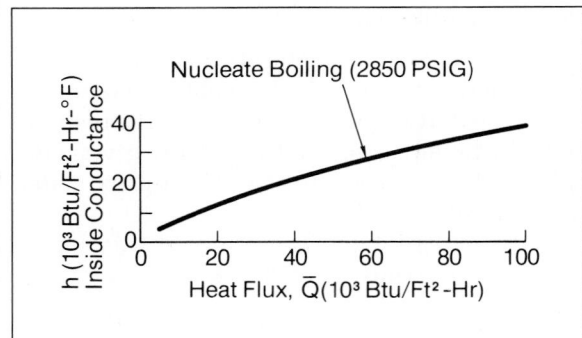

Fig. 39 Correlation between heat flux and nucleate boiling

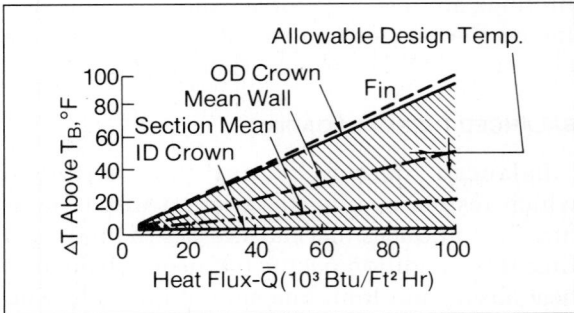

Fig. 40 Temperatures above saturation versus heat flux

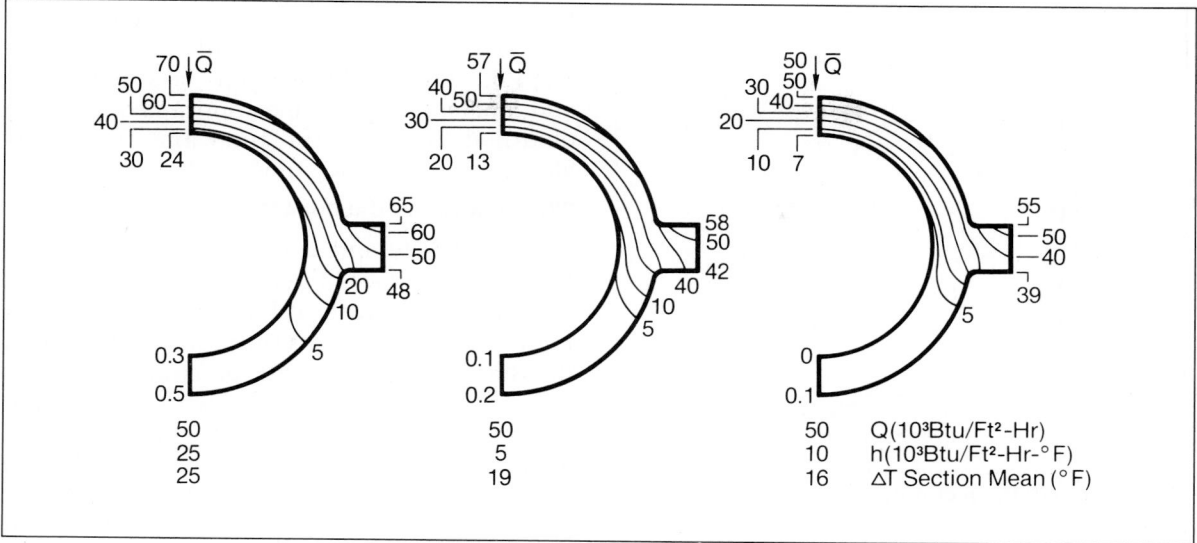

Fig. 41 Temperature distribution in finned furnace-wall tubes

linear Finite Element Program" or "STRUDL" with proper boundary conditions, the computer will calculate thermal stresses and strains in the tube and fin geometry.

The forces and moments from the computer solution of the orthotropic furnace walls (Fig. 37) are applied to the tube and fin geometry as shown in Fig. 43. These forces and moments when used with the finite element model shown in Fig. 42 will generate additional local stresses in the tube and fin geometry. The same finite element model is used to calculate the stresses and strains from the internal pressure applied to the tube. Combining and evaluating these stresses is covered later in this chapter.

UNBALANCED THERMAL LOADING

Unbalanced thermal loading is defined as temperature distribution which varies in either direction along the furnace wall. The variation in heat-absorption rates longitudinally and transversely across the wall is one reason for unbalanced temperaure distribution. Fig. 40 shows how the metal temperatures vary with different heat absorption rates. Unbalances in the flow rate either in neighboring tubes or in different areas of the wall will give unbalanced temperatures. Flow rates in some boilers are very dependent on the heat absorption for a

given tube. Further investigation reveals a very complex interplay between heat absorption and flow rates for some types of furnaces with conditions that can lead to large temperature

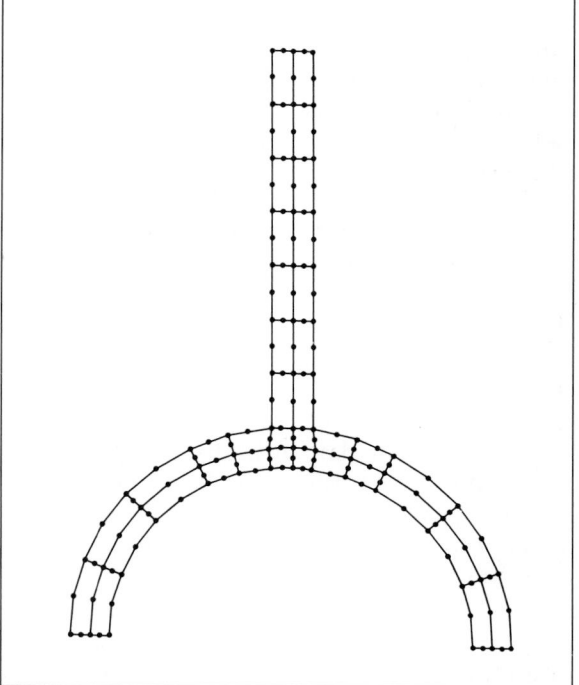

Fig. 42 Finite element model of tube and fin

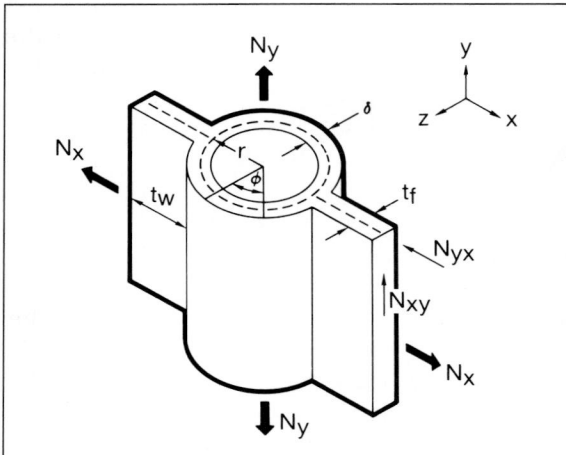

Fig. 43 **Geometrical configuration of tube and fin**

unbalances. Fig. 41 shows temperature distributions for three different inside-film coefficients, which are strongly dependent upon flow rates. A third condition that will result in temperature variations in a furnace wall is that of varying fluid temperatures. In boilers of the once-through type, the fluid temperature continually varies as the fluid progresses from inlet to outlet. In each case there is a very complex interaction between these conditions, and it is the responsibility of the boiler designer to select a suitable set of parameters so that the temperature unbalances will be within reasonable limits.

An experimental technique has been used to determine the stresses from the unbalanced thermal loadings. As an example, Fig. 44 shows a furnace wall panel under test at C-E's Kreisinger Development Laboratory. In this test controlled temperature distributions were imposed on the panel. To identify the influence of each type of discontinuity, electric strain gages placed on the outside surface of the panel, at its end, at its center, at the top bend, and at the soot blower opening determined the resulting strains.[60]

This experimental data has been used to evaluate the procedures for the equivalent orthotropic plate modeling of the furnace walls. Unbalanced thermal distributions are applied

to the furnace model and the resultant forces and moments determined with finite element analysis. Local effects of the resultant forces and moments on the tube and fin geometry are obtained by using the tube and fin finite-element model shown in Fig. 42.

SIGNIFICANCE OF CALCULATED STRESSES

The accurate and detailed determination of the mechanical and thermal stresses in the furnace wall is of little value until the designer knows how these stresses affect wall failure. A measured value of stress or strain has little meaning until it is associated with location and distribution in the panel and with the type of loading which produced it. For example, the average hoop stress in the tube from internal pressure has a different influence on the life of the panel than that of the peak stress at the intersection of the fin and tube from a redundant moment. An excessive hoop stress produces an instability in the tube, which continues to expand and get thinner and thinner until rupture. On the other hand, the peak stress at the notch will relieve itself by yielding and has a bearing only on fatigue life or stress-rupture life.

To help evaluate the different types of stresses that may act on the wall, the various stresses are classified as primary, secondary, and local or peak stresses.

Primary Stress

A primary stress is a normal or shear stress developed by the imposed mechanical loading which is necessary to satisfy the laws of equilibrium of external and internal forces and moments. The basic characteristic of a primary stress is that it is not self-limiting. Primary stresses which exceed the yield strength will result in failure or gross distortion.

Secondary Stress

A secondary stress is a normal or shear stress developed by the constraint of an adjacent part or by self-constraint of a structure. The basic characteristic of a secondary stress is that it is self-limiting. Local yielding and minor distortion can satisfy the conditions which cause the stress. Thermal stresses which produce gross distortions are secondary stresses.

Local or Peak Stress

Local or *peak stress* is a normal or shear stress which is the highest in the region under consideration and is developed by changes in geometry such as a notch. Thermal stresses which are caused by self-constraint are treated as local stresses. The basic characteristic of a local or peak stress is that it causes no significant distortion and is objectionable only as a possible source of either a fatigue failure, brittle fracture, stress-corrosion cracking or stress-rupture failure. These categories also differentiate between a membrane stress (average value across a section) and a bending stress (linearly variable across a section).

COMBINATION OF STRESSES

The characteristic material values on which strength calculations are based are determined under uniaxial tension. Usually this is the common tension test. The furnace walls are subjected to multiaxial stress, and a direct comparison of the stress in the wall with the characteristic strength value of the material is not directly admissible. Only with the aid of a strength hypothesis is it possible to calculate an equivalent stress for comparison with the characteristic strength value as determined by the conventional tension test.

Maximum Shear Theory

The maximum shear stress theory is used as the failure criterion for the furnace walls. The selection of the maximum shear stress theory is based on the use of a good ductile steel for the furnace walls. According to the maximum shear theory, if the principal stresses are σ_1, σ_2, σ_3 and further if $\sigma_1 > \sigma_2 > \sigma_3$, the maximum shear stress is

$$\tau_{max} = \frac{\sigma_1 - \sigma_3}{2}$$

$$(62)$$

The equivalent stress to that of a simple tension or compression test would be

$$S = \sigma_1 - \sigma_3 = 2\tau_{max}$$

Called stress intensity, the equivalent stress is not a single stress quantity. Rather, it represents the combined stress of the three principal stresses or their equivalent. The addition of stresses from different categories must be performed at the component level, not after translating the stress components into stress intensities. Similarly the calculation of membrane stress intensity involves the averaging of stresses across a section and this averaging must be performed at the component level.

The allowable stresses (except for the membrane stress) have been derived from the same considerations that were used in writing Div 2 of Section VIII of the *ASME* Boiler and Pressure Vessel Code. It is an application of limit design theory tempered by engineering judgment and some conservative simplifications.

Allowable Membrane Stress

The first stress limit to consider is primary membrane stress. In this case the value is limited to the values as given in Table PG 23.1 of

Fig. 44 **Furnace-wall test apparatus at C-E Kreisinger Development Laboratory**

the Section I Rules for Construction of Power Boilers of the *ASME* Boiler and Pressure Vessel Code, since the tubing is part of the pressure system in the boiler.

Stress intensities are not determined because this section of the *ASME* Code is based on maximum stress theory.

Allowable Local Membrane or Bending Stress

In the remaining evaluation, the stress intensities are determined for comparison with the stress limits. The next limit is the sum of the primary membrane stress plus the primary bending stress plus any local membrane stress. These stresses are combined at the component level and then the stress intensity determined. This limit is set at 1.5 times the stress limits as given in Table PG 23.1 of Section I of *ASME* Code. In addition, if the temperatures are at the level where stress rupture is the mode of failure, the combined value is limited to those values given in the Table.

Allowable Secondary Stress

In the next consideration, the primary stresses (membrane and bending) are combined with the secondary stresses at the component level and the stress intensities determined. These values are limited to three times

the values from Table PG 23.1 for temperatures below 700°F.

If the temperature is greater than 700°F, the influence of creep strains must be considered. Control of the secondary stresses is to limit deformations and to provide a valid basis for the elastic procedure for the fatigue analysis. When creep alters the stress distribution and increases the strain range, a high-temperature design procedure is required. For the low-alloy ferric materials the temperature range may increase to 850°F without serious creep strain.

Allowable Peak Stress

For the next evaluation all the stresses — primary, secondary and local — are combined for both the maximum and minimum conditions of an operational cycle. From these values the maximum and minimum stress intensities can be determined. Using these stress intensities the alternating stress is calculated. These stresses are usually evaluated on a fatigue basis which will be discussed in the next section.

Where loads are at steady state or of very low cycle frequency and the temperatures are in the realm of stress rupture, the total stresses can be limited by yield stresses. In this region each material has a different behavior as influenced by notches and metallurgical changes. There-

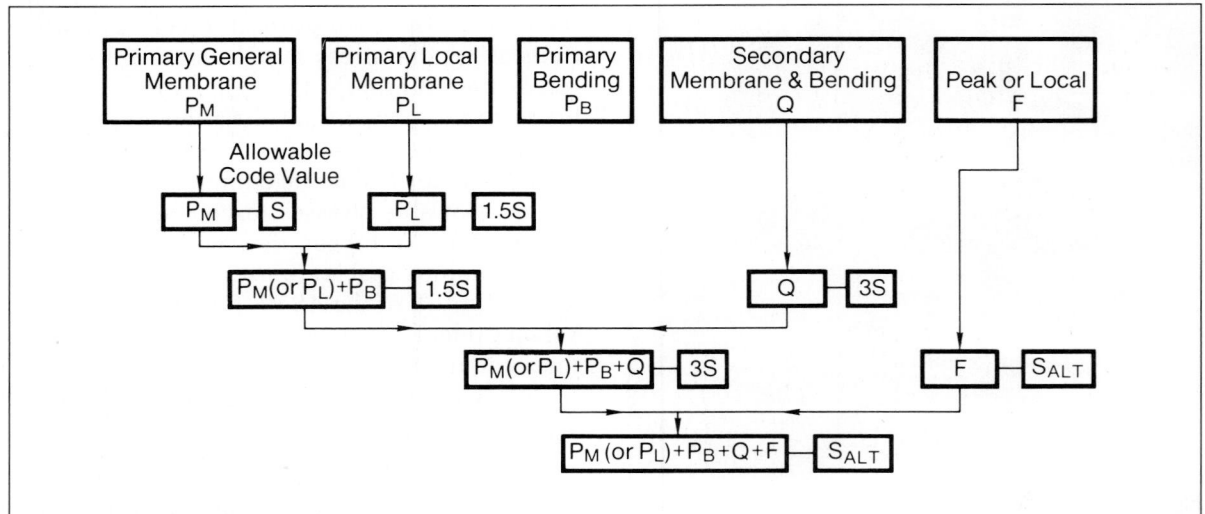

Fig. 45 Stress limits

fore, a survey of the literature on the material should be made as an aid in the final evaluation. Fig. 45 summarizes the stress limits for the various stress categories.

FATIGUE ANALYSIS

The first step in the fatigue analysis is to estimate a stress history for the expected life of the unit. From this stress history, the different operating cycles, such as startups, shutdowns, load changes and other changes which cause the stress level in the wall to vary, can be determined. In each of these cases, the maximum and minimum stress intensities are determined. Alternating stress intensities can then be calculated by

$$S_{alt} = \frac{S_{max} - S_{min}}{2}$$

(63)

If these stresses are significant, they will use up part of the life of the unit. The number of times the significant alternating stresses can be repeated is determined from a design fatigue curve as shown in Fig. 46. In these curves the alternating stress is plotted as a function of the cycles to failure with suitable safety factors. Because safety factors vary considerably and are dependent upon the mechanical behavior of the material, the uncertainty of the assumed loads and the uncertainty in the material (scatter of data), specific values of the ordinate of typical curves are not given. Curves such as these are determined by experimental methods and may be found in the literature. Some fatigue curves for materials normally used in

furnace design can be found in Section III of the *ASME* Boiler and Pressure Vessel Code. These are believed to reflect a conservative design.

In the furnace wall the stresses on both sides of the wall must be determined. The trend of the decrease in fatigue life as the temperature increases is shown in Fig. 47. Thus, because of the temperature influence on fatigue life, the hot face of the tube or fin may fail although the higher stresses may be calculated elsewhere.

CUMULATIVE DAMAGE

The furnace wall is subjected to a wide variety of operating conditions, some of which produce no significant stresses and some of which will produce stresses worthy of consideration. When a design is for either static loads or for a large number of cycles, it is sufficient to design for the condition which produces the highest stresses and to ignore all others. In this case, design is for infinite life, and all stress must be below the damaging level.

Usually the number of operational stress cycles which the furnace wall must withstand is not large, but the value of the operational stress is large unless severe restrictions are imposed.

As soon as the existence of strains beyond the elastic limit is accepted, the design is based on finite life and the damaging effect of all significant strains must be considered. The relationship between the alternating stress and cycles to failure can be determined from the design fatigue curves, and the number of required cycles can be estimated from the anticipated stress history. With these relationships estab-

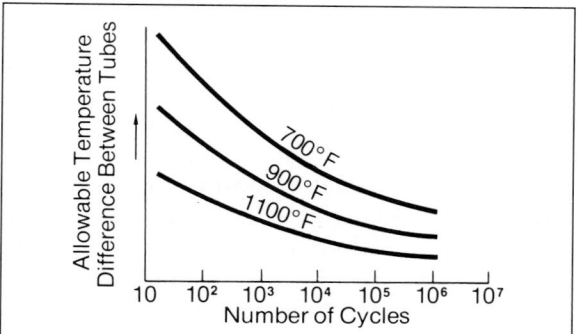

Fig. 46 Design fatigue curve

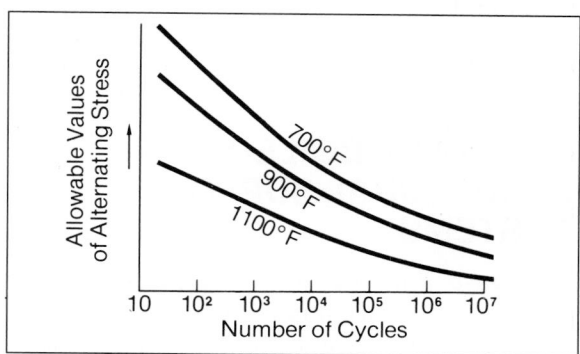

Fig. 47 Decrease in fatigue life

lished, the expected life can be estimated with fair accuracy by adding together the "cycle ratios" produced by each operating condition.

$$\text{Cumulative Damage Index} = \sum_{i=1}^{i=n} \frac{P_i}{N_i} \tag{64}$$

where P_i is the number of times operating cycle will occur and N_i is the number of times alternating stress can be withstood, based on the fatigue design curve.

Fatigue tests have shown that the cumulative damage index can vary over a range of 0.6 to 5. If the most damaging stresses are applied first, the cycles ratio may be as low as 0.6. If the lower stress values are applied first and followed by progressively higher stresses, the cycles ratio can be as high as five. These are extreme conditions. For the instance of random distribution of cycle ratio as for a furnace wall, the cumulative sum will be close to unity. Thus the imposed limiting condition is

$$\sum_{i=1}^{i=n} \frac{P_i}{N_i} = 1 \tag{65}$$

Welded furnace walls must be safe and reliable for a wide range of boiler operating conditions. The analytical techniques and experimental procedures discussed in this section give the boiler designer the necessary tools to make an evaluation of the strength of furnace walls and to assess the suitability of furnace wall design to withstand any of the various operating conditions.

DESIGN OF BOILER STRUCTURAL SUPPORT SYSTEMS

Because the combustion process must be confined within a complex structure capable of supporting many static and dynamic loads, *structural design* is an important sector of overall boiler design. But it is unfortunately true that the fundamental importance of heat transfer and fluid flow tends to obscure the role of the boiler structural designer whose task is to analyze the many internal and external loads and to select suitable structural members.

Structural design of boilers shares much in common with bridge and building design. There is a long history of progressive advancement in this engineering art.[61] Yet there is little evidence of a specialized literature of boiler structural design, and almost no consideration is given to structural problems in the most widely used works on power-plant design.[62]

STRUCTURAL DESIGN CONDITIONS

Structural design of boilers involves the consideration of many factors. The type of boiler, its location, and its method of construction all have a bearing, as do design specifications, codes and ordinances, protective coatings, and type of structure, including choice of members and connections.

Supports for boilers vary greatly, depending upon size and function. Utility boilers and the larger industrial units, including recovery units for pulp and paper mills, are top-supported. Contrarily, shop-assembled boilers including those for waste-heat recovery and high-temperature water are bottom-supported, as are those for marine service. This section, however, will be specifically concerned with the structural design of a top-supported central-station boiler.

LOCATION OF INSTALLATION

Boiler location has various effects upon structural design. It may determine whether indoor or outdoor construction is used. It has a bearing on the specification of domestic or foreign steel shapes. It may require special consideration for climatic conditions, including both natural occurrences and those artifically induced, as in a chemical plant or oil refinery. It will establish wind loads, seismic conditions and soil bearing capabilities for foundations. Local ordinances and regulations may impose special requirements, including restrictions on

materials and limitations of structural size.

Indoor boilers for utility service are a part of the power-plant structure, using the building columns for vertical support. Lateral support is provided by the bracing system of the building, which is completely enclosed and roofed. Boiler platforms for access and maintenance are coordinated with the elevations of power-plant floors.

Although outdoor units may have the boiler completely exposed to wind and weather, many are built with a full or partial roof and a protective enclosure around the fuel-firing equipment. Special consideration must be given to the waterproof design of casings and insulation for outdoor units.

SOURCES OF DESIGN INFORMATION

Many decisions must precede calculations for structural design. For example, will the unit be designed for indoor or outdoor construction? Should the frame be rigid, simple, or semirigid? Related to this is the choice of alternative connections: riveted, bolted, or welded. It is also wise at this time to consider protective coatings and their application.

There are many codes, standards and specifications pertinent to and helpful in making these decisions. One useful publication is *Minimum Design Loads in Buildings and Other Structures*[63] which lists design loads for many materials, gives estimated weights of snowfall and wind loads for practically all geographical areas of the United States, and includes seismic factors relating to the most common occurrences of earthquakes. Detailed information on protective coatings for steel may be found in the *Steel Structures Painting Manual*.[64]

The American Institute of Steel Construction's *Manual of Steel Construction* covers most routine design work. This manual includes the "AISC Specification for the Design Fabrication and Erection of Structural Steel for Buildings" plus an extended commentary.[65]

Structural design of a large boiler involves the solution of many simultaneous equations and can most efficiently be done by using a structural-design computer program such as STRUDL.[66] Such structures are very large, often exceeding 200 feet in height and involving the design and erection of several thousand tons of structural steel.

Top-supported boilers have many hangers which carry the loads of the various component pressure-parts to the structural steel framing. The boiler is allowed to expand downward from the main supports at the top of the structure; both structural-steel members and hanger rods are used. In high-temperature areas, the hangers may be water-cooled or steam-cooled tubing operating at high pressure and designed to the *ASME* Code.

A distinctive feature of boiler structural design is the buckstay system, which protects the pressure parts against transient internal or external forces. The pressure parts, including waterwall tubing and wall radiant surface, are attached to and must therefore deflect with the buckstays. Spacing of buckstays is determined from furnace pressure, tube diameter and spacing, and that part of the allowable stress in the tubing which can be allotted for bending.

REFERENCES

[1] M. K. Drewry, "Economy of Efficient Air Preheating with Extraction Steam," *Transactions of the ASME Journal of Engineering for Power*, 84:1–8, January 1962; also *ASME Paper No. 60-WA-94*. New York: American Society of Mechanical Engineers, 1960.

[2] "Cold-end Temperature and Material Selection Guide for Ljungstrom Air Preheaters," the Air Preheater Company, Inc., Wellsville, NY (request latest edition).

[3] Nsakala ya Nsakala, Ramesh L. Patel, Richard W. Borio; *An Advanced Methodology for Prediction of Carbon Loss in Commercial Pulverized Coal-Fired Boilers*; for Presentation at the 1986 ASME/IEEE Joint Power Generation Conference, Portland, Oregon, October 19–23, 1986, Combustion Engineering's Publication TIS-8211.

[4] Goetz, G. J., Nsakala, N., Patel, R. L., and Lao, T. C., EPRI Final Report AP-2601, September 1982.

[5] Nsakala, N., Patel, R. L., and Lao, T. C., DOE Final Report, DOE/PC/40267-5, March 1985.

[6] Nsakala, N., Patel, R. L., and Lao, T. C., The 189th American Chemical Society National Meeting, Miami, Florida, April 28–May 3, 1985. Combustion Engineering's Publication TIS-7877.

[7] "Steam Generating Units," *ASME Performance Test Codes*, PTC No. 4.1. New York: American Society of Mechanical Engineers, latest edition.

[8] Josef Stefan, "Uber Die Beziehung Zeischen der Warmestrahhung und der Temperatur," *Akademie der Wissenschaft. Sitzungs Berichte*, 79:391–428, 1879.

Ludwig Boltzmann, "Deduction of Stefan's Formula for Radiation from Maxwell's Electromagnetic Theory of Light," *Annalen der Physik und der Physikalischen Chemie*, 22:291–294, 1884.

Max Planck, *Theory of Heat Radiation*, 2nd ed. Translated by Morton Masius. New York: Dover Publications, 1959.

[9] C. M. Vsiskin and E. M. Sparrow, "Thermal Radiation Between Parallel Plates Separated by an Absorbing-Emitting Nonisothermal Gas," *International Journal of Heat and Mass Transfer*, 1(1):28–36, 1960

[10] Thomas H. Einstein, *Radiant Heat Transfer to Absorbing Gases Enclosed Between Parallel Flat Plates with Flow and Conduction*. Report No. NASA-TR-R-154 (N64-11558). Springfield, VA: National Technical Information Service, 1963.

[11] J. R. Howell and M. Perlmutter, "Monte Carlo Solution of Thermal Transfer Through Radiant Media Between Gray Walls," *Transaction of the ASME Journal of Heat Transfer*, 86:116–122, February 1964. Also in: *ASME Paper No. 63-AHGT-1*. New York: American Society of Mechanical Engineers, 1963.

[12] J. R. Branstetter, *Radiant Heat Transfer Between Nongray Parallel Plates of Tungsten*. Report No. NASA-TN-D-1088 (N62-71662). Springfield, VA: National Technical Information Service, 1961.

[13] J. R. Branstetter, *Formulas for Radiative Heat Transfer Between Nongray Parallel Plates of Polished Refractory Metals*. Report No. NASA-TN-D-2902 (N65-27550). Springfield, VA: National Technical Information Service, 1965.

[14] J. D. Buckmaster, *Radiative Heat Transfer and Conduction Between Parallel Walls*. Report No. AD-619 419. Springfield, VA: National Technical Information Service, 1965.

[15] R. Viskanta, "Heat Transfer by Conduction and Radiation in Absorbing and Scattering Materials," *Transactions of the ASME Journal of Heat Transfer*, 87:143–150, February 1965. Also: *ASME Paper No. 64-HT-33*. New York: American Society of Mechanical Engineers, 1964.

[16] H. M. Hsia and T. J. Love, "Radiative Heat Transfer Between Parallel Plates Separated by a Nonisothermal Medium with Anisotropic Scattering." *Transactions of the ASME Journal of Heat Transfer*, 89:197–204, August 1967. Also: *ASME Paper No. 66-WA/HT-28*. New York: American Society of Mechanical Engineers, 1966.

[17] D. K. Edwards and K. E. Nelson, "Rapid Calculation of Radiant Energy Transfer Between Nongray Walls and Isothermal H_2O or CO_2 Gas," *Transactions of the ASME Journal of Heat Transfer*, 84:273–278, November 1962. Also: *ASME Paper No. 61-WA-175*. New York: American Society of Mechanical Engineers, 1961.

[18] Michael A. Lutz, *Radiant Energy Loss From a Cesium-Argon Plasma to an Infinite Parallel Enclosure; Avco-Everett Research Laboratory Report 175*. Report No. AD-431 640 (BSD-TDR-64-6). Springfield, VA: National Technical Information Service, 1963.

[19] Yuan-Siang Pan, *Temperature Discontinuities and Energy Transfer in One-Dimensional Problems*. Report No. AD-618 220. Springfield, VA: National Technical Service, 1965.

[20] T. J. Love and R. J. Grosh, "Radiative Heat Transfer in Absorbing, Emitting, and Scattering Media," *Transactions of the ASME Journal of Heat Transfer*, 87:161–166, May 1965. Also: *ASME Paper No. 64-HT-28*. New York: American Society of Mechanical Engineers, 1964.

[21] John R. Howell, *Radiative Interactions Between Absorbing-Emitting and Glowing Media with Internal Energy Generation*. Report No. NASA-TN-D-3614 (N66-36426). Springfield, VA: National Technical Information Service, 1966.

[22] E. M. Sparrow and A. Haji-Sheikh, "A Generalized Variational Method for Calculating Radiant Interchange Between Surfaces," *Transactions of the ASME Journal of Heat Transfer*, 87:103–109, February 1965. Also: *ASME Paper No. 64-HT-11*. New York: American Society of Mechanical Engineers, 1964.

[23] R. Viskanta and R. J. Grosh, "Heat Transfer in a Thermal Radiation Absorbing and Scattering Medium," *International Developments in Heat Transfer, Proceedings of*

the *1961-62 International Heat Transfer Conference,* Boulder, Colorado, August 28-September 1, 1961 and London, January 8-12, 1962, Part IV, Section B—Radiation, Thermal Properties and Instrumentation, Paper 99, pp. 820-828. New York: American Society of Mechanical Engineers, 1963.

[24] Masayoshi Kuwata and Robert H. Essenhigh, "Comparative Influence of Turbulence and Convention in Combustor Performance" *Proceedings of the Second Conference on Natural Gas Research and Technology,* Atlanta, June 5-7, 1972, Session IV, Paper 3. Chicago: Institute of Gas Technology, ND.

[25] Mehty A. E. Zeinalov, Masayoshi Kuwata, and Robert H. Essenhigh, "Stirring Factors in Combustion Chambers: A Finite-Element Model of Mixing Along an 'Information Flow Path'," *Fourteenth Symposium (International) on Combustion,* University Park, PA, August 20-25, 1972, pp. 575-583. Pittsburgh: The Combustion Institute, 1973.

[26] Satyanarayana T. R. Rao and R. H. Essenhigh, "Experimental Determination of Stirring Factors Generated by Straight and Swirling Jets in Isothermal Combustion—Chamber Models," *Thirteenth Symposium (International) on Combustion,* Salt Lake City, August 23-29, 1970, pp. 603-615. Pittsburgh: The Combustion Institute, 1971.

[27] Robert H. Essenhigh, *A New Application of Perfectly Stirred Reactor (P.S.R.) Theory to Design of Combustion Chambers.* Report No. AD-812 114. (Pennsylvania State University, Department of Fuel Science Technical Report No. FS67-1(U)). (Contract 656(29)). Springfield, VA: National Technical Information Service, 1967.

[28] A. M. Godridge, "Heat Transfer in the Furnace Chamber of Pulverized-Fuel-Fired Water-Tube Boilers," *Institute of Fuel Journal,* 40(318):300-310, July 1967.

[29] V. A. Lokshin, I. E. Semenovker and Yu. V. Vikhrev, "Calculating the Temperature Conditions of the Radiant Heating Surfaces in Supercritical Boilers," *Thermal Engineering* [Translation of Teploenergetika], 15(9):34-39, September 1968.

[30] S. S. Filimonov, et al., "Calculation of Average Temperature and Thermal Resistance of Heating Surfaces in Furnace Chambers," *Thermal Engineering* [Translation of Teploenergetika], 15(9):40-44, September 1968.

[31] E. S. Karasina, "Allowance for Fouling of Radiant Heating Surfaces." *Thermal Engineering* [Translation of Teploenergetika], 15(6):17-22, June 1968.

[32] R. H. Essenhigh, et al., *Furnace Analysis: A Comparative Study.* Pennsylvania State University, Department of Fuel Science Technical Report No. FS/CGSSC-12/69-4. University Park, PA: Pennsylvania State University, 1969.

[33] J. M. Beér and R. G. Siddall, "Radiative Heat Transfer in Furnaces and Combustion," *ASME Paper No. 72-WA/HT-29.* New York: American Society of Mechanical Engineers, 1972.

[34] H. C. Hottell and E. S. Cohen, "Radiant Heat Exchange in a Gas-Filled Enclosure," *ASME Paper No. 57-HT-23.* New York: American Society of Mechanical Engineers, 1957.

[35] H. C. Hottell, "The Melchett Lecture for 1960: Radiative Transfer in Combustion Chambers," *Institute of Fuel Journal,* 34(245):220-234, June 1961.

[36] H. C. Hottell and A. C. Sarofim, *Radiative Transfer.* New York: McGraw-Hill, 1967.

[37] D. B. Spalding, "Mathematical Models of Continuous Combustion," *Emissions from Continous Combustion Systems, Proceedings, Symposium,* Warren, MI, September 27-28, 1971. New York: Plenum, 1972.

[38] William H. McAdams, *Heat Transmission,* 3rd ed. New York: McGraw-Hill, 1954.

[39] S. S. Penner and P. Varanasi, "Simplified Procedures for Estimating Band and Total Emissivities of Polyatomic Molecules," *Eleventh Symposium (International) on Combustion,* University of California at Berkeley, August 14-20, 1966, pp. 569-576. Pittsburgh: The Combustion Institute, 1967.

[40] K. A. Bueters, "Combustion Products Emissivity by F_E Operator," *Combustion,* 45(9):12-18, March 1974.

[41] K. A. Bueters, J. G. Cogoli, and W. W. Habelt, "Performance Prediction of Tangentially Fired Utility Furnaces by Computer Mode," *Fifteenth Symposium (International) on Combustion,* Tokyo, August 25-31, 1974, pp. 1245-1260. Pittsburgh: The Combustion Institute, 1974.

[42] H. F. Mullikin, "Evaluation of Effective Radiant Heating Surface and Application of the Stefan-Boltzmann Law to Heat Absorption in Boiler Furnaces," *Transactions of the ASME,* 57:517-529, 1935.

[43] Frederick P. Boynton and Claus B. Ludwig, "Total Emissivity of Hot Water Vapor-II. Semi-Empirical Charts Deduced from Long-Path Spectral Data," *International Journal of Heat and Mass Transfer,* 14(7):963-973, July 1971.

[44] F. R. Steward and P. Cannon, "The Calculation of Radiative Heat Flux in a Cylindrical Furnace Using the Monte Carlo Method," *International Journal of Heat and Mass Transfer,* 14(2):245-262, February 1971.

[45] H. C. Hottell and R. B. Egbert, "Radiant Heat Transmission from Water Vapor," *AIChE Transactions,* 38(3):531-568, June 1942.

[46] H. R. Hazard, et al., "Field Studies of Slagging in Tangentially Fired Boiler Furnaces—Part I, Labadie Field Trial," *ASME Paper No. 78-Wa/Fu-10.* New York; American Society of Mechanical Engineers, 1978.

[47] W. E. Dalby, "Heat Transmission," *Institution of Mechanical Engineers Proceedings,* Parts 3-4:921-1071, 1909.

W. J. Wohlenberg, et al., "An Experimental Investigation of Heat Absorption in Boiler Furnaces," *Transactions of the ASME,* 57:541-554, 1935.

Osborne Reynolds, "On the Extent and Action of the Heating Surface for Steam Boilers," *Manchester Literary and Philosophical Society Proceedings,* 14:7-12, October 6, 1874.

L. M. K. Boelter, R. C. Martinelli, and Finn Johanssen, "Remarks on the Analogy Between Heat Transfer and Momentum Transfer," *Transactions of the ASME,*

63:447–455, 1941.

W. F. Davidson, et al., "Studies of Heat Transmission Through Boiler Tubing at Pressures from 500 to 3300 Pounds," *Transactions of the ASME,* 65:553–591, 1943.

Max Jakob, *Heat Transfer,* Vol. II. New York: John Wiley and Sons, 1957, pp. 373–381.

[48] Orville L. Pierson, "Experimental Investigation of the Influence of Tube Arrangement on Convection Heat Transfer and Flow Resistance in Cross Flow of Gases Over Tube Banks," *Transactions of the ASME,* 59:563–572, 1937.

[49] E. D. Grimison, "Correlation and Utilization of New Data on Flow Resistance and Heat Transfer for Cross Flow of Gases Over Tube Banks," *Transactions of the ASME,* 59:583–594, 1937.

[50] E. C. Huge, "Experimental Investigation of Effects of Equipment Size on Convection Heat Transfer and Flow Resistance in Cross Flow of Gases Over Tube Banks," *Transactions of the ASME,* 59:573–581, 1937.

[51] Clarence F. Hirshfeld, William N. Barnard, Frank O. Ellenwood, *Elements of Heat-Power Engineering,* Part III; *Auxiliary Equipment, Plant Ensemble, Air Conditioning, and Refrigeration,* 3rd ed. New York: John Wiley and Sons, 1933.

[52] American Society of Mechanical Engineers, Boiler and Pressure Vessel Committee, *ASME Boiler and Pressure Vessel Code.* New York: American Society of Mechanical Engineers, latest edition.

[53] Stephen Timoshenko and S. Woinowsky-Kreiger, *Theory of Plates and Shells,* 2nd ed. New York: McGraw-Hill, 1959.

[54] G. W. Watts and W. R. Burrows, "The Basic Elastic Theory of Vessel Heads Under Internal Pressure," *Transactions of the ASME Journal of Applied Mechanics,* 71:55–73, March 1949.

[55] Miklos Hetenyi, *Beams on Elastic Foundation; Theory with Applications in the Fields of Civil and Mechanical Engineering.* Ann Arbor, MI: University of Michigan Press, 1946.

[56] G. K. Cooper and L. W. Smith, Final Report of PVRC Project of the American Welding Society, Purdue University, August 1952. This report can be obtained from Pressure Vessel Research Council, United Engineering Center, New York.

[57] Julien Dubuc and Georges Welter, "Investigation of Static and Fatigue Resistance of Model Pressure Vessels," *Welding Journal,* 35:329s–337s, July 1956.

[58] G. J. Schoessow and E. A. Brooks, Analysis of Experimental Data Regarding Certain Design Features with Pressure Vessels," *Transactions of the ASME;* 72:567–577, July 1950.

[59] Miklos I. Hetenyi, *Handbook of Experimental Stress Analysis.* New York: John Wiley and Sons, 1950.

A. J. Durelli, E. A. Phillips, and C. H. Isao, *Introduction to the Theoretical and Experimental Analysis of Stress and Strain.* New York: McGraw-Hill, 1958.

Albert Kobayahsi, ed., *Manual on Experimental Stress Analysis,* 3rd ed. Westport, CT: Society for Experimental Stress Analysis, 1978.

Society for Experimental Stress Analysis, *Experimental Stress Analysis; Proceedings of the Society for Experimental Stress Analysis,* V.1–.Westport, CT: Society for Experimental Stress Analysis, 1943.

[60] C. C. Perry and H. R. Lisner, *The Strain Gage Primer,* 2nd ed. New York: McGraw-Hill, 1962.

Frank G. Tatnall, *Tatnall on Testing: An Autobiographical Account of Adventures Under 13 Vice Presidents.* Metals Park, OH: American Society for Metals, 1966.

[61] L. E. Grinter, *Theory of Modern Steel Structures,* revised ed. New York: Macmillan, 1949.

H. M. Westergaard, "One Hundred Fifty Years Advance in Structural Analysis," *American Society of Civil Engineers Transactions,* 94:226–246, 1930.

Stanley B. Hamilton, "The Historical Development of Structural Theory," *Institution of Civil Engineers Proceedings,* 1, Part 3:374–419, 1952.

[62] A. H. Palmer, et al., "Integrated Structural Steel Design Program," *American Society of Civil Engineers Proceedings, Journal of the Power Division,* 89(PO-1):67–82, September 1963.

[63] "Building Code Requirements for Minimum Design Loads in Buildings and Other Structures," *ANSI Standard No. A58.1.* Sponsored by the National Bureau of Standards and approved by the American National Standards Institute. New York: American National Standards Institute, latest edition.

[64] "Steel Structures Painting Manual", Volume 1 covers methods of surface preparation and paint systems. Volume 2 lists all types of important paint systems used in various industries; it also includes a guide with indexes for selection of suitable systems for various structures ad exposures, specifications for surface preparation and pretreatment, paint application and paints. Obtained from the Steel Structures Painting Council, Pittsburgh, PA.

[65] "Manual of Steel Construction", latest edition; American Institute of Steel Construction, N.Y.

[66] Robert D. Logscher, et al., *ICES STRUDL-II: The Structural Design Language Engineering User's Manual,* Vol 1; *Frame Analysis,* 1st ed. Massachusetts Institute of Technology, Department of Civil Engineering, Structures Division and Civil Engineering Systems Laboratory Research Report R68-91. Cambridge, MA: Massachusetts Institute of Technology, 1968.

CHAPTER 7

Central-Station Steam Generators

Most of the fundamentals on boiler design, fuels, ash, coal pulverization, materials selection, manufacture, field construction, and operation in this book are common in whole or in part to the technology of boiler designers and manufacturers worldwide. The engineering information in this chapter departs from that commonality.

This chapter presents the C-E design philosophy as it applies to large high-pressure steam generators for service in electrical-power production. Specifically covered are tangentially fired subcritical- and supercritical-pressure boilers — their design and the ways they are integrated into the overall power plant to obtain maximum reliability and flexibility. The units described are categorized in Chapter 4 as the entrained-flow-reactor, suspension-fired, or open-furnace type; fluidized-bed boilers for power generation are covered in Chapter 9.

Since WWII, manufacturers have developed diverse designs for generating steam at the high subcritical and supercritical pressures. The diversity is particularly apparent in large high-pressure units for which the individual proprietary designs become both the hallmark and *raison d'être* for each manufacturer. Like fingerprints, the arguments made for the superiority of a circulation or firing system characterize and differentiate all boiler designers.

The C-E product line and C-E's desire for excellence extend from the central-station designs to industrial and marine boilers. But it is for the development of its large high-pressure steam generators that C-E has built and staffed some of the finest research laboratories in the world dedicated solely to the development and improvement of equipment for steam generation at the efficient high-pressure levels.

IMPACT OF FUEL ON BOILER DESIGN

The most important item to consider when designing a utility or large industrial steam generator is the fuel the unit will burn. The furnace size, the equipment to prepare and burn the fuel, the amount of heating surface and its placement, the type and size of heat-recovery equipment, and the flue-gas treatment devices are all fuel dependent.

The major differences among those boilers that burn coal or oil or natural gas result from the ash in the products of combustion. (See Chapter 3.) Firing oil in a furnace results in relatively small amounts of ash; there is no ash from natural gas. For the same output, because of the ash, coal-burning boilers must have larger furnaces and the velocities of the com-

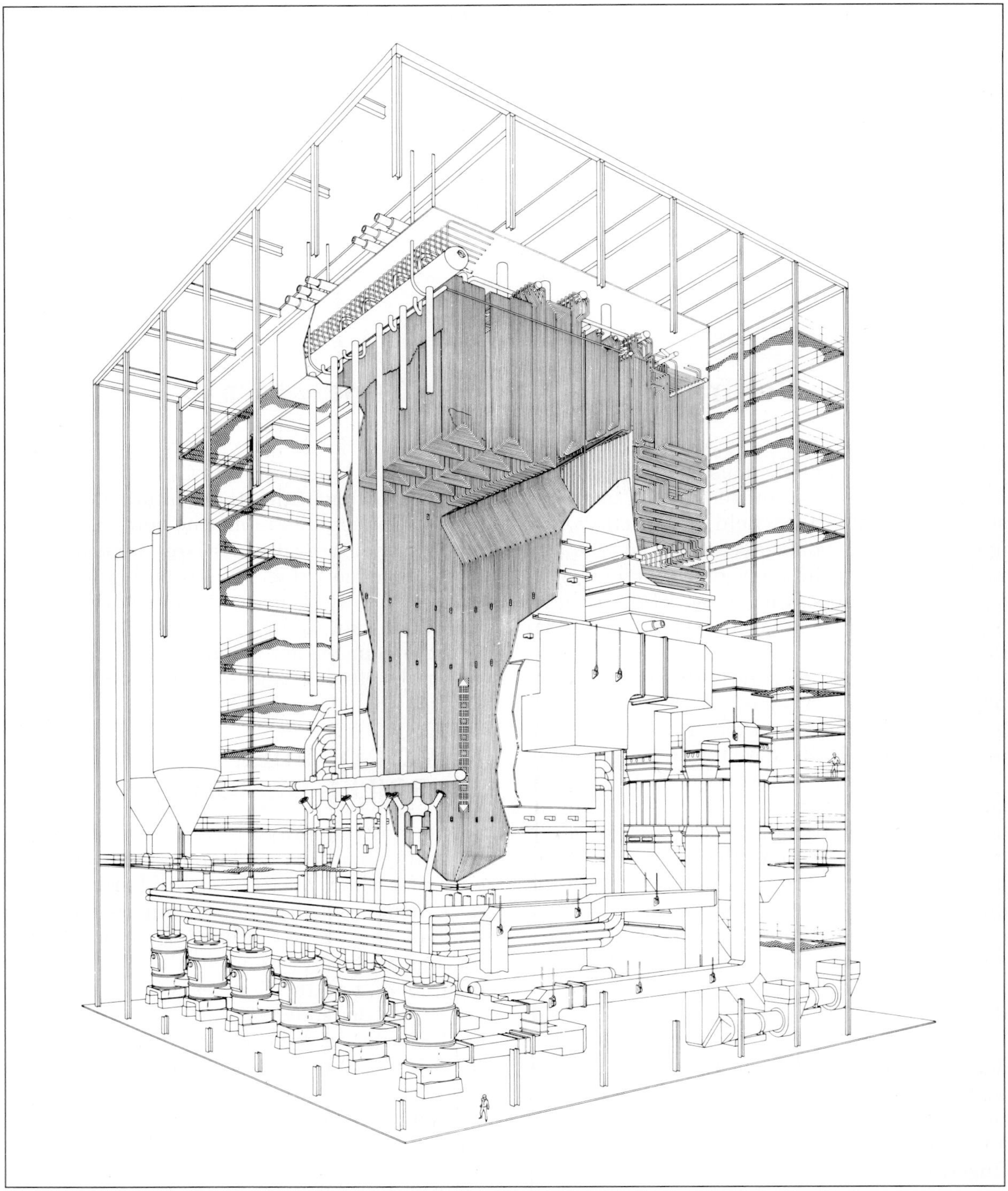

Fig. 1 Typical coal-burning central-station reheat steam generator of the C-E tangentially fired Controlled Circulation® design

bustion gases in the convection passes must be lower. In addition, coal-burning boilers need ash-handling and particulate cleanup equipment that adds a great deal to cost and requires considerable space. (Fig. 1).

Table I lists the variation in calorific values and moisture contents of several coals, and the mass of fuel that must be handled and fired to

Table I. Representative Coal Analyses

	Med.-Vol. Bituminous	High-Vol. Bituminous	Subbitum- inous C	Low-Sodium Lignite	Med.-Sodium Lignite	High-Sodium Lignite
Total H$_2$O,%	5.0	15.4	30.0	31.0	30.0	39.6
Ash, %	10.3	15.0	5.8	10.4	28.4	6.3
VM, %	31.6	33.1	32.6	31.7	23.2	27.5
FC, %	53.1	36.5	36.6	26.9	18.4	26.6
Ash analysis, %						
SiO$_2$	40.0	46.4	29.5	46.1	62.9	23.1
Al$_2$O$_3$	24.0	16.2	16.0	15.2	17.5	11.3
Fe$_2$O$_3$	16.8	20.0	4.1	3.7	2.8	8.5
CaO	5.8	7.1	26.5	16.6	4.8	23.8
MgO	2.0	0.8	4.2	3.2	0.7	5.9
Na$_2$O	0.8	0.7	1.4	0.4	3.1	7.4
K$_2$O	2.4	1.5	0.5	0.6	2.0	0.7
TiO$_2$	1.3	1.0	1.3	1.2	0.8	0.5
P$_2$O$_5$	0.1	0.1	1.1	0.1	0.1	0.2
SO$_3$	5.3	6.0	14.8	12.7	4.6	17.7
Sulfur, %	1.8	3.2	0.3	0.6	1.7	0.8
Fusion (reducing), °F						
Initial Deformation	2,170	1,990	2,200	2,080	2,120	2,030
Softening	2,250	2,120	2,250	2,200	2,380	2,090
Fluid	2,440	2,290	2,290	2,310	2,700	2,200
Fusion (reducing), °C						
Initial Deformation	1 190	1 090	1 200	1 140	1 160	1 110
Softening	1 230	1 160	1 230	1 200	1 300	1 140
Fluid	1 340	1 250	1 250	1 270	1 480	1 200
Btu/lb, as fired	13,240	10,500	8,125	7,590	5,000	6,520
Btu/lb, MAF	15,640	15,100	12,650	12,940	12,020	12,050
Lb Ash/Million Btu	7.8	14.3	7.1	13.7	56.8	9.7
*****Fuel Fired, 1000 lb/hr**	405	520	705	750	1,175	900
*****Ash Fired, 1000 lb/hr**	42	78	41	78	334	57
MJ/kg, as fired	30.8	24.4	18.9	17.7	11.6	15.2
MJ/kg, MAF	36.4	35.1	29.4	30.1	28.0	28.0
kg Ash Million kJ	3.3	6.1	3.1	5.9	24.4	4.2
*****Fuel Fired, kg/s**	51.0	65.5	88.8	94.5	148.0	113.4
*****Ash Fired, kg/s**	5.3	9.8	5.2	9.8	42.0	7.1

*Constant Heat Output, Nominal 600-MW Unit, Adjusted For Effect of Moisture Content on Efficiency

generate the same electrical-power output. These values are important because the quantity of fuel required helps determine the size of the coal storage yard, as well as the handling, crushing, and pulverizing equipment for the various coals; the amount of ash in the coal directly affects the sizing of the ash-handling and flue-gas cleaning equipment.

FURNACE SIZING

As stated in Chapter 6, the most important step in coal-fired unit design is to properly size the furnace. Furnace size has a first-order influence on the size and cost of the convective heat-transfer surface, the structural steel framing, the boiler building and its foundations, as well as on the quantity and length of sootblowers, the extent of platforms and stairways, and the arrangement of steam piping and ductwork.

Three very important parametric influences on furnace sizing are:
- fuel reactivity (discussed in Chapter 2)
- gaseous-emission limitations (particularly those concerning oxides of nitrogen), and
- fuel-ash properties.

Among the ash properties discussed in Chapter 3 and Appendix B, those that C-E considers particularly important when designing and establishing the size of coal-fired furnaces include:
- the ash fusibility temperatures (both in terms of their absolute values and the spread or difference between initial deformation temperature and fluid temperature)
- the ratio of basic to acidic ash constituents
- the iron/calcium ratio
- the fuel-ash content in terms of lbs of ash/million Btu or kg of ash/million joules
- the ash friability.

These characteristics translate into the typical furnace sizes shown in Fig. 2, which are based on the six coal ranks shown in Table I. This size comparison illustrates the C-E philosophy of increasing the furnace plan area, volume, and the fuel burnout zone (the distance from the top fuel nozzle to the furnace arch), as lower grade coals with poorer ash characteristics are fired.

Fig. 2 is a simplified characterization of actual furnaces built to burn the fuels in Table I. Wide variations exist in fuel properties within coal ranks, as well as within several subclassifications (e.g., subbituminous A, B, C), each of which may require a different size furnace. Note that, in general, high-sodium lignites tend also to have high slagging tendencies.

Among the most important design criteria in large pulverized-fuel tangentially fired furnaces are the net heat input in Btu/hr-sq ft or W/m^2 of furnace plan area (NHI/PA) and the vertical distance from the top fuel nozzle to the furnace arch. Furnace dimensions must be adequate to establish the necessary furnace retention time to properly burn the fuel as well as to cool the gaseous combustion products. This is to ensure that the gas temperature at the furnace outlet plane is well below the ash-softening temperature of the lowest quality coal burned. Specifically, the furnace outlet plane is the entrance to the closed-spaced convection surface; the latter is defined as non-platen surface on less than 12 inch (300 mm) horizontal centers. In addition to the tilting of fuel and air nozzles, the heat-absorption characteristics of the walls are maintained using properly placed wall blowers to control the furnace outlet gas temperature by removing ash deposited on the furnace walls below the furnace outlet plane.

To arrive at an appropriate value of NHI/PA and furnace plan area requires thorough evaluation of all fuels expected to be burned in the unit. Depending on the analyses and range of coals to be burned, large pulverized-fuel units have an NHI/PA that generally varies from 1.4 to 2.0×10^6 Btu/hr-sq ft (4.4 to 6.3×10^6 W/m^2).

The distance from the upper fuel nozzle to the furnace arch is a function of furnace width and depth, the upper-furnace superheater and reheater surface arrangement, and, of course, the fuel and ash characteristics.

Widely spaced steam-cooled platen-type heating surface is usually required in the upper furnace to provide further cooling of furnace gases before entering the convection surface. These sections potentially are subject to the

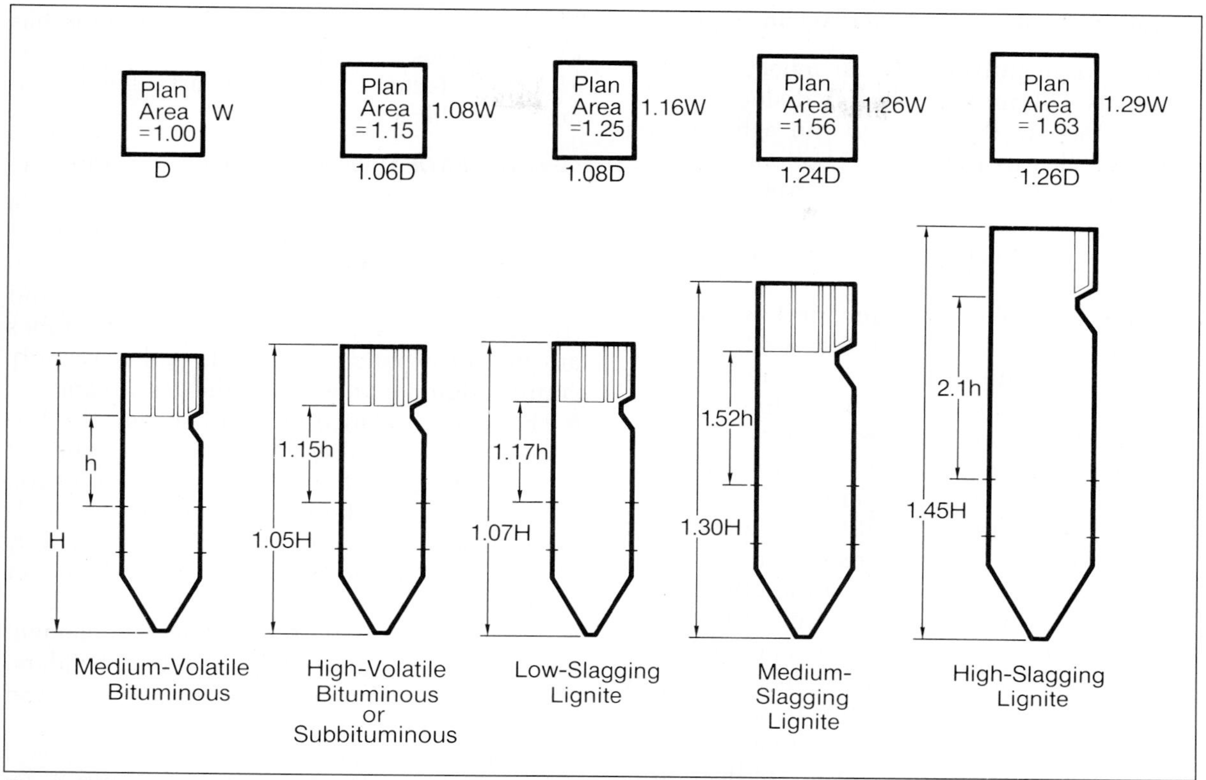

Plan Area = 1.00 W | D
Plan Area = 1.15 1.08W | 1.06D
Plan Area = 1.25 1.16W | 1.08D
Plan Area = 1.56 1.26W | 1.24D
Plan Area = 1.63 1.29W | 1.26D

h | H
1.15h | 1.05H
1.17h | 1.07H
1.52h | 1.30H
2.1h | 1.45H

Medium-Volatile Bituminous
High-Volatile Bituminous or Subbituminous
Low-Slagging Lignite
Medium-Slagging Lignite
High-Slagging Lignite

Fig. 2 Effect of coal rank on sizing of a pulverized-fuel furnace (constant heat output)

most severe fouling conditions and wide spacing helps prevent ash plugging caused by bridging. To perform satisfactorily in this environment, the platens are constructed with the tubes tangent to each other in the direction of gas flow to minimize ash deposition.

A furnace design parameter sometimes specified is burner-zone heat release. This term does not have the same significance in tangential firing as in other firing systems. In tangential firing not all of the fuel is consumed in the windbox zone because of the manner in which the fuel and air are introduced into the furnace. In a unit designed for wall firing, the fuel and air are intimately mixed at the burner throat and consumed directly in front of the burner. With a wall-fired boiler, then, burners can be moved farther apart or more burners can be added to give any value of burner-zone release rate as there is no interaction between rows of burners.

The vertical distance from the lowest coal nozzle to the point where the furnace walls are bent to form the hopper section of the furnace is a function of furnace width and depth and the slagging potential of the fuel. This dimension is generally between 12 and 20 feet (3-1/2 and 6 meters) in a tangentially fired furnace. The hopper section typically slopes at 50° to the horizontal.

With the exception of dry-bottom pulverized-coal units in sizes between approximately 50 and 100 MW, the furnace volumetric heat-release rate (in Btu/hr-ft^3 or W/m^3) generally is not a controlling design parameter. For such units, the release-rate range is from 15,000 Btu/hr-ft^3 for very good bituminous coals to about 10,000 Btu/hr-ft^3 for lignites (150,000 to about 100,000 W/m^3). Similarly, the burner-zone volumetric heat-release rate in Btu/hr-ft^3 or W/m^3 is not a meaningful design parameter for tangentially fired boilers.

DRY-BOTTOM VERSUS WET-BOTTOM FURNACES

All current pulverized-coal reheat units are of the dry-bottom type; the ash dislodged from the furnace walls is below the ash-melting point and leaves the furnace bottom in a substantially dry condition. The wet-bottom or slagging-bottom furnaces are offered today only for special applications, such as in furnaces for gasifying coal.

In the wet-bottom design, the lower part of the flame has to sweep the furnace floor at all loads to maintain the fluidity of the ash. This requirement imposes a definite limitation in the use of slag-tap furnaces. Slag-tap freezing can occur during operation at low load or when fuel is fired only in the upper furnace for steam temperature control. In addition, wet-bottom units have higher nitrogen-oxide production.

The dry-bottom furnace, on the other hand, particularly those designs using tilting fuel nozzles, can provide a wider steam temperature control range and can handle coals with widely varying ash characteristics. It is this latter characteristic, more than any other, that was originally responsible for the greater application of dry-bottom units.

In the usual dry-bottom furnace, the hopper is formed by bending the front- and rear-wall tubes at their lower ends. The tube slope is greater than the angle of repose of the ash, and thus forms a self-cleaning, watercooled hopper. Ash is discharged to the ash receiver through a transverse opening in the hopper, about 4-ft (1.2m) wide on large units.

ARRANGEMENT OF UPPER-FURNACE HEATING SURFACE

Combustion Engineering uses two principal arrangements of upper-furnace heating surface, which can be called the pendant panel and horizontal-surface designs. The reheat steam generators shown in Figs. 1, 6, 11, 17, and 22 of this chapter illustrate the pendant panel design as it is applied to large coal-fired units. The boiler in Fig. 3 is also a large coal-fired unit, but using the horizontal arrangement of superheater and reheater surface in the top of the furnace. Each of these configurations has its advantages and allows for customer preference as a factor in the final arrangement of heating surface.

ADVANTAGES OF THE PENDANT PANEL DESIGN

■ The support elements for the pendant surface are out of the gas stream, above the furnace roof; this eliminates the exposure of load supports and seals to high gas temperatures and flyash erosion. Superheater and reheater tubes are free to expand downward, and have only simple alignment devices in the gas stream.

■ There is no relative motion between the furnace tubing and the superheater or reheater tubes where the latter penetrate the enclosure. Thus, the area of penetration can be seal-welded for maximum gas tightness, thereby eliminating the need for pressurized header enclosures.

■ The above support and sealing arrangement favors shop modularization of tubes, headers, attachments, and supports. Field construction consists of lifting these modules into position, butt-welding adjacent header sections, and seal-welding small areas of skin casing after the horizontal furnace-roof tubes are in place.

■ The lower height of the pendant boiler can be expected to result in lower costs of the boiler building, elevators, platforms and stairways, structural steel framing, and foundations, particularly in high seismic- and wind-load areas.

■ In field erection, major pressure-part construction can be carried out in several areas simultaneously. Also, the pendant panel steam generator is not as tall, as the horizontal type, which reduces construction costs.

■ Any required replacement of, or modifications to, heating surfaces are greatly simplified.

■ Widely spaced panels (6 to 8 feet or 2 to $2^{1}/_{2}$ meters horizontal centers), along with steam-cooled wall sections in the upper furnace, have high radiant-heat absorption, resulting in improved steam-temperature control range.

ADVANTAGES OF THE HORIZONTAL-SURFACE DESIGN

■ The essentially vertical gas flow through the superheater and reheater tube banks minimizes

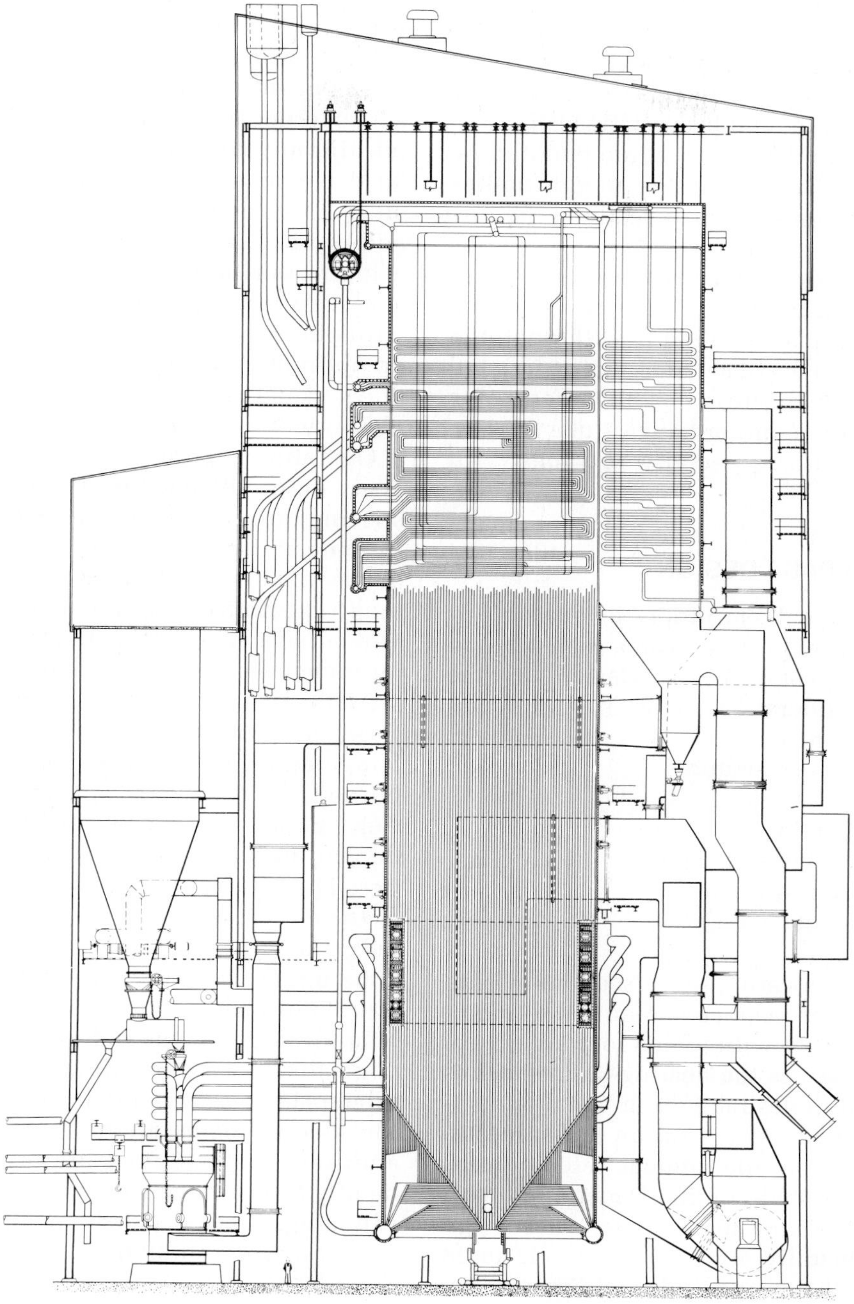

Fig. 3 Large pulverized-coal fired Controlled Circulation® unit with horizontal superheater and reheater surface in top of furnace

the potential for localized tube erosion that might result when gases take a 90° turn into the rear gas pass.

■ Horizontal tubing facilitates designing for drainability, which simplifies freeze-protection procedures, boilout, and hydrostatic testing.

■ Large fused ash deposits that are removed by sootblowers will usually drop through wider-spaced tube sections below, directly to the furnace bottom.

Note that the horizontal arrangement requires that operators take care in start-up to ensure that there is adequate cooling flow through the vertical hanger tubes that support and align the horizontal tube bundles. Thermocouples should be used to monitor hanger-tube temperatures on start-up, especially in tubes with downward flow.

CONVECTION-PASS DESIGN

A number of coal-ash properties have a significant influence on superheater, reheater, and economizer convection-surface design, and the associated considerations of fouling and erosion. Included are:

■ ash-softening temperature
■ base/acid ratio
■ iron/calcium ratio
■ sodium and potassium content
■ silica and alumina content
■ chlorine content
■ ash friability

In sizing and locating these sections within the unit, a proper balance is required to maintain a thermal head with which to transfer heat from gas to steam as the heating-surface use is optimized and undesirably high metal temperatures are avoided. Also, to limit pressure-part erosion from flyash, the flue-gas velocity must not exceed reasonable limits. Depending upon the ash quantity and abrasiveness, the design velocity is generally 40 to 60 ft/sec. (12 to 18 m/s). A boiler that burns coals yielding a heavy loading of erosive ash (usually indicated by a high silica/alumina content) would use the lower velocities. Such velocities are based on the predicted average gas temperature entering the tube section, at the maximum continuous rating (MCR) of the steam generator fired at normal excess-air percentage.

Considering the above factors in furnace sizing and convection-pass design, it is impractical to propose a steam generator capable of burning any kind of coal, nor would such a unit be economically feasible. For example, consider the wide variation in pulverizer size and primary-air-temperature requirements for a high-calorific-value, low-moisture, low-ash coal versus a subbituminous coal or lignite with a high moisture and ash content. Selecting an optimum preparation and firing system for one rank of fuel can result in a much less favorable situation for some other rank of fuel.

Operating experience on units firing certain coals shows the need for wide transverse tube spacing throughout the unit to reduce the fouling rate and possible bridging of ash deposits between adjacent assemblies. This arrangement minimizes serious fouling problems which have an adverse effect on fan power requirements and unit availability.

As the gas temperature is reduced along its flow path to well below the ash-softening temperature of the worst coal expected to be burned, the transverse spacing of the convection-pass tube banks is also reduced— depending on fuel type and the ash characteristics. The tube spacing selected at the entrance to the convection pass depends mainly on the maximum gas temperature possible in this zone under upset conditions and the propensity of the ash to adhere.

In the superheater and reheater sections beyond the ash-adhesion temperature zone, the tube spacing is further reduced to the minimum for effective penetration of sootblower jets. In addition, gas velocities are maintained sufficiently low to prevent erosion. To avoid serious plugging, the transverse spacing in this area must be wide enough to allow pieces of accumulated ash loosened by sootblowers to pass through the tube bank and not bridge the span between adjacent tube rows.

Table II is a guide for establishing transverse clearances for platen and spaced convection

Table II. Transverse Clearance Guide

Temperature Range	Inches of Clearance	
°F	Nonfouling Coal	Fouling Coal
2000–2400	Platen −22	Platen −22
1750–2000	Spaced− 7	Platen −12
1450–1750	Spaced− 3	Spaced− 6
Up to 1450	Spaced− 2	Spaced− 3

surfaces. The guide is based on fouling and nonfouling coals and gas temperatures at maximum boiler load.

COAL PROPERTIES USED FOR DESIGN

Coal properties to be used in a boiler design are established by the purchaser's specifications. It is preferable to show an analysis that represents an actual coal as the performance coal and include low, average, and high analyses to confirm limits for the boiler design. Important items of the analysis are calorific value, moisture and ash contents, grindability index, and the ash analysis. If the fuel specifications are stated as ranges, it is customary to consider the average as the performance fuel but to design the unit to handle fuels with the worst analysis.

If coals from two or more sources may be fired, individual analyses rather than a calculated average are desirable, with "low" and "high" representing the composite. Justification for using individual analyses comes from the necessity to evaluate pulverizer capabilities as well as ash influences on unit performance. The "worst" coal can become the design coal when there is a good probability that the power plant will use a significant amount of the "worst" coal. Finally, if coals are to be blended, information must be provided on both the separate analyses and the possible blends, including values for eutectic fusibility temperatures.

FURNACE DESIGN FOR OIL & GAS FIRING

Selecting a furnace size, its wall tubing, and

its circulating system is a function of two distinct design parameters: the complete combustion of the fuel and the preservation of satisfactory furnace-wall metal temperatures. To this end, furnaces for burning pulverized solid fuels must be configured to prevent the formation of objectionable slag deposits, which can increase the furnace outlet-gas temperature above design values. Generally, the large furnaces and the wall deposits associated with coal firing result in relatively low furnace-wall absorption rates, so that tube-wall metal temperature does not influence furnace size.

When firing oil, the larger amount of fuel reaction surface available for combustion produces complete combustion in a smaller furnace volume. The higher consequent heat-release rate per unit area of furnace-wall surface can cause high localized heat-absorption rates. Furnace sizing for oil burning, then, must account for the effect of high absorption rates on both metal temperatures and circulation, and these factors usually determine the dimensions of an oil-fired furnace.[1]

Gas-fired furnace design is similar to oil-fired, except for the fact that gas firing inherently results in lower localized heat-absorption rates. This allows designing somewhat smaller furnaces with even higher heat-release rates than used for oil firing.

BURNING REFUSE IN UTILITY BOILERS

Chapter 8 describes methods of burning municipal refuse and other cellulosic wastes in industrial boilers. In such units, large percentages of mixed commercial and household refuse are burned; in some, a small startup burner is the only fossil fuel consuming device.

For several important reasons large pulverized-coal units primarily generating electricity should limit refuse firing to 10 to 20 percent of the full-load fuel firing rate. The following are some of these reasons.

■ Even shredded and air-separated refuse is not completely homogeneous, because refuse composition and moisture content vary considerably from season to season, and day to day. This nonhomogeneity will result in much

wider fluctuations in the heating value than that experienced with coal firing.

■ Fired alone, refuse is potentially more corrosive than most coals and fuel oils. It appears that the corrosion potential from burning refuse is nearly linear with the percentage burned. It seems prudent, therefore, to limit the amount of refuse burned in reheat units.

■ Generally speaking, the moisture content of prepared refuse will be more than triple that of bituminous coal (on a pounds-per-million-Btu basis) and will vary widely from day-to-day— as will its heating value. In addition the relatively high moisture content of refuse and the corresponding decrease in fuel-firing efficiency will increase the combustion gas mass compared to the primary fuels being burned in any given plant. Limiting the percentage of refuse burned minimizes the effect on superheater and reheater outlet steam temperatures, and on the air-heater exit gas temperature.

THE DESIGN OF LARGE HIGH-PRESSURE FURNACE-WALL SYSTEMS

The design of the various heat-absorbing equipment described in previous chapters involves a finite pressure drop or resistance to fluid flow. Air heaters, economizers, superheaters and reheaters all have a multiplicity of parallel flow paths or circuits through which the fluid being heated must pass; the uniformity of the flow distribution depends largely upon the magnitude of the pressure drop across the apparatus.

The watercooled furnace, in general, handles the fluid it is heating in many more parallel circuits than are found in the other heat-transfer devices; therefore, the problem of distributing water to each tube and providing adequate circulation for cooling all the tubes is correspondingly more difficult.

The tubes which form the furnace walls are of different lengths and are subject to varying heat-absorption rates. Great care must be exercised to insure that each tube receives a sufficient flow of water to prevent overheating individual waterwall tubes as well as to prevent an excessive temperature difference between adjacent tubes. Flow instability, which can occur in furnace circuits, is considerably more difficult to control when a large number of tubes operate in parallel. The designer must calculate the heat absorption per furnace tube from a forecast of the normal absorption pattern of the entire furnace. This forecast depends on the height and width of the furnace, but most of all, on the manner in which the fuel is fired.

The anticipated pattern of heat absorption might not, however, be reproduced in service; moreover, it will fluctuate from time to time due to load changes, transient conditions, and because the thickness of ash covering the walls will not always be the same. With certain firing systems the number of pulverizers and burners in service also will vary the pattern.

At intervals ash deposits will drop off, or be removed by sootblowing, and this will give rise to further considerable variations. Misjudging the intensity or distribution of heat absorption may cause circulation difficulties that are impossible to correct except by decreasing the rating and operating pressure of the boiler.

As the boiler operating pressure approaches the critical point (Chapter 5) the difference in the density of water and steam to produce circulation approaches zero. Designing tube circuits of low enough resistance to insure adequate velocities and adequate water-to-steam ratios in all tubes at all loads, with all fuels, under varying slagging conditions, and at all firing rates becomes proportionately more difficult at the 2750 to 2850 psig (19.0 to 19.7 MPa gage) drum and waterwall operating-pressure level.

Achieving proper circulation becomes increasingly difficult not only with higher operating pressures, but also with increased unit capacity. In units with high megawatt ratings, the fuel fired per lineal foot of furnace perimeter increases markedly compared to earlier 100 to 200 MW units. This has led the boiler industry to adopt internally rifled tubes for furnace walls

because higher percentages of steam can be tolerated in the steam-water mixture in rifled furnace tubes, at lower mass velocities.

In any case, the difficulties of designing a satisfactory furnace-wall circulating system involve all these considerations:

■ operating pressure level

■ furnace physical arrangement

■ the presence of non-heat-absorbing tubes near the fuel nozzles or burners

■ manufacturing tolerances of commercially available tubing (which affect inside diameter and consequent pressure drop tube-to-tube)

■ allowance for different heat pickup in different portions of the furnace

■ the possibility of internal or external deposits

■ provision for the necessary pressure drop for steam-water separation in the drum

TANGENTIAL FIRING
AND CIRCULATION SYSTEM DESIGN

With C-E tangential firing and its mode of introducing the fuel, the furnace is used as the burner; thus, heat-absorption patterns are much more uniform and more definable than with any other type of firing. As further described and illustrated in Chapters 11 and 12, the fuel and combustion air are introduced in the corners of the furnace, but the actual burning process takes place in the main body of the furnace. Therefore, heat is not liberated in a concentrated form in the area of the fuel admission assemblies, nor is it concentrated on one side of the furnace.

As the load on the steam-generating unit varies, the amount of heat admitted into the furnace is either increased or decreased uniformly, and in such a manner that the heat absorption pattern remains basically unchanged. This is because each pulverizer feeds one horizontal level of fuel nozzles. These two features of C-E's tangential firing, coupled with inherently low flame temperatures, are an asset to the successful design of and operation of the furnace circulation system.

ACQUISITION OF
CIRCULATION DESIGN DATA

The furnace enclosure is one of the most critical components of a steam generator and must be conservatively designed to assure high boiler availability. The *circulation objective* is to assure sufficient cooling of the furnace tubes during all operating conditions with an adequate margin of reserve for transient upsets. Adequate circulation prevents excessive metal temperatures or temperature differentials that could cause failures in the furnace-wall tubes due to overstressing or corrosion.

The design of a furnace-wall circulation system is a procedure requiring knowledge and data obtained both from field operating units and laboratory testing. The field data from operating units is required to understand and characterize the heat-absorption characteristics for a particular type of furnace configuration and firing system, throughout a wide range of load and water-side operating pressures, and the effect that fuel variation has on furnace heat absorption.

Furnace absorption/distribution standards are developed by correlating test data from many operating units. The following important parameters are considered in this work: furnace geometry, fuel, excess air, type of firing, disposition of fuel nozzles, heat input, and furnace-wall cleanliness. By using these standards, it is possible to calculate the vertical absorption-rate profile, the lateral heat-absorption distribution and other furnace heat accumulation patterns to properly engineer the furnace-wall circulation system.

The laboratory data is required for a knowledge of the physical properties of the furnace-wall tubing under actual furnace heat-transfer conditions. In the laboratory, flow versus quality criteria have been established to predict DNB (departure from nucleate boiling) in a furnace tube containing boiling water. This has been done in C-E's high-pressure test loop at the Kreisinger Development Laboratory. The allowable combinations of pressure, mass flow, local bulk quality, and local crown absorption rate have been calculated from correlations based on the laboratory data. Additionally, the derived engineering standards take into account data scatter, dimensional tolerance, instrumentation error, and boiler water chemistry, all to provide sufficient margin to accommodate actual field operating conditions.

C-E researchers have tested both full-size smooth and rifled tubes; this forms the basis for the design engineering of high-pressure boilers using either type of furnace-wall tubing.

FURNACE-WALL CIRCULATION IN SUBCRITICAL-PRESSURE STEAM GENERATORS

Combustion Engineering uses two types of circulating systems for subcritical-pressure applications—thermal (thermosyphonic or "natural") circulation and its Controlled Circulation® system. (Fig. 4). Each is designed to meet specific power-plant requirements and to provide certain advantages.

THERMAL CIRCULATION

Thermally induced circulation, in theory, could be used in the design of a boiler that ap-

proaches the critical pressure of 3208 psia (22.1 MPa abs.). This could be done only by going to extremely large diameter furnace-wall tubes of high-alloy materials and essentially zero pressure-drop downcomer systems, risers, and drum internals. The highest *practical* thermal-circulation pressure that has been designed for is 2800 psig (19.3 MPa gage) in the drum and waterwall circuitry; this limits the available pressure drop for superheater cooling when the required superheater outlet pressure is above 2650 psig (18.3 MPa gage).

As furnace-wall operating pressures and furnace sizes increase, a designer must have substantially greater knowledge of waterwall heat-absorption patterns that result from firing-system characteristics. Predictability and repeatability of such heat-flux patterns must be achievable on a positive, long-term basis. Operator intervention or equipment malfunction must not permit a different distribution of heat input to the furnace walls other than that for which the unit was designed.

CONTROLLED FORCED RECIRCULATION

In Chapter 5, the concept of subcritical-pressure controlled forced recirculation was introduced. It was stated that a boiler with such a circulation system incorporates a recirculating

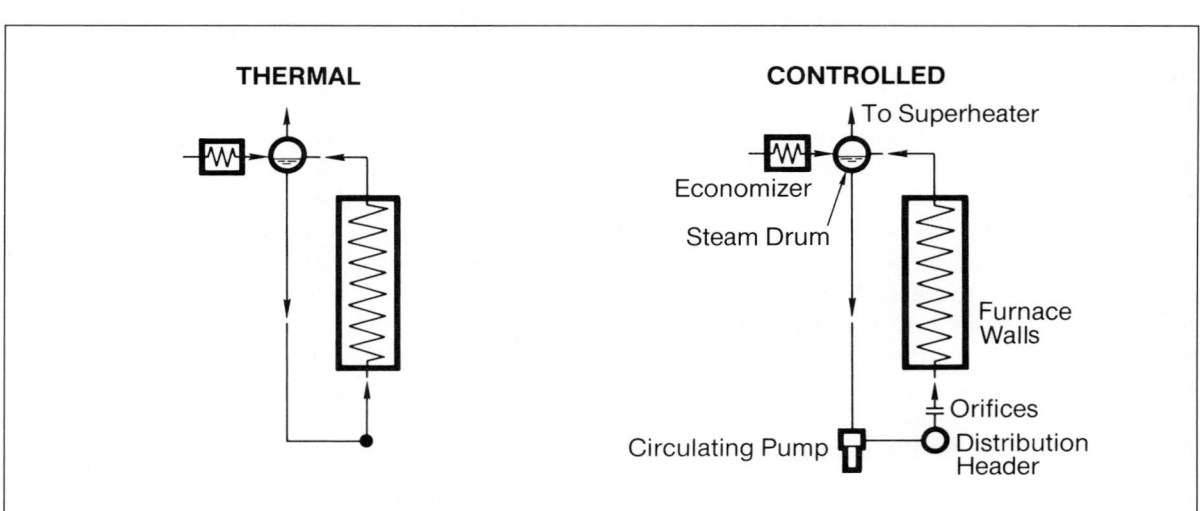

Fig. 4 C-E drum-type subcritical-pressure circulating systems

pump between the drum and waterwalls, thereby freeing the designer from dependence upon the difference in steam and water densities to provide circulating head. Using a recirculating pump means that the designer is assured of positive circulation even before heat is applied. Because the pump helps to overcome friction loss in the waterwall tubes, it is possible to substitute smaller diameter tubes for the larger diameter tubes required with a limited thermal circulation head. The net result is a lower-tube-weight, lower-metal-temperature wall with lower thermal stresses.

Another advantage to incorporating a recirculating pump is the greater flexibility of boiler layout. Also, more freedom exists in the arrangement of boiler heating surface because, with assured positive circulation, horizontal evaporation surface may be used to any extent desired.

The C-E system for controlled forced recirculation at the high subcritical-pressure level is called Controlled Circulation. C-E does not presume to select that pressure which sharply differentiates between the possibility of using thermally induced circulation from the advisability of using a Controlled Circulation system. However, C-E designers believe that

there is a throttle-pressure level above 2200 psig (15.2 MPa gage), particularly in large pulverized-coal units, where a Controlled Circulation system becomes prudent. A 2850-psig (19.7 MPa gage) drum-pressure limit has been established for Controlled Circulation steam generators. This limit permits operation of the 2400-psig (16.6 MPa gage) throttle-pressure cycle at 5 percent overpressure, or 2520 psig (17.4 MPa gage) at the turbine, and as high as 2680 psig (18.5 MPa gage) at the boiler superheater outlet.

THERMAL-CIRCULATION FURNACE DESIGN

Thermal circulation results from a density difference between the mixture of saturated water and feedwater in the downcomers (downtakes) and the lighter steam-water mixture in the furnace-wall tubes. The temperature and corresponding density in the downtakes can be changed somewhat in the initial design of a unit by varying the water temperature leaving the economizer. Fig. 5 shows how the furnace-wall operating pressure and the percent steam

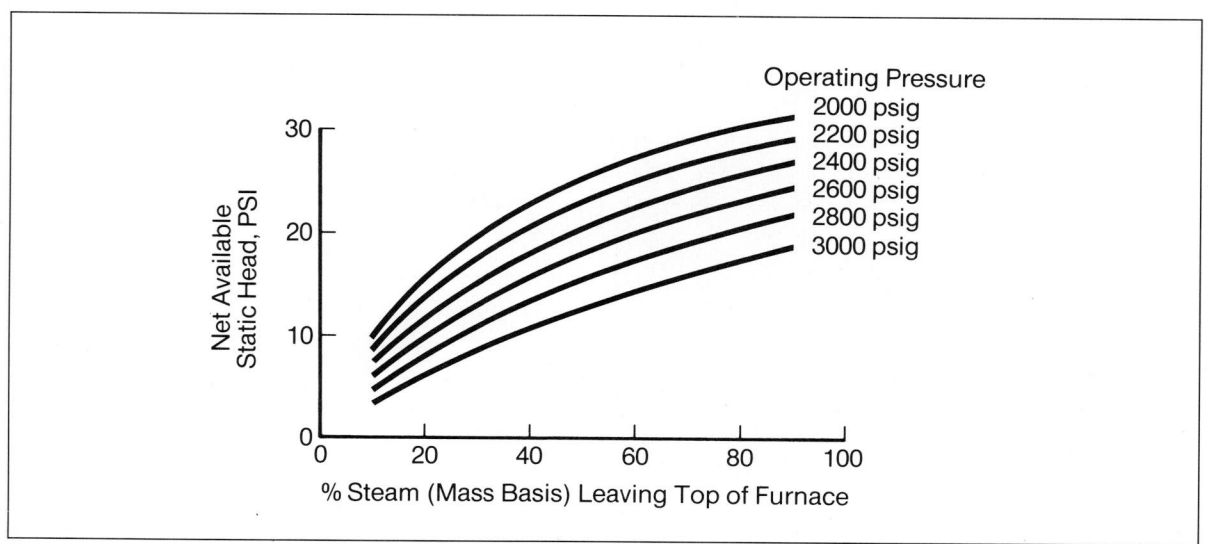

Fig. 5 Typical available static head for circulation based on 200-ft. high furnace; this available head is independent of type of circulation

in the mixture in the wall tube affect the available thermal head for circulation. The available head is the difference between the weight of water in the unheated downcomer and the static head of the steam-water mixture in the heated waterwall tubes.

Fig. 6 shows a typical C-E thermal-circulation reheat steam generator for burning pulverized subbituminous coal, for low to medium sub-critical pressures.

Tangential fuel nozzles are located at five elevations in the four corners of the boiler furnace. Vertical downcomers connecting the steam drum to the bottom waterwall header system supply the boiler water. Lower ends of the front, rear, and side wall tubes connect to the bottom waterwall headers and the upper ends connect to the upper headers. Relief (riser) tubes transport the mixture of steam and water from the waterwalls to the drum, where

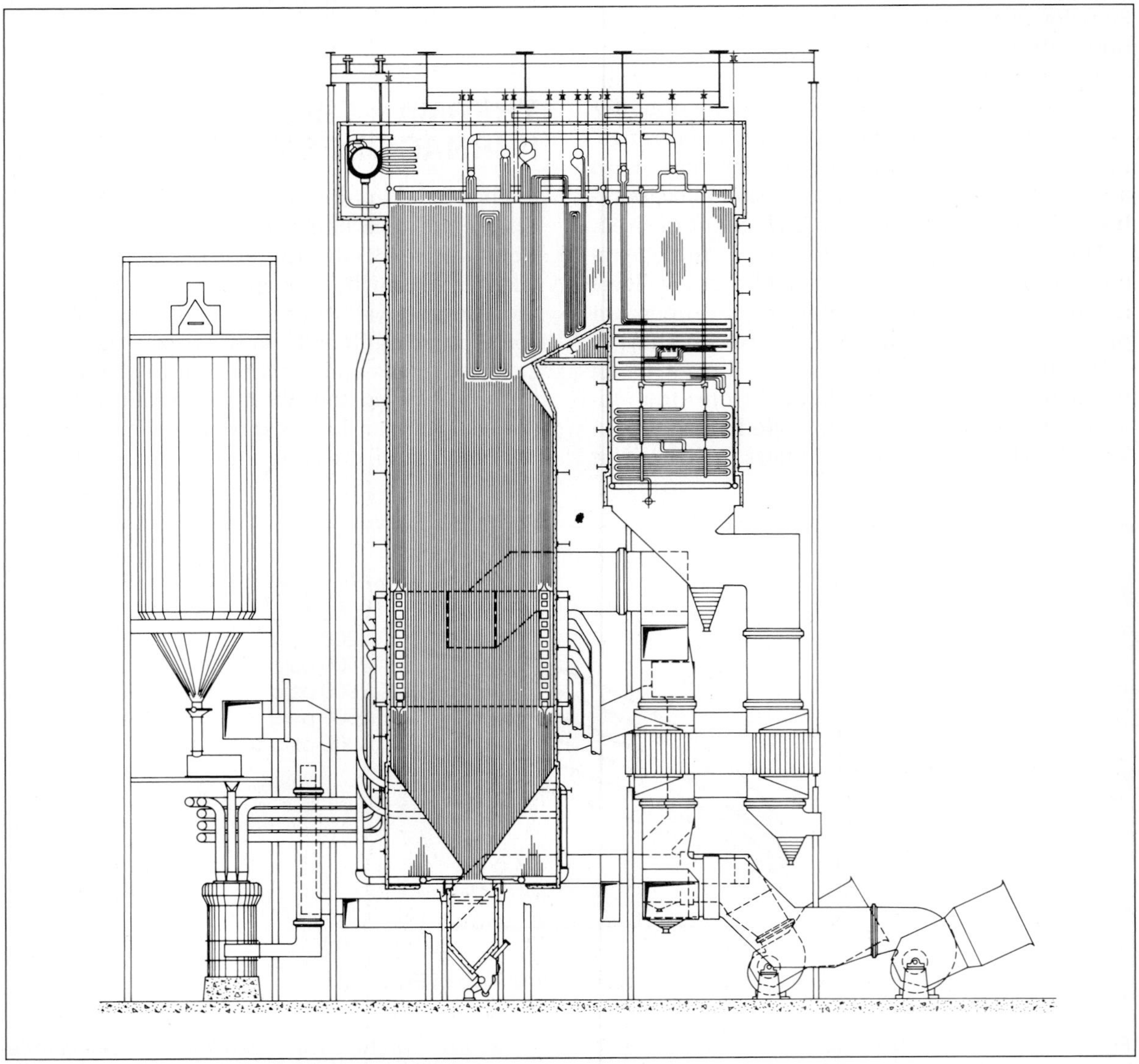

Fig. 6 C-E thermal-circulation tangentially fired pulverized-coal steam generator

the steam is separated from the boiler water. To insure uniform distribution of steam and water, the connecting tubes are spaced evenly along the drum and headers. Steam which has been separated from the steam-water mixture in the drum passes on to the first superheater stage.

The first superheater stage is located in the vertical gas pass to the rear of the furnace. Saturated steam from the drum passes through the roof of the furnace, then down the walls enclosing the rear pass to the lower ring header. From this point the slightly superheated steam flows through the horizontal superheater tubes in a generally counterflow direction to the downward flow of the products of combustion. From the first superheater stage, the steam passes to the finishing or outlet stage which is a platen section located at the gas outlet of the furnace. The reheater is located between the finishing platen superheater and the initial superheater stage. The flow of furnace gas over the reheater is essentially counterflow to the flow of steam.

The economizer is located directly below the first-stage superheater. Water flows upward in the economizer to provide counterflow of furnace gas and feedwater. Outlet tubes of the economizer section extend vertically upward to form the supporting hanger tubes for the horizontal superheater and the economizer. These tubes then connect to a header above the roof; connecting piping conducts the feedwater from this header to the drum.

This boiler is of the balanced-draft type. Pulverizers are pressurized by primary-air fans. Ljungstrom® trisector air preheaters with vertical shafts are located at the economizer gas outlet. The gas ducts connecting the economizer to the air preheater and at the air preheater outlet are arranged with hoppers for ash collection as well as to collect the drain-off wash water.

With the exception of the air preheaters, the entire boiler is supported from steel located at the roof elevation. Expansion joints are provided in the gas duct below the economizer and in the air ducts between the air preheaters and the windboxes to accommodate the downward expansion of the boiler pressure parts in their heated condition.

Shown in Fig. 7 is a cross section of a drum used in C-E thermal-circulation boilers. In this drum, the steam-water mixture from the furnace passes through the centrifugal separators where a spin is imparted. This forces the water to the outer edge of the centrifugal separator where it is separated from the steam. Nearly dried, the steam passes through corrugated plates where, by low-velocity surface contact, the remaining moisture is removed by wetting action on the plates and final screen dryers.

The cross-sectional area of the drum limits the size and arrangement of the internal steam separators and dryers and, consequently, the acceptable rating in terms of mass flow per hour per unit of length. A drum with a large diameter affords more free area for water content at normal water level, and reduced moisture entrainment in the larger steam space. Within the limits of available pressure drop for steam/water separation, higher-capacity separators and increased dryer surface can be provided to achieve a higher rating per unit of length.

For a specific diameter and arrangement of internal components, the drum capacity is proportional to length. The diameter and length, then, are determined by
- unit generating capacity
- operating pressure in the drum
- the practical considerations of the spacing and arrangement of connecting tubes and piping with respect to drum-shell ligaments
- the requirement to insure a uniform distribution of steam and water flow entering and leaving the drum

The percent of boiler-water solids that will be carried over by moisture entrained in the steam leaving the drum is proportional to the boiler-water solids concentration and to the percent moisture carried over. Moisture is removed sequentially by the steam separators, followed by gravitational separation of droplets in the steam space, and finally by the steam dryers, arranged to prevent moisture reentrainment.

A high water level in the drum reduces the steam space available for gravitational separation of moisture. An excessively high level can

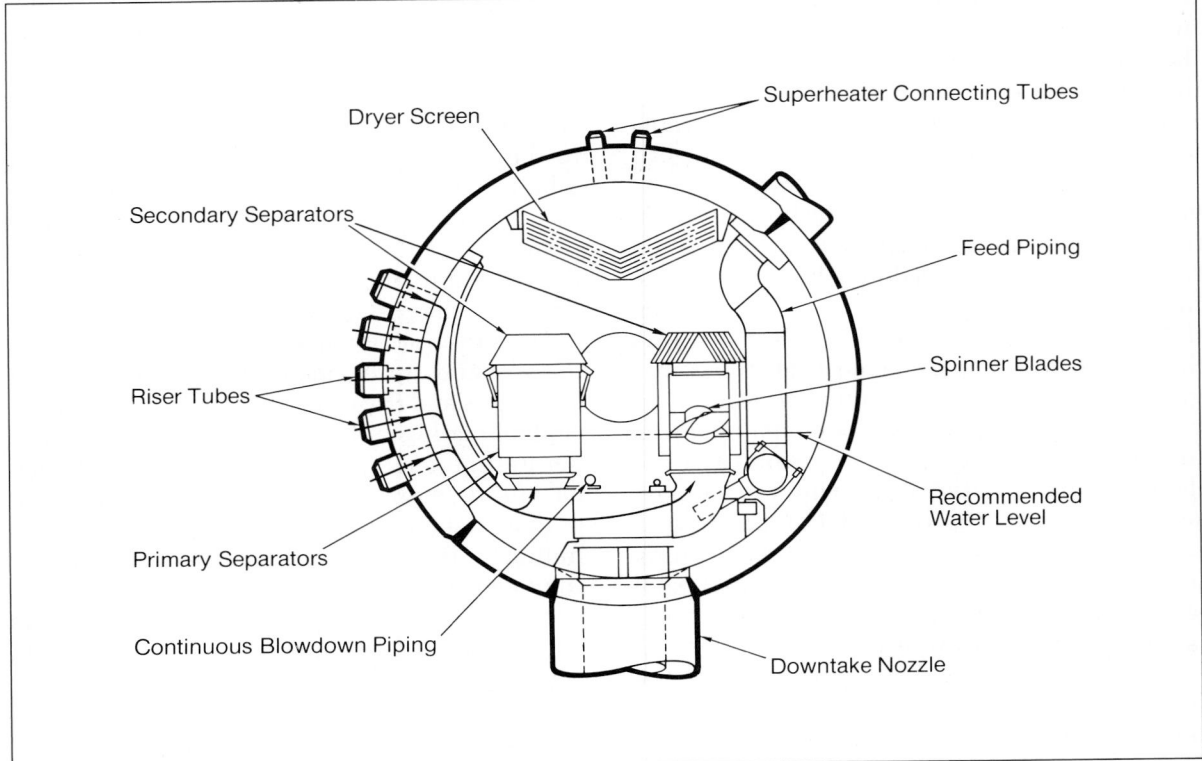

***Fig. 7* Steam-drum internals for C-E thermal-circulation radiant and reheat steam generators**

impair efficiency of primary and secondary separators, resulting in an increased moisture content entering the dryers and a corresponding increase in carryover moisture in which there are entrained solids.

The dissolved solids in the boiler water should be held as low as practicable, consistent with the recommended chemical treatment for boiler-tube corrosion protection. This assures that solids carryover during high water-level excursions will be negligible (see Chapter 20).

Vaporous carryover of certain boiler water constituents, such as silica and sodium, is determined by the respective volatility of these unwanted constituents, or their solubility in steam, as discussed in Chapter 5. They must be controlled by limiting the boiler-water solids concentration at high drum pressures. (See Chapter 20).

CONTROLLED CIRCULATION

C-E's first Controlled Circulation boiler in an American central station went into service in 1942 at the Somerset Station of Montaup Electric Company. Since then, the Controlled Circulation principle has been widely accepted for boilers ranging from waste heat, marine, and industrial units up to the largest central station installations. The reliability and safe operating records of Controlled Circulation boilers have justified their selection in power stations throughout the world.

THE CONTROLLED CIRCULATION SYSTEM

Water flow through the economizer is generally the same with either a thermal-circulation or a controlled recirculation boiler. Water from

the economizer enters the drum, is mixed with the water discharged from the steam-water separators, and is directed to the large downcomers which are evenly spaced along the drum. From this point on, a C-E Controlled Circulation unit (Fig. 1) differs from a thermal-circulation boiler in five respects:

1. Circulating pumps are placed in the downtake circuit to provide sufficient head to insure adequate, positive upward circulation under all operating conditions.
2. Orifices are installed in the inlets of waterwall circuits to assist in obtaining a predetermined, proportioned flow of water to tubes of varying length and heat absorption.
3. Furnace-wall tubing is of smaller diameter and has lower metal temperatures than that for thermal circulation, all made possible by the head available from the pumps.
4. The drums are internally shrouded to provide uniform heating and cooling of the drum shell, for maximum maneuverability during start-up, load changes, and shutdown.
5. Controlled Circulation boilers have an economizer recirculation line which provides a positive flow from the boiler-water circulating pump (BWCP) discharge through the economizer to the drum. Under start-up conditions, with the feedwater valve closed, this feature minimizes any steaming in the economizer and precludes the need for blowdown with its associated heat loss.

The boiler circulating pump suction is taken from a suction manifold, which is supplied by the downtakes. From the boiler circulating pumps the water discharges into the waterwall inlet ring header through the pump discharge lines. In the inlet header, the water passes through strainers and then through orifices feeding the furnace-wall tubes, the extended sidewall tubes and the water-cooled element spacers. The mixture of steam and water leaving the evaporative circuits is discharged into the upper drum, in which the steam is separated from the water.

In the drum of a Controlled Circulation unit, Fig. 8, the basic separation process is similar to that of thermal-circulation drums except that

the available pump head permits more efficient use of the centrifugal devices. One distinct advantage of the Controlled Circulation design is the internal shrouding of the drum. This watertight shrouding directs the flow of steam and water returning from the furnace around the inside surface of the drum, providing uniform heating. This uniform heating effectively eliminates thermal stresses from temperature differences through the thick wall of the drum, between the submerged and unsubmerged portion of the shell, and from end to end. This facilitates rapid start-ups, shutdowns, and cyclic operation.

In operation, a Controlled Circulation boiler has positive circulation established before any heat is introduced into the furnace, which then permits rapid start-up with virtually no time restrictions because of circulation in the pressure-part system.

CONTROLLED CIRCULATION PUMP AND PIPING SYSTEM

Fig. 1 shows the overall arrangement of the boiler circulating pump and downcomer piping system. Downcomers installed on the steam drum carry the recirculated boiler water, mixed with the feedwater, into the circulating pumps. These pumps are connected through a common suction manifold that insures a flow through all the downcomers regardless of the number or location of the circulating pumps in service. This feature minimizes the water level difference along the length of the steam drum and contributes to good performance of the steam/water separation equipment independent of the number of pumps in service.

The downcomers are straight vertical pipes connecting the steam drum and the suction manifold; the circulating pumps are mounted on the suction manifold. Using the boiler as the support for the pumps is a feature that eliminates any need for external supports and prevents undue stresses which might affect pump alignment.

The head developed by the circulating pumps is only that required to supplement the thermal head and, for a typical Controlled Cir-

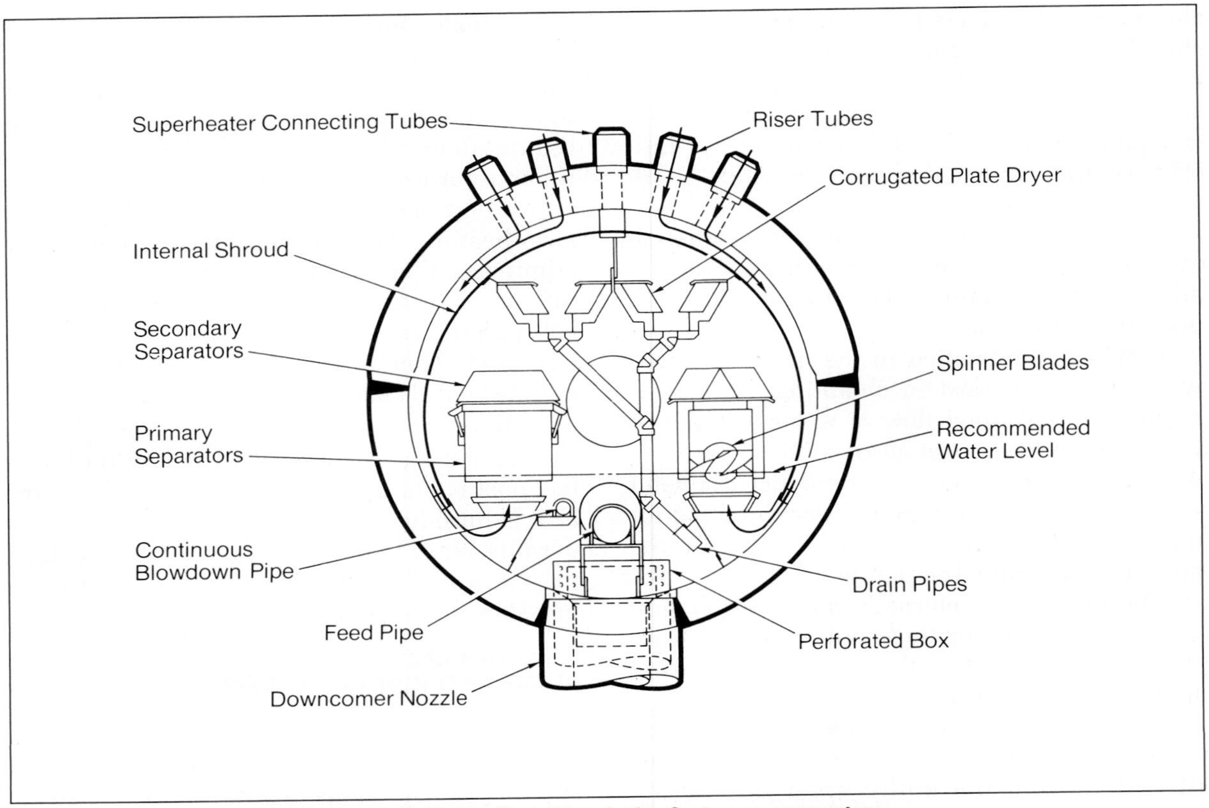

Superheater Connecting Tubes

Internal Shroud

Secondary Separators

Primary Separators

Continuous Blowdown Pipe

Feed Pipe

Downcomer Nozzle

Riser Tubes

Corrugated Plate Dryer

Spinner Blades

Recommended Water Level

Drain Pipes

Perforated Box

Fig. 8 Steam-drum internals for C-E Controlled Circulation® steam generators

culation boiler is 25 psi (175 kPa) or less; for this, a single-stage impeller is adequate. Boiler water circulating pumps and their design are described further in Chapter 14.

ORIFICES FOR FURNACE-WALL TUBES

The orifices used for optimizing flow distribution to the furnace-wall tubes are shop-fitted to adapters welded to the internal header wall. A keying arrangement insures that each orifice is installed in its proper tube circuit once the correctness of the initial installation of the orifice mount adapter has been established. Orifice size varies for different circuits or groups of circuits, depending on their length, arrangement, and heat absorption.

Strainers or screens between the circulating pump discharge and the orifices prevent large particles of foreign material from plugging

the orifices at the entrance to the generating circuits. The strainers are particularly valuable during the initial start-up of a new unit because, during this period of operation, foreign material is brought into the boiler from preboiler-cycle feedwater piping and auxiliary equipment.

DESIGN OF A CONTROLLED CIRCULATION FURNACE-WALL SYSTEM

The orificing and selection of pump head for a Controlled Circulation boiler are best understood by first reviewing a non-pump-assisted thermal circulation unit design.

The designer of a thermal circulation furnace of a given plan area determines the total number of parallel vertical tubes making up the outer walls of the furnace; a common selection is to use 2½-inch OD tubes on 3-inch centers. The next step, based on experience with units of similar physical size and heat duty, is to establish the size and number of unheated downcomers from the steam drum to the lower waterwall-header system. Along with this, it is necessary to establish the arrangement, diameter and number of relief tubes connecting the upper waterwall headers to the drum. An average overall percent steam leaving the furnace is then calculated, based on the heat absorbed by the "average" furnace tube and on the flow resistances in that tube, the downcomers, the relief tubes, the drum internals, and all associated entrance and exit losses.

Within the limits of available thermal head and physical space, adjustments in downcomer and relief-tube size and flow resistances can be made at this time to increase or decrease the average circulation to achieve satisfactory cooling of the furnace-wall.

Tubes in the furnace enclosure have different lengths, dissimilar configurations, and receive varying amounts of heat. Any tube absorbing more heat than the average tube has a *higher* circulating water flow, but probably also a high percent steam leaving; typically, a 50-percent increase in absorption above the average, equally received throughout the length of a tube, produces a 10 to 20 percent increase in water flow entering. For every tube receiving *more* than the average tube, there is another tube receiving *less* water. Each tube must be analyzed on the basis of its individual heat absorption to arrive at its own mass percentage of steam (quality) leaving.

Realize that, in designing a waterwall section—a group of parallel tubes having a common (bottom) inlet header and a common upper (outlet) header, Fig.9—the pressure drop or head loss between the headers has to be identical for every tube. This situation exists irrespective of any pump assistance. Flow in each tube, and percent steam in the mixture leaving, differ depending upon

- heat absorption (overall and local)
- effective inside diameter of tubing (as affected by manufacturing tolerances)
- effective tube length (accounting for bends)

- amount of tubing "shaded" from heat
- location relative to feeder tubes
- orifices (if used)

Tube A, for instance, may be longer, but have a lower overall heat absorption than the "average" tube B; without orificing, it will have a lower flow than B. Tube C, if shorter than B but fully exposed for its entire length to a high heat, will have a higher flow than the average tube (B). But *all* tubes must have the same differential pressure from header to header.

An equilibrium condition will always result in the flow-plus-orifice losses being additive to the weight of the saturated mixture in each tube, at any given moment. This is because the header-to-header differential is comprised of

- the weight of the steam/water mixture in each tube
- the pressure drop due to the resistance to flow of the steam/water mixture in each tube
- orifice pressure drop (if any).

The flow may or may not be enough to cool the tube properly and it may even reverse (from upward to downward, and vice versa) from time to time, dependent upon the variables acting on it.

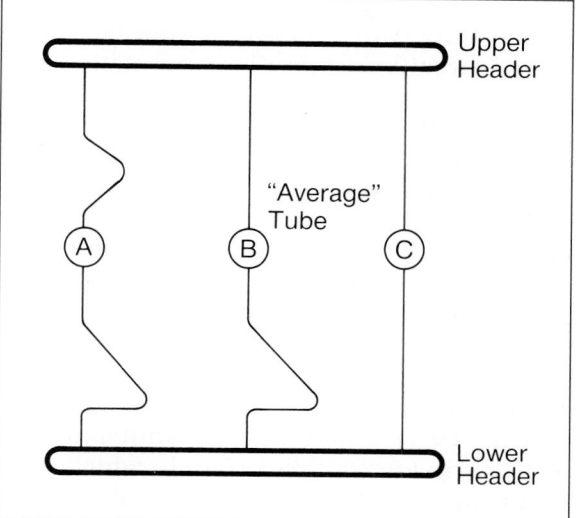

Fig. 9 Schematic of furnace-wall section; head loss between headers is identical for each tube

For unorificed parallel tubes, the flow-related pressure drop equality can be expressed approximately as

$$f\,L_A\,m_A^2V_A = f\,L_B\,m_B^2V_B = f\,L_C\,m_C^2V_C$$

where f is a common friction factor, m_N is mass flow in the Nth tube, L_N is the effective length of any tube, and V_N is the average specific volume of the steam-water mixture in the Nth tube.

In tubes where mass flow is lower because of lower heat absorption, the specific volume is correspondingly higher to compensate for the flow resistance of a high-absorption tube with higher mass flow. For any given overall steam content, water between lower and upper headers redistributes from tube to tube according to the above relationships.

Fig. 10 gives a comparison between thermal-circulation and Controlled Circulation performance. In the upper curve, average mass flow in furnace-wall tubes is shown as a function of design heat absorption. Note that the circulating pumps move the maximum mass of water *before* firing commences, because the water is most dense at that time. The lower curve shows that the designer of a Controlled Circulation unit can preferentially increase the flow in tubes with less favorable configurations to provide more cooling at all loads.

The orifice pressure drop designed into Controlled Circulation circuits is additive to the heated-circuit resistance loss and static head. Orificing individual tubes or groups of tubes provides either adequate mass flow or a desired exit quality under a wide range of postulated operating conditions.

The orifice pressure drop has a dampening effect on the almost random behavior of a purely thermal-circulation unit. Because the orifice drop occurs at the beginning of a circuit, before heat has been added in the furnace, the principal variable is mass flow through the opening, with no effect of differential specific volume. This is a very significant difference between pure *forced* circulation, in which most of the pressure drop is in the heated tube circuits themselves, and C-E's selectively orificed pump-assisted *controlled* circulation.

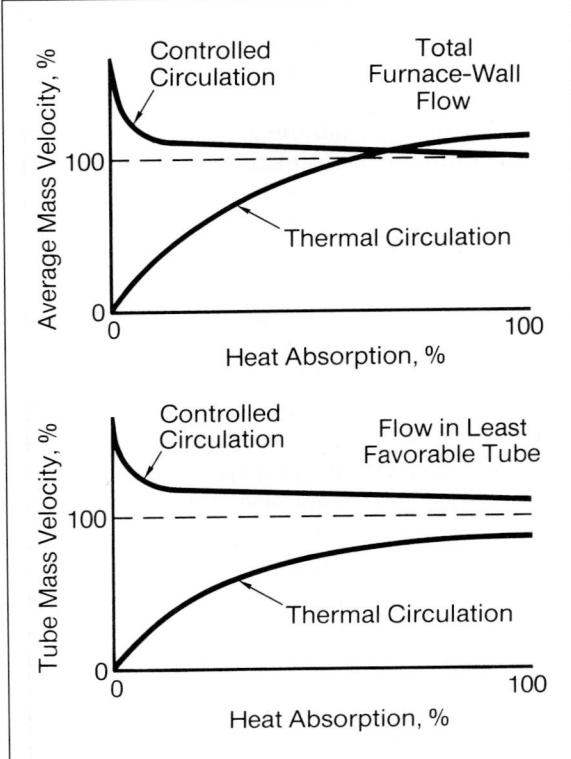

Fig. 10 Comparison of furnace-wall performance — thermal circulation versus Controlled Circulation®

CONTROLLED CIRCULATION STEAM GENERATORS

Fig. 11 shows a Controlled Circulation reheat boiler designed to provide steam to a large high-subcritical-pressure turbine generator. This is a single open-furnace unit. It has a superheater with both pendant and horizontal surfaces and a reheater with both radiant wall and pendant surfaces.

This boiler has superheater outlet conditions of 2620 psig and 1005°F (18.1 MPa gage and 541°C), with single reheat to 1005°F (541°C). It is designed to burn high-volatile bituminous coal and is of the balanced-draft type, with forced- and induced-draft and primary-air fans.

Six C-E pulverizers supply coal to tangential fuel nozzles located in the four corners of the furnace. The supply of water to the waterwalls

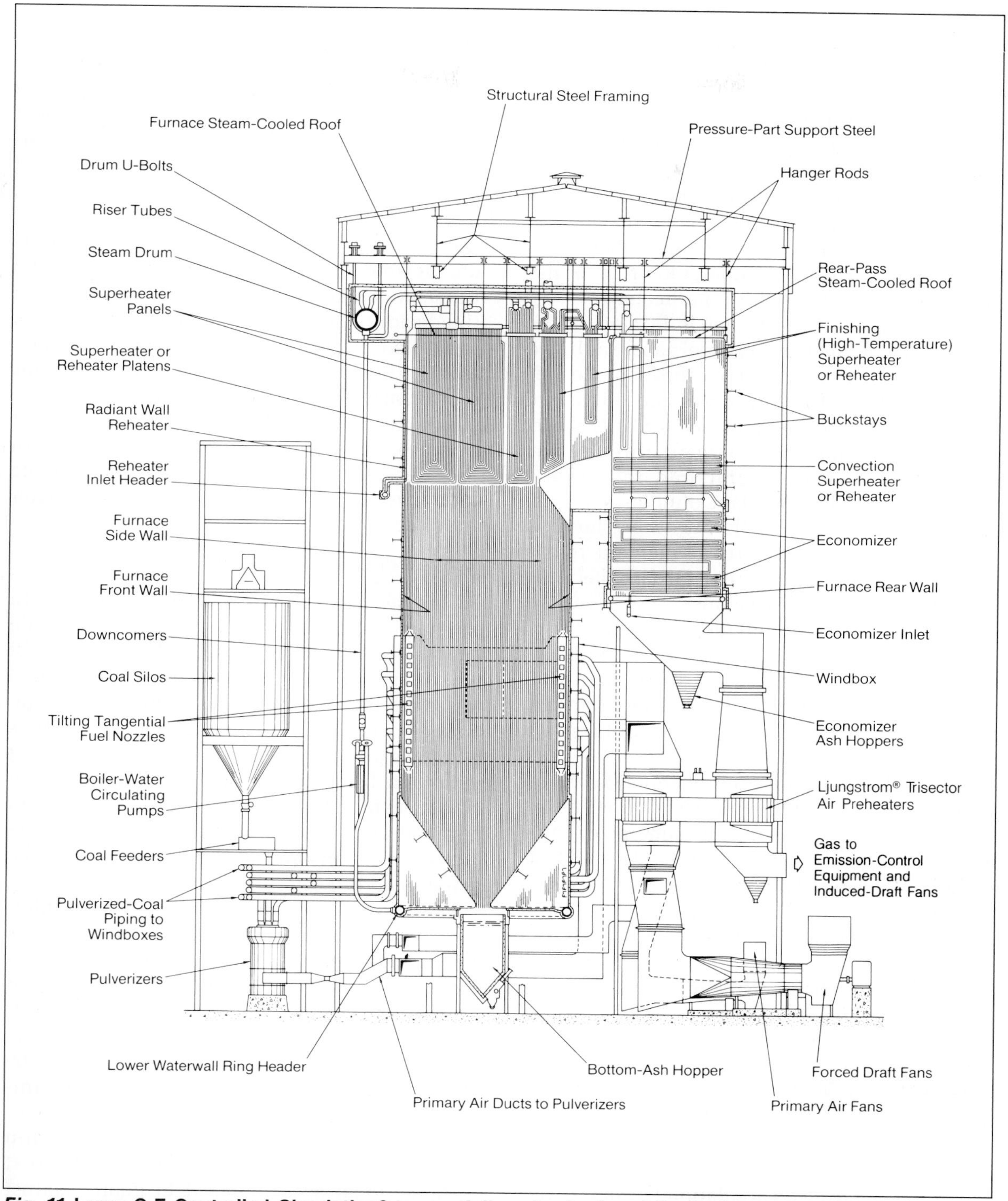

Structural Steel Framing

Furnace Steam-Cooled Roof

Pressure-Part Support Steel

Drum U-Bolts

Hanger Rods

Riser Tubes

Steam Drum

Rear-Pass Steam-Cooled Roof

Superheater Panels

Finishing (High-Temperature) Superheater or Reheater

Superheater or Reheater Platens

Radiant Wall Reheater

Buckstays

Reheater Inlet Header

Convection Superheater or Reheater

Furnace Side Wall

Economizer

Furnace Front Wall

Furnace Rear Wall

Downcomers

Economizer Inlet

Coal Silos

Windbox

Tilting Tangential Fuel Nozzles

Economizer Ash Hoppers

Boiler-Water Circulating Pumps

Ljungstrom® Trisector Air Preheaters

Coal Feeders

Gas to Emission-Control Equipment and Induced-Draft Fans

Pulverized-Coal Piping to Windboxes

Pulverizers

Lower Waterwall Ring Header

Bottom-Ash Hopper

Forced Draft Fans

Primary Air Ducts to Pulverizers

Primary Air Fans

Fig. 11 Large C-E Controlled Circulation® tangentially pulverized-coal-fired reheat steam generator

begins with the steam drum. The water travels down through the downcomers to the suction manifold and into the circulating pumps from which it is discharged to the lower ring header which has flow-distributing orifices. The mixture of steam and water that flows up through the various heated circuits is collected in an arrangement of waterwall outlet headers located just above the furnace roof. The steam-water mixture passes through connecting tubes to the drum, where steam is separated from boiler water and passes on to the superheater.

The first-stage superheater is located in the vertical pass at the rear of the furnace. Saturated steam from the drum passes through the furnace radiant roof, backpass roof and walls, and then enters the lower inlet header of the first stage. The steam flows upward through the first-stage superheater and into the second stage, comprised of pendant double-loop panels located at the top of the furnace. The final superheater is of the vertical pendant-platen type, and is also in the upper furnace, between the panels and the furnace nose.

Reheat steam enters the reheater radiant-wall inlet header and travels upward through the radiant wall tubes. After leaving the radiant-wall outlet header, the reheat steam enters the low-temperature rear pendant, and continues into the finishing reheat pendant.

The horizontal bare-tube economizer is located directly below the first-stage superheater. Vertical outlet tubes from the economizer serve to support both the economizer banks and first-stage superheater. The tubes then terminate in an economizer outlet header above the roof from which connecting piping transports the water to the drum.

Two vertical-shaft regenerative type (Ljung-strom®) trisector air preheaters are located beneath the economizer.

ADVANTAGES OF A CONTROLLED CIRCULATION SYSTEM

The C-E proprietary design of subcritical pump-assisted recirculation boiler, with the registered Controlled Circulation trademark has many features that have made it a preferred choice of electric-power generators throughout the world.

Some of the advantages of Controlled Circulation systems are covered in Chapter 5; however, a primary reason for selecting a steam generator of the Controlled Circulation type is its characteristic of *high dynamic maneuverability*. The fact that furnace-wall flow is positively established, in an upward direction, *before* heat is applied, enables an operator to fire such a furnace at any rate desired—that is, Controlled Circulation furnace walls and steam drums are never a limiting factor in building or changing load. Thus, Controlled Circulation units are inherently better suited to operate under cycling conditions (particularly those demanded by two-shift operation) than other boiler types. This important aspect of the design is covered in depth in the subsection Design Features for Variable Load Operation of large steam generators later in this chapter.

OPTIMUM CHEMICAL CLEANING

Circulating pumps on a boiler facilitate both pre-operational cleaning and operational scale removal. Laboratory testing with circulation at velocities of 0.5 to 1.0 ft/sec (0.15 to 0.30 m/s) or greater has indicated that complete scale removal can be accomplished in 10 to 20 minutes following initial solvent contact. A significant reduction from the usual 4- to 6-hour hydrochloric-acid contact period can be obtained in pumped-circulation boilers where desired velocity conditions and uniform cleaning are assured in all tube circuits.

In a thermal-circulation boiler, solvent interchange or mixing from section to section is slow or does not occur, and significantly different levels of ion concentration can be noted in samples taken from various locations in a unit. A reasonably complete picture of what is occurring in such a unit can be obtained by multipoint sampling. Although multisample monitoring is important during pre-operational cleaning to establish local ion equilibrium, it is of even greater interest in operational scale removal to guard against solvent depletion or hide-out in local areas. In units where positive recirculation can be maintained, uniform solvent concentration is rapidly established in all portions of a unit, and one sample point will give an indication of conditions throughout the unit.

Under a short solvent-contact procedure that is used with Controlled Circulation boilers, then, the solvent contact time is no longer than 30 minutes; with positive circulation, rapid and complete cleaning is assured with a minimum exposure time of pressure parts to the cleaning solvent. This also minimizes boiler downtime and, utilizing the rapid shutdown and start-up capabilities of such boilers, it makes it possible to complete the entire procedure during a weekend outage.

DETECTING DETERIORATION OF FEEDWATER QUALITY

Carryover of metallic oxides and other impurities from the preboiler system may cause deposits on heated water circuits of any boiler. In severe cases such deposits can lead to tube failures.

The Controlled Circulation boiler offers an advantage for detecting deposits without removing the unit from service, by observing the circulating-pump differential pressure between the pump suction manifold and either the discharge manifold or the waterwall lower header. Observation of the differential pressure provides a convenient external means of detecting internal waterwall system deposits and *anticipating* more serious operation difficulties.

The actual differential across the pump at normal full-load operation is established during initial operation, when the boiler is free of deposits. The differential pressure is then checked periodically during full-load operation. If a substantial difference is seen between the baseline and later data, the boiler load may be reduced until an outage presents the opportunity to inspect and clean internal surfaces of water circuits, as described in Chapters 20 and 21.

OPERATION WITH WATERWALL TUBE LEAKS

Because of the available pump head, Controlled Circulation units use smaller diameter tubing than thermal circulation boilers, and incorporate orifices at the entrance to the tubes for optimum flow distribution. The two act to control and reduce the amount of water that can pass through a tube leak of a given size, resulting in a significantly lower damage potential to adjacent tubes than in a thermal unit.

There are many recorded instances in which an operator has kept a Controlled Circulation unit in service through a peak period or until a weekend, when it became convenient to shut down. An additional capability, that of forced furnace-wall cooling with the pumps, allows boiler cooldown with uniform stresses without using the feedwater pump or dumping feedwater to the condenser.

OPERATING-PRESSURE FLEXIBILITY

The curve in Fig. 12 shows that a Controlled Circulation steam generator can be operated through a wide range of pressure at full load. Operation within the permissible area allows great latitude in establishing a desirable sliding-pressure pattern.

WATER-TREATMENT CONSIDERATIONS

In comparison to once-through boilers, Controlled Circulation boilers have numerous ad-

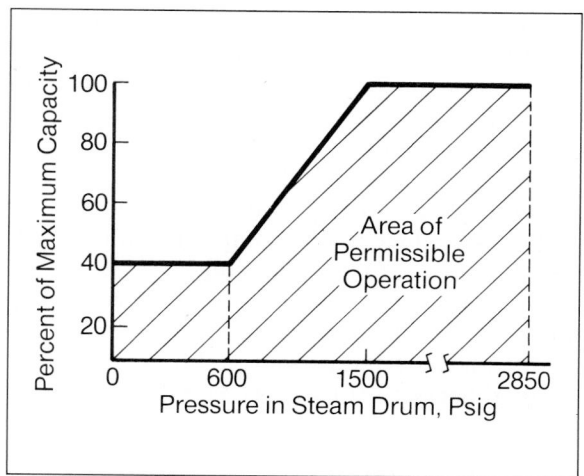

Fig. 12 Maximum permissible capacity as a function of operating pressure in drum, for C-E Controlled Circulation® boilers

vantages in water treatment:

■ Drum boilers can remain "on line" during a condenser leak because the drum gives added protection. In the event of a condenser leak (greater than 0.5 ppm solids in the hot well), desuperheating must be discontinued and steam temperature limited by lowering fuel-nozzle tilt or possibly a drop in load. Under similar hot-well conditions, a once-through unit must be shut down. (See Chapter 20.)

■ During normal operation, polishers may leak small amounts of solids, which may be basic or acidic. In a once-through unit, these solids will continue to the turbine unchanged. There have been cases of turbine corrosion damage from these solids which are concentrated by the turbine. In a drum boiler, chemical treatment of the boiler water renders these solids harmless.

■ Base-loaded drum boilers are not as restricted on start-up as once-through units in regard to feedwater quality. Once-through units should have a cleanup loop associated with the polishers to remove iron oxide and reduce silica in feedwater before start-up. Base-loaded drum boilers do not require this loop.

THERMAL CIRCULATION VERSUS CONTROLLED CIRCULATION

In choosing between thermally induced circulation and a Controlled Circulation system the following must be taken into account:

■ Although thermal circulation does in fact increase in response to heat applied, the increase in flow is not directly proportional to the increase in heat flux. That is, although there is usually a larger amount of water circulated in an individual tube when more heat is applied, the increase may not be completely adequate to provide proper tube protection.

■ Slag can form heavily, and does form unpredictably, on different areas of the furnace at different times; it can be removed by sootblowers, or shed because of its mass, equally unpredictably. It is common to have a major part of the heat input to a furnace-wall tube in its upper half, because of sootblower cleaning or shedding, which can result in a calculable *decrease* in circulation rate in an individual tube or a

group of tubes. Flow reversal can occur, with local overheating a possibility.

■ Controlled Circulation and thermal-circulation boilers have the same thermal circulating head available if they have the same furnace height, the same economizer water-leaving temperature, and the same circulation ratio (Fig. 5). The pumps and orifices incorporated in Controlled Circulation units *do not destroy* the available thermal head or lessen the density difference between furnace-wall tubes and unheated downcomers.

■ To assure that there is adequate available head under all operations of firing and maneuvering the boiler, C-E adds orifices so that all tubes will receive water under *all* conditions and shrouds the thick-walled steam drum to cool it or heat it in a controlled manner under all rates of load change. The circulating pumps complement these engineered modifications to the basic thermal-circulation design, and provide the pressure head to overcome the incrementally increased resistance. A most important point is that flow increases are created with pump assistance that are much greater than those available through the thermosyphonic phenomenon. This point is illustrated (Fig. 13) by comparing the pressure drops in the various components of typical thermal-circulation and Controlled Circulation steam generators.

■ The net power consumption of the circulating pumps is small compared to that in other circuitry of the boiler for accomplishing basically the same objective of equalized flow distribution with variation in load, firing conditions, and heat-absorption patterns. The feedwater-pump power charged to a unit, because of its superheater and economizer pressure drops, is substantially higher than the small power loss of the circulating pumps.

SUPERCRITICAL-PRESSURE STEAM GENERATORS

C-E supercritical steam generators can be classified in three basic design categories according to their operating-pressure regimes.

In the first, for units designed for constant-pressure operation, supercritical pressures are maintained in both furnace walls and superheater over the normal operating range. The furnace-wall arrangement and the use of boiler throttling valves are as described under C-E Combined Circulation® units.

For units that are to have partial sliding-pressure capability, supercritical pressures are maintained in the furnace walls; only the superheater follows a sliding-pressure program. The boiler throttling valve arrangement is modified from the basic Combined Circulation design to allow for increased throttling. For units designed for *full* sliding pressure, the furnace-wall and superheater pressures may vary with load, including operation at subcritical pressure.

Sliding pressure is a highly desirable way to operate central-station steam generators at both high subcritical pressures and in the supercritical-pressure range. A later section of this chapter addresses the subject of sliding (or "variable") pressure operation in detail, and covers the significant advantages to the steam generator and turbine while operating in such a mode.

The three C-E steam-generator designs to meet user requirements for throttle pressures above the 3800 psig (26 MPa gage) level are described in the following sections.

CONSTANT-PRESSURE SUPERCRITICAL-PRESSURE UNITS

The C-E Combined Circulation® system is a once-through steam generator with a superimposed controlled forced recirculation system for the furnace walls. The furnace-wall system is

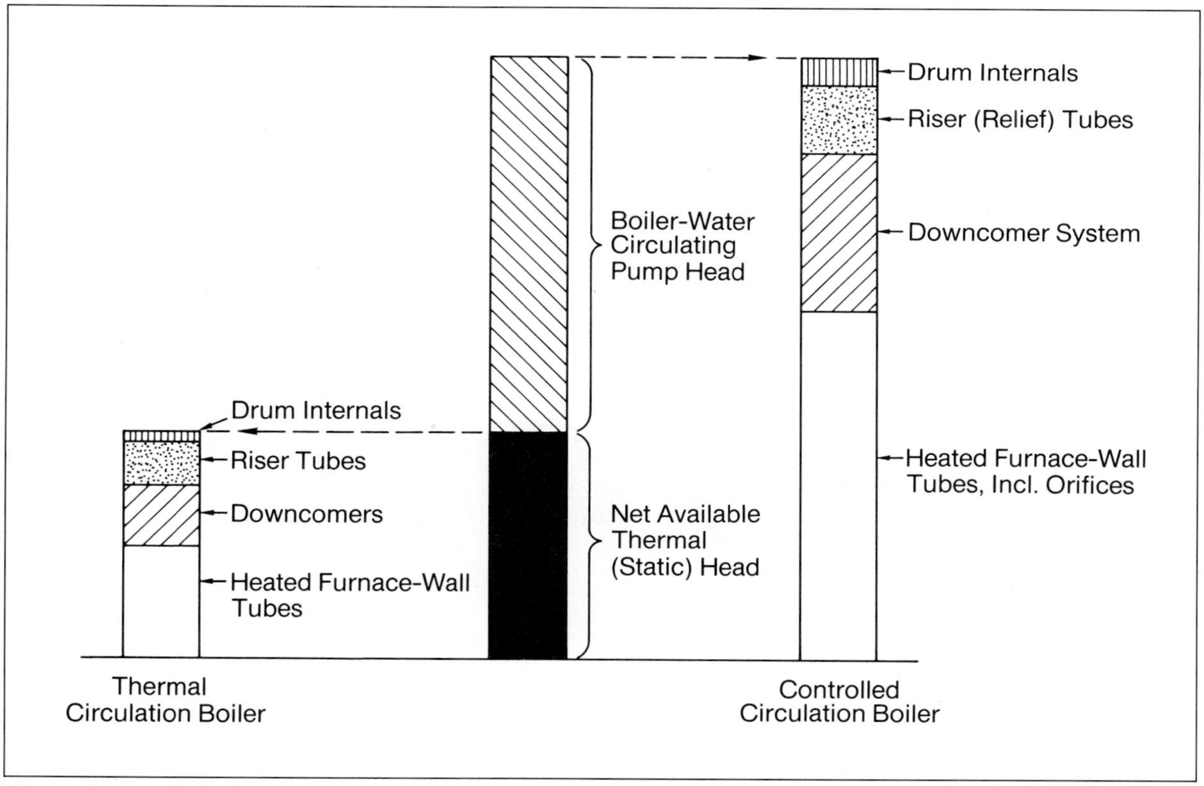

Fig. 13 Comparison of component pressure drops and available static head — thermal-circulation boiler versus Controlled Circulation® boiler

automatically protected by recirculation of fluid with a boiler circulating pump when once-through flow is insufficient at low loads. The recirculation pump eliminates the need for a high-capacity bypass system for furnace-wall protection, while still allowing once-through supercritical flow at higher output. This type of design permits furnace-wall tubes of sufficient inside diameter to maintain an adequate mass flow through the tubes at all operating conditions. The recirculating pump handles both recirculated flow and feedwater flow and thereby assists the feedpump, reducing some of its power requirement.

In the Combined Circulation design, all furnace walls are single-pass upflow, with no mixing headers. Proper design tolerances and the use of tube orifices compensate for flow unbalance or uneven heat absorption. The mixing vessel or sphere is substituted for the conventional drum to properly mix the recirculated flow from the furnace walls with the unit throughput.

The constant-pressure Combined Circulation supercritical units, then, are capable of operating with once-through flow, but cannot be operated with full sliding pressure. The furnace walls must be kept above the supercritical pressure (3208 psia or 22.1 MPa abs.) to avoid the film boiling and tube overheating which can occur at the transition to subcritical pressure. Boiler throttling valves are used at the furnace-wall outlet to keep the wall system pressurized (Fig. 14). The superheater operates in the sliding-pressure mode below approximately 30-percent load. Above that range, turbine load control is accomplished by using the turbine throttle valves.

SYSTEM INTEGRATION

The simplified flow diagram (Fig. 15) illustrates several features that have significantly advanced system integration.

First, the supercritical once-through steam generator closes the steam and feedwater connections of the steam-power cycle in a continuous heat-addition loop without the division established by the water level in the drum of subcritical recirculating units. This fact is recognized by an automatic, non-interacting, feed-forward control system, geared to the dynamic characteristics of the entire plant.

Second, system integration permits starting the turbine and steam generator simultaneously under conditions that provide an optimum of turbine protection for accelerated cold and hot

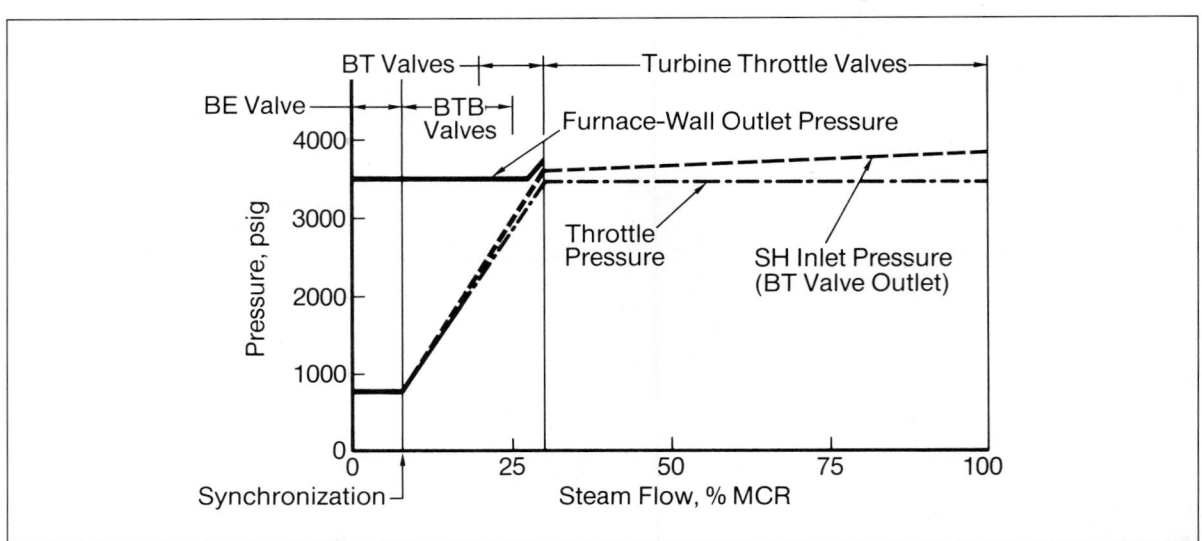

Fig. 14 Constant-pressure program for C-E Combined Circulation® steam generators

starts. This facility is combined with unique features to remove solids and oxides from both preboiler and boiler systems.

Third, the Combined Circulation principle eliminates the basic requirement of other once-through systems for minimum feedwater flow through the furnace. Heat removed from the steam-generating system during any operation is determined only by functions other than furnace-wall protection. The feedwater heating system, therefore, need not be used to reduce start-up heat losses. This in turn avoids contamination of this system by start-up flow and makes the system very simple. Eliminating the minimum flow requirement also greatly facilitates use of the turbine-driven feedpumps for start-up and avoids complications from separate start-up pumps.

BOILER-TURBINE SYSTEM

In Fig. 15, the feedwater enters the steam generator through the feedwater valve (FW). From the economizer the feedwater passes through the recirculating pump and its associated isolating valves to the furnace, where it flows in series through the centerwall and the outerwalls, then through the enclosure walls of the rear gas pass. At this point the steam passes through a system of throttling valves (BTB and BT) and continues through four sections of horizontal and pendant superheater before reaching the steam leads to the turbine through an outlet piping system. Steam is returned to the steam generator for reheating in a two-section reheater, then returned to the turbine. The condensate passes through a demineralizer and a series of low-pressure heaters to an open deaerator. The turbine-driven feedpump then returns the feedwater through a string of high-pressure heaters to the feedwater valve.

The diagram also shows the main components of a simple start-up system within which

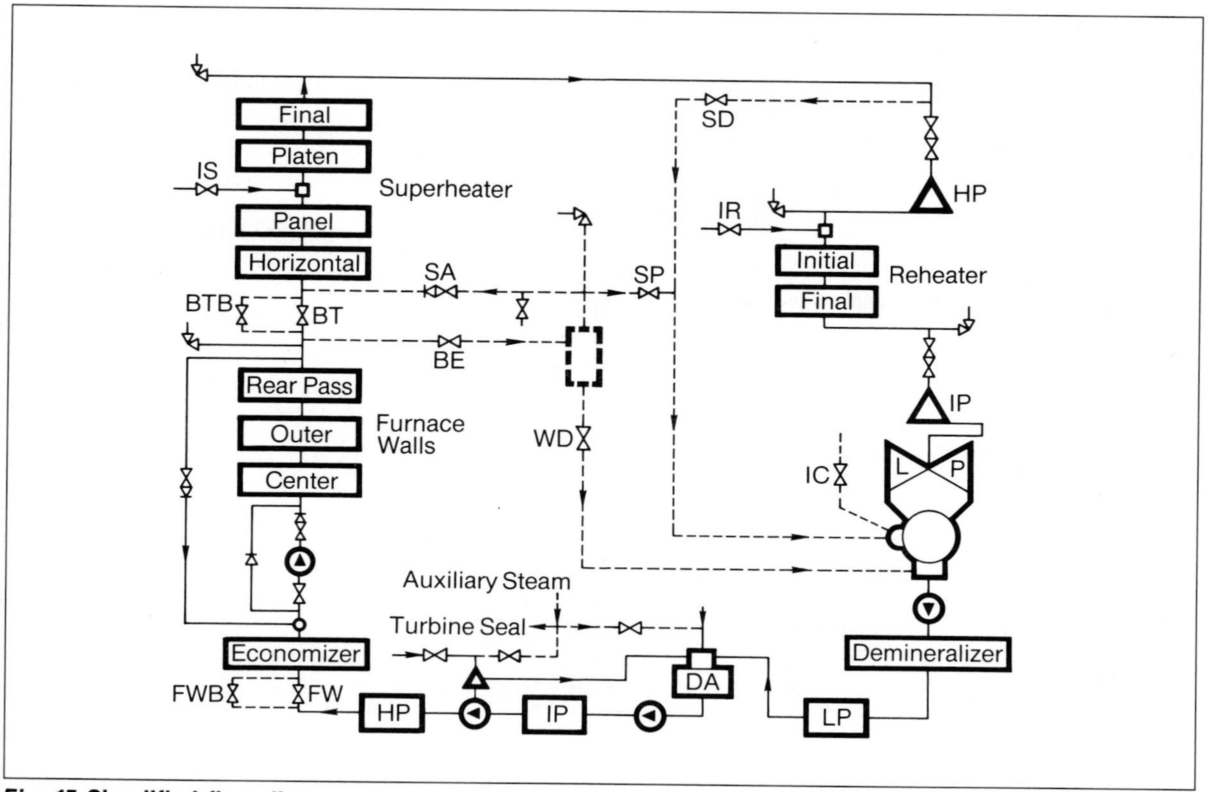

Fig. 15 Simplified flow diagram of C-E Combined Circulation® steam generator

the throttling valves (BT and BTB) belong. Upstream of these valves, piping connects the furnace-wall system with the extraction valves (BE) and the start-up separator. The steam side of the separator is connected to the superheater system through a steam admission valve (SA) and to the condenser through a spill-over valve (SP). Water from the separator is discharged to the condenser through the water drain valve (WD). The steam drain valve (SD), close to the turbine valve chest, discharges the steam for superheater cooling and heating of steam leads to the condenser and an injection valve (IC) regulates desuperheating water flow for condenser protection.

Auxiliary superheated steam is furnished to turbine seals, the deaerator, and the main feed-pump turbine from separate start-up boilers.

COMBINED CIRCULATION RECIRCULATING SYSTEM

The components of the Combined Circulation recirculating system are shown in a simplified isometric view in Fig. 16. Mixing feedwater flow with recirculated flow (as takes place in the drum of a subcritical unit) occurs here in the spherical mixing vessel. A single downcomer brings the fluid to the two recirculating pumps which hang from a symmetrical tee connection on the downcomer. Only one pump is needed for low-load operation, the other being a spare on hot standby. Neither is required at high loads. The pumps discharge to a common header from which supply piping runs directly to the furnace-wall headers.

The subcritical-pressure drum (which functions to separate steam and water) is eliminated at supercritical pressure. The recirculated flow

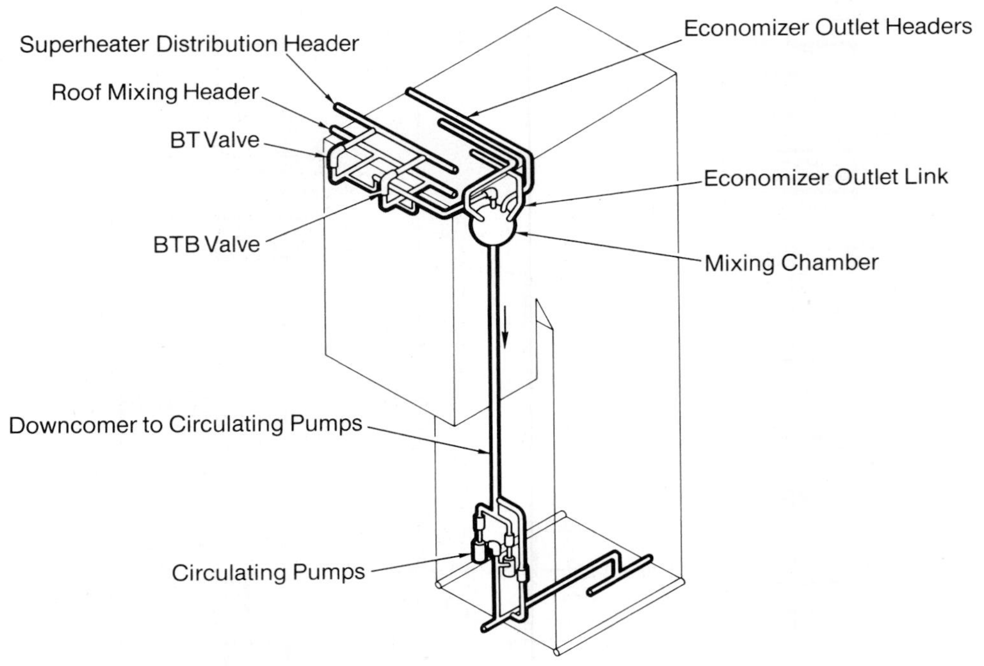

Fig. 16 **Flow diagram of furnace-wall and startup system for Combined Circulation® centerwall unit**

is returned from the outlet of the furnace-wall enclosure to the mixing vessel through a single recirculating line. A stop-check valve in this line automatically prevents bypassing the furnace-wall system when recirculation has ceased at the upper load range. Because downcomer, pumps, and piping to the center-wall all handle mixed or through flow at low temperatures, they are fabricated of carbon steel. The recirculating line and the mixing vessel use chrome-alloy material.

From the furnace-wall outlet headers, the fluid is piped to the backpass, where it cools the walls. From the backpass outlet header, the total boiler through-flow goes through the BT and BTB valve complex to the superheater. During waterwall recirculation conditions, a portion of flow in excess of the steam generator through-flow is recirculated back to the mixing chamber—this flow is extracted ahead of the BT/BTB valve complex.

TYPICAL COMBINED CIRCULATION UNIT

The most common type of supercritical steam generator in operation in the United States is the C-E Combined Circulation design. About 75 percent of the operating units are in the 600 to 900 MW size range, most having single reheat to 1005° (541°C). A Combined Circulation unit in this size range is shown in Fig. 17. It is designed for balanced-draft firing of sub-bituminous A coal, and has superheater outlet conditions of 3590 psig and 1005°F (24.8 MPa gage and 541°C). The reheated steam has an outlet temperature of 1005°F from inlet conditions of 583°F and 676 psig (306°C and 4.7 MPa gage).

Seven C-E pressurized pulverizers supply coal to the seven elevations of fuel nozzles; two centrifugal fans furnish the primary air to the pulverizers.

FURNACE-WALL SYSTEM

The arrangement of the Combined Circulation furnace-wall system has several characteristics:

■ Flow in all furnace and backpass enclosure walls is in the upward direction only.

■ With single-pass flow, all tubes in a welded panel have the same inlet temperature.

■ Centerwall flow and outerwall flow are in series on divided-furnace units.

■ All components of the recirculating system except the recirculating line are in series with the once-through flow.

■ All walls are formed by drainable, welded-tube panels.

These characteristics provide a number of important advantages, among which is the design flexibility available from the presence of a circulating pump with which the flow quantity through the parallel tubes around the furnace periphery can be distributed. Other benefits are improved temperature performance over the entire load range and the elimination of the bypass system as a requirement of the furnace wall.

These advantages will be most clearly understood after considering the flow performance of a typical furnace-wall system, with the volumetric flow at the furnace-wall inlet a function of load (Fig. 18). The through-flow as maintained by the feedpump increases in direct proportion to load. The recirculated flow as maintained by the circulating pumps supplements the through-flow over the low load range in a manner which protects the furnace walls by raising the actual flow to a safe level regardless of feedwater flow.

At low loads the recirculated flow is high, but decreases as the load increases. At about 60-percent load, the pressure drop through the furnace-wall system equals the head produced by the circulating pump; the stop-check valve in the recirculating line automatically closes. The circulating pump then ceases to add to the quantity of furnace flow but continues to contribute its positive head on the once-through flow and so acts as a booster to the feedpump. At this time, the pump may be shut down.

Combined Circulation Performance

Fluid temperatures, both primary and reheat, throughout a typical Combined Circulation supercritical steam generator at maximum con-

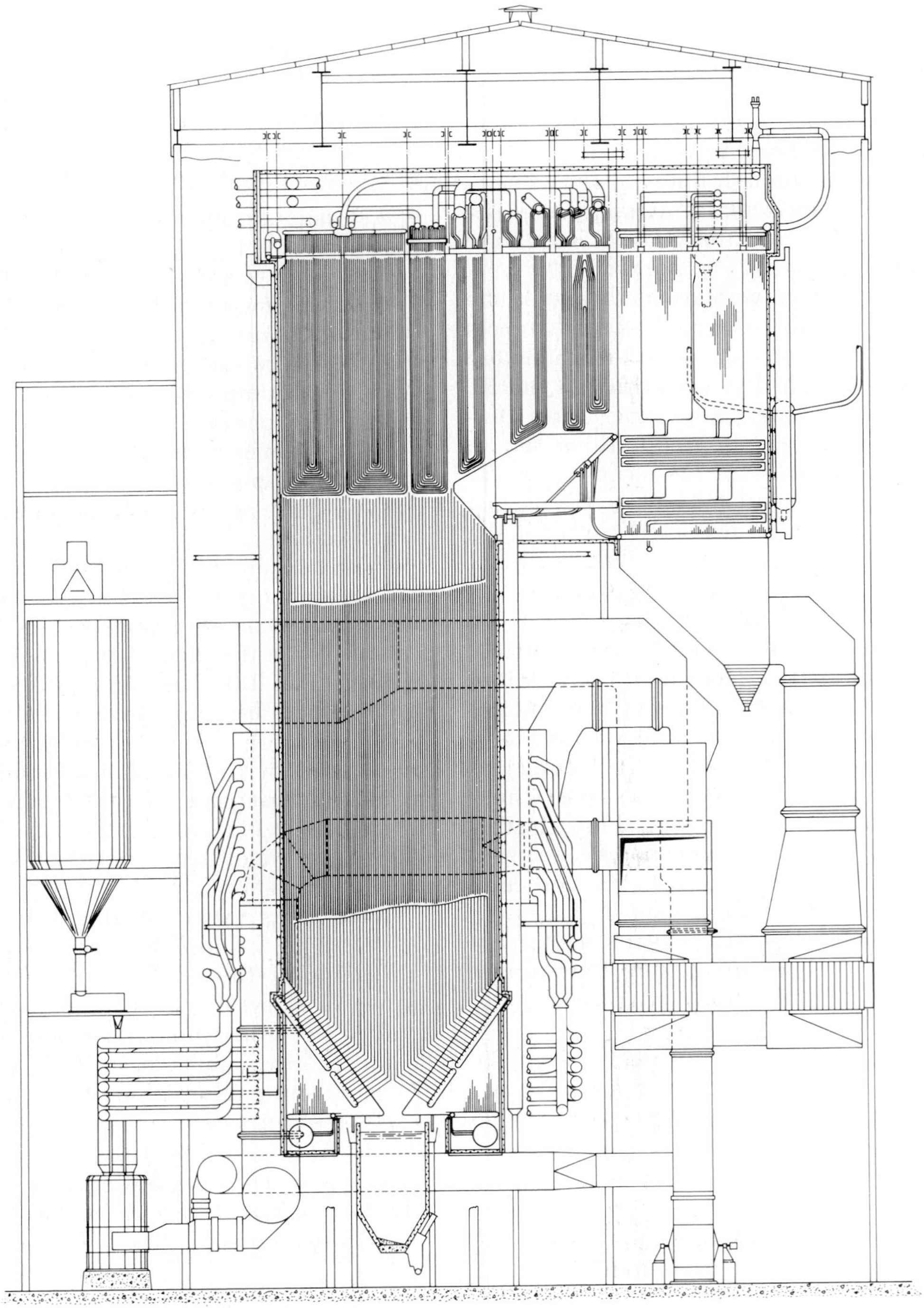

Fig. 17 C-E Combined Circulation® tangentially coal-fired reheat steam generator

tinuous load are plotted against the percentage of heat absorption in Fig. 19.

The primary fluid takes about 83 percent of total output, showing a gradual and continuous rise of fluid temperature from the feedwater inlet to the superheater outlet. The curve of these temperatures follows the characteristic line of supercritical fluid in which any heat absorption or change of heat content is accompanied by a change in fluid temperature.

Desuperheating occurring at the outlet of the panel superheater provides vernier control of average steam temperature. It also adjusts for unbalance of pickup across the width of the unit, as four individually temperature-controlled sections are installed in each furnace. The platen and finishing superheater sections add the balance of superheat.

PARTIAL SLIDING-PRESSURE DESIGN

Combined Circulation supercritical designs with boiler throttle valves (to keep the furnace-wall system pressurized above, say, 3500 psig or 24 MPa gage) can operate in the sliding-pressure mode only below approximately 30-percent rating (Fig. 14). Above this point, turbine load control is achieved through the steam-turbine throttle valves. This causes the usual temperature changes in the first stage of

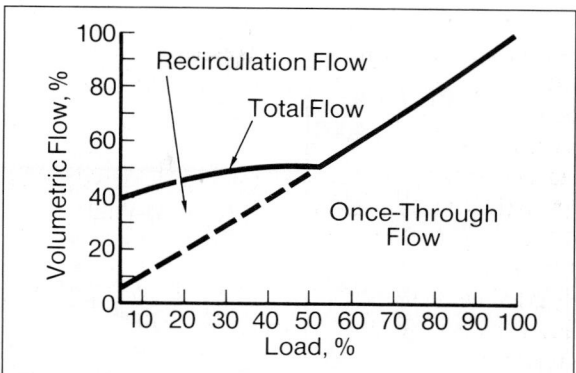

**Fig. 18 Flow in typical Combined Circulation®
furnace-wall system**

the turbine and is a factor in establishing permissible load-change rates. This design is extremely well suited for base-load operation. However, cycling and two-shifting units may require greater flexibility at reduced unit load to protect the steam generator and turbine during the increased number of load swings.

Fig. 20 shows the re-arrangement of a constant-pressure boiler-throttle-valve set to obtain sliding-pressure turbine operation to 80-percent rating while maintaining the furnace walls at supercritical pressure. This design has the usual advantages of operation of the turbine in

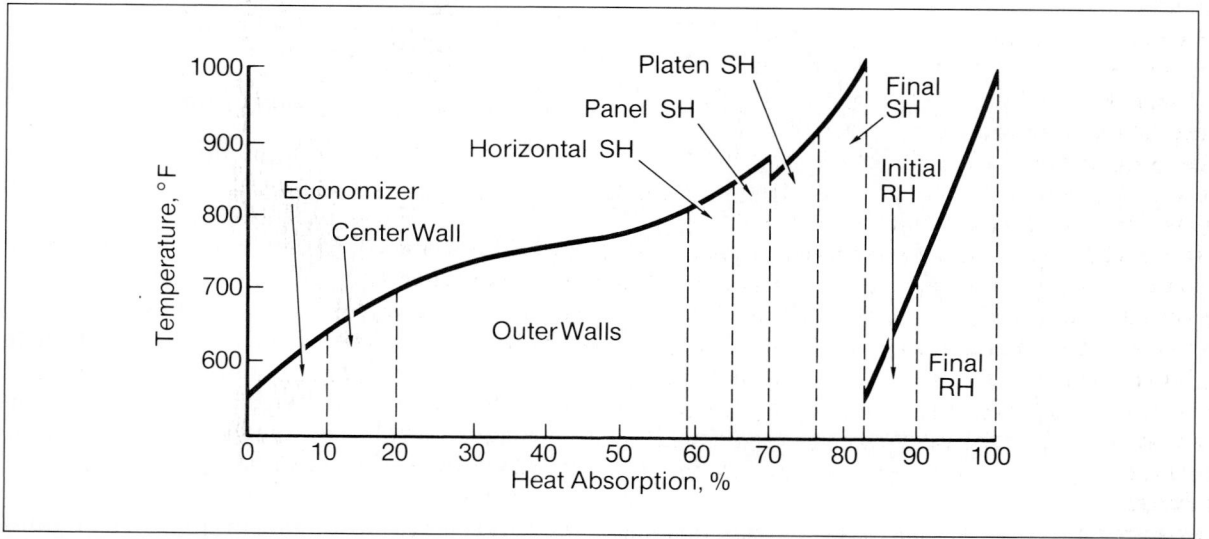

Fig. 19 Fluid temperatures through Combined Circulation® steam generator at full load

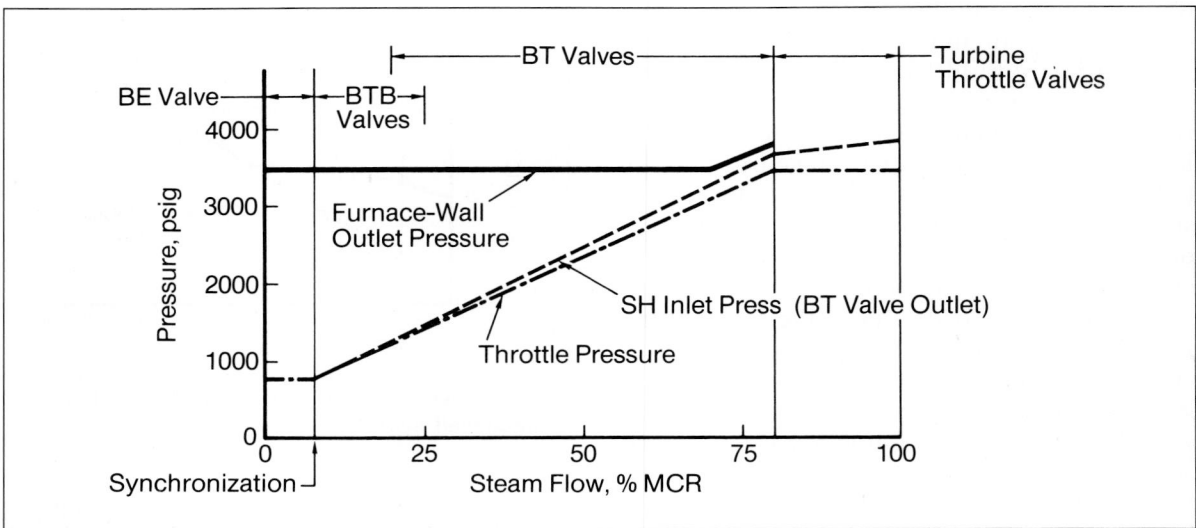

Fig. 20 Sliding pressure operation of C-E Combined Circulation® steam generators

a sliding-pressure mode, including increased reheat-temperature control range, reduced thermal stresses, and improved turbine heat rates as a result of variable-pressure operation with

Fig. 21 Pressure versus load for full sliding-pressure supercritical design

full-arc admission below 80-percent load.

Note that the boiler feed-pump power is the same for this modified design as for constant-pressure operation. However, with the use of boiler throttle valves at all loads below 80-percent, some increased wear and maintenance of them can be expected. This design, then, is often well-suited as a field modification to existing Combined Circulation units which are now required to cycle, or for new units which will cycle a minimal number of times during their operating lives.

FULL SLIDING-PRESSURE DESIGN

The third once-through design eliminates the boiler throttling valves and adopts a full sliding-pressure approach (Fig. 21). The furnace walls are allowed to enter the subcritical pressure range along with the superheater circuits, over the entire load range as shown.

Benefits to plant operation, in addition to those described above, include:
1) Reduced pressure levels at lower loads unload all cycle components pressure-wise, from the feedwater pump to the HP turbine, thereby prolonging life span.
2) Improved overall power plant heat rates when considering power consumption of

7–32

the boiler feed pump and other auxiliaries.

3) Simplified and faster start-up procedures may be employed in the design.

The superheater circuitry for the full sliding-pressure design, as well as the start-up system, the operational controls, and the auxiliary equipment, are essentially the same as for the constant-pressure and modified constant-pressure designs.

For start-up and low-load operation below 30 to 35 percent, the unit utilizes a pumped recirculation system (similar to the Combined Circulation unit) to provide an adequate mass flow through the furnace walls and the economizer. This mass flow is required to provide satisfactory cooling of the furnace-wall tubes and to avoid circuit stability problems. It also avoids flashing and steaming in the economizer. The basic advantage of a pumped recirculation system is that no heat is rejected in the recirculated water, but is all returned to the system.

Fig. 22 shows a side elevation of the C-E supercritical steam-generating unit for full sliding-pressure operation.

The unit has the following features:

■ An integral water separator is placed between the furnace-wall outlet and the superheater inlet. During start-up and low load, the separator operates with a water level, or *wet*. Under these conditions, the excess water from the furnace walls is recirculated through the furnace walls and the economizer.

■ In the once-through mode of operation, the water separator is *dry* and serves as a junction header between the furnace walls and the superheater. The principal advantage of this arrangement is that the water separator remains in the circuit at all times, thereby eliminating the multitude of valves associated with previous designs.

The furnace-wall configuration of a full sliding-pressure supercritical steam generator will vary depending on unit size. Two arrangements are available, the vertical tube-wall and the spiral-wound or helical-tube furnace. Vertical tube-walls are shown in the steam generator in Fig. 22. The spiral-wound design is shown conceptually in Fig. 23. There are design,

performance, and cost advantages to each which must be considered in selecting one or the other for any given application.

VERTICAL VERSUS SPIRAL OR HELICAL FURNACE WALLS

The principal concern with a sliding-pressure supercritical-pressure design is the requirement for once-through operation. The mass flow in the furnace-wall tubes must be sufficiently high to avoid overheating or departure from nucleate boiling (DNB) while generating steam at subcritical pressures, and to avoid excessive metal temperatures and uneven steam outlet temperatures when operating at supercritical pressure at higher boiler loads.

To accomplish these objectives, the spiral-wall design is used for smaller size units. The principle of the spiral- or helical-wall furnace is to increase the mass flow per tube by reducing the number of tubes needed to envelop the furnace without increasing the spacing between the tubes. This is done by arranging the tubes at an angle and spiralling them around the furnace. For instance, the number of tubes required to cover the furnace wall can be reduced to one half by putting the tubes at a 30-degree angle (Fig. 24). Note that the centerline spacing or pitch (P) is made the same as on a vertical wall to prevent fin overheating. Additionally, by spiralling around the furnace, every tube is part of all the walls, which means that each tube acts as a heat integrator around the four walls of the combustion chamber (Fig. 23).

The spiral-wall concept thus addresses two major challenges of the full-sliding-pressure supercritical-pressure boiler:

■ Achieving the required mass flows to avoid overheating and excessive metal temperatures by reducing the number of tube circuits

■ Minimizing differences in tube-to-tube heat absorption by exposing each tube to all four furnace walls

Spiral-wall furnaces have been in operation in Europe for many years and have given satisfactory performance, the majority of them being used with the 2900-psig (20 MPa gage) turbine cycle.

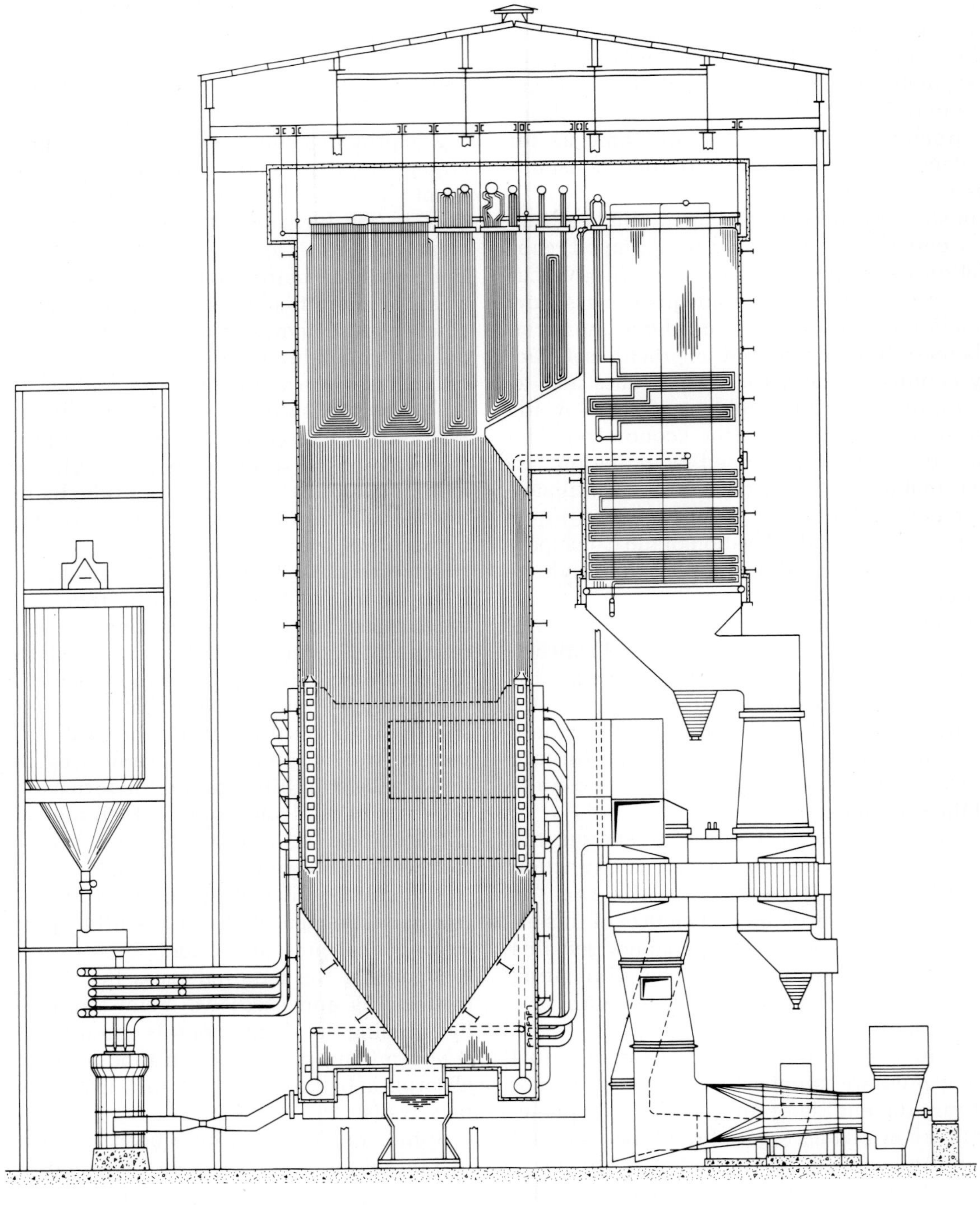

Fig. 22 **C-E supercritical-pressure coal-fired steam generator for full sliding-pressure operation**

There is one major performance penalty with spiral-wound furnace designs. Because of the high mass flow, the pressure drop in the lower furnace walls is generally much higher than for conventional supercritical or subcritical units, which increases the boiler feed-pump power requirements.

Because the furnace-wall tubes are at an angle, the furnace-wall support system is more difficult to build. The load must be transmitted through the fins between the tubes by means of weld attachments and tension strips; consequently, the spiral-wound furnace is more expensive to manufacture and erect than a vertical-tube unit. There are typically four

times as many tube-to-tube butt welds in the furnace walls due to the spiralling arrangement. It is customary with this design to revert to vertical tube construction in the upper portions of the furnace where the heat-absorption rates are lower. This requires the use of an intermediate header or bifurcated/trifurcated sections of tubing, which further multiplies the number of butt welds. Spiral-wall configurations also entail difficult tube routing around all openings for the firing equipment in the lower furnace.

As an alternative to the spiral-wall design for larger-size steam generators, C-E offers a tangentially fired unit with vertical tube walls for ease of fabrication, erection, and maintenance. As described in Chapters 12 and 13, a stable fireball is formed in the center of the furnace with tangential firing, with essentially equal distribution of the lateral heat absorption on all furnace walls. Unbalances are minimized and lateral heat-absorption patterns are predictable over the entire load range.

Rifled tubing is used in the furnace walls as in the Controlled Circulation design, to avoid overheating or DNB at subcritical pressures. As mentioned earlier, C-E has characterized the behavior of rifled tubing for this application by testing different tube diameters and configurations and subjecting full-size tubes to the heat

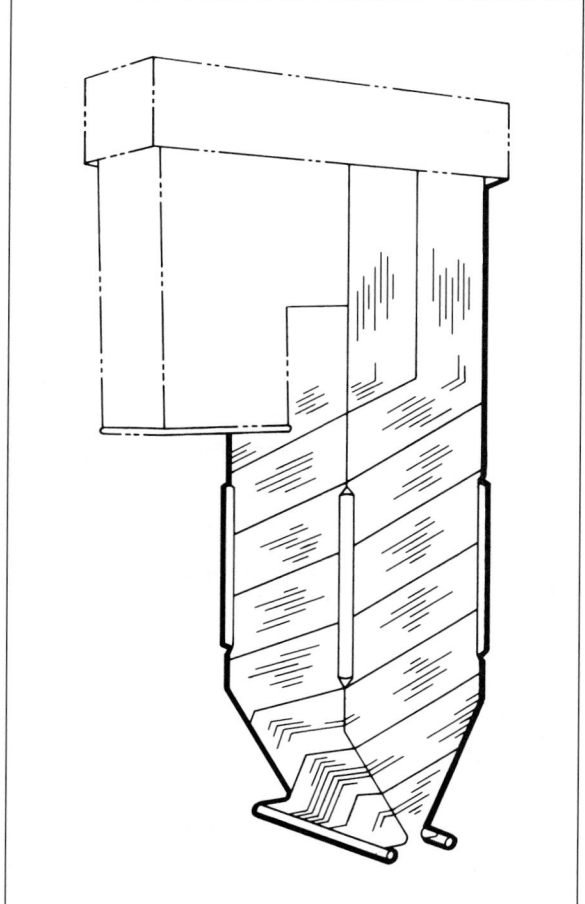

Fig. 23 **Spiral-wall furnace for supercritical pressure**

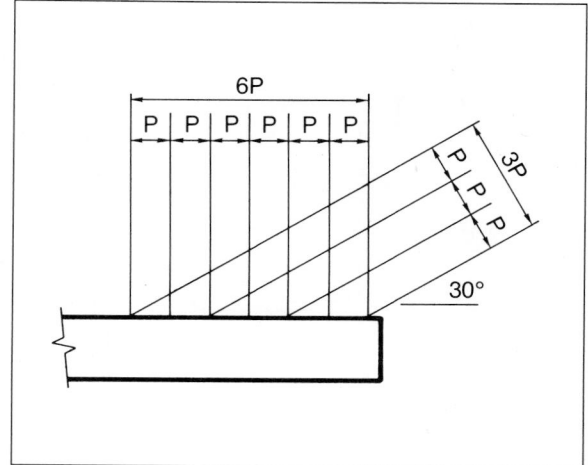

Fig. 24 **Basic principle of spiral-wall furnace**

flux rates, pressures, and mass flows associated with these designs.[2]

The vertical-wall design has individual tube orifices to control the heat absorption in each circuit, in the same manner as has been used in Combined Circulation units. Analyses based on operating experience and design practices for Combined Circulation units have proven that satisfactory temperature differentials throughout the entire operating load range are achieved. Thus, the two problems—of DNB (tube overheating) and waterwall steam-temperature unbalance—can be addressed in a vertical-wall design by employing a tangential firing system and rifled tubing.

Start-up Systems

Both boiler types are started up with sliding pressure and sliding temperature, and dry super-heaters. In the example shown in Fig. 25A, the boiler is equipped with a heat-recovery system effective at start-up and in low-load operation. In this load range, which is run with a minimum feed rate of 30 percent, the surplus water from the separator passes through a start-up heat exchanger into the feedwater tank. During pressureless start-up after a longer shutdown, in the first phase the separated water must be led into an expansion vessel until the pressure in the separator reaches the level in the feedwater tank.

In the second example, Fig. 25B, a recirculation system is used simultaneously as a start-up and low-load system. The surplus water occurring in the separator during start-up, due to the difference between the minimum feed and saturated steam generation or to the water swell, is recirculated to maintain minimum flow for furnace-wall protection.

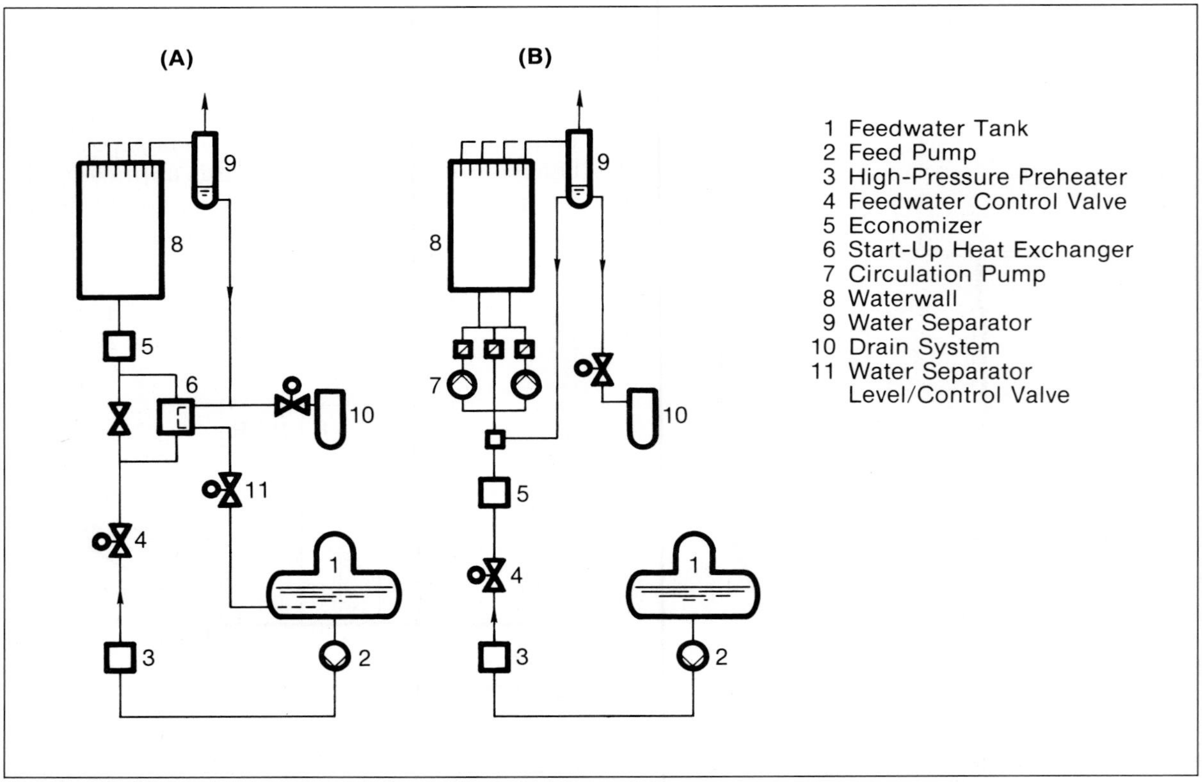

(A)

(B)

1 Feedwater Tank
2 Feed Pump
3 High-Pressure Preheater
4 Feedwater Control Valve
5 Economizer
6 Start-Up Heat Exchanger
7 Circulation Pump
8 Waterwall
9 Water Separator
10 Drain System
11 Water Separator
   Level/Control Valve

Fig. 25 Comparative start-up systems: full sliding-pressure supercritical-pressure units

Though the starting losses on the water side are quite different in the two instances, they are an order of magnitude lower than the starting losses on the steam side (which would arise from turbine bypass operation) and are the same for both examples.

HEAT RATE—THE WHY OF SUPERCRITICAL PRESSURE

In Chapters 1 and 5, we pointed out the advantages in thermodynamic cycle efficiency to be gained by using high steam pressures and temperatures. The data in Table III give some insight into the relative improvement of net plant heat rate with several of the available cycles. Large incremental decreases in plant heat rate (improved efficiency) are very difficult to achieve. Increasing the superheat temperature of a sub-critical unit to 1050°F (565°C) in lieu of 1000°F (540°C) provides only 0.8-percent improvement in the heat rate. Increasing pressure to 3500 psi (24 MPa) at 1000°/1000°F, yields an improvement of 1.5 percent, and the double-reheat ramp cycle improves 4.1 percent above 2400 psi (17 MPa) at 1000°/1000°F.[3]

Many factors affect plant heat rate besides the cycle itself, such as load regime, fuel, condenser temperature, and steam-generator exit-gas temperature. These and other items noted in Chapters 1, 5, and 6 must all be considered when selecting a pressure and temperature.

A detailed history of the development of the once-through steam generator and large supercritical units is given in Chapter 25 of the 1966 edition of this text.[4]

DESIGN FEATURES FOR VARIABLE-LOAD OPERATION

As pointed out in Chapter 1 and elsewhere, it is extremely important for large pulverized-coal-fired steam-generating equipment to have the capability to operate satisfactorily as the load varies widely.

Traditionally, electric utility companies

purchased new boiler additions that were the largest and most efficient equipment available. These new units were base loaded for the first several years of use. But as capital became more difficult to acquire and requirements for emission control equipment increased, this pattern gradually changed.

As more of the older base-loaded plants are relegated to cyclic or low-load-factor operation, it is no longer valid to assume that new, large units are the most cost-effective to operate. On one hand, new units are designed with features for operating in the two-shifting and other cycling regimes. On the other hand, they may have parasitic environmental control equipment that impacts negatively on their overall plant efficiency. In any case, there is a need for maximum system flexibility and using the older equipment for peak loads is not completely reliable because there are strict limits to the cycling capability of any boiler not specifically designed for that kind of service. This section discusses such design limitations.

DEFINITION OF CYCLING

In the past, a conventional cycling unit (either an oil- or natural-gas-fired peaking unit) was described as one designed for rapid rates of load increase and a significantly larger number of start-up and shutdown cycles compared to conventional base load-operation.

A *coal-fired cycling boiler* is one that can operate for long periods of time at reduced capacities ($\cong$ 20 percent of MCR) and reduced pressures (such as 1000 psig) while burning only coal. Such a unit is also capable of base-

Table III. Heat Rate Improvement —Net Plant—

Temperature, °F	2400 PSI	3500 PSI
	——Btu/kWhr——	
1000/1000	9,000	8,860
1050/1000	8,930	8,800
1000/1025/1050	——	8,630

load operation for extended periods. Operation (Fig. 26) is further defined.

■ Base-load mode—The unit is entirely base loaded.

■ Weekend mode—The unit is base loaded during weekdays and removed from service each weekend. In this mode, there is a cold start-up at the beginning of each work week.

■ Cycling mode—The boiler plant operates weekdays at full load during the day and at minimum load ($\cong$ 20 percent MCR) at night. It may be removed from service each weekend. This method of operation may also involve fast load changes. [5]

■ Two-shifting mode—The boiler plant operates weekdays at full load and is taken off line every night for 8 to 10 hours. In addition, the unit is removed from service each weekend. After an 8- to 10-hour nonfiring period, start-ups on weekday-morning are usually categorized as "warm" starts.

To properly consider these operational modes during the design stage, it is necessary to take into account the effect of this type of boiler/turbine loading on furnace sizing, pulverizing and firing systems, operational control systems, turbine start-up systems, and water treatment. [6]

DESIGNING COAL-FIRED BOILERS FOR VARIABLE LOAD

The performance and cost of equipment for boilers dedicated to variable-load service are greatly influenced by the sizing of the furnace and the selection of the pulverizers.

FURNACE SIZING

In recent years, furnace sizing requirements for coal-fired boilers have become very conservative. This trend occurred to permit furnaces to handle the worst possible ash-content coal that might become available at a particular

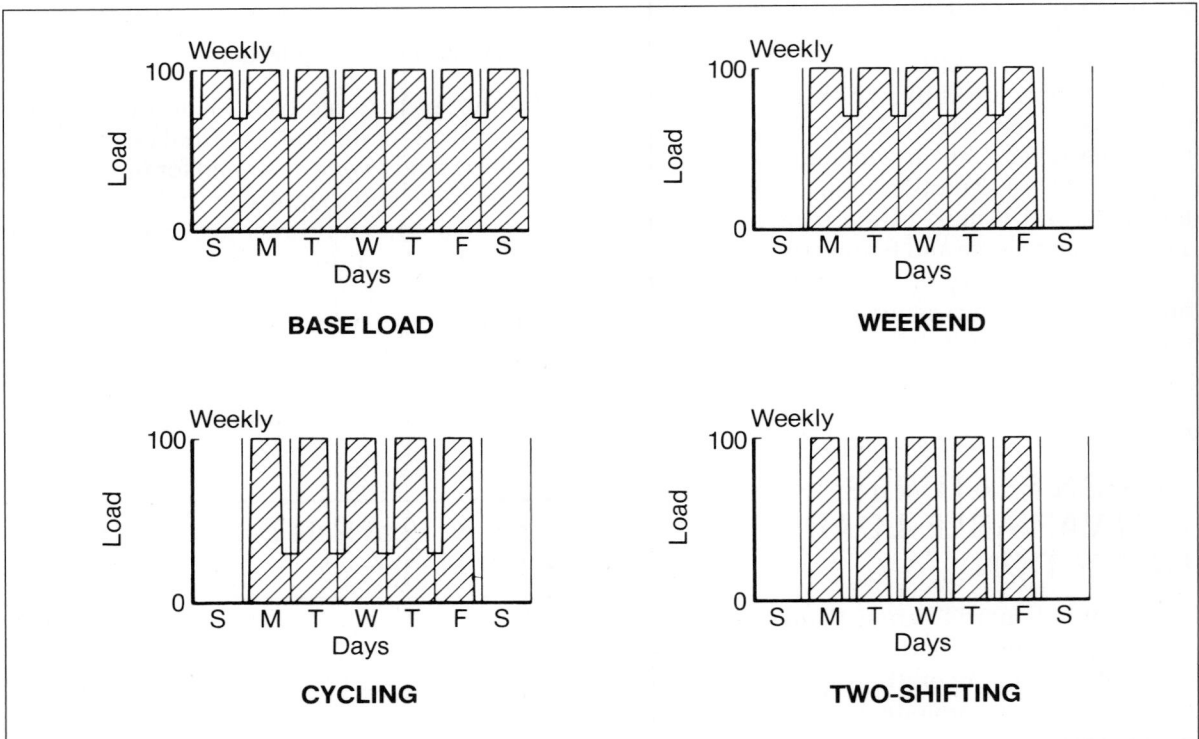

Fig. 26 **Modes of variable-load operation**

site. When a conservatively designed furnace is supplied with coal that produces gas-side conditions much cleaner than anticipated by the design of the furnace, the result can be a lower-than-design steam temperature or reduced temperature control range.

Additionally, several days to several weeks normally are required to "season" the furnace of a coal-fired unit so that the unit can produce design outlet steam temperature(s). When load is reduced substantially, or when the boiler is taken off the line, the furnace sheds its accumulation of ash. After restart, additional time is required for the furnace to season.

PULVERIZER SELECTION

Designing a boiler for cycling operation requires special attention to select and properly size the pulverizing and firing-system components. Unless these components are flexible and reliable over a wide range of conditions, the boiler capability to cycle will be severely limited. Pulverizing equipment must be able to deliver fuel to the unit at low loads, as well as respond to the usual considerations of capacity, fineness and turndown. The alternative is to use large amounts of some stabilizing fuel such as oil or natural gas.

To obtain maximum flexibility and reliability, it is advantageous to use a greater number of pulverizers, of lower capacity, rather than use a smaller number of larger capacity pulverizers, thereby limiting the turndown ratio. The larger number of pulverizers, however, will generally increase initial capital costs.

Primary Air Temperature to Pulverizers

The effect of the hot primary-air temperature on pulverizer performance is well known. Particularly during start-ups and low loads, primary air temperature can be a factor in the operation of the firing equipment. While a problem of low air temperature might be alleviated somewhat on warm restarts following overnight shutdowns, cold starts still are required after weekends and other outages. Therefore, in the selection and arrangement of air-heating equip-

ment, special consideration should be given to obtaining hot air as quickly as possible. This can be done either by directly firing supplemental fuel into the ducts to the pulverizers or preheating the air with steam heaters.

One factor that has a definite effect on the start-up characteristics of the boiler is the purge airflow requirements. Although purge airflow is essential, it acts to cool furnace walls and reduce steam temperature; therefore, every attempt should be made to assure light-off and continued operation after the *first* purge.

UNIT START-UP

Start-up rates for generator-turbine units become increasingly important when system operation departs from base loading. If a unit start-up from a cold condition takes eight hours, but the unit is started only two or three times a year, there is little economic impact from using the more expensive warm-up fuel and the direct labor cost also is minimal. But warm-up and stabilizing-fuel costs can be considerable if the unit is continually two-shifted or cycled at low loads.

The designs described in the following section can assist in matching superheat and reheat temperatures, and pressure ramp, to the turbine start-up requirements.

SUPERHEATER TEMPERATURE AND PRESSURE CONTROL

For a hot restart following an overnight outage, high-enthalpy steam is required. For a cold restart following a weekend outage, low-enthalpy steam is preferred to reduce thermal shock in the first stages of the high-pressure and intermediate-pressure turbines.

Although the steam-temperature control systems as designed for base-load operation generally have been adequate for such cycling, a simple start-up system can provide the necessary flexibility in steam temperature control. Fig. 27 shows an arrangement of valves and piping connecting the backpass ring header of a subcritical-pressure drum-type unit to the condenser. The arrangement is sized to pass ap-

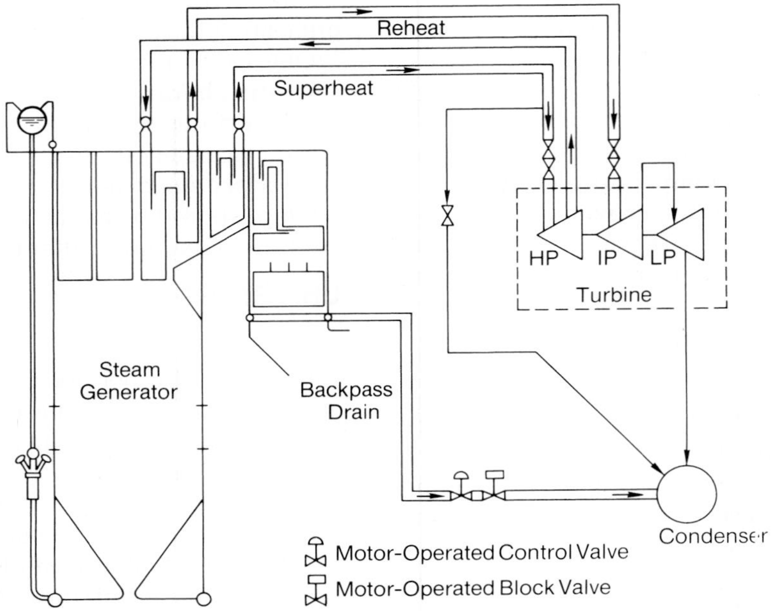

Fig. 27 Backpass drain system for superheater temperature and pressure control, drum-type unit

proximately 5-percent flow at 1000 psig. This system controls pressure rise on start-up, permitting increased firing rates to achieve higher superheat and reheat temperatures. The motor-operated steam-drain bypass valve also preheats turbine leads *before* synchronization. A spray desuperheater in the line from the bypass valve to the condenser provides for thermal protection of the condenser.

REHEAT TEMPERATURE CONTROL

When high primary and reheat steam temperatures are required for hot-start conditions (to match turbine metal temperatures), valved connections from the saturated steam circuits to the reheat outlet can be added to the backpass drain system.

Figs. 28 and 29 illustrate typical boiler start-up conditions for an overnight shutdown of eight hours and a weekend outage of 55 hours, on a high-subcritical-pressure Controlled Circulation boiler. From light-off to synchronization, the unit can be fired at a rate to produce gas temperatures necessary to develop the proper superheat temperature. At the same time the gas temperature at the first section of reheater is limited to approximately 1000°F (or 540°C) until steam flow through the reheater is established. The gas-temperature probe in front of the reheater monitors the gas temperature.

ALLOWABLE START-UP AND LOAD-CHANGE RATES

Disregarding any firing limitations that may exist, the boiler component that limits start-up rate on Controlled Circulation units is the final

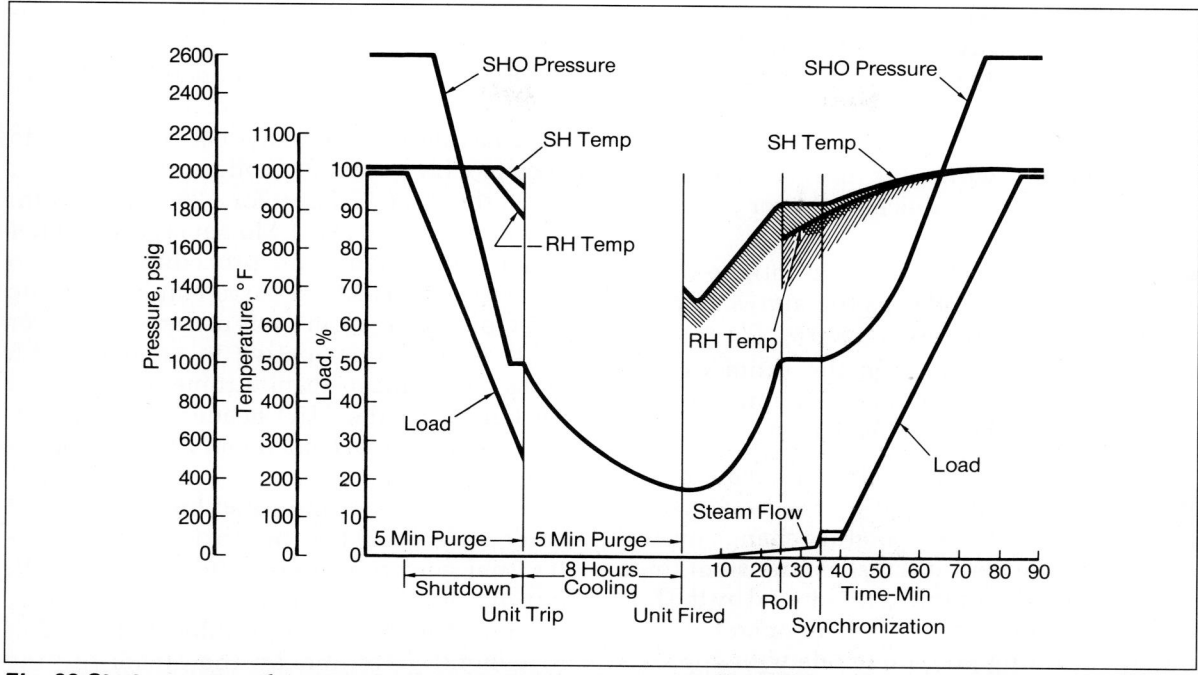

Fig. 28 Start-up curves for operation of a Controlled Circulation® unit following 8-hour shutdown. Hatched areas indicate temperature control achievable with steam-temperature matching systems.

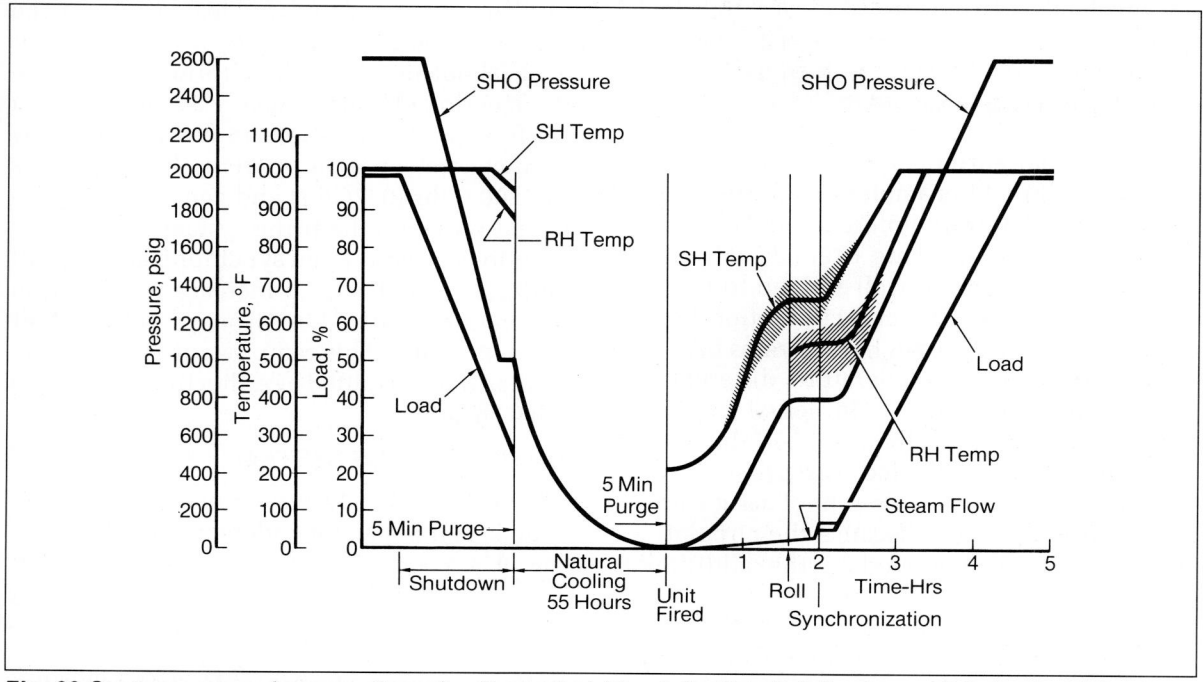

Fig. 29 Start-up curves for operation of a Controlled Circulation® unit following a 55-hour shutdown. Hatched areas indicate temperature control achievable with steam-temperature matching systems.

superheater outlet (SHO) header. There are several reasons why this header, rather than the thicker-wall steam drum, has such an effect on start-up rate.

■ SHO headers experience both fatigue and creep damage, while the drum experiences only fatigue damage due to its lower operating temperature.

■ There is a great difference in fatigue life at 1000°F and 680°F (the temperatures of SHO header and drum plate, respectively).

■ The temperature change the drum experiences on a typical hot start is approximately 400°F less than the SHO header.

Controlled Circulation drums have a nominal ramp-rate limit of 400°F/hr and a fast transient limit of 150°F temperature change during any 15-minute period. This fast transient limit is extremely important, as illustrated by the following example. If the unit is operated in a modified sliding pressure mode with approximately 1500 psig drum pressure at 50 percent load, a rapid increase from 50 percent to 100 percent load will increase the saturated temperature from approximately 600°F to 680°F, or a change of 80°F. During such transients, the drum can tolerate the rapid 80°F change an unlimited number of times.

The internal shroud in Controlled Circulation drums (Fig. 8) promotes heat transfer that is even around the circumference and along the length of the drum. This even heat transfer minimizes through-the-wall and top-to-bottom thermal gradients. Thermal-circulation boiler drums cannot tolerate such ramp rates because of the drum metal temperature differentials that would be generated. In actual practice, they are limited to less than one half the allowable rate of a Controlled Circulation unit. (Thermal-circulation units cannot use an effectively shrouded drum because the circulation system cannot accommodate the additional pressure drop.)

Although the superheater outlet header (SHO) usually is the limiting *boiler* component on Controlled Circulation units, it would be a mistake to assume that the SHO header always limits the boiler/turbine during cyclic operation and start-up. Large reheat turbines have start-up and loading rates compatible with reasonable cyclic life expenditure (CLE).

C-E has developed analytical procedures to generate CLE curves for boiler pressure parts. Fig. 30 shows a CLE curve for a 20″ OD, 4.25 in. wall-thickness, $2^{1}/_{4}$ Cr-1 Mo superheater outlet header. The calculation procedures used to construct the SHO/CLE curve employ a cyclic elastic/plastic creep analysis of the SHO header tee. It is very informative to compare CLE curves for a typical steam turbine (Fig. 31) and CLE curves for the SHO header tee (Fig. 30). The SHO header has much more cyclic capability than the turbine.

The start-up rates indicated for an overnight shutdown (8 hours) are approximately 55 minutes from synchronization to full load (Fig. 28). Start-ups after weekend or holiday shutdowns, in contrast to overnight shutdowns (Fig. 29), are generally limited by the steam-turbine thermal stresses.

When lighting off a drum-type Controlled Circulation unit and raising the steam pressure, the saturated steam temperature increases about 100°F (about 55°C) from first fire to synchronization, a period of approximately 30 minutes for a coal-fired unit. This time is not restricted by steam-generator pressure parts as the ramp rate is only 200°F/hr (95°C/hr) and can be readily increased provided the firing system and turbine conditions permit. From synchronization to full load in 55 minutes is an increase in saturation temperature of 130°F (70°C) or a rate of 142°F/hr (about 80°C/hr). This ramp rate is significantly less than that permitted by either the steam drum or SHO header.

TURBINE BYPASS SYSTEM

The turbine bypass system permits operation of the steam generator independent of the turbine at any rating up to the bypass capacity. This permits turbine start-up with essentially zero temperature mismatch. Bypass systems such as shown schematically in Fig. 32 are usually rated for 50 to 100 percent bypass flow under full-pressure conditions. The major

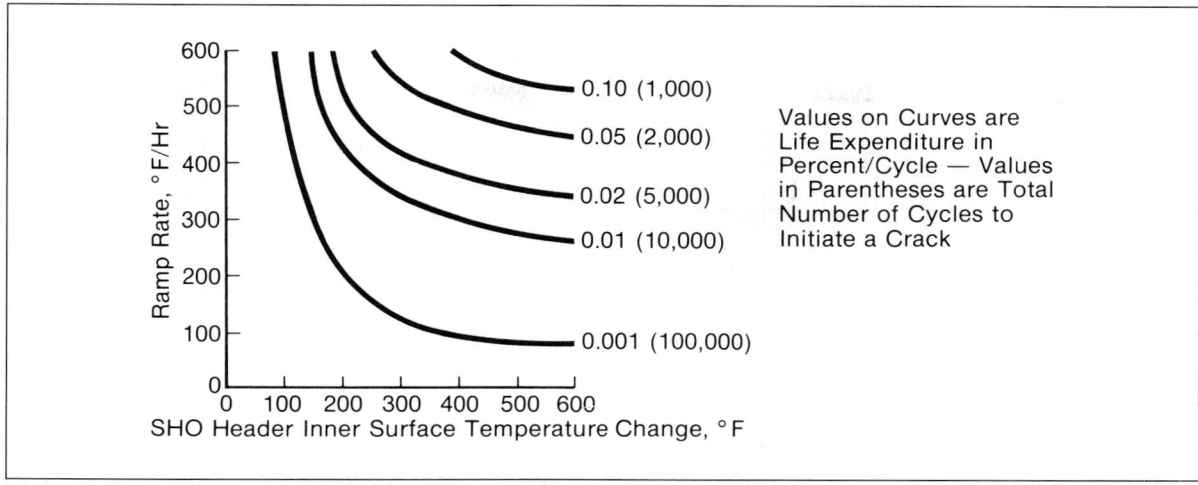

Fig. 30 Cyclic life curves for typical superheater outlet header

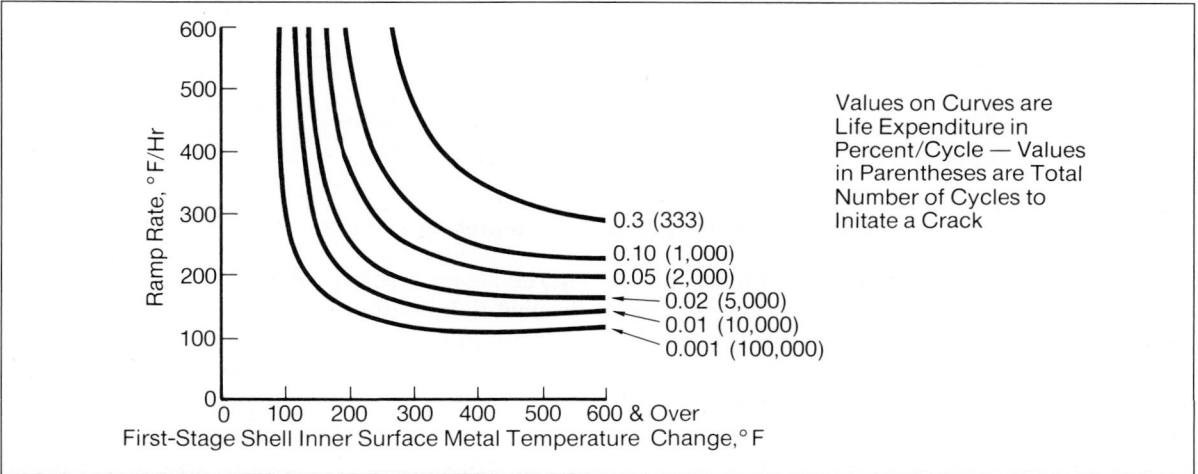

Fig. 31 Cyclic life curves for high-pressure steam turbine

advantages of a bypass system are its

- flexibility of operation
- ability to hold the generator output during start-up without undesirable turbine cooling
- ability to recover following a load rejection before restarting or reloading the turbine
- ability to match turbine metal temperatures on hot restart

Also, there is evidence that exfoliated superheater or reheater material which bypasses the turbine during start-up greatly reduces exfoliation damage to turbine blading.

The disadvantages of a turbine bypass system are

- increased plant cost
- complexity of control
- additional valve maintenance
- possibility of turbine or condenser damage due to malfunction or failure of components in the bypass system
- increases in plant heat rate because of greater condenser heat loss during periods of bypass system operation

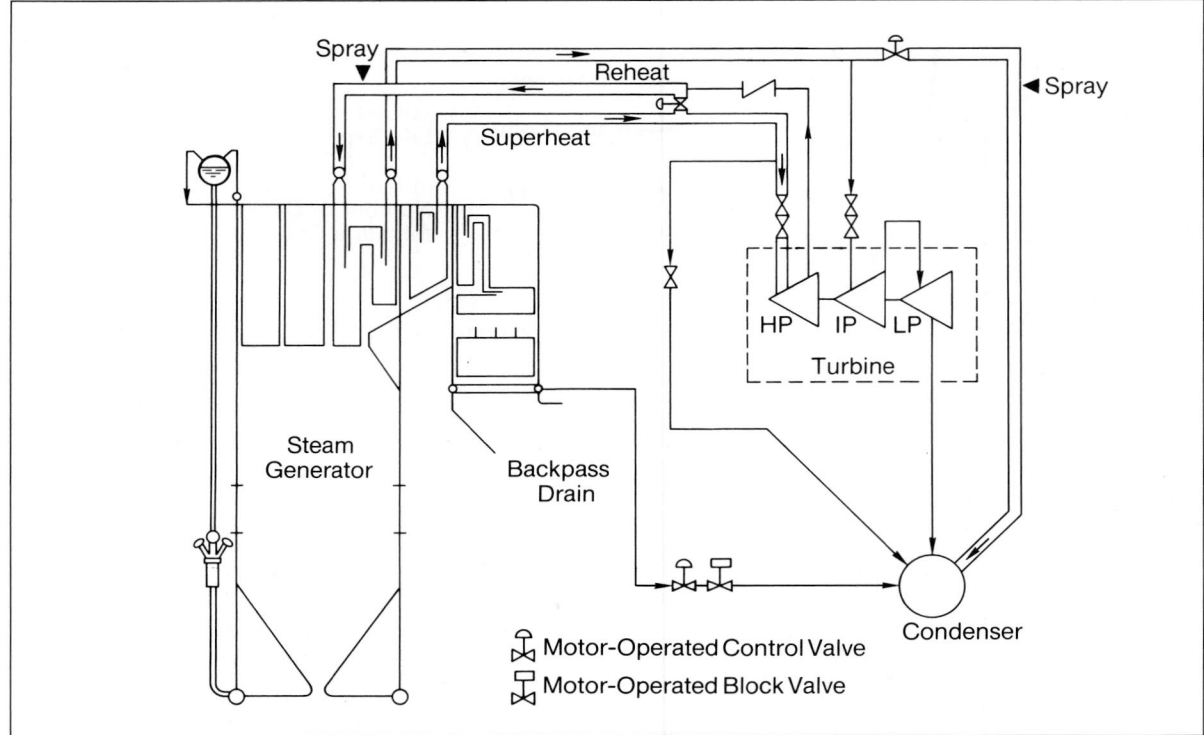

Fig. 32 **Steam-turbine bypass system, drum-type subcritical-pressure unit**

The temperature-matching potential of a bypass system depends upon its sizing; maximum steam temperature capability can be found from the steam temperature curves of the boiler at the maximum rating of the bypass system. If the desired temperature matches between the superheat and reheat steam and the turbine metals can be achieved with other start-up techniques, such as a temperature matching system, no additional saving in start-up time on a hot start can be achieved with a bypass system.

Supercritical units require a start-up system to protect the furnace walls during start-up and low loads. These systems then function as turbine bypass systems to varying degrees depending on their design.

SLIDING-PRESSURE OPERATION

The requirement for daily cycling and/or two-shifting can create undesirable thermal stresses in the steam turbine. One operating regime that is useful in reducing thermal stresses is *variable* or *sliding-pressure* operation.

To control load on a steam turbine, the pressure level in the first stage must be varied. Base-loaded units of traditional design provide constant pressure at the boiler outlet. Throttle valves at the turbine inlet are used to vary pressure and thus load.

This mode of operation generates large differences in temperature inside the first stage of the turbine because of the throttling losses at the inlet and the outlet (Fig. 33). A boiler designed to operate at variable pressure over the load range can match the requirements that the turbine demands. With sliding pressure, the differences in temperature are essentially eliminated because there are no throttling losses.[2]

The principal advantages of sliding-pressure operation are:

1. Minimal variation of first-stage shell temperature of the turbine under varying load condi-

tions. By reducing the temperature difference that the first-stage components must endure during significant load changes, operating constraints are minimized.

2. Improved overall power-plant heat rates considering boiler-turbine, feed pump, and other auxiliaries. Fig. 34 indicates the effect on turbine heat rate of the various operating modes. It can be seen that the two hybrid pressure programs offer better heat rates over the load range than the other alternatives.

3. During start-up, the reduced steam pressure results in smaller heat-transfer coefficients and correspondingly lower thermal stress levels for the same temperature differential between steam and metal components.

4. Extended control range (full steam temperature at lower loads) of both the primary and reheat steam temperatures (Fig. 35).

5. Improved steam flow distribution at lower loads in the superheater and reheater because of the higher specific volume of the steam at lower pressures.

6. The reduced pressure level at lower loads unloads all cycle components between the feed pump and high-pressure turbine, thereby prolonging life span and reducing auxiliary power consumption.

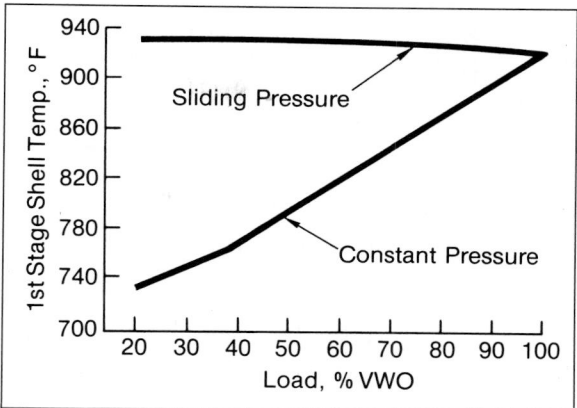

Fig. 33 Typical steam-turbine first-stage shell temperature

As their main goal these various start-up methods protect the turbine. The steam generator must also handle the thermal loads and ensuing thermal stresses that result from temperature differences and gradients imposed by start-up and shutdown, cyclic operations, and the heat-absorption process. As previously emphasized, Controlled Circulation furnace-wall systems provide protection throughout start-up and shutdown, pressure raising, and full-load operation by insuring positive water flow to all furnace-wall circuits independently

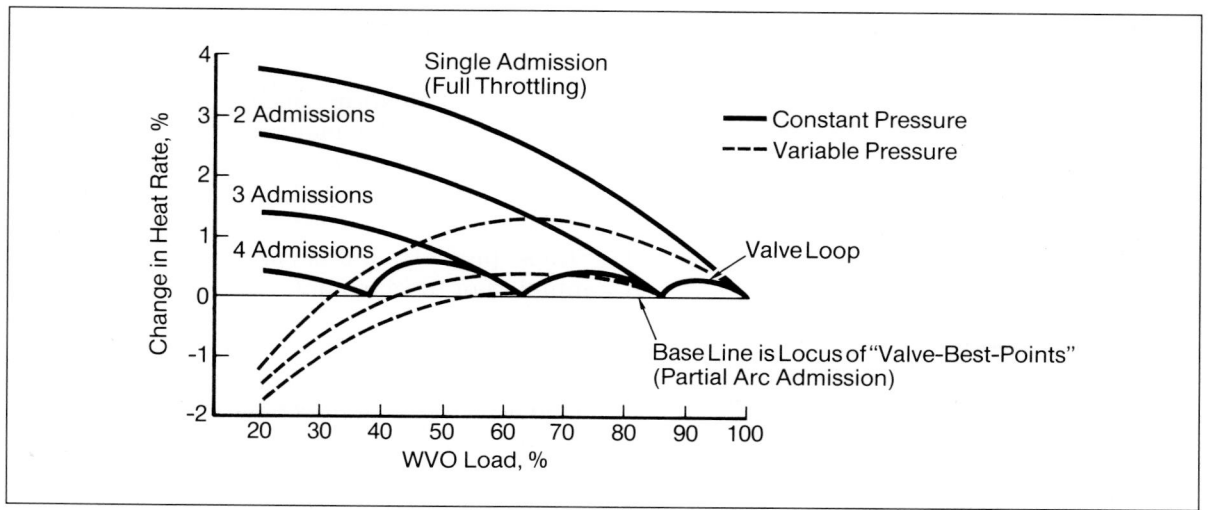

Fig. 34 Effect of admission modes and throttle-pressure programs on heat rate—typical 2400 psig (16.6 MPa gage), 1000°/1000°F (540°/540°C) turbine with 7-heater cycle and turbine-driven boiler feedwater pump

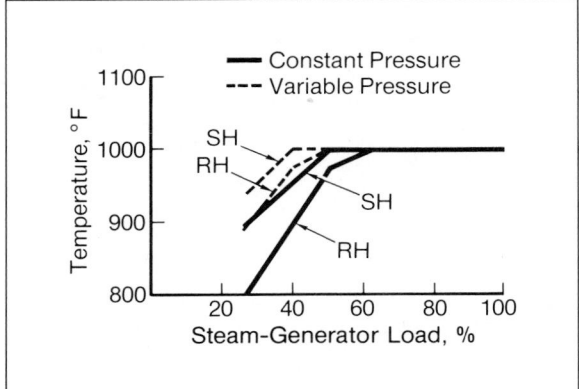

Fig. 35 Typical superheat and reheat temperatures in constant- and variable-pressure operation

of heat-input rate, thereby permitting accelerated start-ups and cool-downs.

Finally, regardless of whether variable pressure is used during normal operation, it can be advantageous for start-up and shutdown by providing better turbine temperature matching, especially with units subjected to cycling and two-shifting. It is normally unnecessary to go below half the maximum operating pressure to obtain the benefits of temperature matching.

**SUBCRITICAL-PRESSURE
SLIDING-PRESSURE OPERATION**

As an example, choose a boiler-turbine combination with nominal steam conditions of

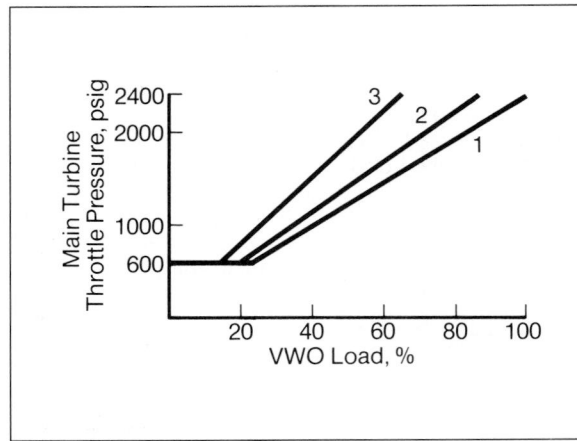

Fig. 36 Sliding-pressure programs for subcritical-pressure drum-type boilers

2400 psig, 1000°F/1000°F (16.6 MPa gage, 540°C/540°C). The turbine has a partial-arc-admission first stage with 4 nozzle groups, each controlled by a separate valve. Three different operating-pressure programs are considered (Fig. 36). They are:

■ full variable pressure, with turbine control valves wide open

■ constant pressure when on control valve No. 4; variable pressure below the point where No. 4 valve closes

■ constant pressure when on valves Nos. 3 and 4; variable pressure below point where No. 3 valve closes

Operating with the first pressure program has the advantage of essentially eliminating first-stage temperature changes inside the turbine—as long as the initial and reheat temperatures are held constant. However, the response of the boiler-turbine combination to changes in load demand is relatively slow and may not satisfy the power-system control requirements. Some operating companies prefer to have the turbine speed governor contributing to system control, and therefore choose a hybrid pressure program such as shown by curves 2 or 3 in Fig. 36. This overcomes the main disadvantages of full variable-pressure operation, while still retaining most of the benefits.

**SUPERCRITICAL-PRESSURE
SLIDING-PRESSURE OPERATION**

The first design condition that must be considered is the mode of variable-pressure operation versus load.

Fig. 37 shows a series of typical variable-pressure programs for supercritical boilers. These programs are determined for a different number of turbine control valves in operation. Operating pressure program Number 1 reflects the maximum number of control valves in operation, and this is essentially constant-pressure operation for a typically base-loaded supercritical unit. Operating program Number 5 shows full variable pressure, with the turbine valves wide open.

Operating in accordance with program Number 5 has the advantage of effectively eliminat-

ing first stage temperature changes inside the turbine, as with subcritical sliding-pressure operation. A hybrid pressure program is shown by Curves 2, 3, and 4 in Fig. 37. As above, this overcomes the main disadvantages of full variable-pressure operation.

Curve Number 3 in Fig. 37 shows a variable-pressure program in which the turbine control valves are throttled above 70 percent of maximum boiler load.

Fig. 21 showed a comparison between waterwall pressure and turbine throttle pressure; below approximately 60-percent boiler load, the furnace walls operate at subcritical pressure, that is, with a mixture of steam and water. Above 60-percent boiler load, the furnace walls operate at supercritical pressure. It illustrated that the potential for departure from nucleate boiling is only at the lower boiler loads where the furnace heat-absorption rates are lower. At higher loads, the furnace walls handle a single-phase fluid and DNB cannot take place. In other words, at the higher loads, the unit operates like a Combined Circulation unit.[2]

Fig. 38 shows the performance characteristics of a once-through sliding-pressure supercritical unit plotting enthalpy vs. load; a standard enthalpy vs. pressure diagram has been used with load substituted for one of the ordinates in order to give a more realistic representation of the performance of the unit.

The following should be noted:
■ The outlet of the economizer has been kept substantially below the saturation condition. Steaming of the economizer would create flow distribution problems entering the furnace walls. This is the reason why the economizer has been included in the recirculation system. For low-load operation (below 35 percent) the increased mass flow through the economizer prevents flashing and steaming, which are common problems in operating such units, particularly under fast-start conditions.
■ The furnace-wall outlet conditions are above the saturation point at subcritical pressures. Therefore, superheated steam is present leaving the waterwalls. As previously stated, this can cause temperature differentials which must be

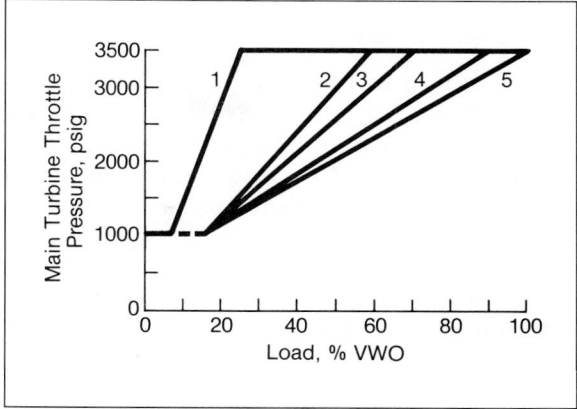

Fig. 37 Sliding-pressure programs for supercritical boilers

dealt with; on the other hand, flow distribution to the back pass and the superheater becomes much more straightforward because a single-phase fluid is present.

■ Below approximately 35 percent, the recirculation system goes into operation and the furnace-wall outlet conditions become saturated. The recirculation system is sized to stay in service, if required, up to 45 to 50 percent of maximum boiler load.

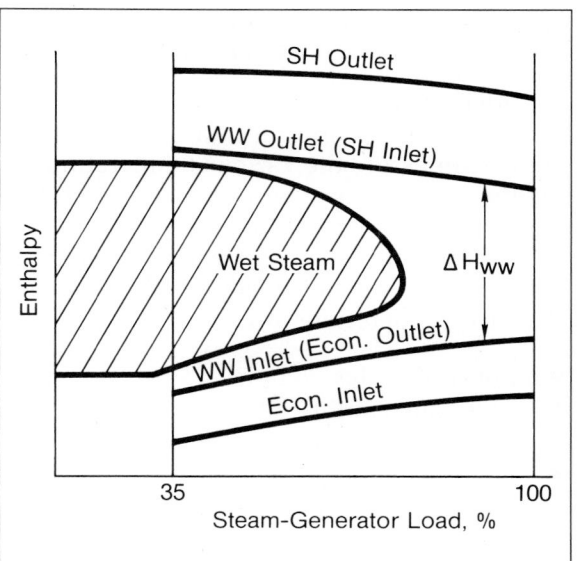

Fig. 38 Performance characteristics of once-through sliding-pressure supercritical steam generator

PRESSURE-PART DESIGN FOR CYCLIC DUTY

In addition to the design and performance requirements for which base-loaded units are selected, two additional requirements must be considered for the cycling boiler. These are rapid rates of load increase and a significantly greater number of start-up and shutdown cycles. The present rules of Section I of the *ASME Code* governing pressure-part design and fabrication do not provide specific guides for dealing with the thermal-stress and fatigue problems encountered in this type of unit. The *Code* does, however, charge the manufacturer with the responsibility of attending to and accounting for these effects.

The first step in an evaluation of this type of operation is making a conservative estimate of the number of operating cycles a plant must undergo. The second step in evolving a good design is to develop and adopt a set of design criteria to allow the intelligent assessment of all types of loading (pressure, mechanical, thermal, and cycling) to insure long-term component integrity. This subject was discussed in Chapter 6.

COMPONENT DESIGN AS INFLUENCED BY CYCLIC OPERATION

Understanding the definitions of cyclic requirements and boiler design criteria, the designer can assess the various temperature and loading conditions in terms of stresses, and can determine the capabilities of the steam-generator components to satisfy system requirements. The major factors affecting these design considerations are

- number of cycles
- heating and cooling rates
- component thickness, diameter, and material
- operating temperature level

Analytical and test methods are used to evaluate specific performance capabilities of a variety of components such as
- drums, including downcomers
- headers, including tees and nipples
- furnace-wall tubes and attachments
- buckstays
- steam/water separators
- pumps and valves
- pulverizer structure and drive train
- duct expansion joints
- hanger rods and supports

Such components are evaluated using various factors, chief being the operating pressures and operating conditions that minimize thermal stresses in the steam-turbine rotor. Similarly, it is necessary to minimize thermal stresses in all boiler components.

Expansion stresses that result from steady-state operation are relatively simple to accommodate in a design, but thermal gradients produced by temperature transients greatly complicate the design. As faster start-up and shutdown rates become necessary, thermal gradients increase and the thermal design of a component may well run counter to the required design for mechanical loads, such as pressure and dead load. Attachments to pressure parts normally are sized by the mechanical loads they carry. To reduce the size of such an attachment (so as to reduce a thermal difference) may invalidate its design. Heavy thick-walled components, such as the SHO tees or steam drums, are sized according to *pressure* requirements. Wall thicknesses, then, cannot be decreased to reduce thermal gradients and thermal stresses.

Pressure and thermal stresses combine to produce a total stress. Fig. 39 shows an ideal operational cycle in which the pressure and steam-temperature curves have the same rate for start-up as for shutdown. The stresses produced in a cylinder with the pressure and thermal loading cycle are shown in Fig. 40. Of interest is the relationship of the thermal and pressure stresses on the inside surface of the cylinder. During the entire cycle, the pressure stresses are tensile. Thermal stresses are compressive during start-up and tensile during shutdown. The pressure and thermal stresses oppose each other during start-up but are additive on shutdown.

The stress range shown in Fig. 40 determines the amount of fatigue damage represented by this cycle. The greater the stress range, the greater the fatigue damage. One possible way of reducing the stress range is to reduce pressure during a cooldown while maintaining the steam-temperature level.

The sliding- or variable-pressure mode of operation allows main steam and reheat steam temperatures to be maintained at a given setpoint lower into the load range. Although it is for this reason and others discussed earlier that variable pressure is used, there is an added benefit. During a cooldown, the pressure stresses are additive as shown in Fig. 41. Therefore, the combined pressure and thermal stresses are lower than the stresses that would have been generated if pressure had been maintained and the temperature dropped.

CONTROL CONSIDERATIONS FOR VARIABLE-LOAD OPERATION

The thermal inertia of steam generating equipment relates directly to its size. It is necessary to understand when designing operational control equipment that, combined with increases in unit size, present practice is to design for variable-load operation. On constant-pressure units, it is only necessary to deal with the energy-storage changes associated with the changes in steaming rate. But, as previously discussed, turbine considerations virtually demand that variable-pressure operation be

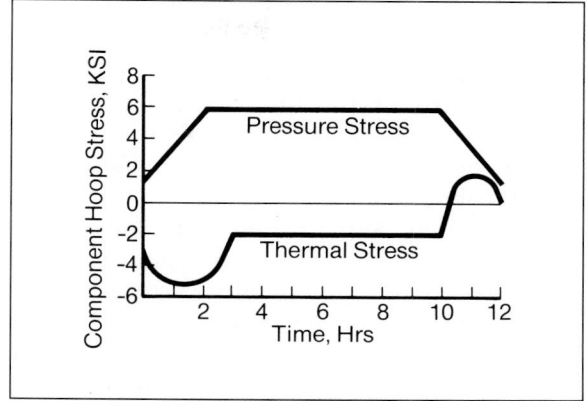

Fig. 40 Pressure and thermal stress versus time

used on cycling units. This requirement places a special burden on the steam-generator control system, with the steam-temperature controls being the most severely affected.

The stored energy in a steam generator (metal temperature and steam enthalpy, both as a function of mass) is higher at full load than at any partial load. Increasing the load, therefore, requires temporarily raising the firing-ramp rate above the load ramp to provide the required increase in stored energy. During load increases on a variable-pressure unit, the slope of the stored energy is substantially steeper than the load output curve.

Overfiring to satisfy the energy storage requirements of the waterwalls generally results in excessive heat pickup in the superheater and reheater, which have lower storage capacity.

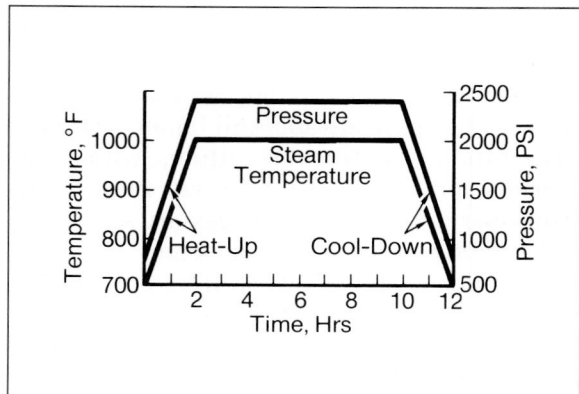

Fig. 39 Idealized operating cycle

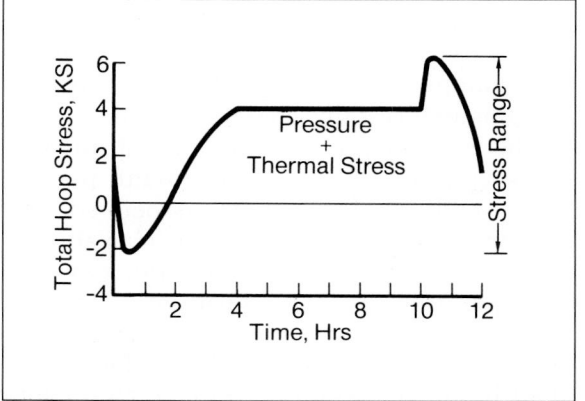

Fig. 41 Combined pressure-/thermal-stress cycle

Further complicating the situation is the fact that the distribution of heat between water-walls and superheater of a typical subcritical-pressure drum-type unit varies significantly with pressure because of changes in the vaporization heat of water and the specific heat of superheated steam as a function of pressure.

Traditional superheater temperature-control systems use cascade-type spray controls, in which the final-temperature error resets the setpoint for the desuperheater outlet temperature. This is a type of feedback control and is inadequate for even moderate rates of load change in a sliding-pressure mode. Any workable steam-temperature control configuration must include both careful timing of any overfiring and a sophisticated anticipation signal in the superheat and reheat control loops. Programming the desuperheater outlet temperature as a function of load and rate-of-change of load is one example of the use of feed-forward temperature control.

In a subcritical unit operating at constant pressure, shrink and swell of drum level occur during a load change because of the change in steaming rate. With sliding pressure, however, there is also a mass-inventory change associated with the pressure change. Feed-forward signals are used to help the drum-level controller distinguish between shrink and swell effects and feedwater/steam-flow mismatches. Regulating the speed of the boiler feed pumps is the most common control of feedwater flow rate. The reduced pressure at low loads with sliding-pressure operation requires that the stable speed control range on the pumps be significantly wider or that supplementary valve control be used.

The specific cycling mode that involves holding minimum load overnight presents some additional concerns associated with keeping supplementary-fuel use to an absolute minimum. Improved flame-scanning systems incorporating higher sensitivity scanners and/or increased quantities of scanners are mandatory. More precise control of fuel and air distribution to maintain the fuel ignition point at each fuel-admission location within the scanner viewing area is also essential.

In a two-shifting mode, the major impact on the control system is the need for more automation. Bringing in additional plant operators to control various subloops manually on every start-up may be acceptable on a base-load unit, but is not likely to be so for a two-shifting unit.

Traditional boiler-following control systems place too much of the burden on plant operators to coordinate both changing conditions of temperature, pressure and flow in the boiler and turbine. The use of a coordinated control system, with its plant-master concept, then, is virtually essential to satisfy the control requirements of cycling operation.

Using boiler start-up systems or turbine bypass systems adds an additional set of control loops that must also be carefully coordinated with the steam-generator and turbine-generator controls.

WATER TREATMENT FOR CYCLING UNITS

Any power plant operation that requires a great number of start-ups, shutdowns, and idle periods will be subject to inleakage of air, especially in the condensate and low-pressure feedwater systems. This increases the potential for preboiler and boiler internal corrosion and over-heating from deposition of corrosion products in internal tube surfaces. The mechanism by which this occurs is oxygen corrosion of feedwater heaters and interconnecting piping, plus the formation of corrosion byproducts such as iron-oxide and copper; these are subsequently transported to the boiler where they will deposit on the inside of the furnace-wall tubes. High oxygen concentrations in the feedwater and boiler water can also produce corrosion pitting, which can directly promote tube failures. Proper precautionary measures are necessary in cycling units to prevent corrosion damage.

Units for 2400 psi two-shifting and cycling operation, and *all* supercritical pressure units should have condensate polishing equipment to control corrosion-products, condenser inleakage contaminants, and make-up water impurities. Its design and operation should keep

feedwater contamination within boundaries of limiting criteria for all modes of operation. Ion exchange to process at least 25 percent of full-load condensate flow is required in two-shift drum-unit operation to process feedwater contaminants and impurities from the turbine and condenser. Cleanup of condensate will also facilitate reaching operating pressure without experiencing undue delays to control silica.

Detailed information on cycling-service pre-boiler cleanup, condensate deaeration, and condensate polishing is given in Chapter 20 and Chapter 21.

REFERENCES

[1] W. H. Clayton, J. G. Singer and W. H. Tuppeny, Jr., "Design for Peaking/Cycling" presented at the *Joint Power Generation Conference*, September, 1970 American Society of Mechanical Engineers Paper No. 70-PWR-9, New York, 1970.

[2] O. Martinez, Jr. and J. A. Makuch, "Supercritical Steam Generator Designs for Sliding Pressure Operation," *Proceedings of the American Power Conference*, 43; Chicago, IL; Illinois Institute of Technology, 1981.

[3] H. E. Burbach and G. F. Shulof, "The Cycle Option," *Proceedings of the American Power Conference*, 42: Chicago, IL; Illinois Institute of Technology, 1980.

[4] G. R. Fryling, "Combustion Engineering," Revised Edition 1966. Library of Congress Catalog Number 66-23939, 1966.

[5] L. W. Cadwallader, E. M. Powell, and J. G. Singer, "A New Look at Peaking Power," *Proceedings of the American Power Conference*, 28:413–421, 1966. Chicago, IL; Illinois Institute of Technology, 1966.

[6] H. E. Burbach, J. D. Fox, T. B. Hamilton, "Steam Generator Design Features for Variable Load Operation," *Proceedings of The American Power Conference*, 41: 561–579. Chicago, IL; Illinois Institute of Technology, 1979.
I. Fruchtman et al., "Cycling of Supercritical Power Plants in the U.S.," presented at the *Joint ASME/IEEE Power Generation Conference*, Miami Beach, FL, October 4–8, 1987; American Society of Mechanical Engineers Paper No. 87-JPGC-PWR-50, New York, 1987.

CHAPTER 8

Boilers for Process Use/Power Production

This chapter is about a wide range of fossil-fueled steam generators often designed to operate at steam temperatures and pressures that are suitable for electrical-power generation by a steam-turbine generator. But, because they can also be used solely for the processes of the several manufacturing industries, co-generatively or not, they have come to be called *industrial boilers*, a term which we use freely in this text. In fact, the versatility of most of these steam generators makes them part of every conceivable type of process-steam, steam-for-power, or cogeneration cycle.

This definition of an industrial power-plant boiler, then, is as arbitrary as previous ones have been and is based on the physical characteristics of the equipment rather than on its ultimate application: it is a stationary (not shipboard) watertube steam generator, without steam reheat; most often, it has a drum operating pressure below 2,000 psig (13.8 MPa gage), is below 100 megawatts in electrical generating capability, and is multifuel fired. Neither cycle or boiler efficiency, sophistication of operational controls, nor the level of government-legislated emission-control systems are pertinent to this definition.

Obviously, this definition does not exclude fluidized-bed boilers which, for the purpose of better delineating the technology, are covered

in the next chapter. Boilers with characteristics particular to the marine industry are found in Chapter 10. Finally, Federal legislation that redefines the relationships between central-station owners and other power generators significantly affects the incorporation of industrial boilers in combined cycles, which we have discussed in Chapter 1, and further in this chapter.

PERFORMANCE REQUIREMENTS OF MODERN INDUSTRIAL BOILERS

Industrial boilers for process-plant service are ordinarily designed to fire many different fuels, each of which influences sizing and configuration of the boiler and its auxiliaries. In addition, the primary fuel may vary over the life of the unit. Also, many manufacturing/process plants are sources of byproducts that are valuable as fuel. The need for industrial boilers to be fuel-flexible, while meeting stringent federal, state, and local emissions standards, requires special design considerations and sophisticated ancillary equipment.

Understandably, there is a preference to design industrial boilers for burning clean liquid and gaseous fuels when such are available at reasonable cost—the boilers and auxiliaries will be of minimum size and cost. Unfortunately, world demand for such prime fuels can be expected to outpace the long-term supply and will result in a decline in oil- and gas-burning steam generators. The eventual shift to solid fuels will require boilers that are not only initially more expensive, but also operationally more difficult. Thus, as more coal and other solid fuels are burned, there will be an increased need for pulverized-coal, stoker, and fluid-bed firing.

Owners and operators of industrial boilers must consider such developments and adapt unit performance to:

■ Multifuel-Burning Capabilities: During the 70's some industrial concerns had to convert boilers from firing coal to gas, to oil, and back to coal as the environmental requirements, fuel costs, and fuel availability changed. To protect an investment in the steam-generating facility, the owner may more frequently request multiple-fuel capability in new units.

■ Waste-Fuel Utilization: Rising fuel costs and the high cost to dispose of waste make refuse, biomass, waste acids, and sludge more attractive as fuels for steam generation; they provide a viable disposal alternative. A growing number of new boiler designs will use waste fuels.

■ Low Operating Costs: The cost of fuel and auxiliary power can in a short time exceed the original cost of a steam generator. Therefore, new equipment must be efficient and have a low auxiliary-power consumption.

■ Cogeneration Potential: Simultaneous generation of electrical power and process steam, as described in Chapter 1, is also common as users strive to extract maximum heat from their fuel. Industrial firms will more frequently select the more efficient high-pressure and high-temperature steam cycles to optimize the amount of electrical power and steam produced.

BOILER DESIGN AS A FUNCTION OF OPERATING PRESSURE

As Fig. 1 shows, the percentage of total heat absorbed in a boiler bank is reduced signifi-

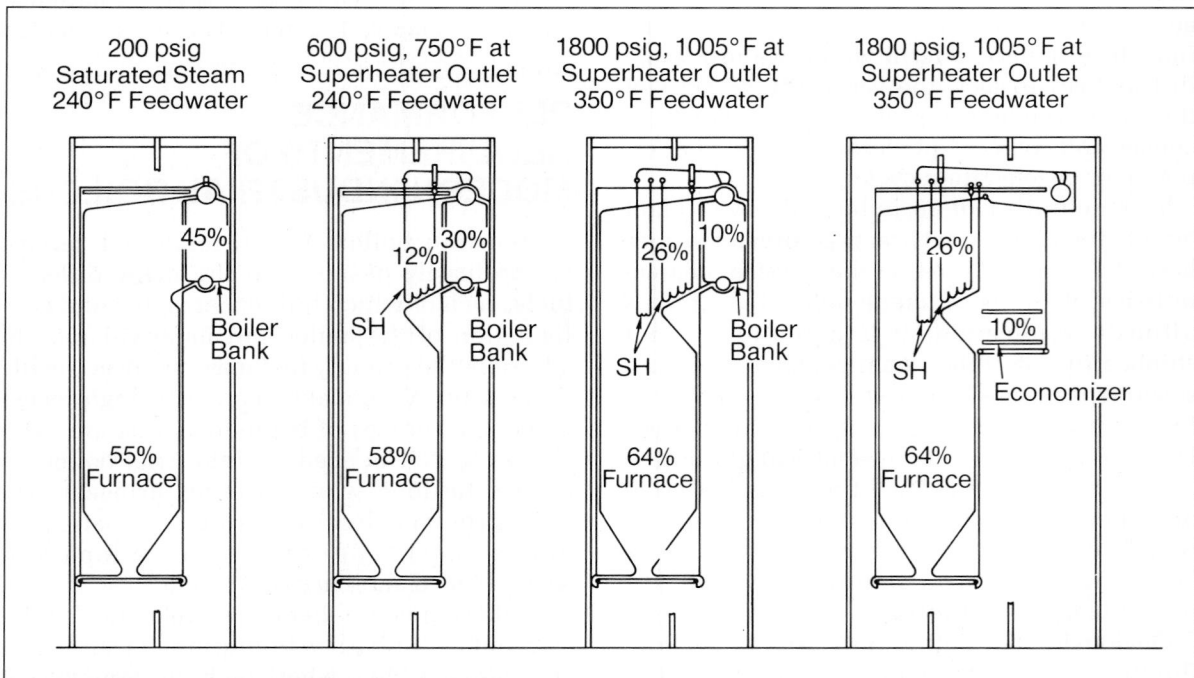

Fig. 1 **Heat-absorption distribution for pulverized-coal suspension-fired boilers**

cantly at higher pressures by two factors:
1. In the superheater, the greater heat absorption required to heat the steam to a higher temperature reduces the gas temperature entering the boiler bank.
2. The water temperature in the boiler bank is essentially at the saturation temperature corresponding to the boiler operating pressure. As the operating pressure increases, the temperature of the water inside the tubes rises thereby reducing the logarithmic mean temperature difference available for heat transfer.

DESIGN CRITERIA FOR INDUSTRIAL BOILERS

Boiler designs are tailored to the fuels and firing systems involved. Some of the more important criteria are

- furnace heat-release rates, both Btu/hr-cu ft and Btu/hr-sq ft of effective projected radiant surface (EPRS)
- heat-release rates of grates
- flue-gas velocities through tube banks
- tube spacings

Table I gives typical values or ranges of these criteria for gas, oil, biomass, refuse-derived fuels, and coal, all fired in suspension or on grates, to illustrate the differences for the fuels and not necessarily to represent specific recommendations. The furnace heat-release rates are important from several viewpoints. The proper Btu/hr-sq ft of EPRS value, along with adequate flame-to-heat-absorbing-surface clearance, will keep maximum local absorption rates within safe limits and avoid "hot spots." Also, with coal and refuse firing, this aspect of design is important to prevent excessive slag accumulations on furnace walls and convection tube surface at the furnace outlet. The furnace heat-release rate in Btu/hr-cu ft directly influences completeness of combustion. Limiting grate heat-release rates to these values minimizes carbon loss, controls visible emissions, and avoids excessive flyash.

The need to limit draft loss controls flue-gas velocities for gas- or oil-fired boilers. On the other hand, with coal and refuse firing the design gas velocities, together with gas temperatures, influence the ability of the boiler to operate continuously without shutdown.

Although convection tube spacing is not critical in boilers designed to fire gas or distillate oil, it is very important when the fuel is a residual oil, municipal refuse, or coal. This is especially so with coals that have low ash-fusion temperatures or high ash-fouling tendencies and with low-rank coals such as subbituminous coals or lignite.

COAL-FIRING SYSTEMS

Many industrial boilers are designed for less than 250,000 lb steam/hr (32 kg steam/s). For many years most of these units used a spreader stoker to fire coal because it was less expensive than installing pulverizers. In this size range, the other benefits of pulverized-coal (PC) did not warrant the increased costs. In today's marketplace, the choice is not as obvious. With the advent of tighter emission regulation, higher fuel costs, and the commercialization of fluid-bed boilers, the advantages of each system must be evaluated thoroughly before selecting a firing system.

EVALUATING STOKER AND PULVERIZED-COAL FIRING SYSTEMS

A reduced efficiency due to the carbon loss is a major factor in comparing a stoker-fired boiler to a PC-fired boiler. A properly designed PC boiler can maintain an efficiency loss due to unburned carbon of less than 0.4 percent. On the other hand, a continuous-ash-discharge spreader-stoker fired unit will typically have a carbon loss of 4 to 8 percent, depending on the amount of reinjection. The PC unit offers a lower carbon loss because of the increased combustion efficiency obtained with the finer coal particles that enter the furnace (normally 70 to 80 percent will pass through a 200-mesh screen). In contrast, the coal particles entering the furnace with a spreader stoker are much coarser. These coal particles have a 3/4-inch (about 20-mm) top size and not more than 50

Table I. Some Typical Boiler Design Parameters

Furnace	Btu/hr-sq ft of EPRS	Btu/hr-cu ft
Natural gas	200,000	N/A
Oil	175,000–200,000	N/A
Biomass	N/A	15,000–20,000
Refuse Derived Fuel (RDF)	N/A	12,000–15,000
Municipal Solid Waste (MSW)	N/A	6,000–10,000
Coal-pulverized	70,000–120,000	15,000–22,000
Coal-spreader stoker	80,000–130,000	20,000–25,000

Stoker, grate heat-release rate, Btu/hr-sq ft	
Continuous ash discharge (CAD), coal	650,000–700,000
CAD – Biomass	900,000–1,200,000
CAD – Refuse-Derived Fuel	750,000–800,000
Dump grate spreader, coal	450,000–550,000
CE/db Pusher Grate (MSW)	270,000–300,000

Flue-gas velocity, ft/sec

	Single-Pass Boiler	Baffled Boiler	Economizer
Gas or distillate oil	100	100	100
Residual oil	100	75	100
Bituminous coal			
Low ash	50–60	50	50–60
High ash	40–50	N/A	40–50
Refuse Derived Fuel	25–30	N/A	30–35
Municipal Solid Waste	20–25	N/A	20–25
Biomass	50–60	40–50	50–60

Tube spacing perpendicular to gas flow, inches clear

	Superheater		Boiler		Economizer
	Front	Rear	Front	Rear	
Gas or distillate oil	2	2	1	1	1
Residual oil	4–6	2	$1^{1}/_{2}$	1	1
Bituminous coal					
Low ash	8	3–6	$1^{1}/_{2}$	1	$1^{1}/_{2}$
High ash	10–16	4–6	2	1–2	2
RDF	6–7	3–6	$2^{1}/_{2}$	$2^{1}/_{2}$	2
MSW	9–10	4–5	N/A	N/A	$2^{1}/_{2}$
Biomass	7	$2^{1}/_{2}$	2	$1^{1}/_{2}$	$1^{1}/_{2}$

percent will pass through a ¹/₄-inch (6-mm) screen (Fig. 2).

The coal itself also affects the total fuel cost difference between these firing methods. Efficient operation of a spreader-stoker-fired boiler requires that the coal has the proper mixture of coarse and fine particles. Normally, double-screened coal is purchased to obtain the proper mixture as run-of-the-mine coal generally does not have the optimum balance of coarse and fine material.

The fine coal particles are burned in suspen-

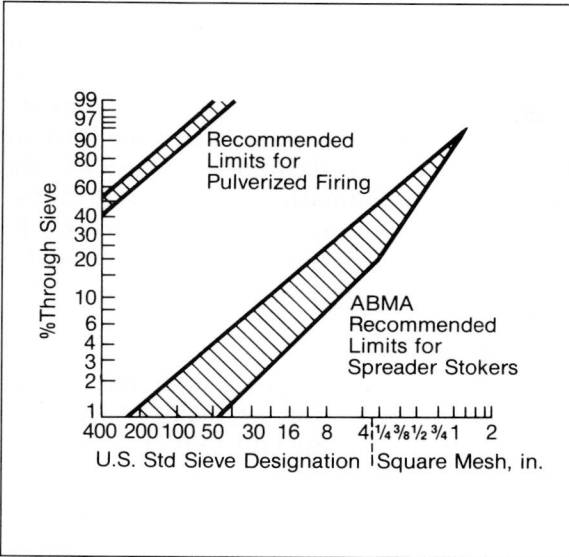

Fig. 2 Coal sizing (as delivered to the furnace)

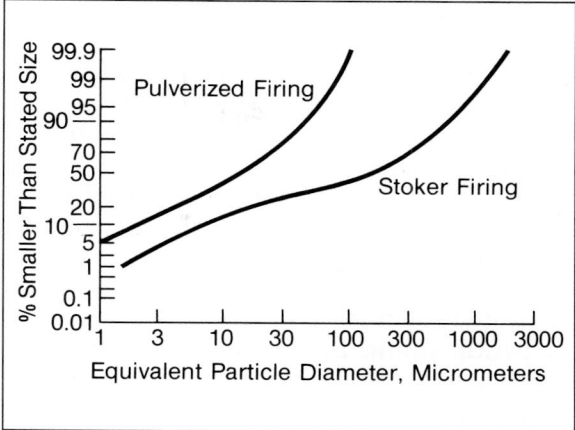

Fig. 3 Size distribution for flyash leaving the furnace

sion. If the amount of fine particles is too great, there will be excessive suspension burning which can produce

- higher stack particulate-matter loading
- furnace pulsations
- possible heat damage to the distributors
- higher carbon loss and visible emissions

If the amount of coarse coal is too great, there will be insufficient burning in suspension, with the potential for loss of flame stability.

If double-screened coal is not used, the day-to-day efficiency of operation will be reduced as the coal sizing varies from optimum. The question of whether the higher cost for double-screened coal offsets the reduced efficiency with unsized coal cannot be answered on a general basis. The answer depends on the degree of coal sizing variation, quality of operators, and boiler or stoker size.

A pulverized-coal-fired industrial boiler requires that the coal entering the mill be 1$\frac{1}{2}$-in. (not over 40-mm) top size. If the proper size is unavailable at the mine, coal of any size can be purchased and crushed at the plant site. Considerable horsepower is necessary to pulverize and transport the coal to the furnace. It is also

necessary to provide hot air to dry the coal to the proper moisture for pneumatic delivery. The need for an air heater, additional ductwork, and a more complicated control system has to be measured against the fuel savings of pulverized-coal firing.

The furnace of a PC-fired system must be designed to prevent ash from slagging on the waterwalls whereas high-ash coals with low ash-fusion temperatures can create a clinkering problem when burned on a stoker. In Ref. 1, Pollock gives a detailed comparison of annual fuel and power costs for pulverized and spreader-stoker coal firing.

EMISSIONS

Control of particulate emissions, sulfur oxides, and nitrogen oxides must also be evaluated in the comparative installed costs of different types of coal-fired equipment.

Formerly, a stoker-fired boiler had a definite advantage in controlling particulate-matter emissions. This type of firing produces large particles that can be removed easily with inexpensive mechanical dust collectors (see Fig. 3). Today, all industrial, commercial, and institutional units firing at or above 100 million Btu/hr (30 MJ/s) must meet strict particulate-emission standards. Compliance will necessitate the use of electrostatic precipitators, fabric filters, or equivalent high-efficiency removal devices regardless of the method by which coal is fired.

Likewise, federally mandated New Source Performance Standards (NSPS) for 100 million and higher Btu/hr (30 MJ/s) sources require sulfur-oxide removal equipment on all coal-fired steam generators. PC boilers with bowl mills can be expected to have less sulfur in the products of combustion than stoker-fired units. During the pulverizing process, the bowl-mill tramp-iron spout may reject up to 40 percent of the pyritic sulfur. Because pyritic sulfur may approach 60 percent of the fuel sulfur, a reduction of nearly 25 percent could be obtained under optimum conditions. With a stoker-fired unit, some pyritic sulfur may remain on the grate and be deposited in the stoker ash hopper.

Both firing systems are equally capable of meeting NSPS NO_x emission levels. In some cases, PC systems may require combustion air to be staged in various configurations to reduce NO_x to these levels, while stoker-fired boilers can use varying proportions of undergrate and overfire air to minimize production of NO_x.

FUEL FLEXIBILITY

Coal-fired boilers can be designed to burn a wide range of coals. For suspension-fired non-reheat units, tilting tangential fuel nozzles and/or spray desuperheating can control the effect of varying furnace heat absorption on the superheater. But during design, coal and ash properties for all of the coals to be fired must be considered. The major factors in designing a boiler for a wide range of coals are

■ Furnace sizing, convective surface spacing, and sootblower coverage: Coals with a lower ash-softening temperature require a larger furnace to assure the ash is in solid form before it enters the convective tube surface. A furnace that is appropriately sized for the poorest ash properties will operate successfully with the better coals. In designing for multiple fuels, it is not uncommon for one fuel to set the furnace size while another dictates the minimum allowable convective-surface spacing and sootblower requirements.

■ Pulverizer requirements: The lowest heating-value coal with the lowest grindability will determine pulverizer size. If the coals vary in moisture, the boiler convective surface (boiler bank, economizer, air heater) must provide the proper air temperature to dry the coal in the pulverizer so it is surface-dry and can be transported pneumatically to the furnace. If, for example, the moisture ranges from 3 to 35 percent, the air temperature required by the pulverizer will vary from 300 to 700°F (150 to 370°C).

■ Stoker requirements: Coals with low ash and/or moisture will tend to have high grate-metal temperatures. For such fuels, maximum allowable undergrate air temperature will be less than 300°F (150°C) and, in the worst case, minimum moisture and ash contents should be specified. High-ash coals or those with low ash-fusion temperatures can clinker and disrupt the proper flow of undergrate air for efficient combusion and grate-bar cooling. The actual physical size of a stoker is, however, far less sensitive to coal properties then is the sizing of pulverizing equipment.

■ Superheater sizing: The specific fuel and ash components determine the heat-absorption characteristics of the furnace and superheater. But generally the best fuel, which produces the lowest gas weight and consequently the lowest heat-transfer rates and LMTD, will set the superheater size. A wide range of fuel moistures will require a spray desuperheater to control final steam temperature when firing high-moisture coal.

■ Fans and gas-cleaning equipment: The highest-moisture coal to be fired in a boiler generates the largest volume of flue gas. The fans and ductwork must be designed for these conditions. If a wide variation in flue-gas volume is anticipated during normal operation, variable-speed motors or turbine-driven fans should be considered to reduce parasitic power consumption (see Chapter 14). Gas-cleaning-equipment size may not necessarily be set by flue-gas volume, but rather by the removal efficiency or the absolute level of emissions in accordance with applicable regulations, whichever is the controlling variable.

PROVISIONS FOR A WIDE RANGE OF COALS

Some special considerations can be made in the original boiler design to accommodate a wide range of coals. With a 3 to 35-percent moisture variation in the coals to be fired, a primary-air preheater may be used to preheat the mill air to 700°F (370°C) from the nominal 550°F (290°C) temperature leaving the secondary-air preheater. When firing low-moisture coals, the primary-air preheater is not used and there is a minimal requirement for parasitic cold tempering air at the pulverizer inlet. Fig. 4 shows a 650,000 lb/hr (82 kg/s) pulverized-coal-fired boiler designed using this principle for various coals.

Although increasing the pulverizer size to handle a wide range of coals is common, it has several drawbacks. The turndown capacity of a pulverized-fuel firing system is principally a function of the mill size. Enlarging the pulverizers has the effect of reducing mill turndown when firing a coal with a higher heating value and grindability than the mill design values. More smaller-size pulverizers may be needed to obtain the required turndown with a system designed for a wide range of coals.

Except for the pulverizers and air temperatures, the design of stoker-fired boilers involves the same considerations for multiple-fuel firing. A stoker-fired boiler requires an air temperature of less than 400°F (200°C) to maintain adequate grate cooling for high-moisture coals and less than 300°F (150°C) for low-moisture coals. Because many high-ash coals with low ash-softening temperatures cause clinkering on the grate, they cannot be used efficiently. To maintain the required bed height, high-ash coals require a faster stoker speed which may increase the unburned carbon content of the stoker ash. Using a large quantity of ash reinjection also aggravates grate clinkering.

BURNING LIQUID, GASEOUS, AND OTHER FUELS

Besides a range of coals, an owner may consider firing other liquid, gaseous or solid fuels (present or future). Because pulverized coal is suspension-fired as are conventional liquid and gaseous fuels, the furnace is already proportioned for this function. Similarly, the required piping, combustion controls, and fuel-firing equipment are in place as part of oil or gas systems for unit warm-up. These fuels can be fired to carry full boiler load with little or no design or equipment modifications.

Up to 20 percent or more of the total heat input can be obtained from well-prepared solid wastes burned in suspension. Biomass, sludge and municipal refuse have all been successfully suspension-fired in a PC-fired boiler. To burn liquid or gaseous fuels in a stoker-fired boiler, the firing equipment must be located relatively high above the grate surface to protect it from the radiant heat produced by the auxiliary burners. For long-term operation, it may be necessary to cover the stoker with refractory. For short-term operation, undergrate air dampers are left partially opened to cool the stoker.

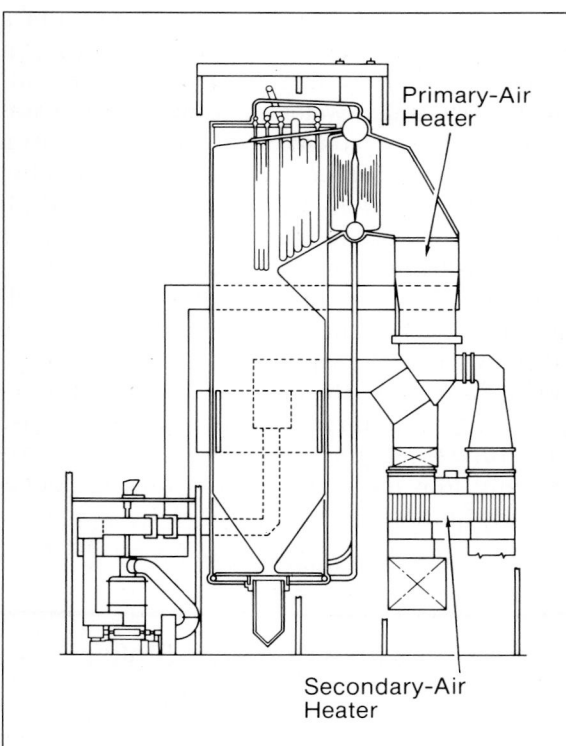

Primary-Air Heater

Secondary-Air Heater

Fig. 4 A pulverized-coal-fired boiler rated 650,000 lb/hr, 1550 psig and 955°F (82 kg/s, 10.7 MPa gage and 513°C) at superheater outlet

The stoker is ideally suited for firing unsized solid fuels other than coal, as shown in Table II. Normally, biomass, sludge, or refuse can supplement the coal with only a minimum amount of additional equipment. Also, in a stoker-fired boiler, the total heat input can come from a larger percentage of waste fuels than in a PC-fired boiler.

THE C-E VU-40 BOILER

The C-E Type VU-40 boiler (Fig. 4) is a field-erected, top-supported, single-gas-pass, thermal circulation, two-drum boiler. Its features make it particularly suitable for the range of steam and fuel conditions of both large industrial and small central-station installations. Boiler applications cover steam capacities from 100,000 to 1,000,000 lb/hr (13 to 125 kg/s), design pressures from 200 to 1800 psig (1.4 to 12.4 MPa gage), and design steam temperatures from saturated to 1005°F (541°C). Virtually any solid, liquid or gaseous fuel can be fired.

DESIGN FEATURES

The VU-40 is produced in sizes and types for every capacity, pressure, fuel, space condition, or method of firing encountered in municipal and industrial power and steam plants. It can be arranged for tangential, horizontal, or stoker firing of single and multiple fuels.

FURNACE

Completely water cooled, the VU-40 furnace uses welded tube panels which are constructed with 2½-in. (64-mm) outside-diameter tubes on 3-in. (76-mm) centers joined by ½-in. (13-mm) wide fins or fusion-welded webs. This construction forms an air- and gas-tight furnace enclosure that eliminates refractory and its attendant maintenance problems.

Complete furnace sidewalls, with inlet and outlet headers shop-welded to the fusion tube panels, are available whenever maximum shop assembly is desired. Also included are seal boxes, miscellaneous attachments and insulation pins.

The drop-bottom furnace of the VU-40 may be designed for any vertical dimension and may include a corner tangential-firing system with either fixed or tilting nozzles, the latter to assist in achieving steam-temperature control. It also may have a horizontal fuel-firing system in the front or rear walls or may be fired by a mechanical stoker either alone or in combination with other types of firing equipment.

CIRCULATION

Fig. 5 shows the circulation of water and steam in the VU-40 boiler. A high furnace and large unheated downcomer pipes provide a thermosyphonic circulation effect resulting in liberal water flow to the furnace waterwall tubes. The steam-water mixture from the waterwall tubes is conveyed to the steam drum through (1) the furnace roof tubes, (2) the relief tubes from the sidewall upper headers and (3) the front-row tubes in the main boiler bank.

Steam-drum sizes vary from 54 to 78 in. (1370 to 1980 mm) inside diameter, with suitable internals of the baffle or centrifugal type (Figs. 6 and 7) to separate the steam from the water. As it leaves the drum and flows to the superheater, or to the main steam line in the case of saturated-steam units, the steam has extremely low moisture content and is consequently quite pure.

Table II. Supplemental Waste Fuels Suitable for Firing with a Spreader Stoker

Fuel	Waste Fuel%	Sizing	Max. Moisture	Limiting Factor
Biomass	100	100% < 6″	55%	None
Shredded rubber	10–20	1″x1″x¼″	–	Smoke carryover, emissions
Dried sludge	20–30	100% < 2″	75%	Fouling, erosion, carryover

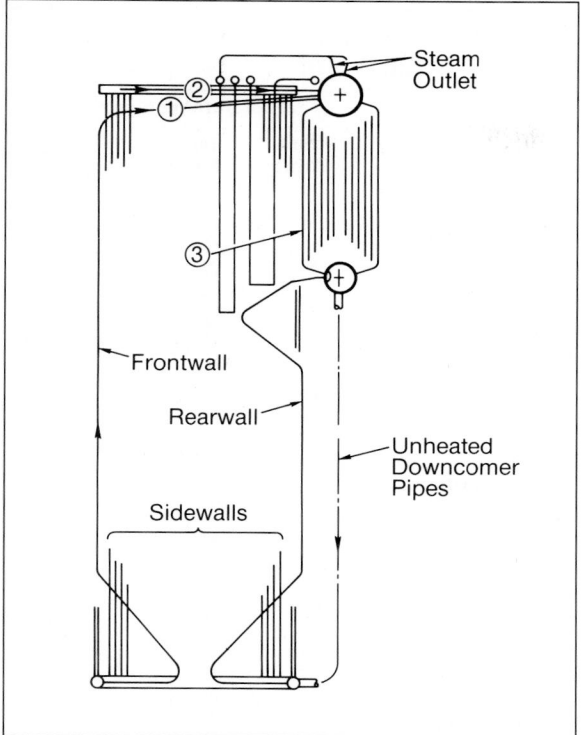

Fig. 5 Circulation system on a C-E VU-40 industrial boiler

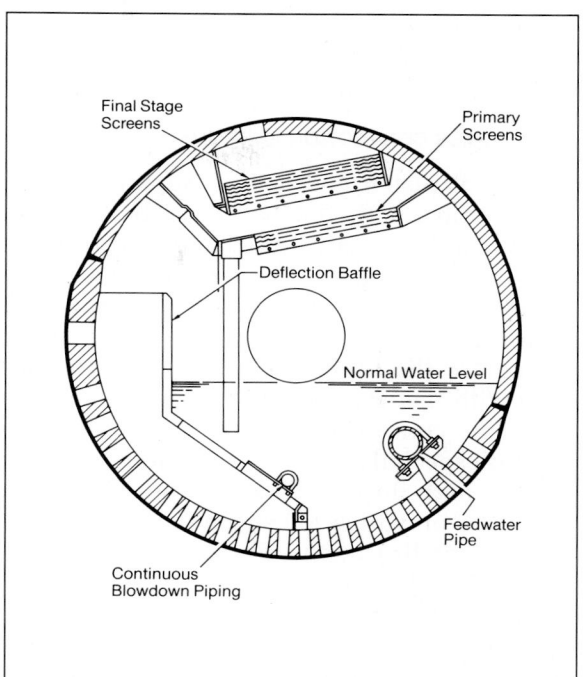

Fig. 6 Steam drum internals — baffle type

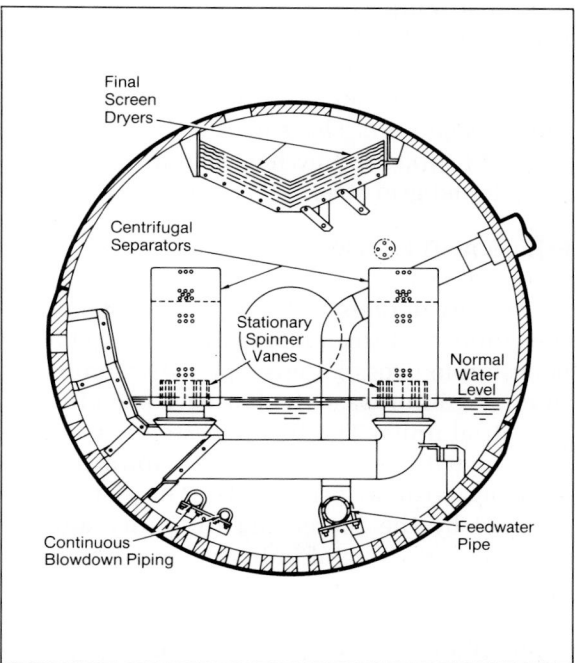

Fig. 7 Steam drum internals — perforated centrifugal type

TOP SUPPORT

VU-40 boilers are top-supported from a structural-steel grid above the unit, with provision for lateral and downward expansion. The furnace section and superheater are suspended from the grid by hanger rods, while the boiler section is hung from two large-diameter U-bolts that cradle the upper drum. The structural steel which supports the grid also supports auxiliary equipment, platforms, and walkways. Suspending the boiler from structural steel permits use of a drop-bottom furnace with a furnace height that is independent of the boiler-bank height.

The boiler size and structural support arrangement necessitate field erection. However, a number of components other than the waterwall panels, such as superheater assemblies, economizers, air heaters, fuel-burning equipment, and ductwork, can be shipped with varying degrees of shop preassembly in order to facilitate erection and to reduce costs.

SINGLE-PASS CONVECTION SURFACES

Pendant platen and spaced superheaters in combination with the baffleless boiler bank result in a single-pass crossflow arrangement of heat-transfer surfaces. The vertical tubes are arranged in-line to facilitate inspection and cleaning. Tube sizes and spacings can be varied to achieve the highest possible heat-transfer rates with the least potential for gas-side fouling. This convection heating-surface arrangement is especially suitable where the flue gases contain high dust loadings. Because there are no baffles on which to concentrate and stratify the dust or flyash, the likelihood of gas-side plugging and tube erosion is substantially reduced as compared to multi-pass convection surfaces. It also has an inherent low draft-loss characteristic because of the lack of gas turns.

Installed across the entire furnace width, superheater surface takes advantage of the uniform distribution of combustion gases as they leave the furnace. Superheater assemblies are supported from their respective inlet and outlet headers located outside the gas stream above the furnace roof. The superheater headers, in turn, are supported by hanger rods attached to the internal grid steel.

FUELS AND FIRING EQUIPMENT

Diversity of fuels and fuel-firing systems typifies industrial steam generation. All grades of conventional fossil fuels—plus numerous byproduct and/or waste fuels—have been fired in industrial units. The list of waste fuels is ever-growing with the decreasing availability and increasing cost of fossil fuels. The list also is affected by the increasing difficulty and expense in disposing of industrial and municipal waste materials in an ecologically acceptable manner. The versatility of the top-supported drop-bottom furnace to accommodate all types of fuel-burning equipment permits the burning of a wide variety of fuels as shown in Table III.

Table III. Fuels and Burning Equipment

Fuels	Front Firing	Tangential Firing	Stoker
Gases			
Natural gas	X	X	
Refinery gas	X	X	
**Carbon monoxide	X	X	
*Blast furnace gas	X	X	
Coke oven gas	X	X	
Hydrogen	X	X	
Liquids			
Oil	X	X	
Tars	X	X	
Pitch	X	X	
*Waste sulfite liquor		X	
Black liquor		X	
*Liquid wastes	X	X	
Solids			
Coal	X	X	X
**Shredded tires			X
Biomass		X	X
**Sludge		X	X
Unsized refuse			X
Sized refuse		X	X

* Usually burned with supplementary fuel.
** Always burned with supplementary fuel.

Fig. 8 illustrates lower-furnace variations for different types of fuel-burning equipment. The water-cooled furnace bottom can be equipped with a dump grate, as illustrated in Chapter 12, to insure complete burnout of relatively large fuel particles of biomass or sized municipal refuse when these fuels are fired tangentially in suspension.

With the C-E tangential firing system, two or more of these fuels can be burned in combination, which is especially desirable if one fuel is a waste material of insufficient quantity to meet steaming requirements or is of a low calorific

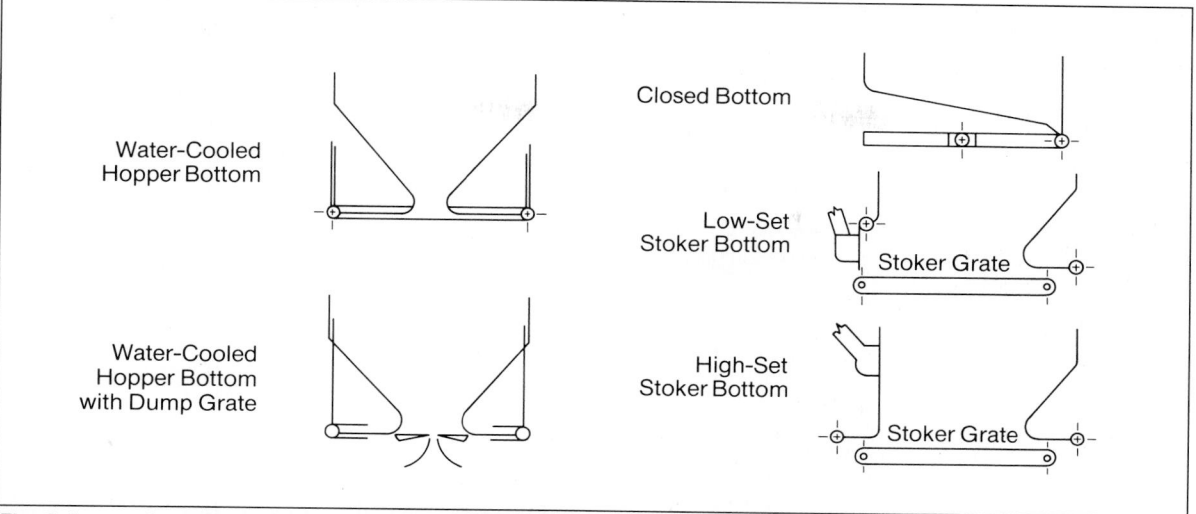

Fig. 8 **Lower-furnace variations of the VU-40 boiler**

value or high-moisture content so as to require supplementary firing.

Fig. 9 illustrates a 500,000 lb/hr (63 kg/s) VU-40 boiler installed in a paper mill. The boiler can fire pulverized coal, heavy oil, and natural gas through the tilting tangential firing system. A row of nozzles located in the front wall directly below the tangential windboxes, introduces a mixture of 65-percent moisture sludge and 50-percent moisture woodwaste for flash drying and partial suspension burning. The firing system also includes a continuous-ash-discharge stoker to burn stoker coal and future woodwaste through combination distributors located on the lower front wall. The tilting tangential fuel nozzles, supplemented by an interstage spray desuperheater, make it possible to achieve the design steam temperature of 900°F (482°C) with many combinations of fuels. The location of the tangential windboxes, positioned high in the furnace, affords protection to the grate should there be occasion to burn heavy oil or pulverized coal alone. Not only does the stoker provide the flexibility to burn stoker-quality coal to 40 percent of full boiler load, but it also provides retention time in the furnace to completely combust any oversized or high-moisture sludge and woodwaste pieces that fall to the bottom of the furnace.

THE C-E RADIANT BOILER

The C-E radiant boiler (Fig. 10) is a field-erected, top-supported, single-gas-pass, thermal circulation boiler designed for industrial and small central-station applications for the 1800-psig (12.4 MPa gage) or higher-pressure steam cycle. The size ranges from 400,000 to over 4,000,000 lb/hr (50 to over 500 kg/s) for non-reheat steam cycles requiring steam temperatures of 1000°F (538°C) at the turbine.

This radiant boiler is designed primarily for pulverized-coal firing, but waste fuels such as sawdust, sewage sludge, or prepared municipal refuse can be fired in combination with the coal. An economizer section is substituted for the boiler bank to facilitate efficient heat transfer when operating at 1850-psig (12.8 MPa gage) or higher superheater-outlet pressure. Welded-wall construction and external pipe downcomers complete the circulation system.

In addition to the above type of boiler, there are many reheat-cycle steam generators producing electricity for large industrial firms, particularly those involved in the production of basic metals. These units are identical to those described in the previous chapter and are found in sizes above 100 MW.

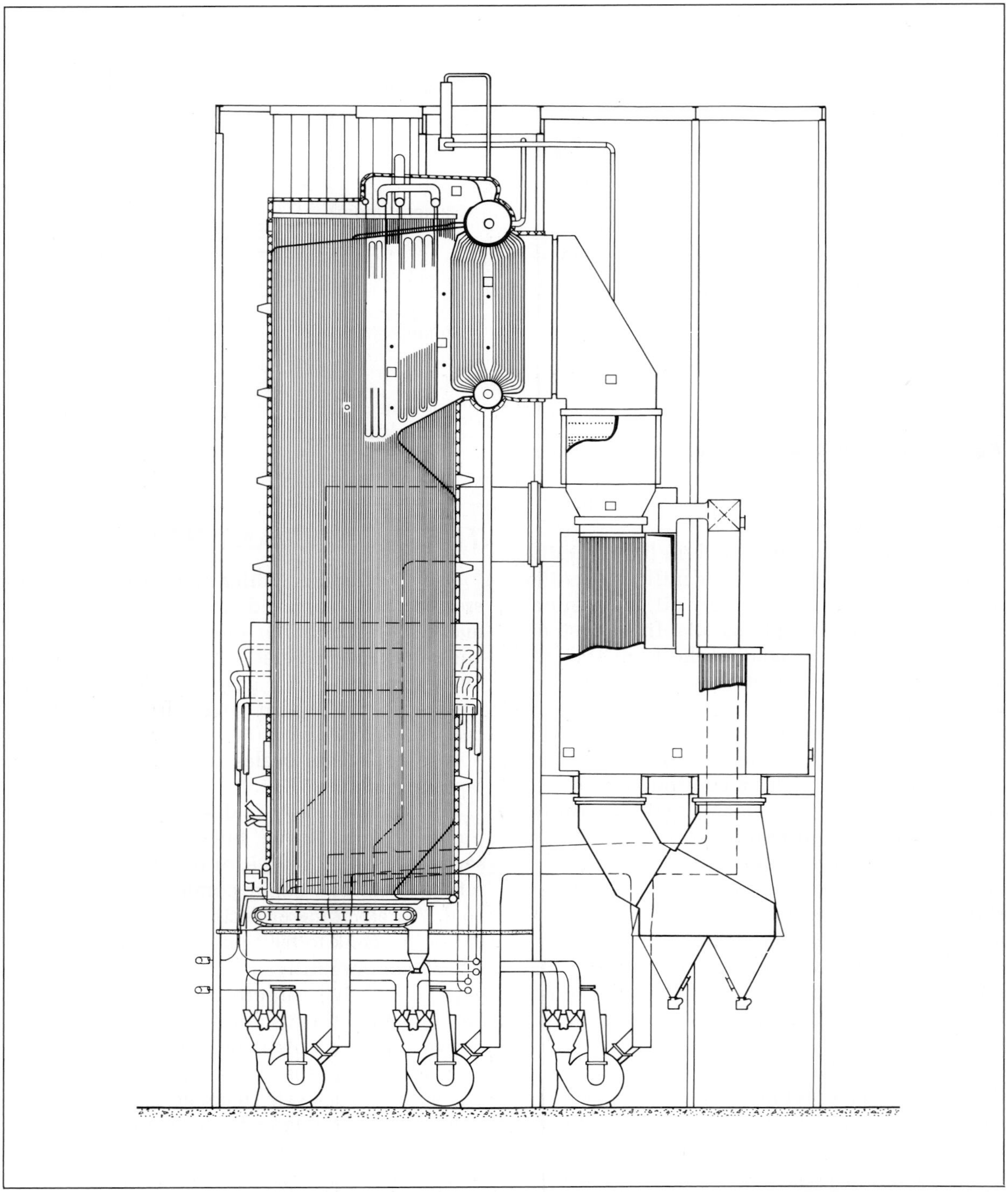

Fig. 9 VU-40 boiler for a paper mill, for pulverized-coal, stoker-coal, heavy oil, suspension-fired wood waste and sludge, future natural gas and stoker wood waste

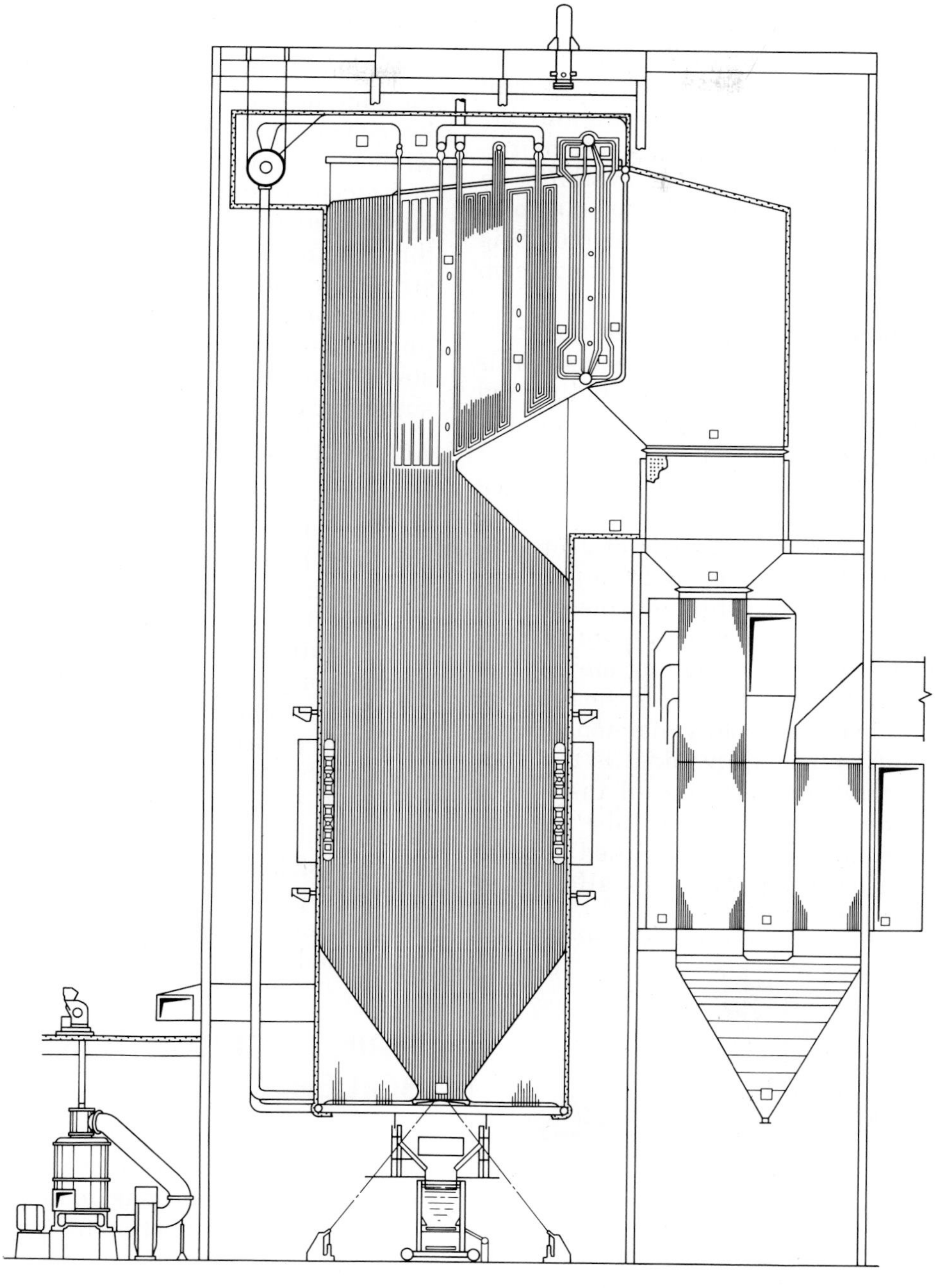

Fig. 10 **Pulverized coal-fired radiant boiler, rated 800,000 lb/hr, 1850 psig and 1005°F (100 kg/s, 12.8 MPa gage and 541°C) at superheater outlet**

THE C-E VU-60 BOILER

The VU-60 is a custom-designed boiler that is built from pre-engineered, modular components (Fig. 11) and is used to fire clean liquid and gaseous fuels and many waste and by-product fuels. This bottom-supported, natural-circulation steam generator is available in capacities from 100,000 lb/hr to 1,000,000 lb/hr (13 to 125 kg/s), design pressures to 1800 psig (12.4 MPa gage), and superheater-outlet steam temperatures to 1005°F (541°C).

Fig. 12 shows the VU-60 arranged for front-wall round-burner firing. Fig. 19, later in this chapter, is an example of a tangentially fired VU-60 unit suitable for firing waste liquid and gaseous fuels in addition to natural gas or oil.

The VU-60 is most often designed for pressurized firing, without an induced-draft fan. The waterwalls consist of panels of 3-in. (76-mm) diameter tubes on 4-in. (102-mm) centers, between which metallic fins are placed and welded, producing a completely tight furnace enclosure. Insulation and lagging are applied directly over the tube panels. This construction virtually eliminates refractory maintenance because the only refractory material is the floor tile and a small amount around the water-cooled burner openings. Also, tube replacement in the welded wall is simplified because the 1-in. fin between the tubes allows cir-

cumferential butt welding without springing adjacent tubes.

SUPERHEATER DESIGN

For steam temperatures above 825°F (440°C), the superheater is arranged in two stages, with interstage desuperheating for control of outlet temperature. Multiloop pendant-type elements and a center connection in the large-diameter outlet header assure even distribution of steam throughout the superheater surface. The construction requires a minimum number of header joints. Superheater elements are supported from headers located out of the gas stream above the furnace roof; these headers are supported by furnace walls and drum.

ACCOMMODATING SOLID FUELS

When solid wastes must be burned along with gaseous or liquid wastes, a modification of the VU-60 modular boiler incorporates a continuous-ash-discharge spreader-type stoker and is bottom-supported (Fig. 13). The VU-60S ("S" signifying a stoker-firing system) has a normal range of steam production between 75,000 and 300,000 lb/hr (9 and 38 kg/s) with operating pressures up to 1550 psig (10.7 MPa gage) at the superheater outlet and total steam temperatures up to 955°F (513°C). Stoker coal, sludge, biomass or other cellulose fuels suitable for stoker firing can be burned. Supplemental oil, gas or other waste fuels are fired with either a tangential-firing system or front-wall round burners.

FIRING REFUSE IN INDUSTRIAL BOILERS

Chapter 2 included a discussion of municipal refuse as a fuel. Its value as an alternate energy source is proportional to the effort spent in "refining" or beneficiating it. And, to a certain extent, the cost of the boiler that can burn a prepared refuse will be inverse to the amount of money spent on the removal of metals and glass and on the size reduction and homogenization of the refuse.[2-13]

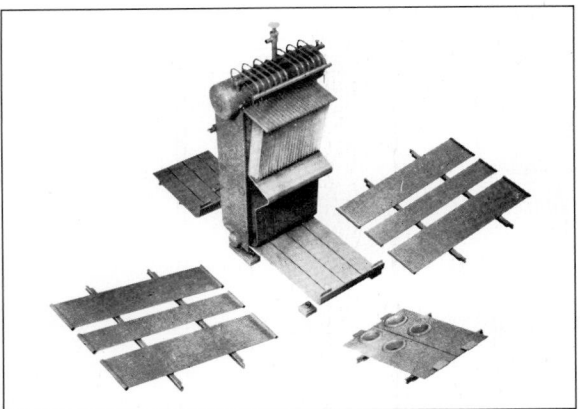

Fig. 11 Pre-engineered modular components of C-E VU-60 field-erected steam generator

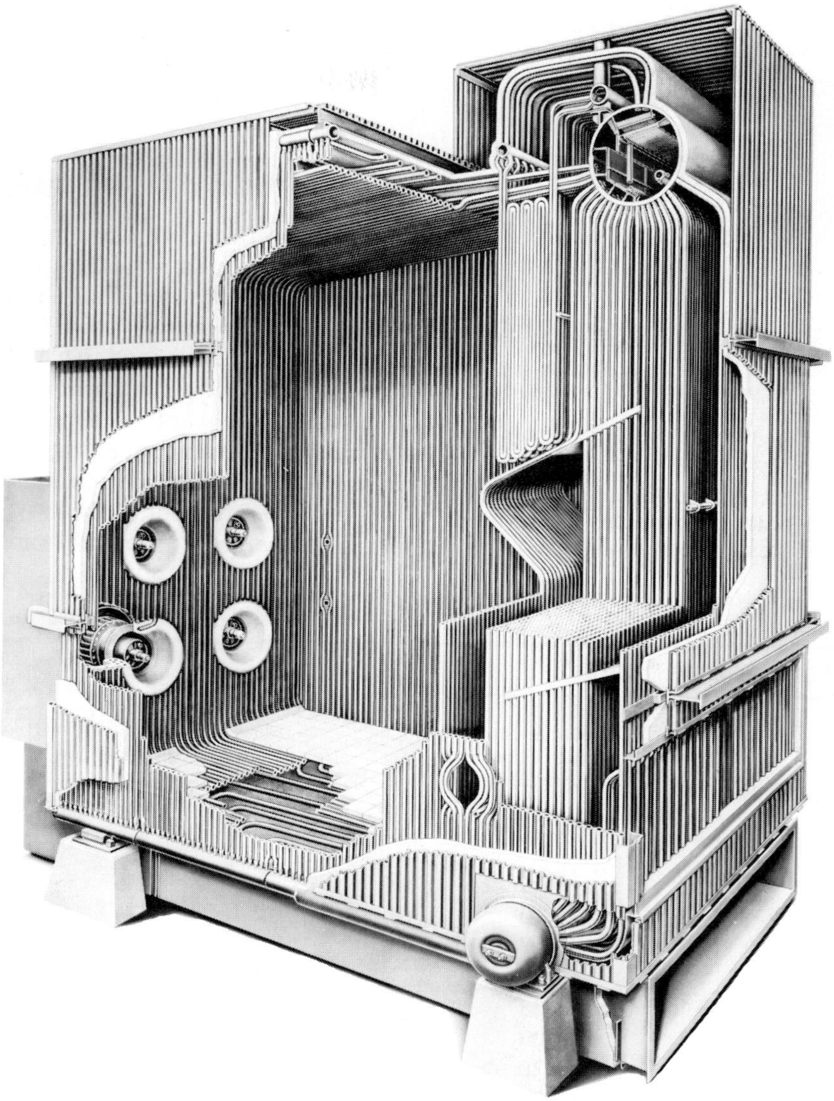

![VU-60 unit cutaway illustration]

Fig. 12 **VU-60 unit with front-wall (round) burners for liquid and gaseous fuels**

The following describes three distinctly different systems for handling and burning refuse, and the types of steam-generating equipment best suited for the specific method. The different systems are:

- mass-burning of unprepared refuse
- spreader-stoker firing of prepared refuse
- suspension firing of prepared refuse

Table IV compares the boiler performance of all three refuse systems when burning 600 U.S. tons/day (545 tonnes/day) of refuse. This performance is for the boiler and emission-control equipment combined. When comparing complete systems, the initial cost and power re-

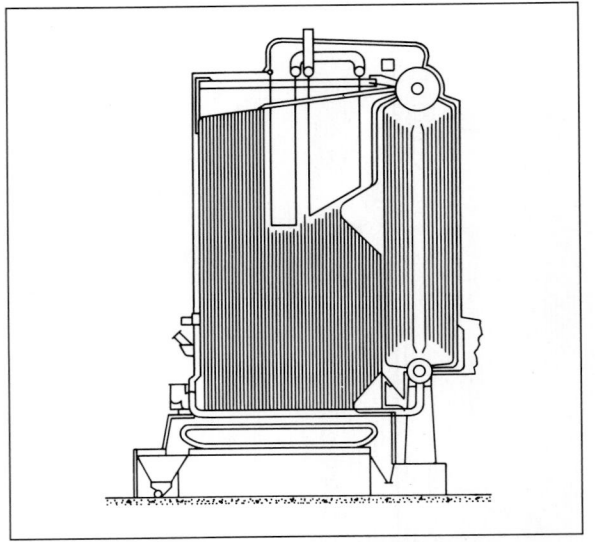

Fig. 13 C-E VU-60S bottom-supported stoker-fired boiler for firing liquid, gaseous, or solid fuels

quirements of all installed equipment must be evaluated. Other factors such as revenue from electrical power generated and products recovered, as well as operation and maintenance costs, must also be considered to determine the true cost-effectiveness of the various systems.

MASS-BURNING OF UNPREPARED REFUSE

A major refuse-firing system (Fig. 14) is that for the mass-burning of unprepared refuse. Usually residential and industrial wastes, this refuse will be referred to as MSW (municipal solid waste), as described in Chapter 2. As compared to conventional biomass fuels, MSW contains quantifiable amounts of heavy metals and an unusually high concentration of sulfur and chlorine. These fuel components dictate that the design requirements of MSW steam generators be extremely different from conventional

Table IV. Systems Comparisons*

	Spreader-Stoker Firing of Prepared Refuse (RDF)	Mass-Burning of Unprepared Refuse (MSW)	Suspension Firing of Prepared Refuse
Fuel Analysis:			
% By Weight			
Carbon	34.00	26.47	34.00
Hydrogen	4.10	3.08	4.10
Oxygen	23.47	17.62	23.47
Nitrogen	0.50	0.45	0.50
Sulfur	0.23	0.15	0.23
Moisture	23.00	38.50	23.00
Chlorine	0.70	0.23	0.70
Ash	14.00	13.50	14.00
Total	100.00	100.00	100.00
HHV, BTU/lb.	5785	4500	5785
Steam Flow, Lb/Hr	205,000	150,000	210,000
Excess Air, %	50	100	30
Efficiency, %	75.0	70.4	76.1
Temp. to FGCE, °F	350	400	350
FD/OFA Horsepower	340	545	335
ID Fan Horsepower**	300	420	295

* 600 U.S. tons/day of fuel fed to each unit; outlet steam conditions: 825°F TST, 900 psig; 300°F feedwater
** Includes flue-gas cleaning equipment (FGCE)

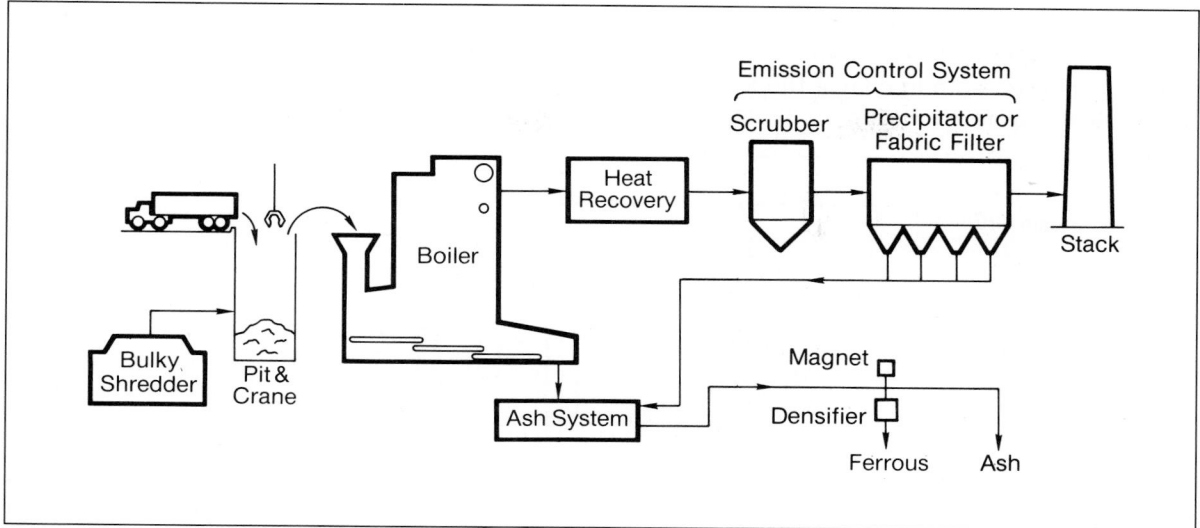

Fig. 14 **Flow diagram for mass-burning of municipal refuse**

fossil-fuel units. The heating value of municipal solid waste can vary from 2,000 to 7,000 Btu/lb (4.7 to 16.3 MJ/kg) with a mean value of 4500 Btu/lb (10.5 MJ/kg). As a general rule, production of a given quantity of steam requires that nearly 3 times as much MSW be burned as coal (by weight). Use of municipal refuse for energy production conserves the remaining supply of fossil fuels and reduces an imposing mountain of garbage threatening to overwhelm landfills. However, to insure an adequate life of a boiler firing MSW, the designer must have an understanding of the impact of this non-homogeneous fuel on the steam generator.

An MSW steam generator has an inherently unique design. The heterogeneous nature of MSW requires long furnace residence time prior to flue-gas contact with heat-transfer surfaces in order to completely oxidize the fuel. Complete oxidation will also reduce the potential for corrosive attack of the heating surfaces. Metal temperatures above 900°F (480°C) accelerate corrosion rates. As a result, the design of the MSW steam-generator heat-transfer surfaces should maintain the metal temperatures below this value. Flyash which deposits on the heat-transfer surfaces as it leaves the furnace initiates a secondary corrosion reaction. The

complex chemicals within the flyash catalyze corrosion reactions involving the sulfur and chlorine present in the flue gases. These sulfation reaction rates are also temperature-dependent. Therefore, the design of an MSW steam generator should attempt to reduce flyash carryover and maintain low metal temperatures.

Combustion Engineering employs a top-supported, multiple-pass, natural-circulation steam generator design to fire MSW (Fig. 15). The first gas pass is the furnace and is located directly above the sloped mass-burning grate. The second, third, and fourth passes contain respectively the evaporator, superheater, and economizer heat-transfer surfaces necessary to achieve the specified steam conditions and heat-absorption requirements. Currently, MSW boilers are available in sizes up to 750 U.S. tons (680 tonnes) of refuse per day. A waste-to-energy plant ordinarily will use two, three, or more such steam-generator units.

Located at the front of the steam generators is a large refuse storage pit common to all of the units. The pit is sized to provide for the proper inventory of fuel, generally three-days storage of the entire plant capacity. Refuse haulers dump the MSW either directly into the pit or onto a tipping floor for fuel sorting before entering the pit. Positioned above the pit is a crane. Normally a clamshell or "orange-peel" type, this crane is used both to charge the feed chute of the steam generators and to remove large non-combustibles (white goods, tree stumps, engine blocks, etc.) from the waste stream if a tipping floor is not used. The crane operator is generally located in an air-conditioned control

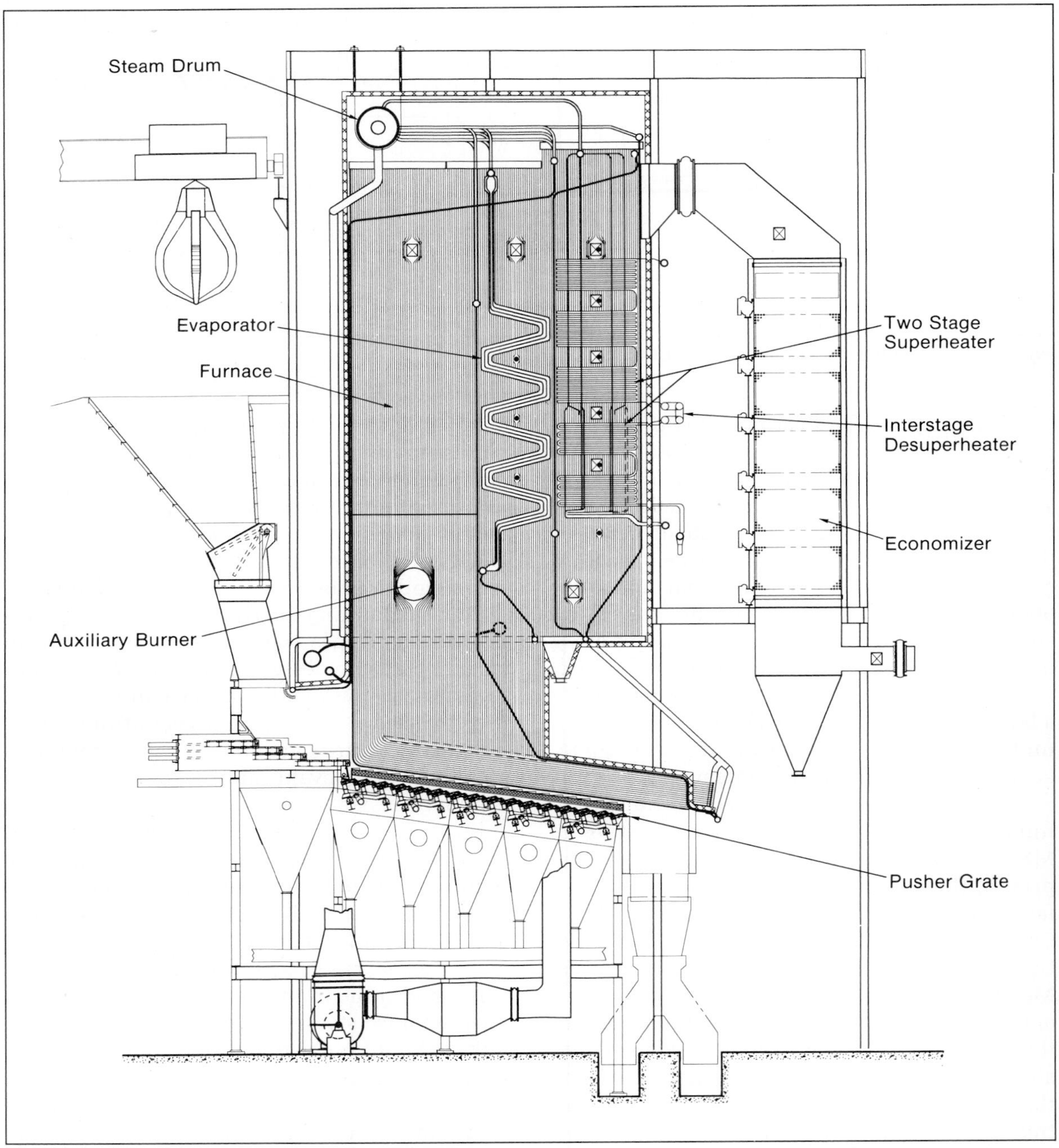

Fig. 15 **Multiple-pass steam generator with CE/db grate for mass-burning of municipal refuse**

booth directly above the refuse pit. This allows the operator to observe the pit area and to visually inspect the fuel sent to the steam generator(s) for non-combustibles and hazardous materials. Since MSW is a highly heterogeneous fuel, the crane also mixes the fuel into a more homogeneous state. During this process, the operator can detect non-combustibles and other unacceptable components. Some operators stockpile paper and cardboard in a corner of the pit to be used if high-moisture fuel is delivered, in order to balance the heating value of the fuel charged to the steam generator.

FUEL-FIRING SYSTEM

As illustrated in Fig. 15, the fuel fed to the charging hopper of the steam generator drops by gravity through the charging chute onto the front of the grate. The charging chute is the transition from the charging hopper to the steam-generator furnace. During normal operation the chute is filled with fuel. This head of fuel provides the seal between the charging hopper and the furnace, preventing air infiltration. The charging chute is water-cooled to provide equipment protection should burning occur within this area. The chute is equipped with a cut-off gate to provide a positive seal during boiler start-up and shutdown. A reciprocating ram feeder located directly beneath the charging chute controls the refuse flow rate onto the grate. This device regulates the quantity of MSW fuel fed to the grate for combustion. The ram feeder function is similar to a fuel-oil flow-control valve. The feeder is operated hydraulically and the stroke length, speed, and frequency can all be controlled and varied independently of the grate controls.

The CE/db grate, described in detail in Chapter 12, is sized for a heat-release rate of 270,000 to 300,000 BTU/hr-ft^2 (850,000 to 950,000 W/m^2) based upon gross heat input. This low heat-release rate is necessary to provide for long residence time of the fuel on the grate, long grate life, and good carbon burnout. The fuel is extremely heterogeneous; at any time it can contain such varied items as cardboard boxes, tires, grass clippings, leaves, clothes, mattresses or carpeting. The varying size, moisture, and combustibility of each item require different lengths of time to burn. Therefore, the MSW firing system must provide sufficient agitation, adequate exposure to oxygen, and long enough time within the combustion zone of the grate to maximize fuel burnout.

Normal operation of an MSW steam generator requires approximately 100-percent excess combustion air. This much air is needed to ensure that the heterogeneous MSW has sufficient air to efficiently oxidize the available carbon and hydrogen. Combustion air is drawn from the pit and enters the steam generator through one of two places. Approximately 40 to 60 percent of the total combustion air is introduced through the grate surface and is called *undergrate air* (UGA). The balance of the combustion air enters through nozzles or ports above the stoker and is called *overfire air* (OFA).

The undergrate air can be preheated to aid in the drying of the fuel on the grate. This is necessary only for extremely wet MSW. A steam-coil air heater after the FD fan is one method used, because it provides the flexibility to preheat the air as the fuel conditions warrant.

The overfire air can also be preheated; however, cold overfire air is more dense than hot OFA and will be able to penetrate deeper into the furnace. OFA nozzles are located in the furnace front and rear walls above the top of the fuel bed; the OFA is pressurized to 20-30″ WG (5 to 7.5 kPa) to ensure air penetration and promote thorough mixing of the air with the volatile combustibles leaving the fuel bed.

FURNACE SECTION

The furnace is composed of water-cooled tubes to form a gas-tight enclosure. Approximately 60 percent of the unit's steam capacity is generated within the furnace section.

Before introduction of overfire air, approximately 15 feet (4.6m) above the grate surface, there exist local reducing environments in the lower furnace. This is because of the incomplete combustion of the volatiles driven off the MSW fuel on the grate. From a thermal standpoint, carbon-steel tubing can be used in the lower furnace since the tube metal temperatures are kept low by choosing operating pressures in the range of 650 to 900 psig (4.5 to 6.2 MPa gage); corresponding furnace wall temperatures are 500 to 550°F (260 to 290°C). However, due to the reducing environment, some type of waterwall protection, such as a layer of castable refractory, is required to physically keep the corrosive gases away from the furnace waterwalls.

The refractory applied to the lower furnace waterwalls is high-conductivity silicon carbide (SiC). This material, which provides excellent heat-transfer characteristics, reduces waterwall slagging and is extremely erosion-resistant. With increased residence time in the furnace and additional oxidation achieved with the OFA injection, the flue gases become more oxidizing, eliminating the need for refractory lining of the upper furnace.

The furnace is sized to provide a low flue-gas velocity and a low exit-gas temperature. A conservative velocity of 18 to 20 ft/sec (5.5 to 6 m/s) will provide long residence times and low particulate carryover. Less particulate carried over into the second pass will reduce corrosion and erosion of the evaporator tubing. A furnace exit-gas temperature of 1450°F or 790°C (under clean furnace-wall conditions) will limit the corrosion of downstream heat-transfer surfaces, since the corrosion reaction rate is very temperature dependent.

Auxiliary fuel-firing equipment located in the furnace can be used to warm the boiler before light-off of the MSW; it can also maintain exiting flue-gas temperatures at necessary levels for proper performance of the emission-control apparatus.

SECOND PASS

As shown in Fig. 15, the flue gas leaves the upper furnace, making a 180° turn through a waterwall screen section before entering the second pass. The watercooled second pass contains convective heat-transfer surface, called an *evaporator*, arranged in serpentine tube bundles. The evaporator surface serves a two-fold purpose. First, it absorbs heat from the flue gas to generate nearly 30 percent of the steam capacity, while lowering the flue-gas temperature to below 1200°F or 650°C (under clean conditions). Second, the evaporator mixes the flue gas, promoting further oxidation and reducing stratification.

The evaporator tubes are constructed of carbon steel and are set on wide transverse spacings, which prevent particulate carried over from bridging between the evaporator assemblies. To prolong the life of the tubing, conservative flue-gas velocities in the range of 15 to 20 ft/sec (4.5 to 6 m/s) are used. Additional tube thickness above ASME Code requirements provides for extra corrosion/erosion protection.

Through normal operation, the furnace walls of the MSW steam generator will become fouled. This causes an increase in the flue-gas temperatures entering the second pass. The evaporator acts as a buffer between the furnace and the superheater, dampening out the temperature increases due to furnace fouling.

SUPERHEATER

The superheater of this multiple-pass boiler follows the evaporator section. The flue-gas temperature entering the superheater should be below 1200°F (650°C) under clean conditions. This low gas temperature entering the superheater provides for low metal temperatures and reduced corrosion potential, while still sufficient for superheated steam temperatures of 800 to 830°F (425 to 445°C). A corrosion allowance is applied to the superheater tubing thickness, and low flue-gas velocities between 20–25 ft/sec (6 to 7.5 m/s) control erosion.

The design of the multiple-pass MSW steam generator combines a number of mechanisms that serve to reduce drastically the erosive and corrosive nature of the flue gas before it enters the superheater. The two 180-degree flue-gas turns and the upstream evaporator section mix the flue gas and minimize CO stratification. The low velocities drop out a substantial portion of the flyash into the hopper under the superheater. Therefore, less flyash deposition will

occur on the superheater tubing. Since the superheater is not exposed to luminous radiation, the surface metal temperatures will be lower.

Experience has shown that unit performance degrades with time. This degradation must be accounted for in the initial boiler design. As the unit begins to "season", the fouled heat-transfer surfaces become less efficient and flue-gas temperatures become elevated; the unit will then have to be shut down for cleaning. Multiple-pass units can run for 4,000 to 5,000 hours before elevated gas temperatures require such shut down. To illustrate this: the flue-gas temperature entering the superheater may be 1150°F (620°C) under clean conditions. After 4000 hours of operation, the temperature will be approximately 1250°F (675°C). The 100°F (55°C) increase must be taken into account when selecting the superheater metallurgy and desuperheater capacity. The advantage of the multiple-pass design is that the evaporator acts as a buffer between the furnace and the superheater, and limits the impact of the fouling on the superheater. This lengthens the time between outages, thereby increasing unit availability and reducing maintenance costs.

ECONOMIZER

The final heat-recovery component of the steam generator is the economizer. On a multiple-pass unit, the economizer is located in a non-watercooled pass following the superheater. The ductwork forming the walls of this pass is of either carbon steel or low-alloy steel depending upon the entering flue-gas temperature. The economizer is constructed of bare, in-line, carbon-steel tubes. The flue-gas velocity entering the economizer ranges from 20 to 25 ft/sec (6 to 7.5 m/s). Generally, the economizer is sized to reduce the exit-gas temperature to about 400°F (200°C), suitable for optimum operation of the flue-gas treatment and cleaning equipment. The large heat duty required of the economizer may result in a steaming condition; this will be determined in the design stage and will affect piping and steam-drum internal arrangements.

REFUSE-DERIVED-FUEL SYSTEM

The refuse-derived-fuel (RDF) waste-to-energy system burns prepared refuse. The degree of refuse preparation may vary from simple removal of bulky material and shredding to an extensive processing system yielding a highly combustible fuel. One type of processing system, as illustrated in Fig. 16, is used in conjunction with the RC-continuous-ash-discharge spreader stoker described in Chapter 12. The resultant fuel from such a system is normally 100 percent less than 6 inches (150 mm), 95 percent less than 4 inches (100 mm), and 50 percent less than 3 inches (75 mm) in size. The high heating value of spreader-stoker-fired RDF can range from 5,000 to 7,000 Btu/lb (11.6 to 16.3 MJ/kg). The ash content ranges from 5 to 15 percent; the moisture content will be from 20 to 30 percent by weight, subject to seasonal variation.

Unlike the MSW steam generator, the RDF steam generator appears similar to a biomass or coal-fired unit with a stoker, as illustrated in Fig. 17. The design features a large single-pass furnace with pendant superheater surface located at the furnace outlet.

The RDF steam-generator system operates in concert with a fuel-processing system. The processing system takes raw refuse and eliminates a portion of the non-combustibles and sizes the resulting fuel. The prepared fuel is then delivered into the furnace for combustion on a stoker.

FUEL-FIRING SYSTEM

Conveyors transfer the RDF from the fuel storage area to the steam-generator area. The RDF metering system is located at the front of the steam generator. A number of devices may be effectively used to meter the fuel. One such device is an auger bin. The bin is equipped with a series of screws at the bottom to regulate the flow of RDF into the chutes leading to the distributors. The distributors are a pneumatic type using high-pressure air to project the fuel into the furnace. Even distribution of fuel on the stoker is essential for efficient operation of the unit.

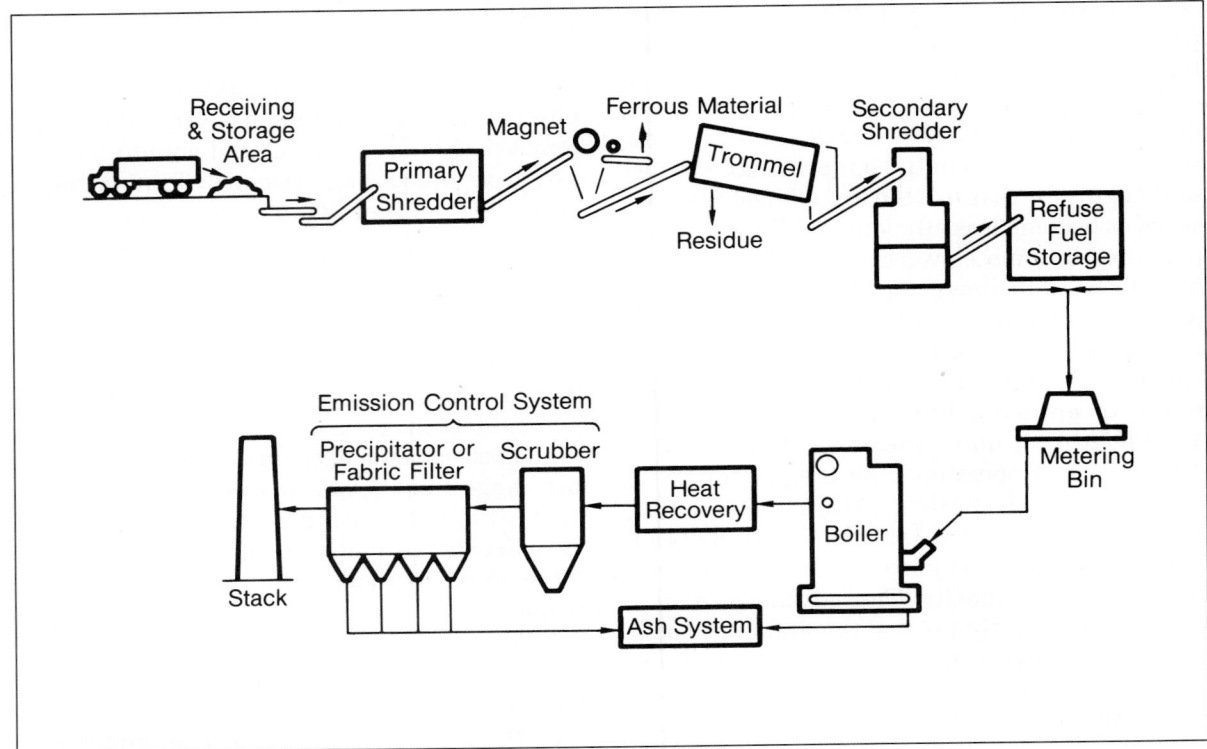

***Fig. 16** Preparation system for spreader-stoker firing of refuse-derived fuel*

The stoker is sized for a 750,000 to 800,000 Btu/hr-ft^2 (2,370 to 2,520 kW/m^2) grate heat-release rate based upon gross heat input. This is slightly more conservative than a normal stoker-fired boiler burning wood because of the potential for clinkering and slagging of the non-combustible ash on the grate. It also results in less slagging on the lower furnace walls.

Combustion air is introduced as either under-grate air (UGA) or overfire air (OFA) as on an MSW unit. Normal operation of an RDF steam generator is with 50- to 60-percent excess air. About 50 to 70 percent of the total combustion air is introduced as UGA with the balance being OFA. One method of introducing the OFA is through four separate windboxes located in the corners above the stoker. The tangential OFA system promotes horizontal mixing and mini-mizes flue-gas stratification thereby improving combustion. Such a system will use an OFA pressure of about 10 to 12″ WG (2.5 to 3kPa) to produce proper penetration into the furnace. The UGA and OFA are normally preheated so that all of the combustion air passes through the air heater to increase unit thermal efficiency.

FURNACE

The furnaces of the first-generation RDF steam generators were designed with the same criteria as conventional biomass units. This re-sulted in units that were far too small for the desired output. When operated at their rated maximum capacities, the units experienced se-vere plugging, erosion, and corrosion. These problems demonstrated that an RDF unit needs its own unique set of design guidelines.

The furnace of the unit in Fig. 17 is com-posed of water-cooled tubes forming a gas-tight enclosure. The tubing is carbon steel, with typ-ical operating pressures in the range of 850 to 950 psig (5.9 to 6.2 MPa gage).

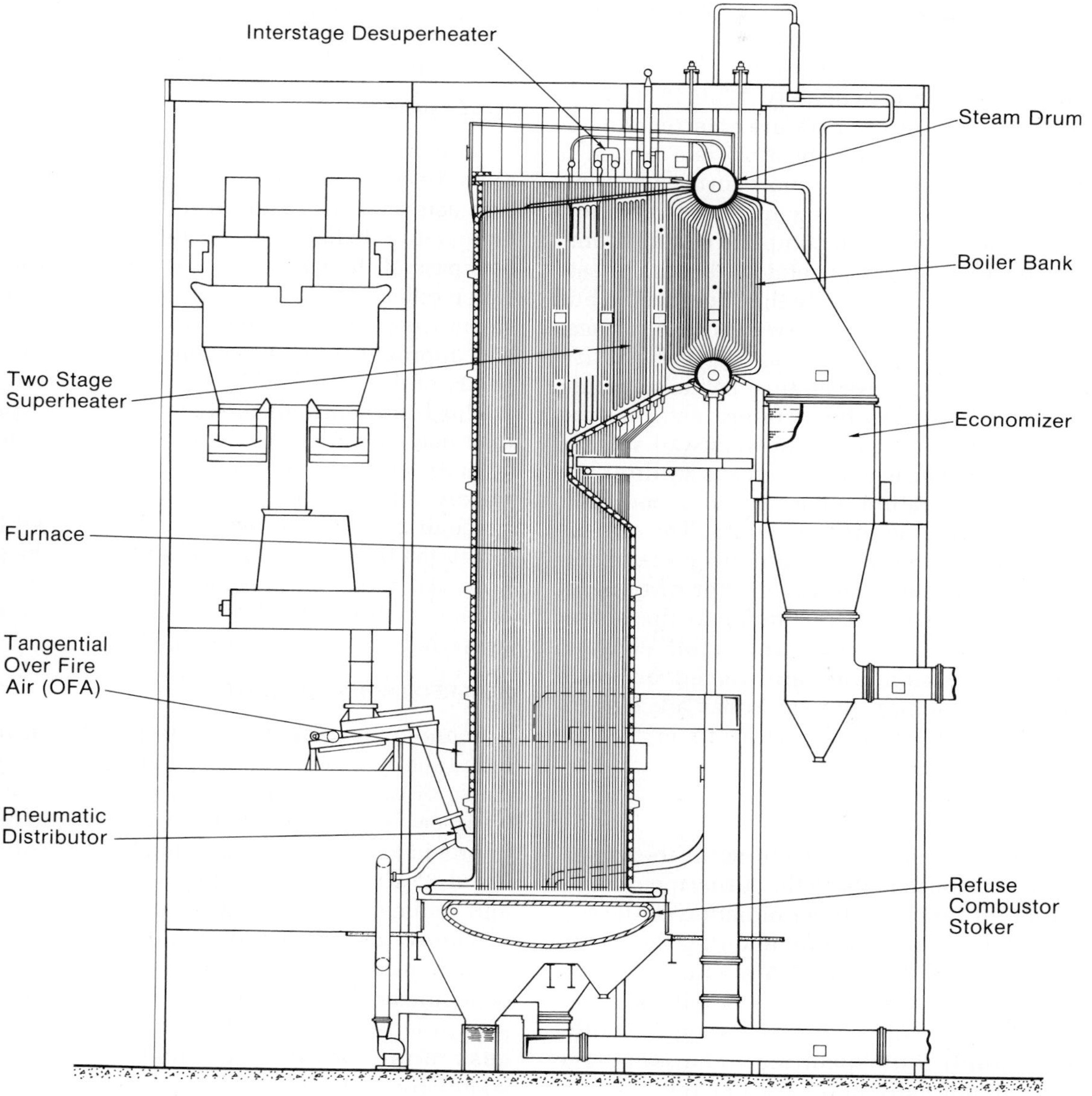

Interstage Desuperheater

Steam Drum

Boiler Bank

Two Stage
Superheater

Economizer

Furnace

Tangential
Over Fire
Air (OFA)

Pneumatic
Distributor

Refuse
Combustor
Stoker

Fig. 17 **Single-pass steam generator with C-E RC grate for burning refuse-derived fuel**

8-23

The furnace is sized for a conservative combustion rate of 12,000 to 15,000 Btu/hr-ft^3 (125,000 to 155,000 W/m^3) and a flue-gas temperature entering the superheater of approximately 1550 to 1600°F (840 to 870°C). These criteria are more conservative than conventional biomass-unit designs with values of 20,000 Btu/hr-ft^3 (205,000 W/m^3) for combustion rate and approximately 1700 to 1800°F (925 to 980°C) gas temperature entering the superheater. The low combustion rate provides approximately 40 percent more furnace volume than a biomass-unit allowing more retention time for gas mixing and complete combustion.

In the RDF steam generator, the first heat-transfer surface is normally the superheater, but may in some cases be waterwall screen surface. The furnace is sized to meet the conservative combustion rate; however, this may not provide the desired flue-gas temperature entering the superheater. The addition of waterwall screens before the superheater is a means to drop the flue-gas temperature entering the superheater without increasing furnace height. The screens also act as a buffer to maintain the gas temperature entering the superheater after dirtying of the furnace takes place, similar to the action of the evaporator tubes in an MSW unit. The screen assemblies are placed on wide transverse spacings and the tubes are tangent in the direction of gas flow to minimize ash deposition.

SUPERHEATER

To reduce metal temperatures in the superheater, it is located above the furnace nose arch to shield it from the direct radiation of the furnace. The flue-gas velocity entering the superheater is about 25 ft/sec (7.5 m/s) to minimize erosion. The superheater may be either a one- or two-stage design depending upon the outlet steam conditions specified. The two-stage design is normally employed for outlet steam temperatures above 725°F (385°C) and incorporates an interstage desuperheater for final steam-temperature control. The two stages may be arranged in either counterflow or parallel flow of steam with respect to flue-gas flow. A parallel-flow arrangement is most often used in the first stage to lower the metal temperatures; a counterflow arrangement will experience higher metal temperatures but will also maximize heat transfer. The decision as to which arrangement to use must be made based upon the particular steam-temperature control range requirements.

BOILER BANK

Following the superheater is the next heat-recovery component, the boiler bank. This section acts as a heat sink to reduce the flue-gas temperature to below 850°F (450°C) so that the backpass of the steam generator does not need watercooled walls. The boiler bank contains a steam-water mixture; the heat it absorbs generates approximately 30 percent of the steam capacity of the unit.

The drum centerline distances and tube spacings may be varied to provide for a maximum velocity entering the boiler bank of approximately 30 ft/sec (9 m/s). A minimum clear space of 2½ in. (65 mm) is recommended between the boiler tubes perpendicular to the gas flow. An erosion/corrosion tube-thickness allowance above the ASME Code-required thickness can be used to extend tube life.

HEAT-RECOVERY EQUIPMENT

Economizers for RDF units are the in-line bare-tube type. Carbon-steel tubing is used because of the relatively low gas temperature. A transverse clear spacing of at least 2 in. (50 mm) is provided to maintain an entering velocity of about 35 ft/sec (10.5 m/s) to minimize erosion and reduce the potential for pluggage. Again, thickness above that dictated by the ASME Code is recommended to prolong tube life. The economizer should be arranged in banks separated by large access cavities for maintenance access and sootblower locations. Rotary sootblowers are used due to the low flue-gas temperature encountered in this area. Counterflow of water to flue-gas maximizes heat transfer. Depending upon the incoming feedwater temperature and the exit-gas temperature specified entering the flue-gas cleaning equipment,

this may be the final heat-transfer surface. Otherwise, a tubular or regenerative air heater will be the terminal heat-recovery device.

SUSPENSION FIRING OF PREPARED REFUSE

For suspension firing, refuse must be reduced to a maximum size of 2 in. x 2 in. (50 mm x 50 mm) so that it burns uniformly in suspension. Such size reduction also allows pneumatic conveying to the furnace.

Fig. 18 illustrates a typical preparation system for suspension firing. C-E uses its tangential system for 100-percent suspension firing. As described in Chapter 12, the refuse and heated air are directed tangent to a circle in the center of the furnace. Fuel and air are mixed in a single fireball, aiding in even distribution within the furnace. The lighter fraction of the refuse quickly burns in suspension. The larger and more dense refuse material will not remain suspended by the air and will fall to the furnace bottom, where a grate is provided to complete the combustion of this material. Several rows of tangentially directed air nozzles are located between the fuel compartments and the stoker. These nozzles maximize the suspension burning of the heavier refuse material as it falls to the grate at the furnace bottom.

For this firing system, boilers able to burn up to 1500 U.S. tons/day (1360 tonnes) of refuse can be designed for steam conditions to 825°F (440°C) total steam temperature. Once again, the total-steam-temperature limitation is imposed to avoid excessive superheater corrosion which increases significantly as the tube metal temperatures approach 900°F (480°C).

The superheater temperature limitation applies to 100-percent refuse-fired boilers regardless of the type of firing systems used. In units deriving 20 percent or less of the total heat from refuse when fired in combination with pulverized coal, the superheater metal temperature may be higher without developing excessive corrosion rates. It is believed that the reduced-corrosion rate with PC firing is a result of the neutralizing effect of the coal ash in combination with the reduced refuse-fuel quantity.

FURNACE AND CONVECTIVE SYSTEM DESIGN

The furnace of a suspension-fired boiler is sized for a gas temperature at the superheater of 1600°F (870°C) to avoid ash build-up and plugging of the convective heating surface. The convective surface must therefore be designed with wide tube spacing so ash build-up between sootblowing cycles cannot bridge from tube to

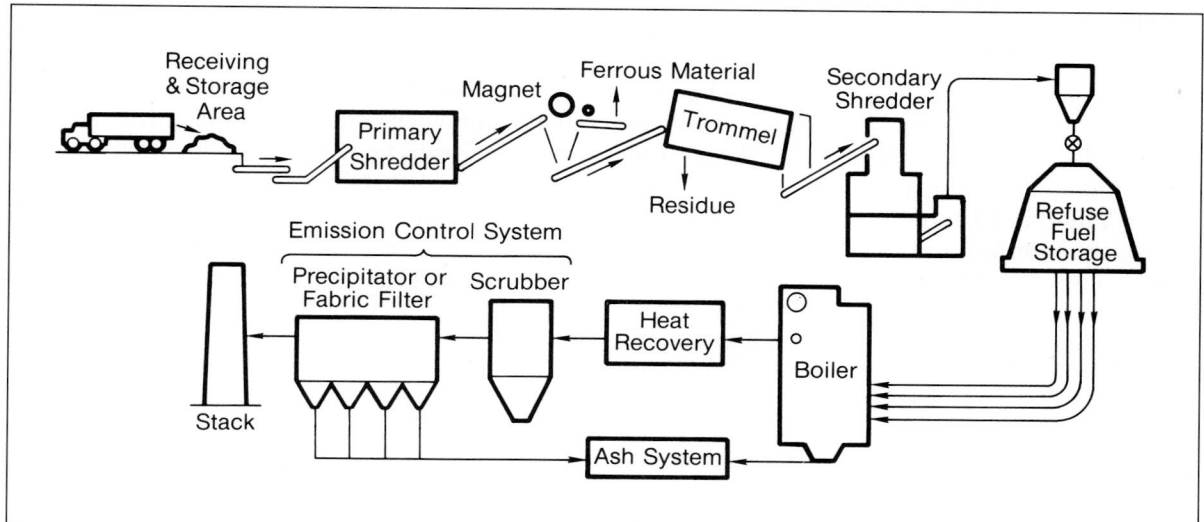

Fig. 18 **Preparation system for suspension firing of municipal refuse**

tube. In-line tube arrangements will assure the cleanability of all tube surfaces using retractable sootblowers.

When it is not possible to lower the flue-gas temperature of the boiler bank to about 600°F (315°C), an economizer cools the combustion gases. The final heat-recovery equipment is an air preheater designed to provide the optimum temperature for the gas-cleaning equipment.

BOILERS FOR FIRING BYPRODUCT FUELS

Many boilers used for process purposes and power production can be designed to fire waste byproduct fuels in order to

- reduce disposal cost
- meet environmental requirements, or
- reduce usage of high-cost prime fuels

The VU-40, VU-60, and shop-assembled boilers (described later in this chapter) have a long history of being used in this rôle. This section covers only a few of the many byproduct fuels and their associated impact on steam-generator and firing-system design. Others are found in Chapter 8 of the 1981 Edition of COMBUSTION: Fossil Power Systems.

CATEGORIZATION OF WASTE AND BYPRODUCT FUELS

Waste gases and byproduct fuels may be characterized as solid, liquid, and gaseous in terms of the state in which they are fired. Data on their chemical composition and calorific value are included in Chapter 2. Solid waste fuels are generally handled and burned in stoker-fired, fluidized-bed, or suspended-fuel-fired arrangements: liquids, in atomized form. Except for special cases, these do not differ appreciably from the handling of coal or oil. Gaseous waste and byproduct fuels, on the other hand, may require special treatment.

Relatively clean, and differing from natural gases mainly in volumetric heat content, refinery gas and coke-oven gas require only minor changes in firing-system design. In general,

fuel gases with heat contents higher than 500 Btu/cu ft (18.6 MJ/m^3) are easily handled by burners of the same type as used for natural gas. With gases of heating value less than 250 Btu/cu ft (9.3MJ/m^3), consideration should be given to firing systems of high volumetric capacity, such as tangential firing.

The large percentages of inert gases or dust carryover differentiate waste gaseous fuels from natural gas and from each other. Most problems involved with these gases are typified in the handling of steel and copper reverberatory gases, blast-furnace and open-hearth furnace gases, catalytic-cracker regenerator gas, and gases from cement kilns. The degree to which the inerts are present, and the quantity, size, and character of the solid, strongly influence the boiler design for a given application.

BOILERS FOR BLAST-FURNACE GAS

Blast-furnace gas is a dust-laden byproduct of the iron reduction process. It is valuable for its high carbon-monoxide content, which varies widely with the furnace charge and affects the heating value of the gas, which can range from 65 to 110 Btu/cu ft (2400 to 4100 KJ/m^3).

The dust carried over is mostly iron oxide which can foul gas mains, firing systems, boiler furnaces, and convection heating surfaces. To extend and improve availability of equipment, the blast-furnace gas may undergo as many as three stages of cleaning. The first stage is a simple mechanical separation; the second, a primary spray washing; and the third, either a more vigorous washing or an electrostatic separation. The dust load after the first stage of cleaning may range from 5 to 8 grains/cu ft. (11 to 18 grams/mk^3). This is reduced to 0.3 to 1.0 grain/cu ft (0.7 to 2.3 g/m^3) in the second stage and, in the final stage, to 0.005 to 0.01 grain/cu ft. (0.01 to 0.02 g/m^3).

Boilers designed for steel-making service must be able to respond to very rapid load changes. When fired by blast-furnace or coke-oven gas, they are further subject to quick changes in the availability of these fuels. For this reason, a supplemental fuel is required. The fuel must rapidly pick up load to maintain continuity of steam-generator operation.

Unlike designs for fossil fuels, the design of a boiler for blast-furnace-gas operation involves many variables. As a first consideration, the quantity of solid carryover in the gas may permit a unit which is as simple and as compact as those used exclusively for natural-gas or oil firing. Conversely, it may require a unit of large size, freely hoppered and provided with as many soot-cleaning devices as a pulverized-coal-fired boiler.

Second, the low calorific value results in 1500 to 1600 lb/hr of combustion products for each million Btu fired (645 to 690 kg/hr per million joules fired). This compares with the 850 to 950 lb/hr of products per million Btu for oil and gas (365 to 410 kg/hr per million joules) and 950 to 1050 lb/hr of products per million Btu (410 to 450 kg/hr per million joules) of coal fired. For a given evaporation, the quantity of products passing through a boiler varies considerably as fuels are changed. This means that the control of the steam temperature for units having high-temperature superheaters becomes a critical consideration. Design conditions are made more difficult because of the erratic availability of the blast-furnace gas, which creates the need to hold steam temperature constant under conditions in which it would tend to swing as much as 100°F (55°C) in seconds. This problem is normally handled by steam-temperature control devices such as spray or gas-recirculation.

There is an additional problem of handling the simultaneous burning of a number of fuels in the same furnace with blast-furnace gas. Tangentially-fired units of the type shown in Figs. 4 and 19 can be adapted easily to fire blast-furnace gas along with several other fuels.

BOILERS FOR REFINERY REGENERATOR GAS

In the petroleum-refining industry, one of the major processes for producing gasoline is catalytic cracking. It is necessary to regenerate the catalyst continuously to maintain optimum output in steady-flow catalytic processes. The process of regeneration involves the removal of coke deposited on the catalyst in the cracking operation. It is accomplished in a regenerator, a large vessel in which compressed air circulates and scours the hot catalyst to burn off the coke and return the catalyst to the reactor for reuse. For this cleaning, large quantities of air must be raised to a pressure as high as 25 psig (170 kPa). At the same time, to minimize compression costs and to keep the temperature within the regenerator to a level which will not destroy catalyst activity, the quantity of air is limited to the smallest amount that will effectively clean the catalyst.

These combustion conditions produce an appreciable amount of carbon monoxide. This gas, which may leave the regenerator at temperatures as high as 1125°F (610°C), contains a large amount of sensible heat, although it may have a heating value below 40 Btu/cu ft (1500 kJ/m^3) with a carbon monoxide content from 4 to 10 percent. Unable to support its own combustion under ordinary conditions, regenerator waste gas, also called "CO gas", requires a supplemental fuel such as refinery gas, oil, or natural gas.

Boilers have been designed to burn this regenerator gas to supply steam at pressures and temperatures suitable for power generation and process use. At the same time, these boilers reduce the potential for atmospheric pollution by oxidizing the carbon monoxide and hydrocarbon content of regenerator waste gas. A tangentially fired VU-60 boiler for burning this gas is shown in Fig. 19.

The main differences between boilers burning regenerator waste gas and conventional fuel-fired units stem from the nature of the gas and the fact that its full flow must be handled by the boiler at all times. A conventional boiler generally maintains a fixed fuel-air ratio throughout its load range and is capable of wide variation within this range. On the other hand, a boiler handling regenerator waste gas begins with a minimum steam output corresponding to the minimum fuel fired for stable and complete combustion. Steam output can be varied from this minimum condition in only one direction, that is, upward.

An important feature of the waste-regenerator gas boiler is the use of *fixed* tangential firing. The waste gas, stabilizing fuel, and air are divided into several streams which are directed from the corners of the lower furnace. Concentration of the windboxes in the lower part of the furnace provides more heat for ignition as well as longer gas travel; location in quiescent corner positions assures positive ignition during all operating modes.

BOILERS FOR SLUDGE

The disposal of various sludges in boilers is receiving increased attention as conventional sludge disposal methods become energy intensive and dumping is banned. There are a

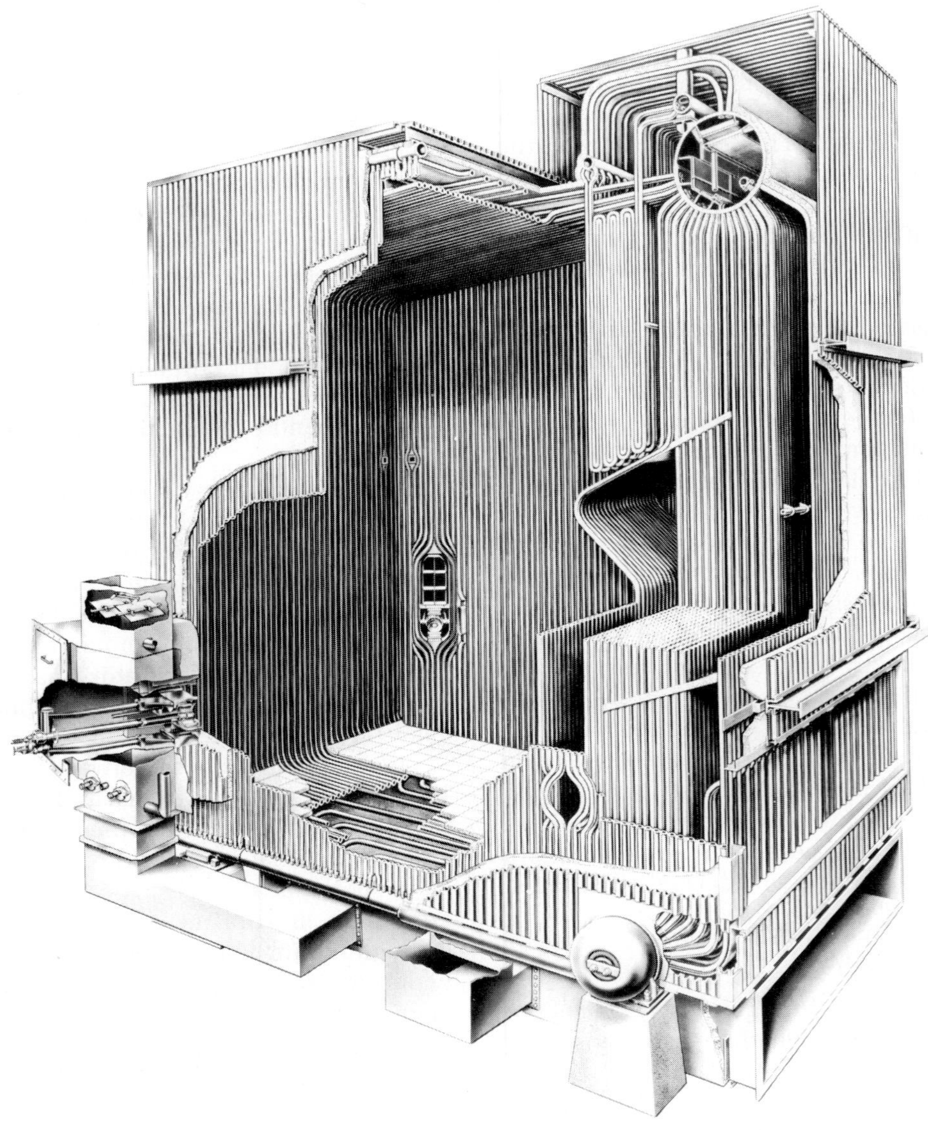

Fig. 19 VU-60 unit with tangential firing system suitable for firing liquid and gaseous prime and waste fuels

variety of sludges produced as process or sewage-treatment residue; they are very high in moisture content, and contain lignins, fibers, combustible organic compounds and ash. It is possible to dispose of most of them in solid-fuel-fired boilers. There can even be a net gain in heat delivered to the boiler; i.e., the chemical heat in the sludge exceeds the latent heat of vaporization required to evaporate the moisture in the sludge. The quantity of sludge that can be disposed of will be maximized by removing the greatest amount of moisture before delivering it to the boiler. If the moisture is removed mechanically, the boiler thermal efficiency will be increased.

Another way to maximize the quantity of sludge to be fired is by the incorporation of a flash-drying system. This system is shown in

Fig. 20. Hot flue gas is taken from the boiler convection-bank outlet and routed to a flash drier where it is used to reduce the moisture content of the sludge. Moisture can be reduced from 80 to 15 percent under optimum conditions. The dried sludge and cooled gas at approximately 300°F (150°C) are separated in a cyclone. The dried sludge is conveyed pneumatically to the furnace for suspension burning and the cooled gas is ducted to the furnace where it is deodorized by the combustion process. The high volatile, low fixed-carbon, and low moisture contents of flash-dried sludge make this an ideal material for suspension burning.

BURNING OF SHREDDED RUBBER

Another category of solid waste fuels that are

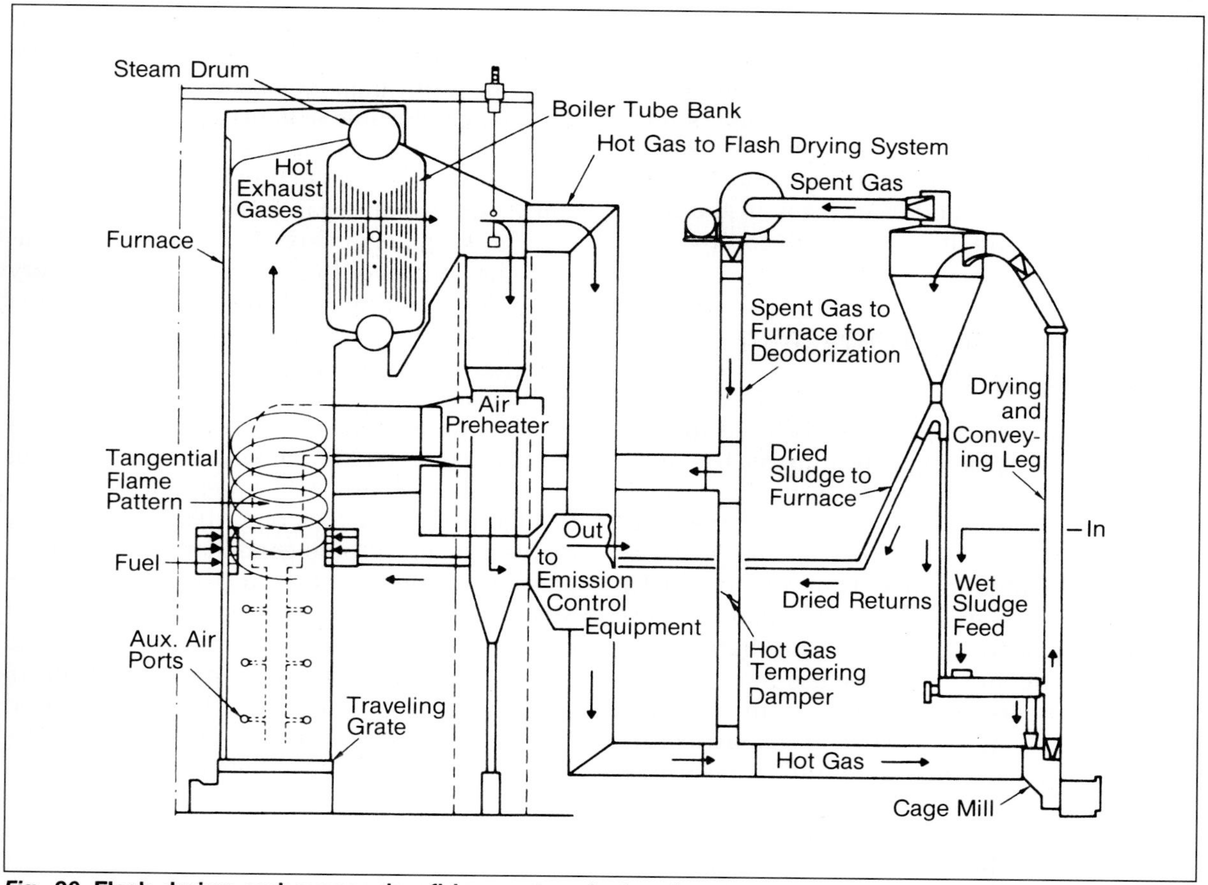

Fig. 20 Flash-drying and suspension-firing system for burning process or sewage sludge

available is shredded rubber from scrap tires. Rubber tires are a nuisance in that they work their way to the surface in sanitary land fills so they present a continual disposal problem. Shredding the tires solves this problem, but the tires still add a considerable volume to our shrinking solid waste disposal areas. It has been estimated that the tire scrappage rate averages 250 million tires a year. With an average heating value of 15,000 Btu/lb (35 MJ/kg), shredded rubber has a considerable potential as a fuel.

Currently, the most common method to burn shredded tires is by mixing them with biomass or coal to be fired on a spreader stoker. The upper limit for tires in this case is 15 to 20 percent on a weight basis.

INCINERATION OF PULP-MILL WASTE STREAMS

To reduce effluent discharge, various waste gases and liquids in pulp mills are often incinerated in the plant's steam generators. These can include total reduced sulfur (TRS) compounds, stripper gas, liquid methanol, and turpentine. Most often, the quantities are small and can be accommodated by modest changes in the fuel-firing system and its associated control and safety system.

STEAM GENERATORS FOR GAS-TURBINE HEAT RECOVERY

Gas turbines have been widely used to provide standby or peaking power for electric utilities, or for unattended service in remote locations. As described in Chapter 1, the thermal efficiency is low because of high exit-gas temperatures (800 to 1000°F, or 425 to 540°C) and high excess-air levels (220 to 300 percent) in the combustion products. The thermal energy remaining in the exhaust gas can be recovered in a heat-recovery boiler to produce additional electricity using a steam-turbine generator. The combined output of electricity from the gas turbine and the steam turbine is 30 to 50 percent greater than that obtained from the gas turbine alone, with no additional fuel input.

Combined-cycle power plants for industrial power-generation applications have much higher thermal efficiencies than conventional steam power plants with the same steam conditions. In general, the high thermal efficiency of a combined-cycle plant can be economically exploited if liquid or gaseous fuels are readily available and the unit can be operated continuously or operated on an interruptible basis at least 50 percent of the time at full power. Chapter 1 identified the four major classifications of combined cycles and their associated heat rates. The two most commonly used cycles for industry employ unfired or supplementary fired heat-recovery steam generators (HRSG's).

Supplementary fired heat-recovery steam generators use firing equipment located in the exhaust gas stream in the boiler inlet transition duct. Since gas-turbine exhaust contains 75 to 80 percent of the oxygen normally found in atmospheric air, fuel may be burned without the need for additional fresh air. By using duct burners, gas-turbine exhaust temperatures can be increased to 1500 to 1600°F (815 to 870°C) with a consequent reduction in the oxygen content of the exhaust gas from 15 percent to 11 percent. Supplementary firing generally doubles the steam output of the heat-recovery boiler by providing a mechanism for varying steam production and matching process-steam demand, independent of the gas-turbine electricity production.

Most applications of HRSG's to gas-turbines of greater than 20 MW generate steam at two or three pressure levels. High-pressure steam (600 to 1800 psig, or 4.1 to 12.4 MPa gage) usually drives a steam-turbine generator. Intermediate-pressure steam (200 to 400 psig, or 1.4 to 2.8 MPa gage) is used for process steam in a plant or is injected into the gas-turbine combustor to reduce NO_x emissions. Low-pressure steam (5 psig to 120 psig, or 35 to 825 kPa gage) is used for plant processes or feedwater heating in a deaerator. An increasing number of installations induce intermediate-pressure steam for additional power recovery in the low-pressure stages of a steam turbine. Certain gas-turbines can accept steam into the power turbine to enhance power output when plant steam demand is low, but electrical demand is high. The heat-recovery steam generator may also incorporate additional water-heating sections for condensate preheating or for high-temperature water for fuel heating or other plant processes.

In some locations, air-quality authorities have imposed very stringent requirements for NO_x emissions from gas turbines. In many cases, these requirements virtually mandate the use of NO_x-reduction catalysts in the turbine exhaust steam. These catalyst assemblies operate in a narrow temperature range that is lower than the turbine exhaust-gas temperature. The presence of an HRSG is an asset in the strategy to control emissions since the NO_x-reduction catalyst may be located in the appropriate temperature zone between sections of heat-exchange surface in the boiler.

STEAM-GENERATOR DESIGNS

The basic principles for selecting heat-recovery steam-generating equipment are similar to those for conventional utility and industrial boilers. However, the designer must be aware of the entire system arrangement in order to integrate the steam generator properly within the overall plant. Although cycle efficiency and economics generally determine the basic cycle conditions, the designer is typically faced with a matrix of conditions which determine the optimum design. These conditions include a wide range of thermal performance parameters, dictated by varying ambient conditions and steam load requirements, limitations on capital cost, and restrictions on available space. In pursuing a solution to the demands of a specific application, three aspects of the boiler design process dominate: (1) extensive use of externally finned tubing for maximum convective heat recovery, (2) an emphasis on low gas-side pressure loss to limit the gas-turbine fuel-rate penalty associated with increased backpressure on the gas turbine, and (3) distribution of heat-recovery surface in multiple sections to achieve optimum heat transfer at each temperature level through the boiler.

BOILER CONFIGURATION

Waste-heat boilers in gas-turbine exhaust service can be configured with gas flow in the horizontal or vertical direction. Vertical gas-flow units permit an arrangement of equipment in the exhaust flow path that occupies less floor space but requires extensive steel support structure. Horizontal gas-flow units generally cover a greater plan area, but afford much better access for maintenance of boiler parts, duct burners, catalyst elements, and other equipment that may be associated with the HRSG.

Boilers with vertical gas flow usually employ horizontal tubes connected by return bends with the tubes supported at several locations along the length of the tube by tube sheets, as illustrated in Fig. 21. Most of these applications require a circulating pump in the steam-generating sections of the boiler. The circulating pump ensures uniform distribution of water to multiple parallel steam-generating circuits. Pumps are usually sized to maintain a circulation ratio of 4 to 1 at the maximum steaming condition.

Boilers with horizontal gas flow use vertical tubes connected to headers at the top and bottom (Fig. 22). The tube and header assemblies may be either top-supported or bottom-supported. Although the tubes are self-supporting in the vertical direction, lateral restraints are required to control gas-flow-induced vibrations. Natural circulation in the steam-generating sections provides high circulation ratios without the use of pumps.

A simplified flow diagram (Fig. 23) for a triple-pressure HRSG illustrates the way in which heat-absorbing sections operating at certain temperature levels are located in the gas stream to minimize the amount of heat-transfer surface required. There are ten discrete heat-exchange sections distributed in descending order based on the gas temperature available and the fluid temperature requirements. Note that the intermediate-pressure generating bank and economizer sections are intermeshed with sections of the high-pressure economizer to optimize boiler performance.

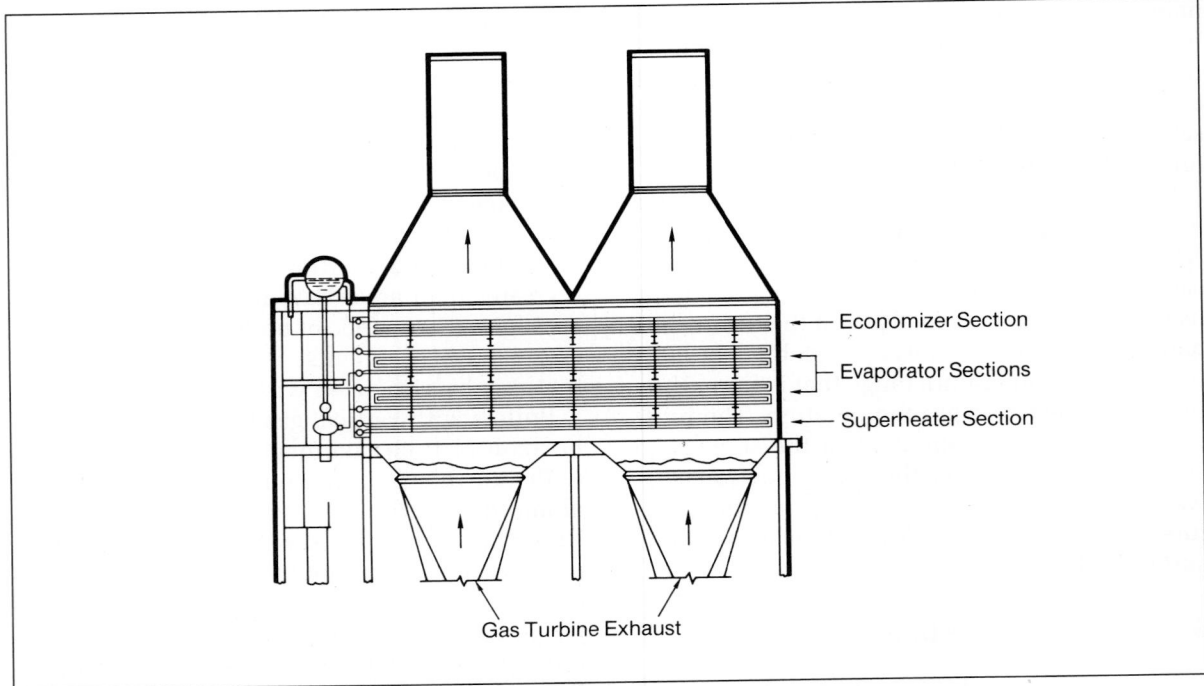

Fig. 21 Unfired steam generator for recovering heat from gas-turbine exhaust

The two critical temperature differences that influence the amount of heat-transfer surface and the overall steam generated at each pressure level are the:

■ Pinch point: The difference between the gas temperature leaving an evaporating section and the temperature at which boiling is occurring (saturated-water temperature).

■ Approach temperature: The difference between the saturated-water temperature in an evaporating section and the incoming feedwater temperature.

The pinch point strongly influences the amount of heat-transfer surface in the evaporating section. Current HRSG designs use pinch points in the 15 to 25°F (8 to 14°C) range. In general, these boilers have 50 percent more surface in the evaporating section than boilers with pinch points of 40 to 50°F (22 to 28°C).

The approach temperature also influences the amount of surface required for an economizer section, with exponentially increasing amounts required for very low approach temperatures. Current HRSG economizers have approach temperatures in the 15 to 25°F (8 to 14°C) range at the design point. Many other operating conditions can occur at off-design points, including start-up. Some conditions will result in steaming at the exit of the economizer, such that it acts as evaporative surface.

Specific provisions to accommodate steaming at levels up to 5 percent of total flow in an economizer include: (1) careful control of water distribution in the last downflow passes of the economizer to cause that portion of the economizer to behave as a forced-circulation evaporator, or (2) configuring the last few passes of the economizer as entirely upflow, with relief by natural circulation into the steam drum.

The triple-pressure HRSG temperature diagram shown in Fig. 23 illustrates the distribution of heat-exchanger sections and the associated temperature differences between exhaust-gas and water and steam temperatures. Pinch points can be observed as a relatively small temperature difference at the right-hand

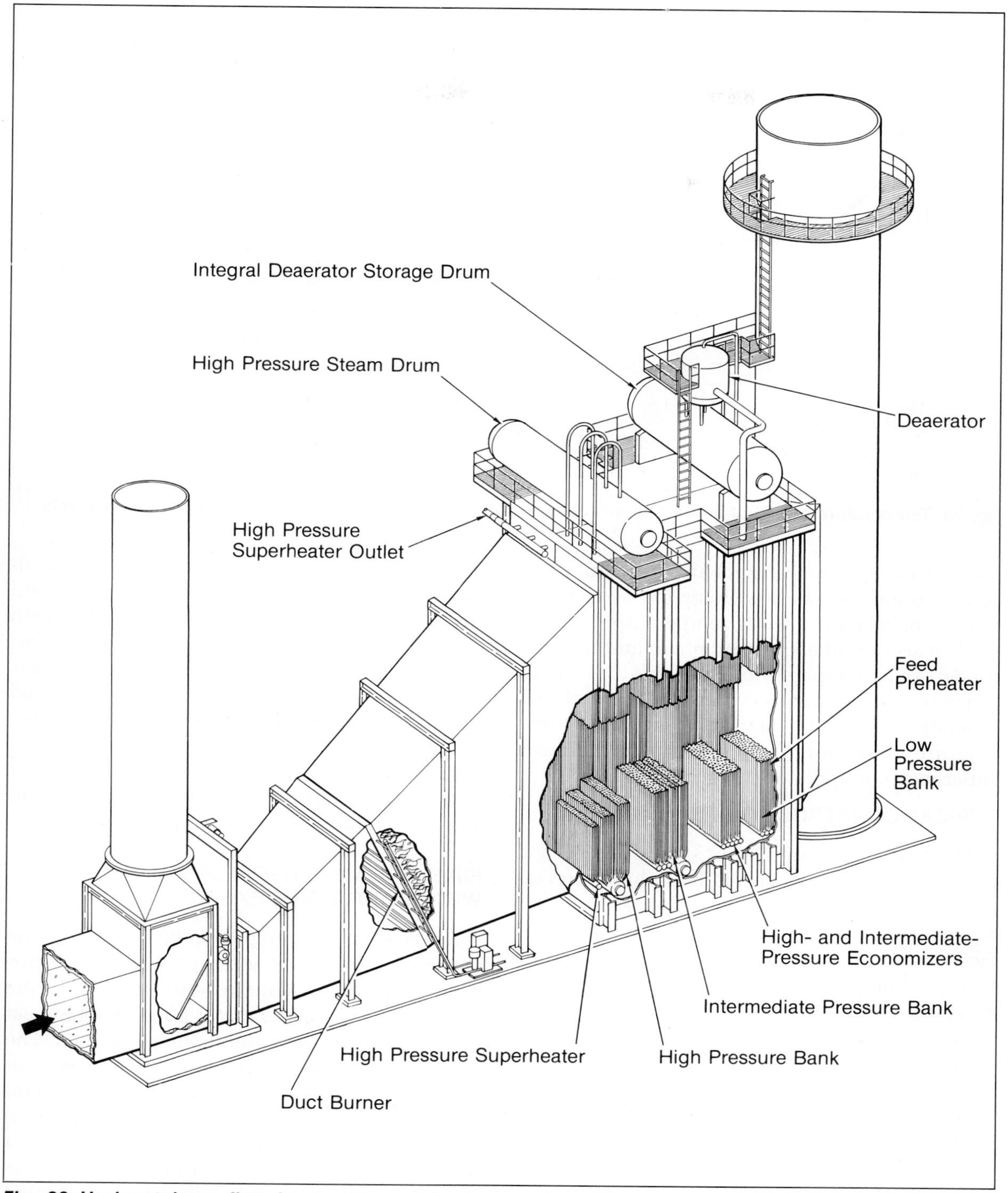

Integral Deaerator Storage Drum

High Pressure Steam Drum

Deaerator

High Pressure
Superheater Outlet

Feed
Preheater

Low
Pressure
Bank

High- and Intermediate-
Pressure Economizers

Intermediate Pressure Bank

High Pressure Bank

High Pressure Superheater

Duct Burner

Fig. 22 **Horizontal-gas-flow heat-recovery steam generator with supplemental firing in duct from gas turbine**

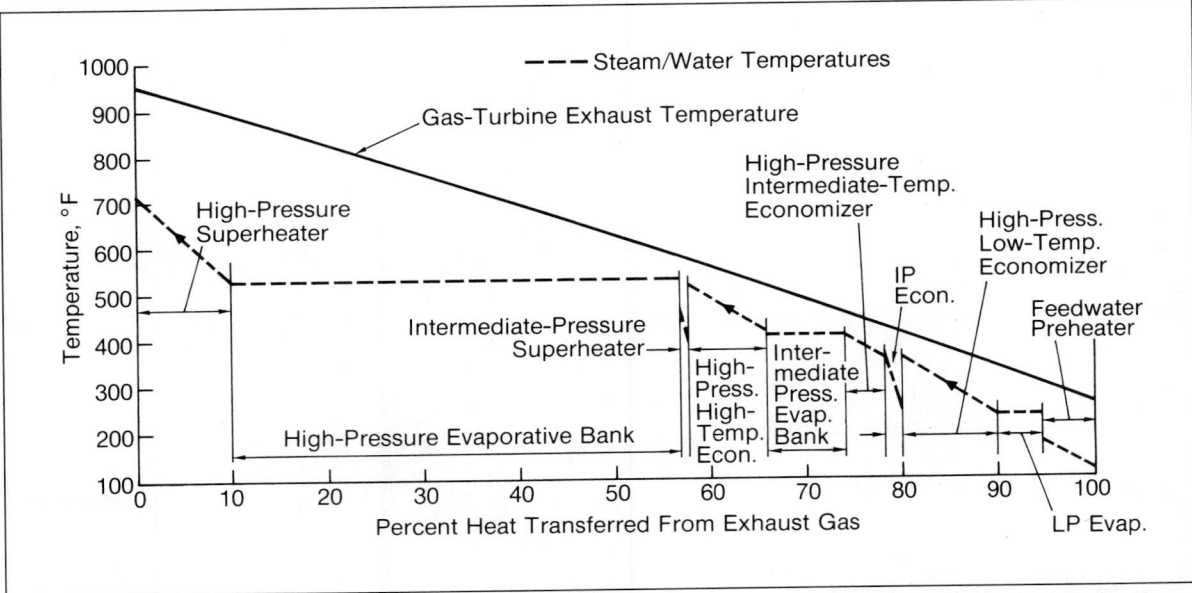

Fig. 23 Temperature profile of unfired heat-recovery steam generator with three operating-pressure levels

side of each evaporating bank section. Approach temperatures are illustrated as the difference between the water temperature leaving the last section of each economizer and the saturated-water temperature. Note that the high-pressure economizer is divided into three separate sections to provide appropriate temperature zones for the intermediate-pressure superheater, evaporator and economizer.

TYPICAL CONSTRUCTION FEATURES

The triple-pressure HRSG shown in Fig. 22 illustrates equipment normally included in the scope of supply of the boiler supplier:
- expansion joint at gas-turbine exhaust interface
- single-blade exhaust diverter valve
- bypass stack with silencer
- inlet transition duct with flow corrective devices
- duct burner
- heat-recovery steam-generator modules
- steam drums
- access ladders and platforms
- exhaust stack

Insulation is placed on the inside of all duct

sections and boiler casing sections, thereby allowing the use of carbon steel casing plate and stiffeners, and minimizing the thermal growth of the overall boiler structure. A system of internal liner plates protects the insulation from gas flow. These plates are segmented for individual thermal expansion and overlapped in the direction of gas flow. All boiler pressure parts are supported in ways that allow complete freedom for thermal expansion relative to the casing and support structure.

FIRED STEAM GENERATORS SUPPLIED WITH GAS-TURBINE EXHAUST

Normally containing 75 to 80 percent of the oxygen found in free atmospheric air, gas-turbine exhaust can concurrently supply to the furnace of a steam generator both sensible heat and oxygen for the combustion of a fuel. The design and operation of such boilers vary considerably, depending upon the ratio of the total exhaust flow to the amount necessary for oxidizing the supplementary fuel needed for a given evaporation. Combustion air preheaters are not used because of the already high level of preheat represented by the 700° to 900°F (370°

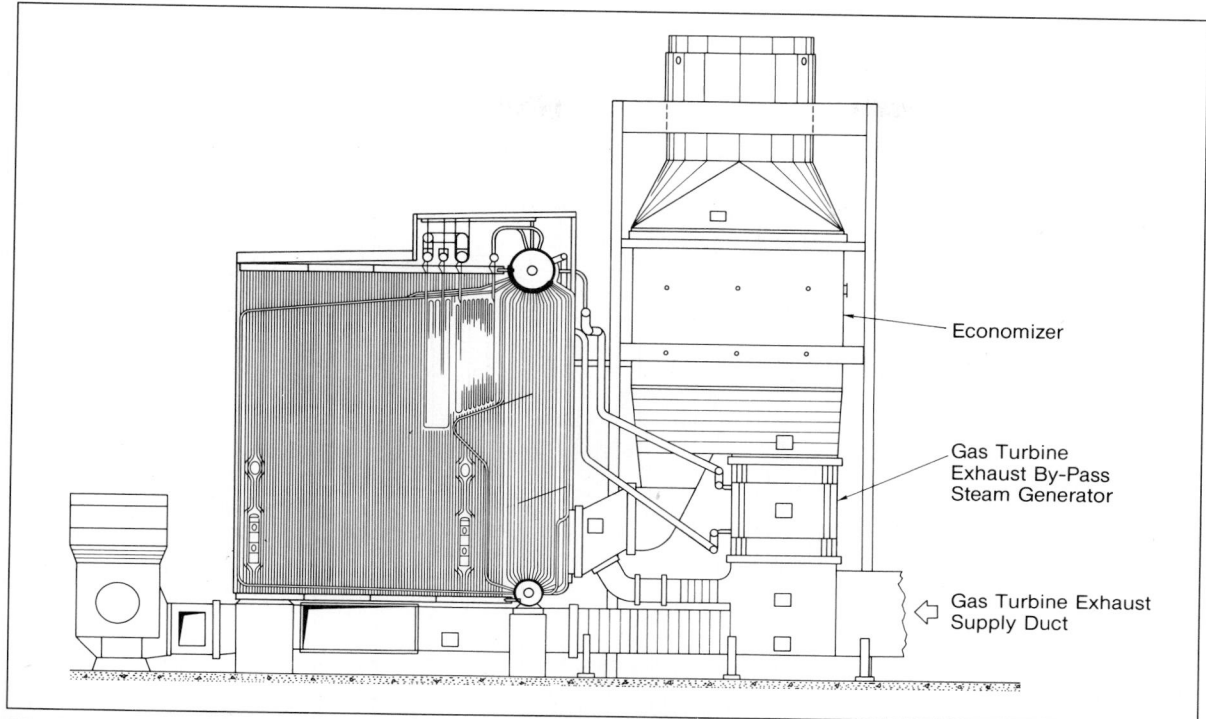

Fig. 24 **Boiler designed for supplementary firing in conjunction with gas-turbine combined cycle**

to 480°C) temperature of the exhaust gases.

Supplementary-fired steam generators (Fig. 24) using most of the oxygen in the turbine exhaust are of the same design and size as units using outside air through forced-draft fans. The stack temperature on such a unit can be dropped economically to within 100°F (55°C) of the incoming feedwater temperature. Since the boiler is sized for a flue gas weight based on fresh-air firing, a portion of the gas turbine exhaust is by-passed. This portion of the exhaust is cooled by passing over a separate steam generating bank, to the same temperature as the gases passing through the boiler, and then proceeds to final heat recovery in the economizer. In such a cycle, gas-turbine and boiler sizes must be matched closely to obtain a high ratio of feedwater flow to gas-turbine exhaust flow.

Because all gas-turbine/boiler applications involve the recovery of sensible heat, the usual concept of boiler efficiency loses its significance. Customary practice, therefore, is to eval-

uate performance of combined-cycle boilers on the basis of stack temperature. The overall station heat balance is determined using the calculated value of fuel fired in the boiler (rather than boiler efficiency as such), in addition to the fuel fired in the gas turbine.

HRSG'S FOR INCINERATOR GAS

Some firing systems for municipal refuse use grates or rotary kilns in conjunction with refractory furnaces. In the past, quench chambers were often used to cool the gases before they entered the particulate-removal device. An alternative is to incorporate a waste-heat boiler with the capability to recover sensible heat in order to generate steam for power or process needs.

Energy-recovery boilers designed for this purpose (Figure 25) include a water-cooled chamber to reduce the flue gas to below its ash-softening temperature. The flue gas from the incinerator will have the same physical and

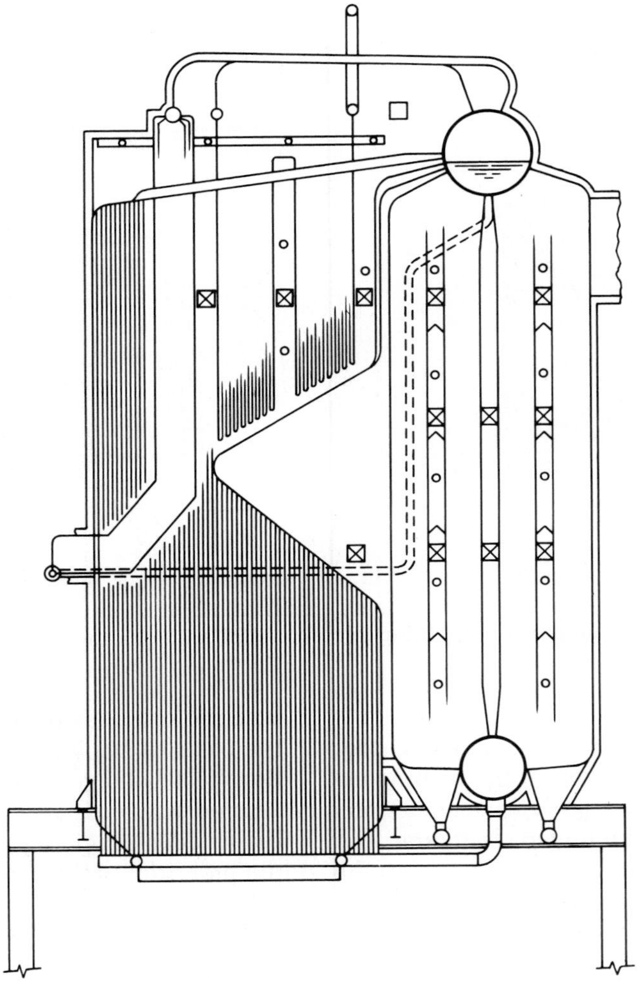

Fig. 25 Heat-recovery boiler with water-cooled furnace for temperature reduction of gases from refuse-fired incinerator

chemical properties as previously discussed for refuse-fired boilers. Therefore, it is important that conservative tube spacings, low velocities, proper erosion and corrosion tube-thickness allowances and other refuse-boiler guidelines be used diligently in the design of these waste-heat boilers.

SHOP-ASSEMBLED BOILERS

The growth of oil and gas suspension-fired shop-assembled boilers since their introduction in the early 1950's has paralleled industry's demand for higher capacities, pressures and temperatures. The evolution of the boiler line

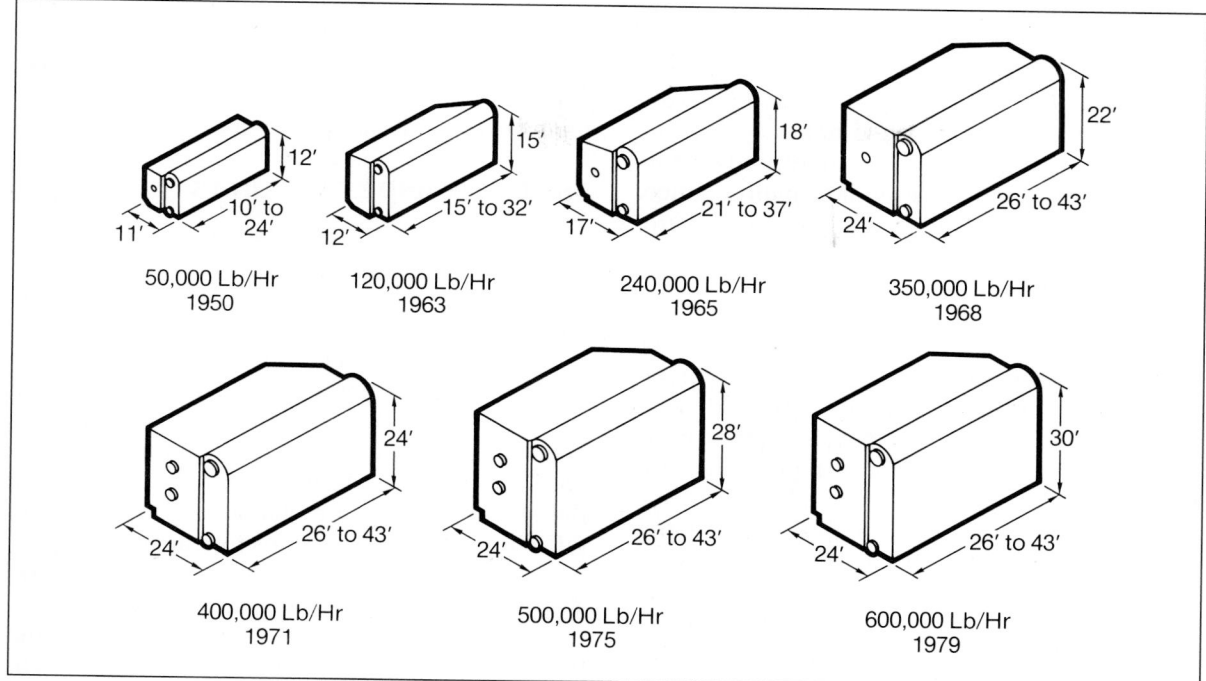

Fig. 26 Size and capacity growth of the VP boiler

has been from 20,000 lb of steam per hour (2.5 kg/s) before 1950 to approximately 600,000 lbs/ hr (75 kg/s) in the late 1970's. As an example, Fig. 26 shows the evolution of the VP (D type) boiler. Steam pressures and temperatures have kept pace with this capacity growth, from the saturated-steam conditions of the 1950's to today's operating pressures of 1650 psig (11.4 MPa gage) and steam temperatures of 950°F (510°C).

As compared with a field-erected unit, the most significant advantage of the shop-assembled boiler is its lower installed cost. This differential comes from the development of standard designs with maximum use of standardized fabrication procedures and minimum field-installation costs.

Shop-assembled boilers can be furnished with integrated auxiliary equipment. The lower capacity units are shipped completely packaged with fuel-burning equipment, safety and combustion controls, and boiler trim. Because of shipping clearance limitations, it is not al-

ways possible or desirable to furnish the higher-capacity boilers in a single package. With higher capacity units it may also be desirable to use heat-recovery equipment which is shipped as a separate package.

Most shop-assembled water-tube boilers use thermal circulation and are designed for pressurized firing. Shipping clearances determine the allowable height and width dimensions of an assembled unit. Usually, the allowable shipping length is greater than can be effectively used. One of the problems of the designer, therefore, is to use the available height and width to best advantage.

The burner/furnace design must be properly coordinated, and burners have been developed specifically for shop-assembled boilers. Having a relatively narrow flame pattern, these burners are designed to burn the fuel completely within the necessarily limited, though sufficient, clearance dimensions from the burner centerline to the furnace walls.

Shop-assembled boilers typically use exten-

sive water-cooled surface. This heat-absorbing surface that "sees" the flame determines the temperature of the combustion gases leaving the furnace. Liberal use of water cooling on the furnace walls, roof, and floor not only reduces the furnace temperature but also minimizes furnace refractory and its attendant maintenance. With a completely welded tube-wall enclosure, the result is highly desirable uniform cubical thermal expansion.

The amount of furnace volume or heat-release rate/cu ft of furnace volume bears no direct relationship to the furnace exit temperatures; the amount of water-cooled furnace surface or the heat-release rate per sq ft or m^2 is the valid criterion for determining furnace conditions. Typically, the furnace heat-release rates when firing oil and gas range from 175,000 to 200,000 Btu/hr-ft^2 (550,000 to 630,000 W/m^2) of EPRS (effective projected radiant surface).

Size and furnace configuration for these package boilers result in a ratio of radiant surface to furnace volume which is greater than for the larger field-erected units. This permits considerably higher furnace liberation rates/cu ft of furnace volume.

The need to limit gas-side draft loss usually determines the flue-gas velocities for gas- or oil-fired shop-assembled boilers. Velocities can go as high as 100 fps (30 m/s) and result in draft losses of 10 to 15″ WG (2.5 to 3.8 kPa).

Convection tube spacing is not critical in shop-assembled units as the fuels generally have little or no ash. However, the designer still must be cognizant of peculiarities in certain fuels such as high-sodium, high-vanadium residual oils which can form tacky deposits at approximately 1100°F (590°C). In this case, tube spacing and sootblower arrangement must be properly factored into the design.

The enclosing casing structure of shop-assembled boilers, including the integral base plate, serves several functions. It is designed to eliminate objectionable gas leakage under maximum pressure conditions. Also, it must have sufficient structural strength for the necessary handling and lifting during manufacture, in transit, and at the installation site. Permanent or detachable lifting lugs are provided to facilitate handling. Many oil- and gas-fired assembled units require only a suitably reinforced-concrete slab or curb for foundation.

THE C-E VP AND A-TYPE UNITS

A typical shop-assembled boiler configuration (Fig. 27) is the C-E VP, which is a D-shape design with a two-drum, vertical bent-tube boiler bank, and water-cooled furnace. These units can be designed to generate as much as 600,000 pounds per hour (75 kg/s) of steam. The VP has welded-wall construction, unheated downcomers, and a combination radiant-convection superheater which produces a flat steam-temperature characteristic. It can be rail-shipped in two sections for capacities up to 240,000 lb/hr (30 kg/s); larger VP units are shipped over waterways on large barges.

Another shop-assembled boiler is the A type (Figs. 28 and 29) which can generate as much as 300,000 pounds of steam per hour (38 kg/s). The A boiler is a three-drum design with one upper (steam) drum and two lower drums. Its symmetry makes the A design ideal for rail shipment because ballast is not needed. This simplifies off-loading and handling.

Both the A and VP designs have boiler banks with a multiplicity of simple tube circuits starting at the lower drum(s) and terminating in the steam drum. Expanded tube joints have proven most practical to connect these relatively close-spaced tubes into the drums. One factor which determines the thickness of drum tube sheets is the width of the ligaments between tube holes, as described in Chapter 6. To achieve the maximum ligament efficiency and minimum drum thickness with these close-spaced tubes, the tube ends are frequently swaged to a smaller diameter in order to reduce the required tube-hole size.

HEAT-RECOVERY EQUIPMENT

Most shop-assembled boilers include equipment to increase overall boiler efficiency and save substantial fuel by recovering heat from the

Fig. 27 C-E VP, a D-type shop-assembled boiler

flue gases leaving the generating equipment. In most cases this heat-recovery equipment can be shipped as a complete, but separate, package.

A counterflow arrangement of flue gas and feedwater for an economizer, or of flue gas and combustion air for an air preheater, results in the most effective use of the heat-recovery surface (see Chapter 6). With high-sulfur fuels, the possibility of corrosion of the cold-end surfaces should be considered, with either an economizer or an air preheater.

Because the gas-side transfer rate in an economizer is relatively low compared with water-side rate, the metal temperature of the cold-end of a counterflow economizer is only slightly above the entering feedwater temperature. With high-sulfur fuels, an economizer is usually not advisable unless the feedwater temperature exceeds 250°F (120°C).

The degree and frequency of partial-load operation are most significant in establishing the optimum air-heater selection. Packaged regenerative air preheaters are designed with reversible and replaceable cold-end layers fabricated

Fig. 28 C-E Type A shop-assembled boiler

Fig. 29 Shop-assembled units like this A boiler are manufactured under controlled indoor conditions

of corrosion-resistant material to minimize maintenance.

SHOP-ASSEMBLED WASTE-HEAT BOILERS

Although applied primarily to fuel-fired boilers, shop assembly is also applicable to heat-recovery steam generators and waste-heat boilers. Unless there is some special consideration, such as auxiliary firing, these units are built without a furnace preceding the boiler convection surface.

The temperature of waste gas and the constituents in gas must be carefully evaluated in establishing the design of waste-heat boilers. The chemical analysis, dust loading, and slagging characteristics all affect arrangement of the boiler surface.

Because most oil- or gas-fired shop-assembled boilers are pressure fired, the forced-draft fan overcomes resistances to gas flow throughout the boiler, whereas with many waste-heat units these resistances are overcome by an induced-draft fan. The usually large volume of gases to be handled makes design for low draft losses desirable to minimize the power requirements of the ID-fan driver.

ECONOMICS OF SHOP-ASSEMBLY

Installation of two or more shop-assembled units instead of a single large field-erected unit aids in load flexibility and scheduling maintenance. Other factors such as limited available headroom for the installation of a field-erected boiler, or much shorter delivery time for a shop-assembled unit, are also important considerations. But the main reason to select a shop-assembled boiler is its low overall installed cost.

Fig. 30 **Type A boiler loaded on Schnabel car**

Standardization is the key to reduced costs: most shop-assembled boilers are pre-engineered. In setting up a line of standard boilers, it is necessary initially to consider all possible variables. Design conditions such as capacity and operating pressure are varied in increments. Physical variables such as terminal connection locations may be varied to some limited extent.

SAVINGS IN MANUFACTURING AND ERECTION

Standarization of design, then, saves time and money prior to actual fabrication of a shop-assembled unit. But the greatest savings with shop-assembled boilers is in manufacturing. Standard shop-assembled boilers are built under controlled conditions which permit a high-quality product at low cost (Fig. 29). The enclosed assembly area is arranged for optimum materials flow both of parts fabricated for individual units and for parts and materials to be taken from stock. Likewise, delays from adverse weather conditions are not encountered.

Aside from the obvious elimination of field erection costs, there are supplementary advantages that further reduce the installation costs and time. Most units require only a simple concrete slab foundation. And, as the boiler is shipped assembled, no erection space or material storage space is required at the site.

SHIPMENT

Smaller shop-assembled units have been shipped short distances by truck or in combination with rail. The development of new shipping methods contributed significantly to the rapid growth of large capacity shop-assembled boilers in the 1970's.

The first VP boilers had a fixed cross section and achieved size flexibility by varying the depth of the boiler. The cross section was limited to standard clearances for American railroads and permitted shipment in standard flat cars to almost any location. Further investigation determined that clearances to some locations were larger. In addition, special railroad cars were available or could be designed, among which is the C-E patented Schnabel (Fig. 30) to accommodate boilers up to 20 ft high x 13 ft wide x 55 ft long (6m high x 4m wide x 17m long).

Barge shipment has paved the way for delivery, to overseas and domestic markets, of shop-assembled boilers with steam flows as high as 600,000 lb/hr (75 kg/s). Any shop-assembled boiler can be shipped by barge, since equipment so delivered is generally not limited by physical dimensions. Boilers up to 30 ft wide x 35 ft high x 70 ft long (9m wide x 11m high x 21m long) weighing up to 950,000 lb (430,000 kg) can be shipped on virtually any existing commercial waterway with available barge equipment.

HIGH-TEMPERATURE-WATER BOILERS

Heating systems using high-temperature water (HTW) became popular after World War II. These systems usually operate in a pressure range from 120 to 300 psig (0.8 to 2 MPa gage) with supply temperatures to the distribution system from 350 to 420°F (180 to 215°C). The higher temperature range affects cost because water-flow rates and resulting pumping costs are reduced and system exchangers can be made smaller. On the other hand, the lower temperatures permit lighter and less expensive materials.

High-temperature water systems are closed systems with makeup required only to restore water that leaks out of the system at valve stems, pump shafts, and similar packed joints. Obviously the average water temperature in the system varies depending on the load, and an expansion drum permits expansion or contraction of the water volume. A small makeup pump automatically adds the necessary makeup to hold the water level in the expansion drum above a predetermined level.

For heat consumers such as unit heaters, radiant panels, and coils of absorption refrigeration equipment, HTW may be used directly. As with steam, it is used indirectly for domestic hot water. Low-pressure steam or low-pressure hot water, if required, may be produced in suitable heat exchangers.

Further description of HTW boilers is found in Chapter 8 of the Third Edition of Combustion: Fossil Power Systems (1981).

PULP AND PAPER INDUSTRY CHEMICAL RECOVERY BOILERS

Although the chemical recovery boiler (Fig. 31) shares its general appearance and many of its physical components with power-boilers, it is unique in the power-generation field in that its purpose is twofold: the first is to recover inorganic chemicals from black liquor to be recycled in the pulping process; the second is to combust the organic constituents in the black liquor to produce valuable steam. The efficient recovery of the inorganic pulping chemicals and the efficient generation of steam from the chemical recovery boiler are both essential elements in the economic and environmental aspects of the kraft pulping process.

THE KRAFT PULPING PROCESS

This is a chemically closed process that produces pulp for paper making from a variety of organic materials. Wood is by far the most common source of pulp used with the kraft process, although bagasse, straw and other nonwood fibers can also be used. The name *kraft*, from the Swedish and German word meaning strength, is a term often applied to sulfate mills, which produce long-fiber pulp for making high-strength paper.

PRODUCING PULP

Fig. 32 illustrates the pulp-making process. As the diagram shows, wood enters the kraft cycle at the digester for pulping. Within the digester, sodium hydroxide and sodium sulfide dissolve the ligneous substances that bind the cellulose fibers of the wood together. The cellulose fibers, which form the pulp, resist the action of the chemicals. In the cooking process, approximately one half of the wood substance is dissolved in the cooking liquor to form what is called black liquor. Two types of digesters are

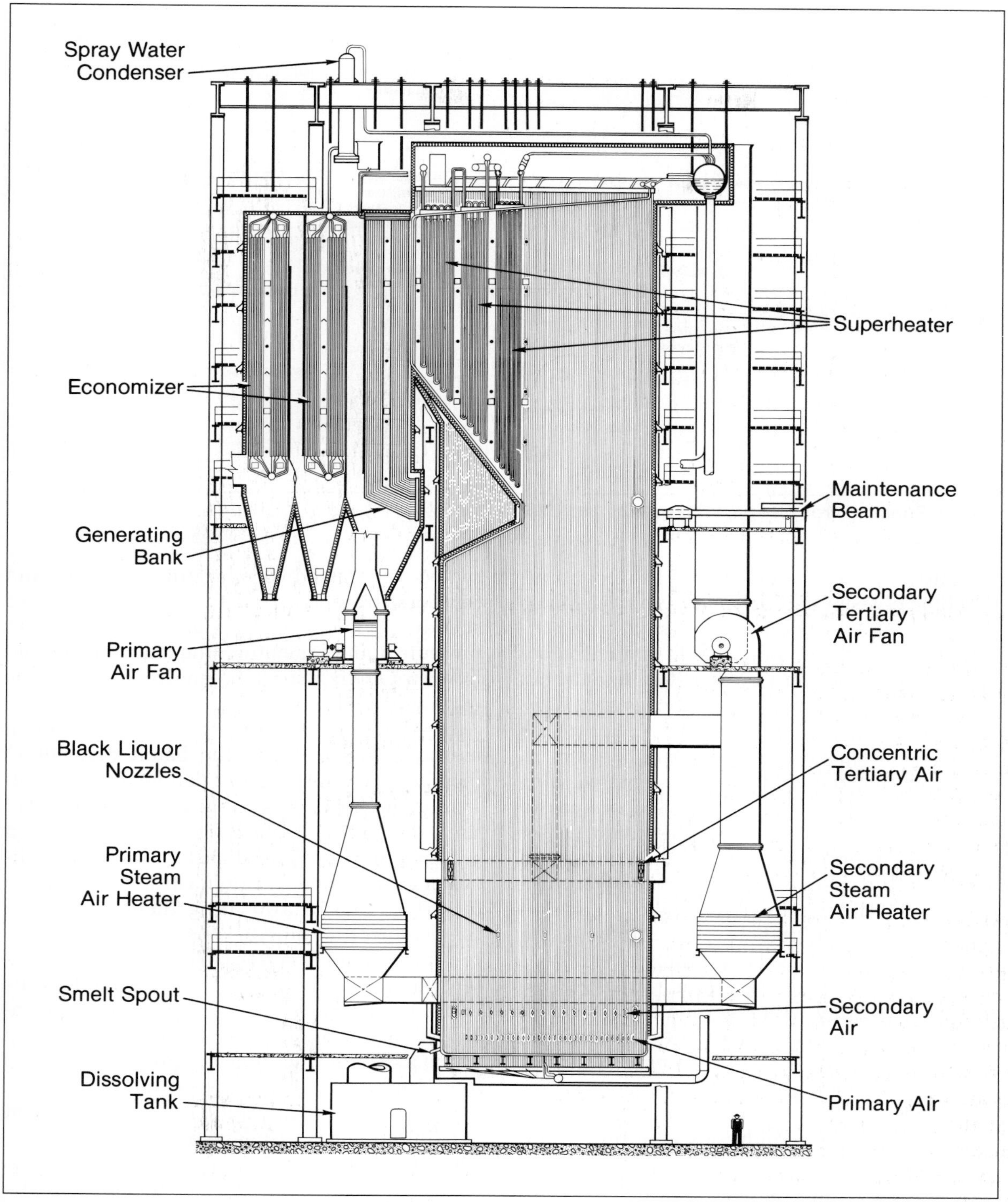

Spray Water
Condenser

Superheater

Economizer

Maintenance
Beam

Generating
Bank

Secondary
Tertiary
Air Fan

Primary
Air Fan

Concentric
Tertiary Air

Black Liquor
Nozzles

Secondary
Steam
Air Heater

Primary
Steam
Air Heater

Smelt Spout

Secondary
Air

Dissolving
Tank

Primary Air

Fig. 31 **C-E chemical recovery boiler**

ABB
ASEA BROWN BOVERI

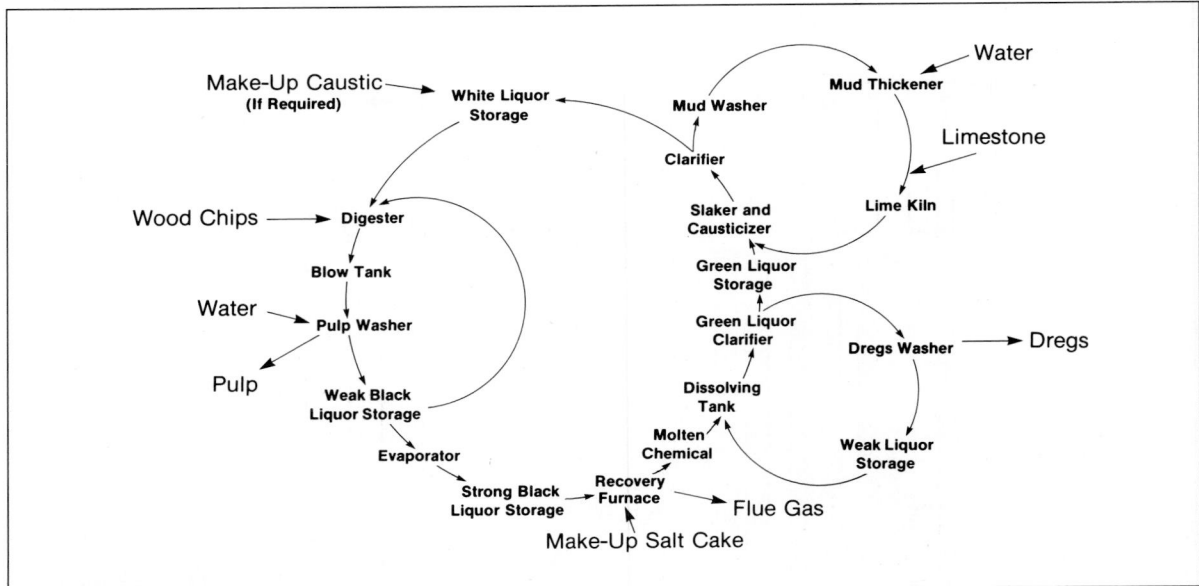

Fig. 32 Principal processes in kraft pulping and chemical recovery

currently in use in the industry: the continuous-type digester and the batch-type digester.

In continuous cooking, the system being installed in most new mills today, the chips are preheated in a steaming vessel before entering the digester. Pre-steaming removes air, non-condensible gases, and volatile constituents such as the terpenes. After entering the continuous digester, the chips are impregnated with cooking liquor at a controlled temperature to ensure uniform penetration of the liquor. After adequate impregnation, the temperature is raised to the cooking temperature of around 330°F (165°C) by indirect heating of circulated cooking liquor and is held there for about one hour. The pulp is then quenched to about 260°F (125°C) with wash liquor. In most cases, diffusion washing which is then carried out in the lower region of the digester removes a considerable proportion of the spent chemicals. The wash temperature in the lower zone of the digester is 175 to 185°F (80 to 85°C). This ensures suitable blow conditions with little or no mechanical damage to the fibers.

In batch cooking, the chips and cooking liquor are charged into the digester which is then sealed and raised to operating pressure and temperature according to a pre-determined schedule. Once the digester is at the proper pressure and temperature, the charge is cooked from two to six hours, depending on the grade of pulp desired.

Direct steam injection is the most common method used to bring the batch digester to cooking temperature, although for some installations the heating is performed in part or entirely by indirectly heating white liquor and recirculating it to the digester. Batch-type digesters using direct steam heating require large quantities of steam to bring them to cooking temperature and pressure. As a result, mills using this type of digester have a heavy intermittent demand from the boilers compared to mills using indirect heating. Batch digesters are installed in groups of at least three and operate in sequence, with each digester preparing 6 to 20 tons (5 to 18 tonnes) of pulp per cook.

After the cook, the mixture of pulp and spent cooking liquor is discharged to the blow tank. The next major step in the process is washing, in which pulp fibers are separated from the black liquor. Most mills use countercurrent rotary vacuum washers in which the pulp is washed with progressively weaker black liquor and finally fresh water. More recently, continuous diffusion washers have been used to complement or replace rotary vacuum washers. After washing, the pulp is further refined and is subsequently sent for conversion into a final product. The weak liquor which was separated from the pulp goes on for further processing.

The kraft process is economically attractive because it is possible to recover and reuse the cooking chemicals, and to generate large quantities of steam and power from the dissolved carbonaceous matter in the black liquor. However, the black liquor leaving the washers contains far too much water to be burned. The next step removes most of this water.

RECOVERING THE SPENT CHEMICALS

Water removal from the weak black liquor is accomplished in large multiple-effect evaporators. Before the black liquor can be fired in the recovery boiler, the concentration of dissolved solids in the black liquor must be increased from an initial 16 to 18 percent to greater than 60 percent. In most mills, the black liquor is fired at a solids concentration of 65 to 75 percent.

Multiple-effect evaporators are used to raise the solids concentration of the black liquor from the initial 16 to 18 percent to approximately 50 percent. An evaporator is a specially designed vertical heat-exchanger in which condensing vapor provides the heat required to evaporate moisture from the black liquor. A multiple-effect evaporator consists of a series of evaporator bodies in which the black liquor concentration is progressively increased in each of the stages or effects. The bodies and heat-exchange surfaces are designed so that the pressure drops successively from the initial steam-supply pressure to a vacuum. The black liquor generally flows countercurrent to the steam. Steam economy is achieved by taking the vapor evaporated from the black liquor in one effect and condensing it in the following effect. Therefore, most of the heat used to evaporate moisture from the black liquor in one effect

is recuperated in the following effect. The vapor from the last effect is condensed and sent for treatment to the foul-condensate disposal or recuperation system. Fresh steam is fed only to the first effect and the clean condensate from the first effect is returned to the feedwater system. Steam economies of up to 7.5 can be achieved with multiple-effect evaporators; i.e., for each pound of fresh steam supplied, 7.5 pounds of moisture can be removed from the black liquor.

Modern installations have concentrators as part of, or following, the multiple-effect evaporator. Concentrators are specially designed evaporators capable of raising the solids content of the liquor to the 65-75 percent level, which is suitable for firing in the recovery boiler. Most recovery boilers installed prior to the early 1970's used a direct-contact evaporator to concentrate the black liquor. The operating principle in a direct-contact evaporator is that hot recovery-furnace flue gases which make direct-contact with black liquor evaporate moisture from the liquor. The evaporated vapors are discharged to the atmosphere along with the combustion gases. Direct-contact evaporators have been almost completely replaced by concentrators in performing the final concentration of the black liquor.

The next step of the kraft chemical-recovery process takes place within the recovery boiler and involves five essential functions in which:

- black liquor is prepared for firing,
- carbonaceous matter is burned out,
- sulfur compounds are converted to Na_2S,
- inorganic salts are smelted for removal, and
- heat is efficiently recovered to produce steam.

The hot smelt (inorganic salts) produced in the recovery furnace flows in a continuous stream into the dissolving tank, where it is quenched and dissolved in weak wash to form "green liquor". This is a solution consisting mainly of sodium carbonate, Na_2CO_3, and sodium sulfide, Na_2S. After the non-soluble materials (dregs) are removed, the green liquor is pumped to the causticizing system in which slaked lime, $Ca(OH)_2$, is reacted with sodium carbonate, Na_2CO_3, to produce a solution of sodium hydroxide, $NaOH$. The sodium sulfide in solution remains unchanged. After this treatment, the liquor becomes "white liquor" and is returned to be reused as cooking liquor in the digester.

The calcium carbonate sludge formed in the causticizing reaction is converted to calcium oxide in the lime kiln and is then hydrolized to calcium hydroxide for reuse in causticizing green liquor to white liquor.

This completes the kraft cycle. The chemicals needed for cooking are recovered and reused. Calcium oxide required for causticizing is reformed, and the dissolved wood substance is burned as a source of heat for steam and power generation.

PRODUCING POWER

In most integrated pulp and paper mills, the waste materials generated by the process furnish a large percentage of the total energy required to produce steam and power. The recovery boiler uses the heat energy from burning the carbonaceous matter in the black liquor, while the bark-fired boiler consumes other waste materials such as sawdust and planer shavings, as well as bark from the debarking drums. When more steam is required than is available from these waste streams, auxiliary fuels can also be fired in the bark or recovery boilers to increase their steam product into the desired maximum continuous rating (MCR).

MILL CAPACITY AND ITS RELATION TO RECOVERY BOILER CAPACITY

The capacity of a mill is based on the tons of pulp that are produced in 24 hours. A mill will also produce a certain amount of spent chemicals, that is, black liquor solids. On an average, 3000 pounds of black liquor solids will be produced for each U.S. ton of pulp produced (1500 kg for each metric tonne). A 1000-U.S. ton (907-tonne) mill will then produce about 3,000,000 lb (1,360,000 kg) of dry solids per day — the actual quantity can vary plus or minus 10 percent. Likewise, the higher heating value of the dry

solids produced can vary from 5,000 to 7,000 Btu/lb (11.6 to 16.3 MJ/kg).

Obviously, the thermal process load which is imposed on the recovery boiler depends on the specific conditions in a particular mill and cannot be related to a nominal mill requirement. Thus, recovery boiler capacity must be related to the following factors:

- dry solids produced per day
- percent solids concentration from multiple-effect evaporator/concentrator
- high heating value (HHV) of the dry solids
- elemental fuel analysis of the dry solids

A typical elemental analysis of black liquor solids derived from a combined hardwood-softwood cook, in percent by weight, is

Carbon	39.4
Hydrogen	3.6
Sulfur	3.4
Sodium	21.2
Inert Mineral Oxides	1.0
Oxygen	31.4
Total	100.0
High Heating Value	6,400 Btu/lb (14.9 MJ/kg)

The above values were derived using the standard as recommended by the Technical Association of the Pulp and Paper Industry (TAPPI); this procedure should be used as others may give varying results.

BLACK LIQUOR FUEL PREPARATION

Before the strong black liquor leaving the multiple-effect evaporators can be introduced into the furnace for burning, the liquor must undergo several important changes.

- If it is not already at the proper firing concentration coming out of the multiple-effect evaporator, the liquor must be concentrated to sufficient strength for firing.
- Ash collected in the precipitator and the boiler hoppers must be returned to the incoming liquor stream.
- Makeup chemicals must be introduced if not added elsewhere in the process.
- The liquor must be brought up to the proper

pressure and temperature required at the spray nozzles.

The final fuel preparation will be done differently depending on whether the liquor is at firing concentration as it leaves the multiple-effect evaporator plant.

BOILERS USING DIRECT-CONTACT EVAPORATION

Direct-contact evaporators are still commonly used on boilers burning very viscous black liquors such as those produced by bagasse pulping. Because of the direct contact of the flue gases with black liquor, the sulfur compounds in the liquor can chemically react with the carbon dioxide in the flue gas to form foul-smelling gaseous compounds which might then be discharged into the atmosphere. To avoid the generation of odorous gases, the liquor in direct-contact evaporator systems is first passed through a black liquor oxidizer in which air or oxygen is bubbled through the liquor (Fig. 33). The oxidizer chemically stabilizes the sulfur

compounds in the liquor so that odorous emissions from direct-contact evaporators are greatly reduced.

After the liquor passes through the oxidizer, it is pumped to the precipitator area. Ash collected in the precipitator is mixed into the liquor stream in either of two ways. The ash can be allowed to fall directly into the liquor as occurs in a wet-bottom precipitator, or the ash can be collected dry and conveyed from the precipitator to a chemical/ash mixing tank, where the ash is thoroughly mixed with the black liquor.

The liquor is then evaporated to a firing concentration of about 68 percent solids in a cascade evaporator by bringing hot flue gases into direct contact with strong black liquor as shown by Fig. 34. The evaporator consists of one or two wheels of rotating tube sheets between which are many horizontal parallel tubes. Part of the tubes are submerged in liquor; the rest are in the gas stream. Those that are in the gas stream are completely coated with wet, dripping liquor. The slowly rotating cascade

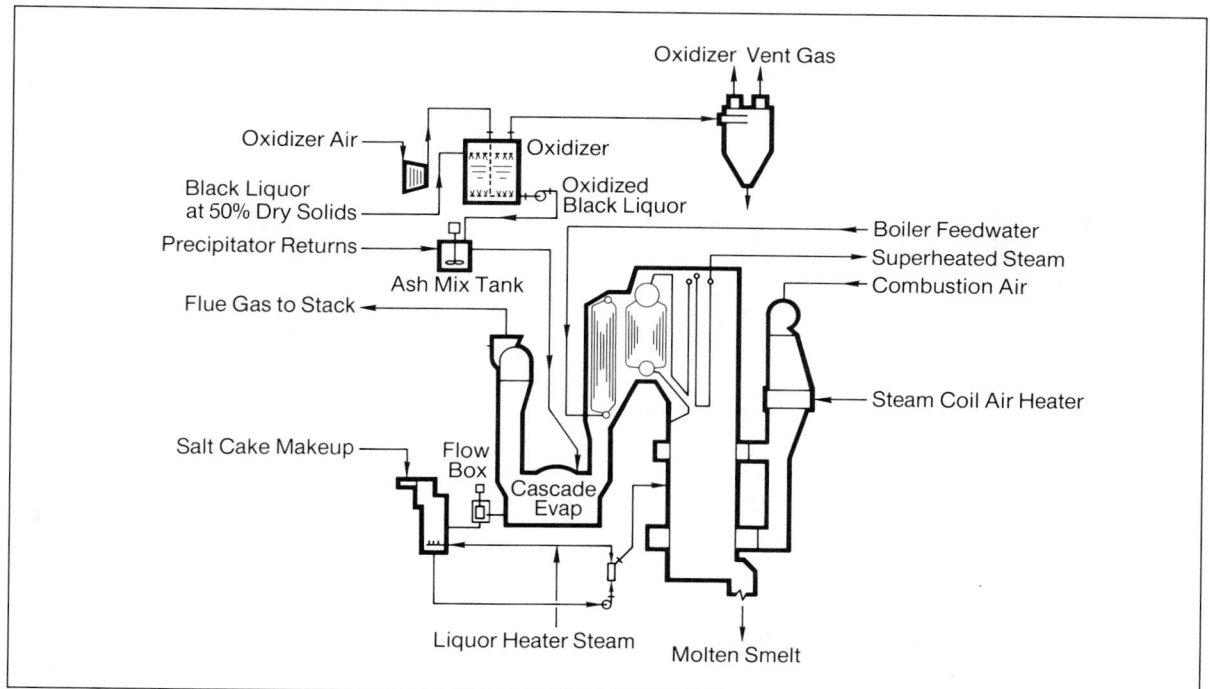

Fig. 33 **Chemical-recovery unit with oxidizer and cascade evaporator**

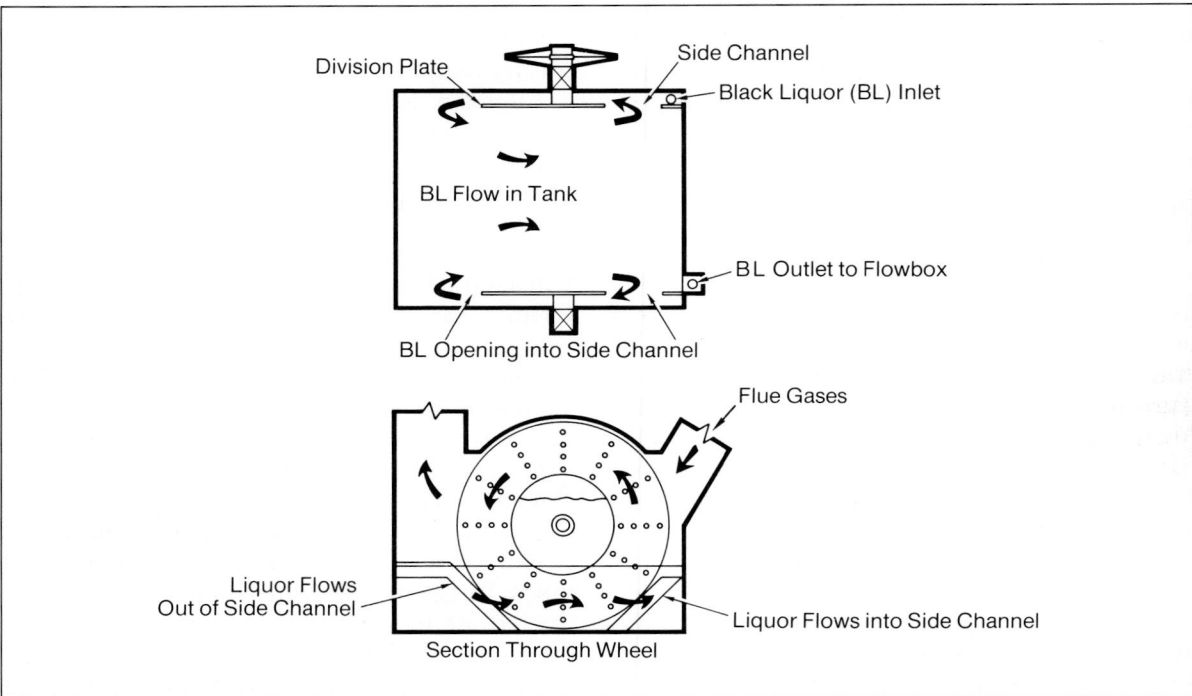

Fig. 34 Pattern of cascade-evaporator liquor flow

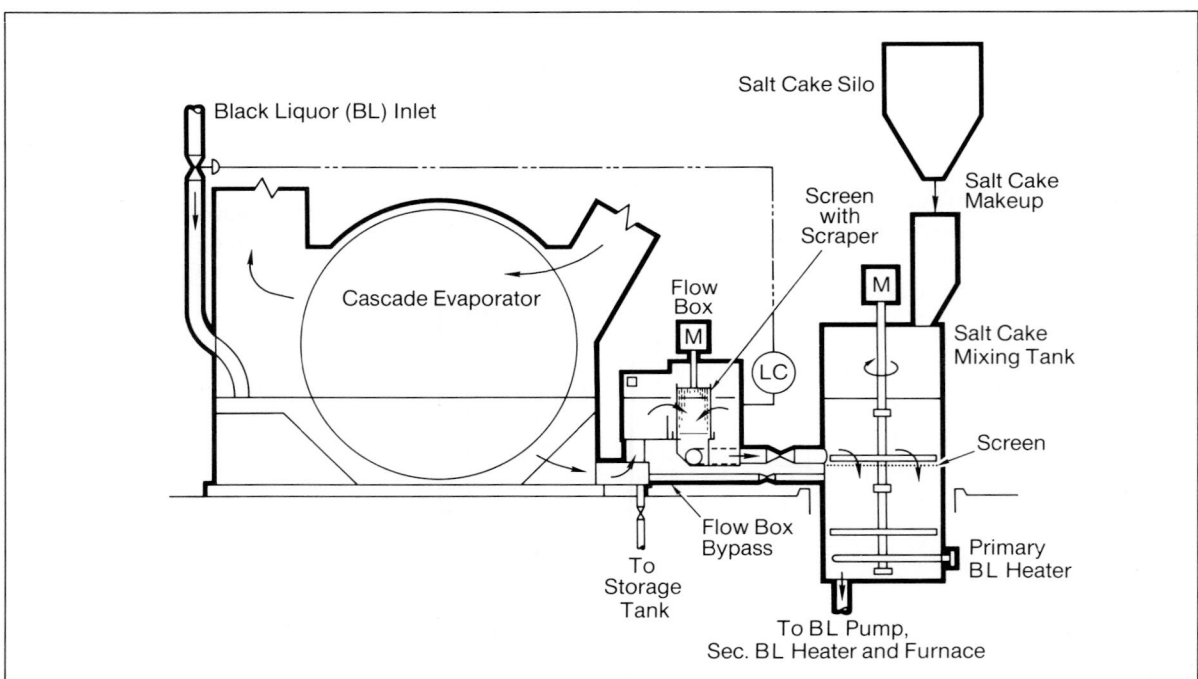

Fig. 35 Black-liquor flow diagram

wheel causes the tubes to be alternately washed in the liquor bath, and exposed to the hot flue gases where water is evaporated. It also induces liquor circulation in the bottom of the cascade which causes the liquor to be well mixed at all times. The tubes never dry out in the gas stream and are, therefore, good collectors of ash particles entrained in the gas stream.

As shown in Fig. 35, the liquor travels from the cascade evaporator to a flow box located next to the cascade tank. Within the flow box is a rotating cylindrical screen through which all the liquor must flow. The screen blocks the larger lumps which are deposited in the box where they dissolve.

From the flow box, the liquor moves into the salt-cake mixing tank where pulverized sodium sulfate can be added to compensate for the chemical losses in the mill. The black liquor is thoroughly mixed with a paddle-type agitator. A horizontal screen within the tank prevents lumps of salt cake from flowing into the black liquor piping. Within the tank a primary black liquor heater raises the liquor temperature to about 210°F (100°C), thereby minimizing

power required by the black liquor pump.

Following the salt-cake mixing tank, the liquor flows to the heavy black liquor pump, and then on to a secondary black liquor heater where the temperature is raised to about 240°F (115°C). From here the liquor flows through the spray nozzles into the furnace.

EXTENDED ECONOMIZER BOILER

The advances in evaporator technology described earlier permitted the development of concentrators capable of producing black liquor at firing strength without the use of a direct-contact evaporator. The better thermal efficiency and reduced-odor emissions of multiple-effect evaporators over direct-contact evaporators caused them to almost completely replace such evaporators on new chemical recovery boilers. To remove the heat from the flue gas that would have normally been picked up in the direct-contact evaporator, the economizer was made larger. Therefore, this type of boiler is referred to as *extended-economizer* or *low-odor system*. Fig. 36 shows a schematic of the liquor preparation system.

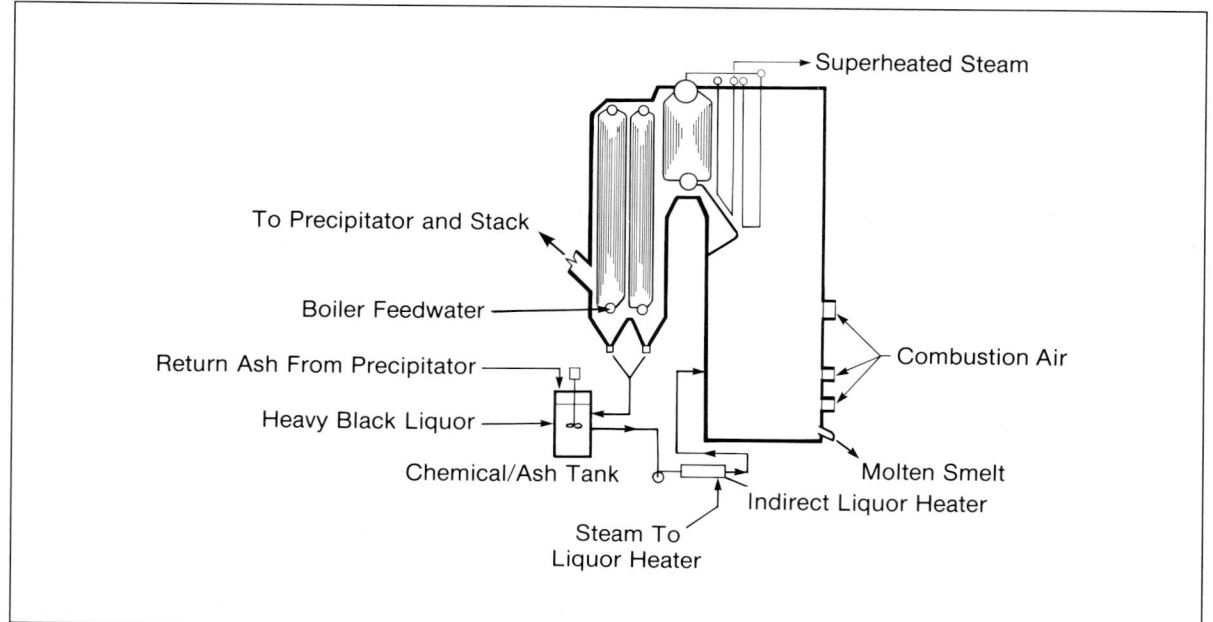

Fig. 36 **Extended-economizer chemical recovery boiler with schematic of black liquor preparation system**

The liquor from the heavy liquor storage tank or the last evaporator effect is sent to the chemical/ash mix tank, where the ash collected in the precipitator and economizer is mixed with the black liquor. In dry-bottom precipitators and economizer hoppers, a dry-ash conveying system removes the ash from the bottom of the hoppers. Dry-ash conveyers can transport it directly to the mix tank, or it can be sluiced with black liquor in sluice bowls, before being sent to the mix tank. An agitator in the mix tank prevents ash settling. In wet-bottom precipitators and economizer hoppers, black liquor flowing through the bottom of the hoppers removes the ash. The liquor leaving the hoppers is returned to the mix tank. Salt-cake makeup can also be added in the chemical/ash tank. In some systems, however, the salt-cake makeup is mixed in a separate, independent mix tank. The chemical/ash mix tank is equipped with a paddle-type agitator and a horizontal screen. It also contains a direct-type heater. However since the strong liquor from storage is already at pumping temperature, it is only used during upset conditions.

From the chemical/ash mix tank, the liquor is pumped to the black liquor heater where its temperature is raised to between 240F and 260°F (115 to 125°C) by indirect heating through a heat exchanger or through direct steam injection. From the heater, the liquor goes to the spray nozzles.

BURNING THE BLACK LIQUOR

The dual purpose of the recovery boiler is:
■ to recover inorganic chemicals from the black liquor in a form in which they can be recycled in the pulping process, and
■ to combust the organic constituents in the black liquor to produce valuable steam.

The chemical recovery efficiency of the boiler is measured in the smelted chemicals which leave the boiler from the bottom of the furnace. This efficiency is measured in terms of the conversion of sodium sulfate (Na_2SO_4) to sodium sulfide evaporated before either the chemical recovery or the combustion process can begin.

As a result of these multiple-process de-mands, the air and liquor delivery systems on the recovery boiler are designed to provide distinct process zones within the furnace, as illustrated in Fig. 37. The black liquor is sprayed into the furnace at an elevation of 17 to 21 feet (5.2 to 6.4 m) above the furnace floor so that it will dry in flight, before reaching the char and smelt bed. This first zone around the fuel entry is, therefore, referred to as the *drying zone*. Once the fuel is dry, pyrolysis of the organic constituents in the liquor will begin; accordingly, the next zone in the furnace is referred to as the *pyrolysis zone*. Char gasification and reduction of the inorganic chemicals take place in the lower part of the furnace and this region is called the *bed zone* or *hearth* of the furnace. The inorganic chemicals are recovered from this zone in smelt form. Finally, the partially combusted gases rise up in the furnace and pass through an oxidizing zone where combustion is completed.

Proper zone control in a chemical recovery boiler is required to achieve a desirable temperature profile similar to the one shown in Fig. 37. Higher gas temperatures in the lower furnace optimize the drying, pyrolysis, char-gasification and chemical-reduction processes. Achieving a lower furnace-leaving temperature will reduce the plugging potential in the convective sections of the boiler.

BLACK LIQUOR DELIVERY SYSTEM

The first priority in achieving optimized zoning in the furnace is to obtain consistently coarse black liquor spray droplets which are evenly distributed across a narrow vertical band that covers the entire furnace cross section. The flat-spray nozzle held in a fixed position is used for this purpose (Fig. 38). The nozzle will produce a flat spray of coarse droplets over a wide range of liquor conditions. A second type of nozzle which can be used to produce a flat spray of coarse liquor droplets is the splash-plate nozzle also shown in Fig. 38. For both nozzles, the angle of the nozzle support can be adjusted to optimize the liquor distribution in the boiler furnace to allow for optimum drying and pyrolysis in flight.

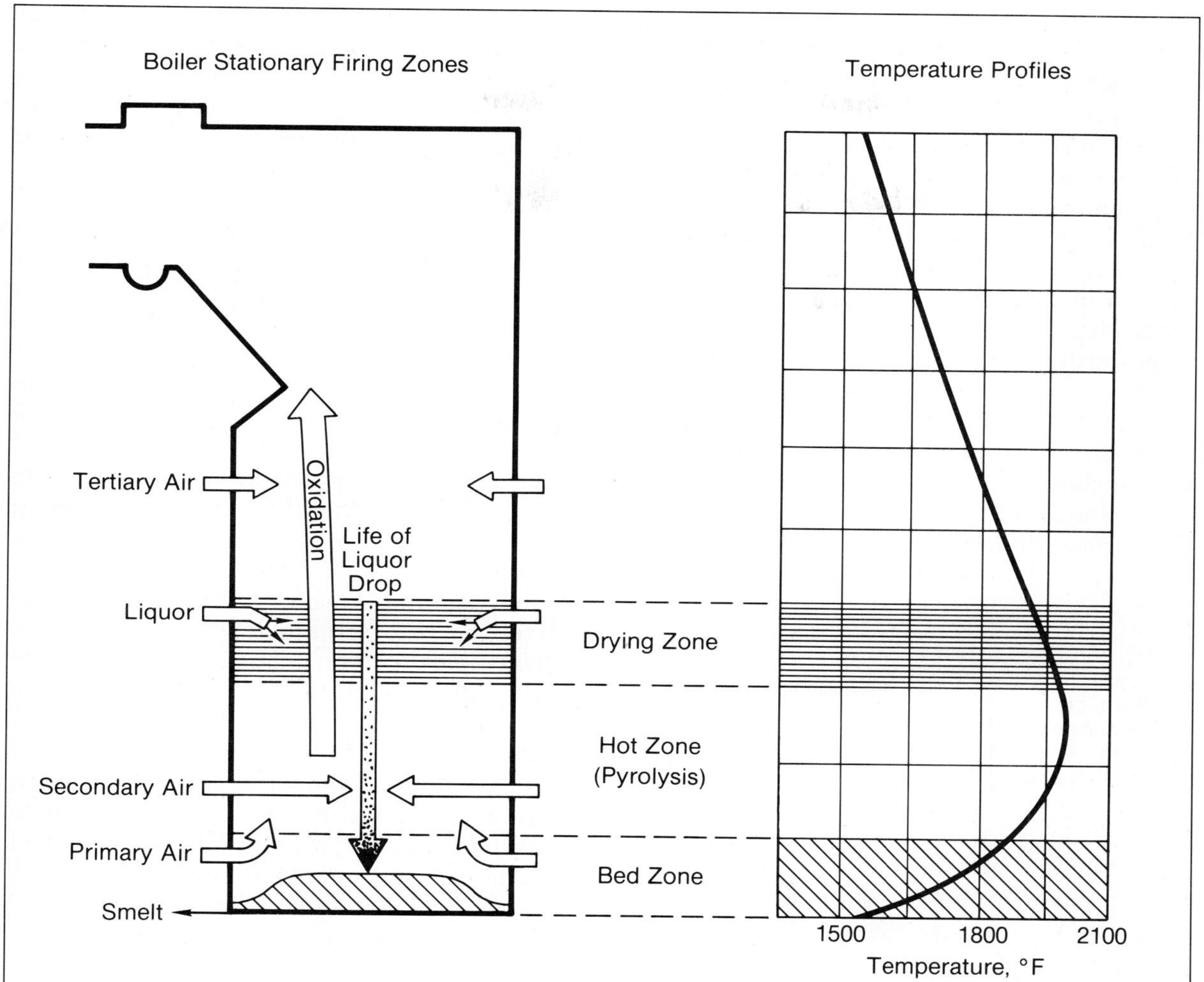

Boiler Stationary Firing Zones

Temperature Profiles

Tertiary Air

Oxidation

Life of
Liquor
Drop

Liquor

Drying Zone

Hot Zone
(Pyrolysis)

Secondary Air

Primary Air

Bed Zone

Smelt

1500   1800   2100
Temperature, °F

Fig. 37 **Process zones in a chemical-recovery-unit furnace**

CHEMICAL RECOVERY BOILER AIR SYSTEM

The three-level air system installed on chemical recovery boilers is designed to optimize the processes occuring in the lower furnace through improved zone control. Each level is independently controlled and monitored. Primary air is admitted at the bed level, secondary air is admitted just above the bed, and tertiary air is admitted above the black liquor spray guns. Each stream serves a distinct function.

Primary air is required to partially oxidize carbon on the bed and thereby liberate heat to maintain the optimum bed temperature, shape, and size. This stream must not disturb the bed as this could result in excessive carryover; however, it must make contact with the bed surface to react with the char. To achieve this, the optimized primary-air stream should enter the furnace through small ports located on all four walls of the furnace. The velocity of this air stream should not be excessive but should be sufficient to provide approximately 4 to 6 feet of penetration (1.2 to 1.8 m) towards the center of the bed. The ports should be equally spaced and directed slightly downward to provide even coverage of the bed around all four walls. Primary air is preheated to 300 to 400°F (150 to 200°C) in a steam-coil air heater to help maintain the high temperatures required to optimize the chemical processes in the lower furnace.

Secondary air is required to supply oxygen to burn some of the combustibles liberated from the bed in support of the requirement for high temperatures in the lower furnace. It also ensures positive smelt bed height control. The secondary-air stream must, therefore, penetrate the furnace to reach all the combustibles to promote chemical-reaction activity. At the same time, it should not disturb the bed or greatly increase the local oxygen concentration on the bed surface. To achieve this, the secondary-air

ports are larger in size than the primary-air ports, are fewer in number, and are angled horizontally into the furnace. This, along with the slightly higher velocity, provides the momentum necessary to penetrate the furnace cross section. The optimum location of the secondary-air ports is 4 to 6 feet (1.2 to 1.8 meters) above the primary-air ports which ensures that the bed surface will not be disturbed. Carryover is minimized and the oxygen concentration at the surface of the bed is not increased so as to adversely affect reduction efficiency. Secondary air is also preheated to 300 to 400°F (150 to 200°C) to help maintain high temperatures in the lower furnace.

Tertiary air is required to supply oxygen to complete combustion of the organics liberated from the bed and to burn any reduced sulfur gases to minimize odors from the unit. A controlled minimum quantity of excess air is required in this stream to ensure that the combustibles (carbon monoxide and total reduced sulfur—TRS) in the flue gas leaving the unit are oxidized. Increased air quantities will reduce the boiler efficiency and lower the total steam generation. The tertiary-air stream must provide the necessary turbulence to generate intimate mixing of air with the gases liberated from the lower furnace. It must also generate an even gas-flow distribution across the width of the furnace at the gas-outlet plane. Ambient air is used for tertiary air because it provides a cooling effect on the flue gases entering the convection zones. It also has a greater momentum and, therefore, greater mixing potential.

Two types of tertiary-air injection systems are used on chemical recovery boilers: concentric injection and interlaced injection (Fig. 39). The concentric arrangement has four windboxes located in the corners of the furnace. The upper compartment of each windbox is aimed tangentially to the large circle. During normal operation, each elevation receives 50 percent of the air at MCR. The interlaced arrangement uses a series of nozzles located on the front and rear walls as shown.

Both the concentric and interlaced arrangements provide an even gas-temperature profile

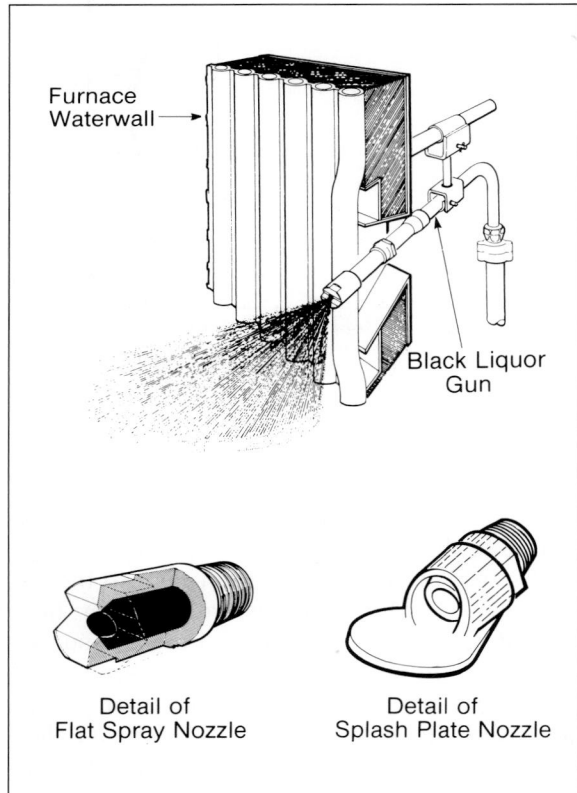

Furnace Waterwall

Black Liquor Gun

Detail of Flat Spray Nozzle

Detail of Splash Plate Nozzle

Fig. 38 **Black liquor firing showing available nozzle types**

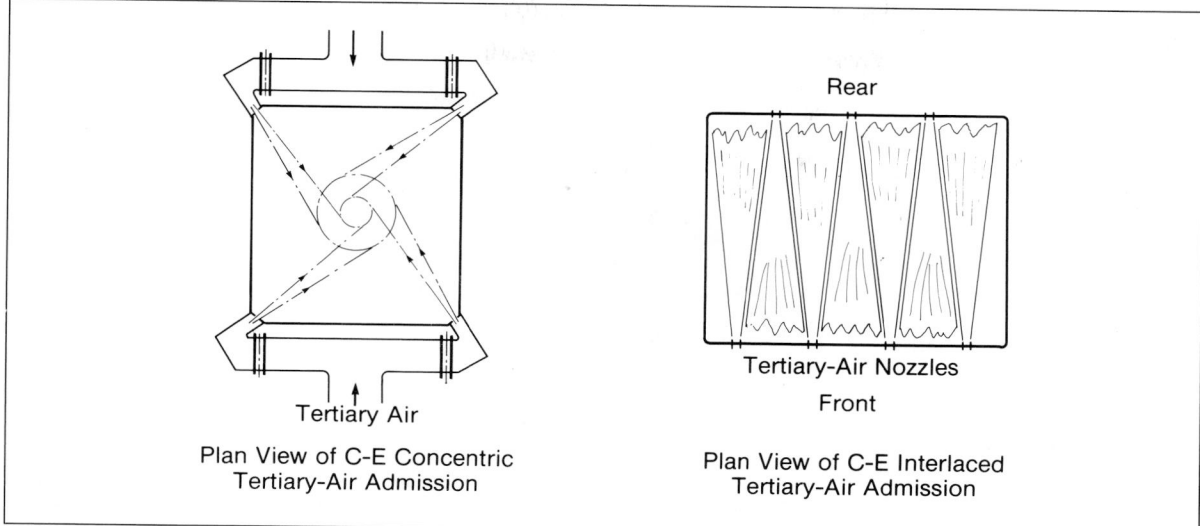

Fig. 39 **Alternative tertiary-air admission systems for C-E chemical recovery boilers**

at the outlet of the furnace. They also provide good mixing of the gases to complete the combustion of the gases and to minimize emissions from the furnace at a relatively low excess-air requirement. Concentric tertiary-air admission is normally preferred for new chemical recovery boilers since load-carrying burners can be included directly into the tertiary-air windbox. With interlaced-air admission, a separate elevation of load-carrying burners and windboxes are required. In addition, unlike the interlaced arrangement, the concentric arrangement uses lower injection velocities to achieve proper mixing. This requirement leads to lower static-pressure requirements at the windbox and lower fan power requirements.

AUXILIARY FUEL FIRING IN RECOVERY BOILERS

Auxiliary fuel is fired in recovery boilers for one of two reasons:

■ to provide the heat required in the lower furnace for warming the unit, stabilizing the char bed during low-load operation or under upset conditions, and burning down the bed during shutdowns. Starting burners are used for these applications, or

■ to raise the steam production from the boiler when the supply of black liquor is not suffi-cient to generate the full-load steam production, or to produce steam when there is no black liquor available. Load-carrying burners are used for this application.

Oil, gas or a combination of both can be used as auxiliary fuel in the recovery boiler.

Starting burners (Fig. 40) provide heat input to the lower furnace. At start-up, the heat from the starting burners dries and ignites the black liquor before a bed is developed. At shutdown, starting burners burn down the bed. At low loads, and when the black liquor is likely to burn poorly, starting burners stabilize the combustion conditions. Four starting burners are provided to fulfill these requirements, one located near each corner about five feet (1.5m) above the floor tubes. They are located close to the bed to minimize auxiliary-fuel requirements while still maintaining the required heat input to the bed, and they are angled to provide good coverage of the bed.

Because of their location close to the floor of the furnace, the starting burners are not suitable for the large heat input required for steam production. Load-carrying burners can be located at the tertiary-air elevation to provide the heat input required to produce additional steam at reduced black liquor flow, or for steam production when no black liquor is available.

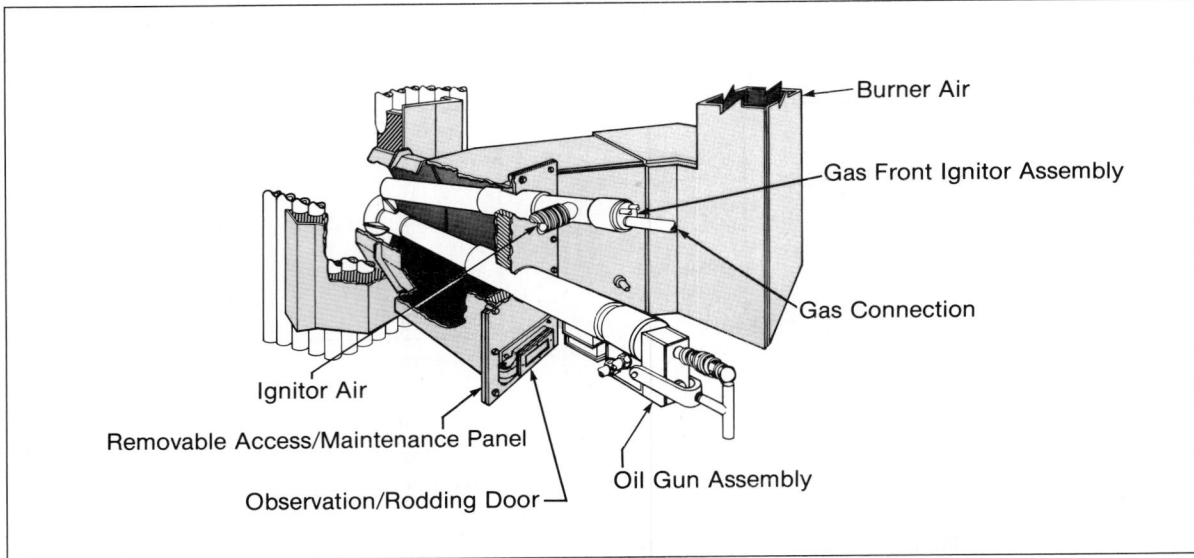

Fig. 40 Recovery boiler starting burner arrangement

DESIGN BASIS FOR RECOVERY BOILERS

Combustion Engineering offers two chemical recovery boiler arrangements: the classic two-drum design and the single-drum design. Typically, the optimum recovery boiler arrangement will be selected based on the operating pressure of the boiler. The two-drum design uses rolled joints for the boiler-bank tube penetrations into the drums. The practical limit for operating pressures using rolled joints is approximately 1500 psig (10.3 MPa gage). Alternatively, the single-drum design features an all-welded construction. With this type of construction, all sub-critical pressures can be accommodated.

The boiler operating pressure also influences the effectiveness of heat-transfer surfaces within the boiler, as described earlier in this chapter. As the pressure increases, the saturation temperature increases and the latent heat of vaporization of the water decreases. Therefore, the need for large evaporative surfaces in addition to the furnace waterwalls decreases with increasing operating pressures. At traditional recovery boiler operating pressures of 600 to 900 psig (4 to 6 MPa gage), the two-drum design provides a large amount of evaporative

surface in the generating bank in a very economical fashion. For higher pressure units, the requirements for evaporative surface is reduced and a single-drum design becomes more economical. There are positive features unique to each arrangement.

The two-drum recovery boiler design provides better access to the boiler tubes and, therefore, is easier to inspect, maintain and repair. The arrangement provides a larger water inventory which makes it less sensitive to water-level fluctuations in the drum and to feedwater-flow upsets. This arrangement also eliminates the need for a hopper and associated ash-removal system below the boiler bank. See Fig. 41.

The single-drum recovery boiler design, shown in Figure 31, has the generating bank located outside the furnace. This location reduces the potential of water entering the furnace cavity from the generating tubes. Locating the steam drum external to the gas stream eliminates the potential for corrosion of the drum metal. The drum also has fewer penetrations and, as a result, thinner drum plate. This thinner tube sheet, coupled with the fact that it is

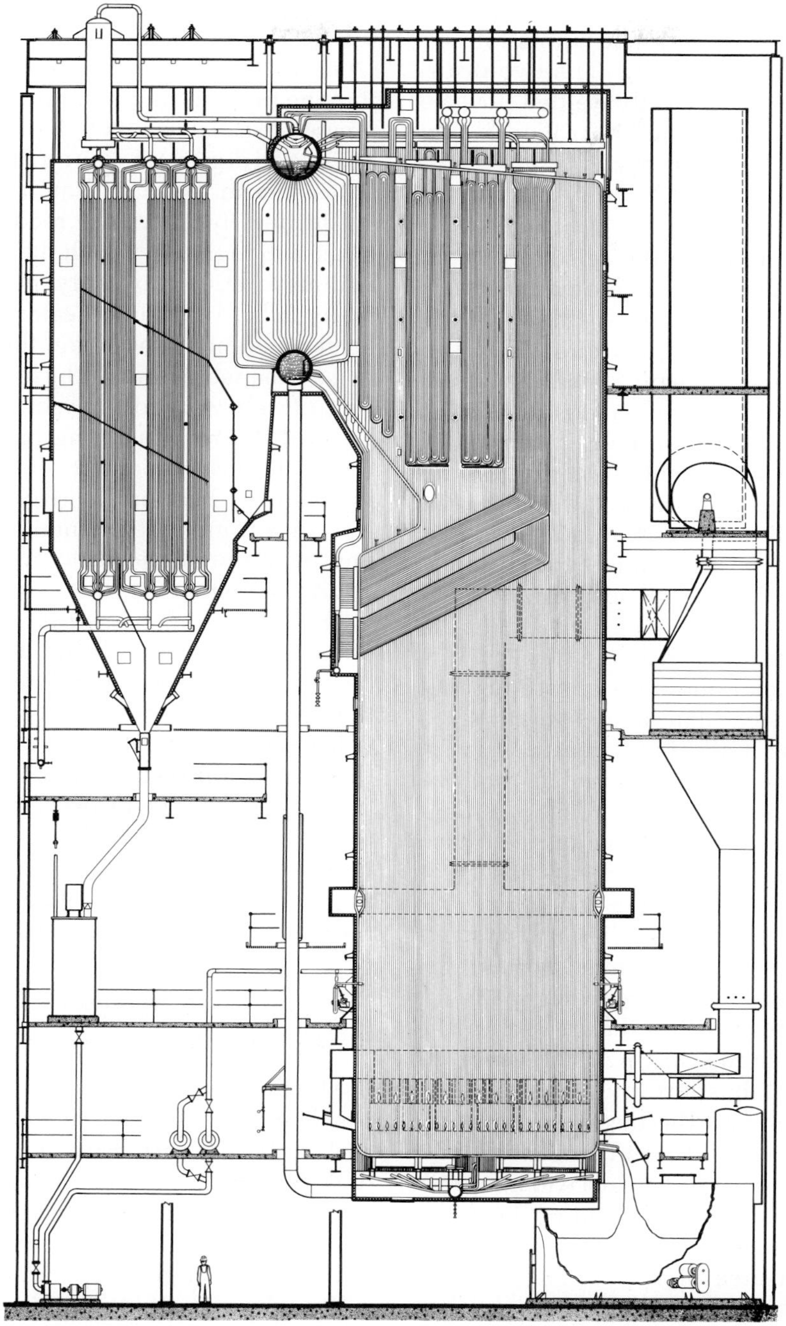

Fig. 41 Two-drum C-E chemical recovery boiler for operating pressures to 1500 psig (10.5 MPa gage)

not in the gas stream, allows faster start-up and shutdown times. Since the single-drum arrangement is a fully welded design, it is not susceptible to weeping at rolled joints following emergency boiler draining. The two-drum design can also be made a fully welded flow arrangement by seal welding the rolled joints in both upper and lower drums.

FURNACE SIZING

Several factors influence the selection of the furnace dimensions for a chemical recovery boiler, including the nature of the fuel, the characteristics of the black liquor, and the chemical recovery and steam generation functions of the boiler. The principal characteristics of black liquor as a fuel are its relatively high ash content (up to 45 percent of the dry solids), its relatively low heating value (HHV between 5,000 and 7,000 Btu/lb, or 11.6 and 16.13 MJ/kg of dry solids) and its relatively high moisture content as fired (25 to 35 percent).

The hearth heat-release rate (HHRR) is the first important factor in sizing the furnace. It is defined as the black liquor gross heat input divided by the furnace plan area. The HHRR is important because it affects the temperature in the lower furnace and it can be used to optimize the reduction efficiency, the combustion stability, pollutant emissions from the lower furnace (TRS and SO_2) and the turn-down stability of the unit. The optimum value for the HHRR must be selected based on the liquor characteristics, the anticipated boiler operating-load variations, and the required future load capacity.

The combustion volume is the second important design variable in selecting the furnace dimensions. The combustion volume establishes the residence time and the degree of combustion in the furnace. Sufficient combustion volume must be present in the furnace to allow for complete mixing and combustion of the gases rising from the lower furnace. Since the HHRR governs the plan area of the furnace, the required combustion volume usually establishes a minimum furnace height.

The third, and possibly the most important variable which influences the furnace dimensions, is the gas temperature entering the convective sections of the boiler. Typical recovery boiler ash deposits have a first melting temperature or "sticky" temperature of 1300 to 1400°F (700 to 760°C); however, studies have shown that the first melting point of the ash can be depressed to as low as 1000°F (540°C) by the presence of impurities such as chloride and potassium in the black liquor or by high liquor sulfidities. In chemical recovery boilers, the transition region in which the ash changes from molten to sticky to a dry powdery form normally falls in the superheater zone. The superheater radiant surfaces are, therefore, designed to be cleanable under these conditions. The first closely spaced convective section in the gas path is the generating bank. In a recovery boiler, the gas temperature entering the generating bank is, therefore, a critical design parameter, since it must be maintained below the first melting point of the ash to prevent the ash from accumulating on the tubes.

The furnace heating surface, based on the required combustion volume, and the superheater surface, based on the required steam temperature control, are not normally sufficient to lower the temperature of the flue gas entering the convection sections to an acceptable value. The designer of the recovery boiler has three alternatives to reduce the gas temperature to these sections:

- Make the superheater larger to increase the heat absorption upstream of the convection section

- Make the furnace taller

- Use radiant furnace panels.

The first option consists of making the superheater much larger than what is required to provide steam temperature control over the specified load range. Steam-cooled screens generally fall under this category since they add superheater surface. This option is usually practical on high-temperature, high-pressure units where the required superheater surface will absorb a large amount of heat and, therefore, the additional superheater surface required is minimized.

The second option is to increase the furnace height to add sufficient heating surface in the furnace to reduce the gas temperature entering the convection sections to an acceptable value without oversizing the superheater. On low-steam-temperature, low-pressure units for which the superheater heat pick-up is not very large, this could lead to extremely tall furnaces. As a result, this option is usually applied in conjunction with a slightly over-sized super-heater and is generally the pre-ferred option for moderate-temperature, moderate-pressure units.

The last option is to use radiant furnace panels to remove heat from the gas before it reaches the superheater. These panels consist of water-cooled tube panels. In addition to reducing the gas temperature, they also serve to protect the superheater from direct furnace radiation. This arrangement is usually the preferred option for low-temperature, low-pressure units since it provides a reliable and economical method of adding heating surface in the furnace.

FURNACE-BOTTOM DESIGN

A major difference between chemical recovery boilers and other types of power boilers is the presence of the char bed and the molten smelt on the bottom of the furnace. The molten smelt flows through the smelt spouts into the main dissolving tank. The floor of the recovery furnace must form a complete seal to prevent smelt from leaving the furnace other than through the spouts. A decanting furnace bottom, Fig. 42, is used for the following reasons:

1. It equalizes the heat distribution in the lower furnace and, therefore, optimizes the reduction efficiency, minimizes the SO_2 and TRS emissions, and optimizes the gas distribution leaving the furnace.

2. It reduces the risk of main dissolving-tank explosions caused by uneven smelt run-off from the furnace.

3. It eliminates the potential for corrosion of the floor tubes by ensuring the continuous presence of a frozen smelt layer on the floor. The frozen smelt layer also minimizes the potential for stress assisted water-side corrosion at the floor tube-to-header attachments by minimizing the operating stresses at these attachments.

4. It favors positive circulation through all the waterwall circuits by reducing the potential of boiling in the floor tubes.

5. It strengthens the floor tubes.

6. It eliminates the complicated floor tube-to-

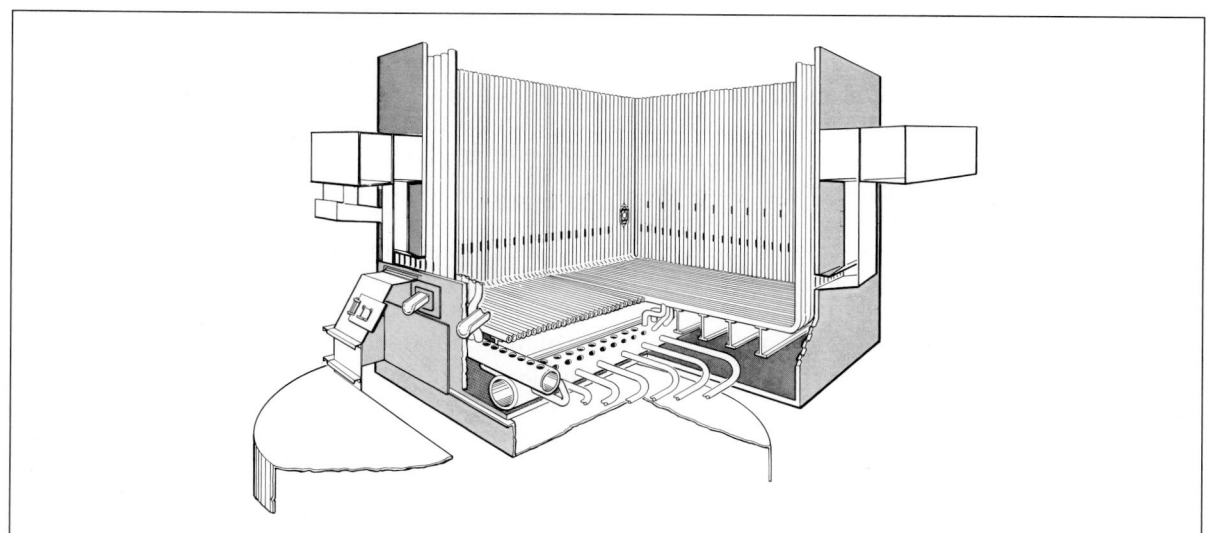

Fig. 42 **Decanting bottom of recovery-boiler furnace**

header design and thereby ensures a positive smelt seal.

7. It reduces the potential for char carryover in the smelt leaving the furnace.

8. It protects the floor tubes from slag falls from the upper furnace.

ARRANGEMENT OF THE HEATING SURFACES

As previously mentioned, the nature of the ash in chemical recovery boilers requires the designer to make special provisions in the design of the heating surfaces to ensure that they will be easy to clean.

Because the gas temperature entering the superheater will usually be higher than the first melting point of the ash, the superheater has to be designed to make it cleanable under that condition. A tangent-tube pendant-platen design on wide spacing is the optimum arrangement to ensure cleanability under sticky-ash conditions. The tangent-platen design prevents the deposits from wrapping around the individual tubes. This situation is similar to the condition at the furnace walls, where the ash deposits build up until they fall off under their own weight. The wide transverse spacing (12 in., or 300mm) prevents the deposits from bridging between elements. These vertical platens combined with judicious spacing of the sootblowers ensure the cleanability of the superheater.

The chemical recovery boiler is designed to ensure that the gas temperature entering the generating bank and economizer is well below the first melting point of the ash. Under normal conditions, the ash in the generating bank is expected to be mostly sodium sulfate in a powdery form, and it is easy to remove from the tubes. However, recovery boiler ash can also become hard through sintering.[14] Sintering is a molecular diffusion process that can occur at temperatures below the deposit first-melting point.

Under suitable conditions, sintering can turn soft deposits which are easily removed into very hard deposits. Deposit control in the generating bank and economizer is achieved through judicious location and operation of the sootblowers,

by adequate spacing between the tubes based on the gas temperature, and by a vertical tube arrangement.

CORROSION

Unprotected pressure-part surfaces in kraft recovery furnaces have historically been sensitive to wastage in specific regions including waterwall and convection section tubing.

WASTAGE LOCATIONS

A pattern of waterwall wastage has been observed in certain kraft recovery furnaces operating above 900 psig (6.2 MPa gage) although it has also been identified in units operating at pressures as low as 600 psig (4.1 MPa gage). In a furnace with unprotected tubing, significant wastage—up to 30 mils/yr (0.75 mm/year)—has been observed in the primary-air port region, and up to 10 mils/yr (0.25 mm/year) in the liquor spray-gun area (Fig. 43).[15] Wastage in the leading tubes of the finishing superheater of 30 to 40 mils/yr (0.75 to 1.0 mm/yr) has been reported.

Wastage rates, a function of both environment and metal temperature, have been observed to increase with operation at higher pressure, more efficient steam cycles.

WASTAGE MECHANISMS

The two most common methods of pressure-part wastage in recovery boilers are sulfidation and liquid-phase sodium hydroxide attack aggravated by the presence of oxygen.

Sulfidation

The primary mode of attack on both upper and lower waterwalls is sulfidation; the unprotected carbon steel conventionally used for construction of waterwalls is vulnerable at metal temperatures exceeding 625°F (330°C).

During furnace operation, tubes in the affected areas are covered with a sheet of frozen "smelt" or wall deposit (Fig. 44). The deposit is generally separated from the tube surface by a thin layer of irregularly spaced porous material, probably dried black liquor. This provides a

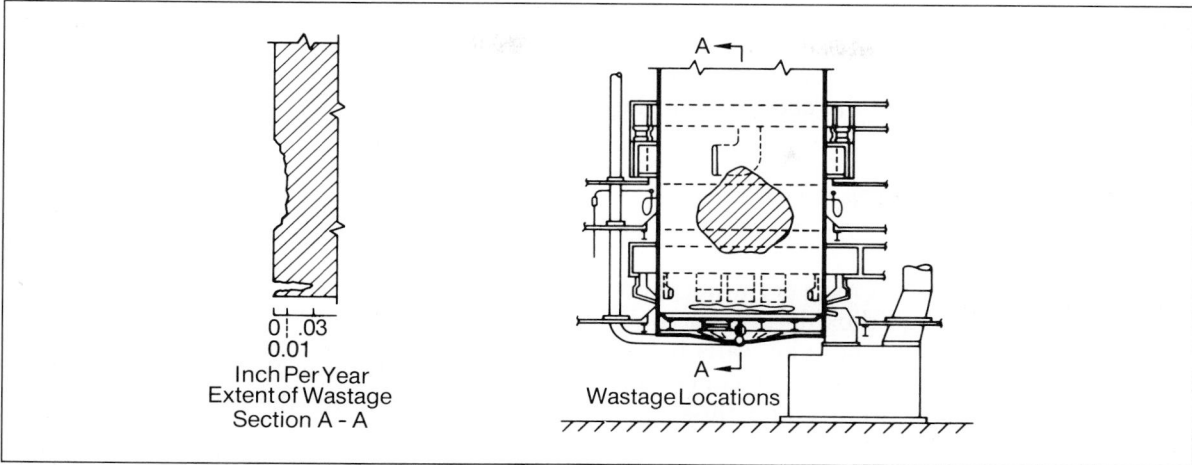

Fig. 43 **Typical recovery-unit furnace-wall corrosion**

space of about 1 mm (0.04 in.) which may serve as a passageway for corrosive gases. On the outer surface of the frozen deposit, there is a pulsating flow of molten smelt. Depending on the ash-bed configuration, this molten outer layer may be in contact with the glowing ash bed and/or bathed in a gaseous stream of widely varying concentrations of primary air and combustion products.

The principal corrosion product is FeS (see Eq. 1). Laboratory and field studies have shown

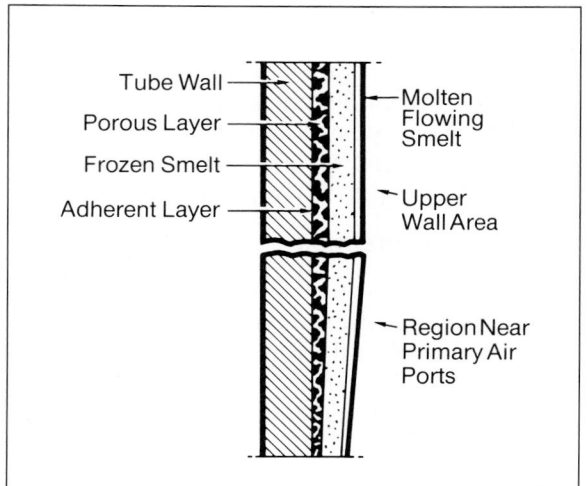

Fig. 44 **Cross section of recovery-unit furnace-wall deposits**

that the corrodent is nascent sulfur formed by the reaction of furnace gases with the frozen smelt on the tube surfaces.

$$Fe + S \rightarrow FeS$$

(1)

A series of laboratory tests has suggested the probable reactions involved in formation of the corroding sulfur.[16] When metal specimens were submerged in powdered sodium sulfide and carbon dioxide was passed through the reaction tube, severe corrosion was observed. Large quantities of sulfur were deposited on the reaction tube and carbon monoxide was detected by gas chromatography. This suggests that, at 700°F (370°C), sodium sulfide is oxidized to sulfur by carbon dioxide:

$$Na_2S + 2CO_2 \rightarrow Na_2CO_3 + S + CO$$

(2)

Increased corrosion in the presence of CO_2 was also reported as one of several significant reactions in papers by the Finnish and Swedish Corrosion Committees.[17,18]

Also studied was the effect of sulfur dioxide on the corrosion of a metal specimen submerged in powdered sodium sulfide. While the

corrosion observed in this test was less than the rate for the carbon-dioxide/sodium-sulfide test, the corrosion was very severe and large quantities of sulfur collected on the cool ends of the reaction tubes.

$$2\,Na_2S + 3SO_2 \rightarrow 2\,Na_2SO_3 + 3S \tag{3}$$

Furnace gas analyses in recovery units show large amounts of carbon dioxide (from 7 to 19 percent) in the area near the tube walls below the primary-air ports. Only small concentrations of sulfur dioxide were found (less than 0.05 percent).

Another way that sulfur may be generated in recovery boilers is by the oxidation of hydrogen sulfide. Fig. 45 shows that the corrosion rate of carbon steel under powdered Na_2S rises rapidly until the stoichiometric oxygen concentration is reached.

$$H_2S + \tfrac{1}{2}O_2 \rightarrow H_2O + S \tag{4}$$

In the absence of H_2S, the effect of the oxygen concentration is less pronounced.

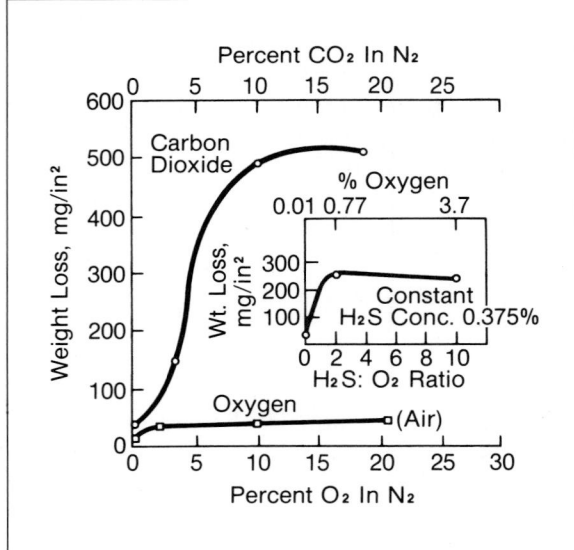

Fig. 45 Effect of furnace gas composition on corrosion of carbon steel

Sodium Hydroxide/Oxygen Attack

A second mechanism of tubing material loss occurs when sodium hydroxide evolved from the bed condenses and reacts with the base metal (see Eq. 5).

$$3O_2 + 4Fe + 4NaOH \rightarrow 4NaFeO_2 + 2H_2O \tag{5}$$

Historically this type of wastage, prevalent in 900 psig (6.2 MPa gage) boilers, was controlled with the advent of fin- and fusion-welded walls which seal out the oxidizing (O_2-in-air) component.

Some higher-pressure boilers (1200 psig or 8.3 MPa gage) equipped with composite stainless tubes that are intended to resist sulfidation attack have been found nevertheless subject to "air-port corrosion" which is observed in crevice areas where hydroxides can condense. At many non-gas-tight locations in the lower furnace (smelt openings, burners, air ports, or observation doors), the stainless layer seems highly susceptible to attack by the liquid-sodium/potassium-hydroxide phase. Since this wastage is confined to a narrow area around the air ports, the present approach is to provide air-tight furnace penetrations so that only high-velocity air enters the furnace and does not meet hydroxides prior to conversion to carbonate and/or sulfate.[19] Some form of lower-furnace waterwall protection is, therefore, necessary for operating pressure above 900 psig (6.2 MPa gage) where saturation temperatures exceed 540°F (280°C) and skin metal temperatures are 600°F (315°C) or greater.

METHODS OF PROTECTION

Several mechanical and metallurgical approaches to protecting recovery-boiler waterwall tubing and convection-pass heating surfaces are successfully applied. They are described in the subsequent sections.

Studs and Refractory

Chrome refractory or frozen smelt is retained by pin studs on the water-cooled tubes. Molten

smelt is thus prevented from contacting the tube metal. Experience with studded tubing on 1250 psig (8.6 MPa gage) C-E recovery boilers both in the U.S. and Finland have indicated that extensive repairs and replacement are required after three to seven years. Because of the stud spacing, monitoring of wall thickness in these units has been extremely difficult. In addition, the labor and fabrication costs can be prohibitive.

Weld Overlay

Protection by weld overlay on the lower 6 to 9 feet (2 to 3 m) of high-pressure units has been satisfactorily employed for several years in Finland, Japan, France and, to a limited extent in the U.S. Austenitic alloys usually containing 18 percent Cr (or more) have been used. There is considerable hand labor involved in this method although satisfactory corrosion protection has resulted.

The weld overlay may be done manually. More recently, some suppliers are using automated systems. Prebending and other special handling is required to produce flat waterwall panels. Surface repair can be done readily in the field, although tube replacement is difficult. Seventeen to twenty years operating experience has been accumulated by users in various countries.

Metallizing

The earliest development of field applied protection was in the form of multi-coat metallizing. Since these coatings are not metallurgically bonded, the most important aspect of flame or plasma spray metallizing is that the tube surfaces be absolutely clean and dry. The tubes should be cleaned by steel or grit blasting (0.005-inch or 0.13-mm size) and spraying should follow immediately. Sand blasting is not recommended because of dust residue left on tubes. More than one blast-cleaning operation may be necessary.

The original multi-coat layers consisted of 0.008 inch (0.20 mm) of 200-series stainless steel adjacent to the cleaned tube surface, fol-

lowed by 0.004 inch (0.10 mm) of aluminum and then a layer of silicate-aluminum sealer. After several years experience, bonding layers were developed to reduce the spalling tendency of these coatings; consequently, a revised coating procedure was developed using the following proprietary products: undercoat layer—0.002 to 0.005 inch (0.05 to 0.13 mm) of Metco 405 nickel titanide adjacent to the grit-blasted tube surface; middle layer—0.007 to 0.009 inch (0.18 to 0.23 mm) of Metco #1 (300 series stainless); outer layer—0.003 to 0.004 inch (0.08 to 0.10 mm) of Metco Superfine (aluminum). A layer of Metco SA silicate-aluminum sealer is sprayed or painted on to fill the naturally occurring pores in the metallized coating.

Annual inspection and repair of spalled areas are required, but it is still not possible to predict where and when coating loss will occur.

More recent trends have been to apply single-layer coatings using plasma-powder spray or wire-arc spray techniques. While these coatings are reported to be more dense and monolithic, they still are porous and are not metallurgically bonded. Although the time of application is significantly reduced with the single layer coatings, they remain subject to spalling and/or undermining by oxide or sulfide formation at the tube interface.

During the time that single-layer coatings have been used, they have apparently not required as extensive touch up or repair during outages as is the practice with the multi-coat process.

Composite Tubing

Composite tubing (bimetallic type 304 stainless over carbon steel) is manufactured by several companies. This development followed successful demonstrations in which C-E participated, where stainless-steel weld overlay was shown to control recovery-unit wastage.

The resistance of type 304 composite tubing to sulfidation is very high; however, recent experience has shown that the more localized air-port attack (attributed to condensation of sodium hydroxide) may not be controllable with this tubing. Fabrication techniques require additional handling and setup. Carbon-steel fins require austenitic stainless overlay. Maintenance or repairs of composite tubing require double scarfing and the use of multiple welding rods. The material cost of the composite tubing is much greater than that of carbon-steel, so that the subsequent cost of boilers with this tubing is considerably higher than with the other forms of protection discussed.

Chromizing

Chromizing is a high-temperature diffusion coating process in which solid-state diffusion produces a chromium-rich metal surface that is resistant to corrosion in a variety of environments. Carbon-steel and numerous alloys can be chromized, including the following steam-generator tubing materials: SA-192, SA213, T-11, T-22 and T-91. Typically, the chromized structure in SA-192 carbon-steel is approximately 10 to 15 mils (0.75 to 0.375 mm) thick, and is up to 20 mils (0.5 mm) thick with SA-123, T-11 or T-22 base metals. This outer layer contains from 30 to 40 percent chromium at the surface, decreasing almost linearly to between 10 and 15 percent at the base-metal/diffusion-zone interface. At the surface, a very thin 0.6 to 1 mil (15 to 25 μm) chromium carbide layer contains 70 to 75 percent Cr. While the total thickness of the coating is important, it is more important that the outer portion of the coating contain at least 20 to 25 percent chromium.

The coating process consists of several steps including placing the part to be chromized in a metal box, covering the part with the chromizing mixture, and heating the box and contents at an elevated temperature. After chromizing, the component is heat treated; the exact procedure is a function of the alloy identity. The process, as developed at Combustion Engineering and also described in Chapter 17 on Materials and Metallurgy, can accommodate large heat exchangers, such as waterwall panels and superheater and reheater assemblies.

Both external and internal surfaces of these tubular components can be chromized. More important, however, the attachments and fins that are fabricated before chromizing can also be chromized, hereby providing protection to the total surface. The enrichment occurs "in" and not "on" the surface; consequently, application of the coating does not change the physical dimensions of the component. Being an integral part of the base metal, the metallurgically bonded, diffused-chromium structure is not subject to peeling or spalling. The 30 to 40 percent chromium in the chromized coating is significantly higher than the 18 to 20 percent chromium level in commonly used austenitic alloys, and thus is considered to be equal to or perhaps more resistant to sulfidation than the 18-8 austenitic stainless steels.

Welding of chromized components is a simple, straight-forward procedure. Butt welds are made after preparatory machining of tube ends, which is done after chromizing. Filler metal is the same as if welding a non-chromized tube. If protection from wastage is desired, an E309 electrode is used as a cover pass. Attachments that must be made to the chromized surfaces use the same procedures that would be specified for welding non-chromized surfaces.

Field and laboratory analyses indicate that chromized carbon-steel and chromized low-alloy tubing is resistant to external corrosion of waterwall and superheater surfaces in chemical recovery boilers and has proven itself to be a cost effective alternative to the use of composite tubing and field applied coatings.[20]

SMELT-WATER EXPLOSIONS

When water contacts molten smelt, the elements exist for a potentially violent physical explosion. Unlike chemical explosions, physical explosions do not involve a reaction between two substances. Rather, a physical explosion results from the very rapid expansion of gas and can be powerful, destructive, and costly. Smelt-water explosions are more violent than typical nondetonative explosions in furnaces which involve conventional fuels such as coal, oil, or gas in air. Whereas the latter develop much lower peak pressures and travel at the speed of sound, smelt-water explosions, like other physical explosions, produce higher pressures, perhaps by a factor of 10 or more, and travel at supersonic speeds. They are, by definition, shock-wave phenomena.

A smelt-water explosion results from water changing suddenly to steam. This rapid change is caused by an extremely high heat-transfer rate that can exist between molten smelt and water. The seriousness of a smelt-water explosion is not related to the amount of smelt or water involved; a teaspoon of molten smelt in a cupful of water can produce a powerful explosion.

Weak black liquor can also produce an explosion when brought into contact with smelt. Concentrations above 58 percent, however, are considered safe to fire into the furnace. Many recovery boilers are equipped with solids meters which continuously monitor the black liquor solids concentration. Should the solids concentration fall to an unsafe level, the liquor is automatically diverted from the furnace.

SAFETY CONSIDERATIONS IN CHEMICAL RECOVERY BOILER DESIGN

Because of the potential for considerable damage and the threat to the safety of operating personnel in the event of a smelt-water explosion, the chemical recovery boiler design includes several safety features. Some special features which minimize such risks are as follows:

■ Corrosion-resistant materials in the lower furnace and very conservative corrosion allowances:

In modern boilers, some form of corrosion-resistant materials, composite tubing with stainless-steel overlay of the fins, or chromized panels are used to eliminate the potential for tube thinning in the critical high temperature zone of the boiler. In addition, the tubes in the rest of the furnace are selected with very conservative corrosion allowances to reduce the potential of water leaking into the furnace.

■ Emergency Drain System:

The recovery boiler is equipped with a rapid drain system to allow the operator to drain the furnace rapidly if it is thought that water may be leaking in the furnace. This system will drain the furnace to 8 feet (2½ meters) above the furnace floor, thereby stopping the leak while at the same time preventing overheating of the lower-furnace tubes.

■ Furnace Framing and Structure:

The furnace-framing system is designed to relieve excess pressure in a safe location such as the upper furnace in the event of an explosion. Zipper corners on the front wall will open to relieve furnace pressure to minimize damage to the pressure parts. Also, the furnace structure has added strength to reduce the risk to operating personnel.

■ Furnace Bottom Design:

The decanting hearth arrangement ensures that a frozen smelt layer constantly covers the floor tubes. This arrangement eliminates the potential for corrosion of the floor tubes, protects the floor tubes from physical damage due to large ash deposits falling from the superheater region, and eliminates hot spots on the floor tubes which assures positive circulation through all circuits. The decanting hearth also adds to the strength of the floor and reduces the risk of main dissolving-tank explosions resulting from uneven smelt run-offs.

RELIABILITY AND MAINTAINABILITY OF THE CHEMICAL RECOVERY BOILER

In addition to safety, reliability and maintainability are also very important aspects of the design of chemical recovery boilers. The recovery boiler often constitutes the bottleneck in the pulp mill production. Reliability is therefore very important since an unscheduled outage usually translates into lost production for the mill. Also, maintainability is important since reduced shutdown time means increased mill production.

Combustion Engineering addresses these issues with conservative materials and equipment selection during engineering of the boiler, and high standards of quality and inspection during manufacturing and erection to ensure the reliability of the unit. Maintainability is improved by providing easy access for inspection and standardizing the equipment to reduce the amount of spare parts required.

REFERENCES

[1] William H. Pollock, "Coal-Fired Industrial Boilers for the 1980's," *Proceedings of the American Power Conference*, 41:835–841, 1979. Chicago: Illinois Institute of Technology, 1979.

[2] L. J. Cohan and J. H. Fernandes, "Potential Energy Conversion Aspects of Refuse," *ASME Paper no. 67-WA/PID-6*. New York: American Society of Mechanical Engineers, 1967.

[3] J. W. Regan, J. F. Mullen, and R. D. Nickerson, "Suspension Firing of Solid Waste Fuels," *Proceedings of the American Power Conference*, 31:599–608, 1969. Chicago: Illinois Institute of Technology, 1969.

[4] J. W. Regan, "Generating Steam From Prepared Refuse," *Proceedings of the Fourth National Incinerator Conference*, Cincinnati, May 17–20, 1970, pp. 216–223. New York: American Society of Mechanical Engineers, 1970.

[5] Joseph F. Mullen, "System for Pneumatically Transporting High-Moisture Fuels such as Bagasse and Bark and an Included Furnace for Drying and Burning Those Fuels in Suspension Under High Turbulence," *U.S. Patent no. 3,387,574*, assigned to Combustion Engineering, Inc., June 11, 1968.

[6] Gerald E. Dreifke, David L. Klumb and Jerrel D. Smith, "Solid Waste as a Utility Fuel," *Proceedings of the American Power Conference*, 35:1198–1206, 1973. Chicago: Illinois Institute of Technology, 1973.

[7] H. Eberhardt and W. Mayer, "Experiences with Refuse Incinerators in Europe: Prevention of Air and Water Pollution, Operation of Refuse Incineration Plants Combined with Steam Boilers, Design and Planning," *Proceedings of the 1968 National Incinerator Conference*, New York, May 5–8, 1968, pp. 73–86. New York: American Society of Mechanical Engineers, 1968.

[8] H. Hilsheimer, "Experience after 20,000 Operating Hours: The Mannheim Incinerator," *Proceedings of the 1970 National Incinerator Conference*, New York, May 17–20, 1970, pp. 93–106. New York: American Society of Mechanical Engineers, 1970.

[9] J. G. Singer and J. R. Mullen, "Closing the Refuse Power Cycle," *ASME Paper no. 73-PWR-18*. New York: American Society of Mechanical Engineers, 1973.

[10] A. P. Konopka, "Systems Evaluation of Refuse as a Low Sulfur Fuel: Part 3, Air Pollution Aspects," *ASME Paper no. 71-WA/Inc-1*. New York: American Society of Mechanical Engineers, 1971. (This information can also be found in NTIS publications PB-209 271 and PB-209 272, Springfield, VA: National Technical Information Service.)

R.E. Sommerlad, et al., "Systems Evaluation of Refuse as a Low Sulfur Fuel: Part 2, Steam Generator Aspects," *ASME Paper no. 71-WA/Inc-2*. New York: American Society of Mechanical Engineers, 1971. (This information can also be found in NTIS publications PS-209 271 and PB-209 272, Springfield, VA: National Technical Information Service.)

R. M. Roberts and E. M. Wilson, "Systems Evaluation of Refuse as a Low Sulfur Fuel: Part 1, The Value of Refuse Energy and the Cost of Its Recovery," *ASME Paper no. 71-WA/Inc-3*. New York: American Society of Mechanical Engineers, 1971. (This information can also be found in NTIS publications PB-209 271 and PB-209 272, Springfield, VA: National Technical Information Service.)

[11] Paul D. Miller and Horatio H. Krause, "Corrosion of Carbon and Stainless Steels in Flue Gases From Municipal Incinerators," *Proceedings of the 1972 National Incinerator Conference*, New York, June 4–7, 1972, pp. 300–309. New York: American Society of Mechanical Engineers, 1972.

[12] F. Nowak, "Considerations in the Construction of Large Refuse Incinerators," *Proceedings of the 1970 National Incinerator Conference*, New York, May 17–20, 1970, pp. 86–92. New York: American Society of Mechanical Engineers, 1970.

[13] Karl-Heinz Thoemen, "Contribution to the Control of Corrosion Problems on Incinerators with Water-Wall Steam Generators," *Proceedings of the 1972 National Incinerator Conference*, New York, June 4–7, 1972, pp. 310–318. New York: American Society of Mechanical Engineers, 1972.

[14] H. Tran, "How Does a Kraft Recovery Boiler Become Plugged?", *1988 TAPPI Kraft Recovery Operation Seminar*; Orlando, FL, 1988.

[15] A. L. Plumley, E. C. Lewis, and R. G. Tallent, TAPPI 49 (1), pg. 72A, 1966.

[16] R. G. Tallent and A. L. Plumley, TAPPI 52 (10), pg. 1955, 1969.

[17] O. Stelling, "What Reactions Cause Corrosion in Soda Plant Boilers," *Annual Nordic Recovery Boiler Conferences*, 1965 – 1967.

[18] O. Moberg, "Recovery Boiler Corrosion," *Pulp and Paper Industry Corrosion Problems*, NACE, 1974, pg. 125.

[19] T. Odelstam, "BLRB Composite Tubes — 15 Years of Experience," *TAPPI Operations Seminar*, 1988.

[20] A. L. Plumley, et al., "Chromizing for Recovery Boiler Corrosion Protection," *BLRBAC*, Atlanta, October 1988; also as Combustion Engineering publication TIS-8417.

CHAPTER 9

Fluidized-Bed Steam Generators

For decades, fluidized-bed reactors have been used in non-combustion reactions in which the thorough mixing and intimate contact of the reactants in a fluidized bed result in high product yield with improved economy of time and energy. Although other methods of burning solid fuels also can generate energy with very high efficiency, fluidized-bed combustion can burn coal efficiently at a temperature low enough to avoid many of the problems of combustion in the other modes. (See the discussion in Chapter 4 that compares the physical mechanisms presently in use for burning coal and other solid fossil fuels in steam generators.)

HISTORICAL PERSPECTIVE

Fluidization and observations of phenomena related to fluidization have been referenced in the literature since the late 1800's. Here, we will try to give the reader a flavor for the origins of fluidized-bed technology and the progress that has been made in the last two decades in applying it to steam generation.

The person most often cited as initiating fluidized-bed technology is Fritz Winkler, who in the 1920's developed the Winkler coal gasifier employing fluid-bed concepts; he received a patent for his work in 1928. With a knowledge of Winkler's work, and a wartime demand for petroleum products, a Fluidized Catalytic Cracking (FCC) process was developed during the early 1940's. This process greatly increased both the quantity and quality of the refined product. Over the years many improvements were made to the original FCC process and, today, various types of fluid-bed processes are used in the refining of petroleum and the production of other chemical feedstocks.

During 1953, patents were assigned to Combustion Engineering for combustion of oil and natural gas in a fluidized bed composed of particles of alumina, as a means to increase heat-transfer rates to boiler heat-absorbing surfaces. C-E also conducted experiments using a fluidized bed to combust coal. However, with low prices for oil and gas during the 60's, there was little incentive to pursue this work in the U.S.

During the 1960's, much of the work of applying fluidized-bed technology to boiler applications was carried out in England by the Central Electricity Generating Board (CEGB) and the British Coal Utilization Research Association (BCURA). The overall emphasis was to develop fluidized-bed combustion (FBC) for industrial steam generation. The FBC technology developed in England found its way to the U.S. in the 1960's. Pilot-plant research work was conducted that was later used to develop concepts for the construction of a 30-MW fluid-bed boiler demonstration plant at Rivesville, West Virginia in 1975. This plant was about 15 times larger than any other FBC facility in operation.

Spurred on by the oil embargoes of the 1970's, FBC technology was considered a potential solution to the energy crisis. Many studies and several demonstration projects were funded in the U. S. by government agencies, with C-E involved in several of them.

In 1976, C-E was awarded a contract by the Energy Research and Development Administration (ERDA), now DOE, to design, construct, and test a 50,000 lb/hr FBC demonstration boiler for industrial applications. The boiler was installed at the Great Lakes Naval Station near Chicago, Illinois, and supplied steam for space heating. A pilot plant was built at C-E's Windsor, Connecticut, facility for process and equipment development.[1] The demonstration plant began operation in 1981 and showed promising results.[2]

In 1977, Combustion Engineering completed a preliminary design study of a large-scale FBC boiler for utility applications, funded by the Tennessee Valley Authority (TVA), involving the design of a 200-MW demonstration unit and an 800-MW commercial unit.[3] Three years later, C-E completed a follow-on study to develop the final design for the 200-MW demonstration unit.[4] In 1984, a contract was received to provide the 160-MW FBC demonstration boiler for installation at TVA's Shawnee Steam Plant in Paducah, Kentucky. This plant began operation in 1988.

Simultaneously with FBC development in England and the United States, Lurgi GmbH in West Germany had developed various fluid-bed processes for roasting of ores and various other materials. In 1965 Lurgi developed a new type of fluid-bed process called a Circulating Fluidized Bed (CFB), as distinct from the more conventional Bubbling Fluidized Bed (BFB). This process was initially applied to processes such as alumina calcination and later to steam generation.

In 1980, C-E was awarded a study by TVA to develop the design of a 200-MW CFB boiler based on the Lurgi technology.[5] This study led to a consortium agreement with Lurgi in 1983 for the joint commercial offering of CFB boilers. Between 1983 and 1987 C-E sold a large number of CFB units for industrial and utility applications ranging in size from 150,000 lb/hr to 1,100,000 lb/hr (with reheat), and firing fuels ranging from bituminous coal and lignite to anthracite waste, wood waste, and other biomass. In 1987, C-E signed an agreement to license the CFB process technology from Lurgi.

This involvement in both BFB and CFB technology, with small- and large-scale units of each type in successful operation, makes C-E unique as a supplier to the boiler industry. C-E can provide the optimum FBC boiler design for any application.

GENERAL DESCRIPTION OF FLUIDIZED-BED COMBUSTION

"Fluidization" refers to the condition in which solid materials are given free-flowing, fluid-like behavior. As a gas is passed upward through a bed of solid particles, the flow of gas produces forces which tend to separate the particles from one another. At low gas flows, the particles remain in contact with other solids and tend to resist movement. This condition is referred to as a fixed bed. As the gas flow is increased, a point is reached at which the forces on the particles are just sufficient to cause separation. The bed then becomes fluidized. The gas cushion between the solids allows the particles to move freely, giving the bed a liquid-like characteristic.

The transition from fixed bed to fluid bed is illustrated in Fig. 1, which plots the gas pressure drop through the bed versus gas velocity. For a fixed bed, pressure drop is proportional to the square of velocity. As velocity is increased, the bed becomes fluidized; the velocity at which this transition occurs is called the minimum fluidization velocity, V_{mf}. V_{mf} depends on many factors including particle diameter, gas and particle density, particle shape, gas viscosity, and bed void fraction.[1,6] At velocities above

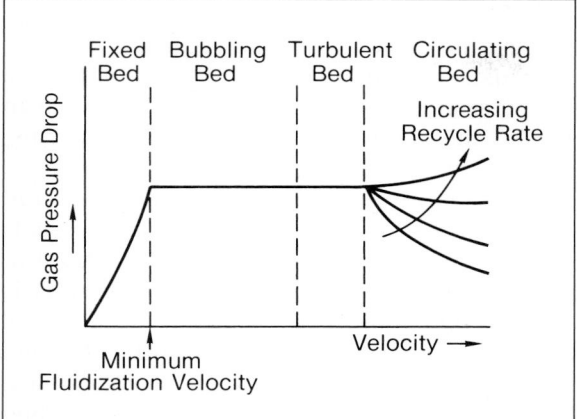

Fig. 1 **Plot of gas pressure drop through a fluidized bed versus gas velocity**

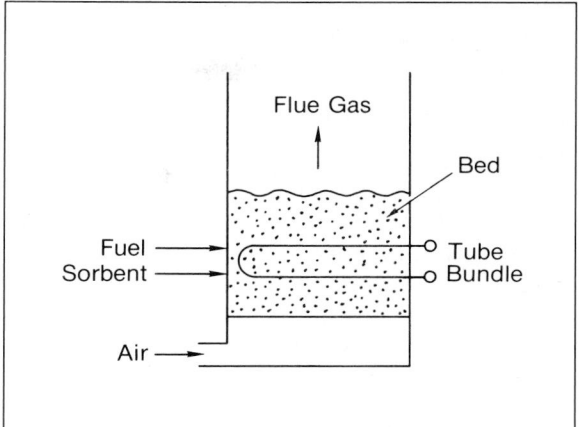

Fig. 2 **Generalized fluidized-bed combustor**

V_{mf}, the pressure drop through the bed remains nearly constant and is equal to the weight of solids per unit area, as the drag forces on the particles just overcome the gravitational forces. Further increases in velocity bring about changes in the state of fluidization, to be discussed later in this chapter.

CHARACTERISTICS OF FLUIDIZED-BED COMBUSTION

Fluidized-bed combustion (FBC) offers the power engineer design versatility for burning a wide variety of fuels, including many that are too poor in quality for use in conventional firing systems. Fuels which contain high concentrations of ash, sulfur, and nitrogen can be burned efficiently, while meeting stringent requirements for the control of stack emissions without the use of flue-gas scrubbers. Fig. 2 shows a generalized fluidized-bed combustor.

In fluidized-bed combustion, fuel is burned in a bed of hot incombustible particles suspended by an upward flow of fluidizing gas. Typically, the fuel is a solid such as coal, although liquid and gaseous fuels can be readily used. The fluidizing gas is generally the combustion air and the gaseous products of combustion. Where sulfur capture is not required, the fuel ash may be supplemented by inert materials such as sand or alumina to maintain the

bed. In applications where sulfur capture is required, limestone is used as the sorbent and forms a portion of the bed. Bed temperature is usually maintained at or near 1550°F (840°C) by the use of heat-absorbing surface within or enclosing the bed, because this temperature is optimal for the chemical processes needed to capture sulfur and control NO_x emissions. It also avoids ash softening in nearly all fuels. At this temperature, efficient combustion can be achieved because of the relatively long residence time of fuel in the bed and the good gas/solids contact there.

The above characteristics lead to the major advantages of FBC:

ABILITY TO BURN LOW-GRADE FUELS

The high thermal inertia of the bed mass provides for stable ignition and combustion of very low grade fuels such as fuels high in ash and/or moisture. Fuels with up to 70-percent ash and 50-percent moisture have been successfully burned in a fluid bed. The high thermal inertia of the bed also provides for good performance when firing low-volatile fuels such as anthracite, anthracite culm, and petroleum coke.

FUEL FLEXIBILITY

Because levels are held below the ash-softening level, the FBC boiler is not sensitive to fuel ash characteristics. A wide range of fuels with varying ash contents and properties can be burned in a single boiler.

LOW NITROGEN-OXIDE PRODUCTION

NO_x emissions are considered to come from two sources: oxidation of nitrogen in the air (thermal NO_x) and oxidation of nitrogen and/or nitrogen components in the fuel (fuel NO_x). At the low temperatures in FBC, thermal NO_x is essentially zero; design features such as staged combustion can significantly reduce fuel NO_x, leading to low total NO_x emissions.

IN-SITU CAPTURE OF SULFUR DIOXIDE

SO_2 emissions are controlled within the combustor by addition of a sorbent material, so a stack-gas SO_2 scrubber is not required. The sulfur sorbent can also react with other fuel constituents such as vanadium, reducing downstream corrosion potential.

TYPES OF FBC SYSTEMS

The state of fluidization in a fluid-bed-boiler combustor depends mainly on the bed-particle diameter and fluidizing velocity. As shown in Fig. 3, there are two basic fluid-bed combustion systems, each operating in a different state of fluidization. At relatively low velocities and with coarse bed-particle size, the fluid bed is dense, with a uniform solids concentration, and has a well-defined surface. This system is called a *bubbling fluid bed* (BFB), because the air in excess of that required to fluidize the bed passes through the bed in the form of bubbles. The BFB is further characterized by modest bed solids mixing rate, and relatively low solids entrainment in the flue gas. While little recycle of the entrained material to the bed is needed to maintain bed inventory, substantial recycle rates may be used to enhance performance.

At higher velocities and with finer bed-particle size, the fluid bed surface becomes diffuse

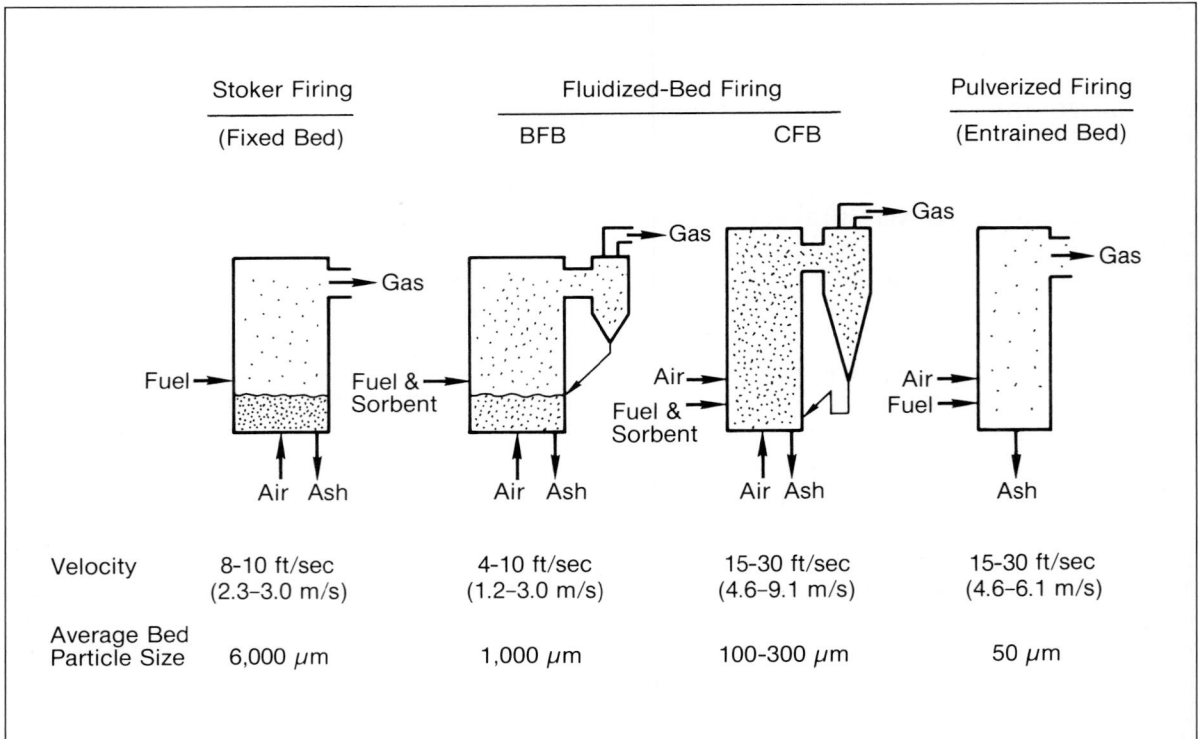

	Stoker Firing (Fixed Bed)	Fluidized-Bed Firing BFB	Fluidized-Bed Firing CFB	Pulverized Firing (Entrained Bed)
Velocity	8-10 ft/sec (2.3–3.0 m/s)	4-10 ft/sec (1.2–3.0 m/s)	15-30 ft/sec (4.6–9.1 m/s)	15-30 ft/sec (4.6–6.1 m/s)
Average Bed Particle Size	6,000 μm	1,000 μm	100-300 μm	50 μm

Fig. 3 Relationships between stoker, fluidized-bed, and pulverized firing of solid fuels

as solids entrainment increases, such that there is no longer a defined bed surface; recycle of entrained material to the bed at high rates is required to maintain bed inventory. The bulk density of the bed decreases with increasing height in the combustor. A fluidized-bed with these characteristics is called a *circulating fluid bed* (CFB) because of the high rate of material circulating from the combustor to the particle recycle system and back to the combustor. The CFB is further characterized by very high solids-mixing rates.

Fig. 3 also shows the typical velocity and bed particle size for BFB and CFB combustors. Also illustrated is the relationship between FBC systems (BFB and CFB), stoker firing, and pulverized-fuel firing. As described in Chapter 4, stoker firing incorporates a fixed bed, having lower velocity and coarser particle size than the BFB. Pulverized firing incorporates an entrained bed having higher velocity and finer particle size than the CFB. The performance differences between BFB and CFB, to be discussed later, will reflect this relationship.

CHEMICAL PROCESSES

Within the bed, several interrelated chemical processes occur, including combustion, sulfur capture, and nitrogen-oxygen conversion.

FUEL COMBUSTION

Even at the relatively low temperatures associated with fluidized-bed combustion, the combustion of fuel in a fluid bed is a rapid process. The combustion rate is a function of the reactivity of the fuel and the fuel surface area available. Solid fuel can be considered to consist of volatile matter and fixed carbon (char) which remains after the volatiles are driven off. Volatile combustible matter generally burns more rapidly than the residual char and can be viewed as a separate process from the char combustion. The concentration of char within the fluidized bed at any given time is typically a small percentage of the total bed material. Because sulfur dioxide is released during the combustion process, fuel-burning characteristics can significantly influence sulfur capture.

The combustible loss from an FBC boiler is predominantly a function of the amount of char that escapes the system without burning. Generally, the loss from unburned volatiles is insignificant. The char particles are entrained (elutriated) from the bed in the flue gas or are drained from the bed in the bottom ash.

In a BFB, the carbon loss by elutriation alone can be on the order of 10 percent. Recycle of the elutriated material to the bed is an effective means for retaining the char within the system long enough for efficient combustion. Combustion efficiencies of 90 to 98 percent for low-reactivity fuels, and 99 percent for reactive fuels, can be achieved with recycle.

In a CFB, there is no distinct bed as in a BFB unit. The conditions within the CFB combustor provide vigorous mixing as a result of the relatively high fluidizing velocity. The very high recycle rate, attained with a high-efficiency cyclone, provides for relatively long solids residence time within the system. Combustion efficiencies of 95 to 99 percent are achievable even with unreactive fuels such as petroleum coke.

SULFUR CAPTURE

The use of limestone as a sulfur-capture sorbent allows sulfur emissions to be controlled within the fluidized bed during the combustion process. Limestone consists of calcium carbonate ($CaCO_3$) and various impurities. Lime (CaO) is formed by calcining the limestone to drive off carbon dioxide (CO_2).

$$CaCO_3 \rightarrow CaO + CO_2 \qquad (1)$$

Sulfur in the fuel is converted to sulfur dioxide (SO_2) during the combustion process. Although nearly all of the sulfur is oxidized, some of the inorganically bound sulfur may be retained in the ash. The sulfur dioxide combines with the calcined lime in the reaction:

$$SO_2 + CaO + \tfrac{1}{2} O_2 \rightarrow CaSO_4 \qquad (2)$$

Eqs. 1 and 2 indicate that a mole of calcium is required to capture one mole of sulfur. Then, defining the Ca/S molar ratio as moles of calcium in the limestone feed to moles of sulfur in the fuel feed, the theoretical minimum Ca/S required for a given level of sulfur removal is 1/1, which assumes 100-percent utilization of the sorbent.

In practical systems, 100-percent utilization is impossible to attain. Because the sulfation process takes place on the surface of the lime particles in the bed, the lime contained in the particle core is generally not utilized. Also, some SO_2 will escape capture if the total sorbent surface within the unit is insufficient. Consequently, Ca/S mole ratios greater than one (1) are necessary.

The porosity of the particle surface formed during calcination is a strong factor in sulfur capture. Slow calcination results in a highly porous particle with an exposed surface larger in area than that of a smooth particle of similar diameter. As it forms, calcium sulfate tends to block the pores. Deep pores provide large surface area but may plug with sulfate before being filled. The optimum particle size provides the maximum surface that can be fully sulfated. The presence of magnesium carbonate ($MgCO_3$) tends to enhance limestone utilization, even though it does not participate in the sulfur-capture process in the bed. This is because, in calcining to MgO, the $MgCO_3$ increases the porosity of the stone.

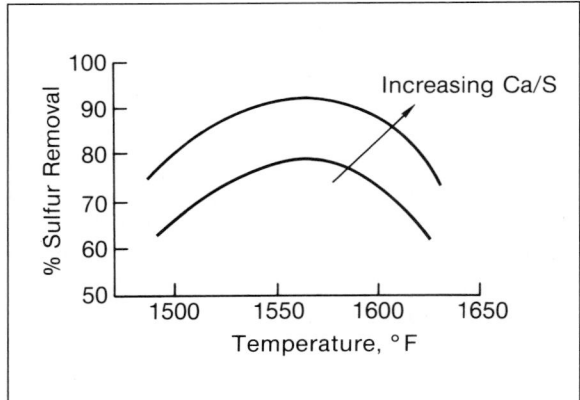

Fig. 4 Sulfur removal versus fluidized-bed temperature

The calcination process begins at around 1300°F (700°C) and, as does the sulfation process, improves with temperature increases. However, the most favorable combination of calcination and sulfation occurs at about 1550°F (840°C). Above this temperature, less-than-optimum porosity forms, limiting the sulfation capacity of the lime particles. Fig. 4 indicates the dependence of sulfur capture on temperature.

There is considerable speculation as to why the sulfur-capture performance falls off so rapidly with increasing temperature. One theory is that sulfate material becomes fluid and coats the particles with a thin, unreactive layer which shields the interior of the particle from further sulfation. Another is the declining rate at which SO_2 is oxidized to SO_3 falls off at temperatures above 1600°F (870°C). The general shape of the sulfation efficiency/temperature curves is common to all types of FBC systems, thus suggesting a fundamental relationship.

Because sorbent particle size is smaller in CFB systems (thus exposing more surface area per unit mass), sorbent utilization is generally better in a CFB than in a BFB. However, BFB units with high recycle rates can closely approach the sorbent utilization of a CFB.

NITROGEN-OXYGEN CONVERSION

NO_x emissions from an FBC boiler are generally less than 0.3 lb/million Btu (0.13 kg/nJ). Values below 0.1 lb/million Btu (0.04 kg/nJ) have been achieved.

Although at the low temperatures typical of FBC no atmospheric nitrogen is converted to NO_x, laboratory data have shown that nearly all of the fuel nitrogen is converted to NO_x during the burning process. For a typical coal containing 1 percent nitrogen, the potential NO_x release is roughly 3 lbs/million Btu (1.3 kg/nJ). Thus, secondary processes are responsible for the low NO_x emissions.

Carbon monoxide (CO) and char are strong reducing agents and appear to be the principal factors in lowering NO_x. These agents strip oxygen from the NO_x in a reduction reaction that produces elemental nitrogen (N_2). In a CFB system, a significant portion of the total air is intro-

duced above the grate. Fuel is normally fed below these air ports, creating a substoichiometric zone in the lower combustor with resulting high concentrations of char and CO. The reducing conditions in the lower combustor enable the attainment of NO_x emissions of 0.1 lb/million Btu (0.04 kg/nJ) and below.

A similar but less extreme condition can be established in a BFB by diverting some of the total air into the freeboard as overfire air. In the bed, the carbon content is around 1 percent and the CO levels are on the order of 5000 ppm. The CO is produced by the burning char in the bed, with burnout completed within the freeboard. Overfire air can be used to redistribute the combustion air, although this must be done with care to avoid the corrosion problems in the bed that would result with substoichiometric firing and yet provide mixing in the freeboard.

There is also evidence indicating that NO_x emissions increase with increasing Ca/S, especially at high SO_2 removal rates.[7] Thus, minimizing Ca/S is important to NO_x emissions as well as to limestone cost.

HEAT TRANSFER IN FLUIDIZED-BED BOILERS

Heat transfer to surfaces immersed in, or bounding, an active fluidized bed occurs by means of three mechanisms acting in parallel: gas convection, radiation, and particle convection. The heat-transfer coefficient to the bed-touched surface is:

$$H_o = H_{gc} + H_{rad} + H_{pc} \qquad (3)$$

where H_o = total outside heat-transfer coefficient
H_{gc} = coefficient of gas convection
H_{rad} = coefficient of particle and gas radiation
H_{pc} = coefficient of particle convection

The gas convection term is the smallest and refers to the transfer of heat from the gas to the surface assuming no solids present. The radiation term is usually the second-largest term and treats the particle "cloud" as a grey body of

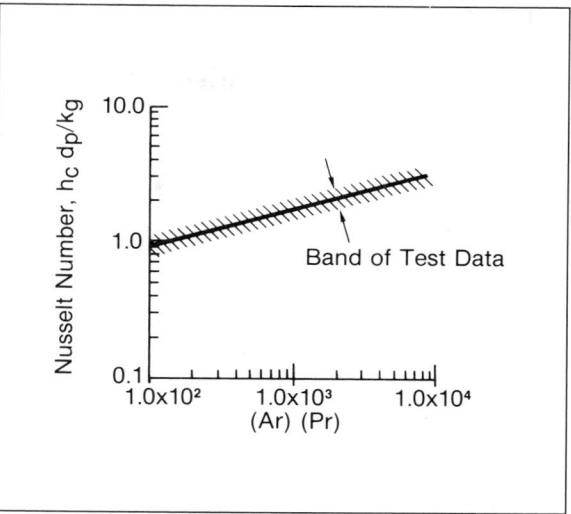

Fig. 5 Correlation of particle convection Nusselt number with Archimedes number

given emissivity radiating to the surface. The particle convection term is usually the largest term, and describes the heat transferred to the surface by particle contact. In practice, the overall heat-transfer coefficient is measured, the gas convection and particle/gas radiation terms are calculated, and the particle convection term is determined by difference. Then, the particle convective component is correlated to certain physical variables for design purposes.

In BFB units, the particle convective component for the active bed area can be correlated reasonably well with the Archimedes number (Ar):

$$Ar = \frac{(d_p)^3 g \rho_g (\rho_p - \rho_g)}{\mu_g^2} \qquad (4)$$

where d_p = mean particle diameter
g = gravitational constant
ρ_g = gas density
ρ_p = particle density
μ_g = gas viscosity, all in consistent units

Fig. 5 is a curve of the Nusselt number versus the product of the Archimedes and Prandtl numbers derived from experimental test results. The curve represents the following correlation:

$$Nu = \frac{h_c d_p}{k_g} = 0.31 \, (Ar)^{0.27} \, (Pr)^{0.33} \qquad (5)$$

where

d_p = the harmonic mean particle diameter
h_c = the total outside heat-transfer coefficient (excluding radiation)
k_g = the gas thermal conductivity
Nu = the Nusselt number
Pr = the Prandtl number, and
Ar = the Archimedes Number

Eq. 5 indicates that the outside convective heat-transfer coefficient is a weak function of particle size and that there is no significant dependence of the heat-transfer coefficient upon the superficial gas velocity through the bed.

In CFB units, work is currently on-going to correlate particle convective component in the combustor with various parameters. The main determinant of combustor particle convection coefficient is the bed density, which varies with combustor height. Factors affecting the density profile include fluidizing velocity, velocity profile, total solids inventory, bed particle size, fuel, and combustor geometry.

Note that CFB units sometimes use a bubbling-bed heat exchanger to cool ash, called a *fluidized-bed heat exchanger* (FBHE) when used in the main ash-circulating loop and a *fluidized-bed ash cooler* (FBAC) when used to cool bottom ash. Work is being done to test the match between the BFB heat-transfer correlation and calculation methods and FBHE/FBAC field data.

Table I indicates the relative magnitude of the various heat-transfer components in and bounding the bed, in typical BFB and CFB steam generators.

The draft loss in the backpass of CFB units is similar to that in stoker or pulverized-fuel boilers; therefore, heat-transfer coefficients are essentially the same. However, for BFB boilers operating at high recycle rates, convective-pass

TABLE I. Relative Magnitude of Heat-Transfer Components in Fluidized-Bed Steam Generators

	BFB	
		Bed
H_{gc}		0–2 Btu/hr-ft²-°F
H_{rad}		8–12 Btu/hr-ft²-°F
H_{pc}		30–50 Btu/hr-ft²-°F
	CFB (same units)	
	Combustor	FBHE/FBAC
H_{gc}	1–3	0–1
H_{rad}	8–12	2–10
H_{pc}	5–20	30–90

dust loading is much higher. Field data indicate that such high dust loading improves the convective-pass heat-transfer rates.

FBC BOILER EFFICIENCY

There are several differences in the calculation of boiler efficiency for a fluid-bed boiler as compared to a stoker or pulverized-fuel boiler. For discussion purposes, the loss method is used, as described in Chapter 6.

An additional loss term is required for the net heat loss or gain from the calcination and sulfation processes in the bed. The calcination loss is that heat lost in calcining $CaCO_3$ to CaO, an *endothermic* reaction. The sulfation heat gain is that from combining the SO_2 with O_2 and CaO to form $CaSO_4$, an *exothermic* reaction. Typically at Ca/S ratios above 2, the calcination/sulfation term is a net heat loss, while at Ca/S below 2, the term is a net heat gain.

The ash sensible heat loss can be large, considering the high-ash fuels frequently fired and the presence of sorbent products in the ash. On a high-ash fuel, the ash sensible heat loss can exceed 5 percent of heat input, as shown in Chapter 6. Therefore, the ash sensible heat loss should be calculated rather than included as an unaccountable loss.

For CFB units, the use of large refractory-lined cyclones increases the radiation loss

above that calculated from the ABMA standard curve in Chapter 6. This loss should also be calculated based on the surface area and skin temperature of the refractory-lined components.

Fan heat credits should also be determined and included in the heat balance. The high fan-discharge pressures used with fluidized-bed boilers result in significant thermal-energy addition to the boilers. Other loss terms are the same for fluid-bed boilers as for other types of boilers, as covered in detail in Chapter 6.

BFB VERSUS CFB

The question of which system, BFB or CFB, is best for a given application is frequently asked but is usually difficult to answer. In certain cases, such as very low SO_2/NO_x emission limits, CFB may be the only choice. Also, there is currently more commercial experience with CFB units on a wide range of unit capacities, steam cycles, and fuels which can influence user selection for new projects. However, for most applications both BFB and CFB are technically feasible. Defining the lowest-cost option requires estimating capital and operating costs for each; the comparison will vary depending on unit capacity, steam cycle, fuel, sorbent, space requirements, and emission limits.

The next two major sections of this chapter describe C-E's fluidized-bed steam-generator designs in more detail, including process design parameters and typical equipment.

BUBBLING FLUIDIZED-BED STEAM GENERATORS

Fig. 6 illustrates a general form of BFB steam generator. Crushed fuel and sorbent are fed to the top or bottom of the bed. Fluidizing air is supplied to the bottom of the bed through a plenum and air distributor. Combustion and sulfur capture (presuming a sulfur sorbent is used) take place in the bed, with the flue gas

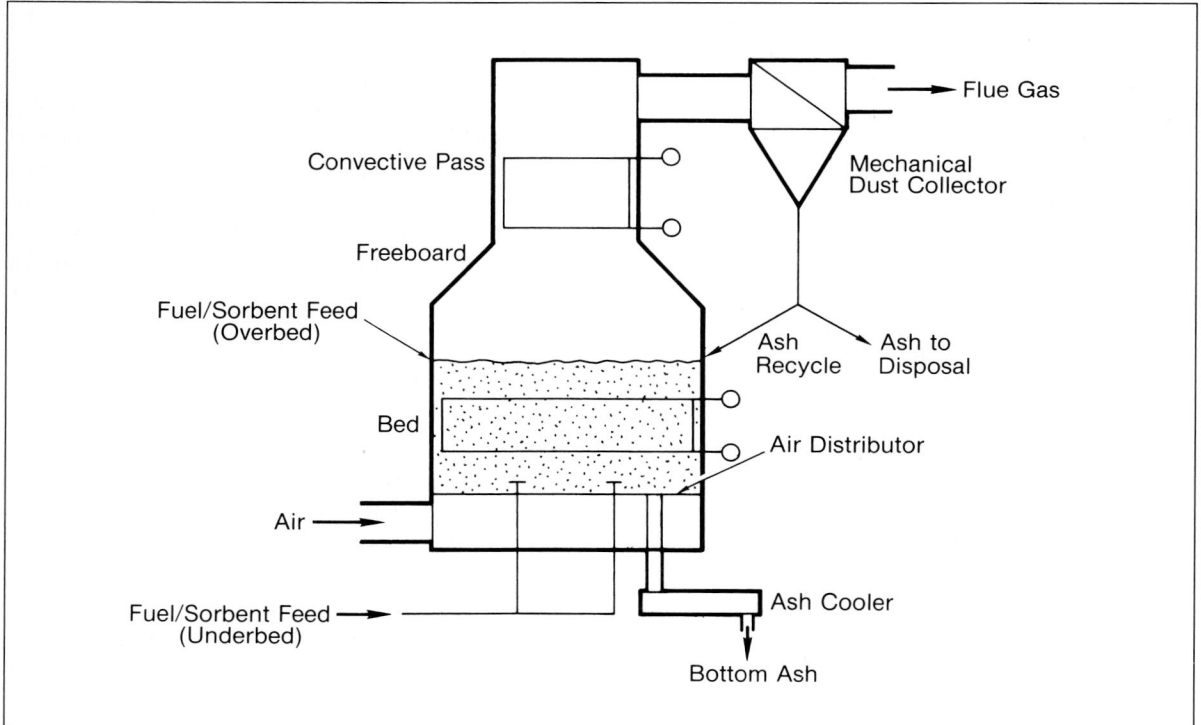

***Fig. 6** Typical bubbling fluidized-bed (BFB) steam generator*

and some entrained solids passing into the section of the combustor above the bed surface, called the *freeboard*. In the freeboard, additional combustion and sulfur capture can take place. From the freeboard, the gas and solids enter the convective pass where they are cooled before entering a mechanical dust collector (MDC). The MDC collects the entrained solids for recycle to the bed or for disposal. Flue gas from the MDC passes to an air heater, fine-particulate collector, and induced-draft fan. Bed temperature is maintained at the optimum for sulfur capture and combustion efficiency, usually by means of the water-cooled walls of the combustor and/or a tube bundle immersed in the bed. Bed level is controlled by draining and cooling an appropriate amount of material from the operating bed.

PROCESS CONSIDERATIONS

At minimum fluidization velocity, a cushion of gas separates the bed particles from one another. Increasing the velocity does not significantly increase the separation distance. Rather, the gas volume in excess of that required for minimum fluidization forms bubbles in the bed. Bubble formation provides a lower resistance path for gas flow. The shape, size, and growth of the bubbles significantly affect bed performance.

The volume of the bubbles causes the bed depth to increase over that of minimum fluidization. The ratio of operating bed depth to the depth at minimum fluidization is called the *expansion factor*. An expansion factor of 1.5 to 2 is typical.

The bubbles provide an important function in mixing the bed. At minimum fluidization velocity, there is very little particle movement. Reactive feed materials such as coal and limestone would not be well dispersed within the bed. The bubbles agitate the bed and induce mixing. On the other hand, the bubbles can provide a path for the bypassing of gases through the bed. If the system is improperly designed, a major portion of the air required for combustion can pass through the bed as bubbles without adequately contacting the fuel. The SO_2 and NO_x

released from the fuel can also escape through the bubbles without fully contacting the sorbent materials in the bed.

The selection of system parameters is important in achieving good mixing without promoting gas bypassing. This is accomplished by designing the bed so that the gases in the bubbles are rapidly exchanged with gases from the non-bubbling regions of the bed. The non-bubbling part of the bed is essentially at the minimum fluidization condition and is referred to as being in the emulsion phase.

The bubbles rise through the emulsion and produce agitation. The effectiveness of gas interchange between the bubbles and emulsion is a function of the bubble size and apparent viscosity of the emulsion phase.

One parameter that can be modified in designing the fluidized bed for efficient gas mixing is the mean bed particle size. At a given gas velocity, the volume of gas in the bubbles depends upon the minimum fluidization velocity, which is affected by particle characteristics. The particle size and density influence the apparent viscosity of the bed which, in turn, affects the average bubble size and rate of bubble growth. With a mean bed particle size of 1000 micrometers (1 millimeter), near-optimum conditions can be maintained over a range of operating conditions with fluidizing velocities from about 2 ft/sec to about 10 ft/sec (0.6 to 3.0 m/s).

For a specific application, it may be desirable to choose a finer or coarser bed-material sizing. For example, a low-reactivity, high-sulfur fuel such as petroleum coke may require a deep bed with a low design superficial velocity; this would serve to increase the residence time for both the fuel-burning and sulfur-capture processes. In such an instance the bed particle size could be reduced while still maintaining desirable bubble conditions. As a secondary benefit, the smaller particle size results in a greater surface area per unit mass for the sorbent bed material. Improved sulfur-capture efficiency is a possible advantage as a trade-off to lower heat release per unit of bed plan area.

In the freeboard, large solids that are ejected from the bed surface separate from the gas flow

and fall back into the bed. The freeboard is considered to terminate at the level at which the gas velocity significantly increases, usually on entering a convective heat-transfer section.

The freeboard can be considered as a second reactor in series with the fluidized bed. The freeboard serves as an important region in which additional combustion and sulfur capture occur. Solids which leave the bed and freeboard are captured in an MDC downstream of the combustor. These solids can be recycled back to the bed, providing additional reaction time for the char and sorbent particles. The amount of recycle is generally described as a recycle ratio, equal to the recycle mass flow divided by the fuel mass flow. For BFB systems burning typical bituminous coals, the recycle ratio is usually selected in the range from 1:1 to 3:1. When firing a typical bituminous coal, with flue-gas flow about ten times coal flow, a recycle ratio of 1:1 gives a solids concentration in the freeboard of approximately 10 percent by weight. This solids concentration provides high sulfur capture and combustion efficiency as the SO_2, char, and combustion air react.

FEED SYSTEMS

The fuel feed system has a major impact on the performance and design of BFB units. Two basic types of feed systems are used, underbed and overbed. Underbed systems include pneumatic as well as mechanical means such as screw feeders and ram injectors. The common feature is that the fuel is introduced at the bottom of the bed. On the other hand, overbed systems use devices such as spreaders to throw the fuel onto the bed surface. Both types of feed systems are shown in Fig. 7.

Underbed feed is the method which takes best advantage of the bubbling-bed dynamics. The fuel is typically introduced by a pneumatic feed system through nozzles located just above the air distributor. For adequate fuel distribution, generally one feed nozzle is provided for each 20 ft^2 (1.9 m^2) of bed area. The nozzles are designed to inject the fuel laterally with a velocity of about 50 ft/sec (15 m/s) to increase dispersion. The fuel is usually limited to a top-size in the range of $1/4$ to $1/2$ inch (6 to 13 mm) to facilitate pneumatic conveying. This size range also provides a large number of fuel particles per unit mass which contributes to a nearly uniform distribution of fuel within the bed.

As compared to overbed, the benefit of underbed feed is improved performance, particularly with regards to sulfur capture. With the fuel introduced upstream of the bed, most SO_2 (especially that from the fuel fines) is released in the bed and so has an opportunity to contact the sorbent in the bed. Thus sorbent utilization is

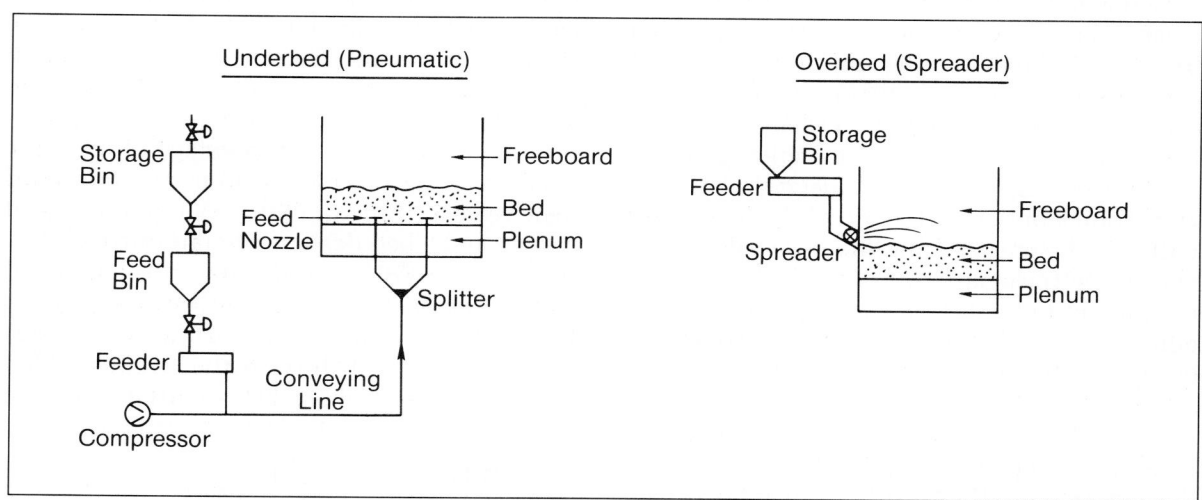

Fig. 7 Fuel/sorbent feed systems for BFB boilers

generally better with underbed feed. Also, underbed feed is more tolerant of fuel fines, which are generally burned in the bed. Further, increasing bed depth can provide acceptable performance for fuels of low reactivity and/or of high sulfur content.

The drawback of underbed feed is its increased complexity. Relatively small pneumatic conveying lines limit the surface moisture of the material being transported to about 6 percent maximum in order to avoid line pluggage. This will usually require a dryer in the fuel-preparation system. Also, the point of fuel introduction at the base of the bed is typically at a positive pressure of 40 to 50 "WG ($1\frac{1}{2}$ to 2 psi, or 10 to 12.5 kPa). Adding to this the pressure lost in the conveying lines and splitter, the feed system must move the fuel from atmospheric pressure to 5 to 10 psig (35 to 70 kPa gage).

The main benefit of overbed feed is that it is much simpler than underbed methods. The fuel is sized as in typical spreader stoker application to 1 to 2" (25 to 50mm) topsize. With the type of equipment involved, this size material is much easier to feed when wet, so fuel drying is not required.

The principal limitations in using overbed feeding arise where high performance is required, especially for sulfur capture. From a process point of view, overbed feed injects the coal downstream of the primary reactor. For combustion, this may not be a major problem because high recycle rates and overfire air can enhance fuel burnout in the freeboard. But the bed is ineffectively used in a system where sulfur capture is a main objective. Larger particles which settle on the bed will liberate volatiles, a large source of SO_2, at the top of the bed. Fines will burn before reaching the bed and liberate SO_2 in the freeboard. All of this SO_2 bypasses the sorbent in the bed. Screening of the coal to eliminate fines, and water sprays to avoid fines burning in the freeboard have been tried to offset this limitation.

Overbed feed also requires closer attention to removal of rock and other inert material fed to the bed. With overbed feed, this material can be quite large and, along with any ash particles formed during combustion, will sink to the bottom of the bed. A layer of large, dense particles can form at the bottom of the bed and disrupt fluidization.

Overbed feed is most appropriate for highly reactive fuels with low sulfur content and for fuels very high in ash. With reactive low-sulfur fuels (such as certain subbituminous coals and lignites) the combustion efficiency can be quite high even with minimal recycle. Also, because the required sulfur capture is low, SO_2 bypassing the bed is not a concern. With high-ash coals and oil shales, the fuel is relatively dense and contains sufficient inerts so that a small particle such as $\frac{3}{4}$ inch (20 mm) or smaller can be uniformly spread. Sulfur-capture efficiency can still be a problem, but often the fuel ash can contain a large amount of calcium. The limestone additive required for sulfur capture may then be negligible or no greater than in the system designed for high-sulfur fuel. Overbed feed is also more appropriate where simplicity rather than performance is the main concern.

DISTRIBUTION OF HEAT RELEASE

Within the BFB combustor, fuel heat is released both in the bed and in the freeboard. Bed temperature is maintained at the desired level usually by heat transfer to furnace-wall tubes and other tubes immersed in the bed. As indicated previously, the heat-transfer rate to these tubes is very high. Thus it is cost-effective to maximize the amount of heat released in the bed. The amount of heat released in the freeboard is expressed as a percentage of the gross heat fired, and is referred to as the *freeboard heat release* (FBHR). With an underbed feed system and a bed designed to minimize gas bypassing, the FBHR is typically in the range of 10 to 15 percent. With an overbed feed system, the FBHR is in the range to 20 to 30 percent, thus shifting heat duty to the freeboard and convective pass where heat-transfer rates are much lower than in the bed.

HEAT-DUTY DISTRIBUTION

The location of evaporator, superheater, and

reheat heating surface within the combustor and convection pass is designed to minimize total cost while providing the proper performance characteristics over the complete range of unit operation.

Usually, the combustor enclosure is composed of welded water-cooled tubing and does evaporative duty. In-bed tube bundles are used for evaporation and for superheater and reheater service. The convective pass contains superheater, reheater, economizer, and air heater surface. With a reheater, a split backpass can be used, with reheat steam temperature controlled by gas-biasing dampers.

PROCESS PARAMETERS

Table II contains a list of typical BFB process design parameters.

Feed size affects both operation and performance. For overbed feed, fines must be limited as they will be entrained in the gases leaving the bed and so not come in contact with sorbent, leading to higher SO_2 emissions. Excessive fines can also lead to more-than-desired burning in the freeboard or localized hot spots near the feeder. For underbed feed, the fuel must be properly sized for pneumatic conveying. High topsize leads to a requirement for excessive transport line velocity and, as a result, potential for erosion. The fuel must have surface moisture less than 6 percent to avoid pluggage in the transport lines. Sorbent size is selected to maximize sorbent utilization, striking a balance between bed solids residence time and surface area.

In the combustor, bed temperature is controlled in the range of 1560 to 1650°F (850 to 900°C), for good combustion efficiency and sulfur capture. The lower temperatures are used with higher sulfur fuels, such as bituminous coal, to maximize sorbent utilization. The higher temperatures are used with high-ash fuels, such as anthracite culm, to maximize burnout. Freeboard temperature is designed to the same temperature range as the bed.

Design bed velocity is usually 8 ft/sec (2.4 m/s). This velocity provides a reasonable plan area, avoids excessive air/gas bypassing via the bubbles, avoids excessive erosion, and retains a margin in fluidization velocity above minimum fluidization for turndown purposes.

Bed particle size depends on the fuel and sorbent sizing and the decrepitation characteristics in the bed. For a given design velocity, bed particle size then determines the state of bed fluidization.

Bed depth is usually set at 2 ft. (0.6 m) in the slumped condition, which results in about 4 ft

TABLE II. Typical BFB Process Parameters

Fuel	Overbed Feed	Underbed Feed
Top size	25–50 mm	10 mm
Moisture	–	<6 percent (surface)
Fines	<25 percent minus 16 mesh	<20 percent minus 30 mesh
Sorbent		
Top size	5 mm	
*d_{50}	800–1400 micrometers or μm	
Combustor		
Bed Temperature	1560–1650°F	
Bed Velocity	4–10 ft/sec	
Bed Particle Size	500–1200 micrometers (μm)	
Bed Depth (slumped/active)	24″/48″	
Bed Pressure Drop	20–50″WG	
Recycle Ratio	0 to 5/1	
Performance		
Carbon Loss	2 to 5 percent or lower	
Ca/S	2.3 to 3.0 (for 90 percent SO_2 capture)	
SO_2	300 ppm or less	
CO	250 ppm	
NO_x	150 ppm or less	

*d_{50} is the diameter at the 50 percent point of a Rosin-Rammler logarithmic probability plot of particle size versus percent passing through a given sieve size.

(1.2 m) active bed height, and an expansion factor of 2. Design bed depth is adjusted (along with bed velocity) to match fuel characteristics. A deeper bed and lower velocity are used with harder-to-burn fuels, or fuels requiring high sulfur capture.

Bed pressure drop is mainly a function of bed depth. Deeper beds, while yielding better performance, also require more fan power. Recycle of entrained material to the bed improves carbon burnout and sorbent utilization. Given the fineness of the entrained material and the multi-cyclone collection efficiency, optimum recycle rates are in the range of 1/1 to 5/1.

PART-LOAD OPERATION

Turndown, or part-load operation, of a BFB is achieved by one or both of two methods. The simplest is called *velocity turndown*. In this method, fuel and airflow are reduced to the entire bed. The bed cooling bundle is designed so that as airflow and, therefore, active bed level are reduced, bed temperature is maintained at or near the optimum. This method alone is usually effective down to 70-percent load only. The other method is called *bed zone slumping*. The air plenum is divided into zones so that the air supply to each zone is separately controlled. As load is reduced, sections of the bed are slumped (defluidized). This allows the maintaining of temperature at the required levels in the active zones. With this method, turndown to 25-percent load and below is easily achieved.

START-UP

Start-up of the unit is accomplished by start-up burners firing oil or gas. These burners can be located underbed in the air plenum, or overbed. Underbed burners heat the bed to solid-fuel ignition temperature (typically 1000°F or 540°C for coal) by first preheating the air to between 1200 and 1400°F (650 and 760°C). Overbed burners fire onto the bed surface, thereby heating the bed material directly.

Because only a portion of the bed is usually heated during start-up, the gas or oil heat-input requirement is minimized. The start-up section may or may not have in-bed surface. If in-bed surface is present, bed level generally is reduced before start-up to avoid having the tube bundles in the active bed. This minimizes both heat losses from the bed and bed heat-up time.

When the start-up section is heated adequately, solid-fuel feed is started. An increase in bed temperature verifies solid-fuel ignition, and fuel and air flows are raised to bring bed temperature up to design. With a multi-zone bed, load is increased by fluidizing a zone adjacent to the start-up zone, bringing the newly fluidized one to the solid-fuel ignition temperature, and then firing additional solid fuel. This process is repeated until all zones of the bed are at temperature on solid-fuel.

SYSTEMS AND COMPONENTS

The following sections cover the major subsystems within the BFB boiler island. Typical equipment is described, along with the major performance criteria affecting design.

FUEL PREPARATION

Selection of fuel-preparation equipment depends on the type of feed system. Overbed feed systems can tolerate relatively high-moisture content fuel, so the preparation system is limited to a primary crusher providing 1 to 2 in. (25 to 50mm) topsize material. Underbed feed systems with pneumatic transport must limit both topsize and surface moisture to avoid pluggage of conveying lines. Thus, the fuel-preparation system contains a crusher supplying 1/4 in. (6mm) topsize material followed by a dryer. The dryer can take several forms. One type is the flash dryer, in which the crusher product is fed to a drying column containing an upward flow of hot gas (typically boiler flue gas). The fuel is dried in the column, separated from the drying gas, and conveyed to the feed system (see reference 8).

An alternate system under development seeks to eliminate the dryer for underbed systems by mixing the wet fuel with recycle material (that material collected by the multi-cyclone for recycle to the bed).[9,10] The hygroscopic components of the recycle material, CaO and $CaSO_4$, react with the water in the fuel to yield a free-flowing material.

SORBENT PREPARATION

Limestone may be purchased with the desired feed-size specification or may be prepared directly at the plant. Smaller installations will typically purchase properly sized limestone and have it pneumatically conveyed into storage tanks. Larger plants will prepare the limestone in a separate yard facility with the system including an impact crusher and size classifiers with the required conveyors. The sized limestone is transported to storage tanks.

FUEL FEED

The fuel feed system introduces the fuel either above the bed (overbed) or at the base of the bed (underbed).

The overbed feed system uses mechanical spreaders to distribute the fuel uniformly across the top of the bed. Underbed systems typically use a pneumatic transport system to distribute the fuel (and often sorbent mixed with the fuel) uniformly throughout the bed. A lock hopper, solids pump, or similar device is required to bring the fuel from atmospheric pressure up to conveying-line pressure.

Fuel feed nozzles must be located every 20 to 30 ft^2 (1.9 to 2.8 m^2) of bed plan area for good fuel distribution. The main-fuel stream from the lock hopper or solids pump must then be split into several streams. For this purpose, C-E uses a fluidized splitter (see Fig. 8) which fluidizes the incoming fuel stream in a low-velocity bed. The fuel overflows uniformly into the individual feed lines within the bed. Splitter fluidizing air becomes the feed-line conveying air.

SORBENT FEED

The simplest way to feed limestone is to the top of the bed through a gravity feed chute. Such a system does not show significantly different sorbent utilization compared with underbed feed.

AIR SUPPLY

A forced-draft fan supplies fluidizing air to a plenum which is usually partitioned to allow control of the airflow to individual sections of

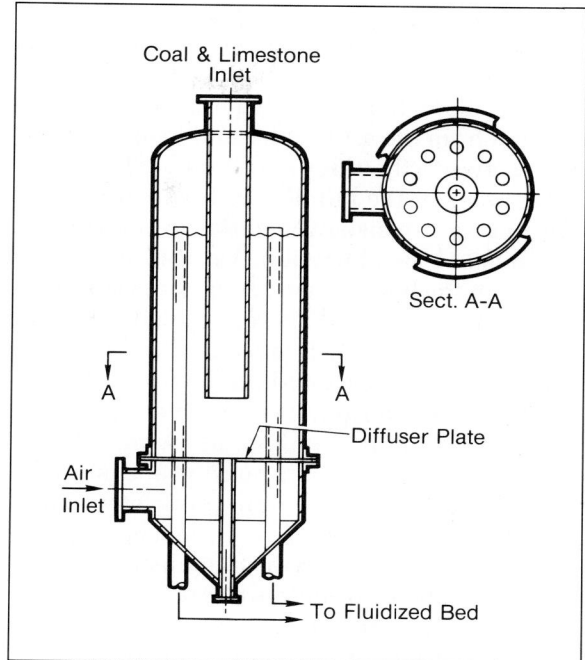

Fig. 8 **Fluidized splitter for dividing fuel flow to bed feed lines**

the bed. The air may be preheated in an air heater. A positive-displacement or rotary blower supplies conveying air at a higher pressure level for pneumatic transport.

Fluidizing-air nozzles, or *bubble caps*, are located at the base of the bed for proper distribution of fluidizing air to the bed. The nozzles are designed to avoid backsifting of bed material into the air plenum. Nozzle pressure drop is selected to ensure adequate air distribution during low-load operation.

COMBUSTOR

The combustor includes the bed and freeboard regions. The enclosure is usually water-cooled and formed from gas-tight finned or fusion-welded panelled tubing, including the bed floor which contains the air nozzles and the fuel/sorbent/recycle feed nozzles (if an underbed feed system is used).

To maintain bed temperature at the required value, horizontal tube bundles are immersed in the bed. Proper design and layout of the tube

bundle and the tube-bundle supports are critical to long-term performance. The bottom tube in the tube bundle is usually located about 2 ft (0.6m) above the fluidizing nozzles, to avoid erosion from the air nozzles and to allow for access between the floor and bundle. The lowermost tubes in the bundle are frequently coated with erosion-resistant material or are studded or finned to avoid erosion. Vertical or inclined tubing in the bundle is especially susceptible to erosion. A C-E developed coating called Extendalloy has proven quite effective in preventing such erosion. The tube-bundle supports must accommodate the gravity and vibrational forces on the tube bundle without allowing differential movement between bundle and support. C-E has developed several proprietary tube-bundle support designs, both mechanical and fluid-cooled, which have demonstrated thousands of hours of trouble-free service.

The combustor can be top-supported, bottom-supported, or a combination of the two. The high mass of the bed material makes bottom support of the bed area attractive.

Circulation Systems

The walls of the combustor are cooled by natural (thermo-syphonic) circulation of the steam-water mixture. In-bed evaporator surface can also have natural circulation, as long as the tubing is inclined. On larger units, inclined tubing is not practical, because the element lengths dictated by the bed depth and the angle of incline are much shorter than large beds require. On such units, the in-bed evaporator bundle uses horizontal tubing and assisted (positive) circulation.

CONVECTIVE PASS

The convective pass is of the same basic design as used in a pulverized-fuel or stoker-fired unit. The enclosure walls are water or steam-cooled tubing, or duct plate, depending on gas temperature. Retractable or rotary sootblowers are used to keep convective surfaces clean, although the ash is powdery with little tendency to adhere to tubing surfaces.

RECYCLE SYSTEM

For units with recycle, a mechanical dust collector collects material entrained from the bed. The device most often used is a multiple-tube cyclone. To provide a substantial recycle ratio, collection efficiency must exceed 90 percent.

Material collected by the MDC is either sent to disposal or recycled to the bed. Like the fuel and sorbent, the recycled material can be fed overbed or underbed. With overbed feed, the recycle material is pneumatically injected from a few locations just above the top of the bed. With underbed feed, the recycle material is elevated to conveying-line pressure by a lock hopper, solids pump, or seal pot (also called a J-valve). For reasonable distribution, recycle feed nozzles must be located every 40 ft^2 (3. 7 m^2) of bed area. Thus, a splitter is also required on the recycle feed system. C-E uses the fluidized splitter for recycle material as well as feed material.

AIR HEATER

The relatively high air/gas pressure differentials affect air-heater selection for BFB applications. Low-leakage designs such as tubular, heat-pipe, and plate-type air heaters have been considered. Also, the development of advanced leakage-control systems for Ljungstrom® air heaters, as described in Chapter 14, has led to use of such on several BFB units.

Use of steam or hot-water preheaters, or a cold air bypass, is required to maintain air-heater average cold-end temperature (ACET) above typical limits. While theoretically the ACET could be reduced for fluid-bed boilers (because of the low SO_2/SO_3 in the flue gas from sorbent addition and unreacted lime in the fly-ash) more operating experience on a wide range of fuels is needed to verify this.

ASH REMOVAL/COOLING

The ash-removal/cooling system includes both the bottom-ash and flyash systems.

Bottom-Ash Removal System

The main function of the bottom-ash removal system is to maintain the desired bed level, or inventory. Bed inventory affects performance in a number of ways, as described earlier. To define bed level, a series of pressure taps are used to locate the active bed surface, while to define bed inventory the total pressure drop across the bed is used. The rate of bottom-ash flow is then adjusted to give the desired value of level/inventory. The bottom-ash system can also control the accumulation of oversize material, which in turn can result in poor fluidizing and mixing, as well as localized defluidization. With an overbed feed system, there is a greater possibility for feed of oversize, and so greater care must be taken to provide a sufficient number of properly located bed drains which can accommodate relatively large material. With an underbed feed system, the opportunity to feed oversize is much less. Providing a bed drain for every 100 to 200 ft^2 (9 to 19 m^2) of bed area is generally adequate for control of bed level/inventory and oversize.

Particularly with an overbed feed system, it may be advantageous to classify the bed-drain material by reinjecting the fines and passing only the coarsest fraction to the bottom-ash system. This arrangement helps remove rocks and such from the bed without excessive ash flow to downstream equipment. The classifier can operate continuously or in batch mode and can also provide some cooling of the ash.

The bottom ash must be cooled from combustor temperature to 250 to 450°F (120 to 230°C) before entering the bottom-ash conveying system. Bottom-ash cooling is accomplished generally by water-cooled screws. Where it is economical to recover the sensible heat in the bottom-ash stream, an air-cooled multi-tube heat exchanger can be used. In this device the ash leaves the bed and flows down into the cooler, which consists of several tubes in parallel over which air is blown. Heat from the ash is transferred to the air which is then mixed with the main fluidizing air.

From the cooler, the ash passes to the bottom-ash conveying equipment for transport to storage. The transport system can be either a mechanical system with flight conveyors or a pneumatic system. Alternately, a mechanical system can be used to transport the material to an intermediate hopper from which the ash can be conveyed penumatically to storage.

Flyash Removal System

Most fluidized-bed boilers have fabric filters for final particulate cleanup. Flyash from the economizer/air-heater hoppers, MDC, and baghouse or ESP is typically handled with a vacuum pneumatic system. No ash cooling is necessary.

The bottom-ash and flyash streams can be stored separately or together. An ash conditioner (see Chapter 16) is used to prepare the ash for transport to disposal.

Chapter 16 describes the complete scope of ash-handling equipment available for use on FBC boilers.

There have been many studies of the possible uses for flyash and bottom ash. Its free-lime content has led to testing of FBC ash as a substitute for cement, with some success. Other uses investigated include: extender in asphalt mix, substitute for gypsum in gypsum board and conditioner for soil.[11] Most often, however, the ash is disposed of as land fill.

INDUSTRIAL APPLICATION

Fig. 9 is an example of a C-E BFB design for industrial application. This type of unit is called the *Fluidized Bed Modular* (FBM) unit as it is designed for shop fabrication and shipment in a minimum number of large sections. The FBM unit can operate over a wide load range while maintaining superheated steam temperatures close to the design value. A boiler firing oil, natural gas, coal, or waste fuel can be designed for capacities from 70,000 to more than 350,000 pounds of steam per hour (10 to more than 45 kg/s). The boiler uses an in-bed superheater, a bottom-support design, and an overbed, underbed, or combined feed system. The large freeboard allows the use of overbed feeding for suitable fuels.

UTILITY APPLICATION

An example of a C-E BFB design for utility applications is the TVA 160-MW demonstration

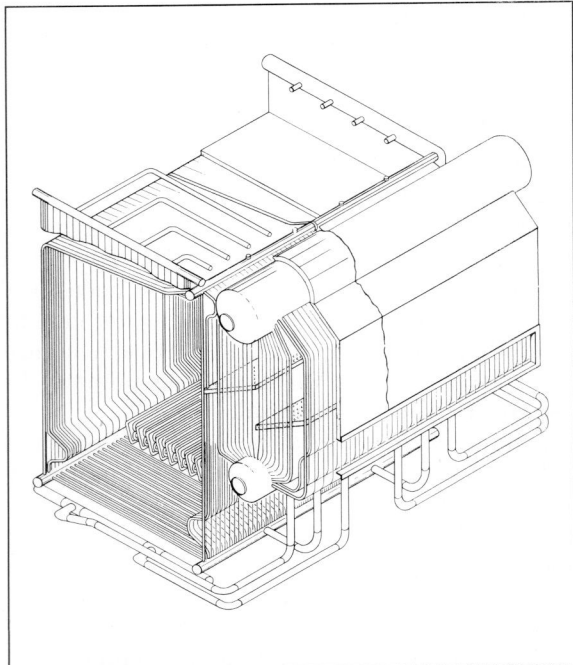

Fig. 9 C-E bubbling fluidized-bed (BFB) boiler for industrial application

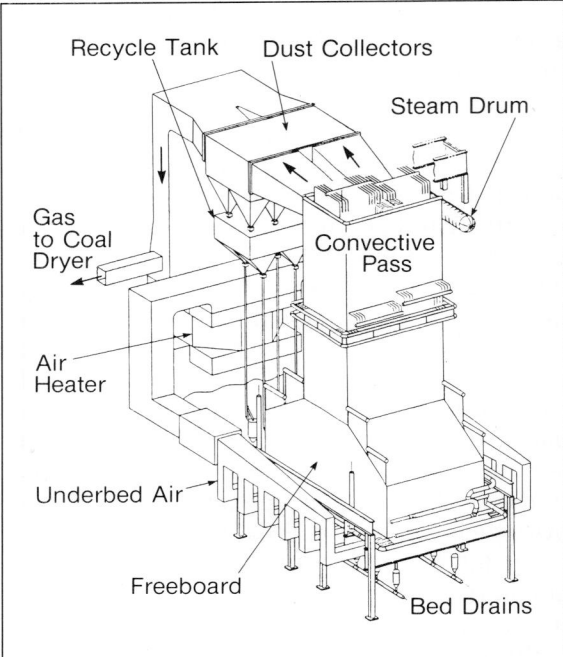

Fig. 10 Arrangement of TVA 160-MW BFB steam generator

boiler, located at TVA's Shawnee Steam Plant near Paducah, Kentucky. The boiler is designed to produce 1,100,000 lb/hr (139 kg/s) main steam at 1005°F (541°C), 1800 psig (12.4 MPa gage), firing a Kentucky bituminous coal. The boiler is the largest fluidized-bed boiler in the world. The plant utilizes the existing steam turbine, condenser, feedwater system, coal yard, fabric filter, induced-draft fan, stack, and flyash transport equipment. Fig. 10 is an isometric view of the boiler.

The combustor consists of a single large bed with freeboard above. The air plenum below the bed is divided into twelve zones, with the airflow separately controllable to each zone. The bed contains several sections of evaporator and intermediate/finishing superheater tube bundles, with mechanical bundle supports. The convective pass which is located above the freeboard in a tower configuration is split, with the reheater on one side. Downstream gas-biasing dampers control reheat steam temperature. An MDC collects flyash for recycle or disposal. A

Ljungstrom air heater is used. Fuel, sorbent, and recycle material are pneumatically fed underbed. Further details of the design are contained in references 12, 13, 14, 15.

CIRCULATING FLUIDIZED-BED STEAM GENERATORS

Fig. 11 shows a typical CFB steam generator. Crushed fuel and sorbent are fed mechanically or pneumatically to the lower portion of the combustor. Primary air is supplied to the bottom of the combustor through an air distributor, with secondary air fed through one or more elevations of air ports in the lower combustor. Combustion takes place throughout the combustor, which is filled with bed material. Flue gas and entrained solids leave the combustor and enter one or more cyclones where the solids are separated and fall to a seal pot. From the seal pot, the solids are recycled to the combustor. Optionally, some solids may be diverted

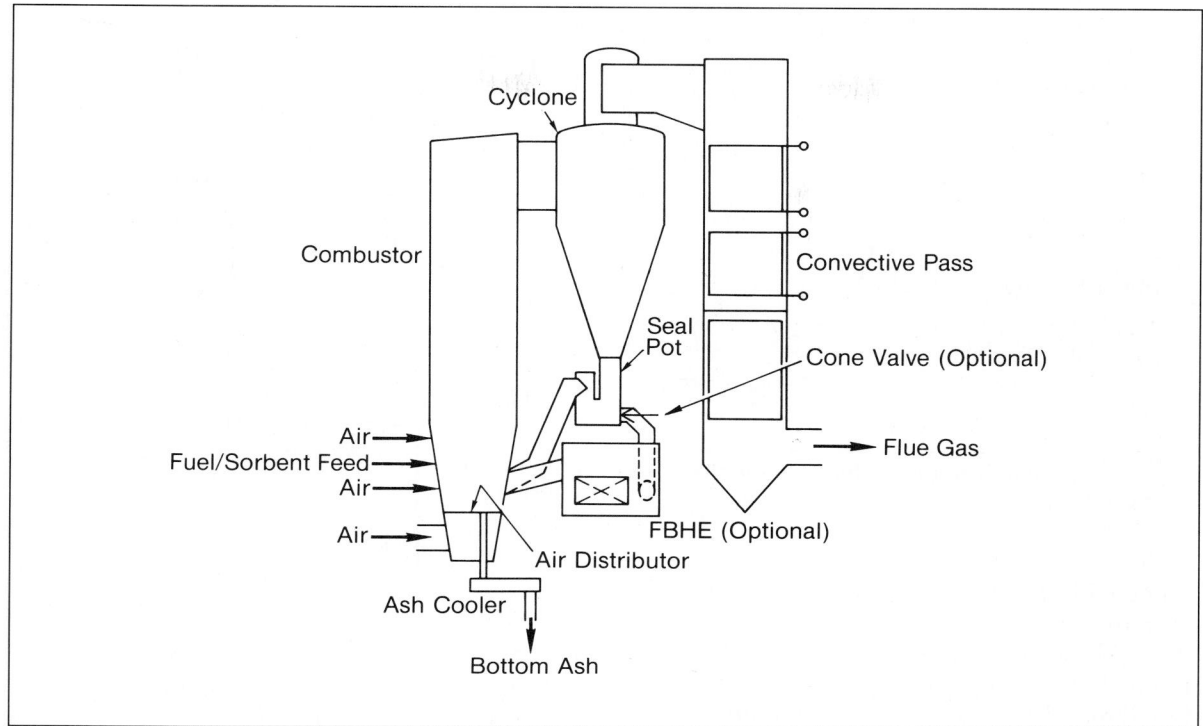

Fig. 11 Typical circulating fluidized-bed (CFB) steam generator

through a plug valve to an external fluidized-bed heat exchanger (FBHE) and back to the combustor. In the FBHE, tube bundles absorb heat from the fluidized solids.

Bed temperature in the combustor is essentially uniform and is maintained at an optimum level for sulfur capture and combustion efficiency by heat absorption in the walls of the combustor and in the FBHE (if used). Flue gas leaving the cyclones passes to a convection pass, air heater, baghouse, and ID fan. Solids inventory in the combustor is controlled by draining hot solids through an ash cooler.

PROCESS CONSIDERATIONS

Circulating fluidized-bed conditions (also called *fast fluidization* or *lean phase fluidization*) are achieved as fluidization velocity is increased past the bubbling regime (see Fig. 1). CFB conditions are generally attained at veloci-

ties greater than 10 ft/sec (3 m/s) with mean bed particle size smaller than 500 micrometers. A large fraction of the bed mass is small enough to be entrained in the gas stream. This material must be collected and recycled to maintain bed inventory. The distinction between bed and freeboard has faded, and bubbles are no longer apparent. The pressure drop from the bottom to the top of the combustor follows a smoothly declining gradient, as illustrated in Fig. 12.

Even though the gas velocity is above the entrainment velocity of most particles in the bed, the entire bed is not entrained out of the combustor. This is because the particles tend to form "clusters" which break-up, reform, and move up and down within the combustor. (Clusters in a CFB are somewhat analogous to bubbles in a BFB.)The gas velocity is below the entrainment velocity of the cluster. The clusters thus allow maintaining considerable bed inventory at usual CFB velocities, and also account

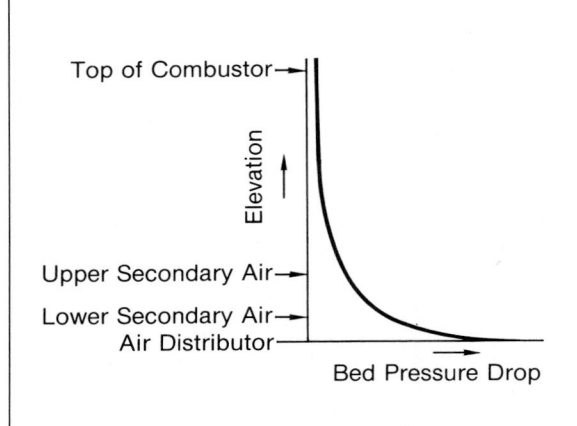

Fig. 12 Circulating fluidized-bed combustor pressure profile

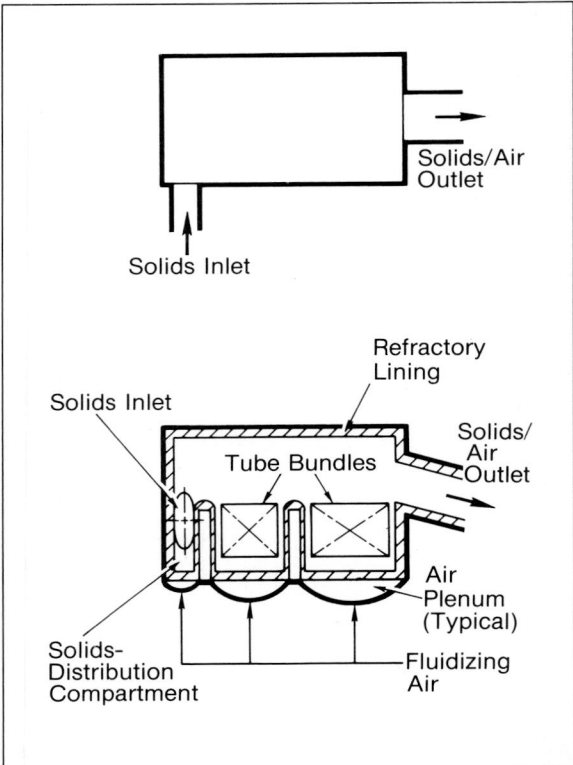

Fig. 13 Typical fluidized-bed heat exchanger (FBHE) with refractory-lined steel-plate enclosure

for considerable internal bed recirculation. The entrained material is of a large enough size to be captured by a cyclone for transport back to the bed. This process results in substantial external recycle and leads to excellent mixing and gas-solids contact with high performance in terms of combustion efficiency and sorbent utilization. Recycle ratios of 10 to 100 and greater are typical and are required to maintain the desired high solids concentration in the combustor. When firing a typical bituminous coal with flue-gas flow ten times the fuel flow, a recycle ratio of 10 provides a solids loading of 1 lb solids/lb gas at the combustor outlet, while a recycle ratio of 100 provides a solids loading of 10 lb solids/lb gas at the combustor outlet. At these solids loadings, the temperature is essentially uniform throughout the combustor and cyclone, and the mixing rates are extremely high. Because of the high mixing rates, only a simple fuel/sorbent feed system with a few feed points is required.

FLUIDIZED-BED HEAT EXCHANGER

For smaller units (less than, say, a steam flow of 500,000 lb/hr or 63 kg/s) with a single design fuel, an FBHE is not needed. For larger units, particularly with reheat cycles or with a wide range of design fuels, an FBHE is highly advan-

tageous.[16] The FBHE is a bubbling-bed heat exchanger with one or more compartments containing immersed tube bundles, as illustrated in Fig. 13.

Additional Heat Duty

The FBHE is an alternate means to remove heat from the hot solids in the primary solids recirculation loop (primary loop), and can reduce the combustor size. Combustor plan area is proportional to unit capacity since design fluidizing velocity is fixed. As capacity increases, the combustor must get taller in order to absorb the required heat in its walls. At some capacity (in the range of 500,000 lb/hr steam, or 63 kg/s), a combustor height of 100 to 110 ft (30 to 33 m) is reached. Combustor heights greater than this offer diminishing returns because heat-transfer rates are low in the upper combustor and additional fan power is required to achieve a desired solids loading at the combustor outlet. C-E normally supplies an FBHE on units above about 500,000 lb/hr (63 kg/s) capacity. The alternative to the FBHE is surface within the combustor bed such as pendant or horizontal surface. Precautions must be taken to avoid excessive erosion of this surface.

Alternate Fuels

Different fuels produce different volumes of flue gas per unit of heat fired. For maximum efficiency, it is necessary to fire each fuel at optimum conditions for sulfur capture and combustion efficiency. This can require a significant variation in the heat absorption required from the primary loop for a wide range of fuels. Table III shows various fuels along with the flue gas each produces. Also shown for each fuel is a term defining the heat in the flue gas leaving the primary loop as a percent of heat input. As indicated, a low-grade, high-moisture fuel such as wood produces large volumes of gas and a relatively large percentage of heat leaving the primary loop (hence lower primary-loop duty) at optimum firing conditions. At the other end of the fuel spectrum, anthracite produces relatively low flue-gas volumes and a relatively small percentage of

heat leaving the primary loop (hence more primary-loop duty) at optimum firing conditions. The FBHE can be used to adjust primary-loop heat absorption when switching fuels, allowing the same combustor to maintain optimum firing conditions on each fuel. For example, when switching from wood to anthracite, solids flow to the FBHE is increased to augment primary loop duty, while combustor temperature and excess air can be maintained at optimum levels. This is illustrated in Fig. 14.

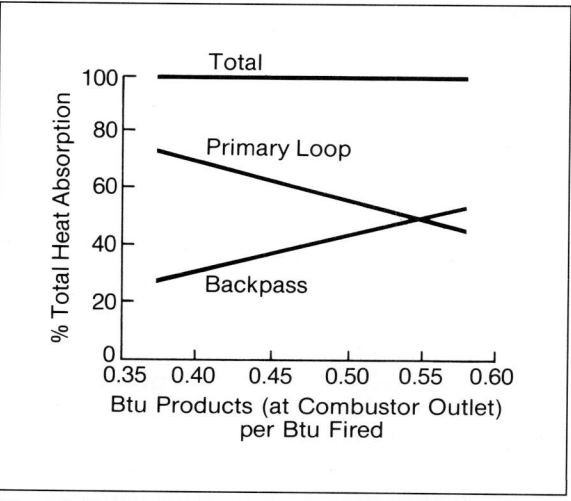

Fig. 14 Total heat-duty distribution for various fuels

TABLE III. Relative Flue-Gas Volumes and Heat-Absorption Patterns for Various Fuels

Fuel	Optimum Combustion Temperature°F	Lb. Combustion Products (Wet) per 10^6 BTU Fuel Fired	BTU Combustion Products* BTU Fuel Fired
Wood Waste	1560	1,329	0.571
Spent Sulfite Liquor	1560	1,181	0.562
Anthracite Culm	1650	1,035	0.447
Anthracite	1650	1,015	0.436
Lignite	1560	1,031	0.431
No. 6 Oil	1560	1,006	0.417
Petroleum Coke	1560	1,004	0.403
Bituminous Coal	1560	992	0.400
Natural Gas	—	925	0.391

*At the combustor outlet

Turndown

Turndown is accomplished by reducing fuel and air flows to the unit. However, it is a characteristic of a CFB that the combustor heat-transfer rate is not reduced proportionally to a drop in airflow. Consequently, without an FBHE, combustor temperature drops as load is dropped. The FBHE can be used to maintain combustor temperature at the optimum level over a wide load range, by reducing solids flow to the FBHE as load is decreased. This feature is especially important when firing certain low-grade fuels. In such situations, maximum combustor temperature must be limited because of the potential for ash softening, thereby leaving only a narrow temperature range for acceptable combustion.

Reheat

Reheater surface can be located in the FBHE, and the flow rate of solids to the FBHE can be used to control reheater outlet steam temperature. The FBHE can also heat air or other process streams, depending on the application.

HEAT-DUTY DISTRIBUTION

The distribution of evaporator, superheater, and reheater duty within the primary loop (combustor and FBHE) and convective pass is set to minimize cost while providing the required performance over the load range. The major considerations in selecting the heat-duty distribution are described in reference 17.

In a unit *without* an FBHE, the combustor walls do evaporative duty, and the convective pass contains the superheater, economizer, and air heater. In a unit *with* an FBHE, the combustor does evaporative duty and the convective pass typically contains the low temperature superheater and economizer plus the air heater. The FBHE then contains the finishing superheater and may contain additional evaporator surface. When fluid-cooled, the walls of the FBHE do evaporative duty.

With a reheat cycle, there are usually two FBHE's. The combustor does evaporative duty,

while the convective pass typically contains low-temperature superheater and low temperature reheater, along with the economizer and airheater. One FBHE then contains the finishing superheater and the other contains the finishing reheater. This arrangement allows control of reheat steam temperature by means of solids flow to the FBHE containing the reheat surface. Either or both FBHE's may contain evaporator surface as well.

The split reheater (part in convective pass, part in FBHE) is designed to put heat into the reheat circuit at low loads (when hot solids may not be available to the FBHE) for turbine steam-temperature matching, without requiring a turbine bypass.

The heat-duty distribution described above minimizes the heating surface and cost in a subcritical drum-type cycle without a turbine bypass (the most common cycle in the U.S.). Other heat-duty distributions may be best for other cycles and to meet certain special performance requirements. One example would be a once-through steam cycle with a turbine bypass, where all reheat surface could be located in one FBHE.

Process Parameters

Table IV shows a list of typical CFB process design parameters.

Proper fuel feed size is extremely important to both operation and performance. If the feed is too coarse, there will be insufficient material in circulation, and reduced burnout, sorbent utilization and combustor heat transfer will result. In the extreme, the solids circulation rate will be low enough to cause large temperature gradients in the combustor. This condition can lead to clinkering and defluidization. Further, the coarse material which resides in the lower combustor adds unwanted pressure drop. If the feed is too fine, excessive material will be entrained from the combustor, thereby producing insufficient material in circulation and a resultant negative impact on performance.

In general, high-ash fuels must be crushed finer than low ash-fuels. This is for two rea-

TABLE IV. Typical CFB Process Parameters

Fuel	
Top size	3–9 μm
d_{50}	500–1000 μm
Sorbent	
Top size	1000 μm
d_{50}	200 μm
Combustor	
Temperature	1560–1650°F
Velocity	18–20 ft/sec
Particle Size	100–1000 μm
Pressure Drop	40–80"WG
Recycle Rate	10 to 100/1
Performance	
Carbon Loss	1–2 percent or less
Ca/S	1.5–2.5 (for 90 percent SO_2 capture)
SO_2	100 ppm or less
CO	100 ppm or less
NO_x	100 ppm or less

sons. First, high-ash fuels tend to decrepitate less and so smaller feed size is required to produce the optimum bed particle size. Second, if high-ash fuels are not adequately crushed, carbon will be encapsulated by ash and the carbon loss will be unnecessarily high. Very high ash fuels such as anthracite culm (which may contain up to 70-percent ash), are typically crushed to 1/8-inch (3mm) topsize, while lower-ash fuels such as bituminous coals are typically crushed to 1/4-inch (6mm) topsize. More reactive fuels such as lignites are typically crushed to 3/8-inch (10mm) topsize.

Proper sorbent size is also important for good performance. Limestone is generally crushed to 1 millimeter topsize, although the optimum sizing depends on the actual decrepitation and sulfur capture characteristics of a given limestone. Occasionally a very fine material, having high surface area but low retention time in the bed, gives the best performance.

The management of the particle size of both fuel and sorbent to achieve adequate solids circulation rates is critical to CFB performance, and must take into account the feed particle size, ash content, decrepitation rates, and cyclone efficiency. The bed particle size shown is that which can be circulated through the system at the design velocity. Larger particles must be drained from the combustor, while finer particles are carried through the cyclone. Fuel and sorbent feed size must be set to provide an adequate inventory of particles in this size range for acceptable combustor performance.[18]

Within the combustor, gas/solids temperatures are typically maintained in the range of 1550 to 1650°F (840 to 900°C). High-sulfur fuels are usually fired at 1550°F (840°C), which is the optimum temperature for SO_2 capture with limestone. High-ash fuels such as anthracite culm are generally fired at 1650°F (900°C) for improved carbon burnout; these fuels are often low in sulfur.

Design combustor velocity is established at 18 to 20 ft/sec (5.5 to 6m/s). This velocity level provides reasonable combustor heat-transfer surface for a given height, low erosion rates, and an acceptable turndown range with adequate bed stability.

Bed pressure drop is generally in the range of 40 to 80" WG (10 to 20 kPa). The indicated level of inventory provides for high internal and external solids recycle rates, and leads, in turn, to good carbon burnout, effective sorbent utilization, and bed stability.

Performance in terms of carbon burnout and emissions is excellent. For high-ash, low-reactivity fuels such as anthracite culm, carbon loss can be as low as 1 to 2 percent, while for high-reactivity fuels such as lignite, carbon loss is typically below 0.5 percent. SO_2 emissions can be reduced below 100 ppm, with 90-percent SO_2 capture at Ca/S of 1.5 to 2.5 (depending on fuel sulfur levels, limestone reactivity, etc.); sulfur capture above 95 percent has been required and achieved in several commercial plants. CO levels are generally in the range of 100 to 200 ppm. NO_x levels below 100 ppm are typical.

PART-LOAD OPERATION

Turndown is accomplished by reducing both fuel and air to the unit. In the process, grate and combustor velocity should be kept above a minimum level in order to produce adequate mixing and solids recirculation for reasonable fuel combustion and to avoid severe temperature maldistribution and backsifting of bed material into the air plenum. This usually implies excess-air levels holding constant as load is decreased until the minimum velocity is reached, after which excess air percentage increases to maintain velocity. Combustor temperature drops, the solid fuel permissive temperature is reached, and the start-up burners must be used.

With an FBHE, combustor temperature can be maintained at high levels through a wide load range. As load is reduced, solids flow to the FBHE is lowered, which adjusts the primary-loop heat absorption to the required level without affecting combustor temperature so that combustor performance is maximized at part load. At reduced load the solids to the FBHE are stopped; below this point combustor temperature will drop with decreasing load.

START-UP

Start-up is accomplished by means of start-up burners located in the lower combustor walls and/or in the primary air duct. Minimum primary-air flow is established and the start-up burners are used to heat the bed material slowly, at a rate dictated by refractory heat-up limits (100 to 200°F/hr or 55 to 110°C/h). When solid-fuel permissive temperature is reached (typically 1000 to 1300°F, 540 to 700°C, for coal), solid fuel is added. Temperature is further increased by adding solid fuel and backing out start-up fuel. At about 30-percent load, the boiler can run on solid fuel alone.

CFB SYSTEM AND COMPONENTS

The following sections describe the major subsystems within the circulating-bed boiler and discuss typical equipment and major performance criteria.

FUEL PREPARATION

Fuel preparation usually consists of one or two stages of crushing, with the system design and layout dependent on such fuel characteristics as moisture and ash content, and required fuel sizing.

Fuels such as anthracite culm generally require two crushing stages. Because the final product must be sized to a topsize of $1/8$ inch (3 mm), or smaller, the material leaving the secondary crushers is typically fed directly to the combustor to avoid hopper pluggage from the fines. Fuels such as bituminous coal permit the material leaving the secondary crushers to be stored in day bins for feed to the combustor.

Many types of crushers, including impact mills, hammer mills, and cage mills, have been applied. Cage mills can limit oversize as the mills wear and so are used with culm firing where oversize fuel is of concern. Feed size is varied by changing mill speed. Hammer mills have changeable grate bars for adjusting fuel sizing. Alternately, air-swept crushers have been very successfully applied in Europe and have been considered for several projects in the U. S. They allow for direct pneumatic feed to the combustor. While air-swept crushers also strictly limit oversize and allow adjustment of fuel sizing, they are usually more expensive than the other types.

SORBENT PREPARATION

Sorbent can be purchased to the correct size specification, or can be crushed on site. Sizing equipment usually consists of an air-swept roller mill with fan, a cyclone collector, and a crushed-product storage bin.

FUEL FEED

The solid-fuel feed system usually consists of a pressurized belt feeder (typically gravimetric) followed by a rotary airlock valve and a fuel chute or pipe leading to the side of the lower combustor. Fuel from the feeder falls by gravity

through the airlock valve, into the combustor. The feeder is pressurized with cold primary air, and the head of fuel in the standpipe of the feeder inlet forms the pressure seal between bin and feeder. At least one completely redundant feed system is recommended.

Alternately, the fuel can be dropped into an air stream and injected pneumatically into the combustor. This approach will help fuel dispersion in the combustor and offers the possibility of using secondary-air ports for fuel feed, thereby reducing the total number of openings required in the walls of the combustor.

High-moisture fuels, such as wood or lignite, are generally fed to the combustor through the discharge side of the seal pot to mix the fuel with hot solids, thereby partially pre-drying the fuel. This feed location also has the advantage of eliminating a separate opening in the combustor for fuel feed, and so the seal-pot feed may be used for any fuel when deemed convenient and economical.

Various pneumatic feed systems can also be used, including air-swept crushers. Fuel from the feeder drops into a pressurized mill, where the fuel is dried, sized, picked up in an air stream, and pneumatically conveyed to the lower combustor. Dense-phase conveying systems are also available. Typically, because of the high internal mixing rates, only a few feed points are required for adequate mixing and dispersion of the solid fuel within the combustor. For smaller units (100,000 to 700,000 lb/hr steam, 13 to 90 kg/s), one or two feed points are usually provided, while for larger units three or more feed points can be required. Of course, fuel heating value can influence the number of feed points. Low-grade fuels with high volumetric flows, such as certain biomass, require more feed points because volumetric flow through certain system components is limited.

Liquid and gaseous fuels for load carrying are fired in lances, fuel-feed pipes which carry only fuel plus an atomizing medium but no combustion air and which are located in the lower combustor. The lance is intended to disperse the fuel within the bed, where it is combusted in the fluidizing air stream. Full load

can thus be obtained on liquid or gaseous fuels with adequate lance capacity. Because these fuels have much shorter bed residence time than solid fuel, they require more feed points for proper fuel distribution and performance.

The lances can be either retractable or stationary. In either case, lance fuel feed can be initiated very quickly on a switch from solid fuel to liquid/gas, or to regain load on a temporary loss of a portion of the solid-fuel feed system.

Liquid and gaseous fuels for start-up are fired in start-up burners, located in the primary/secondary air ductwork and/or in the lower combustor. The burners located in the combustor are retractable.

SORBENT FEED

The sorbent feed system usually consists of a day bin for storing sized limestone, followed by a rotary airlock feeder which drops the sorbent into a pneumatic conveying line for transport to the lower combustor. Gravimetric feeders can be used for a more accurate measurement of limestone flow. To provide the desired number of feed points to the combustor, multiple bin outlets and feed systems can be used or the conveying line from a given feed system can be split. The sorbent can also be mixed with the fuel just before entering the combustor. This is typically done with fine limestone feed, where the limestone is likely to have short residence time in the combustor and so would not be effective if fed far from the fuel.

AIR SUPPLY

Primary and secondary air are supplied to the combustor by separate centrifugal fans, generally arranged in parallel. Either or both of these streams may be preheated in an air heater, depending on the design feedwater and stack temperatures, the economics of air-heater heat recovery versus heat recovery by water or steam heating surface, and whether an air-swept mill is used. An alternate to the above fan arrangement is two fans in series with the second fan supplying the higher-pressure primary air.

With an air-swept crusher, secondary air is generally used for mill air. To overcome the pressure losses through the mill and conveying lines, the air pressure may need to be slightly higher than without such a mill.

Fluidizing air for the seal pot, FBHE, and FBAC is supplied by either positive-displacement or centrifugal blowers. Depending on the flow rates, it may be economical to preheat these air streams in an air heater.

Fluidizing-air nozzles (bubble caps) are provided in the bottom of the combustor, seal pot, FBHE, and FBAC for proper distribution of fluidizing air. These nozzles are designed to avoid backsifting of solids into the air supply system.

COMBUSTOR

The combustor corresponds to the furnace in a pulverized-fuel or stoker-fired boiler. The combustor consists of two zones: lower combustor and the upper combustor.

The lower combustor is that portion containing the fuel, primary-air distributor, secondary-air ports, fuel feed ports, and solids-recycle ports. The density of the bed in this region is relatively high on average, being highest at the elevation of the air distributor and dropping off rather rapidly with increasing combustor height (see Fig. 12). Due to the staged air feed, this region is substoichiometric. Physically, this section is usually rectangular, tapered, formed from finned or fusion-welded waterwall tubing, and lined with refractory to protect the tubing from erosion by the dense bed and corrosion in the substoichiometric atmosphere. The optimum refractory lining is hard (to minimize erosion), thin (to minimize weight), and reasonably conductive (to maximize combustor heat absorption).

The upper combustor, the section above the refractory-lined lower combustor, contains the gas outlet or outlets to the cyclones. The density of the bed in this region is relatively low, and drops off very slowly with increasing combustor height.

Because all air has been fed in the lower combustor, the upper combustor operates under excess-air (oxidizing) conditions. Physically, this section is usually rectangular, straight-walled, formed from finned or fusion-welded waterwall tubing, and unlined to maximize heat absorption.

The air distributor (grate) containing the air nozzles can be uncooled or water-cooled, as can the air plenum below the grate. Water-cooling the grate and plenum provides a seal-welded, gas-tight combustor, and minimizes the size (and thus the maintenance concerns) of expansion joints connecting the primary-air ducts to the combustor.

If necessary on large units, two tapered lower combustors can be used with a single upper combustor, forming a so-called "pantleg" configuration. This configuration improves fuel and air distribution within large combustors.

The combustor can be top-supported or bottom-supported. Top supporting is the more traditional approach but requires that significant differential expansion be taken in the solids-recycle lines connecting the combustor to the cyclones and FBHE, both of which are bottom-supported. Bottom-supporting presumably requires less steel, but often a significant load still must be carried at the top of the structure by the use of constant-load springs, because the number and size of lower combustor openings will not permit the carrying of the entire unit load from below.

Circulation System

The walls of the combustor are cooled by thermo-syphonic (natural) circulation. At high steam/water pressures, the walls of the combustor may incorporate assisted circulation. When provided, an FBHE evaporator bundle can be cooled using pumped (assisted) circulation. Use of natural circulation generally requires inclined tubing and, therefore, less bundle surface per unit of bed plan area. Yet when the total bed plan area required is not excessive, natural circulation is more economical overall. C-E has established the flow requirements for both horizontal and inclined evaporator tubing in laboratory tests.

CYCLONE COLLECTOR

One or more high-temperature cyclones are

used to collect the solids entrained in the gas leaving the combustor. The cyclone is designed to collect essentially all particles with a diameter greater than about 100 micrometers. Given the relatively large particle sizing entering the cyclone, the separation efficiency typically is over 99 percent. When needed, a *vortex finder* (also called a *re-entrant throat*) can be added to the cyclone gas outlet to improve the collection efficiency.

Cyclone Construction

The cyclone is typically constructed of steel plate with a multiple-layer refractory lining. The hot face of the lining is a dense erosion-resistant material, backed up by lighter-weight insulating materials. Proper selection, installation, and subsequent operational care of the refractory materials are essential to ensure long-term lining performance. Alternate construction using water cooling, steam cooling, or air cooling is feasible.

SEAL POT

The seal pot is a non-mechanical valve which moves the solids collected by the cyclone back into the combustor against the combustor back-pressure. Solids flow down on the inlet side, up the outlet side, then back to the combustor. The bottom portion of the seal is fluidized so that material in the seal can seek different levels on each side of the seal, with the difference in level corresponding to the pressure difference across the seal. Then, solids entering the seal inlet displace solids out of the seal on the outlet side.

The seal pot is constructed of steel plate or pipe with a multiple-layer refractory lining. Fluidizing nozzles along the bottom of the seal provide the fluidizing air.

On units with a FBHE, a plug valve is located in the lower portion of the seal pot, to regulate the flow of solids from the seal pot to the FBHE.

FBHE

As described earlier, the FBHE is a bubbling-bed heat exchanger, consisting of one or more compartments separated by weirs and containing immersed tube bundles (see Fig. 13). Hot solids from the seal pot enter the FBHE, where they are fluidized and transfer heat to the heating surface within, and then flow back to the combustor. The tube bundles immersed in the FBHE compartments can be evaporator, superheater, or reheater surface. Here again, proper design of the tube-bundle supports is essential.

Fluidizing velocity is low (1 to 2 ft/sec, approximately 0.3 to 0.6 m/s), the fluidizing medium is air, the particle size is small, and the carbon content of the material is negligible. All these conditions lead to essentially no erosion or corrosion of the in-bed tube bundles. Also, because of the high bed density, heat-transfer rates are very high. Containment can be either refractory-lined steel plate or water-cooled construction.

CONVECTIVE PASS

The convective pass is of the same basic design as used in a pulverized-fuel or stoker-fired boiler. The enclosure walls are usually formed from finned or fusion-welded tubing, steam or water-cooled. Where gas temperatures are sufficiently low, duct plate can be used to form the enclosure.

The convection pass can contain superheater, reheater, boiler bank, and economizer surface. Gas velocities are kept low to avoid erosion from the relatively high dust loading. Retractable or rotary sootblowers can be used to keep heat-transfer surfaces clean.

RECYCLE SYSTEM

Sometimes an MDC is used to collect fine material leaving the cyclone for recycle to the combustor. Such recycle can improve carbon burnout and sorbent utilization. Collection efficiency is usually on the order of 50 percent, given the relatively small incoming particle size and the need to limit collector pressure drop. Devices used include a multi-tube cyclone and a knock-out box (a hopper with chevrons). The collector can be located at any convenient point in the gas pass, such as between the economizer and air heater, or downstream of the air heater.

Provisions should be made for controlled, con-

tinuous (or near-continuous) feed of recycle to the combustor, as well as for transporting collected solids not being recycled to disposal. Ash transport equipment will generally be of the pressure pneumatic type. Lock hopper systems and solids pumps have been successfully used.

AIR HEATER

A prime consideration in selecting an air heater type for CFB applications is the high air-to-gas pressure differential resulting from the high primary-air pressures required. This has resulted in use of low-leakage designs, such as welded tubular air heaters and heat-pipe air heaters (see Chapter 14).

Size of the air heater is then based strictly on economics, since preheated air is not needed by the process. The exception is where an air-swept mill is used and sufficient air preheat is needed for fuel drying.

Generally, two separate air heaters are provided, one for primary air and one for secondary air. These can be arranged in series or in parallel with the gas stream. With large amounts of fluidizing air for FBHE's, it is sometimes economical to provide a separate air heater for fluidizing air. This is because the primary and secondary air heaters become large to achieve a given stack temperature if a significant amount of the total air bypasses these heaters.

With tubular air heaters, the most common design for CFB applications is a gas-over-tube/air-through-tubes design. The dust-laden gas passes over the tubes and, because the tubes are arranged in-line, they can be easily cleaned with sootblowers. Gas-through/air-over designs, though somewhat more difficult to clean, have also been used successfully.

Heat-pipe air heaters, described in detail in Chapter 14, have been used on many CFB projects, due to attractive economics, essentially zero-leakage, and other favorable characteristics. For CFB applications, a maximum of 3 fins per inch on the gas side, retractable sootblowers, and provision to add future surface have been incorporated into the design to accommodate the relatively high dust loading.

Cold-end protection by means of steam pre-

heating, hot-water preheating, or air bypass is usually required to maintain the air-heater ACET above typical limits under conditions of low ambient temperature and/or low-load operation. As mentioned earlier, while there is potential for the ACET to be reduced for fluidized-bed boilers due to low SO_2/SO_3 in the flue gas (from sorbent addition and lime in the flyash), more operating experience on a wider range of fuels is needed to substantiate this approach.

ASH REMOVAL/COOLING

The ash-removal system includes both the bottom-ash and flyash systems.

Bottom-Ash Removal System

The main function of the bottom-ash system is to control bed inventory. Bed pressure drop is the measure of inventory, and bottom-ash flow is adjusted to maintain the desired bed pressure drop. The bottom-ash system can also help control accumulation of oversize material. In a CFB, such accumulation can produce an unfavorable pressure profile with most of the material in the lower combustor and little in the upper combustor, resulting in poor performance. However, the best and most direct way to control oversize accumulation is with proper design of the fuel-sizing equipment to avoid oversize. One or two ash drains per combustor are usually sufficient. A grizzly mounted over the ash drains can keep large material from plugging downstream equipment.

Ash classifiers may also be used, to remove oversize and adjust pressure profile, without requiring excessive bottom-ash flow rates. Such classifiers can operate continuously, in batch mode, and can also cool the ash.

The bottom-ash must be cooled from combustor temperature to between 250 and 450°F (120 to 230°C), before entering the bottom-ash conveying system. On high-ash fuels, the heat in the bottom-ash stream may represent a significant percentage of boiler heat input. Consequently, it can be desirable to recover this heat. Fluidized-bed ash coolers (FBAC's) are generally used for this purpose. The FBAC is a BFB heat exchanger identical in design to the FBHE. Cooling coils immersed in the bed cool the ash and transfer heat to condensate or boiler feedwater. Ash flow from the combustor to the FBAC is controlled by a cone valve, as with the FBHE. The FBAC design must accommodate the accumulation of coarse material which can lead to sintering and/or defluidization. C-E has developed successful FBAC designs, which have been proven on high-ash-fuel applications. On low-ash fuels, there is no incentive to recover heat in the bottom-ash, so water-cooled screws are typically used.

Cooled ash from the ash cooler passes to the bottom-ash handling system for transport to storage. This is usually a mechanical system consisting of flight conveyors, although a pressure pneumatic system can also be used. Alternately, a mechanical system can be used to transport bottom-ash to an intermediate hopper, from which a pneumatic system conveys the material to storage.

Flyash Removal System

As with BFB boilers, fabric filters (baghouses) and electrostatic precipitators (ESP's) are used for final particulate cleanup.

Flyash from the economizer and air-heater hoppers, the MDC, and the fabric filter or ESP is typically handled with a vacuum pneumatic system, although flight conveyors are also used. No ash cooling is necessary. The bottom-ash and flyash streams can be stored for disposal together in the same silo or in separate silos. Ash conditioners then mix in sufficient water for proper handling and transport to the ultimate disposal area.

An alternate to the dust collector and recycle system described earlier is to recycle material to the combustor from the baghouse or ESP. Such a system has the advantage of avoiding an additional dust collector in the system, although care must be exercised not to overload the convective pass and baghouse or to exceed emission levels.

INDUSTRIAL APPLICATION

Two examples of C-E CFB designs for indus-

trial-scale applications are described here, to illustrate major component design and arrangement. The first is a 220,000 lb/hr 955°F, 1255 psig (28 kg/s, 513°C, 8.7 MPa gage) wood-fired steam generator (see Fig. 15). The unit has a single combustor with cyclone and seal pot, but no FBHE. The convective pass includes a superheater, boiler bank, economizer, and heat-pipe air heater.[19]

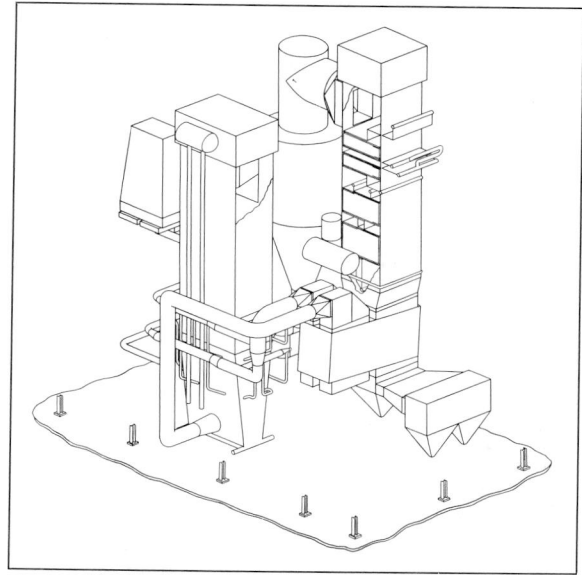

Fig. 15 Small C-E CFB steam generator for industrial/power-production application

The second is an 825,000 lb/hr, 1000°F, 1800 psig (104 kg/s, 540°C, 12.4 MPa gage) boiler, designed to fire anthracite culm (see Fig. 16). The unit has a single combustor, two cyclones with seal pots, and an FBHE containing an evaporative bundle. The convective pass includes a superheater, economizer, and heat-pipe air heater.

UTILITY APPLICATION

One example of a C-E CFB design for utility applications is a 150-MW unit for the Texas-New Mexico Power Company producing 1,000,000 lb/hr (126 kg/s) main steam at 1005°F (541°C), 1800 psig (12.4 MPa gage), with reheat, designed to fire lignite. Fig. 17 is an isometric view of the boiler.

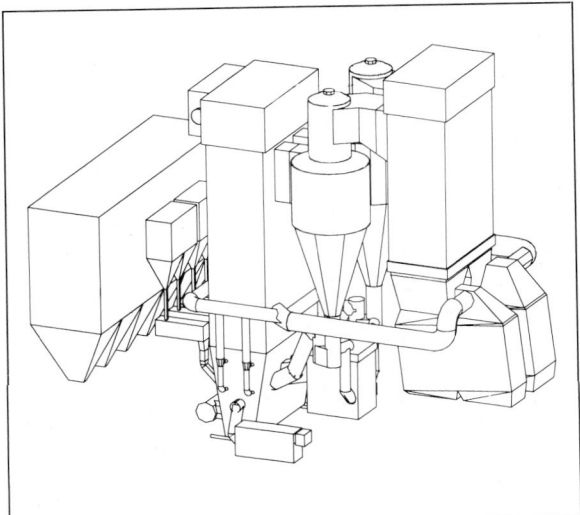

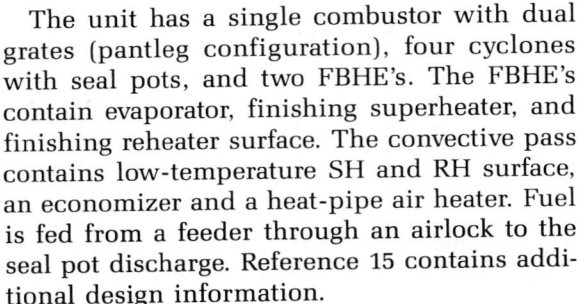

Fig. 16 Large C-E non-reheat CFB steam generator for industrial/power-production application

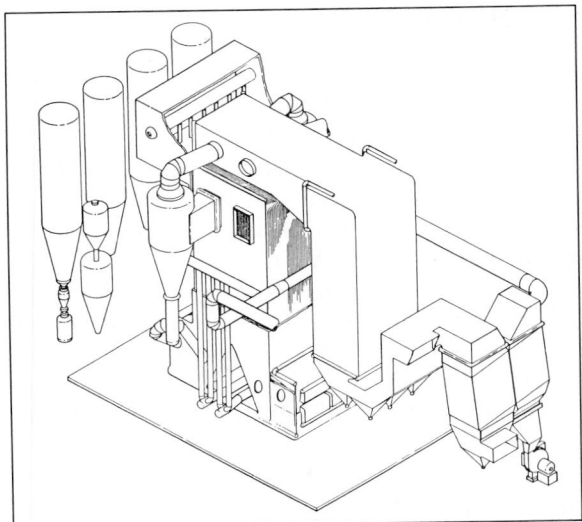

Fig. 17 C-E 150-MW CFB steam generator with reheater

The unit has a single combustor with dual grates (pantleg configuration), four cyclones with seal pots, and two FBHE's. The FBHE's contain evaporator, finishing superheater, and finishing reheater surface. The convective pass contains low-temperature SH and RH surface, an economizer and a heat-pipe air heater. Fuel is fed from a feeder through an airlock to the seal pot discharge. Reference 15 contains additional design information.

PRESSURIZED FLUIDIZED-BED COMBUSTION SYSTEMS

This chapter has introduced all the concepts involved in the design of fluidized-bed steam generators, and has described the implementation of that design for boilers operating at or near atmospheric pressure. Combustion in that regime is referred to as atmospheric fluidized-bed combustion (AFBC). Fluidized-bed operation at pressures 10 to 20 times atmospheric pressure, taking place in large cylindrical or spherical pressure vessels, in combination with axial compressors and gas turbines, is termed pressurized fluidized-bed combustion (PFBC).

In addition to the reduced emissions of sulfur and nitrogen oxides that are possible with fluid-bed combustion, PFBC offers the potential for a significant gain in overall thermal efficiency because of the incorporation of a gas turbine in the cycle. Another advantage of the PFBC system is that all of the equipment operating at the plus-10-atmosphere pressure level is smaller in size than it would be at normal atmospheric pressure, making shop-assembly and barge delivery of components an attractive option.

Two of the many approaches to the pressurized cycle are the PFBC turbocharged cycle and the PFBC combined cycle; even these two have many possible variations, which are being investigated throughout the world, in both design studies and operational units.

In the turbocharged cycle, hot flue gas from a PFBC boiler (at approximately 800°F or 425°C) is expanded through a gas turbine that produces enough power to drive the turbocharging compressor. One version of the higher efficiency power-producing PFBC concept has the gas leaving the pressurized fluidized-bed combustor at about 1600°F (about 870°C). The gas is cleaned in tandem high-temperature cyclones, and is then sent to the gas turbine. The turbine exhaust is further reduced in temperature by passing through heat-recovery equipment, such as evaporative and feedwater-heating surface.

The supercharged furnace-fired combined cycle was introduced in Chapter 1, and is illustrated schematically in Fig. 15 of that chapter. Reference 19 of Chapter 1 and reference 20 of this chapter describe work done by Combustion Engineering and others in the development of this promising technology.

REFERENCES

[1] S. L. Goodstine et al., "Industrial Application of Fluidized Bed Combustion, Phase I, Task 4—Sub-Scale Unit Testing and Data Analysis. Vinal Report, Volume 1," prepared for the U.S. Department of Energy under Contract EX-76-C-01-2473, December 1979; published by the U.S. Department of Energy (NTIS) DOE/ET/10389-T3.

[2] Combustion Engineering, Inc., "Great Lakes Fluidized Bed Combustion, Final Report, Volume 1," prepared for the U.S. Energy Research and Development Administration under Contract EX-76-C-01-247, December 1985. Windsor, CT: Combustion Engineering, 1985.

[3] Combustion Engineering, Inc., "Preliminary Design Study for a Fluidized Bed Demonstration Unit"—Final Report to the Tennessee Valley Authority under TVA contract 45723A, funded by the U.S. Department of Energy under DOE Interagency Agreement EF-77-A-01-6013, June 1, 1978. Windsor, CT: Combustion Engineering, 1978.

[4] Combustion Engineering Inc., "TVA Utility AFBC Project, Phase II, Final Report; Volume 1—200 MW Demonstration Plant Final Design; Volume 2—800 MW Commercial Plant, Proposed Design and AFBC Research and Development Requirements"; prepared under TVA Contract 51863A, January 18, 1981. Windsor, CT: Combustion Engineering, 1981.

[5] Combustion Engineering, Inc., "Preliminary Design Study for a Circulating Fluidized Bed Commercial and Demonstration Unit, Final Report, Revision 1," prepared under TVA Contract 53907A, December 31, 1980. Windsor, CT: Combustion Engineering, 1980.

[6] D. Kunii and O. Levenspiel, *Fluidization Engineering*, ISBN 088275-542-0. Malabar, FL: Robert E. Krieger Publishing Co., Inc., 1969.

[7] B. Leckner and L.-E. Amand, "Emissions from a Circulating and a Stationary Fluidized Bed Boiler: A comparison," *Proceedings of the Ninth International Conference on Fluidized Bed Combustion*, Boston, MA, May 4–7, 1987.

[8] G. C. Dunn, G. D. Jukkola, and R. C. Kunkel, "Fuel Preparation Systems for TVA's 160 MW Demonstration Plant," *Proceedings of the 1988 Joint ASME/IEEE Power Generation Conference*, Philadelphia, PA, September 25–29, 1988.

[9] R. McKinsey et al., "Coal Drying Using Recycle Fly Ash," *Proceedings of the 1986 EPRI Seminar on AFBC Technology—Utility Applications*, Palo Alto, CA, April 1986.

[10] G. D. Jukkola et al., "Application of Coal Drying Systems Using Recycle Material at the TVA Demonstration Plant," *Proceedings of the 1986 EPRI Seminar on AFBC Technology—Utility Applications*," Palo Alto, CA, April 1986.

[11] A. M. Manaker, PH.D. and G. D. Jukkola, "Atmospheric Fluidized Bed Combustion Development at TVA," *Proceedings of the 1988 Joint AME/IEEE Power Generation Conference*, Philadelphia, PA, September 25–29, 1988.

[12] A. Stathoplos et al., "System Developments for Fluid Bed Boilers," presented at the International Symposium on Coal Combustion, Beijing, China, September 1987.

[13] R. V. Jacobs, "Design of the TVA 160 MW Atmospheric Fluidized Bed Steam Generator," *Proceedings of the 1986 Joint ASME/IEEE Power Generation Conference*, Portland, OR, October 19–23, 1986; also as Combustion Engineering publication TIS-8219.

[14] J. W. Regan et al. "Design of the 160 MW AFBC Demonstration Unit at TVA's Shawnee Stea Plant," *Proceedings of the Eighth International Conference on Fluidized Bed Combustion*, Houston, TX, March 18–21, 1985; also as Combustion Engineering publication TIS-8023.

[15] J. W. Regan and H. Beisswenger, "Utility Applications of AFBC." Windsor, CT, Combustion Engineering publication TIS-8244.

[16] S. A. Pierzchala and B. W. Wilhelm, "Design Requirements for Circulating Fluid Bed Units with Reheat," *Proceedings of the 1990 Joint ASME/IEEE Power Generation Conference*, Boston, MA. October 21–25, 1990. ASME Paper No. 90-JPGC/Pwr-70.

[17] E. J. Gottung et al., "Design Considerations for Circulating Fluidized Bed Steam Generators," *Proceedings of the Tenth International Conference on Fluidized Bed Combustion*, San Francisco, CA, May 1–4, 1989.

[18] H. Herbertz et al., "Effects of Fuel Quality on Solids Management in CFB Boilers," *Proceedings of the Tenth International Conference on Fluidized Bed Combustion*," San Francisco, CA, May 1–4, 1989.

[19] R. V. Jacobs and E. Gershengoren. "Design and Operation of a Wood Fired CFB Steam Generator." *Proceedings of the Tenth International Conference on Fluidized Bed Combustion*, San Francisco, CA, May 1–4, 1989.

[20] S. R. Wysk et al., "Technical and Economic Aspects of a Pressurized Circulating Fluidized Bed Combustion System," presented at Second Biennial PFBC Conf., Milwaukee, WI, and 79th Annual Meeting of the Air Pollution Control Association, Minneapolis, MN, June 1986.

BIBLIOGRAPHY

Barner, H. E., Beisswenger, H., and Barner, K. E., "Chemical Equilibrium Relationships Applicable in Fluid Bed Combustion," *Proceedings of the Ninth International Conference on Fluidized Bed Combustion*, Boston, MA, May 4–7, 1987.

Bashar, M. and Czarnecki, T. S., "Design and Operation of a Lignite-Fired CFB Boiler Plant," *Proceedings of the Tenth International Conference on Fluidized Bed Combustion*, San Francisco, CA May 1–4, 1989.

Beisswenger, H., Krittel R., and Ploss, L., "The 95.8 MWe CFB Utility Boiler of the Duisburg Municipal Power Company," Lurgi Publication.

Gendreau, R. J. and Raymond, D. L., "Atmospheric Fluidized Bed Combustion Update – Status and Applications," presented at the 1987 Joint ASME/IEEE Power Generation Conference, Miami Beach, FL, October 4–8, 1987, ASME Paper No. 87-JPGC-FACT-11.

Gottung, E. J. and Sopko, S. J., "Design and Operation of a CFB Steam Generator Firing Anthracite Waste," presented at the 1988 Joint ASME/IEEE Power Generation Conference, Philadelphia, PA, September 25–29, 1988, ASME Paper No. 88-JPGC/Pwr-9.

Patel, R. L. et al., "Reactivity Characterization of Solid Fuels in as Atmospheric Bench – Scale Fluidized – Bed Combustor" presented at the 1988 joint ASME/IEEE Power Generation Conference, Philadelphia, PA, September 25–29, 1988; also as Combustion Engineering publication TIS-8391.

Sainz, F. A. et al., "Chatham Circulating Fludized Bed Demonstration Project," *Proceedings of the First International Conference on Circulating Fluidized Beds*, Halifax, Nova Scotia, Canada, November 18–20, 1985. Toronto, Pergamon Press, 1986.

Wein, W., "Flow Dynamics of Atmospheric Fluid Bed Combustion Systems and their effect on SO_2 capture and NO_x suppression," translated by Lurgi from VGB Magazine, February 1985, pp. 119–123.

Wilhelm, B. W. et al., "100 MW Anthracite Culm CFB Small Power Producer," *Proceedings of the American Power Conference*, Vol. 50. Chicago, IL: Illinois Institute of Technology, 1988.

CHAPTER 10

Marine Boilers

The many limitations and unique requirements applicable to shipboard power plants make the marine boiler a very specialized form of steam generator. Customarily, marine boilers must meet the following overall conditions: high level of dependability, limited space, wide range of steam output capacities, ability to change load rapidly, good accessibility, and limited weight. In addition, because most ships are designed for specific trades, propulsion machinery for a certain ship must be selected to meet the operating conditions encountered in the trade that the vessel is intended to serve.

SERVICE REQUIREMENTS

Marine-boiler design properly starts with a review and study of the service requirements for the particular ship under consideration. Only after an economic evaluation of potential propulsion-plant and cogenerative cycles does the ship designer set the evaporation rate and the required steam pressure and temperature.

The plant heat balances, calculated at various steady steaming conditions, establish certain operating conditions which the boiler must meet. In this connection, it must be recognized that boiler performance is a parameter in the heat-balance calculations. For example,

the boiler efficiency can be set at a given value for any one evaporating rate, but heat-balance calculations at other evaporating rates must be based upon the corresponding boiler efficiencies as set by the performance characteristic of a particular boiler design.

Normal power is a specified characteristic of a given merchant ship and generally corresponds to 100 percent rating of the boiler. For naval vessels, 100 percent boiler rating corresponds to *full power*.

LOAD PATTERN AND BOILER DESIGN

Not only do the conditions set by the quantities which are readily defined by numerical values directly influence boiler design, but operating conditions which will be encountered in service also affect it. Although two ships might have the same shaft horsepower (shp) and the same evaporating rates for the boilers, one ship might be in a trade which calls for prolonged periods of steady steaming, and in contrast, the other ship might be in a service which requires the boilers to be changing load frequently. As examples, the former would be typical of a tanker, and the latter, a dredge or a ferry boat.

Analyses of boiler load variations during a series of voyages make it possible to graph the load pattern for the boilers in a particular trade.

In designing new ships, it is customary to plot these graphs with averages compiled over an extended period from data of ships that have been in the same trade that the new vessels will serve. Fig. 1 illustrates the boiler load pattern typical of a dry-cargo ship during a single round-trip voyage between the West Coast of the United States and the Orient.

Obviously, a dry-cargo boiler has a load pattern which differs from that of a tanker or a destroyer. Although the design of a ship and its machinery must take into consideration possible changes in service conditions, the prime objective is to get the optimum design for the trade in which the vessel will initially serve. Fortunately, most marine boilers have the inherent characteristic of being versatile to a degree. If later service conditions do bring about changes in the boiler load pattern, the boiler usually will be able to meet the new conditions within the limits determined by the other mechanical components of the propulsion plant and the cogenerative cycle.

Not only is the load pattern an important factor in deciding some of the principal characteristics of marine-boiler design, but it will also determine some of the detailed features which must be included. Consideration must also be given to various other service conditions, such as ship sailing schedule and quality of operating personnel. Some of the boiler design features which may be decided entirely, or in part, by service conditions are

- number of boilers per ship
- type of firing system
- furnace dimensions
- overall boiler dimensions
- extent of control systems
- size of steam drum
- type of steam generator
- relative amount of waterwall generating surface

SPACE LIMITATIONS AND BOILER DESIGN

Up to this point there has been no discussion of the extremely basic condition of space

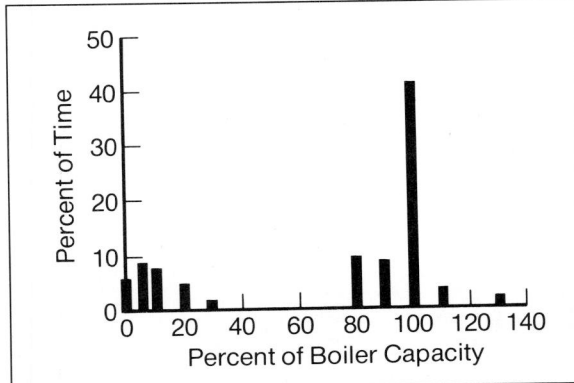

Fig. 1 Boiler load pattern during voyage of dry-cargo ship

restriction which actually dictates the boiler design on many ships. To the boiler designer, it would be ideal if adequate space could be provided in the fireroom so that no limitations on size or shape had to be considered. Table I and Fig. 2 show the principal dimensions of one standard series of boiler designs. Although the ship owner and marine architect usually try to design the ship so that, as nearly as possible, space will be provided for the optimum boiler design, it is almost inevitable that space restrictions will be imposed. From the outset of the preliminary ship design to the time when the boilers are installed and accessibility has been satisfactorily proved, the boiler designer must constantly consider the space problem.

Frequently, the very purpose of the ship establishes limiting dimensions for the fireroom

Table I. Principal Dimensions for Standard V2M8 Marine Boilers

Boiler Design	Capacity* 1000 lb/hr	Dimensions, Ft			
		A	B	C	D
V2M8-6	132	21	24.9	12.7	15
V2M8-7	154	21	24.6	16.3	15
V2M8-8	176	23.6	27.7	14.5	16.5
V2M8-9	198	23.5	27.3	15.9	16.5
V2M8-10	221	26.2	25.2	18.4	19
V2M8-11	243	26.5	26.5	19	19
V2M8-12	265	27.1	28.4	19.8	19
V2M8-13	287	26.2	25.2	23	19

*Varies with cycle conditions

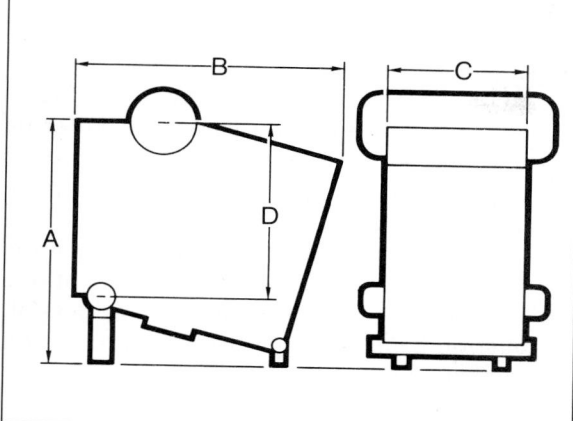

Fig. 2 **Principal dimensions applicable to Table I**

- circulation
- interrelationship of furnace dimensions with main generating-bank dimensions, and
- amount of waterwall surface

The burners must be located in positions that will be readily accessible from the firing aisle. The prescribed minimum burner clearances must be carefully maintained to avoid flame impingement on furnace-wall, floor, or screen generating tubes.

Selection of furnace dimensions must include provisions for the proper flow pattern of the combustion gases within the furnace and into the tube bank. It would be ideal if the gas flow could be controlled to the extent that the mass flow rate and the gas temperature both would be uniform at all points where the gas enters the screen tubes. But achieving this theoretical condition is not generally possible. The boiler designer must, therefore, be guided by the results of past experience obtained during the testing and operating service of designs similar to the one under consideration.

In selecting the preliminary design for a marine boiler, detailed calculations on the circulation of the boiler circuits are usually unnecessary, if the design is reasonably similar to previous designs of the same general type for which complete circulation calculations have been performed. It is important to note that the maximum thermal circulation effect is obtained when the major portion of the heat absorbed is applied to the lower portion of a tube.

Most types of boilers have the arrangement and dimensions of their generating banks related to some degree to furnace dimensions. For example, in early designs of two-drum bent-tube boilers, limitations on the draft loss through the main bank would frequently set the main-bank dimensions, and the furnace height would be fixed accordingly. With some boiler designs, the furnace dimensions can be set at the optimum point, and the main-bank dimensions are essentially independent. This is made possible by having the distributing header located several feet away from the lower drum, as discussed later in this chapter in the section on

space. For example, the train ferry must have as many continuous decks as possible. This feature immediately sets a limit upon the height allowed for the boiler space. For many ship types, the structural members of the ship hull present the most common reason for limiting the boiler dimensions.

SELECTION OF DESIGN

When the service conditions have been properly evaluated, the boiler designer determines the physical design characteristics which will best meet requirements. Ideally, the preliminary design should be completed in time for use by the ship designer in establishing the machinery arrangement and the layout of the main hull structurals passing through the fireroom. Space is generally the first item to be considered. If the optimum boiler design—the one which has the arrangement and physical characteristics most suitable for the service conditions and performance requirements—has any dimension that is definitely in excess of the space allowed, then a compromise has to be made. In designing the boiler furnace, the major factors to be considered are as follows:

- firing-equipment accessibility
- burner clearance to heating surfaces
- gas flow and flame travel

bent-tube boilers. In most instances, boilers are now designed with the maximum practical amount of waterwall heating surface.

DEVELOPMENT OF MARINE BOILERS

In tracing the history of the development and application of marine boilers, it is interesting that some of the basic requirements have always applied, although from time to time the emphasis has shifted from one requirement to another. At the outset, space and weight limitations were not considered to be as important as other factors, and yet at later stages the emphasis on those items brought about the development of new types of boilers. Generally, it has been the economic or military need for increased propulsion power that has instigated significant advances in boiler design.

The earliest types of marine boilers were relatively large pressure vessels, with furnaces located underneath them.[1] But it soon became apparent that such exterior furnaces were far from satisfactory, and they were then located inside the boiler shell. The boiler exterior had now developed into an approximate cube and, no matter how heavily stayed, its flat exterior placed very low limits on the steam pressure that could be carried and also left much to be desired from a maintenance and safety standpoint. The internal furnaces and tubes, however, were cylindrical and entirely satisfactory for the pressure demands of their time. By 1870, marine-boiler pressures had reached only 60 psig; by 1900 they had increased to about 300 psig.

SCOTCH MARINE BOILERS

To overcome the structural weaknesses and at the same time reduce cost of fabrication, the boiler shell was made cylindrical (replacing the earlier box construction) with provision to install one or more combustion chambers between the two flat sides of the shell. Known as the Scotch marine boiler, this boiler as finally developed met with wide and enthusiastic approval by marine engineers. Although minor changes have been made throughout the years, its design has remained much the same and its qualities of ruggedness, reliability, ease of maintenance, and ability to stand abuse made it—until recently—the most popular boiler in the marine field.

The boiler shown in Fig. 3 is coal-fired with a spreader stoker and has two furnaces. A close look at the illustration reveals two separate combustion chambers, each stayed to the other. Tubes used as flues for returning products of combustion from the combustion chambers to the smokebox covering the upper front of the boiler are also shown.

Even though the Scotch marine boiler proved to be popular and well-suited for shipboard installations, the fact that its pressure was limited to 300 psig retarded progress. There was a realization that advanced boiler designs with greater generating capacity, higher pressure, and increased efficiency were needed to meet the changing requirements brought about by the development of turbine propulsion machinery. Stiff commercial competition among merchant fleets as well as various naval powers made these developments most urgent. Marine engineers began to adapt various types of water-tube boilers to fit into the restricted space available for installations on shipboard.

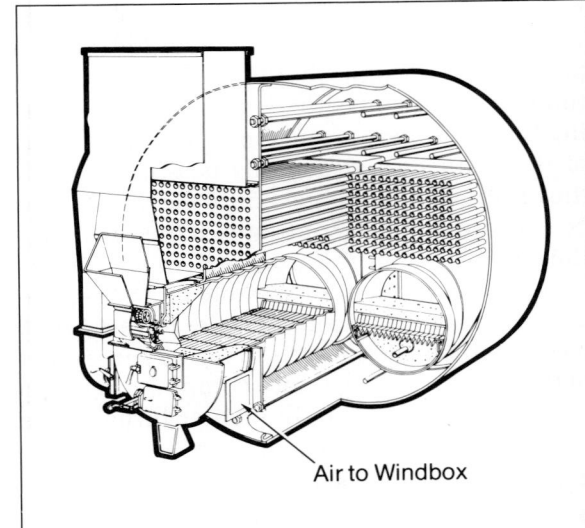

Air to Windbox

Fig. 3 Scotch marine boiler fired by spreader stoker

Some adaptations proved highly advantageous with respect to heat transfer and weight, and offered a variety of designs suitable for pressures in excess of the maximum of firetube types. Thus, new possibilities were presented for the utilization of steam aboard ship.[2]

SECTIONAL-HEADER BOILERS

As an early water-tube boiler, the cross-drum type had generating tubes extending between a row of rear headers arranged vertically and connected to the drum, and a row of uptake headers at the front of the boiler. Vertical waterwall tubes were arranged at both sides of the furnace. At a later stage of development, the downtake headers in the steam drum were relocated to the front of the boiler and that basic arrangement is known today as the cross-drum sectional-header boiler.

The handhole plates in the headers of this type of boiler provide ready access and make it possible to turbine-clean the water sides of the main-bank generating tubes without the need for a person to enter the steam drum. Appreciably heavier than presently available boiler designs, the sectional-header boiler was used during World Wars I and II on many classes of merchant ships. The design was ideally suited for mass production.

The C-E Type SM sectional header boiler as shown in Fig. 4 was designed for capacities up to about 150,000 lb of steam per hour and pressures up to 850 psig. Steel corner columns incorporated into the casing members form a boxlike structure which supports the boiler. An insulated casing encloses the entire unit. Where good accessibility is necessary, the casing panels are removable and are held in place by bolted batten bars. At other locations the casing panels are welded in place. Generally, the side and rear walls of the furnace have vertical waterwall tubes. Superheaters may be either interdeck or overdeck, depending on the steam temperature required.

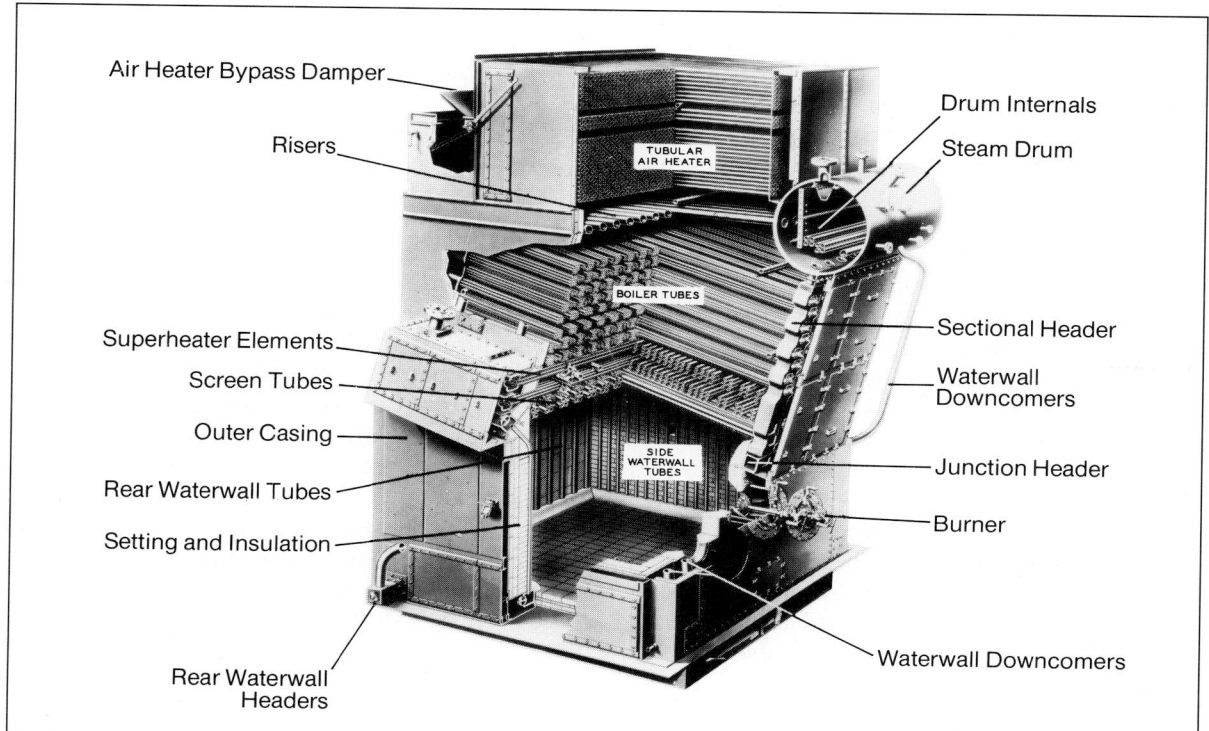

Fig. 4 C-E SM sectional header boiler, principal parts identified

BENT-TUBE BOILERS

Just as the need for increased steaming capacities within fixed space and weight limitations spurred the development of the earlier water-tube-type boilers, so did the same need precipitate the advance to boiler designs of the bent-tube type.

The most compelling reasons for the development of higher capacity boilers came from the naval designers who, at about the end of the nineteenth century, were charged with developing the torpedo boats which were the forerunners of modern destroyer-type vessels.

The term *bent-tube boiler* covers a wide range of boiler designs characterized by the ability to raise steam pressure within a short period of time. This means smaller tube diameters, fewer headers, increased furnace ratings and wider application of watercooled furnaces. The boilers of this category also became known as multidrum type, express type, or by various other names corresponding to the designers of particular arrangements.

With advanced designs available for naval vessels, it was a quite logical step to utilize adaptations of those designs for high-powered passenger ships. Thus during the 1930's, the two-drum bent-tube design was developed for application to merchant ships with power plants ranging up to about 10,000 shp.

After World War II, ship operators and boiler designers became acutely aware that marine-boiler designs would have to be modified to avoid some of the difficulties that were being encountered with the prewar designs. During the war, with the emphasis upon standardization and multiple production, time was not available for the development of improved designs.

Another factor which prompted the development of new boiler designs stemmed from competitive and economic pressures compelling shipowners to look toward higher steam pressures and temperatures, and also propulsion plants of greater power, as a means of achieving more profitable operation. It was also realized that ship boilers would have to be designed to burn efficiently whatever fuel oils would be available. Because of improved refining processes coupled with the natural motivations of profitable marketing, it was to be expected that the crude oil was to be processed to the fullest, and the resulting residual oil which normally goes to bunkers of ships would be progressively of lower quality. The prewar boilers were particularly vulnerable to fireside damage caused by slag deposits from burning such poor-quality fuels.

A comprehensive analysis of these problems led to the development of the C-E V2M Vertical Superheater Boiler. Fig. 5 shows a typical arrangement of this design. By providing a separate drum or distributing header at the lower end of the screen tubes, it is possible to set the floor of the furnace independently of the location of lower drum and main bank. Not only does this allow greater latitude in establishing optimum furnace dimensions, but it also permits the superheater tubes to be parallel to the generating tubes. The access space within the superheater contributes to ease of maintenance and inspection.

V2M8 AND V2M9 DESIGNS

Fig. 6 shows the V2M8 Vertical Superheater Boiler whose major design features are the single-cased welded-wall furnace and the vertical in-line inverted U-loop superheater, fully drainable. The boiler shown is top-fired with resulting improved gas distribution over the entire superheater furnace. Both the main-bank tubes and superheater elements are in-line for improved tube cleaning. Normally, sootblowing equipment includes retractable blowers in the superheater and rotary blowers in the main bank and economizer. Typical heat-recovery equipment consists of a small economizer followed by a regenerative air preheater or, in some cases, a regenerative air preheater only. The welded-wall construction provides a furnace with gas-tight integrity without the need for double casing or heavy refractory bricks behind the tubes. All that is required for insulation behind the walls is blanket insulation.

Some boilers of this type have been built with ''double'' or ''twin'' superheaters. With the

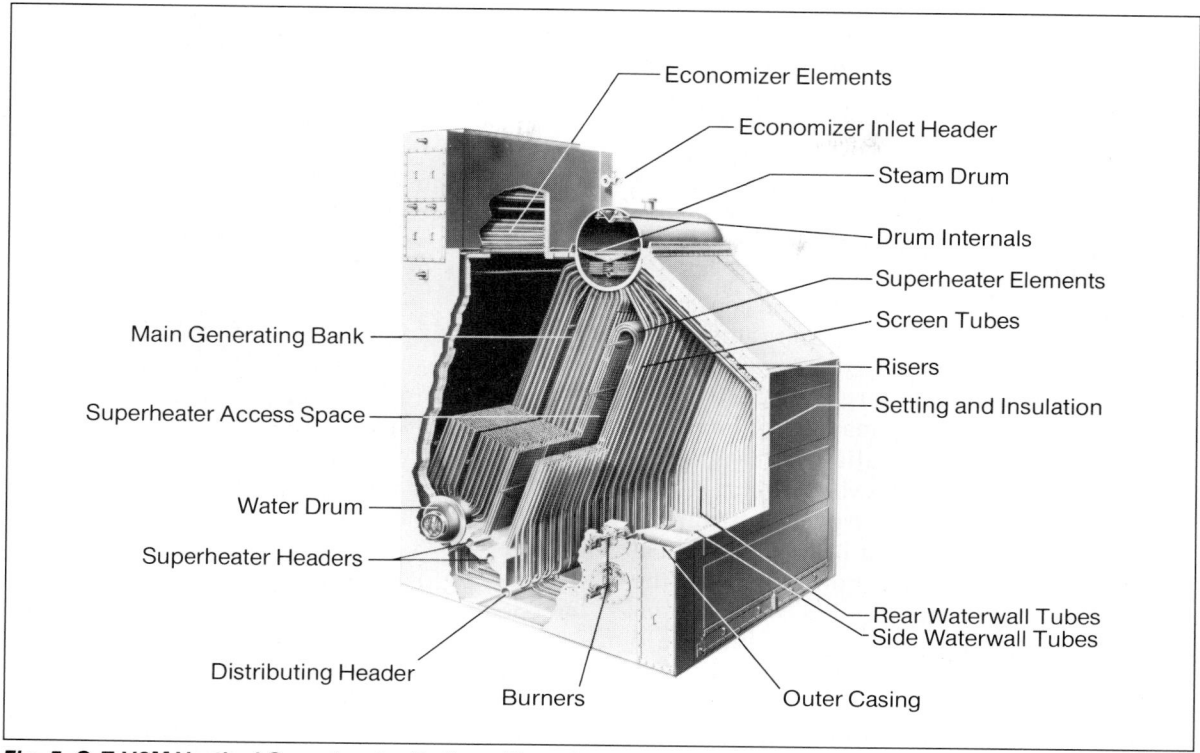

Fig. 5 **C-E V2M Vertical Superheater Boiler with main parts called out**

Economizer Elements
Economizer Inlet Header
Steam Drum
Drum Internals
Superheater Elements
Screen Tubes
Risers
Setting and Insulation
Rear Waterwall Tubes
Side Waterwall Tubes
Outer Casing
Burners
Distributing Header
Superheater Headers
Water Drum
Superheater Access Space
Main Generating Bank

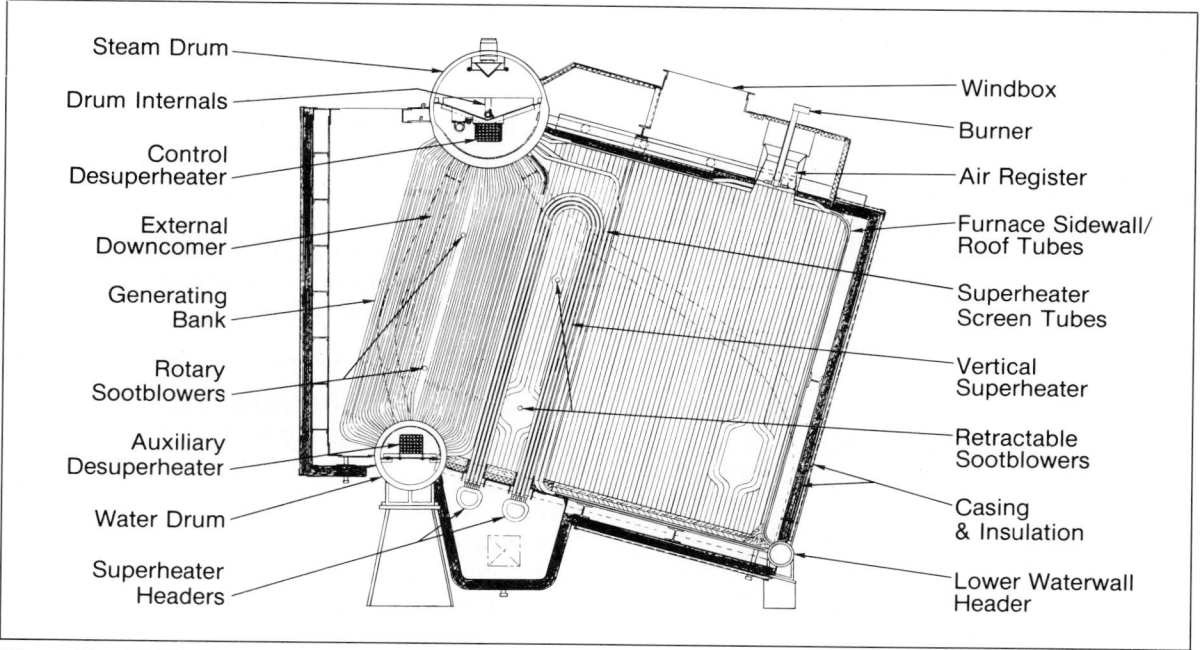

Fig. 6 **The V2M8 Vertical Superheater Boiler design featuring the single-cased welded-wall furnace and the vertical in-line inverted U-loop superheater, fully drainable**

Steam Drum
Drum Internals
Control Desuperheater
External Downcomer
Generating Bank
Rotary Sootblowers
Auxiliary Desuperheater
Water Drum
Superheater Headers
Windbox
Burner
Air Register
Furnace Sidewall/ Roof Tubes
Superheater Screen Tubes
Vertical Superheater
Retractable Sootblowers
Casing & Insulation
Lower Waterwall Header

double superheater, the elements are more widely spaced so that cleanliness is easier to maintain. Tube metal temperatures are also reduced, both because of the lower mass gas flow across the superheater and the location of the secondary superheater, which is furthest away from the furnace. The total surface, overall size, and weight of a double superheater are, however, much greater than that of the single superheater because of the reduced mass gas flow and consequent lower transfer rate.

To generate the much higher evaporations required by increasingly larger ships, the furnaces were required to be made very large. To keep the main bank at essentially the same vertical dimension as the V2M8 boilers, it was necessary to lengthen the furnace in a vertical dimension. The boiler then became a vertical two-drum dropped-furnace boiler. Because the lower part of the furnace (see Fig. 7) was unencumbered, it became clear that the boiler could be tangentially fired. The fuel nozzles are placed in each corner of the furnace, and fire at a target circle in the center of the furnace. This lengthens the fuel-particle residence time in the furnace, and allows for complete combustion and operation at very low excess air. Fig. 7 shows the boiler configuration; Fig. 8A, a diagram of tangential firing. Major design features include single-cased welded-wall construction, tangential firing, and vertical in-line double superheater, inverted U-loop, fully drainable. The boiler is supported at its midpoint, under the water drum and center headers.

In some of the larger designs of single-boiler ships, a nondrainable superheater has been used, because the boiler would be on the line from drydocking to drydocking.

Table II summarizes typical performance data for a regenerative air-heater/boiler.

The registers of a tangentially fired marine boiler are of a straight-throat type with primary- and secondary-air zones. The primary airflow is through the center compartment, where the oil gun and diffuser are also located. Secondary air is above and below the center compartment. See Fig. 8B. Under normal operation, the primary-air damper is fully open at

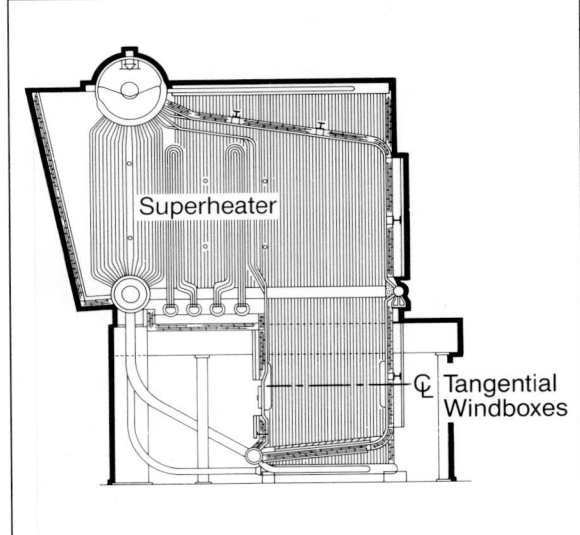

Fig. 7 The C-E V2M9 vertical two-drum dropped-furnace boiler

all times the register is in operation, while the secondary-air dampers are modulated to maintain a constant register draft loss. In this way, good combustion efficiency can be maintained over a wide boiler range.

Fig. 8C shows a typical draft-loss curve for tangential firing. The registers can be operated to cycle on and off for either a two- or four-register operation based on steam flow and drum pressure, or they can operate with four registers on whenever the vessel is under way. The oil guns are of the steam-atomizing type utilizing constant steam pressure at 150 psig with fuel pressure up to about 350 psig. Flame monitoring is accomplished by scanning along

Table II. Typical Performance Data: Regenerative Air-Heater/Boiler

Evaporation	220,000 lb/hr
Total steam temperature	955°F
Feedwater temperature	424°F
Superheater outlet pressure	865 psig
Excess air	3%
Efficiency	90.7%

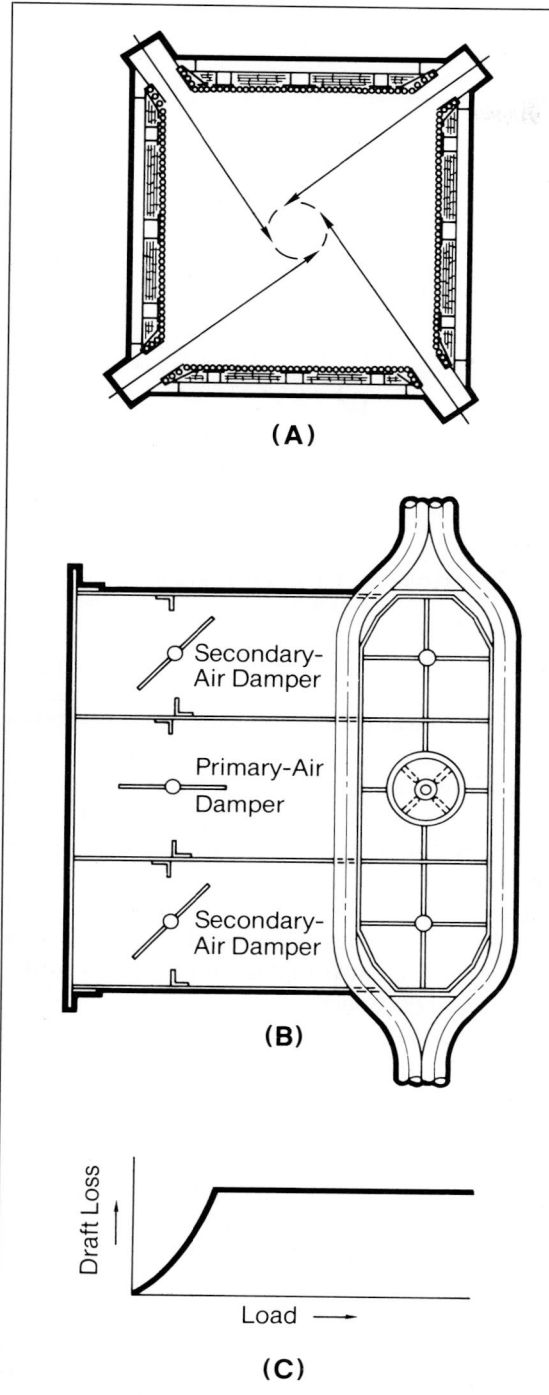

(A)

Secondary-Air Damper

Primary-Air Damper

Secondary-Air Damper

(B)

Draft Loss

Load →

(C)

Fig. 8 Diagram of tangential firing. Fuel nozzles located in each corner of the furnace fire at a target circle in the furnace center.

the axis of the flame and also scanning the flame across the furnace corner, with both scanners showing NO FLAME required to close the fuel-oil solenoid valve. Scanners will scan the individual flame up to about 30 percent boiler rating, after which they will scan the furnace as a whole. If the boiler rating is above 30 percent, the "fireball" in the furnace is large enough to ignite any incoming oil.

STOKER-FIRED DESIGNS

Fig. 9 shows a modern stoker-fired marine boiler designated as a V2M9S boiler. The boiler is a single-cased welded-wall design with single-cased uptakes; some areas around super-heater headers, access doors, and windbox are double cased. Table III compares typical performance data of a coal-fired design with an equivalent fuel-oil-fired design.

By comparing the furnace release rates, it can readily be appreciated that the coal furnace has about 3 times the volume of the oil furnace for the same power output. To minimize furnace and superheater problems, furnace temperature when burning coal is kept lower than the ash-fusion temperature. This temperature can be readily achieved with a totally watercooled furnace by limiting the grate release and furnace heat release. Overfire-air jets and cinder reinjection nozzles are located above the stoker grate, with one line of nozzles serving to reduce the carbon loss by about 4 percent. A cinder hopper is located between the two superheater sections. In addition to the stoker,

Table III. Comparison of Performance Data for Coal- and Oil-Fired Designs

	Coal	Oil
Furnace release rate, Btu/hr-cu ft	25,000	75,000
Grate release rate, Btu/hr-sq ft	700,000	. . .
Furnace absorption rate, Btu/hr-sq ft	37,000	70,000
Furnace outlet temp., °F	1900	2500

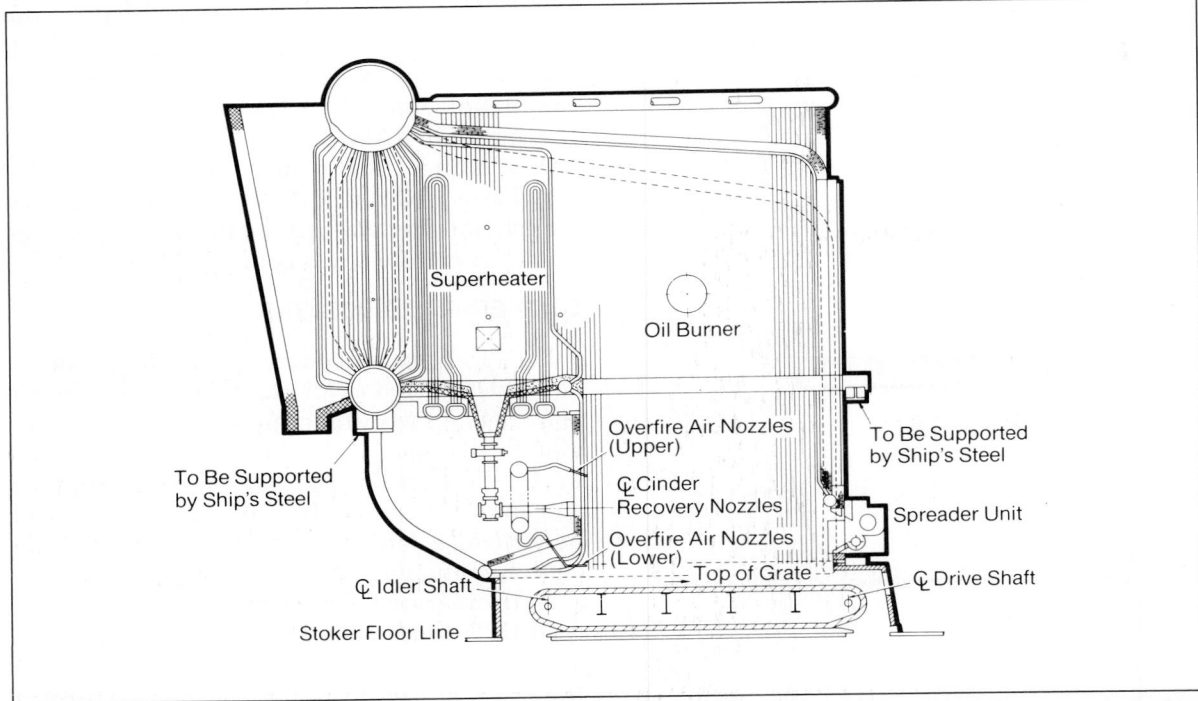

Fig. 9 A V2M9S stoker-fired marine boiler with a single-cased welded-wall design and single-cased uptakes

a fuel-oil burner is incorporated in the upper furnace for in-port firing or take-home purposes. When the burner is in use, a layer of ash is retained on the grate to protect it from overheating. Tube spacings in both the superheater and main bank and the economizer fin spacings are increased over the fuel-oil-fired spacings to keep gas velocities low for minimum erosion.

WASTE-HEAT DESIGNS

Many vessels powered by diesel engines and gas turbines extract waste heat from exhaust gases to create efficient cogenerative cycles. The recovered heat provides shipboard energy in the form of steam for fresh-water production, water heating, fuel heating, cargo heating, galley and laundry services, and, in some instances, to drive a turbogenerator to supply the electrical load of the ship.[3,4,5,6,7] There are various shapes and sizes of waste-heat boilers; Fig. 10 shows a typical waste-heat steam generator composed of horizontal finned tubes connected by return bends. The return bends are installed

outside the gas passage with a gas-tight tube sheet separating them from the gas duct. This type of extended-surface heat exchanger is essentially of the same construction as a marine-boiler economizer. The combination of large-diameter tubes with welded helical fins for extended heat-transfer surface is simple, rugged, and compact.

An important consideration in the design of a waste-heat boiler of this type is to configure a heat exchanger that will produce the required amount of steam within a limited amount of space, with acceptable gas-side pressure drop. A large gas-side backpressure on either a diesel engine or a gas turbine will result in significant degradation of engine performance.

Fig. 11 shows a waste-heat steam generator system as installed in a typical naval vessel. The five modules, each of which is supported on its own foundation, include the steam-generating bank, the drum module, the condenser, the control panel, and the feedwater pump. Each module can be situated to optimize

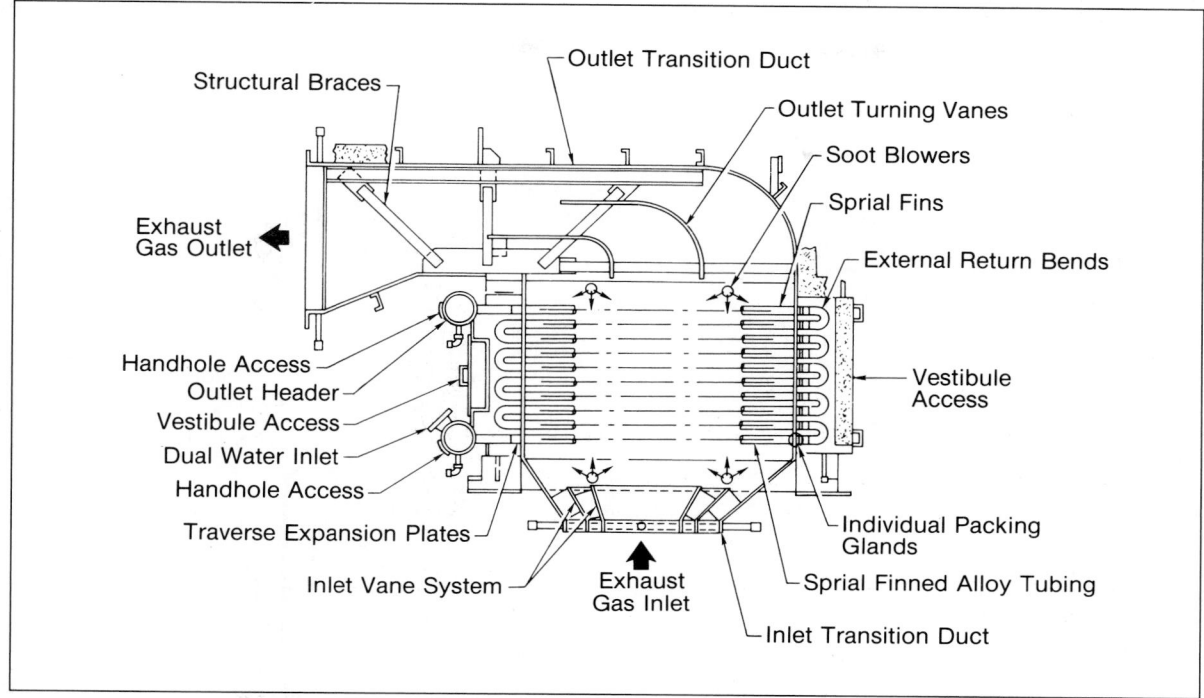

Fig. 10 **Steam generating bank module**

maintenance access on the ship. The following specific design features of the waste-heat boiler system illustrated promote ease of maintenance access.

- Generating bank headers are external to the casing.

- Headers, inspection doors, and access doors are accessible by removing portable insulating pads.

- Full size header handholes are provided, with one handhole for each two tubes.

- Individual generating bank tubes are removable; each tube is easily handled by one person. The entire generating bank need not be removed to replace one tube.

- Individual condenser tubes are removable; each tube is easily handled by one person.

- Large access-way is provided in steam drum for access to drum internals.

The boiler bank shown in Fig. 10 can be used in series with an auxiliary boiler, with water circulated through the waste-heat bank and returned to the steam drum of the auxiliary-boiler. Controls are then arranged to automatically fire the auxiliary boiler so that drum pressure is maintained with varying loads. If the steam generation by waste-heat gas alone is too high, a steam dump to a condenser or gas bypass damper can be installed. If required, a waste-heat bank can also be arranged to run as an independent unit with its own steam drum and circulating-pump system.

The need for oxygen removal on low-pressure auxiliary systems is just as great as in main propulsion systems. The direct-contact feedwater heater is a reliable and effective device for continuously maintaining feedwater at acceptable low levels of oxygen and should be installed. The advisability of using a deaerating feedwater heater has been recognized by not only boiler-water chemistry consultants, but also designers who have experienced problems with corrosion caused by poor water chemistry in waste-heat boiler systems.

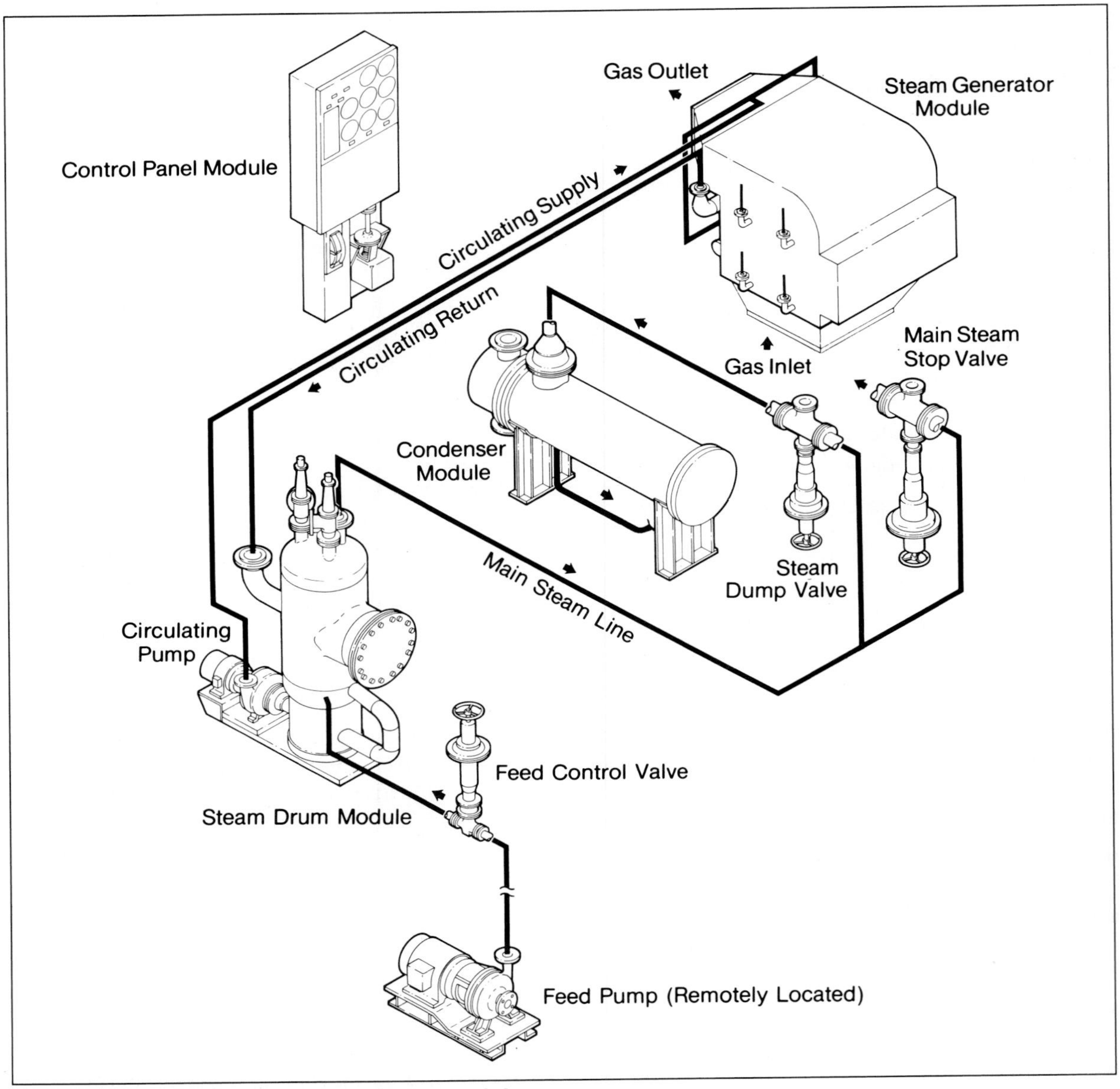

Gas Outlet

Steam Generator
Module

Control Panel Module

Circulating Supply

Circulating Return

Main Steam
Stop Valve

Gas Inlet

Condenser
Module

Steam
Dump Valve

Circulating
Pump

Main Steam Line

Feed Control Valve

Steam Drum Module

Feed Pump (Remotely Located)

Fig. 11 Marine waste-heat boiler system modules

DESIGN PRINCIPLES FOR MARINE-BOILER FURNACES

A number of factors determine the configuration of a marine-boiler furnace. First of all, the type of boiler under consideration sets some of the furnace dimensions. As already discussed, fireroom space limitations, service require-ments, and burner arrangement also influence the furnace layout. The furnace arrangement must also conform to criteria which are pre-requisites for proper thermal and circulation performance. The amount of waterwall sur-face determines the furnace exit-gas tem-perature which, in turn, is most significant in the design of the superheater. Practical re-

quirements must also be considered; for instance, the firing equipment must be located where it can be easily operated.

Two-drum bent-tube boilers are always furnished with watercooled surfaces on the side and roof and, for the majority of designs, both front and rear waterwalls are included.

SUPERHEATERS

Superheaters are essentially a series of tubular elements connected to headers. The saturated steam enters the inlet header from piping connected to the steam drum. In a single-pass superheater, the saturated steam flows from the saturated header through the superheater elements into the superheater outlet header. In the multipass superheater, the headers are diaphragmed to allow the steam to make as many passes as necessary to assure good steam distribution through all elements.

Marine superheater design requires a judicious balancing of economic and practical factors. There must be a sufficiently large temperature difference between the combustion gas passing over the superheater surface and the steam within the surface to result in a superheater of economic proportions. At the same time, there are practical limitations on superheater metal temperatures, and provision must be made to arrange superheater elements to resist slag accumulation. Screen tubes are located ahead of superheaters to obtain reasonable metal temperatures while at the same time allowing a certain amount of direct radiation to be absorbed by the superheater tubes. This results in a flatter steam-temperature curve at the superheater outlet.

Some boiler designs have the superheater behind the main generating bank. Such an arrangement requires a relatively greater amount of heating surface and an elaborate system for controlling superheater outlet temperature. An alternative arrangement incorporates a second furnace for controlling the steam temperature by varying the firing rate of the superheater furnace. Although such a boiler design permits a wide range of control, there are inherent complications in the operating procedures.

Most two-drum bent-tube boilers are designed with the superheater within the generating bank and with either horizontal or vertical superheaters. Positive means must be provided to keep superheaters clear of slag deposits. To permit convenient access for inspection, cleaning, and maintenance, modern designs include access spaces within the superheater banks.

The advantageous features of the vertical superheater, as shown earlier in Fig. 6, result in reduced maintenance and optimum service efficiency. Because the elements are arranged parallel to boiler tubes, clear lanes between generating and superheater tubes present the best arrangement for effective action of the sootblowers. With the superheater tubes in a vertical position, it is difficult for slag to accumulate. Bulky supports are eliminated as it is possible to support each superheater tube from the superheater headers which are outside of the gas pass. Small slip spacers attaching superheater elements and adjacent generating tubes maintain the upper ends of the superheater elements in the correct lateral position.

DESUPERHEATERS

For most modern marine boilers, all of the steam generated by the boiler is passed through the superheater. Auxiliary steam requirements are met by desuperheating the required quantity of auxiliary steam. Because the superheater is located in a relatively high-temperature gas zone and overheating may occur if any substantial proportion of steam were taken directly from the steam drum to the auxiliary steam line, this arrangement is necessary to maintain a flow of steam through the superheater at all times. To provide low-temperature steam for auxiliaries and for other purposes, such as heating, Butterworthing (a special system for steam-cleaning cargo tanks), and evaporators, steam from the superheater outlet is passed through a desuperheater which absorbs some of the heat in the superheater steam.

Desuperheaters can be of the internal or the external type. The internal type contains tubular elements with necessary terminal connections to be installed within the boiler drum. Fig. 12 shows a typical arrangement of an internal-type desuperheater as installed in a boiler drum. Note that the unit is arranged so that it may be passed through the drum manhole as a complete assembly. The external type includes a direct or indirect heat exchanger mounted in the steam line, with feedwater most generally used as the cooling medium.

STEAM-TEMPERATURE CONTROLS

For high-temperature applications, designing for a constant steam temperature over a relatively wide load range may be desirable, as indicated by the superheat characteristic Curve A in Fig. 13, or for nominal temperature applications to limit the steam temperature at infrequent overload ratings, as shown by Curve B. The dashed-line extensions of characteristic curves show expected temperatures without control. Fig. 14 shows a typical arrangement which will provide close control of steam temperature and, in so doing, protect the last passes of the superheater, turbine, and connecting steam piping from high metal temperature. The steam temperature at the superheater outlet actuates an air-operated control valve. Below the set temperature, all steam passes through the orificed line, the control valve being closed. When the tem-

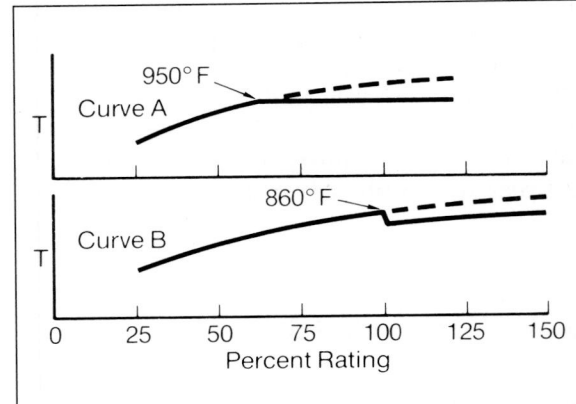

Fig. 13 **Superheater characteristic curves**

perature tends to rise above the set temperature, the control valve opens to permit a portion of the steam to flow to the desuperheater and have its temperature reduced. The control valve has a hand jack for manual operation.

With the exception of manual operation of the automatic control valve, the same general arrangement is used to accomplish the limiting-type of temperature control, as shown in curve B of Fig. 13. At higher boiler ratings, when the temperature starts to exceed the predetermined maximum, the control valve is opened fully, accounting for the temperature drop just beyond normal rating and the continuing rise to maximum rating.

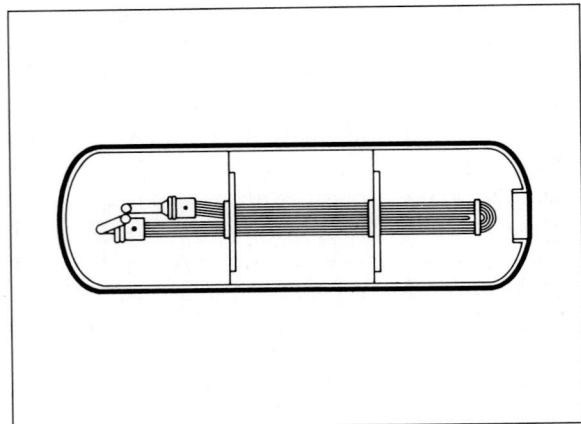

Fig. 12 **Internal-type desuperheater**

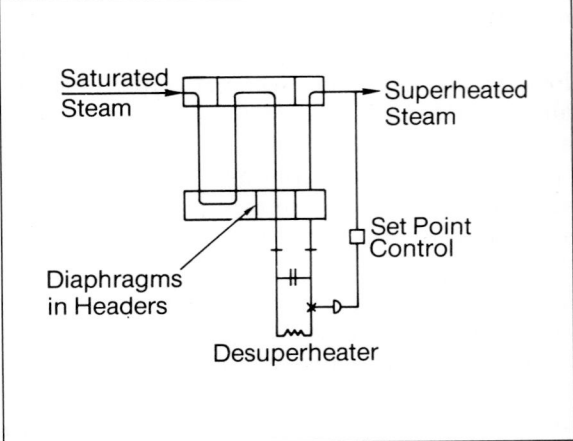

Fig. 14 **Steam-temperature control arrangement for marine boiler**

HEAT-RECOVERY EQUIPMENT FOR MARINE BOILERS

The general classification "heat-recovery equipment" covers heat exchangers which are located such that heat is absorbed from the combustion gases after the gases have passed through the superheater and steam-generating sections of the boiler. Economizers, tubular air heaters, and Ljungstrom® regenerative-type air heaters, are the types most commonly used for marine boilers.

ECONOMIZERS

The economizer is a series of horizontal tubular elements by means of which heat is recovered from combustion gas leaving the boiler and added to the incoming feedwater. The amount of heat absorbed in an economizer depends on the temperature difference between gas and water, the heat-transfer rate, and the amount of heating surface. It is the normal practice to use some form of extended surface on tubes to increase the heating surface per unit of tube length.

Economizers can be categorized as bare tube and extended surface types. The bare tube usually includes varying sizes which can be arranged to form hairpin or multiloop elements.

Because the coefficient of heat transfer is relatively low on the gas side of an economizer tube as compared to that on the waterside, it is desirable to have some form of extended heating surface on the outside of the tube so as to increase the overall rate of heat transfer. Many different configurations of extended surface are used, each with characteristics of a specific standard adopted by a particular boiler manufacturer. In turn, each type of extended surface economizer has a heat-transfer rate that is peculiar to the shape and finish of the surface, the material employed, and the method of fabrication. Fig. 15 shows a widely used type of marine economizer extended surface.

Most marine economizers are designed for counterflow of gas and water, that is, water down through elements and gas up outside of elements. This is done to take advantage of the greater temperature difference between the gas and water. Under these conditions, the heat-transfer rate cannot be increased without incurring a substantial resistance to gas flow, and the use of extended surface to increase the heating surface per lineal foot is desirable. Whatever the arrangement of economizer, ample provision must be made to assure that the heating surface can be kept free of deposits. The coexistence of deposits and moisture can cause corrosion; accordingly, the spacing of the elements, the quantity and effectiveness of the sootblowers, and the design of the breeching above the economizers must all be carefully considered. Moisture can come from a number of sources:

1. Rain or spray down the stack

2. Leaks in economizers or other pressure parts

3. Condensation of moisture in gases because of the gas coming into contact with cold heat-transfer surfaces and cooling to below the acid dew-point temperature. Also, dependent on the sulfur content of the gas, this condensation forms a corrosive sulfuric-acid solution which attacks the metal surfaces.

Although a leak in an economizer pressure part cannot go undetected for very long, substantial corrosion-erosion damage can result quickly if a leak is allowed to persist. Therefore, it is advantageous to design economizers with the

Fig. 15 **Close-up view of economizer spiral steel surface**

minimum of joints and handhole plates consistent with proper provision for replacing an element. Generally, the header tube joint is made up as a full welded joint or, as an alternate construction, the elements are welded to nipples in the headers.

Although the economizer is designed principally for steady steaming conditions, proper consideration must be given to the design of the elements so that steam will not be generated within the economizer during maneuvering. This means that there must be suitable water velocities and pressure drops to assure adequate circulation in each element. Under most conditions, there is sufficient differential between saturation and feedwater temperatures to prevent the generation of steam. But there may be some maneuvering conditions, such as the sequence from stop to full ahead, in which steaming may become a problem. This is particularly true if the water level is high at the start of the maneuver and the boiler control system is of the single-element (water-level) type.

Usually, an economizer is installed at the boiler outlet adjacent to the steam drum with elements parallel to the drum. In most fireroom arrangements, sufficient space is available for a good economizer arrangement. However, to obtain the required economizer heat absorption, the economizer length, width, and number of tubes high may have to be varied to obtain the best arrangement for the space conditions available and to meet thermal performance conditions. In addition, the allowable gas-side pressure loss and water pressure drop must be considered.

Today, it is common practice to design marine economizers so that they can be bypassed if a leak occurs, thereby allowing the boiler to remain in service until the necessary repairs can be made. The economizer then must withstand entering gas temperatures without any feedwater flow through the economizer tubes. With the economizer bypassed, the feedwater enters the boiler directly at a lower temperature, which results in an increased fuel firing rate, increased draft losses, and overall loss of boiler efficiency. Another important consideration is that there is generally an increase of total steam temperature because of the higher fuel firing rate and increased furnace temperature.

AIR HEATERS

Using air heaters to improve the performance of steam-generating units began in marine practice prior to 1880. At that time, practically all the heaters were of the tubular type with various arrangements of vertical or horizontal banks of tubes. In some, the gas passed through the tubes, while in others the tubes served as a passage for the air. Most marine tubular air heaters are of the horizontal type, preferable because of its simplicity and the ease of incorporating it into the boiler arrangement.

Most marine tubular air heaters are located immediately above the boiler gas exit where the unit can be fitted conveniently into the boiler design. For a two-pass heater, air both enters and exits at the rear side of the boiler, presenting an efficient arrangement for transporting air to the fuel burners.

REGENERATIVE AIR HEATERS

Regenerative air heaters for marine service are designed for horizontal or vertical flow of the combustion air and stack gases. In this type of heat exchanger, as described further in Chapter 14, the heating elements are contained in a slowly turning rotor of cellular construction. As the rotor turns, heat is absorbed continuously by the heating elements from the flue gas, while a like amount of heat is released simultaneously to the combustion air, as these fluids flow axially through the rotor. The rotor is enclosed in a gas-tight housing which is fitted at each end to make connections with the air and flue-gas ducts.

Dampered integral air and flue-gas by pass ducts are incorporated in the four corners of the structure. Parallel air by-passes offer a way of controlling cold-end heating-element temperature during operation at reduced steaming rate. The flue-gas bypasses in parallel, together with the air bypasses, provide a means of limiting overall pressure loss at steaming rates above the corresponding to normal power. Fig. 16 shows a typical regenerative air heater for

marine service.

A vertical-flow regenerative air heater can be located directly above the boiler outlet or in the area above the boiler, and connected with conventional ductwork. From the top of the air heater on the gas side, an uptake is led to the stack in the usual manner. Forced-draft fans may be located conventionally and connected to the air side of the heater by means of ductwork. Various arrangements can be utilized to suit the conditions (See Fig. 17).

STEAM AIR HEATERS

Both the boiler operating conditions and the efficiency required of a boiler unit affect the selection of the heat-recovery equipment to be incorporated in the design of a boiler. These are related to the specification of a suitable steam cycle for a marine installation. The best steam cycle for one installation may not be advantageous to another. Therefore, for certain installations, an air heater alone may be optimum. Where feedwater temperature and pressure conditions permit, an economizer or a combination of an economizer and a steam air heater is used to obtain the boiler efficiency required, with no gas-to-air heater.

One method to improve the steam-cycle efficiency is by means of the feedwater heating system, and here the steam air heater uses steam

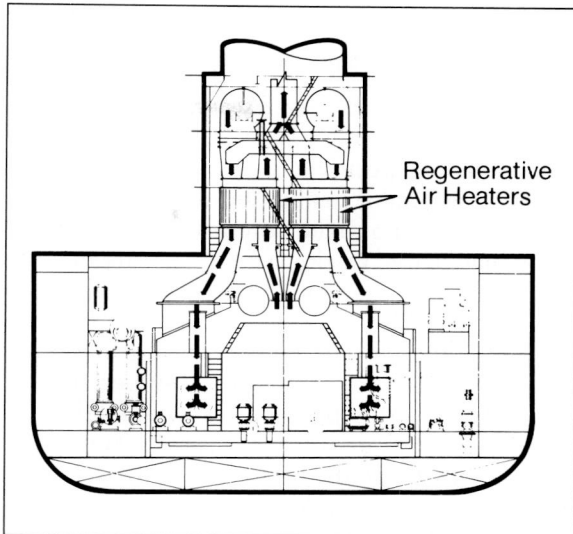

Fig. 17 Arrangement of regenerative air heaters aboard ship

extracted from the feedwater heater or auxiliary exhaust. This reduces the flow of steam to the condenser and improves cycle efficiency. An alternative method is to extract low-pressure bleed steam from the main turbines.

In the condensing of the bleed steam or auxiliary exhaust, heat is transferred to the air used for combustion. Because the condensing steam-film coefficient of heat transfer is high, extended surface is used to advantage. Closely pitched tubes and fins can be utilized, as both the air and heating fluids are clean, eliminating entirely the necessity for cleaning.

In most steam-air-heater installations, bleed steam at 10 to 45 psig is used. Although a feedwater heater uses steam at relatively high pressure, the steam air heater uses auxiliary exhaust or low-pressure bleed steam. Utilizing the lower pressure bleed steam is a good method of improving cycle efficiency.

Another point to be considered is that the auxiliary exhaust can be used to advantage in a steam air heater while boilers are steaming in port, thereby improving low-load boiler operating conditions. Suitable air temperature for good combustion is readily maintained, whereas with a tubular or regenerative-type air heater, a bypass system around the heater is

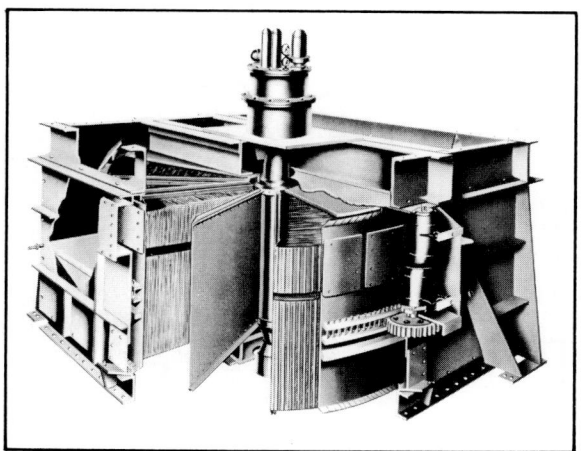

Fig. 16 Regenerative air heater for marine service

necessary to avoid cold-end corrosion.

FLUID REGENERATIVE AIR HEATER

Fig. 18 shows an arrangement of a fluid regenerative air heater. This is an arrangement where the feed pump takes suction from the de-aerating feedwater heater and discharges to the secondary economizer where the 285°F feed-water cools the stack gases to approximately 310°F. In this process, the water is heated to about 340°F. The feedwater is then routed to an air heater where it is cooled by the incoming combustion air to its original temperature of 285°F and, in this process, the combustion air is heated to approximately 310°F. The water then passes through the fourth and fifth-stage heaters, entering and leaving at temperatures of 285°F and 423°F respectively.

From the heaters, the water enters the primary economizer and then goes to the boiler.

The primary and secondary economizer elements are all of spiral steel-fin construction which has been proven by many years of marine service (Fig. 15).

For a 45,000 shp plant, in changing from a steam-air-heater/economizer cycle to a straight regenerative air-heater cycle, there is an overall plant efficiency increase of about 4 percent. Of this 4 percent increase in efficiency, about 1.5 percent is from the decrease in stack temperature, usually from 310°F to about 250°F; the other 2.5 percent is from the additional regenerative feedwater heating of the high-pressure heaters (less heat rejected to the condenser cooling water). From the standpoint of efficiency, the regenerative air preheater is the more attractive. Some shipowners, however, still prefer the economizer/steam-air-heater cycle because of longer life of economizer surface when compared to the cold-end heating

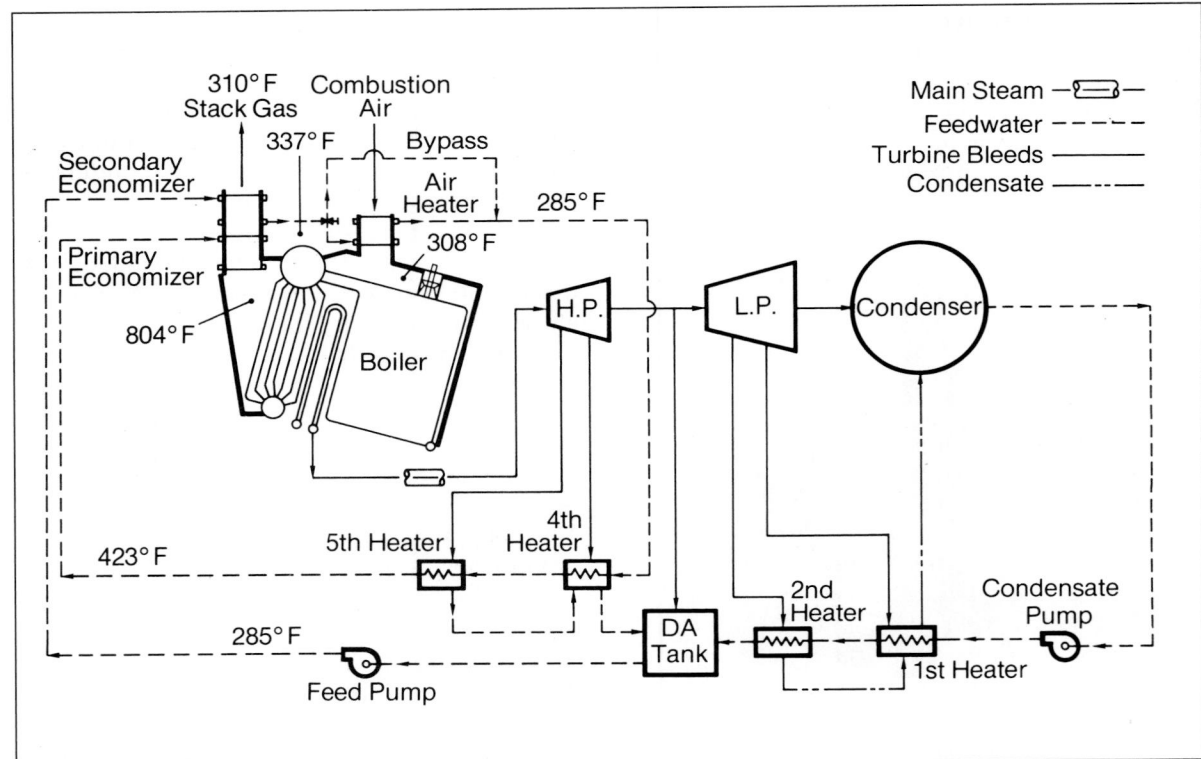

Fig. 18 **Component arrangement of a closed-loop fluid regenerative air heater**

surface of an air preheater.

Under such a constraint, the use of a closed-loop fluid regenerative air preheater can be considered for the following reasons:

1. The plant efficiency will be increased by approximately 2.5 percent because of the use of high-pressure feedwater heaters;

2. The stack gas temperature will remain the same as is presently used in a steam-air-heater/economizer cycle; maintenance, therefore, will remain at the standards that are acceptable today;

3. Gas inerting systems can utilize the cool stack gas after the economizer because there is no air leakage in an economizer system;

4. Forced-draft fans can be selected smaller than in the equivalent regenerative air preheater plant because there is no air leakage in the economizer system.

FUEL BURNING

Oil burners for marine boilers must have features and operational characteristics consistent with the service requirements of the propulsion plant. The burner must also be capable of producting efficient firing over a wide range of oil flow rates and be able to change load rapidly to meet the maneuvering requirements. For ships which maneuver for a substantial proportion of the operating time, it is essential that the boilers be fitted with burners capable of varying the firing rate over a wide range without the need of changing atomizer tips.

Before the fuel oil can be burned, the oil must be changed from a liquid to an atomized condition. Simultaneously, it must be mixed with an ample supply of air to permit combustion. A burner actually consists of two parts; the atomizer or burner barrel which serves to deliver a flow of atomized oil to the furnace (see Fig. 19), and the register which serves to supply a uniform flow of air to the furnace. The burner barrel contains a sprayer plate, or swirl chamber, and an orifice tip which act together to produce the spray of fine oil particles.

The straight mechanical type of oil burner

consists of a burner atomizer which converts the potential energy of the oil pressure into a flow of atomized oil particles. The register introduces and mixes the air with the oil spray produced by the atomizer. The oil flow is regulated by varying the supply pressure. On some installations this is accomplished manually; on others, it is automatically performed. Airflow is regulated by manual adjustment of the forced-draft fan output or by the automatic adjustment of the combustion control. This type of burner has a range or turndown ratio of between 1.5 and 3.0, depending on the pressure of the fuel-oil service system.

The return-flow type of burner functions with a constant-pressure oil supply, and the atomizer barrel contains a passage to permit a portion of the oil to return to the fuel-oil service pump suction. Regulating the control valve in the return line varies the firing rate. This type of burner achieves a much wider range than the straight mechanical burner.

The steam-atomizing-type burner uses a flow of steam which mixes in the atomizer tip with the oil flow. A portion of the energy in the steam serves to break up the oil particles, and better atomization results. This type burner offers a high turndown ratio and the advantage of being able to burn low-quality fuel efficiently.

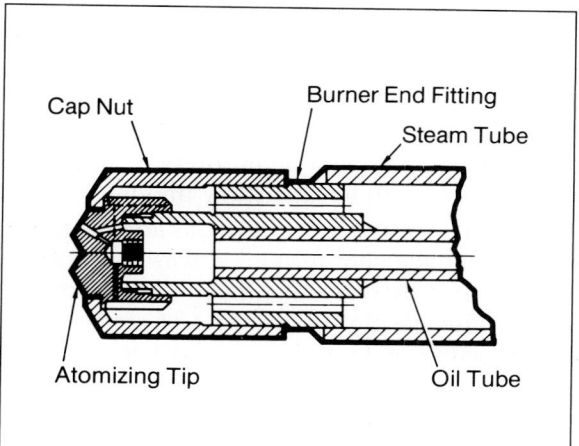

Fig. 19 The atomizer or burner barrel delivers a flow of atomized oil to the furnace

COAL

Although the vast majority of shipboard boilers are fired by oil fuel, there remain some ship trades where economic conditions make it very desirable to burn coal. Modern coal-fired marine boilers use stokers. Although many types of stokers (including chain grate, underfeed, overfeed and spreader types) have been used for marine service, the spreader stoker has proven itself over a long period of operation to be best suited and most efficient in meeting the requirements. It is capable of burning a wide range of bituminous coals and has the ability to respond quickly to load changes. These features make spreader stokers attractive for vessels that will continuously serve in trades which have a readily available source of coal.[8,9] Chapter 12 gives further descriptive material on the spreader stoker for coal firing.

BOILER CASINGS

The primary function of a marine-boiler casing, Figs. 20 and 21, is to contain the products of combustion in such a way that the gases flow from the furnace through the steam-generating, super-heater and heat-recovery tube banks, and out the breeching. Although both single and double-cased designs may be used, the function of containing the products of combustion takes place in that casing which is immediately exterior to the refractory and insulation wall.

Many modern boilers are built double cased. In this arrangement, air pressure in the space between the inner and outer casings is greater than the pressure within the furnace. Accordingly, if there are any open joints in the inner casing, there will be an airflow into the furnace instead of an outward flow of gas and soot. But this advantage should not be overexploited to the extent of disregarding the maintenance of a tight inner casing. If the flow of air through inner casing leaks is allowed to become substantial, boiler efficiency will be impaired and other detrimental effects will accrue.

DESIGN OF INNER CASING

The inner casing, or the only casing in a single-cased boiler, must be designed to accommodate the refractory and insulation. These materials must be arranged to protect the casing from exposure to gases and must insulate the casing from high temperatures. Structural members should preferably be located outside the inner casing. Whenever this is not possible, as may be the case for some supports of furnace insulation and refractory, alloy material should be used if there is any possibility of exposure to high temperature.

The space between inner and outer casings of a double-cased boiler provides a convenient duct system by which the air for combustion can flow to the burners. Because there can be a wide choice of locations for the air inlet connection, the arrangement presents considerable flexibility to the design of the forced-draft system. Those sections of the air space not sustaining the main flow of combustion air are provided with a flow of cooling air which ultimately discharges into the windbox. Thus, all of the inner casing is pressurized. Because the airflow absorbs a large proportion of the heat given off by the inner casing, there is a reduction in boiler radiation losses.

In laying out the arrangement of casing, calculations must be made to determine that the air pressure-drop values are not excessive and

Fig. 20 **Exterior view of marine boiler showing casing arrangement**

that the velocities and airflow pattern in the windbox will be suitable for the even distribution of air to the registers. In determining the dimensions of the air space, allowance must be made for the restrictions imposed by downcomers, structural members and fittings such as sootblower sleeves.

METHODS OF SUPPORT

For most designs of bent-tube boilers, the weight of the pressure parts, and the included water, is carried by the water-drum saddles and waterwall header saddles. Steam and tubular-type air heaters are generally supported by the boiler casing. Heat recovery equipment is usually separately mounted to the structure of the ship.

When it is necessary to obtain a better layout of the breeching when there is a narrow fidley which limits the space directly above the boiler, the heat-recovery equipment may be mounted directly on the boiler. The Ljungstrom air heater is usually located on a deck or platform above the boiler with the ship structure providing support.

For some types of boilers, the casing structure must also support the pressure parts, and this requires that heavier structurals be included at the points of loading. The sectional-header-type boiler has column-type supports built into each corner of the casing, as described earlier.

Fig. 21 Details of boiler casing as fabricated in the assembly plant

The loadings from connecting piping, particularly main-steam piping, can produce excessive stresses in the boiler connections. According to accepted procedure, the boiler manufacturer specifies the maximum allowable forces and moments which can be taken by the connections, and the shipbuilder designs the piping to stay within the limits.

COMMON ATTRIBUTES OF MARINE AND STATIONARY DESIGN PRACTICE

The special considerations required for the design of marine boilers have been discussed in this chapter. Many relate to the differences in operating conditions between land and marine practice. For example, the output of a marine power plant is basically a function of the speed and the displacement of a ship, and there is no exact counterpart to this in industrial or utility power plants. In addition, the majority of marine boilers are integrated into a cogenerative cycle.

Marine boilers operate successfully at very much higher heat release rates and smaller box volumes than field-erected stationary boilers. This is partly accounted for by the nature of marine power-plant operation. For one thing, feedwater is of high quality because, at sea, evaporators furnish makeup. In addition, two or more boilers are generally installed, and although one will operate at rated capacity while the ship is at sea, at least one can be shut down for inspection and maintenance while in port. This is in marked contrast to the operating conditions in some types of industrial and utility power plants which have what amounts to an almost continuous base load. Under these conditions there are rarely periods of low-load demand, and boilers may be operated for sustained periods of a year or longer between shutdowns for inspection and maintenance.

REFERENCES

[1] Robert F. Latham, *Naval Boilers*. Annapolis, Md.: U.S. Naval Institute, 1956.

[2] Otto de Lorenzi, editor, *Combustion Engineering: A Reference Book on Fuel Burning and Steam Generation*. New York, Combustion Engineering Company, Inc., 1947.

See also the 1966 edition, edited by Glenn R. Fryling and published by Combustion Engineering, Inc.

[3] Thomas P. Mastronarde, "Energy Conservation Using Waste Heat Boilers: *The Challenge, Problems, and Solutions*", presented at ASNE Day 1982, sponsored by American Society of Naval Engineers, May 6 and 7, 1982, Washington, D.C. and originally published in *Naval Engineers Journal*, April 1982, also as Combustion Engineering Publication TIS-7039.

[4] Breaux, D. K. and Cdr. K. Davies, "Design and Service of a Marine Waste Heat Boiler," *Naval Engineers Journal*, Volume 90, No. 2 (April, 1978). Pages 165–175.

[5] Graf, T. E. and J. E. Nagengast, "DD-963 Class Waste Heat Recovery System Experience," *ASME Publication* 79-GT-159.

[6] Abbot, J. W., Published Comments on Paper by Dr. Eugene Brady titled "Energy Conservation for Propulsion of Naval Vessels," *Naval Engineers Journal*, Volume 92, No. 3 (June, 1981).

[7] Csathy, D., "Heat Recovery From Dirty Gas," presented at Sixth National Conference on Energy and the Environment, Pittsburgh, Pennsylvania, May 24, 1979.

[8] Fukugaki, A., et al, "Design of a New Generation Coal Fired Marine Steam Propulsion Plant," presented to the *Society of Naval Architects and Marine Engineers* (SNAME) Annual Meeting, New York, November 17th-20th, 1982.

[9] Horlitz, Carl F. and Sabo, Steven E., "Coal Fired Boilers for The 1980's", presented to SNAME at the *Shipboard Energy Conservation Symposium*, September 22nd-23rd, 1980, New York.

CHAPTER 11

Pulverizers and Pulverized-Coal Systems

The main reason why pulverized-coal firing is favored over other methods of burning coal is because pulverized coal burns like gas and, therefore, fires are easily lighted and controlled. Almost any kind of coal can be reduced to powder and burned like gas.

"A change of coal upsets the operation of a pulverized-coal plant to a much smaller degree than it does a stoker-fired plant. Pulverized-coal furnaces can be readily adapted to burn other fuels that burn like gas, and in that respect are capable of burning almost any fuel which is used for making steam."[1] Henry Kreisinger, former Director of Research at Combustion Engineering, made this statement at a 1937 meeting of the American Society of Mechanical Engineers in Windsor, Canada.

Today, over 40 years later, pulverized-coal burning has so dominated the utility market that power generation by stoker firing is no longer a consideration. A major reason for the success of pulverized-coal burning is the ability to adapt operating conditions to all coal ranks from anthracite to lignite. While certain considerations must be made to accommodate fuels with such a wide range of properties, the years of experience since Kreisinger's statement have simply served to prove the versatility and advantages of pulverized-coal firing.

HISTORICAL PERSPECTIVE

One of the most significant engineering achievements of the twentieth century is the commercial perfection of methods for firing coal in pulverized form. In fact, the development is one of the cornerstones making possible the extremely large, modern, steam-generating unit with its high thermal efficiency, reliability, and safety.

Worldwide, practically every coal mined is being burned with complete success in pulverized form. Similarly, many other types of low-grade, waste, and byproduct solid fuels may also be fired economically and efficiently in this manner. As pulverized-fuel firing has contributed to the reduction of labor costs in steam power plants, it has also increased operational flexibility and practicable usage of an extremely wide range of fuels.

Over the years, the concept of pulverized-coal firing has attracted the attention of some of the finest engineering minds. Familiar with early nineteenth century French experiments, Sadi Carnot provided a critical thermodynamic analysis of the pyréolophore, an engine fired by powdered coal, in his 1824 engineering classic, *Reflections on the Motive Power of Fire*.[2]

During the 1890's Rudolf Diesel conducted

his first experiments on the internal combustion engine bearing his name with pulverized coal as the primary fuel. At this time pulverized-coal firing was achieving its first real commercial success in the cement industry.[3] In the early 1900's, Thomas A. Edison made improvements in the firing of pulverized coal in cement kilns, greatly increasing their efficiency and output.[4]

In all fairness, however, recognition must be given also to hundreds of engineers of lesser renown who have made equal or greater technical contributions. Since the first information attributed to the Niepce brothers was published in France in the early 1800's, there have been many examples of engineers whose visions of future developments in pulverized-coal technology have far outreached the materials and technical understanding of their time.

Largely developed as an empirical art, pulverized-coal firing progress has been marked by the efforts of devoted engineers whose success may be attributed to persistence despite many discouraging obstacles. Generally, theoretical understanding has followed rather than preceded practical accomplishment in the field of pulverized-coal firing.

INCENTIVES FOR DEVELOPING PULVERIZING TECHNIQUES

Some elements of engineering reasoning which have stimulated invention and improvement of devices to burn coal in pulverized form include the following: (1) Coal is widely available for combustion purposes. (2) Burning gas appears to be a simpler process than consumption of large pieces of coal. (3) If coal can be finely divided and burned like a gas, it becomes an even more attractive fuel, promising greater boiler efficiency and simplicity of combustion.

In the U.S., the rapid increase in oil prices in the 1890's was the principal incentive for developing the use of pulverized coal for firing cement kilns—the first industrial application to achieve outstanding commercial success. E. H. Hurry and H. J. Seaman of the Atlas Portland Cement Company began a series of experiments relating to the use of pulverized coal in 1894, and in the following year it was successfully applied to a rotary cement kiln. Since that time, pulverized coal has been the dominant fuel in the cement industry.

By the time of World War I, powdered coal—the term then generally used for what is now designated as pulverized coal—had gained sufficient acceptance for the ASME to sponsor a symposium bringing together the accumulated experience in the several fields of application. Reading the record of this symposium will give a clear understanding of the empirical nature of the wide variety of equipment available for various types of pulverized-coal firing.[5]

PIONEERING UTILITY INSTALLATIONS

By the end of the war, pulverized coal still had not achieved its full potential despite an increasing number of applications which spread from the cement to the metallurgical industry, to the steam locomotive and to several stationary boilers. Although all of the elements for outstanding success appeared to be present, someone was needed to integrate the many ideas and to provide a new thrust for pulverized-coal firing in the central-station industry. No one can lay more claim for initiating this impetus than John Anderson, then chief engineer of power plants of what is now Wisconsin Electric Power Company. He effectively enlisted the support and active participation of exceptionally able engineers from his own organization, the public-utility industry, equipment suppliers, and the U.S. Bureau of Mines. Anderson's pioneering efforts resulted in pulverized-coal installations in the existing Oneida Street Station and the bold new concept of Lakeside.[6]

Although American research in this field stems from the establishment of the U.S. Bureau of Mines in 1910 with its extensive program of boiler and equipment testing,[7] there was also earlier work in this country and abroad on the inflammability of dust clouds causing explosions in coal mines. Empirical progress in the art of pulverized-coal firing linked these two areas of research, and brought forth a series of reports and investigations ranging from power-plant tests to studies of particle flow and the thermodynamics of combustion.

Two outstanding test reports stand as engineering classics. Based on experimental work at the Oneida Street and Lakeside Stations, these reports contain important basic information on pulverized-coal firing and establish a pattern for subsequent research, including many of the activities of the ASME Furnace Performance Factors Committee.[8]

Studies on velocities and characteristics of pulverized-coal particles were reported by E. Audibert[9] and John Blizard,[10] who a few years earlier had published a comprehensive study of the state of the pulverized-coal art.[11] Research linking studies of inflammability of coal-mine dust to desired combustion properties appears in an article by Henri Verdinne.[12] W. Nusselt published results of research on coal-particle ignition times in 1924,[13] and P. Rosin reported on studies of heat liberation based on thermodynamic data in 1925.[14] The first of a series of papers on boiler heat-transfer studies at Yale University by W. J. Wohlenberg and his colleagues was also published by ASME in 1925.[15]

Despite the extensive theoretical studies that were made in the 1920's, much of the progress was achieved on an empirical basis of trial and error with boiler installations of ever-increasing size. This was particularly true in the development of pulverizers, where the theory of the underlying principles had not advanced very rapidly. Even today, the laws for crushing materials are subject to much dispute.

Rittinger's law of crushing dates back to a book published in Germany in 1867. It states that the work required to produce material of a given size from a larger size is proportional to the new surface produced. This expression finds more general acceptance than Kick's law, which was first published in 1885 and which states that the energy required to effect crushing or pulverizing is proportional to the volume reduction of the particle. While Rittinger's law is a closer approximation, neither of these laws can be used for comparing efficiencies of

coal pulverizers which are different.

The energy that is required to effect pulverization is dissipated in a number of ways. It cannot be accounted for in the specific manner which is applicable to a boiler or power-plant heat balance. For this reason, both pulverizer design and application have retained many of the elements of an engineering art.

PULVERIZING PROPERTIES OF COAL

To predict pulverizer performance on a specific coal with some degree of accuracy, the ease with which the coal can be pulverized must be known.

GRINDABILITY

A grindability index has been developed to measure the ease of pulverization. Unlike moisture, ash, or heating value, this index is not an inherent property of coal. Rather, it represents the relative ease of grinding coal when tested in a particular type of apparatus. The consistency of grindability test results permits the pulverizer manufacturer to apply the findings to a particular size and, to a lesser degree, type of pulverizer.

Grindability should not be confused with the hardness of coal. (See also Tables IV and V and related text.) The same coal may have a range of grindabilities depending on other constituents in the coal. Fig. 1 gives typical curves for North Dakota lignites and shows the variation in Hardgrove grindability as the moisture content changes. Typically, anthracites and some lignites have at least one point where their grindabilities are very close. Anthracite, however, is a very hard coal whereas lignite is soft, yet both are difficult to grind.

Pulverizing a small air-dried sample of properly sized coal in a miniature mill determines its grindability. Results may then be converted into a grindability-index factor which, with appropriate correction curves, can be used to interpret mill capacity.

The Hardgrove method was developed to measure the quantity of new material that will

pass a 200-mesh sieve. The apparatus for this method, shown in Fig. 2, is extremely simple. A 50-gram sample of air-dried coal, sized to less than 16 and greater than 30 mesh, is placed in the mortar of the test machine along with eight 1-in.-diameter steel balls. A weighted upper race is placed on the ball and coal charge, and is turned 60 revolutions. The sample then is removed and screened.

The quantity passing the 200-mesh sieve is used in the preparation of a calibration chart, from which the grindability of the coal sample is determined in accordance with *ASTM Standards* D 409, Grindability of Coal by the Hardgrove-Machine Method. Four coal samples, obtained from ASTM, standardized especially for this purpose and representing grindability indices of 40, 60, 80, and 100, are used for calibration of each grindability machine and associated apparatus, before the equipment is used to test coals.

Frequently, too much emphasis is placed on grindability while other factors affecting mill capacity, such as moisture, are almost entirely overlooked. Pulverizer capacity is proportional to the grindability index of the coal, but corrections must also be made for fineness of product and moisture of the raw feed.

MOISTURE

Usually a reference to moisture in coal pertains to the total moisture content. This is comprised of what is commonly termed *equilibrium* moisture and *surface* or *free* moisture. Equilibrium moisture varies with coal type or rank and mine location, and would be more accurately called "bed" or "seam" moisture. In reality, surface moisture is the difference between total moisture and bed moisture.

Surface moisture adversely affects both pulverizer performance and the combustion process. The surface moisture produces agglomeration of the fines in the pulverizing zone, and reduces pulverizer drying capacity because of the inability to remove the fines efficiently and as quickly as they are produced. Agglomeration of fines has the same effect as coarse coal during the combustion process, because the surface available for the chemical reaction is reduced. Since in-mill drying is the accepted method of preparing coal for pulverized-fuel burning, sufficient hot air at adequate temperature is necessary in the mill-

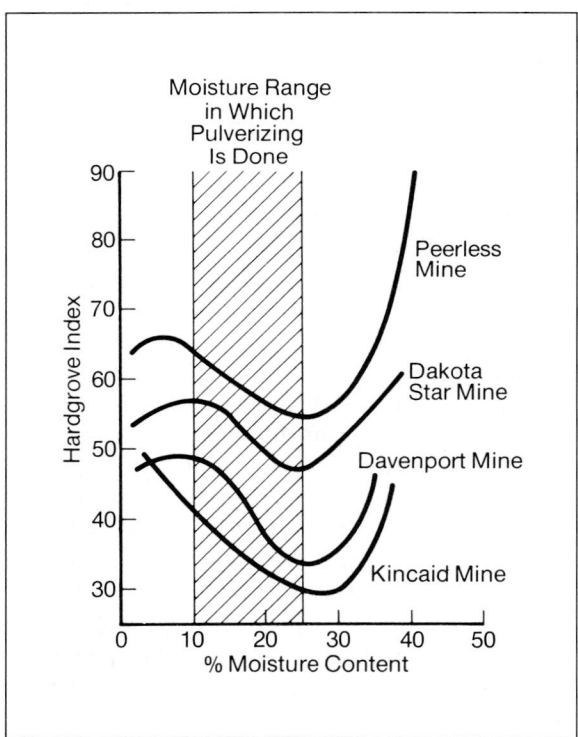

Fig. 1. Variation of grindability index with moisture content, North Dakota lignites (average of standard and corrected values)

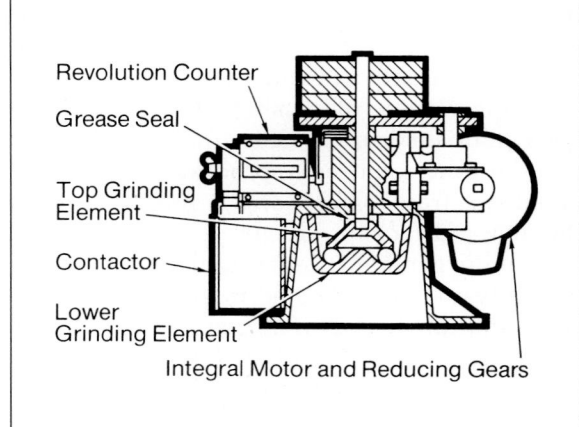

Fig. 2. Hardgrove grindability machine

ing system. Curves such as those shown in Figs. 3 and 4 indicate the air temperature required to dry coal of varied total moistures and coal-air mixtures.

The achieving of rated pulverizer capacity depends upon having sufficient heated air available to dry the coal. If there is a deficiency of hot air, the mill output will be limited to the "drying capacity" and not the "grinding capacity." Thus, it may be possible to obtain more capacity with a relatively dry coal of lower grindability than with a high-moisture coal of higher grindability.

RELATION OF PULVERIZER CAPACITY TO GRINDABILITY

As stated previously, mill capacity is not directly proportional to grindability. Thus, if the actual capacity of a pulverizer with 50 grindability is 10,000 lb per hr, then with 100 grindability it will be about 17,000 lb per hr, and not 20,000 lb per hr. This is because of the

differences between a commercial pulverizer and a grindability test machine which, with no provision for continuous removal of fines, is of the batch type rather than the continuous type. The crushing pressure of the test equipment is also considerably less. As a result, some of its energy is dissipated in deforming the coal particles without breaking them.

Although the test equipment does not indicate a direct proportion of capability between hard and soft materials, the value of these tests is not reduced. Correction factors developed by pulverizer manufacturers on commercial equipment provide for overcoming these discrepancies. Fig. 5 gives correction curves for variations in fineness, grindability, and moisture. As a rule of thumb, for every point the grindability index of a particular coal changes, there will be a corresponding change of 1⅓ percent in pulverizer capacity. Similarly, for every percentage point of change in fineness from the basic design point of a pulverizer,

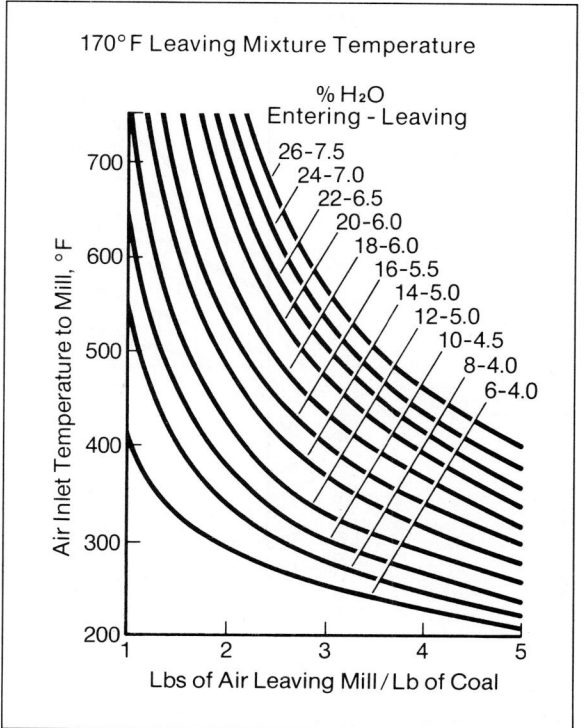

Fig.3 Temperature of air to mill, eastern U.S. coals

Fig.4 **Temperature of air to mill, midwest-U.S. coals**

Pulverizers and Pulverized-Coal Systems

there will be a corresponding change of 1½ percent in capacity.

VARIATION OF LIGNITE GRINDABILITY WITH MOISTURE

The Bureau of Mines at Grand Forks, North Dakota, and others have reported the grindability of lignites at various moistures, and the results show a wide variation, as in Fig. 1. Some feel that such curves are of little value because it seems impossible to select the proper index from them; others feel that they do have significance. With the increased use of lignites, solution of this problem is important.

In setting up the present ASTM code for grindability, the test specifies use of an air-dried sample. In the C-E bowl mill, *all* of the surface moisture and some of the equilibrium moisture are evaporated during pulverization with a hot-air sweep. The moisture content of the pulverized product leaving the mill is con-

ceived to be the moisture level that exists in the grinding zone. Thus grindability vs. moisture indices above the equilibrium level are of little interest. Hardgrove indices therefore have meaning to the pulverizer designer only below the equilibrium-moisture level and in the general range of moisture contents between 10 and 25 percent.

The actual choice of the grindability index for pulverizer design capacity requires consideration of total moisture, equilibrium (bed) moisture, and the selected hot-air temperature.

RELATIONSHIP OF COAL RANK TO REQUIRED FINENESS

Successful pulverized-coal firing depends on recognizing differences in coals and on making whatever modifications are necessary to provide the optimum conditions for efficient combustion. Experience over the years has established that a relationship exists between

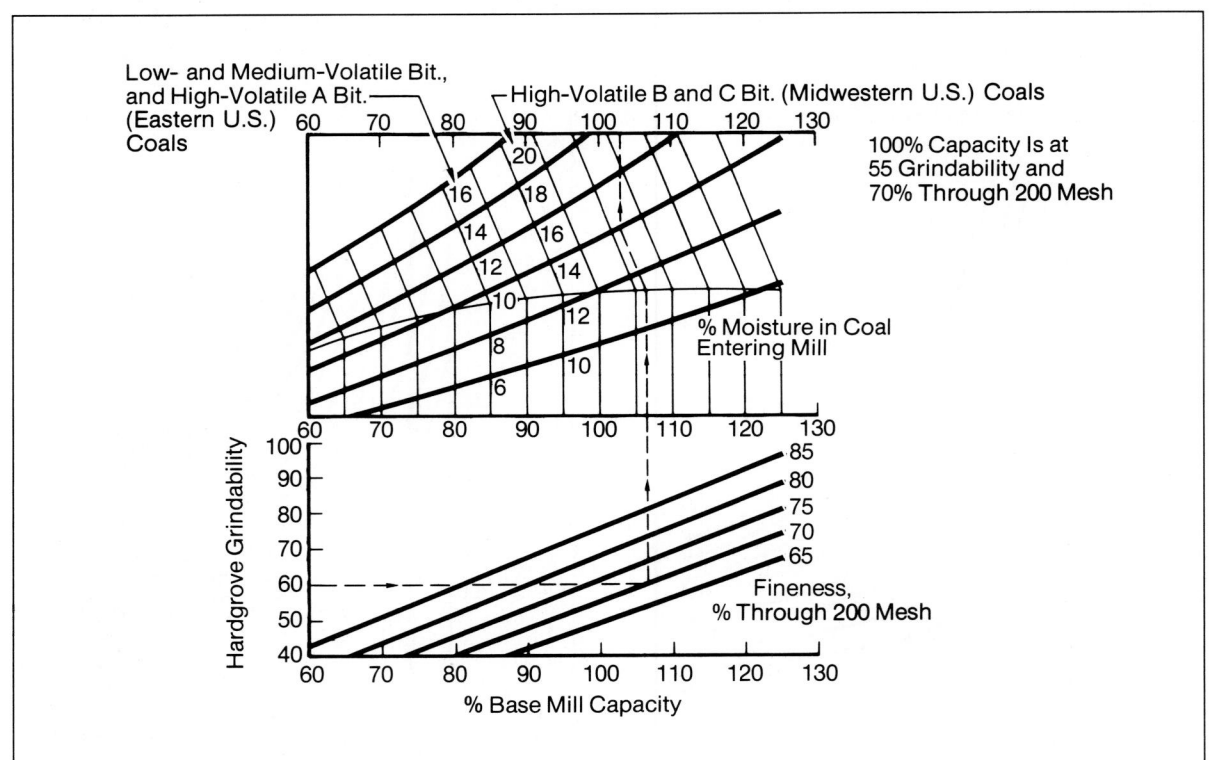

Fig. 5. Mill correction factors for grindability, fineness, coal type, and moisture— bituminous coals with preheated air

rank of bituminous coals and degree of fineness required for successful operation.

To insure complete combustion within the furnace confines for minimal carbon loss, high-rank coals must be pulverized to a finer size than coals of lower rank. Although in part dependent on the type of firing system used, the approximate limits within the following ranges have been established through operating experience.

Coal Rank	Passing 200 Mesh (74 μ) Wt %	Retained on 50 Mesh (297 μ) Wt %
Subbituminous C coal and Lignite	60–70	2.0
High-Vol. Bit. C, Subbituminous A and B	65–72	2.0
Low and Medium-Vol. Bituminous, High-Vol. Bit. A and B	70–75	2.0

When firing certain coals in the low-volatile group in small pulverized-coal furnaces, the fineness percentage may be increased to as high as 80 percent to insure adequate burnout of the carbon content.

SPECIFICATIONS FOR CLASSIFYING COALS

Table I lists the specifications as set by ASTM for classifying coal by rank. Rank classifications are based on varying combinations of volatile-matter content, heating value, and agglomerating properties.

After the predrying step is accomplished in the pulverizing operation, the low-rank subbituminous coals and lignites manifest a higher degree of reactivity than do the higher rank bituminous coals. Recent investigations confirm that this increased reactivity results primarily from the lack of agglomerating properties and increased O_2 content.[16] Data from these investigations are summarized in Table II, Coal Properties.

A comparison of flammability indices with volatile matter and heating value suggests that factors other than these have a great influence on ignition temperature and particle burnout. Because lower flammability temperatures are experienced primarily with lignitic and subbituminous coals, it is reasonable to ask what property is common to lower rank coals. Exclusive of heating value, the most obvious difference between the two groups is that property known as "agglomerating character"—subbituminous coals and lignites do not agglomerate. As applied to coals, agglomeration is the property of particles fusing into a cokelike mass or bonding together into a firm cake when the particles are heated to temperatures of 1000°F or above.

While the determinations of both volatile matter and heating value are well defined, tests for establishing the agglomerating character of coals are less commonly known. *ASTM Standards* D 388, Specifications for Classification of Coals by Rank, describes agglomerating character as: "The test carried out by the examination of the residue in the platinum crucible incidental to the volatile-matter determination. Coals which in the volatile-matter determination produce either an agglomerate button that will support a 500-g weight without pulverizing, or a button showing swelling of cell structure, shall be considered agglomerating from the standpoint of classification."

Since the agglomerating property of coals is the result of particles transforming into a plastic or semiliquid state when heated, it reflects a change in the surface area of the particle. This surface change is manifested by a transformation of the particle from an angular, irregular shape into a spherical or sphere-like particle. Also, the surface character of the particle changes from a porous, irregular, absorptive surface to a glasslike nonporous surface. Thus, with the application of heat, agglomerating coals tend to develop a nonporous surface, while that of nonagglomerating coals becomes even more porous with pyrolysis. This explanation indicates why agglomerating coals require a correspondingly finer particle size to maintain an equivalent surface area for efficient, rapid ignition and burnout.

Table I. Classification of Coals by Rank[a]

Class and Group	Fixed Carbon Limits, % (Dry, Mineral-Matter-Free Basis)		Volatile Matter Limits, % (Dry, Mineral-Matter-Free Basis)		Calorific Value Limits, Btu/lb (Moist,[b] Mineral-Matter-Free Basis)		Agglomerating Character
	Equal or Greater Than	Less Than	Equal or Greater Than	Less Than	Equal or Greater Than	Less Than	
I. Anthracitic							
1. Meta-anthracite	98	. . .	. . .	2	. . .	. . .	nonagglom-erating
2. Anthracite	92	98	2	8	. . .	. . .	
3. Semianthracite[c]	86	92	8	14	. . .	. . .	
II. Bituminous							
1. Low-volatile bituminous coal	78	86	14	22	. . .	. . .	commonly agglomerating[e]
2. Medium volatile bituminous coal	69	78	22	31	. . .	. . .	
3. High-volatile A bituminous coal	. . .	69	31	. . .	14,000[d]	. . .	
4. High-volatile B bituminous coal	. . .	. . .	. . .	. . .	13,000[d]	14,000	
5. High-volatile C bituminous coal	. . .	. . .	. . .	. . .	11,500	13,000	
					10,500	11,500	agglomerating
III. Subbituminous							
1. Subbituminous A coal	. . .	. . .	. . .	. . .	10,500	11,500	nonagglom-erating
2. Subbituminous B coal	. . .	. . .	. . .	. . .	9,500	10,500	
3. Subbituminous C coal	. . .	. . .	. . .	. . .	8,300	9,500	
IV. Lignitic							
1. Lignite A	. . .	. . .	. . .	. . .	6,300	8,300	
2. Lignite B	. . .	. . .	. . .	. . .	. . .	6,300	

[a] This classification does not include a few coals, principally nonbanded varieties, which have unusual physical and chemical properties and which come within the limits of fixed carbon or calorific value of the high-volatile bituminous and subbituminous ranks. All of these coals either contain less than 48% dry, mineral-matter-free fixed carbon or have more than 15,500 moist, mineral-matter-free British thermal units per pound.

[b] Moist refers to coal containing its natural inherent moisture but not including visible water on the surface of the coal.

[c] If agglomerating, classify in low-volatile group of the bituminous class.

[d] Coals having 69% or more fixed carbon on the dry, mineral-matter-free basis shall be classified according to fixed carbon, regardless of calorific value.

[e] It is recognized that there may be nonagglomerating varieties in these groups of the bituminous class, and there are notable exceptions in high-volatile C bituminous group.

Reprinted from *ASTM Standards* D 388, Classification of Coals by Rank.

In addition to the correlation of agglomerating properties with coal reactivity, an equally strong correlation exists between the ultimate-analysis oxygen level of coals and their response to reactivity as reflected in the flammability temperatures. Data in Table II show ranges in oxygen from 4.8 to 14.8 percent for agglomerating coals, moisture- and ash-free (MAF), and from 18.7 to 26.6 percent MAF for the lower rank coals that are nonagglomerating. Seemingly, the higher the inherent or organically bound oxygen content of the coal, the more reactive the coal. These data do not conflict with the observed correlation with agglomerating character. In fact, deliberately varying degrees of oxidation temper or destroy the agglomerating properties of caking coals. Apparently, the breaking point between agglomerating and nonagglomerating coals is a 14–15 percent oxygen level, MAF.[17]

STANDARDS FOR MEASURING FINENESS

When burning solid fuels in suspension, it is essential that the fuel-air mixture contain an appreciable quantity of extremely fine particles to insure rapid ignition. Conversely, to obtain maximum combustion efficiency, a minimum amount of coarse particles in this same fuel-air mixture is desirable. The former condition is usually expressed as percentage through a 200-mesh screen (74 microns), while the latter is designated as percentage retained on a 50-mesh screen (297 microns).

The number of openings per linear inch designates the mesh of a screen. Thus, a 200-mesh screen has 200 openings to the inch or 40,000 per square inch. The diameter of the wire used in making the screen governs the size of the openings. The U.S. Standard and W. S. Tyler are the most common screen sieves. The mesh and opening of these and other international screens are shown in Table III and Fig. 6.

CLASSIFICATION AND SIZE CONSIST

In some reactions, such as setting of cement, surface area is of extreme importance. In the combustion of pulverized coal, however, while it is important to have a proper percentage of

Table II. Coal Properties

	Bituminous					Subbituminous			Lignite
	Low Volatile	Medium Volatile	High Volatile						
			A	B	C	A	B	C	A
Agglomerating character	Agg.	Agg.	Agg.	Agg.	*	Non Agg.	Non Agg.	Non Agg.	Non Agg.
Proximate, %									
Moisture (seam)	2.0	2.0	4.0	7.0	10.0	14.0	19.0	25.0	40.0
Volatile matter, VM	21.1	32.3	38.4	33.8	35.9	35.3	34.5	25.8	25.9
Fixed carbon, FC	68.6	55.8	51.5	47.3	43.3	41.2	37.5	40.9	27.4
Ash	8.3	9.9	6.1	11.9	10.8	9.5	9.0	8.3	6.7
HHV, Btu/lb, As-fired	13,150	13,210	13,410	11,610	10,590	9,840	8,560	7,500	5,940
Flammability index, F	1,010	1,030	950	1,030	990	970	970	990	890
Ultimate (MAF), %									
Hydrogen	5.0	5.5	5.6	4.6	5.5	5.4	5.1	5.6	4.3
Carbon	88.5	84.1	82.5	81.0	74.3	74.2	69.8	66.4	67.0
Sulfur	0.4	1.1	2.5	0.9	4.0	0.5	0.8	0.6	0.9
Nitrogen	1.3	1.7	1.5	1.3	1.4	1.2	1.1	1.3	1.2
Oxygen	4.8	7.6	7.9	12.2	14.8	18.7	23.2	26.1	26.6

* Agglomerating but noncaking

ABB
ASEA BROWN BOVERI

fine particles having a large surface area, it is equally necessary to eliminate the oversize on the coarser screen. Despite the percentage less than 200 mesh (−200 mesh), as little as 3 to 5 percent greater than 50 mesh (+50 mesh) may produce furnace slagging and increased combustible loss, even though combustion conditions are excellent for the finer coal. The small amount of oversize represents very little additional surface if it is pulverized to all −50 mesh and all +200 mesh.

As an illustration, assume a typical screen analysis of a high-volatile bituminous coal sample, pulverized to 80 percent −200 mesh:

- 99.5% −50 mesh
- 96.5% −100 mesh
- 80.0% −200 mesh

This represents a surface area of approximately 1500 sq cm per gram, with over 97 percent of the surface in the −200 mesh portion.

By overgrinding and poor classification it would be possible, on a commercial-sized mill, to have a sample of the following analysis:

- 95% −50 mesh
- 90% −100 mesh
- 80% −200 mesh

Table III. Comparison of Sieve Openings

Mesh	Inches	mm
U.S. Standard Sieve		
20	0.0331	0.84
30	0.0234	0.595
40	0.0165	0.420
50 *	0.0117	0.297
60	0.0098	0.250
100 *	0.0059	0.149
140	0.0041	0.105
200 *	0.0029	0.074
325	0.0017	0.044
400	0.0015	0.037

Mesh	Inches	mm
W. S. Tyler Sieve		
20	0.0328	0.833
28	0.0232	0.589
35	0.0164	0.417
48 *	0.0116	0.295
60	0.0097	0.246
100 *	0.0058	0.147
150	0.0041	0.104
200 *	0.0029	0.074
325	0.0017	0.043
400	0.0015	0.037

*Commonly used screens in pulverized-coal practice for combustion purposes.

Fig. 6. International screen opening comparisons

Screen Opening, mm	USA Tyler No.	USA ASTM/USBS No.	England IMM No.	England Std BESA No.	Germany †Din 1171 No.	France ‡AFN No.	France Std Usuelle No.	Screen Opening, Micrometers
0.03	400	400						30
0.04		325				17		40
0.05	325	270		300		18	300	50
0.06	230	230	200	240	0.06	19	250 / 220	60
0.07	200	200		200	0.075	20	200	70
0.08	170	170	150	170	0.09		190	80
0.09					0.1	21	170 / 150	
0.1	150	140	120	150				100
	115	120	100	120	0.12	22	120	
0.15	100	100	90	100	0.15	23	100	150
	80	80	80 / 70	85			90 / 80	
0.2	65	70	60	72	0.2	24		200
	60	60	50	60	0.25	25	60	
0.3	48	50	40	52	0.3	26	50	300
	42	45	35	44		27		
0.4	35	40	30	36	0.4		40	400
0.5	32	35	25	30	0.5	28		500
0.6	28	30	20	25	0.6	29	30	600
0.7	24	25		22	0.75			700
0.8	20	20	16	18		30	20	800
0.9	16	18	12	16	1.0	31		
1.0	14	16	10	14	1.2	32	16 / 14	1000
	12	14	8	12	1.5		12	
1.5	10	12		10		33		
2.0	9	10		8	2.0	34	10	
	8	8	5	7	2.5	35		
	7	7		6	3.0			
3.0	6	6		5		36		
4.0	5	5			4.0	37		
	4	4				38		
5.0								

* Institute Mining & Metallurgy
  British Engineering Std. Association

† German Industry Norm

‡ Association Français de Norm

This is not a satisfactory grind, because of the high percentage retained on the 50 mesh, even though the surface area is still 1500 sq cm/g.

In the pulverizing process, then, classification plays a major role in matching the particle size to the reactivity of the fuel. Both fine and coarse particles must be controlled within limits by the use of mechanical classification techniques. Careful attention, therefore, must be paid to both the design and the operation of the classification system.

SAMPLING PULVERIZED COAL

It is apparent that product fineness has a considerable bearing on pulverizer performance. Fineness samples should, therefore, be analyzed periodically. In a storage system, this sample may be taken directly from the mill cyclone discharge.

On a direct-fired system, obtaining the sample is more difficult because it must be taken from a flowing coal-air stream. A sampling device, consisting of a small cyclone collector, sample jar, and sampling nozzle with connecting hose, may be utilized. See Fig. 7.

The sample is obtained by traversing the pipe across its diameter, from two points in the same plane and at 90° to each other. The entire pipe diameter must be traversed, and the rate of

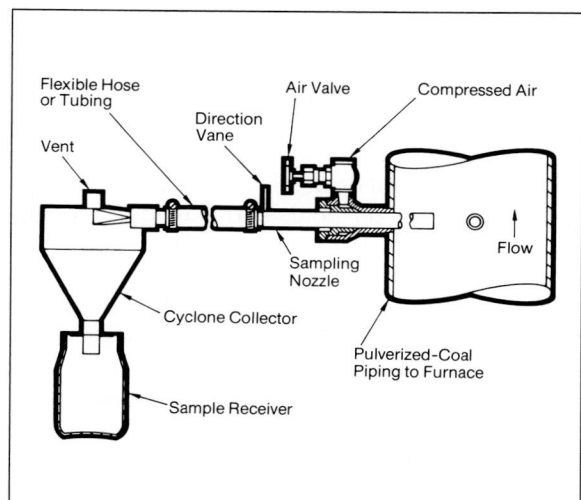

Fig.7 **Pulverized-coal sampling device and aspirating fittings**

movement must be uniform. Samples must be taken in both directions for the same period.

Currently, the industry has further refined these sampling techniques for greater accuracy. In this refinement, the pipe traverse with the coal-sampling device is timed to obtain an isokinetic coal sample. The method uses the proportion of coal-pipe area to the sample-probe-opening area in conjunction with the pulverized-fuel loading in the individual transport pipe. Because pulverized fuel in a transport pipe is not a homogeneous mixture, another sample taken at the same time at a different location in the same pipe may yield different results.

With collection completed, the pulverized-coal samples from each mill are thoroughly mixed. Fifty grams of the sample is placed in the top sieve of a nested stack of 50-mesh, 100-mesh, and 200-mesh sieves. The nest is then shaken either by hand or by a mechanical shaking device until the procedure has separated the coal particles by size. The results of the percentages of coal passing through the individual screens plot as a straight line on a typical sieve distribution chart. (See Fig.8.) *ASTM Standards* D 197, Sampling and Fineness Test of Pulverized Coal and *ASME* PTC 4.2, Coal Pulverizers, give additional information on recommended sampling techniques.

CLOSED-CIRCUIT GRINDING

When a large piece of coal is reduced to a number of smaller ones by any method, a great number of fine particles will be produced simultaneously. Therefore, it is not possible for a pulverizer to produce a product that will pass a 50-mesh screen without also obtaining a large percentage of material finer than 200 mesh. Thus, if a quantity of coal at one stage of pulverization contains 50-percent material that will pass through a 50-mesh sieve, and if this −50-mesh material is removed from the grinding zone, it will contain a smaller percentage of −200-mesh material than if it had been permitted to remain in the grinding zone until the total quantity had been reduced to pass a 50-mesh sieve.

As already noted, an abundance of fine particles is necessary to insure prompt ignition of coal in suspension burning. Substantial energy consumption is required in the production of this fine material. However, when grinding finer than necessary, power is wasted and the pulverizing equipment must be larger than actually required. Removal of the fines from the pulverizing zone as rapidly as they are produced and return of the oversize for regrinding eliminates unnecessary production of fines and reduces energy requirements. Better product sizing and increased capacity result from the removal of the fines, a process called closed-circuit grinding. The pulverizing system component which accomplishes this size control is known as the classifier.

ABRASION

Pulverizing results in an eventual loss of grinding-element material. Balls, rolls, rings, races, and liners gradually erode and wear out as a result of abrasion and metal displacement in the grinding process. Thus, the power for grinding and the maintenance of the grinding elements make up the major costs of the pulverizing operation.

In itself, "pure coal" is relatively nonabrasive; however, such foreign materials as slate, sand, and pyrites, commonly found in coal as mined, are quite abrasive. These are the undesirable constituents that produce rapid and sometimes excessive wear in pulverizing apparatus. The economics of coal cleaning to remove such abrasive foreign materials depends on many variables and must be determined for each individual application.

The resistance of a smooth plane surface to abrasion is called its hardness. It is commonly recorded in terms of 10 minerals according to Mohs' scale of hardness, Table IV. There is no quantitative relation between these, the diamond being much greater in hardness above sapphire than sapphire is above talc. Hardness of selected common materials is shown in Table V; the relatively low hardness of pure coal is compared to the abrasive impurities usually found in the commercial product.

COAL PREPARATION

Coal should be prepared properly for its safe, economical, and efficient use in a pulverizing system. Controllable continuity of flow to the pulverizer must be maintained. Organic foreign materials such as wood, cloth, or straw should be removed. Such materials may collect in the milling system and become a fire hazard, or they may impair material or airflow patterns in the mill. Although many mills are designed to reject, or are not adversely affected by, small inorganic or metallic materials, a magnetic separator should be installed in the raw-coal conveyor system to remove larger metallic objects. If this is not done, these objects may damage the pulverizer coal feeder or obstruct the coal flow.

The raw coal should be crushed to a size that will promote a uniform flow rate to the mill by the feeder. Favorable size consist will minimize segregation of coarse and fine fractions in the bunker, and result in a more uniform rate of feed to various pulverizers being supplied from a given bunker. When mixed with relatively dry lump coal, fine coal with high surface moisture accentuates the segregation problem in bunkers. Crushing by size-reduction of the

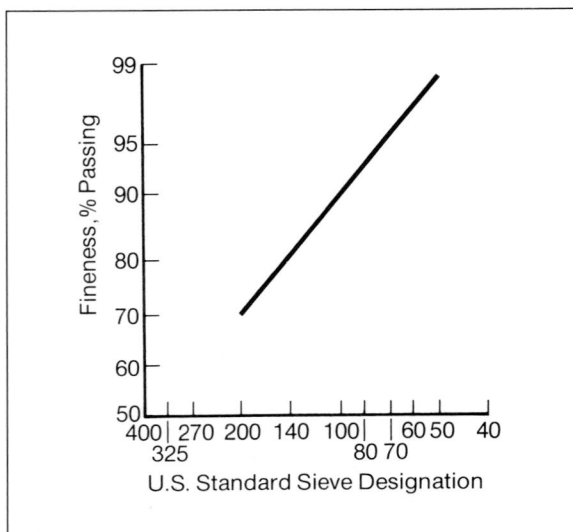

Fig. 8. Typical fineness sample results

dry lumps exposes additional dry surfaces for the adsorption of moisture from the wet fines, thereby producing a more uniform size and moisture distribution in the raw-coal mass.

Most commonly, direct-firing of pulverized coal is the method used for steam generation. In this application, an uninterrupted and uniformly controllable supply of pulverized coal to the furnace is an essential requisite. A steady and continuous flow of raw coal to the pulverizer will insure this.

An ideal feed is one that is closely sized and double-screened; e.g. ¾ in. x ¼ in. Coal of this size will permit excess water to drain off; it will flow freely from bunkers and can be fed uniformly. However, such favorable sizing can be obtained only at a considerable price premium and this usually precludes its use. In most cases, power plants will receive coals classified as run-of-mine or screenings with lumps; therefore, crushing equipment must be installed to provide uniform raw-feed sizing. Generally coal-feed sizing up to 2 in., as sieved through a round screen, is permissible with large pulverizers.

COAL CRUSHERS

Although there are numerous types of crushers commercially available, the type generally used for smaller capacities is the swing-hammer type. This crusher has proved satisfactory for overall use and has demonstrated reliability and economy. The swing-hammer crusher consists of a casing enclosing a rotor to which are attached pivoted hammers or rings. Coal is fed through a suitable opening in the top of the casing and crushing is effected by impact of the revolving hammers or rings directly on, or by throwing the coal against, the liners or spaced grate bars in the bottom of the casing. The degree of size reduction depends on hammer type, speed, wear, and bar spacing. The latter is usually somewhat greater than the desired coal size. These crushers produce a uniform coal sizing and break up pieces of wood and foreign material with the exception of metallic objects. Foreign material that is too hard to crush is caught in pockets.

Roll crushers have been used but are not entirely satisfactory because of their inability to deliver a uniformly sized product. Probably the most satisfactory crusher for large capacities is the Bradford breaker. This design, Fig. 9, consists of a large-diameter, slowly revolving (approximately 20 rpm) cylinder of perforated steel plates, the size of the perforations determining the final coal sizing. In diameter, these openings are usually 1¼ in. to 1½ in. The breaking action on the coal is accomplished as follows: the coal is fed in at one end of the cylinder and carried upward on projecting vanes or shelves. As the cylinder rotates, the coal cascades off these shelves and breaks as it strikes the perforated plate. As the coal drops a relatively short distance, coal crushing occurs with the production of very few fines. Coal broken to the screen size passes through the perforations to a hopper below. Rocks, wood, slate, tramp iron, and other foreign material are rejected. This breaker produces a relatively uniform product and uses very little power.

Table IV. Mohs' Scale of Hardness

1—Talc	6—Feldspar
2—Gypsum	7—Quartz
3—Calcspar	8—Topaz
4—Fluorspar	9—Sapphire
5—Apatite	10—Diamond

Table V. Common Materials and Their Mohs' Hardness

Coal	0.5–2.5
Slate	0.50–6.0
Mica	2.0–6.0
Pyrite	6.0–6.5
Granite	6.5
Marble	3.0
Soapstone	1.0–4.0
Kaolin clay	2.0–2.5
Iron ore	0.50–6.5
Carborundum	9.5

Fig. 9. **Bradford breaker**

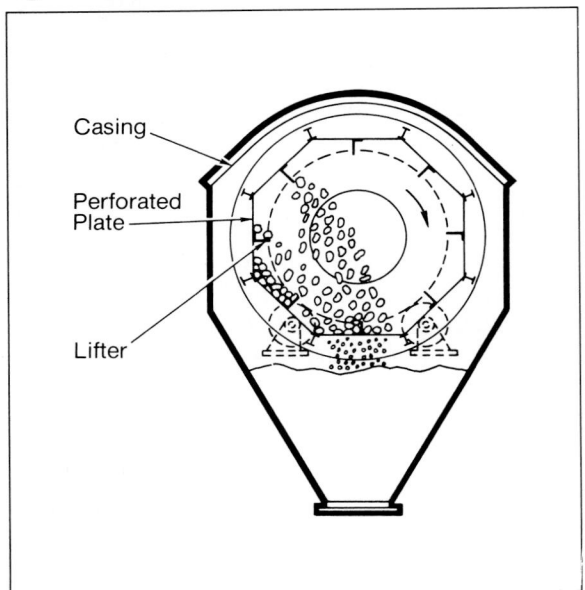

COAL FEEDERS

A coal feeder is a device that supplies the pulverizer with an uninterrupted flow of raw coal to meet system requirements. This is especially important in a direct-fired system. There are several types, including the belt feeder and the overshot roll feeder.

THE BELT FEEDER

The belt feeder uses an endless belt running on two separated rollers receiving coal from above at one end and discharging it at the other. Varying the speed of the driving roll controls the feed rate. A leveling plate fixes the depth of the coal bed on the belt.

The belt feeder can be used in either a volumetric or gravimetric type of application (see Fig. 10). The gravimetric type has gained wide popularity in the industry for accurately measuring the quantity of coal delivered to each

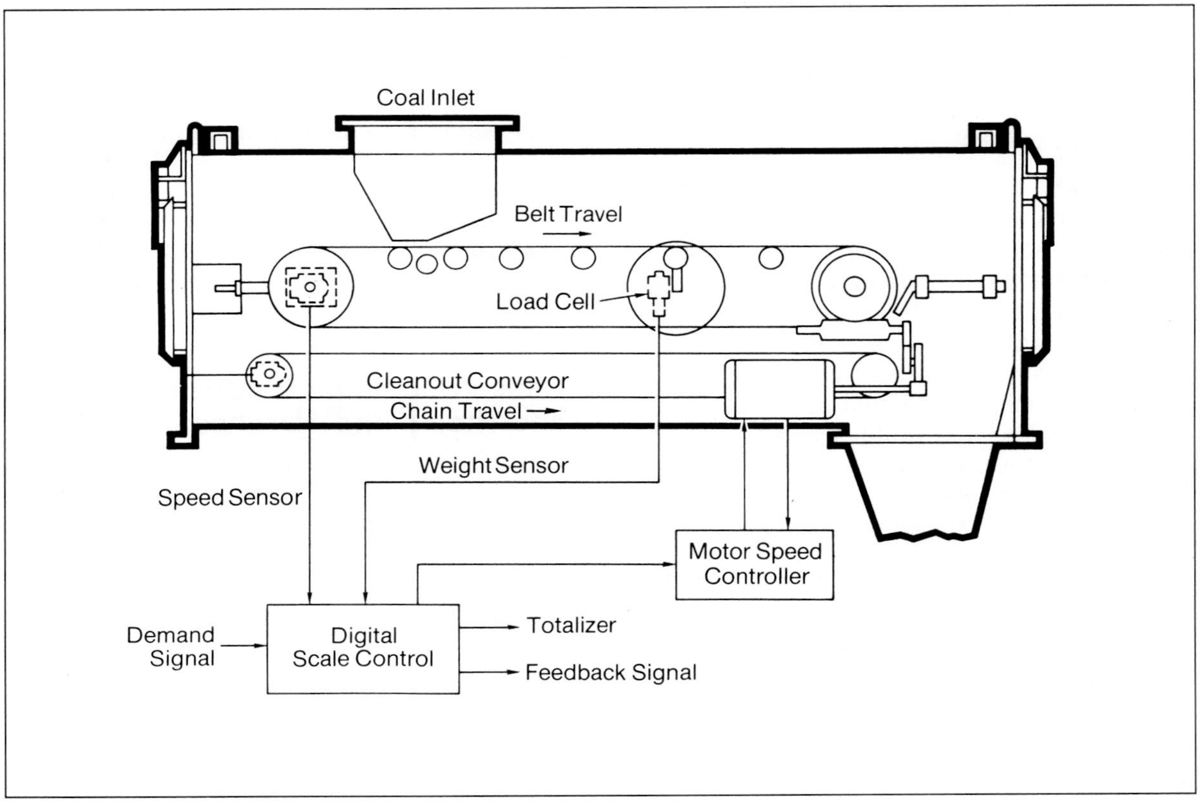

Fig. 10. **Schematic of belt-type gravimetric coal feeder.**

individual pulverizer. Generally, they are applied to steam generators having combustion-control systems requiring individual coal metering to the fuel burners.

There are two accepted methods of continuously weighing the coal on the feeder belt. One method uses a series of levers and balance weights; the other, a solid-state load cell across a weigh span on the belt. Both are very accurate mechanisms and both are well accepted by utilities. This same belt feeder design can also be used for volumetric measurement.

THE OVERSHOT FEEDER

The overshot roll feeder, Fig. 11, has a multi-bladed rotor which turns about a fixed, hollow, cylindrical core. This core has an opening to the feeder discharge and is provided with heated air to minimize wet-coal accumulation on surfaces and to aid in coal drying. A hinged, spring-loaded leveling gate mounted over the rotor limits the discharge from the rotor pockets. This gate permits the passage of over-size foreign material.

Feeders of this type may be separately mounted, or they may be integrally attached to the side of a pulverizer.

The roll feeder and the belt feeder, by virtue of their designs may be considered highly efficient volumetric feeding devices.

METHODS OF PULVERIZING AND CONVEYING COAL

Early coal-pulverizing installations received undried coal and utilized ambient air in the mill system. Because no heat was added to the system, the coal feed was limited to that of very low moisture content; therefore, maximum pulverizer capabilities were not realized. Subsequently, external coal dryers were added to the system. Because of the lack of cleanliness, high initial cost, fire hazard, and space requirements, these dryers were replaced with the now universally accepted in-mill drying.

Three methods of supplying and firing pulverized coal have been developed: the storage or indirect system, the direct-fired system, and the semidirect system. These methods differ on the basis of their drying, feeding, and transport characteristics.

THE STORAGE (INDIRECT) SYSTEM

In a storage system, Fig. 12, coal is pulverized and conveyed by air or gas to a suitable collector where the carrying medium is separated from the coal which is then transferred to a storage bin. The hot air or flue gas introduced into the mill inlet provides for system drying requirements and is vented to rid the system of

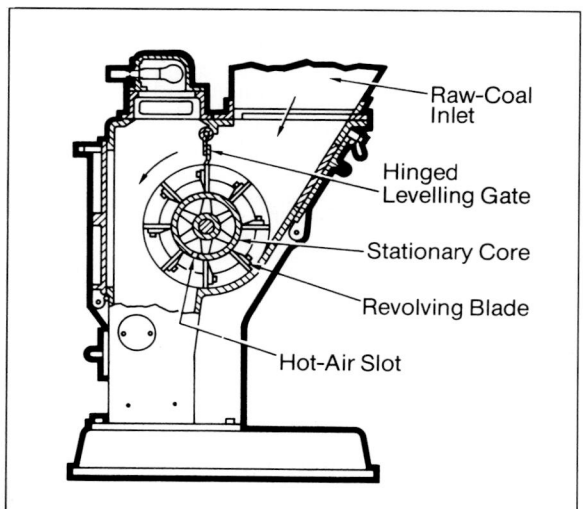

Fig. 11. Overshot roll feeder

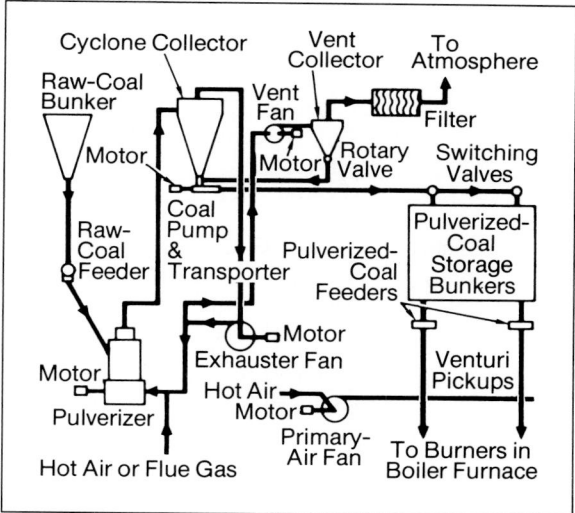

Fig. 12 Storage (indirect) pulverizing system

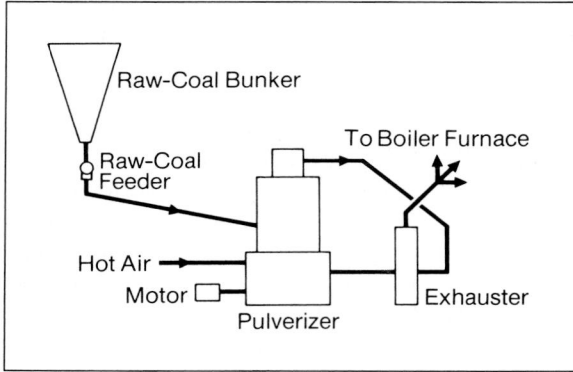

Fig. 13 Direct-fired pulverizing system

moisture evaporated from the fuel. From the storage bin, the pulverized coal is fed to the furnace as required.

THE DIRECT-FIRED SYSTEM

In a direct-fired system, Fig.13, coal is pulverized and transported with air, or air slightly diluted with gas, directly to the furnace where the fuel is consumed. Hot air or diluted furnace gas supplied to the pulverizer furnishes the heat for drying the coal and transporting the pulverized fuel to the furnace. Known as primary air, this air is a portion of the combustion air. As a reduction in oxygen concentration in this primary-air stream affects the rapidity and

stability of ignition, it is necessary, when using hot gas, to draw it from a point of low CO_2 concentration and high temperature. This prevents the use of flue gas for drying in direct-firing.

Fig. 14 illustrates a schematic arrangement of the primary-air system for a pulverizer using a Ljungstrom® trisector air heater. As the name implies, this air heater has three sections: flue gas, primary air (the air that dries and conveys the coal to the furnace) and secondary air (the balance of the air that goes to the furnace). The primary-air section is located between the openings for the secondary air and the flue gas (Fig. 15). With this design, a higher primary-air temperature can be obtained. If there is a large variation in primary airflow, there is relatively little effect on heat recovery, because heat not recovered in the primary section will be picked up subsequently in the secondary section.

THE SEMIDIRECT SYSTEM

In a semidirect system, Fig. 16, a cyclone collector located between the pulverizer and furnace separates the conveying medium from the coal. The coal is fed directly from the cyclone to the furnace in a primary-air stream which is independent of the milling system. The drying medium, therefore, can be the same as in a storage or bin system.

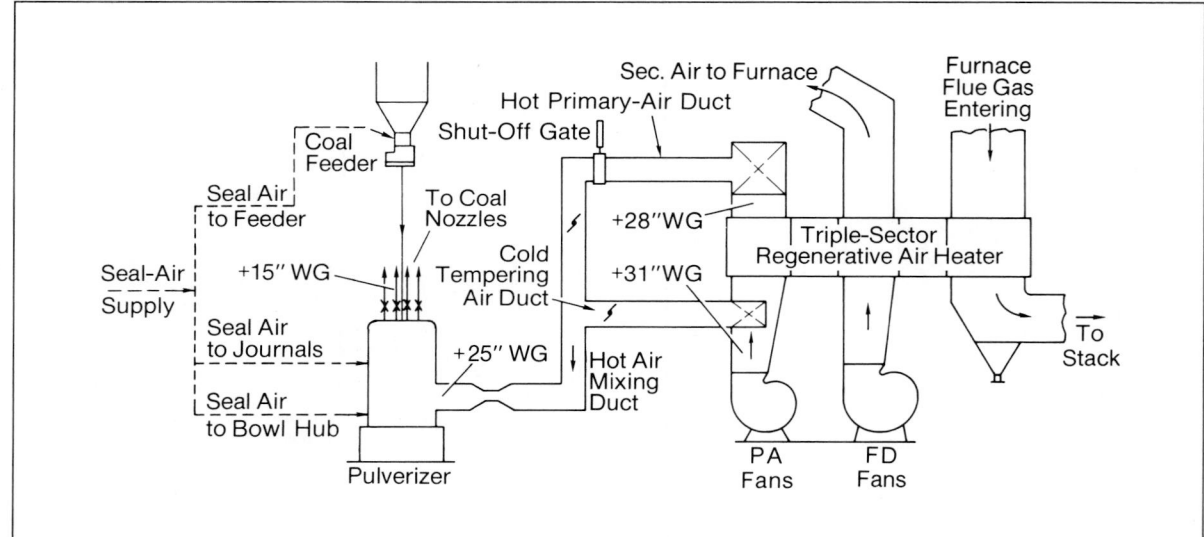

Fig. 14. Primary-air system of balanced-draft furnace. Air pressures shown are illustrative only.

SOURCE OF HEATED AIR

The best source of hot air for mill drying is either a regenerative or recuperative air heater using combustion gas as the source of heat. Those used in connection with large boiler installations usually provide sufficiently high temperature for almost any fuel moisture condition. On small installations, where the moisture in the coal is not high, steam air heaters may dry the coal. For higher moisture conditions a furnace-gas supplement may be necessary. Direct-fired air heaters, properly interlocked and protected, may also be used to supplement air-heating requirements.

All the moisture contained in coal must be evaporated before ignition can take place. For rapid ignition, therefore, surface moisture must be removed before the fuel is injected into the furnace. This same drying process facilitates pulverization. The type of fuel and its surface moisture govern mill-drying requirements. The drying capability of a given pulverizer

design depends on the extent of circulating load within the mill, the ability to rapidly mix the dry classifier returns with incoming raw, wet, coal feed, and the air weight and air temperature which the design will tolerate. C-E pulverizers are designed to operate satisfactorily with inlet air temperatures up to 750°F.

High-rank bituminous coals have a relatively dense structure and appear dry only when containing less than 2 or 3 percent total moisture. The inherent moisture of these fuels varies from 1 to 2 percent. Extremely low-rank fuels, such as lignite, are of a relatively porous or cellular structure and contain inherent moisture of from 15 to 35 percent. These same fuels with 3 percent surface moisture (18 to 38 percent total moisture) still appear dry. For both the pulverizing and ignition processes, it is necessary to reduce the total moisture contained in the fuel to the inherent moisture level.

If a particular pulverizer design requires low airflow for fineness maintenance, wet coal cannot be utilized without a considerable reduction in mill capacity. The design should permit high-temperature incoming air in sufficient volume to maintain a condition of relative humidity below saturation at the mill output temperature. A pulverizer designed for a larger volume of low-temperature inlet air for normal moisture fuel will require the admission of

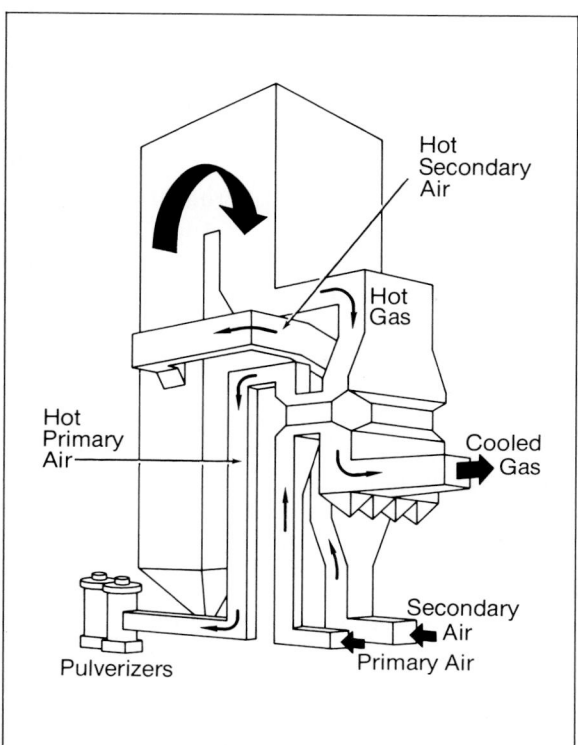

Fig. 15. Ljungstrom® trisector air heater

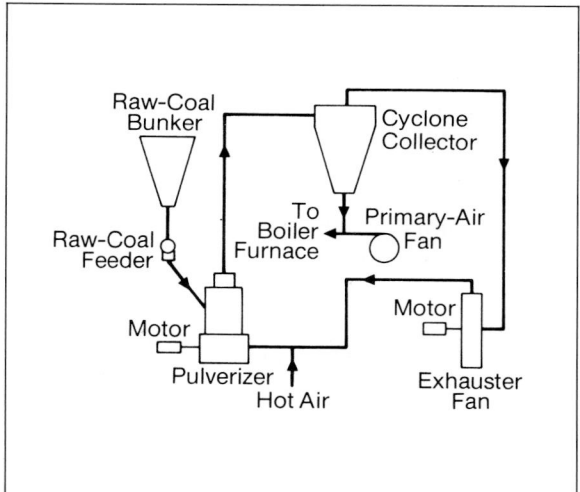

Fig. 16 Semidirect pulverizing system

Table VI. Allowable Mill Outlet Temperatures, °F

System	Storage	Direct	Semidirect
High-rank, high-volatile bituminous	130*	170	170
Low-rank, high-volatile bituminous	130*	150	150
High-rank, low-volatile bituminous	135*	180	180
Lignite	110	150	150
Anthracite	200		
Petroleum coke (delayed)	135	180–200	180–200
Petroleum coke (fluid)	200	200	200

*160°F permissible with inert atmosphere blanketing of storage bin and low-oxygen-concentration conveying medium.

large quantities of cold tempering air when grinding dry coals.

The type of fuel and the kind of system being used will determine the mill outlet temperature. As Table VI indicates, outlet temperatures for storage-systems mills are lower than for direct or semidirect firing, because most coals will not store safely at the temperatures used in direct firing. Storage-bin fires caused by spontaneous combustion of the fuel may result from inadequate mill outlet temperature control. These may be inhibited by maintaining an oxygen-deficient atmosphere, such as flue-gas inerting, over the bin coal level. Oxygen limits for various fuels are shown in NFPA 69, "Explosion Prevention Systems".

Mill capacity is based on coal input, and output on a given coal on a dry basis will decrease with increasing moisture content. Thus, in selecting a mill size, if too much reduction of capacity is experienced with high-moisture fuels, the mill can be too large under normal conditions of moisture or too small under excessive moisture conditions.

PULVERIZING AIR SYSTEMS

All coal-pulverizing systems utilize air or gas for drying, classification and transport purposes. Two methods are utilized for supplying the air requirements and overcoming system resistance. In a direct-firing system, one method utilizes a fan behind the pulverizer, while the other has a fan ahead of it. The former is a suction system and the fan handles coal-dust laden air, while the latter is a pressure system and the fan handles relatively clean air.

With direct firing, the exhauster or blower volume requirement will depend upon the pulverizer size, and is usually fixed by the base capacity of that pulverizer. The pressure or total head requirement is a function of the pulverizer and classifier resistance and the fuel distributing system and burner resistances. These resistances are in turn affected by the system design, the required fuel-line velocities and density of the mixture being conveyed.

In a storage system, the fan is located behind the dust collector and handles only a very small quantity of extremely fine (−200 mesh) dust at a relatively constant temperature (approximately 130°F). These fans are, therefore, designed for high efficiencies and need not be designed for a head higher than their operating temperature requires.

INDIRECT COAL-STORAGE PULVERIZING SYSTEMS

Initial attempts at utilizing pulverized coal as a utility fuel led to the development of the indirect-fired system (Fig. 12). In this system, a cyclone collector separates the coal from the air used in the pulverizer for drying, classifying, and conveying. The pulverized coal is conveyed either

mechanically or pneumatically to storage hoppers or bunkers. Mechanical, controllable feeders at the bunker outlets deliver the required quantity of coal to the fuel lines. At or near these feeders, the coal is reentrained in air called "primary air" in the proper proportions for transport to the furnace.

The indirect-fired pulverizing system is described in greater detail in the previous edition of this book.

DIRECT-FIRING ARRANGEMENTS

Most direct-fired pulverized-coal systems are for furnaces operated under suction (balanced-draft). Fig. 17 shows an arrangement of equipment for a suction-type mill when applied to a furnace of this design; Fig. 14 shows a pressure-type mill for a similar furnace.

With a suction mill, the coal feeder discharges against a negative pressure, whereas in the pressurized mill, the feeder discharges against a positive pressure of 18 to 21 in. WG. No coal feeder can act as a seal; thus, the head of coal above the feeder inlet must be utilized to prevent backflow of the primary air. As a rule of thumb, the height of the coal column above the feeder should equal 1 ft for every 3 in. of primary-air mill-inlet pressure. If the primary-air-fan inlet pressure is 30 in., then the coal-column height will be 10 ft.

There are three basic air conveyance systems used by C-E for direct firing: (1) the suction system in which an exhauster induces airflow through the pulverizer and discharges the coal-air mixture under pressure to the furnace; (2) the pressurized-exhauster system in which the pulverizer is pressurized by the forced-draft fan with both hot and ambient air and discharges the coal-air mixture through an exhauster, acting as a booster fan, to the furnace (this system is used only in conjunction with pressurized furnaces); and (3) the cold primary-air system in which a primary-air fan forces air through an air heater, ductwork, and pulverizer and then forces the coal-air mixture into the furnace.

SUCTION SYSTEM

The suction system has a number of advantages. It is quite easy to keep the area around the pulverizer clean. To control the airflow through the pulverizer a damper is placed in the constant-temperature coal-air mixture just prior to the exhauster entrance. Control of the coal-air mixture temperature is by a single hot-air damper and a barometric damper through which a flow of room air is induced by the suction in the pulverizer. With this control, the fan is designed for a constant, low-temperature mixture and has a low power consumption, even though such material handling fans have a relatively low efficiency of 55 to 60 percent.

The main disadvantage of the suction system is the maintenance required on the exhauster. On the other hand, by using proper design techniques and wear-resistant materials, the maintenance on an exhauster can be minimized. Exhauster maintenance costs are more than offset by the power and capital savings of the system; this justifies the continued use of the suction system on smaller units.

PRESSURIZED EXHAUSTER SYSTEM

To obtain sufficient pressure for firing a pressurized furnace, the pressurized exhauster

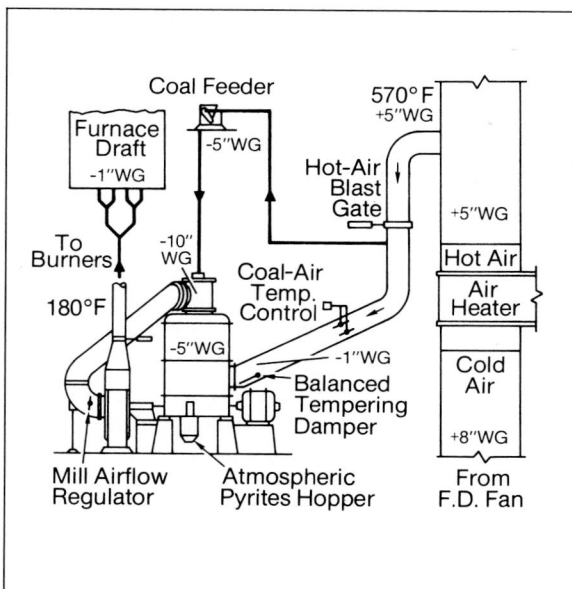

Fig. 17. Balanced-draft furnace with suction mill. Air pressures shown are illustrative only.

system was developed; there are a number of these systems in operation. This system retains the advantages of the suction system in the design of the fan for constant, low-temperature mixture and the relative ease of airflow control. Two dampers, one in the hot-air duct to the mill and one in the cold-air duct to the mill, control the amount of pulverizer airflow. This flow varies with, but is not proportional to, the amount of fuel being fed to the pulverizer. Biasing the hot- and cold-air dampers controls the temperature of the mixture leaving the pulverizer.

One advantage of the pressurized exhauster system is that the low pressures in the pulverizers do not present as severe a problem of sealing the head of coal over the raw-fuel feeder as with pulverizers under direct blower pressure. The disadvantage of the system, as with the suction system, is exhauster maintenance.

COLD PRIMARY-AIR SYSTEM

In this system the primary-air fan handles only ambient air. The fan is located ahead of the air preheater, with a separate primary-air system through the air heater. Although not as simple as the other two systems, its chief advantage is in fan power and maintenance. As the fans handle cold air, they can be smaller, run at higher speeds, and use highly efficient airfoil blade shapes. Inlet vanes can control airflow and further add to fan efficiency.

Some other advantages of this system are that, with high-efficiency fans handling ambient air, design for higher pressure differentials is possible and larger mills with longer fuel-pipe runs are practical. Thus, mills may be located farther from the boiler. Because individual fans for each pulverizer are not necessary, the space requirements for the pulverizer bays can be reduced. Morever, experience has proven that metering airflow on the inlet-air side of the pulverizer is most desirable. With the higher fan head available, the airflow can be quite easily measured by installing a venturi or other metering means.

Even with fewer pulverizer fans, controlling airflow to the various pulverizers is still relatively simple. The total primary air required is a function of the number of pulverizers in operation. This permits the use of a simple control for the airflow requirements. Because of possible variations of load and coal moisture content between pulverizers, it is necessary to control not only the total airflow but the temperature of the air to individual pulverizers. This is accomplished with a system similar to that outlined under the pressurized exhauster system, which uses a hot primary-air duct and a cold primary-air duct with a damper in each for each mill. The airflow requirement for a pulverizer is met by operating both dampers, while controlling temperature by properly proportioning the flow between the hot- and cold-air ducts.

When very wet coals are pulverized in the suction systems described, the exhauster supplies less air at the very time maximum airflow is required for maximum drying. In the suction and pressurized systems with the exhauster located between the pulverizer and the furnace, the high evaporated-water vapor content in the coal-air mixture will reduce the exhauster air-handling capacity. In the pressurized exhauster system only, this can be partially overcome if the forced-draft-fan head capability is adequate. These disadvantages can be corrected by over-designing the fans, but this produces inefficiencies when the unit is operating with normal moisture coal. Conversely, the cold primary-air system may produce a higher capacity at any time it is needed.

The cold primary-air system offers numerous advantages. As compared with the other systems, the total savings with this system, from elimination of exhauster maintenance and reduction in fan power, may be 35 to 40 percent of the cost per unit of coal pulverized. This operating saving is partially offset by the capital charges for additional ductwork, dampers, and controls. With larger units and pulverizers, the use of the cold primary-air system becomes economically favorable.

AIR HEATING FOR THE COLD PRIMARY-AIR SYSTEM

By definition, the cold primary-air system requires independent primary-air heating accomplished in the past by several different

methods. Several systems have been installed using a separate primary tubular air heater with Ljungstrom® air heaters for secondary air, while some have been installed using separate primary Ljungstrom® air heaters. These systems are expensive and become rather complicated in ductwork design. With the cold primary-air system, the trisector Ljungstrom® air heater shown schematically in Fig. 15 is now being used as a basic design. As the name implies, the flow channels of this heater have been divided into three sections, with the primary-air section being located between the secondary-air and flue-gas sections. With this arrangement, primary-air temperatures higher than that of secondary-air can be obtained. Further, the efficiency of heat recovery is not significantly affected by variations in pulverizer hot-air requirements, because the secondary air recovers that heat which is not recovered by the primary air sections.

PRINCIPAL TYPES OF PULVERIZERS

To effect the particle size reduction needed for proper combustion in pulverized-coal firing, machines known as pulverizers or mills are used to grind or comminute the fuel. Grinding mills use either one, two, or all three of the basic principles of particle size reduction, namely, impact, attrition, and crushing. With respect to speed, these machines may be classified as low, medium and high. The four most commonly used pulverizers are the ball tube, the ring-roll or ball-race, the impact or hammer mill, and the attrition type; their speed characteristics are shown in Table VII.

BALL-TUBE MILLS

A ball-tube mill (Fig. 18) is basically a hollow horizontal cylinder, rotated on its axis,

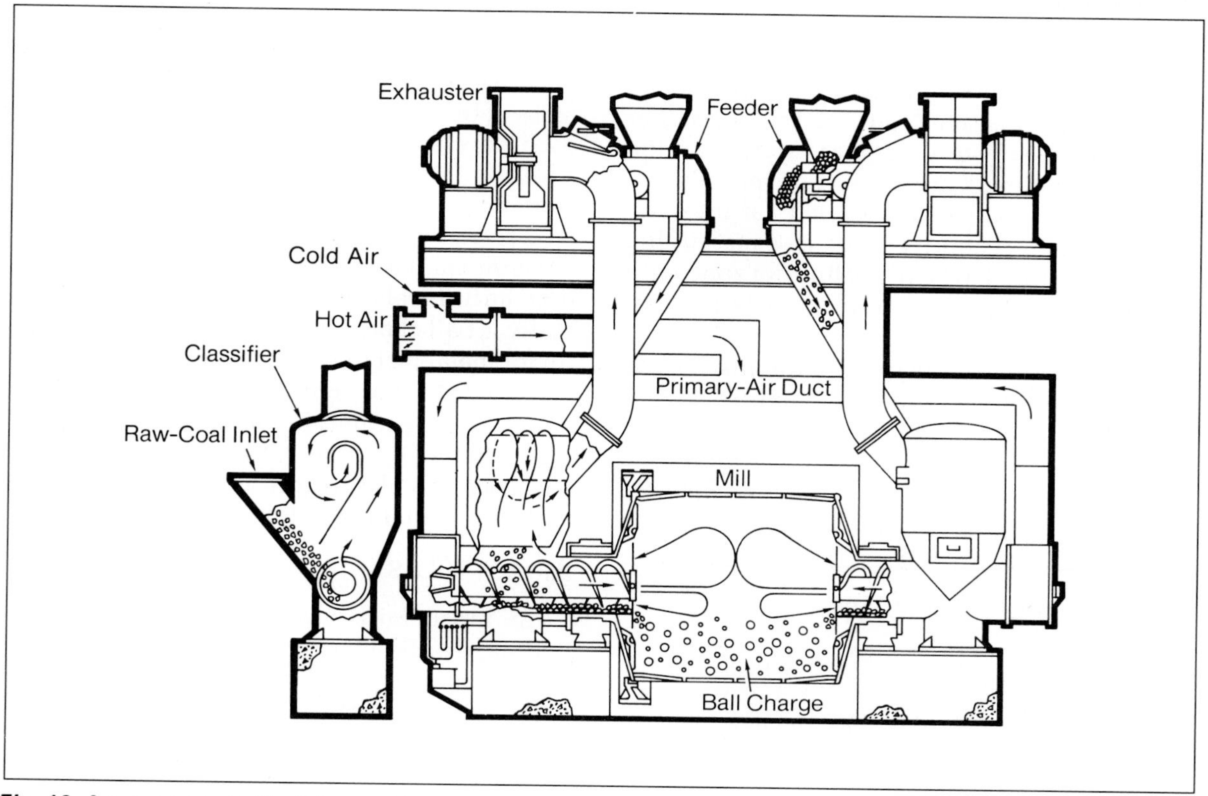

Fig. 18. **Arrangement of ball-tube mill**

whose length is slightly less to somewhat greater than its diameter. Heavy cast, wear-resistant liners fit the inside of the cylindrical shell which is filled to a little less than half with forged steel or cast alloy balls varying from 1 to 4 in. in diameter. Rotating slowly, 18 to 35 rpm (about 20 rpm for an 8 ft diameter mill) the balls are carried about two thirds of the way up the periphery and then continually cascaded towards the center of the cylinder.

Coal is fed into the cylinder through hollow trunnions and intermingles with the ball charge. Pulverization which is accomplished through continual cascading of the mixtures results from (a) impact of the falling balls on the coal, (b) attrition as particles slide over each other as well as over the liners and (c) crushing as balls roll over each other and over the liners with coal particles between them. Larger pieces of coal are broken by impact and the fine grinding is done by attrition and crushing as the balls roll and slide within the charge.

Hot airflow is passed through the mill to dry the coal and remove the fines from the pulverizing zone. In most designs used for firing boilers or industrial furnaces, an external classifier regulates the size or degree of fineness of the finished product. The oversize or rejects from the classifier, sometimes called tailings, already dried in the pulverizing and classifying process, are returned to the grinding zone with the raw coal. This type pulverizer is particularly susceptible to reductions in capacity from surface moisture of the coal. Reducing the average moisture content of the mixture is very important in maintaining a continuous flow of coal through the feed end. The recirculation of dried tailings, then, reduces the tendency for wet coal to plug the feed end.

Table VII. **Pulverizer Types**

Speed:	Low	Medium	High
Type:	Ball-Tube Mill	Ring-Roll or Ball-Race Mill	Impact or Hammer Mill Attrition Mill

During operation, the relatively large quantity of pulverized coal in the grinding zone of a ball-tube mill acts as a storage reservoir from which sudden increases in fuel demand are supplied. The power consumption of ball-tube mills, kW per unit of coal pulverized, is very high, particularly at partial loads. Relatively large physically per unit of capacity, they require considerable floor space. Because of their size and weight, the initial capital cost is quite high. The presence of a high circulating load within the mill results in an overproduction of fines within the mill charge. Using an adequate external classifier permits the removal of the fine product from the grinding zone and reduces the production of extreme fines. The comparatively poor mixing of heated air with the partially pulverized material reduces the drying efficiency of this type of mill. High-moisture coals produce a large reduction in mill capacity.

Ball-tube mills are not well suited to intermittent operation as the large amount of heat stored in the coal and ball charge may produce overheating and fires when the mill is idle. The mass of this mill type makes it necessary to use high-power, high-starting-torque motors. In addition, these mills are very noisy. Most installations require that insulated "dog houses" be erected over each mill for noise attenuation.

Maintenance in the grinding zone of ball-tube mills is relatively easy to perform. Periodically, a ball charge is added to the mill to make up for metal lost in the grinding process. It may take years to wear out the cast liners, but considerable downtime is necessary for their replacement. Over the long run, maintenance costs per unit of coal ground are about the same as for ring-roll type pulverizers.

IMPACT MILLS

An impact mill consists primarily of a series of hinged or fixed hammers revolving in an enclosed chamber lined with cast wear-resistant plates. Grinding results from a combination of hammer impact on the larger particles and attrition of the smaller particles on each other and across the grinding surfaces. An air system

with the fan mounted either internally or externally on the main shaft induces a flow through the mill. An internal or external type of classifier may be used. (See Figs. 24 and 25)

This class of mill is simple and compact, low in cost and may be built in very small sizes. Its ability to handle high inlet-air temperatures, plus the return of dried classified rejects to the incoming raw feed, makes it an excellent dryer.

However, the high-speed design results in high maintenance and high power consumption when grinding fine. Progressive wear on the grinding elements produces a rapid drop-off in product fineness, and it is difficult if not impossible to maintain fineness over the life of the wearing parts. Using an external classifier permits maintenance of fineness, but only at the expense of a considerable reduction in capacity as parts wear. The maximum capacity for which such mills can be built is lower than most other types.

ATTRITION MILLS

No true attrition mill is used for coal pulverizing because of the high rate of wear on parts. A high-speed mill which uses considerable attrition grinding along with impact grinding is, however, used for direct firing of pulverized coal. In this mill, the grinding elements consist of pegs and lugs mounted on a disc rotating in a chamber; the periphery of the chamber is lined with wear-resistant plates and its walls contain fixed rows of lugs within which the rotating lugs mesh. The fan rotor is mounted on the pulverizer shaft. Instead of an external classifier, a simple shaft-mounted rejector type is used. This design utilizes wear-resistant alloy lug and peg facings and casing linings to reduce the wear effect on fineness and extend the periods between parts replacement. This mill type exhibits all the characteristics of the impact mills.

RING-ROLL AND BALL-RACE MILLS

Ring-roll and ball-race mills comprise the largest number of pulverizers used for coal grinding. They are of medium speed and utilize primarily crushing and attrition of particle plus a very small amount of impact to obtain size reduction of the coal. The grinding action takes place between two surfaces, one rolling over the other. The rolling element may be either a ball or a roll, while the member over which it rolls may be either a race or a ring. The ball diameter is generally from 20 to 35 percent of the race diameter, which can be as large as 100″. If the element is a roll, as in the C-E bowl mill, its diameter may be from 50 to 60 percent of the ring diameter, which can be as much as 110″. Its face width, depending upon mill size, will vary from 15 to 20 percent of ring diameter.

When the rolling elements are balls, Fig. 19, they are confined between races. In the majority of designs, the lower race is the driven rotary member, while the upper race is stationary. Some designs also utilize a rotating upper race. The required grinding pressure is obtained by forcing the races together with either heavy springs or pneumatic or hydraulic cylinders. Some additional grinding pressure is obtained from centrifugal force of the rotating balls.

There are two general classes of mill that use rollers as the rolling elements. In one, Fig. 20, the roller assemblies are driven and the ring is stationary, while in another, Fig. 21, the roller assembly is fixed and the ring rotates.

Perhaps the most frequently used application of the first class is the C-E Raymond® roller mill, used in storage systems. Grinding pressure is obtained from centrifugal force resulting from rotation of these rolling elements.

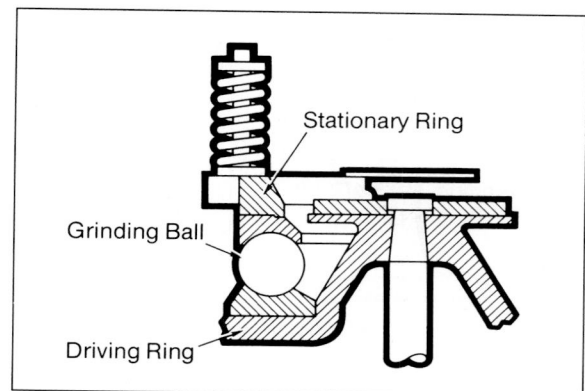

Fig. 19. Diagram of ball-race mill

The other classification of ring-roll mills, namely those in which the ring rotates, constitutes the largest number of mills used for grinding coal; they are manufactured by the major boiler companies. The C-E bowl mill is of this latter type. The vertical driving shaft of these mills operate between 20–70 rpm, with larger mills running at the slower speeds.

Generally, such mills are equipped with self-contained or integral classifiers to regulate the fineness of the finished product. In some cases, this device may be external to the mill itself and is then termed an *instream classifier*.

Primary-air fans or exhausters create a flow of heated air through the mills. This heated air dries the coal, removes it from the grinding zone, carries it through the classifying zone, and conveys it to its point of use, whether this is a direct-fired furnace or the dust collector of a storage system.

When this type of mill is provided with sufficient air at a temperature to produce a satisfactory mill outlet temperature, it can handle very wet coals with only a small reduction in capacity. The high ratio of circulating load (classifier rejects returning to the grinding zone for further size reduction) to output, with the resulting rapid reduction of average moisture content, facilitates the grinding process. These mills require less power than any others.

Physically, these mills are compact and occupy a relatively small amount of floor space per unit of capacity. Some designs of ring-roll mills are extremely quiet in operation. Fineness of the product is relatively uniform throughout the life of the grinding elements.

Several references have been made to the ability of the various types of pulverizing equipment to grind different coals economically and efficiently. This capability is a reflection of the grinding pressures available, the method of application of this force, speed of moving elements, abrasion, and power and size limitations of the particular units. A list of materials capable of being commercially pulverized by the mill types described could be extended indefinitely; therefore, the scope of Table VIII covers combustible materials of primary concern in power generation.

C-E PULVERIZER DESIGN

In the following sections, C-E pulverizers corresponding to the generic types are described. Their detailed design, construction and operational features, and fields of application are given.

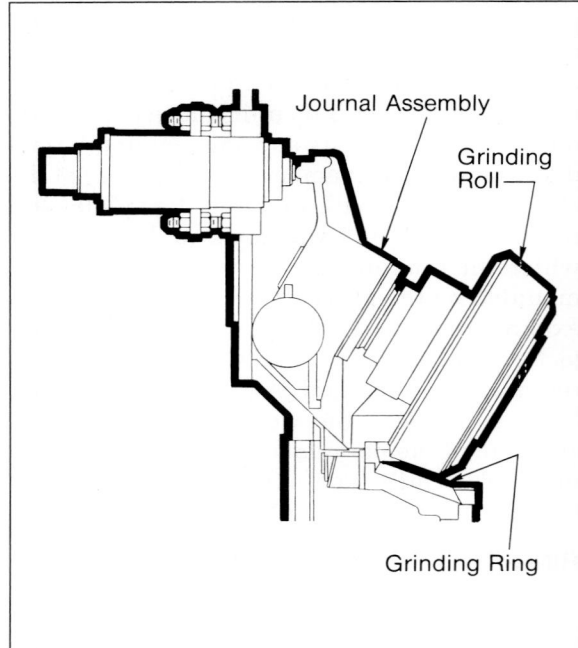

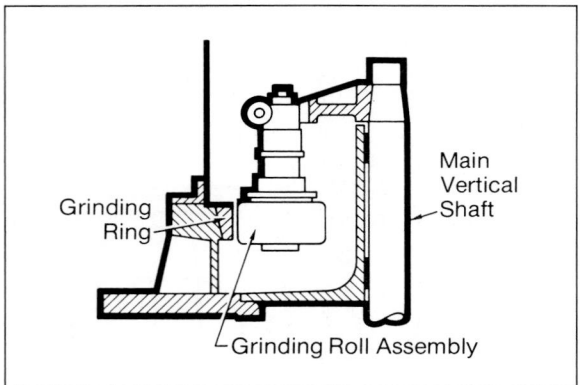

Fig. 20. Diagram of roller-type ring-roll mill journal assembly

Fig. 21 Diagram of bowl-type ring-roll pulverizer journal assembly

C-E RAYMOND ROLLER MILL

The period between 1895, when the first pulverizers were installed in a Pennsylvania cement plant, and 1919, when a successful boiler installation was made at Oneida Street Station in Milwaukee, was a testing time for this type of equipment. Early during this period the C-E Raymond roller mill, Fig. 22, was developed to pulverize coal for storage systems serving metallurgical furnaces and cement kilns. By 1919, several hundred of these mills were being operated successfully. When, after World War I, pulverized coal gained a foothold in the central-station industry, this machine was a fully developed and standardized pulverizer specified for many pioneering installations.

The early mills were built for a maximum capacity of 6 tons per hour of 55 grindability coal when grinding to a fineness of 65 percent −200 mesh. By 1930, the maximum capability of this mill design had been increased to 25 tons per hour. Nearly all used external dryers to predry

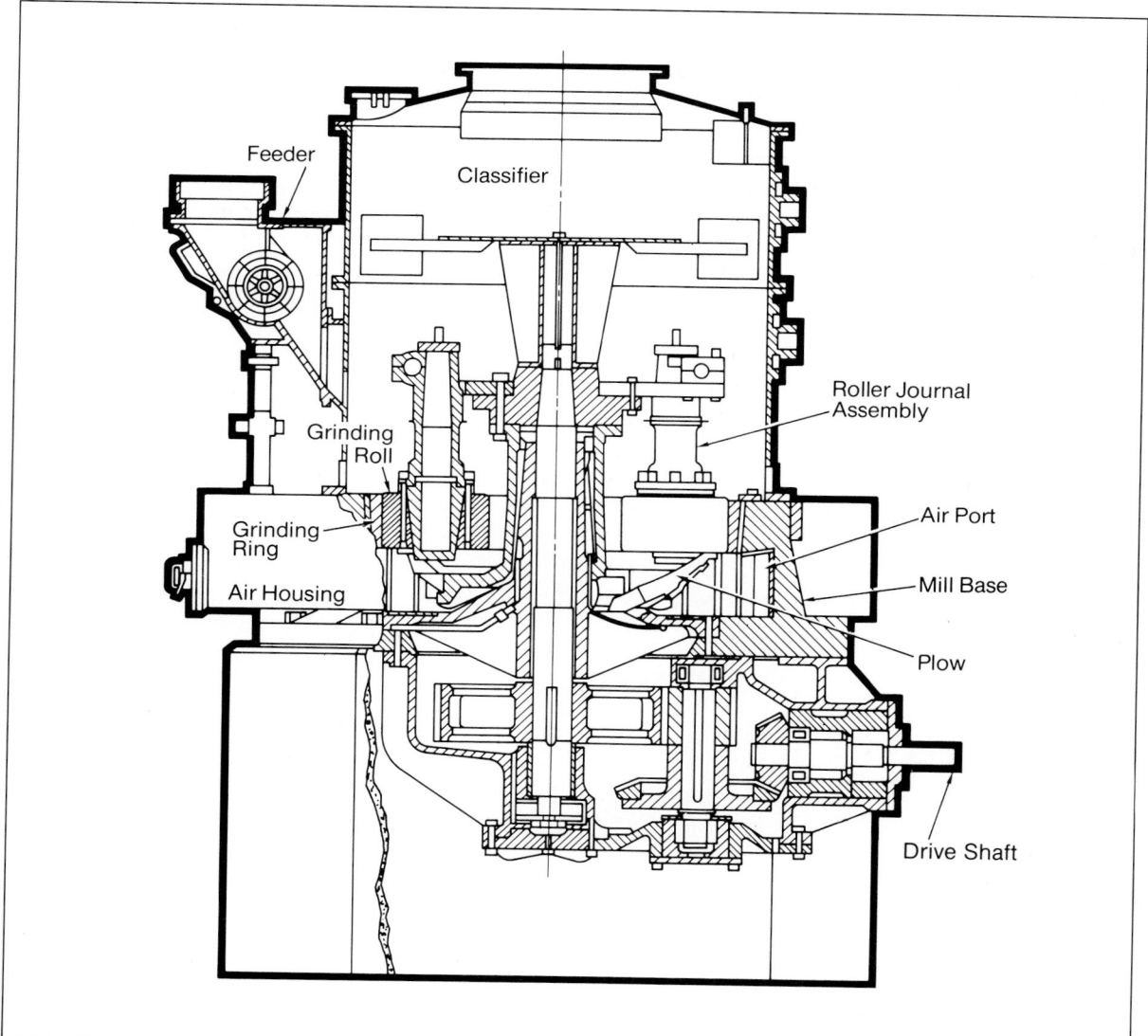

Fig. 22 Sectional view, C-E Raymond roller mill

the coal before use in the pulverizer. The economic disadvantages of these dryers helped to bring about internal mill drying, now the accepted practice. However, the original feature of air-separation and classification inherent in the roller mill has become the accepted standard of ring-roll pulverizer design.

A typical grinding and drying system incorporating a roller mill is shown in Fig. 23.

C-E RAYMOND IMPACT MILLS

The increase in possibilities for direct pulverized-coal firing resulted in the development in 1923 of a series of C-E Raymond impact mills with two basic designs: one for the larger capacities (Fig. 24), utilizing an integral fan but external classifier. The other design (Fig. 25) was for smaller outputs, in which the grinding elements, classifying means, and fan were all mounted on a common shaft rotating at 1800 rpm. Both designs were provided with a mill-housing clearance space or tramp-iron pocket for accumulation and periodic removal of foreign material. The whizzer or rejector blades for fineness control were first used on these mills. These mills are excellent drying pulverizers because high-temperature air can be utilized and there is very violent turbulence of the mixture passing through the pulverizer. They are best suited to soft, relatively nonabra-

Table VIII. Types of Pulverizers for Various Materials

Type of Material	Ball-Tube	Impact and Attrition	Ball Race	Ring Roll
Low-volatile anthracite	x	. . .	. . .	. . .
High-volatile anthracite	x	. . .	x	x
Coke breeze	x	. . .	. . .	. . .
Petroleum coke (fluid)	x	. . .	x	x
Petroleum coke (delayed)	x	x	x	x
Graphite	x	. . .	x	x
Low-volatile bituminous coal	x	x	x	x
Med-volatile bituminous coal	x	x	x	x
High-volatile A bituminous coal	x	x	x	x
High-volatile B bituminous coal	x	x	x	x
High-volatile C bituminous coal	x	. . .	x	x
Subbituminous A coal	x	. . .	x	x
Subbituminous B coal	x	. . .	x	x
Subbituminous C coal	. . .	. . .	x	x
Lignite	. . .	. . .	x	x
Lignite and coal char	x	. . .	x	x
Brown coal	. . .	x	. . .	. . .
Furfural residue	. . .	x	. . .	x
Sulfur	. . .	x	. . .	x
Gypsum	. . .	x	x	x
Phosphate rock	x	. . .	x	x
Limestone	x	. . .	. . .	x
Rice hulls	. . .	x	. . .	. . .
Grains	. . .	x	. . .	. . .
Ores—hard	x	. . .	. . .	. . .
Ores—soft	x	. . .	x	x

Fig. 23 Equipment for a complete grinding and drying system for C-E roller mill

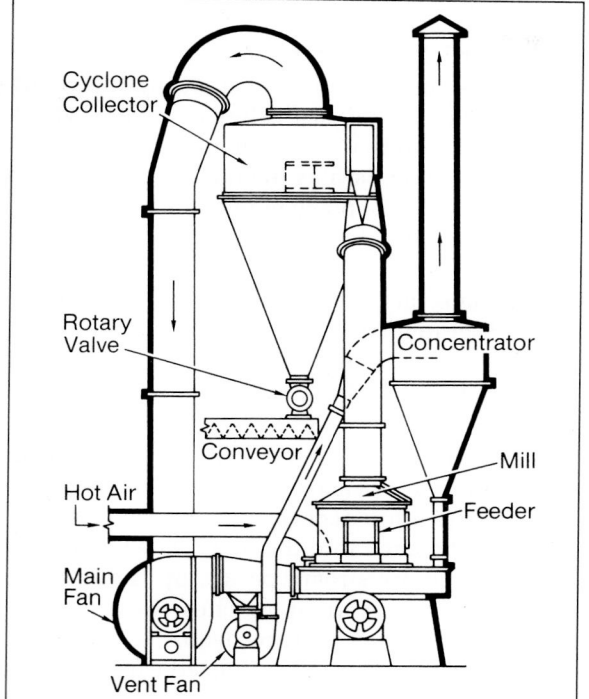

Fig. 24 C-E Raymond impact mill with external classifier and exhauster integrally mounted

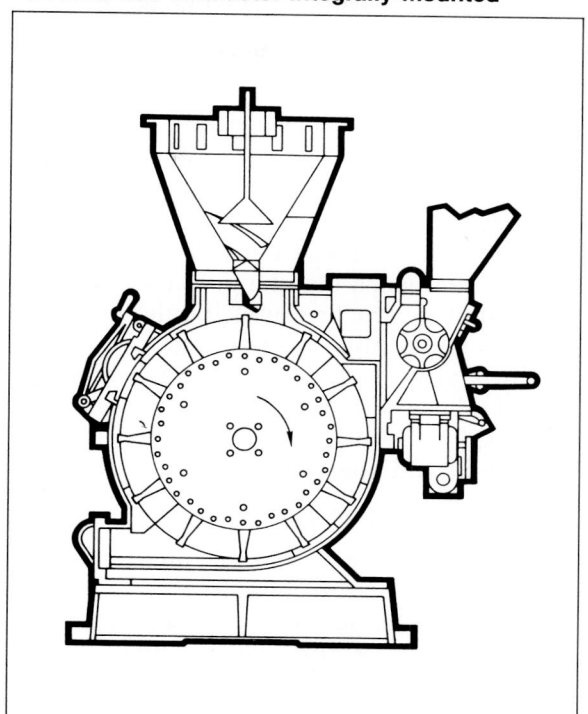

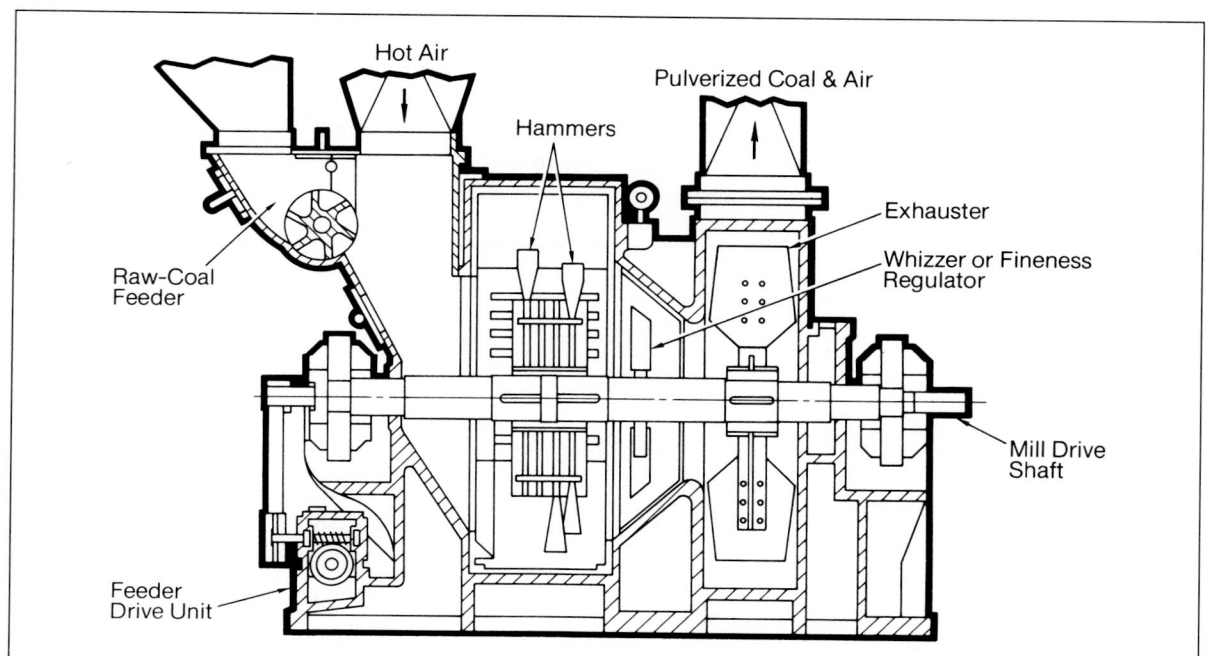

Fig. 25 C-E Raymond impact mill with hammers, classifier, and fan on common shaft

sive coals and for use on small boilers and furnaces. Life of grinding elements is short; maintenance and mill power are quite high.

C-E BOWL MILLS

The positive-pressure type C-E HP ring-roll bowl mill is shown in Fig. 26. It is a medium-speed type, similar in size and performance to the prior C-E "R" series mills. When fitted with an exhauster (such that the pulverizer operates below atmospheric pressure), it is designated an HPS or HPPS mill, a cross-sectional view of which is given in Fig. 27. Table IX shows the type of mill used with different furnaces and air systems, for the two pressure regimes in which these mills can operate.

OPERATION OF THE HP PULVERIZER

When in operation, raw coal enters the center of the pulverizer through a center feed pipe. It falls onto a rotating bowl which has a replaceable wear surface composed of bull-ring segments. Centrifugal force causes the coal to move outward from the center and under the three journal assemblies, where it is crushed by large rolls. To prevent physical contact of the rolls and bull-ring when the mill is run without coal, a stop limits the downward movement of the journal assembly. The force to pulverize the coal is applied to the journal assembly by an external spring. As the journal rotates about its trunnion in response to increasing coal feed, the spring is compressed and the force for grinding is increased.

The partially pulverized coal passes over the rim of the bowl and is entrained by the rising hot-air stream and is flash-dried. The pyrites and tramp iron that enter the mill with the coal follow the same path as the coal until they pass over the rim of the bowl; being denser than coal, they cannot be carried upward by the air stream and fall into the millside. Once there, these rejected materials are swept around by a set of pivoted scrapers until they reach the tramp-iron opening. They then fall into a hopper external to the mill; this rejects hopper can be emptied with the mill in service.

The air-transported partially pulverized coal enters the vane-wheel assembly, where initial size classification occurs, with the heaviest particles falling back into the bowl. The balance of the coal and air stream passes up through the separator body until it reaches the classifier. Here, the coal-air mixture begins to spin in a cyclonic path. Externally adjusted vanes control the amount of spin. Because of the differing mass of the particles and the amount of spin, the oversize particles fall into the cone and slide downward until they mix with the incoming raw coal. In this way, only the desired size of coal leaves the pulverizer. The HP mill has a venturi where the flow is split into four equal streams before exiting through the discharge valves. With

Table IX. C-E HP Pulverizer Application

Mill Type	Furnace Type	Mill Pressure	Air System	Capacity*
HP	Balanced Draft	Positive	Primary Air Fan	16,200 to 200,000 lb/hr
HP	Pressurized	Positive	Primary Air Fan	16,200 to 200,000 lb/hr
HPS	Balanced Draft	Negative	Exhauster Fan	16,200 to 87,200 lb/hr
HPPS	Pressurized	Positive	Exhauster Fan	16,200 to 87,200 lb/hr

*Pulverizer capacitites are based on a 55-grindability coal pulverized to 70 percent through a 200 mesh sieve, having a 12 percent moisture content with low-rank bituminous coals or 8 percent with high-rank bituminous coals.

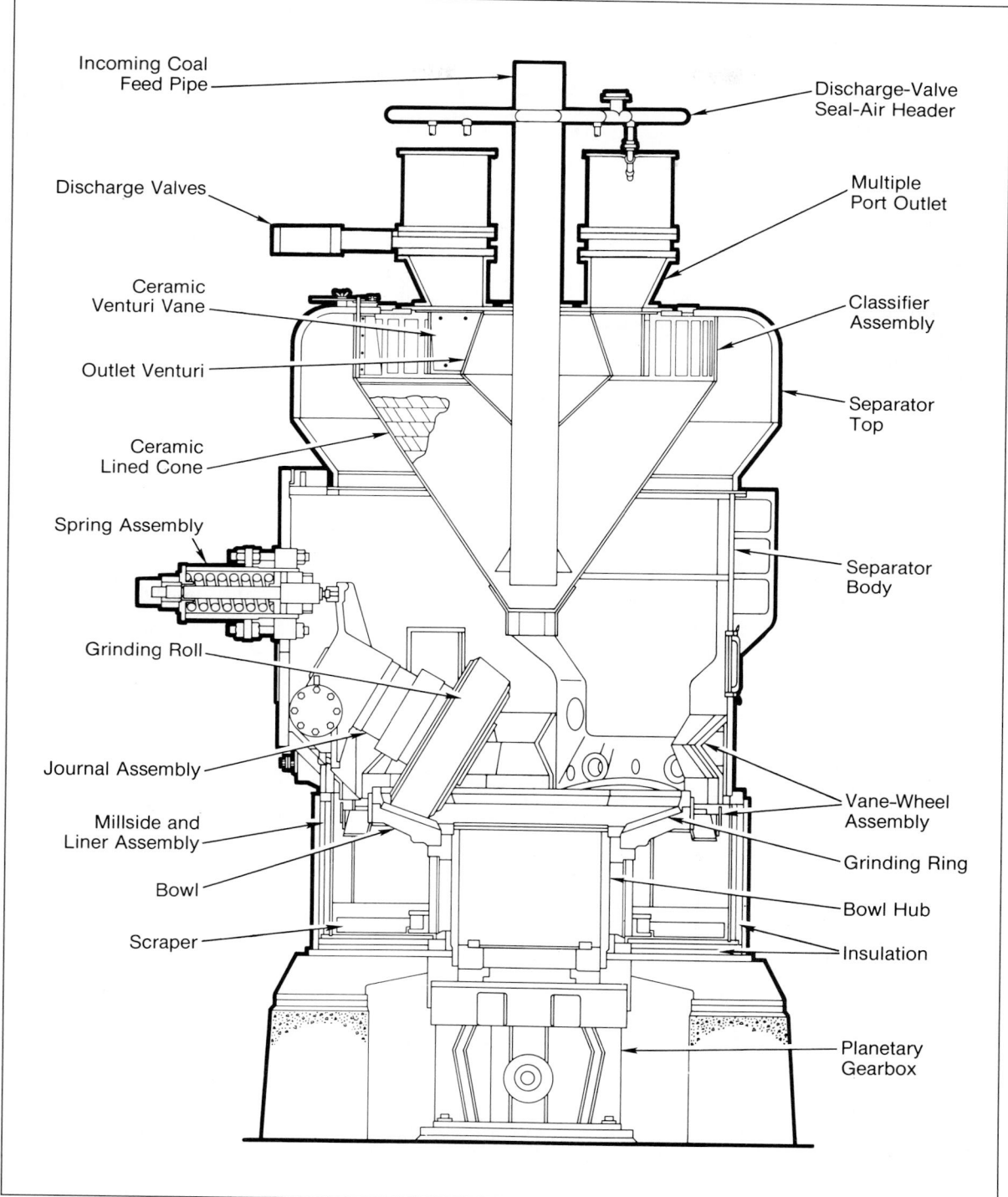

Fig. 26 C-E HP ring-roll bowl mill for positive-pressure operation

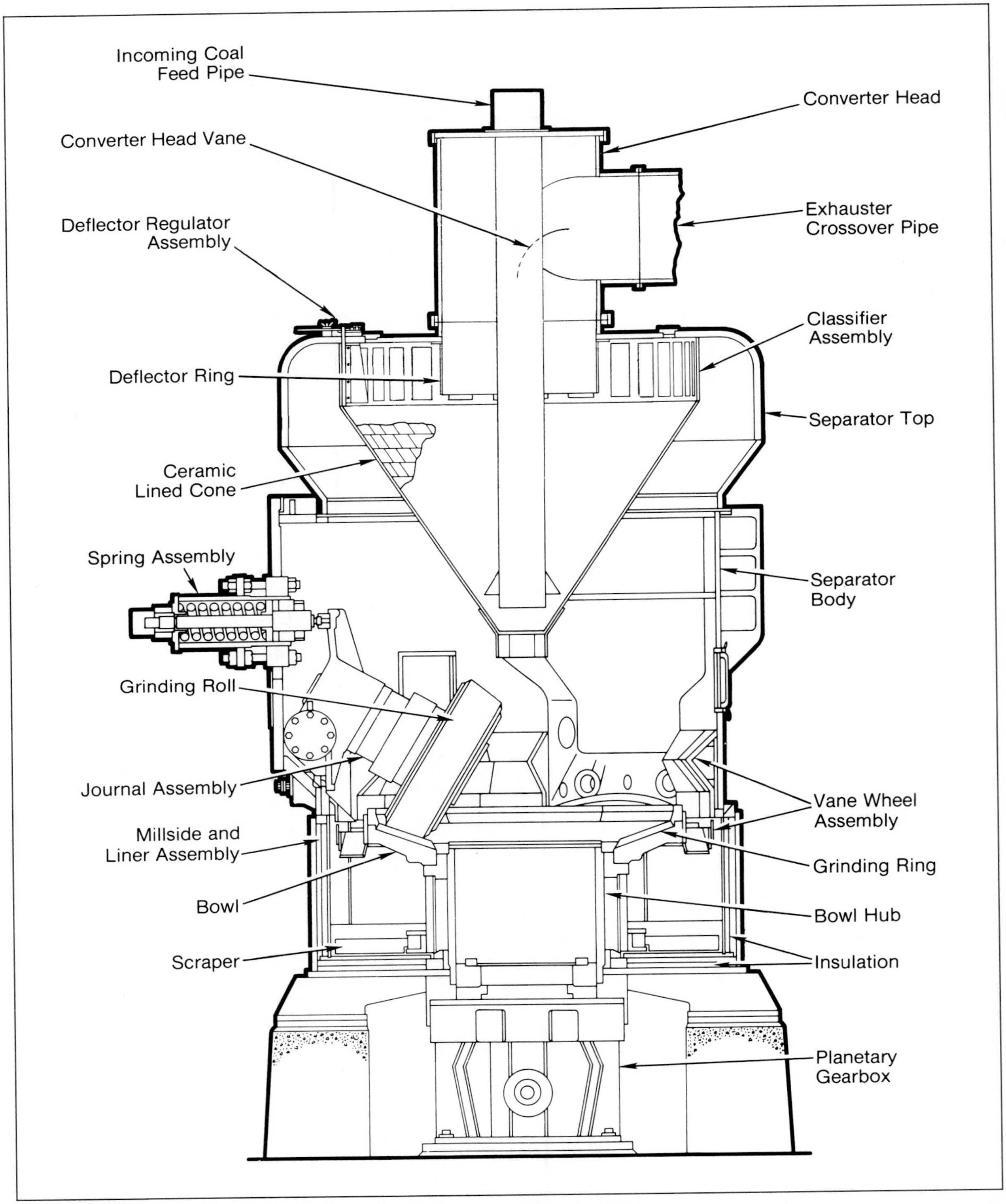

Incoming Coal Feed Pipe

Converter Head

Converter Head Vane

Exhauster Crossover Pipe

Deflector Regulator Assembly

Classifier Assembly

Deflector Ring

Separator Top

Ceramic Lined Cone

Spring Assembly

Separator Body

Grinding Roll

Journal Assembly

Vane Wheel Assembly

Millside and Liner Assembly

Grinding Ring

Bowl

Bowl Hub

Scraper

Insulation

Planetary Gearbox

Fig. 27 **C-E HPS or HPPS ring-roll bowl mill for use with exhauster**

an exhauster mill, there is only one discharge pipe, which conducts the pulverized product to the exhauster. The fineness of the pulverized product leaving this type of pulverizer is affected by the grindability of the coal, the amount of wear that has taken place on the grinding rolls and ring, the air flow and temperature, and the grinding force—both the spring rate and the initial spring compression.

PULVERIZER SHELL DESIGN REQUIREMENTS

All pulverizing equipment containing coal-dust laden air is designed and built according to the recommendations of the National Fire Protection Association (NFPA), their Standard NFPA 85F: Installation and Operation of Pulverized Coal Systems. The pressure-containing shell of the pulverizer, the exhauster, and the rejects hopper are designed to contain a 50-psig pressure. This requirement is independent of any lower design or operating pressure in the equipment.

PULVERIZER FOUNDATION DESIGN

The HP pulverizer exerts three types of loads on its concrete foundation. They are: the static weight of the machine itself; dynamic loads that are the result of the grinding process; and thermal loads from the heating of the pulverizer by the hot primary air, which results in expansion forces on the foundation. The engineer who designs the foundation must take these loads into account in both anchor-bolt and concrete-reinforcement sizing and placement. The mill and its drive motor are both mounted on the same foundation so there is no relative vibration or settling between the two, which could affect component alignment.

REMOVABLE GEAR DRIVE

A distinctive feature of the C-E HP pulverizer is its removable planetary-gear drive (Fig. 28). This drive is lighter in weight yet stronger than similar capacity worm-gear or triple-reduction spiral-bevel helical gear drives. The HP gear unit is independent of the mill housing structure and can be removed for inspection or maintenance, Fig 29. Its size and weight make it practical to move it to a maintenance area away from the pulverizer bay. Mill outage time will be mini-

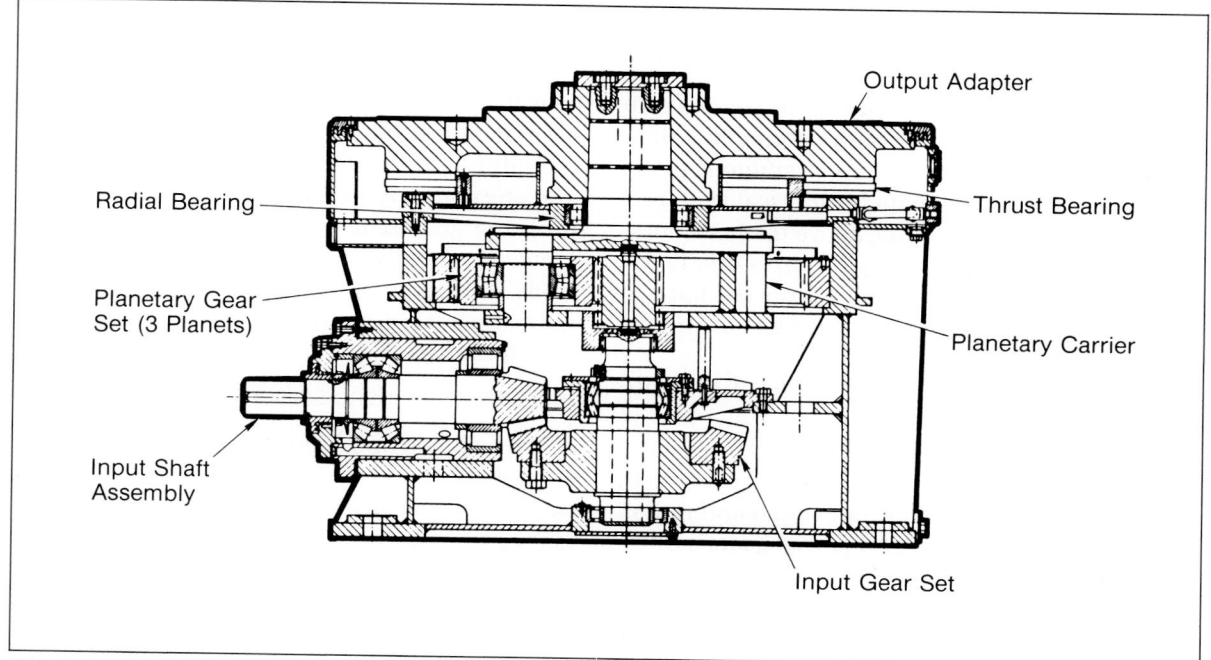

Radial Bearing

Planetary Gear Set (3 Planets)

Input Shaft Assembly

Output Adapter

Thrust Bearing

Planetary Carrier

Input Gear Set

Fig. 28 **Removable planetary-gear drive of C-E HP pulverizer**

Fig. 29 **Shop assembly of C-E HP pulverizer and planetary-gear drive**

mal if a spare gear unit is available at the site.

Since the gear unit does not penetrate the mill housing, it is not exposed to the pulverized coal entrained in the primary air. The input-shaft and output-table seals are of the non-contacting labyrinth type operating at local ambient pressure. The mill-housing penetration seal is on the grinding-bowl support hub, above the gear unit. Any heat load on the gearing is substantially reduced by its physical separation from the hot-air inlet, allowing the gears and bearings to run as cool as possible.

Internally, the gear unit consists of a right-angle spiral-bevel reduction input stage and a planetary reduction output stage. The sun gear is connected to the bevel-gear shaft by a crowned gear-type coupling to allow both axial and radial movement. Floating sun and planet gears assure equalized loads on the meshing teeth; this acts to distribute the total horsepower equally among the three planets. Input-gear, bevel-gear, and planet-gear shaft bearings are all designed for Anti-Friction Bearing Manufacturers Association (AFBMA) B-10 life of 100,000 hours minimum.

Mill grinding forces are carried by a hydrodynamic tilting-pad thrust bearing assembly located above the planetary stage. There are multiple bearing pads, four of which have dual-element sensors to measure pad temperature. The pad temperature is interlocked with mill operation to prevent damage to the thrust bearing. The thrust-bearing pad pitch-circle diameter is the same as the gear-housing outer structural wall, to maximize thrust-bearing support and to transfer grinding forces directly to the foundation without affecting gear meshes.

LUBRICATION SYSTEM

An external lubrication system supplies cooled and filtered oil to the roller bearings, the gear meshes, and the hydrodynamic thrust-bearing pads. All major components of the system are shop-mounted on an oil reservoir tank. An isolation device allows maintenance of the system skid assembly without draining the gear-unit hydrodynamic bearing reservoir, which is at a higher elevation. A duplex filter assembly provides for maintenance of a standby filter while the pulverizer is operating.

After filtering, the oil passes through an oil-to-water shell-and-tube heat exchanger. Electrical temperature and pressure sensors monitor the lubrication system; mill starting, operation, and stopping are all interlocked to prevent running the gear unit without proper lubrication. Low watt-density oil heaters installed in thermowells in the reservoir bring the oil up to minimum operating temperature before mill energization.

MILLSIDE AND LINER ASSEMBLY

The millside and liner assembly (Figs. 26 and 27) is a weldment which supports the upper section of the pulverizer and receives the hot primary air from the combustion-air preheater. It is mounted on sole plates and is secured to the foundation by anchor bolts. Since the hot primary air could produce thermal stresses in the millside and foundation, the millside is internally insulated with ceramic-fiber blankets. Replaceable liners and wear plates cover the insulation.

A tramp-iron spout in the floor of the millside conducts rejected material to the external rejects-conveying system. Doors in the millside give access to the underside of the bowl and provide for a ventilating airflow when maintenance is being done.

BOWL, BOWL HUB, AND VANE-WHEEL ASSEMBLY

Figs. 26 and 27 show these components, located above the gearbox. The bowl and bowl hub carry the skirts, the scrapers, the vane-wheel segments, and the bull ring segments.

The bull ring is subject to abrasive wear as a result of the grinding that takes place on it; replacement is necessary when the wear becomes excessive. C-E furnishes segmented bull rings for ease of handling, and uses high-chrome white iron for the segments for most applications.

The vane-wheel segments are made from heat-treated abrasion-resistant steel plate. Removable air restriction rings provide for adjustment of the pressure differential across the bowl. These rings, welded to the vane-wheel segments, have a weld overlay to protect them from wear.

The pyrites (mill-rejects) hopper is normally opposite the mill drive motor. In this case either the motor or the hopper must be removed to slide the gear unit out. If the hopper is placed to the side of the mill, then the gear box is removed away from the motor. The standard HP mill setting height permits mounting of the hopper and the pyrites handling system without pits or entrenchment.

SEPARATOR BODY AND VANE-WHEEL ASSEMBLY

As described above, initial coal-particle classification is done in the grinding zone by the bowl-mounted rotating vane wheel and the housing-mounted stationary coal/air deflectors. The vane wheel, the primary classifier on HP mills, promotes uniform distribution of the coal and carrier air while it also lessens erosion of mill internals by the coal/air stream. Large coal particles return immediately to the bowl for regrinding before entering the main classifier above.

The rotating part of the vane wheel is con-structed of abrasion-resistant plate, segmented for convenience in assembly and maintenance. Removable liners control the upward airflow past the bowl rim, to adjust pressure drop across the grinding zone and optimize transport of the coal/air mixture to the final classifier.

HP PULVERIZER JOURNAL ASSEMBLY

The roll-assembly bearing system consists of two identical tapered roller bearings in an opposed arrangement (Fig. 30). The system is designed for an AFBMA B-10 life of 100,000 hours minimum, under a severe duty cycle. An oil bath lubricates the grinding-roll bearings. To prevent contamination of the oil, seal air flows through internal roll-assembly ports, then outward through a roll air seal. No parts of the seal-air supply system are exposed to the coal/primary-air stream.

HP mills have a unique grinding-roll tilt-out feature, as depicted in Fig. 30. Each roll assembly can be rotated out of the mill on its trunnion mounting shaft, using a tilting fixture and an overhead mill hoist or crane, as shown in Fig. 31.

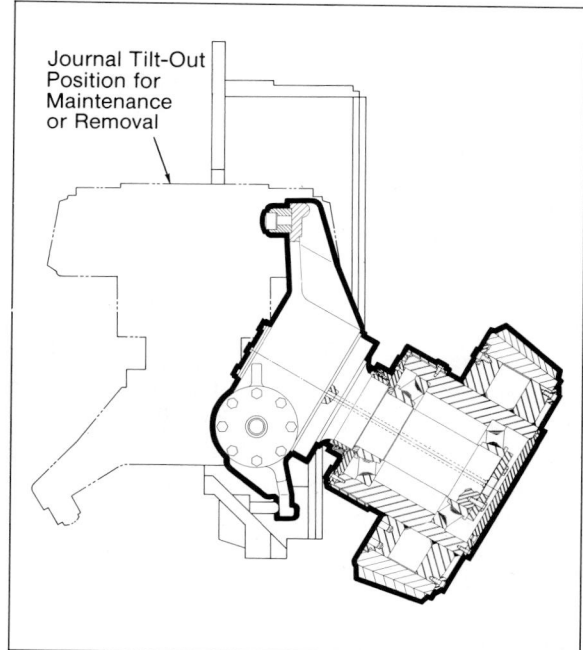

Journal Tilt-Out Position for Maintenance or Removal

Fig. 30 **C-E HP pulverizer journal assembly with tilt-out feature**

Fig. 31 **HP roll assembly tilted-out of pulverizer**

With the roll locked in this vertical service-access position, several inspection and maintenance tasks can be performed without removing the assemblies from the mill: worn grinding rolls can be removed and new ones installed; roll-bearing end play can be inspected and adjusted; oil seals can be inspected or replaced, and oil can be changed; the entire roll stem assembly can be removed and taken to another area for bearing maintenance.[18]

Maximum roll life is a primary goal of the C-E HP pulverizer design. It is accomplished by incorporating large rolls to increase the total volume of wear material available, and by using roll material with high wear resistance. Combustalloy™ wear material, a C-E proprietary weld overlay, provides effective wear life five times greater than standard Ni-Hard.

The externally mounted spring assembly (Figs. 26 and 27) has a major advantage in that maintenance personnel do not have to enter the pulverizer to inspect or adjust springs. Since the assembly is located away from the coal flow, erosion is eliminated. Spring travel can be maximized with an external mechanism, to allow large ungrindable material to pass under the rolls until it is rejected from the mill. A positive spring-preload locking device prevents any change in spring setting during operation. The entire spring assembly is a cartridge type to reduce changeout time, and to allow for stocking of spare assemblies, and convenient inspection and refurbishing.

The spring assembly, as well as the journal assembly and the main bowl-hub air seal, have labyrinth-type seals that use air for sealing between stationary and moving parts. The pressure of the filtered seal air is 8 to 16 in. WG above the mill-inlet pressure, with the flow rate a function of mill size.

CLASSIFIER AND DISCHARGE-VALVE ASSEMBLIES

The top of the mill is made larger than the main body to reduce coal/air stream velocity, to reduce mill aerodynamic pressure drop, and to optimize classifier efficiency.

Plant personnel can adjust the position of the classifier vanes when the pulverizer is operating, by use of two manually operated levers on the mill top, each lever operating half of the vanes. Particle separation is accomplished in a stationary-cone cyclone classifier. As described before, oversized coal particles are returned to the grinding zone through the return spout at the bottom of the cone, where they mix with the incoming raw coal feed to increase drying efficiency.

All areas of the classifier section subject to abrasion wear are lined with cast nitride bonded silicon carbide. The inside of the cone is lined with 90-percent-alumina tile to protect it from wear by the sliding oversize coal particles.

Four knife-gate valves at the outlet of a positive-pressure HP pulverizer isolate it from the fuel piping leading to the furnace, as recommended by NFPA. Each valve has a set of replaceable valve seats, coated to minimize wear. Pneumatic cylinders operating in unison open and close the valves. They are actuated by a single solenoid valve that also controls the supply of purge/seal air to the fuel piping. Fig. 26 shows the seal-air header located downstream of the discharge valves.

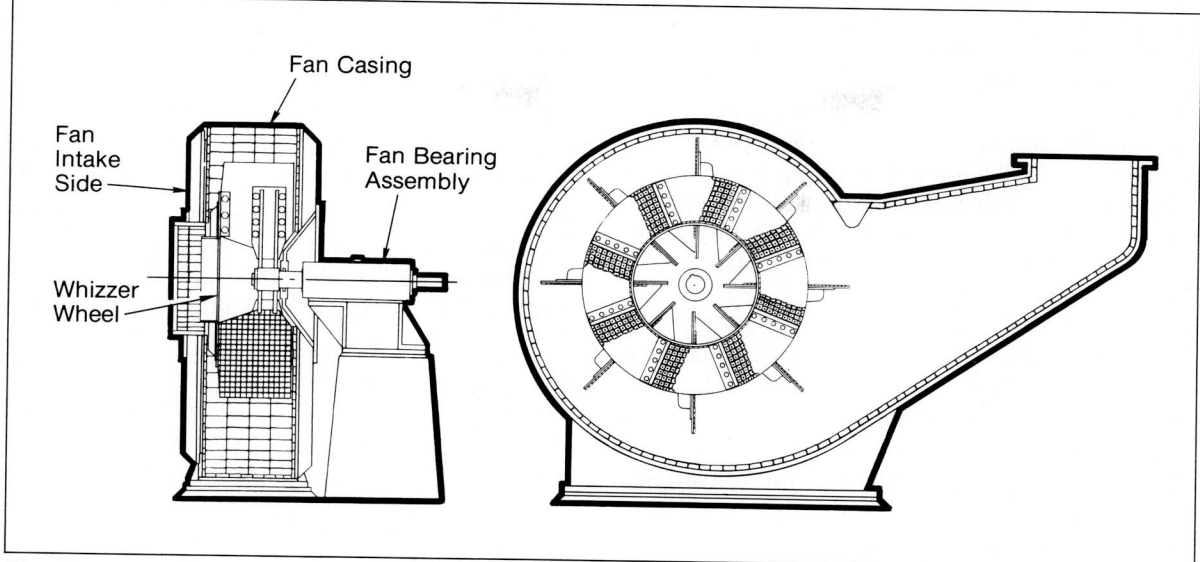

Fig. 32 **Centrifugal material-handling exhauster for C-E suction-type pulverizer**

COAL/AIR EXHAUSTER FANS

For suction-type HP pulverizers, C-E supplies material-handling fans to match the mill and fuel-piping system requirements. These exhausters are horizontal-shaft straight-bladed centrifugal fans (Fig. 32) operating at constant speed and temperatures between 150° and 180°F (65° and 80°C). They are designed in accordance with NFPA standards. There are replaceable liners on the housing scroll and sides. The fan spider is removable if fan blades are to be replaced.

The fan wheel is supported by anti-friction bearings mounted in an external bearing housing located between the exhauster and its driving motor. Exhausters used with HP pulverizers are driven by separate motors. The pulverizer and exhauster motors are interlocked logically to insure coal transport out of the mill. Fan bearing lubrication is independent of the mill lubrication system.

SAFETY AND CONTROLS

The production and handling of pulverized fuels can be hazardous. Because fine particles in suspension or deposition are readily volatilized and become combustible, under certain conditions explosions may take place. Notwithstanding these dangers, the industry has achieved a remarkable safety record since the inception of pulverized-coal firing. To protect property and life, all pulverizing equipment and related auxiliaries, including strength of equipment, valving and inerting, are designed in accordance with NFPA 85F.

Various controls and safety devices are utilized for the correct and proper operation of the equipment. Pulverizer output is controlled by regulation of feed rate in response to load signal. Airflow and air temperature are proportioned to feed rate by automatic control. Also included are permissive interlocks for the proper sequential operation of equipment, flow alarms to indicate cessation of coal flow to and from feeders, and load-limiting devices to prevent overfeeding the mills. Anticipating actions and more responsive feedback may often be included with the above in pulverizer control systems. Control systems and instrumentation are covered in detail in Chapter 13.

PULVERIZER INERTING AND FIRE EXTINGUISHING

To supplement the above operational controls, current practice is to install systems to detect and extinguish mill fires, and to reduce the possibility of destructive positive pressures.

An ideal inerting system continuously purges any combustible volatiles from the pulverizer to the furnace and avoids "bottling up" of the pulverizer; also, it allows a mill to be returned to normal service quickly after the inerting takes place.

C-E's pulverizer inerting and fire-fighting system is designed to provide an early warning of a potentially hazardous situation: it uses readily available plant steam as the inerting medium because steam is less damaging to equipment than other inerting media and makes for easier pulverizer restart. Plant water is used as the fire-extinguishing agent; it is hard-piped to fixed water spray nozzles installed in the pulverizer. The purposes of such a system are

■ to dilute the oxygen content of the mill when there is risk of explosion;

■ to transport pulverized fuel to the furnace by means of an inert medium when transport by air may be hazardous;

■ to extinguish fires in the pulverized-fuel system.

C-E SYSTEM DESIGN FEATURES

The C-E pulverizer inerting and fire-fighting system includes the following features:

■ The pulverizer is automatically inerted when conditions exist for a fire or potential explosion.

■ Steam is used as the primary inerting medium, with CO_2 or other cold inert gas used to cool the pulverizer.

■ The inerting system is capable of supplying steam in such quantities to transport the combustible contents of the pulverizer to the furnace while maintaining an inert atmosphere within the mill.

■ Multiple water-spray nozzles are strategically installed in the pulverizer to provide complete internal fire-extinguishing coverage.

■ The system monitors the entire pulverized-fuel system from the feeders through the fuel piping.

■ Audible and visual alarms are activated in all critical areas and in the control room upon detection of a hazardous condition.

■ Provision is made for interfacing with existing plant control systems.

In Chapter 21 on power-plant operation, we give further information on the efficient and proper operation of pulverizer inerting and fire-extinguishing systems.

REFERENCES

[1] Henry Kreisinger, "Combustion of Pulverized Coal," *Transactions of the ASME*, 60 (Paper No. FSP-60-8): 289–296, 1938.

[2] Nicolas Leonard Sadi Carnot, E. Clayperon, and R. Clausius, *Reflections on the Motive Power of Fire; and other Papers on the Second Law of Thermodynamics*; edited with an introduction by E. Mendoza. New York: Dover Pubs., 1962.

[3] Rudolf Diesel, *Theory and Construction of a Rational Heat Motor*, trans. by Bryan Donkin. London: Spon, 1894.

Friedrich Klemm, *A History of Western Technology.* Cambridge, Mass.: MIT Press, 1964, pp. 342–347.

[4] C. F. Herington, *Powdered Coal as a Fuel.* New York: Van Nostrand, 1918, pp. 68–72.

Frank Lewis Dyer and Thomas Commerford Martin, *Edison—His Life and Inventions*, with collaboration of William Henry Meadowcraft. New York: Harper and Bros., 1929, Vol. II, pp. 953–957.

[5] "Symposium on Powdered Fuel," Spring Meeting, American Society of Mechanical Engineers, St. Paul-Minneapolis, *Transactions of the ASME*, 36: 85–169, 1914.

R. C. Carpenter, "Pulverized Coal Burning in the Cement Industry," *Transactions of the ASME*, 36: 85–107, 1914.

William Dalton and W. S. Quigley, "An Installation for Powdered Coal Fuel in Industrial Furnaces," *Transactions of the ASME*, 36: 109–121, 1914.

F. R. Low, "Pulverized Coal for Steam Making," *Transactions of the ASME*, 36: 123–136, 1914.

"Topical Discussion on Powdered Fuel," *Transactions of the ASME*, 36: 137–169, 1914.

[6] John Anderson, "Pulverized Coal Under Central-Station Boilers," *Power*, 51(9): 336–339, March 2, 1920.

Paul W. Thompson, "Pulverized Fuel at Oneida Street Plant," *Power*, 51(9): 339–340, March 2, 1920.

"Four-Day Test on Five Oneida Street Boilers Burning Pulverized Coal," *Power*, 51(9): 354–357, March 2, 1920.

"Pulverized Coal at Milwaukee," *Power*, 51(9): 341–342, March 2, 1920.

"The New Lakeside Pulverized-Coal Plant, Milwaukee," *Power*, 52(10): 358–360, September 7, 1920.

"Largest Station Using Pulverized Coal," *Power*, 55(16): 604–610, April 18, 1922.

Henry Kreisinger and John Blizard, "Milwaukee's Contribution to Pulverized Coal Development," *Mechanical Engineering*, 61: 723–726 and 737, October, 1940.

F. L. Dornbrook, "Developments in Burning Pulverized Coal—Thirty Year Review of Experience in Milwaukee Plants," *Mechanical Engineering*, 70: 967–974, December, 1948.

[7] Henry Kreisinger, F. K. Ovitz, and C. E. Augustine, "Combustion in the Fuel Bed of Hand-Fired Furnaces," *U. S. Bureau of Mines Technical Paper 137*. Washington: U. S. Bureau of Mines, 1916.

Walter T. Ray and Henry Kreisinger, "The Flow of Heat Through Furnace Walls," *U. S. Bureau of Mines Bulletin 8*. Washington: U. S. Bureau of Mines, 1911.

Henry Kreisinger and Walter T. Ray, "The Transmission of Heat into Steam Boilers," *U. S. Bureau of Mines Bulletin 18*. Washington: U. S. Bureau of Mines, 1912.

Henry Kreisinger and J. F. Borkley, "Heat Transmission Through Boiler Tubes," *U. S. Bureau of Mines Technical Paper 114*. Washington: U. S. Bureau of Mines, 1915.

[8] Henry Kreisinger, John Blizard, C. E. Augustine, and B. J. Cross, "An Investigation of Powdered Coal as Fuel for Power-Plant Boilers—Tests at Oneida Street Power Station, Milwaukee, Wisconsin," *U. S. Bureau of Mines Bulletin 223*. Washington: U. S. Bureau of Mines, 1923.

Henry Kreisinger, John Blizard, C. E. Augustine, and B. J. Cross, "Tests of a Large Boiler Fired with Powdered Coal at the Lakeside Station, Milwaukee," *U. S. Bureau of Mines Bulletin 237*. Washington: U. S. Bureau of Mines, 1925.

[9] E. Audibert, "Etude de l'Entrainement du Poussier de Houille par l'Air," *Annales de Mines*, 1(3): 153–191, March, 1922.

[10] John Blizard, "The Terminal Velocity of Particles of Powdered Coal Falling in Air or Other Viscous Fluid," *Journal of the Franklin Institute*, 197(2): 199–208, February, 1924.

[11] John Blizard, "Transportation and Combustion of Powdered Coal," *U. S. Bureau of Mines Bulletin 217*. Washington: Government Printing Office, 1923.

[12] Henri Verdinne, "The Technique of Powdered Fuel Firing," *Fuel in Science and Practice*, 2: 146–151, 1923.

[13] W. Nusselt, "Der Verbrennungsvorgang in der Kohlenstaubfeuerung (The Combustion Process In Pulverized Coal Furnaces)," *VDI Zeitschrift*, 68(6): 124–128, February 9, 1924.

[14] P. Rosin, "Die Thermodynamischen und Wirtschafthichen Grundlagen der Kohlenstaubfeuerung (Thermodynamic and Economic Bases of Pulverized Coal Firing)," *Braunkohle*, 24(11): 241–259, June 13, 1925.

[15] Walter J. Wohlenberg and Donald G. Morrow, "Radiation in the Pulverized-Fuel Furnace," *Transactions of the ASME*, 47: 127–176, 1925.

[16] R. P. Hensel, "Coal Combustion," presented at the Engineering Foundation Conference on Coal Preparation for Coal conversion, 1st, Franklin Pierce College, Rindge, N.H., 1975. Paper updated in 1978. Combustion Engineering publication TIS-4599.

[17] R. P. Hensel, "The Effects of Agglomerating Characteristics of Coals on Combustion in Pulverized Fuel Boilers," *Symposium on Coal Agglomeration and Conversion*, Morgantown, W.V., 1975. Sponsored by the West Virginia Geological and Economic Survey in Cooperation With the Coal Research Bureau. Combustion Engineering publication TIS-4353.

V. F. Parry, "Production, Classification and Utilization of Western United States Coals," *Economic Geology*, 45(6): 515–532, September–October, 1950.

[18] D. Magnum and P. L. Stanwicks, "HP Series Pulverizer Design, Testing, Maintenance Cost", *Proceedings of POWER-GEN '89 Conference*, Book 2, POWER-GEN '89, New Orleans, LA, December 5–7, 1989.

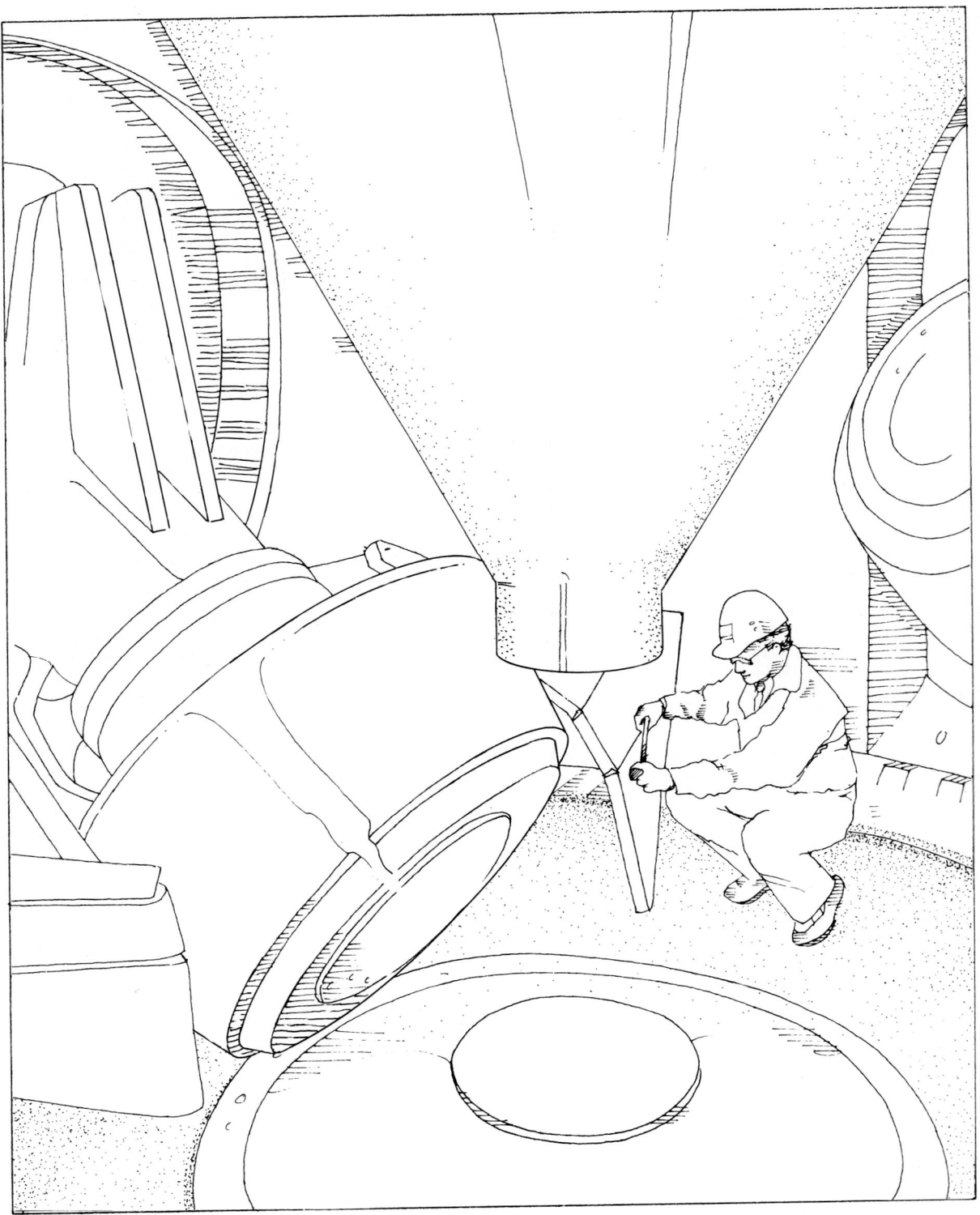

CHAPTER 12

Fuel-Firing Systems

In the process of steam generation, fuel-burning systems provide controlled, efficient conversion of the chemical energy of fuel into heat energy which, in turn, is transferred to the heat-absorbing surfaces of the steam generator. To do this, fuel-burning systems introduce the fuel and air for combustion, mix these reactants, ignite the combustible mixture and distribute the flame envelope and the products of combustion.

An ideal fuel-burning system fulfilling these functions would have the following characteristics:

■ no excess oxygen or unburned combustibles in the end products of combustion

■ a low rate of auxiliary ignition-energy input to initiate the combustion reaction

■ an economic reaction rate between fuel and oxygen compatible with acceptable nitrogen- and sulfur-oxide formation

■ an effective method of handling and disposing of the solid impurities introduced with the fuel

■ uniform distribution of the product weight and temperature in relation to the parallel circuits of heat-absorbing surface

■ a wide and stable firing range

■ fast response to changes in firing rate

■ high equipment availability with low maintenance

In actual practice, some of these characteristics must be compromised to achieve a reasonable balance between combustion efficiency and cost. For example, firing a fuel with the stoichiometric air quantity (no excess above the theoretical amount) would require an infinite residence time at temperatures above the ignition point at which complete burnout of the combustibles takes place. Thus, every firing system requires a quantity of air in excess of stoichiometry to attain an acceptable level of unburned carbon in the products of combustion leaving the furnace. This amount of excess air is an indicator of the burning efficiency of the firing system. (Further discussion of the impact of excess-air percentage on boiler design is found in Chapter 6; included is a table of commonly used excess-air percentages.)

THE COMBUSTION REACTION

The rate and degree of completion of a chemical reaction, such as in the combustion process, are greatly influenced by temperature, concentration, preparation and distribution of the reactants, by catalysts, and by mechanical turbulence. All of these factors have one effect in common: to increase contacts between molecules of the reactants.

Higher temperature, for instance, increases the velocity of molecular movement permitting harder and more frequent contact between molecules. A temperature rise of 200°F at some stages can increase the possible rate of reaction a million fold.

At a given pressure, three factors limit the temperature that can be attained to provide the greatest intermolecular contact. These are the heat absorbed by the combustion chamber enclosure, the heat absorbed by the reactants in bringing them to ignition temperature, and that absorbed by the nitrogen in the air.

The concentration and distribution of the reactants in a given volume are directly related to the opportunity for contact between interacting molecules. In an atmosphere containing 21-percent oxygen (the amount present in air), this rate is much less than it would be with 90-percent oxygen. As the reaction nears completion, the distribution and concentration of reactants assume even greater importance. Because of the dilution of reactants by the inert products of combustion, the relative distribution—and opportunity for contact—approaches zero.

Preparation of the reactants and mechanical turbulence greatly influence the reaction rate. These are the primary factors available to the fuel-burning system designer attempting to provide a desirable reaction rate.

The beneficial effect of mechanical turbulence on the combustion reaction becomes apparent when it is realized that agitation permits greater opportunity for molecular contact. Agitation improves both the relative distribution and the energy imparted. Agitation assumes greater significance if achieved later in the combustion process when the relative concentration of the reactants is approaching zero.

PRACTICAL FUEL-FIRING SYSTEM DESIGN

In the practical application of a burner and fuel-burning system to a boiler, all fundamental factors influencing rate and completeness of combustion must be considered along with the degree of heat-transfer efficiency.

There are two methods of producing a total flow pattern in a combustion chamber to provide successful molecular contacts of reactants through mechanical turbulence. One is to divide and distribute fuel and air into many similar streams and to treat each stream independently of all others. This provides multiple flame envelopes. In contrast, the second process provides interaction between all streams of air and fuel introduced into the combustion chamber to produce a single flame envelope.

The first concept requires that the total fuel and air supplied to a common combustion chamber be accurately subdivided. It also limits the opportunity for sustained mechanical mixing or turbulence—particularly in the early stages of combustion. The necessity of obtaining and sustaining good distribution of fuel and air is a design as well as an operating problem. There must be sufficient opportunity for contact of fuel and oxygen molecules as well as uniform distribution of product temperature and mass in relation to the combustion chamber. On the other hand, the single-flame-envelope technique provides interaction between all streams of fuel and air introduced into a common chamber. It allows more time for contact between all fuel and air molecules, and mechanical turbulence is sustained throughout the chamber. This avoids stringent fuel and air distribution accuracy requirements.

Firing systems representative of these two concepts are horizontally wall-fired systems (characterized by individual flames) and tangentially fired systems (which have a single flame envelope). There are other types and combinations; one such is the vertically fired system, which uses characteristics of both previously described systems.

HORIZONTALLY FIRED SYSTEMS

In horizontally fired systems the fuel is mixed with combustion air in individual burner registers (Fig. 1). In this design, the coal and primary air are introduced tangentially to

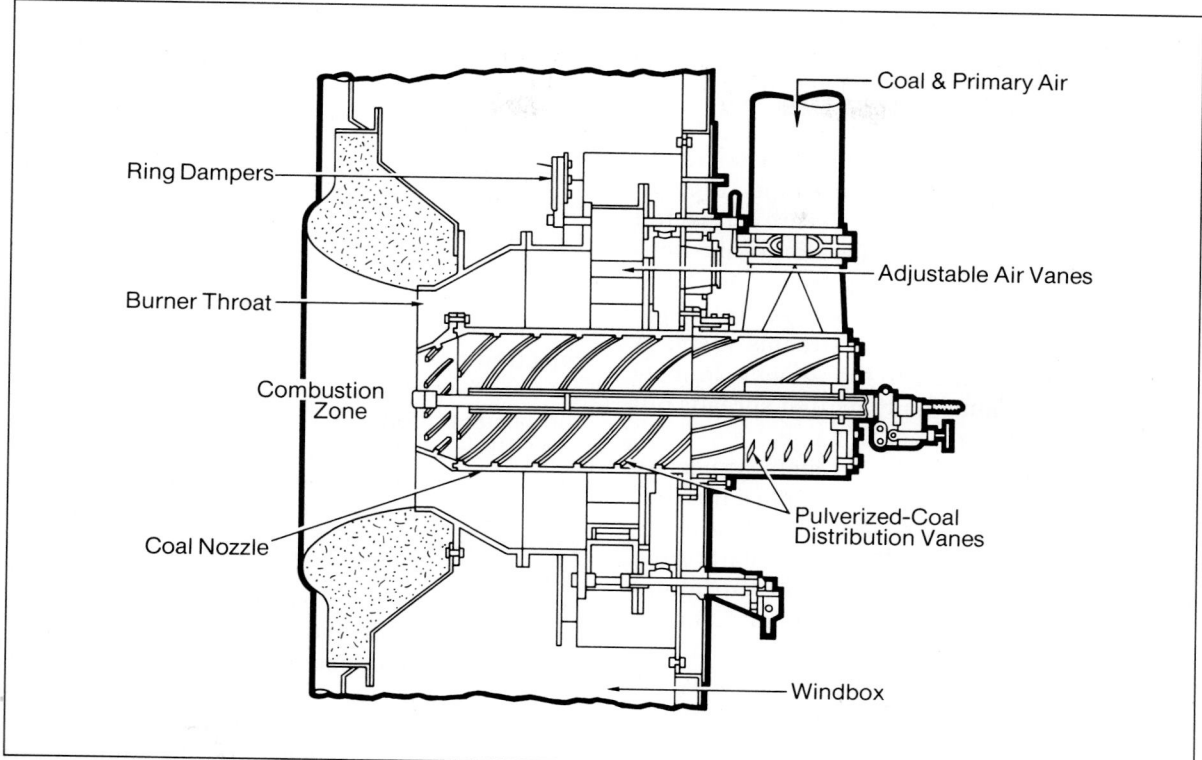

Ring Dampers

Burner Throat

Combustion Zone

Coal Nozzle

Coal & Primary Air

Adjustable Air Vanes

Pulverized-Coal Distribution Vanes

Windbox

Fig. 1. Burner for horizontal firing of coal

the coal nozzle, thus imparting strong rotation within the nozzle. Adjustable inlet vanes impart a rotation to the preheated secondary air from the windbox. The degree of air swirl, coupled with the flow-shaping contour of the burner throat, establishes a recirculation pattern extending several throat diameters into the furnace. Once the coal is ignited, the hot products of combustion propagate back toward the nozzle to provide the ignition energy necessary for stable combustion.

The burners are located in rows, either on the front wall only (Fig. 2) or on both front and rear walls. The latter is called "opposed firing."

Because the major portion of the combustion process must take place within the recirculation zone, it is imperative that the air/fuel ratio to each burner is within close tolerances. The rate of combustion drops off rapidly as the reactants leave the recirculation zone and interaction between flames occurs only after that

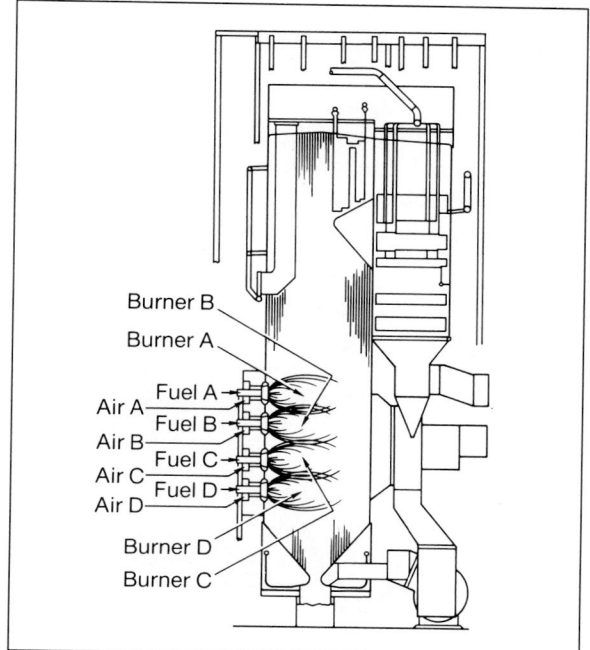

Burner B
Burner A
Air A — Fuel A
Air B — Fuel B
Air C — Fuel C
Air D — Fuel D
Burner D
Burner C

Fig. 2. Flow pattern of horizontal (wall) firing

point. The degree of interaction depends on burner and furnace configurations.

TANGENTIALLY FIRED SYSTEMS

The tangentially fired system is based on the concept of a single flame envelope (Fig. 3). Both fuel and combustion air are projected from the corners of the furnace along a line tangent to a small circle, lying in a horizontal plane, at the center of the furnace. Intensive mixing occurs where these streams meet. A rotative motion, similar to that of a cyclone, is imparted to the flame body, which spreads out and fills the furnace area.

When a tangentially fired system projects a stream of pulverized coal and air into a furnace, the turbulence and mixing that take place along its path are low compared to horizontally fired systems. This occurs because the turbulent zone does not continue for any great distance, since the expanding gas soon forces a streamline flow. The significance of this factor on the production of oxides of nitrogen is discussed later in this chapter. However, as one stream impinges on another in the center of the furnace, during the intermediate stages of combustion, it creates a high degree of turbulence for effective mixing.

The fuel and air are admitted from the vertical furnace corner windboxes (Fig. 4). Dampers which control the air to each compartment make it possible to vary the distribution of air over the height of the windbox. It is also possible to vary the velocities of the air streams, change the mixing rate of fuel and air, and control the distance from the nozzle at which the coal ignites.

The vertical arrangement of fuel and air nozzles provides great flexibility for multiple-fuel firing. It is possible to provide for full-load capability with gas or oil by locating the additional nozzles for these fuels in the secondary-air compartments adjacent to the coal nozzles. In addition, separate nozzles for injecting municipal refuse and other waste fuels are frequently provided in both utility and industrial boilers.

As illustrated in Fig. 5, fuel and air nozzles most commonly tilt in unison to raise and lower the flame in order to control furnace heat absorption and thus heat absorption in the superheater and reheater sections. In addition to controlling the furnace exit-gas temperature for variations in load, the tilts on coal-fired units automatically compensate for the effects of ash deposits on furnace-wall heat absorption.

As wall blowers clean ash deposits from the furnace walls, the furnace exit-gas temperature tends to decrease because of the increase in overall furnace absorption. The windbox nozzles are then automatically tilted upward at a controlled rate, and combustion is completed higher in the furnace. The repositioning effectively reduces the absorption in the lower part of the furnace, and increases the furnace exit-gas heat content to maintain steam at design temperatures.

Conversely, as furnace walls are again gradually covered with ash deposits and furnace heat absorption decreases due to the insulating effects of the ash, the tilts are gradually depressed and combustion is completed lower in the furnace. This exposes the hot gases to a greater proportion of furnace wall surface and effectively controls furnace exit-gas temperature and steam temperature until ash is again removed from the furnace walls.

VERTICALLY FIRED SYSTEMS

The first pulverized-coal systems had a configuration called *vertical* or *arch firing*. They are

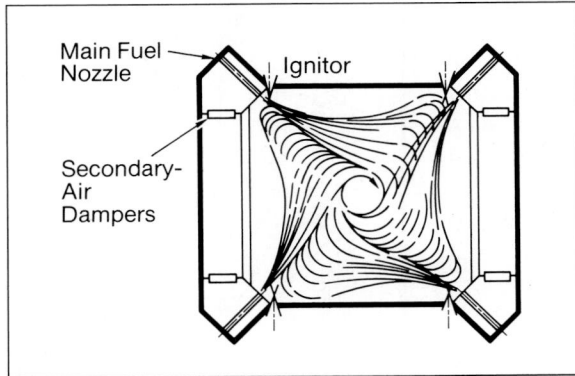

Fig. 3. Tangential firing pattern

Main Fuel
Nozzle

Ignitor

Secondary-
Air
Dampers

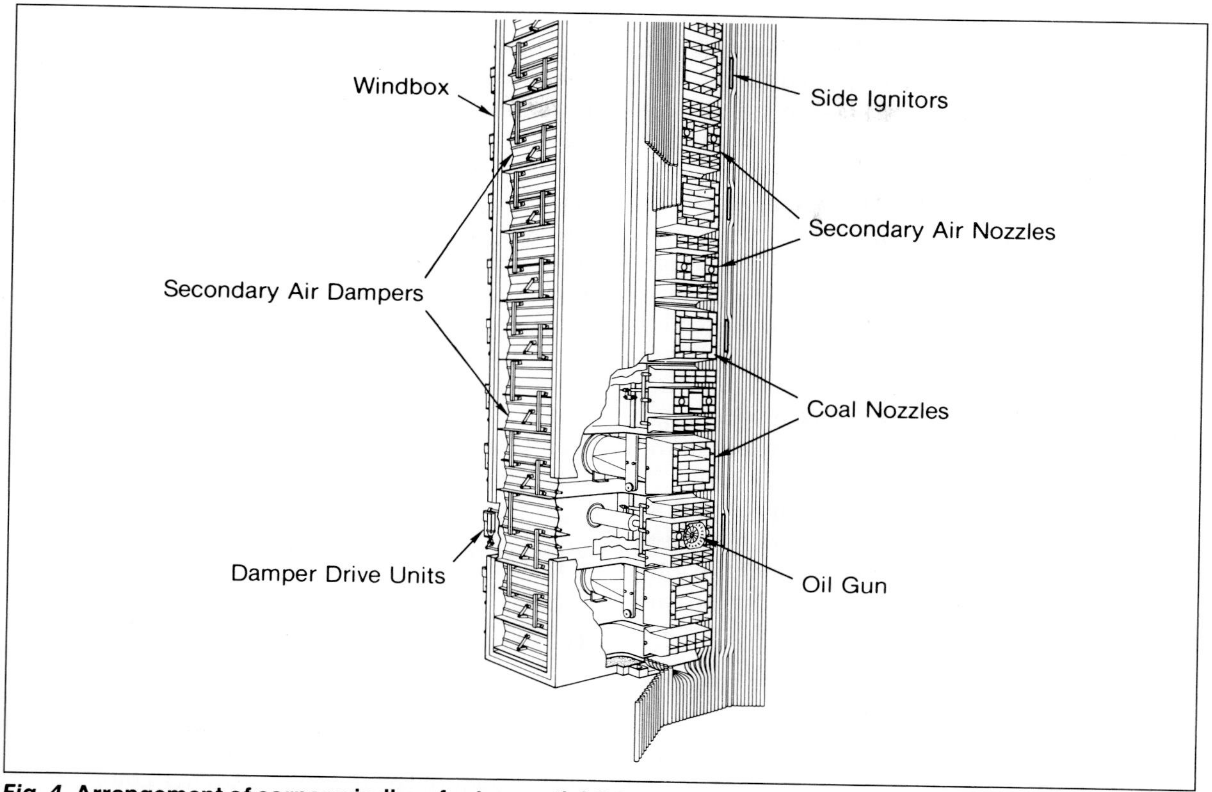

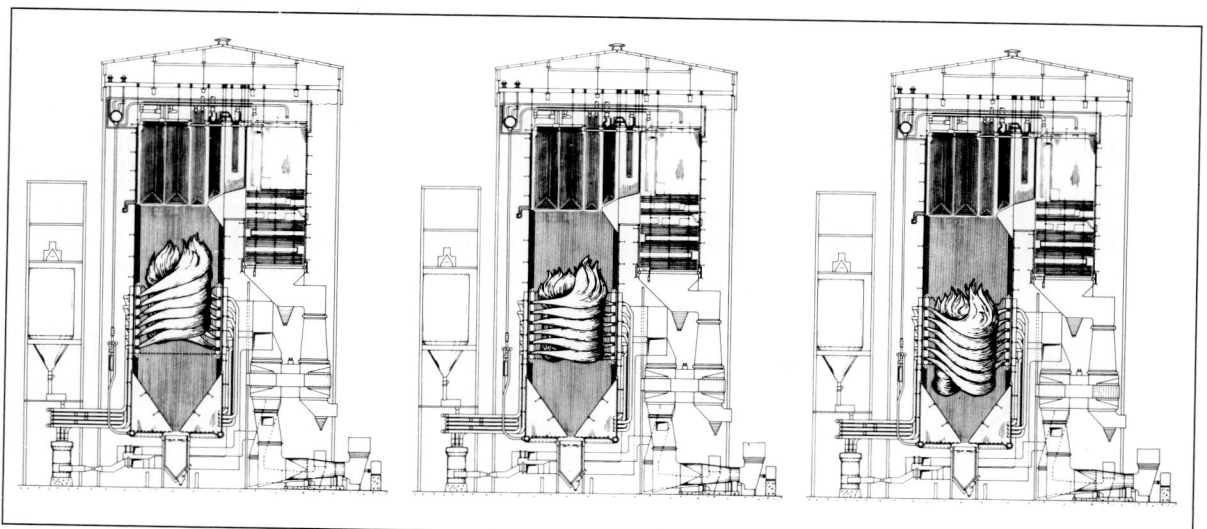

Windbox

Side Ignitors

Secondary Air Nozzles

Secondary Air Dampers

Coal Nozzles

Damper Drive Units

Oil Gun

Fig. 4. **Arrangement of corner windbox for tangential firing of coal**

Fig. 5 **Selective furnace utilization and steam temperature control are accomplished by tilting nozzles in a tangentially fired system.**

now used principally to fire coals with mois-ture-and-ash-free volatile matter between 9 and 13 percent. They require less stabilizing fuel than horizontal or tangential systems, but have more complex firing equipment and, therefore, more complex operating characteristics.

The firing concept and the arrangement of the burners in the arches are shown in Figs. 6 and 7. Pulverized coal is discharged through the nozzles. A portion of the heated combus-tion air is introduced around the fuel nozzles and through adjacent auxiliary ports. High-pressure jets are used to avoid short-circuiting the fuel/air streams to the furnace discharge. Tertiary air ports are located in a row along the front and rear walls of the lower furnace.

The firing system produces a long, looping flame in the lower furnace, with the hot gases discharging up the center. A portion of the total combustion air is withheld from the fuel stream until it projects well down into the furnace.

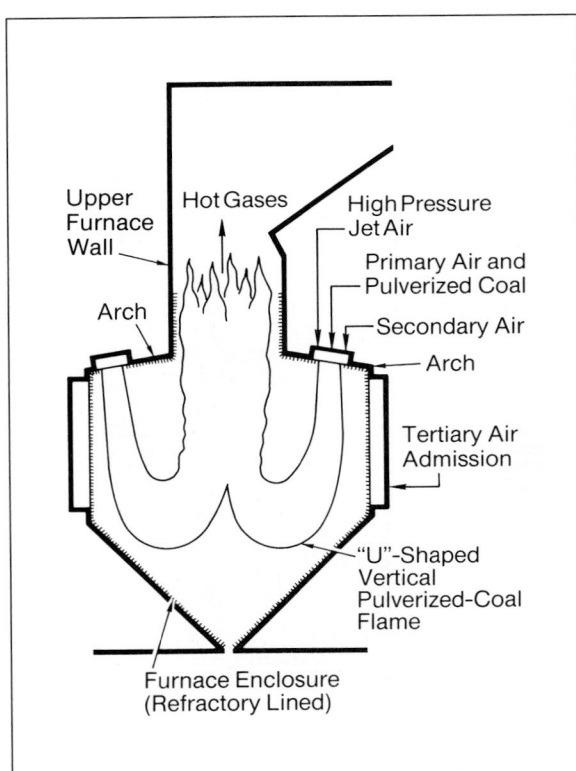

Fig. 6. **Flow pattern of vertical firing**

This arrangement has the advantage of heating the fuel stream separately from a significant portion of its combustion air to provide good ig-nition stability. The delayed introduction of the tertiary air provides needed turbulence at a point in the flame where partial dilution from the products of combustion has occurred. The furnace flow pattern passes the hot product gases immediately in front of the fuel nozzles to provide a ready source of inherent ignition en-ergy which raises the primary fuel stream to ig-nition temperature. The flow pattern also ensures that the largest entrained solid-fuel particles, with the lowest surface-area-to-weight ratio, have the longest residence time in the combustion chamber.

FIRING SYSTEMS THAT MINIMIZE NITROGEN-OXIDE FORMATION

In-furnace firing systems to minimize NO_x formation are designed so that the fuel-bound nitrogen conversion is controlled by driving the major fraction of the fuel nitrogen compounds into the gas phase under overall fuel-rich condi-tions. In this atmosphere of oxygen deficiency, there occurs a maximum rate of decay of the evolved intermediate nitrogen compounds to N_2. Following the admission of the remaining air, the slow burning rate reduces the peak flame temperature, to curtail the thermal NO_x production in the latter stages of combustion.

The final section of Chapter 4 on the forma-tion of NO_x in fossil-fired steam-generating units is a useful treatment of thermal and fuel NO_x, and the design criteria relevant to their control.

TANGENTIALLY FIRED LOW-NO_x SYSTEMS

Early studies of NO_x emissions from all types of steam generators indicated that those from tan-getially fired units were about half the values of those from horizontally fired systems.

The reduced formation of nitrogen oxides results from the relatively low rate of mixing be-tween the parallel streams of coal and second-

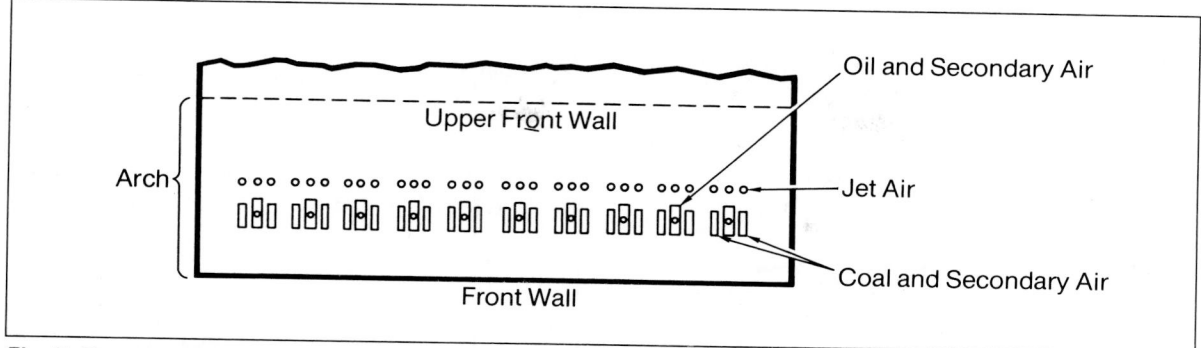

Fig. 7. **Burner arrangement on the front arch of the furnace shown in Fig. 6**

ary air emitted from the corner windboxes. Thus, ignition and partial devolatization occur within an air-deficient primary combustion zone that exists from the fuel nozzle to a point within the furnace at which the stream is absorbed into the rotating mass of gases termed the *fireball*.

The fireball itself is rich with oxygen because it contains all the air required for complete combustion of the fuel. Because the balance of the devolatization occurs after the coal stream enters the fireball, fuel nitrogen-oxide formation is limited.

Two significant modifications in the design and operation of the C-E tangential firing system have resulted in further extension of the oxygen-deficient combustion zone.

TANGENTIAL FIRING WITH OVERFIRE AIR

As shown in Fig. 8, the first modification added air compartments within the windbox above the uppermost coal nozzle. Termed *overfire air* (OFA) ports, these compartments divert approximately 20 percent of the total combustion air to a burning zone above the windboxes. As a result, the fireball at windbox level is at or near stoichiometric air conditions.

Fig. 9 shows the effect on NO_x production of varying quantities of overfire air. Note the effect of total excess air on NO_x levels. This is due primarily to the reduction in available oxygen within the fireball adjacent to the windboxes. With a constant windbox-to-furnace differential, reducing total excess air increases the proportion of overfire air at any given OFA damper

position. This decreases the available oxygen below the OFA ports in even greater proportion to the change in total excess air.

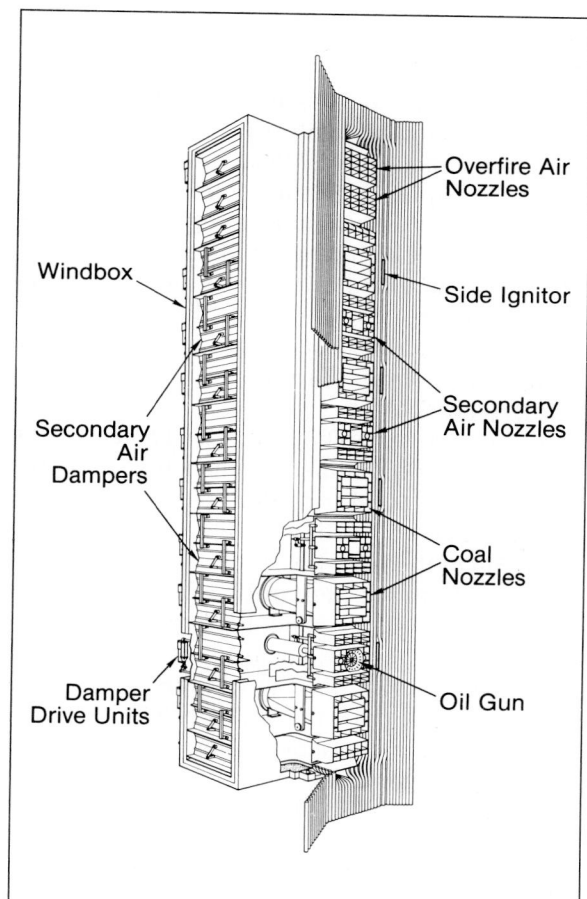

Fig. 8 **Tangential-firing windbox with overfire air (OFA) ports and high energy arc ignitors**

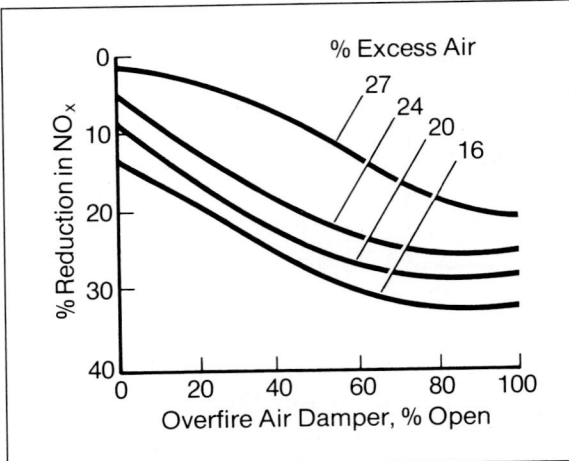

Fig. 9. Effect of overfire air on NO$_x$ production

LOW NO$_x$ CONCENTRIC FIRING SYSTEM

The second modification was the development of a firing-system concept called the *Low NO$_x$ Concentric Firing System* (LNCFS). The LNCFS proportions the secondary airflow through the windbox so as to effect a decrease in the amount of fuel air while increasing the amount of auxiliary air. In addition, the auxiliary air is directed away from the fuel towards the adjacent furnace wall in order to reduce the entrainment of auxiliary air by the expanding primary-air/coal jet.

In the plan view, Fig. 10, LNCFS leaves the fuel and fuel air aimed at the tangent of the small circle central in the furnace. The secondary air, however, is directed 25° away from the direction of the fuel. Like OFA, air is effectively withheld from the fuel; unlike OFA, LNCFS affects the very early stoichiometry of the fuel-burning process. Both techniques are examples of staging, as discussed in Chapter 4. OFA is a type of "vertical staging"; LNCFS can be thought of as a "horizontal staging" technique unique to tangential firing.

As a practical matter, LNCFS affects the "early stoichiometry" for a very limited amount of time. The cross-mixing patterns inherent in tangential firing are massive and the separation of both streams are quickly lost as they penetrate the furnace. Thus, LNCFS requires the use

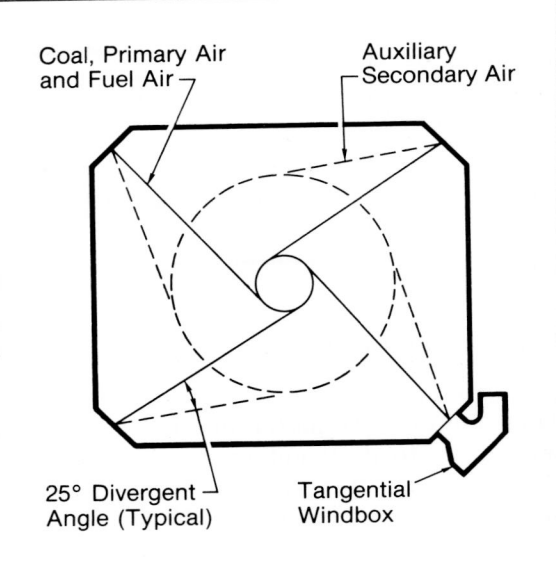

Fig. 10 Furnace plan view — LNCFS

of "flame attachment" nozzle tips to accelerate the devolatization process, Fig. 11.

Fig. 12 presents typical results which indicate an approximate 20% reduction from the standard tangential-firing mode with OFA. The tests clearly show that the NO$_x$ reductions from LNCFS are additional to those already achieved through OFA.

IGNITION SYSTEMS

Modern steam generators are very dependent on their ignition systems for safe start-up and shutdown operations. The need to insure reliable and safe firing conditions has caused the evolution from simple ignitors to highly sophisticated ignition systems. Such advanced systems require the following features:

■ a timed spark-ignition sequence

■ a device to create turbulent fuel/air mixing and sufficient hot gas recirculation to insure flame stability

■ a device to detect and monitor the ignitor flame, and

■ a device to monitor ignitor fuel flow as an indication of ignitor heat-energy release

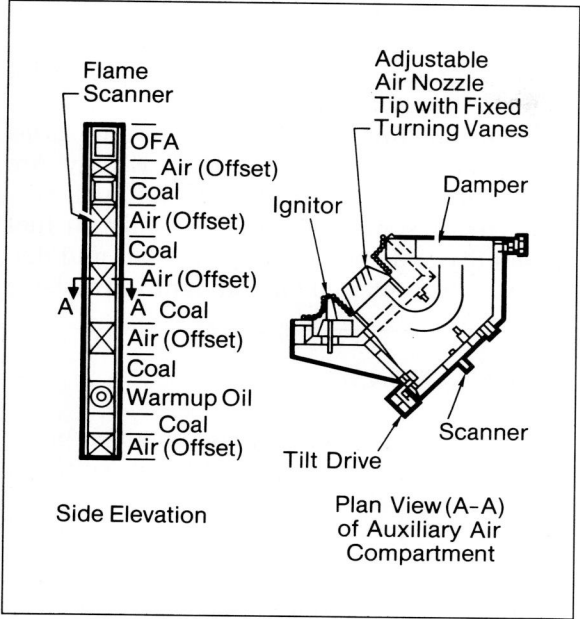

Fig. 11 Air nozzle tip for LNCFS

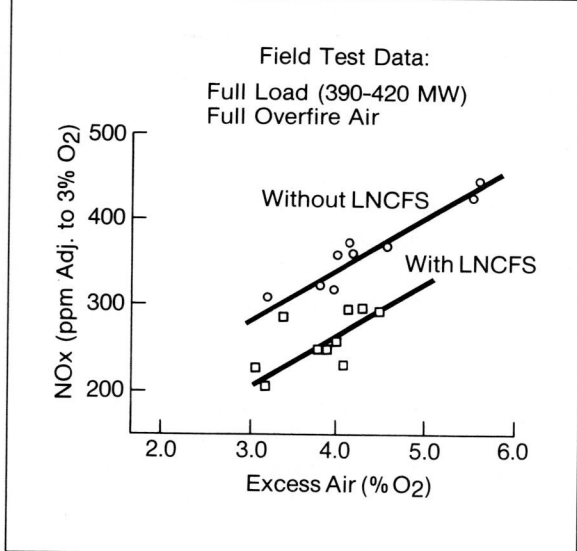

Fig. 12 No$_x$ vs excess air with and without LNCFS

The primary function of the ignition system is to light off the main fuel input to the furnace. Often, the ignition system is designed also to stabilize the main fuel flame under potentially unstable firing conditions. Determining the correct amount of ignition energy to fulfill either function is a challenging task. The correct combination of ignition-energy quantity, quality, and location is affected by many factors, some of which vary with time; what constitutes sufficient ignition energy at one instant may be insufficient the next. From the rate-igniting standpoint, there is never excess ignition energy: the more stable a fire, the more likely that the ignition energy substantially exceeds the minimum necessary to maintain ignition.

Years of experience with various fuels and firing systems have led the National Fire Protection Association (NFPA) to establish practical guidelines for defining ignition-energy requirements based on fuel type and ignition-system function. For NFPA definitions, see the accompanying box Ignitor Functional Definitions.

IGNITION-SYSTEM DESIGN

An electrical spark is the typical initiating source of fuel ignition. High-voltage free-air sparks can reliably ignite high-calorific-value gases and light (distillate) oils. High-energy surface-shunted arcs are used to reliably ignite distillate and heavy (residual) oils. Safety considerations dictate that only a limited quantity of fuel be exposed to the electrical spark because of its relatively low ignition energy. The heat released from this initial fuel input then becomes the ignition energy for a different and/or large fuel input. Essentially, the ignition system provides for a controlled transition from spark to main-fuel firing through an incremental increase in ignition energy.

Not to make too large a step increase in fuel input is also important in order to avoid undesirable furnace pressurization during lightoff. Fig. 13 shows two methods of achieving a step increase in fuel input for tangential firing. In the example on the left, the lowest elevation of ignitors is fired first and, after a suitable time, the elevation of oil warm-up guns is ignited. As furnace temperature stabilizes, the main coal nozzles adjacent to the oil guns are brought into operation. Thereafter, additional ignitors can come into service to light off elevations of main coal nozzles directly.

In the alternative on the right of Fig. 13, ignitors light off successive elevations of warm-up oil guns. Adjacent elevations of main coal nozzles are then ignited by the oil guns.

Experience indicates that firing by these step-input methods will minimize furnace pressurization and unnecessary over-pressure trips.

C-E IGNITION SYSTEMS

Ignitors available for tangential and horizontally fired systems include Ionic Flame Monitoring and High Energy Arc.

These ignitors use a wide variety of gas and oil fuels. The Ionic Flame Monitoring (IFM) ignitor fires medium to high-heating-value gases or distillate oils, while the High Energy Arc (HEA) ignitor fires distillate to residual oils. The type of system selected depends on fuel availability and cost. High-quality gas and distillate oil are preferred for both ignition systems

---

IGNITOR FUNCTIONAL DEFINITIONS

To fully understand the hardware, it may be necessary to review definitions contained in NFPA 85E, "Standard for Prevention of Furnace Explosions in Pulverized Coal-Fired Multiple Burner Boiler-Furnaces, 1985 Edition", from which we quote.

Ignitor Types:

Class 1 (Continuous Ignitor)

"An igniter applied to ignite the fuel input through the burner and to support ignition under any burner light-off or operating conditions. Its location and capacity are such that it will provide sufficient ignition energy (generally in excess of 10 percent of full load burner input) at its associated burner to raise any credible combination of burner inputs of both fuel and air above the minimum ignition temperature."

Class 2 (Intermittent Ignitor)

"An igniter applied to ignite the fuel input through the burner under prescribed light-off conditions. It is also used to support ignition under low load or certain adverse operating conditions. The range of capacity of such igniters is generally 4 percent to 10 percent of the full load burner fuel input. It shall not be used to ignite main fuel under uncontrolled or abnormal conditions. The burner shall be operated under controlled conditions to limit the potential for abnormal operation, as well as to limit the charge of fuel to the furnace in the event that ignition does not occur during light-off. Class 2 igniters may be operated as Class 3 igniters."

Class 3 (Interrupted Ignitor)

"A small igniter applied particularly to gas and oil burners to ignite the fuel input to the burner under prescribed light-off conditions. The capacity of such igniters generally does not exceed 4 percent of full load burner fuel input. As a part of the burner light-off procedure the igniter is turned off when the timed trial for ignition of the main burner has expired. This is to ensure that the main flame is self-supporting, is stable and is not dependent upon ignition support from the igniter. The use of such igniters to support ignition or to extend the burner control range shall be prohibited."

Class 3 Special (Direct Electric Ignitor)

"A special Class 3 high energy electrical igniter capable of directly igniting the main burner fuel. This type igniter shall not be used unless supervision of the individual main burner flame is provided."

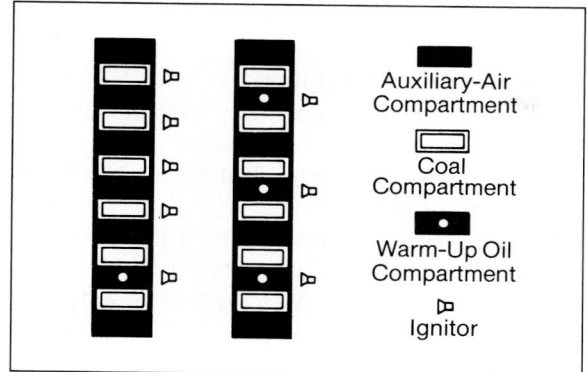

Fig. 13 **Alternative step-ignition processes for tangential firing**

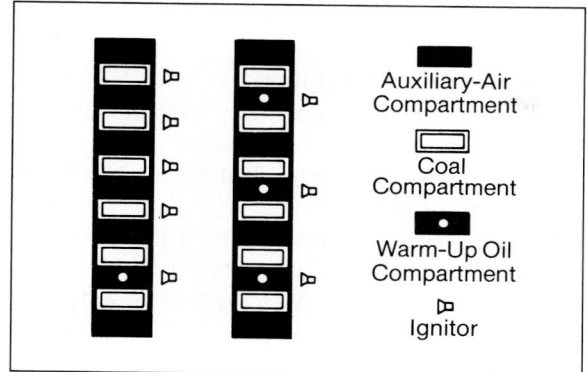 legend:
- Auxiliary-Air Compartment
- Coal Compartment
- Warm-Up Oil Compartment
- Ignitor

and for boiler warm-up; they are easier to handle and cleaner burning in cold furnaces.

IFM IGNITORS

The IFM ignitor is a complete ignition system containing an electrical spark source, a self-stabilizing burner device, flame detection, and fuel-input monitoring. A high-voltage spark or high energy arc source can be used for ignition. Either gases ranging from coke-oven gas to butane or No. 2 fuel oil, is used with this type of system. Flame detection is achieved with the Ionic Flame Monitoring device. Flow switches monitor fuel input. The IFM design (Fig. 14) follows the traditional C-E philosophy of providing both qualitative and quantitative indication of flame.

The IFM system takes advantage of the production of ions and charged particles generated during the combustion of hydrocarbon fuels. A hydrocarbon-fuel flame, then, will conduct electricity. Another characteristic of turbulent flames is that they continuously pulsate at some constant frequency. The quantity of ions and charged particles generated varies as the flame pulsates. Thus, the conductivity of the flame also changes with the pulsation of the flame.

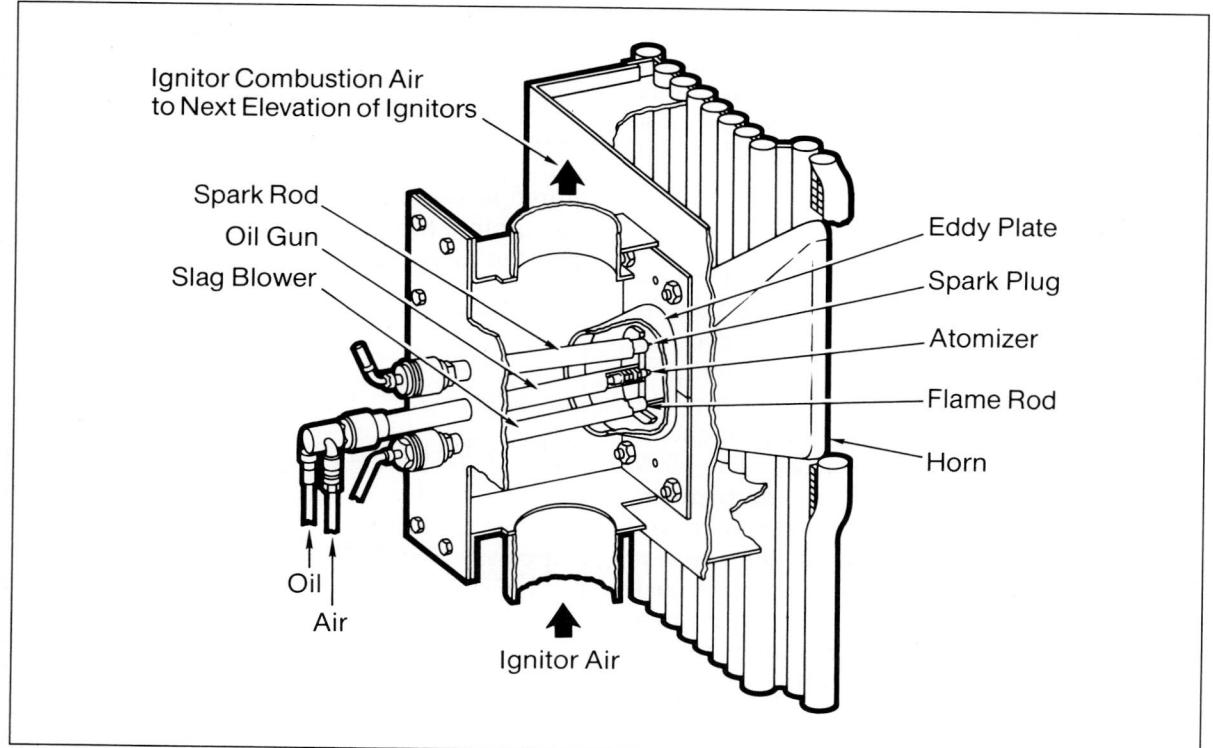

Fig. 14. Ionic Flame Monitoring side ignitor

When a DC potential is placed across a flame, the electric current flow varies at the same frequency as the flame pulsation.

The IFM system operates by imposing a DC potential on an electrode called the *flame rod*, which is in contact with the flame. When there is "no flame", the DC voltage remains at the originally imposed level and no current flows. When there is "flame", the DC voltage drops as current flows, generating what is called an AC feedback signal. This AC signal is filtered, amplified, and modified by the IFM electronics to drive a flame-indication relay. The electronics are designed to be fail-safe. If there is component failure (a short circuit in the flame rod or signal lead wire, or an external AC interference), a "no flame" indication will occur.

The most common application of IFM ignitors is consistent with the NFPA Class 2 definition. IFM ignitors have been used as Class 1, and could be used as Class 3; however, additional burner-control logic and discriminating flame scanners would be required. (Refer to Chapter 13 for more information on flame scanners.) A Class 3 application would be made only when a limited amount of ignition fuel is available, such as with a bottled-gas supply system.

HEA IGNITOR

The High Energy Arc (HEA) ignitor was developed specifically for residual oils, as it eliminates dependence on premium fuels such as natural gas and No. 2 oil for main-fuel ignition and boiler warm-up. The HEA ignitor directly lights off a heavy-oil warm-up gun which, in turn, ignites the main coal nozzles. The HEA ignitor is used with a discriminating scanner, which verifies the operation of the warm-up gun. The warm-up gun is designed for a proven oil flow of about 10 percent of an adjacent coal nozzle, with the discriminating scanner proving the presence of a flame. The combination of an HEA ignitor and No. 6 oil-fired warm-up gun satisfies the NFPA definition of a Class 1 or Class 2 ignitor. The HEA ignitor alone would be a Class 3 ignitor.

The complete HEA Ignition System consists of
- a high energy arc ignitor
- a warm-up oil compartment capable of producing a stable flame at all loads
- a flame-detecting system sensitive only to its associated oil gun
- a control system coordinating all the components and providing unit safety

The High Energy Arc ignitor can ignite warm-up fuel oils ranging from No. 2 to No. 6 and crude oils. The ignitor is a self-contained electrical discharge device which produces a high-intensity spark. Use of a high-resistance transformer, to produce a full wave charging circuit and to control spark rate, enables the sealed power supply to store maximum energy and to deliver a greater percentage of this energy through insulated cables to the ignitor tip. A high spark energy also eliminates coking of the ignitor tip. The HEA ignitor consists of four basic components: the exciter, flexible cable, spark tube and guide pipe, and retractor assembly (Fig. 15).

A key to the successful application of spark-ignition is the presence of a strong recirculation pattern in the primary combustion zone (Fig. 16). The recirculation provides the energy required to vaporize and heat the oil to its ignition point, thus maintaining stable ignition after the spark has been deactivated.

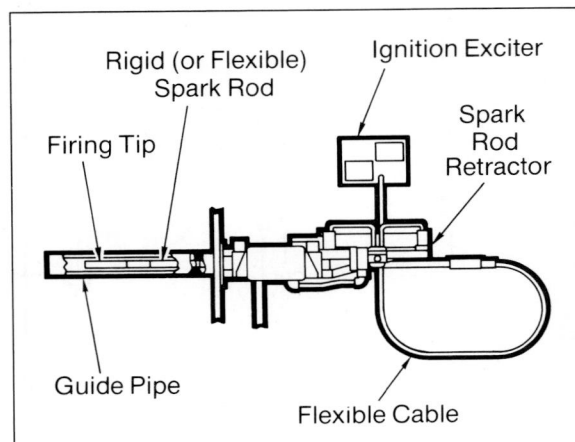

***Fig. 15.* High Energy Arc ignitor**

The discriminating flame scanners operate in the visible-light range, which provides for excellent sensitivity to the presence of a flame. These scanners, therefore, can discriminate between different flames based on both frequency and intensity. This insures safe operation of the overall HEA system.

STOKERS

Stokers are, by modern definition, mechanical devices which feed and burn solid fuels in a bed at the bottom of a furnace. In all cases, the fuel is burned *on* some form of *grate*, through which some or all of the air for combustion passes. The grate surface can be stationary or moving. Stokers are classified according to the way fuel is fed to the grate; the three general classes in use today are underfeed stokers, overfeed stokers, and spreader stokers.

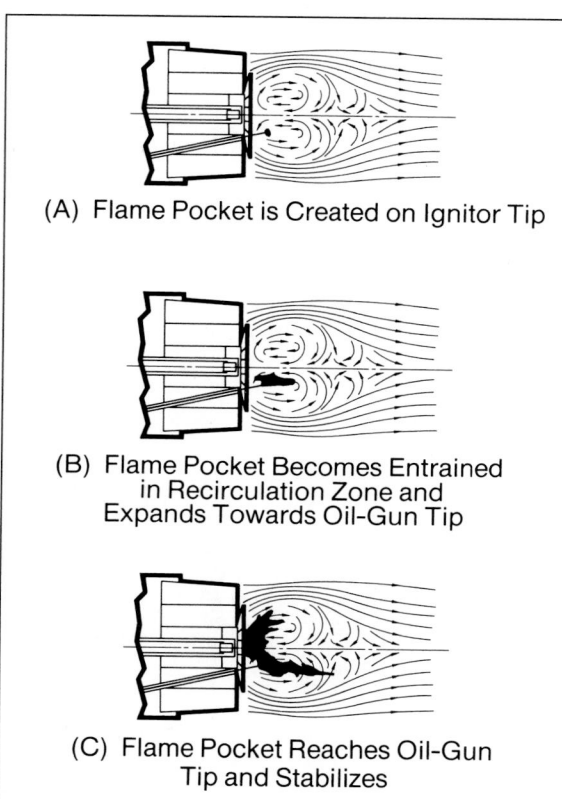

(A) Flame Pocket is Created on Ignitor Tip

(B) Flame Pocket Becomes Entrained in Recirculation Zone and Expands Towards Oil-Gun Tip

(C) Flame Pocket Reaches Oil-Gun Tip and Stabilizes

Fig. 16. Flame pocket is created on ignitor tip

UNDERFEED STOKERS

In an underfeed stoker, the incoming fuel, usually coal, is pushed through one or more troughs, called *retorts*, which are located below the burning fuel bed and air emission grates. Here, moisture is driven off and volatile constituents are distilled, passing up through the incandescent bed. As the retorts are recharged, the char from the previous fill is pushed into and becomes part of the burning bed. On both sides of the retort, the "overflow" burns to completion on air-admitting grates or tuyeres. Ash is removed by dumping grate sections (see Fig. 17). Boilers with underfeed stokers of the single retort design have been built in capacities ranging from 5,000 to 50,000 lb/hr steam flow. Multiple retort stoker units have been built for boilers ranging from 40,000 to 300,000 lb/hr steam flow.

OVERFEED STOKERS

In an overfeed or traveling-grate stoker, gravity feeds the fuel (again, primarily coal) at one end of the air-admitting grate surface (see Fig. 18). The incoming bed depth is adjusted by a gate under which the coal passes before entering the furnace. The rotating grate surface moves away from the feed end through the furnace; combustion is completed as air passes up through the grate. The ash residue is continuously discharged as the grate rotates around to its return run. Traveling-grate stokers can burn every type of coal that is mined with the exception of caking bituminous coal. In addition, by-product and waste fuels, such as coke breeze or anthracite dredged from river bottoms, can be burned effectively. This machine has also been used as a part of chemical process operations which produce coke and carbon dioxide. Steaming rates ranging from 10,000 to 300,000 lb/hr are achievable.

For more detailed discussions about the designs and performance of these first two classes of stokers, the reader is referred to the First Edition of *Combustion Engineering – A Reference Book on Fuel Burning and Steam Generation*.[1]

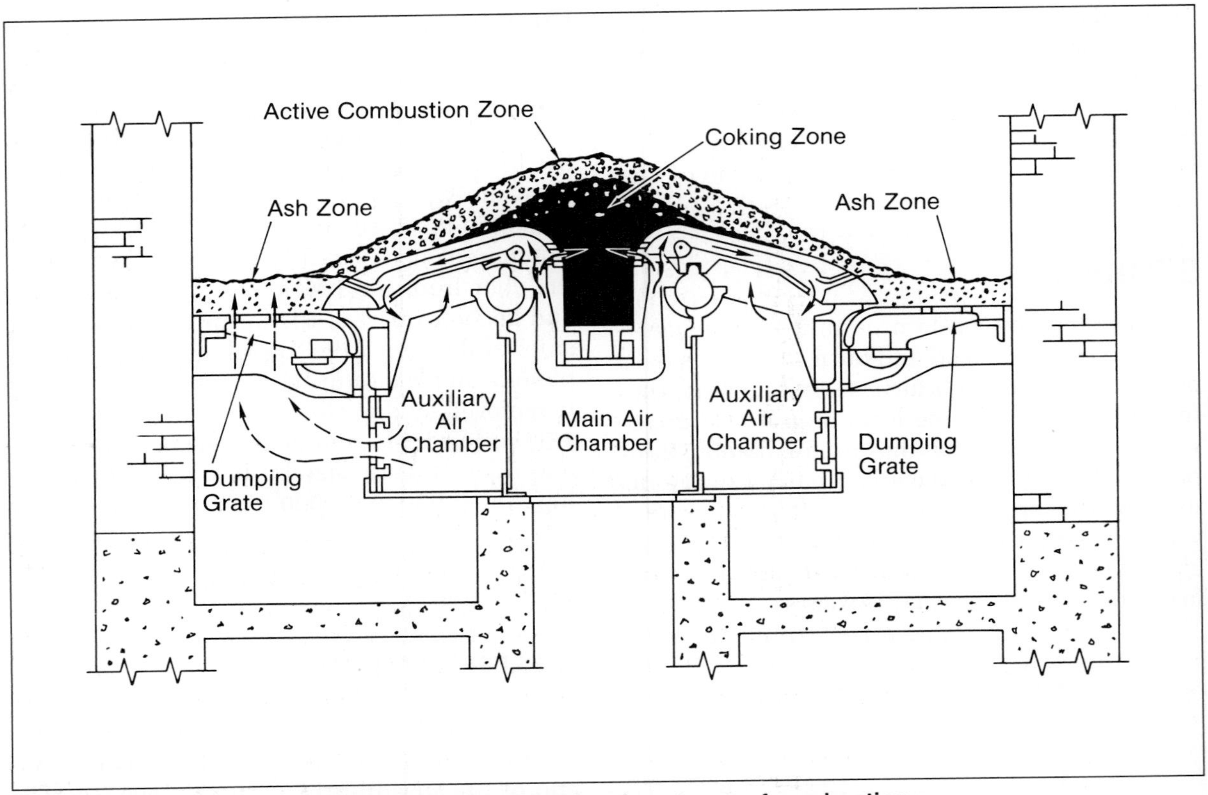

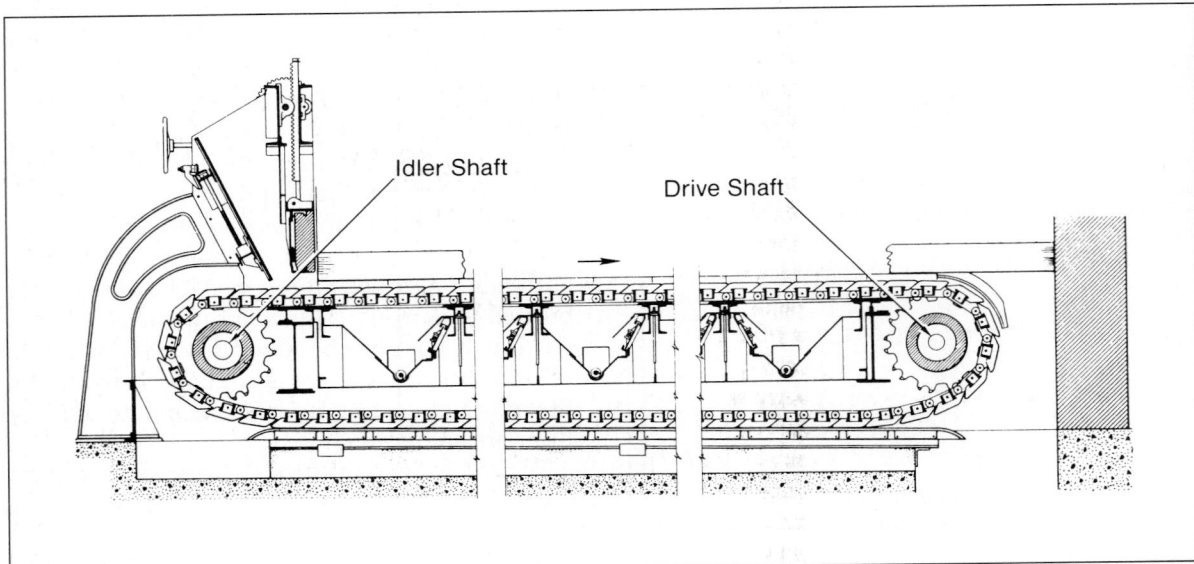

Fig. 17 Cross-section through underfeed stoker showing stages of combustion

Fig. 18 Arrangement of C-E traveling-grate stoker

SPREADER STOKERS

The third major class of stokers, and the machine of choice in modern steam-generation applications, is the spreader stoker (Fig. 19). It combines the principles of suspension burning and thin-bed combustion. Feeder/distributor devices continuously project fuel into the furnace above an ignited fuel bed on the grate. Fines are burned in suspension while larger particles fall and burn on the grate.

The spreader stoker method of fuel firing provides quick response to changes in boiler demand, and generally never has more than a few minutes of fuel inventory on the grate. Flash drying and rapid release of tarry hydrocarbons enable this system to fire caking-type coals without concern of matting or clinkering. The air-cooled, non-agitated ash bed forms few clinkers despite low fusion temperature fuels. Particulate loadings for spreader stokers are higher than for previously described designs in which the fuel quiescently enters the burning zone. Practically all types of coal (except anthracite) and a wide variety of cellulose fuels, including wood wastes, bagasse, furfural residue sludge, rice hulls, and coffee grounds have been successfully burned on spreader stokers.

Two important designs used for spreader-stoker firing are the dump grate and continuous ash discharge (CAD) grate. The dump-grate machine has one or more independent sections which open like venetian blinds to discharge the accumulated ash. When ash builds up on one section, the coal feed is stopped and the fire is burned down. The undergrate air is then shut off to that section only, and the ashes are dumped. After the fire is reestablished, the process repeats in another section. Dump-grate stokers have been built for boilers ranging from 20,000 to 80,000 lb steam/hr.[1]

The most popular type of spreader stoker is that incorporating the CAD grate (Figs. 19 and 20). This machine, unlike the traveling grate stoker, moves toward the fuel feeders, which are throwing new fuel towards the back of the unit. All the fuel is burned before reaching the front end, from which ash is continuously dumped. The return side of the grate carries siftings, which fall through the topside, to the rear and discharges them into a hopper. Grate speed is regulated to maintain an ash bed depth of 2″ to 4″ at the discharge. Typical operating speed ranges from 2 to 20 ft/hr, and is varied via gear or hydraulic drive units.

The C-E CAD stoker grate surface is sectional-

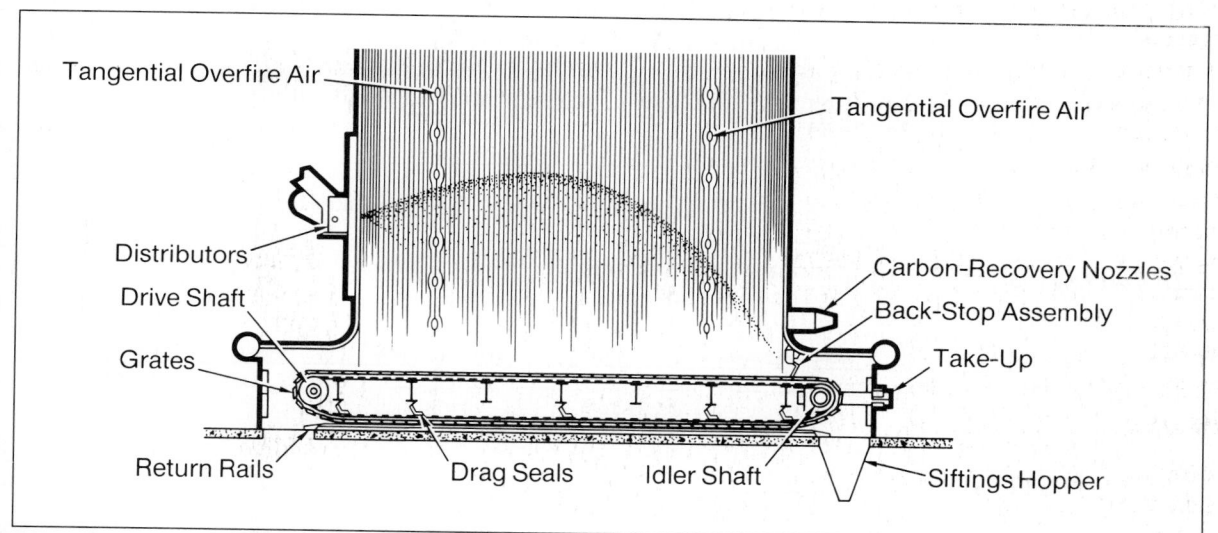

Fig. 19 **Spreader stoker with continuous ash discharge grate**

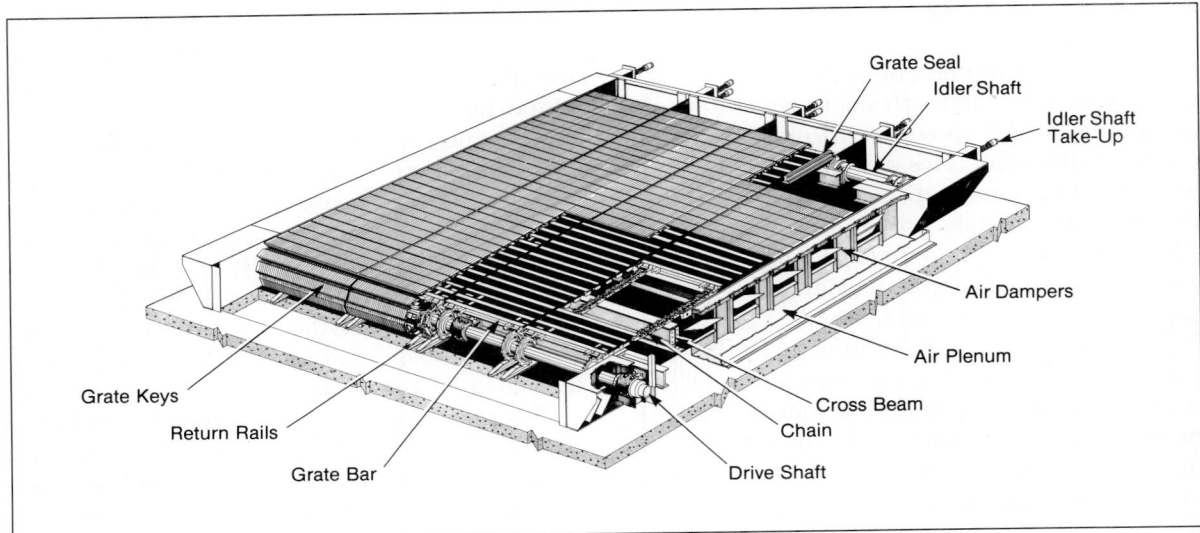

Fig. 20 C-E continuous ash discharge grate

ized. Each section is of the bar-and-key construction (Figs. 21 and 22) and has independent idler shafts. The grate bars are mild-steel fabricated I beams, while the grate keys are the ductile-iron type having a maximum working temperature of 1400°F. Constant load springs maintain proper grate tension. Thermocouples in the stationary castings between the moving-grate sections monitor temperatures and signal overheat conditions that require corrective action.

Current boiler designs that include high-temperature combustion air preclude the use

of conventional grease-lubricated stoker main- and idler-shaft bearings. Therefore, self-lubricating bushings rated at 750°F in an oxidizing environment are used exclusively.

The combustion-air chamber is located below the fuel-supporting surface; the air passes through the keys. Undergrate air is zoned front-to-back by adjusting manual damper assemblies at the entrance to the undergrate compartments. Sealing between compartments is maintained by drag seals and, for cellulose units, by an undergrate sand seal. Thus, under compartmented areas uniform air pressure is

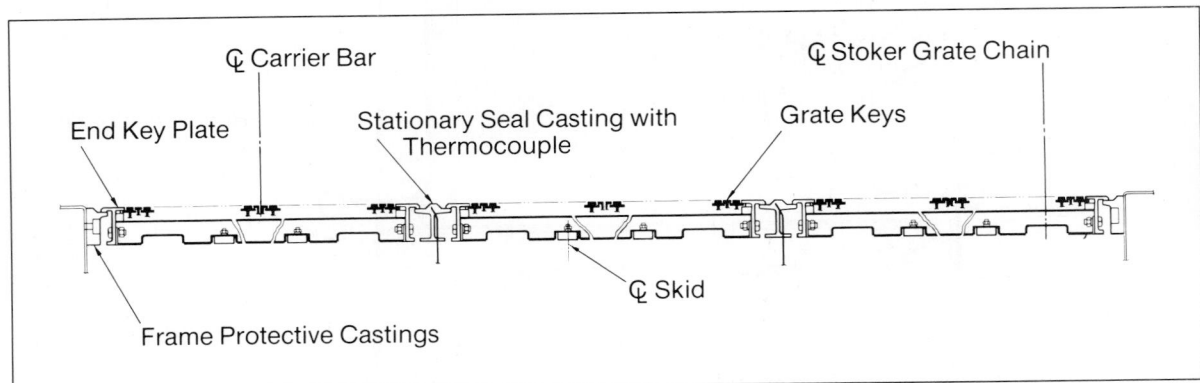

Fig. 21 Sectionalized grate assembly for continuous ash discharge stoker

maintained without air leakage.

One of the outstanding characteristics of spreader-stoker firing is the ability to handle rapid load swings with little or no change in steam temperature and pressure. Turndown of 4:1 is readily achievable. It is not unusual for a spreader-stoker installation to go from 25 percent to 100 percent of rated capacity in less than two minutes. Also, if fuel is shut off, the fire goes out almost immediately.

COAL FIRING ON SPREADER STOKERS

Firing coal with a spreader stoker is accomplished by using one or more coal spreaders (Fig. 23) mounted on a front plate. These include a feeder that regulates the flow of coal in proportion to the load, and a distributor rotor that spreads the coal over the grate.

COAL-SPREADER DESIGN

The design of the coal spreader (major parts of which are identified in Fig. 23), is of utmost importance, because it controls the uniformity with which fuel is supplied to the furnace. When the coal is fine and dry, cascading over the feeder may occur; when it is wet, the particles may cohere or stick to the feeder, causing an erratic supply of fuel to the distributor. For all conditions, including these governing extremes, the rotary feeder provides a positive control of the fuel feed rate. The feeding unit permits positive regulation of the fuel supply over a wide range of operation by either manual or automatic control.

The coal is measured out of the feeder at a rate necessary to carry the boiler load. It then falls in a practically continuous stream into the path of the revolving distributor blades. These blades are usually mounted in rows parallel to the axis of the distributor rotor. The projection of fuel from a single distributor or from a combination of distributors results in uniform dis-

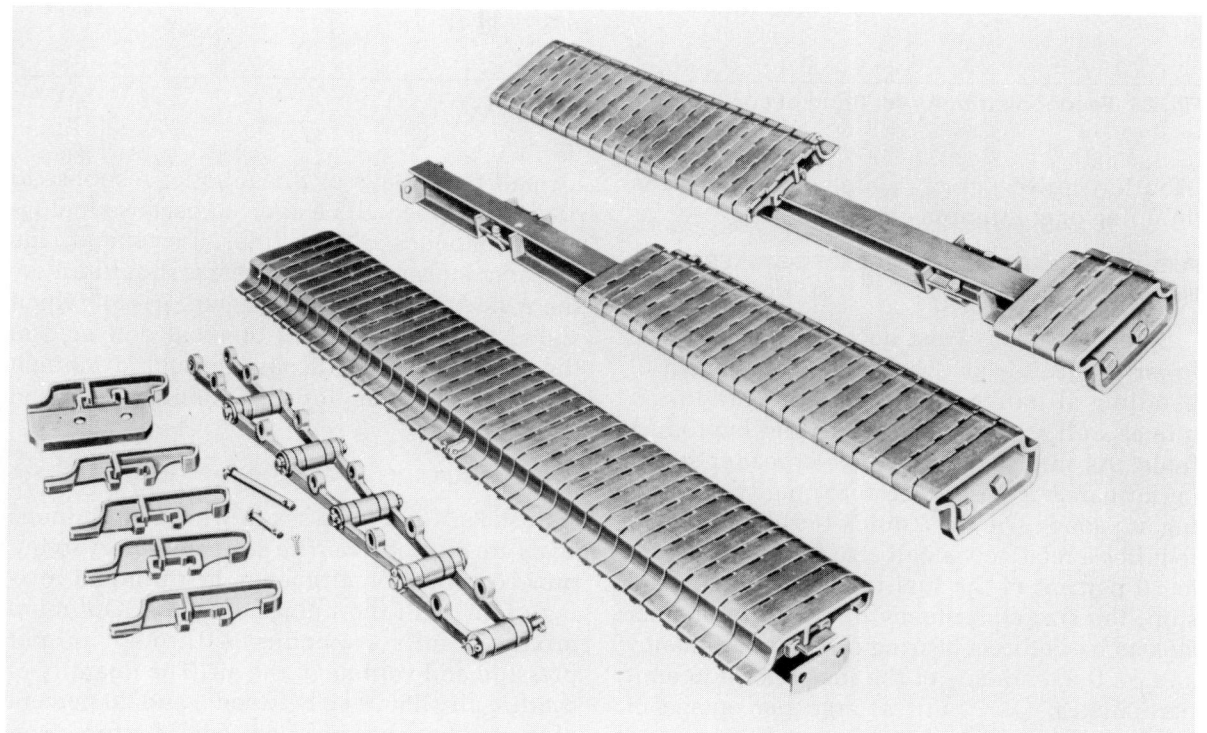

Fig. 22 **Bar-and-key grate construction for CAD stoker**

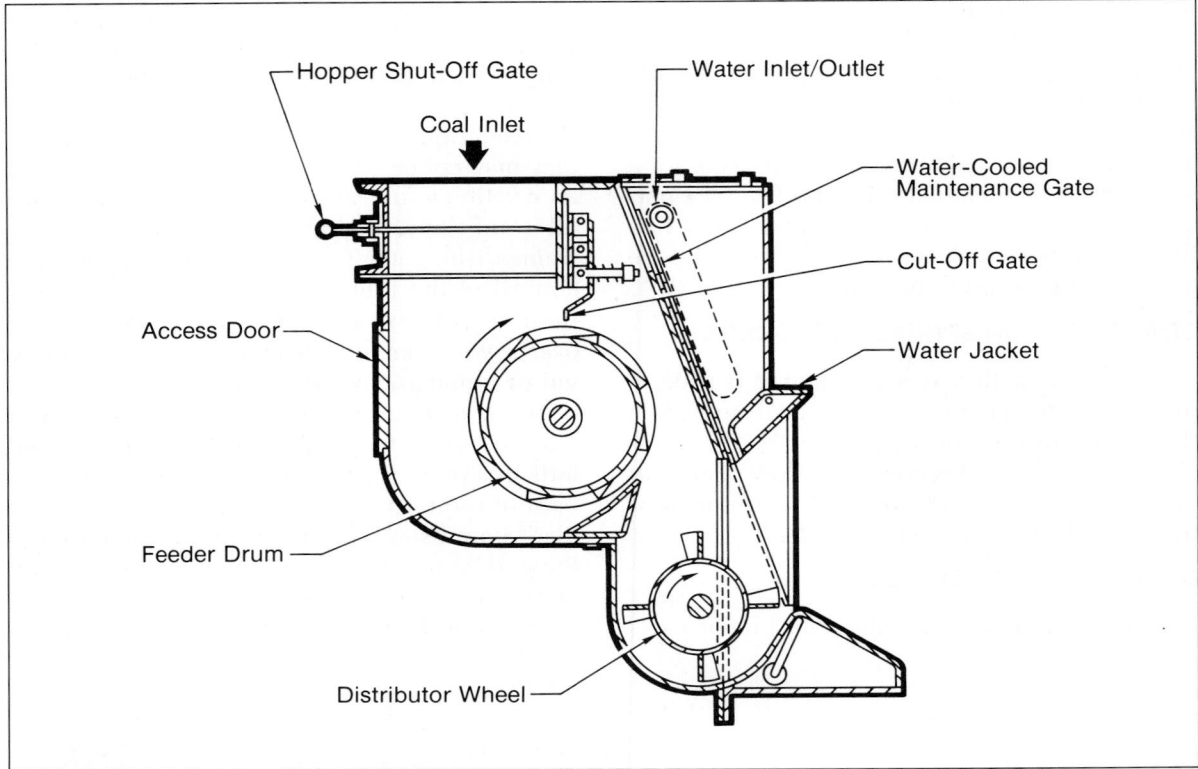

Fig. 23 **Feeder-distributor for firing of coal on a spreader stoker**

tribution on the grate, which may be of the dumping or continuous-discharge type.

FUEL SPECIFICATIONS FOR SPREADER-STOKER FIRING

Spreader stokers were developed to burn the lower grades of coal, but they are capable of handling all ranks from semianthracite to lignite as well as numerous waste and byproduct fuels. As might be expected, spreader-stoker performance is best when fuel quality and sizing are good. The thin, quick-burning fuel bed requires a relatively small size fuel. Because 25 to 50 percent of the fuel is burned in suspension, the size consistency of coal for spreader stokers has a direct bearing on boiler efficiency and on the tendency of the installation to emit particulates. Coal with a large percentage of fines will have high particulate emissions and high carbon loss. If the coal is coarse, with only

a small percentage of fines, boiler response to load variations will be affected because fuel ignition depends on the fines. In general, the spreader-stoker coal consistency should follow the ABMA recommendation per Fig. 24, which shows that 95 percent of the coal delivered to the coal spreader will pass through a 3/4-inch round-hole screen. Top size should not exceed 1 1/4 inches.

OVERFIRE AIR

In stoker-fired furnaces in which bituminous coals are burned, overfire air is necessary to improve combustion efficiency by turbulent mixing of air with the unburned gases. Optimum mixing results are achieved through proper pressure and volume of the air. The quantity of overfire air should be between 5 and 20 percent of the total quantity of air needed for fuel combustion. The amount of overfire air will be a

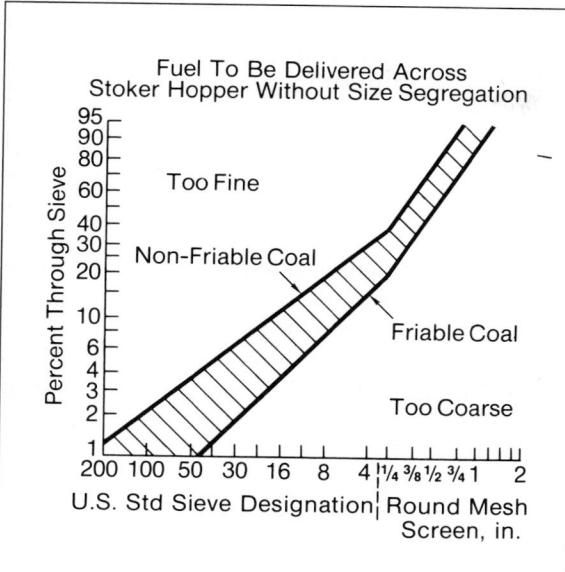

Fig. 24 ABMA-recommended limits of coal sizing for spreader stokers

function of the coal rank and the amount of excess air in the furnace. Relatively small jets of air at pressures up to 25 or 30″ WG are used in most installations for improving combustion in the furnace, and for reducing visible emissions and cinder carryover.

CINDER REINJECTION

A carbon-reinjection system can improve boiler efficiency. The reinjection system reclaims carbon separated from the flyash by using vibrating or rotating screens. Once the carbon has been separated from the flyash, carbon-recovery nozzles reinject it into the high temperature zone of the furnace, just above the fuel bed.

The proper design of a carbon-reinjection system requires that the boiler designer know the percentage of combustibles remaining in various screen-sizes of the coal residue to be reinjected. The combustion potential of this material also is important. This knowledge can only be obtained from test data gathered from operating units. It is probable that the reinjected carbon particles will exhibit properties

quite different from the main coal supply. In some instances, data will indicate that a reinjection system would not be appropriate.

CELLULOSE FUEL FIRING ON SPREADER STOKERS

Spreader-stoker firing systems have been applied to a number of boilers burning a variety of waste products such as bark, refuse, bagasse, and furfural residue. In general, cellulose fuel has a higher moisture and volatile content than coal, and the combustion process involves three phases: drying, distillation of volatiles, and burning of fixed carbon. These phases take place in rapid succession. For complete combustion of the volatile gases formed, 30 to 50 percent of the air supply is furnished as tangential overfire air. This system effectively increases burn-out and reduces particulate carryover. The center lines of the overfire air nozzles are tangent to a firing circle in the center of the furnace. All nozzles are oriented in the same direction to establish the same rotation as that of the tangential windboxes located above the overfire air zone. This continuity of rotation throughout the lower furnace assures a strong vortex, promoting particulate burnout and reducing carryover into the exiting gases.

Distributors for cellulose fuels may be either mechanical or pneumatic. A mechanical distributor (Fig. 25) is essentially a rotating drum with blades placed at an angle to establish uniform fuel distribution over the grate surface. The speed of the rotating drum is adjusted for fuel conditions and the throw required to properly cover the grate.

With the pneumatic distributor (Fig. 26) high-pressure air projects and distributes the fuel uniformly over the grate. The fuel enters the furnace in a uniform, dispersed stream. Most is burned in suspension with the remainder consumed after falling to the grate. As with coal-fired spreader stokers, the grate surface may be of the dumping or continuous-ash-discharge type. A carbon-recovery system with sand-separation screens and steam injection is often incorporated to increase the boiler efficiency.

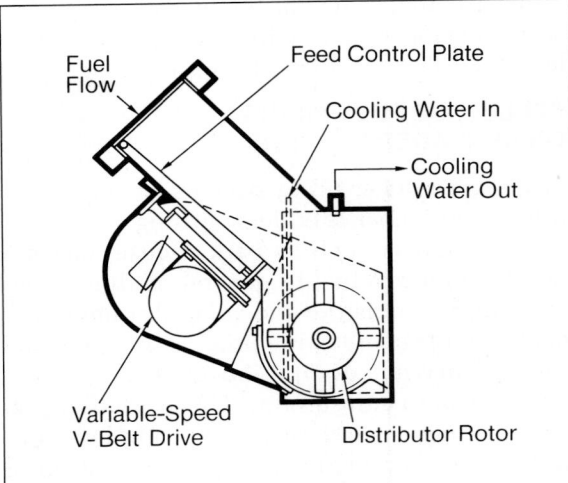

Fig. 25 Mechanical distributor

SUSPENSION FIRING OF CELLULOSE FUELS

Another proven option for burning some cellulose fuels is suspension firing. Here, like pulverized coal firing, the prepared fuel is pneumatically conveyed to the four corners of the furnace and injected into the fireball (Fig. 27). Depending on fuel sizing and moisture, a small dump grate may be affixed at the bottom of the furnace in the coutant throat (Fig. 28). The application of this firing system requires

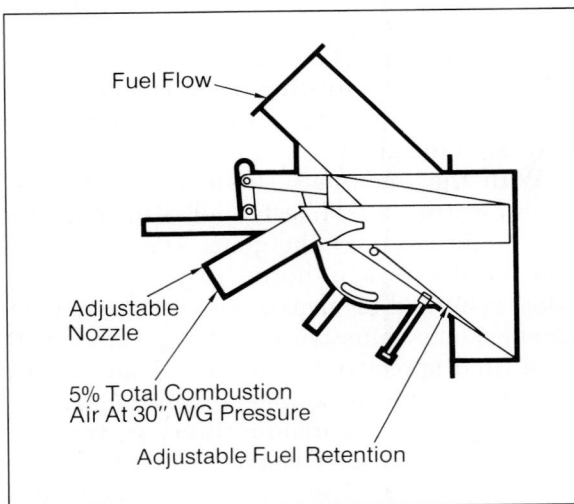

Fig. 26 Pneumatic distributor

that most of the fuel burning occur completely in suspension. In general, low-density materials such as diffuser bagasses, shredded paper wastes, and dry lumber mill waste can be fired at a larger top size or higher percentage of total heat input than higher-density, high-moisture wood waste.

The solid waste fuel is metered to four high-pressure pneumatic transport systems which convey it to tilting nozzles located in compartments of the corner windboxes. The light particles are flash dried and burned in suspension. Larger dense particles are partially dried and burned in the lower furnace before falling to a

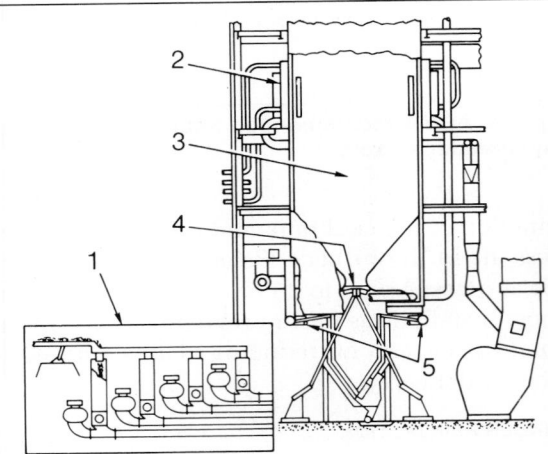

1. **Pneumatic Transport System.** Each of four systems include a positive displacement blower, silencers, air lock, and conveying lines.
   The solid-waste fuel is metered to this system for direct transport to the windbox nozzle.

2. **Windbox Nozzle.** Adjustable nozzles are incorporated in the tilting tangential windboxes. The arrangement permits optimization of firing conditions for the best burn out of the solid-waste fuel.

3. **Tangential Overfire Air.** Multiple elevations of tangential overfire air sustain the suspension burning of the waste fuel.

4. **Dump Grate.** This feature allows burn out of the relatively small percentage of solid-waste fuel which does not burn in suspension.

5. **Grate Air System.** Undergrate air provides most of the air for burning fuel which falls to the grate. Overgrate air is introduced from both ends of the grate to aid in distributing concentrations of fuel on the grate surface.

Fig. 27 Pneumatic firing system

Fig. 28 Dump grates at furnace bottom: closed position on left, open position on right

small dump grate. Burning on the grate adds heat to the lower furnace. This added heat increases the suspension drying and burning of incoming fuel.

The cellulose-fuel nozzle (Fig. 29) is part of the windbox assembly. The bucket at the end of the nozzle can be tilted ±30° along with the adjacent auxiliary air and supplemental fuel nozzles.

The dump grates (Fig. 28) are suspended in the hopper throat formed by the pressure parts. The shafts and bearings are located under the pressure parts where they are protected from the radiant heat of the furnace. Undergrate air is introduced into the ash hopper; overgrate air nozzles are provided at each end of the grate to

blow down any piles that may occur.

Grate castings and support bars are similar to CAD stoker design to maximize component life. Pneumatic cylinder assemblies actuate the grates. This simple design has had a record of very low maintenance and high availability.

C-E TYPE RC SPREADER STOKER

The Combustion Engineering Type RC (refuse combustion) stoker has been designed specifically for prepared-waste-burning facilities that fire refuse-derived fuel (RDF) alone or in combination with coal. The need for such a design grew from operating experience at plants using conventional grates designed for burning woodwaste and coal.

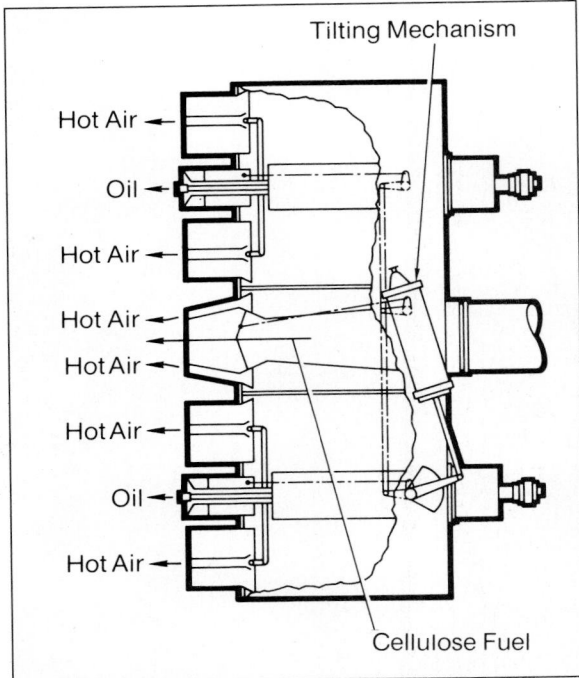

Fig. 29 Tangential corner windbox assembly

TYPE RC STOKER DESIGN DESCRIPTION

Fig. 30 illustrates the RC stoker. It is a continuous-ash-discharge grate design that uses a catenary formed by the return side of the grate to maintain chain tension. The moving grate surface is divided along the grate width to facilitate grate maintenance and to allow for temperature monitoring by means of thermocouples located within the stationary grate-seal castings. These grate surfaces are a bar-and-key design similar in construction and method of assembly to that of the CAD stoker (Fig. 22). The RC key profile, shown in Fig. 31, offers the following distinct advantages:

1. The grate-surface metallurgy is designed with oxidation-resistant keys and carbon-steel grate bars which support the grate keys.

2. Warpage due to high-temperature creep is not a problem since the steel grate bars do not see high temperatures.

3. Grate maintenance is facilitated by key re-

moval from outside the grate and by replacement only of those keys that are broken or oxidized instead of complete assemblies.

The undergrate air is split to each grate half and then into five compartments along the length for a total of ten individual zones. Each compartment has its own damper control to regulate the amount of air to that zone. The air passes around the sifting trough, then up through the free area of the grate surface.

Ash siftings and molten aluminum that fall through the grate are contained within trough assemblies and removed by a compartment screw conveyor which discharges into a common transfer screw. Each transfer screw discharges into the main ash system. While it is recognized that additional moving parts add to the complexity and maintenance needs of the stoker, there are two compelling reasons for the trough/screw conveyor system. First, and most important, is that molten aluminum which flows through the grate will not freeze on the underside of the grate keys, plugging air passages or hindering grate motion. Second, the moving grate surface does not readily allow the siftings to pass through into the siftings hopper.

Both the drive and idler shafts are supported by self-lubricating bushings. Each bearing is sealed around its exposed circumference by graphite fiber packing held in place by a single-piece seal ring bolted to the bearing housing. In addition, lightweight covers enclose the entire housing assembly to minimize sifting contamination. A triple-reduction worm gear is supplied to drive each half of the grate. A 3-HP, 1200-RPM motor is integrally mounted to a speed variator that allows continuous drive motion over a 9:1 speed range.

CE/db MASS-BURNING GRATE

Combustion Engineering is the exclusive licensee for the DeBartolomeis (db) mass-burning pusher grate in North America. This unique grate system, developed by DeBartolomeis over 35 years ago, has been used in Italy, Switzerland, France, Japan, North America and South America.

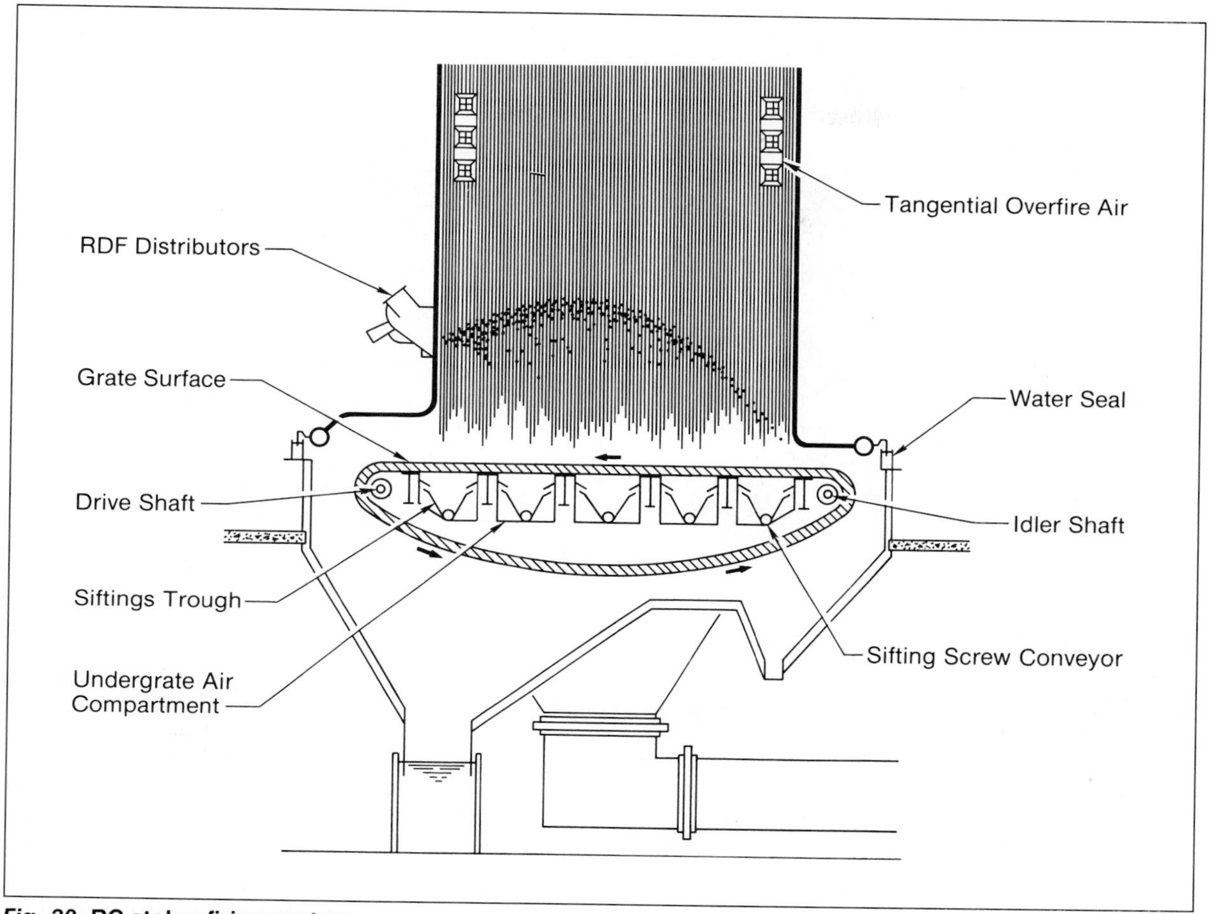

RDF Distributors

Tangential Overfire Air

Grate Surface

Water Seal

Drive Shaft

Idler Shaft

Siftings Trough

Sifting Screw Conveyor

Undergrate Air
Compartment

Fig. 30 RC stoker firing system

CE/db SYSTEM DESIGN

In the CE/db mass-burning system, overhead traveling cranes transfer unprocessed municipal solid waste (MSW) from a storage pit to the furnace-charging hopper of the steam-generating unit. With a clear view of the storage pit, the crane operator selectively removes such bulky items as tree trunks and domestic appliances from the pit.

The MSW falls by gravity into the feeding chute to form a natural seal between the furnace and ambient air and thereby prevent air infiltration. The charging hopper is top supported by the building steel and joins with the bottom-supported feed chute by means of a seal joint to

allow for thermal expansion. A remote-controlled, hydraulically operated, shut-off gate allows the furnace to be isolated from ambient air during normal shutdowns. The walls of the steel feed chute are diverging to prevent bridging and are designed to promote uniform mass flow of MSW to the ram feeder (see Fig. 32).

Thermal protection of the feed chute is accomplished by a water-cooled jacket. Inside the chute is a low-level alarm to advise the operator of the need for more fuel in order to maintain the seal between the furnace and ambient air.

A hydraulically operated ram feeder equipped with three steps is located at the base of the feed hopper. The reciprocating ram feeder moves waste from the feed chute onto a

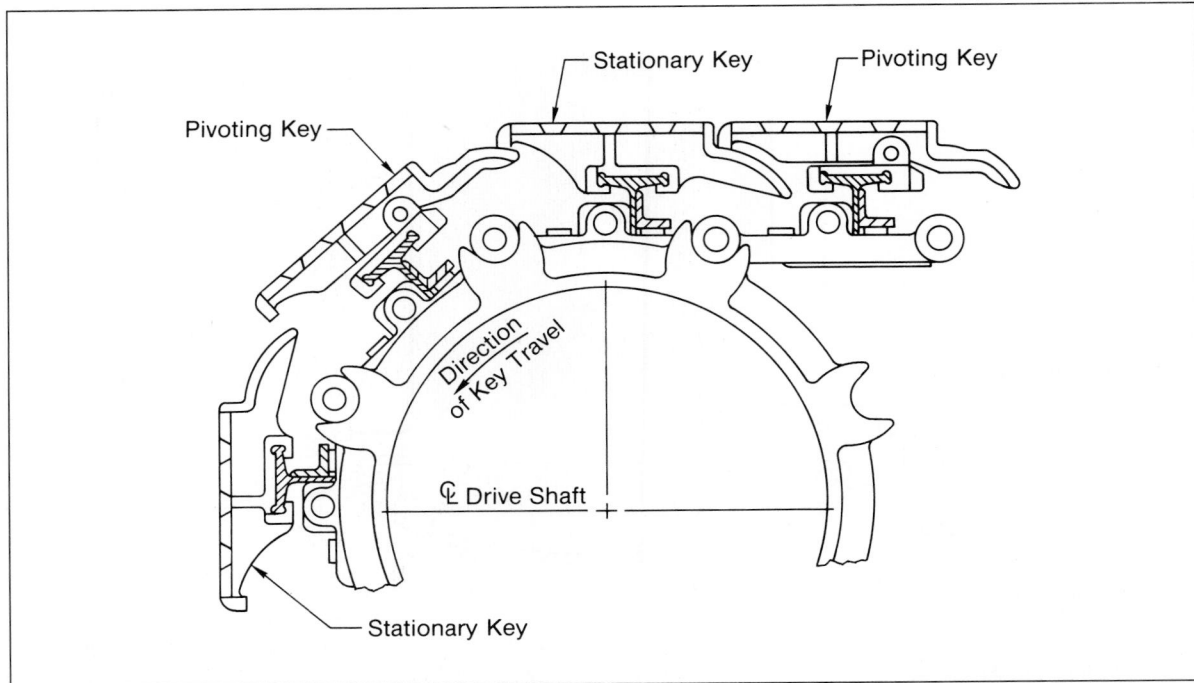

Fig. 31 Design key action at sprocket, C-E RC stoker bar-and-key grate surface

multiple-zone burning grate surface consisting of alternating rows of moving and stationary cast stainless-steel grate bar assemblies. This device follows the steam-flow demand by regulating the quantity of MSW fed onto the grate for combustion.

The pusher grate consists of three or more grate sections inclined slightly downward toward the discharge end. Drying and devolatilization of the MSW occur on the initial portion of the grate, and are followed by active combustion and final burnout on the subsequent sections. The movement of each grate section is separately and independently controlled by hydraulic drive cylinders, allowing adjustment of frequency, speed and stroke. The stroke and speed can be adjusted in place: stroke displacement can vary from 2 to 10 inches, and speed can vary from 3 to 30 feet per minute. Frequency can be adjusted automatically in the control room from one stroke every 30 seconds to one stroke every 5 minutes. With each grate section being equipped with such independent

adjustment, it is possible to vary the fuel-bed height and combustion conditions on the grate to allow for the variable characteristics of the waste. Forward movement of MSW fuel on the grate surface is accomplished by the reciprocating motion of the movable rows of grate bars.

The stationary rows of grate bars for each grate section are directly supported by the steel framework of the grate. The reciprocating grate bars are fastened to a frame which slides on support rollers. Stationary and reciprocating grate bars are exactly alike (see Fig. 33).

The design of the grate bar assembly requires the undergrate air to flow along the underside of the surface plates before entering the MSW fuel bed through the air gaps at the front of the grate bar assembly (Fig. 34). The position of the free area occurs vertically between the nose of the plate and the scraper and provides protection from problems caused by siftings and molten materials which might obstruct the regular flow of undergrate air. The scraper is protected from radiation by the surface plate and acts as a

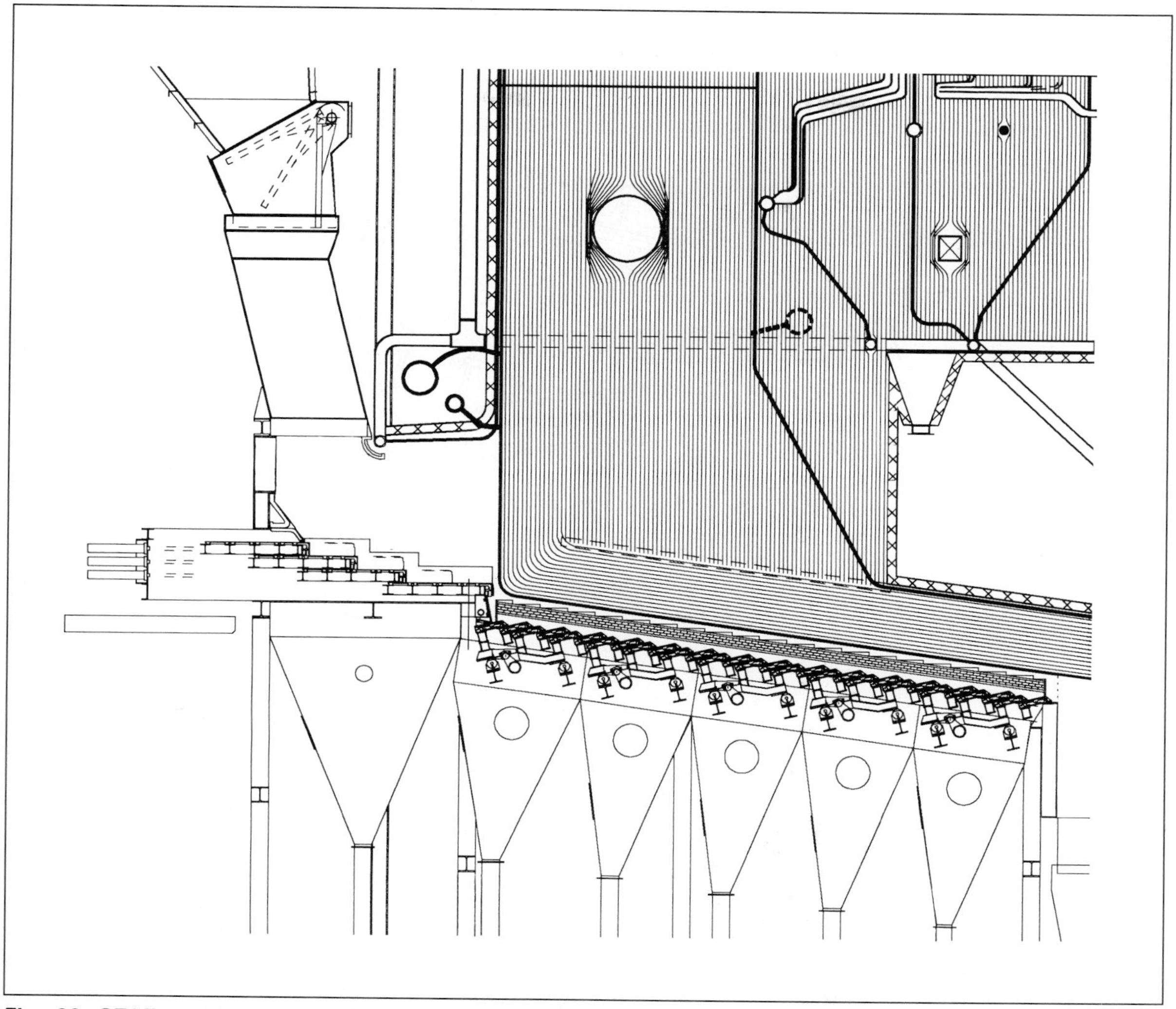

Fig. 32 CE/db stoker in a mass-burning MSW furnace

sealing element against the surface of the lower adjacent surface plate. The free area between the precision-ground cast stainless steel plates and the pivoting scrapers is approximately 2 percent of the total grate surface. The sides of the surface plate contain an overlap/underlap feature minimizing siftings and air leakage between the plates.

CE/db AIR SYSTEM

Approximately 60 percent of the total combustion air is introduced through the grate sur-

face and is called *undergrate air* (UGA). The balance of the air enters through nozzles above the grate and is called *overfire air* (OFA).

The undergrate air can be preheated to aid in the drying of the fuel on the grate. This is necessary only for extremely wet MSW. A steam-coil air heater after the forced-draft fan is one method of preheating which is flexible to varying fuel conditions. Each grate section has its own plenum to provide a controlled quantity of primary combustion air for optimizing combustion of the MSW. Dampers located in the duct-

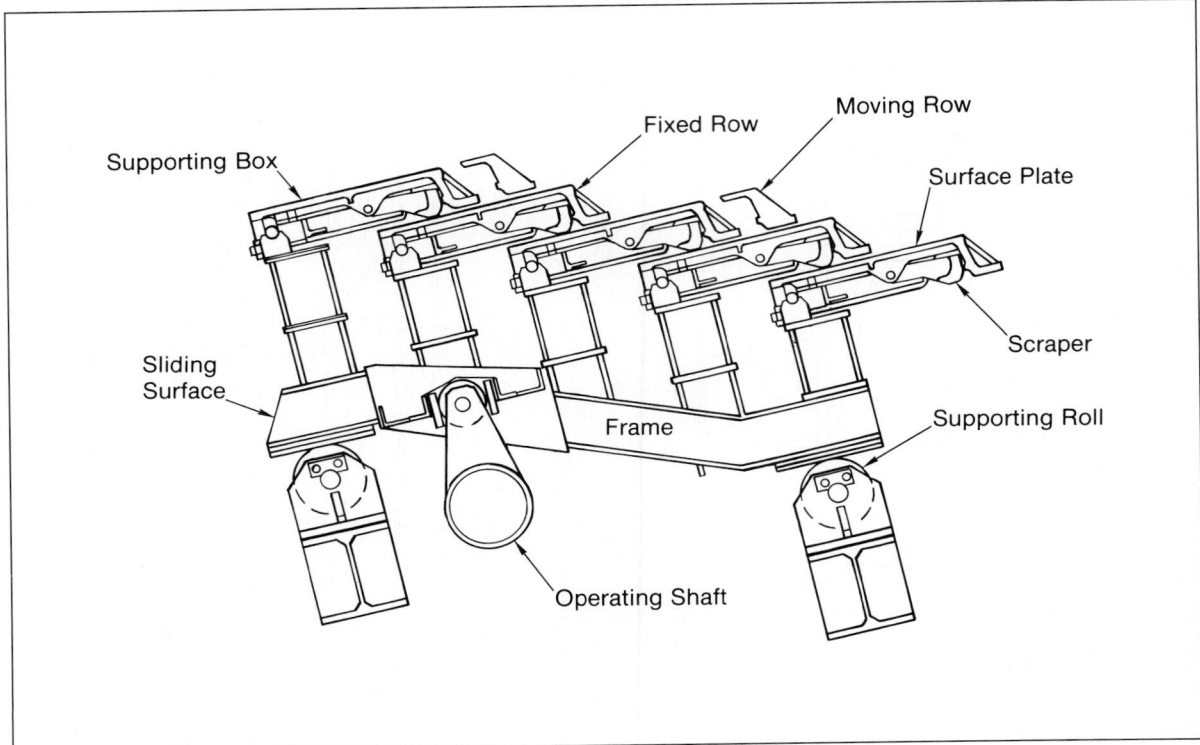

Fig. 33 Typical CE/db grate section

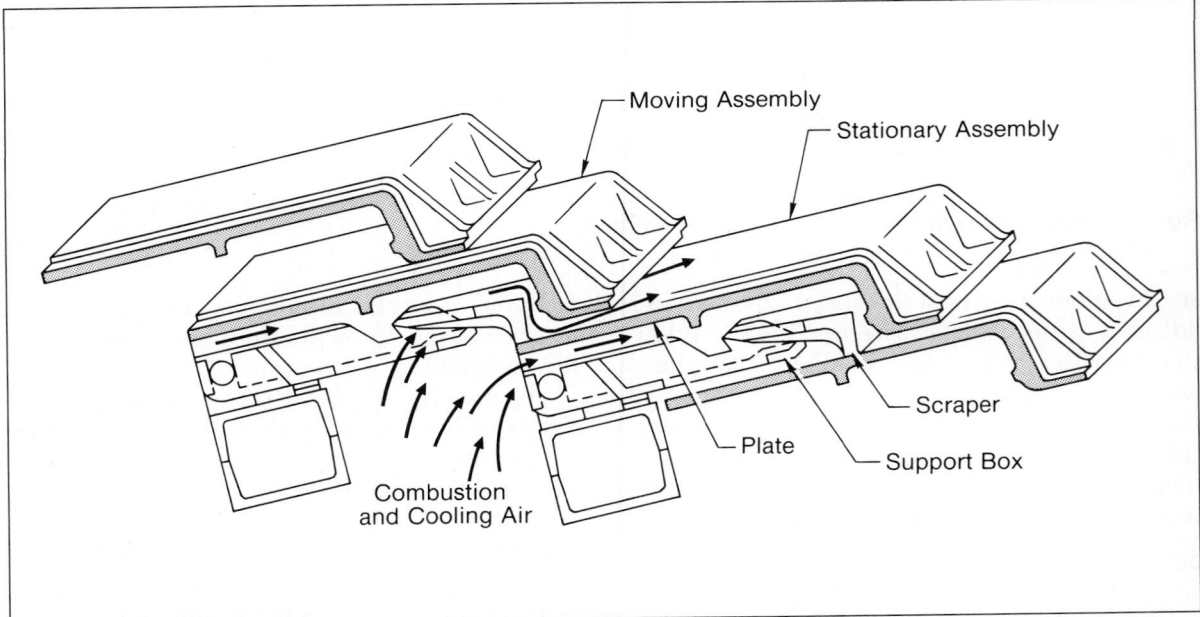

Fig. 34 Undergrate air flow pattern of CE/db grate

work to each plenum chamber provide control for the undergrate air distribution.

Strategically located in the front and rear walls, the OFA nozzles provide turbulence for mixing and ensure complete combustion of the volatile gases. The OFA is at a pressure of 20 to 30″ WG for sufficient penetration.

MORE ON FUEL FIRING

Other aspects of preparing, transporting, firing, and burning fuel in steam-generator furnaces are covered in complementary chapters of this text.

The processes of combustion and the formation of nitrogen oxides are further described in Chapter 4. Chapters 7 and 8 cover the integration of the firing system and furnace in utility and industrial boiler designs. The equipment for injecting fuel into fluidized beds is described in Chapter 9. Firing equipment exclusively oriented to shipboard service is included in Chapter 10, Marine Boilers.

Chapter 11, Pulverizers and Pulverized-Coal Systems, focuses primarily on how coal is reduced in size and transported to furnaces.

Chapter 13, Operational Control Systems, deals at length with safety and flame-detection systems that are so important to the modern firing equipment that has been presented in this chapter.

REFERENCES

[1]Otto De Lorenzi, ed., *Combustion Engineering—a Reference Book on Fuel Burning and Steam Generation*, First Edition, New York: Combustion Engineering-Superheater, Inc., 1947.

BIBLIOGRAPHY

Bartok, William; A. R. Crawford; and G. J. Piegari, "Systematic Investigation of Nitrogen Oxide Emissions and Combustion Control Methods for Power Plant Boilers," R. W. Coughlin, A. F. Sarofin, and N. J. Weinstein, (eds.), *Air Pollution and Its Control*, AICHe Symposium Series, No. 126, Vol. 68, 1972, pp. 66–74. New York: American Institute of Chemical Engineers, 1972.

Bartok, William, et al., *Systems Study of Nitrogen Oxide Control Methods for Stationary Sources. Volume II. Final Report.* June 20, 1968–November 20, 1969. Report No. PB-192 789 (GR-2-Nos-69). Springfield, VA: National Technical Information Service, 1969.

Bueters, K. A., et al., "NO_x Emissions from Tangentially Fired Utility Boilers—A Two Part Paper," Part I, "Theory," by K. A. Bueters and W. W. Habelt; Part II, "Practice," by C. E. Blakeslee and H. E. Burbach, presented at the 66th Annual AICHe Meeting, Philadelphia, November 11–15, 1973.

Chen, S. L., McCarthy, J. M., Clark, W. D., Heap, M. P., Seeker, W. R., and Pershing, D. W., "Bench and Pilot Scale Process Evaluation of Reburning for In-Furnace NO_x Reduction," *Twenty-First Symposium (International) on Combustion*. Pittsburgh: The Combustion Institute, 1986.

Frey, D. J. and M. S. McCartney, "The Formation, Measurement, and Control of Nitrogen Oxides in Existing Fossil Fuel Fired Steam Generators," presented at the 28th Engineering Conference, TAPPI, Boston, October 9–11, 1973; also as Combustion Engineering publication TIS-3598.

Habelt, W. W., "The Influence of the Coal Oxygen to Coal Nitrogen Ratio on NO_x Formation," presented at the 70th Annual AICHe Meeting, New York, November 13–17, 1977; published as Combustion Engineering publication TIS-5140, Windsor, CT: Combustion Engineering, Inc., 1977.

Heap, M. P., et al., "Burner Design Principles for Minimum NO_x Emissions," *Proceedings, Coal Combustion Seminar*, Research Triangle Park, NC, June 19–20, 1973, Report No. PB-224 210 (EPA 650/2-73-021). Springfield, VA: National Technical Information Service, 1973.

Marshall, J. J. and A. P. Selker, "The Role of Tangential Firing and Fuel Properties in Attaining Low NO_x Operation for Coal-Fired Steam Generation," presented at the EPRI NO_x Control Technology Seminar, Denver, November 8–9, 1978; also as Combustion Engineering publication TIS-5623A.

Pershing, D. W. and J. O. L. Wendt, "Pulverized Coal Combustion: The Influence of Flame Temperature and Coal Composition on Thermal and Fuel NO_x," *Sixteenth Symposium (International) on Combustion*, Massachusetts Institute of Technology, Cambridge, MA, August 15–20, 1976, pp. 389–399. Pittsburgh: The Combustion Institute, 1976.

Sarofim, A. F., et al., "Strategies for Controlling Nitrogen Oxide Emissions During Combustion of Nitrogen Bearing Fuels," *AICHe Symposium Series No. 175*, 74, 67, (1978).

Selker, A. P. and R. L. Burrington, "Overfire Air Technology for Tangentially Fired Utility Steam Generators Burning Western U. S. Coal," presented at the 2nd Symposium on Stationary Source Combustion, New Orleans, August 29–September 1, 1977; published as Combustion Engineering publication TIS-5242, Windsor, CT: Combustion Engineering, Inc., 1977.

Taylor, B. R., "Reactions of Nitrogen Species In Fuel-Rich Flames", Sc.D., Department of Chemical Engineering, M.I.T., 1984.

Zel 'Dovich, Ya. B., P. Ya. Sadovnikov, and D. A. Frank Kamenetskii, *Oxidation of Nitrogen in Combustion*. Translated by M. Shelef. Moscow-Leningrad: Academy of Sciences of the USSR, 1947.

CHAPTER 13

Operational Control Systems

A power plant may be thought of as an electricity factory. In that context, modern power plants are among the most highly automated and centrally controlled and monitored production facilities in the world.

As power plants began to grow in size and complexity many years ago, local monitoring and regulation of the plant systems in a timely manner became impossible. Thus, instrumentation and controls in power plants came into being to provide the ability to operate the major plant systems from a central location (Fig. 1). Historically, meters, gages, and lights displayed equipment status to the operator, while recorders made a permanent record of plant performance. Remotely operated air cylinders and electric motors gave plant operators the capability of responding quickly and efficiently to changing plant requirements. More recently, cathode-ray tubes (CRT's) have replaced the panel-board instrumentation, to link the operator with past and present process information through sophisticated microprocessor-based distributed control hardware.

Since the early days of remote operation, the generation of electricity has become a very complicated business. High energy costs demand that as much electricity as possible be extracted from the fossil fuel consumed. Higher availability of equipment is needed to stem rising operating and maintenance costs. Protection of both personnel and equipment must be achieved, and unscheduled shutdowns have to be kept to a minimum.

While obviously instruments and controls cannot of themselves satisfy such concerns, the problems have resulted in a substantially increased requirement for sophisticated instrumentation and automatic control systems.

This chapter deals primarily with the major control and diagnostic systems associated with fossil-fuel-fired utility steam generators. Control systems for the various types of steam generators used in stationary industrial applications are also discussed.

INTERLOCKS AND SAFETY SYSTEMS

The concept of a centralized control room was a big step forward in power-plant control. Nevertheless, as first conceived and executed, remote-manual systems relied entirely upon operator judgment and response.

The multiplicity of operating steps required for safely and properly admitting fuel to a furnace (or the removing of incompletely burned fuel from a furnace) leaves considerable lati-

Fig. 1 **Instrumentation and controls permit operation of major power-plant systems from a central location. Control-room meters, gages, lights, and CRT's display equipment status; recorders permanently chronicle plant performance; and diagnostic systems monitor and evaluate both status and performance.**

tude for operator error, if left solely to her or his judgment. Determining the adequacy of ignition-energy levels is another area that should not be left to operator discretion. Considering that a major furnace explosion can result from the ignition of unburned fuel accumulated in only one to two seconds, it is apparent that human reaction time is inadequate, to say nothing of the need for an instantaneous decision-making capability.

The recognition of these limitations to a completely operator-dependent mode of operation led to the development of automatic protection systems that are designed to minimize the risk of furnace explosions.

FURNACE EXPLOSIONS

Furnace explosions are rare and unlikely. When compared with the total number of unit operating hours, the hours lost because of explosions are minimal. This desirable situation exists because (a) furnaces are supplied with an explosive accumulation only during a small percentage of their operating lives and (b) just a minute part of those explosive charges receive sufficient ignition energy to actually cause an explosion.

In suspension burning, the primary control of the combustion process is the admission rate of fuel and air to a furnace, independently of each other. The dynamic response of the combustion reaction, however, depends on the diffusion of the fuel and air to a flammable limit, and the elevation of this diffused mixture to its kindling temperature. The aerodynamic diffusion of fuel and air results from both the rate and method of admission. This admission flow pattern produces diffusion mechanically by interscrubbing of the fuel and air masses. Molecular diffusion is also present as a result of the elevated temperature level at which the combustion process takes place.

Furnace explosions result from a rapid rate of volume increase of the gaseous combustion products when too great a quantity of fuel and air reacts almost simultaneously in an enclosure with limited volume and strength. Avoiding furnace pressures in excess of furnace enclosure design pressure is, therefore, necessary to prevent furnace rupture.

The basis for any explosion-prevention system must be to limit the quantity of flammable fuel-and-air mixture that can exist in the furnace at any given instant. The rate of maximum pressure rise possible during the reaction is a function also of the available oxygen per unit volume of reactants. The effect of any oxygen-density diluent (nitrogen, increased tempera-ture, decreased pressure, excess fuel, inert gases) reduces the possible explosion pressure.

Furnace-explosion prevention should be aimed at limiting the quantity of diffused flammable fuel-air mixture that can be accumulated in a furnace in proportion to the total volume and the mechanical strength of the furnace.

While fuel and air are being admitted to a furnace, there are only three possible methods of preventing excessive flammable diffused accumulations:

1. Igniting all flammable mixtures as they are formed, before their excessive accumulation.

2. Diffusing all flammable mixtures with sufficient additional air, prior to ignition, to a point beyond the diffused flammable-mixture ratio; and accomplishing this with a sufficient degree of diffusion before a critical percentage of the furnace volume becomes occupied by the flammable mixture.

3. Supplying an inert gas to diffuse simultaneously with the fuel and air, thereby diluting the oxygen content of the mixture below the flammable limit.

Implementation of these preventative methods requires operator action beyond the response, memory, and judgment capabilities of the normal operator controlling a plant in the manual mode. A fireside safeguard system must supervise the flow and processing of fuel, air, ignition energy, and the products of combustion. Satisfactory boiler operation requires that these four ingredients be properly prepared, ratioed, directed and sequenced so that the furnace cannot contain an explosive mixture. At the same time, the combustion process must be supervised to check the results. Combustion must be kept efficient or the unconverted chemical energy may accumulate and subsequently become explosive.

The following factors influence the effective composition change of an explosive charge:

- The facility for mixing
- The inert material in the fuel
- The fuel-air ratio (a near-stoichiometric ratio

develops the highest explosion pressure)

■ The kind of fuel

A furnace explosion requires both sufficient explosive accumulation within the furnace and sufficient energy for ignition. The ignition requirements for an explosive charge are very small, making it impossible to protect against all possible sources of ignition, such as static electricity discharges, hot slag, and hot furnace surfaces. Therefore, the practical means of avoiding a furnace explosion is the *prevention* of an explosive accumulation.

The factors determining the magnitude of a furnace explosion—mass, change in composition, and reaction time—are related in the *explosion factor*.

$$\text{Explosion Factor} = \frac{\text{Mass}}{\text{Furnace Volume}}$$
$$\times \frac{\text{Composition Change}}{\text{Elapsed Time}}$$

(1)

Each furnace has a limiting explosion factor. If the conditions create an explosion factor exceeding this limit, a catastrophic explosion can result. Any lesser reaction will produce a furnace "puff" (a nondestructive explosion) or a temporary upset.

To protect a furnace from an explosion, a safety system must insure a minimum reactive mass accumulation with a minimum available composition change and with a maximum reaction time required. Only control of the composition of the furnace atmosphere offers complete coverage in minimizing the explosion factor. After firing has begun, furnaces always contain sufficient mass to have an explosion and control of the time factor is impossible. Therefore, the composition change must be controlled to prevent furnace explosions with any assurance.

The mechanics of furnace explosions, although defining the actual process, do not de-

scribe the furnace operations which provide the explosive accumulations. Ideal furnace operation continuously converts reactive furnace inputs into unreactive products as fast as the inputs enter the furnace—this precludes furnace explosions. However, in practical furnace firing, unfavorable operations that create explosive situations are difficult to avoid completely.

Several correctly timed events precede a damaging furnace explosion. The furnace explosion event itself is the rapid change in composition of the furnace atmosphere (not the furnace inputs). The change in furnace composition is not spontaneous, and suitable ignition energy, which can be substantially less than that required for continuous furnace-input ignition, must be supplied *after* the explosive composition is attained.

The potentially reactive furnace accumulation must be formed from an earlier buildup process which introduces reactive inputs not converted by oxidation to nonreactive or inert products. This buildup process must continue long enough to create a damaging accumulation. The accumulation composition, which must be within the limits of flammability for that particular fuel, is formed in one or more basic ways.

■ A flammable input into any furnace atmosphere (loss of ignition)

■ A fuel-rich input into an air-rich atmosphere (fuel interruption)

■ An air-rich input into a fuel-rich atmosphere (air interruption)

Furnace firing systems are designed to start up air-rich by introducing fuel into an air-filled furnace. Main fuel is introduced after the integral ignition system has satisfied permissive main-fuel interlocks that it can provide more ignition energy than the main fuel requires to be ignited or to remain ignited. Additional air is introduced around the primary-air/fuel mixture to take it beyond flammable limits, if it has not been ignited and reacted to inert combustion products; this is done to avoid a critical portion of the total furnace volume being occupied by a flammable mixture.

FLAME DETECTORS

In the prevention of furnace explosions, the detection of the absence of flame while fuel is being admitted is the only proper criterion for any control action initiated by a flame-monitoring system.

Flame-monitoring hardware must be reliable, must be sensitive enough to discern the minimum flame envelope, and must have fail-safe characteristics to avoid unnecessary trips. Reliability is improved by fault-detection circuits.

Various methods are used to determine circuit component failures which could cause a false indication of flame. Reaction time of the flame-detecting device must be an absolute minimum to prevent the accumulation of diffused flammable reactants in the furnace following a loss of flame, before the furnace protection system can initiate corrective action to prevent an explosion. Unnecessary trips are avoided by proper flame-safety control logic.

The burning process exhibits many characteristics which can be sensed as indicators of existing flame. Heat sensors, for example, in the form of thermocouples and bimetallic strips have been used successfully for many years in small space-heating systems, though the application of these devices to larger installations has generally not been practical.

Another characteristic of the burning process used for flame detection is the electrical-conduction capabilities of the ionized gases of a flame. An electrical-conduction path is established by a flame rod which extends into the flame envelope; the ionized gases themselves and the boiler tubes serve as the ground return of the system. This flame-rod concept is also in use today on smaller burners fueled by oil or gas, but the erosive nature of suspension-fired pulverized coal makes this type of flame detection impractical for that fuel.

Especially in large suspension-fired burner/furnace installations, the most practical characteristic to sense for proof of flame has proven to be the light emitted by the burning process. The light emission covers a very broad spectrum and it is a continuous, rather than a line, spectrum. The shape of the emission versus wavelength plot for fossil-furnace flames generally resembles the black-body curve, with peak emissions being noted which are characteristic of the fuel being fired.

The light emitted by these furnace flames is further characterized by a fluctuating intensity, commonly called *flicker frequency*, which depends on the area of the flame being viewed, as well as physical conditions at the fuel nozzle. Fuel/air ratio, fuel velocity, air velocity, and the location and effectiveness of turbulence-producing diffusers all influence the frequency of the pulsating intensity.

ULTRAVIOLET DETECTION

In the early 1960's, the use of ultraviolet (UV) detectors for proof of fossil flames became practical when the Geiger/Mueller-tube principle was applied to flame detection. Fig. 2 shows the ultraviolet detecting tube.

The UV detector had a distinct advantage over previous light sensors with cadmium-sulfide or lead-sulfide cells which were responsive to very broad bands of light intensity and

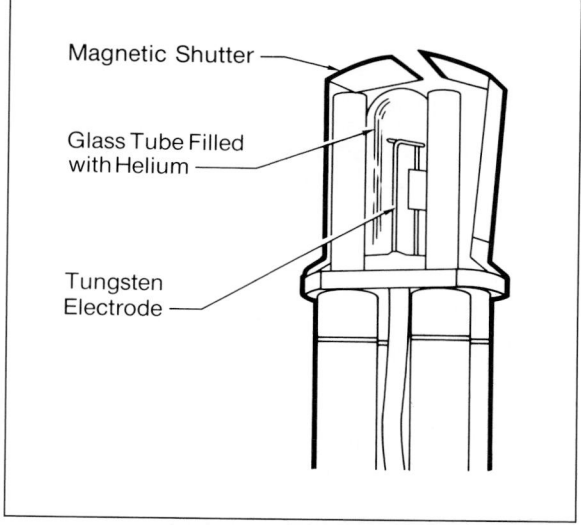

Magnetic Shutter

Glass Tube Filled with Helium

Tungsten Electrode

Fig. 2. An ultraviolet detecting tube. The interior tube atmosphere is helium at low pressure and the electrodes are extremely pure tungsten. Because the special glass envelope has a low attenuation, quartz is unnecessary.

had peak responses in the infrared (IR) range (Fig. 3). The infrared was certainly a characteristic of the flame, but it was also emitted by hot refractory, hot metal, and hot gases, any of which could cause the IR detector to falsely indicate flame intensity present. Contrarily, the UV radiation which is abundant in fossil flames is not emitted in significant quantities from hot bodies at the temperatures encountered in boiler furnaces.

The UV tube works under the photoelectric-emission principle whereby the tungsten electrode will emit an electron when struck by a photon with energy (E_{ph}) in excess of the photoelectric work function (E_w) of tungsten.

Because each photon of light results in only one electron being freed from the electrode, amplifying the electron flow to a detectable level is necessary. The combination of hydrogen and helium gases in the UV tube performs this function. A large voltage difference accelerates photoelectric electrons to the opposite electrode; these electrons collide with molecules of hydrogen and helium, freeing additional electrons that collide with more molecules, setting off a chain reaction. This process is called a *Townsend avalanche*, and the very large current flow it generates is readily sensed with electronic circuits.

UV detectors respond well to natural gas and oil flames. At times, intensity reduction has been needed because of the overabundance of UV generated from natural gas flames as compared to others.

Since a common failure mode of the UV tube is to electrically short out, thereby giving the appearance of seeing a flame, a shutter-check circuit is often used. When a pair of magnetic shutters close off light in front of the detector, there should be no indication of flame.

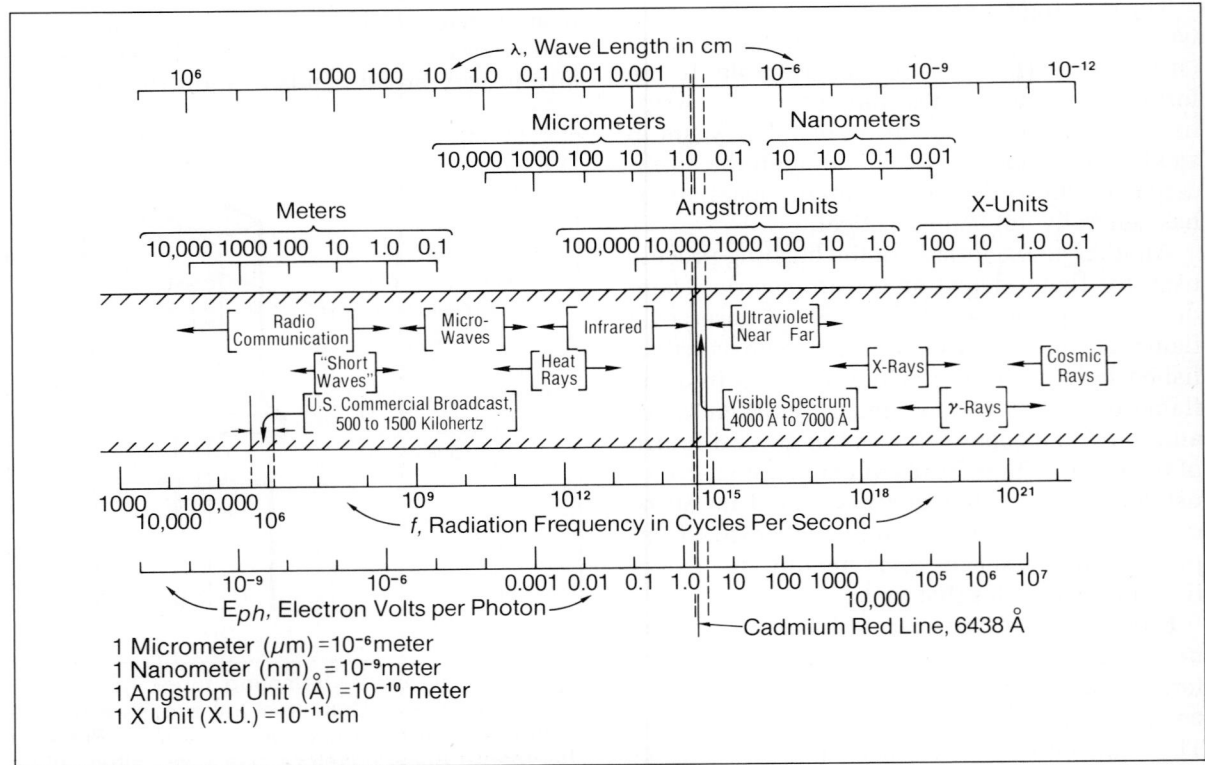

Fig. 3. Total electromagnetic radiation spectrum, logarithmic scales

SOLID-STATE DETECTION

Solid-state light sensors transfer the light energy of the flame to a semiconductor device such as a photodiode or phototransistor. Here, the light energy raises an electron of the semiconductor to the conduction band rather than emitting an electron from an electrode as in the UV tube. Because the electrical potential barrier of the semiconductor is easily varied, a solid-state light detector can be devised which responds to different portions of the light spectrum.

Most commercial solid-state devices have low allowable ambient temperatures (around 70°C or 160°F), whereas burner windbox temperatures can run above 260°C (500°F). For this reason, they are usually mounted in the cooler part (at the back) of the burner. To transmit flame light from the hot burner front to the cooler location, a fiber-optic cable is sometimes used. This transmission distance is generally not more than 6 ft. Commercially available fiber-optic cables exhibit good transmission in the visible and near-IR spectrum. Special high-priced cables are available for the transmission of UV and for the IR spectrum. In those instances where line-of-sight transmission is available, as with fixed burners, fiber-optic cables might not be required. To assist in keeping scanner-head temperatures low and to prevent particulate matter from getting into the scanner, purge or cooling air is often used.

GENERAL SCANNER PERFORMANCE

For the past two decades, ultraviolet flame scanners have been the most widely used flame detectors on large tangentially fired boilers. During the oil crisis of the 1970's, operating coal-fired boilers at low loads without the use of expensive support fuel became necessary. When firing coal at low rates, the emission of UV radiation is reduced; also coal dust and combustion products absorb UV. As a result it is difficult at low firing rates to detect a coal flame with a UV scanner.

As indicated earlier, infrared and visible-light detectors may falsely sense hot material or gases as indicative of flame. To overcome this problem, modern scanners monitor both light intensity and frequency to detect the presence of flame. The frequency of the fluctuations is generally accepted as being between 2 and 600 hertz. The higher frequencies are evident at the base of the flame or the ignition point; lower frequencies originate at the more distant areas or in the "fireball", where fuel streams from many admission nozzles combine and burn in large tangentially fired furnaces.

Reliable flame-scanner design includes self-checking features to assure that sensor failures do not result in false information. This emphasis on assuring the validity of the "flame" or "no-flame" output of the detector is a vital step in preventing furnace explosions.

Proper and accepted proof of flame requires the simultaneous generation of three permissives: (1) no fault, (2) a threshold intensity and (3) a specific flicker characteristic. Fig. 4 shows these operational principles.

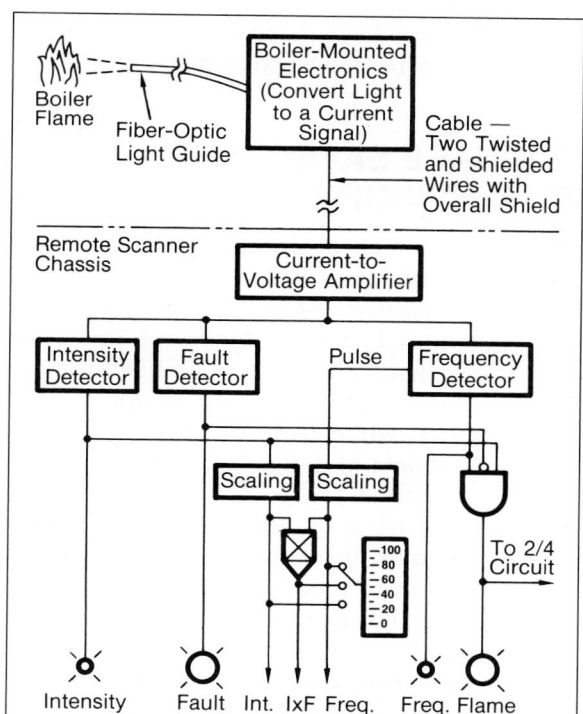

Fig. 4 Furnace corner functional diagram showing principle of operation of SAFE FLAME™ I

A complete scanner system for tilting burners includes a scanner-chassis assembly as in Fig. 5, a scanner-head assembly as in Fig. 6, and a cable to electrically connect the two. A guide tube provides both a cooling-air and a semi-rigid path for the scanner-head assembly.

The scanner-head assembly consists of a light-collecting head, a flexible hose that moves with the tilts of the fuel nozzles, and a housing which contains the scanner-head electronics. The flexible hose contains a fiber-optic cable which passes the flame light from the head back to a photodiode in the scanner-head electronics circuit for signal conditioning.

The scanner-chassis electronics evaluates the flame-signal levels for intensity and frequency, to determine the presence of flame. Also, it contains fault-detection circuitry. Chassis outputs include a flame-relay contact and fault-relay contact as well as intensity and frequency flame-signal outputs. These signals monitor the flame status.

As some of the basics of furnace explosion prevention have been discussed, protection systems can now be described in more detail.

FURNACE SAFEGUARD SUPERVISORY SYSTEM

One of the control systems furnished on almost all modern fossil-fuel-fired steam generators is a protective system which may be called burner control, burner management, fuel-firing safety, or, when supplied by Combustion Engineering, an FSSS® furnace safeguard supervisory system. For a fluidized-bed boiler, the protective system is designated the FBSS® fluidized-bed supervisory system. The main function of such a system is to prevent the formation of an explosive mixture of fuel and air in any portion of a boiler during any phase of operation, including start-up and shutdown.

Preventing damage to the steam generator and/or the firing-system components requires simultaneous, continuous monitoring of a substantial number of parameters and, at times, instantaneous reaction to a hazardous situation.

This task can be performed adequately only by a dedicated protective system such as a furnace safeguard supervisory system.

THE ROLE OF THE SUPERVISORY SYSTEM IN THE STEAM-GENERATING PROCESS

Experience has shown that the furnace safeguard supervisory system cannot be treated as an auxiliary function of the steam-generator process (e.g. combustion, feedwater, and steam temperature) control systems. Rather, the furnace safeguard supervisory system is a separate and distinct system, more closely allied to the firing-system digital actions than to the process controls.

The most important design criterion of a fuel-firing protective system is that it be tailored specifically to the requirements and operating characteristics of the firing system.

SYSTEM ORGANIZATION

The components of the FSSS system can best be divided, for discussion purposes, into four sections (Figs. 7 and 8): (1) the operator's panel, (2) driven devices, (3) sensing elements, and (4) the logic system.

The operator's panel contains all the command devices (panel insert with lights and switches, or CRT and keyboard interface) required to manipulate the firing-system equipment and to monitor the status of this equipment.

The driven devices are primarily those by which fuel and air are admitted to the steam generator. Typical examples are valve operators (oil and gas firing), feeder and pulverizer motor starters (pulverized-coal firing), air-damper drives, and fan-motor starters.

Sensing devices include not only position information on the driven devices, but also such items as fuel pressures, temperatures and flow, and the presence of flame.

The heart of the furnace safeguard supervisory system is the logic system. All operator-initiated commands are routed to the logic system, and the status of all sensing elements is monitored continuously by the logic system. Operator commands are passed on to the driven

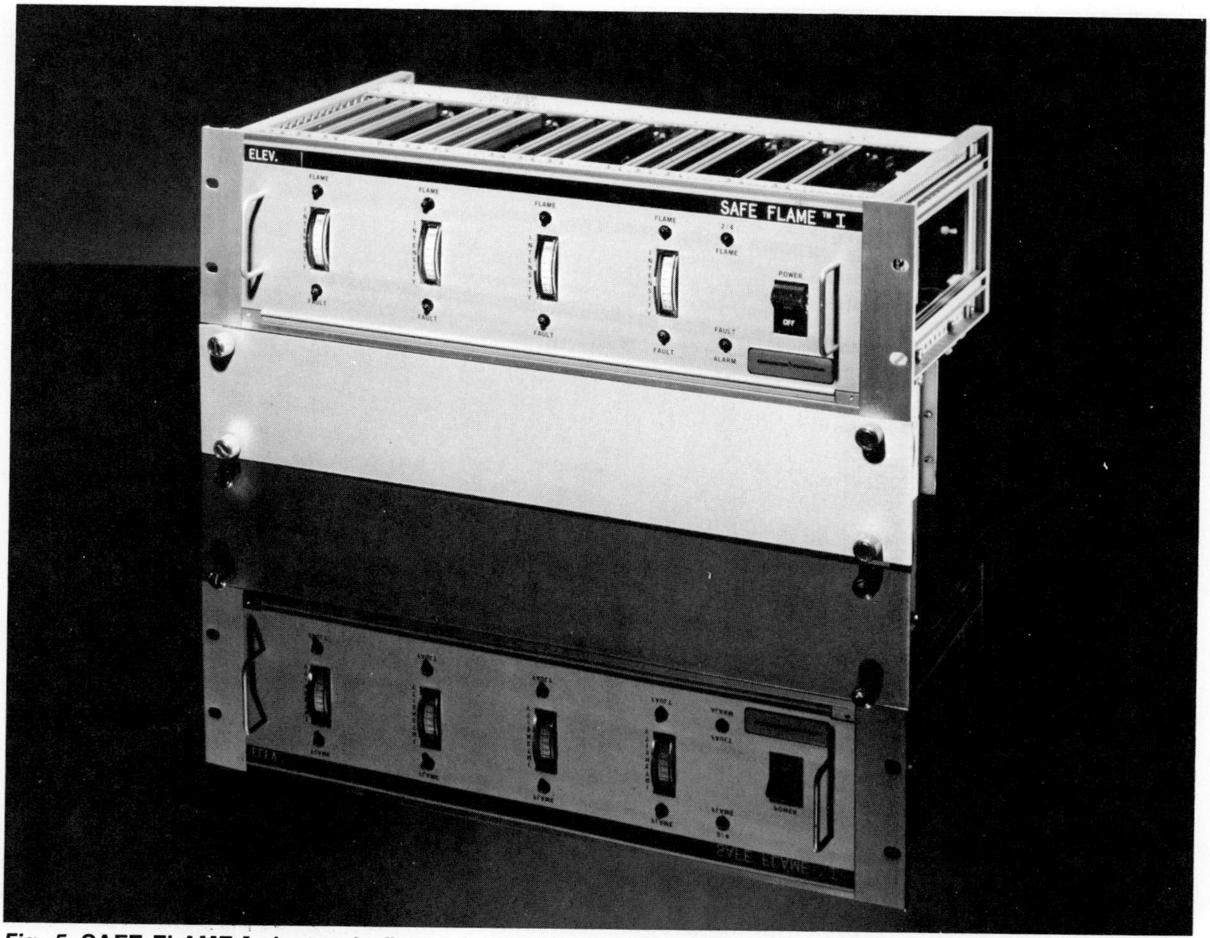

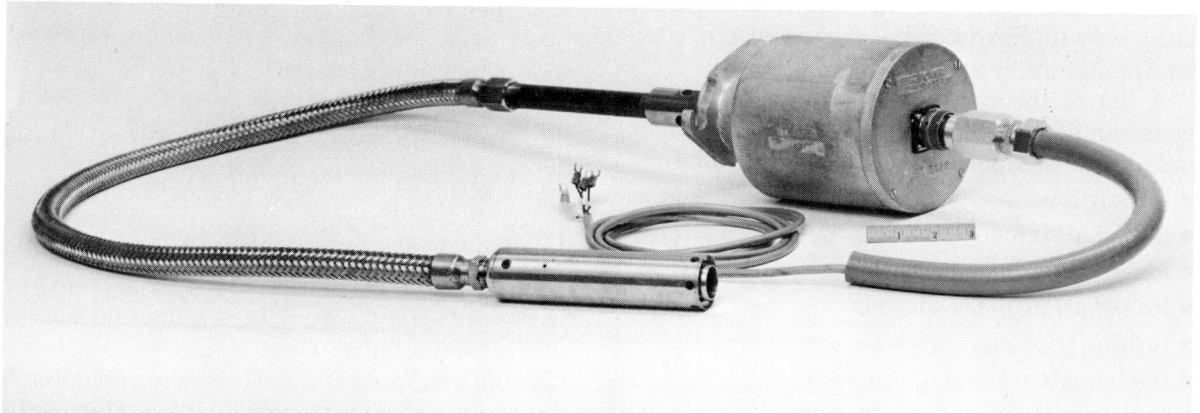

Fig. 5 **SAFE FLAME I electronic flame-detection chassis, with four channels of input, flame indication, and fault-alarm outputs**

Fig. 6 **SAFE FLAME I flame-scanner head assembly for tilting fuel nozzles**

ABB
ASEA BROWN BOVERI

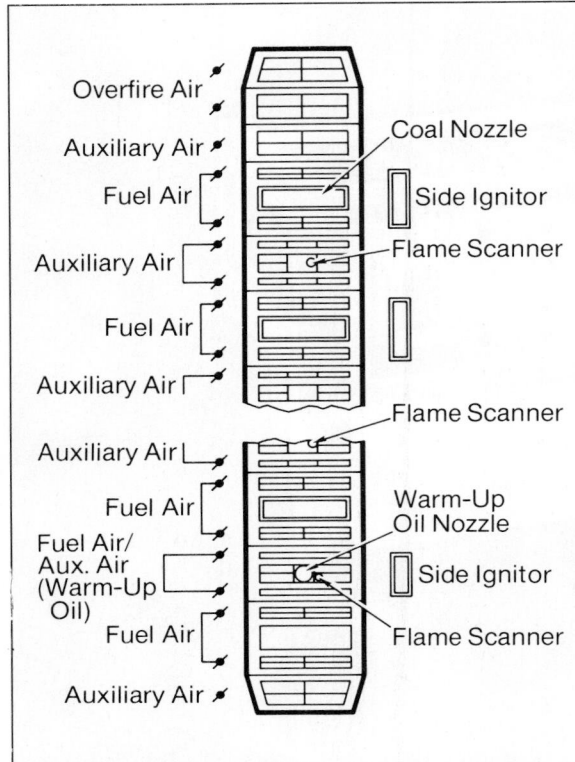

Fig. 7 Typical air and fuel nozzle arrangements within a windbox

devices only if its logic verifies that the proper safety permissives have been satisfied. On the other hand, the logic will shut down equipment automatically whenever continued operation would jeopardize the safety of this equipment or the steam generator.

A list of the typical conditions that would cause an emergency shutdown (trip) of a recirculation-type steam generator is as follows:

- low airflow
- loss of all forced-draft fans
- loss of all induced-draft fans
- loss of all primary-air fans
- turbine trip
- inadequate waterwall circulation
- high furnace-gas-side pressure
- high superheater-outlet pressure

- flame failure
- loss of logic power to FSSS system
- loss of primary and redundant processors (if applicable)
- operator's emergency-trip push buttons depressed

DEGREES OF AUTOMATION

The FSSS arrangement discussed so far is commonly called a remote-manual system. With this type of system, the operator initiates the start-up and shutdown of each individual piece of equipment from a remote operating panel. The system logic insures that the operator commands are performed in the correct sequence and intervenes only when required to prevent a hazardous condition.

Higher levels of automation commonly specified are "automatic" and "automatic with load programming." An automatic system allows an operator to place in service or remove from service a related group of firing equipment (a subloop) in the proper sequence by initiating a single command. A typical example of a subloop would be the feeder, air dampers, and other equipment associated with a single coal pulverizer.

In the automatic mode of operation, a single operator command to start a pulverizer would initiate the following appropriately timed sequence of events (provided all of the required permissive conditions are established):

- associated ignition system placed in service
- pulverizer motor started
- hot-air gate opened
- pulverizer airflow and temperature controls released for automatic operation
- feeder started
- feeder speed control released for automatic operation
- associated windbox dampers released for automatic operation

As the required load on the steam generator is varied over a wide range, firing system subloops are placed in, or removed from, service to

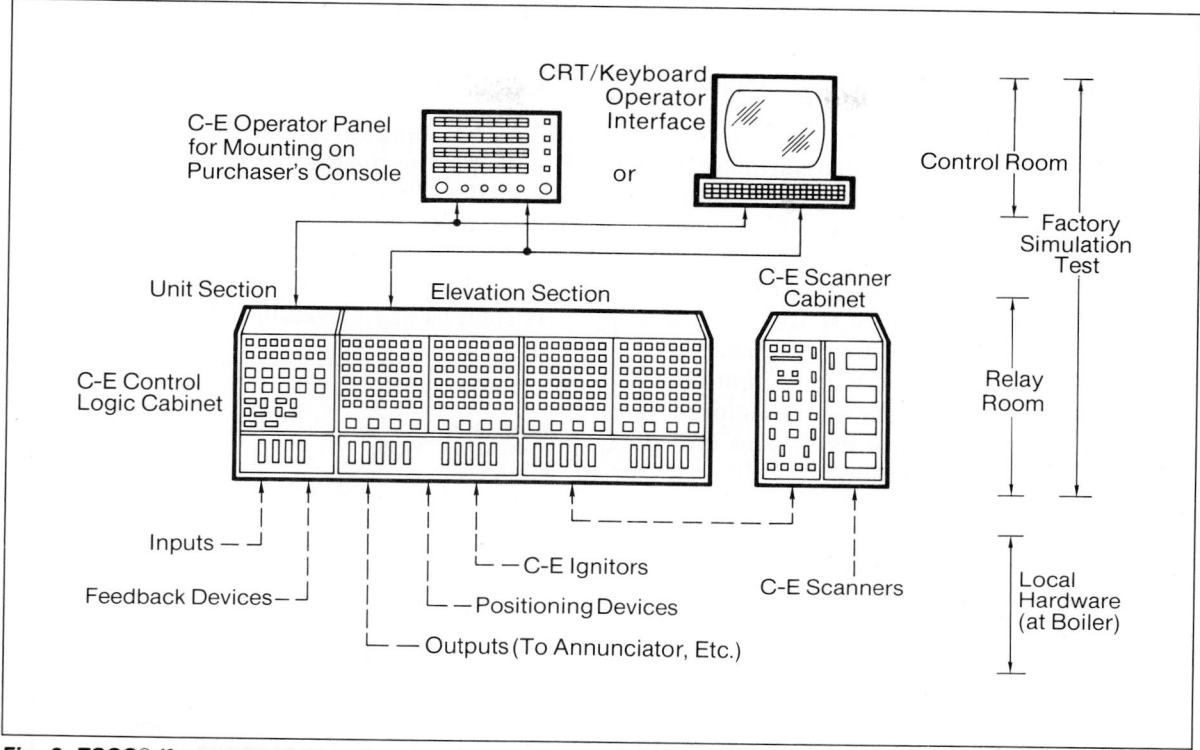

Fig. 8 **FSSS® (furnace safeguard supervisory system) arrangement**

maintain the most stable flame conditions and make the most efficient use of the firing-system components. With the inclusion of the load-programming feature, the FSSS system will take care of this function automatically as steam-generator load demand changes.

BASIC SAFETY FUNCTIONS

Although the furnace safeguard supervisory system does not regulate the fuel and airflow quantities and does not, in most cases, initiate the start-up or normal shutdown of firing-system components, it does exercise authority over both the operator and the process controls through its safety interlocking features.

If, for example, the combustion control should drop the airflow below the minimum value (typically 30 percent) permitted during start-up, the safeguard supervisory system would trip the fuel automatically. Similarly, it will not permit the operator to start equipment

out of prescribed safe sequences and will shut down equipment if prescribed operating practices are not followed. Removing ignitors too early, for example, would result in the shutdown of the associated main fuel.

The specific safety interlocking included in the furnace safeguard supervisory system depends on the physical characteristics of the firing system and type of fuel or fuels being fired. All safety systems of this type, however, are concerned with the following functions:

■ a prefiring purge of the furnace

■ establishment of the appropriate permissives for firing the ignition fuel (i.e., purge complete, fuel pressure within limits)

■ establishment of the appropriate permissives, including ignition permissives, for the main (load-carrying) fuel

■ continuous monitoring of firing conditions and other key operating parameters

■ emergency shutdown of portions or all of the firing equipment when required

■ a postfiring purge of the steam generator

The purpose of the prefiring purge is to be sure that any unburned fuel that may have accumulated in the furnace is completely removed prior to initiating firing. To do this, an airflow is passed through the steam generator at a minimum rating (usually 30 percent of that required for rated steam-generator capacity) for 5 minutes. At the same time, windbox dampers (Fig. 7) are maintained in a particular configuration, the fuel-admission devices are proven closed (or off), and the flame-monitoring devices indicate "no-flame." This combination of conditions will provide the proper velocities and number of air changes through the furnace and convection pass to assure the removal of any fuel accumulations. These purge-permissive requirements also provide a check on the proper operation of the air-damper, fuel-admission, and flame-monitoring sensing devices just prior to firing.

Upon completion of the purge, the steam generator is ready to be fired. The 30-percent minimum airflow requirement is maintained until the steam generator reaches 30-percent load in order to assure an air-rich furnace mixture during the entire start-up phase.

Initial firing is accomplished with a group of ignitors that have the capability of lighting the ignition fuel with an electric spark. A flame detector must be provided with each ignitor to determine the presence or absence of a stable flame. In the case of the C-E ignitor, the flame detector is an integral part of the ignitor, and a fuel-flow measurement is included to insure the proper quantity of ignition energy.

Depending on the choice of load-carrying fuel and the firing-system arrangement, the load-carrying fuel may obtain its ignition energy directly from the ignitors or an elevation of ignition fuel guns may be used as an intermediate step. On coal-fired units, for example, ignition fuel is normally kept in service until two adjacent coal elevations are being fired at 50 percent of their rated capability, thereby assuring sufficient ignition energy to maintain stable

firing conditions during the steam-generator start-up phase of operation.

FLAME MONITORING

The C-E tangential-firing configuration (Fig. 9) operates during start-up with a separate, independent flame emanating from each operating fuel-admission point (nozzle). Each flame is monitored independently by the use of continuous, self-flame-proving ignitors.

As soon as any elevation is fired at a rate ex-

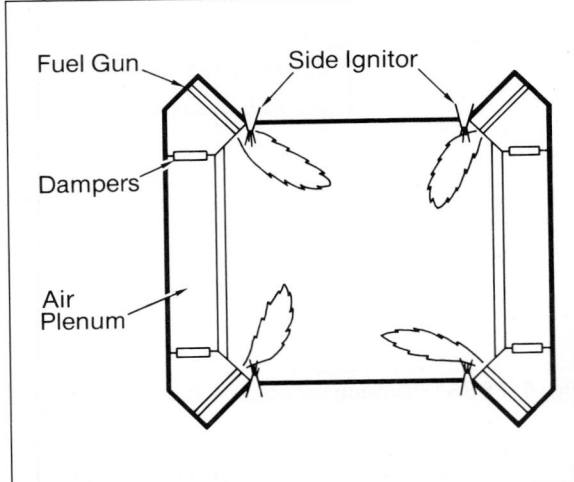

Fig. 9 Tangential-firing pattern at low-firing rates

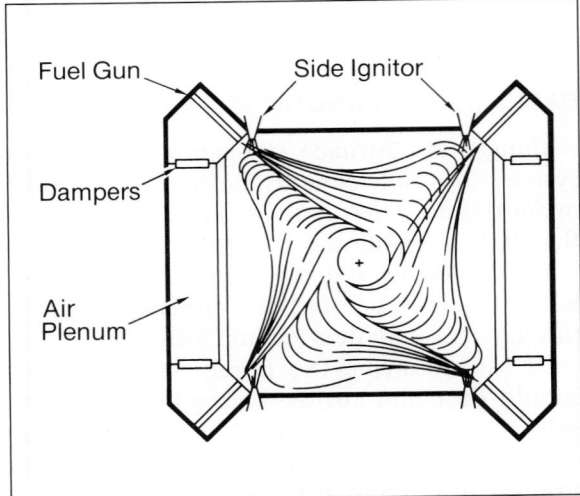

Fig. 10 Tangential-firing pattern: fireball condition

ceeding 30 percent of its design rating, the characteristics of this firing configuration change dramatically (Fig. 10). The thorough mixing of fuel and air produced by the rotating fireball (vortex) permits ignition energy produced at one location in the furnace to support the combustion of fuel admitted at other locations.

The furnace, in effect, becomes a single burner. Under these firing conditions, multiple flame detectors monitor the furnace on a statistical basis. No individual fuel nozzles are shut down because of an adjacent flame detector indicating "no-flame." But should insufficient flame detectors confirm the presence of flame, the entire firing system is shut down.

The transition from start-up firing characteristics to fireball conditions is a gradual transformation as elevation loading is increased. The selection of the 30-percent elevation-loading condition as the point where the logic is transferred to fireball monitoring is based on field experience and assures a completely adequate margin of safety between the two different firing conditions.

During the fireball phase of operation, sufficient ignitors in service at specified locations constitute proof of stable firing conditions. To eliminate the need to keep ignition fuel in service, however, optical flame detectors are used as the normal means of monitoring the fireball.

Front-fired (horizontal-burner) systems exhibit, at all firing rates, the individual burner characteristics which the tangential configuration exhibits at low firing rates; that is, the flame produced at one firing location does not provide ignition energy for fuel admitted at the other locations. The flame-monitoring system for such burners must be designed accordingly. Attempts to use front-fired flame monitoring techniques on tangential-firing configurations and vice versa have met with little success.

LOGIC ACCOMPLISHMENT

The logic portion of the furnace safeguard supervisory system may be implemented with electromagnetic relays, solid-state electronics, or computer-based programmable controllers. All these implementations must be able to tolerate the dust, temperature variations, and electrical noise encountered in a power plant.

The normal method of powering the FSSS logic is to provide redundant AC sources to a transfer switch with the output of the transfer switch going to the system. To prevent a boiler shutdown during operation of the transfer switch, the AC-powered field devices are of the "energize-to-trip" type: that is, they require the presence of power to close. DC power from the plant battery system is also provided to the FSSS system to assure the capability of shutting off the fuel input in the event of a loss of both AC sources.

Electromagnetic relays, which have been the traditional building blocks of furnace safeguard supervisory system logic in the past, are now rarely built. Hard-wired solid-state logic was a successor to electromagnetic relay systems and has been applied to FSSS design for many years, but did not assume a dominant position until the late sixties and early seventies. For a long period, the utility industry continued to judge the traditional advantages of solid-state logic (less power drain, less space required, greater speed, and no moving parts to wear out) to be less important than the disadvantages (higher susceptibility to electrical noise and temperature variations, more complexity, and higher cost).

Fig. 11 shows a typical solid-state FSSS logic cabinet. The built-in simulator provides maintenance personnel with a fast, convenient method to check the operation of the system logic without disturbing field wiring or, inadvertently, operating equipment.

Solid-state logic then evolved such that the electrical noise and environmental problems were mastered. The widespread use of solid-state process controls has resulted in the ability of plant maintenance people to handle solid-state problem-solving easily. Plug-in solid-state modules provide an on-line repair capability not available to relay systems.

The additional expense of a solid-state system versus an electromagnetic relay system results primarily from the need to convert all incoming signals from the sensing-device voltage level to the logic voltage level and back from the logic

Fig. 11 The solid-state-logic cabinet of an FSSS system includes a built-in simulator. The system currently provides protection and automation for an 850-MW unit.

level to the voltage level required by the driven devices. (Logic devices cannot provide either the voltage or the current level required by the field devices.) As systems become more automated, however, the ratio of logic functions (inexpensive) to inputs and outputs (expensive) increases and the differential expense versus relay logic decreases.

PROGRAMMABLE CONTROLLERS

Currently the predominate technique for implementing furnace safeguard supervisory system logic is the microprocessor-based programmable-logic controller (PLC). See Figs. 12 and 13. These devices are minicomputers with a built-in compiler (translator) that makes it possible to program the system directly from logic diagrams. They are also configured to be strong in logic capability and can be packaged ruggedly enough for the power-plant environment.

Although this technology has been in existence for a number of years, further improvements, increased versatility, and wide acceptance of PLC's in coming years are anticipated. As a result of their general popularity, a wide range of data-communication interfaces will be available so the PLC will be easily adaptable to a totally integrated information system.

A distributed, fault-tolerant architecture is used in a microprocessor-based FSSS system. The distributed architecture has a multiplicity of

Fig. 12 **Input/output channels for a programmable system**

Fig. 13 **The central processing units for the programmable system in Fig. 12**

logic controllers in modular fashion to match the functional process modules it serves; for example, pulverizer equipment groups, fuel-firing elevations, or unit functions. In this architecture, each controller and associated input/output module is considered a component with a statistically remote probability of failure. The basic design criterion of this distributed, fault-tolerant architecture is that failure of a single controller, with all its logic capability, or an input/output module, will not jeopardize the steam-generator safety. In addition, a failure will not require the steam generator or even the given functional process module to be shut down. This design criterion is satisfied by the ring-type arrangement shown in Fig. 14.

Fig. 14 represents non-switched fault-tolerant architecture featuring distributed modular redundancy. The ring-type arrangement illustrates the concept of distributed modular redundancy with identical, universal software in each module. Although the primary objective is maximum

safety, maximum availability is also a target, with "on-line" maintenance capability built-in. Fig. 14 is a representative schematic configuration for a coal-fired utility steam generator with four logic controllers in the ring. Each controller serves two pulverizer equipment groups and also contains the logic for overall unit control. (This is a flexible configuration and is adjustable to suit individual plant requirements.)

The functions contained in each processor module consist of two sections. One (shown as white space) is concerned with the control and safety of the associated equipment group. The other section (shown crosshatched) is concerned with the redundant safety of the neighboring equipment group to the left. The arrows in the ring schematic indicate the direction of processor-to-processor redundant coverage, such as 1 providing back-up to 4, 2 to 1, 3 to 2, and 4 to 3.

The total arrangement, in viewing the ring counterclockwise, results in having one processor-control section and two redundant safety sec-

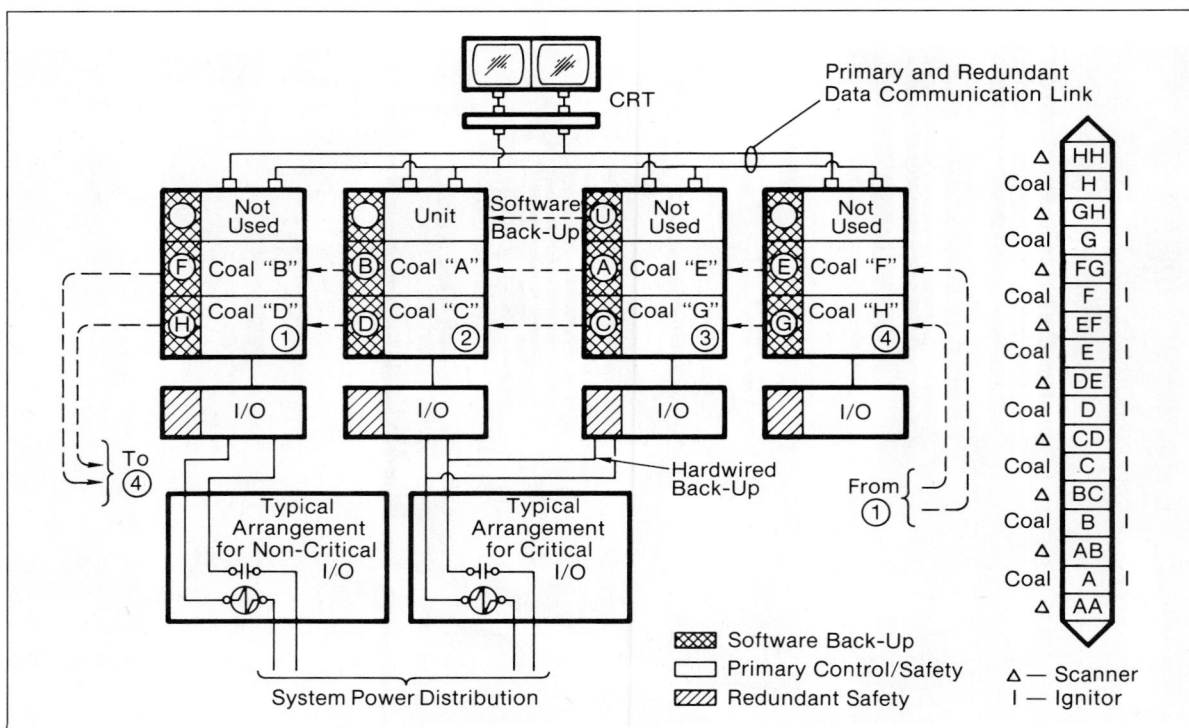

Fig. 14 **Typical ring-type logic-processor architecture, with distributed redundancy**

tions per equipment group distributed in modular fashion.

All combinational logic required to produce the redundant safety action is derived independently. Also, all field-status information to substantiate these safety actions is connected in parallel to and is processed by an independent set of input-conditioning modules. Similarly, an independent set of output drivers processes all safety output command signals. These output drivers act in parallel with the respective output drivers from the other machines to actuate the selected field-driven devices.

The distributed, selective modular redundancy is, therefore, active on a full-time basis. No fail-over circuits are required for switching to backup sections because the backup section is always in operation "on-line," covering safety-related functions. Typically, each pulverizer-equipment group includes a pulverizer and its motor, a coal feeder and its motor, a primary-air supply, a secondary-air supply, associated valves and drives, dampers and drives, and feed-back sensing elements.

The notations A, B, C, D, E, F, G, and H shown on processor sections of the ring configuration of the diagram correspond to the designations given the individual coal-firing physical arrangement. The unit section contained in each processor (but used in only one), has control and safety sections dedicated to serving the overall unit. The corresponding safety section in another processor provides redundant backup for all the critical unit functions. This redundancy will, among other functions, backup the overall flame-failure protection in its entirety, including I/O.

The ringtype architecture (Fig. 14) provides a framework for a very versatile range of redundant coverage options, both processor and I/O, to suit requirements, while maintaining the concept of universal software in each module.

The principle of fault-tolerant partitioning in the example given is applied in comprehensive fashion. Consequently, no common bus exists between the I/O processors of individual controllers. Each processor is electrically and functionally independent and there is also isolation between the low-level internal I/O bus and external I/O bus served by each processor. The processors are, in general, operationally independent of each other and obtain their information independently from their own set of dedicated I/O's. Also, redundant I/O's do not share common connector structure or any other elements having the potential of common-mode failure. A failure of any one processor is localized and does not, therefore, cause misinformation in another processor.

The modular system is designed so that any processor in the ring imposes control on its associated equipment groups only. Furthermore, since a neighboring processor is monitoring the critical safety aspects of the equipment group, an already operating (different) group does not have to be shut down because its processor or associated I/O has failed. Processor and interface maintenance can, therefore, be accomplished with the unit and given equipment group in operation. Should both primary and redundant processors fail, an external monitoring circuit will initiate a trip, independent of the processor logic and confined to the associated equipment only. Emergency-trip pushbuttons can shut down the driven devices from plant DC battery power.

The on-line replacement of a failed processor with a spare unit is accomplished in bumpless fashion using an updating technique which is automatically activated at the time of component replacement. Updating of the "new" processor encompasses all the necessary historical data.

It is important to note that, generally, redundancy structures should be built in simple, straightforward fashion in order to avoid the inadvertent introduction of common-mode failure elements.

OPERATOR INTERFACE

Traditionally, the operator's panel located in the control room contains the command devices, such as switches and pushbuttons for initiating operation of firing-system equipment, and feedback indicators, such as lights or lighted panels, which display the status of the equipment to be controlled. The availability of a communication link allows enhancement of the operator-machine interface with a colorgraphic display. The framework for the CRT displays should be arranged in two categories: Message Summaries and Colorgraphic Animated Schematics.

Message Summaries are primarily dedicated to diagnostic information allowing the operator to identify problem areas blocking orderly start-up or shutdown. The Colorgraphic Animated Schematics (see Fig. 15) allow operator observation of the start-up, shutdown, and operational progress in graphic form, with special techniques for device simulation.

Often for comprehensive visual effect, Message Summaries and Colorgraphic Animated Schematics are combined on the same visual. It is important to specify complete coverage with frames (pages) arranged in functionally modular fashion with clear directions from overview to detail. Process alarms should be grouped separately from hardware alarms.

While the static data for the colorgraphics resides in the graphics processor, the dynamic data resides in the Burner Management Logic processors. The logic processor transmits the dynamic data to the graphics processor by real-time communication links. This data is then superimposed upon the static display image. Limited-scope hardwired control panel inserts

may be specified as a backup feature.

Comprehensive error-checking procedures are utilized. To further assure system integrity, a software buffer zone is inserted into the logic processor application program to provide additional isolation between the base working system and the communication bus. Greater-flexibility is an additional benefit of this feature, which virtually sectionalizes and allows the independent engineering and maintenance of the base control system from the display colorgraphics. The software buffer also helps to reduce traffic on the communication network and prevent overloading.

PROCESS CONTROLS

Plant operators retain the ultimate responsibility for the operation and protection of both utility and industrial steam generators. But as discussed, automatic safety systems are customarily installed to relieve operators of the task of instantaneously analyzing and reacting to a rapidly developing hazardous condition.

Safety systems improve unit availability by preventing improper operating sequences and by automatically shutting down equipment before damage can occur. But acceptable availability can only be achieved by installing

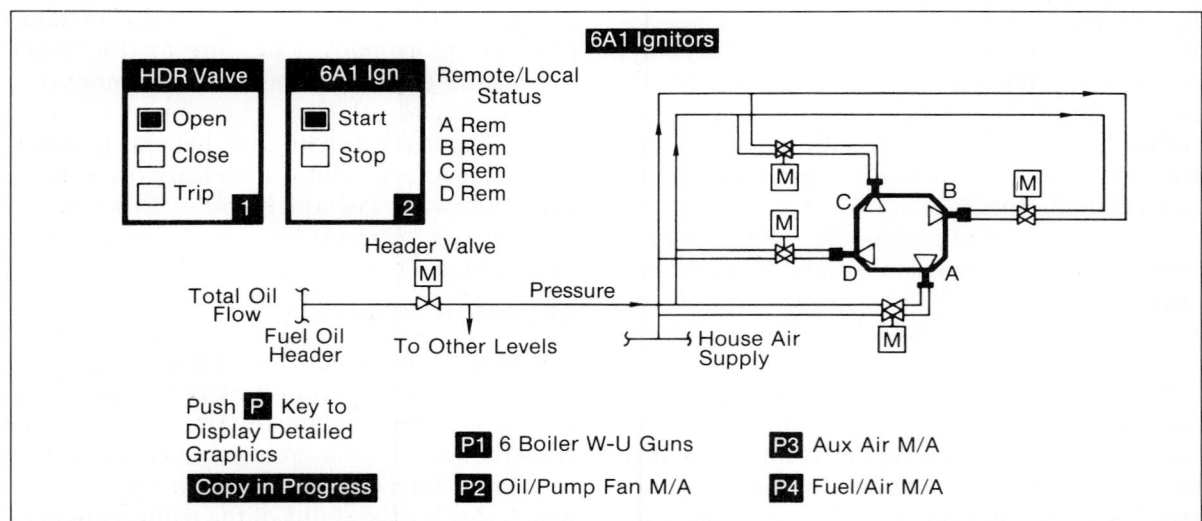

Fig. 15 **Typical FSSS CRT Animated Schematics Screen**

control systems to *prevent* the plant-controlled variables from reaching the deviation from normal values which result in safety system action. The continuous, automatic regulation of controlled variables is the function of the process control systems.

Although it is still common to divide the process controls for a large steam generator into subsystems (such as combustion control, feedwater control, and steam temperature), a *coordinated* approach to process control has proven to be the most effective technique to achieve optimum boiler performance.

BOILER-FOLLOW SYSTEMS

For many years, the most common process control system employed on subcritical drum-type units was the "boiler-follow" system which relies heavily on feedback control principles.

With a boiler-follow system on a subcritical-pressure unit, a change in required electrical generation results in the following events:

1. The turbine control valves are repositioned to establish the new generation level, at the expense of a change in steam-generator energy storage level.
2. Changes in steam flow initiate changes in the flow of feedwater, fuel, and air on a feed-forward basis. Fuel and air are further modified by the feedback of throttle-pressure deviation from setpoint, which is an indication of the energy balance between the steam generator and the turbine. Feedwater is adjusted by drum-level deviation, which is a measure of the feedwater-flow/steam-flow balance.

The large energy storage of drum-type units allows boiler-follow systems to be very effective as long as generation-level (load) changes are small or are accomplished slowly.

COORDINATED-CONTROL SYSTEMS

The advent of the supercritical unit in the 1960's, with its relatively small energy storage, quickly demonstrated the limitations of the boiler-follow system. The solution to this problem was the development of the coordinated-control system.

Ideal for supercritical units, the coordinated-control system is also well-suited for subcritical drum-type units. Its fundamental objective is to operate the steam generator and turbine generator as an integrated unit in order to maintain the process control variables (unit generation, steam pressure, flue-gas oxygen, furnace draft, and steam temperatures) within acceptable limits. The system operates the unit in an integrated (coordinated) manner by developing a unit load demand (ULD) signal in the unit load control (ULC), and transmitting the ULD signal simultaneously to both the steam-generator fuel and air control and the turbine governor-valve control. Fig. 16 shows a typical control-room operator interface for the unit load control in a coordinated-control system.

Fig. 17 shows the functional design arrangement of a coordinated-control system. This design illustrates that the ULD signal, which represents generation demand, goes directly to the steam-generator input variable as well as to the turbine governor. This parallel control arrangement is based on the principle that the only way to permanently change unit generation is to change the energy output of the steam generator.

The presence of any error between required output and actual generation will bias the steam-generator and turbine-demand signals in order to recalibrate the system for cycle changes. Similarly, throttle pressure error is used to recalibrate the balance between steam production and steam usage. Both of these error signals operate on a transient basis to compensate for the difference in response time of the boiler and turbine, allowing the faster responding turbine to minimize generation errors by modifying boiler energy storage levels.

Today's modern coordinated-control systems are designed to accommodate both the U.S. traditional fixed-pressure type of operation and also sliding-pressure type of operation. Sliding-pressure operation has long been popular in Europe because of Europe's smaller grid systems and the resulting need for cycling and two-shift units. Now, in the U.S., because older large units, once base-loaded, are being cycled

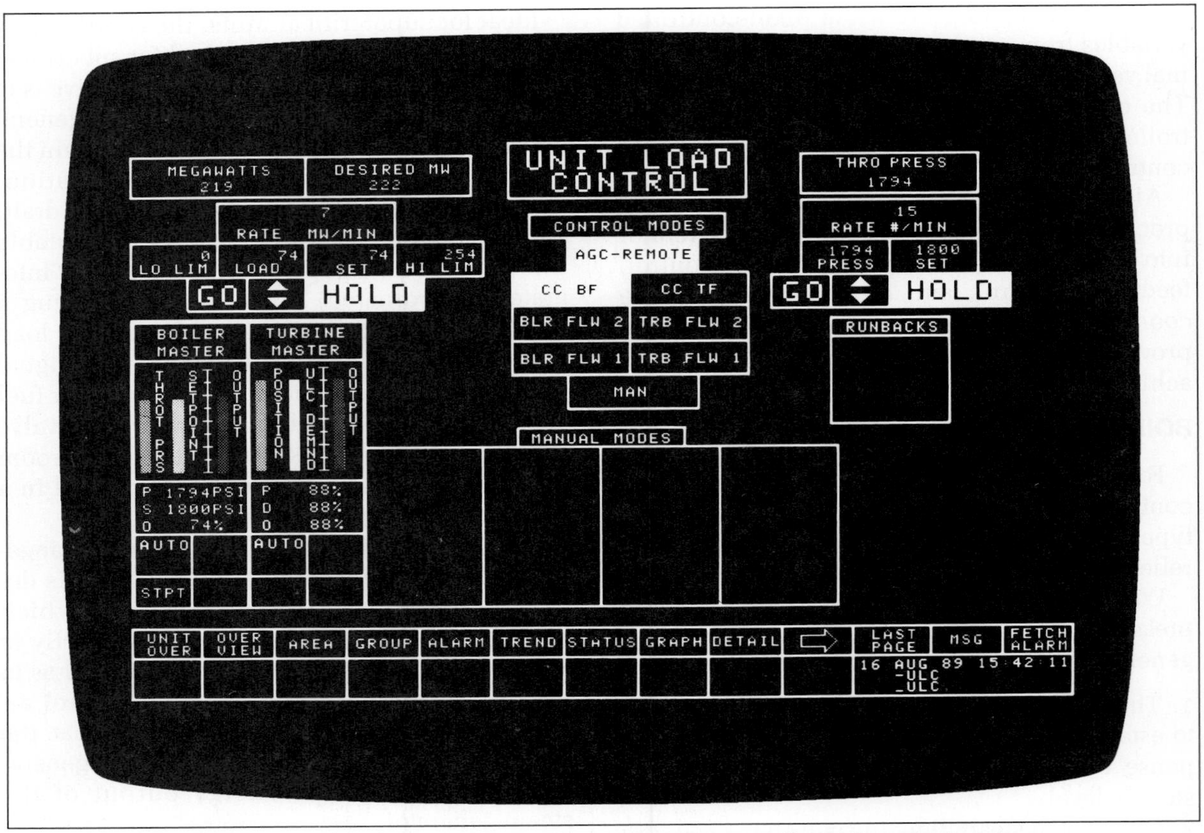

Fig. 16 **Typical operator interface for the unit load control of a coordinated-control system**

and two-shifted, sliding-pressure operation is becoming more common.

In fixed-pressure operation, the coordinated-control system regulates throttle pressure to a fixed setpoint for the entire range of unit load after start-up. The turbine governor valving strokes open as unit load increases. Fixed pressure provides for boiler storage and thus the capability of the unit to more effectively respond during transient load changes without suffering excessive process-variable deviations.

In sliding-pressure operation, throttle pressure setpoint is ramped over most of the unit load range. The turbine governor valving is fixed at a near-open position (normally 90 percent open) throughout the ramp. The 10 percent position reserve allows for moderate unit load-change accommodation. The fixed position of the turbine governor valving is temporarily in-

fluenced during a unit load change by the rate of change of ULD.

THE C-E COORDINATED CONTROL SYSTEM

The preceding discussion summarizes the basic differences between boiler-follow and co-ordinated-control-type systems. The C-E Coordinated Control System (CCS) contains a number of additional features designed to provide several layers of defense between the direct regulation of the variables and the automatic shutdown actions of the furnace safeguard supervisory system.

Each major subsystem, such as fuel, air, or feedwater, contains a flow tie-back loop to insure the process variable will quickly match required output. Redundant transmitters and comparison networks transfer the loop automatically to manual if a transmitter malfunc-

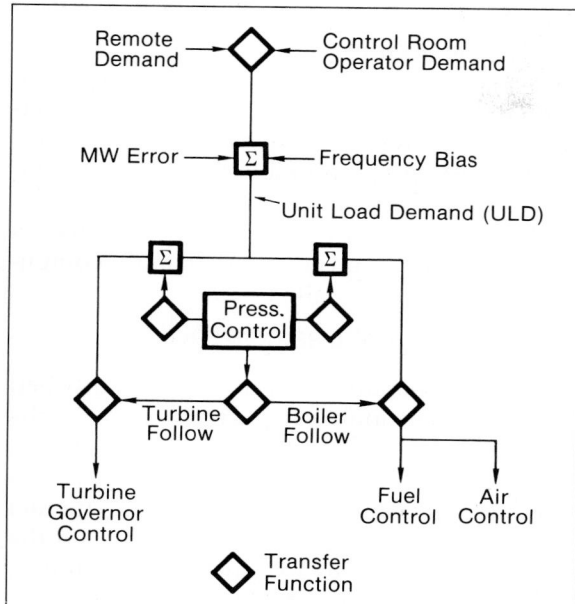

Fig. 17 **Coordinated-control system, signal flow diagram**

tion occurs. Bypass switches permit automatic operation while one transmitter is being calibrated. Ahead of the flow loops are the ratio correction loops. Flue-gas oxygen corrects fuel/air ratio.

When all process loops are well tuned and calibrated, and are functioning properly on automatic, the controlled variables will be maintained well within acceptable limits.

Various limit circuits are provided to take care of the times when this desirable condition does not exist.

In the case of fuel and air matching, a cross-limit circuit is provided which does not permit fuel demand to exceed measured airflow (plus tolerance) and does not permit airflow demand to drop below measured fuel flow (plus tolerance). This circuit is most commonly activated when a malfunction causes airflow and fuel-flow regulating devices to follow demand at different rates.

Referring again to Fig. 17, note that the "output adjustment signal" (the desired generation value established by the operator or the load-dispatch system) must pass through several conditioning

devices before emerging as unit load demand.

The first such device is the operator-set rate-of-change limiter. This device prevents the unit from accepting a load demand change at a rate which would create instability and a resulting unit trip.

Turbine governors automatically change unit steam flow in response to an electrical-grid frequency disturbance. To prevent the CCS from fighting the governor, and to match the boiler demand to the turbine action, the frequency bias conditioner is required. This circuit simply adjusts required output to match the governor action.

Limit actions occur when the CCS detects that one or more process variables cannot follow the output adjustment demand signal. If, for example, a feedwater pump were out of service, the limit circuit would not allow required output to exceed the capability of the pumps in service. Without this limit, firing rate would be allowed to exceed feedwater flow, resulting in loss of drum level. The limit circuits also act whenever such process variables as generation and throttle pressure deviate from their required values by a preset amount. These deviations are indications that the steam generator or turbine, or both, are unable to follow the desired output.

An extension of the limit circuits, which is not shown on Fig. 17, is the run-up and run-down circuits. These circuits act to return the required output within the capability of the unit when a change in boiler-unit conditions occur. For example, if all coal feeders are operating near maximum output and one feeder trips, the unit cannot maintain its existing load. The run-down circuit would automatically reduce load to a value within the capability of the existing feeders.

The run-back actions shown next are simply a more dramatic version of a run-down. The loss of a major piece of auxiliary equipment such as a feed pump or forced-draft fan may reduce the load-carrying capability of the unit to 50-60 percent of maximum continuous rating. When such an equipment loss is detected, the run-back circuit very rapidly reduces the required output signal to a value within the capability of the remaining equipment. Without such a circuit, a unit trip would almost certainly result.

One final protection circuit, which is the last line of defense designed to act before the safety-system limits are reached, is the deviation-limit system. This system continuously makes two

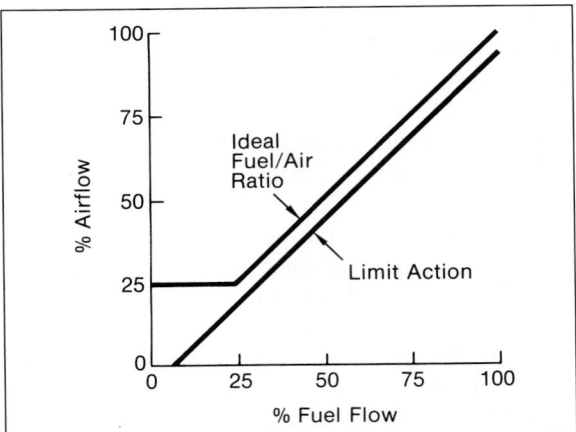

Fig. 18 The deviation-limit system is a protection circuit which continuously monitors fuel/air ratio and, if needed, directly takes action to restore a permissible ratio.

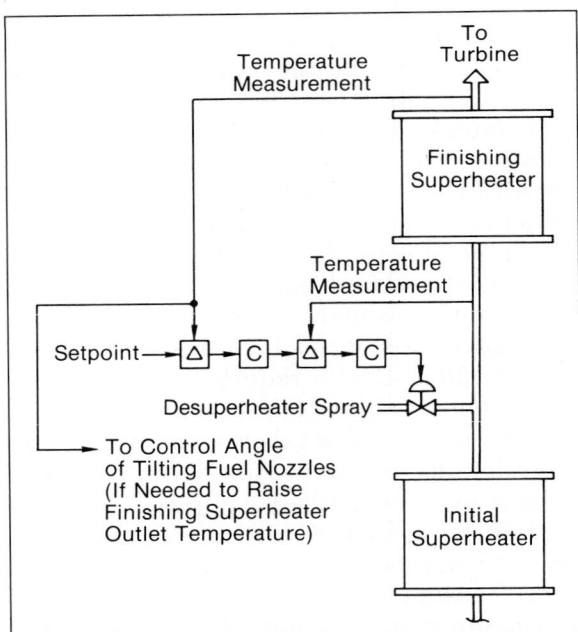

Fig. 19 Coordinated-control system, primary-steam-temperature control

separate and independent calculations of fuel/air ratio. Fig. 18 illustrates the fuel/air ratio monitoring concepts. If both calculations verify that this ratio has exceeded allowable limits, the deviation-limit system directly operates the control loops to restore a permissible ratio. This system, which overrides manual control, would normally activate only when the operator has made a serious mistake or an automatic-control computational element has failed.

STEAM-TEMPERATURE CONTROL

Steam-temperature control is accomplished on C-E suspension-fired pulverized-coal units by a combination of fuel-nozzle tilt positioning and superheater (SH) and reheater (RH) desuperheating sprays. Steam temperature is maintained by allowing fuel-nozzle tilts to respond to the lower of either SH or RH outlet temperatures, with sprays responding to the higher.

Desuperheating spray control is a cascade type control. An "outer-loop" controller associated with final steam outlet temperature sets the setpoint for an "inner-loop" controller, which controls desuperheater steam-outlet temperature. Fig. 19 shows a typical primary steam-temperature control scheme.

Basic boiler design provides for the reheater to be controlled by heat redistribution devices. The several boiler manufacturers have developed numerous mechanisms for this purpose, including flue-gas bypassing, gas proportioning, and excess-air variation. Fig. 20 shows three techniques which C-E uses most commonly: tilting nozzles, desuperheating sprays, and on oil-and-gas fired units, gas recirculation. Only during unusual conditions is reheater desuperheating by water spray used, because of its deleterious effect on heat rate. On coal-fired units, fuel-nozzle tilting is the most common mechanism.

AN OVERVIEW

Primary measuring devices, transmitters, indicators, recorders, annunciators, indicating lights, and CRT graphic displays can be integrated into a system which permits plant operators to continuously and conveniently monitor plant operating conditions. Push buttons,

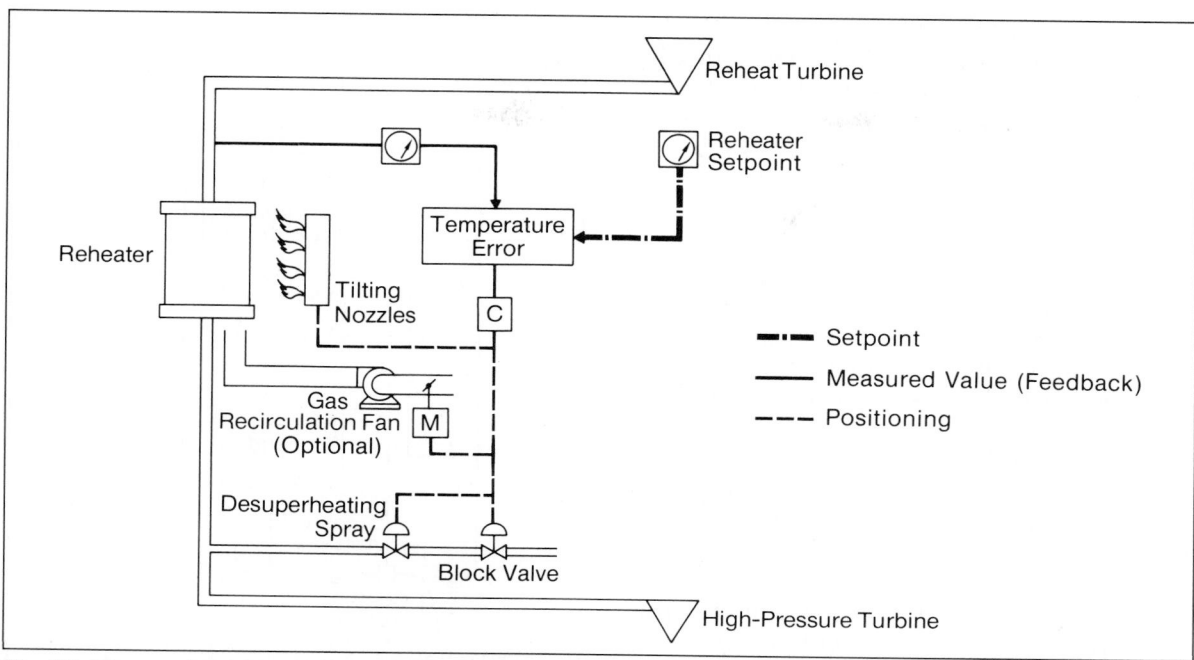

Fig. 20 **Three reheat temperature control techniques commonly used by Combustion Engineering**

switches, manual-loading stations, and CRT keyboards can be combined with pneumatically and electrically powered regulating devices (such as valves and dampers) to provide the capability of starting up, maneuvering, and shutting down a steam generator from a central location.

Using process control systems results in more precise control, thus increasing plant productivity and reducing cycling-life expenditures caused by process variable excursions. When a sophisticated process control is employed, such as the described Coordinated Control System, the steam generator and turbine generator act as a coordinated unit to further improve unit transient response and productivity. The CCS also provides several layers of limit systems designed to improve unit availability by preventing a process deviation from reaching a safety-system limit. Today's modern designs implement the CCS using computerized distributed systems.

Safety systems, such as the FSSS system described earlier, also contribute to plant automation while performing their primary function of protecting both plant personnel and equipment.

A well-designed automation system, with both firing system and process controls, will operate the unit in the same manner that the plant equipment designers would operate it, if they were given the opportunity to do so. It is vital, therefore, that the control-system designers thoroughly understand the process requirements and limitations and that the control systems are tailored to the specific equipment and plant-operational requirements.

INDUSTRIAL BOILER OPERATIONAL CONTROLS

Controls for industrial boilers of the types described in Chapter 8 serve the same overall purposes as controls for utility boilers. Designers of both types of systems are concerned with balancing boiler inputs and outputs of mass and energy, remote operation, precise control, transient response, and operating ease, as well as efficiency, safety and reliability. The principal differences between utility and industrial control systems result from industrial plant steam requirements, types of fuels and firing equipment, and historically divergent design practices. The following sections describe these distinguishing characteristics in greater detail.

PLANT STEAM REQUIREMENTS

Industrial-plant steam requirements serve local needs for direct process use, thermal power, and mechanical power (which may include power to drive electrical generators). The required steam conditions vary from low-pressure saturated steam to high-pressure superheated steam depending on the application.

Process plants, which often have many steam consumers encompassing all of the above uses, supply steam to the various users through a system of steam distribution headers. Steam is generated at conditions to meet the highest-pressure requirements and then is cascaded from the main steam header through pressure-regulating valves to intermediate- and low-pressure headers.

A cogeneration cycle, as described in Chapter 1, uses steam generated at high pressure and temperature for expansion through a back-pressure turbine generator to produce electrical power. The turbine exhausts into distribution headers for plant thermal and mechanical power needs. In effect, the turbine generator acts as a pressure-reducing valve for the plant steam consumers, and the industrial-plant consumers, by using energy that is lost in a typical utility generating station, function as condensers in the steam cycle. In other plants, a condensing turbine with controlled extraction may be the economical choice, depending on such factors as seasonal plant steam demand and prices negotiated with the electrical utility for purchasing the cogenerator's excess electrical power.

Load swings in industrial plants vary according to plant needs and, in some plants, may be quite severe, exceeding 10 percent of maximum continuous rating per minute. Industrial boilers must have greater maneuverability than utility boilers in responding to load changes without great upsets in steam pressure, drum level, and steam temperature.

Due to the multiplicity of steam and/or power

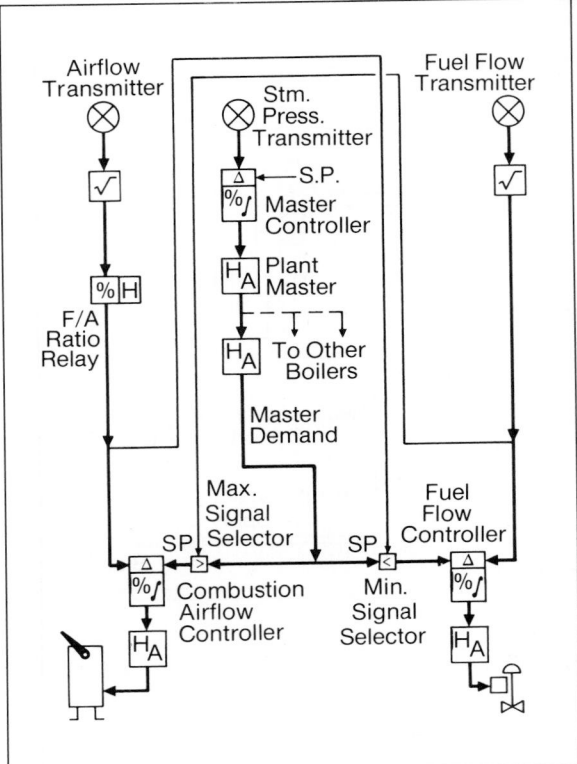

Fig. 21 Cross-limited-metering combustion-control system, single fuel

consumers in an industrial plant, it is usually not practical to feed-forward a load demand to coordinate boiler control with the control of steam consumers. The concept of a coordinated control system is generally not applied to industrial boilers because they are not dedicated to a turbine generator. Figs. 21 through 24 show typical operational-control sub-systems. In the industrial plant, the main steam header pressure is an index of plant load, and a master pressure controller sets the demand for firing rate for boilers supplying steam to the header (see Fig. 21). This is a feedback control loop similar to the boiler-follow system described earlier in this chapter. Note that industrial plants commonly have multiple boilers supplying steam to the same main steam header. In many plants a form of feed-forward control is incorporated in the master controller scheme in which the total steam flow from all boilers is the feed-forward demand.

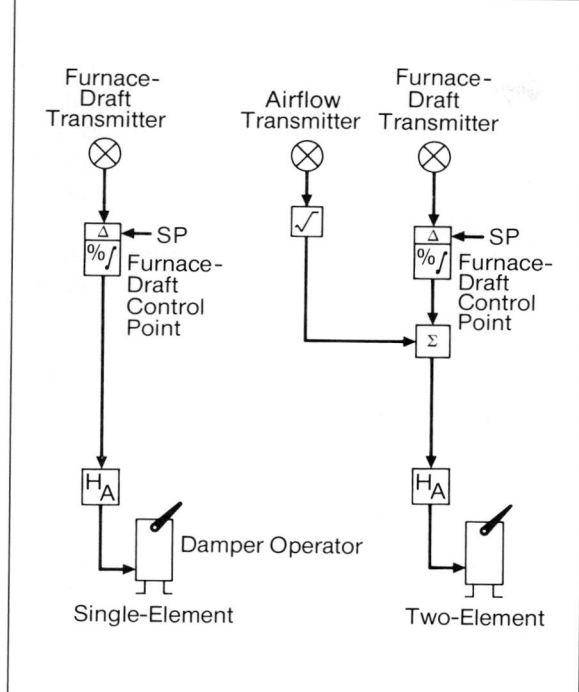

Fig. 22 Furnace-draft control system

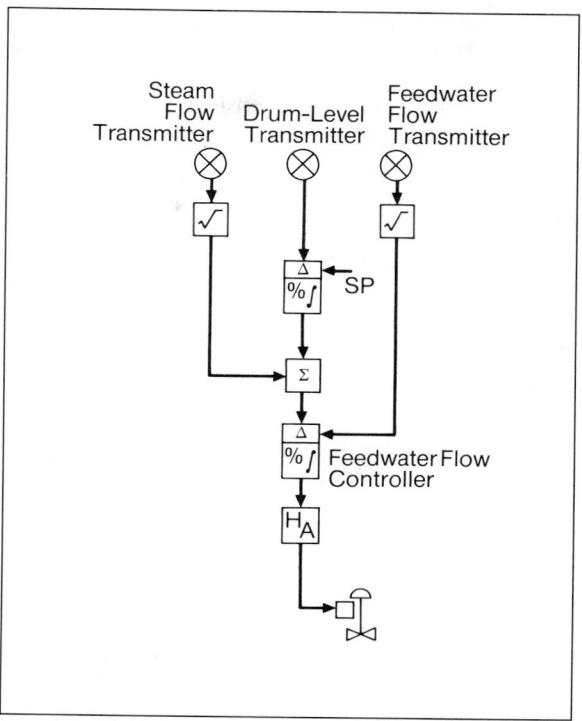

Fig. 23 Three-element feedwater-flow control system

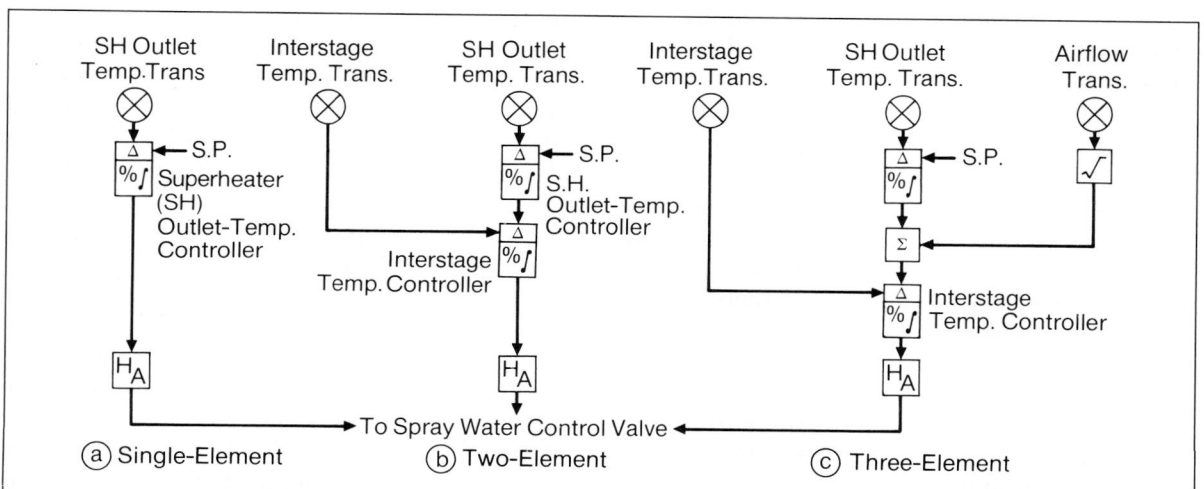

Fig. 24 Typical steam-temperature control systems

In multiple-boiler plants, the operator usually establishes the sharing of load among boilers. There are also several automatic load-allocation schemes in use that vary in complexity up to and including the lowest-operating-cost energy-management scheme.

INFLUENCE OF FUEL
ON CONTROL SYSTEMS

The type of fuel fired in a boiler has a great effect on the combustion-control system. Controls on boilers firing gas, oil, coal, wood or bark are designed to respond to plant load swings. Boilers firing municipal or industrial refuse, black-liquor recovery boilers, and certain waste fuel boilers are usually operated base-loaded to optimize combustion of these difficult fuels. Waste-heat boilers usually are operated to the extent of waste-heat availability.

Gas- and oil-fired boilers are probably the most effectively controlled because the fuels are practically homogeneous and of nearly constant density and heating value as delivered to the plant. Continuous and nearly instantaneous flow measurement of the fuel is possible, and the fuel-firing rate is readily and precisely controlled by control-valve adjustment.

The most commonly used combustion-control system for gaseous and liquid fuels is the cross-limited metering system illustrated in Fig. 21. The fuel and airflow control loops in this system compensate automatically for changes in air- or fuel-system resistance to flow. Continuously active safety constraints automatically limit fuel flow to available airflow. This control system is readily adaptable to automatic adjustment of airflow for fuel/air ratio trim by flue-gas analysis. This basic control system strategy has also been adapted to simultaneous firing of multiple fuels.

When combustible plant-process byproducts can be used as a boiler fuel, burning the byproducts instead of purchased fuels may be economical. In such instances, the combustion-control system may be adapted to fire the byproduct gas or liquid fuels automatically on a preferential basis to the extent of their availability. Such a control strategy has been used in many plants to fire as many as four fuels in the same boiler.

Firing solid fuel presents a challenge because the effective fuel firing rate and the corresponding heat input and combustion air requirements cannot be directly and continuously measured. This difficulty requires that an inferential measurement of fuel-firing rate be made in order to match combustion air and fuel for economical operation. Measuring steam flow and relating it to heat input and combustion airflow are the control strategies commonly used to surmount this difficulty. Controlling steam-flow/airflow ratio approximates fuel/air ratio control, but it should be noted that there are limitations to this method. For example, changes in energy storage during load swings temporarily alter the relationship of heat input to steam flow. Changes in boiler efficiency, feedwater temperature, or boiler outlet steam conditions also affect the relationship. This type of control system is often adapted to adjust the airflow automatically to trim fuel/air ratio by flue-gas analysis.

CONTROLS FOR CHEMICAL RECOVERY BOILERS

Black liquor recovery boilers in pulp and paper mills, described in Chapter 8, are operated to recover spent chemicals from black liquor and to produce steam by burning the carbonaceous matter contained in it. Black liquor is fired in recovery boilers at a rate that is automatically controlled to a value determined by the operator. The black liquor forms a char bed at the furnace bottom and burns in a reducing atmosphere which promotes chemical reactions. To maintain the reducing atmosphere at the furnace bottom and to complete the combustion of the evolved gases higher in the furnace, air is admitted to the furnace through primary, secondary and tertiary air ports. The three airflow streams are each automatically controlled to maintain a fixed ratio to black liquor flow rate. Proper adjustment of each ratio establishes the distribution and quantity of combustion air, which is a critical factor in controlling the chemical-recovery process as well as the combustion efficiency.

It will be noted that, due to the nature of the process, the black liquor flame is not monitored, but an FSSS furnace safeguard supervisory system does monitor the auxiliary fuel burners. Oil- or gas-fired starting burners in the lower furnace of the recovery boiler are fired to initially dry and ignite the black liquor before a char bed is established. These burners are also

fired to stabilize combustion conditions at low load and at times when the black liquor is not burning well. For normal black liquor firing conditions, the starting burners are not in use. However, they are subject to operation on very short notice, if necessary, to stabilize combustion conditions. For this reason, the furnace safeguard system distinguishes between an auxiliary fuel trip and an orderly shutdown of auxiliary fuel burners. Following an orderly shutdown of the auxiliary burners, as long as airflow through the furnace exceeds purging airflow and as long as all operating permissives are within limits, purge credit should be maintained and it should be permissible to initiate an auxiliary burner start sequence when required.

Some recovery boilers include auxiliary-fuel-fired load burners located in the oxidizing zone high in the furnace. The load burners augment steam production when black liquor is insufficient for plant needs or is unavailable for firing. The load-burner firing rate is usually controlled to a setpoint adjusted by the operator, and the recovery boiler continues to be operated base-loaded. Some units have combustion-control systems designed for load burners to respond automatically to load swings as demanded by the plant master controller. The control systems automatically introduce additional air into the oxidizing zone of the furnace in excess of black liquor requirements specifically for combustion of the auxiliary fuel fired at the load burners.

The FSSS flame monitoring and interlocking extend to load burners as well as to starting burners. The recovery-boiler supervisory system is designed to operate these auxiliary burners with continuous ignitors (described in Chapter 12). The interlock system requires proof that adequate ignition is produced by the ignitor as a permissive to admit fuel to the auxiliary burner. Loss of ignition at an ignitor results in immediate closing of the safety shutoff valves at the associated auxiliary burner.

There are some exceptions to the use of continuous ignitors for recovery-boiler auxiliary burners. In response to the scarcity of premium fuels for ignitors in the mid-1970's, some recovery boilers were equipped with burners started with interrupted ignitors, usually direct electric ignitors. For these units the FSSS system allows a short timed trial for ignition, after which the ignitor is turned off and flame detectors monitor the auxiliary burner flame. Failure to detect flame any time after expiration of the time trial results in immediate closing of the associated auxiliary burner safety shutoff valves.

FIRING-EQUIPMENT CONTROL

As has been seen in the previous chapter, some fuels may be burned by more than one type of firing equipment, and the control system must be designed accordingly.

For example, oil, gas, and pulverized coal can be either tangentially or horizontally fired. Flame monitoring of tangentially fired industrial boilers is as described for utility boilers in an earlier section. For multi-burner horizontally fired boilers, however, each burner flame envelope must be monitored individually at all firing rates. The significance of this is that a "no flame" indication by the flame detector will cause individual shutdown of the associated burner. Burners with flame detectors confirming the presence of flame are continued in operation. If the combustion-control system is in automatic operation, it will compensate for the reduced number of burners in operation by increasing fuel heat input at each operating burner. In order to distribute combustion air to the burners in operation, the individual burner dampers or burner vanes must be closed at burners that are shutdown. The FSSS system can be designed to coordinate the operation of the burner air-dampers automatically.

The various means of firing may require control either of the firing equipment or closely associated distinctive fuel-preparation processes. For example, steam-atomized oil firing requires control of oil temperature and atomizing-steam pressure. Black liquor water content, temperature, and pressure must be controlled. Pulverizer coal-feed rate, primary airflow, and mill outlet temperature are regulated. For spreader stokers, feed rate and overfire air are controlled; undergrate air distribution, grate speed, and distributor speed are also adjusted by the operator.

The chosen fuel and means of firing it can also affect the boiler control system by its dynamic responsiveness. Consider the inventory of unburned fuel resident in a furnace in normal operation at any given instant. At equivalent firing rates, it can be seen that pulverized coal (fired in suspension) results in a far smaller inventory than spreader stoker firing over a grate, and several orders of magnitude smaller than for underfeed, traveling grate, and mass-burn refuse firing. The response of the boiler to changes in fuel feed rate to the furnace is much faster for firing systems with a small inventory of finely divided fuel than for firing systems with a large inventory of large pieces of fuel in various stages of drying, devolatilization, and burning. Firing systems with a large fuel inventory are more responsive to changes in combustion airflow than in fuel flow, so it is desirable to apply firing-rate demand changes to the air input first and to retard changes in the fuel input.

CONTROL OF CIRCULATING FLUIDIZED-BED BOILERS

Circulating fluidized-bed (CFB) boilers, which are described in detail in Chapter 9, include many significant features of operation that the control system designer must consider. In its basic elements, the combustion-control system for a CFB unit is a cross-limited metering system similar to that shown in Fig. 21. Primary air and secondary air are individually controlled combustion-air streams that are staged as a function of firing rate for effective distribution. Seal-pot and fluidized-bed heat exchanger (FBHE) fluidizing air streams provide additional combustion air and are measured but not modulated. Oxygen in the flue gas is measured and automatically adjusts fuel feed rate for fuel/air ratio trim. It is important to maintain the combustor temperature within a narrow band. The combustor outlet temperature is controlled by regulating the flow of ash to the FBHE, thereby affecting the distribution of heat absorption in the unit. Sulfur dioxide in the flue gas is measured and controlled by adjusting the rate of alkaline-sorbent feed to the

combustor. The combustor differential pressure is measured and controlled by varying the rate at which ash is drained from the FBHE.

Oil- or gas-fired starting burners operate to increase the combustor temperature gradually and at a low rate in order to protect the refractory during start-up. The starting burners are horizontal burners which are operated with flame monitoring and safety interlocks similar to those required for horizontal burners in conventional boilers. The starting burners are required to raise the combustor temperature to a level exceeding the auto-ignition temperature of the main fuel. This is a key firing permissive for the main fuel interlock in the fluidized-bed FBSS® supervisory system.

The CFB combustor includes oil- or gas-fired burner lances designed to provide a fuel input to the combustor in the event of problems with the solid-fuel feed system. The combustor temperature must exceed the auto-ignition temperature of the boiler lance fuel as a permissive to fire the burner lances.

Key safety functions of the FBSS system are

- a prefiring purge of the furnace,
- establishment of the appropriate permissives for firing the starting burners,
- establishment of the appropriate permissives for firing the main fuel,
- establishment of the appropriate permissives for firing the burner lances,
- continuous monitoring of firing conditions and other key operating parameters, and
- emergency shutdown of portions or all of the firing equipment when required.

DIVERGENT DESIGN PRACTICES

Industrial boilers are usually smaller in scale and less complex than electric-utility boilers. In many industrial plants, an individual boiler may be less critical to the plant operation than a utility boiler is to power-system operation. In the continuing evolution of boiler control designs for industrial and utility boilers, it is understandable that some divergence in design practices has occurred. As might be expected, then, control systems for most industrial boilers

are less complex than for utility boilers.

One of the fundamental differences is the frequent application of "de-energize-to-trip" design to industrial boiler controls. The various control elements to be used in a control system are analyzed to determine their behavior during loss of power and during their most probable mode of failure. The control system is then designed with components selected to position the final elements in a process "safe" position or to shutdown in case of loss of power or component failure.

In many multiple-boiler industrial plants, the shutdown of one boiler does not greatly affect plant operation, so redundant boiler controls and auxiliary equipment have generally not been applied to industrial boilers. It follows that runup, rundown, and runback circuits also have little application in most industrial-boiler control systems.

Nevertheless, each boiler application must be carefully evaluated because, in some cases, the boiler is so critical to plant operation that exceptions to usual design practice are warranted in order to increase availability of the unit. Indeed, for this reason, many industrial-boiler control systems have been designed according to utility-boiler control practice.

FURNACE IMPLOSIONS

Earlier in this chapter, furnace explosions and the techniques to prevent their occurrence were discussed. In the mid-1970's, a new problem, which came to be known as a furnace *implosion*, appeared. Before discussing the impact of the implosion phenomenon on steam-generator control systems, the fundamental nature of the problem should be examined.

THE IMPLOSION PHENOMENON

Two basic mechanisms can cause a negative-pressure excursion of sufficient magnitude to cause structural damage. The first of these mechanisms is well understood.

Simply attach to a boiler an induced-draft (ID) fan that is capable of producing more suc-

tion head than the boiler structure is capable of withstanding. Next, through control malfunction and/or operator error, establish circumstances that result in this pressure capability being applied to the structure. An example would be opening the dampers on an operating ID fan with the forced-draft (FD) fan dampers closed, which can result in a destructive negative pressure with the boiler not being fired.

The second mechanism, the so-called *flame collapse* or *flameout* effect, is not generally understood. As a matter of fact, the unfortunate use of such terms as *flame collapse* has led to many distorted conceptions of the physical reality.

It is vitally important that the following explanation for the negative-pressure excursion which follows a fuel trip and loss of furnace flame be understood, because only through a basic knowledge of this mechanism can pitfalls and preventive techniques be realistically evaluated.

The physical state of the gases in a furnace or other subsystem at any instant in time can be described by the Perfect Gas Law as follows:

$$PV = MRT$$

where: P = absolute pressure
V = volume of system under consideration
M = resident mass (not mass flow)
R = the universal gas constant
T = absolute temperature,
all in consistent units

(2)

Because V is fixed and R is approximately constant, P is directly proportional to the product MT. Thus, for two different conditions of pressure and temperature in a given boiler system, this approximate relationship holds:

$$\frac{P_2}{P_1} = \frac{M_2 T_2}{M_1 T_1}$$

(3)

During steady-state operation, P is held constant at approximately atmospheric pressure

by balancing the resident mass M and the existing temperature T. Furnace temperature in a steam generator is not directly controlled and depends on the thermal balance between the *heat in* (in the form of burning fuel and heated air) and the *heat out* (in the flue gas and heat transferred to the pressure parts). The resident mass is automatically balanced by controlling the flue gas flow out of the boiler to maintain a given furnace pressure.

When the fuel input is terminated, this balance no longer exists. The flue gas being pulled out of the furnace by the ID fan is now being replaced only by preheated air rather than by the products of combustion in the firing zone. As a result of this situation, the average temperature of the gases resident in the furnace (or other subsystem) at any given time following the fuel trip will decrease rapidly. Because of the temperature drop, the pressure in the furnace starts to decrease (Fig. 25).

If the resident mass were to remain constant in quantity, the absolute pressure in the furnace would drop in direct proportion to the absolute temperature drop. A gage pressure of −1.0″ WG is equivalent to 406″ WG absolute. Thus, with a constant resident mass, a 10 percent change in absolute temperature would cause a 10 percent change in furnace pressure; that is, 40.6″ WG.

Fortunately, the natural characteristics of forced-draft and induced-draft fans and their control devices are such that a constant-resident-mass situation is virtually impossible to achieve.

Assume, for the moment, that all FD and ID fan dampers are fixed in position. As the furnace pressure begins to decay because of the

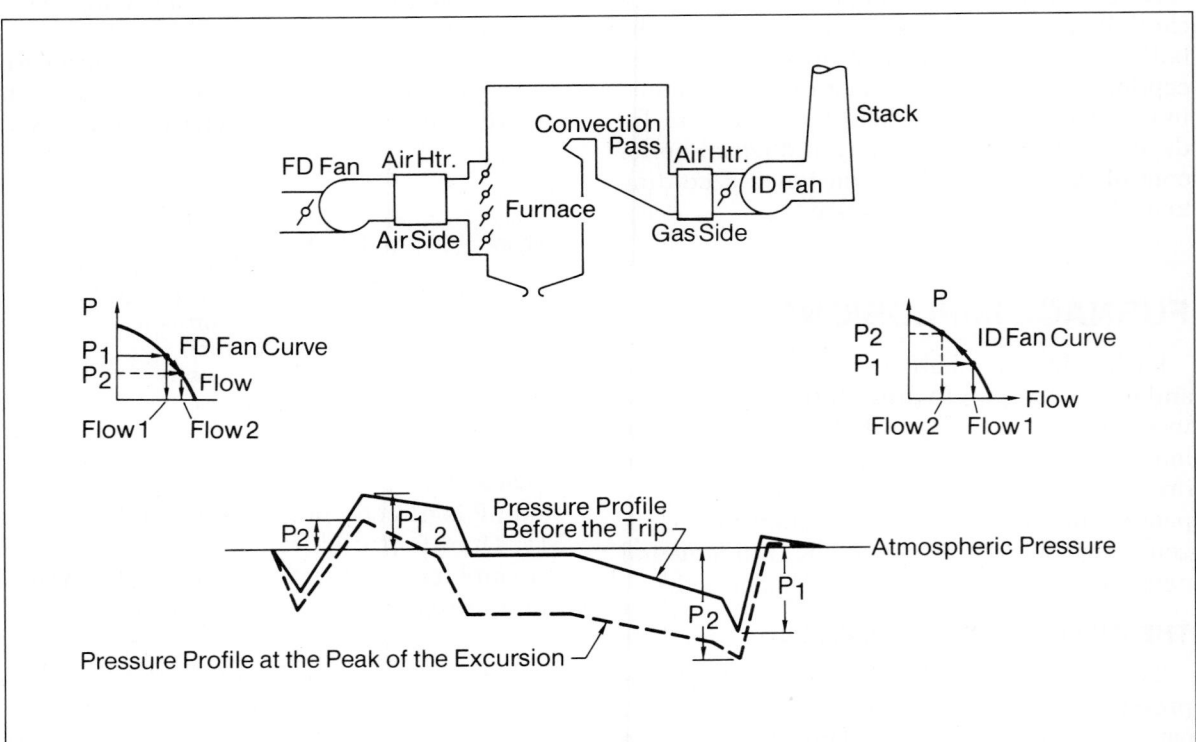

Fig. 25 Pressure profile in a balanced-draft unit before and after trip. Notice that the magnitude of the drop in furnace pressure at the peak of the excursion is higher than those at the FD fan discharge and the ID fan inlet. This is because of the difference in flow conditions at the time of the excursion.

temperature decay, the fans will move to different operating points on their respective performance curves. The FD fan will move in the direction of *increasing* volume; the ID fan, in the direction of *decreasing* volume (Fig. 26).

■ The airflow through the windbox will *increase* because the lower furnace pressure appears to the FD fan to be a reduced system resistance.

■ The flue-gas flow will *decrease* because the lower furnace pressure appears to the ID fan to be an increased system resistance.

The net effect of this increase in airflow and reduction in gas flow is to increase the resident mass in the furnace, which tends to compensate for the temperature decay. This natural corrective action taken by the FD and ID fans is a key ingredient in the ability to contain furnace-pressure excursions to values that are within tolerable limits.

It should be apparent that the Perfect Gas

Law does not recognize the difference between:

■ types of firing systems (corner, front and rear, dry-bottom, or slagging bottom);

■ coal firing, oil firing, or gas firing; and

■ balanced-draft or pressurized operation.

To understand, then, why the most serious implosion problems have been experienced with oil-fired, balanced-draft boilers, it is necessary to keep in mind that the key to holding furnace pressure constant is to maintain the product of resident mass and temperature constant.

On a *pressurized* boiler, the action of the FD fan during normal operation *pushes* the flue gases out of the furnace.

When the combustion process in such a pressurized unit is terminated, the resulting temperature decay causes the furnace pressure to drop rapidly. As soon as the furnace pressure drops below atmospheric pressure (ignoring

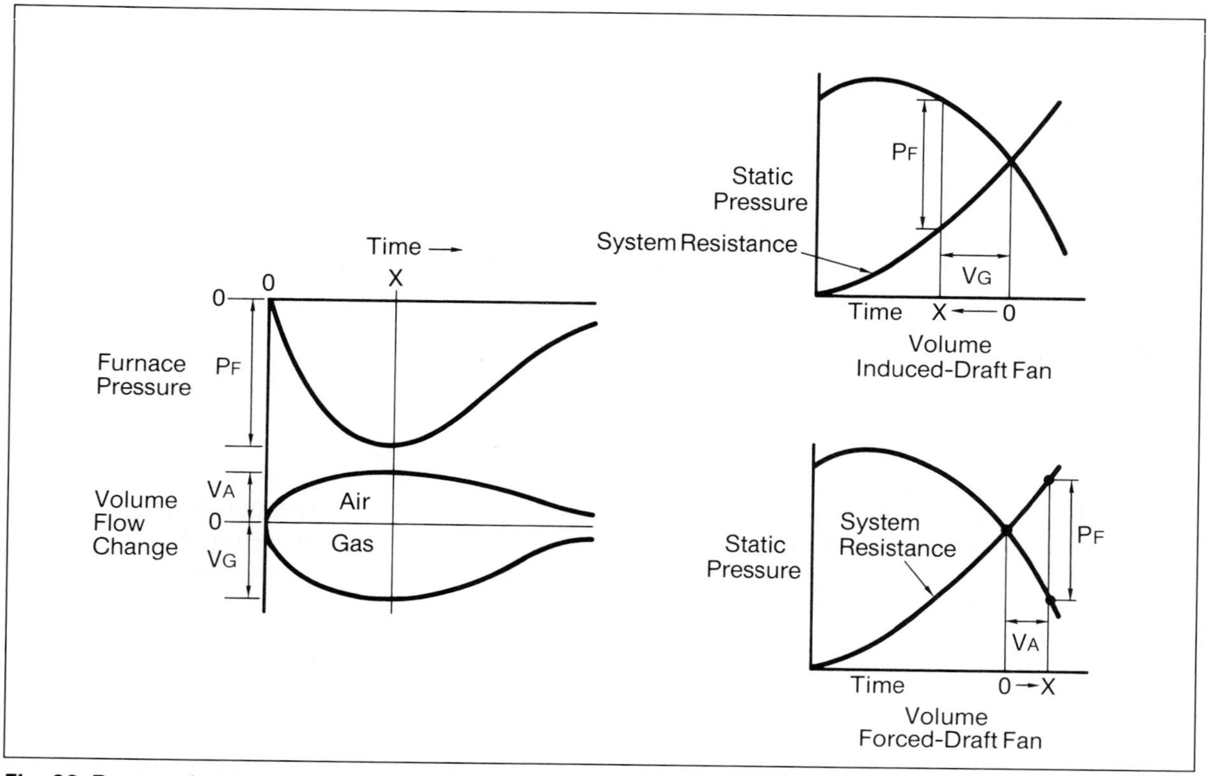

Fig. 26 Destructive negative pressure (DNP) with constant-speed centrifugal fans and dampers locked

stack effect), the flue gas will no longer leave the furnace but the airflow will continue to enter. Moreover, any additional decrease in furnace pressure will cause a reverse flow down the stack, back into the furnace.

When firing is terminated on a *balanced-draft* unit, flue gas (with its mass at furnace temperature) will be pulled out of the furnace until the pressure there reaches a negative value which is directly related to the cutoff pressure of the ID fan corresponding to the temperature entering the fan. The higher the head capability of the ID fan, the larger the pressure excursions that may be experienced.

Thus, although some authors have attempted to treat the ID fan characteristics and the fuel cutoff as two independent effects, they are actually inseparable.

With respect to pulverized-coal and oil firing, the difference is one of the *rate* of temperature change. This rate depends on how fast a given percentage of the furnace volume undergoes a replacement of combustion products with preheated air. Coal-fired units have the following characteristics that produce a slower temperature decay:

1. The fuel cutoff is generally more gradual than with oil firing, because of the pulverized-fuel residual in the pulverizer and fuel piping.

2. The ratio of resident mass to mass-flow rate is higher than for an oil-fired boiler, because coal-fired furnaces are significantly larger in size than oil-fired-only furnaces.

THE "CULPRIT," THE INDUCED-DRAFT FAN

In connection with the implosion phenomenon, it is most important that engineers responsible for draft-system and furnace-framing-system design understand how an induced-draft fan operates, how it is controlled, and how it responds to changes in the inlet pressure that it "sees."

Induced-draft fans are used singly, or in multiples of 2, 3, or 4, to evacuate the products of combustion from a boiler furnace while maintaining the pressure at the top of such a furnace about 0.2" WG below atmospheric pressure.

Centrifugal-type ID fans customarily are equipped with inlet vanes or inlet louver dampers for control of the volumetric flow of gas through them. (See Chapter 14.)

These devices have similar control characteristics (as compared to outlet dampers or variable-speed drives), but can have widely varying time-constants depending upon the type of actuator used to move them.

Fig. 27 shows the effect on the pressure/volume curve of such a fan when either vanes or dampers are closed to reduce the gas through-flow. On this curve, fan static-pressure capability is plotted against rated volume, and a system-pressure curve is shown—a simple square relationship of static pressure to volume (SP $\propto V^2$). The control device is regulated so that the required static pressure to satisfy the resistance of the physical system is developed at any given volumetric flow rate. But, at the same time, for the particular vane position, a fan characteristic curve is established such that the fan is capable of an infinite number of discrete static-pressure values versus any given volumetric flow.

If we consider an operating mode in which the induced-draft-fan control vanes or dampers are capable of infinitely fast response, any change in the furnace pressure of a balanced-draft unit that is sensed by the system will cause the vanes to change position to maintain the required system static pressure exactly, and along the system-pressure curve. For instance, if there should be a reduction in furnace pressure (caused, say, by a decrease in the fuel fired and the corresponding combustion airflow), the vanes will immediately close to reduce the gas flow handled by the fan before the furnace pressure drops any further. With such infinitely fast vane control assumed, even with an instantaneous very large drop in furnace-pressure level (such as there could be if there were a master fuel trip or if the forced-draft-fan dampers were to accidentally close), the system curve will always be followed and the induced-draft fan will make instantaneous corrections to maintain furnace pressure at the level preset (below atmospheric).

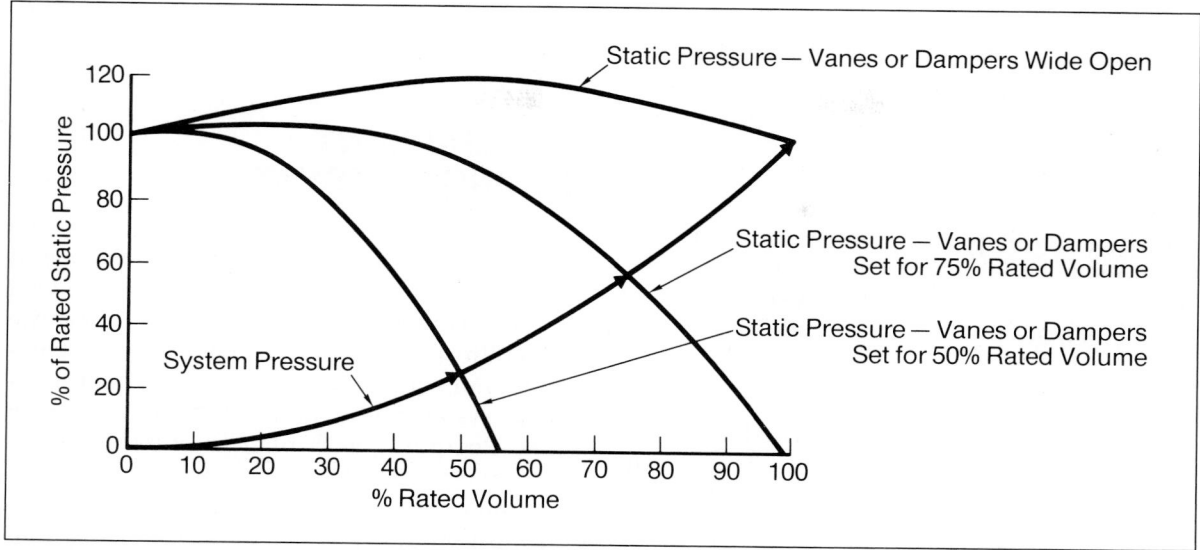

Fig. 27 **The effect on a centrifugal-type ID-fan pressure/volume curve when either vanes or dampers are positioned to reduce the gas through-flow**

Another condition can be postulated, in which the control vanes or dampers are maintained in a fixed position as boiler load varies; this is equivalent to a zero rate of damper response. When boiler load decreases (and fuel flow, airflow, and gas flow decrease correspondingly), the developed fan static pressure increases following a static-pressure versus volume curve sloping upward to the left; furnace suction (pressure below atmospheric) increases rapidly. The obvious limit to the furnace suction that can be developed is the highest static-pressure value on the curve, which for damper settings below 50 percent, appears to be at or about the fan no-load cutoff value (the static head developed by a fan when its volumetric delivery is zero).

Under normal load-changing conditions, if fan vanes are maintained at a given position, either accidentally or purposely, furnace suction will vary, with the maximum static-pressure capability of the fan being the limit. Under these conditions, of course, forced-draft airflow continues, but is automatically varied with boiler firing rate.

Assume that a master fuel trip occurs un-

der this condition of vanes-maintained-in-position. Assume further that the forced-draft fan dampers close quickly, so that there is a rapid decrease in the gas available to the induced-draft fan. The induced-draft fan will immediately ride up along the characteristic curve to the cutoff point; after reaching the cutoff point, reverse flow of gas down the stack will occur (at some indeterminable flow rate) until the furnace suction that has been established by the decrease in the average gas temperature is satisfied.

If the rate of reverse flow is not high enough, the furnace will fail structurally if the framing system is not capable of withstanding the low absolute pressure that results from the fuel trip.

Of course, what happens during a master fuel trip in which equipment is on automatic control is somewhere between the above two postulated situations. Fan dampers or vanes close in an effort to reduce the gas flow leaving the furnace, restricting the magnitude of the furnace pressure-level reduction. With good mechanical response to the control signal (particularly if an "early" signal has been received through the use of a feed-forward arrangement), fan static-pressure capability will match the furnace suction being developed as a result of the reduction in furnace gas temperature, all while the volumetric gas flow through the fan is being reduced to zero. Ultimately, this frustrates any implosion tendency.

The above analysis applies to constant-speed centrifugal-type or vane-controlled axial-flow fans. A similar approach can be used in assessing the implosion potential of variable-speed centrifugal, or variable-pitch axial-flow fans.

CONTROL IMPLICATIONS

Implosion concerns have resulted in many new control-system developments. The search for the ideal implosion-prevention control scheme is still proceeding. Each steam-generator manufacturer has recommendations, and the National Fire Protection Association has issued NFPA 85-G on the subject.

Space does not permit going into the specifics of the control-system designs which are outlined in such recommendations. However, the following must be provided in any furnace-draft and combustion-control system:

1. The airflow to a furnace must be maintained at its pretrip value and must not be prevented from increasing by following natural fan curves; but positive control action to increase airflow is not allowed by NFPA.

2. The flow of combustion products from a furnace must be reduced as quickly as possible following a unit trip.

3. If the removal of fuel from the furnace can be over a 5- to 10-second period (rather than instantaneously), there will be a reduction in the magnitude of the furnace-pressure excursion that follows a unit trip.

Obviously, even the most carefully designed control system will be of limited value if all its components are not completely installed, if it has not been properly checked out or is in "poor tune", or if it has been inadequately maintained. It is essential, therefore, that protective control systems for large, high draft-loss boilers be properly designed, installed, tested,

and maintained, and that plant operators consider these control systems to be vital.

POWER-PLANT DIAGNOSTIC SYSTEMS

The need for performance diagnostic systems and their high cost-effectiveness are discussed in Chapter 24. Although the study referenced in that chapter was aimed at improving the availability and performance of existing steam-generation equipment, the application of diagnostic systems is equally important for new units. Beginning with plant start-up, performance monitoring and diagnosis can provide a consistency of unit operation and can minimize, if not totally avoid, those operating conditions that may have adverse effects on the life of the equipment. Further, because fuel costs constitute the major portion of the cost of electricity, plant heat rate, especially as it is influenced by boiler efficiency, must be a primary consideration in all modes of operation.

SYSTEM STRUCTURE

The purpose of diagnostic systems is to continuously monitor, evaluate, and interpret during operation the thermal, structural, and chemical conditions of the entire power-plant complex in real time. The information obtained is displayed in a meaningful manner to operating personnel and provides them with the opportunity for appropriate action to improve performance and to avoid conditions which may adversely impact equipment life. Further, operating conditions (in the form of data sets) are date- and time-stamped, and archived for off-line recall at a future time. This chronicling allows the opportunity for detailed analysis by results engineers, and the setting of long-term operating strategies consistent with management objectives. In addition, the archiving of data sets with the ability to review operation conditions based on various parameters (such as load, fuel/air ratio, water chemistry, temperature excursions, sootblowing frequency, and the names of operators on duty) can provide insights into long-term degradation of equipment performance or the necessity for additional operator training, which may not be readily apparent during on line-operation.

The sophistication of the installed diagnostic system will, of course, depend on the type and cost of the equipment and software selected. A range of systems is available, from off-line personal-computer-based programs to advanced mini-computer types that allow interactive communication for on-line changes in operating parameters.

DESIGN STRATEGY

The modern power plant obviously will include all the means to generate power; that is, it will have a boiler, a turbine, a generator, primary sensors for monitoring operation and for data acquisition, a data highway, and operational-control equipment. Adding appropriate diagnostic programs, as needed, can improve the efficiency and reliability of power generation. Although the calculated results of on-line analysis can be used to directly affect unit operation, diagnostic systems are better employed to provide adequate feedback information for operators to make intelligent and well-considered decisions to improve unit performance. In order to so improve thermal performance, increase plant availability, and to promote equipment longevity, an integrated system can be built from one of more of the currently available diagnostic modules described below.

■ On-Line Thermal Information System — improves plant thermal-cycle efficiency and optimizes sootblowing frequency.

■ Boiler Stress and Condition Analyzer — extends unit life and improves availability by reducing damage to heavy-walled components.

■ Interactive Chemistry Management System — improves availability by minimizing water and steam chemistry excursions.

■ Acoustic Steam-Leak Detection System — reduces forced-outage time by early detection and location of leaks in boiler pressure parts and feedwater heaters.

■ Tube Temperature Monitoring System — improves availability by reducing superheater and reheater element temperature excursions.

■ OPSIZE — reduces carbon loss, lessens slagging and fouling, and optimizes pulverizer and precipitator performance by measuring and controlling coal-particle fineness.

Although each system functions on a stand-alone basis, they work in combination to complement each other by sharing a common data base. For example, optimizing pulverized-coal fineness can improve thermal efficiency, and so on. Fig. 28 illustrates the integrated concept; it shows how each diagnostic module can communicate with plant data-acquisition systems, the station control room, and engineering work stations, as desired.

Two of the modules above are intended to improve boiler efficiency as well as the overall plant heat rate.

C-E ON-LINE THERMAL INFORMATION SYSTEM (OTIS)

Thermal-performance-optimization software assists operators in selecting controllable parameters of boiler and steam cycles. For coal-, oil-, and gas-fired plants, the controllables include excess air, burner tilt, desuperheating-spray flows, and feedwater heaters in service.

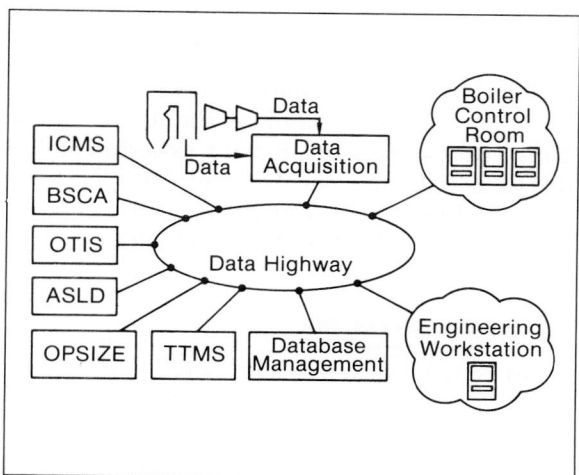

Fig. 28 **Combustion Engineering Total On-Line Performance System**

For coal-fired plants, special importance is placed on sootblowing. Thermal-performance software also assists operators and plant technical staff in diagnosing long-term trends of conditions that can erode heat rate. Examples of such effects are turbine blade erosion, condenser fouling, feedwater-heater leakage or fouling, and air-heater leakage.

Thermal-performance-optimization software is evolving for four distinctly different levels of plant operation. These levels are as follows:

1. Gross plant, boiler, and turbine efficiency.

2. Component-level performance with trending compared to baseline tests. The components analyzed are the boiler, steam turbine, condenser, and feedwater heaters.

3. Subcomponent-level performance, with on-line optimization strategies. The subcomponent-level diagnostics address, for instance, individual heat-exchanger elements and discrete turbine expansion stages in addition to overall component performance. Optimization strategies are based on achieving original design thermodynamic conditions or on iterative searches to optimize overall heat rate. Pattern-recognition techniques are employed to diagnose the causes of long-term heat rate degradation. Predictive modeling allows operators to evaluate alternative control strategies without actually changing operating parameters.

4. Bona fide expert systems employing heuristics to isolate problems and recommend action. These diagnostics are mostly experimental at this time; in them, standard engineering models based on the conservation equations, heat transfer, and thermodynamics interact with a heuristic process.

COAL FINENESS CONTROL

Minimizing carbon loss is widely recognized as an important factor in improving boiler efficiency. Even though a new boiler operates with excellent carbon-conversion efficiencies, older equipment may not do as well. Inadequate coal fineness control is a major contributor to elevated

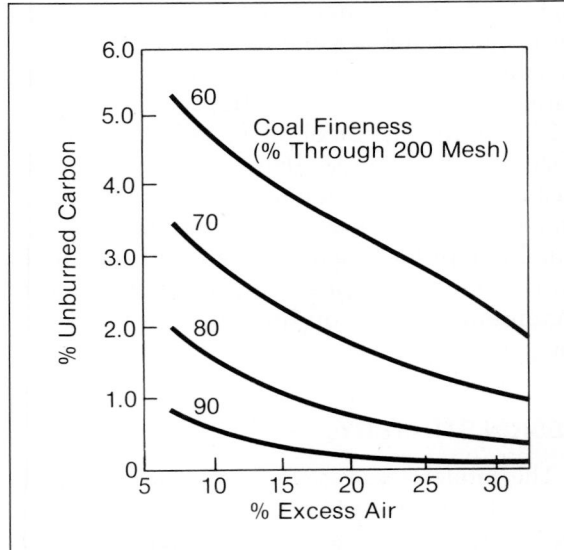

Fig. 29 Sensitivity of carbon loss to coal fineness

carbon loss. Fig. 29 shows the sensitivity of carbon-conversion efficiency to coal fineness for a boiler firing low- to medium-volatile bituminous coal.

The degradation in fineness usually results from one or more of the following problems:

- worn separator body liners
- damage to one or more classifier vanes
- excessive ring-to-roll clearance
- variations in coal grindability resulting from blending or switching sources
- infrequent fineness measurement and adjustment

The C-E OPSIZE system addresses these problems by performing on-line real-time measurement and control of coal fineness.

PLANT AVAILABILITY IMPROVEMENT

Pressure-part failures are generally found to be the single most important contributor to steam-generator forced outages. Interactive chemistry-management systems and acoustic steam-leak detectors address this problem area.

C-E INTERACTIVE WATER-CHEMISTRY MANAGEMENT SYSTEM (ICMS)

This diagnostic tool is designed to improve boiler availability through better control of water and steam chemistry, as described further in Chapter 20 on power-plant water technology. Continuous on-line water-chemistry management assists operators by providing alarms and recommending corrective action as soon as a water or steam chemistry excursion is detected. Fig. 30 shows a schematic diagram of the sampling requirements for a typical water/steam cycle. For feedwater: pH, oxygen, ammonia, hydrazine, and condenser leakage are monitored and displayed; for boiler water: pH, phosphate, total solids, silica, and blowdown; for steam: cation conductivity, sodium, and carryover.

Pattern recognition, trend analysis, and solubility data are used to diagnose phosphate hideout and condenser leakage. Analyses are performed on-line to determine the magnitude of the occurrences, their impact on system chemistry, and the effect of corrective actions.

C-E ACOUSTIC STEAM-LEAK DETECTION (ASLD) SYSTEM

The ASLD system allows plant operators to detect leaks early and locate them more accurately. Both features help minimize the cost resulting from a tube leak. Early detection can minimize the damage that can occur when a leak washes nearby tubes; accurately locating the leak will reduce repair time.

The Acoustic Steam-Leak Detection System consists of a number of tuned listening channels which are strategically located along the furnace walls of a steam generator. Boiler size and configuration determine the required number of listening channels per system. The sound produced by a leak is picked up by the listening channels and is transmitted to a head amplifier which filters and boosts the signal for transmission to the signal-conditioning unit control-room display.

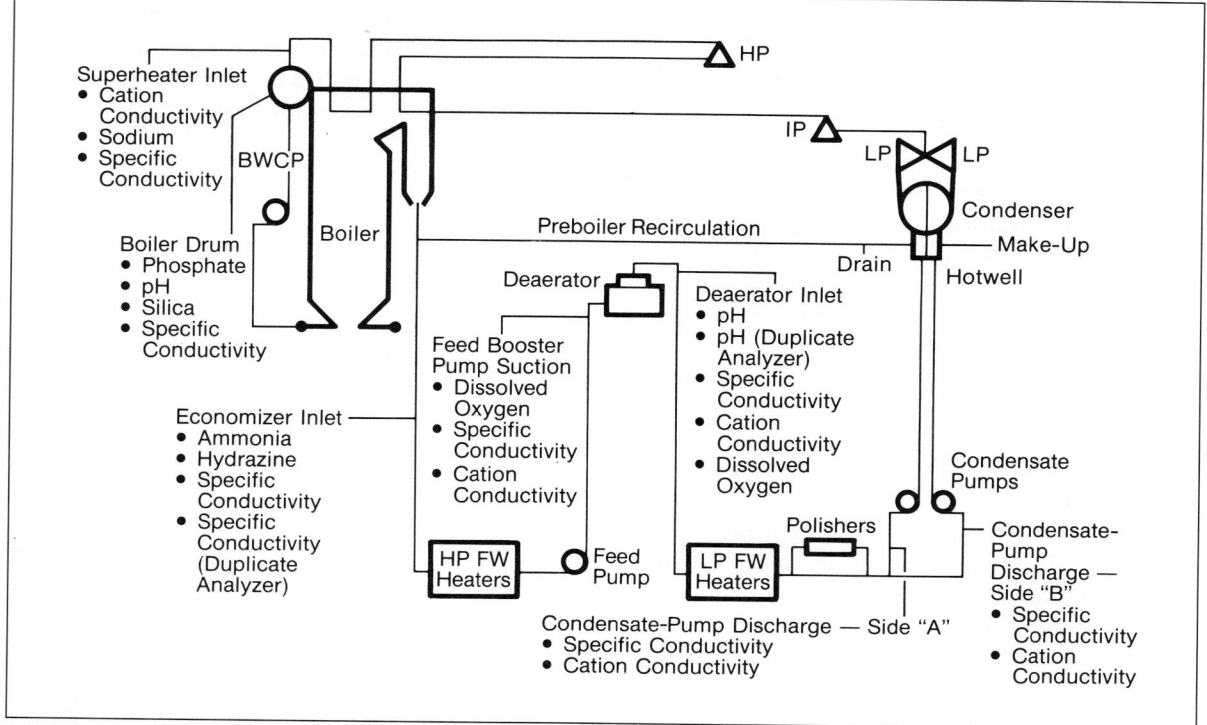

Fig. 30 **ICMS sampling requirements**

LIFE CONTINUATION

A major area of concern in prolonging the life of existing steam-generators is in the avoidance of damage to heavy-walled components. The life of headers, tees, drums, and high-energy piping is critical because of the high costs and long times associated with obtaining replacements for damaged components.

A second aspect of continuing unit life is concerned with preventing superheater and reheater tubing failures, which represent nearly half of all boiler tubing failures.

One of the most cost-effective ways of continuing equipment life is to make it possible for plant personnel to avoid damaging operations. To do this, operators need accurate on-line information, provided in the control room through a stress-and-condition analyzer and a tube-temperature monitoring system.

C-E BOILER STRESS AND CONDITION ANALYZER (BSCA)

The BSCA is an on-line diagnostic system designed to assist operators in avoiding damaging operating conditions. During plant start-up, shutdown, and cycling operation, temperature transients are imposed on the headers, tees, drums, and steam lines. The resulting thermal loadings cause significant stresses which can increase creep damage and cyclic fatigue. The BSCA monitors the unit operation in real time, alerts operators to damaging transients as they occur, and calculates accumulated damage to critical components.

The BSCA continually collects plant operating data including temperature, pressure, and flow data. The effects of pre-existing operational practices on the state of damage in the thick-walled pressure parts are accounted for either by initializing the damage arrays based on historical unit operating data or by a variety of material analysis techniques. Alarm setpoints for differential temperature and overtemperature are established based upon initialized damage arrays, expected mode of unit operation in the future, and optimum component life.

C-E TUBE-TEMPERATURE MONITORING SYSTEM (TTMS)

This system is designed to assist the plant operators and engineering staff in making intelligent operating decisions which will minimize loss of tube life and increase unit availability. The importance of monitoring tube temperatures is illustrated in Fig. 31, which shows stress-to-rupture data for 2-1/4 Cr 1 Mo steel. This data indicates that increasing the temperature of a given superheater element by just 12°F can reduce its time-to-failure from 115,000 hours to 76,000 hours.

The Tube-Temperature Monitoring System monitors the metal temperatures of superheater and reheater tubing. The measured temperatures are compared against predetermined alarm levels, which are based on oxidation limits and stress-rupture limits. Current data, including alarm conditions, are presented to operators on a menu-driven display. Using this information, personnel can modify boiler operation to achieve optimum unit performance consistent with minimizing damage to tubing. In addition, the TTMS maintains a historical data base which the engineering staff can utilize to develop strategies for extended operation and maintenance based on economic criteria.

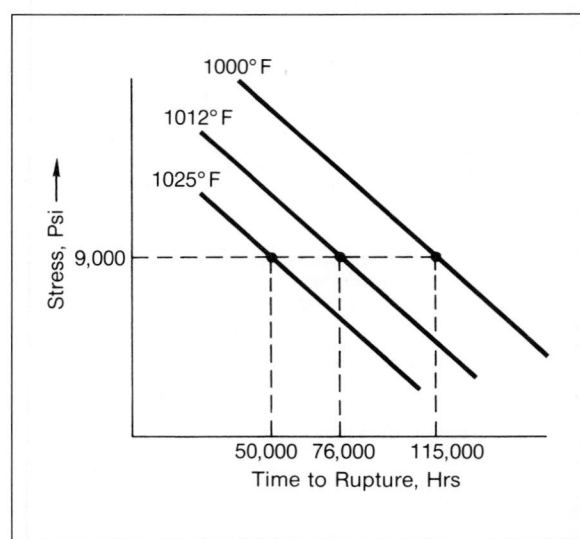

Fig. 31 Minimum stress-to-rupture of 2¼ Cr 1 Mo steel

OTHER POWER-PLANT CONTROL SYSTEMS

This chapter has dealt solely with steam-generator related control and diagnostic systems. Given sufficient space, similar chapters could be written about the controls for flue-gas-desulfurization systems, precipitators, and ash-handling systems, all of which can be in the "steam-generator island." Increased controls sophistication is taking place in such areas, for the same reasons as outlined above.

The power industry cannot be expected to play a significant role in the development of microelectronics. But the industry will take advantage of new developments, particularly in the area of distributed computer-based systems, as they become available.

Increasing demands for higher productivity and availability of steam-generating plants will provide new challenges for designers of operational-control and diagnostic systems. The electronics industry will provide the tools; the power-generation controls engineers must provide the imagination.

CHAPTER 14

Steam-Generator Auxiliary Equipment

This chapter describes power-plant equipment such as fans, air heaters, sootblowers, and the boiler-water circulating pumps. This equipment is usually designated as boiler auxiliaries. Power-plant noise, which applies particularly to these and most other equipment peripheral to the boiler, also is covered. Other such components described in detail elsewhere include pulverizers, Chapter 11; ash-handling equipment, Chapter 16; and structural-steel design and erection, Chapters 6 and 19. Operational control system philosophy and design are included in Chapter 13, and the entire scope of boiler-related emission-control equipment is presented in Chapter 15.

Although boiler auxiliary components are secondary to the design of the boiler itself, they are absolutely essential to its operation. Together, they represent a very sizable portion of the overall steam-generator cost. As differentiated from equipment considered auxiliary to the remainder of a steam power plant, a *boiler auxiliary* is equipment which is an integral part of, or is required for the operation of, the boiler. For example, compressors for air sootblowers and certain sealing requirements, are part of the overall power-plant equipment and are not boiler auxiliaries. Pumps within the boiler circuit are boiler auxiliaries, but feed-water pumps are part of the power-plant cycle. In general, valves required for boiler operation are boiler auxiliaries, but steam stop valves, turbine throttle valves and most other valves related to the steam turbine and feedwater heaters are not.

Equipment is described here to help familiarize the reader with the different components and their specific applications in boiler work. The calculation of volumes handled and static pressures to be encountered with fans are given in Chapter 6. For sootblower selection, the designer must take into account the boiler configuration and the probable dirtying characteristics of the coal ash, as referenced in Chapters 2, 3, 6, 7, 8, and Appendix B. For air-heater sizing, air and gas flows and temperatures have to be calculated, as will allowable static pressure losses; for more information on such calculations see Chapter 6.

It is important to realize that there are instances in which some of this peripheral equipment is purchased separately from the steam-generator contract. However, the boiler manufacturer has to be relied on heavily to provide basic steam generator performance data. The data is necessary for the proper selection and design of the peripheral equipment and other auxiliaries.

POWER-PLANT FANS

Regardless of fuel and method of firing, all boilers for electrical, industrial, or marine power generation use mechanical draft fans (Fig. 1). They supply the primary air for the pulverization and transport of coal to the furnace. They also supply the secondary and tertiary air to the windboxes for completion of combustion. Fans also remove the products of combustion from the furnace and move the gases through heat-transfer equipment and flue-gas desulfurization equipment. Sometimes, gas fans control steam temperature. Numerous small fans are used for sealing and cooling of ignitors, scanners, and other equipment.

Power-station fans are among the largest made: static pressures of 60″ WG and individual fan volumes of 1.5 million cu ft/min are common. And, usually, they are custom-designed with blade configuration and control systems highly sensitive to both owner preference and the evaluated cost of installed and operating power. Applications that require the largest fans (and thereby use the greatest amount of plant power) on a boiler fall into four categories which combined account for more than half of the total boiler power requirements. The four fan categories are: forced-draft, primary-air, induced-draft and gas-recirculation.

FORCED-DRAFT FANS

Forced-draft fans supply air necessary for fuel combustion, as calculated in Chapters 4 and 6, and must be sized to handle the stoichiometric air plus the excess air needed for proper burning of the specific fuel for which they are designed. In addition, they provide air to make up for air-heater leakage and for some sealing-air requirements. Forced-draft (FD) fans supply the total airflow except when an atmospheric-suction primary-air fan is used.

Radial airfoil (centrifugal) or variable-pitch axial fans are preferred for FD service. FD fans operate in the cleanest environment associated with a boiler, and are generally the quietest and most efficient fans in the power plant. They are particularly well-suited for high-speed operation. Most FD fan installations have inlet silencers for noise reduction with screens to protect the fans from any entrained particles in the incoming air.

EFFECT OF AIR TEMPERATURE

Both the air temperature at a power plant and its elevation above sea level affect air density, which in turn has a direct influence on the FD-fan capacity. If steam coils or other means of heating the air *ahead* of the FD fans are provided, the consequent air temperature to the fans must be taken into account. If hot-air recirculation is used for air-heater protection, then both the added volume of air and its higher temperature must be considered. As mentioned in the calculation sections of previous chapters, if air moisture exceeds 2 percent by weight (as can occur in high-temperature tropical installations) the resulting greater volume calls for increases in size of all fans on a unit.

PRIMARY-AIR FANS

Large high-pressure fans supply the air needed to dry and transport coal either directly from the pulverizing equipment to the furnace, or to an intermediate storage bunker. As de-

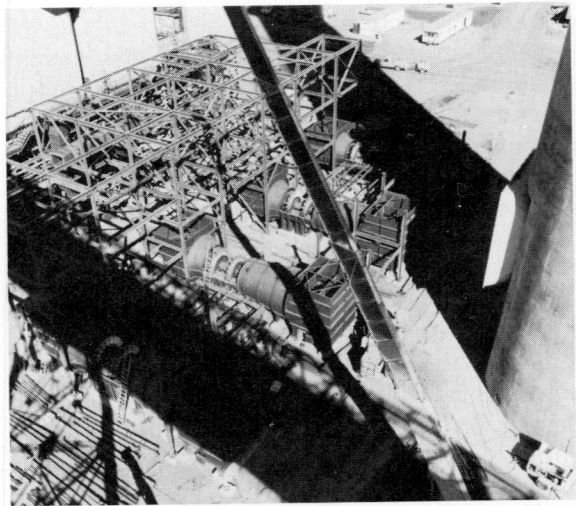

Fig. 1. Construction view of four axial-flow type induced-draft fans serving a large pulverized-coal steam generator

scribed in Chapter 11, primary-air fans may be located before or after the milling equipment. The most common applications are either pulverizer exhauster fans or cold (ambient temperature) primary-air fans.

The mill exhauster fan draws hot air from the secondary-air duct and through the pulverizer. The coal/air mixture from the pulverizer then passes through the fan and discharges into the fuel pipes which carry the mixture to the furnace for ignition. One fan is usually supplied for each pulverizer.

A materials-handling fan of the straight-blade type, the mill exhauster is sized for the maximum airflow needed by the pulverizer. It must develop sufficient pressure at maximum airflow to overcome the resistance of the air ducts, dampers, the pulverizer, and the fuel pipes to the furnace.

Located before the air heater, the cold primary-air fan draws air from the atmosphere and supplies the energy required to force the air through the ducts, air heater, pulverizer, and fuel piping. Usually two fans are supplied for each steam generator.

Cold primary-air fans for ambient-air duty are of the centrifugal airfoil or multi-stage axial type and, like FD fans, have silencers. In situations involving severe particulate concentrations or high temperatures, straight radial or modified radial fans are recommended.

With a cold-air system, the air volume handled by the primary-air fan is subtracted from the amount of air that must be handled by the forced-draft fan. Therefore, if coal is the only fuel being fired, the forced-draft fan may be made smaller because of the amount of air handled by the cold primary-air fan.

INDUCED-DRAFT FANS

Induced-draft (ID) fans exhaust combustion products from a boiler. In doing so, they create sufficient negative pressure to establish a slight suction in the furnace (usually from 0.2 to 0.5″ WG). This condition gives rise to the name "suction firing" or "balanced-draft" operation. These fans must have enough capacity to accommodate any infiltration caused by the nega-

tive pressure in the equipment downstream of the furnace and by any seal leakage in air preheaters. Chapter 6 gives the method of establishing volume and pressure specifications and tolerances.

As ID fans are now typically located downstream of any particulate removal system, they are a relatively clean service fan. In most instances, therefore, radial-tip or solid-reinforced blade centrifugal fans are not required. Typically, airfoil-bladed centrifugal or variable-pitch axial fans are used for this service because of their inherent high efficiency. These fans have high capacities, in excess of 1.5 million cu ft/min, and their airfoil blades minimize turbulence and noise. The blades and center plates may also be fitted with wear pads and replaceable nose sections for greater wear. Structural strength, particularly important in larger sizes, is excellent with these designs.

Where greater wear resistance is needed because of dust burden or where a very conservative approach is desirable, a modified radial or forward-curved, backward-inclined design is used. With some sacrifice in efficiency, these blade shapes minimize dust build-up and reduce downtime for cleaning. They have low noise characteristics, and their relatively simple design allows fabrication in special alloys should they be required for service downstream of a wet scrubber.

The ID fan is sometimes used as a booster fan with flue-gas desulfurizing (FGD) scrubbers. In one such arrangement, ID fans follow the precipitator or baghouse and another set of fans—the booster fans—follows the scrubber. With pressurized scrubbers, the booster fans are placed directly behind the ID's and *ahead* of the scrubbers, acting to "push" the gases through the FGDS.

GAS-RECIRCULATION FANS

Gas recirculation fans draw gas from a point between the economizer outlet and the air-preheater inlet, and discharge it (for steam-temperature control) into the bottom of the furnace. When controlling steam temperature on coal-fired units, a high-efficiency, high draft-loss mechanical dust collector must be installed ahead of the fan. If the recirculation fan on a coal-fired unit is only for standby or emergency oil firing, the dust collector is omitted.

Gas-recirculation duty provides the most severe test of a power-plant fan. The combination of heavy dust loads and rapid temperature changes demands the utmost in rugged, reliable fan design. Particularly important is how the fan hub is mated with the shaft; the often-used shrink-fit may not be adequate. To cope with temperature excursions, fans with an integral hub are preferable. Straight or modified radials or forward-curved, backwardly inclined centrifugal wheels meet these needs the best.

Turning gears are supplied on gas-recirculation fans to turn them at slow speed when the main drive motor is not in operation. If a gas-recirculation or ID fan is exposed to gases above 400°F, the turning gear should be energized to prevent thermal distortion of the motor.

HOW FANS WORK

A fan is a volumetric machine which, like a pump, moves quantities of air or gas from one place to another. In so doing, it overcomes resistance to flow by supplying the fluid with the energy necessary for continued motion. Physically, the essential elements of a fan are a bladed rotor and a housing to contain the incoming air or gas and direct its flow.

ENERGY FACTORS

Because a fan does work it demands energy to operate. The amount of energy depends on the volume of gas moved, the resistance against which the fan works, and machine efficiency. Chapter 6 gives the method of calculation of air and gas volumes. It is important to calculate such volumes at the actual pressure or suction existing at the fan *inlet*.

FAN PRESSURE RELATIONSHIPS

Draft, pressure, draft loss and pressure loss were defined in Chapter 6 which also differentiated between velocity pressure and static pressure: total pressure was stated to be the algebraic sum of velocity pressure and static pressure. Specifically related to fans, total pressure is the air pressure that exists by virtue of the degree of compression and the rate of motion. When applying these definitions to fan performance, there are distinct relationships that exist between each variable.

Fan total pressure is the difference between the total pressure at the fan outlet and the total pressure at the fan inlet. *Fan velocity pressure* is that corresponding to the average velocity at the specified fan outlet area. *Fan static pressure* is the difference between the fan total pressure and the fan velocity pressure. Thus, it is the difference between the static pressure at the fan outlet and the total pressure at the fan inlet. *Static pressure rise*, sometimes mistaken for fan static pressure, is the static pressure at the fan outlet minus the static pressure at the fan inlet. The difference between fan static pressure and static pressure rise is the inlet velocity pressure.

POWER

With the equation for fan work and some basic physical constants, the equation that expresses air horsepower can be developed.

$$Ahp = \frac{V \times H}{6356} \tag{1}$$

where V is the volumetric flow through the fan in cu ft/min and H is the head or pressure difference (in inches of water) across the fan. The air horsepower may also be designated as either static or total. Because the resistance to be overcome in fan application is primarily static pressure, the fan pressure developed is usually referred to in terms of static head. On this basis, the calculated fan power is known as *static air horsepower* (Ahp_s). When the power calculations are based on *total* head, fan power is referred to as the *total air horsepower* (Ahp_t) and is equivalent to the power output.

$$\text{Fan Mechanical efficiency } (\eta_t) = \frac{Ahp_t}{\text{power input}} \tag{2}$$

or, transposing,

$$\text{Power input (brake or shaft hp)} = \frac{Ahp_t}{\eta_t}$$

$$= \frac{V\,H_t}{6356\,\eta_t}$$

(3)

The power input formula assumes that air is an incompressible fluid. But the fact that air is compressible must be recognized when designing for high pressure differentials. The fan power formula then becomes

$$\text{Power input, hp} = \frac{k_c V\,H_t}{6356\,\eta_t}$$

(4)

where k_c can be taken from Fig. 2, which is based on adiabatic compression.

Density

Because they affect gas density, pressure and temperature of the air or gas also influence power output and efficiency. A change in density changes total and static pressure and their subsequent conversion into inches of water at standard conditions. Remember that head and horsepower vary inversely as absolute fluid temperature and directly as absolute

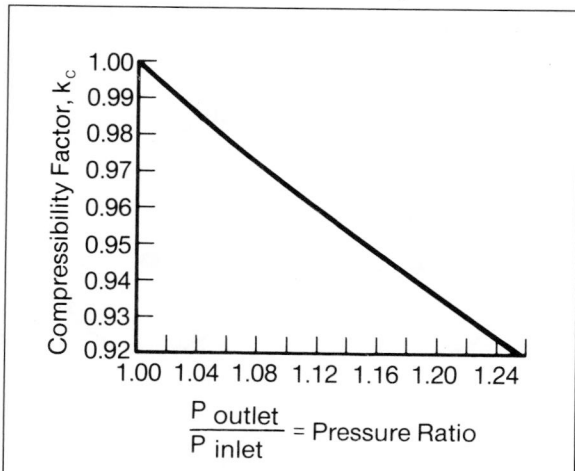

Fig. 2. Compressibility factor for use in calculating fan power consumption, assuming adiabatic compression

fluid pressure (or directly with fluid density)—and that adjustments often must be made for pressure and temperature variations when calculating performance or selecting a fan for a particular application.

SPECIFYING FAN OPERATING CONDITIONS

The problem of specifying the required operating conditions for fans at various load points is usually the responsibility of the boiler, FGD-system, and particulate-removal-equipment suppliers. The fan manufacturer assumes responsibility only for producing the volumetric flow and pressure specified. If the fan and system are incorrectly matched, the fan will not deliver the desired flow or pressure. Once the fan has been installed, it is difficult to increase its capability significantly because of physical limitations such as motor and impeller size. To select the appropriate fan for a given application, the fan vendor needs to know the density, flow and pressure requirements of the system at various points of operation and whether there are any fan speed limitations. If any of these variables change, it can affect the fan selection.

FLOW

The gas flow through the fan is usually expressed as a volumetric flow rate. It is necessary to determine the flow rate in actual cu ft/min (acfm) at the inlet to each fan from the density at the fan inlet and the mass flow rate (lb/hr). Proper corrections for plant elevation and actual conditions of local pressure at fan inlets must be made to the calculated air and gas volumes, as described in Chapter 6.

PRESSURE

On balanced-draft units, the required static head for the FD fan is the sum of all the series resistances in the secondary-air system, including cold-air duct, steam air heater, air preheater, air metering device, hot-air duct, dampers and windbox pressure drops. For pressurized units, the additional loss from the furnace to the stack outlet must also be included in determining total system resistance. ID fans must provide a static head equal to the series resistance from the furnace to the stack outlet, including resistances of superheater, reheater, boiler bank, economizer, air preheater, dust-collecting equipment, scrubbers and all duct-work. For primary-air (PA) fans, static pressure requirements are determined by the resistances through the inlet duct and air preheater, the pressure drop through the pulverizer, and the resistance through the fuel piping to the furnace.

"SPECIFIED" CONDITIONS

The volume flow (acfm) and static pressure ("WG) calculated by the boiler or scrubber manufacturer, as previously discussed, give the actual required fan capacity under ideal operating conditions. Besides the requirements at full load, the boiler or scrubber manufacturer should also calculate performance requirements at several other partial load points to evaluate power consumption, select the control equipment, and assure a fan that will operate at maximum efficiency at the desired normal output of the steam-generating unit. Also, the fan manufacturer must have a clear understanding of the system resistance over its entire range to ensure that a fan will operate at all points along its curve from the point of view of stability, sound, and efficiency.

Usually a margin or safety factor is added to the maximum continuous (MCR) volume and pressure requirements to arrive at maximum design or "specified" rating. Fan tolerances ordinarily used by C-E are given in Chapter 6.

TYPES OF FANS

From the point of view of fluid mechanics, fans represent a class of turbomachines designed to move fluids such as air, gases and vapor against low pressure. From the point of view of mechanical design, fans have a very light casing because inlet pressures are atmospheric or lower. Simplified hydraulic forms and welded steel plate are generally encountered in fans.

Direct-connected drives most often are used in power-plant work with control obtained through variable-speed motors, hydraulic couplings, or variable inlet vanes.

Fans are broadly classed as either centrifugal or axial, according to the flow direction. The centrifugal (radial) fan moves air perpendicular to the rotational axis of the impeller; the axial-flow fan moves air parallel to the rotational axis of the impeller.

CENTRIFUGAL FANS

Centrifugal fans use blades mounted on an impeller (or rotor) which rotates within a spiral or volute housing. Blade design determines fan characteristics, so by using blades of different shapes, a fan engineer can select an appropriate fan design. Basic blade types are the radial, radial tip (forward curved), and backwardly inclined (solid or airfoil). A velocity vector diagram at blade tip (Fig. 3) indicates that backward curved blades produce low velocities for a given tip or peripheral speed, and that forward curved blades give high velocity. Radial blades and radial-tipped blades lie between these two extremes. The backwardly curved blade type, therefore, operates at greater motor speeds than the other types for a given duty and is well adapted to direct drive with motors or steam turbines.

Fig. 4 shows some commonly used blade shapes. In general, blade type limits fan speed. Thus, the backward-blade machines can operate at a relatively higher speed than the for-

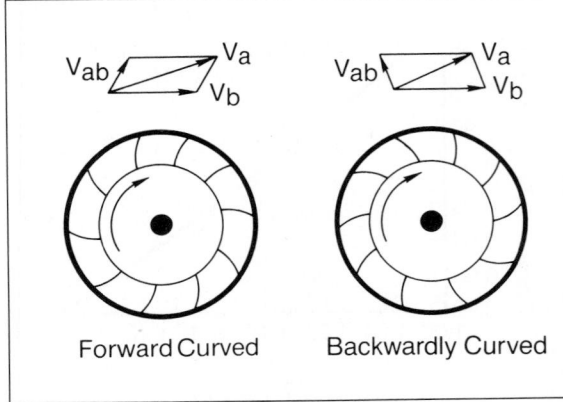

Fig. 3. **Velocity vector diagrams comparing forward-curved and backwardly curved centrifugal-fan blades. At same tip velocity (Vb) each type blade produces different air velocity (Va). Vector Vab is air velocity relative to the blade**

ward-curved design. Blade selection depends on speed limitations, allowable noise levels, efficiency demanded by specified load conditions, and desired fan performance characteristics in the most likely range of operation. In addition, the maximum attainable mechanical efficiencies and tolerance of the blade to corrosion and erosion are selection factors. Table I shows the effect of blade type on both maximum possible efficiency and fan resistance to flyash erosion. Notice that increasing the efficiency of a centrifugal fan sacrifices its erosion resistance.

Once the optimum fan has been determined, the aerodynamic selection process yields width, diameter and speed of the fan wheel. These parameters then become factors for subsequent elements of the fan design. In general, the fan wheel dimensions determine the basic dimensions of the fan housing and inlet boxes.

Centrifugal-Fan Design

A very common centrifugal fan arrangement (Fig. 5) is a single-speed motor drive which controls flow by inlet vanes or inlet louvers. The centrifugal fan in this instance is sized for "specified" conditions and is throttled by the inlet vanes to allow the fan to provide the flow and pressure required at lower operating loads.

As fuel costs and equipment efficiency have become increasingly important, new arrangements for centrifugal fans have developed. Because the welded blades of a centrifugal fan are not easily adjusted, the only other means of controlling flow and pressure is to vary the

speed of the fan. Two-speed motor drives, turbine drives and fluid couplings all have been used with varying success on centrifugal fans, and have helped to increase their efficiency at lower loads. These arrangements do, however, add more moving parts to the fan system, with a potential negative effect on reliability.

The variable-speed and the variable-frequency wound rotor motors are fan drives that allow varying the speed from full to zero in infinite intervals; they improve dramatically low-flow efficiencies. As discussed elsewhere in this chapter, speed variation is the ideal way to change fan operating characteristics; thus, for centrifugal fans, these motor designs are of great interest.

Centrifugal-Fan Construction

Fan scrolls and inlet boxes are of welded steel construction. The housing and the inlet boxes have either removable sections or are split for access and removal of the rotor (Fig. 6). The rotor wheel is of all-welded construction. After final machining, the rotor is usually statically and dynamically balanced and installed on its shaft in the manufacturer's plant (Fig. 7).

AXIAL-FLOW FANS

Axial fans at various boiler loads can maintain far higher efficiencies than constant-speed

Table I. Effect of Blade Type on Erosion Resistance and Efficiency

Blade Type	Typical Max. Static Eff., %	Tolerance to Erosive Environment
Radial	70	High
Radial Tip	80	Medium to High
Backwardly Inclined Solid	85	Medium
Airfoil	90	Low

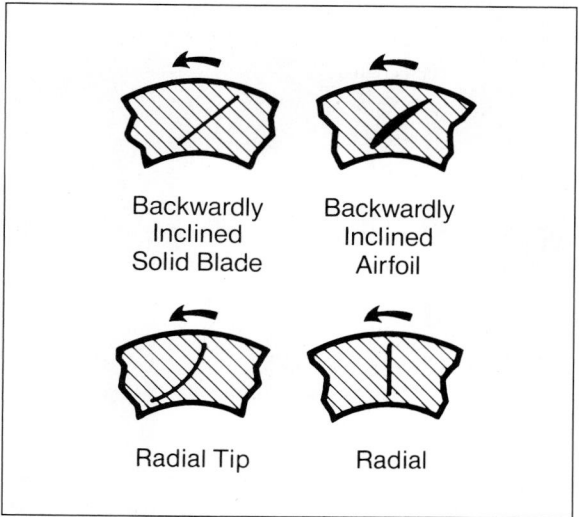

Fig. 4. Centrifugal (radial) fan blade types

Fig. 5. Airfoil-blade centrifugal fan with inlet-vane control

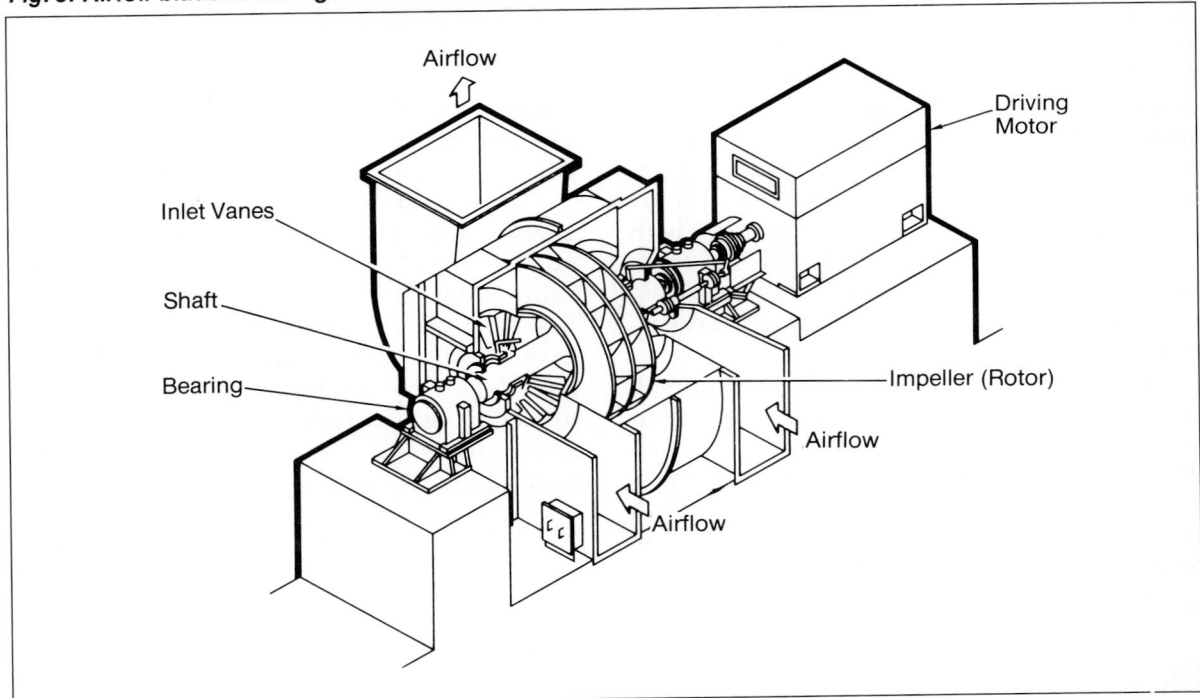

Fig. 6. Airfoil-bladed centrifugal fan showing inlet boxes and housing split for rotor removal

centrifugal fans. As fuel costs continue to rise, their higher initial expense has been offset by operational cost savings over the life of the plant. However, a detailed evaluation is required to show the advantages of using axials in any given situation.

In the most common axial arrangement (Fig. 8), the fan operates at constant speed and the angle of the blades on the hub is adjusted to vary flow; no inlet vanes are required for control. This enables the axial fan to develop, for each point of operation, a unique aerodynamic configuration that is as efficient as possible.

Either a mechanical or hydraulic mechanism adjusts blade pitch while the fan operates at its design speed. Mechanical type pitch-change mechanisms are, however, usually insufficient to properly control fans of the size required for utility applications. Variable-pitch axials are best suited to PA, FD, ID or booster-fan applications on large industrial and utility boilers. As stated earlier, axial fans are not appropriate for small fan requirements or for gas recirculation.

Axial-Fan Construction

Variable-pitch moving blades provide a wide range of flow with satisfactory operational efficiency. Because the fan and its built-in components work under high peripheral speed, and thus great centrifugal force, mechanical reliability is as important as aerodynamic performance. The centrifugal force of a moving blade, for example, can be more than 150,000 lbs.

Fig. 9 shows the structure of a typical two-stage axial ID fan. An overhung hub, with blade-actuating levers and an oil hydraulic piston and cylinder, supports the moving blades. At the free end of the rotating cylinder a control valve is attached.

The variable-pitch scheme shown in Fig. 10, includes an oil-hydraulic control that can be operated both manually and automatically. In automatic operation, an electric drive responds to a signal from the automatic plant controller and moves the control spool in the control valve through a linkage of levers. The control valve consists of a rotating sleeve attached

Fig. 7. Centrifugal-fan rotor undergoing static balancing in manufacturer's shop

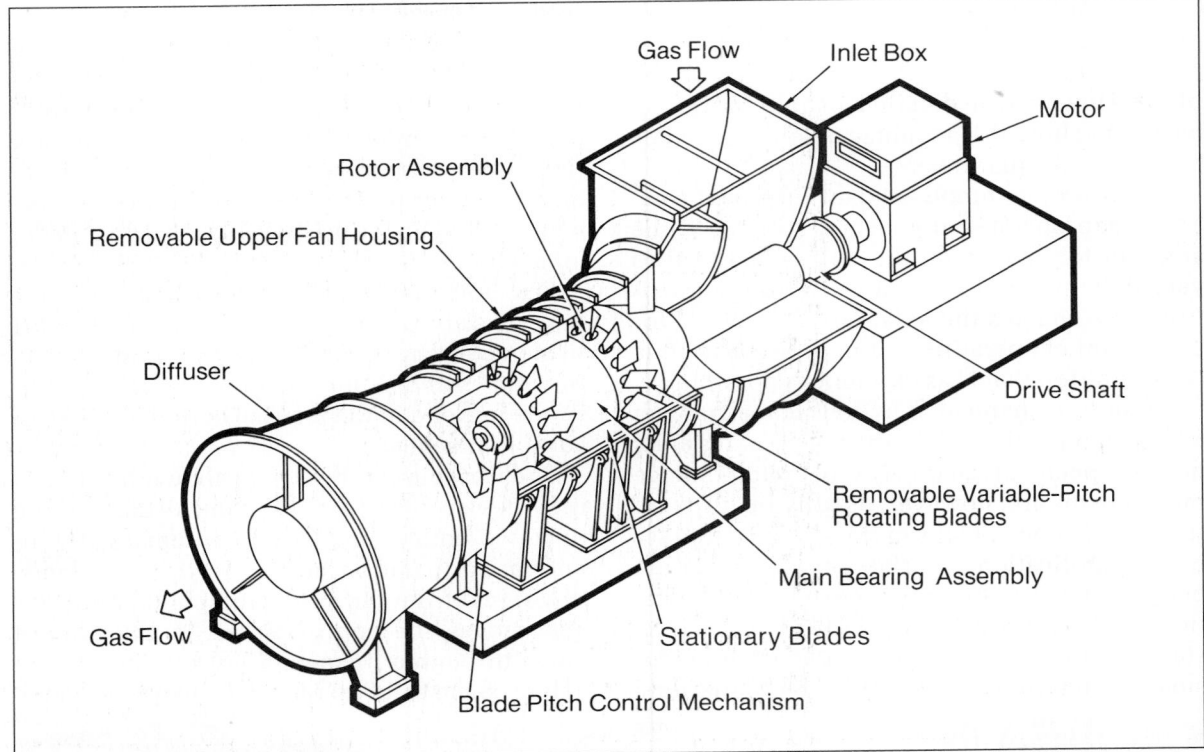

Fig. 8. Two-stage variable-pitch axial-flow fan for induced-draft service

directly to the hydraulic cylinder and a non-rotating spool linked to the external control unit. With the displacement of the spool, pressurized oil flows, accompanying the cylinder movement; the oil is fed back directly to the control valve. Thus, an accurate location of the blade pitch is obtained.

A thrust ball bearing at the blade root supports the blade, and is subject to high centrifugal force. The bearing is one of the most critical elements, as it fluctuates under severe static loading. Such working conditions are most strenuous on such bearings, which require substantial attention in design, with life tests mandatory to simulate actual operating conditions.

FAN CHARACTERISTIC CURVES

Fig. 11 shows a typical set of constant-speed characteristic curves for a centrifugal fan with backwardly inclined blades. Although by applying principles of geometric and dynamic

Fig. 9. Construction view of two-stage axial-flow fan with upper housing not yet in place

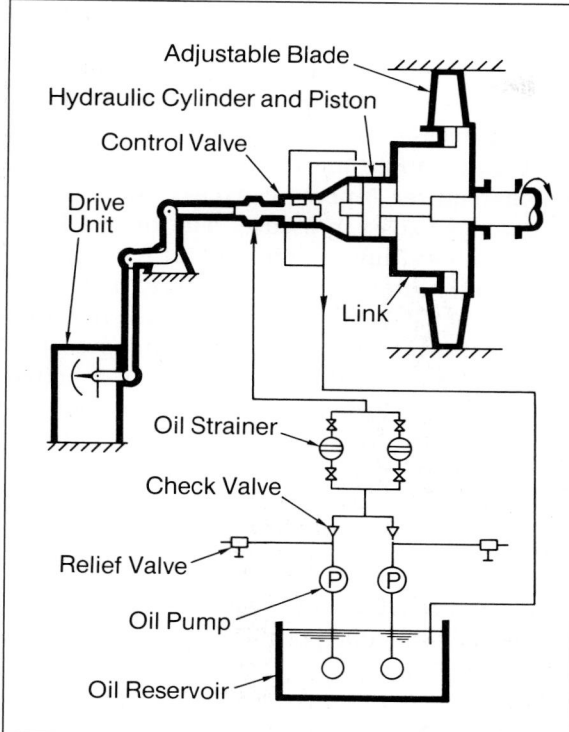

Fig. 10. Schematic arrangement of pitch control for adjustable blade axial-flow fan

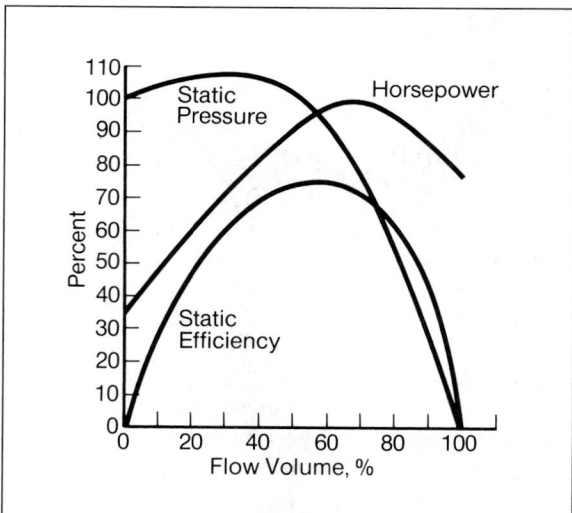

Fig. 11 Chart showing typical constant-speed characteristics for a fan with backwardly curved blades. Pressure decreases as fan capacity goes up; this expands the range of stable operation. Horsepower characteristic is non-overloading.

similarity, the performance of a single fan of a given type may be used to determine the characteristics of a complete line of sizes, each separate type must be tested independently. The curves are for one speed only; when a constant-speed motor drives the fan, it will operate somewhere on the characteristic curves, depending on the resistance imposed by the system through which the air passes. The system resistance, therefore, acts as an automatic control on the fan and will limit the amount of overload that can be developed, despite the flatness of the pressure characteristic. The amount of overload depends also on the shape of the horsepower curve.

To fully appreciate the significance of fan characteristic curves, realize that every fan is restricted to that performance defined by its curve. The fan *must* operate at a point that lies somewhere on the characteristic plot. For example, if the head required for a given volume is less than that specified by the curves, additional resistance must be placed in the system; otherwise, the fan will put out more capacity until it reaches a point on the characteristic curve at which the head matches system resistance. The fan has no choice; it must operate at a point on the curve where head, capacity and system resistance balance.

DEVELOPMENT OF PERFORMANCE CURVES

Fan performance curves are normally developed from base test curves, which, in turn, are developed from test data recorded under controlled laboratory conditions[1]. For continuity throughout industry, as well as for ease of understanding and ready comparison, test data are corrected or adjusted to what is known as standard conditions. Standard conditions for fan design work denote that all flow, pressure and power values are at 70°F and sea level, with air density of 0.075 lb per cu ft.

Normal curve format consists of a graph which has the air or gas volumetric flow numerically on the abscissa or horizontal axis, with both static pressure and brake horsepower numerically on the ordinate or vertical axis. Two separate curves—volume versus sta-

Steam-Generator Auxiliary Equipment

tic pressure, and volume versus power—are plotted on each performance curve.

Several other variables can be incorporated on the constant-speed performance curves. Such things as static or mechanical efficiency, the horsepower or pressure at various temperatures, altitudes, or densities, can all be plotted to illustrate actual conditions. Finally, in rating from a curve, the same fan design must be retained. That is, a radial-blade centrifugal-fan test curve cannot be used to determine the rating of a backwardly-inclined-blade fan.

APPLICATION OF CHARACTERISTIC CURVES

Fig. 12 illustrates the application of typical centrifugal-fan characteristic curves to a fan problem in which points A, B, C, and D are calculated requirements at four load points on a given boiler. The line through them defines the system resistance. The point where this line intersects the static pressure characteristics of any fan, at any given speed, determines the point on the characteristic at which the fan will operate, if both curves are plotted for the same density. However, the fan can operate only on its characteristic curve. If any error has been made in calculating point D—in volume or pressure, or temperature—that point will not fall on the characteristic curve and the fan may not meet the requirements when operating at that particular speed.

For example, if 10 percent more volume is needed at the same pressure, point D will be displaced to the right, but the available pressure at the fan head at the same time drops 14 percent and the fan cannot satisfy the requirements. Similarly, if the volume is correct but 10 percent more pressure is needed, the volume that the fan would deliver at the greater pressure would be only about 90 percent of the requirements, because the fan can operate only on its characteristic curve for a given speed and density condition.

To provide excess capacity it is customary to specify the volume and pressure in excess of the actual calculated requirements and thereby obtain a larger fan. Suppose a portion of the pressure characteristic of this larger fan,

operating at the same speed and density, is represented by line FG in Fig. 12. Then it is apparent that this size fan would be selected by the manufacturer if the purchaser specified 24-percent excess volume. At the same time the fan would satisfy the requirement of point E, which requires 8-percent excess volume and 17-percent excess pressure. The only advantage in attempting to define point E on the extrapolated system resistance curve, instead of point F or G, is that the power requirement given by the manufacturer will then represent a closer estimate of the larger fan under actual operating conditions than if point F or point G had been defined for fan selection. The fan finally chosen can, however, satisfy the requirements of all three points, if sufficient power is available from the drive.

Speed Variations

For a given fan, a family of characteristic curves can be obtained by varying the fan

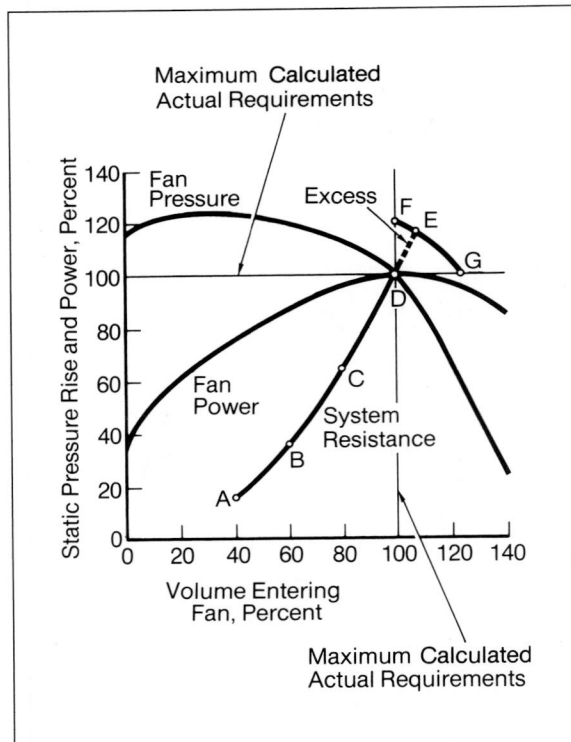

Fig. 12 Use of characteristic curves as applied to a problem of centrifugal fan selection

ASEA BROWN BOVERI

speed. The nature of each curve remains the same because the change in operating speed merely shifts the curve by a proportionate amount. A graph like that of Fig. 13 results if system resistance is plotted on the same grid as the family of curves for different fan speeds.

FAN CONTROL

Very few applications permit fans to operate continuously at the same pressure and volume. Therefore, to meet the requirements of the system, some means of varying fan output becomes necessary. Capacity control of a fan can be achieved in two ways:

- controlling the aerodynamic flow into or within the fan
- controlling the speed of the fan

The first method refers either to altering the flow of gas into the eye of a centrifugal-fan wheel as with inlet vanes, or changing the internal aerodynamics by altering internal geometry as with controllable pitch axial fans. The second method refers to any speed-changing device such as a turbine, fluid drive, multiple-speed motor or an electronically adjustable motor drive connected to the fan.

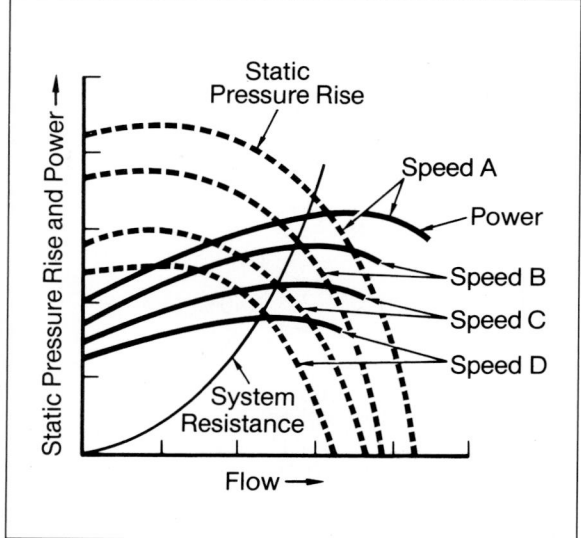

Fig. 13 Fan speed of a centrifugal fan can be varied so that output pressure matches system resistance for desired volumetric flow; this conserves energy.

VARIABLE INLET VANES

Sometimes variable inlet vanes (Fig. 14) are used to control fan performance by providing swirl to the fan impeller, saving significant power. Variable inlet vanes tend to be slightly more effective in saving power than parallel blade inlet box dampers. When furnishing variable inlet vanes for either centrifugal or fixed-pitch axial fans, fan manufacturers provide a complete performance envelope showing the effect of vane position on fan performance and power. Fig. 15 shows the typical percentage range in power reduction along a constant system resistance line as a percentage of design flow for a backwardly-inclined airfoil centrifugal fan.

An inlet-vane-controlled centrifugal fan is selected to produce full specified flow and pressure with no inlet vanes present. Inlet vanes then throttle down this maximum performance capability so that the fan can operate over the range of normal boiler operating load points. Fig. 16 illustrates the effect of this type of control system, which is extremely sensitive at the lower load conditions. Extremely minor changes in inlet vane openings have a dramatic effect on the flow produced by the fan, whereas at higher loads it requires increasingly larger movement of the inlet vanes to have any effect on the flow produced by a centrifugal fan.

AXIAL-FAN BLADE PITCH CONTROL

The axial-flow fan can control flow at constant speed by varying the blade angle of the fan (Fig. 17). The effect is to create a unique aerodynamic configuration for the fan at each point of operation so that the fan is operating at maximum possible efficiency. As the blade angle is adjusted from minimum to maximum position, the flow change is nearly linear (Fig. 16). Another aspect of control rests in the response time of the fan. Most axial-flow fans can move the blades a full stroke, from the maximum open to fully closed position in 30 seconds. This means that under the normal boiler operating range, an axial fan can respond or move from maximum continuous conditions to zero flow in approximately 20 seconds.

Fig. 14. Inlet vanes give an initial spin to air entering a centrifugal fan. By adjusting angle of vanes, the degree of spin and volumetric output are regulated.

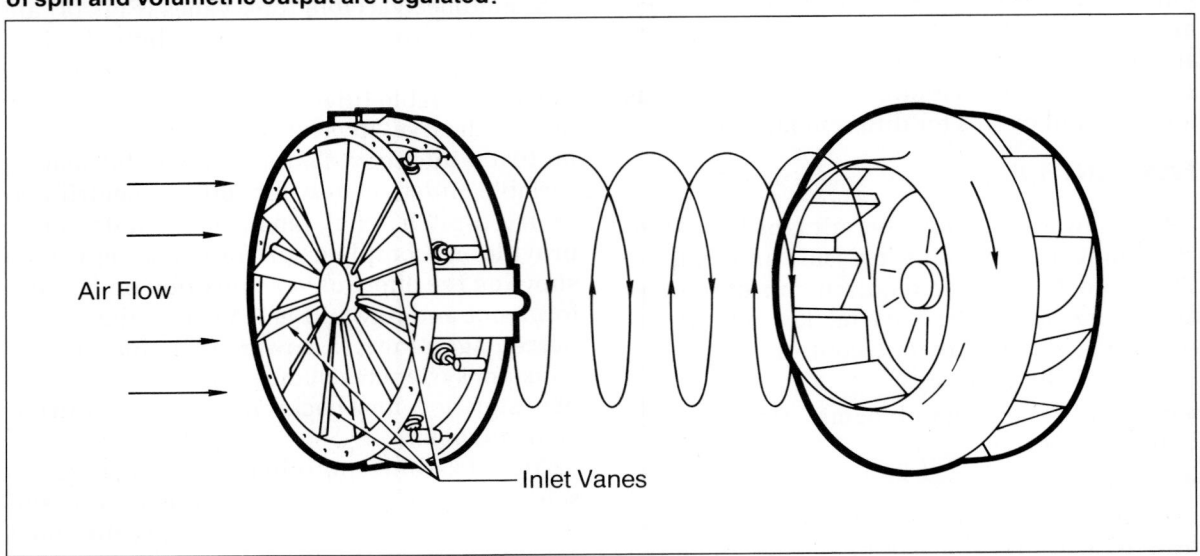

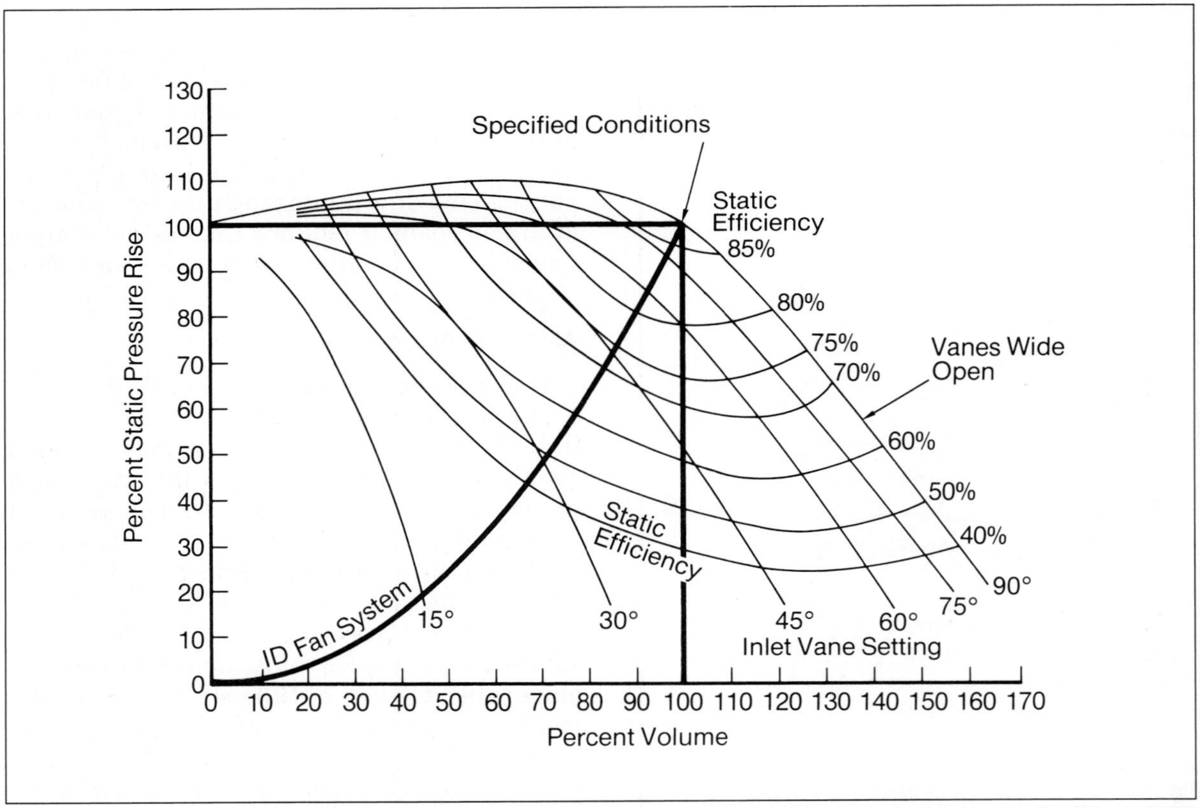

Fig. 15. Typical airfoil blade centrifugal-fan performance characteristics—constant-speed with variable inlet-vane control

SURGE OR STALL LIMIT

The surge limit of a centrifugal fan is that point near the peak of the pressure curve corresponding to the minimum flow rate at which the fan can be operated without instability. The stall limit of an axial fan is that point near the peak of the pressure curve at a particular blade angle corresponding to the minimum flow rate at which the fan may be operated without separation of airflow over the blades. Operation in the surge or stall region for any length of time should be avoided. Such operation can result in a substantial reduction in fatigue life.

An aerodynamic characteristic of all axial fans (just as with an airplane), stall is understood by few. A stall occurs when the angle of attack of the fan blade (or airplane wing) exceeds a certain value in relationship to air velocity. When this angle of attack value is exceeded, airflow becomes separated from the convex side of the blade. Centrifugal force then throws air trapped in this separated portion in a radial direction, to the outer tip of the blade. At this point, pressure builds up until it is relieved through the blade tip clearance. This process creates a very unstable and oscillating pressure force on the blade, and can cause very severe vibrations throughout the entire fan. A reduction in fan flow and head capability also occurs when operating in this mode.

In Fig. 18, the flow separation points occur at the upper end of the several blade-pitch lines, with the dotted lines extending downward and left to the recovery point, where another reversal takes place. At low blade angles, the fan tends to get noisy and somewhat unstable, but it no longer is subject to stalling. (A detailed discussion of stall/air separation is beyond the scope of this text.)

FAN SPEED CONTROL

Controlling the fan speed is potentially the most efficient form of capacity control. The only significant inefficiency that a speed control system can introduce to a fan results from the inefficiencies of the speed control system.

All speed control systems yield certain operational/reliability improvements to large fans in induced-draft service. They include

■ reduction in erosion approximately proportional to the ratio of the squares of the impact velocities

Fig. 17. Close-up of adjustable airfoil blading of an axial-flow fan

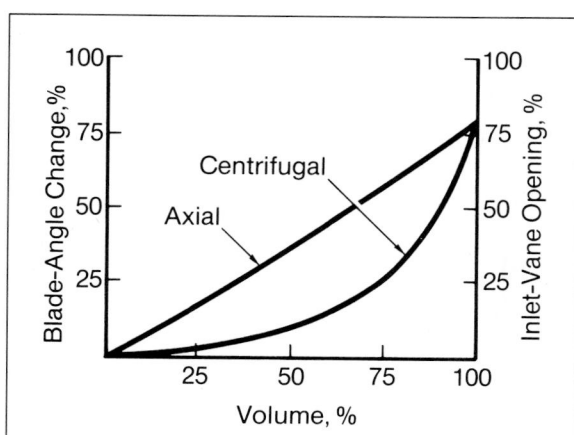

Fig. 16 Comparison of inlet-vane flow control versus blade-angle flow control

■ reduction in mechanical shock at start-up

■ adjustability of speed to any point within the operating speed range

■ reduction in potential fan system-oriented problems (fan/foundation, fan/ductwork, fan noise) at reduced running speeds

■ reduction in electrical power surge on motor start-up

The primary disadvantage of adjustable-speed drives is their cost which, depending on unit size, can be significantly more than a single-speed motor drive with inlet-vane control. Also, adjustable-speed drives add additional components to the drive train which require more careful consideration of system dynamics to avoid torsional oscillations or other undesirable phenomena.

Fluid-Drive Speed Control

For speed adjustment, fluid drives are either hydrokinetic or hydroviscous. In hydrokinetic operation, an impeller attached to the driving motor accelerates oil particles and impinges them against a runner attached to the driven fan. Speed is adjusted by changing the volume of oil in the system and thus the transmittable torque. The hydroviscous method is similar in principle, but relies for speed change upon the increase in torque transmitted by the oil used between alternate rows of driver/driven plates that are forced closer together.

Fluid drives have been used extensively where speed changing was required on large centrifugal fans. They are more efficient than single-speed inlet-vane control as volume output drops below about 75 percent. They are, however, the least efficient adjustable speed method of gas volume control, giving away essentially a percent of efficiency for each percent of speed reduction. They also add another mechanical device in the drive train which increases the cost of installation, requires cooling equipment, and possibly increases initial fan size due to slip at full speed.

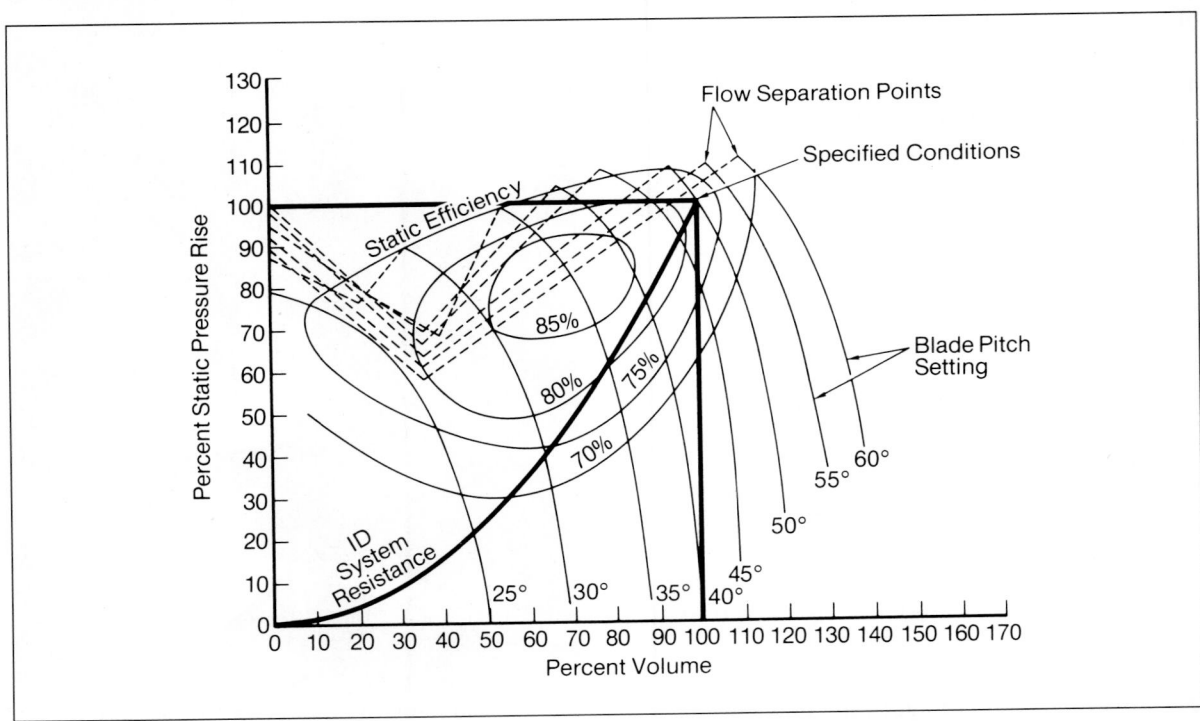

Fig. 18. **Typical axial-flow fan performance characteristics with variable-pitch blade control**

Two-Speed Motor Drives

As a centrifugal fan is vaned or dampered down from the specified point, its efficiency falls off rather rapidly. If, however, at the normal operating point, fan speed is changed to a lower base speed and the vanes or dampers reopened, a system efficiency is achieved which matches the original full load levels. This is the principle of application of two-speed motors to power fans. Many two-speed motors have been applied to fans to gain the overall high efficiencies available from two running speeds. Also, the two-speed motor drive is the least expensive and least complex speed control method. An objection frequently voiced is that operators hesitate to change to the lower speed until far below the designed load limit, which of course negates the planned economy of the system.

Variable-Speed Drives

A variable speed system significantly improves fan efficiency during periods when the boiler is operating at less than its maximum load. A centrifugal fan can be equipped with an adjustable-speed drive, a control and monitoring system. The speed control range can be tailored to a particular installation and drive system costs can be optimized. By equipping a fan with both adjustable speed capability and inlet vanes, the fan output can be tailored to match all possible boiler operating conditions.

Slip-ring (or wound-rotor) motors provide step changes in speed by connecting the motor windings to an adjustable external resistance through slip rings and brushes. By changing rotor circuit resistance, the motor speed/torque characteristics can be adjusted to fit changes in load conditions. With no resistance in the rotor circuit (rings-shorted condition), the wound-rotor motor will operate exactly the same as a squirrel-cage motor.

Another important difference is the starting characteristics of the motor. The wound-rotor motor inrush current can be as low as the magnetizing current—normally about 30 percent of rated current. Most other induction motors draw a very high (600 to 700 percent of rated) current during start-up. This lower current characteristic can provide significant savings in the electrical distribution system. It also eliminates the need for long intervals between motor starts which are needed to dissipate the heat generated by the high starting current. Wound-rotor motor current is essentially proportional to the load torque.

Synchronous motor adjustable-frequency systems use static electronic control equipment to adjust frequency and voltage. The synchronous motor is an AC device with two sets of windings, one fixed in the stator (frame) and the other wound around field poles which are fixed to the rotor in pairs. For a given frequency, the number of poles determines the speed of the motor. To run the synchronous motor, the stator windings are energized by connection to an AC power source. The rotor field pole windings are excited by a DC power source. A magnetic coupling then exists between the stator and the DC field pole windings which are separated only by a small air gap. The windings in the stator when energized with an AC power source set up a rotating magnetic field around the stator. The magnetic DC fields on the rotor couple with the rotating AC field and, thus, mechanically follow the rotating AC stator field. This causes the rotor to turn at the same speed as the rotating AC field.

Sometimes, when a supply of exhaust steam is available, boiler fans can be driven by steam turbines. Such an arrangement is frequently used in industrial plants to achieve an efficient heat balance—as well as drive the fan. Speed variation is feasible down to about 35 percent of rated turbine speed.

FAN SELECTION

Once the boiler or scrubber engineer determines the pressure and flow required by the system at rated load, at the specified condition (with margin) and at various load points, this information can be plotted and the fan selected. A centrifugal fan will have its specified point fall within the area of maximum efficiency. At this point, the fan is operating wide-open with no restrictions. To achieve the lower load points, inlet vanes are closed to throttle the maximum capability of the fan. As the fan performance moves down the system resistance line, it moves through areas of constant efficiency that fall perpendicular to the system resistance line.

The characteristic curve for an axial fan is considerably different from that of a centrifugal fan. Given exactly the same system resistance curve, an axial-fan designer will select a fan that operates at maximum efficiency at boiler full load. If normal fan performance tolerances have been used, the fan will usually have the capacity to achieve the specified requirements. If higher-than-normal tolerances have been applied, a less-than-optimum selection will have to be made, in which full boiler load does not fall inside the maximum efficiency area. The oblong or egg-shaped areas on the axial-fan curve are the areas of constant static efficiency. It is important to note that these areas for the axial fan are oblong in a direction approximately parallel to the system resistance line rather than perpendicular to it, as with the centrifugal fans. Changes in flow requirements along the system curve cause only slight changes in efficiency compared to the significant changes in efficiency for constant-speed centrifugals. The lines that are perpendicular to the system line on the axial curve depict the angle of the blades at that point of operation.

COMPARISON OF FANS AND FAN CONTROLS

Table II summarizes the available fans and operational control equipment; generally, with increasing efficiency of either draft equipment or controls, cost increases. Fig. 19 shows the power consumed by the various designs over the full range of capacity.

In comparing the several types, both capital and operating costs must be established over the anticipated load range. The cost of fans, controls, drives, silencers, foundations, switchgear and other auxiliaries, plus any differences in ductwork, have to be obtained for equal specified capacities and sound-pressure levels. Based upon the relative power-consumption curves, the operating hours at the individual load points for the entire life of the unit must be multiplied by the associated power consumptions of the fans being compared. Maintenance costs also must be considered by analyzing available historic data.

FAN SIZE SCALE-UP

Most fan designs are developed using models of moderate size and input power.[1] The performance obtained from the model provides an information base to calculate the performance of larger fans that are geometrically similar to the model. If only the basic fan laws are used to make these conversions, (with no correction for compressibility of air) the larger fan, in many cases, will perform better than predicted, providing that all geometric, kinematic, and dynamic similarity requirements are satisfied. Geometric similarity requires not only linear proportionality but also requires complete angular similarity without omission or addition of parts.[2]

The three main performance factors of flow, speed, and head are linked in the concepts of specific speed and specific diameter.

Specific speed is that rpm at which a fan would operate if reduced proportionately in size so that it delivers 1 cfm of air at standard conditions, against a 1″ WG static pressure.

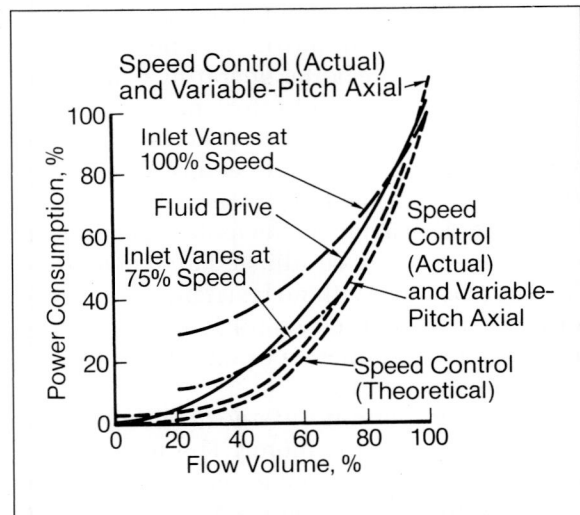

Fig. 19. Comparative fan power consumption versus volumetric flow with various types of control

Specific diameter is the fan diameter required to deliver 1 cfm standard air against a 1″ WG static pressure at a given specific speed. From the fan laws we get these equations:

$$\text{Specific speed } (N_s) = \frac{\text{rpm (cfm)}^{1/2}}{\text{(SP)}^{3/4}}$$

(5)

$$\text{Specific diameter } (D_s) = \frac{D\,(\text{SP})^{1/4}}{\text{(cfm)}^{1/2}}$$

(6)

Table II. Available Types of Fans and Fan-Control Equipment

Centrifugal Fans
Radial
Radial-tip
Backwardly Inclined Solid
Backwardly Inclined Airfoil

Increasing Efficiency Over Normal Operating Range

Axial Fans
Fixed Pitch
Adjustable Pitch
Variable Pitch

Flow Control
Single-Speed Motor
  Outlet Damper
  Inlet Louver
  Inlet Vane
Two-Speed Motor
Fluid Coupling
Variable-Speed Turbine Drive
Variable-Speed Motors

Increasing Efficiency At Partial-Loads

where flow in cfm is at standard conditions, SP is static pressure (in. of water), and D is the diameter of the fan in inches.

Table III shows the parameters for designing to "specified" pressure and volume.

In the case of axial fans, maximum flow can be achieved by proper selection of the rotating-blade tip diameter, so an increase in tip size results in a shift of the entire fan curve along the flow (horizontal) axis to the right. Maximum pressure can be achieved by selection of the proper hub diameter, so an increase in hub size would result in a shift of the entire fan curve along the pressure (vertical) axis upward. The system-resistance requirements given the fan designer remain the same, so by selecting the proper hub and blade sizes, the fan characteristic can be shifted along the system resistance, so that boiler full-load conditions occur in the area of maximum efficiency. A centrifugal fan's characteristic similarly can be shifted along the system resistance line.

FAN LAWS

How a change in any operating condition affects a fan can be predicted by a set of rules known as the fan laws. These are summarized in the box on the next page and apply to fans of the same geometric shape and operating at the same point on the characteristic curve.[3]

FANS OPERATING IN SERIES

Infrequently it is necessary to install two or more fans in series in the same system. Every fan in series handles the same weight of air or gas, assuming negligible air loss or infiltration between stages.

In theory the combined volume-pressure

Table III. Effect of Fan Parameters on Performance Capabilities

	Axial	Centrifugal
To increase volume:	increase blade tip diameter increase rotational speed	increase wheel width increase rotational speed
To increase pressure:	increase blade-root diameter increase rotational speed increase number of stages	increase wheel diameter increase rotational speed

FAN LAWS

1. For a given fan size, system resistance, and air density:

   A—When speed varies,
   - (a) Capacity varies directly as the speed ratio.
   - (b) Pressure varies as the square of the speed ratio.
   - (c) Horsepower varies as the cube of the speed ratio.

   B—When pressure varies,
   - (a) Capacity and speed vary as the square root of the pressure.
   - (b) Horsepower varies as the 1.5 power of the pressure.

2. For constant pressure:

   When density varies, speed, capacity and horsepower vary inversely as the square root of the density; that is, inversely as the square root of the barometric pressure, and directly as the square root of the absolute temperature.

3. For constant capacity and speed:

   When density of air varies, horsepower and pressure vary directly as the air density; that is, directly as the barometric pressure, and inversely as the absolute temperature.

4. For constant mass flow:

   A—When density of air varies,
   - (a) Capacity, speed and pressure vary inversely as the density; that is, inversely as the barometric pressure, and directly as the absolute temperature.
   - (b) Horsepower varies inversely as the square of the density; that is, inversely as the square of the barometric pressure, and directly as the square of the absolute temperature.

   B—When both temperature and pressure vary,
   - (a) Capacity and speed vary as the square root of (pressure × absolute temperature).
   - (b) Horsepower varies as the square root of (cube of the pressure × absolute temperature).

---

curve of two fans operating in series is obtained by adding the fan pressures at the same volumetric flow. (Fig. 20). In practice there is some reduction in volume due to the increased air density in the later stages. There also can be a significant performance loss from non-uniform flow into the second stage fan.

FANS OPERATING IN PARALLEL

Fans commonly operate in parallel in the same system, particularly when large volumes of air or gas must be moved. The combined volume-pressure curve in this case is obtained by adding the volumetric capacity of each fan at the same pressure (Fig. 21). Each fan handles only part of the volumetric capacity. The total performance of the multiple fans will be less than the theoretical sum if inlets are restricted or the flow into the inlets is not straight.

Some centrifugal fans have a pressure-volume curve with a positive slope to the left of the peak pressure point. If fans operating in parallel are selected in the region of this positive slope, unstable operation can occur. Fig. 21 shows the combined volume-pressure curve of two such fans in parallel. The closed loop to the left of the peak pressure point is the result of plotting all the possible combinations of volume flow at each pressure. If the system curve intersects the combined volume-pressure curve in the area enclosed by the loop, more than one point of operation is possible. This may cause one of the fans to handle more air than the other and could cause a motor overload if the fans are individually driven. This unbalanced flow condition tends to reverse readily and fans will intermittently load and unload, with the possibility of damage to the fans, ductwork or driving motors.[4]

When paralleling two variable-pitch axial-flow fans, care must be taken to keep both units out of the stall region. Since this area generally is not near the boiler resistance line, avoiding a stall situation is relatively easy. With two fans

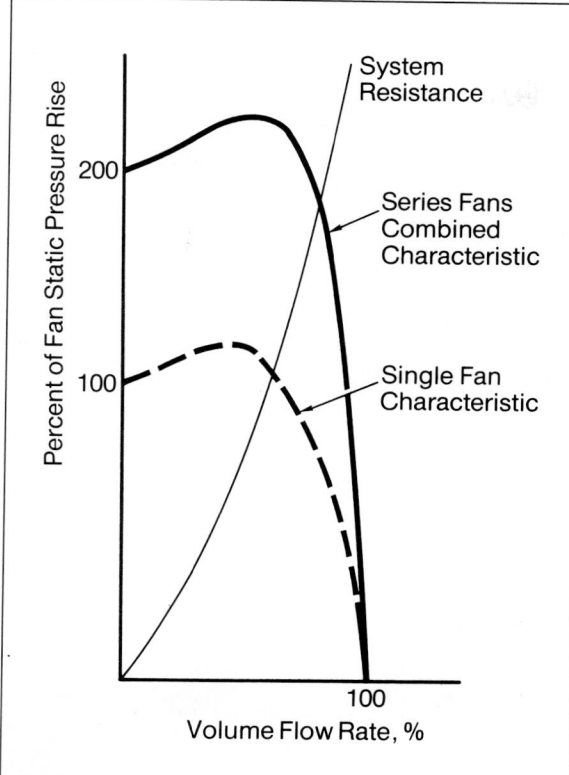

Fig. 20. Typical characteristic curve of two fans operating in series

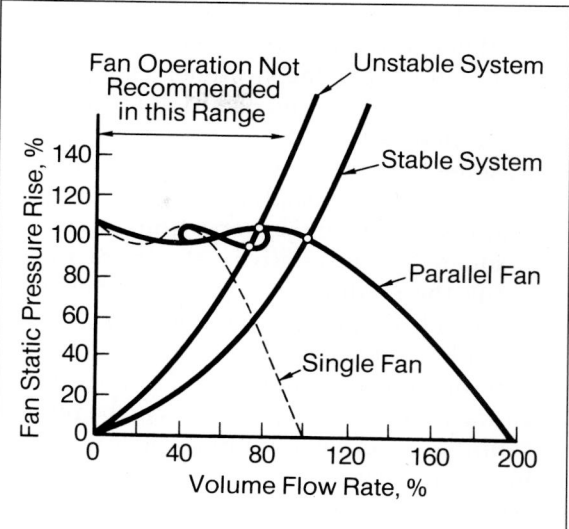

Fig. 21. Parallel fan operation

operating, the resistance line for one fan is influenced by the other fan, as well as by boiler conditions. This means that the two fans together will develop the pressure required to overcome the boiler resistance, but that their volume flow need not be equal.

GAS-FAN EROSION AND CORROSION

Erosion is the loss of fan material through mechanical action of particulate matter in the flue gas impinging on or abrading the fan surfaces during operation.

Corrosion is the loss of material through the reaction of various chemicals in the flue gas coming in contact with the fan.

Erosion is determined not only by tip speed, but also by the properties of the impeller and the particulate matter in the gas. Factors such as ash loading, particle size, composition, density, hardness, shape, surface texture, velocity and impact angle—as well as impeller material, hardness, and use of wear plates—influence erosion rates.[5] Chapter 23 covers the repair of eroded fan blades in detail.

It is significant that fans for induced-draft service have become progressively more efficient as more efficient dust-collection devices have been installed. Airfoil bladed fans could not be used with particulate-collection efficiencies below 95 to 97 percent; axial-flow fans required even higher efficiencies. Now, with dust loadings as low as 0.03 grains/ standard cu ft, the use of more sophisticated equipment is practical. During this same period of development, there had been technical developments on erosion-resistant materials for fan blades, as well as the increased usage of variable speed drivers, which results in further reduction of erosion.

FAN EROSION

The loss of fan wheel material poses two serious threats to fan reliability. First, any change in the rotor mass distribution, especially at the outer extremities, can cause unbalanced forces and high vibrations. Second, a loss of material on highly stressed fan elements such as blades and side plates can cause structural damage and lead to catastrophic failure.

The centrifugal fan designer can consider several alternatives one of which is modifying the fan-blade type and speed. As described earlier, some fan blades are more resistant to erosion than others. Also, the ability of any given fan to resist erosion is proportional to at least the square of the relative velocity of the gas passing the blading. The principal fan wheel areas attacked by erosion are the blades and the center plate. Heaviest erosion occurs in the area where blade and center plate are joined. Erosive wear on these critical parts can be reduced by applying a larger fan running at lower speed which can meet aerodynamic requirements at lower gas velocities.

Another alternative would be to modify the fan inlet conditions where it appears that ductwork design has led to differential erosion of a fan. It is possible for flyash leaving a precipitator to be channeled down a particular leg of a duct and to concentrate in one fan of a multiple-fan installation, or even in one of the inlets of a particular fan.

As a third alternative, repairable liners can be used to control erosive effects on airfoil fans (Fig. 22). This method consists of designing blades with two layers of material, the outer layer being easily repaired in the field with welding. Periodic inspection is required to determine when repairs are necessary. In addition, airfoil blades can be furnished with extra thick nose pieces.

As a fourth alternative, replaceable liners to bear the brunt of erosive forces can be provided on fan blades. It is essential that, in the design of a fan, the added weight of the liner plates be accounted for in the stress calculations for the structural members such as the blade and side plates, as well as in specifying the moment of inertia of the motor.

The axial-fan designer generally uses renewable coatings on blade leading edges and surfaces. In so doing, it is important that the designer consider the strength and durability of the blade base material in terms of both the erosiveness and the temperature of the operating environment. Note that, when axial-fan blades erode, they normally lose material at a greater rate in the tip area than in other areas. This is a less serious situation in terms of overall strength and balance than with a centrifugal fan because the blade actually gets lighter, and blade root stresses reduce.

Because coatings can affect the physical properties of fan structural material, and cracks in coatings can propagate into critical fan members, tests must be performed using proposed spray coats and actual fan structures before such coatings are used.

FAN CORROSION

The problem of chemical corrosion of fans has to be accounted for in the design and operation of ID and booster fans downstream of wet scrubbers. Although early experience with corrosive gases entering fans was limited to chemical process plants, the use of FGD systems has brought the problem to the utility industry.

Fig. 22. Airfoil blades of a centrifugal fan with repairable weld-metal liners

Experience has been excellent with fans of standard carbon-steel or low-alloy-steel construction, where the entering flue gas has been properly dehumidified and heated above the temperature of adiabatic saturation (which is about 130°F). On the other hand, "wet" fans handling scrubber effluent gas without reheating to at least 30°F above saturation temperature have undergone severe corrosion. High-alloy metals used to manufacture such rotating equipment have proven a very expensive alternative to flue-gas drying and reheating.

ACOUSTICS

Controllable-pitch axial fans exhibit a much different noise characteristic than centrifugal fans for similar duty. As shown in Fig. 23, a controllable-pitch axial-fan noise signature contains more acoustic power in the high frequencies due to the large number of rotor blade/stator blades passages in fan operation. This factor must be considered in the design of any noise suppression devices for the axial fan. Because axial fans have a slightly higher total noise sound-power level than centrifugal fans, more acoustical treatment is sometimes needed. Power-plant noise and attenuation are discussed further later in this chapter.

AIR HEATERS

The functions of combustion-air heaters have been described in Chapters 5, 6 and 11. Although justified by the increased efficiency resulting from lower exit-gas temperature, air

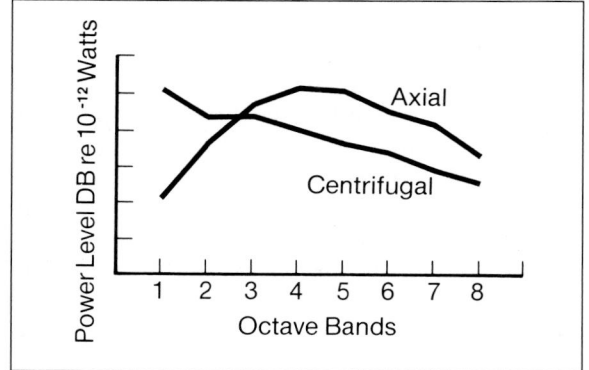

Fig. 23. Fan sound power level comparison

heaters also make pulverized-coal firing practical by providing the drying and transporting medium.

Three principal types are in use: the tubular recuperative, the rotary regenerative (Ljungstrom®), and the heat-pipe (Q-pipe®) air heater. The choice of the size and type of air heater depends on economic and engineering factors. The economic factors include the original cost of the air heater, the maintenance costs, the cost of the fuel, and the cost of the fan power resulting from air-heater draft losses. The engineering factors include the air temperature required for combustion and/or pulverized coal drying, as well as unit reliabilty and installation space requirements.

TUBULAR AIR HEATERS

Fig. 10 in Chapter 8 shows a typical application of a tubular air heater to an industrial boiler. The heater is arranged for vertical gas flow through the tubes. Air flows horizontally across the tubes, usually 2 to 3 inches in diameter, and in a staggered relationship for optimum heat transfer. The air passes over all the sections of the air heater in sequence, the effect of which is to provide counterflow heat transfer. Tube sheets at top and bottom support and guide the tubes. Most frequently, the bottom tube sheet or sheets form the structural support; the upper tube sheet is welded to the outside casing; and the tubes pass through slightly oversize holes in the upper sheet which allows for expansion when the equipment is brought up to temperature. Many designs for use with sulfur-bearing fuels have separated cold-end sections, as shown, to reduce the cost of tube replacement in the event of excessive corrosion.

In contrast to regenerative and heat-pipe designs, tubular or recuperative type air heaters have more severe cold-end corrosion problems. With variations in cleanliness of the tube wall, entering air temperature, and flow intensities on gas and air sides, very low metal temperatures and correspondingly severe corrosion can occur. Although the tube-metal temperature may be considered to be the arithmetic average of the air temperature entering the air heater and the gas temperature leaving, actual field measurements have shown metal temperatures as low as 120°F below the mean of the air and gas temperatures. Such conditions can result in deposits that reduce heat transfer and increase draft loss until they are removed, usually by water washing.[6]

HEAT-PIPE AIR HEATERS

The C-E Q-Pipe® heat-pipe air preheater is a highly efficient heat-transfer device with no moving parts. It consists of many individual heat-pipe tubes formed into an exchanger for transferring heat from flue gases to incoming combustion air. Each heat pipe is a finned steel tube which has been evacuated, partially filled with a heat-transfer fluid, and permanently sealed at both ends (Fig. 24).

MODE OF OPERATION

Heat absorbed from the flue-gas stream vaporizes the fluid within each of the self-contained heat pipes. The vaporized heat-transfer fluid travels up the tube, transporting the heat to the cooler combustion-air side. In the combustion-air end, the working fluid releases its heat energy, is condensed, and returns to the flue-gas side. As long as a temperature difference is maintained externally between those two areas, and the hotter temperature is adequate to vaporize the working fluid, a heat pipe is a self-operating and self-regulating device.

Q-Pipe heat pipes are installed at a slight angle (4 to 7 degrees), with the flue-gas ends lower than the combustion-air section to gravity-assist the return flow of the liquid.

A unique feature of the Q-Pipe heat pipe is its ability to enhance heat transfer by means of a patented internal capillary wick. This wick is a circumferentially spiralled groove that is an integral part of the inner wall surface of the tube; it is continuous along the entire length of each heat pipe. In creating the wick, no material is removed; the metal is merely displaced, as shown in Fig. 25.

The wick serves two different purposes. In the gas section, it distributes the liquid around the entire inner circumference, thereby providing a completely wetted surface for maximum-possible heat transfer. (Without a wick, the fluid would puddle in the bottom portion of the tube, making evaporation less efficient.) In the air section, the wick provides a roughened surface and, therefore, higher heat-transfer coefficients.

ADVANTAGES OF THE HEAT-PIPE AIR PREHEATER

The technology incorporated in Q-Pipe air preheaters offers several benefits. Specific advantages include

- no air-to-gas leakage
- minimum cold-end corrosion
- excellent cleanability
- no maintenance requirements
- compact size
- low weight per unit of heat transferred
- short time and low cost for installation
- design flexibility to suit the needs of a wide variety of applications
- individual operation of each heat pipe, so that the loss of a single pipe has minimal impact on the overall efficiency of the preheater system

THE HEAT PIPE AS A THERMODYNAMIC SYSTEM

The function of the working fluid inside the heat pipe is to absorb the heat energy received at the evaporator end, transport it through the pipe, and release it at the condenser end. It is this process that is called vapor heat transfer. When a liquid vaporizes, two things happen. First, a large quantity of heat is absorbed from the heated area. This takes place because energy is needed to separate molecules that are in contact in the liquid state; the energy required for evaporation at a given temperature is the latent heat of vaporization. Second, as the working fluid vaporizes, the pressure at the evaporator end of the pipe increases; this is caused by the thermal excitation of the molecules comprising the newly created vapor. The higher vapor pressure sets up a pressure difference between the ends of the pipe which causes the vapor, and thus the energy, to move toward

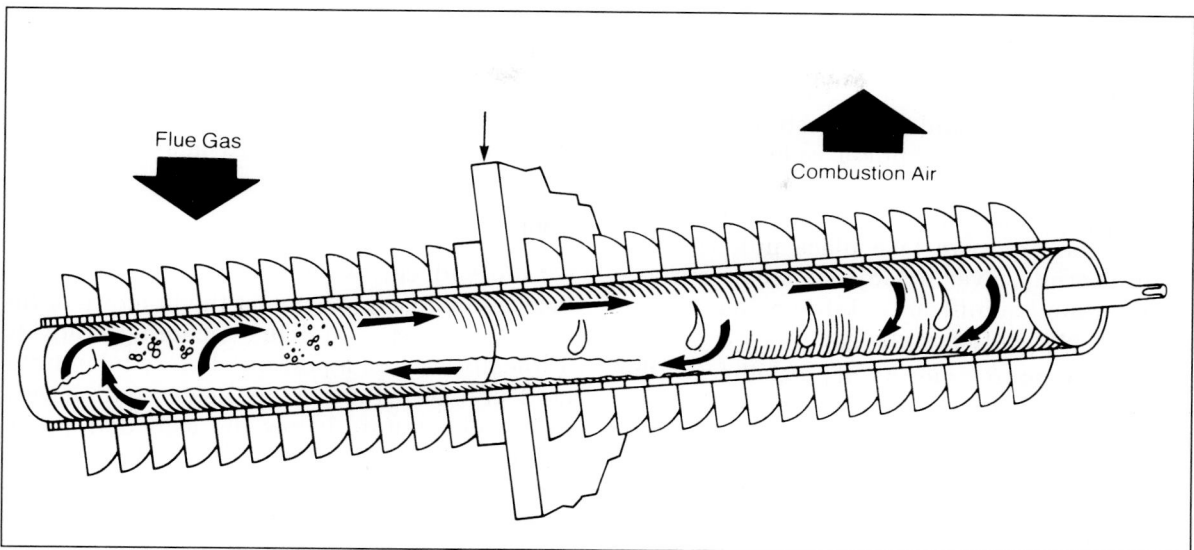

Fig. 24 **Heat-pipe tube for C-E Q-Pipe air preheater**

the condenser end. There, it encounters a temperature lower than that of the evaporator. The vapor then reverts to a liquid and releases the thermal energy stored in its heat of vaporization. In addition, as the fluid condenses, the vapor pressure created by the molecules decreases, so that the necessary pressure differential for continual vapor heat flow is maintained. The condensed liquid returns by gravity to the evaporator to complete the cycle.

It is important to note that the vaporized fluid stores heat energy at the temperature at which the vapor was created, and that it will retain the energy at that temperature until it meets a colder surface. The result is that the temperature along the entire length of the heat pipe tends to remain constant—essentially isothermal. It is this tendency to resist any difference in temperature within the heat pipe that is responsible for its high thermal conductance.

There are four properties of the heat pipe that serve to define the areas for its practical application:

■ Devices that operate on the principle of vapor heat transfer can have a much greater heat-transfer capability than the best metallic conductors.

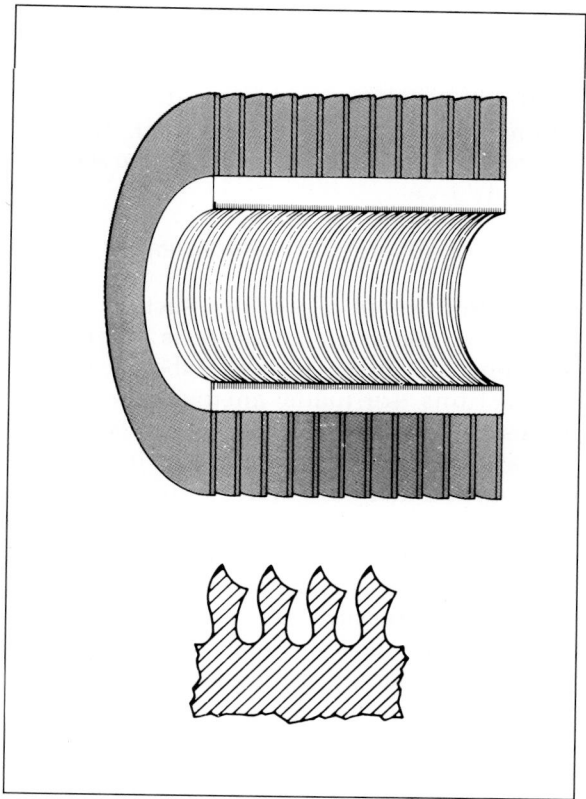

Fig. 25 **Patented capillary wick improves efficiency.**

■ A second property is called "temperature flattening", which provides a uniform temperature over a large surface area.

■ The evaporation and condensation functions are essentially independent operations connected only by the streams of vapor and liquid in the pipe. Thus, the process occurring at one end of the pipe can take place uniformly or nonuniformly, over a large or small surface area, without significantly influencing what is going on at the other end.

■ The heat source can be separated from the heat sink. It is often inconvenient or undesirable to have the heat source and the consumer of heat in close contact.

Although the heat pipe is a highly versatile device, it must operate within certain design limitations. Four limiting factors govern its operation: the maximum total thermal power that can be transferred in a device of a given size; the maximum power per unit of evaporator area that can be handled safely; the maximum and minimum useful temperatures for a given working fluid; and, the extent of operation in a gravitational field or other acceleration. Detailed treatment of these factors is beyond the scope of this text.[7]

COMPARISON TO A TUBULAR AIR HEATER

As described in a previous section of this chapter, a tubular air preheater is constructed by connecting tubes between tube sheets; the connections are made most commonly by means of either tightly or loosely rolled joints. As gas and air flow through the tubular exchanger, the thermal gradients that are generated demand that the various tubes grow in length at different rates and amounts. Either by design or by thermal action, joints loosen and the higher pressure air leaks into the gas stream. Such leakage results in increased pumping costs for both the gas and air streams.

The tube bundles of a Q-Pipe preheater are made up of a number of heat pipes completely seal-welded to the structure dividing the air and gas streams (see Fig. 26). The ends of the heat pipes are free to expand independently on both sides, with sleeves supporting them at the outboard ends. The Q-Pipe preheater is, then, for all practical purposes, a zero-leakage heat exchanger, one that is especially suitable where there are very high air-to-gas-differential pressures.

CORROSION AND FOULING

Cold-end corrosion when handling sulfur-bearing flue gases occurs at the juncture of the cooled-gas outlet and the incoming cold air. Typically, at least a portion of every tube in the cold end of a tubular air heater will have a metal temperature below the acid dew point, leading to corrosion of the carbon-steel tubing.

Because a Q-Pipe air preheater operates totally on counterflow principles and the pipes are isothermal, cold-end corrosion is minimal. Air- and gas-stream temperatures along the row of pipes are essentially uniform, with a much smaller percentage of the total tube bundle at a temperature below the acid dew point. Also the lack of turning baffles (which can add to the thermal stratification) effectively eliminates temperature gradients in the gas and air outlet ducts under all load conditions.

The drier the surfaces of an air heater, the less flyash accumulation throughout the operating range of the boiler. Furthermore, as a deposit begins to form at the cold corner of a tubular heater, the problem accelerates. The insulating effect of the initial deposit extends the area of undesirably cold temperatures, thereby increasing the potential area of flyash fouling.

FLEXIBILITY OF DESIGN OF HEAT-PIPE PREHEATERS

Q-Pipe heat pipes are presently available in lengths up to 42 feet (12.8 meters). The pipes can be manufactured as bare pipes without fins, or with segmented or solid spiral steel fins that are attached by a continuous weld.

The fin spacing can range from one to six fins per inch, depending on the range of fuels to be fired in the boiler. To increase heat recovery while still maintaining ease of cleaning, fins can have a closer spacing on the air side.

If specified, a configuration incorporating removable tubes at the cold end can be provided.

Also, the Q-Pipe counterflow design accommodates space for future heat-transfer surface to increase the gas drop through the preheater if later desired.

CHOICE OF WORKING FLUID

Prior to the charging of the heat-transfer fluid, each heat pipe is evacuated to a pressure of 100 micrometers of mercury or less, at ambient temperature. After charging with an aromatic hydrocarbon, usually toluene or napthalene, the fill tube is closed off by a radially symmetric cold weld, followed by a fusion weld to assure an absolute and reliable seal.

A heat pipe, by design, operates on the saturation curve of its working fluid. The internal pressure, then, is a function of the temperature; the higher the temperature, the higher the pressure. To provide maximum service life, operating-temperature limits of 550°F (290°C) for toluene, and 800°F (430°C) for napthalene, have been established. These have corresponding maximum internal pressures of 405 and 398 psig, respectively (2.79 and 2.74 MPa gage).

The tubing thicknesses selected for air-preheater heat pipes are in strict accordance with the allowable pressures stipulated by the ASME Boiler Code for the material used.

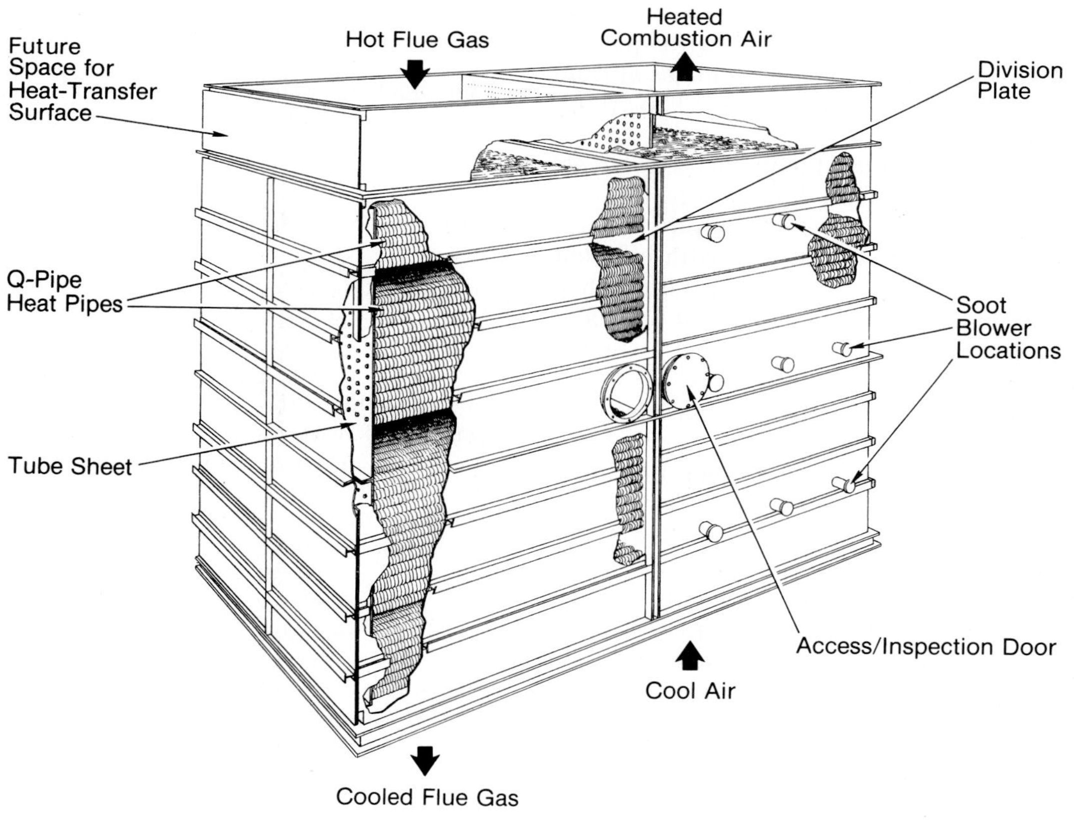

Fig. 26 **Assembly of C-E Q-Pipe zero-leakage air preheater**

14–27

INSTALLATION OF Q-PIPE AIR PREHEATERS

The time required and costs associated with the field installation of Q-Pipe preheaters are substantially less than those of tubular heaters.

Normally, a tubular preheater is shipped as side and end panels, tube sheets, baffle plates, and loose tubes. Each tube must be rolled or welded in place on the job site. A Q-Pipe preheater is compact in size and is shipped in two to six shop-fabricated modules, as governed by shipping limitations. Freight and site installation costs are thus much lower; the modules are simply set in place and seal-welded together.

REGENERATIVE AIR HEATERS

The Ljungstrom air preheater transfers sensible heat in the flue gas leaving the boiler to the combustion air, through regenerative heat-transfer surface in a rotor which turns continuously through the gas and air streams at from 1 to 3 rpm (depending on diameter.) The principle is illustrated in Fig. 27.

Fig. 27 also illustrates the major components of a large vertical-shaft Ljungstrom air pre-heater designed for gas flow downward and air-flow upward. The rotor, packed with efficient heat-transfer surface, is supported through a lower bearing at the cold end of the air preheater and guided through a guide-bearing assembly located at the top or hot end.

Depending on its size, the rotor has either 12 or 24 radial members, which are attached to a center post. The rotor compartments are closed with a shell plate as shown in Fig. 27. The rotor sealing system contains simple leaf-type labyrinth seals bolted to the rotor radial members at both the hot and the cold ends. The radial seals compress against radial plates, again located at both the hot and the cold ends of the rotor. To complete the system, axial seals are positioned at the peripheral end of the radial members of the rotor. There are also leaf-type labyrinth seals used with axial sealing plates. This system effectively separates the air stream from the flue-gas stream. Both the radial and axial sealing plates can be adjusted during operation for maximum effectiveness.

The air preheater design can accommodate any steam-generator duct arrangement. The air preheater ducts and heating surface can be positioned for either vertical or horizontal flow. The vertical-shaft air preheater can be arranged for flue-gas flow upward or downward, while the horizontal air preheater can be arranged for gas-over-air or air-over-gas at the option of the plant designer.

The electric motor provides drive action through a speed-reducer pinion gear which engages a pin rack attached to the periphery of the rotor.

Ljungstrom air heaters are designed with rotor diameters from 7 to 65 feet. The smaller units are completely shop assembled, while the larger utility size units are arranged for convenient field assembly.

HEATING SURFACE DESIGN

The heating surface of a Ljungstrom preheater uses combinations of flat or formed pressed-

---

AIR HEATER TEMPERATURE TERMINOLOGY

Exit-Gas Temperature with Leakage (corrected)—This is the observed or measured exit-gas temperature and includes the dilution effect of leakage through the air-heater seals.

Exit-Gas Temperature, No Leakage (uncorrected)—This is the temperature at which the gas would leave the heater if there were no leakage in the heater. This temperature cannot be measured directly, but is arrived at by accounting for the cooling effect of the leakage air by calculation. (See Chapter 6.)

Gas Drop—Temperature of the gas entering heater minus temperature of the gas leaving heater (not including leakage).

Air Rise—Temperature of the air leaving heater minus that of the air entering heater.

Temperature Head—Temperature of the gas entering heater minus air temperature entering heater.

Gas Side Efficiency, percent =
$$\frac{Gas\ Drop}{Temperature\ Head} \times 100$$

---

steel sheets with corrugated, notched or undulated ribbing. When in combination, they form longitudinal passages of the most desirable contour for the predetermined spacing. The surface design and arrangement provides only point contact between adjacent plates. Although the gas and airflow are turbulent, the smooth path for their travel through the rotor offers low resistance. As an approximate rule, one inch in height of this highly efficient heating surface recovers about as much heat as two feet of surface in a tubular heater with equivalent resistance to gas and air flow. The compact arrangement of light metal sheets permits a large amount of effective heating surface to be placed in a relatively small rotor. The metal sheets are factory packed in containers.

Normally, the heating surface is divided into two or more layers (Fig. 27). Advantages resulting from layering include the ability
■ to provide a field-removable cold-end section, usually 12″ (300mm) in height, that can undergo severe corrosive duty and be first reversed in position, and then replaced;
■ to vary the configuration of surfaces throughout the preheater to satisfy different operating demands; and
■ to make sheets in each layer of different materials, each to accommodate the temperature and other conditions particular to its zone.

Large rotors have 24 self-supporting compartments extending the full length of the heater axially and fabricated as separate modules; the 15° compartments consist of two side plates

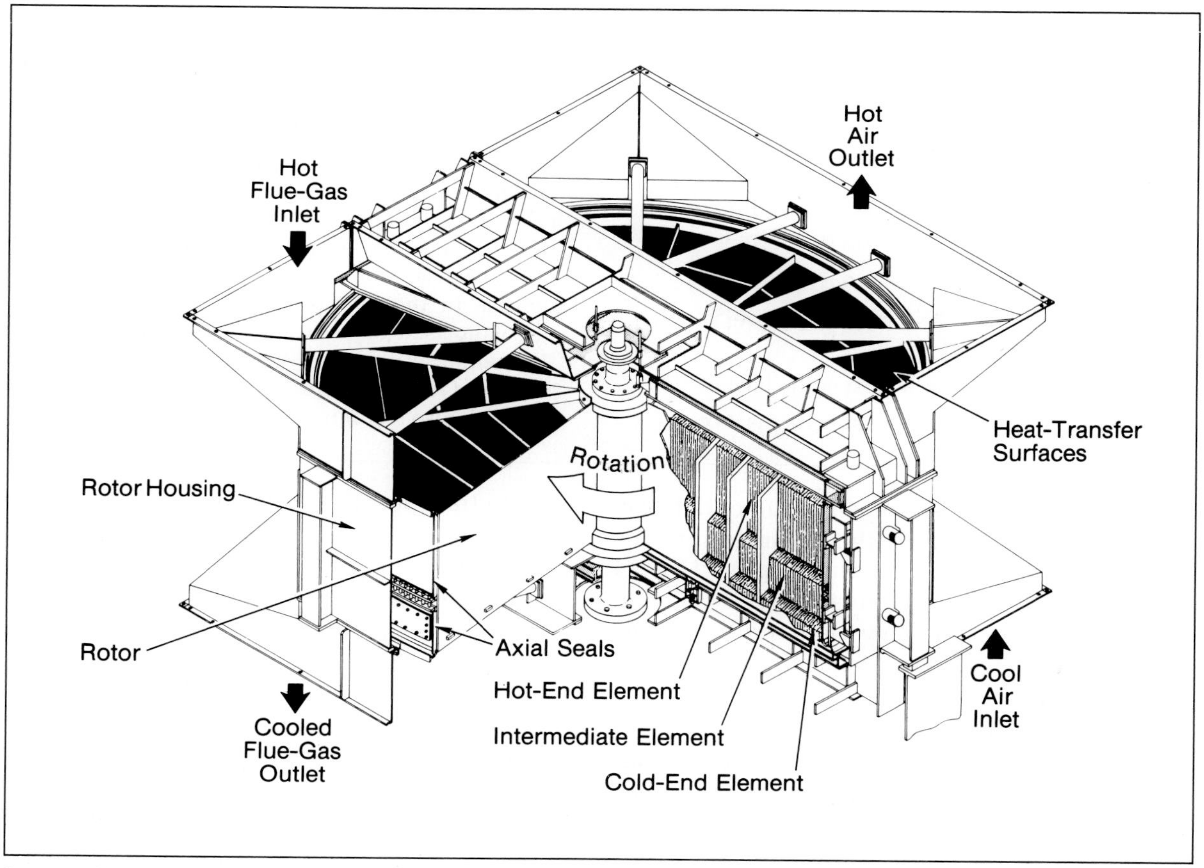

Fig. 27. **Ljungstrom regenerative air preheater, bisector design for vertical air and gas flow**

FACTORS AFFECTING REGENERATIVE AIR-HEATER PERFORMANCE

Entering Air Temperature

A change in entering air temperature will cause the exit-gas temperature to change in the same direction. Changes in entering air temperature result in a change in temperature head which directly affects the drop in gas temperature. For example, if the entering air temperature increases 10°, the exit-gas temperature will increase by 10 × (gas-side efficiency/100)°.

Entering Gas Temperature

A change in entering gas temperature causes the exit-gas temperature to change in the same direction. Changes in entering gas temperature result in a change in temperature head which directly affects the drop in gas temperature. For example, a 10° increase in entering gas temperature will cause the exit-gas temperature to increase by 10 × [1 − (gas-side efficiency/100)]°.

Gas Weight

An increase in gas weight to the heater will result in a higher exit-gas temperature, while conversely, a lower exit-gas temperature results from a lower gas weight entering the unit.

Heat-Capacity Ratio

Also called the "X" ratio, the heat-capacity ratio (HCR) =

$$\frac{\text{mass flow of air} \times \text{avg. specific heat of air}}{\text{mass flow of gas} \times \text{avg. specific heat of gas}}$$

A decrease in the HCR results in an increase in the exit-gas temperature. A 10 to 12 percent change in this ratio may alter exit-gas temperature by as much as 30° to 35°F (15° to 20°C). Factors that affect the HCR include tempering air, overall boiler-system infiltration, and air-heater bypass for cold-end protection.

Pressure Drop Across Air Heater

The air and gas side pressure drops will change approximately in proportion to the square of the air and gas weights through the heater. If excess air is greater than anticipated, the air-heater pressure drop will be greater than expected. A build-up of heating element deposits will result in higher air heater resistances with the consequence of an increase in pressure drop. Pressure drop will also vary directly with the mean absolute temperatures of the fluids passing through the heater as a result of changes in density.

Air Heater Leakage

A change in the temperature of the fluid leaking past the seals will, by reason of density change, have a slight effect on the amount of leakage. Variations in pressure levels between the high and low pressure sides of the heater will, likewise, alter the air heater leakage. An increase in the pressure differential will increase the leakage while a decrease in pressure differential will reduce the leakage. Improper settings of the heater radial and circumferential seals will also result in an increase in leakage.

---

connected by a series of circumferential stay plates and grating. With this rotor design, the rotor post is placed upon its bearings and the 24 rotor modules are then pinned in place, a technique that eliminates any field welding. The basketed heating surface is usually installed in each module before the unit is shipped.

TRISECTOR AIR PREHEATERS

In Chapters 5 and 11, the discussions of the direct-fired pulverized-coal system included the necessity of air heating for coal pulverization. Ordinarily, hot air supplied to the pulverizers furnishes the heat to dry the coal and is used to transport the pulverized fuel to the furnace. This portion of the total air for combustion is called the primary air (Fig. 28).

The trisector air preheater is used on large coal-fired boilers where a cold primary-air fan is desirable. The preheater is designed so that by dividing the air-side of the preheater into two sectors (Fig. 29), the higher pressure primary air and the lower pressure secondary air may be heated by a single air preheater. Some advantages of the trisector over arrangements having separate primary- and secondary-air preheaters include

- no gas-biasing dampers and controls

- because average cold-end temperatures can be maintained using only secondary-air preheater coils, the requirement for primary-air steam coils is eliminated

- fewer sootblowers and water-washing devices are required with one heater in lieu of two

- even with a large variation in primary airflow, there is relatively little effect on overall heat recovery, because heat not recovered in the primary section will be picked up in the secondary section.

The trisector is equipped with an additional radial sealing plate and an axial sealing plate separating the primary and secondary air streams. The trisector heater is readily adaptable to varying coal moisture content. The size of the primary-air opening depends on the coal being fired and its moisture content, with a 35°, 50°, or 72° primary-air opening normally furnished.

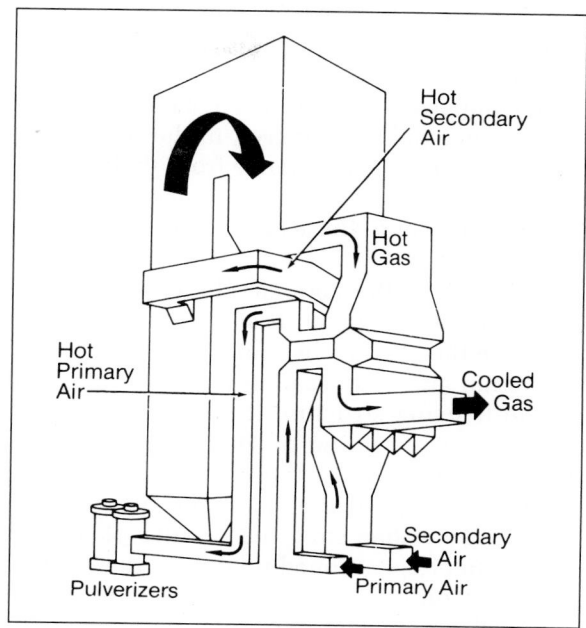

Fig. 28 Trisector air heater arrangement

EFFECT OF THERMAL GRADIENTS ON REGENERATIVE-PREHEATER STRUCTURAL DEFORMATION

The continuous process of heat exchange between cold air and hot gas results in a significant metal temperature gradient through the air preheater which, in turn, leads to a predictable deformation in the shape of the cylindrical rotor. The hot or entering gas/exiting (hot) air side expands radially becoming larger in diameter

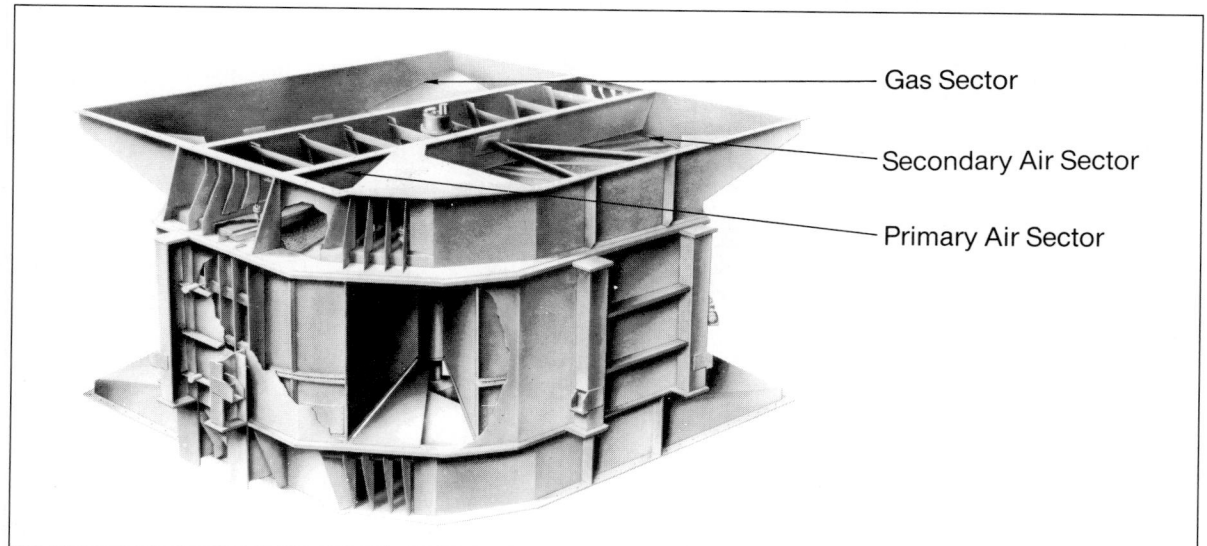

Fig. 29 **Ljungstrom trisector air preheater, vertical- shaft arrangement**

and convex. The cold or entering air/exiting (cooled) gas side end remains smaller in diameter and becomes concave.

To compensate for this thermal distortion, the radial and axial seals are set, in the cold condition, a prespecified distance from their respective sealing plates. During normal operation, in the hot condition, these preset gaps are reduced to near-zero.

The shape of the stationary housing is also affected by the temperature it attains. To further enhance the effectiveness of the sealing system the radial and axial sealing plates are attached to the housing structure through pinned connections. The sealing plates are thereby freed from the effects of the thermal distortion of the housing. The pinned attachments also serve as adjusting mechanisms: the sealing plates can be accurately positioned relative to the rotating leaf seals, on-line if necessary, to achieve an optimum running clearance.

The Ljungstrom Air Preheater Leakage Control System

On larger air preheaters, a leakage control system is furnished. This system acts to minimize the air-to-gas leakage which occurs when the hot-end radial seal clearance increases as the rotor approaches its operating temperature and these seals move away from the sealing plate.

Controlling the hot-end radial seal clearance during operation requires the installation of a drive system to move each sector plate, and a control system to automatically operate the drive. (Fig. 30).

The sector plate drive assembly is comprised of an electric motor, speed reducer, and two mechanical screw actuators mounted to a welded steel base. The electric motor is operated by the control system in response to changes in seal clearance detected by a rotor position sensor.

The rotor sensors are attached to each hot-end sector plate and are actuated upon contact by a sensing lobe that is located on the outer circumference of the rotor.

When the system is in the "auto" mode, the drives independently move the outboard ends of the sector plates approximately 1/16" (1.5mm) per min. toward the rotor. Upon contact with the rotor lobe, the sensor signals the drive to stop and retract the sector plate a small distance from the rotor, leaving a nominal clear-

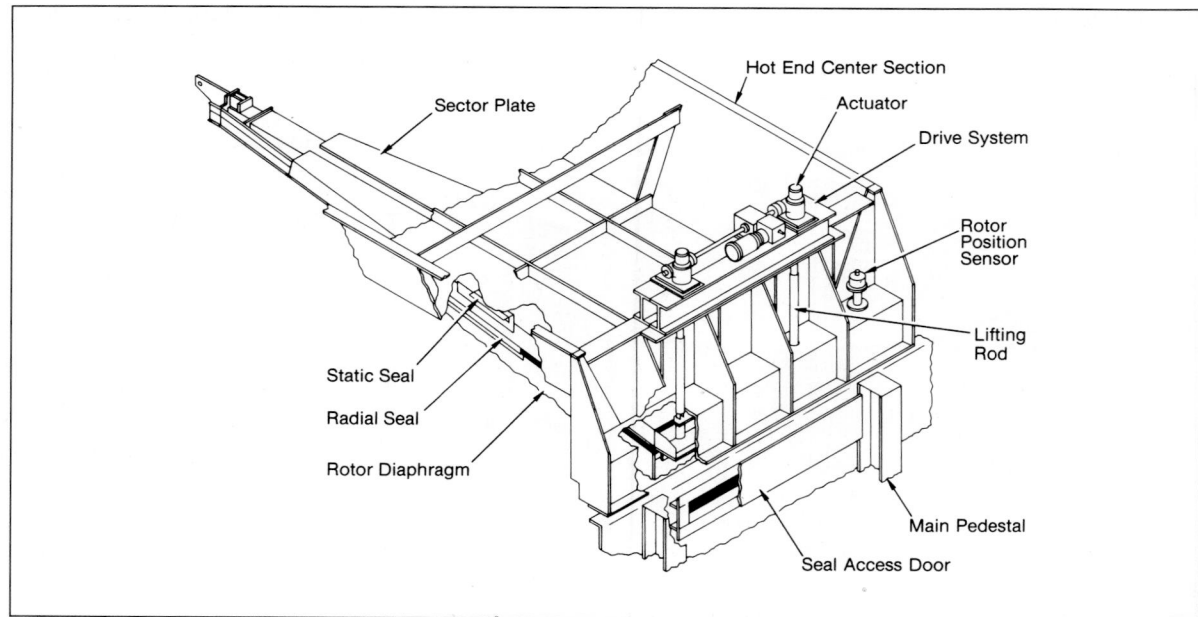

Fig. 30 **Sector-plate drive assembly for Ljungstrom air preheater leakage control system**

ance at the periphery. If the seal clearance changes as a result of temperature change in the rotor, the drive system will automatically maintain an acceptable clearance.

Effect of Rotary Regenerative Air Preheater Leakage

Even with such provisions for minimizing the air-to-gas leakage as described in the previous two sections, such leakage is a design factor that must be taken into account in the sizing of boiler fans. The leakage is defined further as follows.

Direct leakage is that quantity of air that passes into the gas stream between the radial and circumferential seals and sealing surface as a result of the static pressure differential between the air and gas streams. The leakage across the sealing system is directly proportional to the square root of the pressure differential but also depends on the air and gas density.

Entrained leakage is that quantity of air contained in the rotor as it passes from the air side to the gas side and from the gas side to the air side. The quantity of entrained leakage depends on the rotor depth, rotor diameter, and rotor speed.

Whether direct or entrained, leakage has no effect on the heat-transfer efficiency of the Ljungstrom air preheater. There is no difference in the heat transferred to the air stream from the gas stream because of leakage, as discussed in Chapter 6. However, the gas temperature leaving the preheater is decreased by 10°F to 20°F (5°C to 11°C) by the mixture of the cooler air with the hotter gas stream.

CLEANING REGENERATIVE AIR HEATER HEAT-TRANSFER SURFACES

Cold-end deposition occurs when boiler flue gases reach the condensation temperature. The flyash in the flue gas can combine with moisture and sulfur derivatives to form a fine-grained deposit or scale on the cold-end heating surface. Sootblowing can remove and control regenerative air-heater cold-end deposits provided those deposits are not subjected to moisture.

Moisture can be introduced as drainage from water-cooled gas analysis probes, economizer and boiler tube leaks, and unprotected FD fan inlets through which rain can enter the air heater. Leaking steam-sootblower and water-washing shutoff valves add to the problem. But the most frequent source of external moisture is the sootblowing steam. Moisture in this blowing medium can be eliminated by selecting a steam source with controlled pressure and temperature to provide dry steam to the sootblowers at all times.

Air-heater element fouling also can result from the carryover of material from the economizer and the subsequent lodging of the larger particles in the heating surface, particularly at the air-heater hot end.

Regenerative air preheater fouling can be limited by controlling the cold-end temperature level and by the use of proper maintenance procedures and cleaning equipment. The primary requisites for this purpose are sootblowing and washing equipment, a dry blowing medium, an adequate water supply, and a well-engineered drainage system.

Sootblowing

Three types of sootblowing equipment are furnished on Ljungstrom air preheaters. Power-driven sootblowers which have nozzles mounted on a swinging arm are used extensively, while a stationary multi-nozzle type is used on package air preheaters.

Retractable blowers of the same type as used in boiler convection passes (see the next section of this chapter) are installed on air preheaters above 32 ft (10m) in diameter. The equipment is most often located at the gas-outlet side to eliminate flyash from being carried into the windboxes. They are installed either as an integral part of the air-preheater duct or in the gas-outlet ductwork immediately adjacent to the unit.

Air-Heater Sootblowing Media

Superheated steam (approximately 300°F superheat) or dry compressed air is the recommended cleaning medium. Although saturated steam, which contains some moisture, has been used occasionally, superheated steam has been found to be more effective for sootblowing.

Although compressed air is considered to be the premium cleaning medium, its merits do not stem from an inherent cleaning ability, but from its dryness as compared with steam. In fact, under dry discharge conditions, the kinetic energy of a steam jet at 1.4 MPa abs. is approximately twice that of air at the same blowing pressure. But a steam source must be selected to have pressure-temperature conditions which, by proper control measures, may be used to provide dry steam to the air-preheater sootblower.

Steam blowing pressure should be 200 to 250 psig (1.4 to 1.7 MPa gage). A steam-sootblowing piping system should include an automatic drain valve, thermocouple, and an automatic admission valve to the blowers. The automatic drain valve is open to free drain discharge until the temperature-sensing thermocouple indicates steam of adequate quality.

When using air, care should be taken to install a proper line of traps and separators to remove moisture from the blowing medium. Air **at a pressure of 180 psig [1.25 MPa gage] is recommended.**

Water Washing

In cases where sootblowing cannot readily remove residual deposits, it sometimes becomes necessary to water wash the heating surface to maintain acceptable draft losses through the air preheater. In some instances this may be required more frequently than during the scheduled boiler outages.

Most deposits forming on the air-preheater heat-transfer surface are highly soluble in water and, therefore, are easily removed by washing with a sufficient quantity of water. A high-penetration, stationary multi-nozzle device is the standard washing apparatus and is available for all air-preheater types and sizes.

Adequate drainage is necessary before planning to wash an air heater. Washing can be on either the air or gas side, depending upon which has best drainage.

Out-of-service washing is simply washing the preheater in a cold state during periods when the boiler has been shut down. During shutdown is the best time to control the washing operation and to make a thorough inspection of the heating surface, both during and after washing. The rotor can be turned at normal speeds. If it is necessary to restrict the discharge of water to one side of the air heater, however, an auxiliary drive will be required to reduce rotor speed. Surfaces should be examined frequently during the washing process. After the deposits are removed, the unit should be allowed to dry completely before being returned to service.

In-service isolated washing consists of reducing boiler load and isolating one preheater by means of dampers while the boiler remains on the line by using the other preheater(s) and fan(s). After the isolated air preheater is completely washed, the procedure is reversed allowing all the gas and air to flow through the clean preheater while another preheater is isolated and washed in the same manner. The isolated method of washing air preheaters has been found to produce excellent results both from the standpoint of cleanliness and time.

Normally, about 2/3 boiler load can be maintained during isolation of one preheater for washing. Fig. 31 shows the arrangement for the isolated washing procedure. The rotor speed is controlled in a manner similar to that used for out-of-service operation.

On-stream in-service washing is carried out while allowing gas and air to pass through the preheaters. This is only feasible where drains are located to eliminate the moisture entering the dust collectors, precipitators, windboxes, and boiler. The preferred location for operation is in the air side, especially with an electro-static precipitator. On-stream washing should not be done when gas passes are plugged.

The speed of the air-preheater rotor should be reduced to 1/15 rpm by means of an auxiliary drive unit before admitting water to the washing devices. The slow speed permits the wash water to drain from the rotor and heating surface before the wash water enters the other

stream. To insure coverage of all heating surface, the washing cycle must consist of at least one revolution of the preheater rotor.

An intermediate-pressure, high-volume wash is sometimes necessary to remove particles of insoluble material that are carried from the boiler by the gas stream, the larger ones becoming wedged in the flow passages of the hot-end heating surface. These obstructed passages then fill up with flyash, which restricts flow. This type of deposit accumulation is difficult to remove with the usual washing equipment. Special equipment using a high-energy jet has been developed for this purpose. The procedure is similar to that followed in a regular out-of-service washing using standard washing devices. Water nozzles

are moved radially across the surface, allowing sufficient time for the rotor to complete at least one revolution between each movement of the nozzle of the sootblowers.

INFRARED HOT-SPOT DETECTION SYSTEM

C-E Air Preheater's infrared detection system, installed on a rotary regenerative air heater, triggers an alarm if the amount of infrared radiation emitted within the air preheater should ever exceed levels encountered in normal operation. The system operates on the infrared principle, in the 0.9- to 2.5-micrometer wavelength range, where the detector sensitivity is at its maximum.

An infrared system is used because it is easily adaptable to high-speed solid-state electronics.

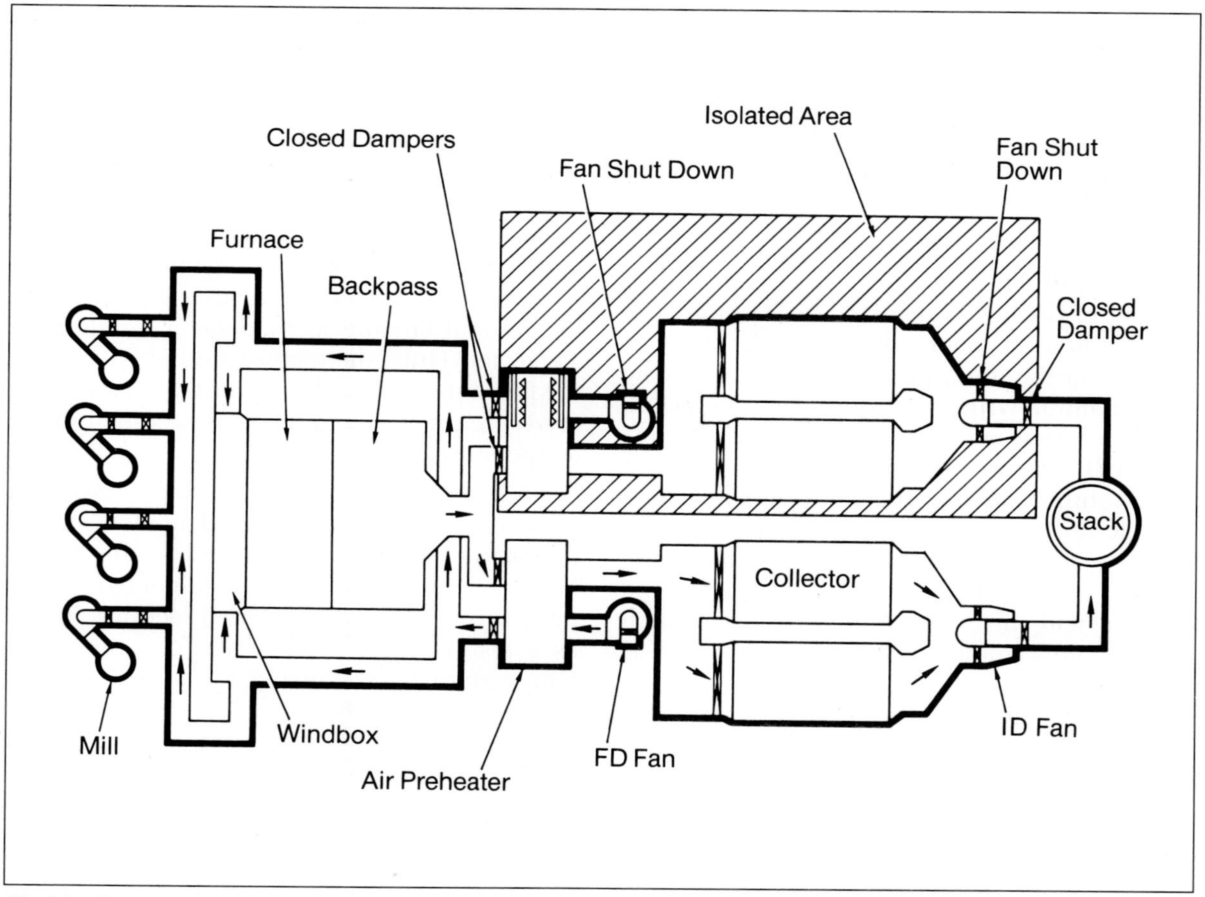

Fig.31 Arrangement for in-service air-preheater washing with one preheater isolated

For example, a six-inch (150mm)-diameter hot spot in the rotating heat-transfer surface will pass a given point in about 150 to 1000 milliseconds; the response time of the detector is less than 100 milliseconds. Also, the lead-sulfide sensing chips used in the equipment are very sensitive to small changes in background temperature. The sensitivity is electronically adjustable and is normally set so that hot spots of 200 to 300°F (90 to 150°C) above the gas inlet temperature will trip the alarm.

During the early stages of deposit heat-up, external effects are not very apparent. The deposit restricts the flow of gas or air so that very little of the heat generated is carried away from the area of origin; most of the heat is absorbed by the metal heat-transfer element nearby. If the condition can be detected during the initial period when temperature buildup is slow, the amount of water needed to reduce the temperatures to a safe level is much less. The C-E infrared system is designed to detect excess heat during this early period.

DESIGNING FOR LOW EXIT-GAS TEMPERATURE

Combustion of sulfur in coal results in the formation of sulfur dioxide, and about 3 to 5 percent of the SO_2 is oxidized to SO_3, depending on the oxygen content, moisture, and temperature of the flue gas. The SO_2 and SO_3 may then combine with moisture in the flue gas to form sulfurous and sulfuric acids. Sulfurous acid will not form above the water dew point temperature, and is seldom a problem. However, sulfur trioxide is hygroscopic and will absorb moisture at temperatures well above the water dew point, resulting in the formation of a sulfuric-acid mist. The temperature at which this acid mist condenses to form sulfuric acid is called the acid dew point.[8]

While boiler efficiency can be improved by adding surface to reduce the air-heater exit-gas temperature, this practice lowers the cold-end metal temperatures, possibly below the acid dew point. Consequently, steel construction materials are subject to corrosion from the sulfuric acid in the flue gas.

The acid dew point varies with the concentration of sulfur trioxide in the flue gas. High-sulfur coals result in the existence of a dew point at a higher temperature, thus exposing more of the air-heater surface to corrosion and fouling. Typical acid dew points for coal are between 280° and 320°F (135° and 160°C).

For coal firing, fouling potential increases as the temperature decreases. Water condensation causes a marked increase in dew-point meter response at 140°F (60°C) or below. This does not necessarily mean as severe corrosion in the water dew-point range, as compared to the acid condensation range, although rapid fouling would likely take place at the water dew point. With low-sulfur coal, the corrosion and fouling potential is low and restricted to the extreme cold end of the air preheater.

A number of means have been developed to minimize the rate of corrosion, as well as provision for replacement of corroded surface. Because corrosion occurs on the lowest temperature surface, air-heater designs have been developed which incorporate replaceable cold-end sections. Other means to minimize corrosion are aimed at increasing the metal temperature. One such arrangement directs a portion of the preheated air to the inlet of the forced-draft fan and recirculates it through the air heater. Thus, the temperature of air leaving the fan and entering the air heater is increased, correspondingly increasing the cold-end metal temperature.

Air bypass around the air heater is used to a limited extent. With reduced airflow, metal temperatures within the air heater are higher because of the influence of the higher gas-to-air ratio. Also, because the overall recovery is less as a result of the reduced airflow, gas outlet temperature rises, causing a rise in the cold-end metal temperature.

The prevalent means of increasing cold-end metal temperature, discussed in detail in Chapter 6, is the use of steam air heaters located in the cold-air duct between the FD fan and the air heater. These increase the temperature of the air entering the heater, correspondingly causing an

increase in the metal temperature. Steam bled from the turbine is used as the heating medium in the steam air heater. In supplying heat to the cold air, steam is condensed and the condensate returned to the appropriate stage of the feedwater bleed heating system.

To assist specifiers and operators in arriving at reasonable cold-end temperatures, average cold-end temperature (ACET) guides for regenerative air preheaters are published. These guides take into account the variables of fuel type, sulfur content, and the effect of excess air. A guide for coal firing was presented as Fig. 4 in Chapter 6.

BOILER SOOTBLOWERS

One of the most important boiler auxiliary operations is the on-line fireside cleaning of heat-absorbing surfaces. Not only is it important for proper heat transfer, but also to prevent sections of the boiler from becoming severely plugged. Plugged sections can restrict gas flow and cause load limitation. Tube erosion due to high local velocities also can occur.

Sootblowing systems are required on coal and oil-fired furnaces. Oil has a low ash content which produces a thin water-soluble deposit on the furnace walls that is normally removed by annual water washing. Furnace-wall sootblowers are, therefore, not required on oil-fired units. In the superheater and reheater sections of an oil-fired unit, ash deposits do accumulate on the tubing surface. This is especially true of high-vanadium oils where additives are used to combat high-temperature corrosion. But with the use of solid-powder additives, ash deposits in the high-gas-temperature areas increase markedly. Ordinarily quite friable, such deposits are easily removed with sootblowers.

Coal-fired units require large complements of permanently installed sootblowing equipment. In the boiler furnace, the concentration of wall blowers depends upon such factors as the ash-fusion temperatures, as described in Chapter 3. The lower these temperatures are, the more likely the ash will adhere to the furnace walls and accumulate. The percentage of ash in the coal is another factor in the design of the furnace wall-blower system, with higher-ash coals requiring greater and more concentrated coverage.

FURNACE-WALL SOOTBLOWERS

A short single-nozzle retractable blower, called a wall blower, removes the ash deposited on furnace walls (Fig. 32). It is a short-stroke lance which penetrates the wall one to two inches. Supported by wall boxes welded to the furnace tubes, it follows both vertical and horizontal movements of the furnace walls as they expand and contract. The single nozzle at the tip directs a supersonic high-energy jet of superheated steam or air parallel to the furnace face of the water-wall tubes, dislodging the slag deposits. The lance rotates through 360° and cleans approximately a five-foot radius; the effective radius depends upon how tenacious the deposit is. Some coals with difficult-to-remove slag require wall-blower spacing to be closer; the maximum cleaning radius may be only 3 to 4 ft. Blowing frequency depends on the rate of slag build-up, but frequencies in the 4- to 8-hour range are common.

CONVECTION-SURFACE SOOTBLOWERS

Superheater, reheater, and economizer sections of large boilers are cleaned with long, fully or partially retractable lances which penetrate the cavities between major heat-absorbing sections. Smaller boilers may use nonretractable blowers with multiple nozzles, which rotate, allowing each nozzle to clean a tube row.

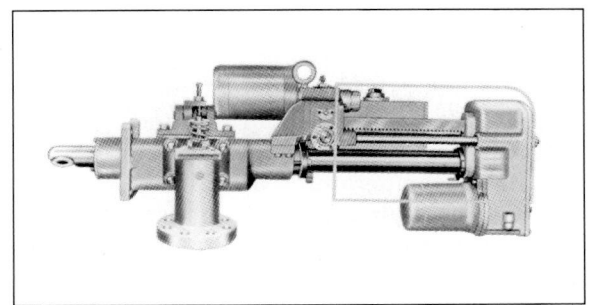

***Fig. 32* Typical retractable furnace-wall sootblower**

The long retractable type sootblower is the most effective way to clean radiant and convective heating surface. It normally uses two 180°-opposed cleaning nozzles at the tip which emit a high-energy jet of superheated steam or compressed air perpendicular to the lance. While the lance traverses the boiler, it rotates, forming a helical blowing pattern (Fig. 33) which effectively cleans the tubes and spaces between tubes in a superheater, reheater, or economizer bank. In widely spaced platenized sections, these nozzles are angled slightly, leading and lagging the perpendicular to gain more dwell time on the tube surface. The effective range of retracts depends upon the gas temperature in the area to be cleaned and the ash characteristics of the particular fuel being fired. Therefore, the maximum effective cleaning radius varies from 4 to 9 feet. It is difficult to relate cleaning radius to blowing pressures because of various nozzle combinations. Blowing pressures depend not only on supplying a flow for cleaning but, in high-temperature zones, supplying an even greater flow for cooling the lance.

Fig. 34 shows a retractable blower on one side of a large utility boiler; this blower penetrates half the width of a 90-ft wide boiler. The blower typically uses a two-point support which allows for boiler expansion. A wall box welded to the tubes supports the front of the lance; the platform structure supports the rear through a slot-and-pin connection.

SOOTBLOWER OPERATION

The type of deposits in the radiant and convection sections of the boiler can vary from very hard tenacious slag to a dry powdery coating. The most important coal properties affecting the severity and rate of ash build-up are ash softening and fluid temperatures, percent ash in the coal, and the percent sodium in the coal ash. Coals with high percentages of ash, low ash-fusion temperatures, and high percentages of sodium in the ash are the most difficult to keep clean. The blowing sequence and frequency of the retract system must be adjusted during initial operation by starting with an assumed sequence of blowing and a frequency of,

say, one complete cycle per 8-hour shift. After observing fouling patterns either through observation doors (on line), or by gas-side inspections during shutdowns, the operator can modify the blowing pattern to that which best suits the boiler for the coal being fired. It is common for certain retracts to be blown more frequently than others, either singly or in groups. Once the pattern is established, it can be programmed in the sootblower system and run automatically.

The air preheaters are also cleaned by sootblowing as described in the previous section on air preheaters. The ash at this point in the system is usually dry and the particles small enough to pass through air preheater elements.

BLOWING MEDIA

The two blowing media are steam and compressed air, with both being equally effective in deposit removal. The utility normally makes the choice based on plant economics. In the case of air, large compressors must be installed with an integrated piping system around the boiler. Steam systems are usually supplied from the boiler through a pressure-reduction station so that, after pressure reduction, dry superheated steam is available at the sootblower nozzle. Steam has the advantage of availability whenever the boiler is in operation.

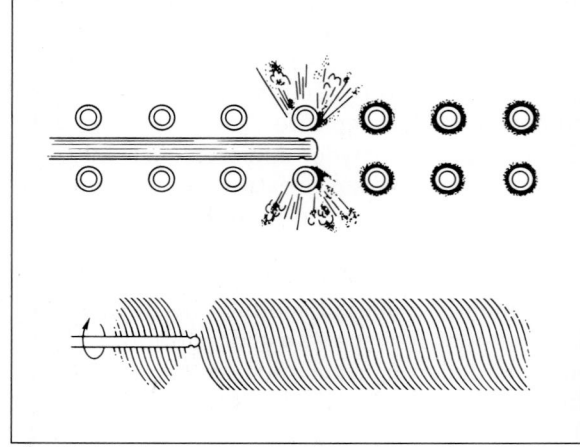

Fig. 33 **Retractable sootblower cleaning pattern**

Fig. 34 Long retractable sootblower for cleaning convection surface of a 750-MW steam generator

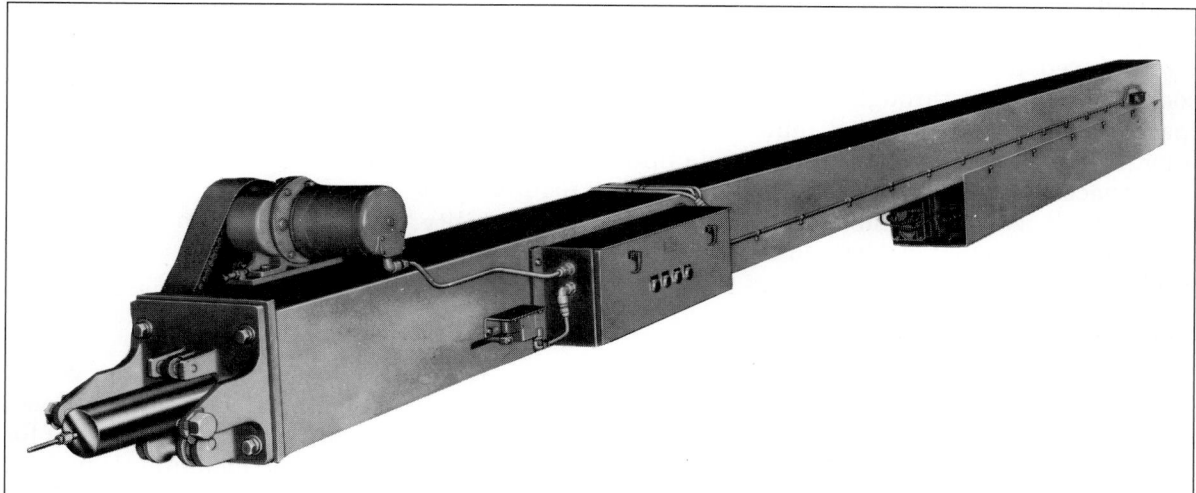

Fig. 35 Retractable probe for measuring furnace gas temperature

ABB
ASEA BROWN BOVERI

With air, there is always the possibility that the blowing medium will be lost if the compressor is out of service.

Water as a medium is not generally recommended for sootblowing and is considered as a last resort for particularly troublesome slag deposits. Thermal shock from water may reduce tube life due to cyclic fatigue.

FURNACE TEMPERATURE PROBES

Fig. 35 illustrates a typical retractable furnace probe for monitoring gas temperatures at the top of a furnace. The probe is inserted during the lighting-off period and remains in the furnace until reheater steam flow has been established. Its design is based on that of the long retractable sootblower, but without the rotating feature. It traverses back and forth across the furnace or convection pass as it is driven by a single electric motor.

As a probe senses temperatures, it also transmits its position. A thermocouple mounted on the end of the lance measures gas temperatures. The standard thermocouple is suitable for temperatures to 1850°F.

For lance extensions below 16 ft and gas temperatures of 1200°F or lower, there is no need for cooling the lance. Automatic retraction from the recorder alarm protects the lance at excessive temperatures. When air cooling is required, the lance is supplied with an inner lining for optimum cooling.

Introduced into the traveling carriage, the cooling air passes through a small annular space between the lance wall and the inner liner. It then is discharged from a nozzle into the furnace. Mounted back from the end of the lance, the nozzle is slanted away from the thermocouple, so that discharged air will have no effect on the temperature readings.

SOOTBLOWER CONTROL SYSTEMS

The mode of sootblower operation depends upon the size and capacity of the boiler unit and the number of blowers installed. Small industrial boilers using few sootblowers cannot economically justify the use of automatic control systems; manually operated blowers are used. On the other hand, large industrial and central-station boilers can justify various degrees of automatic control.

In systems with over 200 sootblowers, programmable controllers frequently are installed so that proper automatic sequential operation of the system can be accomplished after ash-deposition patterns are established under operating conditions. With properly programmed blowing sequence, ash deposits on the furnace walls are minimal and the combustion gases are cooled to the required temperature before entering the convection pass.

Sootblower control panels are now designed for easy modification of the automatic sequential operation. This gives the operator maximum flexibility in maintaining the cleanliness of a unit. Essential parameters for such a sootblower control package include

■ equipment to automatically start each sootblower in the system

■ a method to cancel the operation of any sootblower in the system

■ a way to determine easily which sootblowers have been selected to operate, and their programmed operating sequence

■ the complete capability to monitor and display the operation of each sootblower

■ the capability to monitor all the essentials of the sootblowing system and prevent continued sootblower operation if the system is not functioning properly and abort the operation of any sootblower if a malfunction occurs

■ a method to select and alter various blowing routines as required by the boiler cleaning requirements

■ the ability to operate certain of the sootblowers simultaneously

■ a means of manually overriding the automatic routines

Ideally, a sootblower control system would respond automatically to conditions of load, temperature, pressure and fuel to provide for the most efficient boiler operation. However, because of the number of input variables and the questionable validity of signals and the

complexity of process manipulation, this has not been technically feasible to date. Sequential control with the boiler operator as the decision maker has been used for the majority of sootblower systems. With the advent of more complex blowing operations, the hard-wired solid-state logic systems have taken over some of the decision making. It is extremely beneficial to minimize operator attention and still operate the sootblowers efficiently. Effective performance of any soot-blowing control system depends on its ability to make complex decisions with a minimum of operator input.

BOILER-WATER CIRCULATING PUMPS

Boiler circulation pumps are an integral part of the furnace-wall circulation system on the Controlled Circulation® and Combined Circulation® units. By using pumps in the circulation flow path, the total quantity of water circulated can be apportioned and distributed throughout the furnace heat-absorbing surfaces to suit all the conditions of operation. Boiler circulation pumps also give the operator the capability of insuring positive circulation for a wide variety of pre-operational procedures and off full-load plant cycling conditions.

Most boiler circulation pumps have certain basic similarities. All are vertical, single-stage, centrifugal type with overhung impeller and are designed for constant-speed operation without throttling. The hydraulic end consists of a single vertical suction and either single or double horizontal discharge (Fig. 36).

The pump casing and associated isolation valves are entirely supported by the boiler downcomer piping, thus permitting unrestricted expansion movement during boiler start-up and shutdown. Butt-weld joints between pump nozzles, valves, and piping are designed to withstand the combined moments and forces resulting from dead weight loads and the expansion differential between the furnace and the downcomers. Only the connec-

tions for station services of seal water, cooling water, and electric power supply must be flexible to accommodate pump expansion movements. The pump drives share the basic electrical characteristics of low starting torque, induction squirrel-cage motors, designed for full-voltage across-the-line starting.

DESCRIPTION OF BOILER CIRCULATION PUMPS

Pumping water at or near saturated temperature and at high pressure requires pumps of special design because of the possibility that shaft seal leakage water may flash rapidly into steam. For this service, three types of boiler circulating pumps have been used. The zero-leakage pump is driven by a wet motor, cooled and lubricated by recirculated water at system pressure. The stator windings are immersed

Fig. 36 **Vertical single stage centrifugal pump for boiler-water circulation**

in the high-pressure water. This unit is also known as a glandless submerged motor pump.

The second type employs a canned motor housed within a pressure vessel common to the pump itself. It too is a glandless or zero-leakage type of pump. The third type employs a conventional motor drive. Shaft sealing is accomplished by a stuffing box with packing or a shaft mechanical seal, supplemented by an injection sealing-water arrangement.

ZERO-LEAKAGE BOILER-WATER PUMPS

Controlled Circulation boilers of the type described in Chapter 7 are designed with glandless pumps in which the electric motor and the pump are enclosed in a common pressure casing (glandless meaning that there are no stuffing boxes or mechanical seals around the pump shaft). These zero-leakage pumps are of two main types: those with stator enclosures, but with the rotor immersed in water (known as dry stator or "canned" pumps) and those in which water at system pressure is circulated through the stator windings (known as wet stator or "submerged motor" pumps). In both designs, the high-pressure water in the motor is recirculated through an external heat exchanger and then through the rotor gap and acts as a lubricant for both thrust and journal bearings, and as a cooling medium for removing the heat generated by the motor losses.

SUBMERGED MOTOR RECIRCULATING PUMPS

The motors of these pumps are wound with waterproof insulated cable. The motor power is supplied through pressure-tight cable inlets. As shown in Fig. 37, the internal high-pressure water content is circulated through an external heat exchanger to keep the motor cool and provide bearing lubrication. Circulation is accomplished by an auxiliary impeller, integral with the thrust disc and located at the base of the motor. The design also includes a thermal barrier for minimizing the flow of heat from the pump to the motor. Provision is made for controlling the temperature within the submerged motor by

■ restriction of conduction and convection between pump and motor by reduced cross-sectional areas, and by mounting the pump above the motor

■ heat exchange through a closed-circuit pipe system and external heat exchanger

The electrical leads exit the motor casing through pressure-tight sealed glands to an external terminal box at atmospheric pressure. The main supply cable is connected to the motor phases inside this terminal box.

The entire pumping element—the drive motor, pump shaft, thermal barrier and overhung impeller—can be installed or removed from the pump casing without disturbing the casing or its piping connections.

The pump shown in Fig. 38 is of the diffuser type with a spherical casing. The materials and construction of the pressure-retaining parts— pump casing, heat barrier, forged motor case, and all bolted flanges—are in accordance with *ASME* Section I requirements.

The motor shaft of this pump is fitted with two chromium-plated chrome-steel sleeves running in water-lubricated journal bearings. A double-acting Kingsbury-type thrust bearing locates the rotor assembly axially and takes up any residual thrust. Both the radial journal bearings and the axial thrust bearings are self-adjusting, thus ensuring equal bearing loads.

CANNED-MOTOR PUMP

The canned-motor boiler-circulating pump is a zero-leakage pump designed to circulate boiler water at high temperature and high pressure. No shaft seals are required. The motor and pump components are designed as an integral pressurized unit, sealed by a gasket and studs.

The impeller is rotated by an AC polyphase, squirrel-cage induction motor. The spaces between the stator and rotor and around the bearings are filled with the cooled system water. A heat barrier between the pump and motor limits the conduction of heat from the high-pressure hot water being pumped, to the high-pressure internal motor water. The dry stator winding is completely isolated from the

high-pressure water by a corrosion-resistant "can" lining the stator bore.

The stator assembly consists of the bearing assemblies, the stack of laminated steel punchings, the three-phase insulated copper winding, the outer pressure shell, and stator can.

The rotor assembly contains a shaft, squirrel-cage type rotor body, auxiliary impeller, rotor journals, and thrust runner. The main impeller is mounted on the rotor shaft extension. Secured to the lower end of the shaft, the thrust runner carries the weight of the rotor assembly

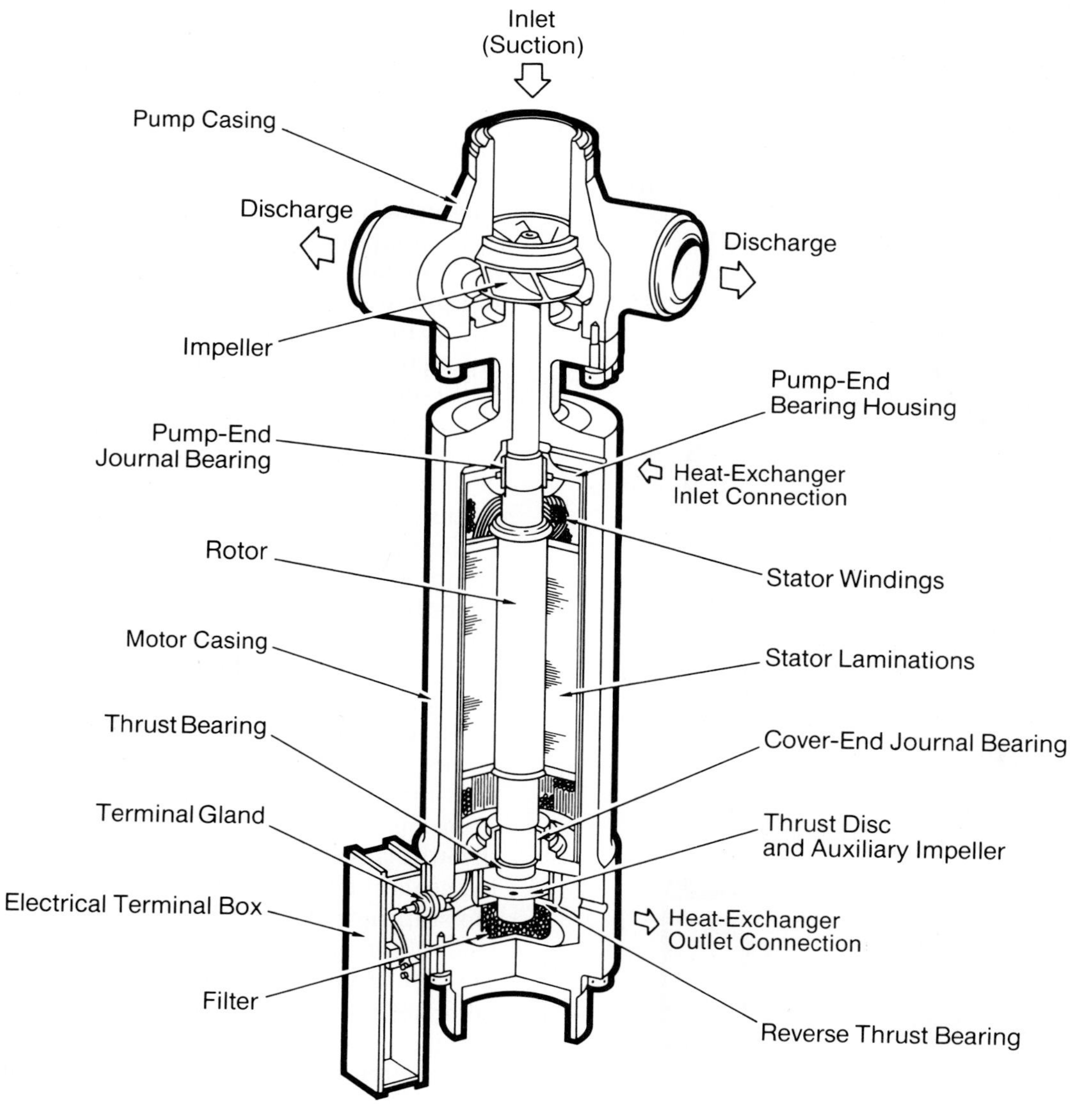

Fig. 37 **Schematic of a typical glandless wet-motor boiler-water circulating pump**

and absorbs the axial hydraulic thrust. The pump motor with its integral pump shaft and impeller can be withdrawn from the pump casing without disturbing the casing or its piping.

In operation, high-pressure cooling water in the motor cavity is circulated by the auxiliary impeller mounted on the rotor shaft. This water is circulated from the auxiliary impeller up through the gap between the rotor and stator, through the upper guide bearing, out to the external heat exchanger, back into the motor through the piping connections to the auxiliary impeller suction in the hollow shaft end. Some of the auxiliary impeller discharge is circulated down through the lower guide bearing and the thrust bearing and then back to the auxiliary impeller suction.

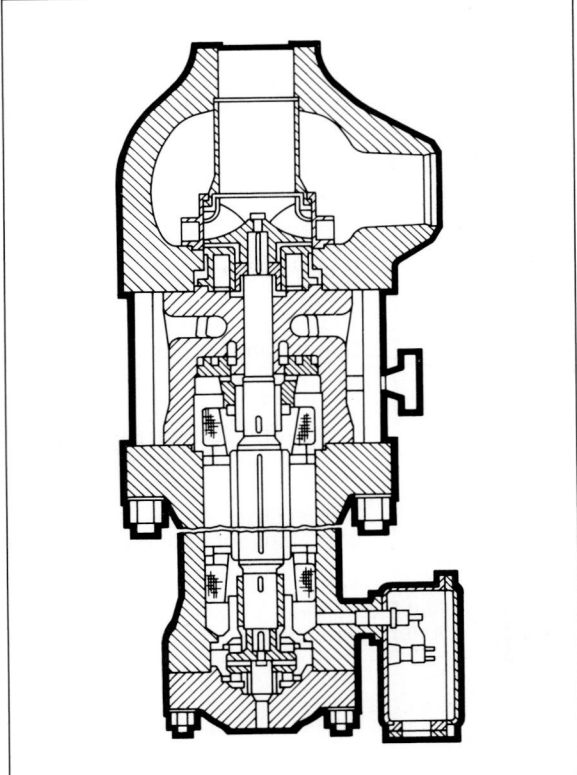

Fig. 38 Single discharge spherical-type pump for supercritical-pressure steam generator. Pump casing design pressure is 4500 psig.

Low-pressure cooling water is circulated through the external heat exchanger. When the pump is idle, but pressurized with high-temperature boiler water in the pump casing, conducted heat is removed by motor-fluid convection flow through the heat exchanger.

INJECTION-SEAL PUMP DESIGN

This controlled-leakage boiler-water circulating pump is mounted vertically with a rigid shaft designed to carry its own thrust. No thrust load is transmitted to the motor, which is entirely supported by the pump. The internal pump materials are selected to accommodate expansion encountered at high temperatures, and to provide corrosion resistance.

Essentially, the pump contains the following main components: casing, stuffing box, pressure breakdown assembly, bearing assemblies and pump shaft with impeller. The pump casing is supported by the suction and discharge piping to which the casing nozzles are welded. A direct-drive coupling connects the pump motor to the pump shaft. The vertical induction motor is of open drip-proof construction for indoor installation, or a weather-protected type for outdoor installation.

The suction pressure and temperature at the pump are virtually the same as the boiler operating conditions. The injection seal reduces the pressure and temperature at the stuffing box packing by injecting cool water at high-pressure into the annulus formed by the shaft running clearance. The injection water serves as a thermal barrier against conducted heat and as a means of keeping the boiler water from entering the close clearances of the throttling sleeve and floating seals. The floating seals and the packing control the leakage at the stuffing box. The injection water, free and clear of foreign particles, can be supplied from the discharge of the boiler feed pump or any other source that can supply a pressure differential of 50 to 100 psi over the discharge of the boiler circulating-water pump. A supplementary booster pump of limited capacity is provided with a takeoff point from the boiler feed-pump discharge to supply the additional pressure necessary for sealing the boiler circulating pump. The booster injection pump has a bypass system for operation at partial loads.

OTHER AUXILIARY EQUIPMENT

Although this chapter has discussed fans, air heaters, sootblowers, and boiler-water circulating pumps, it has only scratched the surface as far as the multiplicity of ancillary equipment involved in a large steam-generator installation. For instance, hundreds of valves for high-pressure steam, water, compressed-air and oil duty are part of the scope of most boiler contracts. The *ASME* Boiler and Pressure Vessel Code and *ANSI* Code B31.1 for power-plant piping govern the design and application of most of the valves.

Because boiler valves experience extremes of temperature and pressure as well as widely varying conditions of operation, the valve manufacturer must give close attention to the materials of construction. For valve parts these may range from low-carbon steel through the low and medium chrome alloys to stainless steel, in both castings and forgings.

Safety valves are regulated by the *ASME* Code. For boilers operating below the critical pressure, superheater safety valves are customarily set to pop before boiler safety valves. This assures maximum possible flow through the superheater, thus protecting the tubes.

Feedwater valves contain stop and check valves near the economizer inlet. Ordinarily these valves are wide open. Blowoff valves are located at low points in the boiler, such as the lower waterwall supply header, for rapid removal of sludge collected at these points. Continuous blowdown and chemical feed valves serve to limit and control concentration of impurities in the water in the main steam drum. Sampling valves provide a continuous record of steam and water purity.

Modern Controlled Circulation boilers have three 50-percent capacity pumps, with outlet stop-check valves. With such an installed spare pump, suction valves are not required at the pump inlets.

Boilers operating at supercritical pressures involve special considerations that differ from *ASME* Boiler and Pressure Vessel Code requirements for subcritical-pressure boilers.

POWER-PLANT NOISE

This final section discusses the problem of noise and its suppression in a large pulverized-coal fired power plant. Generally, the motors, and the rotating equipment they drive, are the source of most noise.

Noise is unwanted sound with a severity that varies with space and time. For example, it is worse at night when the masking effect of background sounds is reduced.

The human ear is more sensitive to high than to low-pitched sounds. Because of this, high frequency noise is normally the greatest offender, particularly if it contains a pure note.

An essential feature of the human sense of hearing is that to perceive a just noticeable difference in volume requires a discrete change of about 10 percent in the subjective noise level on the decibel (dB) scale. Therefore, worthwhile reductions in noise levels should probably involve 5- to 10-fold decreases.

PHYSICAL LEVEL SCALES OF NOISE

The human ear is a frequency meter, a sound analyzer, and an intensity meter. Conveniently, we describe our sensations, sometimes in physical, sometimes in psycho-physical terms, but neither provides terminologies that alone seem suitable for classifying all sounds.

Physically, a pure note is a single harmonic wave. Sound is produced by adding together several, but not too many, pure notes. Too many produce noise. The human ear, to some degree, can separate sound into pure notes, but apparently only if the frequencies involved are well separated. The intensity of specific frequencies then becomes important.

The quantity usually measured in acoustics is the RMS—root-mean-square *sound pressure* (P). The weakest sound pressure perceptible to the average human ear is very small, about 0.0002 microbar. In comparison, the greatest sound pressure perceived without pain is about 1000 microbars. The human sense of hearing covers a wide dynamic range.

The human ear responds to changes in sound pressures, but it is the ratio rather than the absolute difference between two intensities or two frequencies that produces the distinct subjective sensation. In noise measurement, it is practical, therefore, to adopt a relative scale of sound pressure, such as the decibel (dB) scale. The decibel is defined as 10 times the logarithm to the base 10 of the ratio between two quantities of power. (The factor 10 converts bels to decibels.) Thus, as sound power is related to sound pressure squared, the common scale for noise measurement becomes:

Sound Pressure Level (SPL) =

$$10 \log_{10} \frac{P^2}{P_o^2}(dB) = 20 \log_{10} \frac{P}{P_o}(dB)$$

(7)

where P is the actual pressure measured and P_o some reference value, conventionally the threshold of human hearing (0.0002 microbar or 20 micropascals). The use of this equation reduces the range of sound pressures to 0 to 120 dB, 0 dB indicating the reference value and 120 dB an approximate "maximum" level for pain-free stimuli (during short exposures).

NOISE CRITERIA

The dB term, unfortunately, is also used to designate levels of sound power, sound intensity, and in ratings simulating human response such as the *A, B, and C "weightings"* and the *Noise Criteria Curves.* To avoid confusion, reference values should never be omitted.

Some common sound levels are listed in Table IV. Most daily activities take place within ¼ of the scale, from 50 to 85 dB. This covers any sensation between pleasantly quiet and annoyingly noisy. A noise attenuation of a few dB will be most pronounced within this narrow range. At the top and the bottom of the scale, as much as 10 or 15 dB attenuation (*i.e.,* 10% of full-scale) is hardly noticeable.

LOUDNESS AND NOISE NUISANCE

There is no simple relationship between the physical measurement of a sound pressure level and the human perception of sound. In Fig. 39, each curve represents one of constant subjective loudness equal to the physical dB level at a standard noise frequency of 1000 Hz. These "equal loudness contours" are an international standard. Sound-level meters are made with electronic filters which can integrate sound energy in such a way that the filter characteristics (Fig. 40) reflect the subjective response of human hearing. The ensuing noise level (measured in decibels on the A-scale, dBA) is then a good measure of the subjective loudness of the sound provided there are no discrete tones.

Prolonged exposure to noise results in a temporary threshold shift, TTS. It has been suggested that 12 dB TTS (P_o-0.0002 microbar) should be the maximum permissible short-term noise exposure for given noise ratings (N) as a function of TTS at 2000 Hz. For example, exposure to noise rating N-90 should not exceed 2 hours. If the subject becomes exposed for any longer periods, hearing-conservation aids should be provided.

NOISE MEASUREMENT

Most sound-level meters are now equipped with frequency weighting networks to determine overall sound pressure levels at a certain location. However, for abatement purposes, more thorough investigations are often required. Therefore, the next step is to make a frequency analysis of the noise. Usually, the instruments are provided with filters that divide the noise spectrum into 8 to 10 electronically-controlled, well-separated bands. The frequency analysis is then simplified to discrete dB measurements for each band.

In noise studies, the position of observation with respect to the source is of utmost importance, and so is a detailed knowledge of the surroundings. The effects of reflections from wall surfaces and the contribution of the other sources must be known. In fact, to do a noise

Table IV. Typical Sound Pressures and Sound Pressure Levels

P (μ bar)	SPL (dB ref. 0.0002 μ bar)	Source (long time average)	Distance (ft)
2,000	140	Threshold of pain	
	130		
		Pneumatic clipper	6
200	120	Threshold of discomfort	
	110		
		Automobile horn	20
20	100		
		New York subway train	Inside
	90		
		Motor bus	Inside
2	80		
		Traffic on street corner	
	70		
		Conversational speech	3
0.2	60		
	50	Typical business office	Inside
		Quiet residence	Inside
0.02	40		
		Library	Inside
	30		
		Whisper	5
0.002	20		
	10		
0.0002	0	Threshold of hearing	

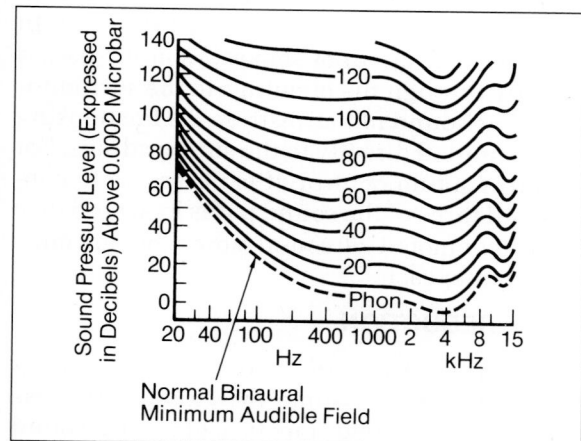

Fig. 39 **Normal equal loudness contours for pure tones**

study properly, one should operate in a free field and, in a standardized way, locate the observation points around the source to account for sound directivity. As measurements usually cannot be made under these conditions, it is extremely important, from the point of view of noise guarantees on equipment, to stipulate not only the noise spectrum, but the exact location at which it applies. Specifying distance from casing or center of the noise source is generally inadequate, as the sound spectrum invariably changes around the source.

In specifying noise ratings, it must be realized that at the present state of the art there are no set regulations, only suggested ones, and

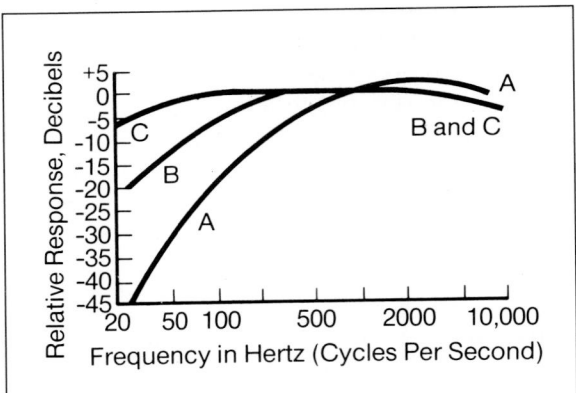

Fig. 40 The standard weighting curve for sound level meters

that each of the several possible ratings has limitations. No sound measurement is that accurate. Plus or minus 2 dB or less in sound power level is insignificant, particularly in the low frequency bands where differences of up to 4 dB should be disregarded.[9]

There seems to be a trend toward the use of sound power level ratings. These indicate the actual acoustic power radiated from a source. From these data, the audible sound pressure levels can be predicted with an accuracy dependent on how well the physics of the surroundings is understood, and one can predict the need for extra abatement measures such as silencers, barriers or plenum chambers.

References 10, 11, 12, and 13 are publications applicable to fan sound ratings, methods of calculating fan sound ratings from laboratory data, and methods of testing sound power from air-moving devices.

NOISE SOURCES

To determine the nature of an offending noise and to make silencing worthwhile, it is necessary to examine the spectral distribution of sound levels at various frequencies. In a power station, nearly every frequency within the audible area is represented. The dominating frequencies differ at different locations within the plant. Where low frequency noise dominates, the offender is likely to be the furnace, fans, reciprocating compressors, or pul-

verizers. High-frequency noises dominate at positions where gas or steam is emitted at high velocity through valves and vents.

There are two principal sources of sound within a fan. The first source is the blade passing tone which is generated by the impeller blades. The frequency of the blade passing tone is equal to the number of times per second a blade passes a stationary object.

$$F_{BP} = \frac{(rpm)\ (number\ of\ blades)}{60}$$

(8)

The loudness of the blade passing tone depends on the width of the fan impeller and the distance between a blade and the cutoff or the stator vanes. The second major source is the turbulence created by the fan while adding energy to the airstream. This sound is generally broad band in character.

There are three primary paths by which sound radiates from a fan; namely, through the inlet, the outlet, and the fan housing. Sound radiated from the fan inlet and outlet is approximately at the same sound power level. Sound radiated by the housing is normally at a lower sound power level than that radiated from either the fan inlet or outlet. If the fan inlet(s) and/or outlet are ducted, some of the sound is radiated into the system while some of the sound is transmitted through the duct wall to the surroundings. In the case of FD fans with open inlets or inlet boxes, sound passes directly through the openings to the surroundings. The sound radiating from the fan housing is transmitted directly to the surroundings. For effective sound control, it is necessary to consider all three of the sound paths in accordance with the amount of sound power being transmitted from each.

NOISE REDUCTION

Basically, there are two ways of abating noise. One is to assume the existing process cannot be modified and to isolate the sound from nearby inhabitants by enclosures, barriers, etc. The other is to examine thoroughly

the nature of the noise and to modify the process or replace the offending component with a less noisy one. In practice, both methods may be required since 10 to 20 dBA reductions are difficult to obtain except early in the design.

In most cases, separation of source and receiver is also only possible during the early design stage. Doubling the distance halves the sound pressure, *i.e.*, reduces the sound pressure level by 6 dB. For this reason, it may be attractive to include a "buffer zone" rather than silencing equipment.

In general, most problem noises are at a fairly high frequency. Therefore, the best results are obtained with double structures (inner steel and outer lead linings with porous glass fibre material between). This method prevents the build-up of reverberant sound levels. However, the structure must be leaktight. With 30 dBA reduction, for example, only one thousandth of the noise energy that hits the inside of an enclosure escapes to the outside. A similar quantity would be emitted through a hole 1/1000 of the wall area. Therefore, small leaks between panels and around openings for supply pipes must be avoided. Also, the wall must be vibrationally isolated from the mechanical sound-source, or, because of its larger area, the noise emission produced by the vibration may be worse than the initial noise.

An example of the noise attenuation that can be achieved by enclosing two 500,000-CFM FD fans for a 500-MW boiler in 30 ft × 45 ft × 35 ft enclosures of 8-in. block masonry is shown in

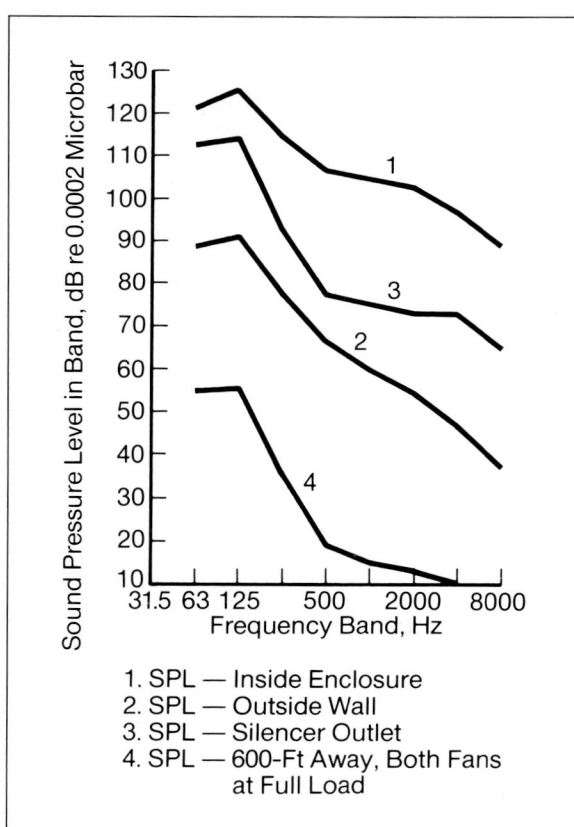

1. SPL — Inside Enclosure
2. SPL — Outside Wall
3. SPL — Silencer Outlet
4. SPL — 600-Ft Away, Both Fans
at Full Load

Fig. 41 Calculated abatement of sound pressure levels for two FD fans for a 500-MW boiler. Each fan is housed in a 30-ft wide × 45-ft long × 35-ft high enclosure of 8-in. masonry.

Fig. 42 Large centrifugal forced-draft fan with silencers mounted on inlet boxes

Fig. 41. At 600 ft (free field assumed), the noise is reduced to a mere whisper.

Fig. 42 shows inlet silencers attached directly to the inlet boxes of an airfoil-type centrifugal FD fan. Such silencers are often used in lieu of acoustical enclosures.

REFERENCES

[1] "Laboratory Methods for Testing Fans for Rating," *ASHRAE Standard 51-75/AMCA Standard 210-74.* New York: American Society of Heating, Refrigerating and Air Conditioning Engineers; Arlington Heights, IL: Air Movement and Control Associations.

[2] *ASME Performance Test Codes:* ANSI/ASME PTC 11-1984, "Fans" (directions for full-scale site testing of fans), New York: American Society of Mechanical Engineers, 1984.

[3] Robert Jorgensen, ed., *Fan Engineering,* Eighth Edition, Buffalo, NY: Buffalo Forge Co., 1983.

[4] Air Movement and Control Association, *Fans and Systems,* AMCA Publication 201. Arlington Heights, IL: Air Movement and Control Association.

[5] R. Kotwal and W. Tabakoff, "A New Approach for Erosion Prediction Due to Fly Ash," *ASME Paper No. 80-GT-96.* New York: American Society of Mechanical Engineers, 1980.

[6] E. F. Rothemich and G. Parmakian, "Tubular Air-Heater Problems," *ASME Paper No. 52-A-124.* New York: American Society of Mechanical Engineers, 1952. Also in: *Transactions of the ASME,* 75:723-728, July 1953.

[7] *Heat Pipe Design Handbook,* prepared for NASA Goddard Space Flight Center by B&K Engineering, under Contract NAS5-23406, U.S. Dept. of Commerce. Springfield, VA: National Technical Information Service, 1979.

[8] Manuel Cadrecha, "Preventing Acid Corrosion in Air Heaters," *Power Engineering,* January, 1980.

[9] Air Movement and Control Association, *Power Plant Fans: Specification Guidelines,* AMCA Publication 801. Arlington Heights, IL: Air Movement and Control Association, 1977.

[10] "Test Code for Sound Rating Air Moving Devices" *AMCA Standard 300.* Arlington Heights, IL: Air Movement and Control Association, 1967.

[11] "Methods of Calculating Fan Sound Ratings From Laboratory Test Data," *AMCA Standard 301.* Arlington Heights, IL: Air Movement and Control Association, 1976.

[12] Air Movement and Control Association, *Application of Sound Power Level Ratings for Ducted Air Moving Devices,* AMCA Publication 303 (Application Guide). Arlington Heights, IL: Air Movement and Control Association, 1979.

[13] "Methods of Testing In-Duct Sound Power Measurement Procedure for Fans," *ASHRAE Standard 68-78.* New York: American Society of Heating, Refrigerating and Air Conditioning Engineers, 1978.

CHAPTER 15

Control of Power-Plant Stack Emissions

This chapter presents the various types of commercially available equipment for the control of particulate and gaseous effluents. No reference is made to national or local legislation or regulations because of the probability of change from year to year. Such regulations are best found in the U.S. Federal Register or obtained from the local environment-regulating agencies in other countries.

ENVIRONMENTAL FACTORS IN POWER PRODUCTION

The installation of highly efficient electrostatic precipitators, fabric filters (baghouses), and flue-gas desulfurization systems (FGDS) has become increasingly necessary; power-plant owners must include them in their planning to obtain permission to start construction of a new facility. Additional controls that are applied to other gaseous emissions such as oxides of nitrogen, and to liquid and solid wastes that result from the processing, handling, and disposal of fossil fuels and their products of combustion add considerably to the complexity and cost of installation, operation, and maintenance of power plants.

Changes in fuel availability as well as air-quality standards have significantly impacted the electric-utility industry. Discounting inflation, the cost of installation of a major facility has more than doubled since the imposition of environmental laws. Because of permits needed to satisfy environmental statutes, and the delay occasioned by debating these in public, the time required to construct a fossil plant has increased by as much as 2 to 3 years.

Besides the emission-control equipment described in this chapter, a coal-fired power plant must have

- very high static-pressure induced-draft fans or additional FGD booster fans to accommodate draft requirements

- FGD solid-waste beneficiation, handling, and disposal systems

- more sophisticated treatment of raw water, makeup water, and waste water

- condenser cooling-water systems that do not add heat to inland or coastal waters.

The auxiliary power requirements of equipment to satisfy environmental regulations will demand a significant percentage of the total installed capacity of a power plant.

POWER-PLANT EMISSIONS

Three classes of emissions from fuel-burning processes are judged significant from an air-quality standpoint: particulate matter, sulfur oxides, and nitrogen oxides. Historically, particulate matter has received the greatest attention because it is easily seen and often labeled a public nuisance. The concern about sulfur oxide comes from its possible health effects and from its potential to damage vegetation and property. Oxides of nitrogen are also significant because they participate in complex chemical reactions that lead to formation of photo-chemical smog in the atmosphere. Additionally, both sulfur and nitrogen oxides have been implicated as precursors to acidic deposition, more commonly called acid rain.

PARTICULATE EMISSIONS

Emissions from coal-fired boilers vary considerably depending on the ash content of the coal and the type of firing. A pulverized-coal-fired unit can be expected to have 60 to 80 percent of the coal ash leaving the furnace with the flue gas. The balance of the ash leaves through the boiler bottom-ash removal system, hoppers under economizers and air heaters, and pulverizer rejects hoppers. Fig. 1 illustrates the exceedingly high efficiency required of particulate-collection equipment for high-ash-con-

tent coals and very low effluent ash levels.

As flyash is nonhomogeneous in properties such as specific gravity and particle shape, size is very difficult to describe in absolute terms. Methods for determining this property include photomicrographs, sedimentation, elutriation and inertial impaction devices. Flyash is the combination of inert or inorganic residue in pulverized-coal particles with varying amounts of carbon or coke particles resulting from incomplete combustion. In general, the inorganic ash particles consist primarily of silicates, oxides, and sulfates, together with small quantities of phosphates and other trace compounds. Particle size varies from below 0.01 micrometer diameter to over 100 micrometers. Fig. 2, based on test data,[1] shows the size distribution between the 2 and 50 μm diameters.

SULFUR-OXIDE EMISSIONS

As fuels burn, most of the sulfur is converted to sulfur dioxide (SO_2) and sulfur trioxide (SO_3). As explained in Chapter 2, the quantity of sulfur varies widely for different coals and may range from less than 0.5 to over 5 percent. In general, 90 percent or more of the sulfur in

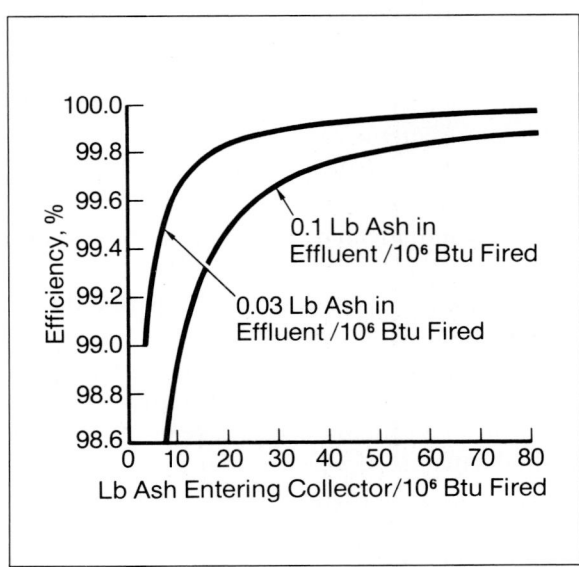

Fig. 1 Required particulate-collection efficiency versus inlet ash concentration

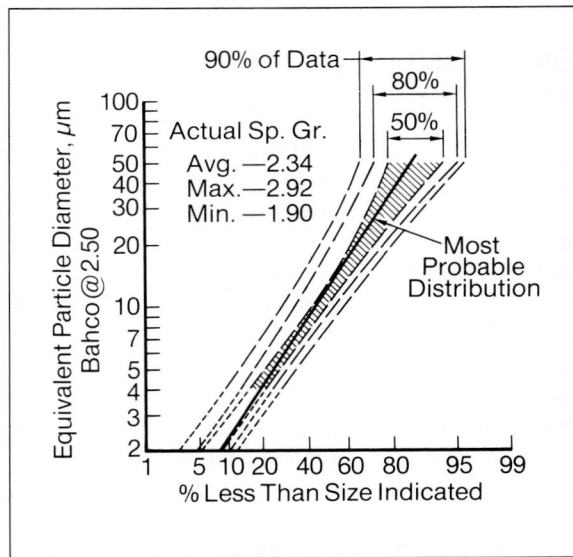

Fig. 2 Flyash particle-size distribution in pulverized-coal-fired boilers

the fuel will oxidize to gaseous sulfur oxides. Typically, concentrations will be 0.05 to 0.30 percent by volume in the products of combustion. The further conversion of SO_2 to SO_3 ranges from 1 to 4 percent. The formation of SO_3 in a boiler is a complex process and is believed to be influenced by the oxidation of SO_2 by molecular oxygen, oxidation of SO_2 in the flame by atomic oxygen, and catalytic oxidation of SO_2.

NITROGEN-OXIDE EMISSIONS

During the combustion process in a conventional steam generator, oxides of nitrogen (collectively referred to as NO_x) form in the high-temperature region in and around the flame zone. The oxidation of both atmospheric nitrogen (thermal NO_x) and nitrogen contained in the fuel (fuel NO_x) is the cause. The rate of formation is influenced by the temperature and the oxygen present. Reducing the flame temperature and the excess air can help control thermal-NO_x formation. Fuel NO_x is related to the available nitrogen in the fuel and is most significantly influenced by the oxygen concentration in the combustion region.

In addition to the information on post-furnace NO_x control in this chapter, refer to Chapter 4 for a discussion of NO_x formation during combustion, and to Chapter 12 for techniques to reduce NO_x in furnaces.

INTEGRATED EMISSIONS CONTROL

For decades, utilities and large industrial power producers in the United States have been routinely purchasing electrostatic precipitators (ESP's) or fabric filters (FF's) to control particulate emissions from their boilers. Since the enactment of the 1970 Clean Air Act requiring control of sulfur dioxide, utilities have also been installing flue-gas desulfurization (FGD) systems. More than 100,000 MW's of FDG systems have been purchased in the United States. Wet FGD systems have accounted for about 80 percent of those purchased; dry-scrubbing FGD systems over 10 percent; and all other types of FGD systems, under 10 percent. Nitrogen-oxide emissions have been almost exclusively controlled in the combustion process, although selective reduction of nitrogen oxides will be applied more widely in the future, to comply with the 1990 CAA.

Fig. 3 illustrates the most common and typical integration of the particulate, nitrogen-ox-

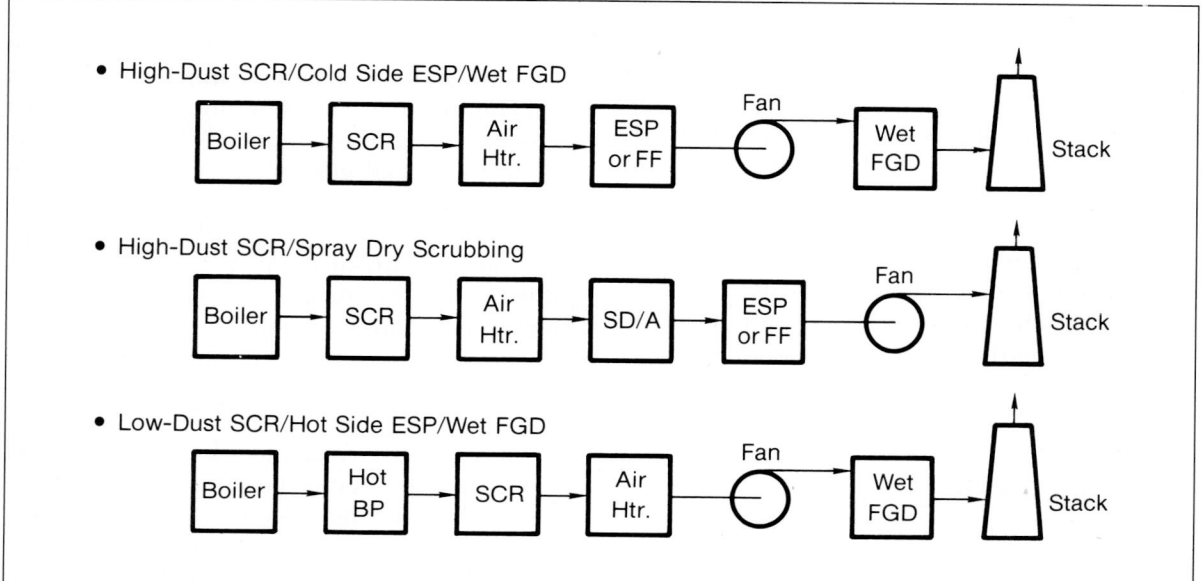

Fig. 3 **Typical power-plant flue-gas cleaning systems**

ide, and sulfur-dioxide control systems. SCR equipment, if used, can be designed for either high or low dust contents, thus dictating the position of the electrostatic precipitator. See reference 2 for a generic description of the various gas-cleaning systems, and other sections of this chapter for detailed discussions of the wet and spray-drying desulfurization processes.

PARTICULATE EMISSION CONTROL

Particulate emission control deals with methods of removing particles dispersed in effluent gases of power, industrial process, and commercial plants. These dispersions have through general use come to include all particles in air or other gases.

Dispersoids are characterized by their physical, chemical, and electrical properties, their particle size and structure, rate of settling under gravity, optical activity, the ability to absorb an electrical charge, the surface-to-volume ratio, and the chemical reaction rate. Particles larger than 100 μm are excluded from consideration because they settle rapidly.

Dispersoids are generally classified as dust, smoke, fumes, and mists. Fig 4A indicates the common dispersoids and the typical particle size for each classification. Fig. 4B shows the devices for collecting the various classifications of dispersoids. Note that methods relying on gravity or centrifugal forces are effective for removal of particles larger than 10 micrometers radius. Efficient collection of 1 μm and sub-micrometer particulate requires devices that depend on electrical force, impaction, interception, or inertial diffusion; these are the capture mechanisms at work in electrostatic precipitators, fabric filters, and high-energy wet scrubbing.

INFLUENCE OF COAL AND COMBUSTION CHARACTERISTICS

The varying coal characteristics described in Chapter 2 have a dominant effect on the re-moval efficiency of particulates. The variation in ash content, ash composition, ash resistivity, and particle size distribution requires an engineering evaluation of collection principles to reduce emissions to required levels.

INFLUENCE OF VOLUMETRIC GAS FLOW

No matter what type of flyash collector is to be installed on a steam generator, it is essential that it be designed for the correct gas volume.

Manufacturers design and size their equipment based upon their experience with this equipment. This experience includes tests on both laboratory or pilot-plant equipment, and full-scale equipment. The joint American Boiler Manufacturers Association (ABMA) and Industrial Gas Cleaning Institute (IGCI) survey has shown there to be a difference in the volume of flue gas that is measured by pitot tube and the volume of gas that is calculated from a stoichiometric balance.[1]

The pitot tube consistently indicates a higher volume. The manufacturer of the dust-collection or desulfurization equipment, therefore, must ascertain if the gas volume for equipment design has been stoichiometrically calculated and be aware that the value may differ from a volume measured by pitot tube in the operating installation. It is vital that proper corrections be made for plant elevation and local negative pressure ahead of the induced-draft fan, to arrive at the actual gas volume. (See Chapter 6 and reference 3.)

Performance testing required to demonstrate compliance with air-quality regulations uses pitot-tube techniques as proscribed by EPA Reference Methods 1, 5 and 16.[5] See Chapter 22 for more on gas-volume measurement.

MEASUREMENT TECHNIQUES

It is difficult to analyze size distribution of fine particulate matter because flyash is non-homogeneous. The varying specific gravity and particle shape make it difficult to describe size in absolute terms. Many methods, including photomicrographs and various sedimenta-

tion and elutriation techniques, have been used to determine size distribution. The ASME Performance Test Code Committee has selected terminal settling velocity to characterize flyash particulate emissions from furnaces.

TERMINAL VELOCITY

Particle terminal velocity is a significant parameter in the design of mechanical separators that use inertial or centrifugal forces to separate dust from a gas stream. It includes the effect of particle shape and specific gravity for each particle. Its determination is important for heterogeneous dusts of varying shape and specific

gravity. Determination of terminal velocity for a centrifugal classifier is included in *ASME Performance Test Code 28*.

Flyash characterization also includes methods for bulk electrical resistivity, in-situ resistivity, and particle size.

BULK ELECTRICAL RESISTIVITY

Particulate resistivity can be determined in a high-voltage conductivity cell in which a sample of known thickness is placed between two oppositely charged electrodes. Guard rings eliminate fringe effects. The potential between the two electrodes is increased and the voltage

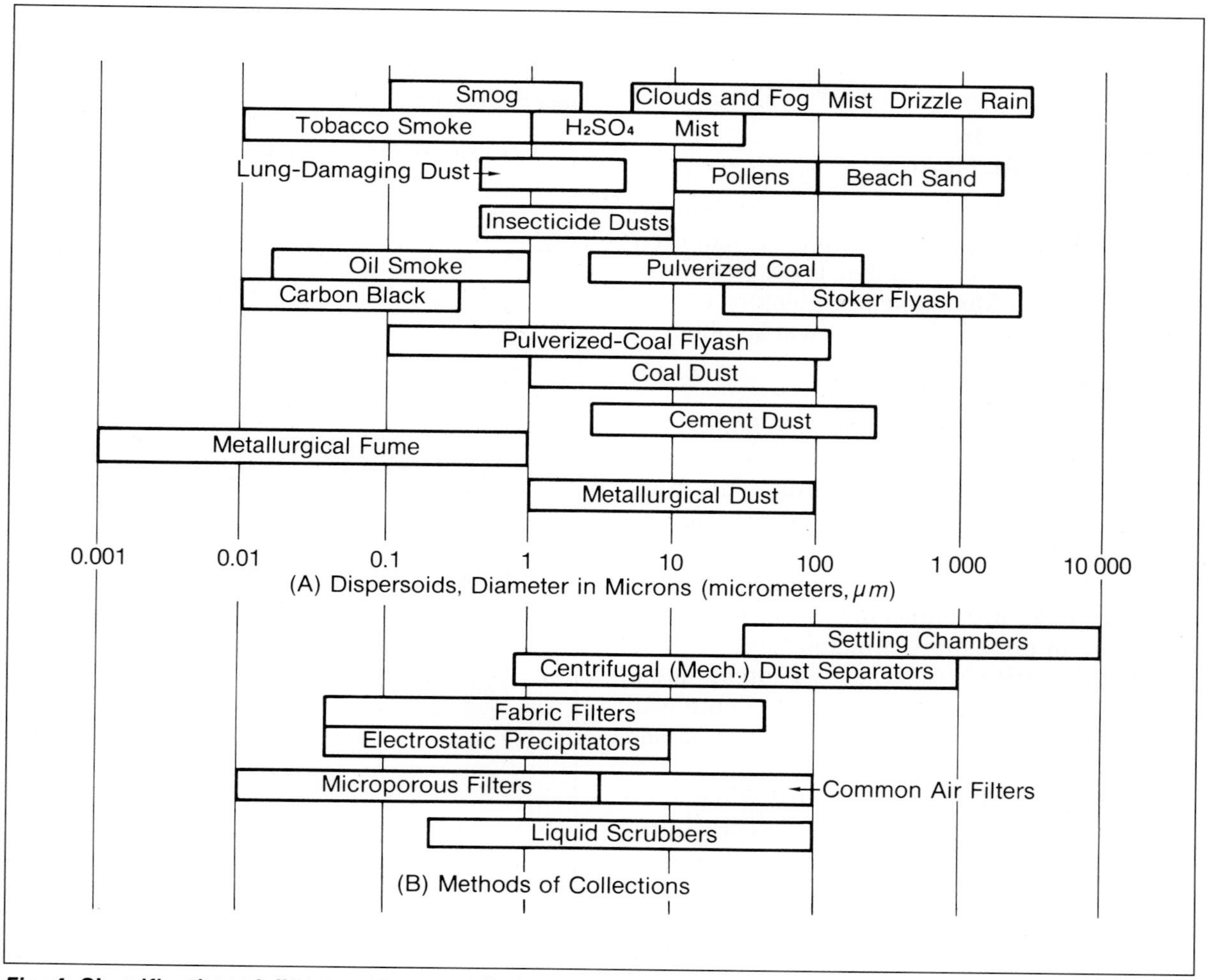

Fig. 4 Classification of dispersoids and methods of collection

and current noted until an electrical breakdown occurs. The resistivity is then calculated and reported in ohm-cm at 85 to 95 percent of the average breakover.

IN-SITU RESISTIVITY

A combination high-voltage and ground-potential probe, inserted in the gas stream, has a negative high-voltage probe consisting of a flat disc concentric about a needle-point electrode. The needle-point electrode precipitates the particulates electrically from the dispersoid. The flat charged disc is separated from the needle point so as not to affect the precipitation of particulate onto a grounded electrode. After a short time, the charged disc is lowered to entrap the precipitated particulate between it and the grounded electrode. Calculation is the same as for bulk electrical resistivity.[6]

DUST COLLECTION
BY MECHANICAL MEANS

Although cyclone-type mechanical collectors have a long history in boiler service, their use is now limited to stoker-fired and fluidized-bed units. Mechanical collectors can be either dry cyclone collectors or wet scrubbers.

MECHANICAL CYCLONE COLLECTORS

These devices achieve particulate removal by centrifugal, inertial, and gravitational forces developed in a vortex separator. The dust-laden gas is admitted either tangentially or axially over whirl vanes (Fig. 5) to create a high velocity in the cylindrical portion of the device. Particles are subjected to a centrifugal force and an oppositely directed viscous drag. The balance between these two forces determines whether a particle will move to the wall or be carried into the vortex sink and be passed on to the clean-gas outlet tube.

The high-velocity, downwardly directed vortex is reversed at the bottom of the cylindrical section. In this reversal, inertia and gravity, as well as centrifugal force, inject the dust into the hopper. The action of the reversed-gas circulation takes place over a small diameter but at a relatively great velocity.

This inner vortex has an axially upward component which carries the spirally flowing gas to the outlet pipe at the upper center of the separating unit. Because these collectors depend primarily on differential inertia, collection efficiencies vary with particle size, particle density, gas temperature, and pressure drop through the apparatus. Efficiencies are very high on material greater than 20 micrometers (20μm) in size, but drop off rapidly for smaller particles. The overall collection efficiency of a cyclone can be estimated when the particle-size distribution and specific gravity of the dust and the allowable pressure drop are known. Fig. 6 relates collection efficiency to particle size. In normal boiler operation, differential-pressure requirements for mechanical collectors are from 2 to 5″ WG.

WET SCRUBBERS

In the analysis of particulate scrubbing, a number of important facts are known. First, the

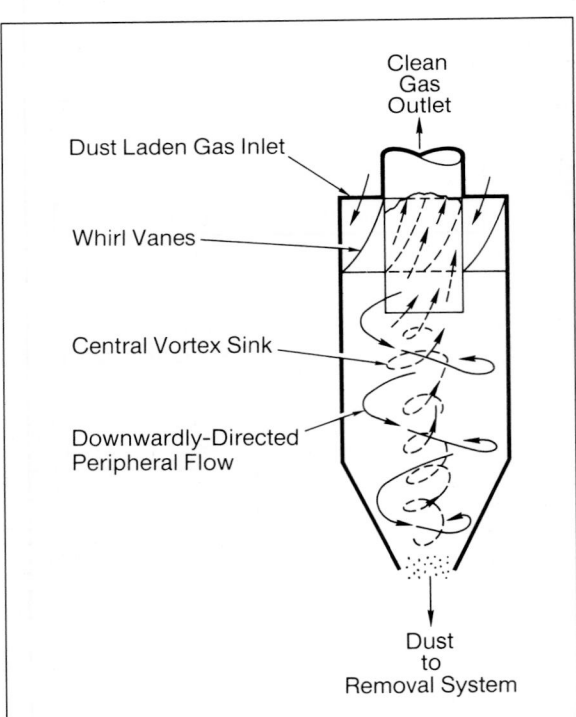

Fig. 5 **Principle of cyclone-separator operation**

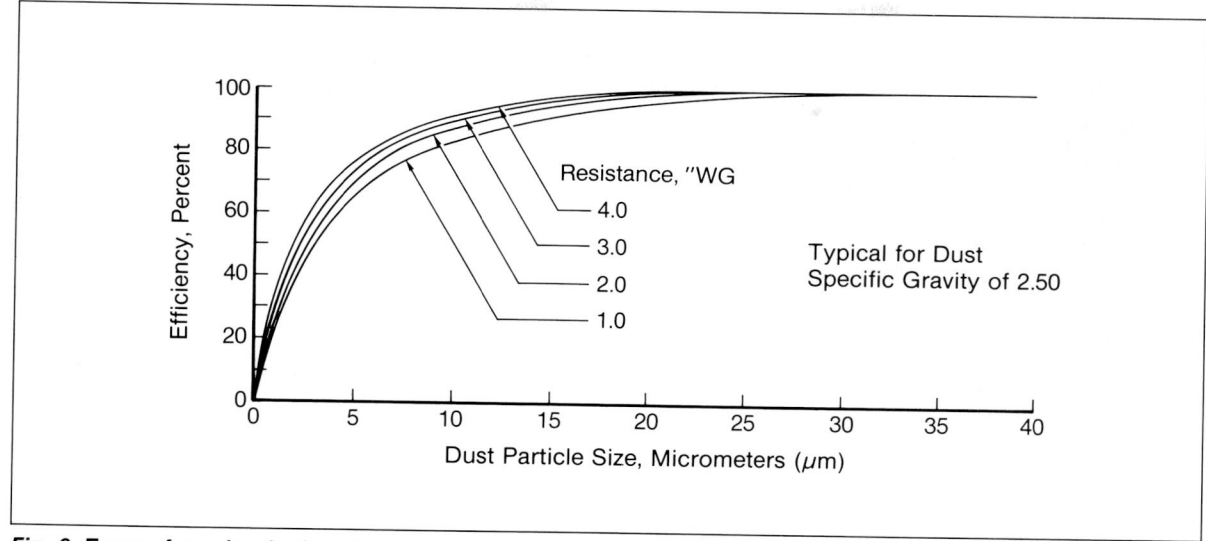

Fig. 6 **Form of mechanical cyclone dust-collector efficiency curve versus particle size**

dust particle must impact on the water droplet. The impaction efficiency is a function of the dimensionless group $\frac{(V_r V_s)}{D g}$ where V_r is the relative velocity between the water droplet and the dust particle, V_s is the settling velocity for the dust particle, D is the diameter of the water droplet in micrometers, and g is the acceleration of gravity. As g and V_s are constant for a dust particle of given size, target efficiency is a direct function of the relative velocity and an inverse function of the droplet diameter. If the collection efficiency depends critically on relative velocity and droplet size, then collection efficiency must also be a function of the power supplied to the unit.

Capture of small particles requires high energy inputs, usually in the form of greater pressure drops across the scrubber. Low-pressure-drop scrubbers, such as spray towers, collect coarse dust in the range of 2 to 5 micrometers. High-pressure-drop venturi scrubbers are effective in removing 0.1 to 1.0 micrometer particles.

Typically, a scrubber operating at 6″ WG pressure drop should capture practically all of the particles greater than 5μm and about 90 percent of the particles in the 1 to 2μm range.

Fig. 7 shows a typical efficiency curve for a wet scrubber operating at 6″ WG pressure drop.

VENTURI SCRUBBERS

In a venturi scrubber, dust-laden gases are wetted continuously at the venturi throat. Flowing at 12,000 to 18,000 fpm, the gases produce a shearing force on the scrubbing liquor due to the initial high velocity differential between the two streams. This shearing force causes the liquor to atomize into very fine droplets.

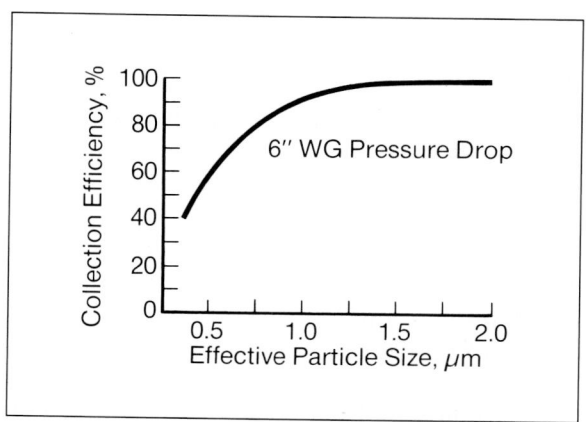

Fig. 7 **Wet-scrubber fractional efficiency curve**

Impaction takes place between the dust entrained in the gas stream and the liquid droplets. As the gas decelerates, collision continues and agglomerated dust-laden liquor droplets discharge through a diffuser into the lower chamber of a separator vessel. Impingement of the stream into the liquid reservoir removes most of the particulate.

A venturi-type scrubber operating in the pressure range of 30 to 40" WG can collect nearly 100 percent of 0.2 to 1.0 μm particles.

ELECTROSTATIC PRECIPITATION OF DUST

In electrostatic precipitation, suspended particles in the gas are electrically charged, then driven to collecting electrodes by an electrical field; the electrodes are rapped to cause the particles to drop into collecting hoppers. This process differs from mechanical or filtering processes in which forces are exerted directly on the particulates rather than the gas as a whole. Effective separation of particles can be achieved with lower power expenditure, with negligible draft loss, and with little or no effect on the composition of the gas.

In the United States, the control of emissions began with preventing or minimizing nuisance smoke. Such control required equipment capable of only 70- to 90-percent efficiency. Later, the need to protect induced-draft fans from erosion by reducing entrained flyash emitted from utility boilers led to precipitators being built for power plants. The first such application was 1923.[7]

The early users of electrostatic precipitators, then installed them to

- recover a valuable product, such as lead, copper, or saltcake
- eliminate a nuisance, either visual or damaging to crops
- protect process equipment, such as induced-draft fans.

As initially applied, precipitators were designed to provide a minimum plate area at low cost; designs used interlocking or opzel collecting plates and hanging weighted-wire discharge electrodes. Roof-mounted gang-rapping vibrators removed particulate from the collecting plates.

To meet a demand for ultra-high efficiency collectors of rugged construction and high reliability, European manufacturers in possession of the basic patents disseminated by Frederick Gardner Cottrell developed the rigid-frame precipitator. Actually, this design more closely approximated Cottrell's original design than did the U.S.-style weighted-wire designs. The term *rigid-frame* refers to the rugged pipe-frame or mast-construction discharge electrode, which largely precludes wire breakage. The basic design incorporates segmented collecting plate configurations and profiles, for close fabrication tolerances over heights up to 52 feet; rigid discharge electrodes; and much greater division of rapping, often with individual rapping for each discharge frame.

ELECTROSTATIC COLLECTION TECHNIQUE

Fig. 8 shows in simple fashion the principle of electrostatic precipitation. The process applies an electrostatic charge to dust particles with a corona discharge and passes them through an electrical field where the particles are attracted to a collecting surface. The basic elements of a precipitator include a source of unidirectional voltage, corona or discharge electrodes, collecting electrodes, and a means of removing the collected matter.

Single-state (Cottrell type) precipitators combine the ionizing and collecting step (Fig. 9A). In the more common plate type, the electrodes are suspended between plates on insulators connected to a high-voltage source. A voltage differential created between the discharge and collecting electrodes develops a strong electrical field between them. The flue gas passes through the field and a unipolar discharge of gas ions, from the discharge electrode, attaches to the particulate matter.

The unipolar discharge of gas ions (normally negative charge) occurs at a critical voltage at which gas molecules are ionized. The ionization is visible as a corona at the discharge electrode. The negative ions move towards the

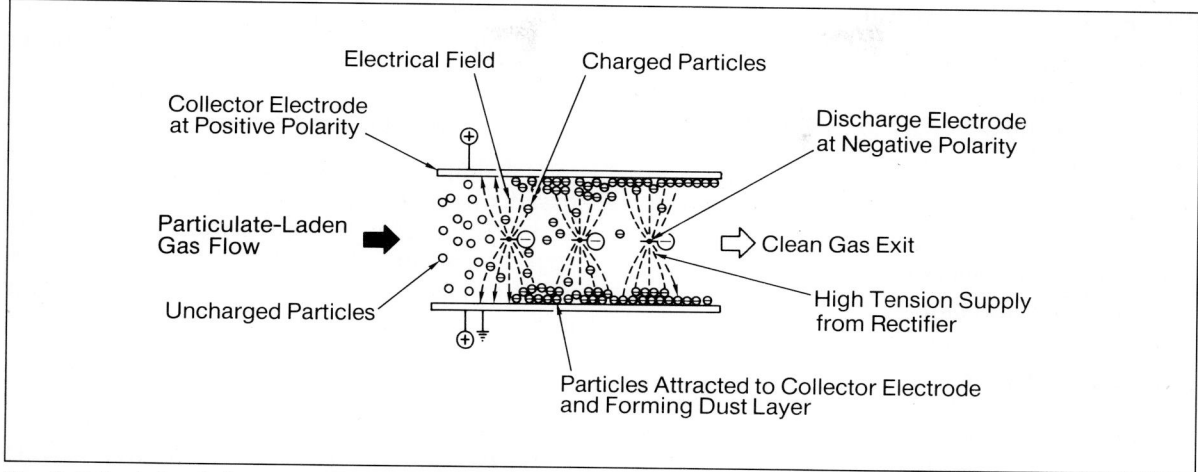

Fig. 8 **Principle of electrostatic precipitator operation**

positive collecting electrode; the positive ions migrate toward the discharge electrode. In this movement, the ions become attached to particles carried by the gas stream. The charge is the force that moves flyash to the appropriate electrode. The particles attached to the positive electrode dissipate their charge and become electrically neutral. The particulates on the collecting electrode are removed mechanically by rapping or washing.

In the two-stage precipitator, the ionizing and collecting stages are separated (Fig. 9B). This design is used for low particulate loadings and minimum ozone generation.

Single-stage precipitators can be either weighted-wire or rigid-frame units. Weighted-wire units work well on high-sulfur coal and easily removed particulates and have economic designs. The rigid-frame precipitator is better suited to applications where high-resistivity ash is to be collected and large collecting areas are needed.

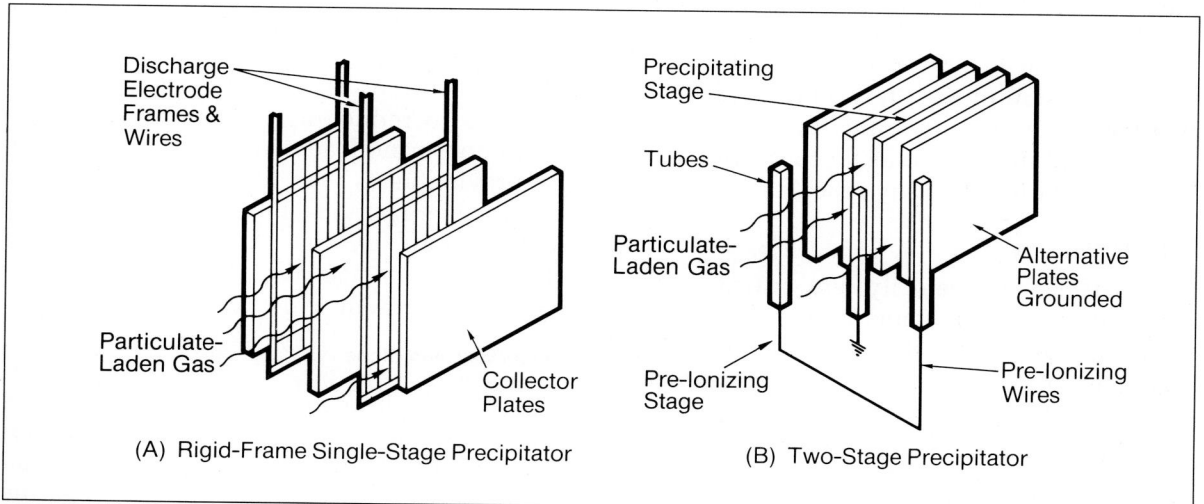

Fig. 9 **Single-stage and two-stage precipitators**

PRECIPITATOR DESIGN FACTORS

Precipitator designs consider
- specific collection area
- treatment time
- gas velocity
- electrode configuration and spacing
- number of fields
- automatic voltage control
- hopper size and slope

Specific Collection Area (SCA)

The SCA relates the size of one precipitator to another in terms of the effective collecting-electrode surface area in square feet per thousand cfm (cubic ft/min) of gas treated. The area considered is the flat projected surface of both sides of the collecting electrodes; the gas volume is the actual volume at the design operating temperature and power-plant elevation. Depending upon the physical, chemical, and electrical characteristics of the coal ash, precipitator SCA's range from 300 to 1,000 for collection efficiencies above 99.5 percent.[8]

Treatment Time

This refers to the length of time a particle will spend in the presence of the electrical field, at design velocity, if it were allowed to traverse the entire length of the precipitator. The treatment length is not the total front-to-rear dimension of the "box," but is rather the horizontal dimension from the plane of the front face of the first collector to the rear plane of the last surface less any space for walkways.

Gas Velocity

Superficial gas velocity is a critical design parameter. It is calculated by dividing the actual gas volume (at design temperature and plant elevation) by the face area of the precipitator immediately in front of the first collecting electrode. Effective plate height is multiplied by the inside face-to-face dimension between the side-wall casings to give the face area.

Designs for low-sulfur, low-sodium coals use velocities between 3 and 4 ft/sec. For high-sulfur, high-sodium coals, velocities on the order of 5 fps are used.

Electrode Configuration and Spacing

The shape and spacing of the electrodes can drastically affect the amount of peak voltage and current achieved in the precipitator before sparking occurs. The selection of transformer/rectifier (T/R) sets with adequate kVA ratings and sensitive controls can affect this voltage and current.[9]

Number of Fields

For reliability, precipitators are divided into fields, each of which is composed of one or more independent electrical sections in the direction of gas flow. The amount of plate area per transformer/rectifier set can affect the peak current and voltage achieved in an electrical section. Systems with a large number of fields (5, 6, or more) are not seriously impaired by the loss of a single field.

Automatic Voltage Control (AVC)

The AVC system keeps the precipitator operating at its optimum voltage. Such control systems can be either analog or digital. Digital control systems are able to maintain higher precipitation voltage potential between the discharge electrode and the collecting plate than analog systems.

Microprocessor-based AVC systems can be designed to reduce overall ESP power consumption during load changes. Such a system can provide users with features such as intermittent energization—blocking selected AC-voltage half cycles to decrease power usage—and automatically modulated ramp rate to maintain present spark rates.[10]

Hopper Size and Slope

The proper design of the collecting hoppers forming the floor of a precipitator is important because a breakdown in hopper outflow that results in overfilling can severely damage the electrical internals. C-E has made comprehen-

sive recommendations concerning the design and heating of precipitator and fabric-filter hoppers, and their flyash-removal systems.[11] These recommendations, some of which are detailed later in this chapter, help the industry avoid costly outages attributable to these portions of the collection/transport complex.

PRECIPITATOR DESIGN ANALYSIS

It is traditional to consider the semi-empirical Deutsch-Anderson equation in the analytical design of a precipitator for efficiencies up to 98 percent. Above that level, performance can be predicted accurately by a modified form of the Deutsch-Anderson formula.

Electrostatic-precipitator collection efficiency is related to the total surface of the collection electrodes per unit volume of gas (the SCA) and is directly proportional to the particle migration velocity. Efficiency, in the Deutsch-Anderson equation, is determined by

$$\eta = \left(1 - e^{\frac{-Aw}{V}}\right) \times 100$$

(1)

where

η = efficiency of collection, percent
e = base of natural logarithms (2.718 . . .)
A = collecting-electrode surface area, ft^2
w = migration velocity, fps, and
V = actual volumetric gas-flow rate, in 1000 ft^3/sec (at temperature and elevation)

Although important to an understanding of the precipitation process, the Deutsch-Anderson equation does not satisfactorily predict performance for a known ash at higher or lower efficiencies than an actual test point. That is because it relies entirely on Stokes law and is properly applied only to average-size particles that are effectively the same size and experience the same force. The lack of correlation of analytical and field-performance data has been found to be specific to the variation in migra-

tion velocity and has led to the development of a modified equation:

$$\eta = \left[1 - e^{\left(\frac{-Aw_k}{V}\right)^k}\right] \times 100$$

(2)

where w_k is now the empirical migration velocity at an observed lower efficiency, and k is a constant, approximately 0.5 but varying between 0.4 and 0.6 depending on the specific ash and application. The other terms are as defined for equation 1.

Migration or Drift Velocity

The drift velocity w, also called the "precipitation rate," is determined by the magnitude of the particle charge, the strength of the electrical field, and by Stokes law for the drag of the particles.[7,12] There is a significant discrepancy between theoretical and practical values of w, with the theoretical being about twice the actual, or more.[13,14] This loss of performance is caused by such factors as uneven gas flow, particle diffusion, electric wind, particle charging time, and loss of particles from collecting surfaces by reentrainment.

Because of these uncertainties, and the effects of performance of concentration variation in SO_3 and sodium in the flue gas and the flyash, it is necessary to rely upon field experience with a variety of precipitator installations.[7,8,13,15,16,17] Where experience is limited, designs are established by means of experimentation and pilot-plant testing.[18, 19, 20, 21]

EFFECT OF ELECTRICAL PROPERTIES

The electrical conductivity of flyash and the dielectric strength of the bulk ash are two properties important to the electrostatic-collection process. For effective operation, a small but definite electric current, in the form of charges carried by gas ions and particles, flows between the high-tension discharge electrodes and the collecting electrodes. This current must pass through the layers of collected ash which normally coat the plates. The ash, therefore, must be able to conduct the ionic current to the grounded metal surfaces of the plates. The electrical conductivity required is very small: about 10^{10} mho. In practice, it is usually more convenient to express this in terms of *resistivity*, which is the inverse of conductivity.

The *dielectric strength* of the collected ash on the plate electrodes is also an important electrical property of the ash. The build-up of charge on the ash layer produces a voltage or potential drop across the layers in accordance with Ohm's law. When the resistivity of the ash is in the critical zone (near 2×10^{10} ohm-cm), the voltage drop across the ash layer can amount to several kV. The existence of this voltage drop across the layer may be sufficient to cause it to break down electrically.

When this happens, a spark flashes through the dust layer and is propagated to the discharge electrode. This is the source of the intensified sparking which occurs with dusts of excessive resistivity. Clearly, the higher the dielectric or breakdown strength of the dust layer, the higher the resistivity which may be tolerated for a given amount of corona current. Thus, dusts of relatively high breakdown strength are less sensitive to resistivity effects than dusts of low dielectric strength.

For flyash, the dielectric strength of the bulk ash in 1/4 to 1/2 in. thick layers ranges from as low as a few hundred V/cm up to as high as 20 kV/cm. The average value of dielectric strength is in the 5 to 10kV/cm range.

Ash Resistivity

Experience has shown that coal ashes with resistivities above 5×10^{11} ohm-cm are difficult to collect. Low resistivity (less than 10^9 ohm-cm) ash tends to suffer from excessive reentrainment, but such low resistivities are rarely encountered. The area between 10^9 and 5×10^{11} ohm-cm usually is normal for satisfactory and predictable precipitator design.

The factors that affect resistivity are the sulfur content of the coal, flue-gas temperature and moisture, and such ash constituents as sodium, potassium, calcium, carbon, alumina, silica, and iron oxide.

Ash resistivity is inversely proportional to the concentration of sulfur trioxide and water in the flue gas and the sodium, potassium, and carbon in the ash. It is directly proportional to the ash constituents of calcium, magnesium, alumina, and silica. Peak ash resistivities occur between 250° and 450°F, depending upon coal-ash and flue-gas characteristics. Above 450°F to 550°F, the ash resistivity is inversely proportional to the absolute temperature, while below 250°F to 300°F, the resistivity is directly proportional to the absolute temperature.

Low-resistivity ash is a problem because the ash easily loses its charge after being collected on the plates. The uncharged particles may then be reentrained in the flue gas. High-resistivity ash does not readily lose its charge when collected on the plates and the agglomerated ash can be very difficult to remove. When a deep enough layer of high-resistivity ash collects on the plate, back-corona develops on the ash surface and the precipitator can no longer collect efficiently.

Back-corona is extremely detrimental to precipitator performance. It occurs when a particle migrates to the collecting surface but fails to dissipate its charge, causing a high potential gradient in the dust layer on the surface of the plate. This layer disrupts the electrical field that induces migration, and repels particles of like charge attempting to migrate to the collecting surface.

Measurement of Dust Resistivity

Dust resistivity, to have any possible value to a precipitator design engineer, must be measured with the dust under the same gas and temperature conditions as those at which the precipitator will operate. Also, the packing density of the dust should be the same as that of the dust layer deposited on the precipitator collectors. It will be appreciated, then, that the measurement of dust resistivity is subject to considerable problems, making it very difficult in practice to relate absolute precipitation efficiency to dust resistivity.

Fig. 10 indicates the in-situ bulk resistivity of a variety of fuels as related to temperature. Although there is a significant difference between many of the ash samples tested at temperatures below 400°F, it appears that, above 600°F, most ash will have resistivities below the level of 5 x 10^{11} ohm-cm.

For this reason some precipitators have been located *between* the boiler economizer and the air heaters, in what is called the "hot" position, instead of downstream of the air heaters, the ordinary "cold" position. Hot precipitators are used to avoid back corona, to minimize the heavy rapping that is sometimes needed at lower temperatures, and to increase the effective migration velocity.[22,25]

The structural and mechanical design of hot precipitators is more critical than that of equipment to be operated at the 300°F temperature level. Designs should minimize thermal gradi-ents and allow for differential expansion if such temperature gradients cannot be avoided.

Gas Conditioning

Migration velocity declines as resistivity increases (Fig. 11), as is characteristic of many low-sulfur subbituminous coals. Adding SO_3 to flue gas can reduce ash resistivity.[26] For some western-U.S. low-sulfur coals, an increase of 30 ppm of SO_3 in the gas has reduced the flyash resistivity from between 10^{11} and 10^{12} to 5 x 10^9.

COHESIVE AND ADHESIVE PROPERTIES OF FLYASH

These properties of flyash are important for collection by electrostatic precipitators, where the ash forms in compacted layers on the electrode surfaces. Typically, these layers are less than 0.5 in. thick, but may in some cases build up to 1 in. or more. Although some ash falls

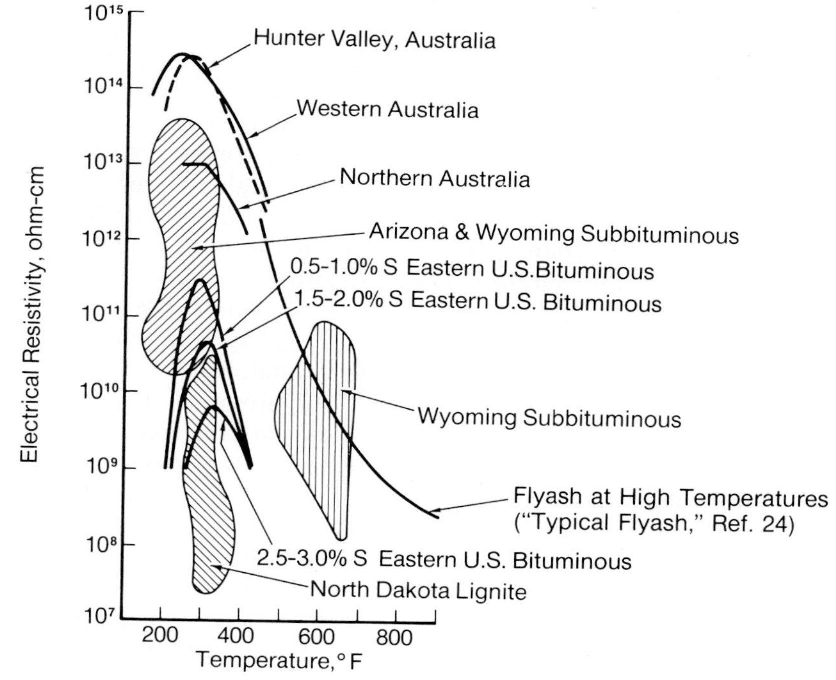

Fig. 10 Composite of in-situ flyash resistivity data versus temperature, from several sources (Refs 18, 22, 23, 24)

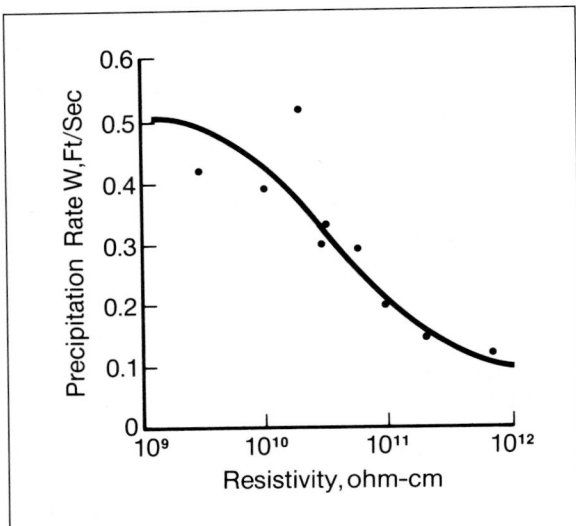

Fig. 11 Precipitation rate versus resistivity for representative group of flyash precipitators

into the hoppers because of either the mechanical vibrations frequently present in plant structures or its own weight, rapping or jarring of the electrodes is usually necessary to effectively remove the collected ash layers. The ash generally breaks away in isolated patches, so that the collecting-plate surfaces present a somewhat spotty appearance.[14]

It is well known that fine solid particles tend to cohere and to form agglomerated masses which have a small but definite stability. The coherence of the particles is caused by molecular attractive forces at the points of contact of the particles. The degree of coherence is appreciable only for relatively fine particles of less than about 20 to 40μm. Coarse or granular particles such as sand, for example, do not cohere in the dry state. On the other hand, fume particles, which typically are less than 1μm in size, usually show a high degree of coherence.

The adhesion of particles to the collecting electrodes is also the result of molecular attractive forces, and, similar to cohesion, the adhesive forces are larger for fine particles. Thus, the adhesion and retention of the collected layers of ash on the electrodes are determined basically by particle size, and increase with

decreasing particle size. But other factors such as particle shape and composition also influence the adhesion and cohesion of particles. The presence of condensed moisture, with possible cementing effects, can also profoundly influence the build-up of particles on electrodes and other surfaces.

In electrostatic precipitators the accumulation of ions on the collected dust layers produces an additional retention force of major importance. These ions tend to charge the collected particles so that they are attracted to the collecting plates by relatively strong electric forces. Further, any particles that do escape from the collected layers of ash tend to be immediately recharged electrically by the corona ions and forced back to the collecting plates. The particle-retaining force of the corona current increases with the electrical resistivity of the particles and with the strength of the current. Thus it is often advantageous to operate precipitators at relatively high-corona current densities—particularly the outlet sections.[14]

Agglomerated masses of particles, removed from the electrodes by rapping, fall into the hoppers by the force of gravity. These agglomerated masses are bound together by molecular cohesive forces. For best performance, these falling agglomerates must be relatively stable and fall with as little loss of particles as possible. If the falling agglomerates have little coherence, some may break up and be redispersed in the gas stream, thus partially nullifying the collecting action of the precipitator.

The presence of coarse particles tends to reduce the cohesive and adhesive forces and, therefore, the agglomerating effect which holds masses of fine particles together. Excessive amounts of coarse, gritty particles in an ash make the ash difficult to hold on the collecting plates and also lead to loose, unstable agglomerates when the plates are rapped. On the other hand, the elimination of all coarse particles above 10 to 20μm may in some cases make removal of the ash difficult. As there also will be a much greater tendency for the particles to build up on the discharge electrodes, a very

fine ash containing no coarse particles presents a more difficult rapping problem.[14]

CONSTRUCTION OF ELECTROSTATIC PRECIPITATORS

The weighted-wire type of precipitator has been well described in references 2, 7, 27, 28 and 29. Features of the rigid-frame design can be found in references 2, 17, 21, and 29.

In the rigid-frame design, the carbon-steel collecting plates are constructed of profiled segments approximately 18 to 19 in. wide and up to 52 ft long. They attach at the top to back-to-back channels spanning the insulator housing, usually on 12-foot centers. At the bottom, anvil or spacer bars maintain plate alignment and act as anvils for plate rapping. The collecting-electrode plates are rapped individually by mechanical hammers indexed on a shaft driven from outside the precipitator shell.

Made from steel tubing, the individually suspended and rapped discharge frames are typically up to 26 ft long and are located one above the other. The discharge wire itself can be star-shaped; thin, flat wire with barbs; or have another profile depending upon the manufacturer. Insulators carrying the high-tension support steel-work are located in the insulator housing. There is no high-tension framing and, therefore, no access above the collecting plates. The discharge frames are individually rapped by mechancial hammers driven from outside the precipitator and are restrained front and rear to maintain electrical clearances. With top suspension for all internals, expansion allowances are provided in the lower alignment steel work.

PRECIPITATOR ARRANGEMENT

After calculating the total collecting area of a precipitator, the application engineer must determine the shape of the box required to house the internals. The beginning point is the computation of precipitator frontal area (width x height) using an acceptable gas velocity to the unit—ordinarily 3 to 5 ft/sec for a coal-fired boiler application. Once field height and number of gas passages have been determined, re-

quired treatment length is figured. Selection is now complete if the equipment fits the allocated space envelope. Should the equipment extend beyond width constraints, alternative selections must be made.

There are four basic arrangements for locating precipitators in a given plot plan (Fig. 12):
- in-line
- cross-flow
- piggyback
- chevron

The preferred arrangement is in-line because of such advantages as:
- lowest capital cost
- good flow balance between chambers
- best gas distribution within a chamber
- lowest system pressure drop
- easy removal of internals in event of damage due to fires, explosion, or corrosion
- ease of erection and shortest time span to install a precipitator system.

In laying out a plant, rigid-frame precipitators have a decided advantage over weighted-wire units because field heights can be 52 ft for the rigid-frame, and a maximum of 36 ft for the weighted-wire. In many instances a rigid-frame unit will fit in-line, whereas the weighted-wire design must be arranged in one of the other, less desirable, ways.

FLYASH RECEIVING HOPPERS

During operation, the flyash accumulated on the collecting surfaces of the precipitator is periodically shaken loose, and dropped into to the hoppers. The level of dust in each hopper will rise until that particular hopper is emptied. If for any reason, emptying the hopper is delayed until the dust level approaches the elevation of the bottom of the discharge electrodes, they will be electrically short-circuited to ground through the mass of collected dust. The affected precipitator section will cease functioning in a normal manner, but will continue to collect some dust by acting as a settling chamber.

If the dust and flue gas entering the precipi-

COMBUSTION
Control of Power-Plant Stack Emissions

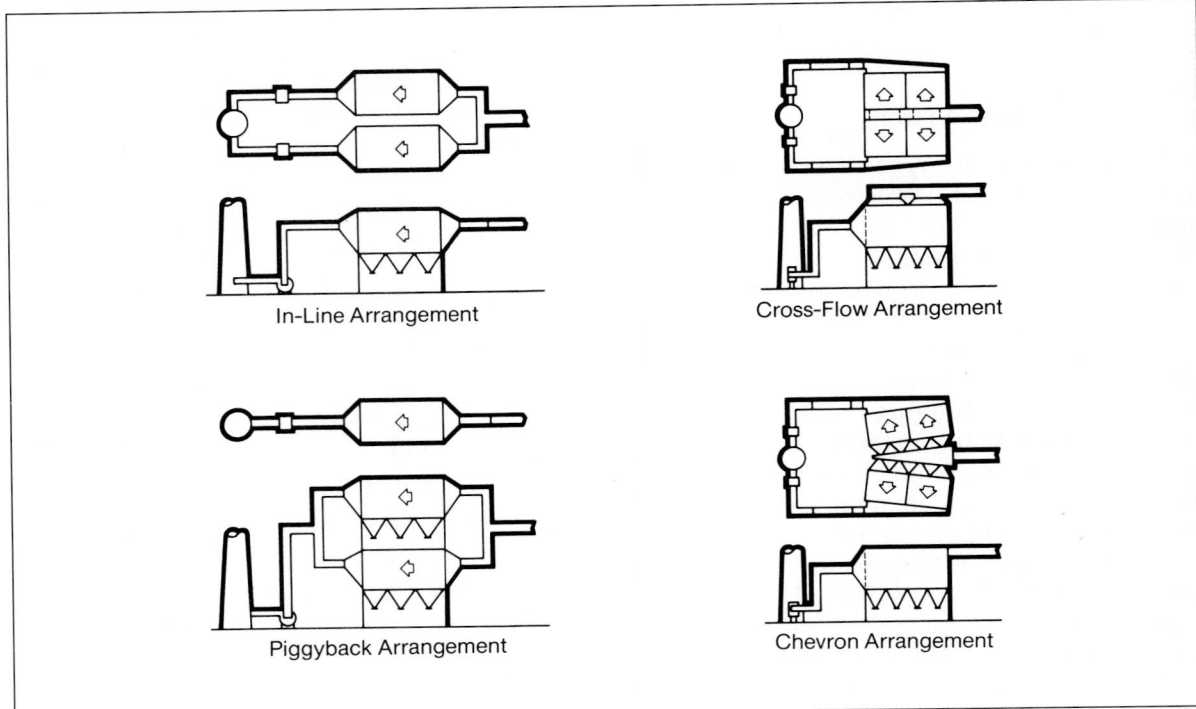

Fig. 12 **Basic precipitator arrangements**

tator are reasonably well distributed, all precipitator hoppers in any row perpendicular to the gas flow will collect substantially the same quantity of dust per unit time. However, much more flyash will be collected in the rows of hoppers closer to the precipitator inlet than in the rows toward the rear of the precipitator. The quantities can be approximated with reasonable accuracy by use of the Deutsch equation. An example is shown in Table I for four precipitators having overall efficiencies of 99.6 percent and from 3 to 6 rows of hoppers in the direction of gas flow. In modern flyash precipitators, then, the inlet row of hoppers can collect from 40 to 100 times as much flyash as does the rearmost row.[30]

FABRIC FILTRATION

Fabric filters, or baghouses as they are commonly called, have a long history of application in the capture of particulate from process

gases. Applications range from nuisance-particulate collection (for the control of silo dust emissions or conveyor transfer station dust control) to process applications where the fabric filter equipment is an integral piece of process equipment used to capture the product (as

Table I. Percent of Flyash Collected In Each Row of Hoppers

Row No.	Total number of rows, front to rear			
	3	4	5	6
1	84.1	74.9	66.9	60.2
2	13.4	18.8	22.2	24.0
3	2.1	4.7	7.3	9.3
4	—	1.2	2.4	3.9
5	—	—	0.8	1.6
6	—	—	—	0.6
TOTAL	99.6	99.6	99.6	99.6

with carbon-black manufacturing), and to strictly particulate capture for environmental compliance (as with coal-fired boiler applications). In general, fabric filters have found increasing acceptance in applications where gas-borne particulate must be efficiently and dependably removed from gas streams.

Fabric filtration is presently considered by many as the best available control technology to control particulate emissions from a gas stream. The development of new and improved fabrics and finishes since the early 1970's has had a dramatic effect on the application potential of fabric filters, which in early years were limited primarily to the natural material cloths available at that time. These new cloths and finishes have greatly extended the allowable operating-temperature window, the expected service life, the resistance to chemical attack, and the cake-release properties.

The preference of fabric filtration over other types of particulate-removal equipment, primarily electrostatic precipitators and high-energy wet scrubbers, is generally because of the superior performance of fabric filters in the following areas:

▪ Outlet emissions are nearly independent of the magnitude of the inlet dust loading.

▪ Special fabrics and fabric finishes can significantly reduce outlet emissions below the capabilities of the electrostatic precipitators and wet scrubbers.

▪ Particulate/gas chemical reactions can occur in the fabric-filter system; thus, in the case of dry flue-gas desulfurization systems, overall sulfur-dioxide removal efficiency is improved by as much as 15 to 20 percent because of the interaction of the flue gas with the fabric-filter dust cake.

▪ Captured particulate and chemical reaction products removed from the filter bags (cake) remain dry for ease of handling and disposal.

Although fabric-filters can be used in a wide variety of applications to control particulate emissions, the emphasis in the following discussion is primarily on those applications in which the fabric-filter controls emissions re-

sulting from the combustion process, i.e., boiler applications.

TYPES OF FABRIC FILTERS

Fabric-filters have a relatively constant collection efficiency and exhibit a varying pressure drop dependent upon the degree of cake thickness at any point of reference. Electrostatic precipitators, on the other hand, have a relatively constant pressure drop, but will vary in overall removal efficiency dependent on inlet loading. Another way of expressing this relationship is to say that the fabric-filter is a constant-emission device (as measured by the mass of particulate emitted per unit of fuel fired), whereas the electrostatic precipitator is a constant-efficiency device (as measured by the percentage removal of inlet loading).

When dirty gas flows through a fabric, the captured particulate matter forms a cake on the surface of the fabric. This deposit increases both the filtration efficiency of the cloth and its resistance to gas flow. Thus, for continuous operation, a fabric-filter must have some mechanism for periodic cleaning of the deposited cake, and the mechanism chosen must be capable of maintaining a reasonable pressure drop, consistent with the operational pressure drop limitations of the system in which it is installed. The cleaning mechanism used frequently represents the generic name of the type of equipment.

In addition to the above, the magnitude of the gas flow will have a bearing on the cleaning mechanism or fabric-filter design selected. Also, the intended service of the fabric-filter will have an additional influence on equipment selection, with more aggressive cleaning mechanisms being used for more-difficult-to-dislodge cake. The designer of the system must address the merits of each filter type and choose the appropriate design for the system requirements.

Shaker Type Fabric-Filters

In this oldest of filter designs, the cleaning mechanism is a vigorous shaking of the filter bag to remove the deposited cake. The shaking action causes the cake to fracture and fall into the collection and disposal hopper (Fig. 13). This method of cleaning has been applied to both inside collectors (those collecting particulate matter on the inside of the individual filter bags) and outside collectors (those collecting the particulate matter on the outside of the individual filter bags).

The cleaning mechanism is a very aggressive system limiting filter-bag cloth selections to those materials and weaves that can withstand the rigors of cleaning without premature cloth failure, and/or where special precautions have been taken to address the forces acting on the filter bags during the cleaning cycle, such as the amplitude of shake, the frequency of shake, and bag tension.

Shaker designs, being the oldest type, have been used in a variety of applications from ambient-condition nuisance-dust collection (those typically seen on bin vent filters and dust-suppression systems on conveyor transfer points) to small industrial-boiler applications (those using the outside-collector type designs) and large utility applications (those using inside-collector-type designs on coal-fired boilers). In all cases, the fabric cloth is carefully selected to

be consistent with the cleaning mechanism, as well as the chemistry and temperature of the filtered gas.

Reverse-Air Type Fabric-Filters

Originally developed to accommodate the relatively fragile fiberglass cloths selected for fabric-filters operating at higher boiler flue gas or process off-gas temperatures, this cleaning method is generally associated with inside-type collection units, although some variation of this design can be found in limited and unique applications. The name *reverse-air* is really a misrepresentation; properly named the cleaning mechanism should be called *reverse-gas*, since the cleaning mechanism is a reverse flow of gas through an isolated compartment to cause an inward collapsing of the filter bag and thus the fracturing of the filter cake (Fig. 14).

Low air/cloth (A/C) ratios characterize reverse-air units, with typical gross ratios (with all compartments in service) ranging between 1.5 and 2.0. The air/cloth ratio given for any operating condition represents the cubic feet of gas filtered divided by the square feet of filter cloth currently in service.

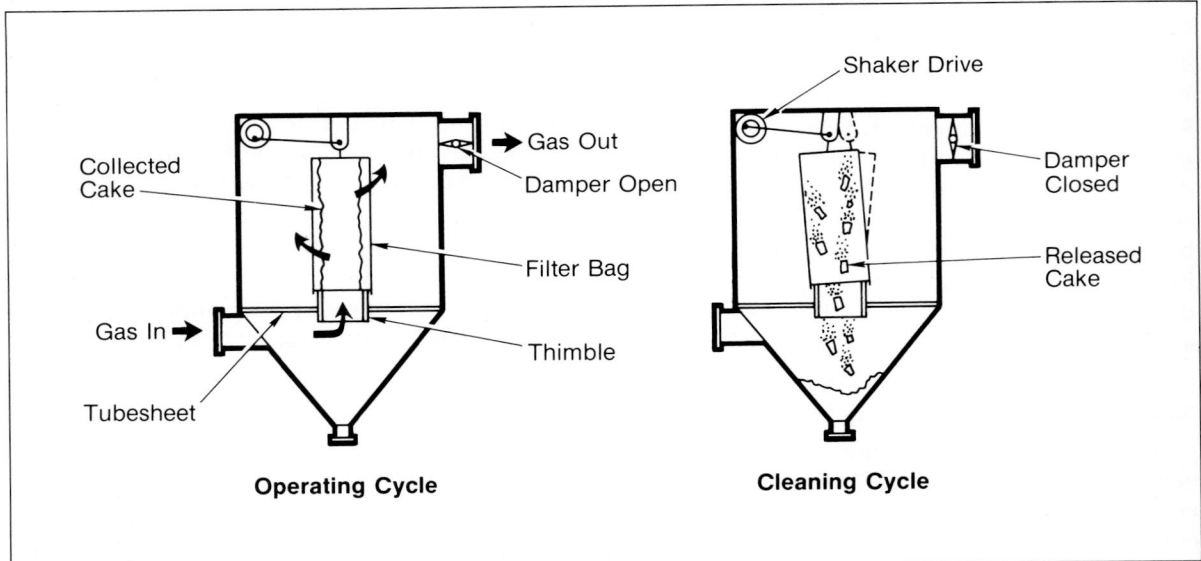

Fig. 13 Schematic arrangement of shaker-type fabric filter

Filter-bag attachment hardware consists of a cap installed in the upper top portion of the bag. This arrangement allows for attachment of the bag to the tensioning device which is hung from the compartment bag hanging frame. An adjustable clamp or a sewn-in band fastens the bottom cuff of the bag to the tube-sheet. Both arrangements allow direct attachment to the tube-sheet thimble. Filter bags for reverse-air service contain multiple rings sewn in the bag length to prevent total collapse during the cleaning cycle (Fig. 14).

Although the bag hardware represents only a small fraction of the total cost of the fabric-filter system, this hardware can cause problems if not properly selected with respect to both design and material. Any component that contacts the cloth needs to be carefully chosen to insure that premature cloth damage will not occur during normal operation of the unit; that is, edges need to be smooth and excessive corrosion should be prevented.

Depending on the inlet grain loading and particle-size distribution, the designer of the fabric-filter unit will adjust the design air/cloth ratio and/or the total number of system compartments required, to insure a reasonable A/C ratio under all operating conditions. Of concern in these calculations is the effective A/C ratio during the cleaning mode, and the maximum ratio in the event of full-load operation with one compartment isolated for any maintenance activity plus one additional compartment in the cleaning mode. Generally, the maximum A/C ratio with two compartments out of service, and with one in the cleaning mode, is limited to 2.5 for these types of units.

The cleaning frequency of the individual compartments will depend, in part, upon the resistivity of the filter cake formed and the inlet grain loading being experienced. As the inlet grain loading increases, all other parameters remaining equal, the frequency of cleaning will also increase, which results in a greater time of operation of the filter unit in the net (less than full complement of compartments operating) condition. Ideally, a properly designed fabric-filter unit should operate in the gross condition the majority of the time as this represents the most efficient filtering mode (lowest A/C ratio) and lowest differential system pressure-drop.

This fabric-filter design using appropriate

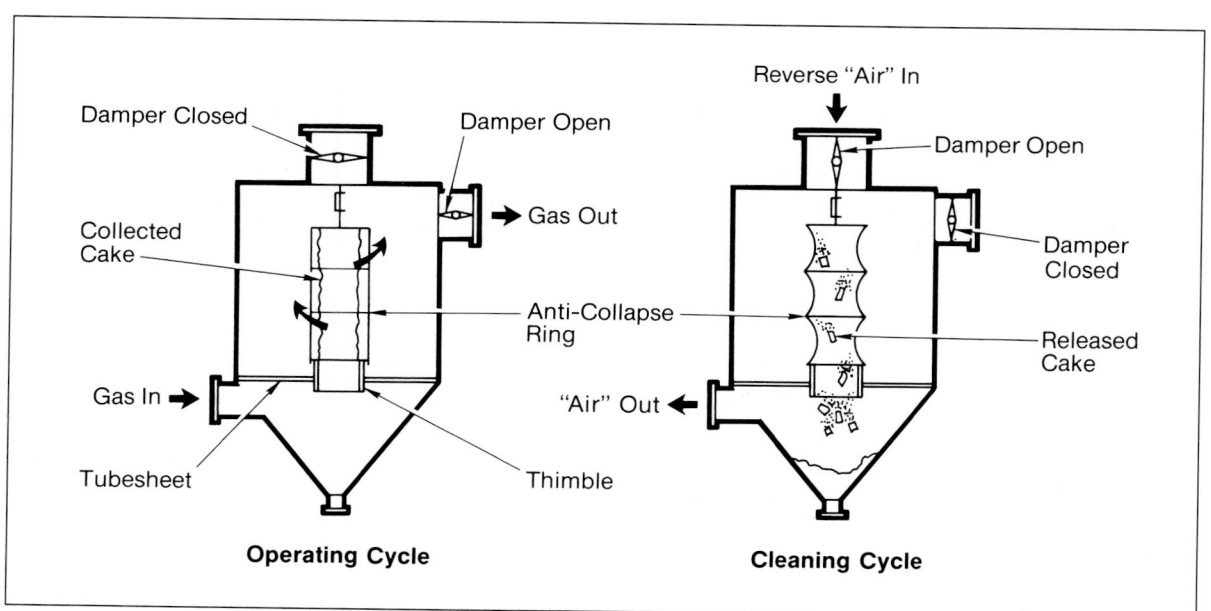

Fig. 14 Schematic arrangement of reverse-air type fabric filter

cloth materials has been applied to a wide temperature range, from near ambient to more than 500°F. An advantage of this particular system is the ability to use fiberglass cloth which can withstand wide ranges in temperature without physical damage. The reverse-air system, being the most gentle method of filter bag cleaning, is compatible with most available bag materials.

Shake/Deflate Fabric-Filters

This type of unit combines the features of the shaker unit and the reverse-air unit in its design. The unit is an inside type of collector which uses shaking and mild reverse gas flow during the cleaning cycle.

An advantage of the shake/deflate unit is its ability to operate at somewhat higher air/cloth ratios than pure reverse-air units due to the more vigorous cleaning action of the shaking mechanism. Following the shaking sequence the filter bags are exposed to a mild reverse-air deflation which further assists in cake removal from the bags. This type of cleaning system is sensitive to cloth selection and the parameters of the cleaning cycle, as previously noted for the pure shaker units, and has generally not demonstrated the bag-life expectancy characteristic of pure reverse-air units.

Pulse-Jet Fabric-Filters

The pulse-jet unit is an outside collector type wherein the particulate-laden flue gas flows from the outside to the inside of the filter bag. Fig. 15 shows a generic pulse-jet collector with the incoming flue gas entering the hopper and flowing upwards and through the filter bags. Wire or mesh-frame cages support the filter bags to prevent their collapse during the filtering period. Cages vary in both design and construction material depending upon the composition of the gas being filtered, the support requirements necessary for the filter-bag construction, and customer preference with respect to ease of handling and the potential re-use of components.

The filtered particulate (cake) is trapped on the outside of the bags while the clean gas flows through the cake and cloth. It travels up to the inside of the bag through the venturi to the clean-gas plenum, and exits through the compartment outlet damper. The individual bags and support cages are installed and removed from the top of the compartment through the clean-gas plenum. The clean-air plenum can be either a full height walk-in type, an extension of the compartment casing, or a top hatch-cover design, which has removable top hatch covers to reach the venturis and bags.

Bag cleaning is by short pulses (50 to 100 milliseconds) of high-pressure air directed downward into each bag in the row being cleaned. Radial acceleration of the fabric and dust cake causes a portion of the dust to be separated from the bags. Gravity allows the released cake to fall down between the bags into the discharge hopper. Within a single pulse-jet unit, or single compartment, rows of filter bags

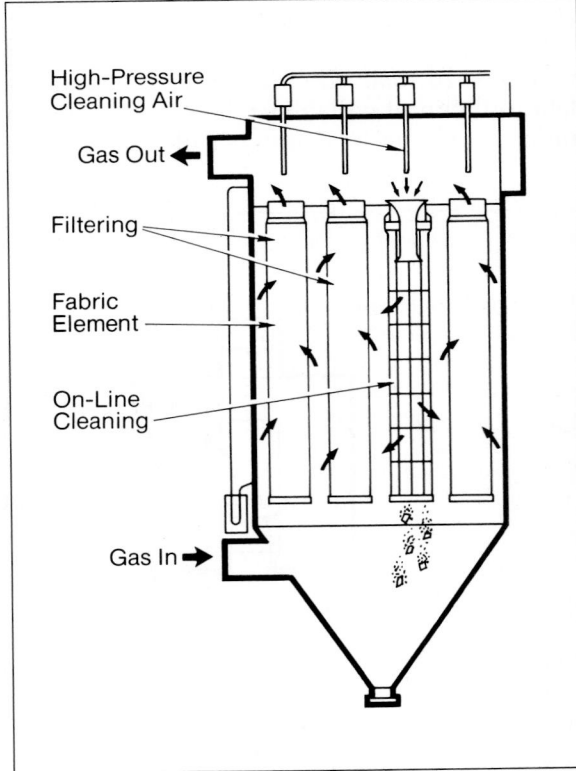

Fig. 15 **Generic pulse-jet fabric filter, with one filter bag being cleaned on-line**

are pulsed sequentially under the control of a solid-state programmer which is usually mounted adjacent to the collector casing.

The original operating concept for all pulse-jet type of collectors was on-line pulse cleaning; that is, the gas flow through the collector was continuously maintained and the rows of filter bags were pulsed sequentially either on a timed sequence or under control of the collector differential-pressure monitor. Although still an established practice in many nuisance-dust collection applications, this cleaning method is a very inefficient cleaning process when compared to off-line pulse cleaning. Consider only these facts:

- The pressure pulse is from 50 to 100 millisecond in duration, and it is only during this period that the gas flow through a row of bags is stopped.
- Dust pulsed from one bag row is immediately reentrained on an adjacent bag row and, therefore, does not reach the hopper.
- Only after repeated pulsing will the captured cake gravitate to the hopper.

The only logical explanation for the success of on-line pulse-jet cleaning is dust agglomeration. The dust particles form agglomerates which are heavy and dense enough to resist attraction to adjacent bags and, therefore, can fall through the rising gas stream into the hopper.

With the advent of large collectors to control emissions from hot processes (such as boilers), the on-line pulse-jet filter has been shown to have several disadvantages:

- On-line maintenance and inspections are not practical.
- Compressed-air requirements are very large, with increased filter area needed to maintain the required pressure-drop.
- The physical properties of fiberglass fabrics required for hot boiler processes are inconsistent with the harsher cleaning mechanism of on-line pulsing at high pulse pressures.

Reverse-air systems, as previously described, are essentially multi-compartment units with off-line cleaning. The carryover of the reverse-air operating scheme to smaller industrial boilers produced the modular reverse-air systems, those that are made up of totally shop-assembled compartment modules; these designs substantially reduced field erection-labor costs.

This same concept was then applied to the pulse-jet units using shop-assembled modular units and off-line cleaning. This allowed for pulse-jet systems composed of totally shop-assembled modular systems including the extra module for off-line cleaning. With this method of operation, the features include:

1. Off-line cleaning, which is more effective as there is little tendency for the dust to migrate to adjacent filter bags. Instead, the cake has an opportunity to fall downward into the collection hopper(s).

2. Pulsing pressure can be reduced as there is no need to overcome the operating-system forward momentum.

3. Lower pulse pressures improve filter bag life without sacrificing fabric-filter performance.

The pulse-jet fabric-filter units, then, can be cleaned in either an on-line mode, where the compartment is not isolated from the system during the cleaning cycle, or an off-line mode, where the compartment to be cleaned is removed from on-line service by closing the compartment outlet dampers prior to and during cleaning. Generally, the off-line mode is the preferred for boiler particulate-control installations, whereas the on-line mode would be commonly used for nuisance-dust applications. In the on-line mode, fewer filter bags are required in the total system. In the off-line mode, multiple compartments are required, including an extra compartment to allow for the removal of one compartment, without affecting the overall system performance, during cleaning.

Pulse-jet module designs have evolved over the years, with numerous design modifications providing improved performance over the generic design of Fig. 15. Some of these major differences are illustrated in Fig. 16.

In place of a dirty-gas inlet in the module

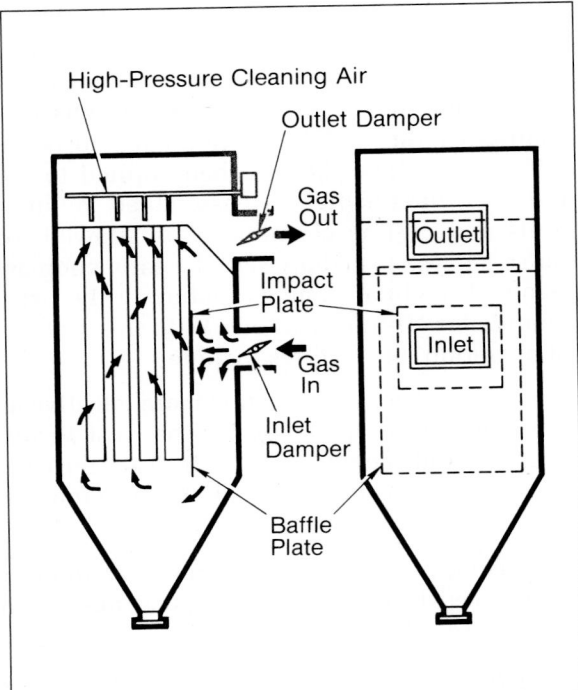

Fig. 16 **Improved design of pulse-jet fabric filter with high side inlet, shown in filtering mode**

hopper, gas entry is into the side of the module casing. An internal baffle and impact plate protect the filter bags from direct impact of the incoming gases and dust-particle stream. Advantages of this arrangement include:

■ Better gas and dust distribution to all module filter bags.

■ A gas-inlet location consistent with a combined manifold design in which the system inlet and outlet manifolds are formed by a single rectangular duct with a diagonal splitter plate forming the two gas paths.

■ Greater hopper storage capacity, which is important on high-inlet-loading systems.

DESIGN OF EQUIPMENT

Overall system designs involve the selection of the cleaning mechanism and type of fabric-filter to be used for a particular service, along with the selection of a suitable cloth and cloth finish to be installed in the unit. To make this selection, the designer needs a complete

knowledge of the process on which equipment is to be installed and a knowledge of the limiting factors of the choices available for application.[31]

Fabric-filter designs have some established criteria such as typically acceptable air/cloth ratios and bag aspect ratios. As noted previously in the discussion of reverse-air systems, the gross air/cloth ratio generally acceptable for the reverse-air system is at or less than 2.0, with a maximum of 2.5 in the net condition of two compartments off-line with one of the compartments in the cleaning mode. In the design of a reverse-air unit, or any off-line cleaning unit, the number of compartments will affect the value of the net condition as, with fewer compartments, the removal of one compartment from service has a greater effect than with a greater number of compartments.

In the design of the reverse-air unit, bag aspect ratios are typically limited to 32 to 1, with a maximum of 35 to 1 under any conditions. Bag aspect ratio is defined as the ratio of the bag length divided by the bag diameter. Here, the concern is the ability to clean effectively the upper section of the bag on this inside-collector-type unit if the length becomes too long. Most designs have used the more common bag lengths of 32 feet in the typical boiler flue-gas application.

Similar design criteria have also developed for the pulse-jet units in defining acceptable ratios or component designs. As an example, typical filter bag lengths range from 12 to 16 feet, although some designs have been up to 20 feet in length. The application of felted cloths on pulse-jet is quite typical; however, the use of woven fiberglass cloth is acceptable especially for the higher boiler flue gas operating temperatures. Filter bag cage materials are commonly of carbon steel with galvanized coating, but other coatings are available, as is alloy construction. High-side inlet collectors have advantages with respect to hopper capacity and hopper reentrainment; however, designs having the gas entering hoppers are acceptable and allow for lower equipment setting heights in overall system design.

The selection of the cleaning mechanism design, the number of compartments needed, the air/cloth ratio necessary, the cloth type and finish, the maximum bag length and a variety of other design considerations, are specific to the service intended and the user's preference.

FABRIC-FILTER MATERIALS

Specification of the weave, thread count, weight, and finish of the cloth for a required filter bag lists only a few of the possible combinations of manufacturing procedures or design that ultimately will describe the finished filter bag for a particular application. Fabric specifications include such properties as tensile strength, abrasion resistance, chemical attack resistance and limitations of operating temperature (Table II). Tensile strength measures the ability of the yarn or fabric to resist breaking when in direct tension. Abrasion resistance is the fabric's ability to withstand externally caused abrasion or that resulting from internal rubbing of the fabric fibers. Some cloths exhibit excellent resistance while others are damaged quite easily. Coatings applied to some cloths not only improve their abrasion resistance, but also their chemical resistance, as in the case of fiberglass cloth, which would self-destruct without a suitable coating.

FABRIC-FILTER PERFORMANCE

Unlike fabric-filters, most other emission-control devices depend upon either pressure drop or collection surface area to reach the required collection efficiencies. Upsets to predicted gas volumes or particulate loadings will typically cause these other types of equipment to have higher emission rates, despite their maintaining the same fractional efficiency. The fabric does not behave in this fashion, because the material collected on the filter bags is the filtering medium that captures the incoming particulate. Although upset conditions may change the flange-to-flange pressure drop, resulting in an excessively fast build-up of cake, these conditions do not materially change the final emission. It can be said, as stated in an earlier section, that a fabric-filter represents a constant-emission device.

The actual performance of a fabric-filter depends on specific items such as air/cloth ratio, permeability of the fabric/cake (particulate bleed-through), the loading and nature of the particulate (irregular-shaped or spherical), particle size distribution (fine versus coarse) and, to some extent, the frequency of the cleaning cycle. By careful selection of design components, fabric-filters can achieve emission levels of higher than 99.9-percent removal of incom-

Table II. Fabric Material Properties

| Fiber | Fiber Property | | Chemical Resistance | | Recommended Operating Temp., °F | |
	Tensile Strength	Abrasion Resistance	Acids	Bases	Continous	Short-time
Cotton	Good	Average	Poor	Excellent	180	225
Polyethylene	Excellent	Good	Excellent	Excellent	190	190
Glass	Excellent	Poor	Good	Poor	500	550
Nylon	Excellent	Excellent	Poor	Excellent	200	250
Dacron*	Excellent	Excellent	Good	Fair	275	325
Acrylic*	Average	Average	Very Good	Fair	260	280
Nomex*	Very Good	Very Good	Fair	Very Good	400	425
Teflon*	Average	Below Average	Excellent	Excellent	450	500

*Trademark E.I. DuPont

ing particulate, or an outlet loading of less than 0.005 grain per dry standard cubic foot.

FABRIC-FILTER OPERATING PROBLEMS

A primary concern in the operation of a fabric-filter is bag life and bag replacement cost. Since the initial cost of the filter bags, excluding installation, can represent 10 percent or more of the total cost of the equipment, a significant operating expense can occur if bag life is shorter than expected. And, since typical bag life is 3 to 5 years (depending on the type of cleaning mechanism selected and the design parameters used in that selection), this operating expense item can represent a significant cost over the life of the equipment. Shortened bag life can be caused by various operating and design problems among which are the following:

■ Bag blinding, usually caused by operating a fabric-filter for frequent or long periods at or below the dew point; such operation tends to plug or blind the cloth. Bag blinding results in unacceptable unit pressure drop which will ultimately become fan- or process-limiting and require filter-bag replacement. The effect on the filter cloth is that material builds up in the passages of the cloth body and restricts the gas passage such that, even after repeated cleanings, the lodged material cannot be removed.

■ Bag erosion, resulting from high-velocity streams of flyash-laden gases where the gas enters the filter-bag or impacts the cloth in any way. Proper selection of the air/cloth ratio and filter-bag length controls the high entry velocity at the bag inlet inside collectors. Higher air/cloth ratios with longer filter bags tend to increase this velocity in the entrance or "neck" area of the bag. Metal thimbles are provided in this area to minimize abrasive action. Velocities are generally limited to less than 300 feet per minute at the bag neck.

■ Filter-bag deterioration, often caused by a poor selection of cloth material or finish, or by abnormal operation of the system, resulting in damage to the cloth or components of construction. Abnormal operation would include oper-

ation outside the design temperature range (either high or low), or operation in excess of the design volumetric flow, or a significant variation in the chemistry of the flue gas being filtered. Any of the above could cause premature deterioration of the cloth selected for service.

FLYASH HOPPER PLUGGING

The reliability and availability of electrostatic precipitators and fabric filters are affected adversely by problems in the dust-receiving hoppers and equipment beneath those hoppers for transporting flyash to disposal. A significant portion of the downtime of such dust-collecting equipment stems from problems with, and malfunctions of, ash evacuation systems.[11]

There is a paradox in the design of equipment for removing the collected flyash from the bottom of precipitators and fabric filters and conveying it away. Suppliers of ash-handling-systems design their equipment for "dry and free-flowing" material at the inlets of the ash-removal equipment. Yet, dust-collecting equipment can have mud or water flowing out of hoppers any time the hopper metal temperatures are below the water dew point (100° to 130°F, depending on fuel type and excess air in the products of combustion). This condition occurs commonly during boiler cold start-up, when many metal surfaces in the gas stream are at ambient temperature.

When sub-dew-point conditions occur, the hygroscopic flyash absorbs the acid or water produced and may agglomerate and cement into large pieces that either cannot pass through the hopper outlets or are too heavy to be conveyed.

EFFECT ON COLLECTING EQUIPMENT

Undetected and/or uncorrected hopper plugging can cause one or more of the following deleterious effects on precipitators:

■ bent or misaligned collecting surfaces

■ discharge-electrode burning

■ distorted or broken discharge-electrode frames

- broken anvil bars
- broken shaft or drive insulators
- shorted high-voltage bus sections
- lowered electrical-power output from bent or misaligned collecting- and discharge-electrode components
- the formation of large, difficult-to-remove clinkers by high-temperature electrical fusing of the ash overflowing a hopper.

With fabric filters, plugging can increase the gas velocities as hoppers fill above the inlet level, resulting in the possiblity of reentrainment of dust. This can result in excessive bag abrasion. With a completely blocked hopper, the compartment will no longer accept gas flow, and the other compartments will have to handle the volume, resulting in increase gas-side pressure drop.

With both precipitators and fabric filters, if a plugged hopper cannot be cleared and discharged through the ash system in a normal manner, the ash must be emptied onto the hopper-room floor or the ground (creating a fire and safety hazard) and be manually removed.[30]

SOLUTIONS TO THE PROBLEM

The problem can be solved by:

1. Removing the ash as continously as possible so that it can remain hot and uncompacted, such that the hoppers are not used for storage of the hygroscopic flyash.
2. Preventing air in-leakage that cools the flyash below the acid or water dew point, by regular inspection and maintenance or replacement of flyash intakes.
3. Modifying hopper design for improved flyash out-flow.
4. Enclosing hopper areas and using adequate insulation thickness to minimize heat losses and cooling of ash-system hardware.
5. Heating hoppers such that the walls do not drop in temperature and have condensation form.

These suggestions, and other recommendations on the design and sizing of flyash hoppers, are discussed further below and in the following chapter on ash-handling systems.

Maintenance of Flyash Intakes

It is estimated that some flyash intakes may open and close nearly a quarter of a million times per year. It is completely reasonable to expect, under such conditions, that solenoids will fail and that valve seats will become worn, resulting in an inflow of air into the hopper. With a pressure-pneumatic system, air can be forced into the hopper at pressures as high as 100 psig (in a dense-phase system). With vacuum systems, the motive force for inducing air into precipitator hoppers or baghouse hoppers is the suction maintained in the precipitator or baghouse by the boiler induced-draft fans, which can create a suction of as much as -20 in. WG. in the precipitator or baghouse. With either a pressure or vacuum system, then, there is a pressure differential that can result in the introduction of large quantities of cool air into precipitator hoppers, which can lead to the condensation of moisture in the flue gas.

Actual field temperature measurements on both "hot" and "cold" precipitators have shown that even very active hoppers can have low (essentially ambient) temperatures. One plausible explanation for such a low internal temperature is leakage of air into the hoppers through the flyash intake valves located at the bottom of the hoppers. Very seldom have such valves been considered "maintenance items" that should be inspected and considered for replacement on some time schedule. But it has become apparent in plant after plant that such valves cannot be expected to last forever and that they are the source of the low-temperature conditions occurring in many hoppers.[11]

Heating of Precipitator and Fabric-Filter Outlet Hoppers

The ABMA/IGCI Committee stated that the heating of hoppers has not been sufficiently emphasized in the purchase, design, and construction of precipitator and baghouse dust outlets. They recommended that, along with a redesign of the dust-conveying equipment, improved hopper heating and insulation has to be accomplished, all with the purpose of facilitating the emptying of the dust outlets.

The principal variables in the application of hopper heating are:

- external (ambient) design temperature
- internal design temperature
- presence of an internal ash layer
- the extent and placement of heating elements
- the design electrical-heat density
- the thickness of insulation outside the heating equipment

Design operating temperature for hopper-heating equipment for boilers firing low- to medium-sulfur coals should be a minimum of 270°F, based on an ash layer inside the hopper. This is 10 to 15 degrees above the normally expected acid dew point. For units firing high-sulfur oils, the hopper plate should be kept above 350°F. The heating equipment must be capable of maintaining such temperatures under the worst expected ambient conditions of temperature and wind velocity (for 97 to 98 percent of the time) and for the lowest anticipated interior gas temperature.

Test reports of temperature measurements in operating precipitator hoppers indicate that gas or air temperatures as low as 90°F sometimes occur inside the hoppers, with the precipitator and boiler operating at essentially full load. Under such conditions, most of the heat loss from the hopper plate is not to the outside atmosphere, but rather to the *inside of the hopper*.[30] The heat-transfer and the mass-transfer situation in such a precipitator can in no way result in the occurrence of such low temperatures by a normal heat loss/transfer mechanism. The greatest probability is that the low temperatures in such hoppers result from the inleakage of relatively cool air from the pneumatic transport system serving the hoppers.

Even when bulk gas temperatures in collecting equipment are above the acid dew-point (240 to 350°F), hopper skin temperatures at the critical throat can be very low, due to the heat-sink effect of the ash-system hardware (strike pads, poke holes and vibrators) and the convective transfer of heat from the bottom of the hopper to the top, by chimney action. This situation is aggravated by severe local weather conditions and exposed hoppers facing a prevailing wind, and can result in localized condensation and wetting of collected ash.[11]

Normally, only the lower one-third or one-half of a hopper is heated to minimize costs and to avoid the problem of heating elements obstructing access to the equipment; also, most plugging is expected to occur in the constricted outlet region. But the hopper plate does not readily conduct heat to other parts of the hopper, especially when the interior gas temperatures are low. If such areas cannot be heat-traced, they must be completely insulated, including doors, poke holes, strike pads, and other protrusions.

Because heater vendors may interpret heating needs very differently, precise heater specifications or hopper heat loads in kilowatts must be made directly to the vendor, both to ensure adequate heating and to prevent the possibility of damage from overheating. Heat loads as high as 32 kW per hopper have been specified on some large coal-fired-boiler precipitators. This is two to three times as much as in many existing installations.

In most current installations, if the flue-gas temperature falls below the acid dew point, or if air or gas in the hopper falls below the water dew point, there is little that can be done by hopper heaters to prevent condensation inside either precipitator or fabric-filter hoppers. This is because their proper function is to heat the steel wall of the hopper and not the air or gas inside.[11]

CONTROL OF GASEOUS EMISSIONS

Sulfur in coal exists in two forms: organic and inorganic. The inorganic compound pyrite, FeS_2, is present as discrete particles within the fuel, and typically accounts for 20 to 50 percent of the total sulfur. Fig. 17 is illustrative of the variety of pyritic sulfur occurring in just one small area of the United States. The coals forming the basis for the plot are all from the

state of Kentucky, which has two separate basins of formation. Pyrite as a percentage of total sulfur is from less than 10 to over 55 percent for the low-sulfur Appalachian basin coal, and from about 25 to 70 percent for the higher sulfur Illinois basin fuel. The milling and classifying operations associated with pulverized-fuel steam generators using pulverizers similar to the C-E Bowl mill separate a substantial portion of the pyritic sulfur from the coal. (See Chapter 11.) An additional amount of inorganic sulfur is retained in the bottom ash and the fly-ash leaving the boiler in solid form.

FORMATION OF SULFUR OXIDES

Contained in the molecular structure of the coal, organic sulfur is oxidized during the combustion process and emitted from the furnace as gaseous oxides of sulfur.

The principal oxidizing reaction leads to the formation of sulfur dioxide.

$$S + O_2 \rightarrow SO_2 \tag{3}$$

In addition to SO_2, lesser quantities of sulfur trioxide, SO_3, are formed during combustion.

$$SO_2 + \tfrac{1}{2} O_2 \rightleftharpoons SO_3 \tag{4}$$

Typically, the ratio of SO_2/SO_3 in combustion gases ranges from 20:1 to 80:1. The chemical reactions that form sulfur dioxide and trioxide are substantially more complex than the overall reactions represented by Eqs. 3 and 4.

Explicit emission standards for the control of SO_3 do not exist. But as SO_3 is highly reactive and extremely hygroscopic compared to SO_2, it is capable of readily combining with water to form sulfuric-acid aerosol. The reaction is enhanced by the presence of fine particles which serve as condensation nuclei; the resulting aerosol is a principal constitutent of visible stack plumes. Thus, opacity and particulate-emission regulations are an indirect method of restricting SO_3 emissions.

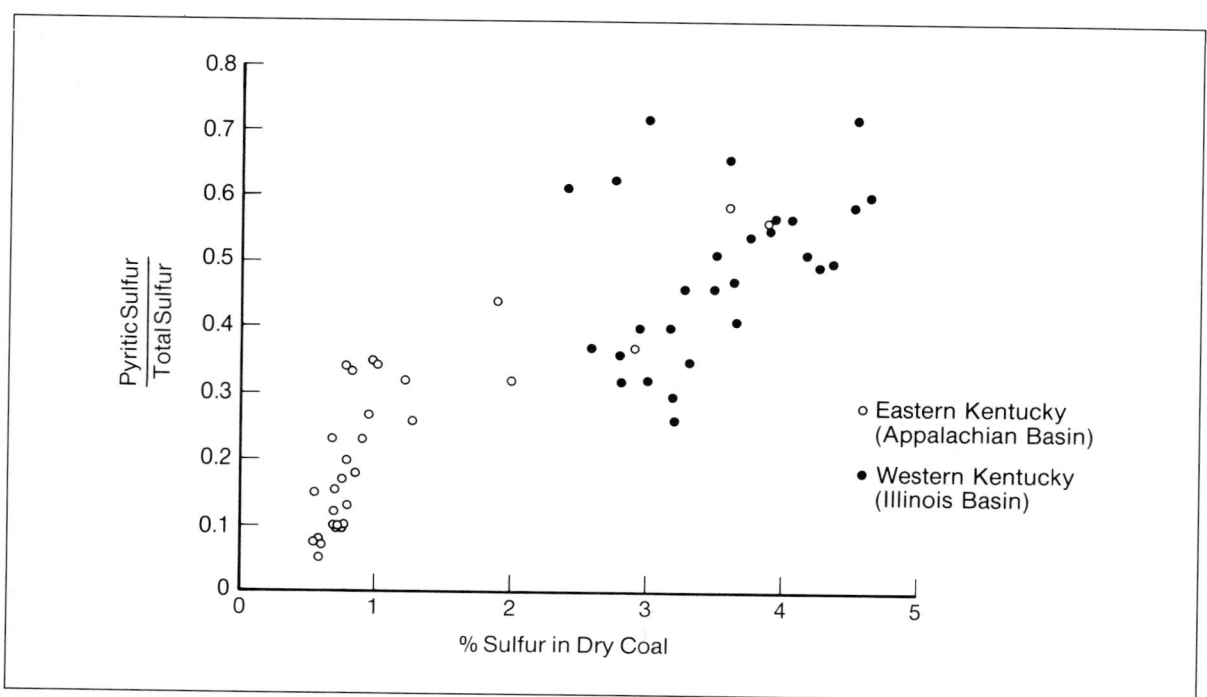

Fig. 17 **Pyrite content of Kentucky coal**

Studies show flue-gas desulfurization systems intended principally for SO_2 control are moderately effective in curtailing SO_3 emissions. Tests indicate that a venturi/spray tower-scrubber removes approximately 60% of the SO_3 in flue gas, with the removal efficiency essentially independent of the SO_3 concentration or scrubber operating variables.[32]

CONTROL OF SO_2 EMISSIONS

Historically, gaseous sulfur-dioxide emissions have been controlled by either dispersion or reduction. Dispersion of SO_2 and other pollutants through tall stacks is the oldest form of control. Under favorable meteorological conditions tall stacks can effectively limit ground-level SO_2 concentration to local ambient air-quality standards. Thermal inversions or comparable short-term meteorological conditions can impair the effectiveness of dispersion as a control method. A further and more critical concern over control by dispersion is the accumulating scientific evidence that oxides of sulfur and nitrogen are the principal precursors to acid precipitation.[33]

Reduction of sulfur emissions has been achieved through one or a combination of the following:

- switching to lower-sulfur fuels
- use of fuel desulfurization methods
- use of flue-gas desulfurization systems

Switching to lower-sulfur fuels is not possible for many large fossil-fired steam-generating plants. Additionally, current air pollution regulations for new power plants require a percentage reduction in SO_2 emissions regardless of the fuel sulfur content.

Fuel desulfurization processes range from conventional coal washing to coal liquefaction and gasification. Coal washing is effective in reducing the pyritic sulfur in the coal. It cannot remove the organic sulfur from the coal matrix, and as such it is limited to sulfur reduction of less than 50 percent.

FLUE-GAS DESULFURIZATION SYSTEMS

Flue-gas desulfurization (FGD) began in England in 1935; the Battersea process used the alkaline Thames River water for scrubbing power-plant flue gas until the visibility of the stack plume caused abandonment of the process early in World War II. The technology remained dormant until the mid-1960's when it became active primarily in the United States and Japan. Since then, over 50 FGD processes have been developed, differing in the chemical reagents and the resultant end products. In the following section, we will describe some of the more prominent, with particular emphasis on those processes which are commercially significant. References 32, 34, 35, and 36 provide a comprehensive overview through the mid-1980's.

In the flue-gas scrubbing process, water evaporates from the scrubbing liquid as the flue-gas is cooled to the adiabatic saturation temperature. Under equilibrium conditions, a quantity of water must be added to the system to make up for the water loss.

In a recirculation type wet-scrubbing system where scrubbing liquid is not bled from the system, the salts in the incoming make-up water gradually concentrate. In any scrubbing system involving solids present in the flue-gas stream as flyash or as solids from the sorption of sulfur dioxide, a stream must be bled off to remove the solids and similarly prevent their buildup to excessively high values in the system. The specific concentration of various chemicals in the scrubbing liquid in any given system depends on the sulfur in the coal, the flyash in the flue gas entering the scrubber, and the type of additive.

The evaporation that occurs in the wet flue-gas scrubbing process, then, results in essentially pure water being discharged to atmosphere, and simultaneously, a concentration of solids in the bleed stream leaving the scrubbing system. Of these solids, the sulfates, sulfites, and chlorides are of the greatest significance.

One approach in categorizing FGD processes is to differentiate between nonregenerable, or "throwaway", and regenerable, or recovery-type, processes. Throwaway systems produce waste products of various sulfur compounds

that must be stored in ponds or treated for use as landfill. Regenerable systems recover the SO_2 in some commercially useful form such as elemental sulfur or sulfuric acid. A further differentiation can be made between wet processes and dry processes: wet processes saturate the flue gas with water vapor and dry processes do not.

NON-REGENERABLE FGD PROCESSES

In the U.S., the overwhelming majority of commercial FGD systems are of the throwaway type. Included in this category are the lime or limestone wet scrubbing, the double or dual alkali, and the lime dry-scrubbing systems.

Lime/Limestone Wet Scrubbing

Lime or limestone scrubbing is the oldest and most common method of flue-gas desulfurization. The first successful closed-loop lime/limestone process is credited to the English firm of J. Howden and Company. During 1931-1933 pilot-plant tests were conducted in which the scrubber effluent was recycled to the absorber; initially, scale formed in the absorber.[37] The critical breakthrough in system chemistry came in 1933 when the problem was identified as one of excessive gypsum ($CaSO_4 \cdot 2H_2O$) supersaturation in the scrubber. The solution was to add a crystallization tank and increase the liquid-to-gas (L/G) ratio. Adding a crystallization tank operating at high solids concentration allows calcium-sulfur compounds to precipitate outside the scrubber. Because of precipitation reactions, the absorbent recycled to the scrubber has a lower concentration of scale-forming gypsum. A high L/G ratio minimizes the increase in supersaturation across the scrubber and thereby further diminishes the potential for scale formation.

In the mid-1960's, Combustion Engineering developed and marketed the limestone-injection scrubbing process. In this variation, pulverized limestone was injected into the furnace simultaneously with the coal. The resulting calcined limestone (CaO) was entrained in the flue gas along with flyash. A water scrubber

was used to remove the flyash, limestone, and SO_2 from the flue gas. Several systems were placed into operation between 1968 and 1970, but persistent difficulties with scaling, poor additive utilization, and plugging of boiler convection surfaces prompted C-E to revert to the older Howden tail-end process.

In a typical lime/limestone wet FGD system (Fig. 18), boiler flue gas enters the scrubber (spray tower) and contacts the absorbent slurry while sulfur dioxide and lesser amounts of oxygen are simultaneously absorbed. The spent absorbent returns to the reaction tank, or scrubber-effluent hold tank, where the dissolved sulfur compounds are precipitated as calcium salts. Fresh limestone or lime slurry is added to regenerate the spent absorbent.

The size of the reaction tank provides sufficient time for precipitation of sulfur compounds and the dissolution of additive. The rate of additive feed is pH-controlled. From the reaction tank, the regenerated absorbent slurry is recycled to the absorber. The slurry typically contains from 5 to 15 percent suspended solids consisting of fresh additive, absorption reaction products, and lesser amounts of flyash. To regulate the accumulation of solids, a bleed stream from the reaction tank is routed to the solid/liquid separation equipment.

Typically, the bleed is routed to a gravity thickener where the suspended solids settle to the bottom and the clear liquid is drawn off the top and returned to the scrubber loop. The thickener underflow containing 25 to 45 percent solids can be pumped to a pond or may be further dewatered through vacuum filtration. Alternative liquid-solid separation equipment, such as wet cyclones and centrifuges, may be used to dewater the bleed stream. The filtrate is returned to the scrubber loop. Alternately, the bleed from the reaction tank can be pumped directly to a settling pond where solids accumulate and clear liquid is recycled to the scrubber. Make-up water is added to the system to replace water evaporated by the hot flue gas and water entrained in the waste stream. The water is added as mist-eliminator wash, and as the additive slurrying medium.

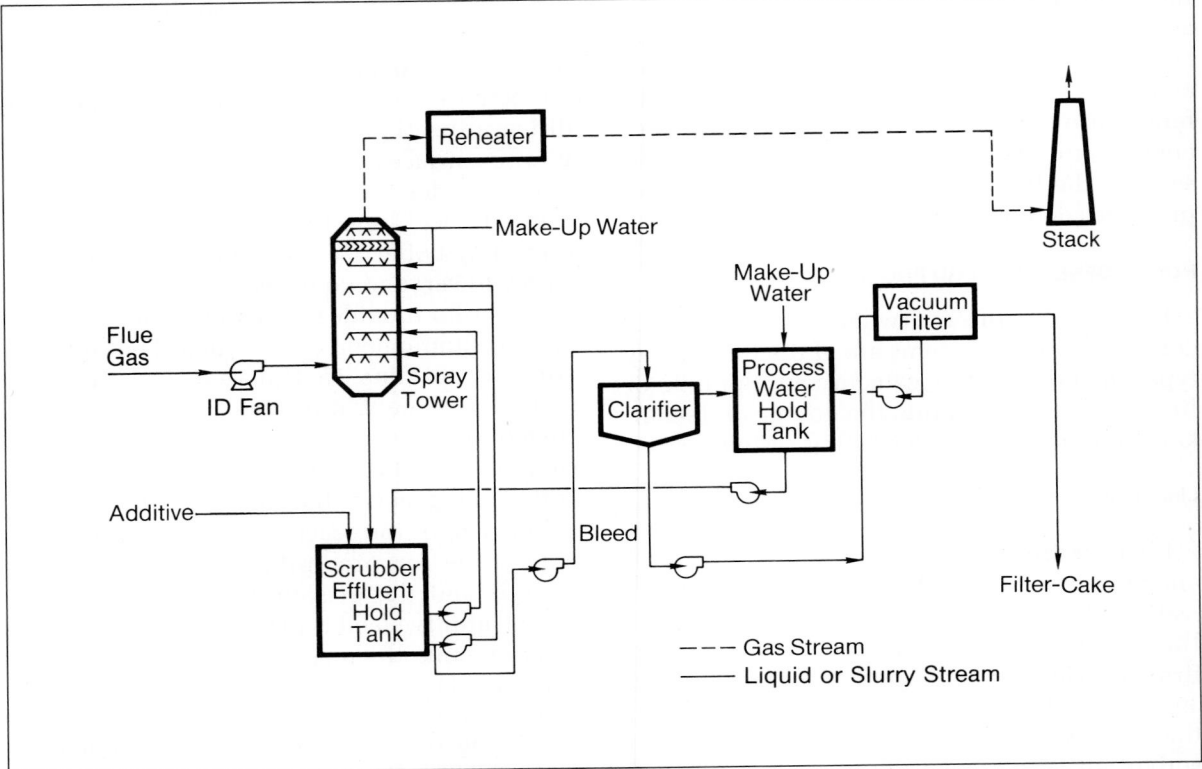

Fig. 18 **Process flow diagram for lime/limestone wet scrubbing**

Double Alkali Systems

Like lime/limestone processes, in double alkali (or dual alkali) processes, lime or limestone is consumed and a calcium sulfite/sulfate wet solid is produced as waste. They differ from the lime/limestone in that SO_2 is absorbed and waste-products are formed in separate components. The absorption and waste generation are separated by using an intermediate soluble alkali such as Na_2CO_3, NaOH, or Na_2SO_3. The separation serves two objectives. First, it permits scrubbing the flue gas with a clear solution, allowing the rate of SO_2 absorption to proceed independently of lime or limestone dissolution rate. Second, it permits better control of the calcium-sulfur precipitation reactions in equipment specifically designed for this function. In theory, this improves the use of lime or limestone.

Double alkali FGD processes developed because they offered significant advantages over lime/limestone slurry processes. First, there are no suspended solids in the absorbent to contribute to scaling and plugging of internal scrubber components. Second, because of a high concentration of dissolved alkali species, there is minimal resistance to liquid-phase mass transfer. The practical implication of these factors is excellent SO_2 removal efficiency achieved with a simple scrubber at lower liquid-to-gas ratios.

Fig. 19 shows a typical double-alkali process. Clear alkali solution is circulated through the scrubber where SO_2 absorption and some oxidation take place. The absorbed SO_2 reacts with sulfite to form bisulfite:

$$Na_2SO_3 + SO_2 + H_2O \rightarrow 2NaHSO_3$$

(5)

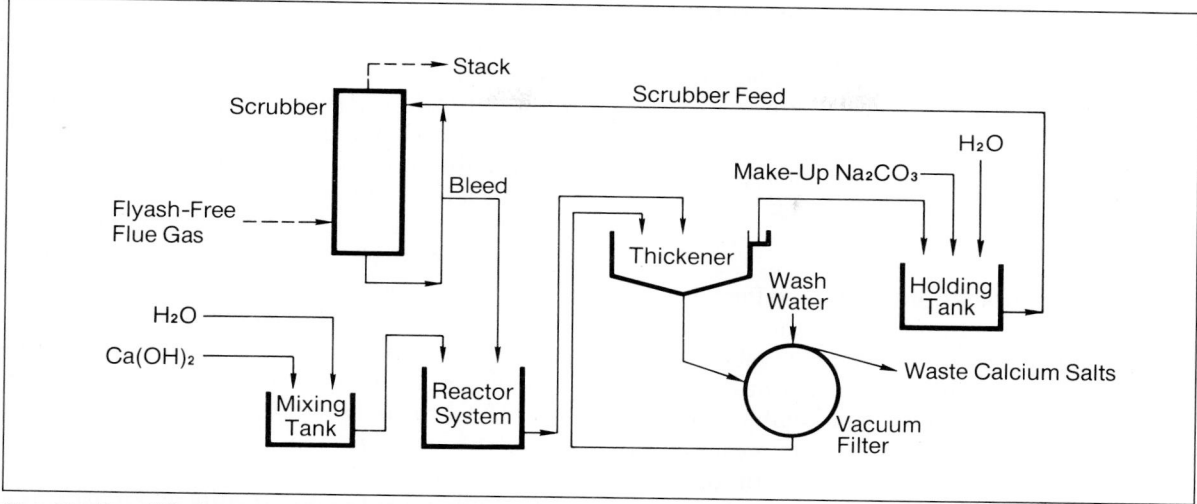

Fig. 19 **Simplified process diagram for double alkali system**

The pH of the absorbent is typically between 6 and 8.5. Scrubber effluent, which contains a lower ratio of sulfite to bisulfite than the scrubber feed, is collected in the absorber sump. A bleed stream from the sump is pumped to the regeneration system where it is treated with lime or limestone. The absorbed SO_2 is precipitated as calcium solids as follows:

$$Ca(OH)_2 + 2NaHSO_3 \rightleftharpoons CaSO_3 \cdot \tfrac{1}{2}H_2O$$
$$+ Na_2SO_3 + \frac{3}{2}H_2O \qquad (6)$$

$$CaCO_3 + 2NaHSO_3 \rightleftharpoons CaSO_3 \cdot \tfrac{1}{2}H_2O$$
$$+ Na_2SO_3 + \tfrac{1}{2}H_2O + CO_2 \qquad (7)$$

The bisulfite is neutralized and active sulfite is regenerated. The sulfite-rich clear solution is separated from the precipitates in conventional solid/liquid separation equipment such as thickeners and vacuum filters. Waste calcium and sulfur compounds are disposed of and sulfite solution is returned to the scrubber loop. Sodium lost as a result of entrainment in the waste solids is replaced with fresh sodium-carbonate addition.

Technically, the most serious difficulty with double alkali processes arises from the irre-versible oxidation of sulfite to sulfate.

$$Na_2SO_3 + \tfrac{1}{2}O_2 \rightarrow Na_2SO_4 \qquad (8)$$

When active sodium sulfite is converted to inactive sulfate, the capacity to absorb additional SO_2 is diminished. To restore dissolved scrubbing capacity, the inactive sulfate must be removed from the system while the sodium remains in solution. A direct purge of sodium sulfate from the system is unacceptable because it contributes to surface and ground water contamination. Additionally, the cost of replacing the lost sodium is further incentive against direct purge.

The methods for removing sulfate depend on the concentration of sulfite in the system. If the sulfite concentration is low, the system is said to operate in the dilute mode. In a dilute double alkali system, sulfate may be removed according to the following reaction:

$$Na_2SO_4 + Ca(OH)_2 + 2H_2O \rightleftharpoons 2NaOH + CaSO_4 \cdot 2H_2O \qquad (9)$$

In a concentrated double alkali system with a high sulfite concentration, this reaction will not proceed appreciably and other techniques must be used to regenerate the inactive sulfate. In Japan, sulfuric acid is added in a separate reactor to force the following reaction:

$$Na_2SO_4 + 2CaSO_3 \cdot \tfrac{1}{2}H_2O + H_2SO_4 + 3H_2O \rightarrow$$
$$2NaHSO_3 + 2CaSO_4 \cdot 2H_2O \qquad (10)$$

This mode of operation is often referred to as low-pH crystallization because the reactor operates at a pH of 2 to 3. Low-pH crystallization often can be combined with simultaneous forced oxidation to yield a marketable gypsum.

Double alkali desulfurization systems in large plants are limited because of the cost of sodium reagents, requirements for large solid/liquid separation equipment, and leaching problems associated with the disposal of wastes containing entrained sodium compounds.

Spray Dry-Scrubbing

A principal characteristic of wet lime/limestone scrubbing is the requirement for a wet purge or bleed stream in order to prevent the accumulation of excessive reaction products. The bleed stream must be treated to separate the water from the solids, to prepare the solids for their final disposition, and to return the water to the scrubbing process.

In the late 1960's, researchers at Rockwell International conceived spray dry-scrubbing as a simple alternative to the bleed stream with its required treatment steps.

Initially, the spray dry-scrubbing process was proposed as an enhancement to dry injection of nahcolite or trona ahead of a fabric filter. As a result, the original absorbents were solutions of soda ash and its related compounds. It was soon discovered that lime slurry also produced very good results when used as a reagent in the spray dryer. The first commercial utility system, the 410-MW Coyote Station near Beulah, North Dakota, used soda-ash solution. Because lime slurry is more economical than soda ash, subsequent installations have used lime slurry for the alkaline reagent.

Fig. 20 illustrates the spray dry-scrubbing process in which flue gas from the air preheater

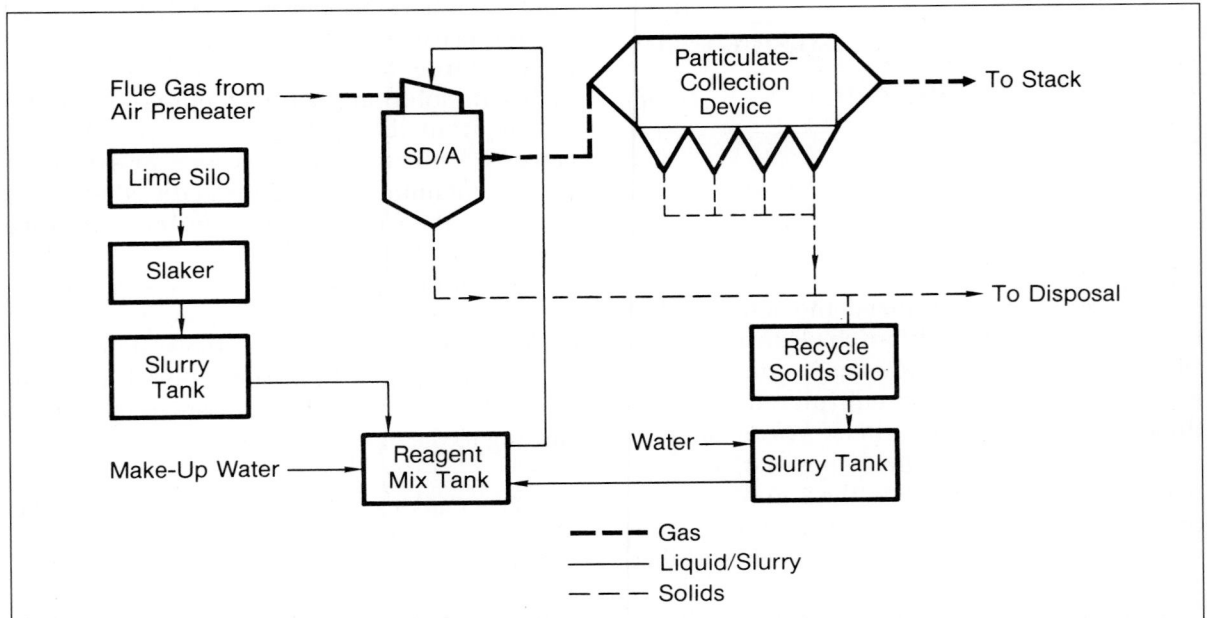

Fig. 20 Flow schematic of spray dry-scrubbing flue-gas cleaning system

is passed into a spray dryer. Spray drying is a unit operation used for more than eighty years in the food and chemical manufacturing industries. It replaces the spray tower in order to simultaneously achieve both absorption of the sulfur dioxide and thorough drying of the reaction product. The amount of reactant slurry introduced to the spray-dryer/absorber (SD/A) is controlled to insure that the reaction products leaving the SD/A will be completely dry. No crystallization tank is required: crystallization of the reaction products occurs quickly in the droplets as they dry and pass through the SD/A. The crystallized reaction products, together with any entrained flyash from the boiler, are collected in a fabric filter or electrostatic precipitator. It is a typical, but not necessary, practice to drop-out some reaction products and flyash in a hopper at the bottom of the SD/A.

The alkaline reactant for the spray-drying process is prepared by slaking pebble lime to produce a slurry of calcium hydroxide as follows:

$$CaO + H_2O \rightarrow Ca(OH)_2 + Heat$$

$$(11)$$

The slaking reaction must take place at temperatures above 170°F, which cause the reaction to occur quickly. The pebbles of CaO "explode", forming a very finely divided milk-of-lime suspension with high particle surface area. Because absorption, dissolution, and recrystallization processes must all occur very quickly in the spray-dryer/absorber, the lime-slurry particle surface area significantly affects how much reagent is required for a given sulfur-dioxide removal. The requirement for relatively fast dissolution of the reagent solids precludes the use of less costly pulverized limestone.

High slaking temperatures are achieved by maintaining the highest practical ratio of pebble lime to water in the slaker reactor. The reactor is usually one of three types: a paste mixer or pug-mill, a detention unit, or a ball mill. The ball mill provides for slaking while simultaneously reducing the size of impurities

which are always present in the pebble lime. Both pug-mill and detention slakers use a screen or settling unit to remove grit or rock-like impurities from the milk of lime. Pug-mill slakers, which allow the highest ratios of pebble-lime to water, can most reliably achieve the high slaking temperatures so important for the spray dry-scrubbing process.

The reagent slurry is prepared by mixing the milk of lime with additional process water as required to achieve a suitable degree of dryness of the solid effluent leaving the spray dryer. Reactant solids from the particulate collecting device may also be mixed into the reagent slurry. The practice of employing recycled solids in the reagent slurry tends to reduce the amount of fresh lime required for a given sulfur-dioxide removal. Most often, the recycled solids are first pre-mixed in a separate slurry tank and pumped to the reagent preparation tank. The reagent preparation tank may be located at ground level with pumps for supplying the slurry to the atomizers or at a high elevation above the atomizers to facilitate gravity feed. Some systems do not use a reagent preparation tank: the milk of lime, recycle slurry, and makeup water are mixed in the line supplying the atomizing device or directly in the device itself.

The most important unit operation in the spray dry-scrubbing system is atomization. Either air-assisted nozzles or rotary atomizers are commonly used. There are two major types of air-assisted nozzles, differentiated by where the slurry is mixed with the atomizing air–inside or external to the atomizer body. Internal-mix nozzles generally are more energy-efficient, although more expensive to fabricate, than the external-mix types. Rotary atomizers are comprised basically of a high-speed rotating atomizer wheel coupled to a drive device. In order to protect the load-bearing components of the drive device, the coupling is usually designed to absorb vibrations induced by imbalanced operation. Generally, the drive device is a two- or four-pole, three-phase electric motor with a speed-increasing gearbox, although in some systems a high-speed electric drive is used without a gearbox.

Rotary atomizers offer some significant advantages over air-assisted nozzles for large industrial and utility systems. They are

- more energy efficient for achieving a uniformly fine degree of atomization
- more capable of handling large slurry flow rates while at the same time producing a uniform spray pattern
- more suitable for atomizing highly abrasive slurries, especially those which contain recycled solids.

Almost as important as the quality of atomization is the mechanism for contacting the atomized slurry and flue gas in the spray-dryer/absorber. Commercial systems can have as many as three spray machines and inlets in one SD/A. Using multiple spray machines reduces the individual power requirements for each machine without appreciably affecting SD/A performance.

REGENERABLE (RECOVERY) PROCESSES

Although regenerable FGD systems can produce marketable sulfur or sulfuric-acid end products instead of essentially valueless wastes, such systems are more expensive and complex. They result in highly sophisticated

chemical plants generally unattractive to electric-power producers.

Limited application of this technology results from several factors. First, regenerable processes are energy-intensive as reducing sulfur compounds to elemental sulfur requires thermal energy. Second, the decomposition processes require substantial amounts of reducing gases including hydrogen sulfide, carbon monoxide, or hydrogen derived from natural gas or other hydrocarbon feed stock.

Aside from the availability problem, there is a reluctance to handle large quantities of potentially toxic reductants like hydrogen sulfide in the power plant. Also, regeneration processes are difficult to justify unless it is impractical to dispose of large quantities of wastes, or there is a secure market for the sulfur. Examples of some regenerable FGD systems follow; others were described in the Third Edition of this text.

Magnesium Oxide

The magnesium-oxide FGD process, a wet-slurry regenerable/recovery system, is shown as a simplified flow diagram in Fig. 21. As with all regenerable processes, the flue gas must have the particulate matter removed to prevent buildup of inert solids and contamination of

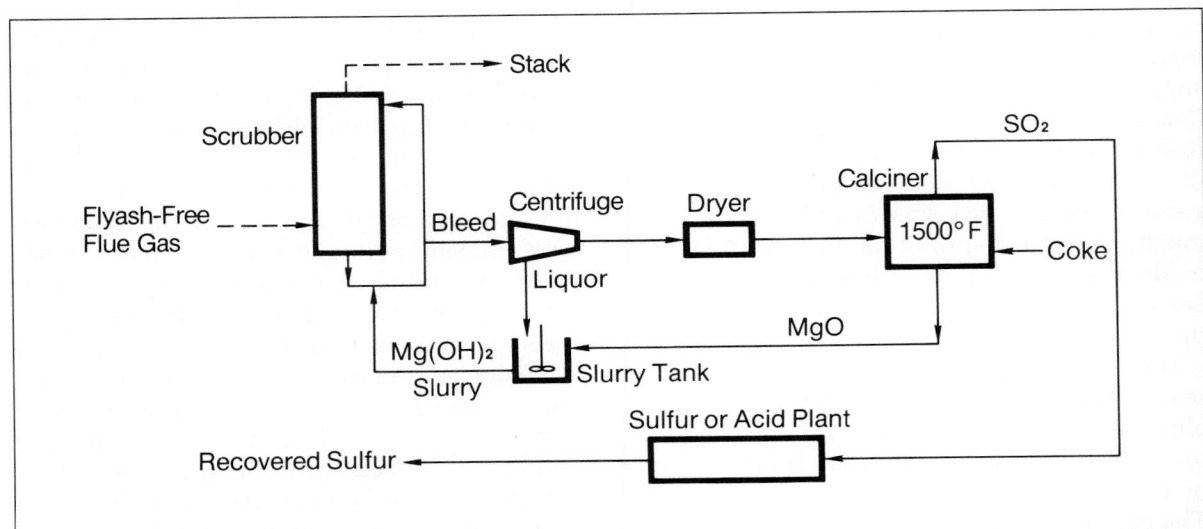

Fig. 21 **Simplified process diagram for magnesium-oxide recovery system**

the final product. Ash-free flue gas is scrubbed with a slurry of magnesium oxide and water in a suitable absorber. Depending upon specific process conditions, sulfur dioxide reacts with magnesium oxide, magnesium sulfite, or magnesium hydroxide to yield soluble magnesium bisulfite.

$$MgO + 2SO_2 + H_2O \rightarrow Mg(HSO_3)_2 \tag{12}$$

$$MgSO_3 + SO_2 + H_2O \rightarrow Mg(HSO_3)_2 \tag{13}$$

$$Mg(OH)_2 + 2SO_2 \rightarrow Mg(HSO_3)_2 \tag{14}$$

In the process, some oxygen is also absorbed into the slurry, an action that forms magnesium sulfate.

The magnesium bisulfite is neutralized with magnesium oxide to produce a magnesium bisulfite precipitate.

$$Mg(HSO_3)_2 + MgO + 11\,H_2O \rightarrow 2MgSO_3 \cdot 6H_2O \tag{15}$$

The magnesium-sulfite/sulfate crystals are withdrawn in a bleed stream from the absorber loop and sent to a centrifuge for dewatering. The product from the centrifuge is thermally dried and the liquor is recycled to the scrubber.

Dried magnesium-sulfite/sulfate crystals are calcined, usually under a reducing atmosphere at approximately 1500°F, to regenerate magnesium oxide and produce a gas stream concentrated in sulfur dioxide. The calciner exit gas is treated for particulate removal and then used as the feed stream to produce sulfuric acid or elemental sulfur. The magnesium oxide is recycled to the scrubber. Commercial magnesium-oxide FGD systems were built at Boston Edison's Mystic station, for an oil-fired boiler, and for coal-fired units at Potomac Electric

Power Company's Dickerson station and Philadelphia Electric Company's Eddystone station.

Sodium Sulfite

The sodium-sulfite process (Fig. 22) is a regenerative FGD process which combines absorption of soluble SO_2 with thermal regeneration of sodium sulfite to produce either elemental sulfur or sulfuric acid.[38] A sodium-sulfite solution absorbs SO_2 from the flue gas, thereby forming sodium bisulfite.

$$Na_2SO_3 + SO_2 + H_2O \rightarrow 2NaHSO_3 \tag{16}$$

A bleed stream is sent to the evaporator/crystallizer where sodium bisulfite is thermally decomposed to form solid sodium sulfite and gaseous SO_2.

$$2NaHSO_3 \xrightarrow{\Delta} Na_2SO_3 + H_2O + SO_2 \tag{17}$$

The sulfur dioxide is stripped with steam to yield a concentrated product stream containing approximately 90 percent SO_2 and 10 percent water. This stream is further processed to recover sulfuric acid or elemental sulfur. Sodium sulfite crystals are dissolved in water and recycled to the absorber. Sodium sulfate formed as a result of oxidation does not decompose in the evaporator and must be purged.

The sodium-sulfite process has been applied since 1971 in Japan on several oil-fired industrial and utility boilers. The only sodium sulfite processes prior to 1979 in the United States controlled SO_2 emissions from Claus sulfur and sulfuric-acid plants. Early in 1980 three systems were installed on coal-fired boilers, all designed to produce elemental sulfur for ultimate use in sulfuric-acid manufacture.

Activated-Carbon Absorption

Activated-carbon processes are dry regenerable FGD processes in which SO_2 is absorbed on activated-carbon granules in either fixed or flu-

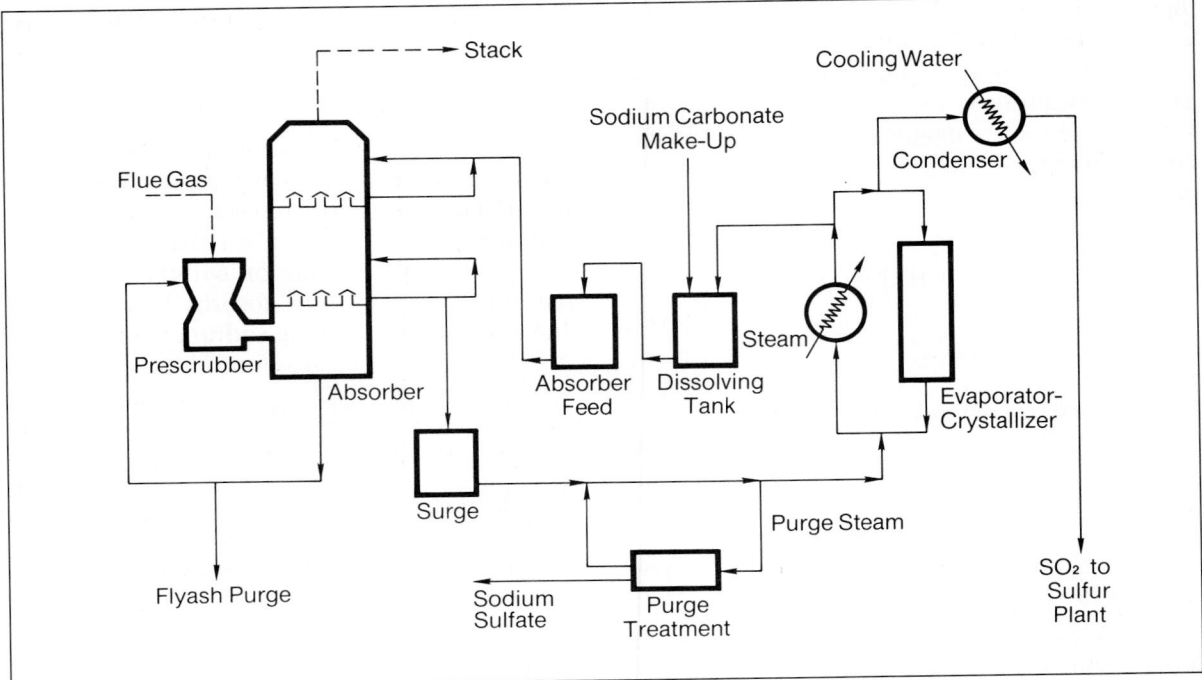

Fig. 22 Simplified process diagram for the sodium-sulfite system

idized-bed reactors. Sulfur dioxide is catalytically oxidized to SO_3 which then undergoes hydrolysis and forms sulfuric acid in the carbon pores.

$$SO_2 + \tfrac{1}{2}O_2 + H_2O \xrightarrow[\text{Carbon}]{\text{Activated}} H_2SO_4 \,(\text{Sorbed}) \tag{18}$$

The carbon, loaded with sulfuric acid, combines with hydrogen sulfide in a separate reactor to yield sulfur.

$$H_2SO_4 + 3H_2S \rightarrow 4H_2O + 4S \tag{19}$$

A third reactor removes the sulfur absorbed upon the carbon by reaction with hydrogen at 1000°F.

$$4S + 3H_2 \xrightarrow[\text{Carbon}]{\text{Activated}} 3H_2S + S \,(\text{Product}) \tag{20}$$

The reactor off-gas, consisting of sulfur vapor and hydrogen sulfide, is cooled to condense out the sulfur.

The ability of the process to consistently remove SO_2 with acceptable carbon degradation has to be demonstrated at the prototype and commercial scale.[39]

WET LIME/LIMESTONE FGD SYSTEMS

In the U.S., the Environmental Protection Agency (EPA) regulations require substantial reductions in gaseous and particulate emissions from power-producing facilities, depending on plant size, age, and the fuel being burned. To meet such standards, and depending on additional local regulations and other site-specific conditions, utilities and large industrial boiler operators have to consider all of the available flue-gas cleaning processes.

The viable options for particulate collection consist of electrostatic precipitation, fabric filtration, and high-energy wet scrubbing. The preferred options for SO_2 removal are lime/

limestone wet scrubbing and lime dry scrubbing. Integrated flue-gas cleaning systems consist of various combinations of particulate and SO_2 removal systems, as shown in Fig. 3.

WET PROCESS VARIATIONS

Lime/limestone wet-scrubbing systems for pulverized-fuel boilers fall into three categories, as described in the following sections.

ESP/Single-Stage Wet Scrubbing

The most common flue-gas cleaning system uses an electrostatic precipitator (ESP) followed by a lime/limestone wet SO_2 scrubber. After the flue gas has been treated in the precipitator, it passes through the induced fans and enters the SO_2 scrubber. If the required SO_2 removal efficiency is less than 85 percent, a fraction of the flue gas can be treated while bypassing the rest to mix with and reheat the saturated flue gas leaving the scrubber.

For higher sulfur fuels requiring SO_2 removal efficiencies of 90 percent or greater, the entire flue gas stream must be treated. Upon leaving the SO_2 absorption section, the flue gas is passed through entrainment separators to remove any slurry droplets mixed with the gas. The saturated flue gas may be reheated approximately 25 to 50°F above the water dew point before it is vented to the stack.

Baghouse/Single-Stage Wet Scrubbing

For certain types of fuels, the size of an electrostatic precipitator to meet current particulate emission standards becomes so large that fabric filters are a technical and economic alternative. The principal advantages of fabric filters (baghouses) are

- very high removal efficiency
- ability to collect fine particles
- insensitivity of collection efficiency to fuel characteristics

This latter advantage is significant because, contrary to the performance of a fabric filter, electrostatic precipitator performance is affected by the characteristics of the flue gas and flyash particles while fabric filter performance

is less sensitive to such changes. Thus, fabric filters give the power generator the option of switching fuels without modifying the particulate collection system. The major disadvantages of fabric filters include gas pressure drops of 4 to 6" WG and more maintenance compared with precipitators.

Two-Stage Scrubbing

Two-stage wet scrubbing, as shown in Fig. 23, can remove not only gaseous emissions but particulates as well. To collect flyash by wet scrubbing, a liquid is atomized to fine droplets which are then dispersed and mixed thoroughly with the flue gas. Atomization takes place by accelerating the gas through a venturi and using the momentum of the gas to shear the liquid into fine droplets. The initial large velocity differential between the liquid droplets and the flyash particles provides ideal conditions for collecting flyash by impaction and interception. Simultaneously with flyash collection, some of the SO_2 is absorbed in the venturi scrubber. A second-stage absorber accomplishes final SO_2 removal.

Two-stage wet scrubbing is particularly attractive when collecting flyash containing appreciable amounts of calcium, magnesium, and sodium. To a certain degree, the alkali content of the flyash reacts with the absorbed SO_2 and reduces lime or limestone consumption.

The primary advantage of particulate wet scrubbers is low initial cost compared to electrostatic precipitators and fabric filters. A significant disadvantage is the high gas-side pressure drop. For very low emission levels, pressure drops in excess of 20" WG are required for most applications.

WET-SCRUBBER DESIGN FACTORS

Lime/limestone-additive scrubber processes commonly use spray towers to absorb SO_2 from the flue gases. Whether operated in the countercurrent, co-current, or cross-current modes, such towers typically have few internals and very low liquid-residence time.

The scrubbers designed by the British in the

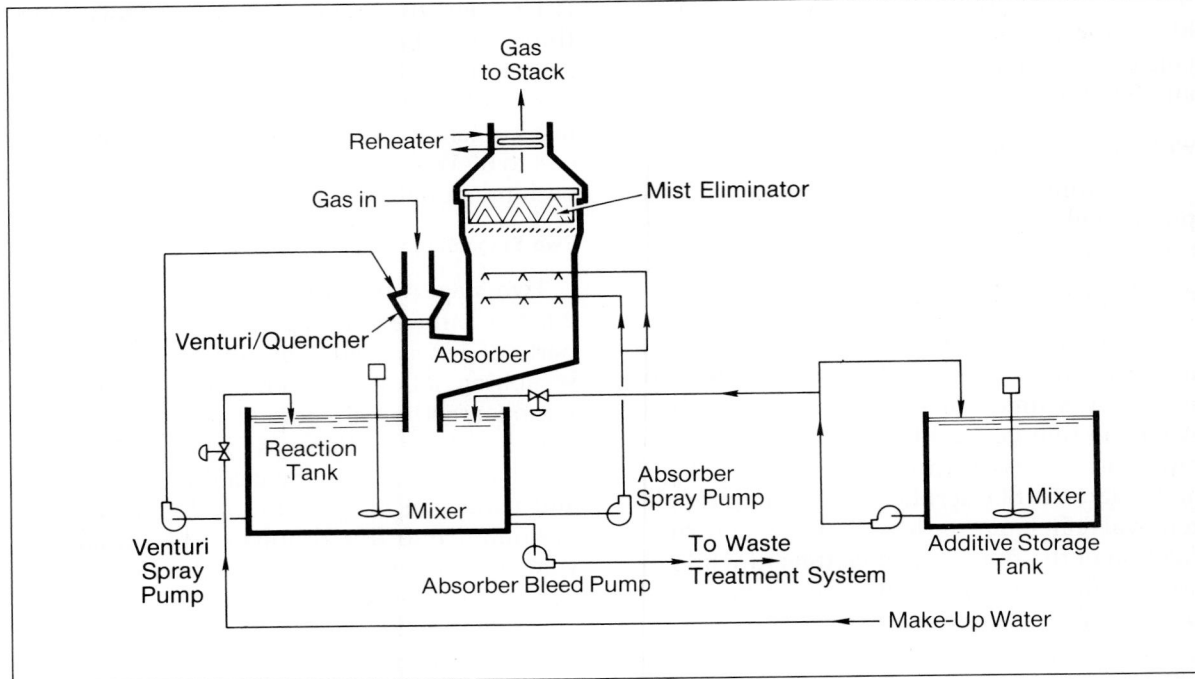

Fig. 23 Two-stage scrubbing flow schematic

1930's reflected the belief that high mass-transfer rates were required to absorb slightly soluble SO_2 gases. Thus, towers were designed with internals to promote liquid hold-up. In the 1960's, U.S. firms, acting on the British experience, adopted packed towers for simultaneous SO_2 and particulate removal.

More restrictive effluent-discharge regulations together with more effluent dewatering of waste solids made it necessary to operate FGD systems at higher concentrations of dissolved solids. In general, as the level of dissolved solids in lime/limestone slurries increases, the potential for scaling also increases. Scale formation in packed scrubbers can shut down a desulfurization system due to excessive gas-side pressure drop. The consequences of scale formation in spray-tower absorbers are considerably less severe.

Spray-Tower Design

Among the advantages of spray towers for lime/limestone FGD processes are

- maximum turndown ratio
- high allowable gas velocities
- low gas-side pressure drop
- low plugging potential
- simple mechanical construction

The most common spray-tower design is the vertical countercurrent type in which the gas enters the bottom of the absorber and flows upward through the absorption/spray zone.

In the typical absorber, multiple stages of atomizing nozzles distribute absorbent slurry into the gas. Designs usually have an optional spare stage. Depending upon the particular FGDS vendor, the nozzles are either hollow or full-cone spray type with capacities that range from approximately 250 to 1000 U.S. gallons per minute in large installations. Nozzles are designed to produce fine droplets over a large area at minimum spray-water pressure.

The sprayed slurry falls by gravity to the bottom of the absorber and is drained to a reaction tank. In the spray tower, above the spray stages, entrainment separators (mist eliminators) separate suspended slurry droplets from the gas stream. Zig-zag baffles are the most common type of entrainment separators because they offer low pressure drop, simple construction, and good cleanability.

Mass-Transfer Concepts

Sulfur dioxide absorption in countercurrent spray towers involves mass transfer from the gas phase to the liquid phase. The design of spray towers and other FGD wet absorbers requires an understanding of mass-transfer.[40]

The rate of mass transfer, N, is determined by the mass-transfer driving force, $y - y^*$, and the resistance to mass transfer, $1/K_gA$. Symbolically, the relationship may be expressed as

$$N = (y - y^*) \div (1/K_g A) = K_g A (y - y^*)$$

(21)

where

N = rate of mass transfer across interface, mol/hr

y = concentration of SO_2 in gas phase, mole fraction

y^* = concentration of SO_2 in gas phase in equilibrium with the existing liquid-phase composition, mole fraction

K_g = gas-phase mass-transfer coefficient, mol/hr-ft²

A = total interfacial area available for mass transfer, ft²

From material-balance considerations, the rate of mass transfer may also be expressed by

$$N = GS \Delta y$$

(22)

where

G = molar gas flux, mol/hr-ft²

S = cross-sectional area of tower, ft²

Δy = change in SO_2 concentration across tower, mole fraction

For any different element of spray-tower height dz, Eqs. 21 and 22 may be equated to give

$$GS dy = K_g (a) (S) dz (y - y^*)$$

(23)

where

dy = differential change in SO_2 gas-phase concentration, mole fraction

dz = differential increment of tower height, ft

a = available interfacial area per unit volume of spray tower, ft²/ft³

Rearrangement of equation 23 gives

$$dz = \frac{G}{K_g a} \cdot \frac{dy}{y - y^*}$$

(24)

where

$K_g a$ = overall gas-phase mass-transfer coefficient, mol/hr-ft³

Eq. 24 may be expressed in integral form as

$$Z = \frac{G}{K_g a} \cdot \int \frac{dy}{y - y^*} = H_{og} \cdot N_{og}$$

(25)

where

H_{og} = height of gas-phase transfer unit, ft

N_{og} = number of gas-phase transfer units

Z = total height of tower, ft

The overall mass-transfer coefficient, $K_g a$, may be evaluated by rearranging Eq. 25 to:

$$K_g a = \frac{G}{Z} \int \frac{dy}{y - y^*} = \frac{G}{Z} N_{og}$$

(26)

and then graphically solving the integral for the appropriate limits of y.

For spray-tower design it is essential to know the relationship between $K_g a$ and design variables such as gas velocity, tower height, number of stages, stage spacing, nozzle characteristics, and L/G. This information usually must be developed through pilot- and prototype-plant experiments. Fig. 24 illustrates how such experiments help.

The operating line, AC is based on test data, where points C and A are the gas-phase inlet and outlet SO_2 concentrations respectively. The slope of the line equals the liquid-to-gas ratio, L/G. The gas-liquid equilibrium line, BD, expresses the equilibrium relations between SO_2 in the liquid and gas phases and is a function of variables such as temperature and liquid composition.

As mentioned, $K_g a$ may be obtained by graphically determining the integral in Eq. 25 to obtain N_{og}. Although details of this integration procedure are not included here, understand that if the operating line AC and equilibrium line BD are close (due to changes in position of either the operating line or equilibrium line), the number of transfer units (N_{og}) will increase and make the removal process more difficult. As seen in Eq. 26, this situation would require an increase in $K_g a$ (G and Z being constant), or an increase in tower height, Z ($K_g a$ and G being constant). If the operating and equilibrium lines are further apart, the opposite relationships will exist.

Through a series of parametric experiments, a mathematical correlation between $K_g a$ and spray tower design variables can be estab-

lished. One such correlation developed by Combustion Engineering[40] is

$$K_g a = K_o (Y_{in})^a (Z)^b (L/G)^c \left[\frac{C_d \Delta P}{D}\right]^d (V_g)^e (Mg, Cl)^f$$

(27)

where

$K_g a$	= overall mass-transfer coefficient, mol/hr-ft^3
Y_{in}	= actual SO_2 concentration at inlet, mole fraction
Z	= spray tower height, ft
L/G	= liquid-to-gas ratio, gpm/1000 cfm (actual) at the outlet
C_d	= coefficient of discharge of the spray nozzle
ΔP	= pressure drop of liquid across the nozzle, psi
D	= Sauter mean droplet diameter of the spray liquid, μm
V_g	= superficial gas velocity, ft/sec
Mg,Cl	= concentration of magnesium and chloride in spray liquid
K_o, a, . . . , f	= correlation constants

Eqs. 26 and 27 are used simultaneously with mathematical models of lime/limestone slurry chemistry to select the optimal spray tower. Other models for prediction of SO_2 removal are available in the literature.[41]

LIME/LIMESTONE
FGDS PROCESS CHEMISTRY

The chemistry of wet scrubbing of SO_2 from flue gas with lime/limestone consists of a complex series of kinetic and equilibrium-controlled reactions occurring in the gas, liquid, and solid phases. Although the overall reactions are commonly expressed as:

$$\text{Lime: } CaO + SO_2 \rightarrow CaSO_3$$
$$CaSO_3 + \frac{1}{2}O_2 \rightarrow CaSO_4$$

(28)

$$\text{Limestone: } CaCO_3 + SO_2 \rightarrow CaSO_3 + CO_2$$
$$CaSO_3 + \frac{1}{2}O_2 \rightarrow CaSO_4$$

(29)

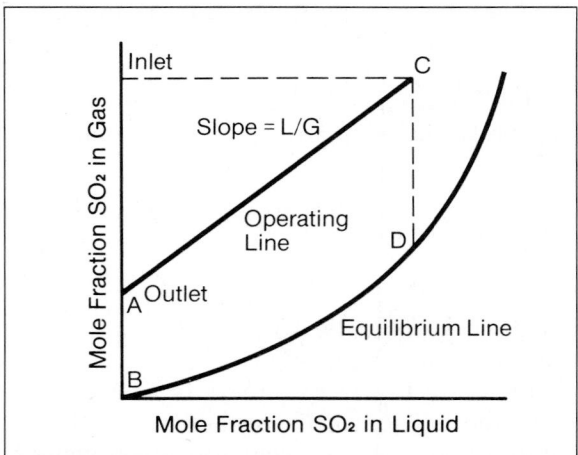

Fig. 24 Determination of $K_g a$

considerably more detail is required to emphasize and understand the effect of several key process variables such as pH and soluble magnesium and chloride concentration on the chemistry of lime/limestone scrubbing.

The chemical reactions occurring in lime/limestone systems can be written in several different forms, depending on which aspects are being discussed. For this discussion, the reactions have been categorized as absorption, neutralization, regeneration, oxidation, and precipitation. Although written sequentially, many of these reactions occur simultaneously in an actual system. Significant reactions are

Absorption:
$$SO_2 + H_2O \rightarrow H_2SO_3\,(HSO_3^- + H^+) \tag{30}$$

Neutralization:
$$H_2SO_3 + SO_3^= \rightarrow 2HSO_3^- \tag{31}$$

$$H_2SO_3 + HCO_3^- \rightarrow HSO_3^- + H_2CO_3 \tag{32}$$

Regeneration:
$$Ca(OH)_2 + 2HSO_3^- \rightarrow Ca^{++} + 2SO_3^= + 2H_2O \tag{33}$$

$$Ca(OH)_2 + 2H_2CO_3 \rightarrow Ca^{++} + 2HCO_3^- + 2H_2O \tag{34}$$

Oxidation:
$$HSO_3^- + \tfrac{1}{2}O_2 \rightarrow SO_4^= + H^+ \tag{35}$$

$$SO_3^= + \tfrac{1}{2}O_2 \rightarrow SO_4^= \tag{36}$$

Precipitation:
$$(m + n)\,Ca^{++} + m\,SO_3^= + n\,SO_4^= + x\,H_2O \rightarrow$$
$$Ca_{(m+n)}(SO_3)_m(SO_4)_n \cdot xH_2O \tag{37}$$

$$Ca^{++} + SO_4^= + 2H_2O \rightarrow CaSO_4 \cdot 2H_2O \tag{38}$$

Replacing Eqs. 33 and 34 with the following regeneration reactions makes it possible to describe limestone scrubbing:

Regeneration:
$$CaCO_3 + HSO_3^- \rightarrow Ca^{++} + SO_3^= + HCO_3^- \tag{39}$$

$$CaCO_3 + H_2CO_3 \rightarrow Ca^{++} + 2HCO_3^- \tag{40}$$

The following text relates this chemistry to key aspects of FGDS design. Specifically covered are the effects of absorbent composition on SO_2 removal, scale control, oxidation, and additive utilization.

Absorbent-Composition Effect

Vapor-liquid equilibrium between SO_2 in the gas and the absorbent liquid governs the amount of SO_2 absorbed from flue gas. If no soluble alkaline species are present, the liquid quickly becomes saturated with SO_2 and absorption is limited.

The neutralization reactions 31 and 32 remove sulfurous acid, H_2SO_3, from solution and allow the absorption reaction to proceed. In a spray tower, the soluble alkaline species in the spray slurry performs most of the neutralization due to the extremely short liquid residence time. If an absorber has greater liquid holdup, additional neutralization can be expected from dissolution of solid calcium sulfite or limestone. Thus, factors affecting soluble alkalinity in the spray slurry impact on the overall SO_2 removal capability of a lime/limestone FGDS. In the development of these processes, it is apparent that the three major factors affecting the soluble alkali level are pH, soluble-magnesium concentration, and soluble-chloride concentration.

Effect of pH

In general, with a lime/limestone FGDS, the higher the pH, the greater the SO_2 removal efficiency. Two factors are involved. First, by definition, a high pH increases the hydroxyl-ion (OH^-) concentration, which acts similar to sulfite and bicarbonate reactions 31 and 32. But at the pH ranges experienced in commercial systems — between 5 and 8 — the impact of hydroxyl-ion neutralization is negligible. So it is the second factor, the impact that pH has on the sulfite/bisulfite and bicarbonate/carbonic-acid equilibrium reactions, that has the primary beneficial effect on SO_2 removal.

Figs. 25 and 26 illustrate the impact of pH on these equilibria. In the range of pH from 5 to 8, increasing pH shifts the sulfite-bisulfite equilibrium in favor of sulfite. Shifting the equilibrium towards the sulfite region by increasing pH is beneficial to SO_2 removal. A similar condition exists with the carbonic acid (H_2CO_3)/bicarbonate equilibrium. Increasing pH favors existence of the bicarbonate ion which can neutralize sulfurous acid (reaction 32).

Effect of Magnesium

Magnesium enters the SO_2 scrubbing system through several mechanisms:
- as a constituent in limestone and lime
- from coal-ash constituents solubilized in the scrubbing liquor
- in the makeup water added to compensate for evaporation and other liquid losses.

The magnesium concentration depends on the total input from the various sources and upon the rate at which liquid leaves the system. In closed-loop operation, the accumulation increases until a magnesium compound, such as magnesium hydroxide, precipitates. In a commercial system some liquid containing magnesium leaves with the solids. Thus, the degree of solids dewatering affects the concentration of magnesium in a wet-process FGDS. Systems with vacuum filters or centrifuges for final solid/liquid separation operate at higher magnesium levels than those with a thickener only.[42]

Commercial operation of FGD systems has shown that soluble Mg in the absorbing slurry significantly improves SO_2 removal efficiency (Fig. 27). The increase is attributable primarily to the higher soluble-alkali levels in the presence of magnesium. In normal lime/limestone spray-tower operation, little of the additive dissolves within the tower because it remains there such a short time. Thus, the amount of sulfur-dioxide absorbed depends principally upon soluble-alkali content. If the alkali concentration in any region of the liquid drops, the rate of the neutralization reactions 31 and 32 also drops. Ultimately, a static equilibrium condition between gaseous and liquid SO_2 concentration is approached.

Where magnesium exists, the soluble alkali level of the absorbent increases primarily be-

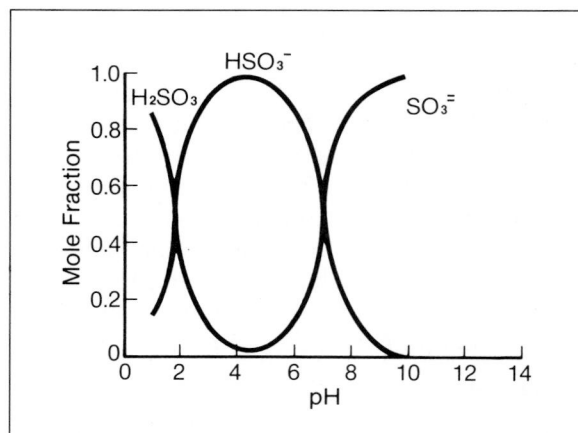

Fig. 25 Bisulfite-sulfite ratio versus pH

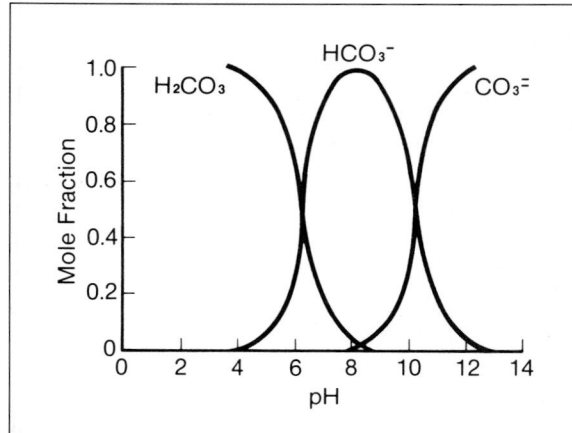

Fig. 26 Bicarbonate-carbonate ratio versus pH

cause of the presence of sulfite and bicarbonate salts of magnesium. As these magnesium alkalies are significantly more soluble than the corresponding calcium alkalies, there is an increase in the sulfur-dioxide absorption capacity of the slurry. The solubility of magnesium alkali compounds depends on pH. In a lime-additive FGDS, therefore, the proper system operating pH is very important to take full advantage of the magnesium. Examination of the sulfite-bisulfite and carbonate-bicarbonate equilibria and the magnesium solubility curve (Fig. 28) indicates the optimum pH to be in the 7 to 8 range. System operation at pH higher than 8 reduces the magnesium solubility and may prevent full use of the additive. Operation below a pH of 7 decreases the concentration of neutralizing bicarbonate and sulfite and the beneficial effect of magnesium on SO_2 removal.

Effect of Chloride

Chlorides generally find their way into an FGD system by one or both of the following: from the chloride present in the coal burned and subsequently dissolved as a hydrogen chloride gas (or as a salt in the scrubbing liquor), or with the makeup water, especially when cooling-tower blowdown or other plant waste water is used.

While magnesium improves SO_2 removal efficiencies by increasing the soluble alkalinity level of the scrubbing slurry, chloride has an opposite effect. Chloride ions combine with available magnesium and displace soluble sulfite and bicarbonate from solution. Many investigations have correlated SO_2 removal efficiency to "effective" magnesium concentration not "actual" magnesium concentration. "Effective" magnesium concentration is defined as:

$$Mg_e = Mg - \frac{Cl}{2.92}$$

(41)

where

Mg_e = effective Mg concentration, mg/ℓ
Mg = actual Mg concentration, mg/ℓ
Cl = actual Cl concentration, mg/ℓ

Using Eq. 41, it can be shown that, in the presence of 10,000 mg/ℓ of chloride, the beneficial effect of 5,000 mg/ℓ of magnesium is reduced to the equivalent of only 1575 ppm magnesium. In the absence of magnesium, it is generally accepted that increasing chloride-ion concentration reduces SO_2 removal efficiency through a pH-lowering effect. Some research-

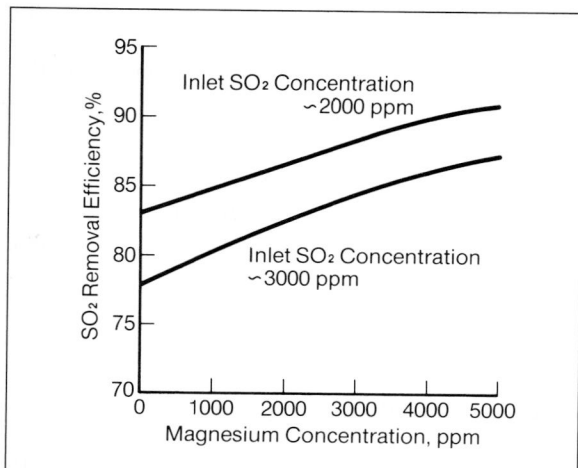

Fig. 27 Effect of magnesium on SO₂ removal efficiency

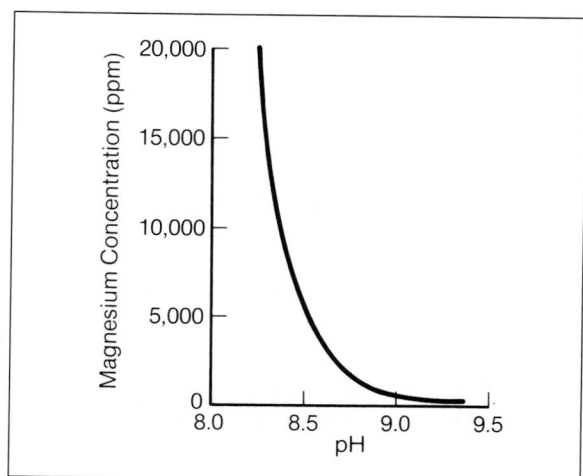

Fig. 28 Magnesium solubility at various pH values

ers, howevers, have failed to observe any SO_2 removal reduction for chloride-ion concentration up to 40,000 mg/ℓ.[43]

Scale Control

The principal compounds crystallizing in a flue-gas desulfurization system are calcium sulfite and sulfate (reactions 37 and 38). The major concern with crystallization of these compounds is the potential for scale formation on scrubber internals. Uncontrolled scale can plug the spray nozzles, mist eliminators, piping, and other equipment. Plugged systems must be shut down and cleaned.

As the crystallization of calcium sulfite is pH-sensitive, abrupt pH increases (especially in lime systems) must be avoided. Control of the spray slurry pH has curtailed sulfite scaling in commercial systems.

Control of calcium-sulfate scaling is more difficult, and the normal 5–8 pH range encountered in operation is not a factor influencing calcium-sulfate saturation or crystallization. Control of calcium sulfate relative saturation depends on seed-crystal concentrations and crystallization time. Any increases in the relative saturation of calcium-sulfate to high values can be conducive to heterogeneous nucleation and the consequent scaling of scrubber components.

Work on scale-control techniques performed by C-E and others in 1974 and 1975 revealed that there was an alternate mode of FGDS operation—subsaturated operation.[42,44] Under some conditions, coprecipitation of sulfite within the calcium-sulfate crystal lattice can occur to such an extent that it removes all the sulfite formed through oxidation. When this occurs, the scrubbing slurry does not become saturated with calcium-sulfate, and scale cannot form. The following sections will discuss various aspects of scale control including relative saturation and scale potential, supersaturated operation, and subsaturated operation.[44]

Relative Saturation and Scaling Potential

As the scrubbing liquor absorbs SO_2 from flue gas, it is in intimate contact with gases containing typically 3 to 5 percent oxygen. The reaction between the oxygen, which is absorbed, and the sulfite/bisulfite in the liquid leads to sulfate formation. Since there may be no accumulation of material at steady-state, sulfate must be continually purged from the system. Although sulfate may theoretically be purged entirely in the liquid phase, this mode is generally unacceptable from an environmental standpoint—large quantities of sulfate would be discharged into the local water.

In practice, sulfate is generally purged from lime/limestone wet scrubbers as relatively insoluble calcium salts which are dewatered and disposed of as solid waste. Thus, it is extremely important to control the reactions of crystallization which form the calcium-sulfate salts. If these reactions are not properly controlled, serious plugging can occur because of a hard scale of calcium-sulfate dihydrate ($CaSO_4 \cdot 2H_2O$, gypsum) that may form.

The equilibrium and kinetic factors in crystallization reactions involving calcium sulfate are complex. Slurries containing high concentrations of salts behave as non-ideal solutions and the relationship between the concentration of calcium sulfate and its equilibrium solubility is best defined in terms of activity coefficients. Activity coefficients are factors which, when multiplied by the concentrations of the respective ions, yield their activities. In a slurry, calcium sulfate does not behave ideally, hence the use of the activity in defining the solubility product, K_{sp}.

$$K_{CaSO_4} \cdot 2H_2O = A_{Ca}A_{SO_4} = 1.9 \times 10^{-5} \text{ at } 50°C$$

(42)

where

$K_{CaSO_4} = A_{Ca}A_{SO_4}$ = solubility product of C_aSO_4
A_{Ca} = activity of calcium ion
A_{SO_4} = activity of sulfate ion

If the activity product of the calcium and sulfate ions is less than the solubility product, the solution is said to be subsaturated. If the prod-

uct is greater than K_{sp}, the solution is supersaturated. The ratio of the activity product to the solubility product is called the relative saturation (RS) and is used as a measure of the saturation level of a solution with respect to gypsum.

$$RS = \frac{A_{Ca}A_{SO_4}}{K_{sp}, CaSO_4 \cdot 2H_2O} \qquad (43)$$

when

RS < 1.0, solution is subsaturated;
RS = 1.0, solution is saturated;
RS > 1.0, solution is supersaturated.

The importance of gypsum relative saturation in designing lime/limestone wet-scrubbing systems for scale-free operation is shown in the following discussions.

Supersaturated Operation—Gypsum Crystallization

Formation of calcium-sulfite and calcium-sulfate scale on scrubber internals can be a severe problem and, while control of calcium-sulfite scaling is possible by avoiding rapid increases in slurry pH, control of calcium-sulfate scaling is not a function of slurry pH.

Considerable effort was made in the early 1970's to establish an expression to describe the desupersaturation rate of calcium sulfate as a function of appropriate system design and operating parameters.[42]

In addition to developing a crystallization-rate equation, it was necessary to determine the level of calcium-sulfate supersaturation that could be tolerated without causing scaling of internal surfaces in any operating scrubber.

A series of experiments to determine the critical supersaturation above which scaling can occur revealed that a high probability of scaling exists if the absorber effluent relative saturation exceeded 1.35. Thus, Fig. 29 shows the supersaturated operating range between relative saturation 1.0—1.35 where scale formation is unlikely and operational difficulties slight.

After establishing a calcium-sulfate relative saturation region for scale-free operation, the industry worked on developing a crystalliza-

tion rate expression.

A crystallization rate equation was developed from bench-tests

$$R = knS \qquad (44)$$

where

R = reaction rate
k = rate constant
n = number of seed crystals
S = driving force

Interpretation of the data collected suggested a driving force term of the form (RS – 1). It also showed that the rate was directly proportional to seed-crystal concentration.

Further work was performed to calibrate the rate expression with data from full-scale, commercial FGD systems.[44,45] The rate expression may be used to predict the seed-crystal concentration required to prevent calcium-sulfate scaling. The rate of calcium-sulfate precipitation in the reaction tank, R, can be computed by material balance if the following information is known: SO_2 absorption rate, fraction SO_2 oxidized to $SO_4^=$, and reaction tank residence time. Sulfur dioxide absorption rate and reaction tank residence time are design param-

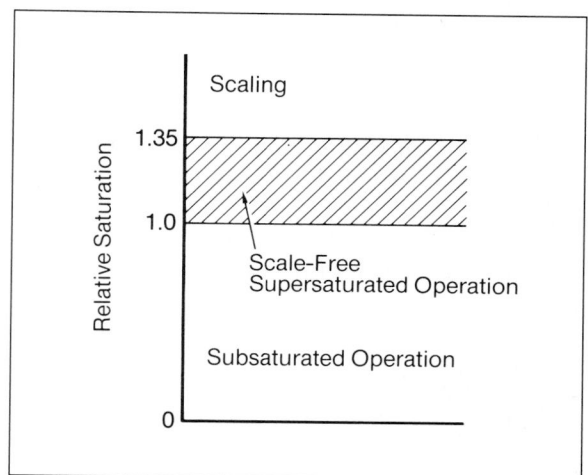

Fig. 29 Relative saturation of gypsum as an indication of scaling potential

eters, and fraction SO_2 oxidized can be predicted from various design models. The maximum relative saturation allowable in the absorber effluent is 1.35; values above this critical level will result in scale formation. By using information on the absorbent composition (such as magnesium and chloride concentrations), rates of oxidation, ionic strength/activity coefficient correlations, and the level of saturation entering the absorber which will result in an absorber effluent saturation of 1.35 can be calculated. This value and the rate of calcium-sulfate precipitation are used in Eq. 44 to predict the seed-crystal concentration required in the reaction tank to maintain the absorber effluent relative saturation below 1.35. Thus, the solids concentration circulated to prevent scaling can be computed and used as an operating point.

Subsaturated Operation—Coprecipitation

Because crystallization is thermodynamically impossible from subsaturated solutions, the potential for scaling can be eliminated by operating the system at a relative saturation less than 1.0. Until recently, the only technique known to accomplish this was liquid blowdown. However, discharging large quantities of process water to the environment is unacceptable. Work by EPA suggested an alternate method for purging sulfate from solutions that are subsaturated with respect to gypsum: coprecipitation of sulfate with calcium sulfite to form a mixed crystal.[46] Although the exact nature of this mixed crystal is not fully understood, several interesting facts have evolved:
- significant amounts of sulfate can be coprecipitated with calcium sulfite from solutions which are subsaturated with respect to gypsum
- the mole ratio of sulfate to sulfite in the mixed crystal can be as high as 0.2—0.3
- x-ray diffraction analysis of samples of the mixed crystal indicates only the presence of calcium-sulfite hemihydrate ($CaSO_3 \cdot \frac{1}{2} H_2O$); no known calcium-sulfate forms are observed.

Bench and pilot scale experiments aimed at studying this phenomenon led to the development of mathematical correlations which related the amount of sulfate coprecipitated to various system parameters. An example of one of the more useful correlations is shown below:

$$MR = KA_{Ca}{}^x \cdot A_{SO_4}{}^y$$

(45)

where
MR = mole ratio of SO_4 to SO_3 in the co-precipitated solid
A_{Ca}, A_{SO_4} = activities of calcium and sulfate ion, respectively
K,x,y = experimental constants

Numerical values for the experimental constants were determined to be
K = 231
x = 0.54
y = 0.76

Fig. 30 is a plot of over 60 bench-scale experiments and several data points taken from full-scale commercial installations. The dependency of mole ratio (MR) on the activities of calcium and sulfate has been demonstrated over a wide range of solution compositions and kinetic conditions. The ranges studied included:

Calcium	60–6,000 mg/ℓ
Sulfate	600–20,000 mg/ℓ
Sulfite	100–6,000 mg/ℓ
Magnesium	0–5,000 mg/ℓ
Chloride	0–14,000 mg/ℓ
Sodium	100–5,000 mg/ℓ
Sulfite precipitation rate	0.15–2.3 millimol/ℓ-min.

The mathematical model can be used to predict the effect of oxidation on the level of gypsum saturation in a subsaturated lime/limestone FGDS.

The left side of Eq. 45 consists of the term, MR, the mole ratio of sulfate to sulfite in the coprecipitated solid. Because nearly all forms of sulfur leave the system in the solid phase, the ratio of sulfate to total sulfur in the solids

equals the oxidation in the system. It can be shown that the following relationship exists between MR and system oxidation:

$$\frac{1}{MR} = \frac{1}{OX} - 1$$

(46)

where
$$OX = \text{fraction oxidized} = \frac{SO_4}{SO_4 + SO_3}$$

The right side of Eq. 45 contains terms for calcium and sulfate activities similar to the definition of RS. Thus,

$$\underset{\substack{\text{(related to} \\ \text{oxidation)}}}{MR} = K \underset{\substack{\text{(related to} \\ \text{relative saturation)}}}{A_{Ca}{}^x \cdot A_{SO_4}{}^y}$$

(45)

Eq. 45 can, therefore, be used to relate relative

saturation to system oxidation. As oxidation can be predicted by other design models, the level of gypsum saturation can be predicted using Eq. 45. Furthermore, the relationship is direct and by increasing the oxidation we increase the saturation.

Maximum Tolerable Oxidation

One particularly useful concept is that of maximum tolerable oxidation (MTO) which is the oxidation level that will cause a lime/limestone FGD system to run saturated with respect to gypsum (RS = 1.0). Maximum tolerable oxidation can be calculated using the proposed design equation. By using an equilibrium program or other experimental or theoretical techniques, the activities of calcium and sulfate for a saturated solution can be calculated. If these values are inserted into Eq. 45, it is possible to calculate the mole ratio of sulfate to sulfite in the coprecipitated solids in equilibrium with this solution. Oxidation can then be calculated

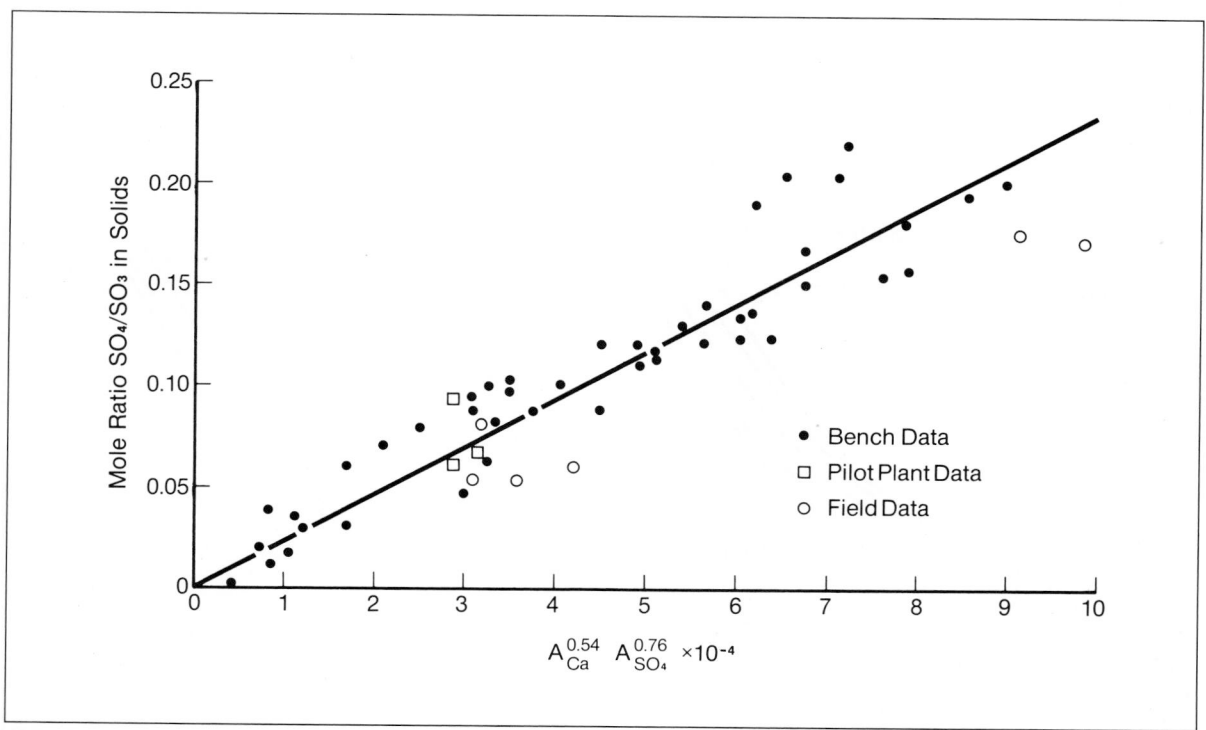

Fig. 30 $A_{Ca}{}^{0.54} A_{SO_4}{}^{0.76}$ versus mole ratio of SO_4 to SO_3

from Eq. 46. For example, in a solution containing no magnesium or chloride, the activities of calcium and sulfate in a saturated solution are approximately 0.0049 and 0.0039 mol/ℓ respectively at 50°C; this translates to approximately 16-percent oxidation (Eq. 46).

If the system operates at 16-percent oxidation, the reaction tank will operate saturated with respect to gypsum, assuming that the precipitation reaction occurs in the reaction tank. Higher oxidation levels will result in supersaturated operation; lower oxidations will result in subsaturated operation. Thus, the MTO for a system containing no magnesium or chloride is 16 percent.

The design equation can be used to predict the relative saturation of gypsum for subsaturated systems as a function of magnesium and chloride concentrations, and temperature. As a consequence, it can also be used to predict whether a system will operate in the subsaturated mode. Calcium and sulfate ion activities can be predicted for solutions of various compositions (i.e., magnesium/chloride concentrations and saturations) and temperatures. Using these activities it is possible to calculate the mole ratio of sulfate to sulfite in the coprecipitated solid and, thus, the oxidation required to produce that mole ratio.

Fig. 31 is a plot of relative saturation versus oxidation as a function of varying magnesium and chloride concentrations obtained by this method. The presence of chloride shifts the curves to the left resulting in high saturations for the same oxidation, while magnesium has the opposite effect. It also can be seen that the presence of chloride diminishes the beneficial effect of magnesium. In addition, the oxidation that corresponds to relative saturation of 1.0 represents the maximum tolerable oxidation for subsaturated operation at these conditions.[43]

Oxidation

In lime/limestone FGD systems, solutions containing sulfite ($SO_3^=$) come in intimate con-

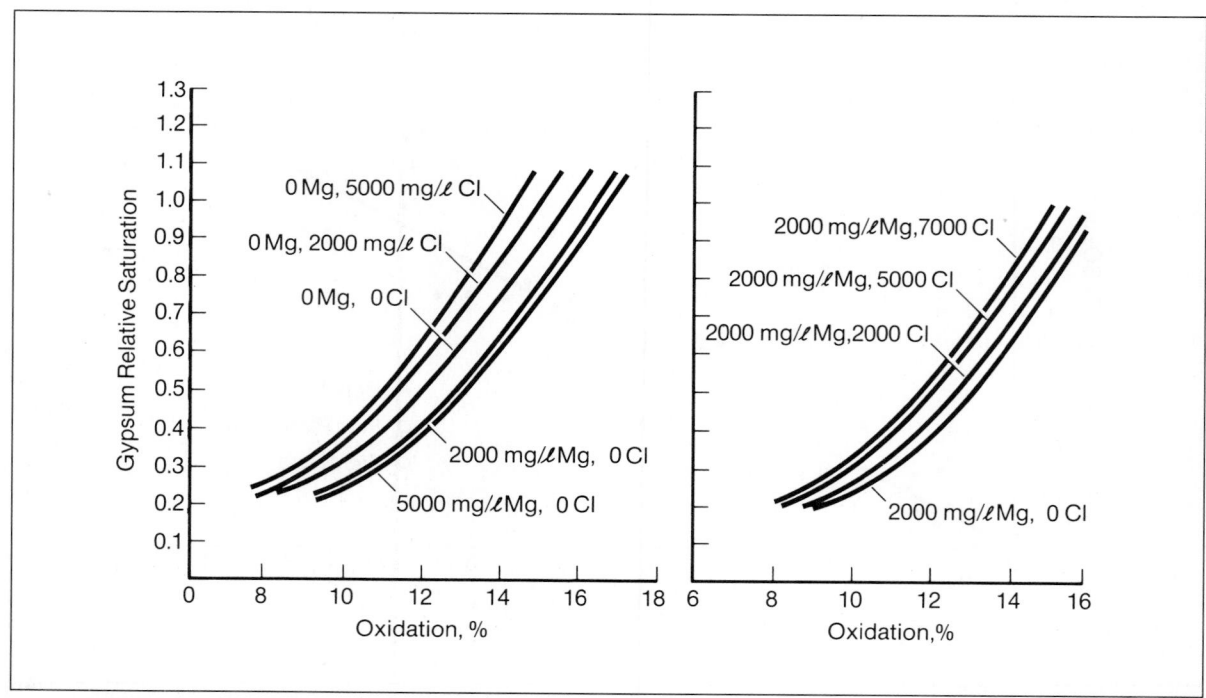

Fig. 31 **Effects of Mg and Cl on gypsum relative saturation as a function of oxidation**

tact with flue gas containing typically 3 to 7 percent oxygen. As a result, significant quantities of $SO_3^=$ (S^{+4}) are oxidized to $SO_4^=$ (S^{+6}). The rate at which sulfate is formed directly determines the required rate of sulfate purge at steady-state which in turn influences the level of calcium-sulfate saturation. The relative amounts of sulfate and sulfite formed in the process also have implications in the ultimate disposal of FGDS waste product. In general, sludges containing high sulfate to sulfite ratios are more easily dewatered and result in a more stable disposal product. Thus, the rate of sulfate ion formation (rate of oxidation) has widespread implications on the performance and reliability of wet SO_2 scrubbing processes. A quantitative understanding of the factors which contribute to sulfate ion formation is essential if lime/limestone wet scrubbing is to have optimum performance.

Two major oxidation process options exist: natural and forced. In natural oxidation, conversion results primarily from a reaction between soluble sulfite in the scrubbing slurry and oxygen in the flue gas. Only a minute amount of oxidation occurs from contact with atmospheric oxygen in open tanks, overflow weirs, and the like.

Several factors influence the rate of oxidation: oxygen concentration in the flue gas, ionic strength and pH of the absorbent slurry, and the liquid-to-gas ratio in the absorber.

In forced oxidation, on the other hand, compressed atmospheric air is injected into the process (usually in the reaction tank) to promote oxidation and achieve oxidation efficiencies of 90 to 99+ percent. Forced oxidation improves the dewatering and structural characteristics of the waste product and consequently reduces disposal costs. The following text discusses factors which influence oxidation in lime/limestone scrubbing systems.

Natural Oxidation

The major source of oxidation in a naturally oxidized FGDS is the absorber in which the oxygen in flue gas reacts with the soluble sulfite in the scrubbing slurry. The reaction between oxygen and sulfite in the aqueous phase is very complex and can be greatly influenced by the presence of catalysts and inhibitors. The rate of the oxidation reaction is nearly always much faster than the rate of oxygen transfer to the liquid phase. Thus, oxidation in FGD systems is generally mass-transfer limited and the rate of chemical reaction can be assumed to be instantaneous. Mathematical models of the oxidation process can then be expressed in terms of classical two-film mass-transfer theory.

An example of the type model which has been developed to describe the rate of oxidation in a spray tower absorber is

$$r_{NO} = \frac{2GC_oRT}{V_AP}\left[1 - \exp\left\{\frac{K_La\,PV_T}{HG}\right\}\right] \tag{47}$$

where

r_{NO} = rate of natural oxidation, mol/min-gal
G = gas flow rate, mol/min
C_o = oxygen content in flue gas, mol/gal
R = gas constant, gal/atm-mol-°R
T = gas temperature, °R
P = total pressure of gas, atm
K_La = overall volumetric mass-transfer coefficient, min^{-1}
V_A = volume of liquid in absorber, gal
V_T = absorber volume, gal
H = Henry's law constant, gal/atm-mol

All variables in Eq. 47, except K_La, are design parameters which are known or may be estimated. The overall mass-transfer coefficient cannot be determined on theoretical grounds except in the most simple cases and must therefore be determined experimentally. C-E has developed mathematical correlations which predict K_La as a function of operating and design parameters for various absorber types. An example of the equation form used to predict K_La for spray-tower absorbers is

$$K_La = a\,(L/G)^b\,(I)^c\,(pH)^d\,(V_G)^e\,(Z)^f\,(C_o)^g \tag{48}$$

where

$$L/G = \text{liquid-to-gas-ratio}$$
$$I = \text{ionic strength of absorbent slurry}$$
$$pH = \text{pH of spray slurry}$$
$$V_G = \text{superficial gas velocity in absorber}$$
$$Z = \text{height of contactor}$$
$$a,b,c,d,e,f,g = \text{experimental constants}$$

With the oxidation rate known, the fraction of SO_2 oxidized can be calculated by dividing the oxidation rate by the SO_2 absorption rate. The fraction oxidized can then be used in the various scale-control models to arrive at appropriate operating conditions, and to predict the characteristics of the disposal product.

Forced Oxidation

Forced oxidation uses atmospheric air injected into the reaction tank of FGD systems. To achieve the best efficiency, the injected air must be finely divided into tiny bubbles to maximize the interfacial area. The dispersion is usually accomplished by using either sparge rings and radial turbine mixers, or jet eductors. With either arrangement, the ultimate problem is to define and predict the impact of various process design and operating parameters on the oxidation rate. This can be achieved in a manner similar to predicting the level of natural oxidation occurring in an absorber. The two-film mass-transfer approach is:

$$r_{FO} = \frac{2G_A C_O RT}{h\,P}\left[1 - \exp\left\{-\frac{K_L a P\,h}{G_A\,H}\right\}\right] \tag{49}$$

where

$$r_{FO} = \text{rate of forced oxidation}$$
$$G_A = \text{airflow rate}$$
$$h = \text{height of reaction tank}$$

Similar to the absorber natural oxidation model, all of the variables except $K_L a$ are known or may be estimated. A correlation describing $K_L a$ as a function of pertinent process variables for a sparge ring/radial turbine forced-oxidation system is as follows:

$$K_L a = K_2 (h)^a\,(I)^b\,(pH)^c\,(V_{air})^d\,(D)^e\,(P_M)^f \tag{50}$$

where

$$K_2 = \text{experimental constant}$$
$$V_{air} = \text{superficial air velocity in tank}$$
$$D = \text{tank diameter}$$
$$P_M = \text{mixer power input to tank}$$

It is generally accepted that oxidation is increased by decreasing pH, increasing mixer power, increasing ionic strength, and increasing tank height.

Note that the total oxidation achieved in a forced oxidation FGDS is a function of the naturally occurring oxidation, for which credit may be taken.

ADDITIVE UTILIZATION IN WET SCRUBBERS

Minimizing the amount of alkali additive required to obtain a desired SO_2 removal efficiency reduces the operating cost of an FGD system. To determine the factors influencing additive use, it is necessary to understand the additive dissolution process and the effect of system operating conditions, design, and additive properties on this process.

In a lime or limestone SO_2 scrubbing system, the addition of alkali is controlled to neutralize the SO_2 absorbed and precipitate the calcium-sulfur salts. Although in theory the amount of lime or limestone required is stoichiometrically equal to the amount of SO_2 removed, the kinetics of the limestone dissolution process dictates the use of excess additive to achieve the desired SO_2 removal efficiency.

Additive stoichiometry is usually defined as the mols of additive fed (excluding impurities) to a scrubber per mol of SO_2 absorbed (although it can be, and sometimes is, defined as the mols of additive fed per mole of SO_2 entering the scrubber). Additive utilization, then, can be expressed by the following relationship:

$$\text{Utilization} = \frac{1}{\text{Stoichiometry}} = \frac{\text{moles } SO_2 \text{ absorbed}}{\text{moles additive fed}} \tag{51}$$

Although not usually significant, the absorption of other acid gases such as HCl can influence utilization of the additive. Utilization then could be more broadly defined as the fraction of additive converted to reaction products in the scrubber. However, unless a detailed material balance is done, it is more straightforward to use Eq. 51 to calculate utilization from system operating data.

From Eq. 51 it can be seen that, to increase utilization, one must either reduce the additive feed rate while maintaining constant SO_2 absorption efficiency or increase the SO_2 removal efficiency for a given limestone feed rate.

The dissolution rate of alkali additives is inversely proportional to the solution pH; therefore, better utilization is possible by operating a system at low pH levels. However, as pointed out earlier, reducing the pH also can reduce SO_2 removal efficiency. Therefore, it is usually more productive to increase the absorber performance for the same system pH, or increase the additive dissolution and utilization at a constant system pH.

Lime Utilization

Lime or calcium oxide (CaO) is obtained from calcination of limestone ($CaCO_3$). The conditions under which the limestone is calcined or "burned" can have a great impact on the reactivity of the lime produced. If the limestone is calcined at too high a temperature, the lime may become dead-burned and unreactive, making it unsuitable for FGDS use.

Because of the high reactivity of most commercially available lime, it is possible to achieve high utilization (as much as 95 percent) with minimum impact on the remainder of the system. The main design parameter in maximizing utilization is the pH of the scrubbing slurry. With pH maintained above 8.5, considerable carbon-dioxide absorption occurs because of the 10 to 15 percent by volume of CO_2 in the flue gas. Carbon dioxide reacts with lime in a manner similar to sulfur dioxide to form calcium carbonate. Calcium carbonate, being insoluble at high pH, precipitates and thus prevents the calcium from reacting with

sulfur dioxide. The following summarize the effect of CO_2 absorption:

$$CaO + CO_2 \rightarrow CaCO_3(aq) \tag{52}$$

$$CaCO_3(aq) \rightarrow CaCO_3(s) \tag{53}$$

Data from a full-scale commercial FGDS indicate that a decrease in utilization from 95 percent to 85 percent may occur as pH is increased from 7.5 to 9.0. To reduce waste of the relatively high-cost lime, the pH of lime systems should thus be held below 8.5.

Limestone Utilization

Limestone is usually less reactive than lime in FGD systems operating under identical conditions. For this reason, system modifications are needed to produce limestone utilizations in excess of 90 to 95 percent.

Limestone dissolution is a function of surface area of the suspended limestone and hydrogen-ion concentration (i.e., pH) of the solutions as follows:

$$r = KA\,[H^+] \tag{54}$$

where

r = rate of limestone dissolution, mol/ℓ-s
K = experimental constant, s/cm^2
A = surface area of suspended limestone, cm^2
H^+ = hydrogen ion concentration (10^{-pH}), mol/ℓ

Use of Eq. 54 suggests two methods of increasing the limestone dissolution rate and, thus, utilization: first, by increasing the surface area, A, and second, by increasing the hydrogen-ion concentration (reducing pH).

Large surface areas can be produced by finely grinding the limestone before it is injected into the system. Limestone is typically ground to 80 to 90 percent through 325 mesh in FGDS applications. Finer grinds are possible, but more

pulverizer power is required. This is many times cost-effective.

Limestone utilization can also be improved by operating at reduced pH's. As mentioned previously, this impairs efficiency—to maintain the same removal at reduced pH, the absorber design must be modified. For example, L/G may be increased (with higher pumping costs) or tower height may be increased (with higher capital costs). The cost-effectiveness of any such adjustments must be examined.

It is also possible to use a two-stage operation in which the primary stage is operated at reduced pH to optimize lime/limestone use, and the second stage is operated at high pH to accomplish the required SO_2 removal. In this operation, limestone is fed to the second stage and the spent slurry from the second stage is fed to the primary stage before being bled off.

A final technique for improving limestone utilization is to increase residence time—at a fixed rate of reaction, the longer a limestone particle remains in the absorber/reaction-tank loop, the greater will be the fraction of the particle that dissolves.

The residence time of the solid particles in the process is a function of the particle inventory and the solids bleed rate. Because the bleed rate is fixed by the SO_2 absorption rate at steady-state, only the inventory can be adjusted. The solids inventory can be made greater by increasing the reaction-tank volume and/or the solids concentration. Because the slurry is abrasive, the solids concentration has a practical upper limit of about 15 percent. The relatively low cost of tanks and low mixer power requirements suggests that, in many situations, it is less expensive to increase the tank size to improve limestone use.

SPRAY-TOWER COMPONENTS

In addition to viable process chemistry, successful spray-tower performance requires the proper selection of system components. These consist principally of the tower shell, the slurry delivery subsystems (pumps, piping and nozzles), the mist eliminator, and the gas reheater. Each must be designed with materials for good reliability and performance in a corrosive and erosive environment.

Tower Shell

Velocity considerations dictate the tower cross-sectional area (Fig. 32). Although higher gas velocities through the absorbing section improve SO_2 removal, the maximum allowable velocity through the close-coupled mist eliminator section (about 10 to 11 ft/sec) sets a practical limit of 12 to 13 ft/sec through the absorbing section. A short transition piece with a gradually expanding area connects the mist eliminator section to the absorber section. The number of absorbing stages required to obtain the desired level of SO_2 removal governs the overall tower height.

The corrosive and erosive environment within the tower requires special materials. Type 316L and 317L stainless steels have been successfully applied in numerous instances. Under certain conditions of scrubber-slurry pH, temperature, and chloride concentration, the use of nickel-based alloys with higher molybdenum and chromium contents can provide superior corrosion protection.[47] The use of such high-grade alloys requires very careful fabrication. In general, failures of high-grade alloys incorporated in the tower shell and internal support members occur primarily because of faulty welding rather than corrosion.

In contrast to the spray-tower shell, the gas ducts upstream and downstream of spray towers represent much more severe corrosive environments. Absorber inlet and outlet ducts, and reheat mixing zones, require special consideration with respect to material selection. Protective linings such as inorganic cements or borosilicate glass blocks over carbon or stainless-steel plate are typically used for the absorber inlet ducting.

Pumps, Piping and Nozzles

Fig. 32 shows independent absorption stages, each consisting of a spray pump, piping, manifold, headers, and spray nozzles. If a spare pump is included, a complete spare absorption stage is also provided. This arrange-

Fig. 32 **Typical spray-tower absorber module**

ment provides maximum flexibility and allows tailoring the spray flow to system requirements by removing absorption stages from operation in steps.

Rubber-lined centrifugal spray pumps provide erosion resistance and protection against unexpected process upsets in pH control. A tall reaction tank provides a large net-positive-suction head to prevent pump cavitation which can damage rubber pump liners.

Fiberglass-reinforced plastic (FRP) spray piping with a abrasion-resistant internal liner provides both abrasion and erosion resistance. The FRP is an inert material used for its corrosion resistance. FRP piping is easily fabricated and, unlike rubber-lined pipe, can be readily repaired in the field.

The manifold and headers are built of FRP for the flow and geometry requirements of each application. Tapered manifolds and headers keep slurry velocities within allowable values (nominally 5 to 10 ft/sec).

The spray nozzles at each absorption stage are arranged to provide full spray coverage over the tower cross-section. The nozzles provide a full-cone, wide-angle spray pattern, and are typically made of a wear-resistant cast refractory containing silicon carbide.

Mist Eliminator

The cleaned flue gas passes through a separation section with two stages: the bulk entrainment separator (BES) and the mist eliminator (Fig. 33). The BES consists of six-inch FRP vanes mounted at a 45-degree angle on two-inch parallel spacing. It extends across the entire face area of the spray-tower absorber.

The mist eliminator located above the BES is made from vee-shaped FRP vanes arranged in a series of chevrons across the gas flow path.

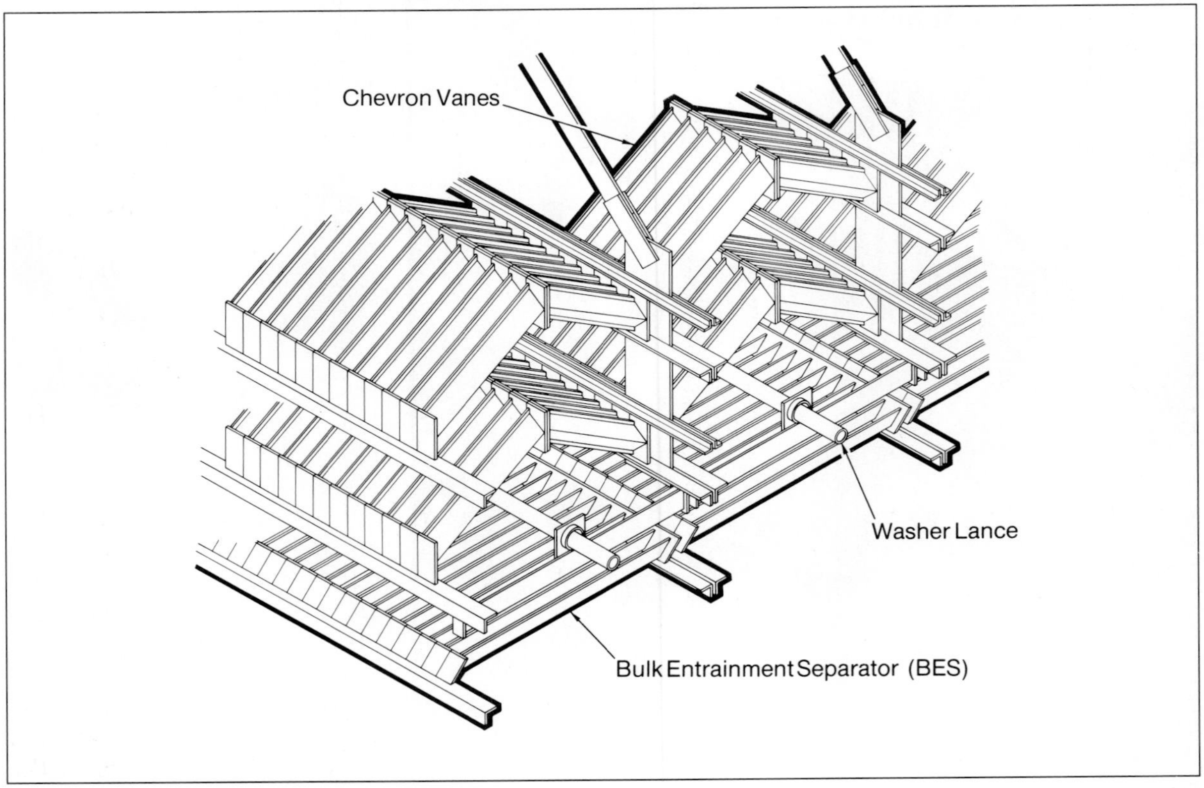

Chevron Vanes

Washer Lance

Bulk Entrainment Separator (BES)

Fig. 33 **Spray-tower mist-eliminator system**

Two rows of chevrons assure droplet impingement and minimize mist carryover.

Retractable lances, which rotate 360 degrees, have pairs of opposed nozzles at the ends and midpoints; they are located between the BES and the lower level of chevrons. These lances provide high-energy jets of water to clean each vane. Efficient cleaning prevents excessive droplet carryover which would result from deposition on the vanes.

Flue-Gas Reheater

Combustion Engineering recommends inline carbon-steel gas reheaters for full-flow heating of treated flue gas before discharge to atmosphere.[48] Compared to austenitic-stainless-steel reheaters, they are the least expensive to install; with proper maintenance, they are also the least costly to operate. C-E uses spiral-wound finned-tube heat exchangers identical to those installed in many high-pressure boilers. Staggered tubes combined with shallow overall depth ensure high heat-transfer efficiency as well as easy cleaning of the reheater.

C-E has had excellent experience with its carbon-steel reheaters placed immediately downstream of the scrubbing equipment. The mist-eliminator system described above exposes the reheater to extremely small quantities of liquids and solids. Any solids that do collect on the reheater are removed easily by conventional air or steam sootblowers.

SPRAY DRY-SCRUBBER DESIGN

The heart of the spray dry-scrubber process is the spray-dryer/absorber (SD/A). Vessels have been built with single inlets and single rotary atomizers, as well as with three inlets and atomizers; from a process standpoint, the three-inlet units perform identically to the single-inlet designs.

The amount of flue gas to be processed dictates the choice of one or three inlets. Single-inlet designs are generally limited to no more than 300,000 cubic feet per minute at the inlet conditions, while a three-inlet vessel can handle three times as much flue gas. Another important factor is redundancy, even though the single-inlet designs have ready spare machines which can be installed in thirty minutes or less after an unplanned spray-machine shutdown. When on-line redundancy is desired, the three machines are usually designed for handling one-half the rated load, thereby allowing one machine and inlet to be isolated if necessary. Three-inlet vessels can even be designed to operate with only one machine and inlet in service, thereby amplifying the turndown capability.

DROP-SIZE REACTOR

Inside the SD/A, the slurry is introduced as a very finely divided spray, often with a mass median diameter below 35 micrometers and a largest droplet well below 150 micrometers. Each droplet, in effect, becomes a miniature reactor. Sub-micrometer particles of lime dissolve according to the reaction

$$Ca(OH)_2 \rightleftharpoons Ca^{++} + 2\,OH^-$$

$$(55)$$

thereby making OH^-, or hydroxyl, ions available in the liquid phase for neutralization of hydrogen ions (H^+). Hydrogen ions are produced as sulfur-dioxide molecules diffuse across the droplet surface, dissolve in the liquid, and dissociate according to the reaction

$$SO_2(g) + H_2O \rightleftharpoons SO_2(l) + H_2O$$
$$HSO_3 + H^+ \rightleftharpoons SO_3 + 2H^+$$

$$(56)$$

The dissolution of sulfur dioxide tends to lower the pH (or raise the hydrogen-ion activity) in the liquid phase, while the dissolution of lime tends to raise the pH. Slaked lime has an equilibrium pH of about 12.0, while sulfur dioxide at 1,000 ppm by volume in the gas phase in equilibrium with pure water will lower the pH to about 2.5. Reactions 55 and 56 characterize the acid-base neutralization which occurs in each droplet as it passes through the spray-dryer/absorber.

Although it is clear how the OH^- ions in reaction 55 are removed through neutralization, it is not immediately apparent what happens to the Ca^{++}. The calcium ions must also be purged from the solution in order to allow more lime to dissolve. One of the most important reactions is the precipitation of calcium and sulfite ions to form calcium sulfite crystals.

$$Ca^{++} + SO_3^= \rightleftharpoons CaSO_3(s)$$

$$(57)$$

It has been observed, as expected, that some of the $SO_3^=$ is oxidized to $SO_4^=$. In turn, $SO_4^=$ precipitates with the calcium ions to form gypsum ($CaSO_4 \cdot 2H_2O$) crystals. Both calcium sulfate and sulfite are relatively insoluble, and in the absence of significant chloride ions, calcium-ionic activity is sufficiently small over the lifetime of the droplet to assure an acceptable rate of lime dissolution.

Several other reactions also significantly impact the performance of the process. Among these are reactions with SO_2 and calcium silicates formed by the chemical combination of lime and flyash. Further, the presence of chloride ions in the droplet not only impacts dramatically the ionic activity of calcium, but also impacts the vapor pressure of water and of the dried reaction products.

MASS AND HEAT TRANSFER

Many interrelated factors govern the rate of SO_2 mass transfer from the gaseous phase to the solid crystal ($CaSO_3 \cdot 1/2H_2O$) or the dihydrate of gypsum ($CaSO_4 \cdot 2H_2O$). In the classical sense, SO_2 must diffuse across a laminar gas-phase boundary layer surrounding the droplet as it is passed through the SD/A. This effect is modeled by an SO_2 diffusion equation

$$dC/dt = A \, \mathcal{D}_{ab}(C_{bulk} - C^*)$$

$$(58)$$

where

dC/dt = the rate of SO_2 mass transfer
  A = the droplet surface area
  $\mathcal{D}_{ab}$ = the diffusivity of SO_2
  C_{bulk} = the concentration of SO_2 in the bulk gas phase, that is, in the well-mixed gas surrounding the droplet
  C^* = the equilibrium concentration of SO_2 that would exist in the gas at the droplet surface

As the droplet first enters the gas phase, the equilibrium SO_2 concentration (C^*) is zero for all practical purposes, because of the high pH in the droplet liquid. In the liquid phase, the droplets quickly become saturated with SO_2 and further mass transfer is limited by the rate of SO_2 removal from the liquid phase, according to Eq. 56.

The mass transfer of SO_2 requires a wet medium. Once the droplet is dry, a different and less understood mechanism influences the rate of mass transfer, and this new mechanism is inferior to the wet mechanism. Consequently, it is important that the droplet remain wet long enough to allow the required mass transfer of SO_2 to take place.

The gas-phase heat transfer to the surface of the droplet and the mass transfer of water molecules away from the surface govern the rate of water evaporation. The rate of heat transfer is modeled, in the classical sense, by

$$dQ/dt = h \, A \, (T - T_s)$$

$$(59)$$

where

  T = the gas-phase bulk temperature
  T_s = the adiabatic saturation temperature of the gas/liquid system, generally assumed to be the temperature at the surface of the droplet
  h = the heat-transfer coefficient
  Q = heat content

The rate of water mass transfer is modeled by

$$dM/dt = \mathcal{D}_{water} \, A \, (H_2 - H)$$

(60)

where

$\mathcal{D}_{water}$ = the diffusivity of water in the gas phase

H = the absolute humidity of the gas

H_2 = the saturation humidity of the gas/liquid system

For a pure water/air system, the heat and mass-transfer driving forces are related by

$$H_s - H = Q_h / \lambda (T - T_s)$$

(61)

where

Q_h = the humid heat of moist air

λ = the latent heat of vaporization of water

This well-known relationship, upon which psychrometric charts are based, has prompted some practical designers to simplify or combine the heat- and mass-transfer Eqs. 59 and 60 to the form

$$dM/dt = K \, A \, (T - T_s)$$

(62)

where K is a pseudo or combined heat- and mass-transfer coefficient. In general, K must be experimentally determined from pilot and full-scale operating data, and depends somewhat upon the numerical method used to integrate Eq. 62. The integration is usually done through a finite range of droplet sizes to compute a value of T along the droplet trajectory. Integration from initial to final moisture content provides adequate time for wet-phase SO_2 mass transfer to occur.

Such integrations provide the system designer with much useful information and afford the serious student of the process a means to grasp a deeper understanding of the observed phenomena. For instance, it can be calculated that, in most practical systems, more than 90 percent of the mass transfer is complete in the first second of gas/slurry contact. It should also be obvious that larger droplets (with more mass) take longer to dry than smaller droplets. It is the largest droplets which principally concern designers of SD/A vessels, because, if these droplets are not dry before the spray cloud intersects the vessel boundaries, then wet deposits can occur.

PROCESS EFFECTS IN THE SD/A

Various process parameters affect the efficiency and economics of the spray dry-scrubbing process: the type and quality of the additive used for the reactant, the degree of ''dryness'' achieved by the spray dryer, the amount of heat available for drying, the relative amount of solids product recycled to the atomizer, and a host of lesser process variables. The impact of these parameters is so important that it is nearly impossible to estimate the process performance and economics without a sophisticated model. Designers thus rely heavily on performance models derived from pilot-plant data. Fortunately, spray dryer pilot-plant droplets look and behave exactly like commercial-scale droplets. For predicting SO_2 removal, it has been determined that process performance can be scaled up greater than 100 to 1 without significant risk. On the other hand, physical scale-up is not as straightforward. Such scale-up has been the source of most difficulties in the commercial application of spray dry-scrubbing. The following section describes the major factors impacting the process performance.

Additive Type and Quality

Any dissoluble base can be used as an additive for the spray dry-scrubbing process; the most important factors are the degree or rate of dissolution and the pH of alkalinity of the dissolved additive. Soda ash and ammonia are both very soluble in water and buffer at relatively high pH values. Correspondingly, these additives have shown to be very efficient when reacting with SO_2 in the spray-dry process. Of course, lime is also an acceptable additive, even though it is only slightly soluble. The rate of lime dissolution accelerates when the pH of the liquid phase decreases; therefore, lime works well in the SD/A, though not as well as soda ash or ammonia. In field tests, lime, soda ash, ammonia, and magnesium hydroxide were evaluated in a 7-foot diameter, 3500-acfm pilot-plant SD/A and pulse-jet fabric filter. As shown in Fig. 34, ammonium hydroxide achieved the highest SO_2 removal at a stoichiometric ratio barely half that of the other additives tested. Soda ash is slightly better than lime at equivalent stoichiometric ratios and approaches to saturation temperature. Magnesium hydroxide, by itself, is a very poor performer.

In most parts of the world, the favorable cost of lime compared to soda ash and/or ammonia has dictated its almost exclusive use as the additive of choice for this process, despite the superior performance of soda ash or ammonia. Soda ash is competitive only where abundant and easily accessible supplies make it commercially viable. Ammonia could be competitive, perhaps, in isolated areas where a gasification plant or other chemical plant is located to provide the ammonia feedstock. Ammonia, however, has the additional drawback of sometimes

producing a white $(NH_4)_2SO_3$ fume which can evade capture in the fabric filter.

Lime is available as either a hydrate powder or an unslaked CaO product called "pebble lime". In general, for moderately sized and large industrial applications, hydrated-lime costs make pebble lime more attractive, even though a slaking facility is required to hydrolyze the CaO to $Ca(OH)_2$. A further advantage of pebble lime is its higher reactivity. Pilot testing indicates that freshly slaked lime exhibits a superior reactivity over powdered hydrate mixed in water; this has been attributed to the much higher surface area of the freshly slaked lime slurry over the hydrate mix.

Stoichiometric Ratio

Additive stoichiometric ratio is the most important parameter affecting the removal of SO_2. Its importance is not because of the magnitude of its effect on process performance, which is substantial, but more to its effect on process economics. In general, control systems of spray dry-scrubbers automatically adjust the additive feed rate (i.e., stoichiometric ratio) to achieve a targeted level of SO_2 removal.

Approach to Saturation

The closer the SD/A operates to saturation temperature at the outlet, the longer the average droplet exists in the SD/A. The behavior of the gas is best understood by examining a conventional psychrometric chart, Fig. 35. The flue gas enters the SD/A at temperature T_{in} and humidity H_{in}. As the water from the evaporating droplets enters the gas phase, the gas temperature decreases and the humidity increases, following the adiabatic operating line. The gas leaves the SD/A at temperature T_{out} and humidity H_{out} with practically all the water from the droplets evaporated. The maximum humidity that the gas could attain is H_s and the minimum temperature is T_s, the saturation values. However, in practice, conventional SD/A's cannot operate at saturation because the droplets will not all be sufficiently dry. In fact, the drying efficiency is directly related to how closely an SD/A can approach saturation at the outlet.

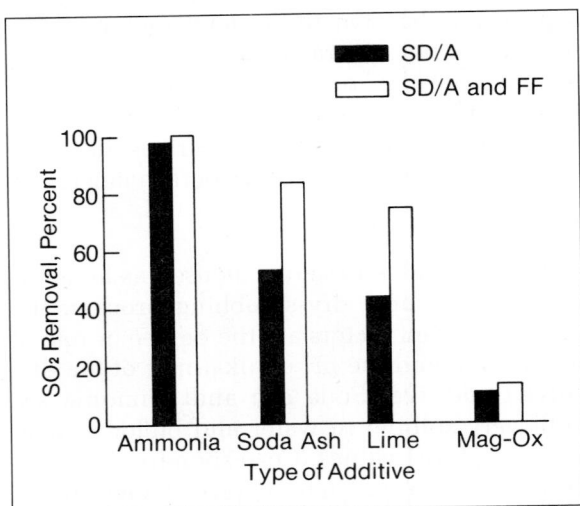

Fig. 34 Relative reactivity of dry-scrubbing additives from test data, also showing the effect of downstream fabric filtration

More important, the SO_2 removal efficiency and the additive utilization are inversely related to how close the SD/A is operated to saturation conditions.

The impact of approach to saturation is dramatic. Fig. 36 shows that, at a lime stoichiometric ratio of 1.4 mols $Ca(OH)_2$ per mol SO_2 entering the SD/A, the SO_2 removal will improve from 62 to 77 percent in the SD/A simply by adding sufficient additional water in the slurry to reduce the outlet approach to saturation (ΔT_s) from 50° to 25°F. Note that the stoichiometric ratio can be decreased from 1.4 to 1.0 at 65-percent SO_2 removal in the SD/A.

Spray-down Temperature

The spray-down temperature is the difference between the SD/A inlet and outlet temperatures. It is directly related to the amount of water supplied to the atomizing system. It is also a key indicator of how much evaporative heat is available in the flue gas. Normally, the spray-down temperature is not a controlled variable unless air-heater flue-gas bypass is available for this purpose. Most often spray-down temperature is set by the requirements for maximizing boiler efficiency—the lower the SD/A inlet temperature, the higher the effective boiler efficiency, but also the lower the spray-down temperature.

Spray-down temperature has a significant and direct effect on the SO_2 removal efficiency. As shown in Fig. 37, there are conditions in which an increase of 50°F in spray-down temperature can mean a decrease in stoichiometric ratio from 1.4 to 1.2 at 75-percent SO_2 removal in the SD/A, all other significant variables being held constant.

Recycle of Solids

The practice of mixing some of the fabric-filter or ESP solids product with the fresh lime slurry has been shown to be beneficial for lime utilization. There are several possible reasons:

- Some of the unreacted lime in the recycled solids is available as it passes through the SD/A again.
- The flyash usually has some alkaline properties which are made more effective when the flyash is dissolved in the slurry.
- The silica in the flyash reacts with the calcium in the lime to form a very reactive calcium-silicate compound; and/or
- The calcium sulfite in the reslurried solids product serves effectively as seed crystals in the droplets, which aid in the precipitation of fresh calcium sulfite (Reaction 57).

All of these mechanisms probably play some

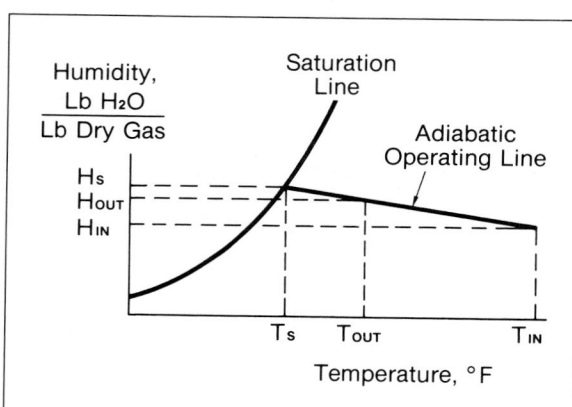

Fig. 35 Temperature and humidity conditions in a spray dryer absorber

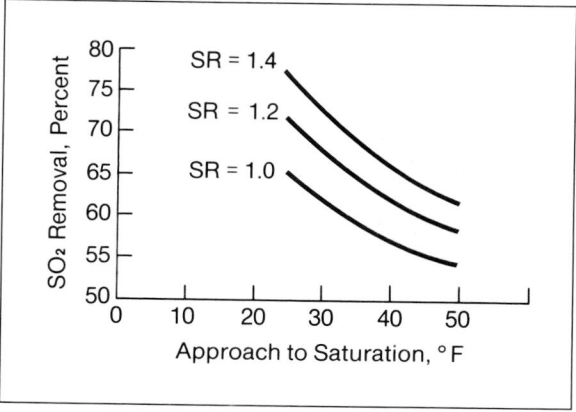

Fig. 36 Effect of approach to saturation temperature and stoichiometric ratio on spray-dryer absorber SO_2-removal efficiency

part in the observed performance improvement seen with the use of recycle. The performance effect of using a ratio of up to 2 lbs of recycle solids for each lb of fresh Ca(OH)$_2$ solids in the atomizer feed is shown in Fig. 38. For a 2.5-percent sulfur coal, the stoichiometric ratio decreases from 1.2 to 1.0 at 74-percent SO$_2$ removal in the SD/A.

Atomization Quality

The quality of atomization affects absorber performance in two ways: the first and most significant is the impact on drying and dryer efficiency, that is, the minimum ΔT_s achievable; the second is the impact on gas-side mass transfer of SO$_2$ and water. Depending upon the shape of the entrained spray cloud and the geometry of the SD/A vessel, along with other operating parameters such as the trajectory time for the largest droplets and droplet mass- and heat-transfer coefficients, an SD/A will have a range of deposit-free operation.

The minimum achievable operating T_s depends on the largest droplet in the spray cloud. For instance, a droplet of 120 μm will result in a safe operating ΔT_s of 20°F above saturation temperature. By comparing Fig. 38 with Fig. 36, it can be seen that the operational penalty associated with poor atomization can have a significant impact on process economics.

Inlet SO$_2$ Concentration

Inlet SO$_2$ concentration has a surprisingly weak effect on overall system performance. Individual pilot test programs often produce conflicting data, usually because the range of testing is necessarily narrow for any single program. Analyzing data from a range of test programs, however, does show some trends. Fig. 39 is a plot of additive utilization from several independent test programs with other effects factored out. There is a slight trend of increasing SO$_2$ removal efficiency in the SD/A and in the combined SD/A and fabric filter as the inlet concentration increases. This is not a large effect, however, considering that inlet SO$_2$ concentration varies by a factor of 5.

Chlorides

The presence of chloride ions in the spray slurry has a dramatic impact on SD/A performance. Chloride ions are not normally present in significant quantities in coal- or oil-fired applications, but they are a large factor in municipal-solid-waste, some types of hazardous-waste, and refuse-derived-fuel applications. Hydrogen-chloride concentrations in the flue gas can range from 300 to 800 ppm by volume for municipal-waste flue gases to greater than 10,000 for chlorine-laden hazardous wastes. Often, both SO$_2$ and HCl are

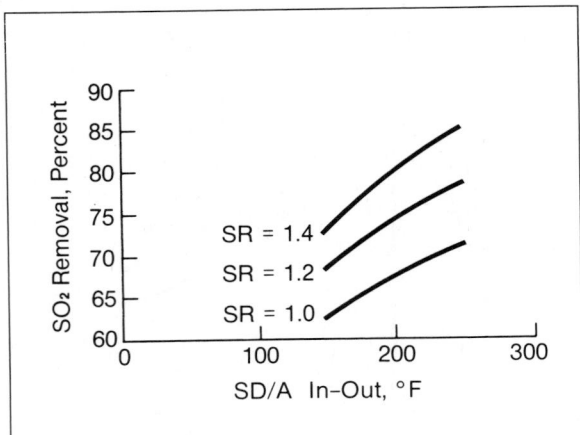

Fig. 37 Effect of spray-dryer/absorber spray-down temperature and stoichiometric ratio on SO$_2$-removal efficiency

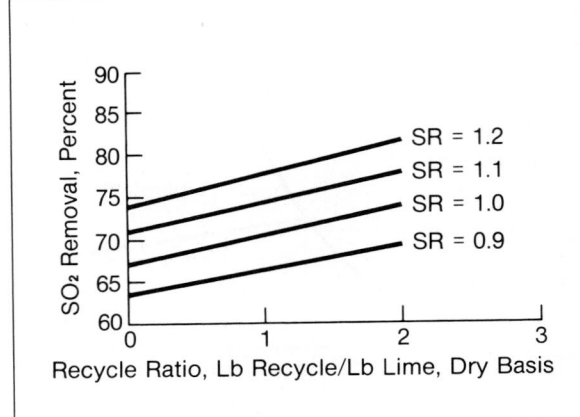

Fig. 38 Effect of recycling of collected solids on spray-dryer/absorber SO$_2$-removal efficiency

present together, with removal requirements established for both species. In other instances, chlorides are introduced through the use of high-chloride makeup water and are concentrated in the feed slurry when solids recycling is employed.

Chlorides in the feed slurry impact operation in two important ways: (1) by changing the apparent saturation temperature of the gas leaving the SD/A, and (2) by increasing the relative solubility of Ca^{++} ions in the slurry liquid.

Spray-dryer/absorbers operating with chlorides present in the slurry respond to a different saturation line than systems with only relatively insoluble salts. This same type of behavior is observed with sodium and ammonium additives which form soluble salt products with SO_2; however, the deviation in saturation temperature from the pure-water system is not as severe as that when both calcium and chlorides are present.

An unfortunate corollary of this situation with chlorides is that the apparent saturation temperature, T_s, depends on the concentration of Cl^- ions in the slurry. Hence, the operating SD/A outlet temperature needed to assure an appropriate approach to saturation temperature (and a corresponding SO_2 removal efficiency) varies with the concentration of chlorides. This concentration varies with the amount of HCl in the gas entering the SD/A relative to other acid-

gas components, as well as with the use of recycle solids to improve efficiency.

PROCESS EFFECTS IN THE PARTICULATE COLLECTOR

The spray-dry process employs a particulate collecting device after the SD/A to collect the dried reaction products and flyash. Through a mechanism not completely understood, the particulate collector is also a reasonable absorber of SO_2 (and HCl, should any escape collection in the SD/A).

The two types of particulate collectors most often applied to spray dry-scrubbing are the fabric filter and the electrostatic precipitator.

Fabric Filter

The fabric filter can be expected to achieve, on average, from 10- to 20-percent SO_2 removal (based on system inlet concentration) during normal operation, with wide variations depending on the operating history immediately prior to a particular measurement.

Increasing the system stoichiometric ratio from 0.8 to 1.3 causes an increase in fabric-filter performance from an average of less than 11 percent SO_2 removal to nearly 14 percent. Of course, the absorber SO_2 is also increasing with increasing stoichiometric ratio, so that the total system removal can easily exceed 90 percent.

The longer the time between cleaning of fabric-filter compartments, the greater will be the SO_2 removal; this is because more reactive cake is available for contact with the flue gas. A differential-pressure signal initiates the cleaning cycle for most fabric filter installations. The value of this signal affects the average length of time between compartment cleaning at a specific plant load.

Electrostatic Precipitator

Precipitators also remove SO_2, although not as well as fabric filters. Test data from a number of sources show an increasing SO_2 removal with precipitator specific collecting area (SCA); for example, system SO_2 removal will increase from an average of 5–6 percent to an

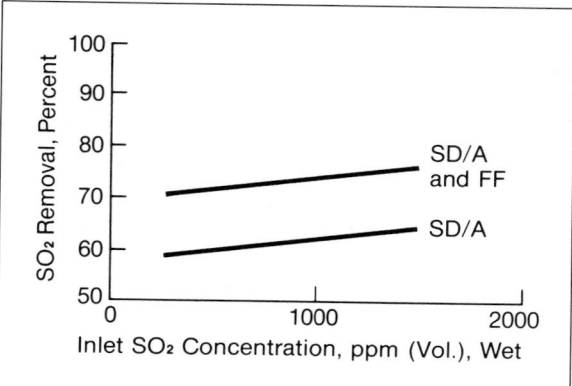

Fig. 39 Effect of inlet SO_2 concentration on spray-dryer/absorber SO_2-removal efficiency

average of 7–8 percent as SCA is increased from 190 to 260 ft^2/1000 acfm.

Since SD/A's were first intended to improve the SO_2 removal of the dry injection process (which historically employed a fabric filter), precipitators have only recently been considered as a competitive substitute for fabric filters. This is because of the large number of existing precipitators on utility and industrial coal-fired boilers; the retrofitting of acid-gas controls on existing boilers makes a significant potential market for SD/A's.

In retrofit applications, it is important not only to be able to predict the SO_2 removal performance of the ESP, but also to project accurately the new migration velocity in the ESP in order to predict the retrofit-system particulate penetration and to determine the need for ESP modifications. Data from a number of SD/A-ESP test programs show that the SD/A tends to have a centralizing effect on particulate size distribution and average resistivity. As a result, ESP's with poor migration velocities before retrofit are improved, while those with reasonably good migration velocities are not improved much, if at all. Consequently, ESP's with good migration velocities but of marginal size will probably require some additional collecting area in order to accommodate the added particulate loading from a retrofit SD/A.

ECONOMIC CONSIDERATIONS

The economics of wet versus dry scrubbing involve a great many factors, with these being the major ones:

- waste-disposal options
- sulfur level in the fuel
- cost of additives
- cost of labor

Waste disposal is becoming the most influential factor in deciding between wet-limestone and dry scrubbing. In many instances, facility planners cannot be assured of adequate land for waste-product disposal for the life of a power plant. When the flyash and scrubber products must be sold as useful byproducts, spray dry-scrubbing often is not feasible. Sometimes, it is possible to find a market for all or a large portion of the spray dry-scrubbing products for concrete aggregate or road bases. Usually, though, the availability of suitable landfill is often the deciding factor.

When disposal options are available for either wet or dry scrubbing products, the economics are driven largely by the other items on the list, the most significant being the sulfur level in the fuel. Since spray dry-scrubbing requires more reactive additives such as lime or soda ash, while the cheaper pulverized limestone can be used for wet scrubbing, the additive costs of wet over dry scrubbing decrease as the sulfur level in the fuel increases. Often, depending on the other cost factors, the two processes can become economically equivalent at a fuel sulfur content of 2 to 3 percent by weight. Below this level, spray dry-scrubbing usually has an economic advantage.[49]

REAGENT PREPARATION SYSTEMS

The reagent preparation systems for flue-gas desulfurization include the handling, storage, grinding, slurrying, and feeding of the additive after it has been delivered to the plant site. Additives such as lime, limestone, magnesium oxide, and sodium carbonate are prepared with such systems. Preparation of the most commonly used additives, lime and limestone, will be described in detail.

LIMESTONE

Limestone is received as crushed rock, typically sized to about $1^{1}/_{2}$″ x 0 or smaller, shipped, unloaded, and stored like coal. The rock is ground to a fine powder (about 90 percent passing through 325 mesh) in wet pulverizers.

A closed-circuit wet-grinding system (Fig. 40) is customarily used to prepare limestone. The system includes classification so that oversized particles are recycled to the mill. The pulverized limestone is slurried and stored in large tanks where the slurry is continuously agitated. Limestone grinding is generally a batch process and the mill operates at maximum capability until the storage tanks are full.

LIME

Lime is generally available as pebble lime produced in $\frac{3}{4}''$ and $\frac{1}{2}''$ mesh sizes by the calcining processes. Commercial lime is typically 85 to 95 percent calcium oxide with 0 to 5 percent magnesium oxide and 0 to 10 percent inerts. The lime is generally fed from the silo to a slaker by a variable-speed gravimetric belt feeder. The slaker controls adjust the water flow rate to provide a 5 to 15 percent solids mixture of slaked lime which is pumped to an additive storage tank for use in the desulfurization system.

REAGENT FEED SYSTEMS

FGD operation requires reagent feeding to the system at a rate proportional to the mass of SO_2 removed. The feed rate should be varied to maximize the additive utilization during partial load operation or changes in SO_2 concentrations in the flue gas. This requires that the additive feed rate be metered and matched to the demand. For process control, a variable slurry feed system is desirable.

The control of a slurry feed does, however, have constraints to consider when designing the feed system. A settling slurry, of lime or limestone, is susceptible to pipe plugging during low flow, thus mitigating against large turndown ratios. High slurry velocities cause excessive component wear, which limits slurry design velocities.

SCRUBBER SYSTEM WASTE DISPOSAL

Depending on the process used, a number of reaction products can result when SO_2 is absorbed from boiler flue gases. Some of these by-products may be marketable while others may be waste products valuable only as landfill. When selecting a sulfur-removal process, it is

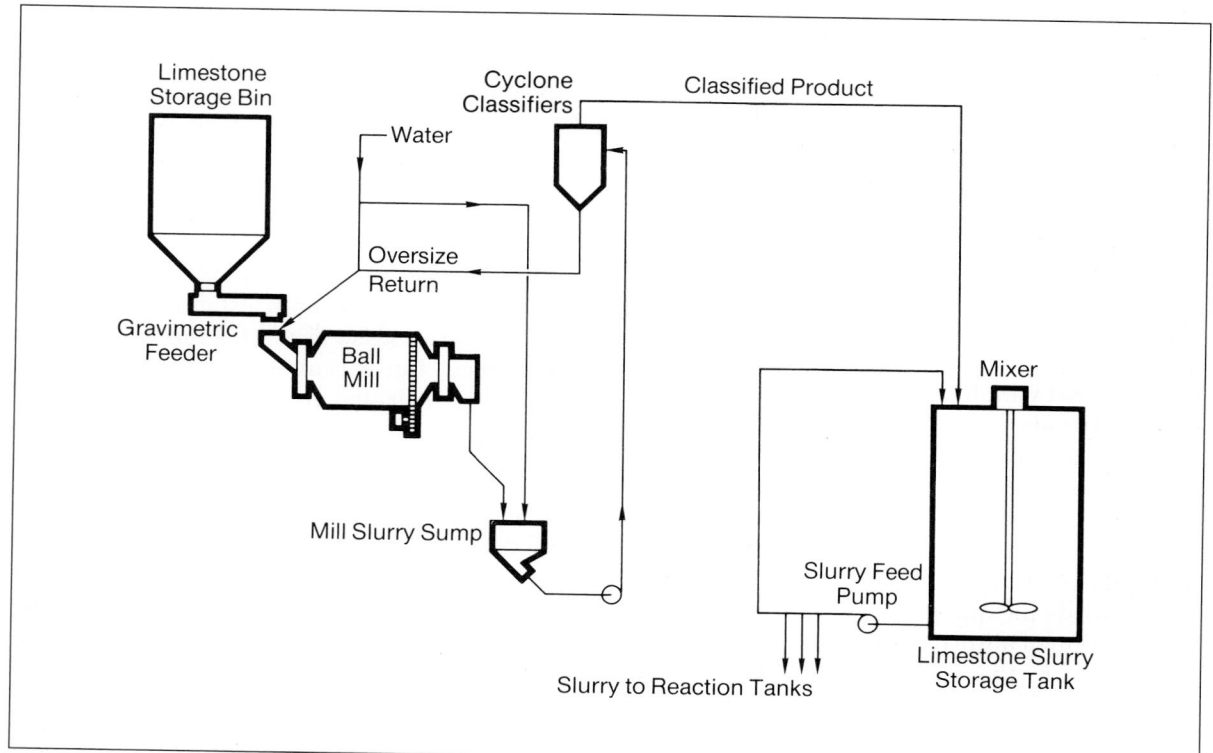

Fig. 40 **Closed-circuit wet-grinding flow diagram**

important to consider local sales markets, applicable disposal regulations, and available land for disposal. Markets for saleable products such as sulfuric acid, sodium sulfate, sulfur, and commercial grade gypsum are usually limited and may be distant from the power plant. Processes that produce a throwaway waste product are relatively simple; and, because the plant must already dispose of ash, it can often accommodate disposal of FGDS byproducts.

CHARACTER OF SOLID WASTE

The solid byproducts of a desulfurization system consist primarily of calcium sulfite, calcium sulfate, and flyash, along with minor quantities of unreacted additive. Calcium-sulfite particles are generally present as thin platelets which often form aggregates with a rosette-like structure. Calcium-sulfate crystals are thick and rod-shaped while flyash particles are predominantly spherical.

The morphology of the waste solids strongly influences activities such as dewatering of wet-process sludge. For example, the thin platelike sulfite crystals settle slowly and tend to retain water in their clustered structure. The sulfite crystalline structure causes a pseudo-thixotropic behavior; sulfite solids which appear dry and stable may reliquefy and flow if vibrated. The rod-shaped sulfate crystals, on the other hand, settle more rapidly and retain less water than sulfite solids. Also, sulfate solids do not exhibit thixotropic properties. Flyash particles appear to respond to dewatering somewhere between that which takes place with sulfite and sulfate crystals.

The exact composition of the waste solids from a particular desulfurization system depends on several operating parameters. The ratio of sulfite to sulfate in the waste solids depends on the degree of oxidation in the FGDS, while the quantity of unreacted additive is a function of additive utilization. Flyash concentrations in the solids may be very low—if produced by a system with an efficient particulate collection device upstream of the FGDS—or very high if produced by an FGDS which removes both particulate and sulfur dioxide.

Variations in waste-solid chemical composition cause corresponding variations in dewatering and handling which affect the design of disposal systems.

DISPOSAL PROCESS SELECTION

Fig. 41 diagrams several waste-disposal options for wet FGD systems. The different alternatives in ponding, landfilling, thickening, and filtering, as well as the structural properties of untreated, physically stabilized, and chemically fixed sludges were presented in detail in Chapter 17 of the Third Edition of this book. Estimates of the relative costs of different options and discussions of the impact of waste-treatment systems on FGS system design are given in references 50 and 51.

When evaluating disposal alternatives, the potential of using waste as raw material in other processes should be considered. Although many uses have been proposed, most are limited by technical or economic factors. The most promising appear to be

- gypsum as a wallboard material or a retarder in cement
- treated wet-process sludge in roads, dams, embankments and the like.

Two determinators are involved in considering waste as a raw material: first, is it possible to use the relatively large volume of waste material produced by a power plant; second, is the product competitive with established sources of the material? Use of wastes has not become widespread, but a limited number of markets may exist in selected areas.

CONTROL OF OXIDES OF NITROGEN

The reduction of NO_x emissions from stationary combustion sources has become a critical issue in most industrialized nations. As a result, the technology associated with the control of nitrogen oxides (NO_x) from fossil-fuel-fired steam generators has matured and expanded significantly.[52,53,54]

The NO_x reduction processes available through both in-furnace NO_x control, e.g., overfire air, gas recirculation, reduced-excess-

air firing, gas mixing, low-NO_x concentric tangential firing, staged combustion, and fluidized-bed firing; and post-combustion NO_x control (primarily selective catalytic reduction, or SCR) provide several alternatives for meeting strict nitrogen-oxide emission levels. Depending on the NO_x emission level required, an optimum NO_x reduction system may result in the integration of several of the above techniques in the overall plant design.

After in-furnace NO_x control has been implemented, as described in Chapter 12, post-combustion controls can result in further NO_x-emission reduction. With dry selective catalytic reduction systems, NO_x reductions of 80 to 90 percent are achievable.

The SCR process was originally developed in Japan where strict NO_x emission requirements dictate the use of post-combustion NO_x techniques. The SCR system was a developmental process in which the catalytic systems were first applied to natural-gas-fired units, then to low- and high-sulfur oil-fired units, and finally to coal-fired boilers. To complement existing C-E NO_x reduction technology, C-E has a license agreement with Mitsubishi Heavy Industries making their low-NO_x technology and products available to C-E for steam-generator applications.

SCR POST-COMBUSTION NO_x REDUCTION PROCESS

The selective catalytic reduction system uses a catalyst and a reductant (ammonia gas, NH_3) to dissociate NO_x to nitrogen gas and water vapor. The catalytic process reactions are as follows:

$$4NO_{(g)} + 4NH_{3(g)} + O_{2(g)} \xrightarrow{\text{catalyst}} 4N_{2(g)} + 6H_2O_{(g)} \quad \textbf{(63)}$$

$$2NO_{2(g)} + 4NH_{3(g)} + O_{2(g)} \xrightarrow{\text{catalyst}} 3N_{2(g)} + 6H_2O_{(g)} \quad \textbf{(64)}$$

Since NO_x is approximately 95-percent NO in the flue-gas of steam generators, equation 63 dominates.

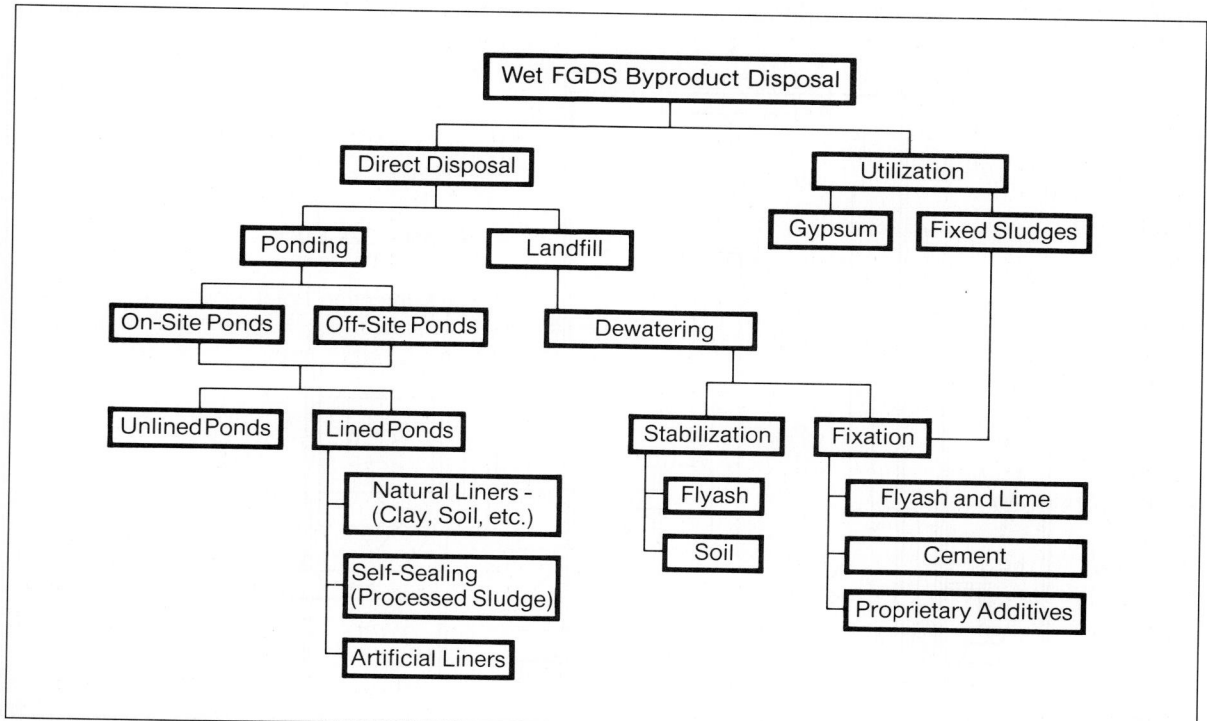

Fig. 41 Wet FGDS waste-disposal alternatives

The SCR catalytic-reactor chamber is typically located between the economizer outlet and air heater flue-gas inlet (see Fig. 42). This location is typical for steam-generating units with SCR operating temperatures of 575 to 750°F (300 to 400°C).

Upstream of the SCR chamber are the ammonia injection pipes, nozzles, and mixing grid. Through orifice openings in the ammonia injection nozzles, a diluted mixture of ammonia gas in air is dispersed into the flue-gas stream. After the mixture diffuses, it is further distributed in the gas stream by a grid of carbon steel piping in the flue-gas duct. The ammonia/flue-gas mixture then enters the reactor where the catalytic reaction is completed.

CATALYST SELECTION

Selection of the proper catalyst for an SCR system is essential not only for economic reasons but also for the reliability of the total system. To reduce operating and maintenance costs, the catalyst should be durable and provide high activity over a long period of time.

Catalyst selection is based on the following criteria:

- resistance to toxic materials
- resistance to abrasion
- mechanical strength against sootblowing
- resistance to thermal cycling
- resistance to the oxidation of SO_2 to SO_3
- resistance to plugging

Each criterion is the subject of a subsequent discussion and involves careful weighing of both economic and reliability characteristics.

Resistance to Toxic Materials

Toxic materials such as alkali metals (specifically sodium and potassium), alkaline-earth metals, and sulfur oxides carried by dust in the flue gas can impair the performance of the catalyst. It is imperative to select catalysts which achieve high NO_x-removal performance without significant deterioration from these harmful substances.

Abrasion Resistance

Erosion of catalyst surfaces from flyash in

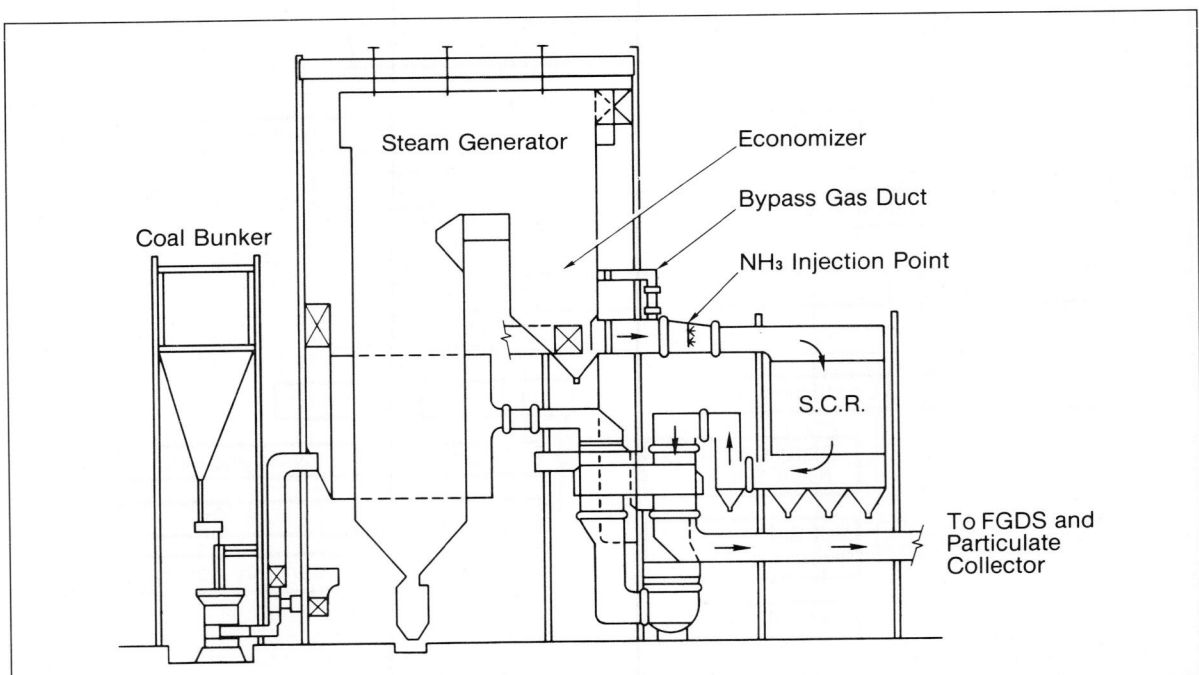

Fig. 42 **General arrangement of an SCR system for a pulverized-coal-fired steam generator**

coal-fired boilers decreases the service life of a catalyst. By using a catalyst with high erosion resistance, as well as a dummy catalyst layer upstream of the active catalyst, erosion problems can be reduced. It is also appropriate to design SCR systems for low flue-gas velocities because erosion potential increases as an exponential function of the gas velocity.

Mechanical Strength Against Soot Blowing

Sootblowers for SCR systems are generally required for coal and high-sulfur oil-fired applications. The catalyst must be designed to withstand the pressures from superheated-steam or compressed-air sootblowing.

Resistance to Thermal Cycling

The catalyst must be selected for thermal stability under all expected operating conditions. The ideal SCR operating temperature range is generally 575 to 750°F (300 to 400°C). When operating temperatures fall below 575°F, the potential for ammonium-bisulfate formation on catalyst surfaces increases. Ammonium-bisulfate formation can cause permanent catalyst activity deterioration when low-temperature operation is maintained over a period of time. Typically, an economizer flue-gas bypass is provided on utility boilers to maintain inlet SCR gas temperatures above acceptable levels during low-load operation. Above 750°F, ammonia gas may dissociate, thus reducing the effectiveness of the SCR process. If temperatures exceed 900 to 1000° (480 to 540°C), the catalyst activity may be permanently impaired, due to sintering of the active catalyst material.

Resistance to Oxidation of SO_2 to SO_3

It is important to select catalysts that minimize the oxidation rate of SO_2 to SO_3. The formation of SO_3 by the catalyst, combined with unreacted ammonia ("ammonia slip") from the SCR system, can cause serious plugging and corrosion problems on such downstream equipment as air heaters, electrostatic precipitators, fabric filters, and ductwork. These problems may occur from the formation of ammonium bisulfate during operation at low flue-gas temperatures. The temperatures at which ammonium bisulfate forms vary with the amount of SO_3 in the flue gas. Typically, if there is 1 ppm of SO_3 present, a minimum operating temperature of 575°F (300°C) is recommended to avoid formation of ammonium bisulfate. As SO_3 and unreacted ammonia concentrations increase, the allowable minimum operating gas temperatures also rise.

Ammonium-bisulfate deposits on catalyst surfaces will cause a reduction in NO_x-removal performance. Usually, this can be avoided by either operating the SCR system above formation temperatures or by halting the injection of ammonia when temperatures are below recommended levels. To minimize the detrimental effects of ammonium-bisulfate formation on downstream equipment, adequate clear spacing for heating surfaces, sufficient sootblowers and washing devices, as well as corrosion-resistant materials, are recommended.

Resistance to Plugging

It is essential to select catalysts with openings that will not become plugged by the flyash in coal-fired combustion gases.

Proper operation of the steam generator as well as careful selection of catalyst material play significant roles toward achieving high NO_x removal and low ammonia slip over a long period of time.

Once the catalyst material is selected, such criteria as the mol ratio of ammonia to NO_x, the volumetric flow of gas to be treated, the entering flue-gas temperature, the inlet NO_x concentration, and the oxygen concentration in the flue gas will affect SCR NO_x concentration, and the oxygen concentration in the flue gas will affect SCR NO_x-removal performance. Each of these parameters must be considered for the proper design of a catalyst system.

CATALYST MAINTENANCE

A routine maintenance program will help catalystic reduction systems to achieve high system availability. Visual inspection of the

reactor chamber is recommended during normal boiler outages to ensure that misalignment of catalyst elements, gas-seal cracks, element clogging by soot or dust, and chipping of element material have not occurred. Another recommendation is the periodic testing of sample sections of the catalyst elements to determine the degree of activity deterioration. When deterioration significantly alters NO_x-removal efficiency, partial catalyst replacement is indicated. Sampling lines of NO_x and ammonia monitoring equipment must also be checked frequently for possible plugging and corrosion to assure the reliability of the ammonia-injection control system.

The catalyst modules may be moved in and out of a vertical chamber by removing cover plates on the side of the chamber. To facilitate handling of the catalyst modules, platforms are provided at each catalyst stage (see Fig. 43). A slow-speed hoist attached to the overhead supporting structure lifts each module to the proper catalyst floor. The module is then placed on a trolley which moves on rails permanently installed in the reactor chamber. A suitable number of soot blowers are provided for each catalyst stage. Each blower is located at a right angle to the direction of module removal.

SELECTIVE NON-CATALYTIC REDUCTION (SNCR)

SNCR post-combustion methods require the injection of a reagent into the flue gas stream of a combustion source to reduce NO_x to elemental nitrogen and water vapor. The predominant SNCR technologies utilize ammonia gas and aqueous urea, $CO(NH_2)_2$, as the reagents.

The ammonia-based process involves the injection of a diluted ammonia gas with a carrier medium such as compressed air or steam. The process is highly dependent upon flue-gas temperature and residence time for achieving high NO_x removal efficiency. An effective temperature window ranging from 1600-2000°F (870-1090°C) is required for these systems. The primary chemical reaction for the process is as follows:

$$4NO + 4NH_3 + O_2 \rightarrow 4N_2 + 6H_2O \tag{65}$$

The addition of hydrogen gas will allow the reaction to proceed effectively down to 1300°F (700°C). If ammonia is injected at too high a temperature (above 2000°F), it will oxidize forming additional NO_x. Injection of ammonia below optimum temperatures will result in increased levels of unreacted ammonia (ammonia slip) exiting the system. Consequently, the location of ammonia injectors is critical in order to obtain

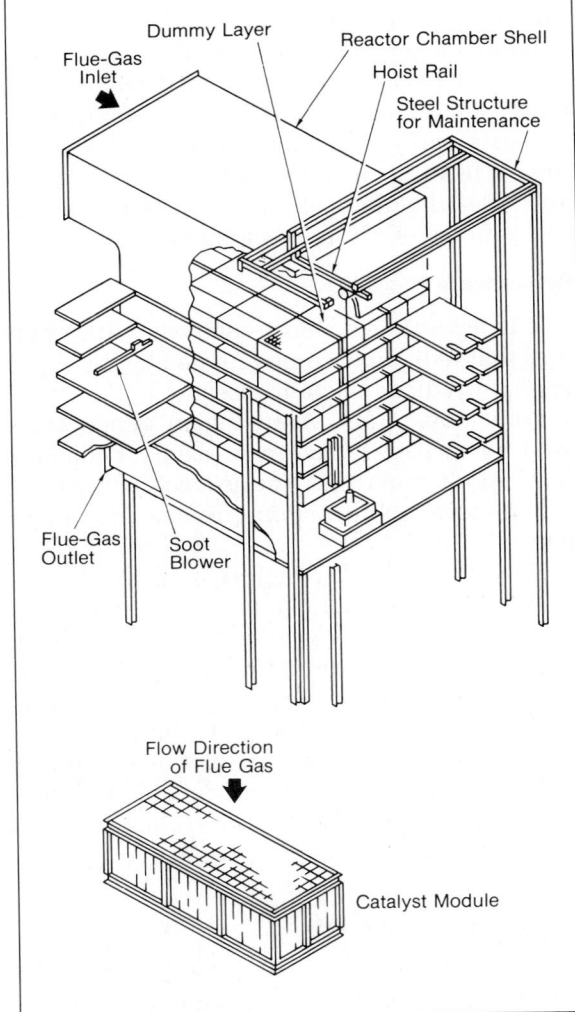

Fig. 43 **Typical SCR reactor chamber**

high NO$_x$ removal efficiency across the expected operating load range.

Similar to the ammonia process, a second method requires the injection of an aqueous urea solution over a defined flue gas temperature range, typically 1600-2000°F. The governing process reaction is as follows:

$$CO(NH_2)_2 + 2NO + 1/2 O_2 \rightarrow 2N_2 + CO_2 + 2H_2O \qquad (66)$$

Enhancers or chemical additives can be used to extend this window down to 1300°F. The effectiveness of this process will be significantly reduced if urea is injected outside this temperature range.

Additional process-design considerations include actual flue-gas velocity gradients, boiler geometry, and flue-gas constituents. Careful evaluation of these parameters allows for the proper location of urea injectors in order to optimize reagent utilization for effective NO$_x$ reduction capability.

FLOW-MODEL TESTS OF EMISSION-CONTROL SYSTEMS

Volumetric gas flow is a fundamental design parameter for emission-control systems (electrostatic precipitators, scrubbers, and fabric filters). It affects not only the system working efficiency but also the engineering design. The purpose of a model study is to investigate the flow characteristics, pressure loss, gas mixing, and temperature distribution. Usually, a three-dimensional model to 1/16th scale or larger is used to obtain design information and to insure that performance predictions are met (Fig. 44). With a flow model, the potential gas-flow problems of a design can be identified and engineering modifications recommended. A flow-model study also can analyze field problems, and is an economic way to decide on modifications quickly and accurately.[55]

MODELING THEORY

Whenever the flow of a fluid through a

Fig. 44 Precipitator flow model for test of flow-control devices

model is studied, there must be some assurance that the flow pattern in the model will closely correspond to the actual flow pattern at full scale. To attain the desired similarity, one should attempt to keep the Reynolds number for the flow in the model equal to the Reynolds number for the full-scale flow. The Reynolds number (Re) is a dimensionless ratio between the inertial and viscous forces existing in the flow pattern and is defined as $\mathrm{Re} = \frac{\rho VD}{\mu}$, where ρ is gas density, V is its velocity, μ its dynamic viscosity, and D is the hydraulic diameter of the model. The flow is usually considered laminar if the Reynolds number is less than 2,000; it is turbulent if the Reynolds number is above 4,000.

The combustion gases flowing through a commercial FGD system normally have a density and a viscosity on the same order of magnitude as density and viscosity for the air in the model. Therefore, to have the same Reynolds number in the scale model as the full-scale unit, the velocity of the air in the model must be 10 to 20 times greater than it is in the actual unit. This causes difficulties and uncertainties in the model study.

Because of the large size of most gas-cleaning systems, the Reynolds number is usually well into the turbulent region. Experience indicates that flow pattern and pressure losses for an FGD system can be reasonably predicted if turbulent (essentially incompressible) flow is maintained in the model.

To predict full-scale pressure losses from model studies, it is important to consider all correlating parameters. The physical flow boundaries and any devices inside the duct work must be modeled to very close tolerance. The flow must be in the incompressible region and, most important, the Reynolds number effect must be determined.

In studying the pressure losses in closed channels, it is helpful to note that the losses generally occur from two effects: loss due to viscous forces and those due to a change in the momentum of the gas. The viscous losses occur in all segments of the duct, while losses due to a momentum change are present only when gas velocity changes magnitude or direction. The boundary layer in the modeled system will, in general, be proportionately larger than the boundary layer in the full-scale system. This tends to make frictional pressure losses in the model higher than full-scale losses. Losses due to momentum changes are similar for both model and full-scale systems because velocity patterns are similar. For these reasons, pressure losses predicted by the model tests are usually conservative.

In model studies, a number of different instruments are used to measure such variables as velocity, static pressure, conductivity and the like. Many of the devices are similar or identical to the measurement equipment described in Chapter 22, and include hot-wire anemometers, pitot tubes and pressure transducers, and thermal-conductivity gas analyzers.

GAS-FLOW CONTROL DEVICES

In precipitator modeling, flow-control-devices such as vanes and perforated plates are studied. Vanes are placed in the turns of the ductwork ahead of the precipitator to smooth the gas flow and lower the pressure-loss coefficient.[56,57] Perforated plates are very effective devices used at the precipitator inlet to reduce turbulence and improve uniformity of gas flow. Basically, they provide the essential transition from large-scale persistent turbulence to small-scale nonpersistent turbulence. Horizontal or vertical vanes or gas-distribution panels may also be installed to obtain a desirable uniformity in flow.

In the final stage of modeling, flow visualization tests that use smoke traces, ground cork, and string tufts are conducted to locate gross flow aberrations, to study dust deposition in the ducts, and to study hopper sweepage. Various configurations of flow-control devices are evaluated until one is found that yields acceptable system flow characteristics. Fig. 44 shows a precipitator flow model with flow-control devices being tested.

Flue-gas scrubbers and fabric filters are modeled using the same techniques as for precipi-

tators except that pressure drops across demisting elements and filter bags are simulated by perforated plates. The experienced modeler will determine optimum turning-vane and baffle arrangements, as well as the size and configuration of gas inlet and outlet plenums. A front view of a fully instrumented fabric-filter model is shown in Fig. 45.

DISPERSION OF EMISSIONS

Control of air pollution is not entirely a question of the quantity of emissions. It is also related to the ability of the atmosphere to assimilate them without adverse effects. Emission control by dispersion requires the optimum combination of such factors as stack height, buoyancy, climate, and topography. Such control requires a study of atmospheric conditions surrounding a power plant to determine airflow patterns and ventilating capabilities of the region. Model studies provide valuable assistance to predict the dispersion of combustion products and thereby maintain particulate or gaseous concentrations to prescribed levels.

Dispersion refers to the movement of parcels of gases, either vertically or horizontally, and their simultaneous dilution with fresh air.

WIND DIRECTION AND VELOCITY

Emissions are dispersed horizontally into the air as the parcel moves with the existing wind. Vertical dispersion results from a stack's discharging a warm parcel upward to mix with fresh air at higher elevations.

Certain phenomena restrict dispersion and must be considered when evaluating plant dispersion capability. One of the most severe natural impediments is thermal inversion of the atmosphere. This atmospheric condition restricts vertical dispersion and, as there is generally little wind during an inversion, it tends to trap and concentrate emissions. Further complicating dispersion is topography such as valleys, where emission concentrations can reach dangerous levels.

To disperse emissions from boilers, very tall stacks have been constructed; their height can be augmented by a high efflux velocity of 75 to 100 feet per second. These factors, coupled with the buoyancy of the hot flue gas, produce an effective stack height substantially greater than the physical stack height (Fig. 46). The effective stack height is the sum of the actual

Fig. 45 A front view of a fabric-filter flow model

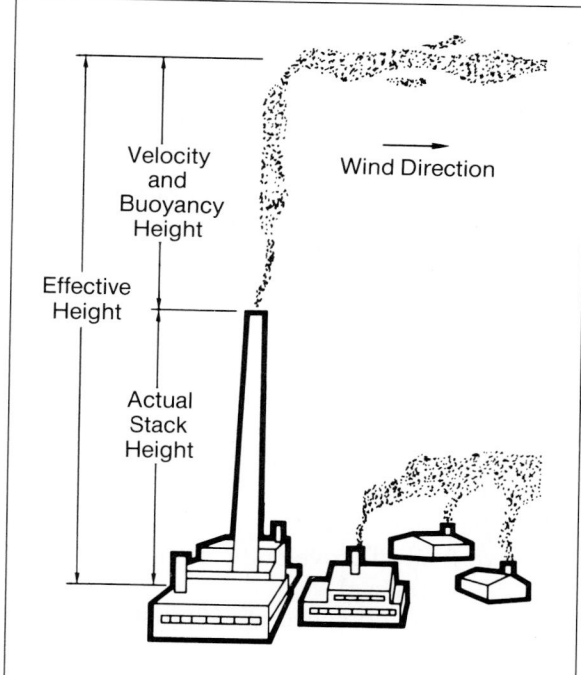

Fig. 46 Effective stack height equals actual stack height plus velocity and buoyancy factors.

stack height plus the height effects due to velocity and buoyancy. If the effective height is great enough, the effluent can "puncture" an inversion to disperse at higher elevations.

"Stack meteorology" has been used to study the factors affecting dispersion of gases from stacks or chimneys. For power plants, the dispersion of particulates and SO_2 must be estimated and compared to federal, state, and local ambient air-quality criteria and regulations.

Three different calculations are generally required to estimate a time-averaged concentration of a gas at a location downwind from stack. First, the plume rise above the stack must be established; second, the dispersion of the gas before it reaches the downwind location must be found; and third, it is necessary to calculate the time average of the ground-level concentration.

DISPERSION CALCULATIONS

The rise of plumes above the top of a stack can be calculated with many formulas, but results vary widely if these equations are applied to sources substantially different from those used in the original testing. Effective plume rise can be obtained from equations such as those given in reference 58.

Small-volume sources having an appreciable exit velocity—but little temperature excess above ambient—are essentially momentum jets in a crosswind and the plume rise Δh may be calculated as:

$$\Delta h = D \left(\frac{V_s}{U_s}\right)^{1.4}$$

$$(67)$$

where, in consistent units

D = diameter of stack

V_s = vertical efflux velocity at release temperature (m/sec)

U_s = mean wind speed at actual stack height (m/sec)

As buoyant plumes have temperatures above that of the ambient air and involve a large vol-

ume release, they require a different calculation treatment. Under stable conditions Δh is

$$\Delta h = 2.9 \cdot \left(\frac{F}{U_s G}\right)^{0.33}$$

$$(68)$$

where

F = buoyancy flux =
$gV_s\left(\frac{D}{2}\right)^2 \left(\frac{P_a - P_s}{P_a}\right) (m^4/sec^3)$

G = stability parameter (sec^{-2})

g = acceleration of gravity (m/sec^2)

P_s = density of stack gas at stack top (mg/m^3)

P_a = density of ambient air at stack top (mg/m^3)

These equations include the plume-rise forces attributable to the thermal buoyancy of the plume and the velocity momentum. The effective stack height (theoretical gas-release height) is the sum of the actual stack height and the calculated plume rise.

Theories based on the assumption that the mean concentration of material dispersed in a diffusion cloud follows a three-dimensional Gaussian distribution have been developed. A Gaussian formulation proposed by Cramer[59] is generally accepted by industry, although other models are available.[60,61,62,63,64]

OPTICAL PROPERTIES OF FLYASH

Although the optical properties of suspended particles do not affect collector performance, they are important because of their relationship to the visual appearance of stack emissions. Fine particles suspended in gases and exposed to a light beam scatter the light in directions other than that of the original beam and obscure or darken the light as viewed toward the light source. The visibility of stack emissions is due to both these effects.

It has been shown both theoretically and experimentally that the light-scattering and obscuring effects increase with the specific surface of the particles and the size or thickness of the stack plume. The specific surface is defined as the total exposed surface area of the particles per unit mass of dust or dispersoid. It is a function of particle size and increases rapidly with decreasing size of particles. For flyash, the specific surface usually will be between about 2,000 and 15,000 cm^2/g. As the fine ash emitted by most pulverized-coal boilers has relatively high specific surface, it is very effective in its light-scattering and obscuring properties. The high-volume gas flows associated with large steam generators also contribute to emissions that have high visibility above the stack.

Removing the coarser fractions of flyash usually will not appreciably reduce the visibility of stack emissions. For example, an inertial collector which removes 75 percent of the ash only slightly improves stack-discharge appearance. The finer particles below about 10 or 20 μm are not collected, and it is these particles with their high specific surface which cause most of the visibility.[14]

OPACITY

Opacity is the percent reduction of light intensity and is defined as:

$$\text{Opacity} = \frac{I_o - I}{I_o} \times 100$$

$$(69)$$

where
  I = intensity of light reaching the observer
  I_o = original intensity of the light

The reduction of light intensity when the light is x feet in length is the following:

$$I = I_o e^{-(bg + a) X}$$

$$(70)$$

where
  b = extinction coefficient
  g = concentration of gases
  a = extinction coefficient for particles (includes dust concentration)

The light path for a flue-gas plume is the plume depth (the stack effluent diameter). The extinction coefficient includes the absorption coefficient, and scattering and diffraction factors that are known for small solid particles, oil mist, and gases.[65]

The required reduction of flue-gas loading due to opacity restrictions becomes the governing requirement when the stack diameter is increased beyond a critical dimension. This is because the light path x in the above equation becomes larger and, with a constant dust concentration, the opacity increases as the stack exit diameter increases. On large units, then, the required degree of dust loading reduction is greater than on small units. Fig. 47 gives the relationships between the Ringelmann number, opacity and transmittance in percent, and optical density.

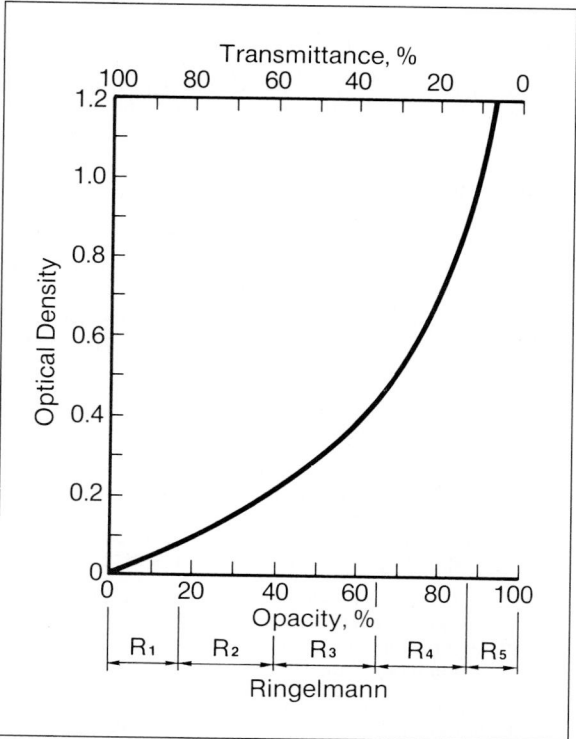

Fig. 47 Relationships between various optical emission-measuring units

REFERENCES

[1] "Criteria for the Application of Dust Collectors to Coal-Fired Boilers: The Results of an IGCI/ABMA Joint Technical Committee Survey," *Proceedings of the American Power Conference*, 29: 572—580, 1967. Chicago: Illinois Institute of Technology, 1967.

[2] Industrial Gas Cleaning Institute, Inc. publications: EP-1, *Terminology for Electrostatic Precipitators*; EP-5, *Factors to Consider When Selecting Electrostatic Precipitators*; EP-7, *Gas Flow Model Studies*; EP-8, *Structural Design Criteria for Electrostatic Precipitator Casings*; EP-9, *Power Consumption Measurement for Electrostatic Precipitators*; F-2, *Fundamentals of Fabric Collectors and Glossary of Terms*; F-3, *Operation and Maintenance of Fabric Collectors*; FGD-1, *Flue Gas Desulfurization Terminology*; G-1, *Gaseous Emissions Equipment: Product Definitions and Illustrations*; M-2, *Cyclonic Mechanical Dust Collector Criteria*; M-5, *Standardized Method of Particle Size Determination and Collection Efficiency*; WS-1, *Wet Scrubber Terminology*; WS-3, *Basic Types of Wet Scrubbers*; WS-4, *Wet Scrubber System – Major Auxiliaries*. Washington, DC, Industrial Gas Cleaning Institute, Inc.

[3] John G. Yost and Philip M. Gerhart, "Field Performance Testing of Large Power Plant Fans Using ASME PTC 11," presented at ASME-IEEE Joint Power Generation Conference, October 20—24, 1985, Milwaukee, WI; ASME Paper no. 85-JPGC-PTC-1.

[4] Several sources, including *Characteristics of Particles and Particle Dispersoids*, Stanford Research Institute Journal, Third Quarter, 1961: Stanford Research Institute, Menlo Park, CA, 1961.

[5] Code of Federal Regulations, Title CFR 40, Parts 53 to 80, "Protection of the Environment," revised as of July 1, 1985. Office of the Federal Register, National Archives and Records Administration, U.S. Government Printing Office, Washington, DC, 1985.

[6] R. E. Bickelhaupt, "A Technique for Predicting Fly Ash Resistivity," U.S. Environmental Protection Agency Report EPA-600/7-79-204, August 1979; Washington, DC, 1979.

[7] Harry J. White, *Industrial Electrostatic Precipitation*. Reading, MA: Addison-Wesley, 1963.

[8] G. W. Driggers, A. A. Arstikaitis, and L. A. Hawkins, "Computer Model Use for Precipitators Sizing," *Proceedings of Fourth Symposium on Transfer and Utilization of Particulate Control Technology*, October 11—14, 1982. Houston, TX; also as Combustion Engineering publication TIS-7252.

[9] Ralph A. Altman, et al., "Technical Evauation of Plate Spacing Effects on Fly Ash Collection in Precipitators," presented at Fifth Symposium on Transfer and Utilization of Particulate Control Technology, August 1984.

[10] B. Phelan and T. Farmer, "Top ESP Performance Demands More Than Opacity Feedback," Power, September 1989, Vol. 133, No. 9; McGraw-Hill, New York, NY, 1989.

[11] Joseph G. Singer, "Design for Better ESP/Fabric-Filter Hopper Operation and Maintenance," presented at Air Pollution Control Association '83, 76th Annual Meeting and Exhibition, June 19—24, 1983, Atlanta, GA; also as Combustion Engineering publication TIS-7402.

[12] Kenneth Wark and Cecil F. Warner, *Air Pollution: Its Origin and Control*. New York: IEP, 1976.

[13] G. W. Penney, "Weaknesses in Conventional Theory of Electrostatic Precipitation," *ASME Paper no. 67-WA/APC-1*. New York: American Society of Mechanical Engineers, 1967.

[14] Harry J. White, "Effect of Fly Ash Characteristics on Collector Performance," *ASME Paper no. 54-A-259*. New York: American Society of Mechanical Engineers, 1954. Also in: *Air Repair*, 5: 37—50, 62, 1955.

[15] Sabert Oglesby, Jr. and Grady B. Nichols, *A Manual of Electrostatic Precipitator Technology. Part I—Fundamentals*. Report no. PB-196 380 (APTD-0610). Springfield, VA: National Technical Information Service, 1970.

Sabert Oglesby, Jr. and Grady B. Nichols, *A Manual of Electrostatic Precipitor Technology. Part II—Application Areas*. Report no. PB-196 381 (APTD-0611). Springfield, VA: National Technical Information Service, 1970.

[16] J. P. Gooch and G. H. Marchant, Jr., "Electrostatic Precipitator Rapping Reentrainment and Computer Model Studies," Electric Power Research Institute Report FP-792, Vol. 3, August 1978; Palo Alto, CA, 1978.

[17] G. W. Driggers, L. A. Hawkins, and R. W. Gray, "Applying Performance Results of C-E Rigid Frame Precipitators over a Wide Range of Coal Characteristics," presented at ASME-IEEE Joint Power Generation Conference, September 25—29, 1983, Indianapolis, IN; also as Combustion Engineering publication TIS-7418.

[18] Kenneth J. McLean, *Survey of Australian Experience in Collecting High Resistivity Fly Ash with Electrostatic Precipitators*. Final Report. Report no. PB-221 139 (EPA-R2-73-258). Springfield, VA: National Technical Information Service, 1972.

[19] Kenneth J. McLean, "Some Effects of High Resistivity Fly Ash on Electrostatic Precipitator Operation," Electrostatic Precipitator Symposium, EPA, Birmingham, Alabama, 1971. Author address: University of Wollongong, P.O. Box 1144, Wollongong, New South Wales, Australia 2500.

[20] Z. Herceg and Kenneth J. McLean, "Efficiency of Electrostatic Precipitators and Relationship to Corona Voltage-Current Characteristics," Presented at the 64th Annual Meeting of the Air Pollution Control Association, Atlantic City, June 27—July 1, 1971, Paper no. 71—178. Pittsburgh: Air Pollution Control Association, 1971.

[21] L. A. Hawkins and H. L. Wheeler, "Characterization of Discharge Electrode Performance: Results of Laboratory and Pilot Plant Experiments," presented at Sixth Joint EPA/EPRI Symposium on Transfer and Utilization of Particulate Control Technology, New Orleans, LA, February 25—28, 1986; also as Combustion Engineering publication TIS-PSG-86-001.

[22] Alan B. Walker, "Operating Experience With Hot Precipitators on Western Low-Sulfur Coals," *Proceedings of the Ameri-*

can Power Conference, 39: 582—594, 1977. Chicago: Illinois Institute of Technology, 1977.

[23] Harry J. White, "Electrostatic Precipitation of Fly Ash; Part II, Section 3: Fly Ash and Furnace Gas Characteristics," *Air Pollution Control Association Journal*, 27(2): 114—120, Pittsburgh 1977.

[24] C. C. Shale, J. H. Holden, and G. E. Fasching, "Electrical Resistivity of Fly Ash at Temperatures to 1500°F.," *U.S. Bureau of Mines Report of Investigations 7041*. Washington: U.S. Government Printing Office, 1968.

[25] W. F. Frazier and W. Borowy, "Evaluation of Utility Cold-Side Electrostatic Precipitators," presented at ASME-IEEE Joint Power Generator Conference, October 20—24, 1985, Milwaukee, WI; ASME Paper no. 85-JPGC-APC-3.

[26] Harry J. White, "Resistivity Problems in Electrostatic Precipitation," *Air Pollution Control Association. Journal*, 24(4): 313—338, April, 1974.

[27] Gilbert G. Schneider, et. al., "Selecting and Specifying Electrostatic Precipitators," *Chemical Engineering*, 82(11): 94—108, May 26, 1975.

[28] Y. Goland, "Mechanical and Structural Design Consideration for Internals in Electrostatic Precipitators," *ASME Paper no. 76-WA/Pwr-7*. New York: American Society of Mechanical Engineers, 1976.

[29] Jason Makansi, "Particulate Control: Optimizing Precipitators and Fabric Filters for Today's Powerplants," *Power Special Report*, December 1986; McGraw-Hill, New York, NY, 1986. (Contains listing of 38 references and papers as suggested reading.)

[30] Joint Technical Committee of the American Boiler Manufacturers Association and Industrial Gas Cleaning Institute, Inc., "Design and Operation of Reliable Central Station Flyash Hopper Evacuation Systems," *Proceedings of the American Power Conference*, 42: 74—85, 1980. Chicago: Illinois Institute of Technology, 1981.

[31] J. A. Hudson, et al., "Design Considerations and Initial Startup of Shawnee Baghouses," *Proceedings of the American Power Conference*, 42, 1980. Illinois Institute of Technology, Chicago, 1981.

Gordon J. Floyd and A. Vandewalle, "Australian Experience with Fabric Filters on Power Boilers," presented at Conference on Fabric Filter Technology for Coal Fired Power Plants sponsored by the Electric Power Research Institute (EPRI), July 15—17, 1981, Denver, CO.

R. C. Carr and W. B. Smith, "Fabric Filter Technology for Utility Coal-Fired Power Plants, Part II: Application of Baghouse Technology in the Electric Utility Industry," Journal of the Air Pollution Control Association, Vol. 31, No. 2, February 1984; includes 53 references.

[32] H. N. Head, *EPA Alkali Scrubbing Test Facility: Advanced Program*. Third Progress Report. September 1977. Report no. PB-274 544 (EPA-600/7-77/105). Springfield, VA: National Technical Information Service, 1977.

[33] "Acid Deposition: Atmospheric Processes in Eastern North America," National Research Council, Committee on Atmos-

pheric Transport and Chemical Transformation in Acid Precipitation, National Academy Press, Washington, DC, 1983.

[34] C. C. Leivo, *Flue Gas Desulfurization Systems; Design and Operating Considerations*. Volume II. Technical Report. March 1978. Report no. PB-280 254 (EPA-600/7-78/030B). Springfield, VA: National Technical Information Service, 1978.

[35] F. A. Ayer, comp., *Proceedings: Symposium on Flue Gas Desulfurization*, Hollywood, FL, 1977. Report no. EPA-600/7-78/058. Springfield, VA: National Technical Information Service, 1978.

[36] G. P. Behrens, et al., "The Evaluation and Status of Flue Gas Desulfurization Systems." EPRI Report no. CS-3322, Project 982-28, prepared by Radian Corporation, January, 1984.

[37] J. L. Pearson, G. Nonhebel, and P. H. N. Ulander, "The Removal of Smoke and Acid Constituents from Flue Gases by a Non-Effluent Water Process," *Institute of Fuel. Journal*, 8: 119—156, 1935.

[38] F. W. Link and W. H. Ponder, "Status Report on the Wellman-Lord/Allied Chemical Flue Gas Desulfurization Plant at Northern Indiana Public Service Company's Dean H. Mitchell Plant." *Proceedings: Symposium on Flue Gas Desulfurization held at Hollywood, Florida, November 1977. Volume II.* Report no. PB-282 091 (EPA-600/7-78-058b), pp. 650—664. Springfield, VA: National Technical Information Service, 1978.

[39] E. F. Aul, et al., *Evaluations of Regenerable Flue Gas Desulfurization Procedures. Volume II*. Final Report. Report no. EPRI-FP-272 (V.2). Palo Alto, Ca: Electric Power Research Institute, 1977.

[40] D. C. Borio, et al., "Design of Spray of Tower Absorbers for Lime/Limestone Wet Scrubbers," *ASME Paper no. 79-WA/Fu-9*. New York: American Society of Mechanical Engineers, 1979.

[41] D. A. Burbank and S. C. Wang, "EPA Alkali Scrubbing Test Facility: Advanced Program—Final Report" (October 1974 to June 1978), EPA-600/7-80-115, PB80-204241, May 1980.

[42] Dante Frabotta and Philip C. Rader, "Lime/Limestone Air Quality Control Systems: Effect of Magnesium on System Performance," *ASME Paper no. 76-WA/APC-10*. New York: American Society of Mechanical Engineers, 1976.

[43] Philip C. Rader, David C. Borsare, and Dante Frabotta, "Process Design of Lime/Limestone FGD Systems for High Chlorides," presented at Coal Technology '82, 5th International Coal Utilization Exhibition and Conference, December 7—9, 1982, Houston, TX; also as Combustion Engineering publication TIS-7293.

[44] Philip C. Rader, M. R. Gogineni and K. Poglitsch, "Coprecipitation: A Method for Gypsum Scale Prevention in Lime/Limestone Flue Gas Desulfurization Systems," presented at the 72nd Annual Meeting of the American Institute of Chemical Engineers, San Francisco, November 25—29, 1979.

[45] Philip C. Rader, et al., "The Role of Crystallization in the Design of Lime/Limestone Wet Scrubbing Systems for Flue Gas Desulfurization," Presented at the Symposium on Crystallization and Energy Systems, 83rd National Meeting of the

American Institute of Chemical Engineers, Houston, March 20—24, 1977.

46 Benjamin F. Jones, Philip S. Lowell, and Frank S. Meserole, *Experimental and Theoretical Studies of Solid Solution Formation in Lime and Limestone Scrubbers. Volumes I and II. Final Report.* Report nos. PB-264 953 (EPA/600/2-76/273a) and PB-264 954 (EPA/600/2-76/273b), respectively. Springfield, VA: National Technical Information Service, 1976.

47 E. C. Lewis, G. W. Driggers, and K. W. Malki, "Laboratory Evaluation of Several Alloys in High Chloride FGD Enviroment—Progress Report," presented at Solving Corrosion Problems in Air Pollution Control Equipment, sponsored by APCA, IGCI, and NACE, Denver, CO, August 11—13, 1981; also as Combustion Engineering publication TIS-6950.

48 R. P. Van Ness and D. C. Borsare, "Design and Initial Operation of High Sulfur Flue Gas Desulfurization Systems with Lime or Limestone at Louisville Gas and Electric's Mill Creek Units 1 and 2," presented at 75th Annual Meeting and Exhibition of the Air Pollution Control Association, New Orleans, LA, June 20—25, 1982; also as Combustion Engineering publication TIS-7140.

49 G. E. Bresowar, D. C. Borsare, and K. W. Malki, "Dry Scrubber Design and Application: The C-E Approach," presented at ASME-IEEE-ASCE Joint Power Generation Conference, St. Louis, MO, October 4—8, 1981.

G. E. Bresowar, P. E. Traccarella, and W. B. Ferguson, "FGDS Selection and Design for Retrofit Applications," presented at Coal Technology 85, Pittsburgh, PA, November 12—14, 1985; also as Combustion Engineering publication TIS-7869.

W. B. Ferguson, Jr., D. C. Borio, and D. L. Bump, "Equipment Design Considerations for the Control of Emissions from Waste-to-Energy Facilities," proceedings of the 1986 Air Pollution Control Association annual meeting; also as Combustion Engineering publication TIS-8189.

M. D. Mirolli, W. B. Ferguson, and D. L. Bump, "Mid-Connecticut Resource Recovery Project," presented at ASME-IEEE-ASCE Joint Power Generation Conference, Portland, OR, October 19—23, 1986; also as Combustion Engineering publication TIS-8194.

50 R. G. Knight, et al., *FGD Sludge Disposal Manual*, Second Edition, EPRI-CS 1515, 1980.

51 Paul E. Traccarella and Nancy C. Mohn, "Disposing of FGD Waste Using Stabilization and Fixation Processes," presented at ASME-IEEE-ASCE Joint Power Generation Conference, Denver, CO, October 17—21, 1982; also as Combustion Engineering publication TIS-7169.

52 Michael S. McCartney and Mitchell B. Cohen, "Techniques for Reducing NO_x Emissions from Coal Fired Steam Generators," *Proceedings of First International Conference on Acid Rain*, March 27—28, 1984, Washington, DC; also as Combustion Engineering publication TIS-7660.

53 Suyama, K., et al, "Operating Experience on a Selective Catalytic Reduction System for Steam Generators," presented at 1985 Joint Symposium on Stationary Combustion NO_x Control, May 6—9, 1985, Boston, MA.

54 "Selective Catalytic Reduction for Coal-Fired Power Plants—Pilot Plant Results," Report no. CS-4386, Electric Power Research Institute, Palo Alto, CA, April 1986.

55 C. L. Burton and R. E. Willison, "Application of Model Studies to Industrial Gas-Flow Systems," *ASME Paper no. 59-A-280.* New York: American Society of Mechanical Engineers, 1959.

56 Lewis F. Moody, "Friction Factors for Pipe Flow," *Transactions of the ASME*, 66: 671—684, New York. 1944.

57 I. E. Idel'chik, *Handbook of Hydraulic Resistance—Coefficients of Local Reistance and of Friction.* Report no. AEC-TR-6630. Springfield, VA: National Technical Information Service, 1966.

58 *Recommended Guide for the Prediction of the Dispersion of Airborne Effluents*, sponsored by ASME Air Pollution Control Division, New York: American Society of Mechanical Engineers.

59 H. E. Cramer, "A Practical Method for Estimating the Dispersal of Atmospheric Contaminants," *Proceedings of the 1st National Conference on Applied Meteorology*, Hartford, CT, October 28—29, 1957, pp. C-33—C-55. Boston, MA: American Meteorological Society, 1957.

60 O. G. Sutton, "A Theory of Eddy Diffusion in the Atmosphere," *Proceedings of the Royal Society of London*, 135, Series A: 143—165, 1932.

61 C. H. Bosanquet and J. L. Pearson, "The Spread of Smoke and Gases From Chimneys," *Transactions of the Faraday Society*, 32: 1249—1263, 1936.

62 H. Moses and M. R. Kraimer, "Plume Rise Determination—A New Technique Without Equations," presented at 64th APCA Annual Meeting, June 27—July 2, 1971, Atlantic City, NJ, Paper No. 71-61.

63 M. E. Smith and I. A. Singer, "An Improved Method of Estimating Concentrations and Related Phenomena from a Point Source Emission." *Journal of Applied Meteorology*, Vol. 5, No. 5, Oct. 1966, pp. 631—639.

64 J. S. Touma, "Dependence of the Wind Profile Law on Stability for Various Locations," *Journal of the Air Pollution Control Association*, Vol. 27, No. 9, Sept. 1977, pp. 863—866.

65 Philip A. Leighton, *Photochemistry of Air Pollution.* New York: Academic Press, 1961.

CHAPTER 16

Ash Handling Systems

Chapter 3 presented both the physical and chemical properties of coal-ash and the effects that ash in its various forms can have on the operation of a boiler. This chapter describes some of the equipment being used for the collection and transport of ash, particularly from pulverized-coal-fired steam generators. Heat losses, power consumption, and water use in operating this equipment also are considered. Only through complete understanding of the design and operation of ash-handling equipment, as well as the comparative features of the many available types, can the large systems currently being installed be made both more reliable and more economical to operate.

There are essentially two types of ash produced in a suspension-fired furnace: bottom ash and flyash. Bottom ash is slag which builds up on the heat-absorbing surfaces of the furnace, superheater, and reheater that eventually falls either by its own weight or as a result of load changes or sootblowing. With low ash-fusion temperatures, a large amount of molten slag can adhere to the furnace walls and subsequently fall through the furnace bottom. (See Fig. 1) Other ash becomes entrained with, and carried away by, the flue gas stream and is collected in economizer or dust-collection equipment hoppers. This is called flyash.

The amount of air-borne ash passing through the furnace depends on the dust-bearing capacity of the combustion gases, on the size and shape of the particles, and on the density of the ash relative to that of the upward flowing gas. Fig. 2 shows the effects of particle size on terminal velocity[1] which is the rate at which particles of various diameter settle in still air.

Fig. 1 **A large pulverized-coal-fired furnace viewed from above. Ash and slag falling from the walls pass through the bottom opening into a bottom-ash hopper or submerged scraper conveyor.**

Coarser particles fall more rapidly than fine. Thus, for a given upward velocity, fine particles will leave the furnace and coarse particles either will drop to the bottom or be thrown from the center of the furnace toward the bounding furnace walls. For example, according to the curve, particles larger than 0.011 in. diameter (not passing through a 50-mesh screen) can be expected to fall to the bottom when the vertical velocity of the gases in a furnace is 10 ft/sec.

In this regard, a situation occurs in the design and operation of large pulverized-fuel furnaces that is of interest. As furnace plan areas increase for a given quantity of heat input (NHI/PA), to accommodate a very poor fuel, for example, the upward gas velocities correspondingly decrease. At the same time, the furnace wall area becomes greater, leading to the possibility of larger accumulations of slag or ash on the walls for a given heat input. The greater ash deposits and lesser upward velocities that result mean that more ash will fall to the furnace bottom.

FACTORS
IN SYSTEM SELECTION

Many factors determine the method of handling and storing coal-fired power-plant ash. They include

■ fuel source and ash content

■ plant siting (land availability, presence of aquifers, adjacent residential areas)

■ environmental regulations

■ steam-generator size

■ cost of auxiliary power

■ local market for ash

■ cementitious character of the ash

Ash quantities and properties, both physical and chemical, determine the type and size of an ash-handling, storage, transport, and disposal system. They also provide an indication of the environmental impact associated with ash disposal, establish overall system constraints, and influence disposal site design. Peak ash production rates (based on 100 percent capacity factor) are used to size ash-handling and transport systems. On the other hand, average ash production rate, which considers the projected plant capacity factor over the operating life of the plant, is used to size disposal sites.

FUEL SOURCE AND ASH CONTENT

A major factor affecting the choice of an ash-handling system is the type of fuel to be fired. Fig. 3 shows the wide variation in ash "fired" into a furnace as a function of fuel ash content and moisture. For example, a boiler firing a 15-percent ash subbituminous coal generates almost three times the total ash as the same size unit burning a 10-percent ash, high-calorific-value, medium-volatile coal.

It is usually more meaningful to define ash content as it relates to energy input rather than on the basis of a weight percentage of the coal alone. The pounds of ash per million Btu fired is arrived at by the formula:

$$\frac{\text{ash (as a \%)}}{\text{HHV (Btu/lb)}} \times 10{,}000 = \text{lbs ash}/10^6 \text{ Btu}$$

(1)

Relating the ash to be handled directly to the required heat input gives a realistic relation-

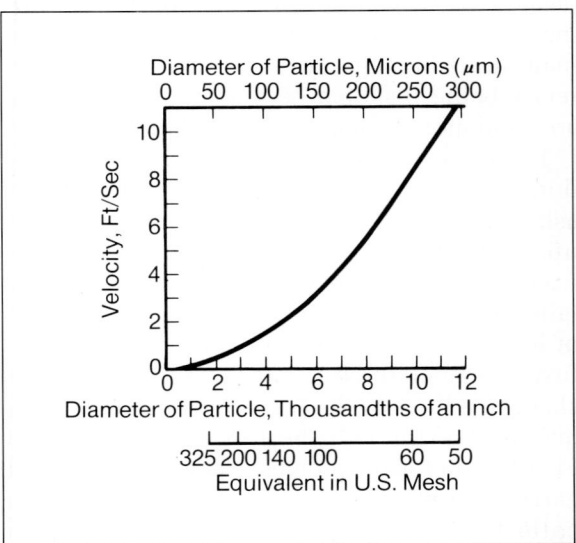

Fig. 2. Velocity of air borne particles (specific gravity 1.2) falling through still air at 70°F

ship to the amount that must be handled as bottom ash as well as that handled as economizer and precipitator ash. It also gives an insight to the potential rate of ash that has to be handled in the furnace and on convective surfaces.

PLANT SITING

The available water supply and locations of an ash disposal site, relative to the power plant, can greatly affect the design and cost of ash disposal. An ample water supply and a great deal of land are prerequisites of many ash-handling systems.

If land for a plant is limited, ash storage bins or small dewatering ponds can be used for temporary holdup. Trucks or rail cars then are used to carry the ash away. If there is insufficient water, the flyash must be collected by one of the available dry systems while bottom-ash is collected in a recirculated water system.

ENVIRONMENTAL REGULATIONS

Federal, state and local effluent regulations probably have the greatest impact on installations today and in the future. For example, Table I lists typical limitations for bottom-ash and flyash transport water.

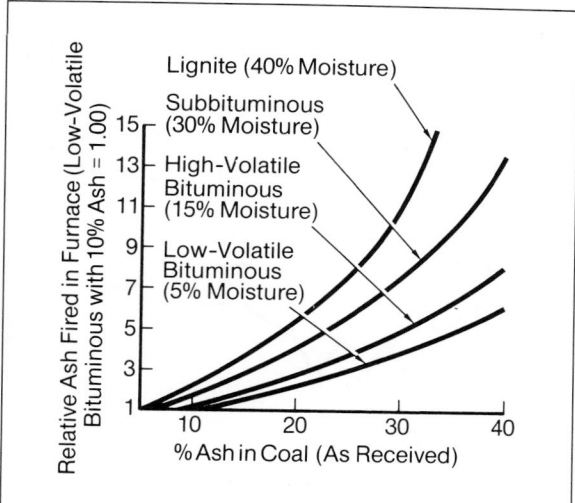

Fig. 3. Relative ash amounts with various types of coals, compared to Eastern bituminous coal with 10% ash as a base (corrected for boiler efficiency). Figures in parentheses are as-received moisture contents (by weight) in the coal

The TSS (total suspended solids) limitations usually are the most difficult to meet, resulting in a trend toward dry flyash systems[2]. But this consideration does not preclude storage ponds to which flyash can be trucked or conveyed pneumatically. Dust is controlled with water sprays and no net effluent is created.

ASH COLLECTION POINTS

After the combustion process in a suspension-fired solid-fuel furnace, the ash collects or is collected in several areas (Fig. 4).

Hoppers or conveyors in troughs are used under the furnace bottom to collect the material falling from the furnace heat-absorbing surfaces. This material can be in either a dry or molten state. If dry, it can be collected in water-impounded or dry receivers. If molten (from wet-bottom furnaces) it must be directed into a water impoundment.

Hoppers also are used under the pulverizer-rejects discharge spout. These collect high-density pyrites and tramp iron separated from the coal during pulverization.

Hoppers under the rear convection pass and air heater(s) of the boiler are commonly called economizer and air-heater hoppers. It is here that coarser particles drop from the gas stream with directional changes in the gas flow.

Flyash hoppers included under the precipitators or baghouses collect the material extracted from the flue-gas stream by these particulate-removal devices.

TABLE I. Typical Effluent Limitations

Bottom-Ash	Oil and Grease	1 mg/ℓ × Flow (max) 0.75 mg/ℓ × Flow (avg)
Transport	pH	6–9
Water	TSS	5 mg/ℓ × Flow (max) 1.5 mg/ℓ × Flow (avg)
Flyash	Oil and Grease	0
Transport	pH	6–9
Water	TSS	0

QUANTITIES OF ASH

Most pulverized-coal-fired steam generators are dry-bottom units in which 20—40 percent of the ash from the coal falls to the furnace bottom. Sixty to 80 percent of the ash leaves the furnace as flyash. Most of the flyash is collected in economizer hoppers and one of several available dust collection systems.

However, with some fuels less than 20 percent bottom ash has been measured, while with certain lignites more than 45 percent bottom ash has been reported. As a guide to establishing the size of the equipment in the total ash-handling system, typical collection quantities (mass basis) are used.

For example, the 20—40 percent figure can vary according to experience with the type of fuel burned. With low- and medium-volatile bituminous coals, a 15—20 percent bottom ash figure is appropriate. With a high-volatile bituminous coal a 20—30 percent bottom ash figure is used. Systems for subbituminous coal and lignite are sized using a 30—50 percent bottom ash figure.

The economizer and air heater ash hoppers each are sized to collect about 5 percent of the flyash while a precipitator, fabric filter or other equipment would be sized to extract 80 to 100 percent of the flyash from the flue gas.

Note that the sum of all the assumed collection-point values can exceed the total ash in the coal by 50 percent or more because of the lack of precision of the individual factors, and to allow for variation in fuel quality.

ASH DENSITY

Flyash density depends primarily on particle size, particle structure, and carbon content. In general, the relatively large, coarse particles containing a high percentage of carbon have a low density. These particles commonly have a porous or lace-like structure and result from the partial combustion of the pulverized coal. It appears that the volatile portion burns out, leaving black coke-like particles. These particles have densities of about 35 to 65 lb/ft^3 (true density). The finer ash particles, which tend to be low in carbon content, have much higher densities, usually ranging from 90 to 180 lb/ft^3 (true density).

In the hot state, freshly collected ash is normally very fluid with a lower density than cold ash—generally between 30 to 50 lb/ft^3 bulk density. The fresh ash is aerated as the individual particles are exposed to the gas. With standing and cooling, de-aeration occurs and the ash compacts. It becomes less fluid and can approach bulk densities as high as 90 lb/ft^3. When

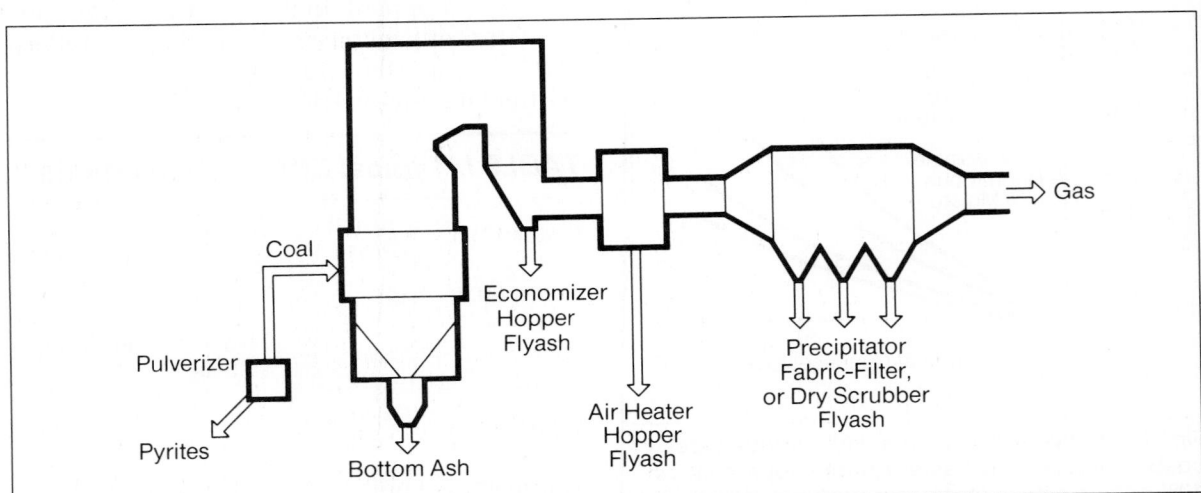

Fig. 4. Ash must be collected and transported from at least the five points shown. In each location, the physical and chemical characteristics of the ash vary

cold, subbituminous coal ash, which is very fine, will usually have loosely packed bulk densities between 65 and 85 lb/ft³. Bituminous coal ash, on the same basis, will have a density between 60 and 70 lb/ft³.

The bulk density of dry bottom ash, as initially dropped into an ash receiver, can be as low as 40 to 45 lb/ft³, but it can compact to above 75 lb/ft³. A value of 55 lb/ft³ bulk density is used for establishing bottom-ash hopper storage capacity. Pulverizer rejects, when principally iron pyrites (iron disulfide, FeS_2), will have a bulk density of at least 135 lb/ft³.

TYPES OF SYSTEMS

The three principal methods of moving ash from the collection points to the on-site disposal or storage location are hydraulic, pneumatic, and mechanical. In a hydraulic system, a stream of water conveys the ash in a closed pipeline or open sluiceway. Commonly, a jet or centrifugal pump provides the motive force. A pneumatic system transports the ash in a stream of air or flue gas induced by either an upstream pressure or a downstream vacuum. The oldest technique for ash removal is the mechanical method, the most rudimentary form being a shovel or hoe, with the more sophisticated type being mechanical scraper or flight conveyors.

Two basic philosophies have evolved throughout the world for removing ash from pulverized-coal-fired boilers:

■ collection, storage and *periodic* removal, usually called intermittent removal systems

■ collection and *continuous* removal, commonly termed continuous removal systems

Ash-handling systems are further differentiated by their conveying frequency and the conveying medium. The major systems are:

1. Bottom ash
   a. Water-impounded hopper with wet pipeline or sluiceway, to a pond or closed recirculation system—intermittent, hydraulic
   b. Dry hopper, dry pipeline—intermittent, pneumatic
   c. Submerged scraper conveyor—continuous, mechanical
2. Pyrites (Pulverizer Rejects)
   a. Hoppers with wet sluicing (pipeline or open trough)—intermittent, hydraulic
   b. Hoppers with manual removal—intermittent, mechanical
   c. Hoppers with dry vacuum removal (infrequently used)—intermittent, pneumatic
3. Economizer
   a. Dry pipeline—intermittent, pneumatic
   b. Dry flight or screw conveyor—continuous, mechanical
   c. Wet pipeline using receiving tanks below economizer hoppers—continuous, hydraulic
   d. Dry pipeline using receiving tanks below economizer hoppers—continuous, pneumatic
4. Flyash
   a. Dry pipeline, vacuum—intermittent, pneumatic
   b. Dry pipeline, pressure—intermittent, pneumatic
   c. Dry/wet pipeline, vacuum—intermittent, pneumatic/hydraulic (water exhausters)
   d. Dry, screw or flight conveyor—continuous, mechanical

BOTTOM ASH SYSTEMS

A wet bottom-ash system collects and hydraulically or mechanically removes ash falling from the furnace heat-transfer surfaces. One type—a water-impounded hopper (Figs. 5 and 6)—receives, quenches, and stores ash from the furnace. In this system the ash is intermittently drained by discharging the water-ash mixture through an outlet gate to clinker grinders that reduce the size of the materials to facilitate transport to disposal by jet or centrifugal pumps.

The principal reasons for water-impounding include the following:

■ to break up large pieces of slag by thermal

Fig. 5 **Water-filled intermittent-removal ash hopper for large pulverized-coal boiler**

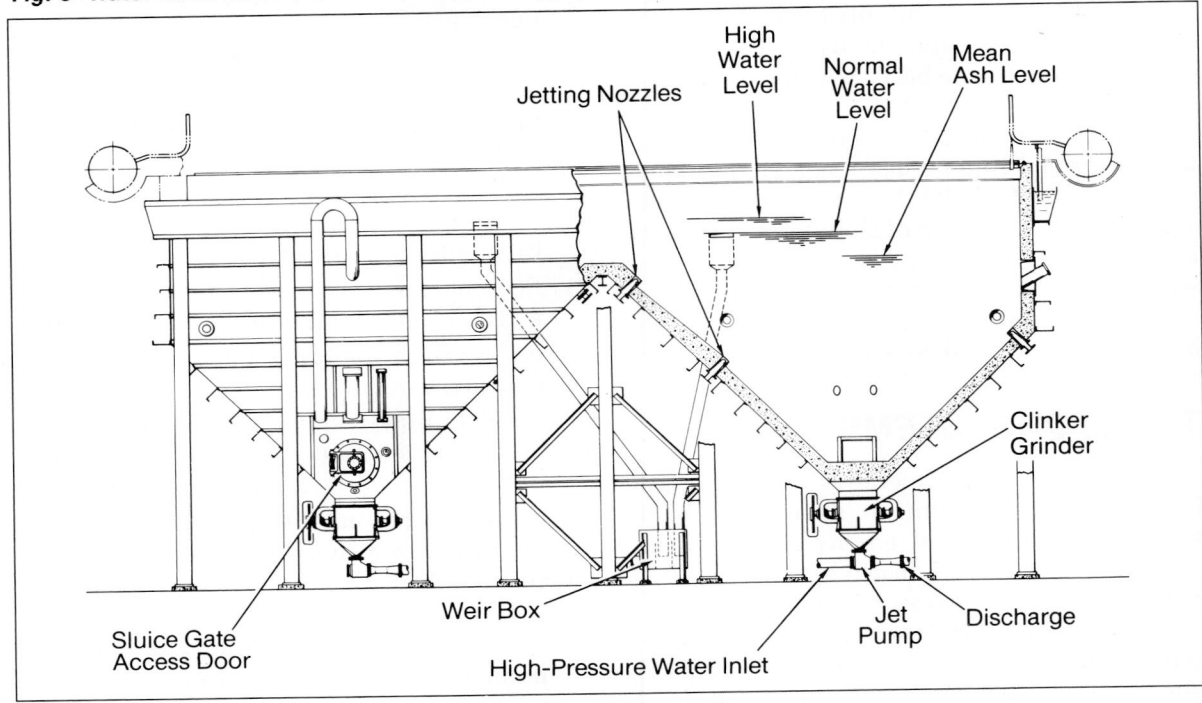

Fig. 6 **Cross section of water-filled ash hopper under a large pulverized-coal boiler**

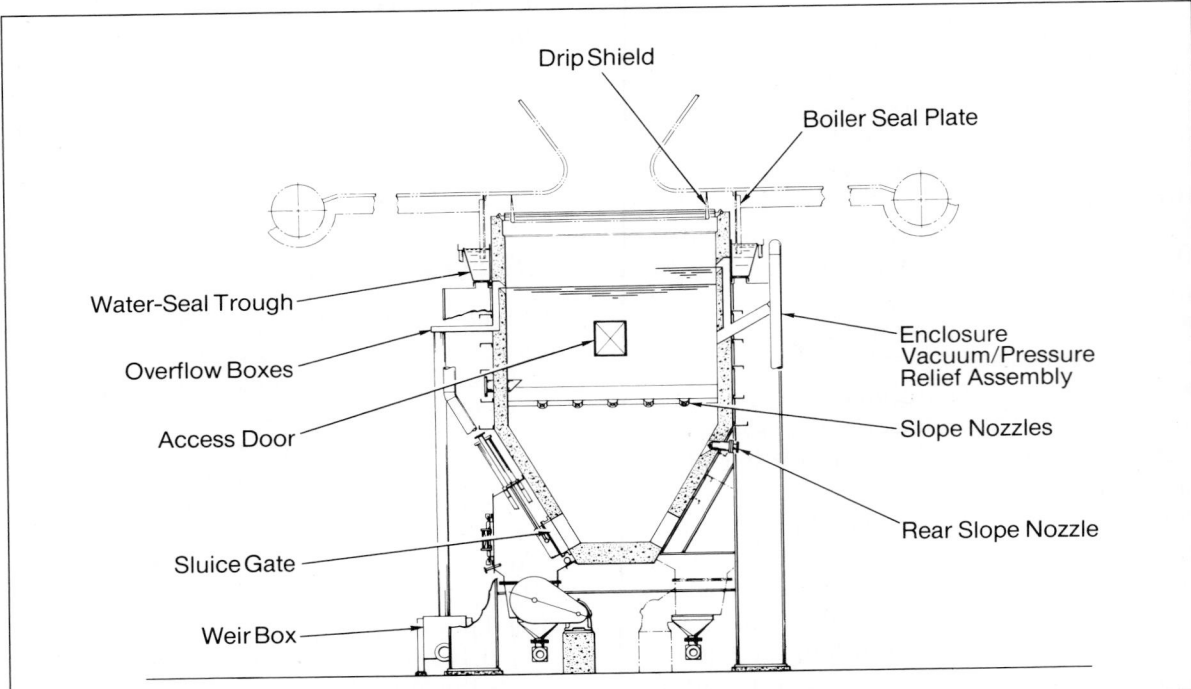

shock as they fall into the pool of approximately 140°F water

- to keep the ash and slag submerged so that they do not fuse into large unmanageable masses that would result if they were exposed to furnace heat
- to provide a resilient medium to decelerate the large pieces of slag
- to assist in the evacuation of ash (from the intermittent-removal type of hopper) by reducing its effective angle of repose

INTERMITTENT-REMOVAL BOTTOM-ASH HOPPER

On most pulverized-coal-fired units above 400,000 lbs of steam/hr, intermittent-removal bottom-ash hoppers are of the V, W, or triple-V design. This design has hopper floors sloping at angles of 35 to 55 degrees allowing gravity to help remove ash from the hopper. During ash removal the high-pressure water to the jet pump is turned on and the clinker grinder is started. The control system opens the hydraulic-cylinder-controlled sluice gate and the ash flows from the hopper through a grinder into the jet pump, then to a transfer tank, dewatering system, or settling pond. Alternatively, a centrifugal materials-handling pump evacuates the hopper and transports the ash to the disposal system.

Operating experience with water-filled hoppers has shown that most of the ash will flow toward and through the sluice gates by gravity; however, large slabs of fused ash do occur that will not move by gravity alone. Their removal requires the assistance of high-pressure (125 psig) water-jetting nozzles located on the slopping floors of the hopper. In addition, jetting nozzles in the wall opposite each sluice gate break up the arching and packing of the ash that may occur at the gate.

Whenever practical the hopper should have a volume corresponding to the ash produced during 12 hours of operation at full boiler load while firing the lowest-heating value and highest-ash-content fuel for which the boiler is designed. This guideline is based on the normal practice in the U.S. of emptying ash hoppers once each 8-hour shift. Depending upon the fuel ash content and the boiler load, the usual time required to evacuate a bottom-ash-hopper is one to three hours. The hopper volumetric capacity is measured below a mean ash level which is usually 12 to 24 inches below the normal water level, depending on the bottom-ash hopper width (see Table II and Fig. 7).

It is important to realize that the ash falling from the furnace walls does not pass through the furnace bottom opening in equal increments along the length of that opening. Fig. 8 shows that, for a large "open" furnace (one without a center wall), all of the ash from each sidewall falls through the first 3 or 4 feet of the bottom opening; the ash falling from the front and rear walls can be assumed to distribute fairly equally along the furnace width. Thus, as shown by the figure, about 25 percent of the total ash to the furnace bottom will go through the 3 to 4 ft portion near the side wall, and less than 2% of the bottom ash will go through each remaining foot of the bottom "slot". It is of utmost importance to take this distribution into account when designing bottom-ash receiving equipment. The normal water level should be at least 30″ below the horizontal plane of the furnace tubes forming the furnace throat opening (A in Figs. 7 and 9); this dimension is with the boiler pressure parts in the hot (expanded)

TABLE II. Determination of Effective Ash Level*

Hopper Width Inside Refractory (See Fig. 7)	Effective Ash Level (EAL) Below NWL, ft
7'-0″	0.8
8'-0″	1.0
9'-0″	1.2
10'-0″	1.4
11'-0″	1.6
12'-0″	1.8
13'-0″	2.0
14'-0″	2.2

*An interpolation can be made for other hopper widths.

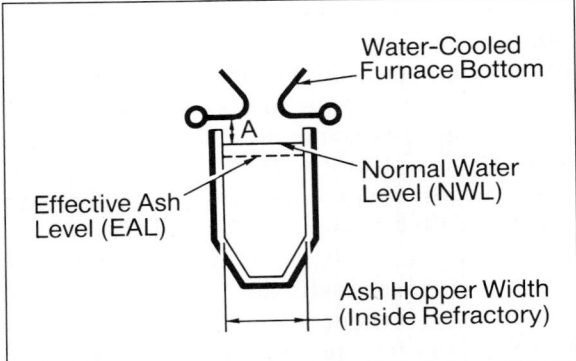

Fig. 7. Effective ash level below normal water level varies as a function of ash hopper width (See Table II)

condition. This distance minimizes the amount of water that may splash on the furnace tubing. The hopper is designed with 5 to 10 feet end-wall submergence to provide a deep pool of water at this critical portion of the hopper.

Depending upon the boiler width and head-room available, the hopper configuration should adequately provide the needed storage capacity. Centerwall boilers must have an odd number of sections. This places the center of a hopper section underneath the centerwall so that there is maximum submergence to collect the ash properly from both sides of the wall.

For ease of maintenance, the ash hopper

should be designed so that the entire hopper and its discharge equipment are located above grade. Pits must be avoided, but if they are used to put the sluice piping below basement floor level, their depth should be limited to a maximum of 3 feet.

OVERFLOW WEIRS

To maintain the normal water level (NWL) in the hopper, an NWL overflow should be provided. The bottom opening of the weir is at the normal water level, Fig. 10. To protect from overfilling the hopper and possibly wetting the boiler tubes, there is a second, independent "high-water-level" overflow. The opening at the bottom of this weir is a minimum of 6 inches above the NWL weir. A conductivity device in the outlet box from the high-level overflow provides an alarm when high overflow occurs.

ASH-HOPPER CONSTRUCTION

Ash hoppers are constructed of carbon-steel plate with structural framing similar to that of the furnace walls. A high-temperature water-resistant gunned or cast refractory, supported by stainless-steel anchors, lines the sides and bottom slopes. Alternatively, on the sloped bottom portions of the hopper, acid-resistant bricks backed by concrete can be used.

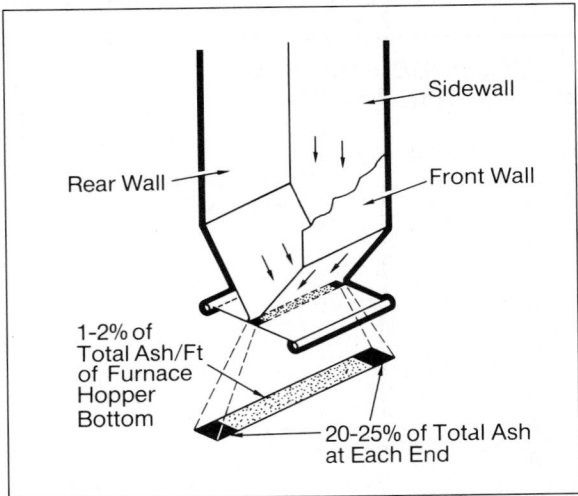

Fig. 8. Approximate distribution of pulverized-coal bottom ash along the furnace hopper opening

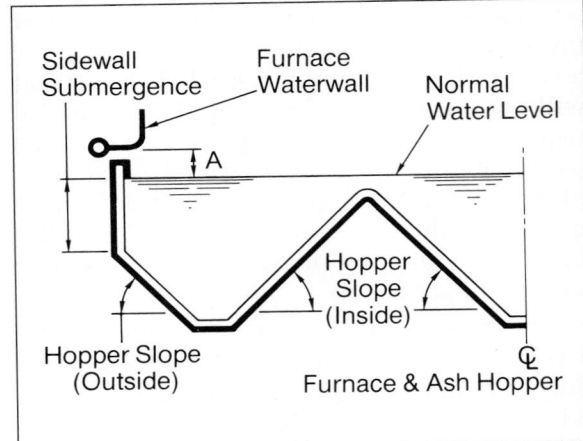

Fig. 9. Schematic arrangement of half of a triple-vee water-impounded bottom-ash hopper. Dimension A is defined with boiler in hot/expanded condition.

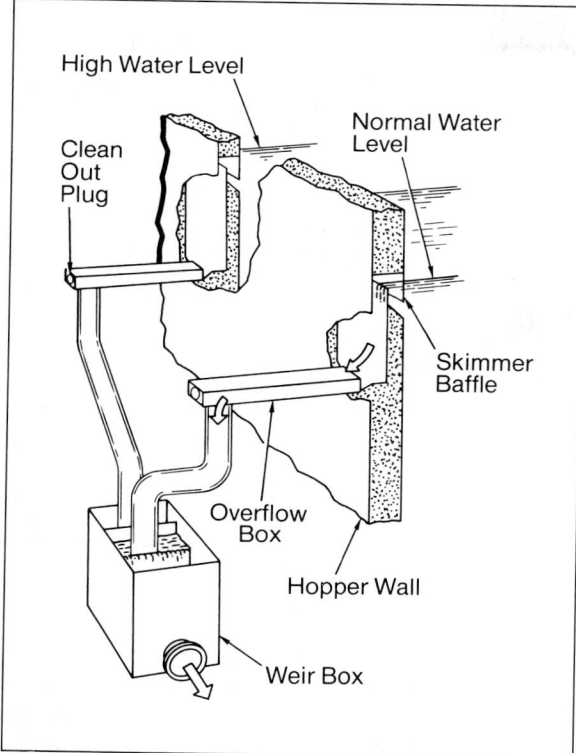

Fig. 10. Arrangement of bottom-ash hopper overflow weirs and piping

A water seal-trough around the upper periphery of the ash hopper, together with the seal plates hung from the boiler tubes, provides a seal against furnace pressure fluctuations and accommodates the downward expansion of the furnace. Nozzles located throughout the seal trough periodically flush accumulated sludge and sediment from the trough bottom.

REFRACTORY COOLING

At the top of the hopper is a water-distribution assembly for cooling the hopper refractory that extends above the normal water level (Fig. 11). The curtain of water flowing over the refractory face also retards refractory deterioration during extended periods when the hopper is drained of water during boiler operation. The water distribution pipe must be equipped with a deflection shield to prevent water from being sprayed on the furnace tubes. Periodic flushing of this pipe with a high-volume water flow through a drain connection in each section insures that pieces of ash do not clog the holes.

Fig. 12, a flow diagram for a single-V type bottom-ash hopper, shows all of the water-flows entering and leaving this hopper. Op-

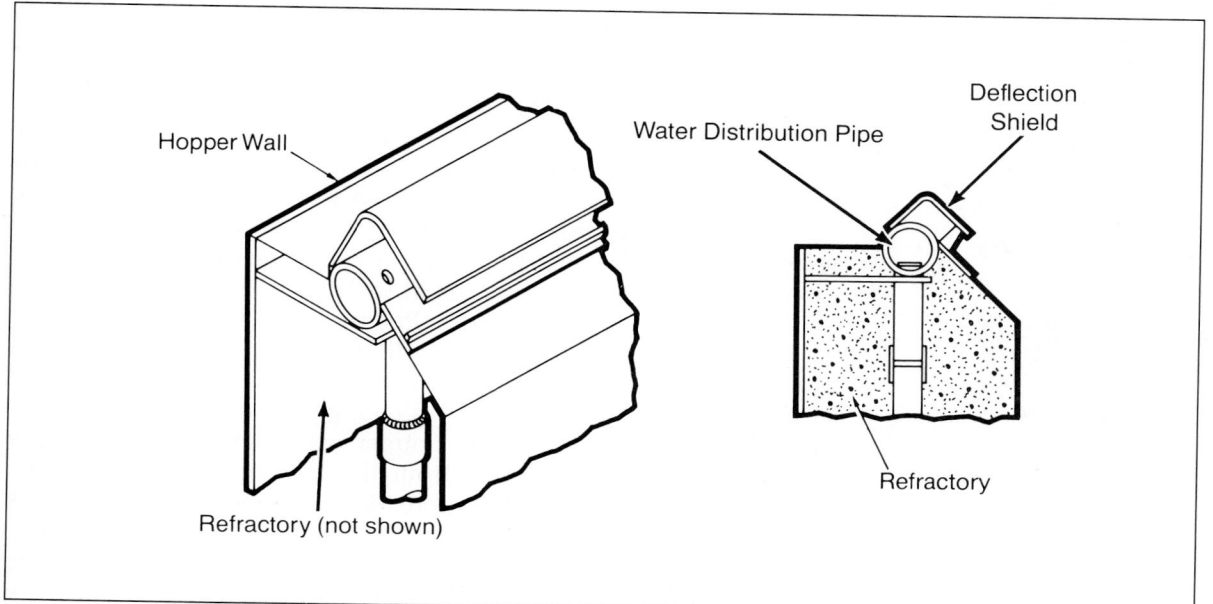

Fig. 11. Distribution-pipe system for refractory cooling water

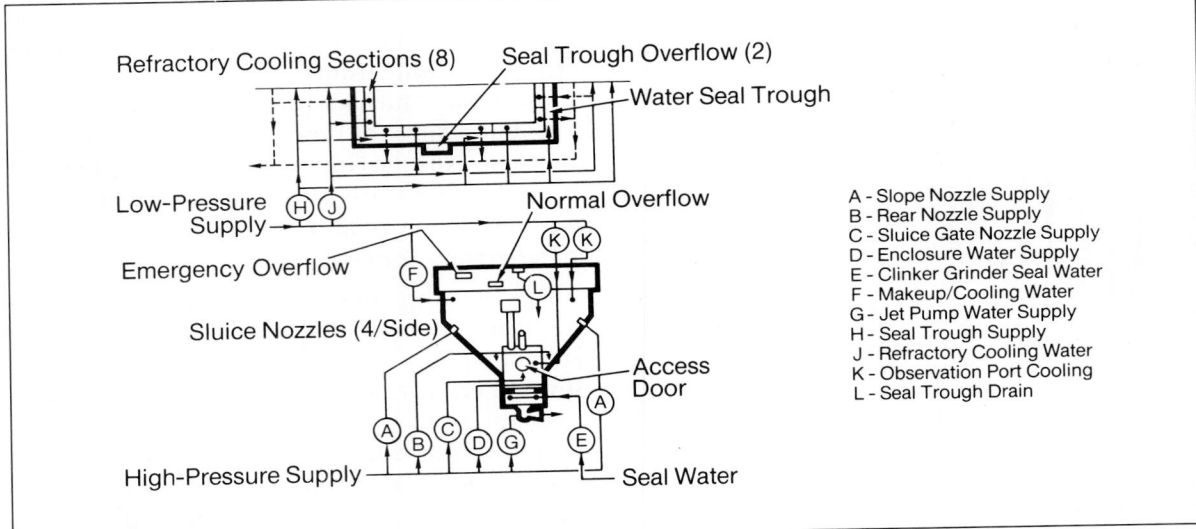

Refractory Cooling Sections (8)   Seal Trough Overflow (2)

Water Seal Trough

Low-Pressure Supply   Normal Overflow

Emergency Overflow

Sluice Nozzles (4/Side)

Access Door

High-Pressure Supply   Seal Water

A - Slope Nozzle Supply
B - Rear Nozzle Supply
C - Sluice Gate Nozzle Supply
D - Enclosure Water Supply
E - Clinker Grinder Seal Water
F - Makeup/Cooling Water
G - Jet Pump Water Supply
H - Seal Trough Supply
J - Refractory Cooling Water
K - Observation Port Cooling
L - Seal Trough Drain

Fig. 12. **Flow diagram of single-vee bottom-ash hopper**

timum operation of a water-impounded hopper can be achieved by maintaining the water level at all times, even while discharging the ash. In this way, the design makes use of all the advantages of water impounding and establishes the maximum possible net-positive suction head on the jet (or centrifugal) pumps at the optimum moment. The only time the water level has to be lowered with this arrangement is when it's necessary to visually inspect the hopper interior or to remove over-sized pieces manually. Each end of the hopper should be equipped with access doors for maintenance.

HOPPER SLUICE GATES

Bottom ash hopper sluice gates (Fig. 13) (usually 24″ × 24″) allow for periodic removal of bottom ash from the hopper. Around the discharge gates are water-tight steel enclosures which flood when ash is being discharged. These enclosures are equipped with access doors, observation ports, and floodlights and are designed to withstand negative and positive pressures. A vent pipe equipped with two vacuum breakers provides vacuum/pressure protection. Positive-pressure protection is achieved by allowing the water in the enclosures to enter the vent pipe to discharge into the bottom-ash hopper. A loop in the vent-pipe above the hopper water level prevents water backflow from the hopper to the enclosure. Vacuum protection is provided by air check valves located at the top of the vent-pipe loop.

The gate is hydraulically lifted by a cylinder designed for water service. Limit switches indicate a fully open or fully closed position.

Operating Cylinder

Ash and Water from Hopper

Wear Liners

Gate Adjustment

Gate

Gate Frame

Renewable Gate Seat

Fig. 13 **Bottom-ash-hopper sluice gate**

High-wear areas in the sluice gate enclosure are often made of ceramics or stainless-steel.

CLINKER GRINDERS

Single- or double-roll clinker grinders are provided at the outlet of each sluice gate. The grinding elements are made from heavy-duty abrasion-resistant materials, such as work-hardened manganese steel or heat-treated high-chrome steel. Ash is ground by a shearing, splitting or crushing action to facilitate hydraulic or pneumatic transport (Figs. 14 and 15).

Because of the severe duty, relatively frequent replacement is required for the wear-prone internals of these crushing machines. It is important to provide easy access for their removal while the steam generator is operating. The design shown in Fig. 14 and 15 provides an easily removed front panel for such access allowing maintenance personnel to replace worn grinding elements without entering the enclosure.

Shaft bearings are spherical roller type with provisions for forced lubrication. Shaft seals have continuous seal-water flow during periods of both grinder operation and shutdown. A close-tolerance throttling bushing impedes the flow of small ash particles into the stuffing box.

JET PUMPS

A jet pump is a venturi-type solids/fluids transport device with no moving parts (Figs. 16 and 17). The pump has an inherent low efficiency but this is offset by its non-overloading characteristic, ease of maintenance, and lack of rotating parts. For these reasons, the majority of hydraulic bottom-ash and mill-rejects systems use jet pumps instead of centrifugal pumps.

To minimize the effects of abrasion, the jet pump is made of abrasion-resistant alloys. The inlet section is heat-treated nodular iron (minimum Brinell hardness of 400); the combining tube (diffuser), high-hardness cast chrome-nickel alloy material, to a minimum Brinell of 550.

A jet pump is supplied at the outlet of each clinker grinder; alternately, a mechanical cen-

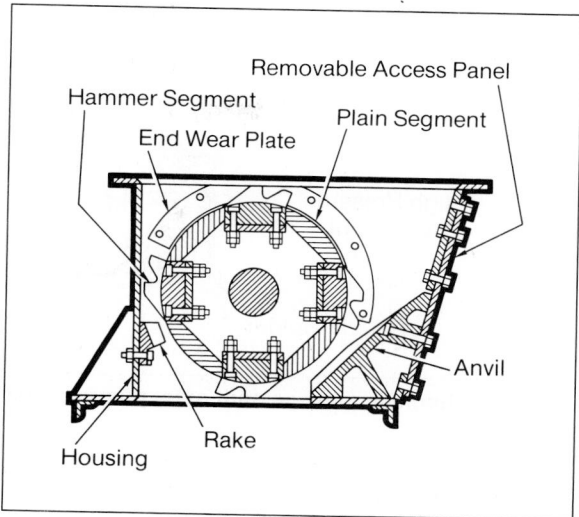

Fig. 14. Single-roll clinker grinder for reducing size of bottom-ash; hammers and anvils are replaceable through removable access panel

trifugal pump can be used. The accompanying box lists the advantages of each.

As more high-ash coal is used for steam-generator fuel—producing greater ash loading on bottom-ash systems—suppliers recommend another sluice gate, grinder and pump for each hopper section. The purpose of this redundant equipment is two-fold: first, to increase the overall system reliability and, second, to in-

Fig. 15 Advanced design of single-roll clinker grinder showing external tripod bearing supports

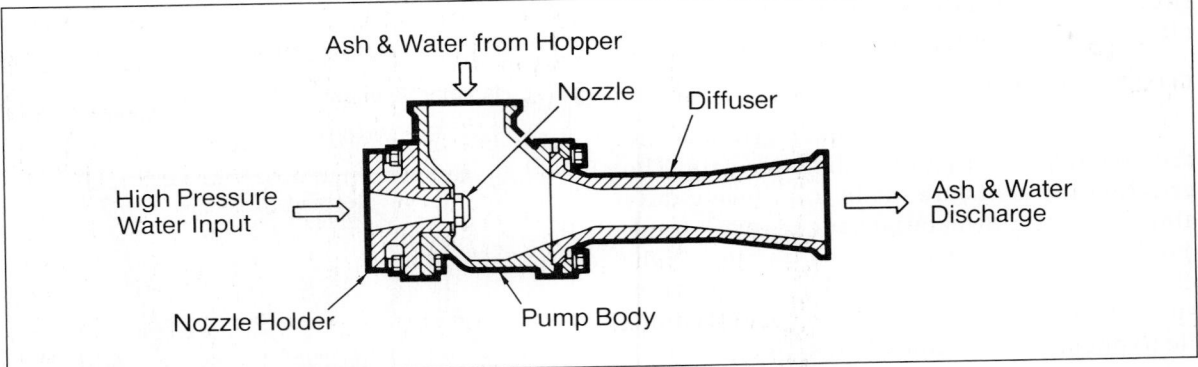

Fig. 16. Schematic arrangement of jet pump for handling abrasive solids

crease the rate of ash-removal when quicker-than-design evacuation is desired.

PIPING, FITTING, AND VALVES

Ash slurry piping is supplied in several forms. Included are
- centrifugally cast chrome-iron alloy with hardness ranging from 280 to 400 BHN (Brinell hardness number)
- surface-hardened carbon-steel piping
- fiberglass-reinforced plastic pipe lined with abrasion-resistant material
- basalt-lined steel pipe

Ash in a water slurry usually travels along the bottom portion of pipes in saltation flow; that is, it alternately "rests" (settles out) and "jumps" (becomes re-entrained in suspended flow). Thus, most wear on slurry piping is on the bottom third (the lower 120° arc) of the pipe. Directly downstream of elbows, piping wear is higher than in straight runs because of the turbulence and impact created at the elbow; this calls for more frequent replacement of downstream runs.

Whenever there is a change in direction in the slurry piping, the fitting located at the change of direction will undergo accelerated wear as a result of turbulence and solids-impact erosion. Thus, fittings, such as 45° and 90° elbows and laterals, are normally constructed from alloy castings that range from 450 to 550 BHN (Fig. 18).

Material hardness alone cannot provide long life at points of change in direction. To add ad-

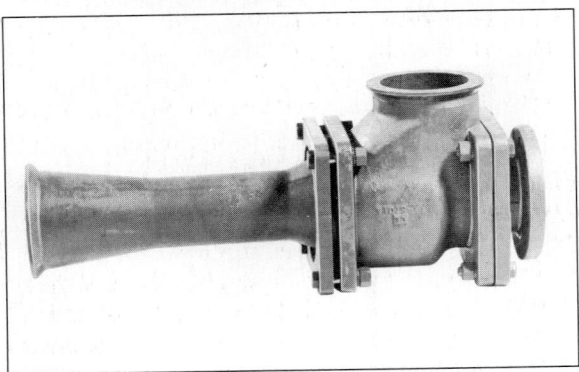

Fig. 17 Jet pump for bottom-ash transport

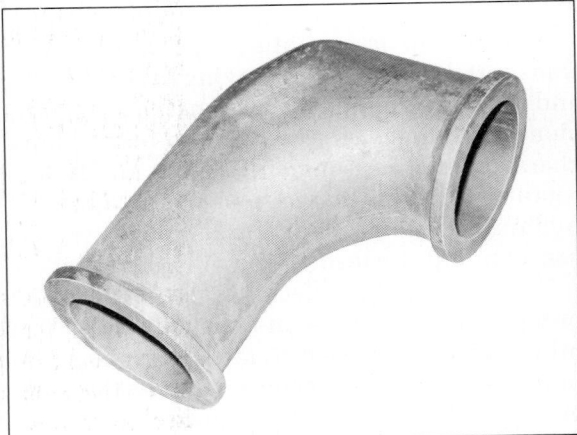

Fig. 18. Chrome-nickel alloy 90° elbow with integral wearback helps reduce wear from solids-impact erosion and turbulence experienced with changes in flow direction

ditional life to the fitting, the impact area has a thicker cross-section than the normal conveying pipeline. This thicker section is commonly termed an *integral wearback*.

Fittings can have connections that are either plain-end or bevelled-end. When plain-end sections are used, piping harnesses prevent the fitting from dislodging due to the reactive loads of the conveying medium. Bevelled-end fittings do not require friction-type couplings to secure the fitting to the pipe.

In conveying water-ash slurries, it may be necessary to divert flow or isolate branch lines when multiple sluice lines exist. Industry practice is to provide heavy-duty knife gate valves for this service because of their profile characteristic. That is to say, when the knife-gate valve is open, its orifice matches the diameter of the sluice pipe in which it is installed. Minimizing the turbulence through the valve reduces erosion. In addition, the insertion of the valve in the line requires no special adaptors or fittings. Also, knife-gate valves can have bonnets to prevent leakage from the packing, and flushing ports for cleaning the seat and the gate before closing. Manual handwheels or pneumatic operators are commonly used to open and close knife gates. Automatic knife-gate systems can have limit switches to provide feedback to the control system.

When the sluice conveying line is made of alloy cast pipe, the couplings between fittings, valves, and plain-end pipe are manufactured by ash-system suppliers. Where two sections of plain-end pipe meet, commercial pipe couplings can be used.

Any slurry piping system must be designed with minimum and maximum velocities in mind. The minimum velocity is that required to convey the particle in the pipeline without settling. This velocity is a function of the specific gravity of the material being conveyed, the pipeline size, and the slurry concentration. Typical ash-slurry pipeline velocities are

- bottom ash—7 to 8 ft/sec
- economizer ash—6 ft/sec
- pyrites—9 to 10 ft/sec
- flyash—4 ft/sec

DISPOSAL OF BOTTOM ASH FROM HYDRAULIC SYSTEMS

Hydraulic ash systems dispose of the ash-and-water slurry in various ways, depending on the overall plant layout, ecological considerations, and available disposal areas.[3] As previously mentioned, the ecological con-

---

COMPARISON OF JET PUMPS VS. CENTRIFUGAL PUMPS

ADVANTAGES OF JET PUMPS

- A jet pump is a simple device. There are no moving parts; it is made up of only three principal components: nozzle, body, and diffuser. This makes maintenance and replacement easier and less expensive than for a centrifugal pump.
- A jet pump can handle air without any deleterious effects; lack of suction head is never a problem.
- A jet pump is self-regulating. The amount of suction material depends on the discharge head. The higher the discharge head, the less suction material will be taken—when there is no suction flow, the only flow through the pump is the motive (clear water) flow. This makes for a self-cleaning action. If the discharge line becomes clogged, the flow of more material into the pump is limited so that the motive flow can flush the line.
- There are no rotating parts subject to erosive ash.

ADVANTAGES OF A CENTRIFUGAL PUMP

- A centrifugal pump can obtain much higher volumetric capacities than a jet pump. A typical jet pump has a practical limit of about 3,500 gpm discharge flow. Centrifugal pumps can produce flows of 35,000 gpm and higher.
- A centrifugal pump has a higher efficiency than a jet pump. Stated in another way, the power required for a centrifugal pump to convey a given flow at a given head is considerably less than the power required to give the jet-pump motive fluid the required head and flow to convey the given mixed water/slurry head and flow.
- No separate motive-water pump is required.

siderations are presently a major influence on decisions made about the means of disposal of ash and slurry.

Most steam-generating plants require either a recirculating system or monitoring (and treatment, if required) of pond blowdown. Recirculating systems frequently include dewatering equipment for the mechanical separation of bottom ash and water.

DEWATERING SYSTEMS

Ash from a conveying sluice line can be deposited into dewatering bins. As the ash settles, it displaces water which overflows into a trough extending around the circumference of the bin top. The overflow water drains by gravity to the recirculation basin or waste drain.

A series of baffle plates, concentric with the outer shell of the bin, prevent ash carryover and undesirable turbulence in the overflow trough. Submerged beneath the water level, the inner baffle inhibits the finer material from reaching the overflow. The second baffle, extending above the overflow trough, creates a barrier to retain the floating material before it reaches the overflow trough. Eventually, floating material settles and is discharged through the sluice gate after the bin is dewatered.

Dewatering bins are installed in pairs. When one bin is filled, the input flow is diverted to the empty bin. After a short period of natural settling, a level of surface water will exist. This surface water is drained off through vertical decanting pipes or floating decanters.

After the surface water is drained, the lower drain valves in the decanting pipes are automatically opened, resulting in a slow, controlled draining of water from the ash in the bin. Depending on the composition of the ash and the desired moisture content, this process generally takes less than eight hours. With proper filtration, settling, and chemical treatment, the overflow water can be recycled to the bottom-ash system in what is essentially a closed loop.

It is important to protect dewatering bins from freezing by providing heaters at the bottom. Also, the lower cone section of the bin normally has vibrators to aid in removing the dewatered ash.

CONTINUOUS-DEWATERING EQUIPMENT

In lieu of the intermittent-dewatering bins just described, continuously operating dewatering systems are available using equipment similar to the submerged scraper conveyor, discussed later in this chapter. With continuous-dewatering apparatus, ash is periodically removed and sluiced from a batch-storage (intermittent-removal) ash hopper to the remote dewatering scraper conveyor. The conveyor dewaters the ash to the same extent as with a dewatering bin and places it on a belt conveyor for further handling. Water from the dewatered ash overflows the conveyor vessel and is treated further as with dewatering bins.

SETTLING AND STORAGE TANK

The discharge water from the dewatering device normally is routed into a settling basin or tank. To separate the ash fines, low velocities are maintained throughout the basin or tank, which are designed for maximum retention time. Any agitation is confined to a small area. Water and ash fines are deposited into the center of the settling tank after passing through a cyclone-type separator and baffles. As the ash fines settle into the tank, the displaced water overflows into a trough which extends around the circumference of the top of the tank. The overflow water drains into the center of the ash-water storage tank. As in the case of the dewatering device, it is important to protect the settling tanks from freezing.

STORAGE OR SURGE TANK

In a recirculating system, an ash-water storage tank provides sufficient volume to absorb the volumetric fluctuations of the hydraulic ash removal system. In addition, this tank provides further separation of fines by maintaining a low velocity. Again, freeze protection must be provided. Centrifugal pumps recirculate the discharge water from the surge tank to supply the jet pumps and other requirements for water in the system.

PONDS OR FILL AREAS

Where geological and siting conditions permit, ash can be pumped to ponds or fill areas. Any ash-receiving pond needs a large capacity to permit the ash to settle by gravity. The design of such ponds can include recovery and reuse of the water in the ash removal system.

DRY BOTTOM-ASH HOPPER SYSTEM

Dry bottom-ash hopper systems for intermittent removal are used when there is no need to provide for water impoundment—and less automation is acceptable. A rule of thumb is that dry bottom-ash hoppers are generally practical on units rated below 400,000 lbs/hr of steam.

Customarily constructed of carbon-steel plate, these hoppers, like the hoppers used in a wet system, have a water-seal trough, complete with flushing nozzles. Unlike water-impounded hoppers, dry hopper slopes must be steep to insure that the dry ash flows to the discharge openings.

The inside of the hopper steel plate is lined with several inches of insulation. Anchors, normally of stainless-steel and welded to the hopper plate, are used to hold 4 to 6 inches of gunned refractory lining forming the inside of the ash hopper.

Ash is periodically removed from the hopper through a series of water-cooled discharge gates. Each gate has a sealed enclosure, a pneumatic operator, an access door, observation port, and floodlight.

At the discharge of the ash gates, grinders may be provided to reduce any large clinkers to a size that can be handled by a pneumatic or hydraulic conveying system. The grinders are similar in construction but normally smaller than the grinders under the wet bottom-ash hoppers. Instead of grinders, sizing grids for manual handling may be placed at the discharge gate outlet.

THE SUBMERGED SCRAPER CONVEYOR FOR CONTINUOUS REMOVAL

Another method of bottom-ash removal uses a flight conveyor of heavy construction submerged in a water trough below the furnace.

With the submerged scraper conveyor (SSC), ash is evacuated mechanically on a continuous basis so there is no long-time storage in the water impoundment beneath the furnace. After discharge from the SSC in a dewatered condition, ash is transported by belt conveyors. The major advantages of the SSC are

- reduced water usage (no transport water)

- reduced power consumption (by eliminating the high-pressure sluicing water required by the jet pumps)

- low boiler setting height

The C-E Combusco submerged conveyor was used in the 1920's for ash removal from stoker-fired boilers. Presently, the C-E/EVT SSC offers similar continuous ash removal from pulverized-coal units.[4] It serves the same functions as the intermittently emptied refractory-lined ash hopper, that is, it seals the furnace bottom, quenches the falling ash and removes it from under the furnace.

The most widely accepted SSC configuration has two separated compartments in which the flights and chains move (Fig. 19). The upper (wet) chamber, containing three to six feet of water, receives the ash falling from the furnace and conveys it up the dewatering slope. At the top of the dewatering slope, the flights reverse direction, dumping the conveyed ash. They then return through the dry chamber below; this has open sides to facilitate inspection while the equipment is operating (Fig. 20).

Generally, the maximum speed of the scraper conveyor is about 20 feet per minute, with a 4-to-1 or 6-to-1 turndown ratio. Operation at low speed, as a function of low boiler load or low ash content, reduces power consumption and abrasive wear on both flights and supporting surfaces.

Travel wheels, when furnished, support the entire conveyor; they ride in tracks or on rails located in the power-plant floor. The dewatering slope on the ash-removal conveyor leads to a transfer chute or hopper, in which a conventional clinker grinder reduces the size of the ash. From the transfer chute, the size-reduced ash is deposited on a belt conveyor.

CONVEYOR CHAIN DESIGN

A common SSC chain is the double-stranded round link or ship's-type chain made from alloy steel, annealed and carburized to a 550–600 BHN hardness. Chain couplings, of the same material as the chain, are designed for ash-handling service. The chain is supplied in matched sets and undergoes stringent quality control and testing to insure a long service life. Useful life is a function of chain lengthening caused by wear of the inner portion of the links. The lengthening can result in a mismatch of the chain to the driving sprocket.

OTHER DESIGN FEATURES

The flights are commercially available structural angles attached to the chains by fittings provided by the chain manufacturer.

The upper trough of the conveyor has overflow boxes to regulate the height of the water above the flights. Cooling water is added to maintain a normal 140°F (maximum 160°F) temperature. In this respect, the submerged scraper conveyor uses exactly the same amount of cooling water as a water-impounded hopper because the heat presented to the equipment is exactly the same.

Dewatering slopes are at an angle between 25 and 45 degrees. The actual physical configuration depends on the unobstructed space available, boiler-house structural-steel design considerations, and the method of handling the ash after it leaves the conveyor.

SETTING HEIGHT OF STEAM GENERATORS AS AFFECTED BY THE BOTTOM-ASH SYSTEM

The operation and maintenance of bottom-ash hoppers and associated auxiliary equipment are easier and better when no equipment is located in sub-basement-floor pits. But, by eliminating such pits it is sometimes necessary to have boiler setting heights up to 35 feet to accommodate the intermittent-removal, water-impounded bottom hopper. (*Setting height* is the vertical distance from the basement floor level to the centerline of the horizontal furnace-wall tubing directly above the ash-removal equipment; this dimension is measured with the steam generator in the "cold" condition, before expansion has taken place.)

Such generous setting heights are not necessary with the submerged scraper conveyor. When SSC's are specified before the boiler is purchased, the setting height can be adjusted to save substantial amounts in the cost of boiler structural steel, piping, and ductwork.

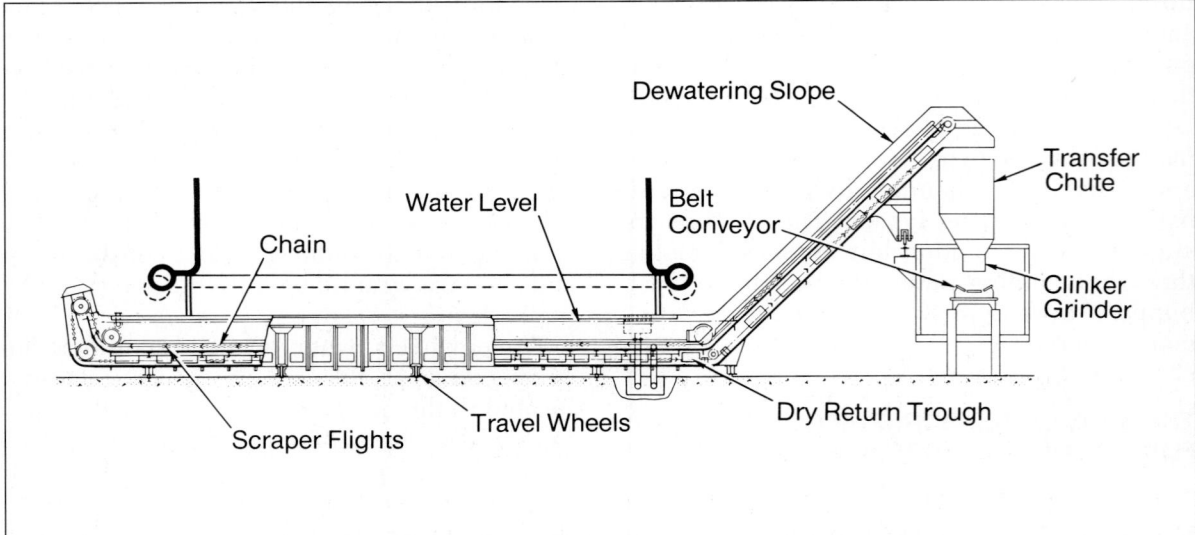

Fig. 19. Submerged scraper conveyor for bottom-ash removal (C-E/EVT SSC)

COMPARISON OF CONTINUOUS- AND INTERMITTENT-REMOVAL BOTTOM-ASH EQUIPMENT

Continuous mechanical removal permits lighter equipment, reduced power consumption compared to hydraulic evacuation, and eliminates the need for conveying water. It's also possible to observe the ash-removal process without drawing down the water-level. Continuous removal provides "auto-dewatering" without the use of dewatering vessels.

In addition, operational control systems for continuous removal are simpler than are those on an intermittent-removal system. Moreover, the reduced height of the boiler above the basement floor has a savings potential in fuel piping, ductwork, structural steel, and building costs. There is also an advantage in being able to move the conveyor to the side at times when the boiler is out of operation. This feature facilitates access to the interior of the furnace through the furnace bottom opening; nor are the high-pressure pumps or water piping of the intermittent hopper (Fig. 12) required.

Because the SSC is designed for *continuous removal*, provision must be made for *continuous receiving* of the ash dropping from the top of the dewatering slope by belt conveyors or other similar mechanical devices. The usual intermittent-removal hopper system in the U.S. is operated only 20 to 30 percent of the time; there is thus a period during which the equipment is not in operation and is available for inspection and maintenance.

Redundancy of mechanical conveying equipment downstream of the scraper conveyor should be considered to assure continuous ash removal. However, despite the fact that ash-storage is not considered in designing an SSC, it can be shut down for short periods to repair downstream equipment.

Fig. 20. C-E/EVT submerged scraper conveyor for 520-MW sub-bituminous-coal-fired steam generator is shown assembled for test operation.

PULVERIZER-REJECTS SYSTEMS

The pulverizer-rejects system collects and transports pulverizer rejects (principally pyrites and tramp iron) hydraulically from the pulverizers (also called mills). A typical system (Fig. 21) uses an enclosed pyrites-reject hopper located next to each pulverizer to receive and store the pulverizer rejects for intermittent removal. Removal and transport from the pyrites hopper is by a jet pump similar to the jet pumps used in bottom-ash sluicing, but smaller in size and conveying capacity.

Although the mill rejects may be pumped directly to a dewatering bin or pond, it is more common to have all the pyrites-removal jet pumps discharge to a local collection or transfer tank (Fig. 21). This arrangement permits emptying more than one hopper at a time.

From this central collection tank, jet or centrifugal pumps remove the pyrites-rejects to the dewatering equipment. Discharging mill rejects into bottom-ash equipment is not acceptable unless a method can be provided to prevent splashing of ash-hopper water onto the boiler tubes above the pyrites injection point. This splashing has caused stress-cracking failure of these tubes. With an SSC, it is practical to introduce mill rejects outside the water-seal plates, obviating the problem.

Plant maintenance and operating flexibility demands that the pulverizer-reject system be operable simultaneously with the main bottom-ash system.

When pressurized mills are used, it is a requirement of NFPA 85F-1988, Para. 2-6.1.4(e) to design the rejects hoppers to withstand 50 psig (Fig. 22). To seal out pulverizer pressure during normal operation, a water-seal overflow box is used. This seal box is sized for 120 percent of the static pressure specified for the primary air fan.

Each rejects hopper must be equipped with a

- level indicator
- steel grate with approximately 1½" square openings to capture large particles for manual removal
- floodlight (for pressurized hopper only)
- inspection port
- access panel
- discharge chute to connect the reject outlet on each pulverizer to the hopper. A pneumatically-operated gate valve is located in the discharge chute for isolation of the hopper. The pyrites slurry piping and fittings are similar to that provided for the bottom-ash systems.

With suction-type mills, the pyrites hopper is an open tank to store and collect the rejects (Fig. 23). On open hoppers, accessories such as floodlights and inspection ports are eliminated. On small steam generators equipped with suction mills from which there will not be a large amount of pulverizer rejects, a simple

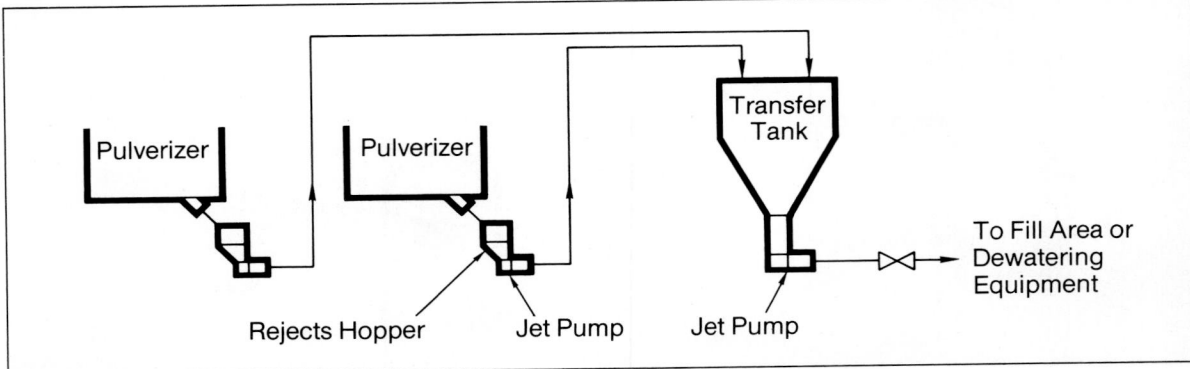

Fig. 21. Mill-rejects (pyrites) hydraulic transport system for mill-rejects (pyrites) uses individual pipelines from each pulverizer to the transfer tank

wheelbarrow may substitute for hoppers.

The amount of rejects from a pulverizer varies greatly as a function of the type of coal, the type of firing system, the mill, and the way the mill is operated.

REMOVAL OF ECONOMIZER-HOPPER ASH

Economizer ash is usually over 700°F, and frequently coarse. It sometimes contains combustible material. This ash can have the physical characteristics of furnace-wall ash and the chemical characteristics of hygroscopic flyash. It is collected in a row of pyramidal hoppers beneath the boiler economizer section, usually at the point of the 90-degree turn in the combustion gas flow.

Formerly, economizer ash was most often handled dry, as an extension of the dry precipitator flyash system; but, exposed to in-leaking air, it could combust and form clinkers making it impossible for the ash to flow through the hopper outlets which usually were only 8 inches in diameter.

Economizer ash should be removed *continuously,* in a method analogous to the continuous removal of bottom ash from a hot furnace. Con-

tinuous removal gets the ash out of the hot environment of the hopper and prevents further combustion of any carbon it may contain. For low-calcium-content bituminous coals, water-filled tanks beneath each economizer-hopper outlet, Fig. 24, should be used. The ash is stored in these tanks for intermittent removal.

Unfortunately, wet holding is not feasible with economizer-hopper ash produced by certain high-calcium lignite or subbituminous coal. Some such ash shows pozzolanic and cementitious properties when dropped into or mixed with water, and may require very frequent evacuation from wet tanks to obviate the possibility of plugging. Utilities burning such "concrete-making" high-calcium coals have used dry transfer tanks below the economizer hoppers to achieve continuous removal without tank or line plugging problems.

It is important, then, to recognize that such flyash has the potential of producing synthetic

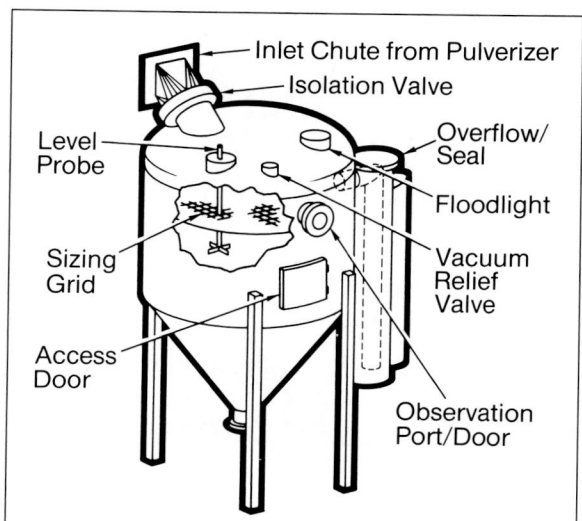

Fig. 22 **Mill-rejects hopper for pressurized pulverizers, in accordance with NFPA 85F-1988**

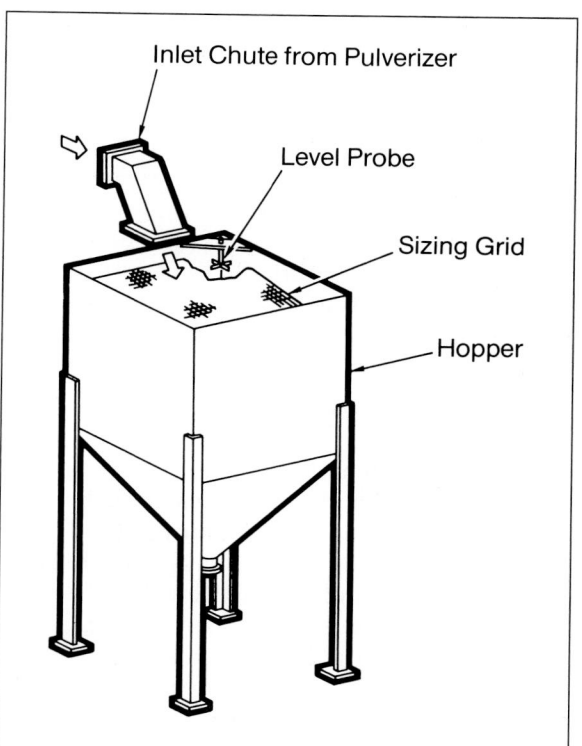

Fig. 23. **Open-top rejects hopper for suction pulverizers**

silicate rocks that cannot be redissolved and flushed out of tanks or transport lines. The ash is in the category of a pozzolan—siliceous or aluminous/siliceous material which in itself possesses little or no cementitious value but which will, in finely divided form and in the presence of moisture, chemically react with calcium hydroxide to form compounds having cementitious properties.

The best solution to the handling of economizer ash having concreting characteristics is to remove it continuously from the hoppers under the high-temperature gas zone into dry holdup/collecting tanks located beneath the economizer hoppers, so as to prevent burning or sintering. The ash can be removed from these tanks on an intermittent basis as a branch of the precipitator or baghouse pneumatic removal system. Alternatively, the flyash can be removed continuously from the hoppers by means of mechanical removal equipment, as discussed later in this chapter.

FLYASH REMOVAL SYSTEMS

Flyash from the hoppers serving air heaters, precipitators, and baghouses is removed intermittently by either vacuum (negative-pressure) or pressure (positive-pressure) pneumatic type systems, or combinations of the two. In addition, several types of continuous-removal systems such as flight conveyors or screw conveyors can be used.

VACUUM SYSTEMS

A vacuum system (Fig. 25) uses a mechanical blower, water exhauster, or a steam exhauster to create a vacuum which removes the flyash from the hoppers. A flyash-intake valve located at each hopper regulates the flow of the flyash. Flyash intakes have carbon-steel or cast-iron bodies and a swing disc which seals against a hardened seat. For maintenance, the outlet of each hopper has a manual isolation gate.

Each flyash intake is actuated automatically by the system logic which controls both the flow rate and quantity of flyash leaving the

hopper to avoid plugging the discharge line. Because the system operates on a vacuum, only one flyash intake and one conveyor branch line operate at any given time. As each hopper is emptied of flyash, the system will step to the next hopper in the same branch line. When all hoppers in a branch line have been emptied, the system will step to the next branch line. The system logic insures the proper sequence of events and positioning of valves.

The intake shown in Fig. 26 uses a microporous stainless-steel filter cloth which passes hot fluidizing air into the body of the intake at a controlled rate. The metallic fluidizing material is used because it does not plug or crack easily, as do some other media.

When a mechanical blower produces the necessary vacuum, flyash is taken to a silo where a cyclone separator and bag filter, in series, separate the air-ash mixture. To protect the blower it is important to collect most of the ash, which is then emptied from the silos into enclosed trucks or railcars.

The water exhauster (Fig. 27) is another method of transporting flyash through the conveying line. High-pressure water supplied to the water-exhauster inlet nozzles creates the

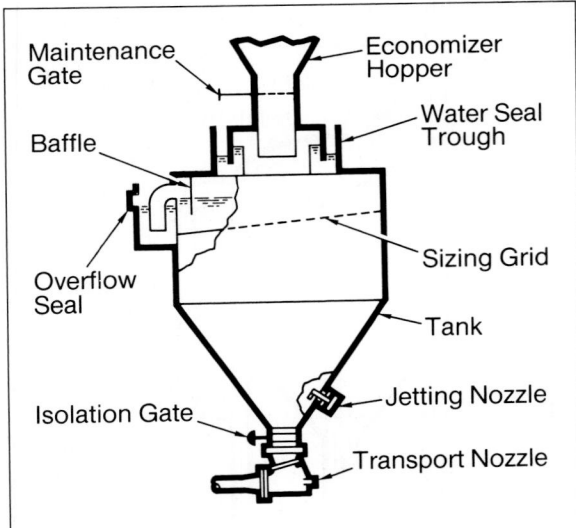

Fig. 24. Water-filled tank beneath economizer hopper receives low-calcium-content ash for cooling and intermittent flushing

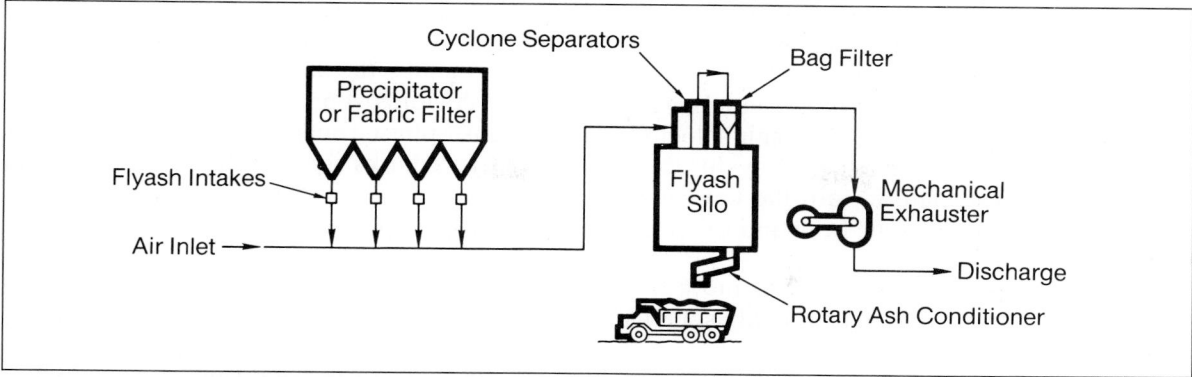

Fig. 25 **Dry pneumatic vacuum flyash system**

transport vacuum; flyash, air, and water are mixed in the exhauster venturi. Water exhausters normally have hardened ductile iron bodies, wear-resistant liners and stainless-steel nozzles. Diffusers are hardened ductile iron.

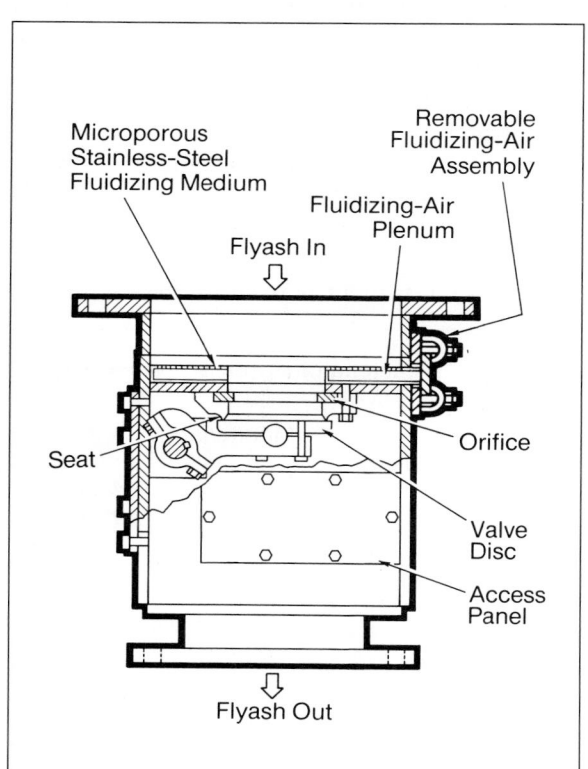

Fig. 26 **Fluidizing-type flyash intake for regulating flyash flow from collection hoppers (fluidizing-air plenum is removable for precipitator washdown)**

Following the water exhausters, an air separator is provided to separate and vent the air from the flyash-water-air mixture (Fig. 28). Separators are made of cast iron or carbon steel with an abrasion-resistant basalt or ceramic liner. The separator discharge is elevated sufficiently to allow the ash-water slurry to flow by gravity to a pond or disposal area. Flyash slurry is never discharged to a dewatering bin because it is very difficult to settle out the fine flyash particles in the dewatering bin; most fine particles would pass over the overflow weir.

PRESSURE SYSTEMS

In a pressure system (Fig. 29), an air-lock feeder transfers flyash from a hopper at a low pressure to a pipeline conveyor at a higher pressure. Compressors or blowers provide the air-flow and pressure to convey the flyash.

Fig. 27 **Water exhauster for vacuum removal of flyash (ash and water are combined in exhauster for transport as a slurry to disposal; upper flange is high-pressure water inlet; dry ash enters left end)**

These feeders are designed with a storage capacity based on the desired conveying rate, ash density and number of hoppers to be conveyed in a given unloading sequence. The volumetric feeders are controlled to empty each selected group of hoppers on a staggered cycle, thus providing uniform loading into the conveyor system. Fluidizing air at the inlet and outlet of each feeder insures ash flow.

Vacuum transport systems usually move flyash in a quite "airy" mixture (dilute phase), but the consistency of the mixture can vary with a positive-pressure system. For example, the mixture might be in a dilute phase, a dense phase (in which there is much more ash than air), or the mixture may begin as a dense phase and gradually change to a dilute phase at the discharge end.

Many low-velocity, dilute-phase systems operate at some time in a dense phase because dense phase occurs whenever saltation takes place in a pipeline.

AIR-LOCK FEEDER OPERATION

It is customary in dilute-phase flyash transport to use the collecting hoppers for storage; the air-lock tank is used only during the actual ash-moving process. Thus, the inlet valve to the pressure tank is kept closed until the trans-

port cycle begins. This exposes the ash to the unheated tank metal for the shortest possible time, but keeps the ash in the collecting hoppers for the longest possible time.

In dense-phase transport, the inlet valves of the pressure tanks are kept open whenever the tanks are not full. The ash, then, leaves the area that is heated by flue gas and drops into the tanks, normally unheated, until the tank level indicator causes the inlet valve to close; the filled tank remains at essentially atmospheric pressure until its turn in the transport "queue" is signalled. The time that the ash remains in the tank before fluidization and transport will depend on the filling/emptying status of any other tanks on the same conveying line.

COMPARISON OF VACUUM AND PRESSURE SYSTEMS

Selecting a vacuum or pressure flyash removal system for a given unit depends on an evaluation of

- conveying rate
- altitude
- number of hoppers to be evacuated
- distance to be conveyed

A vacuum conveying system is normally preferred over a dilute-phase pressure system because less equipment is used under each hop-

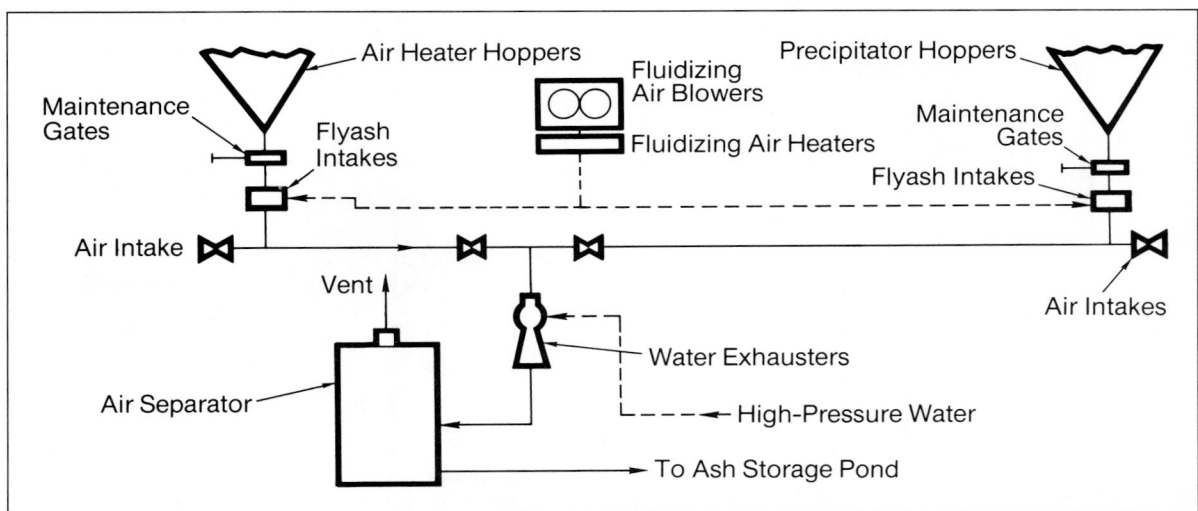

Fig. 28 **Dry pneumatic flyash transport system using water exhausters as vacuum producers**

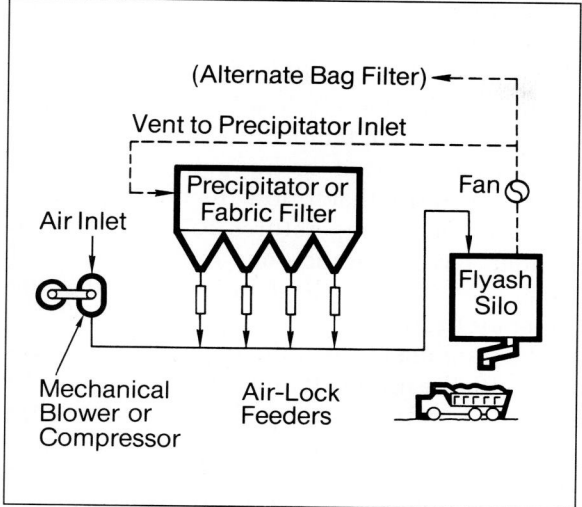

Fig. 29 **Dry pneumatic-pressure flyash system**

per: one flyash intake versus two flyash intakes and an air-lock tank. Also, a vacuum system uses the more positive means of suction to evacuate the flyash from the hoppers whereas a pressure system uses gravity alone to move the ash from the hopper to the feeders. Of course fluidization and rappers are available, but to be effective they must be used properly.

A vacuum system also has the advantage of being able to indicate, from the conveying-line vacuum switches, whether a given hopper is full, empty, plugged or ratholed. This is not possible with a pressure system as multiple hoppers are being evacuated simultaneously.

However, the practicality of a vacuum system is limited by the amount of pressure differential that is available for conveying. The actual pressure drop increases as the conveying rate increases; thus, if all else is constant, higher conveying rates may require a pressure system. (Theoretically, a vacuum system can almost always be used by installing many vacuum-producing devices and pipe-lines in parallel.) When plants are at high elevations, the available pressure decreases (at 7000 feet, approximately 5″ Hg. are lost over that of sea level), thus limiting the application of vacuum systems. In addition, the pressure will drop as conveying distance increases, making a pressure system the more feasible.

To understand a pneumatic transport system, it is important to realize that conveying distance is the total developed length of the pipeline. In pneumatics, the pressure drop through an elbow can easily be the equivalent of 50 feet of pipe; therefore, the optimum system would have a minimum number of elbows.

VACUUM/PRESSURE SYSTEMS

When the number of precipitator or baghouse collection hoppers is relatively high

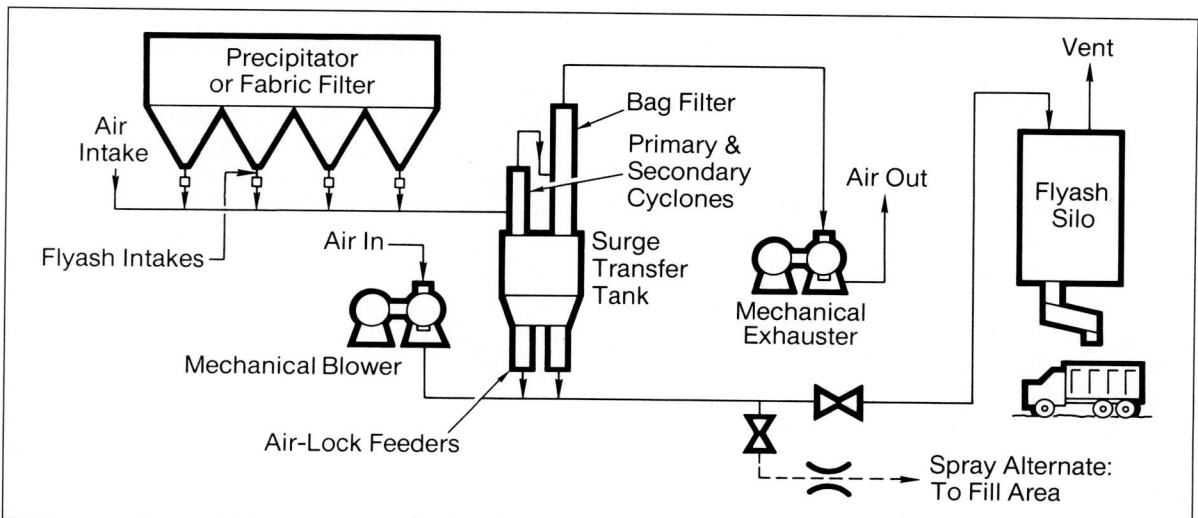

Fig. 30 **Vacuum-to-pressure dry pneumatic flyash system**

(say, 20 to 40, or more) and the conveying distance dictates a pressure system, it is usually economical to provide a vacuum-pressure system, Fig. 30. In this system, a transfer silo is located close to the precipitator or baghouse and the flyash moves by vacuum from the hoppers to the transfer silo. At the discharge of the silo, a set of air-lock feeders forms the pressure system to convey the flyash by pressure to the main ash storage silos. The capital and maintenance cost is reduced because of less equipment under the hoppers (one flyash intake per hopper vs. two flyash intakes and the air-lock tank). This reduction in equipment under the hoppers can more than offset additonal costs associated with the transfer silo and the positive pressure blowers. On the other hand, a vacuum-pressure system uses more power than a straight pressure system.

FLYASH HOPPER DESIGN

Problems have been experienced in the removal of powdery flyash from hoppers, particularly with large precipitators on utility steam generators. Because of the very high ash content found in many coals, and increasingly stringent particulate-collection regulations, precipitators on a large pulverized-coal-fired unit commonly have more than 40 hoppers (sometimes as many as 128). Research has shown that the incidence of hopper operational and maintenance problems is directly related to the number of hoppers on a precipitator.[5]

Pneumatic ash-removal systems *are not designed to handle wet material.* Thus, it is critical in the design and operation of hopper systems to maintain collected material sufficiently above the water or acid dew point to keep it absolutely dry, so that it *will be* free-flowing.

Dry dust in hoppers ordinarily will flow freely by gravity and be transported pneumatically without difficulty. But to do so, it

■ must be kept essentially at the temperature at which it was collected

■ must not be exposed to any moisture

■ must not compact from its own weight to cause bridging above the hopper outlet

■ must have no clinkers formed as the result of oxidation of any combustibles in the ash

In power-plant operation, it is frequently difficult to keep the ash free-flowing. With the boiler at full load, temperatures as low as 90°F have been reported in hoppers of cold precipitators (those that follow the boiler air heaters). Similarly, temperatures 20° to 40°F above *ambient* have been recorded in the hoppers of "hot" precipitators (those located *ahead* of the boiler air heater).

The problem in the industry is even more severe with the high-calcium ash associated with certain subbituminous coals and various lignites. In addition to being highly hygroscopic, such ash shows pozzolanic and cementitious properties when exposed to moisture.

HOPPER SIZING

Hoppers under precipitators and baghouses should be considered only as funnels. Because of difficulties with removing flyash that has been stored for a long time while exposed to flue gas containing moisture and sulfur, recommendations state there be no specified storage time in collecting-equipment hoppers.[5]

Minimum hopper outlet size should be 12 inches (diameter or square). In any case, 6-inch or 8-inch outlets should be avoided wherever possible, unless with identical fuels such sizes proved successful.

The ideal hopper is square in horizontal cross section. Minimum outlet-chute or hopper valley angles should be 55 degrees from the horizontal; 60 degrees if space allows. In any case, hopper sides should have a minimum slope of 60 degrees. In addition, for precipitators, rounded inside valleys are desirable with approximately a 4-inch inside radius. Such rounded corners are not as necessary with fabric filters because the continuous air changes promote ash movement.

Stainless-steel linings at hopper corners or along the hopper's entire lower portion facilitate emptying. Because of size of fields, or where a given row of hoppers serves more than one field, it may be necessary to include baffles in the precipitator. Such baffles should be properly designed and positioned with respect to hopper outlets to prevent bridging. Generally, the baffle plates should terminate 18 inches or more above the hopper outlets.[6]

HOPPER FLUIDIZING DEVICES

A flyash fluidizer, as it is used in a conical or pyramidal hopper, is a porous membrane which allows pressurized airflow through it to be uniformly distributed to the material above, filling the voids between the particles at a slight pressure and changing the effective angle of repose of the material to promote gravity flow.

Properly located fluidizing devices can help evacuate hoppers (particularly precipitators), if these devices are well-maintained and reliably supplied with dry air preheated above dew-point temperatures. If not, fluidizers will only aggravate evacuation by caking the ash, and provide unwanted surface areas for ash accumulation and bridging. Where significant percentages of combustibles are present in the collected flyash, the fluidizing-medium supply must be non-oxidizing to prevent destructive hopper fires.

The fluidizing devices referred to are those located inside the hopper, away from the walls and extending out into the dust stream; they are not those in the hopper walls or those integral with the flyash intake valves.[5]

HOPPER VIBRATORS

When their operation is properly controlled, hopper vibrators can help prevent bridging and ratholing. In the automatic operation of vacuum pneumatic systems, vibrators should be regulated by ash-evacuation controls to insure operation only when an "empty hopper" signal is generated and the evacuation cycle is completed. Routine use of the vibrators during evacuation of damp ash will further compact the ash and make evacuation difficult.[7]

It should be possible to manually operate the vibrators from each hopper to assist maintenance personnel during emergency evacuation.

WEATHERPROOF ENCLOSURES FOR HOPPERS

Many owners add skirts to precipitators or baghouses to keep wind and weather from reducing the hopper metal temperatures. Such enclosures are highly recommended, although reducing the heat loss from the hoppers can be accomplished equally well by judicious use of insulation and lagging.

Skirts also are instrumental in preventing hopper plugging. They

■ keep ash-handling hardware from being chilled to low ambient temperatures

■ provide warmed air from the "hopper room" for pneumatic transport of the flyash, particularly with vacuum systems

■ allow inspection and maintenance under protected conditions

■ provide a plenum for hot air ducted from the top of the boiler house, which then makes heated transport air available without additional energy consumption.[5]

CONTINUOUS REMOVAL OF FLYASH

The ideal flyash-removal system is one that takes ash from the receiving hoppers at the same rate as it enters. The hoppers then effectively become *chutes* and there is no time for cooling, deaeration, or compaction to occur.

There are virtually no such systems in U.S. utility power plants, and only a few in certain industrial boiler or process applications. The majority of multiple-dust-outlet precipitators and baghouses have their removal systems intentionally designed for intermittent removal to minimize the power consumed by the ash-transport system. Such an approach ignores the harmful impact on the reliability of collection equipment when hoppers are used for storage.

Utility plants in Europe report good experience with continuous removal as have some industrial installations in the U.S. In these systems flyash is transported by either mechanical or air-fluidized conveyors, instead of by pneumatic-conveying equipment. It appears that a principal feature of such continuous-removal devices is that, with proper valving, they do not allow any substantial amount of air to leak into the precipitator or baghouse hoppers. This feature maintains the gas temperatures in those hoppers above the dew point.[5]

Air-fluidized conveyors can be used only if the flyash remains fluid under all conditions of start-up, normal load, or shutdown. Although these conveyors are a form of continuous removal, there have been numerous problems in keeping the ash completely dry during start-up. Also, if ash clinkers form, they are impossible to fluidize and transport.

Two types of *mechanical* conveyors have been used for conveying flyash—screw conveyors and flight conveyors. Often prone to wear, screw conveyors are usually used for ash transport only when conveying a relatively small quantity of material. Flight conveyors move the ash in a dust-tight casing using elements linked by a single or double strand of chain.. The units are sized either on a volumetric basis or on an *en masse* basis in which the material is conveyed in bulk, without agitation. Flight conveyors have been successfully operated for many decades on coal-fired units, both in North America and Europe. When properly designed, they are relatively insensitive to problems such as moisture in the flyash and flyash clinkers (which may force outages in other types of systems).[8]

With continuous conveying devices, the system can combine the mechanical conveyor with a pneumatic pipeline. Mechanical devices have the ability to move the ash "uphill" to the top of a storage silo.

FLYASH RECEIVING AND STORING EQUIPMENT

Flyash storage silos and their related accessories form an integral part of ash-handling equipment. Silos are made of concrete or steel, depending upon size, seismic condition, and plant economics.

On vacuum systems, equipment is required to separate the flyash from the conveying air stream before it can be discharged into the silo. This equipment, located on top of the silo, normally has several stages of separation.

AIR/FLYASH SEPARATION EQUIPMENT

Dual-cyclone continuous-separation modules are the primary mechanism to separate the flyash from the conveying air. The module contains a primary and secondary cyclone separator (Fig. 31). The centrifugal action exerted by the cyclone forces the flyash to drop as the lighter airflow is extracted from the top. By using an airlock the storage silo is never under vacuum and the module can function continuously.

The bag filter receives the flow from the separation modules and acts to precipitate the remaining flyash from the conveying air stream. This tertiary separation protects the vacuum pump from unnecessary particulate carry-over. The bag filter usually features reversed-air cleaning with a good air-to-cloth ratio.

The vacuum-producing device (blower, vacuum pump, exhauster) which has pulled the flyash air mixture down the pipeline and through the discharge equipment, is located at the side or underneath the silo.

On a pressure system, somewhat simpler filtration equipment is used. Vent fans and motor drives are mounted on top of the storage silo. The vent fan should be designed to handle not only all the air displaced in the silo by the conveying input stream and silo fluidizing air, but also the static head required to transport it back to the precipitator inlet duct. The vent fan places the silo under a slight negative pressure, eliminating the possibility of blowing dust out of the silo. Vent-fan operation can be synchronized with that of the conveying blowers. Alternately, the vent fan can discharge to a bag filter atop the silo.

SILO FLUIDIZATION

Flyash, when not allowed to agglomerate and when fluidized with heated air, takes on the flow properties of a fluid. To expedite flyash removal, troughs with fluidizing air are provided at the bottom of the silos. Traditionally, the fluidizing element was constructed of canvas or porous stone. But recently, the industry has been using porous laminates of sintered stainless-steel-wire cloth such as used in

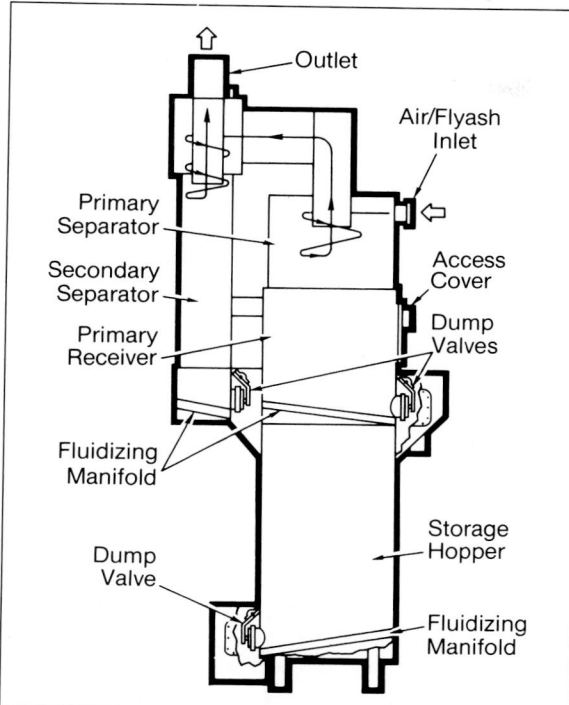

Fig. 31 **Dual-cyclone module for continuous separation of flyash and transport air (used on vacuum-pneumatic systems)**

some flyash intakes. This provides a microporous filter that allows controlled amounts of heated air to pass and fluidize the flyash. The possibility of stress cracking that is common with fluidizing stones is eliminated with this metallic material. Because of the microporous construction, the metallic cloth is highly resistant to blinding; in addition, water will not penetrate into the porous openings. This will allow washing of equipment with water without having to remove the fluidizing element.

Fluidizing-air blowers and associated heaters provide the required air to the fluidizing air troughs. Care must be taken to properly size the blowers to account for a full head of flyash in the silo.

Fluidizing hoods, normally one above each silo discharge opening, minimize the possibility of avalanching, which would adversely affect the operation of the unloading equipment.

Silo discharge equipment is designed so the flyash can be removed dry into enclosed vehicles, or so that it can be moistened (to 12–15 percent moisture by weight) for transport in open vehicles. It is increasingly common for flyash to be sold for use in landfills, roads and the cement industry.

Telescoping ash discharge chutes convey flyash dry from the storage silo to a removal vehicle stationed below. Normally, each chute is capable of extending a full eighteen feet using a motor-driven-winch retract system. A blower and piping in each chute vent, back into the silo, any fugitive dust released during the unloading process. An air-operated gate controls the inlet flow to the chute and is placed just below a manual maintenance gate. A clean-out compartment should be included between the maintenance gate and control gate. This compartment should contain a grid to trap large clinkers that may form in the silo, and an access door for their subsequent removal.

FLYASH CONDITIONERS

Rotary ash conditioners, Fig. 32, add moisture to the flyash discharged from the storage silo into open-top vehicles. The conditioning and moisture prevent any fugitive-dust nuisance. An orifice feeder controls the rate at which the dry material moves from the silo into the mixing drum of the conditioner. A series of atomizing nozzles within the drum spray the material with water as the drum rotates. A longitudinal scraper bar along the full length of the drum plows damp material from the side of the drum, causing it to be mixed thoroughly with the spray water. The continuous tumbling action of the wetted material within the drum results in minimal dust at the conditioner outlet. Stop plates prevent any tendency for dry material to cascade through the drum before being thoroughly mixed with water. Mounted on a central support, the scraper bar and stop plates are readily removed.

OPERATIONAL CONTROLS

Ash-handling controls range from completely manual to completely automatic systems of either the relay, solid state, or

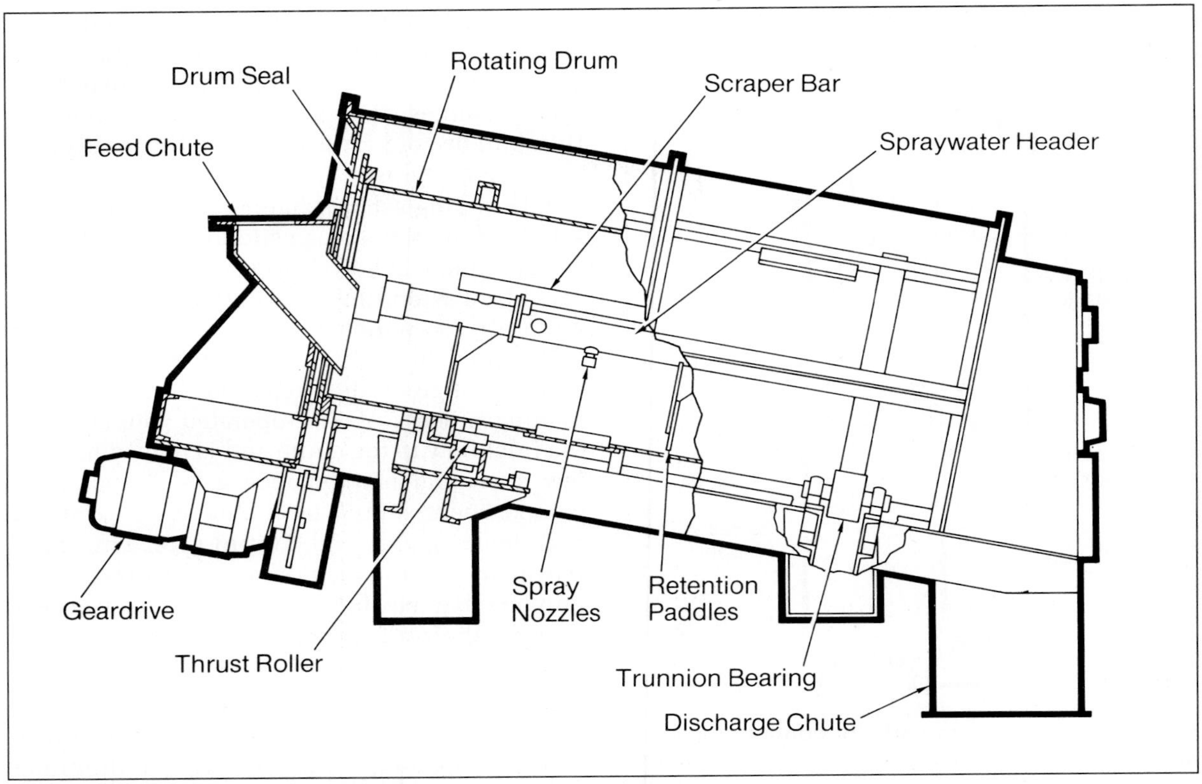

Drum Seal Rotating Drum Scraper Bar Spraywater Header Feed Chute Geardrive Thrust Roller Spray Nozzles Retention Paddles Trunnion Bearing Discharge Chute

Fig. 32 Sectional side elevation of rotary-type flyash conditioner

programmable-controller type.

When selecting the type of control system and the degree of automation that should be employed, many factors should be considered. First, the more automated the plant, the less attention paid by the plant operators. If little attention is paid to ash systems, breakdowns are inevitable and, when breakdowns occur, the operators may lack enough knowledge of the system to take corrective action. For intermittent-removal bottom-ash and pyrites systems many utilities select a push-button automatic system in which the operator, at a local panel, performs the ash-removal sequence. The operator can judge by the crushing sound of the grinder, whether each hopper section has been emptied. If a problem arises, the operator is on hand to analyze it and take action.

In some cases density meters are installed in the discharge line of each bottom-ash hopper section to record the amount of time it takes to empty each section. By comparing the time to empty one section with that of other sections, it's possible to spot signs of trouble.

Similarly, on a vacuum pneumatic system, the operator can make a check by examining the strip chart that records the vacuum in the pneumatic pipeline. As vacuum can be related to the ash flow, the operator can easily determine the removal rate and time required to empty each hopper. Inconsistencies in removal time and/or removal rate per hopper are signs of a problem.

On multifield precipitators, programmable controllers can be used to advantage. As the front fields will collect ash at a much faster rate than the rear fields, the operator may choose to alter the sequence in which hoppers are "visited" by the ash removal system. Such adjustments accommodate unequal distribution across the fields, or compensate for fields being removed from service.

Continuous removal systems obviously require the least sophisticated controls, but on/off switches, indicating lights, and alarms should be provided. Programmable controllers are rarely justified with such systems.[4]

MATERIAL- AND ENERGY-BALANCE CONSIDERATIONS

Water-impounded ash-receiving equipment beneath a furnace, as previously stated, helps to quench the ash as it falls from the furnace and to transport it to a disposal point. The thermal shock to the hot ash as it enters the low-temperature water helps to break up the large pieces, while submerging the ash prevents sintering during the time that it may be stored (with intermittent-removal systems).

Part of the heat released in the furnace is transmitted to the ash hopper by radiation from the burning fuel, but most of the heat input is from the sensible heat given up to the water by the hot bottom ash falling from the furnace. The ash-systems engineer must perform a heat balance to determine the flow of cooling water required to absorb the incoming heat and maintain the water temperature at a predetermined level.

Field experience indicates that the temperature of the impounded water should be optimally about 140°F, and should not be higher than 160°F. Temperatures at this level are effective in rapidly cooling and fracturing the falling ash. Lower temperatures require more cooling water; higher temperature may prove uncomfortable or unsafe in the area around the bottom-ash receiver.

BOTTOM-ASH COOLING SYSTEM DESIGN

Heat is removed from bottom-ash hoppers or submerged conveyors by

■ evaporation of water from the surface of the water pool
■ the heat removed in the water going out the overflow or leakage points
■ the heat removed with the accumulated bottom ash as it is discharged
■ the heat loss by radiation and convection from the outside metal surfaces

The last loss is small and is usually neglected in view of the unknown accuracy of some of the other assumed factors.

Fig. 33 is a simplified heat-balance diagram around a bottom-ash receiver, in this case, a submerged scraper conveyor with continuous ash removal. The heat balance forms the basis for the calculation of the cooling-water flow needed to maintain a temperature of about 140°F (60°C). In Fig. 33

$$Q_{in} = Q_{out} \tag{2}$$

where

$$Q_{in} = Q_{ash\ entering} + Q_{furnace\ radiation} + Q_{inlet\ water} = Q_{ae} + Q_{fr} + Q_{iw}$$

and

$$Q_{out} = Q_{overflow\ water} + Q_{ash\ leaving} + Q_{evaporation} + Q_{external\ radiation\ loss} + Q_{water\ in\ ash\ leaving} = Q_{ow} + Q_{al} + Q_e + Q_{er} + Q_{aw}$$

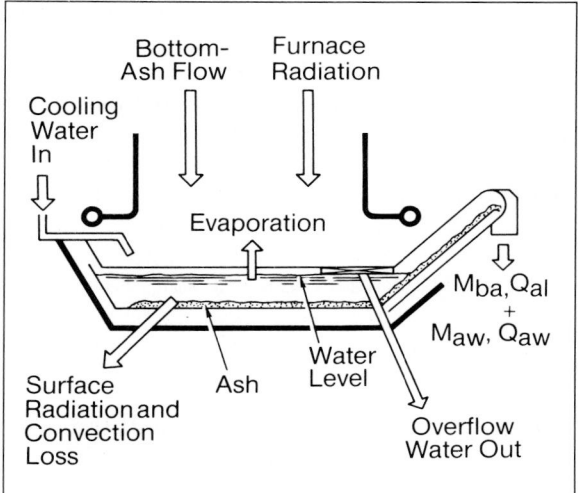

Fig. 33 **Steady-state heat-balance diagram for furnace bottom-ash receiver**

MASS AND ENERGY TERMS

Before equating and simplifying the above inputs and outputs so that they can be mathematically manipulated, it is necessary to define and discuss them further.

HEAT IN THE BOTTOM ASH

The mass of bottom ash, M_{ba}, is the collection rate, lb/hr or kg/hr.

$$Q_{ae} = M_{ba} \times C_{ash_1} \times (T_{hot\ ash} - T_r)$$

(3)

where

C_{ash_1} = specific heat of the ash at the mean temperature between $T_{hot\ ash}$ and T_r

$T_{hot\ ash}$ = the temperature of the falling ash or slag as it enters the water pool

T_r = reference temperature for defining heat contents

The temperature of ash leaving the bottom of a furnace depends on the thickness of the pieces dislodged from the walls, the face-metal temperature of the furnace walls, the temperature of the sootblowing medium, the distance of fall, and the average gas temperature surrounding the material during its fall. For pulverized coal fired in suspension (without the use of furnace-bottom grates), the temperature is estimated to be between 1200°F and 1800°F, averaging 1500°F. Obviously, the validity of the heat-balance calculation depends on this temperature to a great extent.

The heat of the cooled bottom ash exiting the collection device is

$$Q_{al} = M_{ba} \times C_{ash_2} \times (t_{out} - T_r)$$

(4)

where

C_{ash_2} = specific heat of the ash at the mean temperature between t_{out} and T_r

where

t_{out} = temperature of the water and ash leaving

FURNACE RADIATION TO THE IMPOUNDED WATER

Furnace radiation passing through an opening in the hopper bottom varies with firing rate, furnace and firing-system configuration, and the effective center of the fuel-admission nozzles or burners in operation (measured vertically above the furnace-bottom opening). The heat transfer is

$$Q_{fr} = A_{fa} \times R_f \times F_{tr}$$

(5)

where:

A_{fa} = furnace hopper-bottom aperture above the ash receiver

R_f = radiation rate through furnace aperture, Btu/hr-sq ft

F_{tr} = fraction of radiation transmitted to water surface through the space between the furnace bottom and the water, Fig. 34.

A typical value for R_f, the heat radiated through the bottom aperture to the space above the impounded water, is 20,000 Btu/hr-sq ft. Fig. 34 shows that the radiation reaching the water surface decreases as that surface is located farther away from the aperture, as would be expected.

HEAT IN COOLING WATER

The mass of incoming cooling water, from all sources (including any refractory-cooling water or SSC bearing-flush water), is designated M_{iw}.

$$Q_{iw} = M_{iw} \times C_{iw} \times (t_{iw} - T_r)$$

(6)

where:

T_{iw} = incoming water temperature

C_{iw} = incoming water specific heat

The mass of cooling water leaving, M_{ow}, includes any outward leakage, but not the water entrained by the exiting bottom ash; the latter is shown as M_{aw}, the mass of heated water leaving with the ash.

$$Q_{ow} = M_{ow} \times C_{ow} \times (t_{ow} - T_r) \text{ and}$$
$$Q_{aw} = M_{aw} \times C_{ow} \times (t_{ow} - T_r)$$

(7)

where

C_{ow} = specific heat of water leaving

t_{ow} = temperature of the impounded water

HEAT LEAVING IN EVAPORATED WATER

The rate of evaporation from a water pool below a large coal-fired furnace is a complex function of the water temperature, the temperature of the radiating gas, and the other factors involved in the determination of the radiation rate. In general, rates between 4 and 10 U.S. gallons/day-sq ft are to be expected. Large pieces of slag falling into the water can result in additional localized evaporation, which is neither calculable or separable from the overall effect. It is calculated by

$$Q_e = A_w \times R_e \times h_{fg}$$

(8)

where

A_w = total area of ash-receiver water surface

R_e = evaporation rate per unit of surface area

h_{fg} = latent heat of vaporization of water at T_{ow}

SIMPLIFICATION OF THE HEAT-BALANCE EQUATION

If T_r, the reference temperature, is equal to t_{ow}, the water temperature in the water pool, Q_{al}, Q_{ow}, and Q_{aw} go to zero, then,

$$Q_{ae} + Q_{fr} + Q_{iw} = Q_e,$$
$$\text{or}$$
$$Q_{ae} + Q_{fr} = Q_e + [M_{iw} \times C_{iw} \times (t_{ow} - t_{iw})]$$

(9)

Note that the mass of water leaving, M_{ow} must not be less than zero; that is,

$$M_{ow} = (M_{iw} - M_{aw} - M_e) \geq 0$$

(10)

where

M_e = mass rate of evaporated water

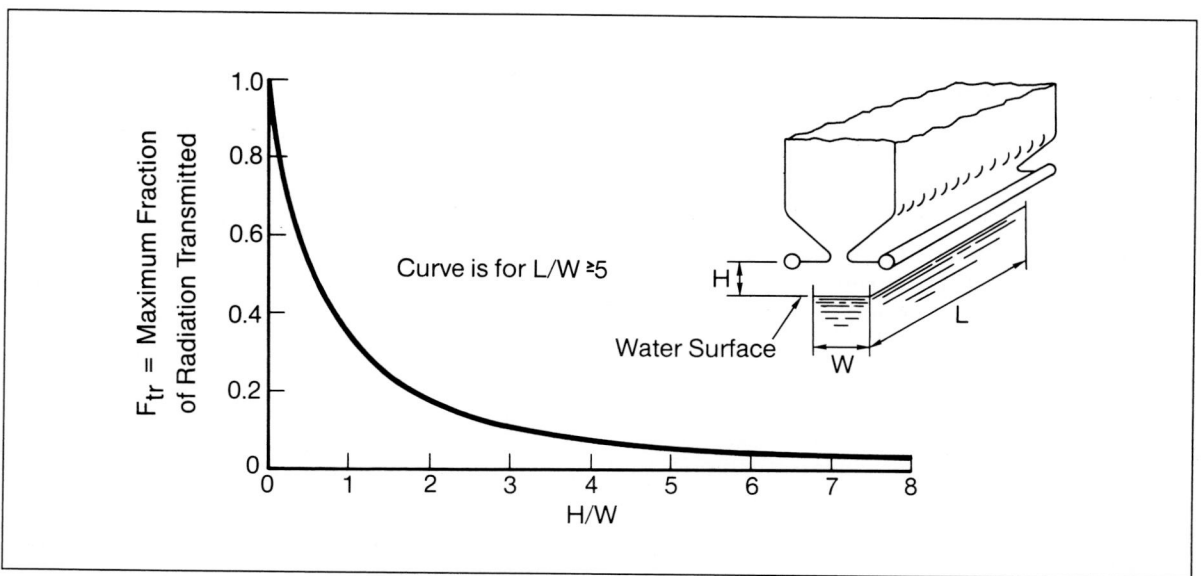

Fig. 34 Curve to determine fraction of radiation transmitted to ash-receiver water surface

In a closed-loop cooling-water system, M_{ow} is the amount of water that must be cooled and recirculated. The extent of physical cleanup and chemical treatment applied to this recirculated water is principally a function of what the system heat-transfer and pumping equipment can tolerate. Generally, the cleanup required for the low-pressure cooling stream of an SSC is much less than that for an intermittent-removal hopper with slope jetting nozzles, which use water at high pressure.

Recirculation equipment size and cost are sensitive to the ash temperature and quantity, the furnace radiation absorbed, and the evaporation rate, as shown above. It is important that economic comparisons of such loops be based on equivalent boundary conditions, which include, in addition, the inlet water and impounded water temperatures.

It is fairly accurate, for these heat-balance calculations, to use 0.25 Btu/lb-°F for the specific heat of ash, 1.0 as the specific heat of water, and 1014 Btu/lb as h_{fg} at 140°F.

COOLING-WATER FLOW VARIATION

Two inputs to the equation can vary widely, greatly affecting the cooling-water flow. First, the bottom-ash collection is a function of

■ quantity of fuel burned (boiler load and heating value of the coal)

■ percentage of ash in the fuel

■ percentage of the total ash that falls to the furnace bottom

Second, the cooling water temperature can vary substantially from summer to winter, and with geographical area.

The designer is faced with the problem of deciding what should be used for ash-collecting rates and the water temperature to the hopper. The common design approach requires excess water because the designer uses *maximum* values for all calculations. Systems so designed will pump and treat excessive quantities of water when the unit is burning a good fuel in winter when the cooling water is cold.

The potential difference in cooling water usage is illustrated in Table III. This data considers only the water necessary to cool the incoming ash (not taking into account the radiation and evaporation heat exchange), to point out how great is the difference between the two postulated conditions.

If the ash-system was designed to provide the maximum cooling water at all times, the pumping power consumption would be considerably greater than needed. Because the purpose of the cooling water is to maintain the ash-hopper water at a temperature such as 140°F, it is logical to regulate the flow on a demand basis. This is done by monitoring the overflow water temperature and modulating a supply valve. The system design has to provide for the worst case (maximum cooling water), but the actual water use will only be as required, thereby saving pumping power and the cost of cleanup of the ash-receiver overflow water under many load conditions.

PROPER INSTALLATION

It is important to state that any ash-removal system must be installed, set up, checked out, and deemed ready for use by a plant startup crew. There have been many reports of damage to boiler pressure parts and precipitator elements, with consequent unit shutdowns caused by well-designed ash-removal systems that were not ready for use.[9]

Also, ash, by nature, is a very abrasive material and wear and replacement of worn components are inevitable. Plant personnel must perform regular inspections of these wear components and make replacements as necessary.

TABLE III. Cooling Water Usage*

	High Use	Low Use
Coal burned, lb/hr	750,000	200,000
Ash content, percent	25	10
Cooling water temperature, °F	90	50
Ash down, percent	50	30
Bottom ash collection, lb/hr	93,750	6,000
Cooling water flow, U.S. GPM	1,275	45

*Approximate, based on absorbing heat from incoming ash at 1500°F only; no radiation or evaporation considered. Ash receiver water temperature 140°F.

REFERENCES

1 E. H. Tenney, "Pulverization and Boiler Performance," FSP-54-7, *Proceedings of the Fourth National Fuels Meeting*, Feb 11 to 13, 1931, pp. 55–65. Chicago, Illinois: The American Society of Mechanical Engineers. 1931.

E. G. Bailey, "Present Status of Furnace and Burner Design for Pulverized Fuel," FSP-50-72, *Transactions of the American Society of Mechanical Engineers*, September to December 1928, p. 177. New York: The American Society of Mechanical Engineers.

2 W. E. Loftus, "Ash Handling, Storage and Utilization," *Proceedings of the American Power Conference*, 38:707–717, 1976. Chicago: Illinois Institute of Technology. 1976.

3 J. G. Singer and A. J. Cozza, "Material and Energy Balances of Ash-Handling Systems," presented at the Joint Power Generation Conference. Dallas, Sept. 10–13, 1978; also as Combustion Engineering publication TIS-5822.

4 J. G. Singer and A. J. Cozza, "Design for Continuous Ash Removal: Alternative Concepts in Ash Handling," *Proceedings of the American Power Conference*, 41: 544–553, 1979. Chicago: Illinois Institute of Technology. 1979; also as Combustion Engineering publication TIS-6211A.

J. E. Horne and A. Bosso, "Southwestern Public Service Company Pioneering in Continuous Bottom Ash Removal," presented at Frontiers of Power Conference, October 11 and 12, 1982, Stillwater, OK; published as Combustion Engineering TIS-7261.

5 Joint Technical Committee of the American Boiler Manufacturers Association and Industrial Gas Cleaning Institute, Inc., "Design and Operation of Reliable Central Station Flyash Hopper Evacuation Systems," *Proceedings of the American Power Conference*, 42:74–85, 1980. Chicago: Illinois Institute of Technology. 1980.

6 J. G. Singer, "Design for Better ESP/Fabric Filter Hopper Operation and Maintenance," presented at Air Pollution Control Association 76th Annual Meeting and Exhibition, Atlanta, June 19–24, 1983; also as Combustion Engineering publication TIS-7402.

7 J. G. Singer and A. J. Cozza, "Ash-Handling Options on Retrofitted and Converted Steam Generators," presented at ASME-IEEE-ASCE Joint Power Generation Conference, St. Louis, October 4–8, 1981; also as Combustion Engineering publication TIS-6869.

8 J. C. Fleming and D. M. Rode, "Ash Removal from Industrial Boilers—the Changing Scene," *Power*, Vol. 126, No. 9, September 1982. New York, McGraw-Hill, 1982.

9 M. B. Caron, A. J. Cozza, J. G. Singer, and J. R. Young, Jr., "Steam-Generator Availability as Affected by Ash-Handling Equipment," *Proceedings of the American Power Conference*, 44:214–225, 1982. Chicago: Illinois Institute of Technology, 1982; also as Combustion Engineering publication TIS-7118.

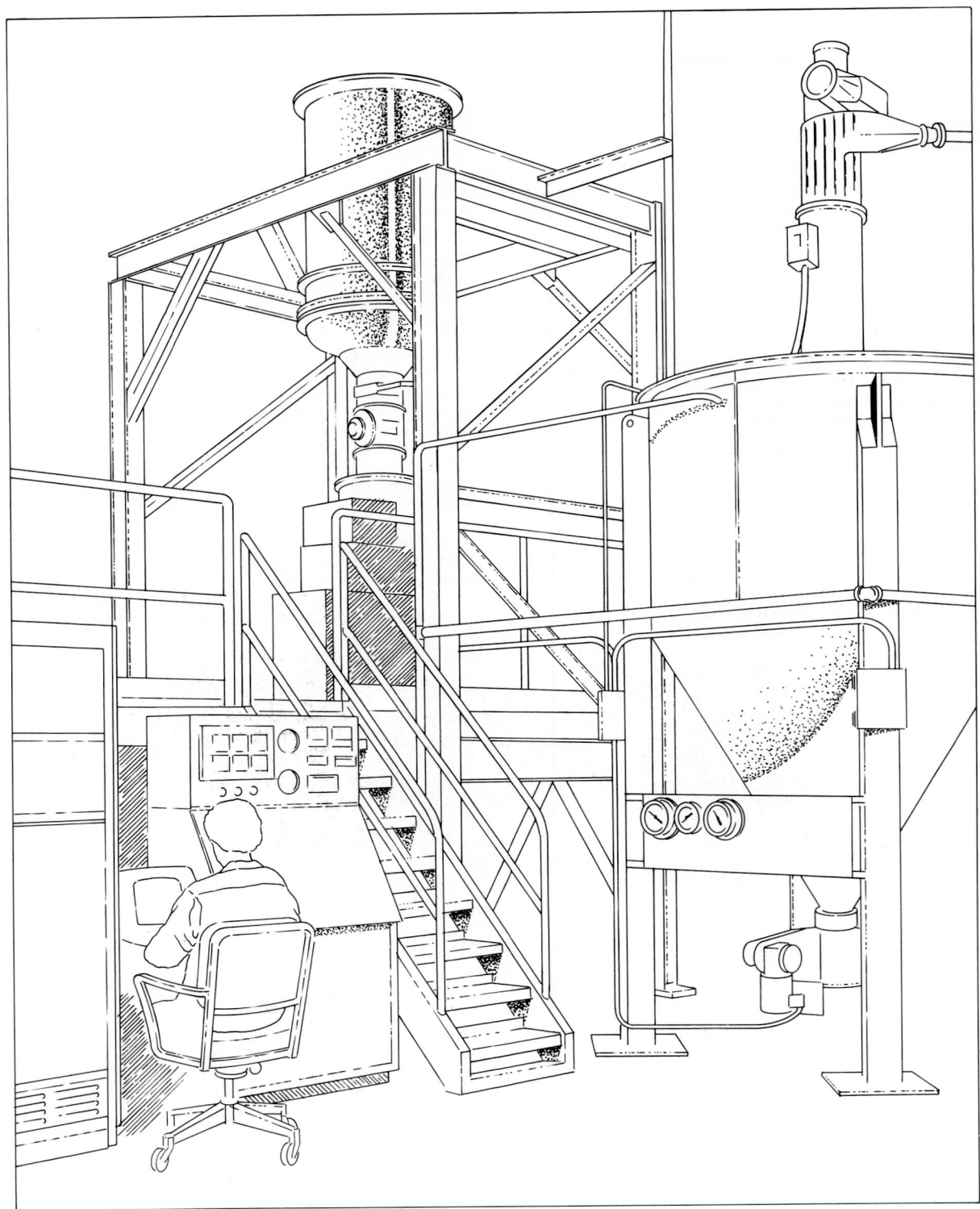

CHAPTER 17

Metallurgy and Materials

The aim of the designer of a large steam-generating unit is to use materials and fabricating methods in the most economical way to produce a unit that will perform properly for its planned useful life. Achieving these objectives requires a fundamental knowledge of material properties and behavior under a wide range of manufacturing and service conditions.

With only a few exceptions, steel is the material used in the construction of pressure parts and support members for boilers. Selection of the appropriate type of steel involves the following considerations:

■ the degree of difficulty or ease with which a particular material can be fabricated or welded

■ mechanical properties of the material such as strength (including tensile, yield, fatigue, creep, and creep-rupture strengths), ductility, and toughness

■ compatibility with the expected service environment

■ effect of fabrication practices on mechanical and corrosion-resistance properties

■ possible degradation of properties caused by long-time service exposure.

Because these material-related considerations are so fundamental to both basic design and details of construction, there needs to be a close interaction between the designer and such material-performance disciplines as metallurgical, corrosion, welding, and materials engineering. All available materials data evolved through experience and research must be used in order to properly select materials and to specify construction details, fabrication procedures, destructive and nondestructive tests, and operating limits and controls necessary for satisfactory service.

This chapter presents some of the many considerations relating to the selection of materials and their operation in service, and describes typical field-service difficulties. Chapter 26 describes the type of laboratory facilities that are used to obtain the materials and metallurgical data required for boiler design.

SPECIFICATIONS AND CODES

The first boiler explosion took place around 1700 in a British mine in which a primitive steam-operated pump had been installed. Over the next 100 years, there was increased use of boilers in stationary installations, but it was the loss of life resulting from marine boiler explosions in the early 1800's that led to a pioneering American investigation by the Franklin Institute of Pennsylvania in 1830.

Many state laws subsequently were enacted to regulate boiler construction and operation, but not until the period 1911–1914 did the *ASME* Boiler Code develop.[1] Originating basically as a safety code to prevent boiler explosions, the *ASME* Code today promotes the acceptance of properly designed equipment on a national and international basis by assisting in the standardization of safety rules.

As formally adopted in 1915 and since administered by the ASME, the Code included many of the earlier regulations and resolved many of the conflicts in the rules. Since that time, a majority of the American states, as well as the Canadian provinces, have adopted boiler laws which make the *ASME* Boiler and Pressure Vessel Code the basis of their legal requirements. The rules of the *ASME* Code are kept up to date by continuing review of the Boiler and Pressure Vessel Committee, a group representing a balance of material suppliers, fabricators, users, insurers, and enforcement jurisdictions. This group sponsors regular revisions of the Code to keep up with technical advancements and responds to questions of interpretation of the rules.

In America, administration of code requirements is the responsibility of the individual state in which the boiler is to be installed, and the National Board of Boiler and Pressure Vessel Inspectors coordinates this effort.

Another technical group, the American Society for Testing and Materials (ASTM), incorporated in 1902, prepares and publishes specifications for purchase, testing, and examination of a variety of materials. This activity, which is directly related to the Society objective of promoting knowledge of engineering materials, is carried on through committees which include representatives of producers, fabricators and consumers.[2] Similarly, Committees of the American Welding Society write specifications for welding consumables.

Such specifications are widely used to procure and acceptance-test materials to meet defined levels of quality, thereby promoting both safety and economy. Groups such as the ASME Boiler and Pressure Vessel Committee also use these materials specifications, either as written or with modifications, to establish recognized grades of materials applying to their code rules for design, allowable stresses, fabrication methods, and inspection.

The American National Standards Institute (ANSI) publishes a variety of standards, many of which apply to boiler and pressure vessel construction or related uses. Standards such as ANSI B31.1 (Code for Pressure Piping), the ANSI B16.5 (Steel Pipe Flanges and Flanged Fittings) and ANSI B16.34 (Steel Valves) are frequently used. Several agencies of the United States government also publish codes and specifications for certain equipment used in their power plants and other facilities.

A basic limitation of all codes is that their rules must be expressed in sufficiently general terms to cover a wide range of applications. Because they define only reasonable minimum standards, there are sure to arise cases in which a power-plant designer must make further investigation and exercise additional effort. A great variety of supplementary information is available from publications of the various engineering societies and from technical bulletins of equipment suppliers.

MATERIALS OF CONSTRUCTION

Fossil boiler equipment in electric-generating service operates at high pressures and temperatures, generally between 2000 and 4000 psig, and 1005° and 1055°F, at the primary-steam outlet. The material grades listed in Table I are widely used in boilers designed in North America; there are extensive data and both shop and field experience with these ferritic and austenitic steels.

RAW MATERIAL FORMS

The product forms of steel available to the designer include plates and sheets, forgings, castings, seamless and welded tubing and piping, and rolled and extruded shapes. All are limited in size by the facilities and capabilities of the various material suppliers. Table I indicates several product forms with specifications

covering the procurement from the *ASME* Boiler and Pressure Vessel Code, Section II, Materials, and from the ASTM specifications.

PLATES

Plates are used in the fabrication of boiler drums and plate-formed headers. The plates used to fabricate the shell and heads of these drums are as thick as 8 inches. Boiler drums are made of carbon-steel plate, usually either SA-515 Grade 70 or SA-299.

During the plate-rolling process in the producing steel mill, discontinuities and heterogeneities in the ingot are elongated in the direction of rolling. Because of this, plates may exhibit what is called directional or anisotropic properties and may be weaker when tensile tested *across* the thickness than when tested parallel or transverse to the direction of rolling. These directional discontinuities are generally not harmful except when they are large enough in size to affect heat transfer or are excessive in quantity so as to affect weldability. The customary practice for heavy plates used in boiler applications is to apply selective magnetic-particle and ultrasonic examinations to determine that the material is free of excessive discontinuities in areas where welding will be performed. See Chapter 18, "Steam Generator Manufacture" for details on drum fabrication and testing.

TUBING

Steam is generated, superheated, and reheated in steel tubing, commonly between 1½ and 3 in. outside diameter. Large units may require as much as 300 miles of such tubing. The tubing is designed to last the life of the unit (up to 30 or more years) even though, in the case of superheated-steam tubing or power-plant piping, it is operated at a visibly "red-hot" temperature sometimes above 1100°F.

From the standpoint of geometrical shape and physical construction, there is no essential difference between pipe and tubing. Pipe sizes are generally designated by their nominal inside diameter for a particular type of service. Tubes, by contrast, are usually specified in terms of outside diameter and minimum wall thickness.

Pipe sizes and wall thicknesses are standardized with consideration of threading ends for joining together lengths with such fittings as flanges, nipples, valves, tees and the like. Such fittings are generally not installed with tubing. Where pressure-tight connections must be made, tubes are welded together, or their ends are expanded into tube sheets.

Selection of the material used depends upon the actual metal temperature to be sustained. The steam-generation tubes are primarily of carbon-steel material in recirculation-type subcritical-pressure units. In low-temperature regions the superheater and reheater tubing is also of carbon-steel analysis. As steam temperature increases to the outlet value, selection then progresses through carbon-molybdenum steel, then low chromium-molybdenum steel, intermediate chromium-molybdenum steel, and finally austenitic stainless steel, depending upon the design metal temperatures in the various areas. Austenitic stainless steel contains sufficient amounts of such alloying elements as nickel and chromium to retain austenite, a solid solution of carbon in gamma iron.

For many years, Grades TP 321 and TP 321H austenitic stainless steel had been used for high-temperature service. But changes in the *ASME* Boiler Code allowable stresses have favored Grades TP 304H and TP 347H, with these latter grades most widely used at present. Although Grade TP 316H is seldom used in the tubing portions of the superheater, it has been and will continue to be used when austenitic stainless piping and headers are required for steam temperatures of 1100°F and higher.

FORGINGS

Boiler drums and similar vessels may use forgings for reinforcing rings around openings and nozzles; some moderately sized high-pressure vessels are made entirely by forging. In most cases, the largest opening required for a boiler drum is an access opening of about 16 in. inside diameter, which is reinforced by a forging welded to the drum head. Forgings are also used for fittings, valves and flanges.

CASTINGS

Castings have not been widely used as boiler pressure parts except in special applications such as valve or pump bodies, where the relative ease in forming thick-walled spherical and other special shapes has been an advantage. Techniques in casting steel have improved, supported by developments in nondestructive testing such as radiography and ultrasonic examination. Some parts and attachments in boiler service are exposed to high temperatures approaching the local gas temperature. Heat-resistant alloys such as 25% Cr –12% Ni (AISI type 309) are required for such items. As is common in many industry publications, steel of this type will be referred to throughout this chapter without the percentage qualifier as 25

Table I. Materials Used in Boiler Construction

ALLOY	Product Form	ASME or ASTM Spec.	Grade	Minimum Tensile Strength ksi	Minimum Yield Strength ksi	C	Mn	P	S	Si	Ni	Cr	Mo
Carbon Steel	Tubes	SA-192	...	(47)	(26)	0.06–0.18	0.27–0.63	0.048	0.058	0.25	...	...	...
Low-Strength	Tubes (ERW)	SA-178	A	...	...	0.06–0.18	0.27–0.63	0.050	0.060	...	...	...	...
	Tubes (ERW)	SA-226	...	(47)	(26)	0.06–0.18	0.27–0.68	0.050	0.060	...	...	...	...
Intermediate	Tubes	SA-210	A-1	60	37	0.27	0.93	0.048	0.058	0.10 Min	...	...	...
Strength	Tubes (ERW)	SA-178	C	60	37	0.35	0.30	0.050	0.060	...	...	...	...
	Pipe	SA-106	B	60	35	0.30	0.29–1.06	0.048	0.058	0.10 Min	...	...	...
	Castings (b)	SA-216	WCA	60	30	0.25	0.70	0.040	0.045	0.60	...	...	...
	Structural Shapes	A36	...	58	36	0.26	...	0.040	0.05	...			
High Strength	Pipe	SA-106	C	70	40	0.35	0.29–1.06	0.048	0.058	0.10 Min	...	...	...
	Plate	SA-299	...	75	40	0.30	0.86–1.55	0.035	0.040	0.13–0.33	...	...	...
	Plate	SA-515	70	70	38	0.35	0.90	0.035	0.04	0.13–0.33	...	...	...
	Forging	SA-105	...	70	36	0.35	0.60–1.05	0.040	0.050	0.35	...	...	...
	Casting (b)	SA-216	WCB	70	36	0.30	1.00	0.040	0.045	0.60	...	...	...
FERRITIC ALLOYS **C–0.5 Mo**	Tubes	SA-209	T1	55	30	0.10–0.20	0.30–0.80	0.045	0.045	0.10–0.50	...	...	0.44–0.65
1 Cr–0.5 Mo	Forging	SA-336	F12	70	40	0.10–0.20	0.30–0.80	0.040	0.040	0.10–0.60	...	0.80–1.10	0.45–0.65
	Tubes	SA-213	T12	60	30	0.15	0.30–0.61	0.045	0.045	0.50	...	0.80–1.25	0.44–0.65
	Pipe	SA-335	P12	60	30	0.15	0.30–0.61	0.045	0.045	0.50	...	0.80–1.25	0.44–0.65
	Plate	SA-387	12 Cl 2	65	40	0.17	0.36–0.69	0.035	0.040	0.13–0.32	...	0.74–1.21	0.40–0.65
	Forging	SA-182	F12	70	40	0.10–0.20	0.30–0.80	0.040	0.040	0.10–0.60	...	0.80–1.25	0.44–0.65
1.25 Cr–0.5 Mo	Tubes	SA-213	T11	60	30	0.15	0.30–0.60	0.030	0.030	0.50–1.00	...	1.00–1.50	0.44–0.65
	Pipe	SA-335	P11	60	30	0.15	0.30–0.60	0.030	0.030	0.50–1.00	...	1.00–1.50	0.44–0.65
	Plate	SA-387	11 Cl 2	75	45	0.17	0.36–0.69	0.035	0.040	0.44–0.86	...	0.94–1.56	0.40–0.70
	Forging	SA-182	F11	70	40	0.10–0.20	0.30–0.80	0.040	0.040	0.50–1.00	...	1.00–1.50	0.44–0.65
	Casting (b)	SA-217	WC6	70	40	0.20	0.50–0.80	0.040	0.045	0.60	...	1.00–1.50	0.45–0.65

ABB
ASEA BROWN BOVERI

Cr – 12 Ni. Many of these parts are cast, but forgings and plate have also been used.

Castings are also advantageous in such boiler components as stokers and pulverizers where parts can be made to desired shapes with little or no machining. Composition of the iron or low-alloy steel may be varied to match the severity of the specific temperature service. For example, gray cast iron is often used in many items of power-plant equipment because of its vibration-damping characteristics. Ductile-iron and steel castings are also frequently used. Where wear resistance is a major requirement, special alloy iron castings and forgings of hard, abrasion-resistant materials are available for liners, grinding rings, rolls and other parts.

STRUCTURAL MATERIALS

A considerable amount of steel in structural and sheet form is used for support and gas-

Table I. Materials Used in Boiler Construction — *Continued*

ALLOY	Product Form	ASME or ASTM Spec.	Grade	Minimum Tensile Strength ksi	Minimum Yield Strength ksi	C	Mn	P	S	Si	Ni	Cr	Mo
2.25 Cr–1 Mo	Tubes	SA-213	T22	60	30	0.15	0.30–0.60	0.030	0.030	0.50	...	1.90–2.60	0.87–1.13
	Pipe	SA-335	P22	60	30	0.15	0.30–0.60	0.030	0.030	0.50	...	1.90–2.60	0.87–1.13
	Plate	SA-387	22 Cl 1	60 (c)	30 (c)	0.17	0.27–0.63	0.035	0.035	0.50	...	1.88–2.62	0.85–1.15
		SA-387	Cl 2	75 (d)	45 (d)								
	Forging	SA-182	F22	75	45	0.15	0.30–0.60	0.040	0.040	0.50	...	2.00–2.50	0.87–1.13
	Casting (b)	SA-217	WC9	70	40	0.18	0.40–0.70	0.040	0.045	0.60	...	2.00–2.75	0.90–1.20
5 Cr–0.5 Mo	Tubes	SA-213	T5	60	30	0.15	0.30–0.60	0.030	0.030	0.50	...	4.00–6.00	0.45–0.65
9 Cr–1 Mo	Tubes	SA-213	T9	60	30	0.15	0.30–0.60	0.030	0.030	0.25–1.00	...	8.00–10.00	0.90–1.10
AUSTENITIC STAINLESS ALLOYS 18 Cr–8 Ni	Tubes	SA-213	TP 304H	75	30	0.04–0.10	2.00	0.040	0.030	0.75	8.00–11.00	18.00–20.00	...
	Pipe	SA-376	TP 304H	75	30	0.04–0.10	2.00	0.040	0.030	0.75	8.00–11.00	18.00–20.00	...
	Plate	SA-240	304	75	30	0.08	2.00	0.045	0.035	1.00	8.00–10.50	18.00–20.00	...
		SA-240	304H	75	30	0.04–0.10	2.00	0.045	0.030	1.00	8.00–12.00	18.00–20.00	...
	Forging	SA-182	F304H	75	30	0.04–0.10	2.00	0.040	0.030	1.00	8.00–11.00	18.00–20.00	...
18 Cr–10 Ni–Ti	Tubes (e)	SA-213	TP 321H	75	30	0.04–0.10	2.00	0.040	0.030	0.75	9.00–13.00	17.00–20.00	...
18 Cr–10 Ni–Cb	Tubes (f)	SA-213	TP 347H	75	30	0.04–0.10	2.00	0.040	0.030	0.75	9.00–13.00	17.00–20.00	...
16 Cr–12 Ni–2 Mo	Tubes	SA-213	TP 316H	75	30	0.04–0.10	2.00	0.040	0.030	0.75	11.00–14.00	16.00–18.00	2.00–3.00
	Pipe	SA-376	TP 316H	75	30	0.04–0.10	2.00	0.040	0.030	0.75	11.00–14.00	16.00–18.00	2.00–3.00
	Forging	SA-182	F316H	75	30	0.04–0.10	2.00	0.040	0.030	1.00	10.00–14.00	16.00–18.00	
	Plate	SA-240	316H	75	30	0.04–0.10	2.00	0.045	0.030	1.00	10.00–14.00	16.00–18.00	2.00–3.00
	Structural Sheet	A167	316L	70	25	0.03	2.00	0.045	0.03	1.00	10.00–14.00	16.00–18.00	2.00–3.00
25 Cr–12 Ni	Casting	SA-351	CH20	70	30	0.20	1.50	0.040	0.040	2.00	12.00–15.00	22.00–26.00	

(a) Single values shown are maximums.
(b) Residual elements not to exceed 1.00%.
(c) Annealed.
(d) Normalized.
(e) Titanium content not less than four times carbon content and not more than 0.60%.
(f) Cb + Ta not less than eight times the carbon content and not more than 1.00%.

enclosure purposes. The basis of selecting the appropriate steel involves design considerations similar to those used for pressure-containing service.

Generally, large utility boilers are suspended from large girders at the top of the unit to allow the unit to expand downward with an increase in temperature. Supporting extremely heavy loads, the girders operate at ambient temperatures which may be quite low, say −40°F. Therefore, they must be designed and fabricated to exhibit adequate toughness for resistance to brittle fracture. The precautions described in the later section on heat treatment are applicable to these structures as they are to the pressure vessels.

In some instances, as with flue-gas scrubber systems, compatibility with a corrosive service environment is of critical importance to the selection of the appropriate grade of steel.

WELD METALS

The metals used in welding drums, headers, tubing, ducts and other platework must be designed to exhibit properties compatible with the base metals. The composition of the weld filler metals and fluxes, when these are required, must be tailored to each different welding process used in the shop and field.

METALLURGICAL FUNDAMENTALS

For purpose of classification, the science of metallurgy is generally divided into two fields: process metallurgy, the science of obtaining metals from their ores, and physical metallurgy, the science concerned with the physical and mechanical characteristics of metal and alloys. These fields, in turn, are part of the basic sciences of chemistry and physics involving the study of the structures of atoms—their sizes and forces of attraction and the arrangements of these atoms in forming molecular structures, grain structures, and grain boundaries of multicrystalline materials.[3]

PHASE DIAGRAMS

Some metals used for engineering purposes

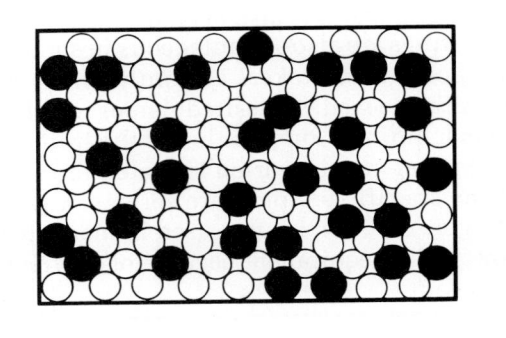

Fig. 1 Solid-solution random distribution of atoms

are commercially pure. This is true of copper for electrical wiring. Deliberately combining elements to form an alloy can enhance mechanical or physical properties. Some combinations result in a single-phase, solid-solution alloy. It is called single phase because the atoms are dissolved in the structure such that visual, microscopic and X-ray observations reveal a single crystal structure. Such a solid solution most readily forms when the two components are similar in atomic size and in electronic structure. The distribution of the atoms may be random (Fig. 1), ordered (Fig. 2), or interstitial (Fig. 3).

Element combinations may result, however, in a mixture of two or more phases. Fig. 4 shows a two-phase structure of sigma phase in type 321 austenitic material. Sigma phase is a hard and brittle intermetallic compound of var-

Fig. 2 Solid-solution ordered distribution of atoms

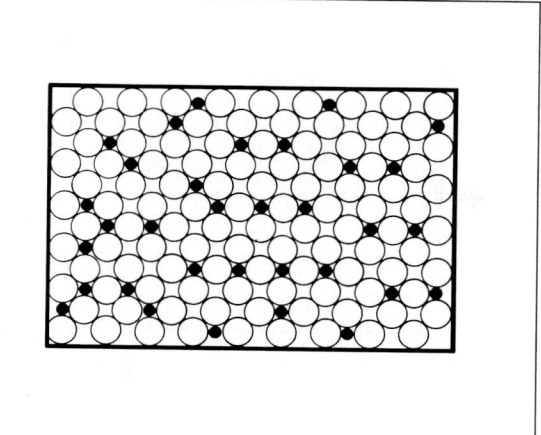

Fig. 3 Interstitial distribution of atoms in solid solution of carbon in iron (face-centered cubic structure)

iable composition that adversely affects the room-temperature ductility of stainless steel.

Fig. 5 shows a two-phase structure of pearlite (alternate lamellae of cementite and ferrite) and ferrite (carbon steel SA-192 of Table I). Cementite is a compound of iron and carbon known as iron carbide of the approximate formula Fe_3C, characterized by an orthorhombic crystal structure; ferrite is a solid solution in which alpha iron is a solvent, characterized by a body-centered cubic crystal structure.

The phase relationships depend upon temperature. Fig. 6 shows a phase diagram for a simple system of lead and tin at equilibrium conditions. This diagram can be used as a map to determine the phases present at any particular temperature and composition. For example, at 50 percent tin and 212°F, the phase diagram indicates two solid phases. Alpha is a solid solution of lead with some dissolved tin; beta is almost pure tin with very little dissolved lead.

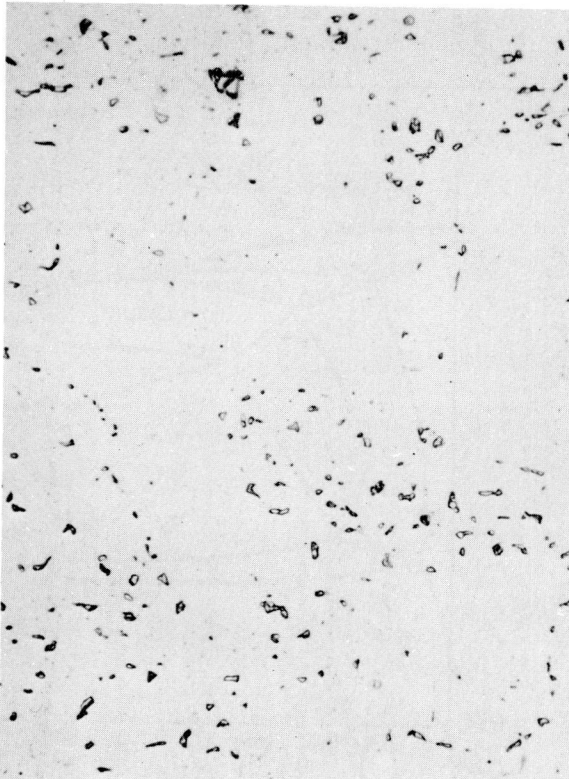

Fig. 4 Two-phase structure of sigma phase in austenitic type 321 material at 500X. Ten-percent potassium hydroxide etch, one half second

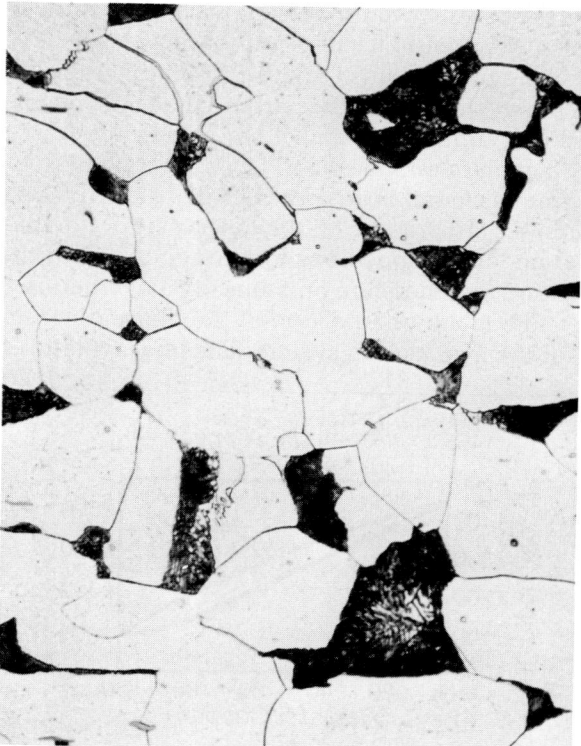

Fig. 5 Two-phase structure of pearlite and ferrite in low carbon steel at 500X. Two-percent nital etch

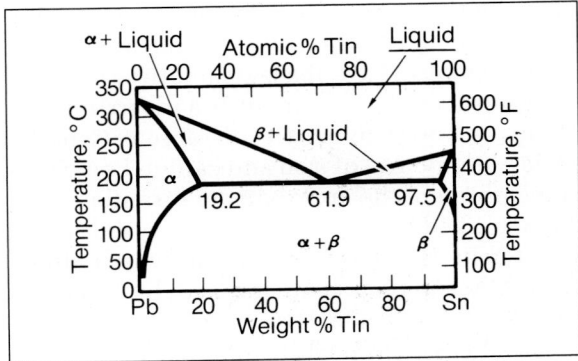

Fig. 6 Equilibrium diagram for a simple system of lead and tin

At about 400°F, an alloy of 10 percent tin and 90 percent lead lies in an area which is entirely in the alpha phase. It is a solid solution of lead with some tin dissolved in it. At the same temperature, but for 30 percent tin and 70 percent lead, the phase diagram indicates a mixture of liquid and solid solution. If this latter composition were heated to a temperature of 575°F, it would become all liquid.

The phase fields in equilibrium diagrams, of course, depend on the particular alloy systems. When copper and nickel are mixed, the phase diagram is as shown in Fig. 7. This phase diagram is comparatively simple, as only two phases are present. In the lower part of the diagram, all compositions form only one solid solution and therefore only one crystal structure.

Fig. 8 shows a portion of the phase diagram for the iron-carbon system. The phase relations

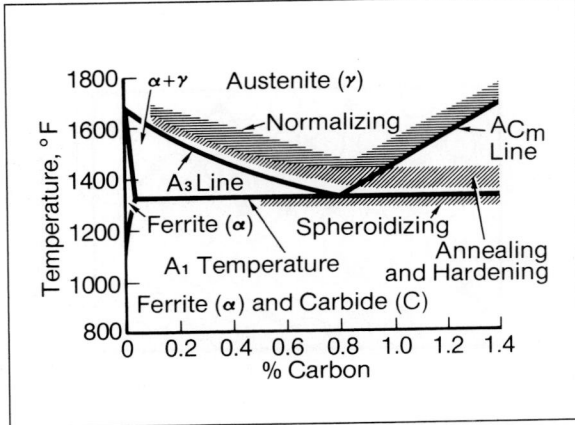

Fig. 8 A portion of the iron-carbon phase diagram showing temperature of importance to the heat treatment of carbon steel

for a ternary alloy become more complex as shown for an iron-carbon-chromium alloy in Fig. 9. Such phase relationships are used in regulating heat treatments in alloys.

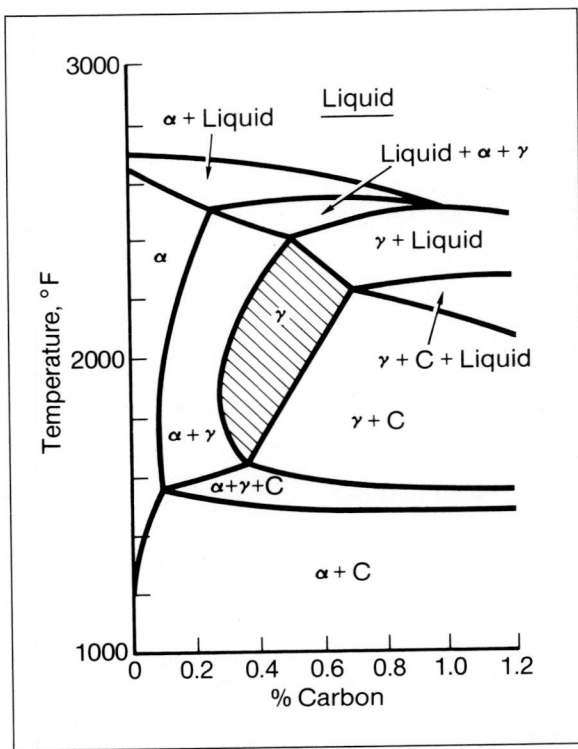

Fig. 9 Iron-chromium-carbon alloy phase diagram (15% chromium)

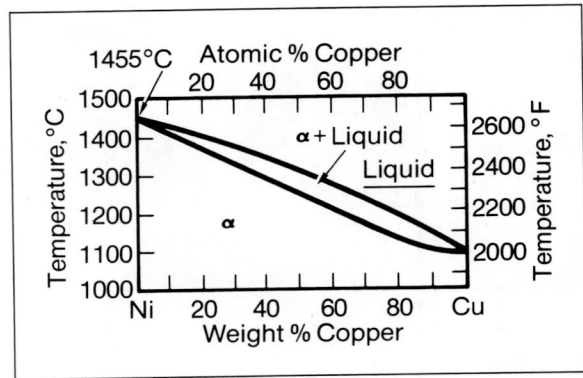

Fig. 7 Simple two-phase equilibrium diagram of nickel and copper

T-T-T CURVES

The phase relations such as shown in Figs. 6, 7, 8, and 9 are representative of equilibrium conditions obtained as a result of slow heating, slow cooling and long time at temperature. Time is required for change from one phase to another. Energy change is involved. Reaction rates are usually faster at higher temperatures except near the temperature of transformation from one phase to another where extra energy is required to nucleate a new phase.

Temperature-time-transformation curves (T-T-T) exemplify reaction rates for phase changes. The Fig. 10 T-T-T curve is for a steel of 0.8 percent carbon content shown in the Fig. 8 equilibrium diagram. It can be seen that rapidly cooling this steel to low temperature allows the gamma phase (austenitic material) to be present down to low temperature for a considerable time period before transformation occurs. The transformation product at this low temperature differs from what it would be if transformation occurred at a higher temperature, and is of higher hardness and strength with sufficiently high carbon content. This transformation product is a hard material called martensite. Such a phenomenon is one of the basic methods of varying properties of steel by heat treatment. Each alloy subject to phase changes with temperature change exhibits a characteristic T-T-T curve.

GRAIN STRUCTURE

A grain is a single crystal usually without a regular external crystalline shape. In a solid, the presence of surrounding grains usually controls its shape. Within any particular grain, all of the atoms are arranged in one pattern, characterized by the unit cell. But at the grain boundary between two adjacent grains, there is a transition zone not directly aligned with either grain. Fig. 11 shows this transition zone.

When a metal is observed under a microscope, although the individual atoms cannot be seen, the grain boundaries can be readily located if the metal has been smoothly polished to a mirrorlike surface and etched with an acid. The grain boundaries will be attacked differently from the body of the grain and become visible. Different phases will also become visible. The grain boundaries of the single-phase austenitic alloy (Fig. 12) shows fine and coarse-grained material (ASTM A-213-TP 321).

Thermal treatment can regulate the grain size. An increase in temperature causes increased thermal vibration of the atoms. This vibration facilitates transfer of atoms across the grain boundary from small to large grains. A subsequent decrease in temperature slows or stops the process, but does not reverse it. Grains may grow through mechanical working, by the use of heat treatment, through phase changes, or a combination of these methods. The only way to refine the grain size is to dis-

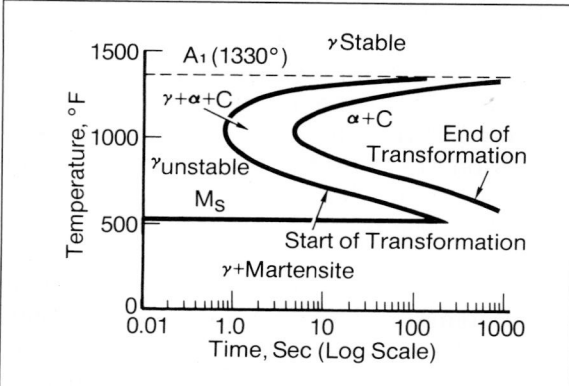

Fig. 10 Time-temperature-transformation (T-T-T) curves for steel in Fig. 8 with 0.8 percent of carbon (eutectoid)

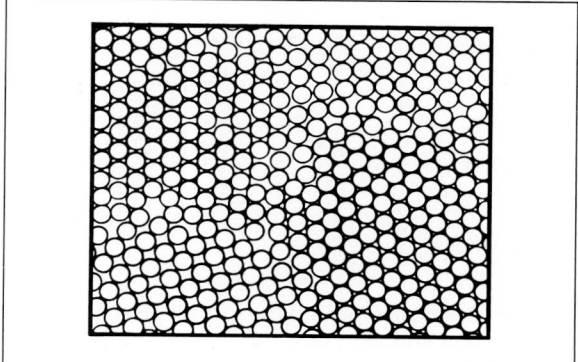

Fig. 11 Atomic transition at grain boundary between two grains

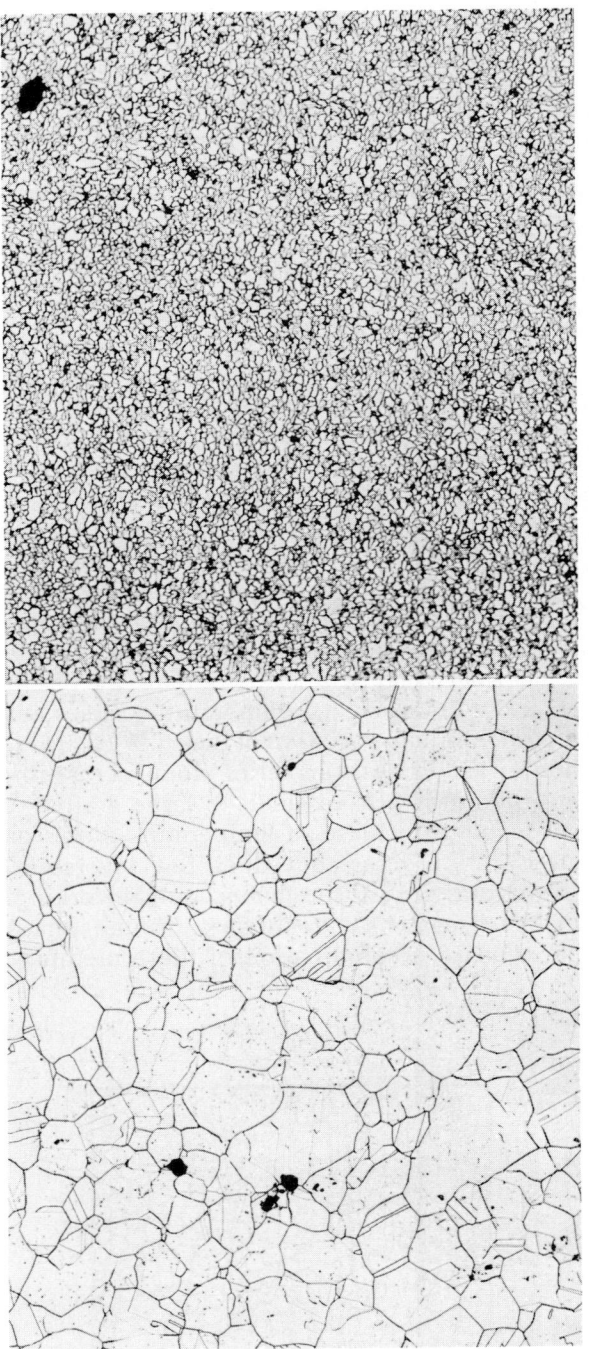

Fig. 12 Grain-size boundaries of single-phase austenitic alloy showing fine and coarse grained material (ASTM A-213-TP321). Grain-size number eight and smaller (top) and four to seven (bottom) at 100X. Sixty percent nitric acid etch

tort, break, or remove the grains which have grown and start new grains. The process depends upon the characteristics of the particular alloy involved. Grain growth will occur only above certain temperatures and depends upon time at temperature. This temperature may be quite low with some materials which have been critically cold-worked.

The ASTM has standardized a grain-size index in which the grain-size number, n, is obtained as follows.

$$N = 2^{n-1} \tag{1}$$

where N is the number of grains observed per square inch at a linear magnification of 100X. For most metals, the grain size ranges from number one to number eight, number one being a large grain and number eight a small grain.

Because the properties of metals vary greatly depending upon grain size, proper control must be exercised.

HARDENABILITY

As discussed previously in the section on T-T-T curves, plain carbon steel is hardened by quenching in water to cool the piece rapidly, thereby avoiding transformation at the fastest reaction rate temperature of 1000°F ("nose" of Fig. 10 T-T-T curve). This is possible if the thickness of material is small enough to cool completely at the required rate. But a large thickness piece cannot be cooled fast enough at the center of the thickness, and transformation will take place more rapidly and at a higher temperature at the center. This results in a softer transformation product.

Alloy additions to this steel retard the transformation rate (shift the curve of Fig. 10 to the right) and allow hardening to greater depths. This also permits less severe quenching and results in lower thermal stress with reduced possibility of cracking. Highly alloyed steels may be hardened by air cooling. Because steels hardened to high values are also brittle as hardened, a reduction of hardness is generally accomplished by reheating to an intermediate temperature. This markedly improves toughness. This process is called tempering.

ASEA BROWN BOVERI

HEAT TREATMENT

Table II[4] helps to understand some of the processes whereby metallic structure can be changed through heat treatment. Thermal treatment can markedly affect ductility and toughness of base metal, weld metal, and heat-affected zones; the type of treatment is selected with knowledge of the material property needs during fabrication and in service.

Thermal treatment is often required to soften zones which become cold-worked in local areas during the forming operations. Whether such treatment is necessary depends on the degree of cold work, the likelihood of subsequent strain aging, and the potential for controlling the degree of cold work developed during processing.

Though careful handling of large parts can sometimes prevent the development of heavily cold-worked surface areas, this is not usual. And heavily cold-worked zones can act as crack starters, causing brittle failures below the nil ductility temperature. On a finished part, such zones can also serve as nuclei for starting cracks at low temperatures under shock conditions. Heat treatment minimizes this effect.

During fabrication, welds may require thermal treatment to eliminate cracking. Heavy plates of carbon steel (more than 5 in. thick) and lighter plates of low-alloy steels have been known to crack if allowed to cool to room temperature after welding. Steels such as these may require heat treatment before cooling from preheat temperatures. This treatment, which may be at a relatively low temperature (600° to 1000°F) for a short time, eliminates cracking. Necessary for certain structures, it is generally termed an *interstage* thermal treatment. This is because it is applied to individual parts of an assembly or after partial welding of subassemblies.

When processing steels for high-temperature operation, it is important to tailor the thermal treatment so that the material will remain stable during the service life. Such custom treatment will avoid the development of serious defects, such as "eyebrow" or "chain-type" graphitization (of carbon and carbon-molyb-denum steels), strain-induced precipitation cracking (particularly possible with the stabilized austenitic steels), aging embrittlement, large grain size (which may cause excessively low ductility) and low stress-rupture strength or creep-rupture ductility.

Some steels are susceptible to very low ductility in a narrow range of elevated temperature. Such a condition, combined with the presence of stress, can produce sufficient strain to exhaust the ductility and result in cracking. This can occur with operation in the low-ductility temperature range, or in passing through the low-ductility temperature range during application of a postweld heat treatment. The latter is known as stress-relief cracking or reheat cracking. The characteristics of such susceptible materials must be established for proper handling.

With steels to be used at low temperature, thermal treatment may be advisable to ensure resistance to brittle fracture. Carbon steels improperly deoxidized are subject to loss of toughness and ductility by quench aging. If not corrected, this deficiency can result in failure during fabrication or operation. Small defects (such as cracks at arc strikes) can exhibit residual stresses as high as the yield point at the defect and may act as crack starters with additional loading. If thermally treated, such areas are generally relatively harmless. For some combinations of conditions, the use of careful nondestructive tests to prove freedom from macroscopic defects may allow the thermal treatment to be eliminated.

MECHANICAL PROPERTIES OF STEEL

Certain mechanical properties of steel affect its fabrication and service. Among these are tensile properties, hardness, toughness, fatigue strength, and high temperature characteristics.

TENSILE PROPERTIES

The standard tension test provides data on yield strength, tensile strength, and ductility. When metals are pulled (stressed) with a uni-

Table II. Heat-Treating Processes

Process	Example	Purpose	Procedure
Annealing	Cold-worked metals	To remove strain hardening[a] and increase ductility	Heat above recrystallization[b] temperature
Annealing	Steel	To soften	Heat into austenitic[c] range and slow cool
Normalizing	Steel	Homogenization and strain relief[a]	Heat into austentic range and air cool
Process annealing and stress-relieving	Steel	To soften and toughen	Heat close to, but below, the eutectoid[d] temperature
Spheroidizing	Steel	To soften and toughen	Heat for a sufficiently long time close to, but below, the eutectoid[d] temperature
Quenching	Steel	To harden	Quench from austenite to martensite[e]. (This is followed by tempering.)
Tempering	Quenched steel	To toughen	Heat briefly at low temperature
Austempering	Steel	To harden without developing brittle martensite	Quench from austenite to a low temperature below the nose[f] of the transformation curve, but above the martensite transformation temperature. Hold until transformation is complete.
Marquenching	Steel	To harden without quench cracking	Quench from austenite to a temperature below the nose of the transformation curve, but above the martensite transforming temperature. Hold until temperature has equalized. Cool slowly to martensite. (This is followed by tempering).
Solution treating	Stainless steel	To produce a single phase alloy	Heat above the solubility curve into a single phase area and quench to room temperature.
Age-hardening	Various ferrous and nonferrous alloys	To harden	Solution treat. Cool to provide supersaturation. Reheat to an intermediate temperature until the initial precipitation starts. Cool to surrounding temperatures.

[a] When a sufficient stress is applied to a metal such that the piece does not return to its original dimensions, the yield strength has been exceeded. Because most metals are weaker in shear than in pure tension, they yield by plastic shear or slip of one plane of atoms over another. This slip occurs most readily along planes containing the greatest number of atoms per unit of area. These parallel planes are more widely

separated than other planes. Plastic movement along these planes causes distortion of the planes because of restraint of surrounding metal and allows added amount of slip to occur less readily. This resultant increase in strength is called strain hardening.

[b] Because plastic deformation at low temperatures causes distorted crystal patterns, the tendency is for the atoms to return to a more perfect unrestrained condition. Heating to a higher temperature increases thermal vibration of the atoms which allows readjustment to take place. As such readjustment also results in decreased hardness, the temperature of marked softening is called the recrystallization temperature. Recrystallization temperature depends upon degree of plastic deformation (cold work), time at temperature, material, and material purity. Generally, it is between ⅓ and ½ of the melting temperature (degrees absolute). Plastic deformation at temperature below the recrystallization temperature is called cold working whereas above the recrystallization temperature is called hot working.

[c] A solid solution in which gamma iron is the solvent; characterized by a face-centered cubic crystal structure.

[d] See Fig. 8. Eutectoid temperature is 1330°F. Eutectoid carbon content is 0.8 percent.

[e] An unstable constituent is quenched steel, formed without diffusion and only during cooling below a certain temperature known as the M_s temperature (see Fig. 10). The structure is characterized by its acicular or needlelike appearance on the surface of a polished and etched specimen. Martensite is the hardest of the transformation products of austenite. Tetragonality of the crystal structure is observed when the carbon content is greater than about 0.5 percent.

[f] See Fig. 10. "Nose" of curve is at approximately 1000°F. The start of martensite transformation on cooling for this steel is approximately 500°F.

---

axial increasing load, the material stretches (strains). Diagrams such as Fig. 13 show the relationship between stress and strain for specific materials. With increasing load, the material strains elastically until it reaches the yield point. During this period of elastic behavior, there is no permanent deformation; strain is directly proportional to the stress. This relationship is described by Hooke's Law which states that stress equals strain multiplied by the modulus of elasticity (Young's Modulus) of the material.

Continuing to increase the loading beyond the yield point results in plastic strains and eventual breakage of the test specimen. Ductility is a measure of the amount of plastic

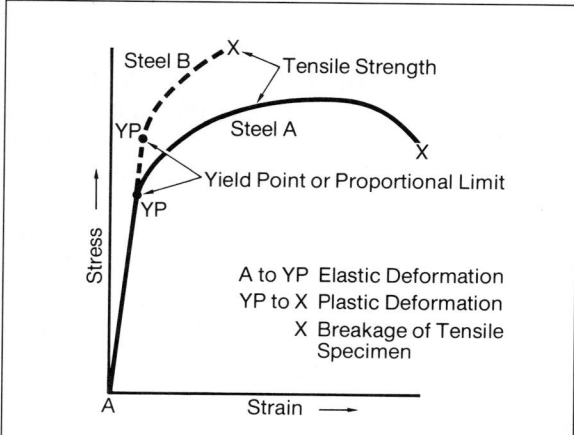

Fig. 13 Stress-strain curve for two steels with different strength and ductility

deformation the steel will sustain before breaking. It is usually expressed as a percentage elongation or reduction in cross-sectional area of the test specimen. How much plastic strain a steel will exhibit is quite variable and is generally inversely related to the tensile strength of the material. High-strength steels typically will exhibit much less ductility than softer, lower-strength steels.

Because the true yield point is difficult to establish with any accuracy, most current tensile testing of steels involves the determination of yield strength rather than yield point. Yield strength, which can be easily established, is defined as the stress necessary to produce a specified value of plastic elongation, usually 0.1, 0.2, or 0.5 percent as measured over the gage length of the tensile specimen.

Most of the specifications covering the steels used in the construction of a boiler require room-temperature tensile testing with minimum values of tensile strength, yield strength and ductility specified. But for design purposes, the tensile and yield strength of each grade of steel used needs to be determined at temperatures up to the creep range of the grade. Various organizations including the producers and fabricators of steel perform such tests and the data from these tests are used by the *ASME* Boiler and Pressure Vessel Code to set maximum allowable stresses.

Some of the many product form specifications—most noticeably those for pipe and tube—require various deformation tests such as flattening, flaring, or bending. These additional tests of ductility provide evidence of the ability of the steel to withstand fabrication or installation operations such as bending, swaging and tube expansion.

HARDNESS

The hardness of a material is a measure of resistance to plastic deformation and is related to the tensile strength of the steel. It is also used as an indicator of the machinability and abrasion resistance of a steel. Usually, hardness is determined by Brinell, Rockwell, or Vickers tests in which a small ball or pyramid-shaped point is pressed into the surface of the metal with a specified force for a specified time. The size of the indentation is measured automatically or by microscope and is expressed as a hardness number. The smaller the indentation, the higher the hardness number.

The use of hardness testing is permitted in some material specifications to approximate tensile strength. ASTM Specification A-370 contains hardness-to-tensile-strength conversion charts for different groupings of steels. In some material specifications, particularly those in which the ductility of the steel is important to its workability, maximum hardness values are specified.

TOUGHNESS

Under most circumstances, steel can tolerate localized stresses above the yield point by plastically absorbing and redistributing these stresses. But under certain conditions, even steels having considerable ductility are subject to a brittle (cleavage) mode of failure when subjected to concentrated stresses at low temperatures. The property of toughness is the ability of a material to resist this failure.

Various impact tests evaluate the property of toughness. One of the more commonly used tests is the Charpy V-notch impact test. In this test, a swinging pendulum strikes a single blow to a notched horizontal specimen supported on both ends. The energy absorbed by the specimen is related to the height of rebound and is expressed as ft-lbs.

The mode of failure in the impact test changes from ductile (shear) to brittle (cleavage) as the temperature is lowered. The temperature range at which this occurs is called the transition range. Material within or below its transition temperature range may crack extensively if subjected to an impact load, or if construction details are such that localized yielding is prevented.

The transition temperature range depends upon the particular metal composition and melting practice as well as the subsequent working and heat treatment. For many types of carbon or low-alloy steels, the transition temperature may be as high as 70°F or above. The possibility of brittle fracture must be considered in the fabrication of materials (bending and forming in various manners), in testing the finished structure, and in any service involving operation below the transition temperature. Care in design, fabrication, inspection and, when necessary, field repair, is required to eliminate conditions which might promote brittle fracture.

Fig. 14 shows the impact strength versus temperature for a material having a high transition temperature. Also shown are the broken surfaces of the impact specimens in which there is a ductility variation with the temperature of testing. The material tested at the lower temperature has broken in a cleavage fracture along crystallographic planes and shows a shiny-grain appearance which at one time led to the mistaken impression the material had crystallized. Actually, what appears in the figure is the progression from a brittle-cleavage fracture to a ductile-shear fracture as the temperature of testing is increased. Corrosion-resisting pressure vessels of this material may be successfully fabricated and used provided sufficient consideration is given to this property. The importance of the degree of stress-concentrating mechanical notch in this type material is shown in Fig. 15.

The examples given are for chromium-iron materials having a particularly high transition

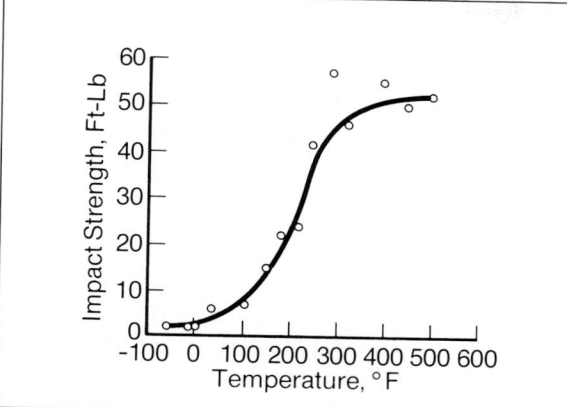

Fig. 14 **Relationship of impact strength to temperature for a material having a high transition temperature. (16% Cr weld metal)**

temperature. Steels used in boilers have transition temperatures considerably below that shown in the example. Metals to be stressed in operation at low temperatures are selected based upon these considerations and purchased to specified impact requirements.

Heat-treatable steels in the quenched and tempered condition may have considerably improved transition-temperature properties. High-nickel steels exhibit low transition tem-

peratures. Austenitic stainless steels and metals such as copper and aluminum, in general, do not change abruptly in toughness as a function of temperature and may be used to temperatures as low as that of liquefied air.

FATIGUE STRENGTH

The stress a material can withstand under repeated application and removal of load is less than that it can withstand under static conditions. The yield strength, which is a measure of static stress a material can sustain without appreciable permanent deformation, can be used as a guide in design only for material subjected to static loading. Dynamic, cyclic loading causes slip and cold-working in minute areas localized at grain boundaries and at stress-concentrating notches of various types.

As sufficient work-hardening develops, microscopic cracking develops and grows until complete fracture results. Fatigue strength is the magnitude of a cyclic stress which a material can resist for a specified number of cycles before failure. But at sufficiently low stress levels, many materials can tolerate an almost infinite number of cycles. Fig. 16 shows the endurance limit for one material.

Areas of peak stress such as changes in section or stress-concentrating notches promote the localized concentration of cyclic strains. Fig. 17 shows the photoelastic distribution of stress in a welded waterwall tube panel when subjected to an in-plane tensile load. Fatigue evaluations of welded walls have included cyclic-loading testing of full-size panels under bending and in-plane loads. With such testing,

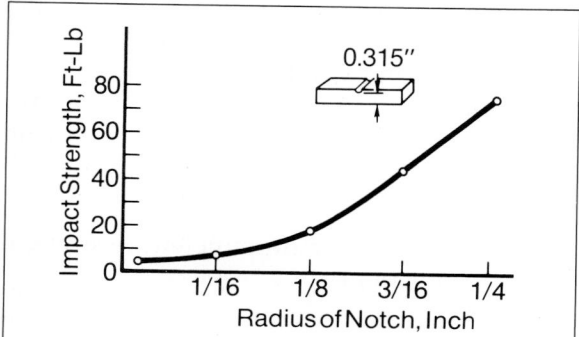

Fig. 15 **Effect of radius of notch (shown by stress concentration) on impact values of 18 percent Cr weld metal**

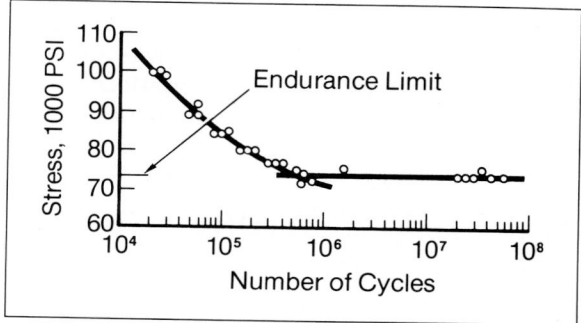

Fig. 16 **Endurance limit for a hot-worked bar stock**

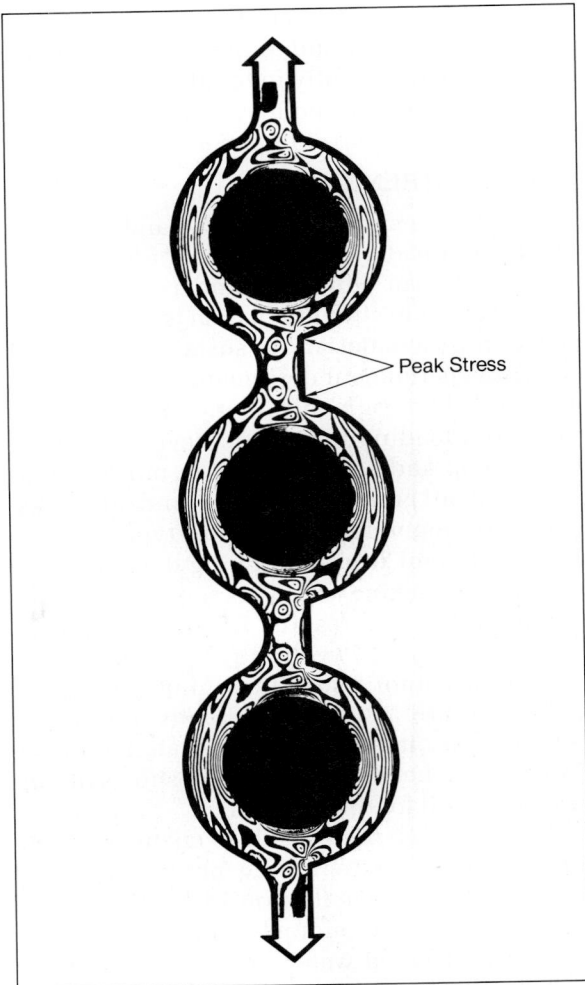

Fig. 17 Photoelastic pattern showing peak stress in welded waterwall tubes

endurance limits for the actual as-built component have been established.

Figs. 18 and 19 present an example of fatigue cracking. The cause of the crack was an unanticipated gas-flow-induced vibration of the superheater tube. Antivibration restraints have been applied to correct this type of problem.

HIGH-TEMPERATURE PROPERTIES

At temperatures exceeding about 650°F, most steels suffer a gradual decrease in tensile and yield strength. At still higher temperatures, the strain in a material is a function not only of the applied stress but also of the time under stress at temperature. In this high-temperature range, the metal will deform (creep) continuously even at stresses much lower than the short-time yield strength. If held for sufficient time under these conditions, the material will rupture.

Because no way has been found to predict this behavior quantitatively from short-time tests, it is necessary to perform tests of creep and stress rupture at several stress levels, temperatures, and over time periods as long as possible. Such tests, and extrapolations as necessary, establish values of creep strength and stress-rupture strength. The creep strength of a metal at a certain temperature is the steady stress that produces a specified low rate of elongation. For long-time service, such as *ASME* Boiler Code applications, a creep rate of 0.01 percent in 1000 hours is used. The stress-rupture strength is the steady stress required at a particular temperature to cause rupture in a specified long period of time.

Fig. 20 illustrates how creep and rupture strengths for a 2¼ Cr-1 Mo steel are established from test data. Creep-rate data for 1000°, 1100° and 1200°F are plotted versus stress. The intersection of these data lines with that for 0.01 percent in 1,000 hours sets the respective creep strengths of 7,800 psi, 5,000 psi and 2,400 psi. The other curves show rupture life versus stress for the same three temperatures. Lines through these data are extrapolated to 100,000 hours to establish rupture strengths of 13,000 psi, 7,000 psi and 3,300 psi respectively.

Paragraph A-150 of Section I of the *ASME* Boiler and Pressure Code lists the following criteria for consideration by the Code Committee in establishing allowable stresses:

■ ¼ of the specified minimum tensile strength at room temperature

■ ¼ of the tensile strength at elevated temperatures

■ ⅔ of the specified minimum yield strength at room temperature

Fig. 18 Fatigue crack in horizontal superheater tube at end of plate support members. Vibration in the length of tube to the right side of the spacer eventually caused the fatigue crack to initiate at the attachment and propagate circumferentially.

Fig. 19 Through-the-wall nature of the Fig. 18 fatigue crack. Typical of fatigue, there is a single, tight trans-granular crack

■ ⅔ of the yield strength at elevated temperatures.

■ 100 percent of the stress to produce a creep rate of 0.01 percent in 1,000 hr

■ 67 percent of the average stress to produce rupture at the end of 100,000 hours, or 80 percent of the minimum stress for rupture at the end of 100,000 hours as determined from the extrapolated data, whichever is lower.

Fig. 21 applies these criteria to establish the allowable stress for a 2¼ Cr-1 Mo steel. For this material, item 1 controls up to 800°F. At 950°F and above, the creep strength and rupture strength, the fifth and sixth items, require a rather sharp decrease in the allowable stress.

The relative influence of the different criteria varies with materials as well as with temperature. Fig. 22 shows the effect of temperature on *ASME* Boiler Code allowable stresses for a number of alloys used for high-temperature service. The carbon steels begin to lose strength above 700°F, and by 850°F are down to about one-half their room-temperature values. The low-chromium ferritic alloys start to lose strength above 800°F and are down to half strength at about 1000°F. The austenitic stainless steels decline somewhat from room temperature to 1000°F because of reduction in yield strength; above 1000°F, creep and rupture strength cause a rapid decrease to half strength or less by 1200°F.

The high-temperature strength and ductility of steel may be strongly affected by grain size, cold working, heat treatment and other variables. Material suppliers and equipment manufacturers must give careful consideration to these factors in their manufacturing and fabrication procedures.

CREEP AND FATIGUE INTERACTION

When metals are exposed to cyclic loadings while operating at temperatures within their creep range, creep effects can reduce fatigue life. Fig. 23 indicates how the fatigue behavior of a steel varies for different testing conditions at elevated temperature.[5] The introduction of

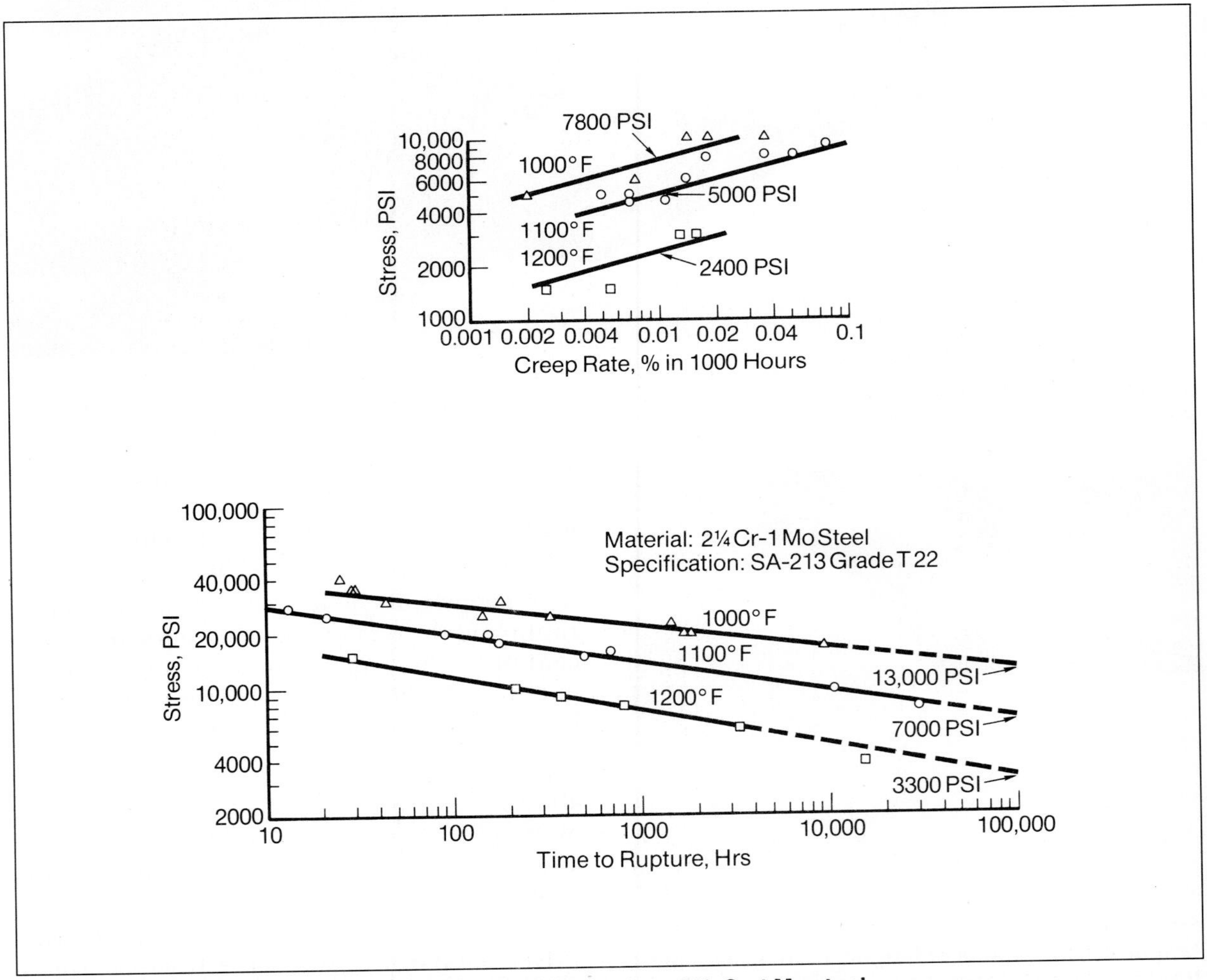

Fig. 20 Creep rate (upper) and rupture strength (lower) for a 2¼ Cr-1 Mo steel

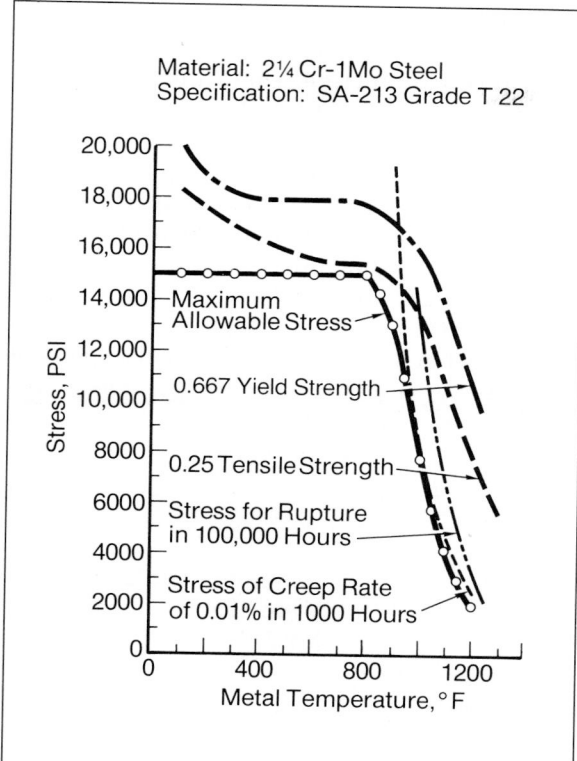

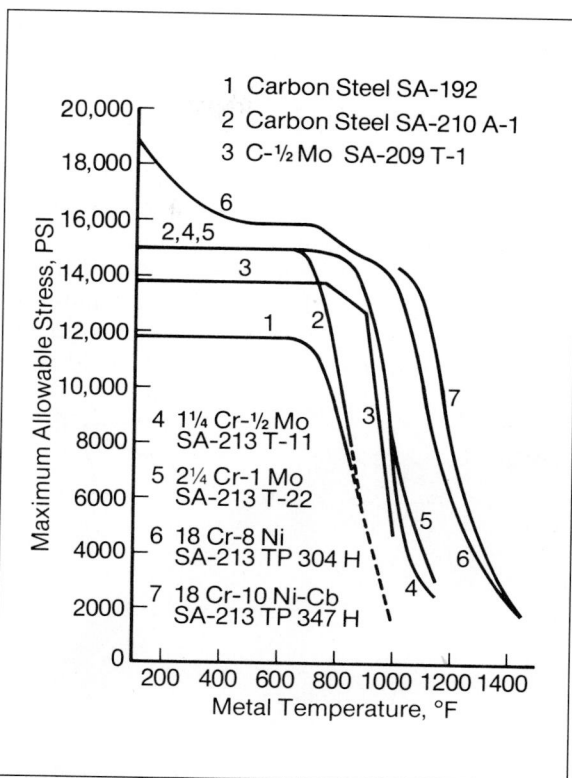

Fig. 21 Use of *ASME* Boiler Code criteria to establish allowable stress for a 2¼ Cr-1 Mo steel

Fig. 22 Effect of temperature on *ASME* Boiler Code allowable stresses for grades of steel tubing

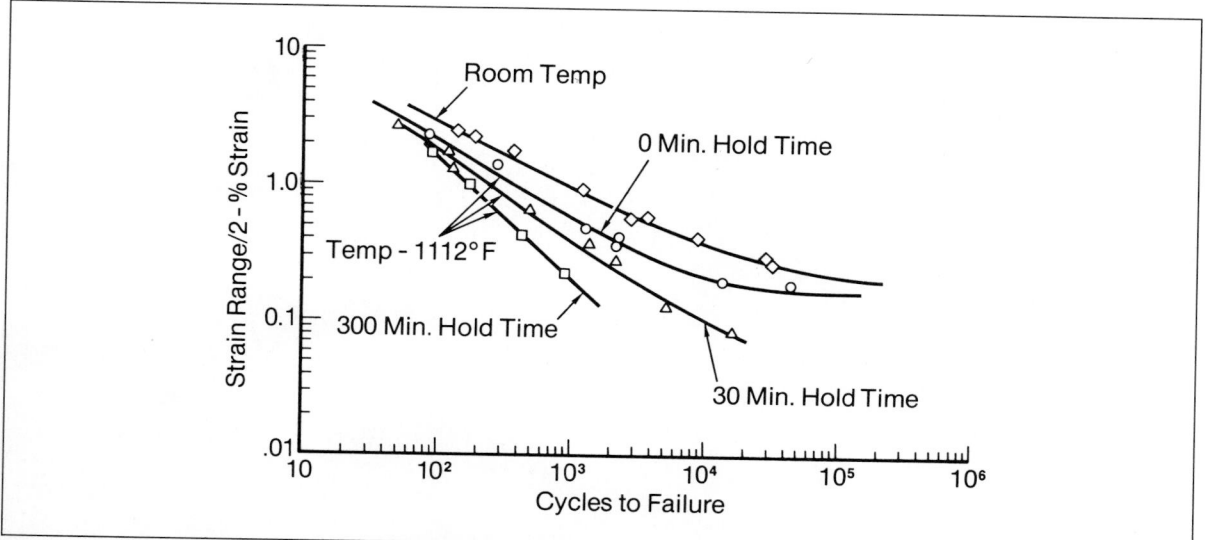

Fig. 23 Effect of temperature and creep-relaxation hold time on number of cycles to failure. Material 2¼ Cr-1 Mo

periods of hold time at temperature, while under maximum strain, causes the slope of the curves to change considerably. The periods of hold time allow creep relaxation to occur to the detriment of fatigue life.

The problem of creep and fatigue interaction is complex both in the formulation of a realistic design basis consistent with proven past practice of successful cyclic operation within the creep range and in the obtaining of material-property data which can be used in design. Attention is being given to the subject worldwide.

In the USA, organizations involved in cooperative research include the Pressure Vessel Research Committee, the Metal Properties Council and the ASME B&PV Committee.

DISSIMILAR WELDS

Fig. 24 is an example of cracking damage because of interaction between creep and fatigue. Welds using type 309 austenitic stainless filler metal have been used to join ferritic tubes to austenitic stainless tubes. The austenitic steels have a coefficient of thermal expansion

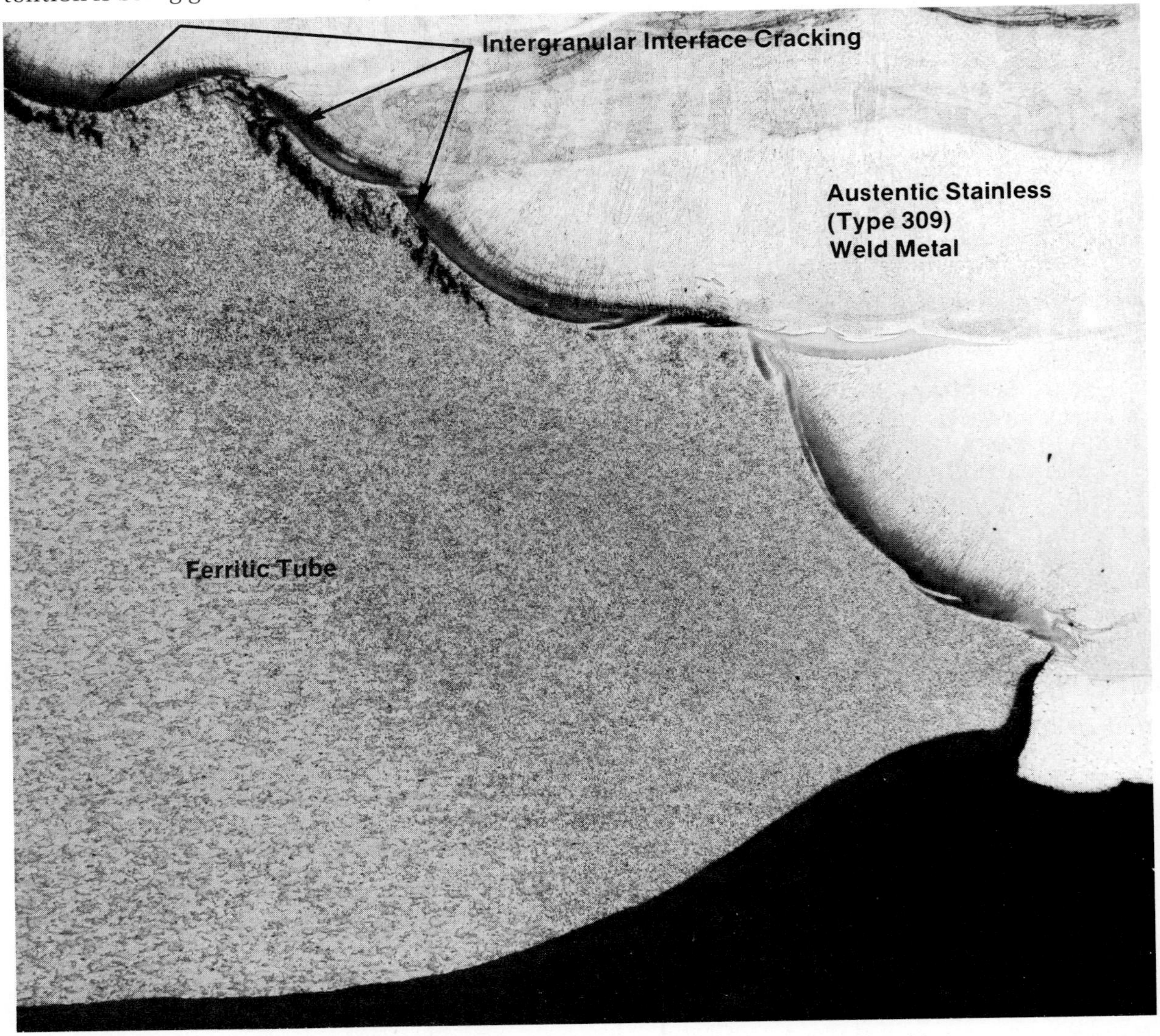

Intergranular Interface Cracking

Austentic Stainless
(Type 309)
Weld Metal

Ferritic Tube

Fig. 24 **Typical interface cracking associated with Grade TP 309 dissimilar metal weld**

approximately 40 percent greater than the ferritic steels. This mismatch results in high shear strains at the interface between the austenitic filler metal and the ferritic tube. With cycling, these strains can cause intergranular cracking within the weaker ferritic material.

Field repairs of type 309 dissimilar-metal welds are made by rewelding with nickel-base filler-metal welds. The nickel-base metals have a coefficient of thermal expansion similar to the ferritic steels, and the expansion-mismatch strains imposed on the ferritic tube are significantly reduced. However, the nickel-base filler-metal welds, while giving better service performance than the type 309 welds, are not totally free of interface cracking problems.

There is some uncertainty as to whether the interface cracking associated with nickel-base filler metals is primarily a function of strain or metallurgical changes. Further research is required to provide an understanding of the factors that promote cracking, such as testing of full-size tubular specimens of various dissimilar metal welds under pressure and temperature cycling conditions (Fig. 25).

Fig. 26 shows another example of interaction cracking from creep and fatigue. Differences in thermal expansion of superheater terminal tubes which are welded to the outlet headers and seal welded at the furnace enclosure have caused circumferential cracking in the tubes at the toe of the tube-to-header weld. Strain cycling caused these cracks to initiate and propagate in a fatigue manner.

Fig. 25 Thirty-two individual creep-rupture tubular specimens, each containing one or two dissimilar welds, before being inserted into furnace on the right. Specimens, which have been thermally aged, are exposed to internal pressures at temperatures somewhat higher than would be encountered in service and cycled, with a 3-day hold time at temperature, to simulate long-time service exposure.

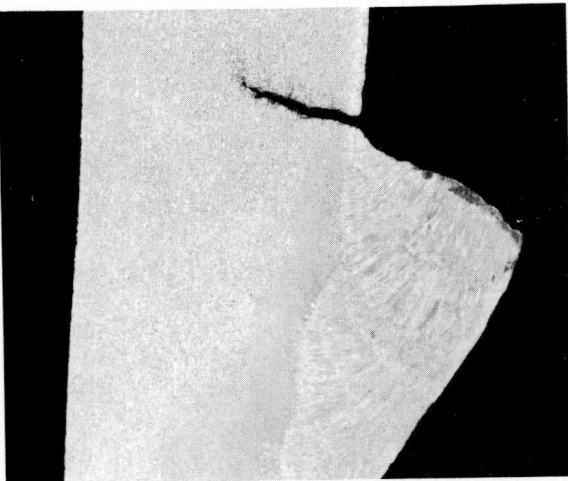

Fig. 26 **Intergranular creep and fatigue cracking at toe of tube-to-superheater outlet-header weld caused by repeated exposure to thermal expansion strains**

Note the similarities in appearance between this cracking and that described in the earlier discussion of fatigue and shown in Figs. 18 and 19. But there is a significant difference in the nature of these two cracks which occurred as a result of fatigue cycling. The vibration-induced crack is transgranular (across grains) which is typical of fatigue, but the cracking in the tube at the superheater outlet header is intergranular (along grain boundaries). Intergranular cracking is typical of that which occurs in longer-time creep or stress rupture,

and low-cycle high-strain loading in the creep range will likely produce intergranular rather than transgranular cracking.

SERVICE EXPOSURE CONSIDERATIONS

Proper selection and usage of materials require knowledge of the expected service conditions and an understanding of how the material will behave under those conditions.

Dependent on its environment, material behavior can differ significantly. Austenitic stainless-steel tubes, for example, are widely used for steam service. In this service, essentially all of the common grades of austenitic stainless can be considered equally compatible with the service environment, and there is no concern about resistance to pitting or intergranular corrosive attack (see the subsection "Sensitization" later in this section). But in the aqueous environment of boiler flue-gas wet scrubbers, only grades of stainless steel containing molybdenum (type 317LMN and other high-nickel alloys) and so forth can tolerate the corrosive environment without excessive pitting. (See Fig. 27.) Also, because of the potential for acidic aqueous conditions in the scrubber, grades of extra-low/carbon stainless type 3161, notably type 317LMN, are used to avoid the possibility of intergranular corrosive attack.

Although type 317LMN stainless is widely used in flue-gas wet scrubbers, austenitic stainless steels are not suitable for the heat-exchange surface which reheats the scrubbed gases to temperatures 25° to 40°F above the adiabatic saturation temperature. In this wet-to-dry service environment, austenitic stainless steels can and reportedly have experienced stress-corrosion cracking. (See "Stress-Corrosion Cracking," this chapter.) Because of this, carbon-steel tubes and fins have been used for the reheaters in scrubber systems.

As the foregoing examples demonstrate, knowledge of the environment and its effect on the material is a key to the proper use of the ma-

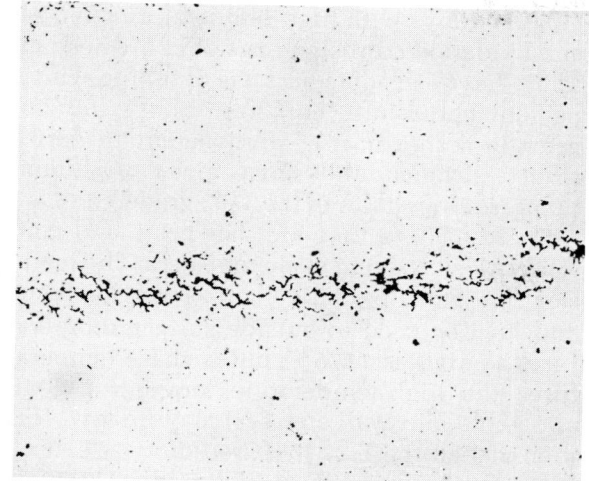

Fig. 27 **Examples of chain-type graphitization at 100X (upper) and random graphite nodules at 50X (lower). Unetched**

terial. The next section presents some of the service exposure conditions that affect the performance of boiler materials.

GRAPHITIZATION

At temperatures above 800°F, carbon steel is subject to graphitization; above 875°F, carbon-molybdenum steel is similarly affected. The carbon normally present in the steel in the form of carbides may transform over a long period of time to graphite. This transformation

may take place preferentially along heat-affected zones of welds, or along stress lines, to form chain-type graphitization, an example of which is shown in Fig. 27.

Carbon-steel pipe and carbon-molybdenum pipe in long-time service, particularly when operated at temperatures well above 800°F and 875°F, respectively, have experienced numerous instances of chain-type graphitization resulting in failure or requiring repair or replacement.[6]

Because of these problems, C-E has for many years avoided the use of carbon molybdenum for piping components and has limited the use of carbon-steel pipe to 800°F. For higher temperature service, the use of steels containing one-half percent or more chromium eliminates the danger of graphitization.

Carbon-steel and carbon-molybdenum tubing has not shown a similar tendency to graphitize as has piping. The graphitization occasionally found in tubing is usually in the form of well-dispersed nodules (Fig. 27) which do not weaken the steel. Carbon-steel and carbon-molybdenum tubing may, therefore, be used in applications when the temperatures reach 850°F and 900°F respectively.

RESISTANCE TO HIGH-TEMPERATURE OXIDATION AND CORROSION

Chromium is the most useful alloying element for imparting oxidation and corrosion resistance to steel. Fig. 28 shows the degree of oxidation scaling versus temperature for a number of steels with a range of chromium from 0 to 18 percent. Similarly, chromium additions improve resistance of a steel to molten-ash corrosion (see Chapter 3). Low levels of chromium, in the range of ½ to 2¼ percent, provide a useful improvement in high-temperature-corrosion resistance. Within this chromium range, high-temperature strength accompanies the trend of somewhat improved corrosion resistance.

For these reasons, there is widespread use of the family of low-chromium alloys, up to 2¼% Cr, for intermediate temperature-range appli-

cations, where carbon steel and carbon molbydenum are not used because of their low high-temperature strength (see Fig. 22) and their tendencies to graphitize. The temperatures at which these alloy grades are used involve considerations of high-temperature strength and corrosion/oxidation resistance. For superheater and reheater tubes in the gas pass, alloy grades SA-213 T-11 (1¼% Cr) and T-22 (2¼% Cr) are limited by C-E to gas-side surface temperatures of 1025°F and 1075°F, respectively. These temperatures, well below the point at which oxidation scaling is significant, were established based on reviews of external-wastage-rate data from units in service experiencing coal-ash corrosion. With these limits, excessive external corrosion metal loss is not expected even when coals with a corrosive tendency are burned.

Generally, the 2¼ Cr-1 Mo steel is the highest alloy used for pressure-part components outside of the gas pass because the metal has good strength and oxidation resistance at outlet steam temperatures between 1005°F and 1055°F. But because some gas-touched portions of the superheater and reheater are exposed to metal temperatures in excess of

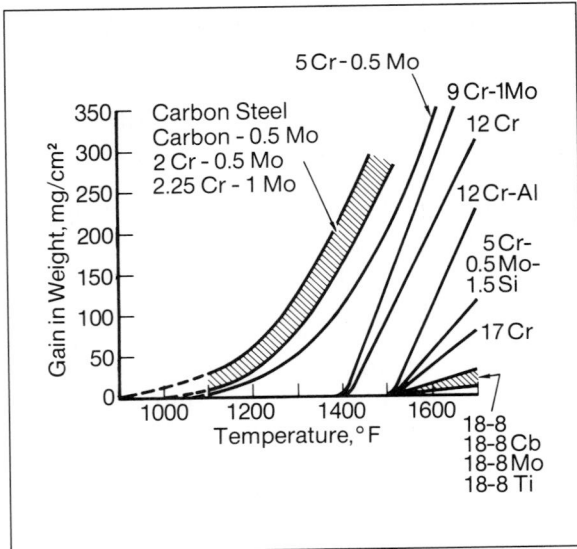

Fig. 28 Amount of oxidation (scaling) of carbon, low-alloy, and stainless steels in 1000 hours in air at temperatures from 1100° to 1700°F

1075°F, alloys with high-temperature strength and oxidation/corrosion resistance are needed. There has been some past use of tubing with 5 percent chromium, but these alloys did not provide a significant improvement in corrosion resistance and have lower high-temperature strength than the T-22 grade.

SA-213 T-9 (9 Cr-1 Mo) has been, and continues to be, used in some reheaters. The material has good oxidation/corrosion resistance and may be used to outside surface temperatures as high as 1175°F, but at these temperatures the high-temperature strength (ASME allowable mid-wall stresses) is quite low, less than that for the T-22 (2¼ Cr-1 Mo) grade.

Austenitic stainless steels (grades TP 304H and TP 347H) are used for most of the tubing operating at surface metal temperatures above 1075°F. Not only do these materials have excellent high-strength properties which are characteristic of austenite, but they also have excellent oxidation resistance because of their 18-percent (nominal composition) chromium content. But it is important to recognize that no tube alloy operating above 1000°F is immune to oil- or coal-ash corrosion. By virtue of their higher operating temperatures, the austenitic stainless steels will experience metal loss if there are molten ash deposits (see Chapter 3).

However, because molten coal-ash deposits do not exist at temperatures exceeding about 1300°F (see bell-shaped curve Fig. 11, Chapter 3, "Properties of Coal Ash"), materials for tube shields, spacer attachments, supports, baffles, and other nonpressure parts normally operating at temperatures above 1300°F can be used without excessive concern for corrosion. Experience has shown that oxidation behavior in normal boiler flue gas is similar to that in air. Table III shows the temperature levels at which oxidation scaling becomes significant for a number of alloy grades.

OXIDE SCALES AND EXFOLIATION

Oxides form on the outside surfaces of boiler components as a result of direct combination with oxygen in the air or in the flue gases. Oxidation also occurs in the internal wetted sur-

Table III. Maximum Temperature Without Excessive Scaling*

Alloy Nominal Analysis	AISI Grade	Maximum Temp°F Without Excessive Scaling
17 Cr	430	1550
27 Cr	446	2000
18 Cr–8 Ni	304	1650
25 Cr–12 Ni	309	2000
25 Cr–20 Ni	310	2000

*ASTM, Data on Corrosion and Heat-Resistant Steels and Alloys—Wrought and Cast, ASTM, May, 1950.

faces of the furnace walls and steam-touched surfaces of the superheater and reheater as a result of iron-water and iron-steam oxidation reactions.

Those oxide scales that form on boiler tube surfaces are thin and protective. In fact, the oxide that forms on wetted surfaces — magnetite — is critical to the compatibility of steel and boiler water. Because of the fundamental importance of magnetite, the half-joking description of a boiler as a thin layer of magnetite supported by steel has been quoted by many. Chapter 20 or water technology describes the boiler-water chemistry control necessary to maintain this protective layer.

Oxides forming on the steam sides of the superheater and reheater tubes inhibit oxidation by reducing the diffusion process. Oxidation rates decrease with time in a manner shown in Fig. 29. This formed-in-place oxide scale should not be confused with deposits on tube surfaces such as salt, debris, or transported oxides. Oxide scale that forms in-situ does not measurably affect heat transfer as deposits can. Under certain conditions, oxide scale that forms on the outside and inside surfaces of superheater and reheater tubes and pipes may break loose and exfoliate. Because of the conservative temperature limits on the usage of these steels adopted by manufacturers such as C-E, exfoliation does not present a problem of excessive metal loss which might lead to rupture of tubes or piping. However, the exfoliated oxide scale

from the inside surfaces of ferritic tubes and pipes is a cause of solid particle erosion (SPE) damage in steam turbines. So the term *exfoliation*, as it is widely used in the steam-power-generation industry, is in connection with the concern for SPE caused by spalled steam-side oxides.

OXIDE SCALE ADHERENCE

In many instances, tubes removed from long-time service exhibit little or no exfoliation or poorly adhering scale. This is typical of most samples removed from superheaters, Figs. 30 and 32. Conversely, ferritic reheater tubes frequently demonstrate severe exfoliation, or large areas of non-adhering scale, Figs. 31 and 33. Examinations of main and hot-reheat steam piping and final outlet headers indicate that these large diameter components tend to exfoliate significantly.

EFFECT OF TEMPERATURE AND SCALE THICKNESS

Although engineering intuition may suggest that exfoliation would tend to be more severe as temperatures and scale thickness increase, examinations of numerous ferritic tubes from service indicate that this is not the case. In fact, the exfoliation tendency seems to be greatest on the thinner scale. (Erosion damage to the turbines, of course, would be more severe with thicker scale. But precise knowledge as to the variables of importance to SPE, including the variable of steam temperature, is lacking.)

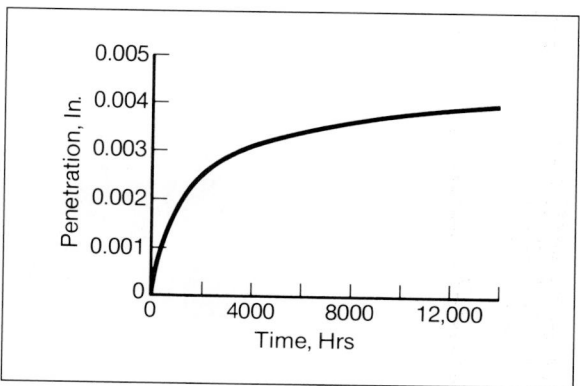

Fig. 29 Oxidation metal loss versus time for 1¼ Cr-1 Mo steel in steam at 1100°F

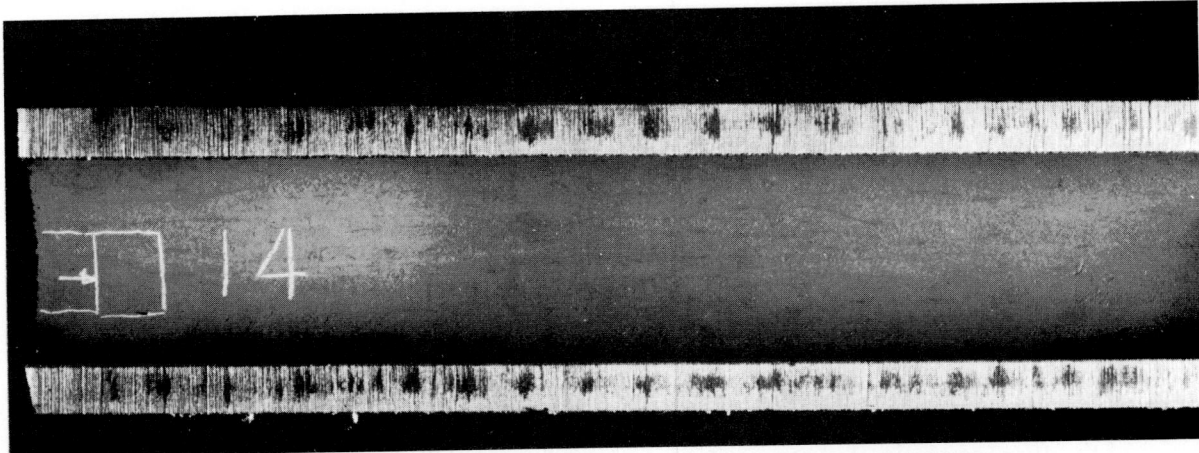

Fig. 30 Superheater tube (2¼ Cr-1 Mo) after 158,000 hours in service. Maximum scale 20 mils

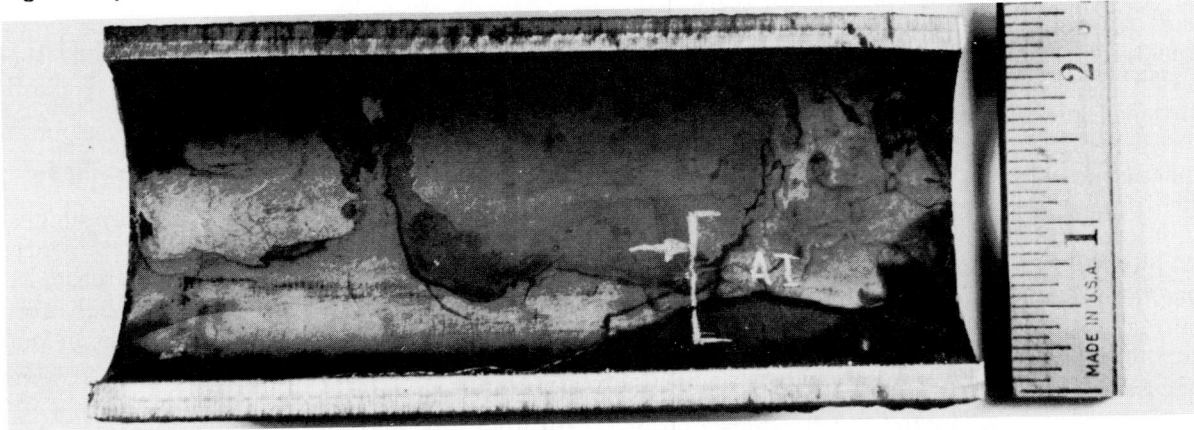

Fig. 31 Severely exfoliated reheater tube (2¼ Cr-1 Mo) after 39,000 hours service. Maximum scale 9 mils

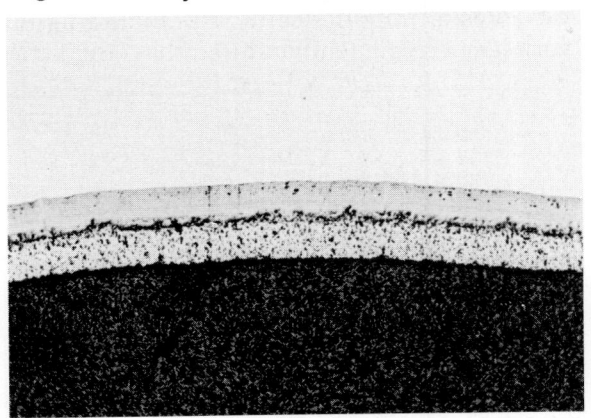

Fig. 32 Tube cross section showing typical two-layer oxide structure adhering well to internal surface of superheater tube (tube base metal is the white portion) at 15X

Fig. 33 Tube cross section showing exfoliated oxides on reheater tube at 15X

POSSIBLE CAUSES OF EXFOLIATION

Several explanations have been offered as possible causes of exfoliation. C-E investigators have postulated a thermal-quench explanation which appears to best fit the presently available facts. During start-up of reheat-type steam generators, there is no cooling steam flow through the reheater. Gas temperatures are limited to 1000°F during start-up, and reheater tube temperatures will approach this limit. When steam is first admitted to the turbine and reheater, its temperature is significantly lower than that of the tubes, and the steam thermally quenches the scale. It appears that the quenching acts to separate the cylinder of internal scale from the tube and to promote separations along planes of weakness within the scale.

This thermal-quench explanation is consistent with the different exfoliation behavior of reheater and superheater tubes. It also fits well with the observation that thin scale tends to have a greater exfoliation tendency. Under equal thermal-quench conditions, the temperature differentials and separation strains between scale and base tube will be greatest with the thinner scale. During hot restarts, with boiler parts still at elevated temperatures, scale on non-gas-touched tubing and piping is exposed to thermal-quench transients similar to that described for gas-touched reheater tubes. Reports from Europe attribute the absence of turbine SPE damage there to the use of turbine bypasses, which allow the boiler to be fired with steam flow passing from the superheater outlet piping directly to and through the reheater to the condenser. In such units, cold- or hot-restart thermal transients which can act to weaken scale-to-tube adherence are avoided. These turbine bypasses may divert any exfoliated scale during start-up directly to the condenser. However, bypass valves and steam distributors in the condenser reportedly do not suffer any noticeable erosion damage, which further supports the view that these bypasses may *prevent* or minimize the exfoliation process.

SOLUTIONS TO EXFOLIATION

Industry has considered, and is exploring further, a range of possible solutions for avoiding or minimizing exfoliation and SPE damage. The following discussion is not intended to provide an all-inclusive listing of these various possibilities, but rather will review what can be done to the steam-touched surfaces in operating units and in units yet to be designed.

EXISTING UNITS

At present, chemical cleaning is the only option available for existing units. With proper care the steam-touched materials can be safely cleaned to remove all oxide scaling. But chemical cleaning, if not correctly performed, can be damaging to the materials, and a decision to chemically clean should only be made with a full understanding of the inherent risks and the precautions to be taken.

A decision of whether chemical cleaning should be performed, and to what extent, involves cost/benefit considerations specific to the individual circumstances. In this connection, it should be noted that particles other than exfoliation scale, such as maintenance repair debris, may cause some SPE damage. So careful assessment of the damage and its likely source is also needed.

The Electric Power Research Institute has funded studies as to the efficacy and feasibility of applying a chromium-rich oxide coating to the inside of tubes following chemical cleaning. Such a chromate treatment process may negate the need for repeat cleanings or may increase the operating time between cleanings.

NEW UNITS

For units not yet designed, steps can be taken to avoid significant ferritic tube and pipe exfoliation. The exfoliation-prone ferritic portions of the superheater and reheater can be upgraded to austenitic stainless steel. There is no established temperature at which exfoliation is not considered a problem, but at 950°F and lower steam temperatures the scale is quite thin even after long-time service. The 1075°F outside-surface-temperature limit that C-E has imposed on the usage of T-22 (2¼% Cr) tubing to minimize external corrosion, also minimizes the quantity of exfoliated scale by reducing the amount of ferritic material.

Austenitic stainless materials can be applied to the inside surfaces of pipes and headers by weld-deposit cladding. C-E and others have extensive experience using these weld cladding techniques in the fabrication of nuclear and chemical-process vessels.

The same internal surfaces of ferritic tubes, pipes and headers which can be replaced or clad with stainless may alternatively be "chromized." This high-temperature diffusion process is described further in the final section of this chapter.

OVERHEATING

The term _overheating_ describes creep-rupture failure. This term can be misleading in the sense that it may convey the incorrect thought that temperature alone was somehow metallurgically damaging to the steel. Actually, overheating describes an exposure to temperatures too high for the operating stress. A T-22 tube, operating at a stress level of 5,000 psi, will give long life at a temperature of 1075°F, but the same tube would fail by overheating at 1000°F if operated at a stress of 15,000 psi.

Overheating failures can be categorized as either short-time or long-time. Fig. 34 shows examples of such failures.

SHORT-TIME OVERHEATING

Such failures result from exposure to temperatures significantly (hundreds of degrees F) above design and indicate a severe abnormality such as a loss of normal water or steam flow. Short-time tube failures exhibit considerable creep elongation which shows itself in large increases in diameter, particularly the inside diameter, and considerable reduction in wall thickness at the fracture surface. Fig. 35 shows a cross section through a waterwall tube which experienced short-time overheating. (A rupture typical of that shown in the lower photo-

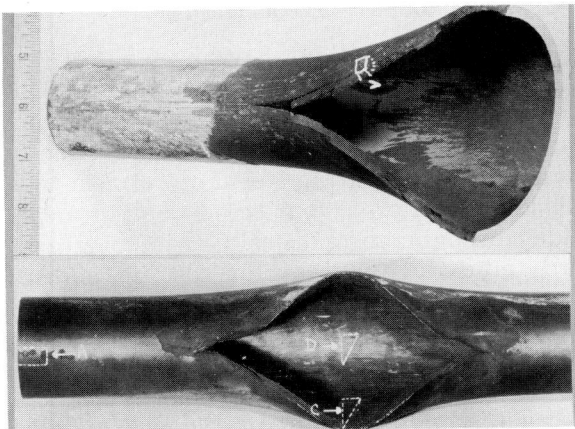

Fig. 34 Stress-rupture failure of carbon steel tube caused by overheating over a long time period (top) and a stress rupture failure because of overheating for only a few minutes (bottom)

graph of Fig. 36 had occurred in the tube a short distance away from this cross-sectional slice.) On the half of the tube exposed to the furnace,

Fig. 35 Cross section of tube exposed to short-time overheating. Furnace side of tube at top has thinned and ovalized as a result of exposure to excessively high temperatures while the tube is under internal pressure.

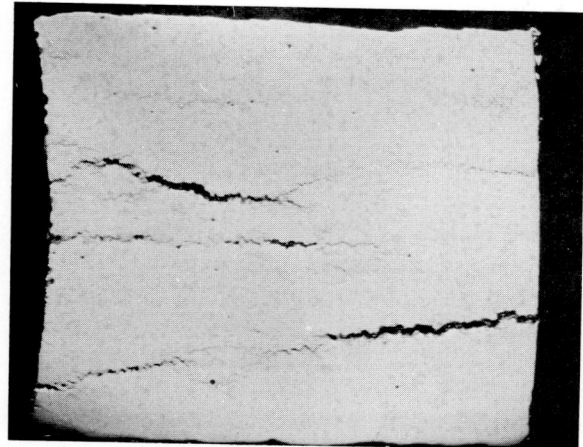

Fig. 36 Macrostructure of a heavy wall carbon-molybdenum steel pipe which has failed because of stress rupture over a long period of time. 2X

there is creep thinning caused by the combination of internal pressure and a temperature too high to resist the pressure forces. The egg-shaped inside of the tube is a typical condition associated with short-time overheating failures. These physical conditions of thinning and diametral growth provide the best and easiest means of assessing the degree of damage sustained. Under short-time conditions, tubes that have not experienced significant thinning or creep swelling have not been measurably damaged as a result of the overheating.

LONG-TIME OVERHEATING

Tubes which accumulate many hours of exposure to temperatures moderately above their long-time strength capability are also vulnerable to creep rupture. Creep elongation effects (swelling and wall thickness reduction) are much less pronounced in long-time overheating failures. The upper photo in Fig. 36 shows a long-time overheating failure and the thick rupture edge typical of this failure mode. Long-time creep-rupture failures usually show evidence of considerable secondary cracking (intergranular) in addition to the main crack associated with rupture. Fig. 36 shows an example of such cracking.

MICROSTRUCTURE

When exposed to temperatures above 900°F, but less than the lower transformation temperature (the A_1 temperature of Fig. 8, which is 1330°F for carbon steel), carbon and low-alloy steels are subject to spheroidization. See Fig. 37. The carbides present in the annealed condition are not in their lowest free-energy state, and exposure to temperature results in the coalescence of these carbides into spheroidal form. This natural change is both time- and temperature-dependent, with less time required to spheroidize as temperatures increase. Because manufacturing heat treatments can cause spheroidization, and in view of its time-temperature dependence, spheroidization is not a good means by which to estimate the service exposure temperature of overheated tubes.

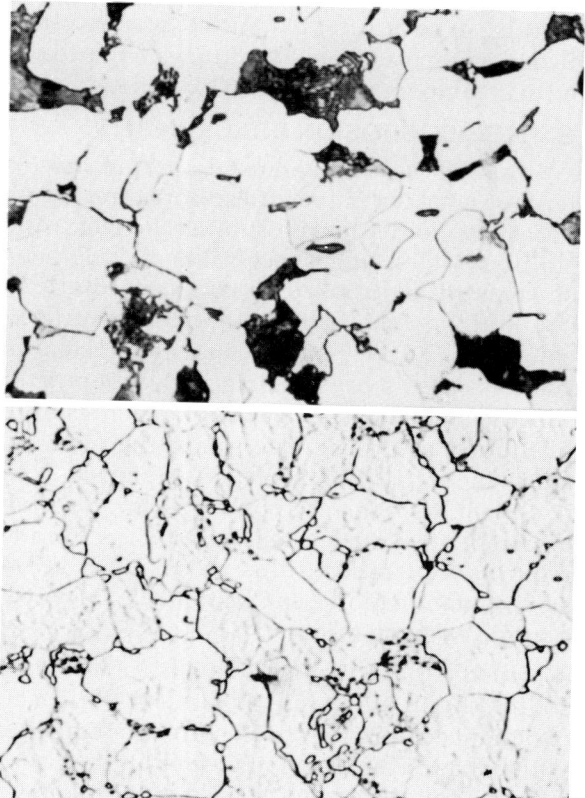

Fig. 37 Photomicrographs showing an annealed structure at top and a spheroidized structure at bottom at 500X. Nital etch

But by comparing microstructures within the tube (for example, the sides exposed to and away from the furnace) spheroidization can be useful as a indicator of differences in, and probable range of, exposure temperature.

When heated to higher temperatures, the ferrite in carbon and low-alloy steels transforms into austenite (see subsection "Phase Diagrams" in this chapter). The temperature at which austenite begins to form is called the lower critical temperature, and the upper critical temperature is that at which there is complete transformation to austenite (A_1 and A_3 temperatures of Fig. 8).

If a tube is heated above the lower critical temperature and then ruptures, the quenching effect of the escaping steam will result in the transformation of the austenite into martensite and other product forms. See Fig. 38. Because of this, exposure temperatures can be established with reasonable accuracy if the lower critical temperature has been exceeded.

STRESS-CORROSION CRACKING

Steels, as well as other metals, are subject to a form of cracking called stress-corrosion cracking. When under tensile strains, the austenitic stainless steels are susceptible to this type of cracking if exposed to aqueous solutions containing caustic or halides, particularly chlorides. See Fig. 39. Because the austenitic stainless steels are used most frequently in superheated-steam environments, there is little likelihood of exposure to damaging environments in service. But during the initial start-up of a plant, the increased potential for exposure to water contaminated with caustic or chloride requires that precautions be taken to avoid such contamination (see Chapter 21).

CAUSTIC EMBRITTLEMENT

Under certain exposures to high levels of sodium-hydroxide, ferritic boiler steels can experience a form of stress-corrosion cracking which has been given the specific name of caustic embrittlement. Leakage of caustic-treated boiler water through crevices such as rolled joints—which have residual stresses introduced during rolling—can concentrate

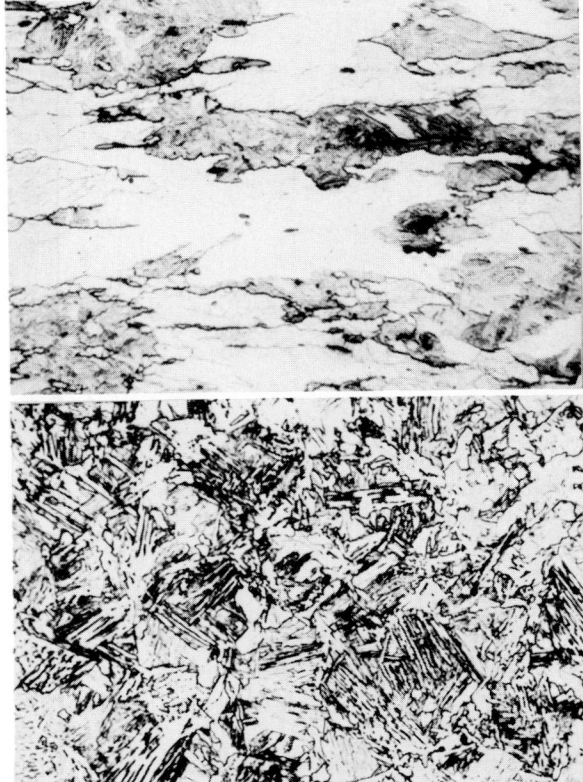

Fig. 38 Microstructure resulting from overheating to between the lower and upper critical temperatures (1330° to 1600° F) for carbon steel tube material (top) and to above the upper critical temperature (bottom) both followed by rapid cooling because of rupture. 500X nital etch

Fig. 39 Very rapid failure of TP 316 austenitic steel through stress-corrosion cracking

caustic and result in cracking of the type shown in Fig. 40. When boilers were of riveted construction and crevices were the rule, it was much more commonly found than in today's welded construction. Proper water treatment can prevent caustic embrittlement.

SENSITIZATION

Austenitic stainless steels, when exposed to elevated temperatures, may lose some of their corrosion-resistant properties at the boundaries between individual grains. (See subsection "Grain Structure," this chapter.) The chromium in the steel has a tendency to combine with the carbon and migrate into the grains, causing a depletion of chromium at the grain boundary. This condition is called sensitization. If sensitized steels are exposed to corrosive aqueous solutions such as acids, the grain boundaries can corrode preferentially resulting in intergranular attack (IGA).

In superheater and reheater service, sensitization which occurs naturally as a result of service exposure presents no problem. But in aqueous applications, such as wet flue-gas scrubbers, IGA is a possibility at heat-affected zones of welded joints, which can become sensitized by the heat of welding.

In these low-temperature applications, either an extra-low-carbon grade of stainless steel, such as type 317LMN, or stabilized grades of

Fig. 40 Caustic embrittlement, a type of stress-corrosion cracking associated with crevices in which caustic can concentrate. It is generally an intergranular attack, although it can also proceed transgranularly.

stainless can provide protection against sensitization and IGA. Those steels greatly reduce the tendency for grain-boundary depletion of chromium. As indicated earlier, wet flue-gas scrubbers also need a type of stainless steel that is resistant to pitting attack; 317LMN satisfies this requirement.

CHROMIZING

Chromium is the most important alloying element in steel for resisting oxidation and the various forms of corrosion encountered in boiler applications; this characteristic of chromium was discussed on page 17-23. Chromizing, one means of providing the levels of chromium necessary to resist such corrosive attack, is a high-temperature diffusion process in which the surfaces to be treated are alloyed with chromium. Chromizing as used for the avoidance of exfoliation of steam-touched surfaces (page 17-28) is also useful for reducing maintenance costs and prolonging the life of other boiler components. C-E's shop-applied chromizing process was specifically developed for pressure parts installed in utility, industrial, and chemical-recovery boilers.

In the diffusion process, an iron-chromium alloy is formed with a composition of 35 to 45 percent Cr, approaching that of AISI 442 stainless steel. The depth and quantity of chromium enrichment is a function of time at temperature. The process is used for chromizing the outside or inside, or both, of superheater or reheater tubing, waterwall panels, piping, or headers, to achieve high levels of Cr enrichment for depths in the 10 to 15 mil range.

Coal-fired utility boilers, especially those of the supercritical-pressure type, have suffered from reducing-atmosphere attack of the fireside of waterwall tubes. A number of owners have installed chromized tubing panels with very favorable results. Chromized tubing, even when located in areas having severe metal loss (up to 60 mils per year) has shown itself to be highly resistant to sulfidation attack.

The firesides of tube with a high front-to-back temperature differential sometimes suffer circumferential cracking. While this cracking is induced by thermal strain, the mechanism of penetration appears to be the result of repeated cracking or degradation of the semi-protective oxidation product formed on the tube surface. The supercritical boilers experiencing sulfidation problems also have corrosion-fatigue cracking in portions of the furnace walls outside the regions of most severe metal loss. Radiant-wall reheater tubes of coal-fired units have also had this type of cracking.

Chromized tubing has proved to be quite resistant to this kind of thermally induced corrosion-fatigue cracking. For example, bare T-11 water-wall tubing in a supercritical coal-fired unit suffered severe corrosion-fatigue cracking while an adjacent chromized panel was found free of such cracks after four years of service.

High-temperature ash corrosion and oxidation in superheaters and reheaters can often be addressed by upgrading the tubing material to the stainless steels. But there are instances where it may be necessary or desirable to continue to use a ferritic alloy such as T-22. In such a circumstance, chromizing will be of benefit in providing resistance to ash corrosion and will significantly raise the effective oxidation limit of the material.

REFERENCES

[1] Arthur M. Greene, Jr., *History of the ASME Boiler Code.* New York: American Society of Mechanical Engineers.

[2] American Society for Testing and Materials, *ASTM Directory.* Philadelphia: American Society for Testing and Materials, latest edition.

[3] Lawrence H. Van Vlack, *Elements of Materials Science: An Introductory Text for Engineering Students,* 3rd ed. Reading, MA Addison-Wesley Publishing Co., 1977.

[4] Lawrence H. Van Vlack, *Elements of Materials Science,* 2nd ed. Reading, MA: Addison-Wesley Publishing Co., 1964, Table 11-1, Common Heath Treating Processes, pp. 308-309.

[5] H. G. Edmunds and D.J. White, "Observations of the Effect of Creep Relaxation on High-Strain Fatigue," *Journal of Mechanical Engineering Science,* 8(3):310-321, Sept. 1966.

[6] Graphitization of Steel Piping, presented at the annual meetings of the American Society of Mechanical Engineers, New York, NY, Nov. 29-Dec. 3, 1943; Nov. 27-Dec. 1, 1944; and Dec. 2-6, 1946; under the auspices of the joint ASTM-ASME Research Committee on the effects of temperature on the properties of metals. New York: American Society of Mechanical Engineers, 1944, 1945 and 1947, respectively. Issued as pamphlets and bound at the end of the *Transactions of the ASME,* Vols. 66, 1944; 67, 1945; and 69, 1947; respectively.

CHAPTER 18

Steam-Generator Manufacture

Manufacturing boiler components requires extensive equipment, machinery, and facilities. Most of these fabrication operations involve the use of steel tubing and plate. As the tubing and plate are built into furnace-wall panels, superheaters and reheaters, drums, headers, pulverizers, ductwork, and structural members, various forms of welding are used. Heavy hydraulic presses are required to form segments of boiler drums and other pressure vessels, and machining likewise plays a vital part in the manufacture of boilers. Heat treatment, materials handling, inspection, and quality control are important adjuncts of the manufacturing operations.

THE INTERRELATIONSHIP BETWEEN PRODUCT DESIGN AND MANUFACTURE

The steam-generator engineer must be continually aware of the interrelationship between product design and manufacture. A design resulting in improved heat-transfer efficiency may be useless if the component cannot be fabricated economically and within the tolerances proposed. On the other hand, savings in manufacturing costs cannot be justified if they result in finished elements which do not meet

design objectives. The personal and computer-driven lines of communication between design and manufacturing functions must be maintained if overall production economy and high-quality products are to result.

New-product design needs have significant effects upon manufacturing processes. Standardized production equipment may not be satisfactory, and new machines may have to be developed. The design of such equipment requires knowledge from many specialized fields, including production practices, component design, materials, metallurgy, and controls.

Since it is obvious that a large utility or industrial boiler must be transported to the jobsite in relatively small assemblies and sub-assemblies, most shipments are made by rail car. A large utility-type unit will have 10,000 tons of boiler parts and equipment and more than 4,000 tons of structural steel supports, and will require about 200 railroad cars and 100 trucks for shipment to the site.

ASME CODE CONSIDERATIONS

All U.S.-installed utility and industrial boilers are designed and manufactured in accordance with the rules of applicable sections of the American Society of Mechanical Engineers Boiler and Pressure Vessel Code, which is the professional standard in the United States and in many other parts of the world. The *ASME* Code is a set of regulations and guidelines incorporating many details for design, manufacturing, and field construction to assure safe construction of pressure parts of boilers. It is not a thermodynamic performance code. There is nothing in it that tells how big a furnace must be to burn a fuel or how large a drum is needed to get dry steam. The Code does concern itself with the materials of construction, the design aspects from a structural or pressure-parts standpoint, the fabrication rules (primarily welding), the rules for nondestructive examination, and certain rules for certification (which include third-party inspection).

This third-party inspection has been a feature of the Code since its inception. It is this which enables it to be a code rather than a standard. Construction of Code-stamped components must be by a manufacturer holding an *ASME* certificate of authorization. One requirement for such a certificate is the manufacturer's arranging for the services of an authorized inspector to verify that the components are constructed in accordance with Code rules. The inspector performs such inspections as he considers necessary in the shops of the boiler manufacturer, and when he is satisfied he so indicates by countersigning a certification sheet (the manufacturer's data report).

The materials in the pressure-retaining parts of the entire boiler structure range from carbon steel for drums, headers, and tubular products to the high-chromium, high-nickel austenitic stainless steels for "fire-exposed" parts of the superheater and reheater assemblies. Each material used in the pressure-retaining parts — drums, headers, tubing, piping, forgings, or castings — conforms to all the requirements of an applicable *ASME* Boiler and Pressure Vessel Code material specification. Furthermore, during fabrication of each material used, the rules relating to welding, material forming, and inspection of materials and the welds in them are dictated by the *ASME* Code. Within the framework of the Code's establishing quality standards there is considerable latitude for selecting materials, welding process and procedures, types of inspection, and thermal treatments of welded assemblies.

FABRICATION OF STEAM-GENERATOR PRESSURE PARTS

Welding is the basic tool for fabricating boiler components. While elements of a skilled art are retained in some aspects, welding is a recognized technical specialty that requires substantial background in the fields of metallurgy, electrical, and mechanical engineering.[1]

In applying the welding process, metallurgical skills are necessary to determine the effects of chemical composition and metal structure on the weld properties. Electrical engineering skills are required in designing the welding power supply, welding controls, and related equipment. Particularly in automatic fusion welding, there must be precise control of the welding arc as well as the welding machine, and such control usually involves electronics. Skills in mechanical engineering are essential to design automatic welding equipment, where the machine designer must have knowledge of mechanics, hydraulics, pneumatics, and heat transfer.

In the sections that follow, the principal fabrication and welding operations for drums, headers, furnace walls, superheaters and reheaters will be described.

STEAM DRUM

The steam drum is a heavy-wall pressure vessel having hemispherical ends, with nozzles and nipples attached for incoming and outgoing water and steam. Fig. 1A illustrates a typical drum for a large boiler. In this figure, the drum is fabricated in three cylindrical sections, called courses, each welded to one another and with a hemispherical head welded to each end of the three-course cylinder. Fig. 1B and 1C show that each cylindrical section is made up of a thick plate (called the tube sheet) and a thin plate (called the wrapper sheet). The tube sheet is designed to withstand the

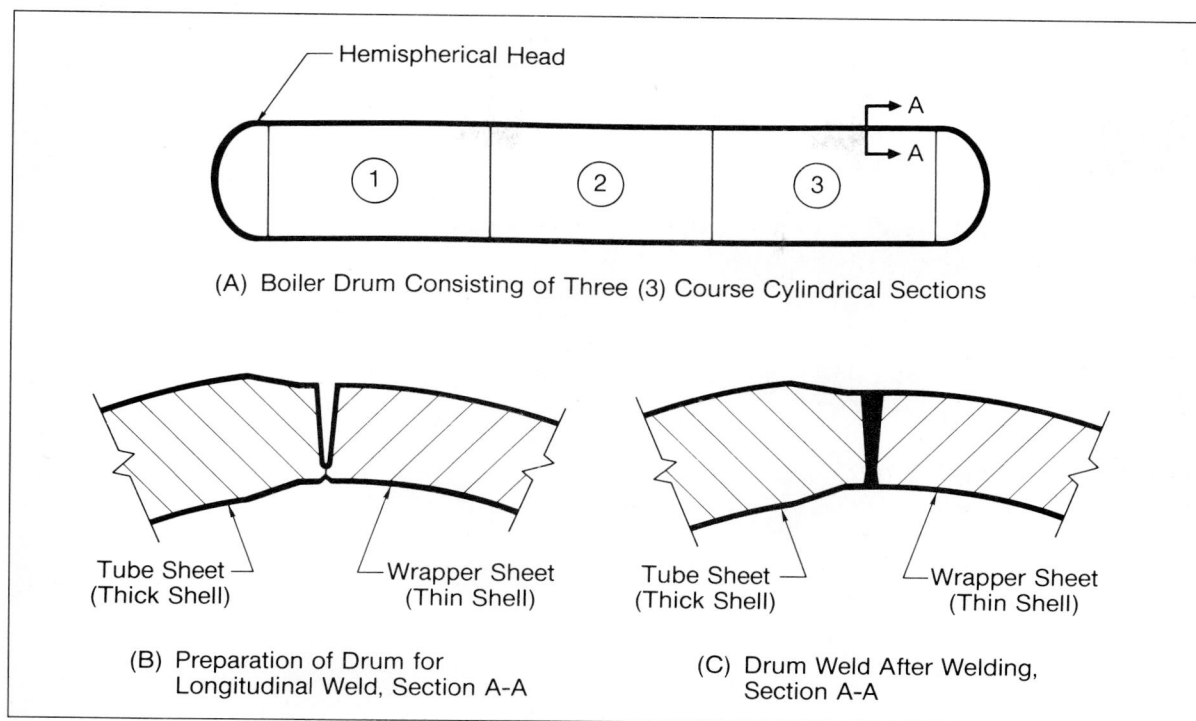

Fig. 1. **Details of a large boiler steam drum**

Fig. 2. **The internals separate water from steam, direct the steam to the outlets leading to the superheater, and provide an exit for the water to be recirculated through the furnace-wall system.**

Fig. 3. Ultrasonic tests are used to inspect plate material. Basically, this form of nondestructive testing uses ultrasonic sound waves to detect flaws in materials and welds.

Fig. 4 Mammoth presses can bend and form steel plate 20 feet wide and 15 inches thick. Here a 6,000-ton hydraulic-beam press is used for cold- and hot-working of plates.

specified pressure and temperature as well as the necessary additional reinforcement to compensate for the weakening effect of the multiple holes drilled for the nipple attachments. The wrapper sheet is designed to withstand the same pressure and temperature parameters but, where openings are required for nozzles, the extra strength or reinforcement is provided by extra thickness in the nozzle forging. This type of steam-drum design provides the most economical use of materials. The internals are welded and installed as shown in Fig. 2. These internals separate water from steam, direct the steam to the outlets leading to the superheater, and provide an exit for the water to be recirculated through the furnace-wall system.

The Drum Plates

The drum plate, which is carbon steel conforming to *ASME* Specification SA 299 or SA 515-70 (both carbon steels), is first ultrasonically tested to assure required quality through the absence of large laminations or seams

(steel-making defects). The ultrasonic test is conducted using a 2.25-megahertz longitudinal wave directed from one plate surface to the other. In this manner, the entire volume of a 12-inch band completely around the plate and a 24-inch-wide strip centered on the longitudinal axis is tested. Fig. 3 shows the ultrasonic-testing equipment being used for examination of a thick drum plate. The forming operation begins only after the plate soundness has been assured by such ultrasonic tests and following a review of the test reports received from the steel supplier.

The plates are heated in a furnace at 1650°F, and each plate is then moved to the 6,000-ton hydraulic-beam press where it is incrementally hot-worked into the required shell-course segment based on the nominal inside diameter of the drum. When the bent plate has cooled to below 300°F, but not less than 72°F, it is then cold pressed to a specified, more precise contour than is possible to attain during the hot-pressing operation (Fig. 4).

Fig. 5. Each of the longitudinal edges of the bent plate is machine weld-beveled.

The next step is to machine the weld bevel on each of the longitudinal edges as shown in Fig. 5. To make up a cylinder or shell course, a series of tack welds join one bent plate with its slightly thinner (wrapper) counterpart. The longitudinal seams are welded using the automatic tandem submerged-arc process to deposit a succession of thin layers from the outside, all with a preheat of 300°F. In the welding process, a blanket of granular fusible material completely covers and shields the welding zone. Although referred to as flux, the material performs functions in addition to those of flux. Visibly, there is no evidence of current passing between the welding electrode and the component being welded. The electrode is not in actual contact with the component. The welding current is carried across the gap through the molten flux which is supplied automatically along the seam to be welded. The entire welding action takes place beneath the flux without sparks, spatter, smoke or flash. When the outer welding is completed, the inner side of the weld is back-gouged to sound metal using the electric arc-air process. Next, the weld is finished on the inside using the automatic submerged-arc process, Fig. 6. As the outside portion of each weld is applied, about 20 percent of the weld is deposited in one seam. The shell is then rotated so that a similar amount of weld can be deposited, with the required preheat, in the other seam; such a procedure helps to control distortion.

Each cylindrical course is fabricated in the same way. After welding, both the inside and outside of each longitudinal seam are ground flush and magnetic-particle inspected. This operation is followed by radiography using a 4-MeV (million electron volts) linear accelerator, Fig. 7. If necessary, repairs are made using low-hydrogen covered electrodes at 300°F preheat. Repairs are re-radiographed until the welding is completely satisfactory.

Drilling for Nipples and Nozzles

Next, each course is drilled for the nipples: first, the holes for nozzles are oxy-gas cut; then, both nozzles and nipples are welded into place using manual metal-arc welding with low-hydrogen covered electrodes.

Joining the Courses

A three-course drum, as we have described, is assembled by first joining two of the courses with a girth seam. Following this, these two joined courses and the remaining course each have a hemispherical head welded to one end. The circumferential welds are ground, magnetic-particle inspected, and radiographed. When the welds have passed inspection, each of the two assemblies is stress-relieved at 1150°F ± 25°F. After the internals are put into place, a final girth weld is made; this is locally stress-relieved, followed by magnetic-particle testing of all welds. Fig. 8 shows a typical steam drum ready for shipment.

HEADERS

Headers may be considered small-diameter versions of the steam drum just described, except that headers do not have steam-separation equipment in them. They collect water or steam from a group of tubes assembled either in panel form or as a group of individual tubes. Headers are made from plate-formed pipe, seamless pipe, or centrifugally cast pipe. The materials usually employed are low-carbon steel, carbon-molybdenum steel and three types of chromium-molybdenum steel.

Fabricating the cylinder or shell from plate for headers differs somewhat from fabricating the drum shell course. To make a "plateformed" header, the plate, after the required inspection, is cut to a predetermined size and the long edges of the plate are bent in a 2,000-ton hydraulic pull-down forming press. This bending operation may be done hot or cold, depending upon the plate thickness and subsequent shell diameter.

After the plate edges have been bent and (depending upon the material) the plate has been heated in a furnace to between 1600 and 1750°F, the plate is pressed into a U-shape and then into a cylinder, slightly larger than the required size. The two longitudinal edges are trimmed and the cylinder is again heated and

Fig. 6. An automatic submerged-arc welder finishes the inside long-seam of the cylindrical course.

Fig. 7. Each cylindrical course is inspected by radiography. Any subsequent necessary repairs are re-radiographed until the welding is satisfactory.

Fig. 8 A typical steam drum is readied for shipment. High-pressure steam drums range in length to over 100 feet and weigh more than 500,000 lbs. Shipment of these drums requires utilization of multiple rail cars.

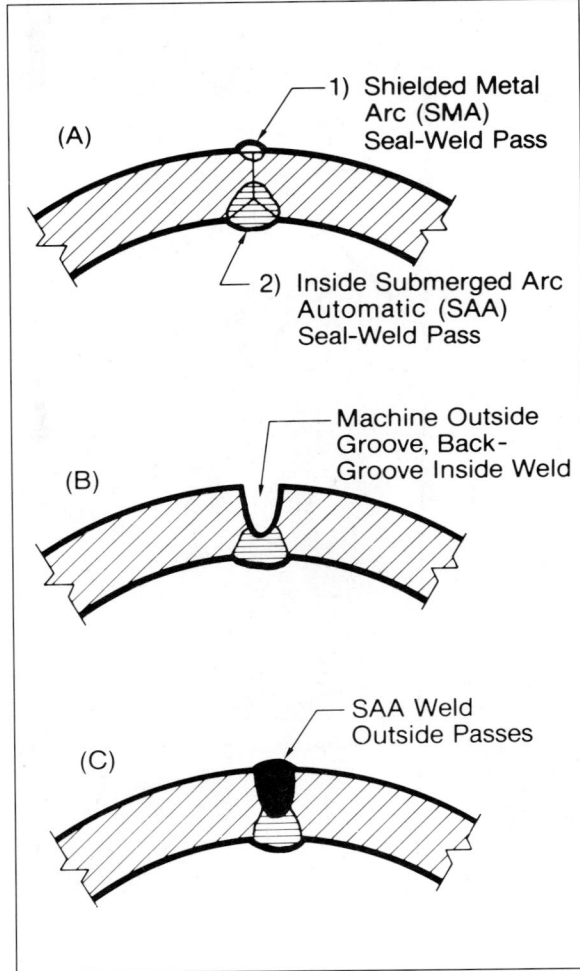

Fig. 9. Header longitudinal welding sequence.

set in dies in a 2,000-ton hydraulic four-post press for final closing and sizing to specified outside diameter. (The long seam has yet to be welded.) The formed shell is then cleaned by grit-blasting and the long seam is fitted-up by depositing a continuous manual weld outside. Next, the shell is moved to the internal longitudinal seam welder, where the inside is finish-welded under the specified preheat conditions, Fig. 9A.

Machining the Shell

The longitudinal seam of the shell is then machined to produce a welding groove on the outside. During this machining, the manually deposited seal weld is removed and the inside automatic weld is machined to clean, sound metal, Fig. 9B. Later, with the specified preheat, the outside portion of the longitudinal seam is completed using the automatic submerged-arc process, Fig. 9C.

Upon completion of the inspection and upon acceptance of the weld quality, the cylinder is machined inside and outside to final required dimensions.

Depending upon the length of the finished header, cylinders and fittings are welded together and the ends may be closed in any one of three ways, by:

1. Butt-welding an elliptical or hemispherical cap on each end,

2. Welding a flat plate into each end of the cylinder assembly,

Fig. 10 Drilling and checking of tube holes in a pressure vessel

3. Forging each end of the cylinder closed (an operation called "wobbling").

The holes for nozzles and nipples are drilled in the headers (Fig. 10) and then the nipples of the desired shape are welded into the sockets previously machined during the drilling operation. After all welding has been completed, the assembly is stress-relieved. Fig. 11 shows a completed header with nipples, ready for rail-shipment.

FURNACE WALLS

C-E furnace walls are constructed of panels made of tubes joined by fusion welding. This construction provides a gas-tight "box" which contains the burning fuel and hot gases. Generally, panels are built up with four tubes, each 75-feet long, welded together simultaneously; single tubes and four-tube subassemblies are added successively to make up panels about 5-feet wide. Details of this fabrication and of the panel-processing welding machines (PPM) are shown in Figs. 12 and 13.

Fusion-Panel Welding

In the PPM, three welding arcs simultaneously deposit molten metal which solidifies to

Fig. 11. Completed header for large boiler is securely loaded on a rail car for delivery to utility destination.

form the web between the tubes. Optimum web widths are in the ⅜ up to ⅝ in. range, and the welder can fabricate panels of ⅞ to 3-½ in. diameter tubes. The machine is capable of welding carbon and alloy tubing at rates up to about two feet per minute.

The fusion panel welder shown consists of a frame, three welding heads, tube drive, pressure rollers, copper chill bars, flux recovery and dispensing unit, filler-wire feed, tube feed-in rack and conveyor, and feed-out and storage racks.

The feed-in rack and conveyor position tubes and panel segments against a gate which aligns the tube ends. The alignment gate is then lowered and the tube drive activated. Held in alignment by contoured pressure and drive rollers, the tubes and panel segments pass under the three welding electrodes where weld wire and filler wire are fed into each arc. A water-cooled copper chill bar, under each welding arc and between the tubes, supports the individual weld puddles until they solidify to form the webs between the tube. At any time

during the welding cycle, the operator may stop or start one, two, or all three welding heads.

After the three webs are welded, the panel moves under automatic scaling hammers which remove the fused flux from the top of the weld. The panel is then discharged from the machine onto the feed-out racks where it is mechanically transferred to storage racks.

Provision for Openings in Tube Walls

It is necessary to provide openings in the tubular walls for sootblowers, insertion of instruments, and access. A special machine called an integral offset-bender makes small openings equal in width to one or two tube diameters. This machine actually cold-distorts tubes after the fusion-weld web joint between

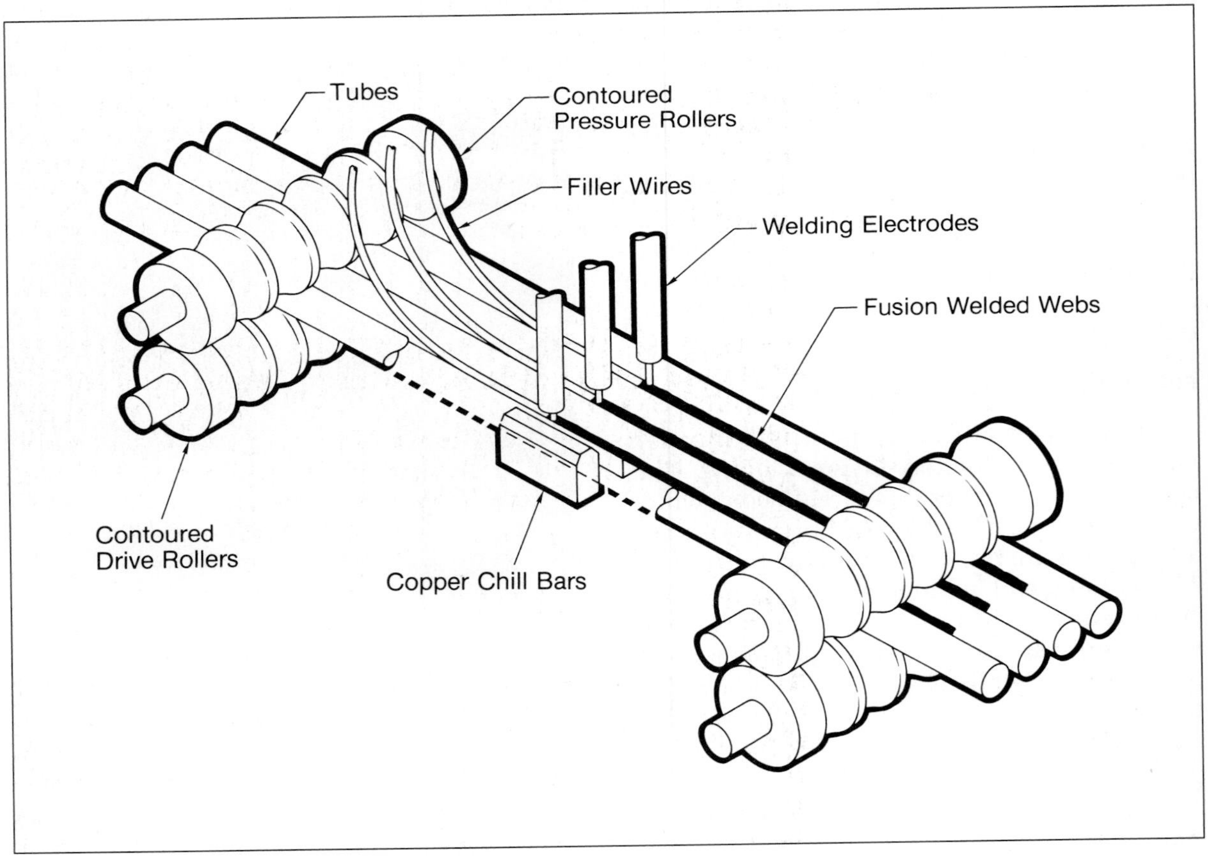

Fig. 12. **Schematic arrangement of fusion panel welder.**

Fig. 13. In panel welding, three welding arcs simultaneously deposit molten metal which solidifies to form the web between the tubes. The machine can weld carbon and alloy tubing at rates up to about two feet per minute.

Fig. 14. An integral offset-bender makes small openings by cold-distorting tubes after the fusion-weld web joint between two tubes is cut.

two tubes is cut, Fig. 14. This pressing-form technique has reduced the use of hand-welded inserts that were a potential source of leaks.

To make larger or wider openings, sections are cut from the panel and then a prefabricated bent-tube assembly is welded back into the opening to permit steam or water flow to continue through the panel but in a path that provides the required opening as shown in Fig. 15. These 75-foot panels, containing attachments for buckstays and other structurals and for attaching insulation, are welded together in place in the field to the required height, width, and depth of the furnace. Generally, the field welds are made by manual shielded-arc welding or tungsten-inert-gas welding, individually or in combination. It should be noted that the welds joining the tubes lengthwise in the panel are intentionally slit in the shop for approximately 18 in. at one end to facilitate the precise alignment of one tube in a panel with its counterpart in the panel above or below, or for joining the tubes to headers or other connecting tubes already in place.

Solid and Peg Fins

Although the fusion-welded panel construction makes a gas-tight furnace enclosure, the space between tubes in the area of the furnace arch and in areas not exposed directly to furnace radiation are filled by either a solid fin or a peg fin. The solid fin is a flat strip of steel welded lengthwise to adjacent tubes where spacing greater than ⅝-inch can be tolerated. The fins are joined to two tubes at one time by a machine having four separate automatic submerged-arc welders. The fins are welded on one side, then turned over; the back side of each previously made weld is again welded to provide a complete penetration weld. Peg fins, carbon steel pieces approximately 1½ in. wide with varying height, are electric-flash-welded to the outside of the tube. The "pegs" provide heat transfer to the tube and fill the space between adjacent tubes to retain refractory.

Rather than by bending individual tubes which are then welded together, welded-wall panels that make up the furnace arch, bottom, and roof section of the boiler are bent after they are formed into panels. Fig. 16 shows how these panel sections are mounted and bent on the shop floor during fabrication.

SUPERHEATERS, REHEATERS, AND ECONOMIZERS

These assemblies are made up of carbon steel, low-alloy steel, and nickel-chromium stainless-steel tubing. Material choice depends upon gas and steam temperatures to which the tubing is exposed; it is selected in accordance with *ASME* Code rules and the boiler manufacturer's design standards.

Shop Fabrication of Superheaters

To use materials effectively and economically, tubes up to 40-feet in length are joined by automatic metal inert-gas arc-welding to form a single tube up to 150-feet long, Fig. 17. The butt welds on this assembly are examined by fluoroscopy and, when acceptable, the tube enters an automatic bending line. Controlled by a previously prepared tape, this equipment bends the long tube at the required location (Fig. 18) and can be programmed to make either cold or hot bends. A superheater or reheater *section* is an assembly of such tubes, each bent to a particular pattern. C-E uses an electric induction pressure-welding process to form butt welds on the long prebent tubes to form element assemblies, Fig. 19. The induction pressure welds are inspected by ultrasonic testing.

In continuous-fin construction, two strips of steel are welded longitudinally to the tube on diametrically opposite sides. Fig. 20 shows the fin detail and the fin welding operation.

Shop Optimization

To save time and labor at the jobsite, panels such as superheaters and reheaters are joined to their respective headers in the shop prior to shipping in a process called "optimization" that is intended to make the most economical assembly for erection upon its arrival at the jobsite. To achieve this, the panels are welded to the tube extensions or nipples on the header by a miniature automatic welding machine that rotates about the two tube ends, the axis of which is horizontal. This is called "orbital" welding; a photo of the orbital welding machine is shown in Fig. 21.

Unquestionably, optimizing shop work on superheater and reheater sections has led to lower forced-outage rates, which directly translates into higher availability. Modules up to 11-feet wide are available that require only field girth welds on the headers themselves. All field welding of tubes to header nipples has been eliminated. Moreover, shipping and erection rigs designed for use with these optimized sections have virtually eliminated any damage during either the shipping or erection stages.

The Economizer Section

The boiler economizer section is another type of tubular assembly, which may have its heating surface extended by fins. The amount of heat transfer and the type of fuel usually dictate the type of heating surface. C-E designs include spiral-finned tubes, continuous-finned tubes, and bare tubes (those without fins). The latter type is fabricated very much like a bare-tube superheater or reheater.

Fig 15 To make larger or wider openings than shown in Fig. 14, sections are cut from the panel. A prefabricated bent-tube assembly is welded back into the opening.

Fig. 16. Entire tube assemblies that make up, say, the furnace bottom or nose section, are bent in panel form rather than by bending individual tubes.

NONDESTRUCTIVE EXAMINATION OF BOILER COMPONENTS

In describing the operations performed in the C-E boiler shops, reference has been made to the several types of nondestructive testing (NDT) for determining the integrity of the fabricated components. By definition, NDT is a means of inspecting an item without destroying its usefulness.

Several factors must be considered prior to selecting an NDT method:

1. Type of imperfection expected to be found: deterioration, cracking, surface, volumetric;

2. Surface condition of areas to be inspected (inside and outside surfaces);

3. Orientation of defects: vertical to wall thickness or horizontal to wall;

4. Permitted severity of imperfections: e.g., 2 percent wall, 5 percent wall, serviceable component;

5. Suspected cause of defect: improper chemical cleaning, water treatment, draft, stress-associated, improper welding.

Once the above conditions have been evaluated, the method of inspection can be determined. Table I lists the most universally used methods. All methods have limits and depend on the ability of the performing inspector. Yet regardless of the inspector's qualifications, *destructive* examination may sometimes be required.

Of the four methods, radiography is the acceptance tool for conformance of pressure parts to *ASME* Code requirements. Ultrasonic, liquid penetrant, and magnetic-particle testing, on the other hand, are used primarily to assure quality in manufacture.

RADIOGRAPHY

Radiographic inspection is the most common volumetric inspection method used in the boiler industry to determine the integrity of welds. Many factors can affect the value of a radiograph as an inspection method; a few are:

Fig. 17. Automatic metal inert-gas arc-welding machines join 40-foot lengths of tubes to form a single tube up to 150-feet long.

Fig. 18. After examination by fluoroscopy, tubes which have passed inspection enter an automatic bending line. Programmable for either hot or cold operation, this equipment bends the long tube at required locations.

Table I. Nondestructive Testing Methods

Method	Definition	Uses	Limitations
Radiographic	Uses electro-magnetic rays (X-rays and gamma rays) to penetrate material, recording on film imperfections in the material.	Used on any metal stock or articles as well as a variety of other materials to detect (and record on film) surface or subsurface imperfections. Film provides a permanent record of the discontinuities.	High initial cost. Requires electrical power source. Potential radiation hazard to personnel.
Ultrasonic	Uses ultrasound to penetrate material, indicating imperfections on an oscilloscope screen.	Used on metal, ceramics, plastics to detect surface and subsurface imperfections. When automated, permanent records are available. Also measures material thickness.	Moderately high initial cost. Interpretation of test results requires highly trained personnel.
Magnetic Particle	Uses electrical current to create a magnetic field in a specimen while magnetic particles indicate where the field is broken by an imperfection.	Used on metal that can be magnetized (ferro-magnetic) to detect surface or subsurface imperfections. Simple to use. Equipment is portable for field testing.	Cannot be used on metal that cannot be magnetized. Requires electrical power.
Liquid Penetrant	Uses a penetrating liquid to seep into a surface imperfection, thus providing a visible indication.	Used on metal, glass, and ceramics to locate surface imperfections. Simple to use and does not require elaborate equipment.	Does not detect imperfections beneath the surface of a specimen.

1. Thickness of the part

2. Radiation emitter physical size

3. Type radiation

4. Radiographic film selection

5. The distance the radiation source is placed from the subject

6. Quality of photographic processing

7. The ability and experience of the person evaluating the radiograph

8. Nature of flaw and degree of acceptance criteria

A radiograph, by definition, is a shadow picture produced by the passage of X-ray or gamma rays through an object onto a film (Fig. 22). Normally, the radiation is produced by one of two methods:

1. Man-made radioactive isotopes, produced by placing an element into a nuclear reactor and bombarding it with radiation; for example, the stable element iridium 191 becomes radioactive iridium 192 after bombardment by radiation in a reactor.

2. X-rays are produced by the sudden stopping of high-speed electrons in a vacuum tube.

ULTRASONIC TESTING

Ultrasonic testing (UT) is based on comparisons of the times it takes ultrasonic waves to travel through materials. After the test operator

Fig. 19. To form element assemblies, an induction pressure welder makes butt welds on the prebent tubes. As part of standard quality assurance procedure, ultrasonic tests check welds for flaws.

Fig. 20. In continuous-fin construction, two strips of steel are welded longitudinally to the tube on diametrically opposite sides.

Fig. 21. Miniature orbital welding machines join superheater elements to header nipples to make optimized superheater modules.

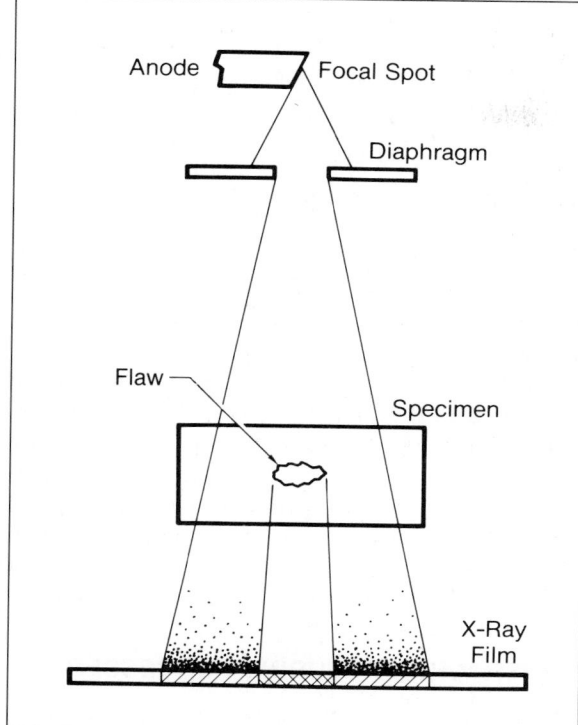

Fig. 22. Schematic diagram showing the fundamentals of a radiographic exposure. The dark region of the film represents the more penetrable part of the object; the light regions, the more opaque. (*From Radiography in Modern Industry,* published by Radiography Markets Division, Eastman Kodak Company.)

establishes the time for an ultrasonic wave to pass through a good specimen of test material, flaws are detected by deviations from the standard time pattern. In Fig. 23, Time A corresponds to the wave entry into the material. Time B corresponds to wave reflection at the extremity of the test specimen. If a void within the component interrupts the sound wave, a portion of the sound is reflected back to the transmitter (transducer) as a mechanical signal. A piezoelectric element in the transducer changes the signal from mechanical to electrical. The electrical signal created by the element is then processed through the instrument and eventually will be produced as an electronic spike on the CRT of the ultrasonic instrument. In Fig. 23, an internal void would cause a wave disturbance to propagate to the surface (and the transducer) sooner than waves reflecting from the extremity at Time B. The electronic spike would show up on the CRT between Times A and B.

Ultrasonic testing is used for volumetric examination of welds and base materials for flaws. When inspecting materials, the sound wave is introduced into the part either normal to the surface or at predetermined angles, providing what is known as angle-beam inspection.

Many factors affect the results of the ultrasonic inspection:

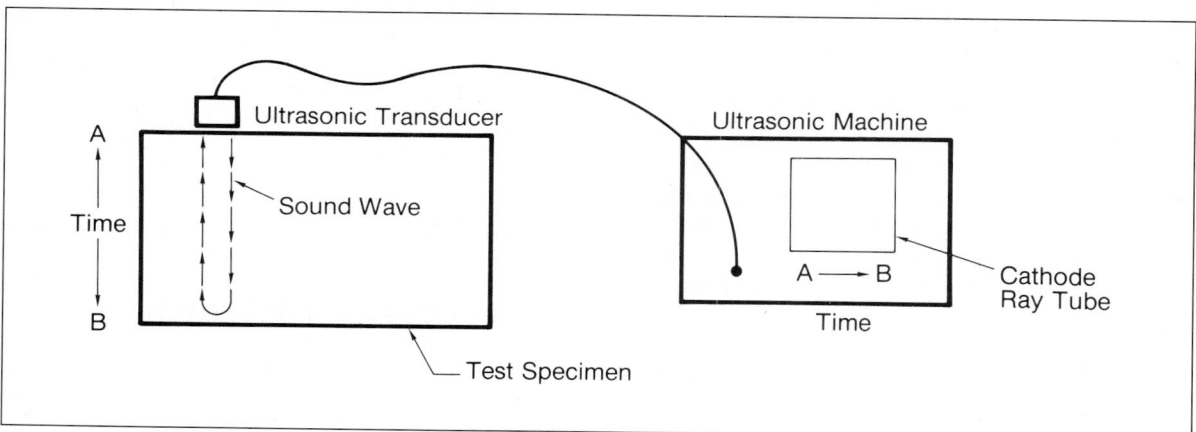

Fig. 23. An ultrasonic transducer changes electrical energy to mechanical energy and mechanical energy to electrical energy.

1. Selection of calibration reflectors
2. Ultrasound transmitter/receiver selection
3. Material composition
4. Surface condition (smooth vs. pitted)
5. Choice of equipment
6. Ability of the inspector

In addition to ultrasonic testing of superheater and reheater tubes and electric-resistance-welded boiler tubes, ultrasonic tests are performed on plate material over four inches thick, induction pressure welds, base material in lifting-lug areas, and nozzle welds (which are tested on a sample basis).

MAGNETIC-PARTICLE TESTING

Magnetic-particle testing is a dry-powder, magnetic-field technique to detect surface defects in

■ welds of fusion-welded panels

■ butt joints on plates 1¼ inch or less in thickness

■ back grooves of multipass submerged-arc welding

■ machined longitudinal weld preparations of plate-fabricated pipe

■ longitudinal and circumferential drum welds after interstage and following final heat treatment

■ nozzle welds, and

■ internal and external attachments

LIQUID-PENETRANT TESTING

Liquid-penetrant testing is a widely used non-destructive test applied to nonporous materials. Its success depends on the ability of a penetrating liquid to enter a surface opening by capillary action, and remain in the opening after the surplus penetrant on the surface is removed. A developer is then used to draw the penetrant out of the opening, back to the surface of the nonporous material.

The visibility of surface imperfections to the operator is increased by the addition of a dye, which may be either a visible dye which can be viewed in normal white light or a fluorescent dye that must be viewed under black light.

In contrast to the magnetic-particle testing method, which normally is used on ferrous materials, the liquid-penetrant method is primarily applied to austenitic stainless-steel and other non-magnetic materials. In providing assistance to manufacturing and quality-control people in a boiler shop, its major field of use is in giving indications of the length of imperfections that are open to the tested surfaces of such materials.

Among the present uses of liquid-penetrant testing are (a) checking of austenitic weld deposits between composite austenitic-ferritic tubes in furnace-wall panels; (b) checking of induction pressure welds where stainless steel is used on one side or both sides of the joint; and (c) as a check method for classification of ghost indications that have been detected by magnetic-particle testing.

OTHER MANUFACTURING PROCESSES

Many aspects of the fabricating process have been omitted in this brief description of a typical steam-generator supplier's pressure-part manufacturing facilities. Details of many types of welding, of heat-treating methods of welded assemblies, and of the inspection and quality-control procedures must, for example, be treated as separate subjects outside the scope of this text. Particularly lacking is reference to the proprietary systems of computer-aided design and fully computer-integrated manufacturing that are being installed by major manufacturers.

REFERENCES AND FOOTNOTES

[1] An authoritative work on this subject is the *Welding Handbook* issued in several volumes and revised periodically by the American Welding Society, United Engineering Center, New York.

CHAPTER 19

Field Construction of Steam-Generating Equipment

This chapter describes the field-construction activities associated with the erection of a coal-burning steam generator and some of the other associated power-plant equipment. The information is of a general nature and does not reflect the detailed design and construction of a specific unit. Rather, it represents common features of many of the large utility units being installed throughout the world. Fig. 1 shows such a large Controlled Circulation coal-fired unit as will be referenced repeatedly in this chapter.

PLANNING AND ORGANIZING FOR FIELD CONSTRUCTION

An electric utility or large industrial organization will prepare the design criteria and scope documents, the detailed engineering specifications, and other technical data needed for a complete plant. This work is accomplished either by a consulting-engineering firm retained to do the architectural and engineering design, or by in-house personnel performing all or part of it.

On most projects, many different vendors supply equipment and services; some of them do their own erection on the plant site. As a result, the owner or engineer issues many specifications to the various suppliers of the plant components—the steam generator, turbine-generator, condenser, emission-control equipment, ash-handling equipment, fans, pumps, piping, and valves. After equipment selection and purchase, the owner frequently will have the engineer-consultant continue as the constructor, supervising the several contractors and subcontractors.

Some of the crafts involved in the erection work are boilermakers, ironworkers, pipefitters, millwrights, electricians, insulators, operating engineers, sheet metal workers, carpenters, teamsters and laborers. These personnel are hired at the local jobsite at the beginning of the project or as the work progresses and are released once the portion of work requiring their craft is completed.

The erection of a utility steam generator is a major portion of the work in the total plant. In many instances, the company that supplies the boiler also constructs it. Early in the project, the erecting company sets up a field organization headed by a construction manager who is experienced and knowledgeable in boiler erection and who holds the authority to make decisions necessary to the operation. The construction manager's staff consists of construction and welding supervisors, field engineers, planning/scheduling and quality-control

19–1

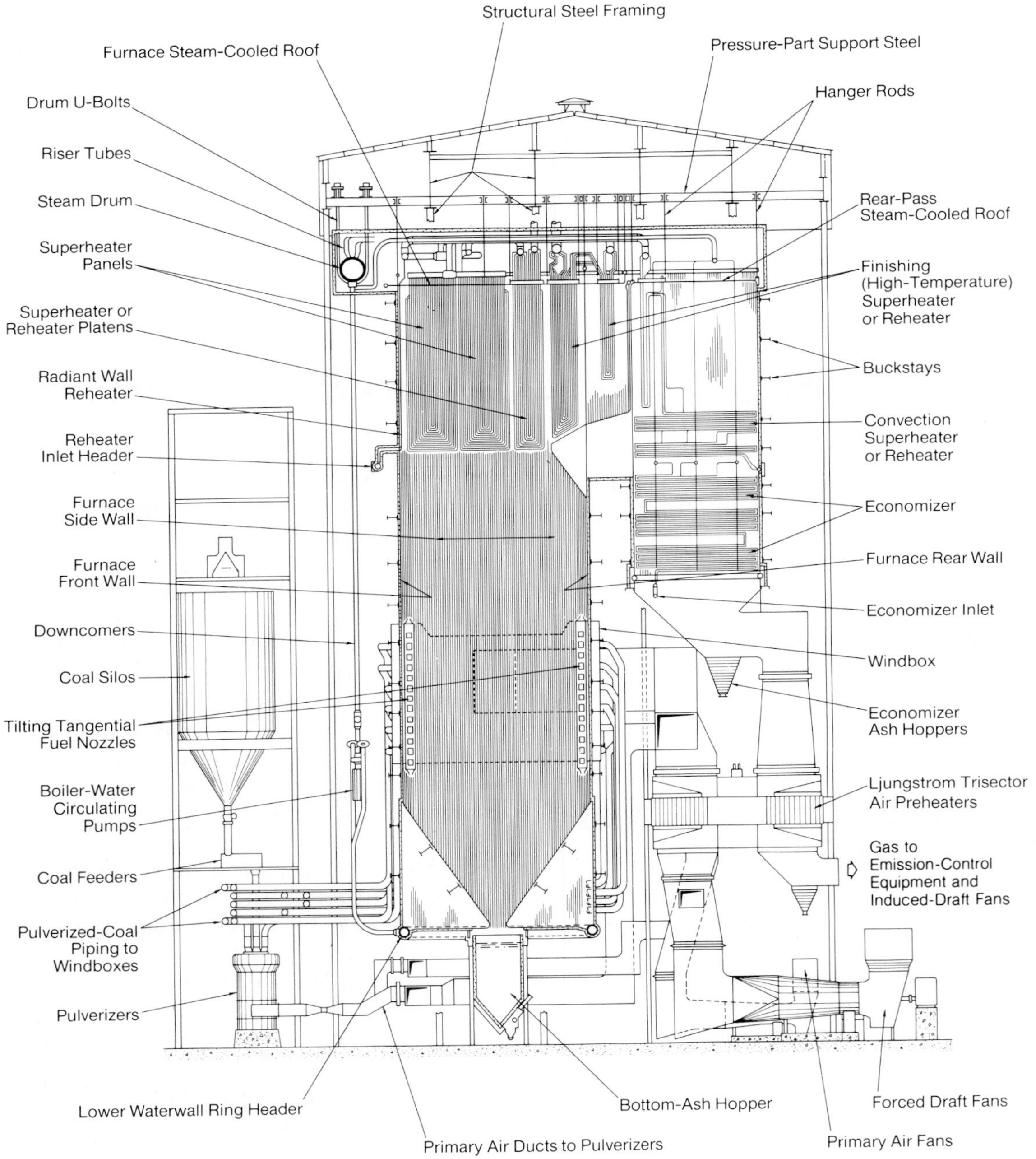

Structural Steel Framing

Furnace Steam-Cooled Roof

Pressure-Part Support Steel

Drum U-Bolts

Hanger Rods

Riser Tubes

Steam Drum

Rear-Pass
Steam-Cooled Roof

Superheater
Panels

Finishing
(High-Temperature)
Superheater
or Reheater

Superheater or
Reheater Platens

Buckstays

Radiant Wall
Reheater

Convection
Superheater
or Reheater

Reheater
Inlet Header

Economizer

Furnace
Side Wall

Furnace Rear Wall

Furnace
Front Wall

Economizer Inlet

Downcomers

Windbox

Coal Silos

Economizer
Ash Hoppers

Tilting Tangential
Fuel Nozzles

Ljungstrom Trisector
Air Preheaters

Boiler-Water
Circulating
Pumps

Gas to
Emission-Control
Equipment and
Induced-Draft Fans

Coal Feeders

Pulverized-Coal
Piping to
Windboxes

Pulverizers

Lower Waterwall Ring Header

Bottom-Ash Hopper

Forced Draft Fans

Primary Air Ducts to Pulverizers

Primary Air Fans

Fig. 1. Side elevation of large coal-fired steam generator for high-subcritical-pressure operation

engineers, field accountants, and clerical personnel. Although all construction personnel are charged with maintaining a safe operation, a safety engineer may also be employed as well as a labor-relations specialist, if deemed necessary by the contractor.

It is the responsibility of this organization to plan and schedule the work in accordance with the overall project requirements; to perform the work as designed in a safe and timely fashion, assuring an adequate flow of materials, manpower, and money to avoid any delay; to verify the attainment of specification and code requirements; and to provide the documents and reports necessary to monitor the work.

FIELD CONSTRUCTION OF A BOILER

One of the first activities of the field-construction phase is the verification of the accuracy of all concrete foundations for the boiler. All column-to-column dimensions, anchor-bolt locations, and concrete elevations are measured and verified against the drawings.

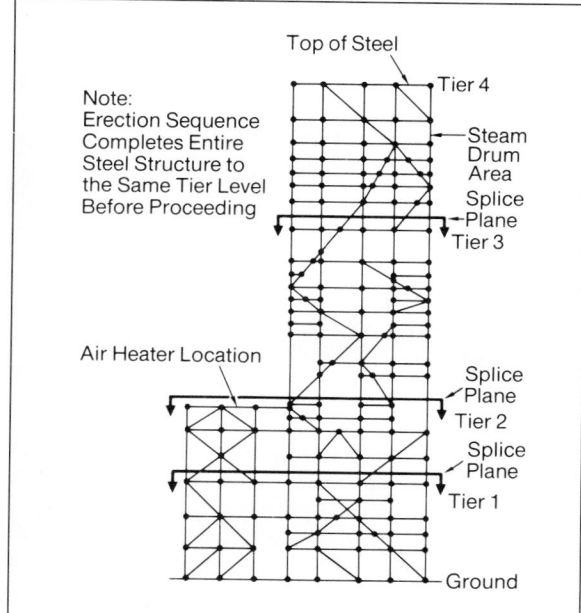

Fig. 2. **Tier method of steel erection**

STRUCTURAL STEEL ERECTION

Since a boiler structure is sometimes 250 feet high, very large cranes are needed for erection. Main cranes with boom heights of 280 ft and lifting capacities of 250 tons or more are common. One crew of ironworkers can erect about 25 tons of steel a day with one such crane.

Because of the large quantity of boiler steel (often greater than 4000 tons for a 600-MW unit), boiler components, and other plant equipment and materials, it is usually impossible or uneconomical to store all the boiler items within reach of the main crane. A multi-acre storage yard and adequate warehouse storage are necessary for the material; such additional equipment as lighter capacity cranes, cherry pickers (small, rubber-tired mobile cranes), and flatbed trucks are required for moving the material to and from the storage areas.

There are basically two ways of erecting boiler steel. One is the tier method, shown in Fig. 2. The other is the cube method. In the tier method, each elevation is completed to the splice point before another elevation is started. In the cube method, sections of the steel are built up to different elevations starting from the turbine deck and progressing back toward the stack. The structure is built in a horseshoe shape, leaving an open area for the crane to back out as work is completed.

The large sizes of some of the boiler components, such as the ductwork, deaerator tank, and sootblowers, require that these components be installed as the building structure is erected. For this reason, close coordination between the material suppliers and the constructor is essential.

It is desirable in many instances for the boiler erector also to erect the boiler support steel. This facilitates the planning of erection sequences, and firm dates for finishing elevations can be planned with a minimum of coordination problems or schedule disruptions. If the steam-generator supplier does not erect the structural steel, it is important that the supplier be consulted so that steel erection and boiler-

Fig. 3 **Erection sequence of a main boiler girder: 3A, main girder being lifted off the ground; 3B, main girder at an angle to the steel; 3C, main girder being set on the support columns; 3D, main girder in place, ready for tie-in steel**

components installation are coordinated, to preclude structural-steel blocking of the component installation.

The steel erection begins with the main boiler columns, usually starting on the turbine and steam-drum sides of the unit. Then the main girders are erected. They support the main boiler sections and the steam drum from the top of the unit. There are usually 5 or 6 main girders spaced about 24 feet apart. Girder sizes range from 6 to 20 feet high, 100 feet long, with weights approaching 160 tons. For structural members of these sizes, shipping to the jobsite becomes a problem, since railroads have height and weight limitations.

MAIN-GIRDER ERECTION

Main-girder erection is one of many rigging problems of the structural-steel erection process. The most accepted method of lifting the heavier girders is to use two cranes in the furnace cavity. Lifting lugs are bolted to the top flange of the girder to attach it to the crane's load blocks. Fig. 3A shows such a girder raised off the ground with the two cranes. Because the girder rests on top of the structure columns, its total length is longer than the open area between the columns. This requires the girder to go up at an angle to fit between the steel. This can be seen in Fig. 3B. Once the girder is raised above the column tops, one of the cranes moves its boom forward while the other goes backward to bring the girder parallel to the front of the boiler. The girder can then be set down on the column tops as seen in Fig. 3C. At this time, the top flange of the girder is very close to the tubular framing on the boom. If the girders begin to swing and touch the boom, the boom could collapse. Once the girder is in place, (Fig. 3D), intermediate steel can be installed to tie all of the upper steel together. Coordinated radio communication, highly experienced operating engineers and rigging supervisors, and first-class equipment are imperative for successful girder erection.

STEAM-DRUM INSTALLATION

The steam drum can be raised into position at the top of the boiler once the main girder steel has been erected. The drum, shown in Fig. 1, is hung just below the main girder steel by means of huge U-bolts that encircle the drum. Steam drums currently being used may be 7 feet in outside diameter, over 100 feet long, and weigh as much as 300 tons.

RAISING A DRUM

As shown in Fig. 4, the drum is often raised at an angle to the ground, because the bay may not accommodate the full drum length. Based on the lifting lugs welded on the drum in the shop, the location of the support steel for the load blocks can be determined before construction is started. A set of "cat heads" (a temporary support for the upper portion of the block-and-tackle arrangement which is used to raise the load) rests on top of the support steel, through which the upper load block is pinned. The block can rotate according to the angle of the drum. The rigging usually consists of 1-inch cable with up to 17 parts of line on each set of blocks.

The steam drum is usually brought inside the furnace cavity on a railroad car and off-loaded with the lifting blocks. In the absence of a railroad track, the cranes used for steel erection can off-load the drum and "walk it" to the lift area. The line-pull two-drum hoist that is used for lifting is usually located on the ground with the two load lines following a column line up to

Fig. 4. Lift of steam drum in boiler cavity. (Note deaerator tank already in place)

the top blocks hanging in the cat heads. Because of the inner dimensions of the structural steel and the location of the cat heads, an out-haul such as that shown in Fig. 4 may be used to keep the drum from drifting into the steel work. Proper location of all rigging is important because of the height of the boiler and the lifting distance, which can be well over 200 feet.

"DRIFTING" A DRUM

On some jobs, it is advantageous to erect the platform steel that is under the steam drum along with the boiler structural steel. This can involve steel in excess of 200 tons that can be put up more efficiently at that time.

When steel has been erected such that the drum cannot be lifted vertically from the ground, the steam drum must be raised up in the boiler cavity to a height just below its final elevation and then drifted into the drum-bay area. This will require 2 sets of cat heads and blocks, and an extra set of lugs on the drum. Fig. 5A shows the drum located in the boiler cavity with the first set of rigging attached to the drum. To the left of the drum is the platform steel that the drum will be drifted over. Fig. 5B shows the drum up to the drift elevation with the second set of rigging being attached to the second set of lugs while the first set still supports the load. The temporary planks on the platform steel provide a work area and also support the hanger rods. Fig. 5C shows the load being transferred from the boiler cavity to the drum bay, with the load equally distributed in all four blocks. In Fig. 5D, the drum is being raised to its final elevation, with the second set of rigging supporting the entire load and the first set slacked off.

PRESSURE-PARTS SUPPORT-STEEL ERECTION

While the steam drum is being erected, the pressure-parts support steel (upper region of Fig. 1) can be installed. This steel, located at the top of the unit, supports all of the weight of the boiler. Unlike conventional structures that are built from the ground up, most modern boilers are built from the top down, with the whole boiler suspended by hanger rods from the main girders. This permits the boiler to expand downward as much as 15 inches as it heats up during the start-up of the plant. The correct setting of the pressure-parts support steel is very important because every part of the boiler is located with respect to this steel. It controls all elevations and is continually used as a reference to align the different components as they are erected.

INSTALLATION OF UPPER HEADERS AND LINKS

Next, the upper headers and links are hung just below the pressure-parts support steel. They may include the main steam lines to the turbine, the crossover links from the different stages of the superheat and reheat systems, the economizer lines to the steam drum, and the riser tubes from the waterwall outlet headers.

By the time these components are raised, the crane used to erect the boiler steel has usually been moved to the back of the unit and cannot reach the areas where these components will be located. Therefore, most are raised from the ground inside the furnace cavity using block and tackle (a set of pulleys in a frame). The components are then hung on their respective hanger rods, which support the different boiler components, or are lashed temporarily to the pressure-parts support steel. The sequencing of component installation requires extra care on units where there are multiple riser tubes to the steam drum from the side-wall headers. These pipe sections are usually 6 inches in diameter and as long as 40 feet. They are very flexible and may contain several compound bends.

Other pipes between the various superheater and reheater headers are often 30 inches in diameter, with sections weighing as much as 10 tons. They require proper sequencing because of the difficulty of moving them once they have been positioned between the existing hanger rods. Fig. 6 shows some completed main steam and reheat piping to the turbine. At this point in the installation of the boiler, no pressure-part final welding has been started. All work has involved the placement of equipment.

Fig. 5. Erection sequence of steam drum being drifted into place: 5A, drum in boiler bay with first set of rigging attached; 5B, second set of rigging being attached to steam drum; 5C, drum being drifted over from the boiler bay with the second set of rigging; 5D, drum raised to its final elevation with the second set of rigging

The downcomers from the steam drum to the suction manifold, as shown in Fig. 1 and Fig. 7, are raised into place after the above-mentioned work is completed. Usually arriving on the jobsite in 60-foot sections, downcomers are about 16 inches in diameter and weigh as much as 7 tons per section. Each downcomer requires about 3 sections. Rigging must be located underneath the steam drum itself, because the downcomers are welded directly to the drum nozzles. Clips are welded on the ends so that threaded rods can be inserted to raise or lower the downcomer a few inches for proper positioning for welding.

As soon as the downcomers are in place, boilermaker welders can begin the fit-up and welding procedures that will ultimately join the downcomers to the drum. A joint can be finished in about 1½ days if a two-person crew works on each weld.

SUPERHEATER AND REHEATER MODULES INSTALLATION

The next boiler components to be installed are the superheater and the reheater pendant modules. Fig. 1 shows their locations within the boiler. These modules, which require some

of the most difficult rigging on the boiler, contain sections of tubes bent in complex configurations and welded together on a common section of header. Each module can measure as much as 10 feet square and 55 feet long and weigh up to 70 tons. They arrive on the jobsite partially encased in shipping rigs that are also used to upend the module to a vertical position.

A large crane and a cherry picker are needed to off-load the modules from the railroad car to a flatbed truck, which brings them into the furnace area. A set of 75-ton capacity blocks are rigged above the final position of the module. The lower block is attached to a special spreader beam, shown in Figs. 8A and 8B. The lower end of the module is picked up with a cherry picker, while the main load blocks raise the header end. The flatbed trailer is removed and the main blocks continue to raise the upper end of the module, while the cherry picker lowers the other end, thus orienting the module in a vertical position for final raising. Rotation to the vertical position is accomplished quite easily with specially designed rods between the high-crown end bar of the module and the spreader beam. The cherry picker and shipping frame are then removed and the module is

Fig. 6. **Main steam and reheat piping to the turbine**

Fig. 7. **Downcomer piping from steam drum**

lifted with the 75-ton blocks (Fig. 9).

Because of the thick-wall tubes required to retain high-pressure steam, the superheater modules are much heavier than reheater modules. When a complete section of superheater modules is in place, the header sections are fitted together and prepared for welding. Girth welds made on the header sections connect the different superheater modules. (Fig. 10).

Depending on the location of the modules in the furnace and the structural steel underneath them, it is sometimes necessary to drift a module into place. To transfer the load to its final position, this lift requires two sets of blocks attached to a steel plate pinned to the module. Generally, this procedure is required when the load is raised or "picked" inside the boiler cavity with the final location of the component in the back pass of the boiler.

COMPONENT LIFTING CONSIDERATIONS

During the erection of the boiler components, hundreds of lifts are made inside the furnace cavity. This necessitates changing the location of rigging almost every time a pick is made. It is often helpful to locate a crawler crane inside the furnace area that has the capacity to boom in and out and swing around to change its location. A crawler is very useful and can be used throughout much of the steam-generator construction.

The elevation of the previously erected pendant sections limits the height of the crane. Upfront planning is necessary so sufficient structural steel is left out to allow access for the crane into and out of the furnace. As the crane is being brought in, the boom is laid all the way down, parallel to the ground. A line is dropped down inside the furnace and attached to the end of the boom. The two lower pins on the front boom section are removed and the section is lifted with the line and rotated around the upper two pins. This puts the top section at a 90-degree angle to the rest of the boom. The rig is driven further into the furnace area for the length of another section and two more pins are removed. The second section is raised to the vertical position and the first two pins are replaced. This is repeated until all of the crane is in the furnace cavity with the boom reassem-

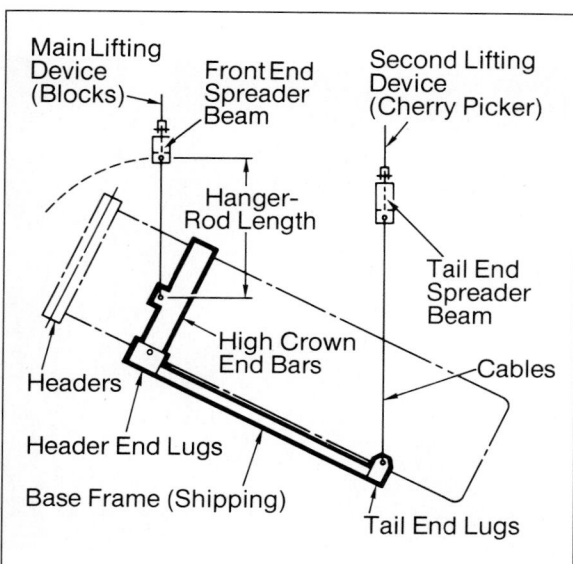

Fig. 8A. **Rigging for superheater and reheater modules**

Fig. 8B. **Raising of superheater and reheater modules**

Fig. 9. Superheater module lift within boiler

Fig. 10. Last module being fit into place. Girth welds will be made between the header sections.

bled in the vertical position. It is removed in the same way. Fig. 11 shows a completed crane within the furnace.

When most of the boiler is completed and there is no more head room available for the crane (about the time that the two lower headers are attached), it is removed.

Another method of lifting inside the furnace is with a guyed derrick located on top of the unit above the main girders. Guyed derricks can rotate 360 degrees on a bull wheel (the lower support for the derrick) and can boom out in any position. A system of guy wires is tied off to boiler steel at different points to support the mast from which the boom is rigged. When the boom is needed on the other side of a guy wire, the boom must be brought up to the vertical position parallel to the mast. Then, by rotating the bull ring, the complete crane is moved to a new position and the boom again lowered.

Other rigs used for lifting in the boiler area are stiff-leg derricks, Sky-horse cranes, ringer and crawler cranes, and travel lifts. All these are used in different ways and in varying capacities.

Some units presently being erected have very tall furnace heights, with top of structural steel in excess of 350 ft, which no conven-

Fig. 11. A completed crane boom within the furnace. Note the waterwall panels enclosing the furnace.

tional crawler crane can reach. These heights sometimes require a "kangaroo" tower crane that sits in the center of the furnace opening and can be used to erect the structural steel and some of the boiler components. The tower rests on a poured concrete base, held down with anchor bolts and tied off to the structural steel in several different places. The tower is a square structural member and, with the crane, can free-stand 160 ft above any tie-offs. The crane is then capable of jacking itself in the air 14 ft and inserting another section of tower for the crane to sit on. Fig. 12A, looking up inside the furnace cavity, shows the tower with its 2 tie-offs to the steel and the crane on top.

Fig. 12B shows the crane after it has erected the main girders and some of the fill-in steel. While the structural steel has been going up, the coal bunkers, ductwork, air heaters and other components have also been erected. The crane has 160 ft of boom and can reach almost all sections of the boiler area. With the crane located on top of the boiler, most of the steel can be erected in the tier method.

BUCKSTAY ERECTION

Once the primary crane has been installed inside the furnace, one of its first jobs is setting buckstays at different elevations. The buckstays are structural shapes or trusses that restrain movement of the waterwalls caused by fluctuation in furnace pressure (see Fig. 1). Buckstays may be all horizontal or may form a system of large vertical steel members combined with small horizontal members.

On boilers with both types, the main vertical buckstays, which are sometimes 3 feet wide and come to the jobsite in 60-foot sections, are hung from the pressure-parts support steel. Between these are smaller horizontal buckstays that are supported by the vertical buckstays.

SOOTBLOWER ERECTION

A crawler crane raises the long-retractable and furnace-wall sootblowers and sets them on the grating at their correct elevations. The type of fuel to be burned determines the number, type, and location of the blowers. Mounted hor-

Fig. 12A. **Tower section and tie-offs for the "kangaroo" crane**

Fig. 12B. **Crane on top of the unit with the main girders in place**

izontally, they have a round tube or lance that is rotated by a driving motor. The sootblowers clean the walls and elements of the steam generator during operation.

One type of sootblower is the wall blower, which extends into the boiler only a few inches to clean the fireside of the waterwalls. A second type is the long retractable blower, which can be inserted up to a distance of either full boiler width, or one half the width of the boiler, depending upon available space and owner preference. These blowers extend between the sections of the superheater, reheater and economizer to clean the convection heat-absorbing surfaces.

WATERWALL ERECTION

Once the sootblowers have been laid on the grating temporarily, the waterwalls which form the sides of the furnace can be erected. The waterwalls are sections of tubes that are fusion welded together into a gas-tight panel. They conduct a mixture of steam and water upward as heat is applied from the furnace side. The walls extend from the base of the boiler (just above the ash hopper) up to the steam-drum level.

Waterwall panels may come from the shop with the upper headers already attached. When two panels are joined in the field, only one girth weld is then needed on the header. This eliminates the need for numerous field welds between header nipples and the panel tubes. Such shop prefabrication can reduce the construction schedule and costs by eliminating field welds, which are often performed under adverse weather or logistic conditions. Other examples of shop prefabrication of waterwall panels include welding of buckstay stirrups, installing insulation pins, and fitting seal boxes on the observation openings.

Another helpful shop procedure is the application of white paint to each panel to brighten the interior of the boiler during erection. This technique, as shown in Fig. 13, improves working conditions.

Typically about 12 feet wide and 80 feet long, waterwall panels are brought into the furnace on a flatbed truck. Because they are so long and flexible, the biggest problem with these panels is upending them. Two lines are needed. The crane in the hole lifts the top line; a cherry picker, the other, in much the same way the superheater modules are upended.

Sometimes, a special lifting device distributes the weight so the panel will not buckle. This requires that several slots be cut in the welded web between the tubes. A bar welded to a small T-beam is inserted and pinned on the back. This device assists in keeping the panel straight along its width during upending process. When the panel is vertical, it is lifted up to its final position.

The desuperheater (or desuperheaters) is

Fig. 13. **Installation of waterwall panels. Vertical and horizontal buckstays are visible.**

Fig. 14. **Installation of roof tubes within furnace cavity**

raised and put in place at this time. Before the superheated steam goes through the finishing superheater assemblies, it passes through the desuperheater. Through a system of spray nozzles that inject water into the steam, the desuperheater controls the temperature of the steam delivered to the turbine.

ROOF-TUBE ERECTION

After the waterwalls are in place, the roof tubes, see Figs. 1 and 14, can be erected usually in either of two ways. One approach uses single tubes that fit between the vertical elements. In other areas with larger openings, full roof-tube panels can be raised from the ground with a set of blocks and a tugger, rather than with a crane.

While the roof tubes are being fitted into place, the intermediate waterwall panels are lifted and fitted below the upper panels. Generally, they are attached to the upper panels by a system of threaded rods and clips that are welded to the welded webs between the panel tubes. When this temporary hanging system is used, final panel spacing for weld-out can be done with the threaded rods.

Fig. 15. **The structural framing between the furnace "nose" and the vertical front wall of the convection pass of a large high-pressure steam generator**

SUPPLEMENTARY FRAMING

Fig. 15 shows the structural framing between the furnace "nose" and the vertical front wall of the convection pass of a large high-pressure steam generator. All of the steelwork shown is supported by the pressure parts; both the pressure parts and these structural members have to be designed to withstand the loads imposed, as well as both positive and negative gas pressures as dictated by codes and owner specifications. The horizontal tubing in the top of the picture forms the floor of the gas pass directly behind the furnace nose.

AIR PREHEATER ERECTION

While the rest of the work is being done on the boiler, the air preheaters, usually two per steam generator, can be erected at the rear of the unit. (See Fig. 1 for their locations.) The air preheaters are heat exchangers made of heat-absorbing metal baskets assembled on a rotor that is turned by a drive. Hot flue gases leaving the boiler pass through one side of the heater. Incoming cold air from the forced-draft fans passes in counterflow through the other side. The baskets are heated by the hot flue gases and rotate into the cold air stream to raise the temperature of the incoming air. Various seals prevent the gas and air flows from mixing.

The air preheater is most often oriented with a vertical shaft. Its main parts are the bearing, bearing supports, baskets, oil circulating system, drive motor, seals, casing and insulation. On some projects, it is economical to fabricate the main sections of the preheater on the ground and raise them into place in the structural framework with a crane from behind the unit. This method is very satisfactory if a large capacity crane, such as the one shown in Fig. 16, is available. The extremely accurate measurements required are performed by millwrights. These include setting the clearances between the rotor and seals, aligning the bearing and support structure and setting the rotor motor and roller pins. Boilermakers erect the main housing and raise the air preheater into place. They also level and set the preheater in its final position in the boiler, as shown in Fig. 17. Once the preheater rotor and other parts are assembled, the heat-absorbing baskets are raised and moved into place.

ECONOMIZER ERECTION

While the air preheater is being erected, the economizer, located just above the air preheater (see Fig. 1), can be erected. This is a heat recovery device designed to transfer heat from the exiting flue gases to the incoming boiler feedwater. The economizer consists of an interconnected horizontal tube array. The upper ends of each economizer assembly connect to the outlet header, which is already in place above the roof tubes.

Two types of economizer heating surface are in common use on coal-fired units: one consists of a pair of longitudinal fins located 180 degrees apart along the top and bottom of the horizontal tube; the second is bare tubing, similar to horizontal low-temperature-superheater surface. Spiral-finned economizers are used on oil- and gas-fired units.

Finned economizer tubes require careful handling because of the heat-absorbing fins. They are grouped together in modules of 3 to 8 elements, each with saddle supports installed to hold them together. They are raised into place from the ground with a tugger rigged in the area of the outlet headers. Pins and rods usually support the assemblies directly to lugs welded to the bottom of the intermediate headers. Because the economizer tubes are often fabricated in a staggered pitch to increase heat transfer and minimize space requirements, fitting the last few elements sometimes presents a rigging problem. It is difficult to pass hoisting falls through the staggered tubes. This often means that the already installed elements must be spread apart and the side walls must be moved out.

WINDBOX-ASSEMBLY ERECTION

After the intermediate waterwall panels are in place, the windbox assemblies, shown in Fig. 18, can be brought into the furnace and positioned for lifting.

The rigging of a furnace windbox assembly is one of the most difficult tasks on the project because of the weight, shape and balance points of the assembly. When the windbox comes from the shop, sections of waterwall tubing are attached to it, with the bends in the tubing prefit around the windbox. These are welded into the already erected wall panels and help support the windbox. Additional hanger rods support the back side of the assembly. The average weight of the assembly is 25 to 30 tons. It is raised in place using the crane in the furnace.

Fig. 16. Assembly of air preheaters on the ground

Fig. 17. Air preheaters in place inside the building

Because the back side of the windbox contains control arms, ductwork and insulated panels, it must be cribbed up on the low-boy truck when it is brought into the furnace cavity.

Special lifting lugs are also installed in the shop to make the windbox assembly hang as close to plumb as possible while it is being raised. To ensure that the tip of the crane boom does not hit the waterwall panel above, special care must be used when attaching the assembly to the hanger rods.

PULVERIZER ERECTION

Coal feeders, coal pulverizers, and the coal piping to the windbox are usually erected as part of the steam-generator contract.

If shipping schedules permit, coal pulverizers such as those shown in Fig. 1 can be set in place in sections as the structural steel is set. On a large unit, there are usually 6 to 10 coal pulverizers which are located on both sides or in front of the boiler at ground level. Generally,

one pulverizer on each side is left out until most of the required components have been moved into the furnace cavity.

A typical bowl mill (Fig. 19) contains about 6 sections, the first of which is the poured concrete foundation with four large anchor bolts. The mill base, which contains the gear case on many mills, is set on top of the foundation. The mill side assembly, where the grinding bowl is located, is installed on top of this. The separator body, which contains the 3 roller journals and covers, goes on top of this. The separator, where the coal is classified cyclonically to the desired fineness, is next. The very top of the mill is the multiple-port discharge assembly. Here, coal is fed into the mill from a central feeder pipe and pulverized coal is discharged to the coal piping through the peripheral ports.

Mills weigh up to 150 tons and are up to 30 feet high. The crane that erects the structural steel can easily set the heavy sections of each

Fig. 18. Windbox assemblies erected in place

Fig. 19. Erection of bowl mills

mill. The lighter internals can be installed later. If the use of a crane is impractical, the different mill parts can be rolled into place on a track and jacked into position. Ironworkers and millwrights do most of the construction work on the mills, with the latter doing all of the alignment and settings because of the close tolerances that are required.

Simultaneously with the work on the bowl mills, hanging and setting of the coal piping to the furnace can also be progressing. Welded joints and Victaulic couplings join sections of the pipe. Fig. 20 shows the final tightening of a Victaulic coupling on an elbow that connects to a windbox assembly.

SUCTION-MANIFOLD ERECTION

The suction manifold, also shown in Fig. 1, is raised into place next. It is the collecting header that distributes the water from the downcomers to the boiler circulating pumps.

Fig. 20. Fuel pipe connections being made to furnace windbox

The suction manifold is hung from the downcomers, which are welded to the steam drum. The downcomer leads join the top of the manifold. Lugs are provided on the downcomers, (and also on the nipples of the manifold), so that threaded rods can be inserted to hold the manifold after it is lifted. These can also be used to adjust the weld gap for proper fit.

The manifold is usually lifted into place using the crane in the furnace cavity. If the crane is not used, blocks must be rigged under the steam drum with the rigging reaching down the full height of the boiler to hoist the manifold into place.

The boiler circulating pump assemblies are usually raised and temporarily rigged before the manifold is raised. As the lift is only about 8 tons for each pump assembly, the crane in the furnace can be used for this operation.

It is often advantageous to fit a number of pump components together on the ground first. The discharge valves can be welded onto the pump casings (Fig. 21) before erection. The pump casing is a volute-shaped chamber into which the pump impeller fits. The pumps force the water from the manifold through discharge links into the lower waterwall headers. Check valves at the pump outlets prevent water flow from backing up into an idle pump.

Fig. 21. Volute and pump assembly with a discharge valve on each side

A rigging beam can be used to install the electric motors and pump rotors. Millwrights perform the critical alignment of motors and pumps.

DUCTWORK ERECTION

Ductwork erection must be completely integrated with boiler erection to avoid problems in fitting ductwork sections into the structural steel framework.

The boiler ductwork, as indicated in Fig. 1, consists of two main systems, the air-supply system, which includes the primary and secondary air and the furnace-windbox ducts, and the flue-gas system, which includes the backpass ash hopper and ductwork to the air preheater, precipitators, absorbers (if included), induced-draft fans, and stack.

The hot-air ductwork from the air preheater to the windboxes, along with the burner connection duct, contains some of the biggest pieces of equipment on the boiler. Because of the sizes involved, ductwork is generally assembled on the ground (Fig. 22) with sections installed simultaneously with the structural steel. Hanger rods and expansion joints in the ductwork allow for thermal expansion. Some ductwork sections are 20 feet by 30 feet in cross section. At this size, temporary internal stiffeners are often needed to provide strength during shipping and erection.

The gas-system ductwork extends from the bottom of the economizer to the exit side of

***Fig. 22.** Ductwork fabricated on the ground*

the air preheater. It also includes the ductwork from the air preheater to the induced-draft fans, to the gas cleaning equipment, and to the stack. Most of this ductwork is installed after the air preheater is in place, usually with the same crawler crane that set the air preheater.

Dampers to isolate equipment or regulate airflow or gas flow to and from the boiler are installed with the ductwork.

LOWER WATERWALL HEADER ERECTION

Typically, the lower waterwall headers are the last major lifts made inside the boiler and are installed after most of the pressure-part welding has been completed. The headers are up to 3 ft in diameter, extend the full width and depth of the furnace, and weigh as much as 60 tons each. During boiler operation, once the water reaches the suction manifold from the downcomers, it is pumped into these headers and up the waterwalls. Because these headers are so heavy, a large cherry picker is brought to the furnace to help the crane make this final lift. Once headers are in position, they are hung with hanger rods from the waterwall structure.

WELDING OF ERECTED COMPONENTS

Thus far, as the boiler erection has proceeded, the different components have been raised and hung by hanger rods or temporary rigging. After a group of components is installed, the fit-up and welding must be performed. Except for the structural steel, which is most commonly bolted together, almost everything else on the boiler is welded.

Several different welding processes are used in the field. One such process is *gas tungsten arc welding* (GTAW), in which a bare filler rod is fed into the molten weld. The tungsten electrode, which is not consumed, provides heat to the workpiece. An externally supplied shielding gas excludes the atmosphere from the weld puddle. Argon is usually the shielding gas, although other gases or combinations of gases may be used. Historically, this relatively slow process is used to put in the root (first) pass in pressure-part butt welds.

The most extensively used process is *shielded metal arc welding* (SMAW), more commonly referred to as the "stick electrode" process. In this process, the flux-covered electrode is consumed in the weld. The flux coating produces the shielding gas which is externally supplied in the GTAW process; it also promotes electrical conductivity across the arc column, adds slag forming materials that help prevent rapid oxidation of the weld metal and, in many cases, adds alloying materials to the weld. This process follows the GTAW root pass and completes the pressure-part butt weld. It is also used to weld ductwork, waterwall seams, and other plate and pipe components.

The third process frequently used during the field construction of boiler components is the *gas metal arc welding* (GMAW) process. In this process, a machine feeds bare of flux-cored (flux-filled tubular) wire from a spool into the weld. The wire is the consumed electrode as in SMAW. The shielding gas may be externally supplied (as with GTAW) or may be supplied by the flux (as with SMAW) if flux-cored wire is used. This process is also used for ductwork, waterwall seams, and other plate welding.

With increasing frequency, machine welding is being performed in the field. In field machine welding, the equipment performs the welding operation under the constant observation and control of a welding operator. Machines which make pressure-part welds are called orbital-welding machines. They usually employ the GTAW process, although the GMAW process is in the early stage of development for use in the field. The main advantage of orbital welding is the higher quality of the deposited weld. The improved quality results from the repeatability of the machine and the decreased fatigue of the operators over manual welders.

Field welds are radiographically examined as required by code, contract, and self-imposed quality control standard requirements. (See Chapter 18 which describes radiography requirements in a steam-generator manufacturing facility.) For this analysis, radiographic film is placed on one side of a weld and a radioactive source is temporarily placed on the other side. (Sometimes the source is placed inside a pipe weld for a panoramic shot and other times the source is placed on the outside of a tube or pipe weld for a double-wall shot.) When the film is developed, discontinuities may be disclosed requiring interpretation by a qualified radiographer. Any discontinuity which is interpreted as a defect must be ground out and rewelded.

Depending on the material specification of the component and the diameter and thickness of the weld, postweld heat treating may be required to reduce the residual stresses induced by the welding. This stress-relieving process involves the placement of electric elements or coils, or gas burners, around the weld area of the component. A band around the weld joint is brought up to a predetermined temperature, typically 1100 to 1300°F, at a controlled rate, and is held there for a time period that is a function of the weld thickness. Thermocouples attached at the outside of the heat band, or temperature-indicating crayons, are used to monitor the temperature. The component is then lowered to ambient temperature at a controlled rate.

ADDITIONAL CONSTRUCTION WORK

When the lower headers have been set in place, the cherry picker and the crane are removed from the furnace. Then the lower waterwall panels are brought in and raised into place, and the downtake system from the suction manifold discharge valves is installed.

Additional work that can proceed simultaneously in other sections of the boiler is attaching the buckstays to the waterwalls, bolting the sootblowers to the waterwalls, connecting the piping for the sootblowers, and welding safety valves to the steam drum and outlet leads to the turbine. At this stage, a "punch list" of work that must be completed before the hydrostatic test should be compiled,

so that all responsible contractors can be aware of unfinished items.

ERECTION OF STEAM GENERATOR AUXILIARY EQUIPMENT

Other equipment that is worked on during boiler erection are precipitators, absorbers, fans and the stack. The precipitator is located downstream of the steam generator proper and removes the solid particulate from the flue gases. It rests on a structural-steel frame which is also tied into the main support members for the precipitator. The hoppers are erected first and lowered into the steel from above with a crawler crane. The hoppers are usually subassembled on the ground and erected as one piece. The main supports, sides and internal bracing are put up next along with the inlet and outlet ductwork. The collecting and discharge electrodes, or other types of internals, are then erected. The inner roof, weather roof, and electrical components finish the main sections. Work on alignment, wiring and checkout is required before the unit is complete. Fig. 23 shows a nearly finished precipitator.

Absorbers, such as the one shown in Fig. 24, are installed as required by environmental regulations necessitating removal of sulfur oxides from the stack gases. Because of the large amount of ductwork and tanks, it is usually easier to erect the absorber at the same time as its structural-steel support framing. Fig. 25 shows the tanks with much of the ductwork in place. Once the main sections are up, the internal piping and spray sections can be installed. Because of the potentially corrosive atmosphere inside the absorber, a noncorroding material, such as austenitic stainless steel, is usually used. Such stainless steel requires special handling and welding requirements not needed on carbon-steel equipment.

On large utility units, there are usually two or more forced-draft fans that move combustion air through the air preheater and into the boiler through the windboxes and two or more induced-draft fans that pull the hot gases from the boiler and direct them to the stack.

Fig. 26 shows two centrifugal induced-draft fans partially completed. Once the support

Fig. 23. **Precipitator during final construction**

Fig. 24. **The erection of a stainless-steel absorber**

foundations are poured, the lower rotor housing, bearing supports, and rotor are installed. Then the remaining inlet chambers and outlet boxes are put together and connected to the supply and discharge ductwork. Millwrights set and align the rotor, a critical job because of the close tolerances required.

Fig. 25. Section of absorber steel and ductwork

Fig. 26. Induced-draft fans under construction

HYDROSTATIC TESTING AND INSULATION

During hydrostatic testing, water is pumped into the complete water and steam circuits of the boiler and raised to the pressure of $1\frac{1}{2}$ times the maximum allowable working pressure (MAWP). The pressure is held for a specified period of time and then lowered to the MAWP and held there for another period. The entire boiler, including piping, valves, wall panels, and drums, is then inspected for leaks. If any are found, the pressure is dropped, the water is drained out, and the leak is repaired. The entire process is then repeated until no more leaks are apparent. A steam-generating unit is "built-in accordance with the ASME Boiler Code" only when it has successfully passed such a hydrostatic test of the entire unit, either in the shop or in the field.

The erector engages the services of a steam-boiler inspection agent who is responsible for verifying the inspection and assuring the integrity of the boiler to the owner. The owner usually has its own insurance representative on the job at the same time to look over the records and ensure that proper inspection and tests are performed.

Fig. 27. Turbine pedestal with most components set in place

Once the hydrostatic test is complete, the insulation and lagging installation can proceed. Insulation blankets or blocks are installed on all piping, waterwalls, drums and ductwork. Metal lagging is installed over the insulation for protection of equipment and personnel. A coating of refractory is troweled on the top side of the roof tubes and a steel "skin" casing is welded over the top.

If the unit is an outdoor unit, the lagging serves to protect the boiler from the weather as well as to provide a satisfactory appearance. If the unit is an indoor unit, the boiler is completely enclosed by an outer building. This is desirable in cold climates, making initial erection and subsequent maintenance easier in the winter months.

ERECTION OF MECHANICAL EQUIPMENT

The installation of mechanical equipment

Fig. 28 Stator lift rig with grip-type jacks used to raise stator from ground level to turbine floor

proceeds while the boiler is being erected and includes the turbine generator, miscellaneous pumps and valves, the water-treatment systems, the boiler feed pumps, and the condenser to name but a few. The turbine generator is a large section of the mechanical work and requires almost the same time span for erection as the boiler. Fig. 27 shows a turbine under construction with most of the components already in place. The overhead gantry crane (upper left corner of Fig. 27), one of the first pieces of equipment erected, is used to raise and set most of the turbine except the stator.

Because the stator is so heavy (200 to 400 tons), a special lifting rig is usually brought to the site just for this one lift. Fig. 28 shows a stator entering through an access opening from the ground. This rig uses a system of grip-type jacks with steel cables to lift the stator from the railroad car and up to the turbine-pedestal floor. Once the stator is up to elevation, special rolling beams are put under the stator. The stator can be lowered a few inches onto special heavy-duty rollers and rolled from the lift frame onto the concrete turbine floor. The stator is transferred to the generator hole and lowered again using the upper section of the same lift rig.

ERECTION-COMMISSIONING INTERFACE

With most of the major work done on the boiler, boilout and acid cleaning of the inside of the tubes can be done. Chemicals are circulated through the boiler for prescribed lengths of time and then drained from the boiler. These procedures remove internal scale from water-bearing systems and assure free water passage in all tubes, headers and drums. (See Chapter 21 for a description of boilout and acid-cleaning of high-pressure steam-generators.)

Installation of boiler-drum internals may now be completed and access openings closed.

The contractual responsibility for the erection phase is usually complete at this point; however, support crafts for assisting test and start-up engineers may be necessary to take care of items that require modification and/or replacement during commissioning.

ASEA BROWN BOVERI

CHAPTER 20

Water Technology

Water is a basic engineering material used in the production of steam for power generation and process use. Effective treatment or conditioning of water has made possible the use of more efficient steam cycles in the subcritical and supercritical pressure ranges. Many industrial processes have special requirements for high quality water, free from objectionable contaminants.

These statements are indicative of the significant role of *water technology* as an interdisciplinary professional field which comprises elements of basic chemistry as well as chemical and mechanical engineering. This chapter will discuss specific boiler applications of water technology, but many of the same considerations hold for various types of process operations. In practically every instance, a careful review of sources of water and its conditioning is required to avoid economic losses because of slowdown in production, damage to equipment, and increased costs of operation.

The sources of water supply to an industrial site differ and can originate from a polluted river or stream, mountain runoff, wells or a lake. The composition of impurities in water varies over a wide range. Water may be polluted with sewage, chemical or organic wastes, bacteria, dissolved gases, suspended mineral matter and dissolved solids of both scaling and nonscaling composition. If water were pure "H_2O" and nothing else, or were of a constant solids composition, the conditioning of water for industrial use would be uniform and simple. Because this is not the case, the conditioning of water contains elements of both a specialized science and an engineering art.

In using water for generation of steam for process or electric power, consideration must be given to water treatment for the prevention of corrosion, scaling, and contamination of steam. Such attention must involve the treatment of the raw water introduced into the cycle and, in many cases, the conditioning of the water present in the preboiler cycle and in the boiler itself. How much consideration is given to the contaminants present in water depends largely upon the end use of the water. The pressure of a boiler is used to classify particular boiler designs. Therefore, the water quality necessary for a particular application is by convention related to boiler pressure, even though it is realized that the corresponding fluid temperature is the real factor influencing the reactivity of chemical and corrosion processes. The quality of the water in a plant becomes more critical as the cycle advances from low-pressure industrial type boilers to high-pressure utility boilers.

As major causes of forced outages, internal corrosion and deposition cost electric utili-

Table I. Common Impurities Found in Water

Constituent	Chemical Formula	Difficulties Caused	Means of Treatment
Turbidity	None—expressed in analysis as units.	Imparts unsightly appearance to water. Deposits in water-lines, process equipment, etc. Interferes with most process uses.	Coagulation, settling and filtration.
Color	None—expressed in analysis as units.	May cause foaming in boilers. Hinders precipitation methods such as iron removal and softening. Can stain product in process use.	Coagulation and filtration. Chlorination. Adsorption by activated carbon.
Hardness	Calcium and magnesium salts expressed as $CaCO_3$.	Chief source of scale in heat exchange equipment, boilers, pipelines, etc. Forms curds with soap, interferes with dyeing, etc.	Softening. Demineralization. Internal boiler water treatment. Surface-active agents.
Alkalinity	Bicarbonate (HCO_3), carbonate (CO_3), and hydrate (OH), expressed as $CaCO_3$.	Foaming and carryover of solids with steam. Embrittlement of boiler steel. Bicarbonate and carbonate produce CO_2 in steam, a source of corrosion in condensate lines.	Lime and lime-soda softening. Acid treatment. Hydrogen zeolite softening. Demineralization. Dealkalization by anion exchange.
Free Mineral Acid	H_2SO_4, HCl, etc. expressed as $CaCO_3$.	Corrosion.	Neutralization with alkalies.
Carbon Dioxide	CO_2	Corrosion in waterlines and particularly steam and condensate lines.	Aeration. Deaeration. Neutralization with alkalies.
pH	Hydrogen ion concentration defined as: $pH = \log \frac{1}{(H^+)}$	pH varies according to acidic or alkaline solids in water. Most natural waters have a ph of 6.0–8.0.	pH can be increased by alkalies and decreased by acids.
Sulfate	$(SO_4)^{--}$	Adds to solids content of water, but in itself, is not usually significant. Combines with calcium to form calcium sulfate scale.	Demineralization.
Chloride	Cl^-	Adds to solids content and increases corrosive character of water.	Demineralization.
Nitrate	$(NO_3)^-$	Adds to solids content, but is not usually significant industrially. Useful for control of boiler metal embrittlement.	Demineralization.
Silica	SiO_2	Scale in boilers and cooling water systems. Insoluble turbine blade deposits because of silica vaporization.	Hot process removal with magnesium salts. Adsorption by highly basic anion exchange resins, in conjunction with demineralization.

Table I. Common Impurities Found in Water — *Continued*

Constituent	Chemical Formula	Difficulties Caused	Means of Treatment
Iron	Fe^{++} (ferrous) Fe^{+++} (ferric)	Discolors water on precipitation. Source of deposits in waterlines, boilers, etc. Interferes with dyeing, tanning, papermaking, etc.	Aeration. Coagulation and filtration. Lime softening. Cation exchange. Contact filtration. Surface-active agents for iron retention.
Manganese	Mn^{++}	Same as iron.	Same as iron.
Oxygen	O_2	Corrosion of waterlines, heat exchange equipment, boilers, return lines, etc.	Deaeration. Sodium sulfite. Corrosion inhibitors.
Hydrogen Sulfide	H_2S	Cause of "rotten egg" odor. Corrosion.	Aeration. Chlorination. Highly basic anion exchange.
Ammonia	NH_3	Corrosion of copper and zinc alloys by formation of complex soluble ion.	Cation exchange with hydrogen zeolite. Chlorination. Deaeration.
Dissolved Solids	None	"Dissolved Solids" is measure of total amount of dissolved matter, determined by evaporation. High concentrations of dissolved solids are objectionable because of process interference and as a cause of foaming in boilers.	Various softening process, such as lime softening and cation exchange by hydrogen zeolite, will reduce dissolved solids. Demineralization.
Suspended Solids	None	"Suspended Solids" is the measure of undissolved matter, determined gravimetrically. Suspended solids cause deposits in heat-exchange equipment, boilers, waterlines, etc.	Subsidence. Filtration, usually preceded by coagulation and settling.
Total Solids	None	"Total Solids" is the sum of dissolved and suspended solids, determined gravimetrically.	See "Dissolved Solids" and "Suspended Solids."

*Adapted from *Betz Handbook of Industrial Water Conditioning* 8th Edition, 1980 Betz Laboratories, Inc., Philadelphia, Pa.

ties millions of dollars in repairs and lost availability. This is both unfortunate and unnecessary; the causes of corrosion are generally known and there are effective means of preventing or reducing damage from corrosion:

■ Maintain recommended water treatment for both the boiler-water and feedwater systems.

■ Control oxygen concentrations in the feedwater to minimize corrosion and formation of pre-boiler corrosion products, which end up as deposits on heat-transfer surfaces in the boiler.

■ Comply with operating procedures during start-ups, shutdown, and outages to minimize corrosion and to avoid the entry of corrosion products into the boiler.

■ Maintain boiler free from significant amounts of deposits by periodic chemical cleaning.

TECHNIQUES OF WATER TREATMENT

The extent of water conditioning or treatment depends both on the original supply source and the ultimate end use of the product. Table I lists common impurities found in water, sets forth some of the resulting problems, and indicates common treatment methods.

The techniques of water treatment cover a wide variety of raw-water preparation schemes. This is covered in detail in established references.[1] Some of the basic chemical reactions involved in the treatment of raw water to produce suitable makeup for boilers will be discussed in a general manner.

Any evaluation of water conditioning must consider the impurities present in power-station raw-water makeup in relation to tendencies toward scaling, corrosion, and deposits. Raw water contains many contaminating elements including (a) mud, clay and silt; (b) oxygen, carbon dioxide and hydrogen sulfide; (c) sewage, bacteria and algae; (d) scale-forming compounds of calcium, magnesium and silica; (e) oil; (f) iron compounds; (g) organic wastes; (h) sulfuric, hydrochloric and other acids; (i) normally soluble compounds, such as sodium bicarbonate, sodium carbonate, sodium hydroxide and sodium chloride.

Table II gives typical analyses of water supplies of the U.S. The river water analyses are of filtered water, and for the most part represent yearly averages. River waters change considerably with the seasons, and the maximum and minimum concentrations may be 50 percent above and 50 percent below the average. Spring and well waters and waters from large lakes have a fairly constant composition.[2]

The usual unit for reporting dissolved and also suspended solids in feedwater and boiler water is *parts per million* (ppm). This is a rational unit which is easily understood and which avoids misunderstanding. In other words, one million pounds of water will contain so many pounds of solids. With high subcritical-pressure and supercritical-pressure boilers requiring very pure water, the term *parts per billion* (ppb) is now in use. One ppb equals 0.001 ppm.

RAW-WATER PROCESSES

Raw water contains various gaseous and solid impurities which must be reduced before the water can be supplied as makeup to a boiler. This is particularly true if the water is taken from a river which may be contaminated with mine washings, gases, organic and chemical wastes as well as silt and other minor impurities. To reduce these substances, the processes of aeration, settling, coagulation, and filtering are commonly used. A brief description of each method follows.

AERATION

This process (Fig. 1) removes such undesirable gases as carbon dioxide and hydrogen sulfide from water by admixing water and air to reduce the solubility of the objectionable gas in water. The removal of gas follows Henry's Law which indicates that the solubility of a gas in water is directly proportional to its partial pressure in the surrounding atmosphere. The partial pressure of a gas such as carbon dioxide is low in a normal atmosphere. Establishing an equilibrium between water and air by aeration results in saturation of the water with oxygen and nitrogen and results in the practical elimination of such gases as carbon dioxide and hydrogen sulfide. Increasing the temperature, the aeration time and the surface area of water improves the removal of gases.

COAGULATION

Adding chemical coagulating materials reduces surface-water contamination in the form of coarse suspended solids, silt, turbidity, color and colloids. The chemicals form a floc which assists in agglomerating impurities. Settlement of the particles permits a clear effluent from the coagulating chamber. Fig. 2 illustrates a typical clarifier. Removal of colloids requires a careful analysis of the impurities to establish the nature of their electrical charge, one of the principal factors contributing to their remaining in the suspended state. Some chemicals used for coagulation are filter alum, sodium

aluminate, copperas, ferrisul, activated silica, and various proprietary organic compounds.

Temperature, pH, and mixing affect the efficiency of coagulation. Some of the reactions involved are

Filter alum (formation of aluminum hydroxide)

$$Al_2(SO_4)_3 + 3Ca(HCO_3)_2 \rightarrow$$
$$3CaSO_4 + 2Al(OH)_3 + 6CO_2$$

Sodium aluminate

$$6NaAlO_2 + Al_2(SO_4)_3 \cdot 18H_2O \rightarrow$$
$$8Al(OH)_3 + 3Na_2SO_4 + 6H_2O$$

Copperas (formation of ferrous hydroxide)

$$FeSO_4 + Ca(HCO_3)_2 \rightarrow$$
$$CaSO_4 + Fe(HCO_3)_2$$
$$Fe(HCO_3)_2 + 2Ca(OH)_2 \rightarrow$$
$$Fe(OH)_2 + 2CaCO_3 + 2H_2O$$

Ferrisul

$$Fe_2(SO_4)_3 + 3Ca(HCO_3)_2 \leftrightharpoons$$
$$2Fe(OH)_3 + 3CaSO_4 + 6CO_2$$

FILTRATION

Filters separate coarse suspended matter from raw water, or remove floc or sludge com-

---

Table II. Analyses of Typical Surface and Ground Waters in the United States

Analysis Number[a]	1	2	3	4	5	6	7	8	9	10	11	12	13
Silica (SiO_2)	2.5	0.4	2.3	8.2	13	8.0	8.4	9.6	16	23	34	12	39
Iron (Fe)	0.03	0.05	.09	.12	0.04		0.15	0.04	0.0			2.1	.09
Calcium (Ca)	5.3	27	32	1.7	72	79	40	3.4	7.2	70	26	72	7.2
Magnesium (Mg)	1.7	7	10	0.4	6.4	28	16	1.5	2.5	24	10	33	4.2
Sodium (Na)	1.4	3[b]	3.5	1.9	41	99	94	5.6	147	12[b]	138	358[b]	7.5
Potassium (K)	0.6		1.0	0.7	4.7	4			0.4		1.6		
Carbonate (CO_3)	0	1	0	0	0	4	0	0	0		0		
Bicarbonate (HCO_3)	10	99	138	3	174	137	46	21	328	179	170	293	50
Sulfate (SO_4)	11	13	17	4.4	138	290	298	3.3	2.6	135	70	560	0.8
Chloride (Cl)	2.6	7	6.5	2.6	9.5	79	14	3.6	51	8	139	195	6.8
Fluoride (F)	0.1		0.1	0.6	0.4	0.4	0.1	0.2	0.8			2.5	0.0
Nitrate (NO_3)	0.3	0.2		0.2	4.0	0.2	3.6	0.7	0.0		0.0	1.1	0.0
Dissolved solids	34	130	171	23	386	661	554	42	392	392	503	1380	90[c]
Total hardness as $CaCO_3$	20	98	121	6	206	315	166	15	28	276	106	316	35
Noncarbonate hardness	6	16	8	3	64	197	128	0	0	126	0	76	0
Specific conductance (micromhos at 25 C)	53.4		263	29.5	575	1040	822	55.5	651		867		
Color	1		3	15	5		6	4	10			0	2.5
pH	6.9	8.1	8.2	5.8	7.7	8.4	7.6	7.0	8.0	7.6	7.9		7.5

[a] Analyses numbers are identified as follows:
1 = New York City, NY, Catskill supply (reservoir—finished).
2 = Detroit, MI, Detroit River (raw).
3 = Chicago, IL, Lake Michigan (raw).
4 = Fitchburg, MA, Pond (finished).
5 = Omaha, NE, Missouri River (raw).
6 = Los Angeles, CA, Colorado River (raw).
7 = Pittsburgh, PA, Monongahela River (finished).
8 = Macon, GA, Ocmulgee River (raw).
9 = Houston, TX, Well 1932 ft deep.
10 = Jacksonville, FL, Well 1064 ft deep.
11 = El Paso, TX, Well 703 ft deep.
12 = Galesburg, IL, Well 2450 ft deep.
13 = Bremerton, WA, Anderson Creek.
[b] Computed by difference in epm and reported as sodium.
[c] Sum of determined constituents.

Values are in parts per million where this unit is appropriate.

ABB
ASEA BROWN BOVERI

Fig. 1 Forced-draft aerator

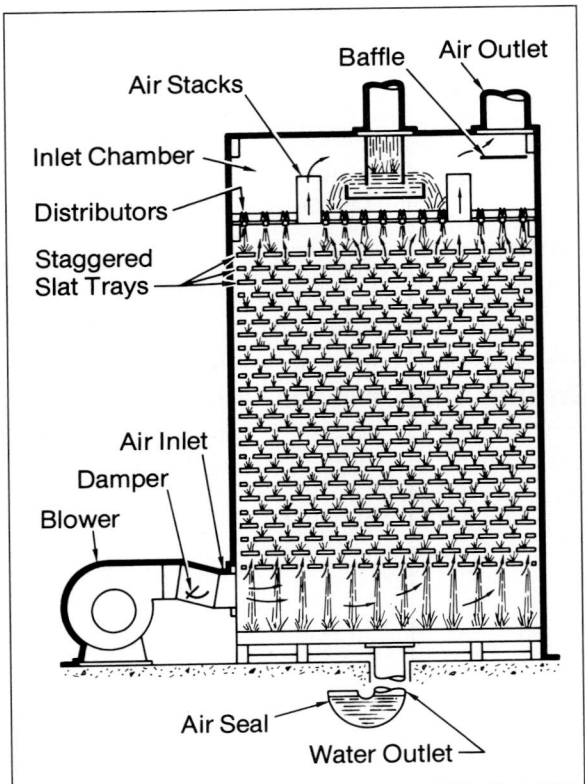

ponents from coagulation or process softening systems. Generally, gravity and pressure-type filters as shown in Figs. 3 and 4 are used for this purpose. Beds of graded gravel or coarse anthracite are the common materials used in the filter bed. Diatomaceous earth and special precoat filters are generally used to remove oil and reduce color in feedwater makeup.

CHEMICAL SOFTENING PROCESSES

Nonscaling feedwater can be obtained with proper pretreatment of raw water. Various chemical combinations can remove hardness, silica and silt from makeup water. Economics and boiler operating conditions dictate the technique selected for the application.

LIME-SODA SOFTENING

To soften water by this process, lime (calcium hydroxide) is added to precipitate the calcium bicarbonate as calcium carbonate and magnesium salts as magnesium hydroxide. Soda ash (sodium carbonate) is added to react with calcium chloride and calcium sulfate to form calcium carbonate. The general reactions are as follows:

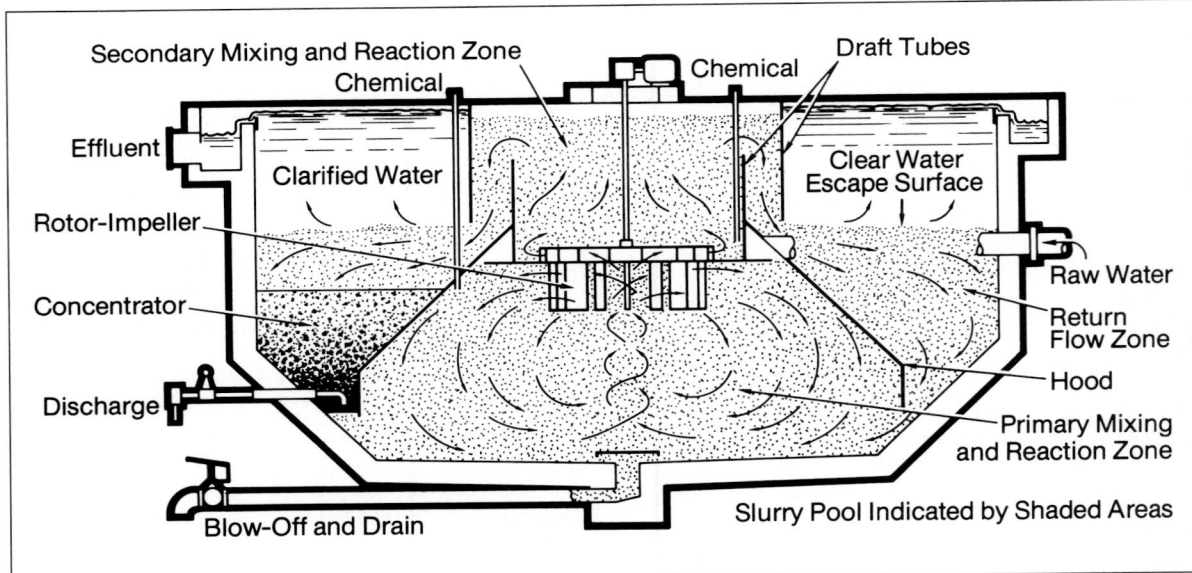

Fig. 2 Solids-contact clarifier

Fig. 3 Gravity filter

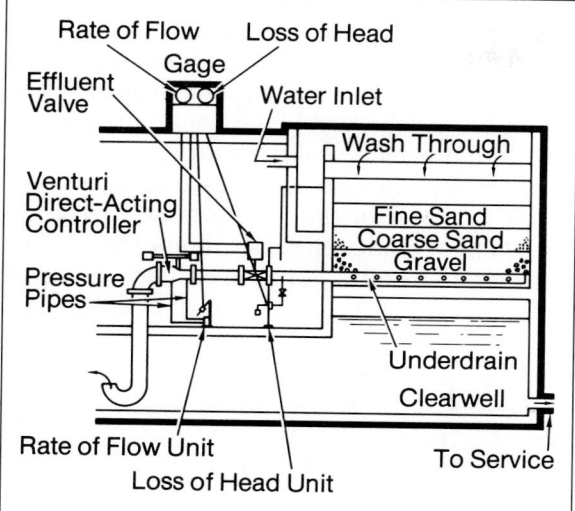

$$Ca(HCO_3)_2 + Ca(OH)_2 \rightarrow 2CaCO_3\downarrow + 2H_2O$$

$$CaSO_4 + Na_2CO_3 \rightarrow CaCO_3\downarrow + Na_2SO_4$$

$$Mg(HCO_3)_2 + 2Ca(OH)_2 \rightarrow$$
$$Mg(OH)_2\downarrow + 2CaCO_3\downarrow + 2H_2O$$

$$MgSO_4 + Ca(OH)_2 \rightarrow Mg(OH)_2\downarrow + CaSO_4$$

Because of the increase in reaction rate and the decrease in solubility of calcium carbonate, this process is more efficient at hot temperatures. The hardness in the effluent of a cold lime-soda process depends upon the excess of soda ash and is generally above 33 ppm. Alkalinity reduction depends on the removal of magnesium and calcium bicarbonates. In the hot lime-soda process, the hardness residual in the effluent ranges from 17–25 ppm.

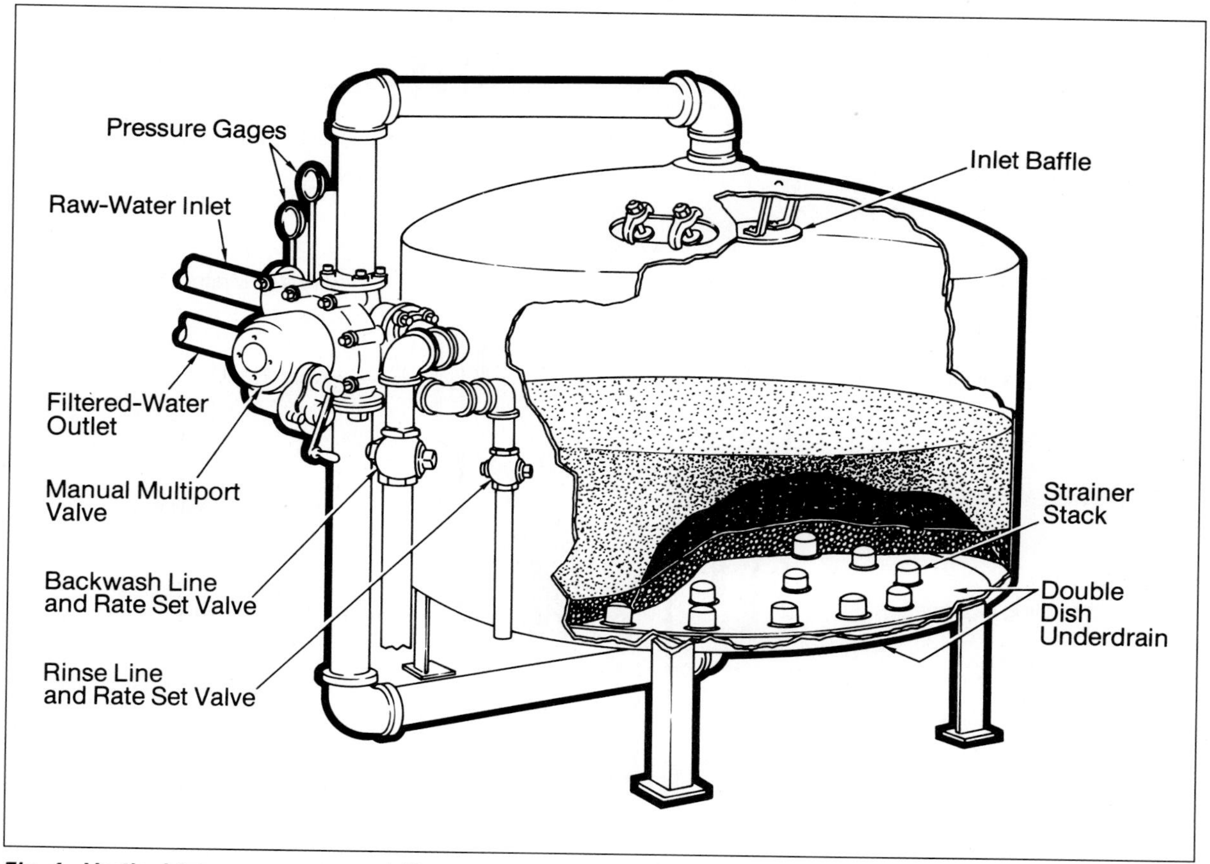

Fig. 4 Vertical-type pressure sand filter

HOT-PROCESS PHOSPHATE SOFTENING

In this process, phosphate chemically precipitates calcium and magnesium salts. The calcium hardness is precipitated in the form of tricalcium phosphate and the magnesium as magnesium hydroxide. Chemical reactions are completed at a temperature of 212°F or above. The reactions can be controlled to reduce the hardness to nearly zero.

Normal low hardness, below 60 ppm, can be eliminated in a single stage. Generally, high-hardness water is handled in a two-stage softener arrangement. The hot-process phosphate softener is particularly suited for turbid waters, which are low in hardness and alkalinity. Softening, turbidity control, deaeration and silica reduction can be combined in a single unit. Fig. 5 shows a typical hot-process settling tank. Usually, surface-acting organic agents are added to stabilize the precipitates formed in the softener. Anthracite-coal filters are best for removal of turbidity from the effluent. Some of

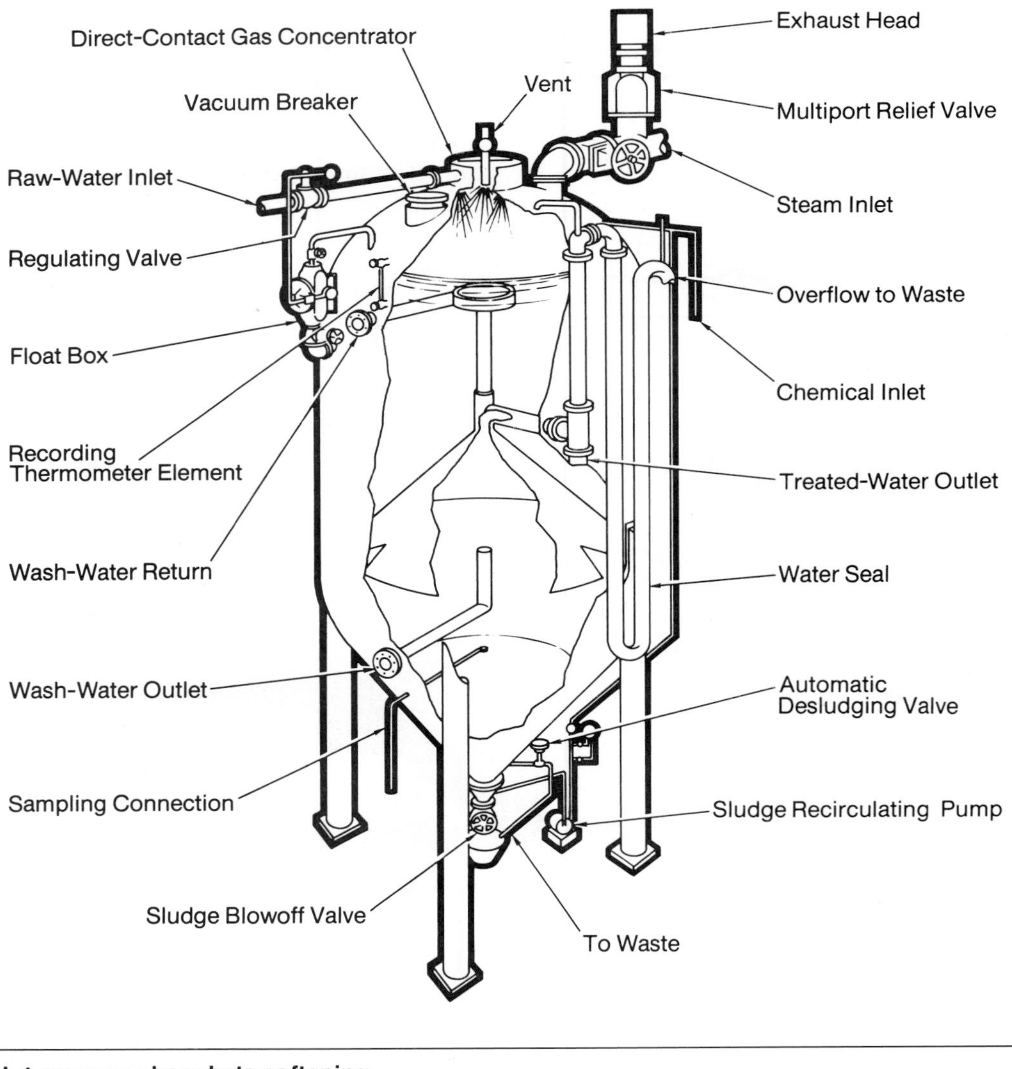

Direct-Contact Gas Concentrator

Vacuum Breaker

Raw-Water Inlet

Regulating Valve

Float Box

Recording Thermometer Element

Wash-Water Return

Wash-Water Outlet

Sampling Connection

Sludge Blowoff Valve

Exhaust Head

Vent

Multiport Relief Valve

Steam Inlet

Overflow to Waste

Chemical Inlet

Treated-Water Outlet

Water Seal

Automatic Desludging Valve

Sludge Recirculating Pump

To Waste

Fig. 5 **Hot-process phosphate softening**

the chemical reactions in the softener to reduce calcium and magnesium hardness are

$$3CA(HCO_3)_2 + 6NaOH \rightarrow$$
$$3CaCO_3\downarrow + 3Na_2CO_3 + 6H_2O$$
$$3CaCO_3 + 2Na_3PO_4 \rightarrow$$
$$Ca_3(PO_4)_2\downarrow + 3Na_2CO_3$$
$$Mg(HCO_3)_2 + 4NaOH \rightarrow$$
$$Mg(OH)_2\downarrow + 2Na_2CO_3 + 2H_2O$$

Addition of magnesium sulfate, magnesium hydroxide, dolomitic lime and magnesium oxide aids in the reduction of silica. Removal of silica from solution is accomplished by adsorption by the presence of magnesium hydroxide formed in the softener.

ZEOLITE SOFTENING

The name *zeolite* refers to a group of water-softening chemicals capable of exchanging ions with which they come in contact. These materials may be natural compounds (green

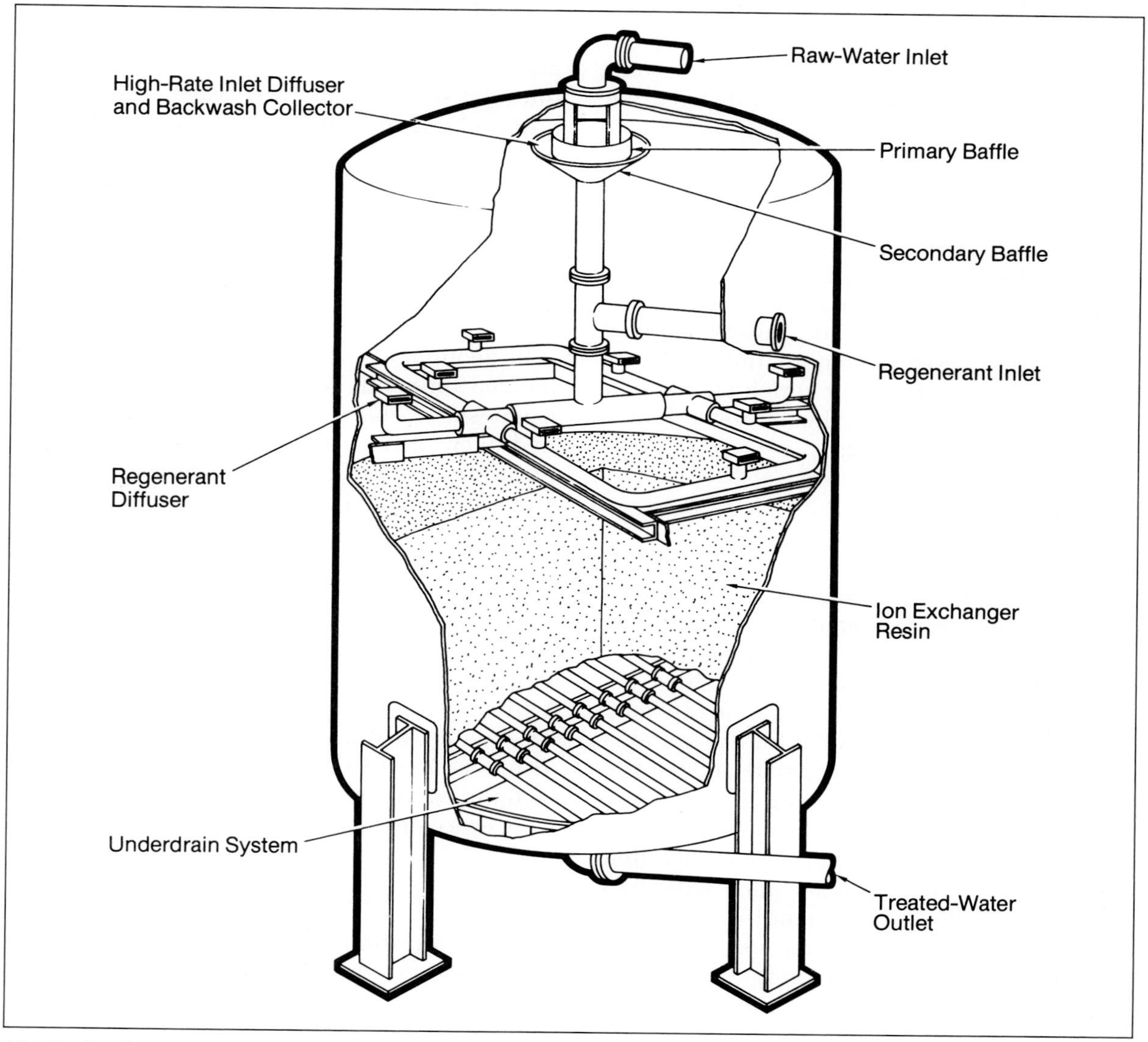

Fig. 6 **Sodium zeolite softener**

sand) or synthetic compounds such as sulfonated coal, phenolic and polystyrene resins. Hard water is passed downward through a bed of sodium-regenerated zeolite contained in a steel pressure vessel. See Fig. 6. As the water passes through the ion–exchange material, the calcium and magnesium ions are exchanged for sodium in the zeolite (Z).

The reactions are

$$Ca(HCO_3)_2 + Na_2Z \rightarrow CaZ\downarrow + 2NaHCO_3$$
$$Mg(HCO_3)_2 + Na_2Z \rightarrow MgZ\downarrow + 2NaHCO_3$$

Regeneration of the zeolite bed is accomplished by passing a salt solution through the softener.

$$CaZ + 2NaCl \rightarrow CaCl_2 + Na_2Z$$

The calcium chloride is passed to waste and the zeolite bed is ready for further softening.

Salt (sodium chloride) or acid can regenerate synthetic zeolites. Acid- and salt-regenerated zeolites can be used in combination to reduce alkalinity in waters having a high bicarbonate hardness. The reaction in the hydrogen zeolite may be written as follows:

$$Ca(HCO_3)_2 + H_2Z \rightarrow CaZ\downarrow + 2H_2CO_3$$
$$Mg(HCO_3)_2 + H_2Z \rightarrow MgZ\downarrow + 2H_2CO_3$$

The mixed effluent from the hydrogen and sodium zeolites is deaerated to remove the carbon dioxide. Generally, the exhausted acid zeolite is regenerated with sulfuric acid.

Zeolite softening of makeup is the most common way to prepare water for use in industrial boilers. It is simple to operate and control. While sodium zeolite softening is very attractive, it must be applied with understanding of its limitations. Turbid waters are unsuited; total solids are not reduced and, with high bicarbonate water, high quantities of carbon dioxide can be expected in the steam. Silica is not reduced by zeolite softening.

Although using hydrogen zeolite reduces solids, the equipment operator must recognize a number of disadvantages: As in the sodium zeolite, turbidity must be low. The cost of the chemicals, acid and salt, may be considerable. Handling of acids introduces a hazard to operators, and corrosion resistant pipe must be used to transport the water to the degasifier.

Synthetic zeolites are used with other water-softening equipment in preparation of makeup water. The hot-lime zeolite softener is a widely used technique in preparing water for intermediate-pressure industrial boilers.

Using chemicals at temperatures about 250°F facilitates reduction of hardness, alkalinity, and silica. The hot-lime softening reduces the carbonate and magnesium hardness as well as the total solids in the raw water. If the magnesium content in the raw water is below the level to reduce silica, activated magnesium oxide or dolomitic lime is added for reaction.

Use of the hot-lime zeolite produces a water of 0–2 ppm hardness, 20–30 ppm alkalinity and 0.5–1.0 ppm of silica.

DEMINERALIZATION

In demineralization, ion exchange removes ionized mineral salts. Cations as calcium, magnesium, and sodium are removed in the hydrogen cation exchanger and anions as bicarbonates, sulfates, chloride, and soluble silica are removed in the anion exchanger.

Synthetic cation and anion exchange resins are used in demineralization of water. Sulfonic, carboxylic, and phenolic hydroxyl compounds are used for cation exchange; amino or quartenary nitrogen, for anion exchange. The cation exchanger is regenerated with acid while the anion exchange material is regenerated with caustic. If the cation resin is designated as Z and the anion material as R, the simple reactions in a two-stage demineralizer may be expressed as:

$$\text{Cation: } H_2Z + CaSO_4 \rightarrow CaZ + H_2SO_4$$
$$\text{Anion: } H_2SO_4 + R(OH)_2 \rightarrow R(SO_4) + 2H_2O$$

Demineralization can yield a pure water, equal or superior to the best evaporated vapor. The anion and cation resins can be arranged in various combinations to produce the best water most economically. Two-, three- or four-bed units or a single mixed-bed demineralizer can be used to accomplish the required result. Fig. 7 illustrates possible combinations of these systems to produce a specific water quality. As strong base exchangers are temperature sensitive, they should not be used above 120°F. Cation exchange resins can tolerate temperatures of 250°F.

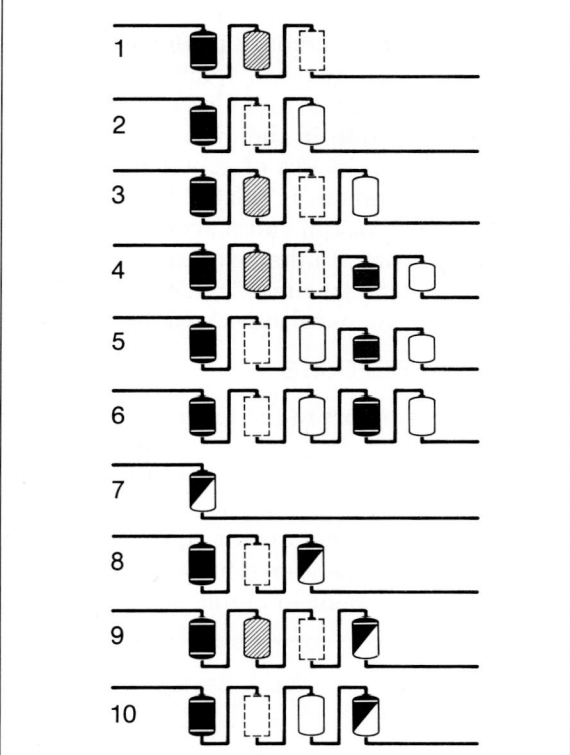

Fig. 7 Ten major demineralizer systems: black, strong-acid cation exchanger; white, strong-base anion exchanger; shaded, weak-base anion exchanger; half black and half white, mixed-bed; broken rectangle, decarbonator or vacuum deaerator. (1) Two-bed with weak-base; (2) two-bed with strong-base; (3) three-bed; (4) four-bed, primary with weak-base; (5) four-bed, primary with strong-base; (6) parallel two-beds, as in system 2, or four-bed as in system 5 (except for size of secondary unit); (7) mixed-bed; (8) cation bed, decarbonator, and mixed-bed; (9) two-bed, as in system 1, and mixed-bed; (10) two-bed, as in system 2, and mixed-bed. The secondary units in systems 4 and 5, which are used only for polishing, may be smaller than the primary, as indicated.

Use of finely sized resins in demineralizers requires a water source low in turbidity and organic matter. Clarification, filtration and chlorination are generally required to reduce organic matter to a low level. Failure to reduce the organic level in the water results in a marked reduction in capacity of the exchangers because of fouling of the resins.

The advance of demineralization to provide a high-purity makeup water may be partly attributed to the development of highly basic anion exchange resins which allow the removal of soluble silica from raw water. Mixed-bed demineralizers, in which the cation and anion resins are intimately interspersed (see Figs. 8 and 9), have successfully provided a high-quality makeup water as well as for polishing purposes in cleaning up impurities in condensates of utility units. Fig. 10 is a schematic of a complete makeup water system for a utility power plant.

WATER TECHNOLOGY FOR BOILERS

Water technology for steam power plants encompasses the production of make-up water

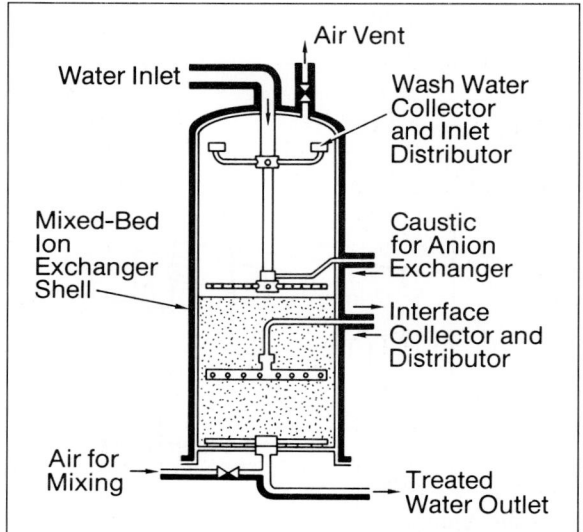

Fig. 8 Mixed-bed demineralizer. It houses mixture of cation and anion exchange resins. Air for mixing enters through bottom distributor.

and the treatment of the water (boilerwater and feedwater) after it enters the cycle. Requirements for make-up water quality and internal treatment may vary depending on many factors such as system design, operating conditions, and type of materials of construction. Modern high pressure power cycles require water of high purity, essentially free of soluble and insoluble solids. As described in previous sections, make-up water of excellent quality can be produced by ion exchange systems at reasonable costs. The internal treatment is then designed to maintain a chemical environment which provides corrosion protection, surfaces free of deposits, and high purity steam.

Plants with boilers operating below 600 psig have less stringent requirements. Therefore, it is often more economical to provide make-up water only partially demineralized (softened) and depend on internal chemical treatment to avoid potential problems due to the increased soluble solids concentration in the cycle.

The greatest incidence of problems in steam generating equipment is related to (a) scale or deposits, (b) corrosion, and (c) carryover of solids with the steam. In the next several sections, various chemical treatment schemes and the corresponding chemical agents will be reviewed.

Fig. 9 Regeneration of mixed-bed, from initial backwash to end of rinse and return to service, takes from 2 to 3 hours.

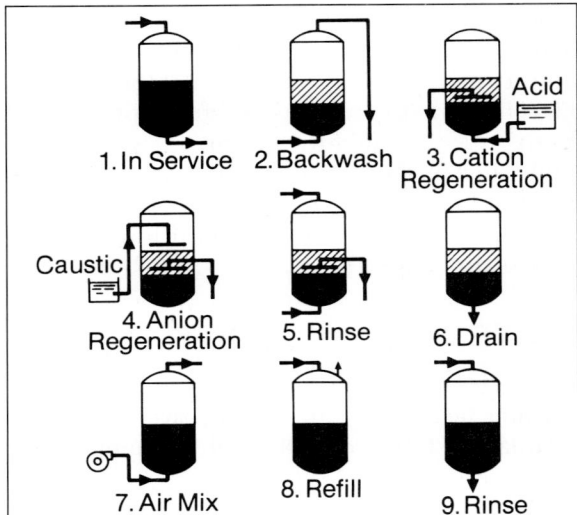

SCALE AND DEPOSITS

Major contributors to the formation of a heat deterrent scale or deposit are (1) contaminating elements present in the makeup water, (2) metal oxides transported to the boiler with feedwater, (3) contaminants from process equipment introduced into the condensate returned to the boiler, and (4) solids present in condenser leakage.

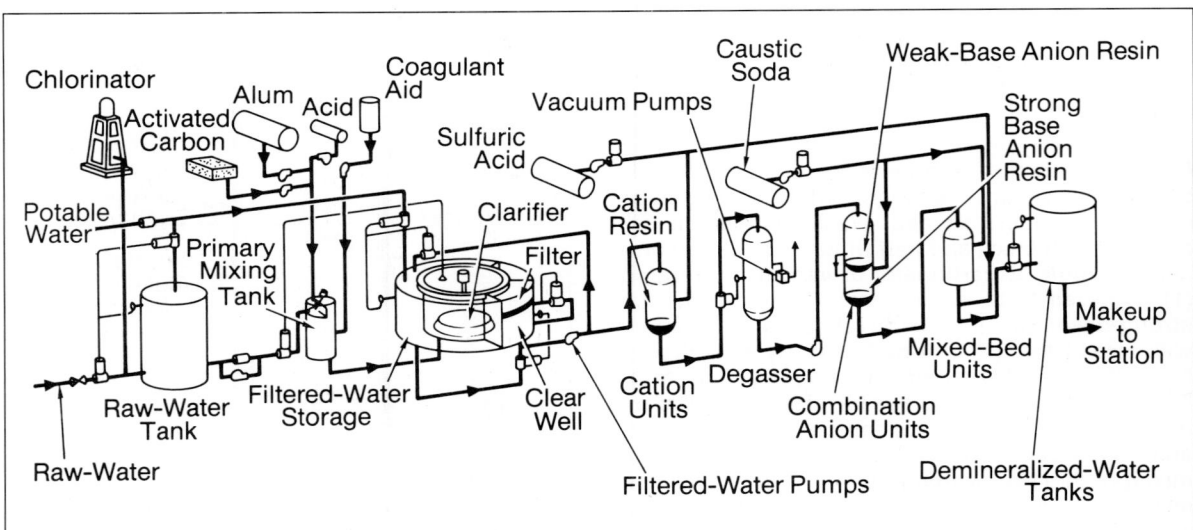

Fig. 10 **Demineralizer and pretreatment flow diagram**

Low-pressure boilers with high makeup and a small heat-recovering preboiler system are more prone to deposits from the precipitation of chemical compounds. Substances such as calcium bicarbonate if not properly removed from the makeup water will decompose in the boiler water to form calcium carbonate as follows:

$$Ca(HCO_3)_2 \rightarrow CaCO_3 + CO_2 + H_2O$$

Calcium carbonate has a very limited solubility and thus, the precipitated particles will agglomerate at the heated surfaces to form a scale. Other substances such as calcium sulphate are also scale producing. In this case, the scaling mechanism is, however, a function of a retrograde solubility or a decrease in solubility in water with an increase in temperature.

These substances have a low thermal conductivity, and if left untreated, even a thin scale will overheat the boiler tube.

USE OF PHOSPHATES FOR INTERNAL TREATMENT

The work of R. E. Hall and associates led to the use of phosphates as an internal boilerwater treatment for converting residual concentrations of calcium and other hardness salts to their respective phosphate compounds. These compounds can be more readily dispersed and removed by blowdown, although if present in large concentrations, these will also form scales on boiler tubes. Several phosphate compounds such as trisodium phosphate, disodium phosphate and monosodium phosphate in conjunction with sodium hydroxide (caustic) can be used for this purpose. Some of the reactions are

$$2Na_3PO_4 + 3CaSO_4 \rightarrow Ca_3(PO_4)_2 + 3NA_2SO_4$$
$$3MgCO_3 + 2Na_3PO_4 \rightarrow Mg_3(PO_4)_2 + 3Na_2CO_3$$
$$Mg(HCO_3)_2 + 4NaOH \rightarrow$$
$$Mg(OH)_2 + 2Na_2CO_3 + 2H_2O$$
$$MgCl_2 + 2NaOH + SiO_2 \rightarrow$$
$$MgSiO_3 + 2NaCl + H_2O$$

The phosphate-caustic treatment is limited to low pressure boilers utilizing softened makeup. It's recognized that for boilers using demineralized makeup, the need for high phosphate concentrations is greatly diminished. Also, the use of caustic is inappropriate because of its corrosivity at high concentrations. With demineralized quality boilerwater, caustic (present either as a treatment chemical or contaminant) can reach high concentrations due to the absence of large quantities of other competing ions.

Whirl and Purcell recognized the ability of sodium phosphate to suitably alkalize and control pH in boilerwater without the negative reaction associated with the use of caustic. They developed the pH-control method termed "the coordinated phosphate pH control". This method is illustrated by the relationship of pH and the phosphate concentration in boiler water shown in Fig. 11. Values noted on the curve represent pH values obtained by dissolving stoichiometrically pure trisodium phosphate (Na_3PO_4). Conditions below the curve represent solutions of trisodium phosphate and disodium hydrogen phosphate (Na_2HPO_4). The area above the curve, which represents solutions of trisodium phosphate and caustic, is to be avoided.

In the coordinated phosphate pH control, specifications can be maintained by adding trisodium phosphate (Na_3PO_4), disodium phosphate (Na_2HPO_4), or monosodium phosphate (NaH_2PO_4). Combinations of trisodium phosphate and disodium phosphate are preferred.

Alkalinity control is attained by the addition of phosphate ions to water to produce a captive quantity of hydroxide (OH) by the reversed hydrolysis reaction as follows:

$$PO_4{}^{\equiv} + H_2O \rightleftharpoons OH^- + HPO_4{}^=$$

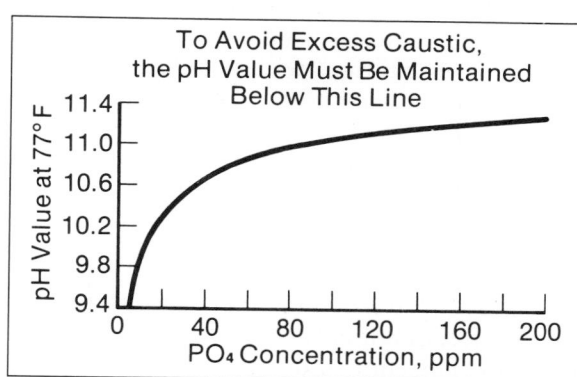

Fig. 11   Coordinated phosphate pH Control

This reaction is complete at pH levels below 11.0. The hydrolysis reaction of dibasic phosphate also proceeds in water as follows:

$$HPO_4^= + H_2O \rightleftharpoons OH^- + H_2PO_4^-$$

The latter reaction is complete at low pH levels (5 to 7) but is less than 0.1 percent at a pH level of 10. Thus, dibasic phosphate neither hydrolyzes nor dissociates in the normal boiler-water pH range. Additions of disodium phosphate have little effect on pH in the range 10 to 11.0.

Monobasic phosphate dissociates to the dibasic form, and the addition of one mol of monosodium phosphate (NaH_2PO_4) is capable of neutralizing one mol of sodium hydroxide. As one mol will increase boiler-water pH equivalent to the addition of one mol of sodium hydroxide, trisodium phosphate assures pH control. Adding monosodium phosphate can reduce excess caustic.

SIGNIFICANCE OF IRON-WATER REACTION

A number of influences characterize the reaction of boiler steel and water: the chemical constituents present in solution, the quantity of dissolved gases, the structural characteristics of the steel, and the rate of heat input to the boiler metal. Although the reaction is understood, in general there is insufficient information to explain the mechanism in exact detail. Many theories have been forwarded to clarify this problem. But despite lack of fundamental data, experience and experimental evidence have permitted the development of controls which satisfactorily contain the corrosive activity. This section will deal with existing techniques developed to understand and control reactions in boilers.

The fundamental reaction of iron and water is to produce iron hydroxide and hydrogen as follows:

$$Fe + 2H_2O \rightarrow Fe(OH)_2 + H_2$$

It is an established fact that the end product of reaction in boilers is magnetic oxide of iron. (Fe_3O_4). The control of corrosion, therefore, is based on the knowledge of the rate-controlling step in the overall reaction. Schikorr established a mechanism which accounted for the production of magnetic oxide of iron. In simplest form, the reactions would be

$$Fe + 2H_2O \rightarrow Fe(OH)_2 + H_2$$
$$3Fe(OH)_2 \rightarrow Fe_3O_4 + 2H_2O + H_2$$

From a consideration of physical chemical relationships, it can be shown that the formation of iron hydroxide is the rate-controlling step in the Schikorr hypothesis. Therefore, the rate of the overall reaction is based on the solubility and stability of this product.

The initial reaction above is pH- or alkalinity-controlled since, by the laws of chemical equilibrium, addition of alkalinity would reverse the reaction to the left. Corey and Finnegan found that iron placed in contact with deaerated and chemically pure water will produce an equilibrium pH of approximately 8.3. Increasing the alkalinity reduces the solubility of the iron corrosion product and inhibits reactivity. The control of this reaction has been well established in the protection of metal surfaces existing ahead of the boiler.[4]

The reaction of water and steel is spontaneous and rapid at high boiler temperatures. The only reason that boiler steel can survive normal operating conditions is that the corrosion end product, magnetite (Fe_3O_4), forms a protective barrier on the metal surface to stifle further corrosion. In the simplest analysis, the function of alkalinity control is to maintain an environment in which the oxide film is stable and protective. The objective of water treatment in boilers is to protect this film against the aggressive action of impurities introduced into the boiler with the feedwater.

The work of Berl and Van Taack (Fig. 12)[5] has been used to relate corrosion of steel over a range of pH values. In interpreting the results of Fig. 12, it may be concluded that the protective oxide is solubilized at pH values below 5.0 and above 13.0. Minimum corrosion is indicated at pH values of 9.0 to 11.0, although corrosion is low over a wide band of pH values. These data are valid, but they were obtained in autoclave tests where boiling, and thus concentration of the alkalinity, did not occur. Unfortunately, corrosion occurs by concentration at the local tube wall and not by concentrations existing in the bulk boilerwater. The attack of the protective oxide film and base metal occurs beneath deposits if the chemical constituents in the boilerwater become corrosive when concentrated.

ROLE OF OXYGEN

The presence of oxygen accelerates the combining of iron and water. Oxygen can react with iron hydroxide to form either a hydrated ferric oxide or magnetite. Generally localized, this reaction forms a pit in the metal. Severe attack can occur if the pit becomes progressively anodic in operation. Oxygen reacts with hydrogen at the cathodic surface and depolarizes the surface locally. This permits more iron to dissolve, gradually creating a pit.

The most severe corrosion action occurs when a deposit covers a small area. The creation of a differential aeration cell about the deposit can lead to a severe local action. The metal beneath the deposit is lower in oxygen than areas surrounding it, becomes anodic and is attacked. Pitting is most prevalent in stressed sections of boiler tubing, such as at welds and cold-worked sections, and at surface discontinuities in the metal.

Power plants employing tight cycles to prevent oxygen infiltration and condenser leakage are generally free from corrosion problems. On the other hand, many cycles are vulnerable to oxygen leakage into the feedwater as a result of design or operation.

Efficient operation of boilers requires the exclusion of oxygen from the feedwater. The normal guaranteed value of oxygen leaving the deaerating heater or a deaerating condenser is less than 0.005 ppm. To achieve low residual, it is necessary to exclude air leakage into the condenser, to judiciously control the addition of undeaerated water to the condensate or feedwater, to prevent the addition of aerated heater drips into the condensate, and to assure the exclusion of air into the feedwater cycle during short outages of the boiler.[7]

One major problem in curtailing corrosion from oxygen is the exclusion of air upon boiler start-up. Normally, pressure in a deaerator is not attained until steam is admitted to the turbine and bleed steam is available for heating. It is possible to introduce more oxygen into the boiler at this time than in several months of normal operation. Admitting auxiliary steam to a deaerator to pressurize the unit to 3–5 psig can prevent much of this problem. In this condition, air is excluded and the feedwater delivered to the boiler is low in oxygen during start-up operation.

REMOVAL OF RESIDUAL OXYGEN

Chemical agents added to the feedwater or boilerwater are generally used to remove small residual quantities of oxygen. Hydrazine (N_2H_4) and sodium sulfite (Na_2SO_3) are the most commonly used "oxygen scavengers".

Sodium sulfite, when added to the boilerwater reacts with oxygen to form sodium sulfate as follows:

$$2Na_2SO_3 + O_2 \rightarrow 2Na_2SO_4$$

Temperature affects the reactivity and stability of sodium sulfite (as well as hydrazine). Generally, a cobalt catalyst is added with the sodium sulfite solution to speed up the reaction at lower

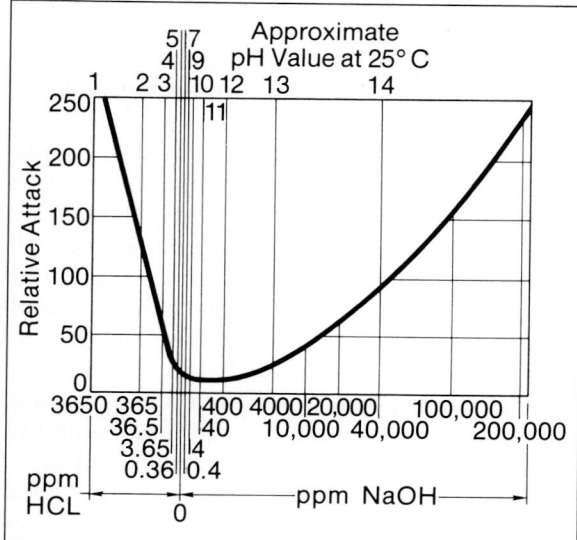

Fig. 12 Relative general corrosion rate of carbon steel versus pH

temperatures. Sulfite decomposition increases with temperature and local concentration in boiler water. Sodium sulfite decomposes in boiler water as follows:

$$Na_2SO_3 + 2H_2O \rightarrow 2NaOH + H_2SO_3$$
$$H_2SO_3 \rightarrow H_2O + SO_2$$

Sulfur dioxide (SO_2) is an acidic anhydride which increases corrosion of metals when it is dissolved in condensate films formed on wetted surfaces of a turbine or condenser.

Concentration of sulfite must be controlled to minimize decomposition at elevated pressures. Recommended limits are shown in Table III. For this reason, sulfite should not be used for reducing oxygen in boilers at pressures above 1200 psig.

Sodium sulfite is an effective reducing agent on boiler water. Besides its reaction with oxygen, the chemical reduces oxides of iron and copper as follows:

$$2CuO + Na_2SO_3 \rightarrow Cu_2O + Na_2SO_4$$
$$3Fe_2O_3 + Na_2SO_3 \rightarrow 2Fe_3O_4 + Na_2SO_4$$

Its consumption in boiler water is a measure of oxygen and oxidized substances added with the feedwater into a boiler. The reduction of cupric oxide begins at about 280°F and is complete at about 400°F. Reduction of ferric oxide begins at about 440°F and is complete at 540°F. Hydrazine reacts with oxygen to form nitrogen and water as follows:

$$N_2H_4 + O_2 \rightarrow N_2 + 2H_2O$$

Hydrazine reacts with oxygen very slowly at temperatures below 350°F. Above 450°F, hydrazine is decomposed rapidly to nitrogen, hydrogen and ammonia. The principal benefit of hydrazine is its ability to reduce the oxidized forms of copper and iron. In this state the general corrosion of the metal surfaces is reduced

Table III. Sodium Sulfite Limits

Boiler Pressure, psig	Concentration, ppm
Below 600	20–40
600–900	5–10

in the preboiler cycle. Copper oxide is reduced with hydrazine at temperatures as low as 150°F. Iron oxide (Fe_2O_3) can be reduced at a temperature of 250°F.

Normally, hydrazine is added to the cycle at the outlet of the condensate pump at a rate to assure a residual of 10–20 ppb (parts per billion) at the inlet of the economizer.

Some of the reactions of hydrazine in the feedwater cycle and boiler are

Decomposition
$$3N_2H_4 \rightarrow N_2 + 4NH_3$$
$$2N_2H_4 \rightarrow N_2 + H_2 + 2NH_3$$

Reduction
$$6Fe_2O_3 + N_2H_4 \rightarrow 4Fe_3O_4 + N_2 + 2H_2O$$
$$4CuO + N_2H_4 \rightarrow 2Cu_2O + N_2 + 2H_2O$$

Reducing agents aid in curbing corrosion but will not prevent metal attack when oxygen is present in the boiler feedwater. At higher temperatures, the reaction rate of oxygen with steel exceeds that of the reducing agents.

FEEDWATER pH CONTROL

Besides the control of oxygen, it's important that the pH of the water be maintained in the proper range. In high pressure units, feedwater is sprayed into superheated and reheated steam to help in controlling final temperatures. Therefore, to assure protection to preboiler equipment, volatile alkaline chemicals have been developed to control the pH of the water without forming solid residue in parts of the boiler where evaporation takes place. Classed as neutralizing amines, the most common of these are ammonia (NH_3), morpholine (C_4H_9NO) and cyclohexylamine ($C_6H_{11}NH_2$). These compounds volatilize with the steam and can react with gases like carbon dioxide to neutralize any acidity in the condensate system. The selection of any of these chemicals are based on obtaining an optimum fit between their chemical and physical characteristics (volatility, solubility, stability, etc.) and the particular application.

Table IV lists the stability characteristics of volatile amines. Both morpholine and cyclohexylamine have temperature stability limits. Ammonia, hydrogen and carbon decomposition products are formed in the dissociation of these amines at high temperatures. Ammonia is stable and has been used at steam temperatures as high as 1200°F. As a result, ammonia is the recommended compound for use in controlling pH of condensate in high-pressure, high temperature boiler systems. However, it's important to maintain ammonia levels generally below 1 ppm to avoid copper alloy attack. A copper ammonium complex is formed in the presence of high ammonia concentrations and the reaction can be further accelerated by the presence of abnormal concentrations of oxygen. Typical reactions are as follows:

$$Cu^+ + 2NH_3 \rightarrow Cu(NH_3)_2{}^+$$
$$Cu^{++} + 4NH_3 \rightarrow Cu(NH_3)_4{}^{++}$$

It has been established that an ammonia concentration of less than 0.5 ppm will assure a protective pH in the cycle without attacking copper surfaces.

Morpholine and cyclohexylamine are preferred to ammonia for pH control in situations where excessive decomposition of these chemicals does not occur and an extensive condensing system or process piping network exist.

USE OF CHELANTS FOR INTERNAL CONTROL

Chelants are weak organic acids with the capability of complexing or binding many cations (calcium and magnesium hardness as well as heavy metals) into a soluble organic ring structure as illustrated in Fig. 13. Chelants are in the neutralized sodium-salt form. They hydrolyze in boiler water to an organic anion. Full hydrolysis depends on a relatively high pH level. The anionic chelant has reactive sites that attract coordination sites on cations. These coordination sites are areas on the ion that are receptive to chemical bonding. In this manner, cations (hardness salts) entering the boiler as contamination from the condensate system, combine with the chelant to form a stable chelate. Deposition of hardness on boiler internal surfaces may therefore be prevented.

Although many substances have chelating properties, *EDTA* (Ethylene Diamine Tetracetic Acid) is the most suitable for boiler feedwater treatment. As with any internal chemical treatment, however, the application and limitations must be clearly understood. Currently, chelants are limited to boiler pressures of less than 1000 psig because of their thermal decomposition at saturation temperatures corresponding to higher pressures. In addition, overfeeding of chelants can lead to corrosion of boiler tubes as chelation of the protective magnetite (Fe_3O_4) film can occur. For a more detailed discussion, see reference 1.

CORROSION DAMAGE IN POWER PLANTS

The major purpose of feedwater and boiler-water treatment is to avoid corrosion and associated tube failures. Most materials that form boiler deposits originate in the preboiler system. Adherence to recommended operating procedures during start-up, normal operation,

Table IV. Stability of Volatile Water-Treating Materials

Chemical	Formula	Pressure psig	Temp °F	Percent Decomposed
Ammonia	NH_3	4270	1202	0
Cyclohexylamine	$C_6H_{11}NH_2$	4270	1202	88
Morpholine	$O(C_2H_4)_2NH$	4270	1202	100
Hydrazine	N_2H_4	4270	1202	100

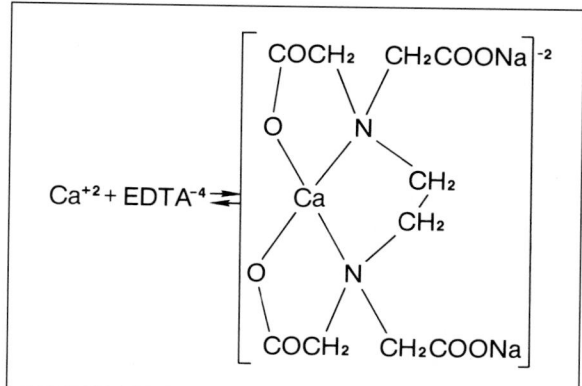

Fig. 13   Calcium bonded with EDTA

shutdown, and outages of a power plant is vital to minimize corrosion and to avoid the entry of corrosion products into the steam generator. Corrosion damage may be in the form of a general attack of the metal surface, pitting, or localized attack of the metal surface beneath internal deposits.

GENERAL SURFACE ATTACK

One type of corrosion is oxygen attack of ferritic materials of construction. Such corrosion may take place throughout the preboiler/ boiler cycle. Corrosion from oxygen can occur during any period of the power-plant lifetime, from erection to operation, as well as standby conditions.

Another type of general overall corrosion is caused by lack of boiler water pH control during system operation. Improper condensate and feedwater pH levels also cause this type of general attack within the preboiler system.

PITTING AND LOCALIZED ATTACK

Pitting of various types can affect the internal surfaces of all tubes and components. Pitting is electrochemical in nature, with the pitted area anodic to the surrounding nonpitted metal surface. If conditions favorable to pitting exist —such as excessive dissolved oxygen in the boiler water/feedwater or an acidic chemical environment—one of several normally innocuous factors can cause severe local attack. Crevices, such as those formed by deposits or minor variations in metallurgical structure, may act to promote localized corrosion. Normal, but higher-than-average, peak stress also can contribute to preferential pitting.

There has been the increased trend in fossil systems toward peaking, variable-pressure, low-load, and two-shifting operations, with frequent shutdowns—all which have contributed to increased numbers of failures associated with peculiar corrosion patterns related to frequent introduction of oxygen during outage and start-up periods. A look at several instances of pitting attack will provide a better appreciation for the need in developing protection when the corrosion potential is high.

EXAMPLES OF PITTING CORROSION

Figs. 14 through 23 show a variety of corrosive attack attributable to pitting. Generally, the term *pitting* denotes localized crater-like attack of a type shown in Figs. 14 and 15. But there are failures in which individual pits join, producing crack-like penetrations of the type indicated in Fig. 16. In the absence of a corrosion or metallurgical term describing this type of corrosion penetration, the condition has been included under the broad category of pitting attack.

The orientation and location of the crack-like indications shown in Fig. 17 can best be explained by pitting preferential to an area of somewhat higher strain than the surrounding metal. Differences in strain promote the formation of anode-cathode cells and the resulting corrosion at the anode is oriented perpendicular to the direction of maximum residual or applied strain.

The pitting attack shown in Figs. 18 and 19 affected the entire tube circumference, but was deepest under the welded fin. In this instance, differences in metallurgical structure may also have been a factor in promoting the electrochemical cell.

This type of pitting attack has also been found in some lower waterwall drums. Fig. 20 shows an example of that attack.

The overall character of these cracklike indications are not typical of fatigue cracking. The bottom of the penetrations show no evidence of cracking and the shape of the void indicates they were formed as a result of corrosion. When tubes exhibiting internal surface cracklike indications (Fig. 21) have been reverse flattened and bent to force fracture along these linear indications, the penetrations show themselves to be a series of aligned corrosion pits (Fig. 22). The evidence strongly suggests that a pitting mechanism can initiate and propagate cracklike penetrations. In some instances the penetrations appear to be a series of pits atop one another, giving the impression of separate time periods when pitting attack has resumed at the bottom of the penetrations as shown in Fig. 23.

Fig. 14 Pitting on internal surface of waterwall tube (mag 3/4X)

Fig. 17 Sectional view across cracklike penetration shown in Fig. 16

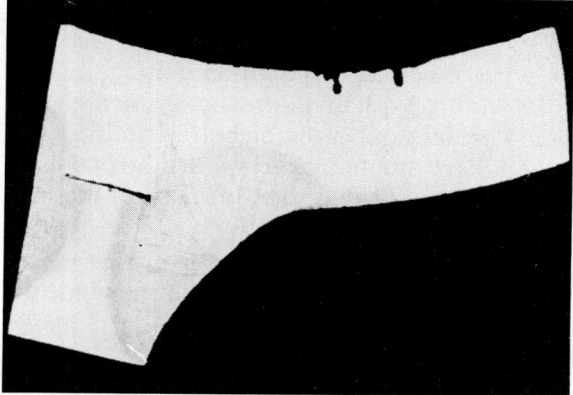

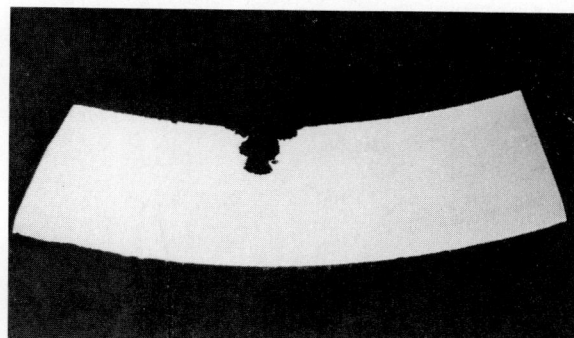

Fig. 15 Sectional view of one pit shown in Fig. 14 (mag 3-1/2X)

Fig. 16 Cracklike corrosion penetrations on internal surface of waterwall tube

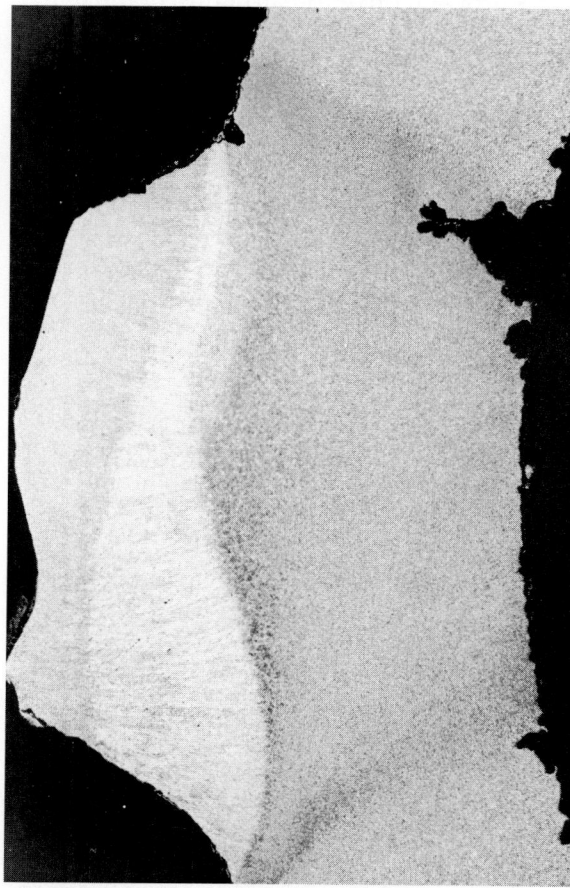

Fig. 18 General pitting attack from inside surface of waterwall tube. Pitting affected entire tube circumference but was more severe under external fin (mag 8X).

ASEA BROWN BOVERI

CORROSION DAMAGE IN BOILERS

Deviations from recommended chemistry limits which result in depressed or elevated pH values promote failures of boiler tubing. Although there are many variations, the majority of failures can be classified into one of the following two categories:

■ Ductile gouging—Normally, irregular wastage of the tube metal beneath a porous deposit characterizes this type damage. The damage progresses to failure when the tube wall thins to a point where stress rupture occurs locally. In this process, the microstructure of the metal does not change and the tubing retains its ductility. Fig. 24 illustrates ductile gouging.

■ Hydrogen damage—This type corrosion damage usually occurs beneath a relatively dense deposit. Although some wastage occurs, the tube normally fails by thick-edge fracture before the wall thickness is reduced to the point where stress rupture would occur. Hydrogen, produced in the corrosion reaction,

Fig. 20 Cracklike corrosion penetration at inside surface of lower waterwall drum at toe of nozzle weld

Fig. 19 Pitting attack similar to Fig. 18 with deep, clearly corrosion-related, localized, cracklike penetration under external fin (mag 8X)

Fig. 21 Internal surface of tube with longitudinally oriented, cracklike indications (mag 3X)

diffuses through the underlying metal, causing decarburization and intergranular micro-

fissuring of the structure. Brittle fracture occurs along the partially separated boundaries, and in many cases, an entire section is blown out of an affected tube. Examples of these failures are shown in Figs. 25 and 26.

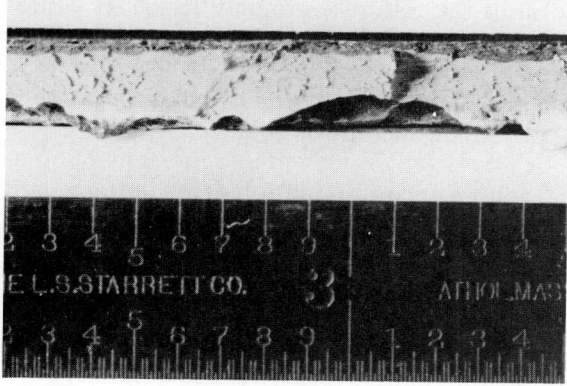

Fig. 22 Sectional through-the-wall view of Fig. 21 tube after flattening and bending to promote fracture along a penetration. Inside diameter (ID) surface is at bottom, and darker craters from the ID surface are pits. (The remaining wall portion is sound metal which fractured as a result of the flattening and reverse bending.) The cracklike penetrations result from aligned pits.

Fig. 23 Crack-like corrosion penetrations halfway through tube wall, strongly suggesting that propagation results from separate time periods of pitting at the root of the penetration.

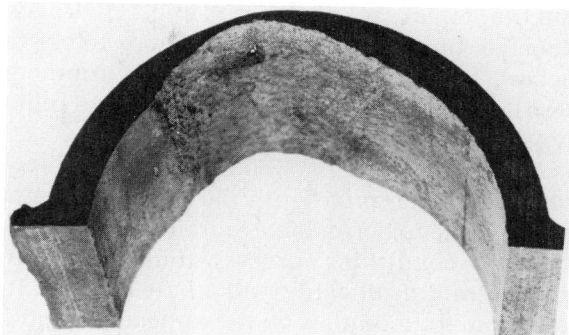

Fig. 24 Ductile gouging

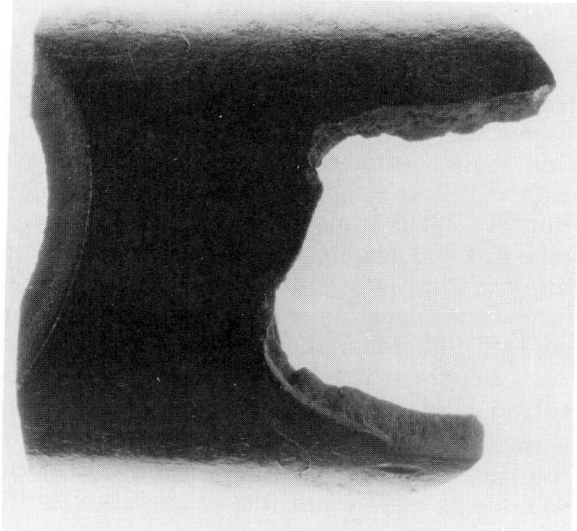

Fig. 25 Thick-edge waterwall tube failure caused by hydrogen damage

Fig. 26 Another typical waterwall failure from hydrogen damage

ASEA BROWN BOVERI

Internal metal-oxide deposits which permit boiler-water solids to concentrate during the process of steam generation accelerate both types of attack. When the boiler water contains highly soluble alkaline treatment chemicals such as sodium hydroxide or potassium salts, ductile attack is more probable. Hydrogen damage, on the other hand, is more likely to occur when a low pH boiler-water environment results from condenser leakage or some other type of system contamination.[9]

According to an ASME/EEI-sponsored laboratory research program[10], clean tubes are not susceptible to corrosion even under high heat-transfer conditions unless unusually high concentrations of acid or alkali are present. In addition, fouled tubes do not appear to be susceptible to corrosion when contaminants are present if proper boiler-water treatment is maintained. These observations can be explained by the "concentrating film" theory.

The Berl and Van Taack curve (Fig. 12) illustrates the relative corrosion of carbon steel under various acid and alkaline conditions. Bulk boiler-water concentrations are always well within the range of low corrosion rates. But, as heat is transferred through the boiler tube wall to a water-steam mixture, a temperature gradient is established, and the internal tube metal temperature must reach a value slightly higher than the bulk fluid temperature. As boiler water is converted to steam, the dissolved solids concentrate in a residual film as shown in Fig. 27. The solids concentration in this film increases until the boiling point of the solution is elevated to the temperature at the tube wall.

It is improbable that such high concentrations can be reached in a clean tube under normal boiling conditions. Theoretically, the bulk of the film-temperature drop is experienced within a laminar film of about 0.001-in. thickness. Free diffusion of ions into the turbulent fluid would prevent the establishment of a concentration gradient. Only when porous internal deposits are formed in areas of heat absorption is it possible to produce very high concentrations. The deposit itself acts as a diffusion barrier and in so doing increases the probability of high local concentrations being reached.

Because porous deposits may exist, boiler-water chemicals should be of a type that will not become highly corrosive when concentrated. Experience indicates that caustic soda and potassium salts are particularly objectionable. Alkalinity can be produced with sodium phosphate (coordinated phosphate treatment) without the risk of extremely high pH values being present upon concentration. Alkalinity control of boiler water may also be achieved by adding a volatile amine such as ammonia and is required for once-through units.

Condenser cooling waters represent a potential source of contamination of the boiler water. Depending on the chemical constituents dissolved in the cooling water, highly acid or alkaline materials may be formed in the boiler environment as a result of condenser leakage.

Because of its poor buffering capability, volatile treatment provides little protection against

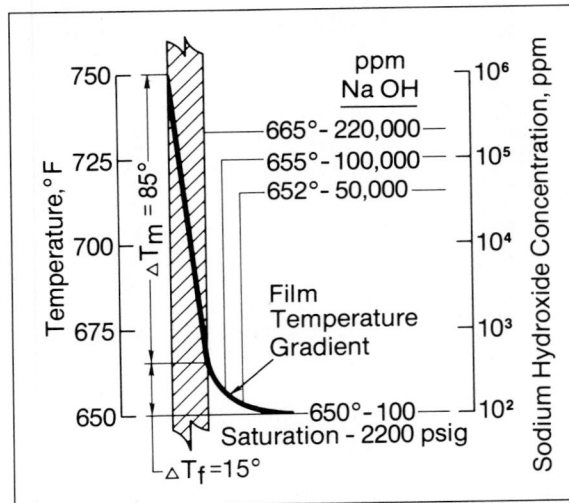

Fig. 27 Illustration of "concentrating film" theory. Film temperature gradient of 15° is developed because of heat transfer. Soluble solids in boiler water are concentrated in the surface film during steam generation to produce "nonboiling equilibrium." Although the caustic concentration in the boiler water is only 100 ppm, the concentration in the surface film must equal 220,000 ppm to elevate the boiling point 15°.

acid attack of boiler tubing. With a stronger alkaline additive, coordinated phosphate treatment results in a more highly buffered boiler water, one in which acid-producing reactions from salt inleakage are better suppressed. Optimum treatment programs based on this method are discussed in the following section.

TUBE FAILURES
CAUSED BY OVERHEATING

While the preceding discussion focused on the role played by internal deposits in the cor-

rosion of boiler tubes, it should be emphasized that deposits also promote the majority of tube failures caused by overheating.

Fig. 28 illustrates the temperature conditions existing in a typical furnace-wall tube. The heat flux of 100,000 Btu/ft² · hr is established by the furnace flame-temperature spectrum and the bulk fluid temperature of 640°F is fixed by boiler pressure. The temperature profile is then established by the ΔT required to drive the 100,000 flux through each segment of resistance. With nucleate boiling on a clean tube, the ΔT across the fluid film is quite small (10°F) and the overall temperature differential (90°F)

Fig. 28 **Furnace wall tube—nucleate boiling**

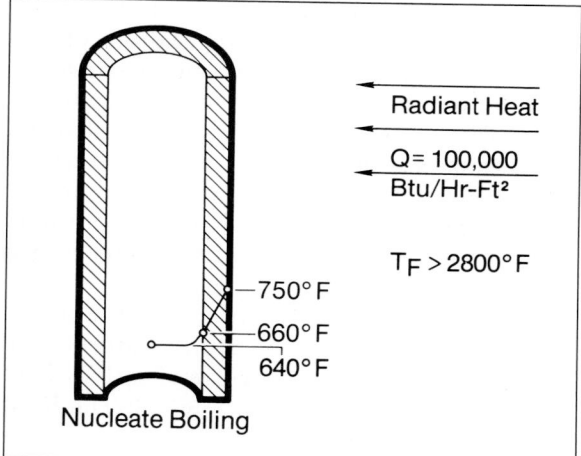

Fig. 30 **Furnace wall tube—film boiling**

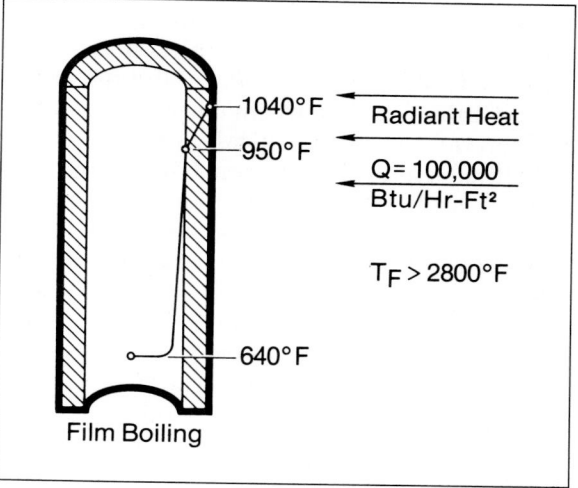

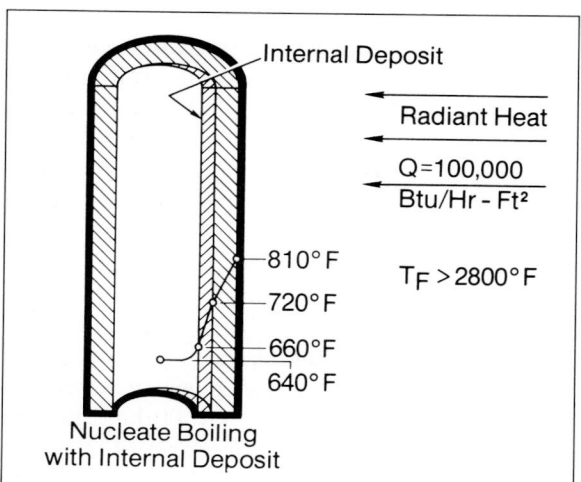

Fig. 29 **Furnace wall tube—nucleate boiling with deposit**

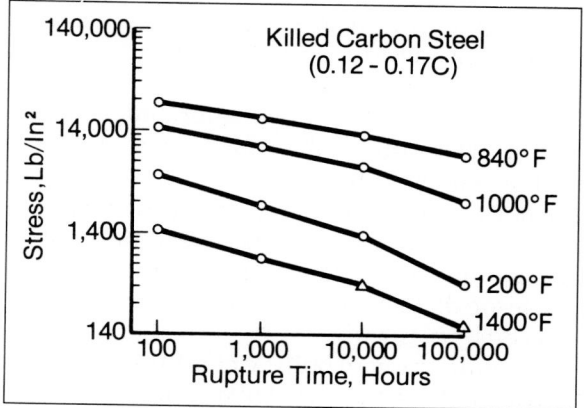

Fig. 31 **Stress and temperature versus time-to-rupture**

results from the resistance of the tube metal.

Fig. 29 shows what happens to the tube-temperature profile if an internal deposit exists. Approximately 60°F is required to drive the 100,000 heat flux through the deposit, so both the inside and outside metal temperatures must rise correspondingly.

Fig. 30 illustrates the temperature profile under a film-boiling condition. The reduced-film coefficient increases the ΔT across the film to 310°F and the metal temperatures rise correspondingly. It should be noted that the flux is constant at 100,000 Btu/ft² · hr for all cases.

STRESS RUPTURE

The strength of carbon steel tubing is nearly constant to about 840°F, at which point it begins to lose strength rapidly. If tube metal temperature is gradually increased beyond this temperature, it will plastically deform (creep) and then rupture. The approximate time-to-rupture is a function of the hoop-stress and the temperature (Fig. 31). The hoop stress (S) equals the product of the internal pressure (P) times the tube radius (r) divided by the wall thickness (x).

$$S = \frac{P\,r}{x}$$

(1)

Simple visual examination of a failed section reveals a great deal about the tube-temperature history. In deposit-caused overheating, local tube temperatures rise gradually so that specific points begin to creep before others (blistering). Deposits tend to form nonuniformly, and even a 20°F difference in temperature can have a pronounced effect on localized creep. Blistering is almost always caused by internal deposition, and conversely, overheating because of deposits will almost always be evidenced by blisters.

When an upset of boiling conditions, such as departure from nucleate boiling (DNB) elevates tube temperature, a temperature excursion will occur. Metal temperature will jump quickly

from the normal condition to a higher value consistent with the upset.

If the temperature excursion is great, the tube may fail very quickly. For instance, if a tube is completely blocked during operation, the tube temperature will rise immediately to the failure-point temperature. The rapid plastic deformation of these very short-term ruptures produces very thin-edged fractures. In the case of major flow interruptions, the tube bulges over an extended length and fails like a balloon bursting.

If a DNB excursion places the tube-metal temperature in the range of 840 to 1020°F, the tube will take a longer time to fail. During this period, intergranular oxidation of the metal will occur and will weaken the structure. Long-term failures, therefore, tend to be thick-edged. The higher the temperature and stress conditions are above the creep limit, the shorter the time to failure and the thinner the metal edge at the fracture. Fig. 32 presents examples of a short, an intermediate, and a long-term failure.

Metallographic analysis is an important tool for analyzing tube-temperature history. The normal pearlitic microstructure of carbon steel boiler tubing can be seen in Fig. 33. The light areas are iron, and the dark areas are pearlite consisting of alternate layers of iron and iron carbide. The normal operating temperature of a boiler tube will not alter this structure. If the temperature of steel is maintained in the range of 930 to 1340°F for an extended period of time, carbon migrates from the pearlite to form spheres of carbon. This process, known as spheroidization, progresses slowly at 930°F and rapidly at temperatures near the lower critical range (1340°F). Fig. 34 shows carbon steel with various degrees of spheroidization.

If the tube is heated to a temperature greater than 1340°F and is then allowed to cool slowly to below 930°F, the normal pearlitic structure will be re-established. But, if the tube is heated above 1340°F and is cooled very rapidly, a martensitic structure (Fig. 35) will be produced. This type of microstructure is likely to be found in very short-term overheating failures in

which the water escaping through the rupture suddenly quenches the overheated tube metal.

RESEARCH STUDY
ON INTERNAL CORROSION

The previously referenced *ASME/EEI*- sponsored laboratory research program on internal corrosion was instrumental in bringing an understanding to the various corrosion mechanisms discussed in the previous sections. The following recap of the more important conclusions includes our interpretation of their practical significance:

1. When heat-transfer surfaces are free of deposits, corrosion is unlikely to occur over a wide range of heat-transfer and boiling conditions regardless of the method of water treatment employed.

2. Corrosion will not occur with volatile treatment of pure boiler water (less than two ppm solids) even when heat-transfer surfaces are fouled and/or conditions of DNB exist.

3. Corrosion will not occur when coordinated phosphate treatment is employed even when heat-transfer surfaces are fouled and/or conditions of DNB exist.

4. Under the conditions of item 3, chemical hideout of sodium phosphate will occur if heat-transfer surfaces are fouled and/or DNB conditions exist.

5. There appears, however, to be no relationship between hideout and corrosion of tube metal.

Fig. 32 Physical appearance of overheating failures

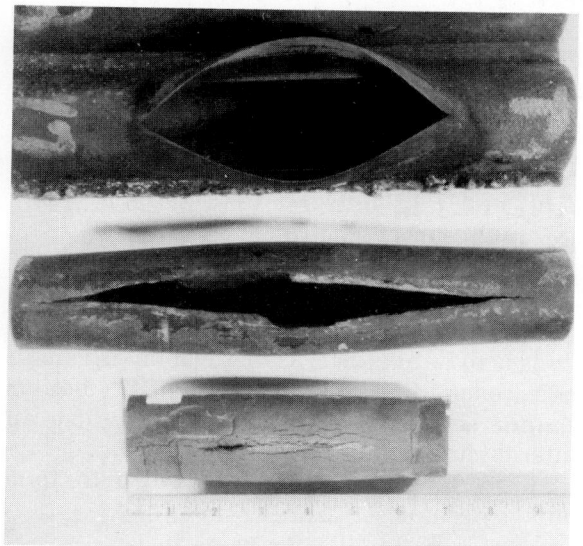

Fig. 33 **Photomicrograph—
normal pearlitic structure of carbon steel**

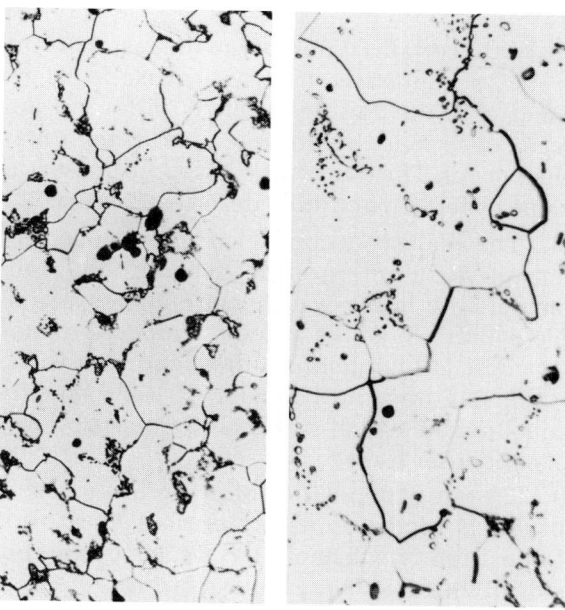

Fig. 34 **Spheroidization: left, mild; right, extensive**

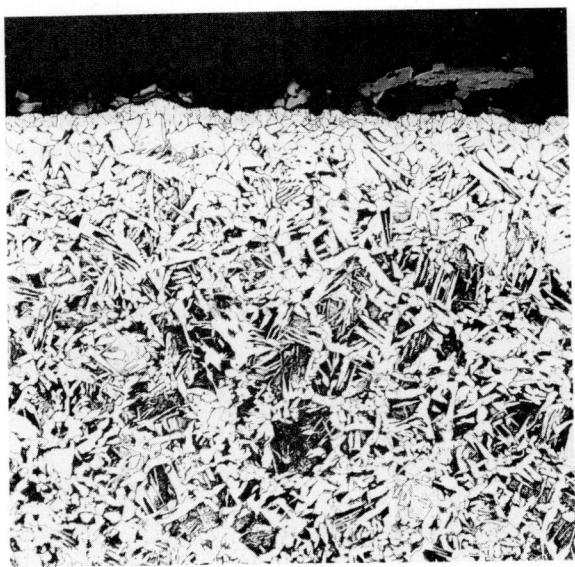

Fig. 35   Martensitic structure

6. Caustic soda concentrations ranging from 10–20 ppm in the boiler water (with a pH of from 10.5 to 10.8) can cause serious corrosion if internal surfaces are fouled with deposits. Corrosion is rapid even under nucleate boiling conditions, and complete penetration of a 5-mm tube wall can occur in several days.

7. With certain types of condenser cooling water, a condenser leak will cause a pH reduction in the boiling water. With fouled surface conditions, hydrogen embrittlement may then occur. The rate of attack is rapid and of the same order of magnitude described in item 6.

8. The analysis of cooled samples can be very deceptive in interpreting behavior in high-temperature boiler-water environment; a sample which is alkaline at room temperature may be acidic at high temperatures, and vice-versa. Understanding the behavior of each specific contaminant which can be introduced by way of condenser leakage is essential.

9. The rate of deposition of preboiler corrosion products increases with increasing heat-flux. Deposition is substantially greater on the hot side of the tube where boiling occurs.

The research program did not conclude that corrosion was the fault of either the operator or the designer. It demonstrated that both chemical and mechanical factors are involved and that they are largely interdependent. A change in one variable will affect the manner in which other variables influence corrosion. Normally, several factors must be simultaneously out-of-line to produce corrosion. As a result, corrosion is a fairly low probability event despite the fairly large number of potentially adverse physical and chemical conditions which temporarily exist.

UTILITY BOILER-WATER AND FEEDWATER TREATMENT IN THE PREVENTION OF CYCLE CORROSION

If extensive corrosion during outage conditions is to be avoided, proper wet lay-up methods are essential. Because large utility boilers cannot be successfully dried to produce an internal atmosphere free of moisture, wet lay-up procedures are recommended even for long outage periods.[11]

PROTECTION DURING OUTAGE CONDITIONS

For any outage longer than four days, the preboiler/boiler cycle should be filled, to the greatest extent possible, with a solution inhibited to prevent corrosion, and pressurized with nitrogen to prevent air inleakage. Excellent results have been obtained with condensate containing appropriate amounts of hydrazine and ammonia for lay-up periods of several years. As both materials are volatile, there is no objection to their use even in nondrainable superheater and reheater sections of the boiler. Where freezing is a problem, it is sometimes necessary to drain portions of the steam generator. This should be accomplished under a positive nitrogen pressure. Nitrogen blanketing is recommended even when the boiler can be maintained full of inhibited condensate.

The success or failure of standby protection also depends upon the proper design and selec-

Table V. Recommended Lay-Up Procedures — C-E Drum-Type Utility Units

Type of Shutdown	Procedure	Notes
Preoperational Period Post Hydro (See Note 1)	With the economizer, waterwalls, superheater, and reheater filled to overflowing pressurize the unit with nitrogen to 5 psig pressure (See Notes 4 & 6).	1. All nondrainable sections to be hydrostatically tested should be filled with demineralized or condensate quality water containing 10 ppm of ammonia and 200 ppm of hydrazine. This should produce a solution pH of approximately 10.0. The superheater should be filled first, to overflow into the boiler drum. Then the economizer and waterwalls can be filled through normal fill connections (See Note 2) with demineralized or condensate quality water, or if not available, any source of clean, filtered water may be used. This water should also contain 10 ppm of ammonia and 200 ppm of hydrazine.
Preoperational Period Post Chemical Cleaning	1. Introduce demineralized or condensate quality water containing 10 ppm of ammonia and 200 ppm of hydrazine into the superheater, reheater, feedwater heaters, (tube side) and associated piping, economizer and waterwalls (Refer to Notes 3 & 4). 2. Nitrogen cap the superheater, feedwater heaters (shell side) and drum. Maintain 5 psig nitrogen pressure (See Notes 5 & 6).	2. Hydrazine and ammonia should be added in a manner that results in a uniform concentration throughout. They may be added to the system in several ways, as for example: a. By pumping concentrated solutions through the chemical feed equipment and blend filling to achieve the desired concentrations. b. If condenser leakage is not a cause for shutdown, concentrated solutions can be introduced directly into the hotwell where they can be mixed to achieve the desired concentrations. If condensate demineralizers are used, they must be bypassed during this operation.
Short Outage— 4 Days or Less	1. Maintain the same hydrazine and ammonia concentrations as those present during normal operation. 2. Establish and maintain a 5 psig nitrogen cap on the superheater and the steam drum (See Notes 5 & 6). 3. Nitrogen cap the shell side of the feedwater heaters.	It is important to have the fluid temperature in the cycle below 400°F before addition of hydrazine. If this temperature is exceeded, the hydrazine will decompose.
Short Outage— 4 Days or Less, Unit Partially Drained for Repairs	1. Drain and open only those sections requiring repairs. 2. Isolate remainder of unit under 5 psig nitrogen pressure where possible (See Notes 5 & 6). 3. Maintain the same hydrazine and ammonia concentrations for water remaining in the cycle as those present during normal operation. 4. Nitrogen cap the shell side of the feedwater heaters.	3. The tube side of copper alloy feedwater heaters should be filled with demineralized water containing 0.5 ppm of ammonia and 50 ppm of hydrazine. 4. If freezing is a problem, the water in drainable circuits can be displaced with nitrogen and the unit laid up under 5 psig nitrogen pressure. Auxiliary heat may be applied to keep the nondrainable sections from freezing.
Long Outage— Longer than 4 Days	1. Fill the superheater and reheater with demineralized or condensate quality water containing 10 ppm of ammonia and 200 ppm of hydrazine. The pH of the solution should be approximately 10.0. Add the fill water to the outlet of the nondrainable sections (See Note 4). 2. Increase the hydrazine and ammonia concentration in the waterwalls, economizer and feedwater heaters (tube side) and associated piping to 200 ppm and 10 ppm respectively (See Notes 2, 3, and 4).	5. If the reheater can be isolated, it should also be capped with 5 psig nitrogen. When the outage is expected to extend beyond 2 months, provisions should be made to isolate and nitrogen cap the reheater, if this was not done previously.

Table V. Recommended Lay-Up Procedures — C-E Drum-Type Utility Units

Type of Shutdown	Procedure	Notes
	3. Establish and maintain a 5 psig nitrogen cap on superheater and drum (Notes 5 & 6). **4.** Nitrogen cap the shell side of the feedwater heaters.	**6.** Nitrogen cap should be applied through the drum vent and superheater outlet header drain/vent, as the unit is cooled, when pressure drops to 5 psig. Admission of air through atmospheric vents should be avoided.
Long Outage— Longer than 4 Days, Unit Partially Drained for Repairs	**1.** Drain and open only those sections requiring repairs. **2.** Fill the superheater and reheater (if not requiring draining for repairs) with demineralized or condensate quality water containing 10 ppm of ammonia and 200 ppm of hydrazine. The pH of the solution should be approximately 10.0. Add the fill water to the outlet of nondrainable sections (Note 4). **3.** Increase the hydrazine and ammonia concentrations in the tube side of the feedwater heaters and the undrained circuits of the economizer and waterwalls to 200 ppm and 10 ppm respectively (See Notes 2, 3 & 4). **4.** Establish and maintain a 5 psig nitrogen cap on the undrained sections of the unit, where possible (See Notes 5 & 6). **5.** Nitrogen cap the shell side of the feedwater heaters. **6.** After completion of the repairs fill the drained sections with demineralized or condensate quality water containing 10 ppm of ammonia and 200 ppm of hydrazine. Cap with nitrogen (See Notes 2, 3, 4, & 5).	

Table VI. Recommended Lay-Up Procedures C-E Combined Circulation Units

Type of Shutdown	Procedure	Notes
Post Chemical Cleaning Preoperational Period	**1.** Introduce demineralized or condensate quality water containing 10 ppm of ammonia and 200 ppm of hydrazine into the feedwater heaters (tube side), economizer, furnace walls, superheater and reheater (Refer to Notes 2 & 3). **2.** Nitrogen blanket the shell side of the feedwater heaters, superheat, and furnace wall sections. Maintain 5 psig nitrogen pressure (Refer to Note 4).	**1.** Hydrazine and ammonia should be added in a manner that results in a uniform concentration throughout. They may be added to the system in several ways as for example: a. By pumping concentrated solutions through the chemical feed equipment and blend filling to achieve the desired concentrations. b. If condenser leakage is not a cause for shutdown, concentrated solutions can be introduced directly into the hotwell where they can be mixed to achieve the desired concentrations. The condensate polishing system must be bypassed. During hydrazine and ammonia introduction, circulation should be maintained through BE and WD valves to insure
Short Outage 4 Days or Less	**1.** Maintain the same hydrazine and ammonia concentrations as those present during normal operation. **2.** Establish and maintain a 5 psig nitrogen cap on the superheater and furnace wall sections (Refer to Note 4). **3.** Nitrogen blanket the shell side of the feedwater heaters.	

Table VI. Recommended Lay-Up Procedures — C-E Combined Circulation Units

Type of Shutdown	Procedure	Notes
Short Outage 4 Days or Less, Unit Partially Drained for Repairs	**1.** Drain and open only sections requiring repairs. **2.** Isolate remainder of unit under 5 psig nitrogen pressure where possible (Note 4). **3.** Maintain the same hydrazine and ammonia concentration for water remaining in the cycle as during normal operation. **4.** Nitrogen blanket the shell side of the feedwater heaters.	adequate mixing in the system. If the superheat section is to be laid up wet, circulation should be maintained through the BE, SA and SD valves. Furnace wall outlet pressure should not exceed 1000 psig while maintaining circulation during introduction of hydrazine and ammonia. It is important to have the fluid temperature in the cycle below 400°F before the addition of the hydrazine. If this temperature is exceeded, the hydrazine will decompose. **2.** All sections to be hydrostatically tested should be filled with demineralized or condensate quality water containing 10 ppm of ammonia and 200 ppm of hydrazine. This should produce a solution pH of approximately 10.0. **3.** If freezing is a problem, the water in the drainable circuits can be displaced with nitrogen and the unit laid up under 5 psig nitrogen. Auxiliary heat may be applied to keep the nondrainable sections from freezing. **4.** If the reheat section can be isolated, it should also be capped with 5 psig nitrogen. When the outage is extended to 2 months or longer, provisions should be made to isolate and nitrogen cap the reheat sections, if this was not done previously.
Long Outage Greater than 4 Days	**1.** Fill the superheater and reheater with demineralized or condensate quality water containing 10 ppm of ammonia and 200 ppm of hydrazine. The pH of the solution should be approximately 10.0. Add the fill water to the outlet of nondrainable sections (Note 3). **2.** Increase the hydrazine and ammonia concentrations in the economizer, furnace walls, and feedwater heaters (tube side) and associated piping to 200 ppm and 10 ppm respectively (Refer to Notes 1 & 3). **3.** Establish and maintain a 5 psig nitrogen cap on the furnace wall and superheat sections (Refer to Note 4). **4.** Nitrogen cap the shell side of the feedwater heaters.	
Long Outage Greater than 4 Days, Unit Partially Drained for Repairs	**1.** Drain and open only those sections requiring repairs. **2.** Fill the superheater and reheater (if draining for repairs is not required) with demineralized or condensate quality water containing to 10 ppm of ammonia and 200 ppm of hydrazine. The pH of the solution should be approximately 10.0. Add the fill water to the outlet of nondrainable sections (Note 3). **3.** Increase the hydrazine and ammonia concentrations in the tube side of the feedwater heaters and the undrained circuits of the economizer and furnace walls to 200 ppm and 10 ppm respectively. (Notes 1 & 3). **4.** Establish and maintain a 5 psig nitrogen cap on the undrained section of the unit, where possible (Refer to Note 4). **5.** Nitrogen cap the shell side of the feedwater heaters. **6.** After repairs, fill drained sections with demineralized or condensate quality water containing 10 ppm of ammonia and 200 ppm of hydrazine. Cap with nitrogen (Notes 1 & 3).	

tion of chemical-feed equipment and the ability to provide an inert gas blanket.[12] Utilities with operating units that are predominantly base-loaded may get along with simple, inexpensive manual systems for introduction of lay-up additives and nitrogen. Systems having load characteristics that result in frequent boiler shutdowns require more elaborate, foolproof, wet lay-up protection systems if problems are to be avoided. Tables V and VI detail recommended procedures to be utilized on all high-pressure utility units.

MINIMIZING CORROSION DURING START-UP OPERATION

During the start-up of a high-pressure boiler, deaeration and pH adjustment must be provided to assure that low feedwater-corrosion-product levels are achieved. Plant start-ups without deaeration will promote continued corrosion of feedwater-system materials of construction.[13,14,15] This will result in significant quantities of corrosion products being carried into, and depositing within, the steam generator. The supply of undeaerated feedwater will also increase corrosion of steel components within the steam generator. Deaeration of condensate during start-up operations can be obtained by use of deaerating condensers or by external deaerating heaters. Since deaeration should be accomplished early in the start-up sequence, auxiliary steam must be available for turbine-seal and deaeration requirements.

CYCLING UNITS

An adequate source of auxiliary steam should be available to the deaerating heater while the unit is starting up. If this is unavailable, pegging-steam directly from the drum should be provided until turbine extraction steam is available. As an alternative, a blowdown flash tank designed for 5 percent of the maximum steaming rate of the boiler can also be used. In units without a deaerating heater, deaeration is achieved solely by the condenser. For the condenser to effectively deaerate cold feedwater, a significant vacuum must be maintained. A reasonable approach is to recycle and deaerate the condensate in the system prior to start-up, with the necessary degree of condenser vacuum and fluid temperature.[14,16,17] (See Fig. 36.) This requires connection of a recycle line from the discharge of the feedwater heaters back to the deaerating section of the condenser to prevent the air-rich condensate in the condenser from being introduced into the boiler. The turbine must be sealed and a condenser vacuum established prior to recycle.

Including a preboiler recycle line has an additional benefit when part-flow condensate demineralization is employed in the cycle. Demineralization permits filtration of corrosion products from the recycling feedwater, and will further permit cleanup of the drains from the shell sides of the feedwater heaters during start-up. Characteristically, these heater drains contain the highest levels of crud in the system. A condensate demineralizer system designed for about 25 percent of full-load condensate flow provides all of the benefits described.[13]

Tables VII and VIII list the recommended feedwater control limits for start-up operation of drum-type cycling units.

HIGH-PRESSURE DRUM-TYPE UNITS

Although base-loaded utility drum-type units suffer from corrosion more during operation than during start-up, attention must still be directed toward procedures to prevent excessive corrosion during start-up periods. It

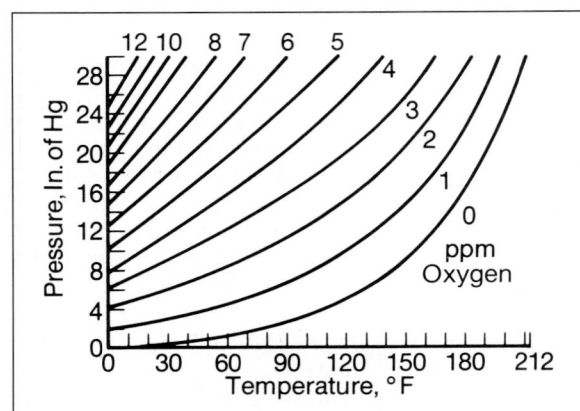

Fig. 36 Solubility of oxygen in water from air saturated with water vapor

has not been found necessary to install special preboiler recirculation lines or condensate demineralization equipment on base-loaded high-pressure drum-type units. Normal start-up sequence which includes the initiation of light steaming rates early in the start-up operation, coupled with boiler drum blowdown, have been found effective in reducing the impact of standby corrosion-product deposition. As already noted, conversion of the base-loaded drum-type unit to peaking or cycling operation can bring with it a multitude of new problems.

SUPERCRITICAL CYCLES: FEEDWATER

Start-up specifications are important in supercritical cycles, and neglect in this area may result in considerable deposit-caused difficulties. Combustion Engineering has placed primary emphasis on the development of start-up/cleanup procedures which assure steam-generator cleanliness over long periods

Table VII. Initial Start-Up and Restart After Long Outages*, Cycling Drum-Type Boiler—see Fig. 37

Phase	Flow Path	Operations	Control Limits	Expected Cleanup Time
	1	Flush to waste at approx. 25% MCR	3 ppm (max.) suspended solids, measured at HP heater outlet	8 Hours
Preboiler Cleanup	2	Pressurize deaerator to 10–15 psig with auxiliary steam, establish condenser vacuum, and place demineralizer in service. Using condensate or feed pump and preboiler recirculating valve, clean up cycle maintaining 25% MCR deaerated flow. Blowdown deaerator storage tank as necessary to prevent recontamination of system.	Total suspended solids 250 (max.) Oxygen 10 Hydrazine 20 pH Copper-alloy system 8.8–9.2 Copper-free system 9.2–9.4 (All concentrations in ppb, measured at HP heater outlet; oxygen measured at deaerator outlet.)	Time dependent on run-in of equipment
System in Service		Close preboiler recirculating valve, place boiler feed pump in service and fire unit for turbine warmup roll, and synchronization. Proceed to normal feedwater control limits. Cascade heater drains to condenser for polishing until unit load exceeds 25%.	Normal Feedwater Limits Total solids 50 (max.) Total iron 10 Total silica 20 Copper 10 Oxygen 5 Hydrazine 10–20 pH Copper-alloy system 8.8–9.2 Copper-free system 9.2–9.4 (All concentrations in ppb, measured at HP heater outlet; oxygen measured at deaerator outlet.)	Normal limits should be obtained before exceeding one-third unit load.

Note: Suspended solids concentration measured by Millipore® filters with appropriate color comparator charts.

*Duration of outage longer than 4 days

of time. A sub-loop cleanup philosophy is utilized in which the crud level in the preboiler system is reduced to a low value before passing feedwater to the steam generator. This procedure assures the delivery to the boiler of feedwater which is low in iron, silica, copper, and oxygen.

Keeping oxygen values below 10 ppb enhances effective and rapid contaminant reduction within the preboiler cycle. Hydrated iron oxide, exposed to high-oxygen values, precipitates readily and delays cleanup time. Conversion of this material by deaeration to black magnetite is essential for fast cleanup because

condensate purification equipment readily removes magnetite. Additionally, the low-oxygen environment prevents further oxide generation.

When performed at a sufficiently high velocity, preboiler recirculation completely removes contaminants prior to furnace-wall cleanup. Based on experience, a velocity of 2 feet per second is recommended for this purpose and is obtained at approximately 25 percent maximum continuous rating (MCR) with currently designed feedwater heaters. When employing the recommended cleanup program, contaminant reduction to specified feedwater quality

Table VIII. Cold Restarts After Short Outages*, Cycling Drum-Type Boiler (Fig. 37)

Phase	Flow Path	Operations	Control Limits		Expected Cleanup Time
Preboiler Cleanup	2	Pressurize deaerator to 10–15 psig with auxiliary steam, establish condenser vacuum, and place demineralizer in service.	Total suspended solids Oxygen Hydrazine pH Copper-alloy system Copper-free system	250 (max.) 10 20 8.8–9.2 9.2–9.4	4 hours if C-E recommended lay-up procedure has been employed.
		Using condensate or feed pump and preboiler recirculating valve, clean up cycle maintaining 25% MCR deaerated flow.			Does not represent a delay; this cleanup may be completed before boiler is available.
		Blow down deaerator storage tank as necessary to prevent recontamination of system.	(All concentrations in ppb, measured at HP heater outlet; oxygen measured at deaerator outlet.)		
System in Service		Place boiler feed pump in service.	Normal Feedwater Limits		
		Fire unit for turbine warmup, roll, and synchronization.	Total solids Total iron Total silica Copper Oxygen Hydrazine pH Copper-alloy system Copper-free system	50 (max.) 10 20 10 5 10–20 8.8–9.2 9.2–9.4	Normal limits should be obtained before exceeding one-third unit load.
		Proceed to normal feedwater control limits.			
		Cascade heater drains to condenser for polishing until load exceeds 25%.	(All concentrations in ppb, measured at HP heater outlet; oxygen measured at deaerator outlet.)		

Notes: 1. No cleanup is required if feedwater quality meets recommended limits.
2. Suspended solids concentration measured by Millipore® filters with appropriate color comparator charts.

*Duration of outage maximum 4 days.

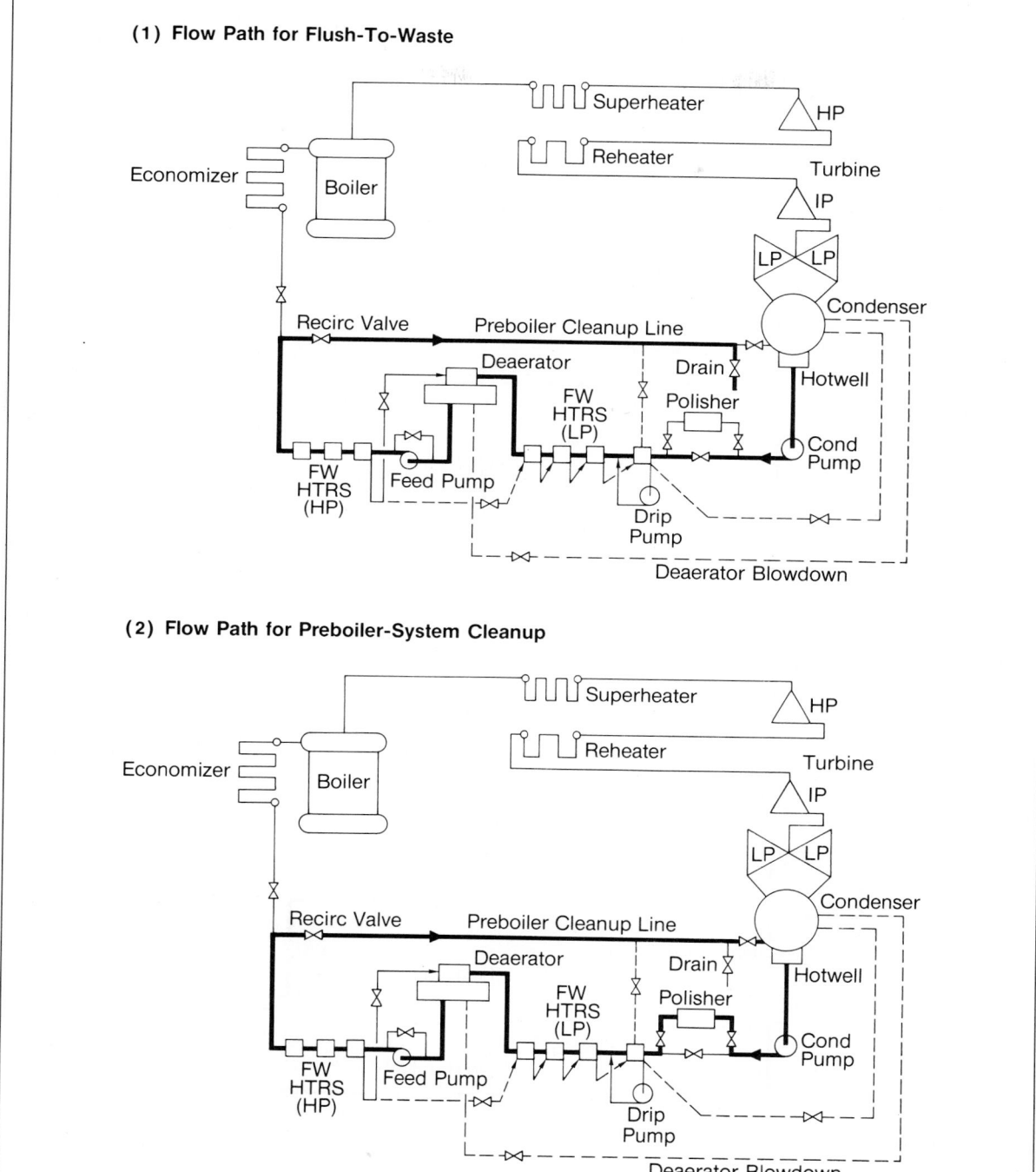

(1) Flow Path for Flush-To-Waste

(2) Flow Path for Preboiler-System Cleanup

Fig. 37 **Procedure for sub-critical unit start-up**

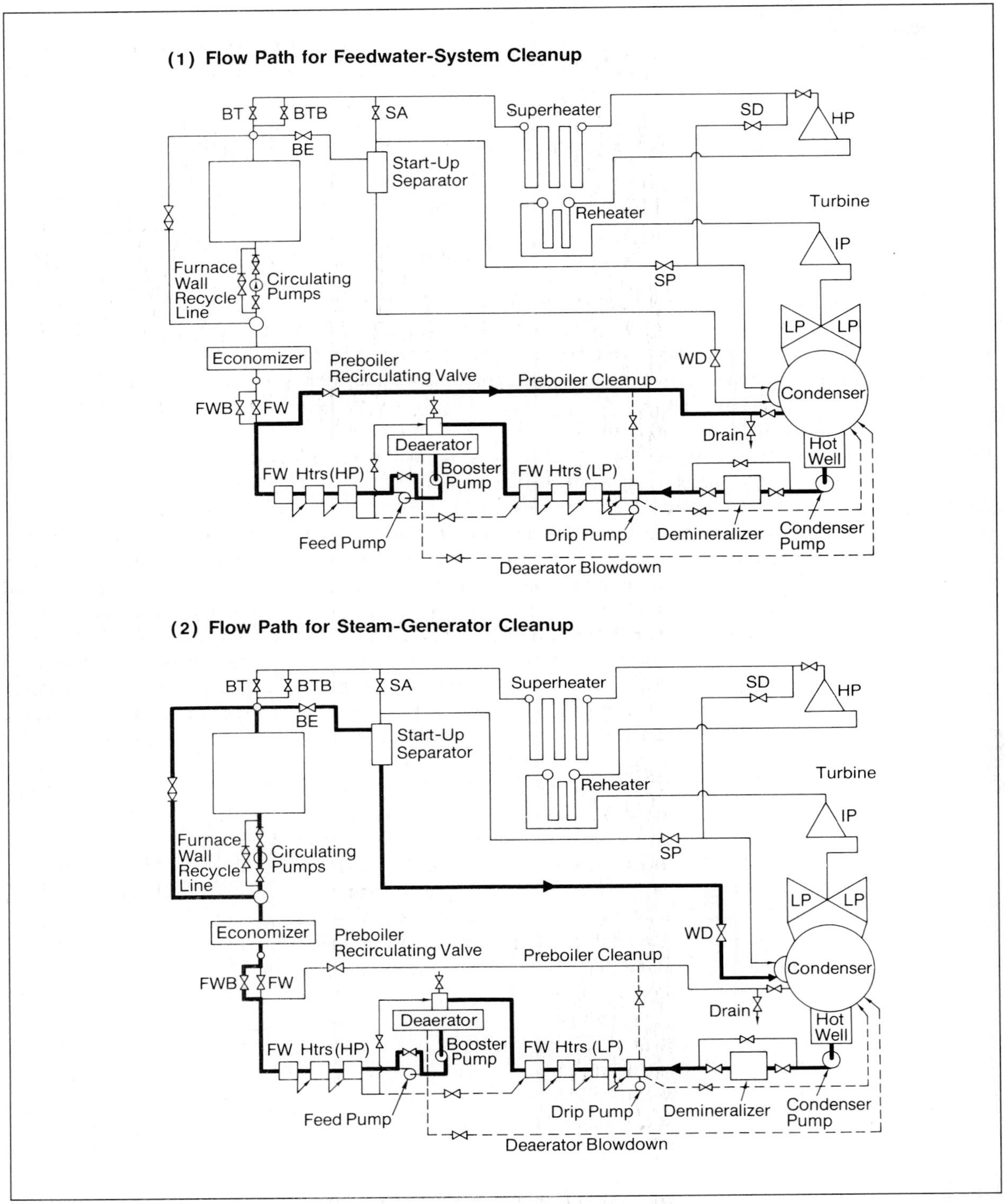

(1) Flow Path for Feedwater-System Cleanup

(2) Flow Path for Steam-Generator Cleanup

Fig. 38 **Procedure for supercritical unit start-up after a short outage**

limits takes less than eight hours on start-up following long outages. Preboiler cleanup following short outages (up to four days) can be accomplished in three to four hours. No preboiler cleanup is required for a hot restart if the quality of the condensate meets the established standards of purity.

SUPERCRITICAL STEAM GENERATOR

Cleanup of the furnace-wall system in a Combined Circulation supercritical unit is performed reliably and quickly because the boiler recirculation pumps provide effective scrubbing velocity. Particulate contamination is maintained in suspension and is easily removed through the start-up/separator system. This blowdown flow through the start-up system rapidly reduces furnace-wall contamination. Experience shows that the boiler furnace walls can be cleaned effectively if supplied with a deaerated feedwater at about 250°F.

Furnace-wall cleanup after long outages will normally take less than three hours. Cleanup operations after outages up to four days will not require a furnace-wall cleanup step if appropriate wet lay-up procedures have been employed. The unit should be capable of being fired as soon as any necessary preboiler recir-

culation is completed. A delay in start-up will occur only when the condensate is badly contaminated; under these conditions, such a delay can always be justified.

A discussion of the complete procedures and feedwater control limits recommended for start-up and normal operation of Combined Circulation units is beyond the scope of this chapter. For illustrative purposes, however, Table IX and Fig. 38 detail the procedure employed for start-ups after short outages. Start-up cleanup operations after long-duration outages require more time and additional flow paths and steps. The effectiveness of this cleanup philosophy has assured that operational chemical cleanings have not been required more often than every three to four years. [18,19]

PREVENTIVE CHEMISTRY DURING OPERATION, DRUM-TYPE UNITS

Two basic types of boiler-water treatment can be successfully utilized in high-pressure drum-type boilers: coordinated phosphate and volatile-based treatment. Volatile treatment does not require the direct addition of any treatment chemicals to the boiler. The ammonia used for feedwater pH adjustment controls boiler-water pH. Fig. 39 illustrates a

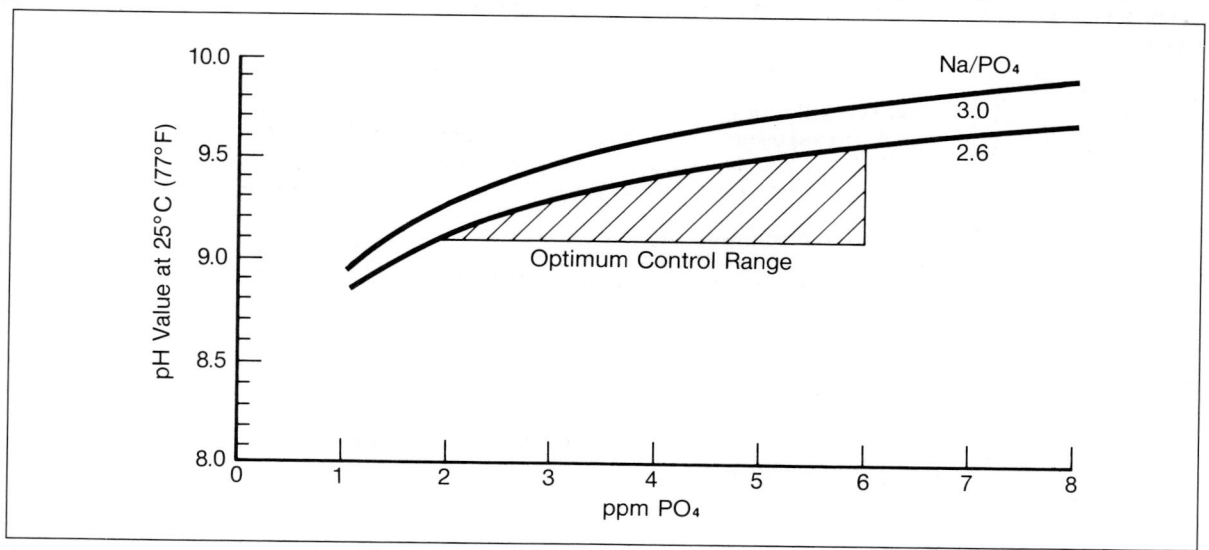

Fig. 39 Recommended coordinated phosphate treatment curve for operation between 2601 and 2900 psig

Table IX. Recommended Start-Up/Cleanup Procedure, C-E Combined Circulation® Units (with Deaerating Heater) — see Fig. 38

Cold Restarts After Short Outages

Phase	Flow Path	Operations	Control Limits		Expected Cleanup Time
Feedwater System Cleanup	1	Pressurize deaerator to 10–15 psig with auxiliary steam, establish condenser vacuum, and place demineralizer in service. Utilizing condensate or booster pump and preboiler recirculating valve, clean up cycle maintaining 25% MCR deaerated flow. Blow down deaerator storage tank as necessary to prevent recontamination of system.	Total Solids Total Iron Total Silica Copper Oxygen Hydrazine pH (All concentrations in ppb.)	250 (max.) 50 30 20 10 20 9.2–9.4	4 hours if C-E recommended lay-up procedure has been employed. Does not represent a delay; this cleanup may be completed before boiler is available.

Transition Period: When control limits are met at HP heater outlet, close preboiler recirculating valve and establish BE and WD through-flow, using condensate or booster pump. To prevent excessive pressure drop across BE valve, furnace wall outlet pressure should not exceed 1000 psig during cold water operation.

Phase	Flow Path	Operations	Control Limits		Expected Cleanup Time
Steam Generator Cleanup	2	Continue BE and WD through-flow until cleanup is complete. Furnace-wall circulating pump on during this phase. Blow down deaerator storage tank as necessary to prevent contamination of system. When control limits are met, proceed to next step.	Total Solids Total Iron Total Silica Copper (All concentrations in ppb, measured at furnace-wall outlet.)	750 (max.) 500 30 20	Less than 1 hour if C-E recommended lay-up procedure has been employed.
System in Service		Place boiler feed pump in service Fire unit for turbine warmup, roll, and synchronization. Proceed to normal feedwater exceeding limits. Cascade heater drains to condenser for polishing until load exceeds 25%.	Normal Feedwater Limits Total Solids Total Iron Total Silica Copper Oxygen Hydrazine pH (All concentrations in ppb, measured at HP heater outlet; oxygen measured at deaerator outlet.)	 50 (max.) 10 20 2 5 10 9.2–9.4	Normal limits should be obtained before exceeding one-third unit load.

Table X. Typical Criteria for Operation of Drum-Type Boilers (2601-2900 psig)

Hotwell Conditions	Operational Limitations	Control Limits		Boiler-Water Control
		Phosphate—pH*	Volatile	
Total Solids <0.05 ppm (Recommended)	None	TS <15 ppm pH 9.1–9.6 PO_4 2–6 ppm	TS <2 ppm pH 8.5–9.0	Normal
Total Solids <0.1 ppm (Acceptable)	Volatile treatment of boiler water not suitable.	TS <15 ppm pH 9.1–9.6 PO_4 2–6 ppm		If on volatile treatment, change to coordinated phosphate-pH control.
Total Solids 0.1–1.0 ppm (Abnormal)	Limited operation. Schedule inspection and repair of condenser as soon as system load requirements permit.	TS <50 ppm pH 9.1–10.1 PO_4 5–20 ppm		Immediately start chemical injection pumps to maintain excess phosphate and pH conditions. Do not continue to operate unit if pH cannot be maintained above 8.0 or the total solids below 50 ppm. Increase blowdown to limit total dissolved solids concentration. Avoid use of desuperheating spray water by (1) permitting reheat temperature to fall, or (2) reducing load.
Total Solids > 1.0 ppm (Excessive)	Emergency operation. 1. Immediately reduce load as necessary to permit isolation of damaged condenser. 2. Prepare for orderly shutdown of the unit if hotwell concentration cannot be quickly reduced below 1.0 ppm.	TS <50 ppm pH 9.1–10.1 PO_4 5–20 ppm		See above. Prepare to establish chemical conditions for wet lay-up of boiler unit and auxiliary equipment during condenser repair.

*Maintain sodium/phosphate ratio $\leq$ 2.6

typical low-level coordinated phosphate curve for use in high-pressure boilers above 2600-psig drum operating pressure. Operation with pH and phosphate values which are below the curve establishes that no free hydroxide exists within the boiler water. The recommended operating chemistry shown in Table X is reflective of this fact. Because free hydroxide does not exist under these specified conditions, ductile tube failures from caustic attack, as discussed previously, will not occur.[20]

Table X illustrates a philosophy of boiler operation with respect to level of contamination entering the cycle. The recommendations and guidelines presented in this table, coupled with recommended feedwater quality, represent the chemistry specifications which will result in long-term steam-generator integrity and cleanliness. It should be noted that coordinated phosphate boiler-water treatment must be employed during periods of condenser leakage. Volatile treatment does not have enough

buffering ability to maintain pH control under these conditions.

Certain deviations from normal chemistry specifications may occur on occasion during plant life, with little or no effect on component integrity. These deviations are designated as abnormal conditions in Table X. Other possible deviations may result in extensive corrosion problems. The serious nature of such deviations requires immediate action, and this is noted where applicable.

FEEDWATER CONTROL

The steam generator acts as a concentrating device during operation. Soluble feedwater impurities which enter the boiler become concentrated in the boiler water because the steam leaving the drum carries off negligible quantities of solids. The recommended feedwater impurity limits given in Table XI cover a number of constituents. The total solids limit is specified to minimize boiler fouling. The silica concentration limit in normal operation is sufficiently low to assure that silica carryover to the turbine by the vapor or steam solubility phenomena does not occur. The oxygen and pH specifications should limit the pickup of iron and copper from the preboiler equipment to those feedwater levels necessary for long-term steam-generator cleanliness.

As long as feedwater total solids (TS) concentration is less than 0.5 ppm, boiler/turbine cycle operation is considered normal. Coupled with this is the requirement that pH value, oxygen level, and copper/iron levels are within the values shown on Table XI. There are no operational limitations, and the boiler-water control is normal.

Operation with feedwater TS concentration between 0.5 and 2.0 ppm should be limited, and condenser repairs should be scheduled as soon as practical. Values of feedwater solids concentration above 2.0 ppm TS are considered emergency levels. Unit load should be immediately reduced, the damaged condenser section isolated, and necessary repair work scheduled. The boiler-turbine cycle should be shut down if feedwater concentration cannot be reduced within eight hours to values below 2.0 ppm.

PREVENTIVE CHEMISTRY IN OPERATION OF SUPERCRITICAL UNITS

Because a supercritical unit requires the ultimate in feedwater quality, a condensate demineralizer must be included in the preboiler cycle to prevent condenser leakage salts from entering the feedwater. The normal feedwater limits for supercritical units are presented in Table XI.

Utility operators who have followed recommended start-up cleanup procedures, as well as recommended lay-up procedures, have maintained a high degree of cycle cleanliness. Virtually all Combined Circulation units are experiencing operational chemical-cleaning frequencies of from three to four years.

The successful operation of a supercritical unit with respect to water technology is coupled to the proper operation of the in-line condensate demineralizer system. Table XII summarizes information on the operation of a supercritical unit during condenser leakage. The methods of limiting operation and controlling the condensate demineralizer presented in this table, if properly implemented, will assure steam generator and turbine cleanliness.

Table XI. Recommended Feedwater Limits, High-Pressure Drum-Type Units

Total solids	50 ppb max.
Total iron	10 ppb max.
Total copper	10 ppb max.
Total silica	20 ppb max.
Oxygen	5 ppb max.
Hydrazine	10–20 ppb residual
pH	8.8–9.2 (copper-alloy system)
	9.2–9.4 (copper-free system)

All measurements made at high-pressure heater outlet or economizer inlet.

Oxygen can be measured at deaerater outlet if cycle is equipped with deaerating heater.

ANALYTICAL METHODS

The implementation of control methods to minimize corrosion-product release, as well as the detection of contaminant inleakage, requires a continuous analytical program. All automatic analytical instrumentation should be maintained operational and calibrated.

If chemical auto-analyzers are not operational, greater emphasis must be placed on laboratory analyses. Membrane-filter testing may provide an adequate indication of water quality with respect to corrosion products. But primary reliance must be placed on laboratory analytical methods for contaminant and corrosion-product-level determinations. The findings of chemical analyses should determine whether, for example, cleanup circulation should continue on a supercritical unit, or whether silica values are low enough in a drum-type unit to allow a pressure increase.

On any high-pressure utility boiler/turbine cycle, the condenser leak detection system is particularly important. Leakage instrumentation should be installed such that, if the condenser has segmented compartments, each compartment can be analyzed for leakage individually. Cation conductivity offers good sensitivity for detecting condenser leaks. Many plants are also being equipped with sodium analyzers for maximum sensitivity. Now, with the availability of computers for powerplant water-chemistry control, monitoring and diagnosing, as well as automatic control, are practical.[21]

Table XIII gives a recommended frequency of analysis for high-pressure utility steam generators if continuous analyzers are not available.

INDUSTRIAL BOILER-WATER TREATMENT

Internal chemical treatment systems are primarily designed to prevent scale formations resulting from residual amounts of calcium and magnesium hardness in the feedwater. External treatment and control must provide a feedwater low in hardness, alkalinity, silica, metal oxides, and oxygen. Because of their extreme solubility, sodium salts are of concern only to the extent of their contribution to the total dissolved solids concentration. However, boiler-tube scale consisting of calcium and magnesium components is not uncommon in units operating below 900 psig. These scales often are the direct result of loss of chemical control, permitting dense calcium precipitates such as calcium sulfate or calcium silicate to form. They may also be a more subtle result of porous-hardness sludge or metal oxides which

---

Table XII. Criteria for Operation of Supercritical Unit During Condenser Leakage

Concentration	Hotwell Salt Description	Leakage Operational Limitations	Demineralizer Control
<0.5 ppm	Normal	None.	Normal
0.5–2.0 ppm	Abnormal	Limited operation. Schedule inspection and repair of condenser as soon as system load requirements permit.	Control demineralizer sequence to assure that tanks will not exhaust simultaneously. If ammonia cycle is employed, override ammonia step.
>2.0 ppm	Excessive	Emergency operation. Prepare for orderly shutdown of the unit. Immediately reduce load as necessary to permit isolation of damaged condenser section. Initiate shutdown if hotwell concentration cannot be quickly reduced below 2.0 ppm.	Place all available tanks in service immediately. If reduced load operation is assumed, transfer resin which is most exhausted and bring in spare charge. Override ammonia regeneration step.

Table XIII. Recommended Frequency of Analysis for High-Pressure Utility Units

Should condenser leak-detection equipment be out of service, manual conductivity readings should be taken at the condensate pump discharge three times daily.

Supercritical-Type Feedwater at Economizer Inlet	Drum-Type Feedwater at Economizer Inlet
pH, per shift O_2, per shift Silica, twice weekly Copper, twice weekly Total iron, twice weekly	Same as supercritical units **Boiler water** pH, per shift PO_4, per shift Total solids, per shift Silica, twice weekly

Table XIV. Typical Phosphate Control Limits for Industrial Units

Pressure (psig)	450–600	600–900
Hydroxide Alkalinity as OH (ppm)	50–100	25–50
Phosphate as PO_4 (ppm)	20–40	5–10
Sodium Sulfite as Na_2SO_3 (ppm)	20–40	5–10
Silica as SiO_2 (ppb)	50 (max.)	10 (maximum)

permit boiler water to penetrate to the tube surface precipitating complex silicate scales.

Formation of scale, below 900 psig, is generally more of a concern with respect to causing overheating failures than promoting corrosion attack. Two basic systems are now available for prevention of internal hardness scales: the conventional phosphate-hydroxide methods and the more recently adopted substitute systems involving chelating agents. Both are proven systems, but the success of either depends on the diligence with which they are applied.

PHOSPHATE-HYDROXIDE METHOD

The phosphate-hydroxide methods can be segregated into one that maintains an excess of hydroxide alkalinity and one that involves no excess or "free" hydroxide content. The former has a long history of use and is still the most prevalent system for low-pressure operation. Table XIV presents typical control limits.

The intent of phosphate control systems is to provide conditions conducive to the precipita-

tion of calcium as a calcium hydroxyapatite, $Ca_{10}(PO_4)_6(OH)_2$ and magnesium as serpentine, $3MgO \cdot 2SiO_2 \cdot 2H_2O$.

In contrast to low-pressure operation, boiler tube failures in units above 900 psig can be caused by corrosion. Accordingly, the coordinated phosphate pH system is used to control boiler-water chemistry as with utility high-pressure boilers.

CHELATION

Chelating agents have received wide acceptance as a substitute system for the conventional phosphate-hydroxide treatment methods. The most common is the sodium salt of ethylenediaminetetraacetic acid (EDTA). As briefly explained in a previous section, these organic agents act with the residual calcium and magnesium in the feedwater to form soluble complexes. Ideally, this should result in boiler surface conditions which are completely free of any hardness deposits. The soluble hardness complexes will be removed through the continuous blowdown line. With conventional phosphate treatment, it is difficult to maintain all of the hardness end products in suspension. As a result, some accumulations normally are found in drums and headers.

Most experience with this treatment method has been limited to operating pressures of 900 psig and below, but some use on 1250- and 1350-psig units has been reported. Economically, justification for its use is limited to feedwaters of low hardness concentrations (1 to 2 ppm). Such evaluation usually considers the savings realized by the elimination or reduced frequency of chemical cleaning.

Chemical additions are made in a continuous mode, with the amount based on the concentration of hardness salts in the feedwater. The chelants must be injected after the boiler feedwater pump to prevent corrosion of the pump components. Earlier attempts to utilize the chemical-feed line to the boiler drum also resulted in severe corrosion of this line. High-oxygen concentrations in the feedwater drastically increase the corrosivity of the chelants. Thus, it is imperative that good deaeration of the feedwater be maintained.

A chelant program can be effective in preventing deposits or damaging boiler tubes depending primarily on the knowledge and experience of the user. Of course, before initiating such a program on any boiler, guidelines, control limits, and analytical procedures must be established with the cooperation of the chelant supplier or water-treatment consultant.

CARRYOVER AND STEAM SAMPLING

The mechanical aspects of steam and water separation are covered in Chapter 5, in which drum-internal design for utility and industrial boilers is described. This section deals with the phenomenon of carryover, methods of steam sampling, and techniques of steam-purity determination. Once again, these subjects cannot be understood solely by rigorous theoretical analysis. Knowledge of laboratory testing procedures and ability to interpret field operating experience are required. Despite intensive research efforts, there is incomplete understanding and much empiricism remains in the techniques for correcting problems resulting from incomplete separation of steam and water in boiler equipment.

CLASSIFICATION OF CARRYOVER

Carryover of boiler water in steam leaving the drum provides a path for introducing solid materials into the steam. Modern separator designs can mechanically reduce the moisture content of the steam to 0.2 percent or less. But in addition to mechanical carryover of boiler water, another mechanism exists that results in the contamination of steam with solid materials.

As operating pressures increase, the steam phase exhibits greater solvent capabilities for the salts that may be present in the water phase. These salts will be partitioned in an equilibrium between the steam and water, a phenomenon known as vaporous carryover. Vaporous carryover will contribute an additional quantity of boiler water solids directly to the steam, independent of the efficiency of steam-water separation components.

Silica was the first material found to exhibit significant vaporous carryover. Silica fouling of turbines was common until it was recognized that successful control of the amount of silica in the steam could only be accomplished by controlling the amount of silica in the boiler water. A similar solution will be required for other solids when operating at high pressures.

Vaporous carryover contributes a significant proportion of total solids in the steam as drum operating pressures increase above 2600 psig. Table XV shows findings of a laboratory study on vaporous carryover.[22] For various salts, the table lists percent vaporous carryover, which is the ratio of the salt concentration in the steam and boiler water. In each case, sodium was measured and then converted to the appropriate salt concentration.

When calculating total solids in the steam, vaporous carryover is assigned a value of 0.1 percent at drum operating pressures above 2600 psig.

Table XV. Summary of Laboratory Results

Pressure (psig)	2600		2800		3000	
Concentration (ppm)	15	500	15	500	15	500
Sodium Sulfate	0.02%	0.03%	0.04%	0.07%	0.28%	0.48%
Disodium Phosphate	0.01%	0.07%	0.03%	0.18%	0.41%	0.74%
Trisodium Phosphate	0.02%	0.11%	0.04%	0.30%	0.35%	1.3 %
Sodium Chloride	0.04%	0.18%	0.09%	0.36%	0.39%	1.2 %
Sodium Hydroxide	0.02%	0.31%	0.08%	0.69%	0.55%	2.2 %

Mechanical carryover may be classified under four headings: priming, spray, leakage, and foam carryover. Each results in troublesome deposits in the superheater or on the steam-turbine blades.

Priming is the development of excessive moisture in the steam because of spouting or surging of boiler water into the steam outlet. This is a rare, easily identified type of carryover. It is usually promoted by the maintenance of too high a water level in the drum, spouting of a submerged riser or sudden swelling of the water in the boiler on a drop in pressure, or sudden increase in rating. Priming is rarely, if ever, associated with boiler-water concentration.

Spray carryover, mist or fog are degrees of atomization of the boiler water. Mist is carried from the drum by the steam as dust is carried by air currents. This carryover is present to a degree in all boilers, and it is the function of drum internals to separate and filter out such spray before the steam leaves the drum. Development of spray carryover indicates failure of the drum internals because of exceeding the velocity limitations of the purification equipment. It is characterized by initial development below the full rating of the boiler and it continues to increase with boiler load. Spray carryover is not sensitive to boiler-water concentration below the foaming range. Improved drum internals are capable of reducing the steam-borne mist to a value as low as a few parts per billion of solids.

Leakage is a general term applied to bypassing of impure steam or boiler water through the drum internals. Normally localized, this form of carryover is directly related to poor design or installation of drum internals. At times, the local contamination may not be sufficient to be reflected in steam-purity measurements of total steam flow. A careful inspection of drum internals will usually reveal this source of carryover. Where the leakage is sufficient to register impurity tests of steam, it will be found that the impurity increases slowly with rating and is relatively insensitive to changes in water level and boiler-water concentration.

Foam carryover is the development of excessive moisture in the steam from carryover of foam from the drum. It is the most common form of carryover in low-pressure units, in which the boiler water may contain high concentrations of dissolved solids, and is the most troublesome and most erratic type.

Foam forms in the steam-generating sections of the boiler when the water-films around the generated steam bubbles are stabilized by the impurities in boiler water. Boiler circulation carries this foam up to the boiler drum where it tends to accumulate at the water level. The foam produced may entirely fill the steam space of the boiler drum or it may be of a relatively minor depth. Although foaming in boilers has been recognized for many years, its causes are not clearly defined and are worthy of further investigation.

The bulk water in the circulating mixture entering the drum is readily separated, but the wet emulsion of very small foam bubbles collects at the water level to a depth largely dependent on the rate of drainage of excess water out of the foamy mass. A considerable amount of moisture is trapped in the foam. When foam carryover occurs, it is frequently sudden and excessive, and the steam sample registers a solids content characteristic of boiler water.

IDENTIFICATION OF CARRYOVER

A systematic field investigation can identify carryover. A variety of factors which may be classified as mechanical, water or operating conditions affects the sources of carryover and the carryover itself to different degrees. As noted previously, foaming in the boiler is the most common type of carryover and is most troublesome and erratic. Special test methods have been devised to demonstrate the presence of foam blankets and for obtaining boiler performance without danger of serious carryover to the superheater and turbine.[23]

Steam flow, water level and boiler-water concentration are the three major factors that can create carryover. By varying these three factors, one at a time, test results can usually be interpreted to determine the specific source of a carryover condition.

■ Steam flow establishes the velocity distribution in the boiler drum. Excessive steam flow can increase steam velocity to a point that entrained mositure can overload the dryer.

■ High water can create spouting and excessive carryover. This can occur at low steaming rates and boiler-water concentrations.

■ Foaming is a characteristic of boiler-water concentration. With water level and steaming rate at recommended values, any carryover which can be precipitated or eliminated by a change in boiler-water concentration can be attributed to foaming.

Fig. 40 illustrates the development of foaming in a drum. The plot indicates purity values of steam samples taken ahead of the steam dryer and at the outlet of the boiler drum for a

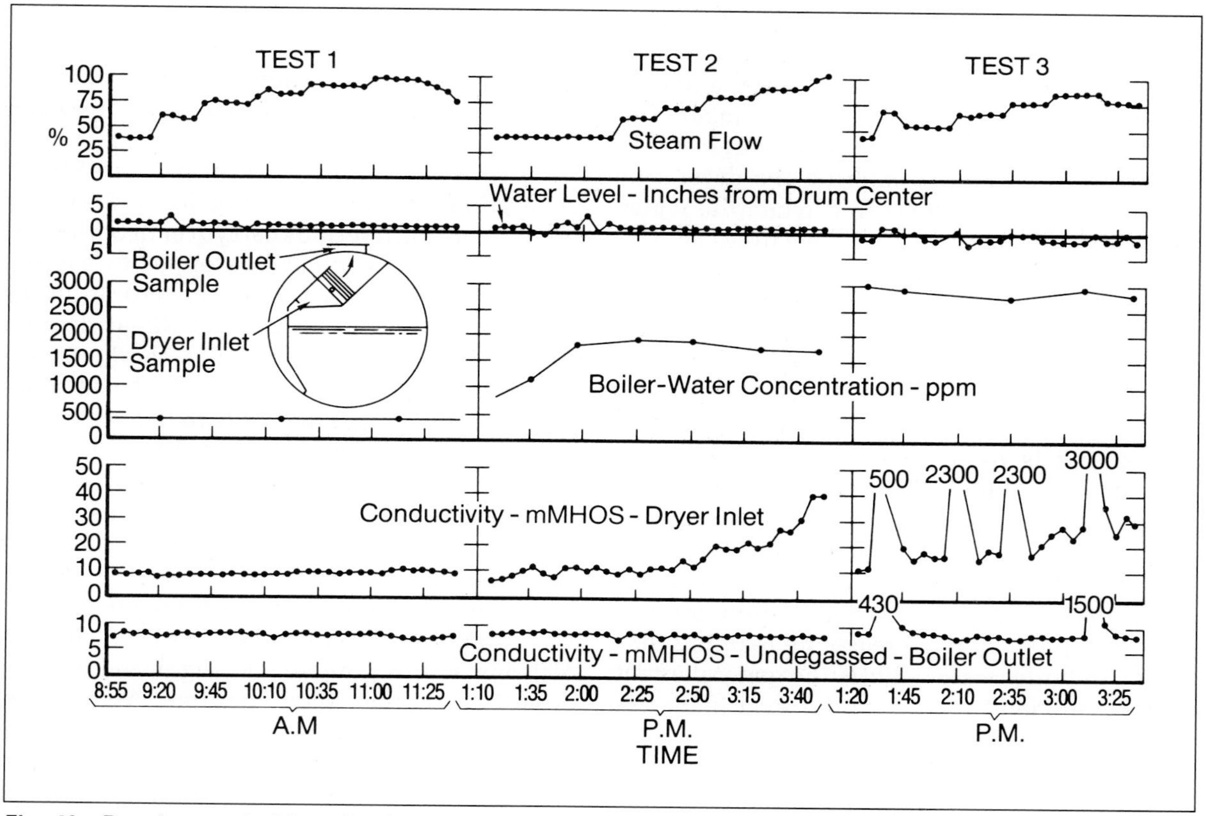

Fig. 40   Development of foaming in a boiler drum

constant level of water in the drum and a typical steam load. The principal change is in the boiler-water concentration.

At a boiler-water concentration of about 550 ppm, the sample ahead of the dryer was about 5 mmhos (1 mmho is approximately 0.5 ppm of solids) and the sample at the boiler outlet was about 7 mmhos. Increasing the boiler-water concentration to about 1800 to 2000 ppm did not alter the purity of the steam leaving the boiler drum. But the sample ahead of the steam dryer increased gradually to a value of about 40 mmhos. Thus, with an established steam flow, and a similar water level, there was a marked increase in solids content of the steam entering the dryer. This is indicative of the presence of a mildly foamy condition in the boiler drum as the only change was in the concentration of boiler-water solids.

Increase of boiler-water solids to about 2800 to 3000 ppm in the boiler water, at the same steam load and water level as in the prior two tests, produced severe foaming in the boiler. The space between the water level and the dryer was practically filled with a foam blanket on the water. This is evident in the high solids content of the sample entering the dryer. Severe foam carryover occurred when the water level in the drum was at or above drum center. This was not a factor in the prior tests. Thus, a small change in water level was sufficient to push the foam blanket into the drum internals creating a severe case of foam carryover.

CAUSES OF FOAMING

Foaming is basically a result of chemical conditions, and boiler-water concentration and composition are important factors involved. High total solids and high suspended solids aggravate the formation of foam. High caustic alkalinity, oil, organic contamination and excess phosphate also increase the foaming tendency in boiler water. While the general effect of a component upon foaming may be anticipated, it is impossible to predict whether foam formation will occur by a cursory examination of boiler water. Extreme cases are on record where excessive foamover occurred with less than 650 ppm in one boiler and, in another case, no carryover troubles developed with concentrations as high as 15,000 ppm. Although these inconsistencies exist, it is necessary to maintain a lower value of foam-producing chemicals in boiler water or add foam dispersing chemicals to the water.

Organic antifoam agents have been developed to dispel certain foams at higher steam pressures. Ordinary tannin and starch compounds are only effective at low pressures. Lignin sulfonates, alkaline polyamides, polymerized esters and alcohols have been effective foam dispersing agents. The function of an efficient antifoam agent is to reduce the number of small bubbles and to confine steam-bubble formation to a small number of large bubbles which will exhibit the tendency to coalesce and grow larger. Under these conditions, the bubbles are unstable and tend to break easily. Antifoam agents are not equally effective with all boiler waters. It is necessary to select an antifoam compatible with the chemical characteristics of the boiler water, and trial of several compounds may be necessary before the foam can be neutralized satisfactorily.

Foam will fill the free surface area of a separating device increasing local velocities and promoting a serious carryover of boiler water. Foam carryover may be stopped by a quick reduction in boiler-water concentration or lowering the drum level. Centrifugal devices have shown a greater ability to handle foamy waters than simple internals. The basic function of a centrifugal device is to dehydrate the foamy emulsion. The dehydrated-foam bubbles can be easily broken up by screens or other simple devices. Foam in this type of separator will decrease the water-separating efficiency of the device.

In general, foam carryover from a boiler can be avoided by keeping boiler-water concentrations within the range suggested by the American Boiler Manufacturers Association (ABMA). These specifications, of course, cannot be a guarantee against foaming which, as indicated previously, is primarily a chemical problem.

Table XVI. Recommended Boiler Water Limits and Associated Steam Purity (at Steady-State, Full-Load Operation)

Drum Pressure psig (Actual)	Range of Total Dissolved Solids[1] in Boiler Water, ppm (Maximum)	Drum-Type Boilers Range of Total Alkalinity[2] in Boiler Water ppm	Suspended Solids in Boiler Water, ppm (Maximum)	Range of Total Dissolved Solids [2,4] in Steam, ppm (Maximum Expected Value)
0–300	700–3500	140–700	15	0.2–1.0
301–450	600–3000	120–600	10	0.2–1.0
451–600	500–2500	100–500	8	0.2–1.0
601–750	200–1000	40–200	3	0.1–0.5
751–900	150–750	30–150	2	0.1–0.5
901–1000	125–625	25–125	1	0.1–0.5
1001–1800	100	Note (3)	1	0.10
1801–2350	50	Note (3)	N/A	0.10
2351–2600	25	Note (3)	N/A	0.05
2601–2900	15	Note (3)	N/A	0.05
Once-Through Boilers				
1400 and above	0.05	N/A	N/A	0.05

Notes:
1. Actual values within the range reflect the TDS in the feedwater. Higher values are for high solids; lower values are for low solids in the feedwater.
2. Actual values within the range are directly proportional to the actual value of TDS of boiler water. Higher values are for the high solids; lower values are for low solids in the boiler water.
3. Dictated by boiler water treatment.
4. These values are exclusive of silica.

Source: American Boiler Manufacturers Association

Concentration limits as a function of pressure are shown in Table XVI.

In both utility and industrial plants, adherence to the ABMA specifications has produced satisfactory operation because of marked improvements in water technology and boiler design. Normally, the steam-purity-specification limit for low-pressure boilers is less than 0.5 percent moisture in the steam. With the use of superheaters and higher pressures, the boilers must deliver a steam product containing less than 1.0 ppm of solids entrained in the steam. Steam purity of high-pressure boilers has been markedly improved and values of less than 0.01 ppm of impurity are achieved.

STEAM SAMPLING METHODS

Steam samples for measuring purity are usually taken ahead of the superheater. These are condensed and cooled. Collecting a true sample that is representative of a large mass of material always presents a difficult problem. The sample size, particle-size distribution and density relationship are some factors which must be considered where there is a question of lack of homogeneity. In a homogeneous sample of fine particle size, sampling is a relatively easy operation.[23]

In sampling steam, the impurities may be solid, liquid, and gaseous. The solid may be in the form of a finely divided sludge particle. Liquid impurity may be in the form of fog or mist in minute droplets, possibly having a solid particle as a nucleus. More adversely, it may be in a form of a surface film on a pipe wall. Moisture itself is not involved in the concept of steam purity except that it may carry solids in solution or suspension.

Sampling impurities in steam is analogous to the difficulty of locating a needle in a haystack.

At a sampling rate of 100 pounds per hour, the impurity content of 1 ppm is represented by the withdrawal of 0.7 grains of solid per hour. Steam lines contain bends, elbows, valves and other fittings which can disturb the flow and segregate the impurities.

B. J. Cross has outlined the assumptions reached in the design of the steam sampling nozzle described in *ASTM Standards* D 1066, Sampling Steam. The velocity front must be reasonably flat, and the density difference of steam and mist or fog carried along with it must be in the same order of magnitude as that of water and steam at the pressure and temperature of the steam in the line. Basic prerequisites for use of the ASTM nozzle design are that the velocity of the steam entering the ports is the same as the line velocity of the steam and that each port of the sampling nozzle shall represent an equal area of the sampling section.[24]

Turns and other irregularities of the steam line influence distribution of solid and liquid impurities. The sample point should be as remote as possible from a source of disturbance. It should also be located where there is a run of at least ten diameters of straight piping. The preferred location with respect to position, in order of decreasing preference, is

(a) vertical pipe, downward flow

(b) vertical pipe, upward flow

(c) horizontal pipe, vertical insertion

(d) horizontal pipe, horizontal insertion

DETERMINATION OF STEAM PURITY

Steam purity is normally determined by measuring either the electrical conductivity or sodium content of the condensed steam sample.

Measuring electrical conductivity is widely used to monitor steam purity in industrial low-pressure boilers. This method is described in *ASTM Standards* D 1125, Electrical Conductivity and Resistivity of Water.

Gases dissolved in a condensed sample affect conductance and indicate an erroneous level of solid impurity. These gases may be removed by degasification. Methods suggested for establishing the content of solids impurity in steam are described in *ASTM Standards* D 2186, Deposit-Forming Impurities in Steam, which provides four alternative techniques. The referee method for establishing the total solids in the steam is by evaporation, as specified in *ASTM Standards* D 1888, Particulate and Dissolved Matter, Solids, or Residue in Water.

Determination of solids in steam by conductance is not sensitive to impurities in the parts per billion range, which is the range required for determination of steam purity in high-pressure utility boilers. Analysis of the steam sample to determine the sodium content in the impurity, by flame spectrophotometric technique, is the most accurate method developed to establish solids content. This technique is described in *ASTM Standards* D 1428, Sodium and Potassium in Water and Water-formed Deposits by Flame Photometry.

REFERENCES

1 Eskel Nordell, *Water Treatment for Industrial And Other Uses,* second edition. New York: Reinhold Publishing Corp., 1961.

Sheppard T. Powell, *Water Conditioning For Industry,* first edition. New York: McGraw-Hill Publishing Co., 1954.

Betz Laboratories, Inc., *Betz Handbook of Industrial Water Conditioning,* seventh edition. Trevose, Pa.: Betz, 1976.

P. Hamer, et al., eds., *Industrial Water Treatment Practice.* London: Butterworths, 1961.

The NALCO Water Handbook, sponsored by The Nalco Chemical Company, Second Edition. New York: McGraw-Hill Publishing Company.

2 E. W. Lohr and S. K. Love, *The Industrial Utility of Public Water Supplies in the United States, 1952: Parts I and II,* Geological Survey Water Supply Papers 1299 and 1300, respectively. Washington: U.S. Government Printing Office, 1954.

3 R. E. Hall, et al., "Phosphate in Boiler Water Conditioning," *Journal of the American Water Works Association,* 21(1): 79–100, January 1929.

4 Richard C. Corey and Thomas J. Finnegan, "The pH, Dissolved Iron Concentration and Solid Product Resulting from the Reaction Between Iron and Pure Water at Room Temperature," *Proceedings of the American Society for Testing and Materials,* Preprint no. 101, 1939.

5 E. Berl and F. van Taack, "Über die Einwirkung von Laugen und Salzen auf flusseisen unter Hochdruck-

bedingungen und über die Schutzwirkung von Natruimsulfat gegen den Angriff von Aetznatron und von Chlormagnesium," (The Action of Caustic and Salts on Steel under Conditions of High Pressure and the Protective Effect of Sodium Sulfate against the Attack of Sodium Hydroxide and Magnesium Chloride), *Forschungsarbeiten*, no. 330. Berlin: VDI, 1930.

[6] S. F. Whirl and T. E. Purcell, "Protection Against Caustic Embrittlement by Coordinated Phosphate-pH Control," *Proceedings of the Third Annual Water Conference*, Pittsburgh, November 9–10, 1942, pp. 45–60. Pittsburgh: Engineers' Society of Western Pennsylvania, 1942.

[7] H. A. Grabowski, et al., "Problems in Deaeration of Boiler Feedwater," *Combustion*, 25(9): 43–48, March 1955.

[8] Samual Glasstone, "Overvoltage and Its Significance in Corrosion," *Corrosion and Material Protection*, 3(6): 15–18, June–July 1946.

[9] H. A. Klein and H. A. Grabowski, "Corrosion and Hydrogen Damage in High Pressure Boilers", Presented at the Second Annual Educational Forum on Corrosion, National Association of Corrosion Engineers, Drexel Institute of Technology, Philadelphia, September 15–17, 1964. Combustion Engineering publication TIS-2652.

[10] P. Goldstein and C. L. Burton, "Research Study on Internal Corrosion of High-Pressure Boilers," *Transactions of the ASME. Journal of Engineering for Power*, 91, Series A: 75–101, April 1969; also in *ASME Paper No. 68-PWR-7*. New York: American Society of Mechanical Engineers, 1968.

[11] J. A. Armantano and V. P. Murphy, "Standby Protection of High Pressure Boilers," *Proceedings of the 25th International Water Conference*, Pittsburgh, September 28–30, 1964, pp. 111–124. Pittsburgh: Engineers' Society of Western Pennsylvania, 1964.

[12] L. H. Vaughn and C. V. Runyan, "Corrosion Protection of Boilers and Associated Equipment During Idle Periods," *Proceedings of the American Power Conference*, 33: 721–729, 1971. Chicago: Illinois Institute of Technology, 1971.

[13] B. T. Hagewood, et al., "The Control of Internal Corrosion in High-Pressure Peaking Units," *Proceedings of the American Power Conference*, 30: 939–948, 1968. Chicago: Illinois Institute of Technology, 1968.

[14] H. A. Grabowski, et al., "Problems in Deaeration of Boiler Feedwater," *Combustion*, 25(9): 43–48, March 1955.

[15] A. F. Kelly, et al., "Modify Base-Load Turbines for Peaking Service," *Power*, 115(4): 62–63, April 1971.

[16] F. N. Speller, "Control of Corrosion by Deactivation of Water," *Franklin Institute Journal*, 193(4): 515–542, April 1922.

[17] F. Gabrielli and W. R. Sylvester, "Water Treatment Practices for Cyclic Operation of Utility Boilers," *Proceedings of the International Water Conference*, Pittsburgh, October 31–November 2, 1978, pp. 193–208. Pittsburgh: Engineer's Society of Western Pennsylvania, 1979; also in

Combustion Engineering publication TIS-5859. F. Gabrielli, N. C. Mohn, and W. R. Sylvester, "Water Chemistry Aspects of Cyclic Operation for Older High Pressure Drum-Type Boilers", *Proceedings of the American Power Conference*, 45:989–999, 1983. Chicago: Illinois Institute of Technology, 1983; also as Combustion Engineering publication TIS-7383.

F. Gabrielli, N. C. Mohn, and B. C. Teigen, "Deposit and Water Chemistry Studies with Rifled Tubing," *Proceedings of the American Power Conference*, 46:973–984, 1984. Chicago: Illinois Institute of Technology, 1984. Also as Combustion Engineering Publication TIS-7530.

[18] J. J. Kurpen and D. L. Dixson, "Operating Experience in Cycle Cleanup for Supercritical Pressure Units," *Proceedings of the American Power Conference*, 30: 883–896, 1968. Chicago: Illinois Institute of Technology, 1968.

[19] K. L. Atwood and G. L. Hale, "A Method For Determining Need for Chemical Cleaning of High-Pressure Boilers," *Proceedings of the American Power Conference*, 33: 710–720, 1971. Chicago: Illinois Institute of Technology, 1971.

[20] H. A. Klein, "Use of Coordinated Phosphate Treatment to Prevent Caustic Corrosion in High Pressure Boilers," *Combustion*, 34(4): 45–52, October 1962.

[21] R. J. Barto, D. M. Farrell, F. A. Noto, and S. L. Goodstine, "Intelligent Chemistry Management System (ICMS)–A New Approach to Steam Generator Chemistry Control", *Proceedings of the American Power Conference*, 48:1025–1031, 1986. Chicago: Illinois Institute of Technology, 1986.

[22] S. L. Goodstine, "Vaporous Carryover of Sodium Salts in High Pressure Steam." proceedings of the *American Power Conference*, 36:784–789. Chicago: Illinois Institute of Technology, 1974; also as Combustion Engineering publication TIS-3973.

[23] P. B. Place, "Carryover Problems and Identification of Carryover Types," *Combustion*, 18(9):29–34, March 1947.

P. B. Place, "Steam Purity Determination. Part I. Evaluation of Test Results," *Combustion*, 25(10): 62–65, April 1954.

P. B. Place, "Steam Purity Determination. Part II. Methods of Sampling and Testing," *Combustion*, 25 (11): 41–44, May 1954.

P. B. Place, "Steam Purity Determination. Part III. Interpretation of Test Results," *Combustion*, 25 (12): 43–46, June 1954.

[24] B. J. Cross, "The Sampling of Steam for the Determination of Purity," *Proceedings of the Eleventh Annual Water Conference*, Pittsburgh, October 16–18, 1950, pp. 71–82. Pittsburgh: Engineers' Society of Western Pennsylvania, 1950.

"Standard Method of Sampling Steam," ASTM Standards D1066. *Annual Book of ASTM Standards*, Part 31: Water. Philadelphia: American Society for Testing and Materials, latest edition.

BIBLIOGRAPHY

Atwood, K. L., "Solvent Selection for preoperational and Operational Cleaning of Utility Boilers," *Combustion*, 42 (3): 16–21, September 1970.

Atwood, K. L. and J. A. Martucci, "The Application of Hydrochloric Acid and Ammonium Bromate for Scale Removal in Utility Boilers," *Proceedings of the 28th International Water Conference*, Pittsburgh, December 11–13, 1967, pp. 167–179. Pittsburgh; Engineers' Society of Western Pennsylvania, 1967.

Brown, R. D. and D. A. Harris, "Large Coal-Fired Cycling Unit," paper presented at the ASME-IEEE-ASCE Joint Power Generation Conference, Portland, OR, September 28–October 2, 1975. Combustion Engineering publication TIS-4558.

Chojnowski, B. and R. D. B. Whitcutt, "Corrosion Failure: One Cause and a Cure in an Operational Boiler," *Combustion*, 48 (5): 28–33, November 1976.

Clayton, W. H., J. G. Singer, et al., "Design for Peaking/ Cycling," paper presented at the Joint Power Generation Conference, Pittsburgh, September 27–30, 1970; also in *ASME Paper No. 70-PWR-9*. New York American Society for Mechanical Engineers, 1970. Also in Combustion Engineering Publication TIS-2927.

Gabrielli, Frank, et al., "Contamination Prevention of Superheaters and Reheaters During Initial Startup and Operation," *Proceedings of the American Power Conference*, 38: 296–310, 1976. Chicago: Illinois Institute of Technology 1976.

Gabrielli, F., et al., "Prevent Corrosion and Deposition Problems in High-Pressure Boilers," *Power*, 122 (7): 85–92 July 1978. also in Combustion Engineering Publication TIS-5271.

Goodstine, S. L. and J. J. Kurpen, "Corrosion and Corrosion Product Control in the Utility Boiler-Turbine Cycle," *Combustion*, 44 (11): 6–18, May 1973.

Greene, N. D. and M. G. Fontana, "A Critical Analysis of Pitting Corrosion," *Corrosion*, 15 (1): 41–47, January 1959.

Kennedy, C. M., et al, "Experience with High Rate Ammoniated Mixed Beds for Condensate Polishing of CIPSCO Coffeen Station." *Combustion*, 39 (9): 19–30 March 1968.

Klein, H. A., "Corrosion of Fossil Fueled Steam Generators." *Combustion* 44 (7): 5–20, January 1973.

Klein, H. A., "A Field Survey of Internation Corrosion in High Pressure Utility Boilers," *Proceedings of the American Power Conference*, 33: 702–709, 1971. Chicago: Illinois Institute of Technology, 1971.

Klein, H. A. and K. L. Atwood, "Chemical Cleaning of Utility Boilers." *Proceedings of the American Power Conference*, 26: 762–778, 1964. Chicago: Illinois Institute of Technology, 1964.

Klein, H. A. and P. Goldstein, "The Effects of Water Quality on the Performance of Modern Power Plants." Presented at the 24th Annual Conference of the National Association of Corrosion Engineers, Cleveland, Ohio, March 18–22, 1968, Paper No. 12; also in Combustion Engineering publication TIS-2891.

Kurpen, J. J., "Externally Regenerated Condensate Demineralization Systems for Once-Through Steam Generators." Presented at the Liberty Bell Corrosion Course, Philadelphia, PA., Sept. 16–18, 1969. Sponsored by Natl. Assn. of Corrosion Engineers and Drexel Institute of Technology; also in Combustion Engineering publication TIS-3071.

Levendusky, J. A. and L. Olejar, "Condensate Purification Applications of the Power Process in High-Pressure Utility Plant Cycles." *Proceedings of the American Power Conference*, 29, 840–856, 1967. Chicago: Illinois Institute of Technology, 1967.

Rivers, H. M., "Concentrating Films: Their Role in Boiler Scale and Corrosion Problems." *Proceedings of the Twelfth Annual Water Conference*, Pittsburgh, Oct. 22–24, 1951, pp. 131–145 Pittsburgh: Engineers' Society of Western Pennsylvania, 1951.

Stocky, D. G. and K. L. Atwood, "Non-destructive Testing for Location of Corrosion Damage in High Pressure Boilers. Presented at the Meeting of the Southeastern Electric Exchange, New Orleans, April 29–30, 1971; also in Combustion Engineering publication TIS-3246.

Ulmer, R. C. and H. A. Klein, "Impurities in Steam from High Pressure Boilers." *Proceedings of the ASTM*, 61: 1396–1411, 1961.

Wages, C. W. and C. W. Smith, "Operating Experience of a Deep-bed Condensate Polishing System." *Proceedings: 38th International Water Conference*, Pittsburgh, PA, Nov. 1–3, 1977, pp. 111–121. Pittsburgh, Engineers' Society of Western Pennsylvania, 1978.

CHAPTER 21

Operation of Steam Generators

The operation of a multi-million-dollar steam generator and its associated power-plant equipment requires the constant exercise of intuitive reasoning and sound engineering judgment. It is in *operation* that all of the factors that went into the design and construction of the facility are put to the test.

A principal objective of proper operation is sustained service between outages while, at the same time, obtaining the highest possible efficiency from all the plant components.

Operation of a steam generator is a balance of inputs to outputs: the better the balance, the smoother the operation. Producing steam from a boiler requires that the weight of water entering the boiler equal the steam leaving, and firing the furnace requires a balance of fuel and air. To equalize these inputs and outputs, one must understand the system, not just the network of hardware that comprises the system. This understanding is the principal ingredient of successful operation.

Too often in recent years, operators have confused operation with control-system management. The operator must realize that a control system is hardware assembled to make operation easier, faster, and safer. All large steam generators require operators, and the control of all major functions can be switched from automatic to operator control. To be effec-tive an operator must know not only what he or she is doing but why it is done and what results from the operator's action.

SAFETY, A PRIME OPERATING CONSIDERATION

The prime consideration for all operation is the safety of people and equipment. Whenever there is any doubt about an unsafe condition, the operator must take immediate action to return the unit to a known safe condition even if it means tripping the unit.

As the loss of a unit even during peak-load requirements is not as important as a human life or the downtime for a major repair, the two most dangerous conditions remain the same today as throughout the history of steam generation: the loss of water or the explosive mixture of fuel and air. Both result from an imbalance: the first from less water than steam produced, even if some of that steam is escaping through a tube leak; the second, from too little air for the fuel present.

Safe operation, then, is a result of comprehensive training programs for operators, well-designed furnace safeguard systems, and an effective preventative-maintenance program.

BASIC OPERATING PRINCIPLES

In this section, certain basic operating guidelines for overall effective operation of a large coal-fired unit are discussed.

STEAM TEMPERATURES

Maintaining desired primary and reheat steam temperatures requires considerable operator attention. Even the best control systems do not anticipate all of the factors affecting steam temperature. Despite the equipment installed for controlling superheater and reheater steam temperatures, certain conditions may produce abnormal steam temperatures.

For instance, with a new coal-fired unit, it may be necessary to operate for a considerable time before normal furnace seasoning allows the unit to make predicted steam temperatures. "Normal furnace seasoning" is often defined as the condition of furnace-wall slag or ash deposits which remains after sootblower operation. Low steam temperatures may also result from:

- insufficient excess air
- higher-than-design feedwater temperature
- reheater inlet temperature lower than specified
- an externally fouled superheater or reheater
- leaking desuperheater spray water
- poorly adjusted controls.

On the other hand, high steam temperature may result from:

- an "over-seasoned" furnace
- too high an excess-air percentage
- feedwater temperature lower than specified
- reheater inlet temperature higher than specified
- irregular ignition or delayed combustion
- poorly adjusted control equipment.

An operating variable with a very great effect on steam temperature is the cleanliness of the radiant and convective heating surfaces. Although all modern coal-fired steam generators are equipped with automatic sootblower systems, the judicious supplemental manual operation of certain blowers can improve overall unit operation. It can save valuable blowing medium and reduce required maintenance by minimizing the number of blowing cycles. To be most effective, a sootblower program requires periodic furnace observations. Based on such observations and performance results, selective sootblowing can lead to better steam-temperature control and reduce the possibility of troublesome accumulations in the furnace and convection passes.

BOILER EFFICIENCY

An effective operator is constantly striving to obtain maximum efficiency from a unit. To do this, he or she must be aware of the effect of all operating variables, and the adjustments required to maximize efficiency.

Two items within operator control that affect boiler efficiency are dry-gas loss and unburned-fuel loss.

DRY-GAS LOSS

Usually the largest factor affecting boiler efficiency, dry-gas loss increases with higher exit-gas temperatures or excess air values. Every 35° to 40°F increment in exit-gas temperature will lower boiler efficiency by 1 percent. A 1-percent increase in excess air by itself only decreases boiler efficiency by 0.05 percent. On most boilers, however, increased excess air leads to higher exit-gas temperature. Consequently, increases in excess air can have a twofold effect on unit efficiency.

Usually, coal-fired units are designed to operate with 20 to 30 percent excess air. To operate a boiler most efficiently, therefore, an operator must have a reliable means of assessing the quantity of excess air leaving the boiler. In-situ oxygen recorders that measure the oxygen at the boiler or economizer outlet are the best information source. They must, however, be checked daily for proper calibration and maintained as necessary. The operator should maintain the required excess air by making sure the controls are in the correct mode or by manual bias of the fuel-to-air ratio.

UNBURNED-FUEL LOSS

On gas- or oil-fired units unburned-fuel loss should be negligible, whereas unburned loss on coal-fired units can be appreciable. The boiler manufacturer will predict unburned-fuel loss and these values can usually be maintained with correct operation. There is no easy way to continuously monitor unburned carbon in the ash. Obtaining such values involves the time and manual effort of laboratory analysis of a flyash sample. The significant point is that the laboratory feed back the information to the operators. If values are consistently high, the plant operations department should develop a program to pinpoint what is causing the high unburns and how to improve the condition.

Usually, high unburns can be traced to the mixing process of the fuel and air in the furnace. Once the source is found, attention must be focused on what corrects it. For example, one cause of poor mixing of fuel and air could be inadequate windbox to furnace pressure. If this is so, monitoring pressure to keep it in line is easier than waiting for the periodic feedback of laboratory results.

There are two other items having an effect on boiler efficiency that an operator can do little about. These are the moisture-in-fuel and hydrogen-in-fuel losses, and the steam-generator radiation loss.

MOISTURE LOSS

Although the moisture loss of the stack gases is considerable, the loss comes from fuel moisture and hydrogen, and the moisture in the air, which are realities not within operator control. The moisture in coal consists of inherent moisture and surface moisture. Although attention to the care of coal from the mine to the coal bunker can minimize the surface-moisture pickup with resultant increase in boiler efficiency, this is seldom within the boiler operator's control. A reduction in exit-gas temperature will decrease the moisture loss as well as the dry-gas loss.

RADIATION LOSS

On large, well-insulated steam generators, the efficiency loss due to radiation is about 0.2 percent at full load. As essentially the same total heat is lost throughout the load range, radiation loss increases with decreasing loads. Considering only radiation losses, operating several units at low load may be less economical than taking one off. But other things must be considered such as sudden additional load demands and the amount of fuel required to return the unit to service.

See Chapters 4 and 6 for further discussion of the above heat-loss items.

AIR HEATERS

As already noted, lower exit-gas temperature is the most positive means for increasing boiler efficiency. The limiting factor is usually the air heater; for purposes of this discussion, observations will be focused on the Ljungstrom® type of regenerative air heater.

For maximum effectiveness the air heater must be kept clean, the baskets must be replaced when acid corrosion has deteriorated enough material to affect performance, and the seals must be adjusted to minimize air and gas leakage. Proper operation of air heaters requires certain instrumentation. Pressure-drop indicators across both the air and gas sides must be available to the operator as well as temperature indicators for gas and air, both entering and leaving. Pressure drops are the best guide on the need to operate the air-heater sootblowers. Once pluggage progresses too far, sootblowers will not remove the deposits and the air heater will have to be removed from service for cleaning. (See Chapter 14.)

Faulty operation causes most air-heater corrosion. The air-heater manufacturer supplies a chart with recommended average cold-end temperatures to keep the metal above the dew point corresponding to the sulfur in the fuel. The "average cold-end temperature" is defined as the arithmetical sum of the temperature of the air entering the air heater plus the gas temperature leaving the air heater, divided by two. Consistent operation below the dew point rapidly corrodes air heater baskets. Steam or water air heaters or bypass ducts control the cold-end temperature. (See Chapter 6.)

High exit-gas temperatures leaving the air heater are often an indication of air bypassing the air heater. This can be from poorly adjusted seals or excess pulverizer tempering air. Pulverizer systems for example, are designed to dry coal of a specified moisture. If the coal has less than designed-for moisture or the mills are at partial capacity, mill tempering air will bypass the air heater and result in higher exit-gas temperatures leaving the air heater.

THE ECONOMIC IMPACT OF OPERATION

The most effective way to achieve maximum boiler efficiency in day-to-day operation is to embark upon an education campaign for plant management, supervisory staff, operators and maintenance personnel. If everyone knows the economic impact of operational variables on fuel costs, this knowledge can lead to significant fuel savings.

Too often campaigns to improve efficiency are carried out by only one group. Tests are performed to collect data rather than to establish "bogies" for day-to-day operation. The people performing the tests do not communicate the results to the rest of the plant, but rather to the company files. It is important, then, to review test results with the unit operators and to establish operating procedures that will take advantage of what is learned from the tests.

Re-starts are expensive. It requires considerable fuel to get a unit up to pressure and bring it on line. Extra maintenance and on-line attention that keeps downtime to a minimum can pay off in fuel savings. Retaining heat in the boiler during a weekend shutdown when there is no demand for steam can save fuel. A boiler can retain a good deal of heat if its isolation dampers are in good condition and are closed tightly. One word of caution—purging requirements both before bottling and prior to light off must be adhered to. The loss of the unit because of an explosion will be infinitely more costly than any heat saving from failing to perform a proper purge.

START-UP
FROM COLD CONDITION: GENERAL

Prior to light off of any boiler, a supervisor of operations should inspect the unit exterior. All doors should be checked, cleared of tags, and then shut. Valves should be correctly positioned for start-up in accordance with the steam-generator manufacturer's valve operating diagram. All areas must be free of debris that will hinder expansion. If repair work was done during the outage, special care must be taken to assure that no permanent ties were made to the furnace structure which will impair expansion. Account for all personnel.

At this point, safety tags can be removed from breakers. As the boiler is filling with treated water, all vents should be open, as noted on the manufacturer's valve operating diagram. On a thermal-circulation unit, the water level should be brought to where it just shows near the bottom of the water glass. On a Controlled Circulation® unit, the water level should be brought near the top of the gage glass; this will prevent the water from dropping from sight when the first boiler-water circulating pump is started. The drainable portions of all steam-circuit headers, connecting links and piping should be drained through lines free from back pressure. Reheater drains and vents are opened so that residual moisture will be boiled off. These reheater drains and vents will have to be closed prior to raising a vacuum in the condenser.

During warm-up and until the unit is carrying load, there will be little or no steam flow through the superheater and reheater. To protect the superheater and reheater metals, the temperature of the gas leaving the furnace should be limited to the manufacturer's recommendation, usually 1000°F. The firing rate must be limited to satisfy this requirement. Thermal-circulation boilers can be warmed up at a rate that does not exceed a saturated-steam temperature rise of 200°F per hour. Controlled Circulation boilers have no saturation steam temperature rise limitation, only the furnace exit-gas limitation previously stated. Most modern boilers have traversing thermocouples to monitor furnace exit-gas temperature during warm-up; older units may use temporary thermocouple probes for each start-up or have con-

servatively established firing limitations based on prior testing.

Before light off, the operator should check all instrumentation and furnace safety systems. Steam-generating units differ too much to be able to give detailed start-up procedures. The operator must be familiar with all manufacturer's instructions and the plant operating procedures.

Usually the equipment will start in this sequence. Air heaters and boiler-water circulating pumps (on positive-circulation units) will be started first. Next the induced-draft fans, followed immediately by the forced-draft fans. On most units, furnace draft will be established and then transferred to automatic draft control. Airflow is raised to at least 30 percent of full load airflow, and the unit purged for at least five minutes to remove any unburned fuel or combustible gases.

During start-up, airflow should be maintained at 30 percent of full load airflow to assure an air-rich furnace mixture and to prevent any settling out of explosive mixtures. Once the unit is purged, oil pumps may be started and gas or oil trip valves opened. At set intervals, the operator should check the proper functioning of any furnace safeguard system. Regardless of how much urgency there is for getting the unit returned to service, no interlock should be jumpered or bypassed.

Never attempt to light off any fuel nozzle without the required ignition-energy source for that nozzle. Any time an operator has doubt about safe combustion in the furnace, he or she should trip the fuel and purge the furnace before relighting.

As pressure is raised, periodic inspections are necessary to assure that the unit is expanding as it should. If oil is the warm-up fuel, air-heater sootblowers should be operated frequently to keep the heating surface clear of flammable deposits.

Once the unit is on line, load may be picked up as swiftly as pulverizers can be brought into service. The usual restriction in most power plants is the warm-up and load-rise limitations of the turbine. (See Chapter 7.)

START-UP FROM COLD CONDITION: DRUM-TYPE UNITS

On drum-type units, maintaining drum water level is of prime importance. Normally, the operator carries out this function in the manual mode until pressure is raised. Before light off, drum water level should be brought in sight. It is best to start firing with a low water level because, as the water starts heating, it will expand. Drum vent valves should be wide open so that the air will vent from the drum. Superheater and reheater should be drained of any condensate wherever possible and then the valves opened or closed according to the valve operating instructions.

For a Controlled Circulation boiler, two of the circulating pumps are started to initiate circulation. To minimize flashing of steam in the downcomers, additional pumps are not started until the boiler water reaches 250°F.

The unit is now fired in accordance with the established furnace safeguards. During the warming period, the economizer recirculating line valves are open. The water will swell as it is heated and the operator will manually blow down the boiler as required to maintain sight of the water in the gage glass.

The firing rate will be controlled to keep the furnace exit-gas temperature below the recommended value, usually between 900°F and 1000°F.

When drum pressure reaches 25 psig, it can be assumed that all air has been purged from the drum and water circuits. The operator can close the drum vent valves. As pressure increases, it is necessary to progressively throttle the superheater drains and vents and to increase the firing rate as required.

If there is evidence of steaming in the economizer by erratic drum control during feeding of water, the operator must be certain that the recirculation line is open. On thermal circulation units it may be necessary to feed more water than required to maintain level in sight and to blow down to control water level.

Start rolling the turbine as soon as the minimum permissible start-up pressure and temperature, specified by the turbine manufacturer, are reached. When bringing the turbine to speed, the firing rate must continue to be controlled to prevent the furnace exit gas from exceeding its temperature limit. Once the turbine is synchronized and minimum load established, this limit can be removed. The gas temperature probe is withdrawn and the firing rate increased as required. Usually by this time all superheater and reheater vents and drains are closed. If drum water level is still under manual control, it should now be placed on automatic control.

LOAD AND PRESSURE CONTROL: DRUM-TYPE UNITS

Opening or closing the turbine control valves increases or decreases the load on a drum-type unit. Changing the firing rate regulates the pressure. In theory, for increasing load the operator will open the turbine control valve. If no other action were taken, the flow to the turbine would increase and the superheater outlet pressure would decrease. Upon seeing the outlet pressure drop, the operator would increase the fuel and air inputs.

Because of the increased steam flow, the water level in the drum would change. Initially, the level would rise due to surging, but shortly it would fall rapidly because more steam is removed than water enters.

A trained operator will anticipate these interactions. As the turbine control valve is opened, the operator will simultaneously increase fuel, air and feedwater flow. A well-designed and executed operational control system will perform the same as will such a trained operator. The more experienced and better trained the operator, and the more sophisticated the control system, the higher the expectation that smooth operation will result.

SHUTTING DOWN DRUM-TYPE UNITS

Time requirements and procedures for shutting down a boiler depend on the nature of the shutdown (normal shutdown to cold, normal shutdown to hot standby, emergency shut-

down) and whether the unit is to be entered.

For a scheduled shutdown, steam pressure can be reduced to the limit of the turbine before the unit is taken off line. Thermal circulation units should not exceed a cool-down rate of 150°F per hour of saturated steam temperature decrease. Controlled Circulation units can be rapidly cooled if the circulating pumps are left in service. Normally, the drum vents are opened at 25 psig and the boiler is not drained until water temperature is below 200°F.

SUPERCRITICAL-PRESSURE BOILERS

Although much of the discussion of drum-type operation pertains also to supercritical-pressure boilers, there are some differences. The supercritical boiler needs an even more precise balance of inputs to outputs, because it does not have the flywheel effect that the boiler drum affords the subcritical-pressure unit.

START-UP

Although the firing system start-up and operation are the same as for a drum-type unit, the fluid system start-up is completely different. The supercritical boiler is furnished with an integral start-up system. The unit is initially fired, warmed up, and brought to partial load on the bypass system.

The Combined Circulation® unit, one type of supercritical-pressure steam generator, isolates the waterwalls from the superheater with boiler throttling (BT) valves (Fig. 1). To gain good control at low flow, there are boiler throttling bypass (BTB) valves around the BT valves. For initial warm-up and supply of steam to the turbine, the BT and BTB valves are closed. To vent fluid, a separator is installed, external to the boiler, with a line from the waterwall to the separator valved off by a boiler extraction valve (BE). Since it is advantageous to start the turbine with low-pressure steam, steam pressure is reduced through the BE valve to the separator. There is a line from the top of the separator to bring the steam to the super-heater, with a check valve to prevent backflow during high-load operation. The check valve is referred to as a steam-admission valve (SA).

A line to the condenser called a water drain valve (WD) removes water and controls water level in the separator tank.

As the boiler is warming up, steam is produced in the separator tank, but very little steam goes to the turbine. A line from the top of the separator tank to the condenser carries the excess steam from the separator. A control valve called the spillover (SP) regulates separator pressure during start-up. For detailed guidelines on the start-up of supercritical units see the adjoining box, "Recommended Supercritical Unit Start-Up Procedure."

LOAD CONTROL

The method of controlling the operation of Combined Circulation supercritical boiler differs from that of a drum-type boiler. The turbine control valves do not regulate the load; rather the load is regulated by flow. To raise or reduce load, the operator varies flow with the boiler feed pump.

PRESSURE CONTROL

The pressure on a supercritical boiler is controlled by the turbine valves, which operate initially as back-pressure valves. The only

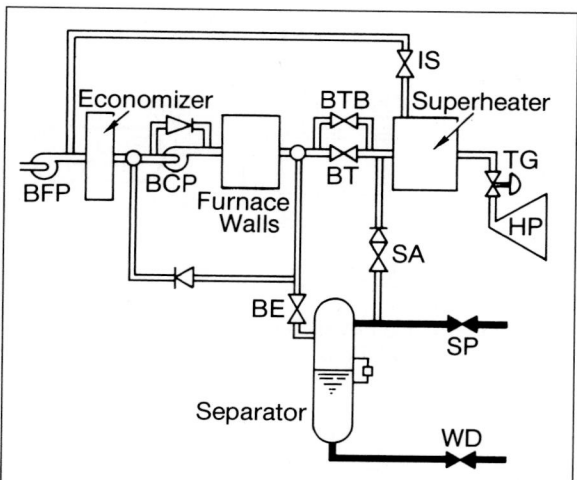

Fig. 1. Schematic diagram of a supercritical boiler

RECOMMENDED SUPERCRITICAL-UNIT START-UP PROCEDURE

1. The feed pump establishes a flow (5 percent of maximum continuous rating) through the economizer and the waterwall. The BT and BTB valves are closed. The flow is brought through the BE down to the separator. The flow continues through the WD valve to the condenser. The WD maintains level in the separator.

2. The boiler circulating pumps are running to provide protection for the waterwall.

3. The boiler is lit off and gradually the waterwall temperature is raised to the point that the turbine can be rolled.

4. As the temperature of the fluid from the waterwall rises, steam will be generated in the separator. Steam will flow through the SA valve and through the superheater. This excess steam is bled off through the SP valve to the condenser. The SP controls the steam pressure in the separator.

5. When the waterwall reaches the proper temperature, the turbine can be rolled and put on the line using steam from the separator. The flow is now through the economizer, the BE valve, the separator, the SA valve, and out through the superheater.

6. The next step is to raise the waterwall temperature high enough to prevent a dip in steam temperature when the boiler is transferred to once-through operation.

7. When the waterwall temperature reaches this point, the flow is transferred from the BE to the BTB valves. The boiler is now operating as a once-through boiler and the separator is no longer in use. The pressure in the waterwalls is controlled by the BTB valves.

8. The turbine valves are set for approximately 30-percent flow. As the load increases, the superheater pressure will gradually rise until it reaches full throttle pressure.

9. When the throttle pressure reaches the design value, the turbine valves are placed on automatic to control boiler pressure.

variable that the turbine valves monitor is throttle pressure, approximately 3500 psig for most operating supercritical units. In normal operation (above 30 percent), these valves are the only boiler pressure control. To raise pressure, the valves close and to decrease pressure the valves open.

STEAM-TEMPERATURE CONTROL

The superheater outlet temperature on a supercritical-pressure boiler is controlled with firing rate, which is increased to raise outlet temperature and decreased to lower it. Although this is an effective means of temperature control, the response time to correct a temperature deviation is too long for smooth operation.

To improve the response, desuperheater water is injected between superheater stages. The injection water is taken from the economizer inlet at full boiler feedwater temperature so there is no cycle efficiency loss. Although the injected water gives fast response, the ultimate temperature control is balancing firing rate to feedwater rate.

The injection water essentially bypasses the waterwalls and economizer. When the superheater temperature is high, the operator or control system will reduce the firing rate and open the spray valves simultaneously until the temperature returns to the desired level.

Usually, for a Combined Circulation unit, adjusting the reheater outlet temperature follows the same procedure as for a drum-type unit. With a C-E tangentially fired unit, the process involves tilting the fuel and air nozzles and maintaining furnace cleanliness by sootblowing.

CLEANUP

Because a supercritical unit has no steam drum to separate the impurities in the waterwalls from the essentially pure steam in the superheater, a waterwall cleanup procedure must precede each start-up. The procedure involves circulating condensate through the polisher, feedwater heater train, economizer, waterwalls, separator, and drain system back to the condenser. The water is monitored for iron concentrations with the flushing continuing until the manufacturer's limits are obtained.

PULVERIZER OPERATION

Although several types of pulverizers are used to grind coal for power-plant service, the majority are of the rotating bowl-type for use in direct-fired systems. Refer to Chapter 11 for a description of this pulverizer.

In operation, four areas must be understood and monitored: lubrication of the gearing and bearings, airflow, mixture temperature leaving, and product fineness.

LUBRICATION

With the exception of the roller journals and exhauster (if so equipped), the mill is completely lubricated from one system in the worm gear housing. The worm-shaft bearings are flood oiled from a bath of oil in the gear case. The circulation is either by an external pumping system or by an internal pump that is a part of the gearing system. The lubricant must meet the manufacturer's specifications and be non-foaming and noncorrosive (see Fig. 2).

Periodic inspection of the gear-housing oil

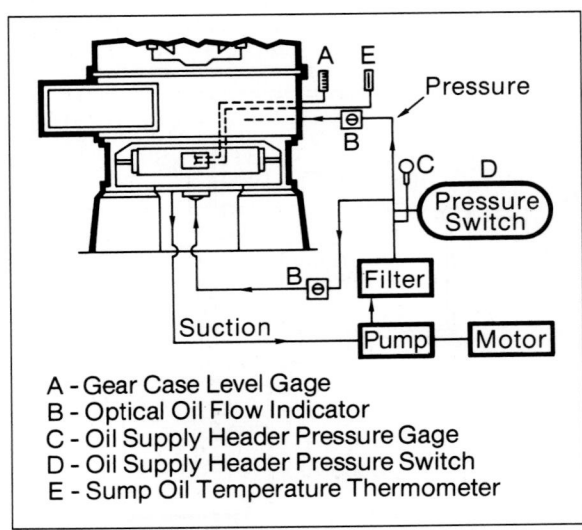

A - Gear Case Level Gage
B - Optical Oil Flow Indicator
C - Oil Supply Header Pressure Gage
D - Oil Supply Header Pressure Switch
E - Sump Oil Temperature Thermometer

Fig. 2. **Simplified pulverizer diagram showing lube-oil system**

temperature is required. For most of the recommended lubricants, oil temperature is maintained at about 150°F. If it is above this temperature, the oil level may be too low; if so, it should be promptly refilled with oil of the same manufacture and type as already in the housing. High oil temperature could also result from a breakdown of the lubricant or sludging. If so, change the oil. It could also be because of scale formation in the cooling coil, insufficient quantity of cooling water, or temperature of the inlet cooling water being too high.

The operator should check the oil in the return glass after each mill start-up, and daily for mills in continuous service.

MILL OPERATION

To place a mill in service, start the mill, allow it to come up to speed, and bring mill outlet temperature up to recommendations before starting the feed of coal. Make sure there is proper ignition energy adjacent to the fuel nozzles supplied by the mill being put in service. Start the feeder with a fairly high feed rate setting. After the mill begins to grind, reduce the feed rate to the desired amount. The feeder drive must be interlocked with the mill motor so that, if the mill power is interrupted, the feeder will shut down.

Mill-outlet temperature should be maintained as high as possible without exceeding the safe limits of the type of coal ground. This temperature may be as low as 130°F with lignites and as high as 180°F for low-volatile bituminous coals. Low mill-outlet temperatures often indicate mill overload which is usually accompanied by other indications such as high motor amperage or high mill differential pressure.

PULVERIZER FIRES

Pulverizer fires can range from nondestructive to highly destructive, dependent mainly on how soon the fire is detected and what action the operator takes. Basic causes of pulverizer fires are excessive mill temperatures; foreign combustible material such as rags, paper or wood; settling out of coal in the pul-

verizers; or excessive accumulations of pyritic material or coal in the mill base area.

Early detection of fires enhances the safety of plant personnel. An automatic system is recommended. Using visual or other sensory means to detect such fires is arbitrary, slow, and worst of all, requires the presence of a person in the immediate hazard area. Temperature-detection devices should be used as the primary indicator of a fire in progress.

There are five types of pulverizer fires which are categorized by where in the fuel-preparation system they occur: feeder fires, above-bowl fires, under-bowl fires, exhauster fires, and fuel-piping fires. Upon detection of a pulverized-fuel system fire, the fire-extinguishing system is operator-activated. A spray of water is introduced into the system at multiple locations; the timing is at the discretion of the operator. However, the pulverizer must be kept inert until the water-injection sequence is over. When the pulverizer is empty of its contents, it can be shut off and isolated. Water injection should continue until all evidence of fire has disappeared. Entry for clean-up is allowable only after the mill and its contents have cooled to ambient temperature. Caution should still be exercised because smoldering pockets of fuel may be present.

Before restart, the entire milling system should be inspected and cleaned of any accumulations. Check the lubricants in the mill base and rolls for any evidence of carbonization. Recheck compression of journal springs by means of a hydraulic jack. If everything is satisfactory, the mill may be returned to service.

EXPLOSIONS IN PULVERIZERS

Pulverizer fires occur more often than pulverizer explosions; if properly handled, they are not overly dangerous. If, however, a fire is not brought under control effectively and expeditiously, an explosive condition can occur. Explosions also can occur without the presence of an obvious fire, if the necessary conditions to support an explosion are present.

All explosions are initiated by fires, but fires do not always initiate explosions. Because of this difference, there are available independently designed and operated systems to handle each condition safely. The C-E explosion-prevention system uses an automatic steam-inerting sequence to reduce the potential for pulverizer explosions when hazardous operating conditions exist. The system can also safely transport the pulverized coal remaining in the mill to the furnace, while maintaining an inert atmosphere inside the pulverizer. The fire-control system detects fires in operating pulverizers and alerts control-room operators. The operators can then initiate fire-extinguishing procedures which include water-spray injection and steam-inerting and transporting. A combination mill inerting and fire-fighting system integrates the two subsystems into one complete package.[1]

SLAGGING AND FOULING

Chapter 3 describes ash characteristics and their effect on slagging and fouling. This section will cover what the operator can do to control slagging and fouling. Successful boiler operation depends to a significant extent on the ability of the operating staff to understand how certain operating variables relate to fuel properties and furnace-sizing criteria. Those operating variables that influence slagging and fouling are unit load, excess air, fuel fineness, and secondary-air distribution.

LOAD

The higher the load, the higher the heat input to the furnace, and the greater the potential for slagging and fouling; therefore, the most direct way to reduce slagging is to curtail load. But this is not always possible, because the rated output may be necessary to meet electric-generation or process requirements. Changes in excess air, fuel fineness, and secondary-air distribution are less-drastic methods for minimizing slagging and fouling.

EXCESS AIR

At high oxidation states, iron compounds in the ash melt at a higher temperature than at lower oxidation states. Therefore, for bitumi-

nous coals, which are frequently high in iron content, there is a significant difference in fusion temperatures measured in reducing (oxygen-starved) and oxidizing (oxygen-rich) atmospheres. Subbituminous coals normally contain less iron and exhibit a smaller difference in melting temperatures produced in oxidizing and reducing environments. This means that, if slagging is a problem with high-iron coals, furnace deposits can be reduced dramatically by increasing the amount of excess air. As a rule of thumb, the higher the fusion temperature, the drier the slag in the furnace, and the easier it is to remove.

FUEL FINENESS

Slagging conditions often can be improved by proper control of pulverizer fineness and classification. Since coarse coal particles take longer to burn, they are more prone to producing slag. High retention on the +50 mesh often increases slagging tendencies. See Chapter 11 for a detailed discussion of the fineness recommended for optimum pulverized firing of various ranks of coal.

SECONDARY-AIR DISTRIBUTION

Because several different types of firing systems are in use, it is difficult to generalize on the subject of air distribution in the combustion zone. Basically, the objective is to provide good mixing of fuel and air so that combustion is efficient and local zones with reducing atmospheres are avoided. In units with tangential firing, for example, slagging sometimes can be reduced by increasing secondary airflow to fuel compartments.

OTHER OPERATIONAL MEASURES

If the coal contains a substantial amount of ash with a tendency to slag and/or foul heat-transfer surfaces, particular attention must be given to equipment capable of cleaning the furnace walls and convection-tube banks. Failure to remove deposits at the proper time may result in a chain reaction of deteriorating events. For example, excessive furnace slagging results from not using the wall blowers at proper intervals. This condition imposes higher gas temperatures in the convection sec-

tions because of the reduced rate of heat absorption in the furnace.

In turn, the higher gas temperature causes the flyash to become sticky, increasing deposition in the convection sections. Depending on gas temperature and ash properties, the retractable sootblowers may not be able to remove these deposits. Ultimately, sections of the convection pass may become plugged. Unless load is reduced at this point, it may not be possible for the induced-draft fans to maintain the proper amount of excess air. This causes additional slagging, and the cycle repeats.

Modern sootblower systems have programming techniques so proper sequential operation of the blowers—on an automatic basis—can be established after ash-deposition patterns are verified during preoperational tests. Through programming, ash deposits on the furnace walls generally can be held to a minimum, and combustion gases cooled sufficiently before they enter the convection pass.

SOOTBLOWER OPERATION

A major guideline to reliable sootblower operation is that plant personnel should not wait until large deposits develop before operating blowers. Waiting too long between operations can seriously hamper the effectiveness of

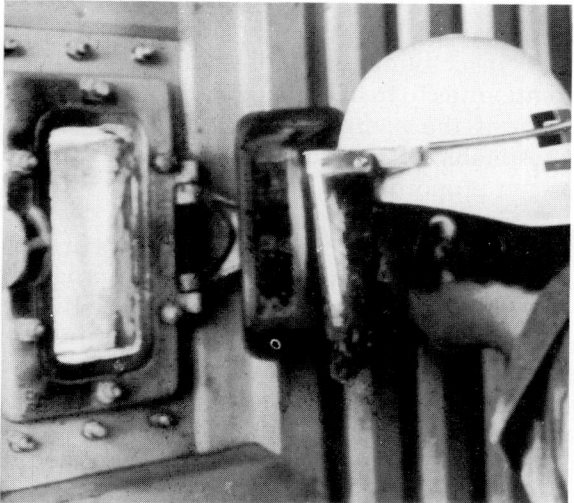

Fig. 3. **Operators must observe fires, evaluate slagging, and listen for tube leaks.**

sootblowers. Observation of furnace conditions at least twice per shift helps eliminate unexpected problems (Fig. 3).

For maximum effectiveness, the blowing sequence must be established to favor those sections of the furnace that foul most easily. Only by careful observation can these areas be identified. If some areas of the furnace are particularly prone to fouling, supplemental remote-manual operation of selected blowers can save valuable blowing medium and reduce system maintenance by minimizing the number of blowing cycles.

Because the sootblower system is so important for reliable, full-power operation of a coal-fired unit, this equipment must be maintained in good operating condition. Realize that blowing pressure can change, especially if valve travel is used for pressure regulation. Thus, air or steam pressure should be checked frequently with blowers of this type, especially when furnace observation shows a buildup in deposits. Sootblowers with adjustable orifices are not so sensitive and will usually retain their settings indefinitely.

FEEDWATER/BOILER WATER

Because internal tube corrosion and deposition are major causes of costly forced outages, operators must be continually alert to the hazard of water neglect. All plants must establish and adhere to a feedwater and boiler-water treatment and control for their system. Chapter 20 describes the chemistry of water treatment in detail. Rather, the following section describes the cautions that must be taken to minimize corrosion and deposition.

OXYGEN CONTROL

Oxygen control is the most important element in feedwater control. Oxygen concentration in the feedwater must be regulated to minimize the formation of preboiler corrosion products which inevitably end up as deposits on heat-transfer surfaces in the boiler.

Oxygen levels are more likely to exceed recommended limits during start-up, shutdown, and low-load operation. At these times low-pressure feedwater heaters and related extraction piping are often under negative pressure, and any leaking valves, pumps or flanges will provide a path into the system.

Excess oxygen is removed from the system by the deaerator, not through chemical additives. Such deaerators have manufacturer's guarantees for levels of oxygen leaving the deaerator.

Oxygen leaving the deaerator should meet these guarantees at all times or the manufacturer should be contacted for his recommendations to bring it back in line. Adequate steam should be available to the deaerator during unit start-up so oxygen is purged from the feedwater. If adequate auxiliary steam is not available because there are no other sources of steam in the station, peg the deaerator with steam from the boiler drum until turbine extraction steam is available.

To minimize the formation of preboiler corrosion products, the oxygen concentration in feedwater should be maintained at less than about 5 parts per billion (ppb) during unit operation. But acceptable feedwater oxygen levels during steady-state operation do not necessarily mean that oxygen concentration is within safe limits. During various phases of operation, conditions can exist that may result in excessive amounts of oxygen. Thus, use of dissolved-oxygen monitors is important, particularly during load swings and start-up operations. And do not be lulled into a false sense of security if oxygen levels are excessive only for a short time. Considerable damage can still occur, a fact that those responsible for plant operation must be aware of. Periodically, plant procedures and controls should be evaluated to ascertain that all sources of oxygen contamination have been eliminated.

pH CONTROL

Of equal importance with oxygen is the control of boiler-water pH. Small deviations from the recommended boiler-water limits will result in tube corrosion. Large deviations can lead to the destruction of all furnace wall tubes in a matter of minutes.

CAUSES OF HIGH AND LOW PH

The primary cause of acidic and caustic boiler-water conditions is condenser leakage. Raw cooling water that leaks into the condenser eventually ends up in the boiler water. The water source determines whether the in-leakage is either acid-producing or caustic-producing. Fresh water from lakes and rivers, for example, usually provides dissolved solids that hydrolize in the boiler-water environment to form a caustic, such as sodium hydroxide. By contrast, seawater and water from recirculating cooling-water systems with cooling towers contain dissolved solids that hydrolize to form acidic compounds.

Strict tolerance levels on condenser leakage should be established for all high-pressure boilers. Set a limit of 0.5-ppm (parts per million) dissolved solids in the feedwater for normal operation; allow from 0.5 to 2 ppm for short periods only. Shut down the steam generator immediately if the surface-condenser leakage produces more than 2 ppm of dissolved solids in the feedwater.

Another potential source of acidic and caustic contaminants is the makeup demineralizer, where regenerant chemicals such as sulfuric acid and caustic may inadvertently enter the feedwater system. Chemicals incorrectly applied during boiler-water treatment also can be corrosive, as for example sodium hydroxide used in conjunction with sodium-phosphate compounds to treat boiler water. Corrosion can occur if the sodium hydroxide and sodium phosphate are not added to the water in the proper proportion.

IMPORTANCE OF WATER ANALYSIS

A comprehensive water-analysis program should be maintained to assure that feedwater and boiler-water chemistry is held within prescribed limits. Continuous, automatic analytical instrumentation is preferred. If automatic analyzers are unavailable or are not operational, conduct water tests daily for pH and oxygen in the feedwater and for pH, PO_4 and total solids in the boiler water. A condenser leak-detection system is of particular impor-

tance in any high-pressure steam cycle. When installing this type of system on multishell condensers, individual analyzers must be provided for each shell.

The operator must remember that many potential tube failures can be avoided by continual attention to the control of the water and steam environment throughout the station.

COMMISSIONING FUNCTIONS

Before a new unit can be put in service, the entire system must be cleaned to remove oil, grease, siliceous material, mill scale, rust, and any other debris. The condensate and feedwater systems are cleaned before the boiler so that none of the debris or dirt is carried into the boiler. These two systems are mechanically cleaned, then given an alkaline flush and sometimes, an acid wash.

The economizer and boiler will be given an alkaline boilout followed by an acid cleaning.

Finally, the superheater, steam piping and reheater will be cleaned by a three-phase scavenging with steam. It is important that each of these processes be conscientiously undertaken and result in as clean a system as is possible. If not, operating problems are sure to develop when the unit goes into service.

PREBOILER CYCLE

All preboiler systems of high-pressure boilers must be thoroughly flushed with a hot alkaline solution to remove oils, siliceous materials and particulate matter which are present following fabrication, storage and erection. It is important that these materials be removed prior to initial operation; otherwise, they will be carried into the boiler. Optimum plant operating conditions will be realized rapidly after start-up if the preboiler equipment is satisfactorily cleaned.

The condensate system, feedwater system and the shell side of all heaters should be included in the alkaline cleaning. This cleaning involves the following basic operations:

1. Manual cleaning of the condenser, all feedwater heaters and the deaerator storage tanks.

2. Gross flushing to waste to remove the bulk of loose material.

3. Preheating of circulation water to increase temperature to 200°F.

4. Circulation of alkaline solution at 200°F (0.5 percent tri-sodium phosphate, Na_3PO_4).

5. Rinse to remove alkaline material.

6. Wet lay-up solution addition and circulation to protect metal surfaces until initial operation.

Cleaning the preboiler cycle normally requires the installation of temporary piping to establish circulation through the system. It is also desirable to install temporary piping to bypass portions of the system, such as the boiler feed pump and deaerator storage tank, which may be damaged by or trap large quantities of loose particulate material. Circulation is normally established with a condensate pump which takes suction from the condenser hotwell, or a special pump of equivalent capacity. Flow is through the condensate and feedwater systems and is returned to the hot-well through temporary piping or through the shell side of heaters. Strainers are placed at the suction of all pumps used during cleaning to protect them from suspended particles. The strainers are checked periodically during the circulation period and cleaned if necessary.

The condenser, condenser hot-well and deaerator storage tank have to be mechanically cleaned to remove loose debris before any chemical cleaning is started. Mechanical cleaning will consist of sweeping and hosing down all surfaces, and removing all loose material by vacuuming, shoveling, or any other convenient means.

A solution containing 5,000 ppm tri-sodium phosphate (5 lbs Na_3PO_4 per 1,000 lbs H_2O) and a detergent are recommended for the hot alkaline cleaning of the preboiler cycle. The cleaning is carried out at about 200°F. Samples are obtained periodically and tested for silica and oil concentrations. The cleaning is continued until the chemical checks indicate no further increase in the silica and oil concentrations. This process is normally accomplished in 6 to 8 hours.

Following the alkaline cleaning, the system is thoroughly rinsed to remove the alkaline material, and refilled with condensate of demineralized water containing 100 ppm hydrazine for wet lay-up of those systems containing copper alloys.

In completely copper-free systems (supercritical boilers), 200 ppm hydrazine and 10 ppm ammonia should be added to the water for wet lay-up. The system should be isolated and stored wet under nitrogen until ready for the next phase of start-up, usually chemical cleaning of the boiler.

CHEMICAL CLEANING OF BOILERS

The internal surfaces of a boiler in contact with water or steam must be kept clean to assure an efficient transfer of heat in the generation of steam. Several cleaning procedures are available to assure a removal of foreign matter introduced into the boiler during the manufacturing process, erection of the equipment and in operation. The general cleaning processes are alkaline boilout and acid washing.

Alkaline boilout removes contaminants commonly found in a boiler following its shop assembly or field erection: lubricants, oil, rust, sand, metal fragments and assorted debris.

Acid cleaning removes scales and deposits formed on internal heat-transfer surfaces in contact with water. This procedure dissolves compounds resulting from contaminants in the feedwater delivered to the boiler. Acid cleaning is also used to remove mill scale and corrosion products.

ALKALINE BOILOUT

The basic reason for an alkaline boilout of a boiler is to remove water- and alkali-soluble and saponifiable compounds from the waterside surfaces of the unit. These compounds may include lubricants used in the erection of the boiler and, in some instances, protective coatings applied to prevent atmospheric rusting following shop fabrication.

Most lubricants used in boiler construction are water soluble and do not offer any difficulty in removal during boilout. Non-water-soluble oils and greases are introduced in small quantities into the boiler from oil-lubricated equipment and workers' clothing. Every effort should be made to minimize the introduction of oil and grease into the boiler because the quantity of these materials determines the length of cleaning and the degree of difficulty in obtaining clean surfaces.

Sand, loose mill scale and corrosion products formed on the tube surfaces during erection and following the hydrostatic test are removed by blowdown during boilout.

Chemicals Used for Boilout

The chemicals used for boiling out a steam generator vary in composition. Generally, some combination of the following chemical compounds is used during an alkaline boilout: caustic soda, soda ash, sodium phosphate, sodium sulfite and sodium nitrate. Sodium sulfite reduces oxygen corrosion and sodium nitrate is added to prevent the possibility of caustic embrittlement. Soda ash and sodium phosphate are most commonly used because of the ease of handling. Potassium salts can be substituted for the sodium form.

Organic detergents are added to improve the effectiveness of the alkaline boilout. These materials must be used with care and according to the supplier's recommendation. Their indiscriminate use may lead to foaming and carryover of chemical to the superheater. The temperature stability of the organic detergent should be ascertained before use in a boiler.

The amount of detergent normally used ranges from 0.05 to 0.1 percent by volume.

The use of sulfite and nitrate is a refinement in the boilout procedures. Using them has not been shown to be a prime necessity. The principal chemical action is the reaction of the alkaline chemicals with non-water-soluble oils and greases. Experience has shown that an effective boilout for drum-type boilers can be attained by any of the following combinations:

1. Soda ash   4000 ppm
   Sodium phosphate   4000 ppm
2. Sodium hydroxide   2000 ppm
   Sodium carbonate or
   Sodium phosphate   2000 ppm

3. Sodium phosphate   5000 ppm
    Caustic soda   500 ppm
4. Sodium carbonate   2000 ppm
    Sodium phosphate   4000 ppm
    Caustic soda   2000 ppm

The first combination is most commonly recommended for boilout of high-pressure boilers. In some instances, a chemical such as sodium silicate is included in the boilout formula. Although this is an effective additive, it is not recommended for use in high-pressure boilers. High silica concentrations have been observed during initial operation when metasilicate was used in the boilout formula.

For once-through boilers, the boilout solution is composed of 5,000 to 10,000 ppm of sodium phosphate. Other formulations and combinations of chemicals can be satisfactorily employed for boilout, as dictated by local water-pollution control requirements. Such revised formulations should be reviewed by the boiler manufacturer prior to use.

Boilout Procedure

Preparatory to boiling out, the chemicals to be added to the boiler should be dissolved completely before introduction into the unit. In thermal (natural) circulation units, these chemicals are most suitably introduced into the boiler by blending them with water as the unit is being filled. This insures a homogeneous concentration throughout the boiler. In positive (pumped) circulation boilers, the boilout chemicals may be pumped into the boiler without concern about proper mixing with the fill water. The boiler-water circulation pumps will insure proper mixing.

Gradual heating of the boilout solution in the boiler is accomplished by the use of ignitors and warm-up burners. This assures a more even heating of the boiler surfaces. Steam pressure is raised to increase the saturation temperature and, thereby, the thermal circulation of the boiler water; this promotes good mixing of chemicals in the boiler circuits. The boiler pressure is raised to about one fifth of the normal operating pressure, or 300 psig, which ever is lower. A pressure of not less than 100 psig is recommended for lower-pressure industrial boilers. Excellent results have been obtained in boiling out at a pressure range of a few pounds to 100 psig with positive-circulation boilers, in which the ability to circulate is not related to pressure.

The quantity of oil and grease found in a boiler determines the duration of the boilout, with boilout periods of 8 to 24 hours being common. During the pressure-holding period, boiler-water solids are purged by blowdown at about 4-hour intervals. A chemical balance is re-established at the end of each purging of boiler water if the chemical concentration decreases to below one-half of the initial value. At the completion of boilout, the boiler is cooled slowly, drained, flushed free of residues, and inspected for cleanliness. If the boilout was unsuccessful and oil and grease are still present, subsequent acid cleaning of the boiler will be ineffective. If an internal inspection still shows oil or grease in the drum, the boilout procedures should be repeated.

ACID CLEANING

As an important part of the commissioning of a new unit, initial acid cleaning is included in this section. But as there will be future needs for acid cleaning to remove operational deposits, the discussion incorporates both initial acid cleaning and future acid cleanings.

Removal of preoperational and operational deposits from the internal surfaces of steam-generator tubing and other components has become an increasingly important maintenance problem in modern equipment. Because mechanical cleaning is virtually impossible in modern boilers, the effective application of chemical-cleaning solvents has become a necessary tool of the power-plant operator.

The primary reasons for boiler chemical cleaning are to prevent tube failures and to improve unit availability. Tube failures in low-pressure boilers are normally the result of creep which occurs when internal deposits produce excessive metal temperatures. A much smaller quantity of deposit will create difficulties in high-pressure boilers. Caustic corrosion and hydrogen damage, which only occur in the presence of deposits, can cause tube failures at temperatures well below the creep limit. Deposits originating both from fabrication and during operation must be considered potential problems.

Boiler manufacturers recommend that boilers operating above 900 psig be acid cleaned prior to initial operation. Also, because of the nature of chemical-recovery-boiler operations, the recommendation dictates that the lower 450 and 600 psig units be acid cleaned initially.

Mill and Operational Scale

All of the pressure parts of a steam generator may be subjected to heat treatment of some sort during fabrication or erection—during forming operations, stress relief, welding, or bending. Whenever carbon or low-alloy steels are subjected to high temperatures in the presence of air, oxidation occurs; the oxide produced is known as mill scale Fig. 4. Mill scale on boiler tubing is normally very thin with the exception of areas near welds and bends. Even where mill scale is initially uniform, its brittleness upon cooling can produce flaking. The resulting non-uniform surface is undesirable from the standpoint of corrosion susceptibility. During operation, mill scale can be readily eroded from the steam-generating surfaces and sub-

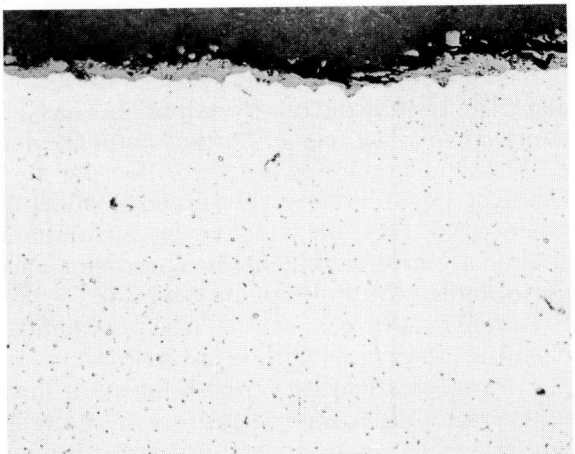

Fig. 4. **Typical mill-scale thickness on tubes installed in modern high-pressure boilers. Scale thickness illustrated 0.5 mils, at 75X**

sequently redeposited in critical areas. Preoperational acid cleaning removes the mill scale and serves to remove atmospheric rust which inevitably accumulates during erection.

The type of operational scale found in steam generators is related to the quality of feedwater supplied to the boiler. In industrial boilers, the principal deposits are calcium and magnesium phosphates, calcium and magnesium silicate, complex silicates as acmite or analcite, quartz, oxides of iron and copper, and organic matter. In utility boilers, the principal deposits are corrosion products, and iron and copper oxides.

Chemical Cleaning Procedures

Only experienced personnel with up-to-date equipment and a detailed procedure to follow should perform chemical cleaning operations.

The procedure should include:

1. The system layout with proper identification of all equipment to be used.

2. A step-by-step description of the functions to be performed. These should be specific for the solvent used for the cleaning.

3. Precautions to be taken against possible inadvertent contamination of equipment not included in the cleaning system.

The boiler operators, cleaning vendor, and boiler manufacturer must cooperate closely for a successful operation. Mutual prior approval of cleaning procedures and assignment of responsibility are desirable. Although respective responsibilities may vary from job to job, they can be generally classified as follows:

1. Normally, the cleaning vendor will supply all of the chemicals and equipment necessary to deliver the solvent to the boiler at a controlled concentration and temperature. Vendor personnel will generally operate the equipment and supervise the overall cleaning. They will perform the necessary chemical analyses during solvent introduction and monitor the spent solvent to determine when the cleaning has been completed.

2. Plant personnel must operate all permanent plant equipment. They are responsible for determining metal temperatures and maintain-

ing proper temperatures throughout the cleaning period. The owner normally supplies the necessary water and steam and sets up the solvent-delivery and waste-disposal systems. An important operator responsibility is to assure that the solvent is not inadvertently introduced to any other part of the steam plant.

3. The boiler manufacturer has the responsibility to provide a boiler that can be cleaned safely and effectively. The boiler must be designed with an adequate filling, draining, and venting capacity. The boiler manufacturer should establish a standard cleaning procedure for the specific boiler that should emphasize the hazards involved and the limitations on the use of specific components. They should be informed of any unusual use of boiler components and be ready to review any cleaning procedures that involve unusual steps or solvents.

The precautions relative to acid cleaning are common to all solvents currently used in practice. The metal temperature of the boiler is raised either by heating water using auxiliary burners or by circulating water which has been increased in temperature by the addition of live steam. The addition of heat by the use of burners is prohibited when the boiler is filled with inhibited hydrochloric acid. This is to prevent a destruction of any inhibitor due to localized application of heat. A similar precaution is required when organic acids are used for cleaning purposes. The only exception is when an organic alkaline solvent (e.g.; such as ammonium EDTA) is used for cleaning the boiler. The boiler is fired to raise the solvent temperature to approximately 275°F.

Selection of Cleaning Solvents

Solvents are selected for their ability to remove boiler deposits. Hydrochloric acid remains the principal solvent used in chemical cleaning. Its wide use is largely related to its lower cost, its availability, and its versatility. It has the ability to remove most of the various deposits normally encountered in boiler tubes even in a stagnant condition.

Although hydrochloric acid is commonly employed for dissolving iron oxide, some organic acids and organic alkaline solvents can also effectively perform the same function. These latter solvents are particularly useful in situations where specific circumstances prohibit the use of hydrochloric acid. Some of the solvents which have been developed, extensively evaluated, and employed in boiler cleanings are (1) ammoniated citric acid (ammonium citrate), (2) formic-hydroxyacetic (glycolic)

Table I. Effect of Velocity on Scale Removal of Various Solvents

Solvents and Cleaning Conditions	Static	Velocity				
		0.03 fps	0.1 fps	1 fps	2 fps	3 fps
Hydrochloric acid (5%) 6 hrs, 160–170°F	C	C	C	C	C	C
Phosphoric acid (3%) 6 hrs, 212°F	C	. . .	C	C	C	. . .
Ammonium citrate (3%) 6 hrs, 200–220°F	U	U	U	C	C	C
Formic hydroxyacetic acid (3%) 6 hrs, 275–280°F	. . .	U	U	C	C	C
Ammonium EDTA (3%) 6 hrs, 275–300°F	. . .	U	U	C	C	C

NOTE: U = Scale not removed (estimated 20–100% of scale remaining)
    C = Scale completely removed (estimated 95–100% of scale removed)
        All samples 5–10 mils scale

acid, (3) ammonium EDTA, and (4) sodium EDTA.

In general, the criteria used to select solvents include:

1. Materials of Construction—The inhibited solvent selected must be compatible with the tube material. For example, hydrochloric acid cannot be used to clean superheaters and/or reheaters because of the possibility of stress-corrosion cracking of stainless-steel materials.

2. Deposit Compositions—Deposit compositions could include iron oxide, copper, zinc, nickel, aluminium, silica, as well as solids from condenser cooling water. Large amounts of silica in the deposit present problems. Ammonium bifluoride is one additional chemical (in conjunction with the solvent) used to remove silica-based materials. Copper complexors must be used with hydrochloric acid to avoid copper plating during the cleaning if small amounts of copper are present. If there is a large amount of copper in the deposit (greater than 10 percent by weight) a multi-step procedure will be necessary. Ammonium bromate has been shown to be an effective solvent to remove deposits with significant amounts of copper.

3. Geometries—Organic solvents are effective under dynamic conditions, usually when velocities are greater than 1 ft/sec. The effect of velocity and circulation on various solvents is illustrated in Table I. Complex circuits such as found in some superheaters and reheaters require special attention to assure removal of all air pockets and positive flow in all circuits.

4. Methods of Disposal—Environmental regulations can greatly affect disposal. The costs and methods of disposing cleaning wastes have a strong influence in the selection of a solvent.

Tube samples should be taken and given to a chemical cleaning vendor to allow determination of the best solvent and cleaning procedure. The thickness, porosity, texture, and composition of the deposit all may affect individual solvent effectiveness and normal cleaning procedures may have to be modified. It is also a good practice to remove additional tube samples after the cleaning to verify that the cleaning was successful.

Determining the Need for Chemical Cleaning

Utility boilers should be cleaned at least once every 3 to 5 years. The empirical relationship given in Table II correlates the amount of deposit on a tube with the cleanliness of the boiler. Tube samples should be taken at yearly intervals from the high-heat-flux areas of the boiler (for example, several feet above the windbox) or other areas that have in the past been prone to deposition.

The information obtained from inspecting the tube samples, in conjuction with operational factors, is used to aid in deciding the need for cleaning. These factors include the number of start-ups, the number of periods with condenser leakage, chemistry deviations, and length of outages.

Table II. Relationship of Analyzed Deposit Quantity to Unit Cleanliness

Boiler Type	Internal Deposit Quantity Limits*		
	Clean Surfaces, mg/cm²	Moderately Dirty Surfaces, mg/cm²	Very Dirty Surfaces, mg/cm²
Supercritical units	Less than 15	15–25	More than 25
Subcritical units (1800 psig and higher)	Less than 15	15–40	More than 40

*All values are as measured on the furnace side of tube samples and include soft and hard deposits.

Note: For all practical purposes, 1 mg/cm² = ~ 1 g/ft².

Chemical cleaning of industrial boilers should also be performed on a periodic basis. The primary purpose for cleaning these units is to prevent buildup of deposits to the point where overheating may occur.

In cases where deposits consist of hardness salts, it is imperative to periodically examine tube samples from the unit to establish the need to clean.

In those units where iron oxide and copper are the main impurities in the feedwater, the information in Fig. 5 can be used as a guide in determining the need to clean. However, good practice would still dictate periodic examination of tube samples to confirm the analytical evaluation.

Acid Cleaning Procedure

In thermal (natural) circulation boilers, no effective circulation can be obtained at the low solvent temperature of 150°F to 170°F. Therefore, the distribution of acid strength and temperature is obtained by blending concentrated inhibited acid and hot water as the solution is injected into the boiler. Superheaters are flooded with condensate prior to the addition

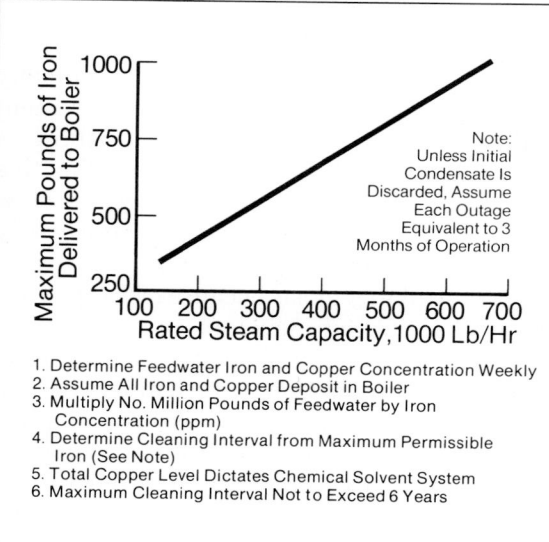

1. Determine Feedwater Iron and Copper Concentration Weekly
2. Assume All Iron and Copper Deposit in Boiler
3. Multiply No. Million Pounds of Feedwater by Iron Concentration (ppm)
4. Determine Cleaning Interval from Maximum Permissible Iron (See Note)
5. Total Copper Level Dictates Chemical Solvent System
6. Maximum Cleaning Interval Not to Exceed 6 Years

Fig. 5. **Operational chemical cleaning guide for industrial-type boilers (based on demineralized or evaporated makeup water)**

of acid to prevent the spillage of the solvent to this section. Thermal circulation boilers are generally cleaned by the soaking process. Samples are taken periodically to check the degree of reaction in the boiler.

Boilers are soaked for a period of 4 to 6 hours. The acid is drained by pressurizing with nitrogen. This step is taken to prevent the oxidation of cleaned surfaces during this time. Water is used to displace nitrogen in rinsing the metal surfaces of acid. The rinse water is subsequently displaced with nitrogen. Two rinses are usually sufficient to attain a pH that is between 5 and 6.

The boiler is then filled with water containing soda ash, 1.0 percent solution. The water is adjusted to the operating level and the temperature is raised to correspond to a pressure of about 100 psig. About 2 hours are required to effect neutralization of the acid and passivation of the metal. The boiler is drained and inspected at the conclusion of the wash period.

Controlled Circulation and Combined Circulation boilers can be cleaned efficiently because the circulation pumps can be used to equalize acid temperatures and concentrations throughout the boiler. The acid is circulated intermittently with one pump during the cleaning, which normally requires 4 to 6 hours.

Organic-type acids are frequently used to clean once-through boilers. At the conclusion of such a cleaning, the boiler is rinsed with water. When the system has then been purged of acid, condensate containing ammonia and hydrogen is circulated to effect neutralization.

Because of the close proximity of the superheater to the drum and the potential of corrosion from the cleaning solvents, particularly on the austenitic alloys, avoiding contamination of the superheater is important. If the superheater is known or suspected to be contaminated, the cleaning should be temporarily halted and the superheater flushed to remove the contaminants. The boiler should never be fired with suspected contaminants in the superheater. For precautions necessary to avoid superheater contamination see the box on the following page.

**AVOIDING SUPERHEATER
CONTAMINATION**

1. The entire cleaning piping layout should be examined to identify all possible areas where any solutions from the boiler could enter the superheater or reheater. Such connections should be eliminated so contamination of the superheater does not occur. Possible leaking valves and pressure levels that might result in backflow must be avoided.

2. Control and monitoring of drum level is of paramount importance.

3. Water used to fill or "backflush" the superheater should be demineralized or of condensate quality.

4. Possibility of preboiler (feedwater train) cleaning solutions "hiding-out" in lines, manifolds or tanks to be used, in, or interconnected to, water fill-lines should be considered before starting the preboiler cleaning operation.

5. Suspicion of contamination warrants a careful assessment of water (or condensate) quality at appropriate locations in the superheater (e.g., locations nearest point of suspected entry). If evidence of contamination exists, firing should be discontinued until a judgment can be made as to the seriousness of the contamination. If doubt exists, the prudent course of action is to water-flush (condensate quality) the superheater and/or reheater.

SUPERHEATER/REHEATER CLEANING AND FLUSHING

Although pre-operational cleaning of the steam-generating section in a high-pressure boiler is strongly recommended, superheater and reheater cleaning is not. The need to eliminate deposit-forming materials is related primarily to their effect on the corrosion problems that are common in waterwalls. We know of no case where the presence of initial mill scale has contributed to failures in superheaters and reheaters or difficulties in turbines. Normal mill scale on tubing is thin, adherent, and not readily eroded by dry steam. Particulate matter and construction debris constitute the major portion of the foreign material present in any superheater or reheater at this point in its life. Because solvents do not efficiently dissolve most particulates, this method of cleaning is not considered completely satisfactory for placing a superheater and reheater in condition for operation. As steam blowing has proven effective in removing particulate matter, debris, loose oxide, and atmospheric rust, it is recommended for any pre-operational cleaning of superheaters and reheaters regardless of other procedures employed.

The need to clean a superheater or reheater at some point in its operating life by using well-engineered programs containing solvent techniques or water flushing cannot be overlooked. Because of the geometry, superheater and reheater cleaning requires special attention. Frequently, these sections contain nondrainable, nonventable sections that are difficult to clean unless the operator has a clear understanding of flow mechanics.

The problem in cleaning superheater or reheater surfaces is to assure that there is positive flow through all the parallel circuits. This is necessary for (1) effective cleaning, and (2) assurance that the solvent can be completely displaced at completion. This requirement applies even when simple water washing is employed for removal of soluble salts. Positive flow can be assured with a relatively low flow of water if the entire section can be initially filled. When a portion of the loop is nonventable, the difficulty exists because of trapped air. Air blockage which develops in nonventable areas will prevent effective cleaning and make it impossible to completely flush unless special procedures are employed to assure complete filling. Fig. 6 illustrates the filling problem. A well-engineered fill-flush program for the individual unit undergoing the cleaning, and the confirming of effective filling prior to cleaning solvent injection, are required if desired cleaning results are to be obtained and damage prevented.

POST-ACID-CLEANING ACTIVITIES

Before the main steam-lines are blown, the steam and lower drums should be inspected. The drums should be flushed of any loose sediment. The internals to the gage glass should be flushed and then blown out. Accessible headers should be inspected and flushed with clear water.

If the primary and secondary separators were not placed in the drum, they are installed. On Controlled Circulation units, each orifice and screen is installed in the lower drum. Each orifice is then checked with a go/no-go gage. Any header handhole caps which were removed or not previously welded are welded in place.

Acid cleaning connections are removed or valved off, and temporary piping removed. The chemical feed and continuous blowdown piping should be flushed and blown out. When all work is completed in the upper and lower drums, they must be inspected to ensure that there is no foreign material remaining. New gaskets should be installed on all drum manholes before closing.

Next, the boiler circulating pumps are prepared for operation as called for by the manufacturer's instructions.

STEAM-LINE BLOWING

The purpose of blowing the main-steam lines and the reheat-steam lines before starting up a new unit is to remove any foreign material remaining in the superheater, reheater and steam piping after erection is completed. Considerable damage can result if such material enters the turbine during initial operation.

On older units, the need for steam-line blowing should be considered following major pressure-parts repairs which introduce the possibility of foreign material into the system.

RESPONSIBILITY

Because prevention of damage to the turbine is the prime concern, the responsibility for determining the effectiveness of the steam-line blowing operation rests with the turbine manufacturer's representative. During this process the unit should be operated in accordance with recommended procedures, with all control systems and protective interlocks functioning.

The design, fabrication and installation of any temporary piping system used for the purpose of steam-line blowing, as well as protection against overpressure or overtemperature, require careful attention by qualified engineers.

GENERAL

Ideally, to obtain optimum cleaning, the flow conditions in the system during steam-line blowing should equal those during normal operation at maximum load. Because it is impossible to exactly duplicate these conditions when blowing through the piping to atmosphere, it is desirable to produce equivalent conditions by using lower pressure steam with a flow rate such that the product of steam flow times velocity will equal that under normal full-load conditions. The determination of the total obtainable flow quantity must be based on flow resistances in the entire system, including the temporary piping. Customarily the designer of the blowing system makes this determination. Most high-pressure units obtain satisfactory results with blowing pressure in the 600- to 800-psig range. The actual steam

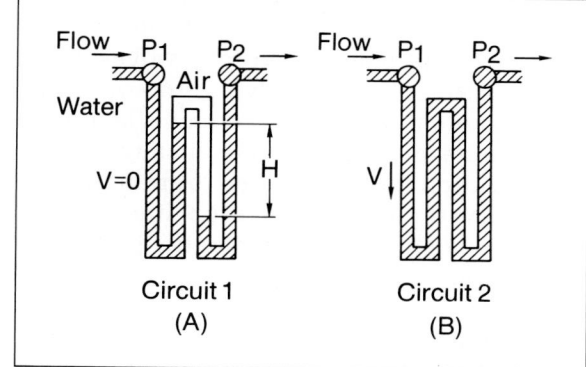

Fig. 6. Typical schematic of a nondrainable, nonventable superheater section where air blockage will occur. (A) The air-blocked circuit(s) will remain stagnant because the unsupported water column H produces a pressure equal to and opposite to the pressure drop between headers P_1 and P_2. (B) Air blockage will be eliminated only when the flush rate is such that the pressure drop between the inlet and outlet headers because of flow exceeds the height of the unsupported water column H.

blow is executed in three phases: main steam, cold reheat and hot reheat.

To prevent foreign material from being transported from one section and deposited in the next, it is important that the system be blown in sections, each section being blown separately. Particular care must be taken when blowing the reheater. If the reheater were to be blown immediately in series with the superheater, material too large to pass through the reheater tubes would remain lodged in the reheater inlet header and cause blockage of reheater tubes, with resulting overheating and failure of reheater tubing when the unit goes into operation. For this reason it is recommended that the temporary piping be arranged such that the main steam lines can be blown to atmosphere first. Similarly, the cold-reheat lines should be blown to atmosphere before they are connected to the reheater inlet header. Only after these precautionary measures should steam be admitted to the reheater for blowing through the hot reheat lines.

The temporary piping is often equipped with shutoff valves which are used as blowing valves. With this arrangement, the superheater and the upstream portion of the temporary piping are maintained at drum pressure at all times when not actually blowing. If the unit is equipped with main-steam stop valves, these may be used as blowoff valves, if the valve design permits this type of service. When more than one steam line is provided in any section, consideration should be given to arranging the temporary piping so that each line may be blown separately. In such situations, the piping system should be investigated to ensure that flow unbalances and expansion stresses at connections do not become a problem.

Impact specimens, installed in the blowoff piping during the final blows, give an indication of the cleanliness of the section. Polished square bar stock suitably mounted and supported has been used effectively for this purpose. The degree of pitting of the specimen surface following a blow determines the end point of the blowing cycle for the particular section.

PRECAUTIONS DURING STEAM-LINE BLOWING OPERATION

1. The process of steam-line blowing imposes abnormal and severe conditions upon the steam generator and steam piping. Large, rapid temperature changes occur during each blowing cycle. This cycling of temperature is far more severe than is incurred in normal operation. Thermal stresses may be excessive in the heavy-wall portions of the system, such as drums, headers and piping. It is prudent to consider this fact when performing the blowing procedure and to limit the number of blowing cycles ("blows") to the minimum consistent with cleaning the system.

2. Firing of all main fuel must be discontinued during all blows. It is permissible to keep ignitors and/or warm up guns in service during the blows in order to speed up reestablishment of the desired blowing pressure after a blow is completed. On Controlled Circulation boilers, circulating pump(s) must be kept in service during the entire blowing process. If a boiler circulating pump loses suction because of low drum water level, it must be shut down. If all pumps are stopped because of loss of suction, the blow must be terminated immediately.

3. Since the temporary steam-line blowing piping may be designed for lower pressure than the steam generator, care must be used to prevent overpressuring this piping during the entire steam-line blowing operation. The pressure in the temporary piping should be monitored continuously and the operators should be alert to prevent overpressure.
It is recommended that a means of overpressure protection be provided in the temporary piping, such as safety valves set at the design pressure of the temporary piping.

4. It is difficult to avoid carryover from the steam drum to the superheater during the steam-line blowing operation. Therefore, to avoid deposits of solid materials into the superheater, boiler water should not be treated with non-volatile chemicals during this process.

OPERATING PROCEDURES

The unit is started in the normal manner, following the cold start-up procedures. All normal recommendations and limitations with respect to pumps, fuel-firing equipment, ash-handling equipment, drains, and vents should be followed, as if the unit were being started for synchronizing the turbine. Before starting the first blow in each section, the economizer recirculating line valves should be closed. Because the steam-blowing operation is the first occasion that the unit is fired at any significant rate, the start-up as well as the steam-line blowing must be conducted with great care.

The unit must be brought up much slower than during subsequent normal start-ups, while all equipment is checked and expansion movements monitored closely. When the unit is fired, the furnace exit-gas temperature limitation must not be exceeded.

During the blows, the drum water level will be subject to extreme fluctuations. As the temporary blowoff valve is opened, the drum water level will rise rapidly and may disappear from sight in the gage glass. As the blow progresses, the drum water level will reappear and may drop out of sight so it is important that the drum water level is established at or slightly above normal operating level before the start of each blow. Feedwater flow must be established as soon as the water level drops back in sight, to prevent excessive low water level with resulting suction loss of the boiler circulating pumps. After the procedure is completed the temporary piping is removed, and final piping connections are made. The unit is now ready for the next commissioning function—setting safety valves.

SAFETY-VALVE SETTING

All safety valves installed on boilers or other pressure vessels should be test-operated before the boiler or vessel is placed in service. This test should involve a check of the proper functioning of the valves as to correct opening pressure, correct blowdown, proper mechanical operation without chatter, and clean closure without leakage.

BLOWING THE MAIN-STEAM LINE— PHASE I

The actual blow is started when the unit reaches the specified blowing pressure, usually 600 to 800 psig. However, the first blow is always at reduced pressure to check out the temporary piping system and its supports and anchors. All main-fuel firing is discontinued. The temporary valve or main-steam valve is opened fully to blow through the superheater, main-steam piping and the temporary blowoff piping to waste.

When the drum pressure drops to about 200 psig, the blowoff valve is closed. The firing rate is increased, and the cycle repeated as often as necessary until cleaning is satisfactory as indicated by inspection of the impact specimen.

BLOWING THE COLD-REHEAT LINE— PHASE II

The blowoff piping is now modified with a temporary connection from the main-steam piping around the high-pressure turbine to the inlet of the cold-reheat piping. Temporary blowoff piping is installed at the reheater inlet. Blowing is executed in the same manner as the main-steam line, but now the blow is through the superheater, main-steam line, cold-reheat line and the temporary blowoff piping to waste. Again, the first blow is at reduced pressure, with subsequent ones from 600 to 800 psig down to 200 psig drum pressure. This cycle is also repeated until the impact specimens are satisfactory.

BLOWING THE HOT-REHEAT LINES— PHASE III

The temporary blowoff line at the outlet of the cold-reheat piping is removed, and the cold-reheat piping is permanently connected to the reheater. A temporary blowoff connection is installed at the hot-reheat piping as close to the turbine as possible and piped to waste. The same blowing procedure is used as in phases I and II. When the specimen results are satisfactory, the steam blows can be considered successful.

In checking opening pressure and blow-down, a calibrated pressure gage must be used and the location of its connection on the vessel, header, or other component must be such that the pressure it indicates is a true indication of the pressure existing at the safety-valve inlet, both when it opens and when it closes. When valves are located on headers or steam lines through which steam is flowing, the test pressure gage should be connected near the valve to eliminate any effect of pressure drop resulting from the flow. In some cases the flow of steam being relieved through the valve itself could produce a significant pressure drop.

For high-pressure boilers (above approximately 1200 psig operating pressure) a manufacturer's representative should initially test and adjust safety valves. Certain local codes require that only licensed people may make adjustments on safety valves. Check with the insurance carrier before making any adjustments on safety valves. The setting or adjustment of safety valves should be done only by a competent person familiar with their construction, operation, and maintenance. A written report should be made of all testing and adjustment of safety valves. This should record the opening pressure, closing pressure and/or blowdown and give an indication of correct functioning of the valves as they are left. In addition, it should include the name and title of the responsible plant observer and the name and title of any insurance inspector, or state or local boiler inspector who witnesses the test and adjustment. The name of the valve vendor's representative making adjustments, if any, should also be included as well as the nameplate data (serial number, set pressure, blowdown, and valve location) of each valve on the unit. The operating company should retain the report as a permanent part of their records.

Before testing safety valves for the first time on new units, the steam lines will have been blown to eliminate as much of the foreign material as possible from the superheater and reheater and from the connecting piping to prevent damage to the safety valves. Damage to the valve seating surfaces from grit blasting by mill scale, weld beads, and other such dirt is common during initial testing. Adequate steam blowing prior to popping values will minimize possible damage. Valves mounted on the dead end of headers are particularly prone to this damage. Inspection and manual clean-

---

PROCEDURES PRIOR TO SAFETY-VALVE SETTING

Before firing the unit for setting the valves, check the following:

1. That hydrostatic test plugs have been removed from the safety valves. If a problem of scheduling a valve vendor's representative for removal of hydrostatic test plugs is involved, the plug from one valve must be removed prior to the first fire. All plugs must be removed before the unit is brought up to full pressure for the first time.

2. That there is no physical interference which would prevent functioning of the safety valves. Expansion of the unit should be considered in this check.

3. That the exhaust stacks from the valves are firmly supported and restrained and that no physical interference which will put external stresses on the valves exists between exhaust stacks and the valves in the cold position, hot position or in between.

4. That the exhaust elbow attached to the valve body is not of excessive length in the horizontal run (maximum recommended— 24″) which may result in abnormal stresses being put on the valve body due to the reaction force when the valve is blowing.

5. That the drain piping from the valve body and drip pans, elbow, etc. is installed and discharges into a location which will prevent injury to personnel when the valve is blowing.

6. That, if flexible hoses are used between the exhaust elbow and exhaust stack, they be of sufficient length and installed so that they do not become "solid" in any position of the valve.

7. That all components of the safety valves are in place and secure, such as manual lifting gear and adjusting ring pins.

ing of the header by vacuum hose and/or magnets through the valve nozzle, with top works removed, is sometimes necessary. With maximum attention given to removing debris before initial popping of the valve, better success will be obtained with initial valve setting.

Leakage of safety valves in operation usually results from one or more of the following:

1. Seating-surface damage

2. Externally imposed stresses on the valve body which distort the seating surfaces

3. Operation at pressure too close to the set pressure of the valve. The valve manufacturers recommend a minimum difference between popping pressure and operating pressure of 5 percent of the popping pressure. With less than 5 percent difference on a continuous basis, problems with leakage and resulting high valve maintenance may be expected.

Safety of personnel and prevention of equipment damage should be the prime concern when testing and adjusting safety valves. Tremendous forces are involved when these valves operate. All precautions necessary to contain these forces and prevent accidents must be taken. For the initial popping of each valve all personnel should be kept at a safe distance from the valve. A rope should be attached to the manual lifting gear so that an uncontrollable valve chatter can be prevented from doing serious damage if it should occur.

Before starting the actual setting, refer to the latest section of the ASME Boiler and Pressure Vessel Code applicable to safety valves.

HYDRAULIC-JACK METHOD FOR SAFETY-VALVE SETTING

The use of a hydraulic jack has made setting safety valves easier. These jacks can be bought or rented from the valve manufacturers. This method was developed originally for setting supercritical pressure valves so that the lift could be limited to reduce damage to the seating surfaces.

The hydraulic-jack system contains a pump, hydraulic piston, yoke, turnbuckle and pressure gage. The hydraulic piston overcomes

some of the safety-valve spring force so that the valve may be set at lower steam pressures.

This method has also been used widely on reheater safety valves as well as with high-pressure valves (drum and superheater) on drum-type boilers. But it must be remembered that this method only tests the popping pressure, not blowdown or the valve action itself. The disadvantages of not testing actual blowdown and the valve operation are outlined in the adjoining box, "Problems Avoided by Blowdown Testing."

Safety-valve manufacturers indicate that they check each valve with steam for blowdown, and that a cold setting of the blowdown rings can be made extremely close if the instruction manual is followed closely. However, C-E recommends the hydraulic jack method for supercritical or reheater applications.

PROBLEMS AVOIDED BY BLOWDOWN TESTING

1. Insufficient blowdown can result in chatter, which can produce extensive damage to the seating surfaces and other valve parts.

2. Excessive blowdown can result in operating problems in getting a valve to close above the normal operating pressure once it pops. The ASME Boiler Code requires that the low-set drum valve shall have no more than 4 percent blowdown and that other drum valves may have longer blowdowns but none shall close lower than 96 percent of the set pressure of the lowest-set drum valve.

3. The valve could fail to reach full lift because of some mechanical problem. This would not be known and hence its capacity might be restricted when it was really needed.

4. The valve could hang open because of a mechanical problem and bleed the pressure completely off the boiler during operation.

5. The exhaust-stack arrangement would not be subjected to a full-flow test before the unit goes on the line. Serious problems and possible damage can result if sizing is insufficient or the supports are not adequate.

Fig. 7. Check each sootblower during weekly walk down.

After the unit has been chemically cleaned, steam lines blown and safety valves set, the boiler is ready for supplying steam to the turbine.

One of the necessities of effective operation is the periodic "walk down" of a steam generator during operation. This requires that the operator keep his or her eyes and ears open for unusual conditions and report any findings (Fig. 7). Potential damage to equipment can be avoided if abnormal conditions are detected in time. The operations department should develop a checkoff list specific to the station.

OPERATION OF FLUIDIZED-BED BOILERS

Many operational aspects of fluidized-bed steam generators are identical to those of other solid-fuel-fired units. This description, therefore, will be limited to the major differences in operation between fluid-bed and other solid-fuel boilers.

START-UP

Cold start-up times are generally 8–12 hours to full load. In a circulating fluid-bed (CFB) boiler, this is because of the refractory lining that is integral to several components, such as the high-temperature cyclones. The rate of refractory temperature change must be limited to approximately 100 to 200°F/hr to avoid cracking and spalling from thermal shock. In a bubbling fluid-bed (BFB) unit, start-up times are determined either by the refractory-lining considerations or by the time needed to heat the bed material, which, in a multi-zone bed, requires the transfer of heat from a start-up zone to adjacent zones. Note that the boiler will not usually delay plant start-up on a cold start, because steam flow is available for turbine warming relatively quickly, and most turbines require many hours of thermal soaking.

Before start-up, if the bed had been drained for maintenance or inspection, a new charge of bed material is needed. This is usually sand, spent bed material (from the ash-disposal silo or a separate bed-material silo), limestone, or a mixture of these. Fluidizing airflow is started and the bed is preheated, using overbed and/or underbed burners, to the temperature required for fuel admission. Bed temperature is the principal permissive for main-fuel firing; with sufficient bed temperature, burners or ignitors are not required to light-off incoming solid, liquid, or gaseous fuels, as the heat of the bed will ensure fuel ignition.

NORMAL OPERATION

In normal operation, a fluidized-bed boiler behaves similarly to other solid-fuel-fired boilers. Firing rate is a function of outlet steam pressure; load swings will result in changes in pressure and, hence, in rates of fuel and air flows. In a CFB, combustor temperature is regulated by variation in excess air, the primary-air to secondary-air ratio, total bed inventory, and (with a fluid-bed heat exchanger) solids flow to the FBHE. Combustor inventory is set by the rate of bottom-ash flow. Superheat and reheat outlet steam temperatures are controlled by desuperheater spray and reheat FBHE solids flow, respectively. In a BFB, bed temperature control is by change in excess air, bed level, and recycle rate. As in a CFB, bottom-ash flow determines bed inventory. Superheat and reheat steam temperatures are controlled by spray desuperheating and by biasing of gas dampers that direct gas flow to the reheat side of a split backpass, respectively.

In general, heat distribution is facilitated in a fluidized-bed steam generator because the combustor temperature (equivalent to the furnace temperatures of stoker- or PC-fired units) can be more easily regulated. Operation is usually very stable because of the large thermal-flywheel effect of the bed mass; thus, wide variations in fuel quality can be tolerated. The bed prevents gas-side transients occasioned by loss or interruption of fuel flow, thereby improving boiler safety. In most instances, there is no concern about furnace-wall slagging, because bed temperatures are held below ash-softening temperatures to ensure fluidization and to optimize sorbent utilization.

The importance of maintaining proper fuel sizing cannot be overemphasized. Fuel size strongly influences bed particle size which, in turn, affects most aspects of boiler performance. The required fuel sizing and its influence on both design and performance of a FBB are discussed in Chapter 9.

SHUTDOWN/RESTART

The shutdown procedure will depend on whether a hot restart is anticipated. If a hot restart is planned, the fuel and air are stopped: bed temperature can remain above the firing permissive level for several hours, such that no purge is required on restarting. On a hot restart, the unit can be brought back to full load within a few hours, depending on the bed temperature level. If no hot restart is planned, the fuel and air flows are gradually reduced, so that the rate of refractory temperature change is kept within the limits of 100 to 200°F/hr.

PLANT MAIN-FUEL TRIP (MFT)

Stored heat in the bed material will continue to generate steam even after fuel flow stops. In a CFB, stored heat in the cyclone refractory contributes to steam generation; precautions must be taken to prevent damage to backpass steam-cooled surfaces from this heat source. Drum level should be maintained by slow feed of water to the drum following an MFT. If cyclone cooling is not possible in any other way, a small steam flow can be induced through the backpass by opening a steam vent to cool the backpass tubing. Steam-cooled surfaces in the combustor and/or FBHE are protected on loss of steam flow in accordance with the manufacturer's instructions, which may call for depressurization or a small cooling-steam flow by means of a steam vent or turbine bypass.

LOSS OF POWER

During a power loss, feedwater flow to the boiler stops. As a result, stored heat can evaporate significant amounts of the water inventory. Whether this will damage the boiler depends on the specific design. The conservative approach is to provide an emergency feedwater pump, which allows water to be fed to the boiler during a power interruption.

TUBE RUPTURE

When limestone is used as the sulfur sorbent, the heated bed material will contain significant amounts of calcined, dehydrated calcium oxide. If this lime comes in contact with sufficient water, such as from a tube leak, it will harden when allowed to dry, and will make cleanup and repair difficult. The leaking water or steam may also damage relatively sensitive refractory linings. Quick action by the operators can minimize outage duration and reduce the extent of cleanup. When a tube leak is detected, it is important to maintain fluidization and to drain the bed material as soon as possible. Acoustic steam-leak detectors are an asset for timely detection of tube leaks.

PROBLEMS IN OPERATION

In spite of adherence to the general procedures and observance of the cautions presented in this chapter, operational problems still occur. Tube ruptures can happen from a variety of reasons. The size and location of the break will determine what action is necessary. On a drum-type unit, the best method for shutting down will be dictated by the ability to maintain normal water level in the boiler and the need for the boiler in service.

If water level can be maintained, the unit can be kept in service until after a peak, or service may be stretched to a weekend outage. There is, however, the ever-present danger of high-pressure water from a break cutting other tubes. If the rupture can be visually observed, to ascertain that it is blowing out into the furnace and not damaging other tubes, the unit may be able to operate for a long time providing sufficient treated make-up water is available.

Unlike a waterwall tubing leak, a leak in a superheater, reheater, or economizer element requires greater attention. Because of the physical arrangement of such surfaces, steam cutting of adjacent tubes can result in making a major repair job out of what might have been a simple and short one. Economizer ruptures, if left unattended, can lead to plugging of the economizer and air heater, as the water mixing with the flyash can set similar to concrete.

A leak in a waterwall of a supercritical unit can result in rapid and extensive damage, not only to the leaking tube, but also to nearby tubes. Thermocouples are installed on individual outlet tubes and representative inlet tubes to alert the operator to such leaks. Any decision to operate a once-through unit with a known waterwall tube leak must be made with the full knowledge of the serious damage that may be incurred.

FURNACE EXPLOSIONS

Furnace explosions usually occur during start-up, shutdown, or low-load operation. Generally, they result from the accumulation of unburned fuel in the furnace because of incomplete combustion, loss of ignition, or fuel-valve leakage. An explosion occurs when the proportion of unburned fuel and air is in explosive range, and some heat source increases the temperature of the mixture to the ignition point. Unburned fuel which causes such fires can accumulate in the furnace in several ways: at leaky fuel inlet valves on idle windbox compartments; when the fire is extinguished and the fuel is not shut off promptly; when the fuel doesn't burn as rapidly as it enters the furnace; or if difficulty occurs in establishing ignition.

EXPLOSION PREVENTION

Because most explosions occur during periods of low fuel input, maintaining a minimum of 30 percent of full-load airflow is important to insure an air-rich mixture and to sweep out any accumulation of unburned fuel. Other preventive measures include

■ Be sure that all liquid or gaseous fuel valves are tightly shut on idle fuel compartments

■ Watch the fires closely at low loads and shut off all fuel immediately if proper combustion is not maintained. Most modern units have flame scanners to trip the unit automatically when poor ignition occurs. The scanners must be properly maintained. They should never be removed from the safety system or their outputs defeated

■ Always use the required ignition energy source when placing any pulverizer in service

■ During low-load operation keep *adjacent* pulverizers in service

■ Regularly check the proper functioning of any furnace safeguard system

■ Never defeat any portion of a safeguard or interlock system

■ On a unit trip, purge the furnace before shutting off the fans

■ Empty the pulverizers of all coal before bottling up the unit.

UNINTENTIONAL FIRES EXTERNAL TO THE FURNACE

One of the most destructive events in steam-generator operation is uncontrolled ignition of fuel in an area external to the furnace. Such fires have taken place in air heaters, ductwork, windboxes, precipitators, hoppers, and fans. In

the immediate vicinity of the boiler, such fires have occurred when oil is being burned.

Generally, fires in air heaters, back-end duct-work, precipitator and induced-draft fans take place when a unit is being brought up to load after a start-up from cold. With inadequate fuel-oil atomization and/or poor mixing of the oil with combustion air, unburned oil distillates will carry to the back of the unit and deposit on the relatively cold back-end surfaces. Later, when load is raised, these deposits will volatilize as temperature increases, and can ignite. Once such a fire starts, extinguishing it is difficult. The metal baskets in an air heater, for example, will continue to burn even after the oil is consumed. The only method of putting out such a metal fire is to flood it with as much water as possible.

Combustion Engineering has always recommended air-atomized light oil or steam-atomized heavy oil for light offs of a cold unit. If mechanical atomization of oil is the only

---

TYPICAL OPERATOR WALK-DOWN CHECKLIST

- Look for unusual traces of either coal dust, oil, flyash or water.

- Look for valve and valve packing leaks.

- Look for any unusual conditions such as discoloration, hot spots on casing and ductwork, or vapor leaking out.

- Open all inspection doors and note any slag accumulations.

- Listen for tube leaks in the furnace. (This is only possible to do on balanced-draft units.)

- Check for unusual noises, overheating, and adequate lubrication of all motors and driven equipment.

- Look for leaks in gage glasses and water columns.

- Check that no sootblowers are stuck in the unit and that there are no leaks in the sootblower lines.

- On tilting tangential units, make sure that the tilt setting is the same on all corners.

- Check the secondary-air damper settings to make sure that all dampers are the same on a given elevation.

- At the firing levels, check for coal, fuel oil or gas leaks. See that warm-up guns are retracted. Note any flyash leaks. Report any oil spills so that they can be cleaned up before a fire occurs.

- Check with a hand touch all vertical coal piping for possible plugging or overheating. A cold pipe on a mill in service is a good indication of plugging.

- Inspect furnace just above ash pit to make sure there is no bridging across bottom.

- Check level of water in ash hopper and bottom seal trough.

- At the pulverizers-

a. Check gear-case oil temperature, flow and level.

b. Check for excessive spillage or malfunction of pyrite system.

c. Look for any indication of mill fires.

d. Listen for unusual noise.

e. Check for coal leaks.

- At the airheater-

a. Check the air-heater sootblower to make sure that it is not leaking when the control valve has closed.

b. Check drive motor, support and guide-bearing lubrication and cooling water.

c. Inspect cleanliness of air side through observation door.

- At least once a week, a more extensive checking should be done such as:

a. Listen for badly leaking safety valves.

b. Check the sootblower cycle by walking down the unit as each blower operates to make sure it is functioning correctly and the packing is tight. Make sure all blowers are blown during a cycle.

c. Put all ignitors and retractable oil guns in service and check to make sure they are operating correctly.

d. Start and stop any idle equipment to make sure it is ready if needed.

means available, special observation of the furnace outlet is necessary when firing with a cold furnace. One method of detecting the carryover of distillates involves air-cooled probes at the furnace outlet which can be periodically extracted and examined.

Anytime there is doubt as to the quality of an oil fire during a cold light off, the air heater and other back-end surfaces should be examined. If oily deposits are found, all firing should cease until the surfaces can be thoroughly cleaned with a hot detergent solution.

Uncontrolled ignition occurring around the windbox area, external to the unit, usually results from oil spills when guns are removed or when valves or gaskets are permitted to leak. Therefore, oil spills should be cleaned and leaky oil valves or gaskets should be corrected immediately. Any plant which tolerates accumulation of spilled oil around the burner area is in danger of a serious destructive fire.

COAL HANDLING IN POWER PLANTS

Improved coal storage and methods of loading bunkers can often result in substantial savings in power-plant operations. Continuous production of steam requires steady flow of coal from the bunkers. Interruption of coal feed to the furnace not only causes loss of production, but can also lead to furnace explosions.

Perhaps the greatest factor impeding continuous coal flow is excessive moisture. Coals containing clays pack easily when moisture content increases; the packing can occur in bunkers, feeders, or pulverizers. In addition, foreign materials such as rocks, metal, slate, wood, and other debris in the coal can block or stall mill or stoker feeders.

Some plant operators learn to live with the problems associated with wet coal and foreign material by fighting stoppages inside the powerhouse with vibrators, manual sledge hammers, poke rods, sluice systems, air cannons, heaters, or coolers. These methods do not stop plugging; they only free plugs. Other operators

keep the problem out of the boiler room by limiting moisture pickup in the coal yard and removing all large debris before the coal arrives at the bunkers.

Coal usually arrives from the supplier reasonably dry and, if properly stored, will remain that way. If a small portion gets wet, it can be moved aside for drying and later use. On the other hand, to minimize fugitive dust, coal is sometimes wetted before shipment from the mine. Additional moisture pickup during transport can result from leaky barges or rain. Stations receiving coal with a substantial moisture increase from mine to plant should consult with the producer and the shipper to see if improvements are possible.

GUIDELINES ON COAL PILING AND RECLAIMING

- Compact the coal; loose coal picks up moisture and encourages fires in the coal pile.

- Pile the coal for maximum run-off of rain; piles should be rounded or pyramid-shaped, with the steepest slopes possible.

- Keep the coal pile free of valleys and pockets; water will collect in them and sink down into the coal.

- Reclaim deep into the pile, not along wide areas of the top. When properly piled and compacted, only the top layers will be high in moisture. This top layer can be reclaimed during drier spells.

- As coal is removed, rework the pile to fill in the reclaimed area and to eliminate gulleys, pockets, and rivulets.

- During reclaiming, do not push coal onto any water deposits or muddy coal in the area of the coal conveyor belt.

- Minimize fine coal, as it absorbs water more readily than coarse coal. Wet, fine coal does not flow easily; it will stick to bunkers and feed pipes. Coarse particles will help to eliminate bunker plugging and keep the coal moving.

COAL–YARD DESIGN AND MANAGEMENT

Good coal yards should be carefully planned. Proper drainage is a must. The yard should be properly graded, and all rocks, wood, and metal removed. A base layer of coal, to a minimum depth of 2 feet, should be spread; this base should never be reclaimed, and should not be included in the stockpile records, even for emergency use. The coal yard should be fenced or otherwise isolated from material storage, scrap, or trash; only coal and equipment needed to handle it should be allowed in the area. Workers should be encouraged to look for foreign material in the yard and to hand-remove it; the screens should not be depended on to remove all debris.

Avoid crushing coal until it is ready to enter the bunkers. Wait until immediately before the boiler is ready to burn coal before filling bunkers. There is a tendency in new plants to fill the bunkers as soon as the coal-handling system is ready; this is understandable, since start-up personnel want as much of the equipment to be operational and ready as possible. However, coal will gradually pack in bunkers and will not flow easily when needed. Also, coal stored in bunkers or silos for long periods can ignite and smolder. Finally, during initial operation, fill only those bunkers that feed the pulverizers required first.

In preparation for a scheduled long-term outage, all bunkers should be emptied, and should not be refilled until the unit is ready for restarting. If a unit has an unscheduled shutdown and it becomes apparent that it will be off line for some time, serious consideration should be given to emptying the bunkers, especially if the coal moisture content is high. Temporary chutes to trucks can be provided in some plants, and the coal can be returned to the stockpile. If major maintenance is to be done on a single pulverizer, its bunker should be run empty before removal for service.

It is important that the coal-yard supervisor become familiar with the needs of the boiler operators, so that his or her responsibility is more than just filling the bunkers; it should include responsibility for the continuous feed of fuel to the mills. In any case, the supervisor should be informed of all coal hang-ups.[2]

OPERATOR TRAINING

A well-trained operations crew means maximum plant availability and optimum unit efficiency. Such a crew is developed by a carefully designed training program which should be a step-by-step learning process. First, there must be an understanding of the fundamentals and principles involved. The individual equipment components and subsystems that comprise the unit must be taught in detail. And, finally, overall generating-unit operation must be discussed fully.

To be effective, a program should be versatile; new operators as well as more experienced personnel should be able to benefit from it. Flexibility is important too; a program that can be used individually or in classroom situations has a distinct advantage. Programs geared to specific equipment should be accurate and to the point. See Figs. 8 and 9.

THE SYSTEMS APPROACH

A well-planned operator-training program looks at the energy supply system as a "system," not just as a network of hardware. Part of that system is the people who manage, operate, and repair the network of hardware and equipment for the owner.

Many operating cost variables can be controlled by plant operating personnel throughout the useful life of the facility. Good operating techniques, for example, extend wearing-part cycles. Fuel usage also depends upon operating technique, as does effective preventive maintenance.

The question of operating-personnel quality is of great concern today, because of the increase in unit size, complexity, and automation. The trend is for fewer personnel to be involved in responsible operation of fossil power plants.

The elements necessary to establish a suc-

Fig. 8. Part of operator training includes classroom presentations covering fundamentals of system equipment and unit operation. A well-planned operator training program looks at the energy supply system as a "system," not just as a network of hardware and equipment joined together.

Fig. 9. Experienced training specialists provide on-the-job training, an essential part of any comprehensive training program.

cessful systems training program include prerequisite conditions such as knowledge of the system, what it is, how it works, who uses it, and what it is for. This is followed by a needs analysis that tailors the program to the specific requirements of the system and the student population. Then the training program must be designed to meet these requirements. Systematic verification of training program effectiveness is vital. Experience has taught that, if the above elements have been successfully applied, it is reasonable to expect demonstration of student operating competence and a high level of student retention of new knowledge and skills.

A prerequisite to a successful program is a cooperative understanding of the objectives and program goals. This is generally accomplished by giving presentations outlining the program goals and methods to station management personnel and officials of the production and maintenance unions. C-E's experience is that these presentations have resulted in excellent dissemination of program information, a high degree of interest, and active participation and cooperation at all levels of plant people. The inclusion of union management has achieved the same resultant support and cooperation and helps to assure cooperation and support for subsequent training activities.

THE NEEDS ANALYSIS

The training requirements analysis is generally based upon data gathered from questionnaires provided to all operating personnel and from a survey of the power station. The survey includes a review of plant configuration, technical documentation, operating procedures and directives, operator job requirements, and, where applicable, review of licensing requirements. These data are analyzed by educational psychologists and training specialists who then prepare a population description of the average student, identify job-knowledge requirements and performance objectives, and identify motivational factors to be considered in training systems design.

The availability of operators for training, with the exception of a new station starting up, represents a significant factor to be considered in a training system design. The complexities of current manning levels, rotating shift schedules, and overtime opportunities complicate the arrangement of an effective training schedule.

Performance measurement criteria applied to training design are:

■ Evaluate the effectiveness of instruction and course material

■ Evaluate the level of each student's achievement and provide necessary additional assistance if required

■ Review and reemphasize important areas of course material

■ Establish procedures for scoring individual students to avoid possible union conflicts

THE RÔLE
OF THE POWER-PLANT SIMULATOR

Training material, equipment, and facility requirements necessary to support the training program should be specified in the design of the program. Of great value to any operator training program is a power-plant simulator designed to demonstrate the concept of a total operating plant.

Simulators allow dynamic demonstrations of overall power-plant systems operation; automatic combustion control system operation; turbine/generator control system operation; and operation of feedwater and fuel control systems. Practice exercises can be provided through the insertion of selected malfunctions and load variations to improve operator ability to diagnose abnormal conditions and initiate corrective action. The application of simulators to the training program is always in the context of specific power-plant operating procedures and as a reinforcement of the training material being taught at the time, Fig. 10.

OTHER SUPPORT MATERIALS

An audiovisual training program is also very effective. Fig. 11. It combines sight and sound which greatly increases retention. Such a program can be used throughout the life of the unit for retraining operators and initial training of replacement operators.

COMPLETENESS

The training program design is complete when the following components have been ac-

Fig. 10 **Simulators provide a highly effective demonstration of plant operational concepts and the application of problem solving logic to determine corrective action to be taken when plant equipment malfunctions.**

Fig. 11 **The equipment used with an audiovisual training program is portable and easy to operate; at any time operators can use it individually for plant system reviews.**

counted for: curriculum, lesson plans, training aids, note-taking guides, simulator, study texts, quizzes, tests, practical exercises (plant walk down and simulator operation), and the teaching team has demonstrated practice-teaching exercises.

PROGRAM VALIDATION

The next step in the process is to provide for program validation. In general, an initial course instruction is provided to a "pilot" class. This is to validate training course content, length of lessons, emphasis, and overall time schedules and objectives. Validation is accomplished by administration of pretests at course start to establish operator entry knowledge, and post-tests at course completion to measure gain in knowledge. Evaluation of training effectiveness through course audit by technical specialists, analysis of test results, and review of student course critiques are additional validation measures. Course material should be revised as necessary based on the results of validation measures.

REFERENCES

[1] S.E. Kmiotek and R.F. Hickey, "Coal Pulverizer Inerting and Fire Fighting System;" Windsor, CT: Combustion Engineering, Inc., publication TIS-8256.

"Explosion Prevention Systems," NFPA 69, National Fire Protection Association, Inc.; Quincy, MA, latest edition.

"Installation and Operation of Pulverized Fuel Systems," NFPA 85F, National Fire Protection Association, Inc.; Quincy, MA, latest edition.

[2] George Thimot, "Want Better Coal Firing? Improve Coal Handling," *Electrical World*, August 1, 1972. New York, NY: McGraw-Hill, Inc., 1972.

CHAPTER 22

Performance Testing of Steam Generators

In the modern power plant, measurements are made for several reasons:
- to monitor such phases of power-plant operation as input and outlet flows, inlet and outlet water and steam temperatures, and power consumption of auxiliaries
- to test individual equipment or systems for initial acceptance
- to verify compliance with federal, state, and local requirements regulating plant emissions
- to check performance following a period of operation.

This chapter concerns the latter types of measurement, many of which are governed by the ASME *Performance Test Codes*[1] and the U.S. Environmental Protection Agency Code of Federal Regulations[2]. Supplemented by additional tests of special character, the ASME *Performance Test Codes* are of considerable interest to manufacturers and users of power equipment, and the results of such tests aid in the evaluation and improvement of designs.

Engineers participating in power-plant testing must become aware of the nature of experimental error, learn how a single error may influence the results calculated from multiple measurements, and gain some knowledge of the effect of instrument selection in minimizing measurement uncertainty. For best results, they must be willing to question why certain procedures and instruments are used and to try to change them if they can find better methods or equipment. Most importantly, engineers must become skillful in interpreting test results and in drawing appropriate conclusions.

POWER-PLANT TESTING AND MEASUREMENT CHARACTERISTICS

Power-plant testing differs in one important respect from product quality control, to which it is sometimes compared: most tests fall into the single-sample category. Unlike the repetitive tests made of product dimensions, most power-plant tests consist, for example, of individual readings taken over a period of time during which some unwanted variations may be introduced. Some well-equipped laboratories may have boilers that can be rigorously tested. But, in general, most large-scale power-plant tests cannot be repeated a sufficient number of times to gain the statistical reliability expected for quality control. Furthermore, the cost of conducting certain full-scale tests can be in the hundreds of thousands of dollars.

UNCERTAINTY ANALYSES

When tests are costly in terms of staffing,

time, and equipment, the uncertainty analyses made in advance help establish procedures and manpower assignments. They can also help determine how uncertainties propagate into the results. In their pioneering article on this subject, "Describing Uncertainties in Single-Sample Experiments,"[3] S. J. Kline and F.A. McClintock show that a given reduction in a large uncertainty is far more important that the same numerical reduction in a small uncertainty. This technique also guides selection of appropriate measuring instruments prior to the start of experimental work. But consideration must also be given to adequate cross checks while the experiment is being conducted.[4]

ADVANCE PLANNING OF FIELD TESTS

Advance planning of field tests is becoming increasingly important. Under some circumstances, it is possible to set forth a fixed objective for overall accuracy and plan instrumentation accordingly.

More often, however, field tests are performed under less than ideal conditions, so the engineer must be capable of making a judicious selection of existing and special measuring instruments. If the objectives and desired end results are well defined, some of the plant control-panel instrumentation may be used, perhaps jointly with special calibration or by checking against standard sources.

Large sums of money may be spent in devising special measuring equipment for a particular part of a test, but these expenditures may be largely wasted if supporting measurements are substantially less precise. Minimally therefore, the engineer should be familiar with the main sources of error and their effect upon the results of the test performed.[3]

By making an error analysis in advance of conducting tests, engineers can determine the crucial points of measurement and select the instruments accordingly.[5] They can also predict effects of propagation of errors on calculated results and obtain a much greater overall consistency in testing activities. In test reports, the combination of an error analysis with interpretation of results adds to the usefulness of tests

as a design tool.

Tests and measurements, as indicated earlier, require a great deal of engineering thought. Their value is greater if engineers are willing to ask searching questions and make all possible analyses before testing is begun. Finally, skill in reporting test results is crucial, particularly in the interpretation of what was observed. Important advances in engineering and science have resulted from skepticism of "obvious" results and perception of new factors that previously went unnoticed.

STEAM-GENERATOR TESTING

When applied to boiler study, testing activities range from daily performance observations using station supervisory instruments to unusual conditions with specialized measuring apparatus. In this latitude, the scope of any single test and the measurement techniques used depend upon the kind of testing information being sought.

Since individual phases of a design problem may range from a completely scientific solution with exact relationships between variables to empirical approaches with variables only generally related, test work for design information is intended either to determine magnitudes and limits of these variables or to establish their cause-and-effect relationships. Although a research laboratory can supply experimental information for basic investigations and observe behavior in many types of models, testing on full-scale units is vitally necessary to verify data thus obtained. Although tests of small industrial boilers may actually be carried out in a laboratory, for large utility-type boilers the power plant itself becomes the testing laboratory. In some cases where unusual fuels are being studied, large quantities of that fuel may be shipped long distances so that tests conducted in a boiler will have the desired firing-equipment and design characteristics.

Although emphasis is on the testing of large central-station boilers, many of the measuring techniques are equally applicable to smaller units installed in marine, industrial, and institutional power plants.

OPERATIONAL TESTING

After a design has been completed and a boiler erected, the initial operating period provides the first opportunity to determine whether it meets performance guarantees. Handled as a normal function of a technical-service organization, this acceptance work may include such items as determination of unit capacity, steam-temperature control range, draft losses and pressure drops, and overall efficiency. If deficiencies exist, some additional testing may be necessary to guide corrective action to meet contractual performance.

When a unit in a central station goes into commercial operation, it becomes an integrated part of a system. It is normal utility practice to determine the unit heat-rate characteristics at this time and periodically later. In these days of high costs, optimum system efficiency is important, and information from these incremental-load cost studies on each unit provides the basis for determining daily load distribution among system plants.

In some central stations, a control-room data-logging computer calculates heat-rate information almost instantaneously. Not only does such rapid information availability permit more detailed analysis of performance, it also supports decision-making for optimal unit operation under such conditions as extreme low loads or loss of a feedwater heater.

In addition to evaluating performance and results, operational testing covers the more detailed aspects of routine operation such as start-up procedures, adjustment and maintenance of unit control systems, and safe-limit monitoring of operating variables.

DETERMINING TEST CONDITIONS

The test objective governs the test condition which can be at either constant or transient load. Historically, the role of the newer, larger units in utility-system steam plants has been one of base loading at design rating, with system load fluctuations being taken care of by the

older and smaller equipment. Consequently, in the past, new unit design emphasized steady-state operating characteristics. In recent years, however, the relationship between required short-term peak capability and off-peak capacity has changed. Because peak load has out-distanced off-peak load growth, there is increasing interest in transient characteristics of the largest and most modern units. Many will, at times, operate at reduced load, and may even be shut down for weekends and other periods of low electrical demands.

Constant-load testing is needed for evaluation of unit heat absorption, heat-transfer parameters, unit efficiency, and heat rates. Transient testing is required for controls-response study and adjustment, and in connection with quick start-up and shutdown procedures.

TESTING MEASUREMENTS

Steam-generator design and operation revolve about three basic premises, any one or all of which may be involved in a test program:
- heat liberation
- heat absorption
- mechanical means to accomplish heat liberation and absorption.

HEAT LIBERATION

Heat liberation results from combustion of a fuel with oxygen in the air. This implies testing to obtain knowledge of fuel properties and combustion characteristics; it leads to establishing the theoretical fuel-air-products relationships from fuel analysis and combining equations and from the actual combustion requirements in an operating furnace. Together with the combustion efficiency, such things as pulverizer (or other fuel-burning equipment) power and capacity requirements, ignition stability, and ash deposits must be evaluated. Testing for heat liberation requires considerable attention to fuel and flue-gas sampling to obtain representative data.

HEAT ABSORPTION

Heat absorbed by unit components reduces

the flue-gas heat level. The reduction can be determined from gas-temperature measurements and specific-heat data. Similarly, the unrecovered heat quantity entering the stack is a loss and must be determined. Solid and gaseous stack emissions are other factors receiving considerable attention because of flyash disposal problems and air-quality considerations. To size fans and exhausters adequately, system resistance appearing as pressure and draft losses must be known. In summary, these are the most common areas of test work in the fuel-air cycle.

For a successful design, heating surface must be properly allocated among furnace, superheater, reheater, economizer, and air heater in order to achieve the desired heat absorption and yield the expected relationships of capacity and steam temperature. Since in modern boilers the furnace represents the most important part of the evaporative surface, its performance is a key factor. The furnace must be sized to be capable of maximum design evaporation yet provide flue gas at the temperature level necessary to superheat and reheat the steam to a desired level over the operating range. Because of the series-type processes involved, miscalculation at any point will upset performance in all the other sections.

TESTING OF MECHANICAL EQUIPMENT

High availability and low maintenance costs are prime requirements for steam-generating equipment. Therefore, materials have to be selected for long-life operation at elevated temperatures and pressures. Since in-service pressure parts and structural-mechanical equipment are subjected to erosion, abrasion, expansion, and vibration, their design requires knowledge of the magnitudes and character of these factors. Only field and laboratory experience will yield this information.

MEASUREMENTS RELATED TO HEAT LIBERATION

Fuel-flow measurement depends upon the fuel state and the desired accuracy. Solid fuels, such as coke and coal, can be measured by scales. Many plants have gravimetric feeders,

suitable for test purposes, between the coal-storage silos and the pulverizers. These scales or feeders must be calibrated before and after tests, in accordance with PTC 19.5, Measurement of Quantity of Materials, Chapter 1.

METHODS OF MEASURING FUEL FLOW

For general test purposes, fuel oils are metered with volumetric instruments of the displacement rotating-disc type. Heavy grades of oil must be heated for proper atomization in burners, and where moderate accuracy is satisfactory, the resultant viscosities permit use of such meters even with these oils. On the other hand, if flow error of less than 4 percent is required, meter calibration is necessary using the intended oil at operating temperature and with a similar piping arrangement. The foaming tendency and volatilization of some oils may prevent their accurate metering by this method.

Using level change in large uncalibrated storage tanks as a flow index is at best an approximate method. For efficiency tests, weigh tanks are the most accurate means of direct measurement, but their use is generally regarded as impractical with large flow rates. Their cost also prohibits widespread use.

For gaseous fuels, orifice-flow measurement is most common. It gives highly accurate results if the actual installation has sufficient approach straight lengths and if it conforms to design of pressure taps and orifice-pipe size relationships.[6]

TECHNIQUES FOR FUEL SAMPLING

In arriving at a representative sample, the inhomogeneity of most fuels requires careful attention to sampling techniques. Accepted standard practices may be found in the latest revisions of *ASTM Standards* D 2013, Preparing Coal Samples for Analysis and D 2234, Collection of a Gross Sample of Coal.

In general, solid-fuel sampling procedures involve collecting small increments at regular intervals over the test period, and reducing their aggregate to obtain laboratory-size samples by successive quartering or riffling. For

some coals, intermediate steps include crushing to definite screen sizes and mixing.

Since surface moisture of coal may vary considerably, separate samples are taken and sealed immediately for moisture-only determinations when high test accuracy is required. This practice is mandatory for efficiency tests and desirable for any high-moisture fuel tests because the normal handling in aggregate reduction can produce significant moisture loss of the sample. Aggregate analysis is then adjusted for the average special moisture value to arrive at the as-fired analysis. The need for such care is readily apparent in weighed-coal tests when the heat input is directly affected by such moisture loss.

Sampling of other solid fuels such as wood and bark should be made with similar care to avoid moisture loss.

With most pumped-liquid and gaseous fuels, the problem is one of possible overall change during the test period rather than inhomogeneity at any instant. Thus, sampling at a single point with fixed extraction rate for the entire test period should be adequate except where known stratification exists as in waste fuel ducts or in residual fuel lines.

ANALYZING FUEL CHARACTERISTICS

In its broadest sense, fuel analysis refers to determination of all physical and chemical properties of a fuel.

Solid Fuels

The basic type of chemical analysis for coal is the proximate, which describes the fuel in weight percent of fixed carbon, volatile material, ash, and moisture contents. In the ultimate analysis, the fixed carbon and volatile material contents are reported in terms of total carbon, hydrogen, oxygen, nitrogen, and sulfur, this form being required for calculation of fuel-air quantities.

Fuel analysis and high-heat-value determination are made in accordance with the Test Code for Solid Fuels PTC 3.2 and *ASTM Standards.*[7] This is a constant-volume determination. Because the fuel is burned at constant

pressure in the steam generator, the high heat value at constant volume as determined in the bomb-calorimeter must, therefore, be converted to a constant-pressure high heat value. Throughout the Codes, this high heat value for constant pressure combustion is referred to as the high heat value. (When testing by the input-output method, only the high heat value and the moisture content of the fuel are required.)

Generally, *fuel sizing* refers to crusher-prepared coal sizing measured with screens of ⅛-inch opening and larger, and is of importance mainly for fuel-bed distribution problems in stoker firing. *Coal fineness* usually means suspension-burning sizes and, for pulverized coal, involves sieving of 50- to 200-mesh (openings per inch) particle sizes. In both cases, a series of screen sizes is used, with the fractions retained or passed by each size being reported. Knowledge of the size consist or overall fraction relationship rather than just one size is required. For example, in pulverized-coal firing, combustion efficiency (carbon loss) can be more affected by larger size consist, whereas grinding power is more closely related to the finer size percentages.

Another physical property related to pulverizer performance is ease of grinding or *grindability index.* This is determined from the amount of power required to pulverize a prepared sample of the test coal in the laboratory. The results are then compared with a standard sample. In the Hardgrove method, the standard is 100, with decreasing numbers indicating progressively harder grinding coals.

Liquid Fuels

Oil fuels are reported in an ultimate analysis together with higher heating value and sample density, viscosity, and flash point. Density can be stated as either standard specific gravity, referred to water, or degrees API, the two terms being readily convertible. Additional properties are pour point, sediment, and the presence of solid impurities such as metallic salts. These properties may influence pumping, storage, slagging, and corrosion problems.

In general, the ultimate analysis is sufficient

to determine fuel-air relationships for by-product liquid fuels such as pulp-mill black liquor, refinery wastes, pitch and others. But operating problems may require special tests in addition to the standard physical property determinations.

Gaseous Fuels

Gaseous fuels are usually combinations of saturated and unsaturated hydrocarbons, and their analysis is reported as a mol or volume percent of these constituents. Common analyses include low-temperature distillation, mass-spectrometer, and adsorption methods. From this constituent analysis, high heating value at standard volumetric conditions and saturated or dry can be calculated by using standard heating values of constituents. Where gas samples of large volume can be provided easily, the classical method of heating-value determination, by calorimeter, is feasible. For instance, recording calorimeters are often found at transmission-line entry points to users' plants for billing purposes.

Fuel-Air Proportioning

Combustion airflow measurement and regulation are requisite to good operation because of the desire to burn a fuel with optimum air quantity. Some air in excess of the theoretical quantity for complete combustion is always required, the amount depending upon the fuel and firing method. Although ideally the lowest excess air is desirable, there are many conditions under which high excess air results in more economical plant operation. If, for example, constant steam temperature is to be maintained at low ratings, increased gas mass flow may be required over heat-transfer surfaces. Increasing excess-air is one way of accomplishing this.

Airflow Measurement

Knowing the quantities of air required and fuel products formed in terms of weight or volume rates is not necessary from an operating point of view. The operator wants to know these quantities relative to the theoretical amounts for the fuel rate being burned. Operating indexes, therefore, are designed to show the percent excess-air or relative percent total-air quantities. On the other hand, for engineering design purposes, these air and gas rates are necessary to properly size fans, ducts, and heat-transfer equipment.

In routine operation, airflow measurement is accomplished through pressure or draft differential sensing of a gas or air-side component, or by means of flue-gas analysis. In either case, field calibration of the metering equipment is ultimately based upon test flue-gas analysis. Flow measurement by pitot-tube techniques is somewhat restricted in boiler air and gas flow problems because duct configuration and low static-impact pressure difference generally limit the resulting accuracy. More importantly, the flue-gas analysis method is superior because of its simple interpretation for overall airflow. However, pitot measurements become necessary for study of air of flue-gas distribution within a system.

Excess-Air Determination by Flue-Gas Analysis

A simple method of volumetric flue-gas analysis is by Orsat apparatus, Fig. 1, in which a known volume of gas is sequentially exposed to absorbent solutions for carbon dioxide and oxygen. Measuring sample volume between stages indicates the constituent volume percentages and the remaining nitrogen. Trace gases such as argon are included with nitrogen, while sulfur compounds such as sulfur dioxide and trioxide are absorbed together with carbon dioxide. For most fuels, their effect on the gas-analysis results is negligible in excess-air determination.

The resultant gaseous products for complete combustion of a fuel using several excess-air values are calculated from the ultimate fuel analysis and chemical combining equations. Plotting the volume percentages of carbon dioxide and oxygen thus formed permits excess-air determination by volumetric gas analysis. As a matter of fact, within certain restrictions, excess-air can be determined without the fuel analysis, because the available oxygen in

products is a direct measure of excess-air and, when compared to the nitrogen, will indicate percent excess-air. For fuel oil or many bituminous coals, oxygen measurement using graphical means such as Fig. 2 is sufficient to indicate excess-air with acceptable accuracy.

The practical uniformity of this oxygen-excess-air relationship over a wide range of fuels makes it extremely useful in both single and combination fuel firing. Moreover, it is responsible for the widespread use of flue-gas oxygen analyzers for operating and test purposes. These devices employ either catalytic combustion of a known gas and the uncombined oxygen in a flue-gas sample within a temperature-sensitive electrical circuit, or magnetic-field distortion due to paramagnetic qualities of the oxygen in the flue-gas sample. In both cases, the effect is proportional to sample oxygen content, and permits calibration of the instrument-output signal in terms of oxygen percent.

Detecting Unburned Combustibles

In modern fuel-burning practice, unburned gaseous combustibles in flue gas are generally nonexistent; however, they can occur with certain operating conditions. Even in these cases, the extremely low quantities to be found (less than 0.10 percent) are usually beyond the accuracy of Orsat work and require detection by chemical reaction-calorimetric methods or laboratory analysis of small samples by mass-spectrometer and adsorption techniques. In either case, to avoid misleading results, the samples must be collected with extreme care to circumvent sample contamination. For example, the sampling probe should be a quenching type since hot-probe materials may act as combustion catalysts in the flue gas, thus indicating no combustibles when they are present.

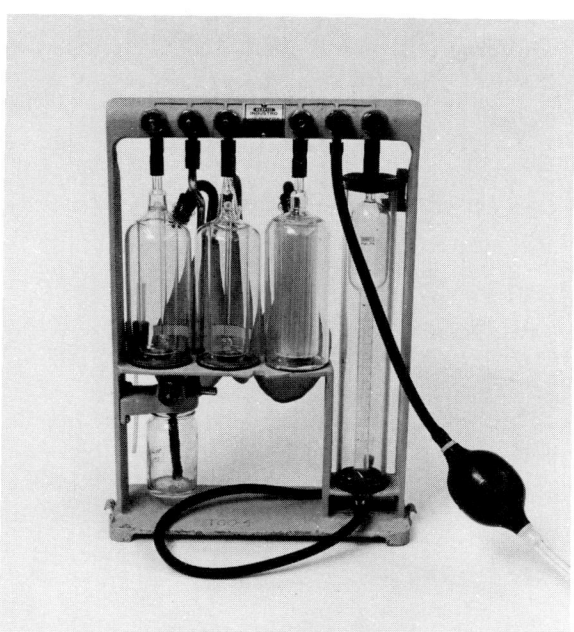

Fig. 1 A basic means of volumetric flue-gas analysis is by Orsat apparatus. In this method, a known volume of gas is exposed to absorbent solutions for carbon dioxide and oxygen.

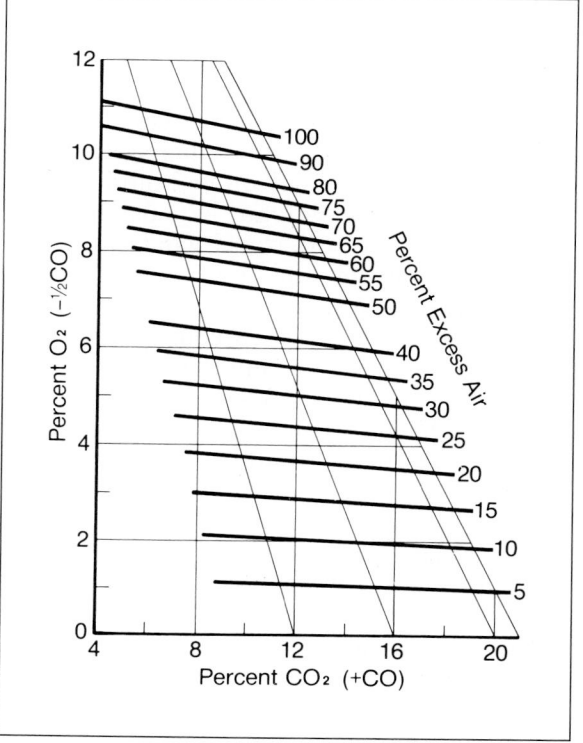

Fig. 2 Percentage relationship of oxygen (O_2), carbon dioxide (CO_2) and excess air. (Orsat analyses for a given fuel should plot on a straight line parallel to the guidelines.)

Conversely, probes and sample containers should be free of oil films, high-vapor-pressure stopcock grease and the like, since these materials may indicate traces of combustibles when there are none in the flue gas.

TEMPERATURE MEASUREMENTS

Temperature-measurement techniques differ depending upon temperature level, fluid stratification, physical accessibility, and accuracy desired.[8]

Generally, two methods are common in steam-generator testing:

1. filled-system thermometers and
2. thermoelectric and electrical resistance effects.

These two methods are direct in that they involve fluid-temperature level attainment by the sensing element. On the other hand, calibration difficulties arising from variations with different fuels and furnace equipment prevent frequent use of indirect techniques, such as utilization of optical or radiation effects accompanying temperature level of the fluid.

The most common of the filled-system thermometers, mercury and other liquid-in-glass thermometers use volume expansion. Their use in gas and air work is restricted to ambient-level temperatures and, when stratification does not exist, in ducts. As a variation of the volume-expansion class, gaseous-bulb type instruments, commonly found in operating instrumentation, also measure accompanying pressure change. Although this type instrument permits remote indication, it is a fixed-position device, usually requiring field calibration by another means. See Fig. 3.

The most versatile temperature-measurement devices are those in the second group. Both thermoelectric and electric-resistance type techniques are readily adaptable to remote measurement, thus enabling monitoring of many points from a single location. Of the two, the thermocouple is more widely used because of its simplicity and low cost.

THE THERMOCOUPLE

Thermocouple measurement employs the thermoelectric effect, a phenomenon whose alternate effects were first observed respectively by Seebeck and Peltier in 1821 and 1834. An electromotive force (emf) is developed if a circuit comprising two different wires has the two junctions at different temperatures. The magnitude of the emf depends on the wire materials and is proportional to the temperature difference between the junctions. Thus, knowing the temperature at one junction and considering the wire materials, the temperature at the other junction can be determined. Special laws govern the effects of intermediate metals and temperatures.[9] See Fig. 4.

Numerous metals are available for thermocouple materials with selection being based on emf developed, expected mechanical life in the atmosphere and temperature involved, calibration constancy, and cost. The most common types are copper-constantan, iron-constantan, chromel-alumel,[10] and platinum-platinum-rhodium; their upper limit temperature ratings for general service are 400, 1100, 1800 and 2700°F respectively.

Several engineering societies[11] have de-

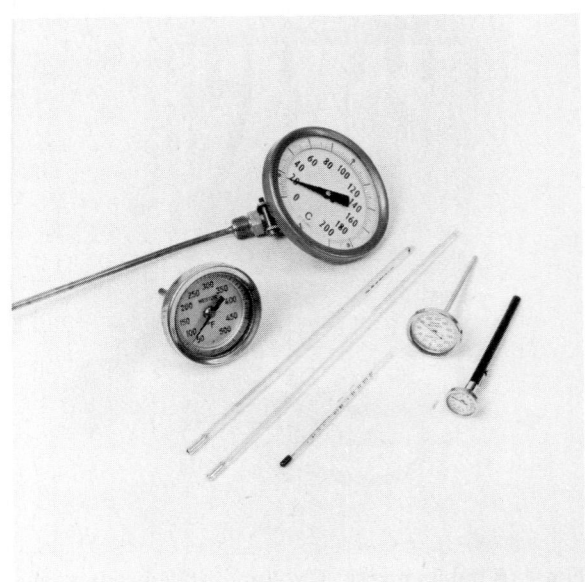

Fig. 3 Mercury and other liquid-in-glass thermometers are used in gas and air work for sensing low- and moderate-temperature levels.

veloped detailed specifications covering temperature-emf relationships and standard accuracy limits. Experience with chromel-constantan thermocouples also indicates suitability of this combination for measurements up to 1600°F with high emf developed.

Thermocouple emf can be measured with either a millivolt galvanometer or a null-balance potentiometer. However, in the galvanometer system, the thermocouple circuit resistance is an additional factor and complicates the measurement, whereas in the null-balance system no current flows in the thermocouple circuit, the measurement being of the opposing emf. Being simpler, the latter system is more widely used.

THE RESISTANCE-BULB METHOD

The resistance-bulb method involves known temperature-electrical resistance change characteristic of a material, usually nickel or platinum, using a Wheatstone bridge type measuring circuit. This method can be made

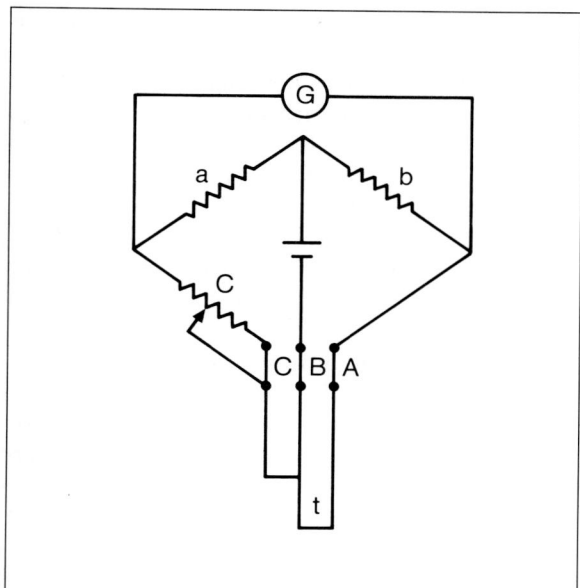

Fig. 4. Resistance thermometers have as their sensing element a resistor usually made of pure-metal wire. Here, a diagram of a typical simple balanced-bridge type of instrument using a Wheatstone bridge is shown.

extremely accurate and is used in temperature standards work. However, the high cost of sensing elements and measuring circuit restricts its use to laboratory calibration work or field test work where high precision is required, such as measurement of fractional parts of degrees.

USING THERMOCOUPLE PROBES

Generally, flue-gas temperature measurement is made with thermocouple probes. Such measurement requires that the hot junction of the thermocouple reaches the fluid temperature level. Thus, the fluid contacting the thermocouple must be flowing at a sufficient rate to supply heat lost by conduction from the hot junction along lead wires and by radiation to colder surfaces. The magnitude of these heat losses will depend upon probe design, including probe temperature and relative orientation of the hot junction, and temperature differential between the hot junction and the heat receiving surfaces.

Thermocouple measurement of gas-air temperature in most ducts is simplified for the following reasons.

1. Temperature level is below 700°F and does not require water-cooling of probes for mechanical strength.

2. Ducts are insulated to prevent heat loss and the interiors of ducts are essentially at fluid temperature.

At a temperature level of 700°F, an error of 15°F is commonly found with bare thermocouples in proximity to economizer heating surface at 500°F.

At gas temperatures higher than 700°F, watercooling of probes is necessary for mechanical strength reasons, and the leadwire conduction heat-loss increases. More significantly, higher temperature levels correspond to areas with heat-transfer surfaces such as economizer, waterwalls, superheater, and reheater, the surface temperatures of which vary between 500 and 1100°F. As a result, the radiation error of a bare thermocouple junction increases severely until at the furnace outlet (with gas temperature of 2000°F) the bare thermocouple error is about 200°F.

To compensate for radiation and conduction heat loss in furnace gas-temperature work, the thermocouple junction is shielded with a thin ceramic cylinder through which a high gas flow is induced. In turn, the shield is subject to the same heat-loss effects and, by increasing the number of annular shield layers, the thermocouple junction can be brought to within ± 5°F of true gas temperature. Usually the fuel determines the limit to the size and number of shield openings. With high-ash coals serious plugging can occur in a short period when shields with small openings are used.

THE ORIFICE PROBE

Another type of furnace gas-temperature measuring device is the orifice probe, in which the hot gas is drawn through two metering orifices in series, with deliberate cooling of the gas occurring between them. By measuring the two orifice-flow rates and cold-gas temperature, the initial gas temperature can be calculated since weight flow is constant and the orifice-flow differentials indicate volume flow, which is a function of temperature. The advantage of this type measurement is that it eliminates thermocouple radiation error.

MEASUREMENTS RELATED TO HEAT ABSORPTION

Heat absorption values are used in developing the surface heat-transfer relationships and are determined from test steam-temperature and flow measurements, together with gas temperatures and flows. Additional performance items in the water-steam cycle are pressures, pressure drops, and quality/purity determinations. Circulation studies in furnace waterwalls and flow distribution between circuits are other examples of water-steam cycle testing.

WATER AND STEAM FLOW

Both steam output and feedwater flow are obtained by pressure-differential measurement across flow nozzle or thin-plate orifice sections. The two flows are equal except for losses such as blowdown and injection leakoff. Of the two, however, feedwater measurement is generally more reliable than steam flow because orifice coefficients are more positively established for the former.

The recording flow meters found in most plants employ commercial orifices or flow nozzles. Overall accuracy is usually plus or minus 2 percent of full scale, unless the installation has been calibrated against a special test orifice. These test orifices are plant primary standards, usually installed in low-pressure, low-temperature condensate points in the feedwater cycle to avoid undue extrapolation of original calibration data. Their calibration is performed with laboratory weigh tanks, and extrapolation of flow coefficients to the condensate conditions of flow, temperature and pressure is usually accomplished on the basis of Reynolds number and other dimensionless criteria.

While a few central stations have reheat steam flow meters, it is more common to determine this quantity from primary steam-water flow, high-pressure heater heat balance and the appropriate turbine gland corrections.

For test of heat-absorption surface characteristics, it is usually sufficient to use the operating feedwater recorder as these characteristics are essentially constant for rating changes within meter accuracy, and the fuel-air-gas flow data are related to this base value by heat balance.[12]

Another method of flow measurement sometimes used is pitot-tube technique. Variations of the standard air-gas type are usually preferred for manufacturing reasons, and these are commonly used in some circulation studies where a low-resistance measuring element is required. These devices are usually not suitable for total unit flow measurement because of structural and flow-pattern uncertainty reasons.

Weigh tanks are rarely found in large utility installations because of unit size and consequent cost. Flow nozzles and orifices are taking

over their functions. However, the weighing facilities are necessary as primary references in standard laboratories where the nozzle calibration work is performed.

TEMPERATURE MEASUREMENT

The common devices for temperature measurement are thermocouples, resistance-bulb elements and thermometers. The general statements on this subject made previously under gas and air measurements also apply to steam and water measurements.

Some aspects, however, present more serious problems than in the gas-air work. Since in this case the fluid pressures are also high, the measuring elements are rarely exposed directly to the fluid stream. Measurement is commonly made by bottoming contact with a metal well which projects into the stream and provides the necessary mechanical strength. Several well designs are shown in the *ASME Performance Test Code*, Supplements on Instruments and Apparatus, PTC 19.3 and in a paper by J. W. Murdock.[13]

Another method of fluid-temperature determination is surface-temperature measurement of the containing steam or water tubing. This method is only possible where there is no heat transfer because of temperature gradient effects from hot gas and steam-film. In these zero heat-flux areas, the tube-wall gradient usually presents no problem in measurement, although even here precautions should be taken against boundary effects. See Fig. 5.

Thermocouples are commonly peened directly into the tube metal or attached by welding the hot junction to the metal. In a well application, spring loading of the element is more practical. In both these cases, the thermocouple is grounded and measurement of a number of points must be by separate circuits since series connections can form unknown loops using boiler metal as a third wire.

PRESSURE AND PRESSURE DIFFERENTIALS

The fundamental pressure and pressure-differential measuring device is the visible liquid column. Although its use is limited to column heights which are practical, the foolproof simplicity of the method is the reason for its retention as a reference. Glass-faced mercury manometers are used for pressure-differential work up to line pressures of 2000 psig, especially for orifice-flow nozzle differential measurements. Some range flexibility exists through use of liquids whose densities are between those of water and mercury. A variation of the visible column, used for pressures over 2000 psig, is a column inside a tube or jacket with a float indicating liquid level through mechanical linkage or magnetic pointer.

Deadweight instruments employ incremental weights acting on a given-sized piston which floats when the liquid and weight pressures are equal. In deadweight gages, the liquid

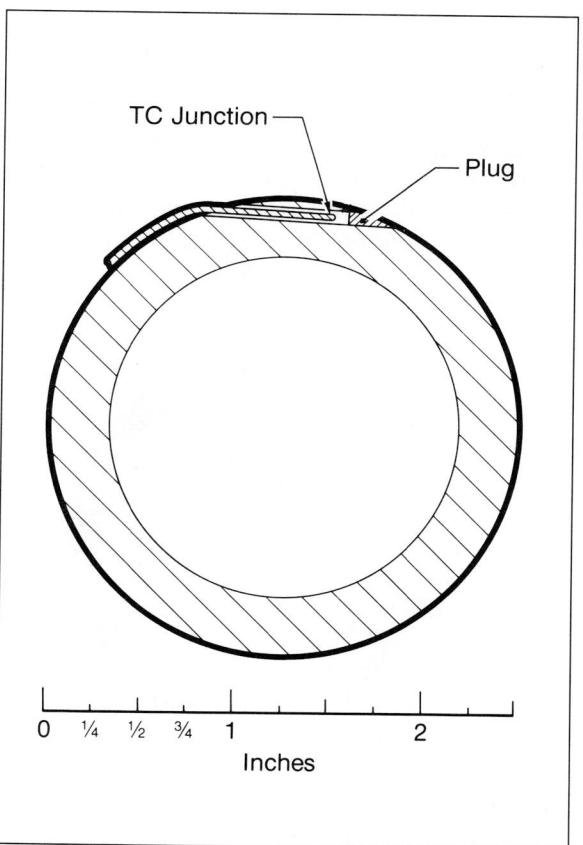

Fig. 5. **Thermocouples are used for surface-temperature measurement of tubing containing fluids at high pressure.**

pressure is line pressure, its value determined from the weight total. In deadweight testers, the liquid is in a closed system with the pressure acting on the instrument to be calibrated and on the weight piston from which the pressure value is determined.

A common steam-water pressure measuring device is the Bourdon gage, Fig. 6, the simplest form of which is a bent flattened tube that tends to straighten itself when internal pressure is applied. Motion of the tube end is transmitted through linkage to a pointer which sweeps a graduated scale. The concept of mechanical motion or distortion due to fluid pressure is common to all pressure elements; the differences occur in the way this force is translated into a signal. Development of this idea is found in diaphragms, bellows and cylinders whose motions are linked to and described by pointers, strain gage wires, varying resistance bridges, differential transformers, and other mechanical or electrical signal generators.

Such combinations of devices are known as pressure transducers. Their sensitivity, linearity, repeatability, and hysteresis are characteristics to be evaluated in selecting a combination for any specific application.

Fig. 6. **In its simplest form, the Bourdon gage, a common steam-water pressure measuring device, is a bent flattened tube that tends to straighten itself when internal pressure is applied.**

ADDITIONAL PERFORMANCE MEASUREMENTS

The steam-generator pressure parts are the means by which heat liberated in combustion of the fuel is absorbed by the steam-water fluid. Since the process is one of heat flow, it must be apparent that temperatures of the various equipment items are the basic criteria for design once the desired pressure is stated.

The thermocouple is by far the most-used piece of test apparatus in steam-generator study. Its adaptability for remote measurement of many widely separated points permits investigations and control otherwise impossible. Usage is not limited to pressure parts but includes structural parts as well.

Aside from temperature considerations, steam-generator equipment is subjected to abrasion and erosion and to vibration forces. The first two are generally long-term processes requiring correlation of metallurgical laboratory studies with prolonged operating experience. The main function of field-test activity in this area is to establish the magnitudes of existing factors such as temperature, gas-flow velocity patterns, ash-fuel character and concentrations.

With the increase in unit size and desire for balance between strength and weight, vibration forces have received considerable attention because of their destructive effects. Standing waves in flue-gas passages induce structural and pressure-parts vibrations. Test activity in this area is required for corrective work and to provide design criteria to avoid susceptible arrangements. Gas vibrations are measured with traversing pressure transducers and oscillographic measuring circuits or mechanical vibration graphing devices, depending upon conditions. The latter are also used for structural and casing members.

Another problem related to temperature factors is equipment expansion between hot and cold positions. These movements are generally measured with trams or dial gages depending upon the application. For instance, since lower waterwall headers on certain units will move

ABB
ASEA BROWN BOVERI

downward as much as 15 to 20 inches, and other parts lesser amounts, unrestrained movement is essential.

GAS-FLOW MEASUREMENT AND SAMPLING TECHNIQUES

Measurement and sampling of the gaseous and particulate products of combustion are essential for verification of compliance with emission regulations; assessment of the performance of emission-control systems; and in evaluating the potential for corrosion, erosion, and deposit formation on internal surfaces such as ducts and stack liners. The principal method of determining the concentrations of particulate material and of the oxides of nitrogen and sulfur in a gas stream is gas sampling. Other gaseous products of combustion that are measured, particularly in municipal-waste-fired plants, include: hydrogen chloride, hydrogen fluoride, heavy metals, and certain organic species. Measurement of the gas volumetric and mass flow-rate is required to assess equipment performance and to determine the total mass emission rates.

TECHNIQUES OF MEASUREMENT

The Environmental Protection Agency (EPA) Methods publications describe those techniques most widely used in the United States for measuring the products of fossil-fuel combustion. These test procedures appear in the Code of Federal Regulations, Title 40, Part 60, Appendix A, and are printed by the Office of the Federal Register, National Archives and Records Administration. Table I lists the principal EPA methods applicable to the products of fossil-fuel combustion. The following sections will cover commonly used EPA procedures for measuring gas flow, particulate matter, sulfur dioxide, sulfuric-acid mist, and nitrogen oxides.

GAS FLOW RATE QUANTIFICATION

Test Methods 1 through 4 detail an approach for measuring the gas mass or volumetric flow-rate using a pitot-static probe, described later in

TABLE I. Principal EPA Methods for Measuring Flow Rate and Sample Products of Fossil-Fuel Combustion

EPA METHOD	Application
1	Layout of sampling/measurement points in a duct or stack.
2	Determination of duct or stack gas velocity and volumetric flow rate using a Stauscheibe pitot tube.
3	Determination of the carbon-dioxide, oxygen, and carbon-monoxide content of a gas sample using an Orsat or Fyrite analyzer.
4	Determination of the percentage moisture in an extracted gas sample.
5	Determination of the particulate concentration of a gas stream. This test method employs a filter enclosed in a heated housing, external to the gas stream.
6	Determination of the moisture-free sulfur-dioxide concentration of an extracted gas sample.
7	Determination of the moisture-free nitrogen-oxide concentration of an extracted gas sample.
8	Determination of the moisture-free sulfur-dioxide and sulfuric-acid concentration of an isokinetically extracted gas sample. A multipoint traverse is employed with this method.
9	Determination of stack-emission opacity through visual observation.
10	Determination of the carbon-monoxide concentration of an extracted sample using a nondispersive infrared analyzer.
17	Determination of the particulate concentration of a gas stream using an in-situ filter.
19	Calculation methods for determining system sulfur-dioxide removal efficiency, and particulate, sulfur-dioxide and nitrogen-oxide emission rates from electric utility steam generators.

the chapter. These methods employ a technique of subdividing the duct cross section into equal-area zones. The number of zones required is a function of the distance of upstream and downstream disturbances relative to the measurement plane. From point gas velocity measurements in each of these equal-area zones, the mean duct velocity can be used to determine the gas volumetric flow-rate as follows:

$$Q = A \times \overline{V}$$

where

Q — gas volumetric flow-rate
A — measurement plane cross-sectional area
$\overline{V}$ — mean duct velocity

Mass flow can be determined from the volumetric flow-rate if the gas composition, absolute pressure, and gas temperature at the measurement plane are known. Methods 3 and 4 detail test procedures for determining the principal products of fossil fuel combustion, including carbon dioxide, carbon monoxide, oxygen, and water.

The equal-area zone technique is also specified by ASME PTC 19.5, Fluid Velocity Measurement. This Code discusses errors produced in non-uniform velocity distribution. Also, several analytical methods of velocity averaging are presented and evaluated.

MEASUREMENT OF PARTICULATE LOADING

Procedures for the determination of the particulate concentration in a gas stream are detailed in EPA Methods 5 and 17. As specified by these test procedures, a sample of gas and particulate is isokinetically extracted, with a sharp edged nozzle, from equal-area zones. Isokinetic sampling simply means that the gas/particulate sample is extracted at a rate equal to the flow rate of gas in that zone. The extraction rate of the gas/particulate sample is measured by a pitot-static tube and must be adjusted, or verified, for each sample point. These methods require that the rate of sample extraction should be within 90 to 110 percent of isokinetic.

Once the gas sample has been extracted, it passes through a glass fiber filter which captures a majority of particulate larger than 0.3 micrometers. The remaining particulate travels through a heated sample line to a series of chilled-water impingers and a desiccant, which remove all remaining particulate and water vapor from the gas sample before it reaches a dry-gas meter and sample pump. The particulate loading in the gas stream is typically reported as grains of particulate per actual or dry standard foot of gas, (gr/acf or gr/dscf). Stack emissions are reported to regulatory authorities as the pounds of constituent per million Btu heat input to the boiler. This may be calculated as follows:

$$E = C_d F_d \frac{20.9}{20.9 - \%O_{2d}}$$

or

$$E = C_d F_c \frac{100}{\%CO_{2d}}$$

where

E = amount of pollutant in an extracted gas sample
C_d = pollutant concentration
O_2 = volume percent oxygen on a water-free basis
CO_2 = volume percent carbon dioxide on a water-free basis
F_d = oxygen fuel factor
F_c = carbon-dioxide fuel factor

Fuel factors can be determined from an ultimate analysis of a fuel sample or an estimated value is provided in EPA Method 19. These equations are applicable in reporting the emission rate of most gaseous or particulate pollutants.

The principal difference between Methods 5 and 17 is the location of the glass fiber filter. Method 17 uses a thimble-type filter located at the end of the probe just upstream of the nozzle, Fig. 7. Method 5, Fig. 8, employs a flat filter enclosed in a heated housing on the end of the probe, external to the gas stream. Method 17 is preferred for high particulate loadings,

such as upstream of a particulate collection device, because a thimble filter can accept much more ash before blinding.

Both Methods require the extraction of at least 30 dry standard cubic feet (dscf) of gas over a period of not less than 60 minutes. The amount of gas sampled, however, is often dependent upon the anticipated particulate loading. For instance, it is not uncommon to extract as much as 60 dscf over 120 minutes at the outlet of a high efficiency particulate-collection device. This provides a larger particulate catch and, therefore, a more accurate result.

An accurate result is also dependent upon

maintaining proper temperature in the sample probe and filter. Insufficient heating of the probe and filter can allow sulfuric acid condensation. Acid which is deposited on the filter will be reported as particulate. Heating the probe too hot can also affect the accuracy of test results. If the probe/filter temperature is too high, the particulate sampled may be volatilized. For this reason, EPA limits the probe/filter temperature to less than 320°F.

MEASUREMENT OF SULFUR-DIOXIDE AND SULFURIC-ACID CONCENTRATION

Procedures for measuring the concentration

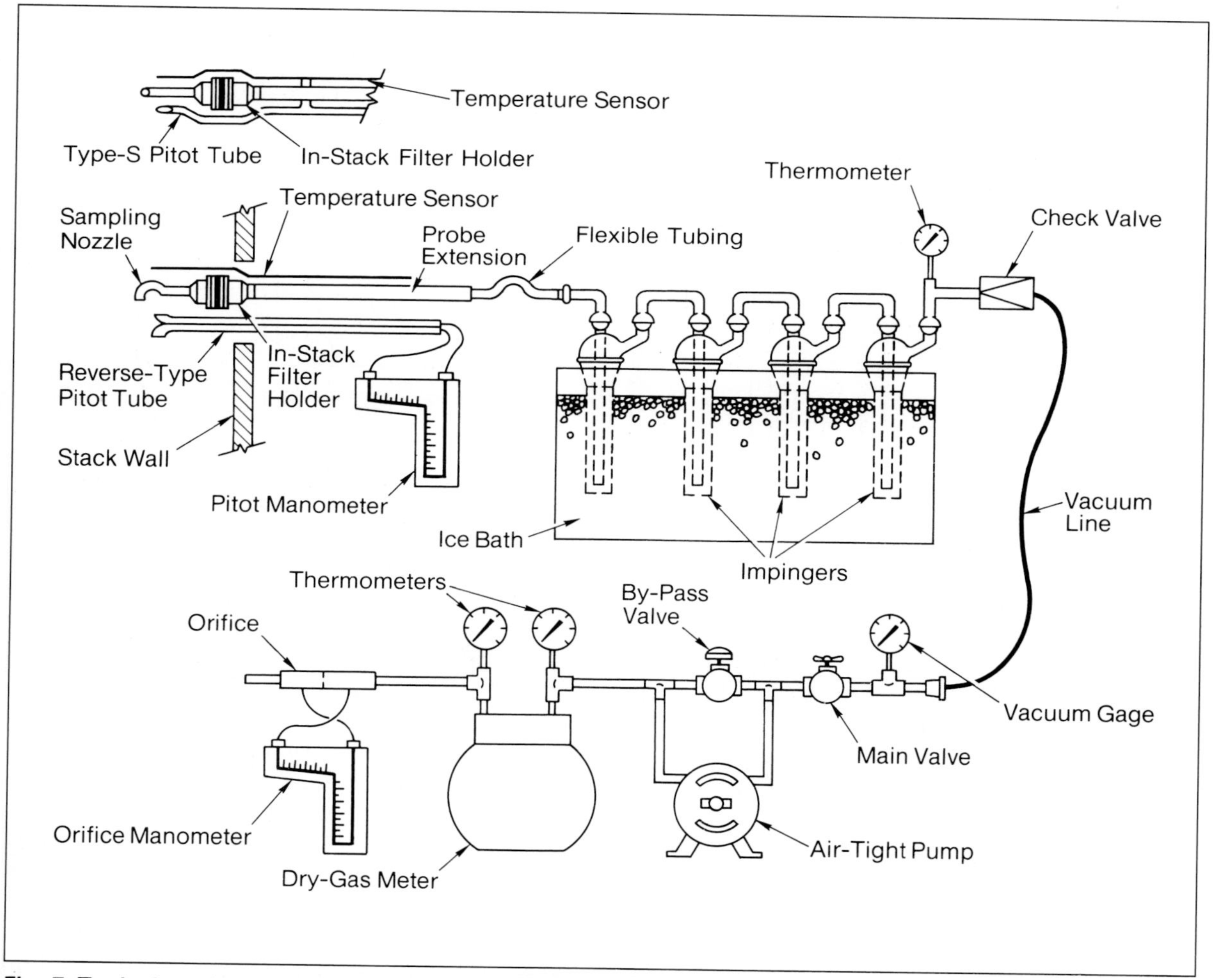

Fig. 7 Typical particulate-sampling train, equipped with in-stack filter.

of sulfur dioxide in a gas stream are outlined in EPA Method 6, Fig. 9. As described in this method, a sample of gas is extracted through a heated probe and sample line and then passed through a series of ice-water-chilled impingers. The first of these impingers contains an 80 percent isopropanol (IPA) solution. The successive two impingers contain a 3 percent hydrogen peroxide solution. Sulfuric-acid mist is preferentially absorbed in the first IPA impinger while the sulfur dioxide is absorbed in the following two peroxide impingers. Following the impingers, the gas sample is further dried using a desiccant and the dry gas volume is me-

tered. Preceding sampling, or in between sample runs, the gas train is purged with air for at least 15 minutes. Gas sample times are at least 20 minutes although longer sample times are desirable when measuring low sulfur-dioxide concentrations.

The sulfuric acid in the gas stream can also be measured with this type of gas train if isokinetic sampling from equal-area zones is used (EPA Method 8). Isokinetic, multipoint sampling is also necessary if sulfurdioxide stratification is present. Stratification is most likely in ductwork runs downstream of the absorber module. Gas entering a flue-gas desulfurization

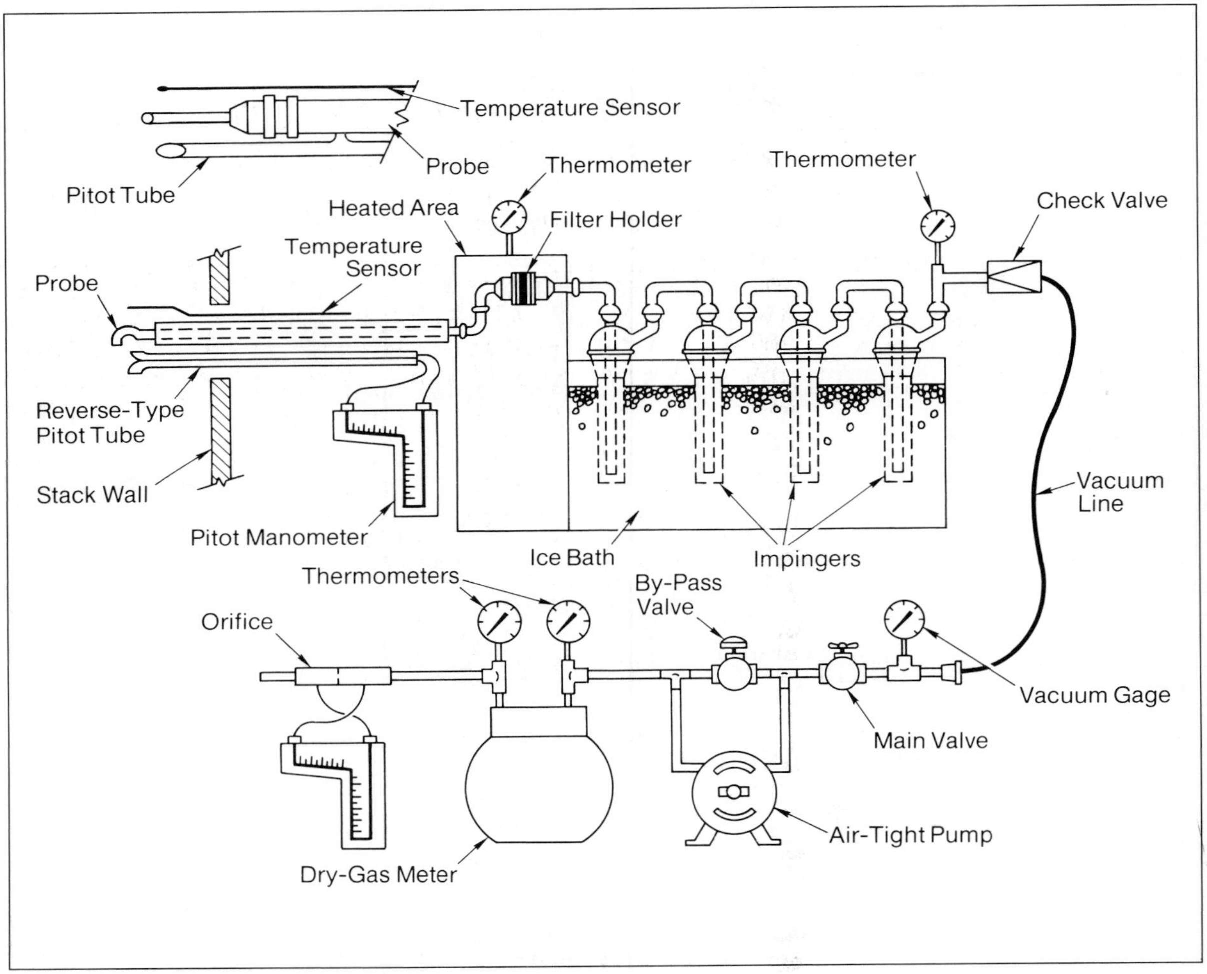

Fig. 8 **Particulate-sampling train**

system is usually well mixed so that a single-point sample, Method 6, is acceptable.

Following extraction of a gas sample, the sulfur dioxide or sulfuric acid absorbed in the impingers is titrated. From the titrant volume required, and the dry gas volume sampled, the concentration of sulfur dioxide and sulfuric acid in the gas stream can be calculated. As with the particulate loading, sulfur-dioxide and sulfuric-acid concentrations are usually reported on a pounds of pollutant per million Btu heat-input basis.

An instrumental procedure for measuring the sulfur dioxide concentration can also be used,

Method 6C. An extracted sample of gas is dried to remove all water, and is then introduced into an analyzer. The analyzer must undergo calibration checks to verify its accuracy preceding and following each sample run. Instrumental methods of measuring gas sulfur-dioxide concentration can provide results in a shorter length of time and can readily identify ductwork sulfur-dioxide stratification.

MEASUREMENT OF NITROGEN-OXIDE CONCENTRATION

Both absorbent and instrumental techniques have been developed by the EPA for measuring

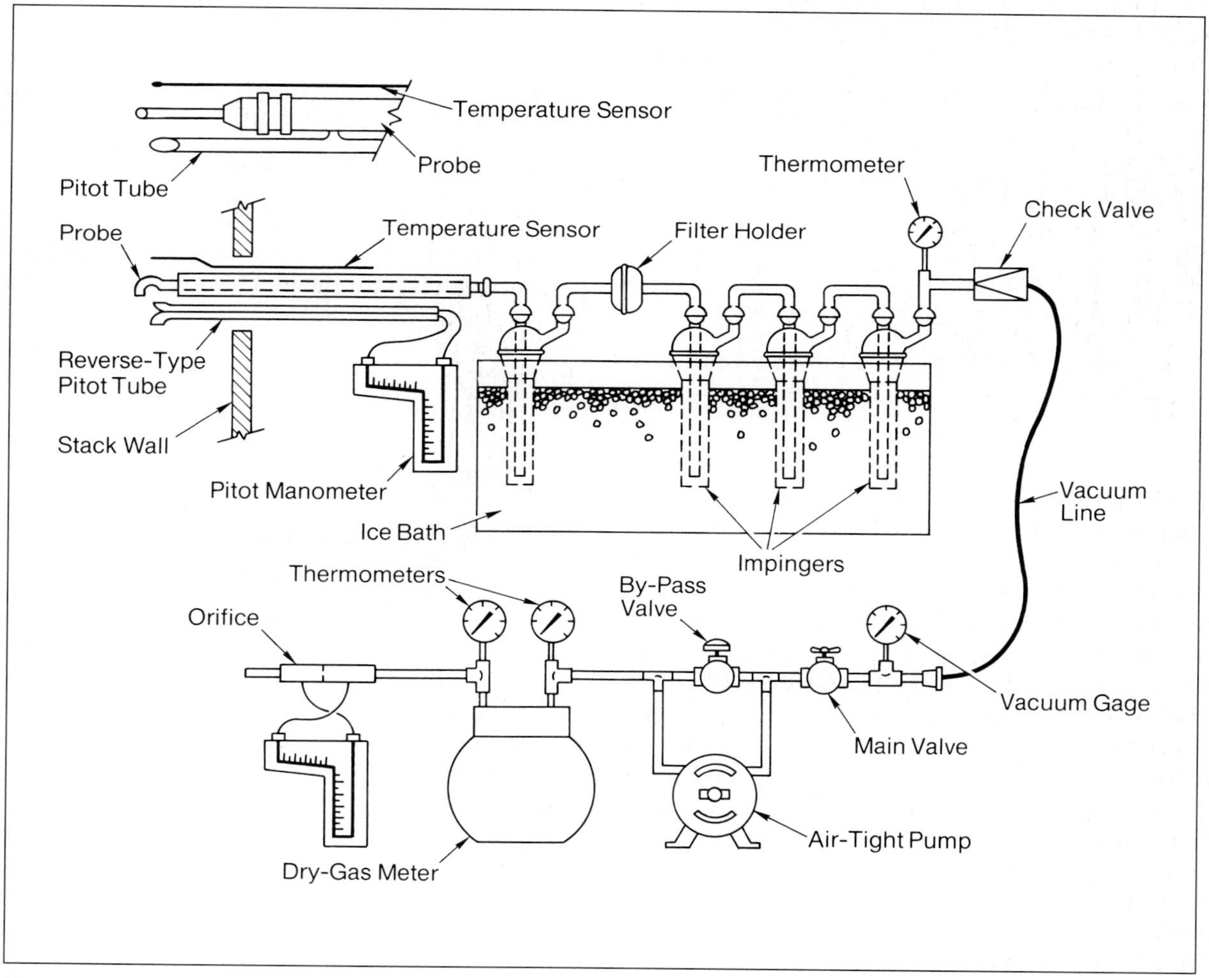

Fig. 9 **Sulfuric-acid mist-sampling train**

the concentration of nitrogen oxide in a gas stream. Instrumental Method 7E has gained the widest use because results can be obtained immediately following gas sampling. Method 7 requires sixteen (16) hours between collection of the gas sample and analysis of the absorbent to determine the nitrogen-oxide concentration.

In both Methods, a sample of gas is extracted through a heated probe. Isokinetic sampling is usually not necessary unless there is reason to suspect NO_x stratification. As specified in Method 7, a sample of gas is collected in a two-liter boiling flask with 25 milliliters of absorbing solution. After contacting the absorbing solution with the gas sample for at least 16 hours, an aliquot of the absorbing solution is analyzed using spectrophotometry. The absorption of NO_2 by the spectrophotometer can then be related to the NO_x concentration of the gas sample.

In Method 7E, the NO_2 in the extracted sample is first converted to NO. A chemiluminescence analyzer is then used to determine the NO_x concentration. As with the SO_2 instrumental method, calibration checks are necessary to verify the accuracy of the indicated results.

THE NEED FOR ACCURATE MEASUREMENT OF GAS FLOW

The information in the final portion of this chapter is included because of the profound effect that the quantity of gas being handled has on the sizing and cost of the gas-treatment equipment.

The determination of *total* quantities of gaseous and solid constituents in the gases leaving a steam generator depends upon an accurate quantitation of the gas flow itself. For instance, in the sampling of dust, the amount of material caught in a sampler is reduced to a dust concentration of, say, pounds of dust per 1000 pounds of gas. The total quantity of dust being emitted by the source, then, is the product of dust concentration and total gas flow by weight.

BOUNDARY-LAYER EFFECTS

Measurement of Solids in Flue Gases by P. G. W. Hawksley defines the approach of the British Institute of Fuel[14]. This publication discusses the effects of nonsymmetrical gas-flow distribution and boundary-layer profiles. It should be noted that the equal-area zone method, as typically used, can easily miss the boundary-layer flow profile. Very small incremental areas would be required for the velocity probing device to "see" all of the boundary layer. However, traversing time can be reduced if the boundary layer is evaluated in detail for one test only and then used to correct coarser subsequent traverses. This approach is justifiable since each flue or stack will have its own characteristic boundary layer which is a function of the specific system geometry and flow turbulence. There have been occasions where boundary layers representing 5 percent of the flue diameter have produced flow errors of some 10 to 20 percent.

VARIATION IN RESULTS USING PITOT-STATIC MEASUREMENT

Pitot-static tube velocity measurements have been coupled with the equal-area technique for many years. The results of total gas flow can be quite varied and misleading, if care is not given to the selection of the measurement plane, or consideration given to the selection of the proper type of flow-measurement device for the flow directionality that is present. Table II[15] presents the results of several total gas-flow measurements using the pitot-static tube and equal-area techniques as called for in the above standards. The results listed are derived from a typical utility steam-generator installation operating at constant load. As noted, the flow as measured can vary from 104 to 150 percent of rated value.

POWER-PLANT FLOW PATTERNS

Industrial ductwork and flues are designed to connect pieces of equipment at the minimum cost and minimum space requirements. The resultant flow patterns quite often deviate from the desired uniform and unidirectional condition required for accurate quantitation of flow.

Table II. Gas-Flow Measurement:

Gas flow measured at different locations in accordance with methods given in *ASME PTC 27* and by using a pitot-static tube. Observe the gross divergence of measured values and the necessity for careful choice of measurement point.

Measurement Point	Gas Flow CFM	Percent Rating
Rated Value	300,000	100
Mechanical-Collector Inlet	383,000	128
Mechanical-Collector Outlet	449,000	150
Induced-Draft Fan Inlet (Precipitator Outlet)	313,000	104
Induced-Draft Fan Outlet	424,000	141

Flow separation and reverse flows, often found in the major runs of flues, are corrected only by major pressure drops such as an air heater or excessively long runs of duct work.

Flow patterns in stacks and fans frequently are highly productive of vortex energy. A basic problem associated with vortex formation is that vortices will persist for many diameters downstream from the source. Vorticity has a very strong effect on dynamic-flow measurement devices.

Occasionally, reverse or negative flow may be encountered. A question then arises as to how to calculate the total volume flow rate from this type of data. Fluid-flow continuity requires that all velocity points, both normal direction and reverse flow, be arithmetically averaged and applied to the total cross-sectional area of the plane of measurement. If this is not done, the recirculated gas flow (the negative values), which can represent 5 to 15 percent of volume throughput, will increase the positive flow values by the same amount and produce a measured flow that is high.

VELOCITY MEASUREMENT

The classical and long-term approach to the measurement of fluid velocity has been by pitot tube. Several references[16-19] are available which describe the developing history of this device. Appendix E of the 1981 edition of this text, Velocity Calculations by Graphical Methods, presents graphs that can be used for quick determination of gas velocities from pitot-tube readings.

THE PITOT TUBE

The typical flow-measurement device is the pitot tube shown in Fig. 10. The impact or total-pressure hole is the most useful part of the probe. Hydraulically, the impact pressure is the true sum of the local static and dynamic (velocity) heads. Also, the impact hole can have a cone of response of some 60 to 70 degrees. Therefore, accurate angular positioning of the probe for total (impact) pressure measurements is not critical. However, positioning of the probe parallel to the flow direction is very critical for static-pressure measurement. The static-pressure-tap reference surface should be parallel to the flow stream-lines. Also, ideally, the flow stream-lines should be straight and parallel. Any flow stream-line

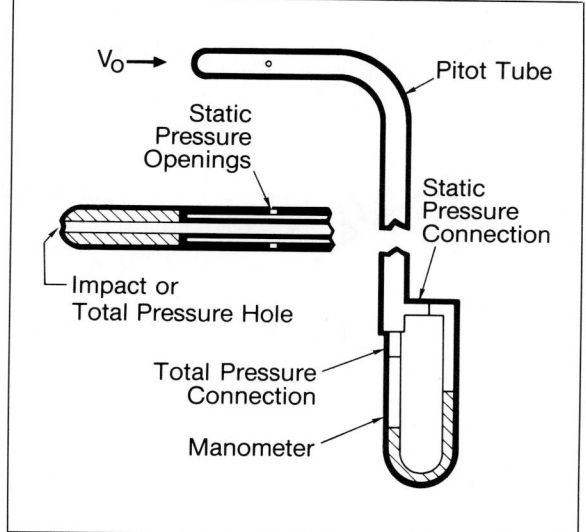

Fig. 10 **A pitot tube for fluid-velocity measurement with pressure connections and readout shown schematically**

which is not parallel to the static surface will induce an impact or reverse-impact, dynamic pressure at the static hole and create an associated error.

One practical problem associated with the pitot tube is the right-angle bend and probe tip length. This configuration requires large holes in the flue, or duct wall, or some awkward probe handling by the user. The bend also makes it difficult to probe the flow at the near side, or access wall, of the flue. Large holes in the flue wall are, in turn, difficult to seal to prevent leakage errors. Another problem associated with the standard pitot tube is the tendency for the small holes to plug in flues with heavy dust loadings.

THE STAUSCHEIBE PITOT TUBE

To eliminate such problems, stack samplers have used the "Stauscheibe" or Type S pitot tube shown in Fig. 11. Because the reverse-impact method is used to read the static pressure, this probe will produce a new velocity head reading h = $KV^2/2g$ where K will be greater than 1.0. A velocity correction factor C_v, where $V = C_v\sqrt{2gh}$, is 0.855. However C_v is not constant, but it is a function of velocity. Cole[18] recommends C_v = 0.870 at 10 ft/sec and 0.885 at 3 ft/sec. C-E has found that C_v is 0.855 only for velocities of 20 ft/sec and higher. Cole shows that there are many variations of C_v with probe manufacture; thus, each Type S probe should be calibrated in a known flow stream and a curve of C_v vs velocity prepared.

THE FECHHEIMER PROBE

A directional pitot tube, see Fig. 12, has been developed using the work of Fechheimer.[19] This probe is basically a reverse impact probe, but makes use of properties of the flow field around a cylinder. Two additional static, or Fechheimer, taps are located on the "forward" face of the cylinder. Two sets of differential readings are required. The probe is rotated in the gas stream until the velocity taps read positive and the two additional static taps are nulled. The probe is then "looking" directly at the flow-velocity vector and the angle can be recorded.

The Fechheimer probe is very sensitive to the problem of flow moving along the length of the stem when the stem is not exactly perpendicular to the flow field. Folsom evaluates this condition, called "stem effects," in reference 19. The probe used in C-E's laboratory places the sensing cylinder at the end of the airfoil stem which will tend to bleed off stem flow before it reaches the pressure sensing holes. This mod-

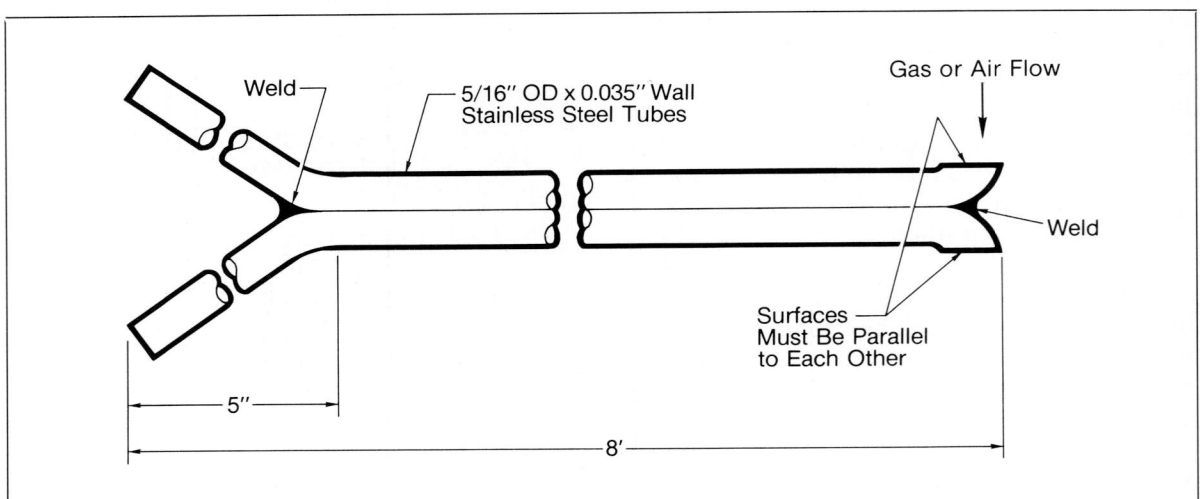

Fig. 11 **The Stauscheibe or Type S pitot tube**

ification minimizes, but does not eliminate, the induced stem errors. The velocity correction factor, C_v, for this probe is not as dependent on small differences in manufacture as is the Type S probe. In this case, the value of C_v is typically 0.78 for probe Reynolds numbers in excess of 100.

THE FIVE-HOLE PROBE

Another type of directional pitot tube is the five-hole or three dimensional probe; see Fig. 13. This probe is similar to the Fechheimer probe; however, it has an additional two pressure-sensing taps which allow measurement of the pitch angle of the velocity vector. As with the Fechheimer probe, the static pressure taps are nulled so that the probe is "looking" directly at the flow velocity vector. The pitch angle is determined from a calibration curve which relates pitch angle to a pitch pressure coefficient, C_v. This coefficient is the ratio of the pitch tap differential pressure over the flow-velocity pressure.

$$C_\phi = \frac{P_4 - P_5}{P_1 - P_2}$$

Each probe head has its own calibration curve which is a function of Reynolds number.

The use of a five-hole probe is most important where there is significant directionality of the gas flow. That is, the flow-velocity vector is not parallel to the duct axis. Most pitot probes are insensitive to flow directionality less than 15 percent. The Fechheimer probe measures directionality only in the plane normal to the probe; a five-hole probe measures directionality in planes normal and parallel to the probe.

If the flow vector at a point is known, then the component of flow normal to the plane of tra-

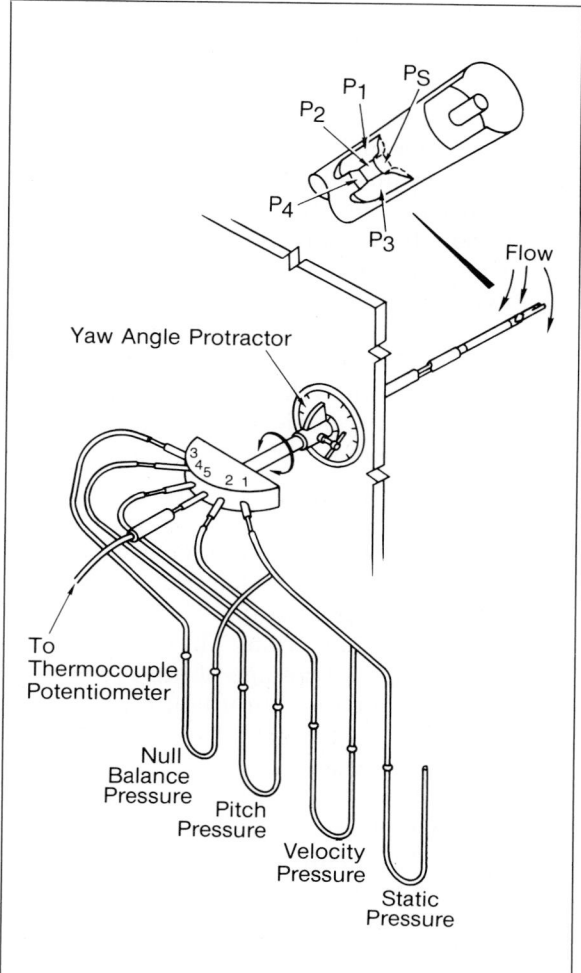

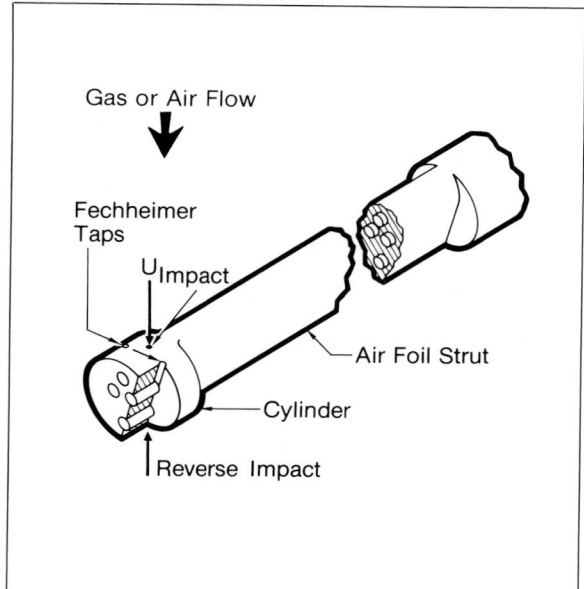

Fig. 12 **A Fechheimer directional pitot tube**

Fig. 13 **Five-hole probe**

verse can be calculated. The duct gas flow can then be expressed as:

$$Q_n = A \frac{1}{n} \sum_{i=1}^{n} V_i (\cos \phi)(\cos \psi)$$

where

Q_n = gas flow rate normal to the plane of travers

A = duct cross-sectional area

V = gas velocity at one point

ϕ = yaw angle of flow-velocity vector

ψ = pitch angle of flow-velocity vector

n = total number of traverse points

i = any one point

As shown by this equation, pitot probes which do not account for flow directionality can indicate gas flow rates which are higher than are actually present. Selection of a probe should be based upon the directionality of the flow stream at the measurement plane. Typically, in most large boiler installations, the stack is the only location where the directionality of the gas flow in both planes is less than 15 percent. However, if a duct has less than this amount of flow directionality along the pitch plane, the Fechheimer probe can be used.

THE HOT-WIRE ANEMOMETER

A device now often used for measurement of industrial fluid velocities is the hot-wire or hot-film anemometer. A single hot-wire system is described in PTC 19.5. The hot-wire anemometer has the advantage of high sensitivity at very low velocities, producing an electric readout and a measurement of turbulence. A disadvantage is its incapability of measuring directional flow in complex flow fields. It is also fragile and can easily be broken in flues with heavy dust loading. This last problem can be overcome by the use of a hot-film anemometer. The hot-film anemometer operates with the characteristics of the hot-wire, but sacrifices the ability to measure very high turbulent frequencies. Since high-frequency response is

typically desirable only in aerodynamic wind-tunnel studies, it is rarely needed for industrial flow evaluation.

INSTRUMENT CALIBRATION

All devices should be calibrated to insure a reasonable level of accuracy. The standard pitot tube will deviate from the theoretical Bernoulli response of $C_v = 1.0$ as a function of wear. The Stauscheibe and Fechheimer probes produce values of C_v which will also vary as a function of velocity and wear. The hot-wire and hot-film response is typically some function of V^n where n is less than 1.0 and will vary with dirtiness. Also, each device is dependent on gas composition. Using these dynamic devices requires that the gas density be known and the hot-wire requires the further evaluation of gas thermal properties. Fortunately, the probes can be calibrated in air and analytically corrected for gas properties.[20]

PROBE DYNAMICS AND TURBULENCE

To simplify energy calculations, flow turbulence is usually considered to be a random, but isotropic process. Typically, flow in industrial flues produces two distinct turbulence characteristics described as rolling and vortex flow.

ROLLING FLOW

Many installations include sharp corners and other abrupt discontinuities which will produce the turbulent roll. This roll is characterized by fluctuating velocity components μ and v usually coplanar with the major flow vector. A third component w is perpendicular to the main-flow vector and is usually of smaller magnitude. Isotropic turbulence, which is defined as $\mu = v = w$, is difficult to find in industrial work.

VORTEX FLOW

Although vortex flow can be produced by stack entrances and fan discharges, it can also be produced by two rolling turbulence patterns intersecting at an angle. It is difficult to characterize vortex flow in linear-velocity vector

notations. Typically, all turbulence is characterized when possible by the combined values of intensity (μ/U), frequency, and scale or size of the major eddies.

EFFECTS ON VELOCITY MEASUREMENTS

Turbulence, when applied to a dynamic device such as the standard pitot tube, will introduce a series of basic errors in the interpreted reading. The first is mathematical in that the reading produced by the pitot tube is a "head" of $V^2/2g$, while the result required is the velocity or $V^{1.0}$. Fig. 14 shows that, when measuring a turbulent or fluctuating velocity, the head measured will be the root mean square (RMS) value of the wave form and will always be higher than the head produced by the "average" velocity. Second, this problem is complicated by the flow dynamics of the resistance of the small-size pressure taps and tubing, the compressibility of air or gas in the tubes (where applicable), and the inertia of the indicating fluid. These effects are additive such that, in turbulent flow fields, when using fluid dynamic devices, the velocity as read will always be higher than the actual velocity being measured.

VELOCITY READOUTS

The problem of flow resistance can be partially corrected by using large sensing holes in

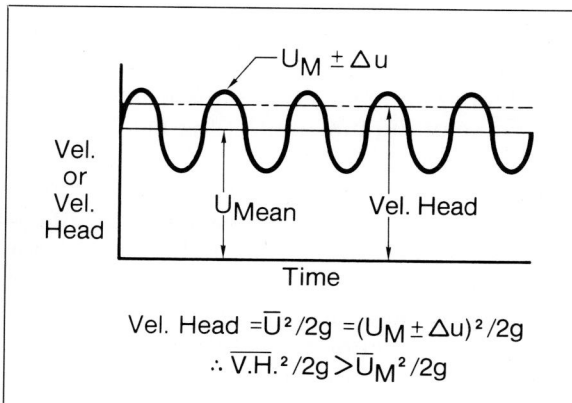

Vel. Head $= \overline{U}^2/2g = (U_M \pm \Delta u)^2/2g$

$\therefore \overline{V.H.}^2/2g > \overline{U}_M^2/2g$

Fig. 14 **Pulsing-flow wave forms obtained in measurements of fluctuating velocities**

the probe, close coupling of the readout device to the probe, and a readout device of low displacement volume and inertia. The readout device should also be capable of producing an electrical output signal. The signal can be recorded, for later analysis of the mathematical RMS effect on the velocity reading.

An approach to the mathematical evaluation of turbulent effects on velocity readings is given in the *ASME Performance Test Code* 19.5 Paragraph 57, which specifies isotropic turbulence and the RMS approach. It is relatively easy to apply this guide if the signal is sinusoidal. However, if the signal is not sinusoidal, turbulence effect is difficult to evaluate. This problem can be resolved by using a true RMS voltmeter to produce a true RMS reading independent of the applied wave form.

The effect of flow-stream vorticity has just the reverse effect on the velocity readout. Vorticity represents a well-ordered flow field of significant curvature. Flow curvature, in turn, produces a very definite radial pressure gradient. If the radius of curvature of the flow is of the same order of size as the measuring device, the device will not be measuring either the correct dynamic or static pressures. Vorticity of small size and high intensity can, therefore, contain a significant amount of dynamic or pressure energy which will not be read if the measuring device is physically larger than the size of the turbulent vortices being produced.

Unfortunately, there is no easy way to evaluate the effect of vorticity on measurements of dynamic flow. Therefore, vorticity should be minimized or eliminated by egg crates or screens if no other flow-traversing station without vorticity is available.

Recommendations on reducing the major sources of error in obtaining total gas-flow rates, based on the above, are given in "Quantitation of Stack Gas Flow" by C. L. Burton.[21]

DRAFTS AND PRESSURE LOSSES (AIR AND GAS)

Draft and pressure losses in steam generators and gas-treatment equipment are recorded in routine and special test operations because they indicate the system resistance which fans must overcome.

Test measurement is usually of draft differential rather than draft alone. Aside from convenience in direct reading, a restriction must be observed in that draft measurement is with reference to some atmospheric pressure. Since atmospheric pressure changes with elevation (approximately one-inch water gage per 100 ft), all draft gages must be at the same elevation or readings corrected for elevation; otherwise, an error of as much as 10 percent can be introduced.

Measurement is with inclined and U-tube liquid manometers. Duct walls and casing connections are usually suitable; the maximum error due to impact would not exceed one velocity head which is less than 0.1-inch water gage in most cases. Since condensation and ash plugging in connecting lines are the most common source of error in draft measurement, the importance of adequate line pitch, drain connections and blowout fittings cannot be overemphasized.

STEAM-GENERATOR EFFICIENCY

The desire to achieve process completion with maximum economy is basic in any technical effort. One measure of performance in steam generators is overall or gross efficiency defined as the ratio of heat output (heat absorbed) to heat input.[22] A heat-balance diagram as shown in Fig. 15 indicates the corollary is efficiency = 100 − percent losses.

The *ASME Performance Test Code* for Steam Generating Units allows efficiency tests that use *only* the chemical heat in the fuel as input. It does not, however, give an explicit definition of boiler efficiency on such a chemical-heat basis where heat credits are to be taken into account. In such practice, the heat diagram and efficiency become as in Fig. 16.

Present day units are large, and the water, steam and fuel quantities associated with them are so great that their measurement with high accuracy is extremely difficult. Weighed water and fuel tests are seldom feasible on these units, and the results with operating instrumentation are subject to commercial meter accuracies. Hence, the input-output method is often unsuitable for an accurate test of very large stationary boilers, because it requires accurate measurement of the quantity and heating value of the fuel, and of the heat absorbed by the steam generator.

Besides costing less, efficiency measurement by the heat-loss method has several advantages over the input-output method. It is more informative since it establishes the individual losses for comparison with expected performance.

Regarding accuracy, the total *losses* are only 10 to 20 percent and fuel sampling and analysis errors affect the end results slightly, whereas in the input-output method these errors are at least four or five times as significant. Also,

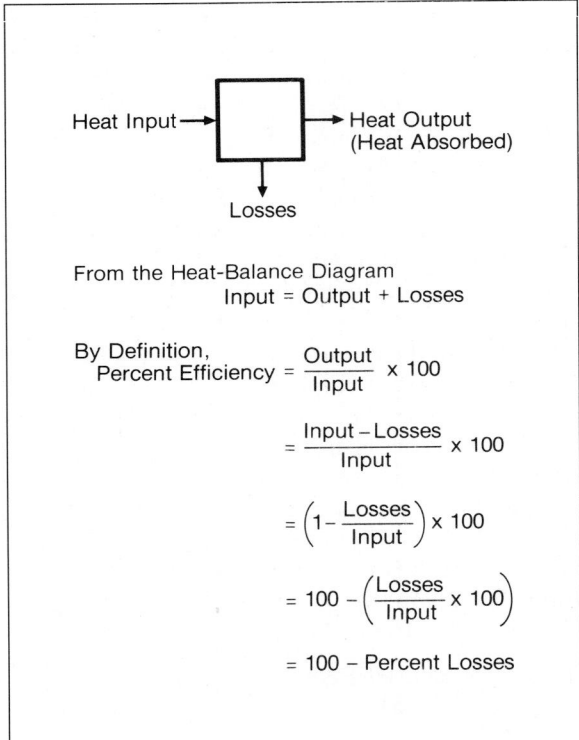

From the Heat-Balance Diagram
$$\text{Input} = \text{Output} + \text{Losses}$$

By Definition,
$$\text{Percent Efficiency} = \frac{\text{Output}}{\text{Input}} \times 100$$

$$= \frac{\text{Input} - \text{Losses}}{\text{Input}} \times 100$$

$$= \left(1 - \frac{\text{Losses}}{\text{Input}}\right) \times 100$$

$$= 100 - \left(\frac{\text{Losses}}{\text{Input}} \times 100\right)$$

$$= 100 - \text{Percent Losses}$$

Fig. 15 Basic heat-balance diagram and efficiency relationships

the basic losses measurements are simple: fuel analysis, exit gas temperature, entering air temperature, and refuse combustible content. Minor items in the heat losses can be obtained from sources such as *ASME Performance Test Code* PTC 4 which contains detailed instructions on test procedures and results calculations.

In order to achieve the objectives of a boiler test in accordance with PTC 4, agreement must be reached, before the test, on the definition of efficiency, the general method (heat loss or input-output), the heat credits and losses to be measured, and the heat credits and losses to be assigned where not measured. In addition, allocation of responsibility for all performance and operating conditions affecting the test must be established, as well as acceptable operating conditions, number of load points, the direction of test runs, and any basis for rejection of runs.

Before a test is started, it must be determined whether the fuel to be fired is substantially as intended. A reliable, accurate efficiency test for purposes of equipment acceptance depends upon the fuel being in close agreement with that for which the steam-generating unit was designed. Significant deviations in fuel constituents and heating value can result in appreciable inaccuracies in heat-loss calculations and resulting efficiencies. Although the magnitude of deviation that is tolerable is difficult to establish, it should be recognized that fuel-analysis variation producing changes in heating value on the order of 10 percent can alter final calculated efficiency about 1 percent.

The Test Code does not include consideration of overall tolerances or margins on performance guarantees; test results are to be reported as computed from test observations, with proper corrections for calibrations. Allowances for errors of measurement and sampling are permissible provided they are agreed upon in advance and are clearly stated in the test report. The limits of probable error on calculated steam generator efficiency are to be taken as the square root of the sum of the squares of the individual effects on efficiency.

Whenever allowances for probable errors of measurement and sampling are to be taken into consideration, the reported test results shall be qualified by the statement that the error in the results may be considered not to exceed a given plus or minus percentage. Table III taken from PTC 4 is included as a guide to show the effect on efficiency of measurement errors (exclusive of sampling errors). The measurement-error range in the table is not intended to be authoritative but conforms approximately with experience. The values in the table are not intended for calculation of test results.

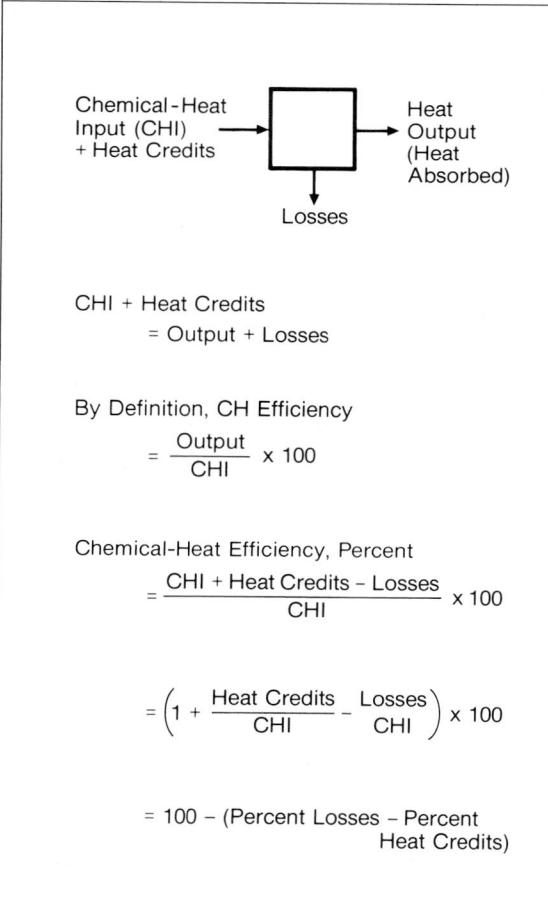

Fig. 16 Chemical-heat input (CHI) is the higher heating value of the fuel fired in the boiler above a stated reference temperature.

Table III. Probable Measurement Errors and Resulting Errors in Efficiency Calculations*

Measurement	Measurement error, percent	Error in calculated Steam Generator Efficiency, percent
Input-Output Method		
(1) Weigh tanks (calibrated scales)	±0.10	±0.10
(2) Volumetric tanks (calibrated)	±0.25	±0.25
(3) Calibrated flow nozzle or orifice including manometer	±0.35	±0.35
(4) Calibrated flow nozzle or orifice including recorder	±0.55	±0.55
(5) Coal scales—batch or dump (calibrated)	±0.25	±0.25
(6) Uncalibrated flow nozzle or orifice including manometer	±1.25	±1.25
(7) Uncalibrated flow nozzle or orifice including recorder	±1.60	±1.60
(8) Fuel heating value (coal)	±0.50	±0.50
(gas and oil)	±0.35	±0.35
(9) Reheat flow (based on heat balance calculations)	±0.60	±0.10
(10) Superheater outlet temperature (calibrated measuring device)	±0.25	±0.15
(11) Superheater outlet pressure (calibrated measuring device)	±1.00	±0.00
(12) Reheater inlet and outlet temperature (calibrated measuring device)	±0.25	±0.10
(13) Reheater inlet and outlet pressure (calibrated measuring device)	±0.50	±0.00
(14) Feedwater temperature (calibrated measuring device)	±0.25	±0.10
Heat-Loss Method		
(1) Heating value (coal)	±0.50	±0.03
(gas and oil)	±0.35	±0.02
(2) Orsat analysis	±3.00	±0.30
(3) Exit gas temperature (calibrated measuring device)	±0.50	±0.02
(4) Inlet air temperature (calibrated measuring device)	±0.50	±0.00
(5) Ultimate analysis of coal (carbon)	±1.00	±0.10
(hydrogen)	±1.00	±0.10
(6) Fuel moisture	±1.00	±0.00

*Taken from ASME PTC 4.

22–26

SUMMARY SHEET
ASME TEST FORM
FOR ABBREVIATED EFFICIENCY TEST

PTC 4.1-a (1964)

OWNER OF PLANT		TEST NO.	BOILER NO.	DATE
TEST CONDUCTED BY		LOCATION		
BOILER MAKE & TYPE		OBJECTIVE OF TEST		DURATION
STOKER TYPE & SIZE		RATED CAPACITY		
PULVERIZER, TYPE & SIZE		BURNER, TYPE & SIZE		
FUEL USED	MINE	COUNTY	STATE	SIZE AS FIRED

PRESSURES & TEMPERATURES

1	STEAM PRESSURE IN BOILER DRUM	psia	
2	STEAM PRESSURE AT S. H. OUTLET	psia	
3	STEAM PRESSURE AT R. H. INLET	psia	
4	STEAM PRESSURE AT R. H. OUTLET	psia	
5	STEAM TEMPERATURE AT S. H. OUTLET	F	
6	STEAM TEMPERATURE AT R H INLET	F	
7	STEAM TEMPERATURE AT R. H. OUTLET	F	
8	WATER TEMP. ENTERING (ECON) (BOILER)	F	
9	STEAM QUALITY % MOISTURE OR P. P. M.		
10	AIR TEMP. AROUND BOILER (AMBIENT)	F	
11	TEMP AIR FOR COMBUSTION (This is Reference Temperature) †	F	
12	TEMPERATURE OF FUEL	F	
13	GAS TEMP. LEAVING (Boiler) (Econ.) (Air Htr.)	F	
14	GAS TEMP. ENTERING AH (If conditions to be corrected to guarantee)	F	

UNIT QUANTITIES

15	ENTHALPY OF SAT. LIQUID (TOTAL HEAT)	Btu/lb	
16	ENTHALPY OF (SATURATED) (SUPERHEATED) STM.	Btu/lb	
17	ENTHALPY OF SAT. FEED TO (BOILER) (ECON.)	Btu/lb	
18	ENTHALPY OF REHEATED STEAM R.H. INLET	Btu/lb	
19	ENTHALPY OF REHEATED STEAM R. H. OUTLET	Btu/lb	
20	HEAT ABS. LB OF STEAM (ITEM 16 – ITEM 17)	Btu/lb	
21	HEAT ABS. LB R. H. STEAM (ITEM 19 – ITEM 18)	Btu/lb	
22	DRY REFUSE (ASH PIT + FLY ASH) PER LB AS FIRED FUEL	lb/lb	
23	Btu PER LB IN REFUSE (WEIGHTED AVERAGE)	Btu/lb	
24	CARBON BURNED PER LB AS FIRED FUEL	lb/lb	
25	DRY GAS PER LB AS FIRED FUEL BURNED	lb/lb	

HOURLY QUANTITIES

26	ACTUAL WATER EVAPORATED	lb/hr	
27	REHEAT STEAM FLOW	lb/hr	
28	RATE OF FUEL FIRING (AS FIRED wt)	lb/hr	
29	TOTAL HEAT INPUT $\frac{(\text{Item 28} \times \text{Item 41})}{1000}$	kB/hr	
30	HEAT OUTPUT IN BLOW-DOWN WATER	kB/hr	
31	TOTAL HEAT OUTPUT $\frac{(\text{Item 26} \times \text{Item 20}) + (\text{Item 27} \times \text{Item 21}) + \text{Item 30}}{1000}$	kB/hr	

FLUE GAS ANAL. (BOILER) (ECON) (AIR HTR) OUTLET

32	CO₂	% VOL	
33	O₂	% VOL	
34	CO	% VOL	
35	N₂ (BY DIFFERENCE)	% VOL	
36	EXCESS AIR	%	

FUEL DATA

	COAL AS FIRED PROX. ANALYSIS	% wt	
37	MOISTURE		
38	VOL MATTER		
39	FIXED CARBON		
40	ASH		
	TOTAL		
41	Btu per lb AS FIRED		
42	ASH SOFT TEMP.* ASTM METHOD		

	COAL OR OIL AS FIRED ULTIMATE ANALYSIS		
43	CARBON		
44	HYDROGEN		
45	OXYGEN		
46	NITROGEN		
47	SULPHUR		
40	ASH		
37	MOISTURE		
	TOTAL		

COAL PULVERIZATION

48	GRINDABILITY INDEX*	
49	FINENESS % THRU 50 M*	
50	FINENESS % THRU 200 M*	

OIL

51	FLASH POINT F*	
52	Sp. Gravity Deg. API*	
53	VISCOSITY AT SSU* BURNER SSF	
44	TOTAL HYDROGEN % wt	
41	Btu per lb	

GAS

		% VOL
54	CO	
55	CH₄ METHANE	
56	C₂H₂ ACETYLENE	
57	C₂H₄ ETHYLENE	
58	C₂H₆ ETHANE	
59	H₂ S	
60	CO₂	
61	H₂ HYDROGEN	
	TOTAL	
62	TOTAL HYDROGEN % wt	
	DENSITY 68 F ATM. PRESS.	
63	Btu PER CU FT	
41	Btu PER LB	

64	INPUT-OUTPUT EFFICIENCY OF UNIT %	$\frac{\text{ITEM 31} \times 100}{\text{ITEM 29}}$

HEAT LOSS EFFICIENCY

		Btu/lb A. F. FUEL	% of A. F. FUEL
65	HEAT LOSS DUE TO DRY GAS		
66	HEAT LOSS DUE TO MOISTURE IN FUEL		
67	HEAT LOSS DUE TO H₂O FROM COMB. OF H₂		
68	HEAT LOSS DUE TO COMBUST. IN REFUSE		
69	HEAT LOSS DUE TO RADIATION		
70	UNMEASURED LOSSES		
71	**TOTAL**		
72	EFFICIENCY = (100 – Item 71)		

* Not Required for Efficiency Testing

† For Point of Measurement See Par. 7.2.8.1-PTC 4.1-1964

Printed in U.S.A. (10/74)

This Test Form (C-36) may be obtained from ASME, 345 E. 47 St., New York, N.Y. 10017

ASME Performance Test Codes: Since its first publication in 1915, *ASME Performance Test Code* 4.1, Steam Generating Units, has been a recognized guide for procedures of boiler testing. The boiler boundary diagram as well as the forms on the following pages are reproduced from the 1964 edition of PTC 4.1.

PTC 4.1-b (1964)

CALCULATION SHEET   **ASME TEST FORM**
FOR ABBREVIATED EFFICIENCY TEST   *Revised September, 1965*

	OWNER OF PLANT	TEST NO.	BOILER NO.	DATE

30 HEAT OUTPUT IN BOILER BLOW-DOWN WATER = LB OF WATER BLOW-DOWN PER HR × $\dfrac{\text{ITEM 15} \;-\; \text{ITEM 17}}{1000}$ = ... kB/hr

24
If impractical to weigh refuse, this item can be estimated as follows

DRY REFUSE PER LB OF AS FIRED FUEL = $\dfrac{\% \text{ ASH IN AS FIRED COAL}}{100 - \% \text{ COMB. IN REFUSE SAMPLE}}$

NOTE: IF FLUE DUST & ASH PIT REFUSE DIFFER MATERIALLY IN COMBUSTIBLE CONTENT, THEY SHOULD BE ESTIMATED SEPARATELY. SEE SECTION 7, COMPUTATIONS.

CARBON BURNED PER LB AS FIRED FUEL = $\dfrac{\text{ITEM 43}}{100} - \left[\dfrac{\text{ITEM 22} \times \text{ITEM 23}}{14,500}\right]$ = ...

25 DRY GAS PER LB AS FIRED FUEL BURNED = $\dfrac{11CO_2 + 8O_2 + 7(N_2 + CO)}{3(CO_2 + CO)}$ × (LB CARBON BURNED PER LB AS FIRED FUEL $+ \frac{3}{8}$ S)

$= \dfrac{11 \times \text{ITEM 32} + 8 \times \text{ITEM 33} + 7\left(\text{ITEM 35} + \text{ITEM 34}\right)}{3 \times \left(\text{ITEM 32} + \text{ITEM 34}\right)} \times \left[\text{ITEM 24} + \dfrac{\text{ITEM 47}}{267}\right]$ = ...

36 EXCESS AIR† $= 100 \times \dfrac{O_2 - \dfrac{CO}{2}}{.2682 N_2 - (O_2 - \frac{CO}{2})} = 100 \times \dfrac{\text{ITEM 33} - \dfrac{\text{ITEM 34}}{2}}{.2682(\text{ITEM 35}) - (\text{ITEM 33} - \frac{\text{ITEM 34}}{2})}$ = ...

	HEAT LOSS EFFICIENCY	Btu/lb AS FIRED FUEL	$\dfrac{\text{LOSS}}{\text{HHV}} \times 100 =$	LOSS %
65	HEAT LOSS DUE TO DRY GAS = $\dfrac{\text{LB DRY GAS}}{\text{PER LB AS FIRED FUEL}} \times C_p \times (t_{lvg} - t_{air}) = \dfrac{\text{ITEM 25}}{\text{Unit}} \times 0.24 \; (\text{ITEM 13}) - (\text{ITEM 11})$ = ...	......	$\dfrac{65}{41} \times 100 =$	......
66	HEAT LOSS DUE TO MOISTURE IN FUEL = $\dfrac{\text{LB } H_2O \text{ PER LB}}{\text{AS FIRED FUEL}} \times$ [(ENTHALPY OF VAPOR AT 1 PSIA & T GAS LVG) − (ENTHALPY OF LIQUID AT T AIR)] = $\dfrac{\text{ITEM 37}}{100} \times$ [(ENTHALPY OF VAPOR AT 1 PSIA & T ITEM 13) − (ENTHALPY OF LIQUID AT T ITEM 11)] = ...	......	$\dfrac{66}{41} \times 100 =$	......
67	HEAT LOSS DUE TO H_2O FROM COMB. OF H_2 = $9H_2 \times$ [(ENTHALPY OF VAPOR AT 1 PSIA & T GAS LVG) − (ENTHALPY OF LIQUID AT T AIR)] $= 9 \times \dfrac{\text{ITEM 44}}{100} \times$ [(ENTHALPY OF VAPOR AT 1 PSIA & T ITEM 13) − (ENTHALPY OF LIQUID AT T ITEM 11)] = ...	......	$\dfrac{67}{41} \times 100 =$	......
68	HEAT LOSS DUE TO COMBUSTIBLE IN REFUSE = ITEM 22 × ITEM 23 = ...	......	$\dfrac{68}{41} \times 100 =$	......
69	HEAT LOSS DUE TO RADIATION* = $\dfrac{\text{TOTAL BTU RADIATION LOSS PER HR}}{\text{LB AS FIRED FUEL} - \text{ITEM 28}}$ = ...	......	$\dfrac{69}{41} \times 100 =$	......
70	UNMEASURED LOSSES **	......	$\dfrac{70}{41} \times 100 =$	......
71	TOTAL	......	......	......
72	EFFICIENCY = (100 − ITEM 71)	......		......

† For rigorous determination of excess air see Appendix 9.2 — PTC 4.1-1964
* If losses are not measured, use ABMA Standard Radiation Loss Chart, Fig. 8, PTC 4.1-1964
** Unmeasured losses listed in PTC 4.1 but not tabulated above may be provided for by assigning a mutually agreed upon value for Item 70.

Printed in U.S.A. (10/74)

This Test Form (C-37) may be obtained from ASME, 345 E. 47 St., New York, N.Y. 10017

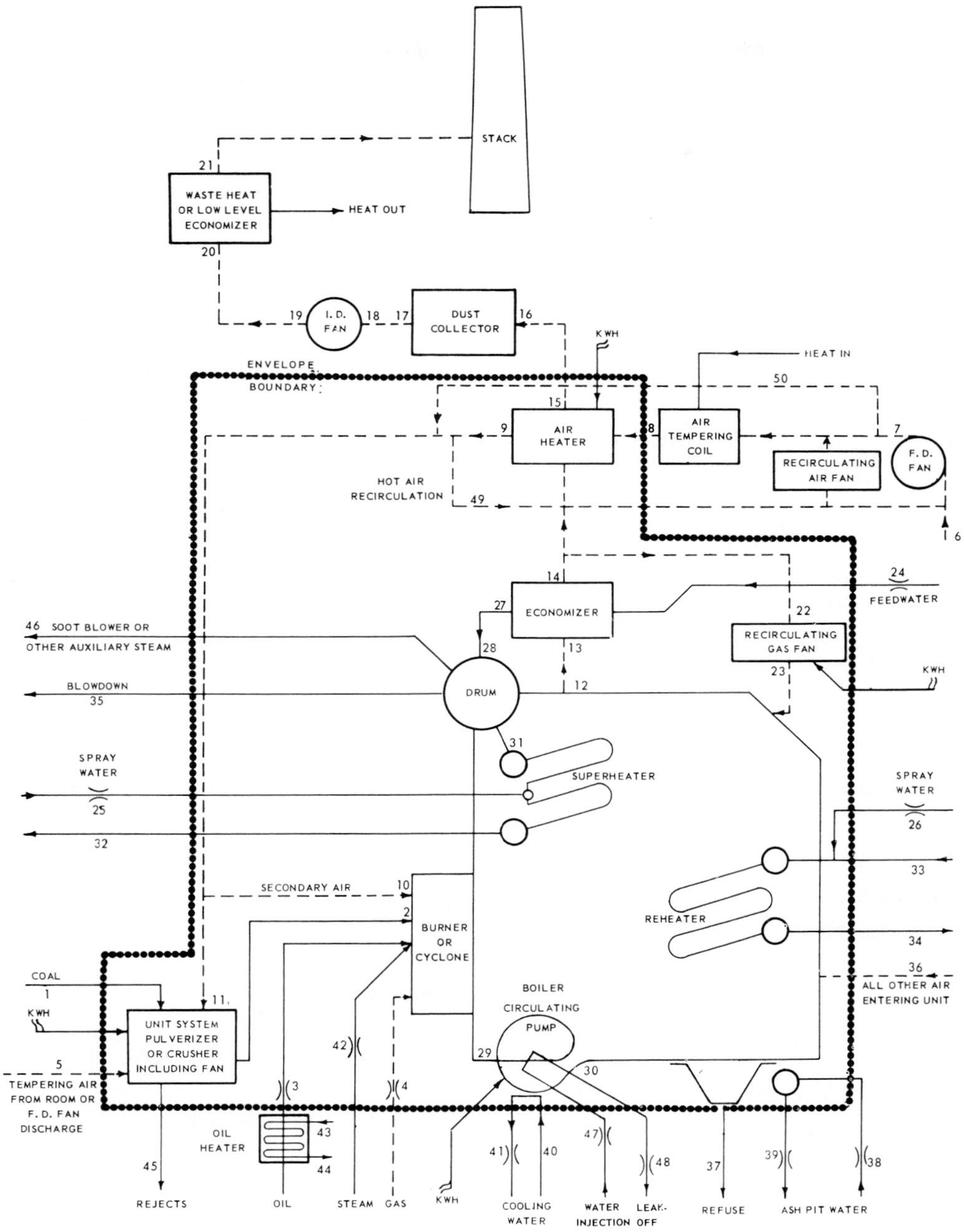

STACK

21

WASTE HEAT
OR LOW LEVEL
ECONOMIZER → HEAT OUT

20

19 I.D. 18 17 DUST 16
FAN COLLECTOR

KWH

ENVELOPE
BOUNDARY

HEAT IN

50

15

9 AIR 8 AIR 7
HEATER TEMPERING
COIL F.D.
FAN

RECIRCULATING
AIR FAN

HOT AIR
RECIRCULATION 49

6

14

27 ECONOMIZER 24
FEEDWATER

22

46 SOOT BLOWER OR
OTHER AUXILIARY STEAM

RECIRCULATING
GAS FAN

28 13 23

KWH

BLOWDOWN DRUM 12
35

31

SPRAY
WATER SUPERHEATER SPRAY
WATER
25 26

32

33

SECONDARY AIR 10

2 REHEATER

BURNER
OR 34
CYCLONE

COAL
1 36
KWH 11 ALL OTHER AIR
ENTERING UNIT

UNIT SYSTEM BOILER
PULVERIZER CIRCULATING
OR CRUSHER PUMP
INCLUDING FAN 42)(

5 29 30

TEMPERING AIR
FROM ROOM OR
F.D. FAN
DISCHARGE)(3)(4

OIL 43
45 HEATER
44

REJECTS OIL STEAM GAS KWH 41)(40 47)( )(48 37 39)( )(38

COOLING WATER LEAK- REFUSE ASH PIT WATER
WATER INJECTION OFF

FIG 1 STEAM GENERATING UNIT DIAGRAM

This diagram for boiler tests serves as a key to the numerical subscripts used throughout *ASME Performance Test Code* **4.1, Steam Generating Units**

REFERENCES

[1] *ASME Performance Test Codes:* PTC 1, "General Instructions," PTC 2, "Code on Definitions and Values"; PTC 3.1, "Diesel and Burner Fuels"; PTC 3.2, "Solid Fuels"; PTC 3.3, "Gaseous Fuels"; PTC 4.2, "Coal Pulverizers"; PTC 4.3, "Air Heaters"; PTC 11, "FANS"; PTC 21, "Dust Separating Apparatus"; PTC 27, "Determining Dust Concentration in a Gas Stream," New York: American Society of Mechanical Engineers, latest edition.

[2] Code of Federal Regulations, Title 40, Parts 53–80, latest edition.

[3] S. J. Kline and F. A. McClintock, "Describing Uncertainties in Single Sample Experiments," *Mechanical Engineering*, 75(1): 3–8, Jan. 1953.

[4] W. A. Wilson, "Design of Power Plant Tests to Insure Reliability of Results," *Transactions of the ASME*, 75: 405–408, May 1955.

L. W. Thrasher and R. C. Binder, "A Practical Application of Uncertainty Calculations to Measured Data," *Transactions of the ASME*, 79: 373–376, Feb. 1957.

S. Baron, "The Effect of Measurement Errors on Plant Performance Tests," *Combustion*, 25(8): 49–54, Feb. 1954.

[5] J. H. Born, Jr., "The Effect of Measurement Errors in the Accuracy of Steam Generator Efficiency Calculators," *ASME Paper 60-WA-301*. New York: American Society of Mechanical Engineers, 1960.

[6] "Orifice Metering of Natural Gas" *ANSI/API 2530. AGA Report no. 3.* New York: American National Standards Institute, latest edition.

American Petroleum Institute, *Oil Pipeline Construction and Maintenance*, 2nd ed. Austin: University of Texas, 1973.

ASME Performance Test Codes, PTC 19.5, "Supplements on Instruments and Apparatus," New York: American Society of Mechanical Engineers, latest edition.

[7] *ASTM Standards*, Part 26, *Gaseous Fuels; Coal and Coke; Atmospheric Analysis.* D 2015, "Test for Gross Calorific Value of Solid Fuel by the Adiabatic Bomb Calorimeter"; D 2961, "Test for Total Moisture in Coal Reduced to No. 8 Top Sieve Size (Limited Purpose Method)"; D 3172, "Proximate Analysis of Coal and Coke"; D 3173, "Test for Moisture in the Analysis Sample of Coal and Coke"; D 3174 "Test for Ash in the Analysis Sample of Coal and Coke"; D 3177, "Test for Total Sulfur in the Analysis Sample of Coal and Coke"; D3178, "Test for Carbon and Hydrogen in the Analysis Sample of Coal and Coke"; D3180, "Calculating Coal and Coke Analyses from As-Determined to Different Bases"; D 3286, "Test for Gross Calorific Value of Solid Fuel by the Isothermal-Jacket Bomb Calorimeter"; D 3302, "Test for Total Moisture in Coal"; Philadelphia: American Society for Testing and Materials, latest edition.

[8] *ASME Performance Test Codes*, PTC 19.3, "Temperature Measurement," New York: American Society of Mechanical Engineers, latest edition.

[9] P. H. Dike, *Thermoelectric Thermometry*, 3rd ed. Philadelphia: Leeds and Northrup Co., 1958.

[10] Chromel and alumel are chromium-nickel and aluminum-nickel alloys respectively. Constantan is a copper alloy.

[11] *ASTM Standards*, Part 44, "Magnetic Properties; Metallic Materials for Thermostatic; Electrical Resistance, Heating, Contacts; Temperature Measurement; Illuminating Standards," Philadelphia: American Society for Testing and Materials, latest edition.

[12] American Society of Mechanical Engineers, *Fluid Meters, Their Theory and Application*, New York: American Society of Mechanical Engineers, latest edition.

[13] J. W. Murdock, "Power Test Code Thermometer Wells," *Transactions of the ASME. Journal of Engineering for Power*, 81, Series A: 403–416, 1959.

[14] P. G. W. Hawksley, et al., *Measurement of Solids in Flue Gases*, 2nd ed. London: The Institute of Fuel, 1977.

[15] Harry James White, *Industrial Electrostatic Precipitation*. Reading, Mass: Addison Wesley, 1963.

[16] W. B. Gregory, "The Pitot Tube," *Transactions of the ASME*, 25:184–211, 1904.

[17] W. C. Rowse, "Pitot Tubes for Gas Measurement," *Transactions of the ASME*, 35:633–703, 1913.

[18] E. S. Cole, "Pitot Tube Practice," *Transactions of the ASME*, 57:281–294, 1935.

[19] R. G. Folsom, "Review of the Pitot Tube," *Transactions of the ASME*, 78:1447–1460, Oct. 1956.

[20] C. J. Fechheimer, "Measurement of Static Pressure," *Transactions of the ASME*, 48: 965–977, 1926.

[21] C. L. Burton, "Quantitation of Stack Gas Flow," Reprint from *Journal of the Air Pollution Control Association*, 22(8): 631–635, Aug. 1972.

[22] *ASME Performance Test Codes*, PTC 4.1, "Steam Generating Units," New York: American Society of Mechanical Engineers, latest edition.

BIBLIOGRAPHY

Buna, Tibor, "Combustion Calculations for Multiple Fuels," *Transactions of the ASME*, 78: 1237–1249, Aug. 1956.

Bostic, J. A., and Long, W. F., "Code Testing of Large Boilers: Input-Output or Heat-Loss Method," *ASME Paper No. 60-PWR-5*. New York: American Society of Mechanical Engineers, 1960.

Cohen, L., and Fritz, W. A., Jr., "Efficiency Determination of Marine Boilers: Input-Output Versus Heat-Loss Method," *Transactions of the ASME. Journal of Engineering for Power*, 84, Series A: 39–43, Jan. 1962.

Vuia, R. E., "Performance Testing of Large Steam Generators," *ASME Paper No. 62-WA-267*. New York: American Society of Mechanical Engineers, 1962.

CHAPTER 23

Maintenance and Repair of Steam-Generating Equipment

A good maintenance program is one of the keys to reliability of any steam generator. To be successful, such a program requires managerial ability, expertise, imagination, planning, training, and the commitment of top management. In addition, maintenance must be closely integrated with both the operation and power-station engineering functions.

MAINTENANCE FORESIGHT

Because plant layout can either facilitate or impede a maintenance program, maintenance personnel should be a part of the initial design team to insure the plant layout facilitates maintenance. Ideally, the person who will be ultimately responsible for plant maintenance should be a member of the design team.

Only during the design stage can truly adequate space be allocated for maintenance. As equipment is located, maintenance personnel must have space to remove the largest replaceable part easily. The steam generator should have enough access doors so that all areas may be entered easily for inspection. Access doors and aisles also must be large enough to accommodate components that must be moved to a maintenance area. Sufficient headroom is a must. Overhead trolleys and cranes for maintenance should be designed *into* the plant rather than retrofitted. All high-maintenance items should have adequate laydown and work areas adjacent to the equipment. Models can help the designer plan for accessibility (Fig. 1).

Designers responsible for piping and electrical work must be aware of areas requiring accessibility for maintenance. Often piping and wiring are field-run from drawings having notes advising construction personnel of special maintenance requirements.

EQUIPMENT MANUALS: AN ASSET TO EFFECTIVE MAINTENANCE

It is important at the time equipment is purchased to specify operation and maintenance manuals which are complete, detailed, and specific to the piece of equipment supplied. Material that does not contribute to an understanding of the design, care, operation, and maintenance of the equipment should be excluded. Manuals should include

- step-by-step disassembly procedures
- step-by-step reassembly procedures
- preventive maintenance and lubrication instructions
- lists of all special tools required and instructions on their use

■ tabulated dimensional data on settings, clearances, and adjustments

DETERMINING THE NEED FOR SPARE PARTS

Often, a complement of spare parts is ordered with the initial purchase. If not, spares should be purchased before the equipment is placed in service. Items critical to plant production should receive special consideration and, for these items, it may be prudent to maintain spare subassemblies. At time of purchase, the owner should ask for complete instructions from the manufacturer for error-proof ordering of replacement parts. Catalogs specific to the equipment and with exploded view drawings help in selecting appropriate spare parts.

In addition, there should be a system for tracking spare parts. Many large plants maintain a replacement-parts inventory on a computer. If establishing an in-plant program is not feasible, equipment manufacturers are prepared to rent computer services for tracking spare parts supplied by them.

Manufacturers can provide guidelines for ordering spare parts based on field experience. Even so, one must realize that all equipment of a given type does not wear at the same rate, because different local conditions can alter wearing life drastically.

As orders arrive, responsible personnel should inspect maintenance material carefully. A defective part may sit in a storeroom for years before the defect is discovered—just at the time of need. Also, to assure correctness and completeness, all the incoming parts should be checked against purchase orders before the parts are stored.

IN-SERVICE MAINTENANCE

To insure that a unit is available on demand requires a constant effort by the maintenance staff. They must implement a day-to-day preventive program while the unit is in operation, establish a well-planned and properly executed inspection program for scheduled outages, and be prepared to handle all forced outages or equipment breakdowns as soon as they occur.

On-line maintenance functions are just as important as those performed during an outage. If properly undertaken, not only do they add to unit reliability, but they also save valuable downtime. On-line maintenance includes lubrication, seal and packing replacement, and repair of any component that has been removed and replaced with a spare.

Proper lubrication is critical for rotating auxiliaries. Usually, the most complete part of any correlated maintenance manual is its lubrication instructions. The manufacturer's recommendations for lubrication should be followed, with substitute oils or greases used only if they meet all specified properties. Each plant must develop and adhere to routine lubrication schedules. Major lubricant suppliers will work with a plant staff to develop a program which allows the most flexibility. Such a program would minimize the types of necessary lubricants, the amount of each kept in the plant, and the best reordering schedules. This lubrication is such an important part of any maintenance program that records of work that has been done should be kept by using checkoff sheets.

Good communications between the opera-

Fig. 1 **Models help lay out a pulverizer lane for optimum maintenance access**

tions and maintenance groups is essential. The day-to-day in-service maintenance program must be closely coordinated with plant operators, who are to inform the maintenance crew of problems and potential problems. For example, maintenance should be informed of slight drops in pressure or temperature which could indicate inefficient operation or deterioration of machinery. In addition to such visual signals, operators inspecting equipment often detect problems by unusual sounds, smells, or tactile sensations.

Similarly, the maintenance crew must inform operators of any equipment being repaired. Often in the case of several sootblowers needing repair, the operator decides which blower is crucial to operation and requests maintenance to give that blower priority.

Fig. 2 **A plant should have three complete journal assemblies stored on transport dollies**

PULVERIZER MAINTENANCE

Although it is beyond the scope of one chapter to describe the maintenance of the various equipment that can be overhauled while a generating unit continues in service, a major component usually overhauled on-line is a pulverizer. The life of mill parts depends on many factors which vary considerably from one plant to another. The main factors are the abrasive characteristics of the coal, operating hours of individual mills, and mill loading as a percent of maximum capacity. Although plants usually install spare pulverizers, two or more of which may be idle at reduced boiler loads, pulverizer outage time for maintenance should be kept to a minimum so that all pulverizers are always ready for service.

Because an effective pulverizer maintenance program must be adjusted on the basis of actual plant operating experience, it is essential to maintain an accurate log of each pulverizer's running time, stoppages for removal of foreign material, maintenance work and inspections. Maintenance must be staggered so that all pulverizers do not require work at the same time.

Rolls and grinding rings are the major replaceable items. (Later in this chapter, there is a further discussion of roll and ring wear.) With good planning and preparation, it is possible to overhaul a pulverizer, renewing the ring and rolls as well as repairing or replacing other wearing parts, in a few days.

As part of its spare-parts complement, a plant should have three complete journal assemblies (Fig. 2). Frequently, such assemblies are interchangeable on all mills in the station and can be stored on transport dollies in the mill overhaul area. Following a mill overhaul, worn journal assemblies that have been removed are rebuilt. The worn grinding rolls may be replaced, or may be resurfaced with weld material to recreate the original shape. A weld material such as COMBUSTALLOY™ can give four to five times the life of the original NI-HARD roll material.

To accommodate transport of worn journals and overhauled assemblies simultaneously, six transport and storage dollies should be on

hand. Storing segmented bowl-ring assemblies on pallets (one complete ring per pallet) permits hoisting to the mill platform in one lift. Frequently used bolts, nuts, and washers are kept in bins in the mill overhaul area.

Once a mill has cooled and been tagged out-of-service, access doors are removed. To further cool mill parts, an air mover often is used to ventilate the mill. When all three journal assemblies have been removed, there is free access to the inside of the mill, and a worn bull ring or worn liners can be removed rapidly. When all obviously worn parts have been stripped from the mill, the remaining parts are carefully inspected for repair or replacement.

When the repair work inside the mill has been completed, the overhauled journal assemblies are installed. All access doors are closed, the mill is cleared of all "hold" tags, and it is started without coal to facilitate adjustment of the grinding-roll-to-ring clearance. When this adjustment has been completed, the mill is released for normal service.

THE ANNUAL OUTAGE

For any power plant, the major part of its unified maintenance program is the scheduled annual maintenance outage. The objectives of the first-annual planned outage are to inspect equipment, recommend repairs, and establish a data base from which to develop a preventative maintenance program. Although subsequent annual outages will concentrate on repair work, a review of these three procedural items must be part of every outage.

Whether accomplished by plant personnel or a maintenance contractor, a well-planned outage program begins as much as a year before the scheduled downtime. The planning and the jobsite organization must include well-defined communication channels (Fig. 3). With an outside contractor, advance communication with the plant staff is crucial to the success of the operation. The plant staff must follow the contractor's work closely so that spare parts, engineering changes, and new material selections are expeditiously handled.

THE PRE-OUTAGE STAGE

Several months before a scheduled outage, the group responsible for the inspection and the work must jointly plan the inspection schedule with plant personnel. In preparation for the outage, personnel assemble tools, scaffolding, spare parts and other required materials. Also during this period, the engineers in charge of work during the outage examine existing unit operating and maintenance records, and discuss with all plant personnel those areas requiring particular attention.

PRE-OUTAGE INSPECTION ITEMS

Usually, while the unit and its auxiliaries are still operating, a complete visual inspection is

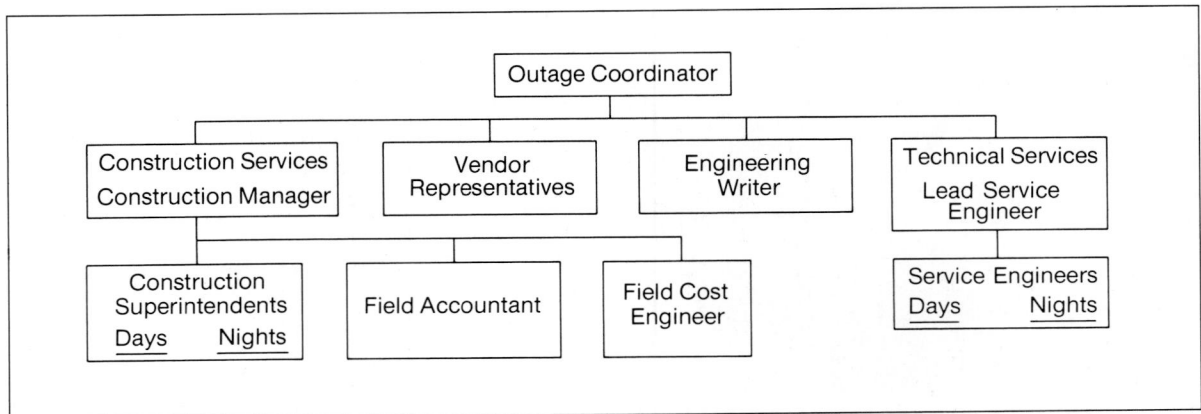

Fig. 3 **A successful outage must include a good organization with well-defined communication channels**

performed to detect any obvious items requiring maintenance or more detailed investigation during shutdown. All items are noted and a list compiled.

Operational checks of certain equipment and systems also are conducted to ensure each is functioning properly—or requires adjustment, repair or replacement. Included are

- fuel-burning equipment such as tilting mechanisms, windbox dampers, coal gates, and oil or gas valves

- furnace-safety-control and ignition systems

- combustion and feedwater-control systems

- sootblowing equipment

In addition, the settings on safety valves are checked. Control and isolation valve performance is verified and any packing leaks recorded. The condition of activators is checked to be sure oil seals are secure and lubricating oil is not contaminated with water or flyash.

All permanent thermocouples, such as those on superheater and reheater tubing and water-wall panel outlets, are checked to isolate any defects requiring repair during the outage.

Readings from expansion trams in important areas should be recorded for comparison with readings taken during the outage when the unit is cold, and after its return to service. These readings are used to confirm free and proper expansion of the unit.

SCAFFOLDING

For outages in which extensive inspection and repairs to the furnace are planned, fixed scaffolding temporarily installed both outside and inside the furnace can save time and money. In the past, craft personnel have experienced considerable difficulty using sky-climbers and have questioned their efficiency and safety. When extensive repairs are planned for furnace and penthouse areas, scaffolding rental fees are more than offset by savings in working-hours. Adapted for quick installation, such scaffolding is modular and made of tubular carbon steel with steel loops at the end of each section (Fig. 4). Fixed-length sections are

Fig. 4 Interior view of unit with modular scaffolding system. Note easy entrance and access to all areas .

safety-locked to one another by inserting steel wedges into the loops.

Basically, the scaffolding for interior access provides a solid platform near the top of the furnace. Typically, ten or more scaffold towers, each with three working platforms, are positioned at the front of the furnace adjacent to each side of the superheater division panels. Scaffolding is also provided across the unit at the front and rear of the superheater or reheater pendent assemblies.

Furnace waterwalls are scaffolded from the lower slopes to a point approximately 6 feet below the lowest element tubing of the superheater division panels. Walkway levels are provided at approximately 7-foot intervals, beginning at the bottom wall-blower elevation and continuing to the top elevation of wall blowers. An extension of the waterwall platform at convenient working elevations provides interior access to the windbox. Exterior scaffolding also can be used at the penthouse front wall near the roof header and superheater

links, at the sootblowers for lance removal, and at the economizer outlet ductwork.

As typical views of scaffolding show (Fig. 4 and 5), maintenance and repair personnel have easy access to the unit during an outage. This provides two significant improvements over previous methods. Repair personnel can enter the interior more safely than by using cable-hung equipment. Equally important, continuous access to all areas is possible and various crafts do not have to compete for access. This significantly reduces the time required to transport personnel and materials to areas to be repaired. Finally, scaffolding allows a more complete and thorough inspection.

THE OUTAGE STAGE

During the outage, every portion of the unit requires inspection. Scheduled repairs are then accomplished by a critical-path analysis to assure all repairs are completed. Key areas to be inspected are the pressure parts, the fuel-firing and transport equipment, boiler casing and

Fig. 5   **View of vertical modular scaffolding system**

structural supports, and such sub-systems as ash-handling and sootblowing equipment.

PRESSURE PARTS: EXAMINATION OF EXTERNAL SURFACES

Within the furnace, external surfaces of tubing are examined for erosion, metal wastage, swelling or developing problems in critical areas such as the firing zone and near sootblowers. Suspicious areas are investigated further by cleaning with sandblasting and then checking wall thicknesses by micrometer, or measuring thicknesses ultrasonically. See Fig. 6. This type inspection also covers

- critical superheater and reheater sections
- economizer tubing
- steam-cooled-wall surfaces and waterwall hanger tubing in the gas passes.

PRESSURE PARTS: EXAMINATION OF INTERNAL SURFACES

The steam drum must be entered and thoroughly inspected for deposits, loose parts, erosion, and areas that may allow internal by-passing of water or steam. Conditions found in the boiler drums may dictate inspection of waterwall headers and other internal surfaces for evidence of corrosion or deposits.

Fig. 6 Tubing is measured ultrasonically for wall thickness

Any lower waterwall headers large enough to have access ways should be inspected. Waterwall distribution orifices and screens are examined closely for deposits. Orifices are checked for proper size with go/no-go gages.

Samples of furnace-wall tubing may be removed from various locations a short distance above the uppermost fuel nozzles. Such samples are examined internally to determine surface condition and to detect any deposits.

FIRING EQUIPMENT

During the internal inspection of the furnace, the condition of the fuel and air nozzles is determined and any serious deterioration (Fig. 7) noted for replacement or repair.

Tilting fuel and air nozzles should be operated through their range while being observed from inside the furnace. Binding, nonparallelism or other malfunctions should be corrected. The tilt mechanism is also inspected from inside the windbox to determine the condition of the linkage.

Windbox dampers are inspected from within to determine that they operate through their normal stroke and that damper-blade movements correspond correctly to external linkage movement. No binding should be evident.

If there are retractable oil guns, the retracting cylinders must be operated to verify proper movement. The clearance of oil guns and condition of the ignitor-horn also should be checked from inside the furnace.

Extensive pulverizer maintenance is not always scheduled during an outage. In any case, a general inspection of the mills must be made. Particularly, inspect all shutoff dampers related to the pulverizers for proper operation.

Coal-pipe elbows in critical locations are examined internally for wear. In particular, the final elbow at the fuel nozzle must be inspected and the wear pattern noted. Defective kicker blocks should be replaced or relocated. Riffle distributors throughout the system are inspected for wear and replaced or rotated as necessary. Coal-pipe orifices are checked for signs of wear.

The feeders must be inspected as recommended by the manufacturer to implement any necessary adjustments and maintenance.

CASING AND ROOF ENCLOSURE

The roof enclosure also is inspected and the condition of pressure parts, hangers, and other pressure-part supports is recorded. Any damage or leakage of the furnace roof casing and seals should be corrected. Condition of insulation and pipe seals should be recorded.

The roof support steel and hanger rods within the air space between the roof deck plates and insulation are checked for corrosion.

Any leaks or damage of the external casing of the entire unit noted during the pre-outage stage are repaired as necessary.

The skin casing and seals within the dead air space behind the deflection arch are inspected as are the casing, seals, and support steel within the dead air space beneath the sloping furnace bottom. All are inspected to determine general conditions and to note any maintenance that should be performed.

OTHER AREAS OF CONCERN

Drip shields and seal plates at the furnace bottom must be inspected and repaired. The bottom-ash hopper and the balance of the ash-handling system are inspected as recommended by the manufacturer.

To the extent possible, all gas and air ducts are examined internally and externally for corrosion, erosion and leakage (Fig. 8).

The entire sootblowing system is checked—any malfunctioning equipment noted during the pre-outage operational check is further investigated. During the internal furnace inspection, sootblower effectiveness and any tube erosion from sootblower operation are noted so that sootblowers can be adjusted as necessary.

Boiler-water circulating pumps are inspected to manufacturer's recommendations.

Air-heater elements, seals, bearings, drives, and lubrication systems are checked.

Malfunctioning or leaking safety valves noted during the pre-outage operational check must be investigated and corrected.

The control system inspection includes

Fig. 7 **Fuel and air nozzles are inspected from inside the furnace**

Fig. 8 **Look for corrosion, erosion or leakage on fan casings**

calibration checks, stroking of drive units, and a general inspection of system components. Any malfunctioning components of the furnace safety system found during the pre-outage operational check are further investigated. Flame-scanner tubes are removed, inspected, and cleaned thoroughly.

All fans must be inspected (Fig. 9) for the condition of the bearings and their lubrication as well as the rotor and the associated dampers and/or inlet vanes. In addition, the induced-draft fans are examined for erosion, corrosion and/or deposits. Finally, all observation ports and access doors are checked for tightness.

THE POST-OUTAGE STAGE

After the unit is returned to service, the repair work is checked under operating conditions. For example, all permanent thermocouples on superheater, reheater, and waterwall individual tubes are observed to confirm proper operation.

As the unit is brought on the line, all control systems should be observed functionally to make any necessary adjustments (Fig. 10).

All valves repaired or repacked during the outage should be observed for proper operation and performance. Re-tighten all packing nuts with extreme care after each valve has reached normal operating pressure and temperature. Do not touch a valve if there is any indication that the valve was not packed properly.

Take an inventory of all recommended spare parts and forward a list of shortages to plant purchasing for ordering.

Fig. 10 Control systems are adjusted as the unit is returned to service

Fig. 9 Fan rotor is removed for inspection

DOCUMENTATION

Documentation is an important phase of any maintenance program. All findings, both good and bad, should be noted in a comprehensive report which covers all pre-outage, outage, and post-outage inspection and repair.

THE PRE-OUTAGE REPORT

The pre-outage-inspection documents contain a review of unit history, visual inspections, a walkdown of the unit while in operation, a checkout of all thermocouples, ignitors, fuel-burning equipment, safety valves, and sootblowers. Any items requiring further inspection or repair are noted.

THE OUTAGE REPORT

The outage report is tangible evidence of all that transpired. It is the basis for any necessary changes in operation, design or repair methods. It is the basis for all subsequent outages. It is the source of spare-part requirements. Outage documentation includes detailed reports, both quantitative and qualitative, describing conditions of both pressure and non-pressure parts of the unit. Fig. 11 shows a typical section of an outage report.

The engineer in charge of the outage should compile the report. An effective report might include sketches, marked prints, or photographs which can be useful where ideas or concepts rather than physical objects are being described for the record.

THE POST-OUTAGE REPORT

The post-outage documents describe conditions immediately after restart, such as thermocouple temperature measurements and expansion data. The final and most important documentation lists the recommendations of

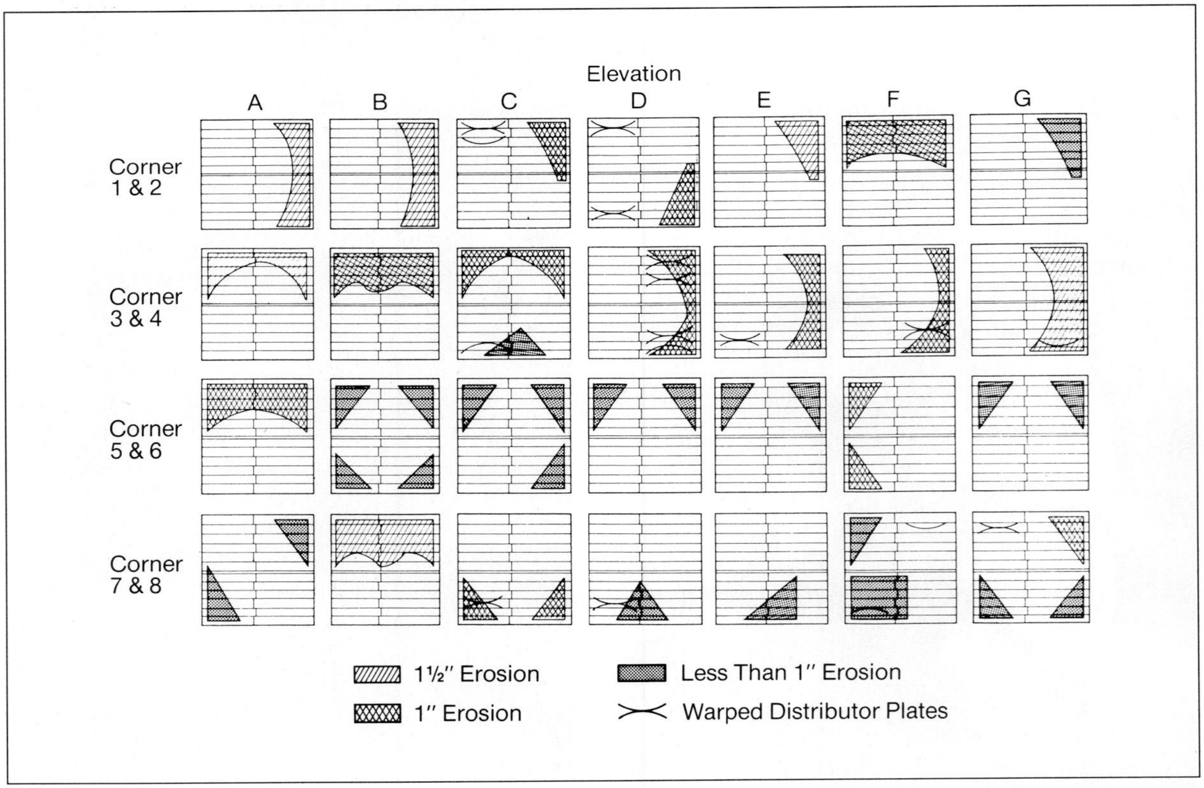

Fig. 11   Typical section of outage report

the inspection team, as well as an engineering department review of those findings. Such recommendations include spare parts to order in preparation for the next outage.

After all the planning, inspection and documentation have been completed, a critique provides all those involved in the outage an opportunity to review their findings and determine improvements for future outages. A frank, open exchange sets the stage for the next outage and also the plan for continuing work throughout the coming operating period.

THE FORCED OUTAGE

A forced outage requires as much or more planning than a scheduled outage, but by its nature, it allows the least time for planning. Consequently, a plant maintenance group must be prepared with an emergency plan formulated before a failure occurs. This plan should be in writing and periodically updated.

The plan should list alternatives such as another division or utility, use of alternate production methods, or purchase of power. Because downtime is critical, plan details and schedule must be worked out immediately. Plant personnel should have lists of other critical work that can be done during an outage.

Pressure-part failure is a major cause of forced outages. Since the unit must be cooled before it can be entered, the cooling time can be used for planning. As the repair of the failure will be the critical path in returning the unit to service, other maintenance must be done only if it does not interfere with the repair.

To save time during cool-down, crews can move parts, tools and scaffolds close to access doors. Usually, it is possible to determine the area of the failure and its general magnitude from the outside and, if a leak is small, sky-climbers, bosun chairs, or two-man scaffolding are generally adequate. If the steam generator has cable holes extending through the roof enclosure, considerable time can be saved because the roof enclosure does not have to be cooled for entry. For safety reasons, it is impor-

tant that the cables be cooled if they are to pass through the hot enclosure; compressed air is the usual cooling medium. If the damage is extensive and requires major repair work, more elaborate scaffolding may be necessary.

The importance of failure records cannot be over-emphasized. The cause of a single failure may not always be determined and an isolated random failure may be of little significance. But several similar failures may indicate serious problems affecting future availability. Good maintenance, then, requires not only that a failure be repaired but also that every effort be made to determine the root cause of the failure.

To retrieve information easily, an effective record-keeping procedure separates the logs for the superheater, reheater, economizer and waterwall. Patterns and trends are easily spotted when presented graphically such as on sketches showing elevation and plan views (Fig. 12). Numerals or dates of the chronological occurrence of a failure can be added.

REPAIR WELDING FOR PRESSURE PARTS

Plant management and maintenance supervisors must bear in mind the legal formalities involved in the repair of boiler pressure parts.

In the United States, the National Board Inspection Code (NBIC) provides guidelines and rules for repairs and alterations to boilers and pressure vessels after they have been placed in service. NBIC rules, including the involvement of the Authorized Inspector (AI), are only mandatory in those states, cities, and provinces which have adopted the NBIC, or where required by the owner or the insurance carrier.

Jurisdictional requirements involving the AI vary widely. Most jurisdictions require the owner to obtain the AI's acceptance before repair or alteration. However, the NBI Code places the responsibility of coordinating the acceptance inspection on the contractor.

Anyone that is required to do pressure-part welding—owner, contractor or manufacturer—must use approved welding procedures done by qualified welders.

GUIDELINES FOR WELDING REPAIR OF LOW-CARBON STEEL TUBES

A damaged tube should be cut out at least 2 in. on each side of the defective area. The minimum replacement tube length should be no less than 6 in.

Backing rings must not be used to weld any heat-absorbing tubes carrying water or a mixture of steam and water. Without a backing ring, the first pass of the weld must be made by gas tungsten arc or oxy-acetylene. The weld passes may be completed by either process, or by shielded metal arc.

Window welds may be used for repair work if access is difficult (Fig. 13). The first pass of a window weld must be made by gas tungsten arc or oxy-acetylene.

Fit-up of the weld joints is important. Although it is difficult to obtain accurate cuts on furnace tubes, it is important to get the existing tube ends squared and correctly chamfered, and to cut the replacement tube to the correct length. A tube-end scarfing tool should be used when possible.

Allow for shrink in welding. Remember that the weld metal and parent metal are melted in the welding process and the molten metal shrinks as it solidifies. (A butt weld in a tube will shorten the total tube length about 1/16 in.)

Use a clamp or guide lug to hold one end of the replacement tube in alignment while the first weld is made. Do not tack weld both ends of the replacement tube, particularly if the existing tubes are rigidly supported. As a general rule, first complete the weld at the lower end of the replacement tube. Do not start welding the upper end of the replacement tube until both the replacement and existing tubes have cooled to ambient temperature.

GENERAL GUIDELINES FOR ALLOY TUBE REPAIRS

If a damaged alloy tube must be replaced, it is always preferable to weld the replacement tube to an existing tube end of the same alloy and the same wall thickness.

Before removing the damaged tube, check the manufacturer's material diagram and locate shop welds used to join the damaged length to tubes of different material or different wall thickness. If at all possible, make the cuts to remove the damaged tube at least 6 in. from the shop weld, thus leaving a "safe end."

CUTTING OUT A SHOP WELD

If necessary to cut out a shop weld joining tubes of different material and/or wall thickness, special attention must be paid, since all

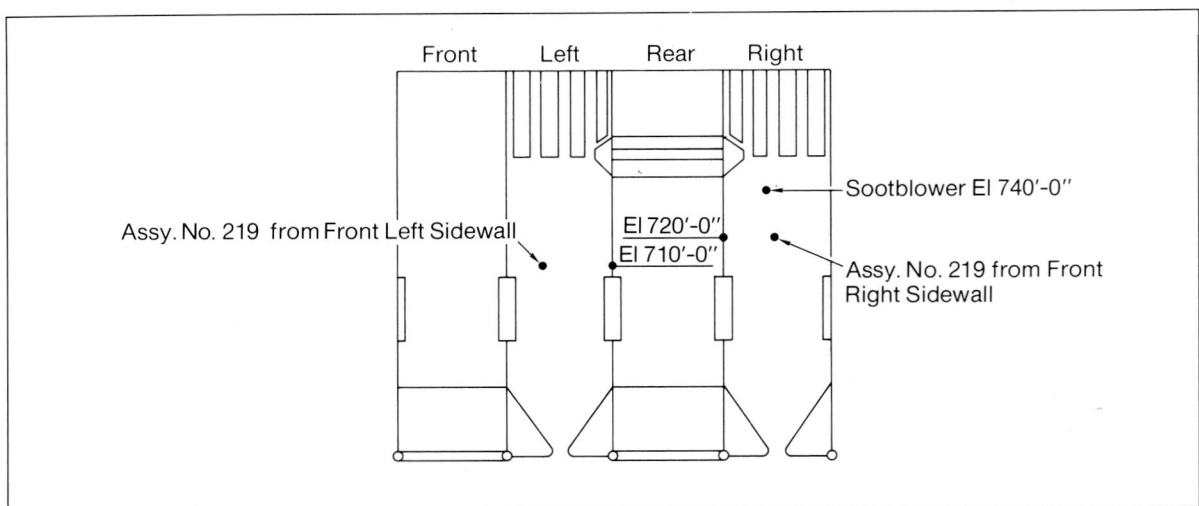

Fig. 12 **Waterwall failure locations (viewed from inside boiler)**

qualified butt-welding procedures require the two tube ends to have the same internal diameter (ID) at the weld root. In some cases, the thicker wall tube may be bored to match the ID of the thinner wall tube as shown in Fig. 14. But the thicker wall tube may be bored only if the strength of the tube, after reducing the wall thickness, is at least equal to the strength of the thinner wall tube at the same operating temperature.

A ferritic-alloy tube must not be bored to match a thinner-wall austenitic alloy tube. The only satisfactory method is to use a connector of austenitic alloy tube having the same wall thickness as the ferritic-alloy tube. One end of the connector is bored to match the wall thickness of the existing austenitic-alloy tube.

FIT-UP AND SHRINK ALLOWANCE

Shrinkage in welding alloy tubes is similar to that for carbon-steel tubes. Allowance must be made for expansion from preheating which will close the root gap slightly.

For shielded metal-arc welding with a backing ring, it is essential that the root-gap opening be sufficient to assure full penetration and fusion with the backing ring during the first pass. For gas tungsten arc welding, a zero root-gap opening is permitted. However, there must be no pressure exerted between the two tubes. It is advisable to allow enough clearance to avoid actual contact at the root-gap opening after the two tubes are preheated.

REPAIR OF TUBES
ROLLED INTO HEADERS OR DRUMS

If a replacement length of tubing requires a weld within 2 or 3 feet of the rolled end, removing the rolled joint and cutting the replacement tube to fit into the tube hole will avoid the stresses from weld shrinkage. This procedure is particularly effective in straight tubes or

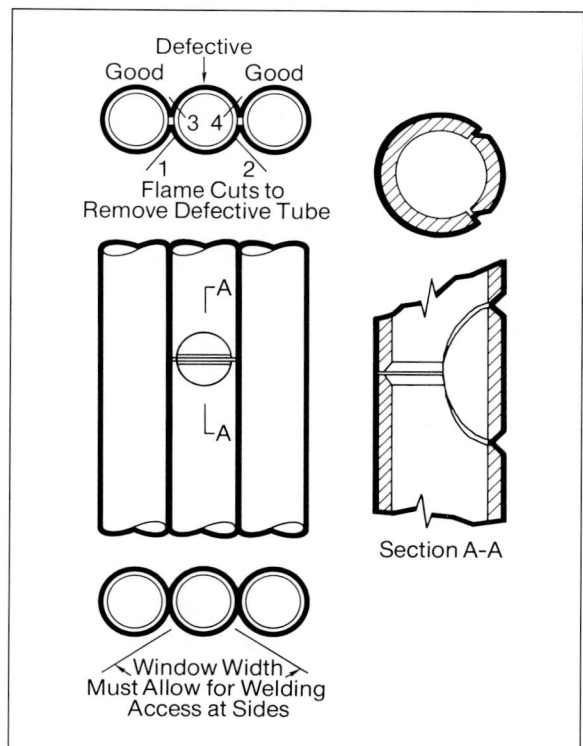

Fig. 13 Window welds may be used to repair damaged tubes when access is difficult

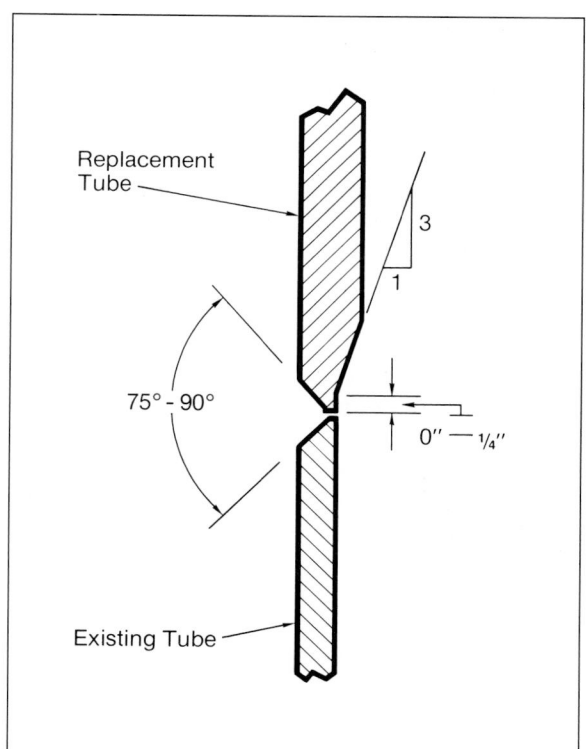

Fig. 14 Fit-up detail to install replacement tube section with wall thicker than existing tube

tubes welded to adjacent tubes. In cutting the replacement tube, allow for the normal projection through the tube hole plus 1/16 in. for shrinkage of the weld. Complete the weld in the replacement tube, allow the tube temperature to cool to ambient, and then roll the replacement tube end into the header.

REPLACING A TUBE LENGTH IN A FIN-WELDED FURNACE WALL

First, the bar or fin along the damaged tube length must be removed by flame cutting. Because it is difficult to make a clean flame-cut in the weld attaching a bar or fin to a tube, it is preferable to make the cut just at the edge of the weld, approximately 1/8 to 1/4 in. from the damaged tube.

Overheating of the tube may have caused a longitudinal crack. In this case, the tube has swollen and the wall thickness reduced. If visual inspection indicates swelling and reduction of the tube-wall thickness at the crack, a complete replacement of the damaged tube length is the best solution. On the other hand, a circumferential crack suggests a failure due to excessive stress applied by expansion restriction, bending, or fatigue. Although such cracks can be repaired by welding, unless the cause of failure is diagnosed and corrected it is possible for a similar failure to occur at or near the original crack.

REPAIR OF TUBE BLISTERS

Internal deposits cause blisters on the furnace wall or boiler tubes. Generally, they occur in boilers operated with a high percentage of make up in the feedwater. A blister forms because an internal deposit increases tube metal temperature until metal creep occurs. As the heated area swells, the internal deposit cracks off and the tube metal temperature returns to normal. The process may be repeated several times before the blister ruptures.

Commonly, a large number of tubes are blistered and not noticed until one of the blisters cracks open. To avoid a massive tube-replacement job, particularly where replacement tubes are not immediately available,

blisters can be worked down to the original tube radius. Follow these general guidelines:

Remove the damaged tube, then carefully cut away enough of the bar or fin to allow chamfering the tube end for welding around the sides of the replacement tube joint (Fig. 15).

After the tube welds are completed, weld the bar or fin to the replacement tube. If the gap between bar or fin is too great for easy bridging, insert a low-carbon-steel welding rod for a filler. The spaces in the bars or fins, at the tube joints, are built up with deposited weld metal. Be sure no cracks exist in these deposits before making the final weld to the tubes.

REMOVAL OF TUBE SAMPLE FOR METALLURGICAL OR CHEMICAL EXAMINATION

When a metallurgical or chemical examination is necessary to determine the cause of tube failure, special precautions are required to remove a defective tube from any type of welded-wall furnace. The defective portion of the tube should not be heated by flame cutting. To avoid such heating, the tube should be

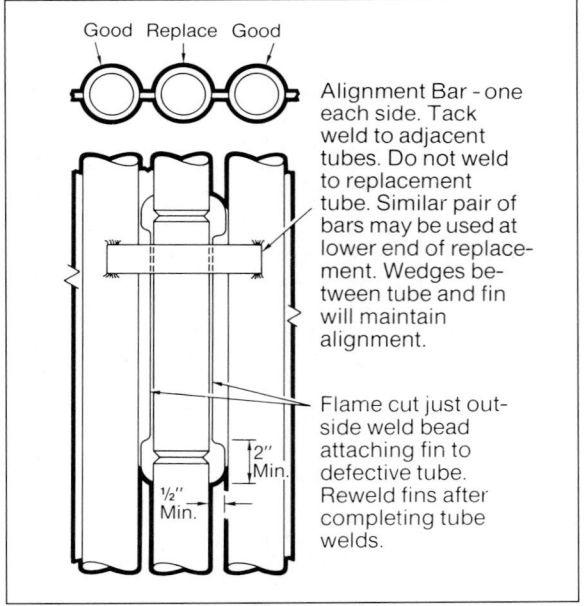

Fig. 15 **Welding detail for replacing a tube length in a fin-welded furnace wall**

flame-cut at a point at least 6 in. away from the damaged area. In a furnace wall with bars or fins welded between tubes, it is preferable to cut the bars or fins with a saber saw or thin disc metal-cutting wheel. If these tools are not available, the bars or fins may be flame cut approximately 1/2 in. from the tube length required for the sample.

After removing the damaged tube from the furnace, the ends may be cut off to reduce shipping weight of the sample. Use a dry hacksaw. Cutting oil will contaminate any deposits and render a useful chemical analysis impossible. Be sure the sample is clearly marked to show which end was at the top and which side was toward the fire (Fig. 16).

COMPONENT MAINTENANCE

Although it is beyond the scope of this treatment to describe the overhaul of all steam-generator components and auxiliaries, a brief discussion of pump, fan, and control-system maintenance is offered for general information.

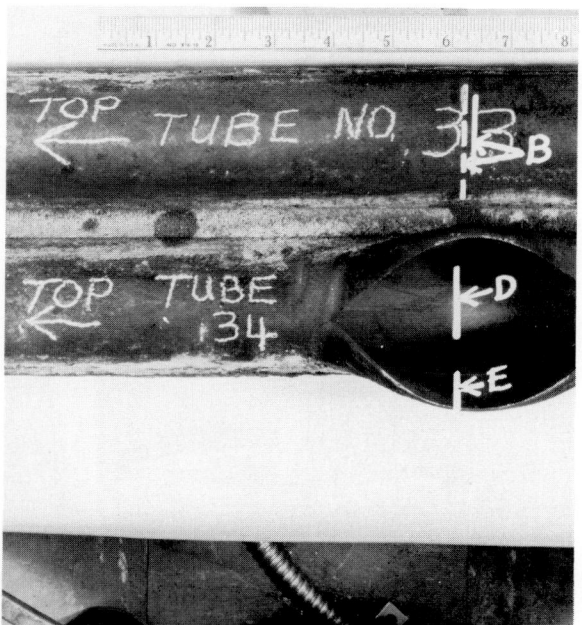

Fig. 16 **Make sure tube samples are clearly marked**

PUMP MAINTENANCE

Various types of pumps are used with a steam generator. For high-pressure recirculation-type boilers, single-shaft centrifugal pumps driven by a wet-stator induction motor move boiler water at pressures and temperatures exceeding 4000 psig and 650°F. Because the motor is enclosed within the pump casing and is filled with boiler water at full system pressure, no seals are used between the pump and motor. A baffle, plus close clearances between the shaft and motor casing, prevents solids in the boiler water and hot water from passing into the motor. Through an external heat exchanger, recirculating water in the motor cavity cools the submerged motor.

Although manufacturers generally recommend an annual inspection for such pumps, local conditions help determine the proper interval for a given unit and installation. Periodic inspection of the pump in service usually reveals when an overhaul is needed. Some indications are an above-normal noise level, high motor temperature, vibration, unusual starting or fluctuating current, reduced head across the pump, leaks from the pressure casing, or a drop in the winding resistance.

At most stations, one of the pumps is dismantled every few years. During disassembly, the motor and pump impeller are lowered as one assembly. The impeller can then be removed from the shaft and the rotor assembly taken from the housing. The assembly is placed horizontally on V blocks at floor level. An inspection is made with as much disassembly of subsections as observation dictates.

Although other pumps on steam generators might handle a cooler liquid at essentially ambient conditions, they may require much higher maintenance if the fluid being handled is more corrosive or erosive than pure boiler water. For example, heavy-duty centrifugal pumps are used to move a slurry solution of limestone and flyash in flue-gas desulfurization systems. The wear rate of these pumps can be substantial and is a function of the severity of the pumping duty and of the abrasive properties of the materials handled. The life of wear parts such as impellers and liners varies greatly. For instance, steel impellers and bodies have had very short life with highly acidic and abrasive solutions. If a new make or different type of pump is used, it should be opened regularly and the parts inspected to establish the life of wear parts.

When handling abrasive fluids, parts other than the impeller and housing are also subject to excessive wear. Shafts and gland seals and even the bearing must be inspected much more frequently. Bearing housings should be opened and inspected at least once a year.

FAN MAINTENANCE

Steam generators require many types and sizes of fans. The largest are used for forced- and induced-draft service, and for moving primary air to pulverizers for drying the coal. These fans can be of either centrifugal or axial-flow design, as described in Chapter 14. Bearing failure, one of the biggest problems with fans, usually is caused by either poor lubrication or excessive vibration. As stated previously, it is important to follow the lubrication requirements specified by the manufacturer.

If it occurs when a fan is initially started, vibration may be the result of inadequate foundations or unsuitable ductwork configuration or design. On the other hand, after operation, rotor imbalance often causes vibration. The imbalance can be a result of flyash accumulation in eroded blades, water, blade erosion, or uneven thermal growth of the wheel. But the possibility of changes in alignment or shifting foundations cannot be discounted.

An outage inspection program permits correction before a dangerous condition exists. The blades and associated ductwork systems should be examined for erosion, corrosion, or cracks (Fig. 17). If cracks are found in any blade, it is wise to magnetic-flux all of the blades for cracks. Fan dampers are as important as the fan itself, for deteriorated dampers can cause poor fan control.

Damper bearings are exposed to the same hostile environment as the fan wheel. Because flyash can penetrate the bearing lubricant, damper bearings must be inspected frequently and cleaned when the lubricant is contaminated. Dampers not in constant use, such as isolation dampers, may freeze in place and become useless. Exercising idle dampers can avoid this condition.

Some erosion of induced-draft fans on coal-fired boilers is to be expected. For centrifugal fans, erosion is generally most severe at the leading edge of the blades and at the joint between the blades and center plate. Axial-flow fans have the heaviest erosion on the front outer tip of the blade. Erosion primarily affects the structural integrity of the fan, but it can also cause imbalance of centrifugal fans.

For severe duty, replaceable wear plates are bolted or welded to centrifugal fan blades. At each outage the wear plates are checked and are replaced when worn halfway through. If wear plates are allowed to erode to a point that damage occurs to the blade material, repair cost and

Fig. 17 **Fan rotor blades with heavy erosion**

outage time can be greatly increased.

Although wear plates can reduce the effects of erosion, it is controlled best by flyash removal upstream of the fans. Again, an effective maintenance program determines and corrects the cause of maintenance sometimes initiated by equipment other than that actually experiencing the problems.

A most important part of any fan repair is the rebalancing and final cleanup. Besides the normal dangers of rotating equipment, fans present an additional hazard in their ability to draw in, not only air or gas, but loose material. Solid objects can pass through the fan and either damage the rotor or discharge the debris as projectiles. Therefore, upon job completion, all ductwork must be inspected to locate any foreign material so it can be removed.

CONTROL-SYSTEM MAINTENANCE

Historically, because of limited manpower and funds, power-plant control-system maintenance has been limited to repair of known deficiencies. With greater emphasis on plant reliability and availability, preventive maintenance programs have grown.

Such a program should begin with a continuous process of recalibration, repair or replacement while the unit is on the line. By relying on a redundant component, transferring a control loop to manual, or by temporarily defeating the component's function, most control system components can be recalibrated, repaired or replaced on line (Fig. 18). Periodic maintenance can prevent unit high or low furnace-pressure trips caused by drifting calibration of transmitters and switches, or by plugged sensing lines between the furnace tap and the transmitters.

A power-plant annual outage offers the opportunity to perform a more complete overhaul of the control system. Usually scheduled well in advance, this work includes the recalibration of field devices such as transmitters; pressure, temperature, and position switches; control valves; speed controllers; damper drives; and thermocouple inputs. If there are known problems in the control system, additional work is done. For example, a dry-run of the furnace supervisory safety system should be conducted if some digital logic is not performing correctly, or a calibration and retuning

Fig. 18 **Periodic maintenance of control hardware can prevent costly trips**

of an analog loop may be in order if poor control is observed during previous operation.

Because the control system is the power-plant component that, as its proper function, trips a unit off line, its role is critical to plant operation. Often, unit trips are attributed incorrectly to control-system malfunctions when, in fact, the control system has sensed an abnormal condition and responded correctly.

EROSION AND ABRASION

On coal-fired units, one of the biggest maintenance requirements is the repair or replacement of parts worn by the erosion or abrasion of coal and ash. Abrasion is the "sandpaper" effect of solid particles moving parallel to, and in contact with, a boundary surface. Abrasion tends to occur at the high spots on the boundary surface and has little effect on the surface matrix. Therefore, a boundary surface containing a small hard-particle matrix will give the best life.

Erosion, on the other hand, occurs from the impact of energy particles. The solid particles strike the wall freely, and cut portions of the boundary or wall material. Hard or erosion-resistant particles in the wall matrix will not protect the less-resistant balance of the matrix. Consequently, many abrasion-resistant materials are not erosion resistant (Fig. 19).

PULVERIZER WEAR

In coal-fired steam generators and auxiliaries, the major wear is from erosion, which occurs in the fuel preparation and delivery system and in the backpass convection section. Although the grinding parts of the pulverizer wear from abrasion, wear in the other parts of the coal milling and transport system is principally due to erosion. The wear in the steam generator itself is almost all by erosion.

As discussed in Chapter 11, the C-E bowl mill pulverizes coal by subjecting it to a grinding force between a rotating bowl and spring-loaded stationary rollers. The bowl has a segmented wear-ring for easy replacement and the rolls are fairly easily removed from their journal assembly for replacement.

The rate of abrasive wear of the grinding parts depends on the type and quantity of impurities in the coal. Considerable laboratory research and field development of materials have reduced the wear and increased the availability of these components. An abrasion-resistant nickel-chromium white cast iron is used for the pulverizer grinding parts. Depending on the abrasiveness of the particular coal being ground, rolls and rings may both have to be replaced at the same time or it may be possible to obtain twice the life from a ring as from a set of rolls (Fig. 20).

For highly abrasive coals, a hard-surface weld overlay which is applied to either new or worn parts has proven effective in increasing roll and ring life. The overlay must be at least $1/2$ in. thick. For used rolls, overlays up to $2 1/4$ in. thick are possible to return the roll to its original diameter.

Other pulverizer parts subject to wear include the various liners fabricated of abrasion-resistant steel plate, high-nickel castings, or ceramic material. These must be checked for material loss at the same time as the rolls and rings.

COAL TRANSPORT PIPING

With more frequent use of highly abrasive coals, pulverized-coal pipeline wear can become a significant problem. The rate of wear generally is a function of

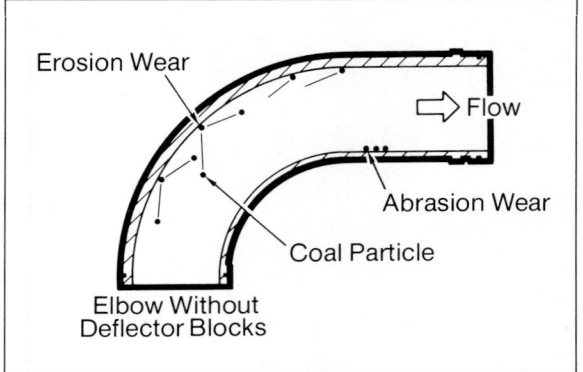

Fig. 19 **Erosion and abrasion in coal piping**

- particle characteristics: hardness, sharpness, and size
- particle velocity
- angle of attack of particles
- properties of the piping material

Fig. 21 shows the standard material selection for the coal-transport piping which provides cast-iron elbows at the pulverizer or exhauster outlet and at the windbox inlet. All straight piping and bends between these points are of commercial carbon-steel pipe. All bends and elbows are designed to a two-pipe-diameter radius, except the windbox inlet elbow, which has a one-pipe-diameter radius.

Rockwell couplings are used on the toggle sections provided to accommodate the vertical expansion of the boiler, and Victaulic couplings for connecting to the cast-iron elbows and orifices. The balance of the piping is welded.

Because erosion potential is greater at and after bends, wear-resistant nodular-iron deflector blocks are positioned in selected elbows to both absorb coal particle impingement and redistribute solids flow concentrations (Fig. 21). Experience dictates initial location of deflector blocks but they can be repositioned in the field to match the elbow wear pattern. System pressure-drop considerations limit the number of deflector blocks that may be installed.

Sometimes, pipeline flow characteristics are so complex that deflector blocks cannot be completely effective. In these installations, the more serious pipeline erosion usually occurs in specific bends and in the first 2 to 3 ft of straight pipe directly following the bends. In such areas, special wear-resistant components are used for highly abrasive coals.

Such wear-resistant materials have been installed in, or in place of, elbows and bends on

Fig. 20 **Checking roll wear**

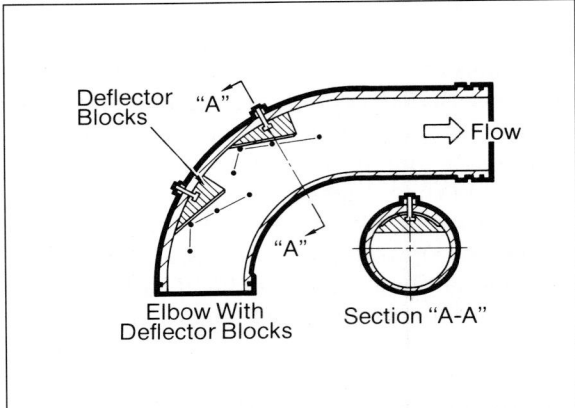

Fig. 21   Coal-piping elbow with deflector blocks

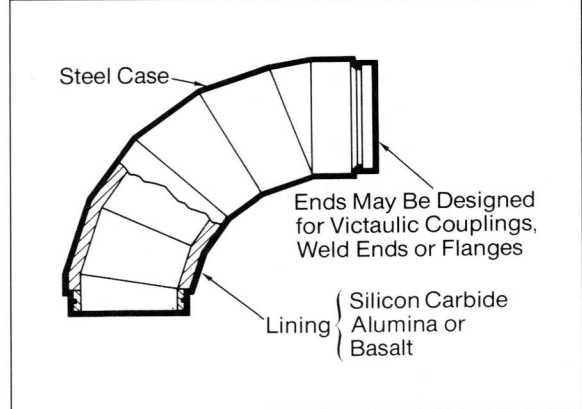

Fig. 22   Typical lined 90-degree steel elbow

operating units and have been evaluated for erosion losses. Some are

- steel elbows lined with alumina
- steel elbows with silicon carbide
- steel elbows lined with basalt
- double wall steel elbows with "lobster-back" wear surface
- heat-treated (carburized) steel elbows

EXPERIENCE WITH WEAR-RESISTANT PIPING MATERIALS

Brittle nickel-chromium nodular cast-iron, previously used to prolong the life of piping elbows and bends, is no longer offered; this is in accordance with the 1988 NFPA 85F standard prohibiting the use of brittle materials in pulverized-fuel piping.

Fig. 22 shows a typical ceramic lined elbow which may be a fabricated mitered elbow or a modified standard steel bend. A steel casing encloses the wear-resistant lining and satisfies code pressure requirements. The lining is secured in place with a high-temperature cement. An advantage of the steel-casing design is that the ends can be prepared for any attachment design including welding, flanging, or machined-end preparation.

The most widely used ceramic for fuel-pipe applications is alumina (Al_2O_3). This material is capable of improving wear-life 5 to 7 times that of carbon steel or nodular cast iron. The alumina is manufactured in wedge-shaped bricks, usually one-inch thick. The bricks are installed in a fabricated mitered elbow row by row. The taper of the bricks is such that the brick in any given row support each other, as in a brick arch. Mortar is used to fill the void between the flat-faced brick and the curved steel shell. The inside diameter of the lined bend is identical to the pipeline I.D., with the O.D. being substantially greater.

Other ceramic materials such as silicon carbide (SiC) have also been successfully used as fuel-pipe bend liners, but alumina brick has been found to be the most economical selection.

A volcanic rock, basalt, is fused and annealed into liners of various shapes and thickness. It has excellent abrasion resistance and is good for straight coal pipes. But as basalt has poor erosive-impact wear resistance, it is not recommended as a lining for elbow or bend installation.

Fig. 23 shows a double-backed elbow with a wear-resistant concrete filling the void between the two surfaces. The steel elbow and the outer containment cover are made from elbow torus fabricating dies. Such elbows have been in service for years with no sign of wear.

Disadvantages of the "lobster-back" elbow are cost and space requirements. It takes two complete elbows and considerable shop labor to construct one wear-back elbow, and results in a component 2 in. greater in outside radius.

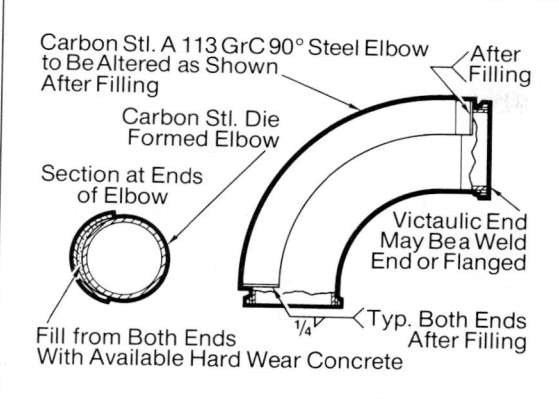

Fig. 23 Double-walled wear-back outlet elbow assembly in steel coal-piping

Two steel elbows carburized to 700 Brinell hardness, installed for test evaluation, showed excellent abrasion and erosive-wear resistance. However, welding the elbows or attachments to the elbows is difficult and flange weld failures have been experienced. Also, the costs of carburizing large elbows appear prohibitive.

EROSION AND ABRASION FROM FLYASH

Approximately 60 to 80 percent of the ash resulting from the burning of coal leaves the furnace with the combustion gases. Gas velocities must not exceed specified limits in order to control erosion from the ash particles. Depending on both the quantity of ash and its composition, the velocity is generally limited to 50 to 60 ft/second.

Two components in flyash that appear to contribute most to flyash erosion are silica (SiO_2) and iron oxide (Fe_2O_3), both relatively hard abrasive oxides. While silica is ordinarily present at higher percentages than iron oxide, iron oxide is more dense and, consequently, produces a greater impact influence.

Flyash erosion is a term that, in general use, includes flyash abrasion. All tubing erodes when subject to a flyash-laden gas stream. As long as the wastage rate (the loss of metal in a given time) is low, the polishing effect is not considered significant. When the wastage rate becomes excessive, premature tubing failures

can occur. Although some erosion is experienced on superheaters and reheaters, most erosion has been on economizers, which usually are horizontal tubes with closer spacing than the superheaters and reheaters.

ECONOMIZER EROSION

In general, predicting where excessive flyash erosion might occur for any given unit is difficult, because such erosion is related to nonuniform flow of flue gas and uneven distribution of flyash in the backpass area associated with centrifugal forces acting on the ash particles during gas-turning.

Many factors affect flyash erosion, which is a very complex process. It is directly proportional to ash loading and is influenced by the relative amounts of abrasive constituents—the silica, alumina, and iron oxide compounds—in the flyash. Most investigators have found that the erosion rate is an exponential function of particle velocity to between the 3rd and 4th power. Gas-side pressure drop (which is a combined function of gas velocity, turbulence, and heating-surface configuration) also influences erosion.

Gas velocity through a tube bank is a function of unit operating load, excess-air level, and the amount of free (net) gas-flow area within the bank. Therefore, unit operation much above maximum continuous rating and design excess air aggravates the potential for flyash erosion. Also, any action that results in the reduction of flow area, such as localized flyash plugging, will increase gas velocity and the erosion potential by concentrating flyash in an area adjacent to the plugged area.

When a unit develops an economizer erosion problem, it is necessary to establish the cause. To add more localized baffling, tube shields, or other devices without regard to the causes, may produce only short-term improvements.

If examination of the unit and its operation indicates that such factors as operating load, excess-air level and presence of plugging are not abnormal, then the addition or modification of baffles or tube shields or other devices is justified.

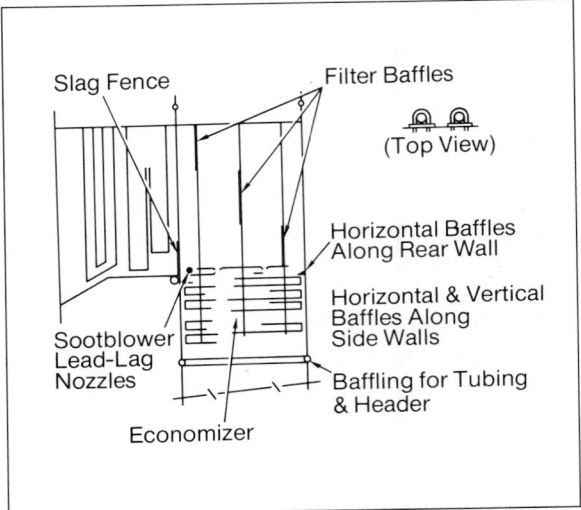

Fig. 24   Erosion reduction methods in economizer

On some economizers where excessive erosion has occurred across the rear of the backpass, a combination of erosion-control methods has been used (see Fig. 24). Collectively, they have abated excessive erosion. Some methods used include

■ adding filter baffles to obtain a more uniform flyash distribution across the economizer

■ adding slag fences to prevent economizer plugging

■ installing plates or screen to block open areas and stop channeling of the flyash

■ adding additional sootblower equipment to maintain cleanliness of the economizer section

■ sequencing the sootblower in the upstream convection sections to blow from top to bottom and rear to front

MAINTENANCE STAFFING

The size of a plant maintenance staff varies with the number and size of units, fuel used, quality of the labor, and extent of outside supplemental help. Large coal-fired stations may have several hundred people employed in maintenance and still hire contractors for major overhauls. Some utilities have roving crews to supplement plant labor when required. Whether to use outside labor or be self-sufficient can only be determined by plant management. But no matter how much outside contract labor is employed, the final responsibility must rest with the station staff. Unless the station staff oversees the work done, there is little chance for a cohesive maintenance program.

TRAINING FOR MAINTENANCE

Training maintenance personnel is vitally important to the successful operation of any plant. Many months before a plant goes into service, maintenance crews should be selected and trained.

Although there are fewer available programs for maintenance people than for operators, the plant should explore all training aids and methods. Many manufacturers have courses and workshops in equipment maintenance and the use of special tools and procedures.

The plant maintenance staff should be used for initial setup of equipment because working with the manufacturers' representatives provides invaluable training. Money spent on vendor servicing can be wasted if no one in the plant knows what is being done. Thus, a plant maintenance supervisor, as well as maintenance mechanics, should be assigned to each vendor representative.

LONG-RANGE PLANNING AND SCHEDULING

Nothing is more important in maintaining a plant than scheduling maintenance intervals. The more crowded and busy the maintenance chief's schedule is, the more the need to take time for long-range planning and scheduling. Such forward thinking includes asking questions which define future needs:

■ What jobs are next in line?

■ On what dates do they have to be started or completed?

■ Which person is going to do what?

■ What equipment and material are needed to get the work done?

■ What problems or roadblocks will they run into?

Equally important, an optimum maintenance program calls for replacing some parts before they are completely worn out so as to take advantage of scheduled outage time. A cost/benefit analysis is one way to decide when to change critical parts; that is, the cost of the life remaining in the component is weighed against the cost of subsequent loss of production if that equipment is out of service again.

For fast-wearing or particularly vulnerable parts, consider substituting new materials or redesigned parts. Improved materials can prolong life and lead to much less maintenance labor. With higher labor costs, materials once considered too expensive are now cost savers. Even plants that seldom rely on manufacturers' representatives for aid in maintenance should periodically ask the manufacturer about new developments in materials and techniques. For instance, high-nickel material or tungsten-carbide overlays on steel have been most effective in prolonging life of steel parts. Ceramics are used in many applications. Initially these were used only on stationary parts but, with improved epoxies, some can now be used on rotating equipment as well.

Wise maintenance supervisors, then, will investigate new materials or other ways to cut maintenance costs. And they will allot some time for discussions with vendors' salespeople to find out what new products may be a solution to difficult maintenance problems.

BEYOND MAINTENANCE AND REPAIR

Maintenance activities, as they have been addressed in this chapter, fall into three overlapping, not-too-well-defined categories. The first, under the general designation of "maintenance and repair", has to do with the day-to-day, month-to-month, and year-to-year actions of inspection, monitoring, and testing (such as of oil or water) to ensure that a steam-generating unit and its auxiliaries will keep running between scheduled annual or bi-annual outages. Strange sounds, visible leaks, higher pressures or temperatures than normal, all tell the alert operator or maintenance person that something has to be done quickly or unit availability will suffer. Unfortunately, this type of maintenance is reactive, but is vital to keeping equipment in service for extended periods of time.

The second type that has been described is anticipatory, or preventive, maintenance—the kind that *prevents* the strange, noises and other undesirable indications. Based on manufacturers' recommendations, in-plant experience, and seat-of-the-pants judgment, preventive maintenance is done at planned intervals while equipment is operating satisfactorily; its purpose is to eliminate unscheduled outages due to component distress. With this type, the timing is arbitrary and consensual; it is not based on rigorous testing, but on an accumulation of favorable experience.

The third type is now called *predictive maintenance*. This has the same aim as preventive maintenance, but employs more sophisticated computer-assisted methods of timing the actions; it is intended to achieve the same or better results at lower cost.

Maintenance and repair activities beyond these three categories lead us into the realm of life continuation, which is discussed in the next chapter. Major efforts to obtain higher availability and indefinite continuation of equipment life are in the area of upper-management decision-making, beyond the proper activity of the maintenance-engineering staff.

CHAPTER 24

Maintaining Availability:
Condition Assessment and Remaining Life Analysis

In many respects, the electric utility or independent power producer is unique. Its product, the day-to-day output of electric energy, is supplied on demand with no opportunity to be inventoried or stored. Thus, the network of generation, transmission and distribution equipment through which the electricity is delivered must have high operating integrity to assure optimum reliability of supply, which is the power industry's highest priority.

In addition, because of the long lead time required to add or contract for new capacity, it is necessary for the industry to plan to build generating equipment years in advance of service dates to supply future electric loads. Further, the power industry ranks among the most capital-intensive in terms of dollars of investment per dollar of revenue. This chapter explores the implications of reliability of supply in the generation of steam-electric power by fossil-fired plants.

AVAILABILITY

The availability of an electric power plant is important to both system reliability and generating-company profit. Improving availability only slightly can save considerably on reserve gener-

ating capacity and the cost of replacement power. As one measure of reliability, published availability statistics are of considerable interest as the power industry emphasizes producing the most energy for the least cost.

The importance of high plant availability has spurred the U.S. Department of Energy (DOE), the Edison Electric Institute (EEI), and the Electric Power Research Institute (EPRI)—the latter acting as the research arm for participating utilities—to pursue programs that identify means of improving plant performance.[1] Some state and federal laws concerning electrical-plant reliability already have been enacted.

In very dramatic terms, the world energy crisis reinforces the tremendous importance of power-generating plant availability. Fluctuations in fuel costs are felt in virtually all energy sources with a significant impact on the political, social, and economic status of the U.S. and most other countries.

PRODUCTIVITY INDICES

The utility industry uses several indices to evaluate power-plant productivity which, in essence, is a measure of the ability of a plant to produce electricity on demand. As defined by the North

24-1

American Reliability Council (NERC), five major power-plant performance indices are:

Net Capacity Factor (NCF), percent =

$$\frac{\text{Total Net (Actual) Generation in MWh}}{\text{Period Hours} \times \text{Net Maximum Capacity}} \times 100 \tag{1}$$

Service Factor (SF), percent =

$$\frac{\text{Service Hours}}{\text{Period Hours}} \times 100 \tag{2}$$

Availability Factor (AF), percent =

$$\frac{\text{Available Hours}}{\text{Period Hours}} \times 100 \tag{3}$$

Equivalent Availability Factor (EAF), percent =

$$\frac{[\text{Available Hours} - (\text{Equivalent Forced} + \text{Planned Derated Hrs})]}{\text{Period Hours}} \times 100 \tag{4}$$

Forced Outage Rate (FOR), percent =

$$\frac{\text{Forced Outage Hours}}{\text{Service Hours} + \text{Forced Outage Hours}} \times 100 \tag{5}$$

where:

Net Maximum Capacity (NMC) = the net sustainable unit capacity when not restricted by ambient conditions.

Available Hours = the period of time during which a unit or major piece of equipment is capable of service whether it is actually in service or not.

Period Hours = the number of clock hours that the unit is in the "active state" (usually taken as one year). The active state includes both the available condition (with the unit operating from zero to full load) and the no-load condition during forced or scheduled outages. It does not include any period in which a unit is on inactive reserve, moth-balled, or retired.

Service Hours = the total number of hours the unit is electrically connected to the transmission system.

Forced Outage Hours = the sum of all hours during which a boiler or other major equipment is unavailable because of a forced outage, which is the occurrence of a component failure or other condition requiring that the unit be removed from service immediately, or up to and including the next weekend.

MEASURING AVAILABILITY

Each equation identifies only a portion of performance measurement for a generating unit. No single index tells the overall performance story for a unit, a fact that is apparent from Fig. 1. Capacity factor, for instance, is not a true indicator of a unit's reliability, because actual generation may be limited by economic or environmental dispatch as well as forced and scheduled outages and deratings. This is particularly true with older units, which typically have higher heat rates (lower thermal efficiencies) than more modern units. The annual availability factor establishes only the percentage of time during the year that the unit was capable of producing power. This factor includes time when the unit was capable of generating power, but was not in service because more efficient units were being used. Thus, the availability factor does not measure the ability of a unit to operate at a specific power level when called upon by the dispatcher. Rather, it measures only the capability of the unit to produce at a power level ranging from 0 to 100 percent of its rated capacity.

The equivalent availability index provides for an adjustment of the availability factor by accounting for the effect of partial deratings (losses in electrical-power output capability) from partial forced and scheduled outages. Essentially, the index is equivalent to the percentage of the year during which the unit was available for operation at full capacity.

What equivalent availability does not indicate is whether an outage is either forced or planned. A forced outage is generally more costly in terms of both replacement power and the ability of the producer to supply system loads. The forced outage rate measures this performance. Net capacity factor is, in effect, the "bottom line" result and also includes discretionary cutbacks in output by a generating company.

Each index, therefore, is limited to defining a specific aspect of power-plant performance. To evaluate the overall performance of a power plant requires subjective and collective review of these and other indices.

AVAILABILITY STATISTICS

Table I is a small sample of the Generating Unit Statistics given by the Generating Availability Data System (GADS) of NERC for the five-year period 1985-1989. The GADS-IEEE Standard 762 and equations and definitions are taken from the GADS short-form summary of its Generating Availability Report.[2] This tabulation illustrates the differences in five performance factors for equipment currently chosen for electrical power

generation in the U.S.

Note that coal-, oil-, and gas-fired steam generators have lower availability factors than jet engines (aircraft-derivative gas turbines), gas turbines (stationary industrial type), and diesel engines. This is because of the extended planned outages taken on such boilers for repair, maintenance, and upgrading, usually in anticipation of major power-generation campaigns. The internal-combustion equipment—which is smaller on average than the boilers—is closer to the ground, often has many replaceable modules, and inherently has shorter preventative-maintenance time; and the less time down for maintenance, the higher the AF.

THE POWER-PRODUCER'S VIEW OF AVAILABILITY

A major reason for improving availability in electricity production is a matter of economics,

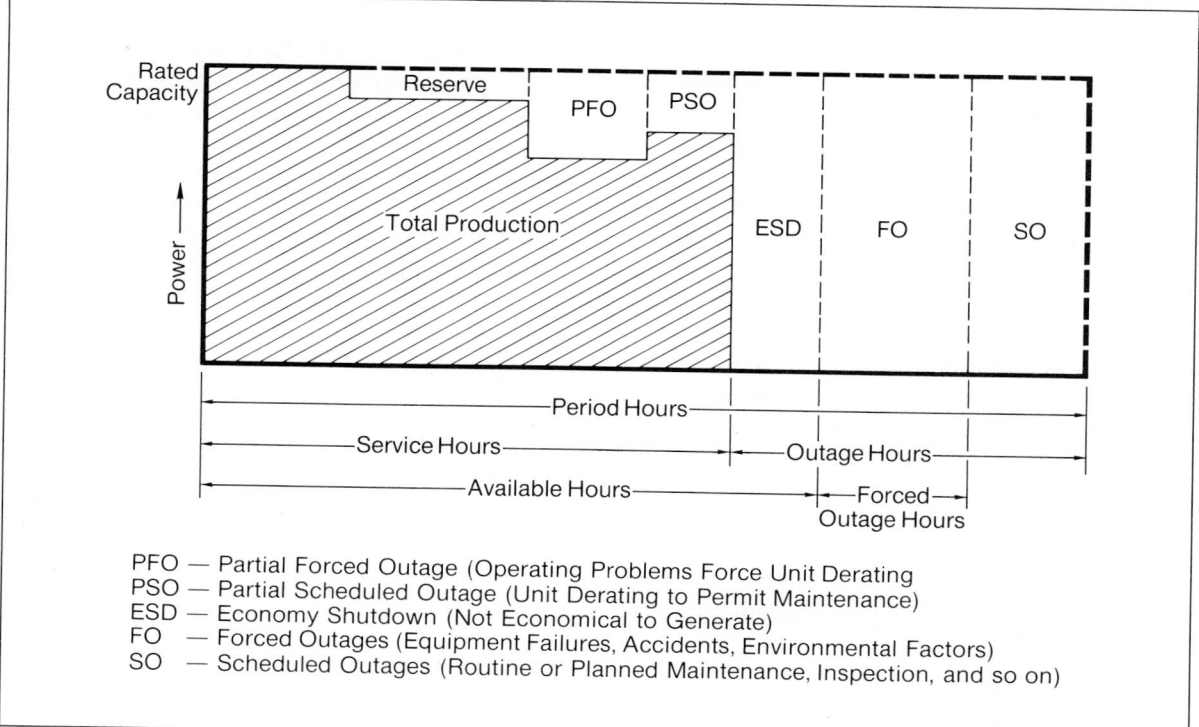

PFO — Partial Forced Outage (Operating Problems Force Unit Derating)
PSO — Partial Scheduled Outage (Unit Derating to Permit Maintenance)
ESD — Economy Shutdown (Not Economical to Generate)
FO — Forced Outages (Equipment Failures, Accidents, Environmental Factors)
SO — Scheduled Outages (Routine or Planned Maintenance, Inspection, and so on)

Fig. 1 **Factors affecting productivity of power plants**[1]

Table I. 1985-1989 Generating Unit Statistics from the North American Electric Reliability Council (NERC) Generating Availability Data System (GADS)

Unit Type	MW Rating	No. of Units	NCF	SF	AF	EAF	FOR
Coal Primary	1-1,000+	887	58.2	70.5	83.1	80.2	6.6
Oil Primary	1-999	244	30.9	47.1	84.4	81.5	7.7
Gas Primary	1-999	472	30.9	42.3	84.8	82.4	7.0
Nuclear (All)	400-1,000+	125	62.1	69.2	69.2	65.7	13.8
Jet Engine	1-20+	347	1.9	2.6	91.8	85.5	55.3
Gas Turbine	1-50+	575	1.8	2.7	91.1	85.1	60.6
Diesel	All Sizes	180	1.9	2.3	95.7	95.6	50.5
Combined Cycle	All Sizes	26	32.5	49.1	85.1	76.8	6.8

which essentially relates to the high cost of constructing power-generating facilities. (The basics of power-plant economics are discussed in Chapter 1.)

Better plant availability could have a significant effect on plans for the construction or deferment of new plants. On the other hand, long-term outages from repeated boiler tube leaks, turbine-blade losses, or other plant equipment problems, could force a utility either to install additional capacity or to purchase replacement power—if it is available—at high cost.

RESERVE MARGINS

One measure of adequate generating capacity is the reserve margin in excess of the peak load demand. This margin offers protection against unanticipated demand growth, forced outages, derating, and other contingencies. In the United States, a reserve margin of about 20 percent is generally considered acceptable. Because of regulatory and/or political factors, certain geographical regions have much less reserve than others, accentuating the requirement for highly reliable generating facilities and adequate transmission interconnections. In all cases, outages of large, highly efficient, base-load equipment increase generating costs.

AN APPROACH TO AVAILABILITY IMPROVEMENT

To help electric utilities improve on unit avail-

ability and obtain maximum reliability from steam generators, C-E at one time formed an Availability Task Force to investigate how best to use EEI, NERC, and other data-gathering systems. A study showed that these data were capable only of providing the component cause for outages and load reduction and were not able to provide information as to why a component failed. Without information on the cause of the component problem, it was not possible to determine whether there existed a deficiency in design, manufacturing, maintenance, or operation. Therefore, C-E established its own program to provide the necessary information.[3]

This program gathered information on the cause of outages and load reductions in nine major equipment categories related to steam generators. Included were

- waterwalls
- superheaters and reheaters
- economizers
- furnace sootblowing/bottom-ash removal equipment
- convection-section sootblowing and flyash-removal equipment
- boiler controls
- fans
- pulverizers
- boiler circulating pumps

This information helped to identify trends in

problems that occur from deficiencies in design, fabrication, operation, or maintenance. Derived from these data is the knowledge that waterwall, superheater, reheater, and economizer tubing leaks account for 80 to 90 percent of all forced outages, whereas coal-pulverizing systems account for approximately 50 percent of equivalent downtime hours in load reductions.

The program was limited to coal-fired units of 390-MW capacity and larger in consideration of the amount of effort required in data recording by the utilities. It helped to improve the performance of the large capacity coal-fired units then operating, and thereby set the pattern for optimal performance of generating capacity in the future.[4]

PROBLEMS AND THEIR ROOT CAUSES

The underlying causes of equipment problems producing forced outages may be difficult to identify, but their correction requires such an identification. Treatment of symptoms will neither cure a power-plant "disease" nor correct a disorder which may, in some instances, be a design or construction deficiency. Chapter 23 emphasized the importance of determining the root cause of incidents causing forced outages.

In the case of waterwall tube ruptures, poor feedwater quality is often the major contributing factor. With high operating pressures, feedwater quality becomes even more critical. The recommendation is a maximum of, for instance, 50 ppm of total dissolved solids in high-pressure drum-type units. Most operators follow this practice, as well as all others given in Chapter 20, Power-Plant Water Technology.

Degradation of coal properties adversely affects the capability of boiler auxiliaries, but its direct effect on availability is difficult to quantify. A valuable study on this subject was done by the Tennessee Valley Authority and is presented in Fig. 2. The capability of components such as pulverizers, fans, economizer, and air heaters frequently limits unit capacity and reduces gen-

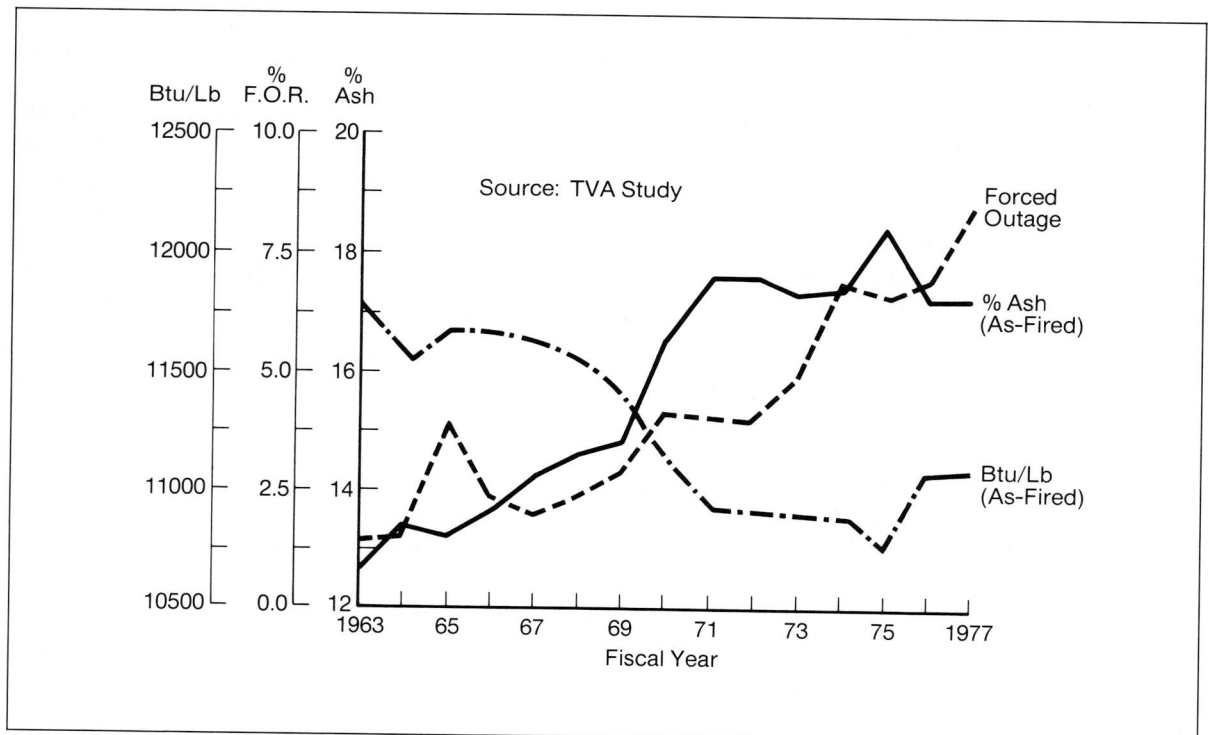

Fig. 2 **Boiler forced outage rate vs coal quality**[5]

erating reserves. If auxiliaries are operated continuously at or near their peak capability without proper maintenance, problems will generally develop.

Furnace-framing structural design to withstand negative pressures can be increased significantly to provide greater protection against implosion.[6]

Damage to fan housings, duct work, and stack liners, resulting from fan-induced duct vibrations caused by pressure pulsations in centrifugal fans, also can be a significant problem. To avert this, C-E has specified equipment that does not emit damaging pulsations. An example is its calling for vane-controlled fans instead of louvered-damper-controlled fans whenever possible. Damper-controlled fans are more susceptible to stalling and consequent large emitted pressure pulsations. A second example is the specification of dorsal fins or an equivalent in all vane-controlled fans. Without this corrective device, some installations experience destructive levels of vibration. Stipulating equipment with these features limits the selection, but markedly reduces fan/duct vibration incidents.[7]

AVAILABILITY/RELIABILITY AND LIFE CONTINUATION

In consideration of many factors, electric utilities and owners of large industrial boilers have often redirected their resources to the improvement and the continuation of life of existing equipment.[8]

For many years, the practical life of power plants was taken to be 30 to 40 years. The rationale for this was based, in part, on the very arbitrary material lives indicated in the ASME Boiler Code. It was felt that major components would be nearing the end of their design lives after such a time, and that the plants would become less economical as their efficiency (relative to newer units), availability, and reliability decreased.

Now, plants over 30 years of age that would have been scheduled for retirement are studied as to the advisability of regeneration to maintain or enhance their availability and to continue their

effective lives for at least 15 years more than originally anticipated. This is because of the very high capital costs and licensing complications of building new plants, and the difficulty of obtaining land-use options. Thus, there is often a genuine interest in keeping older plants in operation, compared to the other available or developable possibilities.

Many large fossil plants were originally designed as base-load units, intended to run steadily with as few starts and stops as possible. But many of these large units have been relegated to cycling duty, which requires an increase in the number of start-ups, shutdowns, and load swings above those contemplated in the original design. Thick metal parts, for example, experience substantially increased thermal stress due to cycling. These stresses, in turn, affect the material properties of the components and make them more susceptible to failure through fatigue, creep, and other conditions. Water-chemistry effects on internal corrosion, external gas-side corrosion, erosion, and other mechanisms can also significantly affect boiler life.

Owners, therefore, need to have qualified engineers examine for the presence of these anomalies on an existing boiler and determine how they might influence the performance of both individual components and the boiler in general. Once this is done, a judgment can be made as to the remaining life of the critical components and whether or not it would be economically feasible to keep the unit operational for some additional period of time.

While there are uncertainties with any life-prediction method, an objective assessment is a prerequisite in any plant life-continuation program. The combination of sound pre-planning, inspection, examination, and the use of the best available life-prediction methodology will permit meaningful conclusions to be drawn as to the current condition of a steam generator and its ancillaries, and of its capability to operate beyond its originally planned retirement date. Combustion Engineering believes that often the most effective approach is a combination of operational and maintenance changes, and material or component modifications. Details of its life-continuation methodology and a differentiation of the several assessment techniques are given by Gelbar and de Mello, reference 9.

MAINTAINING AVAILABILITY

In this book we have already discussed many approaches in design and operation that can contribute to sustaining the performance and availability of large steam-generating units: in Chapters 6 and 7, we went into detail on the design of boiler pressure parts and how the design is affected by cycling operation; in Chapter 13, we emphasized the importance of diagnostic evaluation of boiler on-line performance using modern computer techniques; Chapters 21 and 23 covered the influence that operation and plant maintenance can have on the ability of boilers and their auxiliaries to respond when they are called upon to generate power on demand and to do so efficiently.

In this section, we will go beyond what we have presented in the last chapter, so as to introduce techniques of continuing the intended life of power-plant equipment. As we have intimated, a decrease in the availability of any mechanical or electrical equipment is most often the result of a conscious owner decision to let that happen. The process of life continuation—keeping a steam-electric plant in good condition so that it can continue to operate in active and efficient base-load or peaking service for some undefined length of time—is similarly a management decision to repair, replace, rebuild, or upgrade only as necessary to accomplish that end. Any activities exceeding that represent a judgment to *extend* life to, say, 55 or 60 years, which is greater than the executives originally purchasing the equipment contemplated.

WHY A NEED FOR LIFE CONTINUATION?

For a variety of reasons, the construction of base-load type electrical generating capacity in the U.S. has not kept pace with the projected need for it; this is indicated both by projections of load growth and expected plant retirements.

Currently, power producers have to evaluate the cost of life continuation—which may include the cost of environmental compliance equipment—versus the investment for replacement equipment. The engineering analysis that precedes a life continuation program has to be very thorough and positive in its prediction of remaining life of the components that are *not* to be modified or replaced. The engineers responsible for the replacement or upgrading of pressure parts and other components subject to thermal or pressure stress, or to corrosion or erosion, must assure the owner that the life of the unit will in fact be continued as intended.

It behooves owners to continue in-service, preventive, and planned maintenance on a timely basis, in order to prevent any loss in capacity of power-plant equipment that would result from the deferral of such work. In other words, any reductions in capacity must be avoided lest they are made effectively "permanent" through legal interpretations of environmental regulations. In this context, many owners are understandably reluctant to undertake any actions that could be construed as being "beyond" standard maintenance, to the point that removal of safety hazards and efforts to improve plant reliability may be frustrated.[10]

ASSESSMENT OF REMAINING LIFE OF STEAM-GENERATOR COMPONENTS

Condition assessment, the evaluative process that precedes any actual life-continuation activity, is essentially a review of plant operational and maintenance records plus an inspection that leads to a prediction of expected safe operating life left after the date of the inspection. The process might pin-point the most sensitive component—the one that will probably run out of life first—but the estimate of remanent life is strictly an engineering judgment and is not a calculated value. Any repairs that are indicated, or any upgrading of components or subsystems to the state-of-the-art that are recommended, are only to allow the unit to continue in operation in what is felt to be a safe manner.

A condition-assessment study will identify components having *time-independent* or *time-dependent* characteristics that will affect their service life. Life predictions in the time-independent regime involve knowledge of when effective tube-wall thickness will be reduced to the point of failure. There is essentially no deterioration of material strength properties over time; that is, creep does not occur. The rate of corrosion, erosion, or crack propagation must be established to predict life, taking into account safety factors in the ASME Boiler Code as well as realistic long-term future operating conditions. In instances where observable distress is incident-related (rather than time-related), such as with caustic attack, hydrogen damage, or pitting, life predictions are not practical.

The situation in the time-dependent regime is much different. Because of creep damage along grain boundaries, components in the time-dependent temperature range can rupture with essentially no wall loss. Such components have a finite life because their strength-retaining capability is diminishing with time. Uncertainties in life-continuation studies often relate to predicting metal behavior in the time-dependent regime, especially for thick-walled pressure-containing elements.

DESIGN-CODE ASPECTS

Before discussing what is involved in the remaining-life analysis of thick-walled components, it is important to understand what is *not* involved. Life analyses are not a simple review of the design vis-a-vis the ASME/ANSI Codes and a rubber stamp to say that the design meets the Code and is therefore satisfactory. The fact is that, if the unit met the Codes when it was designed, then, unless there are such effects as thinning present, it will meet the Codes now. The reason for this is that the Codes which thick components are built to, do not explicitly consider the effects of time, nature of service, and fluctuations in load. To understand the reasons for this, it is necessary to consider what these Codes are and some of the philosophy behind their development.

First, the Codes are design tools. They are also, in a very real sense, some of the earliest expert systems. Their purpose is the enforcement of design rules that ensure a low probability of failure by specifying the minimum allowable thicknesses of components made of approved materials fabricated in a stated manner.

Secondly, the philosophy of the Codes is based on such concepts as:

■ Allowed materials are ductile and, therefore, in many cases, forgiving.

■ Life, operational effects, material variability, local stresses and other similar effects are not directly considered but are expected to be accounted for by the conservatism applied in setting the allowable material properties.

■ Design rules are developed by consensus within groups of acknowledged experts and are subject to multiple levels of review. Being developed in this way, there may be no precise physical justification for some of the rules.

The result of this type of approach is that the actual stress in a local region of a properly designed component may exceed the allowable stress given in the Code being used. This does not mean that the component is unsafe, but only that the published values of allowable stresses for the material are sufficiently conservative to ensure that the resulting component *will* be safe, even without considering local effects that may occur in design, manufacture, field construction, or operation.

Unfortunately, the simplifications that make for an excellent design tool destroy the usefulness of the Codes as devices for predicting the life of a component with any precision. The fact is that the life of a component is intimately related to the *actual* local stresses, temperatures, and material properties. Ignorance of any one of these has a significant effect on the precision with which life can be predicted. The Codes ignore all of these and, as a result, while it is possible to manipulate such a Code to obtain a value for the life of a component, any life obtained from such an approach may be in error by orders of magnitude.[11]

FAILURE MECHANISMS

In a well-designed and well-manufactured high-pressure component, failure can be caused by unexpected loading and/or environmental situations. A simple design review and an evaluation of inspection results will generally confirm the quality of the design and manufacturing.

The loading and environmental situations that can adversely affect components are:

1. Water chemistry effects
2. External corrosion
3. External wastage due to erosion
4. Change in thermal conductivity due to scale buildup
5. Service at temperatures and/or pressures above design conditions
6. Changes in the behavior of support structures
7. Abnormal conditions such as temperature and flow maldistributions or defective support systems
8. In-service origination of corrosion sites
9. Non-linear temperature and pressure start-up ramps
10. Faster-than-design start-ups and shutdowns
11. Excessive cycling

Items 1 and 2 are not in general amenable to analytical approaches as discussed in previous chapters, and are beyond the scope of the present treatment. Items 3 and 4, although they can be analyzed, are usually associated with tubular components in the gas pass and, as such, are not generically considered as relevant to thick-walled components. The principal situations to be considered are therefore Items 5 through 11. Some of these, such as thermal transients resulting in cracking of internal scale and exposure of corrosion sites, are not presently capable of being analyzed, and will therefore not be considered here.

The parameters that affect the life of a component are the local values of stress and temperature, and its material properties. Life does not only *depend* on these parameters, it is extremely sensitive to variations in them.

The effects of stress can be appreciated if one considers a component designed such that its thickness is just equal to that required by the Code for operation at 1000°F (540°C). If, in any part of such a component, the local stress is a mere 12% greater than the Code allowable value, then the life of the component will be halved. As an example, consider both unpenetrated and penetrated cylinders designed such that their thicknesses are precisely that required by ASME Section I. As far as the Code is concerned these two components are equivalent. However, a detailed stress analysis shows that the local stresses are significantly different. High local stresses will be somewhat relaxed over time by creep effects and, in fact, the location of the high stress areas will move from the inner surface to the outer surface. However, it is a fact that the lives of the two components are not the same and are probably significantly different.

Assuming the same parameters as in the above example, then a 17°F (9.5°C) increase in local temperature above the design assumptions will also reduce the life of the component by half. This is due to the change in material properties with temperature. As an example, temperature variations along a header can be significant and some allowance is made for this in the design process. However this temperature distribution is not precisely predictable nor is it reproducible between similar units. Aside from the safety aspect, from the user's point of view it is just as bad to underestimate the remaining life as it is to overestimate it. The percentage of deviations from design stress and design temperature required to halve the life of a typical component are shown in Fig. 3 as functions of design temperature.

The third variable involved is the material of the component itself. The properties of any single material specification are extremely variable, particularly those which affect the life of a component.

For instance: at 1000°F, there is a factor of five between the predicted life of a component obtained using minimum properties and that obtained using mean properties. The difference between mean properties and maximum properties would involve an additional factor of five.

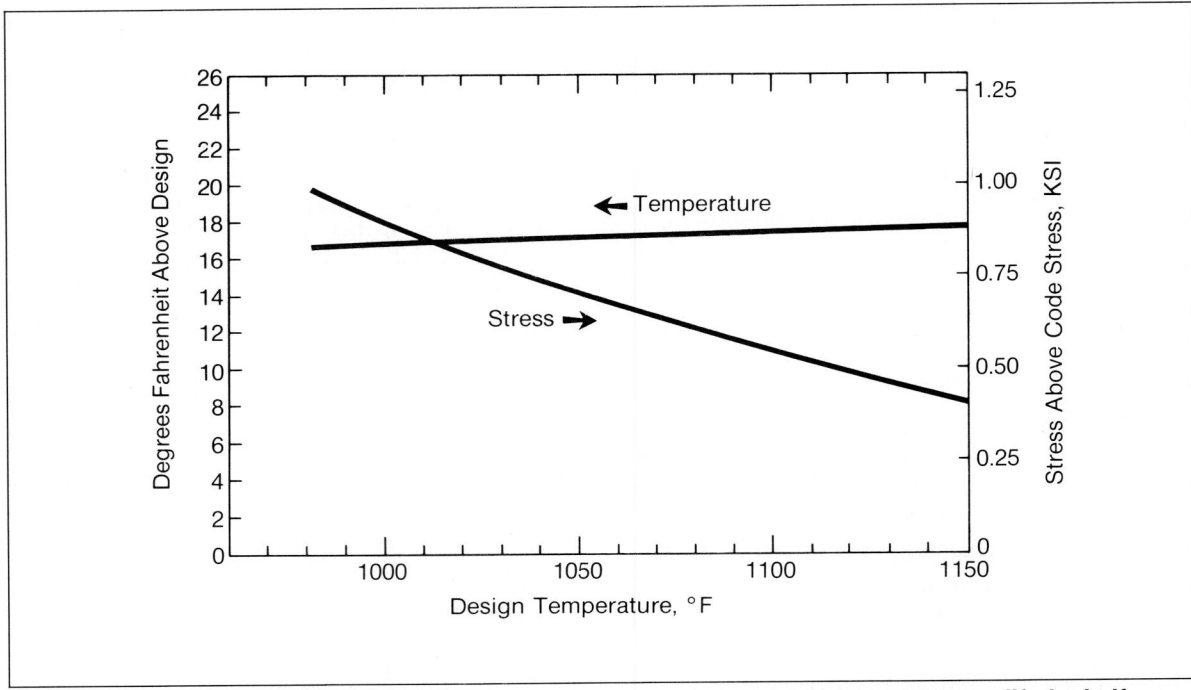

Fig. 3 **Effect of deviations from design for 2¼% Cr, 1% Mo steel, to reduce component life by half**

If one has no information about the material other than its specification, then the minimum material properties have to be used to predict life. Using this approach the predicted life may be underestimated by as much as a factor of 25, purely as a result of the variability of material properties. Fig. 4 shows the minimum and mean lives of a component as a function of operating temperature assuming the actual stress is the allowable ASME Code value.

The above effects apply even if the component is operated purely in steady state; that is, with no variation in thermal, mechanical and/or pressure loadings. In fact, the component undergoes fluctuations in these loadings both in a gross fashion during start-up and shutdown of the system and, in a minor way, during operation. Although the Code does not require it, the manufacturer of a boiler specifies a start-up rate to limit potential damage to the system. However, this start-up is assumed to be smooth and, for stress-analysis purposes, linear. In fact, the change in loadings during start-up can be ex-

tremely non-linear as various subsystems are modulated. In addition, although the shutdown is assumed to be a mirror image of the start-up, the transients in this situation are often even more severe than in the start-up. The effect of transient loadings, particularly in temperature, can be very severe and result in local cyclic plastic deformation which can lead to fatigue failure in a relatively low number of cycles.[11]

As described earlier, there are two principal types of situations that can be considered in the determination of the life of thick-walled components. The first, in the time-dependent regime, is the high-temperature pseudo-steady-state loading where the creep of the material is the phenomenon that leads to failure. The second, in the time-independent regime, is the cyclic plasticity that occurs during start-up and shutdown which can lead to low-cycle-fatigue failures.

CREEP

Creep is the degradation of material properties that occurs with time at temperature. It hap-

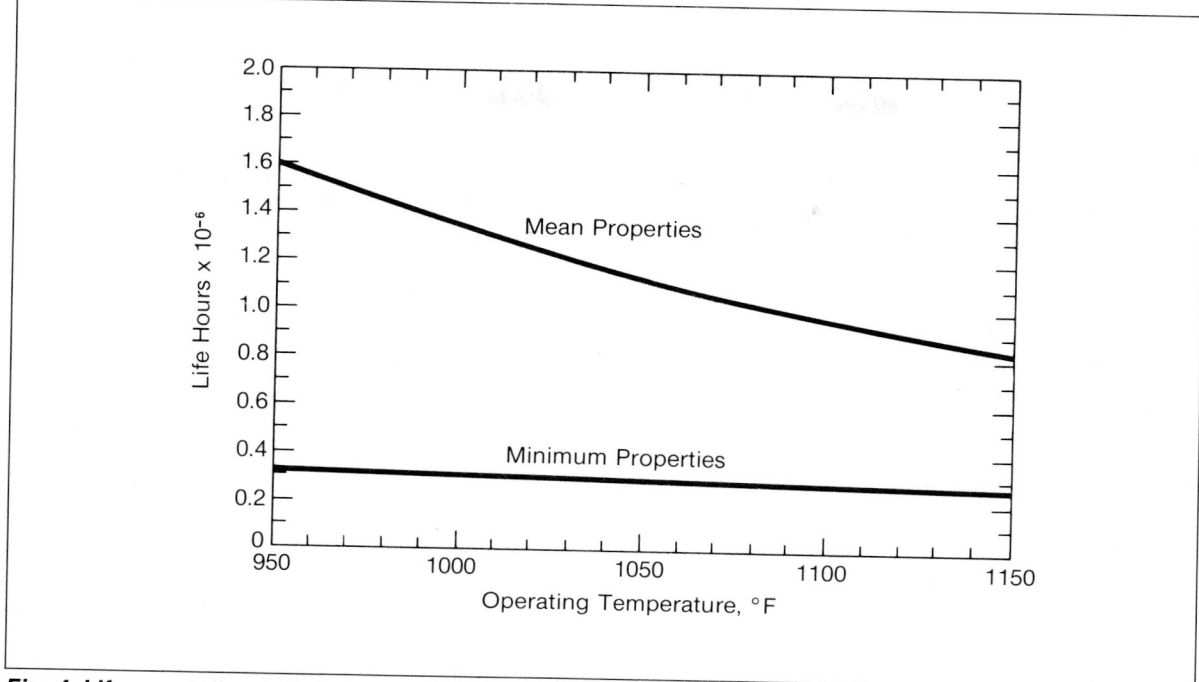

Fig. 4 **Life versus temperature at ASME code allowable stress (material: 2¼% Cr, 1% Mo steel)**

pens even at low temperatures and somewhat differently in the heat-affected zone of a weld than in the base metal. An analysis that does not correctly include the effects of creep may well predict failure in an incorrect location. This is because the damage resulting from creep occurs throughout the life of a component and is exhibited as permanent deformation. This permanent deformation allows a highly stressed area to shed load to a less highly stressed area. Such a redistribution of load can be significant. In the case of a cylinder under internal pressure, if creep is not present the highest stress will be on the inside surface. If the cylinder is operating at creep temperatures, however, the highest stress area will, after time, be on the outside surface and that is where failure will originate. Chapters 6, 7, 17, 20, and 21 cover other design and operational aspects of creep in more depth.

FATIGUE

Fatigue is the process in which materials fail under cyclic loading. For thick-walled compo-

nents of high-pressure boilers, the principal loading that can cause damage is produced by the transient thermal effects that occur during start-up and shutdown. These events are severe but, fortunately, relatively infrequent. When the transient events within a start-up or shutdown are considered, the total number of cycles during the life of a unit tends to be below 10,000. If failure occurs in fewer than 10,000 cycles, it is termed "low-cycle fatigue". It is characterized by local strains in the region of failure that are in excess of the yield point of the material, with the result that the material undergoes plastic flow. (Chapter 6 has a section on fatigue analysis.)

Whenever fatigue failure occurs, it is initiated at a free surface even though this free surface may be at a subsurface flaw. In low-cycle fatigue this initiation is followed by crack growth through a region of previously plastically strained material. Finally, the crack propagates through the rest of the component, through regions that originally saw only elastic strains, and the component fails.

Crack initiation occurs at a very early time in life (<5% of the number of cycles to failure). The surface defect is usually located at a persistent slip band where material extrusions or intrusions occur as a result of cyclic plastic flow. These slip bands occur on planes of maximum principal *shear* stress although the direction of opening will be that of the maximum principal *normal* stress.

Although slip bands occur adjacent to the crack, as soon as the crack starts to open the stress on these adjacent defects is relieved and crack growth is generally limited to a single crack. This is different from the creep situation where gross material degradation occurs throughout the region adjacent to the main crack.

METAL-TEMPERATURE DETERMINATION

Without reasonably accurate estimates of metal temperature, stress-rupture life predictions, whether based on minimum or average strength properties, cannot be accurate. There is research data on the growth of steam-side oxide scales on the low-alloy ferritic steels used in steam generators.[12]

At elevated temperatures, the internal surfaces of pressure parts will slowly oxidize and form an indigenous oxide scale. The rate of oxidation is a function of the internal surface temperature, the type of material, and the time at temperature. It is possible to infer a "historical" tubing metal temperature of a tube sample by measurement of the oxide scale thickness, when there is knowledge of the rate of heat transfer, scale conductivity, scale growth rate, and the length of service. Using such data, the past temperature exposure of a tube can be approximated if the age and thickness of the oxide scale are known. Estimates of this kind of necessity assume unchanging heat input to the tube, and constant steam-side conditions.[12]

DETERMINATION OF CRITICAL COMPONENTS

It is not feasible, for economic reasons, to evaluate in detail the life of every component in a boiler; neither is this necessary. The first step in any life-continuation study, whatever the level of complexity involved, is therefore the determination of what are the critical components.

The initial step in the determination of criticality is an assessment of the consequences of a failure. If the consequences of the loss of a component are significant, then that component meets the first test of criticality.

If the component is one of many similar components with similar loadings and the failure of any one of these components does not result in significant consequences, then it may be that this class of component need not be considered critical. The reasoning behind this is that there will be a distribution in the actual lives of these components and the actual failure rate can be used as the input to the decision as to whether or not it is necessary to replace a system made up of these components. This is usually the case in systems that consist primarily of tubing. The failure of most other components containing high-pressure water or steam would be considered significant.

Design Parameters

The basic parameter to be considered in evaluating the criticality of a component is its design temperature. The fact that a part is designed to operate in the time-dependent regime identifies it as having a finite life and tentatively places it within the class of critical components. Generally, this includes the final superheater outlet header and the hot reheater-outlet header.

Geometric Parameters

Components that are excessively thick may well be sensitive to even relatively slow transients. This is generally not a problem with components fabricated by boiler manufacturers, but there may be a problem with purchased fittings (tees and elbows) and components such as valves and pump casings.[13]

A second geometric consideration is that of configurational complexity. Geometrically simple components can often be eliminated from the list of critical components because of an adjacent one that is more geometrically complex. The reason for this is that, because of the way that design codes handle these, a more complex

geometry will generally have higher local stresses than a simple one.

Experience

The principal guide to determining which of the components that passed the first tests of criticality are in fact critical is experience, either generic or plant-specific.

Generic experience is that knowledge that has been gained from other studies and from observations both within the operating company, the manufacturer's organization, and in the technical community as a whole.

Plant-specific experience involves contractors' personnel and records, as well as those of the owner, and includes such items as knowledge of previous failures and over-temperature operation. It also involves any inspection results; this suggests that the criticality of a component can change following the inspection portion of a life-continuation program. Evidence of distorted or discolored components, cracks in unusual areas, and missing or defective support structures would all cause reassessment of associated components.

Principal Critical Components in Steam Generators

Experience indicates that there are a number of components that should be considered as candidate critical items. The first three of these have experienced difficulties very frequently and should always be considered; an examination of plant records may remove some of the others from the list.

- The final superheater outlet header and the main-steam piping system
- The economizer inlet header
- The high-temperature reheater outlet header and the hot reheat piping system
- Other reheater outlet headers and piping
- Other superheater outlet headers and de-superheater systems, particularly where there are indications of higher-than design levels of desuperheating
- Superheater and reheater inlet headers

APPROACHES TO ANALYTICAL EVALUATION

There are many different ways to evaluate what has happened to a component and what its expected remanent life will be, with the strategy different for pressure vessels and piping. There are five general levels of complexity:
- The design-basis approach
- The operating-records approach
- The pressure and temperature test-data approach
- The material-sampling approach
- On-line monitoring and analysis

As the level of analysis increases, confidence in the results increases, as does cost.

Design-Basis Approach

Manufacturers design boilers to the customer's specifications of pressure and temperature. Each manufacturer has its own design procedure. As with the Boiler Code, this process is an expert system in that it uses constants and coefficients based on experience. Using this procedure, a design is made with the Code used as a design tool to ensure that its requirements are met. In other words, the stresses for the design conditions, calculated in the way that the Code requires, are confirmed to be no greater than the Code-allowable stresses for the specified material at the design temperature. What this means is that the sections will always be at least as thick as the Code requires and often much thicker. In particular, it is frequently standard practice to make the thickness of a cylindrical component equal to the next commonly available size above the precise thickness required by the Code.

Having arrived at knowing what the design criteria were for any given component, and what should be its minimum allowable thickness, we can, with some confirmation from plant records of operational conditions, arrive at a first-order assessment of plant life. In most cases this will say that the majority of high-temperature components are close to, or have passed, their useful life and that the low-temperature components are satisfactory—this despite the fact that there may be physical evidence of damage. In most instances, this approach is just a screening tool at best.

Operating-Records Approach

This level uses the design-basis approach with augmentation by plant records. Because of now-reasonable costs of analytical modelling, we include this type of evaluation although it uses assumed linear ramp rates as input loading data. Minimum material properties and nominal sizes are used.

The analytical tool used is a sophisticated heat-transfer and stress-analysis computer program which includes the effects of creep and plasticity. As such it can overcome the problems of the design-based approach by evaluating the effects of start-ups and shutdowns. However, since these transients are assumed to be linear, this level still ignores their non-linear nature and is therefore non-conservative. This problem can be overcome to some extent by using generic data taken from similar units but, because of the idiosyncrasies of individual units, a high level of uncertainty remains. In general this approach is not recommended because it is frequently assumed, by people not intimately concerned with the process, that, since a sophisticated analysis is performed, a reliable answer is obtained. This is not the case and care should be taken to understand this concern.[14]

The Pressure and Temperature Test-Data Approach

This uses transient pressure, temperature, and flow data, taken during operation, as input for the computer program mentioned above. The data is obtained by installing thermocouples and pressure transmitters on the components under consideration and operating the unit specifically to obtain this data. A more cost-effective approach is to temporarily install a data-acquisition and analysis system so that the evaluation can be done on-line during normal operation. Such an approach avoids the costs of a special test and the associated data handling, and provides the owner with immediate information about what aspects of unit operation are most damaging.

This analysis also uses the measured thickness of components and incorporates the other results of an inspection program.[15] Apart from the uncertainty that the test data may not be representative of past operations, the only other question in this type of approach is the lack of actual material data. As a result, minimum material properties have to be used to provide a lower bound for remaining life.

The use of actual test data allows the prediction of the effects of changes in future operation and also identifies possibly unknown sources of problems such as condensate-flow events.

The Material-Sampling Approach

The next level of sophistication is to take samples of the material present in the components and, after performing accelerated creep tests on these samples, to incorporate these data into the evaluation. This testing is fairly expensive and care needs to be taken to ensure that these results are used correctly.

First, it is part of the requirement for applying accelerated creep test results to actual components, that creep testing be performed at the stress level that the component sees during operation. It therefore introduces errors to perform these tests at stresses that derive from a design approach. The difficulty is that, during operation at elevated temperature, the stress in the component is redistributed by creep relaxation and as a result the operating stress will be significantly different from the design stress. Another aspect of this approach is that, in general, to avoid creating problems with the component, the material sample has to be taken from an area removed from geometrical discontinuities. Unfortunately, in most cases, the critical areas are precisely those areas that have to be avoided during sampling. As a result, unless miniature samples can be taken from the highly stressed areas, it is necessary to translate the results of the accelerated creep testing into the remaining life at the critical areas. This means that, by the use of plant history and analytical techniques, the original creep properties of the material have to be obtained. These can then be applied to the analysis of the critical areas.

Finally, it is unfortunately a fact that high-pressure components are fabricated from substructures made up of different heats of material. Unless the material sample is taken from the critical substructure, the use of accelerated creep test results is inappropriate. In the worst case, if the accelerated creep test results indicate that the material properties are at the top end of the scatter-band, and the critical area is in another substructure where the material properties happen to be at the bottom end of the scatterband, then the life estimate will be non-conservative by a factor of about 25. Here, although an analysis might indicate that the component has many years of remaining life, in fact the critical area may have totally exhausted its useful life.

Despite these cautions, when applied properly this approach does provide the highest level of confidence currently possible, for a single-outage evaluation.

On-Line Monitoring and Analysis

The ultimate level of assessment is similar to that described previously except that on-line monitoring, analysis, and evaluation are accomplished by a device such as the C-E Boiler Stress and Condition Analyzer described in Chapter 13. Such a device monitors temperature, pressure, and flow, and uses this data together with material and geometry information to assess damage on a real-time basis. It makes no assumptions about repeatability of operations. An option to the use of this device is to incorporate "what-if" capabilities such that the effects of changing operational modes can be evaluated.

SUMMARY

Life continuation of a steam-generating system implies an objective of operating the equipment beyond the originally intended life, with these considerations in mind:

- Through rehabilitation or upgrading techniques, the steam generator can be restored to an operational level of availability, reliability, and capacity factor as practically close to "as new" status as possible.

- Currently available "state of the art" component designs can be incorporated in full or in part, to improve the performance level and safety of a steam generator.

- Sophisticated analytical techniques can be applied to estimate the remanent life of boiler components.

- Future operational requirements may change from past or current practice—design changes may therefore be indicated for some of the components subject to rehabilitation.

- Because of aging, or service-related destructive mechanisms, replacement of some parts may be required. Such replacement can be programmed over a period of time based on the calculated remaining life of specific components.

A life-continuation study is not solely a boiler inspection program, or just non-destructive testing of unit components, or only a prediction of remaining life of critical items. Although elements of these are present, timing is also an important issue in the plant life-continuation concept. A proper program is an on-going process in which the owner and the equipment manufacturers work together closely over a long period of time to solve technical and operational problems. Any modifications that are decided on should be phased over as long a time period as possible, for maximum economy. Recommendations for any boiler unit will be owner- and site-specific, and will require a significant degree of owner participation to result in a productive outcome.

REFERENCES

[1] "Availability Patterns in Fossil-Fired Steam Power Plants", EPRI Report No. FP-583-SR, November 1977, Electric Power Research Institute, Palo Alto, CA.

[2] "1985-1989 Generating Unit Statistics", *North American Electric Reliability Council Generating Availability Data System*, Princeton, NJ, 1990.

[3] W. H. Clayton, V. Llinares, and G. C. Thomas, "C-E Availability Data Program", *Proceedings of the American Power Conference*, Vol. 42, Chicago, IL: Illinois Institute of Technology, 1980.

[4] V. Llinares, M. B. Caron, E. J. Schmidt, and F. J. Szela, "The C-E Availability Data Program 1982: Further Developments on MTTR and MTBF Analysis", presented at Ninth Annual Engineering Conference on Reliability for the Electric Power Industry, Hershey, PA, June 16–19, 1982.

V. Llinares and A. B. Lutz, "Long Range Forecast and Availability Improvement Program", *Proceedings of the 1985 Joint ASME/IEEE Power Generation Conference*, ASME Paper No. 85-JPGC-Pwr-9.

[5] "Availability Improvement Program", Division of Power Production, Tennessee Valley Authority, Knoxville, TN, 1977.

[6] S. S. Blackburn and D. E. Lyons, "Design for Availability — An Update", *Proceedings of the American Power Conference*, Vol. 39:349–368, Chicago, IL: Illinois Institute of Technology, 1975.

[7] C. H. Gilkey and J. D. Rogers, "A Summary of Experiences with Fan-Induced Duct Vibrations on Fossil-Fueled Boilers", *Proceedings of the American Power Conference*, Vol. 37:728–734, Chicago, IL: Illinois Institute of Technology, 1975.

[8] "Extending the Lifespan of Fossil Plants", *EPRI Journal*, June 1983, Electric Power Research Institute, Palo Alto, CA.

"Generic Guidelines for the Life Extension of Fossil Fuel Power Plants", EPRI Report No. CS-4778, November, 1986, Electric Power Research Institute, Palo Alto, CA.

[9] D. E. Gelbar and S. J. deMello, "Assessing Boiler Life Continuation Needs", *Proceedings of the 1988 Joint ASME/IEEE Power Generation Conference*, Philadelphia, PA, September 25–29, 1988, ASME Paper No. 88-JPGC-Pwr-37.

[10] J. S. Baylor, "Acid Rain Impacts on Utility Plans for Plant Life Extension", *Public Utilities Fortnightly*, March 1, 1990.

[11] J. D. Fishburn et al., "Approaches for the Determination of Remaining Life in High Energy Piping Systems", *Proceedings of the American Power Conference*, Vol. 50, Chicago, IL: Illinois Institute of Technology, 1988.

T. McColloch, J. D. Fishburn, G. E. Roberts, and G. Hunter, "Evaluating the Structural Integrity of High Energy Piping Systems on Fossil Boilers", presented at the Second EPRI Fossil Plant Inspections Conference, San Antonio, TX, November, 1988.

[12] R. P. Aubrey, B. A. Hawkins, and T. D. Jamison "The Use of Oxide Scale Thickness Measurements in Life Extension Analysis", presented at ASNT 1989 Spring Conference, Charlotte, NC, March 22, 1989.

[13] J. D. Fishburn and R. W. Loomis, "Life Extension of Thick-Walled Components", presented at 11th International Conference of the AMIME, Irapuato, Mexico, November 2–4, 1988.

[14] F. V. Ellis, R. W. Loomis, and S. Tordonato, "Life Extension: The C-E Approach to the Analysis of Thick Walled Components", presented at Conference on Life Extension and Assessment of Fossil Plants, EPRI, Washington, June, 1986.

[15] B. W. Roberts, F. V. Ellis, and R. Viswanathan, "Utility Survey and Inspection for Life Assessment of Elevated Temperature Headers", *Proceedings of the American Power Conference*, Vol. 47:259-301, Chicago, IL: Illinois Institute of Technology, 1985.

CHAPTER 25

Combustion Engineering
Research and Development Facilities

This chapter will describe the laboratory facilities that Combustion Engineering has assembled for conducting research and development in the power generation field. The facilities range from small, sophisticated devices for examining the microstructure of various fuels and materials, to large furnaces which permit testing of full-scale commercial components such as firing systems. The focus of C-E research and development is quite broad, covering such areas as troubleshooting for customers who are experiencing problems, improvement or optimization of present designs, and the investigation and development of new concepts.

A common goal in most of the work conducted in these laboratories is the generation of less-expensive electrical power in an environmentally acceptable fashion. More efficient fuel utilization, greater component availability, and longer component life represent some of the specific R and D objectives which support the overall goal.

Combustion Engineering strongly believes that a well-equipped laboratory which is staffed with highly skilled personnel is mandatory to stay at the forefront of power-generation technology. The C-E laboratory facilities are used for conducting company-sponsored projects and programs sponsored by private organizations, including its customers, as well as those funded by the federal Department of Energy, the Electric Power Research Institute, the Environmental Protection Agency, the Gas Research Institute, and other similar governmental and institutional organizations.

The implementation of most research and development programs requires the integration of efforts from many disciplines within the laboratory complex. For example, if the owner of a large steam generator is contemplating a switch to a new fuel, know-how is required in the areas of fuel combustion and fireside effects, heat transfer, and environmental consequences. The C-E facilities are structured so that the appropriate expertise exists in the following functional groups:

- Fuels Technology
- Fuel Systems Development
- Mechanical Systems Development
- Chemical Systems, including Water Technology
- Chemical Analytical Services
- Materials Technology
- Electrical Systems

The equipment and facilities described in this chapter are located in the C-E Kreisinger Development Laboratory (KDL) in Windsor, Connecticut. The laboratory is named after Henry Kreisinger, Combustion Engineering's first Director of Research. Past accomplishments of KDL have directly led to the development and improvements of the steam-generation, fuel-burning, fluidized-bed, ash-handling, and emission-control technologies that are presented in this book. For example, KDL's work on water-side technology associated with high-pressure steam generators forms the basis for our entire Chapter 20 on power-plant water technology.

FUELS TECHNOLOGY

The Fuels Technology group performs basic and applied research which focuses on the characterization of fuel properties and the impacts of these properties on the design and performance of steam-generating equipment. Work in fuels technology centers on the broad areas of combustion, and the handling and processing of fossil fuels. Typical projects range from bench-scale determination of combustion kinetics to pilot-scale evaluation of boiler fireside effects and flue-gas emissions. Such bench- and pilot-scale evaluations provide detailed quantitative fuel performance data that can be applied to the setting of key boiler design parameters, or that can be used to predict significant operating and performance changes when switching fuels on existing units.

DROP-TUBE FURNACE SYSTEM

Fundamental studies of the reaction kinetics of solid fuels during combustion and gasification processes are conducted in KDL's Drop-Tube Furnace Systems (DTFS), Fig. 1. Two DTFS's are available to provide controlled high-temperature environments representative of suspension-fired commercial systems. A schematic diagram of a DTFS is given in Fig. 3 of Chapter 2. The devolatilization, combustion, and gasification reactions that take place in the furnace can be detected and quantified by analysis of the solid

and gaseous reaction products. The combustion kinetics of a broad range of coals, petroleum coke, coal-derived synthetic fuels, and refuse-derived fuel (RDF) have been studied in the DTFS, as described in Chapter 2 under Char Reactivity. Knowledge gained from this facility has been used in conjunction with C-E's computational boiler models to successfully predict carbon loss in large coal-fired boilers (see Chapter 6). The DTFS has also provided basic data on the formation and control of emissions such as nitrogen oxides, dioxins (from RDF firing), and organic hydrocarbons.

FIRESIDE PERFORMANCE TEST FACILITY

Fuel properties that affect suspension-fired boiler performance are evaluated in the Fireside Performance Test Facility (FPTF), a 4-million Btu/hr (1.2 MW) pilot-scale combustion facility (Figs. 2, 3, and 4). Performance areas addressed include solid-fuel handling and pulverization, combustion, ash slagging and fouling, fireside corrosion, flyash erosion, and gaseous and par-

Fig. 1. **Drop-Tube Furnace System (DTFS)**

ticulate emissions. The facility operates through-out the range of conditions actually found in large central-station boilers, regarding flame and gas temperatures, furnace residence times, and gas velocities. Testing typically focuses on establishing fuel-related operational limits in the various performance areas.

In the radiant section of the test furnace (Fig. 3), a series of simulated waterwall panels collect slag deposits which are studied for their physical and chemical properties, their ease of removal, and their heat-transfer characteristics. A series of probe banks in the convection-pass section (Fig. 4) is used to determine fireside corro-

sion activity, ash-deposit buildup rates, the influence of deposits on heat transfer, and the ease of deposit removal. Flyash erosion rates are measured in a high-velocity test section downstream of the convection test section. An electrostatic precipitator (ESP) follows the erosion section to evaluate flyash collectibility and resistivity; a more detailed description of the ESP is given later in this chapter. A by-pass loop is also available for conducting tests on selective reduction of various regulated gaseous emissions. Continuous-sampling and analysis equipment allows measurement of particulate and gaseous emissions.

The quantitative performance data obtained from the FPTF is used with C-E's computational boiler models to optimize new designs and to predict performance of existing boilers under other-than-design conditions. The FPTF has been used to predict the burning and ash-deposition characteristics of many U.S. and foreign coals, coal blends, synfuels, beneficiated coals, and coal-water mixtures.

FLUIDIZED-BED TEST FACILITIES

Fuel and sorbent performance during fluidized-bed combustion are characterized in a

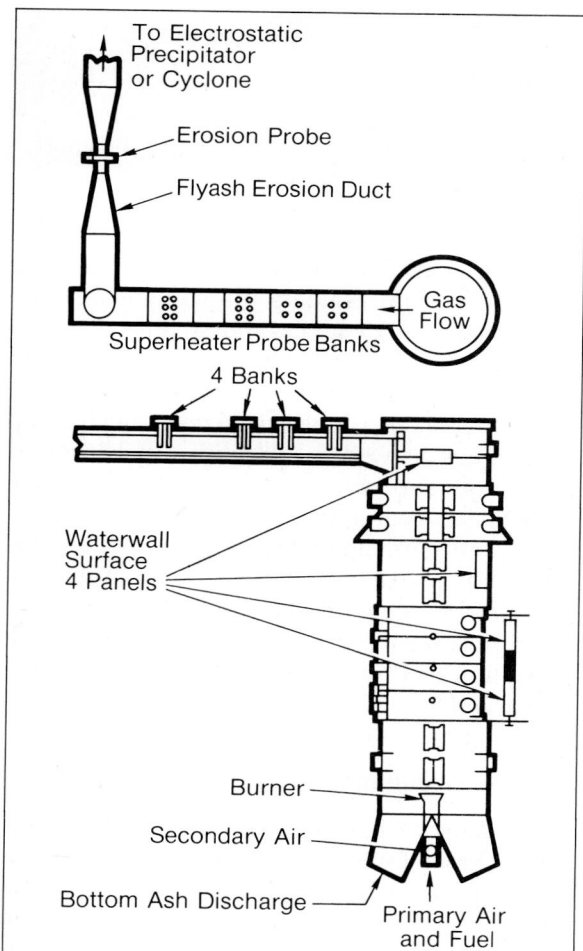

Fig. 2. Fireside Performance Test Facility (FPTF)

Fig. 3. Waterwall test panel in lower furnace of FPTF

Fig. 4. **Upper furnace section and convection-pass sections of FPTF**

4-inch (100 mm) diameter fluidized-bed reactor. This apparatus provides highly controlled conditions for assessment of devolatilization rates, char reactivity, and fragmentation and attrition properties of various solid fuels and additives. The reactivity and attrition attributes of different sorbents for sulfur capture can also be established under actual fluid-bed operating conditions.

FUEL SYSTEMS DEVELOPMENT

Fuel systems development work focuses on large-scale handling, processing, and combustion characterization (including gaseous emissions control) for a wide range of fossil fuels. Typical projects range from designing and assessing oil-atomizer performance to full-scale firing systems development.

FIRING SYSTEMS DEVELOPMENT COMPLEX

This installation, Fig. 5, integrates three commercial-scale test furnaces into a research complex designed to develop advanced firing-system concepts and to evaluate the combustion performance of a wide range of fossil fuels. The support equipment provides maximum investigative flexibility in preparing and handling conventional fuels such as pulverized coal, fuel oil, and natural gas, as well as alternative fuels such as coal-water and coke-oil mixtures; water-in-oil emulsions; coal-derived solids, liquids, and gases; and municipal solid wastes. A pulverization facility which is part of the research complex produces commercial-size pulverized coal for the test furnaces. Located adjacent to the pulverization facility is a pilot plant for producing coal-water fuel (CWF) and for the storage and pumping of CWF.

FULL-SCALE BURNER FACILITY

Commercial scale suspension-firing systems are developed and optimized in this facility. It has a maximum firing rate of 100 million Btu/hr (29 MW_t) heat input on coal and 300 million Btu/hr (88 MW_t) heat input on liquid fuels. Furnace configuration and heat absorption can be

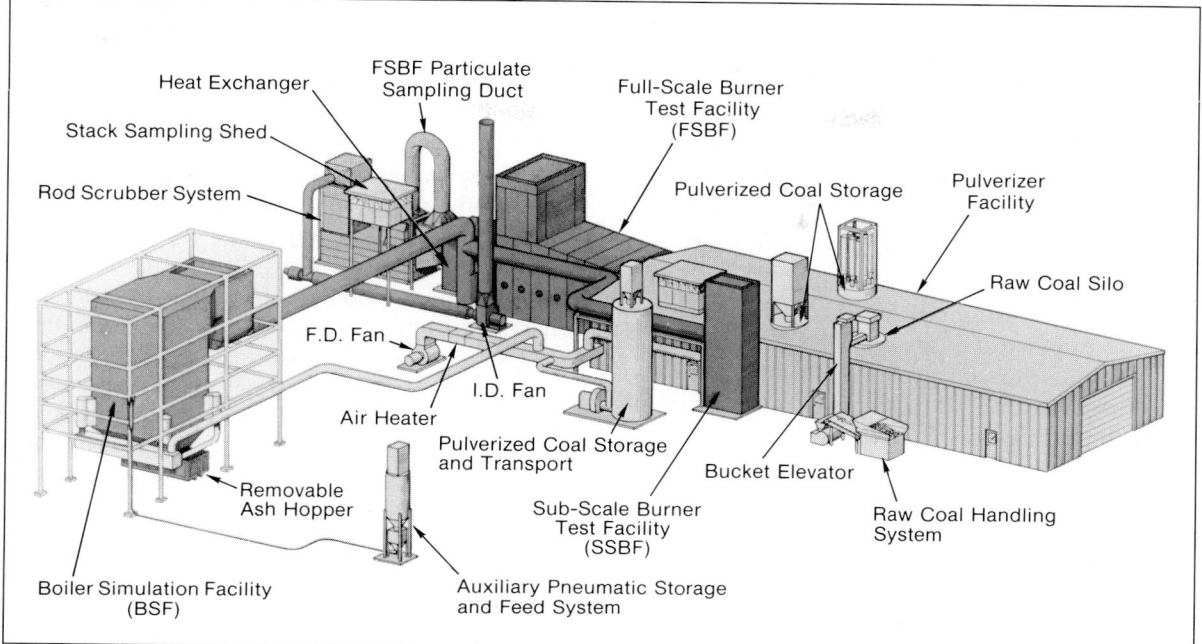

Fig. 5. Arrangement of firing systems development complex

adapted to simulate various commercial furnace applications.

The primary use of this installation is to develop and test central-station scale single-register burners and fuel-admission assemblies. Typical test programs carried out in the full-scale burner facility center on characterizing burner combustion performance in terms of gaseous and particulate emissions, combustion efficiency, flame stability and shape, and other commercially important parameters such as turndown capability. Associated instrumentation provides the means for monitoring both in-flame and downstream flue-gas combustion-related phenomena. New firing systems are optimized through iterative testing and modification, with support often given by other segments of the complex such as the Atomizer Test and the Burner Aerodynamics Facilities discussed below.

SUB-SCALE BURNER FACILITY

This facility (Fig. 6) has essentially the same capabilities as the full-scale burner facility, except for the firing rate, which is 70 million Btu/hr

(20 M_t) on liquid fuels and 25 million Btu/hr (7 M_t) on coal. It is used to develop smaller scale central-station or large-scale industrial ignition and firing systems, and to allow measurement of pertinent combustion-related phenomena.

Fig. 6. Sub-scale burner facility

ATOMIZER TEST FACILITY

This equipment is used to characterize the performance of full scale liquid-fuel atomizers (up to 10 U.S. gallons per minute or 0.6 liters/s capacity) in a non-combustion environment. The facility is uniquely designed to provide optical access to the fuel spray. This facilitates the determination of droplet size distribution, droplet velocity, and droplet trajectory using non-intrusive instrumentation such as double-spark photography and laser-diffraction techniques. This installation has proven to be a valuable tool in designing high-performance atomizers for many liquid and slurry fuels. Atomizer design is further enhanced by the use of proprietary performance-prediction software.

BURNER AERODYNAMICS TEST FACILITY

A key function of wall-fired steam-generator burners, as described in Chapter 12, is to manipulate the inflowing combustion air stream so as to promote ignition and flame stability, and to produce the correct flame shape. One prime method of obtaining burner ignition stability is to employ swirled airflow, which generates internal recirculation of hot combustion gas. The magnitude of swirled combustion air in the burner also has an important bearing on NO_x emissions. A swirler test apparatus is used to measure the angular momentum of the swirled air and to thereby quantify the performance of various designs of swirlers.

The overall aerodynamic characteristics of burners and fuel-admission assemblies are also studied in the burner aerodynamics facility, Fig. 7. Full-scale and partial-scale transparent plastic models facilitate flow measurement and visualization in cold flow experiments. Burner aerodynamics are evaluated first with flow-visualization techniques that are recorded on videotape and film, and second, quantitatively

Fig. 7. **Flow visualization evaluation of a single-elevation fuel-admission assembly**

by three-dimensional velocity- and pressure-mapping.

Aerodynamic data is used in the development of new burners through the interpretation of velocity-vector maps, to calculate the total areas and velocities of near-field recirculated flows, and to compute pressure-loss coefficients.

BOILER SIMULATION FACILITY

This large facility is designed to accurately model the furnace and convective-pass sections of a large suspension-fired steam generator (see Figs. 8 and 9). It can be configured alternately with 3 to 6 elevations of tangential fuel-injection assemblies, for a total of between 12 and 24 assemblies. The unit has a nominal rating of 50 million Btu/hr (15 MW_t) when burning pulverized coal and 100 million Btu/hr (30 MW_t) when burning oil or gas. The firing system can also be arranged to demonstrate a multiple burner array for wall-fired systems, or with a stoker for industrial and refuse-fired applications. The facility also has additive-injection, flue gas recirculation, and reburn-fuel injection capabilities.

The boiler simulation facility is specifically designed for the study of total boiler phenomena. This includes SO_2 capture by sorbent injection, and advanced in-furnace combustion and post-combustion techniques for NO_x emissions control, such as reburning, air staging, or flue-gas treatment using urea or ammonia injection, as described in Chapter 15. The facility will accurately model the time/temperature history, heat release, and heat absorption of a large central-station boiler. It incorporates all the major aspects of such a large unit, including the lower furnace, ash hopper, firing zone, nose section, and upper-furnace. The unit also includes waterwall panels and convective heat-transfer surfaces that duplicate the thermal and aerodynamic conditions of commercial boilers.

COAL-WATER FUEL PILOT PLANT

This facility is used to study the formulation of coal-water fuels (CWF) and to produce CWF test samples for research applications. It has a processing capacity of approximately 1,000 lb/hr (0.13 kg/s) of fuel. The facility can use feedstocks

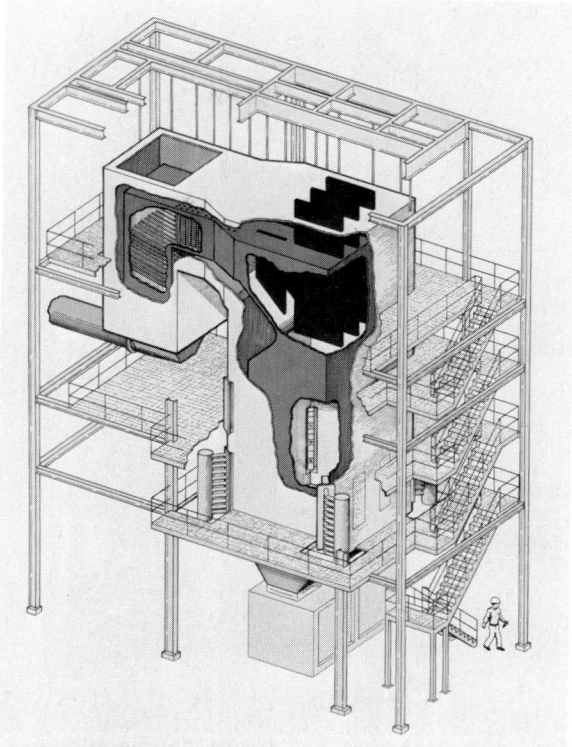

Fig. 8. **Boiler Simulation Facility (BSF) arrangement**

of either dry pulverized coal or wet filter cake. Capabilities exist to produce CWF prepared from coal ground to usual boiler standards (approximately 70 percent through 200 mesh) or to further grind the coal feedstock to produce CWF using micronized coal (with particle size below 10 micrometers).

The pilot-plant CWF production equipment allows evaluation and scale-up of specific fuel formulations developed in the laboratory, as well as assessment of various processing-component design. Process parameters can be optimized with respect to fuel handling and combustion performance. Complete analytical capabilities support the CWF research activities and provide CWF-product quality control. CWF's made in this pilot plant have been used in numerous research and development programs, including those addressing boiler, gas-turbine and diesel-engine applications.

Fig. 9. Boiler Simulation Facility in operation

MECHANICAL SYSTEMS

The mechanical systems group performs basic and applied research in areas of heat and mass transfer, fluid dynamics, thermal hydraulics and two-phase (steam/water) flow, erosive wear, pulverization, bulk-material handling, two-component flow, and particle separation. Customized engineering studies provide cost-effective solutions to design, operational, or maintenance problems at existing boiler plants.

Mechanical-systems research revolves around process modeling, using a combination of experimental and analytical methods. Extensive use is made of physical modeling and statistical techniques, in conjunction with computational modeling. Mechanical design and testing of various power-plant components support programs for steam-generator upgrading and improvement of plant performance.

FLOW MODELING

Flow modeling at KDL uses both physical and computer methods (see Fig. 10). Physical modeling requires the design, construction, and testing of prototypical models using an appropriate fluid. Working fluids can be air, water, or some other medium—including multi-phase fluids or components such as solid particles or liquid droplets—depending on the problem under study. Flow modeling is used for analyzing pulverized-coal and fluidized-bed boilers, heat-recovery steam generators, gasifiers, flue-gas desulfurization spray towers, fabric filters, electrostatic precipitators, and dry scrubbers. Depending on the specifics of the problem, two dimensional water-table models or three-dimensional airflow models are employed.

Qualitative flow visualization plays a dominant role in analyzing many process-equipment situations. This technique is used in conjunction with two- and three-dimensional physical modeling. The two-dimensional technique is a valuable tool in reducing the design effort necessary to achieve optimum equipment performance. Based on qualitative results and the complexity of the

Fig. 10. Large-scale furnace aerodynamic test facility

process configuration, field implementation may follow; alternatively, three-dimensional flow testing is conducted for the most promising solutions. Flow-visualization methods include the use of smoke injection, yarn streamers, or particle seed material. Design changes are first screened visually before quantitative data is collected.

Quantitative component optimization is achieved through evaluation of changes in flow fields due to geometric alterations within process equipment. Flow-field changes are arrived at using pressure and temperature measurements combined with gas sampling. Sampling devices are controlled by one of many host computers located throughout the laboratory. A central data-acquisition system configures the information into permanent files and loads it into the KDL mainframe computer. The principal measurement system is called the Automatic Probe Traversing Device (Fig. 11) and is used for securing quantitative data from 3-dimensional models.

ASH-EROSION TEST FACILITY

In this test stand, furnace-tube material specimens are eroded in crossflow conditions that are both controlled and measured using state-of-the-art instrumentation and controls. From the infor-

Fig. 11. **Automatic Probe Traversing Device**

mation generated, the erosivity of a coal flyash, process solid, or other material can be determined and recommendations made to lessen wastage of pressure-part tubing or other boiler components.

FURNACE AERODYNAMICS TEST FACILITY

This large-scale test facility (Fig. 10) is a focal point in the ongoing study of suspension-fired furnace aerodynamics at C-E. It has been used for many applied research and development programs, including the design of retrofitable low-excess-air firing systems for NO_x reduction and the evaluation of upper-furnace sorbent injection systems for SO_2 removal.

HEAT-TRANSFER AND CORROSION TEST LOOP

This major laboratory installation (Fig. 12) is used to investigate heat-transfer or corrosion on the water or steam side of steam-generator tubing. The pumps, piping, and instrumentation can operate at up to 800°F and 4,000 psig (430°C and 28 MPa gage) in order to test at supercritical pressure and temperature conditions. Heat input is electrically controlled by four separate circuits, with a total capacity of 1.5 megawatts. The loop controls are fully automatic, which permits continuous operation with minimal operator attention. This loop has been used to test commercial boiler tubing for applications to fossil-fuel boilers, nuclear steam generators, and solar-receiver boilers.

ELECTROSTATIC PRECIPITATOR PILOT PLANT

A pilot-scale ESP at the flue-gas discharge of the Fireside Performance Test Facility (FPTF) is used to compare the characteristics of two or more different coal flyashes under similar ESP operating conditions; see Fig. 13. The length, plate spacing, electrode design, rappers, power supply, and specific collecting area (SCA) simulate a commercial ESP, as described in Chapter 15. Since the aspect ratio is significantly different from a full-size ESP, the results from this pilot facility cannot be directly applied to the design or performance of a commercial unit, but must be used in a design program with appropriate scale-up factors.

Fig. 12. **Heat-transfer and corrosion test loop**

The exhaust flue-gas stream from the FPTF passes through a gas cooler to provide a desired inlet-gas temperature to the ESP. Isokinetic samples of the flue gas and flyash are taken at the inlet and outlet to determine dust loading, and gas flow rate, temperature, and moisture content. In-situ ash resistivity and gaseous SO_2 measurements can be made during testing, as well as the electric power consumption of the ESP. After testing, the isokinetic dust sample and dust samples collected in the hopper beneath each ESP field are analyzed for resistivity and particle size distribution. This data is analyzed to determine dust-collection efficiency as a function of the above variables. The efficiencies obtained in the facility are then used to verify or adjust efficiency calculations by utilizing ESP design programs.

HIGH-PRESSURE TEST FACILITIES

C-E has several high-pressure steam-water test facilities for steam-generator component test and evaluation, incorporating a 30-MW_t high-pressure boiler capable of operating at pressures to 3,000 psig (21 MPa gage). Steam-water mixtures from the boiler are used to provide operating environments to evaluate components of fossil and nuclear steam supply systems.

Components tested have included:

- steam-water separators
- pilot-scale nuclear steam generators
- safety and isolation valves
- one-fifth scale nuclear circulating pumps
- Combined Circulation® boiler circulating pumps

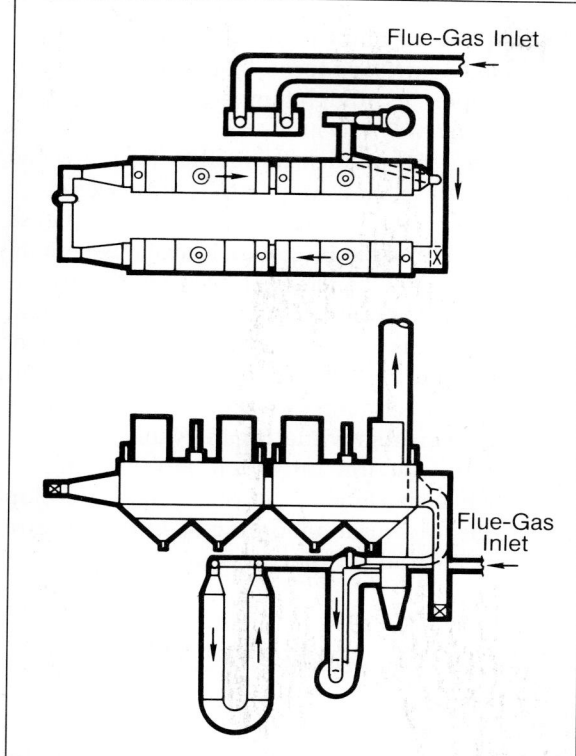

Fig. 13. **Electrostatic precipitator test facility: plan and elevation views**

The facility shown in Fig. 14 was used to test a family of nuclear primary-system safety valves.

CHEMICAL SYSTEMS

Work in this area focuses on water and steam chemistry, internal and external boiler-tube corrosion, and both basic and applied research in the area of environmental control systems. Typical projects may deal with boiler cleaning and pressure-part materials testing, the determination of thermal conductivities of water-side tube scale, and the development of new technologies for flue-gas cleaning. Controlled-temperature probes and integral tubing samples are used to provide design information on external corrosion for alternative pressure-part material selections. These efforts are coordinated with the activities of the metallurgical and materials specialists in the laboratory.

The chemical systems group is responsible for developing water-treatment procedures essential to maximizing fossil-fired steam-generator availability. Their experimental work includes the investigation of ion-exchange technology, operational and out-of-service corrosion protection, chemical cleaning, and water treatment methods. Boiler owners can take advantage of Kreisinger Development Laboratory consulting services to have tube failures analyzed, to resolve operating problems, or to be brought up-to-date on current technology in the above areas.

As described in Chapter 20, steam-generator tubes collect internal deposits during operation. Because these deposits may promote corrosion or cause tubes to overheat, it is necessary to remove the deposits periodically by chemical cleaning. A Chemical Cleaning and Coatings-Evaluation Test Loop is used to evaluate new chemical solvents and to test the effectiveness of standard solvents on unusual deposits. Testing is done under dynamic conditions to simulate actual boiler operation and to find out if a certain chemical-cleaning operation will be effective in the field.

Fig. 14. **High-pressure steam-water test facilities**

FLUE-GAS CLEANING PROCESS DEVELOPMENT

Much of the process development work for the wet-scrubbing flue-gas desulfurization system described in Chapter 15 was done by the chemical systems group in both pilot and bench-scale facilities. Test programs provided information needed to quantify gas-to-liquid absorption, product crystallization, absorbent reactions, and product properties affecting sludge disposal. Interpretation and correlation of the generated data was facilitated by the development of mathematical models which describe unit operations. These models later became part of the design procedure now used for commercial FGD systems.

The process of removal of SO_2 and other acid gases such as HCl and HF in a spray-dry absorber has been described in detail in Chapter 15. A pilot-scale facility (Fig. 15) in KDL has been used to advance the development of dry scrubbing for application to central-station, industrial, and municipal-refuse burning boilers.

Chemical systems development work is also dedicated to advanced concepts for gas cleaning such as the use of permeable membranes to separate SO_2 or H_2S from flue gas, coal-gasification product gas, or natural gas, using electrical potential as the driving force. The electrochemical process operates at high temperature and is similar in many ways to molten-carbonate fuel-cell technology. A bench-scale facility (Fig. 16) has been assembled to investigate this process.

Many of the development tasks or process questions addressed by chemical systems personnel may use such bench-scale apparatus, or information may be obtained from more conventional chemical laboratory equipment. As an example: data on calcination and sulfation of limestone in fluidized-bed combustors were obtained on a standard thermogravimetric analyzer. The information on reaction rates, temperature effects, and particle-size effects was then used to develop equations for overall modeling of sulfur capture in the coal-fired fluidized beds.

CHEMICAL ANALYTICAL SERVICES

This group utilizes a modern chemical laboratory to provide comprehensive chemical and

Fig. 15. Dry-scrubbing pilot-scale facility for flue-gas desulfurization development

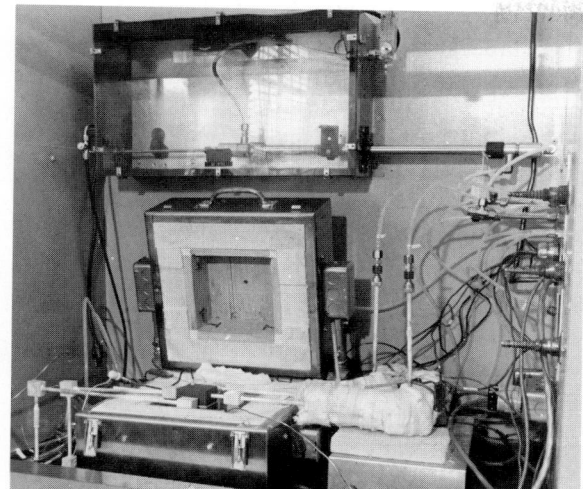

Fig. 16. **Bench-scale facility for investigating electrochemical gas cleaning process**

physical testing of fuels and boiler-related structural materials. The prime functions of this group are to:

- support the fossil-fuel research and development activities

- support field start-up, performance testing, and maintenance activities
- provide analytical and consultative services to C-E equipment owners

The highly qualified staff of chemists and technicians actively participates in *TAPPI* and *ASME* committees, and in *ASTM* committees on coal and water analysis.

The analytical laboratory has a full complement of state-of-the-art instruments to provide accurate data on a whole range of sample analysis, including:

- graphite furnace atomic absorption
- thermogravimetric analysis (Fig. 17)
- differential thermal analysis
- ion chromatography
- x-ray diffraction and fluorescence (Fig. 18)
- inductive-coupled plasma spectrometry (Fig. 19)
- mercury porosimetry
- BET surface area measurement
- particle size analysis

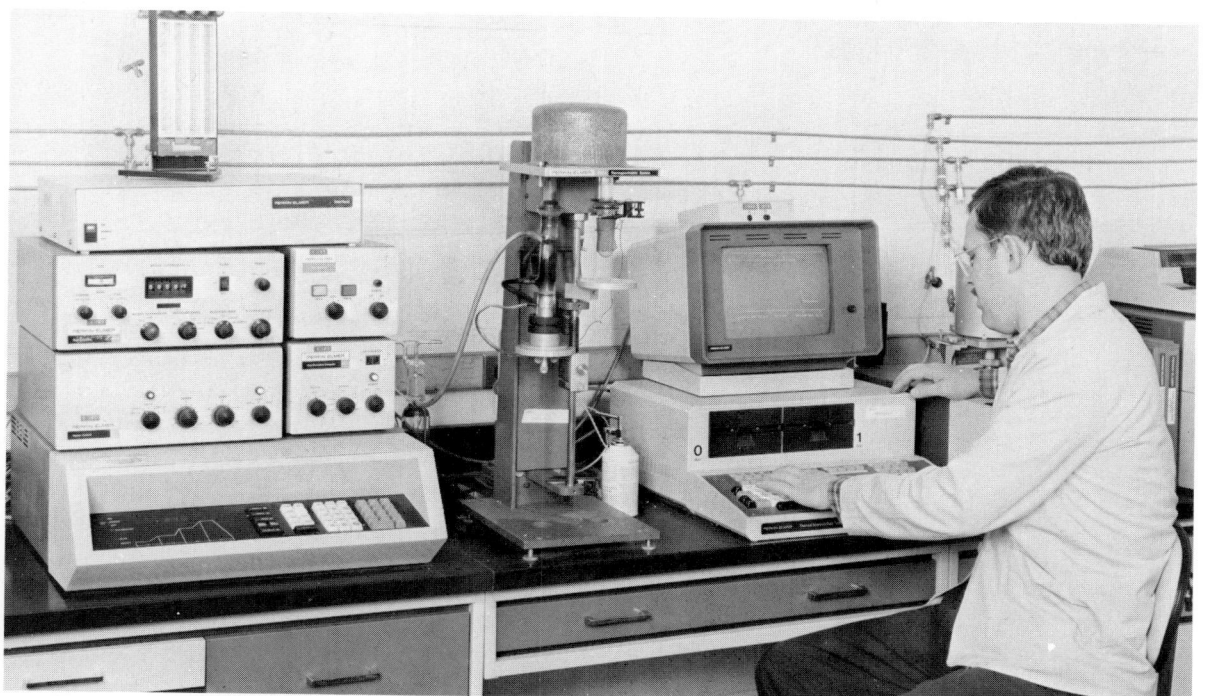

Fig. 17. **Thermogravimetric analyzer**

Fig. 18. X-ray diffractometer for compound identification of crystalline materials

The analytical laboratory is a fully equipped fuels laboratory that can provide a full-spectrum analysis on any type of solid or liquid fuel. It has capabilities in water analysis, boiler tube examination/failure analysis, and metal analysis.

The chemical analytical services group is frequently called upon to support activities in the following areas:

- life-assessment programs on boiler pressure parts
- fuel-blending studies
- pulverizer abrasion
- environmental and industrial-hygiene aspects of steam-generating-unit operation
- analyses of refuse fuels, black liquor, boiler-tube deposits, fuel additives, and chemical recovery unit byproducts, besides the analysis of conventional fuels.

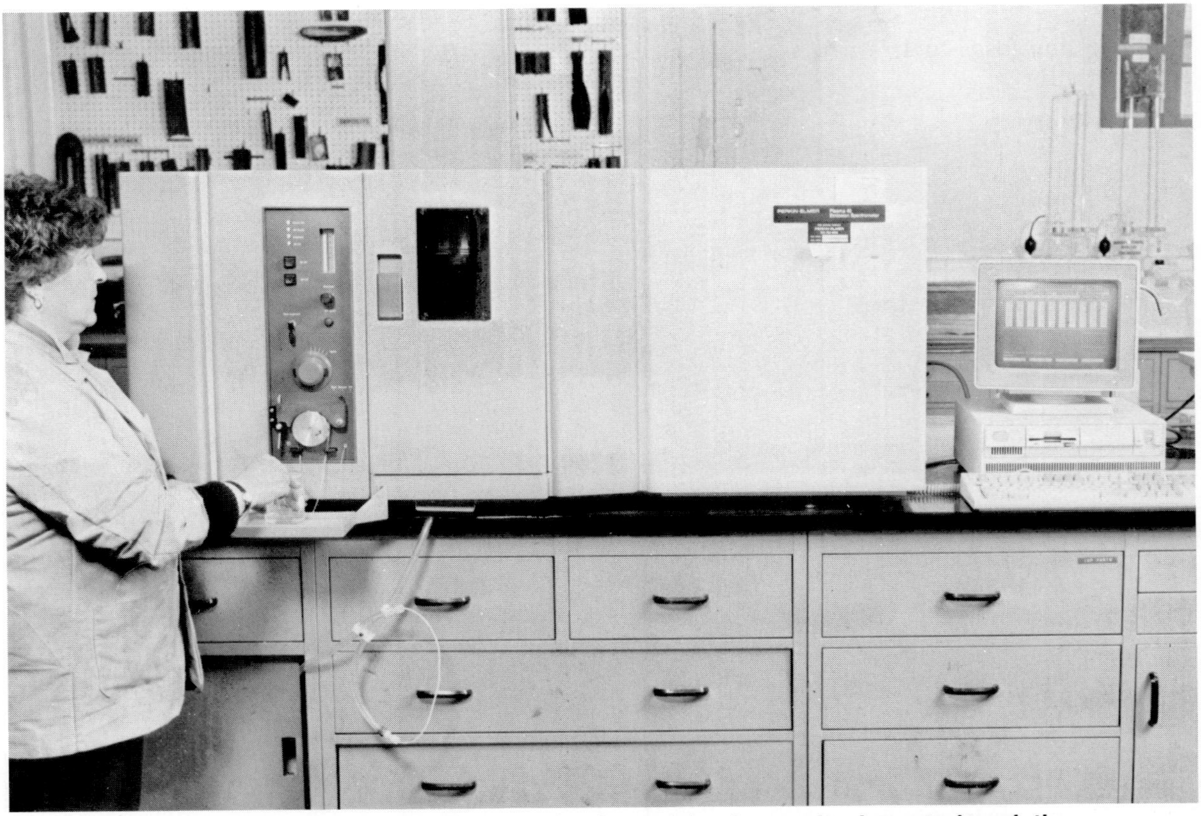

Fig. 19. Inductive-coupled plasma spectrometer for determining inorganic elements in solution

MATERIALS TECHNOLOGY

The close interaction of materials technology with all phases of materials usage is essential for the successful design, fabrication, and service behavior of a boiler and its associated equipment. To attain this end, a laboratory works closely with engineering, manufacturing, quality control, and field services. A continuous process of feedback and information exchange between laboratory and functional departments assures that material-related needs and problems are properly addressed. This process has resulted in the materials technologists at C-E contributing to achievements that include:

■ Production of the first welded boiler drum in America fabricated to ASME Code specifications

■ Development of the low-hydrogen welding electrode, which has since become a universally used electrode in industry

■ Successful welding of the first stabilized stainless-steel pressure vessel

■ Design and fabrication of the first commercial reactor for nuclear power generation

■ Development of the large welded-wall water-wall panel, a design innovation later adopted in boiler manufacture throughout the world

■ Development of the modified 9Cr-1Mo ferritic alloy for high-temperature service, which has been widely adopted for pressure-containment components

■ Formulation and development of a superior hardfacing alloy for use in coal pulverizers

■ Development of surface-modification procedures, chromizing in particular, for protecting metallic boiler components against hostile environments

The laboratory facilities available to C-E materials personnel make the continuation of such achievements possible. Modern equipment and highly skilled specialists in the areas of analytical chemistry, metallography, and mechanical properties determination—all fundamental tools in the study of materials—are continually employed in alloy development, surface protection,

failure analyses, and life-continuation predictions.

ELECTRICAL SYSTEMS

The electrical systems group develops instrumentation and software for the power industry, as well as for internal C-E research and development programs. In addition, it provides engineering and technical service support for KDL's instrumentation, data-acquisition, control, and electrical-power requirements.

LASER LABORATORY

The laser laboratory develops specialized optically based measurement systems. These are intended principally for commercial use, such as the OPSIZE™ in-situ system for measuring the particle size of solids flowing in pneumatic transport pipes (shown in Fig. 20 and described in Chapter 13). Examples of research tools used in other KDL projects include such instruments as a Laser Doppler Anemometer (LDA), a Laser Methane Tracer system (LMT), and a Void-Fraction Measurement system (VFM). The LDA non-intrusively measures velocity in flow-modeling, erosion, and component-development programs. The LMT measures mixing and residence time in flow models, and the VFM determines the void fraction in liquid/air systems.

Fig. 20 **C-E OPSIZE™ in-situ particle size analyzer**

INSTRUMENTATION LABORATORY

This is a multi-functional laboratory used to satisfy several needs within KDL. It is used for the initial bench testing of specialized measurement systems which are not optically based. Instruments developed by Electrical Systems for commercial offering, such as the Mach Therm™ acoustic temperature-mapping system, an enhanced acoustic leak-detection system, and recently, a new digital flame-scanner system, were all initially tested in the instrument laboratory.

This laboratory is also used to test, calibrate, and repair instruments and electronic components used in KDL facilities. Electronic test equipment and calibration standards which are traceable to the National Institute of Standards and Technology (NIST) are maintained by a qualified technical staff for this service. Such standards are maintained in temperature, pressure, force, and displacement. Pressure calibrations from 0.01 in.WG to 12,000 psig (2.5 Pa to 83 MPa gage), including differential pressure at 4,500 psig (31 MPa gage) static pressure, are performed. Temperature calibrations are done for the entire International Practical Temperature Scale (except for the cryogenic range), force to 100,000 pounds (445 kN), and displacement to 20 inches (500 mm).

CENTRAL COMPUTER FACILITY

Electrical Systems designed and installed, and now operates and maintains, KDL's central computer facility, the hub of the laboratory's data acquisition and reduction network. This network is Ethernet-based, with the central computer acting as a file server for all KDL test-facility data-acquisition computers and the many engineering work stations. The central computer simultaneously provides printing services to the various output peripherals such as plotters and printers.

The central computer is also a development machine for several Combustion Engineering Total On-Line Performance System (CETOPS) modules. The Interactive Chemistry Management System (ICMS) used to monitor and control boiler water chemistry was developed in KDL using this facility, as were various parts of the On-Line Thermal Information System diagnostic module (OTIS) and the Boiler Stress and Condition Analyzer module (BSCA), all of which were described in Chapter 13.

ELECTRICAL ENGINEERING SERVICES

Electrical Systems also provides electrical engineering services to KDL; these center around the design, construction, and maintenance of instrumentation, data-acquisition, control, and power systems for the various test facilities. Staff members design these systems, which cover a range from high-speed data-acquisition systems for transient fluid dynamics studies to megawatt-sized electrical heating systems. By having this capability in-house, the Kreisinger Laboratory can respond quickly to facility design changes necessary to satisfy research-program and customer needs.

Appendix A. Coals of the World

This appendix is comprised of two sections. The first has reference data on all the coals of all producing countries of the world; approximate reserves are given, along with some historical and geographical information. Analyses of the coals and their ashes, along with fusibility data, are included to be used in preliminary design of boilers to burn coals from these countries. The data are not intended and are not adequate for the actual design of a steam generator to burn a specific local fuel. Mine or core samples, or detailed fuel analyses, are needed for that purpose.

The International Systems of coal classification are described in the second section. These systems result from a significant effort to categorize in a consistent, rational manner all the different coals (excluding peat) found worldwide. This is vitally important where engineers from one country are involved in the design of fuel-preparation-and-burning equipment, steam-generating equipment, or ash-handling systems for another country where coals may be designated by local custom rather than on a useful engineering basis.

Sources of information for the statistical and analytical data in this chapter have been fuel investigations by the C-E Kreisinger Development Laboratory; publications of the U.S. Bureau of Mines and the Canadian Department of Energy, Mines and Resources; the National Coal Association, Washington, DC; the World Energy Conference Surveys of Energy Resources; the Report of the Commission of Inquiry into the Coal Resources of the Republic of South Africa (1975); and COAL—Bridge to the Future, summary report of World Coal Study.

* * * *

In this Appendix, high heat values are given in both Btu per pound and megajoules per kilogram. Tonnages of coal resources or production are in metric tonnes (2,204.6 pounds). Where a tonne of coal equivalent is shown, it is a metric tonne of coal with a heating value of 12,600 Btu/lb, 7,000 kcal/kg, or 29.31 MJ/kg. Because coals vary significantly in calorific value, much more than 1 tonne of subbituminous, lignitic, or brown coal is required to produce the energy of 1 tonne of coal equivalent (1 tce).

The estimates of economically recoverable coal are at best approximate, and depend just as much on economic and political considerations as they do on exploratory techniques. There are no universally agreed upon definitions of reserves and resources, many of them being a function of quantity of overburden or depth of mining. And, depending upon the source of data and the purpose for which it was gathered, the values may represent undue optimism or conservativeness. In any case, they should be read in the context of this statement by Carroll Wilson (Future Coal Prospects: Country and Regional Assessments—the World Coal Study, 1980): ". . . the use of different fuels is heavily constrained by decisions made in the past and by the fact that new energy facilities being planned now cannot be brought to completion in less than 5 to 10 years. The substantial shift to coal will therefore begin in the mid to late 1980's based on decisions made in the early 1980's. Major effects of this expansion will be felt in the 1985–2000 period . . . The full effects of the switch to coal will, however, not be seen until the early decades of the next century.''

ANALYSES BY CONTINENT AND COUNTRY

Coal is found on all continents and in most countries throughout the world. However, some countries, such as Italy and Sweden, have resources so meager in proportion to their energy demands as to be inconsequential. By far, the largest world coal deposits are located in the Northern hemisphere.

Economically recoverable reserves are those that can be successfully exploited and used within the foreseeable future; they represent about 1/15th of the total world geological resources of solid fuel. The estimated total of such resources for the entire world is in excess of 600,000 million tce; over 95 percent is found in just 13 countries, approximately as follows:

United States	28%
U.S.S.R.	17%
People's Republic of China	16%
United Kingdom	7%
Western Germany	5%
India	5%
Australia	4%
Republic of South Africa	4%
Poland	3%
Canada	2%
Brazil	1%
Yugoslavia	1%
Eastern Germany	1%

The preceding tabulation may very well overstate the resources of such countries as the United States and Western Germany in comparison to the others, because exploration has reached a higher level. The eventual resources, which disregard climate and other factors making recovery difficult, of both Russia and China may be greater than shown. Also, there can be considerable loss in recovering coal, particularly in deep mining—and there can be great differences between run-of-mine coal and a marketable, ash-reduced product.

Further general information on the reserves of all coal-producing countries of the world is found under the individual country headings.

BASIS FOR ANALYTICAL DATA

A most important consideration in the use of coal analyses is the determination of coal type or rank from the data presented. The selection of pulverizing equipment, for instance, depends upon characteristics related to coal rank, as does establishing the heat transfer requirement and size of an air heater.

For the best use of world coal data, the tabulations in this Appendix make it easy to ascertain coal rank. Volatile contents and high heating values are normalized to a moisture- and ash-free basis; ash and moisture contents are given on an as-analyzed basis.

Ash analyses, some of them available from U.S. Bureau of Mines and Canadian Department of Energy, Mines and Resources publications, but mostly done in C-E's laboratory, are given for use in furnace sizing, determining convection-pass fouling potential, and designing ash-transport and storage equipment. All ash fusibility figures are for ASTM reducing atmosphere, as described in Chapter 3.

Regarding the moisture contents shown, realize that there can be no assurance that samples received for analysis have their moisture-sealing intact, so there can be some air-drying before tests occur. In using the moisture contents for pulverizing-system, boiler, and air

Table I. Correlation of Coal Properties (Based on Typical Analyses)

	Bituminous Low VM.	Bituminous Med. VM.	High-Volatile Bit. A	High-Volatile Bit. B	High-Volatile Bit. C	Subbituminous A	Subbituminous B	Subbituminous C	Lignite and Brown Coal
Volatile Matter (VM) MAF	15 to 24	25 to 36	34 to 46	38 to 49	42 to 54	41 to 46	42 to 48	43 to 50	45 and higher
$\frac{C + H}{O_2}$ (Typical)	19.5	11.8	11.2	7.0	5.4	4.3	3.2	2.8	2.7 and lower
Btu/lb, MAF	15,100 to 16,000	15,000 to 15,700	14,600 to 15,400	14,000 to 14,800	13,400 to 14,200	12,900 to 13,800	11,900 to 13,300	11,200 to 12,900	10,200 to 12,700
MJ/kg, MAF	35 to 37	35 to 36.5	34 to 36	32.5 to 34.5	31 to 33	30 to 32	27.5 to 31	26 to 30	23.5 to 29.5

MAF = moisture- and ash-free

heater design, values must be corrected to probable "as-received at the power-plant" moisture. This is particularly important with coals that are transported long distances before analyses, as is frequently the case.

CORRELATION OF COAL PROPERTIES

In addition to the *ASTM Standards* D388, Classification of Coals by Rank (Table II, Chapter 2), a review of the actual coal analyses presented in this Appendix will produce a correlation that is useful in preliminary ranking of coals based on proximate and ultimate analyses. Table I is such a correlation, based on a study of analyses of many typical coals. For such preliminary ranking, refer also to Fig. 1, Chapter 2, which is a graphical presentation of ranking parameters.

COAL DEPOSITS OF NORTH AMERICA

This section describes the coalfields of Canada, the United States, and Mexico, with Alaska being given separate treatment.

CANADA

Canada (Fig. 1) has over 9,000 million tonnes of economically recoverable coal reserves, ranging from anthracite to lignite. Table II shows measured resources (for which tonnages have been computed) of immediate interest for exploration or exploitation activities; generally, for most regions, coal seams have to be at least 5-ft thick to be included in this category. The resources in the table exclude coal that may be present north of 60° North latitude and small occurrences known in Newfoundland and Manitoba.

Ninety-five percent of the recoverable reserves are in western Canada (British Columbia, Alberta, and Saskatchewan), of which about 80 percent can be open-pit mined. As in the United States, coals in western Canada are low in sulfur content, rarely exceeding 0.6 percent on a mass basis, but are high in ash. Eastern Canadian coals generally have high sulfur, particularly those from Nova Scotia.

Approximately 14 percent of the Canadian

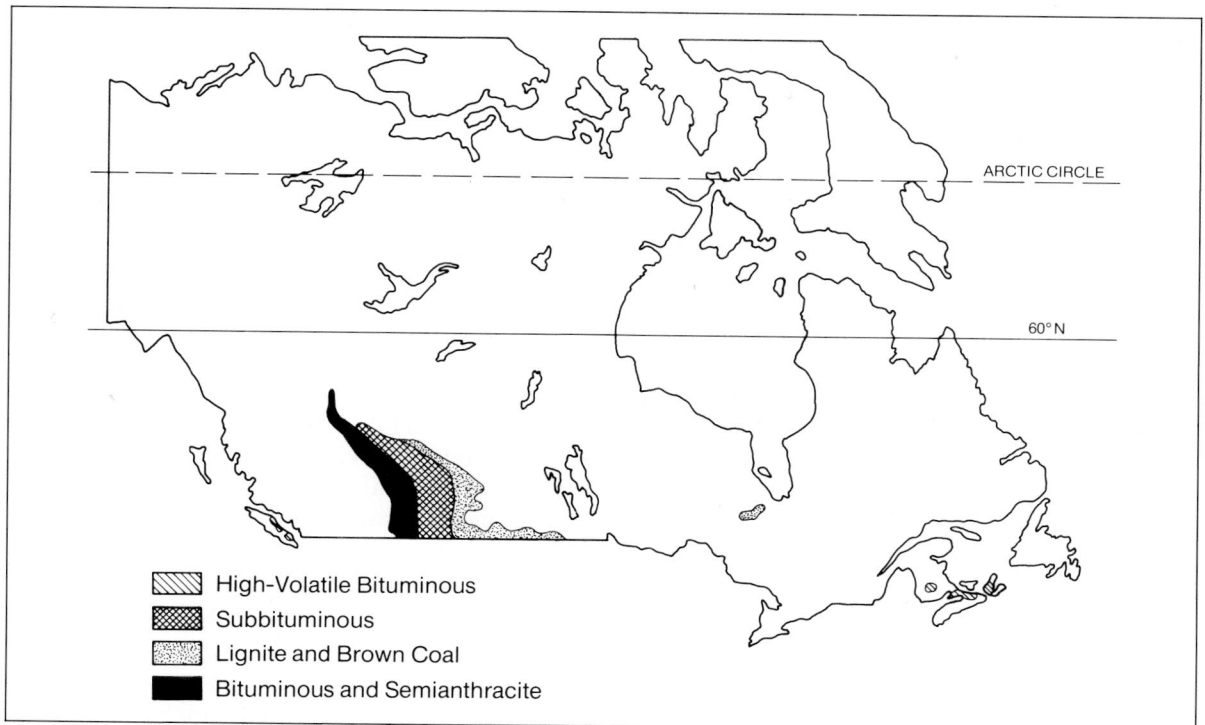

Fig. 1 **Coal Map of Canada**

High-Volatile Bituminous
Subbituminous
Lignite and Brown Coal
Bituminous and Semianthracite

coal produced is bituminous coal from underground mines; the balance is bituminous, subbituminous and lignitic coals taken by strip-mining. Over 70 percent of the domestic consumption is for power generation; most of the balance, for metallurgical uses.

Production of coal in the year 2000 is estimated to be between 75 and 115 million tonnes coal equivalent per year. A steadily growing portion of the production will be used at mine-mouth plants in the West. In Alberta, for instance, coal is planned to supply about 95 percent of the province's electrical needs at the turn of the century. Canada's coal deposits are located in four definable regions.

1. The Mountain Region:
Most of the bituminous coal resources are in the provinces of Alberta and British Columbia (BC), extending in a belt up to 50 km wide from the U.S. border, northwest into the East Kootenay region of southeastern BC, north into Alberta (in the Crowsnest Pass area) and into northeastern BC (Fig. 1). The main center of deposition is located in southeastern BC. Large resources are also being found in the Peace River area to the north.

2. The Foothills Region of Alberta:
Coals from this area are mainly of high-volatile bituminous B and C rank. Mining conditions are not as severe as in the mountain region, but eventually most of the resources will have to be extracted by underground methods.

3. The Plains Regions of Saskatchewan and Alberta:
The four lignite deposits of Saskatchewan are located along the U.S. border and form the northern fringes of the main lignite zone centered in North Dakota. The subbituminous seams of Alberta underlie the southern part of that province from the Rocky Mountains in the west to their outcrops about 250 km farther east, and form an immense arc from the U.S. border up to about 56°N latitude, passing east of the cities of Calgary and Edmonton.

4. The Atlantic Region:
The Cape Breton Island coalfield in Nova Scotia is the most important of the region, although a small quantity of high-sulfur bituminous coal is produced by surface mining in the province of New Brunswick. The Cape Breton field is located on the north shore of the island, facing the Cabot Strait and the Atlantic Ocean. The coal is mostly of high-volatile bituminous A rank, varying from medium to high in sulfur content.

Canadian reserves of peat, which have not been surveyed, are thought to be approximately equal to those of the United States.

Table III gives the analyses of typical Canadian coals. For the ultimate analyses and ash constituents, see Analysis Directories of Cana-

Table II. Measured Canadian Coal Resources of Immediate Interest (1978)

Province and Area	Coal Rank	Millions of tonnes
Nova Scotia		
Sydney	hvb	175
Other	hvb	48
Subtotal	hvb	223
New Brunswick		
Minto	hvb	18
Other	hvb	14
Subtotal		32
Ontario	lig	218
Saskatchewan		
Estevan	lig	310
Willow Bunch	lig	748
Wood Mountain	lig	278
Cypress	lig	162
Subtotal		1,498
Alberta		
Plains	sub	30,000
Foothills	hvb	1,300
Mountains	lvb-mvb	8,000
British Columbia		
Southeastern	lvb-mvb	6,286
Northeastern	lvb-mvb	996
Other	mainly lig some sub-hvb	1,845
Canada total	lig	3,561
	sub	30,000
	hvb	1,555
	lvb-mvb	15,282

lig = lignitic; sub = subbituminous; hvb = high-volatile bituminous; mvb = medium-volatile bituminous; lvb = low-volatile bituminous.

Table III. Analyses of Typical Canadian Coals

Province and Area or seam	Rank	As received % Ash	As received % H$_2$O	% VM	% S	Moisture- and ash-free HHV Btu/lb	Moisture- and ash-free HHV MJ/kg	Ash ST, Red., °F	Hardgrove Grind.
Nova Scotia									
Thorburn	Hvab	14.4	3.0	33.4	0.5	15,180	35.3	2440	60
Broughton	Hvab	14.4	5.2	40.4	7.6	14,560	33.9	2100	59
River Hebert	Hvab	19.3	1.5	41.0	6.8	14,530	33.8	2070	61
Phalen	Hvab	14.4	1.7	34.9	6.6	15,200	35.3	2180	84
Harbour	Hvab	9.3	1.2	40.6	2.7	15,420	35.9	2480	70
Westville	Hvab	15.4	2.8	32.9	0.9	15,020	34.9	2240	64
Sydney	Hvab	10.2	3.8	42.7	6.4	14,660	34.1	2130	63
Joggins	Hvab	15.8	3.7	42.4	6.3	14,460	33.6	2140	60
Inverness	Hvcb	14.2	9.3	44.1	8.0	13,530	31.5	2070	60
New Brunswick									
Minto	Hvab	14.0	3.4	39.0	8.4	15,150	35.2	1980	66
Chipman	Hvab	17.6	2.1	42.2	9.3	14,970	34.8	1920	63
Coal Creek	Hvab	18.3	1.8	41.7	10.3	14,780	34.4	1960	57
Saskatchewan									
Bienfait	Lig A	6.0	34.1	49.9	0.8	12,390	28.8	2300	63
Roche Percée	Lig A	6.3	30.8	46.6	1.0	12,450	28.9	2250	52
Alberta									
Cascade	Sa	8.5	3.1	13.9	0.8	15,580	36.2	2750+	88
Crowsnest	Mvb	10.4	4.6	27.5	0.6	15,200	35.3	2750+	77
Coalspur	Hvcb	10.5	6.3	41.5	0.1	13,680	31.8	2170	52
Lethbridge	Hvcb	10.5	10.2	45.4	0.8	13,590	31.6	2330	40
Pembina	Subc	8.8	19.6	41.3	0.3	12,440	28.9	2450	49
Drumheller	Subb	9.1	17.2	42.2	0.7	13,150	30.6	2270	33
Castor	Subb	5.5	24.8	43.6	0.6	12,470	29.0	2140	45
Edmonton	Subc	13.7	23.5	44.3	0.3	12,870	29.9	2370	38
Taber	Suba	9.9	14.2	44.3	1.8	13,350	31.0	2350	47
Camrose	Subc	5.8	21.4	44.5	0.4	12,710	29.5	2160	31
British Columbia									
East Kootenay	Mvb	9.5	2.4	25.5	0.5	15,350	35.7	2740	88
Vancouver Island	Hvab	12.9	1.4	38.6	1.6	14,910	34.7	2320	57

dian Coals published by the Canadian Department of Mines and Technical Surveys, Mines Branch, Fuels Division.

UNITED STATES

Coal is found in 36 of the states and is currently mined in about 26. Six states contain more than 75 percent of the estimated 245,000 million tonnes of economically recoverable reserves: Montana (28%), Illinois (16%), Wyoming (13%), West Virginia (9%), Pennsylvania (7%) and Kentucky (6%), Fig. 2.

The reserves of peat, not detailed in this text, are estimated to be equal in heating value to about 30 percent of the currently recoverable coal reserves. Peat is located in low-lying coastal areas, in the flatlands of the Great Lakes, and in Alaska.

The electric utilities of the U.S. consume 78 percent of all the coal produced in the states.

The U.S. Geological Survey has divided the coal-bearing areas of the contiguous U.S. into six main provinces designated as (1) Eastern, (2) Interior, (3) Gulf, (4) Northern Great Plains, (5) Rocky Mountain, and (6) Pacific Coast. The provinces are subdivided into coal regions, coal fields, and coal districts. (Alaska is treated as a separate province.)

The Eastern Province

This includes the anthracite regions of Pennsylvania and Rhode Island; the Atlantic Coast region, including the Triassic fields of Virginia and North Carolina; and the Appalachian region extending from Pennsylvania through eastern Ohio, Kentucky, West Virginia, western Virginia, and Tennessee into Alabama. The Appalachian basin contains the largest deposit of the high-grade bituminous and semibituminous coals. The Triassic deposits of central Virginia and North Carolina occur in irregular pockets rather than in seams and are at present of little commercial value, because of expensive mining and rather low-grade quality of coal.

The Interior Province

This province includes all the bituminous coal area of the Mississippi Valley region and the coal fields of Texas and Michigan. This province is subdivided into the Northern region consisting of the coal field of Michigan; the Eastern region, or Illinois basin, comprising the fields of Illinois, Indiana and western Kentucky; and also the Western region embracing the coal fields of Iowa, Missouri, Nebraska, Kansas, Arkansas, Oklahoma, and the southwestern region of Texas. With some exceptions, the bituminous coals of this province are of lower rank and higher sulfur content than those of the Eastern province. Much of the surface-minable coal in the Illinois basin having low chlorine and sodium content has been mined or committed to be mined. Much of the remaining coal, with high chlorine content (above 0.15 percent as fired), will have to be deep-mined and will comprise a large proportion of the future production.

The Gulf Province

This consists of the Mississippi region in the east and the Texas region in the west. The Mississippi region includes the lignite fields of

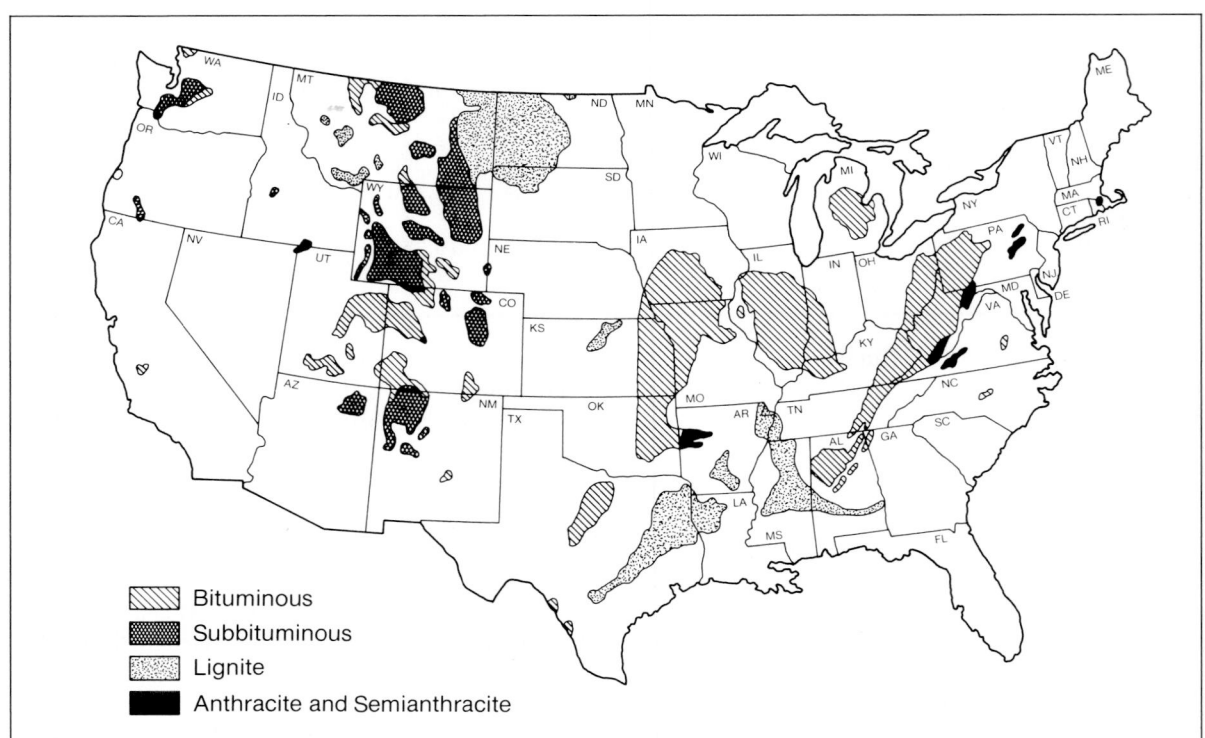

Fig. 2 **Coal Map of United States**

Table IV. Source, Analysis and Classification of U.S. Bituminous Coals

State, County, and Bed	As received % Ash	% H₂O	% VM	% C	% H	% O	% N	% S	HHV Btu/lb	MJ/kg	FSI[A]	A-AD[B]	ASTM[C]	IC No.[D]
Alabama														
Blount, Bynum	7.2	4.6	35.0	85.4	5.4	5.5	2.0	1.7	15,220	35.4	7	120	Hvab	634
Jefferson, Blue Creek	7.1	2.3	29.1	87.2	5.2	4.8	1.8	1.0	15,500	36.0	8½	250	Mvb	535
Jefferson, Mary Lee	10.0	1.9	29.6	87.5	4.9	4.8	1.9	0.9	15,420	35.9	7½	200	Mvb	535
Jefferson, Pratt	6.2	1.6	31.9	87.3	5.4	3.8	1.8	1.7	15,590	36.3	7½	280	Hvab	535
Marion, Black Creek	5.5	4.9	41.5	82.5	5.5	8.6	1.8	1.6	14,750	34.3	3	−27	Hvab	622
Tuscaloosa, Milldale	7.1	5.8	37.6	83.5	5.5	7.4	1.6	2.0	14,900	34.7	5½	30	Hvab	633
Arkansas														
Sebastian, Upp. Hartshorne	7.4	2.4	19.0	90.0	4.5	2.6	1.9	1.0	15,550	36.2	8½	50	Lvb	333
Colorado														
Gunnison, C	3.1	5.8	42.0	81.8	5.7	9.9	1.9	0.7	14,570	33.9	3	−43	Hvbb	721
LaPlata, No. 1	5.8	2.7	41.8	83.7	5.9	7.8	1.6	1.0	15,150	35.2	7	52	Hvab	634
Illinois														
Gallatin, Willis	11.3	3.1	36.3	84.5	5.4	5.1	1.5	3.5	15,170	35.3	5	10	Hvab	633
Knox, No. 6	6.2	18.3	47.9	78.7	5.5	10.9	1.4	3.5	14,130	32.9	3½	−39	Hvcb	821
Vermillion, No. 7	9.0	12.2	49.3	79.2	5.7	9.5	1.5	4.1	14,400	33.5	3½	35	Hvcb	823
Williamson, No. 6	11.7	5.8	43.8	80.5	5.5	9.1	1.6	3.3	14,430	33.6	5	60	Hvbb	734
Indiana														
Knox, No. 5	7.5	9.7	47.0	80.1	5.6	8.5	1.6	4.2	14,530	33.8	4½	70	Hvbb	734
Kentucky														
Bell, Sterling	6.6	4.8	39.5	83.1	5.5	8.4	1.9	1.1	14,870	34.6	4	−5	Hvab	622
Harlan, D	7.0	2.5	39.8	84.0	5.5	7.8	1.7	1.0	15,040	35.0	6	30	Hvab	633
Hopkins, No. 9	11.2	6.1	45.1	80.1	5.5	8.0	1.7	4.7	14,590	33.9	4½	52	Hvbb	734
Hopkins, No. 12	11.2	8.3	46.5	79.0	5.6	10.1	1.7	3.6	14,420	33.5	4	40	Hvbb	723
Muhlenberg, No. 11	8.6	9.0	44.2	79.5	5.6	9.4	1.6	3.9	14,340	33.3	4	25	Hvbb	823
Pike, Alma	3.3	2.5	39.0	85.5	5.5	6.7	1.6	0.7	15,370	35.7	6	100	Hvab	634
Maryland														
Allegany, Pittsburgh	10.2	3.1	20.4	88.5	5.0	3.5	2.0	1.0	15,510	36.1	7½	10	Lvb	433

Table IV. Source, Analysis and Classification of U.S. Bituminous Coals — *Continued*

State, County, and Bed	As received % Ash	% H₂O	% VM	% C	% H	% O	% N	% S	HHV Btu/lb	MJ/kg	FSI[A]	A-AD[B]	Classification ASTM[C]	IC No.[D]
Maryland (Continued)														
Allegany, Tyson	9.1	2.8	21.4	89.0	4.9	2.8	1.9	1.4	15,650	36.4	8½	110	Lvb	434
Ohio														
Belmont, Pittsburgh No. 8	9.1	3.6	45.8	80.9	5.7	7.4	1.4	4.6	14,730	34.3	6	200	Hvab	635
Morgan, Middle Kittaning No. 6	8.5	8.7	45.8	79.4	5.7	10.9	1.5	2.5	14,340	33.3	4½	2	Hvbb	733
Pennsylvania														
Allegheny, Pittsburgh	7.6	2.6	39.5	84.3	5.6	6.7	1.7	1.7	14,870	34.6	8½	180	Hvab	635
Butler, Lower Freeport	8.0	3.7	42.0	82.2	5.6	7.3	1.6	3.3	14,970	34.8	5½	75	Hvab	634
Cambria, Upper Freeport	7.5	1.1	20.0	89.5	4.8	2.9	1.5	1.3	15,660	36.4	9	75	Lvb	334
Clearfield, Lower Freeport	7.9	2.1	35.4	85.3	5.4	3.8	1.6	3.9	15,360	35.7	7	300	Hvab	635
Fayette, Lower Kittanning	8.8	1.7	29.4	87.8	5.2	3.2	1.5	2.3	15,500	36.0	9	310	Mvb	535
Greene, Pittsburgh	8.2	1.8	40.6	84.1	5.6	6.4	1.7	2.2	15,160	35.3	9	250	Hvab	635
Indiana, Lower Freeport	9.7	2.3	24.6	88.5	5.1	3.7	1.5	1.2	15,680	36.5	9	150	Mvb	435
Lawrence, Brookville	10.1	5.2	42.9	81.5	5.7	7.3	1.7	3.8	14,830	34.5	4½	75	Hvab	634
Washington, Pittsburgh	5.1	1.6	40.5	84.2	5.6	7.2	1.6	1.4	15,070	35.0	7½	180	Hvab	635
Tennessee														
Campbell, Jordon	3.8	4.7	38.1	83.2	5.4	8.6	2.0	0.8	14,820	34.5	3	−21	Hvab	622
Utah														
Carbon, D	7.3	3.3	47.2	80.7	5.8	11.7	1.4	0.4	14,330	33.3	2	−45	Hvbb	711
Carbon, Lower Sunnyside	5.6	4.1	42.9	82.1	5.6	9.3	1.8	1.2	14,770	34.3	5	5	Hvab	633
Emery, Lower Sunnyside	6.4	5.2	43.2	79.8	5.6	11.8	1.7	1.1	14,260	33.2	3	−35	Hvbb	721
Washington														
Kittitas, No. 1 (Big)	8.8	3.7	42.4	81.8	6.0	9.7	1.9	0.6	14,710	34.2	4	5	Hvab	623
Kittitas, No. 5 (Roslyn)	9.9	2.9	45.1	82.7	6.2	8.6	2.0	0.5	14,920	34.7	4	35	Hvab	623

Table IV. **Source, Analysis and Classification of U.S. Bituminous Coals — *Continued***

State, County, and Bed	As received		Moisture- and ash-free										Classification	
	% Ash	% H₂O	% VM	% C	% H	% O	% N	% S	HHV Btu/lb	MJ/kg	FSI[A]	A-AD[B]	ASTM[C]	IC No.[D]
West Virginia														
Boone, Alma	9.8	2.3	42.5	83.5	5.7	6.3	1.5	3.0	15,110	35.1	6	120	Hvab	634
Boone, Chilton	7.8	4.2	40.2	83.8	5.5	7.5	1.7	1.5	14,930	34.7	6	10	Hvab	633
Boone, Hernshaw	7.2	2.0	38.0	86.0	5.5	6.0	1.6	0.9	15,320	35.6	6½	140	Hvab	634
Fayette, Fire Creek	6.2	3.0	20.2	90.0	4.7	2.9	1.3	1.1	15,540	36.1	7	10	Lvb	433
Fayette, Lower Eagle	4.8	1.8	33.4	87.7	5.5	4.4	1.7	0.7	15,610	36.3	8½	310	Hvab	635
Greenbrier, Fire Creek	10.2	3.6	24.2	88.5	5.0	3.6	1.6	1.3	15,540	36.1	8½	150	Mvb	435
Greenbrier, Sewell	3.2	3.2	29.1	88.0	5.3	4.3	1.8	0.6	15,540	36.1	9	220	Mvb	535
Kanawha, Winifrede	9.5	4.8	41.5	83.3	5.6	8.6	1.7	0.8	14,880	34.6	4	−5	Hvab	622
Logan, Cedar Grove	9.3	2.4	39.2	84.2	5.5	7.1	1.6	1.6	15,180	35.3	7	150	Hvab	635
McDowell, Beckley	10.3	2.7	18.4	89.8	4.7	3.2	1.6	0.7	15,630	36.3	7½	25	Lvb	333
McDowell, Bradshaw	6.7	1.9	26.5	88.9	5.1	3.4	1.7	0.9	15,690	36.5	8½	240	Mvb	435
McDowell, Pocahontas No. 3	6.2	3.3	18.7	89.4	4.6	4.1	1.3	0.6	15,690	36.5	8½	10	Lvb	333
McDowell, Pocahontas No. 4	7.3	2.8	26.1	89.4	5.0	3.7	1.3	0.6	15,610	36.3	9	170	Mvb	435
McDowell, Douglas	4.0	2.5	25.6	89.3	5.0	3.1	1.7	0.9	15,680	36.5	9	230	Mvb	435
Marshall, Pittsburgh	10.5	4.1	45.6	81.2	5.6	6.8	1.4	5.0	14,790	34.4	7	230	Hvab	635
Mercer, Pocahontas No. 3	7.1	4.4	17.4	90.7	4.6	2.7	1.2	0.8	15,730	36.6	6	5	Lvb	333
Mercer, Pocahontas No. 6	7.9	5.1	20.8	90.3	4.8	3.0	1.3	0.6	15,710	36.5	7	65	Lvb	434
Mingo, Lower Cedar Grove	5.0	1.7	37.5	85.6	5.5	6.4	1.6	0.9	15,390	35.8	8	160	Hvab	635
Mingo, Upper Cedar Grove	9.1	5.1	36.3	85.5	5.5	6.6	1.6	0.8	15,090	35.1	6½	30	Hvab	633
Monongalia, Pittsburgh	6.6	2.3	37.5	85.5	5.5	5.8	1.8	1.4	15,310	35.6	9	250	Hvab	635
Monongalia, Redstone	12.4	2.6	36.6	84.5	5.5	6.3	1.7	2.0	15,090	35.1	7½	180	Hvab	635

Table IV. Source, Analysis and Classification of U.S. Bituminous Coals — *Continued*

State, County, and Bed	As received % Ash	% H₂O	Moisture- and ash-free % VM	% C	% H	% O	% N	% S	HHV Btu/lb	MJ/kg	FSI[A]	A-AD[B]	Classification ASTM[C]	IC No.[D]
West Virginia (Continued)														
Nicholas, Eagle	6.8	4.0	36.2	85.6	5.5	6.4	1.7	0.8	15,270	35.5	7½	140	Hvab	634
Raleigh, Beckley	5.3	1.5	20.0	89.4	4.8	3.5	1.5	0.8	15,660	36.4	8½	60	Lvb	334
Randolph, Sewell	5.4	1.5	31.2	87.4	5.3	5.2	1.4	0.7	15,420	35.9	8	180	Mvb	535
Wyoming, Campbell Creek	2.0	2.2	31.7	86.8	5.4	5.5	1.5	0.8	15,570	36.2	8½	280	Hvab	535
Wyoming, Eagle	3.6	1.9	29.9	88.5	5.3	4.0	1.5	0.7	15,530	36.1	8½	230	Mvb	535
Wyoming, Sewell	2.4	1.5	23.8	89.8	5.0	3.1	1.6	0.5	15,700	36.5	9	170	Mvb	435
Wyoming														
Sweetwater, Nos. 7, 7½, 9, 15	4.1	10.3	47.0	79.3	5.6	12.2	1.8	1.1	13,900	32.3	1½	−22	Hvcb	811

A—FSI = Free Swelling Index
B—A-AD = Audibert-Arnu dilatation index
C—ASTM Classifications: Lvb = low-volatile bituminous; Mvb = medium-volatile bituminous;
   Hvab = high-volatile A bituminous; Hvbb = high-volatile B bituminous; Hvcb = high-volatile C bituminous
D—IC = International Classifications

Alabama, Mississippi, and Louisiana. The Texas region comprises the lignite fields of Arkansas and Texas. The lowest rank coals in the U.S. are found in this province, with equilibrium moisture contents as high as 55 percent. Lignites from Texas can have moisture contents to 40 percent, low grindabilities, and heating values, as received, below 4000 Btu/lb. Some of the Louisiana lignites are very similar to those of North Dakota, having moisture to 36 percent and sodium levels of 5 to 8 percent.

The Northern Great Plains Province

This includes all the coal fields of the great plains east of the eastern front range of the Rocky Mountains. In this province are immense lignite areas of the two Dakotas, and the bituminous and subbituminous fields of northern Wyoming and northern and eastern Montana. Lignites from this province can be of very low rank, having moisture content to 45 percent and sodium oxide as high as 15 percent of the ash.

The Rocky Mountain Province

This comprises the coal fields of the mountainous districts of Montana, Wyoming, Utah, Colorado, and New Mexico. It possesses a great variety of coals, ranging from lignite through subbituminous and high grade bituminous coals to anthracite. The subbituminous coals range in moisture from 10 to 30 percent, with calorific-values (HHV) from 6500 to 10,000 Btu/lb. Several sources in this province have high sodium-oxide in ash, ranging to 15 percent.

The Pacific Coast Province

This area is largely confined to the State of Washington, which contains the largest coal fields on the Pacific Coast. It also embraces the small fields of Oregon and California. The coals in this province range from subbituminous through bituminous to anthracite.

The coals of Oregon and California rank somewhat lower than those of Washington.

Those of California consist mainly of noncoking bituminous coals in the southern part of the State, lignite in the northern, and subbituminous in the center. The fields are scattered and limited in area.

Estimates of U.S. coal production in the year 2000 vary from 1200 to 2000 million tonnes per year, depending upon the amount of nuclear-power generation and the extent of coal exports. Appalachia, which now produces about half of the nation's coal, will decrease to 30 percent; the Midwest will produce 11 percent and the West, 50 percent; the balance will be surface-mined Gulf lignite. Surface mining will predominate at that time and is expected to be over 60 percent of the total output.

Table V of Chapter 2 gives analyses of some U.S. coals. Ash analyses of representative U.S. coals are included in Tables IV, V, and VI of Chapter 3.

Table IV (Page A–10) gives detailed analyses of U.S. bituminous coals, with free-swelling indices, the Audibert-Arnu dilatation, and the ASTM and International classifications.

ALASKA

Coal is distributed widely throughout Alaska in fields differing greatly in size and in geologic environment; much of it is lignitic or subbituminous in rank. Its total reserves are estimated to be 15 percent bituminous coal and 85 percent subbituminous coal and lignite. Only fields close to main lines of transportation have been developed; major reserves occur in more remote parts of the State. The most important fields, in terms of production and known reserves, are the lower Matanuska Valley, northeast of Anchorage, and the Nenana, southwest of Fairbanks. In addition to coal, Alaska has a greater area of peat than all the contiguous states combined.

Alaskan coal deposits are in several regions (Fig. 3). The Northern Alaska and Seward Peninsula regions contain mostly subbituminous and bituminous coal. The biggest Alaskan

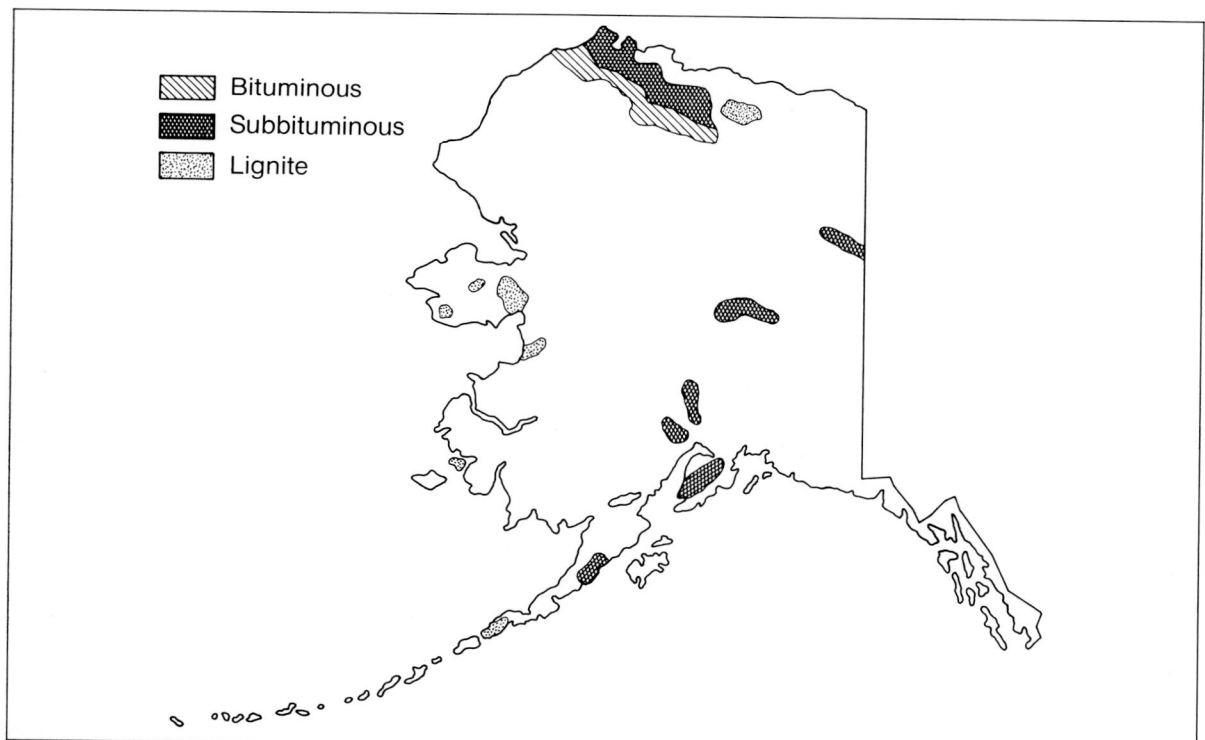

Fig. 3 **Coal Map of Alaska**

Table V. Analyses of Typical Alaskan Coals

| Region, District And Mine | As received | | | | Moisture- and ash-free | | Ash ST, |
| | % Ash | % H₂O | % VM | % S | HHV | | Red., °F |
					Btu/lb	MJ/kg	
Northern Region							
Wainwright	3.2	20.7	41.8	0.4	12,830	29.8	. . .
Yukon Region							
Broad Pass Field							
Costello Creek	13.3	14.3	49.0	0.7	13,080	30.4	. . .
Eagle	24.4	19.8	59.3	0.4	11,240	26.1	. . .
Nenana Field							
Suntrana	6.5	22.3	52.2	0.1	12,130	28.2	2140
Kuskokwin Region							
Nelson Island	18.2	3.9	30.6	0.5	14,690	34.2	2710
Southwestern Region							
Chignik Bay	21.8	7.1	44.3	1.8	13,850	32.2	. . .
Cook Inlet Region							
Matanuska Field							
Chickaloon	10.4	1.9	23.7	0.8	15,120	35.2	2060
Coal Creek	5.8	1.6	24.7	0.7	15,530	36.1	2450
	11.6	4.1	16.4	0.7	14,950	34.8	2240
Eska	10.9	3.7	48.0	0.6	14,530	33.8	2570
Jonesville	21.3	3.5	48.9	0.3	13,900	32.3	2370
Moose Creek	6.4	3.9	44.6	0.2	14,400	33.5	2660
Gulf Region							
Bering River Field							
Katalla	3.8	5.0	15.0	1.0	15,560	36.2	2240
Southeastern Region							
Admiralty Island							
Harkrader	21.4	3.8	47.1	1.7	14,210	33.0	2250

deposits are in the Arctic slope region, in an area 300 miles long (east and west), and 75 miles wide; the coal is low-volatile bituminous to lignite. The Yukon basin coal, lignite and bituminous, occurs in small, scattered areas that have been inadequately explored.

The Nenana deposits are subbituminous high-rank lignite, primarily strip-mined. The Cook Inlet-Susitna region contains some of the most important fields of lignitic and sub-bituminous rank, in thicknesses up to 40 ft. The Matanuska Valley is divided into the lower field, containing the Moose Creek and Eska fields, and the upper field, comprising the Chickaloon and Anthracite Ridge. The Matanuska fields produce high-volatile B bituminous coal from beds 18 to 20 ft. in thickness. Technical Paper 682 of the U.S. Bureau of

Mines (1946), Analyses of Alaska Coals, presents proximate and ultimate analyses, and fusibility data, for coals from all the major Alaskan coal deposits. Table V gives analyses of some typical Alaskan coals.

MEXICO

Mexico has coal deposits in 16 of the 38 Federal States (Fig. 4). Estimated production by the year 2000 is 55 million tonnes. Only three deposits presently have economically recoverable coal reserves: Barrancas basin (Central Sonora), Oaxaca basin and Coahuila basin. In the Coahuila basin, the most significant coal deposits are located within the two largest districts of Sabinas and Rio Escondido.

Brown coals and hard coals are deposited in the Sabinas basin, which is about 35 miles long

and 15 miles wide. The hard coals are of medium rank and have ash contents from 14 to 35 percent. These coals are cokable. High-volatile brown coals are deposited in the Rio Escondido basin. Most of the coals in the Barrancas basin are of high rank, some of these coals are anthracitic.

Table VI gives coal and ash analyses of Mexican bituminous and subbituminous coals.

CENTRAL AMERICA

The coal resources of Costa Rica, Guatemala, Honduras, Panama and Puerto Rico (Fig. 4) comprise subbituminous and lignitic coals. The extent is generally unknown in most countries, and the deposits are not currently of economic importance.

COAL DEPOSITS OF SOUTH AMERICA

Compared to those of North America, South America's coal deposits are very small (Fig. 5). The only Southern countries where deposits of

any economic importance are known to occur are Brazil, Chile, Colombia, Peru, and Venezuela. Argentina has small resources of low-rank coal. Owing to inaccessibility of most of the deposits and lack of transportation, coal mining has not been developed. Lack of capital and skilled labor also limits development.

ARGENTINA

Argentina's only coal deposit is located in the Southwest of the country, in the Rio Turbio Basin, near the Chilean border. The seams are 1 m to 2 m thick. The coal is generally high in ash but, after treatment or blending with imported coal, it is suitable for steam generation. Mixed with high-grade coking coal, Argentinian coal is made available for steelmaking. The Port of Rio Gallegos is used for shipping coal to other parts of the country. Today, the main consumer is the Hydro and Electric Energy Agency. Annual coal production of about 9 million tonnes is anticipated by the

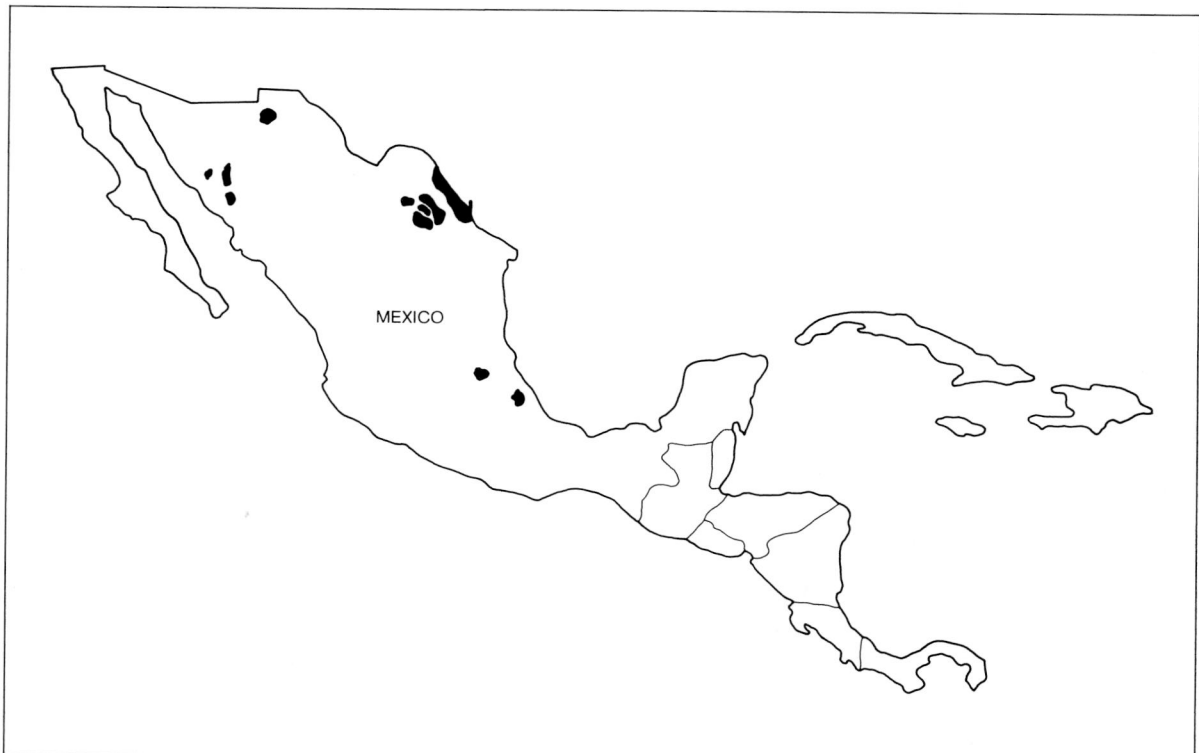

***Fig. 4* Coal Map of Mexico and Central America**

Fig. 5 Coal Map of South America

Table VI. Analyses of Mexican Coals

	Nava District	Rio Escondido	
As received:			
% Ash	30.9	28.3	29.0
% Moisture	4.8	5.2	7.8
Moisture & Ash-Free:			
% Volatile (VM)	50.7	48.3	46.7
% Carbon (C)	76.2	76.5	76.3
% Hydrogen (H)	5.6	5.8	5.7
% Oxygen (O)	14.8	15.0	15.8
% Nitrogen (N)	1.4	1.2	0.9
% Sulfur (S)	2.0	1.5	1.3
HHV, Btu/lb	13,790	13,690	13,260
HHV, MJ/kg	32.1	31.8	30.8
Fusibility Temperature, reducing atmosphere, °F:			
Initial deformation	2700 +	2700 +	2700 +
Ash Composition:			
% SiO_2	60.4	61.7	58.8
% Al_2O_3	25.8	27.7	29.0
% Fe_2O_3	5.0	2.1	2.8
% CaO	3.1	1.4	2.6
% MgO	0.9	0.9	0.6
% Na_2O	0.3	0.5	0.4
% K_2O	0.7	0.8	0.8
% TiO_2	0.8	0.7	1.0
% P_2O_5	0.1	0.1	
% SO_3	1.1	0.8	2.4
Base/acid Ratio	0.11	0.06	0.08
Hardgrove Grindability	58	54	66

year 2000. Analyses of typical Argentinian coals are given in Table VII.

BRAZIL

The principal explored and developed coal deposit of Brazil extends south from southern São Paulo, through the states of Parana, Santa Catarina, and Rio Grande do Sul, to the Uruguay frontier. The seams are thin and lie in flat formations. The coal is classified as bituminous, but is of low-grade because of its high ash content. It is not easily separated from the shale partings and must be crushed to small size to make washing effective. Table VIII gives some proximate analyses of bituminous coals from Brazilian mines.

Large brown coal deposits have been discovered near the border of Peru; also, there are large deposits of lignite in Amazonas, as yet unquantified. Annual production in the year 2000 is expected to be about 19 million tonnes.

CHILE

Chile is the largest producer of coal in South America. The deposits occur in the low-lying land along the ocean, and hence are easily accessible. The average thickness of the main coal seam is 150 cm. The Province of Concep-

Table VII. Coals of Argentina — Rio Turbio Area

	Hvcb	Suba
As Received:		
% Ash	10.9	15.4
% Moisture	8.6	10.3
Moist. & Ash-free:		
% Volatile (VM)	47.1	50.1
% Carbon (C)	77.8	73.2
% Hydrogen (H)	6.3	5.9
% Oxygen (O)	14.2	19.4
% Nitrogen (N)	0.7	1.0
% Sulfur (S)	1.0	0.5
HHV, Btu/lb	14,020	13,490
HHV, MJ/kg	32.6	31.4
Fusibility Temps., reducing at., °F:		
Initial deformation	2020	2190
Softening (H=W)	2280	2300
Fluid	2470	2530
Ash Composition:		
% SiO_2	48.1	53.0
% Al_2O_3	25.6	23.7
% Fe_2O_3	7.0	7.6
% CaO	7.6	6.2
% MgO	1.6	1.6
% Na_2O	0.9	0.3
% K_2O	0.3	0.4
% TiO_2	1.7	2.0
% P_2O_5	0.1	. . .
% SO_3	4.6	5.1
Base/acid Ratio	0.23	0.20
Hardgrove Grind.	49	44

ción and the deposits near the Bay of Arauco have thus far yielded the best coal in Chile. Table IX gives the proximate analyses of the representative coals of these regions. In addition, considerable low-sulfur brown coal deposits exist in the south of the country, in the province of Magallanes.

In 1971 the private Lota-Sweiger Coal Company was nationalized. The Empresa National del Carbon S.A. (ENACAR) was founded to operate the state-owned coal mines. Annual production in 2000 is planned to approach 6 million tonnes.

COLOMBIA

Colombia has the largest coal resources in Latin America, but as yet no comprehensive survey has been made of their extent and quality. Estimates of the total geological resources range from 40 to 80 billion tonnes. The coalfields are situated mainly in the Eastern Cordillera; some coal deposits are located in the Western and Central Cordillera region. All coal ranks are available, from brown coal to anthracite. A considerable portion of the coal has good coking properties. Most of the coking coals have ash contents of less than 10 percent and low sulfur contents of 0.3 to 1 percent, making them suitable for export.

In the deposits of El Cerrejon, open-cast mining produces coal low in ash and sulfur content, and high in calorific value. This coal is well suited for power stations, but can also be used for coking. The deposits producing the largest coal output are situated in the district of Paz del Rio. Further deposits occur in Cundinamarca, Boyaca, the Cauca River basin, and Samaca. Colombia plans to increase its present coal production, ensuring the domestic energy supply, and anticipating future exports up to 30 million tonnes. Table X gives analyses of typical Colombian coals.

ECUADOR

Considerable coal deposits occur in various sections of Ecuador; estimated coal deposits are 13 million tonnes. The bulk of the deposits contain subbituminous coal and low-grade lignite, high in ash. In some localities, coals of better quality, approaching anthracite, have been found. However, the general low quality, high cost of mining, and lack of transportation deter exploitation of Ecuadorian resources.

PERU

Coal deposits have been detected in a number of places along the Andean ranges, mainly in the Oyon area. The coals are both anthracite and bituminous. Two coal deposits in Peru

Table VIII. Analysis of Typical Brazilian Coals

District or Mine	As received % Ash	As received % H₂O	Moisture- and ash-free % VM	Moisture- and ash-free % S	HHV Btu/lb	HHV MJ/kg
Butia	13.6	11.5	42.7	1.7	. . .	. . .
Jacuhy	21.2	9.6	28.0	1.3	. . .	. . .
Tubarão	28.2	1.9	41.6	1.6	. . .	. . .
Santa Catarina	27.0	1.3	43.1	5.3	15,060	35.0
Lauro Müller	18.0	2.0	38.8	2.5	14,500	33.7
Rio Deserto	13.8	2.0	35.6	1.2	15,320	35.6
Rio Negro	39.5	7.0	36.1	2.6	12,600	29.3
Cambiu	18.4	2.5	40.6	. . .	14,600	34.0
Rio De Peixe	15.5	6.3	35.7	. . .	14,710	34.2
São Jeronimo	29.0	7.3	36.3	. . .	13,500	31.4

Table IX. Analyses of Typical Chilean Coals

City or Mine	As received % Ash	As received % H₂O	Moisture- and ash-free % VM	Moisture- and ash-free % S	HHV Btu/lb	HHV MJ/kg
Lota*	15.2	2.6	44.6	1.5	14,650	34.1
Schwager	3.6	2.9	44.2	1.0	15,290	35.6
Lebu	7.8	2.2	48.9	4.4	14,700	34.2
Rios	18.0	2.0	52.6	5.0	14,250	33.1
Arauco	14.3	3.4	50.2	6.6	14,500	33.7
Buen Retiro	7.7	1.7	51.5	6.4	15,080	35.1
Mafil	11.8	12.7	47.1	0.8	13,380	31.1
Quila Coya	13.3	1.7	7.2	1.0	14,820	34.5

* Chemical Composition of Ash From Lota Mine

SiO_2	42.9%		MgO	1.4%
Al_2O_3	26.3%		P_2O_5	0.2%
Fe_2O_3	14.6%		SO_3	4.8%
CaO	7.2%			

Ash Softening Temperature—2280°F

have been developed. Those of Rio Santa mainly consist of anthracite; production ceased in 1962. At present, cokable coals are worked in the district of Cerro de Pasco. Peru plans to bring annual production to about 6 million tonnes by 2000, to meet its future demand for both coking and non-coking coals with no importing required.

The mines at Goyllarisquisga and Quishuarcancha account for the bulk of the Peruvian production. The seams are generally thick, but the quality of coal is low. The analyses of Table XI are of typical Peruvian coals.

VENEZUELA

The main Venezuelan coalfields are in the Eastern Cordillera and in the region of Naricual, in the State of Anzoategui. The coal industry is government-owned. The deposits are in detached basins, some of which cover hundreds of miles. Although most of the coal is of bituminous and lignite rank, some deposits

Table X. Analyses of Typical Colombian Coals

	Hvbb	Subb
As Received:		
% Ash	8.8	7.4
% Moisture	9.0	13.8
Moist. & Ash-free:		
% Volatile (VM)	41.4	47.3
% Carbon (C)	81.0	63.9
% Hydrogen (H)	5.5	6.4
% Oxygen (O)	11.1	27.3
% Nitrogen (N)	1.7	1.4
% Sulfur (S)	0.7	1.0
HHV, Btu/lb	14,360	13,130
HHV, MJ/kg	33.4	30.5
Fusibility Temps., reducing at., °F:		
Initial deformation	2280	2240
Softening (H=W)	2380	2390
Fluid	2575	2500
Ash Composition:		
% SiO_2	61.8	37.2
% Al_2O_3	21.1	31.5
% Fe_2O_3	6.6	7.4
% CaO	2.2	7.8
% MgO	2.1	3.8
% Na_2O	1.1	1.6
% K_2O	2.4	1.2
% TiO_2	0.9	. . .
% P_2O_5	0.2	. . .
% SO_3	1.6	9.5
Base/acid Ratio	0.17	0.32
Hardgrove Grind.	48	40

of semianthracite occur in the Coro district of the State of Falcon.

Of the six coal districts of economic importance, Zulia has the largest coal resources; 1,500 million tonnes. The districts of Labotera, with 100 million tonnes, and of Naricual, with about 50 million tonnes, as well as Unare, Taguay and Tachira, are also of economic importance. By 2000, Venezuela's total coal production will be about 16 million tonnes, which may satisfy domestic needs. Table XII gives analyses of some Venezuelen coals.

COAL DEPOSITS OF EUROPE

Most of the European countries, excluding Scandanavia, possess coal resources. Scandanavia (including Iceland) has substantial quantities of peat, which increases in value as oil prices rise and as there is more uncertainty with regard to other fossil fuels.

In continental Europe, the bulk of coal is found in Germany, Czechoslovakia, Poland, and Yugoslavia (Fig. 6). Many other countries with reasonable reserves have allowed their mines to become idle. French, Belgian, and Spanish coal is expensive to mine. Austria, Denmark, Italy, Luxembourg, the Netherlands, and Switzerland produce little or no coal.

The resources of European Russia are included in the section on the U.S.S.R. and Asia.

AUSTRIA

Coal production from small bituminous coal deposits has been abandoned. Some brown coal deposits, however, have economic impor-

Table XI. Proximate Analyses of Peruvian Coals

Kind of Coal	As received		Moisture- and ash-free		
	% Ash	% H_2O	% VM	HHV BTU/lb	HHV MJ/kg
Goyllarisquisga, Raw	31.5	3.1	54.5	14,710	34.2
Goyllarisquisga Steam, Washed	17.3	4.0	54.8	15,150	35.2
Goyllarisquisga Coking	17.9	4.0	49.9	15,110	35.1
Quishuarcancha, Raw	26.9	2.5	50.7	13,780	32.1
Santa River Valley	7.1	3.9	7.9	15,560	36.2
Paracas Peninsula	9.0	3.1	39.9	15,530	36.1

Table XII. Analyses of Typical Venezuelan Coals

Region	Rank	As received		Moisture- and ash-free		
		% Ash	% H$_2$O	% Vol.	HHV Btu/lb	MJ/kg
Naricual	Hvbb	2.3	2.5	46.5	14,460	33.6
Santa Rosa	Hvab	2.4	1.5	40.1	14,520	33.8
Barcelona	Hvab	3.4	3.0	38.6	14,620	34.0
Simplicio	Mvb	4.3	3.8	27.9	. . .	. . .

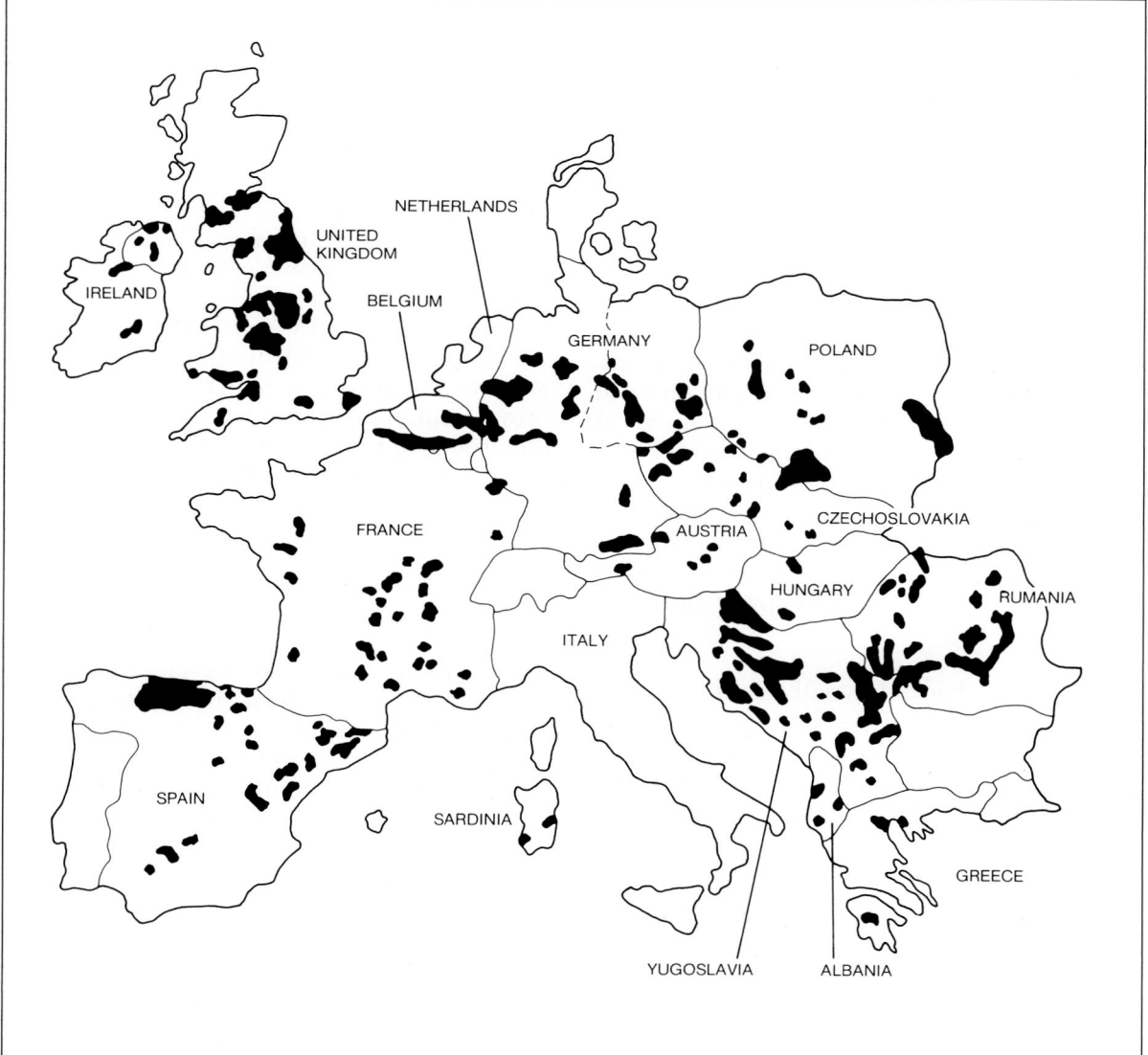

Fig. 6 Coal Map of Europe

tance. There are deposits of glance brown coal as well as of lignite. The most important lignite district is situated near Köflach-Voitsberg, where two 10 to 15 m thick seams are being extracted, partly by open-cast mining, partly by deep mining. There are further brown coal deposits near Wolfsegg-Traunthal and in the Salzach district near Trimmelkam. Apart from local needs, lignite produced in the country is nearly exclusively used in power plants.

BELGIUM

In Belgium, two main bituminous-coal deposits occur, forming part of the northwest European coal belt. The southern coal deposits comprise the districts of Liège and Namur (Bassin de Liège) and Charleroi (Bassin du Hainant). The northern coal deposits include the district of the Campine. There are no brown coal deposits in Belgium.

The South basin extends through the valley of the Sambre and Meuse rivers, running east to the German frontier and west to the French border. The Campine basin, with large reserves, stretches in a westerly direction from the German border, through the Provinces of Limburg and Antwerp. The Belgian coals range from high-volatile bituminous to anthracite.

Belgium produces about half of the coal necessary for domestic consumption; the balance is imported. Only one-fourth of the coal consumed is used for thermal power plants.

BULGARIA

The country has only small bituminous coal deposits. There is a deposit of anthracite coal north of Sofia. Cokable types of coal (cretaceous coal) exist in the small district of Ojabrovo and in the Dobrudsha. The brown coal deposits of Bulgaria are economically more important than the hard coal deposits, being estimated at greater than 2,000 million t.c.e. Deposits of glance coal and of dull brown coal (cannel coal), most of which are extracted by deep mining, occur near Pernik, in the Pirin Mountains, and in Tscherno. Larger lignite deposits in the district of Maritsa can be extracted by open-cast mining. The brown coal mines mainly serve the domestic market, while the lignite mines supply coal mostly to power plants.

CZECHOSLOVAKIA

There are economically important bituminous and brown coal deposits in Czechoslovakia. Technically and economically recoverable reserves exceed 4,000 million t.c.e. The wide Upper Silesian hard-coal deposits in the district Ostrava/Karvina and the Lower Silesian deposit of Walbrzych, near Zaclér, are being mined. In addition to these two large hard-coal deposits, there are the two districts of Plzeň and Kladno/Rakovnik and a small hard-coal deposit near Rosice. In North Bohemia, there are large and economically very important brown coal deposits. The largest district is situated near Teplice and Chomutov on the border of the Krusnéhory, with a further deposit of economic importance in the district of Sokolov. Smaller brown-coal deposits exist in West Bohemia near Cheb, in the south of Moravia, and in Slovakia near Handlova.

The coals from the districts of Plzeň and Kladno are mostly low-rank types. All types of brown coal, from glance brown coal to lignite, are produced in the country; most of them are used for electricity generation.

FINLAND

Finland has no coal reserves and thus no indigenous coal production. But there are large reserves of peat with about one-third of the country considered peatland. Peat, used mainly for industrial steam production and cogenerating district heat and electricity, currently supplies about 3 percent of the fuel for energy consumption in Finland. Production is expected to exceed 2 million tonnes coal equivalent by 2000, corresponding to about 20 percent of the total energy consumption.

FRANCE

French coal deposits consist of a large number of small fields distributed over the northern, central, and southern parts of France. The deposits are generally deep, jagged, and difficult to mine. Usable deposits are between 700 and 1250 meters deep. The most important

Table XIII. Analyses of Typical French Coals

Area	Mining District or Seam	As received % Ash	As received % H₂O	Moisture- and ash-free % VM	Moisture- and ash-free HHV Btu/lb	Moisture- and ash-free HHV MJ/kg	Ash ST, Red., °F	Hardgrove Grind.
Nord and Pas-De-Calais Basin								
Auchel	Marles	26.5	1.2	35.0	15,570	36.2	2100	86
Bethune	Bethune	47.3	1.0	33.1	14,300	33.3	1920	75
	Noeuds-Les-Mines	32.7	1.5	38.1	14,930	34.7	2120	87
Bruay	Bruay	27.9	1.8	39.0	14,700	34.2	1940	83
Douai	Douai	26.0	1.1	24.6	15,250	35.5	2280	95
	Somain	25.3	0.9	18.6	15,340	35.7	2150	106
Hénin-Liétard	Hénin-Liétard	23.5	1.0	12.8	15,300	35.6	2280	79
Lens	Lens IV	17.6	1.0	21.0	15,420	35.9	2010	89
	Lens VII	36.0	0.9	36.0	14,930	34.7	2010	. . .
Oignies	Oignies	24.0	0.9	12.1	15,190	35.3	2280	78
Valenciennes	Anzin	44.9	1.6	17.2	14,360	33.4	2190	71
	Agache	27.6	0.9	11.7	15,310	35.6	. . .	81
Lorraine Basin								
Petite Rosselle	Simon	5.9	3.3	44.2	13,880	32.3	1970	. . .
	St. Charles	34.8	1.8	40.5	13,450	31.3	1970	76
	Wendel	29.5	3.2	40.0	13,710	31.9	2060	57
Saare & Moselle	Merlebach	6.1	2.8	43.0	14,110	32.8	1920	. . .
	La Houve	35.0	3.2	40.6	13,330	31.0	2060	. . .
Faulquemont	Faulquemont	8.6	4.0	41.9	14,750	34.3	2010	. . .

are the district of Lorraine adjoining the Saar in Western Germany, the district of Nord and Pas-de-Calais and some scattered deposits in the district Centre-Midi. In the long term, only coal production in the Lorraine district will be maintained. The smaller brown coal deposits, most of which are in the southwest of the country in Provence and near Arjuzanx, have only local importance for electricity generation.

Most of the French hard coals represented are not cokable. While the mines in the Pas-de-Calais still produce mostly high-rank types of coal and coking coal, the coals from the Lorraine and the Centre-Midi are usually low-rank and only part of them can be used for coking. The low brown coal production is exclusively used in power plants. Table XIII gives analyses of typical French coals.

Economically recoverable reserves of deep-mined hard coal are expected to go from 320 million tonnes in 1985 to 120 million tonnes in 2000; brown coal, from 50 million in 1985 to 20 million in 2000.

GERMANY (EASTERN)

The coal output from the hard coal districts of Zwickau and Lugan--Ölsnitz in Saxony has been stopped because of the properties of the coal deposits. Eastern Germany, however has large lignite deposits in the two principal districts in Lasatia near Cottbus-Senftenberg and near Halle-Merseburg. A third smaller lignite district is situated between Magdeburg and Helmstedt. Germany is the largest lignite producing country in Europe. (See Table XIV.)

Most of the hard coal from the Saxonian districts have unfavorable coking properties and can be used only as a blending coal to the im-

Table XIV. Analyses of Typical Eastern German Coals

District	As received % Ash	As received % H$_2$O	% VM	Moisture- and ash-free % C	% H	% O	% N	% S	HHV Btu/lb	HHV MJ/kg
Sachsen	5–12	2–6	38.0	81.4	5.5	10.4	1.8	1.0	14,690	34.2
Ostelbe	2–5	58–60	56.5	68.2	5.5	24.9	1.1	0.3	11,390	26.5
Westelbe	5–8	52–56	61.0	70.8	6.1	21.3	0.9	0.9	12,110	28.2
Halle-Bitterfeld	5–7	52–56	57.5	72.0	5.5	18.3	0.8	3.4	12,820	29.8

ported coking coal, or as industrial and power-plant coal. German lignites are suitable as power-plant coal, for briquetting and, by employing special methods of coking, they can be used as input coal for coke ovens.

GERMANY (WESTERN)

Western Germany has large deposits of bituminous (stein) coal and brown (braun) coal, the latter being unconsolidated lignite. About 60 percent of the currently economical recoverable deposits have sulfur below 1 percent, as received at the power plant.

The principal districts where bituminous coal occurs in greatest abundance are: the Lower Rhine and Westphalia, Silesia, the Rhenish district and the Saar district. In central north Germany, brown coal occurs in the Rhine Province between Cologne and Aachen (Aix-la-Chapelle). Further economically important brown-coal deposits are in Lower Saxony (Helmstedt) and Lower Bavaria (Schwandorf).

The Westphalian and Rhine district fields include the Ruhr Basin, the most important coal field in Germany. These coal districts are connected with those of southern Holland, Belgium, and northern France. The coal ranges from medium to high volatile, and the ash varies considerably from mine to mine. The bulk of the entire German output comes from the Lower Rhenish Westphalian fields.

The Silesian district extends through the province of Silesia, which, geographically, is divided between Germany, Poland and Czechoslovakia. The bulk of the coal in this district is of first-class coking quality, and is widely used in steel industries. The Saar dis-

trict stretches through parts of Alsace-Lorraine. The coals contain considerable ash, particularly in small sizes, which are usually washed.

The recoverable hard-coal resources of Western Germany amount to about 24,000 million tonnes, with about 85 percent of that concentrated in the Ruhr. The balance comes from the Saar, Aachen, and Ibbenbüren districts. All hard coal is deep mined. Production in excess of 100 million tonnes annually is expected by 2000.

Economically recoverable brown coal totals about 35,000 million tonnes, all from open-pit mining. An annual production of 150 million tonnes is anticipated by the end of the century, practically all of which will be consumed in the country, mainly for power generation. Table XV lists typical analyses of Western German coals.

GREECE

Greece does not have any bituminous coal deposits of economic importance. However, there are important lignite deposits in the districts of Ptolemais and Megalopolis. The production of brown coal was about 18 million tonnes in 1975, and increases steadily to form an important basis for the production of electric energy. The Greek brown-coal mining industry provides 50 percent of the country's energy demand. Analyses of Greek lignites are given in Table XVI.

HUNGARY

Hungary has some bituminous-coal deposits in the districts of Pécs and of Komlo, and several brown-coal deposits of national importance, both totalling about 1,000 million t.c.e.

A large brown-coal deposit is situated in the Budapest district. Glance brown coal is exploited by deep mining in the districts of Tatabánja and Oroszlány. Large lignite deposits, extracted by open-cast mining, exist in the district of Gyöngyös; in the Combinate Borsod, brown coal is extracted by deep mining. A new brown coal district near Bükkabrány projects an annual output of about 20 million tonnes.

Most of the coal produced is used for electricity generation. The use of Hungarian hard coal for coke production is very limited. Table XVII gives analyses of Hungarian brown coal.

IRELAND

There are 8 coalfield areas in Ireland (the Republic of Ireland and Northern Ireland), all of limited size. Only two are currently being

Table XV. Analyses of Typical Western German Coals

District	As received % Ash	% H₂O	% VM	Moisture- and ash-free % C	% H	% O	% N	% S	HHV Btu/lb	MJ/kg
Ruhr (Anthracite)	4–7	3–5	7.7	91.8	3.6	2.6	1.4	0.7	15,440	35.9
Ruhr (Low-Vol. Bit.)	6–9	7–10	10.5	90.8	3.8	2.7	1.7	0.8	15,500	36.0
Aachen (Low-Vol. Bit.)	6–9	8–10	13.8	89.8	4.8	2.8	1.5	0.5	15,410	35.8
Ruhr (Med-Vol. Bit.)	6–9	7–10	24.4	88.7	5.0	4.1	1.6	6.7	15,550	36.2
Saar (Med-Vol. Bit.)	7–9	8–10	32.5	86.9	5.2	5.4	1.3	1.1	15,320	35.6
Ruhr (High Vol. Bit.)	6–7	8–10	33.7	85.9	5.5	6.2	1.6	0.8	15,160	35.2
Saar (High Vol. Bit.)	5–8	3–5	38.2	82.7	5.2	9.4	1.2	1.5	14,220	33.0
Peissenberg (Lignite)	12–20	8–12	52.0	74.0	5.5	14.5	1.4	4.6	12,780	29.7
Helmstedt (Brown Coal)	12–22	42–46	59.4	72.6	5.8	16.7	0.4	4.4	12,790	29.7
Rheinland (Brown Coal)	5–20	50–62	55.0	68.3	5.0	27.5	0.5	0.5	11,340	26.4
Schwandorf (Brown Coal)	6–20	50–58	55.0	63.6	5.0	26.1	1.3	4.0	10,890	25.3

Table XVI. Analyses of Greek Lignites

Region	As received % Ash	% H₂O	% VM	Moisture- and ash-free % C	% H	% O	% N	% S	HHV Btu/lb	MJ/kg
Kimi	5.9	23.0	55.6	...	...	...	...	2.3	11,560	26.9
Aliveri	8.6	33.5	57.7	...	...	...	...	1.4	11,330	26.3
Aegean	18.6	34.9	57.6	...	...	...	...	6.4	10,650	24.8
Attica	15.0	42.0	59.1	...	...	...	...	6.7	11,740	27.3
Ptolemais	6–22	52–60	57.0	65.3	5.3	26.5	1.6	0.5	10,850	25.2
Megalopolis	13–17	60–64	62.0	60.5	6.2	30.6	1.3	1.4	10,510	24.4

Table XVII. Analyses of Typical Hungarian Coals

Region	As received % Ash	% H₂O	% VM	Moisture- and ash-free % C	% H	% O	% N	% S	HHV Btu/lb	MJ/kg
Gyöngyös/Visonta	15–30	46–54	63.0	63.8	4.8	26.8	1.1	3.5	10,670	24.8
Tatabánya	6–12	12–14	52.0	73.0	5.8	17.7	0.9	2.6	13,500	31.4

worked; the Connaught (Arigna) field, which contains semi-bituminous coal with high ash content, provides fuel to an adjacent generating station; the Leinster (Castlecomer) field produces high-grade anthracite for domestic heating purposes. Brown coal occurs in two localities in northeast Ireland, with large reserves in the Thistleborough area.

Peat bogs cover 15 percent of the country; over 3 million tonnes of peat are used yearly for generating electricity and domestic heating. Milled peat powers 8 stations that generate 23 percent of the national electrical demand.

Irish peat in its natural state contains about 95 percent moisture; by long-term air drying, this can be reduced to 20 to 40 percent to produce a commercial fuel. Dried peat has a volatile content between 55 and 70 percent, and ash from 2 to 10 percent. Lower heating value at 30-percent moisture is 6000 Btu/lb (14 KJ/kg); ash softening temperature is about 2100°F.

Current State-owned reserves of peat are equivalent to 40 million tonnes of coal.

ITALY

Italy has no economically important hard coal deposits. The once-mined deposits in Sardinia are to be used again in future. Larger deposits of brown coal and partly also of glance brown coal exist in Toscana, in Calabria, and in the south-west of Sardinia. Outputs are low, about 2 million tonnes annually. The largest open-cast mine is situated in the Valdarno Basin of the district Arezzo. Table XVIII shows analyses of Sardinian coals.

Table XVIII. Range of Analyses of Italian (Sardinian) Coal (As received basis)

Ash, %	15 to 20
Moisture (H_2O) %	12 to 15
Volatile matter (VM) %	32 to 45
Sulfur (S) %	2 to 7
HHV, Btu/lb	9960 to 11,500
HHV, MJ/kg	23.2 to 26.7
Ash softening temp., approx., °F	2000

NETHERLANDS

The most important of the Netherlands coal areas lies in the province of Limburg, where it links up with the Campine basin of Belgium, and extends in a northwest direction into the province of North Brabant. The Netherlands have ample geological resources that are not economically recoverable because of the large Dutch resources of natural gas.

There is an estimated resource base of 100 billion tonnes of hard coal, related to coal seams more than 0.8 m thick and 1500 m deep.

Until 1950 indigenous coal provided more than 70 percent of national coal consumption and about 50 percent of total energy demand, at a production level of some 12 million tonnes per year. Production reached a maximum of 13 million tonnes in 1961 but declined sharply after 1965 as a consequence of the development of world energy prices and low-cost domestic natural gas supply. Coal mining was stopped in 1974, while coal consumption dropped to less than 5 percent of total energy demand.

NORWAY

Economically important hard-coal deposits occur on the arctic island of Spitsbergen; these deposits are currently being worked. The coal produced is used mostly in the north of Norway, for district heating.

POLAND, PEOPLE'S REPUBLIC OF

In Poland, there are several rich and easily accessed hard-coal and brown-coal regions. Hard coals, which are all deep-mined, occur in three basins: the Upper Silesian basin has about 93 percent of the economically recoverable reserves; the Lower Silesian has about 1 percent, found in thin seams (0.4 to 1 m thick) near the towns of Walbrzych and Nowa Ruda. The Lublin basin, with about 6 percent of the reserves, is under development; there, the thickness of the seams ranges from 0.7 to 3.6 m. Most of these coals are high-volatile bituminous in rank, high in calorific value and with sulfur contents about 1 percent.

Brown coal deposits, mined by open-pit methods, occur in the Konin and Adamów dis-

tricts, and near Turów, Legnica, and Belchatów. Brown coal production is solely for consumption inside Poland, primarily for the generation of electricity. (See Table XIX.)

Economically mineable reserves of hard coal are said to be 74,000 million tonnes, to a depth of 1,000 m; of brown coal, 9,000 million tonnes, all to be won by open-cast mining. Production of hard coal is expected to be between 250 and 300 million tonnes by 2000, with an additional 200 to 250 million tonnes of brown coal.

Table XIX. Analyses of Typical Polish Coals

	High-Volatile B Bituminous	Subbituminous A
As Received:		
% Ash	11.5	24.4
% Moisture	8.8	17.4
Moist. & Ash-free:		
% Volatile (VM)	37.1	41.9
% Carbon (C)	82.9	74.2
% Hydrogen (H)	5.2	6.0
% Oxygen (O)	9.9	16.7
% Nitrogen (N)	1.0	0.9
% Sulfur (S)	1.0	2.2
HHV, Btu/lb	14,550	13,020
HHV, MJ/kg	33.8	30.3
Fusibility Temps., reducing at., °F:		
Initial deformation	2160	2280
Softening (H=W)	2210	2460
Fluid	2460	2690
Ash Composition:		
% SiO_2	46.8	55.0
% Al_2O_3	21.8	24.1
% Fe_2O_3	9.6	9.3
% CaO	5.8	3.4
% MgO	3.5	1.5
% Na_2O	0.8	1.1
% K_2O	3.1	1.7
% TiO_2	0.7	1.1
% P_2O_5	0.3	. . .
% SO_3	6.6	2.8
Base/acid Ratio	0.33	0.21
Hardgrove Grind.	49	44

ROMANIA

Romania has numerous small coal deposits, especially along the outskirts of the Carpathian Mountains. Hard coal deposits are mined near Schela, Anina and Cozla. The most important district is situated near Petrosani in the Southern Carpathian Mountains. Here, the sub-bituminous coal seams reach thicknesses of 6 m to more than 10 m deposited in depths from 150 m to 1000 m. Further deposits of brown coal and lignite are situated near Bacau and in the basins of Oltenio, Rovinari Motru and Iilt, most of which are extractable by open-cast mining. Most of the brown coal mined in the country is used in power plants.

SPAIN

Spain has considerable bituminous coal and lignite deposits although, in several deposits, geological conditions render exploitation difficult. The economically important hard coal deposits are concentrated in the province of Asturia in the northwest—the districts of Oviedo and León account for nearly 90 percent of the total Spanish production of hard coal. A second large hard coal deposit is in the district of Mariannica, with individual mines in the provinces of Cordoba and Sevilla. The lignite deposits are in the northeast of the country between Teruel and the Pyrenees. With its present production, Spain can supply more than 75 percent of its demand. (See Table XX.)

SWEDEN

Swedish coal deposits are small and of low quality; technically recoverable reserves have been estimated to be 30 million tonnes. At present, there is no coal being produced except small amounts obtained from clay produced for the ceramics industry. This coal is consumed locally, and production is expected to increase slightly. Sweden has large resources of peat—estimated at about 15 percent of the country's total area—which are yet to be developed.

UNITED KINGDOM

The coal fields of the UK may be grouped into three principal areas designated as the southern, the central, and the northern coal fields.

Table XX. Analyses of Typical Spanish Coals

	Anthra-cite	Low-Vol. Bitum.	High-Vol. Bitum.	Sub-bituminous	Sub-bituminous	Brown Coal
As Received:						
% Ash	31.9	13.2	49.0	12.4	30.9	18.8
% Moisture	3.1	8.7	6.2	25.0	21.5	47.7
Moist. & Ash-free:						
% Volatile (VM)	10.9	20.9	36.8	48.7	48.1	53.1
% Carbon (C)	87.2	. . .	77.8	70.1	67.2	62.4
% Hydrogen (H)	2.5	. . .	4.9	5.0	4.8	5.6
% Oxygen (O)	8.0	. . .	15.6	15.3	16.2	28.4
% Nitrogen (N)	0.9	. . .	0.8	0.7	0.7	1.5
% Sulfur (S)	1.4	2.0	0.9	8.9	11.1	2.1
HHV, Btu/lb	14,090	14,990	13,190	12,640	11,860	9130
HHV, MJ/kg	32.8	34.9	30.7	29.4	27.6	21.2
Fusibility Temps., reducing at., °F:						
Initial deformation	2160	2280	2700+	2070	2040	2120
Softening (H=W)	2420	2370	2700+	2230	2290	2470
Fluid	2640	2510	2700+	2310	2540	2660
Ash Composition:						
% SiO_2	59.0	40.5	53.6	32.3	54.2	54.7
% Al_2O_3	20.5	22.5	29.2	22.4	21.1	21.8
% Fe_2O_3	7.5	14.8	6.1	16.8	15.7	7.3
% CaO	3.5	6.3	1.2	11.0	3.5	5.2
% MgO	1.9	1.1	1.1	1.9	0.8	1.0
% Na_2O	0.9	0.5	0.4	0.3	0.1	0.2
% K_2O	4.0	3.2	4.6	1.2	1.3	0.5
% TiO_2	1.3	. . .	0.7	0.3	0.8	1.1
% P_2O_5	. . .	. . .	1.4	0.2	. . .	. . .
% SO_3	2.0	8.2	1.2	13.2	3.2	5.3
Base/acid Ratio:	0.22	0.41	0.16	0.57	0.28	0.18
Hardgrove Grind.:	54	81	. . .	83	. . .	. . .

The southern area embraces the fields known as South Wales, Forest of Dean, Bristol and Kent. The central area includes the fields of Lancashire, North Wales, Yorkshire, Derbyshire, Nottinghamshire, Staffordshire and Warwickshire. The northern area comprises the fields of Northumberland, Durham, Cumberland, and the fields of Scotland.

Intensive exploration in recent years has revealed large new areas for development, particularly in the Selby district (a major eastward extension of the Yorkshire field) and North-East Leicestershire, with coal of geological disturbance and at relatively shallow depths. Also, over 200 million tonnes of workable coal have been discovered in an area of Coventry and north of Kenilworth, at depths between 1000 and 2500 ft.

The coal of the southern area varies, from bituminous coal through the well-known Welsh steam coal to anthracite of notable purity. The bulk of the central-area coal fields lies in the south of Lancashire. Coals in this field contain 30 to 35 percent volatile matter. Some

Table XXI. Analyses of Typical British Coals

District	— As received —		Moisture- and ash-free							
	% Ash	% H₂O	% VM	% C	% H	% O	% N	% S	HHV Btu/lb	MJ/kg
Durham (Mvb)	6.9	2.6	29.4	87.8	5.3	4.6	1.4	0.9	15,660	36.4
Yorkshire (Hvab)	6.8	2.0	34.4	84.3	5.2	8.0	1.7	0.8	14,980	34.8
West Midlands (Hvbb)	4.9	9.8	39.6	80.5	5.5	11.9	1.4	0.7	14,180	33.0
Scotland (Hvbb)	4.6	13.8	41.5	81.4	5.4	10.3	2.1	0.8	14,580	33.9

Table XXII. Proximate Analyses of Welsh Coals

District	As received		Moisture- and ash-free			
	% Ash	% H₂O	% VM	% S	HHV Btu/lb	MJ/kg
Aberpergwm	4.0	1.3	7.4	0.9	15,820	36.8
Ynisarwed	4.2	1.5	11.3	2.5	15,370	35.7
Bwllfa	4.8	1.1	12.0	0.7	15,560	36.2
Cyfarthfa	4.9	0.8	14.0	0.9	15,430	35.9
Blaenclydach	4.2	0.9	21.3	0.6	15,230	35.4
Meiros	4.4	1.8	36.9	1.2	15,600	36.3

of them yield good coke. In the northern area, the Durham and Northumberland fields are among the most important in Great Britain. The character of the coal varies considerably. It is of exceptionally fine quality for steam, coking, household, manufacturing, and gas-producing purposes. Tables XXI and XXII provide analyses of typical British and Welsh coals.

Coal production in the UK is nearly equal to its domestic consumption, with nearly two-thirds being burned in power-generating stations. Total recoverable reserves are on the order of 45,000 million tonnes; production may be as high as 200 million tonnes annually by the turn of the century, 95 percent deep mined.

YUGOSLAVIA

There is only one hard coal deposit in Yugoslavia, with limited economical importance—on the Istria peninsula in the district of Rasa. Of higher economical importance to Yugoslavia are the brown coal deposits in the districts of Zagorje, Trbovlje, Hrastnik, Kakanj, Jenica, and Nis, with glance brown coals as well as several lignite deposits in Slovenia, such as near Vel-

enje, as well as in Bosnia and Serbia. The economically recoverable deposits, with more than 6.8×10^9 tonnes coal equivalent, are most important to Yugoslavia. Only part of the hard coals of the country are suitable for coking. The vast majority of the coal output is used for electricity generation. Table XXIII gives complete analyses of several Yugoslavian brown coals.

COAL DEPOSITS OF THE U.S.S.R. AND ASIA

The largest countries of this continent in terms of territory, size and population—the U.S.S.R. (Union of Soviet Socialist Republics, also called the Soviet Union or Russia), including the European part, the People's Republic of China, and the Republic of India—are also the countries with the biggest coal deposits. The total technically and economically recoverable reserves are estimated at greater than 200,000 million tonnes coal equivalent; over 10 percent of this is lignite or subbituminous coal, nearly all of which is found in the U.S.S.R. The balance, as described under the individual countries, is mostly bituminous coal.

Table XXIII. Analyses of a Range of Yugoslavian Brown Coals

As Received:				
% Ash	17.8	9.0	19.0	18.7
% Moisture	49.7	49.5	47.8	48.5
Moist. & Ash-free:				
% Volatile (VM)	52.0	51.1	61.0	65.2
% Carbon (C)	61.9	66.0	62.9	62.2
% Hydrogen (H)	4.3	5.1	6.2	4.3
% Oxygen (O)	29.8	25.8	28.3	30.2
% Nitrogen (N)	0.9	1.7	2.0	1.2
% Sulfur (S)	3.1	1.4	0.6	2.1
HHV, Btu/lb	9450	11,110	10,650	10,400
HHV, MJ/kg	22.0	25.8	24.8	24.2
Fusibility Temps., reducing at., °F:				
Initial deformation	2700+	2360	2100	2030
Softening (H=W)	2700+	2390	2420	2090
Fluid	2700+	2400	2450	2150
Ash Composition:				
% SiO_2	21.6	25.3	72.5	36.6
% Al_2O_3	3.1	9.5	12.5	12.9
% Fe_2O_3	6.0	5.7	2.2	7.5
% CaO	45.6	32.5	5.9	25.4
% MgO	2.5	6.6	1.4	3.1
% Na_2O	0.4	2.2	0.6	0.6
% K_2O	0.2	0.6	1.1	1.1
% TiO_2	0.1	0.4	0.5	0.3
% P_2O_5	0.4	0.3	0.1	0.2
% SO_3	15.1	12.8	3.2	11.0
Base/acid Ratio:	1.04	1.35	0.13	0.76
Hardgrove Grindability	A	B		

A: 84 at 35% H_2O; 75 at 30% H_2O; 68 at 22% H_2O; 55 at 16% H_2O; 90 at 7.5% H_2O
B: 58 at 36% H_2O; 47 at 31% H_2O; 48 at 27% H_2O; 45 at 15% H_2O; 53 at 10% H_2O

Fig. 7 is a map of European Russia (that portion of the U.S.S.R. west of the Ural mountains) and Asia, showing the coal-bearing areas.

U.S.S.R.

The Union of Soviet Socialist Republics is divided by the Ural Mountains into European Russia and Asiatic Russia, the latter comprised principally of Siberia. The accessible coal reserves of western and central U.S.S.R. have been largely depleted, while the Siberian and eastern fields are difficult and expensive to develop, and have high transportation costs to the using areas.

Total economically recoverable reserves are estimated at 110,000 million tonnes coal equivalent, of which 75 percent is bituminous coal and the balance lignite and brown coal. Eighty-percent of the presently recoverable resources are located east of the Urals, mostly in northern Siberia. (See Table XXIV.)

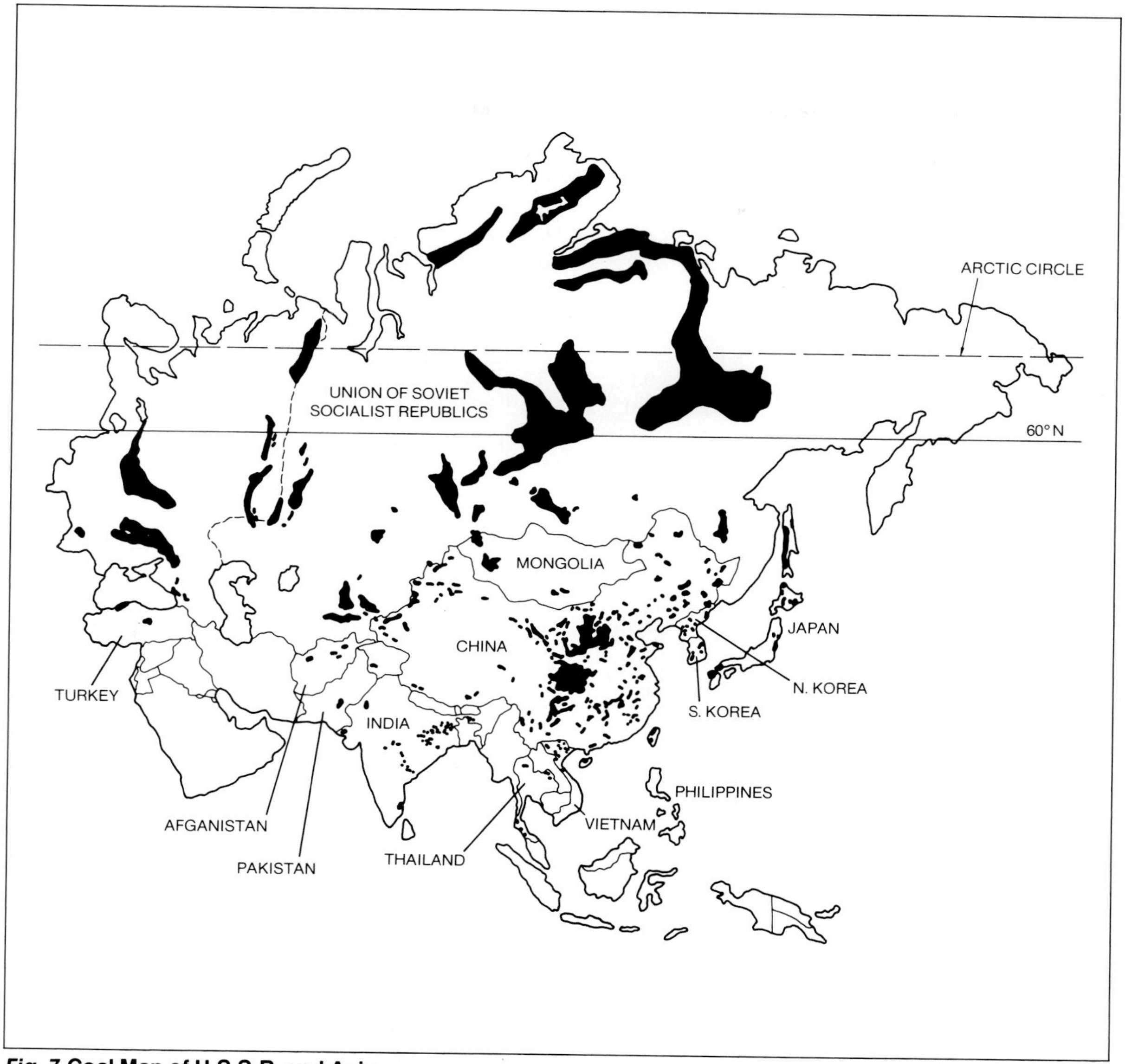

Fig. 7 Coal Map of U.S.S.R. and Asia

European U.S.S.R.

There are three main coal fields in European Russia: the Donetz basin, the Pechora basin, and the Moscow coal fields. All are of limited life. The Donetz basin, which lies directly north of the Sea of Azov, in the Ukraine, has both anthracite and high-grade bituminous coal. The anthracite deposit is extensive, with estimated reserves of about two and a half times

those of the bituminous. The quantity of good coking coal is comparatively small, hence the utilization, on a considerable scale, of anthracite for blast furnaces and boilers. The fact that reserves of coal, iron, manganese, and limestone are all within a short distance creates an unusual economic situation in the Donetz territory.

The Vorkuta field, in the basin of the Pechora

Table XXIV. Analyses of Typical U.S.S.R. Coals

Province or District	Location or Name of Coal	As received		Moisture- and ash-free						HHV	
		% Ash	% H$_2$O	% VM	% C	% H	% O	% N	% S	Btu/lb	MJ/kg
Ukraine	Aleksandrisk	10.8	55.0	57.0	67.6	6.0	22.0	0.8	3.6	12,150	28.3
	Kirovsk	24.8	45.0	56.0	62.5	5.2	27.6	1.0	3.7	10,062	23.4
Grusine	Tkribulsk	26.4	12.0	40.0	78.4	5.9	13.0	1.5	1.2	13,950	32.4
	Tkvarchelsk	27.9	7.0	35.0	81.5	5.9	9.7	1.7	1.2	14,750	34.3
Ural	Cheliabinsk	21.9	19.0	39.0	72.3	5.1	19.5	1.7	1.4	12,520	29.1
	Bogoslovsk	14.0	30.0	43.0	70.0	4.7	23.3	1.5	0.5	11,710	27.2
	Egroshinsk	22.6	6.0	8.0	90.0	3.7	5.1	0.6	0.6	14,780	34.4
	Poltavsk	18.2	9.0	3.5	95.0	0.8	3.5	0.5	0.2	13,860	32.2
Kazachsky	Karagandinsk	19.5	7.0	25.0	85.5	5.2	6.9	1.4	1.0	15,050	35.0
	Berchogursk	37.6	6.0	47.0	77.8	6.6	9.8	1.4	4.4	14,240	33.1
Kirgiz	Kizil-Kia	11.0	27.0	33.0	76.5	4.2	17.1	1.0	1.2	12,600	29.3
	Syliutka	11.9	21.0	29.0	77.7	4.0	17.0	0.8	0.5	12,660	29.4
Siberia	Minusinsky	10.6	12.0	42.0	79.0	5.8	12.5	2.2	0.5	14,050	32.7
	Kansky	10.2	32.0	44.0	73.1	4.9	20.0	1.3	0.7	12,440	28.9
Trans-Baikal	Chernovsky	7.4	33.0	40.0	75.5	5.0	17.5	1.3	0.7	12,960	30.1
	Bukachachisky	14.6	14.0	39.0	80.0	5.5	12.8	1.1	0.6	14,100	32.8
Far East	Kirdinsky	12.7	33.0	41.0	71.0	4.3	23.2	1.2	0.3	11,530	26.8
	Artemovsky	14.1	26.0	49.0	71.5	5.5	20.9	1.5	0.6	12,510	29.1
	Suchansky	27.3	6.0	32.0	86.0	5.0	7.0	1.4	0.6	14,950	34.8
Kuznetsk	Anzhero-Sudzhesky	11.5	4.0	15.0	91.0	4.3	2.1	1.9	0.7	15,490	36.0
	Leninsky	10.3	6.0	39.0	83.0	5.8	7.8	2.7	0.7	14,880	34.6
	Kamerovsky	12.0	4.0	33.0	85.7	5.4	6.5	2.1	0.3	15,030	35.0
	Prokopiersky	10.3	6.0	20.0	89.0	4.6	3.7	2.2	0.5	15,250	35.5
	Kiselevsky	9.3	7.0	18.0	89.3	4.3	3.6	2.3	0.5	15,210	35.4
	Osinovsky	10.2	7.0	30.0	86.5	5.5	4.8	2.7	0.5	15,410	35.8
Donetz	Anthracite	11.3	2.0	3.5	94.6	1.8	1.8	1.0	0.8	14,590	33.9
	Semi-Anthracite	13.4	4.0	12.0	90.0	4.2	2.1	1.5	2.2	15,380	35.8
	Low-Vol. Bitum.	16.0	3.0	16.0	88.0	4.5	2.9	1.5	3.1	15,400	35.8
	High-Vol. A Bit.	19.0	5.0	32.0	83.0	5.1	5.6	1.5	4.8	15,140	35.2
	High-Vol. B Bit.	14.7	8.0	39.0	82.0	5.5	8.5	1.5	2.5	14,600	34.0
	High Vol. C Bit.	19.8	12.0	44.0	77.0	5.6	12.3	1.6	3.5	13,900	32.3

River, contains one of the biggest fields in the U.S.S.R. It is located beyond the Arctic circle, as far north as the 68th parallel, with the almost unaccessible North Ural mountains to the east. The Vorkuta coal is of high calorific value and is easy to coke.

The Moscow field has a large reserve, but the coal is of low-rank and low-grade. It is lignitic in character, and has high moisture, ash and sulfur content. Coal also occurs in the Konban territory and in the Province of Kontais in the Caucasian region.

In the U.S.S.R. there has been considerable development in the use of local low-grade fuels, such as peat and oil shale, for generation of power. The U.S.S.R. has developed technologies for peat dewatering and harvesting, including the milled-peat harvesting tech-

Table XXV. Proximate Analyses of Chinese Coals

Province	District/Mine	As received		Moisture- and ash-free		
		% Ash	% H₂O	% VM	HHV	
					Btu/lb	MJ/kg
Liaoning	Fushun	10.2	6.7	48.0	14,510	33.7
	Sian	1.7	6.8	42.1	15,190	35.3
Jilin	Muling	10.0	3.0	32.2	15,460	36.0
Jehol	Chaoyang	7.0	3.0	35.6	14,350	33.4
	Fusing	7.3	12.3	37.3	14,760	34.3
Heilongjiang	Tangyuan	6.2	2.0	38.2	14,290	33.2
	Lubing	3.7	20.9	48.2	12,660	29.4
Hebei	Lintsing	10.0	1.6	34.1	14,060	32.7
	Wanping	15.0	2.3	9.2	15,360	35.7
Henan	Anyang	11.4	1.1	22.7	15,440	35.9
	Liho	9.9	1.1	9.6	14,790	34.4
	Shen Xien	27.5	4.0	26.9	13,610	31.6
Shandong	Yih Xien	10.0	0.7	30.4	15,500	36.1
	Poshan	8.5	0.3	12.2	15,630	36.3
Shanxi	Dahdong	7.3	3.7	33.7	15,980	37.2
Gansu	Yong Don	3.8	5.6	36.6	15,280	35.5
Hunan	Tzehsin	10.0	2.8	25.7	14,400	33.5
	Siangtan	25.2	3.5	39.8	14,760	34.3
	Ningshian	8.0	6.0	35.6	15,530	36.1
Hubei	Tayei	12.2	1.3	12.0	15,660	36.4
	Tzequai	14.7	4.8	31.9	15,610	36.3
Jiangxi	Po-Loo	10.2	3.1	47.5	13,610	31.7
	Gao Jeng	8.9	1.0	27.1	15,870	36.9
	Jian	6.2	1.5	33.0	12,030	28.0
Anhui	Dahdong	8.9	3.3	41.7	14,600	34.0
	Wheizhou	13.6	0.4	32.1	15,110	35.1
Jiangsu	Xiaohsien	16.2	2.0	26.9	15,550	36.2
Zhejiang	Shangshin	16.6	1.2	44.1	15,530	36.1
Sichuan	Weiyuan	8.2	2.2	33.6	15,610	36.3
	Pahsien	13.2	1.2	21.5	15,940	37.1
Guizhou	Tungtze	12.2	1.0	22.0	15,730	36.6
	Guiyang	15.4	1.8	24.3	14,420	33.5
Yunnan	Shuanwei	12.7	0.9	30.9	15,530	36.1

nique that accounts for 90 percent of the peat won in that country.

Russia consumes 80 million tonnes of peat annually—about 4 percent of the total electrical output is from stations fueled by peat. In the Leningrad district and in White Russia, oil shale and peat are used for generation of power. Oil shale is also used in the Volga Region. The shale contains 50 to 55 percent of ash, which is used as cement for construction purposes. The heating value of the shale is 3200 to 3600 Btu/lb (7.4 to 8.4 MJ/kg) as fired.

Asiatic Russia

The largest coal deposit in Siberia is the Kuznetsk Basin in south central Siberia, which

Table XXVI. Chinese Coals With Ash Analyses

	An-Shan Mine Liaoning Province Hvbb	Guangdong Province Lvb	Shang Shan District South China Lvb
As Received:			
% Ash	1.5	22.5	16.7
% Moisture	2.3	0.5	1.9
Moisture & Ash-free:			
% Volatile (VM)	50.3	13.0	20.1
% Carbon (C)	. . .	. . .	87.0
% Hydrogen (H)	. . .	. . .	4.4
% Oxygen (O)	. . .	. . .	2.9
% Nitrogen (N)	. . .	. . .	0.8
% Sulfur (S)	0.5	1.6	4.9
HHV, Btu/lb	14,290	14,870	15,100
HHV, MJ/kg	33.2	34.6	35.1
Fusibility Temps., reducing atmosphere, °F:			
Initial deformation	. . .	2700 +	2470
Softening (H=W)	. . .	2700 +	2550
Fluid	. . .	2700 +	2700 +
Ash Composition:			
% SiO_2	39.7	52.2	51.2
% Al_2O_3	42.7	34.3	29.2
% Fe_2O_3	6.2	6.6	4.6
% CaO	2.7	0.8	4.3
% MgO	1.6	1.1	0.7
% Na_2O	1.3	0.4	0.7
% K_2O	1.6	2.5	0.8
% TiO_2	1.8	1.2	1.8
% P_2O_5	0.3	0.3	. . .
% SO_3	1.4	0.5	6.4
Base/acid Ratio	0.16	0.13	0.14
Hardgrove Grindability	. . .	95	120

produces 20 percent of the U.S.S.R. total. Next in importance are the Tungus, Kansk-Achinsk, Irkutsk and Minusinsk basins, and the Karagandinsk and the Saghalin deposits. To transport coal to European Russia from many of these fields requires journeys averaging over 900 miles, with some up to 1860 miles.

CHINA, PEOPLE'S REPUBLIC OF

Coal deposits of China are extensive, coal being found in almost every province. One very large area is in northern China, extending over most of the southern part of Shanxi, and one in the south extending over southern Hunan, Guizhou, Yunnan and Sichuan. In variety, Chinese coals range from hard anthracite to lignites of pronounced woody structure. The bituminous coals are of medium and high volatile rank, the medium volatile being rather high in ash. Table XXV gives the proximate analyses of many Chinese coals; XXVI includes ash analyses; and XXVII shows coals presently

Table XXVII. Coals in Use in Chinese Power Stations

	Subc	Hvb	Mvb	Lvb
As Received:				
% Ash	32.8	37.0	29.7	27.7
% Moisture	22.6	3.3	10.3	9.6
Moisture & Ash-free:				
% Volatile (VM)	46.8	39.3	22.7	17.0
% Carbon (C)	74.7	79.6	80.8	83.9
% Hydrogen (H)	4.8	5.4	6.0	4.5
% Oxygen (O)	18.6	12.4	10.7	5.1
% Nitrogen (N)	1.3	1.7	1.4	1.4
% Sulfur (S)	0.6	0.9	1.1	5.1
HHV, Btu/lb	10,890	13,090	13,810	14,030
HHV, MJ/kg	25.3	30.4	32.1	32.6
Fusibility Temps., reducing atmosphere, °F:				
Initial deformation	2070	2700 +	2280	2280
Softening (H=W)	2120	2700 +	2660	2510
Fluid	2280	2700 +	2700 +	2550 +
Hardgrove Grindability	45	44	70	86

in use, categorized by rank.

Most of the coal in China is produced by the longwall retreat method. Hydraulic mining is used in Kailan, with over 3 million tonnes annually being produced by this method, and in Fujian in Shandong province. Most of the coal is found in plains areas rather than in mountain ranges, but have thick overburden. The geological condition of the western part of China is similar to that of the western U.S. Surface production is about 4 percent of the total.

Coal provides two-thirds of China's total energy requirements. Its reserves, estimated to be about 99,000 million tonnes coal equivalent, are exceeded only by those of the U.S. and the U.S.S.R. It ships substantial amounts of steam and coking coal to Japan.

New projects in the design and construction are the 15-to-20 million tonnes/year surface mines in Shanxi province (the An T'ai Pao mine) and in Heilongjiang province (the Yi Min mine).

By the year 2000, Chinese coal production is projected to be on the order of 2,000 million tonnes annually.

INDIA

India's coalfields are extensive and they contain economically recoverable reserves estimated to be between 10,600 and 33,000 million tonnes of hard coal, and from 500 to 900 million tonnes of lower grades. The coals vary greatly in quality, but are generally low in sulfur—less than 1 percent sulfur, as fired, for most of the production. Ash contents are high, ranging from 15 to 45 percent from most mines.

The main coal deposits are located in a triangle, of which the western edge is in the center of India and the eastern corner is about 150 kilometers west of Calcutta. Most of the active mining areas are in Andhra Pradesh, Maharashtra, Madhya Pradesh, Orissa, Bihar, and West Bengal. The oldest coalfields, those of Raniganj and Jharia in the east, produce both high-quality metallurgical and nonmetallurgical coals. The trend now is toward more development in the western part of the Indian coal-belt triangle.

Brown coals are found mainly in the district of Neyveli and in Tamil Nadu. Sixty percent of the brown coal reserves have sulfur contents

Table XXVIII. Analyses of Typical Indian Coals

Area and Rank →	Jharia Mvb	Hvbb	Uttar P. Hvbb	Renusagar Subb	Singrauli Subc	Neyveli Lig A
As Received:						
% Ash	38.9	31.6	28.0	28.6	31.5	4.5
% Moisture	1.1	6.9	10.0	14.9	7.9	53.1
Moist. & Ash-free:						
% Volatile (VM)	25.3	37.2	41.0	45.1	47.4	57.1
% Carbon (C)	83.6	. . .	. . .	74.1	71.9	70.3
% Hydrogen (H)	4.5	. . .	. . .	4.8	5.0	5.2
% Oxygen (O)	9.9	. . .	. . .	18.6	20.3	23.1
% Nitrogen (N)	1.3			1.4	2.0	0.5
% Sulfur (S)	0.7	1.8	0.8	1.1	0.8	0.9
HHV, Btu/lb	14,635	13,500	13,630	12,690	12,230	11,820
HHV, MJ/kg	34.0	31.4	31.7	29.5	28.4	27.5
Fusibility Temps., reducing at., °F:						
Initial deformation	2250	2600	2100	2700+	2700+	2170
Softening (H=W)	2670	2700+	2280	2700+	2700+	2200
Fluid	2700+	2700+		2700+	2700+	2310
Ash Composition:						
% SiO_2	65.9	60.8	66.5	64.3	66.5	41.3
% Al_2O_3	23.7	24.8	26.1	25.2	26.1	12.1
% Fe_2O_3	6.0	6.8	2.8	5.0	2.8	3.3
% CaO	1.1	2.1	0.9	0.8	0.9	12.7
% MgO	0.6	0.5	0.7	0.7	0.7	2.9
% Na_2O	0.1	0.1	0.1	0.1	0.1	0.9
% K_2O	1.4	1.0	0.8	0.9	0.8	0.2
% TiO_2	2.2	. . .	1.5	1.7	1.5	0.4
% P_2O_5			0.2	0.2	0.2	
% SO_3	0.3	1.7	0.4	0.2	0.4	24.3
Base/acid Ratio	0.10	0.12	0.06	0.08	0.06	0.37
Hardgrove Grind.	63	60	50	56	50	*

* Hardgrove Grindability: 98 at 10% H_2O; 134 at 18% H_2O;
135 at 29% H_2O; 148 at 38% H_2O

below 1 percent; the rest can be as high as 3-percent sulfur.

Coal mines in India are generally shallow, with depths to within 300 meters of the surface; the deepest mine is about 800 meters. Between 35 and 40 percent of the future production of bituminous and subbituminous coal will be mined by open-cast methods, the balance being by deep mining. Much of the coal is in thick seams—75 percent appears in seams above 4.8 m in thickness. Production is expected to be between 300 and 400 tonnes annually by the year 2000. Analyses of typical coals from several areas are included in Table XXVIII.

JAPAN

Coal is widely distributed throughout all the islands and territories of the Japanese Empire. The islands of Kyusku and Hokkaido are by far the most important for quantity and quality.

Table XXIX. Analyses of Typical Japanese Coals

| Area | Mining District or Seam | As received | | | Moisture- and ash-free | | Hardgrove Grind. |
		% Ash	% H$_2$O	% VM	HHV Btu/lb	MJ/kg	
Kyushu	Tagawa	35.4	3.9	51.2	13,660	31.8	45
	Iizuka	22.1	2.2	48.3	14,410	33.5	44
	Hiyoshi	24.5	3.0	40.0	14,700	34.2	49
	Kokura	39.1	6.1	53.6	13,160	30.6	37
	Onoura	23.8	2.4	46.2	14,440	33.6	46
	Yamano	51.3	2.2	45.8	13,400	31.2	43
	Meiji-Saga	15.4	2.2	51.7	14,800	34.4	47
	Shinhokusho	23.3	3.6	46.9	14,690	34.2	51
Shikoku	Takamatsu	29.7	4.4	47.6	14,040	32.6	38
Yamaguchi	Sanyo	31.7	2.2	9.5	14,900	34.7	72
Hokkaido	Sunagawa	38.0	1.6	49.7	13,710	31.9	49
	Bihai	15.1	3.6	45.9	14,330	33.3	47
	Yubetsu	21.6	2.4	44.9	14,740	34.3	53
	Akama	30.8	2.6	46.4	14,590	33.9	54
	Hahoro	16.8	12.9	55.8	14,510	33.7	36
	Horonai	8.2	3.4	48.4	14,490	33.7	38

There are two minor deposits on the main island of Honshu. Economically recoverable reserves of coal are about 1,000 million tonnes equivalent, all extractable by deep mining; nearly half of this will come from the Ishikari coalfield on Hokkaido. Production in the year 2000 will be on the order of 20 million t.c.e., which will be less than 10 percent of the national consumption at that time.

The Japanese coals range from anthracite to lignite, but the bituminous outweighs all other in quantity and value. The greater portion of bituminous is of the high-volatile rank containing 6 to 9 percent ash with a high-heat-value range of 12,000 to 13,000 Btu/lb (28 to 30 MJ/kg). Table XXIX gives the proximate analyses of many of the more typical Japanese coals.

KOREA

Although imported bituminous coal is burned in Korea, all of the coal mined in Korea is high-ash-content anthracite. Table XXX gives analyses of several anthracites; as-received sulfur contents range from 0.1 to 0.5 percent.

TURKEY

Considerable deposits of coal and lignite occur in Turkey, the most important so far being those in the northwestern regions of Anatolia. Commercial development, however, has been practically confined to the Erigli Basin. The Erigli coal, when properly cleaned, is suitable for most purposes, and compares favorably wih the bulk of European bituminous. It contains 40 to 45 percent volatile matter, and is utilized by steamships, railways and factories. Table XXXI gives analyses of two coals from Turkey.

VIET NAM

The only known coal deposits in Viet Nam are anthracitic, chiefly from the Nong-Son area. As-received ash content ranges from 15 to 27 percent, with sulfur from 2 to 3.5 percent. Volatile matter (VM) is between 8 and 11 percent, with high heating value (HHV) from 13,800 to 14,600 Btu/lb (32 to 34 MJ/kg), both on a moisture- and ash-free basis. Hardgrove grindability is usually between 32 and 53, with some samples as high as 73.

Table XXX. Analyses of Typical Korean Anthracites

Area or Mine	As received % Ash	As received % H₂O	% VM	Moisture- and ash-free HHV Btu/lb	Moisture- and ash-free HHV MJ/kg	Ash St, Red., °F	Hardgrove Grind.
Kum-Chun	33.4	6.5	8.7	13,580	31.6	. . .	62
Dan-Gok (Hamback Field)	31.1	7.6	7.5	13,280	30.9	. . .	168
Hung-Jun	31.5	5.3	8.1	13,560	31.5	. . .	66
Bang-Je (Hamback Field)	26.0	6.2	6.9	13,880	32.3	. . .	98
Chengsun	21.1	8.7	5.7	13,980	32.5	. . .	49
Cholam	27.3	5.0	6.1	13,750	32.0	2440	45
Dogye	31.5	3.4	7.4	12,980	30.2	2220	59
Yonawol	49.5	2.6	11.7	12,760	29.7	2200	. . .

A representative ash composition is

% SiO_2 -65.0   % MgO - 1.0
% Al_2O_3 - 20.1   % Na_2O - 0.5
% Fe_2O_3 - 5.3   % K_2O - 5.0
% CaO - 1.0   % TiO_2 - 0.9

Table XXXI. Analyses of Typical Turkish Coals

Area or Mine	As received % Ash	As received % H₂O	% VM	% C	% H	% O	% N	% S	Moisture- and ash-free HHV Btu/lb	Moisture- and ash-free HHV MJ/kg
Tuncbilek Subbituminous	14–22	14–24	44.5	76.4	5.8	13.8	2.5	1.5	13,840	32.2
Elbistan Brown Coal	8–24	48–62	67.0	61.4	5.1	29.6	0.8	5.1	10,190	23.7

AFRICA

Both the geological coal resources and the economically recoverable coal reserves are concentrated in the southern part of the continent. They are preponderantly bituminous coal. Technically and economically recoverable reserves are estimated to be in excess of 30,000 million tonnes coal equivalent; the Republic of South Africa has nearly 75 percent of these; Botswana, 10 percent; Swaziland, about 5 percent; and Rhodesia, 2 percent.

BOTSWANA, REPUBLIC OF

Most of the coal reserves are in the eastern part of the country. Coal qualities are poor and there are no existing coking-coal deposits. The present coal output is low, but increases in production have recently taken place at the Morupule Colliery near Palapye. Consumers of coal production are the Botswana Power Cor-

poration and a local nickel-copper mine. The annual production is projected to reach up to 1 million tonnes in the future.

MOROCCO

The coal, mainly anthracite, is mined by the state-owned Société des Charbonnages Nord-Africains (CNA) at Djerada in the east. There are plans to increase the production to over 2 million tonnes by 2000. Nearly 80 percent of the annual production is burned in the power station at Djerada. In future, limited quantities of anthracite could be for export.

MOZAMBIQUE

Since early 1978 the only existing mining company, Companhia Carbonifera de Mozambique (CCM), has been state owned. The annual output is about 700,000 tonnes and consideration is being given to its expansion. A great part

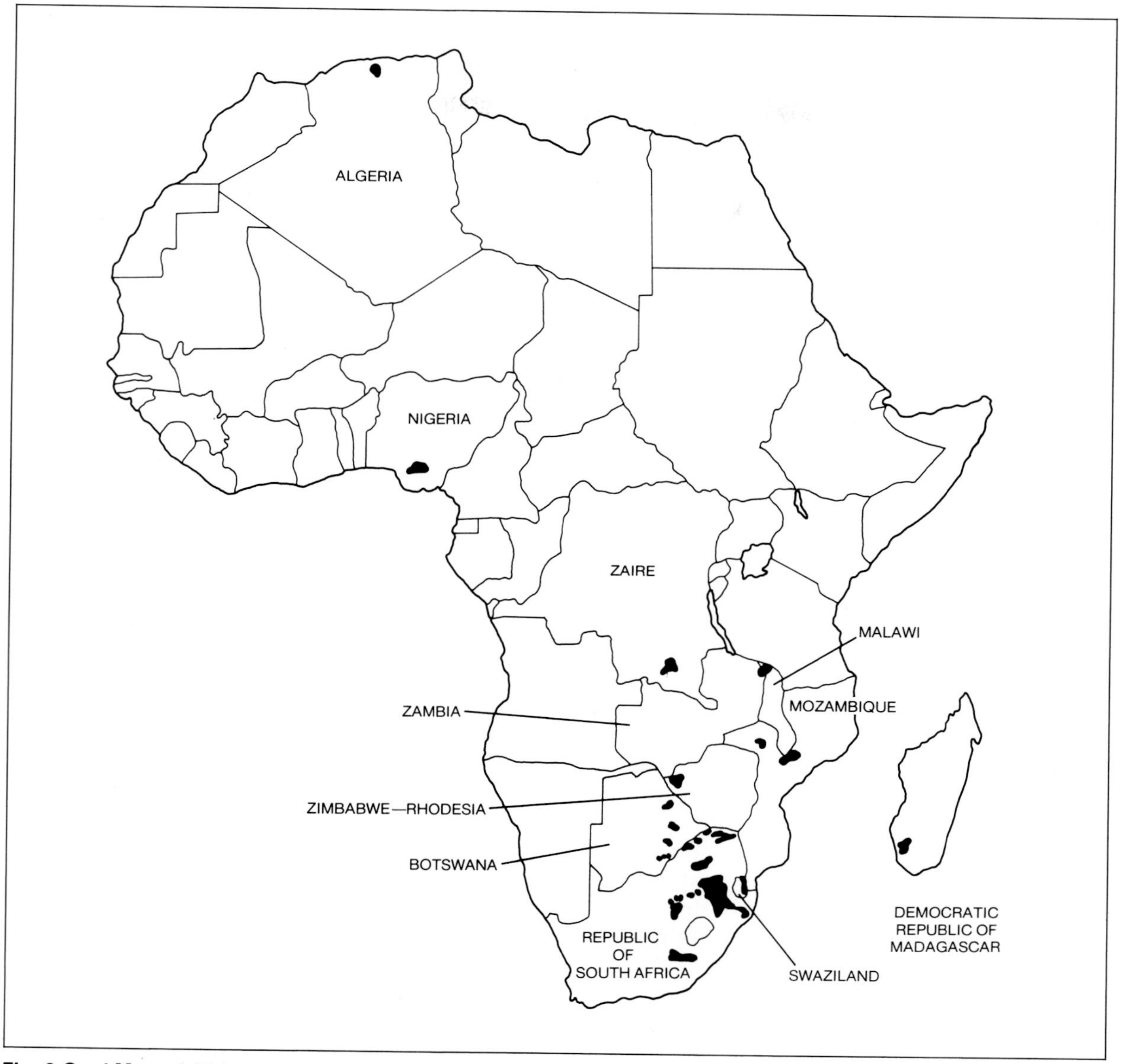

Fig. 8 Coal Map of Africa

of the annual production consists of coking coal. It is possible to increase production up to 7 million tonnes, with a coal export potential of about 5 million tonnes before the end of the century. Today Romania, Japan, Malawi and the German Democratic Republic are importing coal from Mozambique.

NIGERIA

Coal production is now some 300,000 tonnes per year. The coal is produced by the state-owned Nigerian Coal Corporation, with recoverable reserves being on the order of 90 million tonnes. A peak production of 1.5 million tonnes is anticipated by the end of the century. Exports to Ghana are mainly for railways.

SOUTH AFRICA, REPUBLIC OF

The Republic of South Africa (RSA) has about 75 percent of all the coal resources of the

African continent. Most of the coal deposits are situated at shallow depth; more than 10 percent of the RSA coal output is obtained by open-cast mining. In the future, many abandoned "deep" mines will be reworked from the surface with modern heavy equipment.

South African coal production consistently exceeds the country's domestic requirements; it exports both bituminous and anthracite coals. At this time, the Republic of South Africa is one of the 10 largest coal producers and exporters in the world.

The proportion of total coal consumption for the generation of electricity has grown steadily, with approximately 60 percent presently for that purpose. All new thermal stations are located at minemouth, with boilers and pulverizers designed to use run-of-mine coal.

South African coal is deposited in relatively small, isolated basins. The coal seams are thickest and of best quality in the middle of each basin, but pinch out and increase in ash content toward the periphery. The biggest single drawback of the coals of the RSA, particularly for export, is that much of the inert mineral matter present is so intimately associated with the organic matter that its reduction to reasonable levels is generally difficult and costly.

The lowest rank coals in South Africa are found in the Orange Free State; the fields there may be regarded as the southern extension of the Transvaal fields. They have a dry ash-free carbon content of 77 percent; the highest rank coals are the anthracites southeast of Vryheid, with a dry ash-free carbon content of 91.6 percent. The ranks of all the other coals fall between these two limits. Aside from a few minor peat and lignite occurrences, all coals vary in rank from about the lower limit of high-volatile bituminous (or upper limit of subbituminous) coal to anthracite. The higher rank coals are largely confined to Natal, where the fields are essentially an extension of the Transvaal fields.

The coal-bearing areas of the Republic are first subdivided into coal provinces, which in turn are split into coal fields, based either on physical separation of the areas or on conspicuous or slight stratigraphical differences in the coal successions.

The main Witbank field is the center of the coal-mining industry in the RSA; it produces more than half of the coal consumed in the country. Highveld is the biggest of the coal fields in area, but exploitation on a large scale has just begun. Waterberg adjoins the Botswana coal field where the Morupule open-cast is in operation; although it has been intensively prospected and contains 7 seams of workable quality, there are formidable obstacles to economic extraction at this time. Vereeniging-Sasolburg is an important field, with three thick seams. In many of the smaller fields, coal is still being prospected or its mining is yet to be fully developed.

Economically recoverable reserves in the RSA are variously stated to be between one-third and 80 percent of the proven reserves shown in Table XXXII. Estimates of South Africa's coal production in 2000 range from 155 to over 230 million tonnes annually. The quantity of coal to be mined for export is a major variable; an unfavorable political climate toward the RSA will tend to restrict exports below a physically attainable level.

Two bituminous coals from South Africa, together with their ash analyses, are given in Table XXXIII.

SWAZILAND

The Swaziland Collieries Ltd., a subsidiary of Anglo-American Corporation, produces coal for railway, industrial and domestic markets. Only a small quantity is exported. Future prospects are considered to be good as recoverable reserves are in excess of 1500 million tonnes of bituminous coal; production of about 6 million tonnes, nearly 4 million for export, is possible by 2000. Coking coal is exported to Kenya, Japan and Mozambique.

ZAIRE

The coal is produced by Charbonnages de la Luena S.A., Brüssel, a subsidiary of Union Minière. The consumers of the coal are the power stations and the copper industry. A part of the coal is blended with Rhodesian coking

coal for producing a coke suitable for copper smelting. Production of over one million tonnes by 2000 is expected.

ZAMBIA

The coal is produced by Maamba Collieries Ltd., Lusaka, in the south of the country. With capital aid from the African Development Bank and the Federal Republic of Germany, the mines at the Maamba region are to be modernized. The main uses of Maamba coal are in copper smelting, cement and fertilizer production, and for steam generation. By the year 2000 production is anticipated at about 3 million tonnes.

ZIMBABWE (RHODESIA)

Coal is mined by Wankie Colliery Company Ltd., Salisbury, a subsidiary of Anglo-American Corporation. Economically recoverable bituminous-coal reserves approximate 800 million tonnes. The export potential depends on political considerations. Production at present is largely geared to satisfy home consumption, but coal is exported also to Zaire and Zambia. The coal has excellent steaming and coking qualities. The average analysis is:

H_2O	VM	FC	Ash	S	Btu/lb	MJ/kg
0.7	21.0	66.8	11.5	2.8	12,750	29.7

Table XXXII. Proven Resources of Mineable Bituminous Coal (In Situ)

	Millions of tonnes
Witbank, including Springs-Vischkuil	14,186
Highveld	7,231
Waterberg	6,386
Vereeniging-Sasolburg	1,232
Carolina-Wakkerstroom (Eastern Transvaal)	805
Newcastle-Dundee (Klip River)	748
South Rand	600
Other fields, below 600×10^6 tonnes each	1,034
Total	32,222

OCEANIA

Oceania is usually considered to include the lands of the central and south Pacific, including Australia, New Zealand, Tasmania, and sometimes, the Malay archipelago. For convenience, we have included coal data from Indonesia and the Phillipine Islands in this section. Fig. 9 shows the coal fields of Australia, New Zealand, and Tasmania.

Table XXXIII. Analyses of Typical South African (RSA) Coals

	Bituminous Coal and Ash Analyses	
As Received:		
% Ash	13.9	16.6
% Moisture	2.9	6.3
Moist. & Ash-free:		
% Volatile (VM)	33.8	31.8
% Carbon (C)	84.6	82.4
% Hydrogen (H)	4.9	4.4
% Oxygen (O)	7.2	10.6
% Nitrogen (N)	2.2	1.6
% Sulfur (S)	1.1	1.0
HHV, Btu/lb	14,070	14,250
HHV, MJ/kg	32.7	33.1
Fusibility Temps., reducing at., °F:		
Initial deformation	2410	2390
Softening (H =W)	2430	2470
Fluid	2700	2690
Ash Composition:		
% SiO_2	41.5	41.9
% Al_2O_3	30.8	29.4
% Fe_2O_3	4.8	7.2
% CaO	8.7	7.3
% MgO	2.2	1.6
% Na_2O	0.2	0.6
% K_2O	0.3	0.7
% TiO_2	1.0	1.5
% P_2O_5	1.0	
% SO_3	7.6	7.9
Base/acid Ratio	0.22	0.24
Hardgrove Grind.	46	50

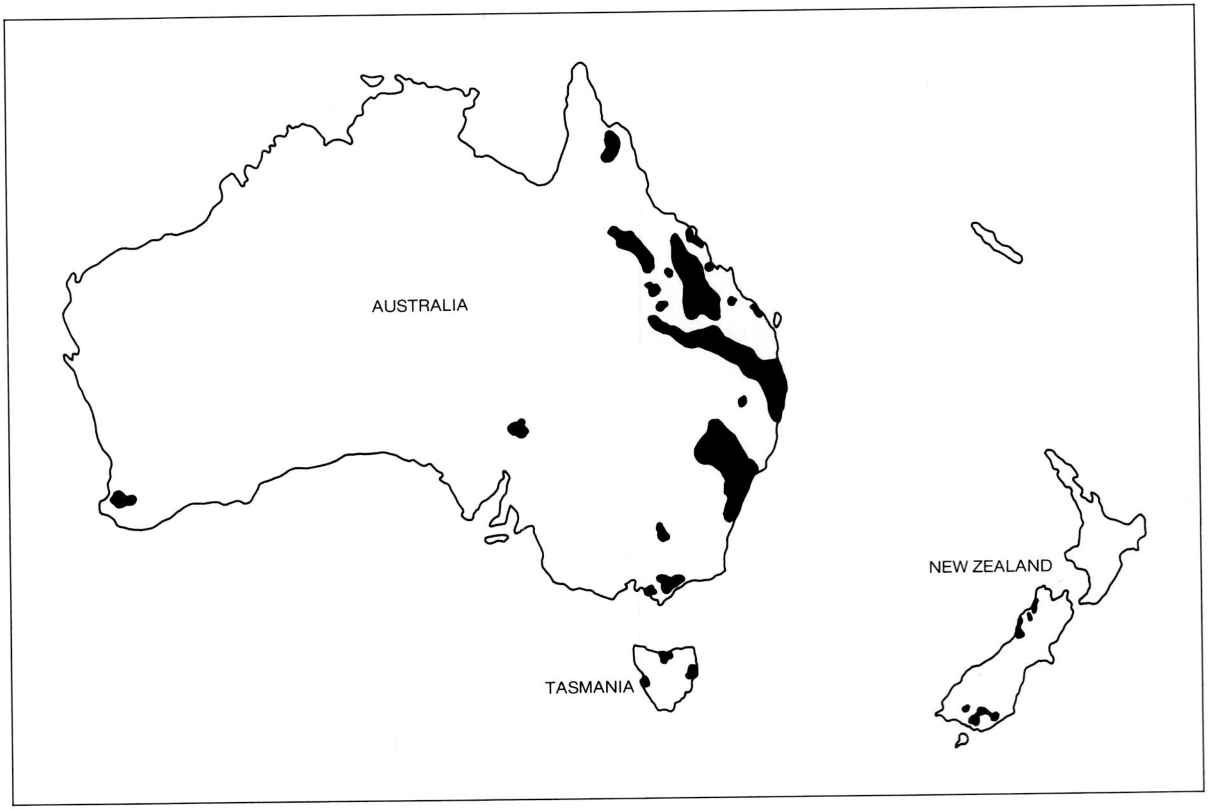

Fig. 9 **Coal Map of Oceania**

AUSTRALIA

Australia's large coal resources are concentrated in the Eastern States. The largest and economically most significant deposits of bituminous coal are found in the Bowen Basin of Eastern Queensland and the Sydney Basin in New South Wales. Nearly half of the country's production is from underground mines in New South Wales; another 30 percent is from open-cut mines in Queensland. The Galilee Basin in Queensland also contains a large resource and the Surat and Clarence-Moreton Basins are of significance. In Australia, a depth of 1,000 meters is generally taken as the current limit for bituminous coal to be regarded as economically extractable.

Major brown coal deposits are located in Victoria where the Latrobe Valley contains proved recoverable reserves of some 35,000 million tons within the limits of a maximum mining depth of 200 meters and a minimum seam thickness of 15 meters. The coal is used predominantly for power generation. The average moisture content is 62 percent and the lower heating value, as received, ranges from 6.0 to 12.5 MJ/kg (2,580 to 5,370 Btu/lb).

The coal resources in other States are relatively small, in most cases comprising isolated deposits of subbituminous coal or lignite. The major use for these coals in South Australia and Western Australia, such as from the Collie basin, is in power stations.

Australia's economically recoverable reserves are in excess of 25,000 million tonnes, about 60 percent of which is bituminous coal and the balance, lower ranked types. About 20 percent of the bituminous coal is capable of being surface-mined. Production of all types of coal is anticipated to be as high as 300 million tonnes coal equivalent in the year 2000.

Table XXXIV. Analyses of Typical Australian Coals

	Farrells C.K.	Bayswater	Liddell Seam	Foxbrook	Puxtrees	Yallourn
As Received:						
% Ash	16.4	25.3	24.7	16.5	26.0	0.8
% Moisture	7.8	6.1	3.3	4.1	7.3	60.2
Moist. & Ash-free:						
% Volatile (VM)	33.5	36.0	39.3	44.7	46.5	60.5
% Carbon (C)	83.4	79.1	80.1	81.2	77.4	. . .
% Hydrogen (H)	4.6	4.8	4.9	6.1	5.5	. . .
% Oxygen (O)	10.6	14.0	12.8	11.0	15.3	. . .
% Nitrogen (N)	1.0	1.7	1.5	1.1	1.2	. . .
% Sulfur (S)	0.4	0.4	0.7	0.6	0.6	0.5
HHV, Btu/lb	14,340	13,650	13,650	14,630	13,490	12,180
HHV, MJ/kg	33.3	31.7	31.7	34.0	31.4	28.3
Fusibility Temps., reducing at., °F:						
Initial deformation	2700+	2700+	2700+	2280	2280	2700+
Softening (H=W)	2700+	2700+	2700+	2500	2480	2700+
Fluid	2700+	2700+	2700+	2700+	2700+	2700+
Ash Composition:						
% SiO_2	55.1	54.9	48.2	56.1	54.5	5.0
% Al_2O_3	31 9	30.6	31.6	25.9	24.0	62.5
% Fe_2O_3	6.6	8.8	7.9	3.8	6.7	2.5
% CaO	0.7	1.1	3.8	5.0	3.7	1.8
% MgO	0.5	0.8	1.5	2.1	2.3	4.1
% Na_2O	0.4	0.5	0.2	0.6	0.3	8.7
% K_2O	1.1	0.8	0.4	0.7	1.0	0.9
% TiO_2	1.3	1.2	1.2	1.1	1.8	2.6
% P_2O_5	0.6		0.5	0.5	0.2	0.3
% SO_3	0.1	0.1	2.9	3.2	3.3	6.6
Base/acid Ratio	0.11	0.14	0.17	0.15	0.17	0.26
Hardgrove Grind.	52	45	50	48	45	

Table XXXIV gives analyses of Australian coals from several mining areas.

NEW ZEALAND

Coal ranging from anthracite to lignite occurs in many sections of New Zealand. The Buller-Mokihinui is the best known bituminous field. The coal is of good quality and makes an excellent steam fuel. It is coking, but the coke produced is not rated high. Table XXXV gives complete coal and ash analyses of two New Zealand coals.

INDONESIA

Since 1972, only two state-owned coal mines are in active production: Ombilin in West Sumatra and Bukit Asam in South Sumatra.

Coal deposits of economic significance are confined to the western part of Indonesia, on the islands of Sumatra and Kalimantan. Estimates of measured reserves of bituminous coal and lignite on these islands are 200 million and 100 million tonnes respectively.

Ombilin coal is a hard bituminous to sub-bituminous coal. It is excellent steam coal with mild coking properties. Detailed explorations carried out in some parts of the Ombilin basin have indicated about 35 million tonnes of mineable coal reserves.

The proximate analysis of a typical Ombilin coal sample is as follows:

Moisture	: 6.0%
Ash	: 1.5%
Sulfur	: less than 0.5%
Volatile Matter	: 46%, m.a.f. basis
Calorific Value	: 14,590 Btu/lb (33.9 MJ/kg)

The coal in the region of Bukit Asam ranges from lignite through bituminous, subbituminous to anthracitic coal and anthracite. The Air Laya opencast mine of Bukit Asam remains the field with the most potential. Recent studies indicate that there is a reserve of close to 100 million tonnes of coal in this field.

The average proximate analyses of the various coals presently produced in Bukit Asam are

Air Laya subbituminous coal

Moisture	: 18.0% to 23.2%
Ash	: 4.4% to 8.9%
Sulfur	: 0.16% to 0.41%
Volatile Matter	: 29.4% to 32.1% as received
Calorific Value	: 10,800 to 11,700 Btu/lb (25.1 to 27.2 MJ/kg), as received

Suban low-volatile bituminous coal

Moisture	: 0.5% to 1.2%
Ash	: 3.6% to 4.4%
Sulfur	: negligible
Volatile Matter	: 15% to 19%
Calorific Value	: 14,400 to 14,760 Btu/lb (33.5 to 34.3 MJ/kg), as received

PHILIPPINE ISLANDS

Coal is mined in eight localities in the Philippine Islands, namely: Bataan, Cebu, Zamboanga, Politti, Masbati, Mindoro, Luzon and Mindanao. The basins are small and discontinuous. The deposits contain lignite, subbituminous and bituminous coals. The bulk of

Table XXXV. Analyses of Typical New Zealand Subbituminous Coals

	Maori Farm	Ohinewai Area
As Received:		
% Ash	5.4	3.0
% Moisture	16.2	22.2
Moist. & Ash-free:		
% Volatile (VM)	48.7	47.6
% Carbon (C)	. . .	73.9
% Hydrogen (H)	. . .	5.5
% Oxygen (O)	. . .	19.4
% Nitrogen (N)	. . .	0.9
% Sulfur (S)	0.4	0.3
HHV, Btu/lb	12,490	12,600
HHV, MJ/kg	29.0	29.3
Fusibility Temps., reducing at., °F:		
Initial deformation	2320	2120
Softening (H=W)	2370	2150
Fluid	2600	2230
Ash Composition:		
% SiO_2	20.1	21.4
% Al_2O_3	7.8	10.8
% Fe_2O_3	9.4	8.7
% CaO	34.9	36.1
% MgO	2.2	4.1
% Na_2O	1.2	2.7
% K_2O	0.2	0.6
% TiO_2	0.3	1.8
% SO_3	18.0	12.8
Base/acid Ratio	1.7	1.5
Hardgrove Grind.	. . .	39

the lignite is black, and seldom displays a woody structure or brown color. The bituminous coals are black, hard and lustrous. They are generally noncoking.

A typical Philippine coal, from Liguan, has an as-received ash content of 11.5%, and moisture of 9.4%. On a moisture- and ash-free basis, volatile matter is 45.3%, sulfur is 0.8%, and HHV is 13,480 Btu/lb (31.3 MJ/kg).

Ash softening temperature of such a coal, in a reducing atmosphere, is approximately 2190°F.

INTERNATIONAL COAL CLASSIFICATION

As the international trade in coal increased following World War II, the Coal Committee of the Economic Commission for Europe established a Classification Working Party in 1949 to develop an international system for classifying coal. The system for classifying hard coals was published in 1956 as a document of the United Nations.[1] In 1958, the U.S. Bureau of Mines published the results of its study of how this classification system would apply to American coals. Two years later the classification system was applied to brown coals and lignites.[2,3] Based on European terminology, the expression *hard coal* as used in the international classification system is defined as coal with a gross calorific value of more than 10,260 Btu per lb (5,700 kg-cal per kg or 23.86 MJ/kg) on the moist, ash-free basis. Coals classified by the American Society for Testing Materials as anthracitic, bituminous, and higher rank subbituminous are included in the hard-coal category. Brown coals and lignites are those coals having gross calorific values below 10,260 Btu per lb, moist but ash-free.

Excerpts from the two referenced classification systems are included in the following text. These excerpts will assist users of various types of coals world-wide, in relating such coals to the American types referred to frequently in this and other engineering references.

CLASSIFICATION OF HARD COALS BY TYPE

Table XXXVI shows the international system of classifying hard coals according to their volatile-matter content, calculated on a dry, ash-free basis, resulting in nine classes of coals. The nine classes of hard coal, based on volatile-matter content and calorific value, are then divided into groups according to their caking properties, as measured by tests when the coal is heated rapidly. The coal groups are further subdivided according to coking properties determined by tests in which the coal is heated slowly.

DIVISION OF HARD COALS INTO CLASSES

Hard coals are first classified in the international system according to their volatile-matter content, calculated on a dry, ash-free basis. These classifications are:

Class No.	VM dry, ash-free %
1 A	3 — 6.5
1 B	> 6.5 —10
2	>10 —14
3	>14 —20
4	>20 —28
5	>28 —33
6–9	>33

As volatile matter is not an entirely suitable parameter for classifying coals containing more than about 33 percent volatile matter, calorific value on a moist, ash-free basis is used for such coals, as follows:

Class No.	Gross calorific value, moist., ash-free basis, Btu/lb	Approximate limits of VM, dry, ash-free %
6	>13,950	33—41
7	>12,960–13,950	33—44
8	>10,980–12,960	35—50
9	>10,260–10,980	42—50

The gross calorific value of coal on the moist, ash-free basis means the calorific value of the coal in equilibrium with air at 30°C and 96 percent relative humidity, calculated to an ash-free basis. For practical purposes, the moisture in the equilibrated sample is considered equivalent to the natural bed moisture of the coal, that is, the coal moisture at freshly exposed bed faces free of visible water.

DIVISION OF HARD-COAL CLASSES INTO GROUPS

The nine classes of hard coal, based on volatile-matter content and calorific value, are divided into groups according to their caking properties. Caking properties, as used in the classification system, are a measure of the behavior of coal when heated rapidly. Either of two methods—the free-swelling test (crucible-swelling test) or Roga test—may be used to

Table XXXVI. International classification of hard coals by type (Ref.2)

Group Number	Free-swelling index (crucible-swelling number)	Roga Index	0	1	2	3	4	5	6	7	8	9	Sub-Group Number	Dilatometer	Gray-King
3	>4	>45						435	535	635			5	>140	>G_8
							334	434	534	634	734		4	>50–140	G_5–G_8
							333	433	533	633	733		3	>0–50	G_1–G_4
							332a / 332b	432	532	632	732	832	2	≥0	E-G
2	2½-4	>20-45					323	423	523	623	723	823	3	>0–50	G_1–G_4
							322	422	522	622	722	822	2	≥0	E-G
							321	421	521	621	721	821	1	Contraction only	B-D
1	1-2	>5-20				212	312	412	512	612	712	812	2	≤0	E-G
						211	311	411	511	611	711	811	1	Contraction only	B-D
0	0-½	0-5		100 A	B / 200	300	400	500	600	700	800	900	0	Non-softening	A

Code Numbers: The first figure of the code number indicates the class of coal, determined by volatile matter content up to 33% VM and by calorific parameter above 33% VM. The second figure indicates the group of coal, determined by coking properties. The third figure indicates the subgroup, determined by coking properties.

Subgroups (determined by coking properties) — Alternative Subgroup Parameters: Dilatometer, Gray-King.

Groups (determined by coking properties) — Alternative Group Parameters: Free-swelling index (crucible-swelling number), Roga Index.

As an indication, the following classes have an approximate VM content of:
Class 6 33-41% VM
7 33-44% VM
8 35-50% VM
9 42-50% VM

Class Parameters

Class Number	0	1	2	3	4	5	6	7	8	9
Volatile matter, VM (dry, ash-free)	0-3	>3-10 (>3-6.5 / >6.5-10)	>10-14	>14-20	>20-28	>28-33	>33	>33	>33	>33
Calorific parameter	—	—	—	—	—	—	>13.950	>12.960-13.950	>12.980-12.960	>10.260-10.980

CLASSES
(Determined by volatile matter up to 33% VM and by calorific parameter above 33% VM)

Note: (i) Where the ash content of coal is too high to allow classification according to the present systems, it must be reduced by laboratory float-and-sink method (or any other appropriate means). The specific gravity selected for flotation should allow a maximum yield of coal with 5 to 10 percent of ash.
(ii) 332a ... >14-16% VM
332b ... >16-20% VM
(iii) Gross calorific value on moist, ash-free basis (30°C., 96% relative humidity), Btu/lb

measure caking properties. The division of classes into groups, according to free-swelling index or Roga index, are:

Group No.	Free-swelling index	Roga index
0	0–½	0–5
1	1–2	>5–20
2	2½–4	>20–45
3	>4	>45

Division of Groups into Subgroups

The hard-coal groups are further subdivided according to coking properties, as measured by tests in which the coal is heated slowly. Coking properties are measured either by maximum dilatation, using the Audibert-Arnu method, or by Gray-King coke type. The division into subgroups follows:

Subgroup No.	Maximum dilatation	Gray-King coke type
0	Nonsoftening	A
1	Contraction only	B-D
2	0 and less	E-G
3	>0 to 50	G_1-G_4
4	>50 to 140	G_5-G_8
5	>140	>G_8

CODE NUMBERS

A three-figure code number is used to express the classification of a hard coal. The first figure indicates its class, the second figure the group, and the third figure the subgroup. An example of the use of the system follows, based on a coal with the following characteristics:

Volatile matter, dry, ash-free, %	37
Calorific value, moist ash-free Btu/lb	14,510
Caking properties:	
Free swelling index	6
Roga index	85
Coking properties:	
Maximum dilatation	60
Gray-King coke type	G_5

As the volatile-matter content on a dry, ash-free basis is greater than 33 percent, the class number is determined by the gross calorific value on a moist, ash-free basis. As the gross calorific value is more than 13,950 Btu the coal

belongs in class 6, which becomes the first figure of the code number. The free-swelling index is 6; alternatively, the Roga index is 85, so the coal belongs in group 3, which becomes the second figure of the code number. As the maximum dilatation is 60, or the Gray-King coke type is G_5, the coal belongs in subgroup 4, which is the third figure of the code number. Therefore the code number of the coal is 634.

CLASSIFICATON OF LOWER-RANK COALS

Table XXXVII shows the international system of classifying brown coals and lignites, those fuels having gross calorific values below 10,260 Btu/lb (5,700 kcal/kg or 23.86 MJ/kg). The parameters for classifying these coals are their total moisture and yield of low-temperature tar. As a correlation exists between the total moisture and calorific values of the lower rank coals, the moisture parameter indicates the value of the coal as a fuel; the tar indicates the value of the coal for chemical processing.

DIVISION OF LOWER-RANK COALS INTO CLASSES

The coals are first classified according to their total moisture content as ash-free coals. Total moisture is that contained in freshly mined coal. The class numbers and corresponding range of moisture values are:

Class No.	Total moisture, ash free, %
10	20 and less
11	>20 to 30
12	>30 to 40
13	>40 to 50
14	>50 to 60
15	>60 to 70

The numbering system starts with 10 to follow consecutively after the classes numbered 0 to 9 in the international classification of hard coals by type.

Division of Classes in Groups

The lower-rank coals divided into classes according to total moisture are subdivided into groups by yield of low-temperature tar, calculated dry and ash-free. Group numbers with corresponding range of tar yields are:

Group No.	Tar, dry, ash free, %
00	10 and less
10	> 10 to 15
20	> 15 to 20
30	> 20 to 25
40	> 25

CODE NUMBERS

A four-figure code number indicates the classification of lower-rank coals. The first two figures identify the class; the last two, the group. For example, if total moisture, ash free, is 35 percent, the coal is placed in class 12; if the yield of tar is 11 percent, the coal is assigned to group 10 and the code number of the coal is 1210.

METHODS OF COAL ANALYSIS AND TEST

The international classification system provides that standard methods of the International Organization for Standardization (ISO) shall be used when such methods become available. Much progress has been made in the work of standardization, although the methods have not been approved in final form. Until standard methods are adopted, it is provided that national standards may be used. The following provisional tolerances are to be used for the parameters pending adoption of international methods of analysis and test:

Volatile matter, VM

20% or less	± 1.0 unit
More than 20%	± 5.0% of value
Free-swelling index	± ½ unit
Roga index	± 5 units

Maximum dilatation:

Subgroups 0, 1, 2, 3	± 5 units
Subgroup 4	± 10 units
Subgroup 5	± 15 units
Gray-King coke type	± 1 type
Gross calorific value	± 110 Btu/lb

The appendix to reference 2 includes methods for determining equilibrium moisture content (moisture-holding capacity) and several of the above indices. The methods are summarized in the following text.

EQUILIBRIUM MOISTURE CONTENT

In the international system, coals with more than 33 percent volatile matter on the dry, ash-free basis are classified according to calorific value on the moist, ash-free basis. The term moist refers to coal containing its natural bed moisture but no visible water on the surface. It is assumed that the natural bed moisture of coal in the ground represents the moisture-holding capacity of the coal when in equilibrium with air at approximately 100 percent relative humidity. Because of the difficulty of equilibrating samples in such an atmosphere, the equilibrium moisture content or moisture-holding capacity is determined at 96 to 97 percent relative humidity, which for practical purposes is considered to represent the moisture in the coal bed.

The test is essential for classifying wet samples of coal in order to bring the coals to their natural-bed-moisture condition. The proposed ISO method for moisture-holding capacity is equivalent to the ASTM method for equilibrium moisture of coal. The method also can restore partially air-dried samples of all coals except lignites to virtually bed-moisture state.

To conduct the test, a sample of coal crushed to pass a number 16 sieve is wetted, drained of excess water, and equilibrated in a reduced-pressure vessel charged with a pulp of potassium sulfate crystals and water. The saturated solution of potassium sulfate at 30°C maintains a relative humidity of 96 to 97 percent in the vessel. After moisture equilibrium is reached, which usually is within 48 hours, the coal sample is covered, removed from the vessel, and weighed. The moisture content of the conditioned sample is then determined.

FREE-SWELLING INDEX TEST (CAKING)

The test for free-swelling index of coal, which is a measure of the coal's caking properties, was originally developed in Great Britain and is called there the crucible-

Table XXXVII. International Classification of Brown Coals and Lignites

Group No.	Group parameter tar yield (dry, ash free), %	Code Numbers					
40	>25	1040	1140	1240	1340	1440	1540
30	>20 to 25	1030	1130	1230	1330	1430	1530
20	>15 to 20	1020	1120	1220	1320	1420	1520
10	>10 to 15	1010	1110	1210	1310	1410	1510
00	10 and less	1000	1100	1200	1300	1400	1500
Class no.		10	11	12	13	14	15
Class parameter total moisture, ash-free, percent		20 and less	>20 −30	>30 −40	>40 −50	>50 −60	>60 −70

Moist, ash-free basis (30°C and 96 percent relative humidity).

Total moisture content refers to freshly mined coal. For internal purposes, coals with a gross calorific value over 5,700 kcal/kg (maf), considered in the country of origin as brown coals but classified as hard coals for international purposes, may be classified under this system, to ascertain, in particular, their suitability for processing. When total moisture content is over 30 percent, gross calorific value is always below 5,700 kcal/kg

swelling test. It is a standard method used in several countries, including the United States.

The test consists of heating a 1-gram sample of pulverized coal to 820°±5°C in a covered crucible and comparing the size and shape of the coke button obtained with the outlines of a set of standard profiles numbered in half units from 0 to 9. The number of the profile most closely corresponding to the cross section of the coke button is the free-swelling index or, in British terms, crucible-swelling number.

ROGA-INDEX TEST (CAKING)

The alternative method for determining caking properties is the Roga test, developed and used in Poland. The test is conducted by carbonizing a mixture of 1 gram of coal and 5 grams of a standard anthracite at 850°C for 15 minutes. The mechanical strength of the resulting coke button is measured by an abrasion test in a special rotating drum. At the end of the tumbling period, the residue is screened on a sieve with 1-mm round openings and the oversize weighed. This process is repeated twice. The index is calculated from the results of the screening test by the following formula.

No information is available on the use of the Roga index for measuring caking properties of

$$\text{Roga index} = \frac{100}{3Q}\left(\frac{a + d}{2} + b + c\right) \quad (1)$$

where:

Q = weight of residue after carbonization,
a = weight of oversize before first screening,
b = weight of oversize after first screening,
c = weight of oversize after second screening, and
d = weight of oversize after third screening.

American coals in combustion processes. As there is an approximate correlation between the results of the free-swelling-index and Roga-index tests, the Roga index probably is about as significant as the free-swelling index for measuring caking properties.

AUDIBERT-ARNU DILATOMETER TEST (COKING)

Better known in Europe than in the United States, the Audibert-Arnu dilatometer test is used in studies of coking properties of coal.

A briquetted pencil of coal is carbonized in a vertical tube topped by a steel rod that slides in the bore of the tube. The pencil, which is formed in a die, is lightly tapered; its diameter averages 6½ mm, and its length is 60 mm. The maximum displacement of the rod measured on an external scale is reported as a percentage of the original length of the pencil. The dilatation is calculated as follows

$$\text{Dilatation, \%} = \frac{\text{Displacement of piston, mm} \times 100}{60}$$

$$(2)$$

GRAY-KING COKE-TYPE TEST

Developed in Great Britain, the Gray-King coke-type test was used at first as a bench-scale, low-temperature carbonization assay for determining yields of coke, gas, tar and liquor. The Coal Survey adopted the test for indicating coking properties. The method was then standardized by the National Coal Board for classifying coals of the United Kingdom.

The test is conducted by carbonizing a 20-gram sample of coal progressively to 600°C in a horizontal tube furnace. The carbonized residue is classified as to volume, coherence, fissuring, and hardness by comparing it with a series of residues. For coals that form powdery to hard coke residues that occupy the same volume as the original coal (standard coke), the type of residue is assigned letters ranging from A to G. For coals that swell to fill the cross section of the tube, electrode carbon is mixed with the coal to obtain a strong, hard coke of the same volume as the original coal-electrode carbon mixture. The coke type is indicated by the letter G with a subscript figure, that is, G_4, G_5, etc. The subscript shows number of parts of electrode carbon needed in the mixture with coal to give a G—type (standard) coke. Reference 2 gives the Gray-King scheme for examining and classifying coke types.

INTERNATIONAL SYSTEM APPLIED TO AMERICAN COALS

The American Society for Testing Materials system of classification of coals by rank was discussed in Chapter 2. The *ASTM Standard Specification of Coals by Rank* D-388 groups coals according to their degree of metamorphism or progressive alteration in the natural coalification series from lignite to anthracite.

The ASTM classification system is based on the parameters of fixed carbon or volatile matter and calorific value calculated on a mineral-matter-free basis. Higher-rank coals are classified according to fixed carbon or volatile matter on the dry basis, and the lower-rank coals according to calorific value on the moist or bed-moisture basis. Agglomerating and weathering indexes are used to differentiate between some adjacent groups.

COMPARISON OF INTERNATIONAL AND ASTM SYSTEMS

The ASTM system provides for classification of all ranks of coal while the international classification is based on two systems: one for the hard coals and the other mainly for brown coals and lignites. The borderline between the two international classifications occurs at 10,260 Btu/lb (5,700 kcal/kg or 23.86 MJ/kg) moist and ash-free and is nearly the midpoint of the subbituminous B group of the ASTM system. Therefore, some subbituminous B coals and all subbituminous A, bituminous, and anthracitic coals of the ASTM system are identified according to the international classification of hard coals by type. The other subbituminous B and all subbituminous C and lignitic coals are covered in the international classification of coals with a gross calorific value below 5,700 kcal/kg.

To divide lower rank coals into classes, the international system uses the total moisture parameter while the ASTM uses the calorific value. Fig. 10 shows that a good correlation exists between total moisture and calorific value of low-rank coals. The figure is based on analyses of exchange samples of coal given in Table XXXVIII.

Fig. 11 compares the class numbers and group boundaries of the international system for hard coals with those of the ASTM classification. As the figure shows, the group boundaries in the two schemes are quite similar; thus, class numbers in the international system are

Table XXXVIII. Source, Analysis and Classification of Exchange Samples of Low-Rank Coal

Source of Sample	As received		Moisture- and ash-free						HHV		Equilib. Moisture,[A]	Classification	
	% Ash	% H$_2$O	% VM	% C	% H	% O	% N	% S	Btu/lb	MJ/kg	%	ASTM[B]	IC No.[C]
Belgium													
Florennes	2.2	63.3	53.8	70.1	5.0	23.5	0.7	0.7	11,910	27.7	39.8	LigB	1510
Canada													
Alberta	7.3	24.2	39.2	76.6	5.5	15.8	1.6	0.5	13,150	30.6	24.4	Subb	1100
Czechoslovakia													
Northern Bohemia	2.7	20.2	49.0	77.6	5.8	14.7	1.0	0.9	13,890	32.3	20.9	Suba	800
	6.4	31.2	50.5	76.0	6.0	15.7	1.1	1.2	13,560	31.5	30.7	Subc	1220
Central Bohemia	5.9	21.1	38.1	79.5	5.0	13.1	1.2	1.2	13,770	32.0	20.4	Subb	900
	15.9	23.2	43.6	77.1	5.2	11.1	1.5	5.1	13,640	31.7	25.9	Subb	1110
Southern Bohemia	7.4	44.6	56.5	70.2	5.4	19.5	0.9	4.0	12,240	28.5	43.5	LigA	1310
Western Bohemia	6.0	39.7	51.7	76.6	6.1	15.2	1.2	0.9	13,680	31.8	37.0	LigA	1330
Slovakia	12.2	36.1	51.3	69.6	4.9	19.5	1.5	4.5	11,870	27.6	39.0	LigA	1300
Federal Republic of Germany													
Rhine Region	3.3	62.2	55.3	69.1	4.9	24.1	0.9	1.0	11,580	26.9	48.4	LigB	1510
Bavaria, Peissenberg	11.9	9.7	52.8	72.8	5.4	12.4	1.6	7.8	13,200	30.7	12.0	Suba	800
Greece													
Athens Area, Peristeri	6.5	22.4	46.9	71.8	4.7	19.8	1.9	1.8	12,140	28.2	27.4	Subc	1100
Island of Euboea, Aliveri	11.3	33.3	56.1	68.0	5.2	23.9	0.9	2.0	11,600	27.0	36.8	LigA	1210
West Macedonia, Ptolemais	12.2	59.0	57.9	65.4	4.3	27.3	1.9	1.1	10,340	24.0	50.0	LigB	1500
Italy													
Lucania, Mercure	10.1	58.8	58.6	65.4	5.2	25.3	1.8	2.3	11,020	25.6	51.7	LigB	1500
Toscana, Pietrafitta	9.9	60.8	57.3	70.5	4.8	19.9	1.9	2.9	11,840	27.5	51.7	LigB	1500
Toscana, Valdarno	5.3	51.9	56.8	66.5	5.2	26.3	1.0	1.0	11,350	26.4	47.2	LigB	1420
Umbria, Spoleto	7.8	31.8	60.4	64.8	5.3	27.3	1.1	1.5	11,140	25.9	34.1	LigA	1210
Federation of Maylaya													
Selangor, Batu Arang	7.5	17.7	48.7	77.1	5.8	15.1	1.6	0.4	13,840	32.2	19.1	Suba	800
Poland	7.3	53.9	60.8	71.4	5.8	20.2	0.6	2.0	12,510	29.1	42.2	LigB	1420
	6.7	17.6	44.0	77.7	5.1	14.3	0.9	2.0	13,540	31.5	21.0	Suba	800
U.S.S.R.													
Chelyabinsk Basin	9.9	17.8	43.6	76.2	5.2	16.2	1.9	0.5	12,910	30.0	18.9	Subb	900
Rhiczechinsk	6.9	37.0	43.7	72.2	4.3	22.1	1.1	0.3	11,910	27.7	34.8	LigA	1200
Borneo-Sarawak													
Upper Rajang Valley	1.3	23.3	48.1	72.4	4.9	21.2	1.4	0.1	12,390	28.8	22.9	Subc	1110
East Germany													
Niederlausitz	2.5	59.8	55.6	68.1	4.7	25.8	0.7	0.7	11,720	27.3	44.7	LigB	1510
Bautzen	3.9	63.4	57.6	70.1	5.3	21.9	0.7	2.0	12,100	28.1	43.3	LigB	1510
Oberlausitz	3.2	58.3	54.4	67.7	5.2	26.0	0.6	0.5	12,020	28.0	47.2	LigB	1510

Table XXXVIII. Source, Analysis and Classification of Exchange Samples of Low-Rank Coal—*Continued*

Source of Sample	As received % Ash	% H₂O	Moisture- and ash-free % VM	% C	% H	% O	% N	% S	HHV Btu/lb	MJ/kg	Equilib. Moisture,[A] %	Classification ASTM[B]	IC No.[C]
Aschersleben	5.0	44.5	72.2	76.4	7.9	12.0	0.4	3.3	14,680	34.1	37.3	LigA	1340
Bitterfeld	5.8	54.6	55.6	71.3	5.1	18.0	0.6	5.0	12,560	29.2	49.3	LigB	1410
Halle	6.0	54.6	58.9	71.6	5.6	17.2	0.6	5.0	12,770	29.7	50.2	LigB	1420
Geiseltal	4.3	55.0	58.0	71.0	5.7	16.5	0.8	6.0	12,920	30.0	48.5	LigB	1420
Borna	4.5	55.1	60.3	72.5	5.9	18.1	0.7	2.8	13,000	30.2	44.2	LigB	1420
United Kingdom	4.0	17.6	43.6	77.6	5.1	13.8	1.4	2.1	13,610	31.7	17.8	Suba	800
United States													
Colorado, Weld County	4.8	25.2	40.8	78.6	5.1	14.1	1.7	0.5	13,630	31.7	24.8	Subb	1110
Montana, Musselshell Co.	7.0	13.9	39.6	80.9	5.1	12.2	1.3	0.5	14,110	32.8	14.1	Suba	800
North Dakota, Burke Co.	4.2	37.2	43.4	74.1	4.7	19.3	1.2	0.7	12,470	29.0	37.5	LigA	1200
North Dakota, Mercer Co.	4.1	38.1	45.3	72.7	4.9	20.8	0.9	0.7	12,330	28.7	37.7	LigA	1200
Texas, Milam Co.	9.4	36.0	49.9	74.7	5.4	16.4	1.5	2.0	12,980	30.2	37.6	LigA	1210
Washington, Lewis Co.	7.1	28.5	50.7	74.6	5.4	18.0	1.3	0.7	12,800	29.8	29.6	Subc	1210
Wyoming, Campbell Co.	5.0	28.2	49.0	74.0	5.6	18.6	0.9	0.9	12,970	30.2	30.8	Subc	1110
Wyoming, Sheridan Co.	3.6	25.9	42.8	75.9	5.1	17.0	1.6	0.4	13,100	30.5	25.3	Subb	1110
Yugoslavia													
Bosnia, Banovici	10.4	17.9	45.6	75.6	5.4	14.7	2.3	2.0	13,310	31.0	19.2	Subb	900
Bosnia, Breza	8.8	17.2	46.9	75.3	5.2	13.6	1.9	4.0	13,330	31.0	18.3	Subb	900
Bosnia, Kakanj	12.5	6.1	49.0	77.2	5.4	11.4	2.3	3.7	13,720	31.9	7.3	Suba	800
Bosnia, Kreka	3.3	39.8	52.0	69.7	5.2	23.7	0.9	0.5	12,110	28.2	37.2	LigA	1310
Serbia, Senjski	10.0	15.6	54.3	71.7	5.6	19.5	1.4	1.8	13,010	30.3	17.1	Subb	900
Slovenia, Trbovlje	7.3	24.9	48.4	71.0	5.0	18.4	1.3	4.3	12,520	29.1	25.9	Subc	1110

[A]—Equilibrium moisture, %, is on an ash-free basis

[B]—Abbreviations are listed in Table II

[C]—IC No. is the International Code Number

nearly equivalent to group names in the ASTM classification. For example,

Class number in international system	Group name in ASTM system
0	Meta-anthracite
1	Anthracite
2	Semianthracite
3	Low-volatile bituminous coal
4, 5	Medium-volatile bituminous coal
6	High-volatile A bituminous coal
7	High-volatile B bituminous coal
8	High-volatile C bituminous coal and subbituminous A coal
9	Subbitminous B coal

The international classification system for hard coals differs from the ASTM classification in that parameters of caking and coking properties are introduced to indicate how the coal can be used in combustion and carbonization processes. This concept is followed in several European national systems of classification. While the ASTM classification of coal was being developed, the feasibility of classifying

Fig. 10 Relationship of Total Moisture and Calorific Value

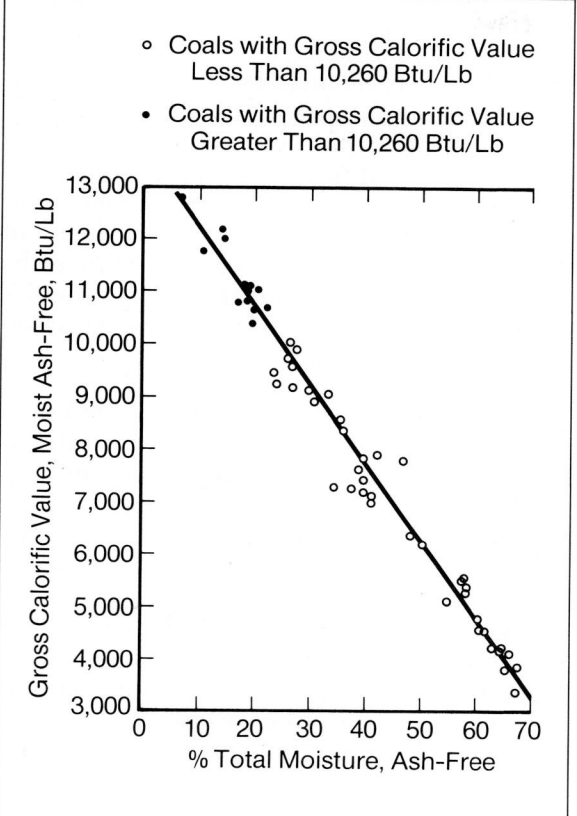

- ○ Coals with Gross Calorific Value Less Than 10,260 Btu/Lb
- ● Coals with Gross Calorific Value Greater Than 10,260 Btu/Lb

coals according to use was studied. It was decided that this was impractical because many coals can be used for various purposes and the type of equipment in which the coal is used probably is as significant as the kind of coal. On the other hand, the caking and coking tests give general information on the characteristics of coals so they should prove of value in considering a coal for a specific use.

INTERNATIONAL CLASSIFICATION OF AMERICAN BITUMINOUS COALS

As little information is available on the coking properties of American coals by the Gray-King coke-type test or the Audibert-Arnu dilatometer test, tests were made on samples of bituminous coal of varous ranks from nearly every coal-producing state. About 80 coals were tested.[2] They were selected to obtain good geographic representation of seams that contain large reserves, are being mined in quantity, or are entering the export market. The Audibert-Arnu dilatometer test was selected for the coking test. Of the 62 code numbers shown in Table XXXVI, American coals occur in 26.

Table IV in the front of this appendix shows the analytical data and the classification of each coal by the ASTM and international systems.

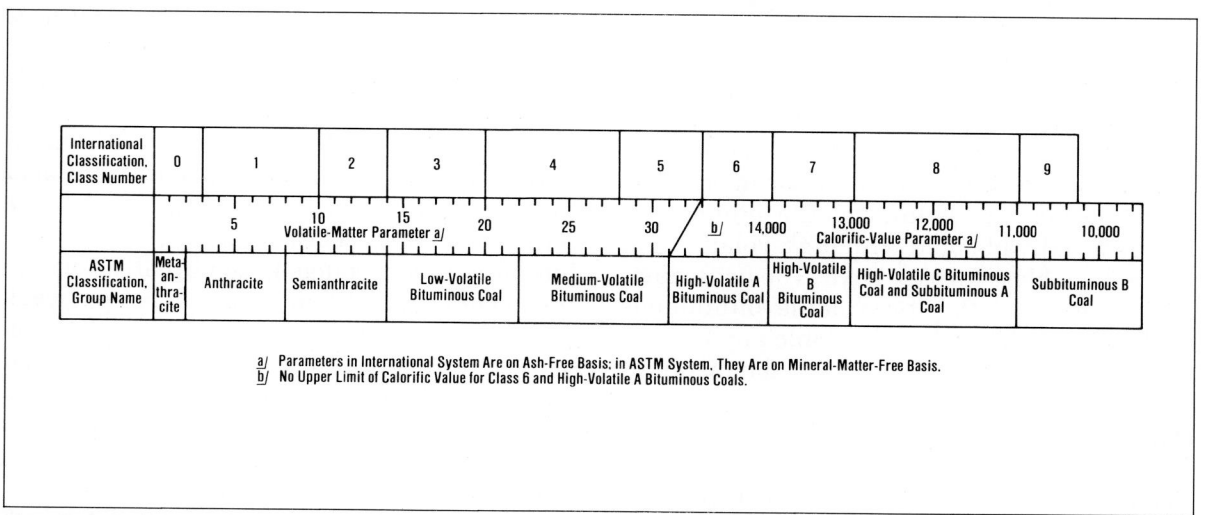

Fig. 11 Comparison of class numbers and boundary lines of International System for Hard Coals with group names and boundary lines of ASTM System

Symbols used in the table for the rank of coal according to the ASTM system are:

Lvb = low-volatile bituminous,
Mvb = medium-volatile bituminous,
Hvab = high-volatile A bituminous,
Hvbb = high-volatile B bituminous,
Hvcb = high-volatile C bituminous.

Based on this data, American bituminous coals are classified in the international system for hard coals as:

Class 0. Although no code number is indicated in the system for coal of this class, coals classified in the U.S. as meta-anthracite would have a code number of 000, as these coals are noncaking and noncoking. The only known deposit of meta-anthracites is in Rhode Island.

Class 1. This class includes most of the anthracites found in the United States, of which Pennsylvania has the most abundant reserves. It includes most of the Sullivan County (PA) coals classified as semianthracites in the ASTM system. The code numbers of the coals are 100A and 100B.

Class 2. The coals included in class 2 are mainly the semianthracites of Arkansas and Virginia. As they are noncaking and noncoking, their code number is 200.

Class 3. This class includes most of the coals classified in the ASTM system as low-volatile bituminous. In general, these coals are strongly coking; if coked alone their coke would be well fused, blocky, and fine-grained, with few fissures. However, they cannot be carbonized alone in commercial slot-type ovens because they expand excessively. In American practice, 10 to 30 percent of low-volatile bituminous coal is usually blended with high-volatile coal to produce blast-furnace coke. The code numbers of most of the low-volatile bituminous coals for which data are available are 333 and 334; coals near the borderline of the medium-volatile group are coded 433 and 434.

Classes 4 and 5. These two classes include all coals classified in the ASTM system as medium-volatile bituminous. If coked alone they form strong, blocky cokes. Their coking properties generally are between those of the low-volatile and the better grades of high-volatile bituminous coals. All coals of these classes that were tested have code numbers of either 435 or 535.

Class 6. This class includes most of the high-volatile A bituminous coals. Many of these coals can be carbonized without blending to produce well-fused, medium- to coarse-grained, though somewhat fingery coke. Usually they are blended with 10 to 30 percent of low-volatile bituminous coal to improve the yield, size, structure, and strength of the coke. Code numbers of coals tested in this class are 635, 634, 633, 623, and 622.

Class 7. This class includes coals classified in the ASTM system as high-volatile B bituminous coals. If carbonized alone they usually produce small, fingery, highly fissured coke with low shatter and tumbler indexes. The relatively poor coke structure is typical of cokes made from high-oxygen coals. If heated rapidly at high oven temperatures, high-oxygen coals yield better coke. In American coking practice, high-volatile B bituminous coals are always blended with more strongly coking coals. The high-volatile B bituminous coals that were tested have code numbers as follows: 734, 733, 723, 721, 711, and 823.

Class 8. This class includes high-volatile C bituminous coals and subbituminous A coals. Coals in this class range from noncoking to fair coking. The noncoking coals include all the subbituminous coals and some of the high-volatile C bituminous coals that are nonslacking on exposure to the weather. Their code number is 800. The other high-volatile C bituminous coals range from virtually noncoking to fair coking. Their cokes range from weak and pebbly to small, fingery, and highly fissured. No American coals in class 8 are coked commercially at present, even in blends with more strongly coking coals. The code numbers of the coals in class 8 that show some caking and coking properties are 823, 821, and 811.

Class 9. This class includes subbituminous B coals with a gross calorific value of more than

10,260 Btu on the moist, ash-free basis. The code number of these noncoking coals is 900.

CLASSIFICATION OF AMERICAN LOW-RANK COALS

To classify American low-rank coals by the international system would require information on total-moisture content and yield of tar. Although a standard method for determining yield of low-temperature tar has not been adopted internationally, it probably will be based on the Fischer-Schrader method[4] used in many countries. Such a method is used in the Pittsburgh laboratory of the U.S. Bureau of Mines, and a somewhat modified procedure in its Denver and Grand Forks laboratories.[5] Data on yields of tar from typical American coals by the Fischer-Schrader[6] and modified method[7,8] have been published. Although precise information is not available on correlating the results by the two methods, a statistical study of the relation of the chemical analysis of the coals with tar yields indicates that the yields are similar enough to be used in a survey for applying the international classification system to American coals.

Table XXXIX summarizes the total moisture and tar yields for 73 low-rank American coals having calorific values of less than 10,260 Btu/lb moist and ash free. The data were taken from the reports listed in references 6 and 7, unpublished data, and supplemental information obtained from American exchange samples tested in developing the international classification system. Included in the table are the international code numbers of the coals based on the total moisture and the tar yields.

The subbituminous B coals for which data on both moisture and tar are available have a moisture content ranging from 22 to 27 percent. The class number therefore is 11. The tar yield is about 7 to 11 percent, which places the coals in groups 00 or 10. The coals with less than 10-percent tar yield have a code number of 1100; those with 10 percent or more are numbered 1110.

For the subbituminous C coals, data were available on several samples from Wyoming

and one from Washington. Coals of this rank also occur in Colorado, Montana, and Oregon. Table XXXIX shows a moisture range of about 27 to 34 percent for the samples tested. The coals are placed in classes 11 or 12 depending on whether the moisture content is less or more than 30 percent.

Tar yield ranges from 6 to 15 percent, which groups the coals into 00 and 10. The table shows that code numbers for the subbituminous coals include 1100, 1110, 1200, and 1210.

American lignites have an ash-free moisture content that ranges from about 35 to 45 percent. Their class number is 12 or 13, depending on whether the moisture is less or more than 40 percent. Lignites of the Fort Union formation in Montana, North Dakota, and South Dakota yield 4 to 9 percent tar, placing them in group 00; code numbers are 1200 and 1300.

Tar yields of the Texas lignites range from 11 to 15 percent; their code numbers would be either 1210 or 1310, depending on total moisture content.

The Arkansas lignites gave a tar yield of about 20 percent; their code numbers are either 1320 or 1330, depending on whether the tar yield is less or more than 20 percent. Because some Arkansas lignites have a total moisture content of less than 40 percent, these coals will have a code number of 1220 or 1230 if their yields of tar are of the same order as for the samples tested. High tar yields from Arkansas lignites probably are due to their relatively large amount of wax and resin.[9]

WORLD CLASSIFICATION OF LOW-RANK COALS

Early in its study the Classification Working Party found little available information regarding the different types of brown coals and lignites occurring in the various countries. Therefore, an extensive exchange of samples was arranged by the laboratories cooperating in the international classification work. According to the plan of the Classification Working Party, the coals selected were to be freshly mined and representative of the various types of brown coals and lignites in each country.

Table XXXIX. International Code Numbers of American Subbituminous B, C Coals and Lignites

State	Number of samples	Range of total moisture, ash-free %	Range of tar yields, dry, ash-free %	Code No.
Subbituminous B				
Colorado	2	22.7–25.2	7.4–8.8	1100
"	1	26.5	11.4	1110
Montana	1	23.7	6.6	1100
Utah	1	21.7	11.2	1110
Wyoming	2	23.5–26.3	7.5–7.9	1100
"	2	23.5–26.9	10.2	1110
Subbituminous C				
Washington	1	30.7	14.8	1210
Wyoming	3	27.4–29.3	6.9–9.0	1100
"	1	29.7	12.2	1110
"	2	30.4–34.2	6.3–7.5	1200
"	1	32.4	12.0	1210
Lignite				
Arkansas	1	41.0	19.8	1320
"	1	41.2	20.5	1330
Montana	1	39.9	6.6	1200
"	5	40.0–42.3	5.6–7.2	1300
North Dakota	12	36.0–39.9	4.3–8.7	1200
"	25	40.0–47.7	4.0–8.4	1300
South Dakota	1	34.4	3.5	1200
"	1	43.6	4.2	1300
Texas	3	37.6–39.7	11.2–12.2	1210
"	5	40.9–45.7	12.0–14.8	1310
Washington	1	40.6	9.8	1300

Also, samples of transition coals were requested that would be close to the borderline of the brown coal/lignite class.

Table XXXVIII shows the source of the coals, some of the analytical data obtained by the Bureau of Mines, and classification of the coals according to the ASTM and International systems. The proximate and ultimate analyses, calorific value, and moisture-holding capacity (equilibrium moisture) were determined by ASTM methods, except that total moisture and

the moisture in the air-dried samples were determined by either xylene or toluene distillation as requested by the Working Party.

Total moisture is used rather than moisture-holding capacity (equilibrium moisture) specified in the classification of hard coals. For practical purposes, moisture-holding capacity is equivalent to natural bed moisture for all ranks of American coals including lignite.[10] However, total moisture and bed moisture are not equivalent for certain European brown

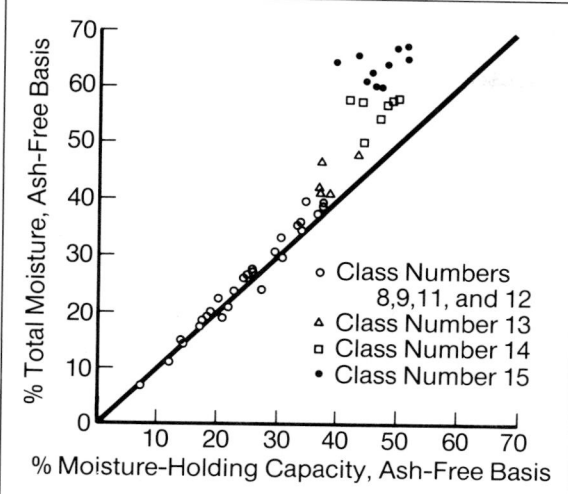

Fig. 12 **Relationship of moisture-holding capacity and total moisture**

coals of higher total moisture content than American lignites. Fig. 12 compares the moisture-holding capacity with total moisture for the exchange samples. The figure shows a linear relationship between moisture-holding capacity and total moisture for coals in the International system with class number 8 to 12 and possibly 13; that is, coals having as much as 40 percent total moisture. For classes 14 and 15, the relationship is poor, with moisture-holding capacity averaging 10 to 15 percentage points lower than total moisture for the coal samples examined.

REFERENCES

[1] United Nations Economic Commission for Europe, *International Classification of Hard Coals by Type.* Document No. E/ECE/247, E/ECE/COAL/110. Publication sales no. 1956 II E.4. New York: Columbia University Press, International Documents Service, 1956.

[2] W.H. Ode and W.H. Frederic, "The International Systems of Hard-Coal Classification and Their Application to American Coals," *U. S. Bureau of Mines Report of Investigations 5435.* Washington: U. S. Government Printing Office, 1958.

[3] W.H. Ode and F.H. Gibson, "International System for Classifying Brown Coals and Lignites and Its Application to American Coals," *U. S. Bureau of Mines Report of Investigations 5695.* Washington: U. S. Government Printing Office, 1960.

[4] Franz Fischer and Hans Schrader, "Crude-Tar Determination with an Aluminum Distillating Apparatus," *Zeitschrift Fuer Angewandte Chemie,* 33 (1): 172–175, 1920.

[5] John B. Goodman, et. al., "Low-Temperature Carbonization Assay of Coal in a Precision Laboratory Apparatus," *U. S. Bureau of Mines Bulletin 530.* Washington: U. S. Government Printing Office, 1953.

[6] W.A. Selvig and W.H. Ode, "Low-Temperature Carbonization Assays of North American Coals," *U. S. Bureau of Mines Bulletin 571.* Washington: U. S. Government Printing Office, 1957.

[7] Manuel Gomez and J.B. Goodman, "Distillation Assays of Missouri River Basin Coals," *U. S. Bureau of Mines Report of Investigations 5009.* Washington: U. S. Government Printing Office, 1953.

[8] J.J. Hoeppner, et. al., "Carbonication Characteristics of Some North-Central United States Lignites," *U. S. Bureau of Mines Report of Investigations 5260.* Washington: U. S. Government Printing Office, 1956.

[9] W.A. Selvig, et. al., "American Lignites: Geological Occurrence Petrograpic Composition, and Extractable Waxes," *U. S. Bureau of Mines Bulletin 482.* Washington: U. S. Government Printing Office, 1950.

[10] W.A. Selvig and W.H. Ode, "Determination of Moisture-Holding Capacity (Bed Moisture) of Coal for Classification by Rank, *U. S. Bureau of Mines Report of Investigations 4968.* Washington: U. S. Gov. Printing Office, 1953.

BIBLIOGRAPHY

V.F. Parry, "Production, Classification, and Utilization of Western United States Coals," *Economic Geology,* 45: 515–532, 1950.

W.A. Selvig and F.H. Gibson, "Analyses of Ash From United States Coals," *U. S. Bureau of Mines Bulletin 567.* Washington: U. S. Government Printing Office, 1956.

R.F. Abernethy and E.M. Cochrane, "Free-Swelling and Grindability Indexes of United States Coals," *U. S. Bureau of Mines Information Circular 8025.* Washington: U. S. Government Printing Office, 1961.

R.F. Abernethy and E.M. Cochrane, "Fusibility of Ash of United States Coals," *U. S. Bureau of Mines Information Circular 7923.* Washington: U. S. Gov. Printing Office, 1960.

"Analyses of Tipple and Delivered Samples of Coal;" Reports of Investigations of the Bureau of Mines, U. S. Dept of the Interior; comprise analytical data showing composition and quality of coal samples collected by the Bureau for nearly every year since 1948. U. S. Government Printing Office, Washington, D.C.

Appendix B. Determination of Coal-Ash Properties

This appendix supplements the data on properties of coal ash included in Chapter 3. Material properties useful in the design of furnaces and ash-handling systems are given. Methods of calculating fusion temperatures and viscosities of coal ash, as well as further descriptions of these properties are outlined, with curves to assist in the determinations.

MATERIAL PROPERTIES OF COAL-ASH SLAG DEPOSITS

There are several properties of slag deposits that are of use in performing such calculations as those of furnace heat-absorption and ash-hopper cooling-water flow.

SURFACE EMISSIVITY

The curves of Figs. 1 and 2 plot surface emissivity of coal-ash slags as a function of temperature. Separate curves are given for fused and particulate states because emissivity differs with the deposit structure. Such emissivities are averaged values of composite data from Australian, Soviet, and U.S. investigations. Heat transfer from the flame to the surrounding surfaces in utility furnaces is predominantly by radiation. It follows that meaningful heat-transfer calculations should include the re-

quirement of a reasonably accurate knowledge of slag-surface emissivities.

THERMAL CONDUCTIVITY

The curves of Fig. 3 show thermal conductivities as a function of temperature. Separate curves are given for fused and particulate states, because thermal conductivity differs with the physical structure of the deposit. The given conductivities are averaged values of composite data obtained from Australian, Soviet, and U.S. investigations.

The particulate deposit referred to is a fine, dry powder that forms an uncompacted layer on the metal surface. The outer deposit layers vary considerably in character from being porous or sintered to being dense, fused, or molten. The conductivity, which will vary with the physical properties of the specific deposit, is expected to increase as the deposit density increases and the deposit becomes more molten.

Heat transfer through a coal-ash deposit to waterwalls is calculated by a conduction model; this requires that the thermal conductivity of the deposits be known.

SPECIFIC HEAT AT CONSTANT PRESSURE

Vargaftik and Oleschuck[1] have measured the specific heat of coal-ash slags at constant pressure with a reference temperature of 60°F. Table I lists these specific heat values.

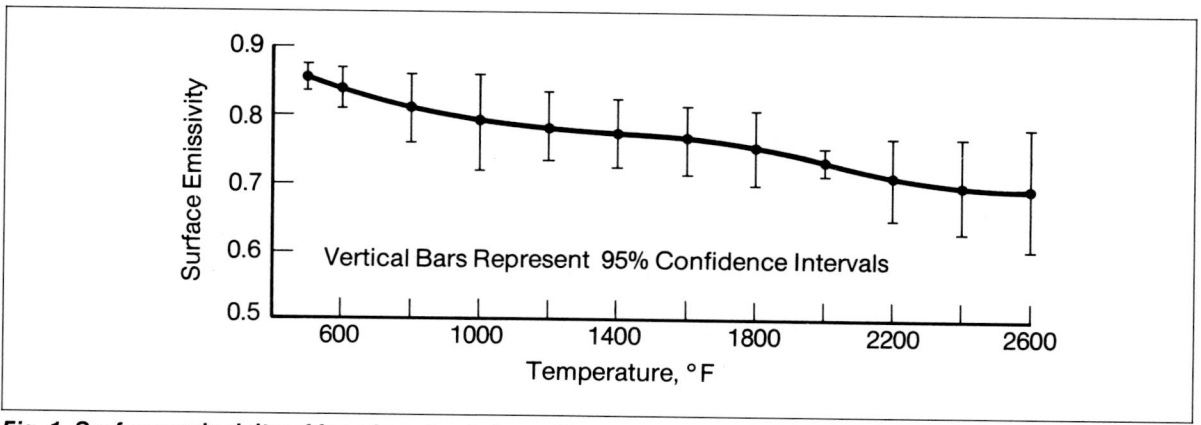

Fig. 1. **Surface emissivity of fused coal-ash deposits**

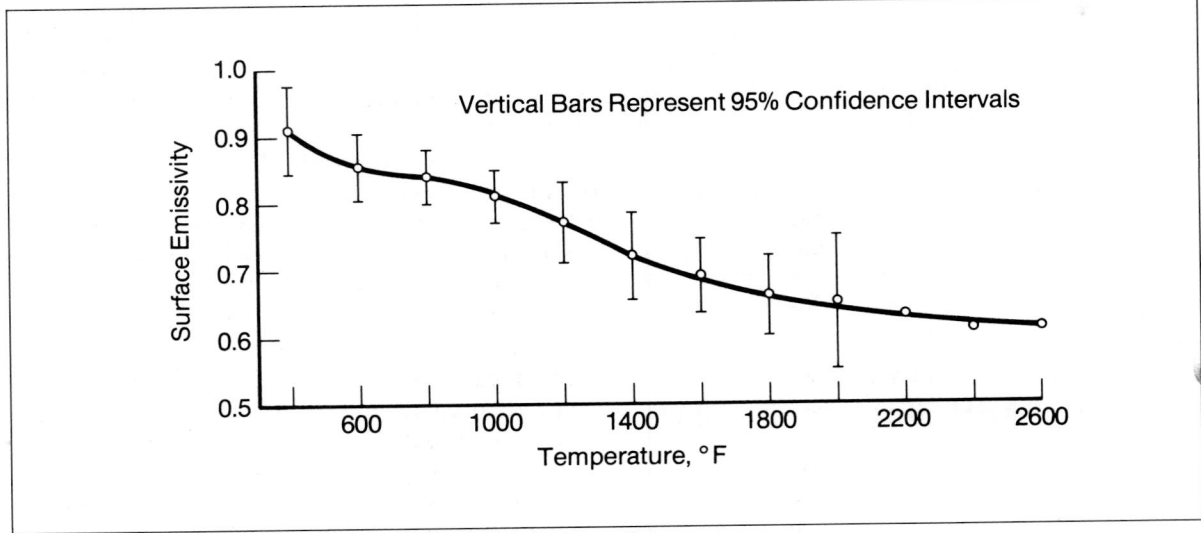

Fig. 2. Surface emissivity of particulate coal-ash deposits

A method for estimating the specific heat of slag from its chemical composition has been developed for blast-furnace slags. Because of the compositional similarity between coal-ash and blast-furnace slags, the method should be applicable to coal-ash slags and is described below. To perform cool-down, ash-hopper evaporation, and transient calculations, knowledge of the slag specific heat is required.

Using a reference temperature of 20°C, Voskoboinikov[2] derived the following empirical equations for estimating the specific heat of blast-furnace slags:

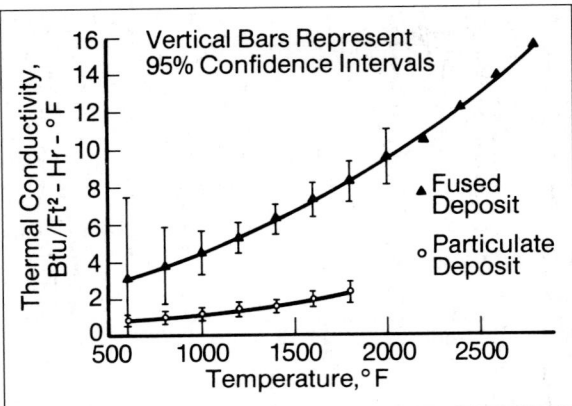

Fig. 3. Thermal conductivity of fused and particulate coal-ash deposits

For the temperature range of 20° to 1350°C

$$c_P = 0.169 + 0.201 \times 10^{-3}t - 0.277 \times 10^{-6}t^2 \\ + 0.139 \times 10^{-9}t^3 + 0.17 \times 10^{-4}t \left(1 - \frac{CaO}{\Sigma}\right)$$

where c_P = specific heat (cal/gm °C)
  t = temperature (°C)
  Σ = $SiO_2 + Al_2O_3 + FeO + MgO + MnO$

(1)

For the temperature range of 1350 to 1600°C

$$c_P = 0.15 \times 10^{-2}t - 0.478 \times 10^{-6}t^2 \\ - 0.876 + 0.016 \left(1 - \frac{CaO}{\Sigma}\right)$$

where c_P = specific heat (cal/gm °C)
  t = temperature (°C)
  Σ = $SiO_2 + Al_2O_3 + FeO + MgO + MnO$

(2)

Foerster and Weston[3] showed that their experimentally measured values for liquid and solid slags agreed well with the calculated values from Voskoboinikov's equation. Experimental values for liquid and solid slags were slightly higher than the calculated values.

The densities of various forms of coal-ash slag deposits were averaged from several investigations and are as listed in Table II.

SURFACE TENSION

Surface-tension forces are involved in many important stages of the slag-deposition process, such as ash fusion, slag adhesion, and slag flow. The methods of estimating surface tension of oxide systems have been developed in the ceramic and glass industry; they should be applicable to some extent to coal-ash slags because of the similarity in their chemical compositions. Surface tension of liquid slags can be estimated according to the following procedure:

1. Surface tension of liquid slags at 1400°C in air can be approximated from the chemical composition using the following equation developed by Lyon.[4]

$$\gamma = 3.24\, SiO_2 + 5.85\, Al_2O_3 + 4.4\, Fe_2O_3 + 4.92\, CaO + 5.49\, MgO + 1.12\, Na_2O - 0.75\, K_2O$$

where γ = surface tension of slag (dyne/cm)

$$(3)$$

2. To estimate surface tension of slags at different temperatures, a temperature coefficient of −0.017 dyne/cm °C is recommended by Parmelee and Harman.[5]

3. To estimate surface tension of slags in different atmospheres, Table III has been derived from data by Parikh.[6]

In general, surface tension of slags does not vary greatly and is within the range of 300 to 400 dyne/cm at 1400°C in air. A lower slag surface tension will usually lead to more rapid fusion and flow properties, and better wetting characteristics. Good wetting is a necessary, but not sufficient, condition for stronger adhesive properties.

RADIATIVE PROPERTIES OF ASH DEPOSITS

Radiative heat-transfer properties of ash deposits are important in determining heat transfer to lower-furnace waterwall tubes when burning coal. The absorptivity of ash deposits plays an important role in determining the heat absorbed by deposit-laden waterwall tubes, while the emissivity of ash deposits determines how much heat will be reradiated by the tubes. The emissivity and absorptivity of ash deposits are a function of the wavelengths of the absorbed and emitted radiation, the surface temperature of the deposits, the physical state of the deposits, and the properties of the coals from which the deposits originated.

THEORY

In Eq. 4, the rate of radiative heat transfer from a flame to a unit area of deposit-laden wa-

Table I. Specific Heat at Constant Pressure (Ref. 1)

Temperature (°F)	C_P (Btu/lb °F)
2010	0.243
2190	0.258
2370	0.272
2550	0.285
2730	0.300
2910	0.303
3090	0.310

Table II. Density of Various Coal-Ash Slag Deposits

Deposit Form	Density (lb/ft³)
Liquid Slag	150–180
Dense, Solid Slag	140–170
Solid Ash	120–160
Loose, Powder Ash	15–35
Fine, Ground Slag	60–90

Table III. Surface Tension of Slags in Different Atmospheres

Atmosphere	Reduction in Surface Tension (%) (Based on Nitrogen Atmosphere)
Nitrogen	0
Hydrogen	1
Air	3
Carbon Dioxide	7
Sulfur Dioxide	27
Water Vapor	35

terwall tube (q/A) depends upon the overall emissivity (ε) and overall absorptivity (α) of the deposit, the flame temperature (T_f) and the deposit surface temperature (T_s). This heat is conducted through the ash deposit of thickness Δx having a temperature drop ΔT_{ash} and a thermal conductivity, k, and transferred to water flowing through the tubes. This simplified energy balance on a unit area of tube (Fig. 4) can be written as:

$$\begin{matrix} \text{Net thermal} \\ \text{energy trans-} \\ \text{fer to water} \end{matrix} = \begin{matrix} \text{Coal flame} \\ \text{radiation} \end{matrix} - \begin{matrix} \text{Deposit} \\ \text{emission} \end{matrix} = \begin{matrix} \text{Conduction} \\ \text{through} \\ \text{deposit} \end{matrix}$$

or,

$$q/A = \sigma \varepsilon_f \alpha T_f^4 - \sigma \varepsilon T_s^4 = k \frac{\Delta T_{ash}}{\Delta x}$$

(4)

where σ is the Stefan-Boltzmann constant and ε_f is the flame emissivity. It should be noted that the above relationship is simplified and does not consider the view factor from the flame to the wall, nor wall reflection.

As in Eq. 5 the overall deposit emissivity (ε) depends upon the monochromatic emissivity (ε_λ) at surface temperature (T_s) and the monochromatic blackbody emission per unit area (W_{s_λ}) at surface temperature (T_s):

$$\varepsilon = \frac{\int_o^\infty \varepsilon_\lambda W_{s_\lambda} d_\lambda}{\int_o^\infty W_{s_\lambda} d_\lambda} = \frac{\int_o^\infty \varepsilon_\lambda W_{s_\lambda} d_\lambda}{\sigma T_s^4}$$

(5)

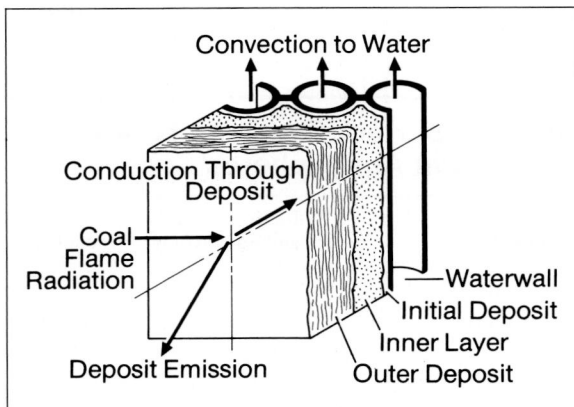

Fig. 4. Section of lower-furnace waterwall depicting heat-transfer mechanism through coal-ash deposit

where σT_s^4 is introduced using Planck's law.

The overall deposit absorptivity (α) depends upon the monochromatic absorptivity (α_λ) at surface temperature (T_s) and the distribution of monochromatic intensities of incident radiation (I_λ):

$$\alpha = \frac{\int_o^\infty \alpha_\lambda I_\lambda d_\lambda}{\int_o^\infty I_\lambda d_\lambda}$$

(6)

where I_λ is dependent on W_{f_λ}, the monochromatic blackbody emission per unit area at flame temperature (T_f) and the monochromatic emissivity (ϵ_{f_λ}):

$$I_\lambda = \varepsilon_{f_\lambda} W_{f_\lambda}$$

(7)

Employing Planck's law and the assumption that a coal flame is nearly black at all wavelengths (that is, $\varepsilon_{f_\lambda} \simeq 1$) gives

$$\alpha = \frac{\int_o^\infty \alpha_\lambda W_{f_\lambda} d_\lambda}{\sigma T_f^4}$$

(8)

When the radiation from the flame is in steady state with the surface, Kirchoff's law ($\varepsilon_\lambda = \alpha_\lambda$) applies and Eq. 8 can be written as:

$$\alpha = \frac{\int_o^\infty \varepsilon_\lambda W_{f_\lambda} d_\lambda}{\sigma T_f^4}$$

(9)

From Eqs. 5 and 9, it is seen that both the overall emissivity and absorptivity depend upon the deposit surface temperature and the monochromatic emissivities; absorptivity is also dependent upon flame temperature. From these equations, it can also be deduced that the overall absorptivity can differ significantly from the overall emissivity, particularly if the monochromatic emissivities vary significantly with wavelength and the flame temperature is significantly different from the deposit surface

temperature. Thus, experimental measurement of only the overall emissivity can result in a significant error in prediction of absorptivity.

The experimental determination of the emissivity and absorptivity of coal ash deposits involves a fair amount of procedural and instrumentational complexity. Early investigations in this field by Agababov,[7] Becker[8] and Mulcahy, Boow and Goard[9] assumed that deposits were "gray bodies." This assumption says that the spectral emissivities and absorptivities ε_λ and α_λ did not vary with wavelength, λ, and were equivalent to the overall emissivity and absorptivity (ε and α) respectively. Hence, they measured either overall emissivity or absorptivity. Later investigations by Khrustalev et al.[10] and Smith, Glicksman[11] proved the "gray body" assumption wrong and employed spectral scans to determine ε_λ. Overall emissivities and absorptivities of ash deposits were calculated by integrating their respective spectral emissivities and their absorptivities.

C-E has developed a laboratory technique employing such spectral spans, and a computational procedure, to determine the emissivities and absorptivities of furnace ash deposits.[12]

Table IV summarizes ash-deposit emissivity and absorptivity results. Here deposits are categorized by physical state (initial deposit, inner layer, and outer deposit), as well as by parent coal source (Eastern and Western U.S.).

On the basis of the limited data, it appears that for initial deposits and inner layers, Western coal-ash emissivities and absorptivities tend to be significantly below those of Eastern

coal ashes. This tendency is expected on the basis of chemical composition; that is, materials composed mainly of oxides of iron, magnesium, and silicon tend to be better emitters and absorbers than those containing high sodium or calcium.[8]

As the physical state of the deposit tends toward a molten condition, emissivity and absorptivity approach 1.0, and differences between Eastern and Western coals would be expected to become less significant. The data also indicate that molten deposits have higher emissivities and absorptivities than powdery deposits. It can be generalized that the emissivity of the ash deposit of a given physical state decreases with increasing surface temperature throughout a wide temperature range.

METHODS OF ESTIMATING ASH-FUSIBILITY TEMPERATURES

Many investigators have attempted to calculate characteristic fusion temperatures of coal ash from its chemical composition. Most of the methods require assumptions that tend to over-simplify the composition of the ash. Also, a common inadequacy among many of the relations is that their validity is usually limited to a certain range of coal-ash compositions. The methods described in this section are those that can be applied to a relatively wide range of coal-ash compositions. The references should be consulted for further information on the accuracy and applicability of the correlations.

Table IV. **A Summary of Emissivities and Absorptivities of Ash Deposits**

Physical state	Parent Coal Region	Emissivity	Absorptivity
Initial deposit	Eastern U.S.	0.75–0.76	0.71–0.76
Initial deposit	Western U.S.	0.37–0.56	0.52–0.61
Inner layer	Eastern U.S.	0.79–0.93	0.67–0.84
Inner layer	Western U.S.	0.68	0.62
Outer deposit	Western U.S.	0.66–0.87	0.62–0.82
Molten inner layer	Western U.S.	0.90	0.91

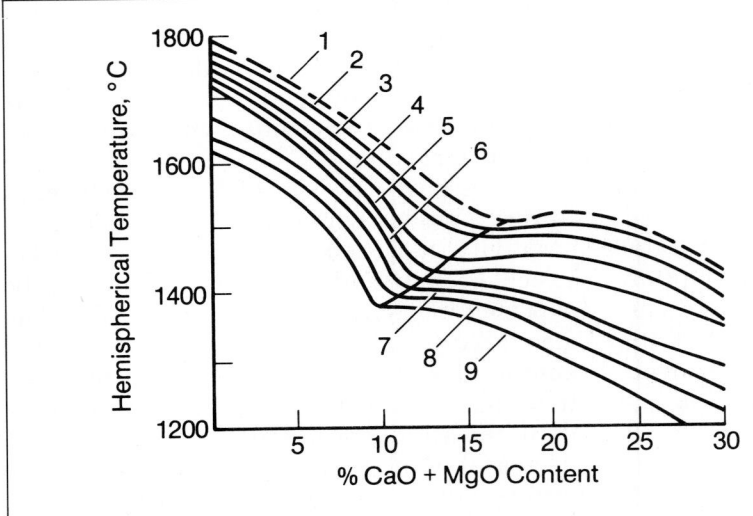

Fig. 5. Hemispherical temperature of coal-ash free of Fe_2O_3, Na_2O, and K_2O

It is very unlikely that calculation of ash-softening temperatures will replace the determination. Nevertheless, using correlations of this type permits conclusions to be drawn as to the relative influence of each component on the ash-softening temperature and the probable effect of addition of reagents.

METHOD OF DUTTA, RAI, AND CHAKRAVORTY[13]

Assuming the British Standard experimental conditions, this method is used to estimate the hemispherical temperature in either a reducing or oxidizing atmosphere. It is based on ash from Indian coals and synthetic ash mixtures. The procedure is as follows:

1. Calculate the $SiO_2/(Al_2O_3 + TiO_2)$ ratio.

2. Obtain the percentage of $CaO + MgO$ based on

$$SiO_2 + Al_2O_3 + TiO_2 + CaO + MgO = 100\% \quad (10)$$

3. From Fig. 5, estimate the hemispherical temperature of the coal ash free of Fe_2O_3, Na_2O, and K_2O.

4. Obtain the percentage of Fe_2O_3 and $Na_2O + K_2O$ based on the following equation

$$SiO_2 + Al_2O_3 + TiO_2 + Fe_2O_3 + CaO + MgO + Na_2O + K_2O = 100\% \quad (11)$$

5. Correct for the presence of Na_2O and K_2O in the ash by reducing the estimated hemispherical temperature by 25°C (45°F) for each percent of $Na_2O + K_2O$ present.

6. Additional corrections for the presence of Fe_2O_3 in the ash are given in the references.

METHOD OF GAUGER[14]

This method, which assumes *ASTM Standards* experimental conditions, is used to estimate the softening temperatures in a reducing atmosphere of ash from Eastern U.S. coals. The procedure is as follows:

1. Calculate S, A, and T, which are defined as:

$$S = SiO_2 + TiO_2 + P_2O_5 \quad (12)$$

$$A = Al_2O_3 \quad (13)$$

$$T = CaO + 0.7\,MgO + 2.25\,Na_2O + 1.5\,K_2O \quad (14)$$

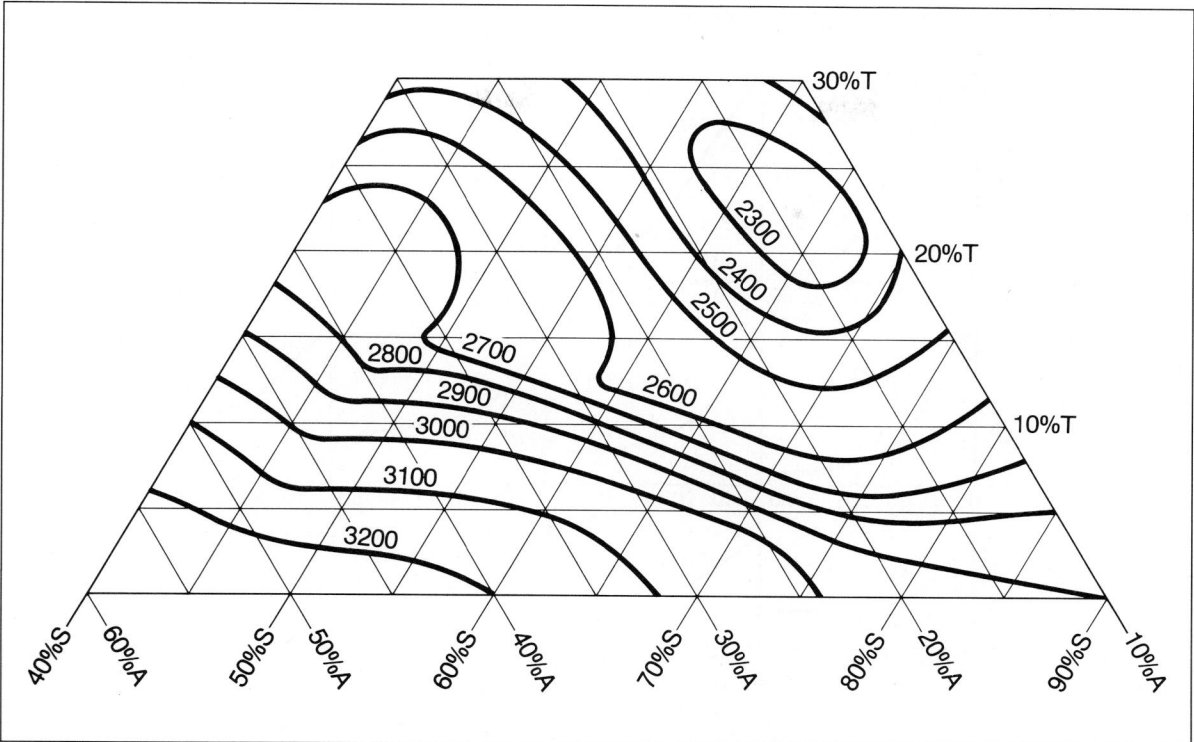

Fig. 6 **Softening temperature of coal ash free of iron oxide, degrees Fahrenheit**

2. Obtain the percents S, A, and T, which are expressed as:

$$\% \, S = \left(\frac{S}{S + A + T} \right) \times 100$$

(15)

$$\% \, A = \left(\frac{A}{S + A + T} \right) \times 100$$

(16)

$$\% \, T = \left(\frac{T}{S + A + T} \right) \times 100$$

(17)

3. From Fig. 6, estimate the softening temperature of the coal ash free of iron oxide, using the above designated percentages.

4. Define I = Fe_2O_3 and calculate the percent I, which is expressed as:

$$\% \, I = \left(\frac{I}{S + A + T + I} \right) \times 100$$

(18)

5. From Fig. 7, estimate the lowering of the softening temperature due to the presence of Fe_2O_3; this will give the softening temperature of the ash.

METHOD OF SCHAEFER[15]

This method, which assumes *ASTM Standards* experimental conditions, is used to estimate the softening temperature in a reducing atmosphere of ash from Eastern U.S. Coals. The procedure is as follows:

1. Calculate R_s, which is expressed as:

$$R_s = \frac{Al_2O_3}{SiO_2} \times \frac{SiO_2 + Al_2O_3}{FeO + 0.6 \, (CaO + MgO + Na_2O + K_2O)}$$

(19)

Determination of Coal-Ash Properties

Fig. 7. Lowering of the softening temperature *from* iron oxide

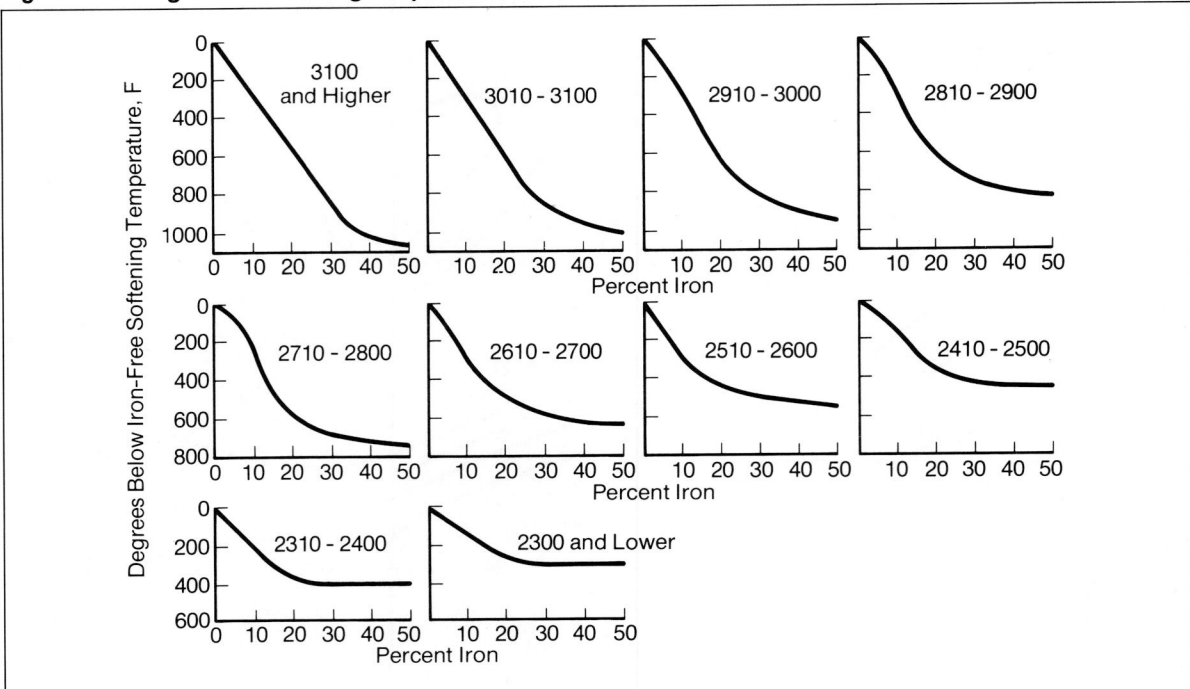

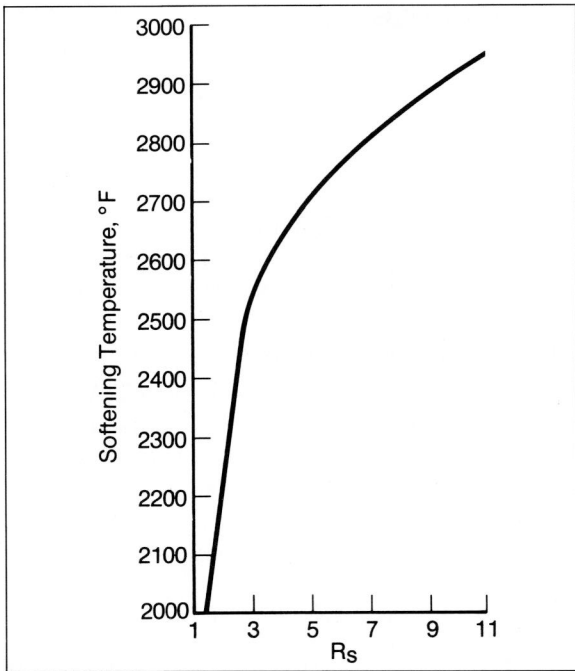

Fig. 8. Softening temperature of coal ash as a function of R_s

2. Using the value for R_s, estimate the softening temperature of the coal ash from Fig. 8.

METHOD OF ESTEP, SELTZ, BUNKER, AND STRICKLER[16]; AND ESTEP, SELTZ, AND OSBORN[17]

This method, which assumes *ASTM Standards* experimental conditions, is used to estimate the hemispherical temperatures in a reducing atmosphere of ash from Eastern U.S. coals and synthetic ash mixtures. If the coal-ash composition contains mainly SiO_2, Al_2O_3, Fe_2O_3, and CaO, then the hemispherical temperature of the ash can be estimated according to the following procedure:

1. Obtain the percentage of CaO based on:

$$SiO_2 + Al_2 + O_3 + Fe_2O_3 + CaO = 100\% \tag{20}$$

2. Typical ternary diagrams are given in Figs. 9 and 10, which are to be used for calcium-oxide contents up to 8.5 percent.

ASEA BROWN BOVERI

Fig. 9. **Softening temperature isotherms (°F) at** *CaO* **= 0−2.5%**

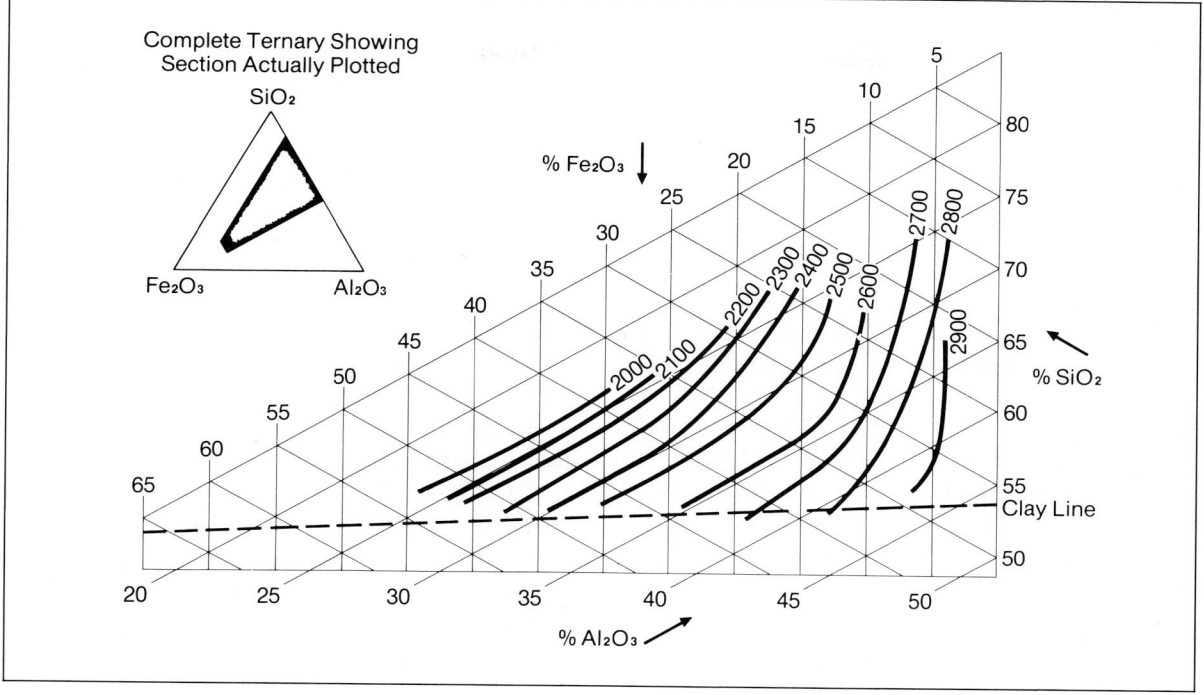

Fig. 10. **Softening temperature isotherms (°F) at** *CaO* **= 7.6−8.5%**

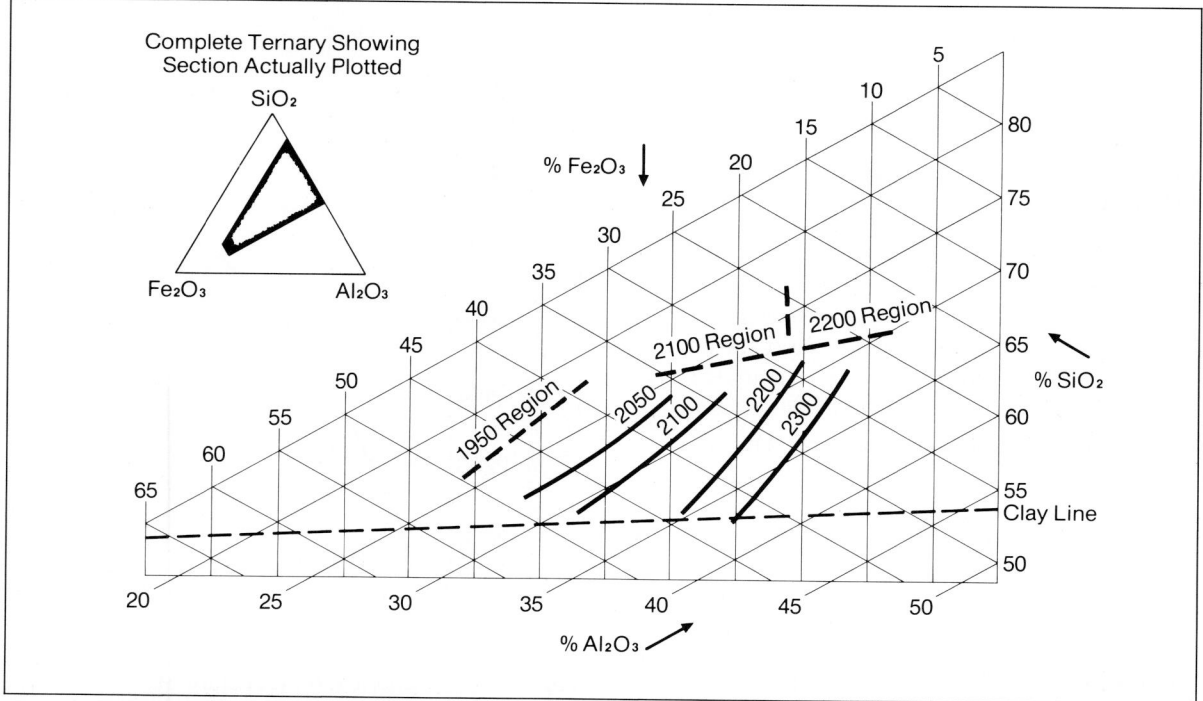

3. Locate the corresponding composition point inside the SiO_2-Al_2O_3-Fe_2O_3 diagram which is based on the following equation

$$SiO_2 + Al_2O_3 + Fe_2O_3 = 100\% \tag{21}$$

This gives the estimated hemispherical temperature of the coal ash.

For ashes with CaO content above 8.5 percent, diagrams showing vertical planes through the prism are presented which give plots of CaO versus Fe_2O_3 at constant SiO_2/Al_2O_3 ratios.

For ashes with substantial amounts of MgO or Na_2O, the following corrections apply:

1. The fluxing effect of MgO is identical to that of CaO up to 10 percent MgO; therefore, CaO + MgO can be considered as a single variable.

2. The presence of Na_2O should be corrected by reducing the estimated hemispherical temperature by 50°F for each percentage of Na_2O based on:

$$SiO_2 + Al_2O_3 + Fe_2O_3 + CaO + MgO + Na_2O = 100\% \tag{22}$$

METHOD OF KOVITSKII, KARAGODINA, AND MARTYNOVA[18]

This method, which assumes Soviet Standard experimental conditions, is used to estimate the softening temperature (t_2) and the temperature of the beginning fused state (t_3) in a reducing atmosphere, of ash from Russian coals. The procedure is as follows:

1. Calculate K_{fu} which is expressed as:

$$K_{fu} = \frac{SiO_2 + Al_2O_3}{Fe_2O_3 + CaO + MgO} \tag{23}$$

2. Estimate the softening temperature and the temperature of the beginning of liquid fused state from the following relations:

$$t_2 = 1094 + 42.5\,K_{fu} \tag{24}$$

$$t_3 = 1139 + 48.6\,K_{fu} \tag{25}$$

where

t_2 = softening temperature (°C)

t_3 = temperature of beginning of liquid fused state (°C).

This relation is claimed to be valid for the following range of K_{fu}:

For t_2: $1.8 \leq K_{fu} \leq 9.9$

For t_3: $1.8 \leq K_{fu} \leq 7.5$

METHOD OF MAJUMDAR, BANERJEE, AND LAHIRI

This method, which assumes British Standard experimental conditions, is used to estimate the hemispherical temperature in a reducing atmosphere of ash from Indian coals. The procedure is as follows:

1. Calculate R_m:

$$R_m = (3.3\,SiO_2 + 1.96\,Al_2O_3)/(2.5\,Fe_2O_3 + 3.57\,CaO + 5.0\,MgO + 3.22\,Na_2O + 3.22\,K_2O) \tag{26}$$

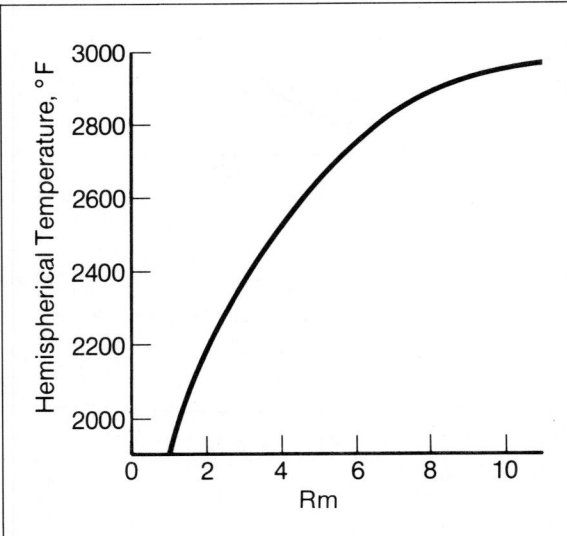

Fig. 11. Hemispherical temperature of coal ash as a function of R_m

2. From Fig. 11, estimate the hemispherical temperature of the coal ash.

This relationship was independently tested on 75 ashes from U.S. coals. The difference between the corresponding estimated and measured values is as follows:

1. In 66.6 percent of the cases the estimated value is within ±100°F of the measured value.

2. The average difference between the two corresponding values is ±84°F.

The original report describing this relation is not available. The curve in Fig. 11 was derived from calculations carried out on ashes from Eastern and Western U.S. coals. The original relation was said to be based on molecular percentages and not on weight percentages. However, the independent tests indicated that the divergence of the estimated values from the corresponding measured values was much smaller when weight percentages were used.

ASH COMPOSITION AND VISCOSITY-TEMPERATURE RELATIONSHIPS

As early as the 1930's, the Bureau of Mines and others conducted studies to measure the flow characteristics of molten ash in wet-bottom furnaces; a method for relating ash composition to viscosity of the molten ash was developed. By placing the ash composition data into such specific arrangements as equivalent ferric oxide, silica equivalent, and calcium oxide percentage, and by developing a series of curves based on relationships of laboratory viscosity to ash composition, viscosity-temperature relationships can be established.

These viscosity-temperature relationships permit the construction of viscosity curves of the ash in both the fluid and plastic states. Although definite predictions are difficult to make from the viscosity curves alone, such plots are particularly useful for predicting the slag characteristics of specific coals by comparing the viscosity-temperature relationships with coals of known performance.

VISCOSITY OF COAL-ASH SLAGS IN THE COMPLETE LIQUID PHASE

The viscosity of coal-ash slag in the complete liquid phase can be measured or estimated at any temperature above its crystallization temperature. Above this temperature range, slag viscosity is mainly a function of temperature and chemical composition. Actual measurement of slag viscosity is usually expensive and difficult; therefore, several methods of estimating slag viscosity from the composition-temperature relationships have been developed. This method should be used only when the slag temperature is above the crystallization temperature, referred to as the temperature of critical viscosity.

METHOD BY REID AND COHEN[19]

The viscosity of coal-ash slags from Eastern U.S. coals in the complete liquid phase can be estimated according to the following procedure:

1. Calculate the silica percentage,

$$SP = \left(\frac{100 \times SiO_2}{SiO_2 + \text{Equiv. } Fe_2O_3 + CaO + MgO} \right) \tag{27}$$

2. Read the viscosity at a given temperature, or the temperature at a given viscosity, from the experimentally derived nomogram shown in Fig. 12. Scale C on the nomogram shows a direct relationship between SP and the liquid viscosity at 2600°F. To find viscosity and any other temperature: (a) connect the 2600°F point on scale A with the desired SP or viscosity on scale C; (b) note the pivot point on line B and (c) draw a line through the desired temperature on scale A, through the pivot point, to obtain viscosity on scale C.

METHOD OF HOY, ROBERTS, AND WILKINS[20]

The viscosity of coal-ash slags from British coals in the complete liquid phase can be estimated according to the following procedure:

1. Calculate the silica percentage,

$$SP = \left(\frac{100 \times SiO_2}{SiO_2 + Equiv.\ Fe_2O_3 + CaO + MgO} \right) \tag{28}$$

2. Estimate the slag viscosity from the following equation:

$$\log_{10}\eta = 4.468\,(SP/100)^2 + 1.265\,(10^4/T) - 7.44$$
$$\text{where } \eta = \text{slag viscosity (poise)}$$
$$T = \text{temperature (°K)} \tag{29}$$

The Hoy, Roberts, Wilkins relationship was based on the following range of coal-ash chemical composition:

SiO_2	31–59%
Al_2O_3	19–37%
Equiv. Fe_2O_3	0–38%
CaO	1–37%
MgO	1–12%
$Na_2O + K_2O$	1–6%
SP	45–75%
SiO_2/Al_2O_3	1.2–2.3

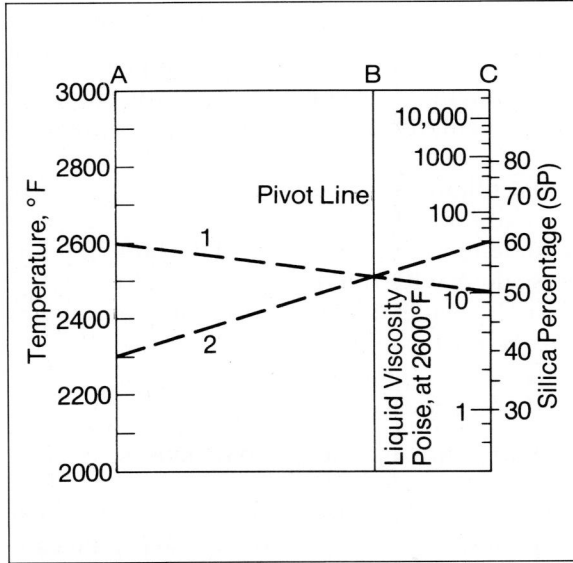

Fig. 12 Nomogram for estimation of coal-ash slag viscosity. See also reference 19.

METHOD OF WATT AND FEREDAY[21]

The viscosity of coal-ash slags from British coals in the complete liquid phase can be estimated according to the following procedure:

1. Normalize the chemical composition of the ash such that:

$$SiO_2 + Al_2O_3 + Equiv.\ Fe_2O_3 + CaO + MgO = 100\% \tag{30}$$

2. Calculate M, defined as:

$$M = 0.00835\,SiO_2 + 0.00601\,Al_2O_3 - 0.109 \tag{31}$$

3. Calculate C, defined as:

$$C = 0.0415\,SiO_2 + 0.0192\,Al_2O_3 + 0.0276\,Equiv.\ Fe_2O_3 + 0.0160\,CaO - 3.92 \tag{32}$$

4. Estimate the slag viscosity from the following equation:

$$\log_{10}\eta = \frac{10^7 M}{(T-150)^2} + C$$
$$\text{where } \eta = \text{slag viscosity (poise)}$$
$$T = \text{temperature (°C)} \tag{33}$$

The Watt-Fereday relationship was based on the following range of coal-ash chemical composition:

SiO_2	29–56%
Al_2O_3	15–31%
Equiv. Fe_2O_3	2–28%
CaO	2–27%
MgO	1–8%
$Na_2O + K_2O$	1.5–5%
SP	40–81%
SiO_2/Al_2O_3	1.4–2.4

FLOW TEMPERATURE

The flow temperature is defined as the temperature at which slag has sufficient fluidity to

allow free flow without difficulty. Normally the flow temperature corresponds to a slag viscosity of approximately 80 poise. For coal-ash slags of known chemical composition, the flow temperature can be calculated by letting $\eta = 80$ poise and obtaining the corresponding T_{80} value. It is used to predict the ease of slag removal in wet-bottom furnaces. It also can be used to estimate the maximum steady-state thickness of furnace-wall deposits, because no accumulation of slag is expected beyond this point.

When estimating slag viscosity in the complete liquid phase, one should consider the existence of the crystallization temperature, which is referred to as the temperature of critical viscosity (T_{cv}). The relation between T_{80} and T_{cv} is as follows:

If $T_{80} > T_{cv}$, then flow temperature $= T_{80}$
If $T_{80} < T_{cv}$, then flow temperature $= T_{cv}$

Prior to the development of the viscometer, the flow temperature was determined manually by stirring the liquid slag with a platinum rod. This stirring helped establish a standard "feel" for the flow temperature. An empirical relation was then developed between the flow temperature of coal-ash slags and the ASTM ash-fluid temperature (FT). For a coal-ash slag with a ferric content of about 10 percent, an approximate relation is expressed as:

$$\text{Flow temperature (°F)} = 1.2\,(\text{FT} - 470) \tag{34}$$

SLAG-REMOVAL TEMPERATURE

The slag-removal temperature is the temperature corresponding to the maximum viscosity at which slag can be tapped from a furnace. This upper limit for fluidity of slag is approximately 250 poise. For coal-ash slags of known chemical composition, the slag-removal temperature can be calculated from the relations described above, by letting $\eta = 250$ poise and obtaining the corresponding T_{250} value. T_{250} can also be estimated from the two graphical methods described later.

Generally, for wet-bottom furnaces, T_{250} should not exceed 2600°F. For dry-bottom units, high values of T_{250} are recommended for easier removal (sootblowing) of waterwall deposits.

When estimating slag viscosity in the complete liquid phase, the existence of the crystallization temperature should be considered. T_{cv}, which is referred to as the temperature of critical viscosity, is found as follows:

If $T_{250} > T_{cv}$, then slag-removal temperature $= T_{250}$
If $T_{250} < T_{cv}$, then slag-removal temperature $= T_{cv}$

Method of Sage and McIlroy[22]

The T_{250} value for slags from Eastern U.S. coals can be estimated from the curves in Fig. 13. The method is said to be valid for bituminous-type ash and for lignitic ash with acidic content over 60 percent.

Method of Duzy[23]

The T_{250} value for slags from Western coals can be estimated from Fig. 14. The method is said to be valid for lignitic ash having an acidic content under 60 percent.

TEMPERATURE OF NORMAL SLAG REMOVAL

The temperature of normal slag removal, t_{ns}, is defined as the recommended temperature for easy slag-tapping from a furnace. Usually, it corresponds to a slag viscosity of 200 poise. This concept has been widely used in Russia and is identical, in principle, to the U.S. criterion of the slag removal temperature. For coal-ash slags of known chemical composition, the temperature of normal slag removal can be calculated from the relations described above by letting $\eta = 200$ and obtaining the corresponding T_{200} value. Alternatively, it can be estimated by the following method.

When estimating slag viscosity in the complete liquid phase, the existence of the crystallization temperature should be considered. The relationship between T_{200} and T_{cv}, the temperature of critical viscosity, is as follows:

If $T_{200} > T_{cv}$, then $T_{ns} = T_{200}$
If $T_{200} < T_{cv}$, then $T_{ns} = T_{cv}$

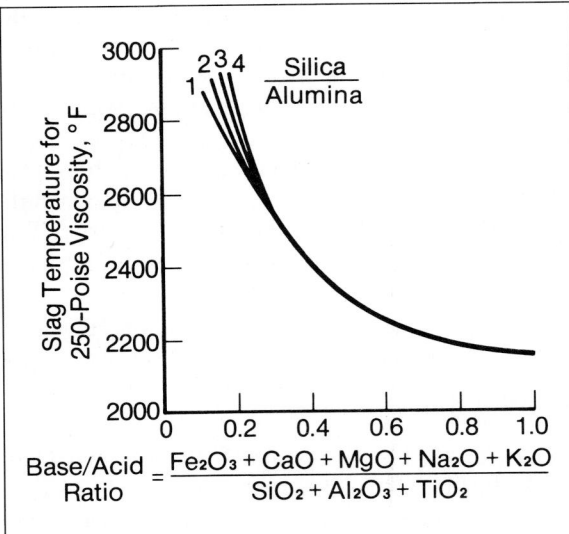

Fig. 13 Estimation of T_{250} value according to base/acid ratio (Ref. 22)

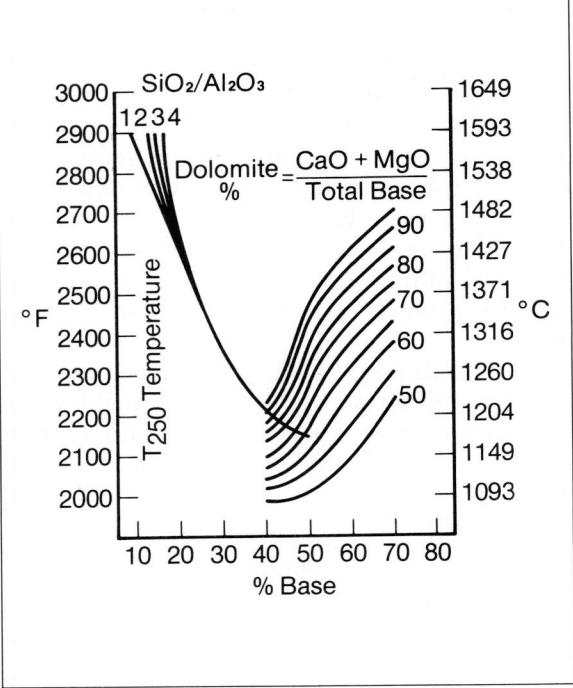

Fig. 14 Estimation of T_{250} value according to dolomite percentage and percent base (Ref. 23)

According to the method developed by Kovitskii, Karagodina, and Martynova,[18] the T_{200} value for coal-ash slags from Russian coals can be estimated by the following procedure:

1. Calculate K_v which is expressed as

$$K_v = \frac{SiO_2 + P_2O_5}{Al_2O_3 + Fe_2O_3 + CaO + MgO}$$

(35)

2. Estimate the T_{200} value from the following equation

$$T_{200}(°C) = 1085 + 314 \, K_v$$

(36)

This relation is claimed to be valid for the following chemical composition range (with K_v between 0.42 and 2.03):

SiO_2	20.9–63.1%
Al_2O_3	14.1–29.0%
Fe_2O_3	3.2–36.3%
CaO	1.2–27.3%
MgO	0.8–7.6%
P_2O_5	0.1–2.9%

EFFECT OF EQUIVALENT Fe₂O₃ AND FERRIC PERCENTAGE

It has been observed that coal-ash slag viscosity in the complete liquid phase can differ between reducing and oxidizing atmospheric conditions. This difference implies the effect of variation in ferric percentage. However, the difference in viscosity is usually small.

Nicholls and Reid[24] initially observed that the entire slag-viscosity profile was raised as the ferric percentage decreased. Sage and McIlroy[22] indicated the opposite; that is, the viscosity profile was lowered as ferric percentage decreased. In either case, the effect of ferric percentage on slag viscosity in the complete liquid phase seems to be small and can usually be neglected. Accordingly, the original Fe_2O_3 content of the ash can be substituted for the equivalent Fe_2O_3 when estimating slag viscosities according to the methods described above.

To obtain results which reflect actual furnace

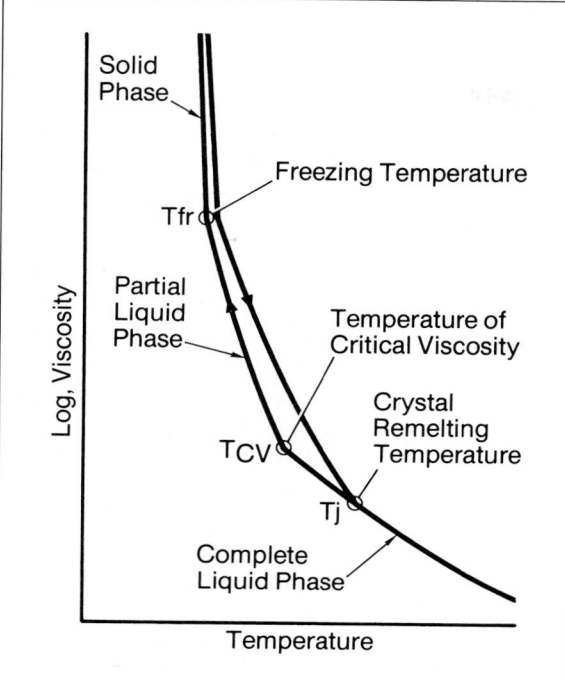

Fig. 15. Effect of cooling and reheating on the viscosity-temperature relation of coal-ash slag

conditions more closely, it is suggested that estimations be carried out at a ferric percentage of 20 percent.

VISCOSITY OF COAL-ASH SLAGS IN THE PARTIAL-LIQUID (CRYSTALLIZATION) PHASE

As coal-ash slags in the complete liquid phase are gradually cooled, the logarithm of viscosity will increase nearly linearly with decreasing temperature. But at a certain temperature, the viscosity departs from this approximate straight line and tends to increase rapidly as temperature decreases. The temperature at which this almost discontinuous transition occurs is referred to as the temperature of critical viscosity, T_{cv}, illustrated in Fig. 15. T_{cv} is believed to be the temperature at which solid phases begin to crystallize and separate out from the liquid and, hence, it can be considered as a crystallization temperature. Actual measurement of T_{cv} is expensive and difficult.

Several methods of estimating T_{cv} from the chemical composition have been developed and are described below.

The temperature of critical viscosity, then, provides the limiting temperature below which slag viscosities cannot be calculated using the methods described earlier. It also allows the prediction of the boundary temperature at which molten slag ceases to flow on furnace waterwall deposits.

Typical cooling curves for four different coal-ash slags are shown in Fig. 16. The explanation of the four curves is as follows:

Curve 1. Represents a true glass which does not have a distinct crystallization temperature. Very few coal-ash slags are of this type.

Curve 2. Represents a slag which is close to glass in behavior. Separation of solids begins after a long cooling period. T_{cv} is usually very low. Slags of this type are frequently referred to as "long" slags.

Curve 3. Represents a slag with a long freezing range. Separation of solids begins at high temperatures at low rates and freezes to essentially a solid at much lower temperatures. T_{cv} is usually high.

Curve 4. Represents a slag with rapid crystallization and freezing characteristics. T_{cv} can be either high or low. Slags of this type are frequently referred to as "short" slags.

METHOD OF SAGE AND McILROY[22]

T_{cv} of coal-ash slags from Eastern U.S. coals at a ferric percentage of 20 percent can be estimated from the following equation:

$$T_{cv} (°F) = HT + 200$$

(37)

where HT is the hemispherical temperature of coal ash (°F).

METHOD OF WATT[21]

T_{cv} of coal-ash slags from British coals can be estimated according to the following procedure:

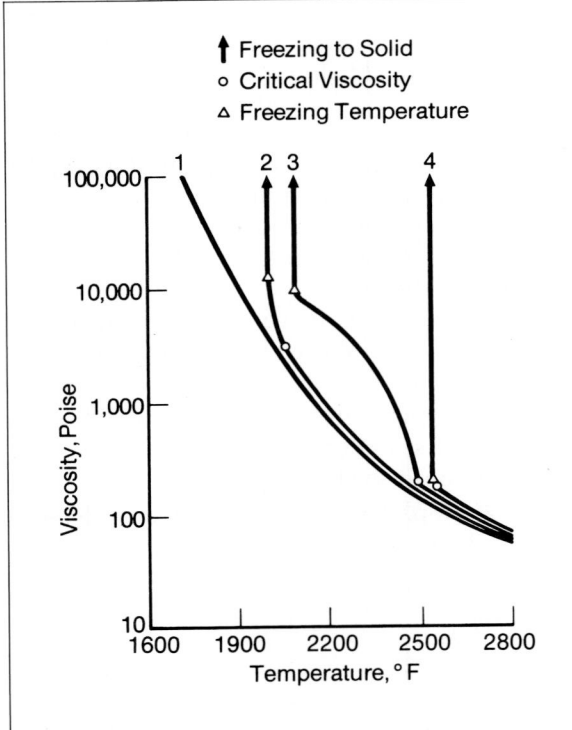

Fig. 16. Typical cooling curves for different types of coal-ash slag

1. Normalize the chemical composition of the ash, such that:

$$SiO_2 + Al_2O_3 + Equiv. Fe_2O_3 + CaO + MgO = 100\%$$
(38)

2. Estimate T_{cv} from the following equation:

$$T_{cv}(°C) = 2990 - 1470 \left(\frac{SiO_2}{Al_2O_3}\right) + 360 \left(\frac{SiO_2}{Al_2O_3}\right)^2$$
$$- 14.7 (Equiv. Fe_2O_3 + CaO + MgO)$$
$$+ 0.15 (Equiv. Fe_2O_3 + CaO + MgO)^2$$
(39)

METHOD OF MARSHAK AND RYZHAKOV FOR SOVIET COALS[25]

The temperature of critical viscosity is referred to as the temperature of the true liquid state, t_o, in the Soviet literature. It has been shown that t_o can be estimated from the following empirical equation:

$$T_o(°C) = 0.75\ t_2 + 480$$
(40)

where t_2 is the softening temperature of coal ash according to the Soviet standard (°C).

VISCOSITY OF COAL-ASH SLAGS IN THE SOLID PHASE

As coal-ash slags are cooled beyond the temperature of critical viscosity, a certain temperature is reached where the movement of the slag is completely terminated. The temperature at which this transition occurs, referred to as the freezing temperature, T_{fr}, is the point of solidification of the slag, as Figs. 15 and 16 illustrate. Only one method of estimation of T_{fr} is available, that by Reid and Cohen[19].

COAL-ASH SLAG DEPOSIT ACCUMULATION AND STRUCTURE

Molten to semimolten coal-ash slag deposits usually will not form on clean waterwall tubes since, upon approaching the relatively cooler tube surface, the slag particles become less adhesive because of rapid cooling in the wall-adjacent area. Accordingly, coal-ash deposition is generally considered to be a two-stage process. A primary layer of deposit first forms on the waterwall tube surface. The resulting rise in the surrounding surface temperatures subsequently allows the adherence of rigid plastic secondary deposits.

MECHANISM OF COAL-ASH SLAG DEPOSITION

The following two types of primary deposits are most commonly observed:

1. Primary deposits that result from the settling of the finer fractions (smaller than 30 microns) of flyash. This type of primary deposit is loose in structure and does not provide strong cohesive or adhesive bonds.

2. Primary deposits that result from the selective deposition of certain reactive components

of the ash (iron, calcium, or alkalies). These components can be present in the deposit in high concentrations as oxides and/or sulfur compounds, leading to the formation of low-melting eutectic mixtures. This type of primary deposit is more dense in structure and has stronger cohesive and adhesive bonds. The properties of the primary layer, which provides the link between the rigid secondary deposits and the tubes, have an appreciable influence on the ease of removal of the deposits by sootblowing.

During the deposition process, there is a transitional stage where the plastic secondary deposits begin to stick on the primary layer. These secondary deposits are strengthened by time and increasing temperatures.

INITIAL SLAGGING TEMPERATURE

The transition from primary to secondary deposits has been shown to be a function of the immediate gas temperature outside the deposit. The temperature at which the secondary deposits begin to form has been referred to as the initial slagging temperature, t_{is}. Methods of estimation of t_{is} are described below.

METHOD BY ALEKHNOVICH, BOGOMOLOV, NOVITSKII, AND IVANOVA[26]

The t_{is} of a coal-ash deposit from Soviet (Kuznetsk) coals can be estimated from the chemical composition of the ash according to the following procedure, using constituents on a weight-fraction (not percentage) basis:

1. Calculate K,

$$K = (Na_2O + K_2O)^2 + 0.048 (CaO + Fe_2O_3)^2 \tag{41}$$

2. Estimate t_{is} from the following equation:

$$t_{is} \ (°C) = 1025 + 3.57 (18 - K) \tag{42}$$

METHOD BY DIK AND SIKHORA[27]

Experimental investigations on deposits of

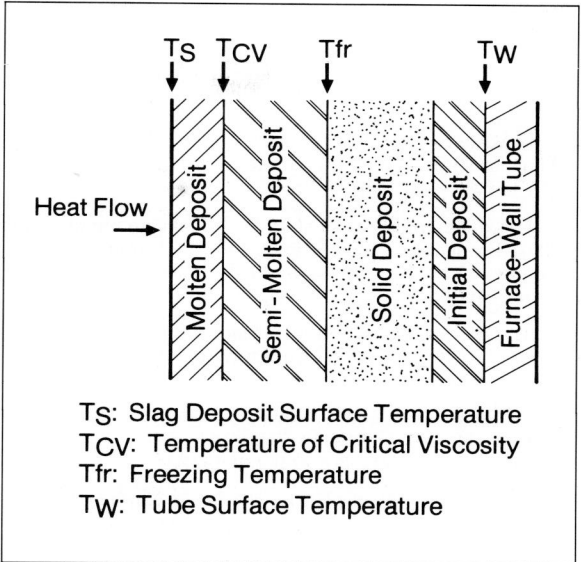

Fig. 17. **Cross section of a typical furnace-wall slag deposit**

coal-ash from Soviet (Kansk-Achinsk) coals showed the following results:

1. t_{is} corresponds to a temperature of approximately 950°C.

2. t_{is} is not strongly affected by the local gas velocity.

DEPOSIT STRUCTURE AND MAXIMUM THICKNESS

Fig. 17 shows a cross section of a typical waterwall slag deposit that has reached a maximum steady-state thickness. The slag surface temperature, T_s, can vary depending on furnace conditions and deposit characteristics. However, in general, T_s should not exceed the flow temperature. From the flow and depositional characteristics of coal-ash slags described in previous sections, it is possible to estimate an approximate structure profile of slag deposits. By knowing the heat flow into the deposit from the furnace flame and the thermal conductivity of the slag, it is also possible to calculate the maximum thickness of a waterwall deposit. This can give an indication of the degree of slagging of different coal types and the effect on heat transfer.

REFERENCES

[1] N. B. Vargaftik and O. N. Oleschuck, "Thermal Capacity of Slags of Various Fuels," *Teploenergetika*, 2(4): 13–17, 1955.

[2] V. G. Voskoboinikov, "The Heat Capacity of Blast-Furnace Slags at High Temperatures," *Teoriya Prabtiba Metallurgii*, 12 (10): 3–5, 1940

[3] E. F. Foerster and P. L. Weston, Jr., "Heat Content of Some Blast-Furnace and Synthetic Slags" *U.S. Bureau of Mines Report of Investigation 6886*, Washington: U.S. Bureau of Mines, 1967.

[4] K. C. Lyon, "Calculation of Surface Tension of Glasses," *Journal of the American Ceramic Society*, 27: 186–189, 1944.

[5] C. W. Parmelee and C. G. Harman, "Effect of Alumina on Surface Tension of Molten Glass," *Journal of the American Ceramic Society*, 20: 224–230, 1937.

[6] N. M. Parikh, "Effect of Atmosphere on Surface Tension of Glass," *J. Amer. Ceramic Society*, 41: 18–22, 1958.

[7] S. G. Agabobov, "Emissivity of Boiler Furnace Slag," *Teploenergetika*, 8, 1958, pp. 56–60.

[8] H. Becker, "The Effect of Ash Fouling on Heat Transfer to Boiler Tubes, Part 2: Thermal Properties of Particulate Brown Coal Ash," Report No. 252, State Electricity Commission of Victoria—Scientific Division, Australia, 1972.

[9] M. F. R. Mulcahy, J. Boow, and P. R. C. Goard, "Fireside Deposits and Their Effect on Heat Transfer in a Pulverized-Fuel-Fired Boiler, Part I: The Radiant Emittance and Effective Thermal Conductance of the Deposits," *Institute of Fuel Journal*, 1966, pp. 385–394.

[10] B. A. Khrustalev, "Spectral Radiation Properties of Some Materials at High Temperatures and Their Influence on the Integral Absorption and Radiation Properties," *Heat Transfer-Soviet Research*, 5, (2) Mar-Apr 1973, pp. 60–64.

B. A. Khrustalev and A. M. Rakov, "Methods of Determining the Integral and Spectral Radiative Properties of Materials at High Temperatures," *Heat Transfer-Soviet Research*, 1, (4), Jul 1969, pp. 163–177.

[11] R. A. Smith and L. R. Glicksman, "Radiation Properties of Slag," Report No. DSR-70332-61, MIT, Feb 1969.

[12] G. J. Goetz, et al., "Development of a Method for Determining Emissivities and Absorptives of Coal-Ash Deposits," *ASME Paper 78-WA/FU-6*. New York: American Society of Mechanical Engineers, 1978; also Combustion Engineering publication TIS-5890.

[13] B. K. Dutta, B. Rai, and K. R. Chakravorty, "Fusion Characterization of Coal Ashes; Softening Temperature of Coal Ashes in Oxidizing Atmosphere," *Journal of Scientific and Industrial Research*, 19B: 206–212, 1960.

B. K. Dutta, et al, "Study of the Fusion Characteristics of Coal Ashes: Softening Temperature of Coal Ashes in Mild Reducing Atmosphere," *Journal of Scientific and Industrial Research*, 21D (2): 44–48, February 1962.

[14] A. W. Gauger, *Progress During the First Twelve Years of Cooperative Research on Bituminous Coals*, Report from Mineral Industries Exp. Station, Pennsylvania State College, 1951.

[15] R. L. Schaefer, *The Relation Between the Chemical Constituents and the Fusibility of Coal Ash*, Ohio State Univ., 1933.

[16] T. G. Estep, H. Seltz, H. L. Bunker, Jr., and H. S. Strickler, "The Effect of Mixing Coals on the Ash Fusion Temperature of the Mixture," *Mining and Metallurgical Investigations*, Carnegie Inst. of Technology, Mining and Metallurgical Advisory Boards, Coop. Bulletin 62, 1934.

[17] T. G. Estep, H. Seltz, and W. J. Osborn, "Determination of the Effect of Oxides of Sodium, Calcium, and Magnesium on the Ash Fusion Temperature by the Use of Synthetic Coal Ash," *Mining and Metallurgical Investigations*, Carnegie Institute of Technology, Mining and Metallurgical Advisory Boards, Cooperative Bulletin 74, 1937.

[18] N. V. Kovitskii, N. V. Karagodina, and M. I. Martynova, "Investigation of the Influence of the Chemical Composition of the Ash of Power Coals in Its Fusibility and Viscosity," *Solid Fuel Chemistry* (Translation of *Khimiya Tverdogo Topliva*), 9(3): 59–63, 1975.

[19] W. T. Reid and P. Cohen, "The Flow Characteristics of Coal-Ash Slags in the Solidification Range," *Transactions of the ASME*, 66, Furnace Performance Factors: 83–97, 1944.

[20] H. R. Hoy, A. G. Roberts, and D. M. Wilkins, "Behavior of Mineral Matter in Slagging Gasification Processes," *Institution of Gas Engineers, Journal*, 5(6): 444–469, 1965.

[21] J. D. Watt and F. Fereday, "The Flow Properties of Slags Formed from the Ashes of British Coals: Part I. Viscosity of Homogeneous Liquid Slags in Relation to Slag Composition," *Journal of the Institute of Fuel*, 42(338): 99–103, March 1969.

[22] W. L. Sage and J. B. McIlroy, "Relationship of Coal Ash Viscosity to Chemical Composition," *Combustion*, 31(5): 41–48, November 1959.

[23] A.F. Duzy, "Fusibility-Viscosity of Lignite-Type Ash," *ASME Paper no. 65-WA/FU-7*. New York: American Society of Mechanical Engineers, 1965.

[24] P. Nicholls and W. T. Reid, "Viscosity of Coal-Ash Slags," *Transactions of the ASME*, 63: 141–153, 1940.

[25] Y.L. Marshak and A. V. Ryzhakov, "Flow of Slag Film When Affected by Gas Flow," *Thermal Engineering* (Translation of *Teploenergetika*), 16(10): 12–16, 1969.

[26] A. N. Alekhnovich, V. V. Bogomolov, N. V. Novitskii, and N.I. Ivanova, "A Study of the Slagging Properties of Kuznetsk Coal Ash," *Soviet Power Engineering* (Translation of *Elektricheskie Stantsii*), 6: 509–513, September 1977.

[27] E. P. Dik, and R. A. Sikhora, "Study of the Conditions of Formation of Ash Deposits When Burning Nazarova Coal," *Thermal Engineering* (Translation of *Teploenergetika*), 16(10): 23–27, 1969.

Appendix C. Combustion Calculations by Graphical Methods

Besides presenting a unique method of accomplishing detailed calculations of combustion products of many fuels, this appendix supplements information from Chapter 2 on the production, transportation, and physical condition of these fuels as they are received in a steam-generating facility. Many of these factors (particularly temperature and water content) influence the calculation of air and gas weights, as well as the boiler efficiency itself. For the reader's convenience, some of the fuel-analysis data of Chapter 2 appear again in this appendix. The order of presentation parallels that of Chapter 2, "Fossil Fuels": first, solid fuels; then, liquid fuels; finally, gaseous fuels.

THE GENERAL METHOD FOR ALL FUELS

Combustion calculations by the designer are one of the first steps in proportioning a steam boiler. For example, the gas weight that results from fuel combustion is required to determine proper arrangement and extent of heating surface in furnace, boiler, economizer, and air heater. The gas weight is also necessary for proportioning gas ducts, dust-collection equipment, induced-draft fans, and stacks. Air weight is used to set the size of air heaters, air ducts, fuel-burning equipment, and forced-draft fans.

This appendix will discuss the following items in detail and will present graphical calculation methods for each.

1. Fuel in products
2. Atmospheric air for combustion
3. Effect of unburned combustible
4. Products of combustion
5. Moisture in the combustion air
6. Moisture from fuel
7. Dry gas content of the combustion products
8. Carbon dioxide in products

In the method outlined here, the first four items are needed for the calculation of gas and air quantities. Items five to seven form the basis of heat-balance calculations in either the design or the testing of a steam-generating unit, when the *ASME Performance Test Codes*, Short Form, is used. The last item, carbon dioxide, and its relationship to excess air, is important in combustion calculations because in test work the CO_2 in the gases is measured and from it the excess air is calculated. The equipment designer works with excess air even though he may appear to have based his estimates on the percentage of CO_2.

The charts make allowance for all the important variables in the analysis. For practical purposes, therefore, they are as accurate as more laborious methods of calculation. *On the other hand, even in working with a complete ultimate analysis and calculating the combustion requirements and products of combustion of each constituent, gross errors may result if, say, the heat value used is not correct for and correspondent to the given analysis.*

The method employed here is based on the concept that the weight of air required in the combustion of a unit weight of any commercial fuel is more nearly proportional to the unit heat value than to the unit weight of that fuel. Consequently, the weights of air, dry gas, moisture, wet gas, etc., are expressed in pounds per million Btu fired. In the case of solid fuels, it is difficult to burn 100 percent of the combustible, so a correction for solid combustible loss must be made.

Before considering in detail the terms found on the charts and in the sample calculations, a quick review of the combustion process of a fuel is necessary. Take, for example, a fuel with no ash; when this fuel burns completely, the weight of the fuel is simply added to the weight of atmospheric air supplied for its combustion to obtain the wet products of combustion or total wet gas. Thus: fuel (F) + air (A) = products (P). But, if some of the fuel is ash, or, if because of incomplete combustion, some of the fuel does not pass out of the furnace with the gases, then F in the above equation will be less and both A and P will also be reduced. This concept of adding F and A to get P is the basis of the method.

FUEL IN PRODUCTS, F

By definition, fuel in products, F, is that portion of the fuel fired which reappears in gaseous form in the products of combustion, as separate elements or in chemical combination with other elements. Since with the method of this appendix all quantities are to be those required for or resulting from firing 1,000,000 Btu, F must also be calculated on that basis. If a fuel contains no ash and if, in addition, it deposits no carbon in the furnace or on other heating surfaces, F is simply obtained by dividing 1,000,000 by the "as-fired" heat value of the fuel in Btu per lb. For the other cases where ash and/or solid combustible loss must be considered, Eq. 1 may be used.

$$F = \frac{10^4(100 - \% \text{ ash} - \% \text{ solid combustible loss})}{\text{fuel heat value}}$$

where:

F = lb per million Btu fired

% ash = percent by weight in fuel as fired

% solid combustible loss = percent by weight in fuel as fired

fuel heat value (HHV) = high heat value as fired, Btu/lb

(1)

Fig. 1 is a graphical solution of Eq. 1, and applies to all fuels.

In the United States and many other countries of the world, the high heat value is the accepted standard and is obtained by calorimetric analysis of the fuel in a laboratory. Producers often sell fuel on the basis of its heat value and users generally check it periodically. In any event, manufacturers of steam-generating equipment need to know it; and, if it is not furnished, they must make an independent test or calculate it from constituents. For solid and liquid fuels, empirical formulas must be used, but for gaseous fuels the sum of the heat value of the various combustible constituents may be employed. Considerable difference may exist between the analysis "as received" and "as fired," even though nothing is intentionally done to the fuel between the time it is received and the time it is burned. Contrarily, certain fuels are purposely dried before they are fired and others take on moisture hygroscopically or are purposely "tempered." In the storage system of pulverized-coal firing, the moisture removed from the coal may actually be vented directly to a stack. In the direct-fired system, the moisture, although removed from the coal, is fired with the coal and therefore does not truly alter the combustion quantities.

Similarly, the heat value and analysis of gaseous fuels are generally reported on the volumetric basis; therefore, the temperature and the pressure conditions of the fuel are an essential part of the analysis. Standard conditions are commonly stated as 60°F and 30 in. Hg.

ATMOSPHERIC AIR, A

All combustion requires oxygen which in commercial practice must be supplied from the atmosphere. The theoretical quantity of oxygen may be calculated accurately from the fuel analysis and simple reaction formulas, such as

$$C + O_2 = CO_2$$

(2)

The corresponding weight or volume of dry air required to supply this oxygen may then be calculated, knowing that air contains approximately 23.1 percent oxygen by weight (20.9 percent by volume).

Natural and manufactured fuels, however, contain varying proportions of several different combustible constituents, such as C, H_2, S, and various hydrocarbons. Some also contain inert constituents, such as ash, N_2, H_2O, CO_2, as well as oxygen in varying quantities. Obviously, determination of air for combustion then becomes a time-consuming task for which shorter methods are desirable if accuracy can, at the same time, be maintained.

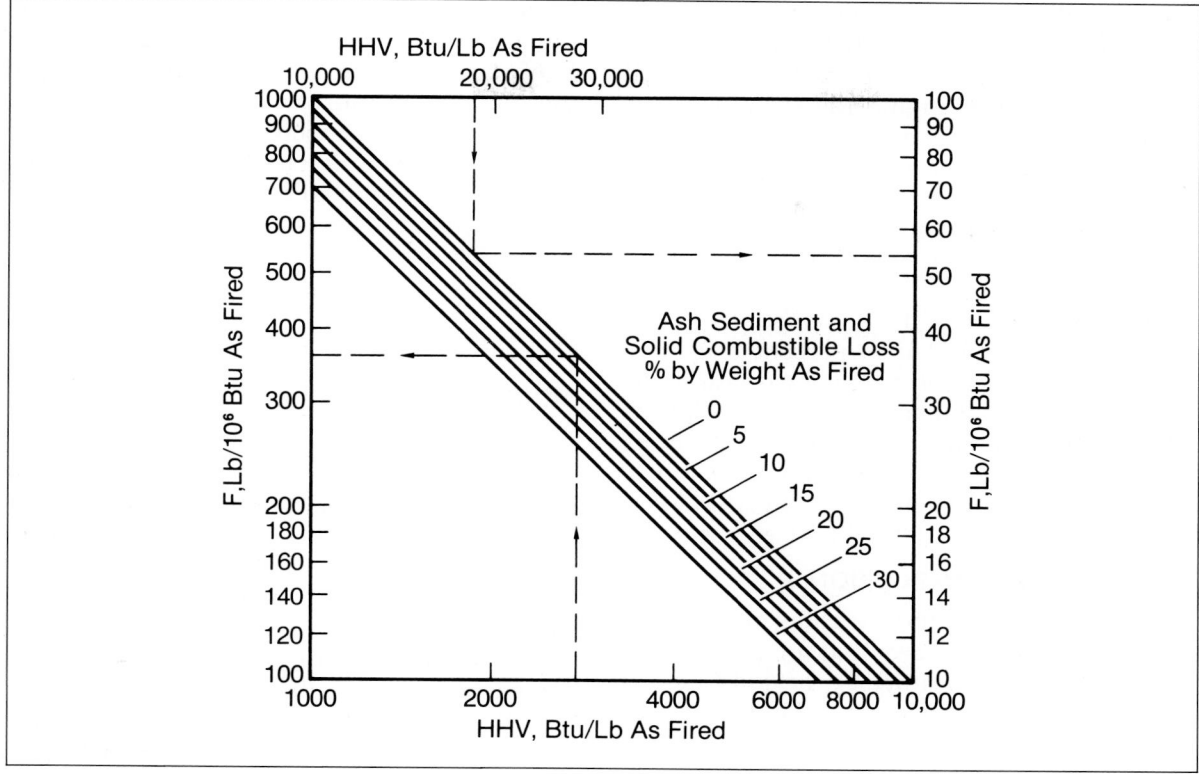

Fig. 1. **Weight of fuel in products of combustion**

Throughout this appendix, the curves labeled "A" give the relation of air quantity to percent excess air calculated for the several different fuels. These curves A are based on "atmospheric air," which means that an arbitrary amount of moisture has been added to the air. The American Boiler Manufacturers Association suggests basing all proposals on 60-percent relative humidity at 80°F, which is equivalent to 0.013 lbs of water vapor per lb of dry air. (To neglect this would introduce an error of approximately 1 percent.)

EFFECT OF UNBURNED COMBUSTIBLE

With a well-designed burner and furnace, it is commercially possible to burn liquid and gaseous fuels completely; no allowance is ordinarily made for loss of combustible in gas and air weight calculations. In the combustion of solid fuels, even in pulverized form, it is not commercially feasible to burn all the available combustible material in the fuel.

The method used to determine the solid combustible loss consists of collecting flyash and refuse from various hoppers. From a dry sample burned in a muffle furnace, weight loss due to combustion is measured. This loss can be expressed as percent by weight of fuel as fired.

In the combustion of a fuel having pure carbon as the only combustible constituent, the air required may be accurately obtained by multiplying a Curve A reading by the factor

$$C = 1 - \frac{\% \text{ solid combustible weight loss}}{100}$$

(3)

because, in this case, air for combustion is directly proportional to carbon in the fuel. If, however, all heat in the fuel does not come from carbon alone (so that the air is not strictly proportional to carbon burned), the factor C will not be exact. For high-carbon, low-volatile fuels it will be nearly exact and will result in only a small error even for fuels low in fixed carbon and high in hydrogen. The error involved by using Eq. 3 in all cases is quite within the limits of accuracy of all other combustion calculations.

For a heat balance the combustible weight loss must be converted to percent heat loss. This can be done conveniently by dividing the percent solid combustible weight loss by the heat value of the fuel as fired and multiplying by 14,500, which is the heat value for combustible in the refuse recommended by *ASME Performance Test Codes*, PTC 4.1.

PRODUCTS OF COMBUSTION, P

Having calculated the foregoing quantities F and A, the gaseous products of combustion may readily be determined by the addition of F and A, as previously corrected. Thus:

$$P = F + CA$$

where:

  P = total gaseous products of combustion, lb/10^6 Btu fired

  F = fuel fired exclusive of ash or solid carbon loss, lb/10^6 Btu fired

  A = atmospheric air required, lb/10^6 Btu fired

  C = combustible loss correction factor

$$(4)$$

MOISTURE IN AIR, W_a

Since the *ASME Performance Test Codes* requires that heat loss due to moisture in air be reported as a separate item in the test heat balance of a steam-generating unit, engineers customarily include it in a predicted heat balance. In an actual test, the moisture in air can be determined from wet- and dry-bulb temperature readings, but for a predicted heat balance it must be assumed. It may vary from day to day even in the same locality, but the American Boiler Manufacturers Association has sug-

gested an arbitrary value for use in preparing proposals. As mentioned, this is taken as 0.013 lb of water per lb of dry air and is included in the atmospheric air, A, as read from the air-weight curve. When required as a separate item for heat-balance calculations, an assumed value from the following equation will be sufficiently accurate.

$$W_a = 0.013A$$

$$(5)$$

MOISTURE FROM FUEL, W_f

This is another item which is separately reported both in an *ASME Performance Test Codes* heat balance and in a predicted heat balance. In the case of some fuels (such as natural and refinery gases), the heat loss due to this moisture may be the largest single item in the heat balance. W_f includes the combined surface and inherent moisture, W_c, from a fuel plus the moisture formed by the combustion of hydrogen, W_h. W_c will vary from zero or a mere trace in fuel oil to over 115 lb per million Btu fired in the case of green wood; W_h will vary from zero or a trace in lamp-black to 100 lb per million Btu fired in the case of some refinery gases.

Special charts, or groups of curves, are used to calculate W_f for each fuel.

DRY GAS, P_d

The need to calculate the "dry gas" or "dry products," in addition to the total products, is due to the *ASME Performance Test Codes*. Dry gas loss is a separate item of the heat balance.

The dry gas may be determined by subtracting the water vapor from the total products, thus

$$P_d = P - (W_a + W_f)$$

where:

  P_d = dry gas, lb/10^6 Btu fired

  P = total products of combustion, lb/10^6 Btu

  W_a = moisture in air, lb/10^6 Btu fired

  W_f = moisture from fuel, lb/10^6 Btu fired

$$(6)$$

CARBON DIOXIDE IN PRODUCTS, CO₂

Operators can use the CO_2 in the gases leaving a furnace or boiler as a guide in adjusting the air supplied, so as not to use too much excess air and in this way decrease the boiler efficiency. Conversely, if the CO_2 is maintained too high, so that the excess air is too low, there will be incomplete combustion of volatile fuel and a higher than normal loss in unburned fixed carbon.

The Orsat apparatus and all other chemical, mechanical, or electrical means for analyzing flue gas, measure the constituents of the gases and not the excess air. However, it is desirable that the excess air be known, because in many calculations the direct use of excess air rather than CO_2 greatly simplifies the work.

RELATION BETWEEN CO₂ AND EXCESS AIR

It is, therefore, convenient to know the relationship between CO_2 and excess air. For bituminous coals, the relationship may be quite accurately expressed in graphical form by means of two curves, one for high-volatile and one for low-volatile fuel. Since in other fuels as oil and gas, there is greater variation in the ratio of carbon to hydrogen, a whole family of curves may be required to cover the full range of analyses. As illustrated by these curves, the wide variation in excess air for a given CO_2, proves that it is unwise to think of CO_2 as synonymous with excess air except when dealing with one particular fuel analysis for which the CO_2-versus-excess-air relationship is known.

The relationship between CO_2 and excess air for a fuel of known analysis can be calculated by methods given in Chapter 4. Some engineers prefer to make calculations on a weight basis even for gaseous fuels, while others prefer the volume or mol system for all fuels.

COAL

Chapter 2 and Appendix A give the properties of the many types of coal found throughout the world. Particularly relevant to the calculation of the combustion products of coal are the methods of laboratory analysis described in Chapter 2, and found in the ASTM Standards referenced there.

PROXIMATE ANALYSIS

This quick and simple method is used to determine the moisture, volatile matter, fixed carbon, and ash of a given coal sample, by weight. The moisture content is obtained by noting the loss of weight in the sample when it is heated to a temperature of about 220°F for an hour. The moisture determined includes surface as well as inherent or hygroscopic moisture. The volatile matter is the loss in weight which follows additional heating of the sample to a temperature of about 1750°F for a half-hour, with all air excluded. The fixed carbon is calculated from the further reduction in weight of the sample which takes place when the coal is completely burned in the presence of air, while the ash is the residue at the end of complete combustion.

Since this laboratory method is approximate in nature, it is well to remember that combustion calculations which are based on it may be considerably in error. For example, it is necessary to know the hydrogen in the coal to evaluate the moisture loss in a heat balance. Fig. 2 shows two curves, taken from DeBaufre's original correlation,[1] which relate the percentage of volatile matter in the proximate

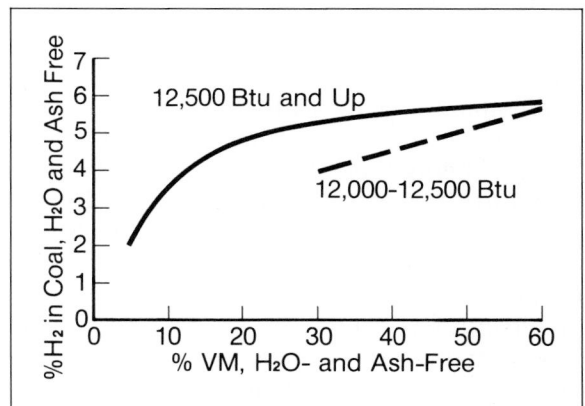

Fig. 2. Determination of hydrogen, moisture and ash free (W. L. De Baufre, "Composition in Heating Value of Fuels," *Combustion,* 2(11): 29–35, May 1931)

analysis to the percentage of hydrogen in the fuel. Actual values of hydrogen for certain coals will differ from these average curves by as much as 1.7 percent. This represents a variation of 12 lbs per million Btu in the water formed by combustion of the hydrogen, which means that the moisture loss estimated from a proximate analysis may be 1.5 percent greater or smaller than it should be, if such loss is based on 500°F. This discrepancy alone may, then, be sufficient to absorb an entire manufacturer's margin and unaccounted-for loss, commonly 1.5 percent.

The character of the volatile matter is not the same for all coals, because some contain more carbon dioxide than others. This fact accounts for the two distinct groups of points occurring on Fig. 3, which is a correlation of the percent CO_2 in the dry products of combustion with the volatile matter in the proximate analysis. Thus, lignites and subbituminous coals form, on the average, about 1 percent more CO_2 than bituminous coals of the same volatile content. The kind of hydrocarbon in the volatile also has a direct bearing on the position of points in Fig. 3, since different hydrocarbons generate different amounts of carbon dioxide. Plotting percent CO_2 against the ratio of fixed carbon to volatile matter does not help to line up the points.

ULTIMATE ANALYSIS

From the foregoing discussion, it is obvious that for accurate combustion calculations it is preferable to have an ultimate analysis of the coal as discussed in Chapter 2. Sometimes, instead of being reported as a separate item, the moisture in the ultimate analysis is divided into 8 parts oxygen and 1 part hydrogen and then added to the oxygen and hydrogen of the combustible in the coal. For combustion calculations, this is not a good practice, because in computing a heat balance, the surface and inherent moisture in the fuel require different treatment from the water formed when hydrogen is burned.

Since an ultimate analysis includes the percent hydrogen, the amount of water formed by

Fig. 3. Variation of theoretical CO_2 with volatile matter (VM)

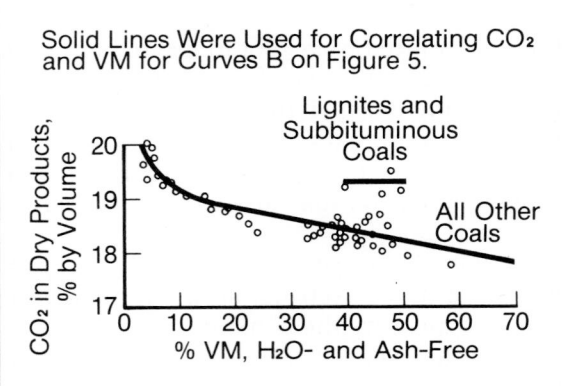

Fig. 4. Variation of theoretical CO_2 with $\left[\dfrac{C}{H_2 - 0.1\,O_2}\right]$

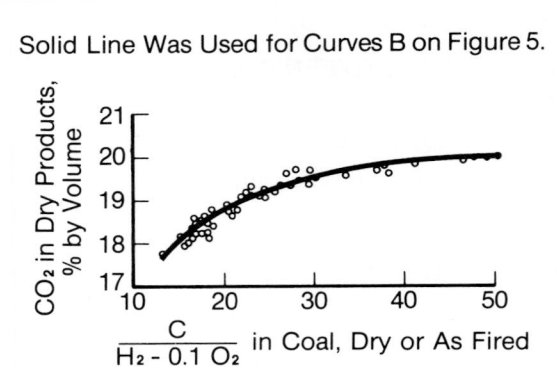

burning it can be accurately calculated. Furthermore, by knowing the ultimate analysis, it is possible to obtain a better correlation between the percent CO_2 in the dry products of combustion and the ratio

$$\left[\frac{C}{H_2 - 0.1\,O_2}\right]$$

for all coals, including lignite and subbituminous. This is shown by Fig. 4 where no CO_2 value plotted deviates from the solid line used in labeling Curves B of Fig. 5 by more than 0.30 percent. These are the principal reasons for preference of the ultimate over the proximate analysis.

Fig. 6 shows the variation of the atmospheric air at zero excess air with volatile matter on a

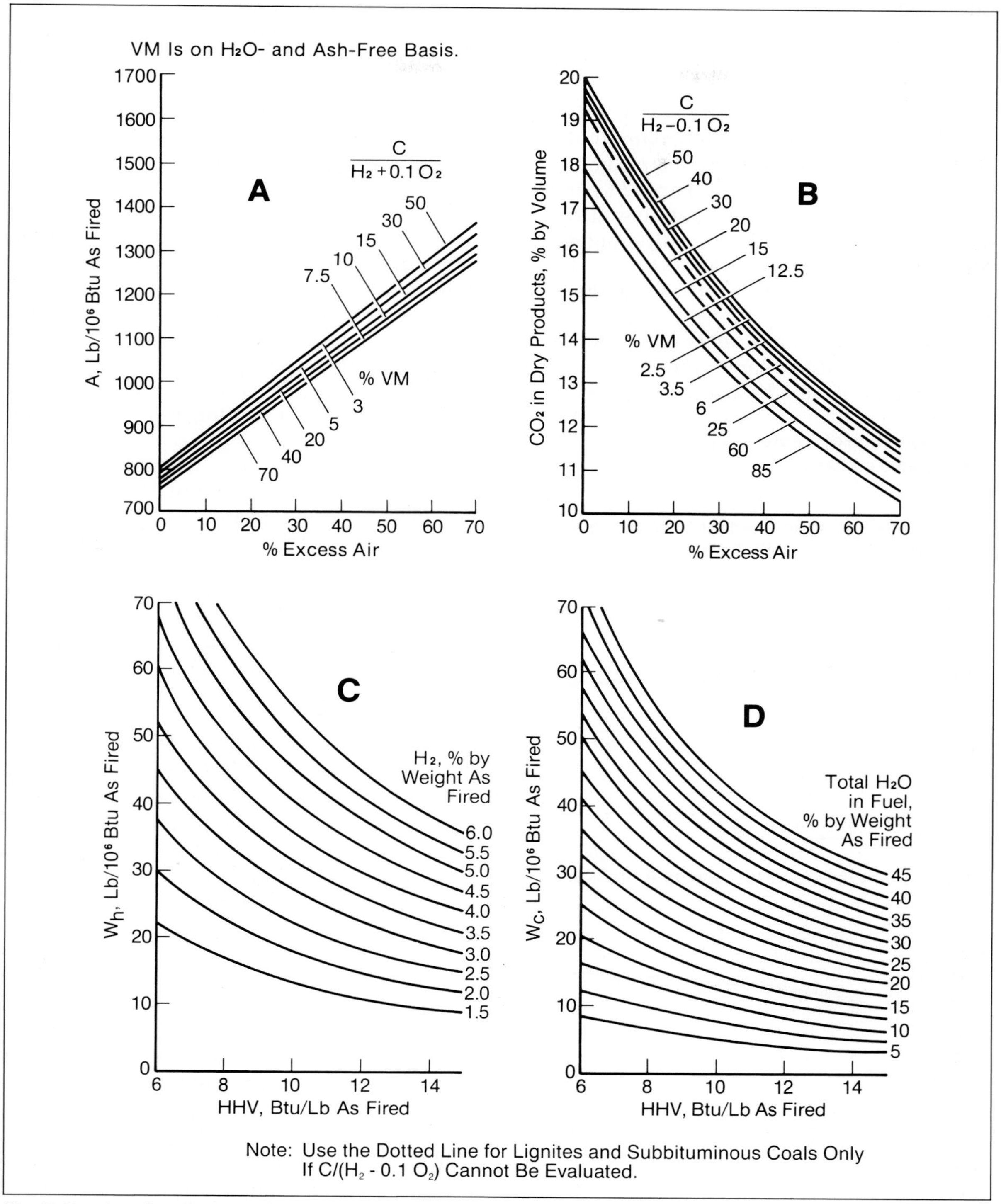

VM Is on H₂O- and Ash-Free Basis.

A

$$\frac{C}{H_2 + 0.1 O_2}$$

A, Lb/10⁶ Btu As Fired

50, 30, 15, 10, 7.5

% VM

3, 5, 20, 40, 70

% Excess Air

B

$$\frac{C}{H_2 - 0.1 O_2}$$

CO₂ in Dry Products, % by Volume

50, 40, 30, 20, 15, 12.5

% VM

2.5, 3.5, 6, 25, 60, 85

% Excess Air

C

Wₕ, Lb/10⁶ Btu As Fired

H₂, % by Weight As Fired

6.0, 5.5, 5.0, 4.5, 4.0, 3.5, 3.0, 2.5, 2.0, 1.5

HHV, Btu/Lb As Fired

D

W_c, Lb/10⁶ Btu As Fired

Total H₂O in Fuel, % by Weight As Fired

45, 40, 35, 30, 25, 20, 15, 10, 5

HHV, Btu/Lb As Fired

Note: Use the Dotted Line for Lignites and Subbituminous Coals Only
If C/(H₂ - 0.1 O₂) Cannot Be Evaluated.

Fig. 5. **Combustion characteristics of coal**

moisture and ash-free basis, while Fig. 7 indicates the change in atmospheric air with the ratio

$$\left[\frac{C}{H_2 + 0.1 O_2} \right]$$

Figs. 6 and 7 reveal that the maximum deviation in the calculated weight of air from the respective average curve is less than ±2.5 percent.

Table I tabulates representative ultimate analyses of various ranks from different states. More extended lists are available in the bulletins issued from time to time by the U.S. Bureau of Mines.

For purposes of reporting or correlation, it is often convenient to express proximate and ultimate analyses on a basis other than as fired.

Fig. 6. Variation of theoretical air with volatile matter (VM)

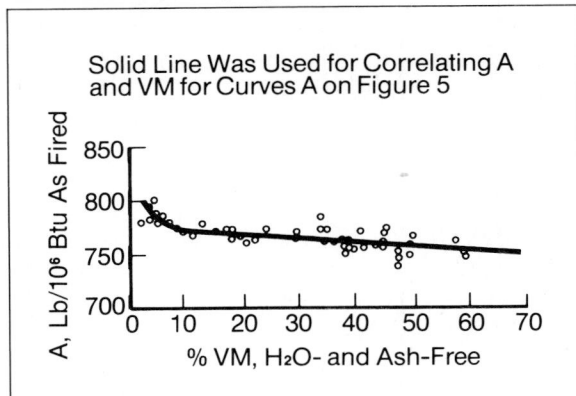

Fig. 7. Variation of theoretical air with $\left[\dfrac{C}{H_2 + 0.1 O_2} \right]$

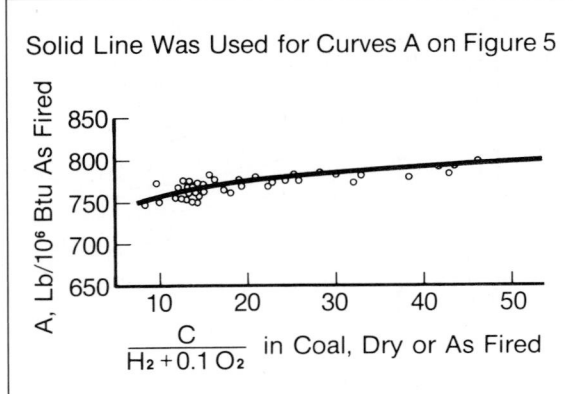

Thus, an analysis on the as-received or as-sampled basis serves to define the condition of the coal at the time it was tested. As *dry*, the fuel is reported with the moisture item omitted; as *ash free*, it is without ash, and as *moisture and ash free*, it has had both moisture and ash eliminated from the anlysis.

HEATING VALUE

It is customary to report the high heating value (HHV) of coal, in Btu per pound, along with the analysis, whether proximate or ultimate, and on the same basis as the analysis. Burning a small sample of coal in a bomb calorimeter immersed in water, and measuring the temperature rise of the water, accurately determine the high heating value at constant volume.

Various formulas have been proposed for evaluating the Btu per pound of coal from its proximate or ultimate analysis. All yield only approximate results because they do not take into account the complex thermodynamic changes occurring when coal is burned. If only the coal rank is known, it is best to assume the heating value from some reasonably close analysis, such as given in Table I.

COMBUSTIBLE LOSS

Combustible loss in coal-fired furnaces results from incomplete burning of either the solid or the volatile combustible in the fuel. How *much* loss is determined by the character of the coal itself, the method of firing it, and the furnace design and operation. With proper furnace design and operation, the loss due to unburned volatile, as determined by the presence of CO in the flue gas, is generally negligible. On the other hand, the loss due to solid combustible in the ash, as measured by the reduction in weight which a sample of ash refuse undergoes on being completely burned, may vary from a negligible amount in burning pulverized coal to 20 percent of the coal fired, or more, with certain sizes of anthracite.

If K_1, K_2, etc., are known or assumed constants representing, respectively, the fractions of the total ash in a pound of coal which is

deposited in the ash hopper, as flyash, etc., W. S. Patterson has shown[2] that the total combustible loss may be expressed by Eq. 7.

$$\% \text{ solid combustible weight loss} = 100\left[\left(\frac{C_{R1}K_1}{1 - C_{R1}} + \frac{C_{R2}K_2}{1 - C_{R2}} \cdots\right) \times \text{ash in coal as fired}\right] \tag{7}$$

where C_{R1}, C_{R2}, etc., are the respective weights of combustible per pound of dry refuse from siftings, ash hopper, flyash, etc. For a given coal burning unit, the sum of all K's is, then, equal to 1.

The results of Eq. 7 may be substituted in Eq. 3 to find the unburned combustible factor, C.

FUEL IN PRODUCTS, F

Knowing the high heating value of the coal, its ash content, and its combustible loss factor C, that portion of the fuel which reappears in the products of combustion, F, is taken directly from Fig. 1.

ATMOSPHERIC AIR, A

For any value of excess air up to 100 percent, the atmospheric air in pounds per million Btu as fired may be taken from Curves A, Fig. 5, after determining either the ratio

$$\left[\frac{C}{H_2 + 0.1O_2}\right]$$

from an ultimate analysis of the coal or the moisture- and ash-free volatile matter from a proximate analysis.

The empirical relation between A and

$$\left[\frac{C}{H_2 + 0.1O_2}\right]$$

is shown by Fig. 7, while Fig. 6 indicates the trend of A with a change in the moisture- and ash-free volatile of the coal.

MOISTURE IN FUEL W_f

As already explained, the loss in weight caused by drying the coal at 220°F constitutes the total moisture reported in an analysis. To this moisture, when converted to pounds per million Btu, the symbol W_c has already been assigned. It is the sum of the surface moisture, which produces "wetness" in the coal, and the inherent or hygroscopic moisture, which being intimately held by the coal is not readily sensed. Table I reveals that, in some high-volatile bituminous coals, the total moisture may be as low as 1.4 percent, while it may increase to 34.8 percent and higher in lignite. The portion of W_c that is inherent moisture, likewise, varies from negligible amounts in super-anthracites to high values in lignites. Curves D of Fig. 5 offer a convenient method of obtaining W_c from the percent moisture and high heating value of the coal as fired.

In addition to W_c, coal will yield water from the burning of its hydrogen content. Curves C, Fig. 5, make it possible to determine this water, W_h, in pounds per million Btu, from the percent hydrogen and high heating value of the coal as fired. If only a proximate analysis of the coal is available, its hydrogen content must first be read from Fig. 2. Special attention is called to the fact that Curves C are based on hydrogen as fired, while the hydrogen from Fig. 2 is on a moisture- and ash-free basis, and must be converted to the as fired condition before entering on Curves C of Fig. 5.

W_f is the sum of W_c and W_h.

PERCENT CO₂ IN PRODUCTS

The percent CO_2 by volume in the dry products of combustion may be read from Curves B, Fig. 5, for excess-air values up to 100 percent of theoretical requirements and for $[C/(H_2 - 0.1 O_2)]$ ratios from 12.5 to 50. The ratio

$$[C/(H_2 - 0.1 O_2)]$$

can, of course, be calculated from an ultimate analysis of the coal. Fig. 6 shows the relation between percent CO_2 at zero excess air and

$$[C/(H_2 - 0.1 O_2)]$$

which was the basis for the CO_2 curves on Fig. 5.

When only a proximate analysis is given, the moisture- and ash-free volatile matter may be

Table I. Analyses of Typical U.S. Coals, as Fired

State	Rank	% Proximate Analysis				% Ultimate Analysis						HHV, Btu/Lb	A at Zero Excess Air, Lb/10⁶ Btu*
		H_2O	VM	FC	ASH	H_2O	C	H_2	S	O_2	N_2		
RI	Ma	13.3	2.5	65.3	18.9	13.3	64.2	0.4	0.3	2.7	0.2	9,313	808
CO	A	2.5	5.7	83.8	8.0	2.5	83.9	2.9	0.7	0.7	1.3	13,720	787
NM	A	2.9	5.5	82.7	8.9	2.9	82.3	2.6	0.8	1.3	1.2	13,340	786
PA **	A	5.4	3.8	77.1	13.7	5.4	76.1	1.8	0.6	1.8	0.6	11,950	791
***	A	2.3	3.1	87.7	6.9	2.3	86.7	1.9	0.5	0.9	0.8	13,540	794
****	A	4.9	3.7	82.2	9.2	4.9	81.6	1.8	0.5	1.3	0.7	12,820	788
AR	Sa	2.1	9.8	78.8	9.3	2.1	80.3	3.4	1.7	1.7	1.5	13,700	770
PA	Sa	3.0	8.4	78.9	9.7	3.0	80.2	3.3	0.7	2.0	1.1	13,450	777
VA	Sa	3.1	10.6	66.7	19.6	3.1	70.5	3.2	0.6	2.2	0.8	11,850	782
AR	Lvb	3.4	16.2	71.8	8.6	3.4	79.6	3.9	1.0	1.8	1.7	13,700	774
MD	Lvb	3.2	18.2	70.4	8.2	3.2	79.0	4.1	1.0	2.9	1.6	13,870	761
OK	Lvb	2.6	16.5	72.2	8.7	2.6	80.1	4.0	1.0	1.9	1.7	13,800	775
WV	Lvb	2.7	17.2	76.1	4.0	2.7	84.7	4.3	0.6	2.2	1.5	14,730	767
PA	Mvb	3.3	20.5	70.0	6.2	3.3	80.7	4.5	1.8	2.4	1.1	14,310	765
VA	Mvb	3.1	21.8	67.9	7.2	3.1	80.1	4.7	1.0	2.4	1.5	14,030	778
AL	Hvab	5.5	30.8	60.9	2.8	5.5	80.3	4.9	0.6	4.2	1.7	14,210	768
CO	Hvab	1.4	32.6	54.3	11.7	1.4	73.4	5.1	0.6	6.5	1.3	13,210	763
KS	Hvab	7.4	31.8	52.4	8.4	7.4	70.7	4.6	2.6	5.0	1.3	12,670	769
KY	Hvab	3.1	35.0	58.9	3.0	3.1	79.2	5.4	0.6	7.2	1.5	14,290	758
MO	Hvab	5.4	32.1	53.5	9.0	5.4	71.6	4.8	3.6	4.2	1.4	12,990	769
NM	Hvab	2.0	33.5	50.6	13.9	2.0	70.6	4.8	1.3	6.2	1.2	12,650	766
OH	Hvab	4.9	36.6	51.2	7.3	4.9	71.9	4.9	2.6	7.0	1.4	12,990	762
OK	Hvab	2.1	35.0	57.0	5.9	2.1	76.7	4.9	0.5	7.9	2.0	13,630	757
PA	Hvab	2.6	30.0	58.3	9.1	2.6	76.6	4.9	1.3	3.9	1.6	13,610	773
TN	Hvab	1.8	35.9	56.1	6.2	1.8	77.7	5.2	1.2	6.0	1.9	13,890	767
TX	Hvab	4.0	48.9	34.9	12.2	4.0	65.5	5.9	2.0	9.1	1.3	12,230	767
UT	Hvab	4.3	37.2	51.8	6.7	43	72.2	5.1	1.1	9.0	1.6	12,990	758
VA	Hvab	2.2	36.0	58.0	3.8	2.2	80.6	5.5	0.7	5.9	1.3	14,510	764
WA	Hvab	4.3	37.7	47.1	10.9	4.3	68.9	5.4	0.5	8.5	1.5	12,610	758
WV	Hvab	2.4	33.0	60.0	4.6	2.4	80.8	5.1	0.7	4.8	1.6	14,350	768
IL	Hvcb	8.0	33.0	50.6	8.4	8.0	68.7	4.5	1.2	7.6	1.6	12,130	766
KY	Hvcb	7.5	37.7	45.3	9.5	7.5	66.9	4.8	3.5	6.4	1.4	12,080	774
MO	Hvcb	10.5	32.0	44.6	12.9	10.5	63.4	4.2	2.5	5.2	1.3	11,300	773
OH	Hvcb	8.2	36.1	48.7	7.0	8.2	68.4	4.7	1.2	9.1	1.4	12,160	762
WY	Hvcb	5.1	40.5	49.8	4.6	5.1	73.0	5.0	0.5	10.6	1.2	12,960	757
IL	Hvbb	12.1	40.2	39.1	8.6	12.1	62.8	4.6	4.3	6.6	1.0	11,480	769
IN	Hvbb	12.4	36.6	42.3	8.7	12.4	63.4	4.3	2.3	7.6	1.3	11,420	758
IA	Hvbb	14.1	35.6	39.3	11.0	14.1	58.5	4.0	4.3	7.2	0.9	10,720	754

Table I. Analyses of Typical U.S. Coals, as Fired — *Continued*

State	Rank	% Proximate Analysis				% Ultimate Analysis						HHV, Btu/Lb	A at Zero Excess Air, Lb/10⁶ Btu*
		H₂O	VM	FC	ASH	H₂O	C	H₂	S	O₂	N₂		
MI	Hvbb	12.4	35.0	47.0	5.6	12.4	65.8	4.5	2.9	7.4	1.4	11,860	762
CO	Sub	19.6	30.5	45.9	4.0	19.6	58.8	3.8	0.3	12.2	1.3	10,130	756
WY	Sub	23.2	33.3	39.7	3.8	23.2	54.6	3.8	0.4	13.2	1.0	9,420	757
ND	Lig A	34.8	28.2	30.8	6.2	34.8	42.4	2.8	0.7	12.4	0.7	7,210	750
TX	Lig A	33.7	29.3	29.7	7.3	33.7	42.5	3.1	0.5	12.1	0.8	7,350	752

*A is the air required for combustion under stoichiometric conditions (no excess air), with 0.013 lb H₂O per lb dry air.

**Orchard Bed, ***Mammoth Bed, ****Holmes Bed, RANK KEY: Ma-Meta-anthracite, A-Anthracite, Sa-Semi-anthracite, Lvb-Low-Vol. Bituminous, Mvb-Med.-Vol. Bituminous, Hvab-High-Vol. Bituminous A, Hvcb-High-Vol. Binuminous B, Hvbb-High-Vol. Bituminous C, Sub-Subbituminous, LigA-Lignite A

used to obtain the CO_2 from Curves B of Fig. 5 for all but subbituminous coals and lignites. Because, as explained before, the volatile in the latter ranks does not possess the characteristics of other coals, the dotted line on Fig. 5 should be used when the ultimate analysis for sub-bituminous and lignitic coals is unavailable.

COKE AND COKE BREEZE

Coke is the fused solid residue left when certain coals, petroleum, or tar pitch are heated in

EXAMPLE

Assume that a Pennsylvania bituminous coal, with the typical proximate analysis of line 25, Table I, is burned with 20-percent excess air and that the expected solid combustible loss is 0.8 percent by weight. Then

1. Fuel, F. The sum of the ash plus solid combustible loss is 9.1 + 0.8 = 9.9 percent by weight. With this sum, and a high heating value from line PA, Table I, of 13,610 Btu per lb, read from Fig. 1, F = 66 lb per million Btu.

2. Atmospheric Air, A. The sum of the ash plus moisture in the coal is 9.1 + 2.6 = 11.7 percent, and the moisture- and ash-free volatile matter =

$$\left[\frac{30}{1 - \frac{11.7}{100}} \right] = 34 \text{ percent}$$

For this value of volatile matter and 20-percent excess air, read from Curves A Fig. 5 A = 920 lb per million Btu.

3. Unburned Combustible Factor, C. Since the solid combustible loss is 0.8 percent, from Eq. 3 obtain

$$C = \frac{1 - \% \text{ solid combustible weight loss}}{100}$$

$$= \left[1 - \frac{0.8}{100} \right] = 0.992.$$

4. Total Products, P. From Eq. 4, P = F + CA = 66 + 0.992 × 920 = 979 lb per million Btu.

5. Moisture in Air, Wₐ. From Eq. 5, W_a = 0.013 × 920 = 12 lb per million Btu.

6. Moisture from Fuel, Wf This is the sum of W_c, the moisture in the coal as fired, and W_h, the water formed in combustion. From Curves D, Fig. 5, for 2.6-percent moisture in the fuel and a high heating value of 13,610 Btu per lb, read W_c = 2 lb per million Btu.

Next, convert the high heating value as fired to a moisture-and-ash-free basis, as follows:

$$\left[\frac{13.610}{1 - \frac{11.7}{100}} \right] = 15,400 \text{ Btu}$$

per lb. and use it, together with the moisture-and ash-free volatile determined for A, to read from Fig. 2, H_2 = 5.4-percent moisture and ash free, or reverting to the as-fired basis, 5.4

$$\left(1 - \frac{11.7}{100} \right) = 4.8\text{-percent hydrogen.}$$

From Curves C, Fig. 5, for a high heating value of 13,610 Btu per lb and H_2 = 4.8 percent, read W_h = 32 lb per million Btu. $W_f = W_c + W_h$ = 2 + 32 = 34 lb per million Btu.

7. Dry Gas, Pd. From Eq. 6, P_d = P − (W_a + W_f) = 979 − (12 + 34) = 933 lb per million Btu.

8. Percent CO₂ in Products. For a moisture-and-ash-free volatile matter of 34 percent and 20-percent excess air, read from Curves B, Fig. 5, CO_2 = 15.3 percent.

(*Note:* Many of the foregoing figures will be somewhat different if the ultimate, instead of the proximate, analysis in line PA, Table I, is used.)

ABB
ASEA BROWN BOVERI

an atmosphere excluding oxygen, so as to expel their volatile content. The process of thus decomposing these fuels into their gaseous and solid fractions is known as destructive distillation or carbonization.

Coke breeze is very small-size coke that is particularly well suited for firing on certain types of stokers. As a rule, it contains a higher percentage of ash than the rest of the coke. Table II presents analysis values of constituents in typical cokes, as fired.

HEATING VALUE

As with coal, the high heating value of coke, in Btu per lb, can be determined by a bomb calorimeter. If not given with the ultimate analysis, it may be calculated approximately from a formula of the Dulong type.

$$HHV = 14,600\,C + 62,000 \left(H_2 - \frac{O_2}{8} \right) + 4050\,S \tag{8}$$

where C, H_2, O_2 and S are weight fractions in the ultimate analysis.

COMBUSTIBLE LOSS

In burning coke or coke breeze in pulverized form or on stokers, a certain amount of the combustible in the fuel is lost as siftings with the flyash, etc., in the manner described in the section on coals. With coke, as with coal, the

$$\% \text{ solid combustible weight loss} =$$
$$100 \left[\left(\frac{C_{R1}K_1}{1 - C_{R1}} + \frac{C_{R2}K_2}{1 - C_{R2}} \cdots \right) \times \text{ash in coke as fired} \right] \tag{9}$$

where C_{R1}, C_{R2}, etc., are the respective weights of combustible per pound of dry refuse from siftings, ash hopper, etc., and K_1, K_2, etc., are constants representing fractions of the ash in a pound of coke found, respectively, in siftings, ash hopper, etc.

FUEL IN PRODUCTS, F

Knowing the high heating value of coke, its ash content, and its combustible loss factor C, that portion of fuel, F, which reappears in the products of combustion, is taken from Fig. 1.

ATMOSPHERIC AIR, A

Analyses of coke are reported on either the proximate or the ultimate basis. The chart, Fig. 8, was prepared so that values of atmospheric

Table II. Analyses of Typical Cokes, as Fired

Kind	% Proximate Analysis				% Ultimate Analysis							HHV, Btu/Lb	A at Zero Excess Air, Lb/10⁶ Btu	CO₂ at Zero Excess Air, %
	H₂O	VM	FC	Ash	H₂O	C	H₂	S	O₂	N₂	Ash			
High-temperature coke	5.0	1.3	83.7	10.0	5.0	82.0	0.5	0.8	0.7	1.0	10.0	12,200	798	20.7
Low-temperature coke	2.8	15.1	72.1	10.0	2.8	74.5	3.2	1.8	6.1	1.6	10.0	12,600	763	19.3
Beehive coke	0.5	1.8	86.0	11.7	0.5	84.4	0.7	1.0	0.5	1.2	11.7	12,527	807	20.5
Byproduct coke	0.8	1.4	87.1	10.7	0.8	85.0	0.7	1.0	0.5	1.3	10.7	12,690	802	20.5
High-temperature coke breeze	12.0	4.2	65.8	18.0	12.0	66.8	1.2	0.6	0.5	0.9	18.0	10,200	805	20.1
Gas Works coke:														
Horizontal retorts	0.8	1.4	88.0	9.8	0.8	86.8	0.6	0.7	0.2	1.1	9.8	12,820	808	20.6
Vertical retorts	1.3	2.5	86.3	9.9	1.3	85.4	1.0	0.7	0.3	1.4	9.9	12,770	809	20.4
Narrow coke ovens	0.7	2.0	85.3	12.0	0.7	84.6	0.5	0.7	0.3	1.2	12.0	12,550	802	20.6

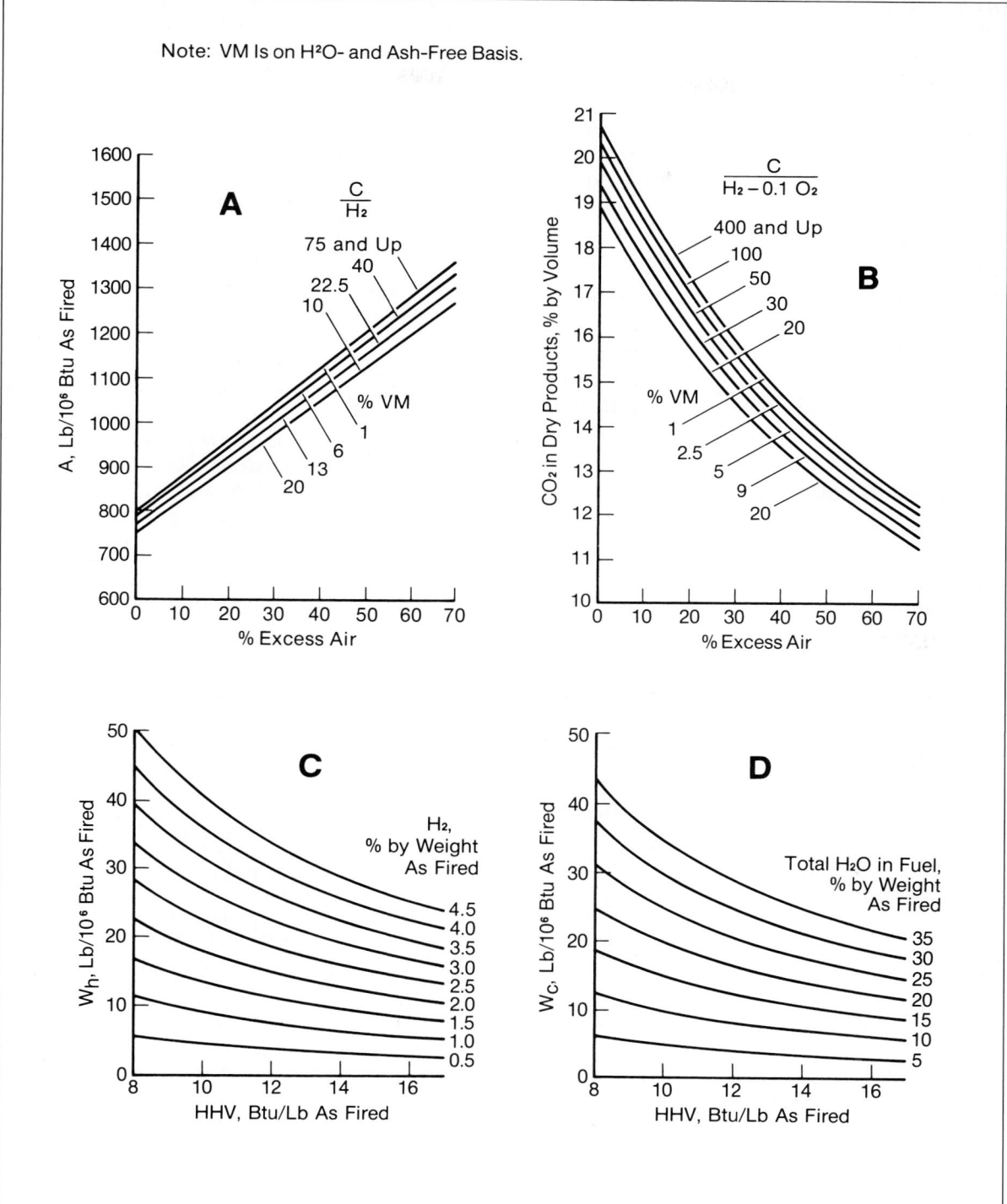

Fig. 8. **Combustion characteristics of coke**

air, A, in pounds per million Btu as fired may be obtained after knowing either the moisture- and ash-free volatile matter in the coke or its C/H_2 ratio.

Fig. 9 shows the empirical relation between A and the volatile matter on a moisture- and ash-free basis and Fig. 10 indicates how A varies with the ratio C/H_2. It will be seen that no calculated point deviates from the solid, average lines on these figures by more than ± 2.0 percent. The solid lines were used in labeling Curves A of Fig. 8.

TOTAL PRODUCTS, P

The unburned combustible factor is obtained from Eq. 3. This factor C is then used in correcting F, the fuel in products, and A, the atmospheric air when determining the total products P from Eq. 4.

MOISTURE IN FUEL, W_f

The moisture reported in a coke analysis, just as for coal, is the loss in its weight when dried at 220°F.

Any moisture present in coke results either from the quenching process or outdoor storage.

W_c, which is this moisture in pounds per million Btu, may, therefore, be a small quantity if the quenching is done rapidly, or may amount to 12 lb or more in cases where storage conditions enabled the coke to absorb moisture, readily because of its porous nature.

Knowing the high heating value and the percent moisture in the coke as fired, W_c may be read from Curves D, Fig. 8.

W_h, the pounds of water per million Btu fired formed by combustion of the hydrogen in coke, may be obtained from Curves C of Fig. 8, provided the hydrogen in the ultimate analysis is given. With only a proximate analysis of coke available, W_h is determined from Fig. 11, after converting the volatile matter to a moisture- and ash-free basis. This curve is based on the graphical correlation shown in Fig. 12, and as such, is of limited accuracy. As in previous sections, W_f is the sum of W_c and W_h.

PERCENT CO_2 IN PRODUCTS

Curves B of Fig. 8 offer a convenient way of determining the percent CO_2 by volume in the dry products of combustion of coke, for any value of excess air from zero to 100

EXAMPLE

Assume that a high-temperature coke breeze, having the typical proximate analysis shown in Table II is burned with 35-percent excess air and that the expected combustible loss is 10 percent by weight. Then

1. Fuel, F. The sum of the ash plus solid combustible loss is $18.0 + 10.0 = 28.0$ percent by weight. With this sum, and a high heating value from Table II of 10,200 Btu per lb, read from Fig. 1, F = 70 lb per million Btu.

2. Atmospheric Air, A. The sum of the ash plus moisture in the coal is $18.0 + 12.0 = 30$ percent, and the moisture- and ash-free volatile matter =

$$\left[\frac{4.2}{1 - (30.0/100)} \right]$$

= 6 percent. For this value of volatile matter and 35-percent excess air, read from Curves A, Fig. 8, A = 1065 lb per million Btu.

3. Unburned Combustible Factor, C. Since the solid combustible loss is 10 percent, from Eq. 3 obtain

$$\frac{1 - \% \ solid \ combustible \ weight \ loss}{100}$$

$$= \left[1 - \frac{10}{100} \right] = 0.90.$$

4. Total Products, P. From Eq. 4, $P = F + CA = 70 + 0.90 \times 1065 = 1028$ lb per million Btu.

5. Moisture in Air, W_a. From Eq. 5, $W_a = 0.013A = 0.013 \times 1065 = 14$ lb per million Btu.

6. Moisture from Fuel, W_f. This is the sum of W_c, the moisture in the coke as fired, and W_h, the water formed in combustion. From Curves D, Fig. 8, for 12 percent moisture in the fuel and a high heating value of 10,200 Btu per lb read $W_c = 12$ lb per million Btu.

Next, with 6 percent moisture- and ash-free volatile determined for A, read from Fig. 11 $W_h = 10$ lb per million Btu. Then

$$W_f = W_c + W_h = 12 + 10 = 22 \ \text{lb per million Btu.}$$

7. Dry Gas, P_d. From Eq. 6, $P_d = P - ((W_a + W_f)) = 1028 - (14 + 22) = 992$ lb per million Btu.

8. Percent CO_2 in Products. For a moisture- and ash-free volatile matter of 6 percent and 35 percent excess air, read from Curves B, Fig. 8, $CO_2 = 14.7$ percent.

(Note: Many of the quantities calculated in the example will be different if the ultimate, instead of the proximate, analysis of coke breeze is used.)

Fig. 9 **Variation of theoretical air with volatile matter (VM) of coke**

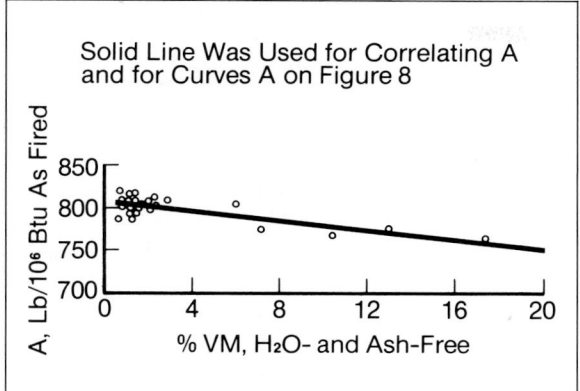

Fig. 11. **Predicted moisture from combustion of coke**

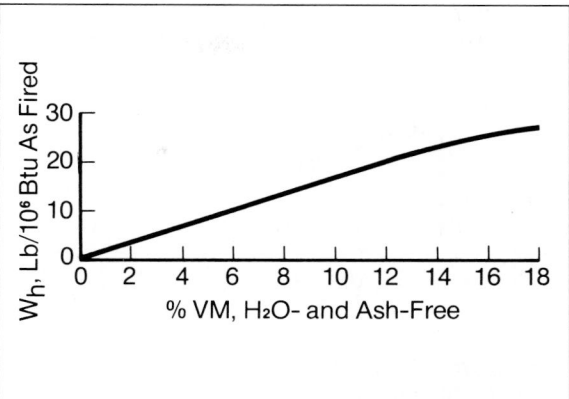

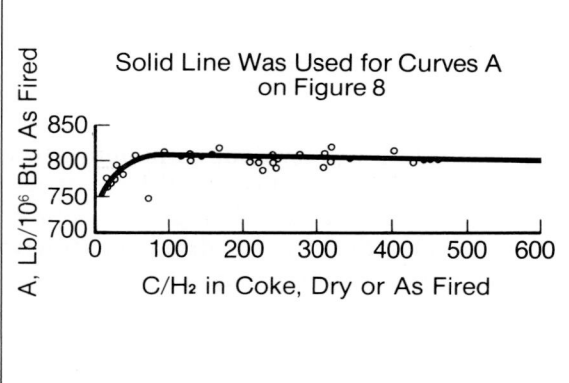

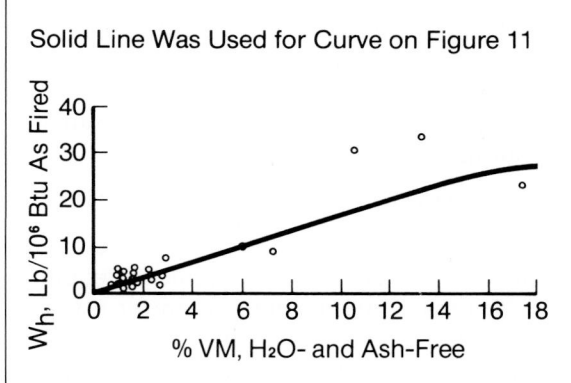

Fig. 10. **Variation of theoretical air with C/H₂**

Fig. 12. **Variation of moisture from combustion of coke with volatile matter (VM)**

percent. As in drawing the lines for A, Curves B are plotted so that CO_2 values may be obtained after calculating the ratio

$$\left[\frac{C}{H_2 - 0.1\,O_2}\right]$$

from the ultimate analysis, or after converting the volatile matter in the proximate analysis as fired to the moisture and ash-free basis.

Comparing Fig. 13, which relates moisture and ash-free volatile matter to the percent CO_2 at zero excess air, with Fig. 14, which shows the variation of CO_2 with the ratio

$$\left[\frac{C}{H_2 - 0.1\,O_2}\right]$$

it is apparent that the latter is a more accurate correlation. Accordingly, in reading Curves B, an ultimate analysis of coke is preferable to a proximate analysis.

WOOD AND BAGASSE

The origin of these two cellulose fuels is described in Chapter 2; typical analyses are shown in Tables III and IV.

Moisture is the most variable single item in the composition of both wood and bagasse, its value ranging from 6 percent in a kiln-dried wood to over 60 percent in green wood or bagasse and up to 80 percent in bark. It is of importance in combustion calculations because it

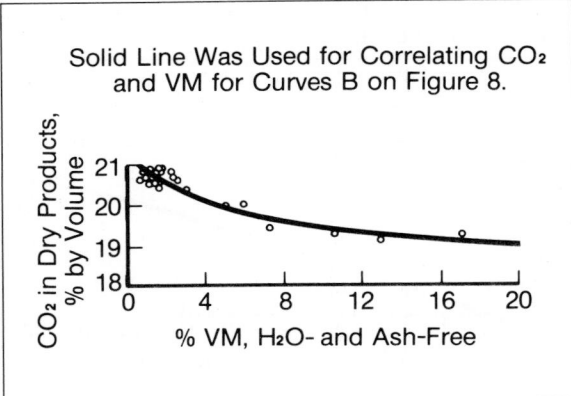

Fig. 13 Variation of theoretical CO_2 with volatile matter (VM) of coke

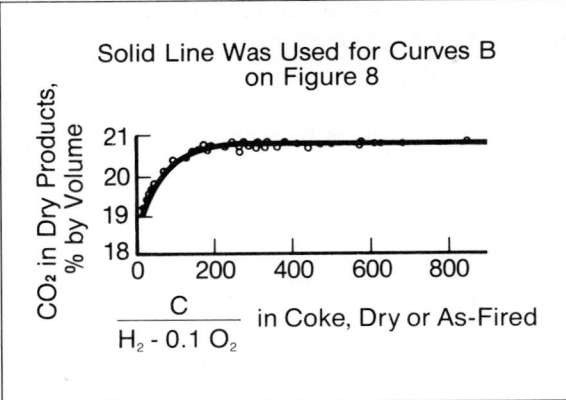

Fig. 14. Variation of theoretical CO_2 with $\left[C/(H_2 - 0.1\ O_2) \right]$

directly influences the heating value of the fuel burned by increasing the proportion of inert matter present. Furthermore, it requires some of the heat released by the dry substance to evaporate and superheat it.

DETERMINATION OF MOISTURE CONTENT

An accurate determination of the moisture content in wood or bagasse is, therefore, necessary. To obtain this, a weighed sample is placed in a steam or electric oven and heated to 212°F until no further loss of weight occurs. The difference between the original weight and that of the dried sample divided by the original weight is the moisture fraction as received. In the lumber industry, it is customary to speak of moisture in wood on an oven-dry basis, whereby the loss of weight in the sample tested is divided by the oven-dried weight rather than the original weight. The percent moisture oven-dry must then be converted to the as-received basis before using it in combustion calculations.

The oven-drying method of determining the moisture content is not accurate with woods such as southern yellow pine, which contain oils that are easily volatilized, because these oils will distill with the moisture. It is then necessary to use the more elaborate distillation method, wherein the water and the oil are measured separately.

Other less reliable, but quicker, ways of judging the moisture in wood consist in measuring its electrical resistance or using a wood hygrometer.

As a rule, kiln-dried woods contain from 6 to 10 percent moisture, although softwoods will show a much higher percentage unless properly handled in the kiln. Air-dried woods have had enough moisture evaporated to be in equilibrium with the surrounding atmosphere. The actual percent present depends on both the temperature and relative humidity of the ambient atmosphere. Fig. 15 shows how the moisture in wood varies with different relative humidities at 70°F. In arid climates, it may be as low as

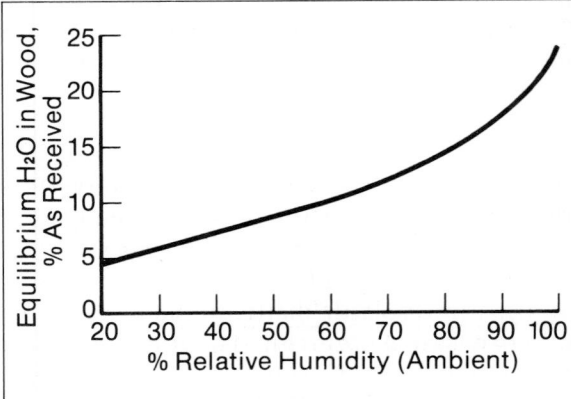

Fig. 15. Moisture content of wood at 70°F ambient temperature

5 percent, whereas in humid regions it may go up to 20 percent or more. Green wood, wood waste, and bagasse contain from 40 to over 60 percent moisture.

HEATING VALUE

The high heating value of wood or bagasse is determined by a bomb calorimeter, as in the case of other solid fuels. From the close similarity of the ultimate analyses of different woods, it might be reasonable to assume that equal weights of dry wood will release the same amount of heat, regardless of species. Actually, as Table III indicates, this is not the case, the heating value of wood depending to some extent on the physical structure and form of the wood tissue, and still more on the presence of resins, gums, tannins, essential oils, or pigments. Resin alone, for instance, may increase the heating value of the same wood as much as 15 percent. Table III shows that the Btu per pound of dry softwoods is a little higher, on the average, than that of hardwoods, due perhaps to the fact that most softwoods are conifers containing resins and oils.

As mentioned, however, the chief factor affecting the heating value of the wood or bagasse as fired into a furnace is its moisture content. In the case of wood refuse, the heating value may also be influenced by the amount of trash and dirt gathered in processing the wood. In harvesting sugar cane, a certain amount of trash,

Table III. Typical Analyses of Dry Wood

	C	H$_2$	S	O$_2$	N$_2$	Ash	HHV, Btu/Lb	A at Zero Excess Air, Lb/10^6 Btu	CO$_2$ at Zero Excess Air, %
Softwoods†									
Cedar, white	48.80	6.37	. . .	44.46	. . .	0.37	8400*	709	20.2
Cypress	54.98	6.54	. . .	38.08	. . .	0.40	9870*	711	19.5
Fir, Douglas	52.3	6.3	. . .	40.5	0.1	0.8	9050	720	19.9
Hemlock, western	50.4	5.8	0.1	41.4	0.1	2.2	8620	706	20.4
Pine, pitch	59.00	7.19	. . .	32.68	. . .	1.13	11320*	702	18.7
white	52.55	6.08	. . .	41.25	. . .	0.12	8900*	723	20.2
yellow	52.60	7.02	. . .	40.07	. . .	0.31	9610*	710	19.3
Redwood	53.5	5.9	. . .	40.3	0.1	0.2	9040*	722	20.2
Hardwoods†									
Ash, white	49.73	6.93	. . .	43.04	. . .	0.30	8920*	709	19.6
Beech	51.64	6.26	. . .	41.45	. . .	0.65	8760*	729	20.0
Birch, white	49.77	6.49	. . .	43.45	. . .	0.29	8650*	712	20.0
Elm	50.35	6.57	. . .	42.34	. . .	0.74	8810*	715	19.8
Hickory	49.67	6.49	. . .	43.11	. . .	0.73	8670*	711	20.0
Maple	50.64	6.02	. . .	41.74	0.25	1.35	8580	719	20.2
Oak, black	48.78	6.09	. . .	44.98	. . .	0.15	8180*	714	20.5
red	49.49	6.62	. . .	43.74	. . .	0.15	8690*	709	19.9
white	50.44	6.59	. . .	42.73	. . .	0.24	8810*	715	19.8
Poplar	51.64	6.26	. . .	41.45	. . .	0.65	8920*	716	20.0

* Calculated from reported high heating value of kiln-dried wood assumed to contain 8-percent moisture.

†The terms "hard" and "soft" wood, contrary to popular conception, have no reference to the actual hardness of the wood. According to the *Wood Handbook*, prepared by the Forest Products Laboratory of the U. S. Department of Agriculture, hardwoods belong to the botanical group of trees that are broad-leaved whereas softwoods belong to the group that have needle or scalelike leaves, such as evergreens: cypress, larch, and tamarack are exceptions.

Table IV. Typical Analyses of Bagasse, Dry

	C	H₂	O₂	N₂	Ash	HHV, Btu/Lb	A at Zero Excess Air, Lb/10⁶ Btu	CO₂ at Zero Excess Air, %
			% by Weight					
Cuba	43.15	6.00	47.95	. . .	2.90	7985	629	21.0
Hawaii	46.20	6.40	45.90	. . .	1.50	8160	687	20.3
Java	46.03	6.56	45.55	0.18	1.68	8681	651	20.1
Mexico	47.30	6.08	45.30	. . .	1.32	8740	646	19.4
Peru	49.00	5.89	43.36	. . .	1.75	8380	700	20.5
Puerto Rico	44.21	6.31	47.72	0.41	1.35	8386	627	20.6

leaves, cane tops, and grass may be included which would result in a lower heating value for the bagasse. As a rule, larger quantities of trash and dirt are picked up when the cane is harvested and loaded by mechanical means than when this is done by hand.

Owing to their complex composition, it is not possible to make use of a Dulong-type formula to predict the heating value of either wood or bagasse. The formula appears to give values of available hydrogen in the factor

$$\left(H_2 - \frac{O_2}{8} \right)$$

which are too low and, as a result, the heating values calculated with it are invariably low.

In sugar-mill practice, bagasse analyses are frequently reported in terms of its fiber, sucrose, glucose, moisture, and ash content. If these are given, the high heating value (HHV) of bagasse in Btu per pound may be computed from

$$HHV = 8550(F) + 7119(S) + 6750(G) - 972(M)$$

where
  F = fiber, lb/lb of bagasse
  S = sucrose, lb/lb of bagasse
  G = glucose, lb/lb of bagasse
  M = moisture, lb/lb of bagasse

(10)

FUEL IN PRODUCTS, F

The combustible loss in burning wood or bagasse is generally assumed to be zero. Therefore, if the high heating value and the ash content of these fuels are known, the portion, F, of the fuel that reappears in the products of combustion may be taken directly from Fig. 1.

ATMOSPHERIC AIR, A

Curves A of Fig. 18 offer a convenient way of obtaining the "atmospheric air," A, in pounds per million Btu fired, for any excess-air value up to 100 percent after calculating the ratio

$$\left[\frac{C}{H_2 + 0.1 O_2} \right]$$

from the ultimate analysis of wood or bagasse. These lines are based on the correlation between this ratio and the theoretical air required shown by Fig. 16.

PERCENT CO₂ IN PRODUCTS

The percent CO_2 by volume in the dry products of combustion may be determined from Curves B of Fig. 18 for any excess air from zero to 100 percent of theoretical requirements. Curves B are based on the solid line of Fig. 17, which shows the change in the ultimate CO_2 with the ratio

$$\left[\frac{C}{H_2 - 0.1 O_2} \right]$$

The correlation of Fig. 17 shows that no value deviates from the solid line by more than ±0.2 percent CO_2. This is the accuracy that may be expected in reading Curves B.

MOISTURE IN FUEL, Wf

The total moisture in the products of combustion of either wood or bagasse has been de-

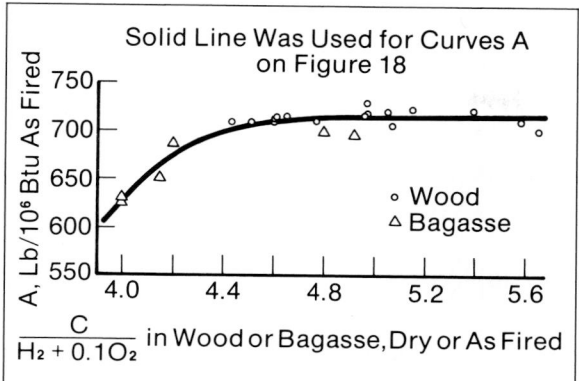

Fig. 16. Moisture of theoretical air

with $\left[C/(H_2 + 0.1\ O_2) \right]$

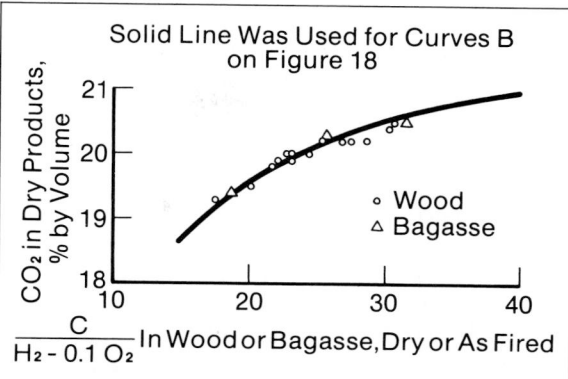

Fig. 17. Variation of theoretical CO_2

with $\left[C/(H_2 - 0.1\ O_2) \right]$

noted by the symbol W_f, pounds per million Btu. It is the sum of W_c, the moisture originally in the fuel and W_h, the water formed from the combustion of hydrogen.

When the ultimate analysis of wood or bagasse as fired is known, as well as the high heating value, W_c may be read directly from Curves D and W_h from Curves C of Fig. 18.

EXAMPLE

Assume that saw-mill refuse from Western hemlock, containing 55-percent moisture as fired and having the ultimate analysis shown in Table III, is burned with 45-percent excess air. Then

1. Fuel, F. Convert both the ash and the high heating value of Table III from the dry to the as-fired condition, or, ash

$$= 2.2 \left(\frac{100 - 55}{100} \right) = 1.0$$

percent as fired and

$$HHV = 8620 \left(\frac{100 - 55}{100} \right) = 3880 \text{ Btu per lb}$$

as fired. With these values for ash and HHV, read from Fig. 1, F = 255 lb per million Btu.

2. Atmospheric Air, A. Calculate the ratio

$$\left[\frac{C}{H_2 + 0.1\ O_2} \right] =$$

$$\left[\frac{50.4}{5.8 + 0.1 \times 41.4} \right] = 5.07$$

and with this ratio read from Curves A, Fig. 18, for 45-percent excess air, A = 1040 lb per million Btu.

3. Unburned Combustible Factor, C. The usual assumption when burning wood in steam boiler furnaces is that the combustible loss is zero; consequently C in Eqs. 3 and 4 may be taken as 1.

4. Total Products, P. From Eq. 4, P = F + CA = 255 + 1 × 1040 = 1295 lb per million Btu.

5. Moisture in Air, W_a. From Eq. 5, W_a = 0.013A = 0.013 × 1040 = 14 lb per million Btu.

6. Moisture from Fuel, W_f. The total moisture in the products of combustion of wood, W_f, is the sum of W_c, the moisture in the wood as fired, and W_h, the water formed by its combustion. Since the wood contains 55-percent moisture as fired and its high heating value as determined for F is 3880 Btu per lb, from Curves D of Fig. 18, W_c = 142 lb per million Btu. Next, convert the percent hydrogen in Table III from the dry to the as-fired basis, or

$$H_2 = 5.8 \left(\frac{100 - 55}{100} \right) = 2.6 \text{ percent}$$

as fired, and with a high heating value of 3880 Btu read from Curves C, Fig. 18 W_h = 60 lb per million Btu. Then
$$W_f = W_c + W_h = 142 + 60 = 202 \text{ lb per million Btu}$$

7. Dry Gas, P_d. From Eq. 6, P_d = P − ($W_a + W_f$) = 1295 − (14 + 202) = 1079 lb per million Btu.

8. Percent CO_2 in Products. Compute the ratio

$$\left[\frac{C}{H_2 - 0.1\ O_2} \right] = \left[\frac{50.4}{5.8 - 0.1 \times 41.4} \right] = 30.4$$

and for 45-percent excess air read on Curves B, Fig. 18, CO_2 = 14.2 percent.

FUEL OILS

The fuel oils ordinarily burned in steam-generating units are derivatives of petroleum comprising the so-called "Bunker" grades, diesel-engine oils, gas oils, kerosene, and gasoline. As they all have a common source, the constituents of their analyses are the same, although the relative quantity of each varies. De Baufre's representative tabulation in his article

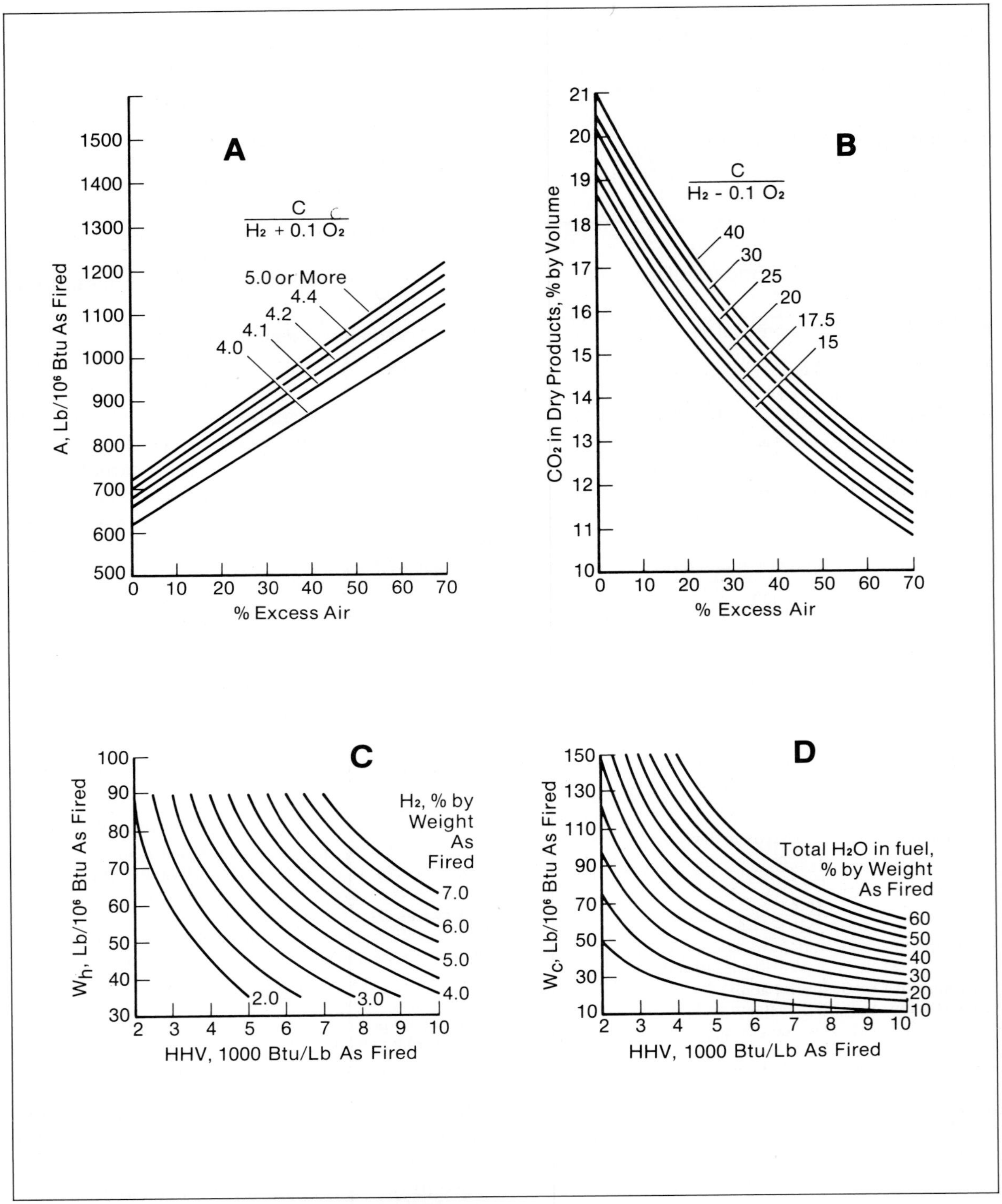

Fig. 18. Combustion characteristics of wood and bagasse

on "Typical Solid and Liquid Fuels"[3] shows that the combustible fraction chiefly consists of carbon and hydrogen with small amounts of sulfur. Negligible percentages of nitrogen and oxygen as well as water and sediment are also frequently found. De Baufre's table is, in part, reproduced as Table V.

The calculated atmospheric air, A, in lbs per million Btu as fired and the percent CO_2 in the products, both for zero excess air, are also tabulated for the different oils.

For commercial furnaces, Bunker C oil is frequently specified. This term has come to be applied to heavier fuel oils having a viscosity from 100 to 300 S.S.F. (Saybolt seconds furol) at 122°F. Bunker C oil roughly corresponds to No. 6 fuel oil in "Commercial Standard for Fuel Oils" by the Bureau of Standards.

FUEL IN PRODUCTS, F

Refer to Fig. 1, or calculate F directly from

$$F = \frac{1,000,000}{HHV}$$

(11)

ATMOSPHERIC AIR, A

The atmospheric air in pounds per million Btu as fired may be taken from Curve A of Fig. 19 for any value of excess air up to 100 percent. The values obtained are averages for a wide range of fuel oils; no chemical analysis is necessary in using Curve A.

TOTAL PRODUCTS, P

The weight of the products of combustion in pounds per million Btu as fired is given by $P = F + CA = F + A$, where C, the combustible-loss correction, is usually taken as 1.

MOISTURE IN FUEL, W$_f$

For fuel oil use curve group C of Fig. 19, where W_f is merely the moisture from hydrogen, and W_h is determined from Eq. 12.

$$W_h = 9 \times H_2 \times \left(\frac{10^4}{HHV} \right)$$

where:

W_h = moisture from combustion of H_2, lb/10^6 Btu fired

H_2 = hydrogen in fuel, % by weight

HHV = high (gross) heat value of fuel, Btu/lb

(12)

PERCENT CO$_2$ IN PRODUCTS

Curves B of Fig. 19 offer a convenient way of determining the percent CO_2 by volume in the dry products of combustion of fuel oil, for any value of excess air from zero to 100 percent. Curves B are plotted so that CO_2 values may be obtained after calculating the ratio C/H_2 from the ultimate analysis of the fuel.

EXAMPLE

Assume a fuel oil with the typical analysis given in Table V burned with 10-percent excess air. Then,

1. Fuel, F. From Fig. 1 for a high heat value of 18,500 Btu per lb and 0-percent ash read F = 54 lb per million Btu.

2. Atmospheric Air, A. From Curve A, Fig. 19, read A = 825 lb per million Btu.

3. Unburned Combustible. In the combustion of fuel oils in stationary boiler furnaces, it is generally assumed that there is no combustible heat loss. Therefore C in Eq. 3 is equal to 1.

4. Total Products, P. From Eq. 4, P = CA + F = 1 × 825 + 54 = 879 lb per million Btu.

5. Moisture in Air, W$_a$. From Eq. 5, W_a = 0.013 (A) = 0.013(825) = 11 lb per million Btu.

6. Moisture from Fuel, W$_f$. Since W_c = 0, from Curves C, Fig. 19, for a high heat value of 18,500 Btu/lb and H_2 of 11.5 percent, read W_f = W_h = 56 lb per million Btu.

7. Dry Gas, P$_d$. From Eq. 6, P_d = P − (W_a + W_f) = 879 − (11 + 56) = 812 lb per million Btu.

8. Percent CO$_2$ in Products. From Curves B, Fig. 19, for 10-percent excess air and a C/H_2 ratio of 7.5, read CO_2 = 14.4 percent.

NATURAL GAS

Characteristics of typical natural gases are given in Table VI, with the volumetric heating value on a moisture-free basis.

From the standpoint of combustion calculations, it is important to remember that the terms *dry* or *wet* as commonly applied to natural gas

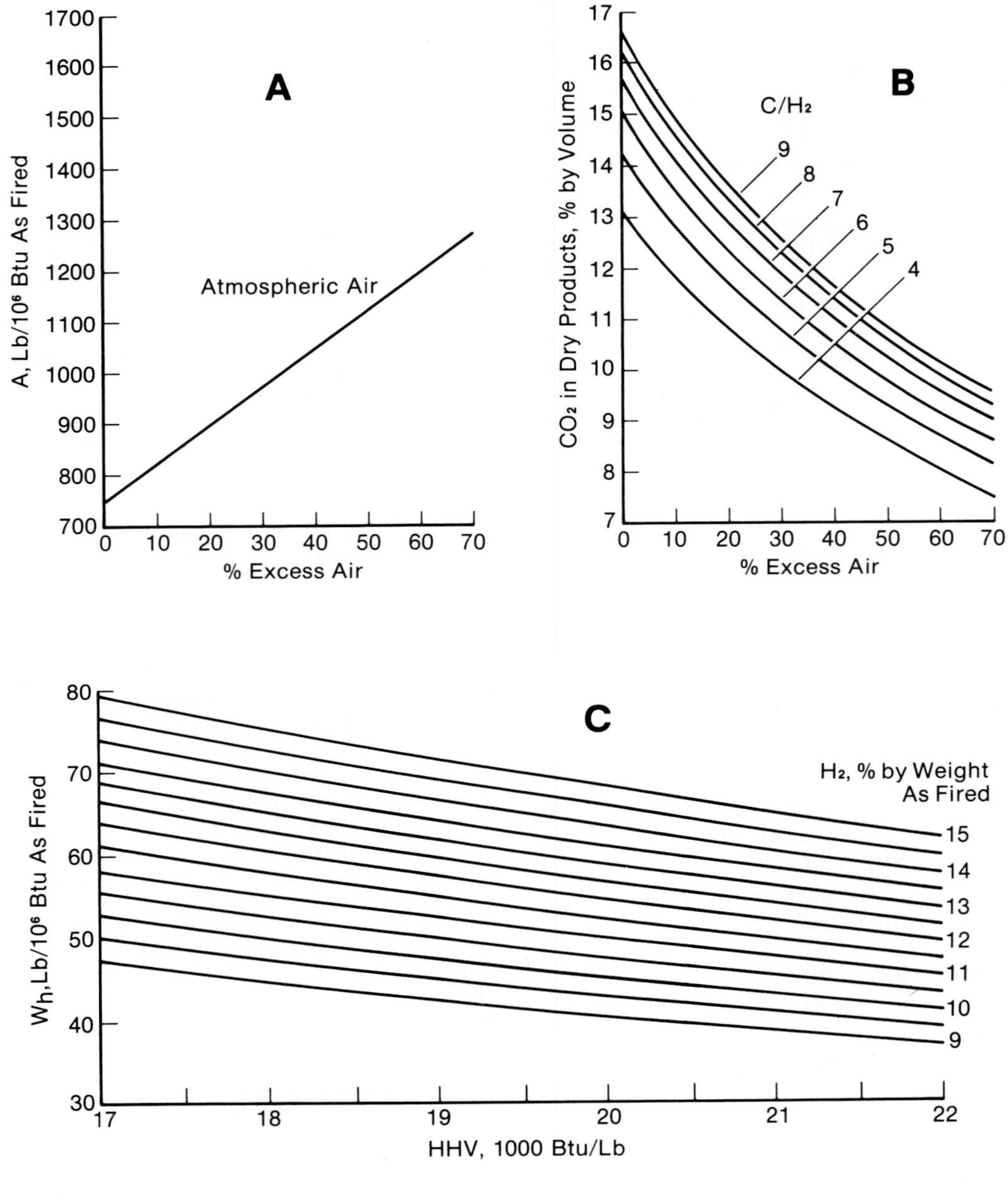

Fig. 19. Combustion characteristics of fuel oil

Table V. Characteristics of Typical Liquid Fuels

	° API	% C	% H_2	% S	% N_2	% O_2	HHV, Btu/Lb	A at Zero Excess Air, Lb/10^6 Btu	CO_2 at Zero Excess Air, %
Gasoline	60	85.0	14.8	. . .	0.1	0.1	20,200	746	14.87
Kerosene	45	85.0	14.0	. . .	0.5	0.5	19,900	742	15.12
Gas oil	30	85.0	12.8	0.8	0.7	0.7	19,300	745	15.48
Fuel oil	15	85.5	11.5	1.6	0.7	0.7	18,500	758	15.90

Table VI. Characteristics of Typical Natural Gases at 60°F and 30 In. Hg, Dry

			% by Volume					Density, Lb/Cu Ft	HHV Btu/Cu Ft**	HHV Btu/Lb	A at Zero Excess Air, Lb/10^6 Btu	CO_2 at Zero Excess Air, %
CO_2	N_2	H_2S	CH_4	C_2H_6	C_3H_8	C_4H_{10}	C_5H_{12}					
5.50	. . .	7.00	77.73	5.56	2.40	1.18	0.63*	0.05622	1060	18,880	738	12.1
3.51	32.00	0.50	52.54	3.77	2.22	2.02	3.44*	0.06610	874	13,220	729	12.3
26.2	0.7	. . .	59.2	13.9	. . .	. . .	. . .	0.06747	849	12,580	732	15.3
0.17	87.69	. . .	10.50	1.64	. . .	. . .	. . .	0.07120	136	1,907	732	6.9
0.20	0.60	. . .	99.20	. . .	. . .	. . .	. . .	0.04491	1006	22,410	732	11.7
. . .	0.60	. . .	. . .	79.40	20.00	. . .	. . .	0.08812	1935	21,960	735	13.3
. . .	0.50	. . .	. . .	21.80	77.70	. . .	. . .	0.11079	2389	21,560	738	13.7

* All hydrocarbons heavier than C_5H_{12} were assumed to be C_5H_{12} for combustion calculations.
** If gas is saturated with moisture at 60°F and 30. in. Hg, reduce by 1.74%.

refer to its gasoline vapor, and not to its moisture content. In fact, the only time natural gas has any moisture at its point of origin is when it is next to salt water. But this is a relatively unusual occurrence.

TEMPERATURE

The temperature of natural gas as it issues from the ground is dependent on the depth of the well and may vary from 32 to 165°F. The general rule among geologists is to assume an increase in temperature of 1°F for every 50 feet in depth, although this rule will not always check with actual measurements taken in the field.

It is also apparent that gas which has flowed through many miles of pipeline will have whatever ambient air temperature prevails in the locality where it is burned, most probably from 40 to 80°F. In combustion calculations for natural gas, it is deemed sufficiently accurate to assume the standard 60°F, when its actual temperature is not reported.

MOISTURE

As already noted, the only time that natural gas can have any moisture on leaving a well is when it previously has lain in contact with salt water. It may then be considered saturated with moisture at its temperature in the well.

However, gas which is delivered from a pipeline has often been "rehydrated," that is, saturated with water vapor by means of steam jets to lower the cost of maintaining pipe gaskets. Since the steam is ordinarily added to the gas in the high-pressure line, when the pressure is lowered for local distribution, the relative humidity of natural gas will also drop. Still another complication is introduced when a wet displacement meter is employed to measure the gas consumption. In this meter, the gas may be saturated with water.

In view of the variable and uncertain moisture content in natural gas, and in the absence of more definite determinations, it has become practice in industry to assume natural gas to be dry at 60°F and 30 in. Hg.

DENSITY

The density of a gaseous fuel may be obtained by adding the weights of the constituents in the fuel. The following is an example of a density calculation for the natural gas listed first in Table VI.

Density of the individual constituents are taken from Table VII, which lists the "saturated" and "unsaturated" or "illuminant" hydrocarbons that are most frequently found in

	% by Vol.	Vol. cu ft constituent /cu ft fuel	Density of constituent, lb/cu ft	Weight lb/cu ft fuel
CO_2	5.50	0.0550	×0.1170	=0.00644
H_2S	7.00	0.0700	×0.09109	=0.00638
CH_4	77.73	0.7773	×0.04246	=0.03300
C_2H_6	5.56	0.0556	×0.08029	=0.00446
C_3H_8	2.40	0.0240	×0.1196	=0.00287
C_4H_{10}	1.18	0.0118	×0.1582	=0.00187
C_5H_{12}	0.63	0.0063	×0.1904	=0.00120
		Total = Density of Dry Gas		=0.05622

gaseous fuels. Saturated hydrocarbons do not unite directly with hydrogen; i.e., they are stable in the presence of hydrogen, while unsaturated hydrocarbons readily take on more

Table VII. Combustion Constants of Dry Gases at 60° F and 30 In. Hg

Gas	Chemical Formula	Density of Dry Gas, Lb/Cu Ft	HHV of Dry Gas Btu/Cu Ft*	HHV of Dry Gas Btu/Lb
Oxygen	O_2	0.08461	. . .	. . .
Nitrogen (atmospheric)	N_2	0.07439	. . .	. . .
Air	. . .	0.07655	. . .	. . .
Carbon dioxide	CO_2	0.1170	. . .	. . .
Water vapor	H_2O	0.04758	. . .	. . .
Hydrogen	H_2	0.005327	325	60,991
Hydrogen sulfide	H_2S	0.09109	647	7,100
Carbon-monoxide	CO	0.07404	321	4,323
Saturated Hydrocarbons;				
Methane	CH_4	0.04246	1014	23,896
Ethane	C_2H_6	0.08029	1789	22,282
Propane	C_3H_8	0.1196	2573	21,523
Butane	C_4H_{10}	0.1582	3392	21,441
Pentane	C_5H_{12}	0.1904	4200	22,058
Unsaturated Hydrocarbons or Illuminants;				
Ethylene	C_2H_4	0.07421	1614	21,647
Propylene	C_3H_6	0.1110	2383	21,464
Butylene	C_4H_8	0.1480	3190	21,552
Pentylene	C_5H_{10}	0.1852	4000	21,600
Acetylene	C_2H_2	0.06971	1488	21,344
Benzene	C_6H_6	0.2060	3930	19,068
Toluene	C_7H_8	0.2431	4750	19,537

* If gas is saturated with moisture at 60°F and 30.0 in. Hg, reduce by 1.74%.

hydrogen. The latter are also called illuminants because they burn with a bright luminous flame, as distinguished from the saturated hydrocarbons which have a blue flame.

Extreme accuracy in calculations of density and heating value of gaseous fuels is not warranted; the precise values of the individual constituents continue to change as more accurate experimental procedures are developed, presenting somewhat of a "moving target."

Figs. 20 and 21 and Eq. 13 may be used whenever it is desired to correct for moisture content or for a temperature higher than 60°F.

$$d_t = \frac{d_{60} + X}{Y} + (D_t + w_c)$$

where:

d_t = density of gas at temperature, t, lb/cu ft
d_{60} = density of dry gas at 60°F, lb/cu ft
X and Y = correction factors read from Fig. 21
D_t = dust content of gas, lb/cu ft at t temperature, °F
w_c = entrained water found only in primary gas, lb/cu ft at t temperature, °F

(13)

Before going to a great deal of refinement, however, remember that the calculated gas density can be no more accurate than the volumetric analysis from which it is derived. A natural-gas analysis which groups all hydrocarbons as CH_4 and C_2H_6, or C_2H_6 and C_3H_8 may indicate a lower density than the actual value.

HEATING VALUE

As with all other gaseous fuels, it is customary to compute the high heating value of natural gas, in Btu per cu ft at 60°F and 30 in. Hg, by adding together the heat evolved by the combustible components reported in the gas analysis. Thus, for a gas such as the first one in Table VI, the procedure indicated below may be employed, using the heating values in Table VII.

The high heating value in Btu per pound is then obtained by dividing the Btu per cubic foot by the density at 60°F and 30 in. Hg. To correct the Btu per pound for the effect of higher

	% by Volume, Dry	Volume cu ft/ cu ft of Gas	Heating Value, Btu/ cu ft	Heat Evolved by Components, Btu/cu ft
H_2S	7.00	0.0700 $\times$	647 =	45
CH_4	77.73	0.7773 $\times$	1014 =	788
C_2H_6	5.56	0.0556 $\times$	1789 =	99
C_3H_8	2.40	0.0240 $\times$	2573 =	62
C_4H_{10}	1.18	0.0118 $\times$	3392 =	40
C_5H_{12}	0.63	0.0063 $\times$	4200 =	26
		Heating Value of Dry Gas	=	1060 Btu/cu ft

temperature and moisture content, refer to Fig. 22 and Eq. 14.

$$(HHV)_t = M(HHV)_{60} + N$$

(14)

The Btu per cubic foot calculated by the foregoing method will generally be lower than that determined by calorimeter because of the arbitrary grouping of hydrocarbons explained before. For the same reason, the calculated density will also be lower than actual. It follows that the Btu per pound, with which we are primarily concerned, is nearer to its actual value when it is figured from the calculated Btu per cubic foot and calculated density than when it is the ratio of the calorimeter Btu per cubic foot and the calculated density of the gas.

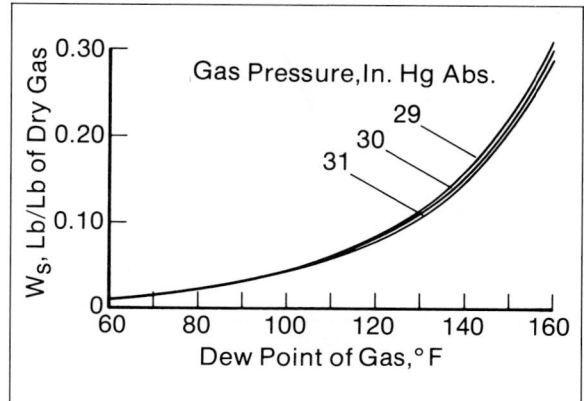

Fig. 20. **Water vapor in gas**

FUEL IN PRODUCTS, F

After determining the Btu per pound (HHV), F, which is that portion of the fuel that reappears in the products of combustion, is easily read from Fig. 1.

ATMOSPHERIC AIR, A

The atmospheric air in pounds per million Btu as fired may be taken from Curve A of Fig. 23 for any value of excess air up to 70 percent. As previously defined, the term *atmospheric air* is used to designate air which contains 0.013 pound of water vapor per pound of dry air, an arbitrary amount equivalent to 60 percent relative humidity at 80°F. The atmospheric air, A, may be corrected for temperatures higher than 60°F using Eq. 15.

$$A_t = \frac{A \, (HHV)_{60}}{(HHV)_t \, (1 + W_s + D_t/d_t)} \tag{15}$$

Note: If t = 60°F, the equation reduces to $A_t = A$, regardless of the moisture and dust content.

No chemical analysis of natural gas is necessary to obtain atmospheric air from Curve A.

TOTAL PRODUCTS, P

The weight of the products of combustion in pounds per million Btu as fired is given by P = F + CA, where C is a factor to correct A for the combustible loss due to imperfect combustion of the fuel. When burning natural gas, C is assumed to be 1 for any properly designed furnace supplied with the correct amount of excess air.

MOISTURE FROM FUEL, W_f

For the general case of a gaseous fuel, W_f is the sum of W_s, the water vapor required to saturate the fuel; W_c, the entrained moisture; and W_h, the water formed in burning the hydrogen compounds in the fuel.

As explained before, natural gas is usually assumed to be dry, and since it seldom carries any entrained moisture, the first two terms generally may be taken as zero. If the gas is not

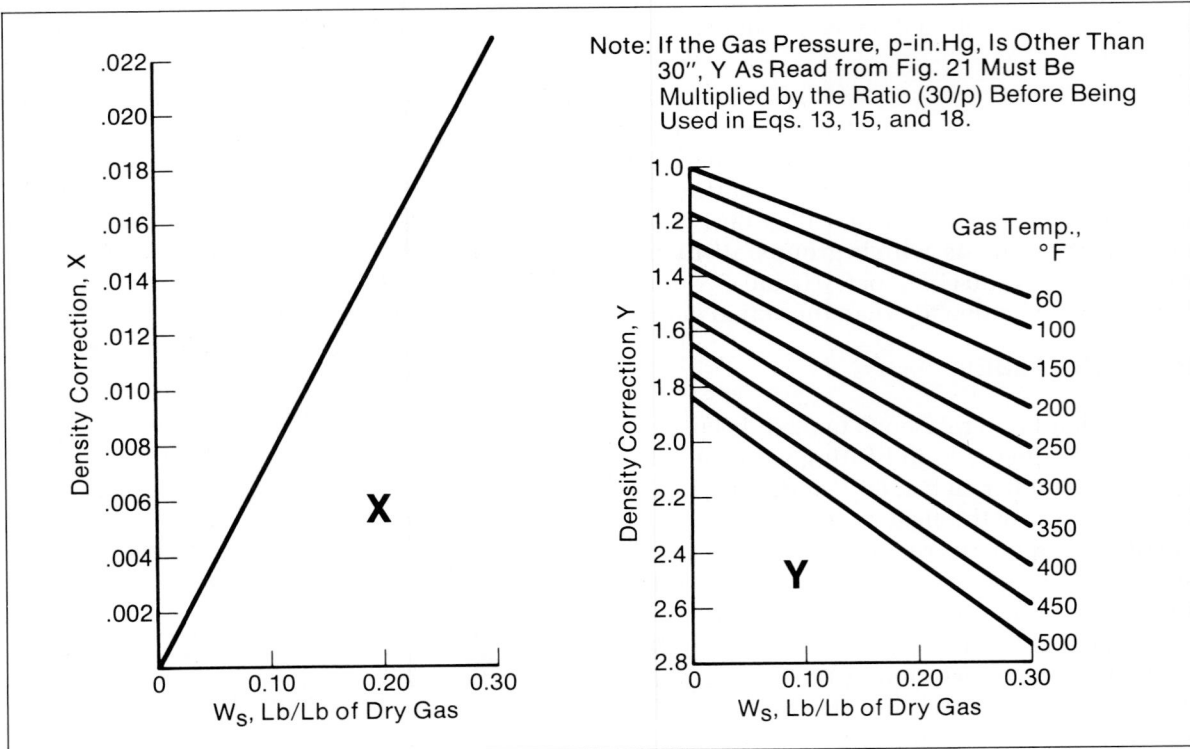

Note: If the Gas Pressure, p-in.Hg, Is Other Than 30″, Y As Read from Fig. 21 Must Be Multiplied by the Ratio (30/p) Before Being Used in Eqs. 13, 15, and 18.

Fig. 21. Density correction for use in Eq. 13

dry, but contains w_s pound of water vapor per pound of dry fuel, the pounds of water vapor per million Btu as fired are given by

$$W_s = \frac{w_s(1 - w_s)10^6}{HHV}$$

(16)

where HHV is the high heating value of the gas in Btu per pound as fired. Fig. 24 is a convenient plot of Eq. 16.

The water produced by the combustion of natural gas is given by equation 17 which follows.

$$W_h = 47,300 \times$$
$$\frac{(H_2S + 2CH_4 + 3C_2H_6 + 4C_3H_8 + 5C_4H_{10} + 6C_5H_{12})}{Btu/cu\ ft\ at\ 60°F\ and\ 30\ in.\ Hg,\ dry}$$

(17)

For dry natural gas at 60°F and 30 in. Hg, W_h may be conveniently read from Curves B of Fig. 23. When it is desired to correct W_h for higher temperature and for moisture content, Eq. 18 may be used.

$$(W_h)_t = \frac{W_h\ (HHV)_{60}}{(HHV)_t\ (1 + W_s + D_t/d_t)}$$

Note: If t = 60°F, the equation reduces to $(W_h)_t = W_h$, regardless of the moisture and dust content.

(18)

PERCENT CO₂ IN PRODUCTS

Curves C, Fig. 23, permit the determination of the percent CO_2 (by volume) in the dry products of combustion for zero excess air. They are a plot of the following approximate equation.

% CO_2 (at zero excess air) =

$$\frac{100}{6.64 + \frac{188 + 0.88(5.5H_2S - 8.5CO_2 - N_2)}{CO_2 + CH_4 + 2C_2H_6 + 3C_3H_8 + 4C_4H_{10} + 5C_5H_{12}}}$$

(19)

Then, by means of guidelines D, the percent CO_2 for any other excess air up to 70 percent may be read. However, it is not advisable to use Curves D for natural gases with a nitrogen percentage of over 40. This is particularly true for high values of excess air.

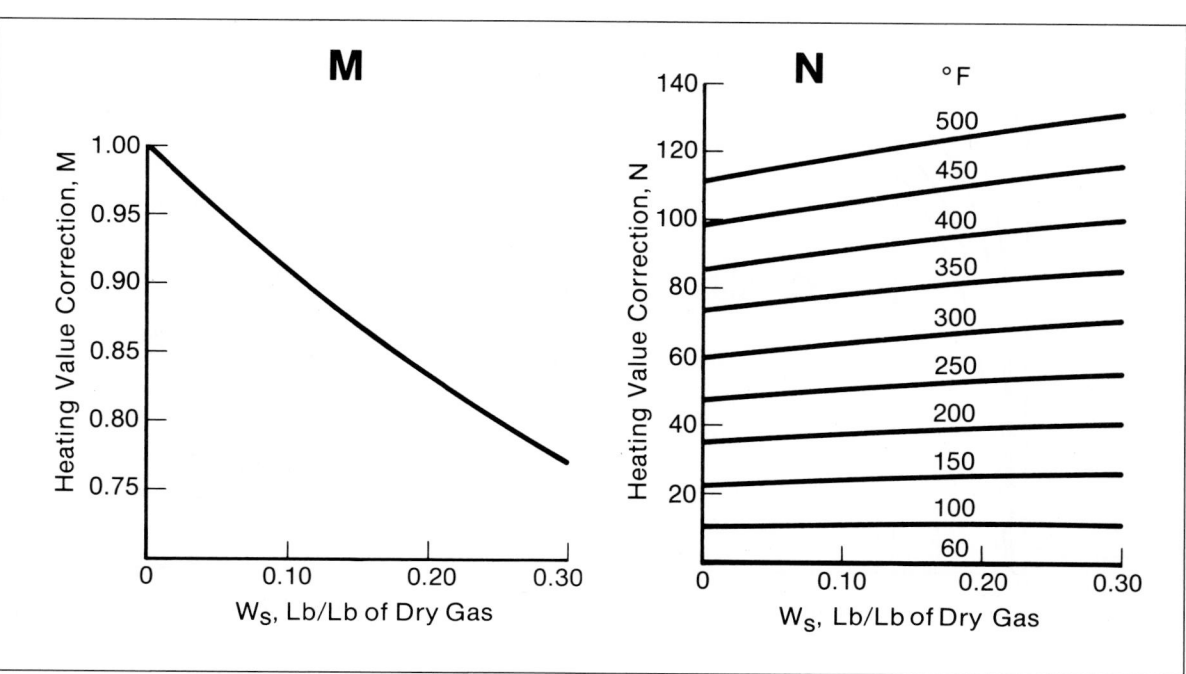

Fig. 22. Heating value corrections for use in Eq. 14

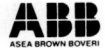

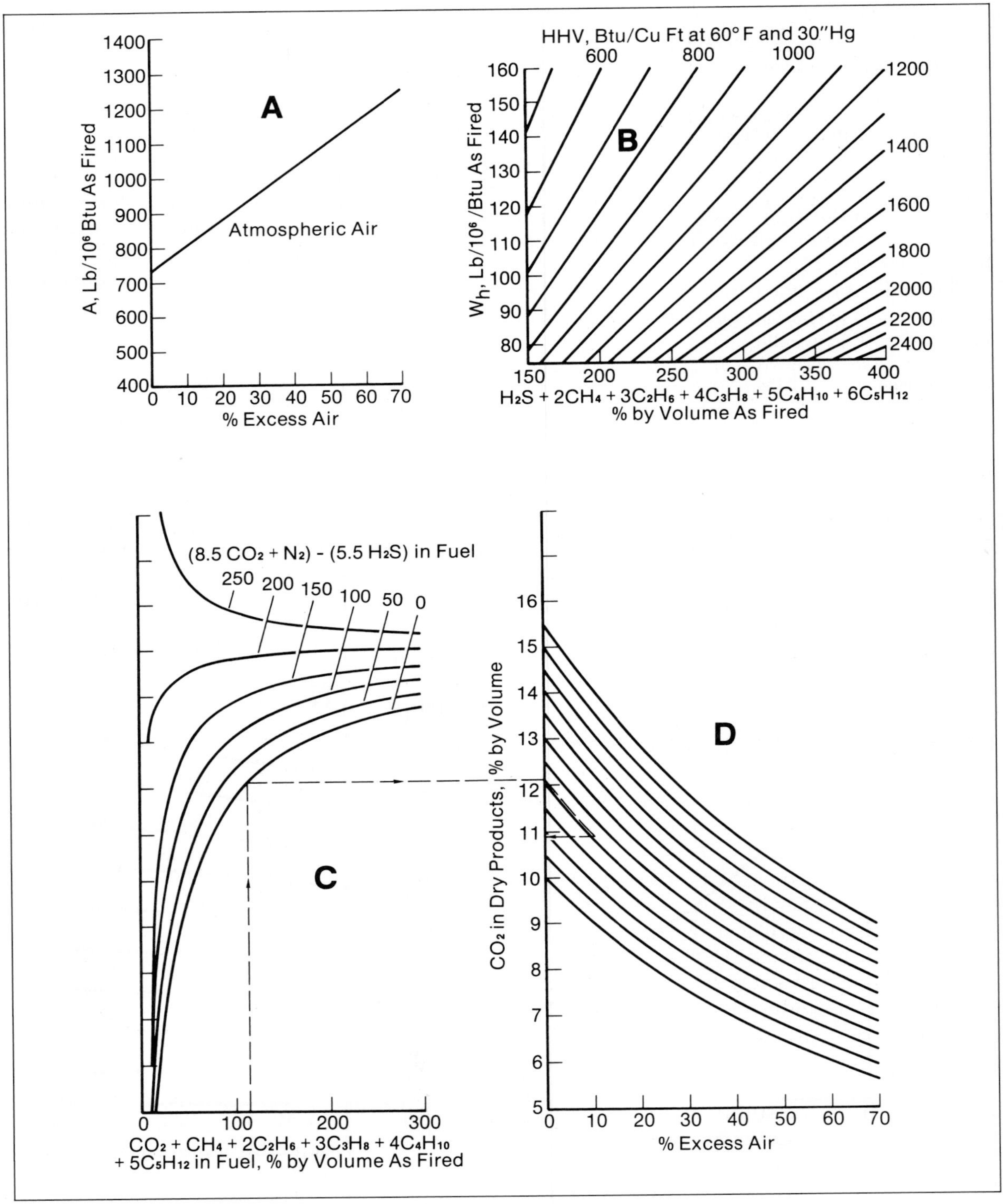
A
Atmospheric Air
A, Lb/10⁶ Btu As Fired
1400
1300
1200
1100
1000
900
800
700
600
500
400
0 10 20 30 40 50 60 70
% Excess Air
HHV, Btu/Cu Ft at 60° F and 30″Hg
600
800
1000
1200
1400
1600
1800
2000
2200
2400
B
Wh, Lb/10⁶ /Btu As Fired
160
150
140
130
120
110
100
90
80
150 200 250 300 350 400
H2S + 2CH4 + 3C2H6 + 4C3H8 + 5C4H10 + 6C5H12
% by Volume As Fired
(8.5 CO2 + N2) - (5.5 H2S) in Fuel
250 200 150 100 50 0
C
0 100 200 300
CO2 + CH4 + 2C2H6 + 3C3H8 + 4C4H10
+ 5C5H12 in Fuel, % by Volume As Fired
D
CO2 in Dry Products, % by Volume
16
15
14
13
12
11
10
9
8
7
6
5
0 10 20 30 40 50 60 70
% Excess Air

Fig. 23. **Combustion characteristics of natural gas**

EXAMPLE

Assume a dry natural gas at 60°F and 30 in. Hg to have the typical analysis listed first in Table VI and to be burned with 10-percent excess air. Then

1. Fuel, F. For a high heat value of 18,880 Btu per lb, refer to Fig. 1 and from it read F = 53 lb per million Btu.

2. Atmospheric Air, A. From Curve A, Fig. 23, for 10-percent excess air read A = 807 lb per million Btu.

3. Unburned Combustible. The general assumption when burning natural gas in stationary boiler furnaces is that the combustible loss is zero. Consequently, C in Eqs. 3 and 4 may be taken as 1.

4. Total Products, P. From Eq. 4, P = F + CA = 53 + 1 × 807 = 860 lb per million Btu.

5. Moisture in Air, W_a. From Eq. 5, W_a = 0.013A = 0.013 × 807 = 10 lb per million Btu.

6. Moisture from Fuel, W_f. Since W_c, the entrained moisture, is zero and W_s, the saturation moisture, is negligible, from Curves B, Fig. 23, for a high heating value of 1061 Btu per cu ft and $H_2S + 2CH_4 + 3C_2H_6 + 4C_3H_8 + 5C_4H_{10} + 6C_5H_{12}$ = 7.0 + 155.5 + 16.7 + 9.6 + 5.9 + 3.8 = 198.5 percent by volume, read $W_f = W_h$ = 88 lb per million Btu.

7. Dry Gas, P_d. From Eq. 6, P_d = P − ($W_a + W_f$) = 860 − (10 + 88) = 762 lb per million Btu.

8. Percent CO_2 in Products. For $CO_2 + CH_4 + 2C_2H_6 + 3C_3H_8 + 4C_4H_{10} + 5C_5H_{12}$ = 5.5 + 77.7 + 11.1 + 7.2 + 4.7 + 3.2 = 109.4 percent by volume and $8.5 CO_2 + N_2 − 5.5 H_2S$ = 46.7 + 0 − 38.5 = 8.2 percent by volume, from Curves C, Fig. 23, read 12.1 percent CO_2 at zero excess air. Following Curves D as guidelines to 10-percent excess air, read CO_2 = 10.9 percent.

REFINERY AND OIL GAS

The range in composition for refinery gas and oil gas is shown in Table VIII. Refinery gas is predominantly made up of saturated gases of low boiling point, such as methane (CH_4), ethane (C_2H_6), propane (C_3H_8), butane (C_4H_{10}), and pentane (C_2H_{12}) with small amounts of illuminants. Minor quantities of carbon dioxide (CO_2), carbon monoxide (CO), and hydrogen (H_2) may also be present with oxygen (O_2) and nitrogen (N_2) introduced to lower the hydrocarbon vapor pressure, in some cracking processes. If the original crudes contained sulfur, the ensuing gases include hydrogen sulfide (H_2S) unless the sulfur compounds have been removed by purifying operations. In Table VIII, the last two analyses represent oil gas, whereas the others are representative of refinery gas. Note that the hydrogen content of oil gas is higher than that of refinery gas, as is also its carbon monoxide content.

DENSITY

In the section on natural gas, a method was outlined for calculating the density of a gaseous fuel, given its analysis by volume. The same method can be used to find the density of refinery gas, or oil gas, either at 60°F or at a higher temperature.

No correction for gas pressure other than the standard 30 in. Hg is required in the calculation of density, because invariably the fuel is burned in a furnace which is at atmospheric pressure. Table VIII indicates that the density of oil gas is considerably lower than that of refinery gas.

HEATING VALUE

The heating value in Btu per cu ft is computed as set forth in the section on natural gas, with appropriate correction factors for temperature and moisture as derived for blast-furnace

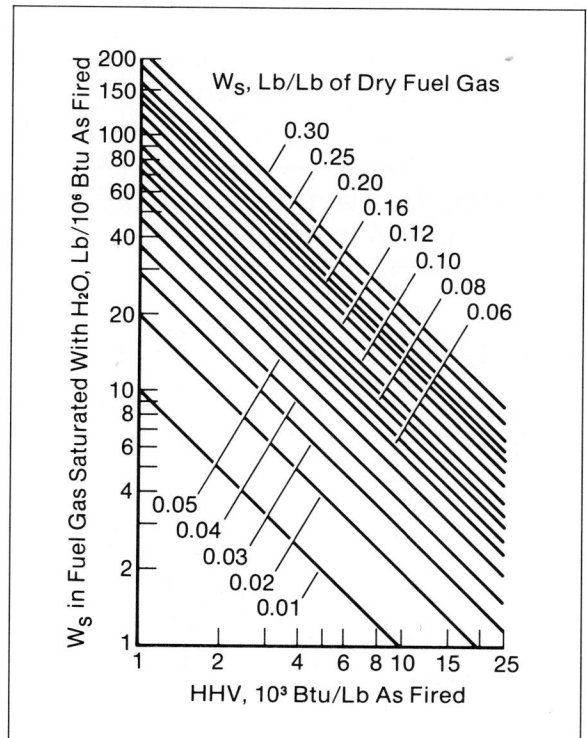

Fig. 24. Water vapor in fuel gas

gas. The Btu per pound is the ratio of Btu per cu ft and the density.

Although there is no doubt that some refinery gases contain little or no moisture, others are saturated with it. The values in the table are given on a dry basis.

FUEL IN PRODUCTS, F

With the Btu per pound (HHV) already determined, it is only necessary to refer to Fig. 1 for the value of F which may also be calculated directly from:

$$F = \frac{10^6}{HHV}$$

(20)

ATMOSPHERIC AIR, A

The air required for combustion may be conveniently read from Curves A of Fig. 26 for amounts of excess air up to 70 percent of the theoretical requirements. Due to the relatively large percentages of carbon monoxide and hydrogen present in some of the oil gases, the magnitude of A in Table VIII varies from 648 to 725 lb per million Btu. The correlation of A at zero excess air with the term $(CO + H_2)$ is brought out by Fig. 25. The average line drawn through the calculated points on this figure was used as the basis for spacing Curves A of Fig. 26.

The atmospheric air as taken from Fig. 26 can be corrected for temperatures other than 60°F and for moisture content by employing Eq. 15.

TOTAL PRODUCTS, P

Knowing A and F, the total wet products formed by combustion of refinery or oil gas can be easily computed, in pounds per million Btu fired, by the addition $P = F + CA$.

MOISTURE FROM FUEL, W_f

The total amount of moisture in the products of combustion, W_f, which is derived from the fuel itself, is the summation of the entrained moisture W_c, the water vapor held by the gas W_s, and the water produced by burning hydrogen, or hydrogen compounds making up the fuel, W_h.

Unless the refinery gas has undergone a scrubbing or washing operation to remove objectionable elements from the distillate, it contains hardly any entrained moisture. Water vapor, however, may be present in the gas not only because of the purifying treatment, but due to water introduced into the system with

Table VIII. Characteristics of Typical Refinery and Oil Gases at 60°F and 30 In. Hg, Dry

| | | | | | | % by Volume | | | | | | Density, | HHV Btu/ | | A at Zero Excess Air, | CO₂ at Zero Excess |
O_2	N_2	CO_2	CO	H_2	H_2S	CH_4	C_2H_6	C_3H_8	C_4H_{10}*	C_5H_{12}†	C_3H_6‡	Lb/Cu Ft	Cu Ft**	Btu/Lb	Lb/10⁶ Btu	Air, %
...	...	...	...	...	2.18	41.62	20.91	19.72	9.05	6.52	...	0.08676	1898	21,880	722	13.3
...	...	...	...	...	...	4.30	82.70	13.00	...	...	...	0.08377	1858	22,170	725	13.4
...	...	...	...	...	...	92.10	1.90	4.50	1.30	0.20	...	0.04845	1136	23,460	723	12.1
...	...	...	...	...	...	5.0	12.0	30.0	34.0	19.0	...	0.13760	2988	21,720	717	13.9
...	3.3	1.5	5.6	...	...	30.9	19.8	38.1	0.6	...	0.2	0.08102	1696	20,930	725	13.4
2.3	8.7	...	...	...	...	30.3	13.4	19.1	14.7	1.8	9.7	0.09232	1844	19,970	715	13.6
0.9	8.4	2.2	14.3	50.9	...	15.9	5.0	...	...	...	2.4	0.03631	519	14,300	648	12.2
0.1	2.7	1.0	6.8	59.2	...	25.4	...	...	...	...	4.8	0.02756	586	21,270	657	10.6

* Includes both iso-C_4H_{10} and n-C_4H_{10}.

† Includes all saturated hydrocarbons heavier than C_5H_{12}, also both iso-C_5H_{12} and n-C_5H_{12}.

‡ Includes all illuminants.

** If gas is saturated with moisture at 60°F and 30 in. Hg, reduce by 1.74%.

the crude oil, or because of steam employed in the distillation process. If the gas contains W_s lb of water vapor per lb of dry fuel, W_s can be determined from Fig. 24.

The water which is formed by combustion, W_h, can be read from Curves B of Fig. 26, which are a plot of the expression

$$W_h = 47,300 \times$$
$$\frac{(H_2 + H_2S + 2CH_4 + 3C_2H_6 + 4C_3H_8 + 5C_4H_{10} + 6C_5H_{12} + 3C_3H_6)}{\text{Btu/cu ft at } 60°F \text{ and } 30 \text{ in. Hg, dry}}$$

(21)

When, for more accuracy, it is required to correct W_h for the effect of temperature and moisture content, Eq. 18 can be utilized.

PERCENT CO₂ IN PRODUCTS

For zero excess air, the CO_2 in the dry products of combustion is found with the aid of Curves C, Fig. 26. These curves were drawn from the approximate equation.

$$\% \, CO_2 \text{ (at zero excess air)} =$$
$$\frac{100}{6.64 + \dfrac{188 - 1.88(N_2 + 4CO_2 + 3CO + C_3H_6)}{CO_2 + CO + CH_4 + 2C_2H_6 + 3C_3H_8 + 4C_4H_{10} + 5C_5H_{12} + 3C_3H_6}}$$

(22)

The guide Curves D of Fig. 26 make it possible to obtain the percent of CO_2 in the products for an excess-air value up to 70 percent.

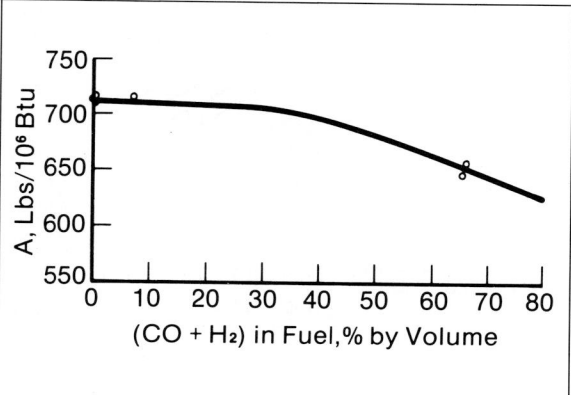

Fig. 25. Variation of theoretical air with (CO + H₂) in combustion of refinery and oil gas

EXAMPLE

Assume a refinery gas of the typical analysis given on the first line of Table VIII to be fired at a temperature of 60°F with 10-percent excess air. Then

1. Fuel, F. With a fuel high heat value of 21,880 Btu per lb, refer to Fig. 1 and obtain F = 46 lb per million Btu.

2. Atmospheric Air, A. On Curves A, Fig. 26, for 10-percent excess air and $(CO + H_2) = 0$, read A = 793 lb per million Btu.

3. Unburned Combustible. When burning refinery gas in boiler furnaces, it is customary to assume that there is no unburned combustible loss, so that in Eqs. 3 and 4 C = 1.

4. Total Products, P. Using Eq. 4, P = F + CA = 46 + 1 × 793 = 839 lb per million Btu.

5. Moisture, in Air, W_a. This item is calculated from Eq. 5, $W_a = 0.013A = 0.013 \times 793 = 10$ lb per million Btu.

6. Moisture from Fuel, W_f. The moisture content of refinery gas itself is generally negligible. The water formed by its combustion is taken from Curves B, Fig. 26. For a high heat value of 1898 Btu per cu ft and $H_2S + 2CH_4 + 3C_2H_6 + 4C_3H_8 + 5C_4H_{10} + 6C_5H_{12} = 2.18 + 83.24 + 62.73 + 78.88 + 45.25 + 39.12 = 311.40$ percent by volume, $W_f = W_h = 78$ lb per million Btu.

7. Dry Gas, P_d. From Eq. 6, $P_d = P - (W_a + W_f) = 839 - (10 + 78) = 751$ lb per million Btu.

8. Percent CO₂ in Products. The CO_2 at zero excess air is determined first. From curves C of Fig. 26, for $CH_4 + 2C_2H_6 + 3C_3H_8 + 4C_4H_{10} + 5C_5H_{12} = 41.62 + 41.82 + 59.16 + 36.32 + 32.60 = 211.40$ percent by volume, and $N_2 + 4CO_2 + 3CO + C_3H_6 = 0.0$ obtain 13.3 percent CO_2. With Curves D as guidelines and 10-percent excess air, then, read $CO_2 = 12$ percent.

COKE-OVEN GAS

Representative analyses of several byproduct coke-oven gases are given in Table IX. Note that the volumetric heating value is based on the gas being moisture free, and that a correction must be made for moisture of saturation, if the gas has been washed and cooled.

It is important to have the correct calorific value at 60°F and 30 in. Hg for a given analysis, since the method employed here relates all combustion calculations to the heat liberated. Unfortunately, many analyses show the heavier hydrocarbons grouped together as "illuminants." DeBaufre[4] and others have suggested taking the illuminants to be propylene, C_3H_6. While this assumption is valid in calculating the Btu per cu ft, it will not give accurate values of H_2O formed by combustion, and

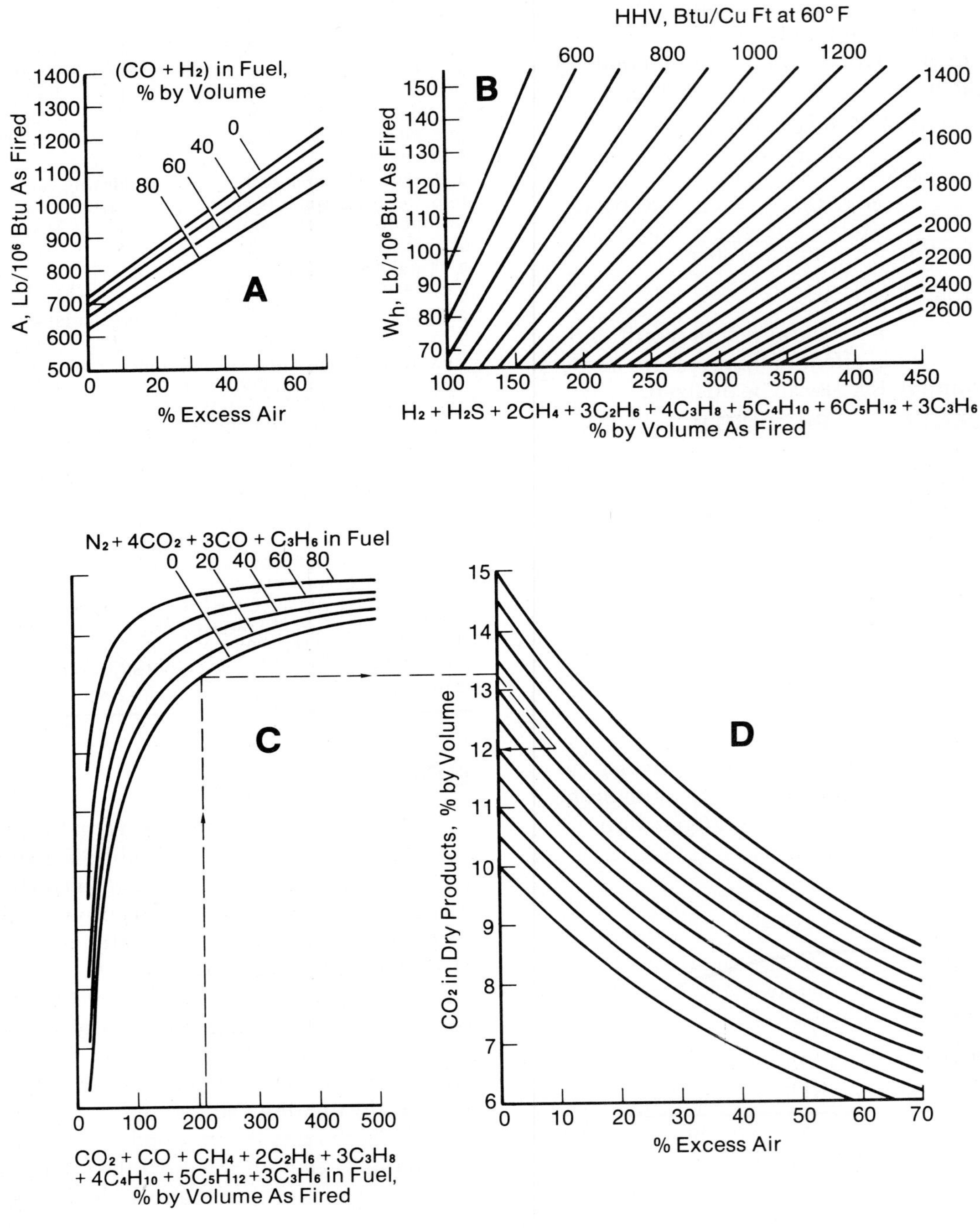
(CO + H2) in Fuel, % by Volume
0
40
60
80
A, Lb/10⁶ Btu As Fired
1400
1300
1200
1100
1000
900
800
700
600
500
0
20
40
60
% Excess Air
A
HHV, Btu/Cu Ft at 60° F
600
800
1000
1200
1400
1600
1800
2000
2200
2400
2600
B
Wh, Lb/10⁶ Btu As Fired
150
140
130
120
110
100
90
80
70
100
150
200
250
300
350
400
450
H2 + H2S + 2CH4 + 3C2H6 + 4C3H8 + 5C4H10 + 6C5H12 + 3C3H6
% by Volume As Fired
N2 + 4CO2 + 3CO + C3H6 in Fuel
0
20
40
60
80
C
0
100
200
300
400
500
CO2 + CO + CH4 + 2C2H6 + 3C3H8
+ 4C4H10 + 5C5H12 + 3C3H6 in Fuel,
% by Volume As Fired
D
CO2 in Dry Products, % by Volume
15
14
13
12
11
10
9
8
7
6
0
10
20
30
40
50
60
70
% Excess Air

Fig. 26. **Combustion characteristics of refinery gas and oil gas**

the resulting heat balance will be in error. A complete gas analysis, including all hydrocarbons, is, therefore, essential. To find the high heating value in Btu per lb accurately, divide the Btu per cu ft by the fuel density.

DENSITY

The density of the gaseous fuel, in turn, may be obtained by adding the weights of the constituents in the fuel. The following tabulation is an example of density calculation for the typical coke-oven gas listed first in Table IX, the density of the individual constituents being taken from Table VII.

	% by volume	Volume, cu ft constituent/ cu ft fuel		Density of constituent, lb/cu ft		Weight, lb/cu ft fuel
CO_2	1.8	0.018	×	0.1170	=	0.00211
O_2	0.2	0.002	×	0.08461	=	0.00017
N_2	3.4	0.034	×	0.07439	=	0.00253
CO	6.3	0.063	×	0.07404	=	0.00466
H_2	53.0	0.530	×	0.005327	=	0.00282
CH_4	31.6	0.316	×	0.04246	=	0.01342
C_2H_4	2.7	0.027	×	0.07421	=	0.00200
C_6H_6	1.0	0.010	×	0.2060	=	0.00206
		Total = Density of Dry Gas			=	0.02977

The saturation pressure of water vapor at 60°F is 0.52 in. Hg. If the gas is saturated with moisture at 60°F and 30 in. Hg, then the partial pressure of all its dry constituents is 30.0 − 0.52 = 29.48 in. Hg. A density correction is made as follows.

$$0.02977 \times \frac{29.48}{30.0} = 0.02925$$

Add to this the density of the water
vapor (from the steam tables)   0.00083
Density of saturated gas   0.03008

It can be seen that the density correction for moisture content will be 1 percent.

FUEL, F

The portion of the fuel which reappears in volatile form in the products of combustion can be read from Fig. 1 when HHV (Btu per lb of fuel) is known. If HHV is not given, it is found by dividing the Btu per cu ft by the fuel density.

No moisture correction is necessary, because F is determined to two significant figures only.

ATMOSPHERIC AIR, A

The atmospheric air required for combustion will vary with the carbon monoxide and hydrogen content of the gas. A correlation of atmospheric air, A, at zero excess air with the term $(CO + H_2)$ is shown in Fig. 27. The atmospheric air in lb per million Btu may be read from the curve family A in Fig. 28. The spacing of these curves is based on the average line drawn in Fig. 27.

TOTAL PRODUCTS, P

The weight of the products of combustion is calculated from $P = F + CA$.

MOISTURE FROM FUEL, W_f

The moisture, W_c, in a fuel gas saturated at 60°F may be neglected as it amounts to less than 2 lb per million Btu as fired. W_h, the water formed by the combustion of hydrogen and the hydrocarbons present in the coke-oven gas, is obtained from Curves B, Fig. 28. Since W_c is neglected, W_f, which is the total moisture in the flue gases derived from the fuel, is equal to W_h.

In the simple case of hydrogen, the volumetric chemical equation for combustion is

$$H_2 + \tfrac{1}{2}O_2 = H_2O \tag{23}$$

Eq. 23 shows that one volume of hydrogen combines with one-half volume of oxygen to produce one volume of water. Similarly, one volume of CH_4 will form, on burning, two volumes of H_2O, etc. Since a coke-oven gas may contain H_2, CH_4, C_2H_4, and C_6H_6, with the aid of Table VII it is seen that the total volume of H_2O evolved by these constituents is proportional to $(H_2 + 2\ CH_4 + 2\ C_2H_4 + 3\ C_6H_6)$. The following exact relation will convert the volume of H_2O to lb per million Btu.

$$W_h = 47{,}300 \frac{(H_2 + 2\ CH_4 + 2\ C_2H_4 + 3\ C_6H_6)}{Btu/cu\ ft\ at\ 60°F\ and\ 30\ in.\ Hg,\ dry} \tag{24}$$

which is also plotted as Curves B, Fig. 28.

Fig. 27. Variation of theoretical air with (CO + H₂) in combustion of coke-oven gas

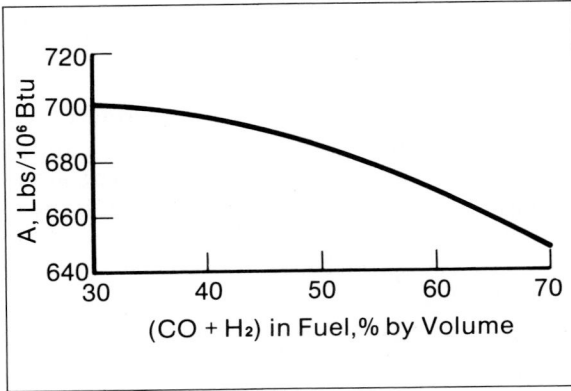

PERCENT CO₂ IN PRODUCTS

There are many constituents in coke-oven gas which on burning will evolve CO_2. A simple equation relating exactly the CO_2 in the dry products of combustion to these constituents cannot be written, but the following equation is sufficiently accurate for analyses which do not contain excessive amounts of N_2 or O_2.

$$\% \ CO_2 \ (\text{at zero excess air}) = \cfrac{100}{1 + \cfrac{0.695 \times \text{Btu/cu ft at } 60°F \text{ and } 30 \text{ in. Hg, dry}}{(100 + C_2H_4 + 5C_6H_8) - (H_2 + N_2)}}$$

(25)

Curves C, Fig. 28, are a plot of Eq. 25, while Curves D serve as guidelines for determining the percent CO_2 in the dry products of combustion for any other value of excess air than zero.

EXAMPLE

Assume a coke-oven gas with the typical analysis of the first gas listed in Table IX and saturated with moisture at 60°F and 30 in. Hg to be burned with 10-percent excess air.

1. Heating Value. The high heat value given as 596 Btu per cu ft must first be corrected for moisture content by deducting 1.74 percent; 596 − 10 = 586 Btu per cu ft saturated fuel. This value is divided by the fuel density (neglecting the moisture correction) to obtain the approximate Btu per lb.

$$\text{HHV} = \frac{586}{0.0298} = 19,700 \text{ Btu/lb}$$

2. Fuel, F. For a high heat value of 19,700 Btu per lb, read from Fig. 1, F = 51 lb per million Btu.

3. Atmospheric Air, A. From Curves A, Fig. 28 for 10-percent excess air and (CO + H₂) = 6.3 + 53.0 = 59.3, read A = 742 lb per million Btu.

4. Unburned Combustible. For stationary boiler furnaces, the general assumption when burning coke-oven gas is that there is no combustible heat loss. Therefore, C in Eqs. 3 and 4 is equal to 1.

5. Total Products, P. From Eq. 4, P = F + CA = 51 + 1 × 742 = 793 lb per million Btu.

6. Moisture in Air, Wₐ. From Eq. 5, $W_a = 0.013A = 0.013 \times 742 = 10$ lb per million Btu.

7. Moisture from Fuel, Wg. Since $W_c = 0$, from Curves B, Fig. 28, for a high heat value of 596 Btu per cu ft and H₂ = 2CH₄ + 2C₂H₄ + 3C₆H₆ = 53.0 + 63.2 + 5.4 + 3.0 = 124.6 percent by volume, read $W_f = W_h = 99$ lb per million Btu.

8. Dry Gas, Pd. From Eq. 6, $P_d = P - (W_a = W_g) = 793 - (10 + 99) = 684$ lb per million Btu.

9. Percent CO₂ in Products. For a heating value of 596 Btu per cu ft and (100 + C₂H₄ + 5C₆H₆) − (H₂ + N₂) = (100 + 2.7 + 5.0) − (53.0 + 3.4) = 51.3 percent by volume, from Curves C, Fig. 28, read 11.1 percent CO_2 at zero excess air. Following Curves D as guidelines to 10-percent excess air, read CO_2 = 10 percent.

Table IX. Characteristics of Typical Coke-Oven Gases at 60°F and 30 In. Hg, Dry

			% by Volume					Density, Lb/Cu Ft	HHV Btu/ Cu Ft*	HHV Btu/Lb	A at Zero Excess Air, Lb/10⁶ Btu	CO₂ at Zero Excess Air, %
CO₂	O₂	N₂	CO	H₂	CH₄	C₂H₄	C₆H₆					
1.8	0.2	3.4	6.3	53.0	31.6	2.7	1.0	0.0298	596	20,010	674	11.1
1.4	0.5	4.2	5.1	57.4	28.5	2.9	...	0.0263	539	20,490	664	10.0
2.6	0.6	3.7	6.1	47.9	33.9	5.2	...	0.0316	603	19,070	676	11.3
3.13	...	...	11.93	42.16	37.14	4.76	0.88	0.0359	663	18,500	684	12.7
0.1	...	2.4	6.8	27.7	50.0	13.0	...	0.0393	829	21,100	700	12.3
0.75	...	12.1	6.0	53.0	28.15	...	...	0.0291	477	16,390	668	9.4

* If gas is saturated with moisture at 60°F and 30 in. Hg, reduce by 1.74%.

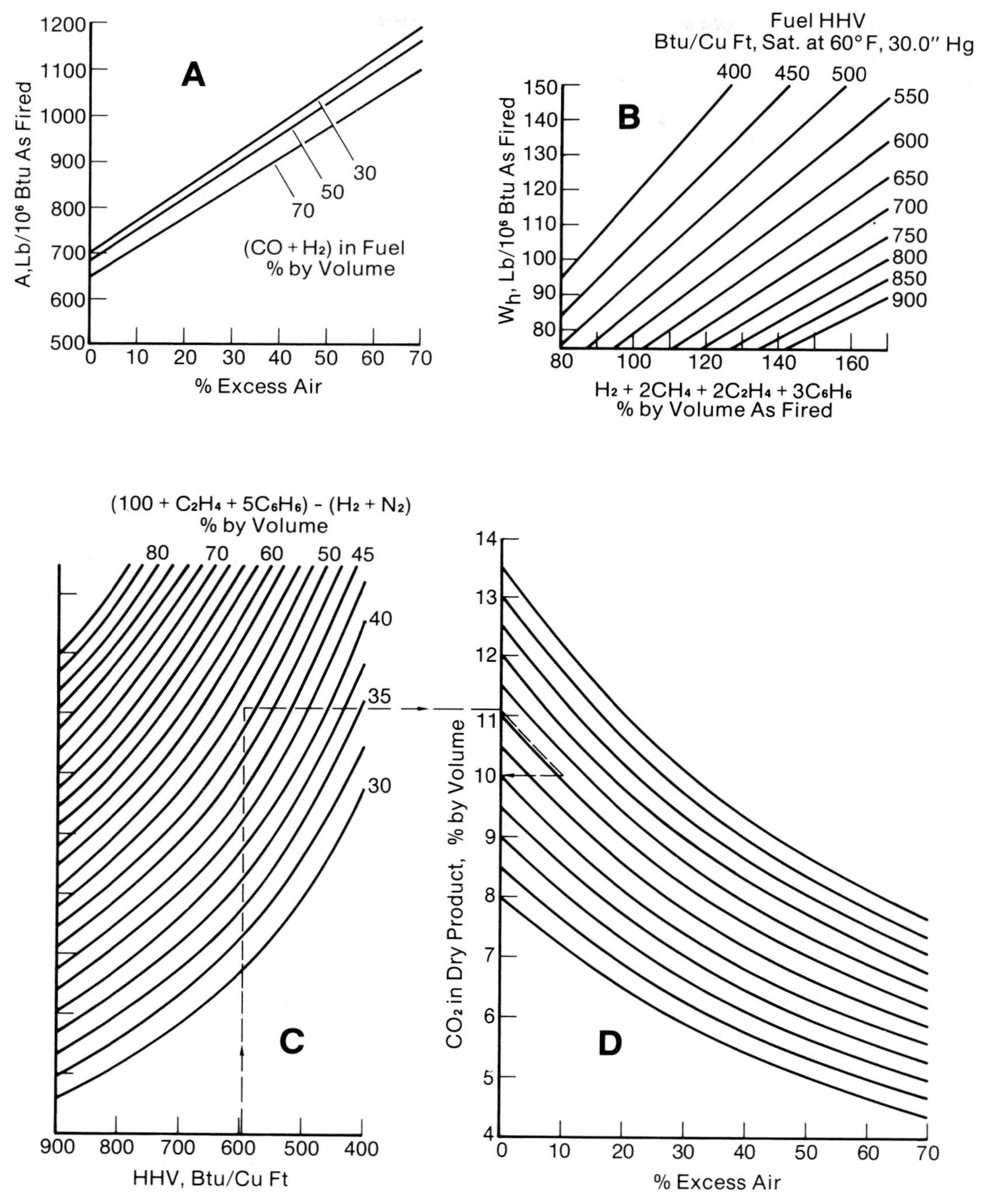

Fig. 28. Combustion characteristics of coke-oven gas

BLAST-FURNACE GAS

The range in composition of steel-plant blast-furnace gas is given in Table X. Note that the volumetric heating value is based on the gas being moisture free, and that a correction must be made for moisture of saturation.

MOISTURE FROM CLEANING

In Chapter 2, blast-furnace gas was classified as raw gas, gas that had a primary cleaning, or gas that has had a final cleaning. For purposes of combustion calculations, however, it is more convenient to speak of *washed* and *unwashed* gas, and these will be the terms most frequently employed in the discussion which follows.

Washed gas is blast-furnace gas that has gone through either primary or final washers. It may be considered saturated with moisture at whatever temperature it leaves the washer because of the intimate contact between cooling water and hot gas. If the gas has only been given a primary wash, it will contain about 4 grains (0.00057 lb per standard cubic foot) of entrained moisture in addition to the water vapor required to saturate it. This entrained moisture is carried along by the gas in the form of suspended droplets which are effectively removed by a final cleaner.

Unwashed gas, on the other hand, may exist as raw gas or as gas that has had a primary, or even a final, cleaning by means of dry separators. Its temperature is much higher than that of washed gas and because of this it is capable of holding much more moisture, since it will contain all the water vapor that it had on leaving the blast furnace regardless of its degree of cleaning. Using a dew point of 140°F has been recommended for calculating the moisture in unwashed gas, but it is obviously preferable to know the actual dew point of a particular gas than to assume it. Fig. 20, then, offers a convenient curve for determining W_s, the saturation moisture, for any dew-point temperature from 60 to 160°F. There is, of course, no entrained water in unwashed gas.

DENSITY

The density of dry blast-furnace gas at 60°F (d_{60}) is easily computed from its volumetric analysis by the method explained in the section on gas, which assumes that the gas is clean. The true density (d_t) is obtained by correcting d_{60} for the actual temperature (t), moisture content, and dust according to Eq. 13.

As an example, assume that the blast-furnace gas in the first line of Table X is unwashed and goes to a burner with 15 grains of dust, at 500°F with a dew point of 140°F and at a pressure of 30 in. Hg. What is its actual density if there is no entrained moisture?

For a saturation temperature of 140°F and a pressure of 30 in. Hg, read from Fig. 20, $w_s = 0.152$ lb per lb of dry gas, and from Fig. 21

Table X. Characteristics of Typical Blast-Furnace Gases at 60°F and 30 In. Hg, Dry

CO₂	N₂	% by Volume CO	H₂	CH₄	Density, Lb/Cu Ft	HHV Btu/Cu Ft*	Btu/Lb	A at Zero Excess Air, Lb/10⁶ Btu	CO₂ at Zero Excess Air, %
14.5	57.5	25.0	3.0	...	0.0779	90.1	1150	576	26.4
13.0	57.6	26.2	3.2	...	0.0771	94.6	1219	575	25.8
15.59	59.28	23.35	1.7	0.08	0.0792	81.4	1021	577	26.7
8.7	56.5	32.8	1.8	0.2	0.0762	113.3	1478	579	25.3
5.7	59.0	34.0	1.3	...	0.0753	113.6	1498	576	24.0
6.0	60.0	27.0	2.0	5.0	0.0734	144.0	1950	631	20.0

* If gas is saturated with moisture at 60° and 30.0 in. Hg, reduce by 1.74%

	% by Volume, Dry	Volume, cu ft/ cu ft of Gas	Density lb/cu ft at 60°F, Dry[2]		Weight lb/cu ft
CO_2	14.5	0.145	× 0.1170	=	0.01697
N_2	57.5	0.575	× 0.07439	=	0.04277
CO	25.0	0.250	× 0.07404	=	0.01851
H_2	3.0	0.030	× 0.005327	=	0.00016
	Density of Clean Dry Gas at 60°F			=	0.07841

obtain X = 0.0115 and Y = 2.29. Then the density of clean gas at 500°F and a dew point of 140°F will be expressed as

$$\frac{d_{60} + X}{Y} = \frac{0.0784 + 0.0115}{2.29} = 0.0393 \text{ lb/cu ft}$$

$$\text{Weight of dust in gas} = \frac{15}{7000} \times \frac{0.0393}{0.0784} = 0.0011 \text{ lb/cu ft}$$

$$d_t = 0.0393 + 0.0011 = 0.0404 \text{ lb/cu ft}$$

HEATING VALUE

The Btu per cubic foot of clean and dry blast-furnace gas at 60°F, may be calculated from its analysis and the heating value of its combustible components as obtained from Table X. The Btu per cubic foot thus determined, when divided by the density at 60°F, will give the Btu per pound $(HHV)_{60}$.

However, in calculating the true heating value $(HHV)_t$ of a gas as fired, we must again take into account its dust content, its temperature, and its moisture content.

The dust in blast-furnace gas is made up chiefly of particles of iron ore, limestone, and coke. Since its structure is sandy or granular, there is reason to believe that iron ore is the main component in this mixture, since its structure is sandy or granular, and is, therefore, more easily lifted by the blast than coke or limestone which comes in relatively large lumps. When the dust burns, the coke in it may be thought of as supplying the heat required to reduce both ore and limestone. In the absence of actual tests showing the calorific value of the dust, it may be assumed that the heat evolved by the coke is just equal to that required to reduce the ore and the limestone. If this is true,

the presence of dust will not influence the heating value of blast-furnace gas.

It should also be clear that any increase in the temperature of the gas above 60°F will enhance its heating value by an amount equal to the sensible heat of the gas above 60°F.

Furthermore, assume, as an example, that an unwashed gas is saturated at 140°F. It will then have 0.152 lb of moisture for each pound of dry gas. As an inert vapor, this moisture will reduce the combustible fraction and, therefore, the heating value of blast-furnace gas, at the same time that it augments the heating value by the sensible heat due to its high temperature.

Having obtained the Btu per pound $(HHV)_{60}$, for the dry gas at 60°F, figure the Btu per pound $(HHV)_t$, at its true temperature, t, and moisture content using Eq. 14.

Knowing the actual dew point of this gas, w_s is read from Fig. 20, and M and N from Fig. 22 for use in Eq. 14.

FUEL, F

After determining the true Btu per pound, F, which is that portion of the fuel that shows up in the products of combustion, may be read from Fig. 1. Any dust in unwashed gas may be treated as ash in coal or sediment in oil.

ATMOSPHERIC AIR, A

The air required for the combustion of blast-furnace gas is a function of the ratio $CH_4/(CO + H_2)$ where CH_4, CO, and H_2 represent the percentage fractions by volume of methane, carbon monoxide, and hydrogen in the fuel.

For clean, dry, blast-furnace gas at 60°F, the atmospheric air, A, in pounds per million Btu, may be read from Curves A in Fig. 29 for any value of excess air up to 70 percent.

This value must be corrected for the dust and moisture content and temperature of the actual blast-furnace gas. After finding A from Fig. 29 and the heating value and density as explained before, the true mass of atmospheric air per million Btu as fired is calculated using Eq. 15.

TOTAL PRODUCTS, P

As with other fuels previously discussed, the weight of the products of combustion is given by P = F + CA lb per million Btu as fired. This equation holds true for either washed or unwashed gas, provided F and A are calculated as explained in the foregoing sections.

MOISTURE FROM FUEL, W_f

As defined in the first section, W_f is the total amount of water vapor present in the products of combustion.

In blast-furnace gas, three different moisture sources eventually become part of the products. First, there is the moisture formed in burning the fuel and previously designated by W_h. Equation 26 shows that for clean gas at 60°F

$$W_h = \frac{47,300\,(H_2 + 2CH_4)}{\text{Btu/cu ft at 60°F and 30 in. Hg, dry}}$$

(26)

Curves B of Fig. 29 are plotted from this equation.

If the actual gas is at a temperature higher than 60°F and contains dust and moisture, the readings from Curves B must be multiplied by the same factor used in correcting the atmospheric air, to obtain the true moisture $(W_h)_t$ in pounds per million Btu as fired. This factor is calculated using Eq. 18.

The next moisture source called "entrained moisture" is carried along by the gas in the form of suspended globules. Designated as W_c, it is considered only when the gas has been given a primary wash and averages 7 lb per million Btu.

The last source is in the water vapor required to saturate the blast-furnace gas at any temperature from 60°F up, as the case may be, and is designated as W_s. For blast-furnace gas saturated with moisture at 60°F, 8 lb per million Btu may be assumed for W_s without serious error. This moisture must be considered separately from W_h and W_c in a heat balance since it exists as a vapor requiring no latent heat of vaporization.

Of course, W_s must be subtracted from the total products, along with W_h and W_c, to obtain the dry gas.

EXAMPLE

Assume a blast-furnace gas with the typical analysis of the first gas listed in Table X and saturated with moisture of 60°F and burned with 15-percent excess air after going through a primary washer.

1. Density. The density given as 0.0779 lb per cu ft for the dry gas must first be corrected for moisture content. The saturation moisture for a dew point of 60°F is read from Fig. 20 as $W_s = 0.02$ lb moisture per lb of dry gas. The values of X and Y are read from Fig. 21 for $W_s = 0.02$ and a temperature of 60°F as X = 0.0015 and Y = 1.03. The entrained moisture in a gas given a primary wash is assumed as $w_c = 0.00057$ lb per standard cu ft.

We can now substitute into Eq. 13

$$d_t = \frac{d_{60} + X}{Y} + (D_t + w_c) = \frac{0.0779 + 0.0015}{1.03} +$$

$$(0 + 0.00057) = 0.0771 + (0.00057) = 0.0777 \text{ lb/cu ft.}$$

2. Heating Value. The high heat value given as 90.1 Btu per cu ft is first corrected for moisture content by deducting 1.74 percent; 90.1 − 1.6 = 88.5 Btu per cu ft saturated fuel. This value is divided by the density to obtain the Btu per lb. HHV ≐ 88.5/0.0777 = 1139 Btu per lb.

3. Fuel, F. For a high heat value of 1139 Btu per lb and a negligible amount of dust refer to Fig. 1 and read F = 878 lb per million Btu.

4. Atmospheric Air, A. From Curves A, Fig. 29, for $CH_4/(CO + H_2) = 0$ and for 15-percent excess air, read A = 662 lb per million Btu.

5. Unburned Combustible. The general assumption when burning blast-furnace gas in stationary boiler furnaces is that the combustible loss is zero. Consequently, C in Eqs. 3 and 4 may be taken as 1.

6. Total Products, P. From Eq. 4, P = F + CA = 878 + 1 × 662 = 1540 lb per million Btu.

7. Moisture in Air, W_a. From Eq. 5, $W_a = 0.013A = 0.013 \times 662 = 9$ lb per million Btu.

8. Moisture from Fuel, W_f. As pointed out in the text for a gas that has passed through a primary washer, $W_c = 7$ lb per million Btu. Curves B, Fig. 29, show $W_h = 16$ lb per million Btu for a blast-furnace gas with a high heat value of 90.1 Btu per cu ft and $H_2 + 2CH_4 = 3 + 0 = 3$ percent by volume as fired. Therefore, $W_f = W_s + W_c + W_h = 8 + 7 + 16 = 31$ lb per million Btu.

9. Dry Gas, P_d. From Eq. 6, $P_d = P - (W_a + W_f) = 1540 - (9 + 31) = 1500$ lb per million Btu.

10. Percent CO_2 in Products. From the fuel analysis, calculate $CO_2 + CO + CH_4 = 14.5 + 25.0 + 0 = 39.5$ percent by volume and $H_2 + 2CO + 8CH_4 = 3 + 2 \times 25 + 8 \times 0 = 53$ percent by volume, and from Curves C, Fig. 29 read 26.4 percent CO_2 at zero excess air. Following Curves D as guidelines to 15-percent excess air, read $CO_2 = 24.8$ percent.

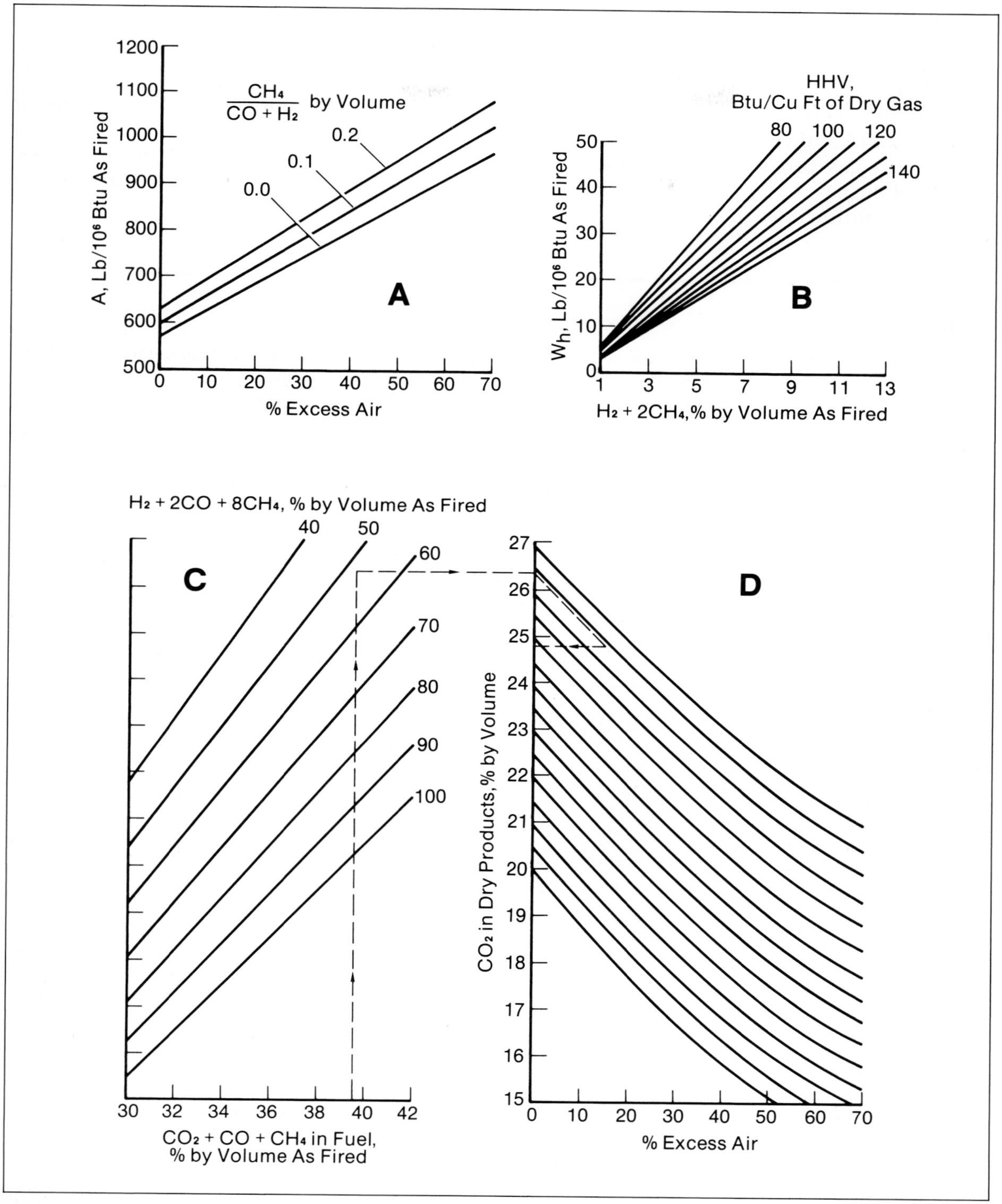

Fig. 29. **Combustion characteristics of blast-furnace gas**

PERCENT CO₂ IN PRODUCTS

Curves C and D of Fig. 29 offer a convenient method to obtain the percent of CO_2 in the dry products of combustion resulting from the combustion of either washed or unwashed blast-furnace gas, provided the dry volumetric analysis of the gas is known. Curves C are plotted from the following equation:

$$\frac{\% \, CO_2}{\text{(at zero excess air)}} = \frac{100(CO_2 + CO + CH_4)}{100 + 0.94(H_2 + 2CO + 8CH_4)} \tag{27}$$

where $H_2, CO_2, CO,$ and CH_4 are the symbols for percent by volume of these constituents in the fuel. This CO_2 must not be confused with the percent carbon dioxide in the dry products of combustion.

REFERENCES

[1] W. L. De Baufre, "Composition and Heating Value of Fuels," *Combustion*, 2(11):29–35, May 1931.

[2] W. S. Patterson, "Calculations of Air Requirements and Combustion Products of U.S. Coals by Simple Graphical Methods," *Combustion*, 3(8): 9–17, Feb. 1932.

[3] W. L. De Baufre, "Typical Solid and Liquid Fuels," *Combustion*, 3(2): 23–31, Aug. 1931.

[4] W. L. De Baufre, "Typical Gaseous Fuels," *Combustion*, 3(6): 26–33, Dec. 1931.

Appendix D. Steam Tables

The following tables are in accordance with those adopted by The Sixth International Conference on the Properties of Steam and published in the *1967 ASME Steam Tables* and the *1977 ASME Steam Tables in SI (Metric) Units* by the American Society of Mechanical Engineers. Combustion Engineering, Inc., has permission to reprint the tables in this form. For a copy of *Steam Tables,* with a convenient Molier diagram included, write to C-E Power Systems, Combustion Engineering, Inc., 1000 Prospect Hill Rd., Windsor, CT 06095.

A thorough historical account of the development of formulations of the thermodynamic properties of steam is presented in Appendix D, *Combustion Engineering,* 1967 edition. It describes the early research begun on steam properties in 1741 and fully covers the proceedings of the Sixth International Conference on the Properties of Steam held between 1958 and 1963.

REFERENCE STATE

The liquid phase at the triple point of water substance is the state for which the specific internal energy and the specific entropy are each made exactly zero.

U.S. CUSTOMARY UNITS

Table I. Saturated Steam: Temperature Table (32 to 705.47°F)

Temp. Deg. Fahr. t	Abs. Press. $Lb_f/$ Sq. In. p	Specific Vol., ft³/lbm			Enthalpy, Btu/lbm			Entropy, Btu/lbmR			Temp. Deg. Fahr. t
		Sat. Liquid v_f	Evap. v_{fg}	Sat. Vapor v_g	Sat. Liquid h_f	Evap. h_{fg}	Sat. Vapor h_g	Sat. Liquid s_f	Evap. s_{fg}	Sat. Vapor s_g	
32.0*	0.08859	0.016022	3304.7	3304.7	−0.0179	1075.5	1075.5	0.0000	2.1873	2.1873	32.0*
34.0	0.09600	0.016021	3061.9	3061.9	1.996	1074.4	1076.4	0.0041	2.1762	2.1802	34.0
36.0	0.10395	0.016020	2839.0	2839.0	4.008	1073.2	1077.2	0.0081	2.1651	2.1732	36.0
38.0	0.11249	0.016019	2634.1	2634.2	6.018	1072.1	1078.1	0.0122	2.1541	2.1663	38.0
40.0	0.12163	0.016019	2445.8	2445.8	8.027	1071.0	1079.0	0.0162	2.1432	2.1594	40.0
42.0	0.13143	0.016019	2272.4	2272.4	10.035	1069.8	1079.9	0.0202	2.1325	2.1527	42.0
44.0	0.14192	0.016019	2112.8	2112.8	12.041	1068.7	1080.7	0.0242	2.1217	2.1459	44.0
46.0	0.15314	0.016020	1965.7	1965.7	14.047	1067.6	1081.6	0.0282	2.1111	2.1393	46.0
48.0	0.16514	0.016021	1830.0	1830.0	16.051	1066.4	1082.5	0.0321	2.1006	2.1327	48.0
50.0	0.17796	0.016023	1704.8	1704.8	18.054	1065.3	1083.4	0.0361	2.0901	2.1262	50.0
52.0	0.19165	0.016024	1589.2	1589.2	20.057	1064.2	1084.2	0.0400	2.0798	2.1197	52.0
54.0	0.20625	0.016026	1482.4	1482.4	22.058	1063.1	1085.1	0.0439	2.0695	2.1134	54.0
56.0	0.22183	0.016028	1383.6	1383.6	24.059	1061.9	1086.0	0.0478	2.0593	2.1070	56.0
58.0	0.23843	0.016031	1292.2	1292.2	26.060	1060.8	1086.9	0.0516	2.0491	2.1008	58.0
60.0	0.25611	0.016033	1207.6	1207.6	28.060	1059.7	1087.7	0.0555	2.0391	2.0946	60.0
62.0	0.27494	0.016036	1129.2	1129.2	30.059	1058.5	1088.6	0.0593	2.0291	2.0885	62.0
64.0	0.29497	0.016039	1056.5	1056.5	32.058	1057.4	1089.5	0.0632	2.0192	2.0824	64.0
66.0	0.31626	0.016043	989.0	989.1	34.056	1056.3	1090.4	0.0670	2.0094	2.0764	66.0
68.0	0.33889	0.016046	926.5	926.5	36.054	1055.2	1091.2	0.0708	1.9996	2.0704	68.0
70.0	0.36292	0.016050	868.3	868.4	38.052	1054.0	1092.1	0.0745	1.9900	2.0645	70.0
72.0	0.38844	0.016054	814.3	814.3	40.049	1052.9	1093.0	0.0783	1.9804	2.0587	72.0
74.0	0.41550	0.016058	764.1	764.1	42.046	1051.8	1093.8	0.0821	1.9708	2.0529	74.0
76.0	0.44420	0.016063	717.4	717.4	44.043	1050.7	1094.7	0.0858	1.9614	2.0472	76.0
78.0	0.47461	0.016067	673.8	673.9	46.040	1049.5	1095.6	0.0895	1.9520	2.0415	78.0

Table I. Saturated Steam: Temperature Table — *Continued*

Temp. Deg. Fahr. t	Abs. Press. Lb./ Sq. In. p	Specific Vol., ft³/lbm			Enthalpy, Btu/lbm			Entropy, Btu/lbmR			Temp. Deg. Fahr. t
		Sat. Liquid v_f	Evap. v_{fg}	Sat. Vapor v_g	Sat. Liquid h_f	Evap. h_{fg}	Sat. Vapor h_g	Sat. Liquid s_f	Evap. s_{fg}	Sat. Vapor s_g	
80.0	0.50683	0.016072	633.3	633.3	48.037	1048.4	1096.4	0.0932	1.9426	2.0359	80.0
82.0	0.54093	0.016077	595.5	595.5	50.033	1047.3	1097.3	0.0969	1.9334	2.0303	82.0
84.0	0.57702	0.016082	560.3	560.3	52.029	1046.1	1098.2	0.1006	1.9242	2.0248	84.0
86.0	0.61518	0.016087	527.5	527.5	54.026	1045.0	1099.0	0.1043	1.9151	2.0193	86.0
88.0	0.65551	0.016093	496.8	496.8	56.022	1043.9	1099.9	0.1079	1.9060	2.0139	88.0
90.0	0.69813	0.016099	468.1	468.1	58.018	1042.7	1100.8	0.1115	1.8970	2.0086	90.0
92.0	0.74313	0.016105	441.3	441.3	60.014	1041.6	1101.6	0.1152	1.8881	2.0033	92.0
94.0	0.79062	0.016111	416.3	416.3	62.010	1040.5	1102.5	0.1188	1.8792	1.9980	94.0
96.0	0.84072	0.016117	392.8	392.9	64.006	1039.3	1103.3	0.1224	1.8704	1.9928	96.0
98.0	0.89356	0.016123	370.9	370.9	66.003	1038.2	1104.2	0.1260	1.8617	1.9876	98.0
100.0	0.94924	0.016130	350.4	350.4	67.999	1037.1	1105.1	0.1295	1.8530	1.9825	100.0
102.0	1.00789	0.016137	331.1	331.1	69.995	1035.9	1105.9	0.1331	1.8444	1.9775	102.0
104.0	1.06965	0.016144	313.1	313.1	71.992	1034.8	1106.8	0.1366	1.8358	1.9725	104.0
106.0	1.1347	0.016151	296.16	296.18	73.99	1033.6	1107.6	0.1402	1.8273	1.9675	106.0
108.0	1.2030	0.016158	280.28	280.30	75.98	1032.5	1108.5	0.1437	1.8188	1.9626	108.0
110.0	1.2750	0.016165	265.37	265.39	77.98	1031.4	1109.3	0.1472	1.8105	1.9577	110.0
112.0	1.3505	0.016173	251.37	251.38	79.98	1030.2	1110.2	0.1507	1.8021	1.9528	112.0
114.0	1.4299	0.016180	238.21	238.22	81.97	1029.1	1111.0	0.1542	1.7938	1.9480	114.0
116.0	1.5133	0.016188	225.84	225.85	83.97	1027.9	1111.9	0.1577	1.7856	1.9433	116.0
118.0	1.6009	0.016196	214.20	214.21	85.97	1026.8	1112.7	0.1611	1.7774	1.9386	118.0
120.0	1.6927	0.016204	203.25	203.26	87.97	1025.6	1113.6	0.1646	1.7693	1.9339	120.0
122.0	1.7891	0.016213	192.94	192.95	89.96	1024.5	1114.4	0.1680	1.7613	1.9293	122.0
124.0	1.8901	0.016221	183.23	183.24	91.96	1023.3	1115.3	0.1715	1.7533	1.9247	124.0
126.0	1.9959	0.016229	174.08	174.09	93.96	1022.2	1116.1	0.1749	1.7453	1.9202	126.0
128.0	2.1068	0.016238	165.45	165.47	95.96	1021.0	1117.0	0.1783	1.7374	1.9157	128.0
130.0	2.2230	0.016247	157.32	157.33	97.96	1019.8	1117.8	0.1817	1.7295	1.9112	130.0
132.0	2.3445	0.016256	149.64	149.66	99.95	1018.7	1118.6	0.1851	1.7217	1.9068	132.0
134.0	2.4717	0.016265	142.40	142.41	101.95	1017.5	1119.5	0.1884	1.7140	1.9024	134.0
136.0	2.6047	0.016274	135.55	135.57	103.95	1016.4	1120.3	0.1918	1.7063	1.8980	136.0
138.0	2.7438	0.016284	129.09	129.11	105.95	1015.2	1121.1	0.1951	1.6986	1.8937	138.0
140.0	2.8892	0.016293	122.98	123.00	107.95	1014.0	1122.0	0.1985	1.6910	1.8895	140.0
144.0	3.1997	0.016312	111.74	111.76	111.95	1011.7	1123.6	0.2051	1.6759	1.8810	144.0
148.0	3.5381	0.016332	101.68	101.70	115.95	1009.3	1125.3	0.2117	1.6610	1.8727	148.0
152.0	3.9065	0.016353	92.66	92.68	119.95	1007.0	1126.9	0.2183	1.6463	1.8646	152.0
156.0	4.3068	0.016374	84.56	84.57	123.95	1004.6	1128.6	0.2248	1.6318	1.8566	156.0
160.0	4.7414	0.016395	77.27	77.29	127.96	1002.2	1130.2	0.2313	1.6174	1.8487	160.0
164.0	5.2124	0.016417	70.70	70.72	131.96	999.8	1131.8	0.2377	1.6032	1.8409	164.0
168.0	5.7223	0.016440	64.78	64.80	135.97	997.4	1133.4	0.2441	1.5892	1.8333	168.0
172.0	6.2736	0.016463	59.43	59.45	139.98	995.0	1135.0	0.2505	1.5753	1.8258	172.0
176.0	6.8690	0.016486	54.59	54.61	143.99	992.6	1136.6	0.2568	1.5616	1.8184	176.0
180.0	7.5110	0.016510	50.21	50.22	148.00	990.2	1138.2	0.2631	1.5480	1.8111	180.0
184.0	8.203	0.016534	46.232	46.249	152.01	987.8	1139.8	0.2694	1.5346	1.8040	184.0
188.0	8.947	0.016559	42.621	42.638	156.03	985.3	1141.3	0.2756	1.5213	1.7969	188.0
192.0	9.747	0.016585	39.337	39.354	160.05	982.8	1142.9	0.2818	1.5082	1.7900	192.0
196.0	10.605	0.016611	36.348	36.364	164.06	980.4	1144.4	0.2879	1.4952	1.7831	196.0

Table I. Saturated Steam: Temperature Table — *Continued*

Temp. Deg. Fahr. t	Abs. Press. Lb/ Sq. In. p	Specific Vol., ft³/lbm			Enthalpy, Btu/lbm			Entropy, Btu/lbmR			Temp. Deg. Fahr. t
		Sat. Liquid v_f	Evap. v_{fg}	Sat. Vapor v_g	Sat. Liquid h_f	Evap. h_{fg}	Sat. Vapor h_g	Sat. Liquid s_f	Evap. s_{fg}	Sat. Vapor s_g	
200.0	11.526	0.016637	33.622	33.639	168.09	977.9	1146.0	0.2940	1.4824	1.7764	200.0
204.0	12.512	0.016664	31.135	31.151	172.11	975.4	1147.5	0.3001	1.4697	1.7698	204.0
208.0	13.568	0.016691	28.862	28.878	176.14	972.8	1149.0	0.3061	1.4571	1.7632	208.0
212.0	14.696	0.016719	26.782	26.799	180.17	970.3	1150.5	0.3121	1.4447	1.7568	212.0
216.0	15.901	0.016747	24.878	24.894	184.20	967.8	1152.0	0.3181	1.4323	1.7505	216.0
220.0	17.186	0.016775	23.131	23.148	188.23	965.2	1153.4	0.3241	1.4201	1.7442	220.0
224.0	18.556	0.016805	21.529	21.545	192.27	962.6	1154.9	0.3300	1.4081	1.7380	224.0
228.0	20.015	0.016834	20.056	20.073	196.31	960.0	1156.3	0.3359	1.3961	1.7320	228.0
232.0	21.567	0.016864	18.701	18.718	200.35	957.4	1157.8	0.3417	1.3842	1.7260	232.0
236.0	23.216	0.016895	17.454	17.471	204.40	954.8	1159.2	0.3476	1.3725	1.7201	236.0
240.0	24.968	0.016926	16.304	16.321	208.45	952.1	1160.6	0.3533	1.3609	1.7142	240.0
244.0	26.826	0.016958	15.243	15.260	212.50	949.5	1162.0	0.3591	1.3494	1.7085	244.0
248.0	28.796	0.016990	14.264	14.281	216.56	946.8	1163.4	0.3649	1.3379	1.7028	248.0
252.0	30.883	0.017022	13.358	13.375	220.62	944.1	1164.7	0.3706	1.3266	1.6972	252.0
256.0	33.091	0.017055	12.520	12.538	224.69	941.4	1166.1	0.3763	1.3154	1.6917	256.0
260.0	35.427	0.017089	11.745	11.762	228.76	938.6	1167.4	0.3819	1.3043	1.6862	260.0
264.0	37.894	0.017123	11.025	11.042	232.83	935.9	1168.7	0.3876	1.2933	1.6808	264.0
268.0	40.500	0.017157	10.358	10.375	236.91	933.1	1170.0	0.3932	1.2823	1.6755	268.0
272.0	43.249	0.017193	9.738	9.755	240.99	930.3	1171.3	0.3987	1.2715	1.6702	272.0
276.0	46.147	0.017228	9.162	9.180	245.08	927.5	1172.5	0.4043	1.2607	1.6650	276.0
280.0	49.200	0.017264	8.627	8.644	249.17	924.6	1173.8	0.4098	1.2501	1.6599	280.0
284.0	52.414	0.01730	8.1280	8.1453	253.3	921.7	1175.0	0.4154	1.2395	1.6548	284.0
288.0	55.795	0.01734	7.6634	7.6807	257.4	918.8	1176.2	0.4208	1.2290	1.6498	288.0
292.0	59.350	0.01738	7.2301	7.2475	261.5	915.9	1177.4	0.4263	1.2186	1.6449	292.0
296.0	63.084	0.01741	6.8259	6.8433	265.6	913.0	1178.6	0.4317	1.2082	1.6400	296.0
300.0	67.005	0.01745	6.4483	6.4658	269.7	910.0	1179.7	0.4372	1.1979	1.6351	300.0
304.0	71.119	0.01749	6.0955	6.1130	273.8	907.0	1180.9	0.4426	1.1877	1.6303	304.0
308.0	75.433	0.01753	5.7655	5.7830	278.0	904.0	1182.0	0.4479	1.1776	1.6256	308.0
312.0	79.953	0.01757	5.4566	5.4742	282.1	901.0	1183.1	0.4533	1.1676	1.6209	312.0
316.0	84.688	0.01761	5.1673	5.1849	286.3	897.9	1184.1	0.4586	1.1576	1.6162	316.0
320.0	89.643	0.01766	4.8961	4.9138	290.4	894.8	1185.2	0.4640	1.1477	1.6116	320.0
324.0	94.826	0.01770	4.6418	4.6595	294.6	891.6	1186.2	0.4692	1.1378	1.6071	324.0
328.0	100.245	0.01774	4.4030	4.4208	298.7	888.5	1187.2	0.4745	1.1280	1.6025	328.0
332.0	105.907	0.01779	4.1788	4.1966	302.9	885.3	1188.2	0.4798	1.1183	1.5981	332.0
336.0	111.820	0.01783	3.9681	3.9859	307.1	882.1	1189.1	0.4850	1.1086	1.5936	336.0
340.0	117.992	0.01787	3.7699	3.7878	311.3	878.8	1190.1	0.4902	1.0990	1.5892	340.0
344.0	124.430	0.01792	3.5834	3.6013	315.5	875.5	1191.0	0.4954	1.0894	1.5849	344.0
348.0	131.142	0.01797	3.4078	3.4258	319.7	872.2	1191.1	0.5006	1.0799	1.5806	348.0
352.0	138.138	0.01801	3.2423	3.2603	323.9	868.9	1192.7	0.5058	1.0705	1.5763	352.0
356.0	145.424	0.01806	3.0863	3.1044	328.1	865.5	1193.6	0.5110	1.0611	1.5721	356.0
360.0	153.010	0.01811	2.9392	2.9573	332.3	862.1	1194.4	0.5161	1.0517	1.5678	360.0
364.0	160.903	0.01816	2.8002	2.8184	336.5	858.6	1195.2	0.5212	1.0424	1.5637	364.0
368.0	169.113	0.01821	2.6691	2.6873	340.8	855.1	1195.9	0.5263	1.0332	1.5595	368.0
372.0	177.648	0.01826	2.5451	2.5633	345.0	851.6	1196.7	0.5314	1.0240	1.5554	372.0
376.0	186.517	0.01831	2.4279	2.4462	349.3	848.1	1197.4	0.5365	1.0148	1.5513	376.0

Table I. Saturated Steam: Temperature Table — *Continued*

Temp. Deg. Fahr. t	Abs. Press. $Lb_f/$ Sq. In. p	Specific Vol., ft³/lbm			Enthalpy, Btu/lbm			Entropy, Btu/lbmR			Temp. Deg. Fahr. t
		Sat. Liquid v_f	Evap. v_{fg}	Sat. Vapor v_g	Sat. Liquid h_f	Evap. h_{fg}	Sat. Vapor h_g	Sat. Liquid s_f	Evap. s_{fg}	Sat. Vapor s_g	
380.0	195.729	0.01836	2.3170	2.3353	353.6	844.5	1198.0	0.5416	1.0057	1.5473	380.0
384.0	205.294	0.01842	2.2120	2.2304	357.9	840.8	1198.7	0.5466	0.9966	1.5432	384.0
388.0	215.220	0.01847	2.1126	2.1311	362.2	837.2	1199.3	0.5516	0.9876	1.5392	388.0
392.0	225.516	0.01853	2.0184	2.0369	366.5	833.4	1199.9	0.5567	0.9786	1.5352	392.0
396.0	236.193	0.01858	1.9291	1.9477	370.8	829.7	1200.4	0.5617	0.9696	1.5313	396.0
400.0	247.259	0.01864	1.8444	1.8630	375.1	825.9	1201.0	0.5667	0.9607	1.5274	400.0
404.0	258.725	0.01870	1.7640	1.7827	379.4	822.0	1201.5	0.5717	0.9518	1.5234	404.0
408.0	270.600	0.01875	1.6877	1.7064	383.8	818.2	1201.9	0.5766	0.9429	1.5195	408.0
412.0	282.894	0.01881	1.6152	1.6340	388.1	814.2	1202.4	0.5816	0.9341	1.5157	412.0
416.0	295.617	0.01887	1.5463	1.5651	392.5	810.2	1202.8	0.5866	0.9253	1.5118	416.0
420.0	308.780	0.01894	1.4808	1.4997	396.9	806.2	1203.1	0.5915	0.9165	1.5080	420.0
424.0	322.391	0.01900	1.4184	1.4374	401.3	802.2	1203.5	0.5964	0.9077	1.5042	424.0
428.0	336.463	0.01906	1.3591	1.3782	405.7	798.0	1203.7	0.6014	0.8990	1.5004	428.0
432.0	351.00	0.01913	1.30266	1.32179	410.1	793.9	1204.0	0.6063	0.8903	1.4966	432.0
436.0	366.03	0.01919	1.24887	1.26806	414.6	789.7	1204.2	0.6112	0.8816	1.4928	436.0
440.0	381.54	0.01926	1.19761	1.21687	419.0	785.4	1204.4	0.6161	0.8729	1.4890	440.0
444.0	397.56	0.01933	1.14874	1.16806	423.5	781.1	1204.6	0.6210	0.8643	1.4853	444.0
448.0	414.09	0.01940	1.10212	1.12152	428.0	776.7	1204.7	0.6259	0.8557	1.4815	448.0
452.0	431.14	0.01947	1.05764	1.07711	432.5	772.3	1204.8	0.6308	0.8471	1.4778	452.0
456.0	448.73	0.01954	1.01518	1.03472	437.0	767.8	1204.8	0.6356	0.8385	1.4741	456.0
460.0	466.87	0.01961	0.97463	0.99424	441.5	763.2	1204.8	0.6405	0.8299	1.4704	460.0
464.0	485.56	0.01969	0.93588	0.95557	446.1	758.6	1204.7	0.6454	0.8213	1.4667	464.0
468.0	504.83	0.01976	0.89885	0.91862	450.7	754.0	1204.6	0.6502	0.8127	1.4629	468.0
472.0	524.67	0.01984	0.86345	0.88329	455.2	749.3	1204.5	0.6551	0.8042	1.4592	472.0
476.0	545.11	0.01992	0.82958	0.84950	459.9	744.5	1204.3	0.6599	0.7956	1.4555	476.0
480.0	566.15	0.02000	0.79716	0.81717	464.5	739.6	1204.1	0.6648	0.7871	1.4518	480.0
484.0	587.81	0.02009	0.76613	0.78622	469.1	734.7	1203.8	0.6696	0.7785	1.4481	484.0
488.0	610.10	0.02017	0.73641	0.75658	473.8	729.7	1203.5	0.6745	0.7700	1.4444	488.0
492.0	633.03	0.02026	0.70794	0.72820	478.5	724.6	1203.1	0.6793	0.7614	1.4407	492.0
496.0	656.61	0.02034	0.68065	0.70100	483.2	719.5	1202.7	0.6842	0.7528	1.4370	496.0
500.0	680.86	0.02043	0.65448	0.67492	487.9	714.3	1202.2	0.6890	0.7443	1.4333	500.0
504.0	705.78	0.02053	0.62938	0.64991	492.7	709.0	1201.7	0.6939	0.7357	1.4296	504.0
508.0	731.40	0.02062	0.60530	0.62592	497.5	703.7	1201.1	0.6987	0.7271	1.4258	508.0
512.0	757.72	0.02072	0.58218	0.60289	502.3	698.2	1200.5	0.7036	0.7185	1.4221	512.0
516.0	784.76	0.02081	0.55997	0.58079	507.1	692.7	1199.8	0.7085	0.7099	1.4183	516.0
520.0	812.53	0.02091	0.53864	0.55956	512.0	687.0	1199.0	0.7133	0.7013	1.4146	520.0
524.0	841.04	0.02102	0.51814	0.53916	516.9	681.3	1198.2	0.7182	0.6926	1.4108	524.0
528.0	870.31	0.02112	0.49843	0.51955	521.8	675.5	1197.3	0.7231	0.6839	1.4070	528.0
532.0	900.34	0.02123	0.47947	0.50070	526.8	669.6	1196.4	0.7280	0.6752	1.4032	532.0
536.0	931.17	0.02134	0.46123	0.48257	531.7	663.6	1195.4	0.7329	0.6665	1.3993	536.0
540.0	962.79	0.02146	0.44367	0.46513	536.8	657.5	1194.3	0.7378	0.6577	1.3954	540.0
544.0	995.22	0.02157	0.42677	0.44834	541.8	651.3	1193.1	0.7427	0.6489	1.3915	544.0
548.0	1028.49	0.02169	0.41048	0.43217	546.9	645.0	1191.9	0.7476	0.6400	1.3876	548.0
552.0	1062.59	0.02182	0.39479	0.41660	552.0	638.5	1190.6	0.7525	0.6311	1.3837	552.0
556.0	1097.55	0.02194	0.37966	0.40160	557.2	632.0	1189.2	0.7575	0.6222	1.3797	556.0

Table I. Saturated Steam: Temperature Table — *Continued*

Temp. Deg. Fahr. t	Abs. Press. Lb$_f$/ Sq. In. p	Specific Vol., ft³/lbm			Enthalpy, Btu/lbm			Entropy, Btu/lbmR			Temp. Deg. Fahr. t
		Sat. Liquid v_f	Evap. v_{fg}	Sat. Vapor v_g	Sat. Liquid h_f	Evap. h_{fg}	Sat. Vapor h_g	Sat. Liquid s_f	Evap. s_{fg}	Sat. Vapor s_g	
560.0	1133.38	0.02207	0.36507	0.38714	562.4	625.3	1187.7	0.7625	0.6132	1.3757	560.0
564.0	1170.10	0.02221	0.35099	0.37320	567.6	618.5	1186.1	0.7674	0.6041	1.3716	564.0
568.0	1207.72	0.02235	0.33741	0.35975	572.9	611.5	1184.5	0.7725	0.5950	1.3675	568.0
572.0	1246.26	0.02249	0.32429	0.34678	578.3	604.5	1182.7	0.7775	0.5859	1.3634	572.0
576.0	1285.74	0.02264	0.31162	0.33426	583.7	597.2	1180.9	0.7825	0.5766	1.3592	576.0
580.0	1326.17	0.02279	0.29937	0.32216	589.1	589.9	1179.0	0.7876	0.5673	1.3550	580.0
584.0	1367.7	0.02295	0.28753	0.31048	594.6	582.4	1176.9	0.7927	0.5580	1.3507	584.0
588.0	1410.0	0.02311	0.27608	0.29919	600.1	574.7	1174.8	0.7978	0.5485	1.3464	588.0
592.0	1453.3	0.02328	0.26499	0.28827	605.7	566.8	1172.6	0.8030	0.5390	1.3420	592.0
596.0	1497.8	0.02345	0.25425	0.27770	611.4	558.8	1170.2	0.8082	0.5293	1.3375	596.0
600.0	1543.2	0.02364	0.24384	0.26747	617.1	550.6	1167.7	0.8134	0.5196	1.3330	600.0
604.0	1589.7	0.02382	0.23374	0.25757	622.9	542.2	1165.1	0.8187	0.5097	1.3284	604.0
608.0	1637.3	0.02402	0.22394	0.24796	628.8	533.6	1162.4	0.8240	0.4997	1.3238	608.0
612.0	1686.1	0.02422	0.21442	0.23865	634.8	524.7	1159.5	0.8294	0.4896	1.3190	612.0
616.6	1735.9	0.02444	0.20516	0.22960	640.8	515.6	1156.4	0.8348	0.4794	1.3141	616.0
620.0	1786.9	0.02466	0.19615	0.22081	646.9	506.3	1153.2	0.8403	0.4689	1.3092	620.0
624.0	1839.0	0.02489	0.18737	0.21226	653.1	496.6	1149.8	0.8458	0.4583	1.3041	624.0
628.0	1892.4	0.02514	0.17880	0.20394	659.5	486.7	1146.1	0.8514	0.4474	1.2988	628.0
632.0	1947.0	0.02539	0.17044	0.19583	665.9	476.4	1142.2	0.8571	0.4364	1.2934	632.0
636.0	2002.8	0.02566	0.16226	0.18792	672.4	465.7	1138.1	0.8628	0.4251	1.2879	636.0
640.0	2059.9	0.02595	0.15427	0.18021	679.1	454.6	1133.7	0.8686	0.4134	1.2821	640.0
644.0	2118.3	0.02625	0.14644	0.17269	685.9	443.1	1129.0	0.8746	0.4015	1.2761	644.0
648.0	2178.1	0.02657	0.13876	0.16534	692.9	431.1	1124.0	0.8806	0.3893	1.2699	648.0
652.0	2239.2	0.02691	0.13124	0.15816	700.0	418.7	1118.7	0.8868	0.3767	1.2634	652.0
656.0	2301.7	0.02728	0.12387	0.15115	707.4	405.7	1113.1	0.8931	0.3637	1.2567	656.0
660.0	2365.7	0.02768	0.11663	0.14431	714.9	392.1	1107.0	0.8995	0.3502	1.2498	660.0
664.0	2431.1	0.02811	0.10947	0.13757	722.9	377.7	1100.6	0.9064	0.3361	1.2425	664.0
668.0	2498.1	0.02858	0.10229	0.13087	731.5	362.1	1093.5	0.9137	0.3210	1.2347	668.0
672.0	2566.6	0.02911	0.09514	0.12424	740.2	345.7	1085.9	0.9212	0.3054	1.2266	672.0
676.0	2636.8	0.02970	0.08799	0.11769	749.2	328.5	1077.6	0.9287	0.2892	1.2179	676.0
680.0	2708.6	0.03037	0.08080	0.11117	758.5	310.1	1068.5	0.9365	0.2720	1.2086	680.0
684.0	2782.1	0.03114	0.07349	0.10463	768.2	290.2	1058.4	0.9447	0.2537	1.1984	684.0
688.0	2857.4	0.03204	0.06595	0.09799	778.8	268.2	1047.0	0.9535	0.2337	1.1872	688.0
692.0	2934.5	0.03313	0.05797	0.09110	790.5	243.1	1033.6	0.9634	0.2110	1.1744	692.0
696.0	3013.4	0.03455	0.04916	0.08371	804.4	212.8	1017.2	0.9749	0.1841	1.1591	696.0
700.0	3094.3	0.03662	0.03857	0.07519	822.4	172.7	995.2	0.9901	0.1490	1.1390	700.0
702.0	3135.5	0.03824	0.03173	0.06997	835.0	144.7	979.7	1.0006	0.1246	1.1252	702.0
704.0	3177.2	0.04108	0.02192	0.06300	854.2	102.0	956.2	1.0169	0.0876	1.1046	704.0
705.0	3198.3	0.04427	0.01304	0.05730	873.0	61.4	934.4	1.0329	0.0527	1.0856	705.0
705.47*	3208.2	0.05078	0.00000	0.05078	906.0	0.0	906.0	1.0612	0.0000	1.0612	705.47*

*The states shown are metastable

Table II. Saturated Steam: Pressure Table

Abs Press. Lb/Sq In. p	Temp Fahr t	Specific Vol., ft³/lbm			Enthalpy, Btu/lbm			Entropy			Abs Press. Lb/Sq In. p
		Sat. Liquid v_f	Evap v_{fg}	Sat. Vapor v_g	Sat. Liquid h_f	Evap h_{fg}	Sat. Vapor h_g	Sat. Liquid s_f	Evap s_{fg}	Sat. Vapor s_g	
0.08865	32.018	0.016022	3302.4	3302.4	0.0003	1075.5	1075.5	0.0000	2.1872	2.1872	0.08865
0.25	59.323	0.016032	1235.5	1235.5	27.382	1060.1	1087.4	0.0542	2.0425	2.0967	0.25
0.50	79.586	0.016071	641.5	641.5	47.623	1048.6	1096.3	0.0925	1.9446	2.0370	0.50
1.0	101.74	0.016136	333.59	333.60	69.73	1036.1	1105.8	0.1326	1.8455	1.9781	1.0
5.0	162.24	0.016407	73.515	73.532	130.20	1000.9	1131.1	0.2349	1.6094	1.8443	5.0
10.0	193.21	0.016592	38.404	38.420	161.26	982.1	1143.3	0.2836	1.5043	1.7879	10.0
14.696	212.00	0.016719	26.782	26.799	180.17	970.3	1150.5	0.3121	1.4447	1.7568	14.696
15.0	213.03	0.016726	26.274	26.290	181.21	969.7	1150.9	0.3137	1.4415	1.7552	15.0
20.0	227.96	0.016834	20.070	20.087	196.27	960.1	1156.3	0.3358	1.3962	1.7320	20.0
30.0	250.34	0.017009	13.7266	13.7436	218.9	945.2	1164.1	0.3682	1.3313	1.6995	30.0
40.0	267.25	0.017151	10.4794	10.4965	236.1	933.6	1169.8	0.3921	1.2844	1.6765	40.0
50.0	281.02	0.017274	8.4967	8.5140	250.2	923.9	1174.1	0.4112	1.2474	1.6586	50.0
60.0	292.71	0.017383	7.1562	7.1736	262.2	915.4	1177.6	0.4273	1.2167	1.6440	60.0
70.0	302.93	0.017482	6.1875	6.2050	272.7	907.8	1180.6	0.4411	1.1905	1.6316	70.0
80.0	312.04	0.017573	5.4536	5.4711	282.1	900.9	1183.1	0.4534	1.1675	1.6208	80.0
90.0	320.28	0.017659	4.8779	4.8953	290.7	894.6	1185.3	0.4643	1.1470	1.6113	90.0
100.0	327.82	0.017740	4.4133	4.4310	298.5	888.6	1187.2	0.4743	1.1284	1.6027	100.0
110.0	334.79	0.01782	4.0306	4.0484	305.8	883.1	1188.9	0.4834	1.1115	1.5950	110.0
120.0	341.27	0.01789	3.7097	3.7275	312.6	877.8	1190.4	0.4919	1.0960	1.5879	120.0
130.0	347.33	0.01796	3.4364	3.4544	319.0	872.8	1191.7	0.4998	1.0815	1.5813	130.0
140.0	353.04	0.01803	3.2010	3.2190	325.0	868.0	1193.0	0.5071	1.0681	1.5752	140.0
150.0	358.43	0.01809	2.9958	3.0139	330.6	863.4	1194.1	0.5141	1.0554	1.5695	150.0
160.0	363.55	0.01815	2.8155	2.8336	336.1	859.0	1195.1	0.5206	1.0435	1.5641	160.0
170.0	368.42	0.01821	2.6556	2.6738	341.2	854.8	1196.0	0.5269	1.0322	1.5591	170.0
180.0	373.08	0.01827	2.5129	2.5312	346.2	850.7	1196.9	0.5328	1.0215	1.5543	180.0
190.0	377.53	0.01833	2.3847	2.4030	350.9	846.7	1197.6	0.5384	1.0113	1.5498	190.0
200.0	381.80	0.01839	2.2689	2.2873	355.5	842.8	1198.3	0.5438	1.0016	1.5454	200.0
210.0	385.91	0.01844	2.16373	2.18217	359.9	839.1	1199.0	0.5490	0.9923	1.5413	210.0
220.0	389.88	0.01850	2.06779	2.08629	364.2	835.4	1199.6	0.5540	0.9834	1.5374	220.0
230.0	393.70	0.01855	1.97991	1.99846	368.3	831.8	1200.1	0.5588	0.9748	1.5336	230.0
240.0	397.39	0.01860	1.89909	1.91769	372.3	828.4	1200.6	0.5634	0.9665	1.5299	240.0
250.0	400.97	0.01865	1.82452	1.84317	376.1	825.0	1201.1	0.5679	0.9585	1.5264	250.0
260.0	404.44	0.01870	1.75548	1.77418	379.9	821.6	1201.5	0.5722	0.9508	1.5230	260.0
270.0	407.80	0.01875	1.69137	1.71013	383.6	818.3	1201.9	0.5764	0.9433	1.5197	270.0
280.0	411.07	0.01880	1.63169	1.65049	387.1	815.1	1202.3	0.5805	0.9361	1.5166	280.0
290.0	414.25	0.01885	1.57597	1.59482	390.6	812.0	1202.6	0.5844	0.9291	1.5135	290.0
300.0	417.35	0.01889	1.52384	1.54274	394.0	808.9	1202.9	0.5882	0.9223	1.5105	300.0
350.0	431.73	0.01912	1.30642	1.32554	409.8	794.2	1204.0	0.6059	0.8909	1.4968	350.0
400.0	444.60	0.01934	1.14162	1.16095	424.2	780.4	1204.6	0.6217	0.8630	1.4847	400.0
450.0	456.28	0.01954	1.01224	1.03179	437.3	767.5	1204.8	0.6360	0.8378	1.4738	450.0
500.0	467.01	0.01975	0.90787	0.92762	449.5	755.1	1204.7	0.6490	0.8148	1.4639	500.0
550.0	476.94	0.01994	0.82183	0.84177	460.9	743.3	1204.3	0.6611	0.7936	1.4547	550.0
600.0	486.20	0.02013	0.74962	0.76975	471.7	732.0	1203.7	0.6723	0.7738	1.4461	600.0

Table II. Saturated Steam: Pressure Table — *Continued*

Abs Press. Lb/Sq In. p	Temp Fahr t	Specific Vol., ft³/lbm			Enthalpy, Btu/lbm			Entropy, Btu/lbmR			Abs Press. Lb/Sq In. p
		Sat. Liquid v_f	Evap v_{fg}	Sat. Vapor v_g	Sat. Liquid h_f	Evap h_{fg}	Sat. Vapor h_g	Sat. Liquid s_f	Evap s_{fg}	Sat. Vapor s_g	
650.0	494.89	0.02032	0.68811	0.70843	481.9	720.9	1202.8	0.6828	0.7552	1.4381	650.0
700.0	503.08	0.02050	0.63505	0.65556	491.6	710.2	1201.8	0.6928	0.7377	1.4304	700.0
750.0	510.84	0.02069	0.58880	0.60949	500.9	699.8	1200.7	0.7022	0.7210	1.4232	750.0
800.0	518.21	0.02087	0.54809	0.56896	509.8	689.6	1199.4	0.7111	0.7051	1.4163	800.0
850.0	525.24	0.02105	0.51197	0.53302	518.4	679.5	1198.0	0.7197	0.6899	1.4096	850.0
900.0	531.95	0.02123	0.47968	0.50091	526.7	669.7	1196.4	0.7279	0.6753	1.4032	900.0
950.0	538.39	0.02141	0.45064	0.47205	534.7	660.0	1194.7	0.7358	0.6612	1.3970	950.0
1000.0	544.58	0.02159	0.42436	0.44596	542.6	650.4	1192.9	0.7434	0.6476	1.3910	1000.0
1050.0	550.53	0.02177	0.40047	0.42224	550.1	640.9	1191.0	0.7507	0.6344	1.3851	1050.0
1100.0	556.28	0.02195	0.37863	0.40058	557.5	631.5	1189.1	0.7578	0.6216	1.3794	1100.0
1150.0	561.82	0.02214	0.35859	0.38073	564.8	622.2	1187.0	0.7647	0.6091	1.3738	1150.0
1200.0	567.19	0.02232	0.34013	0.36245	571.9	613.0	1184.8	0.7714	0.5969	1.3683	1200.0
1250.0	572.38	0.02250	0.32306	0.34556	578.8	603.8	1182.6	0.7780	0.5850	1.3630	1250.0
1300.0	577.42	0.02269	0.30722	0.32991	585.6	594.6	1180.2	0.7843	0.5733	1.3577	1300.0
1350.0	582.32	0.02288	0.29250	0.31537	592.3	585.4	1177.8	0.7906	0.5620	1.3525	1350.0
1400.0	587.07	0.02307	0.27871	0.30178	598.8	576.5	1175.3	0.7966	0.5507	1.3474	1400.0
1450.0	591.70	0.02327	0.26584	0.28911	605.3	567.4	1172.8	0.8026	0.5397	1.3423	1450.0
1500.0	596.20	0.02346	0.25372	0.27719	611.7	558.4	1170.1	0.8085	0.5288	1.3373	1500.0
1550.0	600.59	0.02366	0.24235	0.26601	618.0	549.4	1167.4	0.8142	0.5182	1.3324	1550.0
1600.0	604.87	0.02387	0.23159	0.25545	624.2	540.3	1164.5	0.8199	0.5076	1.3274	1600.0
1650.0	609.05	0.02407	0.22143	0.24551	630.4	531.3	1161.6	0.8254	0.4971	1.3225	1650.0
1700.0	613.13	0.02428	0.21178	0.23607	636.5	522.2	1158.6	0.8309	0.4867	1.3176	1700.0
1750.0	617.12	0.02450	0.20263	0.22713	642.5	513.1	1155.6	0.8363	0.4765	1.3128	1750.0
1800.0	621.02	0.02472	0.19390	0.21861	648.5	503.8	1152.3	0.8417	0.4662	1.3079	1800.0
1850.0	624.83	0.02495	0.18558	0.21052	654.5	494.6	1149.0	0.8470	0.4561	1.3030	1850.0
1900.0	628.56	0.02517	0.17761	0.20278	660.4	485.2	1145.6	0.8522	0.4459	1.2981	1900.0
1950.0	632.22	0.02541	0.16999	0.19540	666.3	475.8	1142.0	0.8574	0.4358	1.2931	1950.0
2000.0	635.80	0.02565	0.16266	0.18831	672.1	466.2	1138.3	0.8625	0.4256	1.2881	2000.0
2100.0	642.76	0.02615	0.14885	0.17501	683.8	446.7	1130.5	0.8727	0.4053	1.2780	2100.0
2200.0	649.45	0.02669	0.13603	0.16272	695.5	426.7	1122.2	0.8828	0.3848	1.2676	2200.0
2300.0	655.89	0.02727	0.12406	0.15133	707.2	406.0	1113.2	0.8929	0.3640	1.2569	2300.0
2400.0	662.11	0.02790	0.11287	0.14076	719.0	384.8	1103.7	0.9031	0.3430	1.2460	2400.0
2500.0	668.11	0.02859	0.10209	0.13068	731.7	361.6	1093.3	0.9139	0.3206	1.2345	2500.0
2600.0	673.91	0.02938	0.09172	0.12110	744.5	337.6	1082.0	0.9247	0.2977	1.2225	2600.0
2700.0	679.53	0.03029	0.08165	0.11194	757.3	312.3	1069.7	0.9356	0.2741	1.2097	2700.0
2800.0	684.96	0.03134	0.07171	0.10305	770.7	285.1	1055.8	0.9468	0.2491	1.1958	2800.0
2900.0	690.22	0.03262	0.06158	0.09420	785.1	254.7	1039.8	0.9588	0.2215	1.1803	2900.0
3000.0	695.33	0.03428	0.05073	0.08500	801.8	218.4	1020.3	0.9728	0.1891	1.1619	3000.0
3100.0	700.28	0.03681	0.03771	0.07452	824.0	169.3	993.3	0.9914	0.1460	1.1373	3100.0
3200.0	705.08	0.04472	0.01191	0.05663	875.5	56.1	931.6	1.0351	0.0482	1.0832	3200.0
3208.2*	705.47	0.05078	0.00000	0.05078	906.0	0.0	906.0	1.0612	0.0000	1.0612	3208.2*

*Critical pressure

Table III. Superheated Steam

Abs Press. Lb/Sq In. (Sat. Temp)		Sat. Water	Sat. Steam	200	250	300	350	400	450	500	600	700	800	900
1 (101.74)	Sh			98.26	148.26	198.26	248.26	298.26	348.26	398.26	498.26	598.26	698.26	798.26
	v	0.01614	333.6	392.5	422.4	452.3	482.1	511.9	541.7	571.5	631.1	690.7	750.3	809.8
	h	69.73	1105.8	1150.2	1172.9	1195.7	1218.7	1241.8	1265.1	1288.6	1336.1	1384.5	1433.7	1483.8
	s	0.1326	1.9781	2.0509	2.0841	2.1152	2.1445	2.1722	2.1985	2.2237	2.2708	2.3144	2.3551	2.3934
5 (162.24)	Sh			37.76	87.76	137.76	187.76	237.76	287.76	337.76	437.76	537.76	637.76	737.76
	v	0.01641	73.53	78.14	84.21	90.24	96.25	102.24	108.23	114.21	126.15	138.08	150.01	161.94
	h	130.20	1131.1	1148.6	1171.7	1194.8	1218.0	1241.3	1264.7	1288.2	1335.9	1384.3	1433.6	1483.7
	s	0.2349	1.8443	1.8716	1.9054	1.9369	1.9664	1.9943	2.0208	2.0460	2.0932	2.1369	2.1776	2.2159
10 (193.21)	Sh			6.79	56.79	106.79	156.79	206.79	256.79	306.79	406.79	506.79	606.79	706.79
	v	0.01659	38.42	38.84	41.93	44.98	48.02	51.03	54.04	57.04	63.03	69.00	74.98	80.94
	h	161.26	1143.3	1146.6	1170.2	1193.7	1217.1	1240.6	1264.1	1287.8	1335.5	1384.0	1433.4	1483.5
	s	0.2836	1.7879	1.7928	1.8273	1.8593	1.8892	1.9173	1.9439	1.9692	2.0166	2.0603	2.1011	2.1394
14.696 (212.00)	Sh				38.00	88.00	138.00	188.00	238.00	288.00	388.00	488.00	588.00	688.00
	v	.0167	26.799		28.42	30.52	32.60	34.67	36.72	38.77	42.86	46.93	51.00	55.06
	h	180.17	1150.5		1168.8	1192.6	1216.3	1239.9	1263.6	1287.4	1335.2	1383.8	1433.2	1483.4
	s	.3121	1.7568		1.7833	1.8158	1.8459	1.8743	1.9010	1.9265	1.9739	2.0177	2.0585	2.0969
15 (213.03)	Sh				36.97	86.97	136.97	186.97	236.97	286.97	386.97	486.97	586.97	686.97
	v	0.01673	26.290		27.837	29.899	31.939	33.963	35.977	37.985	41.986	45.978	49.964	53.946
	h	181.21	1150.9		1168.7	1192.5	1216.2	1239.9	1263.6	1287.3	1335.2	1383.8	1433.2	1483.4
	s	0.3137	1.7552		1.7809	1.8134	1.8437	1.8720	1.8988	1.9242	1.9717	2.0155	2.0563	2.0946
20 (227.96)	Sh				22.04	72.04	122.04	172.04	222.04	272.04	372.04	472.04	572.04	672.04
	v	0.01683	20.087		20.788	22.356	23.900	25.428	26.946	28.457	31.466	34.465	37.458	40.447
	h	196.27	1156.3		1167.1	1191.4	1215.4	1239.2	1263.0	1286.9	1334.9	1383.5	1432.9	1483.2
	s	0.3358	1.7320		1.7475	1.7805	1.8111	1.8397	1.8666	1.8921	1.9397	1.9836	2.0244	2.0628
25 (240.07)	Sh				9.93	59.93	109.93	159.93	209.93	259.93	359.93	459.93	559.93	659.93
	v	0.01693	16.301		16.558	17.829	19.076	20.307	21.527	22.740	25.153	27.557	29.954	32.348
	h	208.52	1160.6		1165.6	1190.2	1214.5	1238.5	1262.5	1286.4	1334.6	1383.3	1432.7	1483.0
	s	0.3535	1.7141		1.7212	1.7547	1.7856	1.8145	1.8415	1.8672	1.9149	1.9588	1.9997	2.0381
30 (250.34)	Sh					49.66	99.66	149.66	199.66	249.66	349.66	449.66	549.66	649.66
	v	0.01701	13.744			14.810	15.859	16.892	17.914	18.929	20.945	22.951	24.952	26.949
	h	218.93	1164.1			1189.0	1213.6	1237.8	1261.9	1286.0	1334.2	1383.0	1432.5	1482.8
	s	0.3682	1.6995			1.7334	1.7647	1.7937	1.8210	1.8467	1.8946	1.9386	1.9795	2.0179
35 (259.29)	Sh					40.71	90.71	140.71	190.71	240.71	340.71	440.71	540.71	640.71
	v	0.01708	11.896			12.654	13.562	14.453	15.334	16.207	17.939	19.662	21.379	23.092
	h	228.03	1167.1			1187.8	1212.7	1237.1	1261.3	1285.5	1333.9	1382.8	1432.3	1482.7
	s	0.3809	1.6872			1.7152	1.7468	1.7761	1.8035	1.8294	1.8774	1.9214	1.9624	2.0009
40 (267.25)	Sh					32.75	82.75	132.75	182.75	232.75	332.75	432.75	532.75	632.75
	v	0.01715	10.497			11.036	11.838	12.624	13.398	14.165	15.685	17.195	18.699	20.199
	h	236.14	1169.8			1186.6	1211.7	1236.4	1260.8	1285.0	1333.6	1382.5	1432.1	1482.5
	s	0.3921	1.6765			1.6992	1.7312	1.7608	1.7883	1.8143	1.8624	1.9065	1.9476	1.9860
45 (274.44)	Sh					25.56	75.56	125.56	175.56	225.56	325.56	425.56	525.56	625.56
	v	0.01721	9.399			9.777	10.497	11.201	11.892	12.577	13.932	15.276	16.614	17.950
	h	243.49	1172.1			1185.4	1210.4	1235.7	1260.2	1284.6	1333.3	1382.3	1431.9	1482.3
	s	0.4021	1.6671			1.6849	1.7173	1.7471	1.7748	1.8010	1.8492	1.8934	1.9345	1.9730

Sh = degrees of superheat, °F
v = specific volume, cu ft/lbm
h = enthalpy, Btu/lbm
s = entropy, Btu/lbmR

Table III. Superheated Steam — *Continued*

Abs Press. Lb/Sq In. (Sat. Temp)		Sat. Water	Sat. Steam	Temperature — Degrees Fahrenheit								
				300	350	400	450	500	600	700	800	900
50 (281.02)	Sh			18.98	68.98	118.98	168.98	218.98	318.98	418.98	518.98	618.98
	v	0.01727	8.514	8.769	9.424	10.062	10.688	11.306	12.529	13.741	14.947	16.150
	h	250.21	1174.1	1184.1	1209.9	1234.9	1259.6	1284.1	1332.9	1382.0	1431.7	1482.2
	s	0.4112	1.6586	1.6720	1.7048	1.7349	1.7628	1.7890	1.8374	1.8816	1.9227	1.9613
55 (287.07)	Sh			12.93	62.93	112.93	162.93	212.93	312.93	412.93	512.93	612.93
	v	0.01733	7.787	7.945	8.546	9.130	9.702	10.267	11.381	12.485	13.583	14.677
	h	256.43	1176.0	1182.9	1208.9	1234.2	1259.1	1283.6	1332.6	1381.8	1431.5	1482.0
	s	0.4196	1.6510	1.6601	1.6933	1.7237	1.7518	1.7781	1.8266	1.8710	1.9121	1.9507
60 (292.71)	Sh			7.29	57.29	107.29	157.29	207.29	307.29	407.29	507.29	607.29
	v	0.01738	7.174	7.257	7.815	8.354	8.881	9.400	10.425	11.438	12.446	13.450
	h	262.21	1177.6	1181.6	1208.0	1233.5	1258.5	1283.2	1332.3	1381.5	1431.3	1481.8
	s	0.4273	1.6440	1.6492	1.6934	1.7134	1.7417	1.7681	1.8168	1.8612	1.9024	1.9410
65 (297.98)	Sh			2.02	52.02	102.02	152.02	202.02	302.02	402.02	502.02	602.02
	v	0.01743	6.653	6.675	7.195	7.697	8.186	8.667	9.615	10.552	11.484	12.412
	h	267.63	1179.1	1180.3	1207.0	1232.7	1257.9	1282.7	1331.9	1381.3	1431.1	1481.6
	s	0.4344	1.6375	1.6390	1.6731	1.7040	1.7324	1.7590	1.8077	1.8522	1.8935	1.9321
70 (302.93)	Sh				47.07	97.07	147.07	197.07	297.07	397.07	497.07	597.07
	v	0.01748	6.205		6.664	7.133	7.590	8.039	8.922	9.793	10.659	11.522
	h	272.74	1180.6		1206.0	1232.0	1257.3	1282.2	1331.6	1381.0	1430.9	1481.5
	s	0.4411	1.6316		1.6640	1.6951	1.7237	1.7504	1.7993	1.8439	1.8852	1.9238
75 (307.61)	Sh				42.39	92.39	142.39	192.39	292.39	392.39	492.39	592.39
	v	0.01753	5.814		6.204	6.645	7.074	7.494	8.320	9.135	9.945	10.750
	h	277.56	1181.9		1205.0	1231.2	1256.7	1281.7	1331.3	1380.7	1430.7	1481.3
	s	0.4474	1.6260		1.6554	1.6868	1.7156	1.7424	1.7915	1.8361	1.8774	1.9161

Abs Press. Lb/Sq In. (Sat. Temp)		Sat. Water	Sat. Steam	Temperature — Degrees Fahrenheit										
				350	400	450	500	550	600	700	800	900	1000	1100
80 (312.04)	Sh			37.96	87.96	137.96	187.96	237.96	287.96	387.96	487.96	587.96	687.96	787.96
	v	0.01757	5.471	5.801	6.218	6.622	7.018	7.408	7.794	8.560	9.319	10.075	10.829	11.581
	h	282.15	1183.1	1204.0	1230.5	1256.1	1281.3	1306.2	1330.9	1380.5	1430.5	1481.1	1532.6	1584.9
	s	0.4534	1.6208	1.6473	1.6790	1.7080	1.7349	1.7602	1.7842	1.8289	1.8702	1.9089	1.9454	1.9800
85 (316.26)	Sh			33.74	83.74	133.74	183.74	223.74	283.74	383.74	483.74	583.74	683.74	783.74
	v	0.01762	5.167	5.445	5.840	6.223	6.597	6.966	7.330	8.052	8.768	9.480	10.190	10.898
	h	286.52	1184.2	1203.0	1229.7	1255.5	1280.8	1305.8	1330.6	1380.2	1430.3	1481.0	1532.4	1584.7
	s	0.4590	1.6159	1.6396	1.6716	1.7008	1.7279	1.7532	1.7772	1.8220	1.8634	1.9021	1.9386	1.9733
90 (320.28)	Sh			29.72	79.72	129.72	179.72	229.72	279.72	379.72	479.72	579.72	679.72	779.72
	v	0.01766	4.895	5.128	5.505	5.869	6.223	6.572	6.917	7.600	8.277	8.950	9.621	10.290
	h	290.69	1185.3	1202.0	1228.9	1254.9	1280.3	1305.4	1330.2	1380.0	1430.1	1480.8	1532.3	1584.6
	s	0.4643	1.6113	1.6323	1.6646	1.6940	1.7212	1.7467	1.7707	1.8156	1.8570	1.8957	1.9323	1.9669
95 (324.13)	Sh			25.87	75.87	125.87	175.87	225.87	275.87	375.87	475.87	575.87	675.87	775.87
	v	0.01770	4.651	4.845	5.205	5.551	5.889	6.221	6.548	7.196	7.838	8.477	9.113	9.747
	h	294.70	1186.2	1200.9	1228.1	1254.3	1279.8	1305.0	1329.9	1379.7	1429.9	1480.6	1532.1	1584.5
	s	0.4694	1.6069	1.6253	1.6580	1.6876	1.7149	1.7404	1.7645	1.8094	1.8509	1.8897	1.9262	1.9609

Sh = degrees of superheat, °F
v = specific volume, cu ft/lbm

h = enthalpy, Btu/lbm
s = entropy, Btu/lbmR

Table III. Superheated Steam — *Continued*

Abs Press. Lb/Sq In. (Sat. Temp)		Sat. Water	Sat. Steam	350	400	450	500	550	600	700	800	900	1000	1100
100 (327.82)	Sh			22.18	72.18	122.28	172.18	222.18	272.18	372.18	472.18	572.18	672.18	772.18
	v	0.01774	4.431	4.590	4.935	5.266	5.588	5.904	6.216	6.833	7.443	8.050	8.655	9.258
	h	298.54	1187.2	1199.9	1227.4	1253.7	1279.3	1304.6	1329.6	1379.5	1429.7	1480.4	1532.0	1584.4
	s	0.4743	1.6027	1.6187	1.6516	1.6814	1.7088	1.7344	1.7586	1.8036	1.8451	1.8839	1.9205	1.9552
105 (331.37)	Sh			18.63	68.63	118.63	168.63	218.63	268.63	386.63	468.63	568.63	668.63	768.63
	v	0.01778	4.231	4.359	4.690	5.007	5.315	5.617	5.915	6.504	7.086	7.665	8.241	8.816
	h	302.24	1188.0	1198.8	1226.6	1253.1	1278.8	1304.2	1329.2	1379.2	1429.4	1480.3	1531.8	1584.2
	s	0.4790	1.5988	1.6122	1.6455	1.6755	1.7031	1.7288	1.7350	1.7981	1.8396	1.8785	1.9151	1.9498
110 (334.79)	Sh			15.21	65.21	115.21	165.21	215.21	265.21	365.21	465.21	565.21	665.21	765.21
	v	0.01782	4.048	4.149	4.468	4.772	5.068	5.357	5.642	6.205	6.761	7.314	7.865	8.413
	h	305.80	1188.9	1197.7	1225.8	1252.5	1278.3	1303.8	1328.9	1379.0	1429.2	1480.1	1531.7	1584.1
	s	0.4834	1.5950	1.6061	1.6396	1.6698	1.6975	1.7233	1.7476	1.7928	1.8344	1.8732	1.9099	1.9446
115 (338.08)	Sh			11.92	61.92	111.92	161.92	211.92	261.92	361.92	461.92	561.92	661.92	761.92
	v	0.01785	3.881	3.957	4.265	4.558	4.841	5.119	5.392	5.932	6.465	6.994	7.521	8.046
	h	309.25	1179.6	1196.7	1225.0	1251.8	1277.9	1303.3	1328.6	1378.7	1429.0	1479.9	1531.6	1584.0
	s	0.4877	1.5913	1.6001	1.6340	1.6644	1.6922	1.7181	1.7425	1.7877	1.8294	1.8682	1.9049	1.9396
120 (341.27)	Sh			8.73	58.73	108.73	158.73	208.73	258.73	358.73	458.73	558.73	658.73	758.73
	v	0.01789	3.7275	3.7815	4.0786	4.3610	4.6341	4.9009	5.1637	5.6813	6.1928	6.7006	7.2060	7.7096
	h	312.58	1190.4	1195.6	1224.1	1251.2	1277.4	1302.9	1328.2	1378.4	1428.8	1479.8	1531.4	1583.9
	s	0.4919	1.5879	1.5943	1.6286	1.6592	1.6872	1.7132	1.7376	1.7829	1.8246	1.8635	1.9001	1.9349
130 (347.33)	Sh			2.67	52.67	102.67	152.67	202.67	252.67	352.67	452.67	552.67	652.67	752.67
	v	0.01796	3.4544	3.4699	3.7489	4.0129	4.2672	4.5151	4.7589	5.2384	5.7118	6.1814	6.6486	7.1140
	h	318.95	1191.7	1193.4	1222.5	1249.9	1276.4	1302.1	1327.5	1377.9	1428.4	1479.4	1531.1	1583.6
	s	0.4998	1.5813	1.5833	1.6182	1.6493	1.6775	1.7037	1.7283	1.7737	1.8155	1.8545	1.8911	1.9259
140 (353.04)	Sh				46.96	96.96	146.96	196.96	246.96	346.96	446.96	546.96	646.96	746.96
	v	0.01803	3.2190		3.4661	3.7143	3.9526	4.1844	4.4119	4.8588	5.2995	5.7364	6.1709	6.6036
	h	324.96	1193.0		1220.8	1248.7	1275.3	1301.3	1326.8	1377.4	1428.0	1479.1	1530.8	1583.4
	s	0.5071	1.5752		1.6085	1.6400	1.6686	1.6949	1.7196	1.7652	1.8071	1.8461	1.8828	1.9176
150 (358.43)	Sh				41.57	91.57	141.57	191.57	241.57	341.57	441.57	541.57	641.57	741.57
	v	0.01809	3.0139		3.2208	3.4555	3.6799	3.8978	4.1112	4.5298	4.9421	5.3507	5.7568	6.1612
	h	330.65	1194.1		1219.1	1247.4	1274.3	1300.5	1326.1	1376.9	1427.6	1478.7	1530.5	1583.1
	s	0.5141	1.5695		1.5993	1.6313	1.6602	1.6867	1.7115	1.7573	1.7992	1.8383	1.8751	1.9099
160 (363.55)	Sh				36.45	86.45	136.45	186.45	236.45	336.45	436.45	536.45	636.45	736.45
	v	0.01815	2.8336		3.0060	3.2288	3.4413	3.6469	3.8480	4.2420	4.6295	5.0132	5.3945	5.7741
	h	336.07	1195.1		1217.4	1246.0	1273.3	1299.6	1325.4	1376.4	1427.2	1478.4	1530.3	1582.9
	s	0.5206	1.5641		1.5906	1.6231	1.6522	1.6790	1.7039	1.7499	1.7919	1.8310	1.8678	1.9027
170 (368.42)	Sh				31.58	81.58	131.58	181.58	231.58	331.58	431.48	531.58	631.58	731.58
	v	0.01821	2.6738		2.8162	3.0288	3.2306	3.4255	3.6158	3.9879	4.3536	4.7155	5.0749	5.4325
	h	341.24	1196.0		1215.6	1244.7	1272.2	1298.8	1324.7	1375.8	1426.8	1478.0	1530.0	1582.6
	s	0.5269	1.5591		1.5823	1.6152	1.6447	1.6717	1.6968	1.7428	1.7850	1.8241	1.8610	1.8959
180 (373.08)	Sh				26.92	76.92	126.92	176.92	226.92	326.92	426.92	526.92	626.92	726.92
	v	0.01827	2.5312		2.6474	2.8508	3.0433	3.2286	3.4093	3.7621	4.1084	4.4508	4.7907	5.1289
	h	346.19	1196.9		1213.8	1243.4	1271.2	1297.9	1324.0	1375.3	1426.3	1477.7	1529.7	1582.4
	s	0.5328	1.5543		1.5743	1.6078	1.6376	1.6647	1.6900	1.7362	1.7784	1.8176	1.8545	1.8894

Temperature — Degrees Fahrenheit

Sh = degrees of superheat, °F
v = specific volume, cu ft/lbm
h = enthalpy, Btu/lbm
s = entropy, Btu/lbmR

Table III. Superheated Steam — *Continued*

Abs Press. Lb/Sq In. (Sat. Temp)		Sat. Water	Sat. Steam	Temperature — Degrees Fahrenheit									
				400	450	500	550	600	700	800	900	1000	1100
190 (377.53)	Sh			22.47	72.47	122.47	172.47	222.47	322.47	422.47	522.47	622.47	722.47
	v	0.01833	2.4030	2.4961	2.6915	2.8756	3.0525	3.2246	3.5601	3.8889	4.2140	4.5365	4.8572
	h	350.94	1197.6	1212.0	1242.0	1270.1	1297.1	1323.3	1374.8	1425.9	1477.4	1529.4	1582.1
	s	0.5384	1.5498	1.5667	1.6006	1.6307	1.6581	1.6835	1.7299	1.7722	1.8115	1.8484	1.8834
200 (381.80)	Sh			18.20	68.20	118.20	168.20	218.20	318.20	418.20	518.20	618.20	718.20
	v	0.01839	2.2873	2.3598	2.5480	2.7247	2.8939	3.0583	3.3783	3.6915	4.0008	4.3077	4.6128
	h	355.51	1198.3	1210.1	1240.6	1269.0	1296.2	1322.6	1374.3	1425.5	1477.0	1529.1	1581.9
	s	0.5438	1.5454	1.5593	1.5938	1.6242	1.6518	1.6773	1.7239	1.7663	1.8057	1.8426	1.8776

Abs Press. Lb/Sq In. (Sat. Temp)		Sat. Water	Sat. Steam	Temperature — Degrees Fahrenheit										
				400	450	500	550	600	700	800	900	1000	1100	1200
210 (385.91)	Sh			14.09	64.09	114.09	164.09	214.09	314.09	414.09	514.09	614.09	714.09	814.09
	v	0.01844	2.1822	2.2364	2.4181	2.5880	2.7504	2.9078	3.2137	3.5128	3.8080	4.1007	4.3915	4.6811
	h	359.91	1199.0	1208.2	1239.2	1268.0	1295.3	1321.9	1373.7	1425.1	1476.7	1528.8	1581.6	1635.2
	s	0.5490	1.5413	1.5522	1.5872	1.6180	1.6458	1.6715	1.7182	1.7607	1.8001	1.8371	1.8721	1.9054
220 (389.88)	Sh			10.12	60.12	110.12	160.12	210.12	310.12	410.12	510.12	610.12	710.12	810.12
	v	0.01850	2.0863	2.1240	2.2999	2.4638	2.6199	2.7710	3.0642	3.3504	3.6327	3.9125	4.1905	4.4671
	h	364.17	1199.6	1206.3	1237.8	1266.9	1294.5	1321.2	1373.2	1424.7	1476.3	1528.5	1581.4	1635.0
	s	0.5540	1.5374	1.5453	1.5808	1.6120	1.6400	1.6658	1.7128	1.7553	1.7948	1.8318	1.8668	1.9002
230 (393.70)	Sh			6.30	56.30	106.30	156.30	206.30	306.30	406.30	506.30	606.30	706.30	806.30
	v	0.01855	1.9985	2.0212	2.1919	2.3503	2.5008	2.6461	2.9276	3.2020	3.4726	3.7406	4.0068	4.2717
	h	368.28	1200.1	1204.4	1236.3	1265.7	1293.6	1320.4	1372.7	1424.2	1476.0	1528.2	1581.1	1634.8
	s	0.5588	1.5336	1.5385	1.5747	1.6062	1.6344	1.6604	1.7075	1.7502	1.7897	1.8268	1.8618	1.8952
240 (397.39)	Sh			2.61	52.61	102.61	152.61	202.61	302.61	402.61	502.61	602.61	702.61	802.61
	v	0.01860	1.9177	1.9268	2.0928	2.2462	2.3915	2.5316	2.8024	3.0661	3.3259	3.5831	3.8385	4.0926
	h	372.27	1200.6	1202.4	1234.9	1264.6	1292.7	1319.7	1372.1	1423.8	1475.6	1527.9	1580.9	1634.6
	s	0.5634	1.5299	1.5320	1.5687	1.6006	1.6291	1.6552	1.7025	1.7452	1.7848	1.8219	1.8570	1.8904
250 (400.97)	Sh				49.03	99.03	149.03	199.03	299.03	399.03	499.03	599.03	699.03	799.03
	v	0.01865	1.8432		2.0016	2.1504	2.2909	2.4262	2.6872	2.9410	3.1909	3.4382	3.6837	3.9278
	h	376.14	1201.1		1233.4	1263.5	1291.8	1319.0	1371.6	1423.4	1475.3	1527.6	1580.6	1634.4
	s	0.5679	1.5264		1.5629	1.5951	1.6239	1.6502	1.6976	1.7405	1.7801	1.8173	1.8524	1.8858
260 (404.44)	Sh				45.56	95.56	145.56	195.56	295.56	395.56	495.56	595.56	695.56	795.56
	v	0.01870	1.7742		1.9173	2.0619	2.1981	2.3289	2.5808	2.8256	3.0663	3.3044	3.5408	3.7758
	h	379.90	1201.5		1231.9	1262.4	1290.9	1318.2	1371.1	1423.0	1474.9	1527.3	1580.4	1634.2
	s	0.5722	1.5230		1.5573	1.5899	1.6189	1.6453	1.6930	1.7359	1.7756	1.8128	1.8480	1.8814
270 (407.80)	Sh				42.20	92.20	142.20	192.20	292.20	392.20	492.20	592.20	692.20	792.20
	v	0.01875	1.7101		1.8391	1.9799	2.1121	2.2388	2.4824	2.7186	2.9509	3.1806	3.4084	3.6349
	h	383.56	1201.9		1230.4	1261.2	1290.0	1317.5	1370.5	1422.6	1474.6	1527.1	1580.1	1634.0
	s	0.5764	1.5197		1.5518	1.5848	1.6140	1.6406	1.6885	1.7315	1.7713	1.8085	1.8437	1.8771
280 (411.07)	Sh				38.93	88.93	138.93	188.93	288.93	388.93	488.93	588.93	688.93	788.93
	v	0.01880	1.6505		1.7665	1.9037	2.0322	2.1551	2.3909	2.6194	2.8437	3.0655	3.2855	3.5042
	h	387.12	1202.3		1228.8	1260.0	1289.1	1316.8	1370.0	1422.1	1474.2	1526.8	1579.9	1633.8
	s	0.5805	1.5166		1.5464	1.5798	1.6093	1.6361	1.6841	1.7273	1.7671	1.8043	1.8395	1.8730

Sh = degrees of superheat, °F
v = specific volume, cu ft/lbm
h = enthalpy, Btu/lbm
s = entropy, Btu/lbmR

Table III. Superheated Steam — *Continued*

Abs Press. Lb/Sq In. (Sat. Temp)	Sat. Water	Sat. Steam	Temperature — Degrees Fahrenheit									
			450	500	550	600	700	800	900	1000	1100	1200
290 (414.25) Sh v h s	0.01885 390.60 0.5844	1.5948 1202.6 1.5135	35.75 1.6988 1227.3 1.5412	85.75 1.8327 1258.9 1.5750	135.75 1.9578 1288.1 1.6048	185.75 2.0772 1316.0 1.6317	285.75 2.3058 1369.5 1.6799	385.75 2.5269 1421.7 1.7323	485.75 2.7440 1473.9 1.7630	585.75 2.9585 1526.5 1.8003	685.75 3.1711 1579.6 1.8356	785.75 3.3824 1633.5 1.8690
300 (417.35) Sh v h s	0.01889 393.99 0.5882	1.5427 1202.9 1.5105	32.65 1.6356 1225.7 1.5361	82.65 1.7665 1257.7 1.5703	132.65 1.8883 1287.2 1.6003	182.65 2.0044 1315.2 1.6274	282.65 2.2263 1368.9 1.6758	382.65 2.4407 1421.3 1.7192	482.65 2.6509 1473.6 1.7591	582.65 2.8585 1526.2 1.7964	682.65 3.0643 1579.4 1.8317	782.65 3.2688 1633.3 1.8652
310 (420.36) Sh v h s	0.01894 397.30 0.5920	1.4939 1203.2 1.5076	29.64 1.5763 1224.1 1.5311	79.64 1.7044 1256.5 1.5657	129.64 1.8233 1286.3 1.5960	179.64 1.9363 1314.5 1.6233	279.64 2.1520 1368.4 1.6719	379.64 2.3600 1420.9 1.7153	479.64 2.5638 1473.2 1.7553	579.64 2.7650 1525.9 1.7927	679.64 2.9644 1579.2 1.8280	779.64 3.1625 1633.1 1.8615
320 (423.31) Sh v h s	0.01899 400.53 0.5956	1.4480 1203.4 1.5048	26.69 1.5207 1222.5 1.5261	76.69 1.6462 1255.2 1.5612	126.69 1.7623 1285.3 1.5918	176.69 1.8725 1313.7 1.6192	276.69 2.0823 1367.8 1.6680	376.69 2.2843 1420.5 1.7116	476.69 2.4821 1472.9 1.7516	576.69 2.6774 1525.6 1.7890	676.69 2.8708 1578.9 1.8243	776.69 3.0628 1632.9 1.8579
330 (426.18) Sh v h s	0.01903 403.70 0.5991	1.4048 1203.6 1.5021	23.82 1.4684 1220.9 1.5213	73.82 1.5915 1254.0 1.5568	123.82 1.7050 1284.4 1.5876	173.82 1.8125 1313.0 1.6153	273.82 2.0168 1367.3 1.6643	373.82 2.2132 1420.0 1.7079	473.82 2.4054 1472.5 1.7480	573.82 2.5950 1525.3 1.7855	673.82 2.7828 1578.7 1.8208	783.82 2.9692 1632.7 1.8544
340 (428.99) Sh v h s	0.01908 406.80 0.6026	1.3640 1203.8 1.4994	21.01 1.4191 1219.2 1.5165	71.01 1.5399 1252.8 1.5525	121.01 1.6511 1283.4 1.5836	171.01 1.7561 1312.2 1.6114	271.01 1.9552 1366.7 1.6606	371.01 2.1463 1419.6 1.7044	471.01 2.3333 1472.2 1.7445	571.01 2.5175 1525.0 1.7820	671.01 2.7000 1578.4 1.8174	771.01 2.8811 1632.5 1.8510
350 (431.73) Sh v h s	0.01912 409.83 0.6059	1.3255 1204.0 1.4968	18.27 1.3725 1217.5 1.5119	68.27 1.4913 1251.5 1.5483	118.27 1.6002 1282.4 1.5797	168.27 1.7028 1311.4 1.6077	268.27 1.8970 1366.2 1.6571	368.27 2.0832 1419.2 1.7009	468.27 2.2652 1471.8 1.7411	568.27 2.4445 1524.7 1.7787	668.27 2.6219 1578.2 1.8141	768.27 2.7980 1632.3 1.8477
360 (434.41) Sh v h s	0.01917 412.81 0.6092	1.2891 1204.1 1.4943	15.59 1.3285 1215.8 1.5073	65.59 1.4454 1250.3 1.5441	115.59 1.5521 1281.5 1.5758	165.59 1.6525 1310.6 1.6040	265.59 1.8421 1365.6 1.6536	365.59 2.0237 1418.7 1.6976	465.59 2.2009 1471.5 1.7379	565.59 2.3755 1524.4 1.7754	665.59 2.5482 1577.9 1.8109	765.59 2.7196 1632.1 1.8445
380 (439.61) Sh v h s	0.01925 418.59 0.6156	1.2218 1204.4 1.4894	10.39 1.2472 1212.4 1.4982	60.39 1.3606 1247.7 1.5360	110.39 1.4635 1279.5 1.5683	160.39 1.5598 1309.0 1.5969	260.39 1.7410 1364.5 1.6470	360.39 1.9139 1417.9 1.6911	460.39 2.0825 1470.8 1.7315	560.39 2.2484 1523.8 1.7692	660.39 2.4124 1577.4 1.8047	760.39 2.5750 1631.6 1.8384

Abs Press. Lb/Sq In. (Sat. Temp)	Sat. Water	Sat. Steam	Temperature — Degrees Fahrenheit										
			450	500	550	600	650	700	800	900	1000	1100	1200
400 (444.60) Sh v h s	0.01934 424.17 0.6217	1.1610 1204.6 1.4847	5.40 1.1738 1208.8 1.4894	55.40 1.2841 1245.1 1.5282	105.40 1.3836 1277.5 1.5611	155.40 1.4763 1307.4 1.5901	205.40 1.5646 1335.9 1.6163	255.40 1.6499 1363.4 1.6406	355.40 1.8151 1417.0 1.6850	455.40 1.9759 1470.1 1.7255	555.40 2.1339 1523.3 1.7632	655.40 2.2901 1576.9 1.7988	755.40 2.4450 1631.2 1.8325

Sh = degrees of superheat, °F   h = enthalpy, Btu/lbm
v = specific volume, cu ft/lbm   s = entropy, Btu/lbmR

Table III. Superheated Steam — *Continued*

Abs Press. Lb/Sq In. (Sat. Temp)		Sat. Water	Sat. Steam	Temperature — Degrees Fahrenheit										
				450	500	550	600	650	700	800	900	1000	1100	1200
420 (449.40)	Sh			.60	50.60	100.60	150.60	200.60	250.60	350.60	450.60	550.60	650.60	750.60
	v	0.01942	1.1057	1.1071	1.2148	1.3113	1.4007	1.4856	1.5676	1.7258	1.8795	2.0304	2.1795	2.3273
	h	429.56	1204.7	1205.2	1242.4	1275.4	1305.8	1334.5	1362.3	1416.2	1469.4	1522.7	1576.4	1630.8
	s	0.6276	1.4802	1.4808	1.5206	1.5542	1.5835	1.6100	1.6345	1.6791	1.7197	1.7575	1.7932	1.8269
440 (454.03)	Sh				45.97	95.97	145.97	195.97	245.97	345.97	445.97	545.97	645.97	745.97
	v	0.01950	1.0554		1.1517	1.2454	1.3319	1.4138	1.4926	1.6445	1.7918	1.9363	2.0790	2.2203
	h	434.77	1204.8		1239.7	1273.4	1304.2	1333.2	1361.1	1415.3	1468.7	1522.1	1575.9	1630.4
	s	0.6332	1.4759		1.5132	1.5474	1.5772	1.6040	1.6286	1.6734	1.7412	1.7521	1.7878	1.8216
460 (458.50)	Sh				41.50	91.50	141.50	191.50	241.50	341.50	441.50	541.50	641.50	741.50
	v	0.01959	1.0092		1.0939	1.1852	1.2691	1.3482	1.4242	1.5703	1.7117	1.8504	1.9872	2.1226
	h	439.83	1204.8		1236.9	1271.3	1302.5	1331.8	1360.0	1414.4	1468.0	1521.5	1575.4	1629.9
	s	0.6387	1.4718		1.5060	1.5409	1.5711	1.5982	1.6230	1.6680	1.7089	1.7469	1.7826	1.8165
480 (462.82)	Sh				37.18	87.18	137.18	187.18	237.18	337.18	437.18	537.18	637.18	737.18
	v	0.01967	0.9668		1.0409	1.1300	1.2115	1.2881	1.3615	1.5023	1.6384	1.7716	1.9030	2.0330
	h	444.75	1204.8		1234.1	1269.1	1300.8	1330.5	1358.8	1413.6	1467.3	1520.9	1574.9	1629.5
	s	0.6439	1.4677		1.4990	1.5346	1.5652	1.5925	1.6176	1.6628	1.7038	1.7419	1.7777	1.8116
500 (467.01)	Sh				32.99	82.99	132.99	182.99	232.99	332.99	432.99	532.99	632.99	732.99
	v	0.01975	0.9276		0.9919	1.0791	1.1584	1.2327	1.3037	1.4397	1.5708	1.6992	1.8256	1.9507
	h	449.52	1204.7		1231.2	1267.0	1299.1	1329.1	1357.7	1412.7	1466.6	1520.3	1574.4	1629.1
	s	0.6490	1.4639		1.4921	1.5284	1.5595	1.5871	1.6123	1.6578	1.6990	1.7371	1.7730	1.8069
520 (471.07)	Sh				28.93	78.93	128.93	178.93	228.93	328.93	428.93	528.93	628.93	728.93
	v	0.01982	0.8914		0.9466	1.0321	1.1094	1.1816	1.2504	1.3819	1.5085	1.6323	1.7542	1.8746
	h	454.18	1204.5		1228.3	1264.8	1297.4	1327.7	1356.5	1411.8	1465.9	1519.7	1573.9	1628.7
	s	0.6540	1.4601		1.4853	1.5223	1.5539	1.5818	1.6072	1.6530	1.6943	1.7325	1.7684	1.8024
540 (475.01)	Sh				24.99	74.99	124.99	174.99	224.99	324.99	424.99	524.99	624.99	724.99
	v	0.01990	0.8577		0.9045	0.9884	1.0640	1.1342	1.2010	1.3284	1.4508	1.5704	1.6880	1.8042
	h	458.71	1204.4		1225.3	1262.5	1295.7	1326.3	1355.3	1410.9	1465.1	1519.1	1573.4	1628.2
	s	0.6587	1.4565		1.4786	1.5164	1.5485	1.5767	1.6023	1.6483	1.6897	1.7280	1.7640	1.7981
560 (478.84)	Sh				21.16	71.16	121.16	171.16	221.16	321.16	421.16	521.16	621.16	721.16
	v	0.01998	0.8264		0.8653	0.9479	1.0217	1.0902	1.1552	1.2787	1.3972	1.5129	1.6266	1.7388
	h	463.14	1204.2		1222.2	1260.3	1293.9	1324.9	1354.2	1410.0	1464.4	1518.6	1572.9	1627.8
	s	0.6634	1.4529		1.4720	1.5106	1.5431	1.5717	1.5975	1.6438	1.6853	1.7237	1.7598	1.7939
580 (482.57)	Sh				17.43	67.43	117.43	167.43	217.43	317.43	417.43	517.43	617.43	717.43
	v	0.02006	0.7971		0.8287	0.9100	0.9824	1.0492	1.1125	1.2324	1.3473	1.4593	1.5693	1.6780
	h	467.47	1203.9		1219.1	1258.0	1292.1	1323.4	1353.0	1409.2	1463.7	1518.0	1572.4	1627.4
	s	0.6679	1.4495		1.4654	1.5049	1.5380	1.5668	1.5929	1.6394	1.6811	1.7196	1.7556	1.7898
600 (486.20)	Sh				13.80	63.80	113.80	163.80	213.80	313.80	413.80	513.80	613.80	713.80
	v	0.02013	0.7697		0.7944	0.8746	0.9456	1.0109	1.0726	1.1892	1.3008	1.4093	1.5160	1.6211
	h	471.70	1203.7		1215.9	1255.6	1290.3	1322.0	1351.8	1408.3	1463.0	1517.4	1571.9	1627.0
	s	0.6723	1.4461		1.4590	1.4993	1.5329	1.5621	1.5884	1.6351	1.6769	1.7155	1.7517	1.7859
650 (494.89)	Sh				5.11	55.11	105.11	155.11	205.11	305.11	405.11	505.11	605.11	705.11
	v	0.02032	0.7084		0.7173	0.7954	0.8634	0.9254	0.9835	1.0929	1.1969	1.2979	1.3969	1.4944
	h	481.89	1202.8		1207.6	1249.6	1285.7	1318.3	1348.7	1406.0	1461.2	1515.9	1570.7	1625.9
	s	1.6828	1.4381		1.4430	1.4858	1.5207	1.5507	1.5775	1.6249	1.6671	1.7059	1.7422	1.7765

Sh = degrees of superheat, °F   h = enthalpy, Btu/lbm
v = specific volume, cu ft/lbm   s = entropy, Btu/lbmR

Table III. Superheated Steam — *Continued*

Abs Press. Lb/Sq In. (Sat. Temp)		Sat. Water	Sat. Steam	Temperature — Degrees Fahrenheit								
				550	600	650	700	800	900	1000	1100	1200
700 (503.08)	Sh			46.92	96.92	146.92	196.92	296.92	396.92	496.92	596.92	696.92
	v	0.02050	0.6556	0.7271	0.7928	0.8520	0.9072	1.0102	1.1078	1.2023	1.2948	1.3858
	h	491.60	1201.8	1243.4	1281.0	1314.6	1345.6	1403.7	1459.4	1514.4	1569.4	1624.8
	s	0.6928	1.4304	1.4726	1.5090	1.5399	1.5673	1.6154	1.6580	1.6970	1.7335	1.7679
750 (510.84)	Sh			39.16	89.16	139.16	189.16	289.16	389.16	489.16	589.16	689.16
	v	0.02069	0.6095	0.6676	0.7313	0.7882	0.8409	0.9386	1.0306	1.1195	1.2063	1.2916
	h	500.89	1200.7	1236.9	1276.1	1310.7	1342.5	1401.5	1457.6	1512.9	1568.2	1623.8
	s	0.7022	1.4232	1.4598	1.4977	1.5296	1.5577	1.6065	1.6494	1.6886	1.7252	1.7598
800 (518.21)	Sh			31.79	81.79	131.79	181.79	281.79	381.79	481.79	581.79	681.79
	v	0.02087	0.5690	0.6151	0.6774	0.7323	0.7828	0.8759	0.9631	1.0470	1.1289	1.2093
	h	509.81	1199.4	1230.1	1271.1	1306.8	1339.3	1399.1	1455.8	1511.4	1566.9	1622.7
	s	0.7111	1.4163	1.4472	1.4869	1.5198	1.5484	1.5980	1.6413	1.6807	1.7175	1.7522
850 (525.24)	Sh			24.76	74.76	124.76	174.76	274.76	374.76	474.76	574.76	674.76
	v	0.02105	0.5330	0.5683	0.6296	0.6829	0.7315	0.8205	0.9034	0.9830	1.0606	1.1366
	h	518.40	1198.0	1223.0	1265.9	1302.8	1336.0	1396.8	1454.0	1510.0	1565.7	1621.6
	s	0.7197	1.4096	1.4347	1.4763	1.5102	1.5396	1.5899	1.6336	1.6733	1.7102	1.7450
900 (531.95)	Sh			18.05	68.05	118.05	168.05	268.05	368.05	468.05	568.05	668.05
	v	0.02123	0.5009	0.5263	0.5869	0.6388	0.6858	0.7713	0.8504	0.9262	0.9998	1.0720
	h	526.70	1196.4	1215.5	1260.6	1298.6	1332.7	1394.4	1452.2	1508.5	1564.4	1620.6
	s	0.7279	1.4032	1.4223	1.4659	1.5010	1.5311	1.5822	1.6263	1.6662	1.7033	1.7382

Abs Press. Lb/Sq In. (Sat. Temp)		Sat. Water	Sat. Steam	Temperature — Degrees Fahrenheit										
				550	600	650	700	750	800	850	900	1000	1100	1200
950 (538.39)	Sh			11.61	61.61	111.61	161.61	211.61	261.61	311.61	361.61	461.61	561.61	661.61
	v	0.02141	0.4721	0.4883	0.5485	0.5993	0.6449	0.6871	0.7272	0.7656	0.8030	0.8753	0.9455	1.0142
	h	534.74	1194.7	1207.6	1255.1	1294.4	1329.3	1361.5	1392.0	1421.5	1450.3	1507.0	1563.2	1619.5
	s	0.7358	1.3970	1.4098	1.4557	1.4921	1.5228	1.5500	1.5748	1.5977	1.6193	1.6595	1.6967	1.7317
1000 (544.58)	Sh			5.42	55.42	105.42	155.42	205.42	255.42	305.42	355.42	455.42	555.42	655.42
	v	0.02159	0.4460	0.4535	0.5137	0.5636	0.6080	0.6489	0.6875	0.7245	0.7603	0.8295	0.8966	0.9622
	h	542.55	1192.9	1199.3	1249.3	1290.1	1325.9	1358.7	1389.6	1419.4	1448.5	1505.4	1561.9	1618.4
	s	0.7434	1.3910	1.3973	1.4457	1.4833	1.5149	1.5426	1.5677	1.5908	1.6126	1.6530	1.6905	1.7256
1050 (550.53)	Sh				49.47	99.47	149.47	199.47	249.47	299.47	349.47	449.47	549.47	649.47
	v	0.02177	0.4222		0.4821	0.5312	0.5745	0.6142	0.6515	0.6872	0.7216	0.7881	0.8524	0.9151
	h	550.15	1191.0		1243.4	1285.7	1322.4	1355.8	1387.2	1417.3	1446.6	1503.5	1560.7	1617.4
	s	0.7507	1.3851		1.4358	1.4748	1.5072	1.5354	1.5608	1.5842	1.6062	1.6469	1.6845	1.7197
1100 (556.28)	Sh				43.72	93.72	143.72	193.72	243.72	293.72	343.72	443.72	543.72	643.72
	v	0.02195	0.4006		0.4531	0.5017	0.5440	0.5826	0.6188	0.6533	0.6865	0.7505	0.8121	0.8723
	h	557.55	1189.1		1237.3	1281.2	1318.8	1352.9	1384.7	1415.2	1444.7	1502.4	1559.4	1616.3
	s	0.7578	1.3794		1.4259	1.4664	1.4996	1.5284	1.5542	1.5779	1.6000	1.6410	1.6787	1.7141
1150 (561.82)	Sh				39.18	89.18	139.18	189.18	239.18	289.18	339.18	439.18	539.18	639.18
	v	0.02214	0.3807		0.4263	0.4746	0.5162	0.5538	0.5889	0.6223	0.6544	0.7161	0.7754	0.8332
	h	564.78	1187.0		1230.9	1276.6	1315.2	1349.9	1382.2	1413.0	1442.8	1500.9	1558.1	1615.2
	s	0.7647	1.3738		1.4160	1.4582	1.4923	1.5216	1.5478	1.5717	1.5941	1.6353	1.6732	1.7087

Sh = degrees of superheat, °F
v = specific volume, cu ft/lbm
h = enthalpy, Btu/lbm
s = entropy, Btu/lbmR

Table III. Superheated Steam — *Continued*

Abs Press. Lb/Sq In. (Sat. Temp)		Sat. Water	Sat. Steam	600	650	700	750	800	850	900	1000	1100	1200
1200 (567.19)	Sh			32.81	82.81	132.81	182.81	232.81	282.81	332.81	432.81	532.81	632.81
	v	0.02232	0.3624	0.4016	0.4497	0.4905	0.5273	0.5615	0.5939	0.6250	0.6845	0.7418	0.7974
	h	571.85	1184.8	1224.2	1271.8	1311.5	1346.9	1379.7	1410.8	1440.9	1499.4	1556.9	1614.2
	s	0.7714	1.3683	1.4061	1.4501	1.4851	1.5150	1.5415	1.5658	1.5883	1.6298	1.6679	1.7035
1300 (577.42)	Sh			22.58	72.58	122.58	172.58	222.58	272.58	322.58	422.58	522.58	622.58
	v	0.02269	0.3299	0.3570	0.4052	0.4451	0.4804	0.5129	0.5436	0.5729	0.6287	0.6822	0.7341
	h	585.58	1180.2	1209.9	1261.9	1303.9	1340.8	1374.6	1406.4	1437.1	1496.3	1554.3	1612.0
	s	0.7843	1.3577	1.3860	1.4340	1.4711	1.5022	1.5296	1.5544	1.5773	1.6194	1.6578	1.6937
1400 (587.07)	Sh			12.93	62.93	112.93	162.93	212.93	262.93	312.93	412.93	512.93	612.93
	v	0.02307	0.3018	0.3176	0.3667	0.4059	0.4400	0.4712	0.5004	0.5282	0.5809	0.6311	0.6798
	h	598.83	1175.3	1194.1	1251.4	1296.1	1334.5	1369.3	1402.0	1433.2	1493.2	1551.8	1609.9
	s	0.7966	1.3474	1.3652	1.4181	1.4575	1.4900	1.5182	1.5436	1.5670	1.6096	1.6484	1.6845
1500 (596.20)	Sh			3.80	53.80	103.80	153.80	203.80	253.80	303.80	403.80	503.80	603.80
	v	0.02346	0.2772	0.2820	0.3328	0.3717	0.4049	0.4350	0.4629	0.4894	0.5394	0.5869	0.6327
	h	611.68	1170.1	1176.3	1240.2	1287.9	1328.0	1364.0	1397.4	1429.2	1490.1	1549.2	1607.7
	s	0.8085	1.3373	1.3431	1.4022	1.4443	1.4782	1.5073	1.5333	1.5572	1.6004	1.6395	1.6759
1600 (604.87)	Sh				45.13	95.13	145.13	195.13	245.13	295.13	395.13	495.13	595.13
	v	0.02387	0.2555		0.3026	0.3415	0.3741	0.4032	0.4301	0.4555	0.5031	0.5482	0.5915
	h	624.20	1164.5		1228.3	1279.4	1321.4	1358.5	1392.8	1425.2	1486.9	1546.6	1605.6
	s	0.8199	1.3272		1.3861	1.4312	1.4667	1.4968	1.5235	1.5478	1.5916	1.6312	1.6678
1700 (613.13)	Sh				36.87	86.87	136.87	186.87	236.87	286.87	386.87	486.87	586.87
	v	0.02428	0.2361		0.2754	0.3147	0.3468	0.3751	0.4011	0.4255	0.4711	0.5140	0.5552
	h	636.45	1158.6		1215.3	1270.5	1314.5	1352.9	1388.1	1421.2	1483.8	1544.0	1603.4
	s	0.8309	1.3176		1.3697	1.4183	1.4555	1.4867	1.5140	1.5388	1.5833	1.6232	1.6601
1800 (621.02)	Sh				28.98	78.98	128.98	178.98	228.98	278.98	378.98	478.98	578.98
	v	0.02472	0.2186		0.2505	0.2906	0.3223	0.3500	0.3752	0.3988	0.4426	0.4836	0.5229
	h	648.49	1152.3		1201.2	1261.1	1307.4	1347.2	1383.3	1417.1	1480.6	1541.4	1601.2
	s	0.8417	1.3079		1.3526	1.4054	1.4446	1.4768	1.5049	1.5302	1.5753	1.6156	1.6528
1900 (628.56)	Sh				21.44	71.44	121.44	171.44	221.44	271.44	371.44	471.44	571.44
	v	0.02517	0.2028		0.2274	0.2687	0.3004	0.3275	0.3521	0.3749	0.4171	0.4565	0.4940
	h	660.36	1145.6		1185.7	1251.3	1300.2	1341.4	1378.4	1412.9	1477.4	1538.8	1599.1
	s	0.8522	1.2981		1.3346	1.3925	1.4338	1.4672	1.4960	1.5219	1.5677	1.6084	1.6458
2000 (635.80)	Sh				14.20	64.20	114.20	164.20	214.20	264.20	364.20	464.20	564.20
	v	0.02565	0.1883		0.2056	0.2488	0.2805	0.3072	0.3312	0.3534	0.3942	0.4320	0.4680
	h	672.11	1138.3		1168.3	1240.9	1292.6	1335.4	1373.5	1408.7	1474.1	1536.2	1596.9
	s	0.8625	1.2881		1.3154	1.3794	1.4231	1.4578	1.4874	1.5138	1.5603	1.6014	1.6391
2100 (642.76)	Sh				7.24	57.24	107.24	157.24	207.24	257.24	357.24	457.24	557.24
	v	0.02615	0.1750		0.1847	0.2304	0.2624	0.2888	0.3123	0.3339	0.3734	0.4099	0.4445
	h	683.79	1130.5		1148.5	1229.8	1284.9	1329.3	1368.4	1404.4	1470.9	1533.6	1594.7
	s	0.8727	1.2780		1.2942	1.3661	1.4125	1.4486	1.4790	1.5060	1.5532	1.5948	1.6327
2200 (649.45)	Sh				.55	50.55	100.55	150.55	200.55	250.55	350.55	450.55	550.55
	v	0.02669	0.1627		0.1636	0.2134	0.2458	0.2720	0.2950	0.3161	0.3545	0.3897	0.4231
	h	695.46	1122.2		1123.9	1218.0	1276.8	1323.1	1363.3	1400.0	1467.6	1530.9	1592.5
	s	0.8828	1.2676		1.2691	1.3523	1.4020	1.4395	1.4708	1.4984	1.5463	1.5883	1.6266

Sh = degrees of superheat, °F
v = specific volume, cu ft/lbm
h = enthalpy, Btu/lbm
s = entropy, Btu/lbmR

Table III. Superheated Steam — *Continued*

Abs Press. Lb/Sq In. (Sat. Temp)	Sat. Water	Sat. Steam		Temperature — Degrees Fahrenheit							
				700	750	800	850	900	1000	1100	1200
2300 (655.89)	Sh v 0.02727 h 707.18 s 0.8929	0.1513 1113.2 1.2569		44.11 0.1975 1205.3 1.3381	94.11 0.2305 1268.4 1.3914	144.11 0.2566 1316.7 1.4305	194.11 0.2793 1358.1 1.4628	244.11 0.2999 1395.7 1.4910	344.11 0.3372 1464.2 1.5397	444.11 0.3714 1528.3 1.5821	544.11 0.4035 1590.3 1.6207

Abs Press. Lb/Sq In. (Sat. Temp)	Sat. Water	Sat. Steam	Temperature — Degrees Fahrenheit										
			700	750	800	850	900	950	1000	1050	1100	1150	1200
2400 (662.11)	Sh v 0.02790 h 718.95 s 0.9031	0.1408 1103.7 1.2460	37.89 0.1824 1191.6 1.3232	87.89 0.2164 1259.7 1.3808	137.89 0.2424 1310.1 1.4217	187.89 0.2648 1352.8 1.4549	237.89 0.2850 1391.2 1.4837	287.89 0.3037 1426.9 1.5095	337.89 0.3214 1460.9 1.5332	387.89 0.3382 1493.7 1.5553	437.89 0.3545 1525.6 1.5761	487.89 0.3703 1557.0 1.5959	537.89 0.3856 1588.1 1.6149
2500 (668.11)	Sh v 0.02859 h 731.71 s 0.9139	0.1307 1093.3 1.2345	31.89 0.1681 1176.7 1.3076	81.89 0.2032 1250.6 1.3701	131.89 0.2293 1303.4 1.4129	181.89 0.2514 1347.4 1.4472	231.89 0.2712 1386.7 1.4766	281.89 0.2896 1423.1 1.5029	331.89 0.3068 1457.5 1.5269	381.89 0.3232 1490.7 1.5492	431.89 0.3390 1522.9 1.5703	481.89 0.3543 1554.6 1.5903	531.89 0.3692 1585.9 1.6094
2600 (673.91)	Sh v 0.02938 h 744.47 s 0.9247	0.1211 1082.0 1.2225	26.09 0.1544 1160.2 1.2908	76.09 0.1909 1241.1 1.3592	126.09 0.2171 1296.5 1.4042	176.09 0.2390 1341.9 1.4395	226.09 0.2585 1382.1 1.4696	276.09 0.2765 1419.2 1.4964	326.09 0.2933 1454.1 1.5208	376.09 0.3093 1487.7 1.5434	426.09 0.3247 1520.2 1.5646	476.09 0.3395 1552.2 1.5848	526.09 0.3540 1583.7 1.6040
2700 (679.53)	Sh v 0.03029 h 757.34 s 0.9356	0.1119 1069.7 1.2097	20.47 0.1411 1142.0 1.2727	70.47 0.1794 1231.1 1.3481	120.47 0.2058 1289.5 1.3954	170.47 0.2275 1336.3 1.4319	220.47 0.2468 1377.5 1.4628	270.47 0.2644 1415.2 1.4900	320.47 0.2809 1450.7 1.5148	370.47 0.2965 1484.6 1.5376	420.47 0.3114 1517.5 1.5591	470.47 0.3259 1549.8 1.5794	520.47 0.3399 1581.5 1.5988
2800 (684.96)	Sh v 0.03134 h 770.69 s 0.9468	0.1030 1055.8 1.1958	15.04 0.1278 1121.2 1.2527	65.04 0.1685 1220.6 1.3368	115.04 0.1952 1282.2 1.3867	165.04 0.2168 1330.7 1.4245	215.04 0.2358 1372.8 1.4561	265.04 0.2531 1411.2 1.4838	315.04 0.2693 1447.2 1.5089	365.04 0.2845 1481.6 1.5321	415.04 0.2991 1514.8 1.5537	465.04 0.3132 1547.3 1.5742	515.04 0.3268 1579.3 1.5938
2900 (690.22)	Sh v 0.03262 h 785.13 s 0.9588	0.0942 1039.8 1.1803	9.78 0.1138 1095.3 1.2283	59.78 0.1581 1209.6 1.3251	109.78 0.1853 1274.7 1.3780	159.78 0.2068 1324.9 1.4171	209.78 0.2256 1368.0 1.4494	259.78 0.2427 1407.2 1.4777	309.78 0.2585 1443.7 1.5032	359.78 0.2734 1478.5 1.5266	409.78 0.2877 1512.1 1.5485	459.78 0.3014 1544.9 1.5692	509.78 0.3147 1577.0 1.5889
3000 (695.33)	Sh v 0.03428 h 801.84 s 0.9728	0.0850 1020.3 1.1619	4.67 0.0982 1060.5 1.1966	54.67 0.1483 1197.9 1.3131	104.67 0.1759 1267.0 1.3692	154.67 0.1975 1319.0 1.4097	204.67 0.2161 1363.2 1.4429	254.67 0.2329 1403.1 1.4717	304.67 0.2484 1440.2 1.4976	354.67 0.2630 1475.4 1.5213	404.67 0.2770 1509.4 1.5434	454.67 0.2904 1542.4 1.5642	504.67 0.3033 1574.8 1.5841
3100 (700.28)	Sh v 0.03681 h 823.97 s 0.9914	0.0745 993.3 1.1373	49.72 0.1389 1185.4 1.3007	99.72 0.1671 1259.1 1.3604	149.72 0.1887 1313.0 1.4024	199.72 0.2071 1358.4 1.4364	249.72 0.2237 1399.0 1.4658	299.72 0.2390 1436.7 1.4920	349.72 0.2533 1472.3 1.5161	399.72 0.2670 1506.6 1.5384	449.72 0.2800 1539.9 1.5594	499.72 0.2927 1572.6 1.5794	
3200 (705.08)	Sh v 0.04472 h 875.54 s 1.0351	0.0566 931.6 1.0832	44.92 0.1300 1172.3 1.2877	94.92 0.1588 1250.9 1.3515	144.92 0.1804 1306.9 1.3951	194.92 0.1987 1353.4 1.4300	244.92 0.2151 1394.9 1.4600	294.92 0.2301 1433.1 1.4866	344.92 0.2442 1469.2 1.5110	394.92 0.2576 1503.8 1.5335	444.92 0.2704 1537.4 1.5547	494.92 0.2827 1570.3 1.5749	

Sh = degrees of superheat, °F   h = enthalpy, Btu/lbm
v = specific volume, cu ft/lbm   s = entropy, Btu/lbmR

Table III. Superheated Steam — *Continued*

Abs Press. Lb/Sq In. (Sat. Temp)		Temperature — Degrees Fahrenheit									
		750	800	850	900	950	1000	1050	1100	1150	1200
3300	v	0.1213	1.1510	1.1727	0.1908	0.2070	0.2218	0.2357	0.2488	0.2613	0.2734
	h	1158.2	1242.5	1300.7	1348.4	1390.7	1429.5	1466.1	1501.0	1534.9	1568.1
	s	1.2742	1.3425	1.3879	1.4237	1.4542	1.4813	1.5059	1.5287	1.5501	1.5704
3400	v	0.1129	0.1435	0.1653	0.1834	0.1994	0.2140	0.2276	0.2405	0.2528	0.2646
	h	1143.2	1233.7	1294.3	1343.4	1386.4	1425.9	1462.9	1498.3	1532.4	1565.8
	s	1.2600	1.3334	1.3807	1.4174	1.4486	1.4761	1.5010	1.5240	1.5456	1.5660
3500	v	0.1048	0.1364	0.1583	0.1764	0.1922	0.2066	0.2200	0.2326	0.2447	0.2563
	h	1127.1	1224.6	1287.8	1338.2	1382.2	1422.2	1459.7	1495.5	1529.9	1563.6
	s	1.2450	1.3242	1.3734	1.4112	1.4430	1.4709	1.4962	1.5194	1.5412	1.5618
3600	v	0.0966	0.1296	0.1517	0.1697	0.1854	0.1996	0.2128	0.2252	0.2371	0.2485
	h	1108.6	1215.3	1281.2	1333.0	1377.9	1418.6	1456.5	1492.6	1527.4	1561.3
	s	1.2281	1.3148	1.3662	1.4050	1.4374	1.4658	1.4914	1.5149	1.5369	1.5576
3800	v	0.0799	0.1169	0.1395	0.1574	0.1729	0.1868	0.1996	0.2116	0.2231	0.2340
	h	1064.2	1195.5	1267.6	1322.4	1369.1	1411.2	1450.1	1487.0	1522.4	1556.8
	s	1.1888	1.2955	1.3517	1.3928	1.4265	1.4558	1.4821	1.5061	1.5284	1.5495
4000	v	0.0631	0.1052	0.1284	0.1463	0.1616	0.1752	0.1877	0.1994	0.2105	0.2210
	h	1007.4	1174.3	1253.4	1311.6	1360.2	1403.6	1443.6	1481.3	1517.3	1552.2
	s	1.1396	1.2754	1.3371	1.3807	1.4158	1.4461	1.4730	1.4976	1.5203	1.5417
4200	v	0.0498	0.0945	0.1183	0.1362	0.1513	0.1647	0.1769	0.1883	0.1991	0.2093
	h	950.1	1151.6	1238.6	1300.4	1351.2	1396.0	1437.1	1475.5	1512.2	1547.6
	s	1.0905	1.2544	1.3223	1.3686	1.4053	1.4366	1.4642	1.4893	1.5124	1.5341
4400	v	0.0421	0.0846	0.1090	0.1270	0.1420	0.1552	0.1671	0.1782	0.1887	0.1986
	h	909.5	1127.3	1223.3	1289.0	1342.0	1388.3	1430.4	1469.7	1507.1	1543.0
	s	1.0556	1.2325	1.3073	1.3566	1.3949	1.4272	1.4556	1.4812	1.5048	1.5268

Abs Press. Lb/Sq In. (Sat. Temp)		Temperature — Degrees Fahrenheit										
		750	800	850	900	950	1000	1050	1100	1150	1200	1250
4600	v	0.0380	0.0751	0.1005	0.1186	0.1335	0.1465	0.1582	0.1691	0.1792	0.1889	0.1982
	h	883.8	1100.0	1207.3	1277.2	1332.6	1380.5	1423.7	1463.9	1501.9	1538.4	1573.8
	s	1.0331	1.2084	1.2922	1.3446	1.3847	1.4181	1.4472	1.4734	1.4974	1.5197	1.5407
4800	v	0.0355	0.0665	0.0927	0.1109	0.1257	0.1385	0.1500	0.1606	0.1706	0.1800	0.1890
	h	866.9	1071.2	1190.7	1265.2	1323.1	1372.6	1417.0	1458.0	1496.7	1533.8	1569.7
	s	1.0180	1.1835	1.2768	1.3327	1.3745	1.4090	1.4390	1.4657	1.4901	1.5128	1.5341

Sh = degrees of superheat, °F
v = specific volume, cu ft/lbm
h = enthalpy, Btu/lbm
s = entropy, Btu/lbmR

Table III. Superheated Steam — *Continued*

Abs Press. Lb/Sq In. (Sat. Temp)		750	800	850	900	950	1000	1050	1100	1150	1200	1250
					Temperature — Degrees Fahrenheit							
5000	v	0.0338	0.0591	0.0855	0.1038	0.1185	0.1312	0.1425	0.1529	0.1626	0.1718	0.1806
	h	854.9	1042.9	1173.6	1252.9	1313.5	1364.6	1410.2	1452.1	1491.5	1529.1	1565.5
	s	1.0070	1.1593	1.2612	1.3207	1.3645	1.4001	1.4309	1.4582	1.4831	1.5061	1.5277
5200	v	0.0326	0.0531	0.0789	0.0973	0.1119	0.1244	0.1356	0.1458	0.1553	0.1642	0.1728
	h	845.8	1016.9	1156.0	1240.4	1303.7	1356.6	1403.4	1446.2	1486.3	1524.5	1561.3
	s	0.9985	1.1370	1.2455	1.3088	1.3545	1.3914	1.4229	1.4509	1.4762	1.4995	1.5214
5400	v	0.0317	0.0483	0.0728	0.0912	0.1058	0.1182	0.1292	0.1392	0.1485	0.1572	0.1656
	h	838.5	994.3	1138.1	1227.7	1293.7	1348.4	1396.5	1440.3	1481.1	1519.8	1557.1
	s	0.9915	1.1175	1.2296	1.2969	1.3446	1.3827	1.4151	1.4437	1.4694	1.4931	1.5153
5600	v	0.0309	0.0447	0.0672	0.0856	0.1001	0.1124	0.1232	0.1331	0.1422	0.1508	0.1589
	h	832.4	975.0	1119.9	1214.8	1283.7	1340.2	1389.6	1434.3	1475.9	1515.2	1552.9
	s	0.9855	1.1008	1.2137	1.2850	1.3348	1.3742	1.4075	1.4366	1.4628	1.4869	1.5093
5800	v	0.0303	0.0419	0.0622	0.0805	0.0949	0.1070	0.1177	0.1274	0.1363	0.1447	0.1527
	h	827.3	958.8	1101.8	1201.8	1273.6	1332.0	1382.6	1428.3	1470.6	1510.5	1548.7
	s	0.9803	1.0867	1.1981	1.2732	1.3250	1.3658	1.3999	1.4297	1.4564	1.4808	1.5035
6000	v	0.0298	0.0397	0.0579	0.0757	0.0900	0.1020	0.1126	0.1221	0.1309	0.1391	0.1469
	h	822.9	945.1	1084.6	1188.8	1263.4	1323.6	1375.7	1422.3	1465.4	1505.9	1544.6
	s	0.9758	1.0746	1.1833	1.2615	1.3154	1.3574	1.3925	1.4229	1.4500	1.4748	1.4978
6500	v	0.0287	0.0358	0.0495	0.0655	0.0793	0.0909	0.1012	0.1104	0.1188	0.1266	0.1340
	h	813.9	919.5	1046.7	1156.3	1237.8	1302.7	1358.1	1407.3	1452.2	1494.2	1534.1
	s	0.9661	1.0515	1.1506	1.2328	1.2917	1.3370	1.3743	1.4064	1.4347	1.4604	1.4841
7000	v	0.0279	0.0334	0.0438	0.0573	0.0704	0.0816	0.0915	0.1004	0.1085	0.1160	0.1231
	h	806.9	901.8	1016.5	1124.9	1212.6	1281.7	1340.5	1392.2	1439.1	1482.6	1523.7
	s	0.9582	1.0350	1.1243	1.2055	1.2689	1.3171	1.3567	1.3904	1.4200	1.4466	1.4710
7500	v	0.0272	0.0318	0.0399	0.0512	0.0631	0.0737	0.0833	0.0918	0.0996	0.1068	0.1136
	h	801.3	889.0	992.9	1097.7	1188.3	1261.0	1322.9	1377.2	1426.0	1471.0	1513.3
	s	0.9514	1.0224	1.1033	1.1818	1.2473	1.2980	1.3397	1.3751	1.4059	1.4335	1.4586
8000	v	0.0267	0.0306	0.0371	0.0465	0.0571	0.0671	0.0762	0.0845	0.0920	0.0989	0.1054
	h	796.6	879.1	974.4	1074.3	1165.4	1241.0	1305.5	1362.2	1413.0	1459.6	1503.1
	s	0.9455	1.0122	1.0864	1.1613	1.2271	1.2798	1.3233	1.3603	1.3924	1.4208	1.4467
8500	v	0.0262	0.0296	0.0350	0.0429	0.0522	0.0615	0.0701	0.0780	0.0853	0.0919	0.0982
	h	792.7	871.2	959.8	1054.5	1144.0	1221.9	1288.5	1347.5	1400.2	1448.2	1492.9
	s	0.9402	1.0037	1.0727	1.1437	1.2084	1.2627	1.3076	1.3460	1.3793	1.4087	1.4352

Sh = degrees of superheat, °F
v = specific volume, cu ft/lbm

h = enthalpy, Btu/lbm
s = entropy, Btu/lbmR

Table III. Superheated Steam — *Continued*

Abs Press. Lb/Sq In. (Sat. Temp)		Temperature — Degrees Fahrenheit										
		750	800	850	900	950	1000	1050	1100	1150	1200	1250
9000	v	0.0258	0.0288	0.0335	0.0402	0.0483	0.0568	0.0649	0.0724	0.0794	0.0858	0.0918
	h	789.3	864.7	948.0	1037.6	1125.4	1204.1	1272.1	1333.0	1387.5	1437.1	1482.9
	s	0.9354	0.9964	1.0613	1.1285	1.1918	1.2468	1.2926	1.3323	1.3667	1.3970	1.4243
9500	v	0.0254	0.0282	0.0322	0.0380	0.0451	0.0528	0.0603	0.0675	0.0742	0.0804	0.0862
	h	786.4	859.2	938.3	1023.4	1108.9	1187.7	1256.6	1318.9	1375.1	1426.1	1473.1
	s	0.9310	0.9900	1.0516	1.1153	1.1771	1.2320	1.2785	1.3191	1.3546	1.3858	1.4137
10000	v	0.0251	0.0276	0.0312	0.0362	0.0425	0.0495	0.0565	0.0633	0.0697	0.0757	0.0812
	h	783.8	854.5	930.2	1011.3	1094.2	1172.6	1242.0	1305.3	1362.9	1415.3	1463.4
	s	0.9270	0.9842	1.0432	1.1039	1.1638	1.2185	1.2652	1.3065	1.3429	1.3749	1.4035
10500	v	0.0248	0.0271	0.0303	0.0347	0.0404	0.0467	0.0532	0.0595	0.0656	0.0714	0.0768
	h	781.5	850.5	923.4	1001.0	1081.3	1158.9	1228.4	1292.4	1351.1	1404.7	1453.9
	s	0.9232	0.9790	1.0358	1.0939	1.1519	1.2060	1.2529	1.2946	1.3371	1.3644	1.3937
11000	v	0.0245	0.0267	0.0296	0.0335	0.0386	0.0443	0.0503	0.0562	0.0620	0.0676	0.0727
	h	779.5	846.9	917.5	992.1	1069.9	1146.3	1215.9	1280.2	1339.7	1394.4	1444.6
	s	0.9196	0.9742	1.0292	1.0851	1.1412	1.1945	1.2414	1.2833	1.3209	1.3544	1.3842
11500	v	0.0243	0.0263	0.0290	0.0325	0.0370	0.0423	0.0478	0.0534	0.0588	0.0641	0.0691
	h	777.7	843.8	912.4	984.5	1059.8	1134.9	1204.3	1268.7	1328.8	1384.4	1435.5
	s	0.9163	0.9698	1.0232	1.0772	1.1316	1.1840	1.2308	1.2727	1.3107	1.3446	1.3750
12000	v	0.0241	0.0260	0.0284	0.0317	0.0357	0.0405	0.0456	0.0508	0.0560	0.0610	0.0659
	h	776.1	841.0	907.9	977.8	1050.9	1124.5	1193.7	1258.0	1318.5	1374.7	1426.6
	s	0.9131	0.9657	1.0177	1.0701	1.1229	1.1742	1.2209	1.2627	1.3010	1.3353	1.3662
12500	v	0.0238	0.0256	0.0279	0.0309	0.0346	0.0390	0.0437	0.0486	0.0535	0.0583	0.0629
	h	774.7	838.6	903.9	971.9	1043.1	1115.2	1184.1	1247.9	1308.8	1365.4	1418.0
	s	0.9101	0.9618	1.0127	1.0637	1.1151	1.1653	1.2117	1.2534	1.2918	1.3264	1.3576
13000	v	0.0236	0.0253	0.0275	0.0302	0.0336	0.0376	0.0420	0.0466	0.0512	0.0558	0.0602
	h	773.5	836.3	900.4	966.8	1036.2	1106.7	1174.8	1238.5	1299.6	1356.5	1409.6
	s	0.9073	0.9582	1.0080	1.0578	1.1079	1.1571	1.2030	1.2445	1.2831	1.3179	1.3494
13500	v	0.0235	0.0251	0.0271	0.0297	0.0328	0.0364	0.0405	0.0448	0.0492	0.0535	0.0577
	h	772.3	834.4	897.2	962.2	1030.0	1099.1	1166.3	1229.7	1291.0	1348.1	1401.5
	s	0.9045	0.9548	1.0037	1.0524	1.1014	1.1495	1.1948	1.2361	1.2749	1.3098	1.3415
14000	v	0.0233	0.0248	0.0267	0.0291	0.0320	0.0354	0.0392	0.0432	0.0474	0.0515	0.0555
	h	771.3	832.6	894.3	958.0	1024.5	1092.3	1158.5	1221.4	1283.0	1340.2	1393.8
	s	0.9019	0.9515	0.9996	1.0473	1.0953	1.1426	1.1872	1.2282	1.2671	1.3021	1.3339

Sh = degrees of superheat, °F   h = enthalpy, Btu/lbm
v = specific volume, cu ft/lbm   s = entropy, Btu/lbmR

Table III. Superheated Steam — *Continued*

Abs Press. Lb/Sq In. (Sat. Temp)		750	800	850	900	950	1000	1050	1100	1150	1200	1250
		Temperature — Degrees Fahrenheit										
14500	v	0.0231	0.0246	0.0264	0.0287	0.0314	0.0345	0.0380	0.0418	0.0458	0.0496	0.0534
	h	770.4	831.0	891.7	954.3	1019.6	1086.2	1151.4	1213.8	1275.4	1332.9	1386.4
	s	0.8994	0.9484	0.9957	1.0426	1.0897	1.1362	1.1801	1.2208	1.2597	1.2949	1.3266
15000	v	0.0230	0.0244	0.0261	0.0282	0.0308	0.0337	0.0369	0.0405	0.0443	0.0479	0.0516
	h	769.6	829.5	889.3	950.9	1015.1	1080.6	1144.9	1206.8	1268.1	1326.0	1379.4
	s	0.8970	0.9455	0.9920	1.0382	1.0846	1.1302	1.1735	1.2139	1.2525	1.2880	1.3197
15500	v	0.0228	0.0242	0.0258	0.0278	0.0302	0.0329	0.0360	0.0393	0.0429	0.0464	0.0499
	h	768.9	828.2	887.2	947.8	1011.1	1075.7	1139.0	1200.3	1261.1	1319.6	1372.8
	s	0.8946	0.9427	0.9886	1.0340	1.0797	1.1247	1.1674	1.2073	1.2457	1.2815	1.3131

Sh = degrees of superheat, °F
v = specific volume, cu ft/lbm
h = enthalpy, Btu/lbm
s = entropy, Btu/lbmR

SYSTÈME INTERNATIONAL (SI METRIC) UNITS

Table I.   Saturated Steam and Saturated Water: Temperature Table

Temp °C T	Press kPa P	Volume, m³/kg		Enthalpy, kJ/kg			Entropy, kJ/kg·K			Temp °C T
		Water v_f	Steam v_g	Water h_f	Evap. h_{fg}	Steam h_g	Water s_f	Evap. s_{fg}	Steam s_g	
0	0.6108	0.0010002	206.31	−0.04	2501.6	2501.6	−0.0002	9.1579	9.1577	0
0.01	0.6112	0.0010002	206.16	0.00	2501.6	2501.6	0.0000	9.1575	9.1575	0.01
5	0.8718	0.0010000	147.16	21.01	2489.7	2510.7	0.0762	8.9507	9.0269	5
10	1.2270	0.0010003	106.43	41.99	2477.9	2519.9	0.1510	8.7510	8.9020	10
15	1.7040	0.0010008	77.98	62.94	2466.1	2529.0	0.2243	8.5582	8.7825	15
20	2.377	0.0010017	57.84	83.86	2454.3	2538.2	0.2963	8.3721	8.6684	20
25	3.166	0.0010029	43.40	104.77	2442.5	2547.3	0.3670	8.1922	8.5592	25
30	4.241	0.0010043	32.93	125.66	2430.7	2556.4	0.4365	8.0181	8.4546	30
35	5.622	0.0010060	25.25	146.56	2418.8	2565.4	0.5049	7.8494	8.3543	35
40	7.375	0.0010078	19.546	167.45	2406.9	2574.4	0.5721	7.6861	8.2583	40
45	9.582	0.0010098	15.276	188.35	2394.9	2583.3	0.6383	7.5277	8.1661	45
50	12.335	0.0010121	12.046	209.26	2382.9	2592.2	0.7035	7.3741	8.0776	50
55	15.741 ·	0.0010145	9.579	230.17	2370.8	2601.0	0.7677	7.2248	7.9925	55
60	19.920	0.0010171	7.679	251.09	2358.6	2609.7	0.8310	7.0798	7.9108	60
65	25.010	0.0010199	6.202	272.03	2346.3	2618.3	0.8933	6.9388	7.8322	65
70	31.16	0.0010228	5.046	292.97	2334.0	2626.9	0.9548	6.8017	7.7565	70
75	38.55	0.0010259	4.134	313.93	2321.5	2635.4	1.0154	6.6681	7.6835	75
80	47.36	0.0010292	3.409	334.92	2308.8	2643.8	1.0753	6.5380	7.6132	80
85	57.80	0.0010326	2.829	355.91	2296.1	2652.0	1.1343	6.4111	7.5454	85
90	70.11	0.0010361	2.3613	376.94	2283.2	2660.1	1.1925	6.2873	7.4799	90
95	84.53	0.0010398	1.9822	397.99	2270.2	2668.2	1.2501	6.1665	7.4166	95
100	101.33	0.0010437	1.6730	419.06	2256.9	2676.0	1.3069	6.0485	7.3554	100
105	120.80	0.0010477	1.4193	440.17	2243.6	2683.7	1.3630	5.9331	7.2962	105
110	143.27	0.0010519	1.2099	461.32	2230.0	2691.3	1.4185	5.8203	7.2388	110
115	169.06	0.0010562	1.0363	482.50	2216.2	2698.7	1.4733	5.7099	7.1832	115
120	198.54	0.0010606	0.8915	503.72	2202.2	2706.0	1.5276	5.6017	7.1293	120
125	232.1	0.0010652	0.7702	524.99	2188.0	2713.0	1.5813	5.4957	7.0769	125
130	270.1	0.0010700	0.6681	546.31	2173.6	2719.9	1.6344	5.3917	7.0261	130
135	313.1	0.0010750	0.5818	567.68	2158.9	2726.6	1.6869	5.2897	6.9766	135
140	361.4	0.0010801	0.5085	589.10	2144.0	2733.1	1.7390	5.1894	6.9284	140
145	415.5	0.0010853	0.4460	610.59	2128.7	2739.3	1.7906	5.0910	6.8815	145
150	476.0	0.0010908	0.3924	632.15	2113.2	2745.4	1.8416	4.9941	6.8358	150
155	543.3	0.0010964	0.3464	653.77	2097.4	2751.2	1.8923	4.8989	6.7911	155
160	618.1	0.0011022	0.3068	675.47	2081.3	2756.7	1.9425	4.8050	6.7475	160
165	700.8	0.0011082	0.2724	697.25	2064.8	2762.0	1.9923	4.7126	6.7048	165
170	792.0	0.0011145	0.2426	719.12	2047.9	2767.1	2.0416	4.6214	6.6630	170
175	892.4	0.0011209	0.21654	741.07	2030.7	2771.8	2.0906	4.5314	6.6221	175
180	1002.7	0.0011275	0.19380	763.12	2013.2	2776.3	2.1393	4.4426	6.5819	180
185	1123.3	0.0011344	0.17386	785.26	1995.2	2780.4	2.1876	4.3548	6.5424	185
190	1255.1	0.0011415	0.15632	807.52	1976.7	2784.3	2.2356	4.2680	6.5036	190

Table I. — Continued

Temp °C T	Press kPa P	Volume, m³/kg Water v_f	Volume, m³/kg Steam v_g	Enthalpy, kJ/kg Water h_f	Enthalpy, kJ/kg Evap. h_{fg}	Enthalpy, kJ/kg Steam h_g	Entropy, kJ/kg·K Water s_f	Entropy, kJ/kg·K Evap. s_{fg}	Entropy, kJ/kg·K Steam s_g	Temp °C T
195	1398.7	0.0011489	0.14084	829.88	1957.9	2787.8	2.2833	4.1821	6.4654	195
200	1554.9	0.0011565	0.12716	852.37	1938.6	2790.9	2.3307	4.0971	6.4278	200
210	1907.7	0.0011726	0.10424	897.73	1898.5	2796.2	2.4247	3.9293	6.3539	210
220	2319.8	0.0011900	0.08604	943.67	1856.2	2799.9	2.5178	3.7639	6.2817	220
230	2798.	0.0012087	0.07145	990.27	1811.7	2802.0	2.6102	3.6006	6.2107	230
240	3348.	0.0012291	0.05965	1037.60	1764.6	2802.2	2.7020	3.4386	6.1406	240
250	3978.	0.0012513	0.05004	1085.78	1714.7	2800.4	2.7935	3.2773	6.0708	250
260	4694.	0.0012756	0.04213	1134.94	1661.5	2796.4	2.8848	3.1161	6.0010	260
270	5506.	0.0013025	0.03559	1185.23	1604.6	2789.9	2.9763	2.9541	5.9304	270
280	6420.	0.0013324	0.03013	1236.84	1543.6	2780.4	3.0683	2.7903	5.8586	280
290	7446.	0.0013659	0.02554	1290.01	1477.6	2767.6	3.1611	2.6237	5.7848	290
300	8593.	0.0014041	0.021649	1345.05	1406.0	2751.0	3.2552	2.4529	5.7081	300
310	9870.	0.0014480	0.018334	1402.39	1327.6	2730.0	3.3512	2.2766	5.6278	310
320	11289.	0.0014995	0.015480	1462.60	1241.1	2703.7	3.4500	2.0923	5.5423	320
330	12863.	0.0015615	0.012989	1526.52	1143.6	2670.2	3.5528	1.8962	5.4490	330
340	14605.	0.0016387	0.010780	1595.47	1030.7	2626.2	3.6616	1.6811	5.3427	340
350	16535.	0.0017411	0.008799	1671.94	895.7	2567.7	3.7800	1.4376	5.2177	350
360	18675.	0.0018959	0.006940	1764.17	721.3	2485.4	3.9210	1.1390	5.0600	360
370	21054.	0.0022136	0.004973	1890.21	452.6	2342.8	4.1108	0.7036	4.8144	370
374.15	22120.	0.00317	0.00317	2107.37	0.0	2107.4	4.4429	0.0	4.4429	374.15

[a] Abstracted from "ASME Steam Table is SI (Metric) Units," The American Society of Mechanical Engineer. Copyright 1977.

Table II. Saturated Steam and Saturated Water (Pressure)

Press kPa P	Temp °C T	Volume, m³kg Water v_f	Volume, m³kg Steam v_g	Enthalpy, kJ/kg Water h_f	Enthalpy, kJ/kg Evap. h_{fg}	Enthalpy, kJ/kg Steam h_g	Entropy, kJ/kg·K Water s_f	Entropy, kJ/kg·K Evap. s_{fg}	Entropy, kJ/kg·K Steam s_g
1.0	6.983	0.0010001	129.21	29.34	2485.0	2514.4	0.1060	8.8706	8.9767
1.5	13.036	0.0010006	87.98	54.71	2470.7	2525.5	0.1957	8.6332	8.8288
2.0	17.513	0.0010012	67.01	73.46	2460.2	2533.6	0.2607	8.4639	8.7246
3.0	24.100	0.0010027	45.67	101.00	2444.6	2545.6	0.3544	8.2241	8.5785
4.0	28.983	0.0010040	34.80	121.41	2433.1	2554.5	0.4225	8.0530	8.4755
5.0	32.898	0.0010052	28.19	137.77	2423.8	2561.6	0.4763	7.9197	8.3960
7.5	40.316	0.0010079	19.239	168.77	2406.2	2574.9	0.5763	7.6760	8.2523
10.0	45.833	0.0010102	14.675	191.83	2392.9	2584.8	0.6493	7.5018	8.1511
15.0	53.997	0.0010140	10.023	225.97	2373.2	2599.2	0.7549	7.2544	8.0093
20.0	60.086	0.0010172	7.650	251.45	2358.4	2609.9	0.8321	7.0774	7.9094
30.0	69.124	0.0010223	5.229	289.30	2336.1	2625.4	0.9441	6.8254	7.7695
40.0	75.886	0.0010265	3.993	317.65	2319.2	2636.9	1.0261	6.6448	7.6709
50.0	81.345	0.0010301	3.240	340.56	2305.4	2646.0	1.0912	6.5035	7.5947
75.0	91.785	0.0010375	2.2169	384.45	2278.6	2663.0	1.2131	6.2439	7.4570
100.0	99.632	0.0010434	1.6937	417.51	2257.9	2675.4	1.3027	6.0571	7.3598
150.0	111.37	0.0010530	1.1590	467.13	2226.2	2693.4	1.4336	5.7898	7.2234
200.0	120.23	0.0010608	0.8854	504.70	2201.6	2706.3	1.5301	5.5967	7.1268

Table II. — *Continued*

Press kPa P	Temp °C T	Volume, m³kg		Enthalpy, kJ/kg			Entropy, kJ/kg·K		
		Water v_f	Steam v_g	Water h_f	Evap. h_{fg}	Steam h_g	Water s_f	Evap. s_{fg}	Steam s_g
300.0	133.54	0.0010735	0.6056	561.4	2163.2	2724.7	1.6716	5.3193	6.9909
400.0	143.62	0.0010839	0.4622	604.7	2133.0	2737.6	1.7764	5.1179	6.8943
500.0	151.84	0.0010928	0.3747	640.1	2107.4	2747.5	1.8604	4.9588	6.8192
600.0	158.84	0.0011009	0.3155	670.4	2085.0	2755.5	1.9308	4.8267	6.7575
700.0	164.96	0.0011082	0.27268	697.1	2064.9	2762.0	1.9918	4.7134	6.7052
800.0	170.41	0.0011150	0.24026	720.9	2046.5	2767.5	2.0457	4.6139	6.6596
900.0	175.36	0.0011213	0.21481	742.6	2029.5	2772.1	2.0941	4.5250	6.6192
1000.0	179.88	0.0011274	0.19429	762.6	2013.6	2776.2	2.1382	4.4446	6.5828
1500.0	198.29	0.0011539	0.13166	844.7	1945.2	2789.9	2.3145	4.1261	6.4406
2000.0	212.37	0.0011766	0.09954	908.6	1888.6	2797.2	2.4469	3.8898	6.3367
3000.0	233.84	0.0012163	0.06663	1008.4	1793.9	2802.3	2.6455	3.5382	6.1837
4000.0	250.33	0.0012521	0.04975	1087.4	1712.9	2800.3	2.7965	3.2720	6.0685
5000.0	263.91	0.0012858	0.03943	1154.4	1639.7	2794.2	2.9206	3.0529	5.9735
7500.0	290.50	0.0013677	0.025327	1292.7	1474.2	2766.9	3.1657	2.6153	5.7811
10000.0	310.96	0.0014526	0.018041	1408.0	1319.7	2727.7	3.3605	2.2593	5.6198
15000.0	342.13	0.0016579	0.010340	1611.0	1004.0	2615.0	3.6859	1.6320	5.3178
20000.0	365.70	0.0020370	0.005877	1826.5	591.9	2418.4	4.0149	0.9263	4.9412
22120.0	374.15	0.00317	0.00317	2107.4	0.0	2107.4	4.4429	0.0	4.4429

Table III. Superheated Steam

Pressure kPa (T_{sat}) °C		v (specific volume), m³/kg				h (enthalpy), kJ/kg			s (entropy), kJ/kg·K		
		Temperature, °C									
		50	100	150	200	300	400	500	600	700	800
1.0 (6.983)	v	149.09	172.19	195.28	218.35	264.51	310.66	356.81	402.97	449.12	495.27
	h	2594.6	2688.6	2783.7	2880.1	3076.8	3279.7	3489.2	3705.6	3928.9	4158.7
	s	9.2430	9.5136	9.7527	9.9679	10.3450	10.6711	10.9612	11.2243	11.4663	11.6911
5.0 (32.90)	v	29.783	34.417	39.042	43.661	52.897	62.129	71.360	80.592	89.822	99.053
	h	2593.7	2688.1	2783.4	2879.9	3076.7	3279.7	3489.2	3705.6	3928.8	4158.7
	s	8.4981	8.7698	9.0094	9.2248	9.6021	9.9283	10.2184	10.4815	10.7235	10.9483
10 (45.83)	v	14.871	17.195	19.513	21.825	26.445	31.062	35.679	40.295	44.910	49.526
	h	2592.7	2687.5	2783.1	2879.6	3076.6	3279.6	3489.1	3705.5	3928.8	4158.7
	s	8.1758	8.4486	8.6889	8.9045	9.2820	9.6083	9.8984	10.1616	10.4036	10.6284
50 (81.35)	v		3.4181	3.8893	4.3560	5.2839	6.2091	7.1335	8.0574	8.9810	9.9044
	h		2682.6	2780.1	2877.7	3075.7	3279.0	3488.7	3705.2	3928.6	4158.5
	s		7.6953	7.9406	8.1587	8.5380	8.8649	9.1552	9.4185	9.6606	9.8855
100 (99.63)	v		1.6955	1.9363	2.1723	2.6387	3.1025	3.5653	4.0277	4.4898	4.9517
	h		2676.2	2776.3	2875.4	3074.5	3278.2	3488.1	3704.8	3928.2	4158.3
	s		7.3618	7.6138	7.8349	8.2166	8.5442	8.8348	9.0982	9.3405	9.5654
200 (120.2)	v			0.9596	1.0804	1.3162	1.5492	1.7812	2.0129	2.2442	2.4754
	h			2768.5	2870.5	3072.1	3276.7	3487.0	3704.0	3927.6	4157.8
	s			7.2794	7.5072	7.8937	8.2226	8.5139	8.7776	9.0201	9.2452

Table III. — Continued

Pressure kPa (T_{sat}) °C		v (specific volume), m³/kg				h (enthalpy), kJ/kg			s (entropy), kJ/kg·K		
						Temperature, °C					
		50	100	150	200	300	400	500	600	700	800
300 (133.5)	v			0.6338	0.7164	0.8753	1.0314	1.1865	1.3412	1.4957	1.6499
	h			2760.4	2865.4	3069.7	3275.2	3486.0	3703.2	3927.0	4157.3
	s			7.0772	7.3119	7.7034	8.0338	8.3257	8.5898	8.8325	9.0577
400 (143.6)	v			0.4706	0.5343	0.6549	0.7725	0.8892	1.0054	1.1214	1.2372
	h			2752.0	2860.4	3067.2	3273.6	3484.9	3702.3	3926.4	4156.9
	s			6.9286	7.1708	7.5675	7.8994	8.1919	8.4563	8.6992	8.9246
500 (151.8)	v				0.4250	0.5226	0.6172	0.7108	8.8039	0.8968	0.9896
	h				2855.1	3064.8	3272.1	3483.8	3701.5	3925.8	4156.4
	s				7.0592	7.4614	7.7948	8.0879	8.3526	8.5957	8.8213
600 (158.8)	v				0.3520	0.4344	0.5136	0.5918	0.6696	0.7471	0.8245
	h				2849.7	3062.3	3270.6	3482.7	3700.7	3925.1	4155.9
	s				6.9662	7.3740	7.7090	8.0027	8.2678	8.5111	8.7368
800 (170.4)	v				0.2608	0.3241	0.3842	0.4432	0.5017	0.5600	0.6181
	h				2838.6	3057.3	3267.5	3480.5	3699.1	3923.9	4155.0
	s				6.8148	7.2348	7.5729	7.8678	8.1336	8.3773	8.6033
1000 (179.9)	v				0.2059	0.2580	0.3065	0.3540	0.4010	0.4477	0.4943
	h				2826.8	3052.1	3264.4	3478.3	3697.4	3922.7	4154.1
	s				6.6922	7.1251	7.4665	7.7627	8.0292	8.2734	8.4997
1500 (198.3)	v				0.1324	0.1697	0.2029	0.2350	0.2667	0.2980	0.3292
	h				2794.7	3038.9	3256.6	3472.8	3693.3	3919.6	4151.7
	s				6.4508	6.9207	7.2709	7.5703	7.8385	8.0838	8.3108
2000 (212.4)	v					0.1255	0.1511	0.1756	0.1995	0.2232	0.2467
	h					3025.0	3248.7	3467.3	3689.2	3916.5	4149.4
	s					6.7696	7.1296	7.4323	7.7022	7.9485	8.1763
3000 (233.8)	v					0.08116	0.09931	0.1161	0.1323	0.1483	0.1641
	h					2995.1	3232.2	3456.2	3681.0	3910.3	4144.7
	s					6.5422	6.9246	7.2345	7.5079	7.7564	7.9857
4000 (250.3)	v					0.05883	0.07338	0.08634	0.09876	0.1109	0.1229
	h					2962.0	3215.7	3445.0	3672.8	3904.1	4140.0
	s					6.3642	6.7733	7.0909	7.3680	7.6187	7.8495
5000 (263.9)	v					0.04530	0.05779	0.06849	0.07862	0.08845	0.09809
	h					2925.5	3198.3	3433.7	3664.5	3897.9	4135.3
	s					6.2105	6.6508	6.9770	7.2578	7.5108	7.7431
10000 (311.0)	v						0.02641	0.03276	0.03832	0.04355	0.04858
	h						3099.9	3374.6	3622.7	3866.8	4112.0
	s						6.2182	6.5994	6.9013	7.1660	7.4058
15000 (342.1)	v						0.01566	0.02080	0.02488	0.02859	0.03209
	h						2979.1	3310.6	3579.8	3835.4	4088.6
	s						5.8876	6.3487	6.6764	6.9536	7.2013

Appendix E. Engineering Conversion Factors

The following table is designed to provide accurate conversion of units between various equivalent representational forms. To convert a unit, locate the unit in the MULTIPLY column of the table, select the desired unit in the TO OBTAIN column, and then multiply the quantity to be converted by the appropriate factor to obtain the desired units.

For example, you have exactly 2.5 acres and you need to know how many square meters. In the table: acre x 4046.87 = square meter. Therefore, 2.50000 acres = (2.50000) × [4046.87] = 10117.18 square meters.

A note should be made regarding scientific notation: 1×10^1, 1×10^2, 1×10^3 are equivalent respectively to 10, 100, 1000. For negative powers, 1×10^{-1}, 1×10^{-2}, 1×10^{-3} represent the numbers 0.1, 0.01, 0.001. Therefore, the number 2.345×10^3 is 2345.0 and 2.345×10^{-3} represents 0.002345.

Accuracy is an important concern with conversion factors. This table presents most conversions with six significant figures; for example, 4046.87 and 39.3701. The actual number of significant figures allowed in a conversion is equal to the number of significant figures in the least accurate of the conversion factors or the unit being converted. For example, in the acre-to-square-meter illustration: if the number of acres was known only to two significant figures, then the conversion to square meters would be (2.5) acres x 4046.87 = 10,000 m^2.

Where conversions involve the International System of Units (le Système International d'Unités, abbreviated "SI"), we show the proper symbols or formulas for the particular SI unit. There are symbols for the recommended SI base, supplementary, and derived units, and for most of their multiples and submultiples. Note that such "metric" units as kilogram-force, calorie, torr, centimeter of water, millimeter of mercury, and metric horsepower are *not* SI units. Also note that, in American Customary Units (ACU), the unit *pound* is used as either a mass unit (pound-mass, or lbm) or as a force unit (pound-force, or lbf); in ACU, *pound* is always pound avoirdupois, not troy pound.

An asterisk (*) after the sixth or higher significant figure indicates that the conversion factor is exact and that all subsequent digits are zero. All other factors are rounded. Where fewer than six significant figures are shown, more precision is not warranted.

As a further explanation of the use of this table, note that: to convert from pound-force per square inch, multiply by 6.894 76 to obtain the number of kilopascals (kPa) means

$$1 \text{ lbf/in}^2 = 6.894 \ 76 \text{ kPa}.$$

Similarly, to convert from inches to millimeters, multiplying the inch value by 25.400 0* to obtain the number of millimeters, means

$$1 \text{ inch} = 25.4 \text{ millimeters (exactly)}.$$

Table 1. Engineering Conversion Factors

MULTIPLY	BY	TO OBTAIN
A		
abampere	10*	ampere (A)
abcoulomb	10*	coulomb (C)
abfarad	10^9 *	farad (F)
abhenry	10^{-9} *	henry (H)
abohm	10^{-9} *	ohm (Ω)

MULTIPLY	BY	TO OBTAIN
abvolt	10^{-8} *	volt (V)
acre (U.S. survey: 1 foot = (1200/3937)m)	4046.87	square meter (m²)
acre foot (U.S. survey)	1233.49	cubic meter (m³)
ampere (A)	1.00000*	coulomb/second (C/s)
ampere/centimeter (A/cm)	2.54000*	ampere/inch
ampere/inch (1959 internat'l inch)	39.3701	ampere/meter (A/m)
ampere/kilogram (A/kg)	0.45359237*	ampere/pound-mass
ampere/meter (A/m)	0.025400*	ampere/inch
ampere/pound-mass	2.20462	ampere/kilogram (A/kg)
ampere/square foot	10.7639	ampere/square meter (A/m²)
ampere/square inch	1550.00	ampere/square meter (A/m²)
ampere/square meter (A/m²)	0.00064516*	ampere/square inch
ampere/square meter (A/m²)	0.09290304*	ampere/square foot
ampere/volt (A/V)	1.00000*	siemens (S)
ampere/volt inch	39.3701	siemens/meter (S/m)
are	100.000*	square meter
atmosphere (kilogram-force/cm²)	98.0665*	kilopascal (kPa)
atmosphere (760 torr)	101.325*	kilopascal (kPa)
atmosphere (760 torr)	29.9212	inch mercury (32°F)

B

MULTIPLY	BY	TO OBTAIN
bar (10⁵ pascals)	100.000*	kilopascal (kPa)
barrel (42 U.S. gallons, liquid)	0.158987	cubic meter (m³)
barrel/ton (U.K. long or gross)	0.156476	cubic meter/metric ton
barrel/ton (U.S. short or net)	0.175254	cubic meter/metric ton
barrel/day	0.00184013	cubic decimeter/second
barrel/hour	0.0441631	cubic decimeter/second
barrel/million std cubic feet	0.133010	cubic decimeters/kilomol
becquerel (Bq) (radioactivity)	1.00000*	disintegration/second
Btu (mean)	1.05587	kilojoule (kJ)
Btu (thermochemical)	1.05435	kilojoule (kJ)
Btu (39°F)	1.05967	kilojoule (kJ)
Btu (60°F)	1.05468	kilojoule (kJ)
Btu (I.T.) (International Table)	1.055056	kilojoule (kJ)
Btu (I.T.)	778.172	foot-pound (force)
Btu (I.T.)/cubic foot	37.2589	kilojoule/cubic meter (kJ/m³)
Btu (I.T.)/hour	0.293071	watt (W)
Btu (I.T.)/hour cubic foot	0.0103497	kilowatt/cubic meter (kW/m³)
Btu (I.T.)/hour cubic foot °F	0.0186295	kilowatt/cubic meter kelvin (kW/m³·K)
Btu (I.T.)/hour square foot	3.15459	watt/square meter (W/m²)
Btu (I.T.)/hour square foot °F	5.67826	watt/square meter kelvin (W/m²·K)
Btu (I.T.)foot/hour square foot °F	1.73074	watt/meter kelvin (W/m·K)
Btu (I.T.)/minute	0.0175843	kilowatt (kW)
Btu (I.T.)/pound-mole	2.32600*	joule/mole (J/mol)
Btu (I.T.)/pound-mole °F	4.18680*	kilojoule/kilomol kelvin (kJ/kmol·K)

MULTIPLY	BY	TO OBTAIN
Btu (I.T.)/pound-mass	0.555556	kilocalorie/kilogram
Btu (I.T.)/pound-mass	2.32600*	kilojoule/kilogram (kJ/kg)
Btu (I.T.)/pound-mass °F	4.18680*	kilojoule/kilogram kelvin (kJ/kg·K)
Btu (I.T.)/pound-mass °F	1.00000*	kilocalorie/kilogram kelvin
Btu (I.T.)/second	1.05506	kilowatt (kW)
Btu (I.T.)/second cubic foot	37.2589	kilowatt/cubic meter (kW/m³)
Btu (I.T.)/second cubic foot °F	67.0661	kilowatt/cubic meter kelvin (kW/m³·K)
Btu (I.T.)/second square foot	11.3565	kilowatt/square meter (kW/m²)
Btu (I.T.)/second square foot °F	20.4418	kilowatt/square meter kelvin (kW/m²·K)
Btu (I.T.)/gallon (U.K. liquid)	232.080	kilojoule/cubic meter (kJ/m³)
Btu (I.T.)/gallon (U.S. liquid)	278.716	kilojoule/cubic meter (kJ/m³)

C

MULTIPLY	BY	TO OBTAIN
calorie (I.T.)	4.18680*	joule (J)
calorie (mean)	4.19002	joule (J)
calorie (thermochemical)	4.18400*	joule (J)
calorie (thermochemical)	0.00396567	Btu (I.T.)
calorie (15°C)	4.18580	joule (J)
calorie (20°C)	4.18190	joule (J)
calorie (kilogram, I.T.)	4186.80*	joule (J)
calorie (kilogram, mean)	4190.02	joule (J)
calorie (kilogram, thermochemical)	4184.00*	joule (J)
calorie (thermochem)/gram kelvin	4.18400*	kilojoule/kilogram kelvin (kJ/kg·K)
calorie (thermochemical)/hour cm²	0.0116222	kilowatt/square meter (kW/m²)
calorie (thermochemical)/milliliter	4.18400*	megajoule/cubic meter (MJ/m³)
calorie (thermochem)/pound-mass	9.22414	joule/kilogram (J/kg)
candela/square meter (cd/m²)	0.291864	foot lambert
candela/square meter (cd/m²)	0.00031416	lambert
centimeter	0.393701	inch (1959 international)
centimeter water (4°C)	0.0980638	kilopascal (kPa)
centipoise	0.001000*	pascal second (Pa·s)
centistoke	1.00000*	square millimeter/second
chain (surveyor or gunter)	20.1168	meter (m)
coulomb (quantity of electricity)	1.00000*	ampere-second (A·s)
coulomb/cubic foot	35.3147	coulomb/cubic meter (C/m³)
coulomb/cubic meter (C/m³)	0.0283168	coulomb/cubic foot
coulomb/foot	3.28084	coulomb/meter (C/m)
coulomb/inch	39.3701	coulomb/meter (C/m)
coulomb/meter (C/m)	0.304800*	coulomb/foot
coulomb/meter (C/m)	0.025400*	coulomb/inch
coulomb/square foot	10.7639	coulomb/square meter (C/m²)
coulomb/square meter (C/m²)	0.0929030	coulomb/square foot
cubic centimeter	0.035195	fluid ounce (U.K.)
cubic centimeter	0.038140	fluid ounce (U.S.)
cubic centimeter	0.0610237	cubic inch
cubic centimeter/cubic meter	1.00000	volume parts/million

MULTIPLY	BY	TO OBTAIN
cubic decimeter (liter)	0.0353147	cubic foot
cubic decimeter (liter)	0.219969	gallon (U.K. liquid)
cubic decimeter (liter)	0.264172	gallon (U.S. liquid)
cubic decimeter (liter)	0.879877	quart (U.K. liquid)
cubic decimeter (liter)	1.05669	quart (U.S. liquid)
cubic decimeter/kilogram	0.119826	gallon (U.S. liquid)/pound-mass
cubic decimeter/kilogram	0.0160185	cubic foot/pound-mass (ft³/lbm)
cubic decimeter/kilogram	0.099776	gallon (U.K. liquid)/pound-mass
cubic decimeter/second	127.133	cubic foot/hour
cubic decimeter/second	543.440	barrel/day
cubic decimeter/second	3051.19	cubic foot/day
cubic decimeter/second	13.1981	gallon (U.K. liquid)/minute
cubic decimeter/second	15.8503	gallon (U.S. liquid)/minute
cubic decimeter/second	2.11888	cubic foot/minute
cubic decimeter/second	0.035315	cubic foot/second
cubic decimeter/metric ton	0.005706	barrel/ton (U.S.)
cubic decimeter/metric ton	0.006391	barrel/ton (U.K.)
cubic decimeter/metric ton	0.268411	gallon (U.S. liquid)/ton (U.K.)
cubic decimeter/metric ton	0.239653	gallon (U.S. liquid)/ton (U.S.)
cubic foot (1959 international)	0.0283168	cubic meter (m³)
cubic foot	28.3168	cubic decimeter (liter)
cubic foot	7.48052	gallon (U.S. liquid)
cubic foot/foot	0.0929030	cubic meter/meter (m³/m)
cubic foot/hour	0.00786579	cubic decimeter/second
cubic foot/minute	0.471947	cubic decimeter/second
cubic foot/pound-mass	62.4280	cubic decimeter/kilogram
cubic foot/pound-mass	0.0624280	cubic meter/kilogram (m³/kg)
cubic foot/second	28.3169	cubic decimeter/second
cubic inch	0.0163871	cubic decimeter
cubic inch	16.3871	cubic centimeter
cubic meter (m³)	1 000 000*	cubic centimeter
cubic meter (m³)	1000.00*	liter
cubic meter (m³)	61 023.7	cubic inch (1959 international)
cubic meter (m³)	6.28976	barrel (42 U.S. gallons, liquid)
cubic meter (m³)	35.3147	cubic foot
cubic meter (m³)	1.30795	cubic yard (1959 international)
cubic meter (m³)	219.969	gallon (U.K. liquid)
cubic meter (m³)	264.172	gallon (U.S. liquid)
cubic meter/kilogram (m³/kg)	16.0185	cubic foot/pound-mass
cubic meter/meter (m³/m)	10.7639	cubic foot/foot
cubic meter/meter (m³/m)	80.5196	gallon (U.S. liquid)/foot
cubic meter/second meter (m³/s·m)	289 870	gallon (U.S. liquid)/hour foot
cubic meter/second meter (m³/s·m)	4022.80	gallon (U.K. liquid)/minute foot
cubic meter/second meter (m³/s·m)	4831.18	gallon (U.S. liquid)/minute foot
cubic meter/metric ton (m³/Mg)	5.70602	barrel/ton (U.S.)

MULTIPLY	BY	TO OBTAIN
cubic meter/metric ton (m³/Mg)	6.39074	barrel/ton (U.K.)
cubic mile (1959 international)	4.16818	cubic kilometer (km³)
cubic yard (1959 international)	0.764555	cubic meter (m³)
curie (Ci) (radiation dosimetry)	3.7000×10^{10} *	becquerel (Bq)

D

degree Celsius (difference) (°C, diff.)	1.80000*	degree Fahrenheit (difference)
degree Celsius (difference) (°C, diff.)	1.00000*	kelvin (difference)
degree Fahrenheit (difference)	0.555556	degree Celsius (difference) (°C, diff.)
degree Kelvin (difference) (K)	1.80000*	degree Rankine (difference)
degree Rankine (difference)	0.555556	kelvin (difference) (K)
	(for temperature, see under "T")	
degree (angle)	0.017453	radian
degree/second (angular)	0.166667	revolution/minute
dyne	0.000010*	newton (N)
dyne	7.23301×10^{-5}	poundal
dyne	2.24809×10^{-6}	pound (force)
dyne	1.00000*	gram centimeter/second squared
dyne/square centimeter	0.100000*	pascal (Pa)
dyne second/square centimeter	0.100000*	pascal second (Pa·s)

F

farad (electrical capacitance)	1.00000*	coulomb/volt (C/V)
farad/meter (F/m)	0.025400*	farad/inch
fathom (6 U.S. survey foot exactly)	1.82880	meter (m)
foot (1959 international)	0.304800*	meter (m)
foot (1959 international)	304.800*	millimeter
foot (U.S. survey)	(1200/3937)*	meter (m)
foot lambert	3.42626*	candela/square meter (cd/m²)
foot of water (39.2°F)	2988.98	pascal (Pa)
foot/degree F	0.548640	meter/kelvin (m/K)
foot/gallon (U.S. liquid) (foot = 1959 international)	80.5196	meter/cubic meter (m/m³)
foot/cubic foot (1959 international)	10.7639	meter/cubic meter (m/m³)
foot/hour	0.0846667	millimeter/second (mm/s)
foot/minute (1959 international foot)	0.0050800*	meter/second (m/s)
foot/second	0.304800*	meter/second (m/s)
foot-poundal	0.0421401	joule (J)
foot-pound (force)	1.35582	joule (J)
foot-pound (force)	0.00128507	Btu (I.T.)
foot-pound (force)/gallon (U.S.)	0.358169	kilojoule/cubic meter (kJ/m³)
foot-pound (force)/second	1.35582	watt (W)
foot-pound (force)/square inch	0.210152	joule/square centimeter
footcandle	1.00000*	lumen/square foot
footcandle	10.7639	lux (lumen/square meter) (lm/m²)

MULTIPLY	BY	TO OBTAIN
G		
gal (galileo)	0.010000*	meter/second squared (m/s²)
gallon (U.K. liquid)	0.00454609	cubic meter (m³)
gallon (U.K. liquid)	4.54609	cubic decimeter
gallon (U.K. liquid)/hour square foot	1.35927×10^{-5}	cubic meter/second square meter (m³/s·m²)
gallon (U.K. liquid)/minute	0.075768	cubic decimeter/second
gallon (U.K. liquid)/ minute square foot	0.00081556	cubic meter/second square meter (m³/s·m²)
gallon (U.K. liquid)/pound-mass	10.0224	cubic decimeter/kilogram
gallon (U.K. liquid)/1000 barrels	28.5940	cubic centimeter/cubic meter
gallon (U.S. liquid)	231.000*	cubic inch
gallon (U.S. liquid)	0.832674	gallon (U.K. liquid)
gallon (U.S. liquid)	0.00378541	cubic meter (m³)
gallon (U.S. liquid)/cubic foot	133.681	cubic decimeter/cubic meter
gallon (U.S. liquid)/foot	0.0124193	cubic meter/meter (m³/m)
gallon (U.S. liquid)/hour square foot	1.13183×10^{-5}	cubic meter/second square meter (m³/s·m²)
gallon (U.S. liquid)/minute	0.0630902	cubic decimeter/second
gallon (U.S. liquid)/ minute square foot	0.00067910	cubic meter/second square meter (m³/s·m²)
gallon (U.S. liquid)/pound-mass	8.34540	cubic decimeter/kilogram
gallon (U.S., dry)	0.00440488*	cubic meter (m³)
gamma (mass)	10^{-9} *	kilogram (kg)
gamma (magnetic flux density)	10^{-9} *	tesla (T)
gauss (magnetic flux density)	0.000100*	tesla (T)
gauss/Oersted	1.25664×10^{-6}	henry/meter (H/m)
gilbert	0.795775	ampere turn
grad	0.0157080	radian (rad)
grain (1/7000 lbm avoirdupois)	64.79891*	milligram (mg)
grain/gallon (U.S. liquid)	17.1181	gram/cubic meter
grain/cubic foot	2.28835	milligram/cubic decimeter
grain/100 cubic feet	22.8835	milligram/cubic meter
gram	0.035274	ounce-mass (avoirdupois)
gram	0.032151	ounce-mass (troy)
gram mole	0.001000*	kilomole (kmol)
gram/cubic meter	3.78541	milligram/gallon (U.S. liquid)
gram/cubic meter	0.058418	grain/gallon (U.S. liquid)
gram/cubic meter	0.350507	pound-mass/1000 barrels
gram/cubic meter	0.00834541	pound-mass/1000 gallons (U.S. liquid)
gram/cubic meter	0.0100224	pound-mass/1000 gallons (U.K. liquid)
gram/gallon (U.K.)	0.219969	kilogram/cubic meter (kg/m³)
gram/gallon (U.S.)	0.264172	kilogram/cubic meter (kg/m³)
gray (Gy)	100.000	rad (radiation dose absorbed)
H		
hectare	10 000*	square meter (m²)

MULTIPLY	BY	TO OBTAIN
hectare	2.47104	acre (U.S. survey)
henry (inductance) (H)	7.95775×10^7	maxwell/gilbert
henry (inductance) (H)	1.00000	weber/ampere
henry (inductance) (H)	1.00000×10^8	line/ampere
henry/meter (H/m)	795 775	gauss/Oersted
henry/meter (H/m)	2.54000×10^6 *	line/ampere inch
horsepower (electric)	0.746000*	kilowatt (kW)
horsepower (hydraulic)	0.746043	kilowatt (kW)
horsepower (metric)	0.735499	kilowatt (kW)
horsepower (U.S.) (550 ft-lbf/s)	0.745700	kilowatt (kW)
horsepower (U.S.)(550 ft-lbf/s)	42.4072	Btu (I.T.)/minute
horsepower hour (U.S.) (550 ft-lbf/s)	2.68452	megajoule (MJ)
horsepower hour (U.S.) (550 ft-lbf/s)	2544.43	Btu (I.T.)
horsepower/cubic foot (550 ft-lbf/s)	26.3341	kilowatt/cubic meter (kW/m³)
hundred weight (U.K., long)	50.8024	kilogram (kg)
hundred weight (U.S., short)	45.359237*	kilogram (kg)

I

inch (1959 international)	25.4000*	millimeter (mm)
inch water (32.2°F)	0.249082	kilopascal (kPa)
inch water (60°F)	0.24884	kilopascal (kPa)
inch mercury (32°F)	3.38638	kilopascal (kPa)
inch mercury (60°F)	3.37685	kilopascal (kPa)
inch/minute	0.423333	millimeter/second (mm/s)

J

joule (J) (energy, work, or heat)	0.737562	foot-pound (force)
joule (J) (energy, work, or heat)	23.7304	foot-poundal
joule (J) (energy, work, or heat)	1.00000*	watt second (Ws) or Newton-meter (N·m)
joule (J) (energy, work, or heat)	9.47817×10^{-4}	Btu (International Table)
joule (J) (energy, work, or heat)	0.239126	calorie (20°C)
joule (J) (energy, work, or heat)	0.238903	calorie (15°C)
joule (J) (energy, work, or heat)	0.238662	calorie (mean)
joule (J) (energy, work, or heat)	0.238846	calorie (I.T.)
joule (J) (energy, work, or heat)	0.239006	calorie (thermochemical)
joule/kilogram (J/kg)	0.108411	calorie (thermochemical)/pound-mass
joule/kilogram (J/kg)	4.29923×10^{-4}	Btu (I.T.)/pound-mass
joule/mole (J/mol)	0.429923	Btu (I.T.)/pound mole
joule/square centimeter (J/cm²)	4.75846	foot-pound (force)/square inch
joule/square centimeter (J/cm²)	0.101972	kilogram (force)-meter/square centimeter (kgf/cm²)

K

kelvin (degree), t_K	1.800 00*	degree Rankin, t_R
kilocalorie (Calorie)	1000.00*	calorie
kilocalorie (thermochemical)	4.18400*	kilojoule (kJ)
kilocalorie (thermochemical)/hour	1.16222	watt (W)

MULTIPLY	BY	TO OBTAIN
kilocalorie (thermochemical)/ hour m²°C	1.16222	watt/square meter K (W/m²k)
kilocalorie (thermochemical)/ kilogram °C	4.18400	kilojoule/kilogram K (kJ/kg K)
kilogram (kg) (mass, by definition)	2.20462	pound-mass (lbm) (avoirdupois)
kilogram (kg) (mass)	0.0685218	slug (mass)
kilogram-meter (kg·m)	0.224809	slug-foot
kilogram meter/second (kg·m/s)	7.23301	pound (mass)-foot/second
kilogram/cubic decimeter	8.34541	pound-mass/gallon (U.S. liquid)
kilogram/cubic decimeter	10.0224	pound-mass/gallon (U.K. liquid)
kilogram/cubic meter (kg/m³)	0.0624280	pound-mass/cubic foot
kilogram/cubic meter (kg/m³)	0.350507	pound-mass/barrel
kilogram/cubic meter (kg/m³)	3.78541	grams/gallon (U.S. liquid)
kilogram/cubic meter (kg/m³)	4.54609	grams/gallon (U.K. liquid)
kilogram/cubic meter (kg/m³)	0.00194032	slug/cubic foot
kilogram/meter (kg/m)	0.671969	pound-mass/foot
kilogram/mole (kg/mol)	2.20462	pound-mass/mole
kilogram/second (kg/s)	7936.64	pound-mass/hour
kilogram/second (kg/s)	2.20462	pound-mass/second
kilogram/second (kg/s)	0.0590524	ton-mass (U.K.)/minute
kilogram/second (kg/s)	0.0661387	ton-mass (U.S.)/minute
kilogram/second (kg/s)	3.54314	ton-mass (U.K.)/hour
kilogram/second (kg/s)	3.96832	ton-mass (U.S.)/hour
kilogram/second-meter (kg/s·m)	0.671969	pound-mass/second-foot
kilogram/second-meter (kg/s·m)	2419.09	pound-mass/hour-foot
kilogram/second-square meter (kg/s·m²)	0.204816	pound-mass/second-square foot
kilogram/second-square meter (kg/s·m²)	737.338	pound-mass/hour-square foot
kilogram/square meter (kg/m²)	0.204816	pound-mass/square foot
kilogram-force (kgf)	9.80665*	newton (N)
kilogram (force)-meter (kgf-m)	9.80665*	joule (J)
kilogram force (kgf) (not SI)	9.80665*	newton (N)
kilojoule (kJ)	0.947817	Btu (I.T.)
kilojoule (kJ)	0.943690	Btu (39°F)
kilojoule (kJ)	0.948155	Btu (60°F)
kilojoule (kJ)	0.947086	Btu (mean)
kilojoule (kJ)	0.948452	Btu (thermochemical)
kilojoule/cubic meter (kJ/m³)	0.0268392	Btu (I.T.)/cubic foot
kilojoule/cubic meter (kJ/m³)	0.00430886	Btu (I.T.)/gallon (U.K. liquid)
kilojoule/cubic meter (kJ/m³)	0.00358788	Btu (I.T.)/gallon (U.S. liquid)
kilojoule/cubic meter (kJ/m³)	2.79198	foot-pound (force)/gallon (U.S. liquid)
kilojoule/kilogram (kJ/kg)	0.429923	Btu (I.T.)/pound-mass
kilojoule/kilogram kelvin (kJ/kg·K)	0.238846	Btu (I.T.)/pound (mass) °F
kilojoule/kilogram kelvin (kJ/kg·K)	0.238846	Btu (I.T.)/pound mole °F

MULTIPLY	BY	TO OBTAIN
kilojoule/kilogram kelvin (kJ/kg·K)	0.239006	calorie (thermochemical)/gram kelvin
kilojoule/kilogram kelvin (kJ/kg·K)	0.239006	calorie (thermochemical)/gram mole °C
kilojoule/kilogram kelvin (kJ/kg·K)	0.239006	kilocalorie (thermochemical)/kilogram °C
kilojoule/kilogram kelvin (kJ/kg·K)	0.00027778	kilowatt hour/kilogram °C
kilojoule/mole (kJ/mol)	0.239006	kilocalorie (thermochemical)/gram mole
kilometer	3280.84	foot (1959 international)
kilometer (km)	0.621370	mile (U.S. statute)
kilometer (km)	0.539957	nautical mile
kilometer/cubic decimeter	2.35215	mile (U.S. statute)/gallon (U.S. liquid)
kilometer/hour (km/h)	0.539957	knot (international)
kilometer/hour (km/h)	0.621370	miles/hour (U.S. statute)
kilomole (kmol)	1000.00*	gram mole
kilomole (kmol)	2.20462	pound-mole
kilomole (kmol)	836.610	standard cubic foot (60°F, 1 atmosphere)
kilomole (kmol)	22.4136	standard cubic meter (0°C, 1 atmosphere)
kilomole (kmol)	23.6445	standard cubic meter (15°C, 1 atmosphere
kilomole/cubic meter	0.0624280	pound-mole/cubic foot
kilomole/cubic meter	0.0100224	pound-mole/gallon (U.K. liquid)
kilomole/cubic meter	0.00834541	pound-mole/gallon (U.S. liquid)
kilomole/cubic meter	133.010	standard ft³/barrel (60°F, 1 atmosphere)
kilonewton (kN)	0.224809	kip (1000 pound-force)
kilonewton (kN)	0.100361	ton-force (U.K., 2240 lbf)
kilonewton (kN)	0.112404	ton-force (U.S., 2000 lbf)
kilonewton meter (kN·m)	0.368781	ton-force (U.S.) foot
kilopascal (kPa)	0.0101972	atmosphere (kilogram-force/ square centimeter)
kilopascal (kPa)	0.00986923	atmosphere (760 torr)
kilopascal (kPa)	0.01000*	bar
kilopascal (kPa)	10.1974	centimeter water (4°C)
kilopascal (kPa)	4.01474	inch water (39.2°F)
kilopascal (kPa)	0.295301	inch mercury (32°F)
kilopascal (kPa)	0.296134	inch mercury (60°F)
kilopascal (kPa)	7.50064	millimeter mercury (0°C)
kilopascal (kPa)	20.8854	pound-force/square foot
kilopascal (kPa)	0.145038	pound-force/square inch
kilopascal/meter (kPa/m)	0.0442075	pound-force/square inch-foot
kilopond force (kgf)	9.806 65*	newton
kilowatt (kW)	56.8690	Btu (I.T.)/minute
kilowatt (kW)	0.947817	Btu (I.T.)/second
kilowatt (kW)	1.34048	horsepower (electric)
kilowatt (kW)	1.34102	horsepower (550 foot pound/second)
kilowatt (kW)	1.34041	horsepower (hydraulic)
kilowatt (kW)	0.2843	ton of refrigeration (12 000 Btu/h)
kilowatt-hour	3.60000*	megajoule (MJ)
kilowatt-hour/kilogram °C	3600.00*	kilojoule/kilogram kelvin (kJ/kg·K)

MULTIPLY	BY	TO OBTAIN
kilowatt/cubic meter (kW/m³)	96.6211	Btu (I.T.)/hour cubic foot
kilowatt/cubic meter (kW/m³)	0.0268392	Btu (I.T.)/second cubic foot
kilowatt/cubic meter	0.037974	horsepower/cubic foot (U.S., 550 ft-lbf/s)
kilowatt/cubic meter kelvin (kW/m³·K)	53.6784	Btu (I.T.)/hour cubic foot °F
kilowatt/cubic meter kelvin (kW/m³·K)	0.0149107	Btu (I.T.)/second cubic foot °F
kilowatt/square meter (kW/m²)	0.0880551	Btu (I.T.)/second square foot
kilowatt/square meter (kW/m²)	86.0421	calorie (thermochemical)/hour cm²
kilowatt/square meter kelvin (kW/m²·K)	0.0489195	Btu (I.T.)/second square foot °F
kip (1000 pound-force)	4.44822	kilonewton (kN)
kip/square inch (force/area)	6.89476	megapascal (MPa)
knot (international)	1.85200*	kilometer/hour (km/h)
knot (international)	1.15080	U.S. statute mile/hour
knot (international)	0.514444	meter/second

L

MULTIPLY	BY	TO OBTAIN
lambert (luminance)	3183.10	candela/square meter
light-year	9.46055×10^{15}	meter
line	1.00000*	maxwell
line	1.00000×10^{-8} *	weber (Wb)
line/ampere	1.00000×10^{-8} *	henry (H)
line/ampere inch	3.93701×10^{-7} *	henry/meter (H/m)
line/square inch	1.55000	tesla (T)
link	0.201168	meter (m)
liter (cubic decimeter)	1000.00*	cubic centimeters
liter (cubic decimeter)	0.0353147	cubic foot
liter (cubic decimeter)	0.264172	gallon (U.S. liquid)
liter (cubic decimeter)	0.219969	gallon (U.K. liquid)
lumen (luminous flux)	1.00000*	candela-steradian (cd-sr)
lumen/square meter	1.00000*	lux (lx)
lumen/square foot	1.00000*	footcandle
lumen/square foot	10.7639	lux (lx)
lumen/square inch	1550.00	lux (lx)
lux (illuminance)	1.00000*	lumen/square meter (lm/m²)
lux (lx)	0.09290304*	footcandle
lux (lx)	0.09290304*	lumen/square foot
lux (lx)	0.00064516	lumen/square inch
lux second (lx-s)	0.09290304*	foot candle second

M

MULTIPLY	BY	TO OBTAIN
maxwell	1.00000	line
maxwell	1.00000×10^{-8} *	weber (Wb)
maxwell/gilbert	1.25664×10^{-8} *	henry (H)
megagram (Mg)	1.00000	ton-mass (metric) or tonne

MULTIPLY	BY	TO OBTAIN
megagram (Mg)	0.984206	ton-mass (U.K. long or gross)
megagram (Mg)	1.10231	ton-mass (U.S. short or net)
megagram/square meter (Mg/m²)	0.102408	ton-mass (U.S.)/square foot
megajoule (MJ)	947.817	Btu (I.T.)
megajoule (MJ)	0.372506	horsepower hour (U.S.; 550 ft-lbf/s)
megajoule (MJ)	0.277778	kilowatt hour (kW-h)
megajoule (MJ)	0.00947813	therm (European Community)
megajoule/cubic meter (MJ/m³)	4.30886	Btu (I.T.)/gallon (U.K. liquid)
megajoule/cubic meter (MJ/m³)	3.58788	Btu (I.T.)/gallon (U.S. liquid)
megajoule/cubic meter (MJ/m³)	0.239006	calorie (thermochemical)/milliliter
megapascal (MPa)	0.145038	kip/square inch
megapascal (MPa)	145.038	pound-force/square inch
megapascal (MPa)	10.4427	ton-force (U.S.)/square foot
megapascal (MPa)	0.0725189	ton-force (U.S.)/square inch
megawatt (MW)	3.41214	million Btu (I.T.)/hour
meter (m)	0.0497096	chain (U.S. survey)
meter (m)	0.546806*	fathom (6 U.S. survey foot)
meter (m)	3.28084	foot (1959 international)
meter (m)	(3937/1200)*	foot (U.S. survey)
meter (m)	4.97096	link (U.S. survey)
meter (m)	0.198838	rod (U.S. survey)
meter (m)	1.09361	yard (1959 international)
meter/cubic meter (m/m³)	0.0929031	foot/cubic foot (1959 international foot)
meter/cubic meter (m/m³)	0.0124193	foot/gallon (U.S. liquid) (1959 international foot)
meter/kelvin (m/K)	1.82269	foot/°F (1959 international foot)
meter/second (m/s)	3.28084	foot/second (1959 international foot)
meter/second (m/s)	196.850	foot/minute (1959 international foot)
mho	1.00000*	siemens (S)
microbar or dyne/square cm	0.100000*	pascal (Pa)
micrometer (μm)	0.0393701	mil
micrometer (μm)	1.00000*	micron (abolished name, 1967)
micron (abolished name, 1967)	1.00000*	micrometer or 10^{-6} meter (μm)
microsecond/foot	3.28084	microsecond/meter (μs/m)
microsecond/meter (μs/m)	0.304800*	microsecond/foot
mil	0.001000*	inch (1959 international)
mil	25.4000*	micrometer (10^{-6} m)
mile (U.S. statute)	5280.00*	foot (survey foot of year 1893)
mile (international)	1.609344*	kilometer (km)
mile (U.S. statute)	1.609347	kilometer (km)
mile (U.S. and international nautical)	1852.00*	meter (m)
mile (U.S. statute)/hour	0.868961	knot (international)
mile (U.S. statute)/U.S. gallon	0.425144	kilometer/liter
mile/hour (international)	1.609 344	kilometer/hour (km/h)
milligram (mg)	0.0154324	grain (mass)

MULTIPLY	BY	TO OBTAIN
milligram/cubic decimeter	0.436996	grain/cubic foot
milligram/cubic meter (mg/m³)	0.0436996	grain/100 cubic foot
milligram/gallon (U.S. liquid)	0.264172	gram/cubic meter (g/m³)
millimeter (mm)	0.0393701	inch (1959 international)
millimeter (mm)	0.00328084	foot
millimeter mercury (0°C)	133.322	pascal (Pa)
millimeter mercury (0°C)	0.133322	kilopascal (kPa)
millimeter/second (mm/s)	11.8110	foot/hour
millimeter/second (mm/s)	2.36220	inch/minute
millimeter/second (mm/s)	0.0393701	inch/second
million Btu (I.T.)/hour	0.293071	megawatt (MW)
million electron volt	0.160219	picojoule (pJ)
miner's inch	1.500	cubic feet/minute
minute (angle)	2.90888×10^{-4}	radian
mole/foot	3.28084	mole/meter (mol/m)
mole/kilogram (mol/kg)	0.45359237*	mole/pound-mass
mole/meter (mol/m)	0.304800*	mole/foot
mole/pound (mass)	2.20462	mole/kilogram (mol/kg)
mole/square foot	10.7639	mole/square meter (mol/m²)
mole/square meter (mol/m²)	0.09290304*	mole/square foot

N

nautical mile (U.S. or international)	1.85200*	kilometer (km)
newton (N)	1.00000*	kilogram-meter/second-squared (kg·m/s²)
newton (N)	1.00000×10^{5}	dyne
newton (N)	0.224809	pound-force, avoirdupois
newton (N)	7.23301	poundal
newton-meter (N-m)	0.737562	pound (force)-foot
newton-meter (N-m)	8.85075	pound (force)-inch
newton-meter (N-m)	23.7304	poundal-foot
newton-meter/meter (N-m/m)	0.0187341	pound (force)-foot/inch
newton-meter/meter (N-m/m)	0.224809	pound (force)-inch/inch
newton/meter (N/m)	0.0685218	pound-force/foot
newton/meter (N/m)	0.00571015	pound-force/inch

O

oersted	79.5775	ampere turn/meter
ohm (electric resistance)	1.00000*	volt/ampere (V/A)
ohm circular mil/foot	1.66243×10^{-9} *	ohm square meter/meter (Ω m²/m)
ohm foot	0.304800*	ohm square meter/meter (Ω m²/m)
ohm inch	0.025400*	ohm square meter/meter (Ω m²/m)
ohm square meter/meter (Ω m²/m)	6.01531×10^{8}	ohm circular mil/foot
ohm square meter/meter (Ω m²/m)	3.28084	ohm foot
ohm square meter/meter (Ω m²/m)	39.3701	ohm inch
ounce-force (avoirdupois)	0.278014	newton (N)
ounce-mass (avoirdupois)	28.3495	gram

MULTIPLY	BY	TO OBTAIN
ounce-mass (troy)	31.1035	gram
		(Note: 12 troy ounces = 1 troy pound-mass)
ounce (U.K. fluid), volume measure	28.4131	cubic centimeter
ounce (U.S. fluid), volume measure	29.5735	cubic centimeter

P

MULTIPLY	BY	TO OBTAIN
parsec	3.08374×10^{16}	meter (m)
pascal (Pa) (pressure, stress)	1.00000*	newton/square meter (N/m²)
pascal (Pa)	10.0000*	microbar or dyne/square centimeter
pascal (Pa)	0.0075064	millimeter mercury (0°C) or torr
pascal (Pa)	2.95301×10^{-4}	inch mercury (32°F)
pascal (Pa)	1.45038	pound-force/square inch
pascal second (Pa·s)	1000.00*	centipoise
pascal second (Pa·s)	10.0000*	dyne-second/square centimeter
pascal second (Pa·s)	0.0208854	pound (force)-second/square foot
pascal second (Pa·s)	2419.09	pound-mass/foot-hour
pascal second (Pa·s)	0.671969	pound (mass)/foot-second
pascal second (Pa·s)	0.0208854	slug/foot-second
picojoule (pJ)	6.24146	million electron volt
pint (U.K. liquid)	0.568262	cubic decimeter
pint (U.S. liquid)	0.473177	cubic decimeter
pint (U.S. dry)	5.50611×10^{-4}	cubic meter (m³)
pint (U.K. liquid)/1000 barrels	3.57426	cubic decimeter/cubic meter
poise (absolute viscosity)	0.100000*	pascal-second (Pa·s)
pound-mole	0.45359237*	kilomole (kmol)
pound-mole/cubic foot	16.0185	kilomole/cubic meter (kmol/m³)
pound-mole/gallon (U.K. liquid)	99.7763	kilomole/cubic meter (kmol/m³)
pound-mole/gallon (U.S. liquid)	119.826	kilomole/cubic meter (kmol/m³)
poundal	0.138255	newton (N)
poundal	0.0310810	pound-force
poundal-foot (1959 international foot)	0.0421401	newton-meter (N·m)
pound-force, avoirdupois (lbf)	4.44822	newton (N)
pound-force	32.1740	poundal
pound (force)-foot	1.35582	newton-meter (N·m)
pound (force)-foot/inch	53.3787	newton-meter/meter (N·m/m)
pound (force)-inch	0.112985	newton-meter (N·m)
pound-force/foot	14.5939	newton/meter (N/m)
pound-force/inch	175.127	newton/meter (N/m)
pound-force/square foot	0.0478803	kilopascal (kPa)
pound-force/square inch (psi)	6.89476	kilopascal (kPa)
pound-force/square inch-foot	22.6206	kilopascal/meter (kPa/m)
pound (force)-second/square foot	47.8803	pascal second (Pa·s)
pound-mass, avoirdupois (lbm)	0.45359237*	kilogram (kg)
pound-mass, avoirdupois (lbm)	1.21528	pound (troy)
pound-mass, avoirdupois	7000.00	grain
pound-mass, troy	5760.00*	grain

MULTIPLY	BY	TO OBTAIN
pound-mass, avoirdupois	0.0310810	slug (mass)
pound-mass/barrel	2.85301	kilogram/cubic meter (kg/m³)
pound-mass/cubic foot (density)	16.0185	kilogram/cubic meter (kg/m³)
pound-mass/foot	1.48816	kilogram/meter (kg/m)
pound-mass/foot-hour	0.00041338	pascal second (Pa·s)
pound-mass/foot second	1.48816	pascal second (Pa·s)
pound-mass/gallon (U.K. liquid)	0.0997763	kilogram/cubic decimeter
pound-mass/gallon (U.S. liquid)	0.119826	kilogram/cubic decimeter
pound-mass/1000 gallons (U.K. liquid)	99.7763	gram/cubic meter
pound-mass/1000 gallons (U.S. liquid)	119.826	gram/cubic meter
pound-mass/hour	0.000125998	kilogram/second (kg/s)
pound-mass/hour square foot	0.00135623	kilogram/second-square meter (kg/s·m²)
pound-mass/minute	0.00755987	kilogram/second (kg/s)
pound-mass/mole	0.45359237*	kilogram/mole (kg/mol)
pound-mass/second	0.45359237*	kilogram/second (kg/s)
pound-mass/second-foot	1.48816	kilogram/second-meter (kg/s·m)
pound-mass/second-square foot	4.88243	kilogram/second-square meter (kg/s·m²)
pound-mass/square foot	4.88243	kilogram/square meter (kg/m²)
pound (mass)-foot/second	0.138255	kilogram-meter/second (kg·m/s)
pound (mass)-square foot (moment of inertia)	0.0421401	kilogram-square meter (kg·m²)

Q

quart (U.K. dry)	1.03206	quart (U.S. dry)
quart (U.S. dry)	0.96894	quart (U.K. dry)
quart (U.K. liquid)	1.13652	cubic decimeter
quart (U.K. liquid)	1.13652	liter
quart (U.K. liquid)	1.20095	quart (U.S. liquid)
quart (U.S. liquid)	0.946353	cubic decimeter
quart (U.S. liquid)	0.946353	liter
quart (U.S. liquid)	0.859367	quart (U.S. dry)
quart (U.S. liquid)	0.832674	quart (U.K. liquid)

R

rad (radiation dose absorbed)	0.010000*	gray (Gy) (joule/kilogram)
radian (rad)	206 265	second (angle)
radian (rad)	3437.75	minute (angle)
radian (rad)	57.2958	degree (angle)
radian (rad)	63.6620	grad
radian/second (rad/s)	0.159155	revolution/second
radian/second (rad/s)	9.54930	revolution/minute
radian/second squared (rad/s²)	0.159155	revolution/second squared
radian/second squared (rad/s²)	572.958	revolution/minute squared
revolution/minute	0.104720	radian/second (rad/s)

MULTIPLY	BY	TO OBTAIN
revolution/minute squared	0.00174533	radian/second squared (rad/s²)
revolution/second	6.28318	radian/second (rad/s)
revolution/second squared	6.28318	radian/second squared (rad/s²)
rod (U.S. survey, 16.5 ft. exactly)	5.02921	meter (m)
roentgen	2.57976×10^{-4} *	coulomb/kilogram (C/kg)

S

second (angle)	4.84814×10^{-6}	radian (rad)
section (1 square statute mile)	2.5898	square kilometer (km²)
section (U.S. survey)	640	acre
siemens (S) (electrical conductance)	1.00000*	ampere/volt (A/V) or mho
siemens/meter (S/m)	0.025400*	ampere/volt inch
slug (mass)	32.1740	pound-mass (avoirdupois)
slug (mass)	14.5939	kilogram (kg)
slug/cubic foot (density)	515.379	kilogram/cubic meter (kg/m³)
slug-foot (1959 international foot)	4.44822	kilogram-meter (kg·m)
slug/foot-second	47.8803	pascal second (Pa·s)
square foot (1959 international)	0.09290304*	square meter (m²)
square foot/hour (thermal diffusivity)	25.8064*	square millimeter/second
square foot/pound-mass	0.204816	square meter/kilogram (m²/kg)
square foot/second	9290.304*	square millimeter/second
square foot pound (mass)/sec²	0.0421401	joule (J)
square inch (1959 international)	645.160*	square millimeter (mm²)
square kilometer (km²)	0.386101	section (1 square statute mile)
square kilometer (km²)	0.386102	square mile (1959 international)
square meter (m²)	10.7639	square foot (1959 international)
square meter (m²)	2.47104×10^{-4}	acre (U.S. survey)
square meter (m²)	1.19599	square yard (1959 international)
square meter/kilogram (m²/kg)	4.88243	square foot/pound-mass
square mile (1959 international)	2.58999	square kilometer (km²)
square millimeter (mm²)	0.00155000	square inch (1959 international)
square millimeter/second (mm²/s)	1.07639×10^{-5}	square foot/second
square millimeter/second (mm²/s)	0.0387501	square foot/hour
square millimeter/second (mm²/s)	1.00000*	centistoke
square yard (1959 international)	0.836127	square meter (m²)
standard cubic foot/barrel (60°F, 1 atmosphere)	0.0075182	kilomole/cubic meter (kmol/m³)
standard cubic foot (60°F, 1 atmosphere)	0.0011953	kilomole (kmol)
standard cubic meter (0°C, 1 atmosphere)	0.044616	kilomole (kmol)
standard cubic meter (15°C, 1 atmosphere)	0.042293	kilomole (kmol)
statampere	3.33564×10^{-10}	ampere (A)
statcoulomb	3.33564×10^{-10}	coulomb (C)

MULTIPLY	BY	TO OBTAIN
statfarad	1.11265×10^{-12}	farad (F)
stathenry	8.98755×10^{11}	henry (H)
statohm	8.98755×10^{11}	ohm (Ω)
statvolt	299.793	volt (V)
stere	1.00000*	cubic meter (m³)
stokes (kinematic viscosity)	10^{-4} *	square meter/second (m²/s)
stokes	$10.7639 \times ^{-4}$	square foot/second

T

temperature (see "degree" for temperature interval or difference)

temperature, degree Fahrenheit, $t_F = 1.8t_C + 32$

temperature, degree Celsius, $t_C = (t_F - 32)/1.8 = 0.555\ 556\ (t_F - 32)$

temperature, degree Kelvin, $t_K = T_C + 273.15$ (by definition)

temperature, degree Rankine, $t_R = t_F + 459.67$ (by definition)

temperature, degree Kelvin, $t_K = 0.555\ 556\ t_R$

temperature, degree Rankine, $t_R = 1.8\ t_K$

temperature, degree Fahrenheit, $t_F = 1.8t_K - 459.67$

temperature, degree Celsius, $t_C = 0.555\ 556t_R - 273.15$

MULTIPLY	BY	TO OBTAIN
tesla (T) magnetic flux density	1.0000*	weber/square meter (Wb/m²)
tesla (T)	10 000.0*	gauss
telsa (T)	64 516.0*	line/square inch
therm (10^5 Btu, I.T., Eur. Comm.)	105.506	megajoule (MJ)
therm (U.S., natural gas)	105.4804*	megajoule (MJ)
ton-force (U.K., 2240 lbf)	9.96402	kilonewton (kN)
ton-force (U.S., 2000 lbf)	8,89644	kilonewton (kN)
ton (force) (U.S.)-foot	2.71164	kilonewton meter (kN·m)
ton-force (U.S.)/square foot	95.7605	kilopascal (kPa)
ton-force (U.S.)/square inch	13.7895	megapascal (MPa)
ton-mass (U.K. long or gross)	2240*	pound-mass
ton-mass (U.K. long or gross)	1.01605	megagram (Mg)
ton-mass (U.K. long or gross)	1.01605	metric ton or tonne
ton-mass (U.K. long or gross)	1.12000	ton-mass (U.S. short or net)
ton-mass (U.S. short or net)	2000*	pound-mass
ton-mass (U.S. short or net)	0.907185	megagram (Mg)
ton-mass (U.S. short or net)	0.907185	metric ton or tonne
ton-mass (U.S. short or net)	0.892857	ton-mass (U.K. long)
ton-metric (tonne)	2204.62*	pound-mass
ton-metric (tonne)	1.00000*	megagram (Mg)
ton-metric (tonne)	0.984206	ton-mass (U.K. long)
ton-metric (tonne)	1.10231	ton-mass (U.S. short)
ton-mass (U.K.)/day	0.0117598	kilogram/second (kg/s)
ton-mass (U.S.)/day	0.0104998	kilogram/second (kg/s)
ton-mass (U.K.)/hour	0.282235	kilogram/second (kg/s)
ton-mass (U.S.)/hour	0.251996	kilogram/second (kg/s)
ton-mass (U.K.)/minute	16.9341	kilogram/second (kg/s)
ton-mass (U.S.)/minute	15.1197	kilogram/second (kg/s)

MULTIPLY	BY	TO OBTAIN
ton-mass (U.S.)/square foot	9.76485	megagram/square meter (Mg/m²)
ton refrigeration (12 000 Btu/h)	3.517	kilowatt (kW)
ton (nuclear equivalent of TNT)	4.184×10^9	joule (J) (defined, not measured)
torr (mm mercury, 0°C)	133.322	pascal

U

unit/foot	3.28084	unit/meter
unit/henry	1.00000*	ampere/weber (A/Wb)
unit/meter	3.28084	volt/meter (V/m)

V

volt (V) electric potential)	1.00000*	watt/ampere (W/A)
volt/foot	3.28084	volt/meter (V/m)
volt/inch	39.3701	volt/meter (V/m)
volt/meter (V/m)	0.304800*	volt/foot
volt/meter (V/m)	0.025400*	volt/inch
volume parts per million	1.00000*	cubic centimeter/cubic meter

W

watt (W) (power)	1.00000*	joule/second (J/s)
watt (W)	3.41214	Btu (I.T.)/hour
watt (W)	44.2537	foot-pound (force)/minute
watt (W)	0.737562	foot pound (force)/second
watt (W)	0.860421	kilocalorie (thermochemical)/hour
watt hour (W·h)	3.60000*	kilojoule (kJ)
watt/inch (international)	39.3701	watt/meter (W/m)
watt/meter (W/m)	0.025400*	watt/inch (international)
watt/meter kelvin (W/m·K)	0.577789	Btu (I.T.)-foot/hour square foot °F
watt/meter kelvin (W/m·K)	6.93347	Btu (I.T.)-inch/hour square foot °F
watt/meter kelvin (W/m·K)	8.60421	calorie (thermochemical)-cm/hour cm² °C
watt/meter kelvin (W/m·K)	0.00239006	calorie (thermochemical)-cm/second cm² °C
watt/square meter (W/m²)	0.316998	Btu (I.T.)/hour square foot
watt/square meter kelvin (W/m²·K)	0.176110	Btu (I.T.)/hour square foot °F
watt/square meter kelvin (W/m²·K)	0.860421	kilocalorie (thermochemical)/hour cm² °C
watt second (W·s)	1.00000*	joule (J)
weber (Wb) (magnetic flux)	1.00000*	volt-second (V-s)
weber (Wb)	1.00000×10^8	line
weber (Wb)	1.00000×10^8	maxwell
weber/ampere (Wb/A)	1.00000*	henry (H)

Y

yard (1959 international)	0.914400*	meter (m)

SUBJECT INDEX

Page numbers for tables, graphs, and illustrations are in italics.

analysis of U.S. water supplies, 20–4, *20–5*
for boilers, 20–11
chemistry management system, ICMS, 13–37
density, 5–26, 7–13
gas (blue gas), 2–39
impurities in, 5–25, 20–1, *20–2, 20–3, 20–4*
oxygen in, 5–10, 20–3, *20–30,* 20–32, *20–38,* 21–11
separation, in boiler drum, 5–25. *See* Drum internals
softening processes, 20–6
and steam mixtures, 5–2, 5–14, *5–19, 5–20,* 5–25, 7–13, 7–19
supplies, 20–1, 20–30
technology, Chapter 20
testing, 20–39
treatment, *7–23, 7–27, 7–50,* 20–4, 20–6, 20–11
vapor in gaseous fuels, 2–41, 6–7, C–36
Water-cooled furnaces. *See* Furnace wall; Waterwalls
Water gas, 2–39, 4–22
Water-impounded hoppers, 6–9, 7–6, 16–5
Water quenching, 17–10
Water-steam cycles, 22–10
Water-tube boilers, 5–5, 10–5
Waterwalls, 5–5, 6–50. *See* Furnace wall; C-E Controlled
    Circulation boiler
  erection, *19–10,* 19–12, *19–12*
  evaporation in, 5–8, 5–16
  flow in, 5–16, 5–18
  steam generation, 5–8
  tube failures, *20–21,* 20–23, 21–15, 24–11
  wastage, 3–27, 8–58, *8–59,* 24–9
Waterwashing, 14–33, 14–34, *14–35*
Watt, J.D., B–12, B–15
Weak-acid analysis, coal, 3–12
Wear resistance in fans, 14–3, 14–21, *14–22*
Weigh tanks, 22–4
Weight of fuel fired, 6–12, 7–3
Weight support in marine boilers, 10–21
Weighted-wire precipitator, 15–8, 15–9, 15–15
Weld-beveling, *18–5,* 18–6
Weld overlay, 8–61
Welded furnace enclosure, 6–49 to 6–55. *See* Fusion-panel
    welding; Furnace wall
Welding,
  boiler components, 18–2
  drum plates, 18–6
  of erected components, 19–17
  flux, 18–6
  hard-face, 8–61

materials, 17–6
orbital, 18–12, *18–16*
repairs, 23–11
solid fins, 18–11
techniques, 18–6, 18–9, *18–10, 18–11,* 18–12
Welds,
  field postweld heat-treating of, 19–18
  field radiographical examination of, 19–18
Weston, P.L., Jr., B–2
Wet-bottom vs. dry-bottom furnaces, 7–6
Wet-motor pump, 14–41, *14–43*
Wet natural gas, 2–36
Wet-process sludge (FGDS), 15–64
Wet scrubbers, 15–6, *15–7,* 15–28 to 15–64
Wet scrubbing,
  effect of chloride, 15–43
  effect of magnesium, 15–42
Wet steam, 5–2
White, David, 2–4
Wilkins, D.M., B–11
Wilson, Carroll, A–1
Windbox-assembly erection, 19–14, *19–15*
Windbox design. *See* Horizontally fired systems;
    Tangential firing systems
Window welds, 23–12
Winkler, Fritz, 9–1
Wobbling of headers, 18–9
Wohlenberg, W.J., 11–3
Wood as fuel, 2–23, *2–24, 2–26,* C–15, *C–16, C–17*
Work, mechanical energy, 1–2 to 1–6, 5–21, 5–23
Work-hardening. *See* Fatigue
World Energy Conference, A–1

Y

Yield strength, 17–13
Young's modulus of elasticity, 17–13

Z

Zeldovich mechanism (thermal NO_x formation), 4–31
Zeolite softening, 20–9
Zerban and Nye, 1–1
Zero-leakage pumps, 14–41, 14–42